Introduction to Nanoscience & Nanotechnology

Introduction to Nanoscience & Nanotechnology

Gabor L. Hornyak • Harry F. Tibbals • Joydeep Dutta • John J. Moore

CRC Press
Taylor & Francis Group
Boca Raton London New York

CRC Press is an imprint of the
Taylor & Francis Group, an **informa** business

This volume consists of the combined books: *Introduction to Nanoscience* (978-1-4200-4805-6) and *Fundamentals of Nanotechnology* (978-1-4200-4803-2).

Cover Title: "Water droplets on lotus leaf nanostructure reflecting technology and nature."

Cover concept and design: Michael Hamers of Lightspeed Commercial Arts, Boulder, Colorado and Anil K. Rao, Professor of Biology, the Metropolitan State College of Denver, Colorado

Cover Permissions: Reflections left and middle are of Zyvex's nProber™ and NanoEffector® respectively; courtesy of Jim Von Ehr, II, CEO of Zyvex Instruments

The butterfly image reflected in the right droplet is *Morpho peleides*, courtesy of Pete Vukusic, School of Physics, University of Exeter, UK.

The lotus flower is courtesy of Gary Lawrence of Westminster, Colorado.

CRC Press
Taylor & Francis Group
6000 Broken Sound Parkway NW, Suite 300
Boca Raton, FL 33487-2742

Printed in the United States of America on acid-free paper
10 9 8 7 6 5 4 3 2 1

International Standard Book Number-13: 978-1-4200-4779-0 (Hardcover)

Visit the Taylor & Francis Web site at
http://www.taylorandfrancis.com

and the CRC Press Web site at
http://www.crcpress.com

CONTENTS

Preface

Introduction to Nanoscience and Nanotechnology

Introduction to Nanoscience and Nanotechnology is a volume that combines the contents of *Introduction to Nanoscience* and *Fundamentals of Nanotechnology*. The combined volume therefore serves as a comprehensive one-year course in nanoscience and nanotechnology. The text consists of a total of 28 chapters—14 that deal with the scientific aspects of nano and the remaining 14 that deal with the technological "nanoengineering" aspects of nano. Each sector is divided into five sections as depicted in below figure.

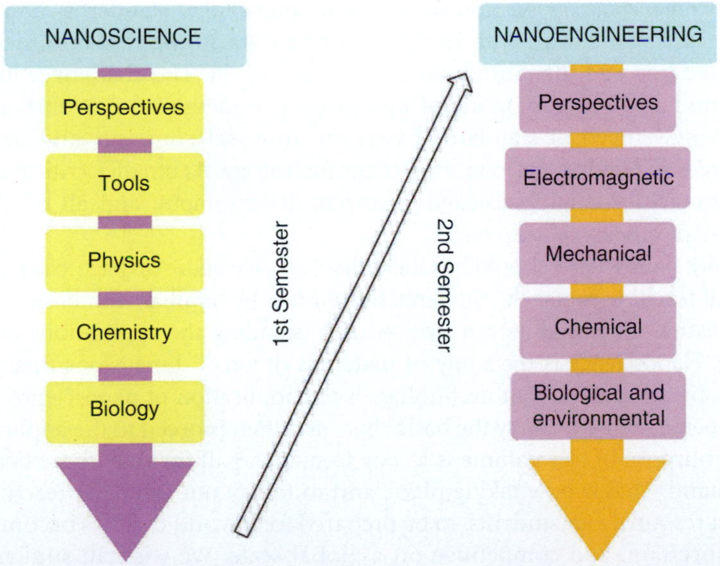

The pedagogical flow and recommended teaching course of Introduction to Nanoscience and Nanotechnology *are depicted. The volume is a combination of* Introduction to Nanoscience *(left) and* Fundamentals of Nanotechnology *(right). Each half of the text is divided into five sections. We believe that the "science" should be taught before the "technology."*

The first half of the volume (nanoscience) is composed of the perspectives, tools, and the monolithic sciences of physics, chemistry, and biology in that order. Clearly, the second half of the volume (technological nanoengineering) is heavier with regard to the technological aspects. In *Nanoscience*, perspectives (introduction and societal implications), tools (characterization and fabrication), physics, chemistry, and finally biology topics related to nanoscience are presented. In *Fundamentals*, we provide perspectives (introduction and metrology), electromagnetics, mechanics, chemical engineering, and biological and environmental topics related to nanotechnology. Each half of the text has its own preface that provides a detailed discussion of respective content and teaching strategy.

In both halves, the "Perspectives" sections provide the foundation for what is to come as well as some insight into factors that are important but exist outside the laboratory—for example, the societal issues. *Introduction to Nanoscience* provides background and historical context to both nanoscience and nanotechnology. In *Fundamentals of Nanotechnology*, a section is dedicated to educate the student in the ways of starting, operating, and exiting a small business. A subsequent section on education and workforce development is provided to guide the student along a potential career path—one that is not necessarily based in the laboratory or on a production line. These kinds of chapters are normally not found in college science texts, but education (and academia in general) and the global economic landscape have changed and are changing continually. Over the past decade, our world has been transformed significantly. Most universities now encourage involvement with business, big and small, as well as international collaboration to levels never seen before. It is prudent therefore, on our part, to help prepare today's students for the future.

In *Fundamentals of Nanotechnology*, nanometrology, standards, and nanomanufacturing are dealt with in detail. All manufacturing relies on metrology and standards and the nanodomain is no exception. The challenges involved are formidable. The very fabric of metrology is undergoing a revolution of its own. A new breed of standards based on universal constants and quantum mechanics is making a strong impact on metrology. As smaller critical dimensions are achieved and exceeded, research, development, and all of industry must adapt quickly to keep pace.

Ideally, nanoscience should be taught first (e.g., see figure on page xxix). This is a practical teaching stratagem. Students should first be familiar with basic concepts of nanoscience before they embark on understanding the applications of nanoscience. Nanoscience is the study of materials of small dimensions that possess remarkable properties. Nanotechnology is the application of nanoscience. It thus makes perfect sense to study the basics first, and then proceed to the applications.

The purpose of this volume is to pay homage to all that has come before, to understand what is now taking place, and to fortify our future by teaching our greatest resource, our students, to be prepared for a world that has become complex, uncertain, and competitive on a global scale. We wish all students and their professors the best of luck as they embark on this most noble of paths. We once again quote the late Nobel laureate Professor Richard Smalley—"Be a scientist. Save the world."

Links to Other Books

www.nanoscienceworks.org/textbookcommunity/introtonanoscience

AUTHORS

Gabor L. Hornyak has an interdisciplinary background in biology, chemistry, and physics. He received his PhD in chemistry from Colorado State University in 1997; BS and MS degrees in biology (human genetics) from the University of Colorado at Denver in 1976 and 1981, respectively; and a BS degree in chemistry from the University of California at San Diego in 1990. Dr. Hornyak worked in the aerospace industry as a senior scientist from 1978 to 1990 in San Diego at the Convair Division of General Dynamics in coatings, adhesives, and corrosion. From 1997 to 2002, he worked at the National Renewable Energy Laboratory on development of chemical vapor deposition synthesis of carbon nanotubes. He has over 15 years of experience in nanotechnology research and development. Dr. Hornyak played a major role in creating an awareness of the promise that nanotechnology brought to the citizens and institutions in Colorado and the surrounding region (2003–2005). He is the editor of the "Perspectives in Nanotechnology" series—a group of books dedicated to bringing topics to the general public that address issues about nanotechnology that are outside the laboratory and production line.

John J. Moore is the trustees' professor and head of department of metallurgical and materials engineering at the Colorado School of Mines (CSM), Golden, Colorado, where he is the director of the interdisciplinary graduate program in materials science, director of the Advanced Coatings and Surface Engineering Laboratory, and director of the Reaction Synthesis Laboratory.

Dr. Moore was awarded a BSc in materials science and engineering from the University of Surrey, Surrey, United Kingdom; a PhD in industrial metallurgy from the University of Birmingham, Birmingham, United Kingdom; and a DEng from the School of Materials of the University of Birmingham, Birmingham, United Kingdom.

Dr. Moore worked as a student apprentice at Stewarts and Lloyds Ltd., United Kingdom, from 1962 to 1966, and as manager of industrial engineering and production control at Birmid-Qualcast Industries Ltd., United Kingdom, from 1969 to 1974. He is currently serving on the board of directors of Hazen Research, Inc., Golden, Colorado (2002–present) and is chairman of the scientific advisory board, XSunX, Aliso Viejo, California (2005–present).

Prior to his appointment at CSM, Dr. Moore served as professor and head, Department of Chemical and Materials Engineering, University of Auckland, New Zealand, from 1986 to 1989; professor of metallurgical engineering at the University of Minnesota, Twin Cities, from 1979 to 1986, and senior lecturer of chemical metallurgy at Sandwell College, United Kingdom, from 1974 to 1979.

Dr. Moore's research interests and activities include physical and chemical vapor deposition of thin films and coatings; synthesis and processing of advanced ceramic, intermetallic, and composite materials using plasma and combustion synthesis techniques; synthesis, processing, and properties of biomaterials; and powder metallurgy processing of advanced materials. He has published more than 620 papers in materials science and engineering journals; holds14 patents; and is the author, coauthor, or editor of 12 books not including *Fundamentals of Nanotechnology*. Dr. Moore is a fellow of the Institute of Materials, United Kingdom; a fellow of American Society for Metals (ASM) International; a fellow of the American Ceramic Society; a fellow of the Institute of Materials, Minerals, and Mining, United Kingdom; and a chartered engineer. He is also an honorary professor at the University of Salford, Salford, United Kingdom and at the University of Auckland, Auckland, New Zealand; he has also been awarded an honorary doctorate from the Moscow State Institute of Steel and Alloys, Russia.

Harry F. Tibbals has served as director of the Bioinstrumentation Resource Center for the University of Texas Southwestern Medical Center since 1997, where he is responsible for providing engineering support to clinical and basic biomedical life science researchers. His funded research includes development of pressure and electrochemical sensors for medical applications, testing and evaluation of life support systems for NASA use in space flight and extravehicular activity, and development of technology for Alcon Research Ltd. for diagnosis of diseases of the eye. Dr. Tibbals's work involves consultation and team leading on a wide variety of analytical, materials, and systems technology in support of medicine and biomedical research. He is frequently called upon to advise on risks and cost benefits for technology decisions.

Prior to joining UTSW, Dr. Tibbals was product line manager, Digital Cardiology Products for Jamieson Kodak, working in Dallas, Texas, and Rochester, New York; and at hospital cardiac catheterization laboratories throughout the United States and Europe. He was responsible for developing and obtaining FDA approval for the first fully digital imaging systems capable of showing the living human heart with medical radiology standards of precision and accuracy. He also consulted for the development of patient anesthesiology systems. As president of Biodigital Technologies, Inc. and board member and consultant to Martingale Research Corporation from 1989 to 1995, he led teams in the development of real-time systems for the identification of bacterial and viral disease agents, and systems for monitoring, analyzing, and reducing environmental hazards. From 1988 to 1991, Dr. Tibbals was product line manager on contract to Inmos and later SGS Thompson for digital signal processing products and applications. Clients for product and development management projects included NEC, Teledyne, Marathon Energy Systems, Coors, Bank One, Shelby Technologies, Innovative Systems SA, Optical Publishing Inc., Gentech, Colorado Medical Physics, and others. He was a consultant and project manager for the development of a production and distribution control system for the world's largest nitrogen fertilizer complex at BASF in Lugwigshafen, Germany.

For most of the 1980s, Dr. Tibbals worked for Rockwell International on trusted and secure systems in government and private areas, serving as principal investigator, project engineer, and systems engineer on projects in the United States and around the world. He was product planner and Rockwell representative on standards committees in the development of the first Rockwell Digital Facsimile Modem systems, which achieved more than 90% market share over a sustained period following its introduction. From 1983 to 1985, he was a principal design engineer for Mostek Systems Technology, heading work on standards, design, and introduction of the VMEbus product line and the development of applications that became dominant in telecommunications, process control, high-end workstations, and scientific and medical instrumentation.

During the 1970s, Dr. Tibbals served on the academic and research staff of Glasgow and Durham Universities in the United Kingdom, where he worked with the Edinburgh Regional Computing Centre, the Digital Cartography Unit, and the Glycoprotein Research Laboratory. He also taught for the Open University and Jordanhill College. He held visiting research and teaching positions at Bogazici University and the University of North Texas, working on instrumentation and systems for environmental monitoring and remediation.

Dr. Tibbals earned his BS degree in chemistry and mathematics at Baylor University in 1965, where he held scholarships in chemistry, English, and Old Testament studies, and was an undergraduate research fellow in electrochemistry. He was awarded his PhD from the University of Houston in 1970 for theoretical and experimental research in nonequilibrium statistical mechanics and kinetics of ion–molecule reactions. He won an SRC postdoctoral fellowship in physical silicon chemistry at the University of Leicester from 1970 to 1972. He has published a number of refereed scientific and technical papers; has received grants and study awards in computer systems architecture, man–machine interfaces, and applications of computers in chemistry; and holds two patents. In 1990, he was awarded a grant from the National Center for Manufacturing Science to visit key technology centers in Japan and study applications of advanced signal processing technology, including fuzzy logic and neural networks. He received a grant from Rockwell to organize a series of three symposia on networks in brain and computer architecture from 1986 to 1988. He served as adjunct professor in the University of Texas at Dallas School of Human Development, and he was a member of the advisory board for the University of Texas at Arlington Advanced Automation and Robotics Center and for the Rutgers Center for Advanced Information Processing.

Dr. Tibbals has served the IEEE Dallas CN group as treasurer, program chair, and board member. He is a member of the American Chemical Society, the Biophysical Society, the Royal Society of Chemistry (chartered chemist), The American Association for the Advancement of Science, The British Computer Society, and the Sigma Xi Society for Scientific Research. He has served on the boards of the Audubon Society Prairie and Cross Timbers Chapter and the Dallas Nature Center. He has served as an advisor to government bodies and companies of information systems and technology for environmental monitoring and improvements. He served on the U.K. Science and Religion Forum and the Commission on Caring for Creation, and he worked with George Dragus and Nicholas Madden on the generation of computerized concordances for early Greek Christian patriarchal writings.

Joydeep Dutta was born on May 5, 1964, and is currently director of the Center of Excellence in Nanotechnology and an associate professor in microelectronics at the Asian Institute of Technology (AIT), Bangkok, Thailand, whose faculty he joined in April 2003. He received his PhD in 1990 from the Indian Association for the Cultivation of Science, India. In 1991 and 1992, he did postdoctoral work at the electrotechnical laboratory (Japan) and at Ecole Polytechnique (France) before moving to Switzerland in 1993, where he was associated with the Swiss Federal Institute of Technology, Lausanne (EPFL), until 2003. From 1997 to 2001, Dr. Dutta worked in technical and managerial qualities in high-technology industries in Switzerland before returning to academia in 2002. He has been a member of the board of two companies working in high-technology electronics and environmental consulting, respectively.

Dr. Dutta has been teaching courses in microelectronics fabrication technology, nanomaterials and nanotechnology, optoelectronic materials and devices, failure analysis of devices, and emerging technologies at AIT. He has also taught nanomaterials (since 1997) at EPFL, and in Uppsala University (2003–2005) and Royal Institute of Technology (2005–present), both in Sweden.

Dr. Dutta is a fellow of the Institute of Nanotechnology (IoN) and the Society of Nanoscience and Nanotechnology (SNN) and a member of several professional bodies, including the Institute of Electrical and Electronics Engineers (IEEE), Materials Research Society (MRS), Society of Industry Leaders, Gerson Lehrman Group Council, the NanoTechnology Group Inc., and the Science Advisory Board—all in the United States; the Asia-Pacific Nanotechnology Forum (APNF), Australia; and the U.K. Futurists Network, United Kingdom, among others. He has reviewed projects of various scientific organizations of different countries (lately in Sweden and Ireland) and has organized a few international conferences and served as a member in several others.

Dr. Dutta's broad research interests include nanomaterials in nanotechnology, self-organization, microelectronic devices, and nanoparticles and their applications in electronics and biology. He recently completed a textbook on nanoscience, and he has over 100 research publications, over 350 citations (Scopus), five chapters in science and technology reference books, three patents (five ongoing applications), and has delivered more than 50 invited lectures.

CHAPTER OPENING CAPTIONS AND CREDITS

Chapter 15 **INTRODUCTION: NANOTECHNOLOGY** 803
NIST's new advanced measurement laboratory (AML) building designed by HDR Architecture, Inc.
Source: G. Porter, National Institute of Standards and Technology, 2004.

Chapter 16 **NANOMETROLOGY: STANDARDS AND NANOMANUFACTURING** 853
An "atomic ruler" uses planes of Si for calibration purposes.
Source: National Institute of Standards and Technology, 2005.

Chapter 17 **NANOELECTRONICS** 923
Synchronized nano-oscillators show interaction of spin waves.
Source: National Institute of Standards and Technology, 2005.

Chapter 18 **NANO-OPTICS** 965
Gallium nitride nanowires emit UV light.
Source: L. Mansfield, National Institute of Standards and Technology, 2006.

Chapter 19 **NANOMAGNETISM** 1003
Magnetic resonances of metallic thin film nanowires.
Source: National Institute of Standards and Technology, 2007.

Chapter 24 **NANOCOMPOSITES AND FIBERS** 1207
Single-walled carbon nanotube polymer composite.
Source: Nanotechnology Gallery, Center for Nanotechnology, NASA Ames.

Chapter 27 **MEDICAL NANOTECHNOLOTY** 1405
Immunohistochemical detection of HER2 biomarker in conjunction with IgY antibody and quantum dots in breast cancer cells.
Source: National Institute of Standards and Technology, 2008.

Part I

Nanoscience

Section 1

Perspectives

Introduction: Nanoscience

I would like to describe a field, in which little has been done, but in which an enormous amount can be done in principle. This field is not quite the same as the others in that it will not tell us much of fundamental physics (in the sense of, 'What are the strange particles?') but it is more like solid-state physics in the sense that it might tell us much of great interest about the strange phenomena that occur in complex situations. Furthermore, a point that is most important is that it would have an enormous number of technical applications. What I want to talk about is the problem of manipulating and controlling things on a small scale.

Richard Feynman, Caltech, 1959 [Fig. 1.0],
"There's Plenty of Room at the Bottom"

Chapter 1

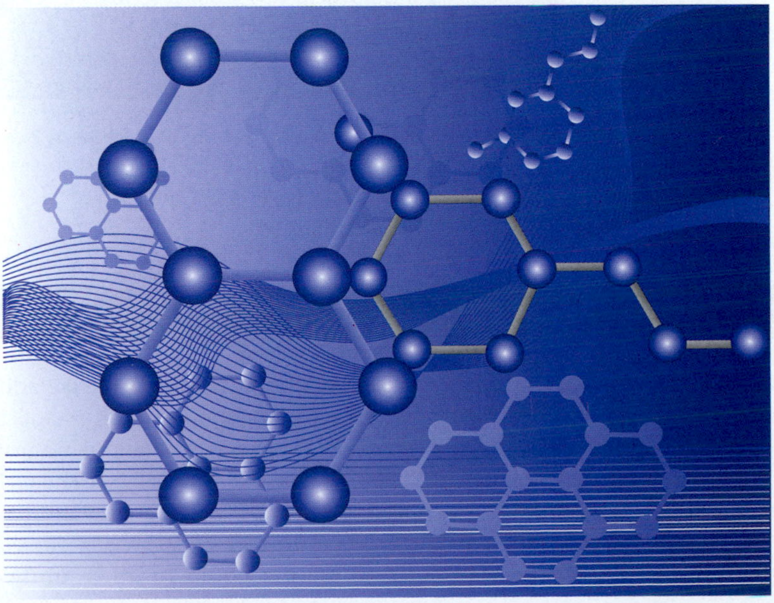

FIG. 1.0 *Nobel Prize laureate Richard Feynman at the blackboard.*

Source: Photograph of Richard Feynman used by permission of Melanie Jackson Agency, LLC. and Caltech Public Relations.

THREADS

"Threads" sections provide perspectives, reinforce general themes and at times serve as a navigational aid—what to expect, where we have been, and what our next objective is. The "Threads" sections provide the site map for the text, a short forum within which we discuss the order of subject matter and emphasize unifying themes. Here we address a fundamental pedagogical theme—specifically, that Part I is about the science of nanomaterials and not about the technology. The technological aspects of nanoscale materials are covered in our sister volume *Fundamentals of Nanotechnology* and also in this book. Another unifying theme of the

Threads (Contd.)

book involves definitions and boundaries. In particular, we construct boundaries only to flagrantly disregard them later. The distinction between science and technology is the first in a series of boundaries that suffers from such apparent nebulosity. We firmly believe that all things converge, forming a continuum. The theme of the integration of everything is a fundamental tenet of nanoscience and nanotechnology. As a consequence, we feel that distinctions between and among things are presented solely for the purpose of perspective (and convenience). Lastly, we attempt to weave the wonders of the natural world into our discussions as often as possible. Nature, the ultimate nanotechnologist, has much to share.

Part I consists of five divisions: I. Perspectives, II. Nanotools, III. Physics: Properties and Phenomena, IV. Chemistry: Synthesis and Modification, and V. Natural and Bionanoscience. Chapters 1 and 2 comprise the perspectives division of the text. Emphasis on our historical scientific heritage is a major theme throughout this book. Chapter 2 delves into societal implications of nanoscience and nanotechnology. Its placement early in the book is strategic because we hope students *always* keep in mind the multifaceted ramifications of past, present, and, hopefully, their own future research. Characterization and fabrication of nanomaterials are the topics of chapters 3 and 4, respectively, and they comprise the nanotools division of the text. The placement of these chapters early in this text is pedagogically motivated. We present images of nanomaterials, analytical data, and synthesis methods throughout the text. Why not understand early on how images and spectra of nanomaterials are acquired? And why not understand early on how nanomaterials are fabricated? In most cases, characterization methods and fabrication procedures predate the formal development of nanotech. The remainder of the text is divided into chapters about properties and phenomena (e.g., physics), synthesis and chemical modification (e.g., chemistry), and, lastly, two chapters that review the remarkable nanomaterials found in the natural world around us (natural nanomaterials and biochemistry).

In the seventeenth century, Sir Isaac Newton provided a fundamental thread that is quite relevant to nanoscience: "If I have seen farther, it is by standing on the shoulders of giants." Newton is credited for making this statement in a letter to Robert Hooke in which he refers to Galileo and Kepler. In the same vein, we all must pay tribute to those who have come before us—those who established the foundation for what we have today—for the benefit of the scientist, technologist, and those properly enlightened with a strong social consciousness.

1.0 Nanoscience and Nanotechnology—The Distinction

We have stated clearly in "Threads" that the focus of Part I is on the science, not the technology. However, especially in the case of "nano," it is often difficult to distinguish between the two. Science involves theory and experiment. Technology involves development, applications, and commercial implications. Both feed on each other (**Fig. 1.1**). We will do our best to adhere to "pure science" without the "applied" component in this text. By necessity, the historical perspectives of both nanoscience and nanotechnology will be presented together, intermingled. Our first boundary therefore already is at risk of "fuzzification."

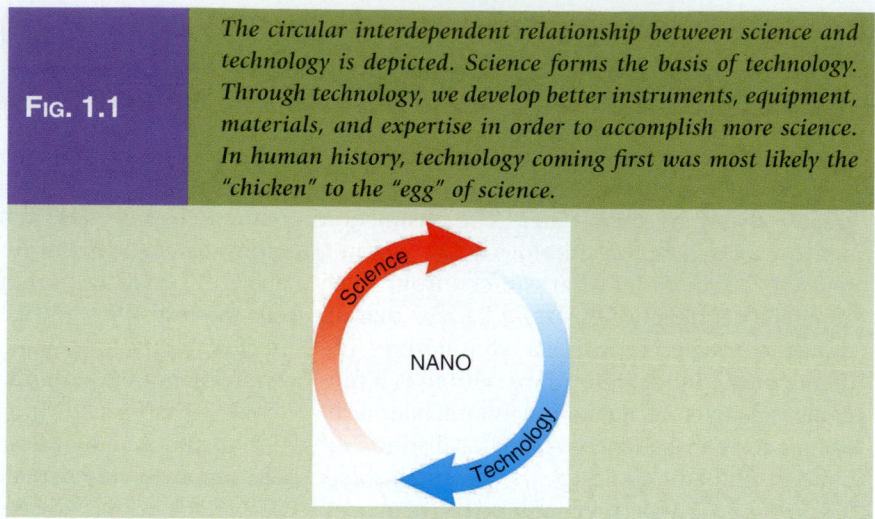

FIG. 1.1 *The circular interdependent relationship between science and technology is depicted. Science forms the basis of technology. Through technology, we develop better instruments, equipment, materials, and expertise in order to accomplish more science. In human history, technology coming first was most likely the "chicken" to the "egg" of science.*

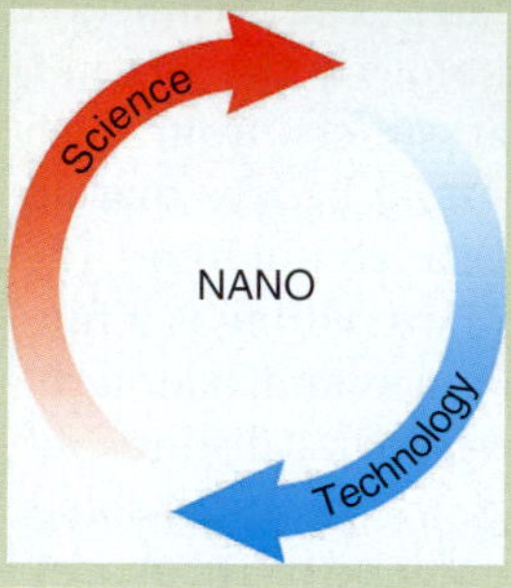

Before we begin to study nanoscience, all of us sincerely hope that you, the student, become inspired by what is to unfold. Pure science is, after all, about truth and the process of digging for the truth—idealistically a good thing for humanity. Democritus is known for saying, "I would rather discover one scientific fact than become king." We unilaterally support Democritus's position. We hope that you are inspired and seriously consider and then pursue a career in nanoscience and technology. Or, we hope that you become involved in a field that is affiliated with nanoscience and nanotechnology outside the laboratory or manufacturing environment. This lofty objective is an underlying goal of this book (both subliminally and overtly stated) with the intent of creating more scientists in a world where technological breakthroughs are needed to help humanity face the grand challenges of the future [1]. The late Nobel laureate Richard Smalley (**Fig. 1.2**) stated directly and sincerely, "Be a scientist—Save the world."

1.0.1 Requisite Definitions

Science requires measurement. Without measurement there is no dissemination of theory and experiment. Civilization requires language. Without language we cannot communicate. Measurement is the language of science. Nanoscience implies a scale of measurement that exists at the level of the nanometer. We begin our journey, then, by submitting fundamental definitions that together yield a sense of the nature of nanoscience and nanotechnology.

The etymological derivation of the prefix *nano* can be traced back to the Latin *nanus* and further back to the Greek root *nan(n)os,* which means dwarf or "little old man." *Nanos* is the foundation of words such as *nana, nanny* (aunt), *nannus* (uncle), and even *nun* [2]. According to dictionary definitions, it is "specialized in certain meanings to mean one-billionth" [3] and has evolved to indicate "extreme smallness" [2]. Nanotechnology therefore is a technology defined by the nanometer. One nanometer is equivalent to one-billionth of a meter (10^{-9} m).

FIG. 1.2 *Nobel laureate and professor Richard E. Smalley of Rice University was a strong proponent of energy development through nanotechnology. He was very passionate about students making a difference in a world that needs technology to help mitigate the challenges that face humanity in the twenty-first century.*

Source: Image courtesy of Rice University Office of Media Relations and Information.

EXAMPLE 1.1 *Nanometer Games*

(a) How many carbon nanotubes 1 nm in diameter can be tightly packed into a cylinder defined by a human hair 100 μm in diameter? Assume packing is done parallel to the long axis and that packing efficiency is not a concern.
(b) Assume that a cubic-shaped transistor in a computer chip has volume of 10 nm^3. How many would fit into a 5-mL drop of water? If currently one billion transistors are fabricated every second, how much time in years is required to manufacture this number of transistors?
(c) Denver is called the Mile-High City. What is its height in nanometers? How far away is our Sun in nanometers?

Solutions:

(a) The area of the cross-section of a 100-μm diameter hair is $A = \pi r^2 = 7.85 \times 10^3\ \mu m^2$. The cross-section area of a nanotube is $7.85 \times 10^{-1}\ nm^2$. The number of tubes that can be packed into this area is $7.85 \times 10^3\ \mu m^2 \div 7.85 \times 10^{-1}\ nm^2 = 1.00 \times 10^{10}$ tubes or about 10 billion tubes.
(b) 5 mL = 5 cm^3. 5 $cm^3 \div$ (1 transistor area · 10 nm^{-3}) = 5×10^{20} transistors. If one billion are made every second today, it would take approximately 16,000 years to make this many transistors. It is safe to assume that this number is greater than that of all the transistors ever made.
(c) Denver is at an altitude of 5280 ft. This is equal to 1.61×10^5 cm, 1.61×10^{12} nm, or approximately 1.6×10^{12} nm. The sun is 92.9×10^6 miles or 1.496×10^8 km from Earth. This is 1.496×10^{20} nm—surely an astronomical number.

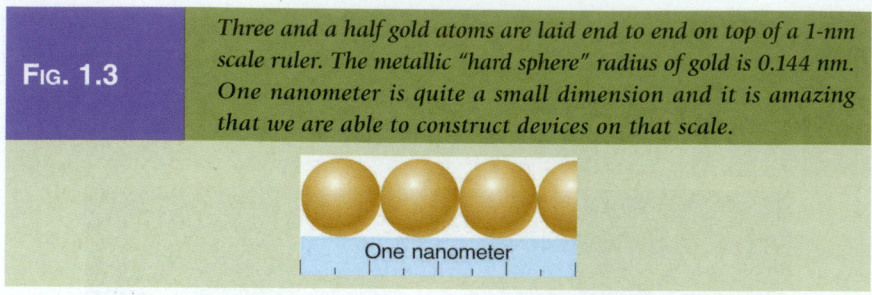

Fig. 1.3 *Three and a half gold atoms are laid end to end on top of a 1-nm scale ruler. The metallic "hard sphere" radius of gold is 0.144 nm. One nanometer is quite a small dimension and it is amazing that we are able to construct devices on that scale.*

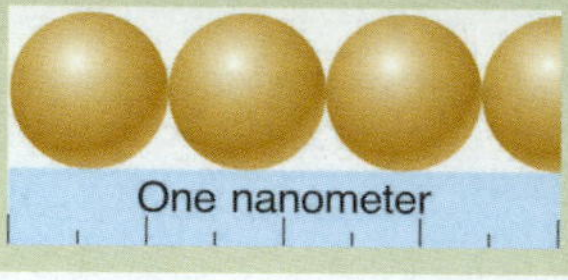

When considering covalent "hard sphere" radii of atoms, 3.5 gold atoms (covalent radius equal to 0.144 nm) placed in a row equal 1 nm [4] (**Fig. 1.3**). A human hair ranges in size from 50 to 100 μm. One nanometer is approximately 1/50,000–1/100,000 as thick as a human hair.

Science (from Latin *scientia*, "knowledge," based on the Greek *skhizein*, "to split, rend, cleave, to separate one thing from another, to distinguish") is defined as "knowledge or a system of knowledge addressing general truths" or the "operation of general laws especially as obtained and tested through scientific method" [3]. Technology (from the Greek *techno*, "art, skill, craft, method, system," based on *tek*, "indicating shape, make" + *logia*, "a study of") is the "practical application of knowledge (science), especially in a particular area" [3]. It is clear to see that science and technology are integrally linked. We will do our best to keep them separate.

1.0.2 Government Line

In February 2000, the National Science and Technology Council (NSTC) Committee on Technology, Subcommittee on Nanoscale Science, Engineering and Technology (NSET) derived the following definition for nanotechnology:

> Research and technology development at the atomic, molecular or macromolecular levels, in the length scale of approximately 1–100 nanometer range, to provide a fundamental understanding of phenomena and materials at the nanoscale and to create and use structures, devices and systems that have novel properties and functions because of their small and/or intermediate size. The novel and differentiating properties and functions are developed at a critical length scale of matter typically under 100 nm. Nanotechnology research and development includes manipulation under control of the nanoscale structures and their integration into larger material components, systems and architectures. Within these larger scale assemblies, the control and construction of their structures and components remains at the nanometer scale. In some particular cases, the critical length scale for novel properties and phenomena may be under 1 nm (e.g., manipulation of atoms at ~0.1 nm) or be larger than 100 nm (e.g., nanoparticle reinforced polymers have the unique feature at ~200–300 nm as a function of the local bridges or bonds between the nanoparticles and the polymer).

A rather cumbersome definition to be sure, but if you are in the process of applying for grant money from the federal government, abiding by these guidelines is a prudent choice. Strict definitions such as these are necessary to filter unwanted solicitations. As with any new technology that is hot and trendy, blatant exploitation of the word is expected. Please recall the buzzword *turbo*. Enter the *nano-pretenders*, those who stretch the concept of nanotechnology beyond any reasonable boundary for the purpose of exploitation. Nano-pretenders are generally business folk or academics that incorporate the prefix nano into a product, company name, or proposal but in reality have nothing or very little to do with it [5]. In fairness to all parties, however, the nebulous nature of nano makes a formal definition rather difficult.

Although this volume is dedicated to nanoscience and not nanotechnology per se, both areas will be addressed early on and used at times interchangeably. Students are encouraged to modify all terms and definitions to suit themselves. Definitions, after all, are basically useful tools that provide a good starting point—a frame of reference. The following obligatory definitions of the nanoscale, nanoscience, and nanotechnology are given to provide a perspective and to lay the foundation for a succinct, meaningful, and luminous 5-min. elevator pitch.

1.0.3 Working Definitions

What is nanotechnology? Nanoscience? **Figure 1.4** illustrates some of the complexity involved in generating a strict definition of such simple words.

FIG. 1.4 *What exactly is nanotechnology? Nanoscience? This schematic illustrates the multiple dimensions, definitions, permutations, and indications of nano above and beyond its relationship to size. The definition of* microtechnology *never underwent such scrutiny or confusion. The definition of* materials science, *when it first appeared on the technological stage several decades ago as a formal academic discipline, was never burdened with such nebulosity or, for that matter, promise.*

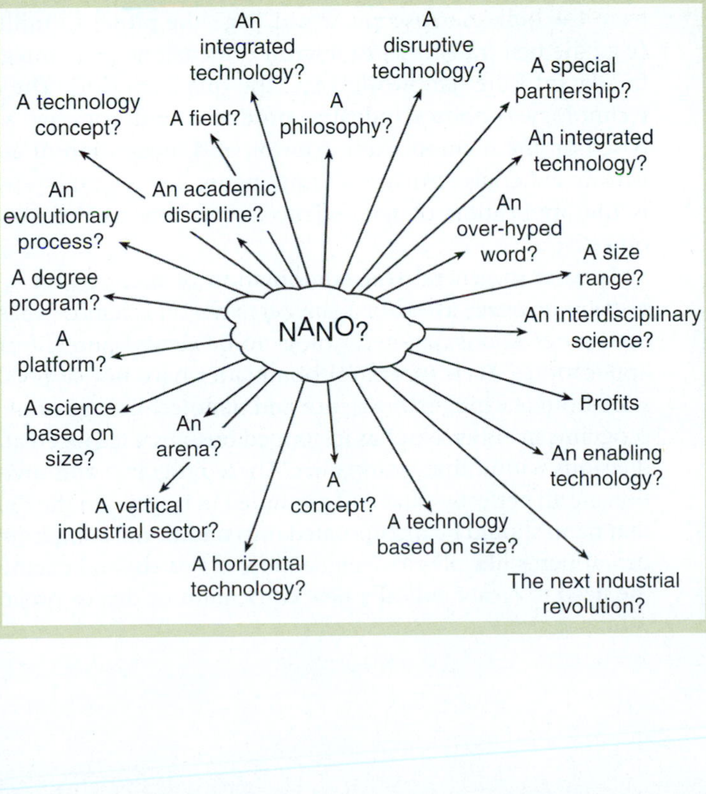

Nanotechnology and nanoscience have different meanings to different people. In the ultimate analysis, a formal definition of nano is not required, but we need to communicate what we discover and manufacture in the language of science.

We begin by defining the nanoscale. By necessity, we must include a broader perspective with regard to size than that given by NSET. Micron-sized particles and those in the submicron realm are generally considered to be bulk materials. In other words, the physical properties of micron-sized materials resemble those of bulk materials; for example, they possess *continuous* (macroscopic) physical properties. It is only when particles assume proportions smaller than ca. 10 nm that the principles of classical physics begin to wobble a bit. We then define the nanoscale to include (1) the size domain less than ca. 10 nm, where materials demonstrate remarkable properties and phenomena apart from the bulk form of that material; (2) larger materials (colloids, thin films) with properties and phenomena between the thresholds of classical physics and the quantum domain; and (3) large molecules that are less than 1 nm in size—and why not include materials that are less than 1 μm in some dimension? We propose the following definition of the nanoscale as it applies to nanoscale science: "The nanoscale, based on the nanometer (nm) or one-billionth of a meter, exists specifically between 1 and 100 nm. In the general sense, materials with at least one dimension below one micron but greater than one nanometer are nanoscale materials."

Nanoscience is the study of nanoscale materials. It is the study of the remarkable properties and phenomena associated with nanoscale materials. Nanoscience is the study of properties and phenomena of materials that are a function of size and size alone. "Nanoscience is the study of nanoscale material—materials that exhibit remarkable properties, functionality, and phenomena due to the influence of small dimensions."

Nanotechnology is the natural progression of technology miniaturization from the bulk macroscopic world (e.g., the plow) to millimeter-sized objects (e.g., the first transistor) to micron dimensions (e.g., integrated circuits), and, finally, into the nanoworld (e.g., the quantum dot). The definition of nanotechnology connotes industry, products, and commerce. "Nanotechnology, based on the manipulation, control, and integration of atoms and molecules to form materials, structures, components, devices, and systems at the nanoscale, is the application of nanoscience, especially to industrial and commercial objectives."

Is nano in general (the combination of nanoscience + nanotechnology) a *platform*, an *arena*, a *field*, or a *frame of reference*? Is nanoscience purely an academic enterprise? And is nanotechnology to be viewed purely from the perspective of applications? Such superficial boundaries have not stopped universities worldwide from teaching nanoscience and nanotechnology or offering special degree programs in either. Nor has it stopped businesses from creating nanotechnology divisions within their companies. These programs and divisions without question are all perfectly valid and grounded in reality. On the flip side, others believe that nano should be incorporated into traditional compartmentalized academic departments like physics, engineering, materials, and chemistry courses without the need to create radically new curriculum or degree programs—once again, a valid position.

FIG. 1.5 *Size scale of nanoscience divisions of physics, chemistry, and biology are expanded to include associated technologies. The graph clearly places the juxtaposition in today's time. From now on, manufacture of very small to very large integrated devices will be made possible by advances in nanoscience and nanotechnology.*

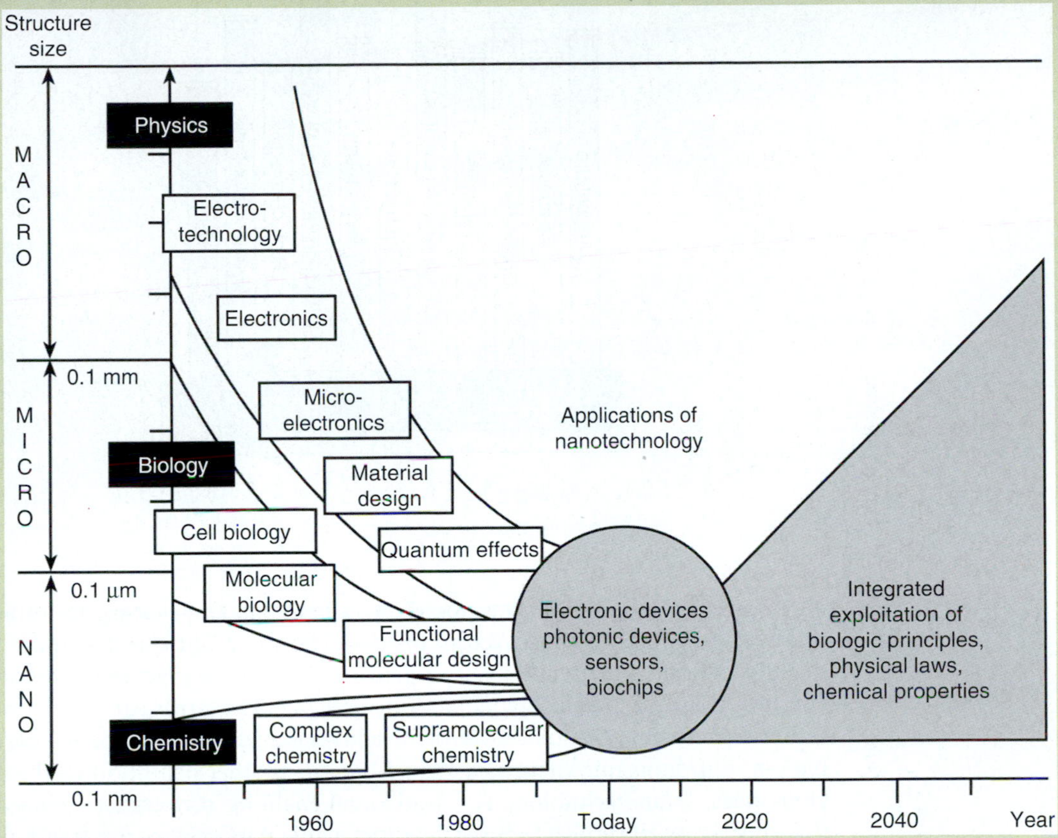

Source: Graph redrawn with permission from VDI-Technology Center, Future Technologies Division—APEC Center for Technology Foresight, Thailand.

The academic platform commonly known as *materials science (and engineering)* comes closest to resembling nanoscience. Although both are inherently interdisciplinary in nature, nanoscience transcends the boundaries of materials science by adding biology and biochemistry into the mix. In addition, materials science, unlike nanotechnology, was never labeled as "the next Industrial Revolution" and societal implications were not a serious part of the materials science equation. A comprehensive graphic relating science, technology, size and a timeline is shown in **Figure 1.5**.

Industrial sectors such as aerospace, biotech, energy, transportation, health care, telecommunications, and information are considered to be vertically oriented (**Fig. 1.6**). Since all industrial sectors depend on materials and devices made of atoms and molecules, by default they can all be improved by application of nanomaterials and nanotechnology. There is no argument with this line

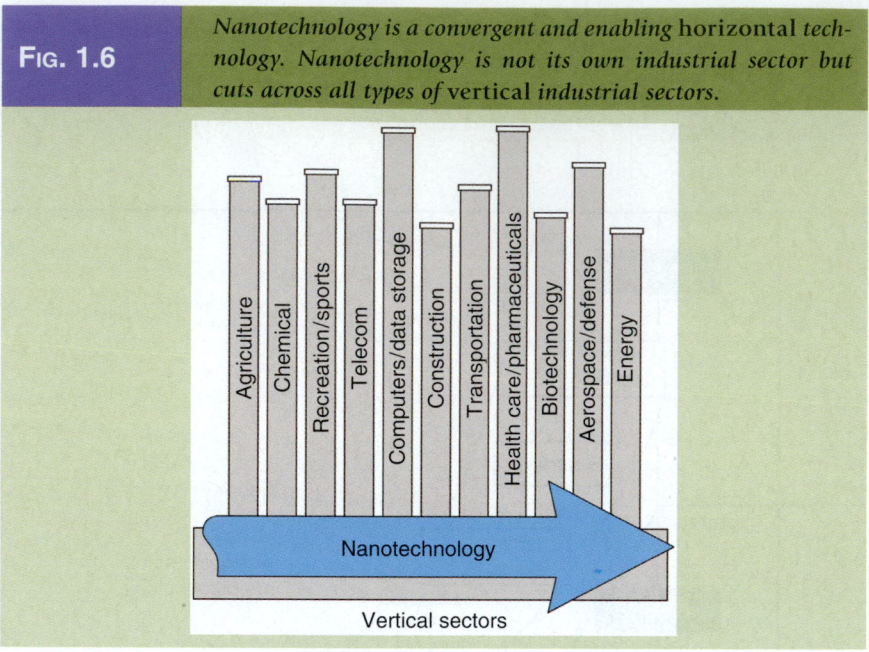

FIG. 1.6 *Nanotechnology is a convergent and enabling* horizontal *technology. Nanotechnology is not its own industrial sector but cuts across all types of* vertical *industrial sectors.*

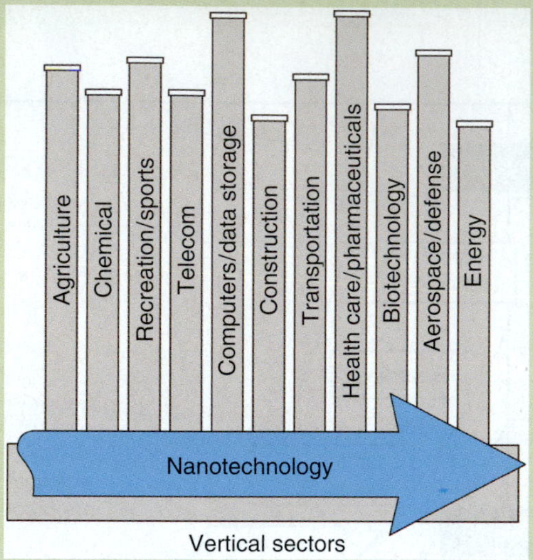

of reasoning. Nanotechnology is, therefore, a *horizontal* and *enabling* technology that will potentially impact ALL industries. The word *convergent* is quite frequently associated with nano. Convergence is the coming together from different directions of previously equal but independent areas of science and technology. The integration of DNA with silicon is an example of convergence of biology with inorganic chemistry. We add, then, another definition to the previous ones: "Nanotechnology is a horizontal-enabling convergent technology that cuts across all vertical industrial sectors, while nanoscience is a horizontal-integrating interdisciplinary science that cuts across all vertical science and engineering disciplines."

Nanotechnology has the potential to impact all products manufactured now and in the future. It has the potential to change the way we all live. This makes nanotechnology a *disruptive* technology. The iron sword of the Hittites, gunpowder, atom bomb, automobile, telephone, penicillin, and computer are examples of disruptive technologies that molded history and changed our lives. These examples, however, are a list of materials and devices rather than a wave of a new type of technology. In this sense, nanotechnology is not a specific entity but rather a generalized form—one that has the potential to disrupt. Nanotechnology, unlike the "dot.com" businesses, requires PhD expertise, significant capital, and a new generation of partnerships to "make it go."

> Nanotechnology is a disruptive technology with a high barrier of entry that will impact the development of materials and devices. Nanotechnology will require that a new genre of partnership be formed among and between business, academe, and government. It will devote study and effort to potential societal implications. Nanotechnology is predicted to significantly impact the wealth and security of nations. Nanotechnology is the next Industrial Revolution.

Societal implications of technology are important now more so than ever. The relationship of nanoscience, nanotechnology, and society will be discussed in more detail in chapter 2. Nanotechnology is expected to exert widespread societal impact. "Nanotechnology is considered, more so than most other technologies, to have great impact on all aspects of culture and society."

We have listed some definitions, platitudes, and phraseology concerning nanoscience and nanotechnology. We have attempted to define the nebulous, the transitory, the ephemeral, and, perhaps, the concrete. Nanotechnology is everything in the basest sense but it is also unique. Nanotechnology has been around for quite some time—thousands of years in our synthetic civilization and billions of years if nature is included. The onus is now upon you, the student, to formulate an impression of this exciting and important field, platform, technology, or business—or word. A Japanese production engineer devoted to accuracy and precision, Norio Taniguchi of the Tokyo Science University, introduced the term *nanotechnology* in 1974. His statement is recorded in the *Proceedings of the International Conference of Production Engineering, On the Basic Concept of NanoTechnology*, 1976. Thus did the word begin.

1.1 Historical Perspectives

Historical transcription of nanotechnology or any science comprises individuals, records, ideas, experiments, tools, applications, and everything outside the science and technology. Historical accounts are a blend of many things. The word *history* itself (from the Greek *historia,* meaning "record, account") is based on the root *histor,* or "wise man." An individual able to grasp all the knowledge of the ancient world was referred to as a *polyhistor*. There are no *polyhistors* of nanotechnology today. As a case in point, a team of contributors was required to compile this textbook. Nanoscience was a part of our heritage well before the formal designation of the concept of nano arrived on the scene. We are at a crossroads, a juxtaposition between the past and present. We ask that you absorb the following historical perspective so that you will be able to stand on the shoulders of the giants and take nanoscience to the next level.

1.1.1 Concept of Atomism

The earliest roots of recorded atomism date back to India 2600 years ago. The *Vaiseshika* philosophical school founded by the Hindu sage Kanada proposed the first record of atomistic theory. The philosophers of the school described atoms as eternal, indivisible, infinitesimal, and ultimate parts of matter [6]. According to the school, if matter could be subdivided infinitesimally, "there would be no difference between a mustard seed and the Meru Mountain" [6]. The size of an atom, from the Buddha biography *Lalitavistara* 2200 years ago, was estimated to be 10^{-10} m [6].

In the West, Anaximander in the sixth century B.C. speculated that the basic element of the universe was an unobservable infinite substance referred to as *apeiron* [7]. Leucippus, a Greek philosopher of the fifth and fourth centuries B.C., invented the concept of atomism (from the Greek *atomos,* uncut: "not" + *tomos,* "cut") [8]. Specifically, he stated that matter is comprised of invisible particles: "imperceptible, individual particles that differ only in shape and position." Democritus (ca. 430 B.C.), Leucippus's student, further developed the atomistic concept and extended the concept of the *apeiron* of Anaximander. Democritus is known for the following atomic maxims [7]:

> Nothing exists except atoms and empty space; everything else is opinion.
> Atoms are indivisible... eternal, unchangeable, indestructible.
> Atoms differ from each other physically, and in this difference is found an explanation for the properties of various substances.

About 100 years later, Epicurus kept Democritus's principle of atomism alive [7].

With the lack of formal experimental methodology, Aristotle and philosophers who followed him adopted other equally intuitive but unfortunately misleading ideas. During the sixth and fifth centuries B.C., Thales and then Anaximenes, respectively, believed that the fundamental elements of the universe are *water* and *air*. Later, Xenophanes and Heraclitus added the elements of *fire* and *earth* to the mix. Empedocles (fifth century B.C.) merged all four and stated that all things are made up of combinations of *earth, air, water,* and *fire* [7]. Certainly this must be the first time in recorded history that a convergence of scientific principles took place—the first grand unification theory! Aristotle adopted and then embellished Empedocles's four-element concept by adding a fifth element, *aether*, described as the ingredient of the heavens [7]. Unfortunately, Democritus's intuitive concept was not to reemerge for two millennia.

In the first century, the Greek engineer Hero postulated that air was compressible and therefore comprised of individual particles separated by space [7]. Unfortunately for Hero, the Aristotelians, who did not advocate experimentation, beat his idea down. It was not for another 1500 years that the principles of Aristotelian thought were finally challenged. Robert Boyle wrote the *Sceptical Chymist* in the 1660s and reintroduced the corpuscular concept of matter [9] **(Fig. 1.7)**: "minute masses or clusters that were not easily dissipable into such particles that composed them."

Because Boyle was privately tutored, he was not influenced by the "didactic Aristotelianism that still victimized most universities" [7]. Boyle reaffirmed Hero's observation that air was compressible: "It must be composed of discrete particles separated by a void." Democritus's concept of atomism emerged once again, but this time it was to prevail.

John Dalton and his "law of multiple proportions" in 1805 reaffirmed that elements are made of atoms [10]. He also stated that atoms of one element are identical to each other and different from atoms of other elements, and that they are capable of combining with other atoms of other elements to form compounds that exist in atomic proportions [10,11]. Lastly, in his treatise he asserted that atoms cannot be divided into smaller particles or destroyed in chemical reactions [11]. The unit of the dalton, represented by "Da," is equivalent to one atomic mass unit. Thus, arguably, began the rise of modern chemistry [10,11].

1.1.2 Colored Glasses

Metal colloids (from the Greek, *kolla*, meaning "glue") provide the best examples of nanotechnology throughout ancient, medieval, and modern times. Metal colloids are the "finely divided" components of catalysts, colored glasses, dyes,

FIG. 1.7 *In his work the* Skeptical Chymist, *Robert Boyle promotes the* corpuscular theory *of matter, thereby laying the foundation for the modern interpretation of the elements.*

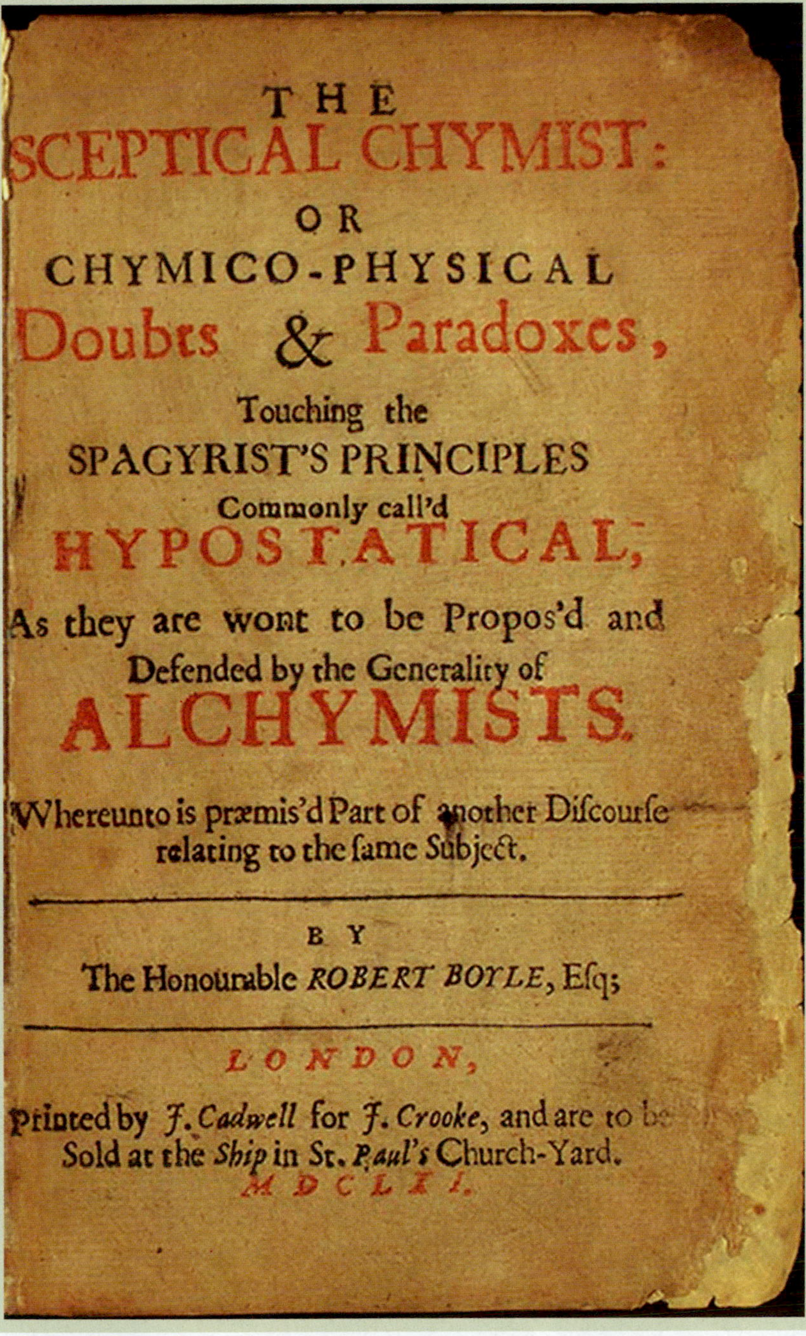

THE
SCEPTICAL CHYMIST:
OR
CHYMICO-PHYSICAL
Doubts & Paradoxes,
Touching the
SPAGYRIST'S PRINCIPLES
Commonly call'd
HYPOSTATICAL;
As they are wont to be Propos'd and
Defended by the Generality of
ALCHYMISTS.
Whereunto is præmis'd Part of another Difcourfe
relating to the fame Subject.

BY
The Honourable *ROBERT BOYLE*, Efq;

LONDON,
Printed by *J. Cadwell* for *J. Crooke*, and are to be
Sold at the *Ship* in St. *Paul's* Church-Yard.
MDCLXI.

and photographic materials, and, eventually, the progenitors of quantum dots. Metal implements fabricated during the Copper, Bronze, and Iron Ages surely were composed of micro- and nanosized metal particles. The same can be said for ceramic materials, dyes, and potions that were developed throughout history. The development of semiconducting colloidal materials also occurred well before the formal embodiment of the *Nano Age*.

A definitive and remarkable piece of Roman glasswork, dating to the fifth century, clearly demonstrates one of the grandest examples of nanotechnology in the ancient world. This magnificent cup, housed in the British Museum, depicts King Lycurgus being dragged into the underworld by Ambrosia. When illuminated from the outside, the cup appears green (**Fig. 1.8a**). When illuminated from the inside, the cup appears crimson red except for the King, who looks purple

FIG. 1.8 *Two versions of the famous Lycurgus Cup, displayed in the British Museum, are shown in this figure. (a) The green appearance is produced by reflection. (b) Red and purple colors are due to transmitted light. The apparent dichroism is due to the interaction of light with gold–silver and copper nanometals and alloys embedded within the soda glass matrix of the cup. King Lycurgus is shown being strangled by the maenad Ambrosia, who was changed into a vine by the goddess Diana. The sad state of King Lycurgus, a Thracian king in 800 B.C., was a consequence of his assault on the god Dionysius and Ambrosia.*

Source: Images reprinted with permission from British Museum Images.

(**Fig. 1.8b**). It was not until 1990 that the specific cause of this color was uncovered after shards of the cup were analyzed with a scanning electron microscope (SEM). Scientists found that the splendid dichroism was due to the presence of nanosized particles of silver (66.2%), gold (31.2%), and copper (2.6%), up to 100 nm in size, that were embedded in the glass matrix [12,13]. The majority of nanometals range from 20 to 40 nm in size. The red color is due to absorption of light by gold at ca. 520 nm. The purple color following absorption is due to larger nanoparticles. The green is attributed to scattering by large silver colloids >40 nm in size. The glass is 73.5% silica, 14.0% Na_2O, 6% lime, and 0.9% K_2O and also plays a role in the optical response of the nanometal-insulator composite glass. The Lycurgus Cup is actually a nanocomposite material.

Following the Romans, medieval craftsman also exploited the addition of metallic constituents to glass to create beautiful stained glass windows. Johann Künckel in the mid-1600s created beautiful ruby-colored glasses from a method to infuse gold into glass [14]. The "purple of Cassius" is a colloidal mixture of gold nanoparticles and tin dioxide in glass [15]. In 1718, Hans Heicher published a complete summary of gold's medicinal uses. In it he discussed how the shelf life of potable gold solutions is stabilized by the addition of boiled starch—perhaps an example of the first ligand-stabilized gold colloids to be synthesized [15,16]. Ancient Chinese cultures are well known for producing beautiful colored glass [17,18]. Chinese porcelain known as *famille rose* contains gold nanoparticles that are 20–60 nm in size. Alchemists have known of colloidal suspensions and potential applications, without providing disciplined experimentation or any intrinsic explanation of them.

Francesco Selmi in 1845 published the first systematic studies of inorganic colloids, particularly those made of silver chloride. He found that a detectable change in the temperature of a solution was absent following coagulation and concluded that such particles could not possibly exist in a state of molecular dispersion [19]. Thomas Graham in 1864, the developer of "Graham's law of diffusion" and considered to be the father of colloid chemistry, discovered that low diffusivity behavior exhibited by colloid solution resulted from their large size (nanometer to micron range) in relation to "ordinary molecules" [20]. Therefore, without having intimate knowledge of nanoparticles, nineteenth century scientists were able to predict the existence of colloids.

It was Michael Faraday (**Fig. 1.9**) in 1857 that first conducted rigorous scientific study concerning the preparation and properties of colloids—those made of gold in particular [21]. The purple color slide he presented at the lecture is shown in **Figure 1.10.** Michael Faraday, in his lecture at the Royal Society of London in 1857, stated that "Gold is reduced in exceedingly fine particles, which becoming diffused, produce a beautiful ruby fluid ... the various preparations of gold, whether ruby, green, violet or blue ... consist of that substance in a metallic divided state."

Not only did Faraday postulate correctly about the physical state of colloids but he also found that the addition of salts transformed ruby-colored gold solutions into blue ones. He demonstrated that such color transformations were prevented by the addition of organic materials such as gelatin. From these experiments and others, Faraday indirectly contributed to organic monolayer chemistry. He also was the first to record the relationship of metal colloids, their surrounding media, and the resulting optical properties.

FIG. 1.9 *A print from the* Illustrated London News *of Michael Faraday at a lecture in 1856. Prince Albert is at the front in the center of the audience.*

Source: Image reprinted with permission from the Whipple Museum of the History of Science, Cambridge University.

In the early twentieth century, Gustav Mie presented his *Mie theory,* a mathematical treatment of light scattering that describes the relationship between metal colloid size and optical properties of solutions containing them [22]. Mie was one of the first to mathematically demonstrate the link between the optical properties of spherical colloids and particle size—one of the prime movers of

FIG. 1.10 *The slide depicting Faraday's gold sol was used in his lecture on nanogold, titled* On the Relation of Gold to Light, *on June 12, 1856. The slide has a reddish-purple color, indicating particles on the order of 20–40 nm in diameter on the average.*

Source: Image reprinted with permission from the Whipple Museum of the History of Science, Cambridge University.

early nanotechnology. In the early 1920s, J. C. Maxwell-Garnett, an outstanding woman scientist who hid her identity from a male-dominated science community, submitted two fundamental papers about the optical properties of metallic colloid solutions. She developed one of the first *effective medium* theories. Effective medium theory relates particle size, shape, orientation, composition, and the dielectric function of the host medium as factors contributing to the optical response of the resulting composite [23,24]. Colored glasses served as one of the first examples of a synthetic nanocomposite material.

1.1.3 Photography

Images were first observed in the *camera obscura*, a phenomenon clearly described by Leonardo da Vinci in 1490 and by Chinese philosopher Mo-Ti (Mo-Tsu) in the fifth century B.C. [26]. Generating an image was one thing; however, recording that image was quite another. Thomas Wedgewood in the early 1800s compiled the observations and work accomplished by numerous scientists throughout the ages to establish the beginning of modern photographic science [25]. His effort demonstrated the first union of interdisciplinary nanoscience followed by a nanotechnology (or was it the other way around?).

Robert Boyle reported that silver chloride blackened upon exposure to the atmosphere; he mistakenly believed that it was exposure to air and not light that caused the reaction. Chemists understood only later that light is able to interact with specific substances. The darkening reaction silver salts undergo when exposed to sunlight has been known since 1585 [25]. In 1727, J. H. Schulze discovered that silver nitrate, when mixed with chalk and placed under stenciled letters, darkened when exposed to light. We should also wonder, as potential nanotechnologists, whether the chalk used in Schulze's experiment was in the form of micro- or nanosized particles. If true, then this would be another example of a human-made nanocomposite. Sir Humphry Davy suggested that silver chloride, a substance more sensitive to light, be used instead [25]. Unbeknownst to the researchers of the time, silver chloride decomposes and is transformed into finely divided silver particles after exposure to light.

In 1839, French chemist Louis J. M. Daguerre was able to piece many aspects of the photographic puzzle together and created the first commercially viable print, the daguerreotype [25]. Several steps were involved in making a daguerreotype print. First, a layer of photosensitive silver iodide was exposed to light. Development of the layer was accomplished over a cup of mercury at 75°C, creating an amalgam. The latent image thus formed was fixed by immersion into a hyposulfite solution. In 1840, William F. Talbot developed a means to coat paper with silver chloride to create the first negative image. This procedure was the earliest way to reproduce images [25]. Ironically, the foundation of photography was based on chemical nanotechnology. We now capture our images with a new form of nanotechnology, as digital nanotechnology displaced the Polaroid camera in quite a disruptive manner.

1.1.4 Catalysis

The ancient Babylonians 5000 years ago stored soap-like materials in clay pots. Writing deciphered on the pots indicated that these materials mere made by boiling fatty substances along with ash. The Egyptians combined animal and vegetable oils with alkaline salts to form soap-like materials for medicinal purposes as well as for cleansing. The word *soap* comes from the Roman period; Mt. Sapo was the site of altar sacrifices where animals were slaughtered and burned. Melted animal fat (tallow) and ashes were washed down into the river. Servants discovered that soapy materials embedded in clay sediments along the riverbank downstream from the mountain aided in washing clothes. The Gauls during the time of the Roman Empire over 2000 years ago are formally credited with inventing soap [27]. Soap is the product of a catalytic process. When boiled in the presence of wood ash (the alkali catalyst), animal fat (the substrate) yields soap and glycerin. The formal investigation of catalytic processes, however, did not occur for another 1400 years. Thus, over the course of several millennia, both catalysts and surfactants were discovered; both materials play a major role in modern nanoscience.

Near the turn of the eighteenth century, Gottlieb Kirschof demonstrated the first controlled catalytic reaction (e.g., the acid hydrolysis of starch to produce glucose [28]). In 1836, J. J. Berzelius coined the word *catalysis* (from the Greek *katalusis,* meaning "dissolution," from *kata,* "down," + *lysis,* "a loosening") to describe the process. Catalysis is the acceleration of a chemical process by a substance known as a catalyst—a material not consumed in the reaction and, ideally, reusable [28]. Heterogeneous catalytic reactions consist of a solid phase catalyst and substrates that exist in liquid or gaseous states. Heterogeneous catalysis is the type of catalysis most important to nanotechnology. The solid phase catalytic particles are usually micro- to nanosized metals made of iron, nickel, or cobalt. Substrates are usually carbon-containing compounds.

Humphry Davy in the early 1800s observed that a heated platinum wire in the presence of air produced water. This is one of the first experiments to demonstrate heterogeneous catalysis [27]. J. J. Berzelius also observed that platinum accelerated chemical reactions without being consumed [27]. Nobel laureate Paul Sabatier discovered the process of catalytic hydrogenation in the early 1900s. He ascertained that catalysts were composed of "finely divided" particles [29,30]—"Finely-divided metal hydrogenation catalysts subsequently formed the bases of the margarine, oil hydrogenation, and synthetic methanol industries."

Heterogeneous catalysis made a significant impact on the chemical and petroleum industries in the mid-1900s. With synergistic breakthroughs in the automobile, oil, and plastic industries to follow, heterogeneous catalyst technology secured a prominent place in the development of nanomaterials to come. We can therefore conclude, with much confidence, that a minitechnological revolution was already intact, thriving, and based on nanotechnology well before the word *nano* became popular.

Interestingly, a pesky by-product of the catalytic decomposition of hydrocarbons also became an unintended contributor to nanotechnology [31]. Carbon fibers, multiwalled herringbone carbon tubes, turbostratic graphite, and amorphous carbons were all considered undesirable by-products of the catalytic decomposition of hydrocarbons. These *soots* typically poisoned the metal catalyst and terminated the primary chemical reaction [31]. How many times have single-walled carbon nanotubes (SWNTs) unknowingly been produced in catalytic decomposition reactions? There is no recorded evidence of nanotubes until the advent of the electron microscope in the 1930s, and in those days the resolution was not good enough to observe a single-walled carbon nanotube. Ironically, the catalytic production of single-walled carbon nanotubes is now a goal and not an undesired byproduct.

Along with carbon nanotubes, the study of catalysis has expanded our knowledge of nanoscience. Properties associated with nanomaterials, such as high surface area, high surface energy, and unique molecular structure, are better understood. We can now, for example, use nanoscopic gold in catalysis. Gold typically is not considered to have catalytic function.

1.1.5 Integrated Circuits and Chips

The drive to minimize electronic components contributed to the onset of the *Nano Age*. J. Bardeen, W. Shockley, and W. H. Brattain invented the transistor on December 23, 1947, and ended the reign of the bulky, heat-generating vacuum tube in electronic amplifiers and other devices. Brattain et al. described their first transistor in their journal [32]:

> Two hair-thin wires touching a pinhead of a solid semi-conductive material soldered to a metal base are the principal parts of the Transistor. These are enclosed in a simple, metal cylinder not larger than a shoelace tip. More than a hundred of them can easily be held in the palm of the hand.

One hundred can be held in the palm of the hand! Amazing! In that time, it was most certainly an amazing accomplishment. Bardeen and his teammates were awarded the Nobel Prize in physics in 1956 for their discovery. Bardeen was awarded another Nobel Prize in physics later for his work on superconductivity.

The first silicon transistor was created in 1951 and the first integrated circuit (IC) was made in 1958. Integrated circuits have shrunk in size over the past 50 years to the level where they certainly qualify as nanotechnology under the definition promoted by NSET that was given earlier. The Intel Corporation marketed a transistor in 2003 that was smaller than the human influenza virus (80 vs. 100 nm, respectively) (**Fig. 1.11**) [33]. Advances in reducing resolving power of lithographic techniques have made it likely that components will become smaller. Lithography will be discussed in more detail in chapter 4, "Fabrication Methods."

Gordon E. Moore helped found Intel (*Int*egrated *El*ectronics Company) in 1968. In 1965, he derived, with input from Douglas Engelbart of the Stanford Research Institute, "Moore's law" [34]. A version of Moore's law is graphically depicted in **Figure 1.12** and is generally summarized: "At our rate of technological development, the complexity of an integrated circuit, with regard to minimum component cost, will double every eighteen months."

The current interpretation of Moore's law states that data density will double every 18 months. In 1965, 30 transistors populated the chip; in 1971, 2000 populated it and ca. 40 million do so today. Reality tracked fairly well with this prediction except that the doubling actually occurred every 2 years rather than every 18 months. How will nanotechnology influence Moore's law? According to the National Nanotechnology Initiative, transistors must be "scaled down" to at least 9 nm. In order to keep pace with Moore's law, downsizing to a 9-nm transistor would result in billions of transistors on a chip by 2016.

FIG. 1.11 *(a) A vintage 2003 transistor fabricated by the Intel Corporation is compared to the human influenza virus. We are now able to make devices smaller than one of the smallest "complete" biological structures. (b) The decreasing trend in transistor size is shown. By 2017, transistors under 10 nm in size are expected to be components in chips.*

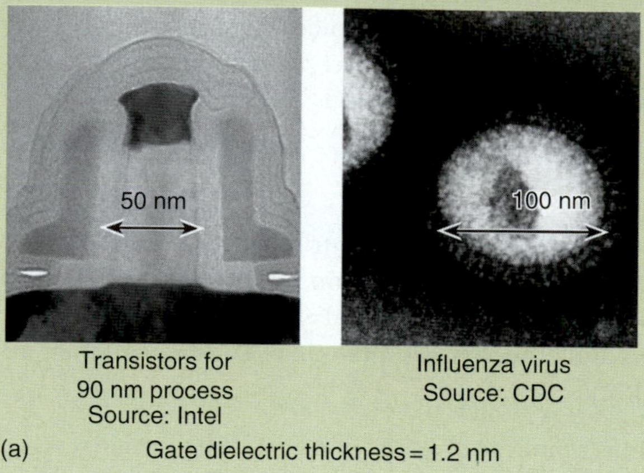

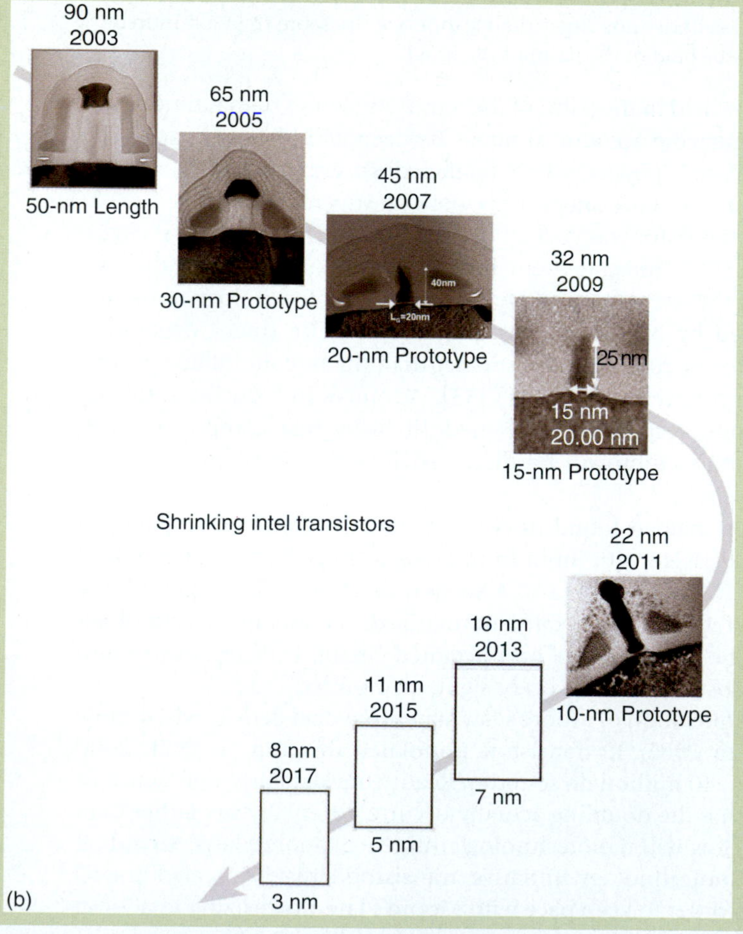

Source: Images reprinted with permission of George Thompson, Intel Corporation.

FIG. 1.12 *An adaptation of Moore's law shows how computing power has increased since the onset of the first mechanical computing devices.*

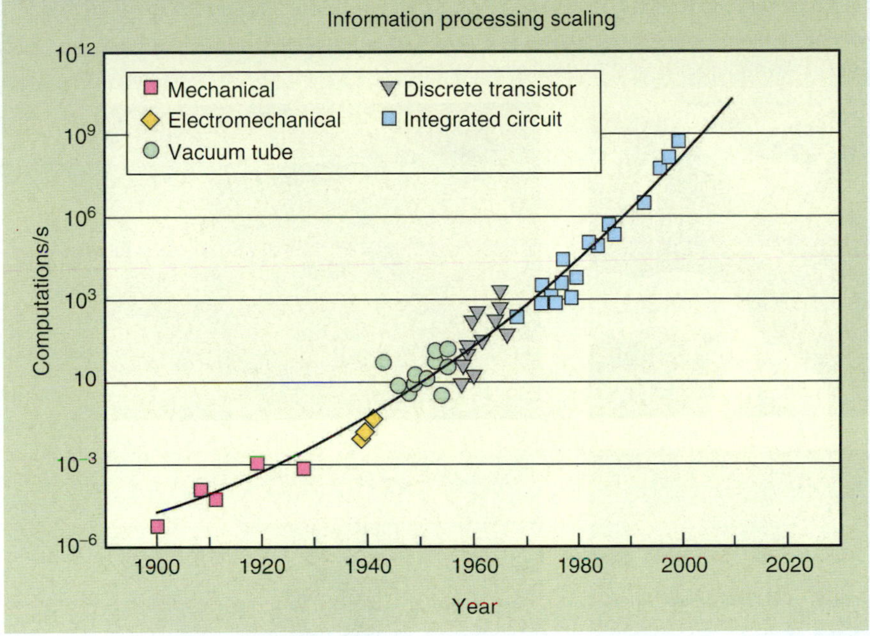

Source: Image courtesy of the IBM Corporation. With permission.

1.1.6 *Microelectromechanical Systems*

Microelectromechanical systems (MEMS) are devices that contain micron-sized components. MEMS devices are fabricated by conventional integrated circuit batch processing techniques. With the onset of ICs and chips and the development of lithography, MEMS production has burgeoned over the past 20 years. Development of micromachined silicon processes led to the development of electronic comb actuators and micropositioning disc drive heads. The automotive and defense industries in particular exploited MEMS technology to produce advanced sensors and actuators for fuel pressure, tire pressure, air conditioner compressors, and exhaust gas sensors. The accelerometer used in airbags is a MEMS device. MEMS devices have long since been applied to medical, environmental, aerospace, electronics, and other industries.

A relatively new area of technology, called NEMS (nanoelectromechanical systems), is an example of the evolutionary trend to make smaller and smaller components. A compilation of MEMS and microlithographically formed devices is shown in **Figure 1.13a–g.**

1.2 Advanced Materials

Examples of nanomaterials are evident throughout the twentieth century. The oxide layer formed on anodized aluminum, for example, is a nanostructured

Fig. 1.13 *Images of microelectromechanical system (MEMS) devices are shown in the figure. (a) Drive mechanism with dust mite on tiny gears, (b) Close-up of drive-gear hub of a microengine, (c) Single-piston microsteam engine, (d) Indexing motor, (e) Grain of pollen and red blood cells on a drive-gear linkage of a microengine, (f) Torsional ratcheting actuator that uses rotationally vibrating (oscillating) inner frame to ratchet surrounding gear, and (g) Six-gear chain capable of 250,000 rpm.*

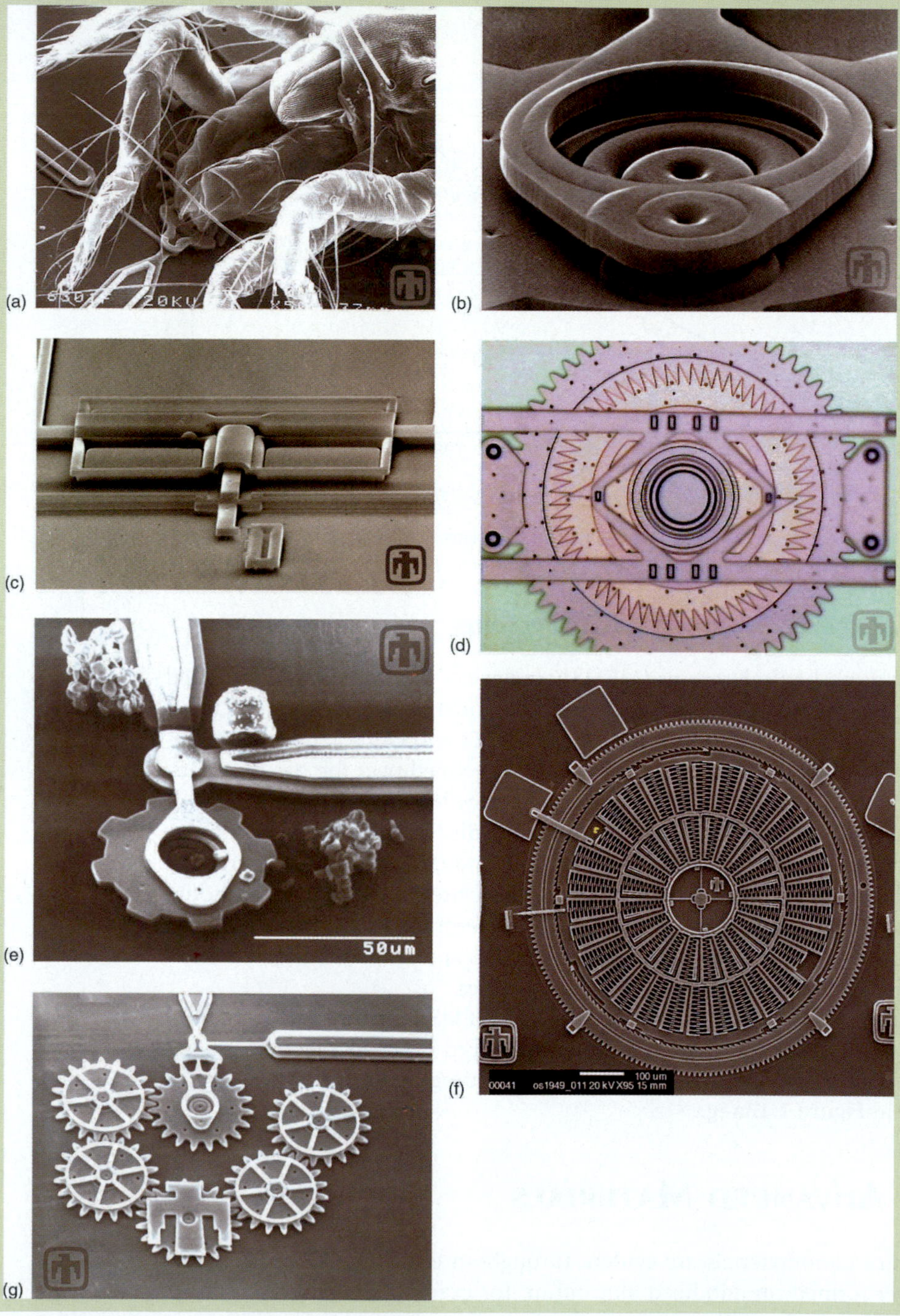

Source: Images courtesy of Sandia National Laboratories, SUMMiT Technologies, www.mems.sandia.gov. With permission.

Fig. 1.14 *Atomic force microscope (AFM) image of a porous alumina membrane surface is shown. The pores are ca. 50 nm in diameter. The film was processed in a 10 wt.% oxalic acid solution at 0°C under 30 V conditions. The hexagonally packed pore channel structure is clearly depicted. Alumina membranes such as this form perfect templates for the synthesis of nanomaterials.*

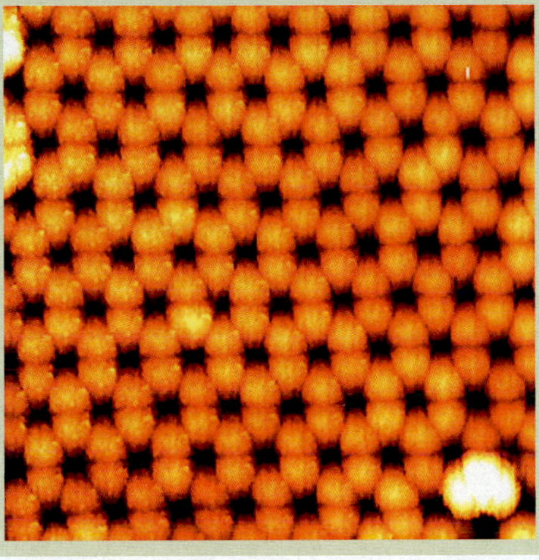

Source: Image courtesy of the National Renewable Energy Laboratory in Golden, Colorado. With permission.

material. The process of anodizing was patented in the United Kingdom in 1926 and then acquired by Alcoa in the early 1930s. Nearly every aluminum panel or part that is used in industry applications has an anodized finish. Anodization is a simple process that provides a protective oxide layer on aluminum to prevent corrosion. Little did the developers know at the time that the anodic layer was composed of hexagonally close-packed channels with diameter ranging from 10 to 250 nm or greater. The aspect ratio (length to diameter) of the channels ranges from 1 to as high as 10,000. Anodically formed porous alumina membranes (**Fig. 1.14**) are used extensively in template synthesis of nanomaterials. Anodic films are discussed more fully in chapter 12 [35].

Other familiar examples include nanoparticles that are found in the rubber component of automobile tires, titanium dioxide pigment found in white paint, chip components in computers, large synthetic biomolecules, polymer monomeric units, nanoparticle slurries for polishing, molecular sieves, ceramic cells and structure, and many more. The recreation industry has already taken advantage of nanomaterials. Multiwalled carbon nanotubes are routinely used as filler material in polymer composites. Single-walled carbon nanotubes (**Fig. 1.15**) can be woven into ropes that exhibit strength greater than that of silk. An image of a *Babolat* tennis racquet reinforced with carbon nanotubes is shown in **Figure 1.16**.

1.2.1 Thin Films

Chemists state, perhaps somewhat ostentatiously, that the practice of nanotechnology originated in chemistry labs long before nanotechnologists claimed to

Fig. 1.15 *Single-walled carbon nanotubes (SWNTs) are arranged in a hypothetical three-dimensional superstructure. Materials reinforced with carbon nanotubes promise to be the strongest known to science.*

Source: Image courtesy of Anil K. Rao, Department of Biology, the Metropolitan State University, Denver, Colorado. With permission.

be nanochemists. Chemists in general manipulate large numbers of atoms and molecules from the "bottom up" to form bulk materials without much consideration of the nanoscale. A major exception to this rule is perhaps found in colloid chemistry, where nucleation and precipitation mechanisms are most certainly nanoscale phenomena. The mainstream chemist designs and directs reactions that require the rupture and formation of strong chemical bonds under kinetic control. The nanochemist, on the other hand, is more likely to conduct synthesis of nanomaterials under thermodynamic control with the formation and dissolution of weaker chemical bonds. Under thermodynamic control, weak bonds are made and broken under mild conditions. More discussion about the distinction between the two types of reactions is presented in chapters 10 through 12.

Thin film technology is a nanotechnology that has been practiced for millennia [36]: "Thin film technology is simultaneously one of the oldest arts and one of the newest sciences. Involvement with thin films dates to the metal ages of antiquity." The ancient Egyptians were capable of producing gold leaf as thin as 3 μm or less nearly 4000 years ago. This was accomplished by physical means such as rolling and beating. Inca goldsmiths practiced "depletion gilding," a process whereby gold is enriched at the surface, over 2000 years ago. The gold leaf

Fig. 1.16 *Recreational applications of nanomaterials include the Babolat tennis racquet depicted in the figure. The racquet's structure at the throat of the frame is reinforced with nanocarbon technology that affords the player better power, control, and stability. "NS" stands for nanostrength technology.*

Source: Image reprinted with permission from Babolat USA.

industry still thrives today and, by means of "machine beating," foils 50 nm thick or less can be attained [36].

Benjamin Franklin in 1773 recorded the unique relationship between oil and water after noticing that the wakes of ships' waters were calmed after cooks dumped cooking grease overboard, as was the custom [37]. John W. Strutt,

Fig. 1.17 *Lord Rayleigh (a.k.a. John William Strutt) was the third Baron Rayleigh. He also happened to be one of the leading scientists of his day. He was capable of "understanding everything just a little more deeply than anyone else."*

Source: Image courtesy of the Nobel Foundation, NobelPrize.org. With permission.

better know as Lord Rayleigh (**Fig. 1.17**), in 1891 studied the surface tension of contaminated waters. He also investigated and described the scattering of light, a fundamental phenomenon responsible for Raman spectroscopy. In 1924, Lecompte du Nuoy, inventor of the de Nuoy ring surface tensiometer, estimated Avogadro's number to be 6.004×10^{23} by application of a thin film of sodium oleate onto water [38]. The amphiphilic molecules formed an organic monolayer on the surface of the water.

One of the greatest influences behind surface chemistry (monolayers in particular) was Irving Langmuir, who, along with Katherine Blodgett, developed Langmuir–Blodgett technology. Langmuir was the first to study monomolecular films and for his effort was awarded the Nobel Prize in chemistry in 1932. While working at General Electric Research Laboratory in Schenectady, New York,

Katherine Blodgett, the first female scientist to be hired by GE, developed the first "invisible glass" in 1938, an application of nanometer thin films. The glass was invisible because it did not reflect light due to the interference effect produced by the 44-molecule-thick film. She also developed a way of measuring thin films by means of interference colors generated by varying the thickness of stearic acid monolayers [39].

In 1980, Jacob Sagiv of the Weizman Institute in Israel found out that octadecyltrichlorosilane would react spontaneously with a glass surface. In 1983, David Allara of Bell Labs found that alkanes containing thiol groups would self-assemble on a gold surface into a monolayer. George Whitesides of Harvard (**Fig. 1.18**) also

Fig. 1.18 *George M. Whitesides was the Mallinckrodt Professor of Chemistry from 1982 to 2004 at Harvard University. He is currently the Woodford L. and Ann A. Flowers University Professor. Dr. Whitesides made significant contributions to nanoscience in general but specifically in the field of molecular self-assembly. The underlying theme of his lab is to* fundamentally change the paradigms of science.

Source: Image courtesy of George Whitesides, Harvard University. With permission.

contributed significantly to the understanding and development of monolayer science and technology. Colin D. Bains, a coauthor of Whitesides's landmark paper in 1989 [40–42], found that "stable surfaces exposing a single organic functional group could be created by adsorption of thiols on gold—for a dozen different groups, from non-polar methyl and CF_3 groups to halides, nitriles and carboxylic acids ... and how these surfaces could be used to predict, tailor, and control wettability."

Thin film chemistry is a central component of nanoscience and nanotechnology and represents a class of fabrication known as "bottom-up" synthesis. Thin films are characterized as two-dimensional materials and their chemistry is introduced in chapter 12; they are also classified as two-dimensional (2-D) quantum structures—a material with length and width but without any apparent thickness (chapter 10).

1.2.2 Fullerenes and Carbon Nanotubes

In 1985, R. E. Smalley, H. W. Kroto, and R. F. Curl discovered a third major form of elemental carbon: the *buckyball*, a molecule consisting of 60 atoms of carbon, C_{60}, assembled in the form of a soccer ball pattern [43]. This new allotrope of carbon, formally called *Buckminsterfullerene*, was named after R. Buckminster Fuller, the architect best known for the geodesic dome (**Fig. 1.19**). Buckyballs were discovered somewhat serendipitously after vaporization of a carbon target with a high-powered laser and then channeling the products into a mass spectrometer (MS). The MS signature of C_{60} and other fullerenes was unmistakable. The primary goal of the Kroto and Smalley group was to create long chain carbons similar to those detected around carbon stars. Little did they know at the time that a Nobel Prize was in the offing or that their discovery would change the face of nanotechnology.

FIG. 1.19 *Buckminster Fuller was an architect known for his geodesic dome design. Carbon C_{60} molecules are called buckminster fullerenes in his honor.*

Source: Image courtesy of the United States Postal Service. With permission.

If one slices a buckyball in half and then extrudes a tube from the hemispherical structure (resembling chicken wire), a single-walled nanotube (a buckytube or SWNT) is formed. Depending on which symmetry axis of the fullerene is chosen—whether projecting from the pentagon apex or the hexagon apex of a C_{60} molecule—different types of SWNTs are formed. Sumio Iijima, of NEC Corporation, reported in the journal *Nature* about his experimental validation of multiwalled carbon nanotubes (MWNTs) [44]:

> Here I report the preparation of a new type of finite carbon structure consisting of needle-like tubes. Produced using an arc-discharge method.... The formation of these needles, ranging from a few to a few tens of nanometers in diameter, suggests that engineering of carbon structures should be possible on scales greater than those relevant to the fullerenes.

In 1993, Iijima described the synthesis of SWNTs [45]. The discovery of carbon nanotubes, just like with fullerenes, was not new. J. R. Rostrup-Nielsen in 1972 reported thermodynamic parameters of carbon fibers from his studies on carbon fibers formed by catalytic chemical vapor deposition reactions [31]. Roger Bacon, a Union Carbide scientist in 1960, determined that samples from a chemical vapor deposition process yielded certain types of MWNTs [46]. Later, in 1979, MWNTs were produced in carbon arc discharge experiments. Theoretical work describing the mechanisms of carbon nanotube formation and structure soon followed [47].

SWNTs are called *quasi*-one-dimensional (1-D) quantum structures or, more accurately, pseudoquantum wires. One-dimensional structures have length but no height or width—a line consisting of infinitely small points (chapter 7).

1.2.3 Quantum Dots

Quantum dots, known as artificial atoms, are called zero-dimensional materials (0-D) (chapter 7). Electrons in a zero-dimensional material are "confined" in all three physical dimensions: the *x*, *y*, and *z*. Quantum dots are quite small, usually less than 10 nm. Quantum dots exhibit properties that are intermediate between the bulk and atoms or molecules. A metal cluster, a type of quantum dot, has characteristic "lowest occupied cluster orbitals" (LOCOs) and "highest unoccupied cluster orbitals" (HUCOs). The electronic properties of colloids, on the other hand, resemble those of the bulk material. Quantum dots are usually smaller than 10 nm and exhibit relatively monodisperse size distribution. Colloids are generally polydisperse with respect to size and shape and are much larger [48,49].

The history of the quantum dot takes us back to the *Stone Ages*, when early researchers developed colloidal pigment elements for wall art. The ancient Egyptians were known to add colloidal constituents into inks to give color and consistency. Medieval artists and craftsmen utilized metal colloids in glasswork. The beautiful stained glasses found in cathedrals certainly were able to impart a sense of other-worldliness due to nanotechnology. Chromium and iron nanoparticles were responsible for Maya blue pigment found in Mayan artifacts and art at the Chichén Itzá site in the Yucatan Peninsula [50].

As we know, the utilization of colloidal metals is well represented in the *Nano Age*. One of the leading nanopioneers in this field is Professor Günter Schmid of the University of Essen (now Duisburg-Essen) in Germany. The development of ligand-stabilized Au-55 quantum dots, transition metal clusters, and bimetallic catalysts offers examples of some of the important developments to emerge from his laboratory [51,52]. Applications of clusters include catalysis, bio-imaging, nano-electronic circuitry [53], and information storage. Both metal and semiconductor clusters have been inserted into porous alumina membranes [54] or secured otherwise into two-dimensional arrays [53]. Ligand-stabilized gold quantum dots comprising 5 to as many as 55 or more gold atoms have been produced experimentally [55]. More discussion will be presented about Au-55 in chapter 12.

Semiconductor quantum dots also possess a rich history. Victorian period glasses that appear red, red-orange, or orange were found to contain nano-crystallites of Zn or Cd sulfides and selenides [56]. Modern era investigations into quantum dots began in earnest in the early 1960s and continued more extensively in the 1970s and 1980s. In 1960, W. D. Lawson and his colleagues wrote a paper titled "Influence of Crystal Size on the Spectral Response Limit of Evaporated PbTe and PbSe Photoconductive Cells" [57]. In the 1970s and 1980s at Bell Labs and other places, A. Efros, L. Brus, M. Bawendi, and P. Alivisatos contributed extensively to the understanding of the role that particle size plays in the manifestation of physical properties [58–62]. In 1982, A. I. Ekimov's paper titled "Quantum Size Effect in Three-Dimensional Microscopic Semiconductor Crystals" describes the importance of size in determining quantum dot properties [62]. In those days, the word *nano* was used sparingly or not at all and most likely in the context of a measurement. An image of CdSe quantum dots synthesized in the laboratory of Professor Zhiqun Lin at Iowa State University is shown in **Figure 1.20.** The vivid fluorescent colors are dependent on the size of the quantum dots in solution.

The drive to develop solar cells also contributed to the development of quantum dots. The Solar Energy Research Institute (SERI, name changed to the National Renewable Energy Laboratory, NREL) was formed in 1978 under the Carter administration. NREL scientists like Arthur Nozik are considered to be pioneers in applications of quantum dots to the field of solar conversion [59]. NREL physicist Alex Zunger (**Fig. 1.21**) contributed significantly to the theoretical understanding of the quantum machinery responsible for the behavior of quantum dots. Michael Grätzel of the Swiss Federal Institute of Technology in Lausanne developed a solar cell based on TiO_2 nanoparticles [63].

1.2.4 Other Advanced Materials

Ferrofluids. In 1960, NASA researchers discovered that ferrofluids could be used in the attitude control devices of spacecraft. A ferrofluid is a liquid containing magnetic nanoparticles that reacts strongly to a magnetic field. Ferrofluids lose their magnetic properties once the field is discontinued. Nanoparticles, however, tend to aggregate (chapter 6). Ligand stabilization of nanomagnetic particles made of iron was prevented from aggregation in polar or nonpolar liquids. Ferrofluids have numerous applications in the electronic, defense, aerospace, metrology, medicine, and automotive industries.

Zeolites. Charles Plank and Edward Rosinski invented a means of converting petroleum base into gasoline by the use of inorganic catalytic

FIG. 1.20 *The wavelength of quantum resonance fluorescence emission of cadmium selenide quantum dots is related to the size of the nanoparticle. The emission of longer wavelengths is indicative of larger size.*

Source: Image reprinted with permission from Professor Zhiqun Lin, Department of Materials Science and Engineering, Iowa State University.

materials called zeolites. Zeolites have been used for decades to filter water out of nonpolar liquids and to separate gases.

DNA–metal inorganic structures. In 1996, C. Mirkin and R. Letsinger of Northwestern University in Chicago attached DNA to gold nanoparticles. This breakthrough helped bring on a new generation of bioinorganic nanostructured materials. The convergence of biology, chemistry, physics, and engineering is apparent in the physical expression of these advanced materials.

1.3 Tools of Nano

Tools (from Old English *tol,* meaning "instrument, implement, prepare") are physical devices that are used to carry out a particular function. The tools of nanotech are the instruments that measure, observe, and manipulate atoms and molecules. Without such tools, there would be no nanoscience, no nanotechnology. The tools of nano are at the heart of nanoscience and nanotechnology. The development of instruments and devices runs parallel to the history of nano and will mold the shape of things to come and perhaps be molded in return.

G. Binnig and H. Rohrer discovered the scanning tunneling microscope (STM) and the atomic force microscope (AFM) in 1982 and 1986, respectively [64,65]. In 1989, courtesy of Donald Eigler and his group at IBM Almaden in

Fig. 1.21 *Alex Zunger of the National Renewable Energy Laboratory (NREL) in Golden, Colorado, made (and is still making) significant contributions to the understanding of quantum dots. Dr. Zunger received his PhD from Tel-Aviv University in Israel and is one of the most cited physicists today. Among his many contributions to physics, Dr. Zunger is known for developing novel theoretical methods for calculating electronic properties of quantum dots.*

Source: Image courtesy of Alex Zunger, National Renewable Energy Laboratory, Golden, Colorado. With permission.

California, 35 xenon atoms were moved about on a nickel (110) surface at low temperature by a homemade STM to spell out the "IBM" logo (**Fig. 1.22**) [66,67]. The "control, manipulation, and integration" of xenon atoms, according to our definition, has been accomplished in spectacular fashion, and the *Nano Age* was formally launched about 30 years after Richard Feynman's bold prediction in 1959 [68]. *Time Magazine* featured the IBM accomplishment in its April 16, 1990, issue [69]. In 2005, in an interview with *Nanooze* (a kids' science

Fig. 1.22 *Thirty-five xenon atoms on a nickel (110) surface at ultralow temperature were placed to spell "IBM" with the aid of an STM by Donald Eigler and his group at the IBM Almaden Research Center. The actual writing took 22 hours to complete. The image was published in* Time Magazine *in 1990 and formally ushered in the* Nano Age.

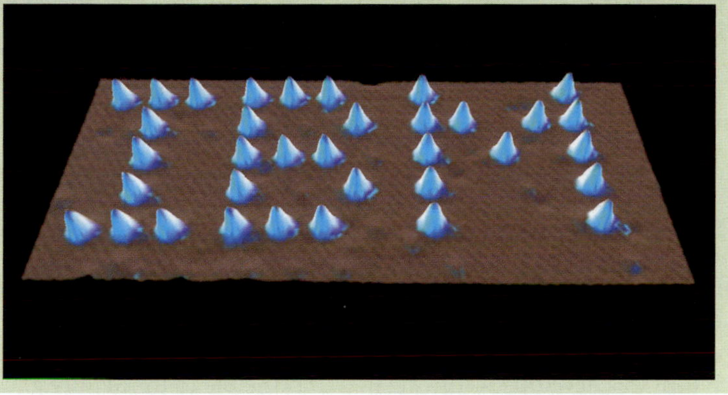

Source: Image reprinted with permission from IBM Research, Almaden Research Center.

magazine sponsored by the National Nanotechnology Infrastructure Network [NNIN]), Donald Eigler said,

> It all depends upon what you call nanotechnology. Rather than get everyone confused about what is or what is not nanotechnology, let's just think about the role that nanometer-scale structures play in our lives and are likely to play in the future. From the transistors in the computer on which I am composing this sentence, to the coatings on the window in front of me, to the drugs and chemicals that are already a part of everyday life, nanometer-scale structures are crucially important. As time goes on, our ability to engineer and fabricate new and useful nanometer-scale structures will only increase. Nanotech will have a profound impact on our lives.

1.3.1 Electron Microscopy

In 1931, Max Knoll and Ernst Ruska invented the transmission electron microscope (TEM), nearly 34 years after the discovery of the electron by J. J. Thomson. M. von Ardenna built the first commercial SEM in 1938, followed soon by the first commercially available TEM, built by Siemens in 1939. The principle of electron beam methods is simple. In order to observe objects with smaller dimensions, the wavelength of the probing source must also acquire comparably smaller dimensions. Erwin Müller, a professor in the Penn State University Physics Department, invented the field-ion EM. He is the first person known to see atoms. The first scientists however to fully exploit electron beam methods were biologists, an interesting irony. Since the resolution of the early scopes was not at the atomic level, it makes sense that relatively large biological structures were first in line to be described by TEM and SEM.

In 1951, x-ray spectroscopic capability was added to TEM and SEM systems with the purpose of conducting in situ elemental analysis. EDS (or EDX, energy dispersive x-ray spectroscopy) systems are included in virtually all EM systems today. In 1970, a high-resolution TEM (HR-TEM) achieved 4-Å resolution. In 2004, Stephen Pennycock of the Oak Ridge National Laboratory (ORNL) achieved 0.6-Å resolution, *transcending the* 1-Å *atomic resolution barrier in the TEM* [70]. Welcome to the *Pico Age*.

Single instruments capable of executing numerous analytical tasks represent another type of convergence. Over the past few decades, development of instruments capable of multiple analyses proved to be a powerful combination in the quest to unravel the mysteries of the nano dominion.

1.3.2 Atomic Probe Microscopes

In 1958, Leo Esaki, a Sony physicist, discovered the phenomenon of electron tunneling and was the first to comment on the special behavior of nanosized materials. His name is attached to the *tunneling diode* and he shared the Nobel Prize in physics in 1973. As a result of Esaki's discovery, another phenomenal research tool was developed—one related in principle to the SEM (e.g., electron emission), but based on the principle of electron tunneling. Gerd Binnig and his colleague Heinrich Rohrer at IBM Zürich introduced a new kind of microscope that is not dependent on wavelength [64].

The STM is capable of creating a topological map of a conductive surface by providing a picture of electron density. The tunneling current is applied under a preset bias between the probe tip and specimen surface. Electrons "jump" from the tip to the surface similar to the way electrons are emitted from the gun of an SEM. Because the size of the analytical current is exponentially proportional to the distance that the tip is from the surface, the topology is mapped by rastering the probe over the surface.

Binnig, Quate, and Gerber invented the atomic force microscope (AFM) 4 years later for the specific purpose of investigating surfaces on the atomic scale that are not conducting [65]. The AFM is a combination of an "STM with a stylus profilometer" [65]. The AFM also is equipped with an atomically sharpened probe tip. The tip is attached to a shiny cantilever that reflects laser light into a detector as the probe is moved across the surface (rastered). Differences in the reflection are recorded in a split photodiode and recorded as changes in topography. Use of the STM is limited to electrically conducting samples. Therefore, the atomic force microscope was developed to image insulating samples. The ability to see atoms and molecules with a device that is purely mechanical is simply amazing. Binnig, Rohrer, and Ruska received the Nobel Prize in physics in 1986 for these incredible inventions—amazing tools that are tailor made for investigating nanomaterials [71].

1.3.3 X-Ray Spectroscopy

German scientist Wilhelm C. Röntgen (**Fig. 1.23**) discovered the x-ray in 1895. He discovered that paper covered with barium platinocyanide fluoresced when placed in the pathway of the radiation emitted from a cathode ray discharge tube [72]. He theorized that x-rays were produced by the impact of the "cathode

Fig. 1.23 *Wilhelm Conrad Röntgen was awarded the Nobel Prize in physics in 1901 for discovering x-rays in 1895 and their application. One of the most useful tools of nanoscience is x-ray diffraction; its roots were laid down at the time of Röntgen.*

Source: Image courtesy of the Nobel Foundation, NobelPrize.org. With permission.

rays" on an object [72]. Max von Laue demonstrated later that x-rays are electromagnetic in nature but with extremely short wavelength [73]. Röntgen's and von Laue's contributions led eventually to the development of x-ray diffraction (XRD). Because the wavelength of x-rays is on the order of distance between crystal planes, XRD is a useful tool in analyzing three-dimensional crystal structure of materials.

The father and son team of William and Lawrence Bragg contributed significantly to this new branch of science [74]. Bragg's law developed instrumental methods to analyze crystal structure based on XRD. The Bragg law (chapters 3 and 5) is the fundamental equation relating interplanar crystal spacing to the x-ray wavelength. XRD methods were expanded to include analysis of powders—another significant contribution to the emerging areas of nanoscience and engineering.

1.3.4 Surface Enhanced Raman Spectroscopy

In 1930, Sir Chandrasekhara Venkata Raman was awarded the Nobel Prize in physics for his discovery of Raman spectroscopy. In 1974, Martin Fleischman et al. reported that metal nanoparticles enhanced the Raman signal of pyridine adsorbed to a roughened silver electrode [75] and erroneously attributed the phenomenon to increased surface area (roughness) alone [75]. In 1977, Richard P. Van Duyne of Northwestern University is credited with explaining the mechanism of surface enhanced Raman spectroscopy (SERS). He found that enhanced Raman signal of the adsorbed analyte species was due to chemical and electromagnetic factors in addition to signal enhancement from increased surface area. Martin Moskovits of the University of Toronto explained the phenomenon in detail in 1985 [76].

The electromagnetic enhancement of SERS is due to induced dipolar and higher order resonance of the surface plasmon on nanometal particles or surface facets. Chemical enhancement, on the other hand, occurs when adsorbed compounds have inherent charge transfer capability (e.g., molecules with lone pairs or π-clouds such as pyridine show the strongest SERS signals). Five or more orders of enhancement using SERS substrates have been reported. Silver is one of the best substrates for SERS use.

SERS techniques have evolved sufficiently to push the detection limit past the attomole (10^{-18} mol) level. This is because SERS is a nanotool that relies on nanoparticles to generate a signal. Specifically, the SERS signal is mediated by the characteristics of the SERS substrate: particle size, shape, and orientation and its composition. The scattering cross-section characteristic of traditional Raman spectroscopy precluded its use for single molecule detection. It is only with the advent of SERS that single molecule Raman detection (SERS-SMD or smSERS) is possible. Single molecule detection of rhodamine-6G on silver nanoparticles was verified by SERS in 2002 [77]. The scattering cross-section of rhodamine-6G was found to be $\sigma = 2 \times 10^{-14}$ cm^2, the largest to date. Although there are skeptics [78], in 2006, Le Ru et al. showed that single molecule signals are very common in SERS [79]. The power of nano!

1.3.5 Lithography

Lithography (from the Greek *lithos,* "stone" + *graphien,* "to write"), a term coined in 1813, has its roots in printing technology. The first lithography was observed in cave etchings dating to the Neanderthal period. Clay Sumerian tablets revealed the earliest evidence of writing. Woodblock printing was developed in China in A.D. sixth century and, in 1040, Pi Sheng invented movable type. Much later, the Koreans also developed movable metal type printing in the thirteenth century. Johannes Gutenburg introduced the printing process to Europeans in 1440. Lithography, however, was not to emerge on the scene for another 300 years.

In 1798 Alois Senefelder of Bavaria invented the modern lithographic process, and in 1819 he wrote a treatise on the subject [80]. Senefelder desperately needed a new method to publish a play he wrote after falling in debt due to problems associated with a "conventional" publisher. He used hydrophobic acid-resistant ink as the *resist* (the positive image) on smooth, finely grained Solnhofen limestone [80]. When the etched plate was immersed in an ink–water mixture, the ink adhered to the positive hydrophobic image and the water adhered to the negative hydrophilic component. He was then able to print from the flat surface of a stone; this was called a *planographic* process. He named it *stone printing* (i.e., lithography).

The photolithography process, a descendent of lithography, is initiated by shining light through a photomask onto a photoresist material. The photoresist material is chemically altered by exposure to the light. Application of a subsequent chemical step etches away the exposed image area (positive resist) or the area around the image (negative resist), depending on the nature of the substrate. Following the etching process, either a positive or a negative of the photomask pattern is produced. Integrated circuits are still produced this way.

Photolithographic processes are limited by the wavelength of the light used. Extreme ultraviolet (EUV) light can be used (wavelength equal to ca. 14 nm), but there are several engineering issues that need to be resolved before its use becomes widespread. There are many more types of lithography, and several will be discussed in more detail in chapter 4.

In 1999, Chad Mirkin of Northwestern University developed the first dip-pen nanolithography system. Nanolithography is lithography at the nanoscale and is capable of producing nanofacets and nanostructures 100 nm or less in size. J. Lyding of the University of Illinois Urbana-Champaign and M. Hersam of Northwestern University developed a variation of lithography called feedback-control lithography (FCL) in 2000. In FCL, an STM is used to build nanostructures on hydrogen-passivated silicon substrates. With the FCL technique, the study of the chemistry of single atoms is possible.

1.3.6 Computer Modeling and Simulation

A valuable nanotool is computer modeling and simulation of atomic and molecular structures, processes, and devices. Although we will not delve into computer modeling and simulation of nanomaterial properties and phenomena in this text, several methods have been developed to probe the behavior of small materials. Computer capability at this time, however, is still limited by computational expense in terms of time and funds to ~100 atoms and timescales on the order of picoseconds.

A versatile and accurate method is molecular dynamic reactive force field simulations based on first principles. Bond order, bond distance, bond energy, and molecular environment dependent charge distribution are examples of parameters that are evaluated by molecular simulation methods. Ab initio Monte Carlo and finite element analysis are also very popular computer simulation methods. Such techniques are analogous to CAD (computer-aided design) programs used by engineers, but operate instead at the nanoscale. Since the design of nanomaterials is usually a costly undertaking, conducting computer modeling beforehand saves time and resources a priori. The nano-engineer–chemist of the future needs to forge good working partnerships with experts in computer modeling.

Phenomena such as hydrogen adsorption on carbon nanotubes, the lotus effect on hydrophobic surfaces, flow through nanochannels, thermal conductivity of carbon nanotubes, diffusion of carbon through (or on the surface of) catalyst particles, light interaction with photonic band gap structures, nanomechanics, and quantum dot optical response all have been modeled with computer simulation techniques. Biological modeling is also routinely done. For example, computer models have simulated the transport of ions via ion channels in membranes [81]. The National Nanotechnology Initiative (NNI) is vigorously funding theory, modeling, and simulation investigations [81].

1.3.7 Molecular Electronics

In 1974, researchers from Northwestern University (Mark A. Ratner) and IBM (Ari Aviram) proposed that electronic circuits be built from the bottom up. Ratner is considered to be the "father of molecular-scale electronics." In 1987, Dmitri Averin and Konstantin Likharev introduced the concept of the "single-electron tunneling transistor" (SET). In 1989, Bell Lab researchers T. Fulton and G. Dolan assembled such a device. The control of the movement of single electrons by a nanodevice was achieved!

1.4 Nature's Take on Nano and the Advent of Molecular Biology

Nature was the master of nanotechnology well before the arrival of the human version (or vision) of nanotechnology. The question to ask nature is, "Why do so many structures and functions exist at the nanoscale?" Nature, of course, does not yield secrets easily. The fundamental building blocks of living things are synthesized from the bottom up. Some remain in the form of individual nanomaterials (e.g., enzymes) and others form hierarchical structures composed of nanomaterials (e.g., connective tissue). The common thread is that all living things without exception are composed of nanomaterials.

The nanoscale represents the size domain above independent atoms and molecules. It also represents the smallest size domain in which there is evidence of functionality. Atoms and molecules per se do not possess functionality; they react and interact according to inherent physical laws, proximity, and probability. It is at the nanoscale, however, where the first examples of function are detected.

1.4.1 Macroscopic Expressions of Natural Nanomaterials

All living matter is composed of nanomaterials that were assembled from the bottom up. The level of complexity (or size) depends on where you look. Following is an example of a hierarchy proceeding from nanoscale materials to assemblies of nanoscale materials. The selected hierarchy starts with the light-absorbing chlorophyll molecule found in plants:

Chlorophyll → Thylakoid → Granum → Chloroplast → Cell → Leaf

Within the chloroplast, the plant cell, and larger structures like leaves, there is a convergence of many types of nanomaterials, structures, and functions (**Fig. 1.24**). Hierarchies such as this can be constructed for any naturally occurring biological material and system, whether structural, metabolic, or even behavioral (e.g., bird feathers). The pads on a gecko's feet, the color of peacock

FIG. 1.24 *A chloroplast is a plastid with a lens-shaped structure about 5 μm in length. It has an inner and an outer membrane. Chlorophyll resides in stacks of* thylakoid *membranes called* grana. *The structure is interconnected with a network of lamellae.*

Source: Image courtesy of Anil K. Rao, Department of Biology, the Metropolitan State University, Denver, Colorado. With permission.

feathers and butterfly wings, the structure of a moth's eye, the silica-based infrastructure of diatoms, the structure of abalone shells, the bone structure of parrot fish teeth, the nanostructured facets of the lotus plant leaf surface, and spider silk comprise a short list of naturally occurring structures composed of nanostructured materials. Natural nanotechnology will be discussed further in chapter 13.

1.4.2 Cell Biology

With the advent of the microscope, cell biologists were the first to investigate micron-sized living things and structures. They also were aware of smaller but irresolvable structures like cellular organelles. The tool of the microworld in the eighteenth and nineteenth centuries was the optical microscope. Dating back to ancient Rome, magnifying lenses are mentioned in the works of Seneca and Pliny the Elder. Spectacles were invented in the thirteenth century (named "lenses" due to their similarity to lentil seeds). Zaccharias and Hans Janssen of Holland in 1590 made the first rudimentary microscope. Galileo in 1609 seized upon that invention and created the first telescope [7]. Robert Hooke invented the compound microscope and in 1665 described microscopic organisms and fossils [7,82]. Hooke is also credited with using the word *cell* (from the Latin *cellulae,* meaning "little rooms") to describe the structure of cork (at 30× magnification). Matthias Schleiden (plant cells) and Theodor Schwann (animal cells) proposed the cell doctrine in 1839 and it remains as the foundation of biology to this day [7].

The "father of microscopy" was Dutchman Anton van Leeuwenhoek [7]. In 1677, with the aid of a microscope capable of 300× magnification, he described entities that dwell at the micron level of measurement: protozoans, red blood cells, and even bacteria. After a 100-year hiatus between major biological breakthroughs, biologists got back on track as Scottish botanist Robert Brown in 1833 coined the word *nucleus* (from the Latin *nucleus,* "kernel," and from the Greek "little nut") after observing cells in the "cellular juice of the orchid" [7]. He is also credited with the first experimental observation of *atomism* after describing Brownian motion.

The discovery of organelles within the cell was documented in the 1800s. Mitochondria and chloroplasts are micron-sized organelles that are composed of nanostructured materials. In 1817, French chemist P. J. Pelletier isolated a nanosized molecule called chlorophyll (from the Greek *khloros,* meaning "pale green" + *phyllon,* indicating "leaf") [7]. Although Julius von Sachs linked the molecule chlorophyll to chloroplasts (an organelle of submicron to micron size) in 1865, Nehemiah Grew is credited with their discovery in 1682. Rudolf Albert von Kolliker described the mitochondrion (from Greek *mitos,* meaning "thread" + *khondrion,* indicating "little granule") in animal cells in 1857 due to improved resolution of optical microscopes. German microbiologist Carl Benda in 1901 named the tiny structures mitochondria. The cellular biologists of the seventeenth through nineteenth centuries went on to investigate and catalogue numerous other intricate microstructures and organelles found in cells. Better microscope resolution would accelerate the study of smaller and smaller components beyond the level of the organelle after the invention of the TEM.

1.4.3 Molecular Biology and Genetics

The area of molecular biology is divided into two parts. One part is biochemical and involves the study and characterization of metabolic processes and syntheses. The study of metabolic pathways fell into the domain of chemists and later biochemists. The other part is the study and characterization of biological nanomaterials like DNA. Description of nanostructured materials in cells became part of the domain of the microscopists. Molecular biology represents the convergence of genetics, biology, and chemistry.

In the 1870s, Walther Flemming discovered chromatin after staining cells with synthetic dyes. He noticed them while observing a cellular division process. Heinrich Wilhelm Waldeyer-Hartz suggested the name *chromosome* (from the Greek *khroma,* "color" + *soma,* "body"). In 1910, Thomas Hunt Morgan linked genetic material to chromosomes of *Drosophila,* the fruit fly. Hermann J. Müller in 1926 postulated that the gene is the basis of life [83]. Unraveling the structure of deoxyribose nucleic acid (DNA) by Watson, Crick, and Franklin in the early 1950s demonstrated that nanomaterials are fundamental to life. Thus was established the *central dogma of biology*—that all life depends on DNA. The recently completed mission of the Human Genome Project (HGP) was to generate a map of all the nucleotides of genes in the human genome.

Diastase, the enzyme responsible for converting starch into sugar, was the first enzyme to be isolated. In 1833, a French chemist, Anselme Payen, purified diastase from a malt extract [7]. The suffix "ase" henceforth has indicated that the substance is an enzyme and performs some metabolic function. He also isolated cellulose from wood, another material that is made of nanosized building blocks, and is credited with introducing the word *ferment* into the lexicon of the early biologists. In 1878, Wilhelm Kühne, a German physiologist, suggested the word *enzyme* (from the Greek word meaning "in yeast," *enzymos*, "leavened") be applied to those substances that "brought about chemical reactions associated with life" [7]. In 1836, Theodor Schwann, who established the *cell theory* along with Matthias Schleiden, isolated pepsin, the first enzyme derived from animal tissue [7]. Biochemist James B. Sumner was the first to crystallize an enzyme, urease, from the extract of black beans in 1926. He was also the first to suggest that enzymes were made of proteins [7].

After he won the Nobel Prize in chemistry in 1902 for his research in purine and sugar chemistry, the great German chemist Emil Fischer (**Fig. 1.25**) joined

Fig. 1.25 *Hermann Emil Fischer was awarded the Nobel Prize in chemistry in 1902 for his work with purines and sugars. He is also the first to propose the* lock and key *hypothesis to describe the relationship between an enzyme (the lock) and its substrate (the key).*

Source: Image courtesy of NobelPrize.org. With permission.

the carboxyl and amino ends of two amino acids together and synthesized the first peptide [7]. Molecular biology became a formal discipline in the 1920s. Warren Weaver, of the Natural Sciences Division of the Rockefeller Foundation, said in 1938, "And gradually there is coming into being a new branch of science—molecular biology—which is beginning to uncover many secrets concerning the ultimate units of the living cell…in which delicate modern techniques are being used to investigate ever more minute details of certain life processes."

The development of a protein synthesizing machine by Bruce Merrifield in 1971 was the first attempt to manufacture proteins in a stepwise synthetic process. The process involved making and breaking strong bonds and therefore required the use of strong acids such as trifluoroacetic acid. Merrifield and other protonanotechnologists were able to anchor amino acids to a board and fabricate protein structures, thus laying down the groundwork for prototypes of "lab-on-a-chip" sensors.

1.5 The Nano Perspective

We have discussed some history, introduced some of the scientists involved, and made mention of some of the significant developments involved in making nanoscience and nanotechnology what they are today. Now we need to delve into the nanoscale itself and gain a feel for the physical perspective. **Figure 1.26** depicts the relative scale of things, depicting images from nature and comparing them to synthetic materials [81]. The electromagnetic radiation spectrum is also shown in order to provide the perspective.

1.5.1 Integration of Everything

We all understand that the major academic disciplines of physics, chemistry, biology, and engineering serve as convenient boundaries to help us make sense of the universe. From the eyes of nature, we can only imagine that the universe is seamless, without demarcation. Intuitively, we sense that all things are connected in some way—that nothing exists independently. The concept of the continuum is referred to often in this text. However, it is difficult to think in terms of "continuum"; otherwise, we would become the polyhistors who know everything. We need boundaries, definitions, and guidelines to sort out our universe. We need to parse data. To cope with the continua abounding in the universe, we invented the major scientific disciplines of mathematics, physics, chemistry, and biology.

Disciplines are created from the top down. For example, chemistry spawned numerous subdisciplines such as organic chemistry, inorganic chemistry, physical chemistry and, later, colloid chemistry and surface chemistry. The list is imposing if we attempt to distinguish more specialized fields such as thin film chemistry, supramolecular chemistry, and, yes, nanoscale chemistry. However, we found that it was not enough to create more specific sub-subdisciplines and sub-sub-subdisciplines complete with journals and conference symposia. As we learned more and more, we needed to cross borders. Hence, subjects such as chemical physics, biophysics, biochemistry, and, yes, even biophysical chemistry emerged.

FIG. 1.26 *The scale of natural things compared to the electromagnetic spectrum are shown in (a), and in (b), synthetic materials of diminishing size are dipicted.*

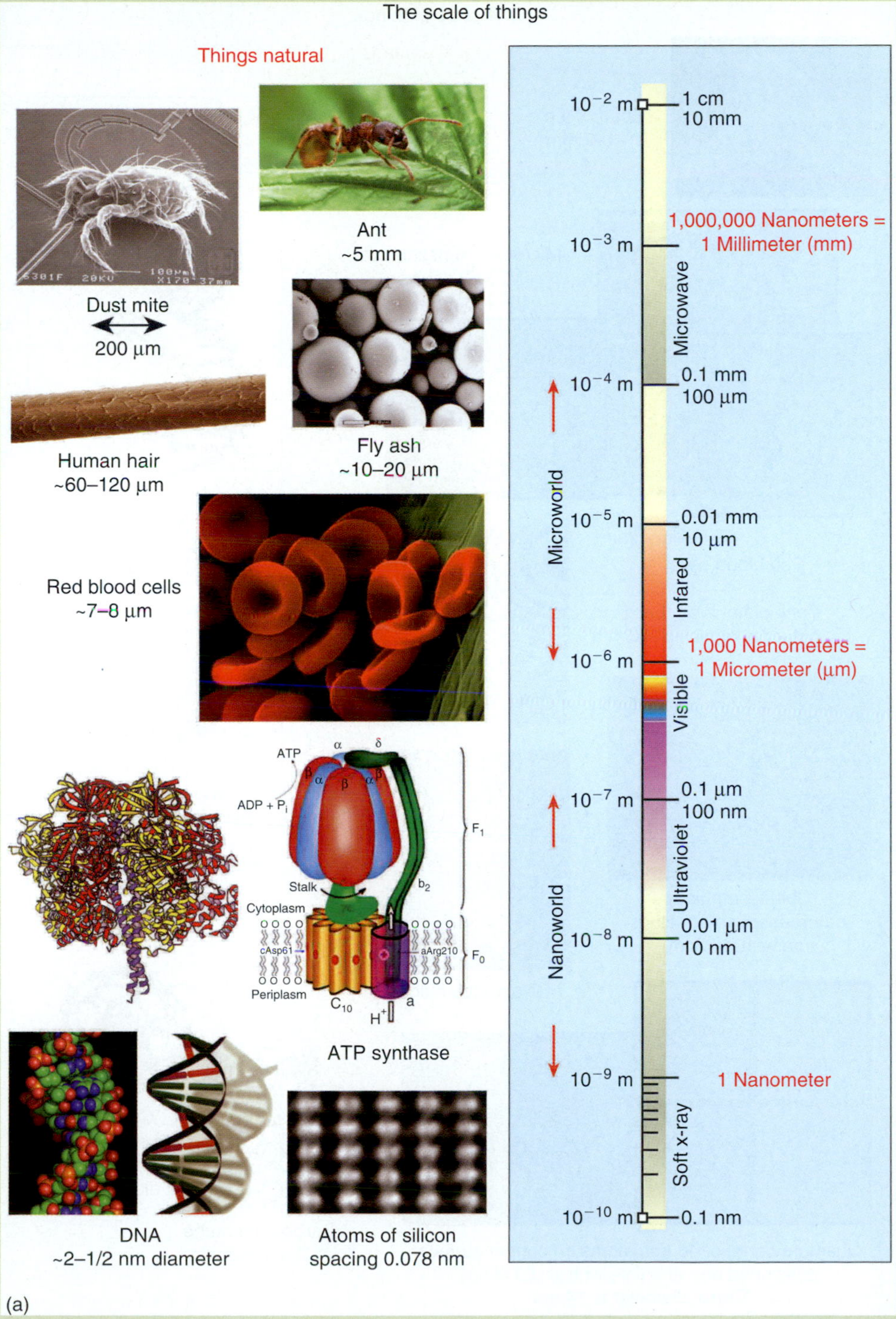

(continued)

Fig. 1.26 (Contd.)

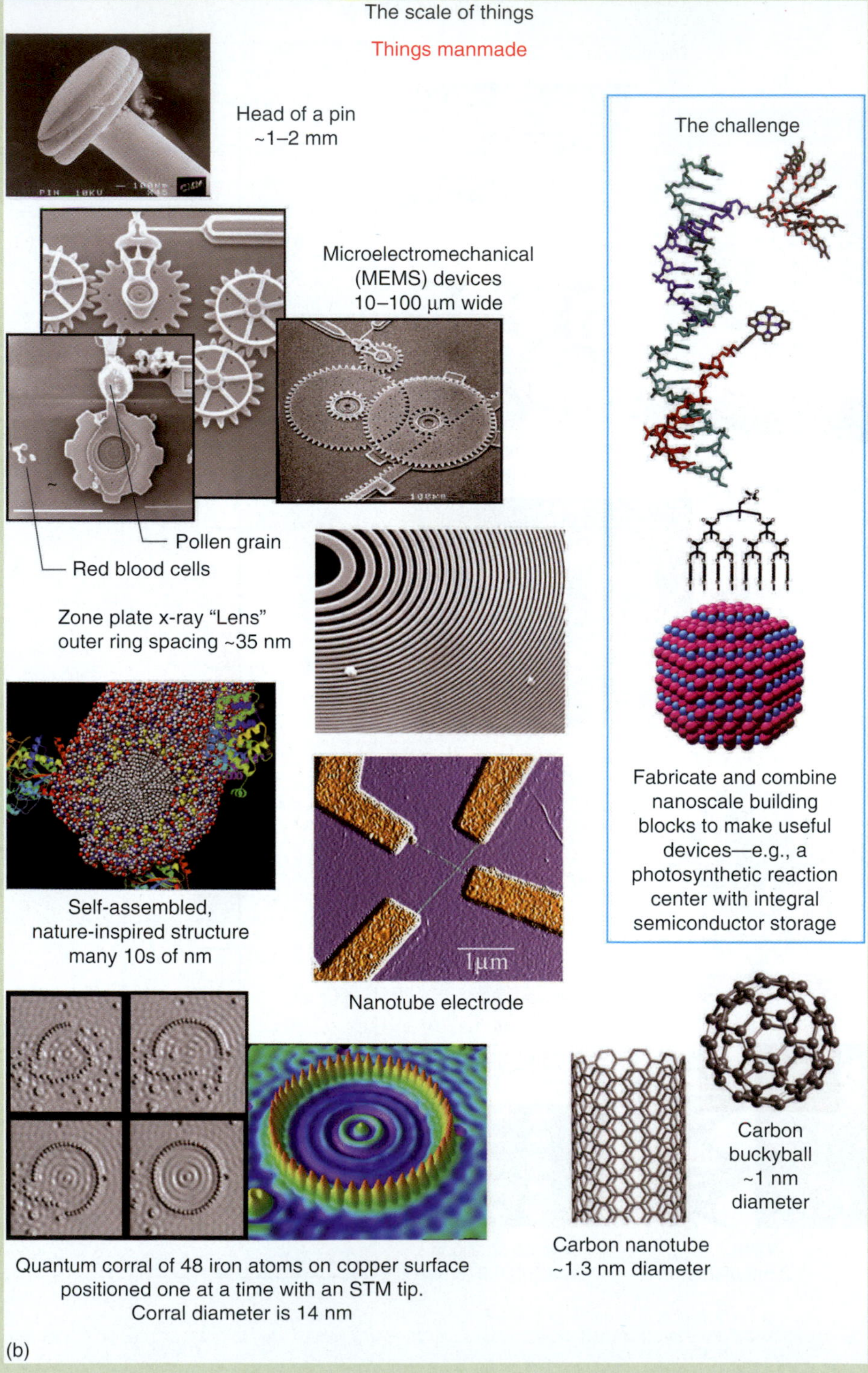

Source: Image reprinted with permission from the Office of Basic Energy Sciences, Office of Science, U.S. Department of Energy.

Since students are not in the habit of reading the preface, we are forced to reiterate some of the salient aspects mentioned there. *Physics* (from the Greek *physikos*, "natural," or *physis*, "nature") is the study of the natural world: the "science that deals with matter and energy in terms of motion and force" [2,3]. The discipline of physics, in the context of this text, is concerned mostly with *properties* and *phenomena* of nanomaterials and is therefore presented first. Properties and phenomena include conductivity, semiconductivity, magnetism, optical response, band gap, heat conduction, quantum phenomena, surface energy, and interaction with electromagnetic radiation and forces in general. Chapters 5–8 comprise the "Properties and Phenomena" (physics) division of the book.

Chemistry (from the Medieval Latin *alchymia*, "the art of transmutation," from the Greek *khenia*, "the art of transmutation as practiced by the Egyptians," and from the Egyptian *Kh'mi*, meaning Egypt) has been in existence for more than four millennia. Chemistry is the science that "investigates the composition, properties and changes of composition, structure, properties and reactions of substances" [2,3]. Within the disciplinary boundaries of this text, chemistry is concerned with *synthesis* of nanomaterials and their modification by chemical methods and is therefore presented after physics. Chemistry also represents the next level of complexity above physics (from a first-principles computational point of view). The content of chapters 9 through 12 comprises the "Chemistry: Synthesis and Modification" division of the text.

Biology (from the Greek *bios*, "life" + *logia*, meaning "study of") is the science of life and life processes, including the study of structure, function, growth, origin, evolution, and distribution of living organisms [2,3]. Biology is where nano begins and is optimized by nature. Within the disciplinary boundaries of this text, the discipline of biology as it applies to nanomaterials is concerned mostly with *describing* and *understanding* natural nanomaterials and is presented after physics and chemistry. Chapters 13 and 14 are in the "Natural and Bionanoscience" division of the book, dedicated to the nanoscience associated with biological and inorganic natural phenomenon. Biology also represents the next level of complexity above chemistry (from a first-principles computational point of view).

Engineering (from O.Fr. *engine*, meaning "skill, cleverness, or war machine" based on the L. *ingenium*, "inborn qualities, talent," *gen* meaning "to produce") is an applied science. The objectives of engineering (chemical, mechanical, and electrical) are to understand the properties of materials and then design devices, circuits, and structures to solve a problem. Engineering is the synthetic analog of biology if materials, structures, and devices are manufactured from the bottom up. Materials science is a subject usually taught in engineering schools. Nanoengineering (e.g. nanomechanics, nanofluidics, nanotribology, nanoelectronics, and the recent fusion of engineering, computer science, and biology via nanotechnology) does place a remarkable spin on the interdisciplinary perspective. Engineering topics are not emphasized in this text.

One more philosophical topic will be discussed and it concerns the convergence of ideas. We have been driven to devise "grand unification theories" ever since earth, air, water, and fire were first merged over 2000 years ago. We know that all academic disciplines are inherently integrated—within an *integration of everything* scenario. We are very interested in the outcome of nanoscopic, microscopic, and macroscopic expressions of nanomaterials from both the synthetic and natural points of view. In many ways we are integrating, unifying, and converging by way of nanoscience and technology. Nanoscience and technology, unlike anything we have seen before, have already raised the bar of interdisciplinary collaboration—proceeding along the path of integrating all scientific and engineering disciplines.

1.5.2 Scale of Things and Timescales

How does one acquire a feel for the nanoscale? One nanometer is 10 × 10 × 10 × 10 × 10 × 10 × 10 × 10 × 10 times smaller than a meter. The distance between cones in the human eye is ca. 2 μm. It is the distance between those cones that limits resolution in the human eye. Robert Hooke in 1673 estimated the resolving power of the human eye to be 70 μm (or 70,000 nm) at the near view limit. This is similar in size to the thickness of a human hair—quite a bit above the nanoscale. Although we cannot see nanoscale materials, we do see the expression of the collective sum of nanomaterials in a macroscopic structure.

Anodized alumina films, ranging in thickness from a few to 20 μm and above, can be easily seen upon close inspection without a microscope, but they appear more like a continuous film without thickness. In order to view the pores of the alumina film, the aid of an electron microscope is required. Light microscopes help to reduce the optical resolution limit to ca. 0.100 μm (or 100 nm), the size of a small bacterium. Electron microscopes, of course, allow us to see atoms 0.0001 μm (or 0.1 nm) and below 1 pm if needed. We are able to see nanomaterials if they are in the form of a powder or in the form of a black soot. We are not able, however, to resolve the detail of the mass without the aid of an electron or atomic force microscope.

The nanoscale is all about size. You will read over and over again in this text how size affects properties and phenomena. The nanoscale is also about time. Emppu Salonen at the Helsinki University of Technology correlated timescales with size. We have added a few extra examples. In terms of seconds, humans live for 2.52×10^9 s (ca. 80 y); red blood cells live for about 1.1×10^7 s (ca. 130 days); *Escherichia coli* can live for $60–8.6 \times 10^5$ s (ca. 10 days); it takes protein about 50–100 s to fold; nanocolloid diffusion happens in 10^{-6} s; surfactant dynamics in 10^{-10} s; molecular collisions in 10^{-12} s; and atomic vibrations every 10^{-14} s in a molecule. The smaller something gets, the faster it reacts and the higher its voice becomes.

1.5.3 Grand Challenges Facing Nanoscience and Nanotechnology

The atomic abacus shown in **Figure 1.27** is an example of manipulation and control of atoms at the nanoscale to form a simple computing system, the abacus. The *beads* consist of C_{60}—otherwise known as a buckyballs—placed in rows of 10 on the steps (*rails*) of a copper substrate [84]. The abacus is manipulated by an STM *finger* probe tip in order to perform simple calculations. The abacus can be operated at room temperature. James K. Gimzewski, leader of the nanoscience project at the Zurich Research Laboratory, states:

> We have made significant progress in handling objects and creating functional units on the nanometer scale at room temperature. Our work demonstrates a

Fig. 1.27 *A nanoscale abacus created in the IBM-Zurich laboratory by Cuberes et al. The beads are actually C_{60} molecules. The rails along which the beads are moved are steps in the copper substrate. Manipulation (calculation) is accomplished with the tip of an STM.*

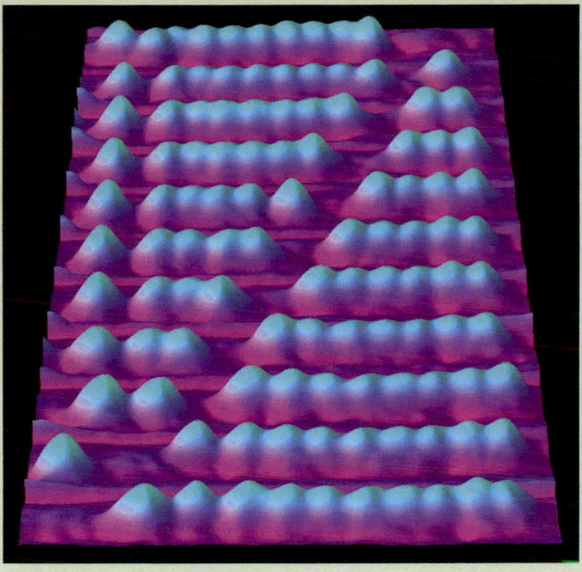

Source: Courtesy of IBM Zurich Research Laboratory. With permission.

> further step in the new and fascinating field of "nano-engineering," where solid-state physics and chemistry merge. We may be able to assemble more complex structures from the bottom up, as nature does, molecule by molecule, and thus break ground for entirely new fabrication technologies with a broad range of applications.

But how practical is a manufacturing process based on this kind of "building one-atom-at-a-time" process? Building nanoscale devices one atom at a time by STM is simply not a reasonable option in today's mass production economy. There needs to be a way to manufacture such devices on a mass scale, simultaneously, and preferably from the "bottom up."

It is also quite straightforward, for example, to fabricate SWNTs. A furnace, some catalyst particles, an inert atmosphere, and a carbon gas source like methane are all that is required. Manufacturing SWNTs that are pure, of the same diameter, of the same orientation, and of the same chirality (chapter 9) is quite another matter. It is somewhat easier to manufacture raw nanomaterials, but it is another matter to manufacture devices that are all the same on a mass scale. George Whitesides of Harvard summarizes some of the challenges that nano will have to address in the next decade [84,85]:

> The age of nanofabrication is here, and the age of nanoscience has dawned, but the age of nanotechnology—finding practical uses for nanostructures—has not really started yet.

> Many interesting problems plague the fabrication of nanodevices with moving parts. A crucial one is friction and sticking .... Because small devices have very large ratios of surface to volume, surface effects—both good and bad—become much more important for them than for large devices .... We will undoubtedly progress toward more complex micromachines and nanomachines modeled on human-scale machines, but we have a long path to travel before we can produce nanomechanical devices in quantity for any practical purpose. Nor is there any reason to assume that nanomachines must resemble human-scale machines.

Within our lifetime, we have seen how technology is able to change the way we live. Nanotechnology, with all of its hype—ranging from eternal life to great wealth—is supposed to change the way we live in revolutionary fashion [86]—or will it? George Whitesides continues:

> None of these things will happen overnight. It is likely that it will take decades ... and nanotechnology will be seen for what it is: another of the wonderful portfolio of technological tools we have available. Not a revolution but an evolution, and one that we don't notice but we take on as another step forward.

Meyya Meyyappan, the director of NASA Ames Nanotechnology Center, states the following about nanosience and its relation to nanotechnology [87]: "lots of nanoscience, very little nanotechnology at this time."

According to Michael Roukes, realistic nanomanufacturing is a long way off. There are two basic types of challenges facing the development of nanotechnology [85]. Communication between the macroworld and the nanoworld is *challenge no. 1*. Starting with the *Heisenberg uncertainty principle*, measuring a system perturbs it. Therefore, any link to a macroscopic detector will change the nanosystem from the ideal state [88]. An interesting permutation of Roukes's proposition is the following: When nanosystems are measured, would the measurement itself serve to stabilize nanocomponents and thereby lose energy (reduce or remove unique properties or phenomena)? *Challenge no.2* involves the nanosurface [88]. The development of an ideal nanocrystal with ultrahigh purity is intended to operate perfectly in ultrahigh vacuum. In any other environment, contact with contaminants would reduce its properties [88].

Some outstanding challenges facing the practical development of nanotechnology are summarized in **Figure 1.28** [85–89].

1.5.4 Next Industrial Revolution

Nanotechnology (as well as nanoscience) has indeed taken center stage in the arena of academic science and is making significant headway into numerous industrial sectors. The number of published scientific literature papers has burgeoned over the past 10 years. The concomitant creation of numerous new journals dedicated specifically to nanoscience and technology soon followed. Hardly any technical conference, whether engineering, optics, chemistry, physics, or biochemistry, is without a symposium on nanoscience and technology. According to government sources, nanotechnology-enhanced products are estimated to contribute $1 trillion to the global GDP and create over two million jobs by 2015 [90].

> While nanotechnology is in the "pre-competitive" stage (meaning its applied use is limited), nanoparticles are being used in a number of industries ... electronic,

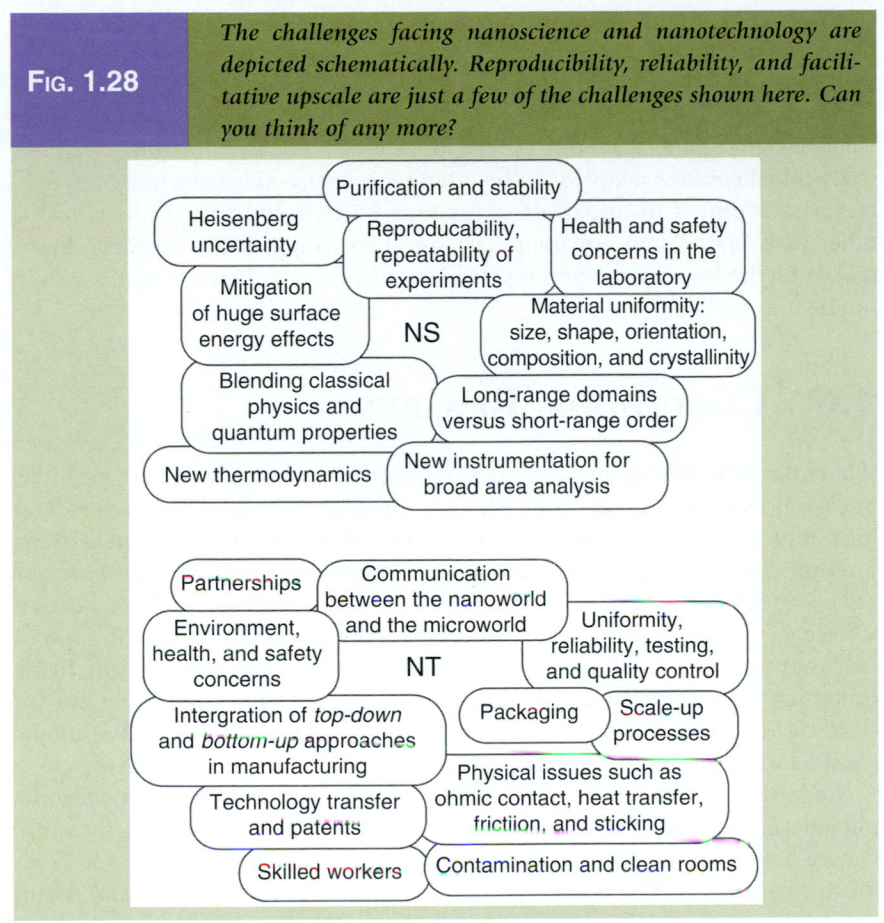

FIG. 1.28 *The challenges facing nanoscience and nanotechnology are depicted schematically. Reproducibility, reliability, and facilitative upscale are just a few of the challenges shown here. Can you think of any more?*

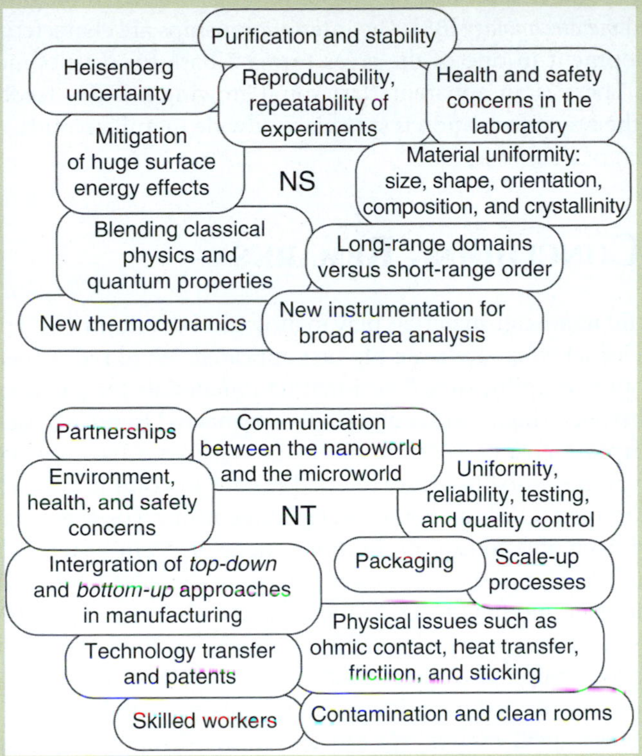

magnetic and optoelectronic, biomedical, pharmaceutical, cosmetic, energy, catalytic and materials applications. Areas producing the greatest revenue for nanoparticles reportedly are chemical–mechanical polishing, magnetic recording tapes, sunscreens, automotive catalytic supports, biolabeling, electroconducting coatings and optical fibers [90].

Basic nanoscience research has been in existence for over 20 years. Since then, many products have made their way into the marketplace. In 2003, for example, Intel introduced the sub-100-nm transistor, smaller than the human influenza virus. Medical diagnostics has exploited the remarkable optical properties of quantum dots. Investigations are taking place now that link quantum dots with molecular recognition systems. Of course, the reliance on nanocatalysts by the oil and gas industries continues unabated. Display technology relies heavily on nanostructured polymer films. Organic light emitting diodes (OLEDs) are being produced with "brighter images, lighter weight, less power consumption and wider viewing angles" [90]. Water purification systems enhanced by carbon nanotube nanotechnology offer inexpensive solutions to filter viruses, bacteria, particulates, and unwanted chemicals. Other products include wear- and chemical-resistant coatings, stain-resistant clothing, inks, dental bonding agents, long-lasting tennis balls, smart golf balls, tennis racquets, cutting tools, special coatings for glasses, waterproof textiles, and catalytic converters.

Be on the lookout for converging technologies, collectively referred to as *nano-bio-info-cogno* (a.k.a. NBIC) [91]. Converging NBIC technologies are collectively called *metatechnology* [88]. The interrelationships are characterized by synergy; development in one of the areas forms a base for developments in the other [91]. There is an apparent "fast spiral-growing positive feedback loop" [91]. With the way information is spread worldwide in milliseconds, this should not be much of a surprise.

1.6 Concluding Remarks

What role did nanoscale materials play in biological development? For example, one would ask at what size scale life first appeared. Would it be, perchance, at the nanoscale? Secondly, were functional nanomaterials the gatekeepers of the threshold between single- and multicellular organisms? Is mitosis itself a process driven by surface area considerations? Large prganisms have a small external surface-to-volume ratio but a huge internal surface-to-volume ratio if all components are summed. Life is all about energy and entropy. For some reason, living things have been able to exploit the sun's energy to form systems that are not clearly favored by entropy (more about this topic in chapter 8). Where do entropic considerations fit into life processes and what role does nano play there?

We have placed a picture of a *Morpho* butterfly on the cover of this text. Its phenotypic expression is a direct result of nanoscale structure and function. Nature has accomplished nano like nobody's business, and we hope you learn much from this great master. We call the study of nature *nanoscience*. From nature's perspective, it is *nanotechnology*.

Due to the accelerated rate of technological development, societal implications of nanoscience and nanotechnology must be considered before, during, and after you write your first research paper. With instant worldwide communication, rapid transport of people and materiel around the globe, and the "accelerated rate of technological development," we can ill afford to neglect consideration of both the fruits or lemons of our nanolabor! As you read and, hopefully, study this text, ask the question, "What is the potential of these things I read about?" Then, as you conduct your own research, ask the question, "What is the potential of these things I am creating?" It all begins with you; where it ends should not be a guessing game.

Acknowledgments

We would like to acknowledge the National Nanotechnology Initiative and its director Mihail C. Roco. He has worked relentlessly for the past several years to make nanotechnology relevant. His emphasis on societal implications in particular has placed such issues at the forefront of an emerging technology like never before. For more information about the NNI, please refer to www.nano.gov.

References

1. R. E. Smalley, Our energy challenge, presentation, *27th Illinois Jr. Science & Humanities Symposium* (2005).
2. *The American heritage dictionary of the English language*, Houghton Mifflin, New York (1969).
3. *Webster's encyclopedic unabridged dictionary of the English language*, Random House, New York (1994).
4. *CRC handbook of chemistry and physics*, 66th ed., CRC Press, Boca Raton, FL (1985).
5. P. McFedries, *IEEE Spectrum Online*/spectrum.ieee.org/apr05/110 (2005).
6. Physics, In *Mechanics of solids*, National Council of Educational Research and Training, India, http://ncert.nic.in/textbooks/, chap. 9 (2005).
7. I. Asimov, *Asimov's biographical encyclopedia of science and technology*, Doubleday & Company, Inc., New York (1982).
8. J. Barnes, Reason and necessity in Leucippus, *Proceedings of the First International Congress Democritus, Xanthi*, 141–158 (1984).
9. R. Boyle, *Sceptical chymist*, Henry Hall, London (1661).
10. J. Dalton, J. Gay-Lussac, and A. Avogadro, *Foundation of the molecular theory: Comprising papers and extracts*, The University of Chicago Press, Chicago, 5–7 (1906).
11. H. E. Roscoe, *John Dalton and the rise of modern chemistry*, MacMillan & Co., New York (1895).
12. D. J. Barber and I. C. Freestone, An investigation of the origin of the color of the Lycurgus cup by analytical transmission electron microscopy, *Archaeometry*, 32, 33–45 (1990).
13. F. E. Wagner, S. Haslbeck, L. Stievano, S. Calogero, Q. A. Pankhurst, and K.-P. Martinek, Before striking gold in gold-ruby glass, *Nature*, 407, 691–692 (2000).
14. F. Mehlman, *Phaidon guide to glass*, Prentice Hall, Englewood Cliffs, NJ (1983).
15. M.-C. Daniel and D. Astruc, Gold nanoparticles: Assembly, supramolecular chemistry, quantum-size-related properties, and applications toward biology, catalysis and nanotechnology, *Chemistry Reviews*, 104, 293–346 (2004).
16. H. H. Heicher, *Aurum Potabile oder Gold Tinstur*, J. Herbord Klossen, Breslau, Lepzig (1718).
17. J. Ayers. In *Ceramics of the world: From 4000 BC to the present*, L. Camusso, and S. Bortone, eds., Abrams, New York, p. 284 (1992).
18. H. Zhao and Y. Ning, Techniques used for the preparation and application of gold powder in ancient China, *Gold Bulletin*, 33, 103 (2000).
19. *Encyclopedia Britannica*, Colloid, Vol. 6, Encyclopedia Britannica, Inc. (1970).
20. T. Graham, On the properties of silicic acid and other analogous colloidal substances, *Journal of the Chemical Society of London*, 17, 318–323 (1864).
21. M. Faraday, Experimental relations of gold (and other metals) to light; the Bakerian lecture, *Philosophical Transactions of the Royal Society of London*, 147, 145–181 (1857).
22. G. Mie, Beiträge zur Optik trüber Medien speziell kolloidaler Metallösungen, Leipzig, *Annals of Physics*, 25, 377–445 (1908).
23. J. C. Maxwell-Garnett, Colours in metal glasses and metal films, *Philosophical Transactions of the Royal Society A*, 203, 385–420 (1904).
24. J. C. Maxwell-Garnett, Colours in metal glasses, in metallic films and in metallic solutions—II, *Proceedings of the Royal Society of London*, 76, 370–373 (1905).
25. *Encyclopedia Britannica*, Photography, Vol. 17, Encyclopedia Britannica, Inc. (1970).
26. photography.about.com/library/weekly/aa010702a.htm
27. http://catalyse.univ-lyon1.fr/0catal.htm
28. *Encyclopedia Britannica*, Catalysis, Vol. 5, Encyclopedia Britannica, Inc. (1970).
29. P. Sebatier, *La catalyse en chimie orgarnique* (first published in 1913, with a second edition in 1920; English translation by E. E. Reid) (1923).

30. P. Sabatier, nobelprize.org/chemistry/laureates/1912/sabatier-bio.html
31. J. R. Rostrup-Nielsen, Equilibria decomposition reactions of carbon monoxide and methane over nickel catalysts, *Catalysis,* 27, 343–356 (1972).
32. W. H. Brattain, Laboratory notebook of December 24, 1947, Press Release, Bell Telephone Laboratories, New York (1948).
33. G. Thompson, Silicon nanoelectronics, INTEL, presentation, Colorado Nanotechnology Summit, Boulder, CO, May (2004).
34. G. E. Moore, Cramming more components onto an integrated circuit, *Electronics,* 38, April 19 (1965).
35. G. L. Hornyak, *Characterization and optical theory of nanometal—Porous alumina composite membranes,* Ph.D. dissertation, Colorado State University (1997).
36. M. Ohring, *Materials science of thin films,* 2nd ed., Academic Press, London (2002).
37. B. Franklin, Letter from Benjamin Franklin to William Brownrigg (1773) www.njsas.org/projects/atoms/monolayers/hist.php
38. P. Becker, History and progress in the accurate determination of the Avogadro constant, *Reports on Progress in Physics,* 64, 1945–2008 (2001).
39. edisonexploratorium.org/bio/blodgett.htm
40. C. D. Bain, E. B. Troughton, Y.-T. Tao, J. Evall, G. M. Whitesides, and R. G. Nuzzo, Formation of monolayer films by the spontaneous assembly of organic thiols from solution onto gold, *Journal of the American Chemical Society,* 111, 321–335 (1989).
41. C. D. Bain and G. M. Whitesides, Formation of monolayers by the coadsorption of thiols on gold: Variation in the head group, tail group and solvent, *Journal of the American Chemical Society,* 111, 7164–7175 (1989).
42. C. D. Bain and G. M. Whitesides, Modeling organic surfaces with self-assembled monolayers, *Angewandte Chemie International Edition, England,* 28, 506 (1989).
43. H. W. Kroto, J. R. Heath, S. C. Obrien, R. F. Curl, and R. E. Smalley, C-60—Buckminsterfullerene, *Nature,* 318, 162–163 (1985).
44. S. Iijima, Helical microtubules of graphitic carbon, *Nature,* 354, 56 (1991).
45. S. Iijima and T. Ichihashi, Single shell carbon nanotubes of one nanometer diameter, *Nature,* 363, 603 (1993).
46. R. Bacon, Growth, structure, and properties of graphite whiskers, *Journal of Applied Physics,* 31, 283–290 (1960).
47. G. G. Tibbetts, *Journal of Crystal Growth,* 66, 632 (1984).
48. R. G. Finke, Transition—Metal nanoclusters. In *Metal nanoparticles: Synthesis, characterization and applications,* chapter 2, D. L. Feldheim, and C. A. Foss, eds., Marcel Dekker, Inc., New York (2002).
49. D. L. Feldheim and C. A. Foss, *Metal nanoparticles: Synthesis, characterization and applications,* Marcel Dekker, Inc., New York (2002).
50. M. Jose-Yacaman, L. Rendon, J. Arenas, and M.C.S. Puche, Maya blue paint: An ancient nanostructured material, *Science,* 273, 223–225 (1996).
51. G. Schmid, Large clusters and colloids: Metals in the embryonic state, *Chemical Reviews,* 92, 1709–1727 (1992).
52. G. Schmid, ed., *Clusters and colloids: From theory to applications,* VCH, New York (1994).
53. G. Schmid and G. L. Hornyak, Metal clusters: New perspectives in future nanoelectronics, *Current Opinion in Solid State Materials Science,* 2, 204 (1997).
54. G. L. Hornyak, M. Kroell, R. Pugin, T. Sawitowski, G. Schmid, J-O. Bovin, H. Hofmeister, and S. Hopfe, Gold clusters and colloids in alumina membranes, *Chem. Eur. J.,* 3, 1951 (1997).
55. J. Zheng, C. Zhang, and R. M. Dickson, Highly fluorescent, water-soluble, size-tunable gold quantum dots, *Physics Review Letters,* 93, 077402 (2004).
56. www.qdots.com (Quantum Dot Corporation)
57. W. D. Lawson, F. A. Smith, and A. S. Young, Influence of crystal size on the spectral response limit of evaporated PbTe and PbSe photoconductive cells, *Journal of the Electrochemical Society,* 107, 206–210 (1960).

58. B. I. Shklovski and A. l. Efros, Interband absorption of light in strongly doped semiconductors, *Soviet Physics Journal of Experimental and Theoretical Physics*, 32, 733–738 (1971).
59. A. Nozik, Photoelectrochemistry applications to solar energy conversion, *Annual Review of Physical Chemistry*, 29, 189–222 (1978).
60. A. I. Ekimov and A. A. Onuschenko, Quantum size effect in three-dimensional microscopic semiconductor crystals, *Journal of Experimental and Theoretical Physics Letters*, 34, 346–349 (1982).
61. L. E. Brus, A simple model for the ionization potential, electron affinity, and aqueous redox potentials of small semiconductor crystallites, *Journal of Chemical Physics*, 79, 5566–5571 (1983).
62. A. I. Ekimov and A. I. Efros, Quantum size effect in semiconductor microcrystals, *Solid State Communications*, 56, 921–924 (1984).
63. M. Graetzel, Artificial photosynthesis: Water cleavage into hydrogen and oxygen by visible light, *Accounts of Chemical Research*, 14, 376–384 (1981).
64. G. Binnig, H. Rohrer, Ch. Gerber, and E. Weibel, Surface studies by scanning tunneling microscope, *Physics Review Letters*, 49, 57 (1982).
65. G. Binnig, C. F. Quate, and Ch. Gerber, Atomic force microscope, *Physics Review Letters*, 56, 930 (1986).
66. D. M. Eigler and E. K. Schwiezer, Positioning single atoms with a scanning tunneling microscope, *Nature*, 344, 524–526 (1990).
67. D. M. Eigler, C. P. Lutz, and W. E. Rudge, An atomic switch realized with the scanning tunneling microscope, *Nature*, 352, 600 (1991).
68. R. P. Feynman, There's plenty of room at the bottom: An invitation to enter a new field of physics, Annual Meeting, American Physical Society/California Institute of Technology, December 29, 1959, California Institute of Technology, *Engineering and Science Magazine*, 23, 22, (1960).
69. *Time Magazine*, 135, 16 (April 16, 1990).
70. M. A. O'Keefe, L. F. Allard, S. J. Pennycock, and D. A. Bloom, Transcending the one-ångstrom atomic resolution barrier in the TEM, *Microscopy Microanalysis*, 12, 162–163 (2006).
71. G. L. Hornyak, S. Peschel, T. Sawitowski, and G. Schmid, TEM, STM and AFM as tools to study clusters and colloids, *Micron*, 29, 183 (1998).
72. W. C. Röntgen, Biography, NobelPrize.org
73. M. von Laue, Biography, NobelPrize.org
74. William Bragg and Lawrence Bragg, Biography, NobelPrize.org
75. M. Fleischman, P. J. Hendra, and A. J. McQuillan, Raman spectra of pyridine. Adsorbed at a silver electrode, *Chemical Physics Letters*, 26, 163 (1974).
76. M. Moskovits, Surface-enhanced spectroscopy, *Review of Modern Physics*, 57, 783–826 (1985).
77. K. A. Bosnick, J. Jiang, and L. E. Brus, Fluctuations in local symmetry in single-molecule rhodamine 6g Raman scattering on silver nanoscrystal aggregates, *Journal of Physical Chemistry B*, 106, 8096–8099 (2002).
78. K. Kneipp, G. R. Harrison, S. R. Emory, and S. Nie, Single molecule Raman spectroscopy: Fact or fiction? *Chimia*, 53, 35–37 (1999).
79. E. C. Le Ru, M. Meyer, and P. G. Etchegoin, Proof of single-molecule sensitivity in surface enhanced Raman scattering (SERS) by means of a two-analyte technique, *Journal of Physical Chemistry B*, 110, 1944–1948 (2006).
80. A. Senefelder, *Vollstandiges Lehrbuch der Steindruckery* (1819) (www.lib.edel.edu/ud)
81. Theory, modeling and simulation, research and development supporting the next Industrial Revolution, Supplement to the President's FY 2004 Budget National Nanotechnology Initiative (2004).
82. R. Hooke, *Micrographia: Some physiological descriptions of minute bodies made by magnifying glasses with observations and inquiries thereupon*, The Royal Society of London, J. Martyn and J. Allestry (1665).

83. J. H. Müller, The gene as the basis of life, *Proceedings of the International Congress of Plant Science*, 1, 897–921 (1926).
84. M. T. Cuberes, R. R. Schlitter, and J. K. Gimzewski, Room temperature—Repositioning of individual C60 molecules at Cu steps: Operation of a molecular counting device, *Applied Physics Letters*, 69, 3016–3018 (1996).
85. G. M. Whitesides, The art of building small. In *Understanding nanotechnology*, Scientific American, Time-Warner Book Group, New York (2002).
86. G. M. Whitesides, The once and future nanomachine, *Scientific American*, 285, 78–83 (September 2001).
87. P. Binks, The challenges facing nanotechnology. Interviewed by R. Williams, *Ockham's Razor*, www.abc.net.au/rn/science/ockham/stories/s1304778.htm (2005).
88. M. Meyyappan, Nanotechnology: The next frontier, presentation, NASA Ames Research Center, Moffett Field, CA.
89. M. Roukes. Plenty of room indeed. In *Understanding nanotechnology*, Scientific American, Time-Warner Books (2002).
90. G. Cao, *Nanostructures & nanomaterials: Synthesis, properties and applications*, Imperial College Press, Singapore (2004).
91. National Nanotechnology Initiative, nano.gov

Problems

1.1 Scientists have made deductions indicating the presence of nanoparticles without actually having seen them. The experiments of Graham and Faraday have done as much. What other examples of such experiments (other than those mentioned in the text) can you think of that imply the existence of nanoparticles before the advent of the electron and atomic microscopes?

1.2 There are numerous examples of nanomaterial-based phenomena all around us, from both technology and nature. Colored materials, for example, have a history of participation of nanomaterials. Can you figure out the basics of the color-generating component of the following (feel free to use the Internet as a research tool)? Not all are based on nanophenomena.

a. Colored titanium jewelry
b. Colored anodized aluminum siding for skyscrapers
c. White/green gold leaf
d. Origin of the color of ruby-colored glass vases in your grandmother's dining room cabinet? What is the origin of the color for the blue-colored glass vases?
e. Colors of bubbles derived from soapy films
f. White-colored titanium dioxide particles
g. Color imparted to a male peacock's feathers
h. White of the edelweiss (*Leontopodium nivale*), an alpine flower that lives at high altitudes, up to 3000 m/10,000 ft, where UV radiation is strong
i. Color of the blue *Morpho* butterfly found in Central and South America
j. Color of emerald, sapphire, and ruby minerals
k. Color of green leaves of plants
l. Color of the sky at noon and at sunset
m. Color of opaline materials like pearls
n. Color of flames

How many of these are based on nanophenomena?

1.3 Without being able to see enzymes, how did chemists and biologist know that enzymes existed and that they performed metabolic functions? How were enzymes discovered?

1.4 What is the nature of the moiety that requires "fixing" in a photographic process? Why is fixing required?

1.5 Why would a nanobrick type of layered structure in the shell of the abalone aid in its survival? The bricks are fused laterally into the shell and there is a biological polymeric adhesive between layers.

1.6 Using only your chemical or physical intuition, provide an empirical explanation for Faraday's observation about the different colors that solutions containing gold particles assume "whether ruby, green, violet or blue." Why would the addition of gelatin or salts prevent the change in color of the colloid solutions?

1.7 What would happen to metabolism in general if all enzymes would suddenly acquire micron-sized dimensions? How would you mitigate the consequences?

1.8 Do you consider nanotechnology to be the next industrial revolution or is it simply a natural progression towards making things smaller?

1.9 What are your views about how nanoscience–nanotechnology should be taught? Should there be degree programs that are anchored in nanoscience and/or technology? Or should the subjects be included in core courses?

1.10 The digital camera all but displaced the photographic film: one nanotechnology displacing another. How many other disruptive technologies have occurred over the years that you are able to recall?

1.11 More nanogames:

a. How many nanotetrahedra can be packed into 1 m^3? Assume that the all sides of the tetrahedra are 10 nm.
b. How many C_{60} would it take to fill a 10-μm long single-walled carbon nanotube of 1.5 nm diameter?
c. Assuming van der Waals radii, how many hydrogen atoms can be squeezed into a length of 1 nm?
d. How many yoctameter-sized cubes can fit into a cubic nanometer?

1.12 What wavelength of probing radiation is required to inspect objects or phenomena on the order of the size of

a. Airplanes
b. Paramecium
c. Gold clusters (~2 nm diameter)
d. Humans
e. Rotation of water molecules
f. Chemical bonds
g. Protons
h. Crystal planes
i. Viruses
j. Chromosomes

1.13 Do you believe that nanobots able to clean plaque from your artery walls will ever be developed?

1.14 Nature's incredible use of nanomaterials and bottom-up synthesis is exhibited all around and within us. Give an example of a process, starting with molecules and ending with a macroscopic structure (e.g., tendon, bone, muscle). Describe the levels of hierarchy.

1.15 Why are nanomaterials inherently unstable? Why are they stable in living systems?

1.16 Which group—chemists, engineers, physicists, or biologists—do you think made the greatest contribution to nanoscience?

1.17 The space elevator project is an effort that intends to place an elevator system grounded on Earth into geosynchronous orbit. Carbon nanotubes are the only material strong enough to make this a reality. What is it about carbon nanotubes that make people think this way? Do you think a space elevator is possible? What are some potential issues?

1.18 Give a 5-min. space elevator speech about nanotechnology.

1.19 How would you teach a course in nanoscience and/or nanotechnology? Is this an important concern?

1.20 List some characteristics of nanomaterials that may be considered to be phenomenal.

1.21 Propose another scenario for the invention of soap.

1.22 Repeat example 1.1 but assume tubes have two-dimensional coordination number of 4. Repeat again with coordination number of 6. What is the number of tubes able to be packed in the hair cylinder in these cases?

1.23 What is grey goo? Green goo?

1.24 What one individual do you think made the greatest contribution to nanoscience?

1.25 List as many women scientists as possible involved in nano in the past.

1.26 How are nanoscience and nanotechnology to be defined? (Question courtesy of Günter Schmid, Uni-Essen.)

Societal Implications of Nano

Addressing societal and ethical issues is the right thing to do *and* the necessary thing to do. *It is* the right thing to do *because as ethically responsible leaders we must ensure that technology advances human well-being and does not detract from it. It is* the necessary thing to do *because it is essential for speeding technology adoption, broadening the economic and societal benefits, and accelerating and increasing our return on investment.*

Philip J. Bond
Former undersecretary for technology,
U.S. Department of Commerce (2003) [1]

Chapter 2

Fig. 2.0 *Former undersecretary of commerce for technology, Technology Administration, Phillip J. Bond. He began his career at the U.S. Department of Commerce on October 30, 2001, and recognized readily the importance of nanotechnology to the prosperity and security of this country.*

Source: U.S. Department of Commerce, Technology Administration, www.technology.gov.

Threads

Definitions, the historical perspective, and the challenges facing nanotechnology from the scientific and manufacturing points of view have been briefly introduced in chapter 1. We now delve into topics that address formidable challenges other than those posed by pure science and technology. In this chapter, we grapple with the complex societal issues that inevitably accompany any new technology, focusing on those that are unique to nano. The world has changed much since the straightforward era of hunting and gathering when choices were ruled by day-to-day-survival. Now, with billions of people living on the planet and communication occurring instantly and globally, we need to be prepared for the ramifications of a new type of technology that is already changing our way of life. In this chapter, we are fortunate to compile the opinions of five experts that present, discuss, and summarize a variety of issues that face the advent of nanotechnology.

The mission of this chapter is awareness. We hope that message stays with you while you read this text—to reflect, contemplate, and formulate and to look into the past, assimilate the present, and project into the future. Societal implications are broad, complex, and highly integrated. For example, specialized groups already protest the potential of nano. Please supplement information gleaned from this chapter with your own outside reading. Stay informed! As the butterfly pictured on the cover of this text flutters its wings, what are the consequences of that action elsewhere? Interestingly, our choices are still ruled by survival.

2.0 Introduction to Societal Issues

This chapter is intended to provide a brief introduction to a broad array of issues that arise as a result of, or in response to, new and potentially disruptive technologies such as nanotechnology and the various other technologies it may enable. Science and technology do not exist in a vacuum separate and apart from the society into which they are introduced. Rather, science and technology influence key dimensions of the larger society, including law, politics, economics and business, public health and safety, national security, education, and popular culture. Science and technology have long challenged and contributed to the evolution of our Western philosophical worldview, altering cultural values, attitudes, beliefs, and practices along the way. Finally, science and technology have become strongly embedded in our assessment of the human condition and in our vision and aspirations for the human future.

As noted in chapter 1, nanoscience and nanotechnology hold the potential for rapid and dramatic changes in our technological capabilities, raising significant practical and moral opportunities and concerns. In this chapter, we focus on the particular dimensions of ethics, law, environment, public perception, and technological forecasting to illustrate the complex relationship between advances in nanotechnology and the larger society. Roco and Bainbridge of the National Nanotechnology Initiative (NNI) stated in 2001:

> Over the next 10–20 years, nanotechnology will fundamentally transform science, technology, and society. However, to take full advantage of opportunities, the entire scientific and technology community must set broad goals; creatively envision the possibilities for meeting societal needs; and involve all participants, including the general public, in exploiting them [2].

Great opportunities are generally accompanied by great challenges and such is the case with nanoscience and nanotechnology. Much has been written about the tremendous potential of nanotechnology; however, barriers exist that go beyond basic science and engineering. The magnitude of scientific advance predicted by Roco and Bainbridge also portends the emergence of a wide range of ethical and societal implications including challenges related to basic moral questions, legal and regulatory responses, environmental concerns, national and global economic and political impacts, public perception, and planning for a range of future scenarios.

2.0.1 *Societal Implications—The Background*

To understand the current context of societal implications research, one must travel back in time to the launch of the Human Genome Project (HGP) in October of 1990 [3]. The HGP was an effort to identify and map all of the human genes, as well as determine the complete sequence of human DNA. Once available, the fully mapped human genome would provide rich new ground for biological study and potential medical advances. However, the project was controversial on many fronts. First, there were many *naysayers* who believed the project was not feasible and amounted to an enormous waste of time and resources. Other critics were concerned with how genetic knowledge would be used, how it might affect human self-perception, and what potential harms could come from the ability to manipulate and control human genetics at its most basic level. In fact, many of the same general societal implications of the HGP pertain to nanotechnology, including matters related to privacy, intellectual property and patent questions, health and environmental concerns, and moral dilemmas related to human bioengineering and commercialization.

In response to these critics, the project identified a key research area that came to be known as ELSI, or ethical, legal, and social issues research. The U.S. Department of Energy (DOE) and the National Institutes of Health (NIH) devoted between 3% and 5% of their total annual HGP budget to the study of ethical, legal, and social issues associated with the availability of genetic information. The ELSI model is now applied around the world with respect to a variety of scientific endeavors. Research continues and the ELSI Website of the HGP has become a clearinghouse for developments in these areas.

Similar to the precedent set by the HGP, and in anticipation of the enormous potential of nanoscience and nanotechnology to impact virtually every realm of human existence, the National Science Foundation (NSF) sponsored a workshop in 2001 from which a report was published. The report, *Societal Implications of Nanoscience and Nanotechnology* [2], is a collection of expert opinions and recommendations on how best to "(a) accelerate the beneficial use of nanotechnology while diminishing the risks, (b) improve research and education, and (c) guide the contributions of key organizations."

The first recommendation reads as follows:

> Make support for social and economic research studies on nanotechnology a high priority. Include social science research on the societal implications in the nanotechnology research centers, and consider creation of a distributed research center for social and economic research. Build openness, disclosures, and public participation into the process of developing nanotechnology research and development program direction.

Additional recommendations address the need to inform, educate, and involve the public; create an infrastructure for interdisciplinary evaluation of the scientific, technological, and social impacts in the short, medium, and long terms; and educate a workforce.

In 2005, the President's Council of Advisors on Science and Technology (PCAST) released the first major progress report on the NNI, entitled *The National Nanotechnology Initiative at Five Years: Assessment and Recommendations of the National Nanotechnology Advisory Panel (President's Council of Advisors on Science and Technology, 2005)* [4]. With respect to societal concerns, 8% of the total NNI budget for 2006, or about $82 million, was requested for societal implication activities, sometimes referred to as SEIN (societal and ethical implications of nanotechnology). Approximately $38.5 million was earmarked for programs working on environmental, health, and safety research and development. The remaining $42.6 million was split between education-related activities targeted at workforce development, as well as public understanding and acceptance, and research on the broad implications of nanotechnology for society, including economic impacts, barriers to adoption, and ethical issues, particularly as related to research priorities [5]. All NNI-funded nanotechnology research must have a societal dimensions component that addresses one or more of the preceding categories.

Presently, the societal dimensions program component area (PCA) of the NNI encourages and funds research initiatives in three broad areas: (1) environmental, health, and safety (EHS) impacts of nanotechnology development and risk assessment of such impacts; (2) education-related activities, such as the development of materials for schools and universities as well as public outreach; and (3) identification and quantification of the broad implications for society. In this last category are included social, economic, workforce, educational, ethical, and legal implications (please consult the National Nanotechnology Initiative website at nano.gov) [6].

In 2007, there was a shift in funding, with approximately $44 million earmarked for EHS programs and $38 million for the combined areas of education and all other societal implications. Among SEIN researchers and other observers, the EHS allocation is somewhat controversial. For example, one might expect environmental, health, and safety concerns to be part of the basic cost of research and development funding as it is in general industry. Critics suggest this assignment of EHS to societal implications effectively reduces the budget for all other societal implications research to far less than what is needed, making SEIN research commitments largely symbolic, geared more towards managing public perception and support than minimizing negative impacts [7].

Regardless of motive, this unprecedented commitment to assessing the societal impacts of new technology does illustrate the breadth and complexity of the issues. At the time of this writing, an infrastructure for interdisciplinary collaboration is beginning to emerge. In reality, there is not much precedent for the idea that scientists and engineers should be working collaboratively with the social sciences and humanities to proactively assess societal impacts and ethical implications. However, National Science Foundation grants have funded large centers at Harvard University, University of South Carolina, Arizona State University, and the University of California at Santa Barbara in an effort to encourage collaboration and coordination of ongoing NSF-funded research initiatives across the country. The overall approach represents a fundamental change in the culture of science and technology, as well as its relationship to both policy making and public dialogue. As scientists and engineers, you will increasingly find yourselves collaborating with ethicists, social scientists, and others to address these larger implications of your work.

2.0.2 Breadth of Societal Implications

As discussed in earlier sections, the range of societal implications is quite broad, reaching into many, if not all, areas of society (**Fig. 2.1**). Berube attempts to define SEIN research as follows [7]:

> SEIN encapsulates research by toxicologists, ethicists, and futurists into a range of issues from environmental impacts, regulatory regimes, workplace and economic dislocations, bionanotechnology convergence, and transhumanism and posthumanism. It involves experts from philosophy, communication studies, law, and political science, as well as fiction studies and art. It also includes the less expert as well as the self-proclaimed technophiles, the social critic, and the crank.

Fig. 2.1 *Societal dimensions of nanotechnology. A single-walled carbon nanotube frames many aspects of nanotechnology within its hexagonal-carbon superstructure.*

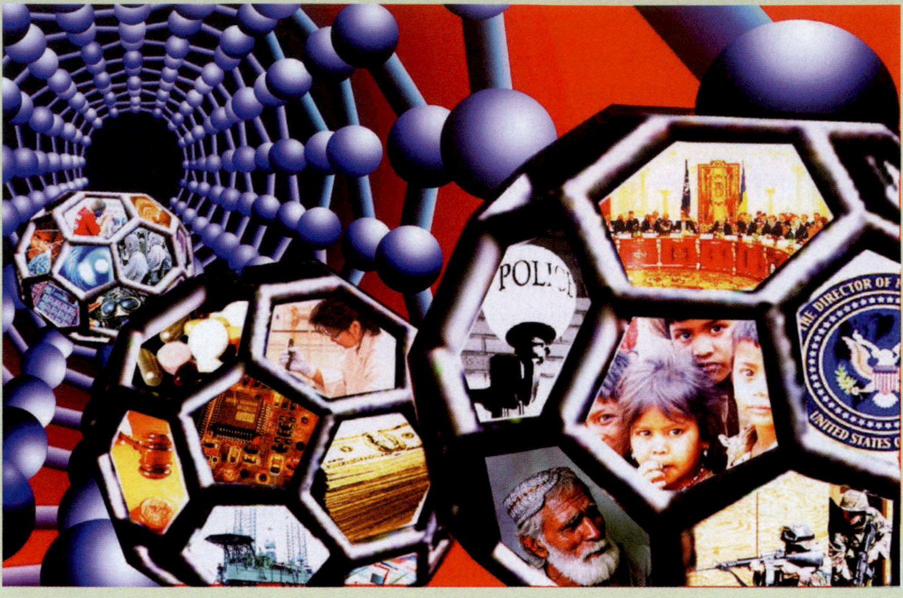

Source: Image reprinted with permission from R. M. Bennett-Woods.

Due to the sheer range of issues, there is no easy way to categorize and describe societal implications. Many, if not all, industries are anticipated to experience significant change in response to discoveries and applications of nanoscience. However, to fully understand the true extent of societal implications, one must also recognize connections branching into nearly every corner of the public sphere; general categories of societal impacts that have been raised as potential concerns range from safety and environmental impacts to workforce and global economic disruptions to controversial applications in medicine.

Any approach to societal implications research is further complicated by the dynamic complexity inherent in any societal shift of the magnitude suggested by nanoscience and its related technological offspring. One example of dynamic complexity is the potential for a relatively small change in one industry to become a large change in another industry, or for changes in the economic sphere to escalate rapidly into the political sphere. For example, nanoscience advances in the biotechnology industry have the potential to alter the very foundation of how medicine is practiced within the health care industry, affecting both the general biotechnology sector and the health care sector. Furthermore, disruptive changes in the health care sector that affect cost or access to services will intensify existing political pressures to reform the entire health care system.

To further illustrate the complex relationships among nano-enabled technological advances, nested within the relationship between the biotechnology and health care industries is the Massachusetts Institute of Technology *MIT Institute for Soldier Nanotechnologies*, which is supporting a number of projects that employ elements of biotechnology and medicine to "enhance soldier survivability" [8]. The combined goals of various projects include improving battlefield triage and the treatment of injuries while still on the battlefield, detection of biohazards, and even augmentation of a soldier's strength and physical agility. Plans are already underway to allow technologies developed for military use on the battlefield to find their way into commercial application within the health care industry and others. At least some of these innovations will likely spark controversy insofar as they raise issues of privacy, cost, access, and human enhancement. Consider, for example, the impact on the sports industry if military applications originally intended to protect soldiers on the battlefield become available as a means to improve endurance or mask injury in sports competitions.

The important point here is to emphasize the complex, highly integrated, and unpredictable nature of societal implications. Some advances will be relatively unproblematic while others will raise a range of concerns that may be difficult to sort out before questionable applications have already made their way into the marketplace. In general, societal implications of nanoscience in most industry sectors raise at least some potential concerns related to health and safety as well as many ethical dilemmas, legal and regulatory issues, barriers to public acceptance, workforce disruptions, education and training needs, and public policy questions.

2.0.3 Meet the Experts

Addressing the full breadth of SEIN is clearly beyond the scope of this introductory chapter. However, we will briefly focus on five specific dimensions of societal impact: ethics, law, environment, public perception, and future forecasting. Experts from these areas will provide an overview of issues that confront scientists, engineers, corporate leaders, policy makers, and community stakeholders involved in research and development of nanotechnology and its potential outcomes.

In the section entitled "Ethical Implications," Professor Deb Bennett-Woods of Regis University (**Fig. 2.2**) introduces a set of concepts from the field of ethics that provide a framework for broadly considering the moral dilemmas raised by nanotechnology in all other dimensions. She suggests that questions based on these concepts can become the basis for collaborative and informed dialogue among scientists, engineers, policy makers, business leaders, and community stakeholders.

"Legal Implications," by intellectual-property attorney Dr. Patrick Boucher (**Fig. 2.3**), a patent attorney for a technology law firm that handles intellectual property, explore the complex legal issues posed by a truly interdisciplinary science and technology [9]. This section takes the approach of illustrating the impact of two fundamentally different approaches to legal doctrine on the development and use of nanoscience. The role of intellectual property in providing inducements to nanoscience developers to share their results for the benefit of all is contrasted with various liability doctrines that attempt to restrain those same developers from prematurely allowing those results to be used without adequate consideration of safety issues. The author suggests a natural legal tension affecting the development of nanoscience that, between factors favoring rapid development and factors favoring a more cautious approach, is reflected in the way these factors interact with existing legal doctrine.

FIG. 2.2

Dr. Bennett-Woods is the chair of the Department of Health Care Ethics and the director of the Center for Ethics and Leadership in the Health Professions at Regis University in Denver, Colorado. She is also a fellow of the Center on Nanotechnology and Society at the Chicago-Kent College of Law. She is the author of Nanotechnology: Ethics and Society *(CRC Press, 2008). Her particular interests are with ethical issues arising from emerging technologies in health care and the convergence of technologies such as nanotechnology, biotechnology, artificial intelligence, cognitive science, and robotics.*

Source: Image reprinted with permission from R. M. Bennett-Woods.

Dr. Jo Anne Shatkin (**Fig. 2.4**), of CLF Ventures, Inc., in the section entitled "Environmental Implications," provides an overview of the process and importance of risk assessment as it relates to nanotechnology. Potential environmental impacts and important questions regarding nanotoxicology are explored.

One specific area of focus for proponents of the NNI has been public understanding and acceptance of nanotechnology, particularly in the face of media hype and science fiction accounts of *gray goo* and *nanorobots* run amuck. In the section entitled "Public Perception," Professor Susanna Priest (**Fig. 2.5**) of the University of Nevada, Las Vegas addresses what the public really thinks

FIG. 2.3

Dr. Patrick M. Boucher holds a PhD in physics (Queen's University, Canada) and a J.D. (Touro College, United States). His technical publications have been in the areas of condensed-matter, nuclear, and astrophysics. For several years, he was associated with Physical Review B, which publishes much of the world's technical nanotechnology research and where he acted as associate editor and managed the journal's scientific editorial staff. He is currently a patent attorney practicing in Denver, Colorado, as a partner of Townsend and Townsend and Crew LLP.

and what, if any, barriers exist to widespread support and adoption of nano-related products.

Finally, in the last section, "The Future of Nanotechnology," futurist Dr. Thomas Frey (**Fig. 2.6**) of the DaVinci Institute in Boulder, Colorado, introduces his brand of forecasting and introduces us to 10 endpoints of nanotechnology that will influence the society of the future.

FIG. 2.4

Dr. Jo Anne Shatkin is managing director of CLF Ventures, a nonprofit group that helps organizations develop complex projects with demonstrable environmental and economic benefits. She provides technical and strategic expertise on projects and is a recognized expert in strategic environmental initiatives, human health risk assessment, technical communication, and environmental aspects of nanotechnology. Before joining CLF Ventures, she managed a human health risk assessment and nanotechnology practice for the Cadmus Group. She brings a wealth of experience in research and application of quantitative human health risk assessment for site redevelopment and remediation, drinking water and air quality, and environmental evaluations of emerging contaminants. Her specialty is the application and communication of innovative science-informed analysis to address complex emerging issues affecting businesses and communities. She has been an active member of the Society for Risk Analysis since 1989, and recently founded the Emerging Nanoscale Materials Specialty Group of the Society for Risk Analysis.

2.0.4 The Nano Perspective

Clearly, the implications for nanoscience and nanotechnology go well beyond the boundaries of the science itself. Renowned science fiction author Asimov once said [10] that "the saddest aspect of life right now is that science gathers

Fig. 2.5

Professor Susanna Hornig Priest is professor in the Hank Greenspun School of Journalism and Media Studies at the University of Nevada, Las Vegas, and is the editor of the journal Science Communication. Previously she served as director of research for the College of Mass Communications and Information Studies at the University of South Carolina and as director of the science journalism master's program at Texas A&M University. Her own research focuses on public and media responses to controversial issues in science and technology, a subject on which she has published many dozens of research articles, book chapters, and unpublished reports. She is also the author of two research methods textbooks for media studies students.

knowledge faster than society gathers wisdom." Take a moment to reflect on what this means for you, as a student of nanoscience and nanotechnology. What wisdom is needed to ensure that scientific advances uniformly guarantee progress and improvements in the human condition rather than making existing social problem worse or creating new ones? And, of equal importance, what is

FIG. 2.6

Dr. Thomas Frey is the executive director and senior futurist at the DaVinci Institute. He is currently Google's top-rated futurist speaker, working with CEOs and executives at top companies from around the world. His talks are known for continually pushing the envelope of understanding, creating fascinating images of the world to come. Before launching the DaVinci Institute, Tom spent 15 years at IBM as an engineer and designer where he received over 270 awards, more than any other IBM engineer.

your responsibility as an agent of scientific and technological change for ensuring that the broader social impacts of your work have been considered? Keep this in mind as you explore the thoughts of various SEIN experts in the following sections.

2.1 Ethical Implications

The term *nanoethics* has begun to appear in the SEIN literature. Whether this term, which translates to *ethics of the extremely small,* catches on or not remains to be seen. However, the broad discipline of ethics does provide a unique and useful lens for framing the diverse range of potential societal effects of nanotechnology with a common language. As such, it is a good starting point for the discussion of societal impacts.

2.1.1 Ethics in the Context of Research and Applied Science

Ethics can be loosely defined as the general study of morals and moral choices. The field encompasses an array of theories, concepts, and methods for conducting moral analysis and providing insight into moral dilemmas. Ethical problems arise when it is not clear what action is the right one to take in particular circumstance. The classic definition of a moral or ethical dilemma is a situation in which *doing something right* also results in *doing something wrong*.

Perhaps the best analogy to the ethical dimensions of nanotechnology can be drawn from the realm of human-subject research in medicine. Modern medicine has come a long way from its primitive roots in which patients were subjected to largely untested, at least by scientific standards, folk remedies, bloodletting, and various cultural rituals. The introduction of modern sanitation, nutrition, and antibiotics has largely shifted the focus of medicine in the developed world from the control of infectious disease to complex management of acute trauma and noninfectious diseases such as diabetes and cancer. However, nearly every major advance in medical science has come as a direct result of research on living human subjects. Such research can often be risky and most subjects never directly benefit from their participation. Abuses in human-subject research, such as the infamous Tuskegee study, in which 400 African American men with untreated syphilis were observed for 40 years in order to document the natural progression of the disease, have led to a set of commonly accepted procedural standards based on ethical principles that safeguard the rights of patients.

Three such principles are formally established in a document entitled the *Belmont Report* produced by the National Commission for the Protection of Human Subjects of Biomedical and Behavioral Research in 1979 [11]. The first principle, *respect for persons*, is the basis for the concept of patient autonomy (self-determination or self-governance) and requires guidelines for obtaining informed consent and the protection of vulnerable populations such as children. The second principle, *beneficence*, obligates researchers to avoid needless harm by minimizing any inherent risks while also attempting to maximize the potential benefits to subjects. *Justice*, the third principle, examines more broadly how likely the benefits and burdens of the research are to be fairly distributed. Using the Tuskegee study as an example, all three principles were violated. The subjects were misled into believing they were being treated and so never gave informed consent; they were needlessly harmed since a cure was known and available; and they bore the full burden of the research with no benefits to themselves or even to science since the stages of syphilis were relatively well documented at the time of the study.

Fascinating, but how does this relate to the societal implications of nanotechnology? Commentators in nanotechnology have actually drawn an analogy between the potential impacts of nanotechnology and human-subject research [12,13]. For example, when exploring the societal implications of nanotechnology, Sarewitz and Woodhouse [12] describe the "unfolding revolution as a grand experiment—a clinical trial—that technologists are conducting on society." The direct implication here is that the considerations and review processes used for clinical trials in human-subject research may also be applicable to the various uncertainties posed by nanotechnology.

Naturally, there are no formal mechanisms in place to accomplish the sort of review and oversight mandated in human-subject research. However, there are key decision points in the continuum of science and technology research and development at which ethical principles can be applied [14]. For example, should we fund new or continued research and development? With increasingly constrained budgets for research and development, we may be obligated to prioritize some research initiatives over others on the basis of a greater good. Should we transfer knowledge developed for one purpose in one industry sector to another? A common example is the transfer of technologies originally developed for military purposes into the private sector. The navigation system in your new car has its origins in military applications of global positioning technology. This raises a legitimate question of fairness and who should benefit from research conducted at the public expense. Finally, do we regulate? This question is often posed only after serious environmental, health, and safety effects have been observed in a product already on the market. Several recent and widely publicized cases of drugs being withdrawn from the market due to serious safety concerns are examples of the value of regulatory assessment and oversight in promoting consumer safety and informed consent in the face of corporate interests.

Although the principles outlined in the Belmont Report are narrowly directed at protecting the individual human subject, they can be expanded to consider broader societal impacts, providing a framework within which ethical analysis and dialogue can occur. The principle of respect for persons can be expanded to the principle of respect for communities, while the principles of beneficence and justice can likewise be respectively broadened to incorporate the common good and a broader conception of social justice [14]. Let us explore each of these individually.

2.1.2 Principle of Respect for Communities

Just as the principle of respect for persons requires that each person be treated with dignity and as an autonomous agent capable of informed decision making in his or her own best interests, so too communities have the ability to act as autonomous, self-governing agents. Naturally, this is much easier said than done, as noted by Sarewitz when he characterizes the "compulsory nature of technological assimilation" as a central cause of social change. In other words, most technology has a tendency to be widely introduced and accepted long before most of us have given much thought to any potential downsides [15].

There are many barriers to a communal informed consent. First, there are few effective mechanisms for timely public engagement in the development and introduction of new technologies. And, even if there were effective forums for public dialogue, informed consent requires that citizens have the information needed to conduct a reasoned deliberation. In a rapidly paced and highly competitive marketplace, the interests of researchers and corporations are not always well served by being too forthcoming about the technical nature of their developments and, in particular, any untoward hazards. Again, even if the information were available, communities marginalized by poverty and illiteracy are not likely to receive it or take advantage of public forums to express their views. Nonetheless, there have been examples of public backlash against the introduction of new technologies, and the NNI's strong emphasis on research initiatives related to public education and perceptions is directly intended to help manage the issue of community consent. Consider the recent rejection of genetically modified organisms (GMOs) by the European Union.

In a nano-related example, let us examine the question of privacy. The development of sophisticated inventory tracking devices, nanosensors for military surveillance, neural implants, and nanobiosensors for detecting biological and environmental threats has raised concerns about individual privacy and related civil liberties [16–18]. The principle of respect for communities would raise a number of basic questions when considering the development of these devices:

- Do potential violations of privacy undermine or enhance human dignity?
- Do these devices have the potential to violate fundamental human rights, including privacy, freedom of conscience, and other basic liberties?
- Are there sufficient information and an effective mechanism available to the public regarding the potential outcomes so that informed and meaningful community consent can occur?
- Are there certain populations that are more vulnerable or under-represented than others, and are there sufficient protections to ensure that vulnerable populations within larger communities have a meaningful voice in the consent process?
- Are there barriers to consent being competent and voluntary?

By periodically posing these questions at those key decision points during the nanotechnology (NT) development process and targeting the most questionable applications for more careful scrutiny and planning, potential problems can more effectively be anticipated and solutions developed or products abandoned. On the other hand, failure to address these issues effectively can lead to widespread public rejection of such technologies, including those that offer great potential benefits to the society as a whole.

2.1.3 Principle of the Common Good

The principle of the common good obligates us to act in ways that respect shared values and promote the common good of communities. On one level, this involves an obligation to promote the well-being of individuals and groups specifically by avoiding harm, minimizing risks, and maximizing benefits. On another level, the concept of a common good is a bit more complicated, but basically encourages actions that respect those shared values that allow us to live effectively in community. In the communitarian traditions of political science and ethics, primary attention is given to a balance between individual liberty and responsibility towards the community, including obligations of sustainability and justice between generations.

Defining the common good is very difficult to pin down in the larger scope of society. The first complicating factor is that, in a complex society, we function as many smaller communities nested within larger ones, often with competing interests and differing values. Second, in order to avoid *harms* you need to know what those *harms* are likely to be—something that is only a matter of informed speculation in the case of emerging technologies. If a technology results in serious environmental degradation or exposes certain populations to a high level of risk, you could argue that the common good has been violated. However, it is not always clear how to identify the greater harm when the impacts are more subtle or complex. How much unintended environmental degradation is acceptable if food production is greatly enhanced? Technology is often referred to as a two edged sword that results in opportunities or solves problems for some, while creating barriers or other problems along the way.

To complicate matters further, when applying this principle, you also need to agree on how to define and weigh various societal benefits with harms, not just in the short term but also in the long term. We often see this dynamic when we aggressively exploit natural resources, such as oil and gas, in the short term for commercial gain and convenience, leaving environmental degradation and limited resources for future generations to confront. And, again, we are faced with the question of how to accommodate vulnerable populations who are not likely to benefit or are actively harmed by the technology in question.

Moving beyond benefits and harms to the context of shared values is even more challenging. What are the shared values and best interests of society and what responsibilities do we owe one another? Where do we draw the line between the ideals of community and individual liberty? In other words, where do my liberties stop and the interests of the larger community kick in? A good example of a nano-related dilemma that may pit the interests of the individual against the common good of the community involves projected shifts in manufacturing and the workforce.

As noted earlier, a primary research focus of the NNI is workforce preparation. Nano-driven advances in business and industry will require a highly trained and specialized workforce that does not currently exist. While this opens the door of opportunity for a segment of the workforce with the means to retool itself, large numbers of manufacturing jobs are likely to be rendered obsolete or displaced [19]. Furthermore, the competition for global dominance in nanotechnology may continue the current trend of any new jobs moving offshore, depending on which countries take the scientific and technological lead in nano-innovation. As corporations position themselves to compete in the global marketplace, what obligations, if any, do they have to the local communities disrupted by rapid shifts in the job market and economic uncertainties? What constitutes an acceptable balance between corporate well-being and individual well-being? In a similar vein, the extraordinary investment in nanoresearch has, by necessity, reduced funding available for unrelated research interests, educational initiatives, and other funding priorities. When faced with scarce resources, every funding decision is a pragmatic choice, to be sure; however, every funding decision is also a statement of what the funding source values and this may or may not align well with the values of the various community stakeholders.

When considering the principle of the common good with respect to a particular issue, several additional questions can be posed:

- How are the values and priorities of communities represented and served?
- How might the values and priorities of communities be violated or undermined?
- What are the potential short-, medium-, and long-term benefits and burdens for individuals and communities?
- What are the most likely outcomes: positive, negative, or neutral?
- What unintended outcomes can be anticipated? What is the level of risk?
- What limitations or safeguards are prudent to prevent negative outcomes?

The principle of the common good essentially boils down to holding all members of the community responsible for the well-being of the larger society and the interests of individual community members. In the face of rapid or unprecedented technological innovation, effective consideration of the common good can allow for a mutually beneficial transition for all members of the larger society.

2.1.4 Principle of Social Justice

The final principle is that of social justice, obligating us to act in ways that maximize the just distribution of benefits and burdens within and among communities. Perhaps more than the first two principles, this principle calls us to think globally. The nature of major technological innovations is that they initially become available largely to those who can afford them. A secondary effect is that related outcomes, such as environmental degradation or political and economic disruptions, tend to result in the least advantaged segments of the global society experiencing the highest level of burden and risk, while receiving little or no benefit. The principle of justice asks us to always give primary consideration to alleviating the suffering of the most vulnerable among us. The sustainability movement provides an obvious example of just such an opportunity.

At least nine biotechnologies related to medicine and the environment have been identified as having the potential to greatly impact health in the developing world [20]. However, there are clear barriers to those technologies becoming available. One barrier is a matter of research priority and investment. The risk profiles of developed and developing countries differ significantly [21]. If research priorities allocate the majority of nanotechnology research and development towards the diseases of the developed world, health disparities may actually increase. A second barrier involves basic market forces and the issue of access. Most companies engaged in NT development seek a solid return on their investment. Increasingly, the partnerships between research universities and the private sector that have been promoted by the NNI have resulted in patenting and restrictive use licensing that directly blocks access to technologies for humanitarian purposes [22].

Assessing the aspect of social justice requires posing one additional set of questions:

- How are benefits and burdens balanced within and between communities?
- How might current social, economic, and political boundaries be enhanced or disrupted?
- How will those communities that are harmed be compensated?
- Does potential lack of access constitute a basic violation of human rights?
- Who is accountable for the fair distribution of benefits?

Taken together, these three principles strive to balance concerns at the levels of the individual, the local community, and the global society.

2.1.5 *You as Moral Agent*

A final word about ethical implications and you as a moral agent needs to be addressed. Individuals all wear multiple hats when it comes to ethical obligations. In the language of ethics, this is termed *conflicting loyalties* or *conflicts of fidelity*. In any particular situation you may find yourself needing to be loyal to your employer, your colleagues, your profession, your family, your community, and yourself. These multiple loyalties often challenge us to consider what we do, and the manner in which we do it, in light of each different stakeholder. For example, the immediate interests of the company you work for may conflict with what you believe to be legitimate concerns regarding public safety. In another example, public funding priorities may violate your sense of how resources should be allocated to serve the greater good. On the flip side, your personal financial well-being could be at stake if funding priorities shift from one direction to another.

We do not have good systems and processes for dealing with such conflicts in modern organizations and social institutions. However, the discomfort you feel when such conflicts arise is evidence that scientists and engineers do have a role and a responsibility to take the societal implications of their work seriously. Using the language and concepts of ethics in a productive and collaborative dialogue, the evolving opportunities of nanotechnology can be realized while also minimizing the inevitable disruptions and harms of technological advancements.

2.2 Legal Implications

The previous section explored some factors that bear on how ethics apply to nanoscience and nanotechnology. Because one of the goals of the law is to provide a framework for promoting morally defensible choices, we now turn to the complex intersection of nanoscience, nanotechnology, and the law. The topics of intellectual property and civil liability are used to introduce the reader to some of the legal challenges raised by nanoscience and technology.

2.2.1 Interaction of Law and Nanoscience

Fundamentally, the purpose of law is to control the behavior of people. No matter what the philosophical underpinnings of different governments in the world might be or what different views those philosophical underpinnings represent about the nature of individual liberty, they all agree there is a need to control aspects of the behavior of those within their borders. Such control can be effected in positive or negative ways—that is, by providing inducements that encourage behavior viewed as socially beneficial or by providing punishments that discourage undesirable behavior. As such, the law insinuates itself into virtually every aspect of human activity, from governing the way children are identified and treated at the moment of their birth to influencing the way they interact with others during their lives to resolving their affairs after death. Nanoscience is one of those rare disciplines that may have the potential to enjoy a similar breadth.

In considering the various ways nanoscience and the law may intersect, it is important to keep in mind that these intersections arise even when laws have not been tailored specifically to address nanoscience. The history of legal doctrine recognizes that human society is constantly in flux and has developed flexibility that aims to have it applied to new circumstances and to new technologies. In this way, legal doctrine attempts to recognize that while circumstances change over time, there are certain constants about the nature of human beings that permit general principles to be applied in circumscribing limits on their behavior.

To be sure, there will be specific issues raised by nanotechnology and nanoscience that will be addressed with targeted legislation. For the most part, this will happen when a perception develops that the more generic legal principles, even when faithfully applied, are not leading to the most desirable result. But in the broad scheme, these targeted laws will be of relatively minor importance in the way nanoscience and the law interact. The vast majority of legal issues that involve nanoscience will draw on legal doctrines that have more general application.

To illustrate how this happens, this section uses two examples of legal frameworks that have been designed with flexibility in mind: intellectual property and civil liability. The first is an example of a set of laws that intends to encourage positive behavior—namely, to provide an inducement for creative individuals to develop and share their creations with the public so that society as a whole may advance more quickly. The second is an example of a set of laws intended to discourage negative behavior by holding those who engage in activities that harm others responsible for their actions. Neither of these doctrines was developed with nanoscience in mind—indeed, they have pedigrees that span centuries before the discipline was even conceived—and yet their principles find ready application to issues of relevance to nanoscience today.

2.2.2 Intellectual Property

Intellectual-property law is a set of legal doctrines that governs the ownership of inventions and creations. It has a number of subcategories—patents, copyrights, trademarks, and trade secrets—that are directed to more specific aspects, with each of these subcategories providing different types of protections to inventors and creators. While those involved in the development of nanoscience have the potential to benefit from any of these doctrines, by far the most important is the ability to obtain patents for inventions in the fields of nanoscience and nanotechnology.

The current patent system has aspects of its origins in laws passed by the city-state of Venice in the fifteenth century. There is a natural tendency among those who invent something useful to want to control the essence of their discovery and thereby to derive a commercial advantage from having exclusive access to it. But such secrecy does little in allowing the discovery to benefit society as a whole. What Venice did, in its years as a thriving center of Middle Ages commerce, was to experiment with a system that would provide a state-sanctioned monopoly on use of an invention in exchange for having the inventor disclose it to the public. The public knowledge that resulted would enable others to make improvements on the invention, and thereby accelerate the pace of innovation. And the inventor would be the beneficiary of certain exclusive rights backed by the power of government. Venice initially experimented with patents in the area of silk-making, but soon expanded to providing protection for other types of devices, including patents for flour mills, cook stoves for dye shops, printing methods, and mills. Because of Venice's commercial importance at the time, its system had the effect of inducing even outsiders to disclose their inventions in exchange for exclusive rights within the city-state, proving the value of the system in putting useful information into the public domain.

The patent system today is predicated on the same basic philosophy. In exchange for disclosure of an idea, the government grants a right to an inventor to exclude others from making, using, or selling that invention unless they have her permission. The right exists for a limited period of time—in almost every country it is 20 years from the date the application is filed with the Patent Office—after which anyone is free to use the invention. One important aspect of this bargain between the inventor and the government is that this right can be exercised even against those who independently develop the same invention. This makes patents especially important in areas like nanoscience, where many researchers are working actively and competitively on developing similar ideas. The first to discover and patent inventions that result from them will have a right of exclusion that can be exercised against all of the others.

The same is not true of trade secrets, which are a form of intellectual property that is in many ways the antithesis of patents. As its name suggests, a trade secret is some kind of information that is withheld from the public. Nothing in the patent system obligates disclosure by an inventor; indeed, when a researcher creates an invention, he still has the choice of disclosing it to the world in exchange for a patent or concealing it to be used as a trade secret. The risk of concealment is that there is no protection provided for independent discovery of a trade secret. If another researcher develops the same idea, there will be no way for the first inventor to interfere with its use by the second inventor and any commercial advantage from being the first to make the discovery will be largely lost. The most significant benefit of a trade secret is that there is no time limit on it. While patents expire in 20 years, trade secrets can be maintained indefinitely, and there are commercial examples of trade secrets that have been maintained for close to a century.

Decisions of whether to patent inventions or to maintain them as trade secrets therefore hinge to a great extent on the likelihood that the invention will be independently developed and on the ease with which its secrecy can be maintained. There is no general prohibition on using reverse-engineering techniques to try to discover how some product that incorporates nanoscience is made. Indeed, the use of such techniques is mostly encouraged as a matter of legal policy since the result is a greater contribution to the public knowledge. Instead, the protection the law affords to those who decide to maintain information as a trade secret is to provide punishments that can be imposed when they are misappropriated. The most dramatic examples of trade-secret theft are sometimes glamorously described as *economic espionage*, but more mundane forms of theft can have similarly negative financial impacts.

Consider, for example, a nanoscience researcher who develops a method of fabricating doped nanotubes that can be easily formed into superconducting wires. In this example, the method uses a catalyst during a reaction that was only accidentally discovered and comes as a complete shock to the scientist who noted it. Should the method be patented or kept secret? The answer is not clear and the decision is a strategic one that requires the exercise of some judgment. Relevant considerations include:

How likely is it that another researcher will stumble on the same discovery?

How difficult will it be to maintain control over the information in production and sales environments?

Is it possible to analyze the superconducting wires and detect use of the catalyst?

The last of these questions is especially important because it relates not only to how easy it is for others to discover the secret but also to how difficult it would be to detect whether there was any actual infringement of a patent.

The other two types of intellectual properties are probably of less direct relevance to nanoscience. Copyright law does have the potential to apply to certain kinds of very small creations—for example, sculptural works created using methods common in nanotechnology. For instance, the Osaka Bull was carved out of resin in 2001 by a group of researchers at Osaka University using a two-photon micropolymerization technique (**Fig. 2.7**).

While one might quibble about whether the bull itself qualifies as "nano," given that its size is on the order of microns, it illustrates the point that creative works are sometimes made using nanoscience and these are entitled to copyright protection. In some ways, an example like the Osaka bull is little more than an isolated curiosity. More practical application of copyright law to nanoscience may arise from ideas currently being formulated for using chemical structures in information processing to produce a *nanocomputer*. The sequence of such chemical structures could easily qualify as a work entitled to copyright protection in much the same way that conventional source code is protected by copyright today. Similar considerations apply to "quantum computing" in which quantum states of particles are used in defining bit states and assembled to produce a form of computational program.

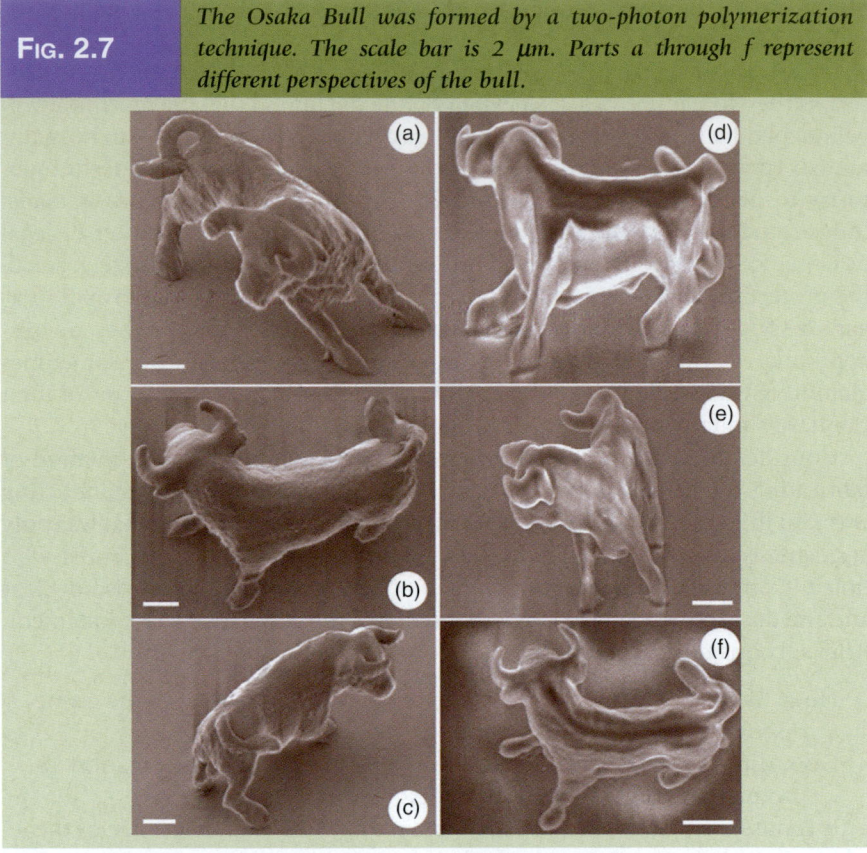

Fig. 2.7 *The Osaka Bull was formed by a two-photon polymerization technique. The scale bar is 2 μm. Parts a through f represent different perspectives of the bull.*

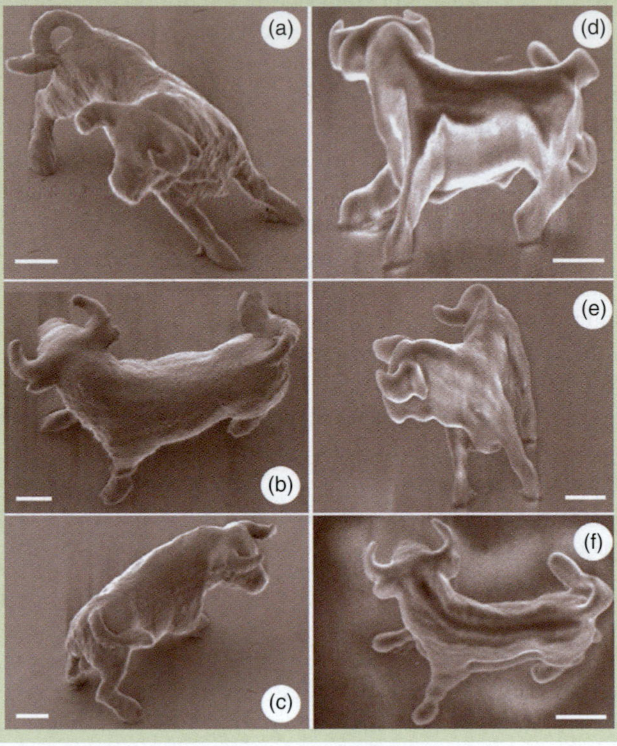

Source: Kawata, S. et al., *Nature*, 412, 697–698, 2001. Image courtesy of the Nature Publication Group. With permission.

Trademarks, the last form of intellectual property, are likely to find application in nanoscience in only the most peripheral way. Trademarks are intended to identify the source of goods or services. Thus, the names of nanotechnology companies could be protected as trademarks to identify the source of the products they produce. And while it is conceivable structures could be included on nanotechnology products as a source identifier, this is relatively unlikely to be done.

2.2.3 Civil Liability Issues

As the discoveries of nanoscience become commercialized, so does the potential for them to act as a source of liability. As discussed in the next section, there are already certain specific concerns about the impact that nanosized structures may have on the environment or human beings, and it is certain that a greater understanding of those impacts will be developed in the future. What the law of liability seeks is a sensible apportionment of financial responsibility when harms occur. For instance, if it turns out that a certain product incorporating nanoscience causes harm, how are the financial and other consequences of the harm to be assigned?

Modern product liability law is based on the idea that there is liability whenever a party introduces a *defective* product into the marketplace. There are three recognized types of defects, the most important of which is a *design defect*. It is not easy to identify when the design of some product is defective since any design necessarily attempts to balance a number of different factors. While the safety of a product is an important factor, economic considerations, the appeal of the product to consumers, the durability of the product, and other factors also bear on the overall design selection. Nonetheless, some objective criteria are needed to evaluate when designs are defective. Two tests have been developed.

The first, the *consumer expectations* test, attempts to discern what levels of safety reasonable consumers anticipate a particular product will have. If the safety level the product provides is incommensurate with what consumers reasonably expect and someone suffers harm as a result of the design, the designer or seller is usually liable for that harm. Difficulties arise in applying this test to *complex designs*—a categorization that applies to many nanotechnology products because of the inherent difficulty for consumers to devise reasonable expectations about the nature of the design. What is more frequently applied to these types of designs is the *risk–utility* test, which seeks to balance the safety risk that a product design has against its utility. This test recognizes more explicitly that making a trade-off with safety risks may often be legitimate to achieve desired functionality. Under this test, a design is defective if the cost of avoiding particular risks is greater than the safety benefit that would result from such avoidance. Particularly relevant in performing such a balancing is whether there are alternative designs for the product that would not entail the same level of risk.

The flexibility of this test can be illustrated by considering the wide variety of products that might incorporate nanoscience and nanotechnology. At one end of the spectrum are consumer products that are entirely traditional except for being augmented using nanoscience to provide some particular benefit. Examples include such things as incorporating nanoparticles in fabrics to enhance resistance to stains or in children's toys to provide antibacterial properties. Most consumers expect such products to be just as safe as their conventional versions, and these conventional versions serve as examples of alternative designs. Even very modest increases in safety risks attributable to the nanotech structures are likely to be seen as outweighing the utility that those structures provide. For instance, improved resistance to staining of fabrics is likely to be seen as having insufficient utility if the risk of illness from wearing the fabrics is even only slightly elevated by the use of nanoparticles. When people are harmed as a result of incorporating nanoscience into consumer products, the producers of those products should expect to have little latitude in avoiding liability.

The same is not true of other types of products that might use nanoscience, with various military applications perhaps illustrating the opposite end of the spectrum. For instance, a very high degree of risk is likely to be acceptable for products that incorporate nano-energetic materials that are intended as explosive devices. The fact that a product is intended to cause harm or to be used in settings where injuries are common is relevant both to consumer expectations and to the balancing of risk and utility.

Where the real issue arises with these kinds of products is under circumstances where the *nature* of the risk is changed by the nanotech structures. When they present a risk that is not conventionally associated with the product, consumer expectations may be quite different from the actual nature of the risk. In such cases, the utility of the nanotech structures when applying the risk–utility test is better considered to be the difference in utility of the product with and without those structures. It is this incremental utility that should be weighed against the excess risk presented by the incorporation of nanotechnology. For example, consider the use of nano-energetic materials for the demolition of structures. There is a certain well-known risk of injury attributable to such demolitions. The mere fact that the demolition is performed with a nano-energetic material would not by itself change the calculus by which liability was determined. It is when the use of such materials produces a new form of risk, such as when the release and inhalation of nanoparticles at demolition sites causes new types of harms, that the evaluation of liability changes. The benefit of using nano-energetic materials as compared with conventional explosives should then be weighed against this new risk. With the test properly applied this way, there is a strong likelihood of liability whenever nanotech structures result in harms that are not well foreseen.

It is worth noting that the law affords special treatment to a class of products described as *unavoidably unsafe*. The doctrine has traditionally found its most effective application in medical contexts and many of the various nanoscience products being developed for medical uses are likely to fall in this category. The doctrine embraces a relatively extreme form of the risk–utility test—that there are certain products that have legitimate uses but are simply of a nature that makes them inherently unsafe. When this is the case and the user of the product is adequately informed about the potential risks, liability for harms caused by the product can be completely avoided.

It is not only in the context of unavoidably unsafe products that the need to inform users about risks arises. Indeed, the principal way to limit liability is to be candid about the nature of all known risks in the form of warnings. A consumer who is adequately warned about the risks associated with a product who still proceeds to use it is generally considered to have assumed the risk and its potential consequences. This illustrates the second type of product defect that is recognized: a warning defect. Manufacturers of products are expected to provide *adequate* warnings of *foreseeable* risks. For a warning to be adequate, it must be clear and specific, and communicate some understanding of the degree of risk to the consumer.

The final recognized product defect—a manufacturing defect—arises when there is a deviation from the standard product design in a particular product. The issues surrounding these kinds of defects are much more straightforward than those related to design defects: If the deviation from the product design resulted in a harm, there is liability, irrespective of almost any other factor that can be conceived.

Each of these liability theories for products is a form of *strict* liability, meaning there is no need to prove any sort of *fault* in order to establish liability. All that is needed is to prove the defect existed and a plaintiff was harmed as a result. There are other theories of liability in which it *is* necessary to establish a form of fault, and while these theories do have potential application to products, they are more usually applied in other contexts. This is specifically true of negligence, which is a legal doctrine that applies when one party owes a duty of care to another party. It may find particular application in medical-treatment settings that use nanotechnology.

Medical malpractice is essentially a specialized form of negligence that considers breaches of the duty owed by physicians or other medical practitioners to their patients. To be liable for medical malpractice, the physician must not only cause harm to a patient, but must also have done so in a way that breaches the duty towards the patient. He must have acted in a manner that is unreasonable according to the standards of the profession. Would a reasonable physician have prescribed the nanoscience treatment for this condition? Would a reasonable physician have administered the treatment in a similar way? If the answers to questions like these are "yes," then there will be no liability for harms caused by the treatment because there has been no breach of the duty. The law of negligence demands only that people behave reasonably; it does not demand infallibility of anyone.

2.2.4 *Evaluation*

These brief examples have been intended to give a flavor of how legal issues may bear on nanotechnology in both positive and negative ways. The rights of intellectual property encourage nanoscience researchers to disclose their results to the public so that all may benefit from developments that are made. The law of liability acts conversely to cause those who introduce nanoscience products or treatments to pause before doing so. The potential to be held accountable for the harms the products cause is intended to ensure that adequate testing is performed, the public is apprised of any risks, and the products are going to be introduced into the market in a responsible way.

There are many other ways in which nanotechnology might be affected by different laws. But ultimately all of those laws will act in one of these two ways. They will either provide an inducement for positive behavior or will provide a penalty to discourage negative behavior. It is in this way that the law acts symbiotically with nanoscience in promoting societal values.

2.3 Environmental Implications

Another aspect of liability is how nanomaterials impact the environment and health. We need not look far into the past to realize the significance of releasing toxic materials into the environment, thus potentially harming people who work with them, use them in consumer products, or are exposed through other environmental pathways such as in their drinking water or food supply. Those who develop these materials and use them in technological applications bear responsibility to ensure they do not harm health or the environment. Nanoscientists and engineers are developing novel materials that did not previously exist, and are using them for broad applications in many economic sectors: medicine, energy, industrial applications, materials science, engineering, electronics, cosmetics, and food science.

The unique behavior of materials 100 nm and below is what makes them attractive for developing new applications. The size raises questions, however, about whether the unique behavior also affects biological systems (people and the environment) differently than other materials. Deoxyribonucleic acid (DNA), the basic building block of life, is made of molecules that are nanosized, and can be called a nanoscale material, since the width of DNA is about 2–3 nm. Because it is not engineered, DNA may not fall into everyone's definition of nanoscale materials; however, a key point is that other nanoscale materials are in the same size range, so they can react with biological systems in ways that larger particles do not necessarily.

Environmental concerns about nanotechnology and nanoscale materials include the use of nanoscale materials in applications likely to result in releases to the environment, and the release of materials throughout the life cycle of a material, from its manufacturing, distribution, use and reuse, recycling, or disposal. Consider sunscreen as an example. First, one must fabricate the nanoscale particles for the sunscreen, which include nanoscale titanium dioxide ($n\mathrm{TiO_2}$) and nanoscale zinc oxide (nZnO). The $n\mathrm{TiO_2}$ and nZnO are packaged and shipped to the sunscreen manufacturer. When the sunscreen is sold, the consumer applies it directly to his or her skin, then swims or sweats off the sunscreen onto a towel or showers off the sunscreen. In each case, the $n\mathrm{TiO_2}$ and nZnO are released to water, a pool, lake, river, or the ocean while swimming, or to wastewater from bathing or laundry. When the container of sunscreen is empty, it either enters municipal solid waste for land or incineration disposal or is recycled. Each of these steps potentially releases the nanoparticles to the environment.

The breadth of potential applications raises questions about the safety of nanomaterials, and whether they have the potential to adversely affect health. C_{60} fullerenes are spherical molecules with a diameter of about 1–2 nm and have been reported to behave as antioxidants, scavenging radical oxygen molecules. Hydroxyl radicals have been associated with aging and stress, and antioxidants are hot market items for skin creams and *nutraceuticals*. Fullerenes are in at least six skin creams currently on the market [26]. C_{60} also appears to have catalytic effects on cell membranes reducing cell viability, and is being investigated as an antibacterial additive for disinfectants. Some early studies on the behavior of C_{60} raise concerns about whether it might cause unintended effects.

In a hospital environment, it is very important to keep surfaces free from contamination by bacteria, and many cleaning and disinfection products are used for washing equipment, hands, floors, and surfaces to help prevent the spread of germs. Using a product containing C_{60} as a disinfectant might mean it would be sprayed, wiped, poured into buckets and on floors, and washed down drains. Where could the C_{60} end up?

Researchers are now detecting bacteriocidal chemicals such as triclosan, commonly found in antimicrobial soaps and cleaning products, in rivers and other water bodies. Populations of bacteria that are resistant to antibiotics are also being found in the environment. Antimicrobial resistance is a big problem, since bacteria are no longer susceptible to the treatments we have for them. Could C_{60} cause antimicrobial resistance in the environment? What other unintended effects could a substance that catalyzes cell membranes cause?

There are hundreds of studies that have measured the effects of nanoparticle exposure on health, but they often suffer from methodological concerns and raise more questions than they answer. Using nanomaterials includes some potential for exposure of those developing the products and those using them to come into contact with nanoparticles. This section defines the field of nanotoxicology, introduces the concepts of environmental health and safety, and addresses larger concerns about environmental aspects of nanoscience.

2.3.1 Nanotoxicology

All materials pose some threat to health. Even water, the sustenance of life, can be toxic. When too much water is consumed, it causes an electrolyte imbalance leading to hypotonia, a condition in which a salt imbalance leads quickly to death. Thus, a key aspect to understanding the toxicity of materials is to know how much of a substance—the dose—can cause harm. Certainly we will not stop drinking water because too much water will kill us. Instead, it is reasonable to assume we will drink adequate, but not excessive, amounts of water. This concept, introduced by Paracelsus in the fourteenth century, forms the foundation of modern toxicology: "The dose makes the poison."

This is the rationale behind alchemy, as well as behind pharmacology, or the use of drugs to treat medical conditions. At some dose level, an effect occurs. In pharmacology, the effect is medicinal to treat some disorder. In toxicology, the effect is often an adverse one; that is, a substance causes toxicity to a cell or an organ system. We all know drugs can have side effects, so one area of toxicology examines the unintended effects of medicines. There are indeed many substances required for health that cause toxicity if taken in excess, including proteins, enzymes, and vitamins. When our bodies are introduced to substances at the wrong dose, or unintentionally, the exposure can result in toxicity.

There are many ways to study the effects of substances in the body. Commonly, substances are tested at different doses in laboratory animals to determine whether adverse effects occur, the types of effects, and the doses at which they occur. While mice and rats are not people, they are mammals and can indicate effects that may occur in people. Alternatives to animal testing include the use of cell culture assays to observe effects at the cellular level. These in vitro tests can be informative, but do not mimic the dynamics of human physiology and thus cannot completely replace animal testing.

More recent developments in toxicology include conducting assays using genetic material and proteins to evaluate specific biological interactions. Termed *-omics,* studies including genomic, proteomic, and metabolonomic assays help scientists understand the fundamental mechanisms by which substances cause toxicity. A genomic test examines how substances interact with genetic material, or DNA. Some substances cause changes to DNA that can lead to the formation of diseases such as cancer. Similarly, metabolonomics observed whether or how substances affect metabolism and can predict toxic interactions in the body. With tools such as micro- and nanoarrays, many tests can be conducted simultaneously. Eventually, there will be databases describing all types of toxicological mechanisms, and we may be able to use these data to predict the toxicology of new nanoscale materials based on their interactions with specific biological molecules.

It is important to keep in mind that we are all a bit different, and some of us are more sensitive than others. One person may be unaffected by an exposure to a substance while another person may suffer an adverse effect. One example of this is allergic reactions. In some cases, genetic differences can be linked to why some people are more sensitive than others to the effects of particular exposures. In other cases, it is unclear why some, but not all, exposed people have a particular reaction.

Nanotoxicology is a new field that evaluates the toxicity of nanoscale materials on the health of organisms. It is the application of toxicology to nanoscience. The term was coined by Oberdörster, Oberdörster, and Oberdörster in 2005 and defined as the "science of engineered nanodevices and nanostructures that deals with their effects in living organisms" [23]. The field of nanotoxicology explores the effects of exposure to nanomaterials.

There are few data available on the toxicology of nanomaterials. Type the search term *nanotoxicology* into the database of environmental health and safety (EHS; work on nanomaterials developed and maintained by the International Council on Nanotechnology [ICON]). How many papers appear in comparison to a search in the ICON EHS database with the term *toxicology*? Most of these studies fall in a few categories: simulations of particles inhaled into the lungs, cellular assays, and short-term ingestion exposure tests. Many are studies of ultrafine articles that are not engineered and are more diverse than engineered particles with a range of sizes and different levels of contamination. Each of these provides different types of information. But with nanoscale materials, the results of the studies are not always easy to interpret. One of the complexities is that some particles behave differently when they are nanoscale than when they are larger. This may be because of increased surface area of the particle, so doses equivalent on the basis of mass are very different on the basis of area.

Alternative explanations for effects from exposure to nanoparticles relate to particle surface charge or shape, among others. Toxicology has not traditionally addressed these complexities about dose, leaving considerable uncertainty about what to measure in studies. For example, what are the key parameters to describe for nanotubes, nanohorns, and aggregate particles? Because of this and similar

Example 2.1 *How to Define the Toxic Dose*

A study by the National Institute for Occupational Safety and Health found that the lungs of mice exposed to single-walled carbon nanotubes (SWNTs) formed "unusual fibrotic responses" compared to mice exposed to other types of carbon particles—specifically, an amorphous particle called carbon black. This response was seen in mice exposed to a dose of 10–40 μg/mouse aspirated into the lungs. The authors conclude that SWNTs are more toxic than carbon black. However, if compared on the basis of surface area instead of mass, 40 μg/mouse of SWNTs is estimated to have a surface area of 1040 m^2/g, while the surface area of the carbon black is 254 m^2/g [24]. In the study, the surface area was determined by Brunauer, Emmett, and Teller (BET) analysis, and diameter was measured by transmission electron microscopy (TEM).

On the basis of surface area, the mice receiving SWNTs received four times as much exposure to carbon particles in their lungs. So, on the basis of surface area, the results showing SWNTs are three to five times more toxic to lungs were actually done at doses four times higher! While this is not definitive evidence that nanotubes are not more toxic to the lung than other carbon nanoparticles, on the basis of surface area, SWNTs may not be more toxic than carbon black when inhaled in the lung.

Another complicating factor in this study is that the SWNTs were found to contain about 0.23% iron, which some have suggested may have contributed to the toxic responses in the lung [25]. Thus, it is not clear whether carbon nanotubes are more toxic than other shapes or sizes of carbon particles, and toxicology research now must consider new factors in the assessment of toxicity.

issues, almost every conference or meeting on nanotechnology includes a discussion of EHS risks. This uncertainty affects decisions about how to work safely with nanoscale materials, since we do not yet understand whether materials can cause toxicity or, if they do, how much is required for an effect to occur. There are many questions about the toxicology of nanoscale materials, and the questions suggest we take a broader look at the potential EHS risks in a larger context.

2.3.2 Nanotechnology Risk Assessment

In section 2.1.3, we asked, "What is the level of risk?" as a question about the nature of a technology. In this section, we discuss environmental risk assessment, a decision-making tool used to analyze and help make decisions about substances and technologies. Risk assessment may be defined as

> a process intended to calculate or estimate the risk to a given target organism, system or (sub)population, including the identification of attendant uncertainties, following exposure to a particular agent, taking into account the inherent characteristics of the agent of concern as well as the characteristics of the specific target system. [27]

More simply, risk assessment allows the estimation of health and environmental impacts from exposure to a substance.

Governmental and private organizations all over the world use risk assessment for environmental and public health decision making for risk management. In the United States, risk assessment is used to understand risks and make management decisions regarding hazardous waste site cleanup, closing municipal solid-waste landfills, setting standards and managing safe drinking water, chemicals' regulation, allowing additives to food and food packaging, and food safety regulations. The European Commission passed legislation in December 2006 to require risk assessments for all chemicals used in commerce in the European Union (Registration Evaluation and Authorization for Chemicals). Companies often perform risk assessments on the substances they manufacture or the products they sell.

You likely conduct risk assessments, too. Is it safe to drink the water from the tap? Is the food in the refrigerator still safe to eat, or has it gone bad? Are you adequately protecting yourself from exposure to materials you handle in the lab? Think of something important for you to succeed in. Will you go for it, or not even try for fear of failing? On an individual basis, we might have different answers to some of these questions. You might be unconcerned about your drinking water, but your classmate may insist on bottled water. If there were an absolute answer, we would not need to conduct an assessment of risk. However, in each case, there is uncertainty that requires some data for our decisions and also means we have to interpret the data, using our judgments. The uncertainty and our individual tolerance for risk affect how we judge the data and the conclusions we reach. The combination of science and judgment constitutes the assessment of risk. Conducting formal risk assessments allows us to apply a process for risk assessment to determine the relative level of risk and judge whether the risk is acceptable or not.

There are four basic steps in a risk assessment. First, we must define the problem or the hazard. Different international frameworks use different terminology for this step, referring to it as problem formulation, hazard characterization, or hazard identification. The hazard identification defines how to conduct the remainder of the assessment. It defines the questions the risk assessment will ask and answer. Hazard identification questions for nanomaterials may not differ from those for other substances; however, the necessary measurements are not widely available.

The second step is to develop an exposure assessment. Exposure assessment considers who might be exposed to the agent and defines the circumstances of those exposures. In human health risk assessment, receptors are those who may be exposed under different scenarios, such as workers in an occupational environment who manufacture a chemical, users of a product, and others who may have incidental exposure occurring as a result of manufacturing, use, or disposal of a product. In ecological risk assessment, the receptors may be specific species, populations, or entire ecosystems.

Substances behave differently in the environment, so it is important to understand whether exposure is likely to occur by a particular pathway. For example, a substance that is a solid would have to somehow be released in the air that we breathe; otherwise, we would not evaluate an inhalation pathway for that substance. For nanoscale materials, it is not clear how to measure exposure as it relates to toxicity (as a concentration, by surface area, or reactive surface area), and there are few analytical techniques currently available that quantitatively measure substances at the nanoscale.

The next step in a risk assessment is to look at how or whether the chemical causes harm. The dose–response assessment identifies the nature of a substance's toxicity by different exposure routes. This assessment relies on data from toxicology studies and, in some cases, from epidemiology studies in populations. The dose–response assessment identifies the effects at different doses and the lowest levels where effects or no effects have been observed. These effect levels become the basis for comparing to the exposure levels. One widely observed effect from exposure to nanoparticles is inflammation, associated with the development of many diseases. It is presently unclear whether the chemical composition, size, shape, or surface characteristics of particles affect the toxicity of nanoscale materials.

The risk characterization step brings together the exposure and dose–response assessments. This is the process of comparing exposure levels to effect levels to see whether the exposures that could occur under different scenarios are significant. The risk characterization also looks at the risks in context of regulations that define how much exposure is allowed under different circumstances and makes comparisons with other types of risks that help to inform how the risks are managed.

Generally, there is a lot of uncertainty associated with this comparison, and the risk characterization step considers these uncertainties. One example of an uncertainty is that we often cannot measure exposure exactly, so we make estimates that rely on assumptions (i.e., assuming a person drinks 2 liters (L; about a gallon) of water every day. Not all of us drink exactly 2 L of water per day; some of us drink 1 L, some drink 3 L, and others drink little if any water. This adds uncertainty to risk estimates because, when we assume some exposure occurs when a person ingests 2 L of water per day, we have simplified reality.

Despite uncertainty, risk assessment is a valuable technique for estimating the potential health and environmental impacts of nanoscale materials used in nanoscience and technology. Even when all of the necessary information is not available, risk assessment can still be helpful to make estimates of potential risk to set research agendas, or make safety decisions about working with or using materials.

2.3.3 Environmental Aspects of Nanotechnology

Many of the current applications of nanotechnology are in consumer products. Maybe some of your personal consumer products use nanotechnology (e.g., an MP3 player or the coating on your cell phone); your laptop screen may use nanotechnology for a stronger, scratch-resistant coating or a more efficient and higher resolution display. You could be wearing antimicrobial socks or static-free pants. Your toilet may have a self-cleaning surface (or perhaps you wish it did). As a student, you may be working with nanomaterials in a laboratory.

How do any of these applications of nanoscience affect the environment? If nanoscale materials are part of a coating on an electronic device, how could the nanomaterials enter the larger environment? When you use your cell phone, how could nanomaterials in the coating of the phone affect the environment? Generally, *exposure* must occur for the environment to be affected. Using a cell phone with a coating including nanomaterials does not result in a release of nanomaterial to the environment, so it is unlikely that it could have an effect.

Working in a laboratory, you might be handling nanoscale materials. You might be making new materials, or testing the properties of a material. Do you touch the materials with your hands? Could the surfaces of your lab have some nanomaterials on them? How could these get into your body? Do you ever put your hands near your mouth or eyes? Is it possible that nanoscale materials are airborne and could be inhaled? When an experiment is finished, where do the materials go? Do they get put in the trash or washed down the drain? If you are working in a clean room or a hood, are particles filtered by an air-circulating system? Are they part of the filter?

These questions are an informal way of conducting an exposure assessment. How materials are used and how they are disposed of influence how and whether they can affect the environment. It is important to consider the entire life cycle of a material to understand the potential for impacts to the environment. What is the life cycle? Some refer to it as "cradle to grave" or "cradle to cradle." Considering the potential for effects throughout the life cycle is an important step in generating new materials.

These questions about whether and how much nanoscale materials associated with nanoscience and nanotechnology enter the environment remain unanswered at the present. It is also unclear whether small amounts released make a difference in terms of health or environmental impact. There are more questions than answers about the fate of nanoscale materials in the environment. Will they enter the air we breathe? Will they contaminate our rivers and our drinking-water supplies? Will they leak out of landfills and enter unexpected places in the environment—for instance, the water we use to irrigate our food crops? Without the answers to these questions, we can only speculate about the behavior of nanoscale materials in the environment. We know nanotechnology is used and useful because of the unique properties at the nanoscale, so we can also speculate that the environmental transport and fate may also be unique from material to material. Considering the life cycle of materials offers an approach to think through the consequences of using nanoscale materials, in the context of their life cycle, both for health and safety and for society as whole.

Whether working in a research lab or for a manufacturer, or consuming products containing nanoscale materials, individual actions will affect whether and how much of the material enters the environment. It is not clear today what the potential impacts are from nanoscale materials in the air, water, and soil. It is not understood whether nanomaterials might enter the food web and become part of what we eat and drink, or whether they can affect our forests and coral reefs and air quality. There is much research to be done to understand the potential impacts. But your actions as a scientist, an inventor, and a consumer make a difference in terms of the extent of impact that occurs.

2.4 Public Perception

Not wanting nanotechnology to be subject to the same sorts of public criticisms that have emerged for controversial forms of biotechnology (such as genetically modified foods, cloning, and stem cell research), considerable effort and expense is being invested in the United States and in Europe in trying to engage the general public early on in considering nanotechnology's benefits and risks. The hope is that later reactions will be better informed if public discussion starts at an earlier point, sometimes referred to as *upstream engagement*.

Ideally, encouraging people to think and talk about emerging technologies at an early stage will also help the futures of those technologies be shaped (in turn) by popular values and priorities. After all, much of the research on these technologies is paid for by tax dollars, so if the public is not "on board" with the directions in which this research is taking us, the odds of negative reactions inevitably rise.

2.4.1 Factors Influencing Public Perception

Many scientists and policy makers alike assume that popular reactions to new technologies are based on superstitions and exaggerated fears when they ought to be based on science. In fact, research has shown that scientific knowledge may not make as much difference as other factors, such as social values (whether the technologies appear to achieve goals that ordinary people consider important) and whether people trust business and government to do the right thing.

Of course, nano is a scale, not a thing! Determining what people think nanotechnology is, let alone whether they think it is good or bad, is not really an easy matter. Focus groups with adults across the United States and Canada, as well as interviews with students at the University of South Carolina in Columbia, have established that, at present, most nonscientists have very little notion of what nanotechnology might be or how it might work. While experts may worry that ordinary people will become unreasonably fearful because of the influence of images of *gray goo* or runaway *nanobots*, so far there is little evidence this is taking place.

Instead, people tend to react to nanotechnology according to their value systems and their experiences with previous technologies. This is not to say that everyone who is opposed to (say) stem cell research or nuclear power is opposed to all other forms of new technology. Rather, people tend to draw from their experience to determine whether the regulatory system for technology is adequate, whether environmental impacts will be difficult to control, whether scientists are conscious of their social responsibilities, and so on.

2.4.2 Nano and Public Opinion Polls

Public opinion polls, while not perfect, are one way we are accustomed to *taking the public pulse* in modern societies. One of the reasons polls are tricky in the very early stages of the development of a technology is because when people have little information, they are very susceptible to small differences in question wording or question order. For nanotechnology, for example, what may seem like minor differences in definitions or the types of examples offered can be quite influential. Also, people may answer questions about their *opinions* even when, in truth, they have not had much opportunity to consider the topic. In fact, some observers would say these people do not really have opinions at all!

Nevertheless, polls are a useful indicator of both initial public reactions and general trends in those reactions. We have developed relatively reliable procedures for conducting polls, even though declining rates of response and shifts away from traditional land-line telephones in favor of cell phones have caused a lot of concern.

One of the first American surveys of public opinion about nanotechnology was conducted over the Internet with over 3900 respondents [28]. As is the case for many Internet surveys, this study was not actually a random or probability sample representative of the U.S. population as a whole. People who chose to answer this survey were likely to have been volunteers and may have been more interested in nanotechnology and more knowledgeable about science and technology generally than the average person. Nevertheless, the results were consistent with other research on popular reactions to technologies in several ways. Women were less optimistic about the risks (a common pattern), whereas people who are generally protechnology were more likely to react favorably.

In this survey, 57.5% of those responding felt that *human beings will benefit greatly* from nanotechnology. This high percentage must be judged in the context of the fact that volunteers in such cases are likely to include a higher proportion of those who are enthusiastic about the technology in question, whatever it might be. However, other surveys also indicate a generally positive public reaction.

In what was probably the first published U.S. phone survey of nanotechnology opinions designed to be representative of the general public rather than composed of volunteers that was conducted a few years later, similar optimism was visible [29]. In this survey of just over 1500 people, respondents felt that the benefits of nanotechnology would outweigh the risks, even though 51.8% had heard "nothing" about nanotechnology! And 80.3% were "not worried" about developments in this area.

In other words, most Americans, in their initial reaction, were not especially concerned about nanotechnology. Treating human disease was identified in this study as the most important of five benefits, clearly reflecting participants' underlying values as well as their overall positive expectations from science and technology. In our proscience, protechnology culture and in the absence of specific evidence that a technology represents cause for concern, this may not be surprising, but it is different from the reactions to some other major technologies we have confronted in recent years, especially biotechnologies. And it is also different from reactions in other parts of the world, to a certain extent.

According to a comparative analysis using data from the "Eurobarometer" (a periodic survey based on in-person interviews with individuals from across Europe—15,000 of them in this particular case) and a phone survey of 850 people in the United States, fully 50% of U.S. respondents believed nanotechnology would improve the quality of life in their country over the following 20 years [30]. Only 29% of Europeans agreed, most likely reflecting cultural differences in technology-related optimism.

However, only small numbers in either location responded that they felt nanotechnology would actually have a negative effect (4% in the United States and 6% in Europe). And in a slightly later U.S. survey of just over 700 people, 74% predicted nanotechnology would help in the area of detection and treatment of human diseases, 64% felt it would help increase national security, and 62% thought it would help in environmental cleanup [31]. A subsequent survey in January of 2005 found 46% of those in the United States foreseeing a positive impact [32]. Nevertheless, only about half the U.S. population seemed to believe that nanotechnology's developers shared their values.

Further, the North American focus group research discussed earlier did suggest some concerns that seemed to "resonate" with the general public [33]. In other words, when small-group discussions turned to certain topics, concerns seemed to be sparked. These topics included issues of the adequacy of control and regulation by government agencies, and also questions of negative environmental and health effects. In addition, more diffuse concerns emerged regarding social disruption, workforce impact, and distribution of benefits. If there are going to be new cures for cancer, for example, people seem to want to know if their insurance will cover these adequately or if these treatments will only be available to the economically well off. They also want to know if some people will lose their jobs or possibly go out of business as a result of technology's evolution.

Why do nanotechnology and biotechnology seem to elicit such different public reactions? Apparently, biotechnology—especially the manipulation of genetic material through recombinant DNA techniques—produces reactions and concerns that are less immediately obvious for nanotechnology. Biotechnology can challenge our cultural notions of individuality (through cloning) and species identification (through genetic engineering that combines genes from different kinds of plants and animals). Stem cell research causes some people pause to reflect on what it means to use human embryos in the laboratory, even for good purposes. Or, as some commentators put it, the *yuck factor* for biotechnology is much closer to the surface than for nanotechnology.

2.4.3 *A Call for Two-Way Communication*

What does all this mean for the science behind nanotechnology, and what do scientists need to do about public opinion? In some ways, nanotechnology has become a giant experiment in new ways of introducing new technologies to society. Because of past difficulties, much more attention is being paid to how the public will react. Increasingly, scientists feel an obligation to become part of the solution here rather than just part of the problem—to help educate the public in the hopes that popular concerns will not mushroom out of proportion. However, communication between scientists and the public—whether it takes place through public meetings, at events at universities or science museums, or through stories in the mass media—will be most meaningful if it is two-way. At the same time scientists are striving to educate the public, they should be striving to understand the public's actual concerns. Ideally, in a democratic society, science policy—the ways we choose to invest in research and development, and the approaches we take to regulation—will evolve in ways that reflect public values.

2.5 Future of Nanotechnology

Professor Kip Thorne of Caltech is a theoretical physicist, known for his visionary understanding of the universe, who was named the 2004 California Scientist of the year. He asks a very probing question: "In an infinitely advanced universe, what things will be possible and what things won't?" The limitations of both physics and humanity are a subject that is not well understood. The future of nanotechnology, as described in this chapter, will be framed by the physical limitations to which Dr. Thorne is referring. Currently some are in the process of being understood and others only imagined.

The study of the future is commonly referred to as a soft science because it often lacks concrete data points around which decisions can be made. However, the most beneficial aspect of the future remains that it is largely unknowable. A known future is demoralizing. Inside the mysteries of a future that we cannot know lie the hope and inspiration that drive humanity forward.

On the following pages we will discuss some of the tools that give us clues about the world ahead, and apply these tools to nanotechnology and its surrounding developments. The tools that will be discussed include cycles and patterns, trend forecasting, attractionary futuristics, and Maximum Freud.

2.5.1 *Cycles and Patterns*

Edward R. Dewey (1895–1978) was an economist who studied cycles and patterns in economics and other fields. Dewey first became interested in cycles while he was chief economic analyst of the Department of Commerce in the 1930s. President Hoover wanted to know what caused the great depression. Dewey found that economists could give no consistent answers on the cause of the depression and he lost faith in the existing economic methods. He then shifted his study to how business behavior occurs, rather than why, and in doing so was able to map business patterns and form a more reliable understanding of the future than economists were able to give.

Dewey devoted his life to the study of cycles, claiming that "everything that has been studied has been found to have cycles present." He carried out extensive studies of cyclicity in economic, geological, biological, sociology, and physical sciences and other disciplines. So far, more than 500 different phenomena in 36 different areas of knowledge have been found to fluctuate in rhythmic cycles. The study of nanotechnology is tied to many different cycles, ranging from harmonic oscillation to circadian rhythms to absorption spectroscopy to standard business cycles.

2.5.2 Trend Forecasting

Similar in some respects to the study of cycles, trend forecasting is an exercise in pattern recognition where a researcher is able to create a high probability trend line, using historic data points, to determine a future event. Commonly adopted methods of trend forecasting include the Delphi method (averaging the opinions of experts), forecast by analogy, growth curves, and extrapolation. Normative methods of technology forecasting such as the relevance trees, morphological models, and mission flow diagrams are also commonly used.

While this list of methodologies makes it sound like trend forecasting has been reduced to some very scientific processes, it really boils down to either rational and explicit methods using hard data or intuitive methods using personal opinion. Moore's law with its prediction of "doubling of the number of transistors on integrated circuits every 18 months" and Metcalf's law, which states that "the value of a telecommunications network is proportional to the square of the number of users of the system" are two examples of how the study of trends can be used to predict future events.

2.5.3 Attractionary Futuristics

An *attractor* is a future event that humanity is being drawn towards. Future events such as putting a man on Mars, finding a cure for cancer, developing a mass energy storage system, or inventing a flying car are all well-known endpoints that have become staples of our culture, and a common theme in media as well as the global conversation. In our daily lives we often discuss these attractors without any awareness that the mere discussion of the topic reinforces its role as an attractor.

A new science that I am involved in developing at the Colorado-based DaVinci Institute, *attractionary futuristics*, is the science of attractors where the effect that a known future event is having on present-day research is being studied. Research is being focused on identifying known attractors, the creation of new attractors, categories of attractors, range of influence, intensity of the attraction, and the directional vectors of these forces. An example of an attractor in the nanotech field is the creation of the ultimate small memory storage particle and the effect this is having on high tech.

Past visionaries have given us the visions we hold today of the future. Sometimes these visions come in the form of illustrations or artwork and at other times in movies or video clips, but very often they start as nothing more than ideas in literature. As an example, Leonardo da Vinci dedicated over 500 drawings and 35,000 words to the concept of flying. He used the tools at his disposal to convey the idea that flying would someday become both viable and practical. His ideas managed to survive the centuries and eventually came to life, first in the form of Joseph and Jacques Etienne Montgolfier's hot air balloon, and later with the Wright brothers' flying machine in 1903 at Kitty Hawk, North Carolina. The da Vinci drawings served as a source of inspiration and as a blueprint of sorts for making it happen.

Other notable visionaries who have influenced our thoughts on the future include Jules Verne's visions of submarines and space travel, Gene Roddenberry's visions of cell phones and teleportation, Arthur C. Clark's visions of talking computers and the space elevator, Philip K. Dick's visions of flying cars and time travel, and George Lucas's visions of robots and space travel.

In 1986, the field of nanotechnology was jolted into existence with K. Eric Drexler's book, *Engines of Creation,* and his visions of what may or may not be possible once we have the ability to work with nanoscale materials. While some of his visions for molecular-scale factories and self-assembling machines remain the source of much discussion and debate, his visions served as a significant turning point for scientific research and commercial development.

Our visions of the future determine our actions today. Very often the images of the future that we hold in our heads have a profound impact on the ways we lead our lives and the decisions we make at work. For this reason it becomes imperative that we continue to improve our visions of the future.

2.5.4 Maximum Freud

Maximum Freud is a technology life-cycle tool that can be used to predict technology endpoints. The topic is best explained with the story in example 2.2.

Clearly this period of time was the end of an era. It was the end of the slide rule era and the beginning of the calculator era. As a society we have not seen

EXAMPLE 2.2 *The Slide Rule*

In 1972, I was a young engineering student at South Dakota State University in Brookings, South Dakota. One of the first courses I was required to take was a short course on slide rules. For those of you who do not know what a slide rule is, first came the abacus, then came the slide rule, and then came the calculator.

This was a time when the real "cool geeks" on campus walked around proudly displaying their black carrying cases for their slide rule attached to their belts. "Brainiacs on parade," as some described them, used their slide rules as a symbol to tell the world how smart they were.

Early calculators were first making their way into stores around 1970, and in 1972 they were still rather expensive. I remember arguing with my teacher about whether or not the slide rule course was necessary and his response was a dismissive "all engineers need to know how to use the slide rule."

Of course his thinking was wrong. Even though I completed the course with flying colors, I have never used a slide rule in my engineering work. Engineers at Hewlett Packard and Texas Instruments who were working on next-generation calculators at the time would have laughed at my teacher's assertion that slide rules were always going to be the centerpiece of the engineer's tool chest.

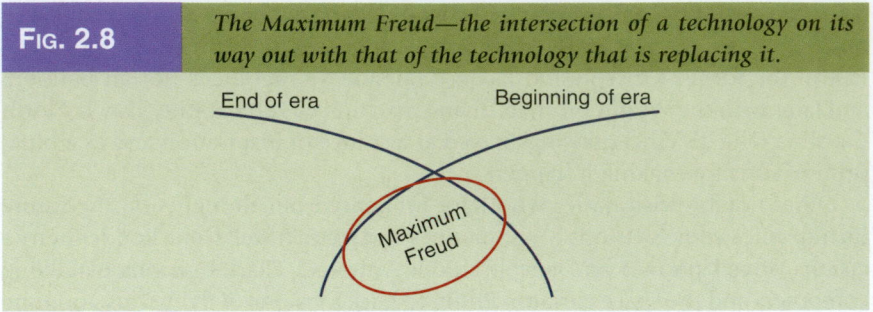

Fig. 2.8 *The Maximum Freud—the intersection of a technology on its way out with that of the technology that is replacing it.*

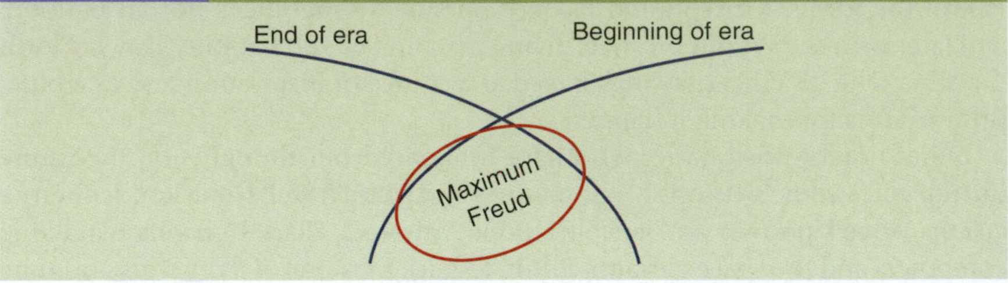

Source: Courtesy of the DaVinci Institute. With permission.

the end of too many eras, but we are on the verge of experiencing many things disappearing in the near future. Most will not be as cleanly defined as the slide rule being replaced by the calculator. Often the soon-to-be-obsolete technology will be replaced by two or three other technologies.

As I sketched out the simple diagram showing the end of one era and the beginning of another, the point where the two eras overlapped caught my attention. This period of time is important to isolate because of the extreme dynamics happening there. It also occurred to me that we did not have a name for this intersection of technology, this collision of business forces. So I came up with the name *Maximum Freud* (**Fig. 2.8**). Yes, it is a rather wacky name, but useful in its description.

As technologies approach Maximum Freud, this is the period when industry players have to spend their time on the Freudian couch to understand what is going on. This is a period of extreme chaos, and also a period of extreme opportunity. But here is the most important part:

> All technologies end. Every technology that we use today will someday go away and be replaced by something else. Every technology will approach its own period of Maximum Freud. So, from the standpoint of making bold predictions, the imminent demise of many of our technologies is a certainty. [34]

Here are just a few examples of technologies that are currently approaching Maximum Freud:

- Checking industry—This is already in decline; the end of the handwritten check is drawing near.
- Space shuttle—This Model T of the space age is long overdue to be replaced by an efficient, low-prep craft that makes space accessible to the common man.
- Sign language—Advances in cochlear implant technology will soon make the need for visual person-to-person sign language unnecessary.
- Fax machine—Museum curators are already dusting off a spot for this once staple of the business world. Already in its twilight, the remaining days of the fax machine are numbered.
- Traditional AM–FM radio—As wireless technologies become more ubiquitous, aspiring radio station owners will find many new low-barrier entry points for commercial broadcast without the need for assigned radio frequency spectrum.

A good example in the nanotech field is the coming demise of the diamond mining industry. Artificial diamonds are on the verge of being created faster, more cheaply, and better than diamonds that nature produces; consequently, the labor-intensive mining industry will soon enter into the chaos of Maximum Freud.

The Maximum Freud concept is a tool we can use to determine when the technologies that we use today will be gone tomorrow, and in the process speculate on their replacements. We live in a very fluid, changing world and each step we take towards the future will enable us to experience life in a new and different way, and nanotechnology will be one of the key agents of change.

2.5.5 Nanotechnology End Points

Tying together this discussion of the future of nanotechnology, we will take a brief look at 10 of the possible end goals, or *attractors,* that are driving the industry. Each of these items can best be described as a disruptive technology with far-ranging business potential:

1. *The ultimate small storage particle.* Eventually Moore's law will end and we will reach a point where we have created the smallest possible particle on which we can store data. Once it is realized that we have created the smallest possible storage particle, the industry will begin to establish standards for information storage, so researchers 200 years in the future will be able to access the data.
2. *The ultimate small motor.* Ultratiny robots will be driven by ultratiny motors—so how small can we make a motor? The Foresight Institute is currently offering a $250,000 Feynman Prize to the person or team that creates a motor no more than 100 nm wide in any direction that is capable of moving atoms around.
3. *The ultimate small flying machine.* The U.S. Department of Defense has funded many projects related to testing the limits of aircraft miniaturization. So far they have been unable to achieve their goal of creating an invisible spy craft that can be flown into an enemy camp without detection. The ultimate small flying machine will, of course, have many other uses.
4. *The ultimate small decision-making (smart) machine.* How small can we scale intelligence and how do we go about building that intelligence into nanoscale machines? The Foresight Institute is currently offering a $250,000 Feynman Prize to the person or team that creates a machine less than 50 nm wide that is capable of adding numbers.
5. *Binary power.* Binary power is a technology where two otherwise harmless beams of energy intersect at a point in space and create power. When imagining systems for powering nanoscale machines with molecule and atom-specific precision, the idea of binary power begins to make sense. This type of power will be very useful for powering nanomotors and nanoscale flying machines.
6. *Molecular alteration.* Much like the medieval dream of being able to turn lead into gold, molecular alteration brings with it the promise of turning inferior grade materials into superior grade materials, useless substances into valuable substances, and problem cells into solution cells.

7. *Robotic nanosubmarine for repairing the human body.* As the ultimate medical device that can be driven to problem areas of the body, the nanosubmarine will come complete with vision tools, cellular diagnostic equipment, and arms, legs, and lasers for effecting change to problem areas of the body.
8. *The ultimate indestructible materials.* Virtually every material known to mankind enters a deterioration process as soon as it is created. The dreams of being able to build a house that lasts forever or roads that never show signs of wear all start with the creation of an indestructible material. However, a truly indestructible material will result in landfills filled with indestructible trash. So the ultimate indestructible material will have a disintegration feature that can be triggered at the end of product life.
9. *Self-repairing cells* (self-healing body tissue). Much like the lizard that loses a leg and is able to grow it back, the idea of creating regenerative cells brings with it the promise of eternal youth and the ability to live forever.
10. *The trifecta particle.* If you can imagine the most useful nanoscale material and the properties it will have, you will have imagined what we call the *trifecta particle*. This particle will have the attributes of being designable, controllable, and intelligent. Computer programming is all about architecting the flow of electrons. In the future, nanotechnology will be all about architecting the flow of matter, and these are the particles that will make that possible.

This list of end goals or "attractors" is not intended to be an exhaustive list of all the possibilities. Rather, its purpose is to stimulate thinking and stretch the imagination. Our visions of the future determine our actions today, and these examples are the first step towards improving your vision.

Although by no means exhaustive, the purpose of this chapter was to help you recognize the tremendous implications nanoscience and technology have for influencing the larger society now and in the future. Proponents of nanotechnology laud the potential for advances at the nanoscale to increase energy efficiency, improve human health, mitigate environmental degradation, and create new economic opportunities. Skeptics are less optimistic and warn that we need to proceed more cautiously to avoid the untoward consequences that inevitably accompany such revolutionary advances. Keep these various issues in mind as you continue through the course. If nanoscience and technology represent *the next big thing,* and broad societal impacts are inevitable, then all of society benefits from thoughtful awareness and responsible action on the part of scientists and engineers.

Acknowledgment

We are extremely grateful to Professor Deb Bennett-Woods of Regis University in Denver, Colorado, for donating her time and effort to organizing and editing this chapter. She also wrote section 2.1, the portion of the chapter that addressed the ethical implications of nanotechnology.

References

1. P. J. Bond, Preparing the path for nanotechnology. In M. C. Roco and W. S Bainbridge, eds., *Nanotechnology: Societal implications—Maximizing benefits for humanity;* Report of the National Nanotechnology Initiative Workshop, Arlington, VA, 16–21, December 2–3 (2003).
2. M. C. Roco and W. C. Bainbridge, eds., *Societal implications of nanoscience and nanotechnology: NSET Workshop Report*, National Science Foundation, 12 (2001).
3. Human Genome Project, U.S. Department of Energy, www.ornl.gov/sci/techresources/HumanGenomne/elsi.shtml (2006).
4. *The national nanotechnology initiative at five years: Assessment and recommendations of the national nanotechnology advisory panel*, President's Council of Advisors on Science and Technology, The National Nanotechnology Initiative (2005).
5. *The National Nanotechnology Initiative: Research and development leading to a revolution in technology and industry, supplement to the President's FY2006 budget*, Nanoscale Science, Engineering, and Technology Subcommittee, Committee on Technology, and the National Science and Technology Council, March (2005).
6. *Societal dimensions*, National Nanotechnology Initiative, www.nano.gov/html/society/home_society.html
7. D. M. Berube, *Nano-hype: The truth behind the nanotechnology buzz*, Prometheus Books, New York (2006).
8. *MIT Institute for Soldier Nanotechnologies*, Massachusetts Institute of Technology, December 24 (2006) web.mit.edu/ISN/resedarch/team02/index.html
9. *Ethical, legal and other societal issues*, National Nanotechnology Initiative, December 15 (2006), www.nano.gov/html/society/ELSI.html
10. J. A. Shulmam and I. Asimov, *Isaac Asimov's book of science and nature quotations*, Grove Press, New York (1988).
11. *The Belmont Report: Ethical principles and guidelines for the protection of human subjects of research*, The National Commission for the Protection of Human Subjects of Biomedical and Behavioral Research, April 18 (1979). www.ohsr.od.nih.gov/guidelines/belmont.html
12. D. Sarewitz and E. Woodhouse, Small is powerful. In A. Lightman, D. Sarewitz, and C. Desser, eds., *Living with the genie: Essays on technology and the quest for human mastery*, Island Press, Washington, D.C., 63–83 (2003).
13. D. Bennett-Woods and E. Fisher, Nanotechnology and the IRB: A new paradigm for analysis and dialogue, *Proceedings of the Joint Meeting of the European Association for the Study of Science and Technology and Society for Social Studies of Science*. Paris, France (2004), www.csi.ensmp.fr/csi/4S/index.php
14. D. Bennett-Woods, Integrating ethical considerations into funding decisions for emerging technologies, *Journal of Nanotechnology Law & Business*, 4 (2007).
15. D. Sarewitz, Science and happiness. In A. Lightman, D. Sarewitz, and C. Desser, eds., *Living with the genie: Essays on technology and the quest for human mastery*, Island Press, Washington, D.C., 181–200 (2003).
16. B. Gordijn, Converging NBIC technologies for improving human performance, *Journal of Law, Medicine & Ethics*, 34, 726 (2006).
17. R. Rodrigues, The implications of high-rate nanomanufacturing on society and personal privacy, *Bulletin of Science, Technology & Society*, 26, 38 (2006).
18. E. Gutierrez, *Privacy implications of nanotechnology*, Electronic Privacy Information Center, www.epic.org/privacy/nano
19. M. C. Roco and W. S. Bainbridge, eds., *Nanotechnology: Societal implications—Maximizing benefits for humanity*, Theme 10: Education and Human Resource Development, Report of the National Nanotechnology Initiative Workshop, 88, Arlington, VA, December 2–3 (2003).

20. A. S. Daar, H. Thorsteinsdóttir, D. K. Martin, A. C. Smith, S. Nast, and P. A. Singer, Top ten biotechnologies for improving health in developing countries, *Nature Genetics*, 32, 229–232 (2002).
21. *Reducing risks, promoting healthy life*, World Health Organization, The World Health Report 2002. Published report, World Health Organization, Geneva (2002), www.who.int/whr/2002/en
22. *Nanotechnology and the poor: Opportunities and risks*, Meridian Institute, published report, Meridian Institute (2005), www.meridian-nano.org/gdnp/paper.php
23. G. Oberdörster, E. Oberdörster, and J. Oberdörster, Nanotoxicology: An emerging discipline evolving from studies of ultra-fine particles, *Environmental Health Perspectives*. Online, July (2005).
24. J. B. Mangum, E. A. Turpin, A. Antao-Menezes, M. F. Cesta, E. Bermudez, and J. C. Bonner, Single-walled carbon nanotube (SWCNT)-induced interstitial fibrosis in the lungs of rats is associated with increased levels of PDGF, mRNA, and the formation of unique intercellular carbon structures that bridge alveolar macrophages in situ, *Particle Fibre Toxicology*, 3, 15 (2006).
25. A. A. Shvedova, E. R. Kisin, R. Mercer, A. R. Murray, V. J. Johnson, A. I. Potapovich, Y. Y. Tyurina, O. Gorelik, S. Arepalli, and D. Schwegler-Berry, Unusual inflammatory and fibrogenic pulmonary responses to single walled carbon nanotubes in mice. *American Journal of Physiology—Lung Cellular and Molecular Physiology*, 289, L698–L708 (2005).
26. Woodrow Wilson International Center for Scholars. *Project on emerging nanotechnologies. A nanotechnology consumer products inventory*. Retrieved April 29 (2007), www.nanotechproject.org/
27. IPCS risk assessment terminology. Part 1: IPCS/OECD Key generic terms used in chemical hazard/risk assessment. Part 2: OPCS Glossary of key exposure assessment terminology; World Health Organization (WHO), January 14 (2007). www.who.int/ipcs/methods/harmonization/areas/ipcsterminologyparts1and2.pdf
28. W. S. Bainbridge, Public attitudes toward nanotechnology, *Journal of Nano Research*, 4, 561–570 (2002).
29. M. D. Cobb and J. Macoubrie, Public perceptions about nanotechnology: Risks, benefits and trust, *Journal of Nanoparticle Research*, 6, 395–405 (2004).
30. G. Gaskell, T. T. Eyck, J. Jackson, and G. Veltri, Imagining nanotechnology: Cultural support for technological innovation in Europe and the United States, *Public Understanding of Science*, 14, 81–90 (2005).
31. D. A. Scheufele and B. V. Lewenstein, The public and nanotechnology: How citizens make sense of emerging technologies, *Journal of Nanoparticle Research*, 7, 659–667 (2005).
32. S. H. Priest, The North American opinion climate for nanotechnology and its products: Opportunities and challenges, *Journal of Nanoparticle Research*, 8, 563–568 (2006).
33. S. H. Priest and H. Fussell, *Nanotechnology: Constructing the public and public constructions*. Paper presented at the Annual Meeting, Association for Education in Journalism and Mass Communication, San Francisco, CA, August (2006).
34. T. Frey, Collision course: Comings and goings of technologies marked by futurist's "Maximum Freud," *Rocky Mountain News*, July 1 (2005).
35. W. Joy, Why the future doesn't need us, *Wired*, 8.04 (2000) www.wired.com/wired/archive/8.04/joy.html?pg=1&topic=&topic_set=
36. D. B. Hughes, Let the nanotech wars begin! KurzweilAI.net, Dec. 14 (2003) www.kurzweilai.net/news/frame.html?main=/news/news_single.html?id%3D2748
37. R. E. Smalley, Of chemistry, love and nanobots—How soon will we see the nanometer-scale robots envisaged by K. Eric Drexler and other molecular nanotechnologists? The simple answer is never, *Scientific American*, 68–69, September (2001).
38. Public policy, definition, www.bitpipe.com/tlist/Public-Policy.html (2007).
39. M. Crichton, *Prey*, Avon Books, Harper-Collins Publishers, New York (2002).

Problems

Introduction to Societal Issues

2.1 Locate an article on a new development in nanoscience or nanotechnology and list the possible societal implications of such a development.

2.2 Using the NNI Website as a resource, review current activities supported by the National Science Foundation. Briefly describe what you believe to be the more important areas of focus. Imagine that $40 million has already been allocated for environmental, health, and safety research and you have been given the job of authorizing the remaining $40 million in miscellaneous SEIN research funds for the coming year. How might you prioritize the allocation of funds and what would be your justification?

Ethical Implications

2.3 Write a paragraph describing the ethical responsibilities of scientists and engineers in nanotechnology.

2.4 One of the most controversial areas of nanotechnology relates to the convergence of nanotechnology, biotechnology, artificial intelligence, and cognitive science, among others. Developments in these areas might not only help restore people who are ill or have been injured to normal functioning, but also may be used to dramatically enhance human performance and perhaps extend the normal human life span. Using the ethics section for reference, pose a list of ethical questions that might be used to consider the societal implications of the use of nanotechnology to enable human enhancement and life extension.

Legal Implications

2.5 Mary is a graduate student working under the supervision of Dr. Brilliant at Prestigious University. The research Mary conducts on the structures of nanotubes leads to a remarkable new process that can produce very long nanotubes with extraordinary efficiency. The most innovative aspects of developing the process are due to Mary, but Dr. Brilliant has consistently overseen how the research was conducted and offered helpful suggestions at many key points in the development. Mary and Dr. Brilliant patent the process and sell it for a very tidy sum to NanoCorp, which becomes the exclusive user of the process. NanoCorp enjoys impressive profits and begins to dominate the nanotube industry as a direct result of using Mary's process. It sells its products in the United States through an exclusive relationship with a single retailer, Bucky Inc. There are no U.S. regulations that restrict it in any way from adopting Mary's process in the manner it deems most useful.

Five years later, physicians begin to report a previously unknown respiratory illness in patients employed by NanoCorp, specifically those who work most actively on implementing Mary's process. Over the next several years, research identifies a biological mechanism that conclusively demonstrates a causal relationship between Mary's process and this illness. Over that same period of time, mortality rates begin to increase among NanoCorp workers and a few reports start end consumers suffering from similar afflictions to appear. Those reports also increase and, after another 10 years, it becomes clear that tens of thousands of consumers have either died or suffered a great loss in the enjoyment of their lives as a result of the process Mary developed. Calls for regulation begin, with many public commentators chastising the U.S. government for taking too little action too late.

Work with a partner. One of you should identify the strongest arguments you can devise for finding each of the following parties liable for the injuries suffered by NanoCorp workers and by consumers:

a. Mary
b. Dr. Brilliant
c. Prestigious University
d. NanoCorp
e. Bucky Inc.
f. The United States

The other of you should respond to those arguments with the best defenses you can devise. In each instance, who do think has the stronger argument? Are there any parties for whom you would like to have additional facts? Which parties and which facts?

After completing the exercise, research the history of asbestos litigation in the United States, particularly the scope of its economic impact. What parallels exist with the hypothetical exercise? What lessons does asbestos litigation provide for those involved in the development of nanotechnology? Summarize your conclusions in a short essay.

2.6 The implementation of the patent system in the United States is frequently criticized for permitting too many obvious inventions to be patented. In some ways, this is reminiscent of an exercise performed several years ago by a major physics journal in which many readers thought too many of the published papers to be of only marginal interest; further investigation found general agreement that about 10–15% of the papers would not be missed if they had never been published, but almost no agreement on which papers were in that group. You are also likely to find that many of the technical ideas discussed in this book will present you with conceptual challenges, but when you review them years from now they will no longer seem at all difficult and the struggles you endured to master them will seem almost quaint; you have probably had similar experiences in the past.

With these issues in mind, explain how you would define an *objective* test for obviousness of a patent application. What are the key factors that indicate something was obvious at the time it was "invented"? Is your test effective at putting the person applying it in the position the inventor found himself in? What safeguards does your test have to prevent a hindsight view from being applied unfairly? Is your proposed test dependent on the technology involved—that is, would you apply the same test to judge the obviousness of a nanotechnology invention as you would to judge the obviousness of a new type of pillow or a new design for a shoehorn? Do you think it is more sensible to have a single objective test applicable across all technology or different tests for different technologies?

2.7 One of the aspects of nanoscience that is commonly lauded is its relevance to an unusually wide range of human activities. When considering regulation of nanotechnology, there are then (at least) two approaches that could be adopted by governments: (1) form a centralized agency that regulates all aspects of nanoscience and nanotechnology or (2) allow existing agencies that are focused on particular areas to regulate nanoscience and nanotechnology as they apply to their own areas. Which approach do you think is better? Defend your view fully.

Environmental Implications

2.8 Let us consider the potential for effects using nanoscale titanium dioxide (nano-TiO_2) in a coating for your cell phone. Here is a brief description of the life cycle of the nano-TiO_2. Titanium dioxide is mined and shipped to a raw material manufacturer. A chorination process converts the raw mineral to liquid titanium tetrachloride, a toxic and unstable substance. The nano-TiO_2 is manufactured from the titanium tetrachloride at high temperature with the addition of water. A coating company buys the nano-TiO_2 and it is shipped in a drum in powder form to the company, which mixes it with a polymer coating. The coating company packages the mixture and ships it to the cell phone housing manufacturer, where it is applied as a coating in the fabrication process and shipped to a company that manufactures your phone. Your phone is shipped to a retailer, where you buy it, and then use it for a period of time. Suddenly, it is time for a new phone, so you dispose of the old phone. If you toss it in the trash, a waste hauler collects your phone, and delivers it to a waste facility—either a landfill, where it enters a pile of trash, or an incinerator, where it is burned. Identify the steps in this scenario in which the nano-TiO_2 in your phone could enter the environment.

2.9 Now, using the example in problem 2.8, let us consider the possibility that when you replace your phone, you send it to an electronics recycling program. Instead of entering the trash, the phone is sent to a company that dismantles the phone and separates it to the original components, from which the metallic components are extracted. The nano-TiO_2 is recovered and placed in a container and shipped to a coating manufacturer. In this case, the TiO_2 is much less likely to enter the environment. Should recycling of such electronics be mandatory? How would this be regulated and enforced? Who is ultimately responsible for the consequences of allowing these materials to enter the environment?

2.10 Finally, again using the example of problem 2.8, let us say you work for a company that makes consumer electronic devices. The company decides to put a coating on the surface of a cell phone to increase its scratch resistance, and protect the underlying paint from fading in sunlight using nanoscale titanium dioxide. You are the scientist in charge of testing the coating. Will your work bring you into contact with the nano-TiO_2? If so, how will it?

When you are finished with your testing, you must dispose of the test material. Your safety officer instructs you to put the test material in the trash. Where does it go then? Your testing involves putting the nano-TiO_2 in a water solution. You have to clean the glassware after the test. Washing the glassware sends the nano-TiO_2 down the drain. The waste water from your lab goes to a treatment facility, where it allows solids to filter out before the water goes to a nearby river. The river eventually reaches the ocean. Where is the nano-TiO_2 now? Should labs be severely restricted from such "dumping" of potentially hazardous materials? What if such restrictions become cost prohibitive to the research? What is your responsibility as a scientist or engineer in the safety of your research or the materials/products you develop? Who, if anyone, should be held liable if such releases result in environmental damage and/or human injuries?

Public Perception

2.11 As nanotechnology becomes better known, how do you think public opinion is likely to change? Will it become more positive, or could it become more negative? What factors will influence this trend?

2.12 How do you feel about nanotechnology and the quality of life? Will it improve our quality of life over the next 20 years, make the quality of life worse, or have no impact one way or another? In what areas might we see improvements? Where might we see new problems?

2.13 Think about the last time you confronted a new technology in your life—whether a new gadget you saw advertised or a new medicine your doctor prescribed. How did you decide whether it was likely to be more beneficial or more risky?

2.14 If you had control over several hundred millions of new government research dollars to invest, where would you invest it? Would it go into nanotechnology or would it go to other areas? What ones and why?

Future of Nanotechnology

2.15 Describe five cycles that apply to nanotechnology.

2.16 List five trends that you see affecting the development of nanoscience.

2.17 Identify two or three additional "attractors" that are related to nanotechnology and describe their probable path of development.

2.18 Select one of the nanotechnology attractors described earlier and discuss its potential impact on society.

Ancillary Problems

2.19 The space elevator is an elevator from Earth that goes beyond geosynchronous orbit (35,786 km above the equator). Make a list of pros and cons that are associated with such a concept.

2.20 Not in my backyard—NIMBY! Discuss the pros and cons related to the placement of a nanotechnology research institute in your neighborhood. What conditions must be agreed upon before such an undertaking gets off the ground? What are the benefits? The risks?

2.21 Bill Joy, the founder of Sun Microsystems, wrote an article titled, "Why the Future Doesn't Need Us" [35]. Read this article, go to a café, and discuss its content with classmates.

2.22 Eric Drexler (self-assembler technology) and Richard Smalley (traditional technology development) have had some heated discussions about nanotechnology. Let the Nanotech Wars begin! [36,37]. The public perception regarding the development of nanobots is a heated issue in terms of funding and in other ways. What do you think?

2.23 Geopolitical and economic forces are in effect in a big way today—forces that are able to change the way you and I live. Will nanotechnology have any impact on the course of these global forces (or vice versa)?

2.24 Public policy is a "system of laws, regulatory measures, courses of action and funding priorities concerning a given topic promulgated by a governmental entity or its representatives" [38]. Discuss how this topic relates to nanotechnology and rate its importance.

2.25 What do you foresee with regard to nanotechnology, education, and workforce development in the United States? In the world? Please answer with short sentences. List as many potential employment scenarios as you can that directly involve nanoscience or nanotechnology and that indirectly involve science and technology.

2.26 We mentioned earlier that nanotechnology requires that new partnerships between and among academia, business, and government be forged. Is this true? Why or why not?

2.27 Michael Crichton's novel, *Prey*, tells a horrific story about nanobots gone berserk [39]. What is your impression of the book? Is its premise feasible? Why or why not?

Section 2

Nanotools

CHARACTERIZATION METHODS

When you can measure what you are speaking about, and express it in numbers, you know something about it; but when you cannot measure it, when you cannot express it in numbers, your knowledge is of a meager and unsatisfactory kind: it may be the beginning of knowledge, but you have scarcely, in your thoughts, advanced to the stage of science.

WILLIAM THOMSON, LORD KELVIN,
Popular Lectures and Addresses (1891–1894)

Chapter 3

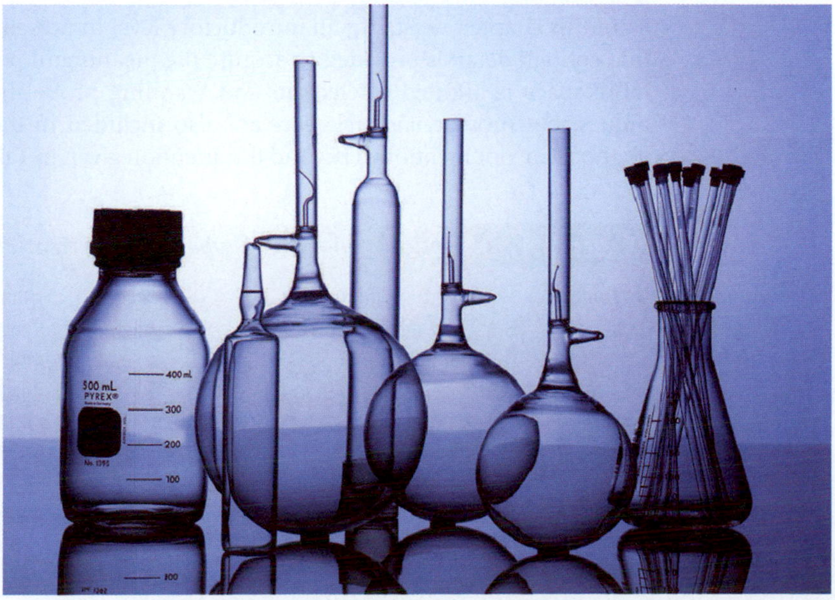

Threads

We are finished with the "Perspectives" division of the text. Hopefully, a generalized impression of nanoscience and nanotechnology has been acquired. Topics brought up in chapter 2, "Societal Implications of Nano," will be referred to throughout the text to reinforce concepts that are important to all that is outside the science. Before we journey on into nanoscience proper, it is timely, and without question obligatory, to familiarize ourselves with the tools of nanoscience. The "Nanotools" division of the text is comprised of two chapters: this chapter and chapter 4, "Fabrication Methods." Because reference is made throughout the text to images of nanomaterials, relevant spectra, and fabrication methods, placement of the two nanotool chapters early in the text is appropriate. These two chapters provide the language of nanoscience and prepare the student for what unfolds.

The purpose of this chapter is to serve as a general catalog for characterization techniques. Following chapters 3 and 4, we enter the physics division of the book, the properties and phenomena, and begin in earnest to delve into nanoscience with the surface of nanomaterials and its importance.

3.0 Characterization of Nanomaterials

The need to sort, name, categorize, catalog, and detail the things, parts, and components of our world has accompanied humanity since the dawn of civilization. Characterization and reporting are fundamental to science. They are the means by which we communicate our scientific achievements. Measurement is characterization, the language of science. Measurement is accomplished with tools: the instruments, machines, equipment, and computer hardware and software. There are many types of characterization methods and most predate the advent of nanoscience and nanotechnology. Some methods have actually evolved alongside nanoscience and helped launch the *Nano Age* itself. The development of novel, integrated methods designed specifically to probe the nanoworld is an ongoing and evolutionary process.

In this chapter, we stay at an introductory level in presentation, but we hope that enough detail is provided to acquire the meaningful perspective. Although more space is allotted to electron and scanning probe methods, other techniques that provide *nano relevance* are also included in the discussion. Many methods are not mentioned beyond the attention given in **Tables 3.1** through **3.6**

Table 3.1 *Optical (Imaging) Probe Characterization Methods*

Acronym	Technique	Analytical value
	Binocular microscopes	Imaging/gross morphology
	Compound microscopes	Imaging/fine morphology
CLSM	Confocal laser-scanning microscopy	Imaging/ultrafine morphology
SNOM	Scanning near-field optical microscopy	Rastered images
2PFM	Two-photon fluorescence microscopy	Fluorophores/biological samples
DLS	Dynamic light scattering	
PCS	Photon correlation spectroscopy	Particle sizing
QELS	Quasi-elastic light scattering	
BAM	Brewster angle microscopy	Gas–liquid interface imaging

Table 3.2 *Electron Probe Characterization Methods*

Acronym	Technique	Analytical value
SEM	Scanning electron microscopy	Raster imaging/topology morphology
EPMA	Electron probe microanalysis	Particle size/local chemical analysis
TEM	Transmission electron microscopy	Imaging/particle size–shape
HRTEM	High-resolution transmission electron microscopy	Imaging structure chemical analysis
STEM	Scanning transmission electron microscopy	Biological samples
FEM	Field emission microscopy	Surface structure/molecular properties
	Electron diffraction	Crystal structure
RHEED	Reflection high-energy electron diffraction	Surface structure
LEED	Low-energy electron diffraction	Surface/adsorbate bonding
EBSD	Electron backscatter diffraction for SEM	Crystallographic information
SAED	Selected area electron diffraction	Local structure information
EELS	Electron energy loss spectroscopy	Inelastic electron interactions
AES	Auger electron spectroscopy	Chemical surface analysis
EBIC	Electron beam induced current	Transport properties in semiconductors

because we assume familiarity with most standard characterization techniques has been achieved by this time. Detailed discussion of characterization methods that measure engineering parameters such as elastic modulus, tensile, hardness and stiffness, etc. is discussed elsewhere. The same is true for biological and medical characterization methods.

3.0.1 Background

Scientists and thinkers over several millennia—Democritus, Hero, Boyle, Dalton, Newton, and many more—have speculated about the nature of matter and light and their interaction. We are now finally able to see atoms and molecules. However, long before the electron microscope could become a tangible tool, the electron had to be discovered. Long before the atomic force microscope became

Table 3.3 *Scanning Probe Characterization Methods*

Acronym	Technique	Analytical value
AFM	Atomic force microscopy	Topology/imaging/surface structure
LFM	Lateral force microscopy	Surface/friction analysis
CFM	Chemical force microscopy	Chemical/surface analysis
MFM	Magnetic force microscopy	Magnetic materials analysis
STM	Scanning tunneling microscopy	Topology/surface structure/imaging
STS	Scanning tunneling spectroscopy	Electronic density of states
APM	Atomic probe microscopy	Three-dimensional imaging
FIM	Field ion microscopy	Chemical profiles/atomic spacing
IAP	Imaging atomic probe	Emitted ions for imaging surface
APT	Atomic probe tomography	Position-sensitive lateral location of atoms
POSAP	Position-sensitive atomic probe	Mass position resolution

Table 3.4 *Photon (Spectroscopic) Probe Characterization Methods*

Acronym	Technique	Analytical value
UPS	Ultraviolet photoemission spectroscopy	Surface analysis
UVVS	UV–visible spectroscopy	Chemical analysis
AAS	Atomic absorption spectroscopy	
AES	Atomic emission spectroscopy	
ICP	Fluorescence spectroscopy	Elemental analysis
FS	Inductively coupled plasma spectroscopy	
SPR	Surface plasmon resonance spectroscopy	Surface/adsorbate analysis
LSPR	Localized surface plasmon resonance	Nanosized particle analysis
PLS	Photoluminescence spectroscopy	Elemental analysis
RS	Raman spectroscopy	Vibration analysis
SERS	Surface-enhanced Raman spectroscopy	Chemical analysis/bond structure
SERRS	Surface-enhanced resonant Raman spectroscopy	SERS coupled with electronic transition
smSERS	Single molecule detection SERS	Ability to probe single molecules
FT-IR	Fourier transform infrared spectroscopy	Asymmetrical vibration analysis
NIRS	Near-IR spectroscopy	Surface/IR analysis
DR-FTIR	Diffuse reflectance FTIR	Surface/adsorbate analysis
ATR	Attenuated total reflection	Adsorbate analysis
XRD	X-ray diffraction	Crystal structure
XRF	X-ray fluorescence	
EDX	Energy-dispersive x-ray spectroscopy	Elemental analysis
WDS	Wavelength dispersive x-ray spectroscopy	
XPS	X-ray photoelectron spectroscopy	Surface analysis/depth profiling
SAXS	Small angle x-ray scattering	Surface analysis/particle sizing (1–100 nm)
EXAFS	Extended x-ray absorption fine structure	Local surface atomic structure
CLS	Cathodoluminescence	Characteristic emission
NMR	Nuclear magnetic resonance spectroscopy	Analysis of odd no. nuclear species
EPR	Electron paramagnetic resonance	Analysis of paramagnetic species
STh-MRM	Scanning thermal microwave resonance microscopy	Thermal detection

a tangible tool, the record player had to be invented. We begin our background discussion with spectroscopy.

Spectroscopy (from the Latin *spectrum,* meaning "an appearance," from *spectare,* "to behold" + "scopy" from the Greek *skopein,* "to view") and other analytical

Table 3.5 *Ion-Particle Probe Characterization Methods*

Acronym	Technique	Analytical value
MS	Mass spectrometry	Material composition
SIMS	Secondary ion mass spectrometry	Composition of solid surfaces
RBS	Rutherford back scattering	Quantitative–qualitative elemental analysis
NSS	Neutron scattering spectroscopy	Chemical structure adsorbate bonding
SANS	Small angle neutron scattering	Surface characterization
NRA	Nuclear reaction analysis	Depth profiling of solid thin films
FReS	Forward recoil elastic spectrometry	Hydrogen deuterium analysis
PIXE	Particle-induced x-ray emission	Nondestructive elemental analysis

TABLE 3.6 *Thermodynamic Characterization Methods*

Acronym	Technique	Analytical value
TGA	Thermal gravimetric analysis	Mass loss versus temperature
DTA	Differential thermal analysis	Reaction heats heat capacity
DSC	Differential scanning calorimetry	Reaction heats phase changes
TMA	Thermal mechanical analysis	Thermal mechanical properties
PnDSC	Parallel nano DSC	Combinatorial analysis
NC	Nanocalorimetry	Latent heats of fusion
TPD	Temperature-programmed desorption	Surface adsorbate properties
TDS	Thermal desorption spectroscopy	Coupled with MS
SCAC	Single crystal absorption calorimetry	Absorption and adhesion energies
TDM	Linear–volume thermodilatometry	Dimensions as a function of °T
TL	Thermoluminescence	Surface states detrapping
BET	Brunauer–Emmett–Teller method	Surface area analysis
MP	Mercury porosimetry	Pore volume, pore size
BJH	Barnett–Joyner–Halenda method	Pore size distribution
Sears	Sears method	Colloid size, specific surface area

techniques possess a long and rich history [1]. It began about 2430 years ago, when Aristophanes demonstrated the first use of lenses. Aristotle speculated that in order to see color, light must exist, and Euclid made observations about the focusing properties of spherical mirrors. Ancient scientists such as Seneca, Cleomedes, and Ptolemy contributed to knowledge about scattering, reflection, and refraction. The rest of the history of spectroscopy is filled with too many milestones to mention here and can be found in any reliable reference. We will focus on the history of electrons, electron probe techniques, and scanning probe characterization methods.

G. J. Stoney in 1874 coined the term electron (from the Greek *electron*, meaning "amber," a substance when rubbed that causes sparks). An electron is a stable subatomic particle with a negative charge equivalent to 1.6022×10^{-19} C and with rest mass m_e equal to 9.1094×10^{-31} kg. In 1895, Jean Perrin determined that cathode rays carry a negative charge. During his study of cathode ray tubes in 1877, J. J. Thomson (**Fig. 3.1**) discovered that the electron was a subatomic particle, but he was not sure exactly what kind of particle: "I can see no escape from the conclusion that [cathode rays] are charges of negative electricity carried by particles of matter.... What are these particles? Are they atoms, or molecules, or matter in a still finer state of subdivision?" He also postulated that electrons were corpuscular (e.g., particulate) in nature. He came to this conclusion after applying a magnetic field to a cathode ray tube and found that the trajectory of an "electron" particle beam is deflected by the magnetic field. He then determined the mass-to-charge ratio (m/e) of the electron by measuring the degree of deflection. Thomson said upon his discovery, "We have in cathode rays matter in a new state." Since electron mass is difficult to measure, $-e/m_e$ is the quotient used by physicists where

$$-\frac{e}{m_e} = -1.758820150(44) \times 10^{11}\ \mathrm{C \cdot kg^{-1}} \tag{3.1}$$

Fig. 3.1 *J. J. Thomson is known for his work with cathode ray tubes and many other notable scientific discoveries. He was awarded the Nobel Prize in physics in 1906.*

THOMSON, Joseph John
Nobel Laureate PHYSICS 1906
© Nobelstiftelsen

Source: Image courtesy of the Nobel Foundation. With permission.

In 1909, R. Millikan conducted the well-known oil-drop experiment and was able to measure the electron's charge by measuring to total charge q on an oil drop of predetermined mass m_{drop}.

$$q = \frac{m_{drop}\, g}{E} \tag{3.2}$$

Where g is the gravitational constant and E is the strength of the applied electric field. By varying the strength of the ionizing x-rays, he found that q was always a multiple of 1.6×10^{-19} C. Louis de Broglie proposed that electrons possess wavelike qualities. In 1924, he theorized that the wavelength of a particle depends upon its momentum. De Broglie was awarded the Nobel Prize in physics in 1929 for his theory of the wave nature of electrons. From the contributions of

Thomson, de Broglie, and others, H. Busch developed the concept of the electromagnetic lens in 1926.

We introduced SEM, TEM, XRD, STM, and AFM methods in chapter 1 and discussed some of the highlights of their unique history. The versatility and power of these nanoscale tools have increased over the past few decades. The control, manipulation, and integration of individual atoms have at last been accomplished due to the advent of these methods. One of the greatest developments in the twentieth century was the application of computer capability to analytical tools. In addition to their value in crunching data, computers also assist researchers with positioning samples, focusing images, monitoring reactions, and data acquisition.

3.0.2 Types of Characterization Methods

Nanoscience and nanotechnology are interdisciplinary in nature. It is fair to say that an equally broad inventory of characterization and analytical techniques applies to nanostructured materials. There are approximately 700 single-signal characterization techniques and approximately 100 that are considered to be multisignal techniques [2]. These numbers are based on the types and combinations of three kinds of physical phenomena: (1) *primary* (1°) analytical probes (electrons, photons, neutrons, ions, etc.) and input stresses (heat, pressure, electric field, magnetic field, electricity, mechanical stress, etc.); (2) the types of measurable *secondary* effects (2°) (electrons, electromagnetic radiation, heat, pressure, volume change, mechanical deformations, etc.); and (3) the monitoring medium of choice (energy, temperature, mass, intensity, time, angle, or phase) [2]. The 1° probe, whether electron or photon, interacts with matter. The matter then responds to regain its equilibrium state. During this process, the 1° probe is modified. Excited electrons, phonons, excitons, or plasmons are some examples of how matter is altered. Alterations of the 1° probe result in the 2° effect—the signal that we measure [2].

Optical Probe Characterization Techniques. The primary probe in optical methods (**Table 3.1**) is visible light of wavelength within the 400- to 800-nm range. The 2° signal from most 1° photon methods is usually another photon. The 2° photon signal arises from many possible physical phenomena, such as elastic scattering, inelastic scattering, or from emission (fluorescence).

Human eyes have great depth of field and depth of focus; however, they are not efficient at peering into nanostructured materials without assistance. Binocular scopes are useful for viewing gross morphology like a clump or large strands of carbon nanotubes. Compound microscopes offer more detail but resolution is limited to a few hundred microns. Confocal, near-field, and two-photon microscopes using visible and near-ultraviolet (UV) 1° photons have managed to push the resolution envelope of optical methods into the sub-100 nm range. Dynamic light scattering is an optical technique that is used to measure particle size, and Brewster angle microscopy (BAM) is used to analyze phase behavior at an air–water interface. Brewster's law is given by

$$\theta_B = \arctan\left(\frac{m_2}{m_1}\right) \tag{3.3}$$

where m_2 and m_1 are the refractive indices of 2 media and is applicable to light that moves between those media. θ_B is the Brewster's angle, the point at which light of one polarization cannot be reflected. Selected optical microscope methods are listed in **Table 3.1.** Optical characterization methods are used primarily for imaging but also are useful in chemical analysis.

Electron Probe Characterization Methods. Electron probe characterization methods use high-energy electron beams. The major 2° effects (signals) of electron interactions with matter are discussed in more detail later in the chapter. Electron beams are used for imaging, chemical analysis, and determination of material structure. The most familiar electron probe methods are scanning electron microscopy (SEM) and transmission electron microscopy (TEM). In SEM, the electron beam is rastered across the analytical surface. A raster (from the German *raster,* meaning "screen," from the Latin *rastrum,* indicating "rake") is a rectangular pattern of parallel scanned lines from which an image is assembled. Electron guns in older televisions produced a picture by rastering electrons over a phosphor grid (the TV screen). Electron probe methods are listed in **Table 3.2.** SEM and TEM instruments are usually coupled to other types of analytical tools, primarily with energy-dispersive x-ray analysis (EDX or EDS).

Scanning Probe Characterization Techniques. All scanning probe microscopy (SPM) methods have two things in common: a finely sharpened probe tip and a system that enables the tip to scan (raster) over the surface of a sample. An image is acquired by moving the probe, or the sample, with a raster action—similar in concept to the SEM but utilizing completely different types of probes. Resolution on the order of atoms is possible with scanning probe techniques. Different types of scanning probe methods are listed in **Table 3.3**.

Spectroscopic Characterization Methods. There are many kinds of 1° photon probe techniques (**Table 3.4**). As in optical imaging, most photon probe methods involve a 1° photon in and a 2° photon out. UV–visible spectroscopy, based on absorption or emission (fluorescence) of photons, is the most common form of spectroscopy. Although the wavelength for UV–visible methods is within the nanoscale (e.g., <1 μm), the dimensions of most nanomaterials are much smaller. UV–visible techniques mainly involve electron transitions. Related parameters such as extinction, intensity, absorption, transmission, reflection, and wavelength are measured during UV–visible spectroscopy. Raman and Fourier transform infrared spectroscopy (FT-IR) are techniques that measure molecular and phonon vibrations of materials. In each case, 2° photon carrier waves are analyzed and information about vibrational states is acquired. Symmetrical vibrations (polarization) are exclusive to Raman spectroscopy, and asymmetrical vibrations (dipolar) reside in the domain of infrared analysis. In surface plasmon spectroscopy, the absorption of 1° photons causes surface electron plasmons to oscillate. The wavelength, λ_{max}, of the absorption is indicative of the surface state of the metal (e.g., adsorbed species).

X-rays form another important category of 1° photon excitation that is important to nanoscale analysis. The wavelength of x-rays ranges from 0.01 to 10 nm, corresponding quite well to the size of nanoparticles. The major 2° effects are photons or electrons depending on the type of analyzer installed in your instrument. The two most *nanorelevant* x-ray techniques are x-ray diffraction (XRD) and energy-dispersive x-ray (EDX), which reveal structural information and chemical analysis, respectively.

Nuclear magnetic resonance (NMR) and electron spin resonance (ESR) are techniques that rely on an applied magnetic field to induce signals. NMR is applicable to molecules that contain odd numbers of protons and neutrons (e.g., have a magnetic moment such as ^1H and ^{13}C). Electron paramagnetic resonance (EPR), commonly referred to as ESR, is a technique that is able to analyze materials that have unpaired electrons. There are numerous excellent sources that describe these techniques in more detail.

Microwave spectroscopy methods are applicable to nanomaterials. Local resolution for microwave spectroscopy by thermal near-field microscopy is less than 100 nm.

Ion-Particle Characterization Techniques. Mass spectrometry (MS) involves the breakup of a parent particle into ionized components. Daughter particles are produced by electron impact (1° electron) and by other techniques. MS used to detect the presence of sodium nanoclusters in particle beams showed that cluster abundance was dependent on size; the most stable clusters possessed a *magic number* of atoms. Smalley and his group used MS to prove the existence of C_{60}. Secondary ion MS (SIMS) is also a popular method. Neutrons and ions are able to modify matter by inducing thermal vibrations, rupturing bonds, and displacing atoms [2]. The 2° signals comprise secondary ions. In addition, 1° ion probes are capable of implantation and inducing ionization and de-excitation [2].

Thermodynamic Characterization Methods. Thermodynamic techniques are characterization methods that involve thermodynamic parameters. The primary probe is not a photon or an electron but rather a thermodynamic parameter such as temperature or pressure. In these cases, changes in temperature, pressure, phase, and volume are considered to be secondary effects. Bond energy, phase change (e.g., melting point), heat capacity, surface adsorption–desorption, volume change, surface tension, and vapor pressure are some effects measured by thermodynamic techniques.

Thermoluminescence is a technique in which a photoluminescence response is generated in a material from heating. Thermoluminescence provides information about surface states—in particular, the ability to detect trapped metastable electron–hole pairs in nanoparticles [3]. Heating the sample induces lattice vibrations that in turn provide energy for electron–hole recombination. Upon recombination, optical photons are emitted [3].

There are methods that have more "nano" relevance than others, but all apply with validity to nanomaterial characterization. The lists provided in **Tables 3.1** through **3.6** are by no means exhaustive. In general, two fundamental types of characterization exist: imaging by microscopy and analysis by spectroscopy [2]. Other conventional techniques, like thermal gravimetric analysis (TGA) and the Brunauer–Emmett–Teller (BET) surface area methods, do not employ electromagnetic radiation per se but can be, in the very broadest of senses, considered to be spectroscopic methods because *x* and *y* axes are involved.

Bulk Engineering Characterization Methods. There are two general types of nanomaterials: (1) bulk materials that are composed of nanomaterials and (2) independent nanomaterials (e.g., a carbon nanotube resonator). Mechanical testing methods are used on both types of nanomaterials. Tensile properties exist at the nanoscale as does friction. Mechanical testing includes abrasion and scratch resistance, hardness, elastic modulus, tensile strength, fracture toughness, stiffness, and fatigue strength. Tensile stress in Pascal,

$$\sigma = \frac{P}{A} \tag{3.4}$$

where P is pressure in Neutons and A is area.

Stiffness K is defined as

$$K = \frac{P}{\delta} \tag{3.5}$$

where P is the applied force and δ is the deflected distance. Electrical testing methods include conductivity, capacitance, and resistivity. Material characterization tests include tensiometry (surface tension) and pycnometry (density). Energy dissipation tests include heat transfer, reflectivity, and thermal expansion. There are many more. All engineering parameters apply to nanomaterials.

3.0.3 Optics and Resolution

Due to the incredible contributions made by TEM and scanning tunneling microscopy (STM) techniques, we are now capable of imaging and detecting individual atoms and molecules! We have indeed achieved one of the holy grails of science by creating images of the "indivisible atoms" about which Democritus and others speculated over the millennia. However, just when we thought we could look no smaller, scientists at IBM Almaden Research Center in California on April 26, 2007, described a new microscope that is able to look *inside* an atom [4]. Some of the challenges associated with imaging smaller and smaller objects are discussed next.

The fundamental relationships that apply to imaging by optical wavelengths also apply to imaging with electrons. The main differences are the source of the 1° probe and the material of the lenses. Lenses in electron probe instruments consist of coordinated magnetic fields.

The relationship between focal length f (the distance from the lens center of a thin lens to the focal point) and the object and image distances is

$$\frac{1}{f} = \frac{1}{o} + \frac{1}{i} \tag{3.6}$$

where o and i are the object and image distances, respectively. The *focal point* is the place on the principal axis of the lens where parallel rays converge to a common point or focus. Graphical depictions of the focal point can be found in many references. Magnification, **M**, is simply the ratio of optical distances.

$$M = \frac{i}{o} \tag{3.7}$$

In the electron microscope technique, magnification is determined by measuring the size of the object displayed on a cathode ray tube (CRT) screen divided by the size of the scanned area. Electron microscopes automatically display magnification and a scale bar for reference. Magnification is simple to calculate in SEM. For example, if the CRT screen is 10 × 10 cm, a magnification of 10× correlates with a specimen area equal to 1 cm^2 and 100× → 1 mm^2; 1000× → 100 μm^2; 10,000× → 10 μm^2; 100,000× → 1 μm^2; and 1,000,000× → 100 nm^2. Magnification is a one-dimensional measure and not the ratio of areas.

Resolution (from the Latin *resolutionem,* "the process of reducing things into simpler forms") is the ability to reduce or separate something into its components (e.g., the minimum distance between two distinguishable objects) or the finest detail that can be distinguished in an image from some set distance. Resolution is related to contrast. *Resolving power* is the ability to measure the angular separation of images that are close together. Counting pixels is another means of calculating resolution. Magnification and resolution, however, are not the same. Prove it to yourself by enlarging an image on your computer screen until you are able to see pixels.

Imaging of samples by optical or electronic means is dependent on the wavelength of the source or probe; the shorter the wavelength of the probe, the greater the resolving power of the instrument. The resolving ability of optical microscopes is ca. 0.35 μm. Resolution can be improved if an oil emulsion is applied to the objective lens to 0.18 μm (180 nm). As a general rule of thumb, the best achievable resolution of an optical microscope is equal to approximately half the wavelength of the illuminating beam. The wavelength of visible light lies roughly between 400 and 700 nm. On the other hand, the practical resolving ability of electron beam microscopes, depending on the applied accelerating voltage, is approximately 0.2 nm.

The quality of the hardware and the operational settings in a microscope also influence the resolving power. The *numerical aperture* of a lens factors significantly into the resolution equation. The numerical aperture (N.A.) of an optical system is a dimensionless unit that indicates the range of angles from which the lens system is able to accept or emit visible light and is defined by

$$\textit{Numerical aperture} = \text{N.A.} = n \sin\theta \tag{3.8}$$

where n is the refractive index of the medium between the lens and the sample and θ is the half-angle that is subtended by the rays entering the objective lens (circular aperture). The resolving power of the lens is proportional to wavelength divided by the numerical aperture (**Fig. 3.2**):

$$R \propto \frac{\lambda}{\text{N.A.}} \tag{3.9}$$

The resolution gets better (e.g., R gets smaller) when (1) the wavelength of the impinging radiation gets shorter (e.g., the λ of blue light is shorter than λ

Fig. 3.2

Three different lens configurations highlighting the numerical aperture of electron microscopes are compared. The closer the aperture is to the sample the larger the angle enclosing the lens becomes. From the equation, N.A. is proportional to the sine of angle θ. For θ = 10° (a long focal length), N.A. = 0.17; for θ = 40°, N.A. = 0.64; and for θ = 60° (a shorter focal length), N.A. = 0.87. Small N.A.s translate into low resolution. High N.A. translates into high resolution.

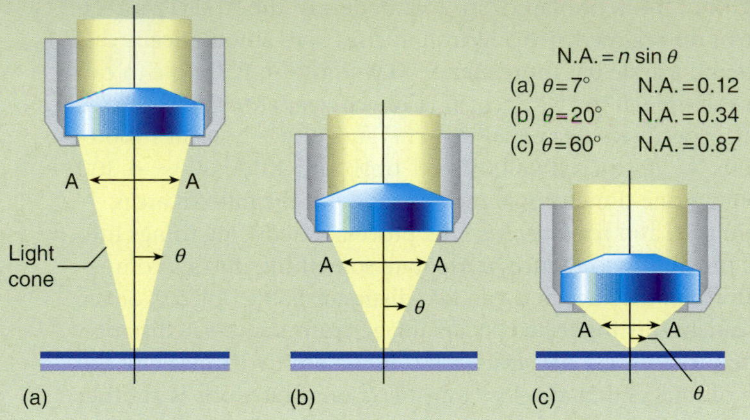

of red light), (2) the refractive index of the connecting medium gets larger (e.g., air with n = 1.0 vs. oil with $n \sim 1.5$), and (3) N.A. gets larger (e.g., the larger the half-angle θ is the bigger N.A. becomes). N.A. also increases as focal length f decreases.

Three phenomena are relevant to resolution. The Abbe diffraction barrier acts to limit resolution that in order to produce a true image, the N.A. must be large enough to transmit the entire diffraction pattern of a point source. The Rayleigh criterion is applied to judge the minimal resolvable detail of an object. An *Airy disk,* named after physicist George Airy, is the projection of light through a circular aperture (e.g., a pinhole) or is formed by focused light impinging upon a point source. The distance between point sources is an indicator of the resolution. The Airy projection appears to be made of concentric rings (orders) with decreasing thickness and increasing spacing between the rings progressing from the center (**Fig. 3.3**). A map of the intensity is shown adjacent to the projection. The diameter of the most intense (the central) Airy disk is inversely proportional to the diameter of the aperture [2]. Because each object in the image is capable of diffracting the impinging electron (or optical) beam, each object is able to form its own Airy disk. If the point images seem to be merged (e.g., unfocused, unresolved), a *disk of confusion* is formed.

The Rayleigh criterion states that two such images are considered to be *just resolved* when the central maximum of the first image falls into the first minimum of the diffraction pattern of the second image. For light transiting through a slit with width equal to d, the Rayleigh criterion is

$$d\sin\theta_R = \lambda; \quad \sin\theta_R = \frac{\lambda}{d}; \quad \theta_R = \frac{\lambda}{d} \tag{3.10}$$

FIG. 3.3 *Three Airy discs are pictured. When light from a point source passes through an aperture, various points of the specimen appear to be a compilation of small patterns (small concentric circles) in the image. These are called Airy discs and are caused by the diffraction of light passing through the circular aperture of the objective. The Rayleigh criterion describes the limits of resolution at which an image can be resolved into separate entities. On the left is an unresolved image; in the middle, the condition of the Rayleigh criterion is depicted; and on the right, a resolvable image is shown.*

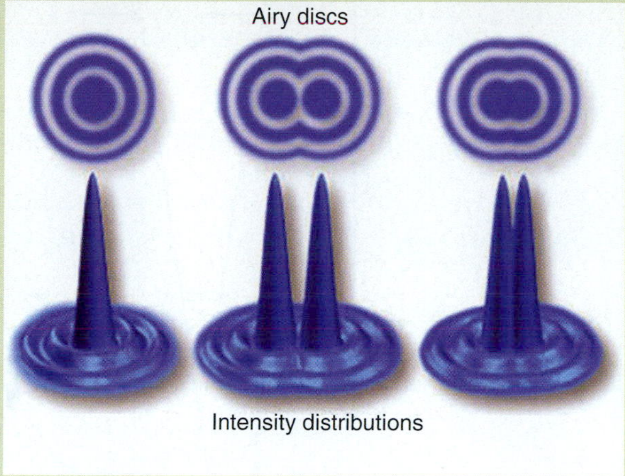

For small angles, the sine of an angle in radians is approximately equal to the angle itself: $\sin\theta_R \sim \theta_R$. Therefore, the Rayleigh criterion minimum angular resolution (in radians) for a circular aperture of diameter D is

$$\textit{Angular resolution} = \theta_R = \frac{1.22\lambda}{2\text{N.A.}} = \frac{1.22\lambda}{D} = \frac{0.61\lambda}{\text{N.A.}} \quad (3.11)$$

where D is the diameter of the circular aperture. Diffraction and resolution are different but related phenomena. It is the diffraction of light that restricts our capability to resolve features that are in close proximity. This factor is also important to optical lithography techniques.

For nanoscientists that use microscopes, the distance between two objects bears more relevance than the angular resolution. Angular resolution is converted into spatial resolution by multiplying by the focal length f. If the distance R is the arc subtended by θ_{min}, then $R = f\,\theta_R$ and

$$\textit{Spatial resolution} = f\theta_R = \frac{(1.22\lambda)f}{D} \quad (3.12)$$

Electron microscopes share similar characteristics with optical microscopes. Mathematical treatments of magnification, resolution, and diffraction for both are essentially the same.

The depth of focus (D_{Focus}) for an electron microscope is about 300 times that of an optical microscope. Why is this? Depth of focus is simply the distance

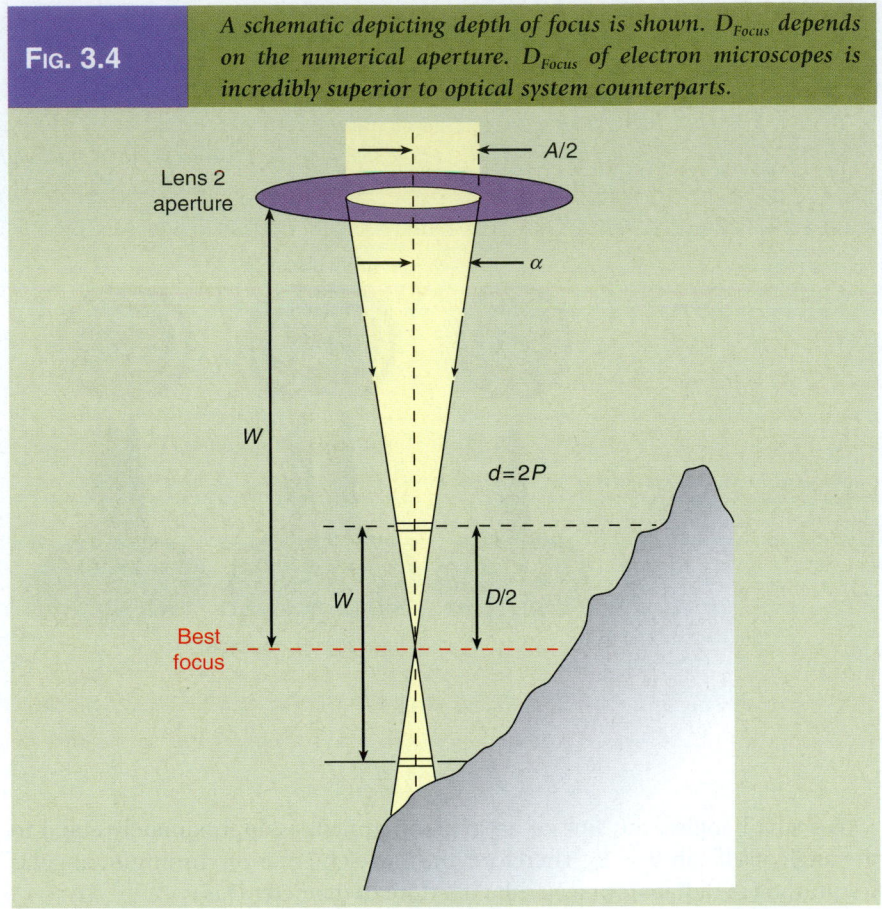

FIG. 3.4 *A schematic depicting depth of focus is shown. D_{Focus} depends on the numerical aperture. D_{Focus} of electron microscopes is incredibly superior to optical system counterparts.*

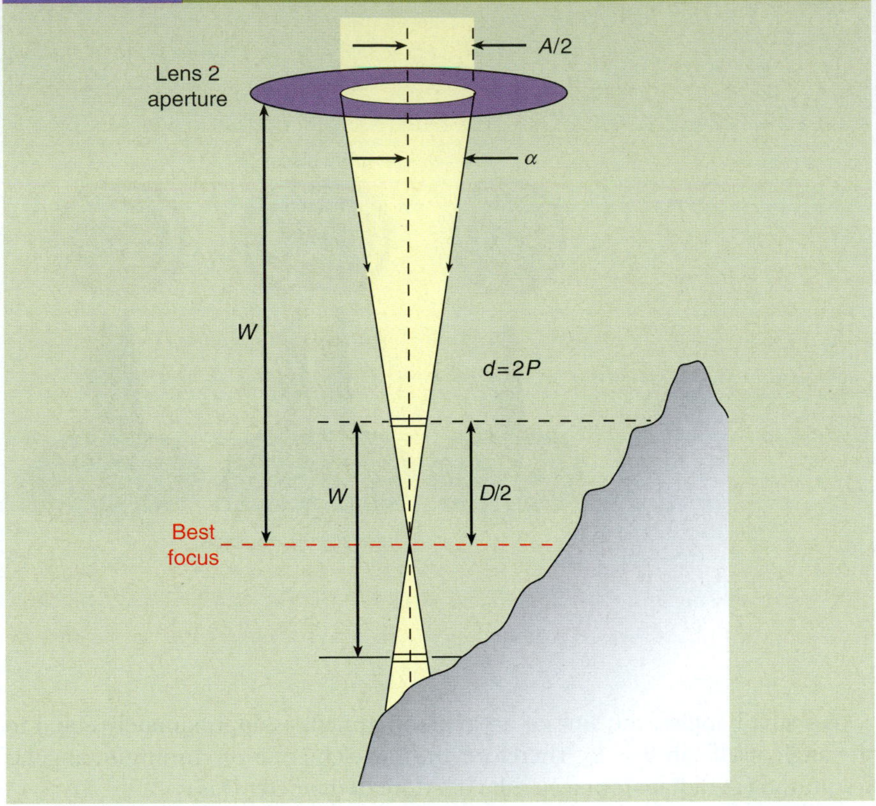

above and below the focal plane in which the object is clearly in focus. Aperture size and working distance (focal length of the objective lens) are two factors that exert influence over the depth of focus. The angle α is the angle formed between the aperture and the focal point on the sample. From **Figure 3.4,** notice that the D_{Focus} becomes larger as α gets smaller. Once again, because α is small we can safely assume that $\tan\alpha \approx \alpha$.

The angle α is the angle of divergence that is dependent on the ratio between the aperture radius and the working distance **W** [5]:

$$\tan\alpha = \frac{R_o}{\mathbf{W}} \sim \alpha \tag{3.13}$$

$$\textit{Depth of focus} = D_{\text{Focus}} = \frac{0.2}{\alpha \mathbf{M}}\,\mu\text{m} \tag{3.14}$$

where **M** is the magnification as before. Another equation that is used to derive D_{Focus} relates the ratio of the wavelength to the square of the numerical aperture [6]:

$$\text{Depth of focus} = D_{\text{Focus}} \propto \frac{\lambda}{(\text{N.A.})^2} \quad (3.15)$$

There are trade-offs, however. Making **W** larger yields a smaller α and thereby a larger D_{Focus}, but an overall decrease in R, the resolution of the image [5]. Increasing D_{Focus} by reducing the aperture diameter results in a decreased signal-to-noise ratio (e.g., a snowier picture). Depth of field (D_{Field}) is not the same as depth of focus. Depth of field is the lateral range within which images remain in focus and is described mathematically by [2]

$$\text{Depth of field} = D_{\text{Field}} = \frac{0.61\lambda}{n \sin\alpha \tan\alpha} \quad (3.16)$$

where α is the semi-angle subtended by the aperture as before.

Marvin Minsky invented the confocal microscope in 1953 but it was not until the advent of lasers that the technique became useful. In a confocal microscope, images (optical sections) are illuminated through a pinhole aperture and a fluorescent signal is channeled via another pinhole near the detector. The image is then compiled from laser scans of the surface. In this way, the instrument is able to screen "out-of-focus" signals. In near-field microscopes, a subwavelength-sized light source (e.g., a glass fiber) hovering a few nanometers above the sample surface is able to acquire optical images on the order of 50-nm resolution. In two-photon microscopy (TPM), the optical section is acquired only from the focal plane where excitation occurs. The Osaka Bull depicted in **Figure 2.4** is an example of two-photon lithography.

The Abbe diffraction limit can be mitigated by application of the *near-field technique*. The basic concept of the procedure is to illuminate a sample through an aperture with subwavelength dimensions held close to the sample (e.g., the aperture-sample distance within half the diameter of the aperture). The resolution in this case depends on the aperture diameter and not the wavelength of the light. In order to acquire a whole image, the aperture is scanned (rastered) over the surface of the sample. Therefore, scanning near-field optical microscopy (SNOM) provided quite a breakthrough in optical imaging.

A *confocal microscope* is able to bypass some of the technical difficulties associated with wide-field optical microscopes. In particular, a confocal laser-scanning microscope (CLSM) utilizes a point-illumination system that allows only the light in the focal plane to be detected. The illuminated volume is on the order of $0.25 \times 0.25 \times 0.5$ μm. Just like with the near-field microscope, rastering is required in order to acquire a larger image. Confocal systems have the capability to block unwanted reflections or emissions. The apertures are designed to eliminate out-of-focus light from entering the detector. Confocal microscopes most commonly operate under the fluorescent mode but are also capable of assimilating reflected light. The resolution of confocal scopes is better than that of traditional optical microscopes (130 nm versus 350 nm, respectively).

The concept of resolution in atomic force microscopy (AFM) and other scanning probe methods is quite different from that of radiation-based microscopy. First of all, AFM imaging is a three-dimensional technique. Resolution limits of STM or AFM samples are quite impressive with resolution on the order of single atoms. The sharpness of the tip, the geometry of the sample surface, and the type of scanning mode (e.g., tapping versus contact) all play a role in determining the value of the resolution of the procedure.

Image Quality Factors. Other factors contribute to the quality of the image. *Contrast* is the difference in intensity between an object and the background and other nearby objects. Because focal length for a specific lens system relies on wavelength, *chromatic aberrations* may occur if polychromatic radiation is used. Lens imperfections also contribute to chromatic aberration. *Spherical aberrations* arise due to imperfections in the lens by causing "rays" to fall away from the point of focus. "Stopping down" the lens (e.g., decreasing the size of the aperture) is a way to mitigate spherical aberration. *Astigmatism* is the result of rays originating from an off-axis to form images at alternative points. It is an aberration caused by unequal focus of vertical and horizontal lines at different points along the optical axis.

Optical Terms. Because terms associated with optics will be used throughout the book, we will list a few of the most common. *Interference* is the superposition of electromagnetic waves. Depending on the structure of a material, electron microscopy (EM) (and particle) waves can add constructively, destructively, or anywhere in between. The colors seen in soap bubble thin films are based on interference phenomena. *Diffraction* is the behavior of a wave when it encounters an obstacle. Once again, depending on the material, interference patterns formed by the probing radiation yield information about its structure. Point objects are able to diffract waves to form Airy disks. *Reflection* of light occurs when an incoming wave is symmetrically bounced off a surface. *Refraction* occurs when impinging radiation is allowed to transit through a material, albeit in bent form. *Absorption* is a process by which electrons of a material interact with EM radiation—the excitation of electronic energy states. Relaxation back to the ground state results in emission that is characteristic of the transition. *Transmission* of radiation through a material is what is left after absorption and reflection have occurred. *Scattering* is a form of reflection and occurs off the surfaces of particles larger than 20–50 nm but smaller than visible bulk materials. Electrons and photons are differentially scattered off surfaces of materials depending on the type of material and the energy of the 1° probe.

3.0.4 The Nano Perspective

Without tools to characterize nanomaterials, we would be blind to the nanoscale. Methods to measure atomic scale phenomena essentially predate the formal designation of nanoscience as a new path in scientific exploration. Of course, techniques that measure macroscopic objects have been around since the beginnings of intellectual thought. One of the major distinctions to keep in mind is the following. Although we are more than able to measure the properties of a macroscopic quantity of nanomaterials—a clump of powder, for example—we are now able to measure the properties of a single atom, a single molecule, a single cluster, a single colloidal particle, and a single nanotube. We now challenge ourselves further to observe, measure, and probe materials that are not much larger than atoms and molecules and to develop integrated techniques that extract information simultaneously, quickly, efficiently, and accurately.

Characterization methods of the past have extracted molecular and atomic information by averaging the bulk response to some stimulation. This kind of methodology also applies to the analysis of nanomaterials. Raman spectroscopic analysis of carbon nanotubes relies on a relatively large amount of sample. However, the focus now is shifting to the characterization of individual nanomaterials and molecules. With regard to this aspect of nanoscience, one must certainly agree that this change in the analytical paradigm is indeed a revolutionary one. Characterization methods have undergone dramatic changes over the past decade and there seems to be no limit of ingenious ways to probe the secrets of individual nanostructures.

3.1 Electron Probe Methods

In this section we focus on characterization methods that rely on the electron as the primary probe. Once electrons interact with a specimen, secondary effects in the form of radiation (photons), electrons, and heat are detected and transformed into images or spectra. Electron probe characterization methods in particular are important to nanoscience due to their capability to analyze the structure of extremely small particles. Of these, the electron microscope is the most significant.

Electrons exhibit particle–wave duality. The de Broglie equation relates the momentum of an electron (a particle phenomenon) to wavelength:

$$p = \frac{h\nu}{c} = \frac{h}{\lambda} \tag{3.17}$$

where

- p is momentum ($m\upsilon$)
- ν is the frequency
- c is the speed of light (2.998×10^8 m·s^{-1})
- h is Planck's constant (6.6262×10^{-34} J·s)
- λ is the wavelength

Converting p into the conventional mass–velocity form, the wavelength of the electron is

$$\lambda = \frac{h}{p} = \frac{h}{m_e\upsilon} \tag{3.18}$$

Relativistic considerations may apply if the energy becomes very high (e.g., as electron velocity approaches the speed of light). The wavelength of the electron must be modified according to

$$\lambda = \frac{h}{p} = \frac{h}{\gamma m\upsilon} = \frac{h}{m\upsilon}\sqrt{1 - \frac{\upsilon^2}{c^2}} \tag{3.19}$$

where γ is the Lorentz factor for special relativity and τ the adjusted speed.

$$\gamma = \frac{dt}{d\tau} = \frac{1}{\sqrt{1-\beta^2}} = \frac{c}{\sqrt{c^2-\upsilon^2}} = \frac{1}{\sqrt{1-\left(\frac{\upsilon^2}{c^2}\right)}} \tag{3.20}$$

β is the relative speed with respect to the speed of light c:

$$\beta = \frac{\upsilon}{c} \tag{3.21}$$

The Lorentz factor is always greater than 1. The relatavistic kinetic energy form is given by

$$\text{KE} = \left[\frac{1}{\sqrt{1-(\upsilon/c)^2}} - 1\right] mc^2 = (\gamma - 1)mc^2 \tag{3.22}$$

The classical form of the kinetic energy of the electron is given by

$$\text{KE} = \tfrac{1}{2} m_e \upsilon^2 \tag{3.23}$$

The wavelength of an electron can then be calculated by

$$\lambda = \frac{h}{\sqrt{2m_e e\text{V}}} \tag{3.24}$$

where V (in $\text{J} \cdot \text{C}^{-1}$), the voltage, is converted to energy in the form of electron volts (eV).

In 1927, American scientists Davisson and Germer diffracted electrons from the surface of a nickel crystal and demonstrated the wavelike character of electrons [7–9]. This experiment validated the wave-particle nature of electrons. Bragg's law exclusively addresses wavelike-interference phenomena (more discussion is presented about Bragg's law in chapter 5):

$$n\lambda = 2d \sin\theta \tag{3.25}$$

Davisson and Germer's experiment showed that interference phenomena also apply to particle waves. Equations (3.26) and (3.27) illustrate the relationship between Bragg's law and wavelength:

$$\frac{1}{\lambda} = \frac{n}{2d\sin\theta} = \left(0.815\sqrt{\text{V}}\right) nm \tag{3.26}$$

$$\frac{1}{\lambda} = \frac{n}{2d\sin\theta} = \frac{p}{h} = \frac{\sqrt{2m_e\text{E}}}{h} = \frac{\sqrt{2m_e e\text{V}}}{h} \tag{3.27}$$

The discovery was accidental—not unlike the results of numerous other breakthrough experiments. The primary intent of their experiment was to measure the energies of electrons scattered from a metal surface. The *Faraday box* detector was rotated along an arc as a pattern emerged. Davisson and Germer also verified a periodic fluctuation in intensity of the scattered electron beam (e.g., wave behavior that conformed to Bragg's law).

3.1.1 Electron Interactions with Matter

Electrons are able to interact with matter in many different ways (**Fig. 3.5a** and **b**). In all cases, electrons serve as the primary probe (1°). An electron impinging on a solid surface may be scattered once, several times, or not at all. Multiple scattering occurs if the dimension of the material is greater than approximately twice the mean free path of the electron. The scattering may be elastic or inelastic and forward or backward. The probability of scattering is determined by the scattering cross-section and the mean free path of the electron—both dependent on the material properties (and size).

When a high-energy beam of electrons is incident upon a thick specimen, a pear-shaped region known as the *interaction volume* (volume of excitation) is formed. The interaction volume, which increases with increasing beam voltage (electron energy) and decreases with increasing specimen density (e.g., generally with increasing atomic number z), is normally between 1 and 5 μm. In general, electrons are able to penetrate solid materials to a depth proportional to $\sim E^2$ and laterally to $\sim E^{1.5}$ (or $\propto V^2$ and $V^{1.5}$, respectively) [2,10].

Types of electron–matter interactions in thick samples (SEM) are listed in **Table 3.7** and for thin samples (TEM) in **Table 3.8**. The important 2° effects that are exploited in the majority of investigations using SEM are secondary

Fig. 3.5 *Electron interactions with matter are shown in the image. On the left, the interaction volume corresponds to materials that have some meaningful thickness. Samples analyzed by SEM are generally thick and demonstrate interaction volume phenomena. On the right, interaction of electrons with a thin sample is shown. The TEM method is dependent on the capability of electrons to penetrate and pass through samples. For this reason, samples must be very thin.*

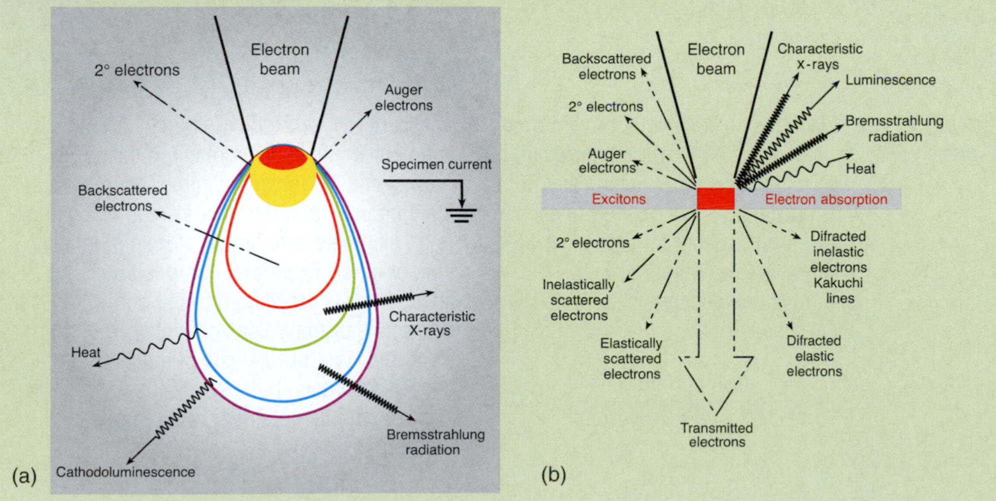

and backscattered electrons (for imaging) and characteristic x-rays (for simultaneous elemental analysis with EDX). The important 2° effects that are exploited in the majority of investigations using a TEM are transmitted electrons (for imaging), elastically scattered electrons (for simultaneous structural analysis by electron diffraction), and characteristic x-rays (for simultaneous elemental analysis with EDX).

TABLE 3.7 *Electron–Matter Interactions Caused by 1° Electron Probe: Thick Samples*

2° Effects	Physical description of effect	Value to analysis	Use
2° Electrons	An electron transits near atom core and releases enough energy to excite a K-shell (2°) electron. A high population of 2° electrons is emitted per 1° electron (electron yield, δ). $$\delta = \frac{\text{Number of } 2^\circ e}{\text{Number of } 1^\circ e} \approx \frac{1}{\cos\beta} \quad (3.29)$$ where β is the incidence angle of the electron beam to the sample. δ values between 0.3 and 0.7 are typical. The energy of 2° electrons is considered to be low and ranges from ca. 5 to 50 *e*V. 2° Electrons are released from a depth of <1–5 nm. Others claim this depth is more on the order of 5–50 nm [10]. The emission of 2° electrons depends on the topography and to a lesser extent on atomic number. The greater the angle between the surface normal and the incident beam is, the greater is the path length of electrons and hence the larger the electron signal becomes (e.g., the contrast).	The SEM image is formed from low-energy 2° electrons after they exit the sample surface and are collected. Surface morphology, topography, and particle shape analysis are commonly performed by an SEM detecting and collecting 2° electrons. Resolution of 2° electrons depends on the diameter of the incident beam, the excitation volume, and the incident angle of the beam. 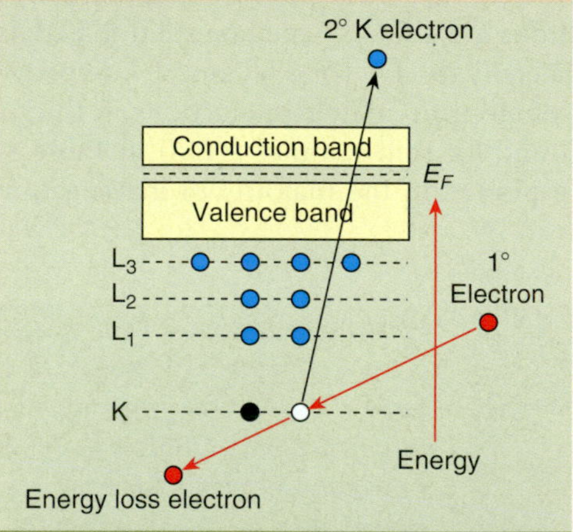 *Formation of 2° electrons. 1° Electrons penetrate a sample and have enough energy to excite a core electron. The 1° electron loses energy in the process.*	SEM EPMA
Backscattered electrons	1° Electrons collide with atoms normal to their path and scattered backward with large deflection (as high as 180°). The scattering is considered to be quasi-elastic. Backscattered electron yield, η, is low compared to 2° electrons and depends heavily on the local chemistry. $$\eta = \frac{\text{BSE current}}{\text{Incident } E - \text{Beam current}} \quad (3.30)$$	Backscattered electrons are highly directional and peak in the plane containing the surface normal and the incident beam. Topographical imaging used to differentiate areas of a sample (e.g., elements with higher atomic number) look brighter. Therefore, compositional analysis is accomplished based on brightness. One of the primary uses of backscattered electrons is in the analysis of mechanically alloyed materials. If the surface is flat, contrast can be obtained from regions differing in *z* by 1.	SEM EBSD

Table 3.7 (Contd.) *Electron–Matter Interactions Caused by 1° Electron Probe: Thick Samples*

2° Effects	Physical description of effect	Value to analysis	Use
	Generation of backscattered electrons relies on specimen atomic number, interaction volume, and accelerating voltage. The larger the *z* of an element is, the larger is the angular deflection. The KE is similar to 1° electrons and sample depth of emission ranges from <10 to 100 nm. Depth on the order of 0.5 μm also is possible. The excitation volume for BSE is large and resolution thereby suffers. If only elastically scattered electrons are selected, resolution is on par with 2° electrons.	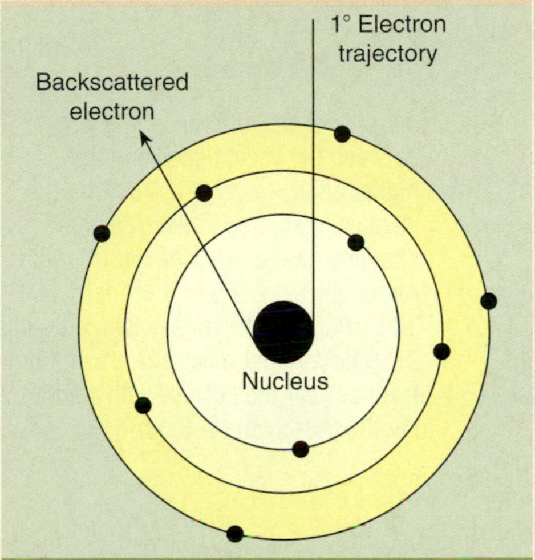 *Backscattered electrons. The trajectory of BSEs is back towards the beam after a slingshot action around the nucleus of the atom.*	
Surface plasmons	Surface plasmons are oscillating waves of electrons on metal. Since surface plasmons are activated by relatively low-energy sources, low-energy electrons (or photons) are responsible for generating surface plasmons on metals undergoing SEM analysis.	Surface plasmon spectroscopy is not performed in SEM analysis. 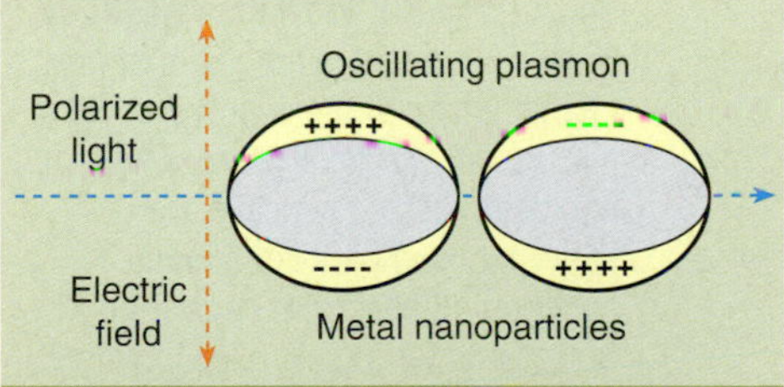 *Surface plasmon oscillations of free electrons on the surface of metals are depicted. A dipolar resonance is established as the free electrons on the metal surface oscillate in response to EM stimulation.*	SPR
Auger electrons	Following release of an outer shell 2° electron, a higher energy electron falls into a vacancy created by the ejected inner shell (usually K-shell electron). Following ionization and relaxation, enough energy is released to emit low-energy outer (surface layer) electrons called Auger electrons. Simultaneous production of visible photons of energy	Auger electrons are used for surface chemical analysis—an electron energy fingerprint. They are characteristic of elements and are also capable of contributing information about the bonding state of that element. Analysis of quantity (intensity, % composition of low *z* elements) and kinetic energy of ejected electrons lead to semiquantitative and qualitative data.	AES SEM

(*continued*)

TABLE 3.7 (CONTD.) *Electron–Matter Interactions Caused by 1° Electron Probe: Thick Samples*

2° Effects	Physical description of effect	Value to analysis	Use
	$$\lambda = \frac{hc}{\Delta E} \quad (3.31)$$ where ΔE is the difference in energy between the vacant shell and the higher energy level [2]. Although the yield is small, Auger electrons are characteristic of the element of origin. Auger electrons are of low energy: 100–1000 eV. Others say it is more like 2–50 keV. Auger electrons are emitted from or near the surface with sample depth <0.3–5 nm (1–3 nm).	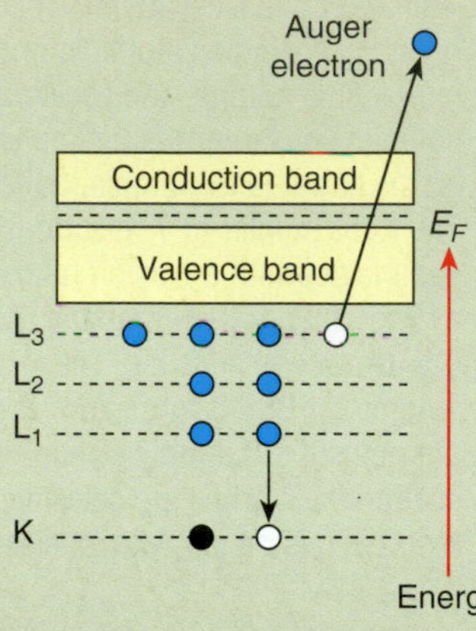 *Formation of Auger electrons. Valence band electrons are emitted following stimulation and relaxation (filling empty core shells). Auger electrons are characteristic of the element.*	
Characteristic x-rays	Following release of a 2° electron, usually K-shell, a vacancy is created. Electrons with higher energy drop down (cascade) to fill the void. In the process, x-rays are emitted that are characteristic of the energy transition and specific to the element. The characteristic x-ray peaks are superimposed on a continuous background spectrum consisting of Bremsstrahlung (see following effect). X-rays are emitted from a sample depth ranging from ~1 to 3 μm. Detection of x-rays is also semiquantitative because x-ray intensity can be converted into percentage composition. ZAF (atomic number effects) correction is applied with the assistance of a computer.	X-rays originating from K-, L-, M-shell x-rays are element specific. The emitted x-ray impinges upon a Li-drifted Si solid-state detector that is able to differentiate between x-rays of differing energy. The x-ray excites electrons in the detector into the conduction band. The greater the energy of the x-ray is, the higher is the number of electrons. Elemental analysis of micron diameter particles and larger materials is routinely accomplished. Silicon detectors are inefficient at low energies and therefore do not detect any element below Na. However, if a wavelength dispersive crystal spectrometer is coupled with the Si detector, elements down to B can be analyzed. 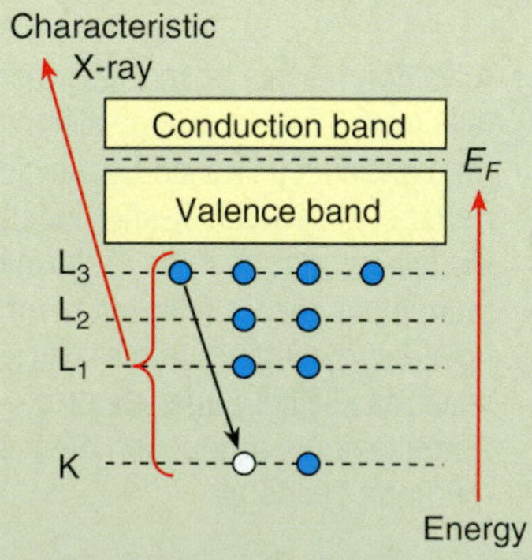 *Formation of characteristic x-rays from electron probe stimulation. Filling a shell vacancy by a higher energy electron results in release of x-rays that are characteristic of the element.*	EDX in SEM EDX in EPMA

TABLE 3.7 (CONTD.) *Electron–Matter Interactions Caused by 1° Electron Probe: Thick Samples*

2° Effects	Physical description of effect	Value to analysis	Use
Bremsstrahlung continuum radiation	When incident electrons undergo inelastic collisions, they slow down due to strong electrostatic forces applied by the nuclei. The lost energy appears in the form of EM radiation. This is the reason it is called "breaking radiation" and covers the entire range of the x-ray region of the spectrum. Bremsstrahlung is the major component of loss for high-energy electrons. Bremsstrahlung radiation is energy emitted by a moving charged particle and is proportional to m_e^{-2}, the rest mass of the particle.	Bremsstrahlung is considered to be background noise upon which the characteristic x-ray spectrum is superimposed.	EDX
Cathode ray luminescence (cathodoluminescence)	Visible light fluorescence following inelastic scattering of 1° electrons is released after ionization of valence shell electrons. In semiconductors, holes are produced after excitation. Recombination of electron–hole pairs results in emission of a photon. UV, visible light fluorescence, IR are produced depending on energy of 1° electron and penetration depth. The sample depth of electron penetration that results in cathodoluminescence ranges from <2 to 8 μm.	Crystal field influence Trace element abundance Semiconductors Geological samples Considered to be a "beam-injection" technique	CL-SEM EBIC
Induced current	Induced current can be generated in a semiconductor sample upon exposure to electrons. Excess carriers are generated with an electron beam and are assembled within an electric field inside the specimen. The electron-beam-induced current induces carriers recombine at defects or are collected at the contact, resulting in a current.	Induced currents typically are in the nano- to microampere range. They are used to locate and analyze *p–n* junctions in semiconductors, as well as to measure the diffusion length of minority carriers from Schottky or *p–n* junctions. Usually a thin evaporated Schottky contact made of aluminum is applied.	EBIC
Heating	Inelastically scattered electrons bounce back and forth within the interaction volume until energy is lost. Phonon excitation of lattice can result. Phonon vibration is then converted into heat. Low-energy continuum photons and low-energy electrons do not escape the lattice and energy is transformed into vibration of bonds and lattice (phonons).	The temperature rise is given by $$\Delta T = \frac{4.8E_0 b_i}{C_\tau d_o} \quad (3.32)$$ where E_0 is the accelerating voltage in keV, b_i is the beam current, C_t is the thermal conductivity of the sample, and d_o is the beam diameter	SEM

Table 3.8 *Electron–Matter Interactions Caused by 1° Electron Probe: Thin Samples*

2° Effects	Physical description of effect	Value to analysis	Use
Transmitted unscattered electrons	No interaction between electrons and sample as electrons transverse the specimen. Acceleration voltage in TEM is on the order of 100 keV or higher. Transmission capability is inversely proportional to thickness. Low atomic-molecular weight materials transmit electrons and appear to be light on the phosphor image screen. Heavy elements interact more strongly with electrons and appear as dark regions on the phosphor image screen.	Light areas equate to more transmission and dark areas to electron-dense areas. The result is a TEM image. Particle size can be measured in TEM. It is good practice to corroborate the measured size with those determined by other methods. TEM allows for direct observation of crystal structure. 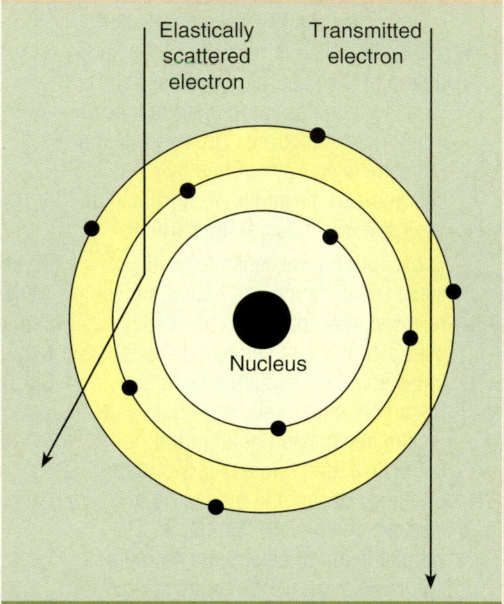 *Elastically scattered electrons in TEM are the bases for the image. Contrast is generated when dense materials absorb electrons.*	TEM HRTEM STEM
Elastically scattered electrons (ESEs)	Elastically scattered electrons are deflected with no loss of energy and transmitted obliquely though a sample. Elastically scattered electrons adhere to Bragg's law of diffraction.	Elastically scattered electrons are used in generating electron diffraction patterns. Information concerning crystal structure and orientation about the sample is acquired from the patterns.	Electron diffraction
Inelastically scattered electrons (ISEs)	Inelastically scattered electrons lose energy by interactions with the sample. Eventually, they too are transmitted through the thin sample. ISEs are transmitted obliquely through sample	Electron loss value typical of element Unique to bonding state (e.g., oxidation) Elemental analysis Kakuchi bands related to atomic spacing are formed. The bands consist of alternating dark and light lines. The inelastically scattered electrons that participate in the formation of Kakuchi lines are related to the atomic spacing of the specimen. The width of the Kakuchi line is inversely proportional to the atomic spacing and can track back to elastically scattered electron diffraction patterns.	EELS TEM HRTEM Electron diffraction
Characteristic x-rays	See **Table 3.7**	See **Table 3.7**	EDX (TEM)

3.1.2 *Scanning Electron Microscopy and Electron Probe Microanalysis*

The main components of the scanning electron microscope are shown in **Figure 3.6.** The electron gun at the top of the electron "optical" column produces the electron beam. The beam is focused to a diameter of ~50 Å at the foot of the column by a series of magnetic lenses and is scanned in a square TV-type raster across the surface of the specimen. The various 2° effects emitted from the surface of the specimen (e.g., secondary electrons, backscattered electrons, x-rays) may be detected and the signals amplified and displayed.

Components. Generally, two types of materials are used to produce electrons: filaments made of either tungsten (W) or lanthanum hexaboride (LaB_6). The tungsten cathode is a wire filament bent into the shape of a hairpin with a

FIG. 3.6 *A schematic representation of a scanning electron microscope (SEM) is shown. A high-powered electron beam originating from an electron gun near the top of the column is accelerated down a column through sets of collimators and magnetic focusing lenses. Detectors are placed near and around a sample. Many kinds of 2° signals are collected and assimilated. If an SEM has EDX capability, in situ elemental analysis is also conducted.*

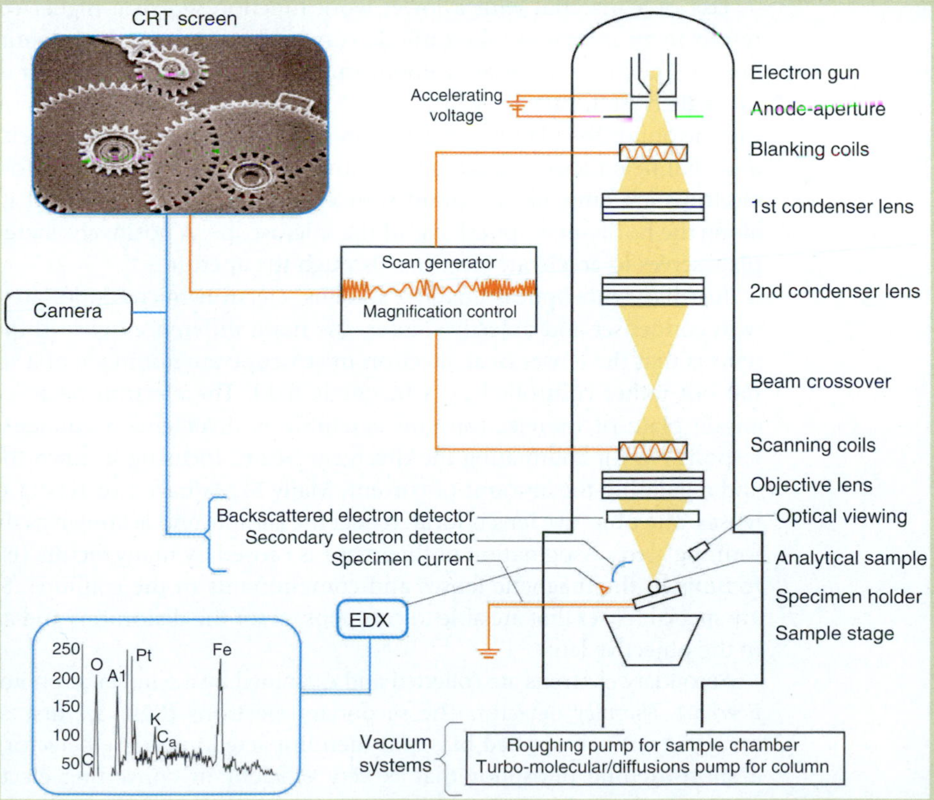

V-shaped tip that has a radius of 100 μm. The cathode is heated directly by a filament current, i_f. Electrons exit the filament with average energy:

$$E_e \sim kT \quad (3.28)$$

The energy necessary to facilitate the ejection of an electron from a material surface is governed by the work function of that material. Metals are the best materials for electron emission due to the presence of free electrons on their surface. The energy required to emit electrons is derived from the heat produced by the filament current. Field emission electron guns add extra energy by application of an electric field to the filament. The temperature of field emission electron guns (a.k.a. cold cathode emission) is much lower than that required for thermionic filaments. Richardson's law (Richardson–Dushman equation) relates the current density J_c obtained by thermionic emission of the filament:

$$J_c = \mathcal{A}_c T^2 \exp\left(\frac{-\Phi_W}{k_B T}\right) \quad (3.33)$$

where

$\mathcal{A}_c$ is a constant that is characteristic of the material
T is the absolute temperature
Φ_W is the work function of the material
k_B is Boltzmann's constant (1.38×10^{-23} J · K^{-1})

Use of a material with a lower work function Φ_W or a higher constant $\mathcal{A}_c$ results in an increase in the cathode current density (e.g., the brightness of the beam). The most important cathode material to be developed so far is sintered, or single crystal, LaB_6 for which $\mathcal{A}_c$ =40 A · cm^{-2} · K^2 and Φ_W = 2.4 *eV*. A Wehnelt cap surrounds the filament and applies a negative potential that repels electrons and channels them through an aperture where a space charge is formed. The electrons are emitted as a point source called a space charge that is centered along the horizontal optical axis of the microscope. A positively charged anodic plate serves to accelerate electrons through the aperture.

Just like with optical imaging systems, electron microscopes are equipped with condenser and objective lenses; the main difference between the two systems is that the lenses in an electron microscope are not made of a solid material but rather controlled by a magnetic field. The electron beam exiting the anode plate of the electron gun assembly is divergent. A condenser lens is responsible for collimating the divergent beam, focusing it down the column and regulating the amount of current. Many SEMs have two sets of condenser lenses. The objective lens is located near the sample and is known as the "probe-forming" lens. Astigmation of the image is caused by many factors (e.g., imperfections in the magnetic lenses and contaminants in the column). Stigmators are special lenses that are able to compensate for the distortions and are housed in the objective lens.

Secondary electrons are collected and examined by a scintillator-photomultiplier *Everhart–Thornley* detector. The secondary electrons (SEs) are first collimated by a grid with an applied bias and then impacted upon the detector surface—a short-lifetime phosphor that is very efficient in converting electrons into ~400-nm ultraviolet photons. The number of electrons that reach the detector is a function of the surface topography. High points and surface features that face the detector produce more 2° electrons and hence a greater signal. Backscattered electrons (BSEs) are detected by a semiconductor array located at the bottom of the column. A current is produced when BSEs strike the semiconductor array. Tapping into the specimen current is another way to acquire an image. Most SEM samples are conducting, and all SEM specimens are fixed to a metal stage with conducting cement or tape. A specimen current is generated by inelastic scattering and absorption of impinging electrons after the SE, BSE, transmitted, and Auger electrons have left the sample. This absorbed current is a function of the atomic number of the elements in the specimen [10].

Image Generation. Image generation occurs by rastering the electron beam (of spot size ca. 50 μm or less) across the sample surface. A reconstructed image is sent to a CRT and viewed. During the rastering process, there is a "dwell time," at which point the beam is paused. During the dwell time, the numerous types of secondary electron effects are expressed. The lifetime of the secondary effects is shorter than that of the dwell time. Before the beam moves to the next segment, the secondary effects have been detected, recorded, and dissipated. Magnification, brightness, and contrast all affect the image quality. Brightness is a function of the entire image and is controlled by the amplitude of the signal. Contrast is the variation of the signal from one rastered point to the next and is expressed by

$$C = \Delta S \cdot S_{\text{Average}}{}^{-1} \tag{3.34}$$

where ΔS is the change in the strength of the signal between two points and S_{Average} is the average signal strength. Cathodoluminescence is an imaging technique used mostly in investigating luminescence from mineral specimens. A camera housed in the column system takes a photograph of the emitted light from the sample.

Operation. The operation of an SEM is quite straightforward, although training is required to protect the high-maintenance instrument and its expensive components. Sample loading and mounting begin by fixing the specimen to a metallic stub with graphite cement or conducting tape. The stub is then attached to the metallic stage and placed inside the sample chamber. The chamber is evacuated with a roughing pump. The column is evacuated to a pressure less than 10^{-6} torr in order to begin the filament warm-up process. A special procedure must be followed to engage the filament (usually of the thermionic variety) and to align the beam. The acceleration potential is stepped up slowly. Once the beam current is detected and adjusted, the sample is introduced into the column and is ready for analysis.

Figure 3.7 depicts a JEOL JSM-7700F field emission SEM [11]. The resolution of this instrument is 0.6 nm at 5 kV and 1.0 nm at 1 kV. Magnification capability ranges from 25× to as high as 2 million×. Accelerating voltage ranges from 0.5 kV to approximately 30 kV. New lens systems are able to correct spherical and chromatic aberration, thereby improving the resolution of objects. This modern SEM is capable of performing high-resolution backscattered electron imaging as well as scanning transmission electron microscope imaging of thin samples [12]. As one can imagine, there is a trend to combine capabilities within one instrument.

FIG. 3.7 *A JEOL JSM-7700F field emission SEM is shown. The resolution limit of this beautiful instrument is less than 1 nm. Magnification is on the order of 2× million. The resolution is 0.6 nm at 5 kV and 1.0 nm at 1 kV. Magnification capability ranges from 25× to as high as 2.0 million. Accelerating voltage ranges from 0.5 to approximately 30 kV.*

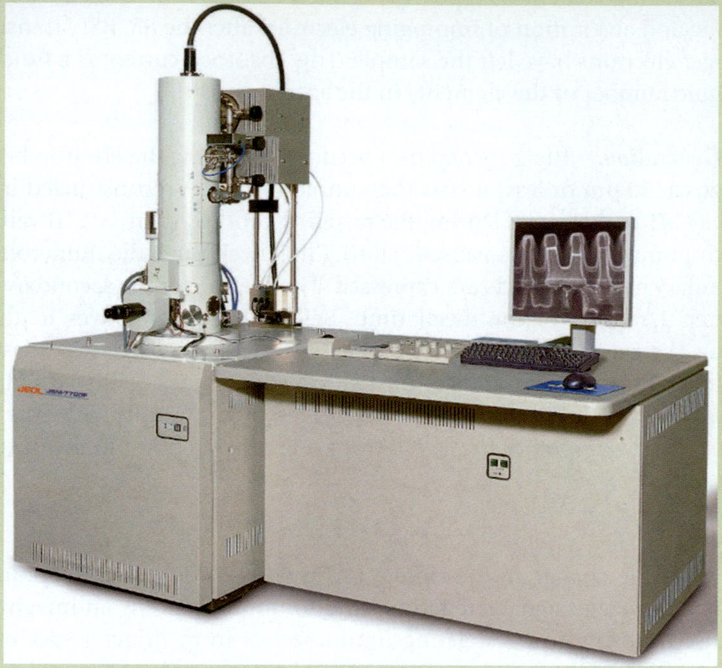

Source: Image reproduced with permission from JEOL, Ltd.

Not only does nano converge with regard to disciplines, nano-instrumentation also converges with regard to capability.

Electron Probe Microanalysis. Henry Mosely in 1913 found that the wavelength of x-rays emitted after excitation was inversely proportional to the square of the atomic number of an element:

$$\lambda \propto \frac{1}{z^2} \tag{3.35}$$

This was the first time the composition of any material (brass in this case) was identified by characteristic x-rays. In 1922, Assar Hadding applied x-ray spectrometry to analyze minerals. The father of electron probe microanalysis (EPMA) is considered to be James Hillier, who patented the process in 1943.

EPMA is designed for nondestructive x-ray imaging and microanalysis of samples. Some special features of the EPMA include capability of ultrafine spot focusing (e.g., down to 1 μm), optical microscope imaging, and enhanced sample positioning [12]. The EPMA has spatial resolution capability of ca. 1 μm

FIG. 3.8

A JEOL JXA-8100/8200 EPMA is pictured. This instrument is capable of analyzing light elements like boron. Elemental characterization is accomplished with wavelength dispersive x-ray techniques (WDS). A comparison between EDX and WDS is given in section 3.3.3, "X-ray Methods." The backscattered electron image detection limit is 6 nm with a working distance (WD) of 11 mm and under 30 kV accelerating voltage conditions.

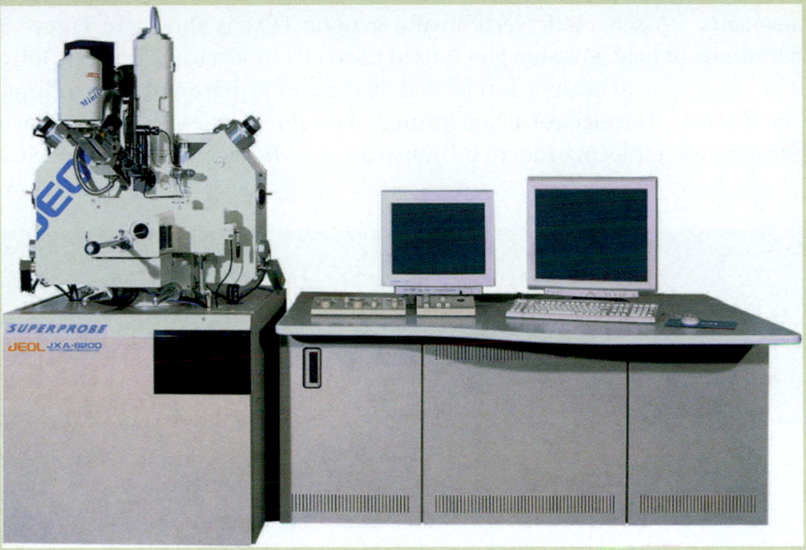

Source: Image reproduced with permission from JEOL, Ltd.

(considered to be excellent), is highly sensitive (on the order of 0.5% for the major elements), and has a detection limit of ca. 100 ppm. Some EPMA have five crystal-focusing wavelength dispersive x-ray spectrometers on board. EDX detectors are also included for quick analysis. X-ray energy in the range of 0.1–15 keV is typically analyzed.

A JEOL JXA-8100/8200 electron probe microanalyzer is shown in **Figure 3.8** [12]. Image acquisition occurs by collection and assimilation of backscattered electrons. The detectable element range begins with very light elements such as $_{5}$B (with $_{4}$Be analysis capability also possible) to very heavy ones such as $_{92}$U. The detectable x-ray wavelength range (by wavelength dispersive spectroscopy, WDS) is from 9.3 to 0.087 nm. Large specimen size can be accommodated with dimensions of 100 × 100 × 50 mm (height) and an analyzable area of 90 mm^2. The probing current ranges from 1 pA to 10 μA. The backscattered electron image detection limit is 6 nm with a working distance (WD) of 11 mm and under 30 kV accelerating voltage conditions [12].

3.1.3 Transmission Electron Microscopy

The TEM functions by the same principles as the SEM except that the detector is a phosphor plate that is able to capture images formed by transmitted electrons. Another major difference between TEM and SEM is that the accelerating voltage in TEM is usually far greater: 300 kV compared to ~50 kV. The wavelength of the electron beam is in the picometer range, 6.13–2.24 pm. The thermionic electron gun consists of a filament, usually made of tungsten or lithium hexaboride. The filament is heated under an applied potential until electrons are produced. Another difference from SEM is that TEMs have a projector lens system at the base of the column. The image that the TEM operator observes on the phosphor screen is the projected image of the sample.

Components. A schematic version of a generic TEM is shown in **Figure 3.9.** A thermionic or field emission gun is used to accelerate electrons between 100 and 400 kV. The electron beam is accelerated by the anode plate and then collimated via an aperture. The electrons pass through a double condenser lens system and focus upon a sample. In order to be transparent to the electron beam, specimens

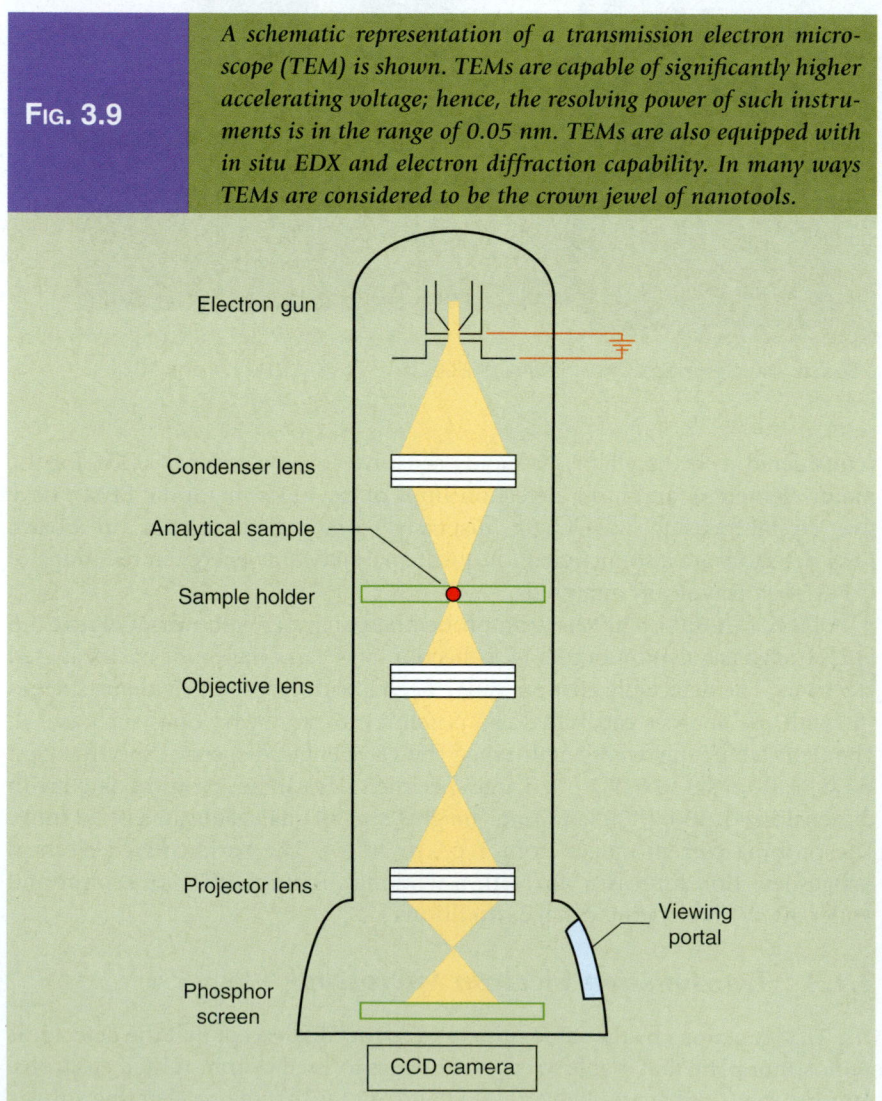

Fig. 3.9 *A schematic representation of a transmission electron microscope (TEM) is shown. TEMs are capable of significantly higher accelerating voltage; hence, the resolving power of such instruments is in the range of 0.05 nm. TEMs are also equipped with in situ EDX and electron diffraction capability. In many ways TEMs are considered to be the crown jewel of nanotools.*

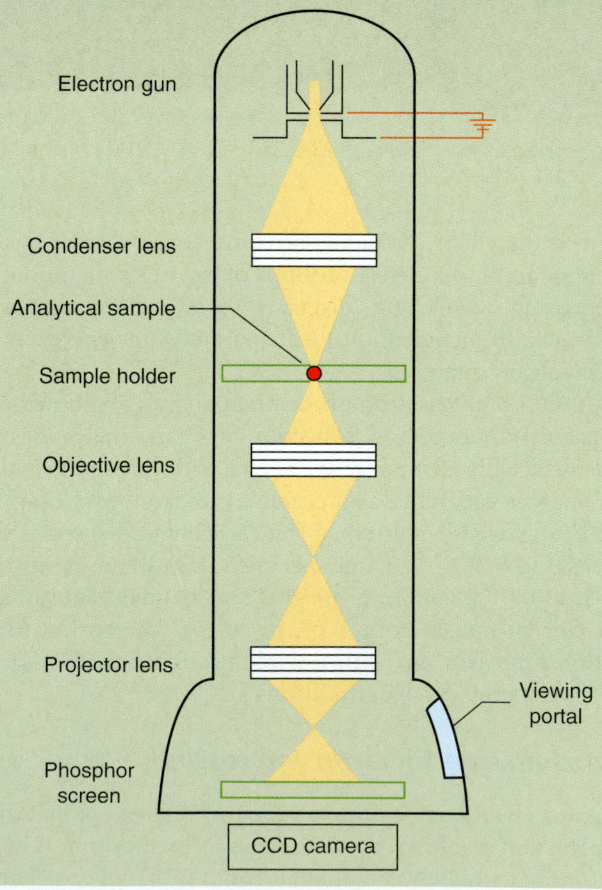

typically are 1000 atoms or less in thickness (e.g., a few nanometers to >100 nm). Electrons that pass through the sample are then focused by an objective lens, channeled through another aperture, and then projected onto the phosphor fluorescent screen by the action of the projector lens for the purposes of viewing. The image is captured with a photographic Polaroid camera (outmoded now) or by a charge-coupled device (CCD) camera.

The CCD camera or photographic paper captures the "shadow image" of the specimen depending on the density of its various components in the sample. CCDs are integrated circuits that are composed of close-packed photodiodes. They function by converting photons into an image. In the TEM, the photons arise from the action of electrons impinging on a phosphorescent screen. When a specimen image is ready to be recorded, the viewing phosphor screen is tilted to open access to the CCD camera. Upon completion of image acquisition, the viewing screen is replaced to its original position.

In 1969, Willard Boyle and George Smith of Bell Labs invented the CCD [13]. They discovered that it was able to interact with electronic charges by way of the photoelectric effect. In 1974, a 100 × 100 pixel CCD prototype device was manufactured and the Sony Corporation began mass production of CCDs for camcorder use. The first CCD camera was used in a TEM in 1982 (the 100 × 100 pixel device directly exposed to 100-keV electrons) [14]. Direct detection, however, presented some serious issues. J. C. H. Spence and J. M. Zhou proposed an indirect detection strategy that employed a scintillation screen with an optical coupler [15]. CCD technology has developed significantly since 1974 with now 4000 × 4000 pixel arrays available for TEM imaging tasks. Thus, the Polaroid method of image acquisition gave way to digital imaging technology once again.

Resolution. The resolution of a good TEM is ca. 0.2 nm, on par with the distance between two atoms and the atomic radii of some heavy metals. Carl Zeiss SMT announced a major breakthrough in TEM resolution in 2005, achieving 0.07 nm with an experimental ultrahigh-resolution 200-kV field emission gun transmission electron microscope (FEG-UHRTEM). This level of resolution is near the theoretical limit of TEM analysis.

Image Generation. There are generally four types of images that are produced by TEM:

1. Transmitted electrons are responsible for generating a *bright field image* of the specimen. When the objective aperture is positioned just below the specimen, only the transmitted electron beam is allowed to pass down the column and form an image of the specimen on the phosphor screen. Any diffracted or inelastically scattered electrons are excluded from passage.
2. To acquire a *low-resolution dark field image,* the objective aperture must be repositioned (e.g., adjusted to one side or another of the main axis of the column) while a metal plate is inserted to block the major beam. Although spherical aberrations are formed in this configuration, a quick identification of specimens that produce diffraction patterns is accomplished.

3. A *high-resolution dark field image* can be formed. In this configuration, electron diffraction patterns are acquired for analysis.
4. If, on the other hand, the aperture is opened to allow many beams, including the direct electron beam, to pass, images are formed by interference of the direct beam and the diffracted beams. These are called high-resolution lattice images and are formed in a high-resolution (HR)-TEM. With HR-TEM, which is a combination of bright-field and lattice imaging, real and reciprocal space can be observed simultaneously.

Operation. Samples are restricted to thickness generally less than 100 nm. The specimen is usually placed on a copper grid that is less than 100 μm thick and 3 mm in diameter. The specimen is prepared in several ways. Initially, for hard inorganic samples, sawing, grinding, and polishing are done to thin the sample. Such mechanical polishing proceeds until a thickness of several microns is achieved. The polished sample is then placed in an ion mill to undergo a finer level of thinning. Argon ions bombard the sample and remove material according to the accelerating voltage of the ion mill system. Milling continues until breakthrough is achieved. The circumferential area surrounding the zone of breakthrough can be extremely thin, less than a few nanometers. Chemical and electrochemical thinning and reactive ion etching are also techniques used to reduce the thickness of samples for TEM analysis.

Another procedure involves embedding the sample in a polymeric resin and sectioning with a microtome by the action of a diamond knife. This is usually accomplished for softer samples—biological samples in particular. The microtome instrument is capable of producing films with nanometer thickness. Following sectioning, the samples float on water and are collected on a copper grid, dried, and placed into the TEM sample holder in the middle of the column. The thickness of sectioned samples is estimated by observation of interference colors. Films that are thinner than a few hundred nanometers appear gray as they float on the water collector. Sections with thickness between ca. 200 nm and 1 μm display all the colors in the spectrum. The phenomenon of interference colors shows how bulk material properties change as size approaches smaller dimensions. Thicker sections look like the parent bulk embedding material. Staining samples, particularly those of biological origin, with heavy metal dyes that contain osmium or lead significantly enhances the structural detail during TEM analysis. The heavy atom sites appear darker because of better scattering or absorption of electrons. Direct deposition of sample materials sonicated in acetone or other volatile solvents onto carbon-coated copper grids without prior preparation is also common. Sonication of the sample beforehand breaks up and disperses the sample.

Limitations of the TEM procedure include the following:

1. Extensive sample preparation is time consuming and this limits the number of samples that can be imaged and analyzed.
2. In addition to the small physical size of samples, the field of view is relatively small as well. For example, the section under analysis may not be representative of the sample as a whole.

3. Sample structure and morphology have been known to change drastically under exposure to electrons with extremely high energy, and damage to biological samples in particular can occur.
4. TEM is a high-vacuum instrument that is costly to operate and to maintain, certainly playing a role in the high barrier entry required to conduct nanoscience research.

However, the advantages of the TEM technique are numerous. Any type of sample, whether electrically insulating, semiconducting, or conducting, is able to be imaged by TEM. The incredible resolution capability allows for atomic level inspection.

An image of a JEOL JEM-3200FS field emission transmission electron microscope is shown in **Figure 3.10** [11]. The TEM is equipped with a 300-kV FEG (a zirconium–tungsten Schottky FEG). It is also capable of conducting electron

Fig. 3.10 *A JEOL JEM-3200FS field emission TEM is pictured. The TEM is equipped with a 300 kV FEG (field emission gun). It is able to conduct EDX, EELS, and electron diffraction analysis. The element in the electron gun is ZrO/W and its point-image resolution capability is 0.17 nm. Beam spot size can be focused down to 0.4 nm in diameter. Magnification ranges from 100× to 1.5 million.*

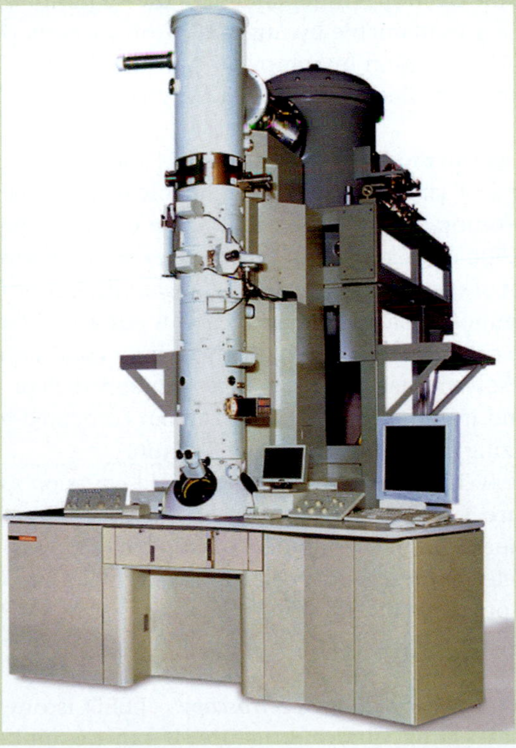

Source: Image reproduced with permission from JEOL, Ltd.

energy loss spectroscopy (EELS), EDS, and electron diffraction. At 300 kV, the TEM is capable of 0.17-nm point-image resolution and energy resolution of 0.9 *e*V. The spot size can be made as small as 0.4 nm. The magnification ranges from 100× to as high as 1.5 million× the size of the specimen structure.

3.1.4 Other Important Electron Probe Methods

Auger Electron Spectroscopy (AES). Lise Meitner and Pierre Auger independently discovered Auger effect in 1923. Although her paper on the procedure predated Auger's, the procedure is known by the name of the latter. Lise Meitner should have also shared a piece of the Nobel Prize in chemistry for her contribution to the discovery of nuclear fission, but Otto Hahn was awarded the distinction in 1944.

Auger electrons, briefly described in **Table 3.7**, are produced from radiation (usually x-rays) or high-energy electron excitation of core electrons. With x-ray excitation, the kinetic energy of the ejected Auger electron is equivalent to

$$E_{\text{KE,Auger}} = h\nu - E_{\text{b}} - \Phi_W \qquad (3.36)$$

where $h\nu$ is the energy of the x-ray; E_{b} is the binding energy of a K-, L-, or M-shell electron; and Φ_W is the work function of the material. Technically, the kinetic energy of an Auger electron is independent of the energy of the excitation source. Each element has a diagnostic Auger electron fingerprint. The kinetic energy of the electron is independent of the excitation energy. Chemical analysis of surfaces is attainable by Auger electron spectroscopy. AES consists of three steps: (1) ionization by removal of a core electron, (2) emission of Auger electrons or x-ray fluorescence, and (3) analysis of the Auger electrons (or x-rays).

AES is highly sensitive for all elements except H and He and is used to monitor surface cleanliness of samples. AES also is capable of semiquantitative analysis of surface composition. Auger depth profiling, in combination with argon ion sputter etching (500 *e*V–5 k*e*V ions), provides semiquantitative compositional data as a function of sample depth. Scanning Auger microscopy (SAM) is capable of providing compositional information about surface heterogeneity. Auger analysis must be conducted under ultrahigh vacuum conditions. SEM provides better surface resolved images; AES is better equipped than XPS to produce depth profiles and its sampling depth is ca. 2 nm compared to EDX, which is capable of analyzing elements at depths of 1–2 μm.

An image of a JEOL FE-Auger JAMP-9500F field-emission Auger microscope is shown in **Figure 3.11** [11]. A Schottky field emission gun acts as the electron source that is capable of delivering 0.5–30 kV accelerating voltage. A Faraday cup serves as a detector for secondary electrons at a resolution of 3 nm (at 25 kV and 10 pA). The probe diameter for Auger analysis is 6 nm (at 25 kV and 1 nA). The range of analytical Auger electrons is 0–2500 *e*V.

Low-Energy Electron Diffraction Spectroscopy. LEED is one of the primary methods of investigating surface structure. It can be applied qualitatively (size, symmetry of adsorbate unit cells) or quantitatively (information about atomic position). The principles of LEED were uncovered accidentally in

FIG. 3.11 *An image of a JEOL FE-Auger JAMP-9500F field-emission (Schottky) auger microscope is shown. Accelerating voltage capability ranges from 0.5 to 30 kV. Resolution is ca. 3 nm. The probe diameter for Auger analysis is 3 nm. Magnification ranges from 20 to 500,000×. This Auger instrument is equipped with an ion gun for depth analysis of samples and is capable of EDX and EBSD analyses.*

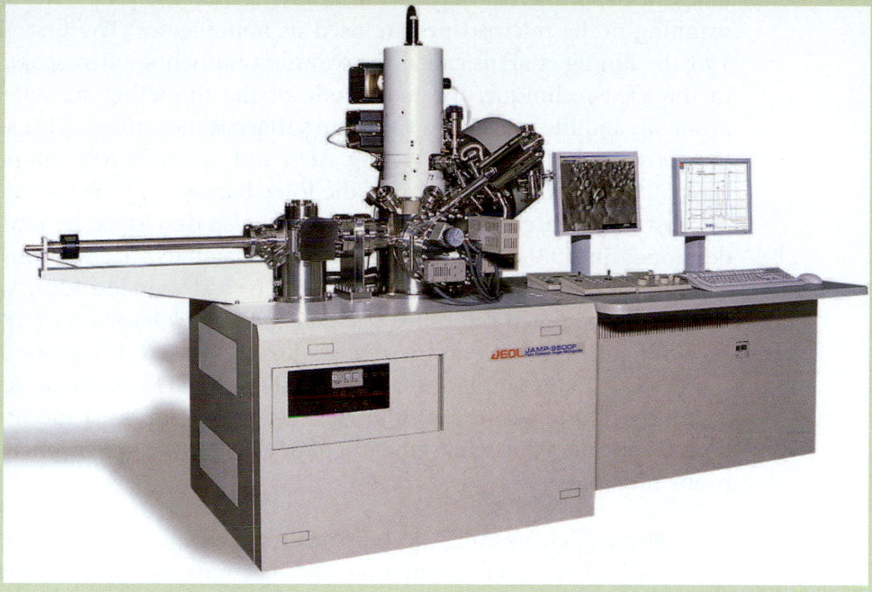

Source: Image reproduced with permission from JEOL, Ltd.

1924 by Davisson and Kunsman during a study of the 2° emission of electrons from a nickel surface. Low-energy electrons (10–500 *e*V) normal to the sample surface are elastically scattered by the electron density surrounding surface atoms [16]. This range of energy corresponds to distance between surface atoms or molecules (de Broglie λ between 0.39 and 0.055 nm) and therefore is able to generate diffraction patterns from the two-dimensional surface structure. The scattered electrons are collected by a retardation grid and analyzed. When analyzing two-dimensional surfaces, one of the Miller indices can be dropped; instead of *h, k,* and *l,* one only need be concerned about *h* and *k*. Miller indices are discussed more fully in chapter 5.

Electron Energy Loss Spectroscopy. Hillier and Baker described the EELS technique in 1944 [17]. It only became popular, however, after better microscopy techniques evolved by the 1990s and with the development of ultrahigh vacuum systems. Spatial resolution of 0.1 nm is obtainable with this method. The EELS technique is based on inelastic interactions of electrons with the specimen. The magnitude of energy loss is measured by an electron spectrometer and is manifested in the form of plasmons, inner shell ionization, inter- and intraband transitions, and phonons. EELS works well for low *z* elements and is therefore a complementary method to EDX but with better resolution.

3.2 Scanning Probe Microscopy Methods

Scanning probe microscopes are able to create detailed three-dimensional images of specimen surfaces with atomic resolution [18]. Two basic types of scanning probe microscopes are used in nanoscience. The first, developed in 1981 by Binnig et al., is called the scanning tunneling microscope (STM) [19]. In the STM technique, the magnitude of the tunneling current between the probe tip and the atoms of a substrate surface is monitored. STM samples must therefore be electrically conducting. AFM and its derivatives comprise a second class of SPMs. In AFM, the size of the force between a probe tip and the atoms of substrate surface is monitored. The AFM also developed by Binnig et al. was developed in 1985 specifically to address materials that were insulating. The AFM is able to measure forces on the order of 1 μN and less [20]. A third type of scanning probe method, called *scanning atom-probe microscopy* (APM), is based on a different approach and is discussed in section 3.2.4 in more detail.

There are many similarities between AFM and STM. Both are equipped with a probe tip fastened to a cantilever, a scanning (motion) mechanism, and a detector system. Four achievements contributed to the development of scanning probe methods:

1. Physical cushioning mechanisms (e.g., bungee cords hanging from a ceiling), motion-dampening eddy currents induced by magnets and copper plates, and nitrogen-gas-regulated suspension systems all serve to isolate the SPM from external vibration sources. Protective enclosures shield the microscope from room drafts. If extremely high-resolution work is required, cryogenic temperatures eliminate atomic and molecular movement.
2. Computerized feedback system control of piezoelectric devices with nanometer level precision guides the probe during its descent to the surface—less than a nanometer above where van der Waals attractions exert force on the probe tip (in AFM) or tunneling current is able to flow (in STM).
3. Computerized control of piezoelectric scanners moves the probe tip across the sample surface with nanometer precision.
4. The fabrication of sharper probes allows for better resolution of surface features. For example, carbon nanotube probes with sharpness of less than 1 nm have shown promise in the past few years.

Scanning Probe Tips. The probe tip of a scanning probe microscope is important to the feature-resolving ability of atomic force and scanning tunneling microscopes. The probe is usually made of silicon or Si_3N_4 for AFM and tungsten for STM. In both cases, the probe is sharpened to a fine tip. Atomic probe methods rely on sharpened probes that are positioned as close to a sample surface as possible. Ultimately, a probe tip sharpened to one atom provides the best resolution (**Fig. 3.12**).

The probe tip has a rich history. The stylus profiler (profilometer) (from the Latin *pro-*, "forth" + *filare*, "to draw out," *filum*, "thread, spin" + "measure") is the progenitor of the atomic force microscope. A profilometer is an instrument designed to measure the topography of industrial material surfaces. Shmalz

Fig. 3.12 *Two probes manufactured by Veeco Instruments are shown. Probe tips are made of many materials that range from silicon to harder materials like tungsten. The resolving power of AFM and STM depends on the sharpness of the tip (the curvature). Sharper tips result in more clarity of image.*

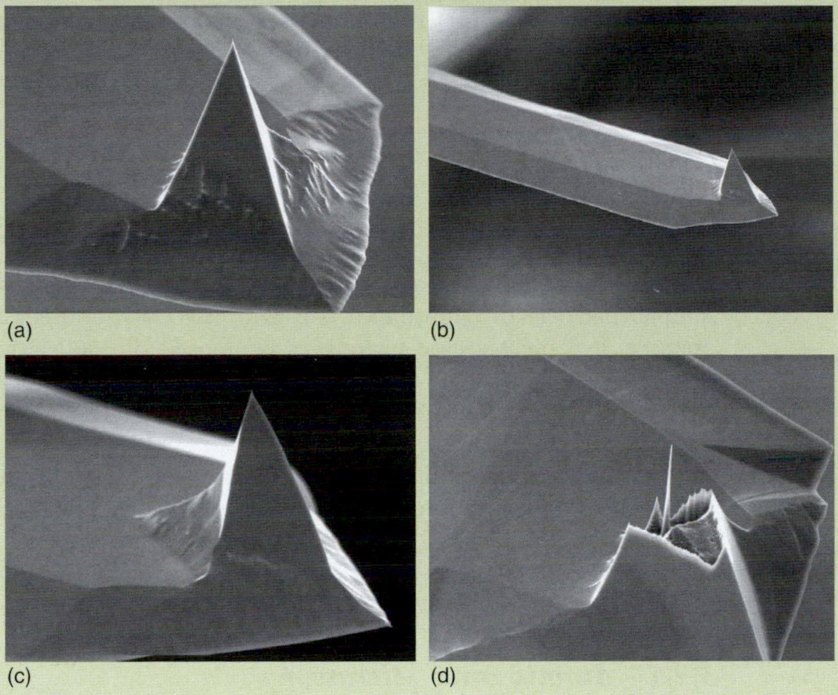

Source: Courtesy of Veeco Instruments, Inc.

invented the first profilometer in 1929 [21–23]. He used an optical lever arm to detect the motion of a sharpened probe attached to a shiny cantilever during a surface raster process. The magnification capability of this instrument was ca. 1000×. Early profilometers were equipped with a diamond-tipped stylus attached to a coil in an electric field. Raster of the profilometer tip across a surface induces a current that is proportional to the surface roughness [24].

A stylus (from the Greek *stylus,* meaning "stake, pillar" adapted to "instrument of writing") is a writing utensil. In ancient times, a small rod with a pointed end was used to scratch letters in wax tablets. Diamond or sapphire-tipped styluses were commonly used in phonographs to replay music. The SPM probe is a stylus that is able to read (and write?) at the atomic level.

Modern profilometers are capable of measuring surface roughness down to 1 nm [25]. In such mechanical probes, the tip is in direct physical contact with the surface of the sample. In 1971, Young developed a noncontact profiler by application of an electron field emission current between the probe and the surface [26]. The AFM evolved from the prosaic profilometer of the 1920s into a powerful and versatile tool of nanoscience. The major difference between the two is the magnitude of the applied force (e.g., the AFM utilizes much smaller forces).

Piezoelectric Materials. Pierre Curie and his brother, Jacques, discovered piezoelectricity in 1883. They showed that electricity was produced when pressure was applied to selected crystallographic orientations. They also demonstrated the converse to be true. Piezoelectricity (from the Greek *piezien*, "to press, squeeze" + electricity) is the induction of electrical polarization in certain types of crystals due to mechanical stress [27]. Piezoelectric devices are made of dielectric components that are able to convert mechanical stresses (e.g., sound waves) into electrical signals and vice versa [27]. Common piezoelectric materials include quartz, various ceramics, and Rochelle salt. Transducers such as phonograph cartridges, microphones, radios, and strain gauges make use of piezoelectric materials.

Electromechanical piezoelectric ceramic transducers are responsible for creating the mechanical motion required to scan AFM and STM specimens. The motion, or change in geometric shape, depends on the type of crystal, its shape, and the strength of the applied electric field [23]. Typically, a standard piezoelectric crystal will expand about 1 nm per AC volt applied [23]. Therefore, to accommodate all the variables of motion required to map the surface of a sample, piezoelectric transducers are comprised of hundreds of layers. For example, if a piezoelectric material is composed of 1000 layers, motion of 1000 nm·V^{-1} is possible. With 50 V applied, motion of 0.05 mm is possible [23]. On the nanoscale, that is a lot of territory. The expansion of a stacked piezoelectric device is shown in **Figure 3.13.**

Depending on the manufacturer, two types of scanning motions are possible. In one, the sample is moved relative to the probe tip. In the other, the sample is kept stationary while the probe is rastered across the surface. For the most common configuration, the piezoelectric device controls the x and y displacement of the sample and the z motion of the probe tip.

3.2.1 Atomic Force Microscopy

One of the most powerful tools of nanoscience is a relatively simple mechanical device that is capable, however, of imaging atoms—a device that requires no special atmosphere, no expensive high-energy radiation or beam source, and can be operated under ambient conditions. Atomic force microscopy relies on the mechanical deflection of a cantilever to relay information about the contour of a sample surface. An atomically sharpened probe tip (50–20 nm or less) descends perpendicularly from the distal end of a cantilever. The tip-to-sample

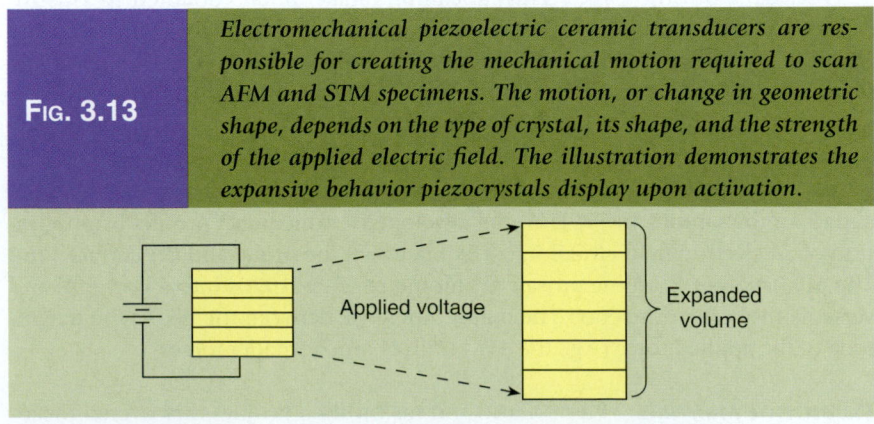

FIG. 3.13 *Electromechanical piezoelectric ceramic transducers are responsible for creating the mechanical motion required to scan AFM and STM specimens. The motion, or change in geometric shape, depends on the type of crystal, its shape, and the strength of the applied electric field. The illustration demonstrates the expansive behavior piezocrystals display upon activation.*

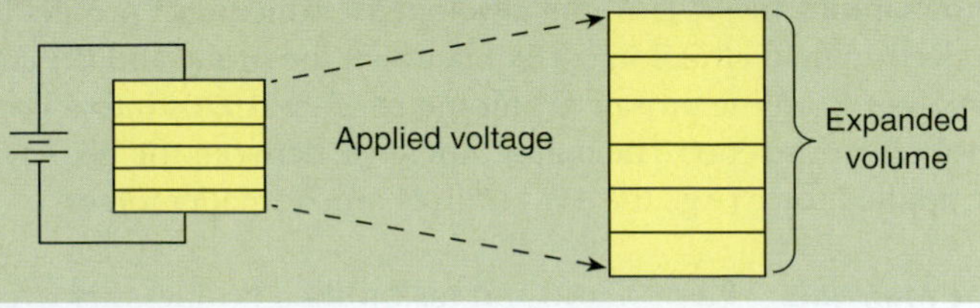

distance is fixed by means of a feedback mechanism that maintains constant force between them (i.e., constant height). A laser beam is focused on the top of the shiny cantilever and is reflected into a photodiode detector (**Fig. 3.14a**). Differences in the reflected beam are recorded by a split photodiode and recorded as changes in topography. The photodetector is able to discriminate motion with accuracy less than 1 nm. The surface is scanned grid-like by the probe to produce a three-dimensional image of the surface. A Veeco Instruments Dimension V Nanoscope multifunctional scanning probe microscope is shown in **Figure 3.14b.**

The AFM technique relies on a balance between attractive van der Waals and repulsive electrostatic forces between the probe tip and the surface. The net force is a function of the distance between the two (**Fig. 3.15**). The force is also a function of the dielectric constant ε of the medium. For example, the forces between the tip and the surface are much weaker if the probe is submerged in a liquid. Such versatility allows for analysis of biological samples by the AFM technique. Hand in hand with atomic-order resolution, the theoretical magnification potential of the AFM technique is on the order of 10^9 [18]. The versatility of the AFM is demonstrated further by its ability to image large objects that are tens of microns or more in size. The tobacco mosaic virus is 300 nm in length and the human hair is 50–100 μm in width. Both have been imaged by AFM. Several modes of AFM operation are listed and described in **Table 3.9.** There are three major modes of operation: contact, tapping, and noncontact.

Different cantilever–probe tip combinations are available for different types of AFM techniques. Cantilevers are fabricated from single crystal silicon. The length of AFM contact and tapping mode cantilevers is between 100 and 500 μm; the width is between 25 and 40 μm, and the thickness is between 1 and 10 μm. The resonant frequency of these cantilevers dwells between 10 and 300 kHz and the spring constant between 0.1 and 50 $N \cdot m^{-1}$. The probe tip diameter ranges from a few nanometers for finer resolution to 20 nm for an average AFM to as large as 50 nm. Noncontact AFM mode cantilevers are made of highly doped single crystal silicon.

The Force in AFM. A sensor measures the force that is generated between the probe tip (cantilever) and the electron clouds of the sample. Hooke's law describes the relationship between the cantilever and the applied force:

$$F = -k\,z \tag{3.37}$$

where F is the force, k is the "spring constant" of the cantilever, and z is the distance of the deflection of the cantilever. Laser light is reflected off the shiny top surface of the cantilever into the photodiode detector. Therefore, the signal from the photodiode detector is proportional to the mechanical displacement of the cantilever [23].

Resolution. Image resolution in AFM is acquired in three dimensions: the x–y plane (or in-plane) typical of optical microscopes and in the z, or perpendicular, direction. The resolving power of an AFM is dependent on the radius of curvature and size of the tip. For example, a thin probe sharpened to a few atoms has better resolving capability than its larger, smoother counterpart. Broadening of features occurs when the tip radius of curvature is on par with the dimension of

Fig. 3.14 *(a) A generic AFM system is shown. A focused laser beam is reflected off the back of the cantilever equipped with a sharpened probe and into a photodiode detector. Any deflection of the beam is translated into a topographical feature. Rastering of the tip over the surface (or vice versa) produces a topographical image. The photodiode is split into two compartments and is able to detect differences in beam position to the level of a nanometer. (b) Veeco Instruments' Dimension V Nanoscope multifunctional scanning probe microscope. The controller for this instrument is able to measure tip-sample/cantilever dynamics. Pixel density is 5120 × 5120. The scan range is 90 × 90 μm and can accommodate a sample size of 150-mm diameter. Modes: contact, tapping, lateral force (LFM), magnetic force (MFM), force–distance spectroscopy, electric force (EFM), scanning capacitance (SCM), scanning spreading resistance (SSRM), tunneling atomic force (TUNA), conductive atomic force (CAFM), scanning tunneling (STM), and torsional resonance mode microscopy (TRmode). Veeco probes include tapping (42 $N \cdot m^{-1}$, 320 kHz, no coating), electrical (0.2 $N \cdot m^{-1}$, 13 kHz, Pt-Ir coating), magnetic (2.8 $N \cdot m^{-1}$, 75 kHz, −400 Oe), conductive (2.8 $N \cdot m^{-1}$, 75 kHz, doped diamond coating), TUNA (2.8 $N \cdot m^{-1}$, 75 kHz, pt-Ir coating), and SSRM (42 $N \cdot m^{-1}$ < 320 kHz, doped diamond coating).*

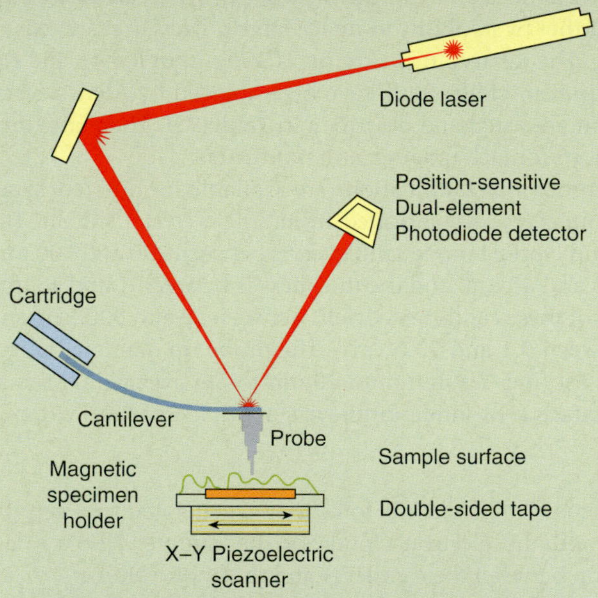

(a)

(b)

Source: Image courtesy of Veeco Instruments, Inc.

FIG. 3.15 *The form of this graph is similar to that of a Lennard–Jones pair interaction potential function and indicates a fundamental behavior characteristic of all physical interactions: There is always attraction and repulsion. Surface profilometers operate in hard contact mode regime of the graph.*

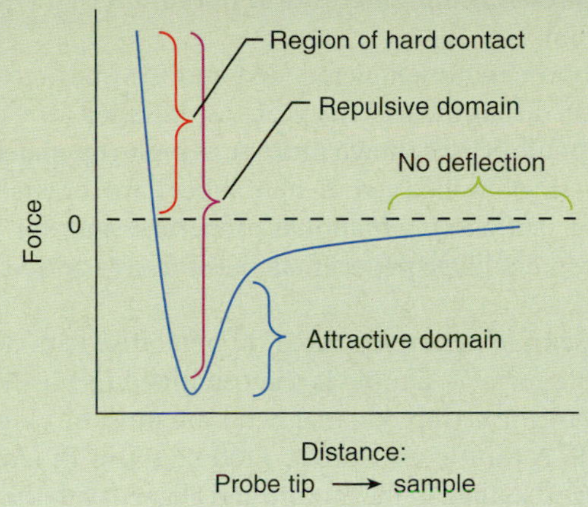

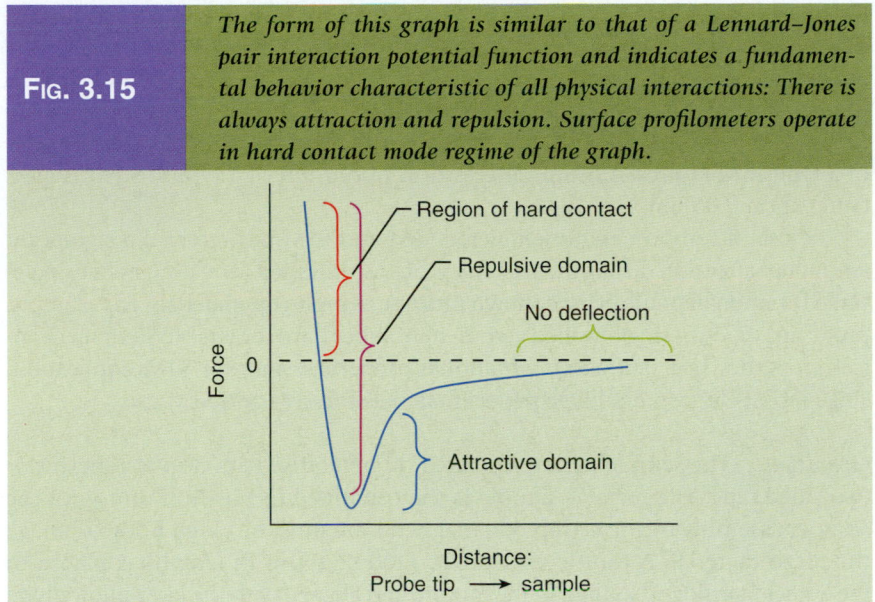

TABLE 3.9 *Atomic Force Microscopy Primary Modes*

AFM mode	Configuration	Capability
Contact mode	The probe-cantilever assembly applies a constant force to the surface; the force constant (or spring constant) is <1 $N \cdot m^{-1}$. The deflection of the cantilever is due to topographical changes characteristic of the surface. The mechanical deflection is translated into an optical signal by reflection of aligned laser light from the top surface of the cantilever into an aligned-calibrated dual photodiode collector. Due to intimate contact with the surface, large lateral forces can influence the probe. Surface contamination, electrostatic forces, and heterogeneous surfaces are able to impact the action of the probe. AFM is capable of imaging nanomaterials in air, vacuum, and liquids.	Samples with hard surfaces are appropriate for contact mode AFM analysis. AFM is capable of imaging insulator, semiconductor, and conductor surfaces. Qualitative information includes three-dimensional visualization and material sensing by phase contrast (chemical contrast of features with the same *z*-displacement). Quantitative information includes topographic mapping, particle and pore size, particle and pore morphology, surface roughness and texture, particle count, size distribution, surface area distribution, and volume–mass distribution [23].
Tapping mode	Strong repulsive regime Oscillating probe (>100 kHz) Intermittent contact with surface Lateral forces reduced significantly Amplitude and phase imaging	Three-dimensional topography Used for soft samples or those weekly bound to surface Biological materials DNA Carbon nanotubes
Noncontact mode	Weak-attractive regime Oscillating probe Can be applied with water layer	Soft surfaces

the object under analysis. The radius of curvature for typical contact mode AFM tips is quite large, on the order of 50 nm [2]. During a scan, the side of the probe makes contact before the tip. Probe tips sharpened to a single atom and close to the substrate surface provide the best configuration for image acquisition. Resolution decreases the farther a tip is from the surface. The lateral resolution of a typical AFM is ca. 1.5 nm. The vertical resolution of a typical AFM is even better at ca. 0.05 nm.

The use of carbon nanotube tips for AFM and STM has received a significant amount of attention during the past few years. Multiwalled (and even better single-walled) carbon nanotubes have shown promise as probe tip materials. For example, probe tips made of single-walled carbon nanotubes have shown excellent mechanical properties (e.g., stiffness), vibrational properties, and, for STM application, single-point electron discharge perfect materials for AFM and STM use.

Operation. The sample material is fixed to metal disc with a spot adhesive or two-sided tape. The metal + sample is then snapped to position on top of the scanner assembly by the action of a magnet. Care must be taken not to damage the piezo material. A cantilever + probe (100–200 μm in length) is placed in the holder (cartridge), secured in the mount, and aligned with the laser-photodiode sensing apparatus. An optical microscope is used to position the probe over the sample surface. The image of the cantilever as it hovers over the sample is visible on a nearby CRT screen. The probe assembly is lowered to the sample surface by an automated feedback control system to a vertical distance of a micron or less. Obviously, we are not able to judge such a small distance and require the use of computerized electromechanical technology to guide the tip down the *z*-axis.

Parameters such as scan area, probe force, and scan rate are input into the system and the scan is put into motion. Successive line scans (rasters) are compiled to generate a three-dimensional image of the surface. There are two generic configurations for scanning. In the first, the sample is mobilized underneath a stationary probe. In the second, the reverse configuration is applied. Numerous computer programs are available for image manipulation.

In contact mode imaging, a soft, deflectable cantilever with a sharp probe tip is brought close to the surface (e.g., in the repulsive regime of **Fig. 3.15**). The cantilever is deflected according to Hooke's law with a spring constant that is typically in the range of 1×10^{-3} to $100\ \mathrm{N \cdot m^{-1}}$. The degree of deflection (usually along the *z*-axis) is recorded via an array of positional photodiodes that capture laser light reflected from the surface of the cantilever. Constant force or distance between the tip and the surface is maintained by a feedback loop. The magnitude of the forces applied to the tip range from 0.01 to 100 nN. Piezoelectric elements hold and position the sample in all three spatial domains—*x*, *y*, and *z*—and participate in the feedback loop process. A three-dimensional rendition of the topography of the surface is achieved by the contact mode of the AFM. Selected AFM images are displayed in **Figure 3.16**.

3.2.2 Scanning Tunneling Microscopy

"The wavelike properties of electrons permit them to 'tunnel' beyond the surface of a solid into a region of space forbidden to them under the rules of classical physics"—so stated Binnig and Rohrer in 1981. Electron (or quantum) tunneling is attained when a particle (an electron) with lower kinetic energy is able to exist on the other side of an energy barrier with higher potential energy, thus disobeying a fundamental law of classical mechanics. Tunneling is the penetration of an electron into a classically forbidden region [28]. Electrons exhibit wave behavior and their position is represented by a wave (probability) function. The wavefunction represents a finite probability of finding an electron on the other side of the potential barrier. Since the electron does not possess enough kinetic energy to overcome the potential barrier, the only way the electron can appear on the other side is by tunneling through the barrier. The principle of quantum tunneling was also used to explain exponential radioactive decay rates (half-lives). In 1928, the great physicist George Gamow showed that alpha decay occurs via a tunneling process. Considering only classical principles, it takes too much energy to pull a nucleus apart; however, in quantum mechanics, a finite probability is allocated to tunneling and hence nuclear decay is allowed. The inversion of the conformation of ammonia is an example of tunneling by a particle, in this case the hydrogen atom [28].

STM relies on an electronic signal to relay information about a sample—the strength of a tunneling current potential that exists between the probe tip and the substrate surface. Small changes in the distance between the probe tip and the substrate surface translate into large changes in the tunneling current. By this phenomenon, atomic scale resolution by STM is possible in the x, y, and z directions. The density of states (DoS) of solid-state materials can be also mapped by the technique called scanning tunneling spectroscopy (STS). Chemical reactions induced and oriented by the STM probe are also available by means of this technique.

There are two kinds of electron microscopes, and we do not mean TEM and SEM because we have already discussed them. The differences between SEM and STM are that the magnitude of current is quite diminutive in STM and that current originates from electron tunneling. Electron tunneling occurs between the conducting sample and the tip of the STM. The tip is very close to the substrate but not in actual physical contact (**Fig. 3.17**).

When an electron interfaces with a finite potential barrier with potential energy U_o that is greater than its own kinetic energy, the electron stays within the "box". Electrons are small enough and "quantum" enough to make tunneling happen. Electron tunneling current between two flat plates separated by a vacuum is given by

$$I \propto Ve^{(-2kW)} \tag{3.38}$$

where I is the tunneling current, W is the distance between the surfaces (the width of the barrier), and k is a term that is related to the potential across the vacuum:

$$k = \frac{\sqrt{2m_e(U - E_e)}}{\hbar} \tag{3.39}$$

where m_e is the electron rest mass as before, U is the potential energy between the surfaces, E_e is the energy of the electron, and $\hbar$ is the Dirac constant ($\hbar = h/2\pi$). Typically, the potential difference, $U - E_c$, is equal to ~4 eV.

Fig. 3.16 *See caption on page 151.*

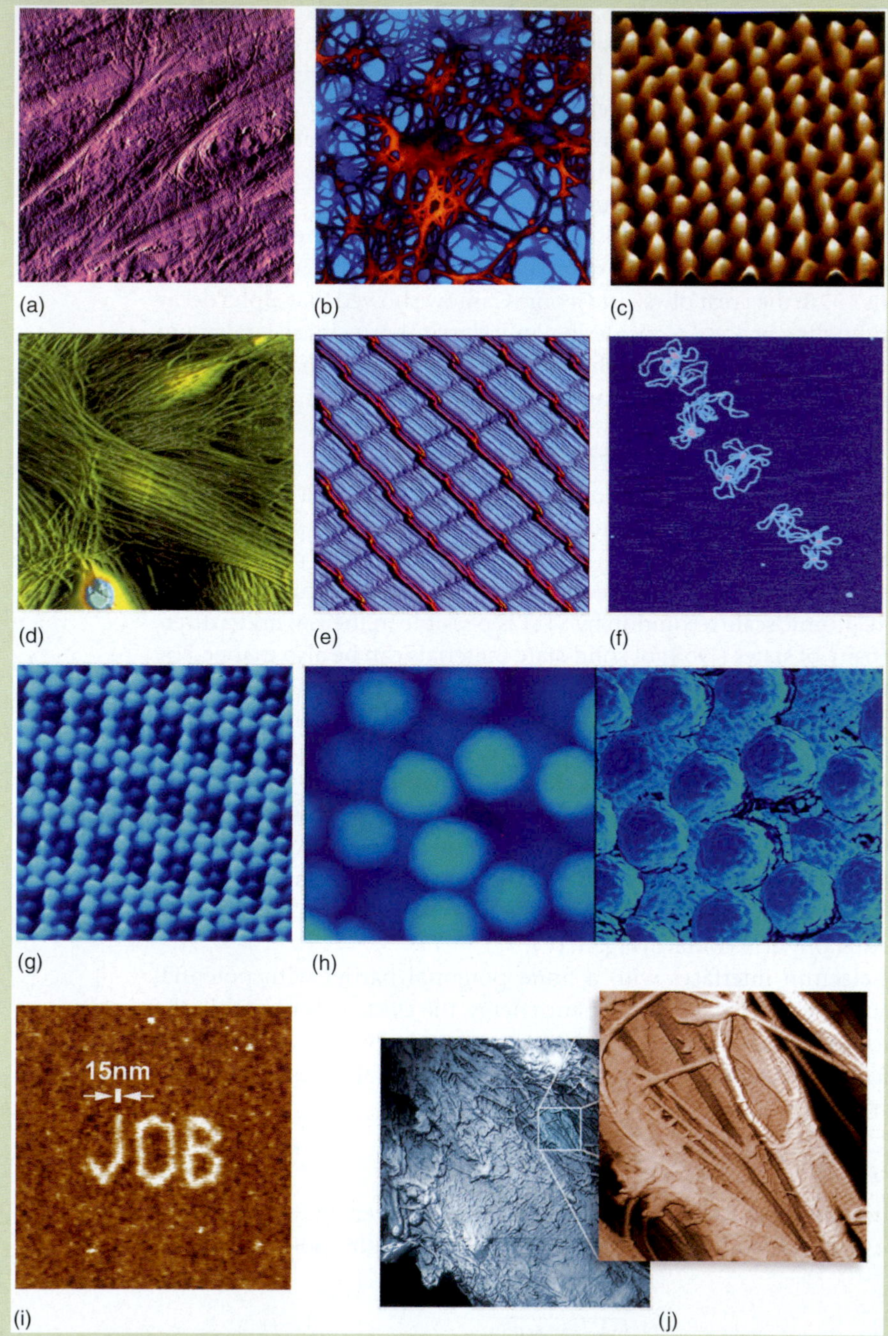

Source: Courtesy of Veeco Instruments, Inc.

FIG. 3.16 *Selected AFM images acquired by various Veeco instruments are displayed. (a) The image shows unmineralized collagen fibrils on the outer surface (periosteum) of trabecular bone. The 67-nm d-banding across the collagen fibrils is visible. The image was acquired by tapping mode with a MultiMode V AFM. Scan size is 8.6 μm. (b) A nanoscale scaffold made of nanofibers fabricated by a template method from a mixture of gelatin and alginate. The structure has potential application as a biosensor. The scaffold has features at both nano- and microscales that mimic the topography of natural extracellular matrices. Image was acquired with a Dimension 31000 AFM. The scan size is 20 μm. (c) A low-pass filtered image of mica surface atoms taken with a BioScope AFM system. The mica surface was imaged in contact mode. A 5-nm scan is shown. (d) Image of live MC3T3F osteobalst cells; networks of actin fibers of the cytoskeleton are portrayed. Scan size is 100 μm. The image was taken with a BioScope II AFM. (e) The periodic three-dimensional structure of the wing elements in* Morpho peleides *butterfly. Image scan size is 10 μm. The structures impart color to the butterfly wing via nanophotonic properties due to the nanostructure of the wing. We have already seen TEM renditions of another* Morpho *species photonic structure on the front cover of the text. (f) DNA strands are imaged with tapping mode. Scan size is 700 nm. (g) STM image of oxygen atom lattice on rhodium single crystal. The enclosed image is the result of a 4-nm scan. (h) Images of poly(methylmethacrylate): AFM height on the left and phase on the right. (i) AFM point-and-click nanolithograpy with the low-noise Dimension CL SPM head (700-nm scan). (j) AFM image of a fibrous structure of naturally aged nineteenth century goat parchment. The uppermost layer was removed by a microtome. The characteristic periodicity of the collagen fibers is distinctly visible. The image was acquired with a MultiMode AFM. Scan size = left: 20 μm; right: 3 μm.*

The current is defined by a modified version of equation (3.38) that takes into consideration the geometry and electronic structure of the probe tip:

$$I = Cp_{tip}p_{sample}e^{-W\sqrt{k}} \tag{3.40}$$

FIG. 3.17 *A schematic rendition of a scanning tunneling microprobe is depicted. The tunneling current is a quantum phenomenon. Direct contact with the surface is avoided. Localized electronic structure (density of states) of a specimen is obtained by scanning tunneling spectroscopy (STS).*

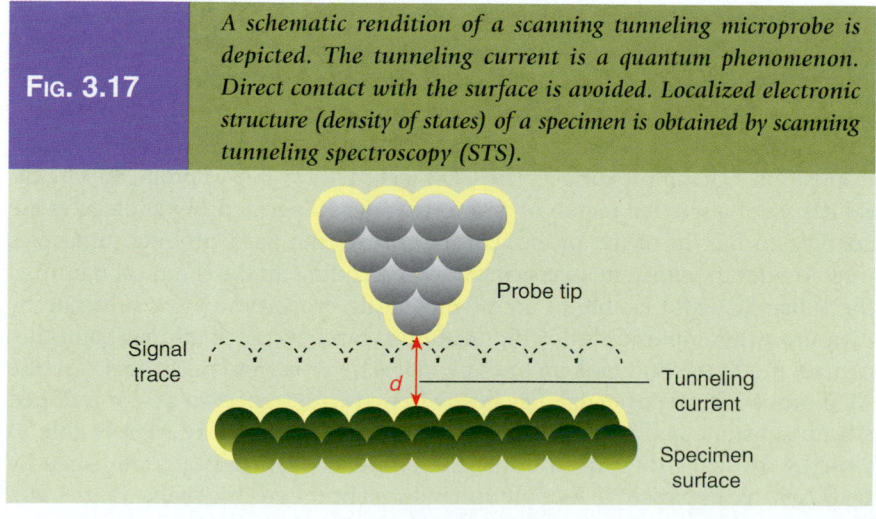

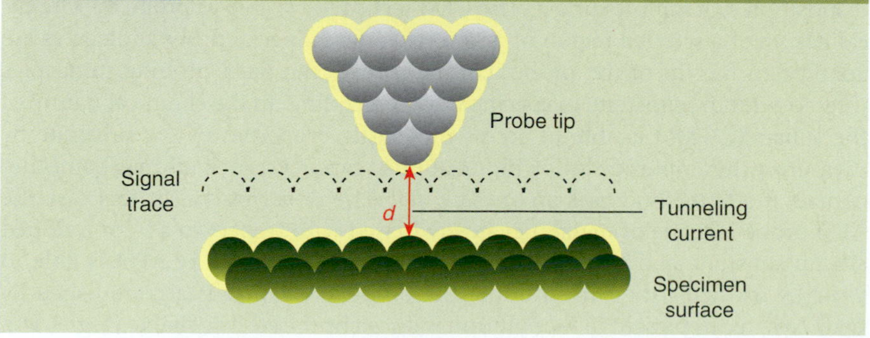

where C is a voltage-dependent constant and the p terms represent the electronic structures (density of states) of the tip and the sample, respectively [29].

The tip potential is biased with respect to the substrate. The direction of electron flow is determined by the direction of the bias. If the tip is negatively biased with respect to the substrate, then electrons will flow from the substrate to the probe. If the reverse configuration is applied, the electrons will flow from the probe to the surface.

Scanning Tunneling Spectroscopy. A complementary method to STM is scanning tunneling spectroscopy. The purpose of STS is to generate a map of the localized electronic structure of a specimen—ideally with atomic level resolution, given that the probe tip conforms to the necessary criteria. The spectrum is a map of the electronic structure (e.g., the DoS). Typically, an STS spectrum is found by measuring the change in tunneling current as a function of voltage yielding an I–V curve of a selected section of the sample. The I–V curve acquired from a bulk metal, for example, is shaped in the form of a smooth continuum. For an atom, its density of states resembles a series of lines. Somewhere in between are the electronic states of nanomaterials, which are expressed as step functions.

The operation of the STM in this technique is straightforward. The tip of an STM is held in place over a region of interest while the bias voltage is manipulated. Other permutations of STS operation exist. In one, current is held constant during a scan while height is recorded. In another, the reverse configuration is applied [2]. Material conductivity (dI/dV) and work function (dI/dz) plotted against applied V are also measured by the STS technique [2].

3.2.3 Other Important Scanning Probe Methods

The development of scanning probe methods has proliferated since the advent of the *nano age*. Offshoots of AFM include *lateral force microscopy* (LFM), *force modulation microscopy* (FMM), *electrostatic force microscopy* (EFM), *magnetic force microscopy* (MFM), *scanning capacitance microscopy* (SCM), *scanning thermal microscopy* (SThM), and *chemical force microscopy* (CFM).

The following three techniques are operated in the contact mode. We understand how the AFM technique measures deflections primarily in the z-direction. On the other hand, LFM is able to detect lateral or torsional deflections of the cantilever. These deflection modes arise from forces that run parallel to the surface of the sample. LFM, therefore, is a useful tool in characterizing the frictional disposition of a sample surface. The CFM technique is able to probe the chemical nature of a selected region of a sample surface. Special probe molecules are adapted to the tip of the probe. During a scan, the fixed probing molecules interact with the substrate in specific ways depending on the chemical nature of the substrate. FMM is able to access the elastic properties of the substrate by measuring the amplitude following an applied cantilever oscillation [2]. Another contact mode method uses an *ambient atomic force microscope*. In this case the AFM probe is made of metal, negatively biased with respect to a p- or n-doped silicon substrate and surrounded by a meniscus of water. The AFM is able to write by oxidizing the silicon under the probe. Oxidation is accomplished by hydroxyl anions present in the solution induced by the electric charge (**Fig. 3.18**).

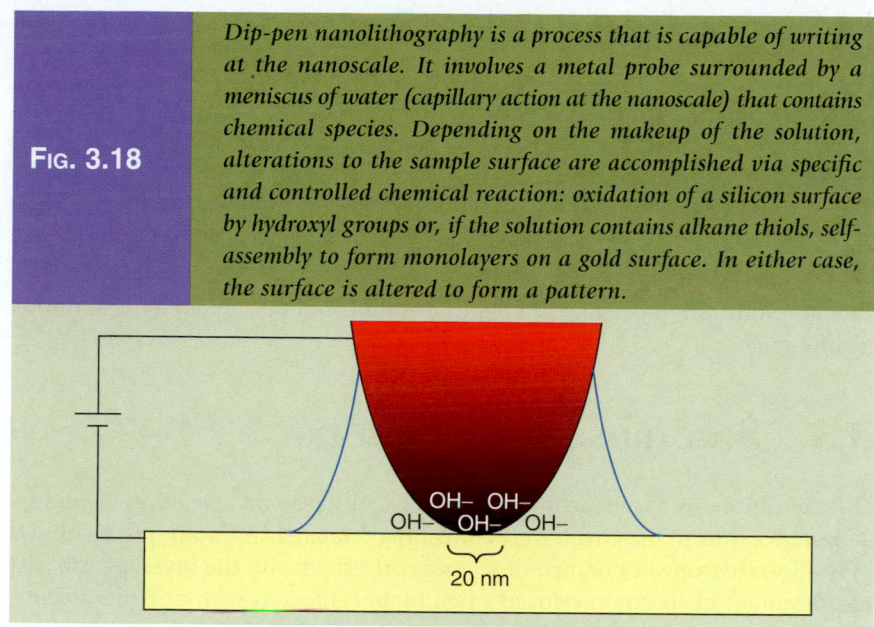

FIG. 3.18 *Dip-pen nanolithography is a process that is capable of writing at the nanoscale. It involves a metal probe surrounded by a meniscus of water (capillary action at the nanoscale) that contains chemical species. Depending on the makeup of the solution, alterations to the sample surface are accomplished via specific and controlled chemical reaction: oxidation of a silicon surface by hydroxyl groups or, if the solution contains alkane thiols, self-assembly to form monolayers on a gold surface. In either case, the surface is altered to form a pattern.*

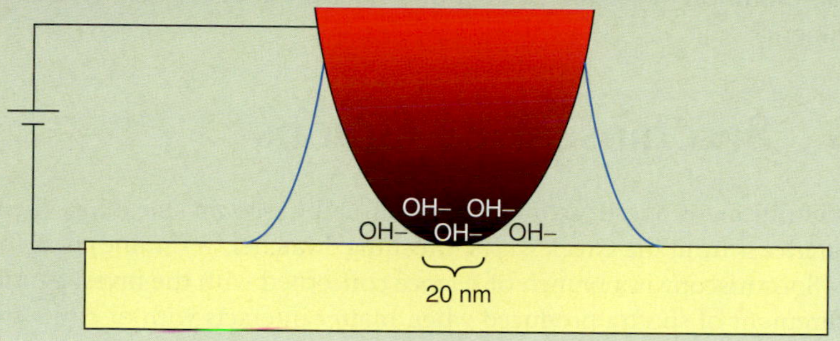

A form of nanolithography, nano-dip-pen lithography, is discussed in more detail in chapter 4.

The following four techniques operate in noncontact mode and rely on changes in the resonant frequency of the cantilever to produce a variable signal. MFM, as its name implies, is a technique designed to measure localized magnetic domains on a material surface. In this technique, a tip coated with a ferromagnetic material is employed. EFM is a technique designed to measure localized charge domains on a material surface. SCM is a technique designed to measure localized capacitance on a material surface. The result is a map of the dielectric outlay of the specimen. The SThM technique uses a heated wire to determine the thermal conductivity of localized regions of a sample surface. Just like with its macroscopic counterpart, thermal conductivity is determined by measuring the change in current as a function of temperature for a specific mass.

3.2.4 Atom-Probe Methods

E. W. Müller, J. A. Panitz, and S. B. McLane invented the atomic probe microscope (APM) in 1967 by combining a time-of-flight mass spectrometer with a field ion microscope (FIM) [30]. APM, like FIM, consists of a movable sharpened tip but is operated in ultrahigh vacuum at cryogenic temperature. Ionization is induced at the surface, and the probe tip repels ions towards a detector aperture. A chemical profile of the surface is obtained in this way with resolution of one-atom spacing. J. A. Panitz went on to develop the imaging atomic probe (IAP) in 1974. The IAP does not require a moving tip. Atom-probe tomography (APT) also uses a detector that is sensitive to the position of the selected area [31]. Tomography (from the Greek *tomos,* "to slice, section") is a technique that is able to analyze cross-sections of a material. This development led to the position-sensitive atom probe (POSAP) invented by A. Cerezo, T. Godfrey, and G. D. Smith in 1988.

Modern atomic probes are capable of providing three-dimensional atomic resolution, compositional imaging, and chemical analysis (similar to SIMS) [32]. Individual atoms of a specimen can be removed and analyzed! Atoms are isolated by a combination of a high electric field with the application of either a laser or voltage pulse. Thin films deposited on the tip itself are readily analyzed. The LEAP 3000X Si, manufactured by Imago Scientific Instruments, has a 200-nm field of view (FOV) and a data collection rate of 5×10^6 ions per minute [32]. Lateral resolution of 0.1 nm and depth resolution of 0.5 nm are typical of this instrument.

3.3 Spectroscopic Methods

The definition of spectroscopy (from the Latin *spectare,* meaning "behold," appearance + from the Greek *skopos,* meaning "watcher," examine) is as follows [33]: "Spectroscopy is a branch of science concerned with the investigation and measurement of spectra produced when matter interacts with or emits electromagnetic (EM) radiation." J. C. Engle, Jr., and S. R. Crouch provide an excellent definition of the spectrum [34]: "A spectrum is a display of the intensity of radiation emitted, absorbed, or scattered by a sample versus a quantity related to photon energy, such as wavelength or frequency."

Spectroscopy, then, in its most basic sense implies the study of phenomena created by the interaction of electromagnetic radiation with matter or, in its purest form, the study of the spectra itself. However, to some, spectroscopy may indicate a broader spectrum of analysis, including those that do not involve electromagnetic radiation. In the context of this chapter, spectroscopic methods are limited to those involving some sort of EM radiation, whether 1°, 2°, or higher order or interaction.

The interaction of radiation with matter assumes many forms. Molecular absorption of radiation is described by

$$E_{\text{Total}} = \sum E_{\text{Interactions}} = E_{\text{Electron}} + E_{\text{Rotation}} + E_{\text{Vibration}} \tag{3.41}$$

For single atoms, these options are limited to electronic transitions. Electronic transitions between energy states are observed by measuring absorption, emission, fluorescence, or luminescence of radiation. Other types of mechanisms include scattering (elastic and inelastic), reflection, refraction, and diffraction. Nonradiative mechanisms such as vibration, rotation, ionization, and heat transfer also factor into many types of EM interactions.

Molecular-level electron transitions exist in several forms. We are quite familiar with the atomic and molecular ground-to-excited state electronic transitions such as $\sigma \rightarrow \sigma^*$, $\pi \rightarrow \pi^*$ and $n \rightarrow \sigma^*$, $n \rightarrow \pi^*$, where σ, π, and n are the sigma-bond, pi-bond, and the nonbond (electron pairs). Nanomaterials also show quantized transitions, especially in small clusters. Larger colloids are able to scatter light but metal quantum dot clusters have HOCOs and LUCOs (highest occupied and lowest unoccupied cluster orbitals, respectively). Nanoscale materials present a new domain of energy transitions: just between the discrete atomic-molecular quantum transitions and continuous transitions of classical bulk domains.

3.3.1 UV-Visible Absorption and Emission Spectroscopy

The ultraviolet–visible spectrum ranges from ca. 350 (ultraviolet) through 770 nm (reds) [34]. Many nanomaterials interact with UV–visible photons to produce a variety of effects: plasmon oscillations on the surface of nanometals (Ag and Au), electron transfer cascades (involving supramolecules like chlorophyll), exciton pairs in semiconductors (nano-TiO_2), particle size-dependent fluorescence (semiconductor quantum dots), diameter-dependent fluorescence (chiral single-walled carbon nanotubes), and interference colors (porous alumina or titania thin films). There are many more examples. We begin by reviewing some basic concepts.

Light can be transmitted, absorbed, or reflected:

$$I_o = I_{Ref} + I_{Abs} + I_{Trans} \tag{3.42}$$

where I_o is the intensity of the incident light and I is the intensity of light after interaction with the specimen. *Transmittance T*, another unitless parameter, is the ratio of radiant power I emerging from a material to the incident radiant power I_o.

$$T = \frac{I}{I_o} \tag{3.43}$$

The Beer–Lambert law is used to describe the phenomenon of *absorbance*, a unitless parameter:

$$A = -\log_{10} T = \varepsilon b c \tag{3.44}$$

where ε is the molar absorptivity ($L \cdot mol^{-1} \cdot cm^{-1}$), b is the path length of the sample in centimeters, and c is the concentration of the compound in the solution. The wavelength of maximum absorption is symbolized by λ_{max}. Emission of light (*luminescence* or *photoluminescence*) after excitation is called *fluorescence* if it happens quickly and *phosphorescence* if it is delayed [3]. The optical response of bulk and nanometals is shown in **Figure 3.19**.

Photoluminescence in Carbon Nanotubes. Bandgap photoluminescence of semiconducting single-walled carbon nanotubes (SWNTs) is due to excitons. Semiconducting SWNTs demonstrate specific electronic absorption, and a strong relationship between optical transitions and SWNT diameter has been demonstrated [35]. On the other hand, nanotubes that are metallic in character do not fluoresce [35]. The optical absorption by π-electrons in carbon nanotubes has been shown to be dependent on the diameter but also the chirality of the nanotubes [36]. Luminescence is related to a phenomenon known as a *van Hove singularity*, a subject that is discussed briefly in chapter 9.

Dipolar Plasmon Resonance in Nanometals. Dipolar plasmon, higher order plasmon excitations, and Mie scattering are phenomena responsible for the varied colors of gold and other metal nanoparticles and colloids. The optical response of gold nanoparticles formed by electroplating into porous alumina

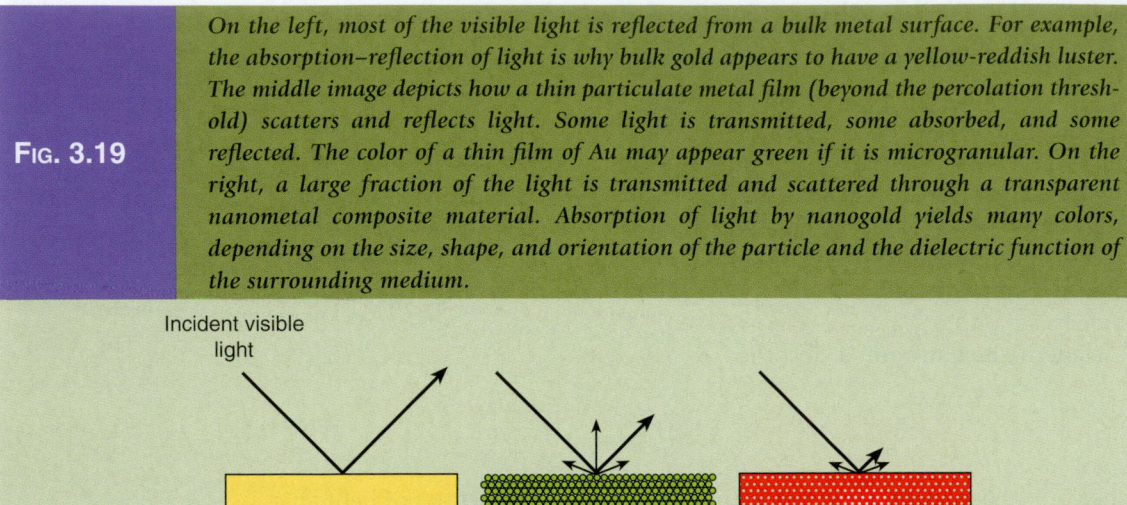

Fig. 3.19 *On the left, most of the visible light is reflected from a bulk metal surface. For example, the absorption–reflection of light is why bulk gold appears to have a yellow-reddish luster. The middle image depicts how a thin particulate metal film (beyond the percolation threshold) scatters and reflects light. Some light is transmitted, some absorbed, and some reflected. The color of a thin film of Au may appear green if it is microgranular. On the right, a large fraction of the light is transmitted and scattered through a transparent nanometal composite material. Absorption of light by nanogold yields many colors, depending on the size, shape, and orientation of the particle and the dielectric function of the surrounding medium.*

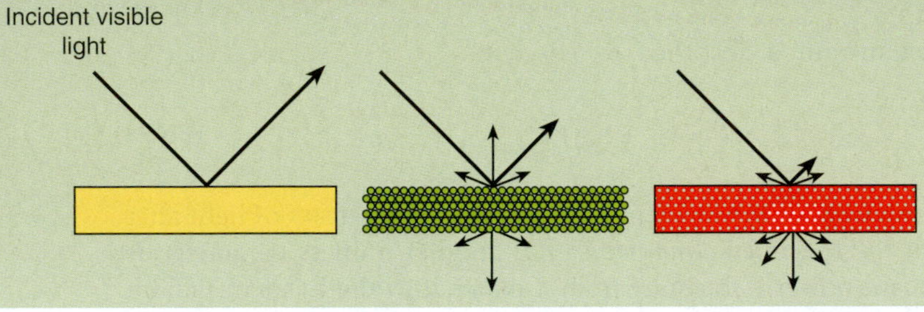

templates shows a definitive size, shape (aspect ratio), and orientation dependence **(Fig. 3.20)** [37]. If the radius of the nanometal particle is approximately 1/100 of the wavelength ($r < 0.01\lambda$), dipolar plasmon resonance is responsible for the optical response. With regard to classical Drude optical theory, smaller particles demonstrate a pronounced surface effect. Specifically, electrons have a better

Fig. 3.20 *Transmission colors produced by gold nanoparticles formed in porous alumina templates. Gold in the form of spherical to cylindrical nanoparticles was plated into the pore channels of anodically formed porous alumina membranes of varying diameters. Particle aspect ratio was controlled by the duration of electroplating. Particle size was controlled by the diameter of the pore channel. A red shift in λ_{max} occurs as particles become larger (increased diameter). In addition to the plasmon resonance, the onset of Mie scattering occurs with larger particles. The image is an approximate reconstruction from physical data [37]. In the figure, particle size increases from bottom to top and particle aspect ratio increases from left to right. The combination of the two independent parameters yields a color array—all consisting of gold nanoparticles.*

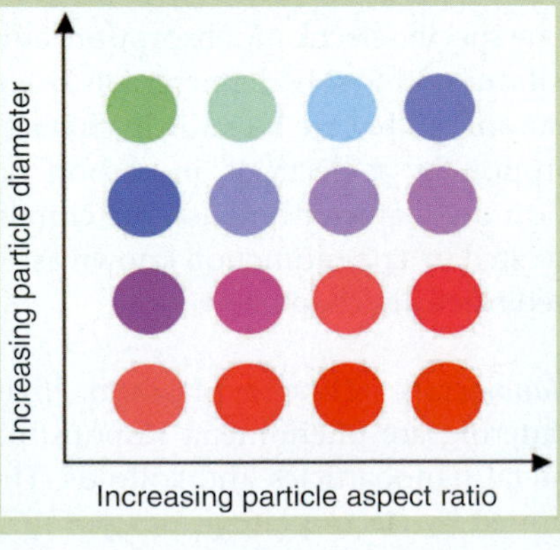

chance of colliding with the surface boundary if the particle becomes smaller. This process increases the energy of the electrons and causes a blue shift in λ_{max}. On the other hand, as the particle becomes larger, scattering and higher order plasmon resonances are excited, resulting in a red shift in λ_{max}.

Composites consisting of a dielectric core coated with a metal shell also demonstrate tunable optical properties. Specific tuning of the optical response was also demonstrated by gold shell–silica–dielectric core composite nanoparticles [38]. Optical resonance of the surface plasmon from the visible to the infrared range was achieved by varying the thickness of the nanoshell and the dimensions of the dielectric silica core material [38].

Quantum Dots. A quantum dot is a nanoscale material within which the motion of electrons is confined in the x, y, and z directions. The definition can be expanded to include the confined motion of conduction band electrons and excitons. Technically, metallic nanoparticles are also considered to be quantum dots—especially upon losing their metallic character with smaller dimensions. The size-dependent emission of visible light from semiconductor quantum dots has fascinated scientists for the past few decades. Quantum dots are usually made of CdSe and other semiconductor materials such as InGaN, InGaAs, CdTe, and ZnSe. A progression in emission color is related to particle size in **Figure 3.21**.

3.3.2 Infrared and Raman Spectroscopy

Far-infrared, infrared, near-infrared, Raman spectroscopy, and surface-enhanced Raman spectroscopy (SERS) are extremely relevant methods used to characterize nanomaterials. These methods provide information about vibrational energy; IR methods rely on asymmetrical (dipolar) vibrations and Raman methods on symmetrical (polarizable) vibrations. In this way, the two methods complement each other.

Raman Spectroscopy. C. V. Raman and K. S. Krishnan in 1928 discovered that light scattered off certain molecules changed the polarization state of the molecules [39]. The Raman effect is a scattering phenomenon that links the vibrational frequencies of the molecule to the energy difference between the incident

FIG. 3.21 *The fluorescent emission from quantum dots of different size made of the same material is represented in this crude reconstruction. It is perhaps the quantum dot that embodies best the spirit of nanoscience—the exhibition of a physical property that is dependent on size and size alone. The figure does not do justice to an actual image of an array of fluorescing quantum dots (Fig. 1.20). The dots range in size from a few nanometers to greater than 5 nm. Excitation is by long-wave ultraviolet and emission ranges from less than 500 to greater than 600 nm.*

and scattered light [34]. However, only molecules with symmetric vibrational (polarization) modes and transitions were amenable to Raman spectroscopic analysis. The SERS process depends on nanometal facets or particles that are able to enhance signal intensity by multiple orders of magnitude—an example of a spectroscopic method enabled by nanotechnology.

Two types of photon scattering exist—elastic scattering and inelastic scattering. Elastic scattering, in which there is no energy loss, is called Rayleigh (Rayleigh–Debye) or Mie scattering. Such scattering occurs when particle dimensions are on the order of or smaller than the wavelength of the incident radiation. Remarkably, then scattering is a true nanoscale phenomenon. If particle size does not conform to this limit (e.g., larger materials), then reflection and refraction also occur [34]. Rayleigh scattering (elastic) involves no change in the polarizability of the molecule [34]. However, a small portion of the incident photons is scattered at frequencies that are different from the incident light. This is called inelastic scattering and the result is exhibited in the form of vibrational, rotational, or electronic energy changes in the molecule. Another form of Raman spectroscopy called resonant Raman spectroscopy (RRS) occurs when there is resonance between the incident radiation and an electronic transition of the molecule [39].

The Raman effect, based on the interaction of a photon with the electronic polarization of the molecule, occurs at energies that are lower than electronic state energies and are known as "excitation to a virtual state." The Raman effect is the inelastic scattering of photons by the sample [40]. Photons that are scattered inelastically by molecules may gain or lose energy depending on the preexisting energetic state of the molecule [40]. Stokes lines are generated when the shifted frequencies are lower than the excitation frequency of the Raman laser. Anti-Stokes lines are formed when the opposite condition exists [40]. Please consult any number of excellent references for further clarification.

SERS is a technique that exhibits incredible sensitivity and has great importance to nanoscience. Some aspects of this method were discussed in chapter 1. Substrates in the SERS technique are noble metals like Au and Ag that happen to be in nanoparticulate form. The primary mechanism of SERS relies on the surface plasmon resonance of the nanometal particulates. An electromagnetic enhancement on the order of 10^6 of the vibrational signal from an adsorbed species is imparted to the spectrum. Analysis of SWNTs by SERS is diagnostic of their presence. Single-molecule detection is also possible with this method.

3.3.3 X-Ray Methods

X-ray methods involve sample excitation by x-rays (creating more x-rays) or by electrons (creating x-rays). X-rays can also be generated from a sample by bombardment by electrons or alpha particles. The energy of emitted x-rays is equal to the difference between the binding energies of the electrons involved in the transition [40]. X-ray fluorescence (XRF) is an elemental analysis procedure that measures 2° x-ray emission. The secondary emission is lower in energy (longer wavelength) than that of the stimulation radiation, and data acquired from XRF can be used in a semi-quantitative manner. X-ray scattering techniques include x-ray diffraction (XRD), small-angle x-ray scattering (SAXS), and x-ray absorption fine structure (XAFS).

X-ray Diffraction. One of the most significant analytical methods to emerge in the late nineteenth and early twentieth centuries is XRD. With XRD, the structure of crystalline materials is revealed and analyzed. Just like with electrons, diffraction of x-rays is a result of scattering from atoms configured in regular arrays. The spacing between atoms and planes of atoms is on the order of the wavelength of the x-rays. Bragg's law forms the foundation for x-ray diffraction:

$$n\lambda = 2d \sin \theta \tag{3.45}$$

The spacing between planes of atoms essentially functions as a three-dimensional diffraction grating [41]. Crystalline solids show long-range periodic structure, exhibited as Bragg diffraction peaks. The Bragg peaks are symmetrically distributed diffraction patterns of focused points [42]. The Cu–K_α (0.15416 nm) source is used for most XRD analyses.

Traditional x-ray diffraction is appropriately used to obtain structural characteristics of bulk crystals. For nanocrystals, traditional XRD is not always appropriate because the coherence length of the structure is limited [42]. Diffuse XRD patterns are formed from bulk materials that have short coherence range and from glasses—materials with no long-range order. Nanomaterials can exhibit a range of periodicity and structural coherence of short size [42]. This also results in diffuse peaks. Extra special techniques need to be applied. High-energy x-ray diffraction (HEXRD) and atomic pair distribution function (PDF) analysis have been shown to be extremely accurate in determining the fine structure of nanomaterials [42]. Nanomaterials in powder form are a popular means of conducting XRD analysis. Nanowire arrays of α-Fe_2O_3 have been analyzed by XRD [43].

Small Angle X-ray Scattering Analysis. SAXS is based on the principle of scattering of x-rays by crystalline or amorphous but uniformly sized small particles. Particle size on the order of 1–100 nm can be measured by this technique. SAXS is applicable to powders in the dry state or suspended in a medium [44]. Particle size can also be estimated by analyzing the width of Bragg peaks in x-ray diffraction spectra [3,45]. Broadening in x-ray diffraction spectra is due to lattice imperfections (microstrain), instrument effects, and crystallite size.

3.4 Nonradiative and Nonelectron Characterization Methods

3.4.1 *Particle Spectroscopy*

Mass spectrometry is an instrumental method in which the mass-to-charge (m/e) ratio of fragmented atomic or molecular clusters is measured. The purpose of MS is to determine the composition of a sample material by analyzing the resultant mass spectrum of the fragmented components of the sample. MS analysis is capable of providing information about isotopic ratio, trace gas analysis structure, and qualitative analysis. E. Goldstein noticed in 1886 that the anode of a cathode ray tube attracted what he eventually named canal rays (Kanalstrahlen). W. Wien later found that the rays could be deflected by an applied electric field and J. J. Thomson in 1913 showed that neon consisted of two isotopes: ^{20}Ne and ^{22}Ne [40]. Thomson described in his book, *Rays of Positive Electricity and Their Application to Chemical Analysis,* how *kanalstrahlen* can be used to analyze chemicals. F. W. Aston constructed the first relatively modern MS in 1919. The first MS to apply electrical detectors was constructed by A. J. Dempster in 1918. The first commercial MS was made available to scientists in 1942 [40].

The technique of MS is straightforward. A sample is vaporized, ionized, and fragmented:

$$M + e^- \rightarrow M^+ + 2e^- \tag{3.46}$$

where M is a molecule of the sample material and M^+ is what is known as the *parent ion*. This reaction is the most common. Other reactions may produce multiple charged positive ions or even negative ions. The parent ions then undergo further fragmentation into lower mass ions. The potential energy of the fragments, electron volts, is converted into kinetic energy according to

$$eV = \tfrac{1}{2}mv^2 \tag{3.47}$$

Deflection of the fragment is caused by an applied magnetic field of strength H (in gauss). If r (in centimeters) is the radius of the curvature of the sector, then the centripetal force experienced by the ions (Hev) is balanced by the centrifugal force of the ions (mv^2/r):

$$\frac{mv^2}{\mathrm{r}} = Hev \tag{3.48}$$

And, substituting $v = H e\mathrm{r}/m$ into equation (3.48), mass-to-charge ratio m/e is calculated from

$$\frac{m}{e} = \frac{H^2\mathrm{r}^2}{2\mathrm{V}} \tag{3.49}$$

The resultant ions are separated by their respective m/e and then detected accordingly. It is easier to vary the applied voltage, inversely proportional to m/e. The semicircular path that each ion undertakes in the analytical sector is a function of H, r, m, e, and the accelerating voltage, V. Each parabolic path is characteristic of some specific ion [40].

There are several types of MSs. The major type of mass spectrometer is the sector MS in which the ion path is deflected. The larger charged and lighter ions are deflected first by an applied magnetic or electric field. The time-of-flight (TOF) MS utilizes an electric field to accelerate ions through an equivalent potential. The time it takes for an ion to reach the detector is then measured. In this scheme, all ions have the same kinetic energy, but the velocity of each individual ion depends on its mass (e.g., ions with lighter mass will arrive at the detector before those with larger mass). The quadrupole MS applies an oscillating electric field that serves to stabilize or destabilize ions in transit [40].

Clusters of extremely small size—a few to several atoms—were first characterized by mass spectroscopic methods. Due to chemical and physical reactions rather than a change in the state of the energy of a molecule, ionization, both positive and negative, forms the basis of mass spectrometry [40]. Fragmented materials in the form of positive and negative species are produced by electron impact from an ion source [40,46].

Forward recoil elastic spectrometry (FReS) is a depth profiling method used to measure the concentration of hydrogen or deuterium in solids. An energetic beam of α-particles is directed 75° from the normal on a sample surface. PIXE is based on particle induced x-ray emission [47]. Inner shell ionization is caused by bombardment with high-energy protons (~MeV), the primary probe material. The protons are produced in an accelerator. The 2° effect of the protons is x-rays. PIXE is similar to EDX but offers much better sensitivity. Characteristic x-rays are produced after activation of the sample with high-energy ions.

3.4.2 *Thermodynamic Methods*

Several thermodynamic methods are used to evaluate nanomaterials. Temperature is the primary independent variable in such methods. Thermogravimetric analysis (TGA) measures the change in mass as a function of temperature under selected environments. TGA data plots, generally shown as percent weight loss versus temperature, reveal information concerning thermal stability, composition (purity), and reaction rates [40]. This procedure is often used to check the purity of single- and multiwalled carbon nanotubes. Differential thermal analysis (DTA) monitors the difference in temperature between a sample and a reference material as a function of temperature. Heats of reaction (endothermic and exothermic), phase transition, and reaction kinetic data are amenable to analysis by DTA [40]. Differential scanning calorimetry (DSC) measures the isothermal differential power between a sample and a reference. Data such as heats of reaction and heat capacity can be analyzed from DSC plots [40].

Recently, nanoscale thermal analysis was accomplished on energetic polycrystalline materials. These substances release stored chemical energy in the form of thermal and mechanical energy [48]. A heated AFM tip was used to initiate local melting, evaporation, and decomposition on materials as small as 100 nm. The temperature range in these experiments was controlled from 25°C–500°C.

3.4.3 *Particle Size Determination*

There are several ways to measure dimensions of nanomaterials. Transmission, scanning electron, atomic force, and scanning tunneling microscopy yield direct data on particle size. Spectroscopic methods based on light scattering phenomena are also used. Each method has its own set of issues when it comes to size determination e.g. artifacts and other systematic errors.

Light Scattering. Although light scattering is not a thermodynamic method per se, the principles behind it, like diffusion, are based on thermodynamics. Dynamic light scattering (DLS), static light scattering (SLS), photon correlation spectroscopy (PCS), and quasi-elastic light scattering (QELS) are analogous methods that measure particle size. Particle size determination by the scattering of laser light has been around for quite some time. The sophistication of the method has increased dramatically over the years as the sizing of particles less than 1 nm is routinely accomplished for micelles, colloids, proteins, and other nanoparticles. The process is based on the principle of light scattering from the surfaces of small particles. If the scattered light is collected as a function of direction, the SLS technique is valid. If the correlation of light scattered intensity is recorded as a function of time from several directions, DLS is appropriate. DLS measures Brownian motion and correlates that information with the size of the particles. Dilute suspensions ranging from 0.0001 to 1% v/v prepared with suitable wetting agents in 40-μL flow cells are analyzed by a 35-mW laser at two scattering angles. The random motion of the particles is correlated with the scattered light intensity fluctuations. From this correlation, the particles' diffusion coefficient is obtained. The equivalent sphere particle size is then calculated from the Stokes–Einstein equation:

$$r = \frac{k_B T}{6\pi\eta D} \tag{3.50}$$

where

r is the van der Waals radius of the molecule in meters
k_B is the Boltzmann constant (1.380×10^{-23} J · K^{-1})
T is the absolute temperature in kelvins
η is the viscosity in pascals per second (or centipoises)
D is the self-diffusion coefficient in square meters per second

Although a spectroscopic method, it is clear from equation (3.50) that DLS has strong thermodynamic roots.

For example, if the self-diffusion constant of 9,10-diphenylanthracene in THF (tetrahydrofuran) at 300 K is equal to 1.04×10^{-9} m^2 · s^{-1} and η is equal to 0.501 mPa · s^{-1}, then the radius of the molecule from equation (3.50) is calculated to be 0.42 nm (the actual van der Waals radius is 0.41 nm) [49].

3.4.4 *Surface Area and Porosity*

Surface area, pore volume, and pore size distribution and pore density measurements are required to characterize nanoparticles, whether porous or otherwise. Seemingly, any and every research paper that involves particles or porous materials includes a section on characterization of surface area and porosity. Several methods are used to acquire information about these parameters. We will present only a few, although every student of nanoscience should become familiar with as many as possible. The most popular methods of surface area calculation are those based on gas adsorption. Other methods include SAXS, small angle neutron scattering (SANS), electron and atomic force microscopy, NMR methods, and mercury porosimetry.

BET Method. Adsorption is the attachment of atoms or molecules to the surface of a solid. The reverse process of adsorption is called desorption. The *adsorbent* is a solid substrate of high surface area upon which the *adsorbate*, a liquid or gas, is adsorbed. Physisorption and chemisorption are two types of adsorption processes. Physisorption is governed by van der Waals forces and usually results in multilayers of adsorbed atoms or molecules. Physisorption occurs at low temperatures with low selectivity, and the heat of adsorption is usually small, $10 < \Delta H_{abs} < 40$ kJ · mol^{-1}. Chemisorption, on the other hand, occurs at higher temperatures with higher selectivity and involves stronger interactions between adsorbates and the surface with $\Delta H_{abs} > 40$ kJ mol^{-1}. The surfaces of nanomaterials are quite large compared to their volume and therefore readily adsorb substances (more discussion to follow in chapters 5 and 6). Surface coverage is defined as the proportion of surface sites on an adsorbent that are partially or completely covered by an adsorbate and is designated as Θ—the fraction of adsorption sites occupied by an adsorbate at equilibrium.

The Brunauer–Emmett–Teller (BET) method to measure surface area was proposed in 1938 [50]. The BET method, however, is based on some broad assumptions [51]. Nonetheless, it is said that this method "serves as a monument to the achievements of imperfection due to its heavy use in determination of surface area of materials."

Regardless of the underlying assumptions, the BET method is extremely valuable in determining relative surface area. The assumptions are that (1) the surface is energetically homogeneous (although most surfaces are rough in nature), (2) only vertical interactions between adsorbed molecules are considered (e.g., any lateral interactions are neglected), and (3) the molecules adsorbed on the substrate demonstrate the strongest energy of adsorption and that the heat of adsorption (ΔH_{ads}) of subsequent layers is the same as the latent heat of condensation (ΔH_{cond}) of the adsorbate gas (e.g., $\Delta H_{ads} > \Delta H_{cond}$). BET is an extremely popular (and useful) means of determining surface area, especially in terms of relative surface area, and we will proceed to discuss its merits.

Specific surface area (m^2 · g^{-1}) is an important parameter in nanoscience research—especially in the study of catalysts, gas separation, and purification materials. Gas adsorption methods enable us to evaluate the surface area, pore size, and pore size distribution of a material. The BET formula is as follows:

$$\frac{P}{V(P_o - P)} \equiv \frac{P/P_o}{V(1-P/P_o)} = \frac{1}{cV_{\mathrm{m}}} + \frac{c-1}{cV_{\mathrm{m}}}(P/P_o) \tag{3.51}$$

where

P is the equilibrium experimental pressure

P_o is the vapor pressure of the adsorbate gas at the experimental temperature

V (m^3 · g^{-1}) is the standardized experimental volume of the adsorbed gas per gram of adsorbant

V_{m} (m^3 · g^{-1}) is the volume of the adsorbate monolayer per gram of adsorbent

c is a constant that relates the heat of adsorption ΔH_{ads} (for the first physisorbed layer) with ΔH_{cond}, the latent heat of condensation (additional layers):

$$c = \exp\left[\frac{\Delta H_{\mathrm{ads}} - \Delta H_{\mathrm{cond}}}{RT}\right] \tag{3.52}$$

Once c is known, the calculation of ΔH_{ads} is straightforward. This relation works well except for cases of high relative pressure, $(P/P_o) > 0.35$, or very low relative pressure, $(P/P_o) < 0.05$. At relative pressure $0.05 < (P/P_o) < 0.35$, equation (3.51) is approximately proportional to (P/P_o) and transforms readily into the equation of a line, $y = mx + b$. A plot of $(P/P_o)/[V(1 - P/P_o)]$ versus (P/P_o) will yield a straight line with slope m equal to $[(c - 1)/(cV_{\text{m}})]$ and intercept b equal to $(1/cV_{\text{m}})$. **Figure 3.22c** shows a sample BET plot consisting of adsorption and desorption isotherms. Specific surface area is calculated by

$$S_{\text{s}} = a_m \left(\frac{P_o V_{\text{m}}}{RT} \right) \mathcal{N}_A = a_m \left(\frac{V_{\text{m}}}{V_{\text{Gas}}} \right) \mathcal{N}_A \tag{3.53}$$

where

S_{s} is the specific surface area in $\text{m}^2 \cdot \text{g}^{-1}$

a_m is the area of solid surface for adsorption of one gas molecule (0.162 nm^2 for N_2)

V_{Gas} is the molar volume of the gas in its standard state (2.24×10^{-2} m^3)

$\mathcal{N}_A$ is Avogadro's number (6.022×10^{23} mol^{-1})

Figure 3.22a and **b** shows an Autosorb-I Series™ surface area analyzer manufactured by Quantachrome Instruments and its schematic of operation. **Figure 3.22c** and **d** shows examples of a BET isotherm (with some porosity) and its slope-form plot, respectively. **Figure 3.22e** and **f** shows plots obtained from BJH analysis (described below) that reveal information about pore volume distribution and size, respectively. Pressure capability ranges from less than 3×10^{-10} to 1000 torr; adsorbate gases, N_2, Ar, CO_2, butane, krypton, ammonia, and water vapor, among others, can be used in the BET procedure. **Figure 3.23** shows SEM images of sample materials that underwent BET analysis.

BJH Method. Porosity is the ratio of void volume to the solid component of a material ($\varepsilon = V_S/V_T$). In addition to the specific surface area, nanoscientists also need to know the overall porosity of a material and the size of pores. Pore size distribution is determined by another thermodynamic method called the Barnett–Joyner–Halenda (BJH) method [52]. The BJH method is based on the Kelvin equation. This method exploits the phenomenon of capillary condensation in mesoporous systems and is valid at P/P_o greater than 0.35. Another mechanism involves capillary filling in micropores (e.g., like those of M-41S type zeolite), and is suitable for BJH at P/P_o between 0.1 and 0.5. Enhancements to BJH by microscopic methods based on statistical mechanics and molecular simulation are applicable for pore size analysis for micro- and mesosystems. Discussions of *density functional theory (DFT)* and *molecular dynamic simulation* methods are beyond the scope of this text.

According to the International Union of Pure and Applied Chemistry (IUPAC), pores are defined according to the following criteria: *Macropores* have pore diameter greater than 50 nm, *mesopores* have pore diameter between 2 and 50 nm, and *micropores* have pore diameter less than 2 nm. Pore channels and cavities also come in various shapes and conformations.

The Kelvin equation relates the equilibrium vapor pressure of a substance above a curved surface to the equilibrium vapor pressure of the same substance over a flat planar surface. With regard to the nanoscale, the Kelvin equation predicts the pressure of condensation or evaporation of an adsorbate in a cylindrical pore that already is coated with a multilayer. Pore size is calculated from the

Fig. 3.22

(a) An Autosorb-I Series™ surface area analyzer manufactured by Quantachrome Instruments and (b) its schematic of operation are shown in the figure. Pressure transducers in the instrument are capable of resolving a minimum of 0.00025 torr with a minimum resolvable relative pressure (P/P_o) of 3.2×10^{-7} torr of N_2 in the 0- to 1000-torr range; 2.5×10^{-6} torr and 3.2×10^{-9} torr in the 0- to 10-torr range; and from 0–1 torr, 2.5×10^{-7} and 3.2×10^{-10} of nitrogen respectively. In (c) and (d), a BET isotherm (with porosity) is shown with its extracted BET slope. The specific surface area is determined from the amount of adsorbate per monolayer. Figures (e) and (f) and derived from BJH and DFT analysis to yield information about pore volume and pore size.

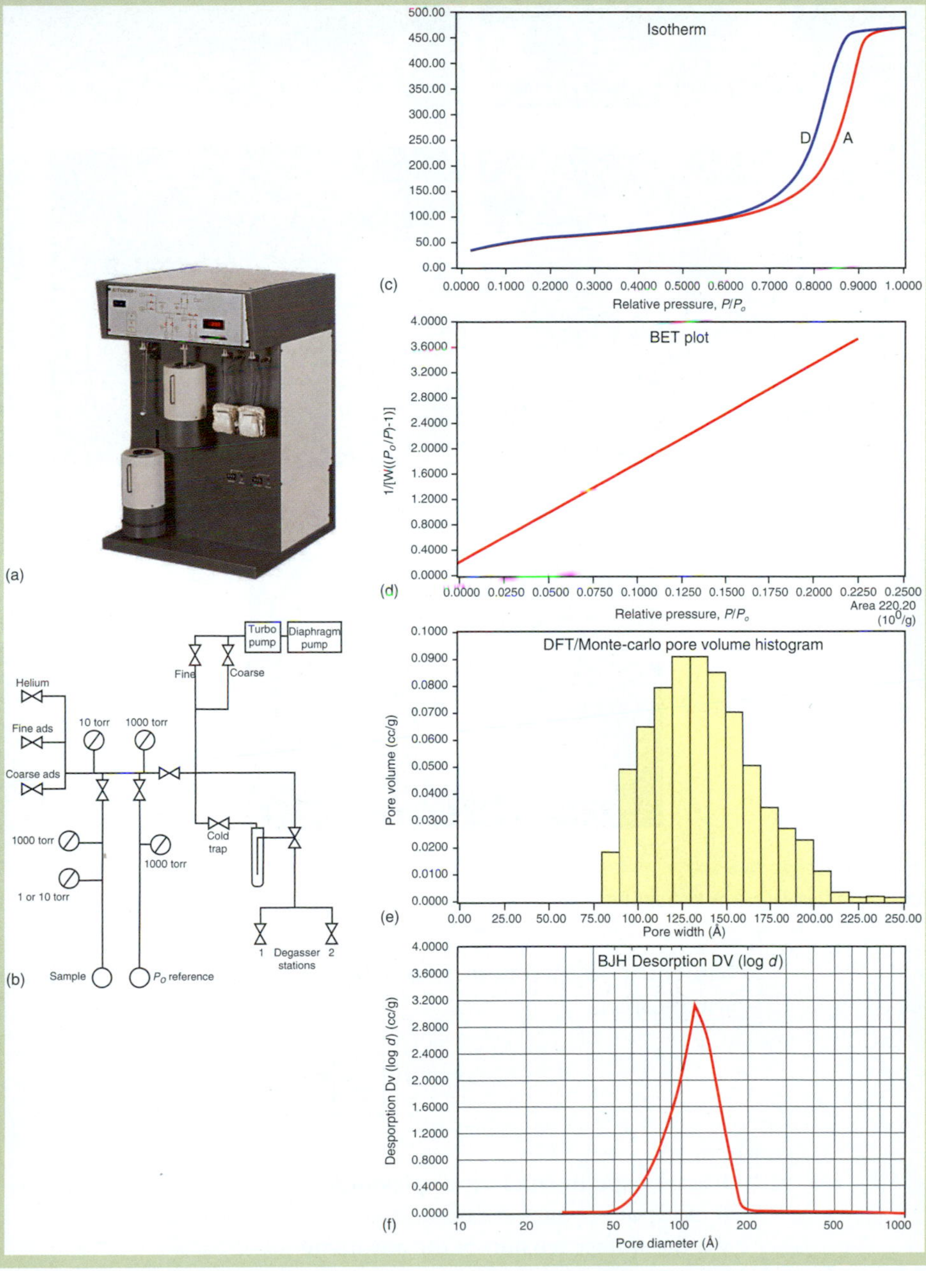

Source: Courtesy of Quantachrome Instruments.

FIG. 3.23 *Carbon black (a), prickly gold (b), radiolarian (c), coal ash (d), and natural zeolite (e) are depicted.*

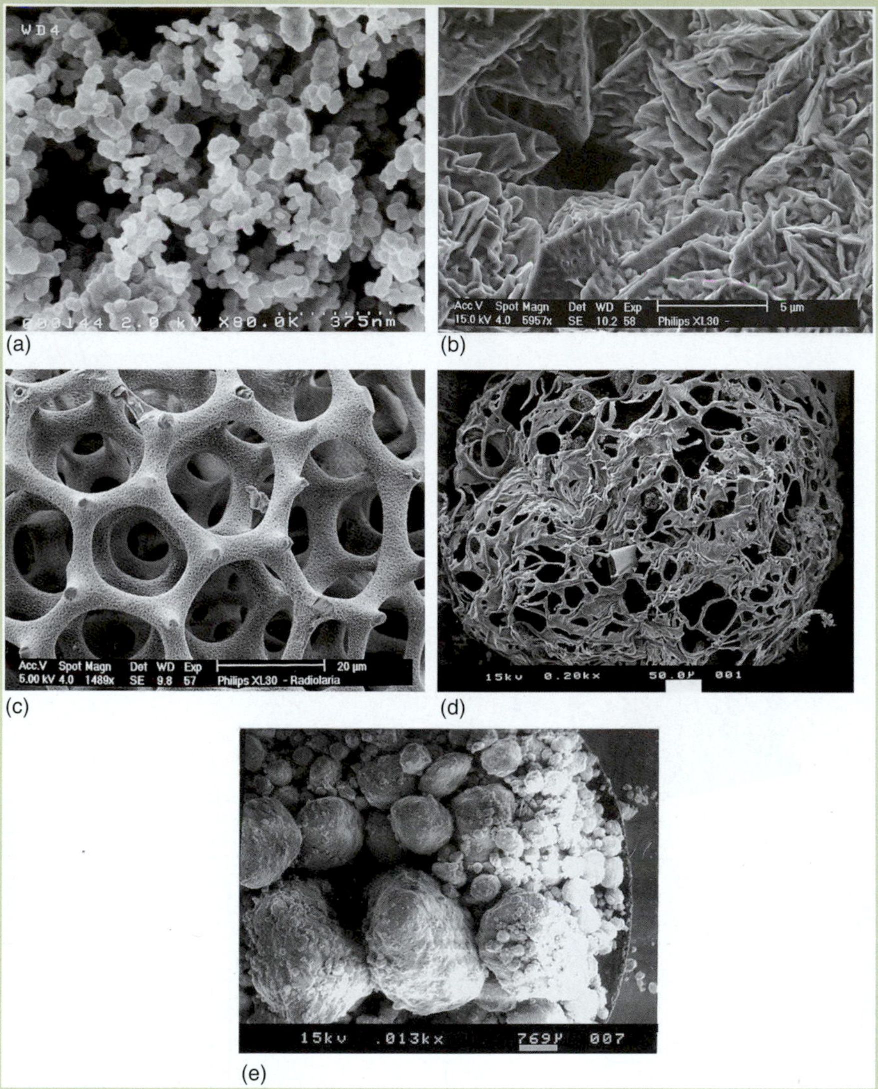

Source: Courtesy of Quantachrome Instruments.

Kelvin equation in conjunction with the statistical thickness derived from a *t-curve equation*:

$$\ln\left(\frac{P}{P_o}\right) = \frac{2\gamma V_{\mathrm{m}}}{rRT} \tag{3.54}$$

where

P is the experimental vapor pressure
P_o is vapor pressure at saturation
γ is the surface tension of the adsorbate

V_m is the molar volume as before
r is the radius of the droplet
RT has its usual meaning

Mercury Porosimetry. Mercury porosimetry is another method by which porous materials are characterized. It is a pore filling method that is complementary to the BJH method because pore dimensions greater than 300 nm can be evaluated. BJH is valid down to ca. 0.3 nm. Pore size distribution is often measured by mercury porosimetry, although the method is not a geometric method per se. It is an indirect technique that works best if correlated to a model or other analytical techniques such as x-ray tomography or magnetic resonance imaging. Mercury porosimetry can be applied to acquire information about the porosity and pore interconnectivity parameters as well as pore size distribution. It is based on the transport and relaxation properties of liquid mercury. Mercury is used because it is nonwetting and has a high surface tension. A sample is evacuated at elevated temperatures and immersed into mercury. An external pressure is applied. As the pressure is increased, mercury fills the pore volume. The amount of mercury that finds its way into the pore volume is a function of the applied pressure—intrusion and extrusion of a nonwetting fluid in a porous solid.

Sears's Method. In 1956, George Sears developed the Sears method to determine the specific surface area of colloidal silica particles by titration with sodium hydroxide (Sears). The surfaces of nonporous particles 5–100 nm in diameter are titrated in a 20% NaOH solution. At *p*H 9, 1.26 hydroxyl groups are adsorbed per square nanometer. The empirical relationship for the Sears method is

$$S_s = 26.4\ (V_t - V_b) \tag{3.55}$$

where S_s is specific surface area as before, V_t is the milliliters of 0.1 *M* NaOH required for titration of 1.50 g SiO2, and V_b is the titration blank in absence of any silica. Particle size d in nm is determined by

$$d = \frac{6}{\rho S_S} = \frac{2.73}{S_S} \tag{3.56}$$

where the density of silica $\rho = 2.2$ g cm^{-3}.

NMR-Cryoporometry. Porous materials can be characterized by determination of the melting behavior of fluids injected into pores [53,54]. Researchers have recently determined that the freezing point depression ΔT_f is a function of surface-to-volume ratio and that melting point shifts (ΔT_m) are influenced by the change in curvature of the pore surface [53,54]. The method used is called NMR-cryoporometry.

3.4.5 Other Important Characterization Methods

Quartz Crystal Microbalance (QCM). A QCM has the capability to measure mass changes as small as a few nanograms per square centimeter, which is sensitive enough to detect monolayers of deposited materials. QCM is used to monitor the thickness of metal accumulating on a surface applied by evaporation or sputtering [55]. The QCM is used in biosensor designs due to this great sensitivity.

We have discussed the importance of vibrating piezomaterials in generating motion necessary for AFM and STM analysis. Another application of vibrating crystals is by the QCM technique. Change in mass can be measured by recording the change in frequency of the piezo crystal. The behavior of a QCM is defined by the Sauerbrey equation:

$$\Delta f = \frac{-2\Delta m f_o^2}{A\sqrt{\rho_q \mu_q}} = \frac{-2\Delta m f_o^2}{A\left(\rho_q \upsilon_q\right)} = -2.26 \times 10^{-6} f_o^2 \frac{\Delta m}{A} \tag{3.57}$$

where f_o is the fundamental resonant frequency of the crystal (600 kHz–30 MHz for quartz and higher for overtones), A is the active area of the crystal between the electrodes, and ρ_q and μ_q are the density ($2.648\ \text{g}\cdot\text{cm}^{-3}$) and sheer modulus ($2.947 \times 10^{11}\ \text{g}\cdot\text{cm}^{-1}\cdot\text{s}^{-2}$) of quartz, respectively. In another form of the equation, υ_q is the acoustic wave speed ($3340 \times 10^2\ \text{cm}\cdot\text{s}^{-1}$).

Most QCMs employ AT-cut quartz (35°15′ across the growth direction of the α-quartz crystal) sandwiched between two electrodes, usually gold. The AT-cut yields a crystal with a shear perpendicular to the crystal face. The sensitivity of a QCM is ca. $1\ \text{ng}\cdot\text{cm}^{-2}$ compared to one of the best electronic microbalances that have sensitivity on the order of 0.1 μg. The QCM is therefore capable of measuring single layers of molecules or atoms.

A KSV Instruments, Ltd. QCM-Z500 (**Fig. 3.24**) has an acoustic sensor that is able to measure the change in mass of surface layers. It is an impedance-based instrument that is a versatile tool of nanoscience and able to monitor antibody (antibody–antigen) interactions, protein binding, lipid bilayer deposition, cell attachment–detachment, vesicle attachment, surfactant adsorption, polymer adsorption, nanoparticle adsorption, self-assembly, surface gels, metal thickness, and corrosion—to name a few applications. The frequency range of this particular instrument is 0.1–55 MHz with a resolution of 0.01 Hz. The mass resolution for a standard 5-MHz AT-cut crystal is $0.177\ \text{ng}\cdot\text{cm}^{-2}$. The maximum mass load and thickness are $\sim 0.1\ \text{mg}\cdot\text{cm}^{-2}$ and ~10 μm, respectively.

In one experiment successive layers of stearic acid ($C_{17}H_{34}O_2$) were deposited on the surface of a gold-coated QCM by the Langmuir–Blodgett technique [56]. Stearic acid is a long-chain aliphatic hydrocarbon with a carboxylic acid end group (**Fig. 3.25**). The frequency change was measured with a dissipative-QCM [55,56] following the deposition of each layer of stearic acid. A dissipative QCM was used because of the built-in capability to gauge another parameter important to monolayers: the softness of the stearic acid film. The correlation between the theoretical and experimental values with the change in frequency is shown in **Figure 3.26.**

Complying with the directive of converging evolution of instrumentation, a combination between QCM and AFM devices was applied to analyze electrodeposition processes [57]. The goal of the researchers was to monitor surface

Fig. 3.24 *The quartz crystal microbalance is a relatively unheralded workhorse for characterizing nanomaterials. Every evaporation instrument has one, but its versatility expands into many fields and applications. A QCM-Z500 quartz crystal microbalance manufactured by KSV Instruments, Ltd. is shown. It is capable of measuring changes in mass and viscoelastic properties of adsorbed or deposited layers. Mass change is proportional to the change in resonance frequency of the crystal and viscoelastic behavior is measured by the change in electrical resistance (impedance). The range of frequency is 0.1–55 MHz with resolution down to 0.01 Hz; mass resolution capability is 0.177 ng·cm^{-3}. The active sensor area is 20 mm^2. The instrument is capable of measuring film thickness as high as 10 μm and a load of 0.1 mg·cm^2.*

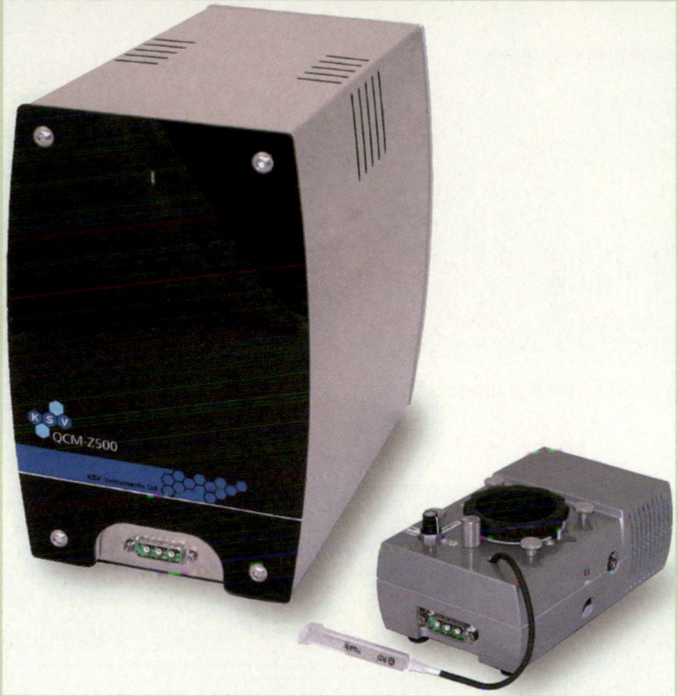

Source: Image and specifications are courtesy of KSV Instruments, Ltd. With permission.

characteristics at the micrometer and nanometer scales. In order to achieve this goal, they combined AFM with two kinds of acoustic wave devices: the QCM and SAW (surface acoustic wave). The QCM–AFM interaction consisted of longitudinal acoustic waves generated by the QCM that were reflected on the AFM cantilever. Sensitivity was calculated by

$$S \equiv \frac{\Delta f}{f_o} = \frac{A}{\Delta m} \qquad (3.58)$$

SAW has greater sensitivity (in the picograms per square centimeter range) than QCM. It uses an interdigitized electrode technology consisting of piezomaterials.

Fig. 3.25 *Stearic acid layers are built on a gold-coated quartz crystal by the Langmuir–Blodgett technique. The frequency change induced following each cycle of deposition was measured with a dissipative QCM. The gold layer was deposited one layer at a time. The stearic acid monolayers were applied in successive deposition steps. Deposition of stearic acid on a gold thin film by the Langmuir–Blodgett process: The gold was first deposited on the surface of a quartz crystal microbalance. The procedure was repeated until 45 layers of stearic acid were deposited. The stearic acid monolayers were deposited with a KSV Minitrough 2 film balance from a 10^{-4} M $MnCl_2$ subphase with pH ~ 6 at a constant trough pressure of 30 mN·m^{-1} [55,56].*

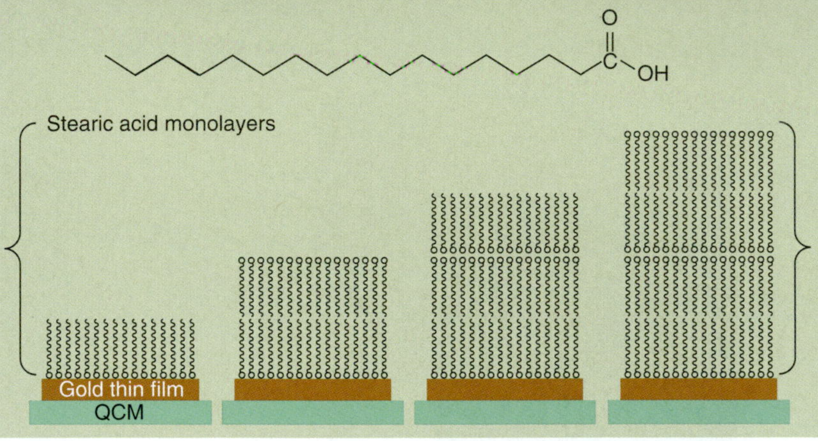

Source: Redrawn with permission from KSV Instruments, Ltd.

Fig. 3.26 *The frequency change of the QCM signal is plotted as a function of the number of layers of stearic acid deposited on the gold surface. The absorbed mass can be calculated from the Sauerbery equation. The theoretical frequency change correlates well with the experimental values. The more material that was added to the monolayer assembly, the lower the resonance frequency became (e.g., the greater the change in frequency, Δf)[55,56].*

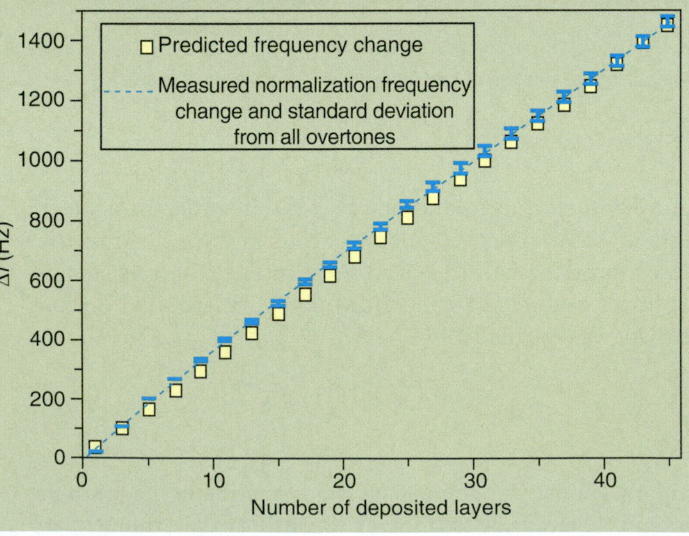

Source: Redrawn with permission from KSV Instruments, Ltd.

References

1. *History of spectroscopy*, College of Charleston, www.cofc.edu/~deavorj/521/History%20of%20Spectroscopy.htm (2007).
2. R. W. Kelsall, I. W. Hamley, and M. Geoghegan, *Nanoscale science and technology*, John Wiley & Sons, Ltd., London (2005).
3. C. P. Poole and F. J. Owens, *Introduction to nanotechnology*, Wiley-Interscience, John Wiley & Sons, Inc., New York (2003).
4. Nanoscience and Technology Institute (NSTI) Conference in Santa Clara, California, April 26 (2007).
5. J. I. Goldstein, D. E. Newbury, P. Echlin, D. C. Joy, C. Fiori, and E. Lifshin, *Scanning electron microscopy and x-ray microanalysis*, 2nd ed., Plenum Press, New York (1992).
6. M. V. Klein, *Optics*, Wiley, New York (1970).
7. C. J. Davisson, Are electrons waves? *Franklin Institute Journal*, 205, 597 (1928).
8. C. J. Davisson and L. H. Germer, Reflection and refraction of electrons by a crystal of nickel, *Proceedings of the National Academy of Science, USA*, 14, 619–627 (1928).
9. C. J. Davisson and L. H. Germer, Reflection of electrons by a crystal of nickel, *Proceedings of the National Academy of Science, USA*, 14, 317–322 (1928).
10. J. H. Wittke, Effects of electron bombardment, nau.edu/microanalysis/Microprobe/Interact-Effects.html, Northern Arizona University (2006).
11. Electron microscopes, JEOL, Ltd. http://www.jeol.com/HOME/tabid/36/Default.aspx (2007).
12. J. J. Donovan, Electron probe microanalysis, In *Scanning electron microscopy and x-ray microanalysis*, 2nd ed., J. I. Goldstein, D. E. Newbury, P. Echlin, D. C. Joy, C. Fiori, and E. Lifshin, eds., Plenum Press, New York (1992).
13. W. S. Boyle and G.E. Smith, Charge-coupled semiconductor devices, *Bell Systems Technical Journal*, 49, 587–593 (1970).
14. M. H. Ellisman, An historical perspective on digital imaging in transmission electron microscopy: Looking into the future, *Microscopy and Microanalysis*, 11, 604–605 (2005).
15. J. C. H. Spence and J. M. Zhou, Large dynamic range, parallel detection system for electron diffraction and imaging, *Review of Scientific Instruments*, 59, 2102 (1988).
16. C. J. Calbick, The discovery of electron diffraction by Davisson and Germer, *The Physics Teacher*, 91, 63–69 (1963).
17. J. Hillier and R. F. Baker, Microanalysis by means of electrons, *Journal of Applied Physics*, 15, 663–675 (1944).
18. B. Bhushan, Scanning probe microscopy—Principle of operation, instrumentation and probes. In *Springer handbook of nano-technology*, B. Bhushan, ed., Springer-Verlag, Berlin (2004).
19. G. Binnig, H. Rohrer, Ch. Gerber, and E. Weibel, Surface studies by scanning tunneling microscope, *Physics Review Letters*, 49, 57 (1982).
20. G. Binnig, C. F. Quate, and Ch. Gerber, Atomic force microscope, *Physics Review Letters*, 56, 930 (1986).
21. G. Shmalz, Uber Glatte und Ebenheit als Physikalisches und Physiologishes Problem, *Zeitschrift des Vereimes Deutscher Ingenieurte*, 1461–1467, Oct. 12 (1929).
22. http://www.pacificnano.com/afm-history_references.html
23. J. Scalf and P. West, Introduction to nanoparticle characterization with AFM, Pacific Nanotechnology, Inc., www.nanoparticles.pacificnanotech.com (2007).
24. D. D. Hsu, *Profilometer*, definition, www.chemicool.com/definition/profilometer.html (1996).
25. High Resolution Surface Profiler/Stylus-Type Profilometer, Ambios Technology, http://www.ambiostech.com/profiler.html (viewed 2007).

26. R. Young, J. Ward, and F. Scire, The Topografiner: An instrument for measuring surface microtopography, *Review of Scientific Instruments*, 43, 999 (1971).
27. Piezoelectricity, *Oxford American Online Dictionary and Thesaurus*, v. 1.0.0, Apple Computer, Inc. (2005).
28. I. A. Levine, *Quantum chemistry*, 4th ed., Prentice Hall, Upper Saddle River, NJ (1991).
29. G. Cao, *Nanostructures & nanomaterials: Synthesis, properties & applications*, Imperial College Press, London (2004).
30. E. W. Müller, J. A. Panitz, and S. B. McLane, The atom-probe field ion microscope, *Review of Scientific Instruments*, 39, 83–86 (1968).
31. M. K. Miller, *Atom probe tomography: Analysis at the atomic level*, Kluwer Academic/Plenum, New York (2000).
32. Product specification, Imago LEAP 3000X-Si, Imago Scientific Instruments, Madison, WI (2007).
33. Spectroscopy, *Oxford American online dictionary and thesaurus*, v. 1.0.0, Apple Computer, Inc. (2005).
34. J. D. Ingle, Jr., and S. R. Crouch, *Spectrochemical analysis*, Prentice Hall, Upper Saddle River, NJ (1988).
35. S. M. Bachilo, M. S. Strano, C. Kittrell, R. H. Hauge, R. E. Smalley, and R. B. Weisman, Structure-assigned optical spectra of single-walled carbon nanotubes, *Science*, 298, 2361–2366 (2002).
36. A. Grünes, R. Saito, G. G. Samsonidze, M. A. Pimenta, A. Jorio, A. G. S. Filho, G. Dresselhaus, and M. S. Dresselhaus, Characterization of nanographite and carbon nanotubes by polarization dependent optical spectroscopy, *MRS Proceedings*, Symposium F, F3.47, Fall (2002).
37. G. L. Hornyak and C. R. Martin, Optical properties of a family of Au-nanoparticle-containing alumina membranes in which the nanoparticle shape is varied from needle-like (prolate), to spheroid, to pancake-like (oblate), *Thin Solid Films*, 303, 84 (1997).
38. S. J. Oldenburg, R. D. Averitt, S. L. Westcott, and N. J. Halas, Nanoengineering of optical resonances, *Chemical Physics Letters*, 288, 243–247 (1998).
39. A. M. Schwartzberg and J. Z. Zhang, Surface-enhanced Raman scattering (SERS). In *Encyclopedia of nanoscience and nanotechnology*, Marcel Dekker, Inc., New York (2006).
40. H. H. Bauer, G. D. Christian, and J. E. O'Reilly, *Instrumental analysis*, Allyn and Bacon, Inc., Boston (1979).
41. J. F. Shackelford, *Introduction to materials science for engineers*, 4th ed., Prentice Hall, Upper Saddle River, NJ (1996).
42. V. Petkov, T. Ohta, Y. Hou, and Y. Ren, Atomic-scale structure of nanocrystals by high-energy x-ray diffraction and atomic pair distribution function analysis: Study of Fe_xPd_{100-x} ($x = 0, 26, 28, 48$) nanoparticles, *Journal of Physical Chemistry C*, 111, 714–720 (2007).
43. X. Wen, S. Wang, Y. Ding, Z.-L. Wang, and S. Yang, Controlled growth of large-area, uniform, vertically aligned arrays of α-Fe_2O_3 nanobelts and nanowires, *Journal of Physical Chemistry B*, 109, 215–220 (2005).
44. A. Trunschke, *Modern methods in heterogeneous catalysis research*. Presentation, AC-FHI, Department of Inorganic Chemistry, w3.rz-berlin.mpg.de/~jentoft/lehre/trunschke_particlesize_031106.pdf (2006).
45. P. Lamparter, *Crystallite sizes and microstrains from x-ray diffraction and line profile analysis*. Power Point presentation, www.imprs-am.mpg.de/nanoschool2004/lectures-I/Lamparter.pdf (2004).
46. A. Markwitz, F. Lucas, J. Rusterucci, J. Kennedy, W. J. Trompetter, M. Rudolph, M. Ryan, V. White, and S. Johnson, A nuclear reaction analysis and optical microscopy study on controlled growth of large SiC nanocrystals on Si formed by low-energy ion

implantation and electron beam annealing, *Nuclear Instruments and Methods in Physics Research Section B: Beam Interactions with Materials and Atoms*, 249, 105–108 (2006).
47. S. A. E. Johansson, J. L. Campbell, and K. G. Malmqvist, *Particle induced x-ray emission spectrometry (PIXE)*, Wiley, New York (1995).
48. W. P. King, S. Saxena, B. A. Nelson, B. L. Weeks, and R. Pitchimani, Nanoscale thermal analysis of an energetic material, *Nano Letters*, 9, 2145–2149 (2006).
49. R. Hoffman, *Diffusion NMR*, Hebrew University of Jerusalem, chem.ch.huji.ac.il/nmr (2005).
50. S. Brunauer, P. H. Emmett, and E. Teller, Adsorption of gases in multimolecular layers, *Journal of the American Chemical Society*, 60, 309–319 (1938).
51. A. Seri-Levy and D. Avnir, The Brunauer–Emmett–Teller equation and the effects of lateral interactions. A simulation study, *Langmuir*, 9, 2523–2529 (1993).
52. E. P. Barrett, L. G. Joyner, and P. P. Halenda, The determination of pore volume and area distributions in pure substances, *Journal of the American Chemical Society*, 73, 373 (1951).
53. O. Petrov and I. Fúro, Curvature dependent metastability of the solid phase and the freezing—melting hysteresis in pores, *Physics Review E*, 73, 011608–011614 (2006).
54. O. Petrov and I. Fúro, Characterizing porous materials through the melting-freezing behavior of pore-filling fluids, 8th International Conference on Magnetic Resonance in Porous Media, September 10–14, Bologna, Italy (2006).
55. *Adsorption sensors: Impedance based QCM-Z500*, KSV, Ltd., Helsinki, Finland, 1–6 (2006).
56. *Langmuir–Blodgett deposition of stearic acid on gold in determining layer mass during repeated LB-film deposition by ex situ measurement of resonant frequency in air*, KSV, Ltd., Helsinki, Finland (2006).
57. J.-M. Friedt, L. Francis, K.-H. Choi, F. Frederix, and A. Campitelli, Combined atomic force microscope and acoustic wave devices: Application to electrodeposition, *Journal of Vacuum Science Technology A*, 21, 1500–1505 (2003).

Problems

3.1 Calculate the de Broglie wavelength of an electron under an acceleration voltage of 150 V.

3.2 Calculate the velocity of an electron moving at this accelerating voltage in problem 3.1. How does this value compare to the speed of light? The speed of light is 299.8×10^8 m·s^{-1}.

3.3 Calculate the wavelength of an electron beam in a TEM with accelerating voltage equal to 100,000 V. What is the velocity of the electron? Should we be concerned with relativistic effects? How would factoring in relativistic effect change your answer?

3.4 Compare the wavelength of a low-energy electron ($\upsilon = 0.01\ c$) to that of a gold atom. Use the de Broglie relation.

3.5 Name some important differences among emission, fluorescence, and luminescence.

3.6 1.8 *e*V of radiation is enough to produce an electron transition. To what wavelength does this correspond?

3.7 Explain the physical principles behind the mechanism of a scintillator. Do the same for a photomultiplier, a collimator, and a CCD camera.

3.8 Review the differences among secondary, backscattered, and transmitted electrons. Which ones are considered to be elastic?

3.9 Why is the wavelength of emitted x-rays following excitation inversely proportional to the square of the atomic number?

3.10 Explain in simple terms the phenomenon of Bremsstrahlung radiation.

3.11 What are Kakuchi lines? Explain this phenomenon.

3.12 What is the work function of a material? Which materials have the lowest work function? Explain the photoelectric effect.

3.13 Why is a phosphor screen used in TEM?

3.14 Explain the Auger equation in more detail. What kinds of analyses are best suited for Auger spectroscopy?

3.15 In Davisson and Germer's experiment, a crystalline Ni surface was exposed to an electron beam $\varphi = 50°$ relative to the normal of the crystal plane. A maximum in the intensity was observed at this angle and the acceleration voltage was equal to 54 V. What is an estimate of the *d*-spacing of the exposed crystal plane?

3.16 What is a Faraday box? Explain its mechanism of detection.

3.17 Interference from sectioned thin films (for TEM) on water: What is the thickness of a thin film of poly(methyl methacrylate) (PMMA) embedding material that looks blue ($\lambda_{incident} = 475$ nm)? The majority of the interference color is due to the PMMA embedding material with refractive index *n* equal to 1.489.

3.18 What is the factor of magnification if a 100-nm square field of view is projected to fill one fourth of a 17-in. diagonal square screen?

3.19 Convert an 850-nm wavelength into units of frequency, electron volts, wavenumbers, joules, and ergs.

3.20 (a) What is the *d*-spacing of a copper plane if it corresponds to x-ray wavelength from erbium-Lγ_3 x-rays (of wavelength 0.13146 nm) at diffraction angle of $2\theta = 50.5°$. Assume first-order diffraction. (b) What is the energy of these x-rays in kiloelectron volts?

3.21 Why would the sharpness of the tip (e.g., the curvature of the tip) play an important role in resolution of a scanned surface?

3.22 Describe the dual-detector system of an AFM in more detail.

3.23 Discuss the piezoelectric effect.

3.24 List the differences between atomic probe methods and atomic force methods.

3.25 The Lycurgus cup introduced in chapter 1 demonstrates dichroic optical behavior. Why? Why is King Lycurgus purple in the transmitted mode?

3.26 Why would the medium (or coating) of a metal affect the position plasmon resonance?

3.27 Before you read the upcoming chapters, why do you think quantum dots made of the same material but of different size demonstrate emission at different wavelengths?

3.28 C_{60} is a new form of carbon with 60 equivalent carbons. How many NMR peaks would you expect for the C_{60} NMR spectrum?

3.29 Describe the mechanism of operation of a time-of-flight MS.

3.30 TGA of a 1-g sample of a porous, anodically formed membrane formed in a 10% v/v sulfuric acid solution was conducted over a range of RT to 1200°C under a nitrogen atmosphere. The *mass versus T* curve is shown in the figure. Assume that the alumina contained water, alumina (as Al_2O_3), and H_2SO_4 only.

a. Using your chemical intuition, speculate upon the types of chemical/physical reactions that occur over this temperature range.
b. Calculate the percent mass loss of each volatile component.
c. What kind of thermal technique would be most suitable for analysis of phase changes?

3.31 DSC is a method that enables the inspection of phase changes. Describe how this is done. Also, discuss the theory behind DSC.

3.32 Calculate the BET surface area for the following generic nanopowder. The isotherm is of BET Type II (multilayer adsorption, no pores). All data are standardized to 1 atm and 273.15 K. Calculate P/P_o and $P/[V(1 - P/P_o)]$.

a. What is the specific surface area in terms of square meters per·gram for this material?
b. Assuming spherical shape, what is the diameter of the nanoparticles?
c. What is the heat of physisorption of N_2 gas on the nanoparticles?

Experimental (P/torr)	Experimental (V/[cm^3 g^{-1}])
060	0.810
100	0.885
150	0.990
190	1.20
230	1.30
295	1.45
450	1.86
560	2.25
620	2.60

3.33 Discuss how BET, BJH, and mercury porosimetry are complementary methods.
3.34 How many gold atoms are contained in a nanogram? A picogram?
3.35 Do you expect the measured frequency to increase or decrease as more material is added to the surface of a QCM?
3.36 The most important methods to determine the size of nanoparticles are TEM and AFM. What are the advantages and disadvantages of the two methods? (Question courtesy of Günter Schmid, Uni-Essen.)

FABRICATION METHODS

Manufacturing takes place in very large facilities. If you want to build a computer chip, you need a giant semiconductor fabrication facility. But nature can grow complex molecular machines using nothing more than a plant.

RALPH MERKLE

Chapter 4

THREADS

Characterization methods have been presented, addressed, and discussed, albeit without providing significant detail. The catalog nature of chapter 3 is deliberately extended into this chapter, which is the last chapter in the "Nanotools" division of the text. Because reference is made continually to various kinds of fabrication techniques throughout the text, it is prudent to place introductory material concerning fabrication early in the book. In this way, the student should be able to establish a level of comfort with, perspective on, and understanding of fabrication methods when the subjects emerge time and time again later in the text. The physics division of the text—chapter 5 through chapter 8—engages the study of nanomaterial properties and phenomena.

Please take note that the fabrication methods listed in this chapter are but a few of the multitude that actually exist. We have tried to categorize in a generic sense the major forms and tried to illustrate the processes with commonly practiced fabrication techniques. Much can be learned about nanomaterials by understanding how they are made.

4.0 Fabrication of Nanomaterials

There is nothing more gratifying, arguably, than holding in one's hand the physical manifestation of an idea, concept, or theory. The link between the idea, concept, or theory and its physical form is the process of fabrication. The fabrication process begins in a laboratory with atomistic simulations, experiments, mock-ups, and prototypes. Eventually, after a battery of testing, the physical embodiment of the idea, concept, theory, simulation, mock-up, and prototype makes it way into a manufacturing facility. We have already listed several characterization methods. It is now time to discuss the fabrication of nanomaterials.

Nanomaterials are made by two generalized processes: *top down* (e.g., subtraction from bulk starting materials) or *bottom up* (e.g., addition of atomic or molecular starting materials). Each scheme has a unique set of advantages and disadvantages. We recommend that you make a checklist of the advantages, disadvantages, limitations, and issues confronting each method as we discuss them through the course of this chapter.

We also add a brief section on molecular modeling, which is a fabrication tool. It is part of the design process. Molecular modeling has become one of the most powerful tools in nanoscale research, development, and material design. There exists a perfect fit between simulation and nanomaterials since atoms and molecules in nanoscale materials are finite and countable, and computer capability in this day and age is still limited with regard to capacity. Depending on the quality of input parameters, molecular simulation is able to generate an accurate rendition of nanoscale material behavior. Low-energy states, structure, dynamical behavior, chemical reactions, fluxes and flows, and more have been modeled with some form of atomistic-molecular simulation.

4.0.1 *Background*

Like anything else that we present in this text, boundaries are drawn for the sake of convenience and clarity, although sharp ones are not always possible.

Boundaries defining fabrication methods are no different; however, we proceed unabated and present the first bifurcation in the road. Like the great baseball player Yogi Berra said, "When you come to a fork in the road, take it."

There are two generic strategies for nanomaterial fabrication: *top down* and *bottom up*. Top-down fabrication methods begin with bulk materials (top) that are subsequently reduced into nanoparticles (down) by way of physical, chemical, or mechanical processes. Bottom-up methods, on the other hand, begin with atoms and molecules (bottom). These atoms or molecules react under chemical or physical circumstances to form nanomaterials (up). Growth proceeds in zero, one, or two dimensions to form dots, wires, or thin films, respectively. There are two generic types of bottom-up procedures: In the first, nanomaterials retain some level of structural and functional independence; in the second, nanomaterials become identical components of a bulk material. An example of the former is an array of gold quantum dots in an electronic device. Examples of the latter case include bulk metals formed from nanocrystallites and the structure of bone tissue.

There is, of course, further blurring of boundaries. Two general kinds of overlap, and possibly a third kind, occur between the two types (bottom up and top down) of fabrication strategies. In one case, a technique may be designated as top down but its microscopic mechanism suggests otherwise. The best example of this is the formation of carbon nanotubes by laser ablation. The starting material is a target made of compacted graphite and catalyst particles—certainly considered to be a bulk material in a compacted form. However, carbon nanotubes form from atoms and molecules via a catalytic process—definitely from the bottom up. Bismuth metal, obtained in bulk form, is melted and subsequently evaporated into atoms that deposit on the surface of a template material. Evaporation of a melted metal source to produce atoms (and perhaps nanoclusters) is a top-down procedure, but the formation of the thin layer of bismuth from those evaporated atoms is certainly from the bottom up.

In the second case, a manufacturing process may consist of both top-down and bottom-up methods. During the course of the fabrication of a computer chip, application of a photoresist material by a process called spin-coating is top down (from a bulk liquid phase). Photolithography is top down; chemical etching of the photoresist or the silicon substrate to reveal features is top down, but chemical derivatization to form a monolayer comprising different materials is bottom up.

Hybrid fabrication technology is a combination of distinct top-down and bottom-up mechanisms that occur simultaneously. This category of fabrication is exclusive to the nanoscale, where top-down and bottom-up techniques converge at the 30-nm size scale [1]. At the 3-nm scale, even hybrid technologies will be challenged by supramolecular and molecular technology that in turn may give way to atomic and nuclear technologies at the subnanometer scale—the realm of the single atom, single electron, single spin, and single photon [1]. These developments will require that we redefine top-down, bottom-up, and hybrid fabrication technologies. In the final analysis, it matters not which designation is assigned to a specific procedure, but for the sake of pedagogical purposes, we will continue to explore many types of fabrication methods and label them as one or the other or both.

Nanofabrication methods, just like characterization methods, have a long history. Fabrication and synthesis processes are the descendents of well-developed chemical and physical techniques developed over millennia. Engineers tend to manufacture components from the top down and then assemble them to make a device. Chemists, on the other hand, have always made materials by reacting atoms and molecules to form chemicals in bulk quantities—from the bottom-up. Chemical synthesis is by definition a bottom-up process. With regard to the biological processes, all structures are formed from the bottom up. Are you able to think of any exceptions to this rule?

The convergent nature of nanotechnology is well represented by fabrication methods. Engineers, physicists, chemists, and biologists respectively bring top-down, top-down, bottom-up, and bottom-up methods to the same table. The future of fabrication will require more cooperation between and among the disciplines, and the design parameters of future *nanofabs* must include such forward thinking in order to accommodate diversity and to enhance interaction among all the participants.

It is not practical to build an automobile engine from the bottom up and, conversely, it is not practical to synthesize aspirin from the top down. However, in nanotechnology, similar structures can be built from either fabrication perspective [2]. Features on a silicon wafer can be produced by a standard top-bottom procedure called lithography (bulk wafer → application of a photoresist layer → mask-UV exposure → etch) or by a bottom-up procedure (bulk wafer → polymer or seed crystals → self-assembly) [2]. Once again, nanotechnology and nanoscience are changing the way we do things and fabrication methods are no exception.

4.0.2 Types of Top-Down Fabrication Methods

We begin our catalog of fabrication methods with top-down methods. Physical fabrication techniques are considered to be mostly from the top down. Top-down methods are extremely diverse. Nanomaterials are formed from the top down by mechanical-energy, high-energy, thermal, chemical, lithographic, and *natural* methods.

Top-Down Mechanical-Energy Fabrication Methods. Cutting, rolling, beating, machining, compaction, milling, and atomization comprise a few examples of mechanical methods used to produce nanomaterials from the bulk. A mechanical method employs a physical process that does not involve chemical change—according to the traditional definition of chemical change (a reaction). Beating metals into a thin film is an ancient mechanical procedure used by the Egyptians and other pre-Hellenistic cultures to make swords, spear tips, and ornamental coatings. Mechanical energy methods such as ball milling operate on the principle of mechanical attrition. Kinetic energy, translated by hard, high-speed pellets, is imparted to samples by collision and friction. Samples are ground into fine powders by this method. An overview of mechanical top-down methods is shown in **Table 4.1**.

Top-Down Thermal Fabrication Methods. In the purest sense, a thermal fabrication method employs a physical process (heating) that does not initiate a chemical change in the sample—according to the traditional definition of chemical change (a reaction). Once again, it has proven difficult to place

TABLE 4.1 *Top-Down Mechanical-Energy Fabrication Methods*

Method	Comments
Ball milling	Production of nanoparticles by mechanical attrition to produce grain size <5 nm [3] High-energy ball milling uses steel balls to transfer kinetic energy by impact to the sample. Highly polydisperse products and contamination are problems.
Rolling/beating	Traditional mechanical methods to minimize material thickness and refine structure. Gold can be beat into a 50-nm thick film [4].
Extrusion; drawing	High-pressure processes of forcing materials in a plastic phase through a die to form high-aspect ratio parts like wires. Bi metal forced through nanopore alumina is an analogous process at the nanoscale and can be considered a thermal–mechanical process.
Mechanical Machining, polishing, grinding, and ultramicrotome	Also known as conventional machining; resolution limit: 5 μm [5] Other techniques analogous to mechanical machining perform the same function with laser beams, focused ion beams, and plasmas. Mechanical grinders/cutters are used to thin TEM samples. These include dimple grinders, diamond saws, ultrasonic disc cutters, and ultramicrotomes (<100 nm sections).
Compaction; consolidation	Metal powder ball milled and compacted. Powders are considered to be bulk materials; therefore, compaction of powders to form bulk material is not considered to be a bottom-up method.
Atomization	Conversion of a liquid into aerosol particles by forcing through a nozzle at high pressure

specific thermal methods into this category. Some of the top-down mechanical methods also involve thermal exchange. During the ball mill process, heat is obviously generated and plays a role in the outcome of the nanomaterial structure. Heat may be deliberately added during ball milling. Several compaction methodologies involve heating of samples during processing. The methods listed in **Table 4.2**, although extremely diverse, involve direct and deliberate heating of the sample during the fabrication process; chemical change may happen or not.

The process of combustion occurs in the presence of oxygen and causes a chemical change. Thermolytic and pyrolytic methods imply a process called "lysis" (from the Greek *lysis*, "a loosening, setting free, releasing, dissolution," from *lysein*, "to unfasten, untie") and usually involve chemical changes to the starting materials. Pyrolysis is the conversion of one material into another material by the application of heat in the absence of oxygen. Thermolysis or thermal reaction is often used synonymously with pyrolysis. Nanomaterials are routinely formed during the combustion of bulk organic materials. In the absence of oxygen, polyaromatic hydrocarbons (PAHs) with nanometer dimensions are formed by chemical reaction in pyrolytic processes. Sublimation, on the other hand, is the process of a solid phase of a material becoming a gaseous phase without experiencing an intermediate liquid phase (top down).

Top-Down High-Energy and Particle Fabrication Methods. High-energy sources such as electric arcs, lasers, solar flux, electron beams, and plasmas are commonly used to produce nanomaterials from the top down. A by-product of high-energy methods is superheating: a desirable or undesirable outcome depending on the objective. Although heat is produced during operation, these are not labeled as thermal methods because the origin of the heat is not a conventional thermal source per se. **Table 4.3** lists several commonly used high-energy methods that

Table 4.2 *Top-Down Thermal Fabrication Methods*

Method	Comments
Annealing	There are two applications of annealing: (1) anneal of bulk materials to form nanocrystallites, and (2) transformation of nanomaterials into another physical phase [6]. Microphase separation to form nanoscopic structures occurs in copolymer bulk materials upon application of thermal anneal above the glass transition temperature.
Chill block melt spinning	Metal is melted with RF coil and forced through nozzle on rotating drum, where it solidifies; strips formed with nanostructure [7].
Electrohydrodynamic atomization (EHDA)	Production of monodisperse droplets; melt or liquid materials at nozzle with electric field between nozzle and surface: cone → thin jet → droplets EDHA + pyrolysis to produce 10-nm Pt nanoparticles [8]
Electrospinning	A high voltage is applied to a polymer melt solution to induce charging. Polymer solutions at room temperature are also used routinely. At an acquired threshold, an electrospun fluid jet emerges from a needle tip to form a Taylor cone. The substrate, held at a lower potential, is covered by the charged polymeric solution
Liquid dynamic compaction (LDC)	Molten stream of metal is atomized by high-velocity pulses of an inert gas and the semisolidified droplets are collected on a chilled, metallic substrate [9].
Gas atomization	Molten metal is subjected to high-velocity inert gas impact that forms metal droplets [7,10]. Kinetic energy is transferred to metal, resulting in small droplets upon solidification to form powders. Powders are then compacted to form high-strength bulk materials.
Evaporation	Evaporation of solid metal or other material samples to form thin films; usually performed under high vacuum (10^{-6} torr). Heat is produced by electrical resistance. If nanoclusters are formed during the evaporation process, it is top down. If atoms or molecules are formed during the evaporation process that recombine to form a thin layer without any chemical reaction, it is a crossover technique.
Electrospinning	The process of electrospinning utilizes electricity to form thin layers of filaments from bulk polymer, composite, or ceramic solutions; fibers with nanoscale diameter can be fabricated [11].
Extrusion	Nanowires by extrusion of bismuth melt by pressure injection into porous template material such as alumina [12]. Parallel Bi nanowires with diameter ~13 nm
Template synthesis + evaporation	Formation of single-crystal Bi nanowires by a vapor-phase technique into porous alumina template—7-nm Bi nanowires [13]; 400–500°C with N_2 trap. Only phase changes are involved in this process.
Sublimation	The physical process of sublimation involves a phase change from a solid into gaseous form without a liquid intermediate phase. If nanoclusters are formed by this process, it is considered to be a top-down process. If atoms or molecules are formed first and then agglomeration into nanoparticles occurs, it is considered to be a crossover technique in which both top-down and bottom-up processes occur nearly simultaneously. Sublimation does not involve a chemical change of the material.
Thermolysis; pyrolysis	Decomposition of bulk solids at high temperature (top-down). These terms are also applied to the decomposition of molecules—nanomaterials are formed after decomposition in a bottom-up way by agglomeration. Because of this crossover, it is hard to place pyrolysis/thermolysis into one category or the other. The most common sense of the terms implies that molecules are simply converted into other molecules. In this sense, pyrolysis and thermolysis are neither top-down nor bottom-up methods. In such reactions (like decomposition), chemical change does occur.
Combustion	Chemical combustion is a top-down process in which there is chemical conversion of bulk organic materials + impurities into molecules like CO_2, H_2O, and nanomaterials such as ash with micron to submicron dimensions. The process of combustion involves oxygen.
Carbonization of copolymers	Spun fibers from polymethylmethacrylate (PMMA)--polyacrylonitrile (PAN) microspheres in PMMA matrix (top down? or bottom up?). Temperature treatment at 900°C removes PMMA and converts PAN into MWNTs [14]. Carbonization is another example of the difficulty encountered in cataloging such processes.

result in nanomaterials. Evaporation, a thermal method based on resistive heating, is considered to be a crossover technique in that a bulk material is converted into small particles (molecules or clusters)—a top-down process—that are then deposited to form a nanomaterial (thin film)—a bottom-up process.

Top-Down Chemical Fabrication Methods. If chemical transformations occur during a fabrication process, we shall designate that process as a chemical fabrication method. Although fabrication (a.k.a. synthesis) methods that employ chemical procedures rightfully reside within the domain of the bottom up, there are several that can be considered to be top down. Combustion is an ambiguous

TABLE 4.3 *Top-Down High-Energy and Particle Fabrication Methods*

Method	Comments
Arc discharge	High-intensity electrical arc discharge directed on a graphite target (anode) + catalyst to produce single-walled carbon nanotubes that accumulate on the cathode Temperature ~4000 K [15,16]
Laser ablation	High-intensity laser beam directed on a graphite target + catalyst to produce single-walled carbon nanotubes; sample warmed to 1200–1500°C by furnace, laser Sample is collected on water-cooled copper collector [17]. This process can be considered to be a thermal and a high-energy method.
Solar energy vaporization	Solar energy focused on graphite target + catalyst to produce single-walled carbon nanotubes Temperature ~3000+K [18]
RF sputtering	Ion bombardment of metal, oxide, or other material targets to form thin film coatings Usually performed under moderate vacuum (10^{-3} torr). Atoms, molecules, and clusters are formed by this process.
Ion milling	Argon ion plasma is used to subtract material from a surface. The purpose is to clean surface or remove (thin) materials for TEM. No change in the chemical nature of the sample happens during this process.
Electron beam evaporation	This is similar to evaporation in **Table 4.2** but uses an electron beam source to heat material. Evaporated material condenses on target substrate. High vacuum is required. Thin-layer antireflection, scratch-resistant coatings are formed by this technique.
Reactive ion etching	Sensitive materials are etched by reactive chemical species in charged plasma. Chemical change of the etched material takes place during this process. The etching process is guided by maskant materials.
Pyrolysis	Pyrolysis can also be considered a high-energy method. Application of high-energy source like fire to bulk hydrocarbon materials (like a steak) in the absence of oxygen creates polyaromatic hydrocarbons (PAHs)—a top-down process (or if considering intermediates—for example, carbon atoms—it can be considered to be a bottom-up process). Pyrolysis of solid refractory nanoscale materials like Si–C–N substrate to form nanotubes at 1500–2200°C is a crossover technique [19]. Large-scale synthesis of multiwalled carbon nanotubes occurs in flame environments by burning carbon sources such as methane, ethylene, or benzene.
Combustion	Combustion can be considered to be a high-energy, thermal, or chemical fabrication method.
High-energy sonication	Ultrasonication uses high-energy sound waves to make nanomaterials from bulk materials. The technique is also used to disperse carbon nanotubes in a suitable solvent. The dispersion of bundles of nanotubes into individual tubes is top down. Probe tips are made of titanium, vanadium, and other metals and alloys. Micron- to nanosized residual tip metal is introduced into solutions during the sonication process.

chemical top-down method, depending on the starting material. The chemical structure of solid constituents is completely altered following a combustion process. Nanosized PAHs and fly ash are by-products of a top-down pyrolysis process, e.g., the burning of coal.

Chemical etching of solid substrates like a silicon wafer (masked or otherwise) is a top-down chemical method. Chemical etching processes, on the other hand, adhere to a slightly different classification criterion—specifically, that chemical alteration occurs only in the layers exposed to, and subsequently removed from, the solid substrate. In other words, although nanofacets or porous structures are formed on or within the solid substrate, the chemical structure of the solid substrate remains intact. Only the surface is altered (passivated, oxidized). The process of chemical alteration is only applicable to substrate material removed during the etching process, e.g., transformation of the solid into a water-soluble oxide.

Anodizing is a chemical etching process that involves electricity (e.g., electrochemical etching). This process is a crossover technique and consists of four parts.

1. Metal is electrochemically removed top down from the surface and released into solution in ionic form, Al^{3+}, during the anodizing of aluminum metal. The cationic products of anodizing are not nanomaterials; they are ions.
2. Hexagonally distributed, monodisperse scalloped structures (nanofacets) are formed on the surface of the aluminum anode during anodizing. The diameter and curvature of individual nanoscale scallops are dependent on the applied anodic voltage. This is a true example of top-down fabrication. The other two parts of the anodize equation are bottom-up procedures.
3. The reaction of metal cations with anions originating from the cathode reaction or with solution anions leads to the formation of nanoscale colloidal oxides that eventually form the porous layer (from the bottom up). Anionic species include oxides, hydroxides, and other negatively charged species (phosphates, sulfates, oxalates, or chromates).
4. The hexagonal porous anodic oxide layer is formed from the bottom up by the electrochemical reaction of Al^{3+} cations with various oxide anions. The scalloped top-down metal surface structures direct the size, orientation, and distribution of the bottom-up pore channels.

Overall, if we had to choose we should probably consider anodizing as a top-down fabrication process. Top-down chemical fabrication methods are listed in **Table 4.4.**

Top-Down Lithographic Fabrication Methods. Many powerful top-down techniques involve some form of lithography. Lithographic techniques are what made the integrated circuit industry what it is today, and it continues to be the most viable method to form nanostructures that actually has widespread applications. The history of lithography was presented briefly in chapter 1. Traditionally, electromagnetic sources ranging from the visible wavelengths are still the most popular—especially in MEMS and circuit fabrication. Ultraviolet and x-ray sources are increasingly in demand as smaller features are required. Electron

Table 4.4 *Top-Down Chemical Fabrication Methods*

Method	Comments
Chemical etching	Standard acid or base solution etching of silicon and other materials, usually guided by maskant materials. Materials with nanometer pore channels are produced by this method. Etching of a metal surface without substantial oxide growth results in nanofacets.
Chemical–mechanical polishing (CMP)	CMP utilizes abrasives with or without a corrosive chemical slurry. Purpose is to thin and flatten samples. Surface roughness depends on size of abrasive. Mirror finishes with nanometer-scale roughness are produced by CMP methods.
Electropolishing	Electropolishing is an anodic method for brightening and smoothing the surface of a metal, primarily used for aluminum. The purpose of electropolishing is to reduce the surface roughness of a metal to nanometer scale. Conditions are extreme: concentrated acids (or bases), elevated temperature, and elevated current.
Anodizing	Anodizing is considered to be a crossover method in that nanofacets are formed on a bulk aluminum surface from the top down that in turn direct the formation of an anodic oxide from the bottom up. Anodizing operates under the same principle as electropolishing except that film growth is favored instead of film dissolution. Conditions are mild in comparison: dilute polyprotic acids, low temperatures (ca. 0°C), and low current.
Combustion	Combustion is an irreversible and dynamic chemical process that is catalyzed by high-temperature flames. Trees burning in a forest fire is a top-down way to form nanoaerosols.

beam sources are used primarily in the manufacture of the masks used in subsequent optical lithographic applications. Direct writing of lithographic patterns by electron and particle beams has proven to be an effective means of pattern generation without the use of a mask. Several low-cost, high-throughput nanolithographic techniques have emerged in the past decade or so (e.g., dip-pen nanolithography, nanosphere lithography, nano-imprinting, and other allied forms).

The new forms of lithography present us with another parsing dilemma. Most lithographic methods are top-down methods. However, if an STM current is involved in sensitizing a surface, should it be called a top-down method even though it is accomplished at the nanoscale from start to finish? If an AFM tip transfers atoms or molecules onto a surface, should this lithographic process be called a bottom-up method? Top-down lithographic techniques are listed and described in **Table 4.5**.

Top-Down Natural "Fabrication" Methods. Both top-down and bottom-up fabrication methods abound in the natural world. Most of these natural processes are quite familiar, and there is no need to allocate any more time or space to them. **Table 4.6** lists selected top-down natural processes to provide another relevant fabrication perspective.

4.0.3 *Types of Bottom-Up Fabrication Methods*

We will not go into significant detail about bottom-up methods in this section because many of them will be discussed more fully in the text in later chapters.

TABLE 4.5 Top-Down Lithographic Fabrication Methods

Method	Comments
LIGA techniques	LIGA is a German acronym for "Lithographie Galvanoformung Abformung," a microlithographic method developed in the 1980s. It was one of the first major techniques to demonstrate the fabrication of high-aspect ratio structures. Beam sources include x-ray, ultraviolet, and reactive ion etching. MEMS devices are fabricated using LIGA techniques.
Photolithography	Light is used to transfer patterns onto light-sensitive photoresist substrates. Photolithography is primarily used in the manufacture of integrated circuits and MEMS devices. The wavelength range of optical lithography techniques ranges from the visible to the near ultraviolet—ca. 300 nm. The resolution of photolithography techniques is ~100 nm [20].
Immersion lithography	Just like with immersion optical microscopy, resolution can be enhanced by 30–40% with application of a liquid medium between the aperture and the sample with higher refractive index. The medium needs to conform to the following criteria: (1) refractive index $n > 1$, (2) low optical absorption at 193 nm λ, (3) immersion fluid compatible with the photoresist and the lens, and (4) be noncontaminating.
Deep ultraviolet lithography (DUV)	Resolution with deep ultraviolet with λ = 248–193 nm, resulting in features on the order of 50 nm
Extreme ultraviolet lithography (EUVL)	Short wavelength ultraviolet, λ = 13.5 nm. EUVL resolution: ~30 nm [20]. The major problem with EUVL is that all matter absorbs EUV and damage to substrates is very likely. High vacuum is required and mask must be made of Mo–Si.
X-ray lithography (XRL)	X-rays are produced by synchrotron sources. XRL is capable of producing features down to 10 nm. Problems include damage to substrate materials.
Electron beam lithography (EBL)	An electron beam source is used instead of light to generate patterns. Although e-beams can be generated below a few nanometers, the practical resolution is determined by the electron scattering of the photoresist material. Just like in SEM, electron interaction volumes are generated during exposure. Line width <20 nm and electron energy: 10–50 keV
Electron beam writing (EBW)	EBW is a direct-writing procedure and, therefore, no pattern masters are required. Direct-write e-beam resolution is ca. 20 nm with lateral dimensions <10 nm [20]. Operation of electron beam parameters and patterning are computer controlled.
Electron beam projection lithography (EPL)	This technique is similar to TEM in that electrons are focused through a lens and projected onto a surface. In this case, however, a pattern is placed near the aperture. EPL is a high-throughput technique. A diamond membrane is used as stencil mask material. The process is not limited by diffraction as are the photolithographic techniques [21].
Focused ion beam lithography (FIBL)	Utilizes a liquid metal ion source (LMIS) with beam size of 10+ nm. FIBL resolution is 30 nm [20]. There is less backscattering than EBL and FIBL resists are more sensitive. FIBL, however, is restricted by limitations in reliable ion sources, difficulty in focusing, shorter penetration depth, swelling of resist, and whimsical ion implantation episodes. FIBL is also more expensive and slower that optical methods.
Microcontact printing methods	The George Whitesides group at Harvard University invented the lithographic method of microcontact printing. A topographical master is created by standard lithographic techniques that employ electron, ion, or electromagnetic beams. A negative replica of the master is made by pouring an elastomeric polymer, usually polydimethylsiloxane (PDMS), over the master. Upon curing, the elastomer is removed and coated with a self-assembled monolayer such as hexadecanethiol. Application of gold then reproduces the master pattern. Sub-100 nm features are possible by this technique [20].
Nano-imprint lithography (NIL)	Nano-imprint lithography is used to fabricate nanometer-scale patterns. It is a straightforward economical process with high throughput and high resolution. Patterns are created by stamping a resist material with a prefabricated stamp. The stamp can be used over and over. There are two types: thermoplastic (TNIL) [22,23] and photo (PNIL) [22,23]

TABLE 4.5 (CONTD.) *Top-Down Lithographic Fabrication Methods*

Method	Comments
Nanosphere lithography (NSL)	NSL is used to fabricate nanometer-scale patterns. It is a straightforward economical process with high throughput and high resolution. It is difficult to categorize this technique as top down or bottom up. Micron-scale latex spheres are often used as the template material. The interstices are nanoscale in size. NSL utilizes nanospherical materials in close-packed configuration as a mask to aid in the fabrication of periodic particle arrays (PPAs). Polymer nanospheres (diameter <300 nm) are in a single or double layer over insulator, semiconductor, metal, inorganic ion insulator, or organic π-electron semiconductor materials. Depending on the sphere diameter, nanoscale facets on the order of 22 nm are easily formed [24].
Scanning AFM nanostencil	An evaporated particle beam source is focused through a hole in an AFM cantilever. The procedure is good for metal deposition. This technique combines the ability to pattern a surface simultaneously with the ability to image the surface with the same cantilever. It is difficult to classify this technique as top down or bottom up (e.g., as it is for the thermal evaporation technique discussed before).
Scanning probe nanolithographies	There exist several forms of scanning probe nanolithographies. Some impart mechanical stress via the probe tip to a sensitized surface, followed by a chemical treatment; others apply an STM current to a substrate to create dangling bonds that react further to produce nanofeatures. These methods can be considered as top down in that nanofacets and features are produced from a solid bulk substrate.
2-Photon polymerization	Photopolymerization causes polymer to solidify to form three-dimensional image. Resolution of ~120 nm, although the laser λ is 780 nm [25]. This is, in the clearest sense of the term, a top-down process.

TABLE 4.6 *Top-Down Natural Fabrication Methods*

Method	Comments
Erosion	Conversion of macroscopic mineral-based materials into micro- and nanoparticles.
Etching	Etching of silicate rocks by carbonic acid from the environment contributes to erosion.
Hydrolysis	The decomposition of organic (and inorganic) matter by hydrolysis is a common way to make nanomaterials in the natural world.
Volcanic activity	Formation of fly ash and other materials by volcanic activity. The dispersion of volcanic byproducts is mostly airborne. Volcanic by-products contribute to the formation of clays like *montmorillonite* (a nanostructured material discussed in chapter 13).
Forest and brush fires	Formation of combustion gases, nanometer scale PAHs, amorphous carbons, and particulates
Solar activity	Radiation degradation of bulk synthetic, inorganic, and organic materials
Pressure and temperature	Formation of diamond crystallites from pressure and temperature processes applied to bulk materials; application to bulk carbon deposits (coal)
Biological decomposition	Decomposition is a process that begins at the bulk, micro-, or nanoscale level and terminates at the nano, molecular, or atomic level. Biological decomposition is mitigated by bacterial and other life forms in addition to inorganic natural processes.
Digestion	Reduction of bulk biological materials into nanometer and subnanometer scale components by the action of acids and hydrolysis; the formation of nitrogenous wastes is a bottom-up procedure, so to speak.

Bottom-up fabrication techniques are divided into four general categories: (1) gaseous phase methods, (2) liquid phase methods, (3) solid phase methods, and (4) biological methods.

Just as with top-down methods, it is difficult to pigeonhole a technique into a general category. Many bottom-up processes are characterized by tandem applications of liquid and gaseous techniques onto solid substrates. There are three generalized states of matter: gaseous, liquid, and solid. The distance d between molecules in a gas is proportional to

$$d = \left(\frac{V}{N}\right)^{\frac{1}{3}} \tag{4.1}$$

where V is the volume and N is the number of molecules. For an ideal gas at standard temperature and pressure (STP), $V = 22.4$ L and $N = \mathcal{N}_A$, Avogadro's number, 6.022×10^{23}. The distance between atoms or molecules, center to center, in an ideal gas is equal to 3.34 nm.

A liquid is a state of matter that has volume but not shape. Although the atoms and molecules in a liquid are compressed as tightly as a solid, the molecules in a liquid are free to move randomly and unfettered. The distance between molecules or atoms in a solid is like that of a liquid, but random movement is severely restricted due to structural factors. Solids, of course, constitute the most condensed form of matter.

A technique is designated as gaseous, liquid, or solid if the process takes place in that appropriate medium or if the active constituent from which nanomaterials are formed is a gas, liquid, or solid. Once again, some difficulty in nomenclature is encountered when more than one phase is present during synthesis, but from a practical point of view, such classification is relatively straightforward.

Bottom-Up Gas-Phase Fabrication Methods. Gases represent a highly dispersed phase of atoms and molecules. Some nanomaterials formed in the gas phase, like clusters, remain in the gas phase. More commonly, gas-phase precursors interact with a liquid- or a solid-phase material. If one of the precursors of a nanomaterial originates from the gas phase or if the reaction takes place in the gas phase, we shall call it a bottom-up gas-phase fabrication method (**Table 4.7**).

Nonbiological Bottom-Up Liquid-Phase Fabrication Methods. Bottom-up liquid methods are numerous and diverse (**Table 4.8**). The choice of solvent is an extremely important parameter in any liquid-based bottom-up fabrication method. The liquid medium can be hydrophilic or hydrophobic, ionic or anionic, or heterogeneous (e.g., for the purpose of phase transfer of product between two immiscible liquids). The new field of supramolecular chemistry is conducted entirely in liquid media. All bottom-up biological fabrication processes occur in liquid media. The liquid phase is also where most chemists feel at home, and it is also going to be one of the prime drivers of nanotechnology. Scale-up of liquid-phase fabrication methods is a relatively straightforward process and it is at the scale-up stage where the chemists turn over the reins of a process to the chemical engineers.

Bottom-Up Lithographic Fabrication Methods. We add a special category for lithography once again, but this time featuring bottom-up lithographic methods.

TABLE 4.7 *Bottom-Up Gas-Phase Fabrication Methods*

Method	Comments
Chemical vapor deposition (CVD)	CVD involves the formation of nanomaterials from the gas phase, usually at elevated temperatures, onto a solid substrate or catalyst. Carbon nanotubes are formed by catalytic decomposition of carbon feedstock gas in inert carrier gas at elevated temperature. Single-walled carbon nanotube production by CVD requires nanoscale Fe, Co, or Ni catalyst plus Mo activator on high surface area support (alumina) at >650°C. Methane gas serves as the carbon source [26].
Atomic layer deposition (ALD)	ALD is an incredibly precise sequential surface chemistry layer deposition method to form thin films on conductors, insulators, and ceramics. The layer formed by ALD conforms to surface topography. Precursor materials are kept separate until required. Atomic scale control pinhole-free layers are formed. Al_2O_3 layers are generated from hydroxylated Si substrate + $Al(CH_3)_{3(g)}$, then H_2O vapor is applied to remove methyl groups. The process is repeated until a target thickness is attained. Layer thickness: 1–500 nm
Combustion	The formation of Si nanoparticles from the combustion of SiH_4 (silane gas) and other silicon-containing gases like hexamethyldisiloxane under low-oxygen conditions produces Si nanoparticles as small as 2 nm. Al_2O_3 and TiO_2 can also be formed by combustion.
Thermolysis; pyrolysis	Solid Si nanoparticles can also be formed by the thermal decomposition of silane gas in the absence of oxygen. The bottom-up decomposition of ferrocene to form Fe nanoparticles is one of the best examples of a bottom-up gas-phase fabrication method.
Metal oxide (MOCVD) Organometallic vapor phase epitaxy (OMVPE)	Chemical characteristics of precursor materials utilize reactive gas-phase-organometallic compounds that decompose to form nanometer-scale thin films or nanoparticles. H_2 carrier gas, group III metal–organic compounds + group V hydrides 500–1500°C at 15- to 750-torr pressure are representative conditions under which MOCVD is performed.
Molecular beam epitaxy (MBE)	MBE is a thin film growth process conducted under high vacuum. A heated Knudsen cell or effusion cell is used to introduce reactants by molecular beams. MBE is able to deposit one atomic layer per application. Examples include alternate layers of GaAs and AlGaAs with each layer of 1.13 nm in thickness and InGaAs quantum dots [27]. The temperature used in MBE is commonly 750–1050°C in H_2 carrier gas.
Ion implantation	This is a tough method to categorize. Nanovoids, for example, can be created by ion implantation of Cu ions into silica and subsequent annealing [28]. It is bottom-up action performed on a bulk material. If the ions come from a bulk source, it has a bottom up component. Once the ions are formed, ion implantation is bottom up.
Gas phase condensation; thermolysis	Formation of Fe nanoparticles by decomposition of ferrocene at 200°C is an example of gas-phase process to form nanoscale Fe. Formation of lithium nanoclusters by decomposition of LiN_3 is another example [7]. Temperature at decomposition depends on the material.
Solid template synthesis	Provides a solid template substrate for gas-phase deposition of materials on the solid substrate. This is considered to be a mixed bottom-up system. Final nanomaterial size, shape, and orientation are predetermined by template parameters.

Bottom-up lithography methods are limited to a few kinds, based on template processes or direct writing (**Table 4.9**).

Bottom-Up Biological and Inorganic Fabrication Methods. Biological processes are overwhelmingly formed from the bottom up (**Table 4.10**). More detail is allotted to this topic in chapter 14.

Table 4.8 *Nonbiological Bottom-Up Liquid-Phase Fabrication Methods*

Method	Comments
Molecular self-assembly	This generic process is supported in liquid media. From some perspectives, supramolecular chemistry is a subset of molecular self-assembly. Almost all molecular self-assembly takes place in liquids. The liquid plays a major role in supporting intermolecular interactions and intermediate metastable species.
Supramolecular chemistry	Supramolecular chemistry, for reasons to be explained in chapter 11, is conducted in liquid media. Weak intermolecular forces are supported in liquids that allow many kinds of intermolecular interactions to take place. All significant biological metabolic processes occur in a liquid medium.
Nucleation and sol–gel processes	Precursor chemicals in a supersaturated state combine by self-assembly or chemical reaction to form seed particles. Thermodynamics drives a nucleation process that forms nanoparticles. The nucleation process depends on prevailing conditions of *p*H, temperature, ionic strength, and time [5]. Due to van der Waals attractions, colloids are formed. Sol–gel methods are irreversible chemical reactions of homogeneous solutions that result in a three-dimensional polymer. Sol–gel methods yield nanostructured materials of high purity and uniform nanostructures formed at low temperatures [5]. Negative replicas of colloidal hierarchical structures, upon drying, yield aerogels or xerogels. Such gels can be back-filled to produce nanocomposites or hybrid materials [5]. These are all pure bottom-up processes.
Reduction of metal salts	Noble metal clusters and colloids are formed by the reduction of metal salts like $HAuCl_4$ and H_2PtCl_6. Common reducing agents come in the form of organic salts like sodium citrate—$Na_3C_6H_5O_7$. By means of phase transfer reactions (consisting of an interface between two immiscible liquids), metal clusters and colloids are stabilized by the addition of organic ligands. For example, phosphine or thiols are adsorbed onto gold-55 to produce a stable cluster [29].
Single-crystal growth	Nucleation process to form single crystals in liquid media
Electrodeposition Electroplating	Electrodeposition is direct deposition of metals from metal salt solutions to form thin layers or films on a solid conducting substrate. Electrodeposition is an electrolytic process that forms thin metal films on the cathode of the cell. The process conforms to Faraday's law.
Electroless deposition	Electroless deposition is the autocatalytic deposition of metals without electrical assistance. It requires metal cation + catalytic (activated) surface + reducing agents like formaldehyde, alkali diboranes, alkali borohydrides, or hypophosphorous acid. Pt, Ni, Co, Au, and numerous other metals can be deposited on many kinds of substrates, including plastics. Electroless deposition has been used to create negative or positive replicas of porous nanostructures [30].
Anodizing	We have already characterized anodizing as a top-down process. We mentioned earlier that anodizing method contains a top-down component (formation of scalloped structure). Here, we focus on the bottom-up formation of the porous alumina. Aluminum metal is made the anode in an electrolytic cell consisting of a polyprotic acid (usually sulfuric, phosphoric, or oxalic). Pore diameter of <5 nm $\rightarrow >200$ nm; with pore density: 20–80+% and film thickness: <1 μm $\rightarrow >100$ μm. Anodized titanium several nanometers thick generates bright interference colors.
Electrolysis in molten salt solutions	Utilization of molten alkali halide salts with graphite electrodes with 3- to 5-A current [31] Erosion at the cathode to form tubes The product is transferred to toluene.
Solid template synthesis	Provides a solid template substrate for electrochemical, chemical, polymerization, and other liquid-phase reactions. Most methods are accomplished in a liquid medium. Final nanomaterial size, shape, and orientation predetermined by template parameters.
Liquid template synthesis	Liquid templates (micelles and reverse micelles) are commonly used to make quantum dots from the bottom up.
Supercritical fluid expansion	Solvent removal under hypercritical conditions forms aerogels and xerogels that contain nanometer-sized voids. Supercritical conditions imply that the medium is in neither liquid nor solid phase.

TABLE 4.9 *Bottom-Up Lithographic Fabrication Methods*

Method	Comments
Nanolithography: Dip-pen methods (DPNL)	Nanoprobe lithography in the form of dip-pen nanolithography was invented by Chad Mirkin's group at Northwestern University in Chicago [32]. DPNL is considered as an AFM-based soft-lithography technique. The operation of this method is quite simple. A water meniscus is formed between an AFM tip and a substrate. The AFM tip, in conjunction with the water meniscus conduit, is able to transfer molecules to the surface. The method has high spatial resolution (<10 nm), has high registration capability (probe can both read and write), and is able to deliver complex molecules such as DNA to a surface [20]. The major disadvantage, like that of STM writing, is low throughput.
Nanosphere template methods	Nanosphere lithography is a template method for fabrication of nanomaterials. Latex spheres are arranged on a substrate surface in various configurations: hexagonal close packed, or into a square array. The interstitial spaces between latex spheres serve as sites through which deposition can occur—a very straightforward, simple process. Although the distribution and placement of the spheres can be considered to be a top-down process, the deposition of material through the interstices definitely occurs from the bottom up.
Nanopore template methods (shadow mask evaporation)	Use of porous alumina membrane templates as templates to form arrays of nanoparticles. The size of the nanoparticles can be controlled from 5 nm to >200 nm. The space between nanoparticles can also be adjusted. Nanoparticle aspects are adjusted by the height of the mask, the pore size, and the direction of evaporation [33]. This technique is good for direct patterning without the need for additional steps such as etching or lift-off. The combinations of masks, materials, and substrates are enormous, and the process allows for straightforward upscale. Arrays have been used in the secondary fabrication of memory devices and carbon nanotubes.
Block copolymer lithography (BCPL)	BCPs applied by spin-coating (top down) self-assemble into an ordered array of nanoscopic domains on a surface. Selective removal of one component yields an etch mask. The substrate pattern is formed by plasma etching. In a specific example: a 35-nm thick polystyrene–PMMA copolymer layer is applied to a Si_3Ni_4-coated Si wafer. Removal of the PMMA leaves an ordered array of polystyrene nanodots. Reactive ion etching (REI) with CHF_3 transfers the pattern to the Si_3Ni_4 layer. The Si_3Ni_4-formed pattern is etched again by REI with HBr. The result is an ordered array of silicon pillars (wires) [34]. Block copolymer lithography was able to produce periodic arrays of 10^{11} holes per cm^{-2} [35]. One problem that faces this procedure is long-range order.
Local oxidation nanolithography	A scanning probe tip (a dynamic AFM tip) is placed a few nanometers above a substrate surface. The environment consists of saturated water vapor. A bias voltage is applied between the tip and the surface. Oxidation of the surface, if silicon, produces lines of silicon oxide. The breadth of the meniscus and the distribution of the electric field within determine the size of the feature [36]. Features as small as 7 nm were produced. One-nanometer projections were formed in the *z*-direction.
STM writing	The IBM logo pictured in chapter 1 was fabricated by a bottom-up method. Starting with xenon atoms, each atom was manipulated by the scanning probe tip into its final position. Other examples of this technique include the *quantum corral*—a circular array of Fe atoms placed on a Cu surface [37]. All scanning probe fabrication methods are hindered by low throughput.

4.0.4 Nebulous Bottom-Up Fabrication Categories

Fabrication of nanoscale materials (structures, domains) within solids is difficult to pinpoint. It is difficult to track the history of an atom or molecule throughout the course of a solid material. Solids contain a number of diverse

TABLE 4.10 *Bottom-Up Biological and Inorganic Fabrication Methods*

Method	Comments
Protein synthesis	Formation of proteins from precursor amino acids by elaborate process of protein synthesis Transfer RNA transports amino acids to ribosomal RNA and link with peptide bonds.
Nucleic acid synthesis	Synthesis of nucleic material (RNA, DNA) from sugars, phosphate, and nuclides (adenosine, guanine, cytosine, and thymine) from the bottom up The processes of mitosis and meiosis are template (replication) methods.
Membrane synthesis	Bottom-up agglomeration of lipids, phospholipids to form organized membrane structures that make life possible
Inorganic biological structures	Mother of pearl (nacre) 95% Inorganic aragonite (platelets 200–500 nm thick) + organic biopolymer Deformable nanograins [38]
Crystal formation methods	Nucleation depends on P, T, concentration, and composition. Flaws reduce surface energy by nucleation. Direction of growth depends on nanostructure.

defects that have nanoscale dimensions. Are these considered to be "nanomaterials" or nanofacets? Or are they merely nanodomains of the bulk type material? Voids formed by ion implantation do agglomerate to form nanovoids from the bottom up. We address this nebulosity in more detail later.

4.0.5 The Nano Perspective

There are many kinds of nanomaterials. When discussing fabrication methods, it is essential that the nature of the end product be understood. For example, some types of nanomaterials retain their nanoscale dimensions (e.g., quantum dots). Others form into components of more complex structures (e.g., one-dimensional, two-dimensional, or three-dimensional arrays of quantum dots). In these instances, the quantum dot retains its identity as a unique nanomaterial. In other cases, nanomaterials form the structure of an integrated bulk material. An example of a bulk material that is composed of nanostructured components is a Cu–Fe alloy in which nanodomains of one or the other metal exist within a bulk material. Steel made of nanosized grains has better mechanical properties than steel made of micron-sized grains.

Silk, collagen, elastin, and keratin tissue found in animals are composed of a hierarchy of increasingly larger structures [39]. The hierarchy begins with sub-nanometer materials and ends with a functional macroscopic material [39,40]. The relationship of nanostructure, muscle fibers, and connective tissue is shown in **Table 4.11.** A similar table can be created for bone tissue and other organ systems in animal bodies. From the purely structural point of view, it is clear that nature begins from the bottom up to build any kind of macroscopic functional material.

Fabrication of inorganic nanomaterials is bottom up, but some well-known methods such as erosion certainly operate from the top down. The construction of a snowflake is a nucleation process that emphasizes eccentricities in the *unit cell* of each snowflake, a bottom-up process. With regard to nanoscience and technology, materials are constructed from the top down, bottom up, or a

TABLE 4.11 *The Nanostructure of Tendons*

Structural component	Dimensions	Description/function
Amino acids	<1-nm	The building blocks of proteins
Collagen	1.5-nm Diameter	Primary structure polypeptide (the protein of connective tissue)
Triple-helix coil (tropocollagen)	1.5-nm Diameter; 300 nm length	Three polypeptide strands form a cooperative quaternary structure.
Microfibrils	<4-nm Diameter	
Subfibril	10–20-nm Diameter	
Fibril	50–500-nm Diameter	Connective tissue called *endomysium* based on collagen subunits that surround muscle fibrils
Fascicle	50–300 μm	Bundle of muscle fibrils (~10) surrounded by *perimysium* (connective tissue) [40]. The *endo-*, *peri-*, and *epimysia* converge to form the tendon.
Tendon	10–50 cm	Attachment of muscle to bone support structure Provides flexibility and strength

combination of both. Although some fabrication categories are placed in one of the two major types, note that several cross borders. Lithographic techniques, when taking in the whole, are a combination of several techniques.

There are numerous challenges facing any kind of nanofabrication technique. According to George Whitesides of Harvard [41], "In almost all applications of nanostructures, fabrication represents the first and one of the most significant challenges to their realization." As with any process, there are advantages and disadvantages. Problems with top-down approaches include alteration of the surface structure during the process [42,43]. Lithography is a method that is capable of causing undesirable changes to the crystal structure and more damage from subsequent chemical etching steps. Reduction in conductivity and generation of excess heat due to surface imperfections in nanowires, for example, could be problematic [43]. Top-down methods can be extremely energy intensive because energy is required to create new surfaces (chapter 6). Nanoscale materials made from bottom-up methods may lack long-range order and structural integrity. There are more disadvantages.

We started our civilization with stone, bone, and wood and then metal and ceramics. These materials are classified as *hard matter*. Biological materials like leather and gut, materials based on biological soft matter sources, were also put to good use. With semiconductors, advanced alloys, and other hard materials, the tradition continued. The advent of plastics, polymers, pharmaceuticals, and other organically based materials ushered in a new era of chemistry: the chemistry of the covalent bond. We have entered the age of *soft matter*.

4.1 TOP-DOWN FABRICATION

Top-down approaches remove, reduce, subtract, or subdivide a bulk material to make nanomaterials. Top-down methods, therefore, are considered to be subtractive. Top-down fabrication methods logically reside within the realm of engineering and physics. Top-down fabrication dominates nanotechnology today, although significant ground has been gained by bottom-up methods [44].

Although tried and true, there are many challenges that confront top-down methods as miniaturization continues unabated towards the nanoscale. Contamination, machine cost and complexity, clean room cost and complexity, physical limits (photolithography), material damage, and heat dissipation are a few of the issues that confront top-down methods. There seems to be a strong link between the cost of a procedure and the size of the intended product. Specifically, it becomes more expensive to make smaller materials and devices. According to pundits, however, once the R&D phase is accomplished and the manufacturing line is in place, the cost of nanomaterial-enhanced products should go down.

A few selected top-down processes will be reviewed in the following sections. There are many we leave out. For the purposes of this course, a representative sample has been compiled that should provide enough insight and information into top-down fabrication methods.

4.1.1 Mechanical Methods (Mechanosynthesis)

Any procedure that involves the action of a bulk implement, tool, or machine on samples made of bulk materials is a top-down mechanical method. Mechanical methods base their action on kinetic energy: a hammer falling, a canister revolving, a roller thinning, a die extruding, a compacter compressing, etc. Beating and rolling methods to form thin metal films with nanometer dimensions and extrusion of soft materials in plastic phase to form wires are widespread industrial practices [5].

Ball Milling. One of the most important mechanical top-down methods is ball milling (and shaker milling), a technique that is able to produce nanoscale materials by mechanical attrition. In ball milling, the kinetic energy of a grinding medium (e.g., stainless steel or tungsten carbide ball bearings) is transferred to coarse-grained metal, ceramic, or polymeric sample materials with the direct purpose of size reduction [3]. Rotation or rapid vibration of a drum or canister imparts kinetic energy to the grinding medium (under controlled atmospheric conditions to prevent oxidation) [5]. During the ball mill process, severe plastic deformation of the sample material initiates the formation of defects and dislocations. Any type of mechanical deformation subjected to high sheer and strain conditions leads to the formation of nanograined material [45]. **Figure 4.1** displays a rendition of a generic ball mill.

The result of the procedure, however, yields nanoparticulate materials peppered with defects with a wide distribution of size. On the upside, mechanical attrition is one of the least sophisticated technical processes and hence the least costly. Although the process has roots in ceramic processing and powder metallurgy for several decades, it is considered to be a rapidly evolving field [3]. Ball milling, first accomplished by J. Benjamin in 1966, produces mechanically alloyed materials. Alloys, metastable phases, quasi-crystalline phases, and amorphous alloys are formed by such mechanical attrition techniques [3].

The principle of mechanical attrition is relatively straightforward. A sample material is placed in a canister filled with ball bearings. The canister is activated and begins to rotate at increasingly higher revolutions per minute. The ball

FIG. 4.1 *There are two ends of the fabrication spectrum: At one end there is the high-priced lithographic equipment that requires a high-vacuum environment and expensive energy sources. At the other end there is the ball mill—a purely mechanical machine that fabricates nanomaterials by mechanical methods. Kinetic energy from a rotating or vibrating canister is imparted to hard spherical materials like ball bearings. The ball bearings in turn reduce bulk precursor materials into nanoparticles.*

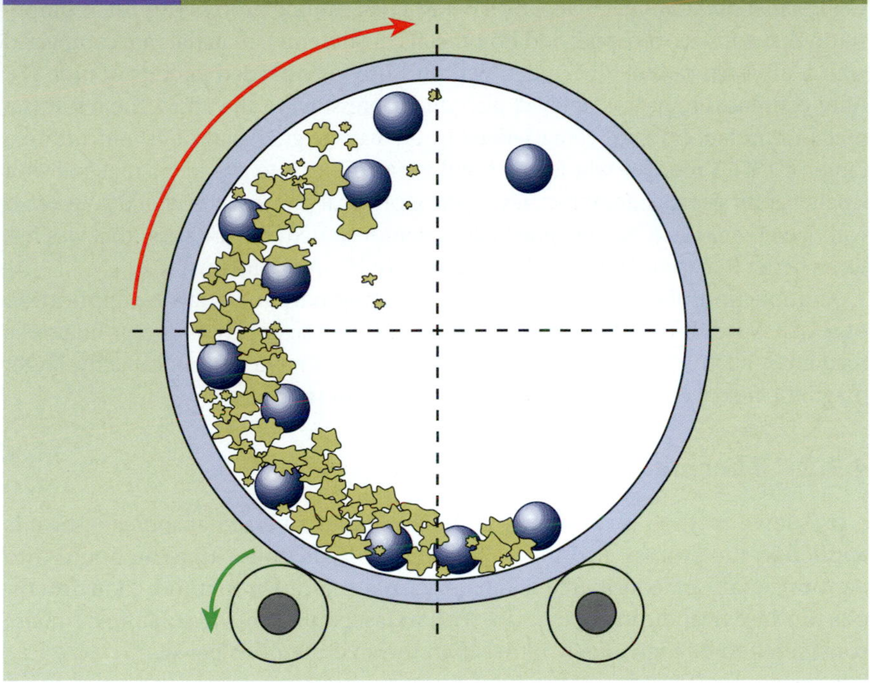

bearings impart significant kinetic energy to the samples, a much softer material. Several processes occur in the following order. The first event to happen is compaction and then rearrangement of particles. Secondly, elastic and plastic deformation and welding occur. Particle fracture and fragmentation further reduce the particle size. Griffith theory describes particle fracture in a mathematical sense:

$$\sigma_F \approx \sqrt{\frac{\gamma E}{c}} \tag{4.2}$$

where σ_F is the stress at which crack propagation leads to catastrophic failure, γ is the surface energy of the particle (joules per square meter), E is Young's modulus, and c is the length of the crack [3]. The tipping point is reached when the stress equals the strength of cohesion between atoms of an isotropic solid. As particles get smaller, due to enhanced surface energy, agglomeration forces (antifracture) predominate. A balance is struck among the stress, increased resistance to fracture, increased agglomeration, and maximum energy that is expended in milling.

There are several types of mechanical attrition devices. Shaker mills are the most popular form used by scientists and are able to produce particles <20 nm in diameter [3]. A back-and-forth high frequency (>1000 cycles · min^{-1}, ball velocity >5 m · s^{-1}) applied to a vial with milling balls ensures that samples pulverize properly. Planetary ball mills are commonly used in laboratories. In this form of mechanical attrition, rotational forces are the source of kinetic energy imparted to the grinding media and the sample.

Compaction and Consolidation. Following a ball milling process (e.g., of a composition that consists of copper and iron metal constituents), materials are compacted with a tungsten-carbide dye under high pressure for extended periods of time [7]. After compaction, heat is applied, also under pressure, to the alloy. The result is a metal formulation that is characterized by an average grain size of 40 nm within a range of 15–75 nm. The whole point of this procedure is to produce a material with smaller grain size that demonstrates superior physical properties to that of a material with larger grain size. Nanograined alloys demonstrated fracture stress that was five times better than pure iron with larger grain size (50 nm–150 μm) [7,46].

Compaction of ceramic and superconducting nanomaterials by application of shock waves limits the grain growth [47]. Ceramic superconductor materials formed by such advanced techniques demonstrate higher current capacity, larger magnetic fields, and no energy loss through resistance.

4.1.2 Thermal Methods

A top-down method is considered to be thermal if an external source of heat is applied to the process. Melting a bulk material and converting the liquid into nanomaterials are considered to be a thermal top-down method. Many methods produce heat during operation, such as laser ablation and solar flux, but are considered to be high energy rather than thermal methods per se.

Chill Block Melt Spinning and Solidification. This is a process that initially applies heat to bulk material with the intent of melting that material and performing an extrusion process. Quick solidification of the metal is induced to freeze the metal into a desired form. An RF (radio frequency) heating source is utilized to create a metal melt. The liquid metal is then forced through a nozzle in the form of a stream that is oriented over the surface of a rotating drum [7]. A bulk alloy material consisting of aluminum nanoparticles, 10–30 nm in size, made by this method demonstrated tensile strength in the gigapascal range.

Gas Atomization. This is another top-down method that is suited for the manufacture of nanoparticulates. In this process, a high-energy stream of some inert gas is directed at a molten metal stream. Just like in the ball milling, kinetic energy is transferred to the sample—this time from the high-energy inert-gas beam. The impact initiates the formation of finely divided metal particles that upon solidification form into a finely divided powder. The nanopowder is then compacted to form a bulk metal with superior mechanical properties.

Electrohydrodynamic Atomization (EDHA). Electrohydrodynamic atomization is an offshoot of electrostatic spray technology and is a subset of liquid disruption processes. The formation of a *Taylor cone* that terminates in a fine-stream jet forms the basic mechanism of EHDA. An electrostatic atomizer causes a net charge to develop on the surface of a droplet that causes dispersion due to coulombic repulsive forces. This process prevents agglomeration of droplets and hence particles are formed. The EHDA process is capable of producing particles as small as quantum dots.

The products of EDHA procedures depend on the flow rate of the liquid, the diameter of the needle orifice, the distance between the needle tip and grounded surface, and the strength of the applied AC field [48]. One of the primary goals of this procedure is to be able to synthesize nanoparticles rapidly and over large areas. The EHDA technique was used to atomize a solution of chloroplatinic acid [$H_2PtCl_6 \cdot (H_2O)_6$] in ethanol. The purpose of the atomization procedure was to produce Pt metal particles. Droplets are sprayed on a Si–SiO_2 substrate and heated at 700°C for a short period of time. The dimensions of the Pt particles were on the order of 10 nm [8].

4.1.3 High-Energy Methods

Arc discharge, laser ablation, and solar vaporization are three high-energy top-down methods that are able to generate nanomaterials by the application of high energy electric currents, monochromatic radiation, or solar radiation to a solid substrate. Each method is capable of forming carbon nanotubes from graphite substrates that contain catalytic Fe, Mo, or Co particles. We consider any process that involves plasma to be a high-energy process. High-energy methods, with the possible exception of the solar version, are not practical to upscale due to the intense investment in energy that is required.

Arc Discharge Plasma Method. The first deliberate attempt to produce carbon nanotubes with an arc discharge method was accomplished with an arc plasma discharge method developed by Y. Ando in 1982 [15,16]. The formation of carbon nanotubes by arc discharge (plasma arcing) process is dependent on the pressure of He, the process temperature, and the applied current. Typical conditions utilize an applied voltage of 20 V, current ranging from 50 to 100 A, and He pressure of 50–760 torr. Two graphite rods are placed millimeters apart (**Fig. 4.2**). The sacrificial anode consists of graphite that is doped with metal catalyst particles. In this configuration, single-walled carbon nanotubes are fabricated. Multiwalled carbon nanotubes are formed if no metal catalyst is present in the graphite. At 100 A, carbon vaporizes in a hot plasma. Carbon cations are formed at the anode and the soot is collected at the cathode. The arc method, although relatively simple, produces an array of unwanted by-products. Samples originating from arc discharge methods often require extensive purification. Basing scientific conclusions on unpurified materials is not a recommended practice.

Laser Ablation of Solid Targets. In 1995, carbon nanotubes were synthesized by pulsed laser method. Graphite rods containing Co and Ni catalyst were heated to 1200°C and then exposed to laser pulses [17]. Heat is, therefore, generated by two means in this process—the furnace and the laser. The vaporized carbon is collected on a cooled finger downstream of the carbon targets. Continuous wave CO_2 (~2 kW) infrared, ultraviolet, or Nd:YAG lasers are the most common types of lasers used in the ablation method. A generic scheme is shown in **Figure 4.3**.

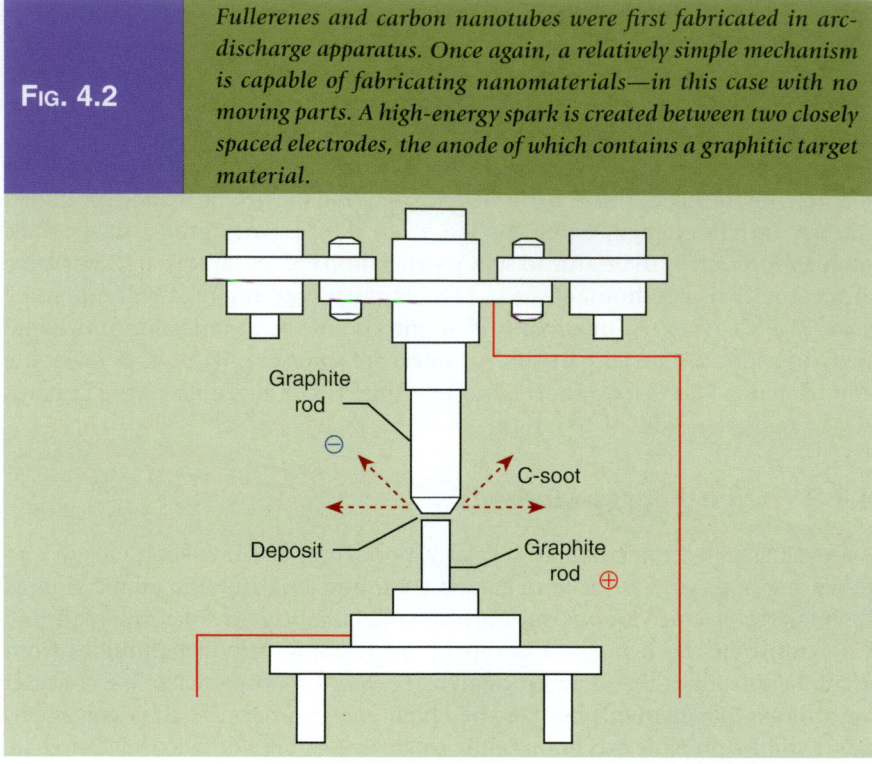

FIG. 4.2 *Fullerenes and carbon nanotubes were first fabricated in arc-discharge apparatus. Once again, a relatively simple mechanism is capable of fabricating nanomaterials—in this case with no moving parts. A high-energy spark is created between two closely spaced electrodes, the anode of which contains a graphitic target material.*

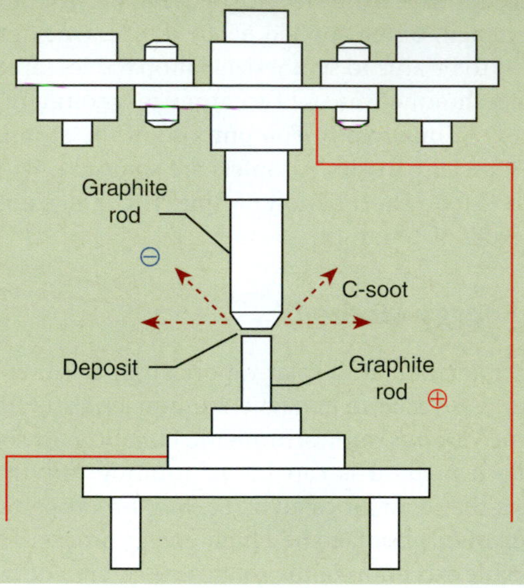

High-Flux Solar Furnace. Solar power has also been used to fabricate carbon nanotubes by a top-down procedure [18]. Since scale-up of the arc discharge and laser ablation methods is problematic, the goal is to increase the power of the solar furnace to a level of 500 kW [49]. At the National Renewable Energy Laboratory (NREL) in Golden, Colorado, researchers were able to produce fullerenes from a 10-mm diameter graphite pellet with a 10-kW high-flux solar

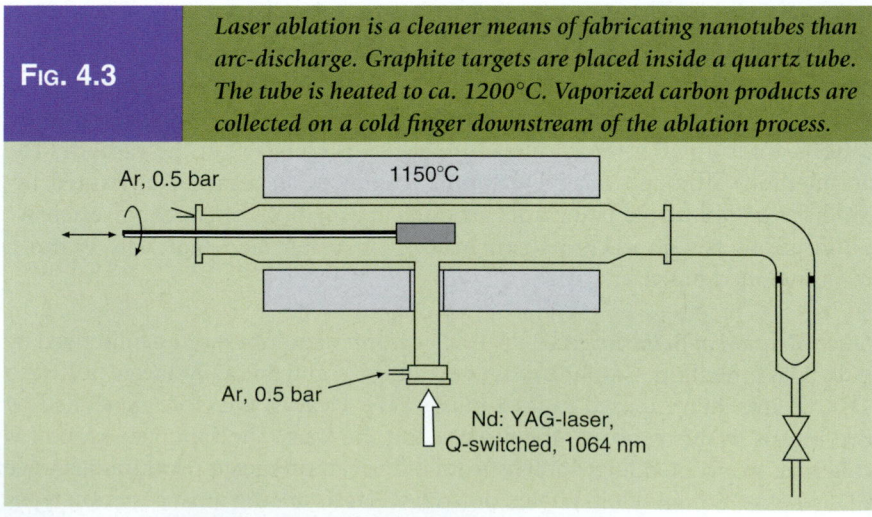

FIG. 4.3 *Laser ablation is a cleaner means of fabricating nanotubes than arc-discharge. Graphite targets are placed inside a quartz tube. The tube is heated to ca. 1200°C. Vaporized carbon products are collected on a cold finger downstream of the ablation process.*

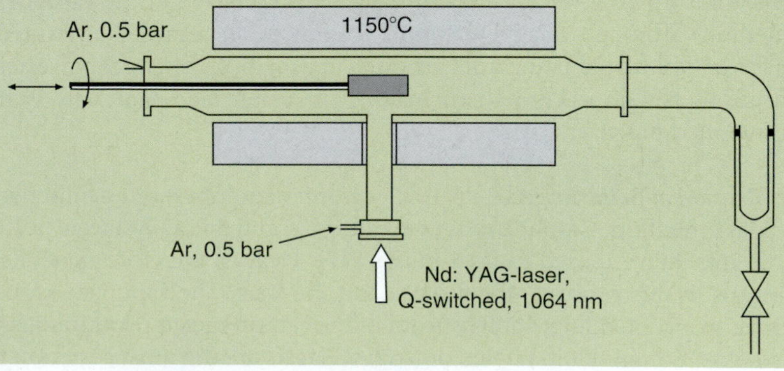

furnace (HFSF) [18]. Temperatures in the range of 3000–4000 K were attained. NREL's high-flux furnace has 25 hexagonal mirrors to concentrate solar radiation that provide flux at 2500 suns or, with adjustments, 20,000 suns—quite impressive.

Plasma Methods. We place ion milling, RF sputtering, plasma cleaning, and reactive ion etching into the category of high-energy methods. Plasma (from the Greek *plasma* or *plassein,* "to mold, to spread") is an ionized gas that is considered to be a distinct phase of matter. Plasmas contain ions and electrons and exist best in a vacuum environment for obvious reasons. Plasmas are electrically conductive and are strongly influenced by electric and magnetic fields. A simple reactive ion etching system is shown in **Figure 4.4**.

FIG. 4.4 *Reactive ion etching (RIE) is an effective means of subtracting material from a substrate—hence, a top-down method. Molecules (usually oxygen, sulfur hexafluoride, fluorine, or other reactive species) are ionized to form chemically reactive plasmas by the action of an applied electromagnetic field (parallel plate configuration) under low-pressure conditions. The apparatus consists of a cylindrical chamber kept under a few millitorr vacuum conditions. Inductively coupled plasma (ICP) produced by RF magnetic fields is another mode of creating RIE plasmas. Combinations of parallel plate and ICP also exist. Since the trajectory of ions produced in RIE is mostly normal to the plane of a substrate, the process is capable of anisotropic etching—as opposed to chemical etching, which tends to act in an isotropic fashion.*

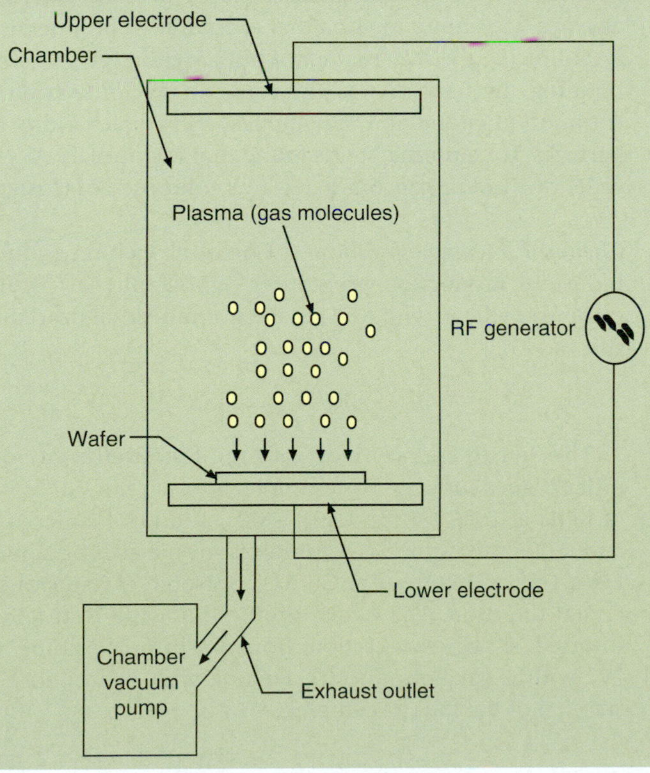

RF sputtering is a physical (as opposed to chemical) vapor deposition (PVD) method. Atoms from a solid target source (hence the top-down designation) are ejected via the process of momentum exchange into the plasma by the action of high-energy ions, usually originating from argon. The ejecta are then deposited on a surface of a sample material to provide a coating. A radio frequency alternating current is commonly used to generate the plasma and a bias voltage is applied to the target to promote acceleration of ions.

Ion milling, another PVD process, is similar to RF sputtering except that no coating is formed. In actuality, the opposite is true. Material is removed to promote thinning or shaping of a sample material (e.g., formation of nanofacets). Thin films with dimensions on the order of a few nanometers for the purposes of TEM preparation are formed after exposure to ion mill plasmas. Reactive ion etching is a chemical process in which a reactive chemical species is added to the plasma mixture. Oxygen, fluorine derivatives, or etchant species that are known to react with targeted substrate materials are commonly used in reactive ion etching (RIE) procedures.

4.1.4 Chemical Fabrication Methods

Combustion of Bulk Materials. Combustion is a top-down chemical method that is capable of producing nanomaterials. However, impurities in bulk carbon materials such as coal and oil contain contaminants that contribute to the formation of fly ash and acid aerosols. Polyaromatic hydrocarbon clusters (PAH) can be produced under incomplete combustion conditions. Pure hydrocarbons produce CO_2 and H_2O under efficient combustion conditions. Combustion also is a bottom-up method that is capable of producing nanomaterials.

Following the combustion of bulk Mg to MgO, a cluster-based nanoparticle bonding mechanism was the cause of agglomeration. This is apparently a common phenomenon that applies with equal validity to titania and alumina particles. For alumina, it was found that the primary Al_2O_3 aggregate was on the order of 1 μm in size, but that it was composed of clusters 10 nm in size [50].

Chemical Etching of Silicon. Chemical etching is important in numerous industrial production procedures, lithography in particular. The anisotropic etching of silicon with KOH is a major industrial procedure. The reaction yields silicates [51]:

$$Si_{(s)} + 4KOH_{(aq)} \rightarrow [SiO_4]^{4-} + 4K^+_{(aq)} + 2H_{2(g)} \tag{4.3}$$

The Si(110) surface undergoes the fastest etch rate of all the primary low-index planes surfaces. For example, the etch rates of Si in a 30% w–w solution of KOH at 70°C for the (110), (100), and (111) surfaces are equal to 1.5, 0.79, and 0.005 $\mu m \cdot min^{-1}$. A common isotropic etching solution used for silicon is HNA ($HF + HNO_3 + CH_3CO_2H$). Isotropic etchants operate independently of crystal direction. The trench profile following isotropic etching looks like an inverted "C" by cross-section; from anisotropic etching, the trench looks like a "V" with a flat bottom [52]. Etching with hydrofluoric acid is driven by the stability of the $[SiF_6]^{2-}$ complex:

$$Si_{(s)} + 6HF_{(aq)} \rightarrow [SiF_6]^{2-} + 2H^+_{(aq)} + 2H_{2(g)} \tag{4.4}$$

As a result of lithographic procedures and subsequent top-down chemical etching, nano- to micron-sized features can be formed on the surface of silicon wafers.

Chemical–Mechanical Polishing. This method is a combination of a chemical etching and a mechanical attrition method. The process of polishing jade with a corundum-based abrasive has been traced back to Neolithic farmers in ancient China 6000 years ago [53]. Grinding is the planar removal of material from a target surface by a fixed abrasive. In polishing, the abrasive is allowed to roll. The surface roughness, determined by profilometers or AFM, is shown to be a few nanometers.

Chemical–mechanical polishing combines the mechanical grinding characteristics of abrasives with the chemical action of an etchant. Pressure is applied on the abrasive and hence on the surface through a conformal pad. This allows for free movement of the abrasive under the pad. The method is important to the lithography industry, where depth of focus (DoF) is ever shrinking with smaller wavelength sources and larger numerical apertures (N.A.). The smoother the surface of a Si wafer becomes, the better is the accommodation of shrinking DoF.

Anodizing and Electropolishing. These two techniques are integrally related and differ only with regard to purpose and conditions. Anodizing is a process that creates an insulating porous oxide layer on a conductive metal anode, usually aluminum, in an electrolytic solution, usually a dilute polyprotic acid. By providing hexagonally packed pore channels that are simple to fabricate and the ability to manipulate pore diameter and length during and after anodizing, the porous anodic film offers a perfect template for nanoscale material synthesis. Anodizing conditions consist of an electrolytic bath made of a polyprotic acid (H_2SO_4, H_3PO_4, $H_2C_2O_4$, or H_2CrO_4) at 0°C with applied voltage of 2–100 V dc. The formation of nanoscale pores with diameters ranging from a few nanometers to several hundred nanometers is the major product of anodizing. The chemical reactions in anodizing are

Anodic reaction $$2Al^{o}_{(s)} \rightarrow 2Al^{3+} + 6e^{-} \quad (4.5)$$

Oxide–electrolyte interface $$2Al^{3+} + 3H_2O \rightarrow Al_2O_{3(s)} + 6H^{+} \quad (4.6)$$

Cathodic reaction $$6H^{+} + 6e^{-} \rightarrow 3H_{2(g)} \quad (4.7)$$

Overall oxide formation reaction: $$2Al^{o}_{(s)} + 3H_2O \rightarrow Al_2O_{3(s)} + 3H_{2(g)} \quad (4.8)$$

Anodizing, however, is a mixed fabrication method. Technically, it contains components that can be classified as top down or bottom up. The top-down component is the electrochemically assisted dissolution of bulk aluminum to form Al^{3+} cations. During this process, nanostructured scallops are formed in the surface of the aluminum metal. Pore diameter is directly proportional to the applied anodic dc potential ($d_{pore} \propto 1.4$ V) and is controlled by the diameter of the scallops on the metal surface. A schematic illustration of an anodic film is shown in **Figure 4.5**.

FIG. 4.5 *Porous alumina membranes formed by an anodic process can be considered to be the ultimate template material. They are insulating, optically transparent, chemically inert (near neutral pH), and thermally stable and possess reasonable mechanical properties. Once again, a relatively "low-tech" method is capable of fabricating nanomaterials (the pore channels).*

Electropolishing involves the removal of metal to form a smooth surface without forming a thick oxide layer. The conditions for electropolishing are rather severe compared to anodizing: elevated temperature (70–90°C), elevated level of current (10–20 A), and concentrated acid or base solutions. Electropolishing often precedes anodizing to prepare a smooth surface.

Hydrolysis Reactions. These reactions can affect inorganic, organic, and biological materials. Hydrolysis occurs by the action of water to disrupt a bond. The bond can be as strong as a covalent bond, ionic bond, or any kind of intermolecular attraction. For example, dissolution of proteins from the top down by acid-catalyzed hydrolytic mechanisms is a common means to regenerate the constituent amino acids. The degree of hydrolysis determines the size of the final product.

4.1.5 Lithographic Methods

A brief history of lithography was presented in chapter 1. Lithographic methods are the most widely utilized industrial process in the high-technology sector. The computer industry, for example, depends heavily on lithography. Integrated circuits, microelectromechanical machines (MEMS), and numerous other applications require lithography during some phase of their manufacture. However, challenges facing lithography today are numerous as well. Fabrication of increasingly smaller features requires sources with smaller wavelength. With increasingly smaller wavelength (e.g., electron beams and x-rays), the resolving power of the procedure is enhanced but the substrate sustains more damage. Fabrication of increasingly smaller features also requires increasingly more expensive equipment. Wavelength-based lithographic techniques, although well established, are rather costly to operate.

Modern optical lithographic techniques utilize radiation sources with wavelength from a few to 300 or 400 nm. Nano-imprint and nanosphere lithography offer cost-effective facilitative alternatives to the high-vacuum, high-energy, high-maintenance processes. Once a few fundamental technical issues in these nanotechniques become better resolved, expect wavelength-based lithography fabrication to start giving way to these nanorevolutionary procedures. With the advent of nanosphere and nano-imprint lithography, both extremely simple methods capable of high resolution, the trend in operation costs may be reversed in the near future.

In general, the underlying operation of lithographic techniques has not changed much since the time of the inventor of the technique, Bavarian author Alois Senefelder, in 1796. Photolithography follows the general procedure of pattern transfer established by Senefelder but employs radiation or particle projection onto a resist material instead of writing on a limestone substrate (**Fig. 4.6**):

Deposition of thin layer on substrate → deposition of photoresist material → *exposure* via mask (the master) by energy source → *development* by etch (positive or negative replica) of excess material → *stripping* of all resist → chemical modification (additive or subtractive)

There are numerous energy sources employed in lithographic processes—visible to ultraviolet radiation and x-rays for photolithography. Electron and ion beams have also been applied in lithographic procedures. Top-down nanolithographic sources consist of photons (UV, DUV, EUV, and x-rays), particle beams (electrons and ions), physical contact printing (nano-imprint methods), and edge-based techniques (shadow evaporation). Bottom-up nanolithographic procedures like dip-pen lithography and self-assembly (surfactant systems and block copolymers) will be discussed in a later section.

There are three primary considerations for any lithographic process: resolution, registration, and throughput. Resolution, first discussed in chapter 3, is defined as the best attainable physical scale of a feature: the smaller the better. Registration

Fig. 4.6 *Lithography is the workhorse of the computer chip industry. It is the most common top-down manufacturing process and it is one nanomanufacturing technique that is widespread. A target material is first applied to the surface of a silicon substrate. Polymeric resist layer is then applied by spin coating. An energy beam, usually in the visible to ultraviolet wavelength range, is shined through a mask that contains a predetermined pattern. Regions exposed to the EM radiation are sensitized (positive resist) or protected (negative) to the subsequent etch step. Following etching, the resist is removed, transferring the pattern inscribed by the mask to the target material. Lithography is a rather expensive process that requires clean room conditions, high-vacuum conditions, and otherwise expensive equipment.*

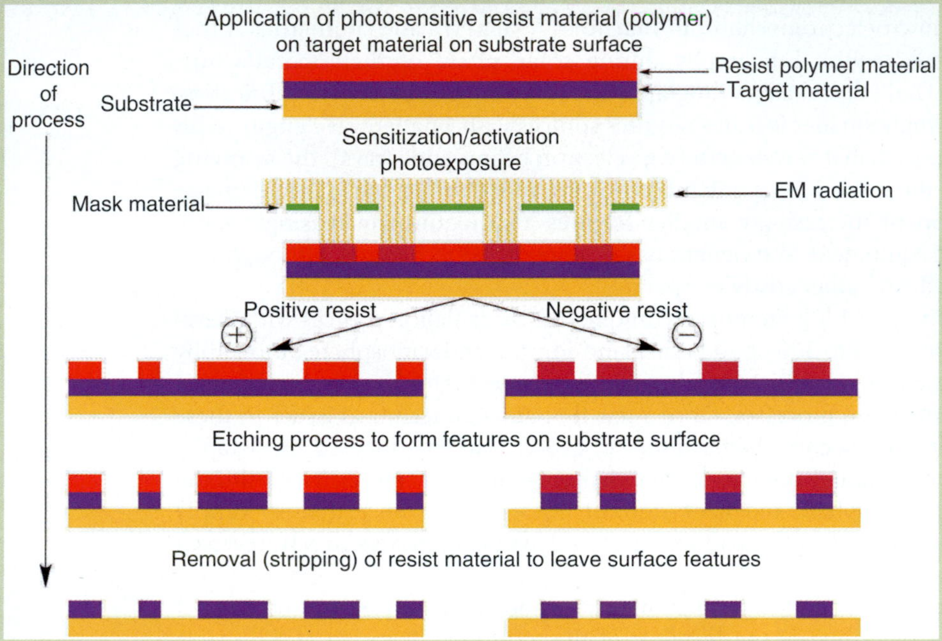

is the process of aligning one layer to another to form an integrated structure. Throughput is a gauge of the balance between cost effectiveness and the rate of production.

Optical Photolithography. Optical lithography employs visible and ultraviolet radiation to transfer a pattern onto a receptive substrate. Ultraviolet radiation (deep ultraviolet lithography, DUV) is the most common kind in use today. Three general methods are used to expose wafers:

1. Contact printing, in which the mask lies on top of the resist (e.g., there is no wafer-mask gap), requires no magnification but resolution is limited (~500 nm). The mask degrades in this configuration resulting in loss of planarity.
2. Proximity printing places the resist in close proximity to the mask. There is no magnification with this configuration and resolution is even lower (~1 μm). Diffraction effects limit the accuracy of the pattern transfer process [54].

3. Projection printing is a widely adopted technique. An image is projected through a mask and reduced by a factor of four to ten times on the resist. Resolution is much better (~70 nm), but equipment is costly and accuracy is limited by diffraction [54].

A computer-generated pattern on a mask (from optical or electron beam generators) is transferred to a chromium surface (~100 nm thick) on fused silica [55]. The mask is then positioned over a substrate, usually silicon, silicon oxide, or a semiconductor material. The substrate is prepped beforehand with a thin layer of oxide, nitride, or other functional material, and then a photoresist material is applied by spin-coating—a photoresist material that is sensitive to the type of radiation used in the lithographic procedure. In optical lithography, the photoresist is illuminated through a mask and is rendered soluble (positive resist) or insoluble (negative resist) during the subsequent developer step. The exposed resist (positive) or the unexposed resist (negative) is removed by etching. For example, in a negative scheme, the exposed resist polymer becomes cross-linked after exposure to the radiation. Cross-linking implies that the resist material is more difficult to dissolve than areas that were unexposed. Following development, an additive process deposits material onto or into the etched areas. In subtractive processes, material may be removed by ion milling through the developed areas. Following these steps, the remnants of the resist are removed.

Resolution in projection lithography is diffraction limited but has improved over the years since the days of the first integrated circuits. Line widths of the late 1960s were on the order of 5 μm [56]. In 1997, this was reduced to 350 nm. Today, sub-100-nm line widths are commonly achieved. Some of the equations present below will look familiar.

For contact style printing, radiation interacts with the sample as a square wave with limited or no diffraction. The near field (or Fresnel diffraction limit), appropriate for proximity printing, and resolution are given by

$$W = k\sqrt{\lambda d_g} \tag{4.9}$$

where d_g is the mask-to-wafer distance (gap), λ is the wavelength of the impinging radiation, and k is a constant that is close in value to 1 and depends on resist material and other technological parameters associated with the process. Fresnel diffraction occurs when

$$\lambda < d_g < \frac{W^2}{\lambda} \tag{4.10}$$

and the minimal resolvable feature is

$$W_{min} \approx \sqrt{\lambda d_g} \tag{4.11}$$

For the projection style of lithography (the most commonly applied form), the optical condition is called *far field* and the mask is called a *mask in the far field*. The optical description of far-field lithography is similar to other types of projection methods, whether optical or electronic. The minimal resolvable feature in a projection lithographic system is

$$W_{min} = k_1\left(\frac{\lambda}{\text{N.A.}}\right) \tag{4.12}$$

where λ is the wavelength of the radiation used for exposure and N.A. is the numerical aperture of the optical lithographic instrument (usually equal to ~0.5). The factor k_1 is a constant for a specific lithographic procedure that depends on the index of refraction and thickness of the photoresist material (0.4–0.8, a quality descriptor). In general, the line width is approximately equal to the wavelength of the incident light.

The resolution limit for optical lithography is given by the following equations:

$$\text{N.A.} = n \sin \theta \tag{4.13}$$

where N.A. is the numerical aperture, n is the refractive index of the medium (if vacuum, $n = 1$), and θ is the half-angle of the cone of light that can enter or exit the lens. Does this look familiar? The numerical aperture is a function of the distance between the lens and the sample and is an indication of the resolving power of the system. The larger the numerical aperture is, the higher is the resolution capability of the instrument. The following equation should look familiar as well; it also applies to lithography:

$$\text{R} = \frac{1.22f\lambda}{d} = \frac{1.22f\lambda}{n(2f\sin\theta)} = \frac{0.61\lambda}{n\sin\theta} = \frac{0.61\lambda}{\text{N.A.}} \tag{4.14}$$

Depth of focus (DoF) (like *depth of field*) becomes a concern as resolution is increased in shorter wavelength tools. DoF is the distance from the objective lens that yields a focused image and it gets worse (smaller) as N.A. becomes larger:

$$DoF = k_2\frac{\lambda}{(\text{N.A.})^2} \tag{4.15}$$

where k_2 is a constant associated with the photolithographic system and is traditionally equal to 1.

Contact, proximity, or projection modes are commonly used photolithography techniques. Contact type of photolithography (or shadow mode) is the case where the mask is right on top of the resists. Resolution is calculated from

$$2b_{\text{min}} = 3\sqrt{\frac{\lambda d}{2}} \tag{4.16}$$

where $2b$ is the grating period of a mask with equally spaced lines and d is the thickness of the resist material. In the proximity method, a gap exists between the mask and the photoresist and its resolution is found from

$$2b_{\text{min}} \approx 3\sqrt{\lambda d_g} \tag{4.17}$$

where d_g is the distance between the mask and the resist.

Particle Beam Lithography (IPL). Because particle beams do not undergo diffraction and scattering is minimal, higher resolution can be achieved with IPL than optical, x-ray, or electron beam methods. Resist materials demonstrate greater sensitivity to ions than to electrons. Ion lithography is mostly used to repair masks in optical and x-ray lithographic procedures.

Extreme Ultraviolet Lithography (EUVL). This technique applies radiation that is as short as 11–14 nm, significantly lower than those used in DUVL [57,58]. Features smaller than 50 nm have been achieved, but theoretically much smaller features are possible (e.g., <25 nm). EUVL is based on multilayer-coated optics that are able to reflect the intense UV radiation. Therefore, reflective coatings are applied on the optics and masks of an EUVL system. EUVL must also be performed in a high vacuum due to absorption of EUV by most forms of matter.

X-ray Lithography (XRL). This method utilizes x-rays produced by a synchrotron source. Electrons are converted into x-rays in a synchrotron. XRL is an extremely expensive method, ca. $25 million for acquisition, setup and operation.

Advantages of XRL include: (1) the wavelength (4 nm) is well matched for nanoscale work, (2) scattering is limited by all materials in contact with x-rays, and (3) resist materials and wavelength can be matched to maximize absorption. XRL is a method that is scaled to the nanometer, minimizing scattering and maximizing resist absorption and image contrast (e.g., absorption without spurious scattering) [59,60]. Disadvantages of XRL include: (1) distortion of absorber (tungsten mask) material due to x-ray-induced internal stresses that result in warping, (2) difficulty in focusing x-rays by conventional lenses requires masks with ultrafine features—a difficult and time-consuming process, and (3) XRL is a very expensive process.

Electron Beam Lithography (EBL). This technique enables patterns to be generated by an electron beam without the use of a mask [61]. Polymethylmethacrylate (PMMA) is sensitive to electron beam exposure. EBL is used to generate nanopatterns on PMMA film supported by a silica substrate. An electron beam at 10 kV with a beam current of 340 pA was used to construct lines on the surface of PMMA [61]. Line widths in the region of 50 nm were obtained [61].

Nano-imprint Lithography (NIL). Nano-imprint lithography, considered to be a *soft lithographic* technique, is an economical method by which nanoscale resolution and high throughput (all desired by industry) are possible. NIL is a clearly defined top-down fabrication method. A stamp pressed into a soft film is responsible for creating a negative replica of the pattern. The film is hardened before the stamp is removed. Structures with resolvable features down to 5 nm have been produced in this way. UV-NIL uses a photopolymerizable thermoresin with a UV-transparent stamp. The liquid resin is easily stamped and then hardened with the application of UV-light. A generalized NIL scheme is depicted in **Figure 4.7**.

Stephen Chou of Princeton University is credited with discovery of NIL in 1996 [22,23]. The basic principle behind NIL is compression molding to create a thickness contrast pattern on a thin resist film on a metal substrate. Anisotropic etching is accomplished for pattern enhancement. The first patterns contained

Fig. 4.7 *Nano-imprint lithography (NIL) was invented in the mid 1990s by Stephen Chou of Princeton University. NIL is a method that is capable of generating features with nanoscale resolution with high throughput. A stamp (top), made by an electron beam lithographic method, is pressed into a soft polymeric material at its glass transition temperature. The film is hardened before the stamp is removed. Anisotropic etching by RIE process removes excess polymer. Metallic pillars are fabricated in the nanofeatures formed by NIL.*

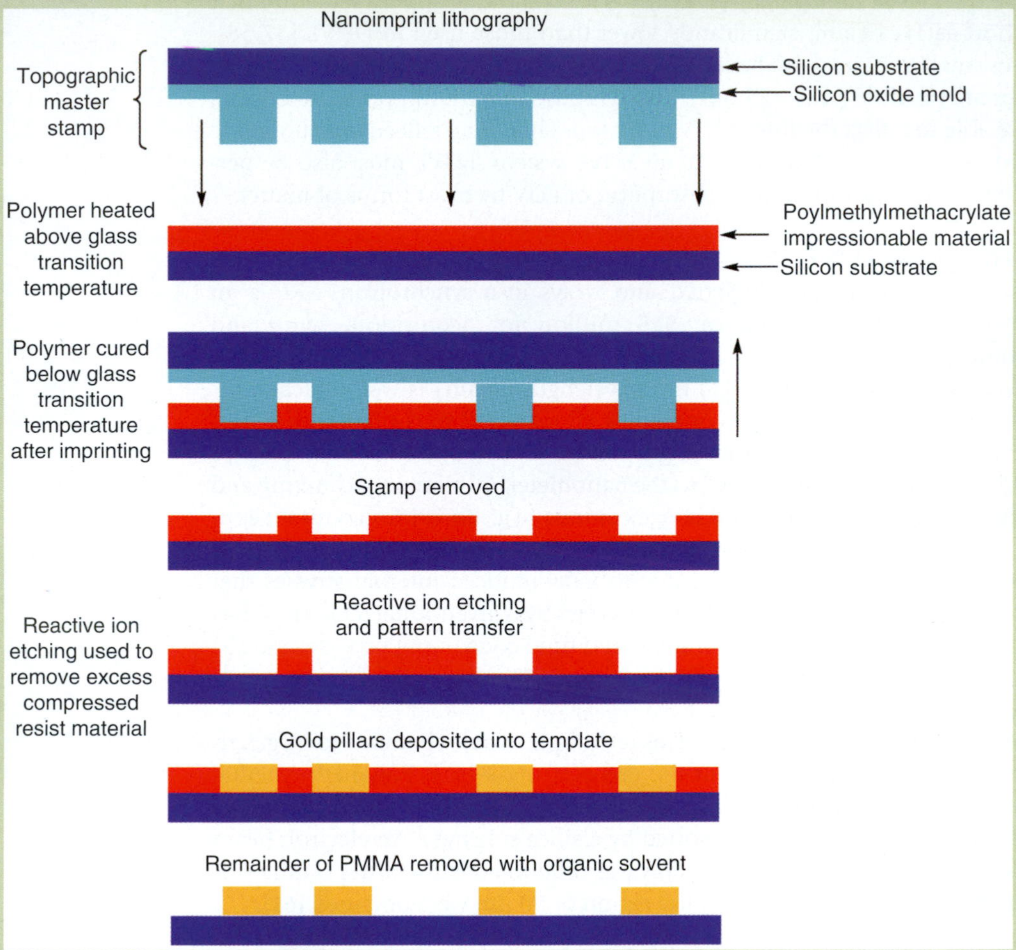

25-nm features spaced 70 nm apart. It is easy to understand why NIL is preferred over wavelength-dependent lithographic techniques: (1) NIL is able to achieve smaller features, (2) NIL takes less time, and (3) NIL is an inexpensive process that does not require ultrahigh vacuum and expensive radiation or electron beam equipment. The biggest problem with NIL is defectivity; although recent methods have driven the defect density to $<0.1\ cm^{-2}$, a practical industrial level of $<0.01 \cdot cm^{-2}$ is desired.

Nanosphere Lithography. R. P. Van Duyne et al. of Northwestern University in 1994 developed the nanosphere lithographic (NSL) process [24,62]. NSL is a straightforward, versatile, high-throughput process that offers nanoscale resolution. Compared to other methods, NSL is quite fast and economical. The NSL process is able to create ordered arrays of differing configurations. In one case, materials are deposited through the open spaces between spheres to form an array (**Fig. 4.8**). In another application, the size of the spheres is

FIG. 4.8

Nanosphere lithography (NSL), like NIL, is another ingenious low-cost, high-throughput method to form nanomaterials and nanoparticle arrays. One simple method utilizes latex spheres that are close packed in a two-dimensional array. Deposition of metal between spheres, the interstitial spaces, forms star-shaped patterns of tetrahedrally formed nano-structured materials. RIE etching, depending on the type of active molecule, is able to reduce the size of the spheres (thereby creating wider gaps among the spherical matrix elements, or etch) in an anisotropic manner, the substrate under the spaces to form pore channels.

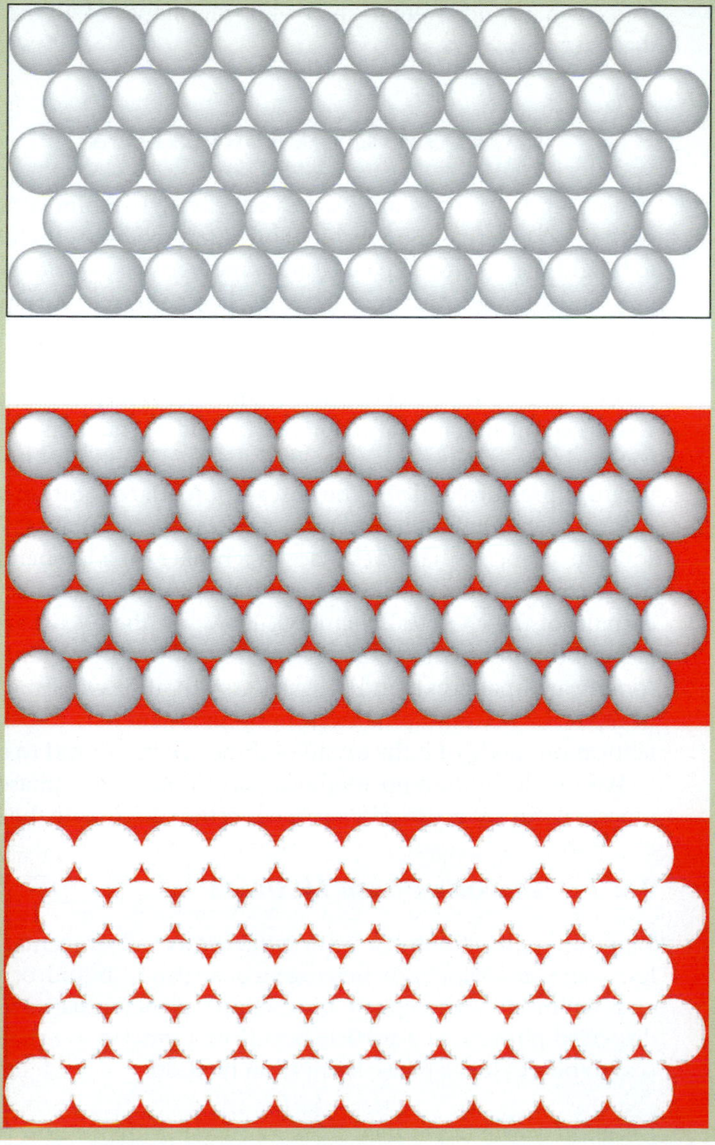

reduced by standard RIE with oxygen [63]. Columnar arrays several microns in height are then fabricated by application of a deep-RIE (Bosch) top-down process.

The NSL is a mixed bag. It is a bottom-up template method during the deposition of material at the base of the interstitial regions of the spheres, but it is a top-down method during the deep-RIE process.

Applications of NSL include its use in manufacturing size-tunable noble metal substrates in the range of 20–1000 nm. The optical response of NSL-formed Ag nanoparticles to their local environment was probed by localized surface plasmon resonance spectroscopy (LSPR) [62]. Results showed that zeptomole-level detection of adsorbed analytes was possible by LSPR spectroscopy [62]. Large-scale fabrication of protein nanoarrays based on NSL was demonstrated by Y. Cai et al. in 2005 [64]. Based on nanospheres with 300-nm diameter, protein islands were formed with ring shapes of 50-nm width and 118-nm diameter [64].

4.2 Bottom-Up Fabrication

Bottom-up fabrication approaches selectively combine atoms or molecules to form nanomaterials. Bottom-up fabrication methods, therefore, are considered to be additive. Bottom-up fabrication methods reside within the realm of chemistry and biology. Nature, of course, has perfected bottom-up fabrication of nanomaterials.

Advantages of bottom-up methods are numerous. Self-assembly processes, for example, occur under thermodynamic control conditions. Because such processes exploit much weaker intermolecular interactions, as opposed to strong covalent bonds, nanomaterials are fabricated under milder conditions of temperature, pressure, and *p*H. The upscale potential of bottom-up methods is enormous. As with any other chemical process, it is "relatively" straightforward to scale up a process that takes place in a beaker on a lab bench (e.g., the domain of the chemist) to a batch production process in a manufacturing line (e.g., the domain of the chemical engineer). However, there exist significant challenges facing bottom-up methods. Overall robustness, long-range order (related to complicated patterns), and directed growth all leave something to be desired. In order for bottom-up fabrication of nanomaterials to become the dominant fabrication mode of industry, all of these concerns need to be overcome.

We divide bottom-up methods according to the phase within which the process occurs. We also add a special section discussing the solid state.

4.2.1 Gaseous-Phase Methods

Vapor phase reactions can be homogeneous (all reactants, products, and catalysts exist as a vapor) or heterogeneous (vapor–liquid or vapor–solid phases exist within the same sphere of reaction). If there exists a vapor (or any highly dispersed phase, e.g., a particle beam) in a process, we shall consider that process to be a gaseous-phase fabrication method.

Chemical Vapor Deposition. Chemical vapor deposition (CVD) is one of the most effective procedures used to produce advanced materials. CVD is the best way to form carbon nanotubes because it is less energy intensive and more control is exerted over products. SiO_2, SiC, Si_3N_4, W, and other materials are routinely deposited on surfaces via CVD methods. In semiconductor industry practice, wafers are exposed to volatile precursor materials that react or decompose on the surface to form thin films. There are many kinds of CVD.

Chemical CVD (CCVD) is used in the fabrication of carbon nanotubes—single-walled and multi-walled operating temperatures range from the low 400°C to produce carbon fibers and multiwalled carbon nanotubes to temperatures >1000°C. The decomposition of methane, ethane, ethylene, propane, propylene, or acetylene or the disproportionation of carbon monoxide—all over catalysts—is an example of some of the carbon source materials (usually in gas form) used in CVD techniques. The decomposition of methane in the presence of catalysts (usually Fe, Ni, or Co) at temperatures of 700°C at atmospheric pressure yields SWNTs.

$$CH_{4(g)} \rightarrow SWNT + H_{2(g)} \tag{4.18}$$

Polysilicon thin films are formed by the decomposition of silanes in a low-pressure CVD (or liquid-phase CVD, LPCVD) chamber at ca. 650°C. If other gases, such as phosphine or arsine, are present in the stream, the silicon can be doped in situ. Silicon dioxide layers are formed by the gas-phase decomposition of tetraethylorthosilicate (TEOS). Since TEOS boils at ca. 168°C, the CVD process is conducted between the boiling point of TEOS and 750°C. TEOS breaks down into solid silica and gaseous diethylether:

$$Si(OC_2H_5)_{4(g)} \rightarrow SiO_{2(s)} + O(C_2H_5)_{2(g)} \tag{4.19}$$

Plasma-enhanced CVD (PECVD) is another bottom-up CVD fabrication method to produce thin films. The plasma is created by radio frequency or direct current discharge between electrodes [65]. Silicon dioxide, from silanes + O_2 or TEOS + O_2, can be formed with the PECVD technique at reasonably low pressure (~100 mtorr). Silicon nitride thin films are also deposited with plasma assistance. An example of a CVD apparatus is shown in **Figure 4.9.**

Metal oxide CVD (MOCVD) utilizes H_2 as a carrier gas, Group-III metal–organic compounds, and Group-V hydrides to make nanometer scale thin films or nanoparticles. Temperatures ranging from 500 to 1500°C at 15–750 torr pressure are representative conditions under which MOCVD is performed.

Atomic Layer Deposition. Atomic layer deposition (ALD; a.k.a. atomic layer epitaxy, ALE) was introduced in 1974 by Tuomo Suntola of Finland with the intent of improving the quality of ZnS films used in electroluminescence displays. After a decade of development, high-quality phosphor layers and dielectric layers were produced, and the process has since acquired major importance to industrial manufacturing. ALD is the process of fabricating uniform conformal films through the cyclic deposition of self-terminating surface half-reactions that allows for thickness control at the level of the atomic layer [66]. ALD is a derivative of chemical vapor deposition (CVD), but one that differs from CVD in several notable ways [67]. The comparison is shown in **Table 4.12.**

ALD is a straightforward synthesis method that exploits specific chemical reactions with the intent of adding one molecular monolayer at a time. The process is

FIG. 4.9 *Chemical vapor deposition, especially in the case of carbon nanotubes, is yet another low-cost, "low-tech" method to form nanomaterials. A carbon source gas (usually methane, CO, acetylene, propylene, or ethylene) is introduced into a chamber (the quartz tube pictured) under reducing conditions. Upon contact with Co, Fe, or Ni catalyst particles, the gases decompose into C and H atoms. Nanotubes nucleate on the catalyst particle and grows out from the particle by either the tip-growth or base-growth mechanism. Typical CVD conditions use 10% methane, 5% hydrogen, 85% argon carrier gas at 700°C, and atmospheric pressure.*

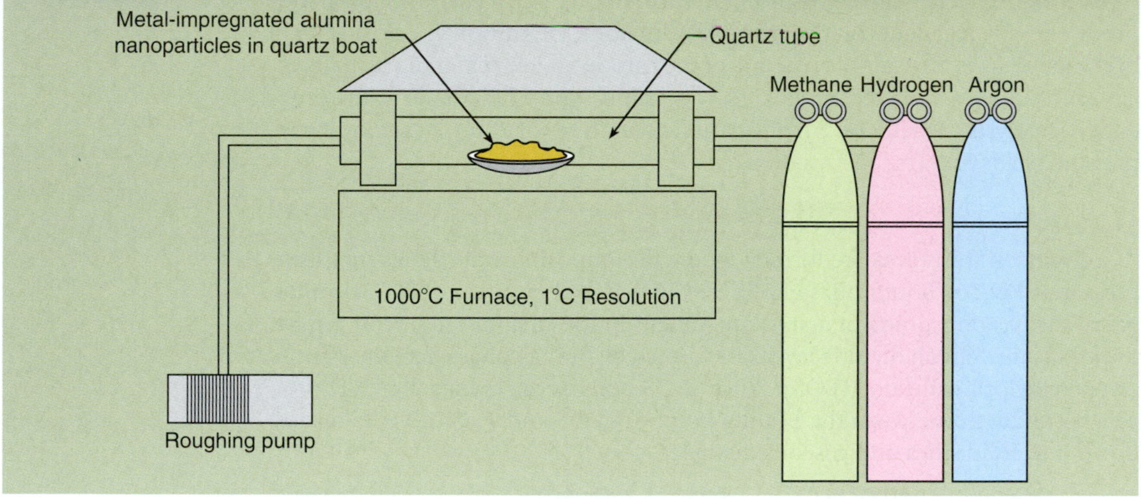

characterized by a binary reaction that is split into two half-reactions applied sequentially. ALD is characterized by the systematic use of self-terminating gas–solid reactions [68]. A self-terminating reaction depends on saturation of available surface sites and that precursors do not react with each other. The ALD process offers a powerful arsenal of properties that are specifically tailored for nanofabrication of thin films. First of all, conformal coatings can be applied to particulates or flat and curved surfaces of bulk materials. Secondly, atomic-scale control of thickness is possible by adding layers with stoichiometric scaling based on a chemisorption–saturation process. The process is broken down into the following general steps:

Surface activation → injection of A → purge → injection of B → purge → injection of A → purge → injection of B → purge → ⋯ → termination

Ultimately a film composed of a structure ABABABA … is formed. The thickness of the film can be estimated instantaneously by counting the steps in the process.

The process is depicted in **Figure 4.10.** In essence, the ADL sequential process alternates between chemisorption and saturation steps. Purging of the process follows each saturation step in the cycle. The ALD film growth process is referred to as *self-limiting* in that a stoichiometric process essentially terminates the reaction upon saturation. Excess reactants and products are purged from the chamber following each step.

Table 4.12 *ALD–CVD Comparison*

Parameter	Atomic layer deposition	Chemical vapor deposition
Precursor reactivity	Highly reactive Self-limiting at saturation	Less reactive Can be autocatalytic
Potential materials	Metals, semiconductors, insulators Wide range of materials	Metal oxides, semiconductors, and carbon compounds
Selectivity	Highly selective	Low selectivity
Surfaces	Layers conform according to surface topography of substrate Surfaces capable of activation (limitation of application)	Layers conform according to surface topography of substrate.
Decomposition at reaction temperature	Reactants and products do not decompose.	Reactants can decompose.
Time of process	Few seconds per cycle	Variable
Uniformity	Saturation mechanism ensures uniformity over a broad area.	Uniformity control by process parameters (partial pressure of reactants, flow, pressure, temperature)—more difficult to execute
Thickness	Controlled explicitly by number of reaction cycles Deposition rate: ~6 $nm \cdot min^{-1}$	Thickness control by process parameters—more difficult to execute
Conditions	Requires vacuum or inert atmosphere Lower temperatures (100–400°C but varies according to application) *P*, *T*, concentration, and gas flow distribution have little effect on the process.	Requires inert atmosphere and higher temperatures (>600°C) *P*, *T*, concentration, and gas flow distribution have significant effect on the process.
Upscale potential	Excellent	Good

The formation of alumina layers on a silicon surface will serve as an example of the ALD process. The first step in the process is the activation of a hydrogen-terminated silicon surface by exposure to water vapor:

$$:SiH + H_2O \rightarrow :SiOH + H_{2(g)} \quad (4.20)$$

After evacuation of the chamber, trimethylaluminum (TMA) is added. The reaction occurs between the lone pairs of the oxygen atom and the empty *p*-orbital of the aluminum atom [66]:

$$Al(CH_3)_{3(g)} + :SiOH_{(s)} \rightarrow :Si\text{-}O\text{-}Al(CH_3)_{2(s)} + CH_{4(g)} \quad (4.21)$$

Following purge of the reactants and products, water vapor is pulsed into the chamber. Water reacts vigorously and completely with the remaining methyl groups attached to the aluminum, replacing them with hydroxyl groups and releasing more methane. New *surface* hydroxyls are formed and aluminum forms bridge-oxygen bonds with its nearest neighbors:

$$:Si\text{-}O\text{-}Al(CH_3)_{2(s)} + 2H_2O_{(g)} \rightarrow :Si\text{-}O\text{-}Al(OH)_{2(s)} + 2CH_{4(g)} \quad (4.22)$$

Oxygen bridges are formed between aluminum elements of the two-dimensional structure. The overall chemical reaction to form layers consisting of aluminum oxide is

$$2Al(CH_3)_{3(g)} + 3H_2O_{(g)} \rightarrow Al_2O_{3(s)} + 6CH_{4(g)} \quad (4.23)$$

Fig. 4.10

Atomic layer deposition (ALD) technique is a straightforward method to manufacture, one monolayer at a time, a two-dimensional functional surface on a substrate. By the use of self-terminating gas–solid chemical reactions, high specificity is achieved. A self-terminating reaction depends on saturation of available surface sites and that precursors do not react with each other. Once saturation occurs over the surface layer, excess reactants are purged in preparation for the next step in the process. In the first step, a silicon surface is activated to generate hydroxyl groups. A highly reactive species, $Al(CH_3)_{3(g)}$, is introduced and one methyl group is readily displaced by a surface hydroxyl to form linkage with the surface. After purging, water vapor is added to initiate cross-linking and activation of the Al for the next round of $Al(CH_3)_{3(g)}$ application.

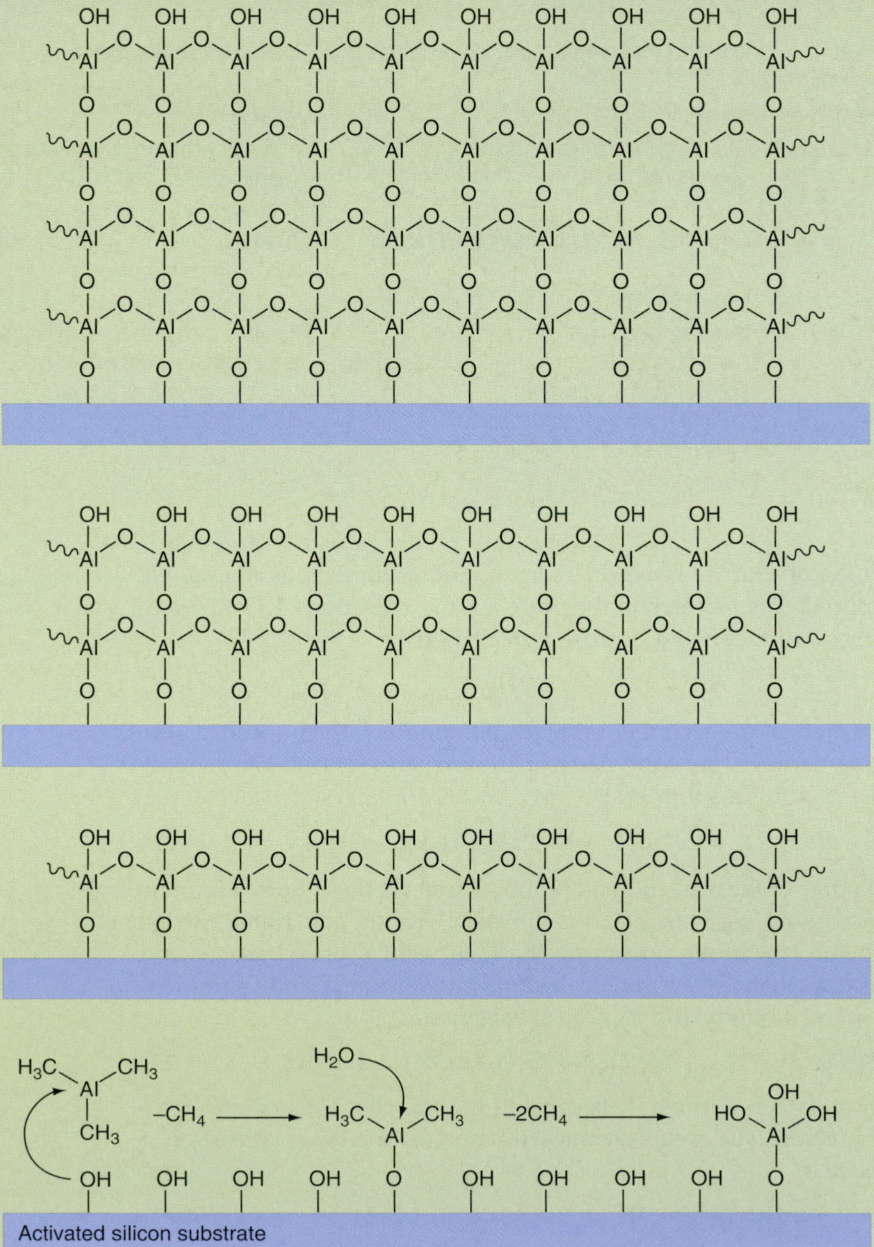

The mechanism of the chemisorption half-reactions was determined by cluster calculation hybrid density functional theory using TMA and water precursors [66]. Both the aluminum and water depositions were determined to be thermodynamically favorable (exothermic) and kinetically uninhibited [66]. STM and ab initio modeling studies of the three-dimensional structure of ultrathin aluminum oxide on NiAl(100) infer that Al is pyramidally and tetrahedrally coordinated [69].

Epitaxy. Epitaxy (from the Greek prefix *epi-*, meaning "upon, placed or rested upon" + *taxis,* based on *taktos,* indicating "to arrange") is the directed growth of a crystalline substance on the crystalline face of a substrate. In this way, the crystal orientation of the deposited material matches that of the substrate. The growth is accompanied by binding of the atoms or molecules to one another to form a two-dimensional crystal. There are two kinds of epitaxial growth: *homeoepitaxy,* in which the layer and the substrate are the same, and *heteroepitaxy,* in which the two materials are different. There are three classes of epitaxial processes: vapor phase, liquid phase, and molecular beam (MBE). We will discuss the MBE method only.

MBE was introduced in the 1970s to make films of high quality—films made of semiconductors, metals, or insulators. The operation principle of MBE is straightforward. Atoms, molecules, or clusters of extremely pure form are produced top down by heating an extremely pure source material (evaporation) in an effusion cell (**Fig. 4.11**). Six to ten effusion cells, each containing a different

Fig. 4.11 *Molecular beam epitaxy (MBE) is a single-crystal film growth technique [4]. MBE is the evaporation of one or more elemental or molecular species onto a heated target substrate material under ultrahigh vacuum [4]. In the figure, five atomic or molecular species are heated in effusion cells and directed towards a substrate. Shutters regulate the timing of release and the level of exposure of the MBE evaporated materials. A built-in RHEED system is able to monitor the development of the epitaxial film. An in situ mass spectrometer monitors vacuum conditions as well as the level of the evaporated species.*

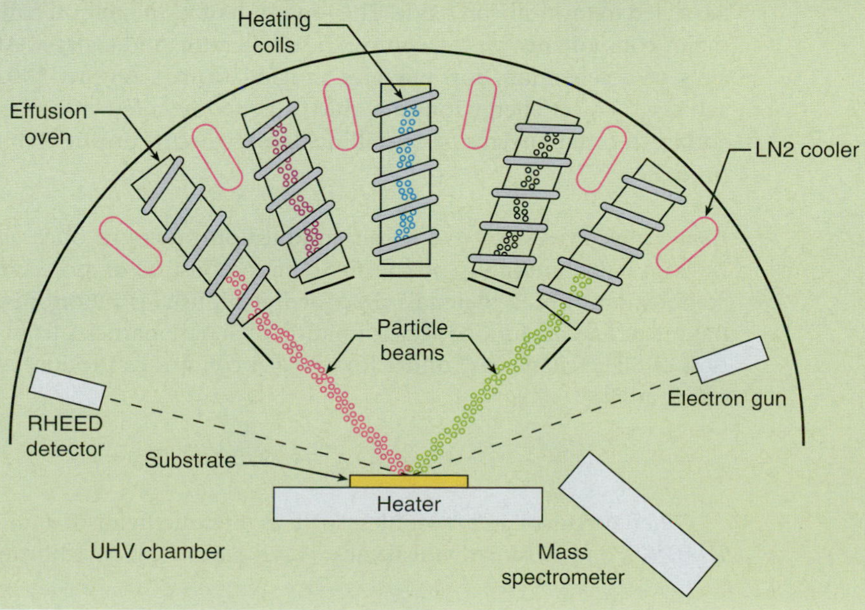

component, are focused onto the sample wafer. The crucibles containing source materials are made of pyrolytic boron nitride that can withstand temperatures of 1400°C; the shutters (that control the flux in the cell) are made of molybdenum or tantalum [70]. The chamber is usually baked at 200°C to remove contaminants for 24 h before use. Cryogenic screens surround the effusion cells to minimize spurious fluxes of atoms or molecules from the walls of the chamber. A RHEED in situ detector is used to monitor the status of the forming film.

The MBE process takes place in an ultrahigh vacuum (UHV), on the order of 10^{-10} torr, to ensure a contamination-free environment. The atomic flux (i.e., the "molecular beam") is directed towards the wafer surface where deposition occurs. The wafer is also heated to enhance diffusion along the surface. Films grow laterally on the surface (e.g., epitaxially). In order to create a film with minimal defects, the MBE process occurs at a slower rate than other deposition techniques like MOCVD. Typical MBE growth rate is on the order of $1\ \mu m \cdot h^{-1}$ [4].

One criterion for MBE operation is that the mean free path λ of atoms, molecules, or clusters be larger than the geometric dimension of the chamber. This prerequisite is fulfilled if a vacuum exists in the chamber. For example, at 1 torr, λ of N_2 is about 10^{-2} cm. At 10^{-11} torr, $\lambda > 10^{8}$ cm, thus exceeding the dimensions of any size chamber [70]. Another criterion for high-quality film growth is that the time it takes for surface-diffusion incorporation be less than the time required to deposit a monolayer [4]. If not, defects in the layer are bound to accumulate due to unincorporated atoms that become buried in the monolayer [4]. Doping of semiconductor thin films occurs with the timely opening of a shutter.

Ion Implantation. Ion implantation is simply a process by which ions of one material are inserted into a solid substrate. Applications of ion implantation are for surface finishing, semiconductor doping, silicon-on-insulator (SOI), or mesotaxy (crystal matching within the bulk). In SOI, an oxide layer is created within a silicon substrate. Oxygen is introduced by ion implantation and then annealed to form silicon oxide. The ion implantation method consists of three major components: an ion source, the accelerator, and a target. At the conclusion of acceleration, ion energies ranging from a few to 500 keV can be achieved. Lower energy ions are able to penetrate a few nanometers into the surface. A certain degree of crystalline damage occurs during ion implantation processes.

Combustion Processes. We revisit combustion processes once again but this time from the bottom-up perspective. The combustion of molecules, unlike the combustion of bulk materials described previously, is a bottom-up process if nanomaterials are formed. The formation of Si nanoparticles from the combustion of SiH_4 (silane gas) under low-oxygen conditions produces silicon oxide nanoparticles:

$$2SiH_{4(g)} + 2O_{2(g)} \rightarrow 2SiO_{2(s)} + 4H_2O_{(l)} \tag{4.24}$$

Other precursor combustibles such as hexamethyldisiloxane ($C_6H_{18}OSi_2$; HMDSO) and hexamethyldisilazine ($C_6H_{19}NSi_2$; HMDSA) burned in air and propane fuel are also capable of forming SiO_2. The SiO_2 nanoparticles are quenched and collected on an aluminum plate [71,72]. Particles ranging from 2.5 to 25 nm were produced [71,72]. Combustion is the dominant pathway for the production of particulate matter into the environment—in particular, carbonaceous nanoparticles that are derivatives of benzene hierarchical structures. Al_2O_3 and TiO_2 nanoparticles can also be formed by combustion of the appropriate metal precursor compounds.

Single- and multiwalled carbon nanotubes can be produced in flames under the proper conditions. Catalysts are entrained into a flame stream by thermal evaporation simultaneously with the introduction of a carbon source gas like CO or ethane. Processes like these have evolved into a major industrial market for MWNTs in nanocomposite materials.

Thermal Decomposition. Solid Si nanoparticles can also be formed by the thermal decomposition of silane gas in the absence of oxygen:

$$SiH_{4(g)} \rightarrow Si_{(s)} + 2H_{2(g)} \tag{4.25}$$

The formation of Fe nanoparticle catalysts in the gas phase by the decomposition of ferrocene is a fundamental ingredient of the HiPCO™ process to manufacture carbon nanotubes—two bottom-up methods in tandem. At temperatures <500°C and at higher temperatures, the cyclopentadienyl decomposes further:

$$Fe(C_5H_5)_2 \rightarrow Fe + 2C_5H_5 \rightarrow H_2 + CH_4 + C_5H_6 + \cdots \tag{4.26}$$

The ferrocene decomposition process is also exploited in a lithographic writing technique utilizing an STM probe tip.

4.2.2 Liquid-Phase Methods

Much of chemistry takes place in solvents (e.g., liquid-phase media). Liquids provide an environment that is able to support molecules, intermediates, metastable materials, and other forms of materials that would otherwise be unable to react, agglomerate, or simply to exist. Biological phenomena, for the most part, are based on a very important liquid: water, which is one of the most important, versatile, and ubiquitous solvents known to science and to the life process itself. There are many kinds of liquid-phase fabrication methods listed in **Table 4.8** and we will discuss selected examples that highlight this diversity.

Molecular Self-Assembly. Molecular self-assembly is one of the most powerful methods to form nanomaterials en masse. Molecular self-assembly is considered to be a *soft-matter* phenomenon. Soft matter abounds all around us. Biological matter is mostly soft, except for bone, shells, teeth, and other hard structures. Organic phenomena like soap bubbles, micelles, polymers, colloids, amphiphiles, and liquid crystals are considered to be soft matter. The chemistry of soft matter is based on a fundamental understanding of intermolecular interactions.

Molecular-based self-assembly is a bottom-up fabrication process in the absolute and most pure sense of the term. The concept of self-assembly, however, does not stop at the molecular level. If a scientist is clever enough to place the proper molecular moieties on the surface of a nanoparticle, it is very likely that the nanoparticles themselves become capable of further self-assembly. What materials can you think of that exhibit this kind of behavior? Proteins involved in the immune system certainly qualify as a suitable answer to this question. Nature operates on the principle of self-assembly, and we are embarking on a similar path; nano is truly revolutionary in that regard.

Intermolecular interactions form the basis of molecular self-assembly, and they are discussed in some detail in chapter 10. Thermodynamic control versus kinetic control is another concept that is introduced in chapters 10 and 11. Examples of molecular self-assembly are all around us. In the basest sense, water achieves its low energy state as a liquid by forming hydrogen bonds. Although a liquid, however, water has structure. When liquid oil and liquid water are mixed, a self-assembly process is automatically initiated without the need for a trigger; it happens spontaneously at mild temperature, pressure, and *p*H. Micelles are another example. In this case, molecules with a hydrocarbon *hydrophobic* end and an inorganic *hydrophilic* end automatically self-assemble into micelles when placed in water (**Fig. 4.12**). This is also an example of self-assembly.

Supramolecular Chemistry. Supramolecular chemistry is a relatively new field of chemistry. It is the chemistry of soft matter and is a bona fide bottom-up method to fabricate nanomaterials. Supramolecular chemistry is based on the *lock and key* hypothesis of Emil Fischer (**Fig. 4.13**).

Fabrication of supramolecular structures requires significant forethought and planning. Just to reiterate, these factors will be discussed more fully later, but we will mention a few salient aspects of supramolecular design and fabrication in this paragraph. To begin with, there exists a vast panoply of intermolecular interactions to choose from: van der Waals forces, hydrogen bonding, hydrophobic interactions, and various species of dipolar interactions, to name a few. Conventional "hard" bonding forms are also important to consider: covalent, ionic, metallic, coordination, etc. and cannot be neglected in molecular design protocol. Complementarity and "host–guest chemistry" are terms often applied in supramolecular chemistry jargon. They are essentially the supramolecular equivalents of the lock-and-key hypothesis, one of the most important and fundamental concepts of supramolecular chemistry. Examples of complementarity can be found anywhere in nature.

Supramolecular design begins with the covalent bond. Synthesis of precursors often requires the making and breaking of strong covalent bonds. This is called molecular chemistry—another bottom-up fabrication method. Once the precursors are made, after due consideration of intermolecular forces, they are allowed to react to form host–guest relationships. The fabrication of DNA, one of the most remarkable of all nanomaterials, is accomplished in this way; the backbone and the nucleotides are covalent bonded structures but the double-stranded DNA is held together with relatively weak intermolecular forces such as π–π interactions, hydrophobicity, and hydrogen bonds.

Sol–Gel Synthesis. Colloids are what make sols. Colloids, a dispersed phase of one substance, exist as discrete entities within a continuous phase, usually water. A *sol* is a colloidal suspension of solid particles within a liquid. A *gel*, on the

FIG. 4.12

Self-assembly is a fast expanding technique that is expected to make significant contributions to bottom-up manufacturing of nanomaterials. Various types of micelles are shown in the figure. Depending on solution conditions, micelles (the spherical structures depicted) or bilayers are commonly formed. Micelles are made of molecules called amphiphiles: single molecules that have both a polar and a nonpolar chemical group. In order for self-assembly techniques to become dominant, problems with self-assembly methods, including lack of long-range order and structural integrity, need to be overcome.

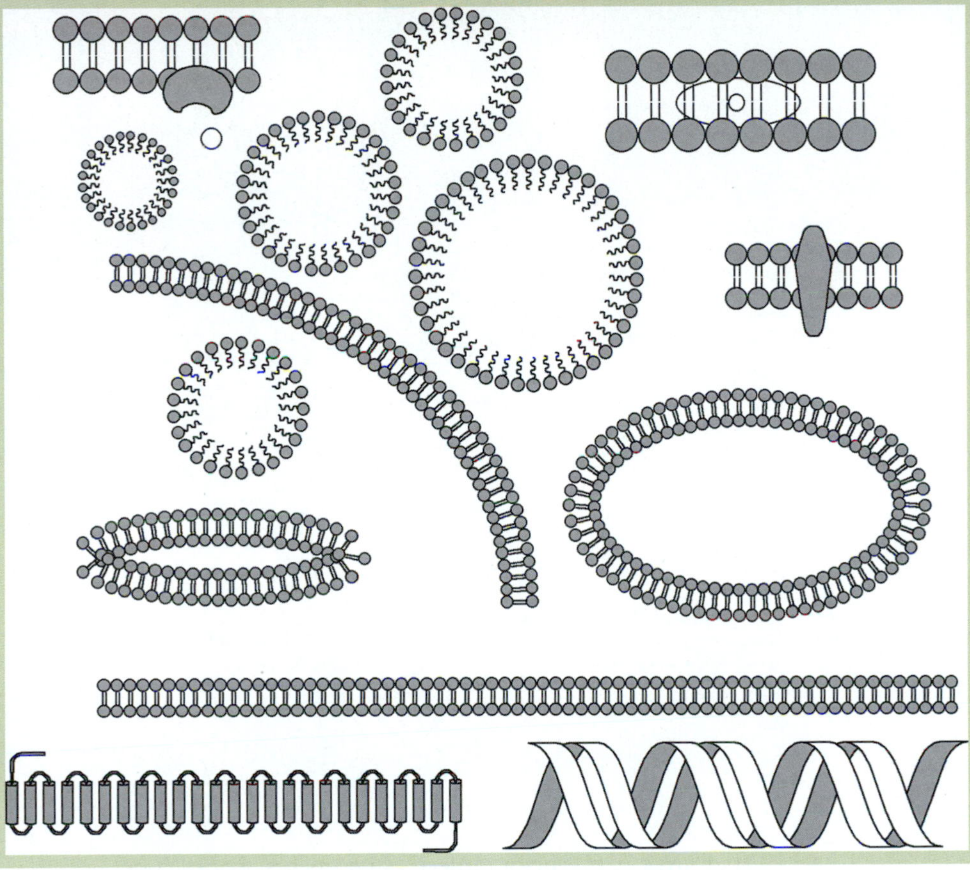

other hand, is a solid that contains liquid within its pore structure. For example, a colloid made from the bottom up begins with the nucleation of appropriate atoms or molecules in a supersaturated solution. When enough colloids are formed, condensation occurs to form three-dimensional structures. The physical conformation of the structure is dependent upon the size of the colloids and the chemical nature of their surface. Supercritical extraction of the liquid and subsequent baking form compounds called *aerogels*. Extraction of the liquid at nonsupercritical conditions forms compounds called *xerogels*. Both forms are built from the bottom up and are bulk materials with nanoscale constituents. There are two general kinds of sol–gel precursor substances: the inorganic salts and the metal alkoxides that are organic materials containing silicon or metal coordinating centers.

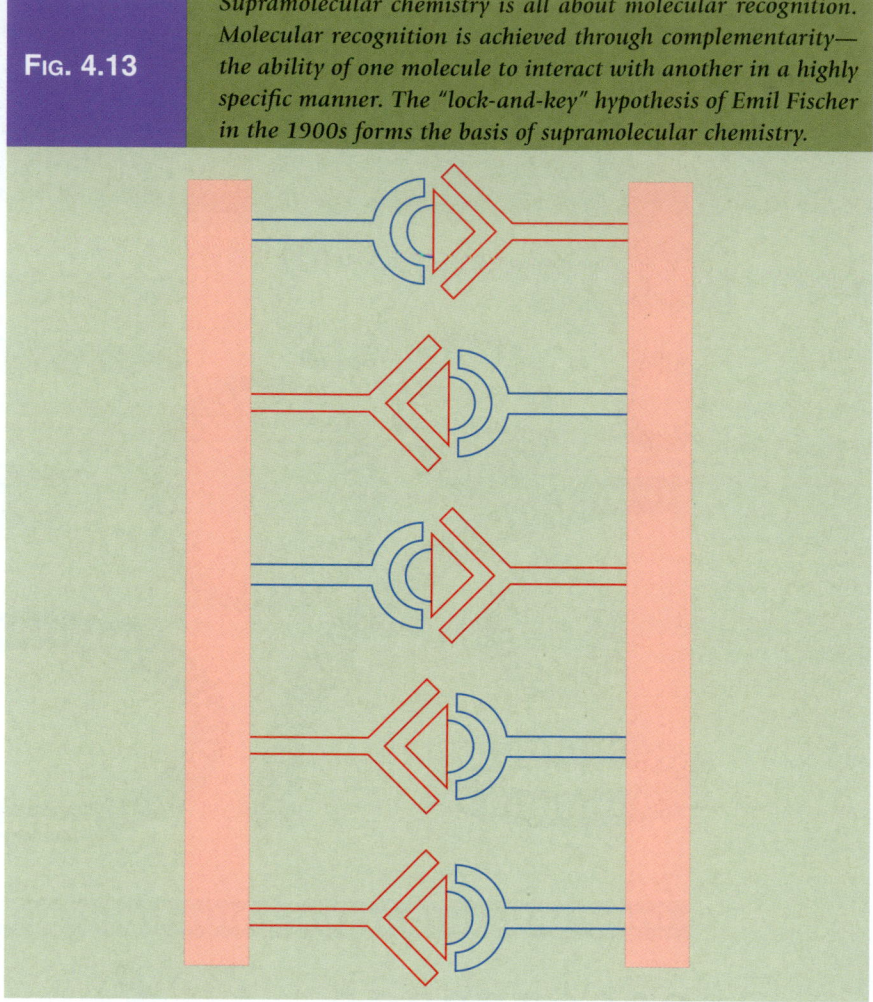

FIG. 4.13 *Supramolecular chemistry is all about molecular recognition. Molecular recognition is achieved through complementarity—the ability of one molecule to interact with another in a highly specific manner. The "lock-and-key" hypothesis of Emil Fischer in the 1900s forms the basis of supramolecular chemistry.*

Hydrolysis of precursor molecules occurs first from TEOS:

$$Si(OC_2H_5)_4 + H_2O \rightarrow Si(OC_2H_5)_3(OH) + (C_2H_5)OH \quad (4.27)$$

Subsequent hydrolysis yields reduced forms of TEOS (an alkoxide precursor):

$$Si(OC_2H_5)_3OH + H_2O \rightarrow Si(OC_2H_5)_2(OH)_2 + (C_2H_5)OH \quad (4.28)$$

The next step is a condensation step that involves the hydrolysis products. The insipient form of the silicon dioxide structure is established at this time:

$$2Si(OC_2H_5)_3(OH) \rightarrow (OC_2H_5)_3Si\text{-}O\text{-}Si(OC_2H_5)_3 + H_2O \quad (4.29)$$

Numerous other processes, many of them without the presence of water, are used to make SiO_2 nanoscale materials. Sol–gel synthesis proceeds with growth of increasingly larger particles via Ostwald ripening and other attractive mechanisms. In Ostwald ripening, larger particles that are energetically favored (due to

FIG. 4.14 *Sol–gel synthesis is an old technology that has incredible potential for nanomanufacturing. Starting from the absolute bottom with molecules and via the process of nucleation and Ostwald ripening, larger and larger particles are grown until the reaction is terminated. Following a sintering process, an array of close-packed spherical particles can be used to form aerogels or xerogels or act as a template to form other nanomaterials.*

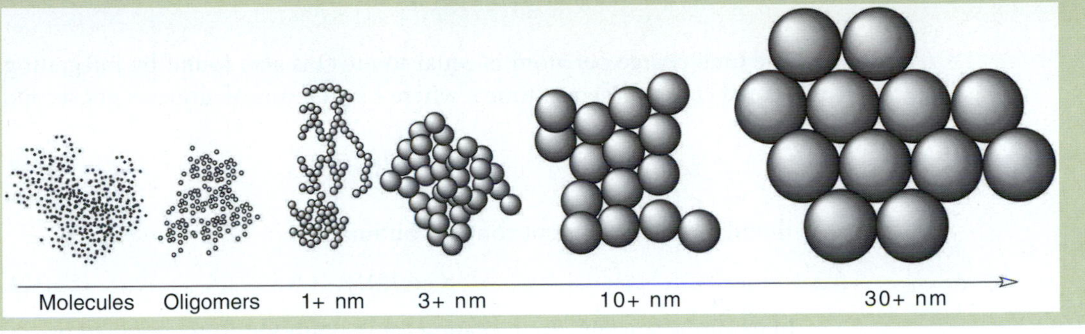

curvature phenomena) grow at the expense of smaller, less stable particles. The process is schematically illustrated in **Figure 4.14**.

One means to obtain colloidal metal oxides is by the hydrolysis of the corresponding metal salt. TiO_2 colloids are formed by the base catalyzed hydrolysis of $TiCl_4$:

$$TiCl_4 + 2H_2O \rightarrow TiO_{2(s)} + 4H^+ + 4Cl^- \tag{4.30}$$

The problem confronting procedures such as those depicted above is agglomeration. Small size is accompanied by high surface energy (chapters 5 and 6) and, therefore, nanosized particles are inherently unstable. Such coagulation can be prevented by addition of chemical agents or adjustment of the solution *p*H and other conditions.

More detail can be found in later chapters; however, the general class of sol–gel synthesis embraces a wide variety of nanoscale fabrication methods that deliver an equally wide variety of nanoscale materials. In addition to aerogels and xerogels, zeolitic structures and inorganic–organic hybrid materials have become increasingly useful.

Electrolytic and Electroless Deposition. These two methods are most appropriate for metal and semiconductor deposition. Thin films or nanoparticles can be fabricated from either method—both originating with atoms to form nanomaterials. Electrolytic deposition is based on Faraday's constant and Faraday's law. Faraday's constant is

$$\mathcal{F} = \mathcal{N}_A \cdot e \tag{4.31}$$

where $\mathcal{F}$ is the Faraday constant (96,485 $C \cdot mol^{-1}$), $\mathcal{N}_A$ is the Avogadro number (6.022×10^{23} per mole), and e is the electron charge (1.6022×10^{-19} $C \cdot electron^{-1}$). The modern form of Faraday's law of electrolysis is

$$m = \frac{Q}{ev}\left(\frac{M}{\mathcal{N}_A}\right) = \frac{1}{e\mathcal{N}_A}\left(\frac{QM}{v}\right) = \frac{1}{\mathcal{F}}\left(\frac{QM}{v}\right) \tag{4.32}$$

where
- m is the mass of metal plated on the cathode in grams
- Q is the total number of coulombs accrued over the duration of the electroplating process
- υ is the charge on the atom (the valence state)
- M is the molar mass in grams per mole

The total charge per atom is equal to $e\upsilon$. Q is also found by integrating the measured current $I(t)$ over time t, where I is in terms of amperes per second:

$$Q = \int_{t=0}^{t=T} I(t)\, dt \tag{4.33}$$

or, if under constant current control conditions,

$$Q = I(\Delta t) \tag{4.34}$$

In its simplest form, electroplating (at the cathode) is represented by

$$M^{n+}{}_{(aq)} + ne_{(aq)} \rightarrow M_{(s)} \tag{4.35}$$

where M is a generic metal and n is an integer that indicates the charge on the metal atom ($n = \upsilon$).

Electroless deposition is, as its name indicates, a plating process for metals that does not require an outside electromotive force. It is a nongalvanic process. Electroless deposition is a purely chemical bottom-up process. The facility of electroless deposition is demonstrated by its ability to plate metal on nearly any substrate—metal, ceramic, plastic, or semiconductor. Electroless plating is an autocatalytic process that requires participation of a metal salt and a reducing agent. The formation of a nanometer scale film of Ag on a substrate is given by a multistep process:

$$AgNO_3 + KOH \rightarrow AgKOH + KNO_3 \tag{4.36}$$

$$AgOH + 2NH_3\ [Ag(NH_3)_2]^{1+} + OH^- \tag{4.37}$$

$$[Ag(NH_3)_2]^{1+} + OH^- + H_2COAg_{(s)} + 2NH_3 + H_2O \tag{4.38}$$

Electroless plating of gold is used extensively in the fabrication of semiconductor devices and metalized ceramic components. A formula for the autocatalytic plating of gold consists of 0.01 M gold chloride hydrochloride trihydrate, 0.014 *M* sodium potassium tartarate, 0.013 M dimethylamine borane, and 400 $mg \cdot L^{-1}$. The *p*H of the solution (adjusted with NaOH) is 13.0 and the temperature is 60°C [73]. If you are a chemist, justify the components and conditions of this formula.

Biological Bottom-Up Fabrication Methods. Last, but most certainly not least, we bring up nature's way to manufacture nanomaterials—yes, from the bottom up. Most natural fabrication processes occur in solution, mainly in aqueous media. Biological processes on the whole occur at mild temperatures, atmospheric pressure, and moderate *p*H. Now that this method has been introduced, please anticipate in-depth treatment to come in chapters 13 and 14.

4.2.3 Solid-Phase Bottom-Up Fabrication?

Solid-State Phenomena. This category is difficult to define. Pure solid-phase bottom-up fabrication methods are "few and far between." By definition, a solid is a close-packed ensemble of atoms and molecules and, therefore, a condensed, bulk state. Any process that converts this bulk state into nanomaterials is rightly considered to be a top-down method. The formation of nanovoids from smaller cavities or vacancies could be labeled as a bottom-up, solid-state "fabrication" method. However, formation of defects, dislocations, twinning, and other material deformations in response to external stresses basically occur from the top down. Atoms do have the ability to diffuse through solids, assemble, and create nanodomains within the solid material. Would this process be considered as a bottom-up process? Defects occur in solids and they play important roles in the properties of bulk materials. It would be expected that such defects play a lesser role in nanomaterials. We will review a few general categories.

Zero-dimensional imperfections (or point defects) occur in homogeneous crystalline materials independent of chemical impurities [74]. One type of point defect is called a *vacancy*—an unoccupied atomic site. A vacancy can also be produced in response to a localized chemical impurity or a nonstoichiometric rendering of compounds within the solid. An *interstitial point defect* is the occupation of a site by an atom that is normally not occupied. A *Schottky* defect is made from a complementary pair of ionic vacancies, and a *Frenkel* defect is a combination of a vacancy and an interstitial point defect [74]. The dimensions of such defects are usually on the subnanometer scale. They are produced from thermal effects, chemical impurities, solid-state diffusion, and other sources of external stress—from the top down? They serve an important function because without such defects, diffusion through metals would be more difficult.

One-dimensional (linear) imperfections cover larger areas. A dislocation is a linear defect in a crystalline solid that is able to impart influence over a nanometer scale. There are *edge* dislocations and *screw* dislocations (a spiral stacking of planes around the dislocation line). Most linear defects consist of a mixture of the two [74]. Mechanical properties of materials are influenced greatly by such material imperfections. However, plastic deformation in metals, for example, is not likely without the presence of imperfections. One-dimensional imperfections are nanometer or greater sized structures. How are they "fabricated"?

Two-dimensional (planar) imperfections involve surface disruption of a crystal. To begin with, the surface is very different from the bulk. *Twin boundaries* (the plane that separates two identical mirror image crystal regions) are formed by deformation or by annealing (top down). The *grain boundary* is a two-dimensional imperfection. Grain size ranges from the nanometer scale to microns. The physical properties of grains in metals are dependent on top-down manufacturing conditions (e.g., quenching rate, temperature). Grain boundaries, like for the surface, represent enhanced areas of material diffusion. Grain boundary dislocations exist between grains with different orientations. Another two-dimensional imperfection, the surface step, has major importance in nanotechnology (discussed in more detail in chapter 6). The most important feature of engineering materials is the grain structure (e.g., grain boundary, size, and distribution) [74]. For example, the mechanical properties of materials containing nanograins outperform counterparts made of micron-sized grains [7].

Three-dimensional imperfections are characteristic of noncrystalline, or amorphous, solids [74]. An example of an amorphous solid is plate glass used in windows. There is no long-range order in amorphous solids. Quasi-crystals represent an intermediate structural state [74]. An offshoot of quasi-crystal research, the study of fractals, has given a new perspective to micro- (and now nano-) scale structures [74]. Some of the benefits of fractal research are the development of new kinds of thin films with tunable properties [74].

Self-Purification of Nanocrystals. Nanomaterials have the capability of self-purification. Because of this, the process of doping quantum dots becomes problematic. Self-purification occurs for several reasons. Because nanomaterials have small dimensions, the diffusion path length of atoms may exceed the physical dimensions of, for example, a quantum dot. Also, impurities can easily find their way to the surface and, once at the surface, impurities easily bind to the surface. The surface energy of any material, bulk or nano, is always higher than the cohesion energy of the volume. The surface energy of a nanomaterial is much higher than its planar bulk surface counterpart. The surface is always seeking ways to become more stable. Binding to complementary materials is one way to achieve stability. The bound atom may be an impurity (e.g., originating from a doping procedure) that has migrated from the interior.

Lastly, the binding energy of volume atoms in very small nanoparticles is lessened due to lowered coordination. The smaller the nanocrystal is, the less binding energy exists and, as result, the difficulty of doping increases. The argument presented here represents one of *antifabrication* rather than fabrication. In order to be able to dope quantum dots, the immediate environment surrounding the dots requires modification [75].

4.2.4 Template Synthesis

This process is perhaps one of the most facilitative in forming nanomaterials from the bottom up. A template (from the Latin *templum,* "plank, rafter," but also "a building for worship") is a material that acts as a gauge, pattern, or mold that is used to guide the manufacture of another piece. According to this definition, DNA serves as a template for the generation of another macromolecule—for example, RNA. This is an example of template synthesis. The mask used in lithography is a template. Porous aluminum oxide films with ordered arrays of pore channels are used as templates and masks. The electroplating (a bottom-up process) of Au into the pore channels of porous alumina results nanoparticles of gold. The aspect ratio and the size of the Au particles are determined by the pore diameter of the channel and the time of electroplating [76,77]. We dedicate a section in chapter 12 to this process.

In its simplest context, template synthesis involves either hard matter or soft matter. Porous alumina fits into the category of a hard matter template. Reverse micelles and DNA are soft matter templates. The basic process is illustrated in **Figure 4.15.**

Fig. 4.15 *Physical dimensions of nanomaterials formed in templates are directed by the template. Template synthesis makes use of a solid or liquid architecture that has the ability to isolate a chemical reaction within its physical boundaries. Templates can contain a cellular array, like the version depicted, or singular form such as that exhibited by micelles. Confinement of chemical reactions results in the formation of materials that would otherwise not exist. The process depicted demonstrates the facility of template synthesis. The pore channels of an anodically formed alumina membrane can act as host to a many kinds of materials.*

4.3 Computational Chemistry and Molecular Modeling

In today's laboratories, R&D firms, and places of manufacturing of nanomaterials and devices, the computer has proven to be an invaluable tool. Computational chemistry integrates theoretical chemical principles into computer programs. The results of computational chemical techniques are quantitative (exact) for hydrogen and assume more qualitative character as molecules become larger and systems become more complex. With increase in complexity, there is a concomitant increase in the level of input parameters, algorithms, and assumptions. Molecular modeling has become highly interdisciplinary as mathematicians, physicists, chemists, biologists, engineers, and computer scientists all work together to produce the desired outcome.

4.3.1 History

The content of this paragraph was obtained from excellent sources found on the Internet: Answers.com, Wikipedia, and the Sci-Tech Encyclopedia site of *Computational Chemistry*, to name just a few. The history of computational chemical methods begins in the early part of the twentieth century. Walter Heitler and German-American physicist Fritz London in 1927 are credited with performing the first theoretical chemistry calculations. The latter is also known for describing in 1930 weak intermolecular dispersion forces that bear his name. Several well-known physicists and chemists of the period contributed to the early development of computational quantum chemistry: Pauling and Eyring, to name a few. When computer technology emerged in the 1940s, solutions to wave equations became practical and, in the early 1950s, semi-empirical atomic orbital calculations were accomplished. The first ab initio Hartree–Fock computations that used a basis set of Slater orbitals were carried out in 1956 at MIT. Polyatomic calculations that employed Gaussian orbitals also were carried out during the 1950s and 1960s. Linear combination of atomic orbital methods (a.k.a Hückel methods) emerged in 1964. The discipline of scientific computational chemistry became a formal discipline around 1979.

The evolution of computational chemistry developed hand in hand with the development of the computer. Computational chemistry officially began in 1962 with the Quantum Chemistry Program Exchange (QCPE). This program helped scientists (chemists in particular) to develop, share, and apply quantum mechanical software [78]. With the advent of small computers, the ability to model real, but very simple, chemical systems began in earnest. In the 1980s, quantum mechanical methods and atomistic simulations were able to predict structures and behavior of small organic molecules and systems of >100 atoms.

Molecular modeling and nanomaterials combine to form a perfect match. There has never been a better, more compatible pair of complementary systems: Computers predict nanomaterial properties and behavior and advanced nanomaterials make computers better and more efficient! Both however are limited by the number of molecules in their structure (or program). Nonetheless, systems of hundreds to thousands of separate constituents (molecules, replicas, or ensembles) are routinely simulated. There are several approaches for computing nanothermodynamic phenomena. Molecular dynamics simulation is regarded as one of the best. In this approach all the atoms and molecules are assumed as "vibrating machines" that are programmed to interact in a predetermined fashion for a period of time under certain conditions to carry out virtual experiments. The results of the simulation are deconvoluted by various kinds of numerical methods.

Why is this important? Design will become one of the most important aspects of nanotechnology development [79]. Prognosticators state that while production costs will diminish significantly due to nanotechnology, the level of design will increase significantly due to the complexity involved [79]. According to Michael Riech at AIFT University of Karlsruhe, "Design will change radically under nanotechnology and for nano-engineers and nano-designers respectively, a broad knowledge will become even more important in the future.... As long as we are still far from the realization of complex nanotechnological applications, nanoengineering and nanodesign almost exclusively take place on computers."

4.3.2 General Types of Molecular Modeling Methods

There are four basic types of simulation methods: quantum mechanical ab initio, Monte Carlo, molecular dynamics, and molecular mechanics methods (**Table 4.13**). Finite element analysis has also become an increasingly popular and powerful tool for analyzing nanomaterial phenomena. Obviously, as stated earlier in this chapter, molecular modeling methods are not trivial in nature. We, therefore, will not offer great detail but rather a qualitative overview of the methods and relevance to nanotechnology.

Ab Initio Methods. Quantum mechanical ab initio (from the Latin *ab*, "from" + *initium*, "beginning") methods strive to simplify the solution of Schrödinger's equation (molecular Hamiltonian) for many particle systems and do not rely on a priori input in the form of empirical or semi-empirical parameters or experimental data. Common techniques include self-consistent field, LCAO (linear combination of atomic orbitals), and density functional methods. The Hartree–Fock version is the simplest in that average electron–electron repulsions are factored into the program. Density-functional theory is applied to calculate molecular electronic structure but empirical data are used to "grease the skids" for this technique. In this method, energy is expressed in the form of electron density and not in the form of a wave function.

The computational investment in ab initio methods is quite large and the technique is confined to the analysis of a few hundred atoms [79]. No dynamical or temporal aspects are considered in ab initio methods.

Molecular Mechanics and Dynamics Methods. Molecular mechanics (static) and dynamic methods, on the other hand, are grounded in classical force field theory [79]. The molecular mechanic method (electrons treated implicitly) is nonquantum mechanical (electrons treated explicitly). The objective of both ab initio and molecular mechanical methods is to develop the lowest energy configuration of the potential energy surface. Molecular mechanics simulations account for bonding energy a priori and the potential energy as the sum of all the forms of electrostatic interactions, stretching, bending, torsion, and van der Waals interactions. To date molecular dynamics (MD) simulations have been successfully applied to a narrow range of nanostructures that are energetically close to equilibrium [79]. Both methods lack information concerning quantum effects. Often, information gleaned from ab initio modeling is input a priori into molecular mechanics and dynamics programs.

Monte Carlo Methods. Monte Carlo methods (random walk, stochastic methods) rely on statistical ensembles based on Boltzmann style distributions. Monte

Table 4.13 *Molecular Modeling Techniques*

Simulation method	Description	Advantages and drawbacks
Ab initio	Solution of Schrödinger's equation for many particle systems Types: Self-consistent field method, linear combination of atomic orbitals, and density-functional method Objective: describe the stable low-energy configuration—the local minima on potential energy surface Result is in the form of quantum effects.	Calculation time is a function of particle number; therefore, a great investment in computer space and time is required. Practical for ca. 10^2 atoms Many approximation algorithms exist to shorten calculation time. QM calculations lack a priori input (no interaction potential, chemical bond, or temperature input parameters). Dynamic aspects are not considered. No temperature data involved in the model. Relatively slow method
Monte Carlo	New particle configurations are created by a step-by-step random process. Systems configurations are sampled according to specific statistical ensemble. Method is grounded in Boltzmann distribution theory.	Calculations are valid for systems close to equilibrium. External input parameters such as temperature are required.
Molecular mechanics	Based on classical mechanics	Able to handle 10^5–10^7 atoms
Molecular dynamics (MD)	Based on Newton's laws of motion The potential energy of all components is calculated using force fields Harmonic, Morse, Gaussian, Lenard–Jones, and other potentials are used in the simulation. Each step is associated with a temporal element. Information: Particle position, velocity, momentum, kinetic and potential energies Utilizes mass points interacting through force fields derived from interaction potentials Describes electron interactions implicitly and considers molecules to be a collection of atoms held together by "sticky forces" Molecular mechanics derives a motionless structure. Molecular dynamics achieves a better model. MD is the choice of most nanotech modeling groups MD is the study of matter in motion. MD is an excellent program with which to study thermodynamics	Require detailed a priori input concerning bond energies; interaction potentials are chosen beforehand As a result, discovery of new molecular systems is not likely by this method. Fitting of potentials to experimental data is required to generate realistic scenarios. Such information is derived from ab initio calculations. No explicit quantum results are offered by these methods.
Finite element analysis	The method does not consider that materials have structure on the microscopic scale or below. Materials are treated like a continuum that has consistent properties when stretched, bent, made to conduct heat, or perturbed in other ways. Molecular level modeling is accomplished in a similar way with atoms and molecules acting as the points, nodes, mesh, etc.	Data depend on the quality and accuracy of applied input parameters, geometry. Properties like Young's modulus have been determined at the molecular scale.

Carlo (MC) methods are based on statistical mechanics rather than the molecular dynamic formulations discussed briefly earlier. These types of simulation techniques require input parameters (e.g., temperature) and, like MD methods, exist near the equilibrium point. Assignment of temporal markers to the evolving configurations is problematic in MC simulations. The five basic types of MC methods are

- classical, where samples are based on a probability distribution with the intent of finding minimum energy configurations and rate parameters;
- quantum, in which stochastic methods are used to predict quantum mechanical parameters based on the Schrödinger equation;
- volumetric, a method that generates numbers from random walk procedures to predict molecular volumes and molecular phase-space surfaces;
- path integral, based on quantum statistical mechanics and used to find thermodynamic properties by application of Feynman's path integral as the starting point; and
- simulation methods like kinetic and thermalization, which use stochastic algorithms to generate initial conditions.

These categories were described on the site of R. Q. Topper at <www.cooper.edu/engineering/chemechem/monte.html> (2002). Monte Carlo simulations make use of the Markov chain principle. The Markov chain is a computational process in which future states are conditionally independent of previous states.

Finite Element Analysis. Finite element analysis (FEA) was first accomplished in 1943 by R. Courant, who conducted numerical analysis and minimization of variational calculus to seek solutions to vibrational systems. FEA methods were first applied to structural mechanics. In 1956, L. J. Topp, R. W. Clough, M. J. Turner, and C. Martin wrote a paper applying FEA to complex structures—in particular, stiffness and deflection relationships. Basically, with regard to engineering systems, FEA is a computer model that is able to analyze engineering parameters such as stress. It is often used in predesign work in the form of computer-aided design (CAD). In FEA, a mesh consists of system points called nodes. Inherent within the mesh are programs that contain structural properties, if engineering applications are the goal, that define how that structure will react to applied stresses. Mass, volume, temperature, strain energy, stress strain, force displacement, velocity, and acceleration are some parameters that are monitored or applied during the simulation [80].

The first step is to divide the structure (mechanical or atomic) into unique sectors called finite elements. Finite elements are joined to form nodes and form the mesh (or grid). Boundary conditions are often applied. Predetermined variables then act over each domain and the results of local equations are assimilated to give the system equation that describes the behavior of the whole system [80].

Examples of finite element analysis of nanomaterials are plentiful in the literature. In one case, the spring constants of AFM cantilevers made of sapphire and tipped with a diamond probe were modeled via nanomechanical FEA [81]. FEA was conducted to analyze quantum mechanical transport in strained quantum

FIG. 4.16 *A Morpho butterfly recreation by finite element analysis methods. The wing structure is made from carbon nanotubes and contains photonic crystals as in the natural form. FEA is a powerful tool in the nano domain; in many ways it can be considered to be a fabrication method. It is a valuable prefab tool involved in the design phase of nanosynthesis and assembly programs.*

Source: Courtesy of Halff Associates, Inc.

dots and wires [82]. It was found that strain is responsible for energy shifts of resonant current peaks and that strain causes additional fine structure in the current peaks. Finite element method (FEM) was used to model the heat generation and distribution and carbon migration in catalysts that form MWNT and SWNTs. From FEM, it was found that growth is mainly driven by a concentration gradient, as opposed to a thermal gradient, with the experimental temperature playing a major role in terms of activating the diffusion process [83].

Computer simulation methods are powerful ways to predict the behavior of nanomaterials. A system that is made of a countable number of atoms or molecules is a perfect candidate for computer-aided simulation techniques. Add computer scientists to the interdisciplinary crew—the continuum of qualified nanoengineers and scientists.

A finite element analysis rendition of the *Morpho* butterfly pictured on the cover of this text is shown in **Figure 4.16**.

References

1. B. K. Teoand and X. H. Sun, From top-down to bottom-up to hybrid nanotechnologies: Road to nanodevices, *Journal of Cluster Science*, 17, 529–540 (2006).
2. S. Price, Top-down and bottom-up processes. Presentation, EE 518, J. Ruzyllo, Pennsylvania State University (2006).
3. C. L. De Castro and B. S. Mitchell, Nanoparticles from mechanical attrition. In *Synthesis, functionalization and surface treatment of nanoparticles*, M.-L. Baraton, ed., American Scientific Publishers, Stevenson Ranch, CA (2002).
4. M. Ohring, *Materials science of thin films*, Academic Press, New York (2002).
5. R. W. Kelsall, I. W. Hamley, and M. Geoghegan, *Nanoscale science and technology*, John Wiley & Sons, Ltd., West Sussex (2005).
6. B. J. Y. Tan, C. H. Sow, T. S. Koh, K. C. Chin, A. T. S. Wee, and C. K. Ong, Fabrication of size-tunable gold nanoparticles array with nanosphere lithography, reactive ion etching, and thermal anneal, *Journal of Physical Chemistry B*, 109, 11100–11109 (2005).
7. C. P. Poole and F. J. Owens, *Introduction to nanotechnology*, Wiley-Interscience, New York (2003).
8. J. van Erven, R. Moerman, and J. C. M. Marijnissen, Platinum nanoparticle production by EDHA, *Aerosol Science Technology*, 39, 941–946 (2005).
9. S. M. Lee, *Dictionary of composite materials technology*, CRC Press, Boca Raton, FL (1989).
10. I. T. H. Chang, Rapid solidification processing of nanocrystalline metallic alloys. In *Handbook of nanostructured materials and nanotechnology*, Vol. 1, H. S. Nalwa, ed., Academic Press, San Diego (2000).
11. D. Li and Y. Xia, Electrospinning of nanofibers: Reinventing the wheel, *Advanced Materials*, 16, 1151–1170 (2004).
12. Z. Zhang, J. Y. Ying, and M. S. Dresselhaus, Bismuth quantum-wire arrays fabricated by a vacuum melting and pressure injection process, *Journal of Materials Research*, 13, 1745–1748 (1998).
13. J. Heremans, C. M. Thrush, Y.-M. Lin, S. Cronin, Z. Zhang, M. S. Dresselhaus, and J. F. Mansfield; Bismuth nanowire arrays: Synthesis and galvanomagnetic properties, *Physics Review B*, 61, 2921–2930 (2000).
14. D. Hulicova, K. Hosoi, S.-I. Kuroda, H. Abe, and A. Oya, Carbon nanotubes prepared by spinning and carbonizing core-shell polymer microspheres, *Advanced Materials*, 14, 452–455 (2002).
15. Y. Ando and M. Ohkohchi, Production of ultrafine powder of p-sic by arc discharge, *Journal of Crystal Growth*, 60, 147–149 (1982).
16. Y. Ando and S. Iijima, Preparation of carbon nanotubes by arc-discharge evaporation, *Japan Journal of Applied Physics*, 32, L107–L109 (1993).
17. T. Guo, P. Nikolaev, A. Thess, D. T. Colbert, and R. E. Smalley, Catalytic growth of single-walled nanotubes by laser evaporation, *Chemical Physics Letters*, 243, 49–54 (1995).
18. C. L. Fields, J. R. Pitts, M. J. Hale, C. Bingham, and A. Lewandowski, Formation of fullerenes in highly concentrated solar flux, *Journal of Physical Chemistry*, 97, 8701–8702 (1995).
19. Y. B. Li, Y. D. Yu, and Y. Liang, A novel method for synthesis of carbon nanotubes: Low temperature solid pyrolysis, *Journal of Materials Research*, 12, 1678–1680 (1997).
20. M. Di Ventra, S. Evoy, and J. R. Heflin, eds., *Introduction to nanoscale science and technology*, Kluwer Academic Publishers, New York (2004).
21. R. S. Dhaliwal, W. A. Enichen, S. D. Golladay, M. S. Gordon, R. A. Kendall, J. E. Lieberman, H. C. Pfeiffer, D. J. Pinckney, C. F. Robinson, J. D. Rockrohr, W. Stickel, and E. V. Tressler, PREVAIL—Electron projection technology approach for next-generation lithography, *IBM Journal of Research and Development*, 45, 615–638 (2001).

22. S. Y. Chou, P. R. Krauss, and P. J. Renstrom, Imprint lithography with 25-nanometer resolution, *Science*, 272, 85–87 (1996).
23. S. Y. Chou, P. R. Krauss, and P. J. Renstrom, Nano-imprint lithography, *Journal of Vacuum Science Technology B*, 14, 4129–4133 (1996).
24. J. C. Hulteen and R. P. Van Duyne, Nanosphere lithography: A materials general fabrication process for periodic particle array surfaces, *Journal of Vacuum Science Technology A*, 1553–1558 (1995).
25. K. Kawata, H. B. Sun, T. Tanaka, and K. Takada, Finer features for functional microdevices, *Nature*, 412, 697–698 (2001).
26. G. L. Hornyak, L. Grigorian, A. C. Dillon, P. A. Parilla, and M. J. Heben, A temperature window for chemical vapor deposition growth of single walled carbon nanotubes, *Journal of Physical Chemistry B*, 106, 2821–2825 (2002).
27. P. Petroff, A. Lorke, and A. Imamoglu, Epitaxially self-assembled quantum dots, *Physics Today*, 54, 46–52 (2001).
28. F. Ren, C. Z. Jlang, Y. H. Wang, Q. Q. Wang, and J. B. Wang, The problem of core-shell nanoclusters formation during ion implantation, *Nuclear Instruments and Methods in Physics Research B: Beam Interaction with Materials and Atoms*, 245, 427–430 (2006).
29. G. L. Hornyak, M. Kröll, R. Pugin, T. Sawitowski, G. Schmid, J.-O. Bovin, G. Karsson, H. Hofmeister, and S. Hopfe, Gold clusters and colloids in alumina nanotubes, *Chemistry, a European Journal*, 3, 1951–1956 (1997).
30. H. Masuda and K. Fukuda, Ordered metal nanohole arrays made by a two-step replication of honeycomb structures of anodic alumina, *Science*, 268, 1466–1468 (1995).
31. W. K. Hsu, M. Terrones, J. P. Hare, H. Terrones, H. W. Kroto, and D. R. M. Walton, Electrolytic formation of carbon nanotubes, *Chemical Physics Letters*, 262, 161–166 (1996).
32. X. Liu, L. Fu, S. Hong, V. P. Dravid, and C. A. Mirkin, Arrays of magnetic nanoparticles patterned via "dip-pen" nanolithography, *Advanced Materials*, 14, 231–234 (2002).
33. Y. Lei, L. W. Teo, K. S. Yeong, Y. H. See, W. K. Chim, W. K. Cjoi, and J. T. L. Thong, Fabrication of highly ordered nanoparticle arrays using thin porous alumina masks, http://dspace.mit.edu/bitstream/1721.1/3662/2/AMMNS009.pdf., ca. 2002, 1–6 (viewed 2007).
34. D. Zschech, D. H. Kim, A. P. Milenin, R. Scholz, R. Hillebrand, C. J. Hawker, T. P. Russell, M. Steinhart, and U. Gösele, Ordered arrays of <100>-oriented silicon nanorods by CMOS-compatible block copolymer lithography, *Nano Letters*, 7, 1516–1520 (2007).
35. M. Park, C. Harrison, P. M. Chaikin, R. A. Register, and D. H. Adamson, Block copolymer lithography: Periodic arrays of $\sim 10^{11}$ holes in 1 square centimeter, *Science*, 276, 1401–1401 (1997).
36. R. Garcia, *Bridging nano and macro worlds with water Meniscii: Attomole chemistry and nanofabrication by local oxidation nanolithography*, keynote address, LITHO 2004, Agelonde-France (2004).
37. M. F. Crommie, C. P. Lutz, and D. M. Eigler, Confinement of electrons to quantum corrals on a metal surface, *Science*, 262, 218–220 (1993).
38. M. Berger, *Nature's bottom-up nanofabrication of armor*, Nanowerk, LLC (2006).
39. T. J. Deming, V. P. Conticello, and D. A. Tirrell, Biocatalytic synthesis of polymers of precisely defined structures. In *Nanotechnology*, G. Timp, ed., Springer-Verlag New York, Inc., New York (1999).
40. D. D. Chiras, *Biology: The web of life*, West Publishing Company, New York (1993).
41. Y. Xia, J. A. Rogers, K. E. Paul, and G. Whitesides, Unconventional methods for fabricating and patterning nanostructures, *Chemistry Review*, 99, 1823–1848 (1999).

42. B. Das, S. Subramanium, and M. R. Melloch, Effects of electron-beam induced damage in back-gated GaAs/AlGaAs devices, *Semiconductor Science & Technology*, 8, 1347–1351 (1993).
43. G. Cao, *Nanostructures & nanomaterials: Synthesis, properties & applications*, Imperial College Press, London (2004).
44. O. Saxl, Opportunities for industry in the application of nanotechnology, The Institute of Nanotechnology, London, nano.org.uk (2000).
45. C. C. Koch, Top-down synthesis of nanostructured materials: Mechanical and thermal processing methods, *Review of Advanced Materials Science*, 5, 91–99 (2003).
46. L. He and E. Ma, Nanophase Fe alloys consolidated to full density from mechanically milled powders, *Journal of Materials Research*, 15, 904–912 (2000).
47. A. G. Mamalis, Advanced manufacturing of advanced materials. In *Innovative superhard materials and sustainable coatings for advanced manufacturing*, J. Lee, N. Novikov, and V. Turkevich, eds., NATO Science Series, II. Mathematics, Physics and Chemistry, 200, 63–80 (2006).
48. K. Sung and C. S. Lee, Factors influencing liquid breakup in electrohydrodynamic atomization, *Journal of Applied Physics*, 96, 3956–3961 (2004).
49. T. Guillard, G. Flamant, D. Laplaze, J.-F. Robert, B. Rivoire, and J. Giral, Towards the large scale production of fullerenes and nanotubes by solar energy, *Proceedings of the Solar Forum 2001: Solar Energy the Power to Choose*, Washington, D.C. (2001).
50. V. V. Karasev, O. G. Glotov, A. M. Baklanov, A. A. Onischuk, and V. E. Zarko, Alumina nanoparticle formation under combustion of solid propellant, www.kinetics.nsc.ru/private_page/en/papers/on2.pdf, Institute of Chemistry Kinetics & Combustion, Russian Academy of Science (2003).
51. Wet-chemical etching and cleaning of silicon, Virginia Semiconductor, Inc., www.virginiasemi.com (2003).
52. W. C. Crone, A brief introduction to MEMS and NEMS. In *Springer handbook of experimental solid mechanics*, W. N. Sharpe, ed., Springer-Verlag, New York (2008).
53. J. Rawson, *Chinese jade from Neolithic to the Qing*, British Museum, London (1995).
54. A. Doolittle, Lithography and pattern transfer, Georgia Tech University, users.ece.gatech.edu/~alan/ECE6450/Lectures/ECE6450L7b-Lithography%20Steps%20for%20a%20CMOS%20Inverter.pdf
55. B. Bushan, ed., Introduction to micro/nanofabrication. In *Springer handbook of nanotechnology*, Springer-Verlag, Berlin (2004).
56. G. L.-T. Chiu and J. M. Shaw, eds., Optical lithography: Introduction, *IBM Journal of Research & Development*, 41, 3–6 (1997).
57. D. Atwood, *Extreme ultraviolet (EUV) lithography based on multilayer coated optics*. Presentation, Virtual National Laboratory, UC Berkeley (2004).
58. V. Banine and R. Moors, Plasma sources for EUV lithography exposure tools, *Journal of Physics D: Applied Physics (London)*, 37, 3207 (2004).
59. B. Braun, *Producing integrated circuits with x-ray lithography*, University of Wisconsin at Madison, http://tc.engr.wisc.edu/UER/uer97/author7/index.html (1997).
60. H. Smith and F. Cerrina, X-ray lithography in ULSI manufacturing, *Microlithography World, Winter*, 10–14 (1997).
61. M. K. Mundra, S. K. Donthu, V. P. Dravid, and J. M. Torkelson, Effect of spatial confinement on the glass-transition temperature of patterned polymer nanostructures, *Nano Letters*, 7, 713–718 (2007).
62. C. L. Haynes and R. P. Van Duyne, Nanosphere lithography: A versatile nanofabrication tool for studies of size-dependent nanoparticle optics, *Journal of Physical Chemistry B*, 105, 5599–5611 (2001).
63. C. L. Cheung, R. J. Nikolic, C. E. Reinhardt, and T. F. Wang, Fabrication of nanopillars by nanosphere lithography, *Nanotechnology*, 17, 1339–1343 (2006).

64. Y. Cai and B. M. Ocko, The large-scale fabrication of protein nanoarrays based on nanosphere lithography, *Technical Proceedings of the 2005 NSTI Nanotechnology Conference and Trade Show*, 1, Chapter 7: DNA, Protein, Cells and Tissue Arrays (2005).
65. Chemical vapor deposition, Wikipedia (viewed 2007).
66. M. Halls and K. Raghavachari, Atomic layer deposition growth reactions of Al_2O_3 on Si(100)-2x1, *Journal of Physical Chemistry B*, 108, 4058–4062 (2004).
67. *Atomic layer deposition tutorial*, Cambridge NanoTech, Inc., www.cambridgenanotech.com (viewed 2007).
68. R. L. Puurunen, Understanding the surface chemistry of atomic layer deposition: A case study for the trimethylaluminum/water process, *Journal of Applied Physics*, 97, 121301–121352 (2005).
69. G. Kresse, M. Schmid, E. Napetsching, M. Shishkin, L. Köhler, and P. Varga, Structure of ultrathin aluminum oxide on NiAl(110), *Science*, 308, 1440–1442 (2005).
70. F. Rinaldi, Basics of molecular beam epitaxy (MBE), Annual Report, Optoelectronics Department, University of Ulm, 1–8 (2002).
71. C. L. Yeh and E. Zhao, Combustion synthesis of SiO_2 on the aluminum plate, *Journal of Thermal Science*, 10, 92–96 (2001).
72. H. K. Ma and C. L. Yeh, The formation of nano-size SiO_2 thin film on an aluminum plate with hexamethyldisilazane (HMDSA) and hexamethyldisiloxane (HMDSO), *Journal of Thermal Science*, 12, 89–96 (2003).
73. J. Henry, Electroless (autocatalytic, chemical) plating. In *Metal Finishing: The Industry's Recognized International Technical Authority since 1903, 54th Guidebook Directory*, 84, 190–191 (1986).
74. J. F. Shackelford, *Introduction to materials science for engineers*, 4th ed., Prentice Hall, Inc., Upper Saddle River, NJ (1996).
75. G. M. Dalpian and J. R. Chelikowsky, Self-purification in semiconductor nanocrystals, *Physics Review Letters*, 96, 226802 (2006).
76. G. L. Hornyak and C. R. Martin, Optical properties of a family of Au-nanoparticle-containing alumina membranes in which the nanoparticle shape is varied from needle-like (prolate), to spheroid, to pancake-like (oblate), *Thin Solid Films*, 303, 84 (1997).
77. G. L. Hornyak, C. J. Patrissi, and C. R. Martin, Optical properties of gold-porous alumina composite membranes: The Maxwell-Garnett limit, *Journal of Physical Chemistry B*, 101, 1548 (1997).
78. *Materials Modeling and Simulation History*, Accelrys Software, Inc., http://www.accelrys.com/technologies/modeling/materials/history.html
79. M. Rieth, *Nano-engineering in science and technology—An introduction to the world of nano-design*, World Scientific Publishing, Hackensack, NJ (2003).
80. P. Widas, *Finite element analysis*, Virginia Tech University, www.sv.vt.edu/classes/MSE2094_NoteBook/97ClassProj/num/widas/history.html (1997).
81. T. Gang and F. Sansoz, *Determination of atomic force microscope cantilever spring constants via finite element modeling for nanomechanical analysis*, University of Vermont, Burlington, www.vtspacegrant.org/ESMD/Travis%20Gang%20URECA_Poster2.ppt (viewed 2007).
82. H. T. Johnson, L. B. Freund, A. Zaslavsky, and C. D. Akyüz, Finite element analysis of quantum mechanical transport in strained quantum dots and wires, *Journal of Applied Physics*, 84, 3714–3725 (1998).
83. C. Klinke, J.-M. Bonard, and K. Kern, Thermodynamic calculations on the catalytic growth of multiwall carbon nanotubes, *Physics Review B*, 71, 035403: 1–7 (2005).

Problems

4.1 Define top-down and bottom-up fabrication. Are there nebulous regions between these extremes?

4.2 Classify the following techniques as top down or bottom up:

a. Plasma etching
b. Epitaxial growth
c. Formation of a hierarchical structure with carbon nanotubes
d. Transformation of amorphous carbon into carbon nanotubes by thermal methods
e. Transformation of a 1-kg diamond into a 1-kg block of graphite
f. Formation of carbon nanotubes in an arc process with a target that contains no catalyst
g. Aspiration of a liquid to form nanoparticulate aerosols that condense to form a bulk material
h. Formation of a very thin layer of paint that contains latex (paint) particulates with 2 μm diameter
i. "Big Bang" formation of the universe and the formation of a black hole
j. Compaction of micron-sized particles
k. Using clay to make a pot

Can you name any other methods employed by nature that are top down?

4.3 What techniques include a combination of distinct top-down and bottom-up methods?

4.4 What is the resolution limit for a projection type of photolithographic system if the incident λ is 365 nm (the i-line of Hg), N.A. = 0.63, and k_1 = 0.6. What is the DoF needed for the best resolution?

4.5 Would etch rate of silicon depend on the type of surface exposed to the etchant? Why or why not?

4.6 Is energy required to reduce a bulk material into nanoparticles? Qualitatively (and relatively speaking), how much energy do you think is required?

4.7 What is the limit on feature size if 600-nm radiation is used in a photolithographic procedure?

4.8 Are there any solid-state "bottom-up" fabrication processes you can think of that result in the fabrication of nanomaterials?

4.9 What is the minimum feature size, W_{min}, possible in a proximity photolithography system with the following criteria:

λ = 320 nm; d_g = 10 μm, 5 μm, 1 μm, or 0.5 μm; and k = 1?

Is there any way to improve the resolution?

4.10 How can you improve the resolution of a projection photolithographic system?

4.11 Trace the bottom-up hierarchy of a biological structure in the human body (akin to the tendon example) starting with any of the three primary precursor molecules (amino acids, lipids, or sugars).

4.12 Can you think of any structures in biology that do not arise from nanomaterials?

4.13 Can thermal top-down methods be considered a unique category or are thermal methods in general components of other top-down methods?

4.14 What is the basic mechanism of RF sputtering? Is it top down or bottom up?

4.15 Anodizing efficiency is defined as the degree of conversion of aluminum into aluminum oxide (Al_2O_3). What factors are able to reduce anodizing efficiency?

4.16 Anodizing is a means of forming nanoporous structures. Is this a top-down method or a bottom-up method?

4.17 What would be the breadth of the features formed by nanosphere lithography (diameter = 1 μm) if the particles' transit is normal to the plane of the spheres? How could you improve the system?

4.18 Self-assembly is a bottom-up method. Do you believe that nanotechnology of the future would be based on self-assembly methods? Why do you think biological structures are based on self-assembly? Are there examples of biochemical processes that are not based solely on self-assembly?

Section 3

Physics: Properties and Phenomena

Materials, Structure, and the Nanosurface

Everything you've learned in school as "obvious" becomes less and less obvious as you begin to study the universe. For example, there are no solids in the universe. There's not even a suggestion of a solid. There are no absolute continuums. There are no surfaces. There are no straight lines.

R. Buckminster Fuller (1895–1983)

Chapter 5

Threads

The nano perspective, an overview of societal implications, a brief foray into the nanoscale world, characterization tools, and fabrication strategies have been inventoried and discussed in a cursory manner. All of these topics dwell within the domain of well-established fields that significantly predate the time nanoscience and nanotechnology were officially anointed the "next big thing." It is time now to get technical and get into the essence of nanoscience. In other words, we are ready now to study the "physics of small" and familiarize ourselves with the properties and phenomena associated with nanomaterials.

Three important aspects of the nanoscale permeate all of nanoscience: the pronounced effects of the surface, the importance of the quantum-bulk boundary, and the inapplicability of bulk scaling laws—three different ways to state a similar message. In chapter 5, we focus on some materials science and the surface and get a feel for the importance of singular and collective surface area to nanomaterial behavior as particles get smaller. In chapter 6, the energetics of nanomaterials is discussed. Scaling laws (and their breakdown) and some basic quantum mechanics are discussed in chapter 7. Chapter 7 also describes zero-, one-, two-, and hierarchical nanostructured materials. Chapter 8 is about nanothermodynamics and potential violations of the well-established thermodynamic laws and with that chapter, the "Physics" division of the text comes to an end, at least in the formal sense of the term.

5.0 Importance of the Surface

Everything has a surface (or interface) and both physical and chemical properties depend on the nature (and breadth) of that surface, regardless of whether it is a bulk material or a nanoscale material. Catalytic activity, electrical resistivity, adhesion, protective coatings, and gas storage are just a few phenomena that depend on the nature of the surface. The surface is important regardless how large or how small the material is. The icy surface of a comet, for example, made of amorphous solid water and known to be efficient at storing and releasing large amounts of gas, has been found to be nanoporous and of high surface area [1]. In upcoming sections, we look into surfaces from a geometric viewpoint and we also look at nanomaterials from a structural viewpoint, once again; no chemical or energetic topics are presented for discussion. Engineering materials are briefly reviewed to provide a perspective for what is to unfold.

5.0.1 Background

Surface science is a discipline that is devoted to the investigation of two-dimensional structures that serve as the interface between physical phases. Surface science involves experimental and theoretical study of the physics, chemistry, and biology of surfaces. The surface of a material is the outermost boundary of any object or substance. The etymology of the word surface (from Old French *sur,* "above" + *face,* from the Latin *facia,* "appearance, form, figure") indicates as much. Irving Langmuir was one of the founders of the discipline of surface science. The advent of surface analysis techniques such as AFM, STM, XPS, Auger electron spectroscopy, LEED, EELS, and SIMS helped establish this field as one of great importance to all of science and technology. Surfaces perform numerous vital functions. They keep things in; they keep things out; or they allow the flow of material and energy across an interfacial structure. Surfaces are also capable of initiating or terminating chemical reactions, as in the case of catalysts.

5.0.2 Natural Perspective

When discussing nanoscience, references made to nature always yield important information and guidelines. Nature, as we have hinted so far, is the ultimate nanotechnologist, at the nano-, the micro-, and the macroscales. At the microscale, single-celled organisms or cells existing within a larger body depend on passive diffusion to obtain nutrients and remove wastes. If cell volume becomes too large, simple diffusion is hindered. Cells developed mechanisms such as pumps to circumvent diffusion obstacles brought about by increases in volume. In the context of evolution, was the size of the cell limited by diffusion kinetics?

On the macroscopic scale, organisms have developed numerous means of increasing surface area to meet metabolic demand. Examples include leaves and the fractal nature of root networks of plants; hair on mammals; microvilli in the small intestine of animals; capillaries in the circulatory system, the liver, and the kidneys; the nervous system; alveoli in the lungs; and so forth. Red blood cells, for example, possess concave surfaces, an adaptation that enhances surface area. At the nanolevel, the level at which all biochemistry occurs, increased surface allows for a greater number of substrates to participate in reactions that are mediated by nanoscale catalytic proteins known as enzymes.

What are the benefits of increased surface area? Surface area in biological terms is a measure of the scale of exposure that a biological entity, whether cell or organism, has with its environment. At the macroscale, small animals have a higher surface-to-volume ratio than larger animals and therefore require a higher metabolic rate to balance heat loss to the environment. Once again, in the context of evolution, did surface-to-volume ratio play a role in limiting the size that a cell can achieve? Is the process of mitosis itself a manifestation of the need for more surface area coupled with the survival advantage that larger size offers? Is it any wonder that most of the cells in an organism are micron sized even though, overall, an organism can be quite large (e.g., the blue whale). The metabolic machinery of nature runs on nano power, function, and structure. The combined surface area of all enzymes must truly be an enormous number. We then ask a fundamental question. Were nanoscale phenomena the gatekeepers at the single- to multicellular threshold?

The surface is just a part of the overall equation. A nanoparticle, like its bulk counterpart, has substance and shape and exists, for the most part, within a specified set of physical dimensions. We begin the chapter with the surface, devoid of references to chemical structure or reactivity and physical properties. The next section introduces (or reintroduces) some basic materials science. Surface-to-volume ratio, specific surface area (surface-to-mass ratio), structure, and orientation follow suit to round out this chapter.

5.0.3 Inorganic Perspective

Surface area is important to inorganic materials, whether they are the size of stars and planets or as miniscule as nanosized catalysts. The surface plays a major role in any science and technology that we investigate or devise.

Example 5.1 *Bulk Physical Example of a Surface Effect*

Sublimation is the conversion of a solid directly to a gas without an intermediate liquid phase. For the purposes of this problem, ice sublimates at a rate of 1 cm·h^{-1} (temperature <0°C and low-pressure conditions). (a) How long would it take a cube of ice with volume equal to 1 m^3 to disappear? (b) for eight cubes derived from the original cube with the same total volume? (c) for a sphere with the equivalent volume?

Solution:

It will take 50 hours to "melt" the large cube. Within 1 m^3 of ice, there exist 50 "boxes" of ice of 1-cm wall thickness, similar to Russian Babuska dolls (2- × 2- × 2-cm cube in the center = 8-cm^3 volume plus 49 additional 1-cm-wall boxes that increase incrementally by 1 cm in the x, y, and z directions).

For the eight smaller cubes, each volume is equal to 0.125 m^3. Each cube has at its center the 8-cm^3 core plus 24 additional 1-cm-walled boxes. The total time to melt each small cube is 25 hours.

For the sphere of 1 m^3 equivalent volume, the radius is 0.620 m. Considering the innermost sphere of radius 1 cm, there are an additional 61 shells with thickness of 1 cm. It would take 62 hours to melt the sphere, a factor of ×1.24 over that of the cube of equivalent volume.

Why do ice cubes assume spherical shape when melting?

Inorganic catalysts are important to industry. The major objectives of catalytic processes are high yield, efficiency, and selectivity. Most inorganic catalysts are nanomaterials. Catalysts reduce the activation energy of a reaction without being consumed. The catalyst is able to impart free energy to a reaction that otherwise would not occur under milder conditions. Since we cannot create energy out of thin air, where does this excess energy come from? We find that the "excess energy" originates from the surface. Heterogeneous catalysis, a state in which the catalyst is a solid phase and the reactant is in the liquid or gaseous phase, is the type of catalysis that is most relevant to nanoscience.

Heterogeneous catalysts have high collective surface area, much like their biological enzymatic counterparts. Collective surface area is the sum of individual surface area of all the particles. Metal nanoparticles, for example, are supported on highly porous oxide materials, usually alumina or silica based. The alumina support also possesses high surface area. High surface area is required to drive the efficiency of catalytic processes, especially if industrial production rates are the goal. The rates of reaction depend intrinsically on the available surface area of the catalyst particles. For some reactions, catalyst size is tailor fitted to a particular reaction (e.g., nanocatalyst size is optimized), another feature of nanotechnology.

Bulk gold, as we all know, is a noble metal that is relatively inert to oxidation and other reactions. Nanoscale gold, on the other hand, is capable of catalyzing chemical reactions—a capability not characteristic of bulk gold or even gold with micron dimensions [2]. Nanosized gold particles on oxide supports and two-monolayer thick gold islands supported on titania exhibited significant catalytic activity towards the low temperature oxidation reaction of CO [2]. Another study showed that gold nanoparticles with low coordination number interacted synergistically with the support material to increase the rate of oxidation of CO at 273 K [3].

Not all implications of small size, however, are beneficial. For example, the inflammatory response in rats upon inhaling titanium dioxide was linked to increased particle surface area—smaller size [4]. The inflammatory response was linear over a range of particle size; 250-nm diameter particles showed 10% response but an over 30% inflammatory response was associated with 25-nm diameter particles. Other work has recently shown a similar dependence of inflammatory response to particle surface area [5]. The inflammatory response also depended on the type of material that was inhaled. A high-activity material such as SiO_2 showed nearly astronomical increases in inflammatory response as a function of increasing surface area.

5.0.4 The Nano Perspective

We are able to determine thermodynamic properties of materials by measuring macroscopic parameters such as pressure, temperature, and volume. The properties associated with any material are macroscopic manifestations of its composition. In addition to molecular makeup and the amount of substance in a material, the properties of nanoscale materials are influenced by size, shape, structure, and orientation. Optical properties in particular are influenced by orientation. Mechanical properties of composite materials depend on the orientation of materials within its matrix. Thermal conduction away from a source into a heat sink is accomplished quite efficiently by single-walled carbon nanotubes-anisotropic conductors of heat.

The past decade has witnessed the integration of biological nanomaterials with all standard classes of engineering materials: metals, semiconductors, ceramics, polymers, and composites. A brief review of these materials follows. All of these engineering materials can be expressed in nanoscale form. "We rebuild it. We make it faster. But we make it smaller."

5.1 Engineering Materials

Material structure and composition are highly integrated. One relies on the other. The structure of any material is based on its atomic and molecular makeup. Metals and alloys; elemental and compound semiconductors; ceramics, powders and glasses; and polymers and composites comprise the traditional types of engineering material categories. Any materials science course instructor will tell you as much, but because this is a course on nanoscience, biological and supramolecular materials must be added to the mix. Special materials based on carbon, like graphite, diamond, fullerenes, and nanotubes—all materials that possess strong directional covalent bonds—also play a major role in the development of advanced materials. A brief overview of the major categories of engineering materials is found in **Table 5.1.**

Delocalized metallic bonds, which involve electron sharing but are nondirectional, predominate in metals. Metals crystallize into three major forms: face-centered cubic (*fcc*), body-centered cubic (*bcc*), or hexagonal close packed (*hcp*). Ceramic materials, the silicates in particular, are held together by strong ionic bonds. Ceramics are more complex than metals and crystallize in many forms. Pure semiconductor materials like silicon crystallize in a diamond-cubic configuration

Table 5.1 Common Engineering Materials

Material type	Bonding structure		Physical properties	Examples
Metals Alloys	Metallic bonds Delocalized (nondirectional) Close-packed structures at minimal energy	Group IB–VIII and heavy elements Crystal structure is *fcc, bcc,* or *hcp* CN: 6–12 Long-range order	Conducting, ductile, readily formable High melting with large thermal expansion Relatively reactive and high strength. Characterized by metallic luster	Cr, Mn, Fe, Co, Ni, Cu, Mo, Ru, Pd, Ag, Cd, W, Os, Ir, Pt, and more
Elemental semiconductors *Intrinsic* *Extrinsic*	Covalent bonds (directional)	Group IVA Diamond cubic *fcc*, tetragonal structure CN: 4	Semiconducting	*Intrinsic:* Si, Ge, Sn (also B, Te) *Extrinsic* (dopants): *n*-type: P, As, Sb, Bi, Se, Te *p*-type: B, Al, Ga, In
Compound semiconductors Oxides	Covalent/ionic	Ceramic-like materials Zinc-blende Rutile, anatase	Semiconducting High melting Some are semiconducting and transparent	AlSb, GaP, GaAs, GaSb, InP, InAs, InSb, ZnSe, ZnTe, CdS, CdTe, HgTe TiO_2 Indium tin oxide (ITO)
Ceramics Ionic solids	Ionic bonds (non-directional) Covalent (directional)	Less densely packed Crystalline Metal + C, N, O, P, or S	Insulating High melting Chemically stable Brittle	Al_2O_3, MgO, SiO_2, silicates, alumino-silicates, ZrO_2, SiC, WC, NaCl, $CaCl_2$, and more
Glasses		Noncrystalline Short-range order	Low melting, soft	Silicates, Na_2O
Carbon-based materials	Covalent bonds (directional) characterized by electron sharing	Group IVA Open structures Crystalline Amorphous	Insulating/conducting Semiconducting High melting	Graphite, SWNT, MWNT, Fullerenes, diamond, fibers, carbon black
Supramolecular structures	Covalent bond backbone precursors, coordination bonds, hydrogen bonds Compounds held together by intermolecular forces	Group IVA–VIIA Light elements C, H, O, N, S, P + minerals Self-assembly capability	Known as "soft matter" Many forms decompose under less than mild conditions. Formed under thermodynamic control conditions rather than kinetic conditions	Biological chemicals like DNA, proteins, lipids. Micelles, self-assembled monolayers Precursors are organic chemical
Polymers	Covalent backbones held together by weak inter-molecular forces	Group IVA–VIIA Light elements C, H, O, N, S, F	Extremely wide range of physical properties. Can be insulating or conducting Polymers are low-melting; some possess high chemical reactivity; others inert. Poor packing	Polyethylene (PE) Polyaniline (PAN) Polyvinyl chloride (PVC) Polymethyl-methacrylate (PMMA) Teflon

TABLE 5.1 (CONTD.) *Common Engineering Materials*

Material type	Bonding structure		Physical properties	Examples
Composites	All classes of bonding are represented in composite materials	All elements are represented in composites. The matrix is usually a polymer resin containing metal/ceramic/carbon fillers	Properties are variable and encompass the entire range of physical properties	Graphite epoxy resins Kevlar Fiberglass Wood Concrete
Biological materials	All classes of bonding are represented in biological materials	Group IVA–VIIA Light elements C, H, O, N, S, P+ minerals	Biomaterials are capable of directed self-assembly Denature easily under nonideal conditions Physical properties span a large range and depend on structure	Proteins Lipids Carbohydrates Nucleic acids Shell, bone, teeth Wood

with covalent bonding, similar to carbon-based systems. Multicomponent semiconductors are similar in structure to ceramic materials. Polymers possess complex structure that may contain a wide range of noncrystalline material and are held together by covalent and weaker secondary forces (intermolecular bonds) such as van der Waals forces. Composite materials, another major category of engineering materials, are more complex due to the contribution of at least two, but usually more, categories of the major engineering materials.

The engineering content of this section was guided by an excellent materials text written by J. F. Shackleford: *Introduction to Materials Science for Engineers* (Prentice Hall, 1996). Although written before the *Nano Age* exerted its impact, it served as an excellent source for the development of this section.

5.1.1 Metals and Alloys

The most dominant metal used in engineering today is iron, mostly in the form of steel. The Earth's core consists mostly of iron. Iron also plays an important role in the fabrication of nanoscale materials. Carbon nanotubes, for example, can be formed from iron catalyst particles. Iron plays a vital role in living things as well in the form of cations. The iron cation is the central metal of hemoglobin, upon which aerobic organisms depend. Therefore, from the atomic level to the nanoscale level to the bulk engineering level and at the geophysical level, iron is an extremely fundamental and important material. Aluminum metal also has widespread use. At the nanoscale, aluminum metal provides the substrate for the fabrication of porous alumina templates. In living things, aluminum is required at extremely low levels but can be quite toxic at higher levels.

Nanometals come in many forms and perform diverse functions. Gold clusters comprising handfuls to thousands of atoms (a.k.a. quantum dots) can be placed in electronic arrays or act in the capacity of catalysts. Nanoshells of Au or

Ag formed on polymer or silica nanoparticles have demonstrated tunable optical response [6–8]. Recently, gold nanoshell materials have proved to be a valuable asset in treating cancer [9]. Gold substrates have shown extreme utility and adaptation in self-assembly processes that include alkanethiols. Nanometals have found their way into tennis racquets, other nanocomposite materials, and thin films. Metals and alloys containing nanograins, for example, exhibit superior physical properties. Fuel cell electrocatalysts made of nanometals have demonstrated efficiency and good performance.

Although metals are characterized by good electrical and heat conductivity, ductility, malleability, reflectance, and strength, many of these traits break down at the nanoscale. If made small enough, the electronic behavior of a metal starts to resemble that of a semiconductor. If made small enough, gold starts to act like a catalyst. The color of nanogold is strikingly different from the color of its bulk form.

5.1.2 Semiconductors

Semiconductors, unlike metals, have a band gap. The band gap existing between the valence and conduction bands, has no electronic states. The electrical conductivity of a semiconductor material lies between that of an insulator (e.g., a large band gap material) and a metal (e.g., a material with no band gap). Semiconductor materials without impurities or defects are called *intrinsic* semiconductors and are found within the metal-insulator boundary on the periodic table. Examples include silicon, germanium, tin, boron, and tellurium. Semiconductor properties, the band gap in particular, are manipulated by addition of dopants: impurities able to donate charge carriers in the form of electrons (*n-type*) or holes (*p-type*). These are known as *extrinsic* semiconductors.

Many semiconductor materials fall into the class of covalent-ionic solids. Examples include titanium dioxide (TiO_2), cadmium sulfide (CdS), and zinc oxide (ZnO). More complex materials such as indium tin oxide (ITO) are "semiconducting materials" that are both conducting and optically transparent. ITO is a mixture of indium(III)–oxide (In_2O_3) and tin(IV)–oxide (SnO_2). When indium is doped with tin, the additional electron becomes "free" and ITO becomes conducting.

Titanium dioxide (titania or titanium-IV oxide, TiO_2) is of great importance to nanoscience, nanotechnology, and industry. Highly porous nanocrystalline TiO_2 sensitized by transition metal or organic dyes is a major component of the Grätzel cell, a photocell capable of ca. 7% efficiency. Thin films ranging from 0.2 to 5 μm in thickness consist of nanosized particles 3–10 nm in size with overall specific surface of 50–150 $m^2 \cdot g^{-1}$ are formed by dip-coating methods. TiO_2 formed by sol–gel methods and sintered under 800°C yields 10- to 50-nm diameter particles with surface ca. 50 $m^2 \cdot g^{-1}$. The photocatalytic activity of TiO_2 is affected by surface area, degree of crystallinity, size, and structure (Li).

Miniaturization of bulk semiconductor materials implies, as for metals, that physical properties undergo change. Confinement, perhaps, is one of the most celebrated examples of the visible change observed in semiconductor quantum dots as particle size is diminished. The size-dependent emission of light by CdSe quantum dots provides a vivid example of this process.

5.1.3 Ceramic and Glassy Materials

Ceramic materials by definition are ionically bound, hard materials that are electrically and thermally insulating. Examples of important ceramic materials include Al_2O_3, Si_3N_4, MgO, SiO_2, Na_2O, CaO, and ZrO_2. Ceramic materials are chemically stable under a wide variety of extreme conditions: chemical, thermal, electrical, and physical. They generally possess high melting points, a characteristic that makes them suitable as refractory materials. Bulk ceramic materials, however, are quite brittle. Thermal expansion coefficient matching between a metal and its oxide coating can be problematic. The strength and fracture toughness of ceramics is enhanced with the incorporation of nanomaterial substituents.

Ceramics exist in crystalline form or noncrystalline form. Zeolites are crystalline ceramics that contain silica, alumina, and alkali metal constituents. Noncrystalline ceramics include the broad family of glasses. The interaction of a host medium, usually a glass, with nanomaterials, usually a metal, gives rise to the dichroism phenomenon of the Lycurgus cup. Ceramics also exist in the nanoform. Silica beads; the cavities in zeolites, aerogels, and xerogels; and the pore channels in anodically formed aluminum oxide are examples of nanomaterials that are made of ceramic materials.

5.1.4 Carbon-Based Materials

Special materials based on carbon form a very important class of engineering materials; as a matter of fact, carbon-based materials are integral components of the materials described in the next three sections. The primary allotropes of carbon are shown in **Figure 5.1.** Most people are familiar with the general forms of carbon, like diamond, graphite, and amorphous carbon. However, an exciting new class of carbon materials has been discovered in the past few decades: the fullerenes and their derivatives. A special section is devoted to these materials as well as to diamondoids and diamond-like materials that have gained significant attention over the past few years.

5.1.5 Polymers

Metals, semiconductors, ceramics, and glasses are inorganic materials. Polymers are made of organic materials and are usually thrown into the category of soft matter. Polymer chemistry is rightly called a spin-off of organic chemistry. As a consequence, most polymeric materials consist of the same elements that define organic chemistry: C, H, O, N, F, Cl, S, P, and Si. The age of modern engineering materials came about because of polymers. Plastics, teflon, lotions, membranes, tires, and thousands of other raw materials and products are based on polymer materials. Thermoplastics (materials that become less rigid upon heating) and thermosetting polymers (materials that become stiffer with heating) form two major classes of polymers.

Examples of nanoscale polymers are numerous. A block polymer is made of one kind of constituent macromolecule (or monomer) that is connected directly by or through a constitutional unit that is not part of the block (e.g., an interlink) [10]. When blocks are made of a different species of polymers, the term *block co-polymer*

Fig. 5.1

Allomorphs of carbon are depicted. (a) Diamond forms in a tatrahedral (sp^3 bonding) arrangement where each carbon is bonded to four others. (b) Graphite is the most stable form of carbon. It forms in planes that consist of hexagonal arrangement of carbon atoms. The bonding in graphite is sp^2. Van der Waals forces hold the planes together. (c) Another form of diamond called lonsdaleite has more hexagonal character to its structure. It is believed to form after a meteor impact that transforms graphite into lonsdaleite while creating the hexagonal form of graphite. (d)–(f) Fullerenes were discovered in 1985 by Richard Smalley and his group. Fullerenes are carbon's representative in the zero-dimensional class of nanomaterials (chapters 7 and 9). (g) Amorphous carbon does not have any crystalline structure (no long-range order). Much of amorphous carbon contains small domains of crystalline graphite (also called turbostratic graphite) with some level of amorphous carbon that holds them together. Coal and soot are amporphous carbons. (h) Carbon nanotubes are called fullerene pipes and comprise (arguably) one of the most remarkable of the carbon allotropic class of materials—or of all materials, for that matter. More will be said about carbon nanotubes in chapter 9. Glassy carbon is a nongraphitizing carbon that is used in electrodes and for high-temperature crucibles. Glassy carbon is a purely isotropic form of graphite. Carbon nanofoams are considered to be an allomorph of carbon that consists of a three-dimensional structure of carbon clusters. Nanofoams were discovered in 1997 by A. V. Rode et al.

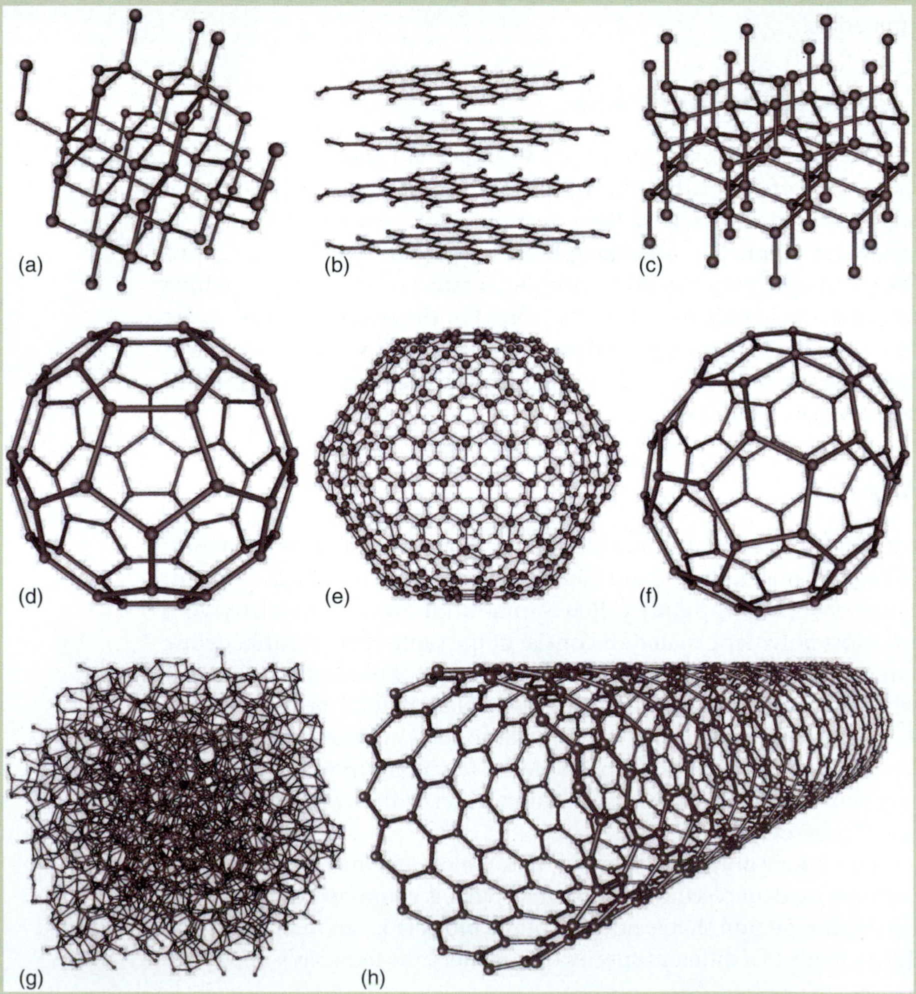

is used to describe the system. An application of block co-polymers—BCPL, is a type of lithography—was introduced briefly in chapter 4. Latex beads made of polystyrene are used in the manufacture of photonic crystals and in nanosphere lithography. Polymeric nanoparticles are used as a template to fabricate gold nanoshells. Membranes made of polycarbonate with track-etched pore channel matrices serve as templates to fabricate devices such as nanoelectrodes [11]. Lastly, polyvinyl alcohol (PVA), polyvinyl chloride (PVC), polystyrene (PS), polymethyl methacrylate (PMMA), polyanaline (PAN), polyethylene (PE), and so many more serve as the polymeric matrix for advanced nanocomposite materials, and all can be fabricated in nanoscale dimensions.

5.1.6 Biological Materials

Nanoscale materials transcend the entire spectrum of biological substances. Macroscale materials like wood and geologically altered materials like coal have been exploited by civilizations for millennia. This is also true for microscale materials like diatoms. DNA and various proteins are relatively new additions to the domain of engineering materials.

All biological materials are fabricated from the bottom up, and it is safe to assume that all macroscopic biological materials are comprised of nanocomponents. Biomimetics is the practice of copying nature. There are many examples of natural phenomena and products that we strive to emulate: polymers (in the substructure of nacre, e.g., the abalone shell); structural elements (wood, ligaments, and bone), electrical conduction (in eels and nervous systems), photoemission (by deep sea fish and glow worms); photonic crystals (butterfly and bird wings); hydrophobic surfaces (the lotus leaf and human skin); chemical sensors (the silkworm moth pheromone detection capability); adhesives (geckos' feet); high tensile strength fibers (spider silk); and artificial intelligence and computing ability (the human brain); the list goes on and on. The most remarkable contributions to engineering materials originate directly from biological nanoscale materials. Biological nanomaterials will be discussed in more detail in chapters 13 and 14.

5.1.7 Composites

Composites are materials that contain selected combinations of two or more of the other major classes of materials previously mentioned in this section: metals, semiconductors, ceramics, polymers, and, now, biological materials. A composite (from the Latin *compositus, componere,* "to put together" + "to place") is a material made of various parts or elements or other recognizable constituents—in essence, all the elements of the periodic table. There are many composites with which we are familiar. For example, fiberglass contains glass fibers embedded in a polymer matrix and collectively exhibits better properties than that of its components taken separately. The glass fibers offer high strength and the polymer material offers ductility. Together, they are considered to be a structural material that makes an efficient (and affordable) insulator. Concrete is an aggregate composite within which rock and sand interact with a silica cement glue matrix to form our bridges, buildings, and roads. Wood is a composite material that exhibits fiber reinforcement of its structure. Composite materials called graphite-epoxy resins are used in aircraft and spacecraft wings, bodies, and components. In addition to high strength, graphite composite materials offer enhanced thermal conduction, radiation ablation resistance, stealth technology, and electrical conduction. The bullet-stopping ability of Kevlar is well known.

An entirely new industry is on the horizon. The industry involves the research, development, application, and marketing of nanocomposite materials. In actuality, nanocomposites have already made an entrance into the global composite market. Multiwalled carbon nanotubes (MWNTs) are used routinely as fillers in nanocomposites and carbon nanotubes in tennis racquets.

5.2 Particle Shape and the Surface

The surface of a material depends on its size and geometric shape. There is no attempt in this section to delve into atomic–molecular details or the chemical or physical nature of nanomaterials. Three factors that relate to surface area will be explored: (1) the effect of subdivision of a parent material on the overall surface area, (2) the effect of particle shape on surface area, and (3) the concomitant increase in surface-to-volume ratio as individual particles assume smaller dimensions.

Nanomaterials have a significant portion of atoms existing at surfaces, of which there are two general types: the exterior surface and the interior surface. Exterior surfaces comprise atoms that exist on the outside surface and edges. The external surface area of a powder, for example, takes into account the roughness of the constituent particulate surfaces. Generally speaking, cavities that are wider than they are deep are considered to contribute to the exterior surface roughness, as opposed to the inverse case in which they are considered to be pores and have internal surface area. Nanomaterials of importance that belong to the category of the exterior nanosurface type include metal and semiconductor clusters and colloids. Interior nanosurfaces refer to the surfaces of pore channels and cavities and the surfaces between grain boundaries of a crystalline or polycrystalline solid. Interior surfaces are represented by inorganic porous oxides found in anodically formed alumina, natural and synthetic zeolites, amorphous carbons, and silicas.

Collective surface area and surface-to-volume ratio are inversely proportional to the size of the particle. We define collective surface area as the sum total of surface area of individual nanomaterials—especially if subdivided from a bulk parent material. Collective surface area is an extensive (additive) property. Surface-to-volume (S/V) ratio is defined as the ratio of the surface area (in square meters) to the volume of the same particle (in cubic meters). As opposed to the collective sense, S/V ratio is relevant to a single entity—e.g., a particle or pore channel. The S/V ratio therefore is an intensive property. Although "surface-to-volume" is technically described in units of reciprocal meters (m^{-1}), S/V (or N_s/N_v) ratios are usually listed as a dimensionless number.

The volume of a sphere increases as the cube of the radius, r, while the surface area increases as the square of the radius. As r is decreased, the proportion of surface units (area or atoms) increases with respect to its volume. With regard to the collective sense, as in the case of a single bulk material divided into multiple clusters, surface area increases geometrically as more new surface is created,

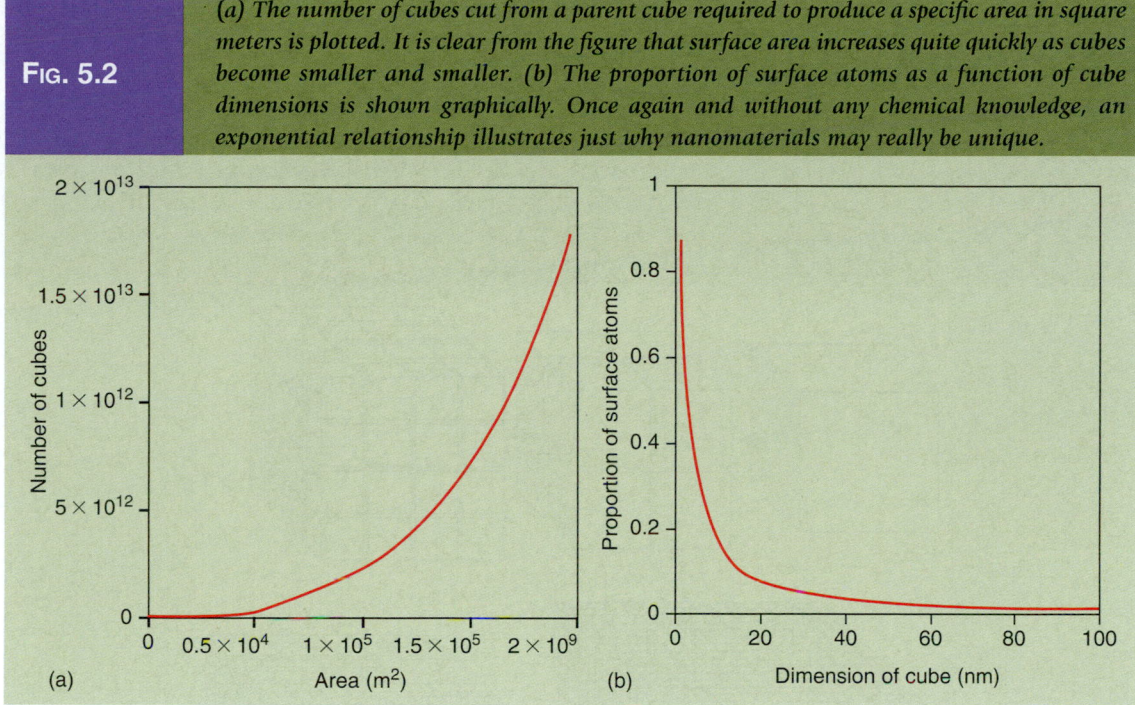

FIG. 5.2 *(a) The number of cubes cut from a parent cube required to produce a specific area in square meters is plotted. It is clear from the figure that surface area increases quite quickly as cubes become smaller and smaller. (b) The proportion of surface atoms as a function of cube dimensions is shown graphically. Once again and without any chemical knowledge, an exponential relationship illustrates just why nanomaterials may really be unique.*

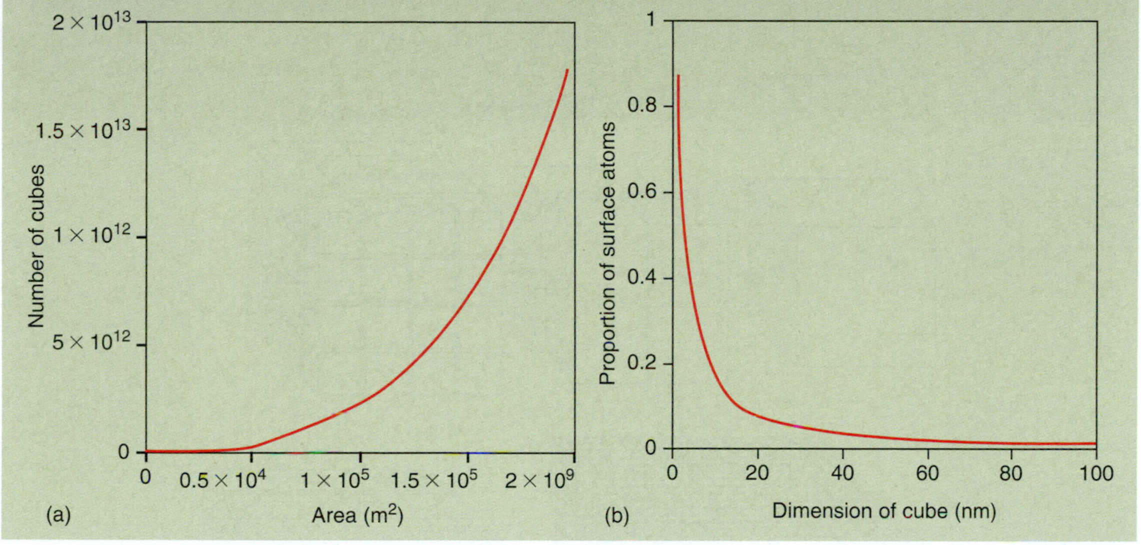

leading to incredibly large values (**Fig. 5.2a**). The trend in surface-to-volume ratio is depicted in **Figure 5.2b.**

5.2.1 Exterior Surface and Particle Shape

Without regard to the details involved in packing atoms or molecules, a cube of some material that is 1 m on each side has surface area equal to 6 m^2. If this cube were broken perfectly into smaller cubes that are 0.1 m on each side, resulting in 1000 additional cubes, the collective surface area becomes 60 m^2. For cubes with 1-cm edges, a million such cubes can fit into a cubic meter, and the collective surface area increases to 600 m^2. For cubes with 1-mm edges, the surface area expands to 6000 m^2. The trend is illustrated in **Figure 5.3.**

Surface area depends on particle shape in addition to size. Cubes, rectangular solids, spheres, spheroids, cylinders, pyramids, and discs are common shapes that nanomaterials assume. Tetrahedrons, icosahedrons, dodecahedrons, cuboctahedrons, tubes, hollow spherical structures, astral structures, and dendritic arrays are more examples of the incredible diversity found in synthetic and naturally occurring nanomaterials (**Fig. 5.4**).

A cube with the same volume as a sphere has a higher surface area. Cylindrical solids, with height h and diameter d, cover a wide range of surface area. The disk ($h \ll d$) and the wire ($h \gg d$) potentially have the highest surface area among the geometric series listed previously (**Fig. 5.5**). The sphere is the most remarkable geometric solid. It has an infinite number of symmetry axes. As result, a sphere

FIG. 5.3 *A visual image of the concept of collective surface area is always helpful in acquiring the proper perspective. As one can readily see, it does not take too many divisions of the parent cube to reach a level of enormous surface area. Simple geometric progression from successive division of a parent cube illustrates the concept of collective surface area. If cubes are divided into eight equivalent cubes with each step, the surface area doubles after each division. For n equal to the number of cubes, surface area proceeds from 6 m² (for n = 1) to 12 (n = 8), 24 (n = 64), 48 (n = 512), and 96 m² (n = 4096). The length of a side of the n = 4096 cube is only 6.25 cm. One can only imagine the total surface area if these cubes are made to be 1 nm on a side.*

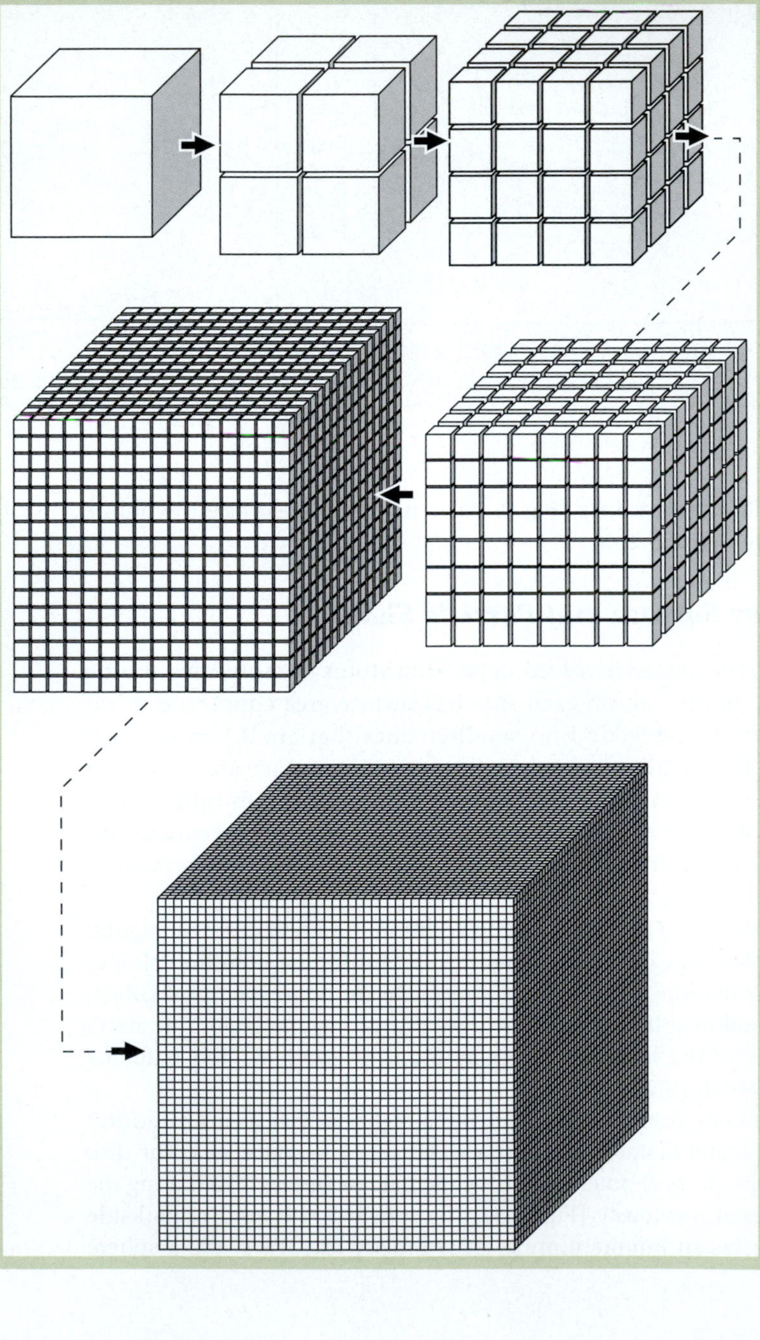

EXAMPLE 5.2 *Determination of the Collective Surface Area of Nanocubes*

How many cubes 1 nm on each side can be carved out of a cubic parent 1 m on each side? Find the collective surface area of the nanometer-sized cubes.

Solution:

This calculation is straightforward. We need to calculate the number of nanometer-sized cubes that are possible from a 1-m sized cube, find the area of one of the smaller cubes, and then sum for the total surface area. The surface area of a 1-m cube is 6 m².

$$\text{Volume of nanocube: } \left(1\times10^{-9}\,\text{m}\right)^3 = \left(1\times10^{-27}\,\text{m}^3\right)$$

$$\text{Surface area of nanocube: } \left(1\times10^{-9}\,\text{m}\right)^2\times6 = \left(6\times10^{-18}\,\text{m}^2\right)$$

$$\text{Number of nanocubes per cubic meter: } \frac{1\,\text{m}^3}{\left(1\times10^{-27}\,\text{m}^3\right)} = 1\times10^{27}\text{ nanocubes}$$

Total surface area:

$$\left(1\times10^{27}\text{ nanocubes}\right)\left(\frac{6\times10^{-18}\,\text{m}^2}{\text{Nanocube}}\right) = 6\times10^{9}\,\text{m}^2 = 6{,}000\,\text{km}^2$$

The total surface area of the nanocubes is a billion times that of the 1-m cube. This is equivalent to more than 2000 square miles—an area bigger than the state of Delaware (1954 sq. mi.). The power of nano!

made of a real material is characterized by an infinite number of degenerate energy states (e.g., all energy levels are equivalent). The sphere represents the lowest energy configuration of solid materials. The sphere also happens to represent the surface minimum in the continuum of geometric solids—a coincidence?

FIG. 5.4 *Nanomaterials can assume any shape imaginable—whether of the two-dimensional variety depicted in the first row, polyhedral shapes depicted in the middle row, or ellipsoidal forms depicted in the bottom row. Spheres, discs, wires, dots, and two-dimensional planar materials abound at the nanoscale.*

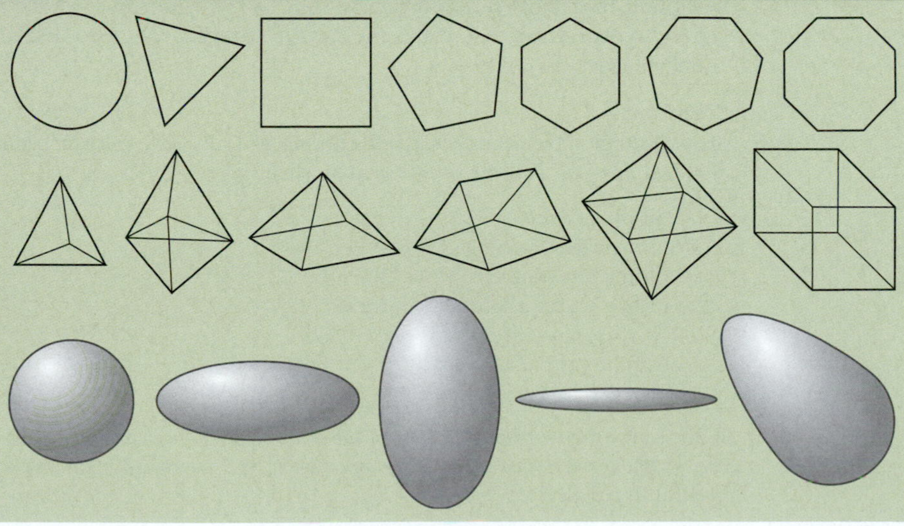

FIG. 5.5 *The ratio between h and d determines whether a shape is like a wire or more like a disc. Many nanomaterials are found in either morphology.*

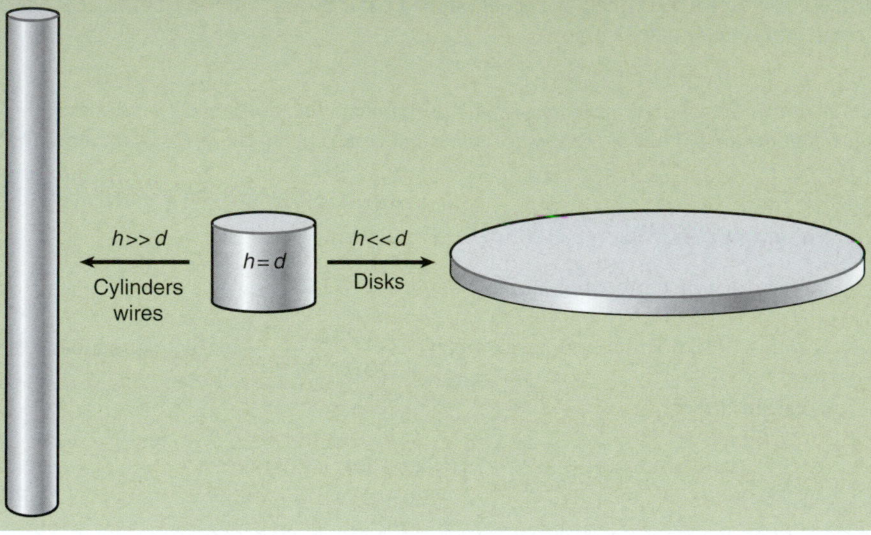

Nanomaterials are found in all of the seven basic crystal systems and consequently in all the Bravais lattices. Crystalline forms derived from Bravais solids are also well represented in nanomaterials. Crystalline nanomaterials can assume the basic shapes of any of those based on Bravais lattices. Bravais lattices are discussed in more detail later.

EXAMPLE 5.3 *Shape Dependence of Surface Area*

Determine the relative surface areas of a cube (a^3) with the same volume as a sphere. Derive an expression that relates the surface area of a sphere to that of a cube of equal volume.

Solution:

Set the volume of the sphere equal to the volume of a cube (a^3); solve for volume in terms of a, and then calculate the surface area for each shape.

Volume of cube: $V_c = a^3$
Surface area of a cube: $S_c = 6a^2$
Volume of a sphere: $V_s = (4\pi/3)r^3$
Surface area of a sphere: $S_s = 4\pi r^2$
$V_{sph} = V_{cube} \rightarrow (4\pi/3)r^3 = a^3 \rightarrow r = 0.62a$
$S_{sph}(a) = 4\pi(0.62a)^2 = 4.83a^2$
Ratio of S_c to S_s: $6a^2/4.83a^2 = 1.24 \rightarrow S_c = 1.24\ S_s$

When comparing a sphere to a cube of the same volume, the sphere will have less surface area. Although a simple geometric phenomenon, this relationship has important consequences in nanoscience.

Example 5.4 *Surface Area of Right-Cylindrical Shapes of Constant Volume*

Determine the surface area of right cylinders of unit volume where $V_{cyl} = 1$. Derive an expression that relates the surface area of the following cylinders normalized to that of a sphere with volume equal to 1:

(a) $d = h$
(b) $d \ll h$ (assume $h = 100$)
(c) $d \gg h$ (assume $h = 0.01$)

Solution:

When the diameter d of a right cylindrical material is much greater than the height h of the cylinder, the shape is a disc. Conversely, if h is significantly greater than d, the shape is equivalent to that of a wire. Both types of shapes are exhibited by important nanomaterials. Use $V_{Cyl} = \pi r^2 h = 1$ *and* $S_{Cyl} = 2\pi r^2 + 2\pi rh$ *and that* $d = 2r$.

$V_{cube} = 1.000$, $S_{cube} = 6.000$
$V_{sphere} = 1.000$, $S_{sphere} = 4.836$
$V_{Cylinders} = 1.000$, $S_{Cyl} = ??$

(a) $h = d$, $r = 0.5419$, $S_{cyl} = 5.535$, normalized to the sphere = 1.14. When $d = h$, this cylinder has the minimum surface area within the family of cylinders with $V_{cyl} = 1$.
(b) $h = 100.0$, $r = 0.5642$, $S_{cyl} = 35.47$, normalized to the sphere = 7.33. In the case of a wire, the second term in the surface area equation $S_{cyl} = 2\pi r^2 + 2\pi rh$ becomes more important as the height (length) of the wire increases.
(c) $h = 0.01000$, $r = 5.642$, $S_{cyl} = 200.4$, normalized to the sphere = 41.4. In the case of a disk, the first term in the surface area equation $S_{cyl} = 2\pi r^2 + 2\pi rh$ becomes more important as the height, h, of the cylinder gets smaller.

5.2.2 Interior Nanoscale Surface Area

According to the International Union of Pure and Applied Chemistry (IUPAC), the definitions of pore size are as follows [12]: *Micropores* have widths smaller than 2 nm, *mesopores* have widths between 2 and 50 nm, and *macropores* have widths larger than 50 nm. We introduced this nomenclature earlier.

Obviously, all of the IUPAC designations could be considered as nanoscopic pores, depending on your point of view. The classification is somewhat confusing because the term "macropores" implies something quite larger in scale. Examples of microscopic pores are exhibited by the zeolites, which have 2-nm sized or less pores; mesoscopic pores are represented by the pore channels in anodically formed alumina that range from a few to a few hundred nanometers. Pores, cavities, and channels are found in crystalline structures (zeolites), assemblies of aggregated materials (ceramics), or in polycrystalline materials like anodically formed alumina.

Some degree of pore structure, whether in the form of cavities, channels, interstices, or grain boundaries, is inherent in most solids. Defects in solids such as voids and vacancies are cavities—albeit very small ones. A porous solid is defined as a solid with pores that are deeper than they are wide. In materials designed overtly to be porous, characteristics such as void volume (or porosity) are important considerations. These include polymers, ceramics, metals, and semiconductors. Pores can be closed or open. Closed pores influence properties such as density, electrical conductivity, and thermal conductivity. Open pores influence transport properties and are able to selectively pass materials based on size, whether gaseous or liquid, through a membrane. Void volume (porosity) ε of a porous material is defined as

$$\varepsilon = \frac{V_{\text{Pores}}}{V_{\text{Material}}} \tag{5.1}$$

Just like with exterior surface area, many chemical reactions and phenomena take place inside pores. Pores are able to confine substances within their boundaries and thereby influence kinetic control over the course of a chemical reaction. Recall that a substrate is confined within the active pocket of an enzyme—a very powerful means of initiating reactions under mild conditions. An entire industry is

Example 5.5 *Surface Area of a Common Porous Material*

Determine the surface area in terms of cm^2 of an alumina membrane (10.0 cm × 10.0 cm × 40.0 μm thickness) with (a) no pores; (b) pore channels with diameter, d_1 = 250 nm, 55% porosity of membrane; and (c) diameter, d_2 = 2.50 nm, 55% porosity of membrane. Consider the pore channels to be right cylinders that are all oriented normal to the surface of the membrane.

Solution:

This example is solved primarily by meticulous bookkeeping.

(a) Surface area of membrane with no pores
Apparent volume of membrane material:
$V_{\text{Material}} = (10.0\text{ cm})(10.0\text{ cm})(40.0 \times 10^{-4}\text{ cm}) = 0.400\text{ cm}^3$
Surface area of the membrane: $S_{\text{Material}} = (10.0\text{ cm})(10.0\text{ cm})(2) + (40.0 \times 10^{-4}\text{ cm})(10.0\text{ cm})(4) = 200\text{ cm}^2$

(b) Total surface area of membrane with 250-nm diameter pores, $\varepsilon = 0.55$
Total porous volume: $V_{\text{Pores}} = \varepsilon V_{\text{Material}} = (0.55)(0.40\text{ cm}^3) = 0.22\text{ cm}^3$
Single pore channel volume: $V_{pc} = \pi r_1^2 T = \pi(125 \times 10^{-7}\text{ cm})^2(40.0 \times 10^{-4}\text{cm}) = 1.96 \times 10^{-12}\text{ cm}^2$
Total number of pore channels: $N = V_p/V_{pc} = (0.22\text{ cm}^3)/(1.96 \times 10^{-12}\text{cm}^3) = 1.12 \times 10^{11}$ pore channels
Single pore channel surface area: $S_{pc} = \pi dT = \pi(250 \times 10^{-7}\text{ cm})(40 \times 10^{-4}\text{ cm}) = 3.14 \times 10^{-7}\text{ cm}^2$
Total surface area of pore channels: $S_p = S_{pc}N = (1.12 \times 10^{11})(3.14 \times 10^{-7}\text{ cm}^2) = 3.52 \times 10^4\text{ cm}^2$
Total surface area of membrane not counting contribution from the pores:
$S_m = (1 - V_p)(2L \times 2) + (T \times L \times 2) + (T \times L \times 2) = 0.45(200) + 4L(40.0 \times 10^{-4}\text{ cm}) = 90.2\text{ cm}^2$
Total surface area of material: $S_T = S_p + S_m = 3.53 \times 10^4\text{ cm}^2$
Obviously, the contribution to the total surface area from the membrane is negligible

(c) Total surface area of membrane with 2.5-nm diameter pores, $\varepsilon = 0.55$
Total porous volume: $V_p = \varepsilon V_{\text{app}} = (0.55)(0.40\text{ cm}^3) = 0.22\text{ cm}^3$
Single pore channel volume: $V_{pc} = \pi r_1^2 T = \pi(1.25 \times 10^{-7}\text{ cm})^2(40.0 \times 10^{-4}\text{ cm}) = 1.96 \times 10^{-16}\text{ cm}^2$
Total number of pore channels: $N = V_p/V_{pc} = (0.22\text{ cm}^3)/(1.96 \times 10^{-16}\text{ cm}^3) = 1.12 \times 10^{15}$ pore channels
Single pore channel surface area: $S_{pc} = \pi dT = \pi(2.50 \times 10^{-7}\text{ cm})(40 \times 10^{-4}\text{ cm}) = 3.14 \times 10^{-9}\text{ cm}^2$
Total surface area of pore channels: $S_p = (1.12 \times 10^{15})(3.14 \times 10^{-9}\text{ cm}^2) = 3.52 \times 10^6\text{ cm}^2$
Total surface area of membrane not counting the pores:
$S_m = (1 - V_p)(L \times W \times 2) + (T \times L \times 2) + (T \times W \times 2) = 0.45(200) + 2(40.0 \times 10^{-4}\text{ cm})(L + W) = 90.2\text{ cm}^2$
Total surface area of the material: $S_T = S_p + S_m = 3.52 \times 10^6\text{ cm}^2$
The surface area of the membrane becomes even less important

What are the respective specific surface areas in terms of square meters per gram if the density of the alumina is 2.0 g·cm^{-3}?

based on the exploitation of the internal surface area of micro- and nanomaterials. Critical biological functions rely on cavities, pores, channels, and active pockets.

5.3 Surface and Volume

By now you, the student, are certainly becoming aware of the power of the nanoscale, at least within the geometric frame of reference presented so far. However, the geometric development has not yielded any clues concerning physical properties and we have advertised at length the remarkable properties that nanomaterials possess based on smaller size and increased surface area. The following surface-to-volume ratio discussion will help pry open that door, if ever slightly. It is actually all about the surface atoms, which are the *first responders* in a reaction and form the interface with the exterior environment. The surface atoms are responsible for chemical behavior of a material. This is true for individual particles that assume new properties as size is reduced or for bulk materials that comprise nanosized grains. An example of the former includes depressed melting points for metal nanoparticles and of the latter enhanced tensile strength in steel. **Figure 5.6** depicts the surface atom fraction of pseudospherical iron particles as size is reduced, ranging from 35 nm down. Particles over 35 nm have a significantly lower percentage of surface atoms.

Fig. 5.6 *Along with the same theme as before, the surface atom percentage (compared to the total volume) of pseudospherical metal particles is shown graphically. Pseudospherical implies that the metal particle is not in the form of a perfect sphere, but rather in the form of a cuboctahedron or other pseudospherical shape. It is apparent from the graph that the ratio of surface atoms increases dramatically as particles become smaller. In many nanomaterials, some actually can be considered to be all surface.*

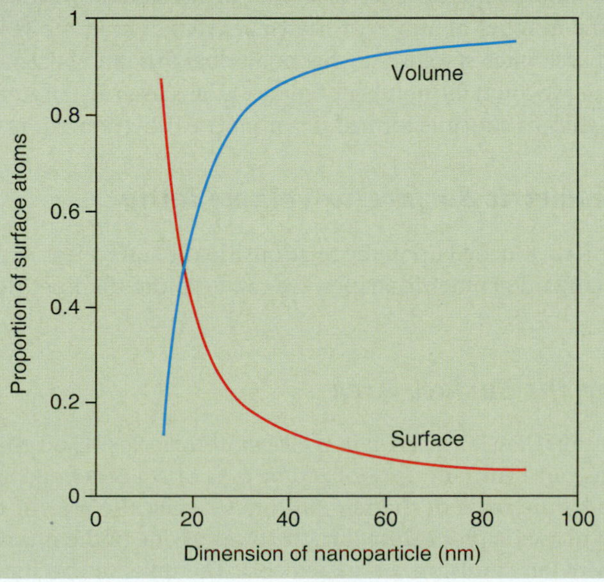

Fig. 5.7 *Gold atoms are arranged in a simple cubic crystal form, where the atoms in each layer are right on top of the atoms in the previous layer (called in register). Brute force implies direct measurement of edges, etc. to derive the ratios we seek with a minimal of sophistication.*

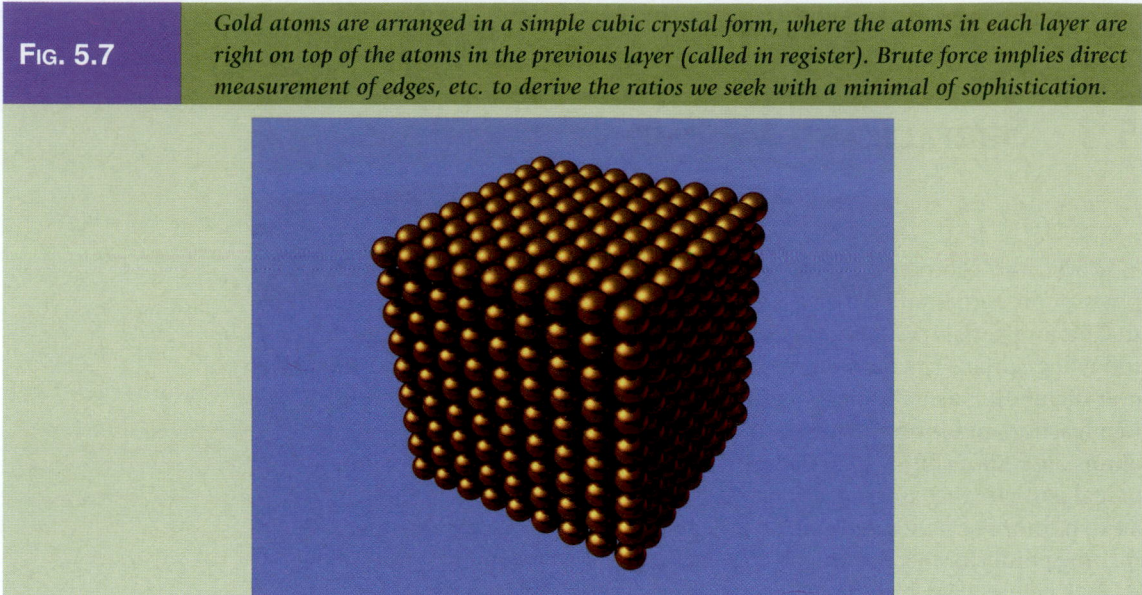

Source: Courtesy of Professor Anil K. Rao, Department of Biology, the Metropolitan State College of Denver.

The drumbeat delivering the message of surface area and its importance in nanoscience continues in this section. In the bulk world, properties such as electrical conductivity and optical response are averaged over a large area or mass. In the world of nanoparticles these properties are subject to change in ways that are quite remarkable (e.g., the Lycurgus cup). Specific surface area (square meters per gram) is an important parameter in the field of catalysis. The effects of particle size on the rate of catalyzed chemical reactions are well documented. The ratio of the number of atoms or molecules on the surface of a material to the number of atoms or molecules that exist in the volume of that material is of keen interest to us. At what point does this ratio become significant? In example 5.6, we count the number of surface atoms in a cube of a material called hypothetical gold—a structure formed in a primitive cubic lattice (**Fig. 5.7**).

5.3.1 Geometric Surface-to-Volume Ratio

Although we have touched upon the relationship the surface has with its volume for several standard geometric shapes, we now explore the finer details of that relationship.

5.3.2 Specific Surface Area

According to the *IUPAC Compendium of Chemical Technology* (2nd edition, 1997), specific surface area (usually symbolized by a, S, or A_s) is defined as the surface area divided by the mass of the relevant phase when the area of the interface between two phases is proportional to the mass of one of the phases (e.g., for a solid adsorbent, an emulsion, or an aerosol). The units of specific surface area

Example 5.6 *Determination of the Percent of Surface by the Method of Brute Force*

This calculation features a material called "hypothetical gold." Calculate the number of surface atoms, N_s, edge atoms, N_e, and volume atoms, N_v, and the surface-to-volume atomic ratio N_s/N_v as size is decreased by order of magnitude increments, starting with a 1-cm^3 chunk of hypothetical gold (AAA...with in register packing throughout—a simple cubic structure—depicted in **Fig. 5.7**). Au: covalent radius = 0.144 nm (Note: Although gold exists in cubic-close-packed (ccp)/face-centered-cubic (*fcc*) structure, for this example we assume that the covalent radius is the same for both the real and hypothetical gold); atomic mass = 196.967 $g \cdot mol^{-1}$, density ρ = 19.31 $g \cdot cm^3$.

Solution:

First calculate the number of gold atoms per meter (and per centimeter, millimeter, micrometer, nanometer, etc.). Take the cube of your answer to find the volume in terms of number of atoms and the square × 6 to find the surface area in terms of the number of atoms. The calculation of N_s/N_v is straightforward.

A sample calculation is given and the remaining values and ratios are summarized in ***Table 5.2***.

$$\text{Number of Au atoms in 1 m: } \frac{1\,\text{m}}{0.288\times10^{-9}\,\text{m}} = 3.47\times10^{9}\ \text{Au atoms}$$

$$\text{Volume of Au atoms: } \left(3.47\times10^{9}\right)^{3} = 4.19\times10^{28}\ \text{Au atoms}$$

$$\text{Surface area Au atoms: } \left(3.47\times10^{9}\right)^{2}\times6 = 7.22\times10^{19}\ \text{Au atoms}$$

$$\text{Surface} - \text{volume ratio: } \frac{7.22\times10^{19}}{4.19\times10^{28}} = 1.72\times10^{-9}$$

are square meters per gram of material ($m^2 \cdot g^{-1}$). Specific surface area is simply the ratio of S/V to the density of a material:

$$S_S = \frac{S_m}{\rho V_m} \tag{5.2}$$

Table 5.2 *Surface–Volume Parameters for Hypothetical Gold of Diminishing Volume*

Cube dimension	Edge atoms (N_e)	Volume atoms (N_v)	Surface atoms (N_s)	Surface–volume ratio (N_s/N_v)
Meter	3.47×10^9	4.19×10^{28}	7.22×10^{19}	1.72×10^{-9}
Centimeter	3.47×10^7	4.19×10^{22}	7.22×10^{15}	1.72×10^{-7}
Millimeter	3.47×10^6	4.19×10^{19}	7.22×10^{13}	1.72×10^{-6}
Micrometer	3.47×10^3	4.19×10^{10}	7.22×10^7	1.72×10^{-3}
500 nm	1.74×10^3	5.23×10^9	1.82×10^7	3.47×10^{-3}
100 nm	3.47×10^2	4.19×10^7	7.22×10^5	0.0172
50 nm	1.74×10^2	5.23×10^6	1.82×10^5	0.0347
25 nm	86.0	6.54×10^5	4.44×10^4	0.0678
10 nm	34.7	4.19×10^4	7.22×10^3	0.172
4.90 nm	17	5.23×10^3	1.81×10^3	0.346
2.88 nm	10	1000	488	0.488
2.02 nm	7	343	218	0.635
1.15 nm	4	64	56	0.875

Note: The ratio of surface atoms to all the atoms of the volume increases from approximately one-billionth to 88%.

Example 5.7 *Determination of Surface-to-Volume Ratio Relations of Selected Geometric Solids [13]*

Derive some generalized formulae of surface-to-volume ratios for the following shapes: cube, sphere, and cylinder with diameter d and height h where $d = h$, $d \ll h$, and $d \gg h$.

Solution:

The units of surface-to-volume ratio are in terms of reciprocal meters (m^{-1}), centimeters (cm^{-1}), or nanometers (nm^{-1}).

Cube with sides equal to a:

$S_{cube} = 6a^2$

$V_{cube} = a^3$

$S_{cube}/V_{cube} = 6/a$

Sphere:

$S_{sph} = 4\pi r^2$

$V_{sph} = (4/3)\pi r^3$

$S_{sph}/V_{sph} = 3/r$, with $2r = d$, $S_{sph}/V_{sph} = 6/d$

Cylinder with $h \gg d$ (the wire):

$S_{cyl} = 2\pi(d/2)^2 + 2\pi(d/2)h \approx 2\pi(d/2)h$

$V_{cyl} = \pi r^2 h = \pi(d/2)^2 h$

$S_{cyl}/V_{cyl} = 2\pi(d/2)h / \pi(d/2)^2 h$, $S_{cyl}/V_{cyl} = 4/d$

Cylinder with $d = h$:

$S_{cyl} = 2\pi r^2 + 2\pi rh \rightarrow 2\pi(d/2)^2 + 2\pi(d/2)d \rightarrow 2\pi[(d^2/2)(1/2) + (d^2/2)]$

$V_{cyl} = \pi r^2 \mathrm{h} = \pi(d/2)^2 d$

$S_{cyl}/V_{cyl} = 2[(1/2) + 1]/d$, $S_{cyl}/V_{cyl} = 3/d$

Cylinder with $d \gg h$ (the disc):

$S_{cyl} = 2\pi(d/2)^2 + 2\pi(d/2)h \approx 2\pi(d/2)^2$

$V_{cyl} = \pi r^2 h = \pi(d/2)^2 h$

$S_{cyl}/V_{cyl} = 2\pi(d/2)^2/[\pi(d/2)^2 h]$, $S_{cyl}/V_{cyl} = 2/h$

where S_s is the specific surface area in square meters per gram, S_m is the measured area in square meters, ρ is the density of the material (grams per cubic·meter), and V_m is the measured volume of the material in cubic meters.

Specific surface area is an important parameter to gauge the effectiveness of catalysts, filtration, and gas chromatographic efficiency, among many other applications. For example, in fish tanks, the specific surface area of the filter (as SSA) indicates how much fish waste can be metabolized by nitrifying bacteria in a 24-h period. In gas chromatography, specific surface area of carbon supports a range from 5 $m^2 \cdot g^{-1}$ for graphitic materials and to 1000 $m^2 \cdot g^{-1}$ for activated carbons. Diatomaceous earth, fire brick (calcined celite), teflon chips, and polymer beads are examples of high specific surface area supports used in chromatographic columns. Catalysis depends on specific surface area. A silicon carbide catalyst, β-SiC (150 $m^2 \cdot g^{-1}$), with 1% Pt doping showed 100% conversion of CO to CO_2 at 175°C [14].

5.3.3 Spherical Cluster Approximation

The spherical cluster approximation (SCA) is based on several fundamental assumptions. First, cluster radius, surface area, and volume are calculated without consideration of packing fraction, coordination number, lattice constants, and other factors associated with structure and the real chemistry involved in its packing [15]. Obviously, the larger the number of atoms N becomes, the more accurate the SCA model becomes. Secondly, the SCA considers, as it name implies,

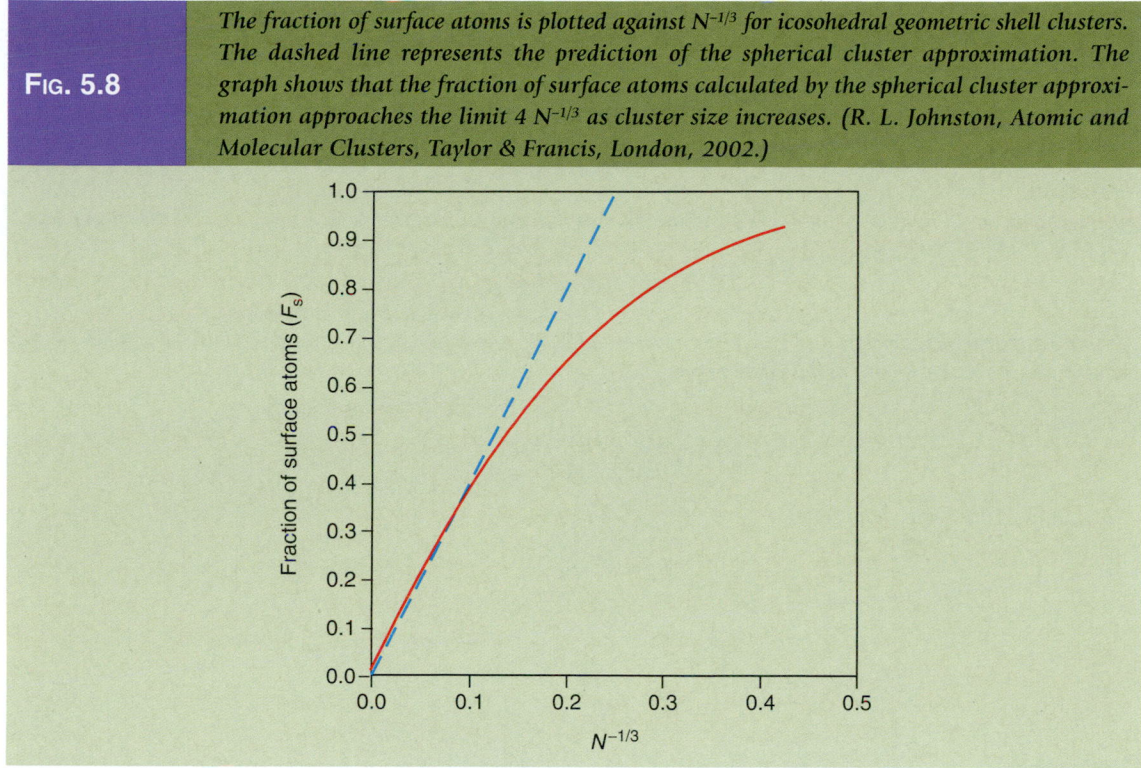

Fig. 5.8 *The fraction of surface atoms is plotted against $N^{-1/3}$ for icosohedral geometric shell clusters. The dashed line represents the prediction of the spherical cluster approximation. The graph shows that the fraction of surface atoms calculated by the spherical cluster approximation approaches the limit 4 $N^{-1/3}$ as cluster size increases. (R. L. Johnston, Atomic and Molecular Clusters, Taylor & Francis, London, 2002.)*

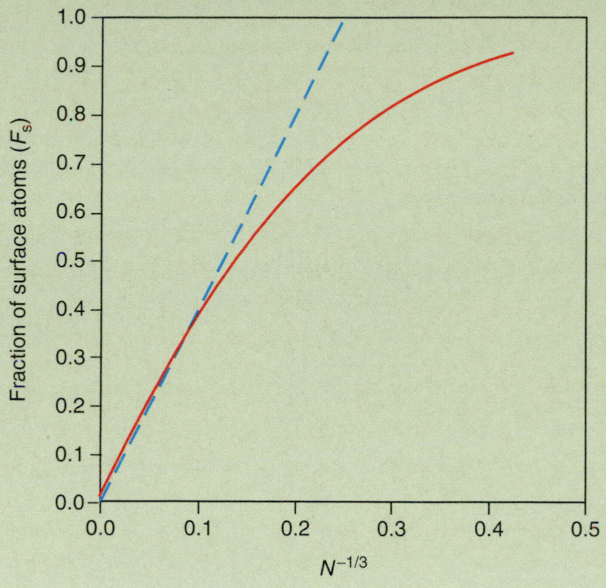

only hard spherical shapes [9]. Finally, in our case, we will designate that the clusters are made of one element: a metal. We begin with a cluster of N_a atoms with individual atomic volume equal to V_a. Our goal is to relate the number of surface atoms to the total number of atoms that comprise the cluster. Please refer to example 5.8.

The fraction of surface atoms is plotted versus $N^{-1/3}$ in **Figure 5.8.** The solid line represents the theoretical limit of the spherical cluster approximation [15]. The limiting value is given by equation (5.8) in example 5.8, $4N_a^{-1/3}$.

5.4 Atomic Structure

Fundamental knowledge of material structure is necessary in order to understand the behavior of nanomaterials. Therefore, a brief overview is presented in this section. Why go through this exercise? All nanomaterials seek a structural configuration that corresponds to minimal energy—highest possible stability. During this process, certain rearrangements within its structure are expected to occur. It is good to get acquainted with the generalized families of crystals to understand the driving forces behind such rearrangement.

5.4.1 *Crystal Systems and the Unit Cell*

Seven geometric crystal (axial) systems and 14 unique Bravais lattices, named after the French physicist Auguste Bravais in 1845, describe the position, direction, and

Example 5.8 *Spherical Cluster Approximation*

Calculate the number of volume atoms, the number of surface atoms, and the percentage of surface atoms N_s to volume atoms N_v for a spherical cluster of N_a atoms. Relate the cluster radius (R_c), surface area (S_c), and volume (V_c) to the radius (R_a), surface area (S_a), and volume (V_a) of an individual atom in the cluster [15].

Solution:

If the volume of an individual atom is V_a and the number of atoms in a cluster is N_a, the volume of the cluster V_c is equal to

$$V_c = N_a V_a \tag{5.3}$$

Assuming that the cluster has a spherical shape with radius R_c, the volume of the cluster equals the volume of the constituent atoms, each with radius equal to r_a:

$$V_c = \frac{4}{3}\pi R_c^3 = N_a\left(\frac{4}{3}\pi r_a^3\right) \tag{5.4}$$

The radius of the cluster then equals

$$R_c = N_a^{1/3} R_a \tag{5.5}$$

and

$$N_a = \sqrt[3]{\frac{R_c}{r_a}} \tag{5.6}$$

The surface area of the cluster, S_c, as it is related to the radius of the constituent atoms, then becomes

$$S_c = 4\pi R_c^2 = 4\pi\left(N_a^{2/3} R_a^2\right) = N_a^{2/3} S_a \tag{5.7}$$

The number of cluster surface atoms is derived from the cross-sectional area of an atom A_a:

$$N_s = \frac{4\pi\, N_a^{2/3} R_a^2}{\pi R_a^2} = 4N_a^{2/3} \tag{5.8}$$

This is the limiting expression for surface atoms as cluster size approaches large dimensions. The calculation of the fraction of atoms, F_s, is straightforward:

$$F_s = \frac{N_s}{N_a} = 4N_a^{-1/3} \tag{5.9}$$

This approximation becomes better as cluster size approaches bulk dimensions [9].

structural planes of most crystalline materials (**Fig. 5.9**). Crystal structure is based on regularly repeating elements that form a pattern in three dimensions (e.g., the unit cell). The unit cell is described by a set of lattice parameters and is the irreducible representation of the crystal structure. For example, knowledge of lattice constants such as unit cell edge length (a,b,c) and crystallographic axis angle (α,β,γ) allows for facilitative quantification of material structure. In other words, translation of structurally equivalent positions of the unit cell over and over again results in a material with long-range order—ultimately, a crystal. In contrast, such long-range order is absent in amorphous materials such as glass, liquids, and gases.

Fig. 5.9 *The seven basic geometric crystal systems are shown with corresponding Bravais lattices. We will just discuss the three cubic systems in this text.*

Crystal system	Definitions	Generic rendition	Bravais lattice
Cubic	$a=b=c$ $\alpha=\beta=\gamma=90°$	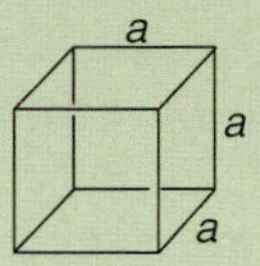	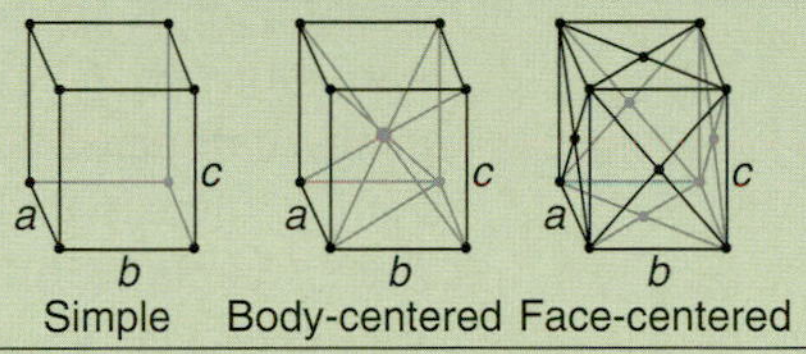
Tetragonal	$a=b\neq c$ $\alpha=\beta=\gamma=90°$	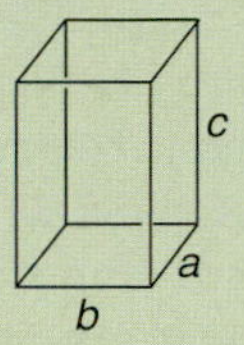	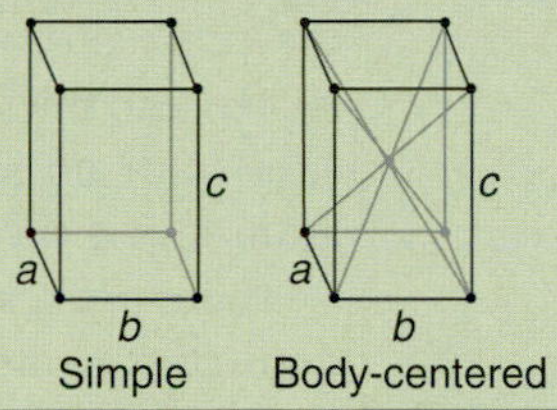
Orthorhombic	$a\neq b\neq c$ $\alpha=\beta=\gamma=90°$	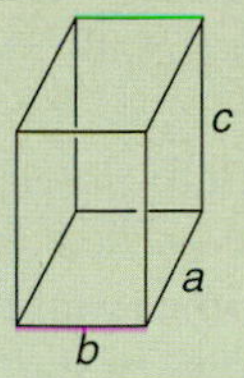	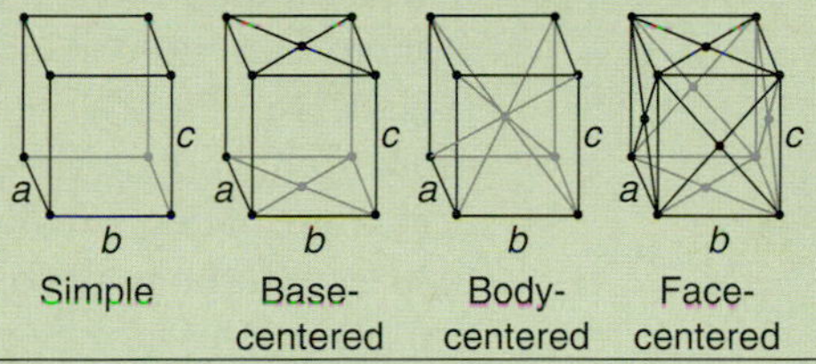
Rhombohedral	$a=b=c$ $\alpha=\beta=\gamma\neq 90°$	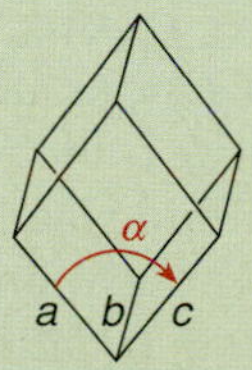	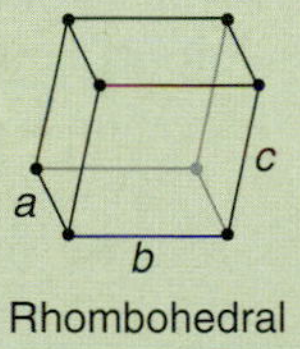
Hexagonal	$a=b\neq c$ $\alpha=\beta=90°$ $\gamma=120°$	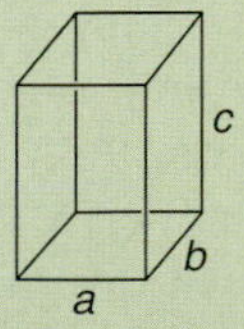	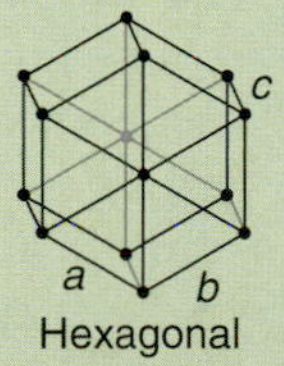
Monoclinic	$a\neq b\neq c$ $\alpha=\gamma=90°\neq\beta$	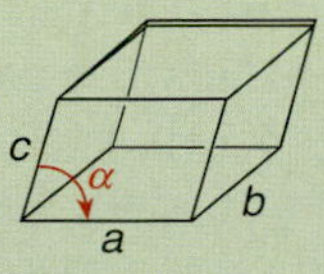	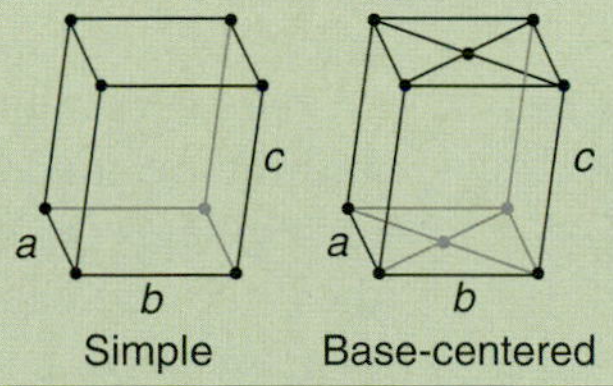
Triclinic	$a\neq b\neq c$ $\alpha\neq\beta\neq\gamma\neq 90°$	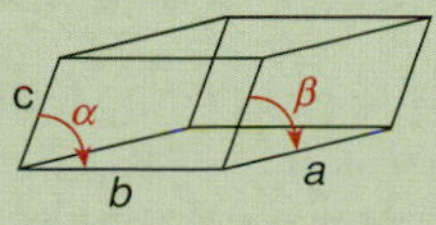	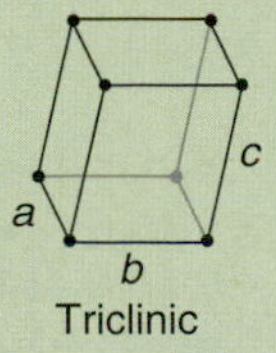

5.4.2 Cubic and Hexagonal Systems

There are three Bravais lattices in the cubic (isometric) system: primitive (*cubic-P*), body-centered cubic (*bcc*), and face-centered cubic (*fcc*). Three cubic systems and the hexagonal close-packed system are depicted in **Figure 5.10a–c**. Metals crystallizing as *fcc* and *bcc* will serve as the models of choice because they are the least complex and easiest to visualize.

Three-dimensional structures have three lattice constants, a, b, and c. Between a and b, there is angle γ; between a and c, there is angle β; and between b and c, there is angle α. In cubic systems, systems that possess the highest symmetry, it follows that $a = b = c$ and $\alpha = \beta = \gamma$. The primitive cube has atoms at each apex of the cube; the *bcc* is the same except that an additional atom is placed in the center, and the *fcc* form has atoms placed in the center of each face of a primitive cube with no atom in the center. The number of nearest neighbors for each atom (a.k.a. the coordination number Z) is 6, 8, and 12, respectively.

There are several ways to pack atoms in a cubic system. As we know, atoms desire the greatest number of nearest neighbors to become stable. For example, in a two-dimensional plane, six atoms are able to completely surround a central atom to form the core of a hexagonally close-packed two-dimensional structure (**Fig. 5.11a**). Such close-packed two-dimensional planar structure has a maximum number of neighbors with the lowest possible void volume. Another identical plane of hexagonally close-packed atoms is placed on top of the first plane to fit into the holes or void volume (gaps) formed by the touching spheres in the first plane. This creates a structure of the AB type (**Fig. 5.11b**). There is another way to place the second plane: right on top of the first plane (on top of the atoms) and not in the gaps. This forms an AA structure (e.g., one that is *in register*) and it comprises the primitive cubic system (*cubic-P*) (**Fig. 5.7**).

We are now ready to place a third plane. Obviously if we stay in register in the AA-base system to create AAA, AAAA, and AAAAA, there are no options and we are finished with its structure (*cubic-P*). However, for the AB-base system, there are two ways to place the third plane, depending on which set of gaps gets covered in the third step. Two stacking arrangements are possible that yield different crystalline arrays: the ABABAB … or ABCABC … type, depending on which gaps get covered. An *hcp* structure ABA is formed if the third layer is placed directly on top of the first layer (**Fig. 5.11c**). The second is an *fcc* (cubic close-packed) structure (**Fig. 5.11d**). In this case, the third layer is placed above the uncovered voids (gaps) of the first layer—the gaps not covered by the original B layer. This forms an ABC structure. The two types are called *polytypes*.

Two types of holes are formed in close-packed structures: the tetrahedral hole and the octahedral hole (**Fig. 5.12**). A tetrahedral hole is a space enclosed by four spheres (three in one plane and the last one on top of the hole formed by the three), and an octahedral hole is a space enclosed by six spheres. The radius of an octahedral site is

$$r_{oct} = (1/4)(2 - \sqrt{2})a = (\sqrt{2} - 1)r = 0.4142r \qquad (5.10)$$

where a is the lattice constant and r is the radius of the sphere. From geometry, then, we see that an octahedral hole can accommodate another atom no larger than $0.4142r$. The number of octahedral holes equals the number of atoms (N_a) in a crystal. Tetrahedral holes are described by

Fig. 5.10

The crystal structures of the basic isometric cubic systems are displayed in point, space-filling, and hard sphere packed forms. (a) Simple cubic (or cubic-P, primitive) is shown on the top. The atoms at the corners are each shared with eight other cubes. Therefore, there is one atom per unit cell $(8)(\frac{1}{8})=1$. Each atom has six nearest-neighbors and therefore the coordination number Z is equal to 6. (b) Body-centered cubic (bcc) has two atoms per unit cell: one in the center and the same eight found in the simple cube at each corner of the cube: $1+(8)(\frac{1}{8})=2$. The coordination number of atoms in bcc crystals is Z = 8. (c) Face-centered cubic (fcc) has one atom on each face (for a total of 6) that is shared by an adjacent cube and the same eight found in the simple cubic and fcc forms. The number of atoms per unit cell then is four per unit cell, one in the center and the same eight found in the simple cube at each point of the cube: $(6)(\frac{1}{2})+(8)(\frac{1}{8})=4$. The coordination number of atoms in fcc crystals is Z = 12.

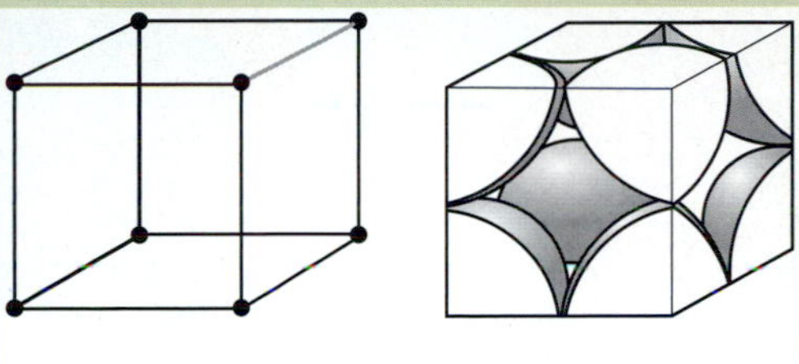

(a) Simple cubic

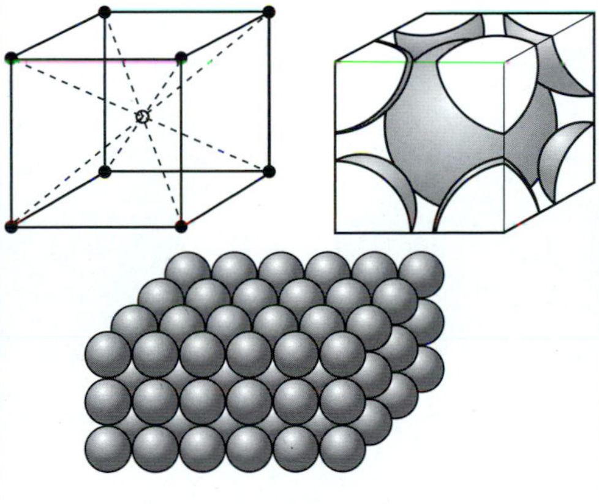

(b) Body-centered cubic (*bcc*)

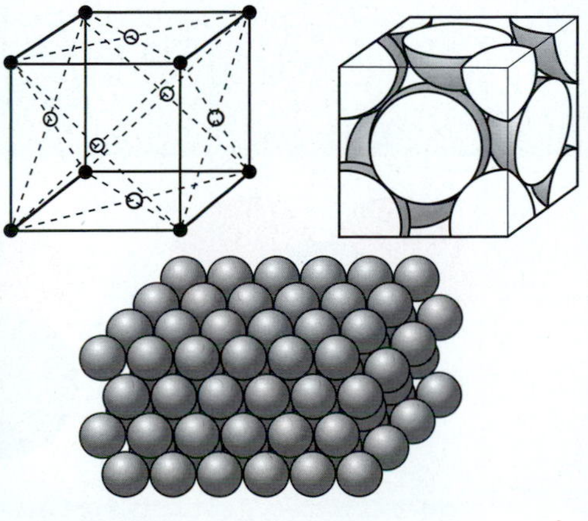

(c) Face-centered cubic (*fcc*)

Fig. 5.11 *Planar packing schemes for close-packed structures are shown. In (a), hard spheres are closely packed into one layer to form a six-coordanate structure (each sphere is surrounded closely by six others). (b) An identical layer is placed in the gaps among atoms of the first layer. Try this at home; can you find two (or more) possible identical configurations (degeneracies) in the placement of the second layer? (c) and (d) show two possible ways of placing the third identical layer. If layer three is placed directly over layer one, then you form the ABAB … polytype characteristic of a hexagonal close-packed (hcp) crystal. Notice that the "holes" are the same holes shown in (b). If the third layer is not placed directly above the first layer, then an ABC … polytype is formed. This is the fcc crystal structure. Notice that in this structure no holes (open gaps or interstitial spaces) are visible.*

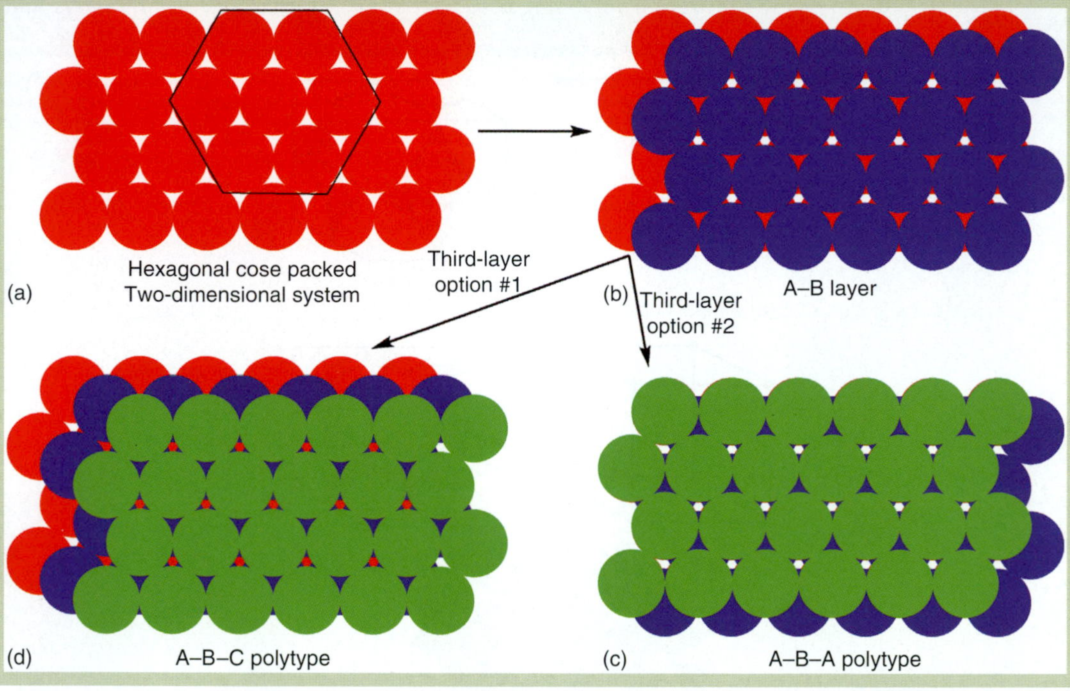

Fig. 5.12 *Two types of holes are displayed. On the left, the tetrahedral hole is illustrated. On the right, the octahedral hole is illustrated. In ABAB … structures, octahedral holes are stacked on top of each other.*

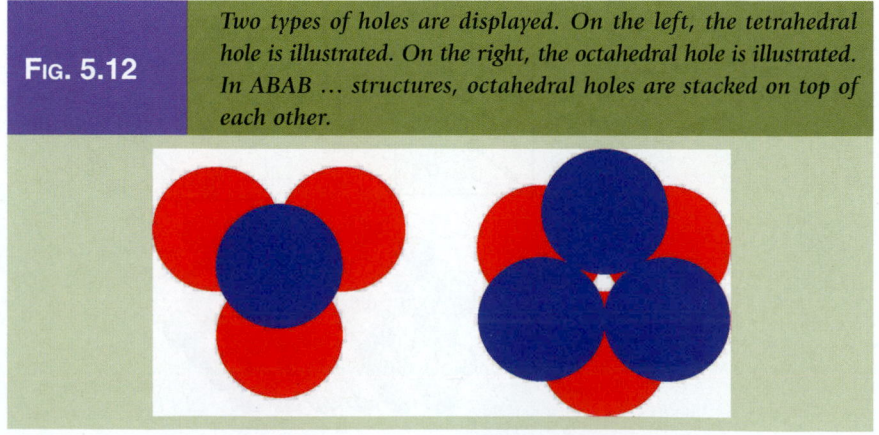

$$r_{tet} = (1/4)(\sqrt{3} - \sqrt{2})a = \left[\sqrt{(3/2)} - 1\right]r = 0.225r \tag{5.11}$$

and there are twice the amount of tetrahedral holes in the close-packed solid than there are atoms ($2N_a$). Note that only smaller atoms are able to fit in the tetrahedral hole.

5.4.3 Packing Fraction and Density

During this brief discussion of packing fraction, we conveniently revert to the application of hard spheres as before. The spherical cluster approximation does not take packing fraction into account. Atomic packing factor or packing fraction, f_p, is a dimensionless quantity always less than one that relates the volume occupied by atoms to the total volume of the material:

$$f_p = \frac{N_a V_a}{V_c} \tag{5.12}$$

where N_a is the number of atoms, V_a the volume of an individual atom, and V_c the volume of the cluster. Void volume, the spaces (gaps) between and among close-packed atoms, constitutes the remainder of the space in a crystal. The first step to determine packing fraction is to calculate the number of atoms in a unit

EXAMPLE 5.9 *Packing Fraction for a Single Element Body-Centered Cubic Structure*

Calculate the f_p for ferrite, the α-iron allomorph that exists as a *bcc* crystal. What is the coordination number? What is the value of the lattice constant *a*?

Solution:

The body-centered cube is depicted in **Figure 5.11b**. *There are eight atoms at the corners of the cube and one atom in the center. Each corner atom shares bonds with the central atom and seven additional atoms within the unit cell and in adjacent unit cells. The coordination number, CN or Z, is the number of nearest neighbors for each atom. In a bcc crystal, each atom has eight nearest neighbors, CN = 8. This holds true for the central atom as well as the corner atoms. The bcc unit cell, as a result, contains just one eighth of the iron atom at each corner and one atom in the center. The total number of atoms in a bcc unit cell is then:*

N_a = *one central atom + (eight corner atoms × one eighth of each corner atom) = two atoms*

α-Iron is therefore not a close-packed cubic structure. The lattice constant, a, is the length of the cell edges. Calculation of the value is accomplished by applying simple geometric relations.

Because the center atom touches both corner atoms along any diagonal, the length of diagonal, d, of a cube in the bcc configuration is equal to d = 4r.

The length of the diagonal in terms of a is $d^2 = 3a^2$ to give $d = a(3)^{1/2}$; therefore,

$$a = 4r/\sqrt{3} \tag{5.13}$$

The packing fraction is calculated by dividing the volume of the atoms by the volume of the cube:

$$f_p = 2 \text{ atoms } [(4/3)\pi r^3]/a^3 = 2[(4/3)\pi r^3]/(4r/\sqrt{3})^3,\ f_p = 0.68 \tag{5.14}$$

The covalent radius of iron is 0.117 nm.

$$\text{The lattice constant } a = 4(0.117)/\sqrt{3};\ a = 0.270 \text{ nm} \tag{5.15}$$

cell. In the primitive cubic system, each corner atom is shared by eight adjacent cells. Thus, there is a total of one atom per unit cell (one eighth of each corner atom × eight corner atoms = one atom).

Calculation of theoretical densities can be found in a straightforward manner by applying

$$\rho = \frac{N_a M}{\mathcal{N}_A V_{uc}} \tag{5.16}$$

where N_a is the number of atoms per unit cell, M is the atomic mass in $g \cdot mol^{-1}$, V_{uc} is the volume of the unit cell in cm^3, and $\mathcal{N}_A$ is Avogadro's number [16].

5.4.4 Structural Magic Numbers

If small materials had the same physical properties as bulk materials, just on a smaller scale, our story would end right now. However, there is a break in the smooth curve of the continuum. Scaling laws break down at dimensions under 10 nm. In actuality, even particles much larger than 10 nm (e.g., colloids) start to display properties that meander away from bulk behavior (the color of gold colloids, for example) but for different reasons. The structure of a bulk material, for all practical purposes, is uniform down to micron-sized particles. When particles get smaller, clusters of similar size may have different structures. One or two atoms may make the difference between a stable cluster and one that is metastable or unstable.

Chemical reactivity of alkali and aluminum clusters is lower relative to open shell counterparts. Small alkali metal clusters are controlled by electronic structure. The reactivity of transition metal clusters, on the other hand, correlates strongly with geometric rather than electronic structure. Alkali and noble metals with an even number of electrons form clusters that are diamagnetic, while those with odd number of electrons form paramagnetic clusters. In nonmetallic clusters such as those composed of the rare gases, magic numbers are due exclusively to structural geometry and not electron considerations.

Example 5.10 *Calculation of the Theoretical Density of Copper*

Calculate the theoretical density of copper from equation (5.16). Copper data: covalent radius = 0.128 nm; M = 63.55 $g \cdot mol^{-1}$; structure = *fcc*.

Solution:

Cu, fcc: Z = 12 and there are four atoms per unit cell, $a = 4r/\sqrt{2}$

Use $\rho = N_a M / V_{uc} \mathcal{N}_A$

Volume of the unit cell, $V_{uc} = a^3 = (4r/\sqrt{2})^3 = 4.74 \times 10^{-2}\ nm^3$; $V_{uc} = 4.74 \times 10^{-23}\ cm^3$

$\rho_{theoretical} = [(4 \times 63.55\ g \cdot mol^{-1})]/[(4.74 \times 10^{-23}\ cm^3)(6.022 \times 10^{23})]$

$\rho_{theoretical} = 8.90\ g \cdot cm^{-3}$

The actual value of the density of copper is 8.94 $g \cdot cm^{-3}$.

Clusters consist of a few to several thousand atoms. A cluster is considered to be a quantum dot. Some clusters exhibit remarkable stability. These stable forms have what is called a *magic number* of atoms. In other words, completeness with regard to physical and electronic structure is achieved by clusters with a magic number of atoms. The concept of magic numbers is not new. Physicist Maria Goeppert-Mayer in 1948 explained why the nuclei of some atoms were more stable than others. Linus Pauling in 1965 added that nuclei with 2, 8, 20, 28, 50, 82, or 126 neutrons or protons were stable and were characterized by extremely long lifetimes. It turned out that the rationalization for nuclear cluster stability was based on a spherical quantum mechanical model called the *jellium model* that was especially suited for describing the stability of alkali metal clusters. Because the jellium model is able to explain both nuclear and electronic phenomena, it is truly one of the fundamental unifying concepts of physics [15].

If atoms are packed in such a way that all possible nearest-neighbor sites are filled to form a complete shell surrounding the central atom or preexisting shell, that cluster is considered to have a *magic number* of atoms. Based on mass spectrometry experiments in the 1980s, a periodic pattern of intense peaks was found in the mass spectra of sodium clusters [17]. The sodium clusters showed intense peaks at 2, 8, 20, 40, 58, and other sums of atoms. Adjacent peaks corresponding to smaller or larger sized clusters on either side of the magic number peak were far less intense [17]. Since MS intensity is proportional to the abundance of daughters of specific size and charge, we can safely conclude that magic number clusters are more stable than others with similar mass. Clusters that have a magic number of atoms have higher abundance, ionization potential, and binding energy. Neutral clusters with closed shells have low electron affinity while those with open shell structures have high electron affinity.

The structure of *fcc* close-packed structures like gold is a good place to begin. Starting with one atom, a maximum of 12 atoms (coordination number $Z = 12$) can be placed around the central atom to form a close-packed structure. The resulting structure is a geometric solid called a *cuboctahedron*—a 14-sided geometric solid with 12 vertices, 8 triangular faces, and 6 square faces **(Fig. 5.13)**.

Thus, the first magic number—$\mathcal{M}^*$, the number 13—is achieved. There are 12 shell atoms and one core atom. There is no more available space to add more atoms, and if an atom is removed, a vacancy is created. Similar to the octet rule for electrons and nuclear closed shells, the magic number is a closed shell geometric configuration for atoms in a cluster and the same result is achieved in all cases: the stable, lowest energy configuration.

The next layer requires that an additional 42 atoms be packed around the first shell. The magic number for the cluster is now 55. The quantum dot cluster Au-55 is a cuboctahedral magic number cluster with 55 gold atoms. For the series of clusters that exist in icosohedron or cuboctahedron form, the appropriate magic number formula for the series is

$$\mathcal{M}^*(K) = 1, 13, 55, 147, 309, 561, 923, 1415, 2057, 2869, \text{etc.} \quad (5.17)$$

Each shell K of the cluster, a pseudoconcentric shell, is described mathematically by

FIG. 5.13 *Cuboctahedral gold clusters are shown. The progression represents a geometric series based on magic numbers: 13, 55, 147, 309, and 561. Notice how the faces retain their integrity regardless of the size of the cluster. As the cluster becomes larger, this figure also illustrates the concept of surface-to-volume ratio: The larger the cluster is, the more volume atoms seem to overwhelm the number of surface atoms.*

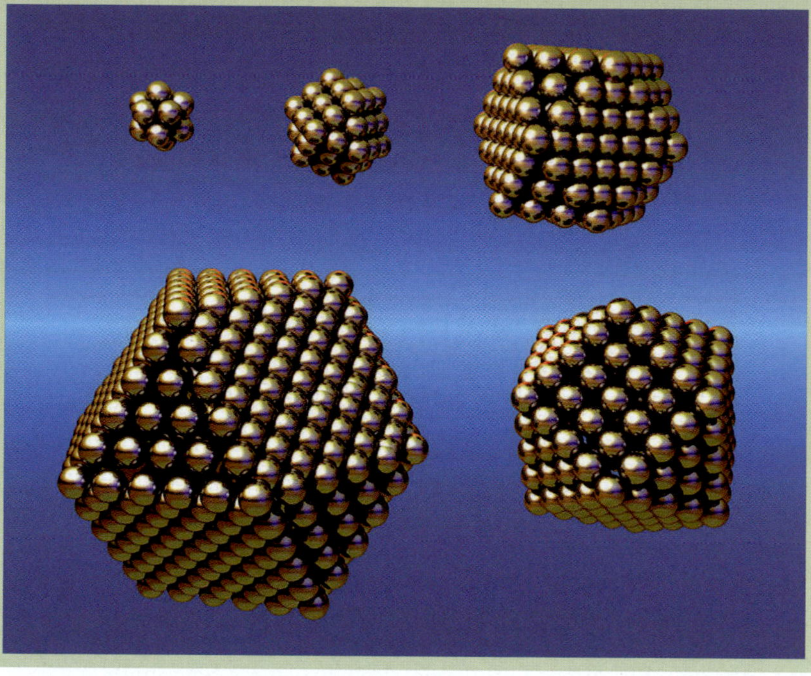

Source: Courtesy of Professor Anil K. Rao, Department of Biology, the Metropolitan State College of Denver.

$$\mathcal{M}^*(K) = (1/3)(10K^3 + 15K^2 + 11K + 3) \tag{5.18}$$

where K is the shell index number. The expression is based on the expansion of

$$\mathcal{M}^*(K) = 1 + \Sigma\,[k = 1 \rightarrow k = K]\,(10K^2 + 2) \tag{5.19a}$$

In other words, every shell consists of $N_s = (10K^2 + 2)$ atoms. If that shell is the last shell to be added, then that shell contains the surface atoms, N_s. Another popular formula for magic numbers is one that includes the central atom as the first shell designated as $K = 1$ [13]:

$$\mathcal{M}^*(K) = (1/3)(10K^3 - 15K^2 + 11K - 3) \tag{5.19b}$$

An expression that includes the central atom as a shell for the number of surface atoms is derived from equation (5.19b):

$$N_s = (10K^2 - 20K + 12) \tag{5.20}$$

We now have another tool with which to calculate surface-to-volume ratio (**Table 5.3**). There are two ways to view surface-to-volume ratios. In one, the surface atoms N_s are related to the whole volume of the cluster (given as $N_s/\mathcal{M}^*$). In the other way, surface atoms are related to only the atoms that exist below the

TABLE 5.3 *Percent Surface Atoms of Cuboctahedral Magic Clusters*

K	$\mathcal{M}^*$	N_s	$N_s/\mathcal{M}^*$
1	1	1	1.00
2	13	12	0.92
3	55	42	0.76
4	147	92	0.62
5	309	162	0.52
6	561	252	0.45
7	923	362	0.39
8	1415	492	0.35
9	2057	642	0.31
10	2869	812	0.28
50	4.04×10^5	2.40×10^4	0.06
100	3.28×10^6	9.80×10^4	0.03
500	4.15×10^8	2.49×10^6	0.006

surface—the true volume atoms (e.g., $\mathcal{M}^* - N_s$). For example, in the Au-55 cluster, only one atom is a true volume atom. We will only be concerned with the former expression.

The diameter of the cluster $2R_c$ is found from [13]

$$2R_c = d[2K - 1] \tag{5.21}$$

where $d = 2r$, the covalent diameter between two atoms for an *fcc* unit cell and is related to the lattice a constant by $d = a/\sqrt{2}$.

Body-centered cubic (*bcc*) clusters are formed as 14-vertex rhombic dodecahedrons. The magic number formula for *bcc* clusters is

$$\mathcal{M}^*(K) = (4K^3 + 6K^2 + 4K + 1) \tag{5.22}$$

Magic number series for *bcc* systems differ from the cubic close-packed *fcc* set (e.g., $\mathcal{M}^*(K_{bcc})$ = 1, 15, 65, 175, 369, 1105, 1695, 2465, etc.) [15]. For the *hcp* system, another set of structural magic numbers is generated: $\mathcal{M}^*(K_{hcp})$ = 1, 13, 57, 153, 321, 581, etc. [13]. All magic number clusters discussed here are of minimum volume, maximum density, and approximately spherical in shape—e.g., pseudospherical [13].

5.4.5 Miller Indices and X-Ray Diffraction

In 1839 British crystallographer W. H. Miller devised a systematic approach that is related to the geometry of the unit cell. The *Miller indices*, as they came to be known, are the fundamental language of crystallographers and engineers and extremely useful tools for predicting structural properties of materials. Miller developed a system of describing the orientation of crystallographic planes inside a crystal lattice by a set of integers. The integers are symbolized by (*hkl*). A similar notation, developed by Bravais and Miller, was invented to describe hexagonally packed systems: (*hkil*).

The index numbers are the inverses of the axial intercepts of the unit cell, e.g., the intercepts that a selected crystal plane makes with the crystallographic axes. Miller indices enclosed in parentheses, like (100), indicate the orientation of a specific plane with respect to a crystallographic axis. Braces denote planes that are related by symmetry. For example, the {100} family contains the (100), (010), and (001) planes and their negative counterparts. Brackets (e.g., [111]), indicate a specific lattice direction—specifically, the direction normal to the (111) plane. This would include the (222), (333), or (*nnn*) planes. The indices [*uvw*] in brackets are used to represent directions.

The Miller index algorithm is created by establishing a Cartesian coordinate system, *x*, *y*, and *z* or *a*, *b*, and *c*, and then placing the unit cell at the origin (0,0,0). The method for obtaining Miller indices is straightforward: (1) Set the origin of the unit sell at 000, (2) determine the value of intercepts of the crystal plane in terms of lattice constants, (3) take the reciprocal of each index, and (4) reduce them to the lowest terms. The use of reciprocals helps alleviate the awkward task of dealing directly with infinite intercepts, the case when a plane is parallel to an axis. **Figure 5.14** displays the fundamental low-index crystal planes of a simple cubic system. The orientation is the same for all cubic crystals. The only difference is the number of atoms and spacing within the low-index planes.

Crystallographic directions (e.g., [*uvw*]) are calculated by repositioning the directional vector through the origin and translating the projections into terms of the lattice constants *a*, *b*, and *c*. The values are then converted into the smallest integer values. Prove to yourself that no reciprocal relations are developed in this calculation. Linear and planar atomic densities are calculated in a manner similar to atom packing fraction. We find the number of whole atoms along a direction or in a plane and then divide that number by the appropriate dimension. Linear density is defined as the number of atoms per unit length centered along a given direction in a crystal. Planar density is defined as the number of atoms centered in a plane per crystallographic plane area.

EXAMPLE 5.11 *Miller Index Calculation*

Determine the Miller indices for the three simplest planes (low index) for any cubic lattice. What does a (111) plane look like? Do not be concerned with the atomic structure, only with the geometric structure of a cube.

Solution:

Consult ***Figure 5.15*** *for assistance with orientation.*

The lattice constants for a cubic crystal are the orthogonal series consisting of $a = b = c$ $(\alpha = \beta = \gamma = 90°)$ *and the coordinate system is set up accordingly.*

The three simplest (low index) planes are those that intersect each major axis of the crystal (i.e., they intersect the designated axis but are parallel to the other two).

The plane (abc) that intersects the a axis at ***a*** *is the* (a, ∞, ∞) *plane. The symbol* ∞ *represents the parallel condition. The reciprocals are* $(1/a, 1/\infty, 1/\infty)$*. With* $\mathbf{a} = 1$*, the Miller indices (hkl) are designated as (100).*

The same logic is applied to the other two orthogonal planes that intersect the b and c axes at the points ***b*** *and* ***c****, respectively, yielding the planes (010) and (001), respectively.*

If the (111) plane intersects each major axis at the midway point, the plane is described by (1/2, 1/2, 1/2). Inverting and reducing to lowest terms, (hkl) = (111).

Example 5.12 *Calculation of Linear Density*

Determine the linear density of atoms in the [110] direction for *fcc* gold (R_{mr} = 0.144 nm, a = 0.408 nm).

Solution:

The [110] vector runs along the bottom-face plane diagonal from coordinate points 0,0,0 to 1,1,0 of the primitive cubic fcc cell (e.g., the diagonal running under the fcc cubic unit cell). The length of the diagonal is equal to four times the metallic radius: $4R_{mr}$. Draw an fcc cubic unit cell. From inspection, there are two whole atoms centered in the line contained by the [111] direction. The linear density is then:

$D_l = 2$ atoms$/4R_{mr} = 1/(2R)$ *or, in terms of the lattice constant a,* $D_l = 2/(a\sqrt{2})$, $D_l = 3.47$ atoms$\cdot$nm^{-1}

In 1895, W. C. Röntgen was the first to detect the arrangement of atoms in a crystal by means of x-ray analysis because the wavelength of the x-rays happened to coincide with the space between atomic planes—e.g., the *d*-spacing. X-ray diffraction (XRD) is important in studying the structure of nanomaterials, because of the need to measure altered *d*-spacing in nanoparticles. XRD depends on the Bragg equation given earlier in chapter 3 ($n\lambda = 2d \sin\theta$).

The *d*-spacing between planes for cubic systems is found from

$$d = \frac{a}{\sqrt{h^2 + k^2 + l^2}} \tag{5.23}$$

where d is the space between adjacent planes and a is a lattice constant.

Fig. 5.14 *The three low-index Miller hkl planes are shown in the figure for a simple-cubic system crystal. Higher index planes of the family of {100}, {110}, and {111} such as (200), (220), and (222), respectively, are not emphasized in this text. There are many hkl planes for every crystal. The (100) plane in the simple cubic system is framed by four atoms; the (110) is also framed by four, but notice how its orientation has been altered. The (111) plane and the (100) plane are displayed gloriously by the cuboctahedrons in Figure 5.13. Can you find the (100) planes in that figure?*

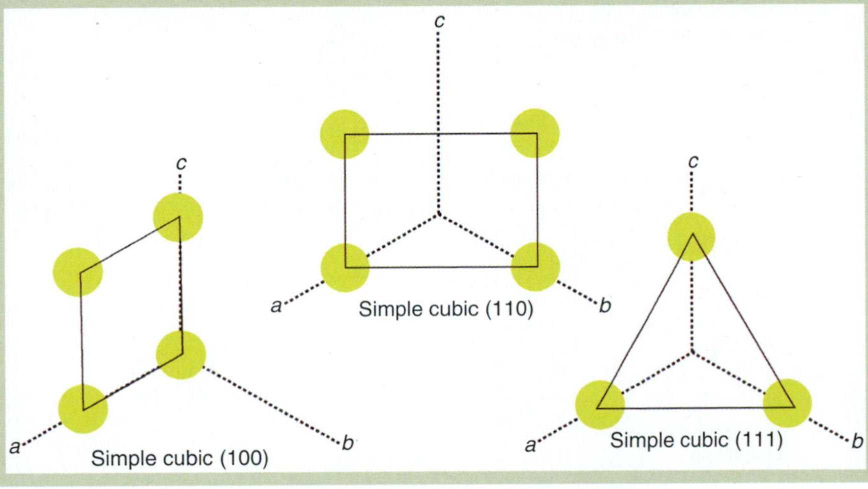

5.5 Particle Orientation

Nanoparticles, particularly those made of noble metals or semiconductors, are routinely placed within dielectric materials such as polymers or porous alumina membranes to form nanocomposite materials. Michael Faraday's experiments investigated solutions containing spherical nanogold particles. Beautiful colors were elicited from the constituent nanoparticles immersed in the surrounding water medium. Once embedded in the dielectric material, the orientation of the nanoparticles is fixed with regard to the plane of the composite. Exposure of the plane of the nanocomposite to impinging radiation (UV–visible light) results in an optical response by the combined effective medium. In other words, the optical response of the nanocomposite is due to a mixture of the dielectric constants of the metal, the surrounding medium, particle orientation, and their relative proportions. An example of such a composite is shown in **Figure 5.15.** The pore channels in a porous anodic alumina membrane (a transparent and insulating host material) were used to house gold nanoparticles of variable aspect ratio and size. The impinging UV-visible radiation was fixed normal to the plane of the membrane during the wavelength scan [18–20]; for example, the oscillating electric field of the EM radiation was oriented parallel to the plane of the membrane and perpendicular to the pore channels (**Fig. 5.15**).

Fig. 5.15 *Orientation of gold nanocylinders is shown with respect to the propagation vector of visible light. The electric field (and the orthogonal magnetic field) in the image is drawn perpendicular to the longitudinal axis of the ellipsoidal nanocylinders. In other words, electrons along the short axis are excited when exposed to the radiation producing plasmon resonances. Porous alumina is a perfect template for fabricating gold nanoparticles. When gold nanocylinders such as these are liberated from the confines of the alumina template and placed in a solution—for example, water—two resonance peaks are observed: one corresponding to the transverse axis (the short axis) and the other to the longitudinal axis. Spectra obtained have two absorption peaks. For ellipsoidal particles, if the particle are spheres, only one resonance is detected.*

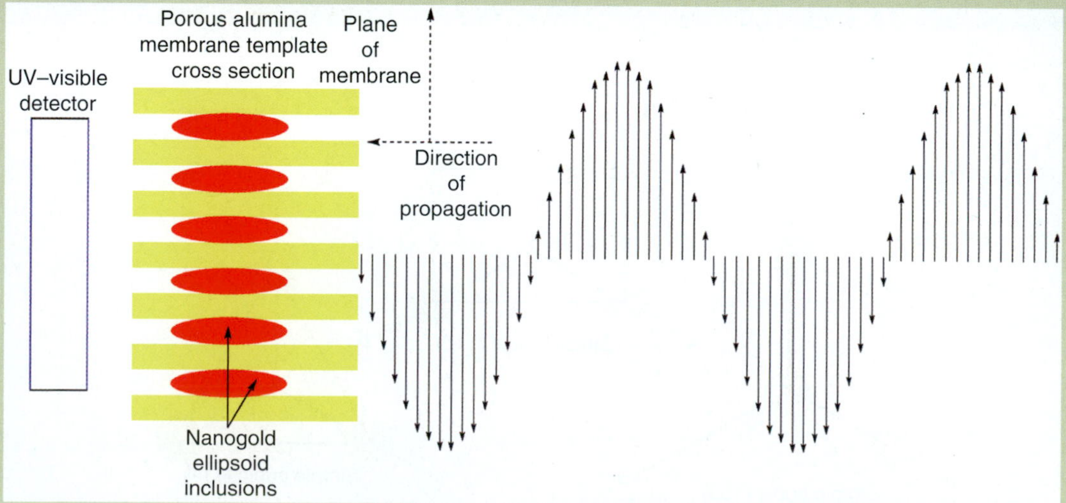

Fig. 5.16 *The family of ellipsoidal shapes originating from the parent sphere is shown. On the right, prolate ellipsoids are shown. For a sphere, a = b = c. Prolate ellipsoids are needle-like in shape with one dimension, say a (the major axis), much greater than the other two (the minor axes): a >> b = c. For oblate shapes, one dimension, say a once again, is much less than the other two axes: a << b = c.*

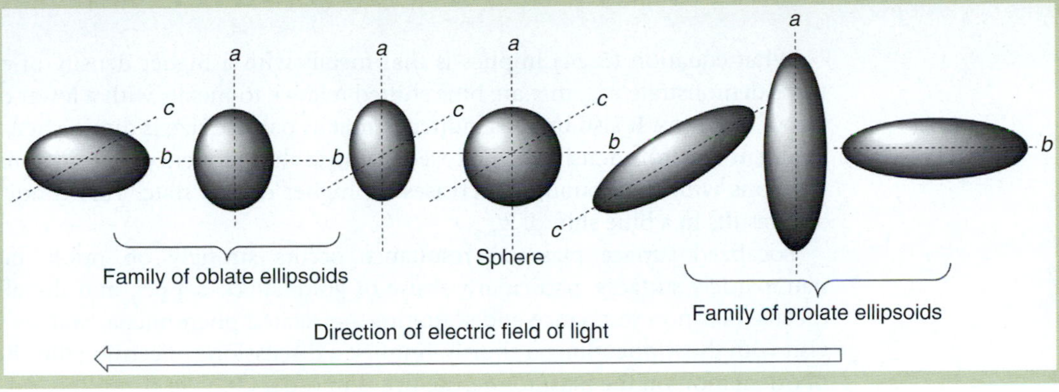

This discussion is confined to ellipsoids of revolution. There exist three types of three-dimensional ellipsoids of revolution: *spherical* ($a = b = c$), *prolate* ($a > b = c$; $b > a = c$; or $c > a = b$), and *oblate* ($a < b = c$; $b < a = c$; and $c < a = b$), where the orthogonal axes a, b, and c comprise a Cartesian coordinate system. Please consult **Figure 5.16** for reference. The prolate forms are similar to wires; the oblate forms are similar to discs. Ellipsoids of revolution are depicted in **Figure 5.16.**

Ellipsoids of revolution were chosen because the electric field that exists within such shapes is uniform throughout and therefore calculable. In this section, atom packing and structural factors are not considered—only the geometric shape is of importance. We are concerned with the maximum absorption λ_{max} of noble metal nanoparticles in the presence of an applied electromagnetic field, specifically in the visible region of the spectrum.

5.5.1 Surface Polarization in Metals

A plasmon (from the Greek *plasma,* "something molded or shaped" [like plastic] + the Latin *mons,* "mountain, body") is a *quasi-particle*—a particle that can be described by a collection of interacting particles—in this case, electrons. Plasmons occur on the surface of a metal, are quantized, and consist of collective longitudinal (or transverse) oscillations of the free electron gas (the conduction electrons). Plasmon oscillations are induced when excitation of the metal surface electrons by visible light or other sources occurs. The location (λ_{max}) and strength (absorption) of the oscillation depend on the wavelength of the incident radiation, particle composition (dielectric constants), size, shape, and orientation.

The plasmon energy is described classically as

$$\omega_p^2 = \frac{N_e e^2}{\varepsilon_o m} \tag{5.24}$$

where

ω_p is the plasmon excitation frequency
e is the electron charge (1.6022×10^{-19} C)
N_e is the electron density in reciprocal cubic meters
m is the mass of the electron (9.1096×10^{-31} kg)
ε_o is the vacuum permittivity of free space (8.8554×10^{-12} F·m^{-1})

What equation (5.24) implies is that metals with a higher density of electrons demonstrate λ_{max} that are blue shifted relative to metals with a lower density of electrons. It also implies indirectly that as particle size is diminished, the apparent electron density is increased because the frequency of collisions of electrons with the boundary increases—a higher energy state. This condition also results in a blue shift of λ_{max}.

Localized surface plasmon resonance occurs strongly on noble metal nanoparticle surfaces, particularly those of gold, silver, copper, and the alkali metals. Plasmon resonance and scattering are related phenomena, and we will visit with those phenomena shortly, but first a discussion concerning the effects of orientation on the plasmon resonance is in order.

An illustration of two ellipsoids of revolution exposed to an electric field is shown in **Figure 5.17.** The applied electric field is designated as $\mathbf{E}_o$. The induced

FIG. 5.17 *The polarization phenomenon of the surface electrons on a metal surface is shown. Upon exposure to the electric field of the light, the surface electrons generate an opposing field that is in turn opposed by an internal depolarization field. The strength of the oscillation depends on the shape and orientation of the nanoparticle (usually a noble metal or copper). The more pointed the nanoparticle is, the stronger the opposing field becomes. This is why molecules situated on the tips of high aspect ratio nanocylinders have significantly enhanced SERS enhancement compared to those placed on spherical particles—a nano-lightning rod effect of sorts.*

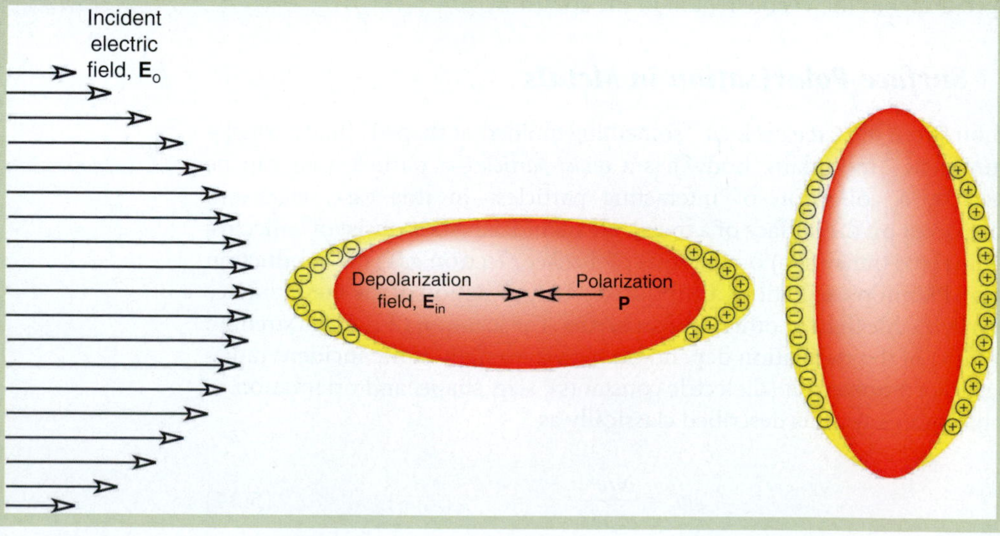

electric field, $\mathbf{E}_{in}$, is opposite in direction to the polarization, **P**, parallel to the applied field. The strength of the depolarization field is dependent on the shape of the nanoparticle.

For spheres, the dipolar resonance, the most fundamental resonance, is described by the expression

$$\varepsilon_o = -2\varepsilon_m \tag{5.25}$$

where ε_m is the real component of the complex dielectric function of the metal and ε_o is the real component (refractive index) of the complex dielectric function of a surrounding insulator material. The factor 2 is typical for spheres—more about that later. The imaginary part of the two materials can be ignored at the wavelength range of interest. This relation is relevant only for particles smaller than 20 nm. For larger particles, higher order resonances and scattering need to be considered—topics beyond the scope of this text. Therefore, this is a case in which the size of a particle causes a change in the manifestation of a physical property.

There are two types of cluster size effects that affect the nature of the surface plasmon. Classical electrodynamic theory using bulk optical constants can be applied safely to describe the optical response of large clusters [21]. The atomic structure of a large cluster is similar to that of its bulk counterpart. This is called an *extrinsic* effect. Therefore, this is not to say that the optical response of large clusters is identical to that of bulk materials because it is not. With smaller dimensions comes the addition of restrictive boundary conditions. Collective electronic excitations known as Mie resonances are important in particles that are the size of colloids or larger. The color of a solution containing large metallic colloids is not the same as the color of the same material in bulk form; colloidal gold sols are purplish, for example.

The optical functions, however, for very small clusters are size dependent. This trend is due to *intrinsic* effects: Changes in the internal structure of the cluster relative to the bulk accompany smaller size. The optical functions (constants like the dielectric constant) do not apply to very small nanomaterials. These changes in electronic and structural properties as size is reduced occur at the threshold between the bulk and quantum domains. Intrinsic effects are not confined to optical properties. Many physical properties undergo radical alteration as size is reduced. They include ionization potential, binding energy, chemical reactivity, crystallographic structure, melting temperature, and optical properties (i.e., the plasmon).

The classification of cluster types shown in the following table for sodium clusters [21].

	Small	Medium	Large
Atoms, N_a	2–20	20–500	500–10^7
$2R_{Na}$ in nanometers	≤1.1	1.1–3.3	3.3–100
N_s/N_v	N_s/N_v inseparable	0.9–0.5	≤0.5

Other classifications exist as well [15] and, as you can see, it becomes a qualitative issue to some extent. In both tables, sodium atoms serve as the model.

	Small	Medium	Large
Atoms, N_a	≤10^2	10^2–10^4	>10^4
$2R_{Na}$ in nanometers	≤1.1	1.9–8.6	>8.6
N_s/N_v	>0.86	0.86–0.19	≤0.19

Surface enhanced Raman spectroscopy (SERS) is a method that exploits the surface plasmon resonance properties of noble metal nanoparticles. The resonance condition is achieved when equation (5.25) becomes very large. Giant electromagnetic enhancements for spheres:

$$G = \chi^{12}\left(\frac{\varepsilon_m - \varepsilon_o}{\varepsilon_m + 2\varepsilon_o}\right)_\lambda^2 \left(\frac{\varepsilon_m - \varepsilon_o}{\varepsilon_m + 2\varepsilon_o}\right)_s^2 \tag{5.26}$$

where $\chi = r/(r + d)$ with r equal to the radius of the spherical metal SERS substrate nanoparticles and d the distance of the analyte molecule from the surface of the metal particle. The λ and s represent the laser and Stokes fields, respectively. Obviously the SERS signal is enhanced significantly as the two denominators approach zero [22].

5.5.2 Particle Depolarization Factor and Screening Parameters

The sphere is a shape that possesses the highest level of symmetry of all geometric shapes. Although electron polarization occurs on spherical metal nanoparticles, the orientation of the spherical particle in an electric field is irrelevant as polarization is infinitely degenerate; that is, the sphere will have the same visible absorption maximum (λ_{max}) regardless of its three-dimensional placement in an electric field. That is why solutions containing spherical gold colloids have one absorption peak. Prolate particles in solution, on the other hand, have two absorption maxima that correspond to the longitudinal and transverse plasmon resonances [23].

We introduce three related terms in this section that help us understand how orientation of nanoparticles affects the optical response of metal nanoparticles. They are, in order of increasing complexity, the eccentricity ζ or f of the ellipsoid, the depolarization factor q, and the screening parameter κ. Eccentricity (from the Greek *ekkentros*, "out of the center," from *ek*, "out" + *kentron*, "center") is the measure of how much an ellipsoid deviates from the shape of a sphere. The eccentricity for prolate ellipsoids is defined as

$$\zeta^2 = \frac{a^2 - b^2}{a^2} = 1 - \frac{b^2}{a^2} \tag{5.27}$$

where (b/a) is the inverse of the aspect ratio of the particle (with b = semiminor axis, a = semimajor axis). The eccentricity of a sphere is $\zeta = 0$; for a prolate particle, $\zeta \rightarrow 1$ as the aspect ratio increases.

The concept of the depolarization factor q (the shape factor) emerged from the concept of the demagnetization factor first developed by J. C. Maxwell and

G. Mie and later embellished by R. Gans in 1915. The magnitude of the depolarization field depends on the shape of the particle and hence on the value of the depolarization factor q. The value of q ranges from $0 \rightarrow 1$ and the depolarization factors along the three coordinate axes of any particle conform to a sum rule

$$\Sigma q_i = 1 \tag{5.28}$$

where i indicates Cartesian coordinates a, b, c. With the electric field along the a-axis, the depolarization factor q_a is found from the eccentricity by

$$q_a = \frac{1-\zeta^2}{\zeta^2}\left[\frac{1}{2\zeta}\ln\left(\frac{1+\zeta}{1-\zeta}\right)-1\right] \tag{5.29}$$

For oblate particles with the electric field along the a-axis, the eccentricity f is equal to

$$f^2 = \frac{b^2-a^2}{a^2} = \frac{b^2}{a^2} - 1 \tag{5.30}$$

and the depolarization factor q_a then is equal to

$$q_a = \frac{1+f^2}{f^2}\left[1-\frac{1}{f}(\arctan f)\right] \tag{5.31}$$

For prolate particles, q_b is found from the sum rule given earlier in equation (5.28). The depolarization factor is conveniently approximated by the following simpler relationship. We take $\boldsymbol{a}$, $\boldsymbol{b}$, and $\boldsymbol{c}$ to be the physical dimensions of the ellipsoid along the a, b, and c axes:

$$q_a = \frac{1/\boldsymbol{a}}{1/\boldsymbol{a}+1/\boldsymbol{b}+1/\boldsymbol{c}}; \quad \text{and if } \boldsymbol{b}=\boldsymbol{c}, \quad q_a = \frac{1/\boldsymbol{a}}{1/\boldsymbol{a}+2/\boldsymbol{b}} \tag{5.32}$$

If the electric field is along the b-axis and from $q_a + 2q_b = 1$, then

$$q_b = \frac{1-q_a}{2} \quad \text{and} \quad \frac{1/\boldsymbol{b}}{1/\boldsymbol{a}+2/\boldsymbol{b}} \tag{5.33}$$

The screening parameter κ indicates the amount of "screening" that takes place in a composite and is a function of the particle shape, size, and orientation of the material. The screening parameter is an indication of transparency. The calculation of κ depends on the depolarization factor q as follows:

$$\kappa = \frac{1-q_i}{q_i} \tag{5.34}$$

The screening parameter assumes any positive value from zero to infinity ($0 \le \kappa \le \infty$) and is highly dependent on particle shape and orientation but not on the composition or distribution of particles. Composition is factored in by use of the complex dielectric constants($\tilde{\varepsilon}$) of a material, and distribution is accounted for by the fraction of metal inclusion. All are displayed in the Maxwell-Garnett equation of effective medium theory [24,25]:

$$f_m\left(\frac{\tilde{\varepsilon}_m-\tilde{\varepsilon}_o}{\tilde{\varepsilon}_m+\kappa\tilde{\varepsilon}_o}\right)=\left(\frac{\tilde{\varepsilon}_c-\tilde{\varepsilon}_o}{\tilde{\varepsilon}_c+\kappa\tilde{\varepsilon}_o}\right) \tag{5.35}$$

where f_m is the volume fraction of the metal inclusion, the subscripts m and o are attached to the dielectric functions of the metal and the oxide, respectively, and subscript c is attached to the dielectric function of the composite (e.g., the effective medium). If the nanostructure of a composite consists of metal particles that are perpendicular to the electric field (e.g., perpendicular to the plane of the membrane) and the impinging radiation is also perpendicular to the membrane (e.g., the electric field of the light is in the plane of the membrane), as shown in **Figure 5.15,** then a screening charge is developed. This charge is able to exclude the photons from the metal and reflect the light back into the transparent medium. It is the screening parameter that we see in equations (5.25), (5.26), and (5.35). $\kappa = 2$ for spheres, and $\kappa = 1$ for wire-like (cylindrical) structures that are perpendicular to the electric field of the light (e.g., parallel to the impinging radiation); $\kappa = 0$, the condition of maximum screening (lowest absorption) in which parallel plates are perpendicular to the electric field (e.g., oblate particles along the c-axis), and $\kappa \rightarrow \infty$, the condition of minimum screening (maximum absorption) with plates parallel to the electric field (e.g., oblate particles along the *a* or *b* axes). Depolarization and screening values as a function of particle shape and orientation are summarized in **Figure 5.18.**

UV–visible spectra of gold nanoparticles of various aspect ratio and size are shown in **Figure 5.19.** The colors produced by gold–alumina composite membranes were illustrated in **Figure 3.32** in chapter 3.

The purpose of this exercise is to demonstrate how particle shape and orientation affect optical properties of nanoparticles. Unfortunately, we had to get at all of this in a roundabout, semi-ellipsoidal manner. We presented an expression for eccentricity and then used that to define the depolarization factor q. From q, the screening parameter κ was found—the term that imparts the effects of shape and orientation into the electrostatic expressions that describe the optical response of nanometals. Because we do not have to address electrodynamic factors (e.g., due to the quasi-static limit), all the working equations boil down to simple algebra. For detailed analysis, the imaginary components of the dielectric functions need also to be factored into the analysis.

5.5.3 Quasi-Static Limit

The quasi-static limit is achieved when the size of a nanoparticle is significantly smaller than the wavelength of the impinging EM radiation. A general rule of thumb is that the radius r of the particle should be on the order of 1/100 of the wavelength λ in order to qualify:

$$r \leq 0.01\lambda \tag{5.36}$$

FIG. 5.18 *The progression of the depolarization factor and screening factor as a function of particle shape and orientation. The condition of the sphere is called three-dimensional isotropic screening (shape and orientation are uniform for a sphere and $\kappa = 2$). With $\kappa = 1$, the condition of two-dimensional isotropic screening is attained. This condition corresponds to the configuration depicted in Figure 5.15. The value κ equal to 0 corresponds to the condition of maximum screening (maximum transparency—e.g., the light is screened into the dielectric medium); κ equal to ∞ corresponds to the condition of minimum screening (maximum absorption).*

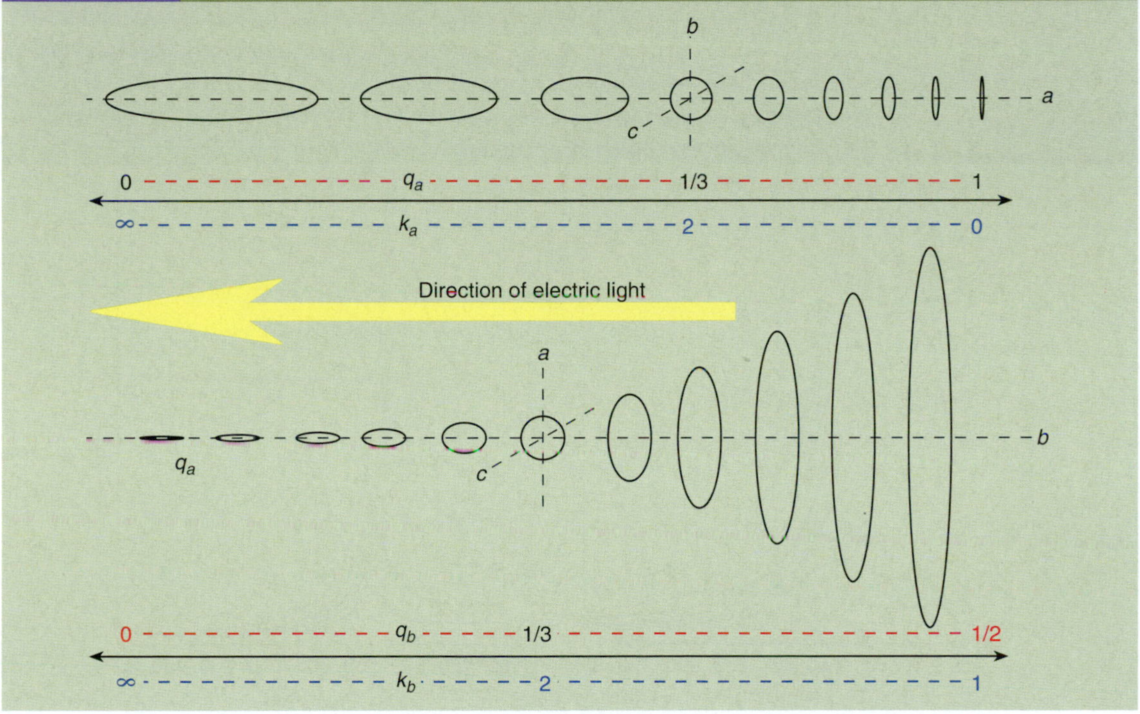

Under these conditions, the oscillations of the time-dependent external field occur slowly relative to the motions of the free electrons comprising the plasmon. As a result, only the dipolar resonance of the plasmon contributes to the optical response. There are no phase shifts (retardation effects), multipole phenomena (e.g., quadrupole scattering and extinction), radiation damping, or dynamic depolarization. This condition lends itself to treatment by *electrostatic* principles, a far simpler method to predict absorption than if an electrodynamic treatment is required. The physical particle size regime that qualifies for quasi-static domain is $r < 10$ nm or so. Larger particles require Mie or Gans models to describe the optical response.

5.5.4 Orientation of Nanometals in Transparent Media

Composite membranes containing cylindrical gold nanoparticles ($a/b \approx 5$) embedded in polyethylene were analyzed with polarized light. In this way,

FIG. 5.19 *Experimental adsorption spectra (left) of gold–alumina composites are compared to simulated spectra (right) derived from Maxwell-Garnett effective medium theory relations. The diameters of the nanoparticles (b = c, the semi-minor axes) were equal to 16 nm in all experiments. The semi-major axis, a, was varied in the experiment and simulations, and the direction of the electric field of the light was perpendicular to the semi-major axis of the nanoparticles (e.g., the electric field of the light was parallel to the plane of the membrane), as shown in Fig. 5.15. The lengths a (in nm) correspond to spectra from top-to-bottom in the figures as follows (aspect ratios are given in parentheses): 165 (10.3), 104 (6.5), 69 (4.3), 39 (2.4), 24 (1.5), and 16 (1.0). The experimental positions of maximum absorption λ_{max} of spectra (in nm), from top-to-bottom in the figures, are as follows (theoretical values are in parentheses): 508 (508), 512 (509), 513 (516), 526 (520), and 539 (533). As aspect ratio was increased, the position of maximum absorption underwent a blue-shift (higher energy, shorter wavelength).*

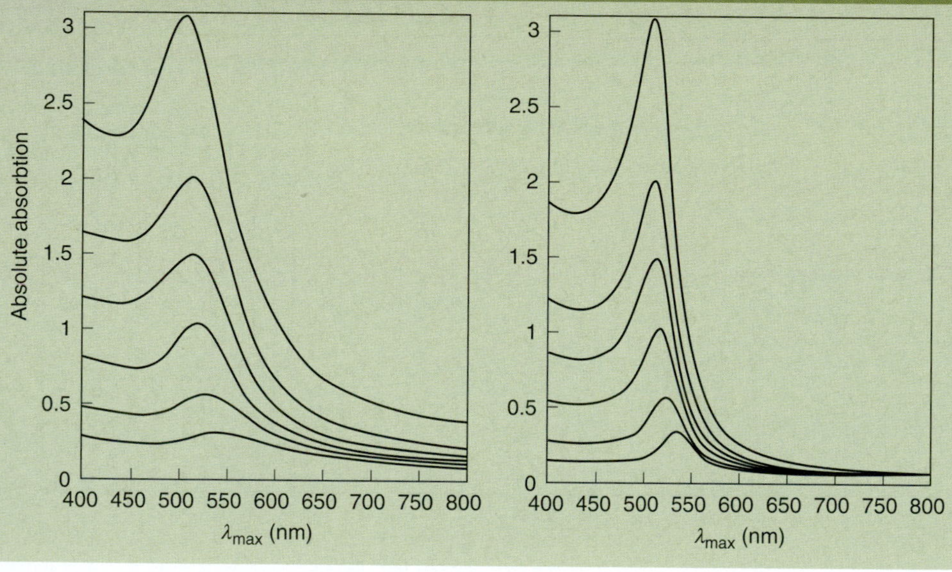

explicit investigation of particle orientation was possible. It is well known that ellipsoids with two perpendicular main axes are capable of generating two distinct λ_{max} [23]. In the case of the sphere, degeneracy exists with regard to plasmon oscillations and only isotropic optical response is expected. The prolate nanogold particles were oriented uniformly in a thin polyethylene film with the semimajor axis parallel to the plane of the film (e.g., the *b*-axis) (**Fig. 5.20a**) [26]. Spectral acquisition was conducted perpendicular to the plane of the nanocomposite film as the sample was rotated in-plane from 0 to 90°. The results are shown in **Figure 5.20b.**

When the incident electric field of the polarized light was held perpendicular to the major axis of the particles ($\theta = 90°$), λ_{max} occurred at 550 nm, yielding a reddish-purple appearance to the composite ($q \approx 1/2$, $\kappa \approx 1$). At the other extreme, with $\theta = 0°$, the case in which the major axis was rotated parallel to the electric field, λ_{max} shifted to an extremely broad peak at ca. 800 nm, nearly a 250-nm red shift, and the composite appeared greenish ($q \approx 1/10$, $\kappa \approx 10$). The condition of dichroism in this case is very different from that observed in the

FIG. 5.20 *(a) Synthesis process of a nanogold–polyethylene composite material. Gold nanocylinders were first formed by electrodeposition inside the pore channels of a porous alumina membrane formed by the anodic process. The alumina was dissolved with NaOH, leaving gold nanocylinders on the surface of the polyethylene. The cylinders were oriented by applying light pressure with the tip of a pencil's eraser. The membrane was placed under a polarizer and rotated 90° (red film) to 0° (green film). The red indicated that the long axis of the cylinders was perpendicular to the electric field of the light and the green color indicated that the cylinders were parallel to the electric field of the light. (b) The corresponding spectra are displayed. The top curve (red, 550 nm) is derived from particles that are perpendicular to the electric field of the light. As the plane was rotated from 90 to 0°, incremental transition of the membrane color from red to green (800 nm) was observed.*

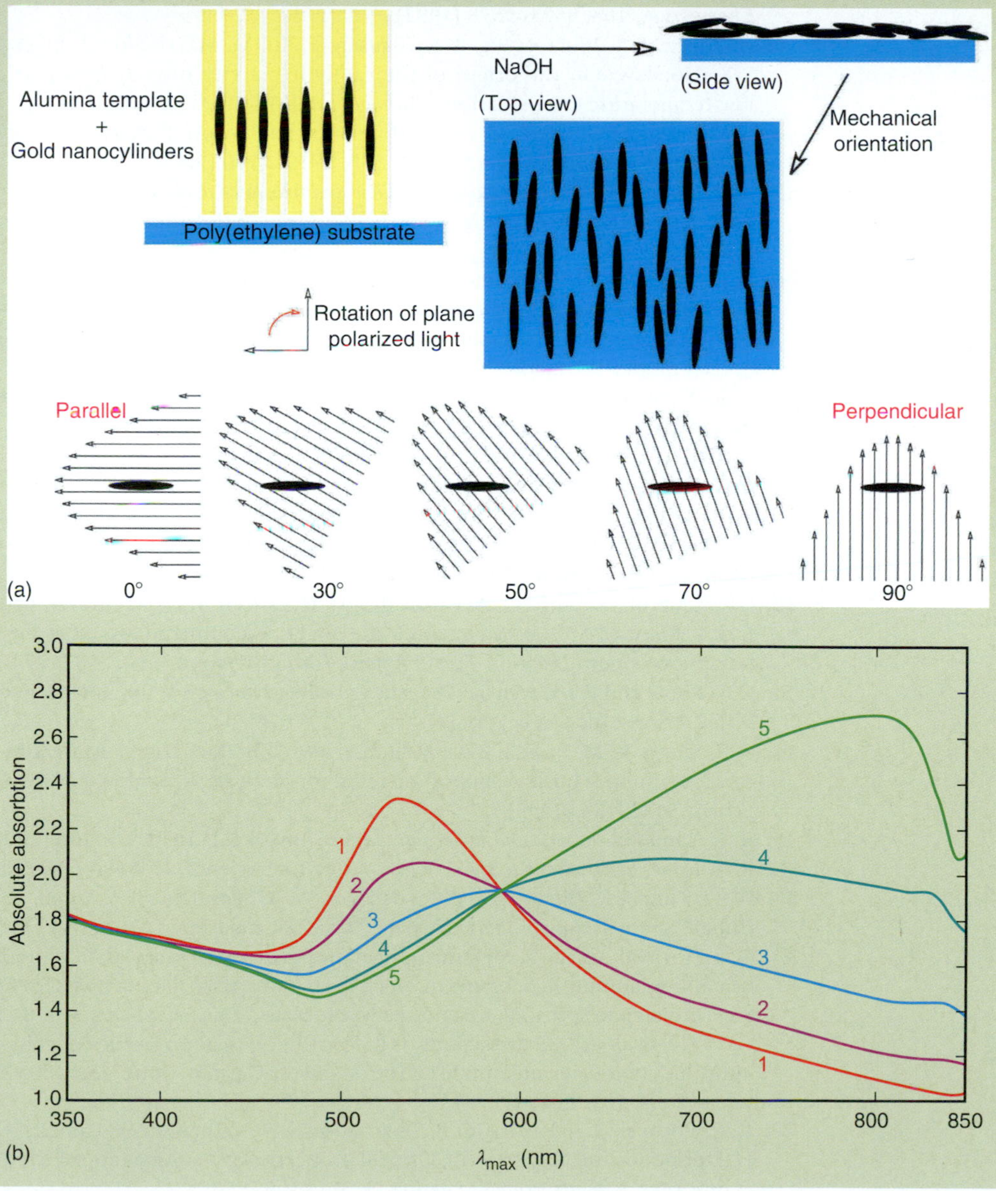

Lycurgus cup. Dichroism in this experiment was due to the orientation of identically aligned particles with respect to polarized light. In the Lycurgus cup, the apparent dichroism was due to absorption, transmission, and reflection phenomena (dependent on where the light source was positioned).

References

1. G. Koller, Study of ice leads to cool new research, Pacific Northwest National Laboratory, *U.S. Department of Energy Research News* (2001).
2. A. Sanchez, S. Abbet, U. Heiz, W.-D. Schneider, H. Häkkinen, R. N. Barnett, and U. Landman, When gold is not noble: Nanoscale gold catalysis, *Journal of Physical Chemistry A*, 103, 9573–9578 (1993).
3. N. Lopez, T. V. W. Janssens, B. S. Clausen, Y. Xu, M. Mavrikakis, T. Bligaard, and J. K. Nørskov, On the origin of the catalytic activity of gold nanoparticles for low-temperature CO oxidation, *Journal of Catalysis*, 223, 232–235 (2004).
4. G. Oberdörster, Toxicology of ultrafine particles in vivo, *Philosophical Transactions of the Royal Society of London Series A*, 358, 2719–2740 (2000).
5. A. D. Maynard, *Nanotechnology: Overview and relevance to occupational health*. Project on Emerging Nanotechnologies for the Woodrow Wilson International Center for Scholars (2005).
6. R. D. Averitt, S. L. Westcott, and N. J. Halas, The linear optical properties of gold nanoshells, *Journal of the Optical Society of America*, 16, 1824 (1999).
7. J. B. Jackson and N. J. Halas, Silver nanoshells: Variations in morphologies and optical properties, *Journal of Physical Chemistry B*, 105, 2743 (2001).
8. S. R. Sershen, S. L. Westcott, J. L. West, and N. J. Halas, An optomechanical nanoshell–polymer composite, *Applied Physics B*, 73, 379 (2001).
9. C. Loo, A. Lowery, N. J. Halas, J. West, and R. Drezek, Immunotargeted nanoshells for integrated cancer imaging and therapy, *Nano Letters*, 5, 709–711 (2005).
10. F. B. Calleja, and Z. Rosianiec, *Block copolymers*, CRC Press, Boca Raton (2000).
11. V. P. Menon and C. R. Martin, Fabrication and evaluation of nanoelectrode ensembles, *Analytical Chemistry*, 67, 1920–1928 (1995).
12. J. Rouquerol, D. Avnir, C. W. Fairbridge, D. H. Everett, J. H. Haynes, N. Pernicone, J. D. F. Ramsay, K. S. W. Sing, and K. K. Unger, Recommendations for the characterization of porous solids, *Pure & Applied Chemistry*, 66, 1739–1758 (1994).
13. C. P. Poole and F. J. Owens, *Introduction to nanotechnology*, Wiley-Interscience, John Wiley & Sons, Inc., New York (2003).
14. S. K. Singh, K. M. Parida, B. C. Mohanty, and S. B. Rao, High surface area silicon carbide from rice husk: A support material for catalysis, *Reaction Kinetics & Catalysis Letters*, 54, 29–34 (1995).
15. R. L. Johnston, *Atomic and molecular clusters*, Taylor & Francis, London (2002).
16. E. A. Irene, *Electronic materials science*, Wiley-Interscience, Hoboken, NJ (2005).
17. W. D. Knight, K. Clemenger, W. A. de Heer, W. A. Saunders, M. Y. Chou, and M. L. Cohen, Spectroscopy of metal clusters, *Physics Review Letters*, 52, 2141–2144 (1984).
18. G. L. Hornyak and C. R. Martin, Optical properties of a family of Au-nanoparticle-containing alumina membranes in which the nanoparticle shape is varied from needle-like (prolate), to spheroid, to pancake-like (oblate), *Thin Solid Films*, 303, 84 (1997).
19. G. L. Hornyak, C. J. Patrissi, and C. R. Martin, Optical properties of gold–porous alumina composite membranes: The Maxwell-Garnett limit, *Journal of Physical Chemistry B*, 101, 1548 (1997).
20. G. L. Hornyak, C. J. Patrissi, E. B. Oberhauser, J.-C. Valmalette, L. Lamaire, J. Dutta, H. Hofmann, and C. R. Martin, Optical properties of Au–Ag nanoparticle alumina composites, *Nanostructured Materials*, 9, 571 (1997).

21. U. Kreibig and M. Vollmer, *Optical properties of metal clusters*, Springer-Verlag, Berlin (1995).
22. A. Campion, Raman spectroscopy of molecules adsorbed on solid surfaces, *Annual Review of Physics*, 36, 549–572 (1985).
23. S. Link and M. El-Sayed, Shape and size dependence of radiative, nonradiative and photothermal properties of gold nanocrystals, *International Review of Physical Chemistry*, 19, 409–453 (2000).
24. J. C. Maxwell-Garnett, Colours in metal glasses and metal films, *Philosophical Transactions of the Royal Society of London A*, 203, 385–420 (1904).
25. J. C. Maxwell-Garnett, Colours in metal glasses, in metallic films and in metallic solutions—II, *Proceedings of the Royal Society of London*, 76, 370–373 (1905).
26. C. A. Foss, G. L. Hornyak, J. A. Stockert, and C. R. Martin, Template synthesis and optical properties of small metal particle composite materials: Effects of particle shape and orientation on plasmon resonance maxima, *MRS Symposium Proceedings*, 286, 431–436, Fall (1992).

Problems

5.1 List 20 examples of systems that demonstrate enhanced surface area, whether at the nanoscale or at the macroscale.

5.2 In example 5.2, we determined the collective surface area of nanocubes derived from a parent cube 1 m on each side. What is the total surface are of spheres inscribed within each cube (diameter = 1 nm)?

5.3 If the parent cube consisted of an *fcc* structure with spheres 1 nm in diameter, what would be the collective surface area of all the individual spheres? How does this compare to your answer in problem 5.2?

5.4 In example 5.6, we introduced hypothetical gold—a material packed in a simple cubic structure (AAA…). What is the density of hypothetical gold (atomic mass = 3.27×10^{-22} g)? How does it compare to the density of real gold, ρ = 19.31 g·cm^3 (an *fcc* structure)? Assume atomic radii are the same.

5.5 Derive a generalized expression of surface area of right circular cylinders of constant volume normalized to the volume of a sphere. What does the graph look like? Hint: derive an expression of S_{cyl} as a function of (h/d).

5.6 Calculate the interior nanosurface area of a hypothetical ceramic material that is configured in an orthogonal network of interconnected tubes 20 nm in diameter. The tubes are spaced 100 nm apart center to center along x–y–z directions. What is the porosity of the material? The pore volume?

5.7 Some quantum dots exist in the shape of a pyramid. Derive a generic expression for the surface-to-volume ratio for such a pyramid. How does the S/V compare to that of a sphere with the same volume? Do the same for a cone.

5.8 Based on the spherical cluster approximation (SCA), calculate the surface fraction of sodium atoms of a 10-nm diameter cluster.

5.9 At what cluster diameter does the number of surface atoms drops below 10% of the total number of atoms in the cluster? (Adapted with permission from R. L. Johnston, *Atomic and Molecular Clusters*, Taylor & Francis, London, 2002.)

5.10 Considering packing fraction (not discussed in our presentation of the SCA), how is the number of surface atoms affected in a tungsten cluster that actually has a packing fraction? Assume packing fraction f = 0.74 for an *fcc* cluster. (Adapted with permission from R. L. Johnston, *Atomic and Molecular Clusters*, Taylor & Francis, London, 2002.) All other

SCA assumptions remain intact, and consider

$$N \cong \frac{V_c}{V_a^*} = \left(\frac{D}{2R_a}\right)^3 f$$

where

V_a* is the effective atomic volume
N is the number of atoms in the cluster
D is the diameter of the cluster
V_c is the cluster volume
R_a is the atomic radius (0.148 nm for W)
f is the packing fraction (0.74 for *fcc* and 0.68 for *bcc*)

5.11 What is the diameter of a spherical *fcc* cluster that has 10^6 atoms with effective $R_a = 0.25$ nm?

5.12 Real clusters are not perfectly spherical. Why do you think this is the case? Would you expect them to become more spherical as the cluster size is increased?

5.13 Tabulate the specific surface areas (square meters per gram) of cubes, spheres, wires, cylinders, and discs made of gold ($\rho = 19.31\ g \cdot cm^{-3}$) using the surface-to-volume relations derived in example 5.7. Beginning with 1 nm and doubling the appropriate dimension until you reach 1052 nm, what trends can you make out from the table? What is unique about the relationship between the cube and the sphere? Is it different from the relationship we derived earlier in example 5.3? Why?

5.14 Using only the C–C bond distance in graphite and simple geometry, calculate (*by brute force!*) the diameter, the surface area, specific surface area, and density of a single-walled carbon nanotube pictured below that has 12 hexagons of carbon along the circumference with the long axis of the hexagon oriented along the long axis of the tube. Do not refer to chapter 9 for vector formulae. Assume that the C–C bond distance in single-walled carbon nanotubes is equal to 0.144 nm. The atomic weight of C is 12.01 $g \cdot mol^{-1}$.

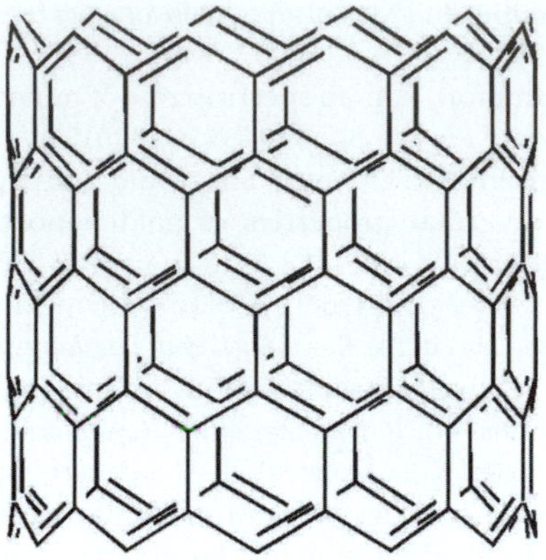

5.15 Calculate the planar density of low-index planes (100), (110), and (111) for an *fcc* crystal.

5.16 With regard to the [*uvw*] crystallographic direction, draw the [111] vector of a cubic system starting at the origin (*a,b,c* or *x,y,z*) = (0,0,0).

5.17 Define crystal density, planar density, atomic density, and linear density.

5.18 Calculate the crystal density of *fcc* gold. How does it compare to the experimentally determined value?

5.19 Calculate the linear density in terms of atom radius along the [110] crystallographic direction of an *fcc* crystal.

5.20 Calculate the planar density in terms of atom radius of the (110) plane in an *fcc* crystal.

5.21 Calculate the *d*-spacing for (111) planes in gold. Correlate these results to experimental data acquired by XRD (do some research).

5.22 Do some research and find out the value of the capacity of the average human lung. What would the size of the lungs be if there were no alveoli in them?

5.23 Concerning the resonance condition discussed in section 5.0, how does the magnitude of the screening parameter κ influence the wavelength of maximum absorption?

5.24 Discuss briefly how eccentricity, the depolarization factor, and the screening parameter are related.

5.25 What is a surface plasmon? Why do you think it is affected by the size of the material? What other factors are able to influence the surface plasmon?

5.26 What kind(s) of hole(s) is (are) found in a NaCl crystal? What is (are) its (their) size?

5.27 Calculate the linear and planar densities of *bcc* α-iron along the [100] direction and the (100) plane, respectively.

5.28 Starting with one cube (the first magic number) and then surrounding it with 26 cubes to generate the second magic number (27), derive a generic equation for magic numbers for such a cubic system. What is the formula for each additional layer?

5.29 Why are smaller nanoparticles usually more active in catalysis than larger ones? (Question courtesy of Günter Schmid, Uni-Essen.)

5.30 Solutions containing nanorods made of gold show 2 absorption maxima. Why is this? If the solution contained only spherical colloids, how many absorption maxima would be expected?

Energy at the Nanoscale

When the attractive power of the solid is greater or less than half that of the liquid, the surface of the liquid must, at its origin, be in the same direction with that of the solid, instead of forming an angle with it, as it often does in reality.... In this manner it may be shown that if the attractive power of the solid be equal to that of the liquid, or still greater, it will be wetted by the liquid, which will rise until its surface acquires the same direction with that of the solid; and that, in other cases, the angle of contact will be greater, in proportion as the solid is less attractive.

Thomas Young, M.D., *A Course of Lectures on Natural Philosophy and the Mechanical Arts* (printed by Taylor and Walton, London, 1845)

Chapter 6

Threads

Surface, geometric, and structural aspects of nanomaterials, particularly of metals, have been introduced and briefly discussed in chapter 5. For the most part, not much discussion about energy or chemistry was presented in the previous chapter. It is time now to investigate the energetic character of nanoscale materials, particularly in the way it correlates with the nano surface. In this chapter, we begin to connect nanostructure with surface energy, and we wrap up the "What are nanomaterials?" and "Why are they interesting?" parts of the text. In chapter 7, discussions about why they have remarkable properties and phenomena begin in earnest. Chapter 7 proceeds to investigate the breakdown of bulk scaling laws at the nanoscale as numerous physical properties and phenomena are individually addressed. In chapter 8, a special section on nanothermodynamics is brought to the forefront—a new way to look at thermodynamics.

6.0 Surface Energy

The structure of nanomaterial surfaces contributes significantly to the energetic character of nanomaterials. Conversely, one can say with equal validity that the structure of nano surfaces is certainly due to energetic considerations. These statements correlating surface structure to energy also apply to bulk material surfaces. However, the net energetic impact of the bulk material is significantly reduced when compared to an equal volume of nanoparticles. This is not to imply that the surface is not important in bulk materials. It most certainly is important. Whether in nanoscale or bulk form, atoms and molecules that exist at the surface or at an interface are different from the very same atoms or molecules that exist in the interior of a material. The difference is manifested in altered structure, enhanced reactivity, and a greater tendency to agglomerate. The surface is the place where periodicity and chemical bonding of a crystal are disrupted and represent a phase that wishes to become something else.

The surface layer (selvage layer) of a bulk solid consists of about three or fewer layers of atoms or molecules, a few angstroms at the most [1]. These surface atoms on average have fewer nearest neighbors. Having fewer neighbors raises the energy and hence lowers the stability of those atoms and, consequently, of the surface itself. We know from our earlier discussion of electron shells, nuclear stability, and magic numbers that closed structures are preferred due to their enhanced stability (e.g., the state of minimal energy). In a bulk metal *fcc* crystal, the optimal coordination number Z for each atom is 12. Anything less than this ideal imparts an inherent instability to the surface (e.g., by way of the *dangling bond*). Dangling bonds spread gloom over the surface—a state of high anxiety. Dangling bonds must find ways to compensate for the apparent deficiencies by whatever chemical or physical means available to them. Please take note that we are only referring to bulk surfaces in this last discussion. If surface atoms do exist in a state of higher energy than fully coordinated volume atoms, we can safely conclude that it must take an investment in energy to create a new surface.

If surfaces were suddenly to become the energetically favored state, there would be driving forces to create more surfaces. That would indeed be a peculiar universe, one that consists entirely of surfaces. In reality, driving forces exist in materials and systems to actually minimize surface area. The smaller the surface-to-volume ratio is, the lower is the energy state of the material. Why then do we have nanomaterials? Nanomaterials are characterized by increased surface area, as we found out in the previous chapter. However, now in the context of this chapter, we try to link these behaviors and phenomena to the surface and hence to the energetic state of the nanomaterial. We have nanomaterials because we have found ways to ensure their presence—as has nature.

We introduced the discipline of surface science in the last chapter and continue here with the same theme. Surface science focuses on the interface between physical phases, whether solid–solid, solid–liquid, liquid–gas, or solid–gas. Therefore, the outermost atomic or molecular layers are involved in surface science investigations. Surface energy is an extrinsic (or *extensive*—more about this term in chapter 8) property of all materials, whether flat or otherwise. In other words, surface energy is an additive quantity. The surface energy of 10 identical nanoparticles is equal to the sum of the surface energy of each individual particle. If these particles were to agglomerate, the overall surface energy is reduced. **Figure 6.1** shows a simple geometric example of agglomeration and how surface energy is reduced overall as larger particles are formed.

EXAMPLE 6.1 *Surface Energy by a Simple–Brute Force Geometric Model*

Devoid of any chemistry or physics, show how surface energy can be reduced by agglomeration of cubes of unit volume to form larger cubic or rectangular solid shapes. Calculate (by brute force) the surface energy of resulting geometric shapes formed by the following agglomeration schemes. Assume that the surface energy of one face of the original cube is equal to γ.

(a) Zero dimensions (a dot): What is the total surface energy of one cube (**Fig. 6.1a**)? What is the surface energy of 27 independent cubes?

(b) One dimension (a wire): Determine the total surface energy of nine cubes agglomerated in one row (**Fig. 6.1b**). What is the surface energy of three rows consisting of nine cubes each?

(c) Two dimensions (thin film): Starting with one cube, completely surround the central cube in two dimensions with additional cubes until you arrive at a square configuration that is 3×3 cubes (**Fig. 6.1c**). What is its surface energy? What is the surface energy of three such squares?

(d) Three dimensions: Starting with one cube, form a larger cube by completely surrounding the central cube to form a large cube that is $3 \times 3 \times 3$ cubes (**Fig. 6.1d**). What is the surface energy?

Solutions:

(a) If we assign the value γ to the surface energy of one face of a cube:
The total surface energy of one cube is $6 \times \gamma = 6\gamma$.
The combined surface energy of 27 cubes is $27 \times 6\gamma = 162\gamma$.

(b) The surface energy of a line of nine cubes is $[(9 \times 6) - (8 \times 2)]\gamma = 38\gamma$. The combined surface area of three such lines is $3 \times 38\gamma = 114\gamma$.

(c) The surface energy of a square of nine cubes is equal to $[(2 \times 9) + (4 \times 3)]\gamma = 30\gamma$.
The surface energy of three such squares is $3 \times 30\gamma = 90\gamma$.

(d) The surface of a cube consisting of 27 cubes is equal to $(6 \times 9)\gamma = 54\gamma$.
Therefore, at constant volume, the surface energy decreases as the level of agglomeration increases:
162γ (independent cubes) $\rightarrow$ 114γ (linear cubes) $\rightarrow$ 90γ (square cubes) $\rightarrow$ 54γ (cubic cubes).

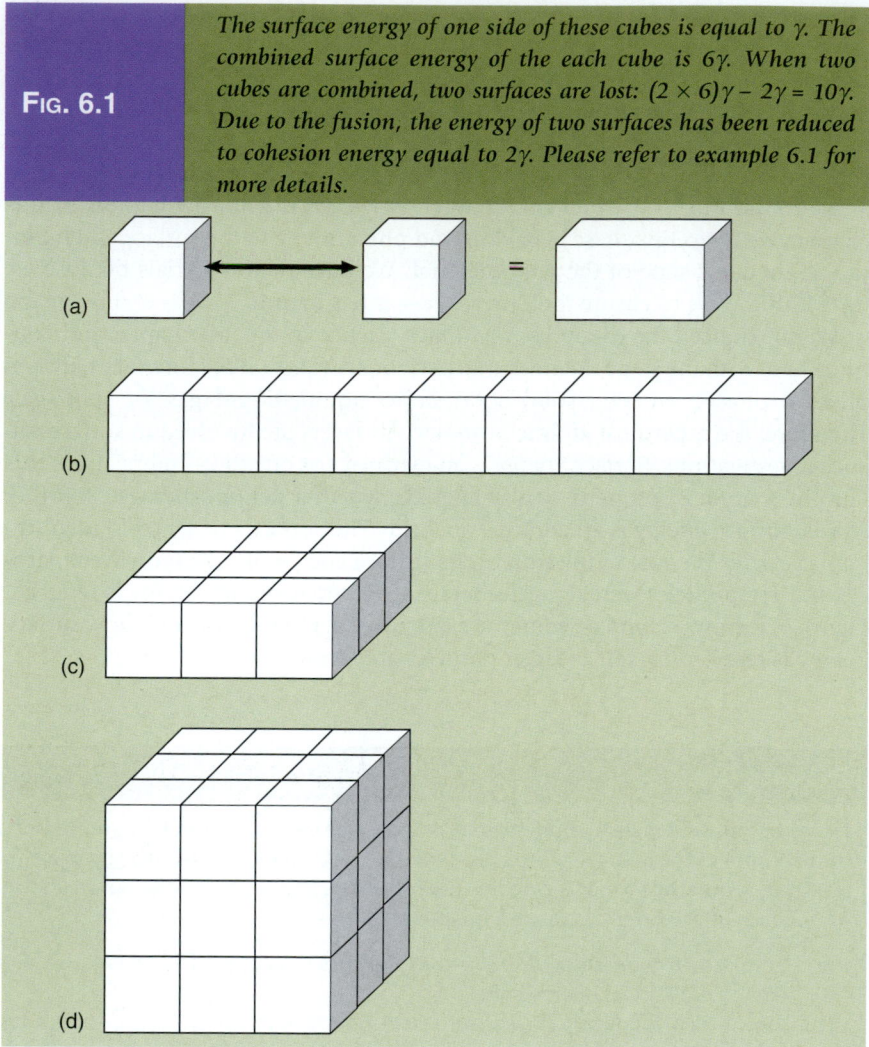

Fig. 6.1 *The surface energy of one side of these cubes is equal to γ. The combined surface energy of the each cube is 6γ. When two cubes are combined, two surfaces are lost: $(2 \times 6)\gamma - 2\gamma = 10\gamma$. Due to the fusion, the energy of two surfaces has been reduced to cohesion energy equal to 2γ. Please refer to example 6.1 for more details.*

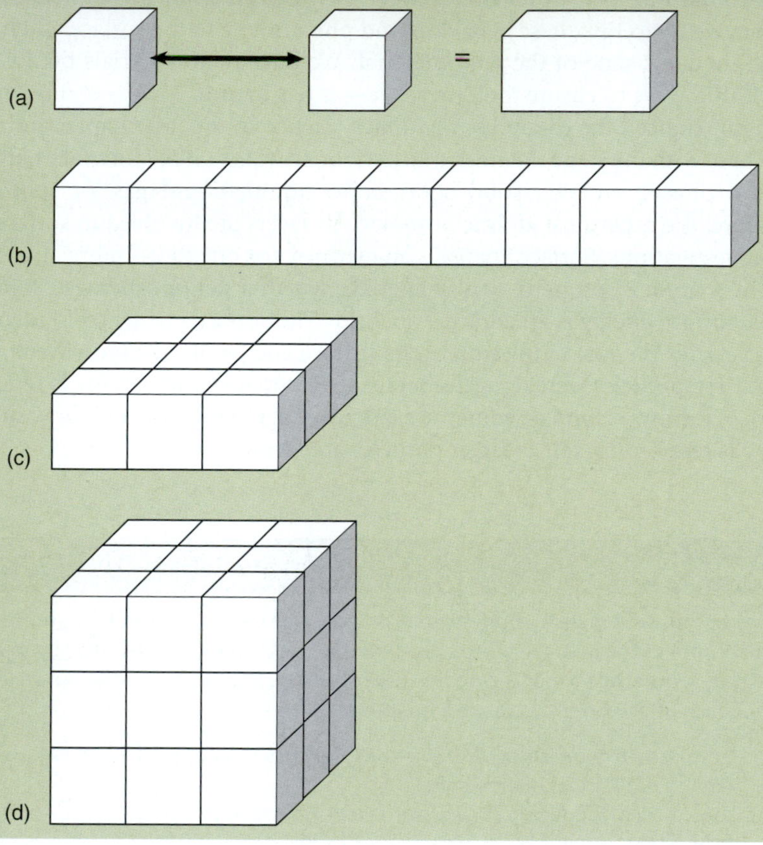

In addition to nanosized liquids and solids, surface phenomena govern much of the physical and chemical nature of colloids (ca. 10^{-6} to 10^{-9} m in size). The chemistry of colloids is really the chemistry of the surface. Surface phenomena are also vital to the functions of living things. Since the living cell is a complex of colloidal systems [2], there must be a significant number of surfaces available in cells to conduct daily business.

The surface is important at all size scales. At the nanoscale, surface phenomena become very interesting. Energy is a part of everything. The nanosurface is another boundary condition—one that actually interfaces with the macroscopic world. It is at this boundary that the currency of energy is exchanged.

6.0.1 Background

The phenomenon of capillarity has been known since the time of Leonardo da Vinci and probably earlier; however, the study of surface chemistry began

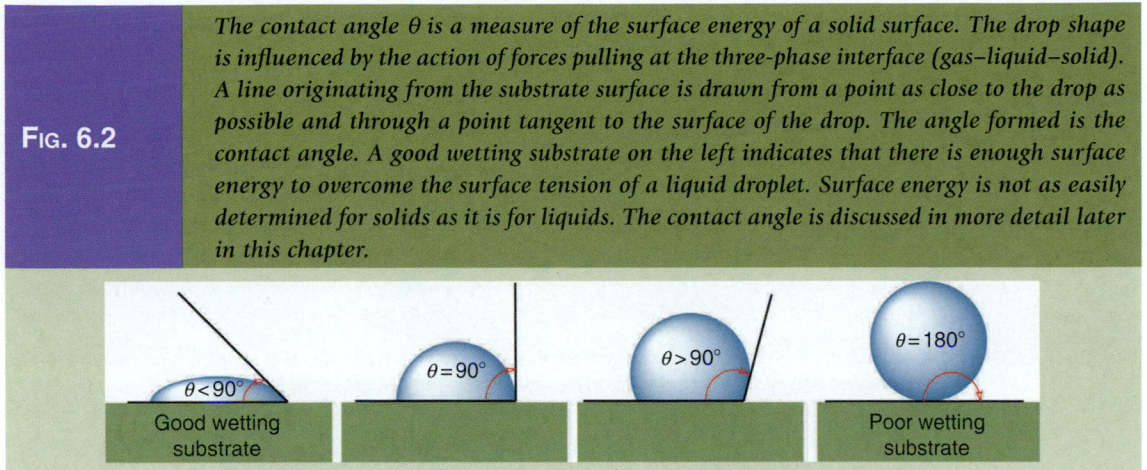

FIG. 6.2 *The contact angle θ is a measure of the surface energy of a solid surface. The drop shape is influenced by the action of forces pulling at the three-phase interface (gas–liquid–solid). A line originating from the substrate surface is drawn from a point as close to the drop as possible and through a point tangent to the surface of the drop. The angle formed is the contact angle. A good wetting substrate on the left indicates that there is enough surface energy to overcome the surface tension of a liquid droplet. Surface energy is not as easily determined for solids as it is for liquids. The contact angle is discussed in more detail later in this chapter.*

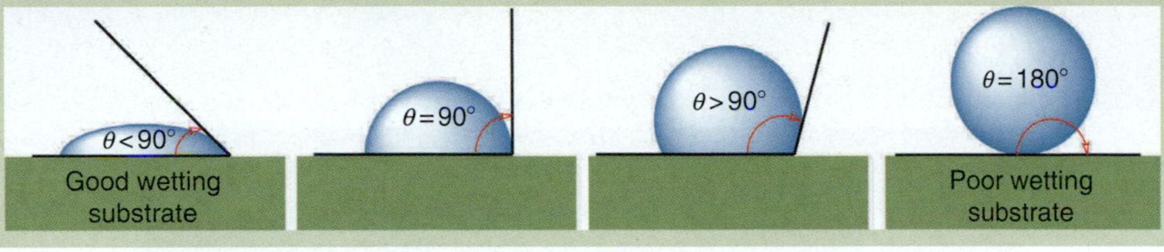

formally when Thomas Young in 1805 and Pierre Simon de Laplace in 1806 first investigated the surface tension (and curvature phenomena) of liquids. Young stated that the contact angle (of a drop of liquid on a flat solid substrate) that exists at the triple phase point of a solid–liquid–gas interface is a physical constant that depends on the materials involved (**Fig. 6.2**). Young is also noted for coining the word *energy* in the context of the "fundamental quantity created by heat which moved particles." Young was considered by some to be the last *polyhistor*, the "last man to know everything" [3].

Johann Carl Friedrich Gauss in 1830 introduced the concept of *surface energy*. He based his concept on the sum total potential of interacting pairs of molecules. In his *Principia Generalia Theoriae Figurae Fluidorum in statu Aequilibrii*, Gauss describes a mechanical theory of heat and claimed that his theory of potentials and methods of least squares connected nature with science. Josiah Willard Gibbs in 1877 developed the classical thermodynamic treatment of surface phenomena and the concept of the *Gibbs dividing surface*. G. Wulff in 1901 is credited with devising "Wulff plots," which depict the equilibrium surface of finite anisotropic solids—the minimum energy surfaces. Significant credit must once again be given to Irving Langmuir who, in the early twentieth century, developed an understanding of heterogeneous catalysis and adsorption kinetics and their associated surface phenomena. John Bardeen advanced the concept of the electric-double layer in 1936. The ability to fabricate a *clean* surface with the advent of ultrahigh vacuum systems enabled researchers to study reproducible crystal surfaces.

6.0.2 Nature

Just like in the discussion of the nanosurface and nanosize in chapter 5, we must always pay tribute to the remarkable examples that nature provides. Macroscopic systems exhibit surface energy phenomena that are based on nanoscale structure. The perfect (and often used) example to illustrate natural surface energy is the leaf of the lotus plant. The *lotus effect* (**Fig. 6.3**), a phenomena described by

Fig. 6.3

The lotus effect is depicted to demonstrate an extreme contact angle: ~180°. Solid surfaces that generate contact angles greater than 150° are considered to be superhydrophobic. For the lotus plant to achieve a contact angle of nearly 180°, hydrophobicity by itself is not enough. The drop is contacted by numerous structures that terminate in a point. In this way, the surface tension of the drop is never overcome and the surface behaves as if it were hydrophobic. Surface roughness of high-energy surfaces (hydrophilic) enhances good wetting. Surface roughness of low-energy surfaces (hydrophobic) enhances drying (e.g., the liquid must expend energy in order to conform to the topology of the surface). Synthetic surfaces called Fakir carpets *are roughened by facets 50 μm wide by 150 μm deep. Spacing between facets is 100 μm. On these surfaces, water droplets are removed without any wetting. If tilted, droplets roll right off.*

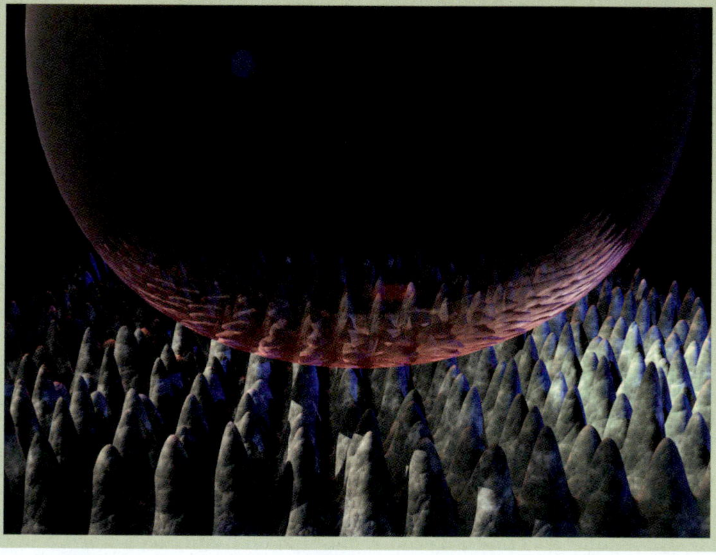

Source: Courtesy of Professor Anil K. Rao, Department of Biology, the Metropolitan State College of Denver.

W. Barthlott in 1990, is a property exhibited by the leaves of the lotus and other similar plants. The lotus leaf is a natural low-energy surface onto which water beads up into a nearly perfect sphere and rolls off the leaf, truly a benefit to a rain forest dweller [4]. Due to the intricate micro- and nanostructure of the leaf surface, enhanced drying and self-cleaning properties are also observed. A detailed presentation of the lotus effect is given in chapter 13, "Natural Nanomaterials." The water strider insect is an example of how another living organism takes advantage of the surface tension properties of water. The legs of the water strider are water repellant [5]. The macroscopic hairs possess nanofilamentous structures that are able to support small air bubbles, thereby allowing the insect to stride over the surface of a pond or stream [5].

Microscale examples of surface phenomena in nature include the cell membrane and any other dividing surface found in eukaryotic cells and bacteria as well as the boundaries that enclose organelles and segregate vacuoles. The formation of hydrophobic or hydrophilic surfaces that are stabilized in complementary environments is a critical component in stabilizing all of life's surfaces. At the nanoscale, protein–surface interactions modulate cell adhesion, migration, proliferation, and differentiation [6]. For this reason, the adsorption of proteins at solid–liquid interfaces is the focus of much scientific endeavor. Depending on which amino acids extend to the surface, proteins can be soluble or insoluble in water.

6.0.3 Introduction to Surface Stabilization

At this introductory stage in the chapter, we assume that nanosurfaces are by nature higher in energy than surfaces of bulk material counterparts. In the next step, we assemble our collective knowledge of chemistry and state that "systems of high energy will strive to attain a state of lower energy, by whatever means possible." Fewer nearest neighbors for surface atoms and molecules result in atomic or molecular species with dangling bonds—bonds that require satisfaction. In **Tables 6.1** and **6.2**, the topic of surface stabilization is summarized along with examples of recourses available to systems to attain a minimal energy condition.

There are many chemical and physical mechanisms available to achieve compensation for bonding deficiencies: higher chemical reactivity, shorter bond length (accompanied by shrinking of interlayer distances), anisotropic structural distortion, coalescence and lowered melting points, and evaporation temperature,

TABLE 6.1 *Physical Modes to Stabilize High-Energy Surfaces*

Physical process	Mechanism
Surface relaxation	Inward forces reduce *d*-spacing. Consequences are reduced lattice constant and shorter bond length with surface atoms. Lateral forces create an asymmetric environment.
Surface restructuring	Formation of highly strained homogeneous bonds that impact crystal growth
Composition segregation	Diffusion of impurities to the surface Phase segregation
Physisorption	Formation of a monolayer (or more) of weakly bound atoms or molecules
van der Waals force	van der Waals forces are exerted at short distances. Does the gecko stabilize a surface by adhering to it?
Sintering	Formation of polycrystalline condensed materials at elevated temperatures. Grain boundary stabilization
Isotropic minimization	Formation of spherical particles to produce the lowest energy geometric solid
Anisotropic minimization	Equilibrium planes that form multifaceted structures. Concave versus convex surfaces
Maximization of packing efficiency	Low-index crystal planes enhance the number of nearest neighbors in three-dimensional solids; elimination of voids, defects, dislocations, etc. in nanomaterials.
Geometric agglomeration	Nonchemical agglomeration in the vapor and liquid phases. In liquid phase, this is called *Ostwald ripening*, where a larger particle grows at the expense of higher energy smaller particles.
Physical confinement	Confinement (kinetic) of metastable species in pore channels to preserve nanoparticle structure. This is a difficult category to precisely define because chemical stabilization may occur from the species in the wall of the cavity.
Artificial environments	Ultrahigh vacuum spaces keep metastable particles free of interaction with contaminants. The development of this technology allows R&D to proceed for high surface energy materials.

Table 6.2 *Chemical Modes to Stabilize High-Energy Surfaces*

Physical process	Mechanism
Chemical termination	Covalently bonded chemical species to surface atoms
Electrostatic stabilization	Formation of an electrical double layer to stabilize reactive species and particles in solution
DLVO theory	Stabilization by van der Waals attraction + electrostatic repulsion in solutions
Supramolecular stabilization	All intermolecular forces serve to stabilize supramolecular species in solution
Stabilization of biological molecules	Environmental controls (mild thermal conditions, protection from radiation); numerous intermolecular forces
Steric stabilization	Polymeric layered materials; ligand formation between sol and polymer chain. The process serves to isolate nanoparticles (e.g., stabilize).
Chemisorption	Chemisorption indicates a weaker chemical bond than a covalent bond. Chemisorbed molecules and atoms can be removed during TPD.
Surfactants	Surface stabilization of metastable phases by hydrophobic–hydrophilic interactions in solution
Kinetic stabilization	Metastable molecules that are thermodynamically unstable exist because there is no pathway available for chemical reaction.

just to name a few. Bulk surfaces, however, are not generally considered to be reactive. How do we reconcile what we just stated with reality? It is true that the surface achieves stability by distortions, passivation, and other processes, but overall, the surface of a bulk material contributes a miniscule amount to the overall stability of the material as a whole. A block of iron sitting in your garage is going nowhere, but it can rust. Once you remove the rusty layer, you basically have not lost much.

Surface energy is minimized in biological systems. One way to minimize the impact of the surface is to grow larger. For example, spherical particles are thermodynamically stable but biologically inefficient, unless they are one-celled organism. Some spherical algal species have evolved into extremely long cylindrical cells. Some of the largest cells in the world belong to algae species. Although extremely long (meters), the thickness of the cell is a micron or so—just about the size of a paramecium.

6.0.4 The Nano Perspective

Nanoscience is, to a great degree, a tribute to those who have been able to fabricate small materials as well as to those who devised ingenious methods that keep them small. These breakthroughs collectively form the basis for nanoscience and technology. Nanoscience is also about how macroscopic phenomena are transposed to the nanoscale and subsequently subjected to interpretation. All macroscopic phenomena retain relevance at the nanoscale; we just have to figure how and why its form has been altered and what it cost us. In this chapter, we discuss surface energy, the energy of adhesion, and the energy of cohesion, and the phenomena of capillarity, curvature, vapor pressure, and the Kelvin effect. These properties and phenomena were developed in the past century and even earlier in the nineteenth century. We now are able to measure and observe them at the nanoscale. Imagine if Thomas Young been able to peer through an electron microscope just once!

6.1 Basic Thermodynamics

Surface tension, γ, is defined as the force per unit length that opposes the expansion of surface area [7]. We will work in some basic concepts and derive surface energy expressions from thermodynamic relations—as always, a good place to start. To break up any condensed phase, whether liquid or solid, work has to be accomplished (energy has to be added) and only then can a new surface be created. Making a new surface involves taking atoms or molecules from the volume of a bulk material where they are stable and forcing them into a surface layer where they are inherently less stable but adapt to achieve stability. This relocation requires positive energy to accomplish. The work is measured in terms of surface energy (appropriate for solids) and surface tension (appropriate for liquids) and has units of joules per square meter or newtons per meter. Both forms are represented by the Greek letter γ:

Surface tension, $\gamma \rightarrow \mathrm{J \cdot m^{-2}}$, $\mathrm{N \cdot m^{-1}}$ (or $\mathrm{erg \cdot cm^{-2}}$, $\mathrm{dyn \cdot cm^{-1}}$) $\leftarrow$ surface energy, γ

The symbols σ and T are also used to represent surface tension. Liquid surfaces tend to contract to achieve a surface of minimum energy (the sphere) due to strong inward attractions. Take note: Once again the sphere makes an appearance to help us explain the nanoscale.

6.1.1 *Derivation of Surface Tension, γ*

Surface tension expressions can be derived entirely from mechanical thermodynamic principles. Application of the Helmholtz (A with V,T constant) and Gibbs (G with P,T constant) functions is required in order to understand the nature of surface tension. The incremental amount of work required to generate an incremental amount of new surface is

$$dW_{rev} = \gamma d\mathcal{A} \tag{6.1}$$

where W is work in joules, $\mathcal{A}$ is the area in square meters, and γ is a proportionality constant that is always positive and equivalent to the surface energy (tension). The corresponding Helmholtz work function is then equal to [8]

$$dA = \gamma d\mathcal{A} \tag{6.2}$$

The magnitude of γ is, of course, characteristic of the chemical nature of the surface material and $\gamma d\mathcal{A}$ is expressed in the form of energy.

To change the surface, clearly some form of pressure–volume work is required in addition to the usual definition of work, W—a term describing the incremental area of new surface ($+\gamma d\mathcal{A}$). With respect to a flat planar interface, the work required to create a new surface also involves changes in volume. The expression of reversible work required to add an increment of new surface is

$$dW_{rev} = -PdV + \gamma d\mathcal{A} \tag{6.3}$$

where $(-PdV)$ is the work required to effect any volume change and P is the equilibrium pressure in the liquid and gas bulk phases. What exactly does surface tension in the form of newtons per meter (or dynes per·centimeter) indicate?

If a force (along the y-direction only) is applied by a piston (with area $\mathcal{A} = l_x l_z$) inside a rectangular container ($l_x \times l_y \times l_z$) and filled with one fluid, the resultant pressure of the system is equal to P, the equilibrium pressure [9]. The incremental amount of work required is simple pressure–volume work and is proportional to the incremental displacement of the piston in the y-direction, dl_y [9]:

$$dW_{rev} = F_{piston}\, dl_y = -P\,(l_x\, l_z\, dl_y) = -PdV \tag{6.4}$$

The expression for work then is adapted to accommodate the contribution of the interfacial layer:

$$dW_{rev} = F_{piston}\, dl_y = -P(l_x\, l_z\, dl_y) + \gamma l_x\, dl_y \tag{6.5}$$

$$F_{piston} = -P(l_x l_z) + \gamma l_x \tag{6.6}$$

$$P_{piston} = P - \gamma / l_z \tag{6.7}$$

In other words, a line of force is in place at the interface of the two phases—one that resists the force ($P\,\mathcal{A}$) of the piston. The surface tension is responsible for the resistance [9] and opposes the force of the piston.

Surface tension and surface energy are related to the differential form of the standard thermodynamic expression for free energy. Since energy is an extensive quantity, we are free to add terms as required. The surface energy requires us to add a surface energy term to the standard free energy expression:

$$dG = -S\,dT + V\,dP + \gamma\, d\mathcal{A} \tag{6.7a}$$

Differentiating with respect to temperature, we get the classical temperature dependence of the free energy (the surface free energy G_s) [17].

6.1.2 Surface Excess

Regardless which convention is used, energy is required to create a new surface and, on the flip side, energy in the form of heat is released when surfaces are merged. This happens when two drops of water merge to become one larger

EXAMPLE 6.2 *How Much Work Is Needed to "Uncontract" (Stretch) the Surface of Water?*

Neglecting the force of gravity, if you have a wire of length 2 cm, calculate the work required to lift a sheet of water held together by surface tension to a height of 5 mm. The surface tension of water is 72.8 dyn·cm^{-1}.

Solution:

This procedure has created a total surface equal to $2 \times l \times h = 2 \times 2\ cm \times 0.5\ cm = 4\ cm^2$. The total work done to create this new surface is $(2\ l\ h) \times \gamma_{water} = 4\ cm^2 \times 72.8\ dyn \cdot cm^{-1} = 291\ dyn \cdot cm^{-1} = 291$ ergs.

drop. This energy is called the energy of cohesion. Obviously, energy of cohesion of a homogeneous fluid is equal to twice the value of the surface tension. Surface energy, then, is also called *surface excess energy*. In other words, there is extra free energy available to drive chemical or physical processes on the surface.

For the reversible process at constant temperature, pressure (and amount of material), the surface energy (tension) is equal to

$$\gamma = \left(\frac{\partial G}{\partial \mathcal{A}} \right)_{P,T} \tag{6.8}$$

where G is the Gibbs free energy in joules and $\mathcal{A}$ is the area in square meters. ΔG is negative, or spontaneous, when surface area is decreased and is positive, requiring energy input, when surface area is increased.

6.1.3 Kelvin Equation

Vapor pressure at the liquid–vapor interface is influenced by pressure. If pressure is applied (e.g., by an inert gas to the surface of a liquid in equilibrium with its vapor), the vapor pressure, p, of the liquid is increased. This is the Kelvin equation

$$p = p_o e^{V_m \Delta P/RT} \tag{6.9}$$

where p_o is the vapor pressure of the liquid under standard conditions and V_m is the molar volume of the liquid. The vapor pressure exerted by a curved liquid surface based on the *Kelvin equation* relates vapor pressure change to surface curvature:

$$\ln\left(\frac{p}{p_o} \right) = \frac{2\gamma V_m}{r_{critical}(RT)} \tag{6.10}$$

where p is the vapor pressure of the particle, p_o is the equilibrium vapor pressure of a planar bulk material, and V_m is the molar volume and $r_{critical}$ the radius of the particle. In other forms of the equations, atomic volume V_a and the Boltzmann constant are used:

$$\ln\left(\frac{p}{p_o} \right) = \frac{2\gamma V_a}{r_{critical}(k_B T)} \tag{6.11}$$

The Kelvin equation in essence gives us a free pass into the nanodomain. The key component is the radius term r in the denominator. As r gets smaller, the vapor pressure of the droplet becomes larger and larger. The Kelvin equation explains the growth of water droplets in the atmosphere and the condensation of cold gases in nanopores by surface area methods such as BET and BJH. The Gibbs–Thomson effect and the Kelvin equation are often used interchangeably. The Gibbs–Thomson effect states that the chemical potential μ and the vapor pressure p of a small particle are related inversely to the surface curvature of a particle, and hence the radius of the particle. Small droplets have high surface

Example 6.3 *Work Required to Create More Surface Area*

A drop of water is forced through an aspirator at 20°C. Calculate the amount of work in millijoules required to increase the surface area of water from that of the 1 cm^3 drop to aerosol particles with 1 μm^3 volume. How many drops are formed from the parent 1 cm^3 drop? What is the new total surface area? What is the specific surface area in square meters · per gram ($\rho_{water} = 1.00\ g \cdot cm^{-3}$)? Ignore any potential Joule–Thomson cooling effect. The surface tension of water is equal to 72.8 $dyn \cdot cm^{-1}$.

Solution:

Calculate the new surface area, find the additional area, and apply equation (6.1), $W_{rev} = \gamma \Delta \mathcal{A}$.
Surface area, $S_{A,drop}$, *of the 1* cm^3 *drop* $= 4.84\ cm^2$
Surface area, $S_{A,aero}$, *of the 1* μm^3 *aerosol particle* $= 4.84 \times 10^{-8}\ cm^2$
Number of aerosol particles formed from parent drop: Aerosols $\equiv 10^{12}$
Total surface area of aerosol particles $= 10^{12}\ (4.84 \times 10^{-8}) = 4.84 \times 10^{4}\ cm^2$
$\Delta \mathcal{A} = S_{A,aero} - S_{A,drop} = 4.84 \times 10^{4}\ cm^2 - 4.84\ cm^2 \approx 4.84 \times 10^{4}\ cm^2$
Energy required to create new area $= 72.9\ dyn \cdot cm^{-1}\ (4.84 \times 10^{4}\ cm^2) = 3.53 \times 10^{6}\ ergs$
$= (3.53 \times 10^{6})(1\ J/10^{7}\ erg) = 353\ mJ$
Specific surface area, $S_s = 4.84\ m^2 \cdot g^{-1}$

curvature and therefore, high vapor pressure. This is because the surface-to-volume ratio is larger than in a bulk material. This is also why depression in freezing point is caused by solutions that contain finely divided particles [10].

6.1.4 Particle Curvature and the Young–Laplace Equation

The Young–Laplace equation relates the pressure difference across the surface of a small particle:

$$\Delta p = \gamma \left(\frac{1}{R_x} + \frac{1}{R_y} \right) \tag{6.12}$$

where R_x and R_y (or R_1 and R_2 or r_1 and r_2) are the radii of curvature along the x- and y-axes parallel to the surface. Because the value of the radius of curvature for a sphere is the value of the radius of the sphere, the Young–Laplace equation simplifies to

$$\Delta p = \frac{2\gamma}{R} \tag{6.13}$$

Selected Δp values for water droplets at different radii in terms of atmospheres are as follows: 0.0014 (r = 1 mm), 0.0144 (r = 0.1 mm), 1.436 (r = 1 μm), and 143.6 (10 nm). A six-magnitude reduction in particle radius results in a five-magnitude increase in the internal pressure of the substance. The power of nano!

The pressure differential across a curved interface, in particular a drop of liquid or a cavity in a liquid with radius r, can be described as follows. Consider a cavity with radius r. The internal pressure, p, balances the tendency of the bubble to minimize its surface area due to contractive forces (e.g., the surface tension γ). The outward force associated with the internal pressure is $(4\pi r^2)$ [8]. The inward pressure bearing from outside the cavity is equal to $p_o(4\pi r^2)$. An incremental change in surface area $d\mathcal{A}$ as the radius changes by an incremental amount dr is

$$d\mathcal{A} = 4\pi\,(r + dr)^2 - 4\pi r^2 = 4\pi r^2 + 8\pi r\,dr + 4\pi(dr)^2 - 4\pi r^2 = 8\pi r\,dr \quad (6.14)$$

Therefore, the work needed to change the area by dr is equal to

$$dW = \gamma(8\pi r\,dr) \quad (6.15)$$

At equilibrium, all the internal and external forces are balanced and

$$p(4\pi r^2) = p_o(4\pi r^2) + \gamma(8\pi r), \quad (6.16)$$

thus simplifying to the Young–Laplace equation for spherical particles:

$$p - p_o = \Delta p = \frac{2\gamma}{r} \quad (6.17)$$

Once again we derive the Young–Laplace equation, albeit from another perspective—one that is purely mechanical. Because all forces sum to zero, $\Sigma F = 0$, this relation accounts for the difference between the internal and external pressure, equivalent to the surface tension times the distance that it is displaced. The Young–Laplace equation shows that the pressure inside a curved surface will always be larger than the pressure outside a curved surface [8]. As radius is increased infinitely, Δp approaches zero and the pressure differential across an infinite plane is achieved, which at equilibrium is equal to zero.

By substituting Δp derived previously into equation (6.9), we get the Kelvin equation for a droplet, equation (6.10). For a cavity surrounded by a liquid,

$$p = p_o e^{-V_m 2\gamma / r(RT)} \quad (6.18)$$

Because the pressure inside a cavity is less than the pressure outside, the only change in this expression from that of the Kelvin equation is the sign of the exponent [8].

Once again, as r increases to infinity, the vapor pressure of the liquid is expected to approach the equilibrium vapor pressure of a bulk planar surface. These relationships tell us that atoms or molecules on the surface of a small liquid droplet are held more loosely than the corresponding atoms or molecules that exist at a flat planar surface of the liquid. Reflecting back to chapter 5, we recall that atoms and molecules at the surface are deficient with regard to coordination number by fellow atoms or molecules. Higher curvature effectively reduces the number of nearest neighbors even further, resulting in higher vapor pressure over the curved surface.

Curvature. As a general rule, liquid materials in droplets with high positive curvature (convex) have a higher vapor pressure than the corresponding bulk material. Liquids in pores demonstrate negative curvature (concave) and have lower vapor pressure than their bulk counterparts. What exactly is curvature? As we progress to smaller dimensions, a parameter known as curvature assumes greater importance. A straight line, or flat bulk material, has no curvature. Curvature is a measure of the degree of deviation from a straight line. The curvature of a circle is the derivative of the angle ϕ (the angle made between some directional vector and the tangent line to the circle) with respect to arc length s (the change in the unit tangent vector per unit distance of the curve). The curvature is a parameter that is equal to the reciprocal of the radius [13]:

$$\kappa(s) = \frac{d\phi}{ds} = \frac{1}{r} \tag{6.19}$$

where κ is defined as the curvature and s is the arc length. Therefore, from equation (6.19), the smaller the radius becomes, the greater the extent of curvature is: $r \to \infty$, $\kappa \to 0$.

It is no surprise by now that physical properties are expected to change as particle size is decreased. In this case we relate how curvature influences physical properties. The pressure differential across the surface of a bubble, as shown previously, is a function of curvature. When a surface is curved, a pressure differential is instantly established across the air–liquid interface. From basic thermodynamics, we know that the vapor pressure of a liquid increases when the external pressure is increased. The Kelvin equation basically states that the vapor pressure at saturation is greater over a curved surface than it is over an infinite flat surface.

This kind of Gibbs surface is valid at the nanoscale if the effective surface free energy (surface tension) is a function of the radius of the nanoparticle [14]. In other words, in this case, classical thermodynamics can be extrapolated to the nanoscale.

6.1.5 Chemical Potential

The chemical potential of physical and chemical processes that involve small particles is dependent on the radius of curvature of the surface. We resort to another form of the Young–Laplace equation:

$$\mu - \mu_o = \Delta\mu = \gamma V_a\left(\frac{1}{r_1} + \frac{1}{r_2}\right) = \frac{2\gamma V_a}{r} \tag{6.20}$$

where V_a is the atomic volume and μ and μ_o are chemical potentials of the curved surface and bulk surface, respectively. Convex surfaces (like the surface of colloids) have a positive chemical potential. To form a convex surface from a flat bulk surface, an input of energy is required [19]. Concave surfaces have negative curvature. Correspondingly, concave surfaces represent low-energy surfaces. The surface of red blood cells is concave. Why do you think that is the case?

6.2 Liquid State

A liquid (from the Latin *liquidus,* "flowing, moist") is a state of matter that has no definite shape, is difficult to compress, and is able to flow. Liquids assume the shape of their container. These definitions can certainly be debated because at the nanoscale drops do assume a definite spherical shape. Liquids lack static long-range order, unlike solids. The atoms and molecules in a liquid are not bound tightly and have fewer neighbors with which to bind. This results in mobility of atoms and molecules that allows restoring forces to react readily in the event of any perturbation of the surface. As a result, liquids are able to respond quickly to any deformation and reassemble in a low-energy, minimum surface area state.

The surface layer is considered to be a few atoms or molecules thick in the absence of ions [9]. The surface of a liquid resembles an elastic skin that is always under tension. In the volume of liquids, balanced cohesive forces tug on atoms or molecules from all directions. In contrast, unbalanced cohesive forces at or near the surface tug on surface molecules and atoms laterally and inwardly (**Fig. 6.4**). In essence, the liquid contracts and compresses until it attains the lowest possible surface area per volume, the sphere. This configuration allows for elastic deformation of the surface. Surface atoms and molecules, therefore, are inherently less stable than their bulk counterparts. This instability is responsible for the enhanced vapor pressure of water droplets over their bulk counterpart.

Surface tension is best visualized as a stretched elastic membrane that tends toward contraction [11]. Drops, bubbles, cavities, and capillarity are all consequences of surface tension. A bubble is a region where gas is trapped by a condensed solid, usually a liquid. A bubble has two surfaces and therefore a factor

Fig. 6.4 *Water molecules near the surface experience different forces. Notice in the figure that surface molecules have fewer nearest neighbors. Surface tension is the direct consequence of surface molecules trying to achieve the lowest energy state. In the bulk, each water molecule experiences the same cohesion forces that result in a balance (net force is equal to zero, $\Sigma F = 0$). Surface molecules are pulled inward by molecular attraction (there is no outward force). Resistance to compression by the liquid balances the inward force. Water is then squeezed until it achieves the lowest surface area possible; for a droplet, this is a sphere. The layer behaves as an elastic sheet that resists any change to its curvature (e.g., the action of a water strider).*

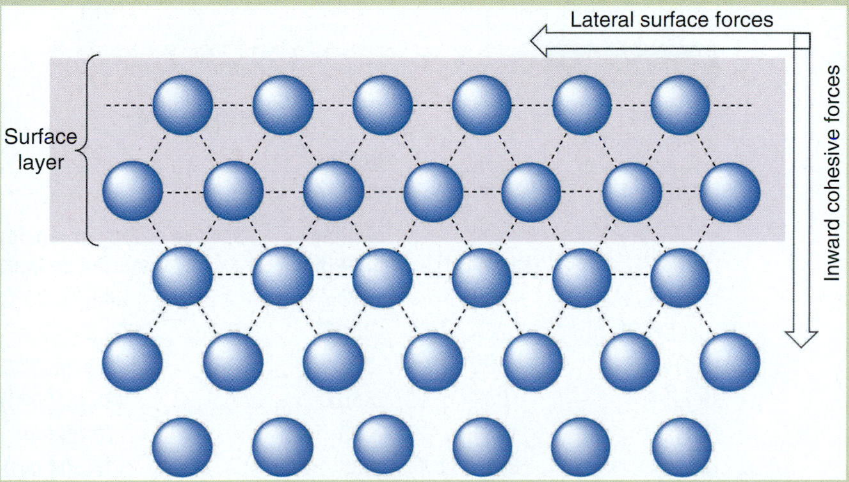

Example 6.4 *Vapor Pressure of Nanosized Droplets*

Calculate the vapor pressure of r = 90 nm water droplets at 20°C. Assume the density of water at 20°C is 0.99823 $g \cdot cm^{-3}$; $\bar{V}_m = 18.047\ cm^3 \cdot mol^{-1}$; $T = 293$ K; $R = 8.314\ J \cdot K^{-1}$; $M_{H2O} = 18.015\ g \cdot mol^{-1}$.

Solution:

Apply the Kelvin equation

$V_m = M/\rho = 18.047\ cm^3 \cdot mol^{-1}$

$2\gamma V_m/rRT = [2(72.25\ dyn \cdot cm^{-1})\ (18.047\ cm^3 \cdot mol^{-1})]/[(90 \times 10^{-6}\ cm)\ (8.314 \times 10^7\ ergs \cdot mol^{-1} \cdot K^{-1})(293\ K)] = 0.01068$; then $e^{0.01068} = 1.0107$

$p = (17.535\ torr) \cdot exp[2\gamma V_m/rRT] = (17.535\ torr) \times 1.0107 = 17.723\ torr.$

of 2 has to be applied in surface energy calculations. A cavity, on the other hand, has just one surface. A drop or droplet is a liquid surrounded by a gas—only one surface. Liquids adopt shapes that tend to minimize surface area. This maximizes the number of atoms or molecules that exist within the volume of the drop.

The *energy of cohesion* is the energy that holds the liquid (or solid) together. It is also the energy that is released in the form of heat when two drops of the same liquid are joined. The energy of cohesion is equal to twice the surface tension energy because, in order to create a surface, the number of nearest-neighbor binding interactions must be halved (example 6.1):

$$E_{co} = 2\gamma \tag{6.21}$$

The cohesive energy density is defined as the ratio of the molar heat of evaporation (or molar heat of sublimation for solids) at standard pressure to the molar volume of the liquid: $\Delta H°_{vap}/V_m$. The molar heat of vaporization is the energy required to convert 1 mol of liquid into gas at its boiling point and is indicative of the forces that hold liquids together. **Table 6.3** lists cohesive energies of several common materials along with their respective surface tensions.

Table 6.3 *Cohesive Energies of Common Materials*

Material	Cohesive energy		
Liquid	~$\Delta H°_{vap}$ ($kJ \cdot mol^{-1}$)	Boiling point (K)	
Solid	~$\Delta H°_{sub}$ ($kJ \cdot mol^{-1}$)	Melting point (K)	Surface energy or tension ($mJ \cdot cm^{-2}$) of materials in liquid form
NH_3 (l)	23.3	239.7	34.5 (at 20°C)
CH_3CH_2OH (l)	26.7	351.5	22.8 (at 20°C)
H_2O (l)	40.7	373.2	72.8 (at 20°C)
Hg (l)	60.3	356.6	485 (at 20°C)
Zn (s)	115.6	693	770 (at mp)
Au (s)	368	1338	1130 (at mp)

TABLE 6.4 *Surface Tension and Interfacial Tension of Common Materials*

Bulk liquid	Surface tension (γ_1)	Interfacial energy (γ_{12}) γ_2, water = 72.9 dyn · cm^{-1}
Benzene	29	35
Chloroform	27	28
Carbon tetrachloride	27	45
Cyclohexanol	32	4
Diethyl ether	17	10.7
n-Octane	22	50.8
n-Octyl alcohol	27.5	8.5
Mercury	475	375

The *energy of adhesion*, on the other hand, is the surface tension of the interface between two nonsimilar liquids and this relationship has the following form:

$$\gamma_{12}\,\mathcal{A} = (1/2)W_{11} + (1/2)W_{22} - W_{12} = \gamma_1\,\mathcal{A} + \gamma_2\,\mathcal{A} - W_{12} \qquad (6.22)$$

where W_{11} and W_{22} are the cohesive energies, proportional to γ_1 and γ_2, respectively, of the pure components and W_{12} is the work of adhesion required at the interface. Just as before, the energy required to create an interfacial surface between two immiscible liquids is called the interfacial energy or interfacial tension. The shape of a material, nanomaterials in particular, depends heavily on the balance between cohesive and adhesive forces. Surface tension and interfacial tension of common materials are listed in **Table 6.4.** The term *interfacial tension* is also used if a surface separates two nongaseous phases.

6.2.1 Classical Surface Tension

Just like with water, the inner structural energy that holds solid materials together is called the *cohesion energy*, E_c. The behavior of bulk materials, considered to be relatively independent of the environment, is attributable to the cohesion energy [12]. We now come to understand that as surface-to-volume ratio is increased, surface energy plays an increasingly larger role in determining the behavior of the material. For the most part, we assume that the intrinsic cohesion energy (bond strength) remains at the bulk value even as size is diminished. Another assumption is that cohesion energy is the opposite of the sublimation energy (although this is true only at 0 K). The size dependence of both the inner cohesion energy and the surface energy is a subject that has received much attention recently [12].

Surface tension depends on long-range forces governed by dispersion, short-range forces such as hydrogen bonding in water, and entropic effects (e.g., the surface is more disordered than the bulk of the liquid) [8]. Surface tension, is a measure of the force required to break a liquid surface of 1 cm length. The surface energy of a solid is estimated by measuring a parameter called the *contact angle*, cos θ. The contact angle is the angle made by a liquid

Table 6.5 *Contact Angles Formed by Selected Common Materials*

Surface	Contact angle
Au	0°
Pyrex glass	ca. 0°
Pt	40°
Polymethyl methacrylate (PMMA)	60°
Teflon	110°
Lotus plant leaf[a]	~180°

[a] Water droplets on the surface of the lotus plant leaf are perfectly spherical.

drop at the three-phase boundary where the liquid, flat solid substrate, and gas phases meet (**Fig. 6.2**). This is considered to be a boundary condition. Surface tension relationships are described by Young's equation (often called the Young–Dupre equation):

$$\cos\theta = \frac{\gamma_{SV} - \gamma_{SL}}{\gamma_{LV}} \qquad (6.23)$$

where γ_{SV}, γ_{SL}, and γ_{LV} are the surface energies at the solid–vapor, solid–liquid, and liquid–vapor phases, respectively. The contact angle is composed of the advancing contact angle θ_a, the receding contact angle θ_r, and the equilibrium contact angle θ_e. Angle hysteresis is said to occur when $\theta_a \neq \theta_r$. We will concern ourselves only with the equilibrium contact angle. The contact angle formed by a drop of water on some common substrates is shown in **Table 6.5.**

The contact angle is an important measure that correlates the surface energy of a solid to the surface tension of a liquid. The contact angle is indicative of a phenomenon known as surface wetting or *wettability*. Surface wetting is the ability of a solvent to spread on a solid substrate, a characteristic that is important, for example, in soldering. Water is a common liquid used in the determination of surface wettability. The magnitude of the spread of a droplet is an indicator of surface energy. The surface energy of a solid can be determined by contact angle measurements against a series of well-characterized wetting agents. If a drop spreads outward after application (e.g., on an unwaxed metal surface of an automobile), the interfacial area between the substrate and the drop is increased. In other words, the substrate surface has enough energy to overcome the surface tension of water and increase the interfacial area of the droplet.

From another perspective, the adhesive forces between the liquid and the solid are comparable in strength to the cohesive forces of the water. If the water forms up in a bead, it is obvious that the cohesive forces—the forces between like molecules—are stronger than the adhesive forces between unlike molecules. For example, if the drop remains spherical or does not spread to any great degree, there is not enough energy emanating from the weak attractive interactions that hold hydrocarbons together, as on a waxed surface, to overcome the surface energy of the water. Good wetting is said to happen if the contact angle lies between $0° < \cos\theta < 90°$. *Dewetting* is the opposite of wetting and is the case when a sheet of liquid is disrupted to form droplets (e.g., on a waxed metal surface of an automobile.

The critical temperature T_c is the temperature at which there is no more distinction between a liquid and its vapor. At the critical temperature, the value of γ becomes zero. The temperature dependence of surface tension of common liquids obeys the following empirical relationship of Guggenheim and Katayama (in E. A. Guggenheim, *Thermodynamics*, 5th ed., North-Holland, 1967):

$$\gamma = \gamma^{o}\left(1 - \frac{T}{T_c}\right)^{n} \tag{6.24}$$

where T_c is the critical temperature at which the liquid phase becomes indistinguishable from the vapor phase, n is equal to ca. 1.22 ≈ 1, and $\gamma°$ is the surface tension of the liquid under standard conditions. Obviously, γ approaches zero as T approaches T_c. Surfactants are also able to lower the surface tension of water. For many compounds, the dependence of surface tension on temperature can be approximated by

$$\gamma = a - bT \tag{6.25}$$

where a and b are constants associated with a particular liquid. For example, for water, a = 24.05 (the actual value at 0°C) and b = 0.0832.

Nanoscale Condensation. The equilibrium vapor pressure of a liquid and the curvature of the liquid–vapor interface are related by the Kelvin equation. The Kelvin equation predicts that unsaturated vapors will condense in nanochannels even into nanochannels. Meniscus radii below 4 nm have been validated for cyclohexane but not verified for water until 1981 when Fisher et al. demonstrated meniscus radii for water as low as 9 nm [15]. Fisher et al. stated in 1979 that "the application of the laws of thermodynamics, and the concept of a bulk surface tension, is valid in principle for such highly curved interfaces."

According to recent findings, the Kelvin effect and capillary rise are different aspects of the same phenomenon [16]. This is a profound demonstration of the validity of classical thermodynamics to nanoscale materials. In chapter 8, we find that classical thermodynamic theory is not enough to explain nanoscale phenomena.

6.2.2 Capillarity

Surface tension is responsible for capillary action. If a liquid has a tendency to adhere to the walls of the tube and spread out to cover more area (e.g., good wetting), energy is minimized. The height h that a liquid is drawn into a capillary tube is described in terms of surface tension:

$$h = \frac{2\gamma_{LV}\cos\theta}{\rho g r} \tag{6.26}$$

where

h is the height of the liquid

γ_{LV} is the liquid–air surface tension

r is the radius of the capillary
θ is the angle of contact
ρ is the density of the liquid
g is the gravitational acceleration constant

On glass, the material of most capillary tubes, water is able to wet the surface completely. In this case the contact angle θ is equal to zero and consequently $\cos\theta$ equals one and the expression falls out of the equation. This relationship is obviously most accurate when complete wetting of the tube material occurs.

If the cohesive forces that hold water together are weaker than the adhesive forces between the water and the glass wall of a graduated cylinder (e.g., there is wetting of the surface), the meniscus of the water forms a concave shape. This occurs when the meniscus contact angle is less than 90° (**Fig. 6.5**). On the other hand, if the liquid is not able to wet the glass surface, as is the case of elemental mercury (e.g., a liquid with extremely high surface tension equal to 485 $\text{dyn}\cdot\text{cm}^{-1}$), there is limited wetting and a convex meniscus is formed. This occurs when the contact angle is greater than 90° (**Fig. 6.5**).

Capillary tube wetting can be described by the following adaptation of the Laplace form:

$$\Delta p = 2\gamma\cos\theta\left(\frac{1}{r_1} - \frac{1}{r_2}\right) \tag{6.27}$$

Capillary condensation is an important factor in determining the mechanical properties of micro- to nanoscopically sized matter, particularly of powders. Capillary condensation depends on the liquid surface tension γ_{LV}, the contact angle θ, and the radius of curvature of the solid surface R. The capillary force is then

$$F_{\text{cap,max}} = 4\pi R\gamma_{LV}\cos\theta \tag{6.28}$$

Capillary forces play an important role with microcontact AFM or STM tips. The principal mechanism forms the basis for several types of dip-pen nanolithography. Effectively, the probe tips act as the nucleation center for condensation. If the radius of curvature of the microcontact probe lies below the critical radius (~ the Kelvin radius), a meniscus is formed. The Kelvin radius is given by

$$r_K = \frac{2\gamma V_m}{RT}\ln\left(\frac{p_o}{p}\right) \tag{6.29}$$

where p_o is the saturation vapor pressure as before. The BJH method to determine porosity and pore diameter is based on the Kelvin equation.

6.2.3 Surface Tension Measurements

Methods to measure surface tension and energy are numerous. There are two principal methods to consider. One, based on a mechanical principle, is called tensiometry (from the Latin *tensio*, "stretching"). Tensiometry is a technique that is able to measure the interaction of a calibrated probe (usually made of a

Fig. 6.5 *The meniscus of mercury (left) is compared to the meniscus of water (right). Mercury has one of the highest surface tensions of any liquid: 485 dyn·cm^{-1}. This is not surprising because metals have high cohesion energy. There is not enough surface energy available on the glass surface to overcome the high surface tension of mercury. The contact angle therefore is very large and the meniscus protrudes upward. For water, the silica glass is able to neutralize the whole of the surface energy of water. The contact angle is nearly 0° and the meniscus therefore is concave. The water is actually dragged upward. If the curvature is small enough, water will rise in the tube (e.g., by capillary action).*

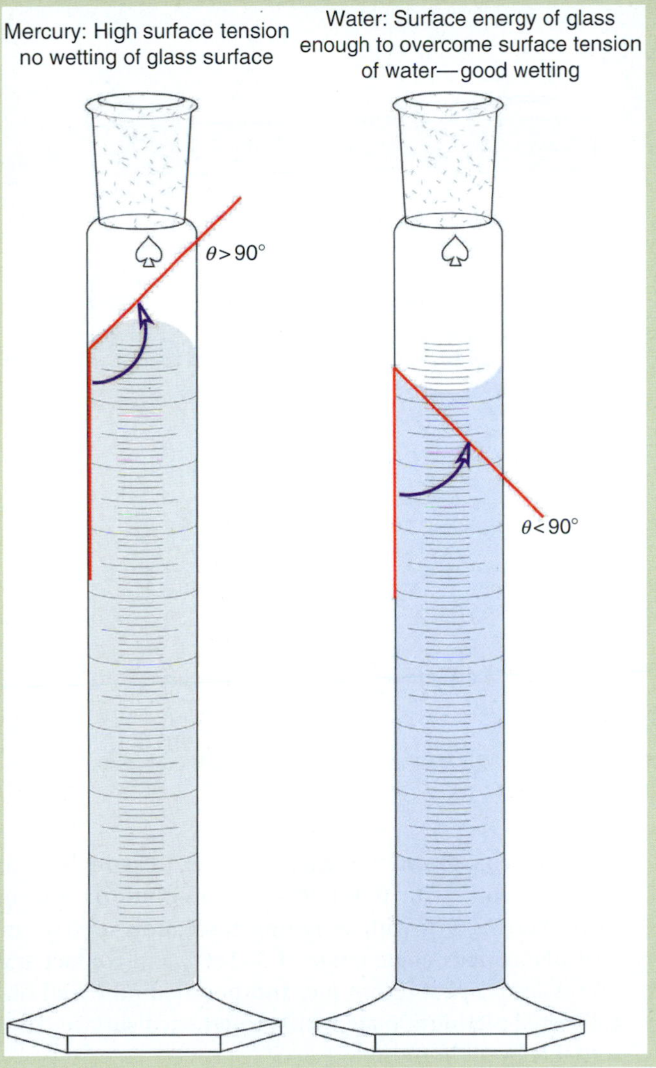

platinum–iridium alloy) of high surface energy with a liquid surface layer. The *Du Noüy ring* method uses a platinum ring that is submerged below the surface of a test liquid (**Fig. 6.6**). The probe is then lifted incrementally until all contact with the liquid is lost. The maximum force during this process is measured and is converted into the dimensions of surface tension. A KSV Sigma-700

FIG. 6.6 *The action of a Du Noüy ring is shown. The ring is made of a wettable metal (as most are)—in this case, platinum, a metal with a very high surface energy. The net force is measured as a function of time along the track of the ring. The graph shows the profile of the measurement step by step: (1)–(4): ring is pushed below the surface of the liquid and the system is allowed to equilibrate; (5)–(7): the ring is pulled upward; the force increases as the surface tension of the liquid exerts its influence. Moving the ring upward results in the creation of a new surface (requiring energy). At (8) the interface breaks as enough force has been exerted to overcome the surface tension.*

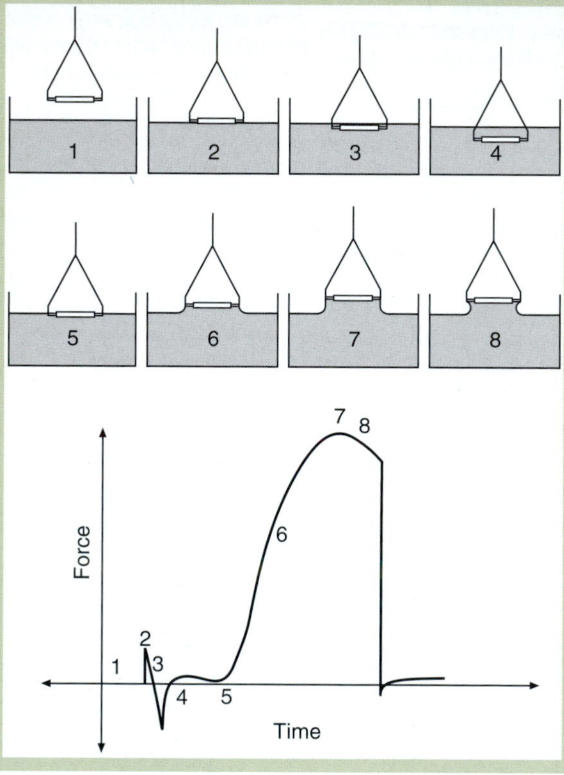

Source: Courtesy of KSV Instruments, Ltd.

tensiometer is shown in **Figure 6.7.** The 701 series is capable of measuring 0.01–1000 $mN \cdot m^{-1}$ with 0.001 $mN \cdot m^{-1}$ resolution. The maximum load for this instrument is 5 g with weighing resolution of 0.01 mg, force resolution of 0.01 μN, contact angle range of 0–180°, and contact angle resolution of 0.01°. The *Wilhelmy plate* technique, another method based on tensiometry, utilizes a calibrated platinum plate with a roughened surface. The surface of the probe is wettable by all liquids.

Another method is based on optical determination of the contact angle and is commonly referred to as a *contact angle goniometry* (based on the Greek *gonia,* indicating angle, + *metron,* "measure"), a process invented by W. Zisman of the U.S. Naval Research Laboratory in the 1950s. Goniometry is a complementary technique to tensiometry and is often used to independently verify results obtained by the latter.

Measurement of the surface energy of solids is not a simple task. The surface tensions against air of water, methanol, benzene, and mercury in air at 20°C are

FIG. 6.7

A KSV Sigma-700 tensiometer is shown in the figure. The 701 series is capable of measuring 0.01–1000 mN·m^{-1} with 0.001 mN·m^{-1} resolution. The maximum load for this instrument is 5 g with weighing resolution of 0.01 mg, force resolution of 0.01 μN, contact angle range of 0–180°, and contact angle resolution of 0.01°. Determination of the surface energy of liquids is quite straightforward.

Source: Courtesy of KSV Instruments, Ltd.

72.9, 22.5, and 485 dyn·cm^{-1} (or mN·m^{-1}), respectively. In methanol, some hydrogen bonding is present and in mercury no hydrogen bonding is present but very strong metallic interactions predominate. The contact angles of a series of liquids with different surface tension values are placed on the solid. Contact angles are recorded and the surface energy of the solid is then calculated indirectly by this means.

6.3 Surface Energy (and Stress) of Solids

Use of the term "surface energy" rather than "surface tension" is appropriate for solids. We stick to homogeneous simple crystal forms in order to illustrate aspects of surface energy. The cubic *fcc* structure always serves as a good base. We also stay with (111), (100), and (110) low-index crystal planes in this section. Topics concerning *surface stress,* another common phenomenon associated with solid surfaces, will not be presented in this section. Liquids do not exhibit surface stress. It is a phenomenon associated strictly with solids only. Solids possess elastic stress up until the melting point [17]. To be complete, both surface energy and surface stress characterization are required to describe the state of the solid surface [17].

Nearest-Neighbor Models. Nearest-neighbor atomistic models are 10–20% accurate but offer a simple physical interpretation of the equilibrium behavior of solids [17]. The N–N model does not consider next-nearest neighbors and higher order interactions. We shall also apply the atomistic perspective in describing the surface energy of solids. We begin our discourse by describing macroscopic condensed systems of bound atoms with pair interaction and their surface energy and then move on to nanoscale materials, *fcc* clusters in particular.

Descartes proposed that solid matter is held together by "hooks and other appendages." His mechanistic approach suggested that external pressure was responsible for solid cohesion. Newton speculated correctly that atoms interacted by attractive forces at long range and repulsive forces at short range. Based on his research of capillarity, Laplace speculated that atomic forces are short range in nature because the rise of a liquid in a capillary tube is independent of the thickness of the capillary wall.

Surface atoms and molecules of solids share similar properties with liquids except atoms and molecules remain relatively stationary. Solids cannot, for example, react spontaneously and change shape to configure a minimum surface. In its most simplistic form, the surface energy E_s of a crystalline solid is the number of surface atoms times the one-half cohesion energy per atom pair:

$$E_s = \tfrac{1}{2}(N_s \varepsilon_b) \tag{6.30}$$

where ε_b is the cohesion (binding) energy of atom pairs in the solid and N_s is the number of atoms that exist in the surface layer. The factor 1/2 is included because making a new surface breaks bonds in two (in a perfect crystal). The assumption that surface atoms have the same binding energy as those within the solid is an approximation at best. The energy of a solid is determined by the strength of the chemical bonds within a crystal and its bonding configuration. The number of nearest-neighbors (and second-nearest, etc.) implicitly contributes to the bonding configuration and stability of the solid.

Phase changes at constant temperature and pressure generally involve a change in enthalpy. The enthalpy of sublimation at 0 K, ΔH_{sub} (or ε_{sub}) is the energy required to convert a solid directly into its vapor form without going through a liquid stage. The enthalpy of sublimation is a convenient approximation of the cohesive energy, or binding energy ε_b. The energy required to melt a solid is represented by the molar heat of fusion, ΔH_{fus}. Melting is an endothermic phase transition process in which energy is absorbed. The processes of heating and subsequent melting reduce the cohesive energy of a solid and, because all bonds are weakened, as a result, the surface energy is also reduced. With melting, the overall coordination of each atom in the bulk is also reduced. The magnitude of the intermolecular interactions in liquids is measured by the latent heat of vaporization, ΔH_{vap}—the energy at constant temperature required to convert liquid into vapor. Recall that at the boiling point of a liquid surface tension is nonexistent.

6.3.1 Interaction Pair Potentials

As any text in physical chemistry will show, there are basically two types of interactions between atoms: short-range repulsive interactions (nearest neighbors only) and long-range attraction interactions [18]. There are many models that illustrate the interatomic potential energy of solids. Hard-sphere potential models treat the atom as hard spheres with the same radius and are appropriate in modeling repulsive potentials. The potential energy between two hard spheres jumps to infinity as soon as the *collision diameter threshold* σ is attained [8]. The *Lennard–Jones (6–12) potential* is one such model that is useful in describing atomic behavior:

$$U(r) = \varepsilon_b\left[\left(\frac{\sigma}{r}\right)^{12} - 2\left(\frac{\sigma}{r}\right)^{6}\right] = \varepsilon_b\left[\left(\frac{r_e}{r}\right)^{12} - 2\left(\frac{r_e}{r}\right)^{6}\right] \tag{6.31}$$

where ε_b is the binding energy between two atoms and σ represents the minimum separation. At equilibrium, the equilibrium radius $r_e = \sigma$ (approximately the sum of the average radii of the two atoms analogous to the lattice parameter a). The restoring force, $F = -dU/dr$, is equal to zero at equilibrium and $U(r_e) = -\varepsilon_b$, the cohesive (binding) energy for the pair interaction. The equilibrium value of the binding energy is equal to the depth of the energy well. When $r \geq 2\sigma$, the potential decreases by a factor of 32 or more and nearest-neighbor interactions can largely be ignored [17]. When $r < \sigma$, the r^{-12} factor dominates the r^{-6} term. When $r > \sigma$, the converse is true. Both attractive and repulsive forces are capable of reducing the potential energy of the system.

6.3.2 Surface Energy of Low-Index Crystals

The energetic state of surface atoms or molecules is one of incompleteness. Atoms with dangling bonds possess extra energy called excess surface energy—useful energy available to accomplish work. Resorting once again to a hard-sphere model, the surface energy of solids cleaved along low-index crystal planes can be approximated by applying the *nearest-neighbor–broken-bond* model introduced earlier [17]. The assumptions in our calculation are as before: that we use a hard-sphere model, consider only nearest-neighbor interactions, and the surface energy be equal to the average of the binding energy of the solid. Other assumptions include a rigid structure with no relaxation, surface strain, or any entropic or pressure–volume contributions [19]. Additional considerations include ignoring any potential surface strain or rearrangement associated with the new surface. Any such restructuring would, of course, contribute to lower surface energy [19]. We will apply the following formulae only to solids and also assume that the resulting surface energy of the newly formed surface is defined according to bulk parameters (e.g., the excess energy the surface atoms have over that of the interior atoms). The *fcc* metal will serve again as our model.

The total energy of a solid is estimated to be equal to the heat of sublimation at 0 K, ΔH_{sub}. The enthalpy and internal energy are considered to be equal. Enthalpy of vaporization (at 0 K), energy of formation, cohesive energy, dissociation energy, and enthalpy of fusion all are important factors with regard to the total binding of a solid. Cleavage of a crystalline solid is somewhat akin to radical sublimation—the abrupt removal of a plane of atoms from its neighbors in the adjacent plane. Calculation of the surface energy of a newly formed surface involves reference to ΔH_{sub} of the bulk material.

The *nearest-neighbor* model [17] estimates that the extensive overall binding (cohesion) energy of 1 mol of atoms E_b is

$$E_b = \tfrac{1}{2} Z\mathcal{N}_A \varepsilon_b \tag{6.32a}$$

where Z is the coordination number as before, $\mathcal{N}_A$ is Avogadro's number, and ε_b is the binding energy between two atoms in a solid. Rearranging and substituting ΔH_{sub} for E_b,

$$\varepsilon_b = \frac{2\Delta H_{\text{sub}}}{Z\mathcal{N}_A} \tag{6.32b}$$

For an *fcc* crystal with $Z = 12$, the cohesion energy of the crystal reduces to

$$\varepsilon_{b,fcc} = \frac{\Delta H_{\text{sub}}}{6\mathcal{N}_A} \tag{6.33}$$

Crystals with faces that have the closest packing of atoms have the lowest surface energy [17].

The energy required to cleave an *fcc* crystal along the (111) crystal plane is calculated as follows [17]. One first counts the number of nearest neighbors in the (111) plane. Three atoms reside above the plane of cleavage and three atoms reside below the plane of cleavage. The result of the cleavage is two additional (111) planes. This means that three bonds are broken for every cleaved atom. The energy of every surface atom is then

$$\varepsilon_s = \frac{3}{2}\varepsilon_b \tag{6.34}$$

The energy of the (111) surface is

$$\varepsilon_s = \frac{3}{2}\varepsilon_b = \frac{3\Delta H_{\text{sub}}}{12\mathcal{N}_A} = \frac{\Delta H_{\text{sub}}}{4\mathcal{N}_A} \tag{6.35}$$

Notice that this value is greater than the cohesion energy of the solid crystal by a factor of 3/2.

Another convenient way to calculate surface energy is to use

$$\gamma = \tfrac{1}{2} N_b \varepsilon_b \rho_a \tag{6.36}$$

where N_b is the number of atoms per unit area and ρ_a is the surface atom density [19]. This form of the calculation relates energy to the lattice constant a of the crystal [19]:

$$\gamma_{(100)} = \frac{1}{2}(4)(\varepsilon_b)\left(\frac{2}{a^2}\right) = \frac{4\varepsilon_b}{a^2} \tag{6.37}$$

The rest of the low-index plane surface energy values are listed in **Table 6.6**.

Table 6.6 *Relative Surface Energy of Low-Index fcc Planes*

Cubic plane	Planar atomic density, σ_{hkl}	N–N broken bonds	Excess energy per atom	Surface energy[a†] γ_{hkl} J·atom^{-1}·a^{-2}	Planar packing density	Surface energy[b] $\gamma_{hkl} \rightarrow \Delta H_{sub}$ J·atom^{-1}·a^{-2}
(100)	$2/a^2$	4	$4\ \varepsilon_b/2$	$4\ (\varepsilon_b/a^2)^{\dagger}$	0.785	$\sigma_{100}\ \Delta H_{sub}/3\mathcal{N}_A$
(110)	$\sqrt{2}/a^2$	5 (6[c])	$6\ \varepsilon_b/2$	$3\sqrt{2}\ (\varepsilon_b/a^2)$	0.555	$\sigma_{110}\ \Delta H_{sub}/2\mathcal{N}_A$
(111)	$4\sqrt{3}/a^2$	3	$3\ \varepsilon_b/2$	$2\sqrt{3}\ (\varepsilon_b/a^2)^{\dagger}$	0.901	$\sigma_{111}\ \Delta H_{sub}/4\mathcal{N}_A$

[a] G. Cao, *Nanostructures & Nanomaterials: Synthesis, Properties and Applications*, Imperial College Press, London (2004).
[b] J. M. Howe, *Interfaces in Materials: Atomic Structure, Thermodynamics and Kinetics of Solid–Vapor, Solid–Liquid and Solid–Solid Interfaces*, John Wiley & Sons, New York (1997).
[c] The sixth broken bond is difficult to find [17].

One would expect the (110) planes to have the highest surface energy because it has the lowest packing fraction of the three low-index crystal planes.

The value of the electronic work function depends on the plane of origin of the electron. **Table 6.7** lists the dependency of electronic work function of the crystal face [1]. Notice that the electron work function is highest for the electron that emanates from the (111) plane. In all three cases, the order of stability is (111) > (100) > (110), similar to the order of surface energy in **Table 6.6.** The close-packed (111) plane contains atoms with the most stability (lowest surface energy) and concentration of surface positive ion cores [1].

The correlation between surface energy γ of the solid–vapor interface and ΔH_{sub} is reasonable. The heat of sublimation, however, is not the best way to calculate the cohesion energy of a metal. A simple empirical means of calculating cohesion energy is to divide the metal melting point temperature by the constant -3.5 ± 0.3 [20]:

$$E_b = T_m/(-3.5 \pm 0.3) \tag{6.38}$$

There is no way to obtain an absolute value of cohesion energy of metals, although better empirical relationships have been attained [20].

The equilibrium surface energy of a solid is the summation of the surface energies of individual facets. A cuboctahedron, for example, has 14 faces: 8 that

Table 6.7 *Electron Work Function Dependency on Surface Plane Face*

Element	Surface plane	Work function, eV
Ag	(100)	4.64
	(110)	4.52
	(111)	4.74
Cu	(100)	4.59
	(110)	4.48
	(111)	4.98
Ni	(100)	5.22
	(110)	5.04
	(111)	5.35

are triangular (derived from {111} planes) and 6 that are square in shape (derived from {100} planes). Starting with a cube of *fcc* close-packed structure, if one were to incrementally chip off the corners (triangular pyramids), the following shape transition progression would be seen (consult **Fig. 6.19** later on):

Cube → cuboctahedron → regular truncated octahedron → octahedron

Without engaging in the complex details, the regular truncated octahedron will have the lowest energy surface of the four major categories of shapes [18]. Why do you think this is the case? Which one resembles a sphere more so than the others? Unless very large clusters are formed, perfect spherical shapes are not allowed in close-packed structures for obvious reasons.

6.3.3 *Surface Energy of Nanoparticles*

Although quantum confinement has not been formally introduced at this stage, for now let us be content with the following definition. Restriction of electrons in one or more dimension translates into quantum confinement. In other words, especially in the case of metals, a continuum of electronic states no longer exists. Quantum confinement in two-dimensional films occurs in the z-direction only.

Thin Films. In bulk material surfaces, the relative surface energy is constant. In nanofilms, the surface energy was shown to fluctuate and that there was a preference of certain thicknesses over others; for example, layers were more stable if they possessed a specific thickness [21]. Layers of Pb were deposited on a Si(111) surface. The thickness of the Pb was varied from 6 to 18 monolayers [21]. Quantum confinement effects are expected to contribute to structural differences such as interlayer spacing. With the utilization of x-rays, information concerning the film thickness, atomic structure, and the interface with the Si surface were obtained. We have already discussed "magic numbers" as they apply to cluster stability. Are there any grounds for phenomena based on "magic thickness"? P. Czoschk, H. Hang et al. state:

> When confinement of the itinerant electrons in the metallic structure is taken into account, oscillations in the surface energy arise, leading to the preference of a certain thickness over others. It is seen that a smooth film represents a metastable state which, upon annealing, evolves through preferred or magic structures to a thermodynamically more stable, highly roughened state.

More Kelvin Effect. The size-dependent evaporation of free Ag nanoparticles showed that the Kelvin effect is verifiable at the nanoscale [22]. Researchers predicted a surface energy value for Ag nanoparticles to be 7.2 $J \cdot m^{-2}$. A linear relationship exists between the onset of evaporation temperature and reciprocal particle size.

Dip-Pen Lithography. There are many examples of surface energy phenomena involving nanoscale materials. Dip-pen nanolithography relies on capillary transport to deliver molecules from an AFM tip to a solid substrate. The capillary filling height and speed of silicon dioxide nanochannels are important in fabrication of wafers [23]. For example, capillary forces drive micropart self-assembly.

6.4 Surface Energy Minimization Mechanisms

Surface tension in a liquid is reduced if the liquid assumes the shape of a sphere, the surface of minimum energy. Liquid surface tension is also reduced by contact with a solid of high surface energy. For example, water wets a gold surface completely as the contact angle of the water is reduced to nearly nothing. With liquids, water in particular, the addition of surface-active agents (surfactants) lowers the surface tension.

The mechanisms to reduce surface energy in solids are abundant. Nanomaterials in particular offer numerous solutions to attain low-energy surface states. Because surface area, both of the collective and singular variety, is an important aspect of nanomaterials, these energy minimization processes are what enable researchers to exert some modicum of control over nanomaterial stability. Part of the emergence of the *Nano Age* is due to this ability.

6.4.1 *Surface Tension Reduction in Liquids*

We focus our attention in this section on the mechanisms of surfactants and how they are able to change the magnitude of the surface tension, specifically of water. Water is a key ingredient of living things as well as the primary medium within which numerous industrial processes depend. There is a constant known as the Hamaker constant, A_H, that describes the dispersion energy of small particles.

A surfactant is a surface-active agent. Many types of chemicals and solid materials can act as surfactants. The seventh century ecclesiastical scholar Bede is known for saying, "Remember to throw into the sea the oil which I gave you, when straightaway winds will abate, and a clam and smiling sea will accompany you throughout your voyage."

In 1757, Benjamin Franklin also noticed that the wakes behind ships that had just dumped their greasy water were smoothed. And when he was at London's Clapham Common pond, Franklin is known for the following experiment:

> The Oil, tho' not more than a Teaspoonful produced an instant calm over a Space of several yards square, which spread, amazingly, and extended itself gradually, until it reached the Lee Side, making all of that Quarter of the Pond, perhaps, half an Acre as Smooth as a Looking Glass.

Surfactants. Surfactants (surface-active agents) are molecules that contain both hydrophobic (nonpolar, water insoluble) and hydrophilic (polar, water soluble) constituents. Long chain hydrocarbons or aromatic moieties contribute to the hydrophobic portion of a surfactant. The hydrophilic moiety can consist of cationic (e.g., amine terminated), anionic (e.g., sulfates, sulfonates, phosphates, or carboxylate terminated groups found in soaps), or non-ionic groups like ethers. When placed in water, the hydrophilic ends of the surfactant interact with the water molecules. On the other hand, the hydrophobic ends seek stabilization at the liquid–vapor interface (e.g., steric acid) or form bilayers in the form of lipid bilayers that are fundamental to biological processes. Packing character of the layer (bilayers) depends on the length of the hydrocarbon chain and the extent of unsaturation.

6.4.2 DLVO Theory

DLVO theory was developed in the 1940s by Boris Derjaguin and Lev Landau and independently by Evert Verwey and Theo Overbeek (hence DLVO). This theory explains the interactions between charged species in liquids. It is especially useful in describing the agglomeration, stability, and phase behavior of colloidal solutions. The stability of particles in solution depends on a total potential energy function, U_T. DLVO theory applies to colloids because most colloids carry an electrostatic charge. DLVO theory is based on a combination of attractive van der Waals forces (U_{vdw}), electrostatic forces (U_{es}), the concept of the electrical double layer, and the potential energy due to the solvent, $U_{solvent}$. The contribution of the solvent is minimal and takes place only within a few nanometers of the particles. An electric double layer (**Fig. 6.8**), initially devised by Helmholtz in 1879, is like a capacitor with the potential at the surface decreasing exponentially as counter ions are adsorbed. The double-layer model also has several assumptions: (1) ions are point charges, (2) only significant reactions are Coulombic, (3) electrical permittivity is constant through the double layer, and (4), the solvent is uniform at the atomic level. DLVO theory describes the interactions between two overlapping double layers.

The curve of a DLVO plot looks very much like one for a Lennard–Jones interaction potential except that the middle curve is positively inclined (**Fig. 6.9**):

$$U_T = U = U_A + U_R + U_{Solvent} \quad (6.39)$$

DLVO theory is another theory applied to the phenomenon of agglomeration. The attractive potential, U_A, is represented by

$$U_A = -\frac{A_H}{12\pi d^2} \quad (6.40)$$

where A_H is the Hamaker constant and d is the distance between colloidal particles. The Hamaker constant determines the effective strength of cumulative van der Waals interactions between colloids. Notice that this potential (attractive potential) is long range ($U \propto d^{-2}$). The repulsive potential, U_R, is given by

$$U_R = 2\pi\varepsilon a\xi^2 e^{-\kappa d} \quad (6.41)$$

where

a is the colloid radius
π is the solvent permeability
ε is the dielectric constant of water
κ is a function of the ionic composition
ξ is the *zeta potential*

There are many assumptions in DLVO theory [19]: (1) flat surface, (2) uniform charge density, (3) constant surface electropotential, (4) static concentration

FIG. 6.8

In 1879, Helmholtz derived a mathematical treatment of a single layer of adsorbed ions on a charged surface. His treatment was based on a model of a capacitor. Gouy and Chapman in 1910–1913 derived a diffuse model in which they showed the potential decreases exponentially going away from the surface. The exponential decrease is due to adsorption of counter ions. Assumptions of this model include: that the ions are point charges, the significant reactions are all Coulombic, and the solvent is uniform with constant dielectric function. In most cases, the particle bears a negative charge. The figure shows a particle surrounded tightly by an inner layer (Stern layer) of cations in which the cations are bound strongly. A diffuse layer, made in this case of anions, is called the slipping plane, in which ions act in a uniform manner. The potential at this boundary is called the zeta potential ξ and decays exponentially away from the boundary. Electric double layers are also found on the surfaces of electrodes.

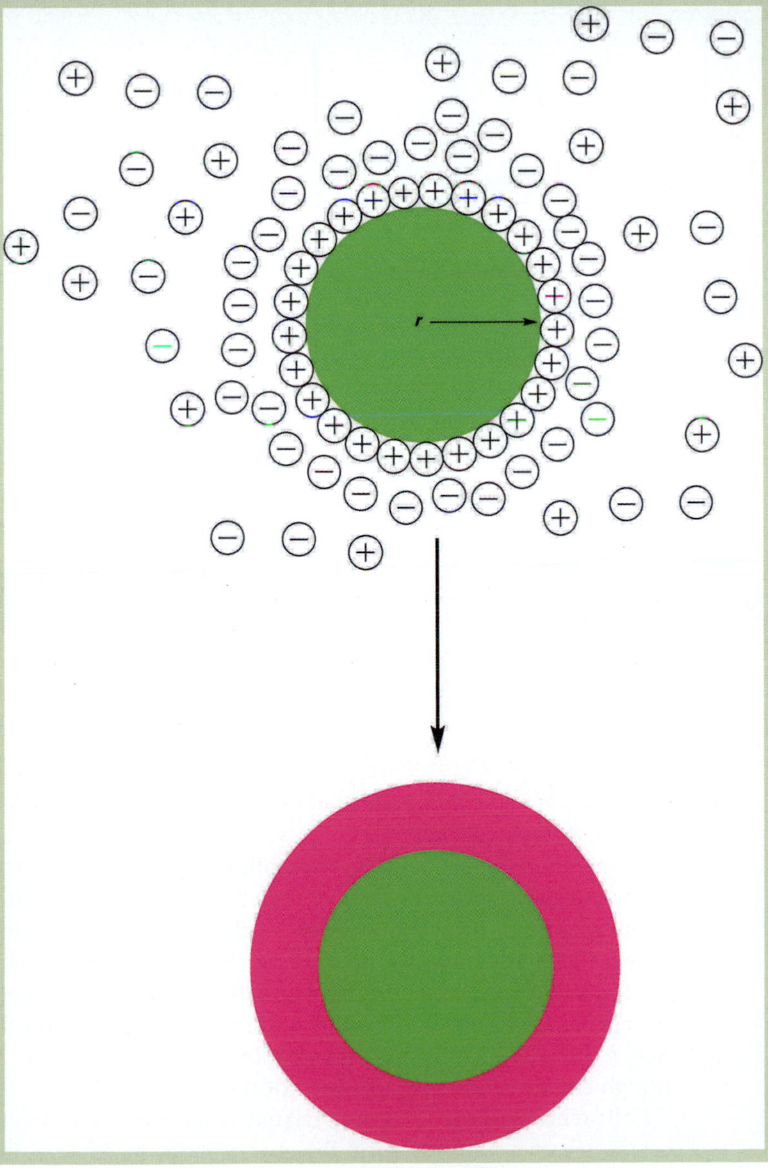

FIG. 6.9 *A generic DLVO potential plot is depicted. The red curve on the left represents the balance between attractive and repulsive forces. DLVO theory integrates the attractive action of van der Waals forces and the repulsive action of electrostatic forces. In this way, it serves as a unifying theory. DLVO theory is actually a good way to describe the interaction between two approaching colloid particles. The maximum depicted in the figure represents the repulsive barrier of the system. If the barrier is greater than $10k_BT$, collisions caused by Brownian motion are not energetic enough to overcome the barrier and, as a result, agglomeration does not happen [19]. The primary minimum occurs after the double layer is totally overcome and agglomeration is the result. U_{max} is a function of the thickness of the double layer; the thicker the double layer is, the higher is the value of U_{max}. The secondary minimum is important in reactions that involve flocculation. The graph on the right represents the approach of two colloids after an increase in the ionic strength of the solution. Apparently, agglomeration becomes enhanced in such an environment.*

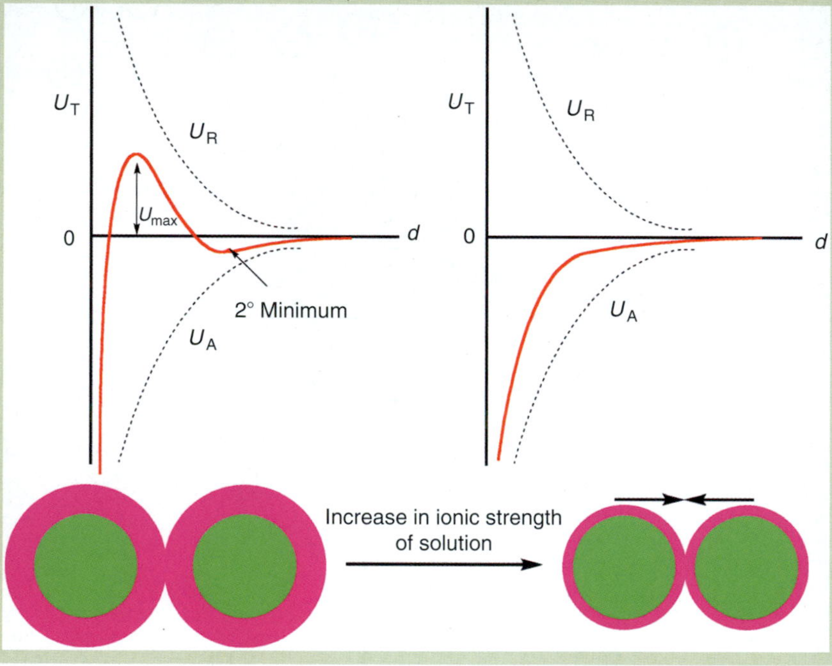

profiles, and (5) dielectric constant only solvent effect. Obviously, these assumptions are quite idealistic.

As the double layer between two colloidal particles starts to overlap, electrostatic repulsion exerts its effect. In such a solution, where colloidal approach is a consequence of Brownian motion, agglomeration occurs when colloidal particles collide with enough kinetic energy to overcome the barrier resulting from the repulsive forces. When this happens, the attractive forces predominate and the particles will agglomerate strongly and irreversibly. If the barrier is greater than $\sim 10k_BT$, then the collisions induced by Brownian motion will not provide enough energy to initiate agglomeration [19].

Cells tend to carry a negative charge. A layer of cations acts to stabilize the surface negative charge to form the electrical double layer. Adhesion of bacteria to cell surfaces can be explained by DLVO theory. Theoretically, bacteria should not be able to approach any closer than 10 nm to the surface of a cell depending on physiological conditions (e.g., approach is encouraged if the ionic strength of the medium is increased). Adhesion in cells is due to the combination of attractive and repulsive forces. At ca. 10 nm, there is some van der Waals attraction. At 2 nm, although the strength of electrostatic repulsion increases, van der Waals interactions start to dominate. At less than 1 nm, cell adhesion is nearly complete.

6.4.3 Polymeric (Steric) Stabilization

Steric repulsion keeps colloidal dispersions intact; that is, there is no agglomeration. The process serves to stabilize individual colloidal particles with additives that inhibit the coagulation of the suspended colloids. Materials used in this process include hydrophilic polymers and surfactants. It is simple to add polymer stabilizer to a solution. The polymer layer acts as a barrier to diffusion and therefore monodisperse solutions of nanoparticles can be synthesized. Polymers can bind tightly to the colloid by forming chemical bonds or form loosely adsorbed assemblies by weak physisorption mechanisms. Steric stabilization is compared to electrostatic stabilization in **Figure 6.10.**

6.4.4 Nucleation

The nucleation and precipitation process is an excellent, and classical, example of surface stabilization of nanoparticles. The process of nucleation is driven by thermodynamic parameters that are related to particle size. The free energy

Fig. 6.10 *Steric stabilization is also known as polymeric stabilization. The form of a steric stabilized particle that is independent of the electrical environment in a solution is compared to one surrounded by an electric double layer in the figure. There are scenarios in which there exist mixed electrostatic and steric stabilization. The importance of steric stabilization is that attached polymeric groups generate a diffusion barrier to further growth. As a consequence, nanoparticles with monodisperse size can be synthesized [19]. The dielectric constant of the solution is important. Obviously, a solvent in which the polymer is soluble is desired.*

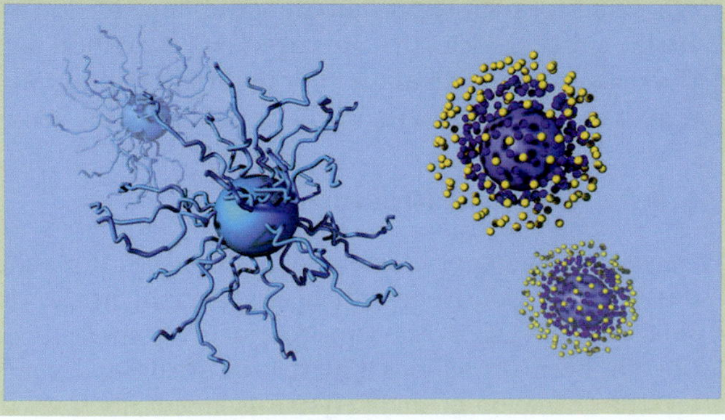

change for spherical particles (like a droplet of water in contact with its vapor) is given by [24]

$$\Delta G = \Delta\mu_V + \mu_S = \tfrac{4}{3}\pi r^3 \Delta G_V + 4\pi r^2 \gamma \tag{6.42}$$

where ΔG_V is the free energy per unit volume. The volume free energy $\Delta\mu_v$ (the volume chemical potential) is the first term on the RHS of equation (6.42) and $\Delta\mu_s$ is the chemical potential of the new surface. This is the classical equation for nucleation of a small particle in three dimensions. From this we can derive another form of the Gibbs–Thomson relation:

$$\Delta G_V = \frac{k_B T}{V_a} \ln\left(\frac{p_V}{p_S}\right) = \frac{k_B T}{V_a} \ln(1+S) \tag{6.43}$$

where S, the saturation, is defined as

$$S = \frac{p_V - p_S}{p_S} \tag{6.44}$$

If $p_V > p_S$, then the solution is supersaturated and there is nucleation and growth. If $p = p_S$, an equilibrium state is attained. If $p_V < p_S$, the solution is not saturated and growth is not likely.

The calculation of the critical nucleus size is dependent on the surface tension—a barrier to nucleation—and relies on the $4\pi r^2\gamma$ term in equation (6.42). The nucleation process is depicted graphically in **Figure 6.11**.

The critical nucleation size occurs when $\partial\Delta G/\partial r = 0$:

$$r_c = -\frac{2\gamma}{\Delta G_V} = \frac{2\gamma V_a}{k_B T \ln(1+S)} \tag{6.45}$$

where V_a is the atomic volume (volume per atom) and S is the saturation as before. The energy barrier to nucleation is [19]

$$\Delta G_c^\ddagger = \frac{16\pi\gamma}{3(\Delta G_V)^2} \tag{6.46}$$

Therefore, the critical nucleus size depends on the degree of supersaturation; r_c is small for high supersaturation and large for low supersaturation [24]. In simple terms, particles that are larger than the critical radius will grow and particles that are smaller than the critical nucleus will shrink. In either case, elimination of the metastable intermediate "small particle" occurs.

6.4.5 *Ostwald Ripening*

Ostwald ripening, or coarsening, is the physical manifestation of the process described previously by which larger nanoparticles grow at the expense of smaller ones (**Fig. 6.12**). Larger crystals are energetically favored over smaller ones for the many reasons that we have discussed during the course of this chapter. The formation of small crystals is kinetically favored because they nucleate

FIG. 6.11 *The free energy profiles of a homogeneous nucleation and growth process as a function of particle size are depicted. The process starts with a supersaturated solution. There exist two thermodynamic contributions to the process: a volume term and a surface term. From the graph it is clear that there is a critical particle size threshold at which stability is achieved. The creation of larger particles following this threshold is a downhill process. The counterbalance in this process is the surface energy. The creation of a new phase and molecules coming together to form a particle with a surface results in increased surface energy. Anything to the left of the threshold then is also a downhill process to revert to the supersaturated state.*

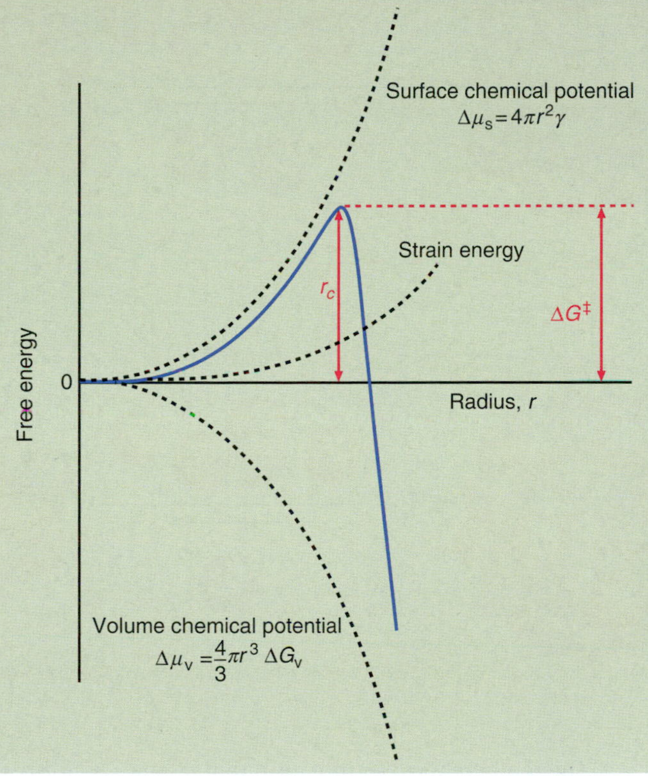

readily; nucleation has to start from the bottom up. Smaller crystals also have greater solubility and mobility in solution. Over a period of time, however, thermodynamic conditions favor the formation of larger crystals.

A dynamic equilibrium exists in solution between the rates of dissolution and precipitation of the dispersed phase to maintain the condition of saturation–solubility [25]. Smaller particles tend to dissolve and larger particles tend to grow. The process of Ostwald ripening is a function of particle curvature (as you might image). Equation (6.47) shows that the chemical potential of the process is dependent on the radius of the particles [19]:

$$\Delta\mu = 2\gamma V_a\left[\frac{1}{r_1} - \frac{1}{r_2}\right] \tag{6.47}$$

FIG. 6.12

Ostwald ripening is the process of forming larger particles at the expense of smaller ones—driven by the reduction of surface energy. Larger particles are inherently more stable than smaller ones. Matter is lost and gained for particles of all sizes. However, the rate at which larger particles exchange matter is much lower than the rate at which smaller particles exchange matter. The net result is the extinction of the smaller particles. Ostwald ripening occurs in the solid, liquid, and gaseous phases and depends on the local environment. Homogenous structures result from this process.

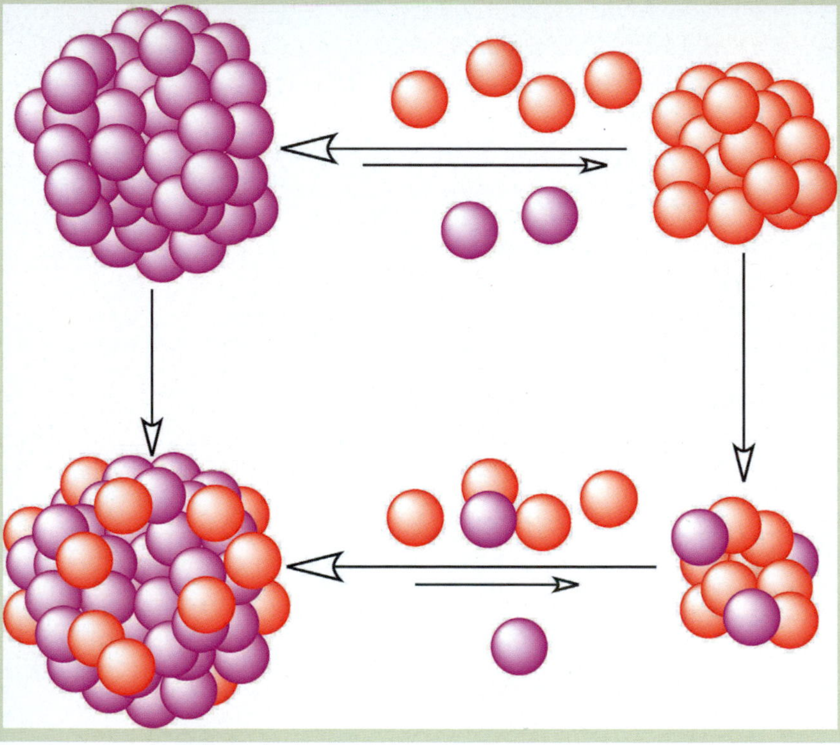

The mathematical formulation that relates solubility to surface curvature is

$$k_B T \ln\left(\frac{S}{S_o}\right) = \gamma V_a \left(\frac{1}{r_1} + \frac{1}{r_2}\right) \qquad (6.48)$$

where S represents the solubility of a particle with a curved surface and S_o is the solubility of the bulk form of the material [19]. Ostwald ripening is a form of sintering that occurs from the evaporation of atoms from one cluster and subsequent transfer to another. It so happens that the rate of "evaporation" is higher for smaller clusters than it is for larger ones—once again due to the numerous reasons we discussed in this chapter so far. The larger clusters then, over a period of time, get larger at the expense of smaller ones [26]. Coalescence implies the merging of similarly sized clusters without the dynamic interchange observed for Ostwald ripening (**Fig. 6.13**). Ostwald ripening occurs in the gas phase, in the liquid phase, and on a surface (surface-mediated Ostwald ripening sintering [SMORS]).

FIG. 6.13 *Coalescence implies the agglomeration of similarly sized particles. In this way, it is different from Ostwald ripening. The dynamic exchange between and among particles is not observed in coalescence and sintering.*

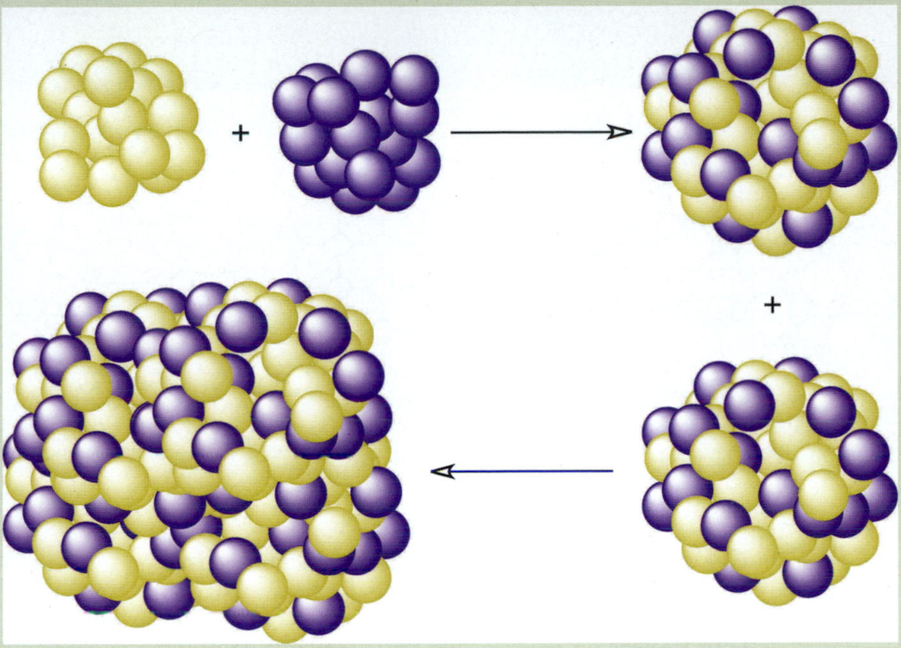

6.4.6 Sintering

The process of sintering is commonly associated with the solid phase, especially with regard to ceramic science. Sintering involves the agglomeration of similarly sized particles to form larger ones (**Fig. 6.13**), as opposed to Ostwald ripening (**Fig. 6.12**). Sintering involves fabrication of materials (usually ceramics, but also powder metallurgy) by heating powder precursors until agglomeration occurs. During sintering, there is a reduction of void volume as material flows into cavities (**Fig. 6.14**). This process results in a decrease in the overall volume of the material. In addition to reduction in porosity, sintering causes material transport from evaporation, condensation, and diffusion [27]. Sintering temperatures exceeding 1000°C are common.

Sintering is a complicated method of fabrication but one that is especially important at the nanoscale. In many cases, sintering is not desired. Sintering involves several stages during the process [19]: (1) diffusion (e.g., surface, volume, or grain boundary), (2) evaporation–condensation or dissolution–precipitation, (3) viscous flow, and (4) dislocation creep. Sintering is used to create dense materials. In sol–gel processes, drying is used to drive out all liquid phases, leaving a skeletal structure (e.g., xerogels). Application of sintering would produce a monolithic dense phase lacking a porous structure.

FIG. 6.14

A generic version of sintering is shown in the figure. Following a compaction process that imparts some low-level order into the structure, the application of temperature converts individual particles into larger sized grains that adhere to one another very strongly. Sintering involves complex diffusion (surface, volume, grain), dissolution–precipitation flows, and evaporation–condensation processes [19]. Sintering is not necessarily desirable in nanomaterial fabrication.

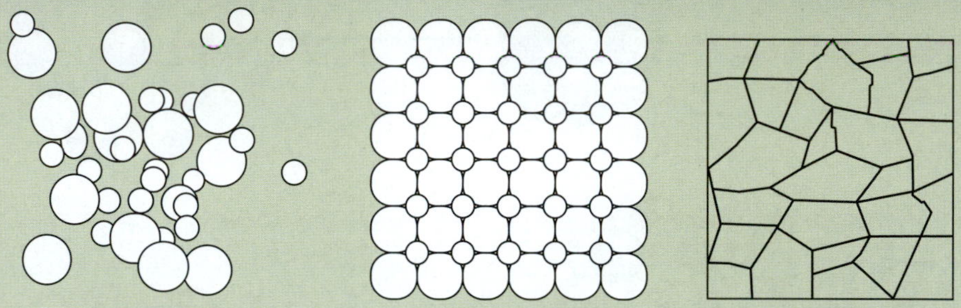

6.4.7 Structural Stabilization in Solids

It is well known that the lattice parameters of metallic nanoparticles contract with decreasing particle size [28]. **Figure 6.15** illustrates rearrangement in a *primitive-C* material. Two mechanisms are shown: contraction of the surface layer into the bulk and lateral shift to increase the number of nearest neighbors. The lattice parameter a decreased in each case [19].

The lattice contraction of nanoparticles as a function of size can be predicted by [28]

$$\frac{\Delta a}{a_{bulk}} = -\frac{1}{1+Kd} \tag{6.49}$$

where Δa is

$$\Delta a = a_{np} - a_{bulk}, \tag{6.50}$$

the lattice parameters of the nanoparticle (np) and the material in the bulk phase. The term K is related to the rigidity modulus G of the bulk material:

$$K = \frac{\pi G}{2\gamma_{bulk}} \tag{6.51}$$

γ_{bulk} is the surface energy of the bulk material and d is the diameter of the nanoparticle. K is in terms of reciprocal meters (or nanometers, angstroms). Plots of $\Delta a/a$ versus reciprocal diameter are linear with negative slope [28]. For palladium, G = 96.2 GPa, γ_{bulk} = 2.046 J · m^{-2}, and a_{bulk} = 3.8904 Å. From these values, K = 240.1 Å^{-1}. For a 2-nm particle, the theoretical percent contraction $\Delta a/a_{bulk} \times 100$ is expected to be

FIG. 6.15 *Restructuring of a simple cubic surface is shown. The coordination of a surface atom is reduced from 6 to 5 (one broken bond forming one dangling bond). Because of this, the surface atom is drawn inward in order to compensate for the lost bond. In this case, the overall structure remains relatively the same. There is only inward shift of the surface atom lattice. Lattice contraction is not just a surface phenomenon, as the whole of the nanoparticle may experience diminishing of its natural lattice constant. If more bonds are broken, as in the case of atoms with higher levels of coordination, the surface is actually restructured. In this case, formation of strained bonds compensates for the loss of coordination. For example, low-index {100} faces of silicon exist at a higher energy level than its low-index {111} sibling. However, restructured {100} faces have a lower surface energy than do natural {111} surfaces [19].*

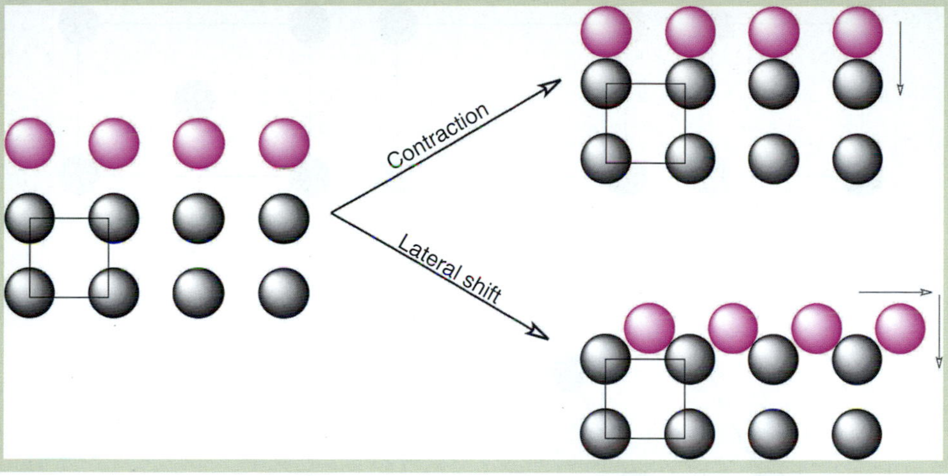

$$\left(\frac{\Delta a}{a_{bulk}}\right) = -\frac{1}{1+(240.1\,\text{Å}^{-1})(20\,\text{Å})} = -0.00021 = -0.02\%\ \textit{contraction} \qquad (6.52)$$

Reconstruction of outer layers also occurs by other mechanisms that involve loss of periodicity [29]. A surface layer that is perturbed is called a *selvedge* layer that perhaps consists of the upper two or so layers of the material. The surfaces of catalysts that accommodate surface carbon diffusion are also called selvedge layers. The selvedge layer may exist in a liquidus state during high-temperature catalytic activity. Atoms in a selvedge layer achieve stabilization by *outward relaxation* rather than contraction. The surface layers move outward and rearrange to preserve the configuration of the bulk atom packing symmetry. This layer is parallel to the layers of the bulk but not normal (original) (e.g., the layer has shifted). In another case, the surface atoms move outward but this time rearrange into a structure that has different symmetry from that found in the bulk of the material. This kind of rearrangement is called *reconstruction*. Following reconstruction, the surface properties of the material can be altered: electrical conductivity, chemical behavior, optical properties, and chemisorption–physisorption characteristics [29].

Surface stabilization is also achieved with the addition of adatoms, which are periodic arrays of an adsorbed monolayer of atoms that are directed by the preexisting surface lattice structure of the solid. Five "Bravais lattices" (instead of the 14 applied in volumetric crystals) characterize surface crystal structure (**Fig. 6.16**). We shall consider the simplest case—that of the simple primitive cubic surface (100) surface (**Fig. 6.17**). If $a = b$ in our simple square surface lattice, then the placement of the adatoms along x and y is related by vectors [28]:

FIG. 6.16 *Five two-dimensional "Bravais" lattices are depicted in (a). A fuller rendition that shows placement of atoms in the two-dimensional crystal is shown in (b).*

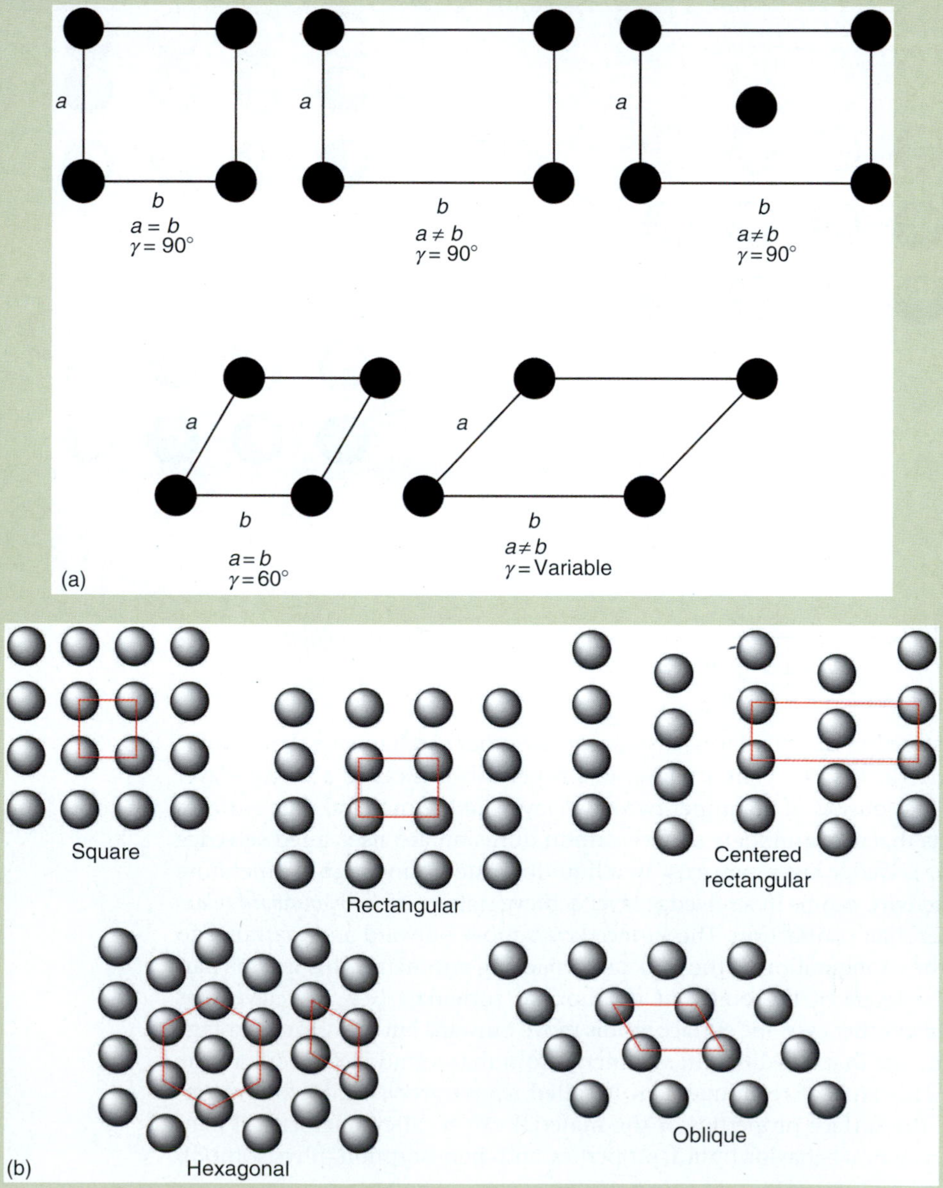

FIG. 6.17 *Adatoms on a simple cubic (100) surface are shown. The structure depicted is referred to as a P(2 × 2) surface.*

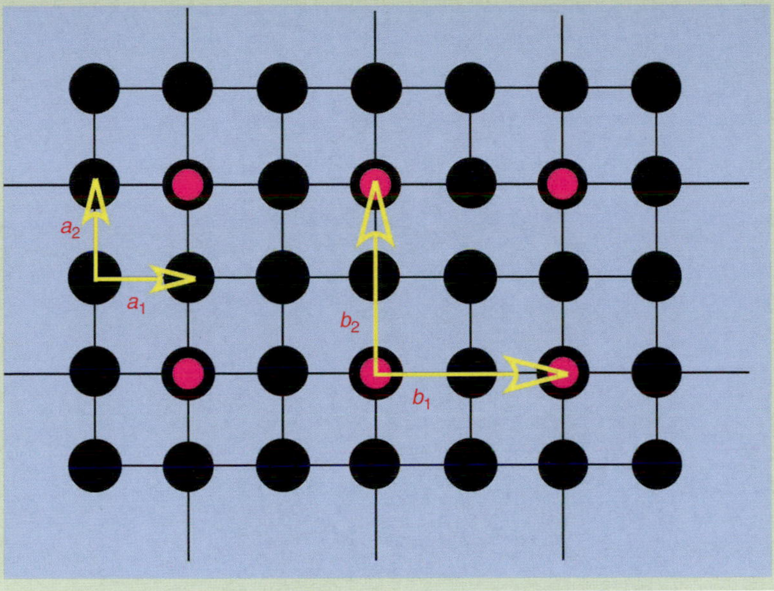

$$b_1 = M_{11}a_1 + M_{12}a_2 \tag{6.53}$$

$$b_2 = M_{21}a_1 + M_{22}a_2 \tag{6.54}$$

where, in the case of the primitive square, $b_1 = 2a_1$ and $b_2 = 2a_2$:

$$M = \begin{vmatrix} 2 & 0 \\ 0 & 2 \end{vmatrix} \tag{6.55}$$

The surface adatom array is designated as $P(2 \times 2)$ relative to the substrate crystal surface. The P stands for primitive.

Nanocrystalline solids are generally not isotropic. In other words, we do not expect nanocrystalline materials to be perfectly spherical in shape. The cuboctahedral structures shown earlier vividly demonstrate that anisotropic surface configurations exist in solid nanomaterials (**Fig. 6.18**). Recall that liquids tend to have isotropic surface energy (e.g., the surface energy is expected to be uniform at any point on the surface). In solids, different crystal planes (facets) on the surface have different surface energy. The individual facets on the surface contribute to the sum total equilibrium shape of the nanoparticle [19]. A general rule with regard to surface energy is that low-index planes—the {111}, {110}, and {100} families of planes—have the lowest surface energy. We have already determined that the {111} family of planes has the lowest surface energy of the group. When considering the packing structure of metals in particular, construction of a perfectly spherical shape requires more energy

FIG. 6.18 *The cuboctahedron structure is shown. The view of the structure through the upper-left vertex is along the [110] direction. The square faces represent {100} planes and the triangular faces represent {111} planes.*

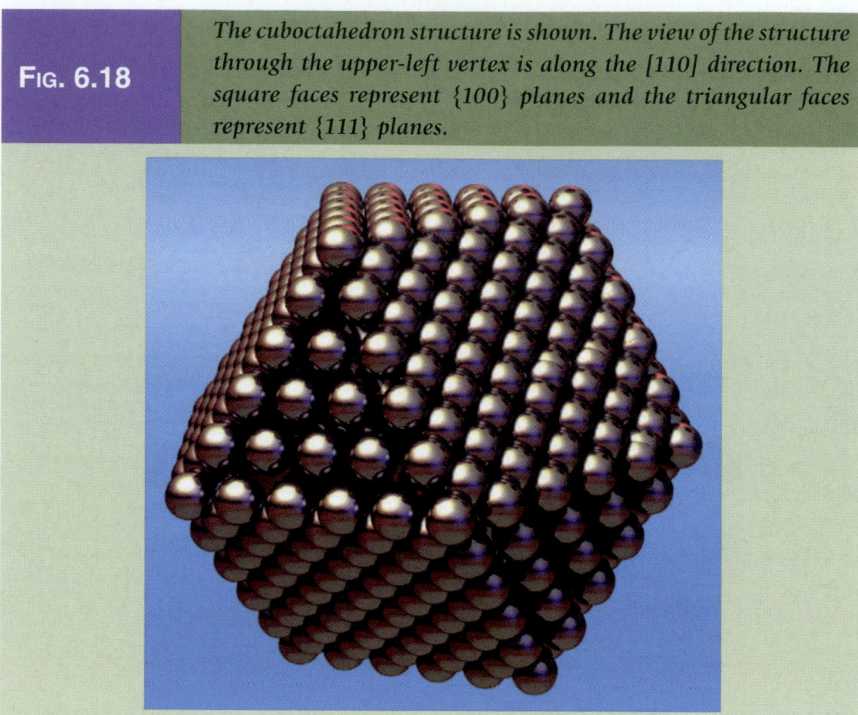

Source: Courtesy of Professor Anil K. Rao, Department of Biology, the Metropolitan State College of Denver.

than constructing one with a pseudospherical anisotropic crystal structure with facets. In this case, the spherical structure would possess higher surface energy.

Wulff plots represent the equilibrium shape of nanocrystalline materials. The general principle behind the Wulff plot is the relationship between the surface energy of a plane and its normal distance ($d_\perp$) from a central point of reference—for example, that the equilibrium distance of the surface facet plane is proportional to the surface energy of that plane:

$$d_\perp = \frac{\gamma_{hkl}}{C} \tag{6.56}$$

where C is a material constant. For example, a (111) plane, of lower surface energy, would be closer to the center of the crystal than would a (100) plane, of higher surface energy. The cuboctahedron is composed of 14 faces: 8 that are triangular (111) and 6 that are square (100). Oxidation of a NiO crystal that exists in the form of a cube with {100} faces goes through several morphological transformations during oxidation (**Fig. 6.19**): (1) cube, (2) truncated-cuboctahedron (contains {110} faces at vertices), (3) cuboctahedron at 75% oxidation with {100} and {111} faces, and finally (4) at complete oxidation into an octahedron consisting of {111} faces [30,31]. The changes in shape were simulated by using Wulff constructions of equilibrium morphologies [30,31].

Fig. 6.19 *During oxidation of cubic NiO, the original cubic form undergoes several morphological transformations. Initially only (100) faces are visible on the surface, forming a perfect cube. After some oxidation of the crystal, the points of the cube are reacted (pyramids are removed) to form flat (110) surfaces and oblique (111) surfaces. As oxidation proceeds, a truncated cubic form and then a cuboctahedron are formed. The cuboctahedron is further oxided to form an octahedron. It also goes through a truncated state before achieving the final octahedron structure.*

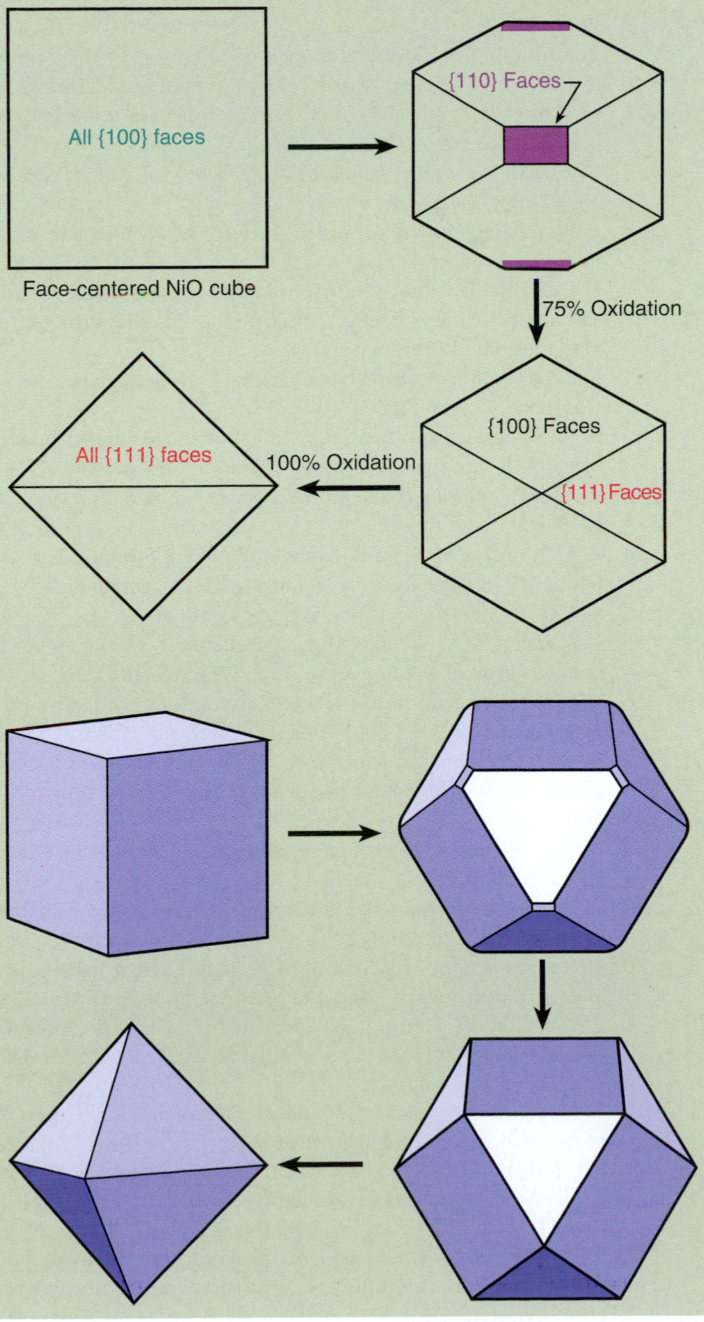

References

1. C. Kittel, *Introduction to solid state physics*, John Wiley & Sons, Inc., New York (1986).
2. W. J. Moore, *Physical chemistry*, 2nd ed., Prentice Hall, Inc., Englewood Cliffs, NJ (1955).
3. A. Robinson, Thomas Young: The man who knew everything, *History Today*, 56, 53–57 (2006).
4. W. Barthlott and C. Neinhuis, Purity of the sacred lotus, or escape from contamination in biological surfaces, *Planta*, 202, 1–8 (1997).
5. X. F. Gao and L. Jiang, Water-repellant legs of water striders, *Nature*, 432, 36 (2004).
6. M. Msksich, A surface chemistry approach to studying cell adhesion, *Chemical Society Review*, 29, 267–273 (2000).
7. J. A. Dean, ed., *Lange's handbook of chemistry*, 13th ed., McGraw–Hill Book Company, New York (1985).
8. P. W. Atkins, *Physical chemistry*, 4th ed., W. H. Freeman and Company, New York (1990).
9. I. A. Levine, *Physical chemistry*, 3rd ed., McGraw–Hill, New York (1988).
10. The Gibbs–Thomson effect, Wikipedia, en.wikipedia.org/wiki/Gibbs-Thomson_effect (updated February 2007).
11. *The American heritage dictionary of the English language*, 4th ed., Houghton Mifflin Company, Boston (2003).
12. M. S. Yaghmaee and B. Shokri, Effect of size on bulk and surface cohesion energy of metallic nanoparticles, *Smart Materials Structures*, 16, 349–354 (2007).
13. A. Shenk, *Calculus and analytic geometry*, Scott, Foresman & Co., Palo Alto, CA (1984).
14. V. M. Samsonov, A. N. Bazulev, and N. Yu. Sdobnyakov, On the applicability of Gibbs thermodynamics to nanoparticles, *Central European Journal of Physics*, 1, 474–484 (2003).
15. L. R. Fisher, R. A. Gamble, and J. Middlehurst, The Levin equation and the capillary condensation of water, *Nature*, 290, 575–576 (1981).
16. S. Siboni, Determination of the Kelvin equation in the presence of an arbitrary gravitational/inertial field, *American Journal of Physics*, 74, 565–568 (2006).
17. J. M. Howe, *Interfaces in materials: Atomic structure, thermodynamics and kinetics of solid–vapor, solid–liquid and solid–solid interfaces*, John Wiley & Sons, New York (1997).
18. B. M. Smirnov, *Clusters and small particles in gases and plasmas*, Springer-Verlag, New York (2000).
19. G. Cao, *Nanostructures & nanomaterials: Synthesis, properties and applications*, Imperial College Press, London (2004).
20. G. Kaptay, G. Csicsovszki, and M. S. Yaghmaee, An absolute scale for the cohesion energy of pure metals, *Materials Science Forum*, 414–415, 235–240 (2003).
21. P. Czoschke, H. Hong, L. Basile, and T.-C. Chiang, Quantum beating patterns in the energetics of thin film nanostructures, *Physics Review Letters*, 91, 226801 (2003).
22. K. K. Nanda, A. Maisels, F. Kruis, H. Fissan, and S. Stappert, Higher surface energy of free nanoparticles, *Physics Review Letters*, 91, 106102-1-4 (2003).
23. S. E. Jarlgaard, M. B. L. Mikkeslsen, P. Skafte-Pedersen, H. Bruus, and A. Kristensen, *Capillary filling speed in silicon dioxide nano-channels*, NSTI Nanotech, Boston (2006).
24. R. J. Hamers, *Materials growth*. Presentation, University of Wisconsin, hamers.chem.wisc.edu/chem630_fall2006/nucleation_and_growth/nucleation_and_growth2.ppt (2006).

25. D. J. Shaw, *Colloid & surface chemistry*, Butterworth–Heinemann, Oxford (1966).
26. M. Bowker, The going rate for catalysis, *Nature Materials*, 1, 205–206 (2002).
27. Sintering, Wikipedia, en.wikipedia.org/wiki/Sintering (2006).
28. W. H. Qi, M. P. Wang, and Y. C. Su, Size effects on the lattice parameters of nanoparticles, *Journal of Materials Science Letters*, 21, 877–878 (2002).
29. M. Ohring, *Materials science of thin films*, 2nd ed., Academic Press, San Diego (2002).
30. S. C. Parker, E. T. Kelsey, P. M. Oliver, and J. O. Titiloye, Computer modeling of inorganic solids and surfaces, *Faraday Discussions*, 95, 75–84 (1993).
31. P. M. Oliver, S. C. Parker, and W. C. Mackrodt, Computer simulation of the morphology of NiO, *Model Simulations in Materials Science and Engineering*, 1, 755–760 (1993).

Problems

6.1 Why are bubbles made of pure water not likely to occur?

6.2 For *fcc* gold, what is the surface energy of a freshly cleaved:

a. (100) Surface?
b. (001) Surface?
c. (010) Surface?
d. (111) Surface?

6.3 Calculate the work required to increase the surface area of mercury from 1 to 6 cm^2 at 20°C.

6.4 A bubble of air exists within a solution comprising acetone at 20°C and 1 atm. Calculate the pressure in a gas bubble if the bubble has a radius of 0.002 mm.

6.5 Find the cohesive energy (E_c), or binding energy, for graphite at 25°C from thermodynamic data (graphite, $\Delta H^\circ_{f,298}$ = 0.00 kJ·mol^{-1}; gaseous carbon, $\Delta H^\circ_{f,298}$ = 717 kJ·mol^{-1}). How does it compare to diamond under the same conditions ($\Delta H^\circ_{f,298}$ = 1.897 kJ·mol^{-1})?

6.6 What is the cohesive energy of bulk water in the liquid state at 25°C?

6.7 Calculate the energy required to atomize one drop of water at 20°C in which the surface area is increased from 1 cm^3 to 1 m^3. Assume that no pressure–volume work is done. The surface tension of water is γ = 72.75 dyn cm^{-1} against air at 20°C. What is the radius of the droplets formed?

Calculate the same problem for atomization of acetone (assume no Joule–Thompson cooling occurs during atomization) with surface tension γ = 23.70 dyn cm^{-1}.

6.8 For a simple cubic structure, what is the coordination number of each interior atom? Each surface atom? Each edge atom? Each corner atom? What amount of energy is required to split a cubic centimeter of the material in half along a (100) plane? Into eight equivalent cubes all along {100} planes?

6.9 For gold, an *fcc* metal, calculate the relative excess surface energy, ε_s in terms of joules per atom, of a freshly cleaved {111} plane. Z = 12 in the bulk. How many nearest-neighbor bonds are broken? What is this in terms of γ (in joules per square meter)?

6.10 Why are heats of vaporization of solids ~10% lower than the heats of sublimation of the same material?

6.11 Neglecting the force of gravity, if you have a wire of length 4 cm, calculate the work required to lift a sheet of acetone held together by surface tension to a height of 100 μm.

6.12 Is surface tension an extensive or an intensive property (consult your thermodynamics text)?

6.13 Would you agree that the Kelvin equation has built-in nano considerations before nanoscience became very popular? Explain your answer.

6.14 Why is the radius of curvature of a sphere equal to the radius of the sphere?

6.15 Young, Laplace, and others developed what is called a mechanical explanation for surface tension. What does this imply?

6.16 Calculate the vapor pressure of $r = 600$ nm water droplets at 20°C. Assume that the density of water at 20°C is 0.99823 $g \cdot cm^{-3}$, $V_m = 18.047\ cm^3 \cdot mol^{-1}$; $T = 293$ K, $R = 8.314\ J \cdot K^{-1}$; $M_{H2O} = 18.015\ g \cdot mol^{-1}$.

6.17 Define the concept of chemical potential. Give some examples of chemical potential.

6.18 What is the difference, qualitatively, between cohesive and adhesive energies?

6.19 Why is the contact angle of a water droplet on Teflon higher than that of the same sized droplet on gold surface?

6.20 What (intuitively) is the meaning of surface energy?

6.21 Make a list of all the factors that contribute to the higher surface energy of nanomaterials.

6.22 Calculate the height of a column of water in a nanocapillary tube (Pyrex™ glass) of diameter of 800 nm.

6.23 List some of the assumptions inherent in the *nearest-neighbor* model to calculate surface energy.

6.24 Why is the work function of a material dependent on the type of surface?

6.25 List as many ways to minimize surface energy as you can.

6.26 Explain DLVO theory.

6.27 Why does the term for the free energy of nucleation include a surface term?

6.28 Would you expect that "surface energy-to-volume energy" scales with surface-to-volume ratio as particles become smaller?

The Material Continuum

As in my conversations with my brother we always arrived at the conclusion that in the case of X-rays one had both waves and corpuscles, thus suddenly— ... it was certain in the course of summer 1923—I got the idea that one had to extend this duality to material particles, especially to electrons.

Louis de Broglie

Chapter 7

Threads

In chapter 5, we introduced topics that described the surfaces of nanomaterials and how size, shape, structure, orientation, and composition influence nanoscale material properties and phenomena. Although devoid of any real discussion of physics or chemistry, we gained a sense of how size and interfacial phenomena can potentially dominate behavior at the nanoscale. Chapter 6 addressed topics involved with the energetic character of nanoscale materials—clusters in particular—primarily from the viewpoint of the chemical bond and the mechanical–thermodynamic treatment of surface tension. In chapter 7, we now prepare to delve into the quantum domain and the effects of size confinement. We reaffirm that nanomaterial behavior does not scale in a linear fashion from bulk material behavior. We also find that nanomaterials, especially those on the scale of a few nanometers, do not necessarily obey the rules of quantum mechanics, although the fundamental relationships still hold true.

Nanotechnologists must be proficient in the basics of quantum mechanics, not only in its mathematics but also in its physical meaning. Indeed, a good intuitive understanding of quantum mechanics is a prerequisite for a thorough understanding of chemistry, physics, and materials science. Why? Because nanotechnology is all about a finite number of atoms or molecules that together are the basic building blocks of everything and so small that they are inherently quantum mechanical. For example, within an atom or molecule, negatively charged electrons are trapped in the neighborhood of one or more (positively charged) atomic nuclei because of the attraction the electrons have for the nuclei and repulsions for other electrons. These electrons only have certain energies that correspond to the 1s, 2p, etc. orbitals we learned in freshman chemistry. For some nanomaterials, electrons can also be trapped and exhibit only certain allowed energies.

Following this chapter, we present a special chapter about nanothermodynamics. We adhere to strict thermodynamic topics and hope to shed light on the energetic relationships of nanoscale materials: Are those relationships worthy of forming the foundation of a new discipline called nanothermodynamics? You decide!

7.0 Material Continuum

The concept of "continuum" (from the Latin *continuus*, meaning "uninterrupted") is ingrained in everything surrounding us. The definition of a continuum is a continuous extent, or whole, no part of which can be distinguished from neighboring parts except by arbitrary division.

Many examples of common continua exist around us—those of language, mathematics, mechanics, time, radiation, and others. The matter–energy continuum implies that all matter and all energy are convertible through $E = mc^2$. The material continuum implies that there is a smooth transition from the bulk world into the quantum domain. This statement is completely true in the sense that all materials are made of atoms and molecules and that nanomaterials simply have progressively fewer of them. However, we have been preaching all along that nanomaterials have neither purely bulk nor strictly quantum properties. We then have to grapple with a dilemma: If there is a continuum of material and energy, then why all the fuss about nanomaterials? In order to come to terms with this apparent difference, it is our desire to understand how properties and phenomena *scale* with regard to diminishing size.

The continuum of interest in this section is that of materials, measured and scaled by size and size alone. The material continuum is

Extremely big things/trees → macroscopic bulk materials/multicellular organisms → micron-sized colloids/crystals/single-celled organisms → ***colloids/nanoparticles/ nanocrystals/clusters/viruses/macromolecules*** *→ molecules/atoms → subatomic particles → subnuclear particles → much smaller things*

7.0.1 Material Properties and Phenomena

Table 7.1 lists properties and phenomena from two different perspectives: the macroscale and the nanoscale.

TABLE 7.1 *Material Properties and Phenomena*

Property/ phenomenon	Specific property	At the macroscale–microscale	At the nanoscale
Structure and electronic configuration	Confinement	No confinement	Confinement in 0D, 1D, and 2D materials
	Surface area	Surface area of bulk materials, although important, is small compared to its volume	Exponentially enhanced Collective surface area can be enormous
	Surface-to-volume ratio	S/V is small; becomes insignificant as objects become larger	Approaches 1 when all atoms are surface atoms
	Lattice spacing	Characteristic of the bulk	Lattice spacing is altered. Spacing near surface contracts due to rearrangement. Ion vacancies larger
	Atom coordination	Coordination saturated except at surface where it is negligible	Coordination undersaturated at surface and in volume
	Electron orbitals	Continuous over the breadth of the material	Quantum: HOMO and LUMO Cluster: HOCO and LUCO Magic electronic numbers in alkali metal clusters
	Quantum mechanics	QM applies at the bulk level: bathtub waves	Nanomaterials exist at the quantum–classical interface
Electromagnetic properties	Radiation: absorption emission	Blackbody radiation Absorption–emission bands broad	Influenced by the Bohr radius
	Optical response	Metals reflect with partial absorption of light Micron-sized particles scatter light and conform to Mie theory analysis Higher order plasmon resonance and plasmon resonance is delocalized	Size-dependent absorption–emission. Environment dependent (effective medium theory) Quasi-static condition ($r \ll 0.01\ \lambda$). Dipolar plasmon resonance Localized surface plasmon resonance (i.e., transverse/longitudinal modes for prolate particles)
	Optical constants, ε, n, k	Bulk values apply and are valid in micron-sized particles.	Bulk optical constants no longer apply below 10–20 nm

(*continued*)

Table 7.1 (Contd.) *Material Properties and Phenomena*

Property/ phenomenon	Specific property	At the macroscale–microscale	At the nanoscale
	Bandgap	Traditional metal, semiconductor, insulator bandgaps. Bandgap independent of size	Bandgap is size dependent. Gold resembles a semiconductor in nanoparticles <2 nm in diameter
	Electrical conduction	Continuous and follows Ohm's law. Conductivity based on band structure Electron mean free path not significant with respect to surface Scattering by lattice defects and thermal phonons	Ohm's law does not apply (classically). Formation of discrete energy levels Coulomb staircase/blockade Tunneling currents are important Ballistic conduction (electron mean free path > dimensions)
	Magnetic memory coercive force	Size independent	Size-dependent magnetic properties. Gigantic magnetoresistance effects possible with stacked magnetic nanoparticle arrays
Thermo-dynamics	Discipline	Classical thermodynamics, statistical mechanics	Nanothermodynamics
	Equilibria	Macroscopic systems—thermodynamic equilibrium capability	Nanosystems subject to environmental fluctuations. Conditions of nonequilibrium/steady state in living systems
	Number of atoms	Thermodynamic infinite limit: $N \to \infty$, $V \to \infty$, N/V constant	N is countable. Thermodynamic limits do not apply
	Intensive properties	Properties independent of amount of material Environment independent	Intensivity not always applicable as intensive properties can change with size. Serious environmental dependency
	Extensive properties	Properties dependent on amount of material	Altered definition of extensivity
	Entropy	Link between micro and macro domains	Violations of the second law. Is a fourth law of thermodynamics required?
	Melting point	Metals relatively high melting	MP drops precipitously below 20 nm Proportional to $1/r$
	Kelvin effect	Function of particle curvature Valid for bulk and nanomaterials	Particles <10 nm conform to Kelvin effect
	Surface tension/ energy	Classical surface tension and surface energy	Surface tension and surface energy a function of size
	Specific heat	Function of elemental makeup	Much higher than bulk counterpart
	Electron affinity	Function of elemental makeup	Electron affinity influenced by magic numbers in clusters
	Work function	Function of elemental makeup	Function of size
	Chemical bonds	The big three: ionic, covalent, and metallic predominate, although all others apply	Intermolecular forces are important: H-bonds, van der Waals, hydrophobic effect, dipole interactions, etc.
Chemical reactivity	Reaction scheme	Kinetic control at higher temperatures, harsh conditions; making and breaking of stronger bonds	Thermodynamic control occurs at lower temperatures, milder conditions; making and breaking of weaker bonds

Table 7.1 (Contd.) *Material Properties and Phenomena*

Property/ phenomenon	Specific property	At the macroscale–microscale	At the nanoscale
	Adsorption	Surface adsorption by chemisorption/physisorption does not result in catalytic activity	Adsorption by small particles can result in catalytic activity
	Solubility	Large particles have limited solubility	Smaller particles have enhanced solubility. This is important in targeted drug delivery systems
	Sintering	Function of elemental makeup. Occurs at high temperatures	Occurs at lower temperatures
	Chemical activity	Function of elemental makeup. Reactivity takes place below the nanoscale	Reactivity of nanoparticles significantly enhanced due to excess surface energy
Mechanical properties	Tensile properties	Bulk materials made of micron-sized grains	Bulk materials made of nanograins have superior properties Tensile properties can approach theoretical limit in carbon nanotubes
Scaling laws Classical continuum models	Density Power Frequency Efficiency Mechanical Dielectrics	Bulk scaling laws and continuum models apply to all materials down to micron-sized particles	Depends on the types of physical phenomena and the size of the particle. Bulk dielectric functions are valid to sub-100-nm particles

7.0.2 Background

Classical mechanics gives the perspective of trajectories of macroscopic particles while wave mechanics gives a view of the discrete energy levels associated with macroscopic particles without actually explaining the trajectories. Sometimes a complete explanation is difficult to obtain, but instead we can take advantage of each view and work on new windows of opportunity to explain different phenomena. Nanomaterials provide us with that new window.

Quantum mechanics assumes importance as material size is diminished. This is especially true in nanoscience; materials comprising a countable number of atoms and molecules are treated mathematically with wave functions. One way to look at the wave function is to consider particles as stationary waves (standing waves), and all that is needed to describe its behavior is this description of the standing wave (e.g., the wave function). We discussed the particle–wave duality in chapter 3, so it should come as no surprise that, at the nanoscale, the dual wave–particle nature of radiation becomes more important. The dual wave–particle nature was first propounded by Louis de Broglie (chapter 3) to address the apparent anomaly observed in black body radiation—specifically, the ultraviolet catastrophe.

In our universe, each particle (whether atom, molecule, or cluster) is characterized by a function called ψ. The only restriction placed on this function is the

Schrödinger equation—the fundamental equation of matter. From this equation, the average values of angular momentum, position, and energy are solvable by the application of mathematical operators (complex conjugates, integration) over all independent variables. Quantum mechanics is a branch of physics that was developed in the early part of the twentieth century. It is extremely effective in describing the behavior of atoms, molecules, nuclei, and, most recently, nanoparticles. Quantum mechanics is based on the concept that matter has wave properties. Quantum mechanics is used to calculate the energies of particles confined to a finite (and small) region of space. The uncertainty principle is a consequence of quantum mechanics, as is the phenomenon of electron tunneling. Quantum mechanics is about probability.

A particle located in a fixed point in space is localized in space with discrete physical properties such as mass. In contrast, a wave is inherently spread out over many wavelengths in space and could have amplitudes in a continuous range with waves superposing and passing through each other. On the other hand, particles are able to collide and bounce off each other. The dual wave–particle nature of radiation is all to do about tiny matter having the ability to behave as a wave as well as a particle (i.e., waves and particles have interchangeable properties). Mathematically, it can be explained by considering that if an object having momentum p has an associated wave whose wavelength is λ, then the momentum can be expressed as

$$p = \frac{h}{\lambda} = \hbar k \tag{7.1}$$

and

$$k = \frac{2\pi}{\lambda} \tag{7.2}$$

Example 7.1 *A Convenient Form of the de Broglie Wavelength*

$$\lambda = \frac{hc}{pc} \tag{7.3}$$

where h is the Planck's constant, p the momentum, and c the velocity. Plugging in the Planck's constant value and the speed of light, $hc = 1239.84\ eV \cdot \text{nm}$ and pc is expressed in electron volts.

This is particularly appropriate for comparison with photon wavelengths since for the photon, pc = E (kinetic energy) and a 1-eV photon is seen immediately to have a wavelength of 1240 nm. For an electron with KE = 1 eV and rest mass energy 0.511 MeV, the associated de Broglie wavelength is 1.23 nm, about a thousand times smaller than a 1-eV photon. (This is why the limiting resolution of an electron microscope is much higher than that of an optical microscope.)

Some examples of physical phenomena that apply to the nanoscale:

Visible light (optical microscope) E ~ 4.1–2 eV): $\lambda \sim$ 300–600 nm
X-rays (synchrotron, E ~ 1 ke$V$): $\lambda \sim$ 1 nm
Electrons (electron microscope, E ~ 100 ke$V$): $\lambda \sim$ 0.004 nm

Ions (He^+ ions, E ~ 100 ke$V$): $\lambda \sim 10^{-5}$ nm where k is the wave vector (also called the *wavenumber*) and

$$\hbar = h/(2\pi) \tag{7.3}$$

the reduced form of Planck's constant. The wave vector is an energy vector of magnitude $2\pi/\lambda$ in the direction of the propagation of the wave.

7.0.3 Nano (Quantum) Perspective

The density of states (DoS) is a physical property of a material that indicates the structure and degree of packing of energy levels in a quantum mechanical system [1]. It is commonly symbolized by N and is a function of $g(\mathrm{E})$ or a function $g(k)$ of the wave vector k. Density of states is a phenomenon usually applied to the electronic levels of a solid. The densities of states for zero-dimensional (quantum dots), one-dimensional (quantum wires), and two-dimensional (quantum wells) materials compared to that of a bulk semiconductor material are shown in **Figure 7.1.**

The purpose of this chapter is to help illuminate the boundary between the quantum domain and that of the nanomaterial and, consequently, to make a link to the domain of the bulk material. The concept of the matter continuum is a valid one; all matter is described by the same equations: quantum, classical, or otherwise. It is simply a matter of complexity. Classical equations break down in many ways (e.g., the ultraviolet catastrophe, to name just one) but they are applied in modified form to describe nanomaterials. If there are enough quantum states, then we have a bulk material. If we have fewer quantum states, then

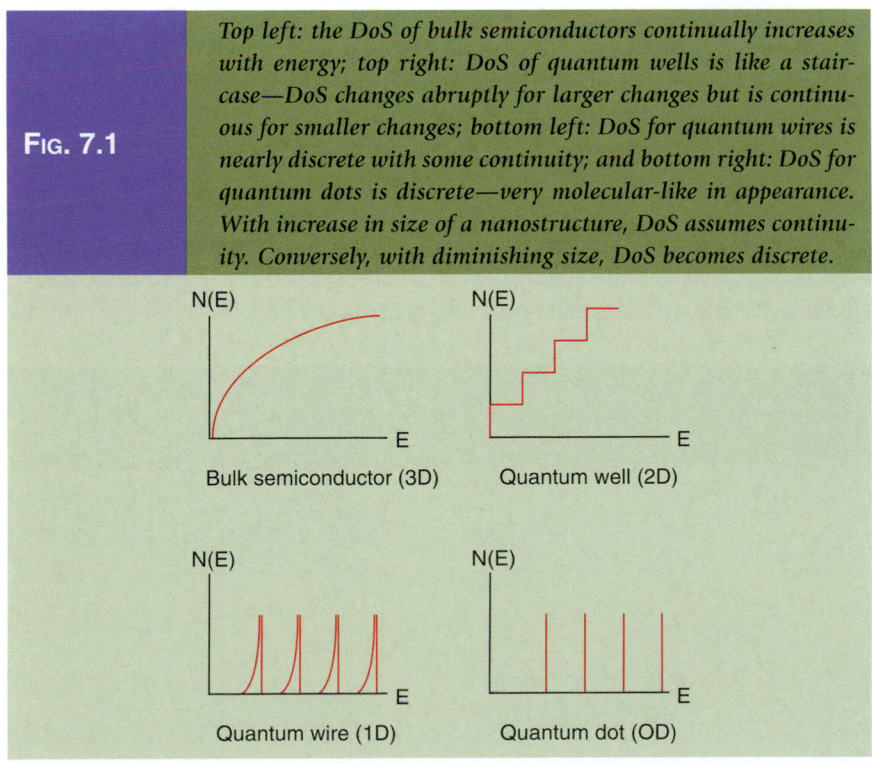

Fig. 7.1 *Top left: the DoS of bulk semiconductors continually increases with energy; top right: DoS of quantum wells is like a staircase—DoS changes abruptly for larger changes but is continuous for smaller changes; bottom left: DoS for quantum wires is nearly discrete with some continuity; and bottom right: DoS for quantum dots is discrete—very molecular-like in appearance. With increase in size of a nanostructure, DoS assumes continuity. Conversely, with diminishing size, DoS becomes discrete.*

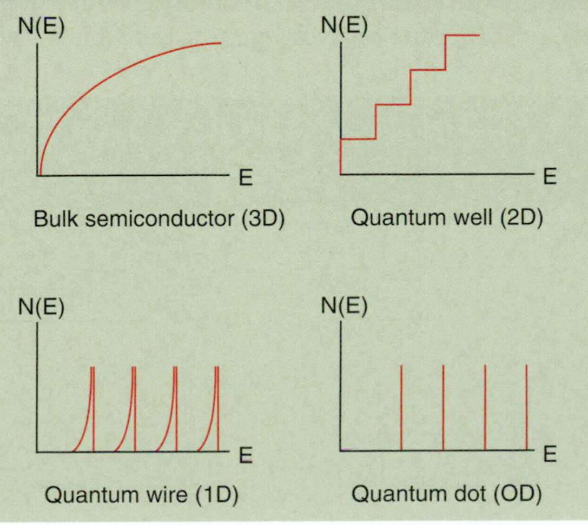

we have a nanomaterial. If we have even fewer quantum states, then we have molecules—a continuum for sure. This process continues to the atom and into the nucleus of the atom in which quantum states also exist. The matter continuum is in every sense of the term a true continuum.

7.1 Basic Quantum Mechanics and the Solid State

Let us begin our description of the solid state by starting with a liquid, by considering a bathtub that is filled with water. If you swish your hand around in a bathtub, you will note that the only waves that persist for a long time are those that can "fit" inside the bathtub; other waves would simply spill water out of the tub. Actually, you will observe that many waves, measuring from crest to crest, fit into the bathtub! These waves are called "standing" waves as they just stand around instead of traveling forward like waves you would see at the beach. In other words, they have their crests and troughs always in the same places, and the nodes (areas midway between the crests and troughs where the water level is undisturbed) are always in the same places. The largest possible bathtub wave that has a crest at each end is shown in **Figure 7.2a.** Another wave with a crest at each end and in the middle is shown in **Figure 7.2b.**

The possible energies of standing waves in the bathtub are limited to certain specific (quantum!) values by the requirement that each wave needs to fit exactly into the bathtub. The human in the tub supplies the energy required to make waves. Now the energies of the bathtub waves can be likened to the wave functions that describe a particle in a box (like electrons trapped inside a quantum dot) that increases with the number of wave crests that are able to fit into the tub. If the water in the bathtub is swished around very gently and slowly, for example, then the only wave we will see would be the wave that has a crest exactly at each end of the tub with a node right in the middle of the tub. If we increase the agitation, we will see waves that have more and more crests in the middle, at least until all the water comes out onto the floor. The wave functions then would resemble the standing waves you observe in the bathtub. In fact, allowed energies are calculated easily by figuring out which waves "fit" into the

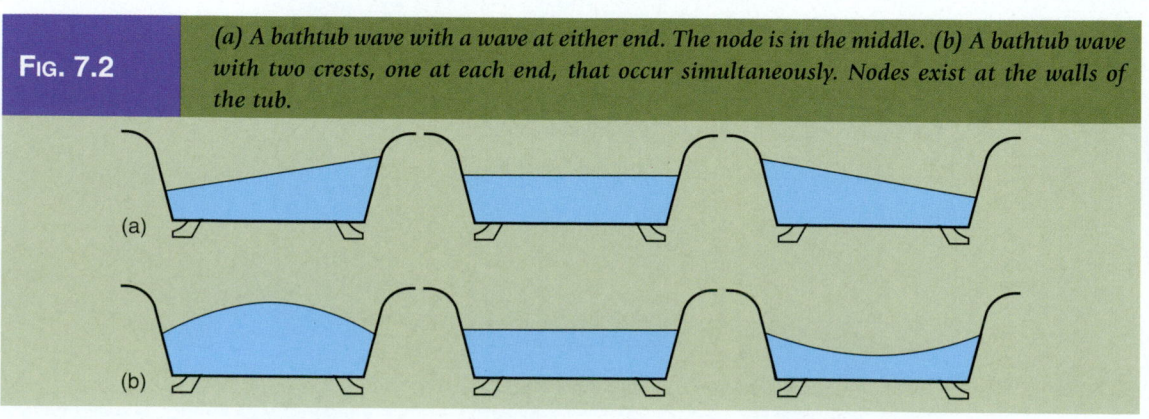

Fig. 7.2 *(a) A bathtub wave with a wave at either end. The node is in the middle. (b) A bathtub wave with two crests, one at each end, that occur simultaneously. Nodes exist at the walls of the tub.*

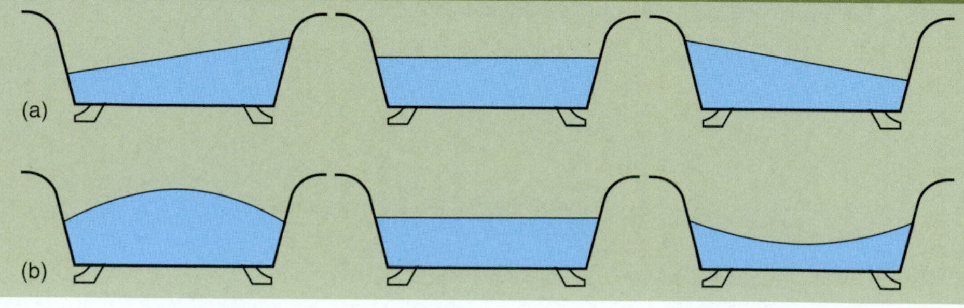

box and which ones do not! The meaning of the peaks and troughs of the wave function is somewhat akin to the waves in the bathtub, in the sense that something is going on at the peaks and troughs but not at the nodes.

7.1.1 Ubiquitous Particle in a Box

One of the simplest quantum mechanical problems is that of the ubiquitous "particle in a box." Solution of this problem is very useful for understanding of many properties of nanoparticles or quantum dots. In a one-dimensional particle-in-a-box problem, we consider a particle that is constrained to move in a single dimension, under the influence of a potential $V(x)$ [or $U(x)$] that is equal to zero for $0 < x < L$ and equal to $\propto$ for all other x. At the walls of the box, $V(x)$ is equal to infinity (hence, the infinite potential well) and the wave function equals zero (**Fig. 7.3a**). The square of the wave function at any given point is equal to the probability that the particle can be found at any point x within the boundaries. Hence, the particle is likely to be found at peaks or troughs but never at the nodes. Classically, if we put a particle in a box, and it is confined to move in only one direction, the particle will reflect off the walls and go backward and forward in the box (**Fig. 7.3b**).

At any time if we open the box we will find that the particle has equal probability of being somewhere in the box. In order to better understand the wave properties from the point of view of the wave equation, let us recall the harmonics of a string held at both ends (like in a guitar; see **Fig. 7.4**).

You already know that waves exist only when the wavelength is an integral multiple of the half wavelengths (**Fig. 7.5**):

$$L = n\frac{\lambda}{2} \quad \text{and therefore} \quad \lambda = \frac{2L}{n} \tag{7.4}$$

In this way, as you can see, the wavelengths of the standing wave of a string are *quantized*. As the particle is confined between two nonpenetrating walls and is confined to moving only parallel to the x-axis (our supposition), its linear momentum (mv, for a mass m, moving with velocity v) remains constant and

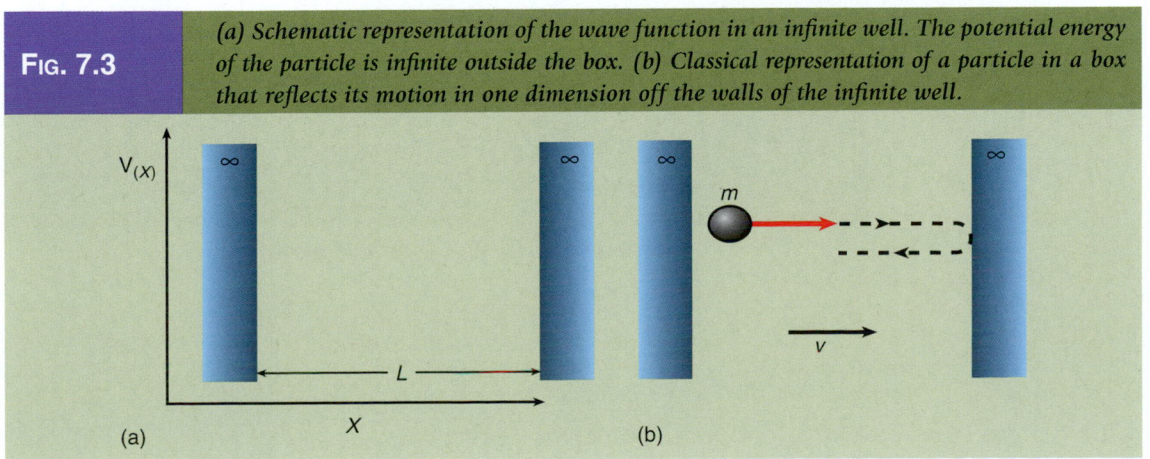

FIG. 7.3 *(a) Schematic representation of the wave function in an infinite well. The potential energy of the particle is infinite outside the box. (b) Classical representation of a particle in a box that reflects its motion in one dimension off the walls of the infinite well.*

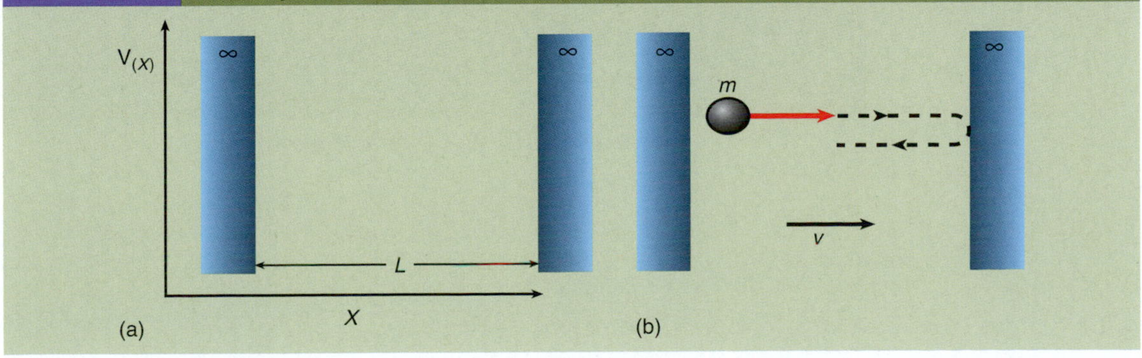

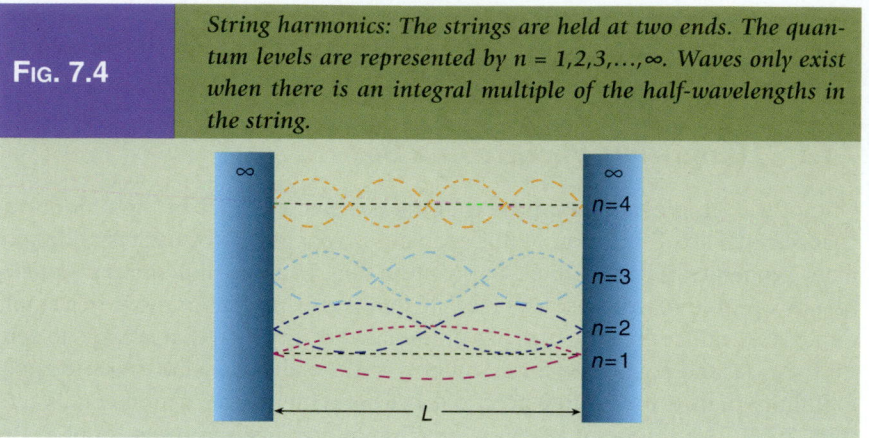

Fig. 7.4 *String harmonics: The strings are held at two ends. The quantum levels are represented by n = 1,2,3,...,∞. Waves only exist when there is an integral multiple of the half-wavelengths in the string.*

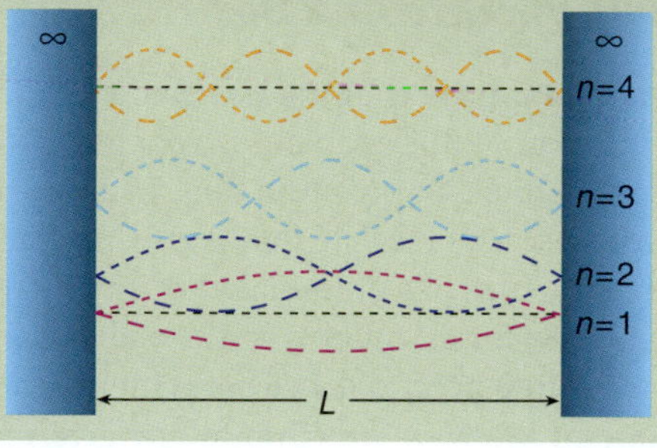

its kinetic energy (KE) is a constant as well. If we consider the wave to be described by a simple function for a standing wave:

$$y(x) = A\sin(kx) \tag{7.5}$$

where A is a constant that defines the amplitude. If we modify the value of k, the wavenumber is then

$$\text{If } k = \frac{2\pi}{\lambda} \text{ and } \lambda = \frac{2L}{n}, \quad \text{then } k = \frac{n\pi x}{L} \text{ and } y(x) = A\sin\left(\frac{n\pi x}{L}\right) \tag{7.6}$$

1n 1926, Erwin Schrödinger proposed the wave equation that is used to describe how waves of matter (or the wave function) propagate in space and time. The wave function is similar to our identity card and contains all of the information that can be known about a particle. The wave function of a free particle moving along the x-axis is given by

$$y(x) = A\sin\left(\frac{2\pi x}{\lambda}\right) = A\sin(kx) \tag{7.7}$$

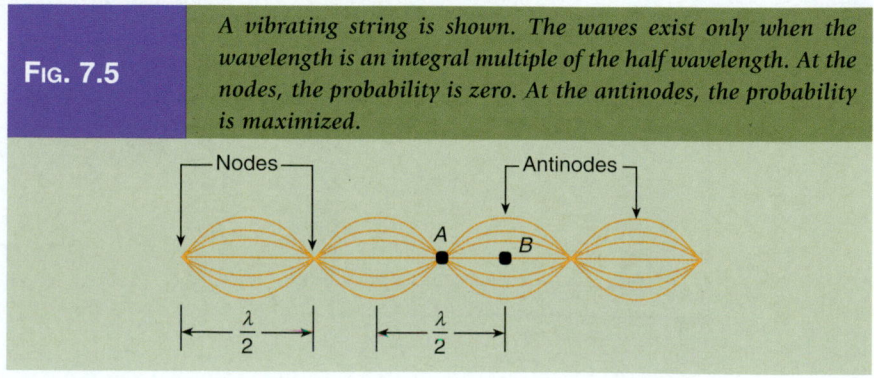

Fig. 7.5 *A vibrating string is shown. The waves exist only when the wavelength is an integral multiple of the half wavelength. At the nodes, the probability is zero. At the antinodes, the probability is maximized.*

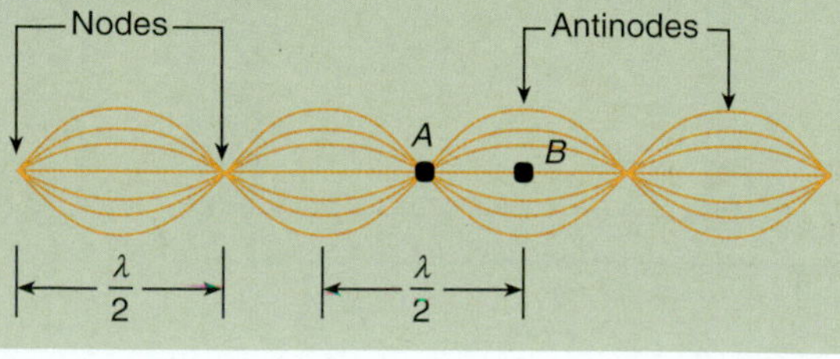

The waves, like those shown in **Figure 7.6**, represent a snapshot of a wave function at a given time. The probability of finding a particle in a position inside the box is given by

$$|\psi(x)|^2 = A\sin^2\left(\frac{2\pi x}{\lambda}\right) = A\sin^2(kx) \tag{7.8}$$

The probability is always positive and the sum total probability equals one. Because the de Broglie wavelength is quantized, the momentum is also quantized. Higher energy levels of the particle correspond to higher potential energy in the box as it climbs the walls of the infinite well. The rest of the energy of the particle is the kinetic energy, the energy due to its motion. The kinetic energy is also quantized:

$$E_n = \frac{1}{2}mv^2 = \frac{p^2}{2m} = \frac{(nh/2L)^2}{2m} = \left(\frac{n^2h^2}{8mL^2}\right); \quad n = 1, 2, 3, \ldots \tag{7.9}$$

EXAMPLE 7.2 *Electron Confined within Atomic Dimensions*

Consider an electron in an infinite well of size 0.1 nm (typical of an atom). What is the ground state energy (n = 1) of the electron? What energy is required for the electron to jump to the third energy level (n = 3)?

Solution:

The electron is confined in an infinite potential well; therefore, its energy is given by

$$E_n = \left(\frac{h^2n^2}{8mL^2}\right); \text{ for } n = 1 \text{ and } L = 0.1 \text{ nm} \tag{7.10}$$

$$E_1 = \frac{(6.626\times10^{-34}\,\text{J}\cdot\text{s})^2(1)^2}{8(9.110\times10^{-31}\,\text{kg})(1\times10^{-10}\,\text{m})^2} = 6.025\times10^{-18}\,\text{J} \; or \; E_1 = 37.6\; eV \tag{7.11}$$

The frequency of the electron associated with this energy is

$$v = \frac{E}{\hbar} = \frac{6.025\times10^{-18}\,\text{J}}{1.055\times10^{-34}\,\text{J}\cdot\text{s}} = 5.71\times10^{16}\text{ radians}\cdot\text{s}^{-1} \text{ or } v = 9.092\times10^{15}\,\text{s}^{-1} \tag{7.12}$$

The third energy level (E_3 at n = 3) is

$$E_3 = E_1n^2 = (37.6\; eV)(3)^2;\; E_3 = 338.4\; eV \tag{7.13}$$

The energy required to take the electron from 37.6 to 338.4 *eV* is 300.8 *eV*.
This energy can be provided by a photon of exactly that energy—no more and no less. Since the photon energy is $E = hv = hc/\lambda$

$$\lambda = \frac{hc}{E} = \frac{(6.626\times10^{-34}\,\text{J}\cdot\text{s})(2.998\times10^{8}\,\text{m}\cdot\text{s}^{-1})}{(300.8\,eV)(1.6022\times19^{-19}\,\text{C})} = 4.12\,\text{nm} \tag{7.14}$$

which is the energy of an x-ray photon

Example 7.3 *Particle-in-a-Box Application*

Consider a linear butadiene molecule. What is the length of the box? Calculate the energy of an electron confined in this molecular box? What is the expected absorption of butadiene CH_2=C–C=CH_2 [2]?

Also what is the energy of a dye molecule that is 0.8 nm in length with an electron $n_1 \rightarrow n_2$ transition?

Solution:
The C–C bond length is 1.54 Å and that of the C=C bond is 1.35 Å. The length of the box is then: $L = 2 \times 1.35 + 1.54 = 5.78$ Å

$$E_n = \frac{n^2h^2}{8mL^2}, \quad n = 1,2,3,\ldots \tag{7.15}$$

Since each state holds two electrons and there are four π-electrons, the $n = 1$ level is full and the $n = 2$ level is full. The promotion of the electron then must proceed as $n_2 \rightarrow n_3$.

Fig. 7.6 *The particle-in-a-box model is applied to the four π-electrons of the conjugated butadiene molecule. In reality, butadiene is not a linear molecule. The length of the box is equivalent to the length of the molecule that consists of two double carbon–carbon bonds and one single carbon–carbon bond. The absorption of butadiene is at 217 nm. The particle-in-a-box model derived value is 220 nm.*

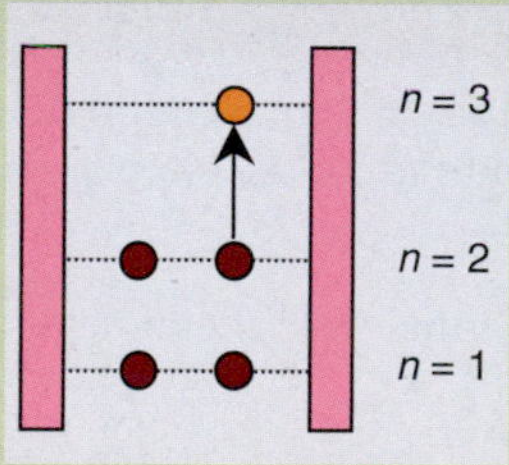

$$E = \frac{h^2}{8mL^2}\left(n_3^2 - n_2^2\right) = \frac{h^2}{8mL^2}\left(3^2 - 2^2\right) = \frac{\left(6.626\times10^{-34}\,\mathrm{J\cdot s}\right)^2}{8\left(9.110\times10^{-31}\,\mathrm{kg}\right)\left(5.78\times10^{-10}\,\mathrm{m}\right)^2}(5) = 9.02\times10^{-19}\,\mathrm{J} \tag{7.16}$$

This energy corresponds to wavelength through

$$E = h\nu = \frac{hc}{\lambda};\ \lambda = \frac{hc}{E} = \frac{\left(6.626\times10^{-34}\,\mathrm{J\cdot s}\right)\left(2.998\times10^{8}\,\mathrm{m\cdot s^{-1}}\right)}{9.02\times10^{-19}\,\mathrm{J}}\left(\frac{10^9\,\mathrm{nm}}{\mathrm{m}}\right) = 220\,\mathrm{nm} \tag{7.17}$$

The experimental absorption of butadiene is 217 nm—not a bad match from a particle-in-a-box calculation. A dye molecule with box length $L = 0.8$ nm has energy 2.8×10^{-19} J. This corresponds to $\lambda = 709$ nm.

The kinetic energy of the particle at any point inside the quantum box is equal to the relative magnitude of the curvature of the wave function (this is true for real waves as well—like in the sea wave, where the water moves fastest where the wave is most sharply curved at the very tip of the crest). One is able to observe that higher energy wave functions are sharply curved to give shorter distance between successive wave peaks. Hence, more wave peaks can fit inside

Example 7.4 *Confinement of Proton and Electron in a Quantum Well*

Consider a proton confined in a one-dimensional box of width 0.200 nm. (a) What would be the lowest energy of the proton? (b) What is the lowest energy of an electron confined in the same box?

Solution (a)

The energy of a particle of mass m in a 1D box of length L is

$$E_n = \frac{n^2\pi^2\hbar^2}{2mL^2} = \frac{n^2\pi^2(h/2\pi)^2}{2mL^2} = \frac{n^2h^2}{8mL^2} \quad (7.18)$$

The lowest energy ($n = 1$) of a proton with $m_p = 1.67 \times 10^{-27}$ kg in a box with $L = 2 \times 10^{-10}$ m is

$$E_{1,p} = \frac{h^2}{8m_pL^2} = \frac{(6.626\times10^{-34}\,\text{J}\cdot\text{s})^2}{8(1.67\times10^{-27}\,\text{kg})(2\times10^{-10}\,\text{m})^2} = 8.22\times10^{-22}\,\text{J}\left(\frac{eV}{1.6022\times10^{-19}\,\text{J}}\right) = 5.13\times10^{-3}\,eV \quad (7.19)$$

Solution (b)

The lowest energy of an electron in a similar box is

$$E_{1,e} = \frac{h^2}{8m_eL^2} = \frac{(6.626\times10^{-34}\,\text{J}\cdot\text{s})^2}{8(9.11\times10^{-31}\,\text{kg})(2\times10^{-10}\,\text{m})^2} = 1.506\times10^{-18}\,\text{J}\left(\frac{eV}{1.6022\times10^{-19}\,\text{J}}\right) = 9.40\times10^{-3}\,eV$$

Note that the electron energy is much larger than the proton energy because of the difference in the mass: $m_p \approx 2000\ m_e$.

the box. This reflects the physical condition that increasing the energy of the particle in the box increases its kinetic energy. The average kinetic energy of the particle for any of the allowed total average energy is just the difference between the horizontal line and the floor of the box.

The curvature of a function is just the *change in slope* of the function as one moves from left to right through any point. Since the slope of a function is expressed as its first derivative, the curvature must be the second derivative. That means if we represent the graph of the wave function by a function $\psi(x)$, then the curvature of the graph is given at any point x by

$$\frac{d^2}{dx^2}\psi(x) \quad (7.20)$$

and the relative magnitude of the curvature (the magnitude of the curvature divided by the magnitude of the wave function) is

$$\frac{d^2}{dx^2}\psi(x) \Big/ \psi(x) \quad (7.21)$$

The expression gives the kinetic energy of the particle if it happens to be at the point x. To get the average kinetic energy of the particle we need to multiply by the probability that the particle actually is at the point that we are measuring (we already know that the probability is given by the square of the wave function).

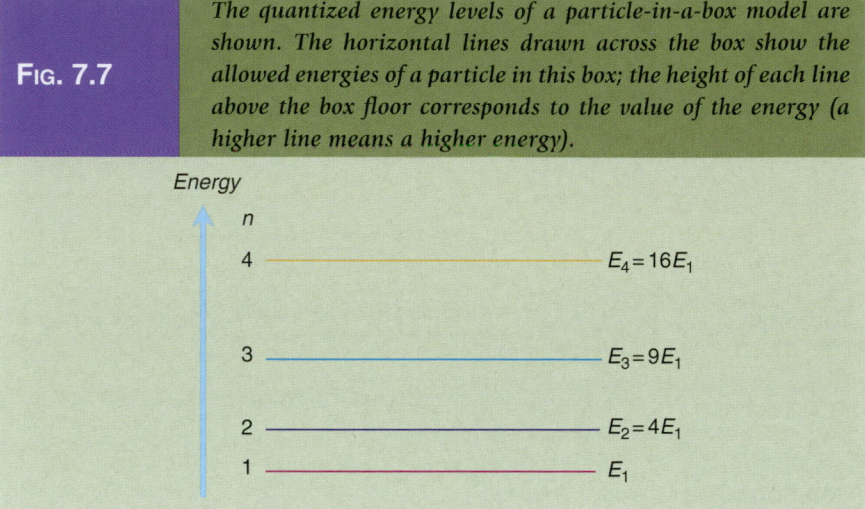

FIG. 7.7 *The quantized energy levels of a particle-in-a-box model are shown. The horizontal lines drawn across the box show the allowed energies of a particle in this box; the height of each line above the box floor corresponds to the value of the energy (a higher line means a higher energy).*

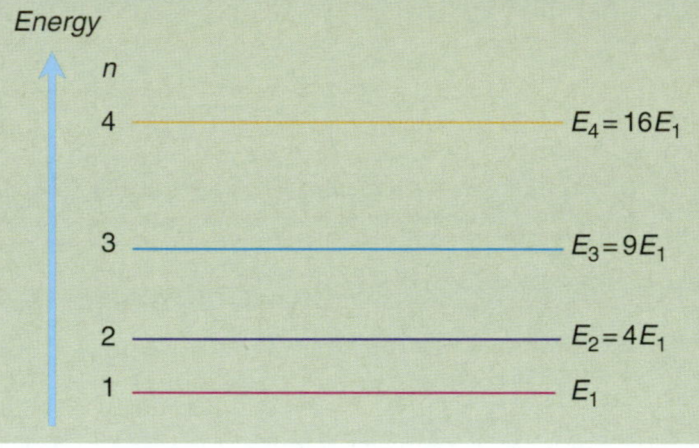

Hence, the average kinetic energy of the particle is given by multiplying the preceding expression by $\psi^2(x)$, like

$$KE(x) = \frac{1}{2}\psi(x)\frac{d^2}{dx^2}\psi(x) \tag{7.22}$$

7.1.2 Two-Dimensional Quantum Systems

The allowed energy levels inside the box are quantized and are shown in **Figure 7.7.** If the particle is electrically charged, it can emit a photon when it jumps from a higher state to a lower state. It can also jump from a lower to a higher state by absorbing a photon. The horizontal lines drawn across the box show the allowed energies of a particle in this box; the height of each line above the box floor corresponds to the value of the energy (a higher line means a higher energy). The lines on the figure show the wave functions that correspond to each allowed energy state.

The particle-in-a-box problem is extended from one dimension into higher spatial dimensions (e.g., two and three dimensions). Since the potential inside the box is always zero, the Schrödinger equation can be easily separated into two or three separate equations dependent on the spatial coordinates x, y, and z that have the exact same form as for the equation in the one-dimensional space. The solutions are the same for each individual dimension, and the total wave function is then a product of the individual wave functions, while the energy is a sum of the corresponding individual energies.

$$E_{n_1,n_2,n_3} = \frac{h^2}{8m}\left(\frac{n_1^2}{a_x^2} + \frac{n_2^2}{a_y^2} + \frac{n_3^2}{a_z^2}\right) \tag{7.23}$$

Because the walls are impenetrable for $x \le 0$ and $x > L$, the wave function bouncing between the walls is $\psi(x) = 0$. The particle is never found outside the

Example 7.5 *A Special Two-Dimensional Case*

Consider a rectangular infinite square well. The potential is zero inside the rectangle of dimension L_x by L_y. The potential is infinite outside the rectangle (**Fig. 7.8**). How do you go about deriving an expression for the energy of a particle in this well?

Fig. 7.8 *The classically allowed region exists inside this rectangular two-dimensional box. The same principles apply in this case except that now the particle is free to move in two dimensions.*

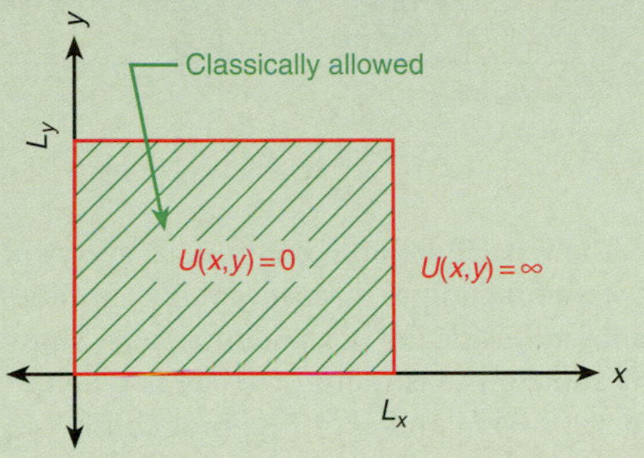

As before, we establish a boundary condition that y be zero in the infinitely disallowed region outside the rectangle. Inside the rectangle, where the potential is zero, we have the Schrödinger equation

$$-\frac{\hbar^2}{2m}\left(\frac{\partial^2}{\partial x^2}+\frac{\partial^2}{\partial y^2}\right)\psi(x,y)=E\psi(x,y) \tag{7.24}$$

Since the Hamiltonian is the sum of two terms with totally independent variables, we try a product wave function $\psi = X(x)Y(y)$ where

$$-\frac{\hbar^2}{2m}\left(\frac{\partial^2 X(x)}{\partial x^2}\right)=E_xX(x) \text{ and } -\frac{\hbar^2}{2m}\left(\frac{\partial^2 Y(y)}{\partial x^2}\right)=E_yY(y) \text{ and } E_{Total}=E_x+E_y \tag{7.25}$$

The following differential equations have the usual solution:

$$X(x)=A_x\sin(k_xx)+B_x\cos(k_xx);\quad E_x=-\frac{\hbar^2k_x^2}{2m} \text{ and } Y(y)=A_y\sin(k_yy)+B_y\cos(k_yy);\quad E_x=-\frac{\hbar^2k_y^2}{2m} \tag{7.26}$$

Application of boundary conditions gives

$$\begin{aligned}\psi(x=0)=0 &\Rightarrow B_x=0 \qquad y(x=L_x)=0\Rightarrow k_xL_x=n_x\pi\\ \psi(y=0)=0 &\Rightarrow B_y=0 \qquad y(y=L_y)=0\Rightarrow k_yL_y=n_y\pi\end{aligned} \tag{7.27}$$

The resulting normalized wave functions are

$$\psi(x,y)=\frac{2}{\sqrt{L_xL_y}}\sin(k_xx)\sin(k_yy);\ k_x=\left(\frac{\pi n_x}{L_x}\right) \text{ and } k_y=\left(\frac{\pi n_y}{L_y}\right) \text{ and } E=\frac{\hbar^2\left(k_x^2+k_x^2\right)}{2m}=\frac{\hbar^2\pi^2}{2m}\left(\frac{n_x^2}{L_x^2}+\frac{n_y^2}{L_y^2}\right) \tag{7.28}$$

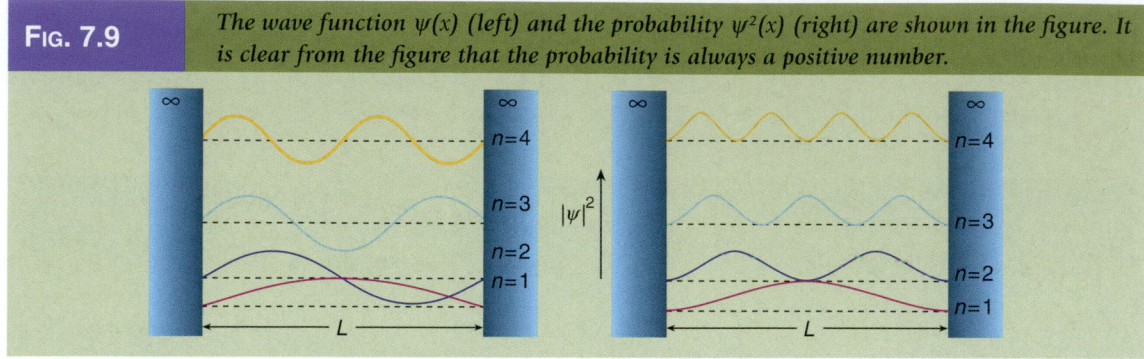

FIG. 7.9 *The wave function $\psi(x)$ (left) and the probability $\psi^2(x)$ (right) are shown in the figure. It is clear from the figure that the probability is always a positive number.*

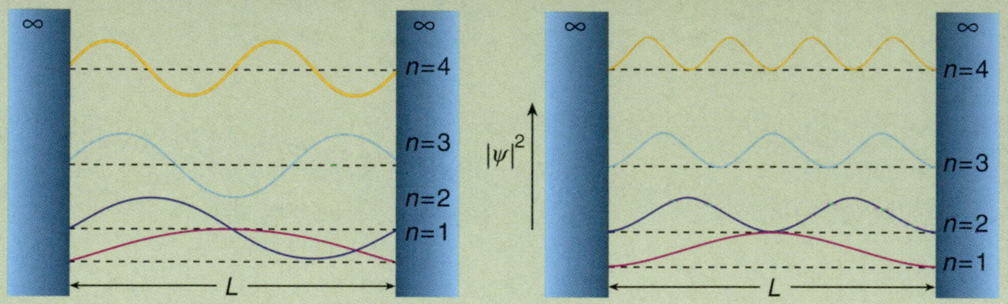

box. The wave function is continuous everywhere—that is, for $\psi(x) = \psi(L) = 0$. The wave function that satisfies these conditions will be the only one that will be allowed inside the box and must have sinusoidal form. For $n = 1$, the probability function maximum of $\psi^2(x)$ is when $x = L/2$. For $n = 2$, the maximum occurs at $x = L/4$ and $3L/4$. The probability is zero for $x = L/2$. For the first three stationary states for the particle in a box ($n = 1,2,3, \ldots$), the wave function will be shown in **Figure 7.9** (left) while the probability for finding the particle in the box is shown in **Figure 7.9** (right).

7.1.3 Schrödinger Equation

The complete average energy comprises the sum of the average kinetic and average potential energies of the particle at some point x. If the potential energy is V(x), then the average potential energy is $V(x)$ by the probability that the particle is at the point x:

$$V(x)\psi^2(x) \tag{7.29}$$

The total average energy including the kinetic energy of a particle at x is given by

$$\frac{1}{2}\psi(x)\frac{d^2}{dx^2}\psi(x) + V(x)\psi^2(x) \tag{7.30}$$

To find the all the energy at all x, we integrate over all x. E is

$$E = \int dx\left[\frac{1}{2}\psi(x)\frac{d^2}{dx^2}\psi(x) + V(x)\psi^2(x)\right] \tag{7.31}$$

This is a form of energy in terms of the wave function—the *variational principle*. It is also considered to be the second law of thermodynamics in disguise in that the wave function will adjust itself until the lowest possible energy minimum is achieved. This equation gives the relationship between the curvature of the wave function at any point x and the potential energy at that point. Another familiar form of the *time-independent* Schrödinger equation is

$$-\frac{\hbar^2}{2m}\frac{d^2}{dx^2}\psi(x)+V(x)\psi(x)=E\psi(x) \tag{7.32}$$

$$H\psi_n = E_n\psi_n \text{ and that } H\,|\psi_n\rangle = E_n\,|\psi_n\rangle \tag{7.33}$$

where H is a mathematical operator called the *Hamiltonian,* $|\psi_n\rangle$ are a set of *eigenstates,* and E_n is the *eigenvalue* of the Hamiltonian.

Boundary Conditions and Quantization. The solution of the Schrödinger equation is quite straightforward if we know where the value of the wave function is known. We know, for example, that the probability of a particle being inside the left wall of the box is equal to zero. Then we pick any slope for the wave function and use the Schrödinger equation to calculate the curvature. At this point, we find how the curvature behaves. We then calculate the slope and subsequently the wave function.

With Schrödinger's equation we can generate a wave function for any energy. In order to ensure that quantization of the energy occurs, the wave function must attain a certain value at the *boundary condition* that is forbidding of all but specific values of the energy. It is to these boundary conditions that the wave function owes its quantum nature, just as the walls of the bathtub enforce the quantization of waves within its porcelain boundaries (the open sea would not). In the case of the square well, our boundary condition is that the wave function will be zero as soon as it reaches the right-hand wall. If we require this, it turns out we can only choose certain specific values for the initial slope and curvature in slope of the wave function at the left-hand wall. This, in turn, means we can only use wave functions corresponding to certain energies.

Quantization of the Energy! Classically Forbidden Zones. When you raise the energy of the particle in a box high enough, the top energy level exceeds that of the allowed energy levels. Then a particle having this allowed total energy is forbidden in classical (nonquantum) mechanics from being inside the box since the potential energy at a point in this region is greater than the allowed total energy at this point. Hence, the particle, if it were here, would have to have a negative kinetic energy. This makes no sense classically because we cannot have kinetic energy less than that implied by zero motion (which is zero). But, from the quantum mechanical perspective, this makes good sense! Quantum mechanics only requires that the total average energy summed up over all possible positions of the particle be equal to the total average energy and hence it is a possibility to have a negative kinetic energy at some point as long as it is balanced by a positive contribution somewhere else. Thus, the quantum particle is allowed to visit *classically forbidden* regions of the box! The wave function in this region will, however, be greatly damped and more so if the level of the particle energy is raised further, indicating that the probability of the particle being in a classically forbidden region is greatly reduced.

This gives rise to the famous phenomena of *quantum tunneling*. Suppose that a particle is trapped in a well that has well-defined boundaries. If the classically forbidden region has finite width and there is a classically allowed region on the other side (as there is in this system, for example), then a particle trapped in the first allowed region can be found on the other side, having apparently traversed a region of space in which it was *not allowed*. This is called *tunneling*, so named because one imagines the particle tunneling through the barrier, instead of going over the top; it does not have enough energy to get over the top.

Erwin Schrödinger expressed the hypothesis of particle–wave nature in 1926. The continuous nonzero nature of its solution, the wave function which represents an electron or particle, implies an ability to penetrate classically forbidden regions and a probability of tunneling from one classically allowed region to another. In 1928, Fowler and Nordheim explained the main features of electron emission from cold metals by high external electric fields on the basis of tunneling through a triangular potential barrier. Conclusive experimental evidence for tunneling was, however, found in the 1960s by L. Esaki in 1957 and by I. Giaever in 1960. Esaki's tunnel diode had a large impact on the physics of semiconductors that led to important developments such as the tunneling spectroscopy and to increased understanding of tunneling phenomena in solids. They won a Nobel Prize along with B. Josephson in 1973.

With the ability to fabricate horizontal structures lithographically with dimensions of 50–100 nm, as well as the self-organization of different structures, it becomes possible to provide confinement in any or all three dimensions, leading to a variety of quantum plane, wire, and dot configurations. Because of the quantum effects, the number of electrons in the well of a quantum dot is quantized to an integer number of electrons, and even if the number is as large as tens to hundreds, single-electron changes can be observed. Combining quantum dots with very thin insulating layers, single-electron tunneling transistors have been implemented, and the possibility of single-electron logic is on the horizon. Research in this promising discipline has been active since its formulation in the mid-1980s. Even quantum molecules—that is, several closely spaced quantum dots interacting via quantum mechanical tunneling—have been explored.

7.1.4 Bohr Exciton Radius

A zero-dimensional material, called a quantum dot, is a material in which the crystallite size is negligible in all three spatial dimensions. The motion of electrons and holes, therefore, is also restricted in all three spatial dimensions. Thus, a quantum dot can be defined as a structure where all dimensions are comparable to the *Bohr exciton radius* (**Fig. 7.10**). The Bohr exciton radius is the distance between an electron in the conduction band and the hole it left behind in the valence band. As motion of electrons is restricted in all directions (like a particle in a box), it is called quantum confinement, and electrons exist exclusively in quantized levels of energy. A quantum dot is like a custom designed atom with a countable number of electrons. It is a semiconductor particle of usually 2–8 nm in diameter. The energy states of the electrons and holes in quantum dots are discrete and given by

$$E_n = \frac{3\hbar^2 n^2}{2m^* a^2}; \qquad \text{where } n = 1, 2, 3, \ldots \tag{7.34}$$

where m^* is the effective mass of the electron (a is analogous to L for the linear box described previously).

Fig. 7.10 *The Bohr exciton radius is shown in the figure. When an excited electron is promoted to a higher energy level in the conduction band, a hole is created in the valence band. In semiconductors, the bandgap is the region between the conduction and valence bands. The Fermi energy corresponds to a region inside the bandgap. The Fermi surface is the surface that separates the occupied level from the unoccupied level in metals. Technically, there are no states inside the bandgap of intrinsic semiconductors. Therefore, a semiconductor does not technically have a Fermi surface (or a Fermi energy). It is more accurate to refer to the "Fermi energy" of a semiconductor as its chemical potential.*

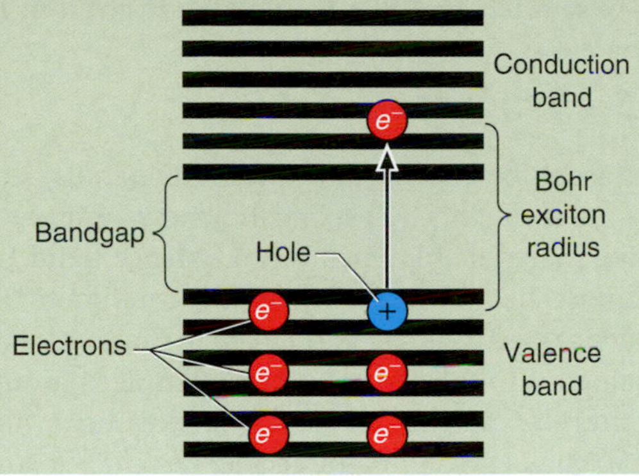

Source: N. W. Ashcroft and N. D. Mermin, *Solid State Physics*, Saunders College Publishing, Harcourt Brace College Publishers, New York, 1976. With permission.

The wave functions of the electrons and holes vanish on the particle surface, and the particle acts as a potential well with infinitely high walls. Each energy level is then associated with an atomic orbital. Two electrons can have the same energy in the orbital at the same time if they have different spins, but no more than two electrons (spin up and spin down) are able to occupy an energy level at the same time (*Pauli exclusion principle*).

Quantum confinement occurs as the semiconductor particle size, a, is reduced to that of the excitonic Bohr radius, which is the radius of the sphere defined by the three-dimensional separation of the electron–hole pair. The excitonic Bohr radius is given by

$$a_{exciton} = a_o \left(\frac{\varepsilon}{m^*_{exciton}} \right) \tag{7.35}$$

where ε is the dielectric constant of the semiconductor, a_o is the Bohr radius, and $m^*_{exciton}$ is the effective exciton mass given by

$$m^*_{exciton} = \frac{m^*_{electron} + m^*_{hole}}{m^*_{electron} m^*_{hole}} \tag{7.36}$$

The Bohr radius should not be confused with the Bohr exciton radius. The Bohr radius is the distance between an electron and its nucleus in the lowest energy level (smallest possible orbit) of the hydrogen atom and is given by

$$a_o = \frac{4\pi\varepsilon_o \hbar^2}{m_e e^2} = 5.29\times10^{-11}\,\text{m} = 0.53\,\text{nm} \tag{7.37}$$

TABLE 7.2 *Bohr Exciton Radii of Selected Semiconductor Materials*

Material	Si	Ge	GaAs	GaP	InP	CdSe	InSb
$a_{exciton}$ (nm)	4.9	17.7	14	1.7	9.5	3	69

where ε_o is the permittivity of free space, m_e is the electron rest mass, e is the elementary charge, and $\hbar$ is the reduced form of Planck's constant. The Bohr exciton radii of selected crystalline materials are given in **Table 7.2**.

7.1.5 Bandgaps

There are three general categories of materials based on their electrical properties: (1) conductors, (2) semiconductors, and (3) insulators (**Fig. 7.11**). In conducting materials like metals, the valence band (consisting of electrons in the outermost shell of the atom) and the conduction band overlap. This means that all valence electrons can participate in the conduction process. The valence band is analogous to the highest occupied molecular orbital (the HOMO) in a molecule, and the conduction band is analogous to the highest occupied molecular orbital (the LUMO) in a molecule, except, of course, that there are many more in the solid. The energy separation between the valence and conduction band is called E_g. The ability to fill the conduction band with electrons and the energy of the bandgap determine whether or not a material is a conductor, a semiconductor, or an insulator. In metals, thermal energy is enough to stimulate electrons to move into the conduction band.

The bandgap in semiconductors is a few electron volts. If an applied voltage exceeds the bandgap energy, electrons jump from the valence band to the conduction band, thereby forming electron–hole pairs called *excitons*. Insulators have large bandgaps that require an enormous amount of voltage to overcome the threshold. These materials therefore do not conduct electricity.

FIG. 7.11 *The picture illustrates the relative positions of the valence and conduction bands in the three kinds of materials: left: an insulator; middle: a semiconductor; and right: a conductor.*

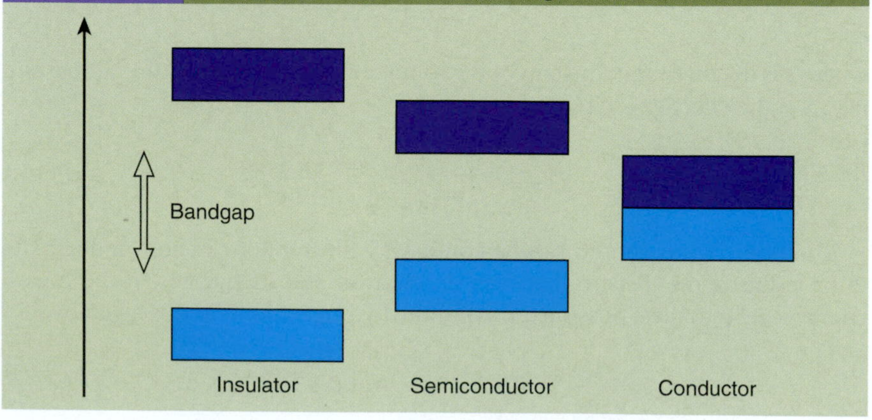

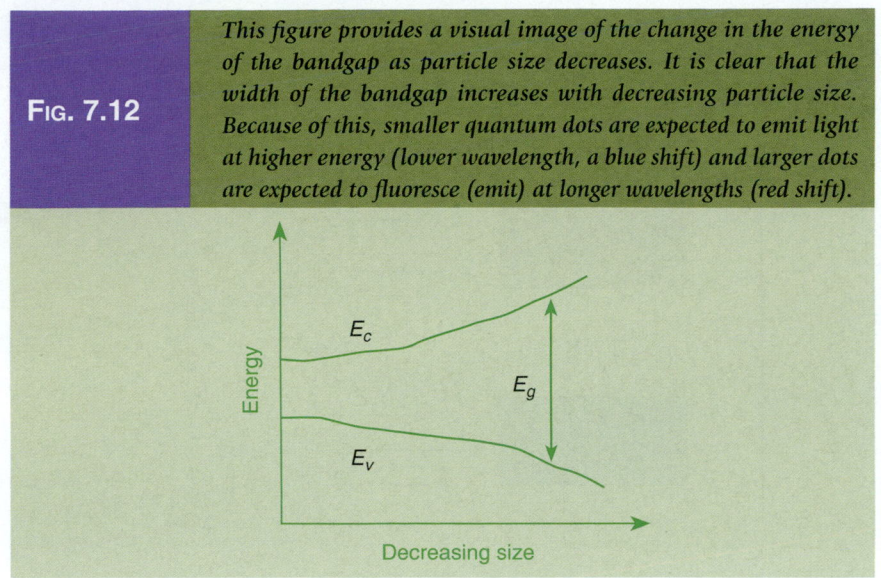

Fig. 7.12 *This figure provides a visual image of the change in the energy of the bandgap as particle size decreases. It is clear that the width of the bandgap increases with decreasing particle size. Because of this, smaller quantum dots are expected to emit light at higher energy (lower wavelength, a blue shift) and larger dots are expected to fluoresce (emit) at longer wavelengths (red shift).*

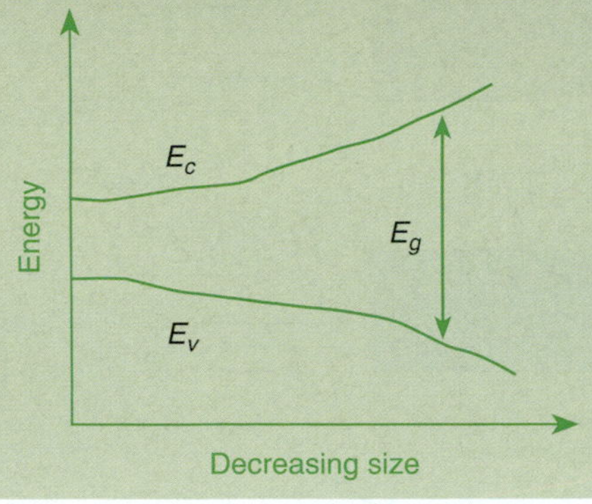

The increase in the effective bandgap energy is caused by quantum confinement effects; the photon energy is regarded as the sum of the bulk bandgap and the quantum confinement energy. At *very small dimensions* when energy levels are quantified, band overlap disappears in metals during transformation into a bandgap (**Fig. 7.12**). Normally, in the case of semiconductors, the bandgap increases with smaller dimensions due to quantization. Some states disappear from the bulk material entirely, including the highest levels of the valence band and the lowest of the conduction band. In the case of a metal, the widening of the gap actually causes the overlap to become a gap. This means that quantum dots can be made from materials that are either semiconductors or conductors as bulk materials.

The energy of the bandgap of a nanoparticle differs from the bulk by

$$E_{g,nanocrystal} = E_{g,bulk} + \Delta E_g = E_{g,bulk} + \frac{h^2\pi^2}{2mr^2_{nanocrystal}} \tag{7.38}$$

As a particle becomes smaller, its bandgap increases. The relationship of shrinking size to the bandgap is shown graphically in **Figure 7.13** from semiconductor to a quantum dot to the atom.

Following absorption of energy (voltage or light source), excited electrons are able to jump from the valence band to the conduction band. If the energy of the light is equal to or higher than the bandgap, absorption occurs and an electron–hole pair is generated. When recombination of electrons and holes happens, energy is emitted. In some cases, energy is emitted in the form of phonons (vibrations in the material causing temperature to increase—also called nonradiative emission) or in the form of photons (electromagnetic waves—also called radiative emission). Some semiconducting materials have a direct bandgap (e.g., no states exist in the gap). Other materials have an indirect bandgap (e.g., one or more states exist in the bandgap).

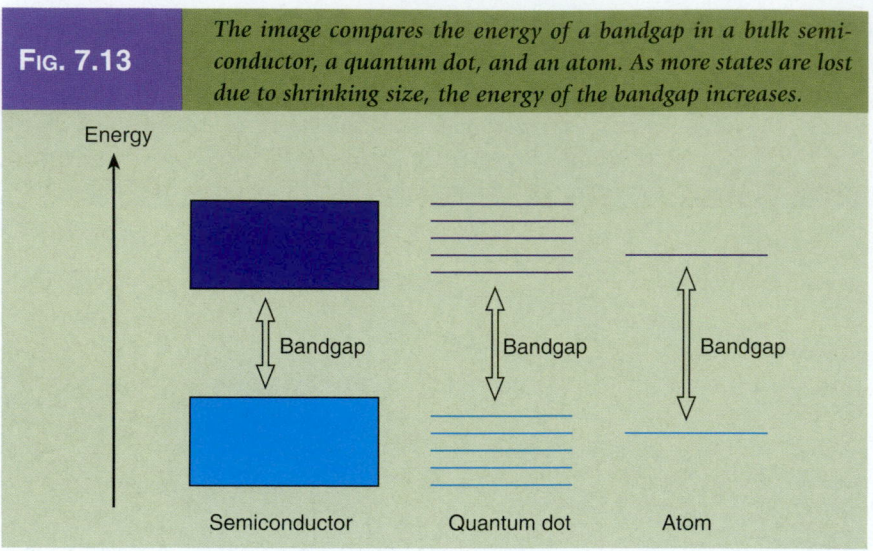

FIG. 7.13 *The image compares the energy of a bandgap in a bulk semiconductor, a quantum dot, and an atom. As more states are lost due to shrinking size, the energy of the bandgap increases.*

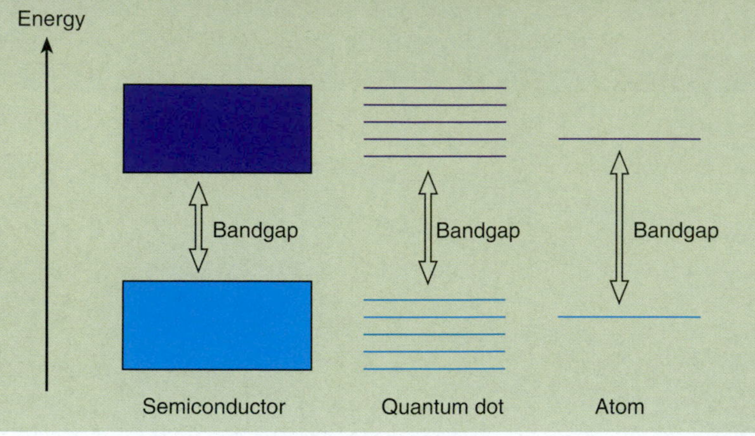

In order for an electron to be excited in an indirect bandgap material, the presence of both a photon and a phonon is essential because intermediate states, in addition to different energy from the valence band, also have different momentum. These materials can simultaneously emit both nonradiative and radiative emissions. One of the emissions is from the transition between the conduction band and the intermediate state and the other emission is from the transition between the intermediate state and the valence band. If light is absorbed and then electron–hole transition causes emission of another wavelength of light, the process is referred to as photoluminescence.

If the bandgap increases due to quantum confinement, the wavelength of light needs to be of higher energy (shorter wavelength) in order to be absorbed by the bandgap of the material. The wavelength of the fluorescent light emitted from the material also changes in the same manner (e.g., a blue shift in fluorescence is observed for smaller quantum dots). We therefore have a method of tuning the optical absorption and emission properties of a semiconductor over a range of wavelengths by control of the crystallite size. If the semiconductor in question is undoped and is a direct bandgap material, then both absorption and luminescence occur at the energy band edge. We can also introduce trap states within the bandgap to act as luminescent centers. A shallow trap can be described as a mobile charge orbiting the trap in a large 1*s* orbital, resulting in a susceptibility to quantum confinement effects. Deep traps have highly localized orbiting charges and are therefore relatively immune to confinement effects.

Semiconductor Quantum Dots. The energy gap of a quantum dot depends on the size of the dot. The larger the size is, the lower is its absorption and fluorescence energy (red shift). The smaller the dot is, the higher is its absorption and fluorescence energy (blue shift). In addition to quantum size effects, doping nanoparticles with different materials is a means of controlling the dimensions of the bandgap of nanoparticle quantum dots. Doping introduces altered energy levels and defect sites within the bandgap of the semiconductor that can serve as electron "resting sites" during a transition. Therefore, semiconductors that normally absorb only UV light are now able to emit visible light. ZnS quantum dots, when doped with manganese, absorb in the UV range and emit orange light. In another study, Ge quantum dots on Si(100) showed narrow band luminescence as a function of dot size and proximity [3].

7.2 Zero-Dimensional Materials

A zero-dimensional materials exhibits quantum confinement in all three spatial dimensions—a quantum sphere as it were or, more correctly, a quantum dot (**Fig. 7.14**, left). In other words, no matter which direction one views the spherical nanoparticle, the electronic behavior along that dimension is quantized. In **Figure 7.14** (right), images acquired by TEM show gold nanoparticles.

7.2.1 *Clusters*

Bulk matter is made of atoms and molecules, and as such, they have been widely classified and their properties satisfactorily explained. An ensemble of atoms or molecules, the so-called *clusters*, is far from being well understood. Elemental clusters are held together by various forces depending on the nature of the constituent atoms: (1) inert gas clusters are weakly held together by van der Waals interactions, for example, $(He)_n$; (2) semiconductor clusters are held together by directional covalent bonds; and (3) metallic clusters are held together by fairly strong delocalized nondirectional bonds.

Fig. 7.14 *Left: the electrons in a zero-dimensional material (a quantum dot) are confined in all three dimensions. In other words, quantum properties are expected to be observed in this material regardless of particle orientation. Right: a TEM image of gold nanoparticles on the order of 20-nm diameter. The optical absorption spectra of such particles are shown in Figure 7.16.*

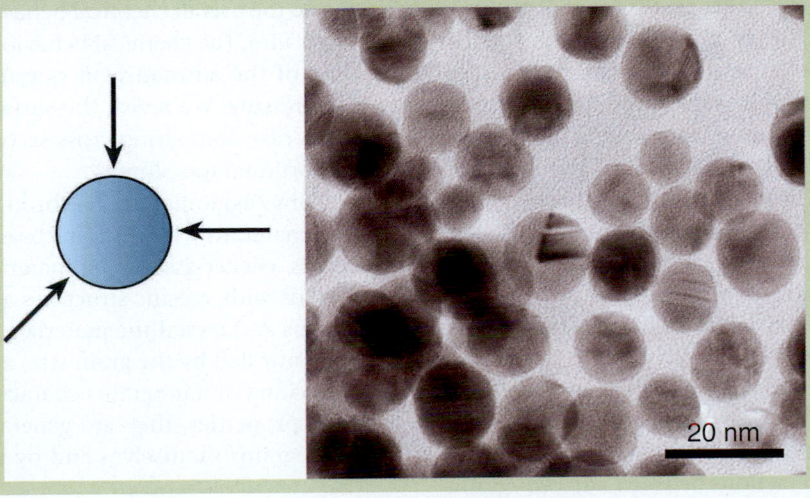

TABLE 7.3 *Cross-Section, Mass, N, and Surface Fraction—Revisited*

Size (nm)	Cross-section (10^{-18} m^2)	Mass (10^{-25} kg)	No. of molecules (N)	Fraction surface molecules (%)
0.5	0.2	0.65	1	
1	0.8	5.2	8	100
2	3.2	42	64	90
5	20	650	1,000	50
10	80	5,200	8,000	25
20	320	42,000	64,000	12

Elemental clusters, or a mixture of clusters of different elements, constitute the vast expanding field of materials sciences called *nanomaterials*. One has to be clear right away that clusters are not a fifth state of matter, as some may believe, but rather are simply intermediate between atoms on the one hand, and solid or liquid state of matter on the other, with widely varying physical and chemical properties. The number of atoms forming the cluster determines the percentage of atoms that are exposed on the surface of the cluster, with decreasing number of surface atoms with increasing size of the cluster, as we discovered in chapter 5. Some make a distinction among molecular clusters, nanoparticles, and nanomaterials. Physicists working with molecular beams have extensively studied agglomerations of a few atoms. Today, the mystery related to larger ensembles of atoms (in other words, *nanomaterials*) is fading due to active research being carried out across the world over the last few decades.

In **Table 7.3,** we see that smaller particles contain only a few atoms that are practically all at the surface. As the particle size increases from 1 to 10 nm, cross-section of the cluster increases by a factor of 100× and the mass of molecules by a factor of 1000×; concomitantly, the proportion of molecules at the surface falls from 100% to just 25%. For particles of 20-nm size, a little more than 10% of the atoms are on the surface. Of course, this is an idealized hypothetical case. If particles are formed by macromolecules (that are larger than the present examples), the number of molecules per particle will substantially decrease and their surface fraction increase accordingly. The electronic properties of such ensembles of atoms or molecules are the result of mutual interactions, so the overall chemical behavior is entirely different from individual atoms or molecules. The chemical behavior is also different from the macroscopic bulk state of the substance in question under the same conditions of temperature and pressure. We revisit the surface-to-volume relationships of chapter 5, but in this case comparing cross-section (which is proportional to r^2) to kilograms (proportional to volume).

The idea of tailoring properly designed atoms into agglomerates has brought in new fundamental work in the search for novel materials with uncharacteristic properties. Among various types of nanomaterials, cluster-assembled materials represent an original class of nanostructured solids with specific structures and properties based on structures between amorphous and crystalline materials. In fact, in such materials, the short-range order is controlled by the grain size, and no long-range order exists due to the random stacking of nanograins characteristic of cluster-assembled materials. In terms of properties, they are generally controlled by the intrinsic properties of the nanograins themselves and by the interactions between adjacent grains.

Cluster-assembled films are formed by the deposition of clusters onto a solid substrate and are generally highly porous with densities as low as about one half of the corresponding bulk materials densities. Both the characteristic nanostructured morphology and a possible memory effect of the original free cluster structures are at the origin of their specific properties. From recent developments in the cluster source technologies (thermal, laser vaporization, and sputtering) [4,5], it is now possible to produce intense cluster beams of any materials, even the most refractory or complex systems (bimetallic, oxides, and so on) for a wide range of size from a few atoms to a few thousands of atoms.

Cluster Particle in a Box. The de Broglie wavelength of an electron is found from de Broglie's relationship. The electron possesses a characteristic wavelength due to the particle duality exhibited by all matter. If a cluster's size is on the order of this wavelength, quantization of the energy of the electrons is likely. **Figure 7.15** illustrates the particle-in-a-box analogy for clusters. A ligand shell

FIG. 7.15 *The n = 1 and n = 2 levels are shown in a ligand-stabilized gold cluster. The cluster is the "box" and the ligands are the "walls." The de Broglie wavelength of the cluster is ca. 1.4 nm, corresponding to the diameter of the cluster. The complete particle is 2.1 nm in diameter [(2 × 0.35 nm = 0.7 nm) + 1.4 nm = 2.1 nm] [6,7]. If clusters are placed in a row to form a quantum wire, electrons are able to tunnel through the ligand barrier and conduct. In order for this to occur, the wavelengths need to match up constructively between clusters. The length of a two cluster wire is ca. 4.2 nm [6,7].*

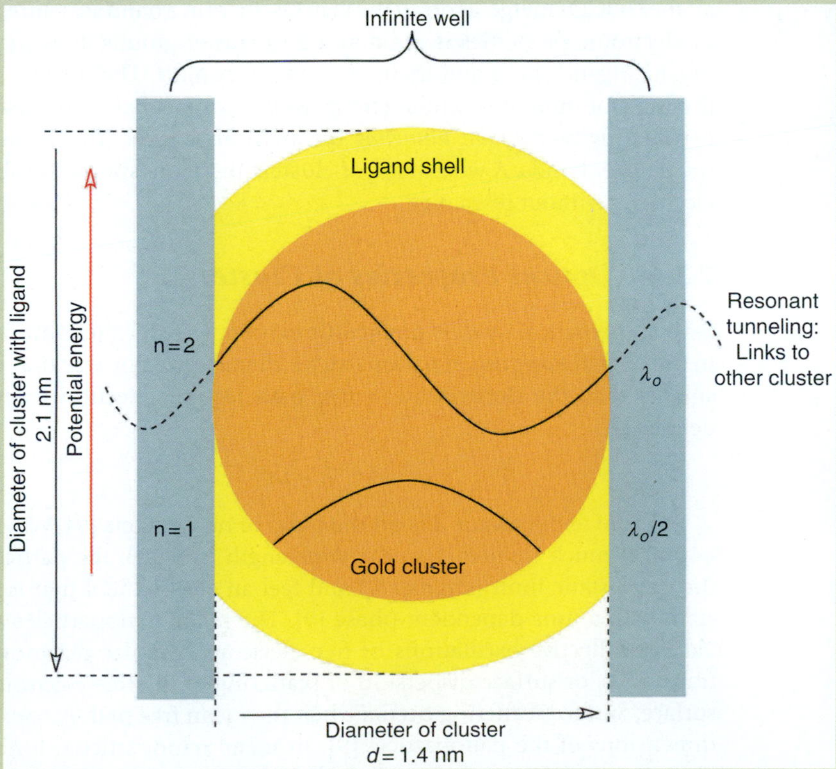

surrounds the cluster and in essence provides a barrier for the electronic wave functions of the cluster. Clusters in contact with one another (via the ligand shells) communicate electronic information by the phenomenon of tunneling (chapter 3).

7.2.2 Metal Clusters and the HOCO–LUCO

HOCO–LUCO. A gold nanoparticle that is 1 μm^3 in volume has 10^{10} atoms and is still considered to be a bulk material [6]. In bulk metals, energy levels are numerous and essentially continuous. If the diameter of the particle is reduced to some integral multiple of the de Brogile wavelength, λ_o, of an electron in the ground state of that material, quantum effects begin to emerge. If the size is reduced further to integral multiples of $\lambda_o/2$ (half the de Broglie wavelength), then electron energy levels in clusters become quantized (discrete). The ground state of the ligand-stabilized Au_{55} cluster has two electrons; the diameter of the cluster, $d_{cluster}$, is equal to $\lambda_o/2$ = 1.4 nm and the diameter of the cluster plus the ligand sphere is equal to ~2.1 nm (the walls of the infinite well, 0.35 nm thick) [6]. In the quantized electronic state, very small metal particles lose their ability to conduct electrons due to the existence of clearly defined highest occupied molecular orbitals (HOMOs) and the lowest unoccupied molecular orbitals (LUMOs) [7]. The HOMO in a cluster is aptly renamed the HOCO—the highest occupied cluster orbital. Concomitantly, the cluster LUMO is renamed the LUCO [7].

If naked clusters (on the order of $\lambda_o/2$) are in contact with one another, agglomeration occurs, particle size increases, and quantum properties are lost; all the HOCOs merge, as do all the LUCOs, to form a band structure. Quantization of electronic properties is maintained in cluster groups, however, if they have organic ligand shells and then are made to contact. The organic shells serve as the walls of infinite potential energy wells. Contact between clusters, therefore, occurs only by *resonant tunneling* through this barrier. This leads to electronic intercluster bands. A wire made of clusters therefore should be able to conduct electricity without resistance.

7.2.3 Optical Properties of Clusters

Several theoretical models predict the existence of so-called intrinsic size effects in the optical response of nanoparticles characterized by size that is significantly smaller than the electron mean free path [8]. The electron mean free path is defined by

$$\lambda = v\tau \tag{7.39}$$

At room temperature, λ is on the order of nanometers [9]. When particles are <40 nm (much smaller than the wavelength of light), the particles experience the quasi-static limit (chapter 5) and feel an electric field that is spatially constant with a time-dependent phase [9]. The metal nanoparticles absorb energy via the collective oscillations of free electrons (dipolar plasmons), interband transitions, or surface dispersion or scattering of the free electrons on a metals surface. Surface scattering occurs when the mean free path is comparable to the dimensions of the nanoparticle [9]. In metal nanoparticles, limitations of the electron mean free path as well as damping by the chemical interface increase the damping rate of the surface plasmon resonance, leading to shorter dephasing times. Due to these intrinsic size effects, the dielectric permittivity of a nanoparticle differs from that of the bulk metal because of the additional surface damping contributions. Experimental studies on ensembles of metallic nanoparticles revealed the existence of such effects. However, a quantitative description of these effects is difficult to acquire for ensembles, mostly because of inhomogeneous broadening in spectra.

The electronic structure of nanoparticles is unraveled by studying its interaction with electromagnetic radiation, specifically in the range of optical wavelengths. In absorption, the frequency of the incoming light wave is at or near the energy levels of the electrons. After absorption, the energy is (1) remitting by a photon or (2) retained and distributed by nonradiative mechanisms. Apart from absorption of light by objects, scattering of light also takes place. Gustav Mie developed a generalized theory of light scattering by a spherical particle at the turn of the nineteenth century. We discussed the quasi-static limit in chapter 5. The absorption scattering cross-section of a particle is proportional to r^3, while that of the scattering cross-section is proportional to r^6. If particle size is significantly less than the wavelength, then absorption is more important than scattering for very small particles ($2\pi r << \lambda$) [8]. Scattering becomes more important than absorption when the circumference of the particle is comparable to the wavelength of light ($2\pi r \approx \lambda$). Scattering does not occur for very small particles.

Dipolar Plasmon Resonance. In metal nanoparticles, absorption by plasmon resonances arises from the large electron density. The collective particle plasmon mode strongly interacts with optical waves [8]. Plasmon resonance is observed in particles or planar surfaces that are metallic (e.g., silver or gold) nanoparticles or nanolayers that scatter optical light elastically with remarkable efficiency because of a collective resonance of the conduction electrons in the metal. The excitation of surface plasmons by light is denoted as a *surface plasmon resonance* (SPR) for planar surfaces or *localized surface plasmon resonance* (LSPR) for nanometer-sized metallic structures.

The position, width, and amplitude of the surface plasmon wave depend on the nature of the nanoparticles. The dipolar resonance was introduced in chapter 5. In local surface plasmon resonance, incident light strikes the surface of metal nanoparticles to stimulate the conduction electrons. This action creates a dipole that oscillates in phase with the electric field of the light and generates a restoring force in the metal nanoparticle as it tries to compensate for the change. The result is a unique resonance wavelength. The character of the dipolar plasmon is also influenced by the nature of the support material (or the surrounding dielectric medium). The position of the absorption, therefore, depends on particle size and shape, the type of metal, its orientation, and the surrounding medium. The plasmon is, not surprisingly, a quantized entity.

Noble metals of small size have a very strong visible absorption due to the resonant coherent oscillation of the free electrons in the conduction band. The brilliant ruby color in stained glass windows of cathedrals built in the seventeenth century is due to the strong absorption of gold nanoparticles. If the particle is not spherical (e.g. rod shaped), the optical (absorption and fluorescence) and physical properties of these gold nanorods become quite different from those of spherical ones.

In case of gold nanorods that contain nearly 10^5–10^6 atoms, there exists significant polarizability. In-phase excitation of the collection of these electrons with an incident light generates an electric field on the particle surface that is proportional to the dielectric of the metal and its volume. This enhanced field is a new property—one that is absent in the bulk gold.

Figure 7.14 showed gold nanoparticles of diameter of around 20 nm. Depending upon the size of the nanoparticles, the absorption of UV–visible light is different and, as such, gold nanoparticles show different colors. The optical spectra of those gold nanoparticles are shown in **Figure 7.16.**

7.2.4 Other Physical Properties and Phenomena

Surface Energy. The trend in surface energy as a function of size is tabulated in **Table 7.4.** The surface energy of bulk calcite, $CaCO_3$, is 0.23 $J \cdot m^{-2}$ [10]. It is clear from the table that the materials we discussed in chapter 6 are relevant to nanomaterials. The energy of the edges exceeds the surface area.

Another stellar example of the concept of surface energy is given by a 1-g cube of NaCl. Starting with 1 g of material, the parent is subdivided into smaller cubes with increasingly larger surface area. The density of NaCl is 2.17 $g \cdot cm^{-3}$. The average energy of surface atoms is 2×10^{-5} $J \cdot cm^{-2}$ and the average energy for edge molecules is 3×10^{-13} $J \cdot cm^{-1}$. Notice that edge molecules have significantly higher energy than do surface atoms. From **Figure 7.17,** we see that the surface molecular density of NaCl per unit cell is equal to 2 molecules per a_o^2 or 6.27 molecules per square nanometer. The linear density is equivalent to 1 molecule per a_o or 1.77 molecules per nanometer. From this information, the average surface energy per molecule is calculated. One would expect that edge atoms (and corner atoms) have the highest surface energy (**Table 7.5**).

Thermal Properties. One of the most popular illustrations of nanoparticle physical properties is that of the dramatic deviation of melting point from the bulk (**Fig. 7.18**). Recall our previous discussion concerning the coordination number of surface atoms—specifically, that surface atoms are undersaturated with regard to nearest neighbors. Although this physical condition is also true for bulk materials, the ratio of surface atoms is so small compared to the bulk volume that any discussion concerning surface effects, especially when it involves the melting of solids, is inconsequential.

The melting point change is inversely proportional to the radius r of the particle

$$\Delta\theta = \frac{2T_o\gamma}{\rho r(L)} \tag{7.40}$$

where

$\Delta\theta$ is the change in melting point from the bulk temperature
T_o is the melting point of the bulk material in kelvins
γ is the surface tension of the solid–liquid interface in joules per square meter
ρ is the particle density in kilograms per cubic meter
L is the latent heat of fusion in joules per kilogram [13]

Fig. 7.16 *(a) The optical spectra of increasingly larger spherical gold clusters (colloids) are shown. The transmission color of gold solutions is shown at the top left. The absorption spectra correspond to: clockwise from top right: <20 nm diameter, $\lambda_{max} \approx 518$ nm; $\lambda_{max} \approx 530$ nm; and $\lambda_{max} \approx 560$ nm. (b) The variation of size-dependent emission from quantum dots is graphically depicted in the figure. The larger the semiconductor dot is, the lower energy is the emission band. (c) The orange luminescence of manganese-doped zinc-sulfide is shown.*

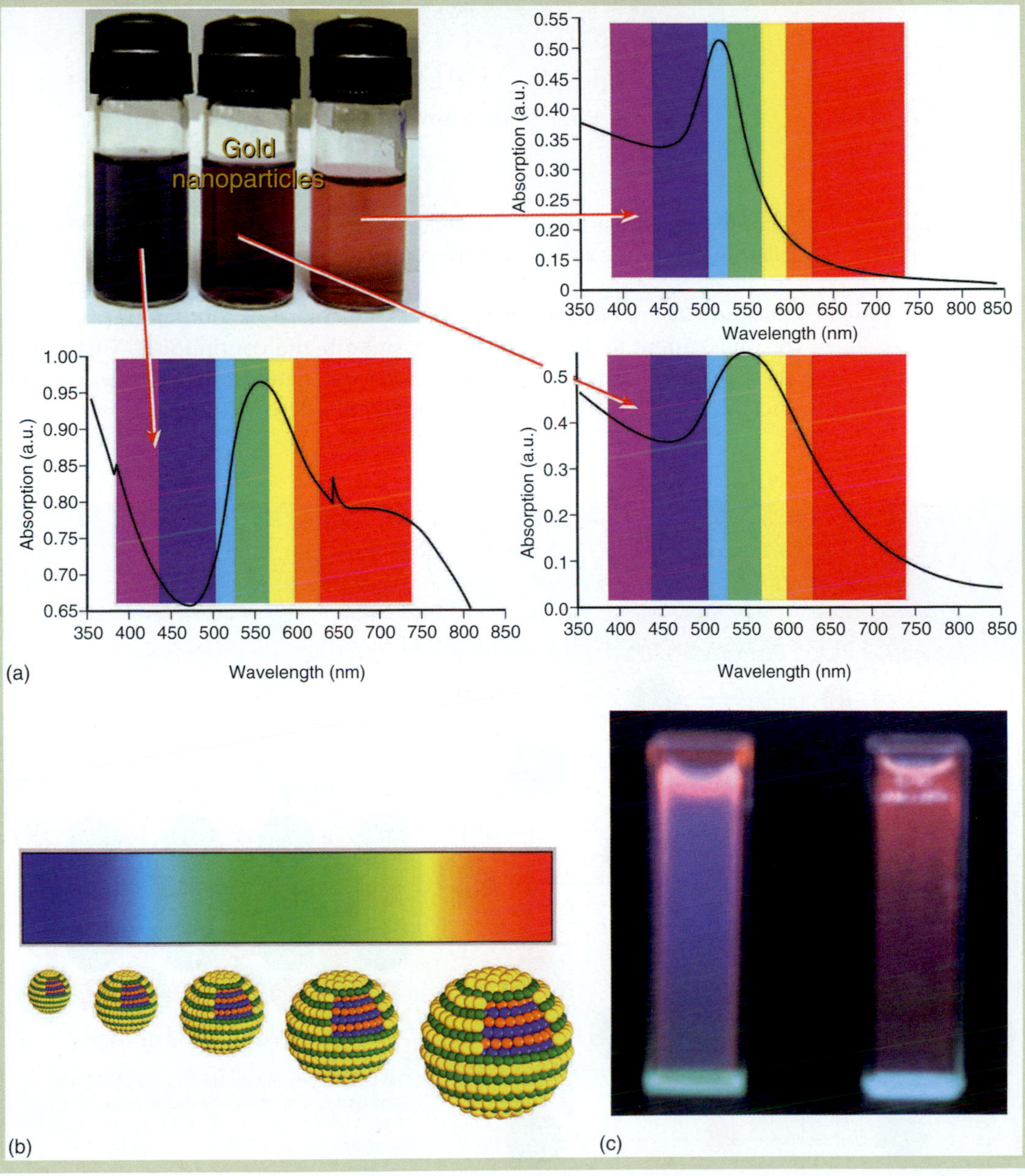

TABLE 7.4 *Surface Energy of Calcite Particles*

Size (nm)	Surface area (m·mol⁻¹)	Surface energy (J·m⁻²)
1	1.11×10^9	2.55×10^4
2	5.07×10^8	1.17×10^4
5	2.21×10^8	5.09×10^3
10	1.11×10^8	2.55×10^3
20	5.07×10^7	1.17×10^3
100	1.11×10^7	2.55×10^2
1000	1.11×10^6	2.55×10

Source: K. Kamiya and S. Sakka, *Gypsum Lime*, 23, 163 (1979).

The smaller the particle is, the greater the range of melting differential can be. Nanomaterials are able to undergo sintering at lower temperatures.

Conductivity. Electrical conductivity in bulk metals is continuous (e.g., the curve relating current to applied voltage is smooth and continuous). This is because there are an enormous number of electronic states in the conduction band of metals (**Fig. 7.19**, left). Electron mobility μ is described by [6]

FIG. 7.17 *The fcc structure of a sodium chloride crystal is depicted. The surface energy depends on the planar face of the molecule that is exposed. The (100) surface and the linear density in the [010] direction are shown. Edge atoms, due to their relative undersaturation of nearest neighbors compared to the surface atoms, should have higher energy. Many ligands attach to edges and corners (vertices) of metal clusters for this reason.*

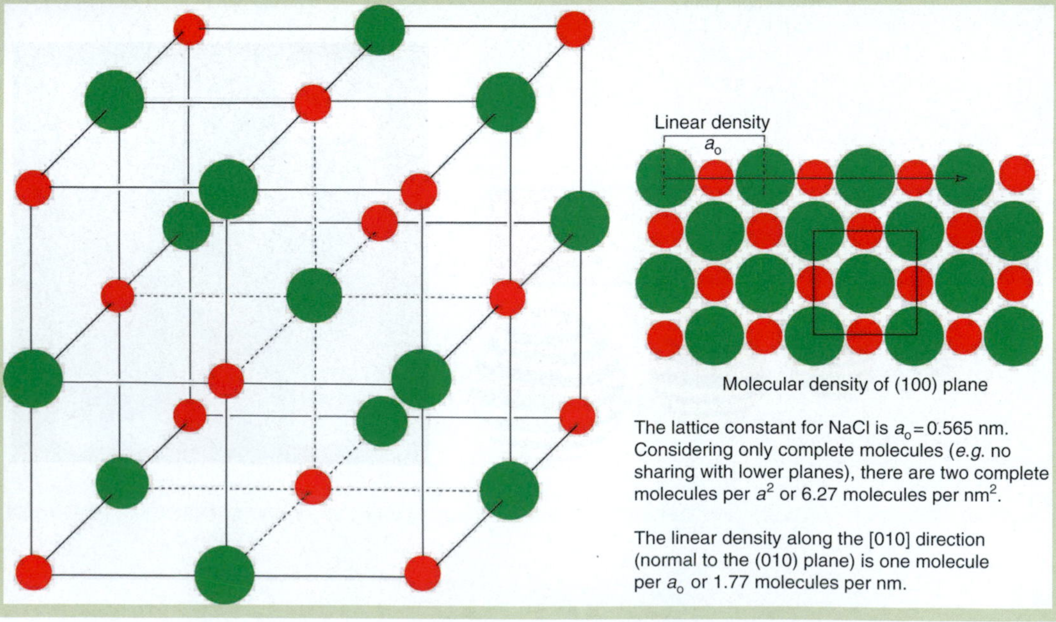
Linear density
a_o
Molecular density of (100) plane
The lattice constant for NaCl is $a_o = 0.565$ nm. Considering only complete molecules (*e.g.* no sharing with lower planes), there are two complete molecules per a^2 or 6.27 molecules per nm^2.
The linear density along the [010] direction (normal to the (010) plane) is one molecule per a_o or 1.77 molecules per nm.

TABLE 7.5 *Surface and Edge Energy*

Size (nm)	Collective surface area (cm^2)	Collective edge (cm)	Surface energy ($J \cdot g^{-1}$)	Edge energy ($J \cdot g^{-1}$)
1 nm	2.8×10^7	5.5×10^{14}	560	170
1 μm	28,000	5.5×10^8	0.56	1.7×10^{-4}
1 mm	2,800	5.5×10^6	0.056	1.7×10^{-6}
1 cm	280	55,000	0.0056	1.7×10^{-8}
1 dm	28	550	0.00056	1.7×10^{-10}
7.7 dm	3.6	9.3	0.000072	1.7×10^{-12}

Sources: G. Cao, *Nanostructures & Nanomaterials: Synthesis, Properties & Applications,* Imperial College Press, London (2004); A. W. Adamson and A. P. Gast, *Physical Chemistry of Surfaces,* 6th ed., John Wiley & Sons, New York (1997).

$$\mu_e = \frac{e\lambda}{4\pi\varepsilon_o m_e \upsilon_F} = \frac{e}{4\pi\varepsilon_o m_e}\tau \qquad (7.41)$$

where

e is the electron charge in coulombs
λ is the mean free path of the electron
ε_o is the permittivity of space
m_e is the effective mass of the electron
υ_F is the Fermi velocity of the electron
τ is equal to λ/υ_F and is the time between collisions

FIG. 7.18 *The melting curve of a generic metal as a function of particle size is shown. Below ca. 10 nm, the melting temperature takes a nosedive. The melting point of gold is given at the top of the graph to provide a reference.*

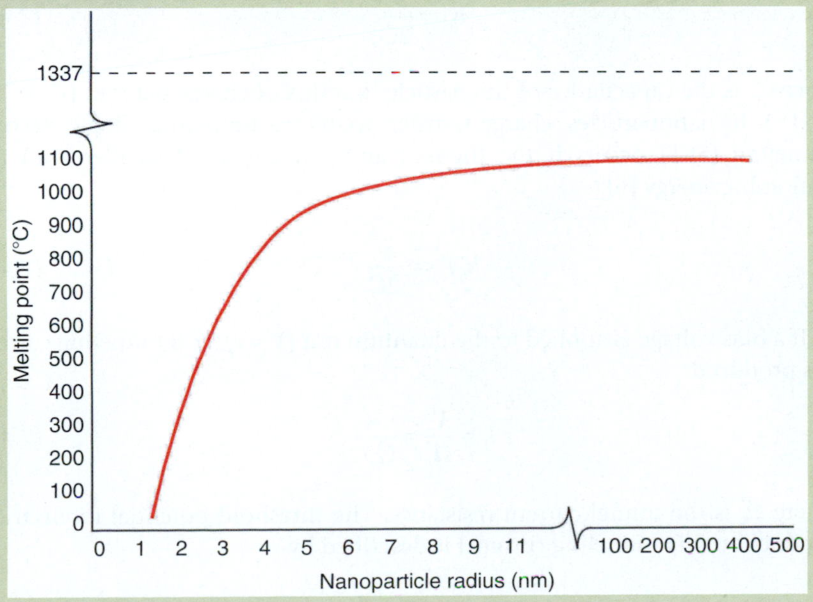

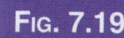

FIG. 7.19 *Generic (and idealistic) current–potential relationships are shown for a bulk metal (left) and a material capable of single-electron transport (right) [6]—like a gold cluster. The steps correspond to jumps in the current after a threshold level of voltage is applied. This image is also known as a Coulomb staircase or a Coulomb blockade.*

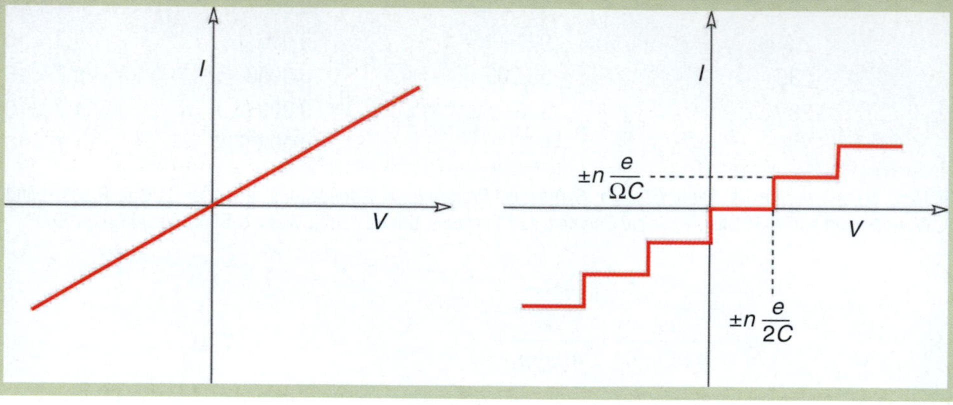

Scattering occurs in metal lattices due to electrons colliding with lattice defects (contaminants, vacancies, surfaces, grain boundaries, dislocations, etc.) and due to phonon vibrations [6]. For Cu, υ is 1.6×10^6 m·s^{-1} and λ is 43 nm and t is 2.7×10^{-14} s [13].

Ohm's law states that electromotive potential is proportional to the product of current and resistance:

$$V = I\Omega \tag{7.42}$$

The magnitude of the coulombic energy E_C involved in electron transfer is

$$E_C = \frac{e^2}{2C} \tag{7.43}$$

where C is the capacitance of the particle in terms of charge per volt ($C \cdot V^{-1}$ or $C^2 \cdot J^{-1}$). In nanoparticles, charge transfer occurs via tunneling. Single-electron tunneling (SET) occurs if the thermal ambient energy is smaller than the coulombic energy [6]:

$$k_B T \ll \frac{e^2}{2C} \tag{7.44}$$

If a bias voltage is applied to the quantum dot ($V = e/C$), a tunneling current I_t is produced:

$$I_t = \frac{V}{\Omega_t} = \frac{e}{\Omega_t C} \tag{7.45}$$

where Ω_t is the tunnel current resistance. The threshold potential to electrical conduction (via tunneling current) is described by

$$V_{Threshold} = \pm\frac{e}{2C} \tag{7.46}$$

This relationship describes the phenomenon known as the *Coulomb blockade* (**Fig. 7.19**, right). The current-voltage transient shows no activity until this threshold is reached, at which point the electron is transferred. The staircase step height

$$\Delta I \propto e/\Omega C \tag{7.47a}$$

and

$$\Delta V \propto e/2C \tag{7.47b}$$

Catalytic Behavior of Nanoparticles. We have mentioned early on in the text how small gold nanoparticles demonstrate enhanced catalytic activity. Gold is not considered to be a catalyst, even in the domain of nanomaterials. Nanoparticles have catalytic properties due to the increased level of surface atoms—atoms that are eager to seek stability. Inherent high curvature is not a desired state and therefore the smaller the particle is, the more reactivity it tends to demonstrate. This does not always mean that smaller particles lead to increasingly faster reaction rates [14,15]. The dissociation rate of CO on Rh island aggregates peaks at 1000 atoms [14,15]. Smaller islands down to 100 atoms show decreased dissociation rates. The reaction rate of hydrogen gas with Fe nanoparticles was highest with 10 atom hydrogen-clusters; 17 atom clusters showed lower reaction rate, as did 7 atom clusters [15]. In another example, the turnover frequency of cyclohexene to cyclohexane was shown to increase with smaller Rh particle size [16].

The challenge involved in application of small catalyst particles is maintaining surface area and to keep agglomeration at a minimum. Ligand-stabilized clusters, although reducing overall surface area (that regulates the spacing of catalyst particles), are able to influence the selectivity in catalytic processes [6]. The semihydrogenation (and subsequent *trans*- vs. *cis*-directing capability) of alkynes by 3- to 4-nm Pd clusters serves as an example. The activity and selectivity of TiO_2-supported Pd clusters stabilized by two different kinds of ligands (1,10-phenanthroline vs. 2-*n*-butylphenanthroline) are strikingly different [6]. Application of the unsubstituted phenanthroline ligand yielded 95% of the *cis*-product after 75 min, but immediately transformed into the *trans*-product, which is not desirable. On the other hand, although the reaction took longer to complete (500 min), the *cis*-product comprised 100% of the product. Conversion to the *trans*-form occurred over a long period of time (ca. 2000 min) [6,17].

7.3 One-Dimensional Materials

One-dimensional materials are those in which crystallite size is negligible in two dimensions but is not restricted in the third direction; that is, electron and hole motion are confined in two spatial directions while free propagation is allowed in the third direction. Perhaps one of the best examples of nanowires is the carbon nanotubes (**Fig. 7.20**). These are also called *quantum wires*. Thus, a quantum wire can be defined as a structure where two dimensions are comparable to the exciton Bohr radius (**Fig. 7.21**).

In other words, a quantum wire is an electrically conducting or semiconducting wire in which quantum effects control the transport properties in two dimensions. Due to the confinement of conduction electrons in the transverse direction of the wire, their transverse energy is quantized into a series of discrete values E_0 ("ground state" energy, with lower value), $E_1, \ldots, E_n$. One consequence

FIG. 7.20 *TEM images of single-walled carbon nanotube bundles and "wires" taken with a Philips CM-30 operated at 200 kV. Left: a global image of a tangled mass of SWNTs is shown. The dark regions of the image correspond to catalyst particles used in their synthesis. Right: A close-up image shows the bundling tendency of the SWNTs. SWNTs are actually known as "pseudoquantum wires."*

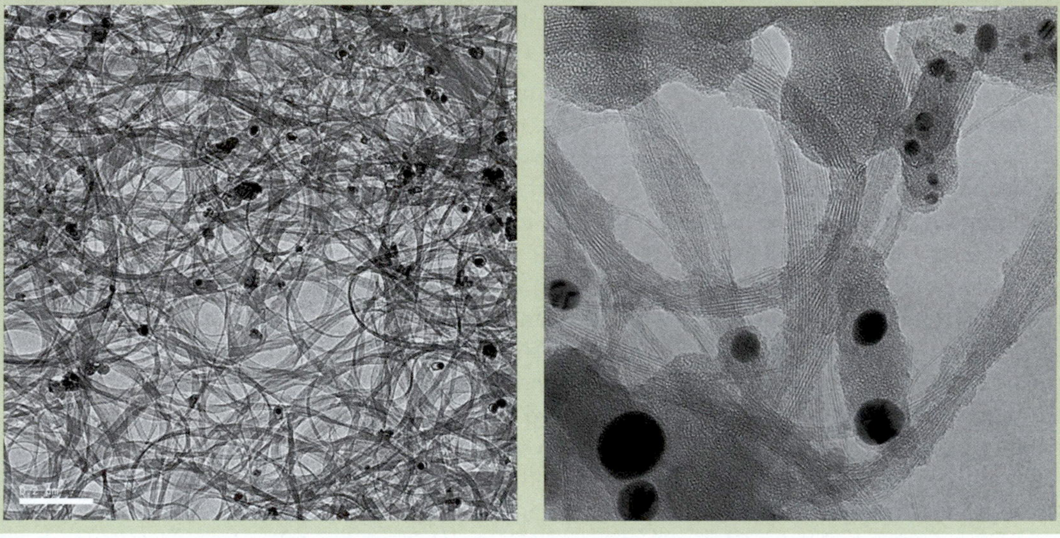

Source: Image printed with permission from Kim M. Jones, staff scientist in the Measurements and Characterization Division at the National Renewable Energy Laboratory (NREL) in Golden, Colorado.

of this quantization is that the resistance of a quantum wire cannot be deduced from the classical formula:

$$\Omega = \rho \frac{L}{A} \tag{7.48}$$

where ρ is the resistivity, and L and A are length and cross-sectional areas of the wire, respectively. For such quantum wires, an exact calculation of the

FIG. 7.21 *A one-dimensional material is a high aspect ratio nanoparticle with electron confinement along its diameter (e.g., confinement in two dimensions). On the right, ZnO wires are shown. Note: The diameter of these wires is rather large—not exactly a one-dimensional material.*

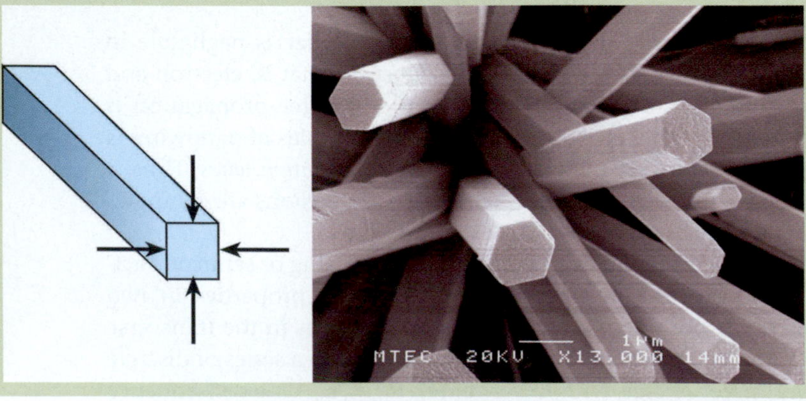

transverse energies of the confined electrons has to be accomplished to find the resistivity of the wire. Following from the quantization of electron energy, the resistance is, naturally, also quantized. The strength of the influence of quantization is inversely proportional to the diameter of the nanowire for any given material. For different materials, the quantized conductivity–resistivity is dependent on the electronic properties of the specific material—especially on the effective mass of the electrons. Stated simply, it will depend on how conduction electrons interact with the atoms within a given material. In practice, semiconductors show clear conductance quantization for wires with relatively large transverse dimensions (~100 nm). The energy of a simple square nanowire can be approximated by a two-dimensional potential well with wave function

$$\Psi_{n_x,n_y} = \sqrt{\frac{4}{L_x L_y}} \sin\left(\frac{n_x \pi x}{L_x}\right)\left(\frac{n_y \pi x}{L_y}\right) \tag{7.49}$$

in which the quantized energy is derived from

$$E_{n_x,n_y} = \frac{\hbar^2 \pi^2}{2m}\left[\left(\frac{n_x}{L_x}\right)^2 + \left(\frac{n_y}{L_y}\right)^2\right] \tag{7.50}$$

7.3.1 Types of Nanowires

The dimensions of nanowires made of metal, semiconductor, carbon, oxides, and organic materials range from a few nanometers to roughly 100 nm in diameter and a few to over 100 μm in length. Nanowires are made by single crystal growth, vapor–solid (VS), vapor–liquid–solid (VLS), vapor–solid–solid (VSS), laser ablation, chemical vapor deposition (CVD), metal oxide CVD (MOCVD), and molecular beam epitaxy (MBE) (chemical beam epitaxy [CBE]) procedures. Si nanowires surrounded by an oxide casing are formed from Si–Fe substrate at 1200°C. GaN nanowires (whiskers) can be grown epitaxially from sapphire surface by VLS of a nickel catalyst. The diameter of the catalyst dictates the dimensions of the nanowire and homogeneous nucleation events enable the control of wire length. Silicon nanowires were also grown by catalysts in the presence of silane and hydrogen gas. GaAs nanowires formed from Au catalyst with 50 nm diameter. Gold is not usually used in the role of a catalyst. Solution phase synthesis of ZnO vertical nanowire arrays using textured ZnO seeds generated wires 15–65 nm in diameter with 250–400 nm in length. Carbon nanotubes are discussed in more detail in chapter 9.

7.3.2 Physical Properties and Phenomena

Optical Properties. Nanowire optical properties are expected to be different from those of spherical clusters. Spherical particles are highly symmetric and their optical response is isotropic, regardless of the direction of the probing radiation. Orientation however is an important aspect of nanowire properties and allows them to be aligned in electric or magnetic fields. In 1912, Gans proposed that small ellipsoids abide by the dipole approximation—however that the surface plasmon mode splits into two distinct modes [18]. The curvature of the surface is responsible for this phenomenon due to its effect on the restoring force (or depolarization field) that affects the "confined" population of electrons [18]. The degree of the effect is a function of the aspect ratio of the particle. Spectra of solution containing gold nanorods reveal two absorption maxima: one for the longitudinal plasmon mode and the other for the transverse plasmon mode; both modes are independent of one another [19].

Surface modification of metal nanowires is able to alter the response. The optical response of metal particles is affected by the surrounding medium. Application of an organic material, essentially an insulating dielectric material, is expected to shift the wavelength of maximum absorption (λ_{max}) to lower energies. The dielectric material is able to diffuse the polarization and thereby cause the red shift in absorption.

Mechanical Properties. There is much interest in the use of nanowires as connectors in electrical devices. Nanomechanical measurements using an AFM tip found that in Au nanowires, Young's modulus is independent of diameter and that the yield strength is the largest for the smallest tubes with strengths up to 100 times better than that demonstrated by bulk materials [20].

Ballistic Conduction. Ballistic conduction occurs when the length of the conductor is smaller than the mean free path of the electron [11,21]. In other words, electrons are allowed to flow without collisions with phonons, impurities, etc. No energy is dissipated during the electrical conduction process [11]. The lack of impurities that cause elastic scattering in the conduit material is imperative for this to occur; loss of quantization results if this happens. Frank et al. first demonstrated ballistic conduction in carbon nanotubes [11,22].

7.4 Two-Dimensional Materials

Two-dimensional materials are those in which the crystallite size is negligible in one dimension and is not restricted in the other two directions; electron and hole motion is confined in only one spatial direction, whereas free propagation is allowed in the other two spatial directions. These are also called *quantum wells*. Thus, a quantum well can be defined as a structure where one dimension is comparable to the exciton Bohr radius.

Since particle motion is confined to two dimensions, the geometric correlation is to that of a plane with no thickness. When the quantum well thickness is comparable to the de Broglie wavelength of the carriers (i.e., electrons and holes), just like with zero- and one-dimensional materials, quantum confinement takes place in that material in just one dimension: that normal to the surface of the film (**Fig. 7.22**).

A simple approximation of the energy of square quantum well with thickness L that is comparable to the de Broglie wavelength of the carriers (generally electrons and holes) is

$$E_n = \frac{\hbar^2 \pi^2}{2mL^2} n^2 \tag{7.51}$$

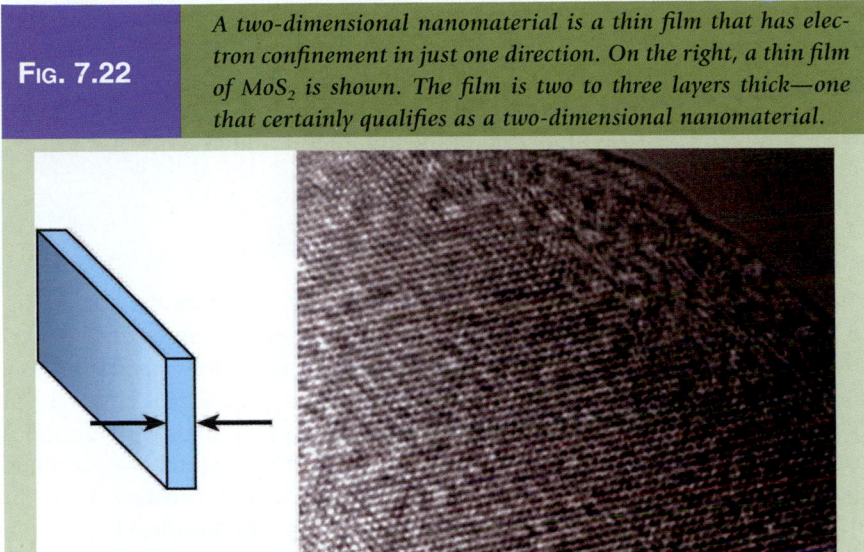

FIG. 7.22 *A two-dimensional nanomaterial is a thin film that has electron confinement in just one direction. On the right, a thin film of MoS_2 is shown. The film is two to three layers thick—one that certainly qualifies as a two-dimensional nanomaterial.*

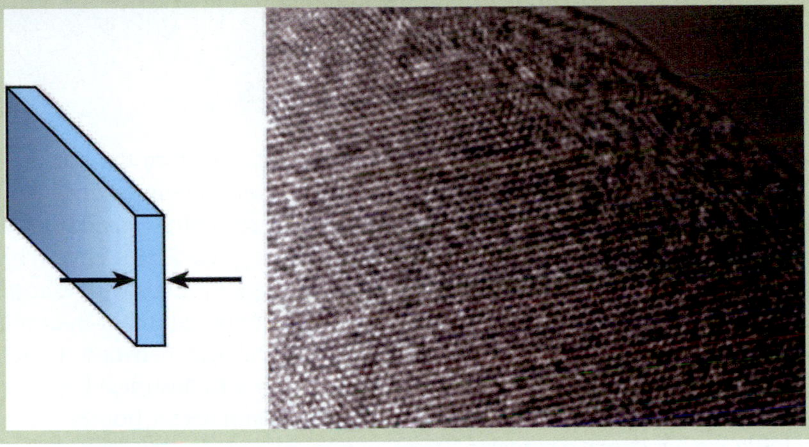

Source: Image courtesy of Kostya Novoselov, School of Physics & Astronomy, University of Manchester. With permission.

7.4.1 Types of Thin Films

Two-dimensional materials are covered extensively in this text and therefore, we provide a brief overview of such materials. There are many ways to fabricate two-dimensional films: electrodeposition, evaporation, wet chemical methods, CVD, electroless deposition, and many, many more. More discussion of thin film types, fabrication, and modification ensues in later chapters.

7.4.2 Physical Properties

Optical Properties. Optical properties depend on several parameters: (1) film composition, (2), film thickness, and (3) film structure (and orientation). These, in turn, affect properties such as interference, reflectivity, absorption, and the refractive index. All of these characteristics also depend on the kind of deposition mechanism. The films can be crystalline, polycrystalline, columnar, amorphous, or lacunar (fractal). If the film substructures are small enough, the film does not scatter light. Films can be hydrophilic or hydrophobic. Nanostructured thin films are anisotropic, usually in the direction normal to the surface. Films can be polarizing or neutral.

The optical behavior of *anatase* (TiO_2) thin films can be altered by addition of $ZnFe_2O_4$. Apparently, the zinc compound acts as a photosensitizer that enhances the photoresponse, and hence the photoactivity, of the TiO_2. The addition of $ZnFe_2O_4$ encouraged a strong photoluminescence emission due to enhanced localization of impurity- and defect-trapped excitons [23].

Mechanical Properties. Mechanical properties of thin films differ from bulk material counterparts because of their unique nanostructure (and microstructure) and because they possess inherently large surface-to-volume ratio [24]. Recall that in chapter 4 we discovered that disc-shaped materials have the potential of largest surface-to-volume ratio. Thin film integrity is influenced strongly by its substrate. The substrate (thick and underlying) therefore plays a key role in determining the physical properties of the film that it supports.

7.5. Hierarchical Structures

Hierarchical structures are structures that are synthesized (from the bottom up or top down—usually a combination) at different levels of complexity by altering growth conditions. Simply stated, hierarchical structures gradually grow from one parent structure into a more complex form (**Fig. 7.23**). Hierarchical nanostructures are three-dimensional materials. Biology abounds with examples of hierarchical structures. There are numerous examples of natural or man-made materials with hierarchical structures, such as bone or collagen networks. Bones in animal bodies are light and stable. That is because they are constructed optimally from the smallest to the largest levels. Their smallest elements are bound to fibrils that fold together to make lamellae. These, in turn, organize themselves into girders that form scaffolding—an inspiration, for sure, to structural engineers.

7.5.1 *Importance of Hierarchical Materials*

It is an exhilarating process to develop new hierarchical materials that will find applications that are important to humanity. Nanosized hierarchical structures

Fig. 7.23 *A generic version of a hierarchical structure is depicted. Hierarchical structures are typically found in biological materials.*

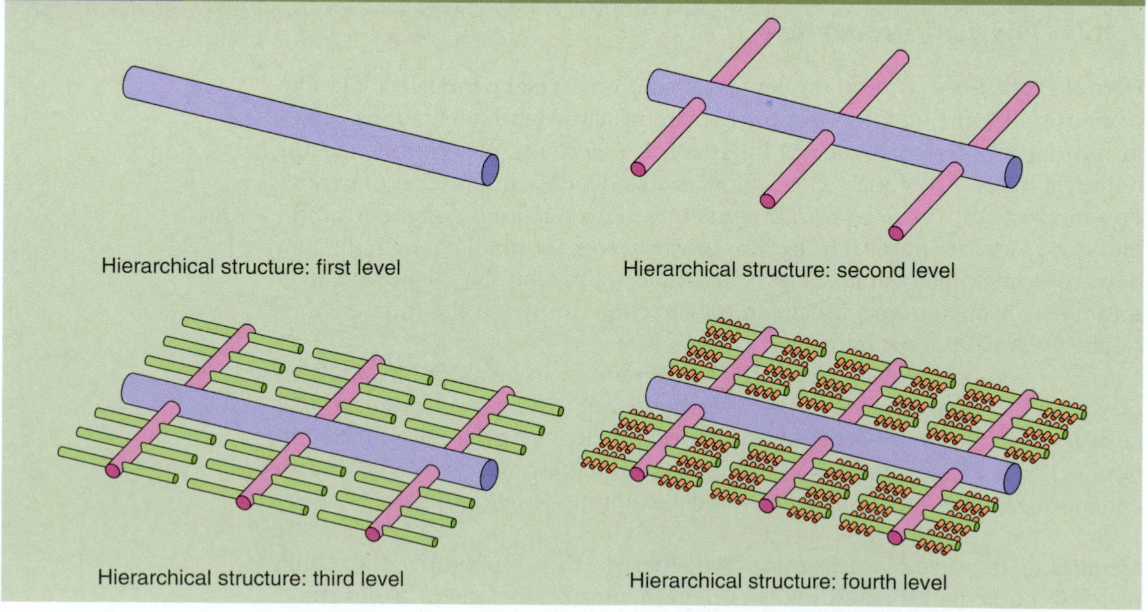

are good candidates for use in medical applications, environmental greening, and renewable sources of energy. Various materials like ZnO, TiO_2, C, Au, and Ag have been successfully applied in hierarchical structures. The morphology of ZnO nanostructures can be significantly altered by modifying the composition of the source materials. Nanowire ribbon arrays, for example, consist of a central axial nanowire surrounded by a radial distribution of nanobranches. The growth of the structure is in two stages: (1) rapid growth of the ZnO nanowire, and (2) nucleation and epitaxial growth of nanoribbons.

7.6 Quantum Size Effects and Scaling Laws

Specific Heat. Specific heat is the amount of heat required to raise the temperature of a sample of mass m by ΔT [13]:

$$C = \frac{\Delta Q}{m\Delta T} \tag{7.52}$$

In 1912, Peter Debye proposed a quantum mechanical model for estimating the phonon contribution to the specific heat. A classical version was originally proposed by P. L. Dulong and A. T. Petit in 1819. They stated that the specific heat of a crystal is due to lattice vibrations equal to $3R \cdot M^{-1}$. (R is the gas constant and M is the molar mass in kilograms per mole). The model fails at low temperatures and is appropriate for solids with simple crystal structure [24]. The Debye model transforms the particle-in-a-box principle to phonons—"phonons in a box." The resonant modes of sonic phenomena of a cube of sides L have wavelength equal to

$$\lambda_n = \frac{2L}{n} \tag{7.53}$$

where n is an integer. The energy of a phonon is

$$E_n = h\nu_n = \frac{hc_s}{\lambda_n} = \frac{hc_s n}{2L} \tag{7.54}$$

where c_s is the speed of sound. This equation relates the energy of phonons to the size of the box.

The specific heat of crystalline nanoscale materials is higher than those of the bulk. C_v (the specific heat at constant volume) for bulk Pd is $25\ \mathrm{J \cdot mol^{-1} \cdot K^{-1}}$ (for a 6-nm nanocrystal, $C_v = 37\ \mathrm{J \cdot mol^{-1} \cdot K^{-1}}$). For 8-nm Cu, the C_v increases from 24 to $26\ \mathrm{J \cdot mol^{-1} \cdot K^{-1}}$, and for 6-nm Ru nanoparticles, C_v increases from 23 to $28\ \mathrm{J \cdot mol^{-1} \cdot K^{-1}}$ [13].

Magnetism. Magnetic properties are size dependent in materials like Fe, Co, Ni, and Fe_3O_4. The coercive force behind magnetic memory required to reverse an internal magnetic field within a particle is size dependent. The strength of a particle's internal magnetic field is also size dependent.

Structural Fluctuations. Independent clusters in particular have the ability to alter their structure rather quickly—a trait not characteristic of bulk materials. At high temperatures, these fluctuations cause breakdowns in symmetry that may result in the formation of a liquidus state in the particle [14].

7.6.1 Scaling Laws

Scaling laws offer us a means to place our universe into perspective. The following discussion is gleaned from an essay written by Chris Phoenix of the Center for Responsible Nanotechnology (CRN) and summarizes the importance of scaling to smaller sizes. Scaling laws tell us, for example, that small things are not as affected by gravity as are larger things [25]. Most importantly, scaling laws are important when it comes to the manufacture of small things. We are familiar with the scale-up process where a prototype is developed and placed into the hands of manufacturing engineers with the prospect of mass production in mind. This is the scale-up of a process from small to large. Scaling laws start from the large and predict properties of smaller things or vice versa.

Scaling laws with positive exponents become more important as things get larger (e.g., r^3). Scaling laws with negative exponents (e.g., r^{-1}) get more important as things get smaller. In nature, the muscles of elephants and fleas are very different. *Strength* is proportional to cross-sectional area ($\propto L^2$). The *weight* of the muscle is proportional to its volume ($\propto L^3$). Strength versus weight is proportional to L^2/L^3 or L^{-1}. This ratio gets 10 times larger as organisms get 10 times smaller—a scaling law. At the nanoscale, ca. 10^6 times smaller than a flea, there is no concern at all about weight [25].

Timescale considerations affect potential productivity. With regard to time, the speed of a bulk manufacturing process may be a few units per second, per hour, or per day depending on its size and complexity. An enzyme is able to transform substrates into products at a rate of thousands to millions times per second. The wingbeat frequency of dragonflies (~40 Hz) like Zygoptera (a larger species) is half as fast as that of Anisoptera (a smaller species) [26]. The smaller species is capable of quicker acceleration.

Power density is also important at any scale. Power density (force × speed) scales with L^2, just like strength. If a cubic engine 10 cm on one side is capable of producing 1000 W of power, then an engine of 1 cm on a side (10 times smaller along one dimension) should produce 1/100 of the power of the larger engine [25]. However, the volume of the smaller cube is 1000 times smaller than that of the larger cubic engine and 1000 of those smaller engines would end up producing 10,000 W of power—10 times as much power as the original 10-cm cubic engine. Power per volume scales with $1/L$. Therefore, scaling laws tell us that by manufacturing 1000 more parts that are 1000 times less in volume, 10 times the output of the original cubic engine is achievable with the same total mass [25]. Add to this that smaller things run at higher frequencies and, according to temporal scaling laws, the small cubic engines would be able to operate at a rate that is 10 times faster than the parent engine—the power of small!

Functional density is proportional to L^{-3}. This is simply the number of devices that can be crammed into a unit of space; the smaller the device is, the more of it one can pack into the usable space. The computer industry serves up the best example of the importance of this scaling law; each year, more and smaller transistors squeeze into the same universal-sized wafer. The result of such miniaturization is more memory and faster processing.

Forces operating on small things differ widely from those operating at the macroscopic level. Van der Waals forces originating from the setae of gecko feet are able to hold the creature to a ceiling. Electrostatic forces assume more importance at the nanoscale; at our scale, it is referred to as "static cling" of clothing materials [25]. As motors are miniaturized to the level of cells or molecules, power density increases, with nanoscale electric motors having a power density on the order of $10^6\ \mathrm{W \cdot cm^{-3}}$ [25].

Friction at the nanoscale dwells in an altered reality. Friction scales with L^2 in the macroworld, indicating that frictional power is proportional to power: The more power that is exerted, the more wear due to frictional forces is expected to occur in large things. Wear life is proportional to L (thickness of a film or coating). Despite this drawback (e.g., thinner films are worse and therefore lead to lower operational lifetime), the intervention of *nonscaling mechanisms* saves the day. There should be no wear and tear as we know it at the nanoscale due to strong chemical bonds able to withstand localized heating forces.

7.6.2 Classical Scaling Laws and the Nanoscale

The content of this section was summarized from *Nanosystems*, chapter 2, "Classical Magnitudes and Scaling Laws," by Drexler [27]. The *magnitudes* of properties that characterize bulk materials, by definition, are not directly applicable to nanomaterials. Macroscale magnitudes, part of the classical continuum, are adjusted by the application of scaling laws, but they have a tendency to lose validity due to the effects of atomic structure, mean free path effects, and quantum mechanics [27]: "When used with caution, classical continuum models of nanoscale systems can be of substantial value in design and analysis. They represent the simplest level in a hierarchy of approximations of increasing accuracy, complexity and difficulty."

Scaling laws focus on some parameter that has an inherent characteristic of "scalability." This seems like a circuitous argument, but nonetheless it is an accurate one. For example, if we wish to predict the properties of an object that is larger than the one used in the model, we conduct the calculations based on increasingly larger values of some scale. If material strength and stress are considered to be constant, then several scaling laws relate to a dimension L [27]:

$$\text{Total strength} \propto \text{force} \propto \text{area} \propto L^2 \tag{7.55}$$

If one assumes a state of constant density, then

$$\text{Mass} \propto \text{volume} \propto L^3 \tag{7.56}$$

This is straightforward. The mass of a block of material with density $\rho = 3.5 \times 10^3\ \mathrm{kg \cdot cm^{-3}}$ has mass equal to $3.5 \times 10^{-24}\ \mathrm{kg^{-1}}$, assuming that the density remains constant at that scale—a nonscaling parameter. For a more complicated property like acceleration,

$$\text{Acceleration} \propto \text{force/mass} \propto L^{-1} \tag{7.57}$$

A nanometer-sized object is capable of accelerating to a level of 3×10^{15} m · s^{-1}. Other relations are given: *power* $\propto$ *force · speed* $\propto L^2$; *power density* $\propto$ *power/volume* $\propto L^{-1}$; *frequency* $\propto$ *speed/length* $\propto L^{-1}$; *speed* $\propto$ *acceleration · time = constant*; *time* $\propto$ *frequency*$^{-1} \propto L$, and *frequency* $\propto$ *acoustic speed/length* $\propto L^{-1}$—all good scaling laws for bulk materials. According to Drexler [27],

> The accuracy of classical continuum models and scaling laws to nanoscale systems depends on the physical phenomena considered. It is low for electromagnetic systems with small calculated time constants, reasonably good for thermal systems and slowly varying electromagnetic systems, and often excellent for purely mechanical systems provided that the component dimensions substantially exceed atomic dimensions. Scaling principles indicate that mechanical components can operate at high frequencies, accelerations, and power densities.

7.6.3 Scaling Laws for Clusters

The following discussion is summarized from the book *Atomic and Molecular Clusters,* by Johnston [28]. Ionization energy, electron affinity, melting temperature, and cohesive energy vary with cluster size [27]. It is possible to model cluster size effects (CSEs) by application of the spherical cluster approximation (SCA) presented in chapter 5 in the case where an N-atom cluster is a sphere with radius r. Interpolation-based scaling law formulae that incorporate a scaling factor proportional to the power law of the cluster radius (like L before) are able to smoothly approximate the behavior of nanoclusters. A scaling law that incorporates the radius as a factor is

$$G(r) = G(\infty) + ar^{-\alpha} \tag{7.58}$$

where $G(\infty)$ is the value of the specific property G at the limit of the bulk material. Or, if it is by the number of atoms that potential scaling is focused (since properties at the nanoscale depend on the number of atoms and that surface fraction $F_s \propto N^{-1/3} \propto r^{-1}$),

$$G(N) = G(\infty) + BN^{-\beta} \tag{7.59}$$

The value of the exponent α in the previous equation is equal to 1; for β, the exponent is equal to 1/3.

Ionization energies of potassium clusters provide a good example to illustrate this process. The IE can be fitted with great accuracy to the following interpolation formula:

$$\frac{\mathrm{IP}^K(N)}{eV} = 2.3 + 2.04\,N^{-1/3} \quad \text{or} \quad \frac{\mathrm{IP}^K(r)}{eV} = 2.3 + 5.35\left(\frac{r}{\text{Å}}\right)^{-1} \tag{7.60}$$

Figure 7.24 plots the result of the interpolation scaling of the ionization potential versus cluster size. The fit with experimental data is excellent [27].

This process also works well with melting temperature of clusters. The melting point of gold, T_m^{Au}, decreases with decreasing size (as we pointed out before) and follows the r^{-1} rule. A good fit between experiment and theory is given by the expression

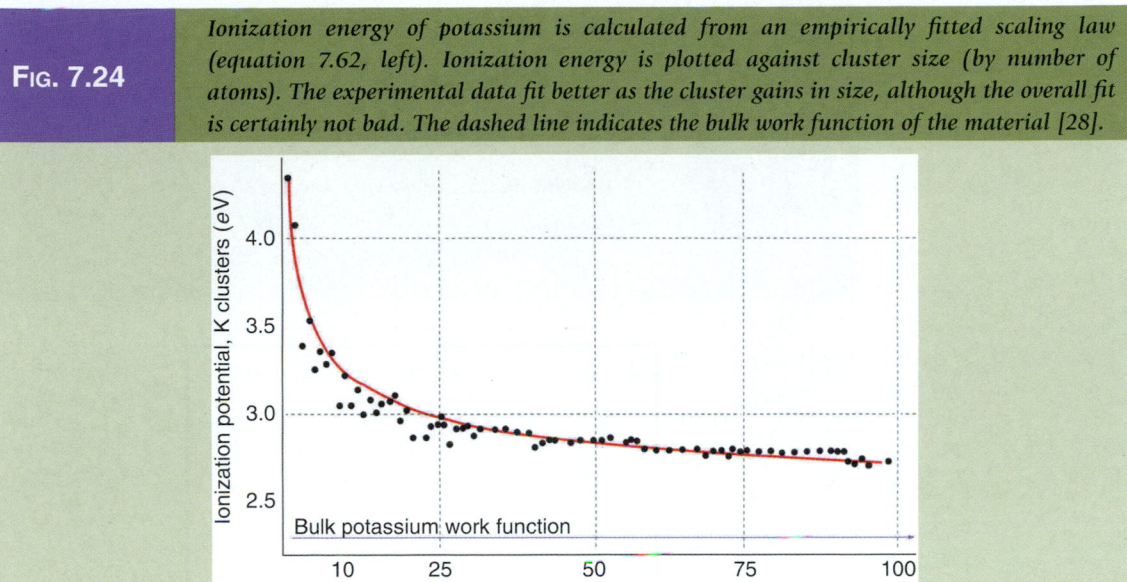

Fig. 7.24 *Ionization energy of potassium is calculated from an empirically fitted scaling law (equation 7.62, left). Ionization energy is plotted against cluster size (by number of atoms). The experimental data fit better as the cluster gains in size, although the overall fit is certainly not bad. The dashed line indicates the bulk work function of the material [28].*

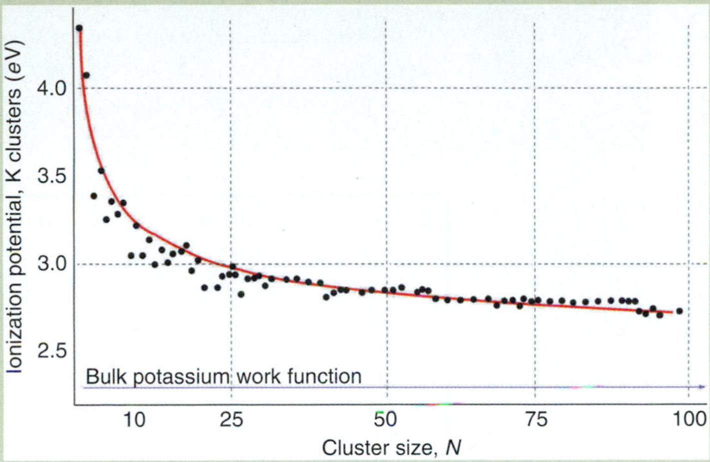

$$\frac{T_m^{Au}(r)}{K} = 1336.15 - 5543.65\left(\frac{r}{\text{Å}}\right)^{-1} \tag{7.61}$$

Large clusters show good agreement with the scaling law but smaller ones deviated from the theoretical line (**Fig. 7.25**).

Fig. 7.25 *The melting temperature of gold is represented by an interpolation formula (equation 7.63) in the figure. Apparent oscillatory behavior in the melting point is observed as cluster size is decreased. The oscillations are due to quantum and surface (geometric) effects. The dashed line represents the melting temperature of the bulk material [28].*

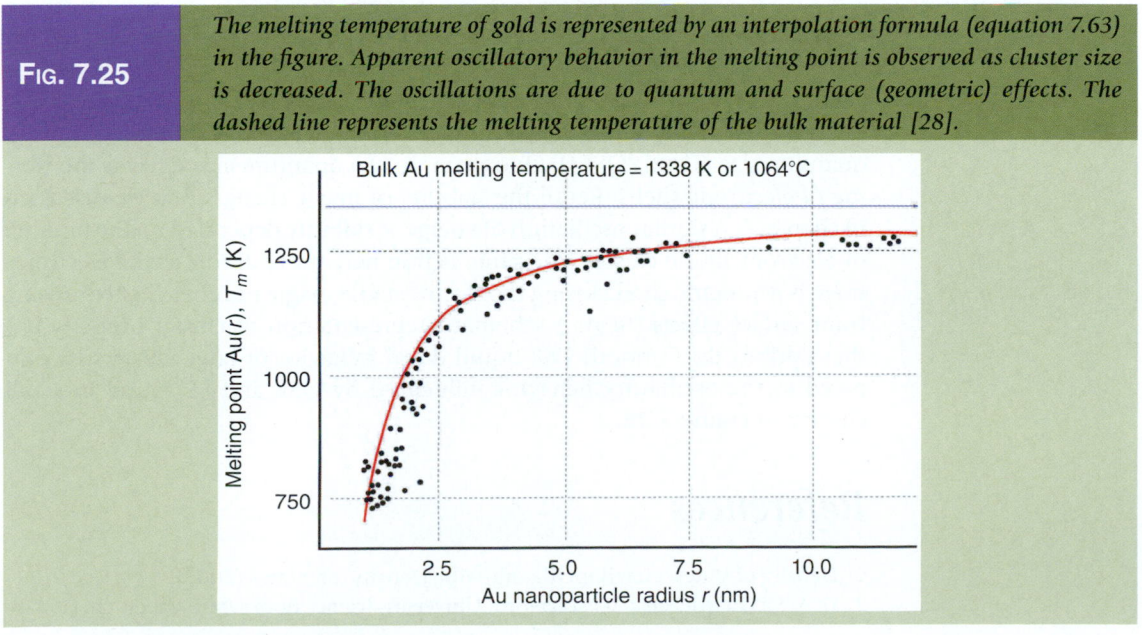

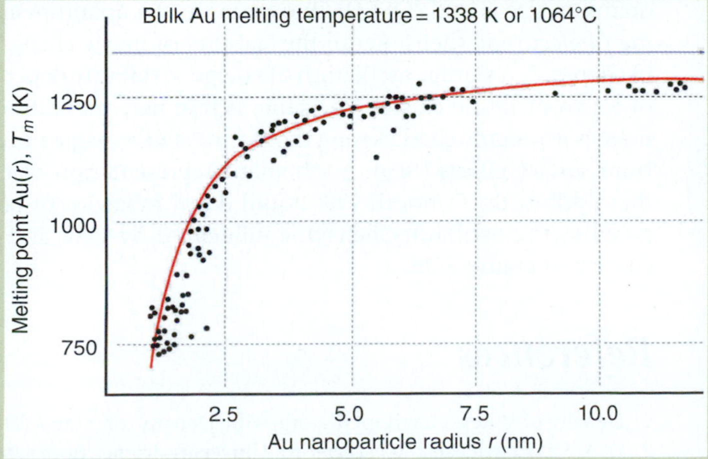

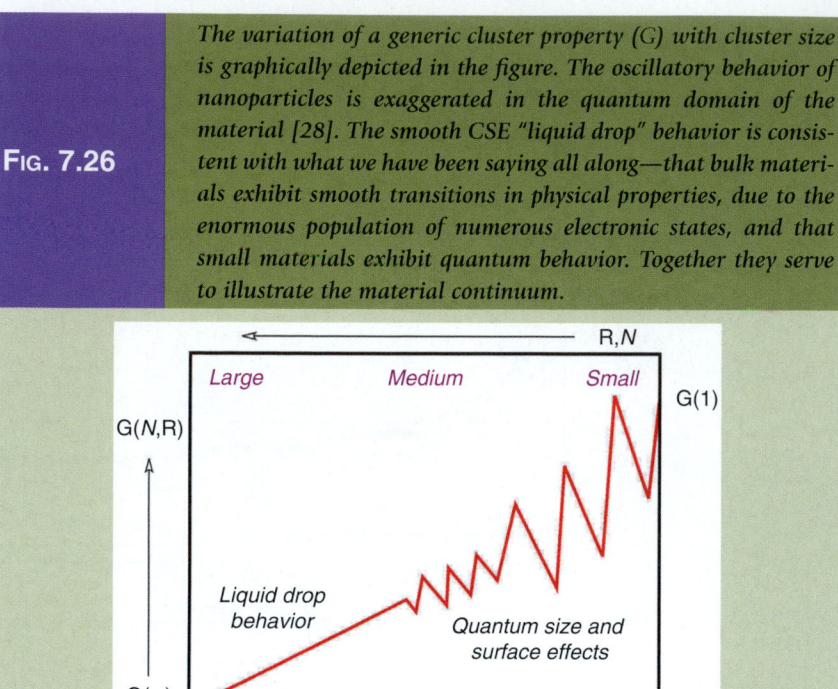

Fig. 7.26 *The variation of a generic cluster property (G) with cluster size is graphically depicted in the figure. The oscillatory behavior of nanoparticles is exaggerated in the quantum domain of the material [28]. The smooth CSE "liquid drop" behavior is consistent with what we have been saying all along—that bulk materials exhibit smooth transitions in physical properties, due to the enormous population of numerous electronic states, and that small materials exhibit quantum behavior. Together they serve to illustrate the material continuum.*

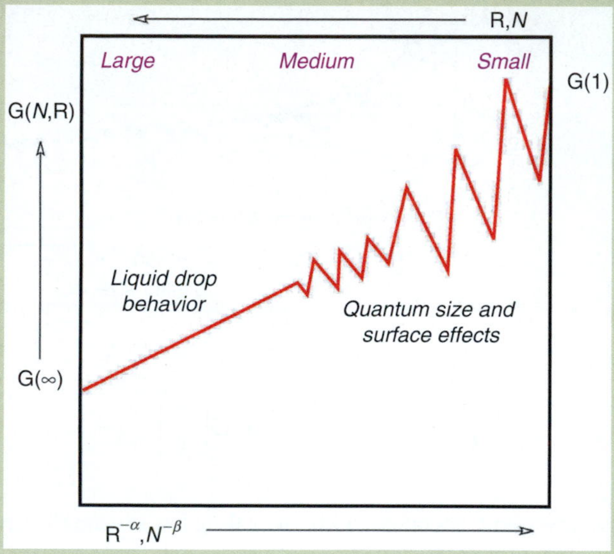

What are the reasons for these deviations? If one inspects the graph carefully, one sees that oscillatory behavior around the smooth CSE power-law curve is observed in the small cluster regime. These oscillations are typically observed for many kinds of phenomena [27]. These deviations are due to quantum size effects (QSEs). One example of a quantum size effect is the closing of electronic shells. Recall the stability of magic clusters that we discussed in chapter 6. A similar oscillation of energetic stability depended on the number of electrons in the cluster. The same is true here. In addition to the electrons, there is geometric shell closing that forms stable magic number clusters arising from surface effects (SEs). A schematic representation of cluster property (G) that exhibits the "smooth CSE liquid drop" behavior of large clusters is compared to the oscillatory behavior influenced by QSE and SE found in small clusters in **Figure 7.26.**

References

1. Density of states, en.wikipedia.org/wiki/Density_of_states (2007).
2. D. A. McQuarrie, *Quantum chemistry*, University Science Books, Mill Valley, CA (1983).

3. X. Ma et al., The size dependence of the optical and electrical properties of Ge quantum dots deposited by pulsed laser deposition, *Semiconductor Science Technology*, 21, 713–716 (2006).
4. P. Melinon, V. Paillard, V. Dupuis, A. Perez, P. Jensen, A. Hoareau, M. Broyer, J. -L. Vialle, M. Pellarin, B. Baguenard, and J. Lemme, From free clusters to cluster-assembled materials, *International Journal of Modern Physics B*, 9, 339–397 (1995).
5. H. Haberland, M. Moseler, Y. Qiang, O. Rattunde, T. Reiners, and Y. Thirmer, Energetic cluster impact (ECI): A new method for thin film formation, *Surface Review and Letters*, 3, 887–890 (1996).
6. G. Schmid, Metals. In *Nanoscale materials in chemistry*, K. J. Klabunde, ed., Wiley-Interscience, John Wiley & Sons, Inc., Hoboken, NJ, chap. 2 (2001).
7. G. L. Hornyak, S. Peschel, T. Sawitowski, and G. Schmid, TEM, STM and AFM as tools to study clusters and colloids, *Micron*, 29, 183–190 (1998).
8. U. Kreibig and M. Vollmer, *Optical properties of metal clusters*, Springer-Verlag, Berlin (1995).
9. C. Noquez, Surface plasmons on metal nanoparticles: The influence of shape and physical environment, *Journal of Physical Chemistry C*, 111, 3806–3819 (2007).
10. K. Kamiya and S. Sakka, Formation of calcium carbonate polymorphs, *Gypsum Lime*, 163, 243–253 (1979).
11. G. Cao, *Nanostructures & nanomaterials: Synthesis, properties & applications*, Imperial College Press, London (2004).
12. A. W. Adamson and A. P. Gast, *Physical chemistry of surfaces*, 6th ed., John Wiley & Sons, New York (1997).
13. M. Meyyappan, *Introduction to nanotechnology, II. Nanoscale properties*. Presentation, Center for Nanotechnology, NASA Ames research Center (viewed 2007).
14. C. P. Poole and F. J. Owens, *Introduction to nanotechnology*, Wiley-Interscience, John Wiley & Sons, Inc., Hoboken, NJ (2003).
15. R. L. Whetten, D. M. Cox, D. J. Trevor, and A. Kaldor, Correspondence between electron binding energy and chemisorption reactivity of iron clusters, *Physics Review Letters*, 54, 1494–1497 (1985).
16. G. W. Busser, J. G. van Ommen, and J. A. Lercher, Preparation and characterization of polymer-stabilized rhodium particles. In *Advanced catalysis and nanostructured materials*, Academic Press, San Diego (1996).
17. G. Schmid, V. Maihack, F. Lantermann, and S. Peschel, Ligand-stabilized metal clusters and colloids: Properties and applications, *Journal of Chemical Society, Dalton Transactions*, 589–595 (1996).
18. J. Pérez-Juste, I. Pastoriza-Santos, L. M. Liz-Marzán, and P. Mulvaney, Gold nanorods: Characterization and applications, *Coordinated Chemistry Reviews*, 249, 1870–1901 (2005).
19. G. L. Hornyak, *Characterization and optical theory of nanometal/porous composite membranes*, Ph.D. dissertation, Colorado State University (1997).
20. B. Wu, A. Heidelberg, and J. J. Boland, Mechanical properties of ultrahigh-strength gold nanowires, *Nature Materials*, 4, 525–529 (2005).
21. S. Chappel and A. Zaban, Nanoporous SnO_2 electrodes for dye-sensitized solar cells: Improved cell performance by the synthesis of 18 nm SnO_2 colloids, *Solar Energy Materials & Solar Cells*, 71, 141–152 (2002).
22. S. Frank, P. Poncharal, Z. L. Wang, and W. A. der Heer, Carbon nanotube quantum resistors, *Science*, 280, 1744–1746 (1998).
23. Y. Jin, G. Li, Y. Zhang, Y. Zhang, and L. Zhang, Photoluminescence of anatase TiO_2 thin films achieved by the addition of $ZnFe_2O_4$, *Journal of Physics: Condensed Matter*, 13, L913–L918 (2001).
24. Dulong–Petit law, en.wikipedia.org/wiki/Dulong-Petit_law (2007).

25. C. Phoenix, Scaling laws—Back to basics, CRN Science & Technology Essays-2004, Center for Responsible Nanotechnology, crnano.org/essays04.htm#Scaling (2004).
26. G. Rüppell, Kinematic analysis of symmetrical flight maneuvers of Odonata, *Journal of Experimental Biology*, 144, 13–42 (1989).
27. E. K. Drexler, *Nanosystems*, John Wiley & Sons, New York (1998). Foresight Institute, www.foresight.org/Nanosystems/Ch2/chapter2_1.html
28. R. L. Johnston, *Atomic and molecular clusters*, Taylor & Francis, London (2002).

Problems

7.1 Why does classical blackbody radiation treatment break down at high energies?

7.2 Describe the photoelectric effect and provide the generalized equation. Why is it considered to be a quantum phenomenon?

7.3 Werner Heisenberg stated that two complementary properties of a system (like position and momentum) can never be simultaneously known. Is this principle applicable to nanoparticles? Use $\Delta p \Delta x \geq h$ as a frame of reference.

7.4 A nanoparticle with mass 5×10^{-27} g exists in a 1-nm, one-dimensional box. What is the wavelength of radiation that is emitted when the nanoparticle loses energy from the $n = 3$ level to the $n = 2$ level?

7.5 What does the term degeneracy imply with regard to energy levels?

7.6 Explain in the simplest of terms the phenomenon of quantum tunneling.

7.7 Define the terms wave, particle–wave duality, vibration, rotation, and momentum. Which possess quantized characteristics?

7.8 Why there are differences in bandgap energies between insulators, semiconductors, and metals?

7.9 Explain the presence of the zigzag lines in the material continuum trace shown in **Figure 7.26**.

7.10 Is there a material continuum or does the concept break down in the quantum domain?

7.11 Bulk gold is yellow, and gold nanoparticles are light red, purple, or blue. What is the reason for the change of color? (Question courtesy of Günter Schmid, Uni-Essen.)

7.12 What does the specific color depend on? (Question courtesy of Günter Schmid, Uni-Essen.)

7.13 Why is the plasmon resonance visible only for Cu, Ag, and Au as opposed to other metals? (Question courtesy of Günter Schmid, Uni-Essen.)

7.14 What is a "Coulomb blockade" and what do its properties depend on? (Question courtesy of Günter Schmid, Uni-Essen.)

7.15 How can Coulomb blockades be measured? (Question courtesy of Günter Schmid, Uni-Essen.)

7.16 When are quantum confinement effects observed in nanocrystallites?

7.17 What is meant by quantum confinement in a semiconductor quantum dot? What are the implications of this for the energy levels?

7.18 In how many dimensions free electron motion is restricted in quantum wires?

7.19 What do you understand by the wave–particle duality in nanotechnology?

7.20 A proton is confined to moving in a one-dimensional box of width 0.200 nm.

(a) Find the lowest possible energy of the proton.
(b) What is the lowest possible energy of an electron confined to the same box?
(c) How do you account for the large difference in your results for (a) and (b)?

7.21 An electron is confined to a 1 μm thin layer of silicon. Assuming that the semiconductor can be adequately described by a one-dimensional quantum well with infinite walls, calculate the lowest possible energy within the material in units of electron volt. If the energy is interpreted as the kinetic energy of the electron, what is the corresponding electron velocity? (The effective mass of electrons in silicon is 0.26 m_0, where $m_0 = 9.11 \times 10^{-31}$ kg is the free electron rest mass).

7.22 Estimate the minimum velocity of an apple of mass 100 g confined in a crate of size 1 m.

NANOTHERMODYNAMICS

I see evidence of the second law of thermodynamics all over my room. Are you trying to tell me that if I made my room much much smaller, it would clean itself?

UNKNOWN STUDENT, 2007

Chapter 8

Threads

The physics division continues with a special chapter on nanothermodynamics. It is recommended to students with at least college level physical chemistry. For those who do not have such a background, we recommend that you at least read the chapter and discuss it with your instructor and classmates. Much of the chapter is nontechnical/nonmathematical and should acquaint you with the basic terminology and the concepts. In many ways the chapter will serve as a primer on thermodynamics, but familiarization with the subject matter, particularly as it pertains to living things, should provide for interesting reading.

So far we have investigated nanomaterial surface and structure and discussed the energetics of nanoscale materials in chapters 5 and 6, respectively. Chapter 7 showed us how scaling laws break down at the nanoscale as properties and phenomena acquire more quantum character with diminished size. We now present another viewpoint of the nanoscale—a view from a purely thermodynamic perspective. We begin by reviewing some obligatory fundamentals of classical thermodynamics and statistical mechanics and then discuss selected theories that apply specifically to nanomaterials; these theories essentially bring into question the validity of applying classical thermodynamics to small systems without considering them as small systems.

Special sections on fullerenes and carbon nanotubes are presented in chapter 9. With chapter 9 we begin the chemistry division of the book and we move on to intra- and intermolecular chemical bonding in chapter 10.

Although an in-depth discussion of nanothermodynamics is beyond the scope of this text, we feel strongly that the subject matter must be mentioned and introduced at this time. As a result, we limit our discussions to qualitative perspectives without much quantitative treatment. It is up to you the student to absorb, learn, and assimilate what you feel is interesting, necessary, or relevant and develop your own unique perspective on this complex subject.

8.0 Thermodynamics and Nanothermodynamics

Nanothermodynamics, like atomistic simulation, is nontrivial, and understanding nanothermodynamics requires some tweaking of preexisting thermodynamic paradigms. We have referred to nanoscience and nanotechnology as disruptive to society. In this chapter, we assert that thermodynamic properties of nanomaterials need to be viewed in a different light as well.

Adhering to our historical theme, many of the concepts that support nanothermodynamics were developed well before the word "nanotechnology" became popular. Adhering to our continuum theme, we vigorously assert that nanothermodynamics is a part of disciplines belonging with the energy neighborhood. Lastly, adhering to our "nature is still the best teacher" theme, we once again place focus on nature's mastery of the thermodynamics of small things.

8.0.1 Background

Thermodynamics (from the Greek *thermos,* meaning "hot" and *therme,* indicating heat, + *dynamikos,* meaning "powerful") is a branch of science that tries to bring understanding and order to chaos—a somewhat paradoxical definition in that one of the basic drivers of thermodynamics is chaos. Thermodynamics is the study of energy and its interconversion. It is the study of heat and its relationship with physical parameters like temperature, pressure, density, and other macroscopic measurable properties of matter. Thermodynamics is a science that is grounded in experiment and theory and centered on the concept of equilibrium. Thermodynamics, along with statistical mechanics and the kinetic theory, provides tools that enable us to understand the relationships between a system and its surroundings.

Historically, thermodynamics developed from the need to increase the efficiency of early steam engines. The two Roberts, Boyle and Hooke, initiated an insipient form of thermodynamics when they created an air pump to measure the pressure–temperature–volume correlations and eventually to describe a relationship that we now know as Boyle's law. Nicolas Léonard Sadi Carnot in the early 1800s added significant contributions to early thermodynamic theory. Carnot, who is known as the father of modern thermodynamics, invented the Carnot engine and developed the working theory of the Carnot cycle. Concepts such as engine power and efficiency soon followed. James Joule coined the word *thermodynamics* in the mid-1800s. While working with steam engines, he strove to understand the relationship between heat, power, and mechanical work and later on with electricity.

Kelvin, Clausius, and Rankine took Boyle's law to the next level and the first and second laws of thermodynamics emerged by the mid-nineteenth century. Josiah Willard Gibbs was one of the most productive experimentalists and theoreticians of the mid- to late nineteenth century (**Fig. 8.1**). He published a fundamental

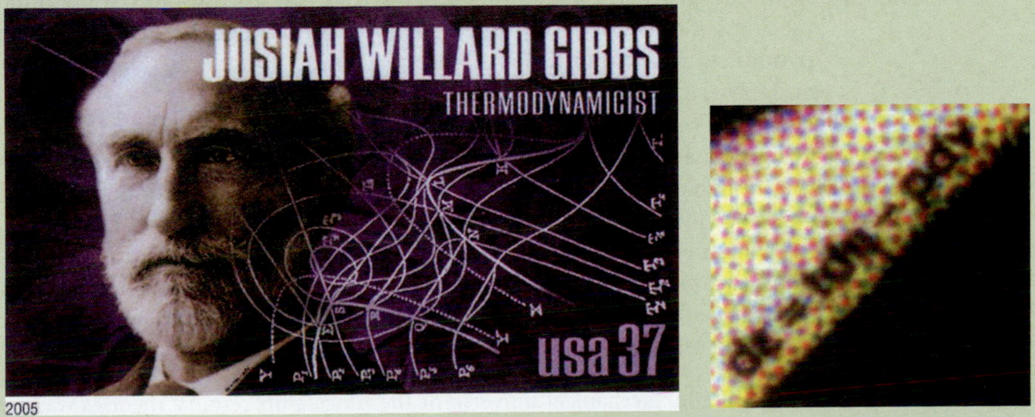

(a) (b)

FIG. 8.1 *(a) The U.S. Postal Service released this stamp in May 2005 as a tribute to Josiah Willard Gibbs. Gibbs was the first American to receive a doctorate of engineering in 1863. His contributions to the first and second laws of thermodynamics and the Gibbs equation are fundamental to thermodynamics. The background is made of thermal coordinates. (b) An interesting subtlety exists in the image. If one inspects the collar of his shirt, his famous equation will be found there:* $dE = Tds - PdV$ *(U.S. Postal Service and Zolti Spakovsky, Cambridge, Massachusetts, in* Mechanical Engineering, *8 April 2006). The equation is reproduced exactly as it is found in Gibbs's paper "Graphical Methods in the Thermodynamics of Fluids," published in the* Transactions of the Connecticut Academy *in 1873.*

paper in 1873 detailing geometric representations of thermodynamic surfaces. By the age of 34, he changed the face of science. He is credited with formulating the concept of thermodynamic equilibrium of systems in terms of energy and entropy. His book, *Elementary Principles of Statistical Mechanics*, published in 1902, linked thermodynamics to statistical analysis. Gibbs was a fundamental force behind the development of nanothermodynamics, albeit without prior knowledge of the term. T. L. Hill, one of the pioneers of nanothermodynamics, gave credit to Gibbs for adding chemical potential terms to the basic energy–heat–work equations. It is from the chemical potential term, μ, that Hill's *subdivision potential term* to describe nanothermodynamic systems is derived.

The concepts of system, surroundings, internal energy, free energy, entropy, and enthalpy gradually matured in the early twentieth century and were fundamental not only to engine design but also to a wide range of various applications such as mechanical engineering, chemical engineering, aerospace engineering, materials science, and biology. The work accomplished by these early pioneers laid the foundations of macroscopic *classical thermodynamics*—an approach that lacks any obvious atomic (or nano) interpretation and assumes in its interpretation a state of equilibrium.

8.0.2 The Nano Perspective

Thermodynamics is defined within the framework of a continuum that is centered on the transfer of energy. Once again we have to draw lines to help us understand its meaning. We invented *classical thermodynamics* to understand how engines behave and then, from the top down, to understand how molecules behave, from the perspective of average properties. We invented *statistical mechanics* to illustrate how thermodynamics is constructed from the bottom up, thus adding a complementary component to the continuum. We invented *quantum mechanics* to describe the behavior of atoms and molecules knowing that all thermodynamics is based ultimately on quantum mechanics. These are "integrated into everything" within the greater space of thermodynamics, although each component may seem to be extremely segregated in its expression: classical thermodynamics, kinetic theory, statistical mechanics, quantum mechanics, pseudoequilibrium thermodynamics, quasi-equilibrium thermodynamics, nonequilibrium thermodynamics, nonergodic thermodynamics, and, now, nanothermodynamics.

Nature once again gives us an amazing number of paradoxes to ponder. We all understand that macroscopic behavior can be described by using classical thermodynamic principles. However, these principles do not seem to apply clearly to living systems. Although metabolic processes ultimately follow thermodynamic laws when considering the system as a whole, much of what goes on in the cell seems to be quite mysterious—quite nonthermodynamic—as if a *fourth law of thermodynamics* is needed to explain what actually transpires inside the living cell [1]:

> The sterile, mechanical universe of the nineteenth century theories was conceived as a closed, isolated system tending to equilibrium. As predicted by the second law of thermodynamics, it will inexorably expire as available energy is

> spent doing work and converted to entropy ... But a recent revision is underway by which life has become known as an open system infused and organized by a flow of energy and information .... An effort to articulate a "fourth law of thermodynamics" counter to the second law is now in progress.

We have forged strong links between nanotechnology and nature throughout the text. The preceding statement is another example of the revolutionary imperative of nanotechnology, once again grounded within the intricate machinations of nature.

Replication, for example, exists in biological and nonbiological systems. The process of replication is more a kinetic process than a thermodynamic one [2]. On the other hand, nonreplicating systems have more *thermodynamic* (thermostatic?) character and less kinetic character. Proponents of this train of thought believe that kinetic stability and kinetic selection are forces that drive replication processes in biology [2]. Unfortunately, this makes thermodynamic interpretation of such complex processes even more difficult.

Before we begin our quest, let us propose a few simple questions. When did biological nanomaterials start to display their special functions and what thermodynamic driving forces made this possible? How do they accomplish those functions and how do they fit into the complexity that characterizes all biological systems? And, why nanomaterials? What is it about the special relationship nanomaterials have to energy transfer that makes life work?

8.1 Classical Equilibrium Thermodynamics

We begin this section with a brief recapitulation of classical thermodynamics. We will strive to be as succinct as possible in our presentation.

8.1.1 *Extensive and Intensive Properties and State Functions*

Extensive and Intensive Properties. There are two basic categories of thermodynamic properties: *intensive* and *extensive*. Intensive and extensive properties and *intensivity* and *extensivity* are terms that describe physical parameters that are fundamental to thermodynamics and to nanothermodynamics, albeit in altered form. Properties and phenomena that do not rely on "how much" is involved are called *intensive properties* and are independent of the size of a system. Examples of intensive properties include density ρ (grams per cubic centimeter), pressure P (newtons per square meter), temperature T, and chemical potential μ. For example, the temperature of a system in equilibrium is the same throughout that system no matter how large or small. And, conversely, in the absence of a gradient, thermal equilibrium is achieved when all parts of a system, no matter how large or small, have the same temperature.

Extensive properties are those that have linear dependence on the amount of matter or size involved. Internal energy (U), enthalpy (H), Gibbs free energy (G), Helmholtz free energy (A), volume (V), area ($\mathcal{A}$), mole number or number (n or N), and entropy (S) are scaled by the amount of material on hand. Energy released during a forest fire depends on the number N_{Trees} that burned. The total energy of the fire is equal to the sum of the energy of each component (tree). The relationship between extensive and intensive variables is accomplished through the theory of homogeneous functions (in which all terms are of the same degree, e.g., polynomials with the same exponent).

From a mathematical perspective, thermodynamic functions are either homogeneous to the first degree ($\lambda = 1$) and are called *extensive* functions or homogeneous to the zeroth degree ($\lambda = 0$) and are called *intensive* functions. The straightforward mathematical representation is

$$f(x_1,x_2,\ldots,x_n) \Rightarrow f(hx_1,hx_2,\ldots,hx_n) = h^{\lambda} f(x_1,x_2,\ldots,x_n) \quad (8.1)$$

where $f(x)$ is a smooth continuous function with the following characteristics: (1) the arguments of the function are scaled by h, (2) the function f is scaled by the factor h^{λ}, and (3) the function is homogeneous to the degree λ [3,4]. Euler's theorem of homogeneous functions is expressed as a differential equation of the form

$$x_1\left(\frac{\partial f}{\partial x_1}\right) + x_2\left(\frac{\partial f}{\partial x_2}\right) + \cdots + x_n\left(\frac{\partial f}{\partial x_n}\right) = \sum_{i=1}^{n} x_i\left(\frac{\partial f}{\partial x_i}\right) = \lambda f(x_1,x_2,\ldots,x_n) \quad (8.2)$$

where λ is called the Euler exponent and $x_k\left(\frac{\partial}{\partial x_k}\right)$ is called the Euler operator. Stated in more complex terms, homogeneous functions are *eigenfunctions* of the Euler operator, with the degree of homogeneity as the *eigenvalue* [4].

The entropy, S, for example, is completely described by three extensive variables: internal energy, volume, and number: $S = S(U, V, N)$. If we state that the change in internal energy U is a function of extensive variables only, $U = U(S, V, N, X)$, a general case relating U and the extensive variables to the Euler exponent scaling factor λ is

$$U(S, V, N, X) \rightarrow U(\lambda S, \lambda V, \lambda N, \lambda X) \rightarrow \lambda U(S, V, N, X) \quad (8.3)$$

where X symbolizes all other types of extensive variables. Differentiating U with respect to λ yields

$$U = S\left(\frac{\partial U}{\partial S}\right)_{V,N,X} + V\left(\frac{\partial U}{\partial V}\right)_{S,N,X} + N\left(\frac{\partial U}{\partial N}\right)_{V,S,X} + X\left(\frac{\partial U}{\partial X}\right)_{S,V,N} \quad (8.4)$$

Breakdown of individual terms yields intensive terms T, P, μ, and xt, respectively, where x is a generalized intensive parameter that corresponds to the extensive variable X:

$$T = \left(\frac{\partial U}{\partial S}\right)_{V,N,X}; \quad -P = \left(\frac{\partial U}{\partial V}\right)_{S,N,X}; \quad \mu = \left(\frac{\partial U}{\partial N}\right)_{V,S,X} \quad \text{and} \quad x = \left(\frac{\partial U}{\partial X}\right)_{S,V,N} \quad (8.5)$$

The extensive variable X is planted at this time as a teaser to be explained later, and the form of the final term in equation (8.5) is explored further in section 8.4.3. The integrated form of U from equation (8.4) is given as

$$U = TS - PV + \mu N + xX \tag{8.6}$$

Notice the additive nature of extensive–intensive terms in classical thermodynamic formulations and that ratios of extensive terms yield intensive terms.

For the Gibbs free energy, $G = G(N, P, T)$, the extensive dependence of G is only on N (P and T are intensive properties). Therefore, $G(\lambda N, P, T) = \lambda G(N, P, T)$. G is a homogeneous function of degree 1 with its extensive argument as N. From Euler's theorem,

$$G(N, P, T) = N\frac{\partial G}{\partial N} = \mu N \tag{8.7}$$

Equation (8.7) is an important equation and serves as a starting point in our discussion of nanothermodynamics.

Although we will not go into any more detail, the description of a thermodynamic system requires at least one extensive function and at least one intensive function. It is also true that the ratio of two extensive quantities results in an intensive quantity. Entropy per mole number (S/N) is an intensive property based on the ratio of two extensive properties. Density, an intensive property, is defined by the ratio of mass to volume—both extensive quantities. The Gibbs–Duhem equation

$$SdT - VdP + Nd\mu = 0 \tag{8.8}$$

is a combination of extensive and intensive variables. The intensive variables are not independent of one another. Knowledge of the values of two of them will yield the value of the third.

Conjugate variables consisting of an intensive property and an extensive property yield information about a thermodynamic system—for example, pressure-volume (mechanical parameters), temperature–entropy (thermal parameters), and chemical potential–particle number (material parameters). The terms *extrinsic* and *intrinsic* are used in physics and thermodynamics too but should not be confused with the extensive and intensive defined previously. In the general sense, the term extrinsic implies components, phenomena, and conditions outside a system while the term intrinsic implies the opposite.

State Functions. The value of a property in a state of thermodynamic equilibrium is called a *state function* [5]. Properties that are independent of the path taken to reach a final state from an initial state are called state functions and are represented by ΔU for internal energy. Path-independent state functions contain exact differentials where i represents the initial state and f the final state (**Table 8.1**):

$$\Delta U = \int_i^f dU = U_f - U_i \tag{8.9}$$

The first law of thermodynamics (discussed more fully in the next section) is represented as $U = q + w$, where q and w are heat and work, respectively. The variables q and w are dependent on the path taken and therefore are not state functions. Therefore, the parameters dq and dw are not exact differentials (e.g., the integration of q or w does not result in a Δq or a Δw [6]).

TABLE 8.1 *State Functions—Summary of the Five Kinds of Energy*

Name	Euler form	Differential form	Environmental variables
Internal energy	$U = q + w$ (8.10)	$dU = TdS - PdV + \mu dN$ (8.11)	S, V, N
Enthalpy	$H = U + PV$ (8.12)	$dH = TdS + VdP + \mu dN$ (8.13)	S, P, N
Helmholtz free energy	$A = U - TS$ (8.14)	$dA = -SdT - PdV + \mu dN$ (8.15)	T, V, N
Gibbs free energy	$G = H - TS$ (8.16)	$dG = -SdT + VdP + \mu dN$ (8.17)	T, P, N
Grand potential energy[a]	$\Omega = A - \mu N$ (8.18)	$d\Omega = -SdT - PdV - Nd\mu$ (8.19)	T, V, μ

[a] Process at constant temperature and chemical potential. The grand potential energy parameter is a measure of nonchemical work (e.g., mechanical work).

8.1.2 The System, Its Surroundings, and Equilibrium

Thermodynamic equilibrium implies that only infinitesimal reversible changes of thermodynamic parameters are allowed. These infinitesimal changes occur in the forward or the reverse direction (a.k.a. a reversible process). A system is in thermodynamic equilibrium when temperature, pressure, and material are in equilibrium. For example, Gibbs free energy for a system is considered to be in thermodynamic equilibrium when its overall thermodynamic potential is minimized. Mathematically, the equilibrium state of Gibbs free energy at constant temperature and pressure (T, P) is represented by

$$\Delta G_{T,P} = 0 \tag{8.20}$$

or when all chemical potentials are equal (*diffusive* or material equilibrium)

$$\mu_1 = \mu_2 = \mu_3 = \cdots = \mu_n \tag{8.21}$$

where μ represents the chemical potential for a substance.

The macroscopic compartment in which a reaction takes place is called the system. Everything else belongs to the surroundings. If there is no exchange of mass, it is called a *closed system*. Conversely, if matter is exchanged, it is considered to be an *open system*. If there is no exchange of matter and energy with the surroundings, then it is called an *isolated system*. A *reversible process* is one that is in equilibrium and causes no change to the system or its surroundings. An *irreversible process* causes the system to permanently deviate from its original state—a characteristic of nonequilibrium systems. Factors that cause irreversibility are friction, mixing, convection, turbulence, and heat transfer.

If there is no exchange of heat, it is called an *adiabatic system*. An *isothermal system* implies constant temperature, and if the pressure remains constant, the system is called *isobaric*. If volume is held constant, it is called an *isochoric process*. A *steady-state process* is one in which the internal energy is kept constant. If the process is reversible and adiabatic, entropy is considered to be *isentropic*—a system at constant entropy.

Parameters that describe a thermodynamic state consist of combinations of intensive and extensive variables: P, V, T, N, $\mathcal{N}$, μ, $\mathcal{E}$, U, S, H, A, and G. The term $\mathcal{E}$, the subdivision potential, will be defined later in the chapter. Therefore, a thermodynamic state may be thought of as the manifestation of a system with a set number of variables held constant within its intrinsic and extrinsic environment.

8.1.3 Laws of Thermodynamics

The Zeroth Law of Thermodynamics. If two thermodynamic systems are separately in thermal equilibrium with a third, they are also in thermal equilibrium with each other (**Fig. 8.2**). This law helps to compare the systems. The zeroth law is needed for the development of thermodynamics.

The First Law of Thermodynamics. If it were not for Carnot's untimely death of cholera in 1832 at age 36, he would have certainly written both the first and second laws of thermodynamics. Unpublished notes reveal that he understood the concepts quite well: "Heat is simply motive power, or rather motion, which has changed its form ... power is, in quantity, invariable in nature; it is... never either produced or destroyed."

The change in the internal energy of a closed thermodynamic system is equal to the sum of the amount of heat energy supplied to the system or created by the system and the work done on or by the system. It is about the conservation of energy:

$$E_{\text{Total}} = KE + V + U \tag{8.22}$$

where E_{Total} is the total energy, KE and V are the macroscopic kinetic and potential energies, respectively, and U is the internal energy. For a closed system, the first law is

$$\Delta U = q + w \tag{8.23}$$

and its differential form is

$$dU = dq + dw \tag{8.24}$$

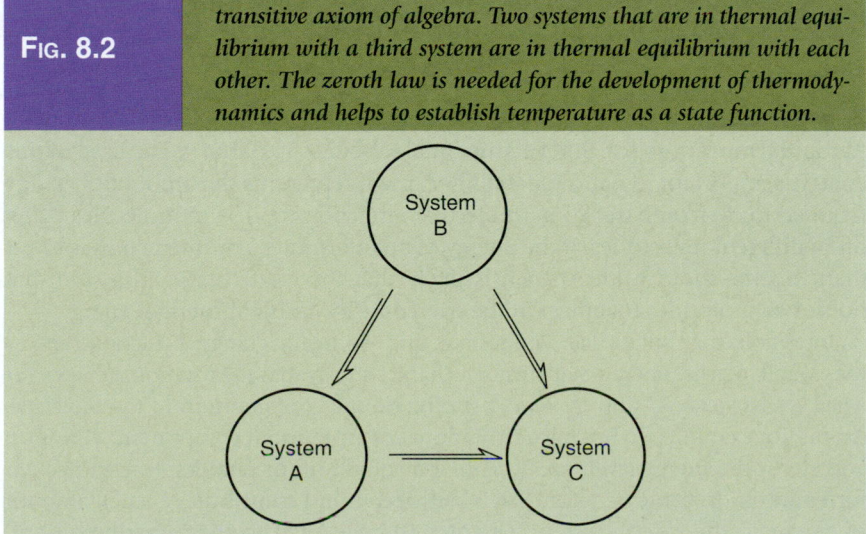

FIG. 8.2 *Illustration of the zeroth law of thermodynamics–similar to the transitive axiom of algebra. Two systems that are in thermal equilibrium with a third system are in thermal equilibrium with each other. The zeroth law is needed for the development of thermodynamics and helps to establish temperature as a state function.*

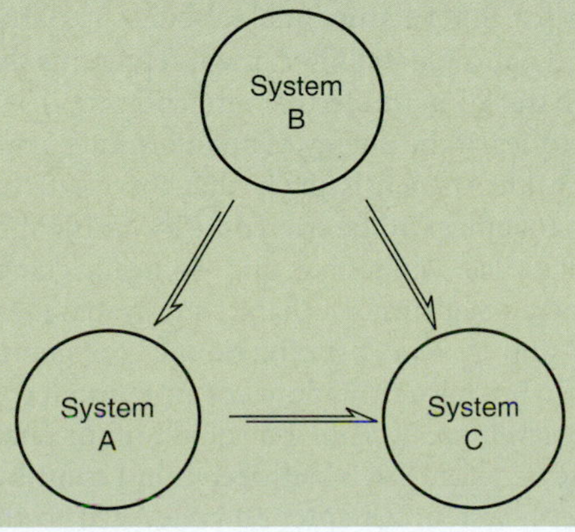

where q and w are heat and work, respectively. The following equation also presents the first law from another perspective:

$$\Delta E_{\text{Total-System}} + \Delta E_{\text{Total-Surroundings}} = 0 \tag{8.25}$$

In most cases in nanoscience, K and V equal 0. Because U is a state function and there is no consideration of the path, the internal energy of a system in equilibrium is

$$\Delta U = 0 \tag{8.26}$$

Specific heat (or specific heat capacity) is the amount of heat (energy) required to raise the temperature of 1 g of a substance by 1°C:

$$C = \frac{Q}{m\Delta T} \tag{8.27}$$

where Q is the heat content, m is the mass of the substance, and ΔT is the change in temperature. Heat capacity is expressed in joules per kelvin, the same units as for entropy S. If it is measured at constant pressure, heat capacity is called *heat capacity at constant pressure* (C_p); if is measured at constant volume, then it is called the *heat capacity at constant volume* (C_v):

$$C_v = \left(\frac{dq}{dT}\right)_V \tag{8.28}$$

and

$$C_P = \left(\frac{dq}{dT}\right)_P \tag{8.29}$$

Due the higher proportion of surface atoms in nanoparticles, both volume and surface atoms make different contributions to the specific heat capacity [7].

The molecular nature of internal energy comes in many forms. Molecules have kinetic energy equal to

$$KE = \tfrac{1}{2}mv^2 \tag{8.30}$$

At temperature K, $KE = {}^3/_2RT$ where R is the gas constant ($R = 8.314\ \text{J}\cdot\text{mol}^{-1}\cdot\text{K}^{-1}$). At room temperature (ca. 300 K), this equals about 3.8 kJ · mol^{-1}. The level of this energy is important in nanomaterials because it represents the amount of energy required to overcome weak intermolecular attractions and send molecules flying off in different trajectories. If the energy of an intermolecular interaction is lower than this threshold value, then it is likely that the electrostatic attraction that holds two molecules together will be overcome by ambient thermal energy.

Intermolecular, rotational, vibrational, and electronic energy also contribute to the overall internal energy. Statistical mechanics shows that the rotational energy for a gas lies between RT and ${}^3/_2RT$, depending on the configuration of the molecule. Atoms, for example, do not have a rotational or vibrational component. Vibration and electronic energy levels are described by quantum mechanics. Intermolecular forces are electrostatic in nature. Each hydrogen bond contributes a small amount of energy to the overall system. Chapter 10 is devoted to the presentation and discussion of intramolecular and intermolecular bonding. In summary [5],

$$U_{Total} = U_{Trans} + U_{Rot} + U_{Vib} + U_{Elec} + U_{Rest} \quad (8.31)$$

where U_{Rest} is the *molar rest mass energy* of electrons and nuclei. Without exception, all of these forms of internal energy apply to the combined internal energy of nanomaterials.

The Second Law of Thermodynamics. William Thomson (Lord Kelvin) in the 1840s and Max Planck later are credited with the formulation of the second law. In the mid-1800s, Rudolf Clausius published his fundamental thesis on the second law, called the *Clausius statement*. J. Willard Gibbs stated that modern thermodynamics began with the Clausius statement. Clausius is also credited with coining the term entropy. Gibbs was the first scientist to apply the second law of thermodynamics to experiment and was awarded the Copley Medal of the Royal Society of London in 1901 for his work and accomplishment. Gibbs's award was for "the exhaustive discussion of the relation between chemical, electrical and thermal energy and the capacity for external work."

The essence of the second law is that, for an isolated system, conditions proceed to a state of maximum entropy. It also implies that 100% efficiency is impossible and that the existence of perpetual motion machines is equally impossible. The second law is about entropy and equilibrium. According to P. W. Atkins in his book, *Physical Chemistry,* the second law exploits the parameter of entropy to "identify the *spontaneous changes* among the *permissible changes*" (e.g., permissible with regard to the first law) [6]. In other words, irreversible changes increase the entropy of a system. Reversible reactions, by definition, do not generate changes in entropy. Can you think of any system in the real world that is ideal in this way?

Entropy is defined as a measure of order within a closed system. Entropy is also defined as the thermal energy unavailable to do useful work (please keep this statement in mind for later discussion). The total entropy of any isolated thermodynamic system tends to increase over time,

$$\frac{dS}{dt} \geq 0 \quad (8.32)$$

where t is time, S is entropy (joules per kelvin) and approaches a maximum value. Equation (8.32) encapsulates the concept of the second law and signifies the irreversibility of most natural processes. Many experiments and simulations of nanomaterials focus on the validity of the second law—in particular, violations of the second law by small systems under short timescales [8].

The Third Law of Thermodynamics. The entropy of a perfect crystal approaches 0 K at absolute zero temperature. The third law quantifies entropy in the following form:

$$\Delta S \rightarrow 0 \text{ as } T \rightarrow 0 \text{ K} \quad (8.33)$$

The third law was developed by Walther Nernst in the early part of the twentieth century. A modified version of the third law was proposed by M. Randall and G. N. Lewis. They added that not only does the change in entropy approach zero, but also the value of entropy itself becomes zero. According to the Boltzmann relation ($S = k \ln \mathcal{W}$, where $\mathcal{W}$ is the number of ways energy can be distributed), when $W = 1$ (e.g., only one energy state), then $S = 0$. J. P Abriata and D. E. Laughlin in 2004 proposed that solids in equilibrium at 0 K exist in the form of pure elements or atomically ordered phases [9]. However, apparent violations of the third law have been exhibited by one-dimensional classical systems [10].

Application of the third law was the driving force behind imaging the physical structure of carbon nanotubes [11]. Obviously, colder samples provide for better scanning tunneling microscope (STM) images. At colder and colder temperatures, molecular vibrations diminish. Scanning tunneling microscope imagery of a bundle of single-walled carbon nanotubes was accomplished at 78 K and is shown in **Figure 8.3.**

The Fourth Law of Thermodynamics. Although there is no clear declaration of the fourth law of thermodynamics, many researchers have tried to implement

FIG. 8.3

Three carbon nanotubes imaged by STM are shown in the figure. The image was acquired under the following conditions: constant current STM taken in UHV at 78 K, with a sample bias −1.5 V, tunneling current of 0.2 nA, and a scan size of 17-nm image of three carbon nanotubes on Au (111) surface. Recall that we discussed surface reconstruction in chapter 6. A herringbone pattern reconstruction on the Au (111) surface is also visible. The Au (111) reconstruction consists of partial dislocation ridges due to a uniaxial contraction (4.2%) along the [110] direction. There are 23 atoms for every 22 sites. The reconstruction results from atoms in bridge sites at elevated positions and other regions where atoms occur in the hollows of fcc ABCABC … and hcp ABABAB … crystal structures. The point of the image in this text, however, is to show how the third law of thermodynamics is demonstrated during the acquisition of this image. Quite impressive!

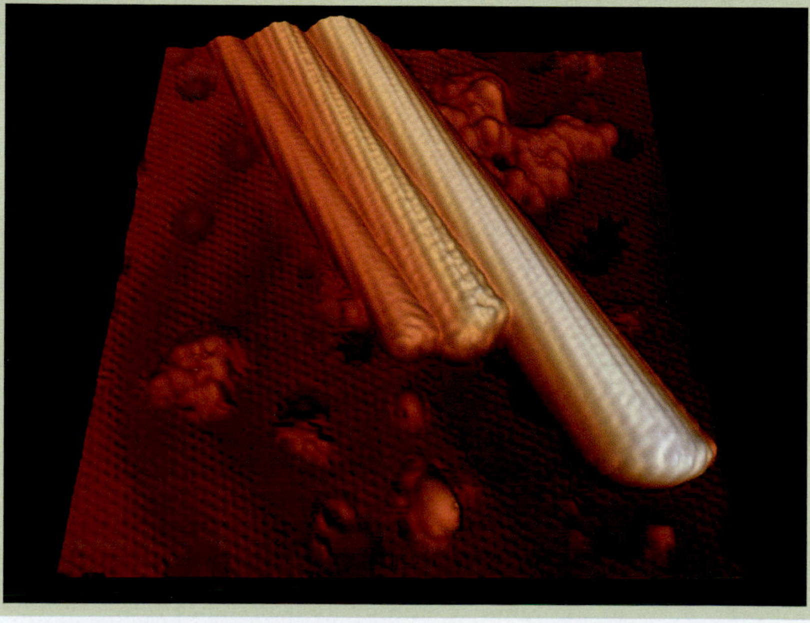

Source: Image is courtesy of Ping Xie, Lieber Group, Harvard University. With permission.

one. Keep yourself posted on the developments. Efforts to generate a fourth law of thermodynamics (and even fifth and sixth laws) have been ongoing since the late nineteenth century. Ludwig Boltzmann proposed that "the fundamental life-struggle in the evolution of the organic world is available energy" [12]. He was aware of a special thermodynamics that applied to living things.

The Onsager reciprocal relations, an attempt to apply the laws of thermodynamics to nonequilibrium systems, were developed by Nobel laureate Lars Onsager of Norway in 1931. His formulation was compelling to the extent that many scientists refer to it as the fourth law of thermodynamics [13]. Any detailed discussion of Onsager relations is way beyond the scope of this text. They will, however, be displayed later on in the chapter. The *maximum power principle* of A. Lotka in 1922 implied that various versions a fourth law exist [13,14]. These fourth laws were all based on the notion of reconciling thermodynamics with life processes.

8.1.4 Fundamental Equations of Thermodynamics

Pressure (P), volume (V), temperature (T), the number of moles or amount of material content (n or N), energy (E), and work accomplished (W) are familiar parameters. Terms like specific heat (C), chemical potential (μ), enthalpy (H), entropy (S), and free-energy (G) are also fundamental to thermodynamics.

Internal energy is the heat content of the system under constant volume and temperature. *Enthalpy* (H) is the heat content of the system under constant pressure and temperature. *Entropy* (S) contributes to the spontaneous change in the system and is the measure of disorder at the molecular level. The tendency towards greater disorder is a driving force in reactions. *Free energy* is defined as the amount of useful work that can be obtained from a system. There are two major types of free energy: the *Gibbs free energy*, G, and the *Helmholtz free energy*, A. The Gibbs form is valid within the frame of constant pressure and temperature; the Helmholtz form is valid at constant volume and temperature. Basic thermodynamic equations are listed in **Table 8.2** for reference without further ado and without further elaboration.

TABLE 8.2 *Thermodynamic Parameters*

Parameter—constant	Common form		Comments	
Ideal gas law	$PV = nRT$	(8.34)	$R = 8.314\ J \cdot mol^{-1} \cdot K^{-1}$	(8.35)
Work	$w = \int F_x dx + \int F_y dy + \int F_z dz$	(8.36)	$w = -\int PdV$	(8.37)
First law internal energy	$\Delta U = q + w$	(8.38)	$dU = TdS - PdV$	(8.39)
	$\Delta U = q - \int PdV$	(8.40)		
Enthalpy	$\Delta H = \Delta U + P\Delta V$	(8.41)	$dH = TdS + VdP$	(8.42)
Entropy	$\Delta S = \frac{1}{T}\int dq$	(8.43)		
Heat capacity	$C_P = \frac{dq_P}{dT} = \left(\frac{\partial H}{\partial T}\right)_P$	(8.44)	Constant P	

(continued)

Table 8.2 (Contd.) *Thermodynamic Parameters*

Parameter—constant	Common form	Comments
Heat capacity	$C_V = \frac{dq_V}{dT} = {}_P\left(\frac{\partial U}{\partial T}\right)_V$ (8.45)	Constant V
Joule–Thomson effect	$\mu_{JT} = \left(\frac{\partial T}{\partial P}\right)_H$ (8.46)	Constant H
Second law	$\Delta S \geq 0$ (8.47)	No such thing as 100% efficiency
Gibbs free energy	$\Delta G = \Delta H - T\Delta S$ (8.48)	$dG = -SdT + VdP$ (8.49) Constant P, T
Helmholtz free energy	$\Delta A = \Delta U - T\Delta S$ (8.50)	$dA = -SdT - PdV$ (8.51) Constant V, T
Chemical potential	$\mu_i = \left(\frac{\partial G}{\partial n_i}\right)_{T,P,n_j}$ (8.52)	Constant T, P and n_j; intensive property Gives free energy after adding or removing material from an open system
Third law	$\text{Lim}_{T \to 0}\, \Delta S = 0$ (8.53)	by Walther Nernst
Standard equilibrium constant	$\Delta G^o = -RT\ ln\ K_P^o$ (8.54)	For chemical reactions
Gibbs equation	$dG = -SdT + VdP + \sum_i \mu_i dn_i$ (8.55)	The fundamental equation of thermodynamics
Clapeyron equation	$\frac{dP}{dT} = \frac{\Delta H}{T\Delta V}$ (8.56)	Phase changes for a one component system
Clausius–Clapeyron equation	$\frac{d\ln P}{dT} = \frac{\Delta H}{RT^2}$ (8.57)	Phase equilibria
Gibbs–Helmholtz equation	$\frac{d}{dT}\left(\frac{\Delta G^o}{T}\right) = \frac{-\Delta H^o}{T^2}$ (8.58)	Used to calculate the changes in Gibbs energy as a function of temperature
van't Hoff equation	$\frac{d\ln K_P^o}{dT} = \frac{\Delta H^o}{RT^2}$ (8.59)	Good way to find ΔG^o, ΔH^o, and ΔS^o if K_p, the equilibrium constant, is known from slope of ln K versus $1/T$
Kelvin equation	$\ln\frac{P}{P_o} = \frac{2\gamma V_m}{rRT}$ (8.60)	Vapor pressure of liquid over curved surfaces
Raoult's law	$P_i = x_i P_i^o$ (8.61)	Molecular view of vapor pressure depression—for liquids (usually volatile)
Henry's law	$P_i = K_i x_i$ (8.62)	Partial vapor pressure of a solute in an ideally dilute solution
Ideal solution chemical potential	$\mu_i = \mu_i^o + RT\ln x_i$ (8.63)	$x_i = \exp\left[\left(\mu_i^{id} - \mu_i^o\right)/RT\right]$ (8.64)
Gibbs–Duhem equation	$\sum_i \mu_i dn_i = 0$ (8.65)	The increase of μ of one substance causes the decrease of μ for another Constant T, P
Excess Gibbs functions	$G^{XS} = G - G^{id}$ (8.66)	Deviation from ideality
Surface tension	$\gamma = \left(\frac{\partial G}{\partial A}\right)_{P,T}$ (8.67)	Units of $J \cdot m^2$

8.1.5 Equilibrium Constant and Reaction Kinetics

The link between thermodynamics and kinetics for a generic reaction given here between species A and species B to produce species C,

$$A + B \underset{k_{-1}}{\overset{k_1}{\rightleftharpoons}} C, \tag{8.68}$$

is given by the following relation involving the equilibrium constant and the rate constants (k):

$$K = \frac{\Pi[\text{Products}]}{\Pi[\text{Reactants}]} = \frac{[\text{C}]}{[\text{A}][\text{B}]} = \frac{k_1}{k_{-1}} \tag{8.69}$$

The link between the equilibrium constant to Gibbs free energy under standard conditions is

$$\Delta G^{\circ} = -RT \ln K_{\text{P}}^{\circ} \tag{8.70}$$

The activation energy, E_a, of a generic chemical reaction can be found by the Arrhenius relationship:

$$\ln k = \ln A - \frac{E_a}{RT} \tag{8.71}$$

where k is a rate parameter of the reaction and A is the Arrhenius pre-exponential factor. A is approximated by classical collision theory:

$$A = \mathcal{N}_A \sigma \sqrt{\frac{8kT}{\pi\mu}} \tag{8.72}$$

where $\mathcal{N}_A$ is the Avogadro's number, σ is the cross-section, and μ is the reduced mass. Activation energy E_a is found from the slope of (ln k) versus ($1/T$) equal to ($-E_a/R$).

Nanoparticle catalysts are used in many applications where the Arrhenius relationship is applied. Catalytic systems do not represent ideal classical thermodynamic systems due to the presence of steady-state and nonequilibrium factors. The catalytic synthesis of single-walled carbon nanotubes from nanosized catalysts is not adequately explained by invoking classical thermodynamic principles alone. We will discuss this in more detail shortly.

Reaction kinetics implies some degree of nonequilibrium character (e.g., steady-state reactions and irreversibility depending on the type of reaction). The steady-state approximation is derived from kinetic theory. One of the characteristics of steady-state processes is that the concentrations of intermediates remain constant throughout the reaction:

$$\frac{dC_i}{dt} = 0 \tag{8.73}$$

where C_i is the concentration of the intermediary component i.

Catalysts, a broad class of nanomaterials, are able to influence the rate of a chemical reaction. Ideally, catalysts are not consumed in the reaction and are able to cycle indefinitely. Catalytic reactions are often discussed from a kinetic point of view. Following activation and induction, catalytic processes endure a steady-state phase until the end of the run, when the feed material is stopped, the catalyst is deactivated, or the temperature is turned down. We are quite familiar with the catalytic function of enzymes in living things—at body temperature nonetheless! Once again, life is able to impart the kinetic, steady-state aspect to the nano perspective. What systems exhibited by living things fall in the domain of nonequilibrium processes, at least as defined in this section?

8.2 Statistical Mechanics

We are familiar by now with macroscopic properties such as heat capacity, internal energy, enthalpy, entropy, conductivity (heat, electrical), viscosity, surface energy, dielectric behavior, magnetism, and many more. On the other hand, molecular properties involve intramolecular and intermolecular forces and phenomena such as vibrations, rotations, geometry, and structure [5]. Statistical mechanics couples knowledge of atomic/molecular energy levels with the macroscopic expression of properties of bulk materials and serves as the link between quantum mechanics and classical thermodynamics. It is the bridge between classical bulk thermodynamics and molecular properties [6].

James Clerk Maxwell and Ludwig Boltzmann established the basis for statistical mechanics with their development of statistical distributions and the kinetic theory of gases in the mid- to late 1800s. The Maxwell–Boltzmann probability distribution function forms the foundation of the kinetic theory of gases. They proposed that the occupation (distribution) of energy states is proportional to $\exp(-\Delta E/k_B T)$. In 1889, Gibbs, applying classical thermodynamic principles, began to work in earnest on the subject of statistical mechanics. He had published much of that work in books in 1902, that provided a foundation for the later development of quantum theories. Albert Einstein also made contributions to statistical mechanics at the same time [5].

There is a field called *equilibrium statistical mechanics* and, not surprisingly, one called *nonequilibrium statistical mechanics*. Nonequilibrium statistical mechanics involves mathematical treatment of phenomena such as transport and chemical reaction rate. Due to its complexity, this field of nonequilibrium statistical mechanics is not as well developed [5]. We shall adhere to the former in this section. Our objective is to simply get a feel of the discipline without too much detail.

8.2.1 Microstates and Macrostates

The term "microstate" refers to the quantum state of a system. Ludwig Boltzmann, perhaps one of the most brilliant physicists ever, derived his famous equation, the *Boltzmann principle,* in the latter part of the nineteenth century. It links entropy (S) with probability and serves as the basis for statistical mechanics:

$$S = k \log W \tag{8.74}$$

where $\mathcal{W}$, weight, represents the number of microstates (system energy levels) that have a significant probability of being occupied [5]. Another popular form of the equation uses the natural log rather than $\log_{10}$ ($S = k \ln \mathcal{W}$). $\mathcal{W}_i$ is often used to denote the degeneracy (distinct states with the same energy) of energy level E_i. Regardless which form is used, the equation states that the systems energy is proportional to the probability of population of energy states. As $\mathcal{W}$ increases, the system's entropy also increases. An inscription of the formula is carved into Boltzmann's gravestone in Vienna (**Fig. 8.4**).

The weight, $\mathcal{W}$, of a configuration comes from probability theory and is represented by

$$\mathcal{W} = \frac{N!}{n_0!\, n_1!\, n_2!\, \ldots} \tag{8.75}$$

where N is the total number of molecules and n represent individual microstates (specific sets of populations of molecules). Each molecule of population n_x is represented by an energy ε_x. Note that $n_0! = 1$ [6]. The ability to predict microscopic properties is the basic advantage of statistical mechanics over classical thermodynamics. Both theories are linked through the second law of thermodynamics by way of the concept of entropy. However, entropy in classical thermodynamics can only be known empirically, whereas, in statistical mechanics, it is a function of the distribution of the system on its microstates. The essence of statistical mechanics is known as the "equal a priori probability postulate": "Given an isolated system in equilibrium, it is found with equal probability in each of its accessible microstates."

This postulate is a fundamental assumption in statistical mechanics and indicates that any system in equilibrium will have an equal probability for any of its available microstates. In the thermodynamic limit of $N \to \infty$, the dominating configuration is overwhelmingly the most probable [6].

8.2.2 Canonical Ensembles

An *ensemble* is an infinite number of noninteracting (e.g., chemically noninteracting) systems of differing microstates that comprise the same macrostate [5]. According to some, an ensemble is composed of interacting (e.g., energetically) molecules that all have the same temperature [6]. A *canonical ensemble* (from the Latin *canon,* indicating "measuring line, rule, collection") is associated with constants N, V, and T (e.g., all systems have the same temperature); a *microcanonical ensemble* with N, V, and E (or U) constants (e.g., all systems have the same energy); and the *grand canonical ensemble* with constants μ, V, and T (e.g., it is an open system). We only address equilibrium statistical mechanics in this section.

For the canonical ensemble, one of the underlying principles of such a system is that, at constant T, V, and N, all quantum states with the same energy have an equal probability of occurring [5]. The canonical (molecular) partition function is

$$Z = \sum_j e^{-\beta E_j} = \sum_j e^{-E_j/kT} \tag{8.76}$$

Fig. 8.4 *Ludwig Boltzmann's grave in Vienna. His famous equation, $S = k \log W$, is carved into the stone. Sadly, there was much controversy over his theories and perhaps that contributed to his suicide. Truly, Boltzmann was one of the most brilliant minds of all time—someone who understood the major driving force of the universe: entropy.*

Source: en.wikipedia.org/wiki/Image:Zentralfriedhof_Vienna_-_Boltzmann.JPG.

where Z is the sum over all states, β is equal to $1/k_B T$ (the thermodynamic beta), and E_j is the "quantum-mechanical energy" level of the macroscopic system [5]; $e^{-E_j/kT}$ of state j is proportional to the probability p_j of the state j:

$$p_j(E_j) = \frac{e^{-E_j/kT}}{Z} \tag{8.77}$$

This is the probability that a system of fixed volume, temperature, and composition is in quantum state j of energy E_j [5]. And for the system that has energy of E_i,

$$p_i(E_i) = \mathcal{W}_i \frac{e^{-E_i/kT}}{Z} \tag{8.78}$$

where $\mathcal{W}_i$ is the number of degenerate quantum states with energy E_i. Statistical mechanics is absolutely grounded in probability theory.

The following statistical mechanical relations are used to express pressure, internal energy, entropy, the Helmholtz function, and chemical potential. Notice how the mathematical expressions are tied into classical thermodynamic forms but with the addition of a statistical component:

$$P = kT\left(\frac{\partial \ln Z}{\partial V}\right)_{T.N} \tag{8.79}$$

$$U = kT^2\left(\frac{\partial \ln Z}{\partial T}\right)_{V.N} \tag{8.80}$$

$$S = \frac{U}{T} + k \ln Z \tag{8.81}$$

$$A = -kT \ln Z \tag{8.82}$$

$$\mu = -RT\left(\frac{\partial \ln Z}{\partial N}\right)_{T.V,N} \tag{8.83}$$

According to I. N. Levine's *Physical Chemistry* [5], the steps needed to calculate macroscopic thermodynamic parameters are as follows: (1) solve the Schrödinger' equation to obtain quantum mechanical energies (E_j) for the system, (2), evaluate the canonical partition function Z, and (3) use ln Z to calculate the system's macroscopic thermodynamic properties [5]. Z is approximately equal to

$$Z = \mathcal{W}e^{-U/kT} \tag{8.84}$$

and

$$\ln Z = \ln \mathcal{W} - \frac{U}{kT} \tag{8.85}$$

Thus, the Boltzmann principle (equation 8.74) can be derived by combining equations (8.81) and (8.85) [5]:

$$S = \frac{U}{T} + k \ln Z = \frac{U}{T} + k \ln \mathcal{W} + k\left(-\frac{U}{kT}\right) = k \ln \mathcal{W} \tag{8.86}$$

An interesting paradox to ponder is that entropy only has significance for a large number of atoms and molecules, yet it is derived from molecular properties [5]. How is this apparent paradox reconciled? We discover later that it is possible to estimate the entropy for a single molecule.

8.2.3 Energy (Molecular) Partition Functions

A small case z is used to represent molecular partition functions and the small case Greek epsilon is used to represent molecular energies. The molecular energy is the sum of all the forms of energy of a molecule presented in a manner similar to equation (8.31):

$$\varepsilon_{\text{Molecular Energy}} = \varepsilon_{\text{Trans}} + \varepsilon_{\text{Rot}} + \varepsilon_{\text{Vib}} + \varepsilon_{\text{Elec}} \tag{8.87}$$

and z, the molecular partition function, is the product of all the probabilities of the individual molecular partition functions:

$$z = (z_{\text{Trans}})(z_{\text{Rot}})(z_{\text{Vib}})(z_{\text{Elec}}) \tag{8.88}$$

or, given in the usable form,

$$\ln z = \ln z_{\text{Trans}} + \ln z_{\text{Rot}} + \ln z_{\text{Vib}} + \ln z_{\text{Elec}} \tag{8.89}$$

Relating Z to the molecular partition functions yields

$$\ln Z = N \ln z_{\text{Trans}} + N \ln z_{\text{Rot}} + N \ln z_{\text{Vib}} + N \ln z_{\text{Elec}} - N(\ln N - 1) \tag{8.90}$$

For example, internal energy, from equation (8.80) is then represented by

$$U = kT^2\left(\frac{\partial \ln Z}{\partial T}\right)_{V.N} = NkT^2\left[\left(\frac{\partial \ln z_{\text{Trans}}}{dT}\right)_V + \left(\frac{d\ln z_{\text{Rot}}}{dT}\right) + \left(\frac{d\ln z_{\text{Vib}}}{dT}\right) + \left(\frac{d\ln z_{\text{Elec}}}{dT}\right)\right] \tag{8.91}$$

There are many excellent sources available if you wish to pursue this fascinating field on your own. The purpose of the section on statistical mechanics is simply to show how thermodynamic properties can be derived from the bottom up using the tools of statistical mechanics.

8.3 Other Kinds of Thermodynamics

Near-equilibrium, nonlinear, nonequilibrium, steady-state, and pseudoequilibrium thermodynamics are terms you will encounter often—especially if you choose to study nanomaterials. There are also *quasi-equilibria*. Most nonequilibrium thermodynamic experiments are conducted at near-equilibrium conditions (conditions not too far off equilibrium). All types of thermodynamics are part of the *thermodynamic equilibrium continuum*. The discriminator among them is the level of complexity involved in describing their respective systems. Pseudoequilibrium (*pseudo-*, from the Greek meaning "false, feigned or erroneous") thermodynamics is a bit more difficult to define. The term implies that there is some kind of equilibrium apparent in the system but one that does not quite conform to the criteria required for a true equilibrium. Therefore, it is a false equilibrium! It may refer to a system that is kinetically confined but still demonstrates some degree of reversibility—perhaps called pseudoreversibility.

If one were to step back, contemplate, and pick one word to describe conventional thermodynamics, that word would probably be *static*. Mostly, classical thermodynamics is about macroscopic evaluation of a system in equilibrium by the use of extensive and intensive parameters. According to M. Agrawal of Stanford University in his 2005 article "Basics of Irreversible Thermodynamics" [68]:

> The dynamic part is a stretch indeed when the most dynamic happening is whether or not there is spontaneity between the initial and final state or if a change is so slow that the system is evolving through a series of quasi-equilibrium states.... The dynamics is certainly evident in non-equilibrium situations where demonstrable evolution is taking place.

Nonequilibrium thermodynamics implies irreversibility, fluctuations, dissipation, open systems, and "time independent (steady-state) thermodynamic systems" [14]. Nonequilibrium at the extreme implies convection, flow, and perhaps even turbulence. Steady-state thermodynamics is a configuration in which there is no accumulation of heat or matter within the system and extensive properties within the system are time independent. Steady-state thermodynamics and nonlinear thermodynamics are considered to be a subset of nonequilibrium thermodynamics. According to PBS.org, nonlinear, nonequilibrium thermodynamics is

> A branch of physics developed in the latter half of the twentieth century that deals with systems of particles far from the near equilibrium, conditions studied in classical thermodynamics and which are governed by complex, non-linear forces. Significant attempts have been made to extend this theory into the realm of living (self-replicating) organisms.

In the early 1990s, the development of micromanipulation methods allowed researchers to study the energy fluctuations of small systems [15]. In macroscopic thermodynamics, behavior is predictable and therefore fluctuations (e.g., deviations from ideal behavior) are averaged over the whole system and therefore rendered insignificant. In a nonequilibrium small system, fluctuations can dominate as materials acquire smaller dimensions [15].

8.3.1 The Onsager Relations

Onsager equations are in actuality fourth law interpretations of thermodynamic potentials, forces, and flows. These relations define systems out of equilibrium but suggest that there exist local equilibria. For example, evaluation of thermal conductivity by equilibrium methods is problematic and is based on $\Delta T \to 0$, although a real steady-state temperature gradient is in place. The best way to analyze this process is to quantify the rate of change of the entropy [16]. One fundamental expression in thermodynamics relates entropy to other extensive properties like internal energy and volume and shows how the change in entropy is a function of intensive properties such as temperature, pressure, and chemical potential:

$$dS = \frac{1}{T}dU + \frac{P}{T}dV - \sum_{i=1}^{n}\frac{u_i}{T}dn_i \tag{8.92}$$

The corresponding thermodynamic forces are the gradients $1/T$, P/T, and μ/T [16].

$$\frac{\partial S}{\partial t} = \sum_i \mathbf{J}_i \cdot \mathbf{X}_i \tag{8.93}$$

where **J** is a heat flux and **X** is a force. The force in this example is supplied by a temperature gradient, ∇T. The entropy is defined as a function of all extensive quantities where $\mathbf{I}_i$ represents a conjugate intensive variable involving the entropy (Ext_i) [17]:

$$\mathbf{I}_i = \left(\frac{\partial S}{\partial \text{Ext}_i} \right) \tag{8.94}$$

J is defined by a linear matrix of coefficients that is expressed by an Onsager equation:

$$\mathbf{J}_i = \sum_j \mathbf{L}_{ij} \cdot \nabla \mathbf{I}_i \tag{8.95}$$

This expression is valid in the regime of small forces and slow variation. Quantification of entropy is a key factor in nanothermodynamics.

Onsager's PhD dissertation, titled *Onsager Reciprocal Relations,* was rejected by the faculty at the Norwegian Institute of Technology due to "insufficiency." His second dissertation, a solution of the *Mathieu equation*, was labeled as "incomprehensible." He was fired from his position at Johns Hopkins in 1928 for "incomprehensible teaching" and for the same reason at Brown University in 1933. When he was hired at Yale, a controversy erupted because he did not have a PhD He finally obtained a PhD in 1935 and then went on to win the Nobel Prize in chemistry in 1968.

8.3.2 Nonequilibrium Thermodynamics

We are all familiar with irreversible reactions. An example of an irreversible reaction of a material that is kinetically stable is the explosion of nitroglycerin. Nitroglycerin is thermodynamically unstable but kinetically stable unless you start shaking the liquid or drop it. The reaction is, for all practical purposes, irreversible. Fullerenes are thermodynamically less stable than graphite but are kinetically stable. It all depends on the environment. The thermodynamic behavior of nanomaterials is very much dependent on their immediate environment. Does this imply that nanomaterials possess inherently nonequilibrium characteristics?

A subtle example of kinetic stability is provided by the diamond–graphite equilibrium. Diamond is only stable at very high pressure. Graphite is stable at atmospheric pressure. Although diamond exists at STP conditions, the transformation of diamond into graphite is extremely slow (a kinetic process). A way to accelerate the process is to provide heat and more pressure.

The bond strain in C_{60} is about the highest known in carbon materials, yet the material does not spontaneously decompose at room temperature. Kinetic stability is related to the threshold energy required to start a reaction. The entropy produced in an irreversible reaction is overall positive:

$$\Delta S_{\text{Irreversible}} > 0 \tag{8.96}$$

Kinetic stability is also demonstrated by colloids that are thermodynamically metastable or unstable with respect to their bulk counterparts because they do tend to exist for quite some time. Why is that? Obviously some form of environmental support must be in place to ensure that colloids persist in their nano- to microscale form.

An automobile engine is an example of a steady-state system. Material is added at a constant rate (unless you have a lead foot) and material is released in the form of gases—CO_2, H_2O, and others, also at a relatively constant rate. Heat is dissipated at a constant rate and the temperature of the system is held constant by transfer of heat to water (unless your car has bad hoses). Once the power cycle is terminated, the system goes to equilibrium with its surroundings and assumes the same temperature. In living systems, this would mean *cellular death*. Combustion itself is a highly nonequilibrium thermodynamic condition that is regulated by steady-state systems such as the automobile engine. The amount of energy released by the gasoline and the equilibration of the engine temperature with the environment are thermodynamic processes. The rate at which gasoline is burned and the scrubbing of exhaust gases by catalysts are kinetic parameters. The operation of the engine is an irreversible nonequilibrium steady-state process.

Use of the steady-state approximation (equation 8.73) simplifies rate equation calculations, as in the case of intermediates in catalytic processes. Rates in catalytic reactions are expressed as $(\text{mol} \cdot \text{min}^{-1} \cdot \text{g}^{-1})$ of catalyst or $(\text{mol} \cdot \text{min}^{-1} \cdot \text{M}^{-2})$ per surface area of catalyst. Steps in the catalytic reaction process, in order of occurrence, include diffusion of reactant to catalyst surface, adsorption of reactant on the catalyst surface, the targeted chemical reaction, desorption of products from the catalyst surface, and diffusion of the products away from the catalyst. Particle size and temperature affect the progress of the catalytic reaction. Although the change in energy between the starting material and the product is considered to be a state function and, as such, dwells in the domain of classical thermodynamics, it is obvious that catalytic processes are characterized by far more complexity. In other words, the path assumes a high level of importance in catalytic processes, especially if commercial factors are involved. Many steady-state processes exist in cellular processes.

Consider the nonequilibrium convection pattern mechanism of a Bénard cell, an inorganic system comprised of water and copper plates (**Fig. 8.5**). A thin layer of water is trapped between two petri-dish-sized Cu plates. As heat is added, molecular equilibrium is first established until a critical level is reached, at which point convection of water begins. A nonequilibrium self-organized structure, known as a *dissipative* structure, is formed. Upon elimination of the heat source, the self-organized patterns disappear and the equilibrium condition is restored. The capability to perform this kind of self-organized work is driven by the steady-state input of heat energy. Self-organization is an interesting consequence of nonequilibrium states. Biological entities are considered to be self-organizing, self-replicating, dissipative structures [18–20].

FIG. 8.5 *A Bénard cell is a flat, circular dish filled with a thin layer of water that is uniformly heated from below. If the water layer is of appropriate thickness, heat is dissipated rapidly through the self-assembly of hexagonal patterned convection cells (energy gradients) rather than turbulent boiling. The Bénard cell is a dissipative structure. Does the Bénard cell provide an adequate model for living systems that are also dissipative in nature? One criterion to consider is that of the enduring structure. Living systems are characterized by enduring structures that are capable of accomplishing work. Is a Bénard cell an enduring structure capable of accomplishing work beyond that of maintaining its cellular pattern? Living systems, for example, are able to create less entropy from ordered structures. "The 2nd Law cannot be violated, but it can be stalled" (interesting discourse from J. Fournier,* Evolution, Entropy and Work, *www.geoman.com/jim/entropy.html).*

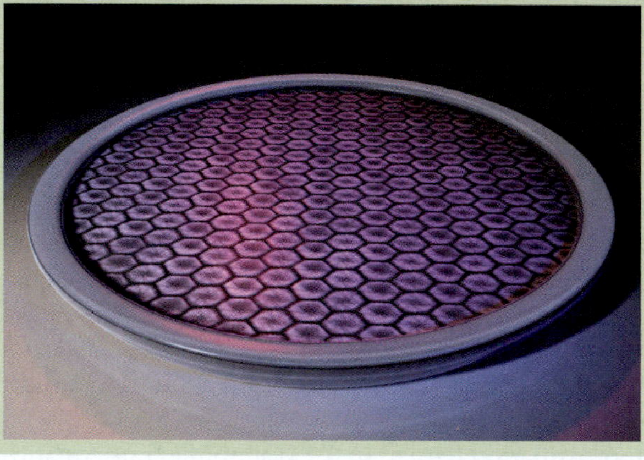

Source: Courtesy of Professor Anil K. Rao, Department of Biology, the Metropolitan State College of Denver.

8.3.3 The Concept of Pseudoequilibrium

According to one definition, *pseudoequilibrium* is a state of a system in which the distribution of a component of interest exists between the solid and liquid phase (perhaps solid and gaseous phase) and does not change upon further equilibration [21]. The time of equilibration required to reach this state of pseudoequilibrium depends on experimental conditions. Pseudoequilibria phenomena are often encountered in sorption and catalytic studies that are eventually uncovered in subsequent kinetic analysis. In a similar vein, we can say that pseudoequilibrium implies a metastable state of a solid in which the content of free energy is higher than that characteristic of the equilibrium state. A pseudoequilibrium rate constant can be considered to be associated with a process that is constantly unbalancing the equilibrium.

Pseudoequilibrium is observed in the conditions of carbon plasmas for diamond deposition [22]. Pseudoequilibrium components are detected in the conversion of methanol to CO_2 and H_2 in the pores of a catalyst after analysis of the equilibrium constant in which significant deviations were found to occur. Kinetic confinement of the catalytic reaction and thermal gradients in the pores contributed to the pseudoequilibrium state [23]. The conversion of methane to single-walled carbon nanotubes under CVD conditions by Fe–Mo catalysts is reported to closely approximate pseudoequilibrium conditions [24]. The

Michaelis–Menten constant K_M in enzyme kinetics is considered to be a dynamic or pseudoequilibrium constant.

Protein adsorption phenomena are often associated with a pseudoequilibrium state. The protein system is considered to be a self-contained small thermodynamic system complete with associated water molecules. Such small systems are susceptible to thermal fluctuations in the immediate environment. Fluctuations in internal energy, volume, and conformational changes of the protein are therefore likely to occur when proteins are exposed to thermal fluctuations. Introduction of a solid substrate like silicon (e.g., a new environment) causes transient conformations to become stable at the solid–liquid interface during adsorption [25]. The protein is able to adsorb in many different configurations. The adsorption can be rapid and reversible but dependent on prevailing environmental and kinetic conditions.

One can safely conclude that true equilibrium most likely never exists in vivo in plasmas, aquatic environments, and other systems of great complexity. Ideal thermodynamics, in the natural world, is not the rule for dynamic systems but more like an exception. The closest a real system comes to the thermodynamic ideal is in the form of a dilute ideal gas or when two blocks of purely homogeneous materials are in thermal equilibrium.

8.3.4 Cellular and Subcellular Systems

We have already discussed terms such as irreversibility, nonequilibrium and kinetic stability, and pseudoequilibria and have yet to be introduced to concepts such as fluctuations, dissipation, nonextensivity, and nonintensivity. We have also indicated on several occasions how nanotechnology—nanomaterials in particular—is fundamental to life and its development. Living processes and structures cannot be explained by classical thermodynamics even though the specter of classical thermodynamics permeates all of biology. The fourth law of thermodynamics may be coming soon, but until then, let us take a brief look into nonequilibrium thermodynamics and apply some of its concepts to biological phenomena.

Stephanie E. Pierce of the Department of Biological Sciences at the University of Alberta stated in 2002 that the organization of information systems in living things follows the second law in that speciation occurs when these systems become complex (e.g., addition by mutation) and disorganized and are subjected to environmental forces [18]. She states that the "entropic drive to randomness underlies the phenomena of variation and speciation" [18]. How does this tie in with nonequilibrium thermodynamics? Stephanie Pierce goes on to say

> They take in a[nd] give off energy from the environment in order to sustain life processes and in doing so function in a state of non-equilibrium. Although biological organisms maintain a state far from equilibrium, they are still controlled by the second law of thermodynamics. Like all physiochemical systems, biological systems are always increasing their entropy or complexity due to the overwhelming drive towards equilibrium. But unlike physiochemical systems, biological systems possess "information" that permits them to self-replicate and continuously amplify their complexity and organization through time.

Fritjof Capra in his book, *The Web of Life,* states [26]:

> We emphasize that life is at its very center. This is an important point for science, because in the old paradigm, physics has been the model and source of metaphors

> for all other sciences. "All philosophy is like a tree," wrote Descartes. "The roots are metaphysics, the trunk is physics, and the branches are all the other sciences." Deep ecology has overcome this Cartesian metaphor...physics has now lost its role as the science providing the most fundamental description of reality. However, this is still not generally recognized today...Today the paradigm shift in science, at its deepest level, implies a shift from physics to the life sciences.

Kinesin. The biomolecular machine called kinesin serves as an excellent example of a nonequilibrium steady-state thermodynamic system. Most cellular processes function by random diffusion. Others require directed action. Molecular motors are able to direct cellular functions to produce desired outcomes. There are three types of molecular motors: myosins (muscle), dynesins (inward draggers), and kinesins (outward draggers). Members of the family of nonequilibrium steady-state systems include the steam and automobile engines, the motion of *E. coli* swimming in water, the smallest artificial motor, a 1-μm bead dragged in water, and the action of kinesin [15]. Because these systems require an input of energy and then dissipate energy continuously, they are considered to be in a state of nonequilibrium [15]. The function and operation of the molecular motor kinesin offers clues to unraveling some secrets of nonequilibrium thermodynamics at the nanoscale.

Kinesin is a microtubule-based ATPase motor that performs tasks like organelle transport (**Fig. 8.6**). Kinesin is involved in replication, transcription, translation, and repair of DNA/RNA and operates along microtubule tracks (protein filaments). It obtains energy from hydrolysis of ATP (adenosine triphosphate) at each step and is able to exploit energy from thermal fluctuations. Kinesin takes one 8-nm step every 10–15 ms while dragging its cellular load outward away from the nucleus [15].

The distinction between macroscopic nonequilibrium steady-state systems and nanoscale nonequilibrium steady-state systems is profound. Molecular machines like kinesin, unlike their big counterparts, are able to utilize thermal fluctuations to power motion. By a process known as rectification, kinesin is able to channel captured thermal fluctuation energy into motion only in the forward direction.

FIG. 8.6 *The action of kinesin is depicted in the figure. Kinesin is shown hauling a subcellular object away from the center of the cell. Its "feet" are able to extract energy from ATP as well as from thermal fluctuations in the surrounding cytoplasm.*

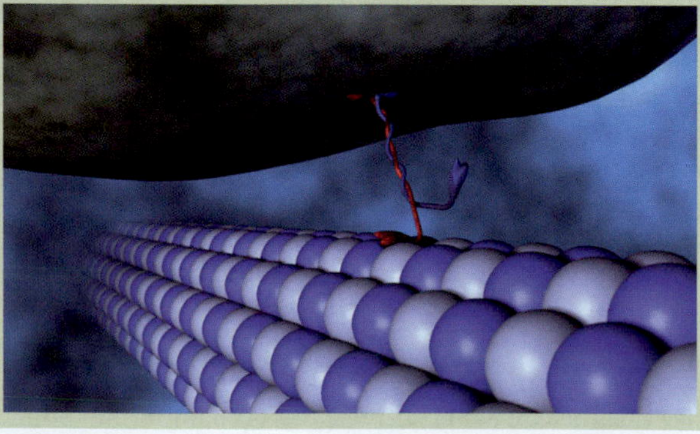

Source: Courtesy of Professor Anil K. Rao, Department of Biology, the Metropolitan State College of Denver.

Example 8.1 *Efficiency and Heat Dissipation of Kinesin Action*

What is the efficiency of a kinesin motor and how much heat does kinesin dissipate in one second of work? Assume 20 k_BT of energy in kilojoules per · mole is released during the hydrolysis of ATP and that the motor does about 12 k_BT of work with each step [15]. The Boltzmann quantity is equal to 1.38×10^{-23} J · K^{-1} and body temperature is equal to 310.15 K.

Solution:

Efficiency ε is defined as

$$\varepsilon = \frac{W_{Max}}{\text{Energy Input as Heat}} = \frac{w}{q_h} = 1 - \frac{q_c}{q_h} = 1 - \frac{T_c}{T_h} \tag{8.97}$$

$$\varepsilon = \frac{12k_BT}{20k_BT} = 0.60 \text{ Efficiency}$$

If 8 k_BT are released as heat per step and the average time between steps is [(10 + 15) ÷ 2 = 12.5], then the energy dissipated per second is

$$\textit{Energy Dissipated} = \left(\frac{1000\text{ s}}{12.5\text{ s}\cdot\text{step}^{-1}}\right)\frac{8k_BT}{\text{step}} = 640\,k_BT \rightarrow 4.28\times10^{-21}\text{ J} \rightarrow 2.57\text{ kJ}\cdot\text{mol}^{-1} \tag{8.98}$$

The kinesin moves by extracting heat energy from the surroundings and uses the energy obtained from ATP to ensure that only forward fluctuations are utilized. This is called energy rectification of thermal fluctuations towards directed motion [Bustamante]. The kinesin-microtubule system is an example of a nonequilibrium steady-state system that requires a constant energy input to operate and is characterized by the constant dissipation of energy. Life itself is an assembly of nonequilibrium systems.

From the nano perspective, and again reflecting on the origins of life and the role nanotechnology played, this concept of nanothermodynamics is most certainly an intriguing one.

Notice that the calculations performed in this example were all based in classical thermodynamics on principles one would find in any generalized physical chemistry text. There is nothing exclusively nano about this treatment, although the acquisition of thermal fluctuations to power forward movement happens at the nanoscale. That form of energy acquisition is not addressed by classical methods.

Therefore, in addition to obtaining energy from ATP, kinesin is able to cultivate energy from thermal fluctuations—not a process explained adequately by classical thermodynamics. Rectifying behavior is behavior allowing heat (or electricity) to flow only in one direction.

According to C. Bustamante et al., RNA polymerase moves along a DNA strand to transcribe a complementary strand of RNA. During this process, the polymerase moves by extracting energy, like kinesin, from a thermal bath, and then uses the bond hydrolysis energy released by ATP to ensure that only forward fluctuations are captured (rectification). The motion is described as a nonequilibrium time independent steady-state system.

Thermal fluctuation forces are considered to be random, but in small systems, their effects can be significant and therefore "arbitrarily large" [15]. Local thermal fluctuations will cause fluctuations in ΔU, *q*, and *w*. For a macroscopic system, the ratio of the probabilities of absorbing heat and releasing heat in an equilibrium system is equal to one (e.g., there is no net thermal fluctuation).

$$\frac{P(+Q)}{P(-Q)} = 1 \tag{8.99}$$

The average amount of heat $\langle Q \rangle$ is correlated with an average amount of entropy in a steady-state system:

$$\langle S \rangle = \frac{\langle Q \rangle}{T} \tag{8.100}$$

The entropy production, σ, is synonymous with the rate of heat exchange with a surrounding bath ($\sigma = Q/Tt$). Reformulation of the probability distribution yields

$$\lim_{t \to \infty} \frac{k_B}{t} \ln \frac{P_t(+\sigma)}{P_t(-\sigma)} = \sigma \tag{8.101}$$

This equation favors the probability of a steady-state system delivering heat ($+\sigma$) to the bath over that of the probability of the system absorbing heat ($-\sigma$). This is not surprising because nonequilibrium systems are dissipative. For large extensive systems, the probability of heat absorption by the system is insignificant [15,27]. For small systems, like molecular motors, the probability of absorption becomes significant—an apparent violation of the second law. Although molecular motors deliver heat to the "bath" (e.g., they too are dissipative), they also move by rectifying thermal fluctuations (e.g., absorption of heat from the bath) [15].

Times are changing and nanotechnology is a great of part of the driving force. The purpose of this section is to provoke thought—to alter your paradigm and perhaps instill a sense of curiosity. Nanomaterials are fundamental to living things. Nanothermodynamics, a topic discussed more fully in the next section, is not equivalent to our classical preconception. However, when evaluating the sum total of the energy transformations, as ΔS is always greater than zero, at the nanolevel, the self-organizational level, this postulate itself may deviate from the accepted norm—that ΔS on a local scale may be less than zero.

When one ponders the entropic aspects of living things, one realizes that order is made from disorder to make functioning living things. Although the overall entropy in the universe is constantly increasing and the energy from the sun is the external source that drives our steady-state engines, local entropy within living things is decreased or held in check. This also makes one wonder about the aging process. Is aging a process wherein entropy incrementally reasserts itself in living things due to the breakdown of regulatory "ordering" mechanisms that are genetically controlled? And why do organisms display such a wide range of lifetimes? What is the relationship of environment, information, and entropy?

8.4 Nanothermodynamics

The thermodynamics of nanomaterials, existing somewhere between the bulk and the atomic, is a key component in the energy transfer continuum we refer to often in this text. It is also a key link between bulk thermodynamics and the apparent thermodynamics of living things, another theme of this text. Research over the past few years has revealed that nanomaterials do not follow the strict laws established by classical thermodynamics. As we stated before, classical thermodynamic approaches assume that a system is in a state of equilibrium. This assumption does not always apply to all nanomaterials because not all nanomaterials represent stable phases, but yet many persist. Some metastable phases are stable long enough for engineering applications. One reason for this is that a great amount of energy is spent on synthesizing nanomaterials; this is true especially for top-down fabrication methods.

Nanomaterials in metastable phases fabricated from bottom-up methods are done so under moderate temperature and pressure conditions, as opposed to the corresponding bulk version of the material where high temperature and pressure often exist or if top-down methods are employed. The theory of Wang and Wang suggests that size effects are responsible for the formation of metastable phases and that the capillary effect due to particle curvature is enough to drive metastable phases into new stable phases [28]. In other words, nanomaterials have higher free energy content than their bulk counterparts, as we learned in chapter 6.

Nanomaterials have higher levels of energy than their bulk counterparts, yet they can be stable. Since the goal of classical thermodynamics is to lower energy, how does this happen and why does it persist? Such a condition is possible if there exists a large proportion of disordered structures in crystals. Most of classical thermodynamic calculations do not take these into account. For example, the strain energy associated with a crystal lattice plays a crucial role in the thermodynamics of these materials. In fact many processes of synthesis of nanomaterials are based on introducing more defects into the structure of crystals like equal channel angular pressing (ECAP) and severe plastic deformation (SPD). Arresting a structure formed at a high temperature by suddenly lowering the temperature converts thermal energy into the crystals' defects. This technique is accomplished by CVD, PVD, and bulk metal glass (BMG) procedures.

Trapping chemical energy inside nanomaterials is also accomplished by wet chemical methods. The trapped energy is sequestered in crystal defects like twins (a signature of highly stressed crystals). Thus, the thermodynamic principles are modified from their classical appearance to accommodate factors such as strains and defects in order to explain the properties and phenomena of nanomaterials. Hence, there are several new approaches to compute the thermodynamic aspects of nanomaterials. The most common of them takes the grain size or the particle size and the defect concentrations into account. More sophisticated approaches employ higher orders of atomic interactions.

Nanosystems may not be large enough to assign extensive properties to them [29]. Particles can be considered to be polymorphs of the bulk material without statistical homogeneity, one of the tenets of the Euler theory [29]. In nanosystems, intensive properties can change in an instant, depending on the environment—a state of affairs not applicable to macroscopic equilibrium systems; the term equilibrium implies that a certain amount of built-in inertia is in effect. Therefore, transformation of one equilibrium state into another is an infinitely slow process that is time independent [30]. Transformations of nanosystems, on the other hand, are time dependent. The parameter λ, the Euler exponent, must deviate

Table 8.3 *Comparisons between Macro- and Nanothermodynamics*

Parameter	Classical thermodynamics	Nanothermodynamics
Number of particles, N	N is infinite	N is countable
Number of independent variables (degrees of freedom)	Fewer degrees of freedom	More degrees of freedom
Surface-to-volume ratio	Miniscule	Significant
Thermodynamic limit	Yes	No
Heisenberg uncertainty	Not important	Important to set bin size for computation
Environment	Not important	Important
Size	Not important	Important
Geometry	Not important	Important
Fluctuations	Small	Large
Extensivity	Homogeneous to degree one ($\lambda = 1$) λ is the Euler exponent	Loses its significance—upon scaling, extensive parameters may change disproportionately
Intensivity	Homogeneous to degree zero ($\lambda = 0$)	Loses its significance due to large fluctuations—may not be retained
Materials of interest	Any bulk material	Droplets, bubbles, clusters, confined spaces, edges, defects, large molecules, kinesins, enzymes
Heat transfer	Heat dissipation	Heat dissipation and rectifying behavior

from unity in nanothermodynamics or even assume negative values. What all this means is that thermodynamic parameters like entropy for small systems depend on a small number of atoms and molecules, their environment, and temporal factors. Note that all of these characteristics deviate from the macroscopic factors that affect entropy (**Table 8.3**).

Another aspect to keep in mind while reading this chapter is that the second law of thermodynamics implies that it is impossible to extract heat from a reservoir and convert it to accomplish work [31]. We have noted in the last section that extraction of "unusable heat" is possible at the nanoscale, at least theoretically. Because of these fundamental distinctions, the classical thermodynamic approach may not be accurately transposable to nanomaterials. A new branch of thermodynamics, called *nonextensive thermodynamics,* offers some innovative theories that explain nanoenergy transfer [32]. We now are challenged to explain the entropic properties of small systems that depend on a countable number of atoms.

Fluctuations. What are fluctuations? Fluctuations are all about the second law of thermodynamics. In general, fluctuations are perturbations in a system—equilibrium or otherwise. Fluctuations are variations in extensive or intensive properties. With an infinitely sized ensemble, fluctuations are ignored. For microscopic systems, fluctuations cannot be ignored. Fluctuations have significant impact near phase transition boundaries where pseudoequilibrium conditions prevail. When a system is far removed from a phase transition, fluctuations are not important and it is possible to define the system with thermodynamics. Also, the larger the system becomes, the impact of phase boundaries (two-dimensional surfaces) diminishes. In small systems of less than a few thousand atoms or molecules, the boundaries become more important but are no longer clearly defined. Small metal clusters, for example, do not possess a clearly defined melting point because at the so-called "phase boundary" it is hard to distinguish whether the cluster is in liquid or solid form. This is the reason the parameter called the melting point is a macroscopic property.

Fluctuations within a large system that has an extremely large number of constituents tend to average out to zero. The energy and material in a subsystem may fluctuate by $\langle \Delta U \rangle$ and $\langle \Delta N \rangle$, respectively, but the statistically derived average values of fluctuations for a large system at the thermodynamic limit average out to zero.

$$\langle \Delta U \rangle = 0 \text{ for internal energy and } \langle \Delta N \rangle = 0 \text{ for material} \qquad (8.101b)$$

Statistically, one is able to grasp the idea that fluctuations in small systems assume greater weight. The best way to probe the internal structure of small systems is to apply entropic considerations. Expressions involving entropy are directly able to scale the probability of fluctuations involved.

Emergent Properties. The concept of *emergent properties*, applied universally to any system—not just ones based in thermodynamics—is quite intriguing. Insect groups like bees and ants, for example, rely on a few key chemical signals to trigger actions that are responsible for nearly all their behavior. An emergent property is based on simple properties that, when taken together and provided with the proper stimuli, result in a new, higher level of complexity. Such systems are characterized by a finite probability of pathways that may disobey the second law of thermodynamics that are capable of extracting order and information from the environment. The system may be open (material flow) and not isolated (energy flow) from the environment. Depending on the ambient conditions, atoms, molecules, and particles are able to self-organize to form phases. This forming of phases and boundaries is classified as an emergent process. The Bénard cell discussed earlier can be considered to be an emergent process, although no transfer of information involved.

If you are interested in acquiring a better understanding of nanothermodynamics, please refer to one of the best books on the subject matter, written by G. A. Mansoori and called *Principles of Nanotechnology: Molecular-Based Study of Condensed Matter in Small Systems*, World Scientific Publishing, 2005 [30].

8.4.1 Background

In the early 1960s, Terrell L. Hill wrote a book on thermodynamics of small systems related to his interest in colloidal systems, polymers, and macromolecules [33,34]. The notion of nanothermodynamics soon followed. His first book addressed the thermodynamics of metastable states in microenvironments [34]. A year later, Hill followed up with a book on the thermodynamic considerations of metastable nanostructured materials at microscopic levels [35]. R. S. Rowlinson at the 1983 Faraday lecture summarized that, for particles smaller than a few nanometers [36], "thermodynamics and statistical mechanics lose their meaning."

Although the field of nanothermodynamics, once again as with other things nano, has been around for a little while, it must be considered to have a spark of revolutionary element. T. L. Hill went on to introduce the term "nanothermodynamics" in 2001. Hill's abstract from the article follows [37]:

> Gibbs initiated his main contribution to thermodynamics by adding new chemical potential terms to the basic energy–heat–work equation. Nanothermodynamics is initiated, as a next step, by adding (at the ensemble level) a further subdivision potential term to the chemical potential terms of Gibbs. The basic equations of nanothermodynamics can then be deduced efficiently from this starting point.

Hill stated in 2002 that the "thermodynamics of a small system will usually be different in different environments" [38]. In other words, material properties may depend on the surrounding environment and not just on size [36]. Also in 2002, Hill, with R. V. Chamberlin, adapted classical thermodynamic principles to small particles by taking into account surface effects studied in the form of "small system ensembles" [38]. Modern nanothermodynamics began with Constantino Tsallis in 1988, who proposed the concept of Tsallis entropy and Tsallis statistics in his paper, "The Possible Generalization of Boltzmann–Gibbs Statistics" [39]. From this seminal work, the foundation for nonextensive thermodynamics was laid.

Latex Beads and Thermal Fluctuations. In 2002, scientists at the Australia Northern University found that the second law can be broken for short periods of time in microscopic materials [8]. In some recent experiments, micron-sized latex beads were suspended in water and infrared lasers were used to track movement and actually drag the beads through the water for short distances; the IR laser beams acted as tweezers that were able to push the bead along a trajectory [40]. They found that the system showed negative entropy over short time frames (the latex beads absorbed energy from random movement of the water molecules, a.k.a. thermal fluctuations), but quickly returned to the normal positive entropy state when measured over longer time frames. Measurement of the motion of the latex beads and extrapolating the level of forces applied to it by the water demonstrated that the bead was "kicked" by the water molecules. This suggests that energy was transferred from the water (the thermal bath) to the bead and that energy in the form of heat was transferred from the water to the bead [40]. Denis J. Evans of the Research School of Chemistry at the Australian National University States [40]: the results imply that the fluctuation theorem has important ramifications for nanotechnology and indeed for how life itself functions where disorder can suddenly become order.

Microscopic systems are capable of becoming more ordered for short periods of time [31]. Because of this and related findings, researchers feel that limits could be placed on miniaturization because "nanoscale devices will not be a simple scaled-down version of bulk counterparts" in that they could possibly operate backwards [41]. Recall that macroscopic thermodynamics is a statistical rendition of the behavior of bulk materials. When the sample size is reduced to such a great degree, macroscopic statistics may not apply.

Nanoparticle Structural Transformation. In 2003, it was shown that water is able to drive the structural transformation of nanoparticles [42]. The thermodynamic behavior of small particles is different from the bulk by the additive quantity $\gamma\mathcal{A}$, where γ is the interfacial surface energy and $\mathcal{A}$ is the interfacial surface area. It was reported that a nanoparticle system is able to exhibit structural changes as a function of the surface environment rather than just particle size [42]. Earlier results have shown that polymorphs of the same material with different interfacial free energies can cause phase stability changes when particle size is reduced. In experiments, ZnS nanoparticles with average diameter equal to 3 nm exhibited reversible structural transformation (to a minimum energy configuration) upon removal of the synthesis solvent (methanol). Immersion into water revealed that the structure of the ZnS particles was substantially altered from the previous configuration. For example, the particles exhibited significant reduction in surface distortions. Zhang et al. concluded that structure and reactivity of nanoparticles depend on both particle size and the surrounding environment [42].

8.4.2 Application of Classical Thermodynamics to Nanomaterials

The machinery of classical thermodynamics can be extended successfully to nanomaterials, without consideration of small systems, and derive quantitative data. One major criterion is that the sample size be relatively large. Allotropes of carbon were shown in **Figure 5.1**.

Free energy of formation of selected carbon allotropes is compared to graphite and diamond in **Figure 8.7**. The energy data were acquired from calorimetric studies (a tried and true classical thermodynamic procedure), experimental CVD data, and atomistic simulations. Since fullerenes are "highly strained molecules," they are expected to be less stable (higher energy) than graphite. Therefore, it is expected that fullerenes acquire more stability as additional carbons are added to the structure. Hence, the ΔH^{o}_{f} values of fullerene-like compounds approach the value of graphite in asymptotic fashion as more carbon is added [43,44]. Fullerenes and nanotubes exist in a kinetically stable *metastable phase* and are formed under *pseudoequilibrium* conditions. Single-walled carbon nanotubes (SWNTs), multiwalled carbon nanotubes (MWNTs), graphitic fibers and tubes, and fullerenes all exist at higher energy than graphite [44].

Single-Walled Carbon Nanotube Growth Mechanism. Classical thermodynamic principles can be extended to explain the energy associated with graphitic structures. A single-walled carbon nanotube is nothing more than a sheet of graphite (called graphene) rolled to make a cylinder. Multiwalled nanotubes are a series of concentric SWNT tubes of increasing diameter. However, an excess amount of energy is required to get this rolling accomplished. The trade-off in energy is a balance between forming an open structured planar sheet of graphite of limited size (more accurately, a graphene sheet that has numerous dangling bonds) and forming a closed structure with bond strain but with no dangling bonds. As we find out, the size of the catalyst particle has great influence over these two parameters.

Fig. 8.7 *Relative enthalpies of formation of various carbon allotropes are graphically depicted. All values are relative to the enthalpy of formation of graphite (which is the most stable form of carbon, $\Delta H_f = 0.0$). "A" indicates a type of nanotube called the armchair nanotube (more detail provided in chapter 9). "Z" stands for zigzag carbon nanotube. "C" indicates a fullerene. The number in parentheses indicates the radius of the nanotube in nanometers. The asterisk (*) indicates that the enthalpy of formation of the fullerene was determined by calorimetry from references 43 and 44. ESWN is an experimental estimate of enthalpy of nanotubes. (Unpublished results, G. L. Hornyak et al., National Renewable Energy Laboratory, 2002.) All other results (all A, Z, and C data points) are from reference 48 and were estimated by quantum molecular dynamic (QMD) simulations.*

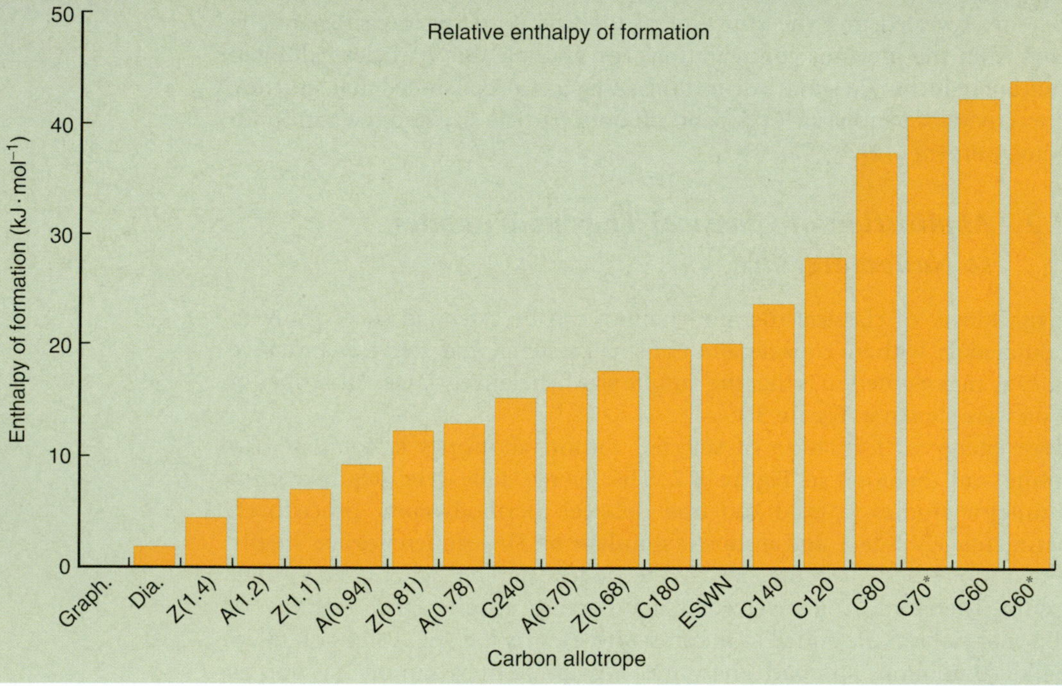

Richard Smalley et al. in 1999 proposed a growth mechanism of single-walled carbon nanotubes called the *yarmulke* mechanism [45]. Fe metal clusters with diameter of 0.7–1.4 nm (50–200 Fe atoms) are formed from the gas-phase decomposition of ferrocene. Upon the addition of a carbon source gas like CO, solid carbon is formed on the surface of the Fe catalyst from disproportionation of CO gas. Carbon nucleates on the surface of the Fe catalyst and then aggregates to form a hemifullerene cap. The cap, a structure that resembles a yarmulke (a skull cap), lifts off when additional carbon atoms are added, and a nanotube is formed. The diameter of the tube correlates with the diameter of the Fe catalyst particle. The 0.7-nm diameter tube is the smallest practical limit for SWNTs, although a tube of diameter of 0.4 nm was detected inside MWNT structures—constituting the theoretical and experimental limit [46]. The 0.7-nm value also happens to correlate with the diameter of a fullerene [45]. Larger catalyst particles formed *nanocages*, structures that encapsulate the metal catalyst. The process of SWNT propagation is by a base-growth mechanism (e.g., the catalyst remains anchored to a substrate, see chapter 9). From the kinetic perspective, SWNT growth depends on the number of active sites on the catalyst surface and the adsorption rate [47].

The growth dynamics of a single-walled carbon nanotube were monitored in situ with an ultrahigh-vacuum transmission electron microscope at 650°C [47]. M. Lin et al. found that SWNTs preferentially grow on catalyst particles less than 6 nm in diameter. They identified three distinct growth domains: (1) incubation (nucleation), (2) growth, and (3) passivation [47]. Larger particles generate *nanocages* that encapsulate (and therefore, inactivate) the catalyst. Following decomposition of a carbon source gas on the surface, two possible paths are available for carbon diffusion: along the surface (lower energy of activation due to lower coordination number) or diffusion through the bulk of the catalyst particle. Both thermodynamic and kinetic effects are responsible for the nucleation and growth process [47]. In smaller sized catalyst particles, however, the probability of diffusion through the catalyst increases due to three factors: (1) path length of diffusion on the order of the dimensions of the catalyst, (2) reduced coordination number within the volume of the catalyst, and (3) the phase of catalyst may be no longer solid but rather in the form of a liquidus—a factor that is highly related to the level of coordination.

Ab initio calculations have determined that growth of a graphene sheet is more likely (more stable) than an aggregate of isolated carbon atoms. In other words, a level of order is preferred over a loose agglomeration of carbon atoms that are seeking coordination. This result is not surprising in that the driving force to reduce the number of dangling bond sites is quite powerful. Therefore, the future fate of the graphene sheet relies on the size of the catalyst particle.

To review, there are two possible types of carbon products: a hemispherical graphene cap or a graphene sphere (carbon nanocage) that encapsulates the catalyst. The path that is selected is dependent on the catalyst size. The system seeks to find a balance between surface energy (a consequence of curvature and dangling bonds) and strain energy (a consequence of C–C bonds out of their normal orientation) [47]. Larger catalyst particles have less curvature, and the resulting graphene sheet that forms on its surface has lower torsional bond strain. Nanocage formation is favored in this configuration.

For smaller catalyst particles, bond torsional energy dominates the surface energy of the graphene layer and the hemifullerene cap is formed into a tube. As the tube is formed, the overall bond strain energy is relaxed as more carbon is added. The bonds in a tube are strained in just two dimensions rather than the three for a spherical structure of the same diameter.

Calorimetry. Calorimetry, a tried and true classical procedure, is the measure of the amount of heat released or absorbed in a chemical reaction, usually that of a combustion reaction. Heats of formation in particular are calculable by calorimetric methods. The heats of formation of fullerenes have been determined in an *isoperibolic aneroid microcalorimeter* [43,44]. They were found to be $\Delta H^o{}_f(C_{60}) = 42.5\ \mathrm{KJ \cdot mol^{-1}}$ and (of carbon atoms and not a mole of fullerenes) above that of graphite (**Fig. 8.8**) [44]. This was accomplished with no consideration of smallness—just the result of a standard calorimetric study. The goal

FIG. 8.8 *Chemical structure of polyhedral oligomeric silsesquioxane (POSS) and pentacene.*

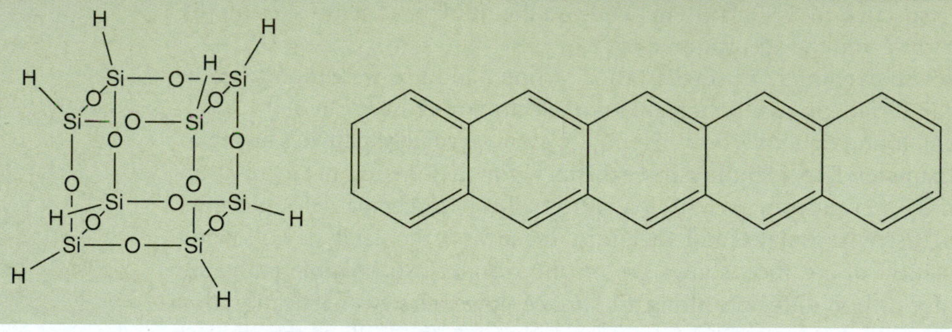

was to obtain the heat of formation of a macroscopic amount of fullerene a nanomaterial.

Molecular Modeling Method to Calculate ΔH_f^o. The equilibrium energies of fullerenes and carbon nanotubes were also estimated by first principles (ab initio) quantum molecular dynamics (QMD) calculations [48]. The single input parameter in the simulations was the *planarity* of the bonds in the fullerene. Planarity is defined by the π-bonding angle, ϕ_π. The interaction between parallel π-orbitals is directly proportional to cos ϕ_π. The definition of planarity was then refined to indicate the average value of cos ϕ_π. Therefore, fullerenes that exhibit more curvature naturally had more bond strain. The curvature is a function of particle diameter and the number of carbon atoms along a specified circumference. Data obtained for fullerenes and tubes by the QMD method are shown in **Figure 8.7** along with calorimetric, CVD, and other determinations.

Classical Thermodynamics and Carbon Nanotubes. Classical equilibrium thermodynamic analyses have been conducted on graphitic fibers formed over a nickel catalyst [49,50]. The studies were conducted to optimize conditions that reduced coking of catalysts. Ironically, the cause of the catalyst deactivation was growth of carbon fibers—an unwanted consequence of carbeurization reactions [49,50]. The fiber morphology was either multiwalled carbon nanotubes or solid fibers of graphite. Later it was discovered that the thermodynamic analyses of these early experiments were flawed. Specifically, the work of Rostrup-Nielsen and others measured the equilibrium conditions of an intermediate species (e.g., surface metal carbides) and not that of the carbon fibers [24]. Because carbon activity in the catalyst particle is not constant, the gas phase carbon activity is in *pseudoequilibrium* with the surface carbide rather than with the filaments [24]. The mechanism to form fibers and MWNTs occurs by a tip growth mechanism, in contrast to SWNTs that form by a base-growth mechanism. The question to pose while reading the remainder of this section is simply this: Are classical thermodynamics principles enough to explain all that is occurring during the SWNT growth process?

Recent studies adopted similar classical procedures. SWNTs are formed by a chemical vapor deposition (CVD) process. Methane–hydrogen mixtures at temperatures from 600 to 1000°C were reacted with iron–molybdenum catalysts. Presence of SWNTs was verified by Raman spectroscopy. The endothermic SWNT formation process based on the decomposition of methane is illustrated in the following equations:

$$CH_4 \rightarrow C_G + 2H_2 + \Delta H_G(T) \tag{8.102}$$

$$C_G \rightarrow C_{SWNT} + \Delta H_C(T) \tag{8.103}$$

$$CH_4 \rightarrow C_{SWNT} + 2H_2 + \Delta H_{SWNT}(T) \tag{8.104}$$

$$\Delta H_C(T) = \Delta H_{SWNT}(T) - \Delta H_G(T) \tag{8.105}$$

$$K_{SWNT}(T) = \frac{\left(p_{H_2}\right)^2}{\left(p_{CH_4}\right)} \tag{8.106}$$

Example 8.2 *Decomposition of Methane to Form SWNTs*

Calculate the enthalpy of reaction of SWNTs (ΔH_{SWNT}) from the decomposition of methane (Equations 8.102–106). Temperature and the equilibrium constant data are given below. Compare the enthalpy of decomposition of SWNTs to that of the decomposition of methane to form graphite.

Data:

T/K	892	913	958	1005
K_P	0.713	1.281	2.766	4.053

The onset temperature of SWNT formation was chosen to be the thermodynamic threshold of interest. The onset temperature, and hence the K_P of the decomposition–growth process, of SWNT formation depended on the partial pressures of H_2 and CH_4.

A van't Hoff plot of $-\ln K$ versus $1/K \times 1000$ is required to determine the ΔH of the reaction.

Transformed data:

$1/K \times 1000$	1.121	1.095	1.044	0.995
$-\ln K_P$	0.3387	−0.2478	−1.0175	−1.3995

Calculation of ΔH_{SWNT}:

These data show how the equilibrium constant depends on the reaction temperature. The reaction enthalpy can be extracted from the slope of the line plotting $-\ln K_P$ versus $(1/K) \times 1000$. The van't Hoff equation assumes that ΔH does not change much over the temperature range in question.

The equation of the line is

$$y = -15.107 + 13.655x \tag{8.107}$$

The slope of the relationship is proportional to $\Delta H/R$

$\Delta H_{SWNT} = (+13.655 \times 10^3\ K) \times 8.314\ J \cdot mol^{-1} \cdot K^{-1} \div 1000 = +114\ kJ \cdot mol^{-1}$

The decomposition of methane to form graphite is $\Delta H_G = 78.8\ kJ \cdot mol^{-1}$.

The excess enthalpy required to form SWNTs is ca. 25 $kJ \cdot mol^{-1}$ more than required to form graphite, the most stable allotrope of carbon. This value seems about 7 $kJ \cdot mol^{-1}$ greater than values predicted using QDM.

What factors may contribute to deviations from predicted values?

ΔG and ΔS are also calculable from the preceding data using $\Delta G = -RT \ln K_P$ and $\Delta G = \Delta H - T\Delta S$.

where ΔH_C is the difference in enthalpy at constant pressure and temperature between graphite and SWNTs and p is the partial pressure of the gaseous reactants and products. Plots of ln K versus reciprocal temperature are linear (assuming that ΔH does not change significantly over the experimental temperature range—a.k.a. a van't Hoff plot—and that the system is an equilibrium system. If ($-\ln K_P$) is plotted against $[(1/K) \times 1000]$, the slope of the line is proportional to $\Delta H/R - \Delta H = \text{slope} \times R \times 1000$. If there is a positive slope, the reaction is endothermic. A negative slope is indicative of an exothermic reaction.

The calculation of the Gibbs free energy also proceeds via classical methods.

$$\Delta G_{\text{SWNT}}(T) = -RT \ln K_{\text{SWNT}}(T) \tag{8.108}$$

$$\Delta G_C(T) = -RT \ln \frac{K_{\text{SWNT}}(T)}{K_G(T)} \tag{8.109}$$

Once ΔH is calculated, the values of ΔG and ΔS can be found from equation (8.108) and the equilibrium constant and $\Delta G = \Delta H - T\Delta S$, respectively. The decomposition of CH_4 over Fe–Mo catalyst is endothermic overall. The disproportionation of CO over nickel catalyst is exothermic. So far this treatment has been entirely from the classical perspective.

Gibbs Free Energy Estimate of SWNT Growth. L. M. Wagg et al. of the National Renewable Energy Laboratory found the Gibbs free energy of formation for SWNTs relative to the formation of graphite to be between 16.1 and 13.9 $\text{kJ} \cdot \text{mol}^{-1}$. Gas feed composition (methane and hydrogen) and reaction temperature (700–1000°C) were varied to determine the thermodynamic threshold for nucleation and growth of SWNTs [24]. Experimental parameters for efficient growth of SWNTs by the decomposition of methane have been reported earlier [51].

Are explanations of the SWNT nucleation-growth process based only on classical thermodynamic principles valid? The answer to the question is an unequivocal "no." It is not appropriate to interpret the SWNT growth process by classical thermodynamic principles and methods alone. The authors claim that the concept of *pseudoequilibrium* needs to be invoked in order to explain the processes involved [24]. For one thing, the process resembles a steady-state process where there is a (material in $\Leftrightarrow$ material out) scenario with constant internal energy. There are potential nonequilibrium domains that exist below 700°C that, according to classical analyses, were detected at p_{H2} less than 10%. The purported "decomposition of methane" in these cases actually exhibited an exothermic response as opposed to the normal classical endothermic nature of methane decomposition at equilibrium conditions (higher temperature, higher p_{H2}) [52]. Extreme driving forces, nonequilibrium mixing of gases and thermal gradients may be responsible for exothermic character of the low hydrogen partial pressure domain [52].

Intermediates formed throughout the reaction are complex and rely on kinetic parameters as well as upon thermodynamic ones. As a case in point, iron carbides formed within the catalyst particles may be mistaken for the thermodynamic product (tubes) if weight gain–loss methods of tracking the reaction coordinate are employed [24]. Regardless of the ideality, this approach does provide a good starting point, but extreme caution must be practiced when interpreting experimental data based on purely classical principles.

8.4.3 Small System Thermodynamics (the Theory of T. L. Hill)

The work of J. M. Rubi et al. [53,54], H. J. F. Jansen [55], R. M. Baigi [56], G. A. Monsoori [3,30], A. K. Rajagopal [33], and J. -R. Roan [57] (and others) was extremely helpful in putting this section together by providing some of the best overviews of nanothermodynamics to be found in the literature. Two factors wrap themselves around this new thermodynamics: (1) modification of the terms *extensivity* and *intensivity* and (2) the dependence of nanomaterials on their immediate surroundings (e.g., their immediate environment).

A good place to start is with the Gibbs–Thomson effect and its description of the nucleation and growth of nanoparticles and nanodroplets:

$$\Delta G = \frac{4}{3}\pi r^3 \Delta G_V + 4\pi r^2 \gamma \tag{8.110}$$

If this equation looks familiar it is because we covered it in chapter 6. It describes the free energy change of a small particle in which ΔG_V is the free energy change per volume, $4\pi r^3\gamma$ is the free energy change per unit surface area, and γ is the surface tension (joules per square meter) as before. Notice that the surface tension (energy) term γ is assumed to be independent of size. In actuality, the surface energy increases dramatically once the radius dips below 3 nm [58]. What this relation indicates is that a smaller particle will have a higher relative energy than a larger one. It also illustrates the concept of additivity. The volume term alone is not able to describe the change in free energy for the nucleation process. However, the addition of a surface term provides the necessary correction to the energy. G, r, and V are extensive terms. G_V and γ are intensive terms. As with any classical thermodynamic parameter, the relation can be expressed in terms of entropy.

As experimental techniques were improved, researchers found that classical nucleation theory was incomplete and did not account for the temperature dependence of the rate of nucleation. Several theories were proposed to explain the variance: Hill theory (given next), Tsallis entropy, and a new approach proposed by Wang and Wang [28].

Hill Theory for Small Systems. Thermodynamic limits apply in classical thermodynamics and statistical mechanics: $N \to \infty$ and $V \to \infty$ where N/V is a constant. If volume V is doubled, then energy, entropy, and all other extensive properties are also doubled. Intensive properties remain the same. Rephrased, it means that thermodynamics is only valid if a system contains an infinite number of particles distributed in an infinite volume with a constant density [53]. This viewpoint guarantees existence of extensivity without fluctuations (independent of the environment). Statistical mechanics is a means by which the thermodynamic limit is approximated in an asymptotical manner. But what if there is no thermodynamic limit?

When particles assume small dimensions, as in the case of biological and colloidal systems, the concept of extensivity breaks down. The best example of this is provided by surface energy. We know from chapter 6 how the behavior of surface atoms differs significantly from that of the bulk. In the bulk, the proportion of surface atoms is small and the contribution of surface energy to the energy of the system as a whole is miniscule. However, in small materials, the surface assumes greater influence. A modification of macroscopic thermodynamics is in order. T. L. Hill presented this equation of the Gibbs energy for a small cluster [59]:

$$G = uN + aN^{\beta} \tag{8.111}$$

where μ is the chemical potential, a is an intensive parameter constant, and β is less than 1. As $N \rightarrow \infty$, the second part of the equation becomes less significant, but for small particles like clusters, both terms are important. In the former scenario, $G \propto \mu N$ yields the energy per particle—an additive extensive parameter (e.g., G for the system is the sum of G of all the parts). However, if the second term in equation (8.111) (the exponential) becomes important, then the concept of extensivity for G is no longer valid [54]. A mathematical fix that restores extensivity is

$$G = \hat{\mu}N \text{ where } \hat{\mu} = \mu + aN^{\beta-1} \tag{8.112}$$

$$\hat{\mu} - \mu = aN^{\beta-1} \tag{8.113}$$

What this states is that for small systems, an ensemble of constituent equivalent replicas becomes a large system that has the usual thermodynamic infinite limit properties [54]. An ensemble such as this provides a macroscopic base from which to study the nanodomain. A consequence of this, according to Hill, is that thermodynamic parameters of different ensembles may not be equivalent [53,59].

The differential form of the internal energy and the other forms of the total system by combining the first and second laws of thermodynamics is

$$dU = TdS - PdV + \mu dN \tag{8.114}$$

To start, let us visualize an ensemble of $\mathcal{N}$ equivalent components (identical, independent replicas of small systems) of a one material system. For a macroscopic system, $\mathcal{N} = 1$. The total energy, entropy and volume, etc. are a function of the number of replicas $\mathcal{N}$ in the system at large:

$$U_{\text{Total}} = \mathcal{N}\bar{U}; \quad V_{\text{Total}} = \mathcal{N}\bar{V}; \quad S_{\text{Total}} = \mathcal{N}S \quad \text{and} \quad N_{\text{Total}} = \mathcal{N}N \tag{8.115}$$

where the barred terms represent average values per replica [54]. Another term called the *subdivision* potential, $\mathcal{E}$, is a type of chemical potential that includes the additive contributions of small systems to the total energy. Therefore, for the set of small N systems for constant T, P, change in the energy of the ensemble is given by

$$dU_{\text{Total}} = TdS_{\text{Total}} - PdV_{\text{Total}} + \mu\, \mathcal{N}\, dN_{\text{Total}} + \mathcal{E}\, d\mathcal{N} \tag{8.116}$$

Obviously, if we deal with a macroscopic system, $\mathcal{N}=1$ and the $\mathcal{E}$ subdivision potential term would vanish. For this ensemble of small systems, $N \neq 1$ and $\mathcal{E} = 0$. Now, keeping T, P and N constant, integrating equation (8.116) yields

$$U_{\text{Total}} = TS_{\text{Total}} - PV_{\text{Total}} + \mathcal{E}\mathcal{N} \tag{8.117}$$

Dividing equation (8.117) by $\mathcal{N}$ yields

$$\bar{U} = TS - P\bar{V} + \mathcal{E} \quad (8.118)$$

Identifying $\mathcal{E}$ with G gives [3]:

$$TS - P\bar{V} + \mathcal{E} \rightarrow TS - P\bar{V} + G \quad (8.119)$$

Recalling that $G = \hat{\mu}N$,

$$\bar{U} = TS - P\bar{V} + \hat{\mu}N \quad (8.120)$$

Rewriting, we get [3]

$$\bar{U} = TS - P\bar{V} + \mu N + (\hat{\mu} - \mu)N \quad (8.121)$$

At the thermodynamic limit, G equals μN and the last term of equation (8.119) disappears (as $\hat{\mu} \rightarrow \mu$) [54].

This is the *Gibbs–Hill* adaptation to small system thermodynamics. In other words, it is an adaptation of macroscopic thermodynamics with additive terms that address small system effects. This brand of thermodynamics does not wander too far off equilibrium. There are, of course, a handful of recent interpretations that differ from this approach.

Example 8.3 *Evaluation of $\mathcal{E}$*

Derive a form of the subdivision potential $\mathcal{E}$ from the linear homogeneous Euler form of U_{Total} at T, P. U_{Total} for a small one-material system is

$$U_{Total} = TS_{Total} - PV_{Total} + \mu N_{Total} + \mathcal{E}\mathcal{N} \quad (8.122a)$$

Solution:

Division by $\mathcal{N}$ small systems in the ensemble yields the average quantities:

$$U = TS - PV + \mu N + \mathcal{E} \quad (8.122b)$$

Differentiating gives:

$$dU = d(TS) - d(PV) + d(\mu N) + d\mathcal{E}$$

$$dU = T\,dS + S\,dT - [P\,dV + V\,dP] + \mu\,dN + N\,d\mu + d\mathcal{E}$$

From the definition, $dU = T\,dS - P\,dV + \mu\,dN$

$$T\,dS - P\,dV + \mu\,dN = T\,dS + S\,dT - [P\,dV + V\,dP] + \mu\,dN + N\,d\mu + d\mathcal{E}$$

Therefore:

$$d\mathcal{E} = -S\,dT + V\,dP - N\,d\mu \quad (8.123)$$

Notice from equation that $(-S\,dT + V\,dP) = dG - \mu\,dN$

$$d\mathcal{E} = (dG - \mu\,dN) - N\,d\mu = dG - d(\mu N)$$

$\mathcal{E} = G - \mu N$ *and from equation 8.112,* $G = \hat{\mu}N$*, restoring extensivity to the expression*

Example 8.3 (Contd.) *Evaluation of $\mathcal{E}$*

$$\mathcal{E} = (\hat{\mu} - \mu)N \tag{8.124}$$

Deviations from large systems are due to $(\hat{\mu} - \mu)$, which becomes insignificant at the thermodynamic limit. In the form below, it is easy to see why $\mathcal{E}$ vanishes as N gets infinitely large and $\hat{\mu} \to \mu$.

$$\hat{\mu} = \mu + \frac{\mathcal{E}}{\mathcal{N}} \tag{8.125}$$

The subdivision potential $\mathcal{E}$ is a form of the chemical potential. Surface, edge, and other small system effects contribute to $\mathcal{E}$. In different environments, $\mathcal{E}$ assumes different forms. Macroscopic thermodynamics is recovered as $N \to \infty$ (the thermodynamic limit) and $\mathcal{E}/N \to 0$. Therefore, $\mathcal{E}$ is the limit for small system thermodynamics. Another form that you may find for $\mathcal{E}$ is given by

$$\mathcal{E} = \left(\frac{\partial U_{Total}}{\partial \mathcal{N}}\right)_{S_{Total}, V_{Total}, \mathcal{N}_{Total}} \tag{8.126}$$

where $\mathcal{E}$ is an "intensive parameter" in the form of a chemical (subdivision) potential that represents all other extensive variables that can be measured and specified to determine the thermodynamics of the system.

8.5 Modern Nanothermodynamics

In 1876, physicist Josef Loschmidt, in response to the prevailing thought that the universe will suffer "heat-death" as a result of the second law of thermodynamics, submitted the "reversibility paradox" to the scientific community: "If the motion of individual particles is considered to be reversible, why then is their collective behavior irreversible?" This critique encouraged Boltzmann to create his statistical concept of entropy shown in equation (8.74). Equations of thermodynamics are based on the random motion of many Avogadro's numbers of particles (in Germany, this value was referred to as Loschmidt's number). Heat always flows in one direction towards the cold pole and entropy always increases. Right? Boltzmann, upon reformulation of his classical version of the second law into the widely accepted statistical form, understood the importance of statistical fluctuations.

8.5.1 Nonextensivity and Nonintensivity

Tsallis Formulation and Entropy. C. Tsallis proposed a new definition for entropy in 1989 [3,39]. His concept is summarized in equations (8.127) and (8.128). The idea that thermodynamics of small systems have nonextensive and nonintensive components is quite an out-of-the-box viewpoint. Its intent was not to replace Gibbs–Boltzmann statistics, but rather to offer explanations to anomalous systems characterized by nonergodicity or metastable states (fluctuations). We have seen, by way of classical thermodynamic precepts, that extensivity and intensivity explain thermodynamics and cooperate with differential and integral formulations. This is evidenced by corrections such as $G = \hat{\mu}N$ with the purpose of restoring extensivity into classical formulations.

The mathematical formalism of Tsallis's new definition of entropy is given by

$$S_q = k_B \frac{1 - \sum_{i=1}^{W} p_i^q}{q-1} \tag{8.127}$$

where p_i is the probability of microstate i and where $0 \leq q \leq 1$ [3,39]. The entropy for two independent systems A and B with independent probability of occurring is

$$S_q(A+B) = S_q(A) + S_q(B) + \frac{1-q}{k} S_q(A) S_q(B) \tag{8.128}$$

The entropy, as defined in the preceding equation, is nonextensive (nonadditive) for $q < 1$. As $q \to 1$ and when $q = 1$, the entropy according to this relation is reduced to the Boltzmann–Gibbs form and becomes extensive [3]. Others have shown that Tsallis and Hill thermodynamics can be reconciled—specifically, that the nonadditive property of the Tsallis entropy forms the basis of the subdivision entropic potential of Hill [60]. This last statement provides some level of support for the continuum theory of energy we espouse.

Non-Tsallis Formulations. However, Mohazzabi et al. in 2005 showed, by way of molecular dynamic simulations, that nonextensivity and nonintensivity can be explained within the framework of Boltzmann–Gibbs formalism without consideration of Tsallis's thermodynamics [3]. They demonstrated that internal energy and entropy are nonextensive and that temperature and pressure are nonintensive [3].

In the study of Mohazzabi et al. that we mentioned in section 8.4.3, systems of variable number of identical particles in three dimensions were simulated by a molecular dynamic program [3]. The parameters of the simulation were as follows: (1) argon-like particles of mass m (the unit mass) were made to interact according to a pair-wise Lenard–Jones (6–12) interatomic potential energy function potential: $u(r) = 4\varepsilon[(\sigma/r)^{12} - (\sigma/r)^6]$, where ε (the unit energy) is the depth of the well and σ (the unit length) is the "hard core radius", (2) the number of particles within each system varied from 8 to 1000: $n(2,3,4,\ldots,10) = n^3(8,27,64,\ldots,1000)$ that were distributed in a square lattice, (3) the particles were randomly assigned velocity v and equations of motion were solved, and (4) internal energy was calculated from kinetic and potential terms, temperature from ε/k_B, and pressure from the change in momentum of particles per unit time per unit area after collision with the container walls.

The unit of entropy was the Boltzmann constant k_B. The computation of entropy was not straightforward [3]. Entropy was not calculated from classical thermodynamic procedures that assume a finite number of microstates, that all microstates in an ensemble are equally likely, and that entropy is an extensive parameter where $S(N, V, U) = k \ln \mathcal{W}(N, V, U)$. Calculation of S from particle trajectories for continuous microstates (e.g., from positions and velocities) cannot be accomplished by classical methods. Therefore, entropy was calculated from particle trajectories by choosing a "bin size" for each coordinate of momentum (e.g., setting a limit on bin size) [3]. This example demonstrates the level of thought and expertise required to produce a molecular dynamic simulation. Obviously, the results depend on the number and quality of the input parameters.

The molecular dynamics (MD) simulation consisted of small systems interacting via Lenard–Jones type potential energy functions ranging in size from a few argon-like particles to 1000 interacting argon-like particles. The authors found deviations from macroscopic thermodynamics even in systems as large as 1000 particles [3]. Conclusions from this work based on MD simulations are [3]:

- Plots of internal energy versus N were not linear (to be linear, slope = U per particle), indicating a small nonextensivity. U was found to be slightly subextensive (more negative).
- With regard to internal energy, the nonextensive behavior of simulated particles is related to the surface-to-volume ratio. This implied that as surface-to-volume ratio increased, nonextensivity became more pronounced.
- $U_2 \neq 2U_1$. This means that U is nonextensive in that two particles did not result in twice the energy. Extra energy was obtained from particle interaction—converting external potential energy into internal energy.
- Entropy was shown to be nonextensive. Because it increased for smaller systems, entropy was classified as superextensive.
- Neither pressure nor temperature demonstrated intensivity and they are therefore nonintensive properties.
- All parameters approached extensive or intensive states as N was increased.

According to preceding discussion, extensivity and intensivity for small systems exist in an altered state from that of the standard macroscopic state. The smaller the system was, the more pronounced became the deviation [3].

Extension of the Young–Laplace Equation. Wang and Wang proposed in 2005 that a universal quantitative thermodynamic nanoscale model based on the extension of the Young–Laplace equation is sufficient to explain deviations from classical nucleation theory. The theory is based on computer simulations of diamond and C–BN (amorphous carbon boron nitride) nanocrystal nucleation processes. Correlation of Young's equation with equilibrium phase diagrams was used to describe the thermodynamics of metastable phase nucleation [28]. According to the authors, no adjustable parameters in simulations were required to explain the phenomena and simple extrapolation of the phase equilibrium was successful in predicting nanothermodynamic behavior from macroscopic thermodynamic data [28]. Simulations were also accomplished for one-dimensional structures formed under vapor–liquid–solid mechanisms. From their method, Wang and Wang were able to predict thermodynamic and kinetic size limits of nanowires as well as nucleation thermodynamics and diffusion parameters of catalyst nanoparticles at the tip of the nanowire [28].

8.5.2 *Nanothermodynamics of a Single Molecule*

It is inevitable that we ask this question: Is thermodynamics valid at the single molecule level? We stated earlier that entropy only applies for a great number of atoms and molecules and is meaningless for a single one. J. M. Rubi et al. in 2006 studied this phenomenon [54]:

> We show how to construct non-equilibrium thermodynamics for systems too small to be considered thermodynamically in a traditional sense... We show in particular that the Gibbs equation, when formulated in terms of average values of the extensive quantities, is still valid, whereas the Gibbs–Duhem equation differs from the equation obtained for large systems due to the lack of a thermodynamic limit ... The potentials of mean force and mean position correspond respectively to our Helmholtz and Gibbs energies. The results show that a thermodynamic formalism can indeed be applied at the single-molecule level.

8.5.3 *Modeling Nanomaterials*

Examples of computer simulations of nanomaterials are abundant in the literature. Selected examples of nanomaterial modeling gleaned from the literature are presented next.

Polyhedral Oligomeric Silsesquioxane. The bottom-up design of self-assembling nanoparticles consisting of polyhedral oligomeric silsesquioxane (POSS) (**Fig. 8.8**) was demonstrated by a "multiscale computer simulation method" that included four types of modeling systems: (1) ab initio quantum mechanical calculations, (2) molecular dynamics simulations with classical and reactive force fields, (3) Monte Carlo simulations, and (4) coarse-grained mesoscale simulations [61,62]. The purpose of the study was to optimize optical, thermal, and mechanical properties of nanocomposite materials that contain the functionalized POSS constituents. Thermal, vibrational, mechanical, and structural properties were first studied by ab initio and MD simulations. Results were then compared to preexisting experimental data. It was found that the POSS cages have a unique rhombohedral packing that implies parallel packing in the crystal form. The consequences of such packing are higher charge carrier capability when functionalized with acene type molecules (pentacene) (**Fig. 8.8**). Evaluation of the electronic properties of the composite organic–inorganic hybrid semiconductor simulations revealed similarities to acenes but predicted superior thermal and mechanical bulk characteristics [62].

Another study involved the modeling of POSS dissolved in hexadecane at 400–1000 K. Radial distribution functions, potentials of mean force, and self-diffusion coefficients were determined by molecular dynamic simulations [63]. Canonical (@ N, V, T) molecular dynamics simulations were used to compute the thermodynamic and transport properties of the POSS. Substitution parameters included temperature and atomic replacement or removal. The simulation showed that the mechanical properties of synthetic polymers are enhanced with the incorporation of nanoscale particulate materials.

Dodecanethiol SAMs. In another example, equilibrium structures and thermodynamic properties of dodecanethiol self-assembled monolayers were investigated by MD simulations [64]. It was found that compact passivating monolayers were formed on (111) and (100) Au surfaces. At lower temperatures, the passivating molecules organize into parallel bundles. When the temperature is raised, the bundles melt and cover the gold in a uniform way [64]. The "melting temperature" is much lower than that found in a conventional self-assembled monolayer of dodecanethiol on a flat gold surface. Phase transitions such as the one described in this example are commonly studied with MD simulations.

Nanovoids. Nucleation and melting of materials containing nanovoids were also investigated by MD simulations [65]. It was shown that the behavior of four melting stages in nanomaterials differs significantly from the bulk form. The melting in each stage depends on the interactions among thermodynamic mechanisms that arise from changes in the interfacial free energy, the curvature of the interface, and the elastic energy induced by the density change during melting [65]. Ultimately, melting was shown to depend on the internal defects of a material (e.g., the nanovoids). The parameters of the simulation were as follows: (1) zero external P and constant T, (2) standard Lenard–Jones potentials, where $\phi(r) = 4\varepsilon–[(\sigma/r)^{12} - (\sigma/r)^{6}]$, where $\varepsilon = 119.8\ k_B$ potential well (where $\varepsilon/k_B \approx 120$ K) and $\sigma = 3.405$ Å length parameter, (3) cutoff distance was 2.5σ, (4) *fcc* structure lattice, (5) three different size systems: 20 × 20 × 20, 30 × 30 × 30, or 40 × 40 × 40 unit cells, (6) periodic boundary conditions applied in all directions, (7) spherical voids created by removing atoms from the center, (8) system relaxed to $T = 0.2\ \varepsilon/k_B$, (9) initial void radius ranges = 0.58–0.6.62 nm, and (10) void sample was heated and held isothermally for 200,000 time steps and observed nucleation and growth processes of melting from the nanovoid [65].

Lipid Bilayer Property Modeling. We also add a biological example. Classical MD simulations of atomistic models of interactions between combustion-formed carbon nanoparticles and lipid bilayers were accomplished to determine structural, dynamical, and thermodynamic effects on biological membranes [66]. The authors found that the membrane acts as a discriminator of particle shape and structure, resulting in altered solvation, mobility, adsorption, and permeation [66]. For example, from simulations, it was found that spherical particles tend to stay near the surface of the biological membrane, whereas higher aspect-ratio particles sequester themselves, although not trapped, within the hydrophobic tails of the hydrocarbon region of the membrane [66]. Particle size was also modeled and found to be a factor in the strength of adhesion to the membrane (e.g., larger particles adhered more strongly to the membrane). The dynamical aspects of this study were quite intriguing. Combustion-generated nanoparticles, when inside the membrane, perturbed the structure of the natural lipid and, when coupled with thermal fluctuations, induced pore formation in the membrane [66].

Monte Carlo and MD methods were employed for this study. All dangling bonds of the combustion-generated carbon moieties were terminated with hydrogen in simulations. Applied force fields were obtained for generic nonmetallic main group elements. A preprogrammed lipid bilayer model was used in the MD simulations; 64 lipids were agglomerated to form a membrane 34 Å in thickness and 200 water molecules were included. The longest carbon nanoparticle was 17 Å.

8.5.4 Modern Non-nanothermodynamics?

No theory, however, is embraced and accepted fully. S. Gheorghiu et al. state that application of Occam's razor is the best policy when it comes to the explanation of the *nonergodic* theories proposed earlier [67]. William of Ockham, a fourteenth century logician and Franciscan friar, stated (simply) that "the explanation of any phenomenon should include as few assumptions as possible, eliminating those that make no difference in the observable predictions of the explanatory hypothesis or theory," or, "all things being equal, the simplest solution tends to be the best one." This philosophy is also expressed as the "principle of parsimony."

Gheorghiu et al. stated emphatically that phenomena characterized by *power-law distributions* are plentiful in nature and in the applied sciences [67]. They go on to state that "complicated arguments based on long-range correlations or non-ergodicity are often incorrect or misleading in explaining many naturally observed power laws, in particular those described by the Student distribution." The Student's distribution (t- or Student's t-distribution) is a means of quantifying deviations from the mean without knowledge of the standard deviation. Its objective is to estimate the mean value of a normally distributed population when the sample size is small. The Student's t-test was derived by W. S. Gossett in 1908 while employed by the Guiness Brewery in Dublin.

These claims in effect challenge the assertions of Tsallis et al. and the theory of nonergodic systems in general—that nonextensivity exists for small systems. The authors claim that the Boltzmann factor itself has a Student form and that the t-distribution provides an accurate description of power-law statistics associated with polydisperse heterogeneous systems—for example, that the representation is completely thermodynamically classical without the need for spatial or temporal correlations [67].

Concluding Thoughts. Please recall that this was a special chapter. Its purpose was to introduce concepts in thermodynamics that differ (perhaps drastically) from the well-established paradigms with which we are all so familiar. The purpose was to bring awareness and to provoke thought. Although we authors do not claim to be experts in this new, dynamic field, rest assured that due diligence was practiced during the acquisition of references, consultations, and the writing. Therefore, any feedback from students and professors would be greatly valued. In future editions, we hope to present a formal treatment of the subject matter complete with a challenging set of problems.

The message to take away with you after completing this chapter is a simple one: Thermodynamic behavior of materials at the nanoscale may not be the same as thermodynamic behavior of the same materials at the macroscopic scale. Nanomaterials are subjected to their environment, especially to thermal fluctuations, which may not be considered large on the macroscopic scale but are significant at the local level. Such behavior is expressed as nonextensive and/or nonintensive—definitely anathema to thermodynamics at the big level. Terms such as nonequilibrium, pseudoequilibrium, quasi-equilibrium, and metastable are associated with nanomaterials and a deeper level of understanding of those basic concepts is required in order to understand what transpires at the nanoscale.

Lastly, the link between thermodynamics and living things was explored from yet another frame of reference. Throughout the text we have emphasized the importance of nanomaterials to living things. We believe that this chapter, more than any other, has helped to frame that special and fundamental relationship in a proper perspective.

References

1. E. J. Chaisson, *Cosmic evolution—The rise of complexity in nature,* President and Fellows of Harvard College, Boston (2001).
2. A. Pross, Stability in chemistry and biology—Life as a kinetic state of matter, *Pure & Applied Chemistry,* 77, 1905–1921 (2005).
3. P. Mohazzabi and G. A. Mansoori, Nonextensivity and nonintensivity in nanosystems: A molecular dynamics simulation, *Journal of Computational Theoretical Nanoscience,* 2, 1–10 (2005).
4. Euler's theorem on homogeneous functions, PlanetMath.org, planetmath.org/encyclopedia/EulersTheoremOnHomogeneousFunctions.html (2007).
5. I. N. Levine, *Physical chemistry,* 3rd ed., McGraw–Hill, Inc., New York (1988).
6. P. W. Atkins, *Physical chemistry,* 4th ed., W. H. Freeman and Company, New York (1990).
7. B.-X. Wang, L.-P. Zhou, and X.-F. Peng, Surface and size effects on the specific heat capacity of nanoparticles, *International Journal of Thermophysics,* 27, 139–151 (2006).
8. G. M. Wang, E. M. Sevick, E. Mittag, D. J. Searles, and D. J. Evans, Experimental demonstration of violations of the second law of thermodynamics for small systems and short time scales, *Physics Review Letters,* 89, 050601(2002).
9. J. P. Abriata and D. E. Laughlin, The third law of thermodynamics and low temperature phase stability, *Progress in Materials Science,* 49, 367–387 (2004).
10. S. Lovesey, Condensed matter theory, The Rutherford Appleton Laboratory, http://www.isis.rl.ac.uk/ISIS97/theory.htm (1997).
11. P. Xie and C. M. Lieber, *Carbon nanotubes on gold (111) surface,* Lieber Group, Harvard University (2007).
12. Tentative fourth law principles. In laws of thermodynamics, Wikipedia, en.wikipedia.org/wiki/Laws_of_thermodynamics (2007).
13. Lars Onsager, Biography, Answer.com, www.answers.com/topic/lars-onsager?cat=technology
14. Nonequilibrium Thermodynamics, Wikipedia, en.wikipedia.org/wiki/Non-equilibrium_thermodynamics (2007).
15. C. Bustamante, J. Liphardt, and F. Ritort, The nonequilibrium thermodynamics of small systems, *Physics Today,* 58, 43–48, July (2005).
16. G. D. Mahan, *Many-particle physics,* 3rd ed., Physics of Solids and Liquids Series, Kluwer Academic/Plenum Publishers, New York, (2000).
17. Onsager reciprocal relations, Wikipedia, en.wikipedia.org/wiki/Onsager_reciprocal_relations (2007).
18. S. E. Pierce, Nonequilibrium thermodynamics: An alternate evolutionary hypothesis, *Crossing Boundaries,* 1, 49–59 (2002).
19. D. R. Brooks and E. O. Wiley, *Evolution as entropy: Towards a unified theory of biology,* University of Chicago Press, Chicago (1986).
20. D. R. Brooks, D. D. Cumming, and P. H. LeBlond, Dollo's law and the second law of thermodynamics: Analogy or extension? In *Entropy, information, and evolution,* Weber, Depew, and Smith, eds., The MIT Press, Cambridge, MA (1988).
21. M. Kosmulski, *Chemical properties of material surfaces,* Surfactant Science Series, 102, CRC Press, Boca Raton, FL (2001).
22. H. Zhu, S. F. Webb, and R. Blumenthal, The chemical composition of diamond plasmas, *Journal of Vacuum Science Technology A,* 14, 952–959 (1996).

23. T. A. Brabbs, Catalytic decomposition of methane for onboard hydrogen generation, NASA technical paper 1247, 1–48, June (1978).
24. L. M. Wagg, G. L. Hornyak, L. Grigorian, A. C. Dillon, K. M. Jones, J. Blackburn, P. A. Parilla, and M. J. Heben, Experimental Gibbs free energy considerations in the nucleation and growth of single-walled carbon nanotubes, *Journal of Physical Chemistry B*, 109, 10435–10440 (2005).
25. J. D. Andrade, V. L. Hlady, and R. A. Van Wagenen, Effects of plasma protein adsorption on protein conformation and activity, *Pure & Applied Chemistry*, 56, 1345–1350 (1984).
26. F. Capra, *The web of life: A new scientific understanding of living systems*, Anchor Books, Doubleday, New York (1996).
27. G. Gallavotti and E. G. D. Cohen, Dynamical ensembles in nonequilibrium statistical mechanics, *Physics Review Letters*, 74, 2694–2697 (1995).
28. C. X. Wang and G. W. Wang, Thermodynamics of metastable phase nucleation at the nanoscale, *Materials Science and Engineering: R Reports*, 49, 157–202 (2005).
29. D. R. Baer, J. E. Amonette, and P. G. Tratnyek, Small particle chemistry: Reasons for differences and related conceptual challenges, Interagency Grantees Meeting/Workshop—*Nanotechnolgy and the environment: Applications and implications*, Pacific Northwest National Laboratory, September 15–16 (2003).
30. G. A. Mansoori, *Principles of nanotechnology—Molecular-based study of condensed matter in small systems*, World Scientific Publiching Co. Pte. Ltd., London (2005).
31. P. F. Schewe, B. Stein, and J. Riordon, Nanothermodynamics, physics news update, *The American institute of physics bulletin*, 598, July 17 (2002).
32. G. R. Vakili-Nezhaad, Euler's homogeneous functions can describe non-extensive thermodynamic systems, *Journal of Pure & Applied Mathematical Science*, 1, 7–8 (2004).
33. A. K. Rajagopal, C. S. Pande, and S. Abe, Nanothermodynamics—A generic approach to material properties at the nanoscale. Invited presentation, Indo-U.S. Workshop on *Nanoscale materials: From science to technology*, Puri, India, April 5–8 (2004).
34. T. L. Hill, *Thermodynamics of small systems, I*, W. A. Benjamin, New York (1963).
35. T. L. Hill, *Thermodynamics of small systems, II*, W. A. Benjamin, New York (1964).
36. R. S. Rowlinson, Molecular theory of small systems, Faraday Lecture, *Chemical Society Reviews*, 12, 251–265 (1983).
37. T. L. Hill, A different approach to nanothermodynamics, *Nano Letters*, 1, 273–275 (2001).
38. T. L. Hill and R.V. Chamberlin, Fluctuations in energy in completely open small systems, *Nano Letters*, 2, 609–613 (2002).
39. C. Tsallis, Possible generalization of Boltzman–Gibbs statistics, *Journal of Statistical Physics*, 52, 479–487 (1988).
40. P. Weiss, Law and disorder: Chance fluctuations can rule the nanorealm, *Science News*, 162, 51 (2002).
41. M. Chalmers, The second law of thermodynamics "broken," *New Scientist*, 09, 21, July 19 (2002).
42. H. Zhang, B. Gilbert, F. Huang, and J. F. Banfield, Water-driven structure transformation in nanoparticles at room temperature, *Nature*, 424, 1025–1029 (2003).
43. H.-D. Beckhaus, C. Rüchardt, M. Kao, F. Diedrich, and C. S. Foote, The stability of buckminsterfullerene (C_{60}): Experimental determination of the heat of formation, *Angewandte Chemie International Edition in English*, 31, 63–64 (1992).
44. H.-D. Beckhaus, S. Verevkin, C. Rüchardt, F. Diedrich, C. Thilgen, H.-U. ter Meer, H. Mohin, and W. Müller, C_{70} is more stable than C_{60}: Experimental determination of the heat of formation of C_{70}, *Angewandte Chemie International Edition in English*, 33, 996–998 (1994).

45. P. Nikolaev, M. J. Bronikowski, R. K. Bradley, F. Rohmund, D. T. Colbert, K. A. Smith, and R. E. Smalley, Gas-phase catalytic growth of single-walled carbon nanotubes from carbon monoxide, *Chemical Physics Letters*, 313, 91–97 (1999).
46. H. Y. Peng, N. Wang, Y. F. Zheng, Y. Lifshitz, J. Kulik, R. Q. Zhang, C. S. Lee, and S. T. Lee, Smallest diameter carbon nanotubes, *Applied Physics Letters*, 77, 2831–2833 (2000).
47. M. Lin, J. P. Y. Tan, C. Boothroyd, K. P. Loh, E. S. Tok, and Y.-L. Foo, Direct observation of single-walled carbon nanotube growth at the atomistic scale, *Nano Letters*, 6, 449–452 (2006).
48. G. B. Adams, O. F. Sankey, J. B. Page, M. O'Keefe, and D. A. Drabold, Energetics of large fullerenes: balls, tubes and capsules, *Science*, 256, 1792–1795 (1992).
49. I. Alstrup, A new model explaining carbon filament growth on nickel, iron and Ni–Cu alloy catalysts, *Journal of Catalysis*, 109, 241–251 (1988).
50. J. R. Rosrtup-Nielsen, Equilibria of decomposition reactions of carbon monoxide and methane over nickel catalysts, *Journal of Catalysis*, 27, 343–356 (1972).
51. G. L. Hornyak, L. Grigorian, A. C. Dillon, P. A. Parilla, and M. J. Heben, Temperature window for optimized single walled carbon nanotube synthesis by CVD, *Journal of Physical Chemistry B*, 106, 2821–2825 (2002).
52. G. L. Hornyak, Thermodynamic parameters of SWNT formation from the decomposition of methane over Fe–Mo catalysts, unpublished results (2002).
53. J. M. G. Vilar and J. M. Rubi, Thermodynamics "beyond" local equilibrium, *Proceedings of the National Academy of Science USA*, 98, 11081–11084 (2001).
54. J. M. Rubi, D. Bedeaux, and S. Kjelstrup, Thermodynamics for single-molecule stretching experiments, *Journal of Physical Chemistry B*, 110, 12733–12737 (2006).
55. H. J. F. Jansen, Thermodynamics, Dept. Physics, Oregon State University, www.physics.oregonstate.edu/~tgiebult/COURSES/ph642/Jansen.pdf, December 14 (2003).
56. R. M. Baigi, *Nanothermodynamics: A subdivision potential approach*. Presentation, Computational Physical Sciences Research Laboratory, Department of Nano-Science, IPM. Viewed (2007).
57. J.-R. Roan, Introduction to biophysics I. Presentation, Fall 2004, *Bioenergetics*, Chapter3, Energy, 140.120.11.15/vedio/biophysics/Chinese/teachning/Fall%202003/Lecture_Notes.files/Lecture4_Bioenergetics.ppt (2004).
58. C. T. Campbell, S. C. Parker, and D. E. Starr, The effect of size-dependent nanoparticle energetics on catalyst sintering, *Science*, 298, 811–814 (2002).
59. T. L. Hill, *Thermodynamics of small systems*, Dover, New York (1994).
60. V. Garciá-Morales, J. Cervera, and J. Pellicer, Correct thermodynamic forces in Tsallis thermodynamics: Connection with Hill nanothermodynamics, *Physics Letters A*, 336, 82–88 (2005).
61. T. C. Ionescu, F. Qi, C. McCabe, A. Striolo, J. Kieffer, and P. T. Cummings, Evaluation of force field for molecular simulation of polyhedral oligomeric silsesquioxanes, *Journal of Physical Chemistry B*, 110, 2502–2510 (2006).
62. J. Kieffer, Simulation of self-assembly of functionalized silsesquioxane molecules, http://meetings.aps.org/link/BAPS, MAR.H14.3 (2005).
63. A. Striolo, C. McCabe, and P. T. Cummings, Effective interactions between polyhedral oligomeric silsesquioxanes dissolved in normal hexadecane from molecular simulation, *Macromolecules*, 38, 8950–8959 (2005).
64. W. D. Luedtke and U. Landman, Structure and thermodynamics of self-assembled monolayers on gold nanocrystallites, *Journal of Physical Chemistry B*, 102, 6566–6572 (1998).
65. X.-M. Bai and M. Li, Nucleation and melting from nanovoids, *Nano Letters*, 6, 2284–2289 (2006).

66. R. Chang and A. Violo, Insights into the effect of combustion-generated carbon nanoparticles on biological membranes: A computer simulation study, *Journal of Physical Chemistry B*, 110, 5073–5083 (2006).
67. S. Gheorghiu and M.-O. Coppens, Heterogeneity explains features of "anomalous" thermodynamics and statistics, *Proceedings of the National Academy of Science USA*, 101, 15852–15856 (2004).
68. www.stanford.edu/rmukul/tutorials/irreversible.pdf

Problems

8.1 Does the concept of continuum apply to thermodynamics?

8.2 List as many intensive properties as you can. Do the same for extensive properties.

8.3 Define a state function.

8.4 A great thermodynamic review is achieved by deriving a thermodynamic relation from first principles. Do this for the Clausius–Clapeyron equation.

8.5 Explain to a classmate the phenomena of reversibility and irreversibility.

8.6 Explain in more detail Boltzmann's famous equation: $S = k \ln W$.

8.7 Construct a case for the existence of non-equilibrium thermodynamics in living things.

8.8 List some examples of the successful application of classical thermodynamics to nanosystems.

Section 4

Chemistry: Synthesis and Modification

Carbon-Based Nanomaterials

Chemists are a strange class of mortals, impelled by an almost maniacal impulse to seek their pleasures amongst smoke and vapour, soot and flames, poisons and poverty, yet amongst all these evils I seem to live so sweetly that I would rather die than change places with the King of Persia.

Johann Joachim Becher,
Physica Subterranea (1667)

It is disconcerting to reflect on the number of students we have flunked in chemistry for not knowing what we later found to be untrue.

Robert L. Weber,
Science with a Smile (1992)

Chapter 9

Every attempt to employ mathematical methods in the study of chemical questions must be considered profoundly irrational and contrary to the spirit of chemistry... if mathematical analysis should ever hold a prominent place in chemistry—an aberration which is happily almost impossible—it would occasion a rapid and widespread degeneration of that science.

Auguste Comte,
Cours de Philosophie Positive (1830)

Threads

This special chapter is devoted to carbon-based nanomaterials. Carbon is the basic element of life. Carbon is the basic element of organic chemistry. Why should not carbon take the spotlight, albeit in a shared capacity, in nanoscience as well? Arguably, there are many equally fascinating nanomaterials and who is to say that quantum dots are not more fascinating? In answer, quantum dots are rather one dimensional (please pardon the pun), whereas carbon-based nanomaterials have a wide range of applications. We understand how size affects physical properties. That concept is rather clear by now—how a material is transformed into a semiconductor as small dimensions are achieved. We now ask you to believe the unbelievable: Materials made of one component (carbon), all of the same relative size with similar structure (e.g., a tube), can act as an insulator, a conductor, or a semiconductor. How can this be? How is this possible with an element that makes the strongest bonds and one that has the ability to bond in so many ways? How can materials that consist entirely of carbon be so different?

With chapter 9, we begin the chemistry division of the text. This chapter represents the perfect transition between the physics and chemistry divisions of the book. Although there is plenty of material describing properties and phenomena (a.k.a. the physics of carbon nanotube structure and properties), the chapter includes plenty of material describing synthesis and chemical modification. The chapter then also serves to introduce intermolecular interactions without actually addressing them in detail; that is reserved for chapter 10. Chapters discussing the chemical bond (chapter 10) and supramolecular chemistry (chapter 11) follow; chapter 12 is dedicated to chemistry of nanomaterials in general. These four chapters comprise the chemistry division of the text. Biological topics are presented and discussed in chapters 13 and 14.

9.0 Carbon

The term carbon (from the French *charbone,* "glowing coal," based on the Latin *carbo*) was coined in the 1780s by French chemist Lavoisier. J. J. Berzelius of Sweden in 1824 also played a role by designating the letter "C" for carbon in the periodic table. Carbon is the most important element found in living things. Without carbon, there would be no life (at least as we know it). Carbon is special because of its ability to bond to many elements in many different ways. Its four-coordinate bonding capability allows for the synthesis of unique structures. Although silicon also is capable of four-coordinate bonding, its versatility is limited compared to carbon. Carbon is smaller than silicon. Ironically, silicon is the most important element in the semiconductor industry and a quite important player in polymer chemistry and sol–gel synthesis.

Potential applications of inorganic carbon materials are well known: lead pencils, graphite fibers, highly ordered pyrolytic graphite (HOPG), diamonds, amorphous carbon filters, chromatographic column packing materials and sieves, hardeners in steel, moderator material in nuclear reactors, plastics, paints, and lubricants. Organic chemistry has given us the polymer and its multitude of applications. Organic carbon, in the form of dinosaurs, Paleozoic and Mesozoic plants, and plankton, was transformed over the ages into oil that powers our civilization.

Inorganic carbon in the form of CO_2 is converted into high-energy sugars by the action of enzymes in plants. The sugars in turn are broken down during metabolism back into CO_2. Such inorganic carbon is extremely important on the global scale due to its strong ability to absorb infrared radiation. One can state with great confidence that all of supramolecular chemistry, an offshoot of biochemistry and organic chemistry, is a chemistry that relies heavily on the carbon atom. Biological materials are of course based on carbon; all life is carbon based. Starting with amino acids, lipids, proteins, nucleic acids, and a host of biological vitamins and cofactors all possess a carbon backbone. Because biological materials are nanomaterials—at least materials that are made of nanomaterials—carbon also is important to nanoscience. Many kinds of inorganic nanomaterials are based on carbon, which is the topic of this chapter.

The abundance of carbon in the universe is ca. 0.5 ppm; it is the sixth most abundant element in the universe. Its atomic mass is 12.011 a.m.u. Its electron configuration is $1s^22s^22p^2$. Carbon oxidation states are ±4 and ±2 with the valence state of 4 the most common. Its covalent radius is 0.77 Å, its ionic radius is 0.91 Å, and its atomic volume is 4.58 $cm^3 \cdot mol^{-1}$. The density of C is 2.62 $g \cdot cm^{-3}$, but of course this value varies significantly depending on the type of carbon.

Sir Harold W. Kroto of the University of Sussex and Richard E. Smalley of Rice University were awarded the Nobel Prize in chemistry, along with Robert Floyd Curl, Jr., also of Rice University, in 1996. In 1985, they discovered physical evidence for the existence of fullerenes [1]. Experimental mass spectrometry of molecular beam products showed discrete peaks that corresponded to 60 carbons. Higher order carbon peaks, corresponding to larger fullerenes, were also detected. Smalley was a molecular beam expert, specializing in the synthesis of laser vaporization techniques applied to semiconductor materials as well as carbon. Kroto was interested in stellar processes and how the vaporization of graphite may provide clues to their origin. The combination helped to kick start the *Nano Age*. The archetypal fullerene is C_{60}, a spherical structure that is composed of 60 carbon atoms configured as a soccer ball (**Fig. 9.1**).

Carbon nanotubes have been manufactured for centuries on Earth by natural pyrolytic processes, but they were formally discovered by Japanese electron microscopist Sumio Iijima of NEC Laboratories in Japan in 1991 [2]. Iijima and Toshinari Ichihashi (and, independently, Donald Bethune of the IBM Almaden Research Center) discovered single-walled nanotubes (SWNTs) in 1993 [3,4]. Multiwalled tubes have been around for decades, an undesirable by-product of catalytic processes.

9.0.1 Types of Carbon Materials

We are interested in four types of carbon materials in this chapter: diamond (specifically diamondoids and thin layer diamond-like films), graphite, fullerenes, and carbon nanotubes. Diamond is found in cubic and hexagonal forms. In cubic diamond, each carbon atom is linked to four others by sp^3 bonds to form a tetrahedral array. The bond length of each C–C in diamond is 1.544 Å, almost 10% longer than the C–C bond length in graphite. This, of course, makes sense because double or partially double carbon bonds are shorter than single bonds. Allotropes of carbon were displayed in **Figure 5.1**. There are many kinds of carbon materials: (1) diamond, Lonsdaleite, and diamondoid materials; (2) graphite (Bernal and others), turbostratic graphite, highly ordered pyrolytic graphite

FIG. 9.1 *The fullerene C_{60} is depicted—a truly magnificent molecule.*

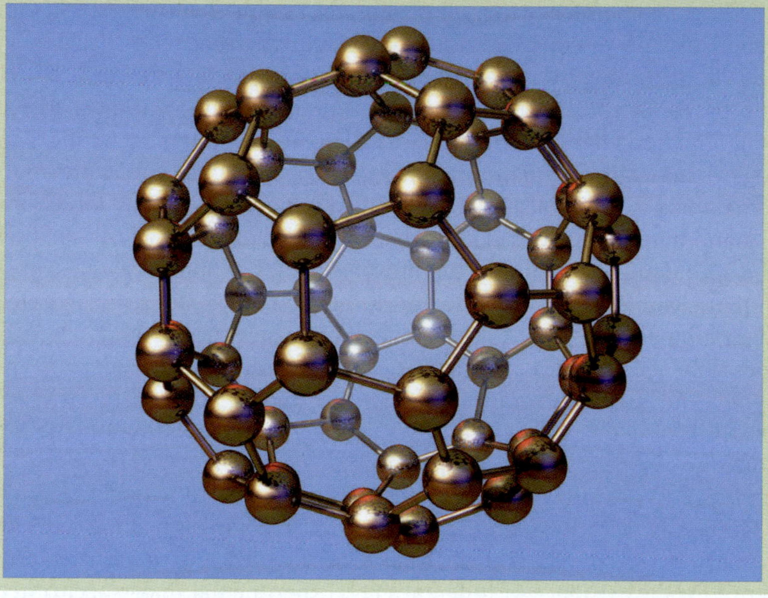

Source: Image courtesy of Professor Anil K. Rao, Department of Biology, the Metropolitan State College of Denver.

(HOPG), and lampblack; (3) fullerenes (spherical and elongated), nanobuds, nanohorns, and carbon nanotubes (MWNTs, SWNTs, and DWNTs); (4) glassy carbons; and (5) amorphous carbons. Carbon allotropes span a wide range of properties, in many cases representing the extremes of physical properties.

Graphite and Diamond. No two materials made of the same element can be more different than graphite and diamond or more extreme in their differences. Graphite is one of the softest materials; diamond is one of the hardest. As a result, diamond is used as an abrasive and graphite is used as a lubricant (interestingly, graphite has recently been shown to be a poor lubricant in space—no intercalated moisture?). Diamond is also a good lubricant that is used in the space shuttle on bearings. Graphite is electrically conducting and diamond is electrically insulating; however, diamond can be made semiconducting by adding dopants (*p*-doping only). Diamond is optically transparent and graphite is optically dense. Diamond crystallites are members of the isometric (cubic) system; graphite crystals are members of the hexagonal system. Diamond has an isotropic structure, whereas graphite structure is anisotropic. Diamond conducts heat extremely well (one of the best in that capacity), but graphite, taken as a whole, does not ($graphite_{||}$ conducts heat very well "in plane"; interplane conduction $graphite_{\perp}$ is poor). Interestingly, graphite is the stable form of the two. Diamond is constantly undergoing transition into graphite, albeit quite slowly; the activation energy for this conversion is nearly equal to the lattice energy of diamond. The fact that diamond was of made of carbon was verified in 1796.

Abraham Gottlob Werner in 1789 coined the term "graphite" (from the Greek *graphein*, "to write") because the material was used commonly in pencils. The structure of graphite was elucidated by John Bernal in 1924 and was henceforth

TABLE 9.1 *Physical Properties of Graphite, Diamond, and C_{60}*

Physical property	Graphite	Diamond	C_{60}
Crystal structure	ABABAB plane stacking called Bernal stacking In-plane nearest neighbor distance = 1.421 Å ABCABC stacking is rare	Zinc blend, *fcc*, diatomic basis Hexagonal diamond is called Lonsdaleite (Wurzite crystal form)	Fullerene molecules crystallize in an *fcc* structure C_{60}–C_{60} nearest approach is 3.1 Å [5] Octahedral site radius: 2.07 Å
Lattice constant	$a = 2.461$ Å $c = 6.708$ Å Interplanar distance, $d = 3.34$ Å	$a = 3.567$ Å	$a = 14.15$ Å
Bond length	1.42 Å	1.544 Å Lonsdaleite = 1.52 Å	C–C = 1.455 Å C=C = 1.391 Å Cage diameter: 7.11 Å
Bonding	Trigonal bonding network is present in graphite—sp^2 with delocalized π-network Interlayer bonding due to overlap of π-orbitals and not van der Waals forces. Others claim that weak van der Waals forces are responsible for interplanar interaction due to the fourth electron in valence shell [6]	sp^3 hybridization to form tetrahedral bonds	Bonds vary with length and the type of associated polygon. Hybridization is an admixture of sp^2 and sp^3 Bond angles vary from 108 to 110°
Standard heats of formation	0.0 kJ · mol^{-1}	1.67 kJ · mol^{-1}	37.99 kJ · mol^{-1}
Band gap	No band gap, $E_g = 0.0$ *eV* (~ −0.04 *eV*)	Wide band gap, $E_g = 5.47$ *eV*	1.8 *eV* (1.7 *eV*)
Atomic density	1.133×10^{23} cm^{-3}	1.77×10^{23} cm^{-3} Diamond has the highest atomic density	Molecular density: 1.44×10^{21} cm^{-3}
Gravimetric density	2.26 g · cm^{-3}	3.52 g · cm^{-3}	1.65 g · cm^{-3} (1.72 g · cm^{-3})
Melting point	4200 K (3800 K)	4500 K (3700 K)	Sublimes at 800 K
Electrical conductivity	Semimetal	Insulator	Semiconductor
Thermal conductivity	In plane graphite (‖) has very high conductivity ~2500 W · m^{-1} · K^{-1} or better	Diamond has the highest thermal conductivity ~2500 W · m^{-1} · K^{-1}	
Hardness	Graphite is a relatively soft material: Mhos hardness 1–2	The hardest material in nature: Mhos hardness ~10	

known commonly as Bernal graphite (consisting of an ABABAB structure). Graphite is essentially a planar material with each carbon atom bonded in a planar configuration to three others by σ-sp^2-bonds. The physical properties of graphite, diamond, and C_{60} are compared in **Table 9.1.** A graphic rendition of the unit cell of graphite is shown in **Figure 9.2** (left). On the right in **Figure 9.2,** the relationship between two planes of graphite, as found in the native material, is illustrated. The lattice parameters, bond length, and interlayer *d*-spacing are shown in **Figure 9.3.**

Fig. 9.2 *The unit cell of graphite (containing two atoms) is shown on the left. On the right, the relationship between two planes of graphite is depicted.*

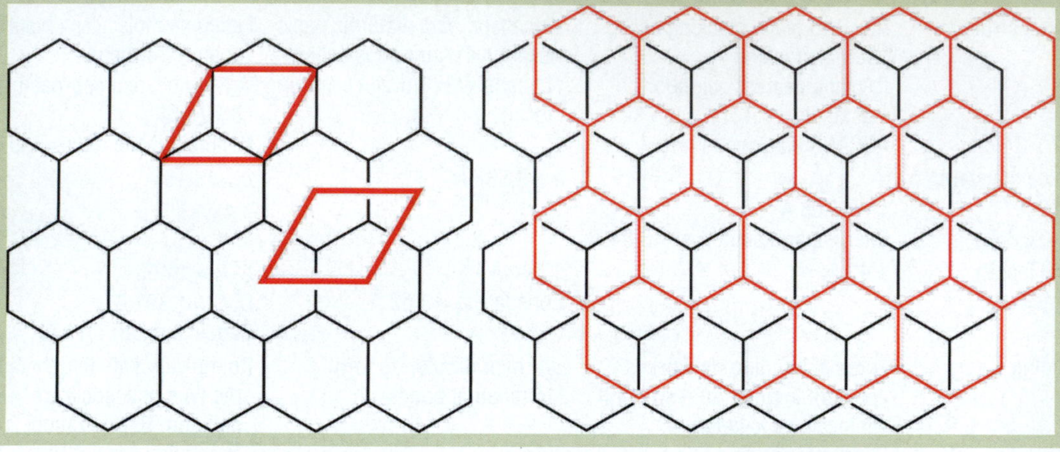

9.0.2 Bonding in Carbon Compounds

Carbon has four electrons with which to form bonds: two 2*s* electrons and two 2*p* electrons; all four are in the $n = 2$ quantum energy state of the atom. Depending on the chemical circumstances, these electrons hybridize to form

Fig. 9.3 *Three layers of graphite are shown. The d-spacing between layers is 3.34 Å. The lattice parameter a of the unit cell is equal to 2.46 Å. The C–C bond length is 1.42 Å.*

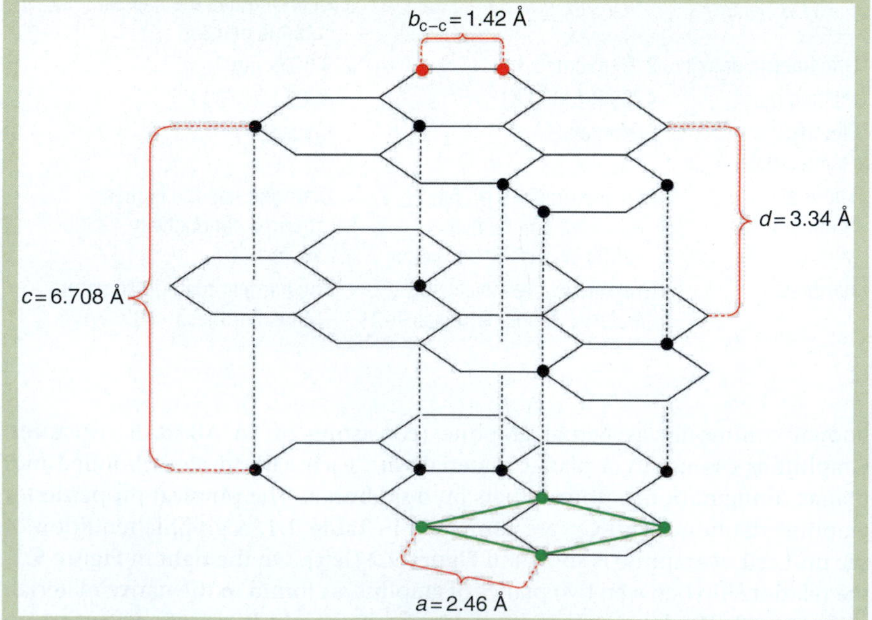

FIG. 9.4 *Three kinds of hybrid orbitals of carbon are shown. Tetrahedral carbon has four equivalent sp^3 orbitals with which to bond four substituents (e.g., methane, CH_4). The bond angle of sp^3 hybrid orbitals is 109.5°. The bond angle of sp^2 hybrid orbitals is 120° (e.g., ethylene, $H_2C{=}CH_2$). The three sp^2 orbitals per carbon form a single plane. The orientation of the single available p-orbital is perpendicular to the plane of the sp^2 orbitals and is available to form one π-bond. Finally, the sp hybrid orbital is shown (e.g., acetylene, C_2H_2). The two sp orbitals form a 180° and are therefore linear. The two available p-orbitals are orthogonal to each other and are able to form two π-bonds.*

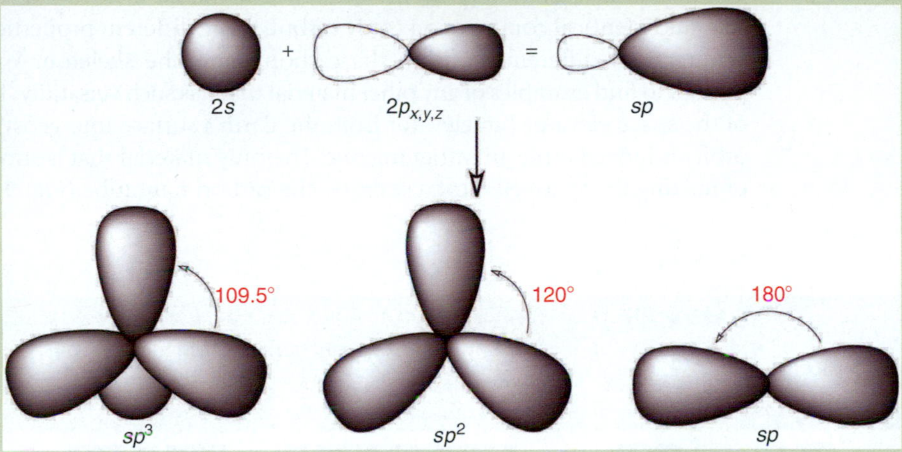

4 sp^3 (no *p*-orbitals available for bonding), 3 sp^2 (one *p*-orbital available for bonding), or 2 *sp* orbitals (two *p*-orbitals available for bonding). In the sp^2 and *sp* configurations, the remaining *p*-electrons go into forming π-bonds (**Fig. 9.4**). Carbon with sp^3 orbitals form tetrahedral structures (bond angle = 109.5°) is like that found in methane or in methylene groups of a hydrocarbon chain; sp^2 orbitals result in planar structures (bond angle = 120°) and are found in compounds like ethylene with π-bonding orbitals perpendicular to the plane of the molecule; and *sp* structures result in linear molecules (bond angle = 180°) like acetylene with capability to form two orthogonal π-orbitals. Hybrid orbitals are represented by the generalized wavefunction ψ:

$$\psi = s + \chi p_i \tag{9.1}$$

where χ equals 1 for an *sp* orbital, 2.5 for an sp^2, and 3.5 for an sp^3. There are gradations of χ in molecules that have strained angles such as in fullerenes and carbon nanotubes. Carbon nanotubes, for example, have more sp^3 character than do unstrained materials like ethylene or graphite. Obviously, having more sp^3 character enhances three-dimensional flexibility in bond angle when compared to the planar sp^2 bond.

9.0.3 The Nano Perspective

Fullerenes were discovered recently and became important enough to award a Nobel Prize to their discoverers. Fullerenes are the perfect nanomaterial. The smallest fullerene, C_{60}, contains 60 carbon atoms formed into the structure of a soccer ball. They are stable (kinetically) and provide great versatility in chemical reactions and applications. The soccer ball shape is indicative of the high level of symmetry of C_{60}. C_{60} is a monodisperse material: All C_{60} are identical. Synthetic batches of C_{60} have 99.99% + purity; no other nanomaterial can make that claim—not quantum dots, not gold clusters, not carbon nanotubes, not diamondoids, or anything else.

Carbon nanotubes are found in the form of semiconductors, conductors, and insulators even though all three kinds have the same relative shape, relative size, and identical composition (only carbon). The different properties arise due to structural differences within the carbon nanotube skeleton. We are hard pressed to find examples of any other material that has such versatility. The concept of the space elevator (an elevator from the Earth's surface into geosynchronous orbit) is indeed quite an intriguing one. The only material that is strong enough of making the space elevator a reality is the carbon nanotube (**Fig. 9.5**).

Fig. 9.5 *The concept of the space elevator is shown. An elevator anchored on Earth (on the ocean surface) at one end and placed in geosynchronous orbit at the other is indeed a "far out" concept that stretches the imagination to its ultimate limit. Carbon nanotubes are the only material that has the necessary strength to pull this off.*

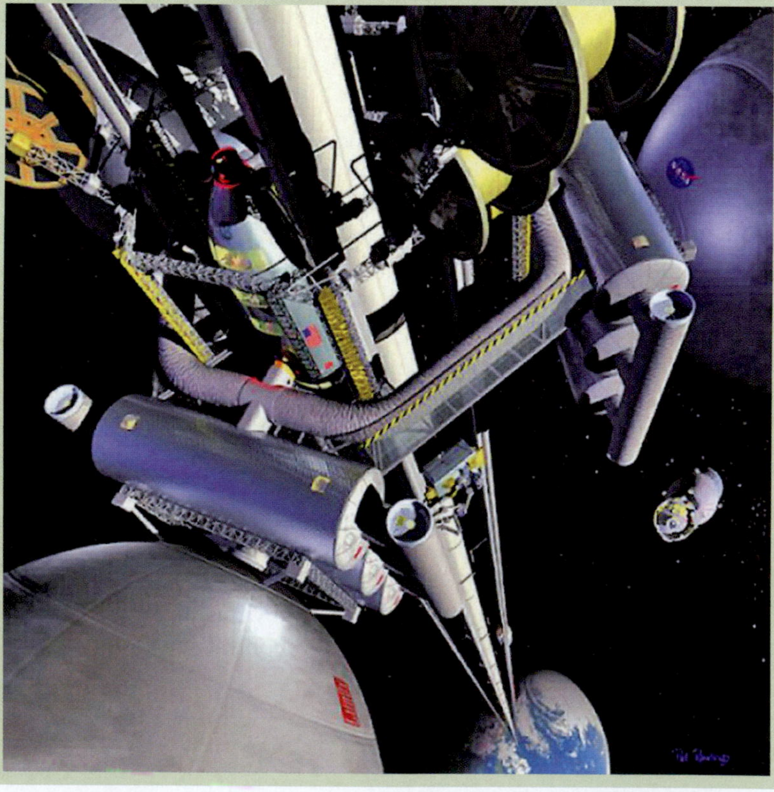

Source: Image courtesy of Michael J. Laine, The Liftport Group. With permission.

Diamondoid materials are also remarkable in the sense that monolayers of carbon in diamond form can be applied to surfaces and are manufactured under relatively mild conditions. Diamondoid materials are made of carbon that contains small amounts of hydrogen, oxygen, nitrogen, or sulfur. Because diamond has excellent physical properties (if hardness, thermal conductivity, and strength are important to you), it is expected to be one of the most valuable construction materials of the future. The strength-to-weight ratio, just like for carbon nanotubes, is expected to be 50 times better than any conventional material in use today. Some nanotechnology pundits already refer to the coming "diamond age" [7]. Applications of diamondoid materials (both current and potential) include: cages for drug and gene delivery, molecular capsules and probes, positional assembly, shape-targeted nanostructures, supramolecular synthesis and host–guest chemistry, fluorescent molecular probes, diamond monolayers, and seed molecules for dendrimer structures. Diamondoids are the only semiconductor materials that show a negative electron affinity [8].

9.1 Fullerenes

Fullerenes (buckyballs) were discovered in 1985 by researchers at Rice University and the University of Essex in England [1]. The general class of fullerenes includes carbon nanotubes as well as buckyballs and derivatives. In this section we focus mainly on the spherical or ellipsoidal forms of fullerenes: C_{60} and its allies. The presence of fullerenes has been detected in the outer bounds of our solar system in various nebulae. Buckyballs form under conditions of extreme temperature and pressure. Large fullerenes consisting of C_{60}, C_{70}, C_{100}, and C_{400} and many species in between have been found in the 4.6 billion-year-old carbonaceous chondrite Allende meteorite in Mexico [9]. There is also evidence that outer space gases are trapped in buckyballs. Australia's Murchison meteorite was found to enclose extraterrestrial helium gas within its cage. Some red giant stars emit a wavelength centered on 21 μm that is caused by interaction with large, complex molecules or solid materials [10]. Fulleranes, a hydrogenated from of fullerenes, are also associated with meteoric masses [11].

Carbon clusters comprise many interesting metastable species [12,13]. Clusters up to 10 atoms are stable in the form of linear chains, whereas 10–30 carbon atoms prefer to form into an annular configuration [14]. P. Eaton in 1964 synthesized a cubic form of carbon known as *cubane* [C_8H_8] [15]. $C_{20}H_{20}$, a dodecahedron, was synthesized in 1983 by L. Paquette [15]. The bond angles in cubane are 90° (highly strained), and in the dodecahedron, bond angles range from 108 to 110°; both deviate slightly from the ideal 109.5° but are characteristic of many "saturated" species [15]. Mass spectrometer data revealed that for carbon number N_C less than 30, clusters ranging from 2 to ca. 30 are well represented [15,16].

Molecular orbital theory calculations showed that linear structures form when N_C is an odd number (with *sp* hybridization) and cage structures like cubane when N_C is even. Linear or ring clusters with $N_C = 3, 11, 15, 19$, and 23 have tetrahedral bond angles and are therefore more abundant in the mass spectrum [15]. Between 30 and 40 carbons, clusters do not form metastable entities, but above with N_C greater than 40 atoms, enclosed fullerene structures are preferred [1]. The mass spectrum of ionized carbon materials shows a very large peak at $N_C = 60$. Because cage structures appear to possess inherent stability, one would expect that the stable forms of fullerenes also contain an even number of carbon atoms (e.g., C_{60}, C_{70}, C_{76}, C_{78}, C_{80}, C_{84}, C_{90}, etc.). These species are relatively more abundant than neighboring species that have odd numbers of carbon atoms [15].

9.1.1 Fullerene Properties

C_{60} is an extremely stable molecule (in the kinetic sense) with a carbon atom placed at each of the 60 vertices of the structure. C_{60} is not a stable molecule in the thermodynamic sense due to extreme bond strain. The structure is formally known as a truncated icosahedron. C_{60} has 90 edges and 32 faces and, of course, 60 vertices, one at each carbon atom. Of the 32 faces, 12 are pentagons and 20 are hexagons. Euler's theorem of polyhedra relates faces, edges, and vertices of a geometrical solid:

$$f + v = e + 2 \tag{9.2}$$

where f, v, and e are the number of faces, vertices, and edges of the polyhedron, respectively. In a single layer of graphite, the C–C bond distance is on the average equal to 1.42 Å. However, C_{60} is not like graphite in that it contains pentagonal carbon structures intermingled with hexagons. Pentagons exhibit no double bonds, per se, and hexagons have alternating double bonds. Single bonds (existing on the pentagonal moieties) have bond length equal to 1.46 Å, while double bonds are shorter at 1.40 Å [17]. The diameter of C_{60} is 7.10 Å [1]. Each carbon atom forms a trigonal relationship with three other carbon atoms, just like in graphite. Every carbon in C_{60} is identical (evidenced by ^{13}C NMR spectrum that reveals just one peak at 142.68 ppm).

A more accurate description of the bonds in fullerenes is one that includes some sp^3 character that results in an sp^2–sp^3 admixture. Higher fullerenes are formed by the addition of hexagonal rings around their middles. The limiting case for a fullerene therefore is a carbon nanotube (or vice versa). If carbon nanotubes were to extend from the hemifullerene caps shown in **Figure 9.6**, an *armchair tube* is formed from the cap (a pentagonal at the apex) on the left and a *zigzag tube* is formed from the cap (the hexagonal apex) on the right. There is more discussion about carbon nanotubes to follow.

Fullerenes are resistant to pressure and reclaim their original shape even after experiencing over 3000 atm [18]. Bonding between fullerenes is accomplished via intermolecular interactions (noncovalent) and by covalent bonding. In general, fullerenes interact via weak van der Waals bonds, similar to those experienced between layers of graphite.

We discussed calorimetry briefly in chapter 8 [20]. The difficulty in applying calorimetric techniques to carbon nanotubes is that uniform samples are hard to obtain. This is not an issue with fullerenes, especially the popular ones like C_{60} and C_{70} for which very pure samples can be obtained. The energetic relations among various forms of carbon are graphically illustrated in **Figure 8.8**. The $\Delta H_f^\circ (C_{60})$ and $\Delta H_f (C_{70})$ of these basic fullerenes were found to be 42.5 kJ · mol^{-1}

FIG. 9.6 *Two different views of the same fullerene are shown. Looking down the pentagonal axis, armchair carbon nanotubes are consistent with the pentagonal hemifullerene cap on the left. Zigzag tubes are consistent with the hexagonal hemifullerene cap on the right.*

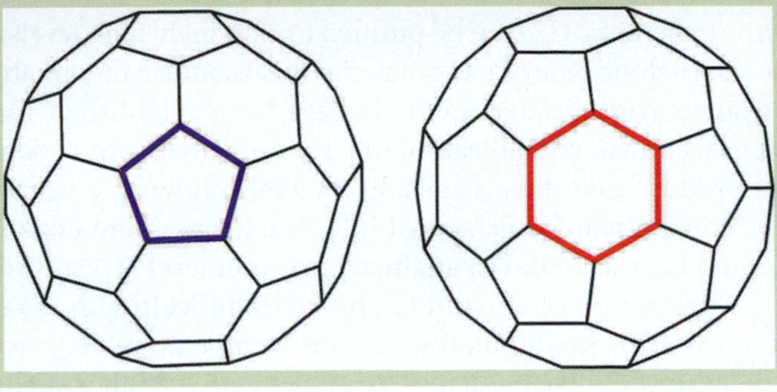

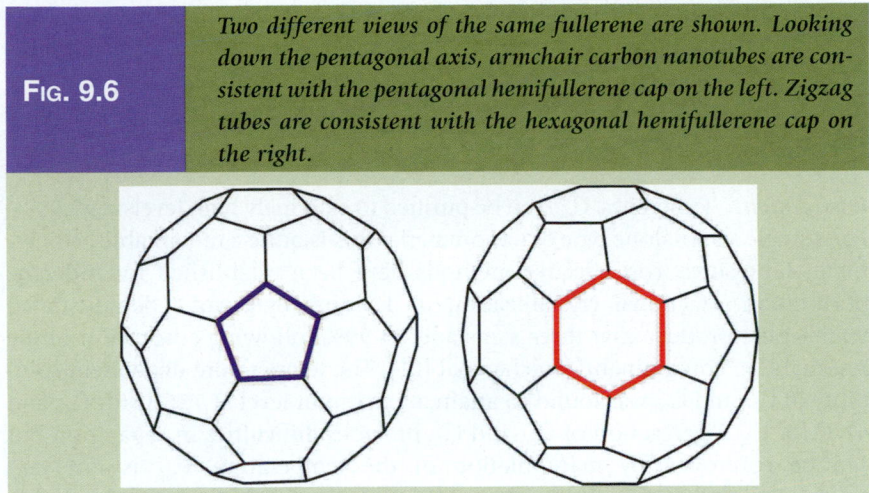

and 40.4 kJ · mol^{-1} respectively—quite a bit higher than the ΔH_f° (graphite), which is equal to 0.0 kJ · mol^{-1}.

9.1.2 Fullerene Synthesis

Fullerenes are routinely produced in nature by high-energy mechanisms such as lightning strikes and combustion processes. There is evidence that links fullerenes to the Permian extinction well over 250 million years ago. Apparently, fullerenes were produced either upon impact or were already part of the chemical makeup of the suspected asteroid. We discussed some methods to fabricate fullerenes and carbon nanotubes in chapter 4 that included arc-discharge, laser ablation, and flame methods. Electron beam methods produce higher order fullerenes with $N_C > 70$.

A few high-energy processes are capable of producing fullerenes like C_{60}. The electric arc discharge method is one of the most popular and was discussed briefly in chapter 4, "Fabrication Methods." In essence, the surface of a graphite rod is vaporized by exposure to an electric arc discharge under high-current and low-pressure conditions in the presence of an inert gas. If metal catalyst particles are integrated within the graphite target, carbon nanotubes are formed. The products resulting from this procedure usually have a high level of contamination and therefore require significant effort to purify. Fullerene products are collected on water-cooled surfaces (cold fingers) in the reactor. The resulting soot contains ca. 15% fullerene products, mostly C_{60}. Constituents are separated from the parent soot by mass by liquid chromatography (with toluene solvent) and other separation techniques.

Fullerenes and *endohedral fullerenes* (fullerenes that contain atoms, ions, or clusters enclosed within their cavity) are synthesized by laser vaporization. During laser ablation, a graphite target is exposed to extremely high temperatures. The best way to make fullerenes, however, is by combustion of a hydrocarbon fuel under low pressure. The fullerene products produced by the combustion method are significantly cleaner than those produced by arc discharge and laser ablation

(>95%). In addition, the combustion method requires significantly less energy. In this method, the oxidation region of the flame is decoupled from the post-flame region. This gives the operator control over equivalence ratio, temperature, and pressure in the primary zone. In a secondary zone, further residence time is allotted to recycling reactants.

Purification. Fullerenes, C_{60}, can be purified to extremely high levels: +99.99%. For a single stand-alone type of nanomaterial this is quite a remarkable achievement. Innovative, cost-effective methods have been established for fullerene purification. Fractional crystallization in 1,3-diphenylacetone demonstrated 99.5% pure product after three steps and 99.99% following adsorption of the residual C_{70} "contaminant" on charcoal [21]. The temperature-dependent solubility of C_{60} and C_{70} was found to attain a maximum level at 136°C for C_{60} and 41°C for C_{70}. Aggregation of C_{60} and C_{70} provides difficulties in separation but can be controlled by manipulation of the temperature. C_{60} is not very soluble in polar solvents like methanol ($S = 0.0$ mg · mL^{-1}) but is slightly soluble in ethanol ($S = 0.001$ mg · mL^{-1}), isooctane ($S = 0.026$ mg · mL^{-1}), and cyclohexane ($S = 0.036$ mg · mL^{-1}) [22,23]. Solubility is enhanced in solutions containing larger molecules and aromatics, such as benzene and toluene ($S = 2.8$ mg · mL^{-1}), 1,2-dichlorobenzene ($S = 8.5$ mg · mL^{-1}). Even better solubility is obtained in 1-methylnaphthalene ($S = 33.0$ mg · mL^{-1}), 1-phenylnaphthalene ($S = 50.0$ mg · mL^{-1}), and 1-chloronaphthalene ($S = 51.0$ mg · mL^{-1}) [22,23].

9.1.3 Physical and Chemical Reactions of Fullerenes

Carbon forms stable bonds in a few preferred geometric structural configurations: tetrahedral (109.5°, hybridization = sp^3, e.g., alkanes), planar (120°, hybridization = sp^2, e.g., alkenes), or linear (180°, hybridization = sp, e.g., alkynes). The preferred geometric configuration for aromatic compounds is also planar. Any deviation from the ideal geometric configuration results in bond strain, higher energy, and enhanced reactivity. Bond strain relief is the driver for fullerene addition chemistry. Any addition to C_{60} results in bond strain relief for all remaining fullerene carbons [24].

Fullerenes are considered to be *soft electrophiles,* which are molecules that are ready to accept electrons from a donor molecule during the first step in a potential chemical reaction. Fullerenes usually are able to accommodate three additional electrons from donors such as hydrogen, methyl groups, and amines. There are three general ways to modify fullerenes: (1) methods that enclose a metal or other small species (e.g., *endohedral* fullerenes); (2) chemical modification of the surface of the fullerene (e.g., to form *exohedral* complexes); and (3) assembly of fullerene one-, two-, and three-dimensional arrays by modification methods 1, 2, or intermolecular interactions. Of course, there do not seem to be any chemical or physical restrictions on combining all three types of fullerene modification methods to form superstructural nanomaterials. *Fullerites* (or fullerides), for example, are crystalline structures of fullerenes composed entirely of C_{60} blocks. The C_{60} form an *fcc* structure held together by weak intermolecular forces with lattice constant $a = 14.17$ Å and the C_{60}–C_{60} bond distance equal to 10.02 Å. Interestingly, the C_{60} molecules comprising the crystal are able to rotate isotropically (freely) at room temperature (**Fig. 9.7**) [17].

FIG. 9.7 *A fulleride crystal is displayed. The crystal shown is an fcc type within which van der Waals gaps separate the constituent fullerenes. Intercalation of metals like lithium between and among fullerenes occurs to form stable complexes.*

Source: Courtesy of Professor Anil K. Rao, Department of Biology, the Metropolitan State College of Denver.

Exohedral Fullerenes. Fullerenes can be modified by chemical methods. Monoadducts form when one type of chemical group is bonded to the surface [25]. An interesting procedure to form monoadducts that involves many types of bonding is shown in **Figure 9.8.** A hierarchical structure is achieved by utilizing all the types of bonding available on a single adduct molecule: thiol linkage to gold surface; lateral hydrogen bonding between adduct molecules; and π–π stacking between fullerenes (**Fig. 9.9**). Combination of multifunctional fullerenes is another approach to forming arrays. In this case, the required functionalities are all located on a single fullerene. For example, four addends are added around the equator of the fullerene and a tethering group is attached to one of the poles [25]. If the four equatorial addends have different length, expanded forms of patterning are possible.

Two- and three-dimensional arrays of fullerenes can be produced by various polymerization methods. For example, dimers result if double bonds between adjacent fullerenes are broken by exposure to laser or UV light and subsequently connected via a cyclobutane derivative—by a [2 + 2] cycloaddition mechanism—forming four-membered carbon rings (**Fig. 9.10**). The cell parameter ($a = 14.17$ Å) of the C_{60} *fcc* lattice is slightly decreased by the formation of these linkages ($a_{rxn} = 14.05$ Å) [26,27].

Rhombohedral, linear orthorhombic, tetrahedral, and other two-dimensional lattice configurations are possible. The arrangement depends on the level of modification, the placement of the functional groups on the fullerene, and length and chemistry of the adduct species (**Fig. 9.11**).

Fig. 9.8 *Formation of single adduct (single functional) fullerenes: Each fullerene has 30 π-bonds, many of which are available for supramolecular synthesis, which involves intermolecular interactions. All functionality provided for a chemically modified fullerene in this figure occurs at a single addend attached to the fullerene with covalent bonds. The purpose of the addend is to modify the fullerene with a chemical moiety that provides: (1) an anchor substituent between the fullerene and a metal surface, and (2) a means to interlink with other fullerenes derivatized in the same way or in different ways. The monoadduct contains thiol groups for bonding to a gold surface and carboxylic acid groups to provide intermolecular bonding (via hydrogen bonds) between fullerenes and the π–π interaction capability.*

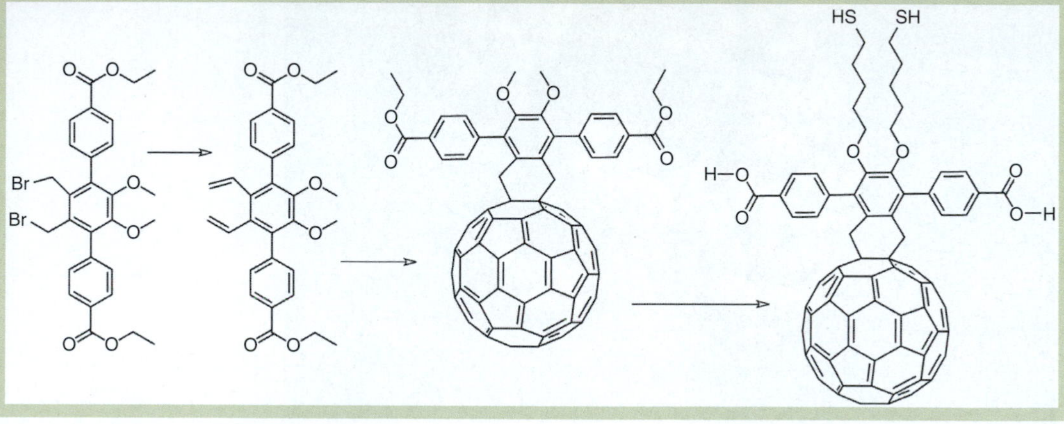

Source: Image redrawn with permission from Glen P. Miller, University of New Hampshire, Center for High-Rate Nanomanufacturing.

There are several types of fulleride–polymer networks. It is relatively easy to bond C_{60} together to form such networks to yield many kinds of one-dimensional, two-dimensional, and layered structures. The Li_4C_{60} polymer serves as an excellent example of a mixed-bonding configuration that yields a close-packed structure.

Fig. 9.9 *A two-dimensional array of fullerenes (modified as pictured in Fig. 9.8) on gold substrate is shown. Along the lateral (x) axis, modified fullerenes are held together by hydrogen bonds between carboxylate functionalities. Along the longitudinal (y) axis, fullerenes are attracted to each other via π–π interactions. Molecular models have confirmed the viability of such a system.*

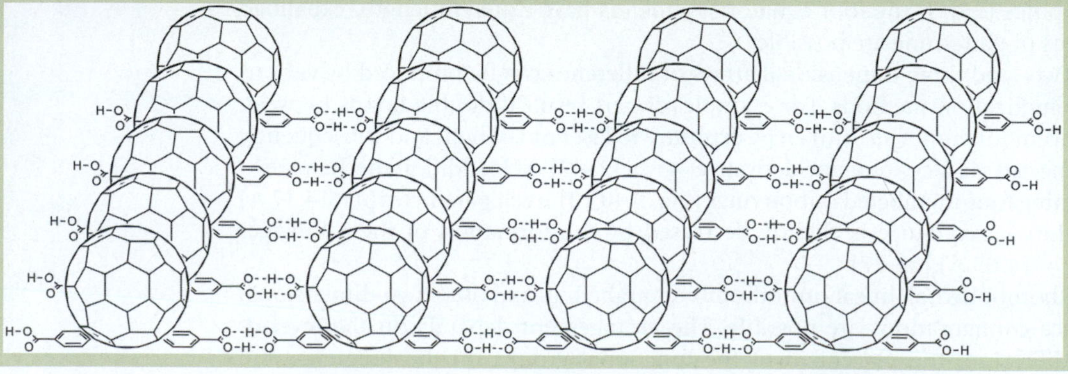

Source: Image redrawn with permission from Glen P. Miller, University of New Hampshire, Center for High-Rate Nanomanufacturing.

FIG. 9.10 *C_{60} dimer formed by [2 + 2] cycloaddition. The structures are formed during prolonged exposure to UV or laser light. The dimers (and higher forms) are insoluble in solvents that buckyballs normally disperse. The dimers revert to individual C_{60} molecules by heat treatment under ambient pressure conditions.*

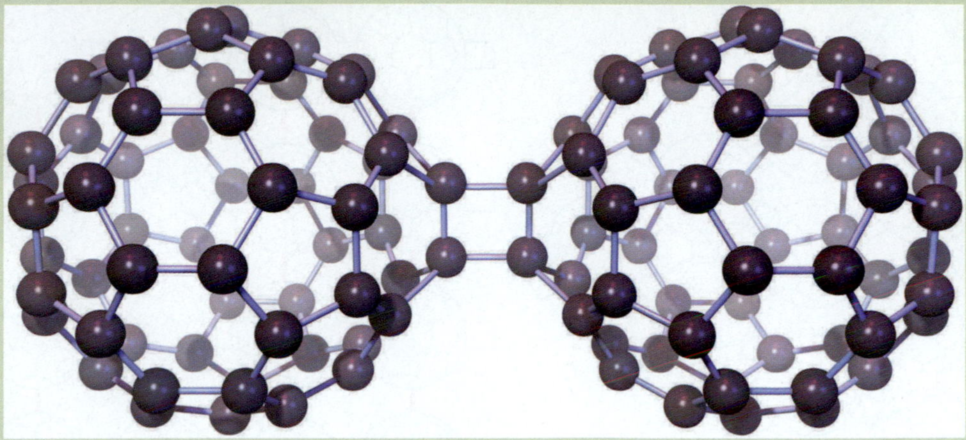

Source: European Synchrotron Radiation Facility, http://www.esrf.eu/UsersAndScience/Publications/Highlights/2005/Materials/MAT11; Image courtesy of Professor Anil K. Rao, Department of Biology, the Metropolitan State College of Denver. With permission.

XRD of a powder sample revealed that the anisotropic fulleride polymer structure was body centered monoclinic with lattice parameters $a = 9.3267$ Å (involving the C–C lateral bonding scheme); $b = 9.30499$ Å (involving the cycloaddition scheme); $c = 15.03289$ Å and $\beta = 90.949°$. Center-to-center dimensions were found to be approximately 9.33 Å along a and 9.05 Å along the b lattice parameter. Lithium ions reside in pseudotetrahedral and pseudo-octahedral holes formed between the layers [26,27].

On the other hand, Na_2RbC_{60} fullerides are bridged by C–C bonds only. Na_4C_{60} fullerides are bridged with four C–C bonds. Factors that influence the mode of bonding include the charged state of the fullerene and the steric influence of the alkali ion. Two Li^+ actually fit into the pseudo-octahedral site of the Li_4C_{60} fulleride. The charge on the fullerene is less than −4 if Li is applied and approximately equal to −4 if Na is incorporated. All caged carbon structures from fullerenes to nanotubes are excellent electron acceptors.

Endohedral Fullerenes. Metal atoms inserted within the fullerene cage during energetic high temperature synthesis form endohedral fullerene structures (a.k.a. "dopyballs") and are designated by the notation $X@C_{60}$ where X represents the enclosed metal. Lithium atoms, for example, can be inserted within a fullerene cage by bombardment with 13-*e*V Li atoms. Radioactive isotope ^{133}Xe gas ions were implanted into fullerenes by an "isotope separator online" (ISOL) instrument [28]. Fullerene targets were made by vacuum evaporation of C_{60} r C_{70} onto Ni foils. Radioactive ^{133}Xe atoms were ionized and accelerated to keV. The Xe ions were separated by mass from the stable ^{129}Xe isotope orehand. The energetic Xe ions penetrate the fullerene by expanding one of

Fig. 9.11 *Generic C_{60} arrays are shown: Top left: rhombohedral; top right; tetrahedral; and bottom: linear orthorhombic polymers. Red lines represent chemical linking groups. Blue squares represent fullerene dimer linkage. Three-dimensional arrays are also possible.*

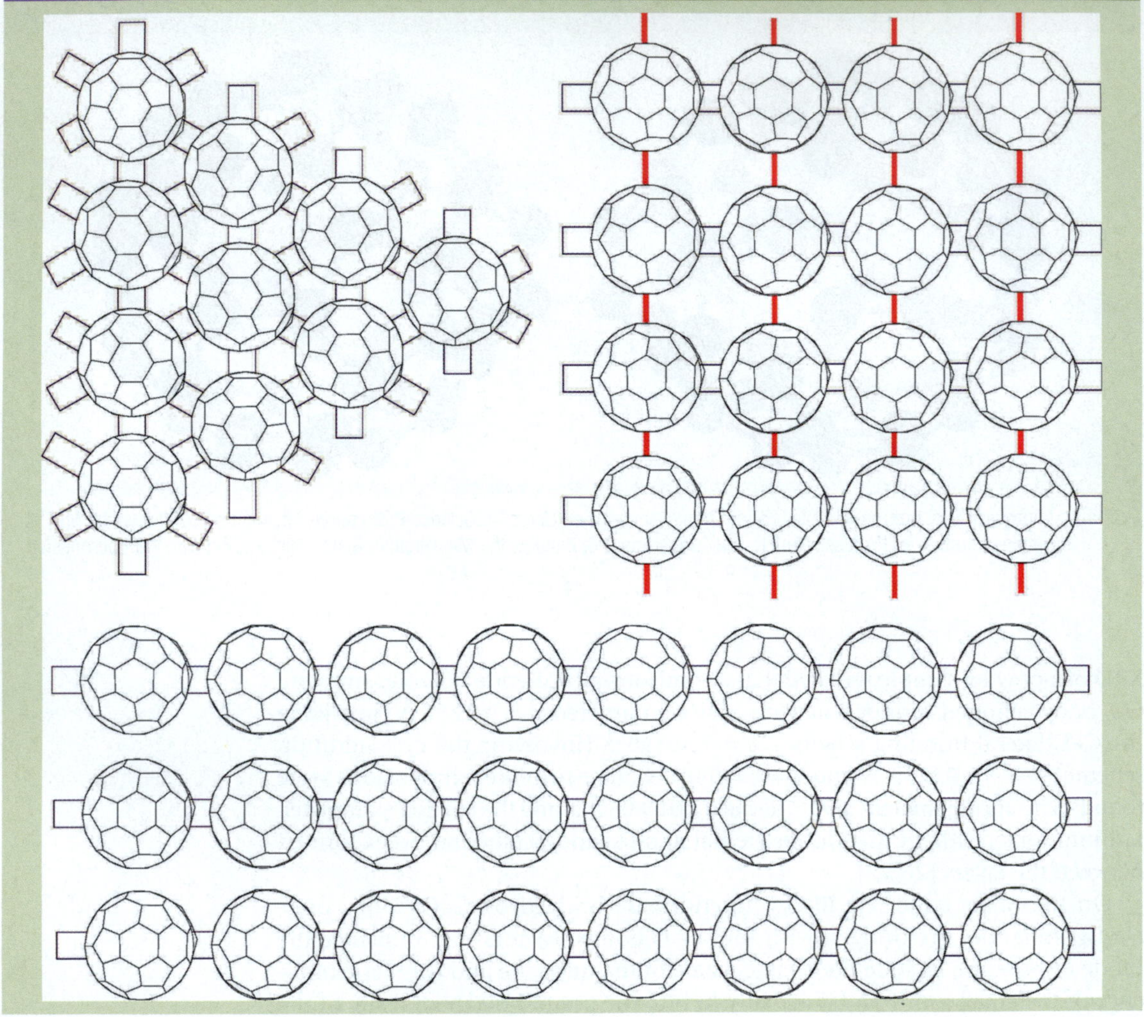

Source: European Synchrotron Radiation Facility, http://www.esrf.eu/UsersAndScience/Publications/Highlights/2005/Materials/MAT11. With permission.

the six-membered rings until the center of the fullerene cavity has been achieved (**Fig. 9.12**). Following implantation, the fullerenes were removed from the Ni target and dissolved in *o*-dichlorobenzene and injected into an HPLC chromatographic column (Cosmosil 5PYE) at a flow rate of 1 mL·min^{-1}. The desired product was detected with UV spectroscopy. These endohedral radioisotope fullerenes (RIF) have application in radiopharmaceutical therapy. Subsequent chemical modification is able to render the RIF hydrophilic with molecular recognition to target cancer cells [28].

The endohedral $Er_3N@C_{60}$, for example, has interesting magneto-topic activity as a solid-state quantum information processing system. This proced

FIG. 9.12 *Endohedral fullerene. Xenon is inserted into the fullerene cavity by the technique of ion implantation.*

Source: S. Watanabe et al., *Journal of Radioanalytical and Nuclear Chemistry*, 255, 495–498, 2003; Courtesy of Professor Anil K. Rao, Department of Biology, the Metropolitan State College of Denver. With permission.

involves the fabrication of an open-grid endohedral fullerene surface array by template-assisted assembly on an oxide-crystal surface. The procedure applied is as follows: (1) substrate preparation: Ar^+ sputter of 0.5%-wt, Nb-doped $SrTiO_3(001)$ single crystal surface in ultrahigh vacuum; (2) anneal of substrate: $c(2 \times 4)$ reconstruction was generated by annealing the sputtered surface at 1090°C; repeated annealing forms; (3) application of endohedral fullerenes: thermally stable $Er_3N@C_{60}$ is deposited on the surface from the vapor phase at 480°C from a Createc Knudsen cell for 20–30 minutes. A 0.15- to 0.20-monolayer coverage per deposition is achieved by this process. Result no.1: close-packed surface ordering of fullerenes is achieved on a $c(4 \times 2)$ surface reconstruction. Result no.2: the heights of the fullerene islands are ca. 1 nm in the *z*-direction .

Fulleranes. A less well-known form of fullerenes is the family of fulleranes—hydrogenated versions of the fullerenes. Fulleranes can be prepared from various solvents using Zn–HCl as the reducing agent [11]. The molecule $C_{60}H_{36}$ is not stable in air in the presence of light. Oxidation of this species yields hydroxyl and ketone groups with concomitant cage breakdown. The presence of ozone accelerates the breakdown of the cage. The absorption band of $C_{60}H_{36}$ matches that of several infrared emission bands detected from astrophysical objects such as nebulae [11]. Other fulleranes such as $C_{280}H_{120}$, $C_{70}H_{60}$, $C_{100}H_{60}$, $C_{120}H_{72}$, and $C_{336}H_{144}$ were modeled by computer simulations. Icosahedral fulleranes such as $C_{80}H_{80}$ and $C_{180}H_{180}$ have shown remarkable stability [29].

9.2 CARBON NANOTUBES

Nanomaterials are remarkable materials. Carbon nanotubes, especially single-walled carbon nanotubes, are one of the most remarkable of nanomaterials. Their physics and chemistry are both well understood. Their synthesis is closing the gap between short tubes and long tubes, polydisperse and monodisperse batches, chiral and achiral forms, metallic and semiconducting, and pure and highly pure. Their potential is enormous, although it remains to be seen how many commercially viable applications are developed over the next decade.

Their electrical, mechanical, chemical, and thermal properties are, in no uncertain terms, quite fantastic. The discovery of fullerenes, in addition to adding another allotrope to known forms of carbon, ignited the imagination. We now present our discussion on carbon nanotubes—elongated versions of the fullerenes.

We do not focus our effort on multiwalled tubes in this section although manufacture of multiwalled nanotubes (MWNTs) has achieved the status of economic viability—mostly in the capacity of fillers for polymeric matrix materials. The first nanotubes discovered were MWNTs; their discovery preceded that of SWNTs by several years.

9.2.1 Structure of Single-Walled Carbon Nanotubes

Two types of high symmetry SWNTs are considered to be archetypal: the zigzag SWNT and the armchair SWNT. All other kinds of SWNTs are called *chiral* nanotubes and have structure somewhere in between these two limits. Each chiral nanotube, as the term implies, has a mirror image. The two archetypal tube types and a generic chiral form are shown in **Figure 9.13a.** In the figure, a small planar segment of graphite is laid out to form a plane—an unrolled carbon nanotube. Just like graphite, carbon atoms exist in a hexagonal configuration. In zigzag tubes, hexagons of carbon are arranged around the circumference so that an apex of the hexagon is oriented parallel to the longitudinal axis (the long axis) of the SWNT (shown at the right in the figure). Armchair SWNTs, on the

FIG. 9.13 *(a) An unrolled carbon nanotube (called a graphene sheet) is displayed. Three types of single-walled carbon nanotubes are enclosed in red rectangles. Bottom right: a zigzag tube SWNT in which the hexagons point along the long axis of the tube; top left: an armchair tube SWNT in which the flat side of the hexagon lies along the long axis of the tube; and bottom left: a chiral nanotube in which the configuration lies between the two extremes. (b) Single-walled carbon nanotubes are shown schematically from the perspective of the tube axis. Left: armchair SWNT; middle: chiral SWNT; and right: zigzag SWNT. The vector renditions represent lines through the long axis (the pointing end) of the hexagones.*

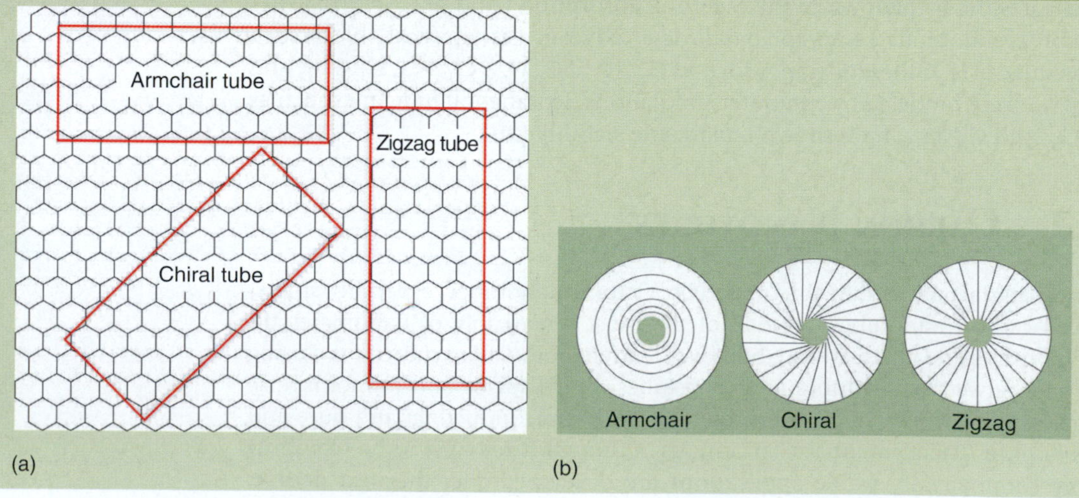

other hand, have the apex oriented along the circumference of the tube and a C–C bond oriented along the longitudinal axis of the tube. Another way to visualize the two is to imagine that the tubes are derived from the hemispherical fullerenes shown in **Figure 9.6.** The end cap of a zigzag tube corresponds to the hemifullerene depicted on the right in the image (e.g., the hexagonal derivative). The cap of an armchair tube, on the other hand, corresponds to the fullerene on the left (e.g., the pentagonal derivative). Front-end views of the three basic kinds of fullerenes are represented schematically in **Figure 9.13b.** The physics of SWNTs is extremely rich and the mathematical description of SWNTs is quite well defined.

Vector Notation for SWNT Structures. Information for this section was gleaned from excellent textual and Internet sources—in particular, *Physical Properties of Carbon Nanotubes* by R. Saito, G. Dresselhaus, and M. S. Dresselhaus (Imperial College Press, London, 2004) and *Carbon Nanotubes and Related Structures* by P. J. F. Harris (Cambridge University Press, Cambridge, 2001). Most nanotubes have chiral form [30]. Specification of the structure is accomplished via the chiral vector **C**. The origin of the vector intersects two equivalent points on the graphene sheet. There are many ways to produce degeneracies (folding the sheet in different ways that produce the same nanotube). Just as in group theory, there are irreducible representations. The sector bounded by two chiral vectors is shown in **Figure 9.14** and enclose the irreducible region. This wedge is 1/12 of the graphene lattice [30].

FIG. 9.14 *Illustration of nanotube structure and the chiral vector. The dark arrows represent the tube axes of zigzag and armchair tubes. The arrow in the middle represents the tube axis for chiral tubes, the vector sum of the others.*

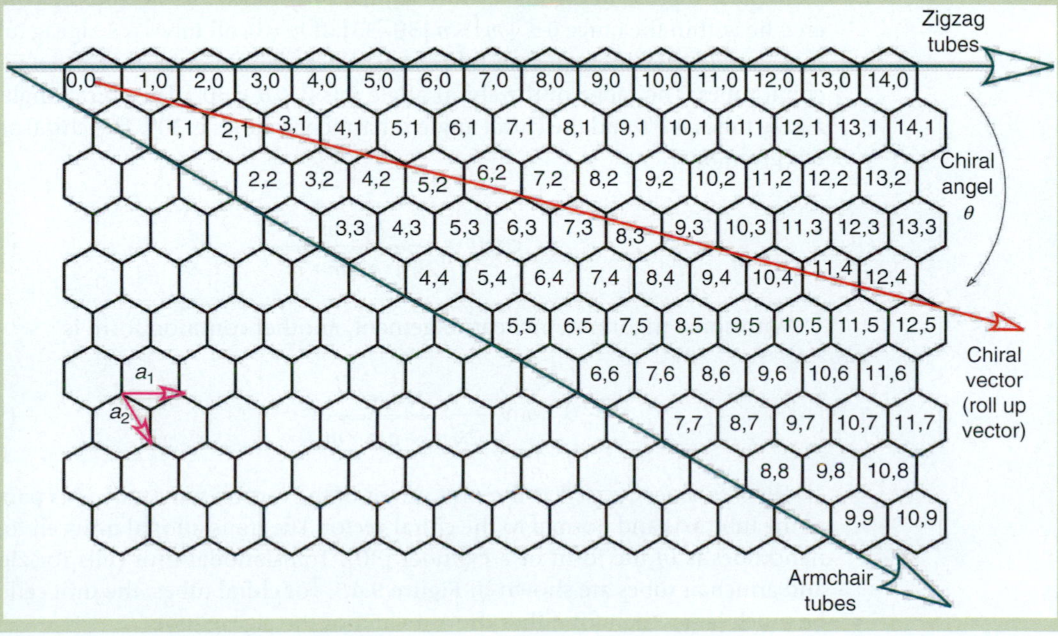

The chiral vector is described by

$$\mathbf{C} = n\mathbf{a}_1 + m\mathbf{a}_2 \tag{9.3}$$

where $\mathbf{a}_1$ and $\mathbf{a}_2$ are the unit base cell vectors of graphite (e.g., in the form of the graphene sheet) and $n \geq m$ [30]. The unit cell of the graphene sheet is not the same as the translational unit cell of the nanotube along the tube axis (more about that later). The magnitude of the unit cell vectors $\mathbf{a}_1$ and $\mathbf{a}_2$ of graphite is

$$|a_1| = |a_2| = a = 0.246 \text{ nm} \tag{9.4}$$

The measured C–C bond length is 0.142 nm. Derivation of the bond length from the lattice constant is given by

$$b = \frac{a}{\sqrt{3}} = 0.142 \text{ nm} \tag{9.5}$$

The diameter of the nanotube is also given by the ratio of the circumference of the tube, L, to π:

$$d_{\text{SWNT}} = \frac{L}{\pi} = a\sqrt{n^2 + nm + m^2} \tag{9.6}$$

or, in terms of n and m and b, is given by

$$d_{SWNT} = \frac{|\mathbf{C}|}{\pi} = \frac{b}{\pi}\sqrt{3\left(n^2 + mn + m^2\right)} \tag{9.7}$$

where b is the C–C bond length in graphite as before. These equations assume that the C–C bond length is 0.142 nm, when in actuality it is longer.

Because there is hexagonal symmetry, the only tubes that need to be considered lie within the range $0 \leq |m| \leq n$ [30–33]. If $m = 0$, all tubes are zigzag tubes. If $m = n$, all tubes are armchair tubes. Any other combination of m and n yields chiral tubes. The range of the chiral angle θ is $0 \leq \theta < 30°$. The chiral angle for zigzag tubes is 0°, and the chiral angle for armchair tubes is 30°. The chiral angle θ is given by

$$\cos\theta = \frac{2n + m}{2\sqrt{n^2 + nm + m^2}} \tag{9.8}$$

With some trigonometric rearrangement, another common form is

$$\theta = \sin^{-1}\frac{\sqrt{3}m}{2\sqrt{n^2 + nm + m^2}} \tag{9.9}$$

The *translation vector* **T** is the unit vector of the carbon nanotube. **T** is parallel to the tube axis and normal to the chiral vector. The translational unit cell for all nanotubes is in the form of a cylinder [30]. Translational unit cells for zigzag and armchair tubes are shown in **Figure 9.15.** For chiral tubes, the unit cell will be much larger than for either the armchair or the zigzag tube.

FIG. 9.15 *The translational unit cell of a (10,0) zigzag tube is shown above. The width of the cell is equal to √3a, where a is the unit vector for the graphene lattice. The translational unit cell for a (6,6) armchair tube is shown below. The width of this unit cell is equal to the unit vector a of the graphene lattice [30]. Chiral tubes that have lower symmetry are characterized by larger cells than the two depicted in the figure [30]. The magnitude of the chiral vector is the circumference of the tube. The magnitude of the translational vector is the width of the cell. The chiral vector (along the circumference) and the translational vector (along the tube axis) are orthogonal to one another.*

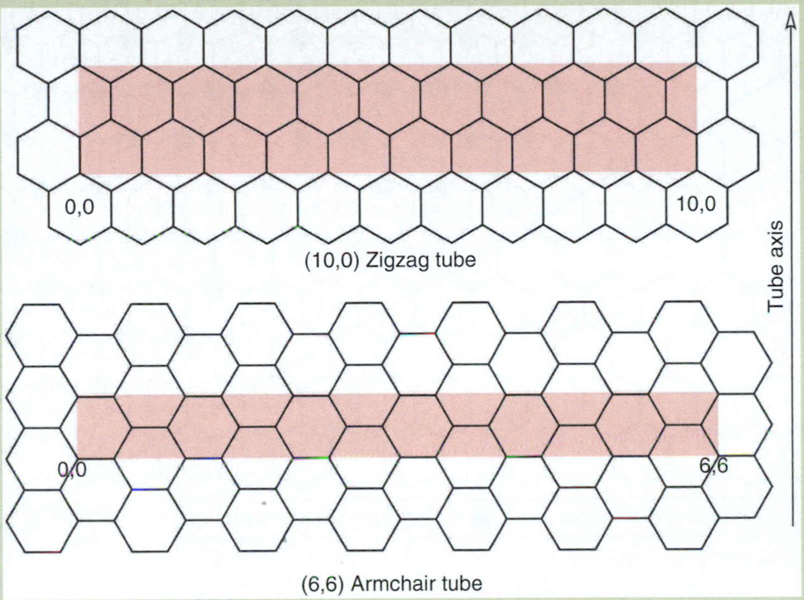

Vector Notation and Electrical Conductivity. Not all carbon nanotubes conduct electricity. Some kinds are semiconducting. The distinction between the two kinds is found in the vector notation. Tubes are conducting if $(m - n)$ or $(n - m)$ is a multiple of 3: $n - m = 3q$, where q is an integer. Therefore, all armchair tubes are metallic because $n - m = 0$, all zigzag tubes with n a multiple of 3 are conducting—for example, (6,0), (9,0,), (12,0), etc.—and all chiral tubes that fall into the general category of $n - m = 3q$ are conducting as well (e.g., (6,3), (9,6), (12,9), etc.). In **Figure 9.16**, an unrolled nanotube graphene grid depicting the kinds of nanotubes is utilized again to show which tubes are semiconducting or metallic.

How to Make a Carbon Nanotube. A good exercise for students is to print out a hexagon matrix of carbon atoms onto a piece of transparency film. Print a graphene lattice on the transparency and fold the tube into various configurations. A sample is shown in example 9.1 (**Fig. 9.17**).

For more information concerning vector relationships and symmetry in carbon nanotubes, consult the text *Physical Properties of Carbon Nanotubes* by R. Saito, G. Dresselhaus, and M. S. Dresselhaus (Imperial College Press, London, 2004).

Fig. 9.16 *SWNTs are either electrically semiconducting or electrically conducting (metallic). Semiconducting SWNTs are represented by solid green dots and conducting tubes by hollow purple dots. Tubes are conducting if (m – n) or (n – m) is a multiple of 3: n – m = 3q, where q is an integer. Therefore, all armchair tubes are metallic, all zigzag tubes with n a multiple of 3 are conducting, and all chiral tubes that fall into the general category of n – m = 3q are conducting as well.*

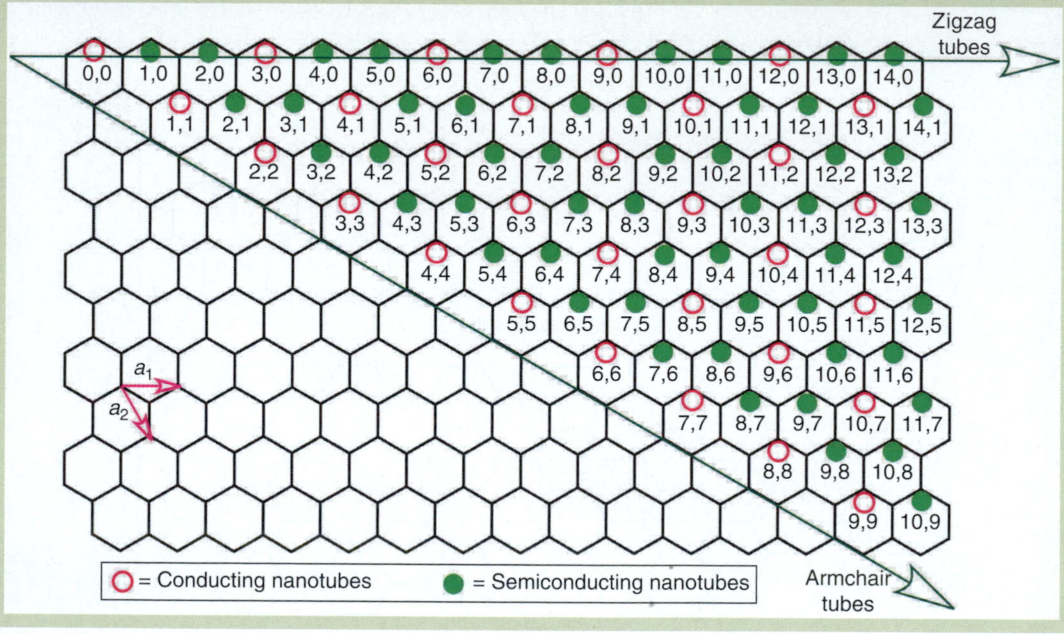

9.2.2 *Physical Properties of Single-Walled Carbon Nanotubes*

We have been saying all along that SWNTs have some incredible physical properties. We now discuss a few of the well-known ones.

Mechanical Properties. CNTs in general are some of the strongest and stiffest materials known to science. As opposed to other materials that consist of melded

Example 9.1 *Making a Carbon Nanotube*

The best way to understand the geometrical relationships within a carbon nanotube is to construct one. Start by printing out a graphene sheet on a clear 8.5- × 11-in. slide (transparency film, used by overhead projectors and very rare nowadays). A good number is ca. 15 hexagons to fill the width of your page.

The simplest tube to make is a zigzag tube. Let us start with the (10,0) tube. Take the (0,0) atom and superimpose it on the top of the (10,0) atom. You now have a (10,0) zigzag tube. A (5,5) armchair tube is also constructed in ***Figure 9.17****.*

Chiral tubes are made in a similar fashion. Notice that the zigzag and armchair tubes have a high degree of symmetry. You will also notice that when the sheet is rolled up, the ends of the chiral vector should intersect. The magnitude of the chiral vector then is the circumference of the newly formed tube.

FIG. 9.17 *Making (10,0) and a (5,5) carbon nanotubes by folding a graphene transparency.*

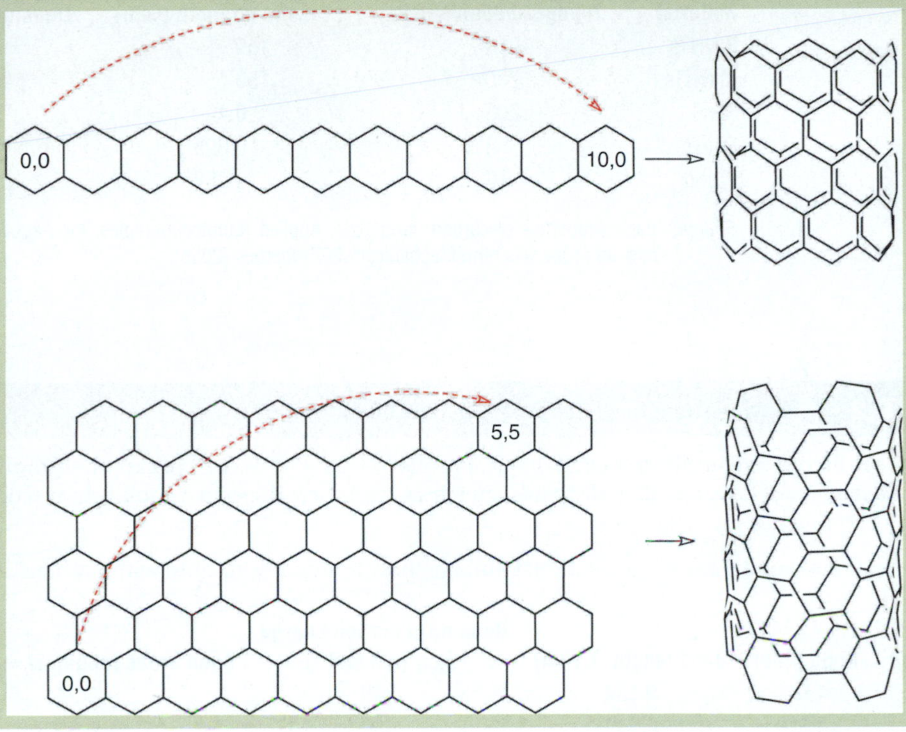

grains and micron-sized substructure, the constituents holding nanotubes together are carbon bonds—the strongest in nature. The breaking point therefore of the C–C or C=C bond must be an enormous number. The theoretical breaking point of carbon nanotubes then must be proportionally enormous and a function of the number of C–C bonds that contribute to the circumferential structure (the translational unit cell). The covalent sp^2–sp^3 bonds that hold carbons in place in a tube are the basis for the strength of the material. A multiwalled carbon nanotube demonstrated experimental tensile strength on the order of 63 GPa compared to high-carbon steel that measured in at a mere 1.2 GPa [34].

The specific strength of carbon nanotubes is on the order of 48.5×10^3 $kN \cdot m \cdot kg^{-1}$. The specific strength of steel is 1.54×10^2 $kN \cdot m \cdot kg^{-1}$ [35]. The combination of high tensile strength with low density in a material is an engineer's dream—definitely those of aerospace engineers. The tubes undergo plastic deformation under extensive tensile stresses. Under compression, tubes are not very strong due to their hollow structure and they tend to buckle (**Table 9.2**).

A good exercise is to calculate the diameter of a megabundle of SWNTs that would be required to pick up the Empire State Building (weight = 365,000 tons). Information about the type of nanotube, the geometrical packing area, and the density of a bundle of SWNTs would be required. For the sake of simplicity, assume the tubes are zigzag (10,10) tubes. Under mechanical tensile stress, the tube undergoes ~0.35% axial stretch and ~0.45% radial stretch [37,38].

TABLE 9.2 *Mechanical Properties of Materials*

Material	Young's modulus (GPa)	Tensile strength (GPa)	Density ($g \cdot cm^{-3}$)
SWNTs	1054	150	1.33
MWNTs	1200	150	2.6
Steel	208	0.4	7.8
Epoxy	3.5	0.005	1.25
Wood	16	0.008	0.6

Source: Basic properties of carbon nanotubes, Applied Nanotechnologies, Inc., applied-nanotech.com/cntproperties.htm#Mechanical%20Properties (2005).

EXAMPLE 9.2 *Tensile Strength of Carbon–Carbon Bonds*

(a) What is the theoretical tensile strength σ_S in gigapascals ($1\ GPa = 10^9\ N \cdot m^2$) of carbon–carbon bonds of different order [Gilkes]? Assume that all bonds stretch to 2.5× their respective equilibrium length before breaking.

(b) How much mass under Earth's gravitational pull are these bonds able to hold without breaking?

Type bond	Bond order	Bond length, L (nm)	Bond dissociation energy, E_{BDE} ($kJ \cdot mol^{-1}$)	Bond break length, $d = 2.5L - L$
C–C	1.0	0.154	348	+0.231
C⩦C	~1.33	0.142	480[a]	+0.213
C=C	2.0	0.134	614	+0.201
C≡C	3.0	0.120	839	+0.180

[a] Hypothetical value based on a bond order ≈1.33 of graphite, bond length ≈0.142 nm.

Solution:

The bond dissociation energy for one C–C bond: $$E_{Carbon\ Bond} = \left(\frac{E_{BDE}}{\mathcal{N}}\right)\left(\frac{10^3\ J}{kJ}\right) \tag{9.10}$$

where $\mathcal{N}$ is Avogadro's number

The force required to break one C–C bond: $$F_{Carbon\ Bond} = \left(\frac{E_{Carbon\ Bond}}{d}\right) \tag{9.11}$$

The covalent radius r_C of a carbon atom is 0.077 nm—the cross-sectional area $\mathcal{A}_C$:

$$A_C = \pi r_C^2 = (3.1416)(0.077\times10^{-9}m)^2 = 1.86\times10^{-20}\ m^2 \tag{9.12}$$

Tensile strength σ_S ($1\ Pa = N \cdot m^{-2}$) of one C–C bond: $$\sigma_{Carbon\ Bond} = \frac{F_{Carbon\ Bond}}{A_C}\left(\frac{1GPa}{10^9 Pa}\right).$$

The maximum force F_{Max} that one C–C bond is able to hold without breaking is the product of the tensile strength and the area of the material A_M (e.g., the diameter of a cable)

If it is just for one C–C bond, then $A_M = \mathcal{A}_C$: $$F_{Max} = A_M\sigma_S = A_C\sigma_{Carbon\ Bond} \tag{9.13}$$

Example 9.2 (Contd.) *Tensile Strength of Carbon–Carbon Bonds*

The maximum mass m_{Max} one C–C is able to hold without breaking: $$m_{Max} = \frac{F_{Carbon\ Bond}}{g} \quad (9.14)$$

Worked example for the single C–C bond:

$$E_{C\text{-}C} = \left(\frac{348\ \text{kJ}\cdot\text{mol}^{-1}}{6.022\times10^{23}\ \text{bonds}\cdot\text{mol}^{-1}}\right)\left(\frac{10^3\ \text{J}}{\text{kJ}}\right) = 5.78\times10^{-19}\ \text{J}\cdot\text{bond}^{-1};$$

$$F_{C\text{-}C} = \left(\frac{5.78\times10^{-19}\text{J}\cdot\text{bond}^{-1}}{0.231\times10^{-9}\text{m}}\right) = 2.50\times10^{-9}\text{N}$$

$$\sigma_{C\text{-}C} = \frac{2.50\times10^{-9}\ \text{N}}{1.86\times10^{-20}\ \text{m}^2}\left(\frac{1\text{GPa}}{10^9\text{Pa}}\right) = 134\ \text{GPa}; \quad F_{Max} = (1.86\times10^{-20}\ \text{m}^2)(134\ \text{GPa})\left(\frac{10^9\text{Pa}}{\text{GPa}}\right) = 2.50\times10^{-9}\ \text{N}$$

$$F_{Max} = (1.86\times10^{-20}\ \text{m}^2)(134\ \text{GPa})\left(\frac{10^9\text{Pa}}{\text{GPa}}\right) = 2.50\times10^{-9}\ \text{N}; \quad m_{Max} = \frac{F}{g} = \frac{2.50\times10^{-9}\text{N}}{9.801\ \text{m}\cdot\text{s}^{-2}} = 2.54\times10^{-10}\ \text{kg}$$

Type of carbon bond	Bond dissociation energy ($\text{J}\times10^{-19}$)	Force, F ($\text{N}\times10^{-9}$)	Tensile strength, σ_S (GPa)	(b) Mass, m_{Max} ($\text{kg}\times10^{-10}$)
C–C	5.78	2.50	134	2.54
C⩦C	7.97	3.74	201	3.82
C=C	10.2	5.07	272	5.18
C≡C	13.9	7.74	416	7.90

Clearly, the tensile strength increases with the bond strength. Therefore, theoretical limit of the tensile strength of one of the strongest bonds in nature is quite high. If the mass of two carbons in the bond is 1.99×10^{-26} kg, these bonds can hold about 10^{15}–10^{16} times their own mass.

Electrical Conduction. Nanotubes are one-dimensional nanomaterials. Many refer to them as *pseudo-one-dimensional nanomaterials*. Due to confinement, electrical conduction is considered to be quantized, and because the mean-free path of electrons is on the order of the dimensions of nanotubes, electrical conduction is also considered to be ballistic. Because resistivity in nanotubes is constant, nanotubes are excellent materials for high current applications [39]. S. Frank et al. in 1998 showed, by scanning probe microscopic methods, that jumps in conductance were exhibited by SWNTs in contact with mercury [40]. D. Tománek et al. confirmed these results in 1999 [41].

Metallic single-walled carbon nanotubes, in theory, should be able to carry electric current density greater than 1000× that of silver or copper [42].

Thermal Properties. Along the tube axis, carbon nanotubes have excellent thermal conductivity to the tune of ~6000 $\text{W}\cdot\text{m}^{-1}\cdot\text{K}^{-1}$ at room temperature [42]. On the other hand, copper, an excellent conductor of heat, is valued at 385 $\text{W}\cdot\text{m}^{-1}\cdot\text{K}^{-1}$

[42]. Along the diameter of the tube, however, carbon nanotubes are insulating. The thermal stability of carbon nanotubes is very high (ca. 3100 K in vacuum). In the presence of oxygen, carbon nanotubes are easily oxidized at ~900 K.

Optical Properties. Metallic tubes, as for any other metal, have no band gap (0.0 *e*V). Metallic carbon nanotubes, as we found out earlier, arise when the condition $n - m = 3q$ is satisfied. Semiconductor tubes, on the other hand, do have a bandgap that is a function of the diameter, another example of a nanomaterial property that relies on size . The bandgap ranges in energy from 0.4 to 0.7 *e*V [39]:

$$E_g = \frac{2wb}{d_{\text{SWNT}}} \tag{9.15}$$

where E_g is the band gap in electron volts, w is the tight-binding overlap energy (2.7 ± 0.1 *e*V), and b is the bond length as before.

The difference is best illustrated in a *density of states* (DoS) plot. DoS is obtained by scanning tunneling spectroscopy (STS). STS is obtained by an STM parked over a predesignated region of the sample while the bias voltage transmitting from the tip is scanned from plus to minus. The DoS is a function of the number of quantum states in the energy range between E and $E + dE$ divided by the product of dE and the volume of the sample. Examples of generic DoS plots of a metallic and a semiconducting nanotube are shown in **Figure 9.18.** In a metal, the Fermi level is populated and therefore the DoS shows a positive population in that area. For a semiconductor, the Fermi level (technically, the chemical potential) is empty (the bandgap), and the DoS in that region has zero population (no states). As energy is increased on either side, peaks called van Hove singularities appear. These regions are responsible for the optical properties of carbon nanotubes (e.g., transitions between the van Hove singularities) [39].

FIG. 9.18 *Optical properties of single-walled carbon nanotubes. Van Hove singularities are shown for (a) (9,9) metallic conducting tubes and (b) (11,7) semiconducting tubes. Note that there is no band gap in the metallic tube (all states populated). The semiconducting tube has the region in the center with no populated states. Optical transitions occur between the van Hove peaks.*

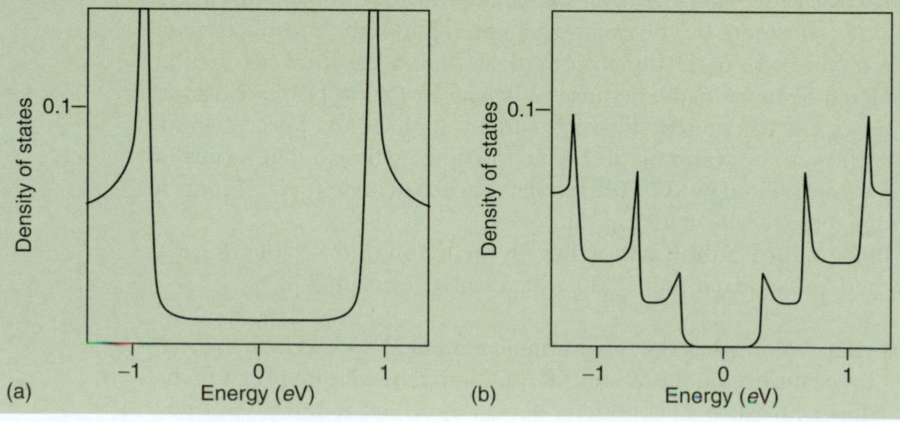

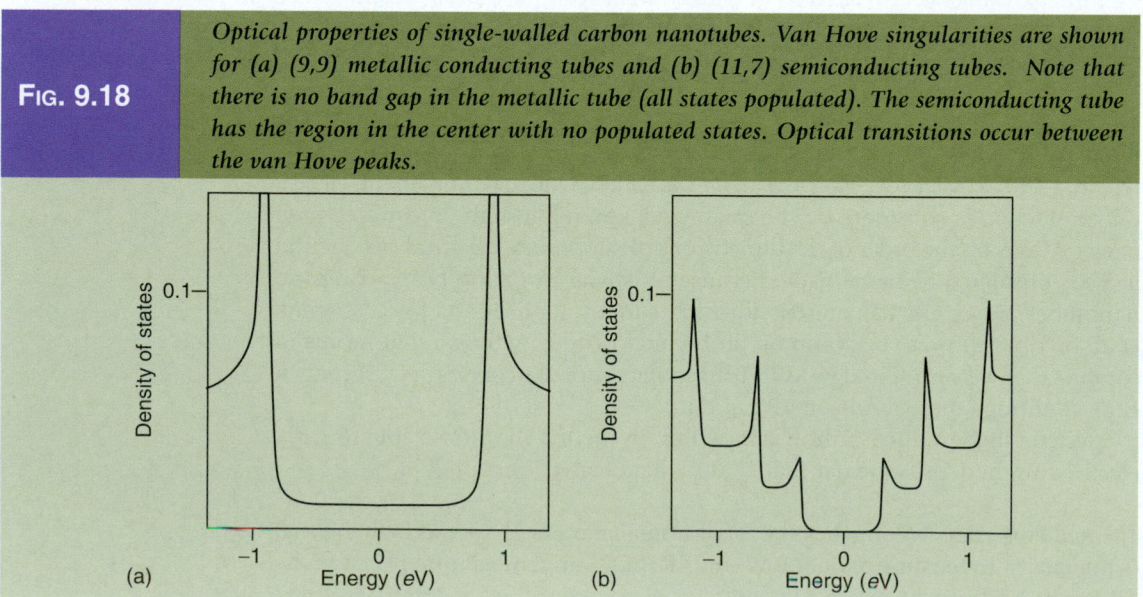

Source: www.pa.msu.edu/cmp/csc/nanotube.html

Fig. 9.19 *A generic Raman spectrum of single-walled carbon nanotubes is shown. The spectrum is unique to SWNTs because of the radial breathing modes found at lower energy. The location of the RBMs is a function of the SWNT diameter; the larger the diameter is, the lower energy the breathing mode has. The intense peak at ca. 1600 is the tangential E_{2g} mode (a.k.a. the G-band) that reflects the electronic properties of the SWNT. The D-band is an indication of the level of disorder in the batch.*

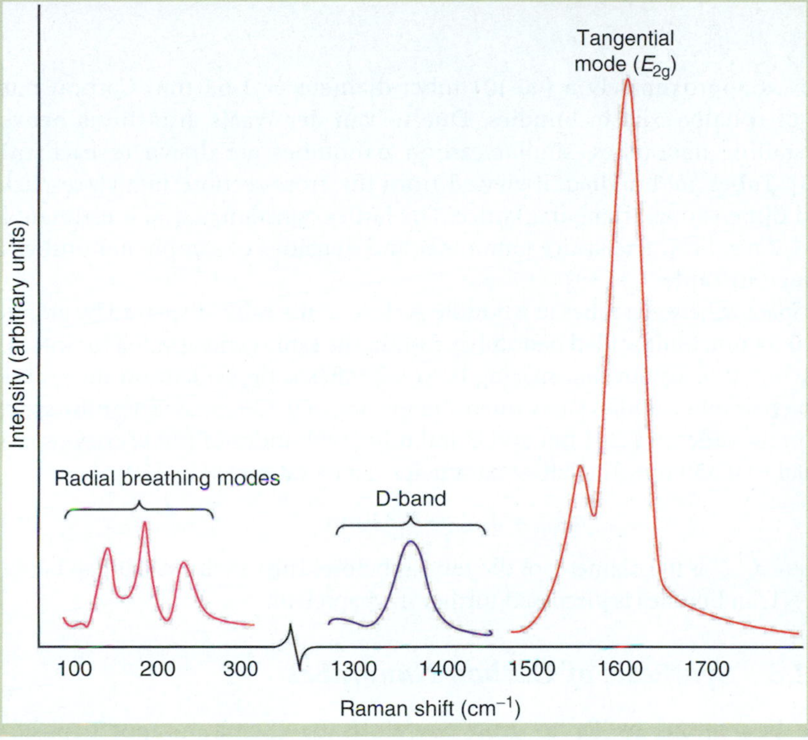

Raman Spectra. Raman scattering spectra provide a clear and unambiguous indication of the presence of SWNTs. Due to unique symmetrical phonon modes of SWNTs, such as the radial breathing mode (RBM) and the tangential mode, Raman spectra acquired from SWNTs are diagnostic. Three features are apparent in Raman spectra of SWNTs in **Figure 9.19**: (1) the intense band on the right in the figure corresponds to tangential symmetric vibrations (the G-band or E_{2g} tangential mode) around 1590 cm^{-1} and is related to the strain in the tubes and semiconducting properties; (2) the D-band (ca. 1340 cm^{-1}) in the middle represents the level of disorder in the tube/bundle; and (3) the radial breathing modes (140–300 cm^{-1}) are indicative of the diameter of the SWNTs. Another band, not shown, is the G′ band, an overtone of the D-band, that provides information about the electronic structure of the nanotube (e.g., whether it is metallic vs. semiconducting).

Carbon Nanotube Bundles. A common example put forward by nanotubers is that of the 1-nm diameter nanotube (inner diameter). If the wall thickness is assumed to be 0.34 nm, then the diameter of the SWNT overall is 1.68 nm.

Table 9.3 *SWNT Bundle Parameters*

Nanotube index (n,m)	Lattice parameter, a_{SWNT} (nm)	Density (cm^{-3})
10,10	1.678	1.33
17,0	1.652	1.34
12,6	1.652	1.40

Source: T. A. Adams and D. Tomànek, Physical properties of carbon nanotubes, The Nanotube Site, pa.msu.edu/cmp/csc/nanotube.html (visited 2007).

This is approximately a (10,10) tube; diameter = 1.65 nm. Carbon nanotubes usually exist in bundles. Due to van der Waals attractions between crystalline nanotubes, similar carbon nanotubes are drawn to each other [43]. Tubes are bundled, if viewed from the cross-section, in a close-packed two-dimensional triangular lattice. The lattice constant, a_{SWNT}, is estimated to be 1.7 nm [44]. The lattice parameter and densities of sample nanotubes are shown in **Table 9.3** [39].

Spacing between tubes in a bundle is close to the value displayed by graphite: ca. 0.34 nm. Multiwalled nanotubes exhibit the same general value for intertube spacing. It turns out that spacing between tubes is dependent on the chirality. Armchair tube bundles show intertube spacing of 0.338 nm; zigzag tube spacing is on the order of 0.341 nm and chiral tubes with index of $(2n,m)$ have spacing equal to 0.339 nm. The lattice parameter can be estimated by [39]

$$a_{SWNT} = d_{SWNT} + 0.34 \text{ nm} \tag{9.16}$$

where d_{SWNT} is the diameter of the tube as before. The van der Waals gap between SWNTs in bundles is discussed further in chapter 10.

9.2.3 Synthesis of Carbon Nanotubes

The best means by far to grow SWNTs is by chemical vapor deposition. High-energy methods such as laser ablation and electric arc are energy intensive and therefore make scale-up problematic. Extreme temperature conditions (1000–2000°C or greater) make it impossible to exert any control over tube purity, uniformity and orientation. The product is usually in the form of a tangled mass with nanotubes projecting in all directions. Flame methods to produce MWNTs, on the other hand, have achieved a certain level of sophistication with regard to bulk processing and purity. MWNTs have found a niche as fillers in the polymeric materials industry and have become an economically viable process. Flame synthesis is a subset of CVD fabrication methods.

Chemical vapor deposition was discussed briefly in chapter 4 and we will not elaborate much more on the process itself. Instead we will focus on recent developments that have enhanced CVD methods to the point that it has become the "method of choice" for researchers. Although its commercial potential is still in its infancy, the future bodes well for CVD methods—especially for fabrication of SWNTs. Commonly used carbon sources include methane [45], acetylene [46], ethylene [2,3], alcohols [47], CO [48], propylene [45], hexane [49], benzene [50], and others.

Synthesis of carbon nanotubes, especially SWNTs, is remarkably easy; it is ironic that so many nanomaterial synthesis processes are relatively simple, cost effective, and straightforward with high throughput. The catalyst is generated by any number of techniques. Physical methods include RF sputter and thermal evaporation; chemical methods include aqueous incipient wetness impregnation (AIWI) and template synthesis [45]. Substrates like silicon, silicon oxide, alumina, and mica serve as the support for catalyst nanoparticles.

Process Improvements. In general, current CVD methods produce CNTs that are polydisperse with regard to diameter, chirality, and length; randomly oriented; defect laden; and impure, for the most part. Nanotube length has been limited to a few hundred microns, although nanotubes in the millimeter to centimeter range have been reported [49]. However, mechanical properties of SWNT bundles 20 cm in length failed to live up to expectations [49]. Although fabricating a 20-cm bundle is quite an achievement, researchers concluded that the strands were composed of shorter nanotubes, hence the poor mechanical properties.

The work of Hata et al. demonstrated highly pure 2.5-mm SWNTs made from ethylene and water vapor in the CVD atmosphere [51]. The function of the water vapor is to extend catalyst lifetime by oxidizing amorphous carbon by-products formed during the CVD process. The purity of the SWNT product was 99.8% and the SWNT/catalyst ratio was reported to exceed 50,000 [51]. The application of water to scrub catalyst particles was reported by Cao et al. in 2001 [52]. Other mild oxidizers such as oxygen and CO_2 have also been applied during the CVD process [53,54].

Other improvements have been made over the past few years. The use of alternate carbon sources was reported by Maruyama et al. [46]. High-quality SWNTs were produced at 550°C, a very low temperature for a CVD process. Alcohols are partially oxygenated hydrocarbons and may provide a built-in means of scrubbing catalyst particles and removing unwanted CVD by-products. The use of ohmically heated substrates has demonstrated an efficient and cost-effective means of forming CNTs by CVD [55]. MWNTs grown from Ni, Co, or Fe catalyst particles were supported on resistively heated carbon paper. Aligned two-dimensional arrays of CNTs can be produced with the application of an electric field [56,57]. CNTs formed in RF plasmas are aligned by the applied field and thereby it is possible to fabricate aligned vertical arrays of carbon nanotubes in this way.

Seidel et al. showed that SWNTs are efficiently synthesized between catalyst-covered Mo pads [58]. The purpose was to develop new methods to fabricate electronic device circuitry. Evaporation of a thin layer of alumina anchors the catalyst particles at predetermined locations on the substrate. The thin alumina layer also prevents the diffusion of the catalyst into the Mo substrate [58,59]—a tribute to the versatility of nanoengineering! G. S. Duesberg et al. showed growth of single strands of MWNTs from lithographically fabricated nanoholes in SiO_2 [60]. The diameter of the catalyst particle and hence the diameter of the MWNT were directed by the diameter of the nanohole—a template synthesis process. Catalysts were implanted into the nanoholes by "spin-on" deposition or by application of a high-precision ion coater. SWNTs were also grown in nanoholes [60].

Although technology is not stressed in this text, many recent developments are demonstrating the incredible utility of carbon nanotubes. Two-centimeter-long nanotube aspect ratios exceeding 900,000 were produced by V. Shanov and

M. Schultz of the University of Cincinnati [61]. Apparently an innovative means to keep the catalyst alive during the CVD process was discovered (details not available).

A nanotube-based synthetic gecko tape was developed by Ge et al. [62]. Micropatterned carbon nanotube arrays were transferred onto a flexible polymer tape patterned after gecko setae (chapter 13). The tape supported a shear stress of 36 $N \cdot cm^{-2}$, approximately four times better than that of the gecko foot and it was able to stick to many different kinds of surfaces, even Teflon [62]. The setae were represented by nanotube bundles and the *spatulae* (on the order of 200 nm in geckos) were represented by individual carbon nanotubes. The stickiness is due to van der Waals forces and offers a means to provide reversible adhesion that also has the capability to conduct electricity [62].

Purification and Separation. More often than not, carbon CVD products require purification and separation in order to extract nanotubes. Reflux of sooty products in 10% HNO_3 at 150°C for 16 h removes most of the amorphous carbon and graphitic debris [30] by oxidation and then, depending on the type of catalyst, most of the metal. The process is not particularly effective for graphite encapsulated metal catalysts. Following rinsing in surfactants and washing in toluene, the nanotubes undergo further graphitization at temperatures exceeding 1500°C to remove structural defects (heptagons and pentagons). Other purification methods apply equally harsh conditions.

Removal of the protective fullerene cap on nanotubes is also desired in many applications (e.g., hydrogen storage and functionalization). Defect-rich nanotubes in particular are easy to oxidize, albeit at great cost: Most of the product is lost in the process [63].

Purification is just the first part of the process. Separation is the other. To date, no one has been able to synthesize exactly one kind of nanotube, although exclusive fabrication of SWNTs, MWNTs, or DWNTs (double-walled carbon nanotubes) is possible by controlling reaction conditions (pressure, gas composition, catalyst composition and size, and temperature) [38]. Most as-formed product requires postsynthesis separation because, in addition to numerous carbon byproducts, the batch includes a mixture of tubes with polydisperse diameter, length, and chirality and both semiconducting and metallic tubes.

Many types of separation procedures have been applied to CNTs—some effective and some not effective [38]: chromatography (size exclusion, HPLC, gel permeation, and ion exchange); electrophoresis (capillary, ac dielectrophoresis, gel, dielectric field-flow fractionation); fluid-based methods (flow-field fractionation [FFF], nematic liquid crystal extraction); destructive methods (current-induced oxidation, fluorination, and annealing); chemical methods (amino acid–amine adsorption, photoelectrochemistry, density-gradient centrifugation). Capillary electrophoresis has been applied to separate CNTs based on length [21]. Flow-field fractionation [64] has proven to be an effective means of separating CNTs. One means of separating semiconducting tubes from conducting (metallic) tubes is by passing a large current through an SWNT mass [65]. This method, however, results in the destruction of the metallic tubes due to defects caused by the large current that lead to exclusive oxidation of the conducting tubes [65]. Hongie Dai et al. of Stanford utilized low-temperature methane plasma to selectively destroy metallic tubes and nanotubes with diameter in the range of 1.0–1.3 nm [38,68].

Researchers at Rice University recently developed a means to separate semiconducting nanotubes from metallic ones based on size [67]. The separation procedure is based on the diameter-dependent dielectric constant of individual nanotubes. Dissolved nanotubes were pumped into an electrified chamber in which metallic nanotubes became trapped. Semiconducting tubes, depending on their chirality and size, floated at different levels in the chamber. The smaller diameter tubes possessed larger dielectric constants and were found lower in the system. As the speed of the flow was varied, with upper level currents traveling faster, the Rice scientists were able to separate tubes [67].

9.2.4 Growth Mechanisms

R. E. Smalley et al. proposed a straightforward explanation of nanotube growth [68]. The theory explains the size and structure of two kinds of graphitic materials: SWNTs and onions (graphite encapsulated catalysts) (**Fig. 9.20**). The model,

FIG. 9.20 *The yarmulke mechanism of SWNT growth is depicted. It is energetically better for the growing fullerene to remain in the form of a tube rather than spread out over the particle surface if the curvature of the particle is extreme (such as the case for very small particles)—the trade-off between dangling bonds and bond strain. Encapsulation of larger catalyst particles is depicted below. Depending on process conditions, large particles are either encapsulated or grow multiwalled tubes. The onion configuration is schematically depicted at the bottom right of the figure. The structure resembles graphite with interlayer spacing on the order of 0.34 nm. Once a particle is encapsulated it loses its catalytic function.*

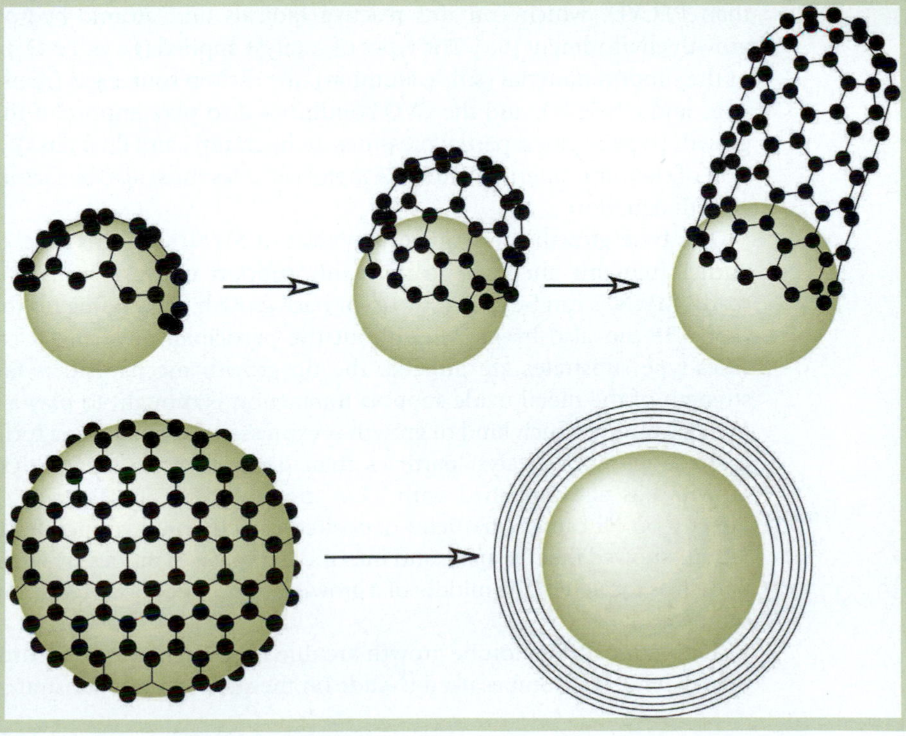

called the *yarmulke* mechanism, is based on the formation of a hemispherical fullerene cap on the surface of the iron nanoparticle. What happens after the formation of the cap depends on the size of the catalyst particle. Experimental results demonstrated the gas-phase synthesis of SWNTs from iron catalysts formed by the decomposition of ferrocene. Following decomposition and nucleation, the catalyst grows by collision with other clusters. Depending on reaction conditions, clusters in the range of 0.7–1.4 nm are formed (50–100 iron atoms). CO gas, the carbon source, disproportionates on the cluster surface according to the Boudouard reaction:

$$CO + CO \xrightarrow{\text{Catalyst}} C_{(s)} + CO_2 \tag{9.17}$$

Classical thermodynamic principles can be extended to explain the energy associated with graphitic structures. A single-walled carbon nanotube is nothing more than a sheet of graphite (called a graphene sheet) rolled to make a cylinder. Multiwalled nanotubes are a series of concentric SWNTs of increasing diameter. However an excess amount of energy is required to get this rolling accomplished. The trade-off in energy is a balance between forming an open structured planar sheet of graphite of limited size (more accurately, a graphene sheet that has numerous dangling bonds) and forming a closed structure with bond strain but with no dangling bonds. As we find out, the size of the catalyst particle has great influence over these two parameters.

The growth mechanisms of carbon nanotubes were discussed briefly in chapter 8. In this chapter, we will touch upon on the distinctions between base growth and tip growth and other factors involved in CNT growth. The types of synthesis process—e.g., CVD, arc discharge, laser ablations, and PECVD (plasma-enhanced CVD), affect the growth mechanism of CNTs. CVD is a cleaner process than PECVD, which contains reactive radicals and atomic hydrogen in the growth environment [68]. The types of catalyst applied (Fe vs. Fe_2O_3), the nature of the support material (silica, alumina), the carbon source gas (methane, acetylene, and ethylene), and the CVD conditions also play important roles in CNT growth (type of gases, partial pressures, temperature, and flow rates). The formation of chemical intermediates like metal carbides must also be factored into the growth equation.

The base growth mechanism is typical of SWNTs. In this case, the catalyst particle remains anchored to the oxide support material, usually aluminum oxide. MWNTs can be grown by tip or base growth, depending on the substrate (MWNTs can also be grown without the participation of metal catalysts). If silica-type substrates are utilized, the tip growth mechanism is favored. The strength of the metal oxide support interaction is thought to play a major role in determining which kind of growth is expressed—e.g., alumina forms stronger interactions with catalyst particles than does silica. In situ analysis of CNT growth was accomplished with TEM methods [69]. CNTs were synthesized directly on Ni catalyst particles deposited directly on a copper TEM grid [69]. Results showed that tip, base, and intermediate growth modes (where the catalyst particle is located in the middle of a growing nanotube) occurred simultaneously on the grid.

Limitations to nanotube growth are due to several factors. According to one theory, whole nanotubes need to slide on the surface of the substrate [70]. Once the tube becomes longer, van der Waals forces between the nanotube and the substrate become too strong and overcome the force required to move the nanotube during growth. Thus, growth is terminated. This apparent difficulty can be overcome by application of an electric field (from RF sources). In this case, the tubes would align with the field during growth. This is not a problem with the tip growth mechanism. Another theory proposes that mass transport limitations account for termination of nanotube growth. The flow rate of carbon source gases to the catalyst becomes restricted as the nanotube forest is grown [70]. Also, SEM and AFM showed that most tubes nucleate via base growth and thereby the restrictions placed on mass transport apply [70].

In general, carbon source materials like methane or acetylene adsorb onto the surface of the catalyst and decompose. The use of saturated hydrocarbons like methane and propane initialize endothermic conditions whereas the use of unsaturated gases like acetylene or ethylene produce exothermic conditions—e.g., $C_2H_2 \rightarrow 2C_{graphitic} + H_2$, $\Delta H_{rxn} = +262.8\ kJ \cdot mol^{-1}$ [71]. The liberated carbon diffuses into the metal catalyst or upon its surface. The activation energy for surface diffusion is obviously smaller than diffusion through the bulk of the metal and is therefore the preferred path, at least in the formation of SWNTs. The surface diffusion argument, however, does not adequately explain the growth of MWNTs [71]. Because certain amount of carbon must be dissolved in the catalyst before nanotube growth is possible, other methods of carbon transport may also affect CNT growth.

Klinke et al. applied thermodynamic calculations and finite element analysis to CNT growth based on hydrocarbon precursors at elevated temperatures [71]. Temperature gradients were thought to drive the diffusion of carbon during the growth phase but more recently, a concentration gradient is more likely the cause of the carbon flow with regard to the catalyst [71]. Carbon precipitates at the part of the catalyst known as the depletion zone [71]. The temperature however plays a major role in activating the diffusion process. Although much progress has been achieved, there is much about CNT growth mechanisms in general that remain to be clarified.

9.2.5 Chemical Modification of Carbon Nanotubes

This field of nanochemistry has burgeoned in recent years, especially as chemically modified nanotubes have become increasingly significant components in nanocomposites. The applications of carbon nanotubes in polymers are very broad—electrical conductors and heat transfer conduits to structural elements. The extreme aspect ratio of SWNTs, ranging from 1000 to 10,000 and more, makes their use as fillers in polymeric matrices very attractive to nanocomposite technology. Chemical challenges facing nanotube chemistry include dispersion and solubility in the polymer, a fact exacerbated by bundling. Dispersion and dissolution is problematic with SWNTs due to strong intertube attractions (bundling) and the fact that they are large molecules. Interfacial adhesion between the polymer and the "smooth" nanotubes is also problematic. In order to reinforce, enhance, or add properties to composite materials, carbon nanotubes need to be modified chemically. Although physical techniques such as sonication are able to temporarily disperse CNTs, once the power is turned off, the tubes reagglomerate into bundles. SWNTs, however, require special treatment to enhance

FIG. 9.21 *Two growth mechanisms of multiwalled carbon nanotubes are shown. On the left, the base-growth mechanism is depicted. Catalyst particles that are adsorbed onto alumina support materials facilitate base growth of carbon nanotubes. Tip growth, depicted on the right in the figure, is the case in which catalyst particles do not adhere to the surface of the substrate and occurs on silica surfaces. Chemisorption between catalyst particles and alumina appears to be strong enough to anchor the catalyst to the surface. The carbon source shown above is methane. Methane decomposes following adsorption onto the surface of the metal catalyst.*

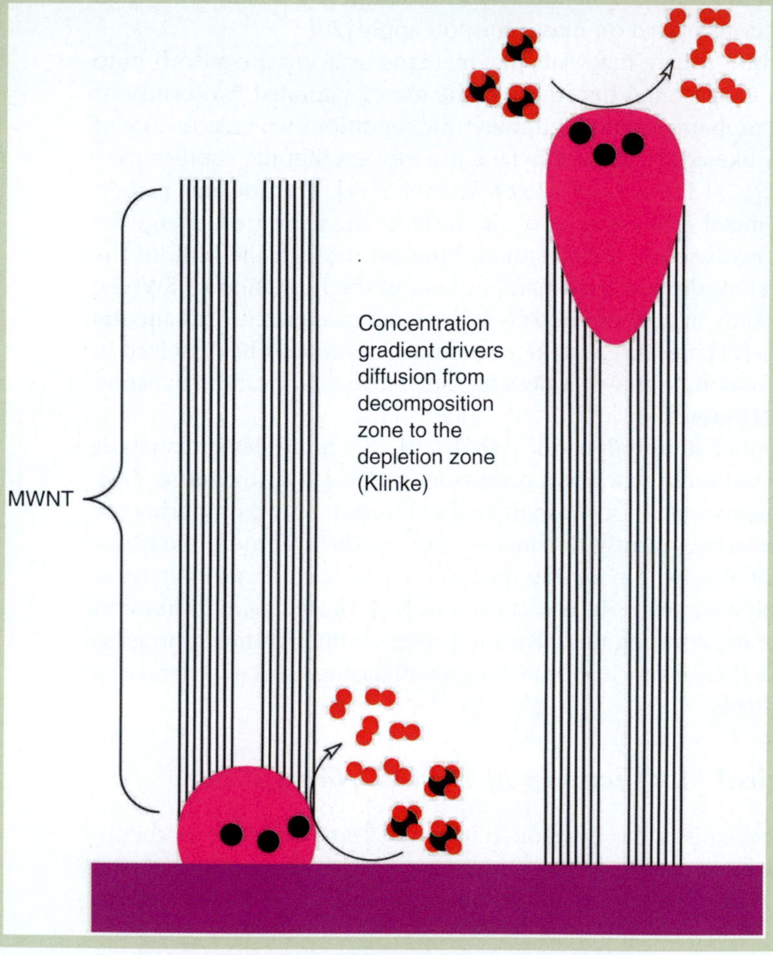
MWNT
Concentration gradient drivers diffusion from decomposition zone to the depletion zone (Klinke)

their chemical compatibility, dispersion, dissolution, and interfacial adhesion properties. Although significant strides have been made in recent years, fabrication of CNTs with uniform length, diameter, and chirality; inexpensive mass production; and purity plague advanced nanotube development.

SWNTS are all surface, both inside and outside. Every single atom is a surface atom. Carbon nanotubes are supermolecules that are held together by some of the strongest bonds known to nature, the carbon–carbon bond; there are no dangling bonds since the ends are technically capped by hemispherical fullerenes.

One would expect that SWNTs in particular would be extremely reactive due to high surface energy. However, eventhough carbon nanotubes have no bulk atoms, they are relatively inert—not because of thermodynamic considerations but rather because they have no chemical functional groups. It is possible, however, to disturb the conjugated system of CNTs by attack of strong acids such as HNO_3 and/or H_2SO_4 to form carboxylates [24]. Applications of heat and/or ultrasonication along with acid treatment result in the formation of dangling bonds that render the tubes reactive [24]. The apparent stability of SWNTs is demonstrated by the low degree of functionalization, even under such harsh conditions.

Under the right conditions, the intrinsic curvature of chemical bonds (thermodynamic properties) making up nanotubes provides the basis for chemical modification. Modification of the outer surface of SWNTs reduces intertube attraction because the perfect Van der Waals interactions no longer overlap exactly. Chemical functionalization also enhances interaction between the SWNT or MWNT and a potential polymer matrix material. The design of modified nanotubes to suit a specific purpose is a high-priority research endeavor.

Single-walled carbon nanotubes are made of two components: the high-aspect-ratio tube and the end caps. The end caps, hemispherical fullerenes, naturally exhibit fullerene behavior and are therefore more susceptible to oxidation and addition reactions. There are two generic types of chemical modification that can be performed on nanotubes: covalent modification and noncovalent modification [72,73].

Covalent chemical modification involves breaking old covalent bonds and forming new covalent bonds. Processes such as oxidation and fluorination are examples of covalent modification. Noncovalent modification implies that carbon nanotubes are functionalized by means of intermolecular bonding processes, some forms of which were discussed in previous sections of this chapter. By its very nature, covalent modification perturbs the structure of the nanotube and introduces defects into the carbon nanotube that disrupt the aromatic structure. Defects alter the mechanical strength, electrical conductivity, and thermal conductivity of the nanotube. On the flip side, depending on the chemical nature of the modifying group, the nanotube is capable of linking to its host polymeric matrix. Such cross-linking, always an important aspect of polymer design, serves to strengthen the polymer as a whole. A balance must be achieved between the degree of chemical modification of the nanotube and the degree of enhancement of the strength of the polymer as a whole: What is the load-transfer efficiency from the polymer to the nanotube [73,74]?

Noncovalent types of modification are governed by thermodynamic parameters. This form of chemical functionalization is accomplished by means of intermolecular attractions between an adduct molecule and the nanotube. In this way the backbone structure of the nanotube is not altered and properties such as mechanical strength, electrical conductivity, and thermal conductivity are compromised, relatively speaking. The problems of solubility, dispersion, and separation of native tubes are addressed by noncovalent chemical functionalization. As a result, subsequent processing of nanotubes with the desired outcome is more likely.

The potential routes to nanotube chemical modification are summarized in **Table 9.4**.

TABLE 9.4 *Chemical Modification of Carbon Nanotubes*

Major category	Type	Comments/examples
Noncovalent methods	π–π Stacking	Increased solubility in aromatic solvents, aromatic surfactants, and aromatic polymers Immobilization of chemical and biological molecules by polynuclear aromatic compounds (anthracene, pyrene, etc.) Increase solubility of SWNT in aqueous media by application of ionic pyrenes [75] Immobilize enzymes on a surface via pyrene-derivative modified CNTs
	Polymer wrapping	Poly(styrene sulfonate), polyvinyl sulfate, polyvinyl alcohol, polyethylene glycol, and many others wrapping around SWNT driven by thermodynamic propensity to eliminate interface between hydrophobic surface of the SWNT and aqueous medium [76]
	Surfactants	Application of surfactants aids in dispersion but forms no permanent structures
Covalent methods	Side wall chemistry	Electrophilic nature of SWNT electronic structure
	End and defect chemistry	Most reactive region of SWNT (highest bond strain)
	Grafting to	Polymers are grown first and then grafted to the SWNT. Poly(styrene), PS, with nitride functionality was synthesized by atom transfer free radical polymerization (ATRP) followed by end-group transformation ($PS–N_3$) and then grafted onto the SWNT [77]
	Grafting from	Polymers are grown from immobilized groups on SWNTs. Immobilization of initiators onto tubes followed by surface-initiated polymerization (e.g., formation of polymer brushes) [77,78]. ATRP styrene with 2-bromopropionate immobilized on SWNT. Methly-2-bromopropionate was added as initiator. The PS was attached covalently to the SWNT
	Fluorination	Fluorination destroys electrical conductivity of nanotubes [79]. Oxidation first, then fluorination under F gas to produce routes to further chemical derivatization [79]
	Carboxylation	Addition of carboxylic groups + sonication shortens the length of the nanotubes and thereby enhances dispersion
	Electrochemical reduction	Reduction of aryldiazonium salts [80]

Noncovalent Modification. This form of chemical derivatization is relatively nonspecific and is based on intermolecular interactions that do not involve strong bonds. The electrical conductance of SWNTs, for example, relies strongly on its immediate environment and modification by noncovalent methods tend to preserve the basic structure although electronic properties (e.g., the bandgap) may be influenced [72].

R. E. Smalley et al. reported in 2001 about the derivatization of SWNTs by a process called *polymer wrapping* [76]. The process entailed association of the SWNT with water-soluble linear polymers such as polyvinyl pyrrolidone (PVP) and polystyrene sulfonate (PSS). The process was successful in disrupting the hydrophobicity of the nanotube as well as the intertube attraction to form bundles [76]. Unwrapping of the nanotubes occurred by changing the supporting solvent. Chemical modification by this process also allowed for analysis by solution-phase techniques such as chromatography and electrophoresis [76].

According to the authors, a thermodynamic driving force strives to eliminate the hydrophobic interface between the SWNT and the water solvent [76].

Although noncovalent modification technically does not affect the structure of nanotubes, it is capable of influencing the bandgap of the material [81]. Metallotetraphenyl porphyrins are electron donors; SWNTs are electrophiles. In one experiment, the wavelength and absorption intensity of the SWNT were red shifted and increased, respectively, relative to the undoped state [81]. Electron donation into the *p*-doped valence band increased the electronic density while simultaneously decreasing the transition energy of the unoccupied conduction band [81].

A. Noy et al. accomplished layer-by-layer electrostatic assembly of polyelectrolyte nanoshells on individual suspended carbon nanotubes [82]. The goal of the study was to develop a robust strategy for noncovalent modification. The purpose of the nanotube was to serve as a bridge and template. The first step involved exposing the CNTs to pyrene derivative (1,3,6,8-pyrenetetrasulfonic acid tetrasodium salt and 1-pyrenepropylamine hydrochloride or $PyrNH_3$) followed by layer-by-layer deposition of polyelectrolyte macro-ions. The polymers used in the study were poly(diallyldimethylammonium chloride (PDDA) and polystyrene sulfonate sodium salt (PSS). The entire process was based on stepwise self-assembly among the constituents. A generic version of carbon nanotube decorated with pyrene groups is shown in **Figure 9.22**.

Covalent Modification. Covalent modification strategies rely on the curvature of the tube and on the electronic properties [81]. Functionalization of SWNTs to

FIG. 9.22 *The planar pyrene molecule reacts with nanotube surfaces by π–π interactions—intermolecular interactions that are not as strong as covalent bonds. They serve as nonspecific anchoring constituents. Reactive groups attached to the pyrene enable further chemistry to take place.*

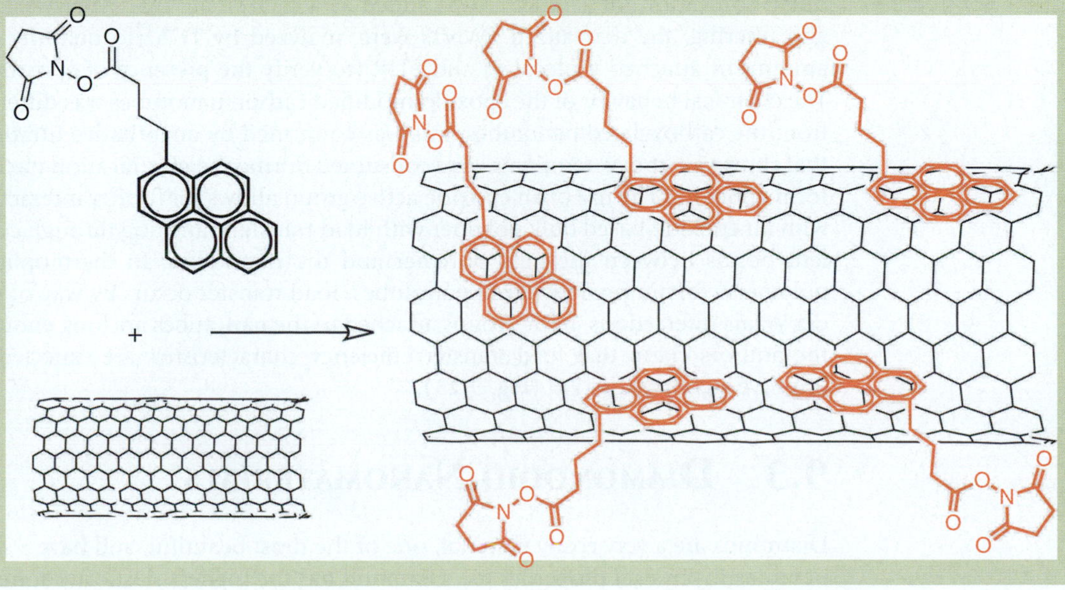

generate $-CO_2H$ decorations takes place in a rather severe chemical environment. SWNTs are first refluxed in 2–3 *M* nitric acid for 12–48 h at 115°C. Chemical modification can cease at this point if carboxylic acid groups are the desired decoration. The product is washed and dried. If further modification is required, carboxylated SWNTs suspended in dimethylformamide (DMF) are exposed to thionylchloride ($SOCl_2$). This process transforms the $-CO_2H$ group into a $-COCl$ chloric acid derivative. This transformation renders the SWNTs reactive to further modification by long-chain amines such as dodecylamine ($C_{12}H_{27}N$, a.k.a. *f12*) and octadecylamine ($C_{18}H_{39}N$, a.k.a. *f18*). A surfactant, NaDDBS (sodium dodecylbenzenesulfonic acid), was used to disperse the mixtures [83].

Evidence of carboxylation is provided by FTIR and Raman analysis. The presence of sp^3-hybridized carbon, a consequence of carboxylation, is indicated by the increase in the Raman disorder mode (D-band) at 1292 cm^{-1}. The G-mode, also called the tangential mode, is shifted to higher energy, an indication of the electron withdrawing affect of the carboxylate groups [83]. FTIR analysis shows the diagnostic carbonyl stretch located at 1727 cm^{-1}. Following treatment of the carboxylate moiety to form the chloric acid derivative, an amide linkage is formed between the *f12* or *f18* and the SWNT [83]. One means of measuring the degree of functionalization, measured as the percentage of carbon atoms that are actually derivatized, is by TGA. Alkyl chain mass losses account for up to 48% of the total mass of the derivatized SWNT system. This yields a grafting density of approximately four alkyl chains per nanometer of a (10,10) SWNT. Overall, chemical modification of the SWNT surface was less than 5% [83].

In another application, SWNTs were first carboxylated in a sulfuric-nitric acid (3:1%-vol) and then sonicated for 3 h at ambient temperature [74]. Following dilution with distilled water (1:5), the mixture was forced through a PTFE filter (10-μm pore size) and washed until no residual acid remained. The success of carboxylation was verified by FTIR [74]. Di-epoxide-terminated molecules were then attached to the CO_2H-modified SWNTs by adding EPON 828 and subsequent sonication for 1 h. KOH was added as a catalyst at 70°C. After washing and filtering, the derivatized SWNTs were analyzed by TGA (to quantify the amount of attached molecules) and FTIR (to verify the presence of epoxides). The chemical behavior of the epoxide-modified carbon nanotubes was different from the carboxylated nanotubes; this was confirmed by quantitative titrations that showed carboxyl terminals were consumed during the modification track to form epoxides. The use of an epoxide active group allows for further interaction with an epoxide-based bulk polymer with load transfer transiting through covalent bonds between the bulk polymer and the nanotubes. In thermoplastic polymers containing derivatized nanotubes, load transfer occurs by way of van der Waals interactions. If the groups attached to the nanotubes are long enough, the authors claim that load transfer efficiency characteristics are expected to improve significantly [74] (**Fig. 9.23**).

9.3 Diamondoid Nanomaterials

Diamonds are a very costly material, one of the most beautiful, and have a slew of fantastic physical properties [6]. Diamond has the highest elastic modulus of

FIG. 9.23 *Covalent surface modification involves making bonds between the carbon framework of the nanotube and the modifying moiety. A multiwalled carbon nanotube underwent carboxylation by application of a sulfuric-nitric acid (3:1) mixture while under sonication. Following rinsing and purification, the carboxylated MWNTs were reacted with di-epoxide-terminated molecules (Epon 828) in the presence of KOH catalyst. The terminal epoxide is available to undergo further chemical reactions. If the derivatized nanotubes are immersed into a bulk epoxy polymer matrix, strong links with the bulk polymeric material are formed. Such linking allows for efficient load transfer from the bulk polymer to the MWNTs. In this way, the MWNTs are able to enhance the mechanical properties of the polymer composite material.*

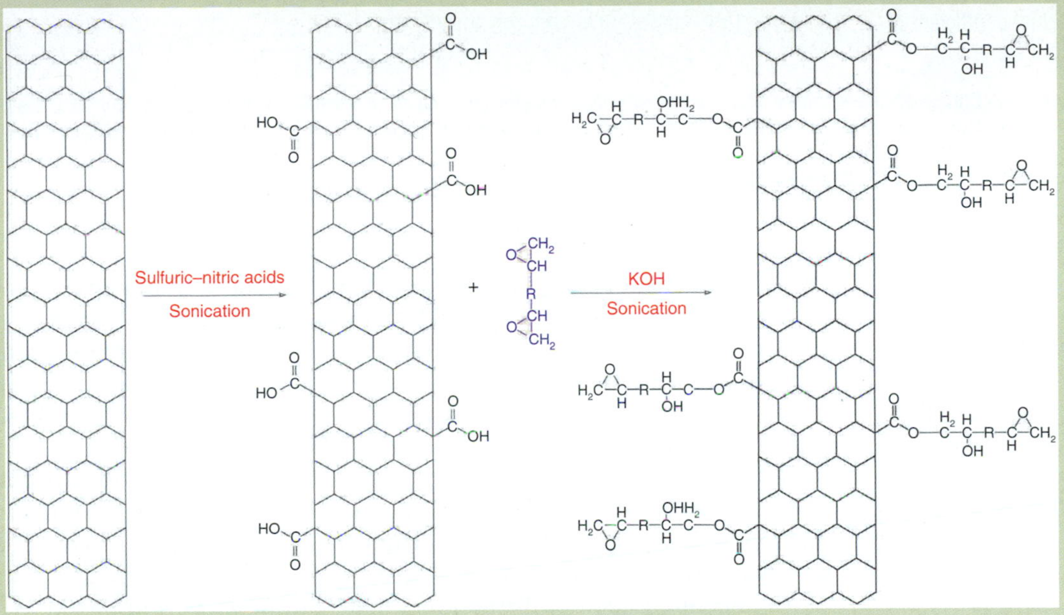

Source: Image redrawn with permission from Linda S. Schadler, Materials and Science Engineering, Rensselaer Polytechnic Institute, Troy, New York.

any material (1050 GPa), one of the highest electrical resistivities (~10^{16} $\Omega \cdot$ cm), high thermal conductivity (2000 $W \cdot m^{-1} \cdot K^{-1}$), and low thermal expansion (1.2 $\times 10^{-6}$ K^{-1}). Diamond is for the most part resistant to chemicals; transparent to visible, IR, and microwave radiation; and insulating (high band gap) [6]. The fabrication of diamonds has always been one of the holy grails of science. The feat was accomplished in 1954 by the General Electric Corporation; diamonds were formed under high-temperature and -pressure conditions (HPHT) [6]. These diamonds have numerous defects but are useful in industrial processes. Bulk diamonds are metastable forms of carbon. Diamond nanoparticles and thin films are also metastable, perhaps more so.

Diamondoids were first isolated in 1933 from Czechoslovakian crude oil and named adamantane (from the Latin *adamentum,* "the hardest metal," and the Greek *adamas,* "unconquerable"). Deposition of diamondoids is an undesirable occurrence during the oil refining process and subsequent transportation of

Table 9.5 *Comparison of Diamond, Diamondoid, and Amorphous Carbons*

Physical property	Diamond	CVD diamondoid	Amorphous carbon (diamond-like)
Crystal structure	Cubic, $a = 3.567$ Faceted crystals	Cubic, $a = 3.561$ Faceted crystals	Mixed sp^2 and sp^3 Smooth to rough Domains of microcrystalline diamonds
Hardness (H_v)	7000–10,000	3000–12,000	900–1200
Density ($g \cdot cm^{-3}$)	3.51	2.8–3.5	1.2–2.6
Electrical resistivity ($W \cdot cm$)	$\sim 10^{16}$	$\sim 10^{13}$	10^{6}–10^{14}
Thermal conductivity ($W \cdot m^{-1} \cdot K^{-1}$)	2000	1100 (2100)[a]	(1700)[a]
Thermal coefficient of expansion (K^{-1})	1.2×10^{-6}	1×10^{-6}–2.0×10^{-6} [86]	—

Sources: M. Ohring, *Materials science of thin films: Deposition and structure*, 2nd ed., Academic Press, San Diego, 2002; R. C. DeVries, *Annual Review of Materials Science*, 7, 161, 1987; H. C. Tsai and D. B. Bogy, *Journal of Vacuum Science and Technology*, A5, 3287, 1987.

[a] J. Norley, *Electronics Cooling*, 7, 50–51 (2001).

products like natural gas, gas, and petroleum products. Diamonds are made entirely of carbon that crystallizes in two forms: diamond and Lonsdaleite.

Diamondoids are materials that resemble diamond with regard to hardness, strength, stiffness, structure, and density. Diamondoids are covalently bonded sp^3 tetrahedral materials consisting of light elements with valence of three or greater. Diamond and sapphire are examples of diamondoid materials, but the definition is sometimes extended to sp^2-hybridized structures that form in planes like graphite. Therefore, fullerenes and nanotubes are sometimes referred to as diamondoid materials. We will not include them in the general category of diamondoid materials (**Table 9.5**).

Another application of diamondoid material thin films is to provide a surface with the capability of radiation resistance (called radiation hardness)—a required feature for the development of radiation detectors and linear colliders that undergo severe radiation environments.

9.3.1 Diamondoids

Adamantane is a diamond-like single cage structure with chemical formula [$C_{10}H_{16}$]. Two or more linked cage structures are called diamantanes, triamantanes, and, in general, polyamantanes—up to a dozen or so adamantane groups that include isotetraamantane (18 faces and four isomers) [$C_{22}H_{28}$], cyclohexamantane [$C_{26}H_{30}$], and superadamantane [$C_{35}H_{36}$] (**Fig. 9.24**). These structures are similar to diamond and have hydrogen terminal groups that stabilize the surface (just like diamond).

We are aware that nanomaterials have size-dependent properties. Ideally, we all want to study the quantum confinement properties of ideal materials. In reality, we usually deal with materials that are polydisperse or charged with nonuniform surface reconstruction or termination [8,87,88]. Diamondoid materials offer the opportunity to study electronic properties of the smallest possible single cage unit of the diamond lattice. Each dangling bond is terminated with hydrogen

FIG. 9.24 *Left: adamantane; middle: triamantane; and right: pentamantane that has been chemically modified with a thiol group—at the locus of bridging groups. Although these structures are considered to be diamondoids, they are really just simple hydrocarbons that are able to form into cages. The thiol group is able to bind the adamantane to a gold surface (or a gold nanoparticle) forming a diamond-like layer.*

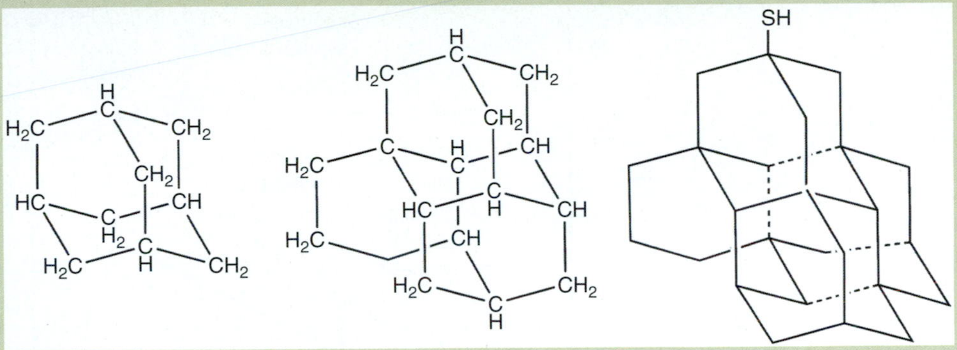

and each diamondoid is capable of adding another cage to form semiconductor clusters [8]. Bostedt et al. in 2004 monitored the size-dependent changes in electrical properties of individual diamond clusters [8]. Diamondoids (adamantane to hexamantane) were brought into the gas phase and analyzed by carbon K-edge x-ray adsorption spectroscopy. Data revealed that diamondoids absorb x-rays at lower energies and that the terminal hydrogens influence the density of states of the nanoparticles. Adamantane absorption resembles the x-ray profile of cyclohexane while hexamantane begins to show the characteristic profile of bulk diamond [8].

9.3.2 Thin Diamond Films (and Other Ultrahard Substances)

There are three hard substances known to science: diamond, boron–nitride, and carbon nitride. Interestingly, silicon carbide is not on this list. All three can be deposited by CVD. The deposition of perfect diamond lattices by CVD has been accomplished in the past couple of decades. In general, hydrocarbon sources such as CH_4 are activated over a solid substrate surface (720–1200°C) by application of a hot filament or electrical methods (DC arc discharge, RF, or microwave). In this way, true diamond lattice structures are formed—or diamond-like structures that are mostly amorphous with microcrystalline phase inclusions [86,89]. Room temperature thermal conductivity of such diamond-like carbon thin films is 1700 $W \cdot m^{-1} \cdot K^{-1}$. Values for CVD diamond as high as 2100 $W \cdot m^{-1} \cdot K^{-1}$ have been achieved [86,89].

CVD of metastable diamond occurs over a narrow range where there is a very small free energy difference between diamond and graphite: 2.1 $kJ \cdot mol^{-1}$ (**Fig. 9.25**) [6]. Kinetic factors can predominate in this small domain and the probability that both phases of carbon are produced is finite [6]. Kinetic control can be inserted to prevent the formation of graphite or its removal as it forms. This is accomplished by generating a superequilibrium (supersaturation) of atomic

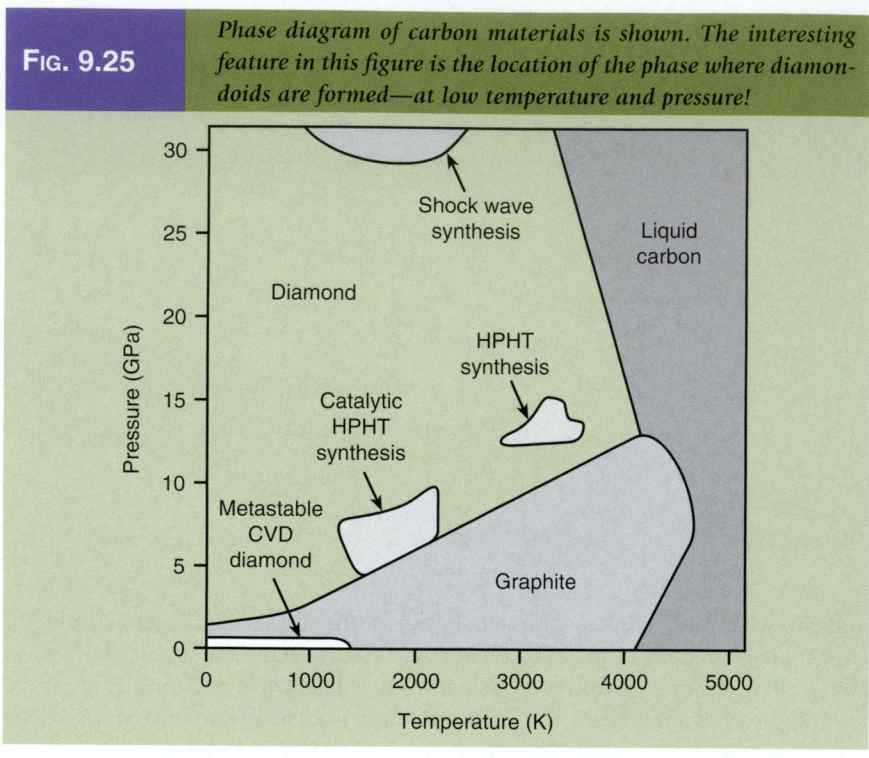

FIG. 9.25 *Phase diagram of carbon materials is shown. The interesting feature in this figure is the location of the phase where diamondoids are formed—at low temperature and pressure!*

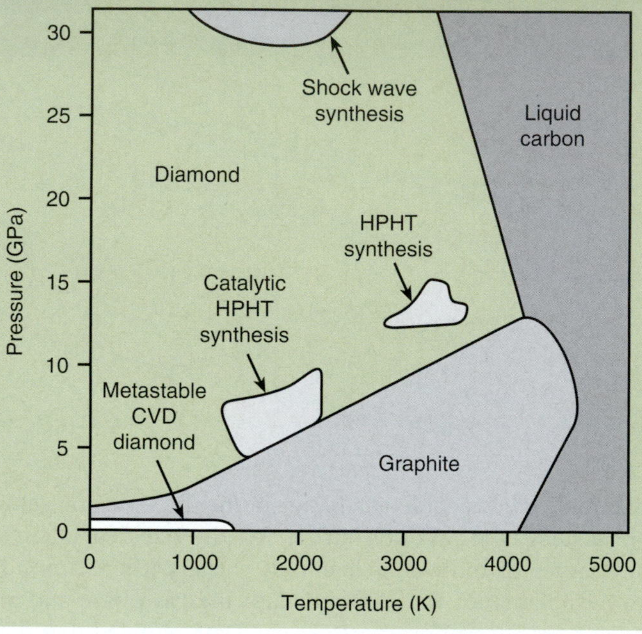

hydrogen radicals [H ·]. In this way, the atomic hydrogen radical prevents the formation of graphite by dissolving it during its nucleation [6]. Typical conditions to form microcrystalline diamond films are as follows: 0.5–2% CH_4 in H_2 at 50–100 torr; 2.45-GHz microwave plasma or hot filament; and a roughened silica substrate held between 800 and 900°C. Polycrystalline diamond was produced at a rate of 2 $\mu m \cdot h^{-1}$ with grain size equal to 1–10 μm. Nanocrystalline films are characterized by diamond grain size between 5 and 50 nm. CVD conditions are altered by significant dilution of reactants achieved by the presence of a 90% argon fraction.

9.3.3 Chemical Modification of CVD Diamond

The growth of covalently bonded nitrophenyl layers on atomically smooth boron-doped crystalline diamond surfaces was successfully accomplished by Uetsuka et al. in 2006 [90]. The purpose was to fabricate a biosensor that has long-term chemical stability and inherent biocompatibility. Chemical modification of diamond surfaces, because of its remarkable properties, is problematic, although reactions with atomic hydrogen, fluorine, and chlorine have been reported [90]. In 2000, Takahashi et al. demonstrated that covalent layers were possible on diamond surfaces by photochemical chlorination–amination–carboxylation of hydrogen terminated CVD diamond surfaces. Subsequent photochemical methods were able to modify CVD diamond with alkenes as well [90,91]. Attachment to ultrananocrystalline diamond (UNCD) has also been accomplished [90,92].

Boron-doped single-crystal diamond films (1 μm thickness) were formed homoepitaxially on diamond substrates (900°C) by microwave plasma-assisted (1200 W) CVD at 50 torr in 0.6% CH_4 in H_2 in the presence of B_2H_6 [90]. The boron concentration in carbon was 1.6×10^4 ppm. Hydrogen termination was achieved by application of pure hydrogen plasma for 5 min. One goal of the preparation was to form oxygen-terminated groups. This was accomplished by exposing the as-formed diamond film to oxygen plasma (13.56 MHz, 300 W) at 20 torr for 2.5 min. The degree of success was determined by wettability experiments; the contact angle θ approached 0° upon rendering of the surface to a hydrophilic state. Prior to this step, the contact angle was >94°, suggesting strong hydrophobicity of the H-terminated surface. The resulting surface had surface roughness < 1 Å. Nitrophenyl groups were attached by reduction of 4-nitrobenzene diazonium tetrafluoroborate. Attachment was initiated by cyclic voltammetry or constant potential attachment (**Fig. 9.26**).

In another study (R. J. Hamers et al.), researchers were able to attach DNA to CVD-formed diamond films [94,95]. DNA or other biological entities such as antibodies are chemically stable when they are bonded covalently to diamond surfaces. The use of electrical signals to direct chemical modification of diamond surfaces has become a very effective means of generating electrodes that have biosensing capability. The procedure to form the biological active electrodes is as follows: a Si wafer was coated with 300 nm of silicon nitride. Photolithographic masks were used to transfer a pattern of Ti–Mo derivatized contacts to the surface. Nanocrystalline diamond powder was introduced followed by lift-off of the photoresist. Nanocrystalline diamond was applied on Ti–Mo contacts in a 2.45-GHz microwave plasma reactor that contained methane, hydrogen, and B_2H_6. Grain size by this method was 20–100 nm. The film was continuous and pinhole free. The surface layer was H-terminated as discussed before. Forty millimolars of 4-nitrobenzene diazonium tetrafluoroborate in 1% sodium dodecyl sulfate (SDS) surfactant produced aryl-nitro groups attached to the surface. Reduction of the nitro groups to primary amines proceeded via cyclic voltammetry, followed by reaction with bifunctional linking groups that

Fig. 9.26 *Reduction of 4-nitrobenzene diazonium tetrafluoroborate to form a reactive radical and subsequent attachment to a CVD diamond film is depicted. The nitro-moiety allows for further attachment of functional groups such as DNA or the creation of additional layers.*

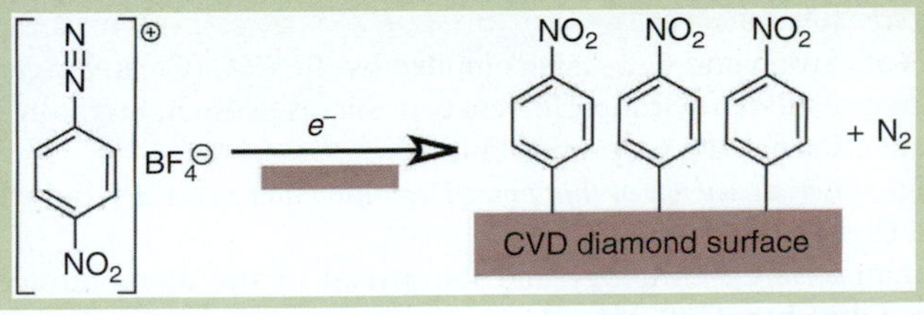

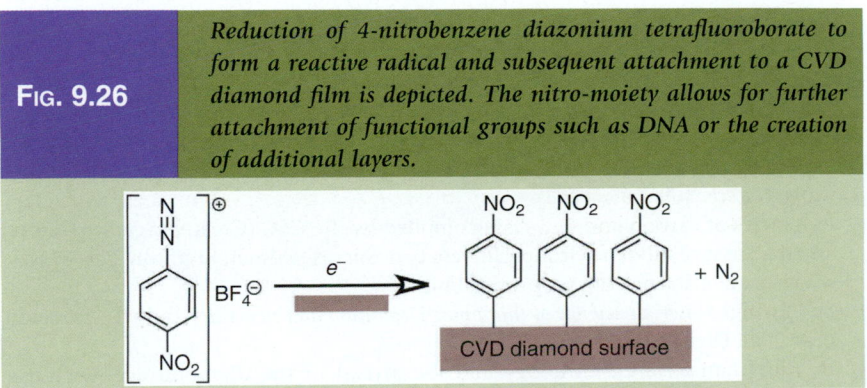

Source: J. F. Bahr et al., *Journal of the American Chemical Society*, 123, 6536–6542, 2001. With permission.

FIG. 9.27 *Attachment of DNA to a CVD diamond nanocrystalline thin film [94]. These DNA-modified diamond films show extraordinary stability under conditions that would denature natural DNA (elevated temperatures, repeated hybridization–denaturation cycles).*

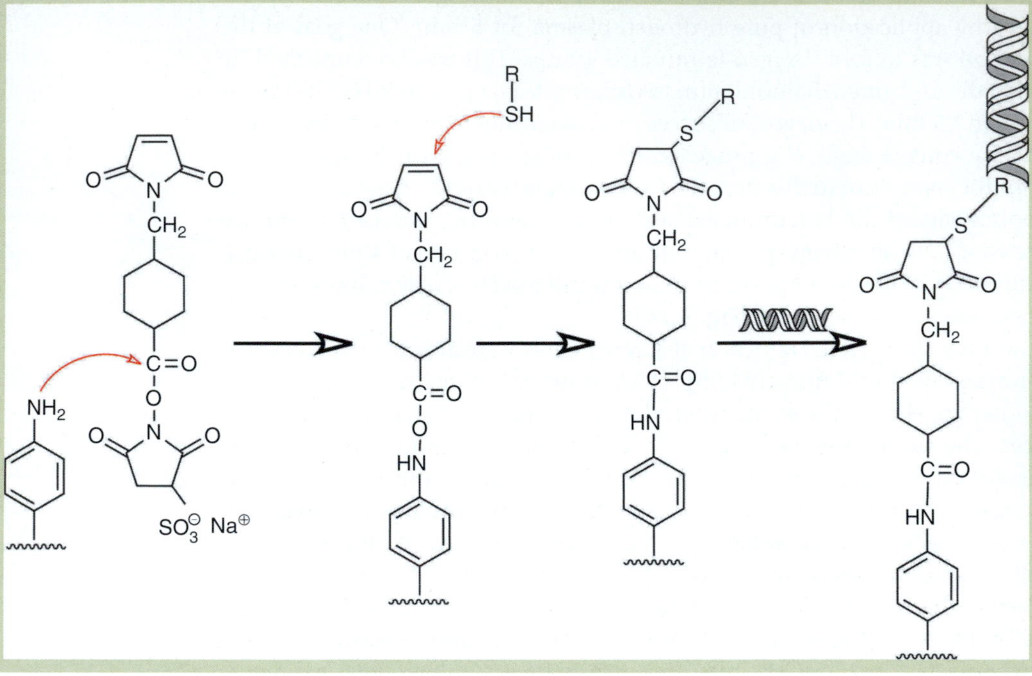

Source: Image redrawn with permission; R. J. Hamers, Department of Chemistry, University of Wisconsin.

are able to covalently attach to DNA—a heterobifunctional cross-linking molecule called sulfo-succinimidyl 4-(*N*-maleimidomethyl)cyclohexan-1-carboxylate (or SSMCC) [94] (**Fig. 9.27**).

References

1. H. W. Kroto, J. R. Heath, S. C. O'Brien, R. F. Curl, and R. E. Smalley, C60: buckminster fullerene, *Nature*, 318, 162 (1985).
2. S. Iijima, Helical microtubules of graphitic carbon, *Nature*, 354, 56 (1991).
3. S. Iijima and T. Ichihashi, Single-shell carbon nanotubes of 1-nm diameter, *Nature*, 363, 603 (1993).
4. D. S. Bethune, C. H. Kiang, M. S. De Vries, G. Gorman, R. Savoy, J. Vasquez, and R. Beyers, Cobalt-catalyzed growth of carbon nanotubes with single-atomic-layer walls, *Nature*, 363, 605 (1993).
5. Properties of carbon and C_{60}. Data compiled by the CSC (Creative Science Center) and the Sussex Fullerene Group, University of Sussex, Falmer, Brighton, East Sussex; www.creative-science.org.uk/propc60.html
6. M. Ohring, *Materials science of thin films: Deposition and structure*, 2nd ed., Academic Press, San Diego (2002).
7. H. Kluytmans, Nanotechnology and the arrival of the diamond age, www.dse.nl/~hkl/e_nano1.htm (1997).
8. C. Bostedt, Electronic properties of diamond clusters (diamondoids, nanodiamonds), www.physik.tu-berlin.de/cluster/diamondoids.html (2004).

9. L. Becker and T. E. Bunch, Fullerenes, fulleranes and PAHs in the allende meteorite, *Meteoritics*, 32, 479–487 (1994).
10. A. Hellemans, Labs hold the key to the 21-micrometer mystery, *Science*, 284, 1113 (1999).
11. F. Cataldo, Fullerane, the hydrogenated fullerene: Properties and astrochemical considerations, *Fullerenes, Nanotubes and Carbon Nanostructures*, 11, 295–316 (2003).
12. P. Mauron, *Growth mechanisms and structure of carbon nanotubes*, inaugural dissertation, Universität Freiburg, Department of Physics (2003).
13. E. A. Rohlfing, D. M. Cox, and A. Kaldor, Production and characterization of supersonic cluster beams, *Journal of Physical Chemistry*, 81, 3322–3330 (1984).
14. H. S. Carman and R. N. Compton, Electron attachment to c_n clusters (n#30), *Journal of Physical Chemistry*, 98, 2473–2476 (1993).
15. C. P. Poole and F. J. Owens, *Introduction to nanotechnology*, Wiley-Interscience, John Wiley & Sons, Inc., Hoboken, NJ(2003).
16. S. Sugano and H. Koizuni. *Microcluster physics*, Springer-Verlag, Heidelberg (1988).
17. G. Cao, *Nanostructures & nanomaterials: Synthesis, properties & applications*, Imperial College Press, London (2004).
18. P. Holister, C. Román Vas, and T. Harper, Fullerenes, Technology White Papers, NR.7, Científica, www.scientifica.com (2003).
19. H.-D. Beckhaus, C. Rüchardt, M. Kao, F. Diedrich, and C. S. Foote, The stability of buckminsterfullerene (C_{60}): Experimental determination of the heat of formation, *Angewandte Chemie International Edition in English*, 31, 63–64 (1992).
20. H.-D. Beckhaus, S. Verevkin, C. Rüchardt, F. Diedrich, C. Thilgen, H.-U. ter Meer, H. Mohin, and W. Müller, C_{70} is more stable than C_{60}: Experimental determination of the heat of formation of C_{70}, *Angewandte Chemie International Edition in English*, 33, 996–998 (1994).
21. R. J. Doome, A. Fonseca, H. Richter, J. B. Nagy, P. A. Thiry, and A. A. Lucas, Purification of C_{60} by fractional crystallization, *Journal of Physics and Chemistry of Solids*, 58, 1839–1843 (1997).
22. N. Sivaraman, R. Dhamodaran, I. Kalliappan, T. G. Srinivasan, P. R. Vasudeva Rao, and C. K. Matthews, Solubility of C_{60} in organic solvents, *Journal of Organic Chemistry*, 57, 6077–6079 (1992).
23. R. S. Ruoff, D. S. Tse, R. Malhotra, and D. Lorents, Solubility of fullerene (C_{60}) in a variety of solvents, *Journal of Physical Chemistry*, 97, 3379–3383 (1993).
24. S. Niyogi, M. A. Harmon, H. Hu, B. Zhao, P. Bhowmik, R. Sen, M. E. Itkis, and R. C. Haddon, Chemistry of single-walled carbon nanotubes, *Accounts of Chemical Research*, 35, 1105–1113 (2002).
25. M. Jazdzyk, W. Jia, and G. P. Miller, *Patterning surfaces with functionalized fullerenes*, Poster, Emerging Technologies and the Chemical Sciences, NERM 2006, American Chemical Society (2006).
26. Mixed interfullerene bonding motifs in C60-based polymers, European Synchrotron Radiation Facility, http://www.esrf.eu/UsersAndScience/Publications/Highlights/2005/ (2007).
27. S. Margadonna, D. Pontiroli, M. Belli, T. Shiroka, M. Riccò, and M. Brunelli, Li_4C_{60}: A polymeric fulleride with a two-dimensional architecture and mixed interfullerene bonding motifs, *Journal of the American Chemical Society*, 126, 15032–15033 (2004).
28. S. Watanabe, N. S. Ishioka, T. Sekine, A. Osa, M. Koizumi, H. Shimomura, and H. K. Yoshikawa Muramatsu, Production of endohedral fullerene—^{133}Xe by ion implantation, *Journal of Radioanalytical and Nuclear Chemistry*, 255, 495–498 (2003).
29. M. Linnolahti, A. J. Karttunen, and T. A. Pakkane, Remarkably stable fulleranes: $C_{80}H_{80}$ and $C_{180}H_{180}$. *ChemPhysChem*, 7, 1661–1663 (2006).
30. P. J. F. Harris, *Carbon nanotubes and related structures: New materials for the twenty-first century*, Cambridge University Press, Cambridge (2001).

31. M. S. Dresselhaus, G. Dresselhaus, and P. C. Eklund, *Science of fullerenes and carbon nanotubes*, Academic Press, San Diego (1996).
32. M. S. Dresselhaus, G. Dresselhaus, and R. Saito, Physics of carbon nanotubes, *Carbon*, 33, 883 (1995).
33. R. Saito, M. S. Dresselhaus, and G. Dresselhaus, *Physical properties of carbon nanotubes*, Imperial College Press, London (1998).
34. M.-F. Yu, Strength and breaking mechanism of multiwalled carbon nanotubes under tensile load, *Science*, 287, 637–640 (2000).
35. P. G. Collins, M. S. Arnold, and P. Avouris, Engineering carbon nanotubes and nanotube circuits using electrical breakdown, *Science*, 292, 706–709 (2001).
36. Basic properties of carbon nanotubes, Applied Nanotechnologies, Inc., applied-nanotech.com/cntproperties.htm#Mechanical%20Properties (2005).
37. A.N. Kolmogorov, V.H. Crespi, M.H. Schleier-Smith, J.C. Ellenbogen, Nanotube-substrate interactions: Distinguishing carbon nanotubes by the helical angle, *Physics Review Letters*, 92, 085503 (2004).
38. M. D. Taczak, Controlling the structure and properties of carbon nanotubes, MITRE Corporation, the MITRE Nanosystems Group, www.mitre.org/tech/nanotech (2007).
39. T. A. Adams and D. Tomànek. Physical properties of carbon nanotubes, The Nanotube Site, pa.msu.edu/cmp/csc/nanotube.html (visit 2007).
40. S. Frank, P. Poncharal, Z. L. Wang, and W. A. de Heer, Carbon nanotube quantum resistors, *Science*, 280, 1744–1746 (1998).
41. S. Sanvito, Y.-K. Kwon, D. Tománek, and C. J. Lambert, Fractional quantum conductance in carbon nanotubes, *Physics Review Letters*, 84, 1974 (2000).
42. P. G. Collins and P. Avouris, Nanotubes for electronics, *Scientific American*, December, 62–69 (2000).
43. A. Thess, R. Lee, P. Nikolaev, H. Dai, P. Petit, J. Robert, C. Xu, Y. H. Lee, S. G. Kim, A. G. Rinzler, D. T. Colbert, G. Scuseria, D. Tománek, J. E. Fischer, and R. E. Smalley, Crystalline ropes of metallic carbon nanotubes, *Science*, 273, 483 (1996).
44. G. Gao, T. Cagin, and W. A. Goddard, III, Energetics, structure, mechanical and vibrational properties of single walled carbon nanotubes, *Nanotechnology*, 9, 184–191 (1998).
45. G. L. Hornyak, L. Grigorian, A. C. Dillon, P. A. Parilla, and M. J. Heben, A temperature window for chemical vapor deposition growth of single walled carbon nanotubes, *Journal of Physical Chemistry B*, 106, 2821–2825 (2002).
46. S. Maruyama, Y. Murakami, Y. Miyauchi, and S. Chiashi, Generation of single-walled carbon nanotubes from alcohol and generation mechanism by molecular dynamic simulations, *Journal of Nanoscience and Nanotechnology*, 4, 360 (2004).
47. L. Ci, S. Xie, D. Tang, X. Yan, Y. Li, Z. Liu, X. Zhou, W. Zhou, and G. Wang, Controllable growth of single wall carbon nanotubes by pyrolyzing acetylene on the floating iron catalysts, *Chemical Physics Letters*, 349, 191 (2001).
48. H. Dai, A. G. Rinzler, P. Nikolaev, A. Thess, D. T. Colbert, and R. E. Smalley, Single-wall nanotubes produced by metal-catalyzed disproportionation of carbon monoxide, *Chemical Physics Letters*, 260, 471 (1996).
49. H. W. Zhu, C. L. Xu, D. H. Wu, B. Q. Wei, R. Vajtai, and P. M. Ajayan, Direct synthesis of long single-walled carbon nanotube strands, *Science*, 296, 884–886 (2002).
50. N. Hatta and K. Murata, Very long graphitic nano-tubules synthesized by plasma-decomposition of benzene, *Chemical Physics Letters*, 217, 398 (1994).
51. K. Hata, D.N. Futaba, K. Mizuno, T. Namai, M. Yumra, and S. Iijima, Water-assisted highly efficient synthesis of impurity-free single-walled carbon nanotubes, *Science*, 306, 1362 (2004).
52. A. Cao, X. Zhang, C. Xu, J. Liang, D. Whu, and B. Wei, Aligned carbon nanotube growth under oxidative ambient, *Journal of Materials Research*, 16, 3107 (2001).

53. S. C. Tsang, P. J. F. Harris, and M. L. H. Green, Thinning and opening of carbon nanotubes by oxidation using carbondioxide, *Nature*, 362, 520–522 (1993).
54. P. M. Ajayan, T. W. Ebbesen, T. Ichihasi, S. Iijima, K. Tanigaki, and H. Hiura, Opening carbon nanotubes with oxygen and implication for filling, *Nature*, 362, 522–523 (1993).
55. O. Smiljanic, T. Dellero, A. Serventi. G. Lebrun, B. L. Stansfield, J. P. Dodelet, M. Trudeau, and S. Desilets, Growth of carbon nanotubes on ohmically heated carbon paper, *Chemical Physics Letters*, 342, 503 (2001).
56. Y. Zhang, Electric field directed growth of single-walled carbon nanotubes, *Applied Physics Letters*, 79, 3155 (2001).
57. T. Ono, E. Oesterschulze, G. Georgiev, A. Georgiev, and R. Kassing, Field-assisted assembly and alignment of carbon nanofibres, *Nanotechnology*, 14, 37 (2003).
58. R. V. Seidel, *Carbon nanotube devices*, Ph.D. dissertation, Technischen Universität Dresden (2004).
59. R. Seidel, M. Liebau, G. S. Duesberg, F. Kreupl, E. Unger, A. P. Graham, W. Hoenlein, and W. Pompe, In-situ contacted single-walled carbon nanotubes and contact improvement by electroless deposition, *Nano Letters*, 7, 965–968 (2003).
60. G. S. Duesberg, A. P. Graham, M. Liebau, R. Seidel, E. Unger, F. Kreupl, and W. Hoenlein, Growth of isolated carbon nanotubes with lithographically defined diameter and location, *Nano Letters*, 3, 257–259 (2003).
61. V. Shanov and M. Schultz, The longest carbon nanotubes you've ever seen, *Nanotechnology Now*; www.nanotech-now.com/news.cgi?story_id=22540; Posted May 10 (2007). The breakthrough was presented in April 2007 at the *Single Wall Carbon Nanotube Nucleation and Growth Mechanisms* workshop organized by NASA and Rice University.
62. L. Ge, S. Sethi, L. Ci, P. M. Ajayan, and A. Dhinojwala, Carbon nanotube-based synthetic gecko tapes, *Proceedings of the National Academy of Science USA*, 104, 10792–10795 (2007).
63. T. W. Ebbesen, P. M. Ajayan, H. Hiura, and K. Tanigaki, Purification of carbon nanotubes, *Nature*, 367, 519 (1994).
64. N. Tagmatarchis, A. Zatton, P. Rescgiglian, and M. Prato, Separation and purification of functionalized water-soluble multi-walled carbon nanotubes by field-flow fractionation, *Carbon*, 43, 19984–19989 (2004).
65. P. G. Collins, M. S. Arnold, and P. Avouris, Engineering carbon nanotubes and nanotube circuits using electrical breakdown, *Science*, 292, 706–709 (2001).
66. G. Zhang, P. Qi, X. Wang, Y. Lu, X. Li, R. Tu, S. Bangsaruntip, D. Mann, L. Zhang, and H. Dai, Selective etching of metallic carbon nanotubes by gas-phase reaction, *Science*, 314, 974–977 (2006).
67. Rice Develops Method to Separate Nanotubes Based on Size, brightsurf.com/news/headlines/25097/Rice_develops_first_method_to_sort_nanotubes_by_size.html; based on information acquired from H. Schmidt, Carbon Nanotechnology Laboratory at Rice University (2007).
68. M. Meyyappan, Growth: CVD and PECVD, Chapter 4 in *Carbon nanotubes: Science and applications*, M. Meyyappan, ed., CRC Press, Boca Raton (2005).
69. S. Huang and J. Liu, Direct growth of single walled carbon nanotubes on flat substrates for nanoscale electronic applications, Chapter 4 in *Applied physics of carbon nanotubes: Fundamentals of theory, optics and transport devices*, S. V. Rotkin, S. Subramoney, eds., Springer-Verlag, New York (2005).
70. S.-J. Eum, H.-K. Kang, and C.-W. Yang, Electron microscopy investigation at the initial growth stage of carbon nanotubes, *Journal of the Korean Physical Society*, 42, S727–S731 (2003).
71. C. Klinke, J.-M. Bonard, and K. Kern, Thermodynamic calculations on the catalytic growth of multiwall carbon nanotubes, *Physics Review B*, 71, 035403:1–7 (2005).

72. R. J. Chen, S. Bangsaruntip, K. A. Drouvalakis, N. W. S. Kam, M. Shim, Y. Li, W. Kim, P. J. Utz, and H. Dai, Noncovalent functionalization of carbon nanotubes for highly specific biosensors, *Proceedings of the National Academy of Science*, 100, 4984–4989 (2003).
73. P. Liu, Modifications of carbon nanotubes with polymers, *European Polymer Journal*, 41, 2693 (2005).
74. A. Eitan, K. Jiang, D. Dukes, R. Andrews, and L. S. Schadler, Surface modification of multi-walled carbon nanotubes: Toward tailoring of the interface in polymer composites, *Chemistry of Materials*, 15, 3198–3201 (2003).
75. H. Palmoniemi, T. Aarital, T. Laiho, H. Liuke, J. Lukkari, and K. Haapakka, Noncovalent functionalization of single-wall carbon nanotubes to improve water-solubility, University of Turku, Department of Chemistry, Poster, Segovia, Abstract_Poster_PaloniemiHanna.pdf (2004).
76. M. J. O'Connell, P. Boul, L. M. Ericson, C. Huffman, Y. Wang, E. Haroz, C. Kuper, J. Tour, K. D. Ausman, and R. E. Smalley, Reversible water-solubilization of single-walled carbon nanotubes by polymer wrapping, *Chemical Physics Letters*, 342, 265–271 (2001).
77. S. Qin, W. T. Ford, D. E. Resasco, and J. E. Herrera, Functionalization of single-walled carbon nanotubes with polystyrene via grafting to and grafting from methods, *Macromolecules*, 37, 752–757 (2004).
78. X. Lou, C. Detrembleur, V. Sciannamea, C. Pagnoulle, and R. Jérôme, Grafting of alkoxamine end-capped (co)polymers onto multi-walled carbon nanotubes, *Polymer*, 45, 6097–6102 (2004).
79. H. Aziz, Routes to carbon nanotube solubilization and applications, Department of Chemistry, Duke University: www.lib.duke.edu/chem/chem110/papers/Hamza%20Aziz.thm (2003).
80. J. F. Bahr, J. Yang, D. V. Kosynkin, M. J. Bronikowski, R. E. Smalley, and J. M. Tour, Functionalization of carbon nanotubes by electrochemical reduction of aryl diazonium salts: A bucky paper electrode, *Journal of the American Chemical Society*, 123, 6536–6542 (2001).
81. D. R. Kauffman, O. Kuzmych, and A. Star, Modification of the semiconducting single walled carbon nanotube valence band through non-covalent attachment of tetraphenyl metalloporphyrins, Department of Chemistry, University of Pittsburgh, www.nanofab.ece.cmu.edu/avs/abstracts/DouglasKauffman.doc
82. A. B. Artyukhin, O. Bakajin, P. Stroeve, and A. Noy, Layer-by-layer electrostatic self-assembly of polyelectrolyte nanoshells on individual carbon nanotube templates, *Langmuir*, 20, 1442–1448 (2004).
83. R. Haggenmueller, F. Du, J. E. Fischer, and K. I. Winey, Interfacial in situ polymerization of single wall carbon nanotube/nylon 6,6 nanocomposites, *Polymer*, 47, 2381–2388 (2006).
84. R.C. DeVries, Synthesis of diamond under metastable conditions, *Annual Review of Materials Science*, 17, 161–187 (1987).
85. H. C. Tsai and D. B. Bogy, Characterization of diamond like carbon films and their application as overcoats on thin film media for magnetic recording, *Journal of Vacuum Science and Technology*, A5, 3287–3312 (1987).
86. J. Norley, The role of natural graphite in electronics cooling, *Electronics Cooling*, 7, 50–51 (2001).
87. T. M. Willey, C. Bostedt, T. van Buuren, J. E. Dahl, S. G. Liu, R. M. K. Carlson, L. J. Terminello, and T. Möller, Molecular limits to the quantum confinement model in diamond clusters, *Physics Review Letters*, 95, 113401 (2005).
88. J. E. Dahl, S. G. Liu, and R. M. K. Carlson, Isolation and structure of diamondoids, nanometer-sized diamond molecules, *Science*, 299, 96 (2003).
89. J. Rantala, Diamonds are a thermal designer's best friend, *Electronics Cooling*, electronics-cooling.com/articles/2002/2002_february_techdata.php (2002).

90. H. Uetsuka, D. Shin, N. Tokkuda, K. Saeki, and C. E. Nebel, Electrochemical grafting of boron-doped single-crystalline chemical vapor deposition diamond with nitrophenyl molecules, *Langmuir,* 23, 3466–3472 (2007).
91. W. Yang, S. E. Baker, J. E. Butler, C.-S. Lee, J. N. Russwell, L. Shang, B. Sun, and R. J. Hamers, Electrically addressable biomolecular functionalization of conductive nanocrystalline diamond films, *Chemistry of Materials,* 17, 938–940 (2005).
92. J. Wang, M. A. Firestone, O. Auciello, and J. A. Carlisle, Diamond films by electrochemical reduction of aryldiazonium salts, *Langmuir,* 20, 11450 (2004).
93. J. F. Bahr, J. Yang, D. V. Kosynkin, M. J. Bronikowski, R. E. Smalley, and J. M. Tour, Functionalization of carbon nanotubes by electrochemical reduction of aryl diazonium salts: A bucky paper electrode, *Journal of the American Chemical Society,* 123, 6536–6542 (2001).
94. W. Yang, O. Auciello, J. E. Butler, W. Cai, J. A. Carlisle, J. E. Gerbi, D. M. Gruen, T. Knickerbocker, T. L. Lasseter, J. N. Russell, L. M. Smith, and R. J. Hamers, DNA-modified nanocrystalline diamond thin-films as stable, biologically active substrates, *Nature Materials,* 1, 253–257 (2002).

Problems

9.1 Carbon is fundamental to life. Why? Why would silicon not be as good?

9.2 Calculate the diameter of the following SWNTs:

a. (6,6)
b. (10,10)
c. (12,12)
d. (12,0)
e. (9,3)
f. (10,2)
g. (11,7)

9.3 Which SWNTs are conducting? Which ones are semiconducting?

a. (4,3)
b. (4,4)
c. (11,5)
d. (5,1)
e. (9,0)
f. (9,9)
g. (12,4)

9.4 Calculate the tensile strength of a (12,12) carbon nanotube. How many are required to pick up the Empire State building? Use the dissociation energy of a 1.5 C–C bond (between energy between that of a C–C and a C=C bond). The Empire State building weighs 365,000 tons.

9.5 Explain the van Hove singularity in more detail.

9.6 Explain the differences between Raman and infrared techniques.

9.7 Name the fundamental Raman modes of SWNTs and describe briefly their energy and source in the nanotube.

9.8 Please research the following: The translation vector that describes SWNTs is given by

$$T = t_1 a_1 + t_2 a_2$$

where

$$t_1 = \frac{2m+n}{d_H} \quad \text{and} \quad t_2 = \frac{-(2n+m)}{d_H}$$

where d_H is the greatest common divisor.

a. Show that $T = \dfrac{\sqrt{3}C}{d_H}$ if $n - m \neq 3rd_H$.
b. Draw the unit cell for a (5,5) SWNT. Determine the area of a (5,5) nanotube unit cell (use: area in square nanometers $= |C \times T|$)
c. Draw an image of the graphene sheet representing this tube.

9.9 Using a transparency, construct a nanotube that is

a. (10,10)
b. (5,5)
c. (5,0)
d. (5.3)
e. (10,0)

Scotch tape the ends of the transparency together to hold the tube together.

9.10 Do some research and find an expression that reveals the number of carbon atoms per unit cell.

9.11 Determine the chiral angle in the following nanotubes:

a. (4,3)
b. (4,4)
c. (11,5)
d. (5,1)
e. (9,0)
f. (9,9)
g. (12,4)
h. (10,10)
i. (5,5)
j. (5,0)
k. (5.3)
l. (10,0)

9.12 Some carbon nanotubes are insulating. Why do you think this could be?

9.13 SWNTs are often found in bundles. Why do you think this occurs?

9.14 How many peaks do you expect in a ^{13}C NMR spectrum of C_{60}? Of C_{70}?

9.15 Explain briefly the mechanism of formation of SWNTs.

9.16 Why is *base-growth* mechanism favored over *tip-growth* from catalysts on alumina substrates?

9.17 Would it be possible to determine nanotube diameter from Raman spectra? Explain your answer.

9.18 Can you think of any other pure material (one element) that can exist as a conductor, a semiconductor, or an insulator based solely on its structure and size?

9.19 Boron–nitrogen materials also form solid sp^2 structures. How are they different from or similar to graphene? Boron (3 e^-) and nitrogen (5 e^-) are on the left and right side of carbon, respectively, in the periodic table.

9.20 Check Euler's rule of polyhedra, $f + v = e + 2$, for a tetrahedron (where f = faces, v = vertices, e = edges); for an octahedron; and for a cuboctahedron. In addition to metal clusters, the formula also works well for sp^2-bonded carbon materials. Does it apply to C_{60}? How about C_{70}?

9.21 Theorize as to why small carbon clusters (~20 atoms or less) prefer to exist in chains. (Consult M. Di Ventra et al., *Introduction to Nanoscale Science and Technology*, Kluwer Academic Press, the Netherlands, 2004.)

9.22 How are the energetics of fullerenes related to curvature of the nanostructure? Is there a relation to $1/R$ or $1/R^2$ (where R is the curvature)? (Consult M. Di Ventra et al., *Introduction to Nanoscale Science and Technology*, Kluwer Academic Press, the Netherlands, 2004.)

Chemical Interactions at the Nanoscale

Even the formal justification of the electron-pair bond in the simplest cases ... requires a formidable array of symbols and equations.

Linus Pauling, 1931

Chapter 10

Threads

The chemistry division consists of four chapters. We started with chapter 9, "Carbon-Based Nanomaterials." These materials form an extremely important class of materials and consequently the topic was awarded a special chapter. Materials based on carbon contribute to zero dimension (fullerenes), one dimension (carbon nanotubes), two dimensions (diamond thin films), and three dimensions (hierarchical structures). Carbon contributes to a wide class of applications including conductors, semiconductors, insulators, and structural components. The importance of nanocarbon materials in biology, biotechnology, and medical technology cannot be understated.

In this chapter, we present and discuss many kinds of intermolecular chemical interactions. Although the chapter is all about chemistry, the topic unavoidably includes a hefty dose of physics. Supramolecular chemistry is the focus of chapter 11. All of supramolecular chemistry takes place at the nanoscale. Chapter 12 delves into chemical modification of nanomaterials, template synthesis, and other nanorelevant topics. With chapter 12, we conclude the chemistry division of the text.

Since numerous subdisciplines of chemistry infringe upon the domains of physics at one end and biology at the other, we will try our best to stay within the boundaries of synthesis-oriented topics as they apply to nanomaterials. Chapters 13 and 14 are allocated to the biological nanoscience division; they focus on natural nanomaterials and biochemical nanoscience, respectively.

It is time to refresh our memories once again concerning chapter 2 on the societal implications of nano. As you proceed through this chapter and the rest of the text, always ask the question, "What are the consequences—good, bad, or indifferent—of these materials?"

10.0 Bonding Considerations at the Nanoscale

Chemical and physical interactions between atoms and molecules comprise two sides of the same coin. This apparent dichotomy is once again a product of convenience, generated to feed our instinct to catalog. Two atoms held together by attractive forces form a chemical bond that in turn yields a molecule (from the Latin *moles,* "small unit of mass" or "mass barrier" based on the Greek *molos,* "exertion"). Attractive forces between atoms form strong *intramolecular* bonds as a result of chemical reactions. We have briefly introduced types of *intramolecular* bonding such as the covalent bond, the ionic bond, and other types like the metallic bond in chapter 5. We now expand our discussion to the *intermolecular bond*—a type of interaction that exists between two or more molecules.

We define physical processes as those that involve no change in the chemical structure of molecules. We define chemical processes as those that do involve changes in the chemical structure of molecules. If we apply these definitions, in the strictest sense, to intermolecular bonds, they should be classified as a physical interaction because the structure (chemical nature) of the precursor molecules is not altered (significantly). However, any type of bonding, regardless of how strong or weak, causes perturbations to the integrated molecular orbital of any preexisting molecular system—perturbations that can be construed to be chemical in nature. Although no bonds are made or broken (except in the case of disulfide linkages), is the folding of a protein a chemical or a physical process? The result of the folding yields a molecule, albeit a quite large one, that is bestowed with altered chemical properties.

Interactions between atoms or between molecules, regardless of our need to classify them, belong to a continuum of bonding. It is just a matter of degree with regard to bond strength and the type of bond that is formed between specific chemical species that allows us to classify them as one type or another. All bonding is due to some kind of electrostatic attraction or repulsion or a combination of both and therefore belongs to the "electrostatic bonding continuum." Although bonding forces involved in forming intramolecular and intermolecular bonds have many different labels and are assigned numerous levels of strength, they all fall under the generalized umbrella of Coulombic (electrostatic) interactions. Electrons and their relationship with positively charged nuclei account for all types of bonding that occur between and among atoms and molecules. The distinction, therefore, between chemical (intramolecular) and physical (intermolecular) interactions is somewhat arbitrary but certainly one that helps us to derive a perspective and to seek an understanding of macromolecular systems. In an analogous manner, the distinction between chemisorption and physisorption is also worth a second look with respect to this context. In the former, a chemical bond is formed between an atom or molecule and a substrate; in the latter, a physical interaction is responsible for the attachment. Physisorption, therefore, can be considered to be one of the weakest types of chemical bonding.

Two or more molecules that are held together by attractive forces other than the previously mentioned "big three" also result in a generalized chemical bond. These bonds assume many forms and range in strength from almost nothing to approximately kT to several tens or more kilojoules per mole, but they are usually much weaker than those found in ionic, covalent, and metallic materials. The formation of covalent bonds, for example, requires energy between 80 and 400 times kT (200–1000 $kJ \cdot mol^{-1}$) [1]. These bonds are generally stable to fluctuations in their environment. Intermolecular bonding, as we shall soon discover, falls under another set of parameters and can be significantly influenced by the immediate environment.

Molecules held together by intermolecular forces are characterized by the following generalized parameters:

- Primary structure of molecules that are bound to each other by intermolecular bonds is not altered (e.g., the covalent backbone of each molecule remains unchanged although the molecule exists in a newly combined state).
- Strength of intermolecular bonds is relatively weak compared to the big three.
- Intermolecular bonds are generally formed under thermodynamically controlled conditions.
- Solvent is important during synthesis of macromolecules and their subsequent maintenance.
- Both enthalpic and entropic conditions are important to overall ΔG.
- Although individually relatively weak, the energy of intermolecular bonds summed over the whole molecule or system can be quite formidable.

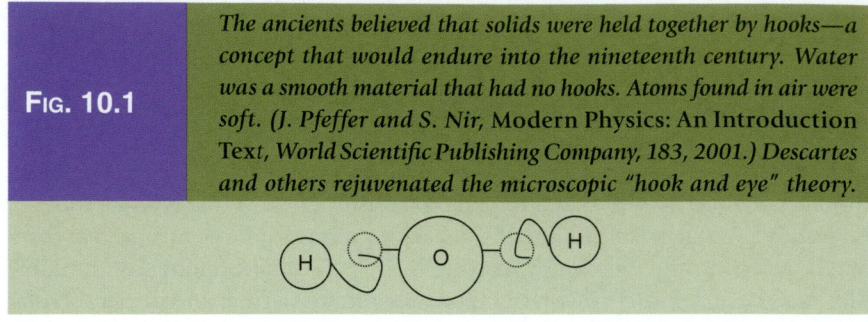

FIG. 10.1 *The ancients believed that solids were held together by hooks—a concept that would endure into the nineteenth century. Water was a smooth material that had no hooks. Atoms found in air were soft. (J. Pfeffer and S. Nir,* Modern Physics: An Introduction Text, *World Scientific Publishing Company, 183, 2001.) Descartes and others rejuvenated the microscopic "hook and eye" theory.*

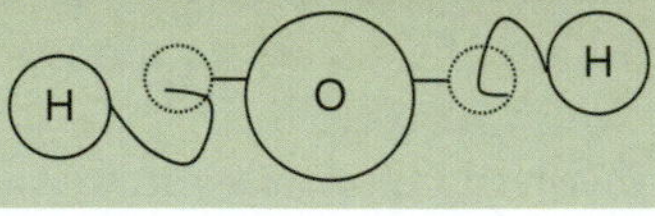

10.0.1 Background

The concept of chemical bonding dates back to ancient times when Asklepiades of Prusa (ca. 100 B.C.) theorized about clusters of atoms [2]. In ancient Rome, Lucretius speculated in his *De Rerum Natura* (*On the Nature of Things*) that atoms were tiny spheres held together by fishhooks and that atoms bonded when these hooks became entangled [2]. Robert Boyle in 1661 (*The Sceptical Chymist*) rejuvenated the idea of the cluster and that chemical reactions resulted in rearrangement of the clusters. His corpuscular theory of matter helped construct the platform of modern chemistry. Later, in the early seventeenth century, René Descartes proposed that "some atoms were furnished with hook-like projections, and others, with eye-like ones" (**Fig. 10.1**). He affirmed that two atoms combined when the "hook of one got caught in the eye of the other."

The idea of "bonded combinations of atoms" was broadened into the concept of chemical affinity in 1718 by French chemist E. F. Geoffroy, who developed one of the first periodic tables: the Affinity Table (**Fig. 10.2**).

William Higgins in 1789 proposed the concept of *valency* based on "ultimate particles" that had an associated strength of force between them. It was a few years later in 1803 that John Dalton proposed the law of simple proportions and established the basics for modernistic atomic theory (**Fig. 10.3**).

In 1811, a paper by Amadeo Avogadro titled "Essay on Determining the Relative Masses of the Elementary Molecules of Bodies" was published. In it Avogadro claimed that "atoms are united by attractions to form a single molecule." M.-A.-A. Guadin of France correctly proposed molecular formulae (e.g., H_2O) in 1833. Friedrich Kekule in 1857 claimed that carbon was tetrahedrally bonded to four other constituents. In 1861, Joseph Loschmidt in his book *Chemischen Studien I* introduced the double-bonded carbon structure. The first stick-and-ball model of molecules was fabricated by August W. von Hofmann in 1865. Alexander C. Brown expanded the concept of valency. Emil Fischer in 1894 introduced the idea of "lock-and-key," specifically within the context of enzyme action. In 1916, G. N. Lewis generated the first *Lewis structures* and the *octet rule* by using dots to represent shared electrons.

It was not until 1931, following the discovery of the electron and the development of quantum mechanics, that Linus Pauling and colleagues established the modern understanding of the chemical bond, reported in his landmark work titled *The Nature of the Chemical Bond* [3]. Hybridization theory and

FIG. 10.2 *E. F. Geoffroy of France in 1718 developed a theory about chemical affinity. He created a table that described "certain alchemical forces that draw components together." It never ceases to amaze how early scientists were able to speculate about the fundamental nature of matter without the ability to see atoms.*

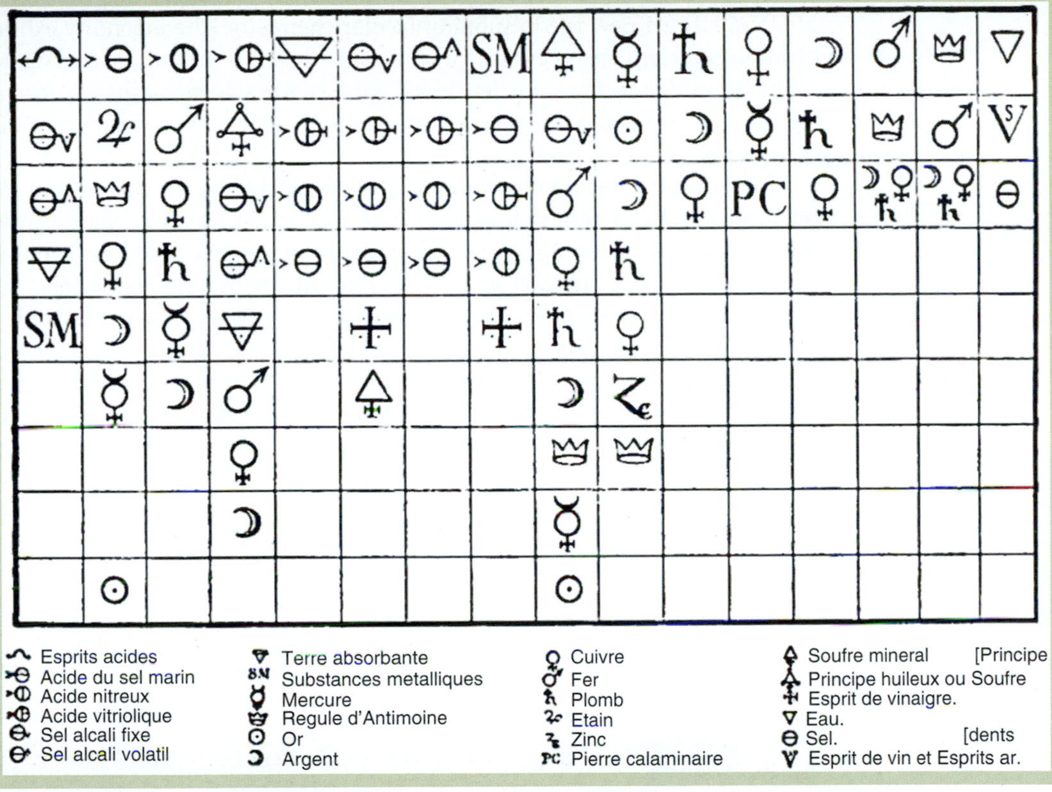

FIG. 10.3 *John Dalton also believed in the "hook theory" of atomic combination. Regardless of his understanding of the exact mechanism of bonding, hooks, or others, he was able to compile an impressive table that describes the various combinations of atoms to form common molecules.*

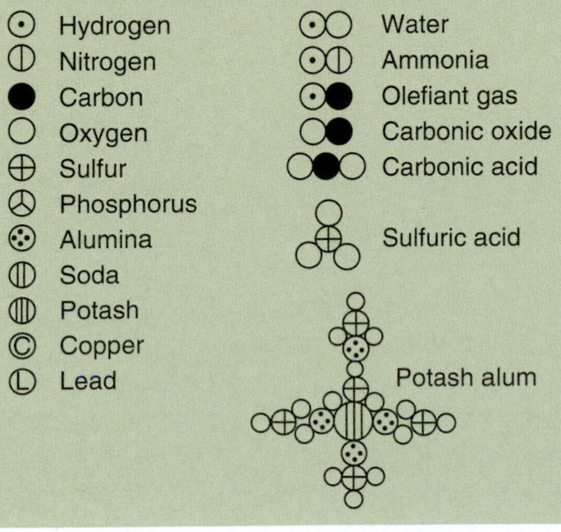

electronegativity were just a few of the phenomena described in this fundamental work. Details concerning the nature of the *intermolecular* bond also emerged about this time. K. L. Wolf in 1937 described the hydrogen bond, the first intermolecular bond to be designated as such, in *supermolecular* systems. In 1953, Watson and Crick's (and Franklin's) landmark work illuminating the structure of DNA led to a new age in supramolecular chemistry. The ingenuity exhibited by all these scientists is certainly inspirational if not outright astounding. Most of the bonding models were developed without knowledge of the electron and, for the first few thousand years, without the benefit of any clear proof of the existence of the atom itself.

One can just imagine the exhilaration experienced by the first person to see the atom—a distinction that goes to Professor Erwin Mueller (the inventor of the field-emission, atomic probe, and field-ion microscopes) of Penn State University in 1955. He used an electron microscope that he developed: "It was a sticky day in August 1955 that I became the first person to see an atom."

The first truly atomic-scale resolution image of an atom was achieved by University of Chicago graduate student Joe Wall in 1969 with an STEM (Discover Magazine.com/2007). Of course, as we know so well by now, the development of scanning tunneling microscopy (STM) allowed for manipulation of single atoms and a close-up look at crystal structure and molecular bonding.

10.0.2 Intramolecular versus Intermolecular Bonding

We proceed now to reinforce our previously developed dichotomy in bonding: the differences between the intramolecular bond and the intermolecular bond. The three major kinds of intramolecular bonds are summarized in **Table 10.1**.

TABLE 10.1 *The "Big Three" of Intramolecular Bonding*

Type of bond	Comments/examples	Bond strength ($kJ \cdot mol^{-1}$)
Ionic bonds	Nondirectional bonds between ions of opposite electronic charge Bond strength is proportional to the magnitude of the charges and inversely proportional to distance between them	Lattice energies (calculated and thermochemical)
	NaF	910–923
	NaCl	769–786
	KF	808–821
	KCl	701–715
	CsF	744–740
	CsCl	657–659
	Al_2O_3	15,916
	Ga_2O_3	15,590–15,220
	Ti_2O_3	14,702
	Na_2O	2,481
	MgO	3,795–3,791
	CaO	3,414–3,401
	$NaNO_3$	755–756

TABLE 10.1 (CONTD.) *The "Big Three" of Intramolecular Bonding*

Type of bond	Comments/examples	Bond strength ($kJ \cdot mol^{-1}$)
Covalent bonds	Highly directional (with well-defined angles), hybrid bonds and sharing of outer-orbital electrons	Strong bonds ranging from 100 to 1000 $kJ \cdot mol^{-1}$
	H–H	436
	O–H	463
	C–H	412
	C–C	348
	C=C	690
	C≡C	839
	C–O	360
	C=O	690
	C≡O	1072
	C–Cl	338
	C–Br	276
	C–I	238
	C–N	305
	N≡N	942
Metallic bonds	Nondirectional delocalized electrons exit as "electron gas" based on free electrons of metal. Transition metals are the most common metals	Energy of cohesion, ΔH_{sub}
	Au (*fcc*)	368
	Ag (*fcc*)	284
	Cu (*fcc*)	336
	Al (*bcc*)	327
	Hg (*rhomb*)	60.3
	Os (*hexagonal*)	788
	W (*bcc*)	848
	Mo (*bcc*)	718
	Fe (*tetragonal*)	413
	Pb (*fcc*)	196
	Mn (*hexagonal*)	282
	Po (*monoclinic*)	145
	Pt (*fcc*)	564

In order for atoms to stay in close proximity, the strength of the bond that holds them together must be able to overcome the thermal kinetic energy of an average molecule

$$KE_{\text{Molecule}} > \frac{3}{2} k_B \text{T} \qquad (10.1)$$

where k_B is the Bolztmann constant (1.3806×10^{-23} $J \cdot K^{-1}$) and T is the temperature in Kelvin. At room temperature, the kinetic energy of an average molecule is ca. 3.72 kJ · mol^{-1}. If bond strength is below this value, atoms run the risk of vibrating off the bond and going on separate trajectories. It is clear from **Table 10.1** that intramolecular bonds are quiet strong and run no risk of dissociation at k_BT. However, many ionic compounds, however, seemingly dissolve in water quite easily. What is the basis of this apparent paradox? The heat of solvation, whether exo- or endothermic, is the difference between the lattice enthalpy and the enthalpy of solvation. Obviously, other factors besides thermal energy are at work. For example, enough free energy is gained by the interaction of the ions in a crystal with the molecules of the solvent to allow for dissolution of the solid.

Although covalent interactions are always part of the bonding scenario, they take a backstage role for now as we direct our attention to intermolecular attractions that combine to form the primary actors (attractors) of this chapter and the next: supramolecular chemistry. Chemistry of the nanoscale is very much about exploitation of weak interactions: some types that exist below the kinetic energy of the average molecule. So long as temperature, *p*H, and pressure remain relatively constant (made possible by confined or insulated systems and proper choice of environment), supramolecular structures, including the major subset of biomacromolecules, remain in stable form. Intra- and intermolecular bonding are compared side by side in **Table 10.2**.

10.0.3 Types of Intermolecular Bonding

In this section we set the table of intermolecular bonding by providing a brief overview. All intermolecular interactions arise from the relationships between

Table 10.2 *Intermolecular—Intramolecular Bond Comparison*

Parameter	Intermolecular	Intramolecular
Molecular structure of constituents	Molecular structure remains relatively intact after intermolecular bonding episode Example: amino acids linked by hydrogen bonds that are involved in the 2° (folding) structure of a protein. The 1° structure of the protein is not altered	New molecules formed with chemical properties that may be radically different from precursors Amino acids linked by condensation reactions form the 1° structure of a protein
Bond strength	Bond strength in general ranges from nearly nothing to less than 100 kJ · mol^{-1}	The bond strength of the CO triple bond is 1072 kJ · mol^{-1}
Reaction type	Thermodynamic control	Kinetic control
Importance of solvent and immediate environment	Extremely important during formation and maintenance (stability) of macromolecular systems (e.g., biological systems)	Solvent important during synthesis to stabilize intermediates and affect solubility
ΔG	Both ΔH and $T\Delta S$ are important	ΔH is important
Overall bond strength	Even weak bonds between molecules summed over all molecules can influence physical properties (e.g., the hydrogen bond in water) The gecko's ability to cling to a ceiling is another example of overall bond strength	Overall enthalpy is a straightforward extensive property of the material
Self-repair	Yes	No
Degrees of freedom	Greater flexibility	Limited

and among the electrons, the nuclei, and the solvent if in liquid phase. The basic types of attractive intermolecular interactions are listed and discussed below.

Ion–Ion and Ion–Dipole Interactions. Ion–ion interactions are purely electrostatic in nature. Apart from those found in ionic crystals, ion–ion bonds are technically characterized as individually weak. For example, the interaction between a charged species (e.g., an anionic macromolecule) and a cation (either in atomic or molecular form) falls into this category. Obviously, ion–ion reactions that are relevant to nanoscience take place in solvents, although gas phase ion–ion reactions also exist. Electron transfer from one atom to another is typical in such reactions, although subtle exchanges between (+) and (−) charges are also observed. To be purely intermolecular, no electron transfer should occur (e.g., the discrete character of the molecule is retained during the interaction). These kinds of interactions are generally classified as nondirectional (delocalized), nonspecific, and lacking unique stoichiometry. Although many ion–ion configurations incorporate other selectivity criteria (e.g., size exclusion), in general, ion–ion interaction reactions are difficult to assimilate into a molecular design process. Ion–dipole interactions are also considered to be purely electrostatic (Coulombic); the difference arises only in the strength of the charge on the dipole. The strength of ion–dipole interactions is reduced compared to ion–ion interactions. The strongest interaction occurs when the proper alignment of ion to dipole occurs.

Van der Waals Interactions. Van der Waals interactions include a wide array of electrostatic interactions based on polarity and electric charge. The *dipole–dipole* electrostatic interaction (the Keesom van der Waals interaction) is based on Coulombic attraction between two permanent dipoles, each characterized by a unique dipole moment μ. The angle of approach is a critical factor that influences the strength of the intermolecular bond. Rotations around bonds influence the strength of the dipole interaction and can range from attractive to repulsive. A unique subset of dipole–dipole interactions is the hydrogen bond—a category that deserves its own section in this text.

Polarization interactions contribute two major subsets of the van der Waals interaction category: *ion-induced dipole, dipole-induced dipole,* and *induced dipole–induced dipole* interactions. Ion-induced dipole interactions are technically not considered to be a member of the general class of van der Waals forces. Dipole–dipole interactions occur in two ways: head on head, similar in form to the hydrogen bond, and in the case where the polar ends of molecules overlap simultaneously. Dipole-induced dipole interactions (the Debye form of the van der Waals interaction caused by the formation of instantaneous dipoles) are caused by polarization of a previously nonpolar molecule by a nearby dipolar molecule. The last category, induced dipole–induced dipole (London dispersion forces), is the most common form of van der Waals interactions. London dispersion forces are characterized by charge fluctuations (e.g., dynamic electron cloud fluctuations of participating molecules). All electron clouds of all elements and molecules are subject to polarization from outside influences. Once polarized, molecules are able to remain in combined form due to these weak electrostatic interactions, depending on the strength of perturbations arising from the local environment.

Hydrogen Bonds and Allies. Hydrogen bonds are dipole–dipole interactions in which a relatively strong bond is formed between substituents. It occurs between two molecules that contain highly electronegative (and small) atoms like O and N and a shared hydrogen. H-bonds are directional and can therefore be quite specific (e.g., the interaction between nucleotide base pairs to form DNA). Their utility in supramolecular design, unlike ion–ion, ion–dipole, or other forms of dipole–dipole interactions, is highly valued. Cα–H-bonds and halogen bonds, are similar in character to hydrogen bonds and will be discussed later in the text.

Hydrophobic Interactions. Hydrophobic interactions occur between nonpolar molecules in aqueous solvents. Nonpolar molecules do not form hydrogen bonds with water but are nonetheless "organized" by water molecules (hydrated). Additional hydrophobic constituents in the solution cause release of organized water (entropy driven) and eventual coalescence of nonpolar constituents. Hydrophobic interactions are not classified as true bonding. They are nondirectional and nonspecific. Although extremely important in biology, hydrophobic interactions are difficult to account for in the molecular design process. *Attractive entropic elasticity* is a characteristic of macromolecular systems that accounts for an attractive–recoiling force produced by deformation due to extension of the molecules in a polymer.

Repulsive Forces. *Steric repulsion* occurs from overlap of electron clouds and is based on the Pauli exclusion principle (no two electrons can have the same quantum number). Repulsion forces arising from the (+) charge of nuclei also contribute to electrostatic repulsion between two atoms or molecules. Repulsive interactions are described by Lennard–Jones potential functions and quantum mechanics. *Born repulsion* is the electrostatic repulsion between two entities with the same electrical charge. Subtle forms of repulsion also occur between two hydrophilic molecules. In particular, *hydrophilic repulsion* occurs in aqueous solutions between polar molecules to maximize the formation of hydrogen bonds with water. *Repulsive entropic elasticity* is the opposite of attractive entropic elasticity. In this case, repulsive forces arise due to compression of polymeric or other macromolecular systems. *Repulsive osmotic forces* arise from diffuse counter-ion electrical double layers between two charged surfaces and from steric repulsion in polymers originating from solvent translational entropy and interchain excluded volume.

10.0.4 The Nano Perspective

All types of bonding are important in nanomaterials. In chapter 6 we probed the various methods by which nanomaterials seek stabilization. Distortion of the lattice structure of metals held together by metallic bonds, for example, is an effective means of providing relief for the high energy and strain of the surface. Agglomeration, of course, is a straightforward means of alleviating instability. The adsorption (chemisorption or physisorption) of chemical moieties is fundamentally one of the most effective means of stabilizing nanomaterials. Chemisorbed ligands formed by dative bonds to the surfaces of nanometals serve to stabilize clusters. All systems seek the condition of minimal energy. Intramolecular metallic bonds hold together the Au-55 cluster, but intermolecular bonds stabilize the cluster through the action of ligands. Ionic bonds hold together TiO_2 nanoparticles, but dye molecules attached to the surface via intermolecular bonds serve to stabilize and functionalize. Carbon nanotubes, although kinetically stable, are held together by strong covalent bonds but can be chemically modified by covalent, ionic, and intermolecular reactions such as π-interactions.

Intermolecular bonds are well represented by nanoscale materials that we shall now designate as *soft matter* [4]. The term soft matter applies to materials like polymers, colloids, amphiphiles, and liquid crystals and, of course, almost all the biological materials. Colloid chemistry is an often neglected form of soft chemistry [4]. Well, not anymore! The soft matter of nature exhibits a level of bonding expertise that is by far the most impressive. The functionality of nanoscale biological systems relies on the timely manipulation of numerous forms of chemical bonding—often simultaneously. Supramolecular chemistry, our chemistry of the nanoscale, is discussed in chapter 11. Supramolecular chemistry is chemistry beyond the single molecule and is the chemistry of the intermolecular bond [5–7].

Several tables listing the types of intermolecular interactions follow. Please view these lists somewhat open mindedly and consider that the distinctions between and among all types of bonding are fuzzy at best. Numerous opportunities will become available throughout the text to sort out and evaluate unique bonding character. We certainly understand that all intermolecular bonds are derived from generalized Coulombic interactions and that there is significant intermingling of one type of bonding category into another type—especially in cooperative interactions. It is all a matter of degree, albeit one that forms yet another continuum comprising bonding in matter.

We also ask that the relationship of bonding to the perspective of nanomaterials be placed at the back of your mind as well as within the frontal lobes during all the forthcoming discussion. Bonding between and among entities that are larger than atoms or molecules, such as between materials with nanoscale proportions, is the sum total of the individual "bonding elements" integrated across the entire volume and surface of the constituents. The sum total energy in these cases may be quite large.

10.1 Electrostatic Interactions

Although all bonding is electrostatic, we especially direct our attention to interactions in which there is a clear case for physical charge separation between entities that have an extra electron or two or three or are missing electrons. Permanent dipoles are not charged but exhibit charge separation within their structure. The permanent dipole is due to the difference in electronegativity of its atomic constituents. All chemical bonding is due to interactions between electronic charges that arise from manifestations of Coulomb's law, given below in its force and energy forms:

Fig. 10.4 *A commemorative stamp issued by the French postal service to honor the work of Charles-Augustin Coulomb is depicted. He is best known for the formulation of Coulomb's law—that "the force between two electrical charges is proportional to the product of the charges and inversely proportional to the square of the distance between them" (Geocities.com).*

$$F = \frac{1}{4\pi\varepsilon_o\varepsilon_r}\left(\frac{q_1q_2}{r^2}\right) \tag{10.2}$$

$$E = \frac{1}{4\pi\varepsilon_o\varepsilon_r}\left(\frac{q_1q_2}{r}\right) \tag{10.3}$$

where F is the force experienced between two elementary charges q, ε_o and ε_r are the dielectric permittivity of vacuum and medium, respectively, and r is the distance between the two charges. E is the potential energy between two point charges and is inversely proportional to the distance between them. The value of the vacuum permittivity constant ε_o is equal to 8.854×10^{-12} $C^2 \cdot J^{-1} \cdot m^{-1}$ and that of the elementary charge is 1.6022×10^{-19} C. Ion–ion, ion–dipole, and dipole–dipole interactions are electrostatic interactions between atoms and/or molecules that possess relatively permanent electrostatic charges. Charles-Augustin Coulomb is pictured in **Figure 10.4.**

Repulsive Interactions. Lest we place too little value on repulsive interactions, we shall provide a cursory overview of a few kinds of electrostatic repulsive interactions that exist in materials. Without repulsive interactions, all matter would collapse and there would be no investigators and nothing to investigate. Pauli exclusion is responsible for repulsive forces. Repulsive interactions provide the balance to attractive interactions at the distance of equilibrium position. Interaction potentials function based on distance between two charged species and like the *Lennard–Jones 6–12* potential, mentioned in chapter 6, incorporate both repulsive and attractive forces that additively combine to generate the curve shown in **Table 10.3.** In 1903, Gustav Mie developed an expression for a generalized interaction pair potential (before definitive evidence for atoms and molecules existed). The overall energy potential V_{total} (or U_{total}) is

TABLE 10.3 *Repulsive Interactions*

Description	Example	Repulsion energy ($kJ \cdot mol^{-1}$)
Steric repulsion	Steric repulsion results from filled orbitals that cannot be involved in bonding; such excluded volume contributes to a kinetic effect Reduces strength of interaction important in *lock and key* scenarios [8] Molecular crowding in cells Equilibrium constants inside small cavities or cells are different than of standard equilibrium constants determined in a beaker	Variable
Steric hindrance (steric resistance) Steric protection Steric shielding (of charged groups)	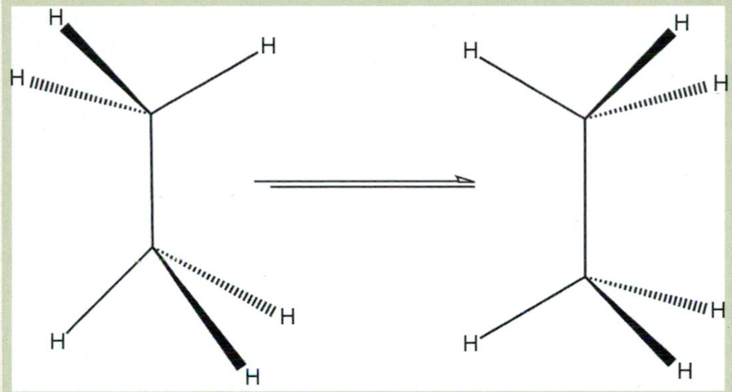 *The staggered version of ethane (left) is more stable than the eclipsed version (right). Unwanted reactions can be prevented by steric protection (t-butyl groups are often used).*	12.5 $kJ \cdot mol^{-1}$
Lennard–Jones 6–12 potential	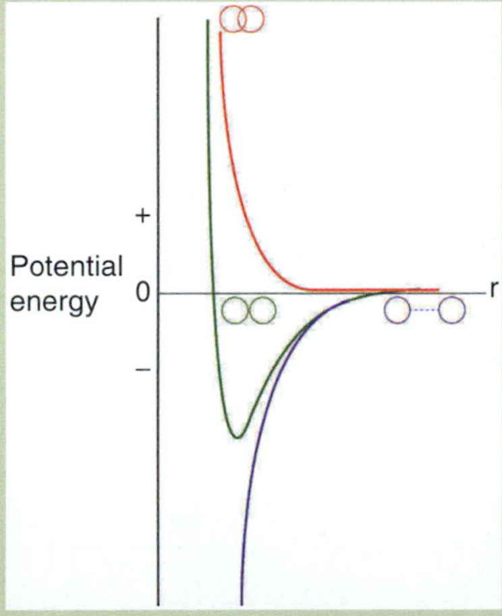 *The Lennard–Jones potential. The green line represents the average of the attraction (red) and repulsion (blue) curves. Circles represent atomic pairs. Repulsion due to Pauli exclusion and repulsive forces, therefore, is close range. Potential energy is inversely proportional to r^{12}. Repulsion potentials are positive and attraction potentials are negative.*	Interaction energy is a function of distance

$$V_{\text{total}} = -\frac{C_1}{r^n} + \frac{C_2}{r^m} \tag{10.4}$$

where C_1 and C_2 are constants and r is the interatomic or intermolecular distance. You should be able to recognize the form of equation (10.4) and that exponents n and m are similar in form to the Lennard–Jones 6–12 exponents that we encountered in chapter 6. The Lennard–Jones potential is shown in **Table 10.3.**

Other electrostatic repulsive species exist that are equally as important, especially with regard to macromolecules and biological molecules. The steric (repulsive) character of rotamers such as ethane requires about 12 $kJ \cdot mol^{-1}$ to overcome the barrier to rotation. Steric hindrance in the form of *t*-butyl groups is able to protect chemically active species (**Table 10.3**).

10.1.1 Ion Pair Interactions

Ion–ion interactions are nondirectional. The strength of the interaction, the bond energy, is directly proportional to the magnitude of the charges involved and indirectly proportional to the distance between them. Expressions that describe the ion-attractive type interaction are extensions of Coulomb's law. Ionic solids are characterized by their lattice energy. The lattice energy is the change in enthalpy between an ionic solid and its ionic forms in the gas phase. The lattice energy is a good approximation of the strength of the ionic bond in the solid. An estimate of lattice energy for ionic solids is given by

$$E_a = \frac{\mathcal{M}\, z_1 z_2 e^2}{4\pi\varepsilon_o d_o} \tag{10.5}$$

where $\mathcal{M}$ is the Madelung constant for the crystal, z_1 and z_2 are the respective integral charges on the ions, d_o is the equilibrium interatomic spacing, and e and ε_o assume their usual values. The Madelung constant $\mathcal{M}$ is a scaling constant that factors in the geometric arrangement of ions in a crystal to the *n*th level of coordination—nearest neighbors and beyond. Sample Madelung constants for a few crystals are NaCl (1.748), CsCl (1.763), CaF_2 (2.519), TiO_2 (2.408), and Al_2O_3 (4.172). A better approximation of E_a is obtained if a repulsive expression (r) is added to the lattice energy approximation. The total energy E_L of the lattice then is given by

$$E_L = E_a + E_r \tag{10.6}$$

$$E_a = \frac{\mathcal{M} z_1 z_2 e^2}{4\pi\varepsilon_o d_o}\left(1 - \frac{d^*}{d_o}\right) \tag{10.7}$$

where d_o is the equilibrium spacing as before and d^* is typically equal to 0.345 Å. The ion–ion energy depicted in this form of the equation yields energy in terms of electron volts (of the single ion pair interaction). It is difficult to determine lattice energy solely by experimental means; however, these calculations yield fairly good approximations of bond strength in ionic compounds.

The sodium chloride crystal is the universal example of a material based on ion–ion interactions. Each sodium cation is surrounded by six adjacent chloride anions in a face-centered cubic close-packed configuration. Each chloride, in a

FIG. 10.5 *The dissolution of NaCl is shown. Dissolution is driven by the ability of water to stabilize sodium cations and chloride anions. Hydration spheres with specific structure are able to solvate the ions.*

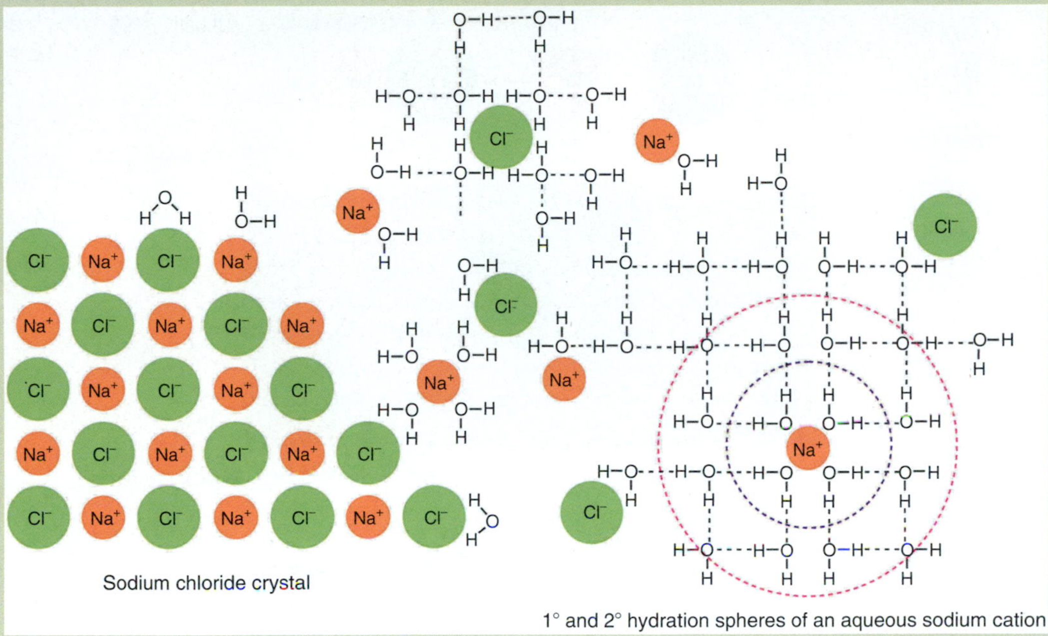

complementary fashion, is coordinated by six sodium cations. The dissolution of a NaCl crystal is shown in **Figure 10.5**. Sodium chloride, although possessing relatively high lattice energy, will dissolve readily in water.

NaCl is technically not a supramolecular compound because it possesses only strong bonds and its precursors are all atoms. It is, however, an example of the simplest kind of donor–acceptor relationship—one that exhibits self-organizational behavior [9]. Ionic chelation compounds such as *tris*-diazobicyclooctane and cynanides serve as better examples of supramolecular ionic compounds that are able to complex metal anionic clusters such as $[Fe(CN)_6]^{3-}$ [9]. Other well-known chelating agents such as EDTA (ethylene-diaminetetraacetic acid) sport two lone pairs and four negative charges (Y^{4-} species), depending on the solution *p*H, and are readily able to complex metal cations. The coordination of biological metals (cofactors) by organic supramolecular moieties relies on ion–ion interactions as well as other types of bonding species (e.g., salt bridges). Ionic compounds—inorganic, metallic, and organic—all contribute in a profound way to supramolecular and biological systems (**Table 10.4**).

10.1.2 Solvent Effects

Although lattice energies of ionic solids are relatively high (**Table 10.1**), many ionic compounds dissolve readily in water. We now address the paradoxical statement proposed earlier. If we inspect a Born–Haber style cycle for NaCl, starting from ions in their gaseous state, we see that the lattice enthalpy for NaCl

Table 10.4 *Ion Pair Interactions*

Description	Example	Bond strength (kJ · mol−1)
Ion–ion interactions $E_{Ion-ion} = \frac{q_1 q_2 e^2}{4\pi\varepsilon_o\varepsilon_r r}$ (10.8)	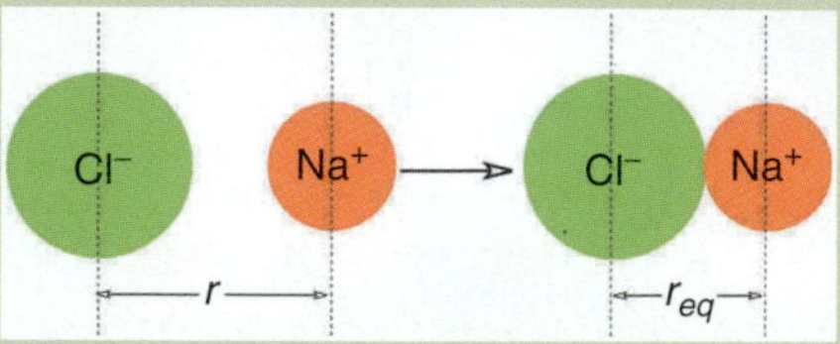 *Ion–ion attractions are responsible for the formation of NaCl crystals. Long-range attractive or repulsive forces have energy proportional to $1/r^2$; repulsive energy is proportional to $1/r$ (Table 10.3). Ion attractions are nondirectional and depend on the dielectric constant of the medium. The local minimum depicted in Table 10.3 corresponds to the equilibrium distance between the two charges.*	50–300 (100–350) Usually >190
Salt bridge	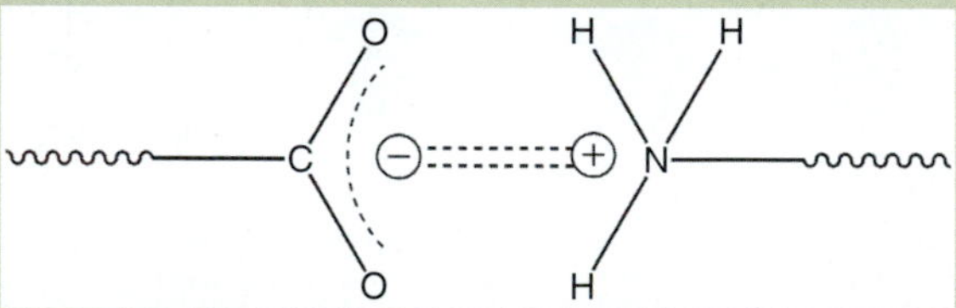 *Strength of salt bridges can be as high as 25 kJ·mol−1. Salt bridges are commonly found in protein complexes, especially in the active pocket of enzymes (usually between +Arg, +Lys, +His, and −Asp, −Glu).*	4–25

Example 10.1 *Lattice Energy of Ion–Ion Compounds*

(a) Calculate the lattice energy (ΔH°_L) of the NaCl crystal, at 25°C and standard pressure, from the Born–Haber cycle method ($\Delta H^\circ_f = -411$ kJ · mol−1; $\Delta H^\circ_{sub} = +108$ kJ · mol−1; BE(Cl_2) = +242 kJ · mol−1; $I_{Na} = 5.13$ eV; $E_{Cl} = 3.68$ eV; the Faraday constant, $\mathcal{F} = 96{,}484.56$ C · mol−1.

(b) Calculate the answer using e–e potential energy functions and the Madelung constant (equation 10.7) and compare to the value derived in (a).

Solutions:

(a)

$$\Delta H^\circ_L = -\Delta H^\circ_f + \Delta H^\circ_{sub} + \Delta H^\circ_{diss} + \Delta H^\circ_{ion}, Na + \Delta H^\circ_e\text{–}aff, Cl$$
$$= -\Delta H^\circ_f + \Delta H^\circ_{sub} + [BE(Cl_2)/2] + I\mathcal{F} + \mathcal{E}\mathcal{F} \quad (10.9)$$

The enthalpy of formation of NaCl is negative and the enthalpy of sublimation is positive. The energy of the Cl–Cl bond is 242 kJ·mol−1, the ionization energy associated with $Na \rightarrow Na^+ + e^-$ is 5.14 eV; the electron affinity E of $Cl^\circ + e^- \rightarrow Cl^-$ is 3.68 eV.

$\Delta H^\circ_L = -(-411\ kJ\cdot mol^{-1}) + (108\ kJ\cdot mol^{-1}) + [242/2\ kJ\cdot mol^{-1}] + [(5.13\ eV)(98{,}484.56\ C\cdot mol^{-1})/1000]$
$-\ [(3.68\ eV)(96{,}484.56\ C\cdot mol^{-1})/1000];\ \Delta H^\circ_L$
$= [411 + 108 + 121 + 495 - 355]\ kJ\cdot mol^{-1} \approx 780\ kJ\cdot mol^{-1}$

Example 10.1 (Contd.) *Lattice Energy of Ion–Ion Compounds*

(b) In general, the potential energy function, PE, is given by Coulomb's law: PE $\propto$ $[q_1q_2/d]$ where q is equal to e (1.6022×10^{-19} C) and d is the distance between the two charges. The total electronic interaction of a single Na^+ and the chlorides in the crystal is mediated by the Madelung constant M that equals 1.748 for NaCl. Directionality is not a factor.

The NaCl crystal comprises two intermingled fcc structures. The distance from Na^+ to Cl^- ions from all possible directions within its realm of coordination comprises the Madelung constant used below.

The sum of the ionic radii, d, of the closest Na–Cl pair is [0.97 + 1.81 Å] = 2.78 Å, but the actual lattice constant of NaCl is 2.825 Å. Note that d is equal to 1/2 the lattice constant a of a NaCl crystal (a = 5.65 Å).

The ionic potential energy relationship for the NaCl crystal is

$$\Delta H_L = \frac{\mathcal{N}_A \mathcal{M} z_{Na} z_{Cl} e^2}{4\pi\varepsilon_o d_o}\left(1 - \frac{d^*}{d_o}\right)$$

where $\mathcal{N}_A$ is the Avogadro number, $\mathcal{M}$ is the Madelung constant for NaCl equal to 1.748, and d is a constant equal to 0.345 Å. The expression [1 – d*/d] is the repulsion term.*

$$\Delta H_L = \frac{(6.022\times10^{23})(1.748)(+1)(-1)(1.6022\times10^{-19}\,\mathrm{C})^2}{4\pi(8.854\times10^{-12}\,\mathrm{C^2\cdot J^{-1}\cdot m^{-1}})(2.78\times10^{-10}\,\mathrm{m})}\left(1 - \frac{0.345\,\mathrm{n}}{2.78\,\mathrm{n}}\right)$$

$$= -\frac{(1.389\times10^{-4}\,\mathrm{J\cdot m\cdot mol^{-1}})(1.748)}{(2.78\times10^{-10}\,\mathrm{m})}(0.876)(1\,\mathrm{kJ}\cdot 1000\,\mathrm{J^{-1}}) = 765\,\mathrm{kJ\cdot mol^{-1}}$$

Please recall that the lattice energy is altered for compounds that have high surface area to volume ratio, such as for nanomaterials.

is 788 kJ · mol^{-1}. For this solid to dissolve into its constituent ions, there must be quite a hefty investment in energy. This energy comes from the formation of new bonds formed by hydration of the ion (ΔH_{Hyd}). The formation of new coordinative water bonds to Na^+ and Cl^- species is 784 kJ · mol^{-1} (other estimates claim that the hydration of ions from the gaseous state, ΔH_{Hyd}, is Na^+ = –406 kJ · mol^{-1} and ΔH_{Hyd}, Cl^- = –363 kJ · mol^{-1} = 769 kJ · mol^{-1}). Based on enthalpic considerations alone, ΔH_{Sol} is equal to +3.88 kJ · mol^{-1}, an endothermic discrepancy between the lattice energy and the energy of hydration—certainly not enough to drive the dissolution of NaCl in water. What other factor must we consider?

$$\Delta H_L \text{ (always endothermic)} + \Delta H_{Hyd} \text{ (always exothermic)} = \Delta H_{Sol} \text{ (endo- or exothermic)} \quad (10.10)$$

If we consider all of the entropic trade-offs in the process of dissolution and hydration, we find that the overall free energy, ΔG, of the dissolution reaction is negative because most dissolution reactions involving ionic compounds occur spontaneously. There are three entropic factors to consider: (1) completely replacing the crystalline order of the lattice by independent ions decreases overall order and ΔS is positive; (2) rearranging the host solvent (creating cavities) increases order and ΔS yields a negative contribution to the overall entropy; and (3) incorporating the ions within relatively ordered hydration cages (hydration spheres) causes more order and ΔS is also negative in this case. For the dissolution of NaCl in water, the value of the entropy for each step is $+240\ \mathrm{J \cdot mol^{-1} \cdot K^{-1}}$, $-20\ \mathrm{J \cdot mol^{-1} \cdot K^{-1}}$, and $-180\ \mathrm{J \cdot mol^{-1} \cdot K^{-1}}$, respectively. The overall free energy, ΔG, is found from

$$\Delta G = \Sigma(\Delta H) - T\Sigma(\Delta S) \tag{10.11}$$

and

$$\Delta G = (+788 - 784)\ \mathrm{kJ \cdot mol^{-1}} - [298\ \mathrm{K}\ (+0.240 - 0.020 - 0.180)\ \mathrm{kJ \cdot mol^{-1} \cdot K^{-1}}] \approx -8\ \mathrm{kJ \cdot mol^{-1}}.$$

The overall free energy is less than zero and the dissolution process occurs spontaneously at 298 K. Because ΔH_{Sol} is positive, the solution becomes cooler during dissolution of NaCl.

The purpose of this exercise is to bring attention to solvent effects. The solvent, as we shall find out, plays a major role in nanochemical processes. Dissolution, dispersion, stabilization, surfactants, ligand exchange reactions, and hydrolysis all figure into the grand scheme of supramolecular chemistry. The ion–ion example is perhaps one of the easiest concepts to visualize. Unfortunately, it only gets more complex from this point.

The Dielectric Constant. The dielectric constant is based on the phenomenon of permittivity. Permittivity is an intensive quantity that gauges the interaction of an electric field within a medium. The dielectric constant ε_r of a specific material is the ratio of the permittivity within a material to the permittivity of free space:

$$\varepsilon_r = \frac{\varepsilon_s}{\varepsilon_o} \tag{10.12}$$

where ε_s is the specific permittivity of a material. The dielectric constant of air is approximately 1. The dielectric constant of water is 78.85. The relative polarity (or nonpolarity) of a solvent is scaled by the dielectric constant. The strength of ion–ion attractions, as well as any other type of electrostatic interaction, is mitigated by the medium enclosing the charges. As we will learn later, the design of host–guest complexes must take into account not only the type of bonding interaction but also the immediate environmental conditions, such as temperature and solvent.

Dielectric Constant and Solvents. In biology, ionic bonds serve an important function. The tertiary (3°) structure of proteins is stabilized by the participation of ionic bonds to form salt bridges between the carboxylate ($-COO^-$) moiety of aspartate or glutamate and the amine ($-NH_3^+$) group of lysine. In the enzyme aspartate aminotransferase, the salt bridge between Lys-68 and Glu-265 plays a vital role in the kinetic mechanisms of the enzyme action [10]. The surfaces of proteins are usually enveloped in water (e.g., polar groups on the protein face outward into the aqueous medium). The interior environment of proteins, on the other hand, especially the active sites, is often hydrophobic in nature. Proteins are able to sequester single molecules of water in the confined space of an active pocket and utilize water's polar character to influence substrate-specific reactions—all within the hydrophobic pocket of an enzyme. The dielectric nature of water, or any physiological fluid, is an important factor in enzyme function and protein structure.

10.1.3 Ion–Dipole and Dipole–Dipole Interactions

Ion–Dipole Interactions. Ion–dipole interactions comprise the next level of complexity along the electrostatic interaction continuum as weaker interactions are accompanied by greater complexity of mathematical expressions. The strength of ion–dipole interactions is significantly less than that of ion–ion interactions and ranges from 50 to 200 $kJ \cdot mol^{-1}$ [9]. A dipole (from the Greek *dyo*, "two" + *polos*, "pivot") is exactly what the term implies—that regions of opposite charge (poles) exist within a single molecule.

In the case of the dipole, we need to introduce another electrostatic term known as the *dipole moment*. The measure of dipole strength is the electric dipole moment, μ—the product of charge (in coulombs) and the distance between the charges (meters). The *debye* is the unit of measure for the dipole moment—1 debye, D = 3.336×10^{-30} $C \cdot m$ and is symbolized by ⟷ with the positive charge stationed at the start of the arrow. Permanent dipoles are formed when a molecule is composed of two or more atoms that have substantially different electronegativity. The presence of a molecular dipole depends on the symmetry of the molecule and the electronegative nature of its constituent atoms. For example, carbon tetrachloride (CCl_4) does not have a dipole moment even though four very electronegative chlorine atoms are attached to a significantly less electronegative carbon atom. Chloroform ($CHCl_3$), on the other hand, does possess a dipole moment. Quadrupole, octupole, and higher orders of charge distribution also are found within molecules.

An ion–dipole interaction is an interaction between a charged ion, in elemental or molecular form, and a polar molecule (e.g., one that contains a dipole). The implicit existence of ion–dipole interactions was mentioned in the previous section during our discussion of solvent effects. Na^+ coordinated by water molecules serves as an excellent example of an ion–dipole interaction. The ion–dipole attraction between water and Na^+ contributes to the overall energy required to overcome the lattice energy of a NaCl crystal. The energy between ion–dipole attractions is inversely proportional to the square of the distance between them, r^2. The coordination of Na^+ or K^+ by crown ether complexes is a good example of an ion–dipole interaction found in supramolecular organic and biochemical systems. The energy of an ion–dipole attraction is given by the following expression:

$$E_{Ion-dipole} = \frac{ze\mu\cos\theta}{4\pi\varepsilon_o\varepsilon_r r^2} \tag{10.13}$$

where z is the unitless magnitude of the charge on the ion, μ is the dipole moment in D, $\cos\theta$ is the angle of approach of the ion to the center of the dipole (ranges from 0° to 180°). Obviously, the strongest interaction occurs when θ equals zero.

In biology, ion–dipole interactions are extremely important. Metal cations are coordinated in the center of porphyrin ring complexes and include the binding of Fe^{2+} or Mg^{2+} by heme, cytochrome-C, and chlorophyll. A cobalt cation is bound within the corrin ring system of vitamin B_{12}. These kinds of interactions are classified as dative bonds—bonds that coordinate metal cationic species within an organic framework. The distinction between metal cation–dipole interactions and dative or coordinative bonds is rather blurry. In some instances, the metal cation serves as a template around which a supramolecular species is synthesized.

Another form of ion interactions involves negative ions (anions). Anion binding is prevalent in biological systems. This is important because nearly two thirds of all enzyme substrates are anions. An ion–dipole interaction between a phosphate moiety and a *dipolar group transfer receptor* provides overall electrostatic balance [11]. Anion recognition is a fundamental mechanism in the development of antibacterial agents. Anions form nondirectional interactions with neutral, positively polarized, and cationic groups. Since anions are heavily solvated, the binding energy of any potential host must overcome this high solvation energy barrier—another factor to consider in supramolecular host–guest system design.

The action of ionophores depends on an ion–dipole mechanism. An ionophore is a lipid-soluble structure, channel, or carrier that transports metal ions through natural or synthetic lipid membranes. They range in size from small

Example 10.2 *Ion–Ion Interaction: Effects of the Liquid Medium Ion–Dipole Interactions*

What is the ion–ion potential energy of Na^+ and Cl^- ions that are 1 nm apart in water? in methanol? in trimethyl amine? ε_r of water, methanol, and trimethyl amine at 25°C are 78.85, 32.63, and 2.44, respectively.

Solutions:

Permittivity of free space, $\varepsilon_o = 8.854 \times 10^{-12}\ C^2 \cdot J^{-1} \cdot m^{-1}$, $e = 1.6022 \times 10^{-19}\ C$

For water:

$$= \frac{(+1)(-1)(1.6022\times10^{-19}\,\text{C})^2}{(4\pi)(8.854\times10^{-12}\,\text{C}^{-2}\text{J}^{-1}\text{m}^{-1})(78.85)(1\times10^{-9}\,\text{m})}$$

$$= -\frac{(8.988\times10^{9}\,\text{J}\cdot\text{m}\cdot\text{C}^{-2})(2.567\times10^{-38}\,\text{C}^2)}{(78.85)(1\times10^{-9}\,\text{m})} = -2.926\times10^{-21}\,\text{J}$$

$$= -(2.926\times10^{-21}\,\text{J}\cdot\text{bond}^{-1})(6.022\times10^{23}\,\text{bonds}\cdot\text{mol}^{-1})(1\,\text{kJ}\cdot 1000\,\text{J}^{-1}) = -1.76\,\text{kJ}\cdot\text{mol}^{-1}$$

For methanol:

$$= -\frac{(8.988\times10^{9}\,\text{J}\cdot\text{m}\cdot\text{C}^{-2})(2.567\times10^{-38}\,\text{C}^2)}{(32.63)(1\times10^{-9}\,\text{m})} = -7.07\times10^{-21}\,\text{J}$$

$$= -(7.07\times10^{-21}\,\text{J}\cdot\text{bond}^{-1})(6.022\times10^{23}\,\text{bonds}\cdot\text{mol}^{-1})(1\,\text{kJ}\cdot 1000\,\text{J}^{-1}) = -4.26\,\text{kJ}\cdot\text{mol}^{-1}$$

For trimethyl amine:

$$= -\frac{(8.988\times10^{9}\,\text{J}\cdot\text{m}\cdot\text{C}^{-2})(2.567\times10^{-38}\,\text{C}^2)}{(2.44)(1\times10^{-9}\,\text{m})} = -9.46\times10^{-20}\,\text{J}$$

$$= -(9.46\times10^{-20}\,\text{J}\cdot\text{bond}^{-1})(6.022\times10^{23}\,\text{bonds}\cdot\text{mol}^{-1})(1\,\text{kJ}\cdot 1000\text{J}^{-1}) = -56.9\,\text{kJ}\cdot\text{mol}^{-1}$$

EXAMPLE 10.3 *Ion–Dipole Interactions*

Calculate the interaction energy between a potassium cation and urea in carbon tetrachloride. The dipole moment, μ, of urea is equal to 4.56 D. Assume that the ion is 2-nm radius from the urea at an angle of 45° and that the solvent is chloroform (ε_r =2.238 for carbon tetrachloride). What would the interaction energy be if the K^+ cation were in line with the carbonyl group of the urea?

Solution:

$$E_{Ion-dipole} = \frac{ze\mu\cos\theta}{4\pi\varepsilon_o\varepsilon_r r^2}$$

$$= \left[\frac{(+1)(1.6022\times10^{-19}\,\text{C})(4.56\times3.336\times10^{-30}\,\text{C}\cdot\text{m})\cos 45}{4\pi(8.854\times10^{-12}\,\text{C}^2\cdot\text{J}^{-1}\cdot\text{m}^{-1})(2.238)(2\times10^{-9}\,\text{m})^2}\right]\left(\frac{6.022\times10^{23}}{\text{mol}}\right)\left(\frac{1\,\text{kJ}}{1000\,\text{J}}\right)$$

$$= -1.04\ \text{kJ}\cdot\text{mol}^{-1}$$

The minus sign indicates that all attractive interactions are negative. If the cation interacted head-on with the urea, the interaction energy would be greater ($\cos 0° = 1$) $= -1.47\ \text{kJ}\cdot\text{mol}^{-1}$.

molecules that complex the metal ion and shield its charge to larger structures that are able to sequester the ion completely. Structures called *ionophore channel-formers* insert hydrophilic pores into membranes.

Dipole–Dipole Interactions. Dipole–dipole interactions are weaker forms of electrostatic interactions and therefore exhibit yet more mathematical complexity:

$$E_{Dipole-dipole} = -\frac{\mu_A\mu_B}{4\pi\varepsilon_o\varepsilon_r r^3}(2\cos\theta_A\cos\theta_B - \sin\theta_A\sin\theta_B\cos\phi) \qquad (10.14)$$

If dipoles react in a head-on configuration, θ is equal to zero and the expression simplifies to

$$E_{Dipole-dipole} = -2\left(\frac{\mu_A\mu_B}{4\pi\varepsilon_o\varepsilon_r r^3}\right) \qquad (10.15)$$

Please refer to **Table 10.5** for an illustration of the dipole–dipole interaction. In dipole–dipole interactions, the energy of interaction is inversely proportional to the cube of the distance between the dipoles, r^3. Dipole–dipole interactions also involve rotations (ϕ). Rotation of dipoles acts to diminish the overall strength of interactions in a solution due to averaging as the system alternates between attractive and repulsive domains (**Table 10.5**).

The dipole–dipole interaction is the basis of the hydrogen bond. It is without question one of the most important types of bonding—whether inside or outside the biological domain. From the macroscopic perspective, hydrogen bonding in ice makes life possible under frozen lakes. From the nanoscopic perspective, hydrogen bonds act collectively to impart structure to DNA. The hydrogen bonds between and among water molecules are strong and flexible enough to

Table 10.5 *Ion–Dipole and Dipole–Dipole Interactions*

Description	Example	Bond strength (kJ · mol^{-1})
Ion–dipole	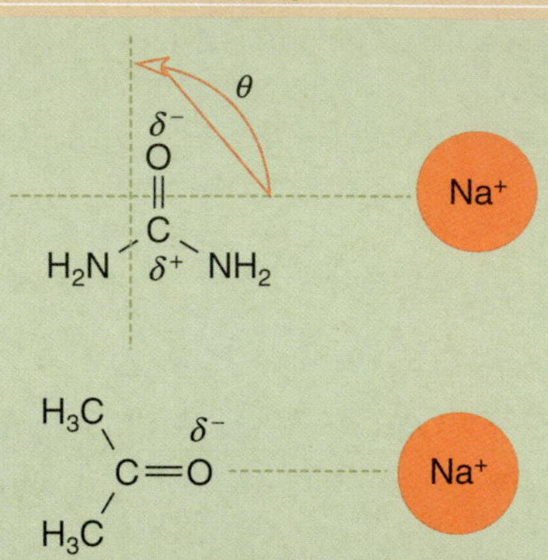 *Ion–dipole interactions. Two approaches of the sodium cation are shown. Ion–dipole reactions can be attractive or repulsive. The strength of the interaction has $1/r^2$ dependence.* 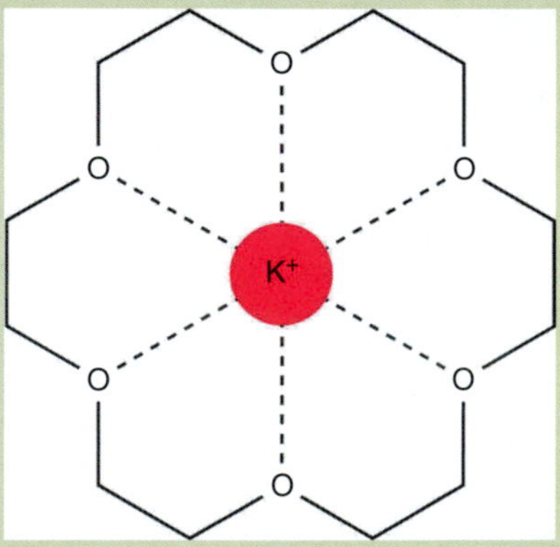 *Crown ether ethers also are able to undergo ion–dipole interactions. A potassium crown ether [K(18-Crown-6)] molecular complex is displayed.*	50–200 (40–120)
Dipole–dipole	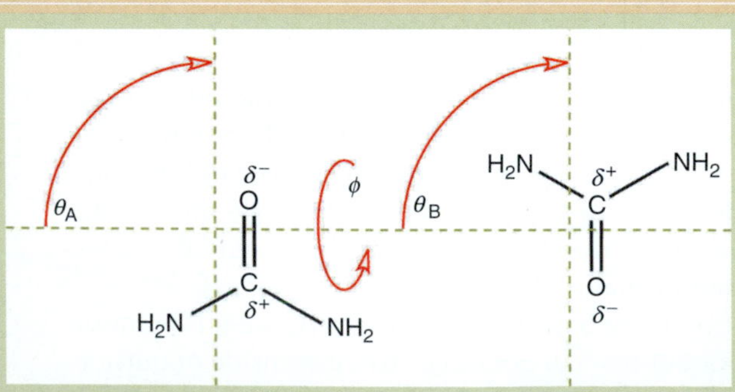 *A strong interaction between two dipoles requires the proper alignment.*	5–50 (5–40)

TABLE 10.5 (CONTD.) *Ion–Dipole and Dipole–Dipole Interactions*

Description	Example	Bond strength (kJ · mol^{-1})
	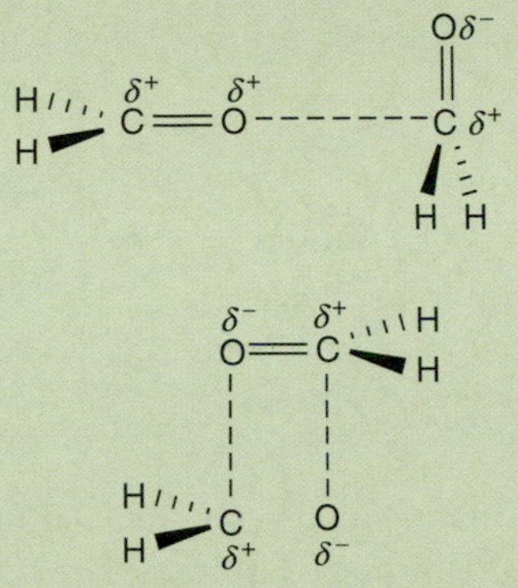 *Dipole–dipole interactions between two formaldehyde molecules. The strongest interactions require maximal overlap of electrostatic potentials. On the top is a* Type I carbonyl *type interaction; on the bottom, a* Type II carbonyl *interaction is shown.*	5–50 (5–40)

enable living systems to adjust to prevailing circumstances. More about hydrogen bonding is presented in section 10.2.

10.1.4 Dative Bonds

The stabilization of metal ions by ligands that donate electron pairs to vacant metal outer orbitals is called dative bonding. The dative bond (from the Latin

EXAMPLE 10.4 *Dipole–Dipole Interactions*

Apply the same conditions that were given in example 10.3 to the following dipole–dipole example. Assume that the head-on dipole configuration is intact. What is the value at 0.4 nm approach?

Solution:

$$E_{Dipole\text{–}dipole} = -2\left(\frac{\mu_A\mu_B}{4\pi\varepsilon_o\varepsilon_r r^3}\right)$$

$$= -2\left[\frac{\left(4.56\times3.336\times10^{-30}\,\mathrm{C\cdot m}\right)^2}{4\pi\left(8.854\times10^{-12}\,\mathrm{C^2\cdot J^{-1}\cdot m^{-1}}\right)(2.238)\left(2\times10^{-9}\,\mathrm{m}\right)}\right]\left(\frac{6.022\times10^{23}\cdot\mathrm{mol^{-1}}}{1\,\mathrm{kJ}\cdot1000\,\mathrm{J^{-1}}}\right)$$

$$= -0.031\,\mathrm{kJ\cdot mol^{-1}}$$

This is a much smaller value than that for the ion–dipole interaction of example 10.3. The minus sign indicates that all attractive interactions are negative.

At 0.4 nm, the dipole–dipole interaction is much stronger: $E_{Dipole\text{–}Dipole} = -1.96\ kJ\cdot mol^{-1}$.

FIG. 10.6 *The first metal coordination complex was synthesized by Alfred Werner in 1913. Six-coordinate complexes assume different colors depending on the ligands: $[Co(NH_3)_6]^{3+}$ = yellow-orange; $[Co(NH_3)_5Cl]^{2+}$ = purple; $[Co(NH_3)_5H_2O]^{3+}$ = red; and $[Co(NH_3)_4Cl_2]^{+}$ = green. What is the point of this? Ligands influence the physical (and chemical) properties of complexes.*

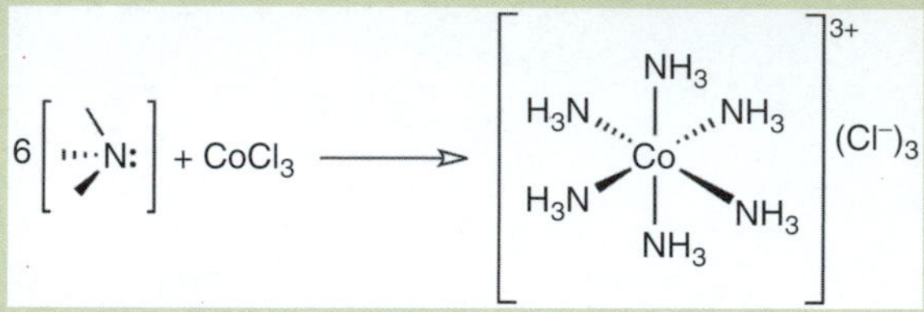

datives, "to give, donate") involves a metal (usually a transition metal) that forms neutral, cationic, or anionic complexes. The coordination of metal ions can vary from 2 to 12 ligands, but the usual number of coordinating ligands is 6. Complexation of metals is relevant in biology, pharmacology, environmental chemistry, and supramolecular chemistry.

The process of binding metals to form macrocyclic rings is called chelation (from the Greek *chela,* "claw"). Morgan and Drew coined the term "chelate" in 1920 [12]. Alfred Werner invented the field of coordination chemistry, for which he won the Nobel Prize in 1913. He is known for the reaction depicted in **Figure 10.6**.

A dative bond is like a covalent bond in which electrons are shared, but in the case of dative bonding, both electrons are usually donated by one atom. Dative bonds are also known as "coordination or dipolar bonds" (ion–dipole bonds). Once again we are up against the chemical bonding continuum, trying to define a type of bonding that can be characterized in several ways. Coordination bonding comes in many forms: ionic, covalent, dative, or mixed. The major types of metal–ligand binding are listed in **Table 10.6.** The binding (stabilization) of metal clusters is a purely nano phenomenon. As we found out in chapter 6, clusters are inherently unstable due to the high associated surface energy. Ligands provide avenues to coordinate (and therefore stabilize) surface atoms or individual metal cations—both are deficient with regard to "nearest neighbors." The use of ligands to stabilize clusters also opens new routes to hierarchical nanomaterial synthesis.

Ligand-Stabilized Clusters. One of the significant achievements of nanocluster science is the stabilization of metal complexes like Au_{55} with organic ligands to

EXAMPLE 10.5 *Binding Constant of Zn^{2+}*

Calculate ΔG of formation of the EDTA–Co^{2+} coordination complex at 25°C in a buffered solution containing EDTA and Co^{2+} if the overall formation constant K_f of the EDTA–Co^{2+} complex is equal to 2.0×10^{16}.

Solution:

$$\Delta G = -RT \ln K_f = -[8.314 \times 10^{-3}\ kJ \cdot mol^{-1} \cdot K^{-1}\ (298\ K)\ \ln (2.0 \times 10^{16}) = -93\ kJ \cdot mol^{-1}$$

TABLE 10.6 *Metal–Ligand Interactions*

Description	Example
Dative bonding	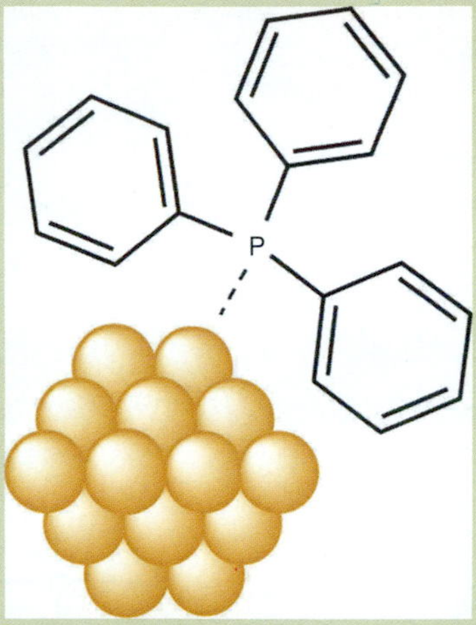 *$Au_{55}(PPh_3)_{12}Cl_6$ cluster is an example of a ligand-stabilized system [13–16]. The size of the constituents is not in scale and the Au-ligand structure is not exactly displayed; there are actually 12 phosphine ligands surrounding the gold cluster. Coordination of metals by ligands occurs by donation of a lone pair of electrons—e.g., dative bonds.*
Metal chelation	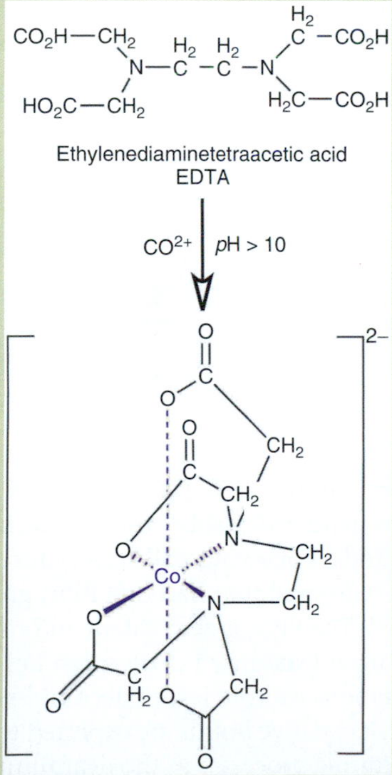 *EDTA is one of the best chelation agents in industry and research. It has six ligands with which to bind metal cations. The mechanism is based on Lewis acid–base reactions (dative bonds).*

(*continued*)

TABLE 10.6 (CONTD.) *Metal–Ligand Interactions*

Description	Example
Typical metal coordination by monodentate ligands	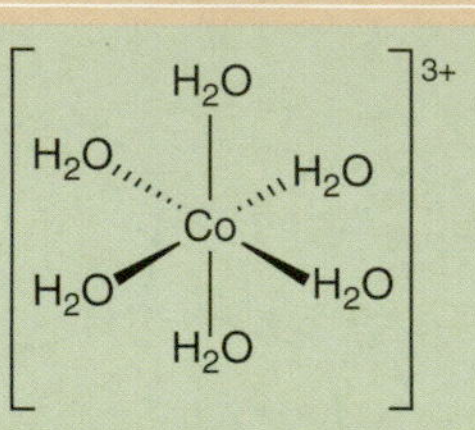 *A hexaaquocobalt(III) complex is depicted. The water molecules form dative bonds to the Co^{3+} metal cation via lone pairs on oxygen atoms.*
Agostic interactions (α, β, γ C–H bonding)	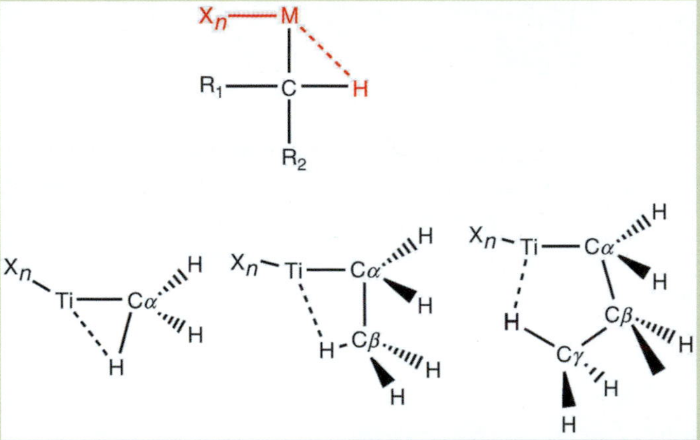 *Agostic bonds involve the attraction of a C–H bond to a metal. Their structure (top) was elucidated in 1983 by M. Brookhart and M. L. H. Green (M. Brookhart and M. L. H. Green, Journal of Organometallic Chemistry, 250, 395–408, 1983.) M = metal; X = H, halogens or other substituents. Three kinds of organometallic titanium complexes are shown: left: α-agostic; middle: β-agostic; and right: γ-agostic. (I. Vidal et al., Organometals, 25, 5638–5647, 2006. With permission.)*

form the $Au_{55}(PPh_3)_{12}Cl_6$ cluster [13–16]. More recently, a tetrahedral $Au_{20}(PPh_3)_8$ was rendered stable when coordinated to eight PPh_3 (triphenyl phosphine) ligands. Following collision-induced dissociation, stable $Au_{20}(PPh_3)_4^{2+}$ species were formed that had one PPh_3 group bonded to each corner of a tetrahedron [18]. The PPh_3 is zero valent and forms dative bonds to the gold cluster. Batches of ligand-stabilized clusters can be dried, packaged, and shipped to far away destinations without fear of decomposition or agglomeration. In terms of molecular design, dative bonds are expected to contribute significantly to molecular manufacturing processes in the near future [19]. B. Olenyuk et al. state [19]

> A recent novel synthetic protocol in the construction of organized nanoscopic assemblies from multiple building blocks in a single step, namely self-assembly, relies on critical information about the shape and properties of the resulting

> structure being preprogrammed into each individual building block. Although this approach was initiated by artificial mimicking of natural receptors that utilize weak hydrogen bonds, it has now resulted in an entirely different "unnatural" strategy, molecular architecture that employs transition metals and dative bonds to achieve structurally well defined, highly ordered assemblages.
>
> This approach relies on the fact that fewer metal–ligand bonds may be used in place of several hydrogen bonds owing to their greater strength. Another advantage lies in the existence of a large variety of transition metals with different co-ordination numbers [sic], thus facilitating the building of diverse nanoscopic entities, with tremendous variations in shapes and sizes.

Binding modes in coordination complexes goes beyond donation of neutral lone electron pairs. Ionic constituents, primarily in the form of anions, also form complexes with metals. The mechanism of chelation involves the use of both types of ligand donors (ionic and neutral), sometimes simultaneously to bind the same metal cation. Dative bonds are, of course, important to nanoscience, and the formation of well-ordered organic monolayers has great potential for use in sensors and other devices. For example, the adsorption of pyrimidine onto a Ge(100) surface was found to have a "double dative bond configuration"—confirmed by STM, TPD, and density functional theory (DFT) [20]. Specifically, dative bonds are formed between the nitrogen of pyrimidine (Lewis base) and a Ge-dimer (the *buckle-down* atom). In this configuration, electron transfer occurs from one member of the Ge-dimer to the other and is accomplished without the loss of aromaticity in the rings of the pyrimidine molecules [20].

Chelates. Chelation is a specialized form of metal coordination binding that involves a molecule with at least two binding sites or ligands (from the Latin *ligamentum,* "to bind, tie"). Monodentate (from the Latin *dentis,* "tooth"), bidentate to multidentate chelates are named according to the number of ligands that a single molecule can provide to bind a metal cation. The more ligands that a molecule can provide to bind a metal or cation, the better is the entropic trade-off and the more spontaneous the binding becomes. Hence, one of the driving forces behind the chelate effect is the increase in entropy afforded by polydentate ligands that release monodentate counterparts into solution. On the other hand, the enthalpy of complexation for a polydentate chelate is about the same as that of an equivalent group of monodentate ligands.

The exothermic character and high formation (complexation) constant of chelate binding processes is due in part to the *chelate effect*—the enhanced stability exhibited by metal–ligand complexes that contain chelate rings (e.g., polydentate ligands) compared to similar configurations that contain fewer or no rings (e.g., monodentate ligands) [8]. The magnitude of the effect depends on the free energy change, the basicity of the ligand, the *p*H of the solution, and the type and concentration of the chelate and substrate [21]. Multidentate chelate complexes remain stable even under extremely dilute conditions, whereas monodentate ligand complexes are able to dissociate under dilute conditions.

The coordination sphere of a metal–ligand complex is composed of the number of ligands directly attached to the metal. These are called inner-sphere complexes. Complex cations, however, are able to associate with anions without displacement of inner-sphere ligands. These are called outer-sphere complexes and are characterized by short lifetimes and nonspecific geometry. Coordination complexes consisting

Example 10.6 *Free Energy of Chelation of EDTA*

Calculate ΔG of formation of the EDTA–Co^{2+} coordination complex at 25 °C in a buffered solution containing EDTA and Co^{2+} if the overall formation constant β of the EDTA–Co^{2+} complex is equal to 2.0×10^{16}.

Solution:

$$\Delta G = -RT \ln\beta = -[8.314 \times 10^{-3}\ kJ \cdot mol^{-1} \cdot K^{-1}\ (298\ K)\ \ln(2.0 \times 10^{16}) = -93\ kJ \cdot mol^{-1}$$

of multidentate ligands invest a large covalent contribution to the inner-sphere structure. These species are long-lived and have a definitive geometry.

The free energy of complexation is given by $\Delta G = -RT \ln \beta$ where β is the overall equilibrium complexation constant). The stability constant for adding a succession of monodentate ligands is as follows [8,9]:

$$ML_{n-1} + L \rightarrow ML_n\text{: } K_n = \frac{[ML_n]}{[M_{n-1}][L]} \tag{10.16}$$

and for the cumulative addition of ligands:

$$ML + nL \rightarrow ML_n\text{: } \beta_n = \frac{[ML_n]}{[M][L]^n} = \prod_1^n K_n, \tag{10.17}$$

where $\beta_n = (K_1)(K_2)(K_3)(K_4) \ldots (K_n)$

Example 10.7 *Free Energy and Formation Constants of Chelation Reactions*

Calculate ΔG and the formation constants K_f (and log K_f) for the following coordination reactions at 298 K and discuss the reasons for the differences. M = metal cation. In reaction a, four monodentate ligands coordinate a single metal cation. In reaction b, two bidentate ligands coordinate a single cation.

a. $M^{2+}_{(aq)} + 4NH_{3(aq)} \rightarrow [M(NH_3)_4]^{2+}_{(aq)}$ $\quad \Delta H° = -53\ kJ \cdot mol^{-1}$, $\Delta S° = -42\ J \cdot mol^{-1} \cdot K^{-1}$

b. $M^{2+}_{(aq)} + 2(NH_2{-}CH_2{-}CH_2{-}NH_2)_{(aq)} \rightarrow [M(NH_2{-}CH_2{-}CH_2{-}NH_2)_2]^{2+}_{(aq)}$
$\Delta H° = -56\ kJ \cdot mol^{-1}$, $\Delta S° = 10\ J \cdot mol^{-1} \cdot K^{-1}$

Solution:

In all cases, use $\Delta G° = \Delta H° - T\Delta S°$ *and* $\Delta G° = -RT \ln K_f$ *(where* $\log K_f = \ln K_f/2.303$*)*

a. $\Delta G = -53\ kJ \cdot mol^{-1} - (298\ K)(-0.042\ kJ \cdot mol^{-1} \cdot K^{-1})$
$\Delta G = -41\ kJ \cdot mol^{-1}$
$\ln K_f = \Delta G/[-RT] = -41\ kJ \cdot mol^{-1}/[-0.00834\ kJ \cdot mol^{-1} \cdot K^{-1} \times 298\ K] = 16.5$
$\log K_f = \ln K_f/2.303 = 7.2$
$K_f = 1.5 \times 10^7\ M^{-1}$

b. $\Delta G = -56\ kJ \cdot mol^{-1} - (298\ K)(+0.010\ kJ \cdot mol^{-1} \cdot K^{-1})$
$\Delta G = -59\ kJ \cdot mol^{-1}$
$\ln K_f = \Delta G/[-RT] = -59\ kJ \cdot mol^{-1}/[-0.00834\ kJ \cdot mol^{-1} \cdot K^{-1} \times 298\ K] = 23.8$
$\log K_f = \ln K_f/2.303 = 10.3$
$K_f = 2.3 \times 10^{10}\ M^{-1}$

There is a three order of magnitude increase in the complexation constant of the metal dication by the bidentate ligand. The difference is due to the ca. +52 $J \cdot mol^{-1} \cdot K^{-1}$ *increase in entropy afforded by the bidentate chelation agent.*

Agostic Interactions. Agostic (from the Greek "to hold close") interactions involve weak bonding interactions between transition metals (usually second row metals) and the alkyl hydrogen of a C–H bond; specifically, the alkyl hydrogen is "held close" to the metal atom. The strength of the interaction is about 42–63 $kJ \cdot mol^{-1}$ [17]. In 1983, M. Green and M. L. H. Brookhart introduced the term *agostic* to the chemical bond community. Other forms involving transition metals and groups with more polar character than alky groups, more polar groups such as R-S–H and R-N–H, are also included into the category of agostic bonding [17]. Although its exact nature is still not clear, agostic bonds have been described as either van der Waals type or "three-center-two-electron" bonds. Agostic bonds are not a special case of hydrogen bonds but should be considered more along the line of a real bond.

10.1.5 π-Interactions

π-interactions are another important type of electrostatic interaction (energy 0–50 $kJ \cdot mol^{-1}$) [9]. For example, DNA, in addition to being held together by hydrogen bonds, is held together by π-interactions between aromatic constituents of the nucleotide bases. Interplanar π–π interactions are responsible for the three-dimensional structure of graphite and for its lubrication properties. Intercalation phenomena, aromatic packing in crystals, the 3° structure of proteins, complexation in host–guest configurations, and stacking of porphyrins are all dependent on π–π interactions [22,23]. DNA, graphite, and molecular tweezers all rely on intercalation to provide additional functionality. The basic forms of π-interactions are shown in **Table 10.7.**

C. A. Hunter et al. in 1990 proposed that aromatic interactions rely heavily on geometrical factors in order for bonding to occur [22]. The σ-framework is an important component in facilitating π–π interactions because attractions between π–σ systems (via van der Waals forces) are strong enough to overcome direct π–π repulsive forces. The authors claim that the geometrical point of intermolecular contact rather than the overall molecular oxidation–reduction potential is what dominates the interaction. If true, this implies that conventional electron-donor~electron-acceptor models are not applicable to π–π interactions [22,23]. The Hunter–Saunders model is based on competing π–π and van der Waals electrostatic interactions. Van der Waals attractions arise between the electrons of the π-system and the positive-leaning σ-framework of another aromatic molecule [9,22,23]. Stacking of porphyrins in nanostructured crystals happens by means of a coplanar-offset geometry. This example demonstrates how cooperative interactions are most likely responsible for binding in complex systems. Molecular tweezers that consist of two aromatic surfaces are able to sandwich adenine with a "high degree of face-to-face complementarity" [25]. The binding energy for this complex is on the order of 25.1 $kJ \cdot mol^{-1}$ in deuterated chloroform [25].

Neutral Guests. The binding of neutral guests raises certain issues with regard to selectivity and stability [9]. However, hydrogen bonding, dative bonding, hydrophobic effects, and π–π interactions and mixed forms are all ways of pulling neutral molecules into a reactive zone. The basic driver in chemical binding is to maximize the amount of orbital overlap. The electronic configuration of an

Table 10.7 *π-Interactions*

Description	Example	Bond Strength (kJ · mol^{-1})
π–π Interactions	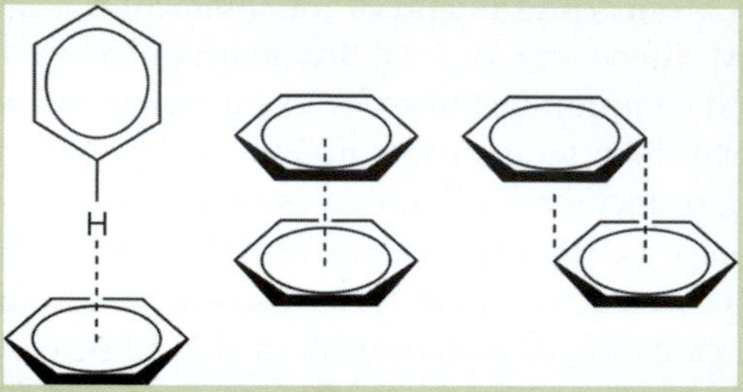 *Attraction between the electron-rich interior of the aromatic ring with the electron-poor exterior is the most likely configuration. The interaction is considered to be electrostatic and van der Waals. The configuration in the middle is least likely. The interaction found in graphite is shown on the right.*	0–50 (10–15 Face) (15–20 Edge)
d-Block–π-interactions	Ferrocene $Fe(C_5H_5)_2$ 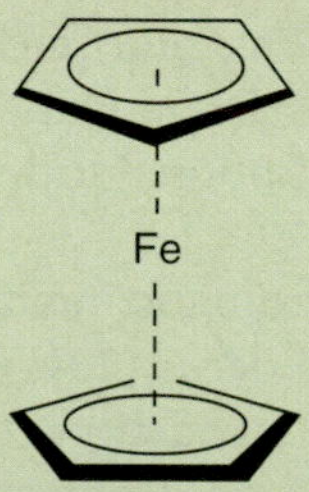 *Coordination of iron between two cyclopentadyl rings yields the compound ferrocene—one of the most important in the synthesis of carbon nanotubes. The preparation and organometallic chemistry of ferrocene was established in 1951 (G. Wilkinson, Journal of Organometalic Chemistry, 100, 273–278, 1975. With permission.)*	5–80
Cation–π-interactions	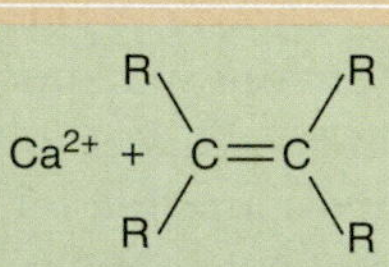 *Metal addition to alkenes to form strong covalent bonds—a superdative bond of sorts.*	

unsuitable bonding partner, like a neutral molecule, can be altered by a cooperative effort. The binding of *p*-benzoquinone with hydrogen bonds and π–π interactions demonstrates the concept of cooperativity between (or among) different types of molecules. A π–π interaction acting in concert with a hydrogen bond is able to alter the electronic properties of the neutral guest and render it bondable by changing its polarity or tweaking its symmetry [9]. This example demonstrates how such intermolecular interactions can influence the chemical nature of the guest molecule [8]. The best examples of cooperative bonding exist in the active pockets of enzymes.

10.2 Hydrogen Bonding

Of all the kinds of intermolecular bonding, chemists and biologists alike are most familiar with the hydrogen bond (H-bonds), a special subset of dipole–dipole interactions. H-bonds are mostly directional bonds that consist of a shared hydrogen atom between small electronegative elements (usually O, N, F, Cl) that possess at least one lone pair of electrons. Carbon, however, is another element capable of forming hydrogen bonds, especially if an electron-withdrawing group is nearby on its backbone structure [9,26]. Physical properties like boiling point, melting point, and surface tension are influenced by hydrogen bonds. For example, the boiling point of Group V, VI, and VII hydrides is anomalously higher for NH_3, H_2O, and HF than their heavier counterparts. **Table 10.8** lists some of the relevant features of hydrogen bonds and those of other compounds that make hydrogen bond-like bonds.

10.2.1 Standard Hydrogen Bonds

Although formed from dipole–dipole interactions, the bond energy of H-bonds can be much stronger than those of standard dipole–dipole bonds: ~120 versus ~50 $kJ \cdot mol^{-1}$, respectively. Hydrogen bond length ranges from 2.5 to 3.5 Å. This is somewhat more than the length of a normal covalent bond between hydrogen and other electronegative atoms (<2.0 Å). H-bond length depends on the electronegative constituents of the molecules involved. For example, an H-bond between water and chlorine, a larger (softer) and therefore less electronegative atom, is generally longer and weaker than those between water and fluorine [9]. The strength of hydrogen bonds also depends on the immediate environment: the medium (type of liquid or gas phase), chemical composition, *p*H, and temperature. The hydrogen bond is responsible for the physical properties of water, the binding in DNA, and protein folding to form 2° structures.

Even though the hydrogen bond is relatively weak compared to covalent bonds, the summation of all hydrogen bonds over the volume and surface of a system results in compounds or phases that are quite stable. In the liquid state, each water molecule is involved in an average of 3–3.5 hydrogen bonds—even as hydrogen bonds in water are formed and broken every 10^{-12} s. Molecules of water pack together to form short-lived hydrogen-bonded tetrahedral lattice transients called *flickering clusters* [$8H_2O \leftrightarrow 2(H_2O)_4$] caused by fluctuations in thermal energy. In addition to hydrogen bonding interactions, water molecules experience dispersion and polarization. Polarization effects in water mutually act to strengthen the hydrogen bond—an example of another cooperative effect.

For the pure octomeric form of water, ΔH_{vap} is equal to 44.3 $kJ \cdot mol^{-1}$. Many kinds of interactions contribute to the heat of vaporization: hydrogen bonding (35 $kJ \cdot mol^{-1}$), dipole interactions (1.7 $kJ \cdot mol^{-1}$), van der Waals interactions (5.0 $kJ \cdot mol^{-1}$), and the RT (or k_BT) contribution (2.5 $kJ \cdot mol^{-1}$) [30]. In ice, each water molecule is coordinated by four hydrogen bonds to form a tetrahedral

TABLE 10.8 *Hydrogen Bonding and Allies*

Description	Example	Bond strength (kJ · mol⁻¹)
Standard hydrogen bonding	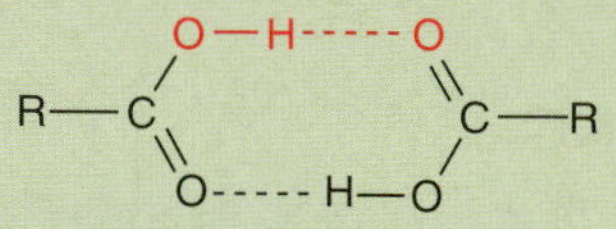 *Hydrogen atoms are shared between electronegative oxygen atoms. H-bonds are directional; bond character is electrostatic (90%) + some covalent (10%) and has short-range attraction (0.25–0.35 nm). (O. Markovitch and N. Agmon, Journal of Physical Chemistry A, 111, 2253–2256, 2007. With permission.)*	4–120 [9] (15–40 strong, 5–15 moderate, <5 weak)
	O–·····HO	21
	O–H······N	29
	N–H······N	13
C–α–H·······O/N C–α–H·······π Hydrogen bonds	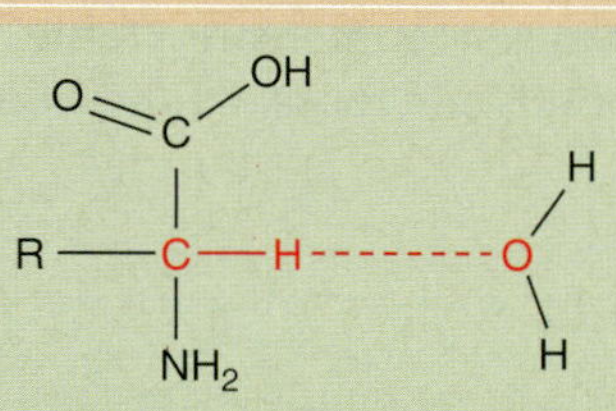 *Cα–hydrogen bond in amino acids. These bonds are commonly found in protein structures. (S. Aravinda et al., Biochemical and Biophysical Research Communications, 272, 933–936, 2000. With permission.)*	The archetypical weak hydrogen bond of weak acid–strong base or soft acid–hard base is one kind. Hydrogens also donated to π-acceptors like C=C and phenyl rings are included in this category
Halogen bonding	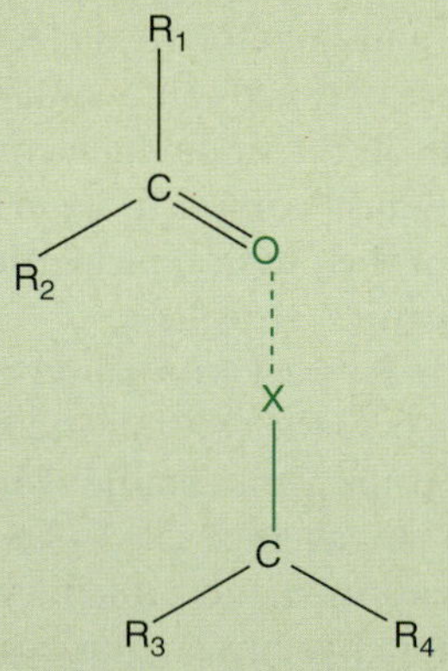 *Halogen bonding is analogous to hydrogen bonding. The substituent X is F, Cl, Br, or I or oxygen-containing groups like carbonyls (pictured), carboxylates, hydroxyls, phosphates, or sulfates. (P. Auffinger et al., Proceedings of the National Academy of Sciences, USA, 101, 16789–16794, 2004. With permission.)*	Weak to strong, just like with H-bonds Some bond strengths exceed H-bonds Bond strength dependent on environment

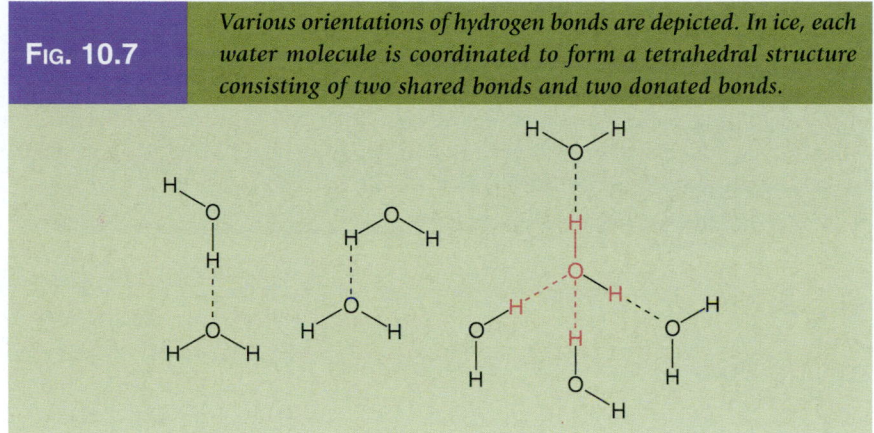

FIG. 10.7 *Various orientations of hydrogen bonds are depicted. In ice, each water molecule is coordinated to form a tetrahedral structure consisting of two shared bonds and two donated bonds.*

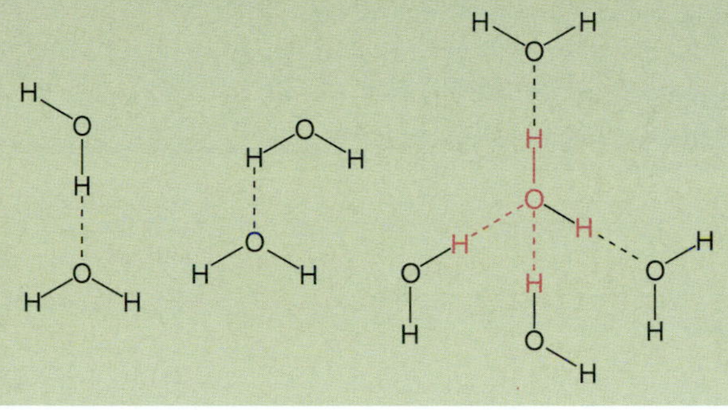

geometry (two shared, two donated). Such factors make H-bonding by itself a cooperative effort [31]. The structure of water is a function of the concentration of ions and other constituents. Hydration shells are another aspect of organization in water. Hydration around an ion changes the structure of the hydrogen bond as H_2O reorients itself to accommodate the ion. Depending on the charge of the ion (+ or −), water will direct its opposite polarized charge to create the hydration sphere. Various orientations of hydrogen bonds in water are shown in **Figure 10.7.**

The structure of water in confined spaces, as in the active pockets of enzymes and in carbon nanotubes, is a vital area of current research. Electrodes consisting of nanolayers of chemically modified with osmium bipyridine, poly(allylamine) and poly(vinyl)sulfonate were found to alter the hydration number of certain ions [32]. Molecular dynamic simulations have shown that the hydration number of Na^+ (4.5 in bulk water) is reduced to 3.5 when confined in 0.90-nm diameter nanotubes [33]. In 0.82-nm diameter tubes, the hydration number is reduced even further [33]. Solubility of ions in water is a function of the ability to create hydration spheres around the ion. If space is restricted, the solubility of ions goes down. The solubility of NaCl decreases by a factor of 2 if the solvent is confined in 0.8-nm diameter pores [34]. Confinement effects are able to influence the course of chemical reactions—the case once again in the active pockets of enzymes.

Hydrogen bonding also plays a role in solids. Approximately 20–25% of the lattice energy of dihydrogen phosphate salts is due to hydrogen bonds [36]. The hydrogen bond also contributes to the packing and crystal geometry of proteins. Supramolecular chemistry (chapter 11) is highly dependent on the hydrogen bond. Large organic supramolecules based on melamine and cyanuric acid (**Fig. 10.8**) or extremely large supramolecules like DNA all rely on hydrogen bonding to maintain the integrity of their structure.

Just as with the naming of disciplines and subdisciplines of chemistry, boundaries that distinguish between and among the types of H-bonding are also fuzzy. Hydrogen bonds have both covalent and ionic character, are strongly directional, or have angular components. Bonds can be short or long. Because of these and other effects imparted by the solvent and other solutes, a complete description of the bonding scenario is difficult to achieve (**Table 10.9**).

FIG. 10.8 *Hydrogen bonding is important in supramolecular complexes such as the melamine–cyanuric acid complex shown. It is clear from the image that hydrogen bonds are able to direct supramolecular geometry with high specificity. The same is true for DNA; we are all quite familiar with its specificity and overall structure. The rosette structure on the left is a kind of supramolecular structure that creates a cavity in the center. A ribbon is formed by control of the size of the R-groups. Such nanomaterials are capable of self-assembly due to the complementarity character of hydrogen bonds.*

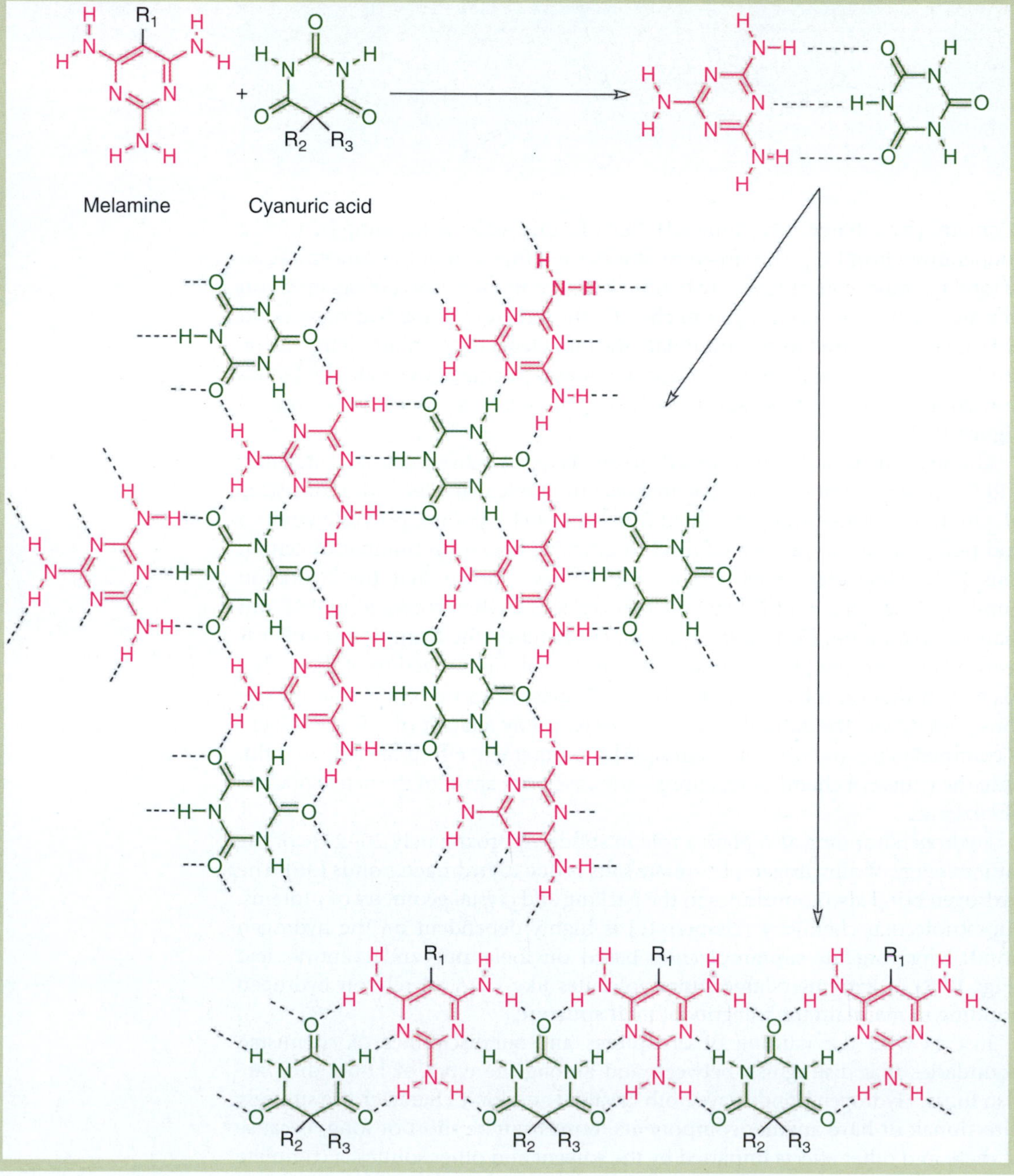

TABLE 10.9 *Hydrogen Bond Energies of Some Common Molecules*

Molecule	Hydrogen bond strength (H_nX--H) (kJ · mol^{-1})	Covalent bond strength (H_nX······H–XH) (kJ · mol^{-1})	Energy ratio % hydrogen/covalent
H_2S	7	363	1.9
NH_3	17	386	4.4
H_2O	22 (Pure water: 10.2) [35]	464	4.7
HF	29	565	5.1

Note: X is S, N, O, or F.

Technically, the hydrogen bond is the result of an insipient proton transfer reaction [37]. Dissociation energies of hydrogen bonds range from 0.8 to 170 kJ · mol^{-1} depending on the ambient conditions [37]. The hydrogen bond is characterized by broad transitions that merge continuously with covalent bonds, ionic interactions, van der Waals phenomena, and π-cation interactions. Interestingly, it was shown by simulation that the strength of hydrogen bonds is stronger in ordered clathrate solvation shells of nonpolar groups (18.5 vs. 17.4 kJ · mol^{-1}) and that the correlation time of hydrogen bonds persisted significantly longer near hydrophobic groups [31]. If cations like Na^+ are nearby, hydrogen bonds in the hydration sphere are weakened by the strong electric field of the cation. The Na^+ cation is capable of causing perturbations to the alignment of the water dipoles around hydrophobic groups, distorting the hydrogen bond [31]. Others claim that water at hydrophobic surfaces demonstrates weaker hydrogen bonding but stronger orientation effects [38]. Selective probing of the molecular structure of CCl_4–H_2O and hydrocarbon–H_2O interfaces with vibrational methods found that H-bonding was weakened near hydrophobic surfaces [38].

The Born–Haber method of determining the bond strength is an approximate method at best because it is not able to discriminate among the numerous types of bonding that may be involved. Many methods have been applied to determine the bond strength of H-bonds. For example, the hydrogen bond strength was determined for a fluorescent solute, neutral red, in water by measuring the temperature dependence of the ratio of the intensity of the dye at 625 nm in water to its intensity in benzene (at 530 nm). The H-bond energy was determined to be 10.1 kJ · mol^{-1} [35].

Analytical Methods. Infrared methods such as FTIR rely on shifts (in reciprocal centimeters) and the width of the –OH stretching bands to detect H-bonds. Hydrogen bonding in general results in decreased frequency and a broadening of the absorption band [39]. H-bonding in alcohols, for example, exhibits this phenomenon. FTIR spectra of alcohols reveal two types of –OH stretching frequencies: a "free" hydroxyl stretch (no hydrogen bonding) located at ca. 3620–3640 cm^{-1} and a broader bound hydroxyl stretch (with hydrogen bonding) at 3250–3450 cm^{-1} [40]. Polarized infrared absorption spectra acquired from silicate minerals (pectolite and serandite) showed that H-bonds are short, strong, and asymmetric [41]. FTIR and XRD were applied in tandem to analyze the active-site structure role of Tyr^{34} residue of manganese superoxide dismutase. Results indicate that the phenolic hydroxyl of Tyr^{34} is hydrogen bonded and acts as a proton donor to an adjacent H_2O in the active pocket [42].

Example 10.8 Calculation of the Hydrogen Bond Strength of a Crystalline Solid

A hypothetical material that contains a small electronegative anion, X^- forms a crystalline solid with an organic carbonyl, R, of the formula MXR where M is some metal. Using the Born–Haber cycle, calculate the strength of the hydrogen bond if:

ΔH_L MX·R = +725 kJ·mol^{-1}
ΔH_{Sol}, MX·R = –2.9 kJ·mol^{-1}
ΔH_{Vap}R = +19.7 kJ·mol^{-1}
ΔH_LMX = +798 kJ·mol^{-1}
ΔH_{Sol}, MX = +34 kJ·mol^{-1}

Solution:

Using the Born–Haber cycle:

$MX_{(s)} + R_{(l)} \rightarrow M^+_{(g)} + X^-_{(g)} + R_{(l)}$ $\quad \Delta H_L = +798\ kJ \cdot mol^{-1}$ *[lattice energy of $MX_{(s)}$]*

$R_{(l)} \rightarrow R_{(g)}$ $\quad \Delta H_{vap} = +19.7\ kJ \cdot mol^{-1}$ *[energy of vaporization of $R_{(l)}$]*

$\mathbf{MX_{(s)} + R_{(l)} \rightarrow M^+_{(g)} + X^-_{(g)} + R_{(g)}}$ $\quad \mathbf{\Delta H_{rxn} = +817.7\ kJ \cdot mol^{-1}}$ ***[overall energy on left side]***

$MX_{(s)} + R_{(l)} \rightarrow M^+_{(l)} + X^-_{(l)} + R_{(l)}$ $\quad \Delta H_{sol} = +34\ kJ \cdot mol^{-1}$ *[solvation energy of $MX_{(s)}$]*

$M^+_{(l)} + X^-_{(l)} + R_{(l)} \rightarrow MX \cdot R_{(s)}$ $\quad \Delta H_{sol} = -2.9\ kJ \cdot mol^{-1}$ *[solvation energy of $MX \cdot R_{(s)}$]*

$MX \cdot R_{(s)} \rightarrow M^+_{(g)} + X^- \cdots\cdots R_{(g)}$ $\quad \Delta H_L = 725\ kJ \cdot mol^{-1}$ *[lattice energy of $MX \cdot R_{(s)}$]*

$M^+_{(g)} + X^- \cdots\cdots R_{(g)} \rightarrow M^+_{(g)} + X^-_{(g)} + R_{(g)}$ $\quad \Delta H_{H\text{-}bond} = H\text{–}B\ kJ \cdot mol^{-1}$ *[unknown energy of H–bond]*

$\mathbf{MX_{(s)} + R_{(l)} \rightarrow M^+_{(g)} + X^-_{(g)} + R_{(g)}}$ $\quad \mathbf{\Delta H_{rxn} = +761.9\ kJ \cdot mol^{-1} + H\text{–}B}$ ***[overall energy on right side]***

Balance
817.7 = 761.9 + H–bond
H–bond energy = 55.8 kJ · mol^{-1} $\quad \Delta H_{H\text{-}bond} = -55.8\ kJ \cdot mol^{-1}$

Glass transition temperature analysis by differential scanning calorimetry (DSC) was used to determine the behavior and extent of hydrogen bonding in phenolic polymers. The thermodynamic competition between inter- and intramolecular hydrogen bonding was evaluated by molecular mechanics modeling [43]. Analysis of water by NMR methods revealed an up-field (lower frequency) chemical shift of the proton involved in hydrogen bonding, and the more hydrogen bonding there was, the greater the up-field shift became. At higher temperatures (100°C), the shift was even greater; better shielding occurred due to the reduced strength of the hydrogen bond [43].

Molecular modeling methods have proven to be extremely useful. Ab initio molecular orbital theory methods were employed to determine the stability of hydrogen bonds in methyl substituted adenine and uracil [44]. The results agreed quite well with experimentally determined binding constants [44]. Bond strength, depending on the substitution constituent, ranged from 40 to 60 kJ · mol^{-1}.

10.2.2 C-α-H···O Hydrogen Bonds

Weaker forms of hydrogen bonding are exhibited by C-α-H······O compounds [45]. These are of the form C–H······O/N or C–H······π. These weaker forms of hydrogen bonding were all discovered in the 1930s [45]. Validation of their existence was accomplished in the 1960s [45]. By the time the 1990s rolled around, the C–H·····O and π-types of hydrogen bonds were fully accepted—even by the crystallographers who grudgingly released their stamp of approval. Intermolecular O–H·····π hydrogen bonds were discovered in solutions of R–O–H donors in solvents with π-acceptors like benzene and toluene. This weak form of the hydrogen bond plays a role in supramolecular chemistry. Quinones in particular, like 1,4-benzoquinone, can be formed in a planar configuration that depends on an extensive network of C–H·····O bonds [45]. Weak H-bonds also exist in purine and pyrimidine interactions and in collagen and polyglycine [45].

In 2000, researchers detected an unusual C-terminal conformation that was due to a strong C–H·····O bond between Ala(4)-CαH and D-leu(9)CO [28]. Ab initio quantum calculations validated the presence of C–H···O hydrogen binding and showed that peptide CαH groups can be good hydrogen donors to water molecules with the binding energy in the range of 8–11 $kJ \cdot mol^{-1}$ for nonpolar and polar amino acids, respectively [46]. This relatively new type of hydrogen bond is considered to be important in many kinds of molecular complexes and crystals [47,48]. C–H·····O bonds are important to biological systems with nontrivial bond energies. **Table 10.10** lists simulated FTIR and NMR results for several amino acids as proton donors to water [48].

10.2.3 Halogen Bonds

An interesting analogy to H-bonds is the halogen bond. The interactions are fundamentally electrostatic but polarization, dispersion, and charge transfer also contribute to the bond—just like with hydrogen bonds [29]. The Br·····O distance is fairly long, ~3 Å, 12% shorter than the van der Waals radii sum (data acquired from an aldose reductase complex with halogen inhibitor) [14]. In

TABLE 10.10 *Interaction of Proton Donors with Water*

Proton donor	Interaction energy ΔE ($kJ \cdot mol^{-1}$)	FTIR stretch shift: C–H Δν (cm^{-1})	NMR isotropic shift of bridging proton $\Delta\sigma_{iso}$ (ppm)	Anisotropic shift of bridging proton $\Delta\sigma_{aniso}$ (ppm)
H–OH	−18.9	−31 (O–H)	−2.6	10.9
F_2HCH	−10.6	26	−1.3	6.1
Gly [H]	−10.6	14	−1.4	5.9
Ala [CH_3]	−8.8	51	−1.4	6.8
Val [$CH(CH_3)_2$]	−8.4	56	−1.5	6.7
Ser [CH_2OH]	−9.6	22	−1.7	7.5
Cys [CH_2SH]	−7.9	51	−1.6	7.7
Lys^+ [$(CH_2)_4NH_3^+$]	−20.6	6	−1.7	6.7
Asp^- [CH_2COO^-]	+5.98	70	−1.5	6.7

other compounds, bond lengths on the order of 80% of the van der Waals (*vdw*)-radii sum are not uncommon [14]. The thyroid hormone utilizes halogen bonds in molecular recognition. Halogen bonds are short-contact charge-transfer bonds that involve a negative charge from oxygen, nitrogen, or sulfur (well-known Lewis bases) and a polarizable halogen such as bromine (Lewis acid) [29]. The most celebrated naturally occurring halogen-bonded macromolecules are the class of thyroid hormones. Short I · · · · · O contacts between tetraiodothyroxine and its transport protein transthyretin play a role in recognition. There are greater than 3500 halogenated metabolites, including antibiotics [29]. The authors of this paper state:

> It is stunning to see how Nature has exploited all possible intermolecular interactions, even the most "exotic" ones, to design very specific and efficient recognition systems.
>
> Halogen atoms can be involved in electrostatic-type interactions that are strong enough to compete with hydrogen bonds, in addition to their better documented abilities to serve as electron withdrawing substituents or their supposed "hydrophilic" properties, would contribute to the design of ligands by providing a framework for the use of this interatomic interaction.

This type of bond has already been utilized to form supramolecular assemblies. It is capable of inducing conformational perturbations and therefore, with proper preorganization, design parameters of supramolecular assemblies can be expanded.

10.2.4 Hydrogen Bonds and Living Things

The effect of hydrogen bonding on living systems is fundamental to life itself [49]. The body temperature of humans is 36.1–37.8°C. Reduction in H-bond strength would bring about the following ramifications (at body temperature): Reduction in hydrogen bond strength (by 29%) would cause water to boil at 37°C; by 18%, proteins would denature; and by 11%, the potassium ion becomes *kosmotropic* (or *chaotropic*, the ability to destabilize H-bonding and hydrophobic interactions—the more negative ΔG_{hyd}, the more kosmotropic the salt becomes); and a decrease in H-bond strength by 7% would result in increased pK_w by a factor of 3; by 5%, CO_2 would be 70% less soluble.

Increase in H-bond strength would effect the following changes (at body temperature): an increase of 2% would cause significant metabolic changes; by 3%, viscosity increases by 23% and diffusivity is reduced by 19%; by 5%, O_2 would be 270% more soluble; by 5%, CO_2 would be 440% more soluble; by 7%, pK_w would decrease by 1.7; by 11%, Na^+ becomes chaotropic; by 18%, water freezes; and by 51%, many proteins cold denature [48]. There is, of course, a wealth of information about the hydrogen bond but due to the limited space, we must by necessity move on to the next category of bonding. We recommend that you independently pursue more information about the hydrogen bond and its importance in chemistry and biology—and to skating on frozen lake surfaces [50].

10.3 Van der Waals Attractions

In 1873, J. D. van der Waals (**Fig. 10.9**), a Dutch scientist, noticed that the pressure of a real gas was lower than the value predicted by the ideal gas relation. He

Fig. 10.9 *Johannes Diderik van der Waals was awarded the Nobel Prize in physics in 1910 for his contributions to understanding equations of state for gases and liquids—the continuity of the gas and liquid state. Indeed, van der Waals showed in 1893 that there is a material continuum among the states of matter. In his famous equation, he showed that van der Waals forces need to be considered in describing the physical state of a gas (e.g., its deviation from ideality). In his law of corresponding states he developed a general form of the equation of states. Van der Waals died in 1923.*

surmised that this was due to attractive and repulsive forces between atoms and molecules of the gas—a condition not addressed by the ideal gas laws. Van der Waals was awarded the Nobel Prize in physics in 1910 for his work on the equation of state for gases and liquids. His equation of state for homogeneous substances consists of the usual pressure, volume, and temperature correlations, but also includes constant factors, characteristics of a specific gas, that offer corrections to the ideal gas law (e.g., nonideality due to deviations from the action of intermolecular forces) and that atoms and molecules have nonzero volume. These weak forces are known as van der Waals forces and they constitute an important category of intermolecular attractions, especially if summed over many molecules. The generalized van der Waals equation is an expanded version of the well-known $PV = nRT$ form. The term n is the number of moles of gas, and subtraction of the product nb from V corrects for intermolecular repulsion [51].

$$\left[P+\left(\frac{an^2}{V^2}\right)\right](V-nb)=nRT \tag{10.18}$$

The term an^2/V^2 is a measure of the intermolecular attraction in the system [51]. Familiar forms of this equation employ a molar volume V_m term. The constants a (units of $L^2 \cdot atm \cdot mol^{-2}$) and b (units of $L \cdot mol^{-1}$) are typical of a specific gas, and the equation applies appropriately to spherical atoms or molecules. Van der Waals attractions arise due to the polarization of an electron cloud of an atom or molecule by a nearby nucleus that results in an electrostatic attraction that is <5 $kJ \cdot mol^{-1}$ in strength [9]. *VDW* interactions are weak intermolecular or particulate forces with $1/r^6$ dependence. Transient fluctuating dipoles are *VDW* interactions that contribute to stabilize hydrophobic guests in hydrophobic pockets [8].

10.3.1 Contributions to the van der Waals Interaction

Van der Waals forces can be broken down into three contributions: (1) the Keesom-*vdw* attraction is a dipole–dipole angle-averaged orientation that involves electrostatic interactions between charges such as ions, dipoles, quadrupoles, and other permanent multipoles; (2) the Debye-*vdw* interaction involves the process of attraction by induction (polarization). In this case, a molecule with a permanent charge or dipole induces a dipole (or multipole) in another molecule within proximity; and (3) the final type of *vdw* attraction is the London-*vdw* (or dispersion) force that relies on transient dipoles for interaction. London dispersion forces are the only attraction force experienced by the inert gases. Except for the case of the inert gases, all of these *vdw* forces depend on the shape (iso- or anisotropic charge distribution) and orientation of the participating molecules (**Table 10.11**).

The Keesom type of *vdw* interaction involves electrostatic forces that can be attractive or repulsive depending on the rotational configuration of the participating molecules. Due to thermal motion, electrostatic forces based on the Keesom form are thermally averaged due to molecular rotations that limit the strength of the interaction. Keesom and Debye interactions abide by classical electrostatic (Coulombic) and Bolztmann distribution laws [52]. London dispersion phenomena require quantum mechanical explanations [52]. The Keesom form for two dissimilar molecules is given by

$$\mathrm{V_P}(r) = -\frac{\mu_1^2\mu_2^2}{3(4\pi\varepsilon_o)^2 k_B \mathrm{T} r_{12}^6} = \frac{C_P}{r^6} \tag{10.19}$$

where μ is the dipole moment of the species and C_P is the Keesom dipolar orientation interaction coefficient. The Debye dipole–induced dipole interaction is expressed as

$$\mathrm{V_I}(r) = -\frac{\mu_1^2\alpha_2 + \mu_2^2\alpha_1}{(4\pi\varepsilon_o)^2 r^6} = \frac{C_1}{r^6} \tag{10.20}$$

where α is the polarizability of the species and C_1 is the Debye-induced interaction coefficient.

London dispersion is represented by

$$\mathrm{V_d}(r) = -\frac{3\alpha_1\alpha_2}{2(4\pi\varepsilon_o)^2 r^6}\left(\frac{I_1 I_2}{I_1 + I_2}\right) = \frac{C_\mathrm{d}}{r^6} \tag{10.21}$$

Table 10.11 *Van der Waals Interactions*

Description	Example	Bond strength ($kJ \cdot mol^{-1}$)
Keesom forces	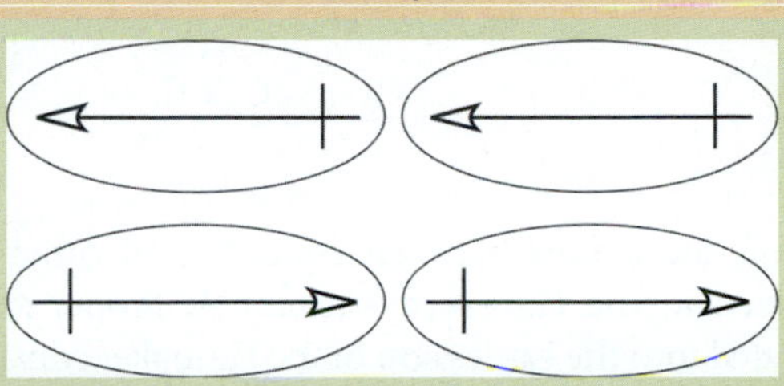 *Keesom type van der Waals alignment. Permanent dipole–permanent dipole interactions are capable of repulsive interactions upon rotation.*	0.4–4.0
Debye forces	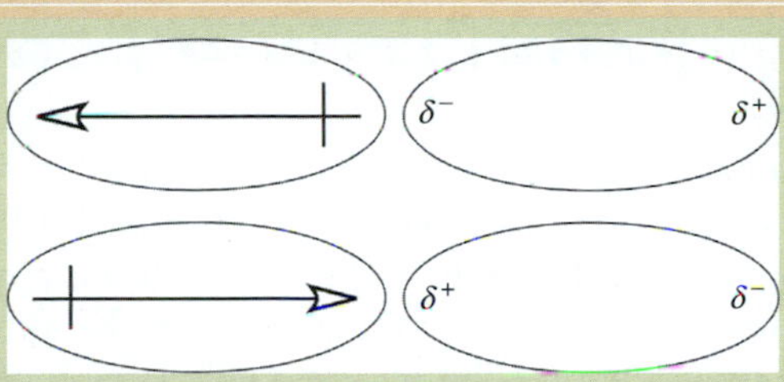 *Debye type van der Waals permanent dipole–induced dipole interaction.*	(<5)
London dispersion forces	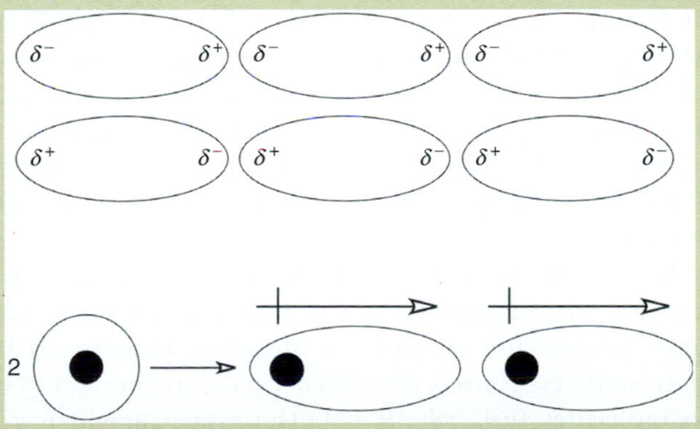 *London type (dispersion) van der Waals induced dipole–induced dipole (above) and transitory induced dipole (below) interactions.*	<<5
Hamaker constant	Hamaker constants are combinations of all van der Waals forces that are used to evaluate colloid–colloid, colloid–surface, and similar interactions. The Hamaker constant is shape dependent and specific for materials. Most Hamaker constants are on the order of 10^{-19}–10^{-20}J.	10^{-19}–10^{-20} J

where I_n represents the ionization potential of the species. The link to quantum mechanics is made through the ionization potential term. The total van der Waals interaction is the sum of the three contributions:

$$V_W(r) = \frac{C_W}{r^6} = \frac{C_P + C_I + C_d}{r^6} \tag{10.22}$$

The *vdw* attractions between colloids and other solubilized materials are summarized in the Hamaker constant *A*. Proper shape-dependent terms are incorporated into the expression of the Hamaker constant. A complete derivation of the Hamaker constant is beyond the scope of this text. A generalized relation appropriate for spherical particles is [53,54]

$$V_W(r) = \frac{A}{6}\left[\left(\frac{2R^2}{d^2 + 4Rd}\right) + \left(\frac{2R^2}{d^2 + 4Rd + 4d^2}\right) + \ln\left(\frac{d^2 + 4Rd}{d^2 + 4Rd + 4d^2}\right)\right] \tag{10.23}$$

All Hamaker constants range from 10^{-19} to 10^{-20} J. The Hamaker constants of water (4.4×10^{-20} J), toluene (6.3×10^{-20} J), metals ($25–40 \times 10^{-20}$ J), quartz (8.7×10^{-20} J), and CCl_4 (5.5×10^{-20} J) serve as a few examples [52].

Of course, we are not interested per se in the attractions between gases but rather of molecules in solution, especially large molecules. Depending on the size, shape, proximity, and orientation of one molecule to another, physical properties are expected to be affected, most notably the boiling point of liquids and the stability of monolayers. Since all atoms and molecules are impacted by van der Waals forces, it is considered to be a universal factor in intermolecular bonding. In nature, the humble gecko (chapter 13) has mastered the ability to cling to ceilings, whether wet or dry, due to capillary forces or van der Waals forces, respectively. At the nanoscale, van der Waals forces assume a greater level of importance than they do at the bulk scale; small molecules or clusters are influenced greatly by van der Waals forces and must be considered in the design of devices.

The Lennard–Jones potential relationship gives us a good feel for van der Waals interactions and it serves as a good first approximation of the potential energy between any two atoms or molecules that approach each other. Van der Waals energy is inversely dependent on the sixth power of the distance between two molecules (e.g., $vdw \propto r^{-6}$). The same dependency is exhibited by the attractive force component in the Lennard–Jones potential treatment. The van der Waals radius of atoms can be calculated from the generalized *Lennard–Jones 6–12* equation. Some van der Waals radii are listed in **Table 10.12.** Reading towards the right in the periodic table, *vdw* radii decrease. As we proceed down a group, *vdw* radii in general increase in magnitude. As a first approximation for molecular dynamic simulations, atoms cannot approach each other closer than their *vdw* radius allows. The van der Waals radius is the equilibrium distance-closest approach between two atoms or molecules. Generic hard-sphere van der Waals radii are shown in **Figure 10.10.** Space-filling models generated by computer programs do a great job in approximating the van der Waals surface. Selected examples of space-filling models of large molecules are shown in **Figure 10.11a** and **b.**

TABLE 10.12 *Van der Waals Radii of Some Common Elements*

Atom	Radius (Å)	Atom	Radius (Å)
H	1.20	Te	2.06
C	1.70	As	1.85
N	1.55	Sb	2.2
O	1.52	Cl	1.75
F	1.47	Br	1.85
S	1.85	I	1.98
Se	1.90	P	1.80
Si	2.10	Ni	1.63
Na	2.27	Pb	2.02
K	2.75	Xe	2.16
Au	1.66	Zn	1.39

Source: A. Bondi, *Journal of Physical Chemistry*, 68, 441–451, 1964.

10.3.2 Van der Waals Radius

Van der Waals forces involve the interaction between closed-shell molecules [55]. They include interactions between partially distributed electrical charges of polar molecules and the repulsive interactions (Pauli exclusion) as molecules approach within the van der Waals radius. Although repulsive forces are indeed important (very important), we focus primarily upon the attractive variety of van der Waals interactions. The induced dipole is the fundamental electrostatic entity involved in van der Waals interactions, and there are many forms of van der Waals forces. Strictly speaking, we will refer to the generalized van der Waals force that arises from the polarization of the electronic space of a molecule by a nearby nucleus that results in fluctuating dipoles, multipoles, octupoles, etc. [9].

10.3.3 Physical Property Dependence

Van der Waals forces are roughly proportional to the number of atoms, and hence the number of electrons, incorporated into a molecular structure. As we

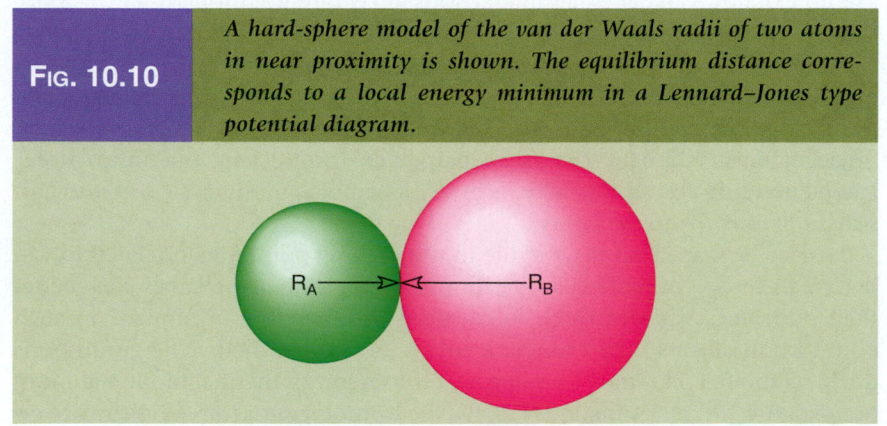

FIG. 10.10 *A hard-sphere model of the van der Waals radii of two atoms in near proximity is shown. The equilibrium distance corresponds to a local energy minimum in a Lennard–Jones type potential diagram.*

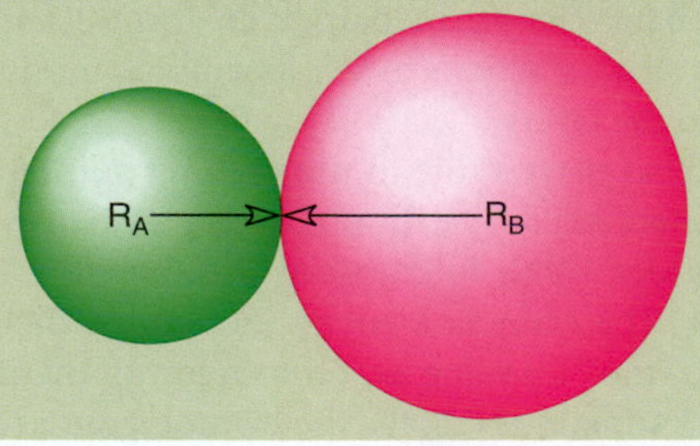

FIG. 10.11 *The structure of vitamin A is shown by two illustrations: (a) by a succinct chemical formula and (b) in the form of a space-filling model. Computer models are also adept at showing the van der Waals surface—a somewhat better rendition of the molecular orbital that envelopes the entire molecule.*

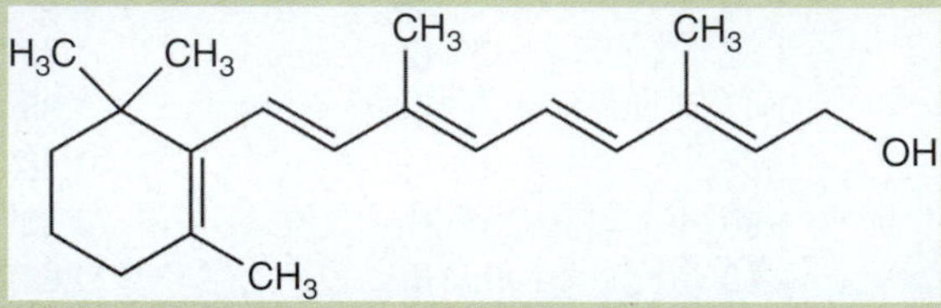

(a)

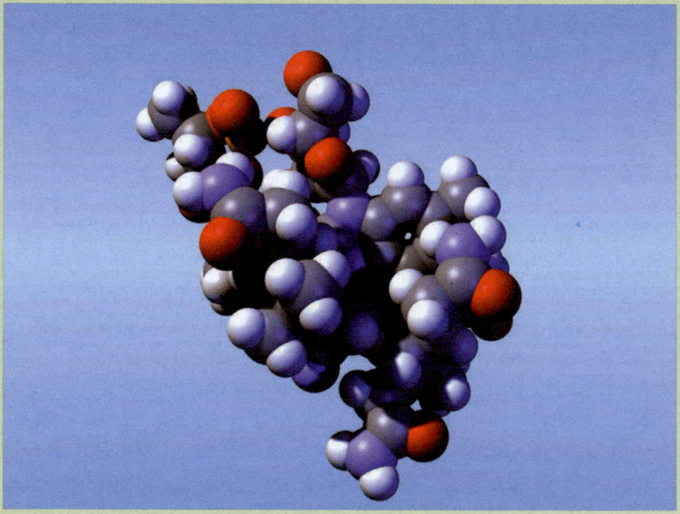

(b)

demonstrated earlier with the boiling point of small hydrides, the boiling points of linear methane series (paraffin series) alkanes (C_nH_{2n+2}) serve as an excellent example of how van der Waals interactions affect physical properties [57]. Since these molecules have extremely limited polarity, van der Waals attractions are expected to contribute to most of the interbonding potential energy. The boiling point of *n*-alkanes will be higher than their branched-chain counterparts due to the greater common surface area (better fit) shared among linear molecules. Adjacent linear aliphatic chains, in particular, form strong *vdw* interactions. Lateral bonding via *vdw* interactions in self-assembled monolayers serves to give structural integrity to the layers.

Molecular shape also plays a role in the melting point (again related to van der Waals forces) of materials. Specifically, even numbered alkanes pack better than odd ones. Van der Waals interactions are therefore maximized in such even-carbon alkanes. The estimated energy per added methylene group boils down to about 3.5 $kJ \cdot mol^{-1}$, a value near the energy of thermal motion at room temperature. Others estimate ΔH°_{vap} for the homologous series of alkanes to be 4.9 $kJ \cdot mol^{-1}$ per added methylene group [58] (**Table 10.13**).

Example 10.9 Van der Waals Radius Calculation

Calculate the VDW radius, r_{VDW}, for the water molecule and compare to that of carbon tetrachloride. What is a major assumption of this equation and what are the major limitations of this method? Which is a better approximation?

Solution:

Use the following relationship:

$$\frac{4}{3}\pi r_{VDW}^3 = f_P\left(\frac{M}{\mathcal{N}_A \rho}\right) \tag{10.24}$$

where M is the molecular weight, $\mathcal{N}_A$ is the Avogadro number, ρ is the density of water, and f_P is an estimated packing fraction of liquid water (equal to 0.7405 assuming 12 nearest neighbors—chapter 5).

The equation assumes a rather hard-sphere model and close packing.

For water:

$$r_{VDW} = \sqrt[3]{0.7405\left(\frac{3}{4\pi}\right)\left(\frac{M}{\mathcal{N}_A \rho}\right)} = \sqrt[3]{0.7405\left(\frac{3}{4\pi}\right)\left(\frac{18\,\text{g}\cdot\text{mol}^{-1}}{6.022\times10^{23}\text{mol}^{-1}\times1.00\,\text{g}\cdot\text{cm}^3}\right)} = 1.74\,\text{Å}$$

For carbon tetrachloride:

$$r_{VDW} = \sqrt[3]{0.7405\left(\frac{3}{4\pi}\right)\left(\frac{M}{\mathcal{N}_A \rho}\right)} = \sqrt[3]{0.7405\left(\frac{3}{4\pi}\right)\left(\frac{153.82\,\text{g}\cdot\text{mol}^{-1}}{6.022\times10^{23}\text{mol}^{-1}\times1.5867\,\text{g}\cdot\text{cm}^3}\right)} = 3.05\,\text{Å}$$

CCl_4 is more purely tetrahedral than water and by default better approximates a sphere. In both cases the VDW radius is overestimated by ca. 30%. The molecular diameter of water is about 2.75 Å and from the preceding equation, the derived molecular diameter is 3.48 Å. Also recall that both water and carbon tetrachloride are bonded with strong covalent bonds. This also helps to explain the serious deviation from experimental results.

Better packing parameters are available for water (0.63 → 1.65 Å) and carbon tetrachloride (0.53 → 2.74 Å).

Table 10.13 Effect of Size on Physical Properties of n-Alkanes

n-Alkane	Molecular weight	Melting point	Boiling point	ΔH_{vap}
Methane	16.04	−182	−164	8.91
Ethane	30.07	−183.3	−88.6	15.6
Propane	44.11	−189.7	−42.1	19.0
Butane	58.12	−138.4	−0.5	24.3
Pentane	72.15	−130	36.1	27.6
Hexane	86.18	−95	69	31.9
Heptane	100.21	−90.6	98.4	37.4
Octane	114.23	−56.8	125.7	38.6
Nonane	126.26	−51	150.8	43.8
Decane	142.29	−29.7	174.1	45.7
Undecane	156.32	−25.6	196	48.0
Dodecane	170.34	−9.6	216.3	49.6
Tridecane	184.37	−5.5	234.4	54.4
Tetradecane	198.40	5.9	253.7	57.5
Pentadecane	212.42	10	270.6	61.2

We understand that van der Waals forces are relatively weak and only exert their influence at very short distances (proportional to r^{-6}) for closed-shell molecules.

Van der Waals forces are active in stabilizing the lateral attractions experienced by nitrogen molecules in BET analysis. They are not so important arguably, with regard to supramolecular design because of their inherent lack of directionality [9]. Van der Waals interactions are a factor, however, in molecular cavities that trap small organic molecules such as the van der Waals inclusion complex toluene by *p-tert*-butylcalix[4]arene [9]. In ever-complex supramolecular clusters, it would be nearly impossible to predict the sum total of all van der Waals forces. For other systems, the van der Waals contributions are well known. Van der Waals interactions bundled into the Hamaker constant serve as a valuable tool in predicting the physical properties of colloidal systems. Discussion of the shape-dependent forms of the Hamaker constant is beyond the scope of this text. An example of a van der Waals inclusion complex molecule is shown in **Figure 10.12.**

Van der Waals forces, although not of the highest priority in supramolecular design, are important in polymer chemistry. Van der Waals interactions are important in the chemistry of macromolecules and self-assembled molecular assembly, but they also are important in the stabilization of fullerene clusters (and carbon nanotube bundles). Both fullerenes and nanotubes are formed in an uncharged state. The surfaces of both are kinetically stable. Single-walled carbon nanotubes in particular are extremely difficult to dissolve, especially if they are bundled (e.g., they behave like a *supermolecule*). Large molecules in general are more difficult to solvate than smaller ones. A bundle of SWNTs may

FIG. 10.12 *A van der Waals inclusion complex (p-tert-butylcalix[4]arene) is depicted. This guest molecule is able to accommodate, for example, toluene within its cavity. Van der Waals attractions are nondirectional and therefore a high level of specificity between guest and host is not expected.*

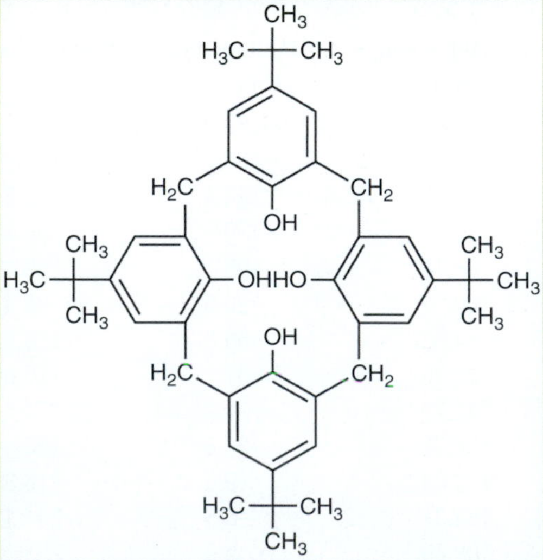

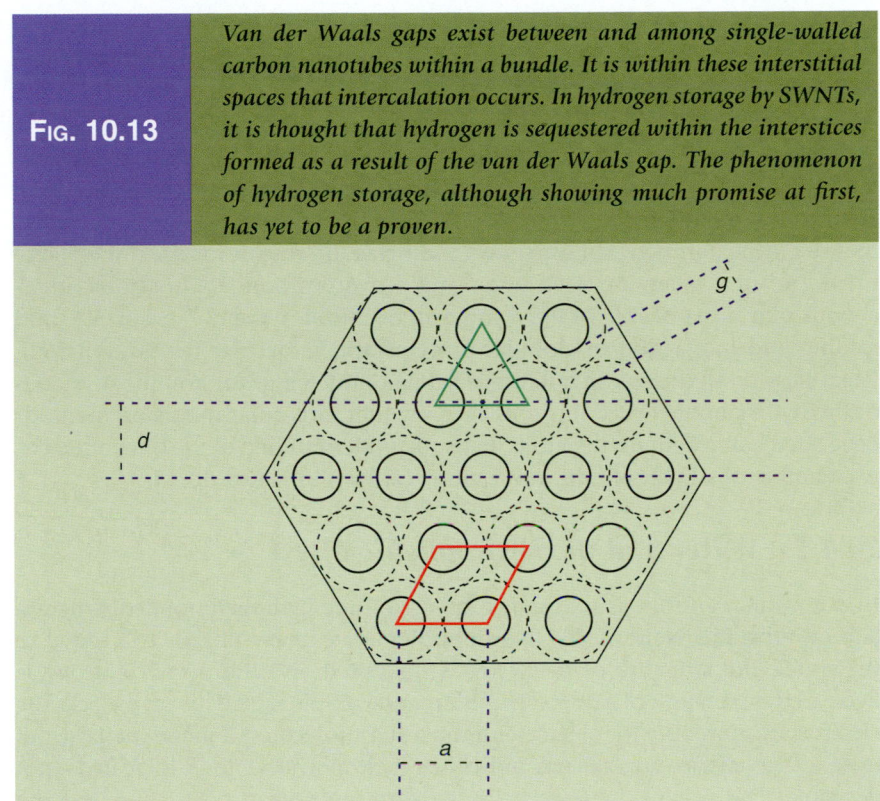

FIG. 10.13 *Van der Waals gaps exist between and among single-walled carbon nanotubes within a bundle. It is within these interstitial spaces that intercalation occurs. In hydrogen storage by SWNTs, it is thought that hydrogen is sequestered within the interstices formed as a result of the van der Waals gap. The phenomenon of hydrogen storage, although showing much promise at first, has yet to be a proven.*

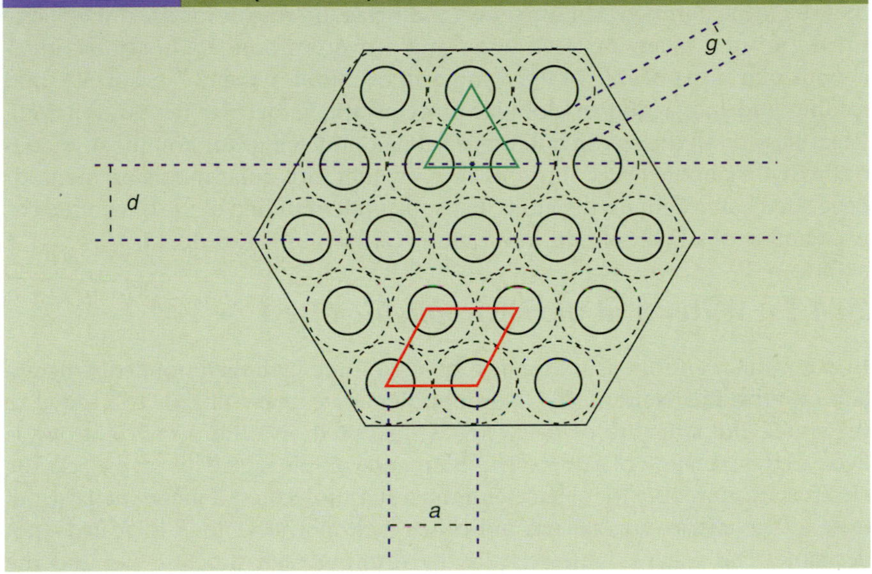

consist of tens to hundreds of carbon nanotubes that are many microns in length (even millimeters and centimeters in length). Trying to disperse the bundle into individual nanotubes is a formidable task that requires extra energy to accomplish. A phenomenon known as the *van der Waals gap* is commonly applied to the distance between fullerenes in the solid state and bundled SWNTs. The van der Waals gap is equal to 3.15 Å in both cases. Fullerenes crystallize in a three-dimensional *fcc* close-packed structure while SWNTs bundle in a two-dimensional trigonal lattice (**Fig. 10.13**). The van der Waals gaps within the nanotube bundles serve as sites for intercalation.

10.4 Hydrophobic Effect

The hydrophobic effect is one of the least understood but a very significant noncovalent interaction—especially for biological materials. Entropy-driven processes play a major role in forming hydrophobic-coupled structures.

10.4.1 Background

G. S. Hartley first described the hydrophobic effect in 1936:

> The antipathy of the paraffin chain in water is, however, frequently misunderstood. There is no question of actual repulsion between individual water

molecules and paraffin chains; nor is there any very strong attraction of paraffin chains for one another. There is however, a very strong attraction of water molecules for one another in comparison with which the paraffin-paraffin or paraffin-water attractions are slight.

Tanford in 1973 coined the phrase hydrophobic effect [59,60]. A little bit later, Hildebrand asked if there was such a thing as a hydrophobic effect [61]. Tanford essentially agreed with Hartley's observation given previously. He stated that the attraction between water and itself was the reason for the unfavorable interfacial free energy between water and a hydrocarbon. Hydrophobic interactions play a major role in the formation of micelles, membranes, DNA, and folding and interaction in globular proteins and molecular recognition [62]. Regardless of all the research that has been conducted to unravel the mysteries of the hydrophobic effect, it is only recently that molecular modeling methods have shed some light on the molecular scale understanding of the phenomenon. An example of a hydrophobic interaction is shown in **Figure 10.14.**

10.4.2 Water and the Hydrophobic Effect

In pure water, entropy is maximized by its unique hydrogen-bonded structure and electrostatic relations. If a hydrophobic molecule or substance is added to the water, the structure of the water is disrupted and the overall entropy is decreased by creation of a cavity. This is an unfavorable state. This cavity lacks the electrostatic potential to interact with the water molecules. If many such cavities exist, a high surface area of such individual hydrophobic entities is created—not desirable. The water counteracts this by forming cage structures around the

FIG. 10.14 *A hydrophobic interaction with host–guest specificity is depicted. When a hydrophobic surface is exposed to an aqueous medium, water molecules form an organized layer on the surface. Water molecules undergo tangential organization of dipoles on the hydrophobic surface. (D.T. Bowron, J. L. Finney, and A. K. Soper, Journal of Physical Chemistry, 102, 3551–3563, 1998.) Displacement of the water molecules by another hydrophobic entity raises the entropy of the systems—the increase in entropy gained by the removal of hydrophobic surface area from ordered solvating water. This phenomenon and water's intrinsic attraction to other water molecules act as drivers in hydrophobic organization. Hydrophobic interactions are one of the most important noncovalent intermolecular forces in the folding of linear polypeptides.*

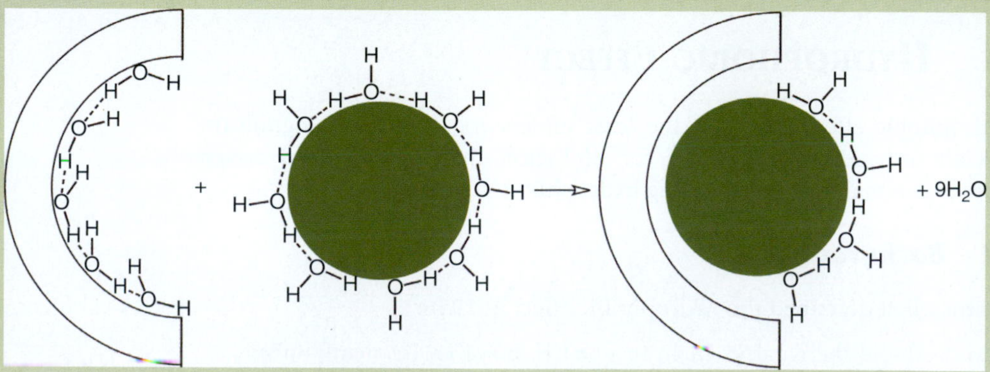

Table 10.14 *Hydrophobic Effects*

Description	Example	Bond strength (kJ·mol^{-1})
Micelles Lipid bilayers Hydrophobic pockets	Micelles Lipid bilayers 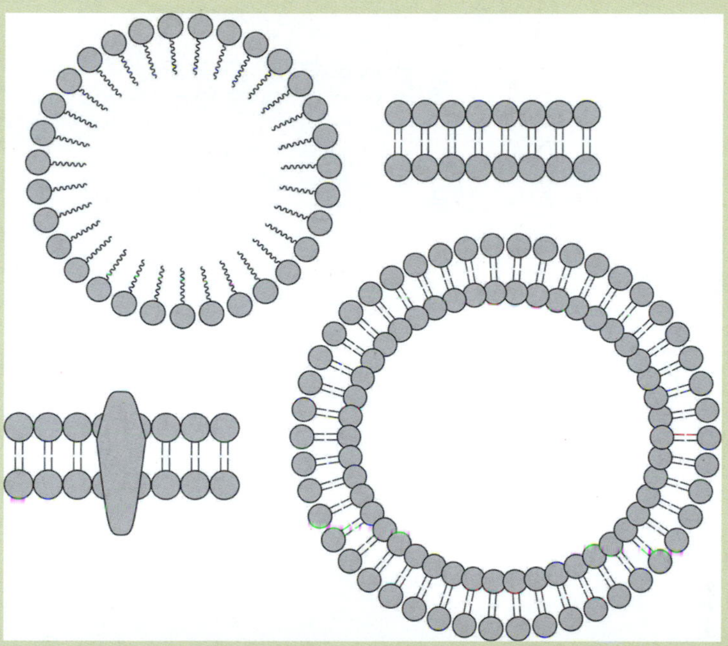 *Attraction of amphiphiles in aqueous media forms micelles (from the Latin mica, "crumb") and lipid bilayers. Ordered water in a hydrophobic pocket or surface when released by a suitable guest raises the entropy of the system and serves as a driving force to form micelle structures. (P. D. Beer et al., Supramolecular chemistry, Oxford Scientific Publications, Oxford University Press, Oxford, 2003. With permission.)*	0–0.2

hydrophobic molecules or inspiring bubbles to coalesce (reduce collective surface area). This action lowers the overall surface area (**Table 10.14**).

Recent work has shown that small and large hydrophobic solutes are hydrated in different ways [62]. Pressure, temperature, and additives all affect the hydrophobic interaction, regardless of the size of the hydrophobic solutes. However, the behavior of the solute–water system is a function of the length scale of the hydrophobic moiety [62] (e.g., small solutes have different hydration thermodynamics than do large solutes). For example, small solutes exhibit entropic effects while large solutes exhibit enthalpic effects. Apparently, according to simulations, there is a crossover state of nanoscopic size that consists of both molecular and macroscopic hydration thermodynamics [62]. This is important because biological and supramolecular assemblies include the integration of molecules as small as amino acids and as large as proteins. Chaotropic substances such as urea and guanidinium chloride increase the solubility

Table 10.15 *Hydrophobic Character Scales of Amino Acids*

Amino acid	Mol. wt.	Rating				Amino acid	Mol. wt.	Rating			
		[64][a]	[65][b]	[66][c]	[67][d]			[64][a]	[65][b]	[66][c]	[67][d]
ALA[1]	89	1.80	0.42	8.11	−2.09	LEU[1]	131	3.80	1.80	9.54	−7.45
ARG[2]	174	−4.50	−1.56	−83.3	12.6	LYS[2]	146	−3.90	−2.03	−39.8	12.6
ASN[3]	132	−3.50	−1.03	−40.5	0.837	MET[1]	149	1.90	1.18	−6.19	−5.44
ASP[4]	133	−3.50	−0.51	−45.8	12.6	PHE[1]	165	2.80	1.74	−3.18	−10.5
CYS[5]	121	2.50	0.84	−5.19	−4.18	PRO[1]	115	−1.60	0.86	0.00	0.00
GLN[3]	146	−3.50	−0.96	−39.2	0.837	SER[3]	105	−0.80	−0.64	−21.2	1.26
GLU[4]	147	−3.50	−0.37	−42.7	12.6	THR[3]	119	−0.70	−0.26	−20.4	−1.67
GLY[1]	75	−0.40	0.00	10.0	0.00	TRP[5]	204	−0.90	1.46	−24.6	−14.2
HIS[2]	155	−3.20	−2.28	−43.0	−2.09	TYR[3]	181	−1.30	0.51	−25.6	−9.62
ILE[1]	131	4.50	1.81	9.00	−7.45	VAL[1]	117	4.20	1.34	8.32	−6.28

Note: [1] neutral, nonpolar; [2] basic, polar; [3] neutral, polar; [4] acidic, polar; [5] neutral, slightly polar.

[a] Value determined by hydropathicity.

[b] Value determined by HPLC.

[c] Value determined by hydration potential (kJ·mol^{-1}).

[d] Value based on hydrophilicity.

of hydrophobic particles in aqueous media. These materials have the capability to destabilize aggregations of nonpolar solute particles and micelles and to denature proteins [63].

10.4.3 Amino Acids and Proteins

There are numerous *hydrophobicity scales* that catalog amino acids with regard to their hydrophobic and, conversely hydrophilic, natures (**Table 10.15**). The purpose of the scales is to be able to predict or determine the hydrophobicity of proteins assembled of those amino acids. The hydrophobicity values for the scale of Kyte and Doolittle (1982) were derived from experimentally measured amino acid solubility in various organic solvents [64]. More detail will be afforded to the topic of amino acids and proteins in chapter 14.

The hydrophobicity of proteins can then be acquired by compiling an average hydrophobicity over all amino acids that comprise the protein [68]. Advanced calculations of *vdw* volume are used routinely in developing drug compounds [69]. The hydrophobic effect and cooperativity are figured in together in the calculation of binding constants [70].

There is no question about the importance of van der Waals interactions in all materials—especially in nanomaterials like colloids, proteins, and pharmaceuticals. Van der Waals forces are not the most significant interaction at the macroscale (unless you are a gecko), but they assume huge importance at the nanoscale. If only van der Waals himself could see how his fundamental concept has been developed [71].

References

1. H. Y. Erbil, *Surface chemistry of solid and liquid interfaces*, Blackwell Publishing, Oxford (2006).
2. Chemical bond—History, *Science Encyclopedia*, Vol. 2, http://science.jrank.org/collection/2/Gale-Encyclopedia-Science.html (2007).
3. L. Pauling, *The nature of the chemical bond*, Cornell University Press, Ithaca, NY (1960).
4. I. W. Hamley, *Introduction to soft matter: Polymers, colloids, amphiphiles and liquid crystals*, John Wiley & Sons, Ltd., West Sussex (2000).
5. J.-M. Lehn, *Supramolecular chemistry*, Weinheim, Germany: VCH (1995).
6. J.-M. Lehn, *Organic chemistry: Its language and its state of the art*, Proc. Centenary Geneva Conf., M.V. Kisakürek, VHCA, Basel; VCH, Weinheim (1993).
7. J.-M. Lehn, Cryptates: Inclusion complexes of macrocyclic receptor molecules, *Pure Applied Chemistry*, 50, 871–892 (1978).
8. P. D. Beer, P. A. Gale, and D. K. Smith, *Supramolecular chemistry*, Oxford Scientific Publications, Oxford University Press, Oxford (2003).
9. J. W. Steed and J. L. Atwood, *Supramolecular chemistry*, John Wiley & Sons, Ltd., Chichester (2000).
10. E. Deu, K. A. Koch, and J. F. Kirsch, The role of the conserved Lys68* Intersubunit salt bridge in aspartate aminotransferase kinetics: Multiple forced covariant amino acid substitutions in natural variants, *Protein Science*, 11, 1062–1072 (2002)
11. N. K. Vyas, M. N. Vyas, and F. A. Quiocho, Crystal structure of M tuberculosis ASB phosphate transport receptor: Specificity and charge compensation dominated by ion–dipole interactions, *Structure*, 11, 756–774 (2003).

12. G. T. Morgan and H. D. K. Drew, Researches on residual affinity and coordination. Part II, acetyl acetones of selenium and tellurium, *Journal of the Chemical Society*, 117, 1456–1465 (1920).
13. G. Schmid, Large clusters and colloids: Metals in the embryonic state, *Chemical Reviews*, 92, 1709–1727 (1992).
14. G. Schmid, ed., *Clusters and colloids: From theory to applications*, VCH, New York (1994).
15. G. Schmid and G. L. Hornyak, Metal clusters: New perspectives in future nanoelectronics, *Current Opinion in Solar State Materials Science*, 2, 204 (1997).
16. G. L. Hornyak, M. Kroell, R. Pugin, T. Sawitowski, G. Schmid, J-O. Bovin, H. Hofmeister, and S. Hopfe, Gold clusters and colloids in alumina membranes, *Chemistry: A European Journal*, 3, 1951 (1997).
17. I. Vidal, S. Mechor, I. Alkorta, J. Elguero, M. R. Sundberg, and J. A. Dobado, On the existence of α-agostic bonds: Bonding analyses of titanium alkyl complexes, *Organometals*, 25, 5638–5647 (2006).
18. H.-F. Zhang, M. Stender, R. Zhang, J. Li, and L.-S. Wang, Toward the solution of the tetrahedral Au_{20} cluster, *Journal of Physical Chemistry B*, 108, 12259–12263 (2004).
19. B. Olenyuk, A. Fechtenkötter, and P. J. Stang, Molecular architecture of cyclic nanostructures: Use of co-ordination chemistry in the building of supermolecules with predefined geometric shapes, *Journal of the Chemical Society, Dalton Transactions*, 1707–1728 (1998).
20. J. Y. Lee, S. J. Jung, S. Hong, and S. Kim, Double dative bond configuration: Pyrimidine on Ge(100), *Journal of Physical Chemistry B*, 109, 348–351 (2005).
21. J. J. da Silva and R. Frausto, The chelate effect redefined, *Journal of Chemical Education*, 60, 390–392 (1993).
22. C. A. Hunter and J. K. M. Saunders, The nature of π–π interactions, *Journal of the American Chemical Society*, 112, 5525–5534 (1990).
23. P. A. Brooksby, C. A. Hunter, A. J. McQuillan, D. H. Purvis, A. E. Rowan, R. J. Shannon, and R. Walsh, Supramolecular activation of *p*-benzoquinone, *Angewandte Chemie International Edition*, 33, 2489–2491 (1994).
24. G. Wilkinson, The iron sandwich. A recollection of the first four months, *Journal of Organometallic Chemistry*, 100, 273–278 (1975).
25. M. Kamieth, U. Burkert, P. S. Corbin, S. J. Dell, S.C. Zimmerman, and F.-G. Klärner, Molecular tweezers as synthetic receptors: Molecular recognition of electron-deficient aromatic substrates by chemically bonded stationary phases, *European Journal of Organic Chemistry*, 1999, 2741–2749 (1999).
26. G. A. Jeffrey, *An introduction to hydrogen bonding*, Oxford University Press, Oxford (1997).
27. O. Markovitch and N. Agmon, Structure and energetics of the hydronium hydration shells, *Journal of Physical Chemistry A*, 111, 2253–2256 (2007).
28. S. Aravinda, N. Shamala, A. Pramanik, C. Das, and P. Balaram, An unusual C–H···O hydrogen bond mediated reversal of polypeptide chain direction in an synthetic peptide helix, *Biochemical and Biophysical Research Communications*, 272, 933–936 (2000).
29. P. Auffinger, F. A. Hays, E. Westhof, and P. S. Ho, Halogen bonds in biological molecules, *Proceedings of the National Academy of Science, USA*, 101, 16789–16794 (2004).
30. J. Barciszewski, J. Jurczak, S. Porowski, T. Specht, and V. A. Erdmann, The role of water structure in conformational changes of nucleic acids in ambient and high pressure conditions, *European Journal of Biochemistry*, 260, 293–307 (1999).
31. H. Xu, H. A. Stern, and B. J. Berne, Can water polarizability be ignored in hydrogen bond kinetics? *Journal of Physical Chemistry*, 106, 2054–2060 (2002).
32. M. Tagliazucchi, D. Grumelli, and E. J. Calvo, Nanostructured modified electrodes: Role of ions and solvent flux in redox active polyelectrolyte multilayer films, *Physical Chemistry Chemical Physics*, 8, 5086–5095 (2006).

33. H. Liu, S. Murad, and C. J. Jameson, Ion permeation dynamics in carbon nanotubes, *Journal of Chemical Physics*, 125, 084713–084726 (2006).
34. A. A. Malani, Structure and dynamics of water under confinement, presentation, Indian Institute of Science, Bangalore (viewed 2007).
35. M. K. Singh and G. E. Walrafen, New method for determining H-bond energy: Fluorescence from neutral red in water compared to fluorescence from neutral red in benzene, *Journal of Solution Chemistry*, 34, 579–583 (2005).
36. C. B. Aakeröy, K. R. Seddon, and M. Leslie, Hydrogen-bonding contributions to the lattice energy of salts for second harmonic generation, *Structural Chemistry*, 3, 63–65 (1992).
37. T. Steiner, The hydrogen bond in the solid state, *Angewandte Chemie International Edition*, 41, 48–76 (2002).
38. L. F. Scatena, M. G. Brown, and G. L. Richmond, Water at hydrophobic surfaces: Weak hydrogen bonding and strong orientation effects, *Science*, 292, 908–912 (2001).
39. R. S. Drago, *Physical methods for chemists*, 2nd ed., Saunders College Publishing, Hartcourt-Brace-Jovanovich College Publishers, New York (1992).
40. A. Streitwieser and C. H. Heathcock, *Introduction to organic chemistry*, 3rd ed., Macmillan Publishing Company, New York (1985).
41. V. M. F. Hammer, E. Libowitsky, and G. R. Rossman, Single-crystal spectroscopy of very strong hydrogen bonds in pectolite, $NaCa_2\ Si_3O_8(OH)$, and serandite, $NaMn_2\ Si_3O_8(OH)$, *American Mineralogist*, 83, 569–576 (1998).
42. I. Ayala, J. J. P. Perry, J. Szczepanski, J. A. Tainer, M. T. Vala, H. S. Nick, and D. N. Silverman, Hydrogen bonding in human manganese superoxide dismutase containing 3-fluorotyrosine, *Biophysical Journal*, 89, 4171–4179 (2005).
43. C. L. Aronson, D. Beloskur, I. S. Frampton, J. McKie, and B. Burland, The effect of macromolecular architecture on functional group accessibility: Hydrogen bonding in blends containing phenolic photoresist polymers, *Polymer Bulletin*, 53, 413–424 (2005).
44. S.-I. Kawahara, K. Taira, M. Sekine, and T. Uchimaru, Evaluation of the hydrogen bond energy of base pairs formed between substituted 9-methlyadenine derivatives and 1-methyluracil by use of molecular orbital theory, *Nucleic Acids Symposium Series*, 44, 237–238 (2000).
45. G. R. Desiraju and T. Steiner, *The weak hydrogen bond*, Oxford University Press, Oxford (2001).
46. S. Scheiner, T. Kar, and Y. Gu, Strength of the CαH····O hydrogen bond of amino acid residues, *Journal of Biological Chemistry*, 276, 9832–9837 (2001).
47. G. R. Desiraju, Crystal gazing: Structure prediction and polymorphism, *Science*, 278, 404–405 (1997).
48. T. Steiner, The influence of C-H···O interactions on the conformation of methyl groups quantified from neutron diffraction data, *Journal of Physical Chemistry A*, 104, 433–435 (2000).
49. M. F. Chaplin, Water's hydrogen bond strength. In *Water of life: Counterfactual chemistry and fine-tuning in biochemistry*, R. M. Lynden-Bell, S. C. Morris, J. D. Barrow, J. L. Finney, and C. L. Harper, eds., manuscript in preparation (2007).
50. A. H. Nissan, The hydrogen bond strength of ice, *Nature*, 178, 1411–1412 (1956).
51. I. A. Levine, *Physical chemistry*, 3rd ed., McGraw–Hill Book Co., New York (1988).
52. H. Y. Erbil, *Surface chemistry of solid and liquid interfaces*, Blackwell Publishing, Oxford (2006).
53. G. Cao, *Nanostructures & nanomaterials: Synthesis, properties and applications*, Imperial College Press, Singapore (2004).
54. P. C. Hiemnz, *Principles of colloid and surface chemistry*, Marcel Dekker, New York (1977).
55. P. W. Atkins, *Physical chemistry*, 4th ed., W. H. Freeman & Co., New York (1990).

56. A. Bondi, van der Waals volumes and radii, *Journal of Physical Chemistry*, 68, 441–451 (1964).
57. R. C. Wiest, ed., *CRC handbook of chemistry and physics*, CRC Press, Boca Raton, FL (1985).
58. J. D. Dunitz and A. Gavezzotti, Attractions and repulsions in molecular crystals: What can be learned from the crystal structure of condensed ring aromatic hydrocarbons? *Accounts in Chemical Research*, 32, 677–684 (1999).
59. C. Tanford, *The hydrophobic effect*, Wiley, New York (1973).
60. C. Tanford, Interfacial free energy and the hydrophobic effect, *Proceedings of the National Academy of Science, USA*, 76, 4175–4176 (1979).
61. J. H. Hildebrand, Is there a hydrophobic effect? *Proceedings of the National Academy of Science, USA*, 76, 194 (1979).
62. S. Rajamani, T. M. Truskett, and S. Garde, Hydrophobic hydration from small to large length scales: Understanding and manipulating the crossover, *Proceedings of the National Academy of Science, USA*, 102, 9475–9480 (2005).
63. S. Moelbert and P. De Los Rios, Chaotropic effect and preferential binding in a hydrophobic interaction model, *Journal of Chemical Physics*, 119, 7988–8001 (2003).
64. J. Kyte and R. F. Doolittle, A simple method for displaying the hydropathic character of a protein, *Journal of Molecular Biology*, 157, 105–132 (1982).
65. R. Cowan and R. G. Whittaker, Hydrophobicity indices for amino acid residues as determined by HPLC, *Peptide Research*, 3, 75–80 (1990).
66. D. Ring, Y. Wolman, N. Friedmann, and S. L. Miller, Prebiotic synthesis of hydrophobic and protein amino acids, *Proceedings of the National Academy of Science, USA*, 69, 765–768 (1972).
67. T. P. Hopp and K. R. Woods, Prediction of protein antigenic determinants from amino acid sequences, *Proceedings of the National Academy of Science, USA*, 78, 3824–3828 (1981).
68. C. M. Roth, B. L. Neal, and A. M. Lenhoff, van der Waals interactions involving proteins, *Biophysics Journal*, 70, 977–987 (1996).
69. Y. H. Zhao, M. H. Abraham, and A. M. Zissimos, Fast calculation of van der Waals volume as a sum of atomic and bond contributions and its application to drug compounds, *Journal of Organic Chemistry*, 68, 7368–7373 (2003).
70. D. H. Williams and B. Bardsley, Estimating binding constants—The hydrophobic effect and cooperativity, *Perspectives in Drug Discovery and Design*, 17, 43–59 (2004).
71. J. D. van der Waals, *Over de Continuiteit van den Gas*—en. Vloeistoftoestand, thesis, University of Leiden (1873).
72. J. Pfeffer and S. Nir, *Modern physics: An introduction text*, World Scientific Publishing Company, Hackensack, NJ, 183 (2001).
73. M. Brookhart and M. L. H. Green, Carbon–hydrogen transition metal bonds, *Journal of Organometallic Chemistry*, 250, 395–408 (1983).
74. D. T. Bowron, J. L. Finney, and A. K. Soper, Structural investigation of solute–solute interactions in aqueous solutions of tertiary butanol, *Journal of Physical Chemistry*, 102, 3551–3563 (1998).

Problems

10.1 What is the ion–ion potential interaction energy for Na^+ and Cl^- if the ions are in water at 25°C at a distance of 3 nm? 2 nm? 1 nm?
Permittivity of free space, $\varepsilon_o = 8.854 \times 10^{-12}$ $C^2 \cdot J^{-1} \cdot m^{-1}$, $e = 1.6022 \times 10^{-19}$ C, $\varepsilon_{r,H_2O} = 78.85$.

10.2 How close must two attractive ions be in order to satisfy the 50–300 $kJ \cdot mol^{-1}$ range?

10.3 What ways can you think of to alleviate entropic factors in systems that have increased level of organization?
a. Hydrophobic entities in water?
b. Micelles?

10.4 The dipole of diethyl ketone is in line with a potassium cation at a distance of 0.5 nm. What is the energy of the interaction in water ($\varepsilon_r = 78.85$)? In methanol ($\varepsilon_r = 32.63$)? In trimethyl amine ($\varepsilon_r = 2.44$) at 298 K?

10.5 Which molecules have a permanent dipole moment? Why?

a. CH_4
b. CO
c. N_2
d. H_2O
e. CO_2
f. *n*-Heptane
g. HF

10.6 List, in order of lowest to highest boiling point, the following hydrides and explain.

a. HF, CH_4, H_2O, NH_3
b. H_2S, H_2Se, H_2Te, H_2O

10.7 Molecular shape is also an important consideration in van der Waals bonding. How would the boiling point of a solution filled with nonhomogeneously shaped molecules compare to one that is homogeneous?

10.8 How would the melting and boiling points of *n*-butane and methylpropane compare? Is there any correlation between mp, bp, and thermodynamic stability? The $\Delta H°$ of the reaction $CH_3CH_2CH_2CH_3 \rightarrow (CH_3)_2CH_2CH_3$ is -8.58 kJ·mol^{-1}.

10.9 What kinds of interactions do you expect between the following ion or molecular pairs? List them in order of increasing strength:

a. Dichloromethane and methanol
b. Octane and heptane
c. Bromide anion and octane
d. Na+ and water
e. Dimethyl ketone and heptane
f. Ethanol and water

10.10 Calculate the cohesion energy (binding energy) of a KBr crystal from a Born–Haber cycle. How does the value compare to that of NaCl? Why the difference? Experience the Born–Haber cycle for KBr by writing out all the equations involved. Feel free to use references.

10.11 Calculate the density of a fullerene crystal made of C_{60} that is fcc-close-packed structure. What is the packing fraction of the solid-state form of the C_{60}?

10.12 Rank the following ionic compounds in order of decreasing lattice energy and explain the trend:

a. NaCl—NaF—NaBr—Na_2O—NaI—$NaNO_3$
b. Al_2O_3—MgO—KCl—KBr—KI

10.13 Using your newly acquired knowledge of chemical bonding, why does acetone have a lower surface tension than water?

10.14 Calculate ΔG and the formation constants K_f (and $\log K_f$) for the following coordination reactions at 298 K and discuss the reasons for the differences. M = metal cation:

a. $M^{2+}{}_{(aq)} + 4NH_{3(aq)} \rightarrow [M(NH_3)_4]^{2+}_{(aq)}$
$\Delta H° = -53$ kJ·mol^{-1}, $\Delta S° = -42$ J·mol^{-1}·K^{-1}
b. $M^{2+}{}_{(aq)} + 4CH_3\text{–}NH_{2(aq)} \rightarrow [M(CH_3\text{–}NH_2)_4]^{2+}_{(aq)}$
$\Delta H° = -57$ kJ·mol^{-1}, $\Delta S° = -67$ J·mol^{-1}·K^{-1}
c. $M^{2+}{}_{(aq)} + 2(NH_2\text{–}CH_2\text{–}CH_2\text{–}NH_2)_{(aq)} \rightarrow [M(NH_2\text{–}CH_2\text{–}CH_2\text{–}NH_2)_2]^{2+}_{(aq)}$
$\Delta H° = -56$ kJ·mol^{-1}, $\Delta S° = 10$ J·mol^{-1}·K^{-1}

10.15 Calculate ΔG and the formation constant K_f (and $\log K_f$) for each of the following coordination reactions at 298 K in water and discuss the reasons for the differences. M = metal cation. How does the value for ethylenediamine (en) compare to the answers you derived in problem 11.3? What would account for the difference in the K_f between the metal chelated in problem 10.14?

a. $[M(H_2O)_6]^{2+}{}_{(aq)} + 2NH_{3(aq)} \rightarrow [M(H_2O)_4(NH_3)_2]^{2+}_{(aq)} + 2H_2O$
$\Delta H° = -46$ kJ·mol^{-1}, $\Delta S° = -8$ J·mol^{-1}·K^{-1}
b. $[M(H_2O)_6]^{2+}{}_{(aq)} + (NH_2\text{–}CH_2\text{–}CH_2\text{–}NH_2)_{(aq)} \rightarrow [M(H_2O)_4(NH_2\text{–}CH_2\text{–}CH_2\text{–}NH_2)]^{2+}_{(aq)} + 2H_2O$
$\Delta H° = -54$ kJ·mol^{-1}, $\Delta S° = 23$ J·mol^{-1}·K^{-1}

10.16 Answer the following questions with short sentences:

a. How do enzymes function? Give an example and describe the chemical processes (and bonding) involved.
b. What is a cofactor and what is its purpose? How is it bound? What is a coenzyme and what is its purpose. How is it bound? Give an example of each.
c. What is a heme group and how is it coordinated? What is the name of the generic type of bioorganic compound that coordinates Fe in living systems? How does it bind oxygen?
d. What is a vitamin? Please supply an example and describe its generic structure.
e. Of what is the tobacco mosaic virus composed? Is it a supramolecule? What kinds of bonds hold it together?
f. Would other metals besides magnesium function properly in the chlorophyll supramolecule? Why or why not?

10.17 Calculate the lattice energy of a KCl crystal by the Born–Haber method if $\Delta H^\circ_{f,\text{KCl}} = -438$ kJ·mol^{-1}, $\Delta H_{\text{sub,K}} = +89$ kJ·mol^{-1}, $\Delta H_{\text{ion,K}} = +425$ kJ·mol^{-1}, $\Delta H_{\text{diss,Cl2}} = +244$ kJ·mol^{-1}, $\Delta H_{\text{e-aff,Cl}} = +355$ kJ·mol^{-1}. How does the value compare to that found for NaCl? Why the difference?

10.18 In a nanotube bundle of uniform SWNTs, what force holds the bundle in place? What factors contribute to the integrity of the bundle?

10.19 Are all interactions between atoms and molecules electrostatic in nature?

10.20 We know that hydrogen bonds when taken collectively are quite strong. Is it possible to calculate the strength of H-bond from its ability to rupture strong containers when water freezes? How would you go about setting a value in this problem?

10.21 Hydrogen bonds are fundamental to life. Give as many examples as possible.

10.22 A linear amphiphile has a cross-sectional area of 50 Å^2. How much area of water surface, assuming tightest packing, would 8.0×10^{-2} mol of this species cover? How many moles would be required to cover Lake Michigan?

Supramolecular Chemistry

Chemistry is an ever expanding "universe" at the microscopic level requiring mastery of the invisible . . . at ever increasing levels of complexity.

Nicholas J. Turro,
Columbia University, 2005

Chapter 11

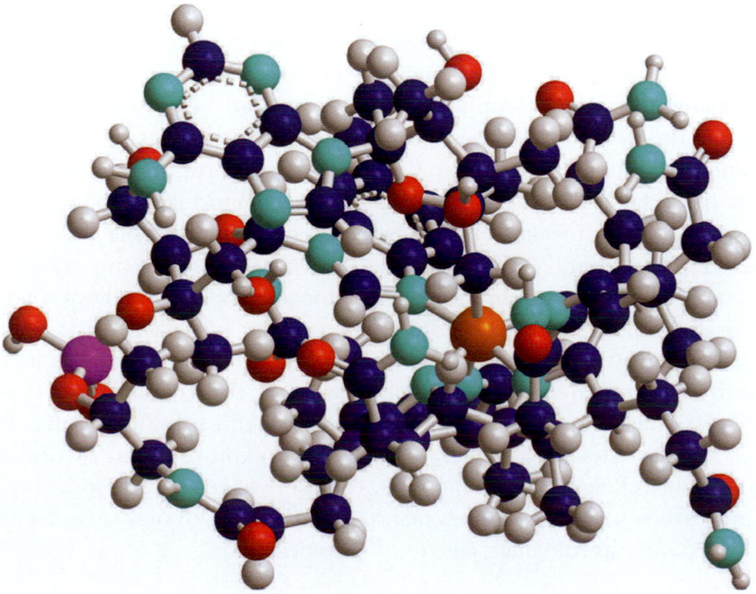

Fig. 11.0 *Nicholas Turro of Columbia University, renowned supramolecular chemist.*

Source: Image courtesy of Dr. Nicholas Turro.

Threads

Chapter 11 is about chemistry and synthesis at the nanoscale. How different would you expect the subject matter in this chapter to be from any other found in traditional chemistry textbooks? Chemistry, after all, is just chemistry. Traditional chemistry has always involved some semblance of the nanoscale, the best example of which is given by colloid chemistry. Chemistry at the nanoscale involves every kind of traditional chemistry. As we have stated before, human beings have dabbled in the chemistry of nanoscale materials for the past millennia or so, but we have had limited success in controlling the size, shape and orientation of nanomaterials—atleast until now. In the latter part of the twentieth century to the present, major breakthroughs have been accomplished in that regard. The key ingredient is to keep nanomaterials from morphing into larger materials or decomposing into smaller ones.

Supramolecular chemistry is a new field. Biochemistry is an old field. Colloid chemistry is an old field. Chemical functionalization of nanomaterials is new because the substrates are relatively new. Polymerization is an old technique. Nanocomposites, quantum dots, and carbon nanotubes are new materials. C, N, H, O, F, S, and P are old materials.

Our understanding of chemical bonding has expanded significantly in the past few years, and we plan deliberately, consciously, subliminally, and subconsciously (hopefully, not unconsciously) to make reference to chemical bonding and intermolecular interactions throughout the chapter; it is unavoidable to do otherwise. Chapter 12 takes on more chemistry topics: cluster synthesis, self-assembled monolayers, template synthesis, and more chemical modification of nanomaterials.

We can never, especially at this stage, sidestep our analogies to nature and learn from this ultimate tutor. Nature has incorporated, by the most ingenious of means, the entire continuum of chemical bonding into its wondrous biological products with the utmost in efficiency and under the mildest synthesis conditions imaginable—just how perfect is that? From nature we continue to learn how this is accomplished so that we can construct our own version of macro-, super-, and supramolecules. In chapters 13 and 14, we explore nature's world in more detail. The closer we get to biological topics, the more relevant societal issues seem to become. Keep them in mind.

11.0 Chemistry of Nanomaterials

Every discipline of chemistry applies with equal validity to nanoscience. Elements of organic, inorganic, organometallic, coordination, physical, bio, surface and interfacial, colloidal, electro, and even cosmo chemistry have ties to nanoscience. This chapter is about chemistry of the nanoscale and focuses on molecules and synthesis and in particular supermolecules, supramolecules, and their design. The focus is not so much upon properties and phenomena as in previous chapters, although we do not pledge to stay clear of such topics. However, we have limited time and limited space and must settle for this precursory format.

Chemistry of the nanoscale is all of traditional chemistry plus chemistry that is conducted on nanomaterial synthesis and the subsequent chemical derivatization on nanomaterial substrates. The chemistry involved in building hierarchical structures is also presented and discussed briefly. Nanoscale chemistry is a convergence of disciplines at the molecular level that includes an infusion of biochemistry and chemical engineering to create designer molecules. This chapter could easily be expanded into additional chapters that address supramolecular chemistry in more detail, self-assembly, and chemical functionalization. Unfortunately, we must be rather succinct in our presentation.

We start our journey with a generalized introduction to nanoscale chemistry, list types of chemical synthesis, and discuss thermodynamic versus kinetic control of reactions. Then we tackle the burgeoning field of supramolecular chemistry. Supramolecular chemistry is based on complementary synthesis of thermodynamically stable compounds that are held together by relatively weak intermolecular interactions. We have just had a taste of intermolecular interactions in the previous chapter, and we understand, at least at the introductory level, how the different types of chemical bonds work together to form macromolecular systems.

First, supramolecular chemistry is the chemistry of the intermolecular chemical bond. Secondly, reactions are designed to synthesize supramolecules proceed via thermodynamic control rather than kinetic control, a distinction that leads to thermodynamically stable products. Lastly, factors such as solvent effects and entropic considerations fundamentally have greater impact on the course of a reaction and overall product stability. We have had a small dose of each of these topics in previous chapters. There is no question that this exciting new field has roots in biology and organic chemistry, and from that unifying principle, we shall proceed to take our first steps.

In section 11.3, we provide a few outstanding examples of hierarchical supramolecular fabrication to demonstrate the interplay between intramolecular and intermolecular bonding. Traditionally, the chemistry of self-assembly is a subset of supramolecular chemistry but one that is mostly restricted to two-dimensional synthesis strategies (e.g., one monolayer at a time). The synthesis of self-assembled monolayers, however, does not rely exclusively on intermolecular bonds and in that regard does not fall completely under the supramolecular umbrella. Chemical functionalization is addressed more fully in the next chapter. Chemical functionalization involves the chemical modification of nanomaterials or the application of nanomaterials as building blocks with which to form three-dimensional hierarchical structures.

11.0.1 Background

German chemist Friedrich Wöhler synthesized urea (carbonyl diamide $[CO(NH_2)_2]$) in 1828. This accomplishment was the first experimental debunking of the *Vital Force* school of thought by demonstrating that organic compounds can be synthesized from inorganic starting materials (e.g., that synthesis of biological molecules is no longer exclusive to nature). The beginnings of *molecular chemistry* are attributed to the work of Wöhler. *Macromolecular chemistry* is the chemistry of larger molecules and has its roots in the early part of the twentieth century. Hermann Staudinger received the Nobel Prize in chemistry for his discoveries in the new field of macromolecular chemistry. He also was fundamental in laying the foundation for polymer chemistry by stating that "polymers were comprised of long-chains of repeating molecular units."

In 1894, Emil Fischer, another Nobel laureate, proposed the rigid enzyme *lock and key* model for enzyme action. The *lock* in the analogy is the enzyme and the *key* represents the substrate, usually a small molecule. He also contributed to understanding of the synthesis of sugars and purines. Paul Ehrlich developed the first antibiotic and popularized the terms *chemotherapy, magic bullet,* and *blood–brain barrier* and predicted the term *autoimmunity.* He investigated the affinity of chemical substances for use against antigens and laid down much of the foundation of modern immunology. In 1908, Paul Ehrlich shared the Nobel Prize in medicine with Ilya I. Mechnikov for their work on immunity. In the same period, Alfred Werner invented the field of coordination chemistry, for which he won the Nobel Prize for chemistry in 1913.

Irving Langmuir and Katherine Blodgett, mentioned briefly in chapter 1 and sporadically throughout the text, developed the science of self-assembled monolayers in 1938. Not surprisingly, the process is named after them; the Langmuir–Blodgett process involves the synthesis of monomolecular layers on the surfaces of water, metal, or glass. Blodgett later went on to develop a coating on glass that consisted of 44 monolayers of barium stearate. The result was a glass that is 99% nonreflective, appropriately named "invisible glass"—one of the first deliberate applications of nanomaterials. The development of invisible glass helped to mitigate distortion in eyeglasses and other optical equipment. Katherine Blodgett also used interference colors to measure the thickness of each layer as it was applied. In retrospect, Katherine Blodgett should have been awarded a Nobel Prize for the breakthroughs she ingeniously delivered to the world of science.

James D. Watson, Francis H. C. Crick, and Rosalind Franklin uncovered the structure of DNA in 1953. Watson and Crick received the Nobel Prize for their work in 1962. Franklin did not win the prize due to her untimely death in 1958 of cancer, coupled with the policy that the Nobel Prize is not awarded posthumously. DNA (aside from its fundamental role in life) is perhaps the most celebrated and vitally important supramolecule in the brief history of the field. DNA has also become one of the most interesting and useful nanomaterials.

Supramolecular chemistry officially began in the second half of the twentieth century with the work of Woodward, Pederson, Cram, and Lehn. Robert Woodward was awarded the Nobel Prize in chemistry in 1965 for his work in organic synthesis. In 1987, Charles Pedersen (crown ethers), Donald Cram (host–guest chemistry), and Jean-Marie Lehn (cryptands) were awarded the Nobel Prize in chemistry for development of complexes that demonstrated high selectivity and structural specificity. They are considered to be the founders of modern supramolecular chemistry. More recently, Nicholas Turro of Columbia University is widely recognized as a pioneer in the field of supramolecular chemistry (a.k.a. host–guest chemistry) [1,2].

Although the existence of fullerenes and their derivatives (the carbon nanotubes) was unveiled in the latter part of the twentieth century, they cannot technically be considered as supramolecules, at least not according to our upcoming definition. Rather, these large molecular structures must be designated as *supermolecules*: big versions of the same repeating small molecular (or atomic) unit that are held together throughout with strong covalent bonds. Colloids, sodium chloride, and gold clusters are not supramolecules either. These three materials are relatively homogeneous and are held together with covalent, ionic, or metallic bonds, respectively. The distinction, however, is another boundary that we shall certainly violate.

Colloid and surface scientists were the first researchers to be actively involved in supramolecularity and self-assembly [3]. Whereas colloids are not considered to be *designed* (not a fair assessment according to some), supramolecular structures are designed to assemble in a specific way that yields unique structure and function [3]. Colloid chemists and surface scientists laid down much of the foundation for nanoscience and deserve all the credit that is due. It was their discovery and work that established much of the theory and practice (supramolecular chemistry, self-assembly, the Hamaker constant to nucleation, and other relevant phenomena) with which we now are so familiar.

Self-assembly is a phenomenon we that is self-explanatory. The phrase implies that things, components, constituents, molecules, etc. all come together spontaneously without the input of energy or design. Some input energy is usually required, of course, but it is expressed at new levels of subtlety. The driving energy may be sequestered in the surface of a nanoparticle, in the form of a molecular recognition couple or within an excited state of a molecule. Rest assured, it is there. From what we have gleaned in previous chapters, we know that the making and breaking of relatively weak intermolecular bonds are controlled by entropic trade-offs and small ΔH_{Rxn}, thus giving the appearance of self-assembly. The following statement embodies the importance of self-assembly to nanoscience [3]:

"Self-assemblies are not merely beautiful structures (although this is certainly a source of their appeal); they allow alternative solutions to problems that have hitherto been addressed only at the single-molecule level." This is certainly well stated. We now have developed new routes to synthesize materials that were essentially "unsynthesizable" in the past. These new routes go through the zone of nanoscience.

Chemical functionalization techniques are not new. Many, if not all, synthesis techniques to produce and modify nanomaterials have roots in inorganic, organic, polymer, colloid, and biochemistry. For example, carbon nanotubes, in addition to DNA, are some of the most remarkable materials known to nanoscience, or science in general for that matter. Neat carbon nanotubes without chemical modification do not enhance the mechanical properties of nanocomposites. Scientists had to find ways to incorporate them into polymer matrices and other technological materials without loss of their remarkable properties. Various chemical modification schemes have been developed to do just that. Attachment of proteins to solid surfaces and modifying targeted groups is the basis of biological sensor technology.

In chapters 5–8 we addressed physical properties and phenomena, thermodynamics, and chemical bonding without much fanfare about synthesis. It is time for synthesis, from the bottom up, to carry the baton of nanoscience in the forward direction. Bottom-up chemical routes are becoming increasingly practical as better understanding of the forces that drive the assembly of nanomaterials is acquired. Although top-down fabrication will always be important and bottom-up chemical methods are still in the minority, bottom-up chemical synthesis methods are gaining ground fast. Atoms and molecules can be added batch-wise to surfaces to stabilize nanoparticles and to create entities with highly specific structure and functionality. Nature does, after all, accomplish wondrous things in this way. The importance of chemical methods cannot be understated.

Supramolecular chemistry is chemistry "beyond the molecule" [3–6]. It is the chemistry of the intermolecular bond [3–6]. The term *supramolecular chemistry* was introduced in 1978 by Nobel laureate Jean-Marie Lehn [6]: "Just as there is a field of molecular chemistry based on the covalent bond, there is a field of supramolecular chemistry, the chemistry of molecular assemblies and of the intermolecular bond."

11.0.2 Types of Chemical Synthesis

Although every type of chemical synthesis approach applies with validity to nanomaterials, nanoscale synthesis must be considered to be a specialized field—one that takes seriously the effects of small dimensions and the importance of forces that are considered to be relatively weak on larger scales. Although catalysts, colored glasses, and photoemulsions have been studied and used in industrial processes for centuries, the conscious and deliberate use of nanomaterials as precursor materials, substrates, or moderators in reactions to synthesize new materials is a relatively new undertaking. Today, many routes to nanomaterials and hierarchical structures go through the supramolecule or nanomaterial phase. Just as with any table that involves a catalog, the divisions, although not entirely arbitrary, are susceptible to interpretation. For the purposes of the text, consider them to be pedagogically oriented and susceptible to change over time. **Table 11.1** is directed towards students who have not had the blessing of a rigorous background in chemistry.

The purpose of this table is to help to construct a perspective about the vast field of nanochemistry. It is also provided to illustrate the extreme interdisciplinary nature of nanoscience. As you proceed in this text, try to correlate the subject matter with one or more of the disciplines listed in the table. Once again, it was Nobel Prize winner Jean-Marie Lehn who said in 1993 [5]:

> Definitions have a clear, precise core but often fuzzy borders, where interpenetration between areas takes place. These fuzzy regions in fact play a positive role

TABLE 11.1 *Chemical Synthesis—All Bottom-Up Methods*

Type of synthesis	Discipline[a]	Comments
Traditional synthesis-oriented fields of chemistry	Inorganic chemistry Coordination chemistry	Noncarbon chemistry. Basic coordination chemistry contributed to the founding of supramolecular chemistry. The use of ligands to stabilize metals also applies in nanoscience.
	Organometallic chemistry	Organometallic chemistry is a hybrid field. Supramolecular chemistry in nature and in the laboratory involve organometallic complexes that have functionality.
	Organic chemistry	The chemistry of carbon compounds. Synthesis of naturally occurring compounds by organic methods began over 100 years ago. For the most part, organic chemistry involves breaking and making covalent bonds and kinetic control of the reaction. Organic chemical synthesis is used to make precursors to supramolecular species.
	Polymer chemistry	Polymer chemistry is an offshoot of organic chemistry. It is the use of monomeric constituents to form a bulk material with specific properties. Nanoscale polymeric materials contribute to drug delivery systems. The field of nanocomposites is an offshoot of polymer chemistry.
	Biochemistry	Bottom-up synthesis with roots in biology is the primary focus of biomimetic technology. Supramolecular chemistry owes much of its existence to biochemistry and models many of its reactions on biochemical phenomena.
Specialized fields of traditional chemistry	Colloid chemistry	From several nanometers to several microns, colloids occupy a broad stage in science and nanoscience. Concepts of surface area and surface energy were first developed in colloid chemistry. Colloid chemists do not get the credit that is due them.
	Surface science	Surface science is the study of the physical chemistry of surfaces and interfaces, primarily those involving a solid–liquid or solid–gas interface. Surface science of nanomaterials is a new field.
	Catalysis	Although micro- and nanoscale catalysts have been on the scene for decades (centuries), the use of surfactant-stabilized catalysts of extremely small size is a relatively new development for hydrogenation and oxidation reactions. Materials not thought to be good catalysts, such as gold, have proven to be otherwise.
Recently developed fields of chemistry	Supramolecular chemistry	Supramolecular chemistry is the convergence of organic, inorganic, polymer, and biochemistries. It is the chemistry of more than one molecule—chemistry beyond the atom or molecule. It is the chemistry of the intermolecular bond.
	Metallo-supramolecular chemistry	Developed in the 1990s, metallo-supramolecular chemistry involves metal-directed synthesis and metal-directed self-assembly in the formation of discrete to infinite structures.
	Chemical functionalization of nanomaterials	Nanomaterials are the starting materials. Chemical modification of nanomaterials offers stability and function. Due to the high surface energy of nanomaterials, reactivity is the rule rather than the exception. Functionalization allows for carbon nanotubes to be inserted into polymers and in the stabilization of nanometals.
	Template synthesis	Kinetic confinement of nanoscale reactions to directly produce nanomaterials that are confined in small spaces. The use of porous alumina membranes, zeolites, block copolymers, and micelles to form nanomaterials that would be impossible to form by other methods is gaining momentum.
	Physical–chemical synthesis methods	ALD, MBE, and CVD are physical–chemical methods to form nanomaterials—for the most part, 2-D nanomaterials, although 0-D and 1-D materials are also possible by sputtering, evaporation, and templates.

(*continued*)

Table 11.1 (Contd.)	*Chemical Synthesis—All Bottom-Up Methods*	
Type of synthesis	**Discipline[a]**	**Comments**
	Nanomaterial chemistry[b]	Use of catalysts, colloids, carbon nanotubes, or other nanomaterials to form new nanomaterials with greater complexity, hierarchical nanostructured materials, or bulk materials with nanoscale components
	Nanochemical engineering	Serious challenges face the scale up of nanomaterials to industrial proportions. Experiment repeatability from lab to lab let alone to the production line has not been realized in bottom-up methods. The interdisciplinary nature of nanotechnology offers its own set of issues.
	Chemistry in nanomedicine	The pharmaceutical industry has always had a chemical, supramolecular basis. Now more than ever, nanomaterials have made inroads into medical imaging, cancer therapy, structural prosthetics, Alzheimer's disease mitigation, and numerous other applications.
Biochemical nanoscience	Bionanotechnology	Biomimetics and bionano are already big industries. The pharma industry has been nano for decades. Trying to mimic nature's incredible arsenal of nanoscale materials and phenomena is burgeoning.

[a] There are many more subdisciplines, of course.

[b] Nanomaterials chemistry in the context of this table implies the use of nanoscale materials as a starting material or substrate in a chemical reaction or process.

since it is often there that mutual fertilization between areas may occur. This is certainly also true for the case at hand, the case of supramolecular chemistry and its language.

11.0.3 Thermodynamic versus Kinetic Control and Selectivity

There are two general types of chemical synthesis: (1) those that are conducted under kinetic control conditions and (2) those that are conducted under thermodynamic control conditions. Once again we create a boundary for the purposes of understanding. In actuality, both factors are always present in any reaction and in every kind of synthesis. The important aspect to remember is that we are able to manipulate experimental conditions to favor a preferred product whether it is the kinetic product or whether it is the thermodynamic product.

According to IUPAC (the International Union of Pure and Applied Chemistry), thermodynamic control of product composition indicates that "… conditions that lead to reaction products in a proportion governed by the equilibrium constant for their interconversion of reaction intermediates formed in or after the rate-limiting step." And, according to IUPAC, kinetic control of product composition indicates that "… conditions (including reaction times) that lead to reaction products in a proportion governed by the relative rates of the parallel (forward) reactions in which the products are formed, rather than by the respective overall equilibrium constants."

Table 11.2 lists some of the differences between the two aspects of chemical reaction control. The universal time-honored and often used example of

TABLE 11.2 *Thermodynamic versus Kinetic Control*

Parameter	Kinetic control	Thermodynamic control
Field of use	Organic synthesis	Supramolecular synthesis
General descriptor	Rate of product formation	Relative stability of products
Strength of bonds	Strong	Strong or weak
Types of bonds	Covalent, ionic, metallic; the "big three"	Interactions: covalent + hydrogen bonding, van der Waals, hydrophobic, π–π
Driving force	Product with lowest activation energy (competing rates of product formation) Reverse reactions not favored	Formation of thermodynamically stable products (equilibrium thermodynamics) Reverse reactions are possible (if allowed)
Thermodynamic parameters	Although entropy and enthalpy are important, kinetic methods can be made to overcome thermodynamic disadvantages	Entropy and enthalpy important in nanoscale chemistry and are used to drive reactions
Reversibility	Generally irreversible for kinetic product to predominate	Generally reversible Thermodynamic product predominates
Equilibrium constant	The equilibrium constant is the same for both. Product proportion governed by kinetic factors ("Le Chatalier's principle" manipulation, confinement, and other kinetic tricks)	Reaction product proportion governed by equilibrium constant
Types of reactants	Covalent, ionic, and metallic substrates	The same three plus supramolecules and nanomaterials
Stability of intermediates	More stable transition state with lower activation energy resulting in faster reactions	Less stable transition states with higher activation energy in covalent reactions. Self-assembly process occurs at low temperature to form thermodynamic products.
Stability of products	Products energetically less stable Products are kinetically stable (no path available for decomposition or reaction)	Products energetically more stable
Purity of products	Kinetic products as a rule require purification (there are exceptions, of course)	Thermodynamic nanomaterials are monodisperse and pure.
Number of products	Numerous products are possible	Thermodynamically favored (usually one product favored)
Molecular recognition	Molecular recognition requires a long gestation period. Kinetic reactions by definition usually occur quickly	Allows manipulation of free energy of product. Desired compound will dominate the mixture at the expense of undesired products
Temperature	Favors reaction path of lowest activation energy—hence lowest temperature	Reactions proceed at higher temperatures in covalent chemistry; low temperature (over long periods of time) in intermolecular domain.
Biology	Interestingly, kinetic reactions are prominent in the biological world—living things' need for reactions to happen quickly. Energy to overcome activation barriers provided from external sources	Thermodynamic factors were probably more important in the prebiotic world. Underlying thermodynamic self-assembly and other reactions are governed by kinetic factors.

thermodynamic versus kinetic control is the addition of hydrogen bromide (HBr) to the double bond of butadiene. The 1,2-addition to form 3-bromo-1-butene (kinetic product) is faster than the 1,4-addition to form 1-bromo-2-butene (thermodynamic product) (**Fig. 11.1a**). The reaction coordinate for both types is depicted in **Figure 11.1b.**

Fig. 11.1 *(a) The addition of HBr to the alkene 1,3-butadiene is pictured. Two products are formed: On the top, the kinetic product is formed faster by the 1,2-addition; on the bottom, the thermodynamic product is formed by the slower (thermodynamically stable) 1,4-addition. 1-Bromo-2-butene is the thermodynamically favored product (blue curve) with higher activation energy and overall more negative ΔG. (b) Graphical comparison of thermodynamic versus kinetic control of a reaction. Kinetic reactions are essentially irreversible and are controlled by the rates of formation of constituents. The transition state has lower activation energy. Thermodynamic control implies that equilibrium conditions are in place. Right: The universal example of kinetic versus thermodynamic control is depicted. It is the reaction of 1,3-butadiene with HBr. The thermodynamic product 1-bromo-2-butene is depicted in red. The overall change in free energy is depicted on the right. The thermodynamic product is expected to be more stable than the kinetic product. Left: The transition state, K, is formed at a lower energy than the transition state, T, of the thermodynamic product.*

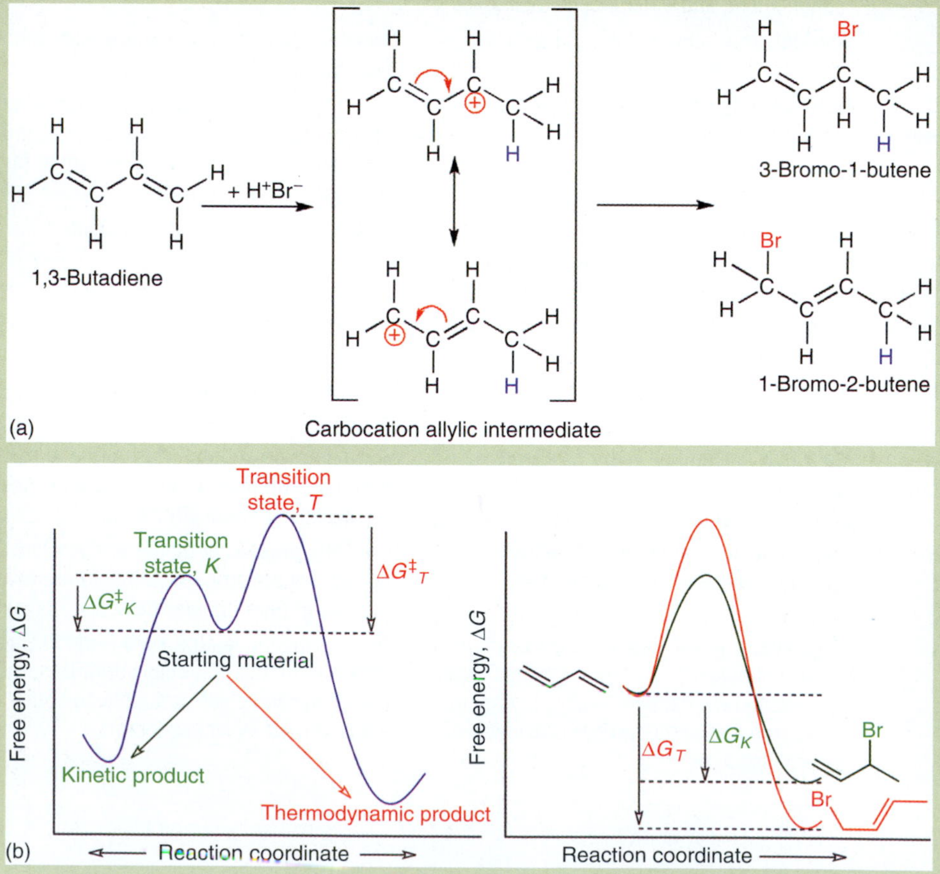

At low temperatures, the less stable kinetic product is favored because the intermediate in the reaction goes through a pathway that requires lower energy (e.g., lower activation energy). At elevated temperatures, there is enough energy available for the reaction to proceed through another intermediate: the one with higher activation energy that leads to the thermodynamically stable product. With enough time or if the temperature is raised, the kinetic product is transformed eventually into the thermodynamic product. Therefore, at higher temperatures and longer reaction times, the thermodynamic product is favored. The example of the transformation of diamond into graphite illustrates this process. There is a distinction to be made at this point between reactions like the one discussed in **Figure 11.1** and those that we consider for supramolecular chemistry. Specifically, in the HBr reaction, strong covalent bonds (intramolecular) are made and broken. In supramolecular chemistry, weak intermolecular bonds are made and broken and they fall mostly under the umbrella of thermodynamics.

A phenomenon called the *template effect* is a good example of a kinetic parameter that is able to influence the outcome of a reaction by control of the steric pathway. Templates in general exert kinetic control over a chemical reaction. In actuality, the template effect was fundamental in giving supramolecular chemistry its start; supramolecular chemistry may never have existed had it not been for the participation of some good kinetically based luck [7]. A kinetic pathway due to the template effect was responsible for the formation of the historic crown ether, dibenzol[18]crown-6, synthesized by Nobel laureate Charles Cram. Although thermodynamically less stable, the crown ether was favored over the thermodynamically more stable polycondensation product (**Fig. 11.2**) [7].

The *kinetic template effect* in this historic reaction was due to the ability of the K^+ cation to organize the assembly of a stable intermediate into an octahedral geometry. By this way, the templating action of the potassium ion was able to place reactive moieties on either end of the precursor molecule into close proximity [7]. Often you will here how a nanomaterial is *metastable* or *kinetically stable*. These terms imply that the material is not thermodynamically stable but that it is able to exist nonetheless. Kinetically metastable materials will exist when: (1) there are no other molecules around with which to react; (2) the environmental conditions are able to support their stability (e.g., low temperature); (3) the space remains confined; or (4) it becomes chemically stabilized by another reaction (e.g., a shell of ligands). The template effect witnessed in confined spaces is able to bring about reactions that normally would not occur in other environs, somewhat akin to catalytic processes that serve to reduce the activation energy of a reaction by forming kinetically stable intermediates.

In the case of the *thermodynamic template effect*, the reaction equilibrium shifts to favor a metal-stabilized product with high yield [8]. Without the presence of the templating metal, other products would participate in the product mixture. The metal then is able to exclude other reactions from occurring by forming a thermodynamically stable product. A metal cation's ability to "select" the best ligand to form an energetically favored complex in a reaction mixture alters the equilibrium in favor of the thermodynamic product [7].

Thermodynamic factors play a major role in selectivity, molecular recognition, self-assembly, and self-replication. Several processes consist of numerous steps involving both thermodynamic and kinetic factors. For example, thermodynamic

FIG. 11.2 *Synthesis of the crown ether [18] crown-6 by the template action of a potassium cation. In the presence of KOH, the oxygen atoms of the precursors form dative bonds with the potassium cation. Notice that in this configuration, the distal ends of the linear molecules are placed within close proximity of each other—a by-product of the kinetic templating effect of the metal cation [7]. In the presence of an organic base, no direction occurs and the polymer condensate consisting of precursors is formed. The kinetic pathway leads to the formation of one of the first crown ethers. Conducting this reaction in the condition of high dilution enhances the formation of the macrocycle rather than the polymer.*

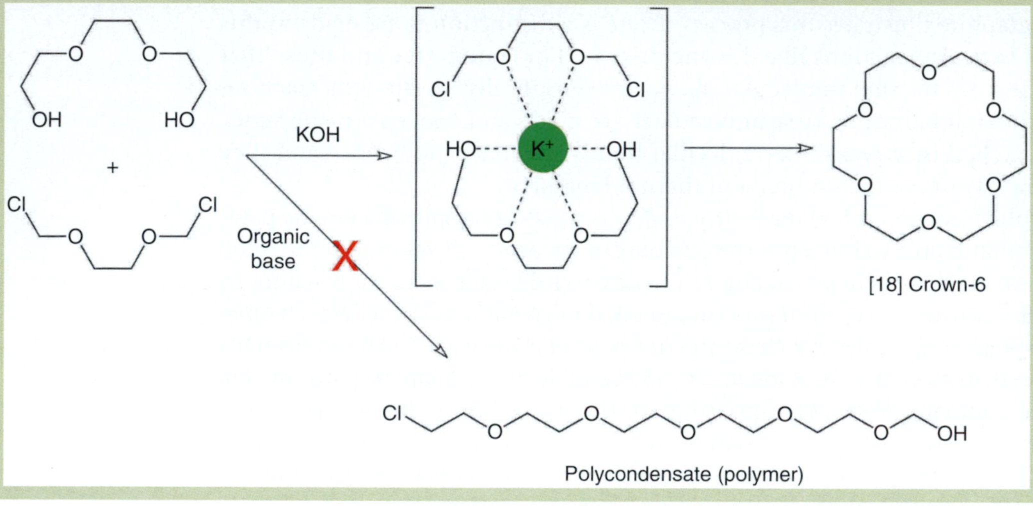

self-assembly may initially involve π–π stacking, while subsequent covalent reactions involve kinetic parameters [7]. Another example involves hydrogen bonding—an intermolecular attraction characteristic of thermodynamic self-replicating systems—to form precursor molecules that are covalently modified later in the process. The action of enzymes involves both thermodynamic and kinetic controls. A potential substrate is lured into an active pocket by means of intermolecular factors and then modified under kinetic conditions to form a completely new molecule. Ripping a hydrogen atom from a substrate during an oxidation process taking place in the active pocket of an enzyme requires the kinetic confinement of the substrate and the application of strain to the bond in question. Once all factors are in place, the hydrogen is removed under physiological conditions.

11.0.4 Introduction to Supramolecular Design

Selectivity in the host–guest process is an important aspect of supramolecular chemistry and other chemistry of the nanoscale. As chemists, we design reactions in ways that favor specific products—preferably in high yield. Such prudent strategy saves time and expense in necessary post facto purification, further processing, and proper functioning. Selectivity is highly desired in any chemical reaction, whether it is chromatography, polymerization, or enzyme–substrate interactions. With regard to supramolecular chemistry (and biochemistry), selectivity is the ability of a potential host molecule to distinguish between and among potential guest substrates and be able to bind that guest molecule for some predetermined length of time. The host must prefer a certain substrate or class of substrates over others in order to accomplish any meaningful function.

Distinguishing between a structural (static) element and a functional (dynamic) element is important to any kind of molecular design. For example, if one plans to design a structural element that fits into a larger supramolecular system, it would be prudent to consider a reaction that forms a complex with a high binding constant. On the other hand, if the purpose of your supramolecular system is more centered on chemical processing, such as turning over a substrate in an enzymatic system, a prohibitively high binding constant may not be what you want. The rate of complexation of a potential guest influences the outcome of a host–guest relationship. Biological enzymes, for example, depend on kinetically selective conditions. The formation of an enzyme–substrate complex that has a prohibitively large binding constant would slow down the rate of the primary reaction, which is certainly not desirable in metabolic processes [7].

The ability to accomplish a chemical task at a facilitative, moderate, or deliberate pace relies on both the kinetic and thermodynamic parameters of the chemical process. Depending on the application, whether it has a metabolic, catalytic, or structural mission, the timescale of the process, its product quality, its yield and rate of turnover mean the difference in investment of time, energy, capital, and personnel. In the case of nature, life and death itself depend on host–guest relationship parameters. The best example of the criticality involved in such metabolic processes is the binding of oxygen by the heme complex (e.g., life based on carbon monoxide simply would not work under the current metabolic hierarchy).

11.0.5 The Nano Perspective

The material presented in this chapter falls under the category of *soft matter* as opposed to hard matter that consists of, for example, minerals and metals. Soft matter lacks long-range crystalline order and is held together by intermolecular bonds. Members of the class of soft matter include living tissue, polymers, micelles, colloids, amphiphiles, liquid crystals, gels, and supramolecules [9,10]. The strength of many kinds of intermolecular attractions between and among components of soft matter is on the order of magnitude of k_BT where k_B is the Boltzmann constant. All of the materials considered to be soft matter get their start at the nanoscale. A comparison between soft and hard matter is given in **Table 11.3.** However, hard matter makes great for substrate materials upon which soft matter can be manipulated.

We spoke briefly about the threshold between single- and multicellular organisms in chapter 5 and what roles increased surface area and size may have contributed to crossing that threshold. Before the existence of eukaryotic cells there was the prokaryote and before that the protocell—a cell that is not considered to be life per se but possesses some of the rudimentary characters of living things. The protocell is thought to have existed within a thermodynamic envelope that was separate from its environment and within which it was able to replicate itself (grow?) by exploiting nutrients and energy sources that existed outside its

Table 11.3 *Soft Matter versus Hard Matter*

Parameter	Soft matter	Hard matter
Examples	Assembly of large molecules [9]: micelles, colloids, amphiphiles, liquid crystals, self-assembled monolayers, polymers, gels, sols, supramolecules, biological materials, composites	Assembly of atoms and small molecules [9]: metals, minerals, ionic solids, semiconductors, ceramics, composites (concrete), ice, graphite, carbon nanotubes, quantum dots, and fibers
Bonding and energy	Intermolecular interactions on the order of 1–10 k_BT Energy proportional to k_BT; entropy measured in units of k_B	Intramolecular interactions of covalent, metal, and ionic bonds Energy $>> k_BT$
Structure	Micro- and nanostructure between 1 nm and 1 μm. Although nanomaterials self-purify, long-range order is absent in 2-D soft matter nanomaterials. Order is achieved on the mesoscopic scale	Contains micro- and nanostructure but may be homogeneous on the bulk scale Crystalline materials have long-range order.
Reaction control	Thermodynamic and kinetic	Thermodynamic and kinetic
Behavior	Viscoelastic	Rigid structures prevalent

Sources: J. G. E. M. Fraaijie, Soft matter. Presentation ISM-Series, University of Leiden, wwwchem.leidenuniv.nl/scm/Presentations/ISM/ISM09/ISM09.pdf (viewed 2007); I. W. Hamley, *Introduction to soft matter*, John Wiley & Sons, Ltd., Chichester (2000).

envelope [11]. The protocell required a minimal lipid structure to form an exclusion membrane and the vesicles within (a.k.a. the liposomes). These structures must have formed spontaneously (negative ΔG) via self-assembly by the action of large molecules (amphiphiles) that have hydrophobic and hydrophilic ends. How fortuitous was this scenario? Transfer of nutrients and energy perhaps first occurred in such structures.

The theory of Morowitz, Heinz, and Deamer states that self-assembled membrane vesicles made of rare organic amphiphiles were responsible for the beginnings of life. A prebiotic foundation for the photosynthetic process was most likely established by an amphiphilic pigment molecule that became incorporated into a lipid bilayer [11]. It is not as far-fetched as one might think to imagine a system in which a source of energy was spontaneously incorporated into a system that required energy, thus developing a relationship between structure and function in living things. For example, why would any prebiotic structure require energy in the first place? What driving force is there that requires materials to seek energy? Is the answer "to make more of themselves"? we then ask, why make "more of themselves" in the first place?

Proteinoid microspheres (protein, from the Greek *protos,* "first quality"), formed from a mixture of amino acids at ca.150°C, have been shown to spontaneously assemble into spherical structures (in the laboratory) [12]. This work was based on Stanley L. Miller's landmark paper in 1953 that described the synthesis of primitive α-amino acids under prebiotic conditions [13,14]. DNA and RNA are also supersized molecules that have built-in self-assembly mechanisms. Initially, there may have been a non-DNA–RNA means to transfer information—a fact contrary to the "central dogma" of biology. Prions, for example, are considered (arguably) to replicate (reproduce?) without the aid of nucleic acids. Prions have the ability to influence (infect?) the normal form of the amyloid protein with the identical amino acid sequence by initiating refolding into an abnormal shape that eventually sets up a chain reaction (self-assembly?) [15–17]. Interestingly, prions are able to resist proteases better than normal counterparts as well as higher levels of heat and radiation—conditions that may have existed in early Earth history. Prions may be the most primitive form of protein. Certainly these are interesting aspects to ponder. Prions have been determined to be the active agent in the cause of Alzheimer's disease.

The message implied in the previous paragraph is that supramolecular nanoscale materials were made of natural inorganic and organic elements and not necessarily from biological sources. Later, biological forces were able to exploit processes that require little input of energy, such as the molecular self-assembly process, to form supramolecular molecules and structures. Whether DNA came first or some kind of prion preceded the advent of DNA is a debate better left to *paleobiologists, paleochemists,* and perhaps nanoscientists to sort out in coming years. The rest is history and we anticipate many exciting developments in this extremely relevant field in the very near future. The *nano perspective* of supramolecular chemistry is one of great size and breadth. It is a field that will render unimaginable impact upon our technological world. The contributions of pharmaceuticals have already improved the quality of life for humans. What wonders will we be able to contribute to civilization and with equal importance—give back to the natural world?

11.1 Supramolecular Chemistry

Supramolecular chemistry (from the Latin *supra,* "above, beyond," + chemistry) has its roots in organic, colloid, coordination, polymer, and biochemistries. It is defined in a general sense as any chemistry that addresses synthesis "beyond that of the atom or molecule" and is widely known as the "chemistry of the intermolecular bond" [18–20]. The study of materials that have two or more molecules, in the narrowest sense, conforms to this definition. This relatively new field of chemistry involves design, synthesis, and characterization of large molecules. It is the chemistry of molecular assemblies and the intermolecular bond—a generalized coordination chemistry that goes beyond the simple binding by ligands of transition metals. An important aspect of supramolecular chemistry is that molecules that are part of a supramolecular complex retain their original structure without too much modification, although their chemical nature may be modified.

11.1.1 The Host–Guest Relationship

Supramolecular chemistry is indeed very much about the *lock and key* (or guest–host) hypothesis first proposed by Emil Fischer in 1894. Within the context of nanoscience, supramolecular chemistry focuses on noncovalent interactions between molecules. Fabricating a *supermolecule,* on the other hand, is akin to making the same molecule that is held together with strong bonds simply bigger. Many nanomaterials are supermolecules. Many kinds of nanomaterials are supramolecules. This is not to say that nanogold has different properties apart from the bulk and atomic forms because it certainly does; it just is not due to supramolecular chemistry. Another strongly bonded material, NaCl, is capable of extremely primitive "atomic recognition" and "self-assembly," but it is not the same self-assembly characteristic of soft-matter materials. Carbon nanotubes are precursors in supramolecular chemical reactions, but are not considered to be supramolecules, although they are certainly supermolecules. In the carbon nanotube, there are no precursor molecules, just carbon atoms that are covalently bonded (very strongly) to form a very long molecule—a supermolecule. Although nomenclature and categories are convenient tools, distinguishing between terms *supra* and *super* should not provide any cause for consternation.

Supramolecular chemistry is focused on the relationship between a host molecule and a potential guest atom, molecule, or substrate. **Figure 11.3** describes in minimal detail the fundamental concept of supramolecular chemistry: the host–guest relationship.

Aside from the obvious conclusion that supramolecular structures can be rather large, each molecule has the ability to accommodate other complementary pairs until the precursor material is exhausted—a version of the *ying–yang* scenario of the molecular world. This general principle forms the basis for

FIG. 11.3 *A hypothetical host–guest interaction leading to a supramolecular structure is displayed. The "guest" hexagons (blue) "fit" perfectly in the grooves formed by a complementary "host" structure (red). The process continues as long as material transport is unhindered, precursor supply is abundant, or the complex remains soluble. Structures such as the one depicted come together due to two factors: molecular recognition and molecular self-assembly. Host–guest chemistry is about weak intermolecular forces that hold a guest and host together. Host and guest molecules, individually, are held together by covalent bonds. The bond to the substrate is often of covalent character. Each individually is a molecule. Together they comprise a supramolecule.*

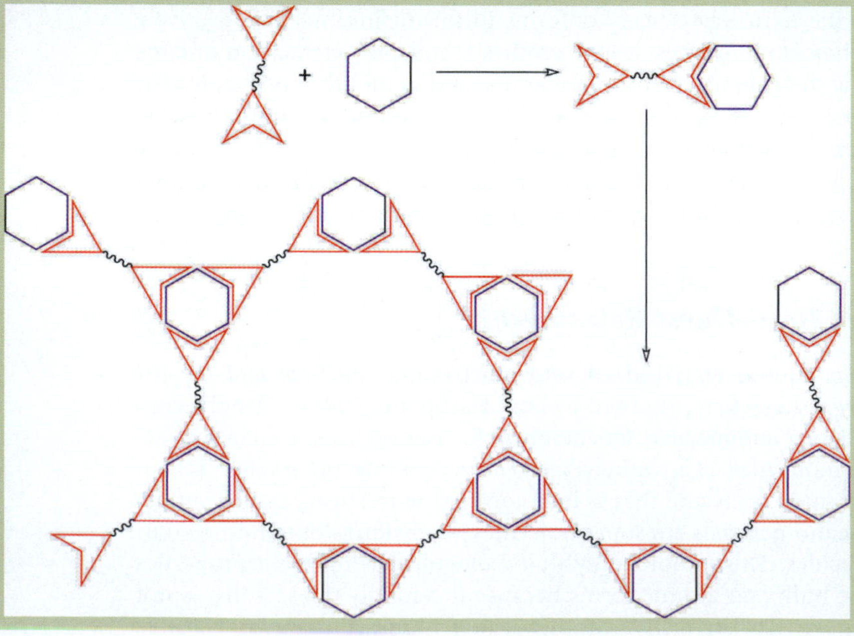

coordination compounds (that involve a ligand and a metal), the enzyme–substrate specificity, and the antibody–antigen relationship that is fundamental to the immune system. *Complementarity* is perhaps the best term to describe the host–guest concept [8]. Hosts can be biological, organic, or inorganic; guests are usually smaller entities like ions, metals, and neutral chemicals. Hosts can be anionic, cationic, or neutral or combinations of anionic–neutral or cationic–neutral and even zwitterionic. Substrates are also found in anionic, cationic or neutral states.

11.1.2 Molecular Recognition

There is no better way to introduce a subject than by quoting one of the founders of the field—Jean-Marie Lehn:

> Molecular recognition is defined by the energy and the information involved in binding and selection of a substrate(s) by a given receptor molecule; it may also involve a specific function. Mere binding is not recognition, although it is often taken as such. One may say that recognition is binding with a purpose. It implies a pattern recognition process through a structurally well-defined set of intermolecular interactions. Binding of σ to ρ forms a complex or supermolecule characterized by its (thermodynamic and kinetic) stability and selectivity, i.e., by the amount of energy and of information brought into operation.

According to this quotation, we understand that there is something special about this field, above and beyond the normal boundaries of chemistry with which we are so familiar. Supramolecular chemistry implies purpose, design, and engineering, as well as an inherent beauty. As a case in point, DNA, the premier supramolecule, is all about information—a molecule with a purpose. It is no wonder with the advent and development of supramolecular chemistry that we are ready to knock on nature's door itself—ready to insert a key, albeit still on the small scale, into its mysterious lock. "Binding with a purpose!" Molecular recognition, after all, provides the foundation for replication mechanisms.

Molecular recognition represents a process that defines specific interactions between two or more molecules (or supramolecules) by means of noncovalent forces. Host–guest chemistry, and hence supramolecular chemistry, in general is based on molecular recognition. However, recognition in its broadest sense is responsible for colligative properties such as melting point, freezing point, and boiling point—relationships that arise between and among atomic or molecular constituents of a material. Atoms are able to recognize each other and bond with each other "knowing" that the minimal energy state will result from such binding. But there is no information or function associated with such colligative properties, only "purposeless" stochastic interactions that are driven towards a state of minimal energy. Supramolecular chemistry, in another broad sense, is chemistry with a purpose.

Moving on from sodium chloride, the *fitting* together of molecules implies a higher level of recognition called molecular recognition. Is *information* exchanged in all molecular recognition processes? If we label one bonding episode as "1" and a nonbonding episode as "0," then our answer is in the affirmative. It would be a stretch, however, to claim that any sort of information exchange is actually involved. We know that DNA is capable of undergoing supramolecular interactions that code for information transfer processes. Although no information may be exchanged per se in many supramolecular reactions, the foundations for information exchange are there to be exploited when the conditions are appropriately defined.

Supramolecules come together to form larger entities with the same goal that drives metals to agglomerate: achieving that state of lower energy. Because supramolecular reactions involve intermolecular interactions, the reactions are reversible. Reversibility (a condition characterized by small reaction constants, K_{rxn}) is a fundamental property of biological phenomena. If RNA coding were not reversible, then there would be no protein synthesis, at least in the way with which we are familiar.

The aim of early molecular recognition processes was to bind metal cations [21]. C. J. Pedersen in 1967 helped to launch the science of molecular recognition after his synthesis of crown ethers [22,23]. Crown ethers are able to selectively bind alkali metals [22,23]. The binding of metals by chelates is a generalized form of molecular recognition within the general bounds of the definition given in this chapter. EDTA, for example, is able to bind many kinds of metals. Is this a case of molecular recognition? The answer is, "Yes, it is, but in a less specific way." Other chelating agents offer better specificity. Iron is coordinated in heme by a porphyrin ring to form a near planar complex. Is the binding of iron by porphyrin a true case of molecular recognition or more like recognition by a molecular species of a cation?

The biological realm offers numerous examples of specific molecular interactions. DNA, RNA, structural proteins, antibodies, cellular receptors, and enzymes are a few examples of materials that combine by the action of noncovalent bonds. This is not to say that DNA, RNA, and proteins are without covalent bonds. Covalent bonds are well represented in the backbone structures of each material. However, higher order noncovalent interactions also occur between DNA and protein (histones), RNA and ribosomes, sugar and lectin, antibody and antigen, and between magnesium or iron and porphyrins (chlorophyll and heme, respectively). Molecular recognition forms the fundamental basis of immunology and of life itself.

Taking back the clock to the timeless period of prebiotic protocellular matter, insipient stages of molecular recognition must have been linked, albeit tenuously, to some specific function or just some statistically based process that found its way to achieve the state of lowest energy and stability. The purpose of life is to replicate. The function of molecular recognition is to make life possible. Just think for a moment about the attraction that results in romance and the chemistry in relationships—in actuality nothing more than a glorified version of molecular recognition with the underlying end goal of replication.

Types of Molecular Recognition. There are two basic kinds of molecular recognition. *Static* molecular recognition is similar to the *lock and key* hypothesis proposed by Emil Fischer in the later nineteenth century. It involves complexation between one host and one substrate: a one-on-one relationship. *Dynamic molecular recognition,* based on dynamic equilibrium, is more complex and involves more hosts (or sites) and more guests. In one host, several binding sites may exist simultaneously. If a guest is bound by one site, the thermodynamic propensity to associate by another site may be increased (*positive allosteric effect*) or decreased (*negative allosteric effect*). One of the consequences of dynamic molecular recognition is the capability to regulate substrate binding, a useful characteristic of sensors [24,25]. Unless we are dealing with an isolated host and guest, all integrated supramolecular systems of two or more molecules exert influence on one another and the whole.

Preorganization and Complementarity. Right along with the terms molecular recognition, host–guest, and lock and key are the terms *preorganization* and *complementarity*. Preorganization is an engineering term—a design term involved in the selection of host and guest so that a thermodynamically stable product is formed during a process. The term complementarity describes a state in which one component complements the structure, the electronic character, and function of another component. It is the "hand-in-glove" term fundamental to supramolecular chemical compounds and synthetic procedures. Complementarity in physics implies that the particle–wave duality of the electron explains certain phenomena associated with electrons. Complementarity in chemistry means a good fit. The best example of a good fit is that between an enzyme and its substrate. The most commonly used example is that of the enzyme carboxypeptidase-A (CPA) (**Fig. 11.4**). We shall conform to the trend and use it ourselves to illustrate this important point.

CPA is a peptidyl-L-amino-acid hydrolase (metalloexopeptidase) that exists in the small intestine and is made up of 307 amino acid residues. Its job (function, purpose) is to cleave the terminal amino acid from the carboxyl terminal (C-terminus) with the assistance of the metal cofactor zinc. A single activated water molecule and the zinc dication work together to hydrolyze the peptide bond. The hydrophobic pocket is there to provide the proper chemical environment for groups such as the phenyl depicted in **Figure 11.4**. The *active site* of CPA pictured in **Figure 11.4** is a static version frozen in time. The process is actually quite dynamic. CPA is formed in the pancreas and once in the small intestine it is activated by trypsin, another peptidase, to open its active site. The entire process becomes even more remarkable when the turnover rate is considered (in the range of 100–1000 substrates per second).

The DNA double helix is sine qua non when it comes to biological complementarity—a complementarity based on hydrogen bonds between heterocyclic nucleotide base purines (adenine and guanine) and pyrimidines (cytosine and thymine) (**Fig. 11.5a**). The backbone of the DNA consists of covalent-bonded phosphate and deoxyribose sugar molecules that provide the framework (substrate) for linked nucleotides. The primary structure is composed of a nucleotide base-pair sequence that is able to code information in groups of three that consist of various permutations of the four base pairs adenine, guanine, cytosine, and thymine (ACGT). "A" only binds with "T" and "C" only binds with "G." The coupled base pair trios run up and down the length of the double helix to form a DNA strand that makes up the chromosome—indeed, a hierarchical extension of complementarity. Informational exchange is linked to the order of the base pairs. Information coded by DNA exists at a very high and sophisticated level of complementarity.

Therefore, beginning with a simple complementary molecular recognition system consisting of mainly hydrogen bonds, an intricate information data system is constructed. Overall, complementary hydrogen bonding, enthalpic π–π stacking interactions, and entropy-driven hydrophobic interactions

Fig. 11.4 *A schematic rendition of carboxypeptidase-A (CPA) and substrate (in blue). The primary function of the enzyme carboxypeptidase-A (CPA) is to hydrolyze the carboxy-terminal (C-terminus) peptide bond of proteins. By this action, the last amino acid in a peptide is removed. The interplay between and among hydrophobic pocket, hydrogen bonds, salt bridge, anionic and cationic moieties, metal cofactor, and a water molecule is certainly mind boggling. The action of CPA is considered to be nonselective (e.g., it acts on nearly all C-terminus peptides). The products of the reaction depicted in the figure are the amino acid phenylalanine and the rest of the protein. The squiggly red line indicates the site where hydrolysis of the peptide bond takes place. Carboxypeptidases A_1, A_2, and B are made in the pancreases and are involved in the breakdown of food stuffs in the small intestine. CPs also assist in the synthesis of other proteins like insulin and many other processes. CPA is a metallo-CP. CPs that prefer amino acids that contain phenyl groups (shown here) or branched hydrocarbons are know as CPA (where A stands for aromatic or aliphatic). CPA is converted into its active form by another enzyme called enteropeptidase. By this mechanism, the digestion of CPA is prevented.*

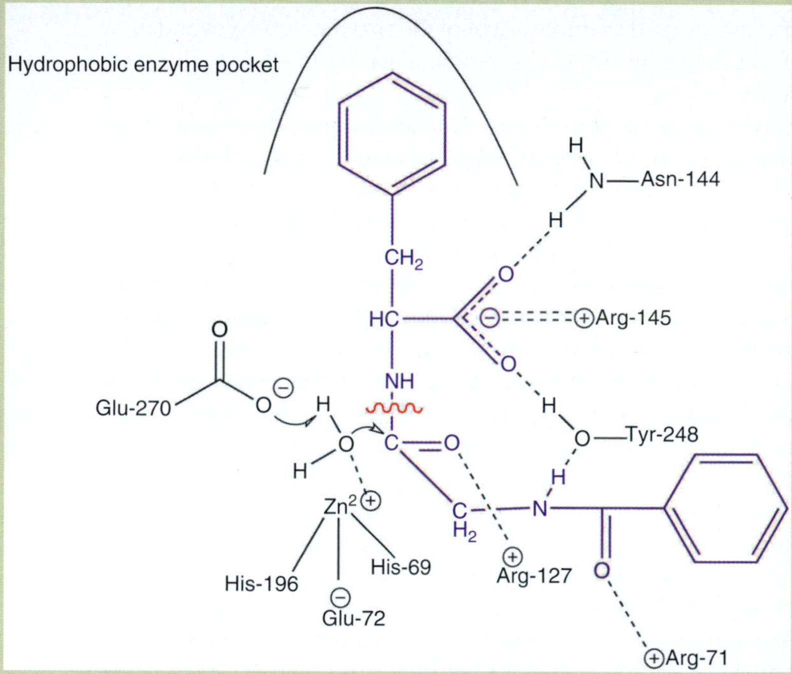

contribute to the self-assembly of DNA into the α-helix configuration (**Fig. 11.5b**) [26].

Molecular Recognition in Nature. Metal binding by macrocyclic biological molecules is another type of molecular recognition exhibited by living systems. This kind of binding is classified as molecular recognition between a metal cation and an organic chemical. It is the least complex form of molecular recognition. Potassium in the form of a monovalent cation electrolyte is responsible for maintaining the membrane potential of cells and also serves as a cofactor for some enzymes (pyruvate kinase, Na–K-ATPase). Aside from its apparent complementary relationship with crown ethers, potassium also has a special relationship with a biological macrocyclic molecule called valinomycin $[C_{54}H_{90}N_6O_{18}]$ (**Fig. 11.6a**). Valinomycin is a depsipeptide that consists of six alternating amino acid and organic acid precursors (**Fig. 11.6b**). It is a natural compound that also happens to be an effective antibiotic. Valinomycin specializes in the transport of the K^+ cation (and Rb and Cs cations) and acts specifically to increase the permeability of membranes to potassium cations [27]. The antibiotic function arises from its ability to alter the permeability of K^+ across cellular membranes in bacteria, mitochondria, and erythrocytes. Valinomycin was isolated in 1955 by H. Brockmann and G. Schmidt-Kastner [28].

Valinomycin synthesis, structure, and function are summarized in four steps: (1) Covalently bonded precursors combine to form a repeating unit (a monomer) consisting of four precursors (**Fig. 11.6b**); (2) three monomer units are covalently bonded to form a macrocyclic compound; (3) hydrogen bond interactions stabilize a conformation that allows tight binding of the potassium cation (e.g., the ring is puckered to form a three-dimensional octahedral enclosure around the potassium cation); and (4) after capture and a change in its conformation, a hydrophobic perimeter is generated that enables the valinomycin, with its sequestered cationic cargo, to become soluble within the lipid of a membrane, thereby allowing passage. This is an excellent example of a simple function at the nanoscale—a function that is critical to life. No information is coded into this process. Thermodynamic (and perhaps kinetic) driving forces are the only factors involved in its function. However, living things created the environment within which valinomycin operates—an environment that did

FIG. 11.5 *(a) DNA consists of complementary sets of purine–pyrimidine base pairs connected to a phosphate–deoxyribose backbone—all formed into a double-helix structure. The purine adenine pairs only with the pyrimidine thymine; the purine guanine pairs only with the pyrimidine cytosine. Molecular recognition is achieved with highly specified hydrogen bonding. This is a remarkable complementary situation that is able to store data, regulate cellular function, and make replicas.*

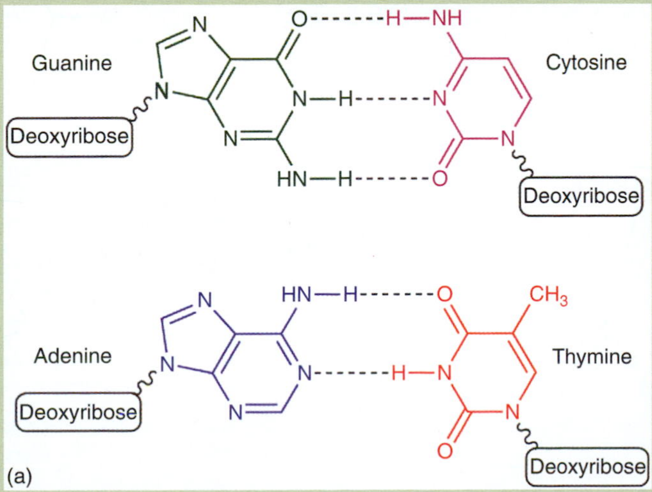

(*continued*)

Fig. 11.5 (Contd.) *(b) The double helix is the result of straightforward hydrogen bonding between purine and pyrimidine nucleotide base pairs, π-stacking interactions between the aromatic constituents of the nucleotides, and complex hydrophobic factors. Red: thymine; blue: adenine; green: guanine; purple: cytosine. The code in this segment is TAGCT on the left strand. Additional carbons in the backbone and the sugar are not shown for purposes of clarity. Covalent bonds hold the nucleotides to the backbone structure. Hydrogen bonds between the nucleotides are unzipped during replication. It is difficult to show on this image a rigorous rendition of the π–π interactions that occur between the linkage groups.*

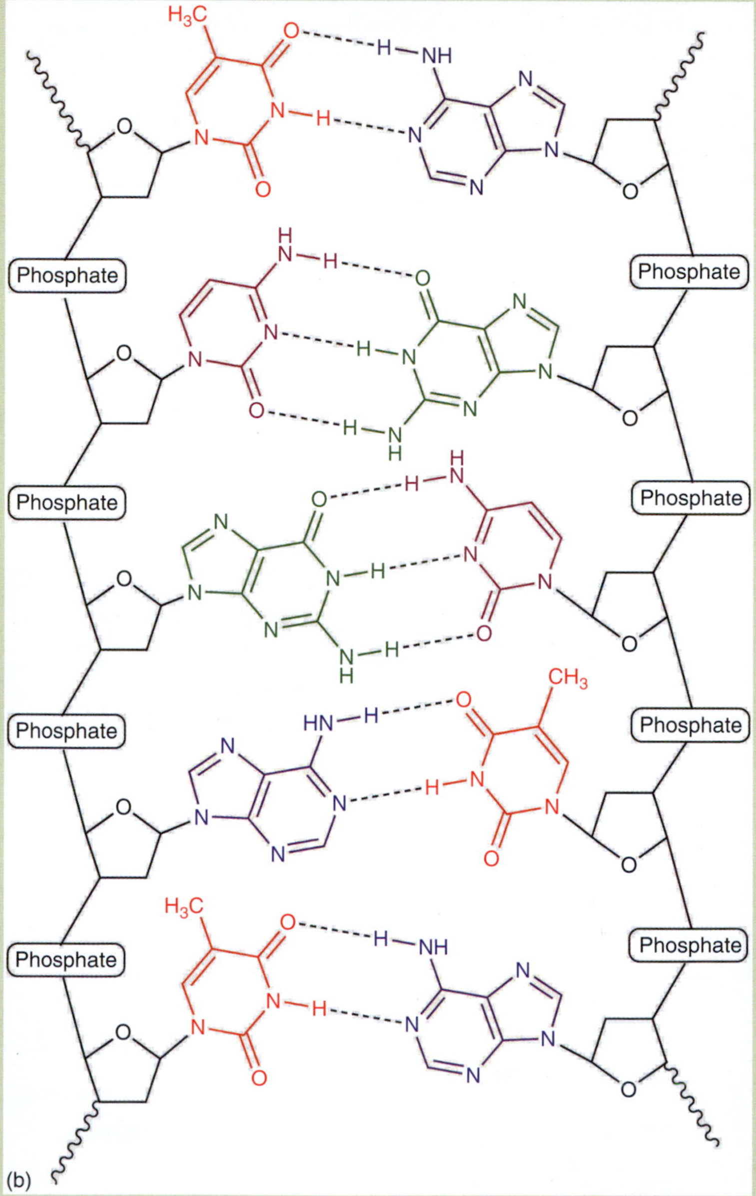

Fig. 11.6 *(a) The biological macrocycle valinomycin is a mobile carrier of K^+ across phospholipid bilayer membranes. Hydrogen bonding between –[N–H···O=C]– enhances the folding of the ionophore into an octahedral shape around K^+. Such action is termed preorganization. Valinomycin's antibiotic potential relies on its ability to disrupt the transport of potassium across bacterial membranes. (b) Precursor molecules of valinomycin (from left to right): orange = lactic acid; blue = valine; green = hydroxy-isovaleric acid; red = valine. These precursors are bonded together with strong covalent bonds, thus synthesizing a supermolecule. It is only the metal cation guest that is bonded with labile intermolecular bonds.*

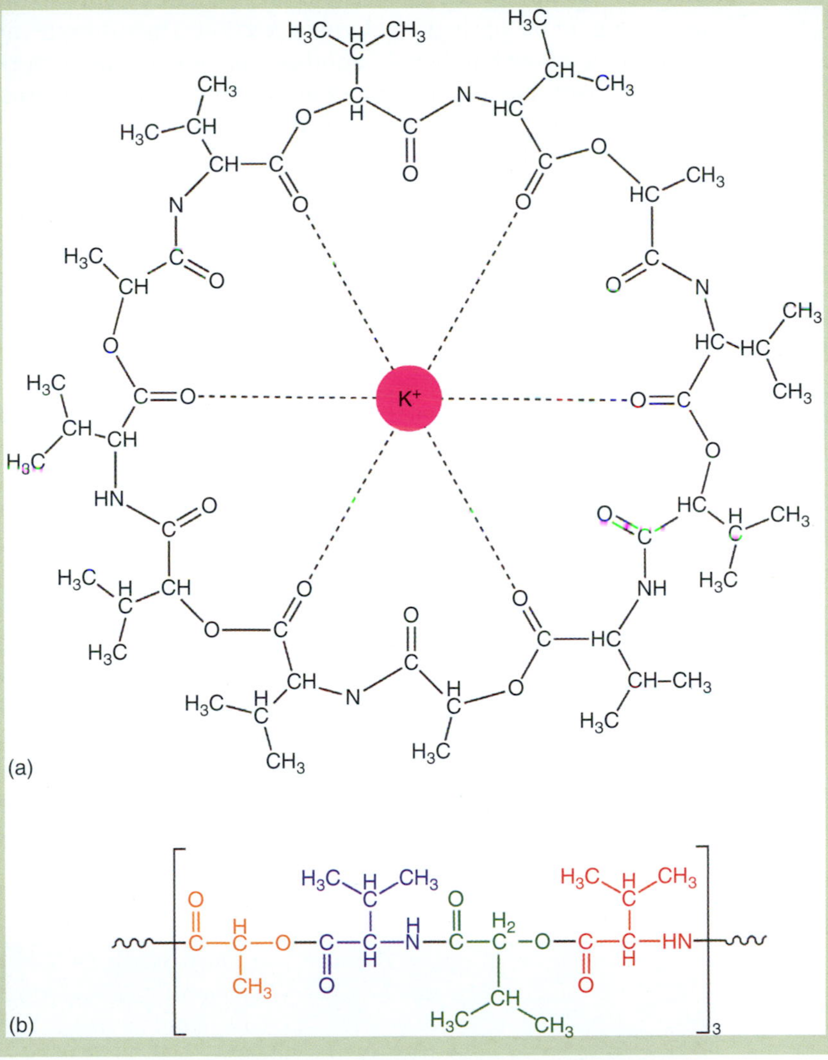

Source: F.F. Nachtigall et al., *Journal of the Brazilian Chemical Society*, 13, 295–299, 2002.

require information, translation, and synthesis, which all existed in a state of lowest energy specially designed for valinomycin to accomplish its task. Perhaps the earliest forms of "living things" did not rely on coded information. If true, this makes the valinomycin–potassium cation relationship one of the oldest in the life evolutionary timeline. Nature exploited the special relationship between valinomycin and the potassium cation and created an environment for it within which to function. Once inside the cell, the potassium is released in another environment designed just to accomplish that special task. Human science has found a way to apply valinomycin in the capacity as an ion-exchange component in potassium selective electrodes.

A more advanced form of molecular recognition is illustrated by the humble silkworm moth, *Bombyx mori*. It is a form of molecular recognition between a molecule and a protein receptor. Although still rather simplistic, in some ways this relationship can be considered as the ultimate expression of molecular recognition in terms of single molecule detection. The substrate molecule is a sex pheromone called bombykol—another type of naturally occurring supramolecule precursor. Pheromones are examples of precursor molecules that participate in extremely sensitive and selective reactions.

Bombykol [$C_{16}H_{30}O$] (**Fig. 11.7**) is released by the female silkworm moth and is detected by pheromone-binding protein receptors that exist in the nano- and microstructure of the *sensilla* of the male moth [29,30]. Sensilla are capable of detection down to a single molecule of bombykol [29,30]. Nature has achieved the level of single molecule detection! Considering the extreme distances between the pheromone and its molecular recognition receptor (often measured in kilometers) and the dilution factor involved, this process is nothing short of remarkable.

Jean-Henri Fabre, a French naturalist, discovered pheromones in 1870 when a female peacock moth attracted numerous male suitors within moments after emerging from its cocoon. Adolf Butenandt, a German biochemist, is credited with naming the silkworm moth pheromone bombykol in 1959. The word "perfume" has its roots in the word pheromone (from the Greek *pherein*, "to carry," + *hormon*, to "impel, urge on"). Biological entities such as cells are considered to be a *collection of sophisticated nanomachines* that self-assemble to interact with complex chemical and physical networks [31]. The example of bombykol is just one great example of natural molecular recognition mediated by nanostructures that are integrated into complex chemical and physical networks. Perhaps the highest form of "dynamic molecular recognition" is exhibited by the immune system. The immune system is an example of a dynamic molecular recognition system that is able to scan antigenic invaders, construct a chemical surface that is specific to the antigenic surface and subsequently neutralize it,

FIG. 11.7 *Bombykol is a relatively simple molecule; it is not very much like a supramolecule at all but more like a large precursor molecule. It obviously plays the role of the guest to the host protein receptors of the sensilla of the male moth.*

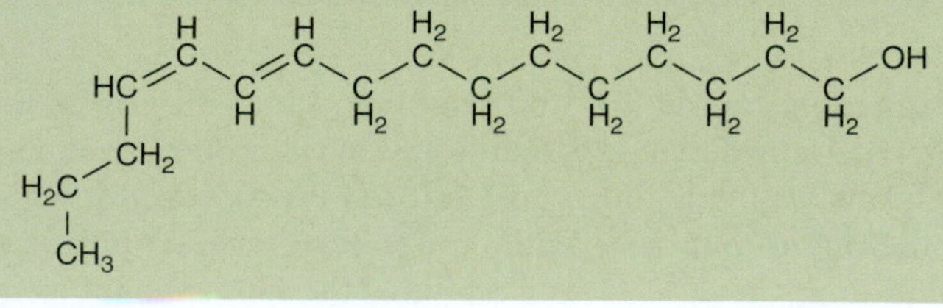

and make it soluble to hydrolysis or excretion. We know that it works quite well most of the time.

Molecular recognition is one of the greatest challenges faced by the developers of pharmaceuticals. The overall objective is to create a drug with selective action, rather than one that has systemic effects. Drugs that are able to target only sick cells via molecular recognition would certainly spare the host the agony of systemic chemotherapy treatment. Cancerous liver cells, for example, have abundant folic acid receptors on the surface. Recently Z. Tang et al. have proposed a system based on *aptamers* that are able to characterize cancer cells based on molecular recognition [32]. They describe a means to synthesize molecular probes for specific recognition of cancer cells. Aptamers are short single strands of DNA or RNA (oligonucleotides) or peptides with capability to bind target molecules. Aptamers are created by statistically based techniques from a large pool of candidates and are touted to be the "chemist's antibody" that can function as molecular probes for many kinds of biochemical applications [32]. "The strategies used here will be highly useful for aptamer selection against complex target samples in order to generate a large number of aptamers in a variety of biomedical and biotechnological applications, thus paving the way for molecular diagnosis, therapy, and biomarker discovery."

Aptamers based on nucleic acids have the capability of molecular recognition similar to that of antibodies, but "antibodies" that can be easily modified chemically to extend lifetime in fluids and bioavailability [33]. Standard antibodies are generally not modifiable because they lose specificity during the process. Antiviral aptamers are actively being developed as a mode of therapy against tenacious virus systems [33]. Small RNA is capable of folding into three-dimensional structures that are able to bind to proteins and hence deactivate them in a manner similar to other protein antagonists [34]. Natural decoy RNAs in the recent past have been shown to slow down the replication of the AIDS virus [34,35]. An aptamer product that targets vascular endothelial growth factor (cause of age-related macular degeneration) is in Phase II/III clinical trials. Another aptamer that targets thrombin is used as an anticoagulent that is reversible upon application of an antidote [34].

11.1.3 Synthetic Supramolecular Host Species

The number of supramolecules is vast and it would be impossible to list them all. The number of supramolecular precursor species is equally as vast. We will catalog only a few of the well-known supramolecular precursor organic host species. Biological supramolecular building blocks and complexes are numerous as well. Supramolecular building blocks (precursors) are covalently bonded macromolecules that either act separately, as in the case of ionophores, or in concert with others, as in the case of proteins, coenzymes, chlorophyll, columnar structures, and DNA.

The shape of substrates is also an important consideration, especially to the supramolecular receptor design engineer (and chemist!). Substrates assume varied geometries: spherical, linear, trigonal-planar, tetrahedral, and octahedral, to name just a few. Recognition between guest and host occurs by means of the many intermolecular interactions discussed in the previous chapter: electrostatic, hydrogen bonding, Lewis acid–base relations, and hydrophobicity. Inorganic anion substrates include atomic species such as fluoride, chloride, bromide, and iodide; linear molecules such as nitride, cyanide, thiocyanide, and hydroxide; trigonal species like carbonates and nitrates; tetrahedral moieties that include phosphates, sulfates, vanadates, molybdates, and manganates; and octahedral entities such as ferricyanide and other octahedrally coordinated metal cyanides [8]. Approximately 70–75% of biological substrates are anions. To become familiar with the multitude of organic species, the supramolecular precursors are presented in **Table 11.4** in alphabetical order. There is no way we can discuss each and every one in great detail, but a few selected hosts are provided with a brief discussion.

Host–guest complexes come in a variety of forms [4,7,8]. There are several kinds of host–guest relationships: (1) *encapsulation*, in which the guest molecule is sequestered within the host; (2) *nesting*, where the guest sits on the surface of the host; (3) *perching*, in which the guest is associated with the edge of the host; (4) *nonpolar surface interaction* in which the guest is placed near the host; (5) the *sandwich* relationship, akin to that of the iron in ferrocene; and (6) the *wrapping* configuration, in which the guest is surrounded by the host with an exit portal available for decomplexation [7]. At higher levels of complexity, host materials are able to complex with other host materials to form superhosts or bind with complex guests. Some hosts are able to bind more than one guest. Binding in hosts often occurs by means of dative style bonds (Lewis acid–base relationships) if the guests are metal cations, but all kinds of intermolecular interactions are capable of binding guest species. Cooperative binding alliances work in concert to attract and then hold guest species and release them when needed.

Hosts are able to accommodate neutral species, anions, and cations or, in some cases, both anions and cations, simultaneously, or zwitterions. In addition to atomic species and molecular anions, substrates like alcohols, amino acids, and peptides are able to fill the role of guest species in enzymatic reactions. Supramolecular design requires diligent analysis of potential host–guest relationships. In order to achieve the level of a tangible nanomaterial, more than a simple dative bond with a metal is often required. A nanomaterial with supramolecular ancestry is the sum of all its host–guest reactions. There are few generalized categories of host–guest relationships. They consist of: (1) the same kind (homogeneous) in which the nanomaterial is assembled in one step; (2) the same kind, but in which the nanomaterial is assembled in multiple steps by the same process; (3) multiple tandem steps consisting of different processes (different host–guest relationships); or (4) simultaneous reactions consisting of heterogeneous processes akin to what was most likely found in the primal soup.

With regard to nomenclature the suffix "and" indicates the synthetic host species (e.g., cryptand, carcerand, and catenand). The suffix "ate" (also "plexes") indicates that the host species exists in a complexed form (e.g., cryptate, encarcerate, and catenate or caviplexes, carceplexes, and spheraplexes). As one can see from the table, the list of supramolecular hosts is quite imposing. We will discuss a few of them in more detail. Several host species are depicted in **Figure 11.8a–p**.

Chelates. Chelates and the chelate effect were introduced and defined in the last chapter. Chelates represent a broad range of compounds that are vital to biology and extremely useful in supramolecular assembly. Chelates are able to

TABLE 11.4 Catalog of Supramolecular Host Species A–Z

Host species	Description	Species	Description
Amphiphiles Surfactants	Molecules with hydrophilic and hydrophobic groups (e.g., soaps, phospholipids). Surfactants are able to self-assemble in aqueous solutions and change the properties of water (surface tension).	Intercalates	Layered solids, intercalates can be considered as clathrates. Examples include graphite, cationic/anionic clay minerals, and metal phosphates. Organic urea clathrates form solid-state clathrates. Urea clathrates form helical host channels.
Antennae complexes	Complexes capable of harvesting light (e.g., metal polypridyls)	Ionophores	Organic ion carriers in biological systems that provide binding and shielding to metal cations—K^+ in particular. The purpose of ionophores is to transport metal ions through membranes.
Calixarenes	Macrocycle with –OH centers + phenyl rings (e.g., *p-t*-butly-calix[4]arene). Useful in cation complexation. Calixarenes are cavitands.	Katapinands	Macrobicyclic cavity hosts that complex halides. One of the first examples of anion binding
Carcerands Carceplexes Hemicarcerands	Permanently trap guest species unless covalent matrix of host ruptures. Hemicarcerands have a measurable activation barrier for releasing guest. Contain a concave inner surface—a molecular capsule. Prominent in neutral guest binding with *vdw* forces	Lariat ethers	A crown ether with appendages that enhance metal cation complexation by providing another dimension to binding
Catenanes Catenands Catenates	A host consisting of two or more rings that are interlocked in a mechanical way (e.g., [2]catenand). Although linked mechanically, the rings are not covalently bonded. Complexes are stabilized by H-bonds/charge transfer π–π interactions.	Lewis acid hosts	Lewis acid hosts are capable of binding anions. Organoboron compounds are electron deficient. These hosts are opposite to chelates that are Lewis bases. Binding usually occurs at the periphery of the molecule or between two molecules.
Cavitands Cavitates Caviplexes	Host with intramolecular cavities. Host–guest complexes are stable in both liquid and solid state. A molecular container with an enforced concave surface. Calixarenes are cavitands. Cavitands are neutral molecules; cyclodextrins are cavitands.	Ligands	A ligand is an organic molecule that is able to complex with a metal. The Au–triphenylphosphine is an example of a metal–ligand complex. Water is a monodentate inorganic ligand that is able to form complexes with metals.
Chelates	The binding of a single metal atom with two or more binding sites on the same ligand. In the general sense, most hosts that bind metals can be considered to be chelates. Chelates can be charged or neutral. Recognition is by Lewis acid–base interaction.	Micelles Lipid bilayers Reversed micelles	Micelles (3-D) and lipid bilayers (2-D) are made of amphiphilic molecules that are able to self-assemble in aqueous solutions. Reversed micelles self-assemble in hydrophobic solutions.

(*continued*)

Table 11.4 (Contd.) *Catalog of Supramolecular Host Species A–Z*

Host species	Description	Species	Description
Clathrands Clathrates	Hosts with extramolecular cavities—between two or more host molecules. Water forms polyhedral networks (hydrates) around dissolved species. Zeolites are inorganic solid-state clathrates. Clathrates are a broad category that includes intercalates, helical tubulands, inclusion polymers and zeolites, and more.	Molecular clefts Molecular tweezers	Two guest-binding domains are positioned to form a binding site.
Coordination polymers	Metal–ligand directional coordination reactions to produce crystalline polymeric architecture. Guest bound by Hoffman type clathrate	Molecular knots	Molecular knots are intricate forms of catenanes formed with multiple metal helicates and other substructures.
Corands Corates Coraplexes Azacorand	Crown ethers belong to this general class. A corand is a closed monocyclic heteroring. If containing nitrogen and/or oxygen it is called an azacorand or azacrown.	Molecular squares Molecular boxes	Squares, triangles, and cubes with metal coordination at corners and linked with aromatic species. Molecular boxes, etc. are assembled with metal coordination.
Crown ethers Anticrowns Azacrowns Heterocrowns	The roots of supramolecular chemistry are traced to the crown ethers of C. J. Pedersen. Crown ethers are oxygen-containing heterocycle corands. They are 2-D heterocycles that contain oxygen. Azacrowns contain nitrogen. Crowns are suitable for cation binding.	Podands Podates	Open chain hosts. They are acyclic hosts with pendant binding sites. Open crown ethers form podants. Because they are open structures, they do not have the affinity for cations as do the crown ethers due to unfavorable enthalpic and entropic effects.
Cryptands Cryptates Hetero-cryptands Cryptophanes	Cryptands are 3-D bicyclic or oligocyclic systems. Crown ethers that have 3-D components are cryptands. Cryptands have high stability constants because the cryptand cavity is poorly solvated and there is less reorganization required to bind metals.	Rosettes	Hydrogen bonded structures consisting of melamine and cyanuric acid form planar circular structure with a central cavity.
Cyclophanes Paracyclo-phanes	A cyclophane is an organic host molecule with a bridged aromatic ring. Cyclophanes contain multiple bridged benzene rings. The bridges are usually in the form of an aliphatic chain. Cyclophanes have parallel aromatic walls with a well-defined cavity. Cyclophanes are good hosts to hydrophobic apolar guests.	Porphyrins	Porphyrins are classical coordination compounds consisting of planar pyrrole rings that bind by classical chelate and macrocycle effects. They are tetrapyrrole macrocycles. Binding of Fe by heme is a well-known example.
Dendrimers	Dendrimers are cascade molecules with highly branched 3-D structure. The dendritic core is porous and capable of hosting a variety of materials. By *site isolation*, a dendrimer can be constructed from a core guest. *Guest inclusion* is the case in which the dendrimer is already formed and a guest is invited in.	Rotaxanes Pseudorotax-anes	Rotaxanes are long molecules that exist in the center of macrocyclic rings in needle–thread fashion. These are permanent structures, especially if the linear molecule is terminated with a bulky group. Pseudorotaxanes allow the central molecule to slip out. Rotaxanes are similar to catenanes.

Table 11.4 (Contd.) *Catalog of Supramolecular Host Species A–Z*

Host species	Description	Species	Description
Fullerenes	The cavity within a fullerene is a perfect place to sequester metal or other species. Metals are introduced by ion implantations and chemical techniques. Space inside carbon cage ranges from 0.4 to 1.0 nm for C_{60}–C_{240}.	Schiff's bases	Macrocycles formed from Schiff's condensation (amine + aldehyde–water = imine) to complex metals
Helicands Helicates Helical tubulands	These clathrates can coordinate metal ions or anions like Cl^-. They are usually made of organic hosts. Helicate assembly occurs with templated metal assist of organic threads that resemble DNA—transition metal-directed assembly. Only helices of strands of identical length would self-assemble. The metal ion sits in the center of periodic twists in the helix.	Sepulchrates Sarccophagines	Sepulchrates are cryptands which are made from cobalt(III) template synthesis of *tris*(chelate) complexes of bidentate ligands. As with other cryptands, the metal is sequestered. Sarcophagines contain C-bridged species.
Hoffman inclusion compounds	Belong to the family of clathrates. Lattice voids in inorganic coordination compounds with general formula: $M(NH_3)_2M'(CN)_4{\cdot}2G$ where M = Zn, Cd; M′ = Ni, Pd, Pt and G = small aromatic molelcule	Siderophores Siderands	Siderophores are naturally occurring ligands like myobactin and enterobactin that mobilize iron. Siderands are synthetic analogs. Siderophores form a 6-coordinate complex with Fe.
Hourglass inclusions	Formed by the nonstoichiometric inclusion of water-soluble organic dye molecules within an organic lattice like K_2SO_4. The colored and colorless regions overall look like an hourglass.	Spherands Spheraplexes Hemispherands	Spherands are macroscyclic cation hosts. They are based on *p*-methylanisole. Crowns and cryptands are flexible. Spherands are rigid. Due to this, spherands have high selectivity.
Hybrid hosts	Hybrid hosts form by combining two or more kinds of hosts: crown ethers, cryptands, spherands and podands, etc. A wide variety of hosts are created by mixing host types.	Speleands Speleates	A speleand is an amphiphilic receptor that combines two or more forms of guest recognition. A synergistic enhancement of binding results. Polar or charged sites can be combined with a hydrophobic residue. The active sites of enzymes certainly display speleand character.
Hydroquinone Phenol cages Hexahosts	Hydroquinone and phenol form hexagonal cyclic hydrogen-bonded rings with bulk substituted groups alternating in up–down pattern. Cavities are formed between rings. These are a kind of clathrate.	Zwitterions	Neutral molecules with both a positive and a negative charge. Anionic binding proteins and enzymes are zwitterionic.

Sources: J.-M. Lehn, *Supramolecular chemistry: Concepts and perspectives*, VCH Verlagsgesellscahft GmbH, Weinheim (1995); J. W. Steed and J. L. Atwood, *Supramolecular chemistry*, John Wiley & Sons, Ltd., Chichester (2000); P. D. Beer, P. A. Gale, and D. K. Smith, *Supramolecular chemistry*, Oxford Chemistry Primers, Oxford Science Publications, Oxford University Press, Oxford (2003).

FIG. 11.8 *(a) Chelate: $[Ru(2,2'\text{-bipyridyl})_3]^{2+}$ hexadentate chelate; (b) crown ether corand: [24]crown-8; (c) azacrown ether analog; (d) MCM-41 zeolite pore structure: 1.5–10 nm, surface area: 400–1700 $m^2 \cdot g$; pore volume limit: 1.1 $cm^3 \cdot g$; pentasil zeolite ZSM-5 0.54 × 0.56 nm; (e) cryptands are three-dimensional analogs of the crown ethers. A [2.2.2]cryptand is on the top and a [2.2.1]cryptand is on the bottom; (f) benzol[18]crown-6crown ether; (g) fullerene hosting a xenon gas atom; although it is a host molecule, would you consider a fullerene a supramolecule? (h) A podand: pentaethyleneglycol dimethylether if R = Me (EG5); (i) p-tert-butylcalix[6]arene, a cyclophane, is pictured on the bottom; a spherand is pictured on top; (j) [2]catenane; (k) α-cyclodextrin [6-sugar ring]; (l) dendrimer made from amine groups—poly(propyleneimine); (m) trimesic acid (1,3,5-benzentricarboxylic acid) tubuland network precursor; (n) helicate precursor 2,2′:6′,2″:6″,2‴:6‴,2⁗:6⁗,2′′′′′-sexipyridine; (o) and (p) molecular squares $[(en)Pd(4,4'\text{-bipy})]_4(NO_3)_8$, and $[Cd(4,4'\text{-bipy})]_4(NO_3)_8$.*

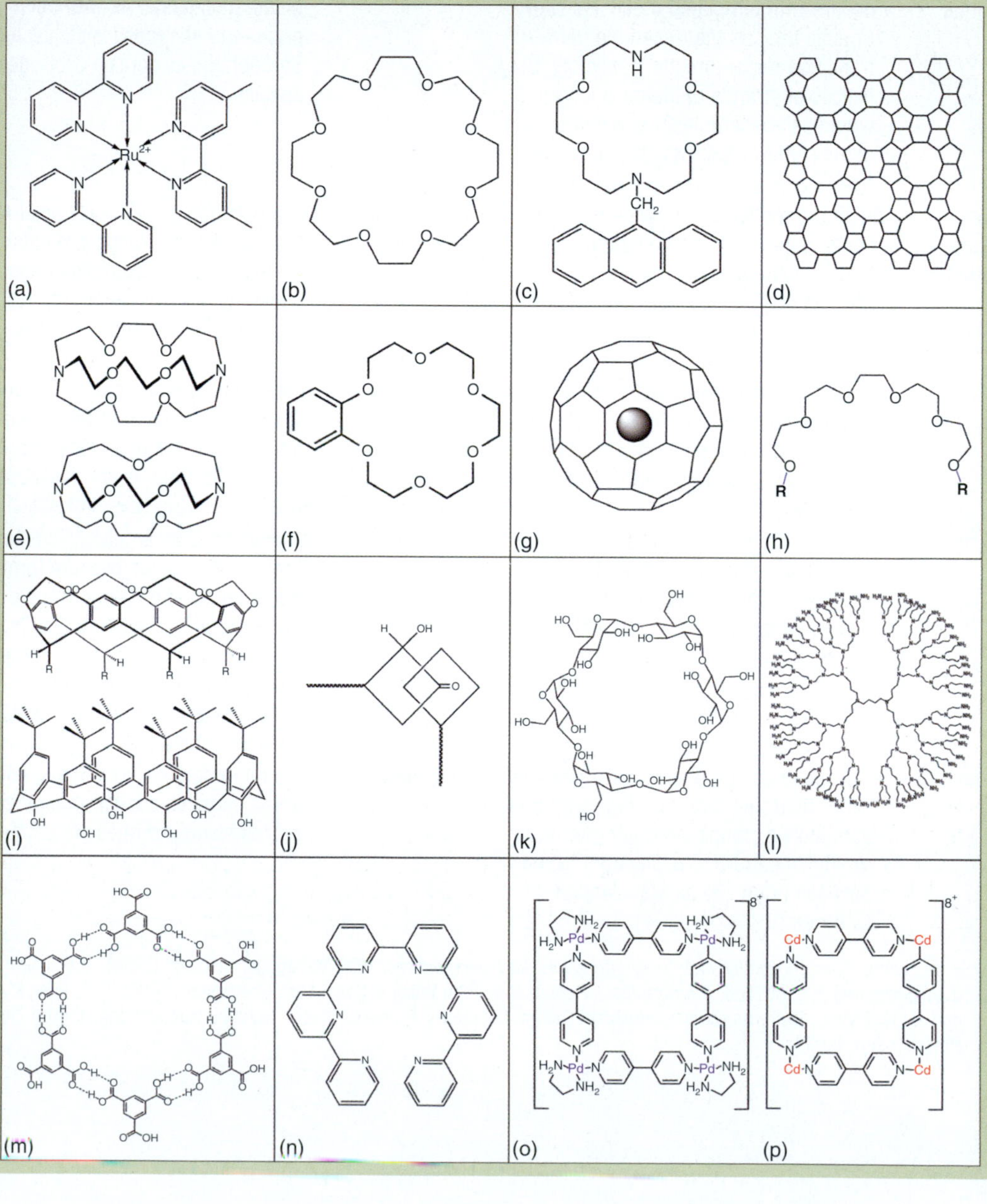

complex with metal ions to form very stable structures and bind metals on the basis of Lewis acid–base recognition—dative coordinative bonds that collectively contribute to form a complex (**Fig. 11.8a**). In the broadest sense, many host species can be considered to be glorified chelates.

Clathrates. Clathrates (and intercalates) form the largest and most diverse class of host materials. Inorganic, organic, biological, mineral clays, graphite, and polymers all have representative clathrate compounds. Clathrates (from the Greek *klethra,* "bars") are chemical substances that consist of a lattice that is able to sequester or trap guest species. A clathrate hydrate (a gas hydrate) is pictured in **Figure 11.9**.

Crown Ethers. Supramolecular chemistry began with the crown ether (**Fig. 11.8b, c, f**). Crown ethers have oxygen atoms that link aliphatic ethyl bridges to form a circular macrocycle (heterocycle). Crowns that contain nitrogen-bearing ligands are called azacrowns. Crown ethers belong to a class of host molecules called *corands*. The cavity diameter of crown ethers can be tailor fitted to accommodate

Fig. 11.9 *A clathrate consisting of a water cage enclosing one methane molecule is depicted. This structure is called a methane clathrate, methane hydrate, or even methane ice. Its physical form is that of solid water that contains a significant amount of methane. Naturally occurring methane clathrates are found under sediments on the ocean floor. The clathrate remains stable up to 18°C. There is usually 1 mol of CH_4 per 5.8 mol of water with a density of 0.9 $g{\cdot}cm^{-3}$. The basic structure consists of 20 water molecules to form a dodecahedral shape. If oceans were to warm further, the release of methane would significantly contribute methane to the atmosphere and further exacerbate global warming.*

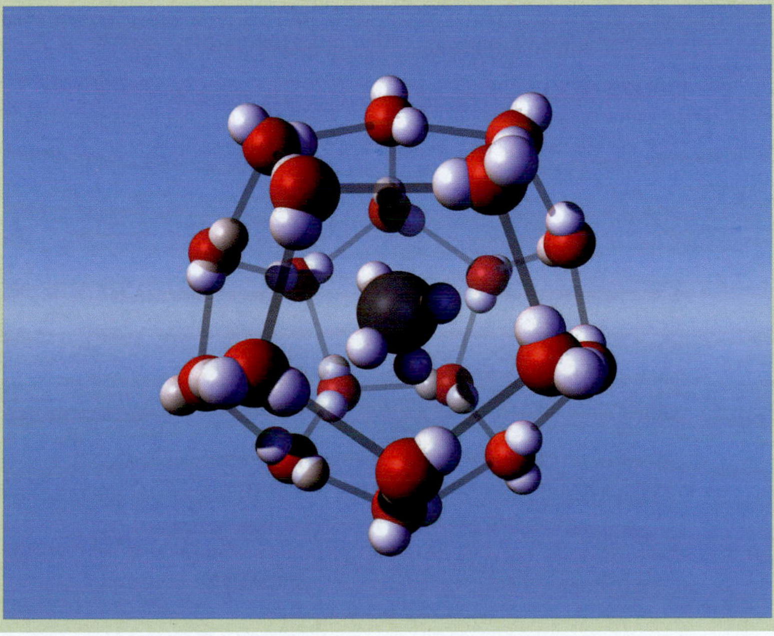

Source: Image courtesy of Anil K. Rao. With permission.

cations of differing sizes. The [12]crown-4 (four oxygens with four aliphatic ethyl group bridges) has a cavity size of 1.2–1.5 Å, perfect for a Li^+ cation. The [21] crown-7 possesses a cavity that is 3.4–4.3 Å, perfect for the accommodation of a Cs^+ cation [8].

Cryptands. Cryptands are cage-like bicyclic molecules that are three-dimensional analogs of the crown ethers [8]. Mixed cryptands contain other atoms besides oxygen, usually nitrogen (**Fig. 11.8e**). Amine-based cryptands have an affinity for binding the ammonium cation, alkaline earths, and lanthanides. The cavity of cryptands, as the name implies, is deep within the molecule. Therefore, once a metal is bound within the cavity, it tends to remain there.

Cyclophanes. Cyclophanes consist of benzene rings held together by aliphatic or amine bridges (**Fig. 11.10**). A positive charge imparted to cyclophanes is able to enhance its selectivity towards anionic-hydrophobic guests. In this way, the positive charge attracts the anion and the hydrophobic pocket provides for the binding specificity [8]. The size of the cavity depends on the number of benzene rings in the macrocycle.

Dendrimers. Dendrimers are incredible molecules with a fractal-like quality very much like the roots or branches of a tree [36]. A fractal is a structure that does not exist in integer space but rather a space governed by fractions—the dimensionality of a fraction [36]. It is a structure that is subdivided but one in which each subdivision resembles the structure as a whole—a property known as *self-similarity*. Dendrimers are cascade molecules with highly branched three-dimensional structure and are often spherical in shape (**Fig. 11.8l**) [36]. The subunits of dendrimers are relatively simple amines or other branchable molecules. There are three general ways to synthesize dendrimers: divergent, convergent, or by site isolation. Site isolation technique begins with a core "guest" from which the dendrimer is constructed. In divergent methods, the dendrimer is assembled from the center to the periphery. In the convergent method, the opposite strategy is employed. Each layer (or generation) is built step by step. The dendritic core is porous and capable of hosting a variety of materials. For

FIG. 11.10 *Cyclophanes are organic host molecules that contain bridged aromatic rings that may contain a cavity [7]. A [2,2]-cyclophane is depicted. With appropriate positioning (preorganization), cyclophanes are able to contribute steric and electrostatic complementarity to a potential guest [7]. Calixarenes and hemicarcerands are also cyclophanes. Cyclophanes are able to bind neutral molecules and cations.*

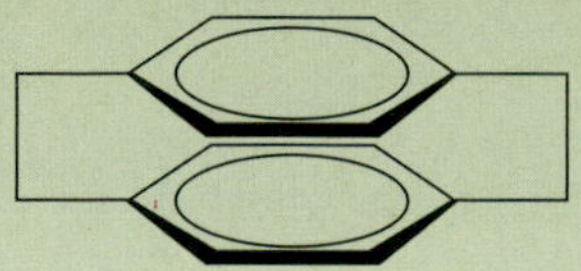

Fig. 11.11 *A dendrimeric synthesis scheme is depicted. 1-Bromopentane undergoes nucleophilic substitution with triethyl sodiomethanetricarboxylate. Lithium aluminum hydride is then applied to reduce the ester groups to alcohols. Functionalization of the terminal groups into good leaving groups (tosylates) prepares the process for another generation of growth. The process continues until a dendrimer is synthesized.*

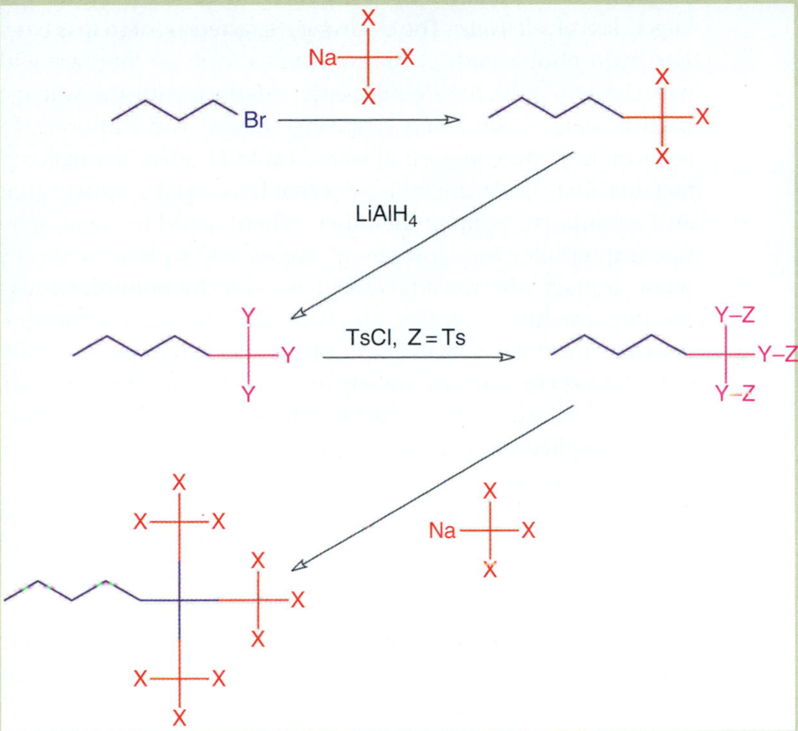

example, a dendrimer can be constructed from polyethylene glycol arms attached to an iron–porphyrin core to form a dendritic water-soluble heme [7].

The first dendrimer (the *Newkome* dendrimer) was synthesized in 1985. Dendrimers are synthesized by a series of repetitive growth-activation steps (**Fig. 11.11**). First, 1-bromopentane [$CH_3(CH_2)_3CH_2$-Br] underwent nucleophilic substitution by triethyl sodiomethanetricarboxylate [$(CO_2Et)_3Na$] in DMF (dimethlyformadide) and benzene at 80°C. Lithium aluminum hydride [$LiAlH_4$] in diethlyether was used to reduce the ester groups to alcohols. The terminal groups were then transformed into tosylates (good leaving groups) by the application of tosyl chloride for further activation to complete the first generation of the dendrimer. Nucleophilic substitution was then performed again. The process was repeated for each generation of the dendrimer structure.

Recently, a Swedish research team headed by Michael Malkoch found a way to eliminate half the number of steps involved in dendrimer synthesis [37,38]. In the new method, two different monomers are used alternatively without the need for an activation step. B. Sharpless, of the Scripps Research Institute in La Jolla, stated in 2007 that "Malkoch and colleagues have provided a beautiful new dendrimer synthesis. It is simple and efficient, and to the best of my knowledge, represents the state of the art."

11.1.4 Surfactants and Micelles

Surfactants and Micelles. Surfactants are surface-active agents. They are part of the larger class of *soft matter*. The cleansing characteristic of soap is based on surfactants. The hydrophobic ends of the surfactants form an interface with insoluble dirt particles while the hydrophilic ends interface with the water; this renders the particle water soluble and cleansing ensues. The addition of surfactants also reduces the surface tension of water. **Table 11.5** lists the major categories of surfactants. *Amphiphiles* are molecules that have a polar moiety attached to one end and a nonpolar group at the other. When placed in water, self-organization of the amphiphiles into low-energy forms takes place as the hydrophilic ends retain contact with the aqueous phase and the nonpolar ends agglomerate by creating a middle where water is absent as is the case for micelles. At a gas–liquid interface, the polar ends of amphiphiles stay in the water while the tails interface with the gas (usually air), thereby forming a film across the surface. Depending on the chemical nature of the amphiphile, many different structures are possible, from spheres to cylinders to planar bilayered structures, giving supramolecular design using amphiphiles great flexibility.

The simplest amphiphilic structure is the micelle. Although micelles are homogeneous in the sense that they consist of the same molecule, they must be considered as supramolecular structures. Why? The molecular components of micelles are held together by intermolecular forces. Assuming that all amphiphiles in a solution are identical, predictions about their structure are made by application of the *critical packing parameter* [37]. The *cpp* is given by

$$cpp = \frac{V}{A_{\text{cap}} L_{\text{cap}}} \tag{11.1}$$

where V is the volume of the amphiphile, A_{cap} is the area of the head-group, and L_{cap} is the length of the head-group. Packing shapes vary from cone like to cylinder like to wedge like. For single tail surfactants that have relatively large head-groups (e.g., sodium dodecyl sulfate, SDS), *cpp* is equal to 1/3 and the packing shape is conical with the tails petering out into a point. Spherical micelles conform to these criteria as well. Single-tailed amphiphiles with small head-groups (e.g., hexadecyltrimethylammonium bromide, CTAB) pack into a truncated cone with *cpp* between 1/3 and 1/2 and show incipient cylindrical tendencies. Double-tailed amphiphiles with large heads (e.g., phoshpatidic acid) form truncated cones (*cpp* = 1/2–1) that in turn form flexible bilayers. Double-tailed amphiphiles with small heads (e.g., phosphatidyl serine) form cylinders (*cpp* = 1) that settle eventually into planar bilayers (like membranes) or into inverted truncated cones (wedges). In the cases where *cpp* > 1, inverted micelles (reverse micelles) are formed under the proper solution conditions (e.g., phosphatidyl ethanol amine) [37].

Each micellular system is characterized by a unique *critical micelle concentration* (CMC). Surfactants exist as independent molecules up until a limiting condition at the point of the CMC. It is one of the most important surfactant

TABLE 11.5 *Types of Surfactants*

Type of surfactant	Comments	Example
Anionic	An example of an amphiphilic molecule with hydrophobic tails and hydrophilic head groups	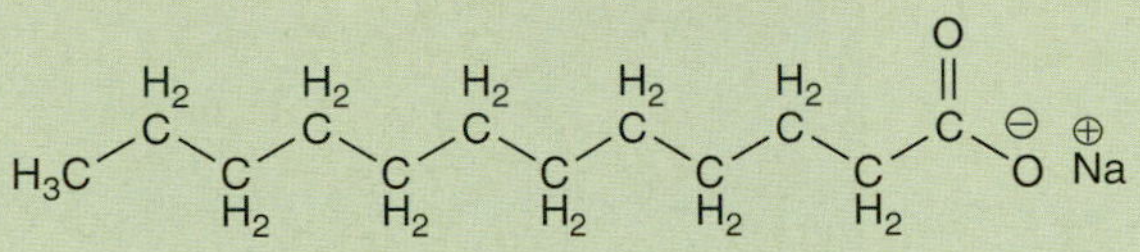 *Sodium laurate [$CH_3(CH_2)_{10}(CO_2)Na$] anionic surfactant commonly found in shampoo. (Image redrawn with permission from Tapani Viitala, KSV Instruments, Ltd.)*
	Double-tailed anionic surfactant	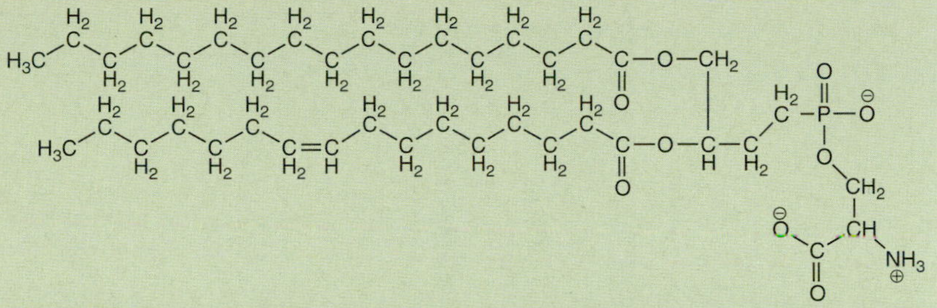 *Dipalmitoyl phospahtidyl serine is a double-tailed anionic surfactant.*
Cationic	Used as a topical antiseptic against bacteria and fungi. It is also used in a buffer solution for extracting DNA.	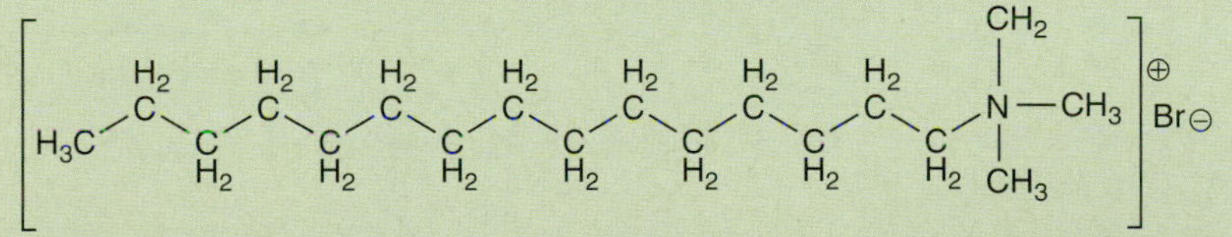 *Cetrimonium bromide is a cationic surfactant.*
Non-ionic	Non-ionic surfactants are used in detergents that are able to dissolve proteins. Also act as emulsifiers in beverages and food products	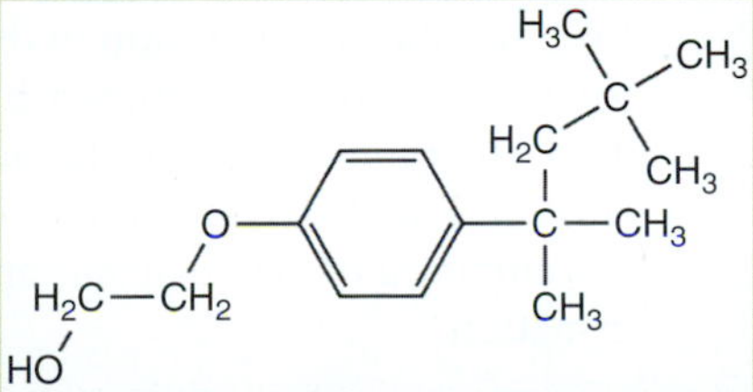 *Triton X-100 [$C_{14}H_{22}O(C_2H_4O)_n$] is a non-ionic surfactant. The hydrophilic contribution is from the poly(ethylene oxide) or from a sugar. Nonionic surfactants are used in detergents and are able to solubilize membrane proteins.*
Zwitter-ionic	Zwitterions have both acidic and basic functional groups on the same molecule.	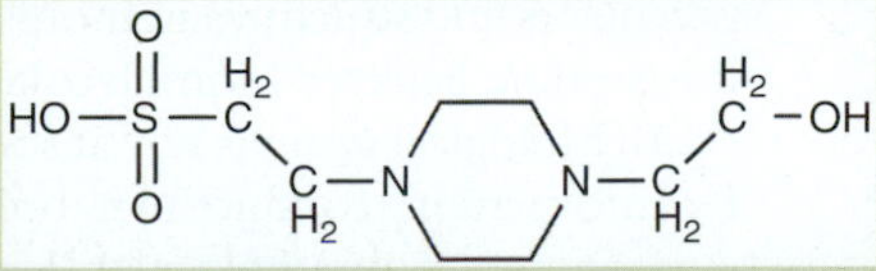 *HEPES[4-(2-hydroxyethyl)-1-piperazineethanesulfonic acid] is a zwitterionic surfactant that contains both anionic and cationic groups. The molecule exists in the zwitterionic form at certain pH.*

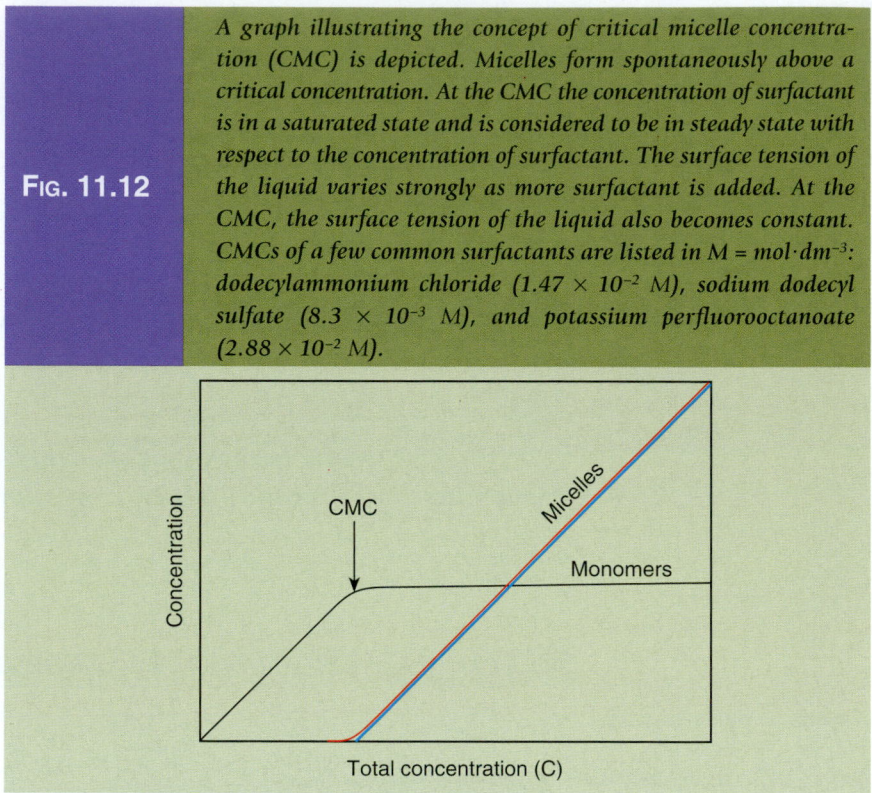

FIG. 11.12 *A graph illustrating the concept of critical micelle concentration (CMC) is depicted. Micelles form spontaneously above a critical concentration. At the CMC the concentration of surfactant is in a saturated state and is considered to be in steady state with respect to the concentration of surfactant. The surface tension of the liquid varies strongly as more surfactant is added. At the CMC, the surface tension of the liquid also becomes constant. CMCs of a few common surfactants are listed in M = mol·dm^{-3}: dodecylammonium chloride (1.47 × 10^{-2} M), sodium dodecyl sulfate (8.3 × 10^{-3} M), and potassium perfluorooctanoate (2.88 × 10^{-2} M).*

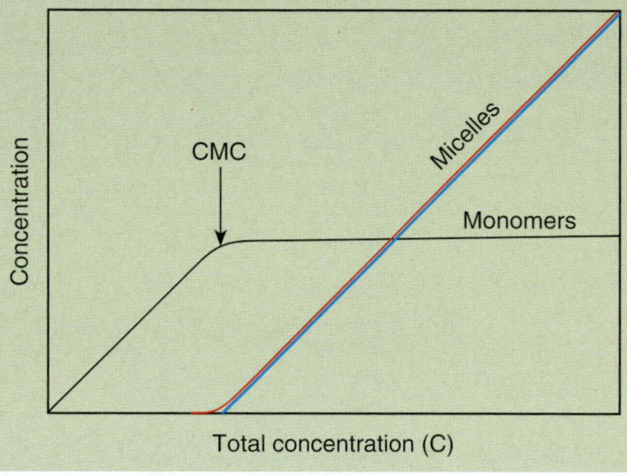

Source: Image redrawn with permission from Tapani Viitala, KSV Instruments, Ltd.

properties to chemical engineers. In plots describing the concentration of surfactant and micelles, micelles form when a critical concentration is exceeded. At this point the concentration of the micelle increases linearly, but the concentration of surfactant monomers achieves a steady-state concentration (**Fig. 11.12**). The CMC of anionic surfactants is $\sim 10^{-2}$ *M* and that for nonionic surfactants $\sim 10^{-4}$ *M*. For SDS, CMC = 8.3×10^{-3} *M*. This means that, at this molarity, the concentration of SDS plateaus and solution dynamics is dominated by micelle formation.

11.1.5 Biological Supramolecular Host Species

We are all quite familiar with nature's arsenal of biological host molecules and host nanomaterials. Chapters 13 and 14 will address natural supramolecular precursors and structures in more detail. We therefore will not dwell on them in this section. Selected biomolecular hosts are shown in **Table 11.6.**

All biological systems rely at some level on supramolecular structure to transfer information, conduct metabolism, build structures, and interface with the environment. Supramolecular hosts in the form of enzymes, DNA, heme, ionophores, neuroreceptors, and chemical receptors accommodate diverse substrates like sugars, intercalation materials, transition metals ions, alkali metal ions, acetylcholine, and hormones, respectively. In the case of heme, once iron is

TABLE 11.6 *Biological Supramolecular Structure Precursors and More*

<table>
<tr><th>Biological supramolecule</th><th>Comments</th><th>Examples</th></tr>
<tr><td>Enzyme pockets</td><td>70–75%
Substrates and cofactors are anions.
Substrates include phosphites, phosphates, sulfates, chloride</td><td>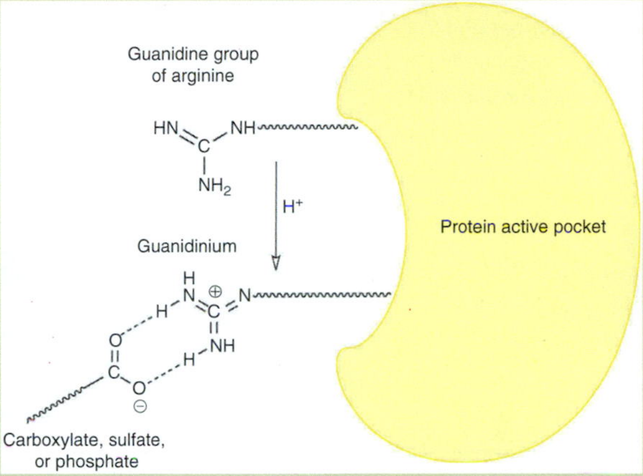

Generic rendition of an enzyme pocket is shown. Many cooperative intermolecular and intramolecular (bond breaking) reactions occur in such pockets.</td></tr>
<tr><td>Ionophores</td><td>Membrane transport of cations.
Lipophilic carriers of cations
Valinomycin K^+/H^+
Nonactin and enniatins have less selectivity for cations than valinomycin and nonactin.</td><td>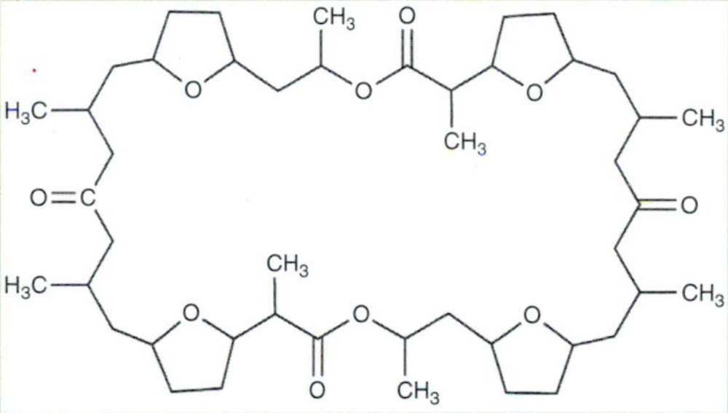
The ionophore nonactin is depicted. Its action is similar to that of valinomycin. The ionophore is specific to K^+.</td></tr>
<tr><td>Coenzymes</td><td>NAD^+ and NADH
Electron transfer in oxidoreductase enzymes
Molecule is all covalent bonded.
Action is by intermolecular attractions.</td><td>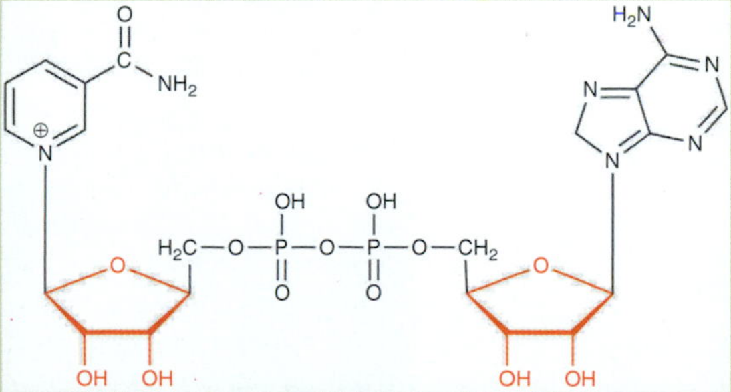
Nicotinamide adenine dinucleotide (NAD^+) is a coenzyme that provides the vehicle to carry electrons. It therefore is a major participant in redox reactions. NAD^+, an oxidizing agent, is reduced to NADH during metabolic processes. NAD is a coenzyme that is able to transfer electrons in the form of hydrides.</td></tr>
</table>

(*continued*)

Table 11.6 (Contd.) *Biological Supramolecular Structure Precursors and More*

Biological supramolecule	Comments	Examples
Natural chelation agents	Synthesized by bacteria (*Streptomyces*) Used as antibiotic	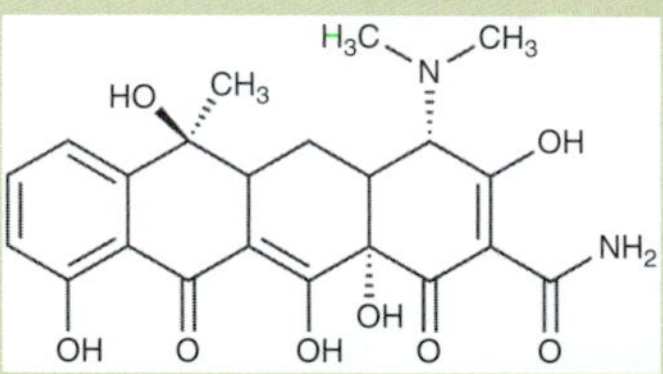 *Tetracycline is a natural chelation agent that is produced by the bacterium Streptomyces. As an antibody, its function is based on the ability to inhibit the action of ribosomes in procaryotic bacteria by binding with t-RNA. In other words, it has the capability to inhibit protein synthesis. Is it the host or is t-RNA the host?*
Porphyrins	Chlorophyll Porphyrins Tetrapyrroles Vitamin B_{12}	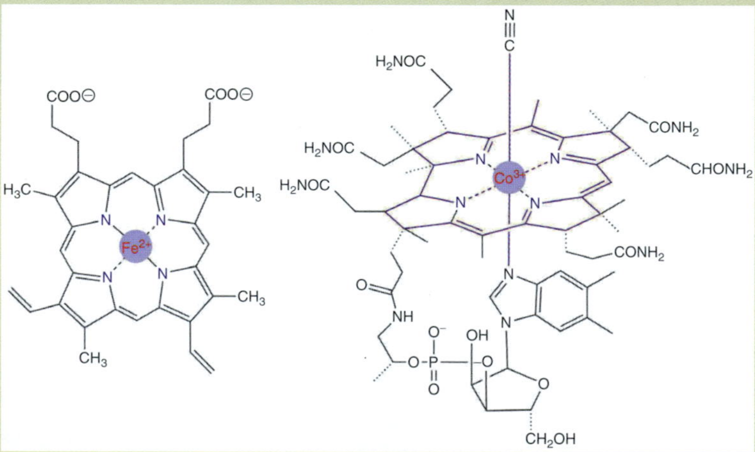 *Left: A porphyrin ring guest coordinates an iron cation. In unoxygenated form, the ground state Fe cation has a +2 charge and exists in the Fe(II) state. It is paramagnetic with four unpaired electrons. In this configuration, the heme complex has a domed shape [7]. In the bound state, electron transfer occurs from Fe(II) to the oxygen to form an Fe(III) species. The system is diamagnetic and the porphyrin ring is flattened out. This example illustrates an intricate relationship between the host (the porphyrin ring), the guest (the iron cation), and a substrate (the oxygen molecule). The binding of O_2 is influenced by the allosteric effect in that as soon as one O_2 is bound, the affinity for oxygenation increases until all sites in hemoglobin are filled. Right: vitamin B_{12} is depicted. Cobalt(III) is coordinated by six dative bonds to anchor the core of the vitamin.*

TABLE 11.6 (CONTD.) *Biological Supramolecular Structure Precursors and More*

Biological supramolecule	Comments	Examples
	Chlorophyll-*a*	
	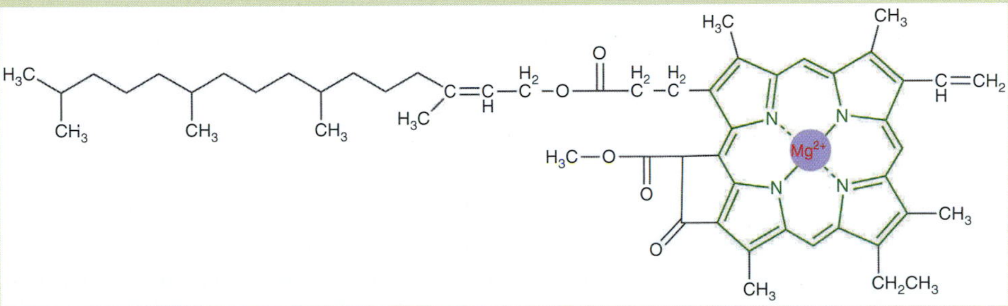	
	Chlorophyll-a is a conjugated π–system that is characterized by low-energy π–π electron transition capability [7]. Perhaps it is one of the most amazing of the supramolecules; it captures photons and begins the process of making energy that drives most forms of life on this planet. The magnesium tetrapyrrole complex is anchored by the Mg^{2+} cation. The long aliphatic (phytyl) chain anchors the molecule into a phospholipid membrane of the chloroplast. Light harvesting antennae produce an electron on excitation that is transferred away from the excited state center to a chemical reduction center. The hole left behind is utilized to oxidize water into oxygen. The superstructure consisting of chlorophylls, electron transfer moieties, and all other infrastructure must be considered to be the premier accomplishment of natural supramolecular chemistry—all existing at the nanoscale nonetheless.*	

incorporated during a "primary level host–guest synthesis process," it becomes the guest of a quaternary protein structure to become the integral functional part of hemoglobin. All of this is in place so that hemoglobin can perform the vital function as host to molecular oxygen.

Molecular recognition is the foundation of biological materials and of life itself. Life began with the earliest forms of recognition: Two molecules combined to form a lower energy, lower entropy but restless state. Molecular recognition operated at the simplest level: the binding and transport of metal cations by simple molecules and the self-assembly of lipid bilayers. For us, it now operates at the highest level, where we rely on molecular recognition to guide our senses, our neural networks, and our chemical receptors to cope with the pressures and stresses of the outside world. Adrenaline, acetylcholine, dopamine, vitamins, and many more all have the appropriate receptors in our bodies to help us survive—where other supramolecules like DNA can make sure that we continue.

11.2 Design and Synthesis of Selected Supramolecular Species

"The Chemist as Engineer and Architect" is a title that should be applied to this section. Although chemists have been designers of chemicals since making the scene in ancient times, and especially in the glory days of organic and polymer chemistry and later for pharmacological synthesis, this field offers as much as or more in terms of creativity, engineering, and design. The supramolecular chemist is interested in systems of great complexity, adding molecule after molecule to make species that can be quite large. Supramolecular chemistry, the chemistry of self-assembly and modification of nanomaterials, offers new sets of challenges as its golden age is upon us with no perceivable limit in sight.

The transformation of materials from precursors and substrates into hierarchical nanostructures is reviewed in **Figure 11.13.** Precursor molecules, synthesized by traditional chemical methods, are groomed to become hosts (and guests) and then undergo transformation into supramolecules. Once formed, supramolecules react, again primarily via intermolecular interactions, to form integrated structures. Although we are not emphasizing applications in the text, some are listed to provide perspective.

Supermolecules are large "molecules" that are held together by covalent, ionic, or metallic bonds; others provide alternative definitions. Supermolecules are largely homogeneous (e.g., they are made of repeating units of the same molecule or atom) and are not in the strictest sense considered to be hosts or guests. Obviously, carbon nanotubes, gold surfaces, and latex beads are members of this class of materials. Supermolecules therefore logically serve as substrates for other chemical species. Substrates are activated chemically and bonds are formed with surfaces via chemisorption (covalent) or physisorption processes. Therefore, supermolecules are not supramolecules, but they serve an important role as substrates that are able to anchor interesting species. Many types of substrates are nanomaterials.

The major components involved in a supramolecular design process are summarized in **Figure 11.14.** A simple but rather obvious rule applies to supramolecular synthesis: "The larger the system is, the more complicated its synthesis becomes" and the more aspects need to be considered. We start once again with precursors—more specifically, the selection of precursors that have the capability to serve as a host species in a potential supramolecular synthesis. Most precursor species are already available for commercial consumption—obviously, the best-case scenario.

The design process involves the following key elements: (1) selection of precursor molecules (starting materials); (2) analysis of thermodynamic and kinetic variables (this includes consideration of solvent effects, intermolecular interactions, and reaction schemes); (3) selection or design of receptors and substrates (another way to say "host–guest"); (4) design of the target hierarchical structure; and (5) process termination (passivation) schemes. Integral to every step of the process are: (a) the synthesis conditions, (b) purification steps (if required), and (c) characterization of products (and, of course, abiding by all safety protocols!). One is compelled to ask what the difference is between this and any other usual form of chemical synthesis. The major difference is stated in (3): the host–guest relationship.

Jean-Marie Lehn, one of the founders of the field, provides the following strategy [4]:

Precursors → receptor synthesis → host–guest reaction → supermolecule

Lehn uses the term supermolecule (not by our definition). There is more:

FIG. 11.13 *The synthesis process is tracked through a material perspective. Starting with precursors, supramolecular synthesis proceeds through a succession of more complex interactions until an end product is obtained—whether a device or a natural product. Supermolecules, according to our definition in this text, are held together with covalent, ionic, or metallic bonds and are not perceived as host species, although good arguments can be made for the opposite case (e.g., carbon nanotubes and polymer wrapping). Substrates in general can be considered to be supermolecules. Even in nature, supramolecules rely on supermolecules for support, like the phosphate–deoxiribose backbone in DNA. Are peptides supermolecules? Are they precursors? According to our definition, they are precursors (or substrates). Proteins, on the other hand, are supramolecules because intermolecular interactions fold them into large molecules with unique conformation (and function).*

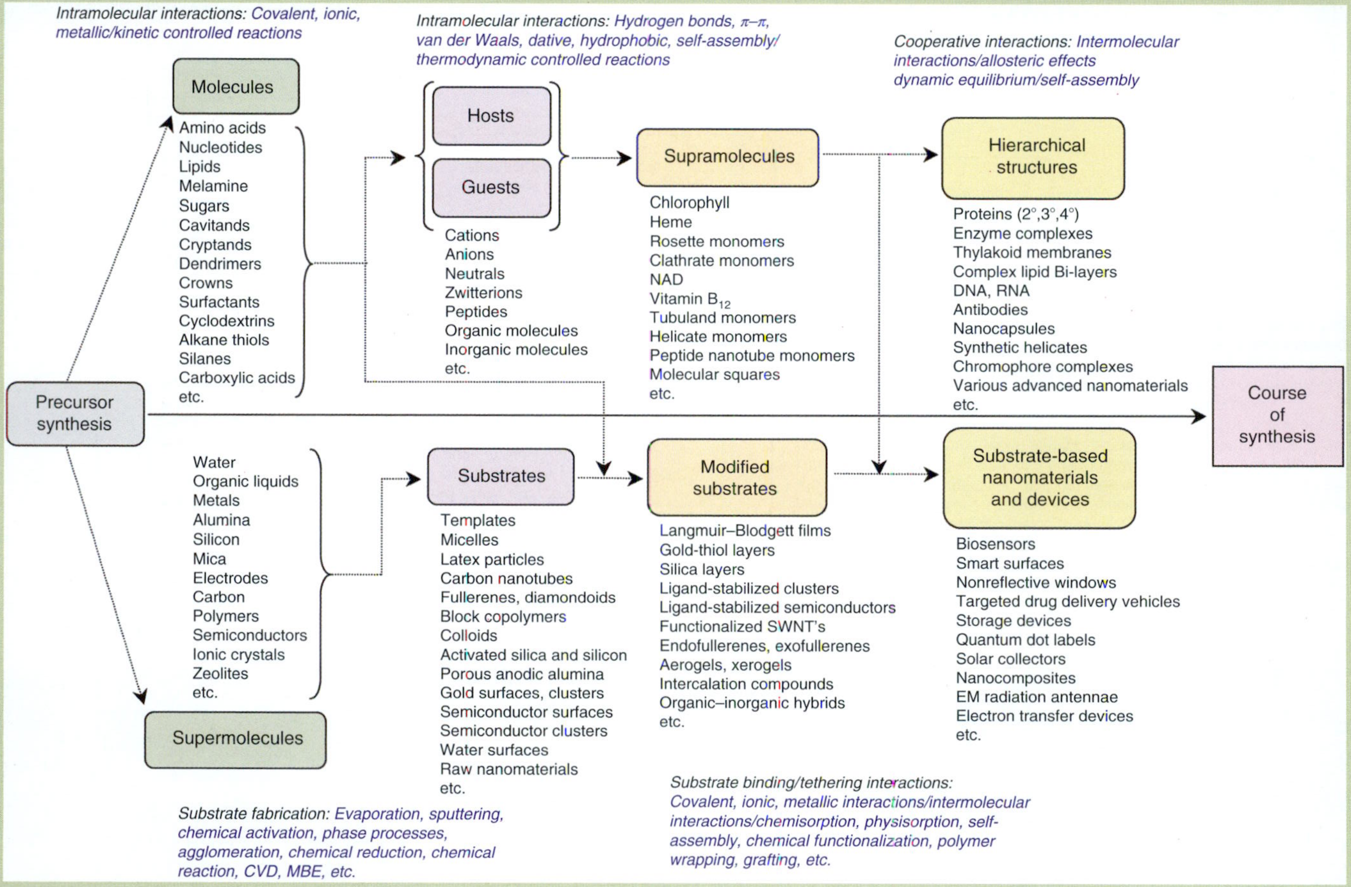

FIG. 11.14 *One form of a supramolecular design process is summarized. Preorganization implies that the host molecule does not undergo significant conformational changes upon accommodation of a guest species. Complementarity is achieved after preorganizational deliberations are accomplished. Complementarity simply implies a good stable fit between guest and host. Obviously, thermodynamic and kinetic considerations permeate all of supramolecular chemistry. There are many factors that affect the thermodynamics and kinetics of reactions and a few are summarized in the figure. The solvent is another crucial variable in these reactions. When all is in place, activation takes place and the reaction concludes with the formation of a host–guest complex. With due foresight and consideration, hierarchical structures consisting of supramolecular subunits self-assemble into a functioning nanomaterial.*

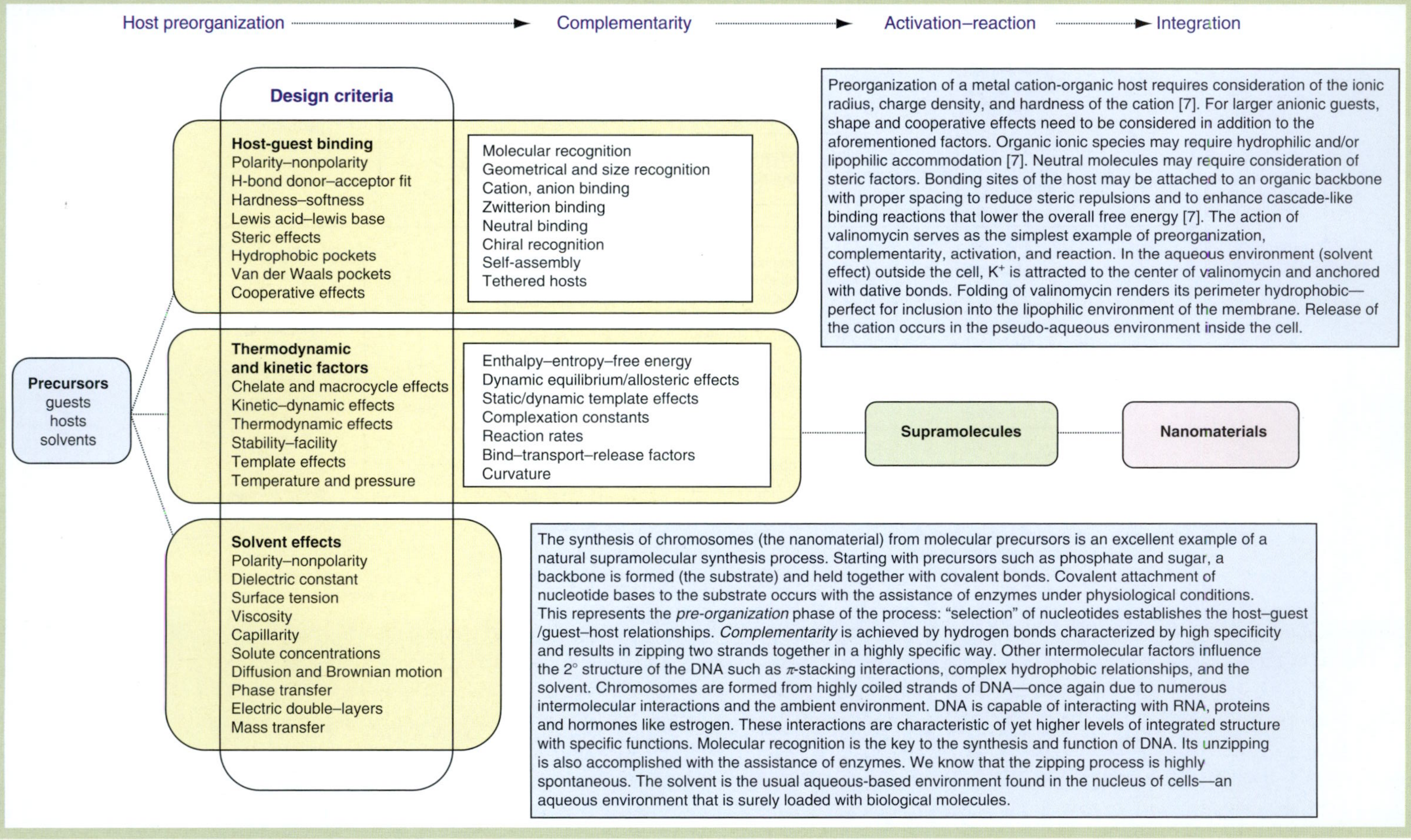

Supermolecule → recognition, transformation, translocation → polymolecular self-assembly/self-organization + functional components → molecular and supramolecular devices

Jonathan W. Steed and Jerry L. Atwood [7] place emphasis on the following parameters:

Chelate–macrocycle effects + preorganization–complementarity + thermodynamic–kinetic selectivity + intermolecular–supramolecular interactions → HOST DESIGN

Not all of these categories are independent parameters. Accordingly, host *preorganization* is perhaps the most important aspect of the design process—the host–guest relationship: Is there *complementarity* between the host and the guest? Do electronic and steric interactions, physical size, molecule spacing, and potential conformational changes of the host match those of the guest? Electronic compatibility includes polarity and charge, hardness–softness, H-bond alignment, and Lewis acid–base complementarity, to name just a few [7].

Successful *preorganization* results in hosts that exhibit little conformational changes after binding with a guest [7]. This is an important factor in host–guest design because the most stable complexes are achieved with preorganized hosts [7]. Interestingly, dramatic (and dynamic) conformational changes occur routinely in enzymatic hosts during and after binding. The conformational changes serve to strain bonds in substrates once drawn into the active pocket of an enzyme. It is much easier, for example, to break strong covalent bonds that are artificially lengthened or bent in this way. The coordination of substrates with one water molecule is a kinetic effect that facilitates the dissociation of substrate bonds by providing a sources of hydrolysis of polarity required for orientation. In an ideal synthetic world, we wish our supramolecular structures that comprise the bulk of the nanomaterial to be static once formed, unless the purpose is that of sensing. In sensors, allowing for guests to "check out" ensures that the sensor is able to continue with its purpose.

11.2.1 Thermodynamic and Kinetic Effects

Thermodynamic and kinetic factors are major players on the supramolecular design stage. The biological domain offers remarkable and numerous examples of selectivity. It also offers examples of nonselectivity (e.g., the oxidation–reduction of many kinds of substrates by oxidoreductases implies there is some flexibility in choice of substrate). One of the prime directives of supramolecular chemistry is selectivity because stable structures are preferred to those that accommodate more than one species. Obviously it would be problematic to build a uniform structure from a pool of similar but structurally varied materials. Therefore, analysis of binding constants and kinetic parameters of potential guests beforehand is a prudent path to take.

Selectivity and stability of the resulting host–guest complex are interrelated. The "right guest" usually tends to form the most stable H–G complex. Chelate effects, macrocycle effects, intermolecular interactions, solvent effects, concentration profiles, and reaction rates all influence the thermodynamics and kinetics of a potential H–G couple. Selectivity by itself is not enough, however.

The guest molecule must also form a thermodynamically stable complex with the host—in some instances, one that is able to release the guest if necessary (as in living things). For synthetic purposes, irreversible complexes are usually desired except in the case of sensors where reversible binding is preferred [7]. The magnitude of the binding constant (a.k.a. complexation, association, or formation constant) reflects the degree of thermodynamic stability of a host–guest complex:

$$H + G \leftrightarrow HG \tag{11.2}$$

$$K_{f,1} = \frac{[HG]}{[H][G]} \tag{11.3}$$

where K_f is the equilibrium constant, and H and G represent the concentrations of the host and guest, respectively. Selectivity with regard to the equilibrium constants of two or more potential guest molecules is the ratio of the respective binding constants [7]:

$$Selectivity = \frac{K_{G,1}}{K_{G,2}} \tag{11.4}$$

The proper choice of guest (lock and key), preparation of the host (preorganization), and complementarity (host–guest fit) all contribute to a successful supramolecular design process [7,8].

The Chelate and Macrocycle Effects Revisited. Binding constants are estimated by experiment and by simulation. Molecular dynamic modeling in particular has proven to be extremely useful in describing H–G energy and structures. Numerous experimental methods, some direct and some indirect, have been employed to determine binding constants and other H–G-related parameters. Calorimetry, solubility, titration, potentiometry, spectroscopy, x-ray crystallography, and the vast array of nanotools discussed in chapter 3 have all provided useful information concerning substrate selection and binding.

Although binding constants have a favorable enthalpy due to the chelate or macrocycle effects, the entropy of their formation is usually unfavorable [8]. The association constant of a 1:1 (monodentate) H–G reaction was shown in equation (11.3). For metal binding, chelates with a higher number of binding sites demonstrate dynamic molecular recognition as the binding constant changes with the addition of another guest:

$$G + G \leftrightarrow HG_2; \quad K_{f,2} = \frac{[HG_2]}{[HG][G]} \tag{11.5}$$

and the overall reaction (with equation 11.2):

$$H + 2G \leftrightarrow HG_2; \quad \beta_2 = \frac{[HG_2]}{[HG][G]^2} \tag{11.6}$$

With $\beta_1 = K_{f,1}$, β_n overall is [8]

$$\beta_n = \prod_{1}^{n} K_{f,n} \tag{11.7}$$

The macrocycle effect is similar to the chelate effect. Macrocycles are more stable than their respective open chain analogs [8]. Increased stability of macrocycle complexes is because; (1) macrocyclic hosts are less solvated (therefore require less energy to desolvate); (2) macrocycles are rigid to begin with and therefore potential ΔS upon complexation is smaller than for a flexible system; and (3) macrocycle complexes are relatively inert chemically compared to their acyclic analogs [8]. Although having one more binding site in a single host is better in terms of complex stability, larger cycles do not necessarily translate into more stable complexes [7]. Formation constants range from 10^2 to 10^4 for podates (single chain complexing agents) and 10^4 to 10^6 for corates (cyclic complexing agents) due to the macrocyclic effect. The *macrobicyclic effect* demonstrates even higher K_f. A macrobicycle host is characterized by two macrocyclic rings linked at the poles and is represented by the cryptands. The formation constant macrobicycle complexes like cryptates range from 10^6 to 10^{12} [7].

Recall from earlier discussions that chelates and templates exert a kinetic effect on processes. The ability to bring reactants into close proximity results in species that may not ordinarily form. This is a kinetic effect. Kinetic studies are based on reaction rates, concentration, and temporal coordinates.

A study by Nachtigall, Lazzarotto, and Nome provides an excellent example of a complex binding study [38]. The host is calix[4]arene (**Fig. 11.15**).

EXAMPLE 11.1 *Coordination Thermodynamics and the Chelate Effect*

Calculate ΔG and the formation constants K_f (and log K_f) for the following coordination reactions at 298 K and discuss the reasons for the differences. M = metal cation.

(a) $M^{2+}{}_{(aq)} + 4NH_{3(aq)} \rightarrow [M(NH_3)_4]^{2+}_{(aq)}$ $\Delta H° = -53$ kJ·mol^{-1}, $\Delta S° = -42$ J·mol^{-1}·K^{-1}
(b) $M^{2+}{}_{(aq)} + 4CH_3\text{-}NH_{2(aq)} \rightarrow [M(CH_3\text{-}NH_2)_4]^{2+}_{(aq)}$ $\Delta H° = -57$ kJ·mol^{-1}, $\Delta S° = -67$ J·mol^{-1}·K^{-1}
(c) $M^{2+}{}_{(aq)} + 2(NH_2\text{-}CH_2\text{-}CH_2\text{-}NH_2)_{(aq)} \rightarrow [M(NH_2\text{-}CH_2\text{-}NH_2)_2]^{2+}_{(aq)}$
$\Delta H° = -56$ kJ·mol^{-1}, $\Delta S° = 10$ J·mol^{-1}·K^{-1}

Solution

In all cases, use $\Delta G° = \Delta H° - T\Delta S°$ and $\Delta G° = -RT \ln K_f$ (where $\log K_f = \ln K_f/2.303$)

(a) $\Delta G = -53$ kJ·mol^{-1} $-$ (298 K)(-0.042 kJ·mol^{-1}·K^{-1}) $= -41$ kJ·mol^{-1}
$\ln K_f = \Delta G/[-RT] = -41$ kJ·mol^{-1}/[-0.00834 kJ·mol^{-1}·K^{-1} × 298 K] = 16.5
$\log K_f = \ln K_f/2.303 = 7.2$; $K_f = 1.5 \times 10^7$ M^{-1}
(b) $\Delta G = -57$ kJ·mol^{-1} $-$ (298 K)(-0.067 kJ·mol^{-1}·K^{-1}) $= -37$ kJ·mol^{-1}
$\ln K_f = \Delta G/[-RT] = -37$ kJ·mol^{-1}/[-0.00834 kJ·mol^{-1}·K^{-1} × 298 K] = 15
$\log K_f = \ln K_f/2.303 = 6.5$; $K_f = 3.1 \times 10^6$ M^{-1}
(c) $\Delta G = -56$ kJ·mol^{-1} $-$ (298 K)($+0.010$ kJ·mol^{-1}·K^{-1}) $= -59$ kJ·mol^{-1}
$\ln K_f = \Delta G/[-RT] = -59$ kJ·mol^{-1}/[-0.00834 kJ·mol^{-1}·K^{-1} × 298 K] = 23.8
$\log K_f = \ln K_f/2.303 = 10.3$; $K_f = 2.3 \times 10^{10}$ M^{-1}

The formation constant is higher for multidentate ligand in (c) by several orders of magnitude. Notice that the entropic consideration for (c) is positive—a definite plus if spontaneity is desired in a reaction.

FIG. 11.15 *Calixarene host species are shown. Left: A p-t-butylcalix[4]arene molecule binds a sodium cation in its lower tier structure forming an endo-calix complex. Middle: Bonding of toluene guest is shown in the upper tier binding complex. Bonding is accomplished primarily with van der Waals interactions. Right: Binding of an aliphatic amine to form an exo-calix complex is shown. In the experiment of Nachtigall et al. [38], a calixarene without para-substituted groups was used.*

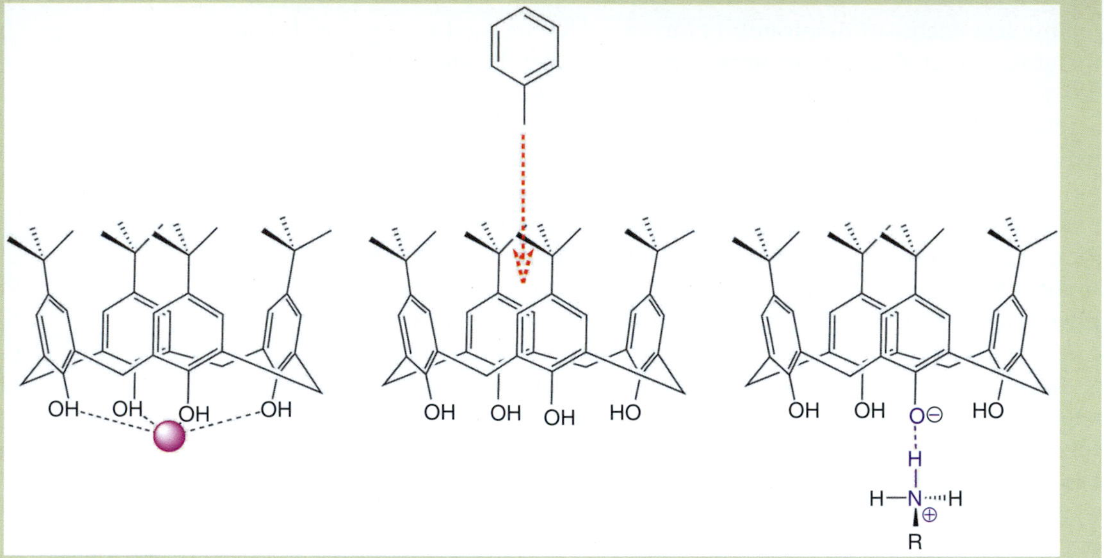

Calixarenes are members of the family of cyclophanes because they possess bridged aromatic groups [7]. They are formally known as substituted [1,1,1,1] metacyclophanes. Macrocyclic calixarenes are synthesized from linked phenol groups to form a four-membered macrocycle. *p*-Methylcalix[4]arene was synthesized in 1956 by R. F. Hunter and B. T. Hayes. Calixarenes are able to bind cations, anions, and neutral molecules.

Calixarenes offer more than one means to complex guest species. If the guest is a monovalent cation, the lower rim (acidic) hydroxyl groups are able to collectively bind, with dative bonds, the cation by the hydroxyl groups. This is referred to as the *endo* mode of binding. On the other hand, the binding of aliphatic amine complexes was found to proceed via an *exo* mode of binding in which the bound guest is outside the lower rim phenolic substituents—akin to a perching mode mentioned earlier [38]. The complex is called an *exo*-calix-ammonium complex. The major intermolecular interaction responsible for binding is the hydrogen bond between N^+–H · · · · · O^-–*Phe*. The calixarene is also capable of binding toluene and other similar guests within the hydrophobic upper rim region of the molecule to form an inclusion compound. In this case, van der Waals forces and steric and hydrophobic interactions need to be considered during the molecular design process.

Nachtigall et al. studied the binding between calix[4]arenes (without substituted *para* groups) and aliphatic amines by conductance, spectroscopic, and NMR methods [38]. Titrations were carried out in acetonitrile by hexylamine at constant temperature (298.0 ± 0.1°C). UV absorbance (310 nm) versus

FIG. 11.16 *(a) A typical Job plot relates the concentration of the complex to the ratio of the concentrations of the calixarene to the total concentration of calixarene + amine (e.g., the host and guest). In this figure, the absorbance is a function of the ratio. They have similar form because concentration of the complex $[C_o]$ is directly proportional to absorption. (b) A plot showing the results of a spectrometric titration (310 nm) of the H–G calixarene–amine complex is shown. The shape of this plot is typical for methods that use titration to determine relative concentrations of host–guest species.*

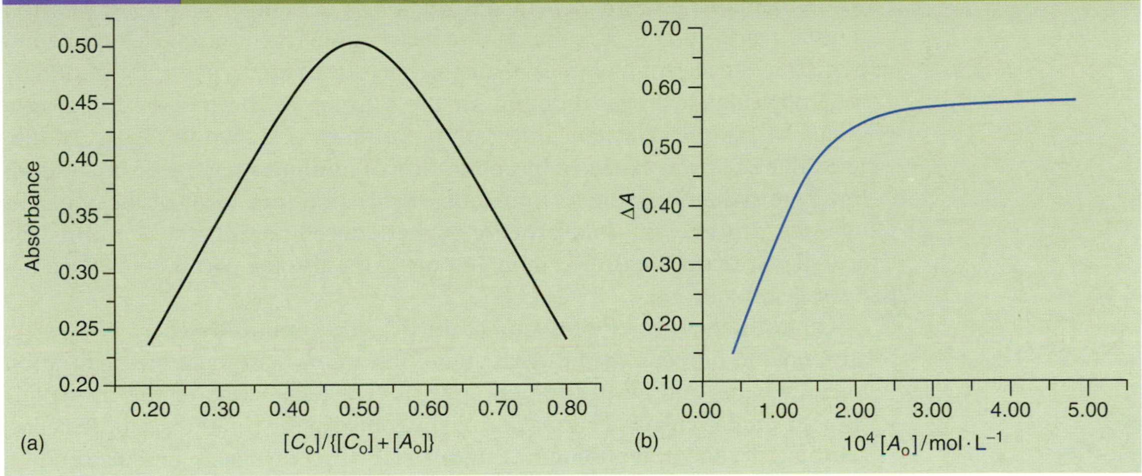

Source: F. F. Nachtigall et al., *Journal of the Brazilian Chemical Society*, 13, 295–299, 2002.

$[C_o]/\{[C_o] + [A_o]\}$ indicated 1:1 stoichiometry, where $[C_o]$ is the total concentration of the calixarene and $[A_o]$ is the total concentration of the amine. $[C_o] + [A_o] = 3.0 \times 10^{-4}\ M$ (**Fig. 11.16a**). The concentration relationships between one-on-one host–guest complexes are often portrayed by Job plots (**Fig. 11.16a**). The concentration of the complex [C] is plotted against the ratio of the host concentration to the total concentration of host and guest: $[H]/\{[H] + [G]\}$. From the figure it is obvious that, in a 1:1 ideal circumstance, the maximum in complex concentration is achieved when each component of host and guest has the same concentration—an idealistic case that does not consider higher order complexation and numerous other factors [7].

The equilibrium between the calixarene and the amine conformed to the form of an acid–base reaction:

$$\text{Calix-OH} + NR_3 \rightarrow \text{calix-O}^- + HNR_3^+ \tag{11.8}$$

and K_p, the equilibrium constant, is equal to

$$K_p = \frac{[C^-][A^+]}{[C][A]} \tag{11.9}$$

Rewriting using total concentrations,

$$0 = \left(1 - \frac{1}{K_p}\right)[C^-]^2 - \langle [C_o] + [A_o][C^-] + [C_o][A_o] \rangle \tag{11.10}$$

The concentrations are related to the Beer–Lambert law by

$$\Delta Abs = \varepsilon\left(\frac{[C_o]+[A_o]-\sqrt{([C_o]+[A_o])^2-4[C_o][A_o]\left(1-\frac{1}{K_p}\right)}}{2\left(1-\frac{1}{K_p}\right)}\right) \quad (11.11)$$

with ε the molar absorptivity equal to 3954 mol · L^{-1} and $K_p = 18$ for the proton transfer reaction. The pK_a value for the hexylamine reaction was determined to be 18.26. The pK_a for phenol is higher at 26.6 due to the lack of stabilization by intramolecular hydrogen bonding in the calixarene. The plot of absorbance versus $[A_o]$ of the calixarene–hexylamine complex is shown in **Figure 11.16b.** The value of K_p was extracted by application of nonlinear regression analysis. At lower concentrations, the K_p form (the dissociation into ions) of the apparent two-step process was dominant. At higher concentrations, the K_a form (the acid–base reaction) assumed more importance—also the part that involved the hydrogen bond [38].

Other titrant species were applied in the study, and differences in K_p were attributed to differences in the basicity of the amines. Conductimetric titration measurements and NMR titration analysis (chemical shifts of the ^{1}H signal) showed similarly shaped curves during the course of the titrations. Fluorescence, calorimetric, and potentiometric titrations are other means to measure binding constants and other parameters. Extraction methods, in which there is a partition of a guest between two immiscible phases, are a means by which macrocycle selectivity can be determined [7].

Thermodynamic Control in Natural Systems. The self-assembly of a double helix consisting of adenine and uracil nucleic acids will serve us as a perfect example of the concept of dynamic equilibrium [39]. Dynamic equilibrium is defined as a dynamic state of balance between forces—a characteristic of many metabolic processes. After some back and forth, dynamic equilibrium will ultimately drive the process to favor the thermodynamic product [26]. In **Figure 11.17**, the free energy of complexation is tracked along the course of nucleation and propagation of two complementary strands of RNA (comprising only of adenine and uracil nucleic acid bases). Although ΔH is negative from the onset (nucleation), ΔG is positive due to the unfavorable $T\Delta S$ term, and it becomes clear from the graphs that a minimal number of base pairs are needed before a double helix can be formed by self-assembly. After four base pairs are zipped together, ΔG becomes negative and the reaction proceeds spontaneously. During propagation, the entropic factor plateaus and self-assembly fall into place to zip together the remainder of the complementary base pairs [26,39]. The bottom line is that assembling larger supramolecular assemblies becomes easier as more noncovalent bonds are involved in the process [26,39]. The activation energy required to overcome the entropic penalty of the initial step is probably overcome by participation of an enzyme.

The tobacco mosaic virus (TMV) shown earlier illustrates another example of thermodynamic influence over a self-assembly process [26]. Recall that the virus is a cylindrical structure about 300 nm in length with a diameter of 18 nm. It is

Fig. 11.17

The free energy of nucleation and propagation of RNA zipping (consisting only of adenine and uracil) is tracked in the figure. Two components are contributing to the overall free energy: enthalpy and entropy. Initially, ΔG is positive due to the unfavorable TΔS term. This makes sense because zipping two strands together contributes inherent entropy hits to the overall free energy. From the graph it is clear that a minimal number of bases have to be zipped together before the enthalpy term dominates the process. The entropic component plateaus as expected because each increment provides the same level on negentropy. This phenomenon is an example of dynamic equilibrium. Once a complex is established, the affinity for establishing more complexes increases and it is driven by increasingly favorable enthalpic factors. In the very broadest sense, this process resembles an allosteric effect *(a phenomenon characteristic of enzymatic systems) in that ligands (the nucleotides) are able to affect the progress of complexation and exhibit a degree of cooperativity.*

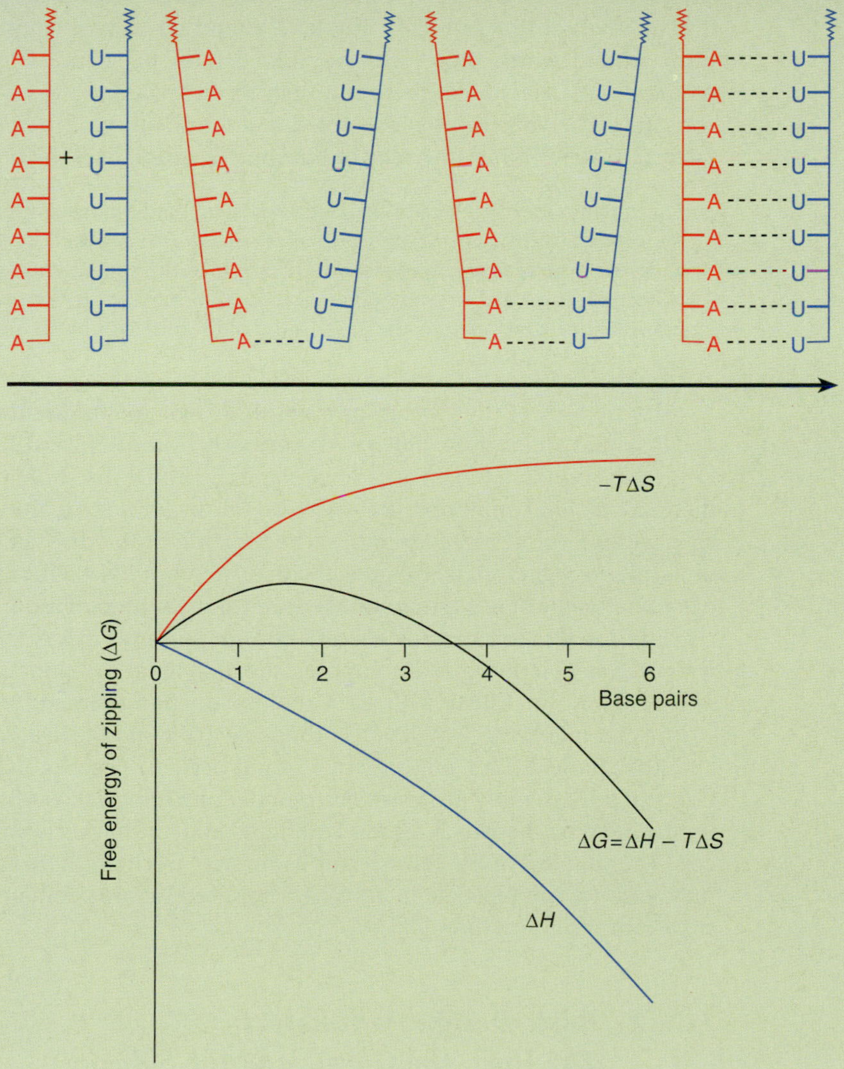

Source: M. E. Craig, D. M. Crothers, and P. Doty, *Journal of Molecular Biology*, 62, 383–401, 1971. With permission.

made of 2130 identical protein subunits wrapped around a single strand of RNA in a helical fashion. The RNA contains ca. 6400 nucleotide bases. During the assembly progresses of the virus, the rush to completion gains momentum as more units are self-assembled into the body of the virus.

During the synthesis of this structure, several nanoscale interactions are evident: (1) preorganization, complementarity, and molecular recognition; (2) thermodynamic control and the dynamic equilibrium effect; (3) self-assembly of a hierarchical structure; (4) template effects (of the RNA and the preformed protein wedges); (5) self-correction (due to thermodynamic factors); and (6) efficiency (due to high selectivity). In the first step, a wedge-shaped protein monomer is synthesized. The monomers self-assemble (by molecular recognition) to form disc-shaped subunits. Once a disc is formed, it binds to a single strand of RNA that induces a conformational change in the disc's geometry into that of a helical structure [26]. The RNA strand serves as a templating mechanism around which more discs are added until the entire virus is formed. As more units are added, there is a rush to completion, a cascading effect of sorts, until the virus is fully assembled and termination of the process occurs. Trent H. Galow of the University of Edinburgh summarizes the process:

> Equilibrium means reversibility. Intermediates that do not represent the thermodynamic minima can progress in a forward or reverse direction—relatively easily to overcome the activation barrier. Once the lowest energy aggregate is formed, it is hard to overcome the activation barrier for the reverse stage. Equilibrium continues to the right-hand-side until complete. Inherently, a self-correcting process.

Self-assembly of the virus is an example of dynamic molecular recognition and dynamic equilibrium, illustrated previously by the RNA and TMV examples. Once a few building blocks are in place and the initial threshold overcome, the thermodynamic conditions that exist at the beginning of the process have changed in favor of more assembly. The function of the disc, an expression of molecular recognition, is to interact with a strand of RNA. Once the RNA is in place, the energy of self-assembly is reduced and addition of more discs is facilitated; this is akin to a cascade effect (one with purpose) that gains momentum with each step. If the virus decomposes, the individual components are capable of reassembly into a TMV with all its natural function restored [26].

The example of the TMV involves thousands of molecules, scores of proteins, and 2130 protein discs, but only one strand of RNA. These molecular recognition and self-assembly processes occur under physiological conditions. Backing up the TMV synthesis is a central command system that orchestrates this process. The simplest of life forms exhibits mastery of chemical processes that we only wish we can replicate in a beaker; however, we are getting closer, step by step, with each breakthrough.

11.2.2 Basic Design Parameters: The Host, the Guest, and the Solvent

Precursor Molecules. Precursor molecules are defined as rather small molecules that require making and breaking of covalent bonds during their synthesis. The

FIG. 11.18 *Condensation reaction between two amino acids to form a small dipeptide is depicted. Covalent bonds are broken and formed in this reaction. In living things, enzymes mediate such processes under conditions that are considered to be rather mild. Peptides should be considered as supermolecules held together by covalent bonds with relatively similar composition.*

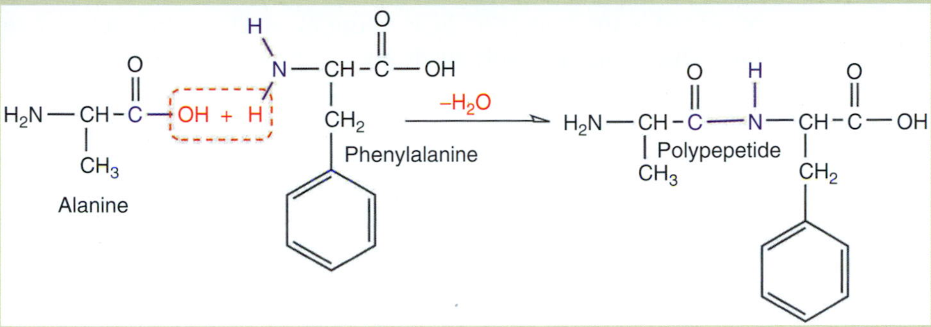

precursor molecule or molecules will form the basis of the next step: designing the receptor and substrate. Proteins, for example, are the result of a condensation reaction between two amino acid precursor molecules **(Fig. 11.18)**. In polymer chemistry, precursor molecules are considered to be monomers. The backbones of polymers are analogous to those of proteins in that covalent bonds hold them together. The interactions between polymeric chains, however, can be covalent in nature, but many are considered to be intermolecular. Interactions between proteins (or within one protein) can be intermolecular or covalent (the disulfide linkage). Are proteins supermolecules or supramolecular molecules?

Once proteins are folded and provide a specific function, do they then become supramolecules? How about a compromise? Let us call peptides superprecursors. It is easier to define synthetic supramolecular precursors. All synthetic precursor molecules consist of covalent bonds that are made by hard chemistry techniques (under kinetic control conditions). The synthesis of the first crown ether shown earlier is a typical example.

Receptors and Substrates. Molecular receptors (hosts) and substrates (guests) are generally classified as organic chemicals (precursor molecules) held together by strong covalent bonds [4]. The key to designing receptor (and substrate) molecules is the ability to convert organic molecules into ones that have potential for molecular recognition [4]. This paradigm does not apply to all precursor molecules. For example, the precursor molecules that combine to form the rosette (melamine and cyanuric acid) already possess a high degree of molecular recognition via their complementary H-bonding structure. But we get the point nonetheless. There are many possible kinds of substrates (guests): cationic, anionic, zwitterionic, or neutral species. These species originate from inorganic, organic, or biological sources.

Another challenge faced by designers of drugs is estimation of binding constants of drugs to receptors [40]. Molecular recognition mechanisms bring the parties together, but what happens afterwards is also important. Partitioning free energy of binding or association according to the type of bonding (hydrogen bonds, π–π interactions, salt bridges, or the hydrophobic effect) is one approach, but that would ignore the cooperative effect that intermolecular attractions are known to exhibit. The strength of the hydrophobic effect, a measure that depends significantly on the type of method applied, must be measured in the context of the medium in which it occurs and not in isolation [40]. This brings us to solvent effects.

Solvent Effects. Macromolecular interactions with water are one of the most important factors to consider when designing a supramolecular system. Structure, function, specificity, stability, and the dynamical potential of supramolecules are all influenced by solvation with water if taking place in aqueous media. For example, proteins rely on water for their three-dimensional structure [41]. Water also plays a vital role in protein–nucleic acid recognition by mediating bridging hydrogen bonds between functional groups [42]. Water acts as a lubricant that facilitates hydrogen bonding. Proteins have *conformational flexibility* and in solution exhibit a high degree of hydration states (e.g., the existence of ordered water in the form of a hydration shell that is denser than bulk water and facilitates proton transfer [41]. The energetic price tag for

EXAMPLE 11.2 *A Simple Demonstration of a Solvent Effect*

The dipole of α-alanine ethylester (μ = 2.09) is in line with a potassium cation at a distance of 0.5 nm. What is the energy of the interaction in water (ε_r = 78.85)? in methanol (ε_r = 32.63)? in trimethyl amine (ε_r = 2.44) at 298 K? The

Solution

$$\text{Use } E_{Ion-Dipole} = \frac{ze\mu\cos\theta}{4\pi\varepsilon_o\varepsilon_r r^2}$$

Solution:

Permittivity of free space, $\varepsilon_o = 8.854 \times 10^{-12}\ C^2 \cdot J^{-1} \cdot m^{-1}$, $e = 1.6022 \times 10^{-19}\ C$

$$E_{Ion-Dipole}(water) = \left[\frac{(+1)(1.6022\times10^{-19}\,\mathrm{C})(2.09\times3.336\times10^{-30}\,\mathrm{C\cdot m})\cos 0}{4\pi(8.854\times10^{-12}\,\mathrm{C^2\cdot J^{-1}\cdot m^{-1}})(78.85)(0.5\times10^{-9}\,\mathrm{m})^2}\right]\left(\frac{6.022\times10^{23}}{\mathrm{mol}}\right)\left(\frac{1\,\mathrm{kJ}}{1000\,\mathrm{J}}\right) = -0.31\ \mathrm{kJ\cdot mol^{-1}}$$

Reducing all nonvariable components to 24.18 (the variable component is the dielectric constant of the solution):

$$E_{Ion-Dipole}(methanol) = \frac{24.18}{32.63} = -0.74\ \mathrm{kJ\cdot mol^{-1}}$$

$$E_{Ion-Dipole}(trimethylamine) = \frac{24.18}{2.44} = -9.91\ \mathrm{kJ\cdot mol^{-1}}$$

The solvent matters. If you are designing a host species that is trying to attract an ion with a dipolar appendage, then choice of solvent is very relevant.

hydrating a new surface or removing water from a surface is between 3.6 and 63 $kJ \cdot mol^{-1} \cdot Å^{-2}$ [42].

Hydrophobic effects influence the three-dimensional folding of proteins. Entropic considerations arise when a hydrophobic surface is in contact with polar water molecules. Quasi-crystalline organization of water occurs on the surface of the hydrophobic entity, but H-bonding saturation (four bonds) with other water molecules does not occur and this phenomenon results in water that is denser than it is in its natural state. As a result, slight negative entropy is obtained when water molecules are released from the pocket into the bulk aqueous solution as each H_2O is bonded tetrahedrally to other water molecules.

Solvent effects were covered in some detail in chapter 10 but we will elaborate upon this important parameter further in this section. The solvent plays a role in the recognition process—particularly with regard to the degree of solvation. Going back to our discussion of hydration spheres in chapter 10, removing the hydration sphere around a cation would have a positive (unfavorable) enthalpy component, although entropy would increase as a result (favorable). When designing a host–guest relation that takes place within a solvent, a priori inspection of the solvation properties of potential guest molecules and the thermodynamic trade-offs between the host and guest must be evaluated.

In addition to the solvation interactions and the electrostatic factors discussed in example 11.2, other solvent evaluation criteria that weigh into the design process include the fact that the hydrophobic effect (from chapter 10) affects the formation (complexation) constant of H–G reactions. If a H–G complex is rendered insoluble in the parent solvent, then the tendency to bind to the guest is enhanced. The binding of anion guests depends strongly on the type of solvent. For example, binding constants of small inorganic anions are better in carbon tetrachloride than it is in water [7].

The design of supramolecular systems is not trivial and requires detailed analysis of precursors, hosts and guests, and the complexes themselves once formed. Consideration of thermodynamic and kinetic effects as well as the contribution of the solvent, all in concert, is a formidable task that may require the assistance of a computer program. Nature, of course, does this on a routine basis—the result of hundreds of millions of years of trial and error. Perhaps we can devise a combinatorial chemistry scenario that is able to accelerate the replication (molecular recognition) process, at least to the level of the simplest primitive, most rudimentary organism.

11.3 Extended Supramolecular Structures

Most chemical systems are fabricated by mixed bottom-up methods. Mixed methods involve several kinds of materials, intra- and intermolecular interactions, solvents, and reaction conditions. Each step requires meticulous forethought and design. We show one remarkable example: the formation of monolayers of gold clusters, molecular squares, and surface tethered groups.

11.3.1 Golden Molecular Squares

The synthesis of molecular squares is straightforward (**Fig. 11.19**) [43]. It is based on the transition metal–ligand interaction and it has recently been applied to construct metal macrocycles and cages [43]. Molecular squares from osmium tetroxide, olefin, and bispyridyl ligands come together to form an electrically neutral octahedral coordination center around osmium, making them amenable for hydrogen bonding without interference from a charged species. Mixtures contained small to large macrocyclic osmate esters (COEs). It was found that higher ordered structures were obtained when aromatic diols were added to the mix. Apparently, π-stacking and hydrogen bonding self-assembly occur between the COEs and the diols. The osmium(VI) is capable of directing the synthesis of COEs and higher structures by coordination, chemical reaction, hydrogen bonding, and aryl stacking. Hydrogen bonds form between the alcohol and the available osmium oxygens. Aryl stacking was achieved between 24 individual components. The size of the diol plays a role in the size of the supramolecule [43]. This strategy is an interesting route to synthesize integrated supramolecular structures.

Lahav et al. in 1999 developed ordered arrays of supramolecular structures for sensing, electronic and photochemical purposes [44]. Fujita et al. in 1994 synthesized molecular squares with the goal of fabricating new types of host–guest cavities [45]. The squares consist of bipyridine units linked by transition metal ions such Pd^{2+} and Pt^{2+} [45]. Au-molecular square structures are depicted in **Figure 11.20** and illustrate a multistep fabrication process involving numerous kinds of chemistry—a trait very typical of modern nanoscale engineering and chemistry. Precursors such as 3-aminopropyltriethoxysilane are synthesized according to traditional organic chemical routes and serve as the precursor molecules. An indium–tin-oxide (ITO) supermolecular surface is activated to generate hydroxyl groups. The ethoxy group on the silane is exchanged by a surface hydroxyl and the silane is anchored to the surface. The monolayer is further stabilized by lateral interactions between adjacent silanes. The amino-terminus is made positive by a subsequent protonation step. Prefabricated supermolecular gold colloids of ca. 12 nm diameter are introduced to the positively charged surface and allowed to equilibrate. The gold colloids were made by a reduction of gold salts in a citrate solution. Molecular squares are then introduced onto the Au colloid modified surface. Au and cross-linking molecules are alternatively added until the desired number of layers is achieved. Amazingly, all the reaction sequences occur at room temperature under mild conditions. This type of synthesis illustrates the utility and practicality of self-assembly processes.

The typical plasmon absorption of Au colloids of that size is $\lambda = 518$ nm, but another absorption was noticed at 650 nm, suggesting the presence of aggregates due to interparticle coupling [44]. The molecular squares carry positive charges and were thus able to cross-link with the Au nanoparticles by "multisite" ion pairing [44]. The ratio of Au:Pd was 30:1; the ratio of Au:Cd was 100:1. As a sensor, the cross-link molecular squares act as π-receptors for π-donor substances such as dialkoxybenzene derivatives. This is an excellent example of a self-assembly process following covalent interactions to synthesize

FIG. 11.19 *Molecular square supramolecular complex anchored by osmium(VI) is depicted. The repeating units are capable of forming hexagonal and larger structures.*

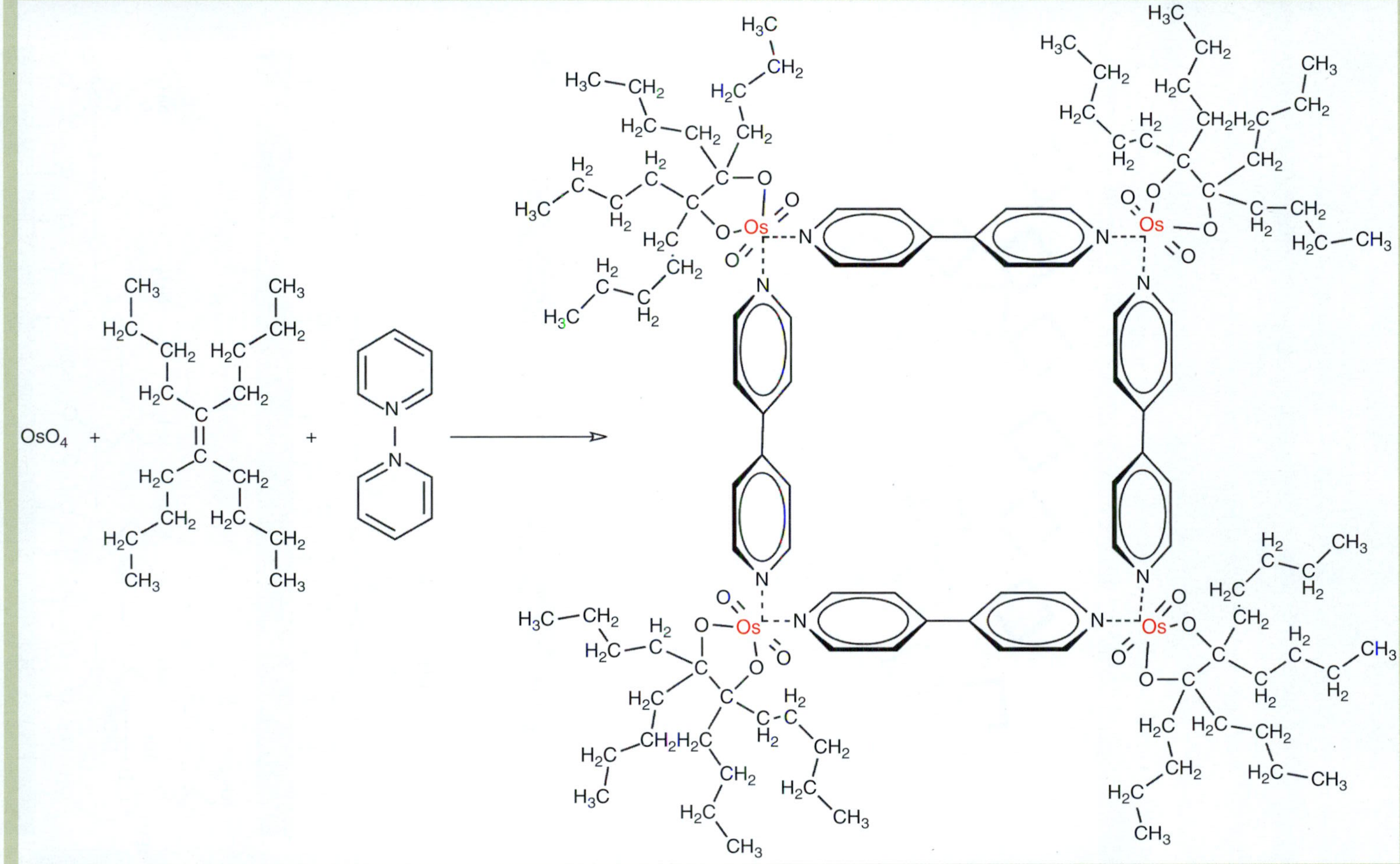

Source: Y. L. Cho et al., *Journal of the American Chemistry Society*, 123, 1258–1259, 2001. With permission.

Fig. 11.20

Synthesis and structure of a multilayer golden molecular square complex is depicted [44]. The figure shows a process in which covalent bonding, electrostatic interactions, and π-processes are in effect to form the structure. Fujita et al. developed a series of cyclophane-like molecular square precursors that consist of bipyridines linked by metal ions (e.g., molecular squares) [45] (Fig. 11.8o and p). Gold colloid (represented by gold spheres) and molecular square (represented by blue squares) were assembled into an integrated structure. The assembly was formed by noncovalent cross-linking of 12-nm Au colloids with molecular square transition metal complexes. An activated ITO surface was reacted with 3-aminopropyltriethoxysilane by covalent interactions. The amine groups yielded a positively charged surface. The chemical-sensing layers were fabricated by alternating applications of citrate stabilized Au colloids molecular squares. The molecular squares served to cross-link

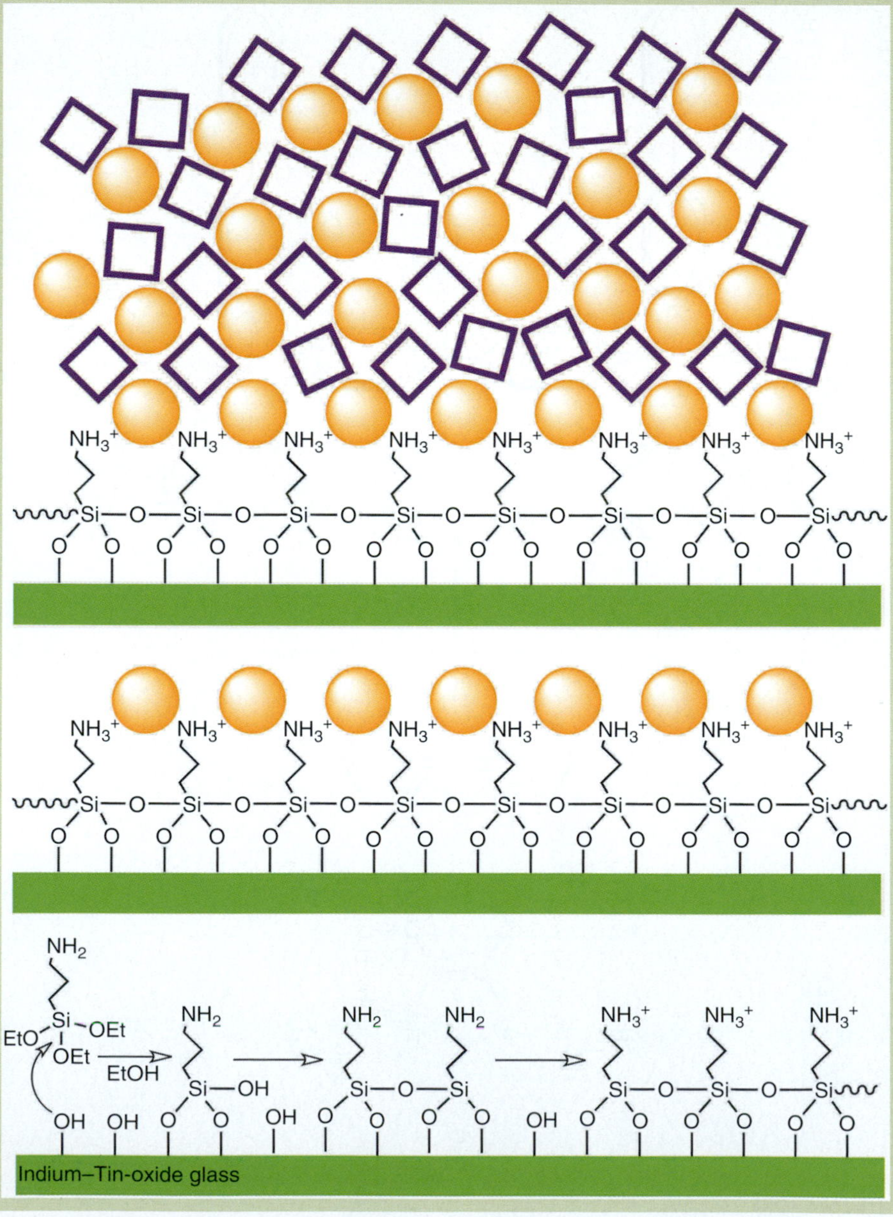

Source: M. Lahav et al., *Chemical Communications*, 1937–1938, 1999.

FIG. 11.20 (CONTD.) *the Au nanoparticles. Molecular squares are a new generation of hosts that contain cavities suitable for binding guests. For example, the squares can be used to sense (complexation) anions. An ITO substrate is first activated to generate accessible hydroxyl groups. A reactive silane (3-aminopropyl-triethoxysilane) was applied to the surface to generate a positively charged monolayer. Citrate-stabilized Au colloids (12 nm) were then applied to the monolayer. Palladium squares* $[(en)Pd(4,4'\text{-}bipy)]4(NO_3)_8]$ *or cadmium squares* $[Cd(4,4'\text{-}bipy)]4(NO_3)_8]$ *adhered to the gold colloids and fabrication of the "first receptor layer" was completed. Alternating steps of gold colloid and square addition led to the structure depicted at the top of the figure. The plasmon band absorption of gold colloids of this size (12 nm) is 518 nm. Interestingly, an additional plasmon absorption that was significantly red shifted (to 650 nm) was detected [44]. The shifted peak is attributed to interparticle plasmon coupling—proof that particles are in contact with each other via the molecular squares. The absorption at 650 nm increased as the number of layers was increased. Electrical conductivity through the layer was also enhanced as more layers were added.*

a supramolecular structure. It also serves as an excellent example of a potential sensor application based on π-interactions.

11.3.2 Synthesis of Benzocoronene Complexes

Benzocoronenes are a versatile class of nanographitic precursors that are relatively easy to modify chemically and are able to form aryl stacks by self-assembly. Benzocoronene materials have promising applications in electronics. The fabrication of highly ordered two- or three-dimensional polymolecular architectures consisting of benzocoronene discotic liquid crystals involves intramolecular, intermolecular, and interfacial forces. Replacement of the R-groups is possible with a variety of species. Benzocoronenes are graphite analog precursor molecules synthesized by conventional organic chemistry by making and breaking covalent bonds (**Fig. 11.21**). Chemical modification of benzocoronenes is accomplished by organic chemical means as well. Ethylene oxide residues can be attached to the modifiable regions of the molecule to alter properties (**Fig. 11.21**). Once synthesized, the precursors self-assemble to form columnar structures called liquid crystals (**Fig. 11.22**).

The structure of liquid crystals (LCs) is somewhere between that of a crystalline solid and a liquid [46]. There are two kinds of liquid crystals: thermotropic (in which phase transitions are observed as a function of temperature) and lyotropic (in which phase transitions occur as a function of concentration). Liquid crystals were discovered in 1888 by the Austrian botanist Friedrich Reinitzer after he noticed that cholesteryl benzoate exhibited two melting points. At 145°C it melted into an opaque liquid and at 179°C, it melted but this time into a clear liquid. The term liquid crystal was credited to Georges Friedel in 1922.

LCs are found in two phases: *smectic* and *nematic*. Smectic phases are characterized by positional order (along one direction) and orientational order (along the layer normal or tilted away from the layer normal). Nematic phases exhibit no positional order but are characterized by long-range orientational order; all

FIG. 11.21 *A planar hexa-peri-benzocoronene is depicted. The diameter of the molecule is 0.15 nm. Left: Synthesis of hexabenzocoronene from an alkene is depicted. Right: Chemical modification of R-groups yields many kinds of hexabenzocoronenes.*

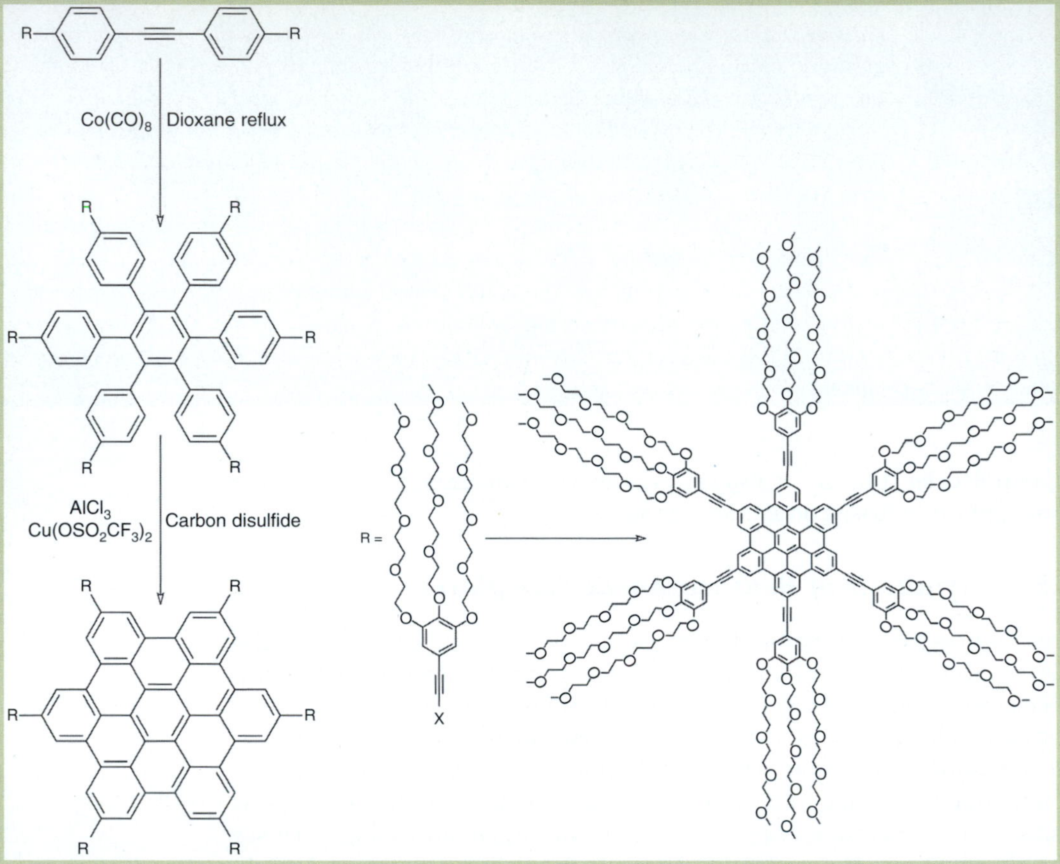

Source: Redrawn with permission from Professor Klaus Müllen, Max-Planck Institut für Polymer Forschung.

crystals are oriented ($a = b << c$) in the same direction within their respective domains. Nematic LCs can be aligned by external magnetic fields.

Benzocoronenes form uniaxial discotic (nematic) columnar liquid crystals. The columns can be further organized into rectangular or hexagonal arrays. The structure following mild heating is shown in **Figure 11.22** [47,48]. A benzocoronene species with $-C_{12}H_{25}$ substituted constituents showed LC phase at 106.5 and 124.4°C. A single LC phase was demonstrated at ~400°C [49]. In another study, hexa-peri-hexabenzocoronene (also a discotic liquid crystal) was combined with perylene dye to form thin films with vertically separated perylene–hexabenzocoronene domains with extended interfacial surface area [50]. The structure was applied to a photodiode device in which a photovoltaic response on the order of 34% efficiency was achieved at 490 nm [50]. The photovoltaic response was due to charge transfer between vertically aligned dye and the hexabenzocoronene π-systems. Over 50 different cores have been created from which about 3000 discotic LCs have been developed [51].

FIG. 11.22 *A schematic illustrating the transformations of benzocoronenes. Left: Self-assembly in water; middle: herringbone structure following self-assembly; right: liquid crystal columnar discotic structure.*

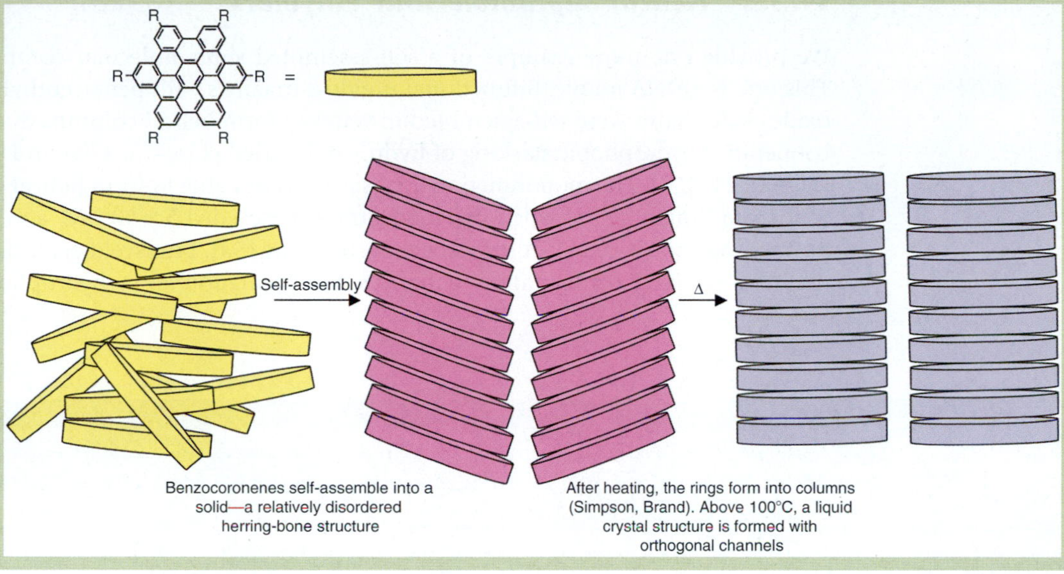
R
R
R
R
R
R
=
Self-assembly
Δ
Benzocoronenes self-assemble into a solid—a relatively disordered herring-bone structure
After heating, the rings form into columns (Simpson, Brand). Above 100°C, a liquid crystal structure is formed with orthogonal channels

Source: C. D. Simpson et al., *Journal of Materials Chemistry,* 14, 494–504, 2004; A. M. van der Craats et al., *Advanced Materials,* 10, 36–38, 1998. With permission.

Graphitic nanotube analogs can be made from chemically modified benzocoronenes. Mixing R-groups (**Fig. 11.23**) to create an amphiphilic benzocoronene results in helical self-assembly in THF or THF–water mixtures [52]. Self-assembly in THF results in a tighter helix.

The aspect ratio of the tubes was ca. 1000 and the diameter of the opening was 14 nm. The wall was 3 nm in thickness and consists of a helical array of π-stacked graphene molecules. The exterior and interior of the molecules were covered

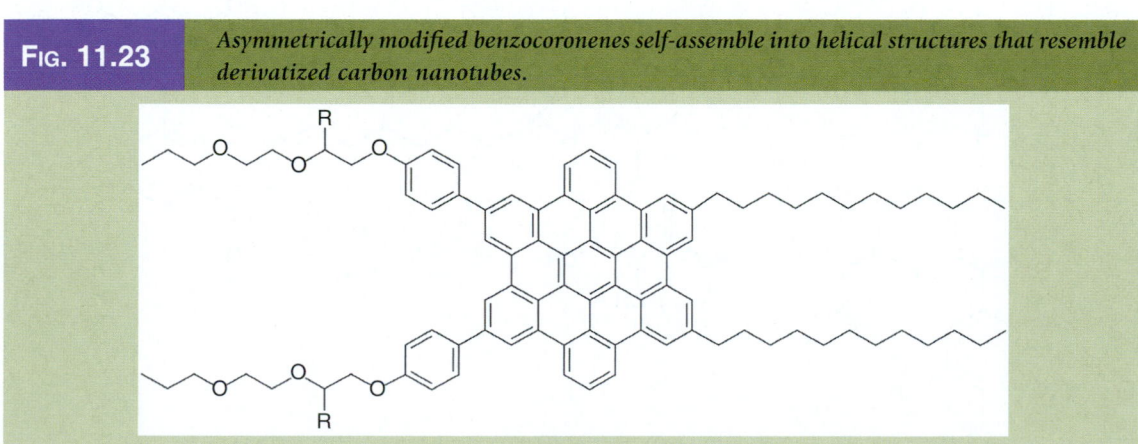

FIG. 11.23 *Asymmetrically modified benzocoronenes self-assemble into helical structures that resemble derivatized carbon nanotubes.*

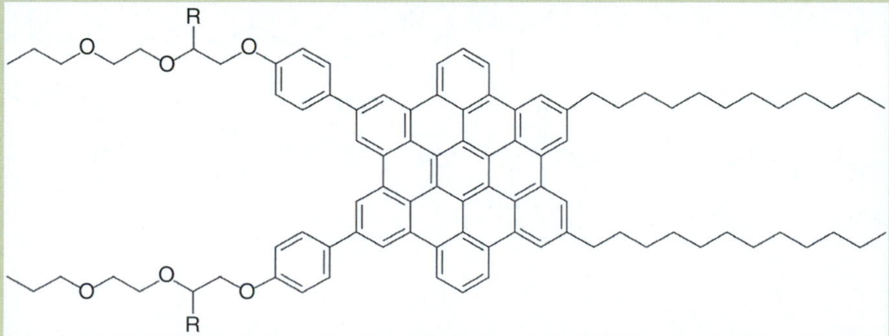
R
O
O
O
O
O
O
R

Source: J. P. Hill et al., *Science,* 304, 1481–1483, 2004. With permission.

with hydrophilic triethylene glycol residues. The tubes were electrically conductive [52]—and all by self-assembly of a precursor molecule.

11.3.3 Helical Supramolecular Polymers

We provide one more example of a self-assembled supramolecular complex. This one is a DNA mimic. Bifunctional ureido-*s*-triazines with penta(enthylene oxide) side chains were self-assembled in water to form helical columns due to cooperative hydrophobic stacking of hydrogen bonded pairs—just like in DNA (**Fig. 11.24**) [53]. The monofunctional version was not able to form helical columns. Stacking occurred when the concentration exceeded 5×10^{-6} *M*, and the process was monitored by fluorescence spectroscopy [53]. Brunsveld et al. claim that the stacking configuration enhances the formation of a hydrophobic

FIG. 11.24 *Ureido-s-triazines modified with ethylene oxide side chains are depicted in the figure. These materials appear to be nucleotide mimics and it is no wonder that they are able to form helical structures in water. The structures dimerize via four-fold hydrogen bonding and are able to stack aromatic cores by aryl–aryl intermolecular interactions. The core is protected from the solvent by the creation of a hydrophobic interior—all by self-assembly. The column radius was found by small angle neutron scattering (SANS) to be 0.16 nm. Interestingly, columns formed by the molecule on the left in the figure had discs that were able to rotate. There was no disc rotation in columns formed by the entity on the right due to the hexamethylene spacer between dimers. This experiment demonstrated the formation of highly ordered aggregates by cooperative intermolecular interactions: hydrophobicity, π–π stacking, and hydrogen bonding—the same three that are responsible for DNA's conformation.*

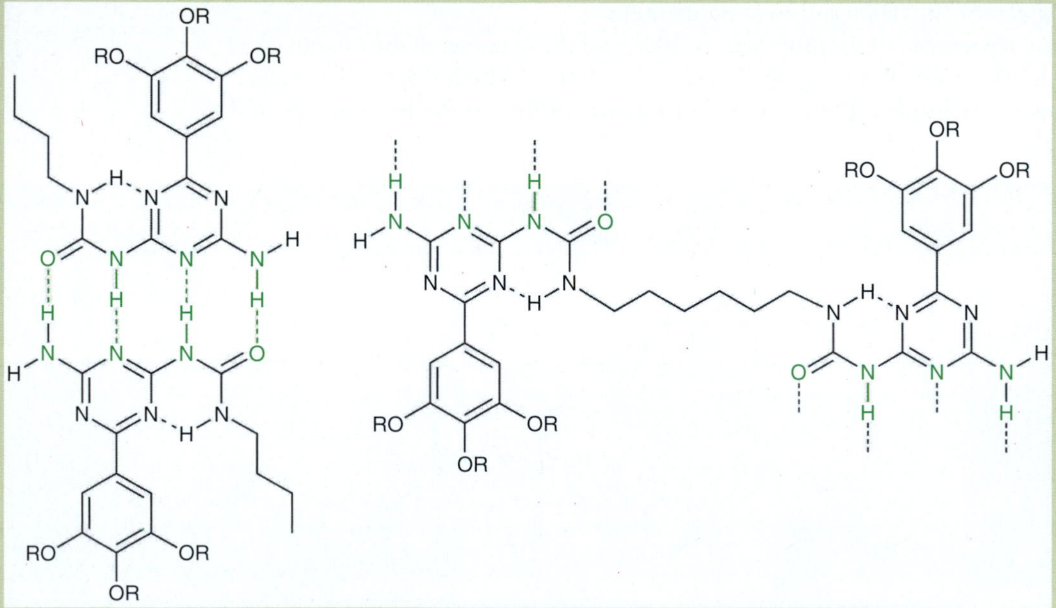

Source: L. Brunsveld et al., *Proceedings of the National Academy of Science USA*, 99, 4977–498, 2002. With permission.

"microenvironment" that promotes intermolecular hydrogen bonding to occur at higher concentrations due to shielding from water. Addition of side chains skews the helicity of the structures. Specially designed precursors allow for control over the design of chiral nanotubes—nanoscience in action! The R-groups are either long chain aliphatics or chainlike derivatives of ethylene oxide.

References

1. N. J. Turro, Supramolecular organic photochemistry: Control of covalent bond formation through non-covalent supramolecular interactions and magnetic effects, *Proceedings of the National Academy of Science USA*, 99, 4805–4809 (2002).
2. N. J. Turro, From molecular chemistry to supramolecular chemistry to superdupermolecular chemistry. Controlling covalent bond formation through non-covalent and magnetic interactions, *Chemical Communications*, 20, 2279–2292 (2002).
3. F. M. Menger, Supramolecular chemistry and self-assembly, *Proceedings of the National Academy of Science USA*, 99, 4818–4822 (2002).
4. J.-M. Lehn, *Supramolecular chemistry: Concepts and perspectives*, VCH Verlagsgesellscahft GmbH, Weinheim (1995).
5. J.-M. Lehn, Organic chemistry: Its language and its state of the art, *Proceedings Centenary Geneva Conference*, M. V. Kisakürek, VHCA, Basel; VCH, Weinheim (1993).
6. J.-M. Lehn, Cryptates: Inclusion complexes of macrocyclic receptor molecules, *Pure Applied Chemistry*, 50, 871–892 (1978).
7. J. W. Steed and J. L. Atwood, *Supramolecular chemistry*, John Wiley & Sons, Ltd., Chichester (2000).
8. P. D. Beer, P. A. Gale, and D. K. Smith, *Supramolecular chemistry*, Oxford Chemistry Primers, Oxford Science Publications, Oxford University Press, Oxford (2003).
9. J. G. E. M. Fraaijie, Soft matter. Presentation ISM-Series, University of Leiden, wwwchem.leidenuniv.nl/scm/Presentations/ISM/ISM09/ISM09.pdf (viewed 2007).
10. I. W. Hamley, *Introduction to soft matter*, John Wiley & Sons, Ltd., Chichester (2000).
11. J. H. Morowitz, B. Heinz, and D. W. Deamer, The chemical logic of a minimum protocell, *Origins of Life and Evolution of Biospheres*, 18, 281–287 (1988).
12. A. Pappelis and S. W. Fox, Domain protolife, *Journal of Biological Science*, 20, 129–132 (1995).
13. S. L. Miller, A production of amino acids under possible primitive earth conditions, *Science*, 117, 528–529 (1953).
14. D. Ring, Y. Wolman, N. Friedmann, and S. L. Miller, Prebiotic synthesis of hydrophobic and protein amino acids, *Proceedings of the National Academy of Science USA*, 69, 765–768 (1972).
15. B. Commoner, Unraveling the DNA myth: The spurious foundation of genetic engineering, *Harper's Magazine*, February, 39–47 (2002).
16. B. Commoner, Biochemical, biological and atmospheric evolution, *Proceedings of the National Academy of Science USA*, 53, 1183–1194 (1965).
17. S. B. Prusiner, The prion diseases: One protein, two shapes, *Scientific American*, 272, 48–57 (1995).
18. J.-M. Lehn, *Supramolecular chemistry: Concepts and perspectives*, Weinheim, Germany: VCH (1995).
19. J.-M. Lehn, *Organic chemistry: Its language and its state of the art, Proceedings Centenary Geneva Conference*, M. V. Kisakürek, VHCA, Basel; VCH, Weinheim (1993).
20. J.-M. Lehn, Cryptates: Inclusion complexes of macrocyclic receptor molecules, *Pure Applied Chemistry*, 50, 871–892 (1978).
21. J. Rebek, Jr., "Some got away, but others didn't...," *Journal of Organic Chemistry*, 69, 2651–2660 (2004).

22. C. J. Pedersen, Cyclic polyethers and their complexes with metal salts, *Journal of the American Chemical Society*, 89, 7017–7036 (1967).
23. C. J. Pedersen, Cyclic polyethers and their complexes with metal salts, *Journal of the American Chemical Society*, 89, 2495–2496 (1967).
24. Molecular recognition, Wikipedia, en.wikipedia.org/wiki/Molecular_recognition (2007).
25. S. Shinkai, M. Ikeda, A. Sugasaki, and M. Takeuchi, Positive allosteric systems designed on dynamic supramolecular scaffolds: Toward switching and amplification of guest affinity and selectivity, *Accounts of Chemical Research*, 34, 494–503 (2001).
26. T. H. Galow, Non-covalent synthesis, University of Edinburgh. Presentation, www.chem.ed.ac.uk/teaching/chem4-5/resources/course_d.pdf (viewed 2007).
27. P. Bhattacharyya, W. Epstein, and S. Silver, Valinomycin-induced uptake of potassium in membrane vesicles from *Escherichia coli*, *Proceedings of the National Academy of Science USA*, 68, 1488–1492 (1971).
28. H. Brockmann and G. Schmidt-Kastner, Valinomycin 1, XXVII, Mitteil Über Antibiotica aus Actinomycetes, *Chemische Berichte*, 88, 57–61 (1955).
29. Z. Syed, Y. Ishida, K. Taylor, D. A. Kimbrell, and W. S. Leal, Pheromone reception in fruit flies expressing a moth's odorant receptor, *Proceedings of the National Academy of Science USA*, 103, 16538–16543 (2006).
30. F. Gräter, W. Xu, and H. Grubmuller, Pheromone discrimination by the pheromone-binding protein of *Bombyx mori*, *Structure*, 14, 1577–1586 (2006).
31. J. C. Love, L. A. Estroff, J. K. Kriebel, R. G. Nuzzo, and G. M. Whitesides, Self-assembled monolayers of thiolates on metals as a form of nanotechnology, *Chemistry Review*, 105, 1103–1169 (2005).
32. Z. Tang, D. Shangguan, K. Wang, H. Sui, K. Sefah, P. Mallikratchy, H. W. Chen, Y. Li, and W. Tan, Selection of aptamers for molecular recognition and characterization of cancer cells, *Analytical Chemistry*, 79, 4900–4907 (2007).
33. D. H. J. Bunka and P. G. Stockley, Aptamers come of age—at last, *National Review of Microbiology*, 4, 588–596 (2006).
34. E. Smyth, RNA in therapy, Horizon Symposia, Connecting Science to Life, Nature Publishing Group, nature.com/horizon/rna/background/therapy.html (2003).
35. B. A. Sullenger and E. Gilboa, Emerging clinical applications of RNA, *Nature*, 418, 252–258 (2002).
36. C. P. Poole and F. J. Owens, *Introduction to nanotechnology*, Wiley-Interscience, John Wiley & Sons, Inc., Hoboken, NJ (2003).
37. J. N. Israelachvili, *Intermolecular & surface forces: With applications to colloidal and biological systems*, 2nd ed., Academic Press, San Diego (1992).
38. F. F. Nachtigall, M. Lazzarotto, and F. Nome, Interaction of calix arene and aliphatic amines: A combined NMR, spectrophotometric and conductimetric inverstigation, *Journal of the Brazilian Chemical Society*, 13, 295–299 (2002).
39. M. E. Craig, D. M. Crothers, and P. Doty, Relaxation kinetics of dimer formation by self-complementary oligonucleotides, *Journal of Molecular Biology*,62, 383–392 (1971).
40. D. H. Williams and B. Bardsley, Estimating binding constants—The hydrophobic effect and cooperativity, *Perspectives in Drug Discovery and Design*, 17, 43–59 (2004).
41. B. Barbiellini, Water interaction with proteins, Northeastern University, stardec.ascc.neu.edu/~bba/RES/PROT/WATER.htm
42. J. Barciszewski, J. Jurczak, S. Porowski, T. Specht, and V. A. Erdmann, The role of water structure in conformational changes of nucleic acids in ambient and high pressure conditions, *European Journal of Biochemistry*, 260, 293–307 (1999).
43. Y. L. Cho, H. Uh, S.-Y Chang, H.-Y. Chang, M.-G. Choi, I. Shin, and K.-S. Jeong, A double-walled hexagonal supermolecule assemble by guest binding, *Journal of the American Chemical Society*, 123, 1258–1259 (2001).

44. M. Lahav, R. Gabai, A. N. Shipway, and I. Willner, Au-colloid—"Molecular square" superstructures: Novel electrochemical sensing interfaces, *Chemical Communications*, 1937–1938 (1999).
45. M. Fujita, F. Ibukuro, H. Hagihara, and K. Ogura, Quantitative self-assembly of a catenane from two preformed molecular rings, *Nature*, 367, 720–723 (1994).
46. P. J. Collings and M. Hird, *Introduction to liquid crystals*, Taylor & Francis, Bristol, PA (1997).
47. C. D. Simpson, J. Wu, M. D. Watson, and K. J. Mullen, From graphite molecules to columnar superstructures—An exercise in nanoscience, *Journal of Materials Chemistry*, 14, 494–504 (2004).
48. A. M. van der Craats, J. M. Warman, K. Müllen, Y. Geerts, and J. D. Brand, Rapid charge transport along self-assembling graphitic nanowires, *Advanced Materials*, 10, 36–38 (1998).
49. S. Ito, M. Wehmeier, J. D. Brand, C. Kubel, J. P. Rabe, and K. Müllen, Synthesis and self-assembly of functionalized hexa-peri-hexabenzocoronenes, *Chemistry A European Journal*, 6, 4327-4342 (2000).
50. L. Schmidt-Mende, A. Fechtenkötter, K. Müllen, E. Moons, R. H. Friend, and J. D. MacKenzie; Self-organized discotic liquid crystals for high-efficiency organic photovoltaics, *Science*, 293, 1119–1122 (2001).
51. S. Kumar, Self-organization of disc-like molecules: Chemical aspects, *Chemistry Society Reviews*, 35, 83–109 (2006).
52. J. P. Hill, W. Jin, A. Kosaka, T. Fukushima, H. Ichihara, T. Shimomura, K. Ito, T. Hashizume, N. Ishii, and T. Aida, Self-assembled hexa-peri-hexabenzocoronene graphitic nanotube, *Science*, 304, 1481–1483 (2004).
53. L. Brunsveld, J. A. J. M. Vekemans, J. H. K. K. Hirschberg, R. P. Sijbesma, and E. W. Meijer, Hierarchical formation of helical supramolecular polymers via stacking of hydrogen-bonded pairs in water, *Proceedings of the National Academy of Science USA*, 99, 4977-4982 (2002).

Problems

11.1 In your spare time, create an academic tree that relates all the disciplines and subdisciplines of chemistry. Create one with a timeline as well.

11.2 Define in your own words the concepts of thermodynamic and kinetic control. Are they one and the same (a part of a continuum) or are they distinctly different phenomena?

11.3 Design a supramolecule from the bottom up. Consider solvent effects, substrates, and precursors. Will your system provide a function?

11.4 Give the IUPAC name for valinomycin.

11.5 Name all of the types of natural or synthetic examples of self-assembly that you witnessed before becoming a chemist, biologist, or physicist. Is the formation of the gritty layer on the surface of a bathtub after one takes a bath an example of self-assembly?

11.6 Calculate the thickness of a layer of 2 mL of a large *n*-alkane spread out over the surface of water. How much water is covered, assuming a tightly packed layer?

11.7 Summarize the differences between hard and soft matter in one paragraph.

11.8 Define the terms host–guest, complementarity, and preorganization.

11.9 What is molecular recognition? Define the terms lock and key, dynamic molecular recognition, positive allosteric effect, and negative allosteric effect.

11.10 Describe the mechanism of a dehydrogenase enzyme (e.g., alcohol dehydrogenase). Is the enzyme able to oxidize and reduce substrates? Under what conditions is it able to do so? What are these types of enzymes called?

11.11 Would you consider valinomycin a molecule with a specific function or simply a molecule that reacts in a stochastic manner?

11.12 Is there a difference between a supramolecule and a supermolecule?

11.13 Summarize the relative bond strengths of all the types of intermolecular interactions in this chapter.

11.14 Is there an entropic factor involved in this reaction?

$$M(H_2O)_6^{2+} + 6NH_3 \rightarrow [M(NH_3)_6]^{2+} + 6H_2O$$

How about this reaction?

$$M(H_2O)_6^{2+} + 3\,[H_2N\text{-}CH_2CH_2\text{-}NH_2]$$
$$\rightarrow [M(H_2N\text{-}CH_2\text{-}CH_2\text{-}NH_2)_3]^{2+} + 6H_2O$$

Which one of the two is favored and why?

11.15 a. We know there is a difference in reactivity between linear bidentate (and greater) chelating agents and monodentate ligands like NH_3. Is there a difference between linear polydentate chelates (left) and macrocycle chelates (right) containing the same amount of ligand moieties? What are those differences and which kinds are entropically favored in a reaction? Why?

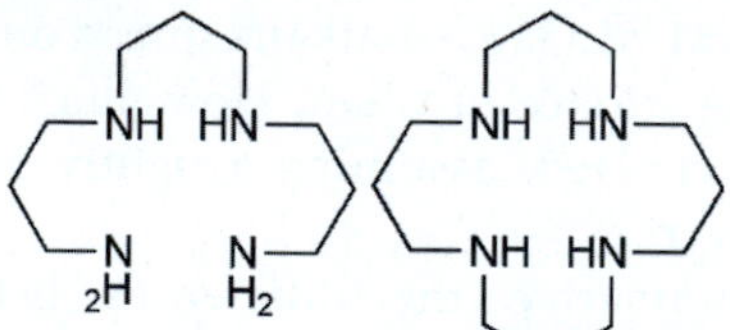

b. Is there a limit on the number of ligands a macrocycle can have before it encounters unfavorable energetics? Why or why not?

11.16 Donating and accepting electron pairs are important solvent effects. Which of the following are good donors? Good acceptors? Good donors and acceptors? Neither?

(a) $R-S(=O)-R$

(b) H_2O

(c) R–OH

(d) R–C≡N

(e) 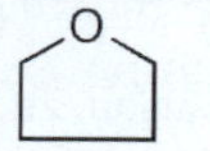

(f) 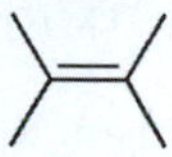

11.17 Research: Look into *Gutmann donor* and *acceptor numbers*. Create a table of molecules that lists donor number, acceptor number, and the dielectric constant of each. What trends do you see?

11.18 A recent study showed that stability constants of Ag(I) and Cu(II) metal cations with selected macrocycles were higher in perchlorate than in nitrate solutions. Stability constants were also higher in DMF rather than DMSO solutions. Can you explain these results? (A. Sil et al., *Supramolecular Chemistry*, 15, 451–457, 2003.)

11.19 Cation complexes (Ca^{2+}, NH_4^+, K^+, or Na^+) with crown ring size of 18 atoms seem to demonstrate the highest stability constants. Why do you think this is the case?

11.20 Give examples of a *chelate effect*, a *macrocycle effect*, and a *macrobicyclic effect*. Which kind generally demonstrates the highest binding constants?

11.21 What is the difference between the *thermodynamic template effect* and the *kinetic template effect*?

11.22 Alkali metals (Li^+, Na^+, K^+, Rb^+, and Cs^+) are bound by a variety of hosts in the following order of decreasing $-\Delta G$ (J. W. Steed et al., *Supramolecular Chemistry*, Wiley, New York, 2000): spherands > cryptaspherands > cryptands > hemispherands > corands > podands > solvents.
Explain this trend.

11.23 Metal chemical hardness versus softness is also a factor in supramolecular design, as are hard acids and hard bases.

Devise a scenario that involves these parameters.

11.24 What is the *synergic effect*? (J. W. Steed et al., *Supramolecular Chemistry*, Wiley, New York, 2000.)

11.25 What is *chiral recognition*? Is it important in biological systems? Name some important chiral biological molecules and their corresponding receptors.

11.26 What is an *amphiphilic receptor*? List a few examples.

11.27 What is a *π-acid ligand*?

11.28 Anion coordination chemistry is important in biology. Because anions are larger than cations, what kinds of receptors are able to accommodate such species? Name a few examples. Why do anions have higher solvation ΔG (more spontaneous) than cations of similar size?

11.29 Anions assume a variety of shapes: spherical, linear, planar, tetrahedral, and octahedral. Give examples of each. Also, provide an appropriate receptor for each.

11.30 Design a nanoscale molecular light conversion device. What components are required to transmit a signal?

11.31 Design a nanoscale electrochemical sensor. What components are required to transmit a signal?

11.32 What kinds of molecules would serve well as molecular wires? Molecular rectifiers? Molecular switches? Molecular machines?

11.33 Design a problem that involves a bipyridine species as the chromophore. Use the Beer–Lambert law. Find a species that absorbs and then luminesces. Calculate the quantum efficiency.

11.34 What host species can act as enzyme mimics? (Consult J. W. Steed et al., *Supramolecular Chemistry*, Wiley, New York, 2000.)

11.35 With classmates, discuss the types and structures of liquid crystals. Design a system that performs a function.

11.36 Much of what we have covered previously seems to be more related to organic or biochemistry. Although important, we also need to ask some nano-oriented questions. One good question to ask concerns the size of supramolecules. In general, what happens to solubility as size is increased?

11.37 Many enzymes are extremely large. How do they coexist with their immediate surroundings?

Chemical Synthesis and Modification of Nanomaterials

If a metal is downsized to a critical dimension on the nanoscale, the classical physical laws are no longer valid to describe its physical properties but instead, quantum mechanical rules must be applied. Particles of that size are called nanoclusters, nanoparticles or, related to their special physical behaviour, quantum dots. Quantum dots are a fundamental part of nanoscience and will play a decisive role in future nanotechnology. They will influence our life dramatically, since applications from information technology up to diagnosis and therapy in medicine on a molecular level will become possible.

Günter Schmid, University of Essen (2006)

Chapter 12

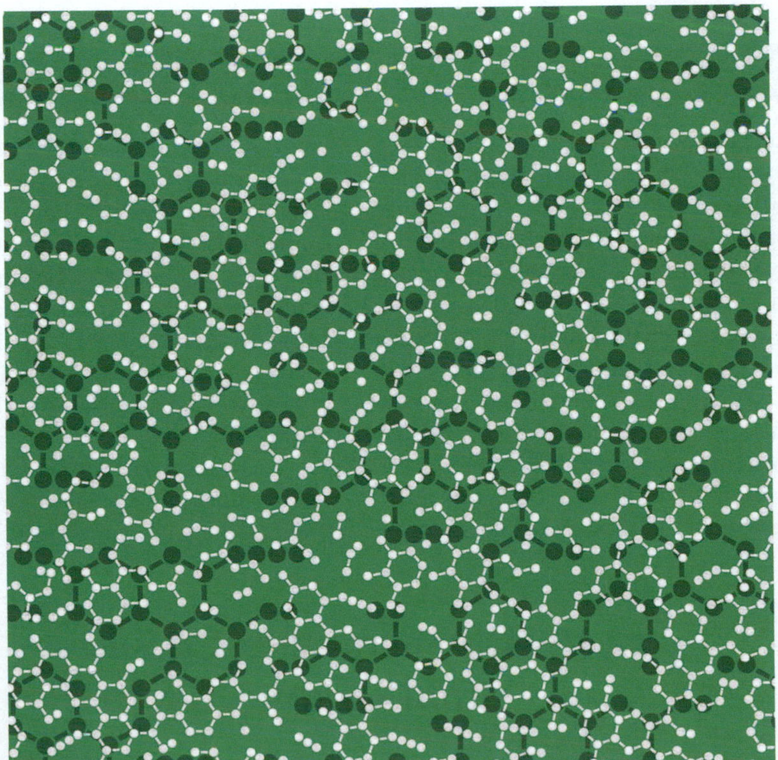

Fig. 12.0 *Professor Günter Schmid of the University of Essen (non Essen–Duisburg) has made major contributions to modern cluster chemistry and catalysis. The synthesis of Au-55 quantum dots, their physical and chemical properties, and applications have been described in great detail by Schmid.*

Source: Image courtesy of Professor Dr. Günter Schmid of the University of Essen in Germany. With permission.

Threads

Supramolecular chemistry revolves around intermolecular interactions and host–guest relationships. The precursors to supramolecules are manufactured by mostly traditional means: kinetically controlled reactions involving the making and breaking of covalent bonds. We have shown how dative bonds play a major role in biological and supramolecular complexes—in both their synthesis and function. In this chapter, we continue with intermolecular interactions but also review chemical reactions that are more traditional in their makeup that involve breaking old bonds and making new ones. Polymer chemistry is reviewed in the context of nanomaterial synthesis. Polymers are projected to make a significant contribution to the science of drug delivery by providing substrates and capsules for active components. Self-assembled monolayers have been in existence for several decades but are proving to be a significant material for sensors and electronic and optical devices, and they are proving to be facilitative in terms of their scale-up potential. With this chapter we round out the chemistry division of the text; the two chapters of the biological division remain. Our progress through perspectives, tools, properties and phenomena, chemical synthesis, and the biological domain is nearly completed.

12.0 Chemistry and Chemical Modification

Supramolecular chemistry is the chemistry of the molecule and the intermolecular bond. Chemical modification of precursor molecules, supramolecules, and nanomaterials in general proceeds with an arsenal of techniques, including covalent, ionic, and metallic interactions and with all the intermolecular forms we have delved into in previous chapters. In many ways, this chapter is focused more on synthesis than on describing types of molecules. The bottom-up method of fabricating nanomaterials is a core ingredient of every process.

12.0.1 *Types of Synthesis Processes*

Chemical synthesis strategies are numerous and varied. We assemble another catalog to describe them and classify them. Although there is much overlap, we attempt to place each kind of synthesis method within a convenient perspective. One convenient means of drawing boundaries is to place a synthesis method in one of the major classifications of nanomaterials: zero-, one-, and two-dimensional systems. This makes some sense, but we all know that several types of synthesis methods cross the boundaries of material geometry. For example, the process of forming a silica layer on a surface is also relevant for zero-, one-, and two-dimensional materials or on anything that has a surface, even one with a nano-surface. A special section is devoted to template methods of synthesis. Template methods offer a kinetically controlled environment that is able to alter the course of reactions that may otherwise not happen. In the last section, we cover the enormous domain of polymer chemistry and its relation to nanoscale materials. **Table 12.1** lists a few major chemical synthetic methods. Once again boundaries are placed for convenience; for example, we are aware that many template methods involve the sol–gel process.

12.0.2 *Introduction to Molecular Self-Assembly*

In the broadest sense, molecular self-assembly (MSA) implies an effortless construction of an object—a fabrication that simply, quickly, and automatically "falls into place." Molecular self-assembly has a similar meaning but it occurs at the level of the nanoscale with molecules. Molecules spontaneously fall into place without significant energy or direction from an outside source. MSA is influenced by chemical reactions, electrostatic interactions, and physical phenomena such as capillarity [3]. Just like with any reaction, minimization of chemical potential is the overall driver. With nanomaterials and surfaces, the driving force behind self-assembly of monolayers is to reduce the high energy associated with nanoparticles and reduce the high energy of surfaces to form a low-energy stable structure.

There are two general kinds of molecular self-assembly. *Intermolecular self-assembly* is the spontaneous aggregation (or reaction) of molecules. The molecules may collectively react with a substrate via covalent bonds (e.g., chemisorption) or by means of intermolecular forces (e.g., physisorption). *Intramolecular self-assembly* occurs within large molecules or complexes like proteins. Proteins are able to assume their 2° structure (folding to make helices or pleated sheets) due to the interactions of hydrogen bonds. The 3° structure of proteins (higher level of folding) is due to covalent interactions in the form of disulfide linkages and intermolecular interactions like hydrogen bonding, and/or hydrophobic interactions in the presence of water. Protein quaternary structure is due to proteins self-assembling into complex structures—mainly in the form of

TABLE 12.1 *Types of Chemical Synthesis*

Type of synthesis	Description
Template synthesis	Template synthesis involves a solid (or liquid) material that is capable of directing the geometrical development of a nanomaterial during its chemical or physical fabrication of nanomaterials to resemble its own image (or negative image). If pores or channels are small enough, like they are inside carbon nanotubes or zeolites, the template is capable of exerting kinetic confinement effects during a chemical process.
Sol–gel methods	Sol–gel methods have been in existence for several decades. It is a process whereby monodisperse particles can be generated in large quantities. Most sol–gel systems are based on silicate chemistry but others exist as well, such as the formation of monodisperse hematite (α-Fe_2O_3) from ferric hydroxides/ferric oxyhydroxides [1]. Sol–gel chemistry is exploited to form silica nanoparticles as drug delivery vehicles.
Metal reduction	Reduction of metals, particularly gold salts like hydrogen tetrachloroaurate [$HAuCl_4$], by organic bases such as sodium citrate is a very old procedure. As the solution becomes saturated, Au nanoparticles nucleate and start to precipitate. Control of solution parameters such as *p*H, concentration of reducing agent, and potential stabilizing ligands lead to control over particle size [2].
Molecular self-assembly	Molecular self-assembly is the spontaneous assembly of precursors to form metastable or stable nanoparticles. Molecular self-assembly is a process that is driven by favorable free energy to form zero-, one-, and two-dimensional nanomaterials.
Atomic layer deposition	ALD is a precise means of applying layers of select materials onto any surface that is capable of chemical modification. The method was discussed in some detail in chapter 4. Formation of covalent bonds and intermolecular bonds occurs during ALD deposition. ALD is based on one-to-one stoichiometry that is highly specific (e.g., any reactants that are allowed to react with a substrate react completely and all excess reactants are purged before the next step in the process).
Emulsion polymerization	Emulsion polymerization is a method that involves an emulsifier with a monomer species and a surfactant. This method is used to form latex particles that have applications as a vehicle for targeted drug delivery. In addition, the solution may contain comonomers, initiators, stabilizers, buffering agents, and preservatives.
Block copolymerization	Block copolymers are formed from two or more monomeric species. The net polymer is composed of blocks that contain one kind of polymer that is linked covalently or otherwise to another block made of another kind of polymer. In the majority of cases, a patterned array is formed that permeates the structure of the polymer—at least to the extent of forming domains with good order.
Electrodeposition electroless deposition	Electrodeposition is a means to apply thin metallic layers on conducting surfaces. Electroless deposition techniques are applied to form metallic layers on the surfaces of any material that is capable of chemical modification. Formation of three-dimensional nanomaterials is possible by this technique.

globulin proteins like hemoglobin. These interactions are more along the lines of glorified intermolecular self-assembly. Self-assembly is a subset of the general class of chemical modification. If breaking old covalent bonds and making new ones happens under mild conditions quickly, effortlessly, and spontaneously, it should be labeled as a self-assembly process. For example, the nucleophilic substitution of ester-modified silanes by surface hydroxyl groups occurs rather spontaneously, to completion, and with relatively small equilibrium constant—surely a bona fide self-assembly episode.

Molecular self-assembly is a characteristic of supramolecular reactions. Natural processes, in which energetic considerations are always at a premium, readily exploit such reactions to accomplish many metabolic and physiological tasks. Although nature is quite efficient, any error in the transfer of information, the code itself, or any environmental perturbation of the process could lead to defective products that can become detrimental to the organism. For example, it is known that certain amyloid proteins in Alzheimer's patients have flawed structure due to improper folding of the protein and that those forms tend to self-assemble. The anomalous proteins (prions) are rendered insoluble and tend to agglomerate (self-assemble) into structures called plaques in the brain. What happened to these proteins during the self-assembly (intramolecular) folding process that resulted in such devastating consequences to the organism?

According to George Whitesides et al. [4]:

> Molecular self-assembly is the spontaneous association of molecules under equilibrium conditions into stable, structurally well-defined aggregates joined by non-covalent bonds. Molecular self-assembly is ubiquitous in biological systems and underlies the formation of a wide variety of complex biological structures .... Self-assembly is also emerging as a new strategy in chemical synthesis, with the potential to generate non-biological structures with dimensions of 1 to 100 nm.

and

> Molecular synthesis is a technology that chemists use to make molecules by forming covalent bonds between atoms. Molecular self-assembly is a process in which molecules (or parts of molecules) spontaneously form ordered aggregates and involves no human intervention: the interactions involved usually are non-covalent. In molecular self-assembly, the molecular structure determines the structure of the assembly. Synthesis makes molecules: self-assembly makes ordered ensembles of molecules (or ordered forms of macromolecules). The structures generated in molecular self-assembly are usually in equilibrium states (or at least in metastable states).

George Whitesides was mentioned in chapter 1 and is one of the leaders in the field of molecular self-assembly. Interestingly, self-assembly is not the exclusive domain of nanomaterials. Micromaterials and larger materials are also able to fall into place under the proper conditions [4]. We discussed in chapter 8, "Nanothermodynamics," how self-assembly processes are characteristic of dissipative systems that are fueled from outside sources (the Bénard cell). The concept of self-assembly actually can be extended to include galaxies! However, within the frame of this text and the prevailing definition, let us adhere to molecular processes that require little input of energy or management.

12.0.3 Introduction to Chemical Functionalization

Nanomaterials tend to aggregate, react chemically, or decompose if not treated properly. Nanomaterials are metastable materials that need to find ways to address the inherently unhappy state of high surface energy. There are many means to stabilize nanosurfaces and render them chemically receptive to further modification (e.g., attachment of ligands with good leaving groups or of ligands that can be easily displaced by ones that make stronger interaction with the nanoparticle). All kinds of nanomaterials are capable of undergoing chemical functionalization. Attaching molecular recognition moieties to the surface of ligand-stabilized quantum dots, adding polymer linking groups to the surface of single-walled carbon nanotubes, and attaching photo-sensing charge transfer dye molecules to titanium dioxide nanoparticles to initiate the exchange of solar energy to PV devices are all means of modifying nanomaterial substrates for further chemical processing.

There are many ways to chemically modify nanomaterials. Chemical modification methods involve addition, subtraction, or exchange of chemical groups on the surface of nanomaterials or internal restructuring of the molecule itself. One objective of chemical modification is to enhance the chemical behavior of nanomaterials without altering (at least drastically) the physical properties of the underlying molecule. Nanocomposite physical properties such as mechanical strength and electrical and thermal conductivity benefit composites immensely if reinforced with carbon nanotubes. However, pristine tubes are ineffective in the matrix of the polymer without chemical modification of the surface (e.g., not much is able to stick to the surface of carbon nanotubes). Bonding chemically labile species to the surface of nanotubes enhances cross-linking between nanotubes and with the materials of the polymer matrix. The degree of cross-linking affects the mechanical properties of the composite but also the intrinsic physical properties of the carbon nanotube. This trade-off needs to be considered during the design phase of the nanocomposite—an issue for the mechanical and chemical engineers to address. The terms *functionalization, modification,* and *derivatization* are used interchangeably in this chapter.

12.0.4 The Nano Perspective

Bottom-up methods will assume more importance in nanotechnology as the years go by. Obstacles such as the apparent lack of long-range order will diminish as better precursors are developed, techniques are refined, and atomically smooth substrates become commonplace. Formation of undesirable side products becomes less important as supramolecular chemistry and site-specific chemistry become better understood. Whole-scale manufacturing of nanomaterials by bottom-up techniques at the nanoscale is not a reality at this time, although its upside is enormous (in particular, its high throughput and inexpensive capital needs). However, there are many examples of well-developed (and profitable) manufacturing processes to make nanoparticles of all sorts from the bottom up. Latex particles that have nanometer to micron dimensions have been around for decades. In the computer industry, bottom-up techniques are just starting to make their presence known. Industrial fabrication of multiwalled carbon nanotubes from the bottom up by flame techniques for use as polymer fillers is acquiring an increasingly bigger share of the global market.

Efforts to copy nature (biomimetics) are ongoing as more and more of nature's secrets are unraveled. The key to success must lie in the ability to fabricate incredibly complex materials at body temperature and neutral *p*H—easier said than done.

12.1 Self-Assembly Revisited

We have been directly and indirectly alluding to the concept of molecular self-assembly in this chapter and previous chapters. MSA occurs between and among mobile molecules and between molecules and surfaces. The surface can be flat (planar) or with high curvature (colloids, cavities, nanowires, edges, etc.). Molecular self-assembly is the method of choice of bottom-up chemical synthesis of nanostructure materials for many reasons—not all of them fully realized at this stage of its development. Some forms of MSA require high specificity, as in the case of supramolecular recognition, and others not so much, such as in the case of monolayers and metal surfaces. Self-assembly implies that materials come together in a rather spontaneous way without the need for excess energy. The term self-assembly also implies that chemical reactions are reversible (e.g., relatively small but positive reaction equilibrium constants, K_{rxn}).

In addition to intermolecular and intramolecular MSA, there is also static and dynamic MSA. *Static self-assembly* is the embodiment of the equilibrium condition (e.g., in that $\Delta G = 0$) and the process operates under the conditions of thermodynamic equilibrium. *Dynamic self-assembly* processes exist away from equilibrium conditions, dissipate energy, and stop existing when the steady-state flow of energy ceases. What kind of system resembles a dynamic self-assembly process? The answer is that living organisms are full of processes that resemble dynamic self-assembly.

We have now encountered a small dilemma. All along we have been stating that nature relies on self-assembly processes to accomplish its multiple purposes and that these processes are spontaneous, directed by molecular recognition, and occur under mild conditions. These are all true but the bottom line is that all self-assembly processes in living things are solar powered. This implies that MSA in living systems is a dynamic process and not a static one per se. Recall that the general state of thermodynamic equilibrium is only achieved when the organism ceases to exist—the condition of cellular death. Are there examples of static equilibrium in natural systems? In the beaker, molecular self-assembly occurs at moderate temperatures. If the system achieves equilibrium, it is called static MSA. There is no dilemma because the energy required to drive it is supplied by a few integral portions of $k_B T$. After all, very few self-assembly processes are conducted in a refrigerator. For our synthetic purposes, we wish that reactions terminate (achieve thermodynamic equilibrium or kinetic stability) so that we can use the products to accomplish other tasks. MSA for us implies equilibrium; for nature, it implies nonequilibrium.

The formation of macromolecules and macromolecular systems from inorganic starting materials that become integrated into living things is the biological foundation for self-assembly. As with most biological processes, a change in state is effected from that of a disorganized system to one that inherently demonstrates a higher level or organization. Entropy therefore must be accounted for in any self-assembly process. Driving forces for self-assembly include chemisorption of the potential layer to the substrate, surface effects, hydrophobic–hydrophilic factors, intermolecular forces, and capillary forces [3].

Self-assembly processes tend to produce products with relatively short-range order. Long-range order, although extremely desired, breaks down due to the

FIG. 12.1

Image of an n-alkane thiol (derived from tetradecane) with a 2° amine capping (terminal) group is shown in the figure. The thiol group (yellow) chemisorbs to gold surfaces to form a monolayer. The alkane tails of the SAM interact with each other via lateral van der Waals forces. The capping group (the amine, blue) undergoes numerous organic reactions to form chemically modified monolayers.

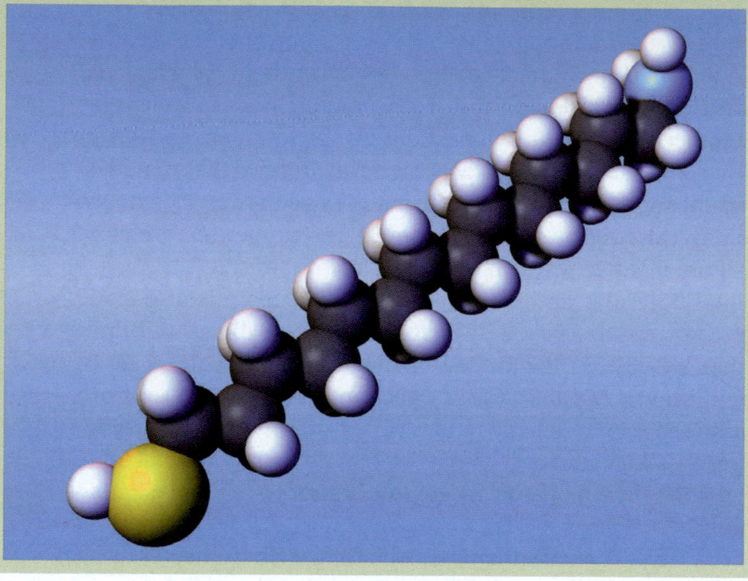

occurrence of any number of possible aberrations during the synthesis process or perturbations transferred to the process by surface defects. The goal of achieving long-range order is one of the holy grails of molecular self-assembly research. Interactions that result in forming labile bonds and molecular recognition leading to selectivity are key ingredients of MSA.

A generic molecule that is suitable for self-assembly on a surface consists of a surface-active head group, a cylindrical body that usually is an aliphatic chain, and a capping (or terminal) group (**Fig. 12.1**). The head group is either an electron-rich moiety like sulfur that is willing to donate electrons to a transition metal surface or one that contains labile ester groups willing to exchange with surface hydroxyls. The body of the typical surface-active molecule is usually a long-chain alkane that experiences lateral van der Waals interactions to form a tightly bound monolayer once bound to the surface. The capping group is usually capable of further derivatization with the purpose of adding additional layers or highly specific host molecules (**Table 12.2**).

Self-Assembled Monolayers. Self-assembled monolayers (SAMs) form a distinct class of molecular self-assembled materials. In the spirit of this chapter, we shall continue with the premise that intermolecular attractions and bonds form the core of the structures of interest, although a strong case can be made for covalent "self-assembly" interactions as well. Yes, a self-assembled monolayer is a coating on a substrate by a single layer of molecules that fall together spontaneously, but covalent bonds can be and are formed between the monolayer entity and an

Table 12.2 *Types of Self-Assembly Methods and Materials*

Generic type of SAM	Description	Applications substrates
Langmuir–Blodgett films on liquid surfaces	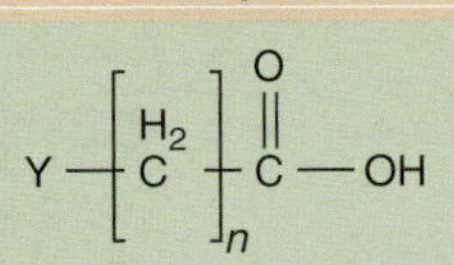 *A generic alkyl acid is depicted: Y = functional group, usually $-CH_3$, $-OH$, or $-CO_2H$*	Hydrophobic or hydrophilic liquid substrates Electrostatic interactions: Charge–dipole Dipole–dipole
Self-assembled monolayers on solid surfaces	Thiols 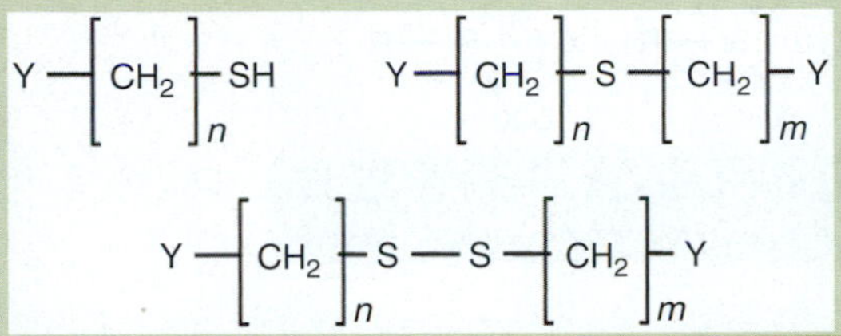 *Sulfur-containing SAMs bond well to metals. Left top: n-alkane thiol; top right: two-tailed thiol; bottom: a disulfide is depicted. Disulfide linkages play a major role in determining protein structure*	Neat, structured Au, Ag, Cu surfaces Covalent *d*–*d* interactions ~180 $kJ \cdot mol^{-1}$
	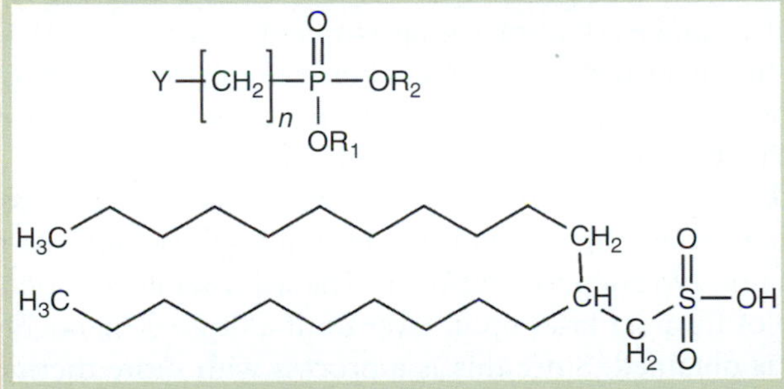 *Phosphonates and fatty acids. R = –H, $-CH_3$, or other hydrocarbon groups*	Glass, SiO_2, Al_2O_3, AgO, or other polar surfaces SiO_2, Al_2O_3, TiO_2 Ionic/electrostatic interactions ~50 $kJ \cdot mol^{-1}$
	Other SAM ligand systems on thin films, bulk substrates, nanomaterials [5] R–OH on iron oxides, Si–H, Si $R-CO_2^-$ and $R-CO_2H$ on alumina, iron oxides, Ni, Ti, and TiO_2 $R-NH_2$ on FeS_2, mica, stainless steel, CdSe R–CN on Ag, Au R–SH on Ag, AgS, Au, CdTe, CdSe, CdS, Cu, Ge, Hg, InP, Ir, Ni, PbS, Pt, Ru and Zn, ZnS R-S-S-R′ on Ag, Au, CdS, Pd R–SeH on Ag, Au, CdS, CdSe $P-R_3$, $P(Ph)_3$ on Au, FeS_2, CdS, CdSe, CdTe $R_3-P=O$ on Co, CdS, CdSe, CdTe $R-PO_4^-$ on Al_2O_3, Nb_2O_5, Ta_2O_3, TiO_2 $RHC=CH_2$ on Si RCCH on Si(111):H	

(*continued*)

TABLE 12.2 (CONTD.) *Types of Self-Assembly Methods and Materials*

Generic type of SAM	Description	Applications substrates
Silanization	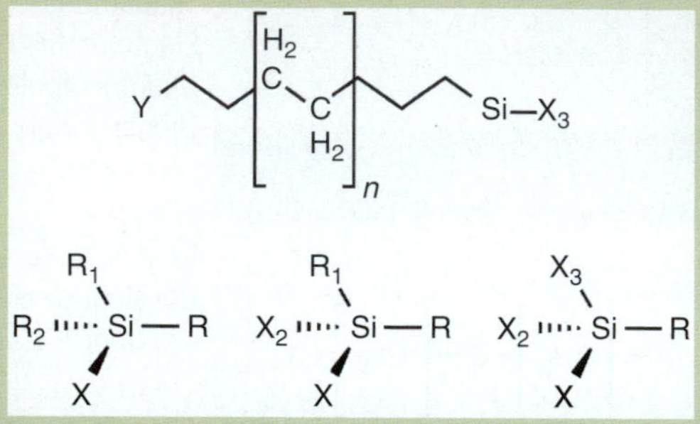 *Alkylsilanes. X = halogen, ester, hydroxide, amine (1°, 2°) or other reactive group; R = –H, $-CH_3$, or hydrocarbon moiety*	Hydroxylated surfaces (in general) and carbon Covalent Si–O ~440 $kJ \cdot mol^{-1}$ Covalent Si–C ~300 $kJ \cdot mol^{-1}$

activated surface. Yes, the interactions involved in spreading a monolayer of stearic acid on the surface of water is due to weak interactions, such as dipole–dipole, van der Waals, and hydrophobic effects, but many kinds of self-assembled monolayers form strong covalent (chemisorbed) bonds with the substrate. In other words, it is perhaps better to classify the self-assembly process as occurring between a superprecursor (the monolayer moiety) and a supermolecule (the metal) that results in the formation of a stable layer.

Regardless where we draw the line, the application of a self-assembled monolayer onto a surface is accomplished by liquid-phase and gas-phase methods (e.g., by atomic layer deposition [ALD]). The self-assembled monolayer technique is capable of forming layer upon layer until a desired hierarchical structure or thickness is obtained. Since this is a process with more thermodynamic than kinetic character (once again, where do we draw such lines?), the driving force to form SAMs is the reduction of the overall chemical potential [3]. Monolayers of special molecules can be formed on surfaces of bulk materials composed of metals, metal oxides, semiconductors, ceramics, polymers, and various liquids. The layers can be organic or inorganic. Covalent bonds are generally formed between the layer and the substrate as well as at the terminal (solution) end, the reactive site at which an additional layer can be added.

Substrates. Substrate preparation is accomplished by well-established techniques [5]: physical vapor deposition (PVD), evaporation (thermal and electron beam), and electroless and electrodeposition, to name a few. Silicon wafer surfaces (in the natural oxidized form) are commonly used as substrates. A 1- to 5-nm layer of Ti, Cr, or Ni (or Ta) is applied to provide adhesion for noble metals. The thickness of Au films after deposition ranges from 10 to 200 nm. Such metal films form polycrystalline domains (on the order of 10 nm–1 μm) usually of the (111) orientation for *fcc* metals. Grain size and surface structure can be modified by manipulation of temperature, composition, deposition rate, annealing, chemical treatment, and mechanical methods. Cleaved mica substrates provide "pseudosingle crystal surfaces." Terraced Au(111) layers by epitaxial growth on mica(100) surface are prepared by thermal evaporation of gold (@ 0.1–0.2 $nm \cdot s^{-1}$) onto mica surface at 400–650°C. Interestingly, self-assembled layers on liquid mercury have shown high levels of organization [5]. Liquids tend to form "perfect" surfaces, whereas solid surfaces are never perfect unless it is a single crystal material.

SAM Techniques. Several techniques take advantage of SAM technology. Metal-coordinating pyridines serve as gate molecules to link electrodes. The link is formed by SAM mechanism on the gold surfaces of either electrode. Nano-imprinting methods are able to produce patterned surfaces by application of hexadecanethiol onto a gold layer. Silicon is coated with a thin Ti layer upon which Au is applied. Microcontact printing (μCP) technique is able to form patterns of self-assembled monolayers with the alkanethiol SAM as the ink (**Fig. 12.2**). A mask is first prepared by optical, x-ray, or e-beam lithography and its pattern is subsequently transferred to stamp made of silicon elastomer (poly[dimethylsiloxane], or PDMS). The patterned PDMS is then dipped into the ink (the *n*-alkanethiol) and the substrate is stamped accordingly [6], thus transferring the ink to the substrate. Etching and applying another kind of thiol (e.g., one terminating in carboxylic acid groups that also reacts with the metal surface) yields a surface that is completely covered in SAMs but that has patterns. The regions coated with thiol are hydrophobic and thereby resistant to wet chemical etching.

12.1.1 Langmuir–Blodgett Films

In 1891, Agnes Pockels, an "uncredentialed" physicist, submitted a letter to Lord Rayleigh. The letter summarized work she had accomplished in her kitchen sink between times she washed dishes and conducted other household chores. She did not know at the time that her work would lay down the foundation for a new field of chemistry and that the "trough" she used to study the effect of substances on water would essentially form the basis for the trough used by Langmuir and Blodgett in their landmark discovery. The great Ostwald in 1932 paid tribute to her work [7]: "Every colleague who is now engaged on surface layer or film research will recognize that the foundation for the quantitative method in this field ... has been laid by observations fifty years ago."

The definition of a Langmuir–Blodgett (LB) film, according to KSV Instruments, Ltd. is

> Lipids, polymers or other water insoluble atoms or molecules can form ultrathin and organized monolayers at the air/water interface, i.e., Langmuir films. These films can be deposited on solid substrates to form highly organized regular multilayer stacks called Langmuir–Blodgett films. The LB films are prepared by successively dipping a solid substrate up and down through the monolayer at a constant molecular density or surface pressure. In this way multilayer structures of hundreds of a few nanometer thick monolayers can be produced.

Langmuir–Blodgett films are two-dimensional nanostructured materials. They are molecular monolayers that have length and width but no apparent thickness.

FIG. 12.2 *Microcontact printing (μCP) developed by the Whitesides group of Harvard University. This technique is an inexpensive process (except for the single top-bottom lithographic step involved in making the mask) with high potential throughput. After application of the first monolayer, the wettability of the surface (one way or another) is altered. This allows for specific treatment of the remaining surface with other types of SAMs. The patterned surface is characterized by domains with different chemical and physical properties. The sample surface depicted in the figure is a hypothetical pattern and is not grounded in any form of reality.*

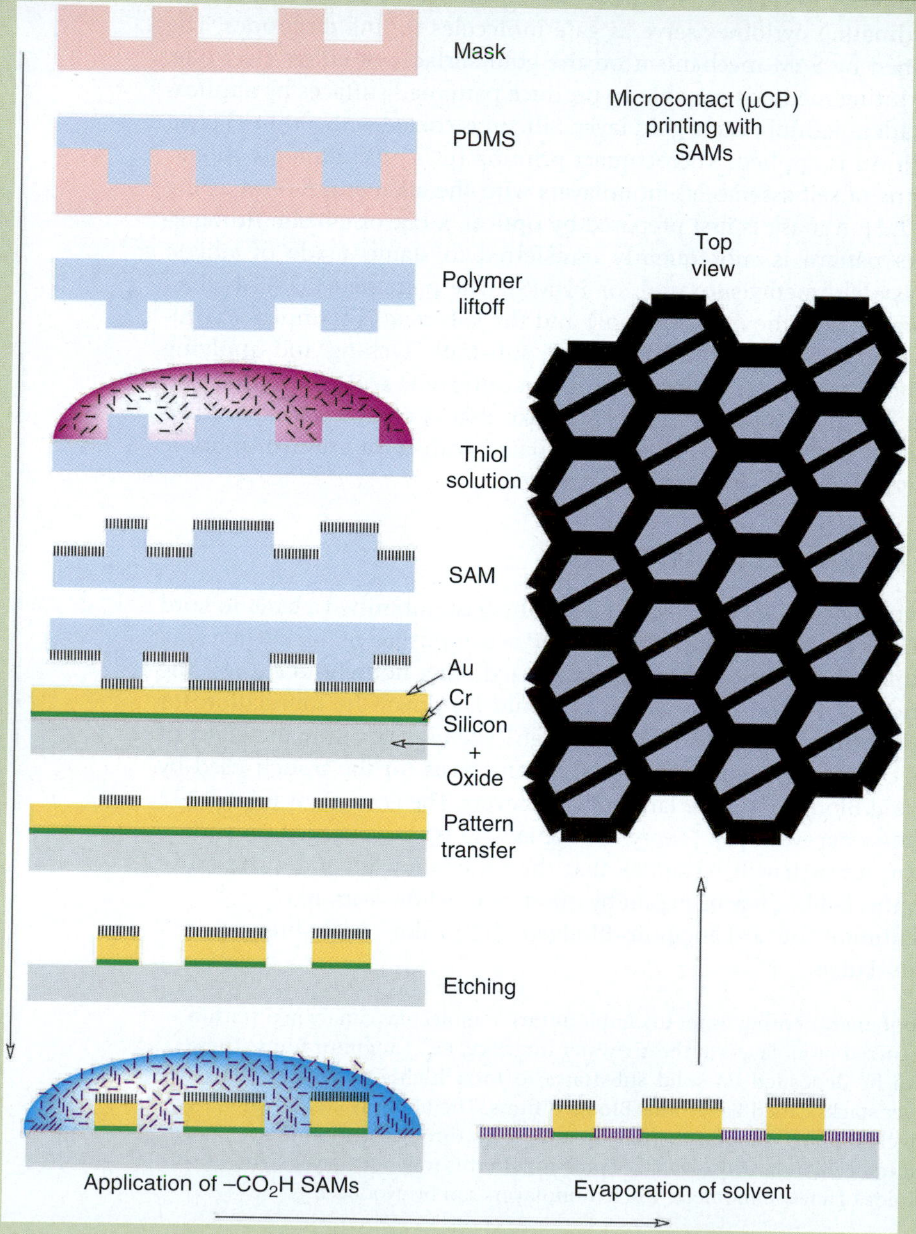

If linking groups are attached to the caps, more layers can be applied to form contiguous robust films. Additional layers can be added and held by simple hydrophobic–hydrophilic interactions but without the integrity afforded by covalent linkage between layers. In many ways, LB films are a chemist's dream because they are easy to prepare; are reproducible; possess excellent molecular organization, orientation, and structural properties; and are easy to characterize by a multitude of analytical techniques. Molecules that are not soluble in the liquid medium (usually water) qualify as candidates for LB application. Gibbs layers, on the other hand, are formed from molecules that are soluble in the liquid medium. Micelles and similar structures are members of the family of Gibbs materials.

Langmuir–Blodgett films were discussed briefly in chapter 4, "Fabrication Methods." The substrate is lowered normal to the surface of the liquid phase. Langmuir–Schaefer deposition, a related technique, involves lowering a solid substrate in a horizontal orientation to the liquid phase. Self-assembled/organized deposition of amphiphilic molecular layers at a liquid–air interface (or solid–liquid–air) forms the basis for the Langmuir–Blodgett film technique. The formation of thin organic films is accomplished by transferring monolayers of amphiphiles (surfactants) from the surface of water, where they are entrained, to a solid substrate. The observations of Benjamin Franklin in 1774 and Lord Rayleigh later about the spread of an oil monolayer at an air–water interface—in this case, that of a pond—have already been mentioned.

What is the importance of monolayers? To begin with, monolayers have the ability to modify the surface energy of a solid or liquid. A surface that was previously hydrophilic (e.g., hydrated alumina) can be rendered hydrophobic after the application of a long-chain carboxylated hydrocarbon such as stearic acid. The surface tension of solutions is also altered by application of a monolayer, as astutely observed by Benjamin Franklin. Surfactants have an affinity for interfaces and the reduction in surface tension results in the reduction of the overall free energy of the interfacial system. Monolayers are able to alter the optical properties of a surface. The best example of this is the invisible glass developed by Katherine Blodgett that consisted of 44 monolayers. Monolayers provide a means of electrical communication between a conducting substrate such as gold and sensing moieties tethered to capping groups. Because the thickness of the layer can be controlled and *preorganized* molecular arrangement is achievable, Langmuir–Blodgett films support a wide variety of molecular, electronic, and bioelectronic devices. Such films have already achieved widespread commercial success and therefore must be considered as one of the first applications of nanoscience to make an impact.

All of us chemists have estimated Avogadro's number by determining the molecular cross-section of stearic acid on the surface of water. It is incredibly ironic how thoughts of nanotechnology never entered our heads during our freshman laboratory endeavors. All we were interested in that lab was to determine Avogadro's number, the thickness of the stearic acid monolayer, calculating the molecular volume, and passing the class (right?). The spread of stearic acid on the surface of water is a pure example of MSA. It involves only intermolecular interactions; each stearic acid molecule retains its chemical identity and a condition of static MSA in equilibrium is attainted. The polar group of stearic acid interacts with polar water molecules via hydrogen bonding. The linear long-chain aliphatic components of stearic acid attract each other by van der Waals interactions to provide an evenly spaced, hexagonally distributed contiguous layer.

The Langmuir–Blodgett Technique. The Langmuir–Blodgett technique and a schematic of the mechanism are shown in **Figure 12.3a** and **b.** KSV Instruments, Ltd. Series 5007 instrument and an LB-substrate, are illustrated in **Figure 12.4a** and **b,** respectively [8]. A flat planar material with an appropriately treated surface (e.g., clean and chemically compatible with the chemistry of the monolayer) is thrust vertically downward into the liquid phase supporting the monolayer. Following equilibration, the plunger is withdrawn slowly as the monolayer adheres to the surface. Meanwhile, a moving barrier applies pressure laterally on the film to ensure maximum contact among the amphiphiles in the monolayer. The LB technique is a very simple technique that is capable of fabricating two-dimensional nanomaterials. Film thickness is linearly proportional to the number of LB steps [9].

As molecules are squeezed together in the trough, the incipient monolayer undergoes phase changes that range from the gaseous (the highly dilute phase)

Fig. 12.3

(a) Successive layers of amphiphiles are added to a surface. A solid substrate is dipped into a liquid (usually water) that has a monolayer of a selected amphiphile on its surface. In the figure, the solid surface is an activated silica that is decorated with hydroxyl groups. The surface therefore is hydrophilic and attracts the hydrophilic moiety of the amphiphile (the spherical structures of the molecule). For the second application, the reverse process occurs; specifically, it is now the hydrophobic parts of the amphiphile that interact with the hydrophobic parts of the first monolayer. The process continues in this way until enough layers are built to satisfy the specification. (b) Graphical depiction of the Langmuir–Blodgett apparatus and operation. The SAM molecules are introduced onto the surface of the liquid (usually water) with a syringe. Lateral pressure is applied to the monolayer from the barrier plunger apparatus. It is amazing that Agnes Pockels conducted the first LB experiments in her kitchen sink.

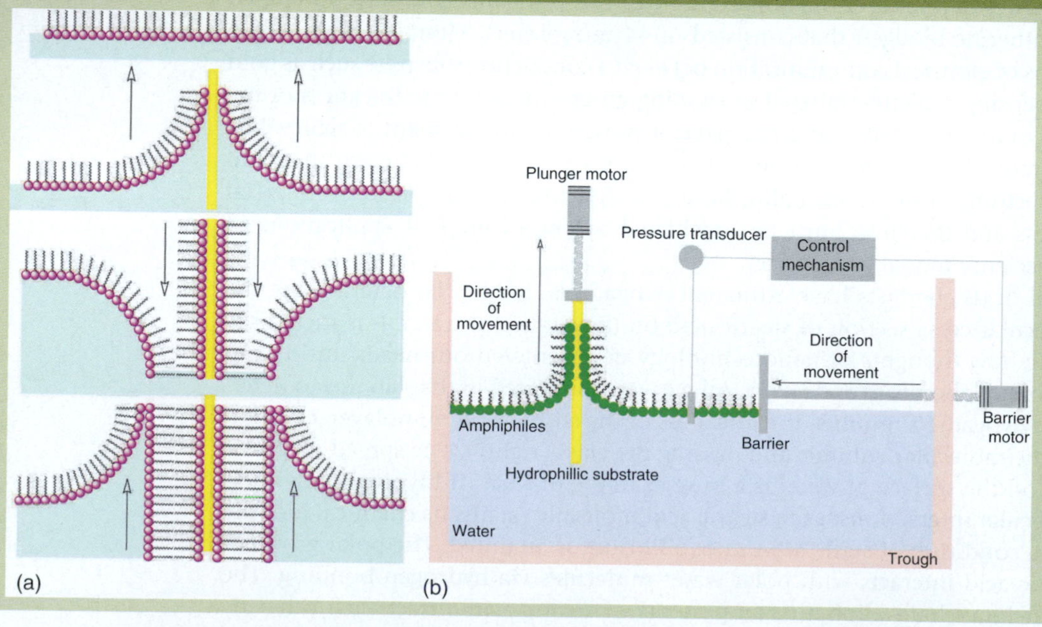

Fig. 12.4 *(a) Image of a KSV Langmuir–Blodgett 5007 Series instrument provide by KSV Instruments of Finland. Compare the components in this image with the schematic shown in Figure 12.3. (Image courtesy of Tapani Viitala of KSV Instruments, Inc.) (b) Image of a substrate undergoing the LB procedure is shown. The surface is that of a silicon wafer with a native oxide coating. The type of surface and its rate of entry and withdrawal affect the quality of the final SAM on the surface.*

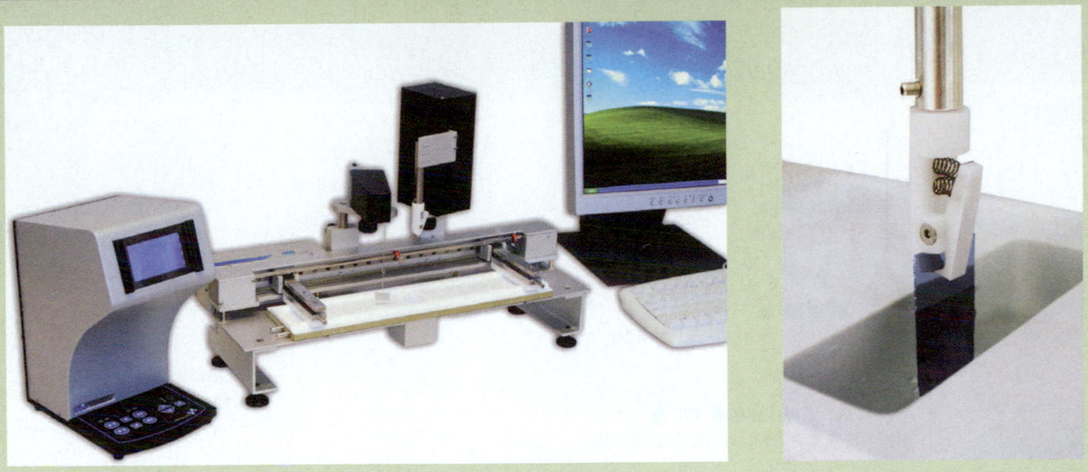

Source: Image courtesy of Tapani Viitala of KSV Instruments, Inc. With permission.

to liquid phases (expanded to condensed) and finally into the solid phase. A generic pressure–area isotherm profile is shown in **Figure 12.5.** Many factors affect the quality of LB films:

- Most LB substrates are made of hydrophilic materials like quartz, glass, silica, and metals. The computer industry employs silicon wafers covered with a thin layer of silicon oxide. Substrates with extremely high surface energy like meat silver and gold make monolayer transfer problematic. Metal surfaces are also easily contaminated [3].
- The chemical nature and structure of the amphiphile also influence the character of the LB process and the resulting monolayer. Careful selection of chain length and polar head strength is important. Head groups that interact strongly with the water may not be suitable for LB processing. If the strength of the head group is on the weaker side, then it may be difficult to form an LB film due to lack of interaction with the substrate [3]. For C_{16} amphiphiles, the most stable films are formed by surfactants with head groups that terminate (in general) with alcohols, esters, carboxylic acids, cyanides, and amides. Weaker films result with ethers and peroxides, and very weak LB films are generated if halogens are employed. Strong soluble end groups include the class of sulfate-based anions and ammonium (1, 2, and 3°) cations [10].

Fig. 12.5 *A typical pressure–area isotherm is shown in the figure. Upon compression, the system proceeds from a gaseous (highly dilute) phase depicted on the far right through a liquid–gas phase and then into transitions between expanded liquid and condensed liquid phases—all distinct regions on the graph. When the solid phase is achieved, the distance between molecules is minimized and the molecular area equals the area of the molecule. If the monolayer is compressed beyond the solid phase boundary, the film collapses into a three-dimensional structure. The surface tension of water is also shown as a function of surface coverage—it decreases as molecules pack the surface. The monolayer provides the requisite interactions with undercoordinated water molecules at the surface.*

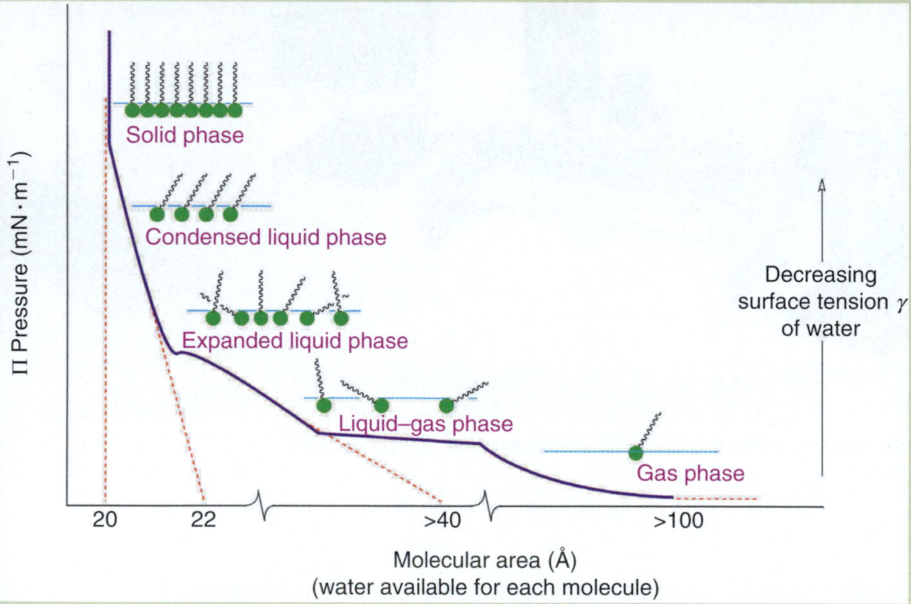

- The kind of solvent (mostly water), its *p*H, temperature, and the relative concentration of amphiphiles all influence the LB process to some degree.
- The cleanliness of the substrate surface, the rate of harvesting, and the rate of compression also influence the course of an LB process.

Zwitterionic Phospholipids. A recent study by L. Cristofolini et al. utilized LB layers of zwitterionic phospholipids as a method to immobilize DNA. An LB film consisting of dipalmitoylphosphatidylcholine (DPPC) (**Fig. 12.6**) was formed in a KSV Instruments 5000 Langmuir trough. All monolayers were formed on a DNA subphase with *p*H 6.0 containing 5 m*M* $CaCl_2$ and 10 m*M* NaCl. The DNA covered between 20% and 30% of the surface area. Compression was accomplished at a rate of 10 mm · min^{-1} [11]. The purpose of the study was to find a way to transport DNA into mammalian cells by nonviral gene therapy. The authors found that the immobilized DNA retains its double-helix structure when stabilized by the zwitterionic lipid monolayer [11]. An incredible representation of the tiered structure is shown in **Figure 12.7**.

FIG. 12.6 *Dipalmitoylphosphatidylcholine (DPPC), a zwitterionic phospholipid, is a two-tailed SAM molecule.*

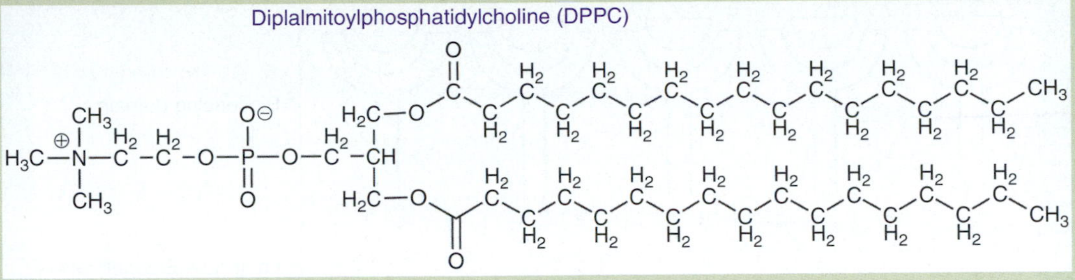

PET–Gold Clusters. Langmuir–Blodgett monolayers of nanoparticles of phenylethylthiolate (PET)-passivated gold were fabricated at various Π (surface pressures) [12]. The diameter of PET–gold nanoparticles $[Au_{38}(Ph\text{-}C_2S)_{24}]$ was approximately 1.6 nm [13]. A monolayer on water was prepared by addition of the PET–Au onto the surface of water in an LB trough and then deposited on an interdigitated electrode array. Measurement of interparticle spacing was accomplished with transmission electron microscopy (TEM) and scanning tunneling microscopy (STM). The surface pressure in the trough is related to the interparticle distance; π–π stacking of the phenyl groups of neighboring particles occurred during the self-assembly of

FIG. 12.7 *The DPPC monolayer–DNA interaction is depicted. The model is based on x-ray reflectivity data. The formation of the zwitterionic DPPC–DNA–Ca^{2+} system (the calcium cations are represented by the red spheres) at the air–water interface demonstrates the versatility of the LB technique. Ca^{2+} is a necessary ingredient for this structure to form. Because DNA did not denature in this configuration, its utility as a method for DNA immobilization (hence transport) offers inroads to developing advanced gene-therapy vehicles.*

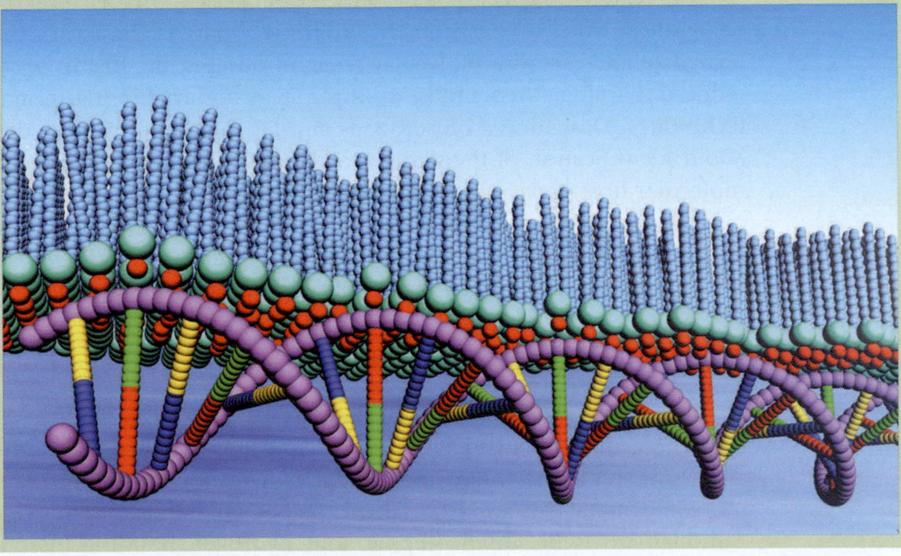

Source: Image reprinted with permission of Luigi Cristofolini of the Department of Physics at the University of Parma, Italy.

FIG. 12.8 *A general schematic of a biosensor is depicted. Contact between a host and guest initiates a signal transfer process. In these cases, tight permanent bonding with a substrate is not desired.*

the LB film. Current-potential profiles indicated that interparticle coupling occurred through the aromatic ring overlap of neighboring particles.

Electron transfer occurred through bonds and through space via van der Waals contacts between interdigitated ligands. The authors concluded that the ligand van der Waals interactions are enhanced by the π–π stacking of the phenyl groups from adjacent particles. By way of computer modeling, they found that the edge-to-edge separation of particles was 1.1 nm when the phenyl groups completely overlapped [12].

Biosensors with LB Films. A biosensor is a molecular sensing element coupled to a transducer (**Fig. 12.8**). A biosensor operates at a higher level of molecular recognition with emphasis on the "cognitive" part of the term—the ability to recognize and record a molecular recognition episode. In this way, sensing is reduced to a single step, unlike multiple steps involved in traditional analytical techniques. Designing a biosensor is similar to supramolecular design and is about identification of the target molecule (the guest), selection of a suitable molecular host species, tethering strategies to fix biosensing elements, and, of course, the associated electronics that are required. The design of such sensors involves chemists, biochemists, engineers, and electronics specialists. The detection limit and resolution of such devices should be better that 1 nmol.

12.1.2 Gold–Thiol Monolayers

The formation of self-assembled monolayers on metals is not considered to be supramolecular chemistry, technically speaking. It is the reaction between an organic molecule (not considered generally to be a supramolecule) and an inorganic metal surface (the supermolecule?) that is often made of gold. Although no ionic, dipole–dipole, and hydrogen bonding interactions take place, a combination of dative bonding (chemisorption) and van der Waals forces, as well as hydrophobic interactions, is in effect. The chemisorption of a monolayer upon a metal substrate implies the formation of stronger bonds (e.g., bonds between electron-donating moieties like sulfides and disulfides to an electronegative metal like Au). One of the driving forces for chemisorption (and physisorption) on metal or metal oxide surfaces is reducing the surface excess energy (chapter 6). Surface metal atoms in particular are undersaturated with regard to bonding and are willing to seek electrons from other sources for stabilization—especially from electron-rich elements like sulfur.

The synthesis of Au–thiol SAMs is relatively straightforward. The result is a monolayer system that is one of the most fundamental, flexible, and useful in nanotechnology; gold is the perfect substrate due to its high conductivity and relative chemical inertness, and that sulfur loves gold. The ability to chemically modify the tail group of the attached moieties allows for additional layers and specific chemical functionalization in later steps. According to George Whitesides et al. of Harvard, SAMs (1) are easy to prepare; (2) conform to the topography of kind of shape; (3) are able to modulate information between the environment and its substrate metal structure via electrical I–V response, surface plasmon resonance, and refractive index; and (4) are the nanostructured component of macroscopic "interfacial phenomena" like surface wetting, adhesion, and friction [5].

The tethering of alkylthiolate monolayers (ca. 1–3 nm) on gold occurs by dative bonding between sulfur and gold. The strong affinity of S for Au is one factor that plays a role in the stability (and hence popularity) of Au–thiol systems, but intermolecular van der Waals forces are also important; they are responsible for the lateral packing of the monolayer. Lateral packing provides uniformity and robustness to the layer. The strength of the lateral interaction, dependent on chain length, was found to be ca. 4–8 $kJ \cdot mol^{-1}$ per methylene group [14], but 6–8 $kJ \cdot mol^{-1}$, according to others [15]. The chemisorption energy of dimethlydisulfide adsorption was found by the TPD method to be 117 $kJ \cdot mol^{-1}$ [16]; others report a larger bond strength of ~180 $kJ \cdot mol^{-1}$ [15]. The bond between S and Au is formed by a chemisorption mechanism. The generic reaction to form such a monolayer is as follows:

$$2R(CH_2)_nSH + 2Au^o \rightarrow 2R(CH_2)_nS^- + 2Au^{1+} + H_{2(g)} \qquad (12.1)$$

The capping R-group represents species that are either active (available for further chemistry) or terminal (allowing no further chemistry). The hydrophobic aliphatic tail is optimally 2–3 nm in length and tilts at an angle of 30° from the normal to the gold surface. This tilt angle maximizes the interchain van der Waals attractions of the hydrocarbon chain and hence lateral continuity. The Au(111) surface provides an optimal substrate and directs the thiols to be laid out in a hexagonal pattern (**Fig. 12.9**).

Metal surfaces are characterized by high surface energy. In other words, if a droplet of water is placed on the surface, the contact angle is expected to be very small. Chemical modification of the gold surface by an alkane–thiol monolayer (depending on the chemical nature of the capping group) is able to drastically alter the wettability characteristics of the gold surface. If the capping group is a terminal methyl group, the contact angle of water on a hexadecane thiolate SAM-modified surface was found to be 115°. If, instead, a hydrophilic substituent is the capping group, such as [$HS(CH_2)_{16}$–OH], the contact angle is less than 10°. Special surfaces can be fabricated to suit the wettability requirements with mixed species of monolayer [17].

Fig. 12.9 *Left: alkanethiol distribution on Au(111) surface is depicted. The strength of the Au–S bond is relatively strong (ca. 184 kJ·mol^{-1}) and the intermolecular attraction of the alkane chains of the thiol groups is approximately 6–8 kJ·mol^{-1}. Right: idealized rendition of thiol layer on gold substrate. Red dots represent tethered functional groups at the end of the thiol aliphatic chains. Alkane chains with 12 or more methylene units form dense, well-ordered monolayers. The sulfur of the thiol attaches to the three-fold pocket (pinning site) formed by adjacent gold atoms. Distance between pinning sites is ca. 5 Å. This provides an area equal to 21.4 Å^2 with which to accommodate a chemisorbed molecule. The van der Waals radius of the thiol is only 2.3 Å (area = 17 Å^2); therefore, it cannot completely fill the space provided by the gold interstitial surface area. The chain therefore tilts 30° to enhance van der Waals interactions between and among the monolayer molecules.*

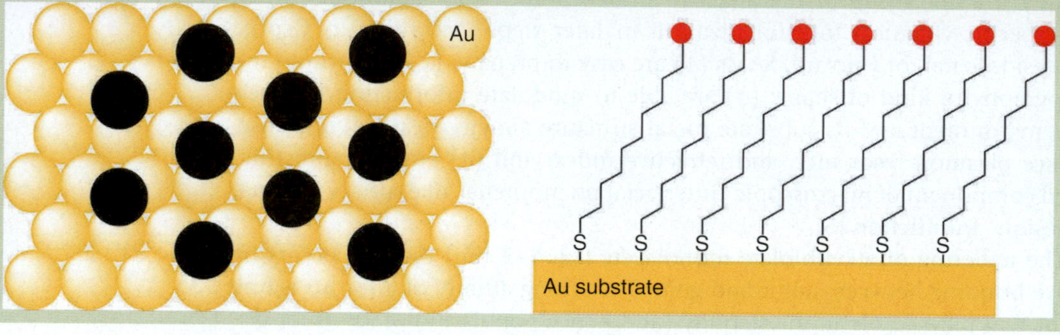

Source: A. Ulman, *An Introduction to Ultrathin Organic Films: From Langmuir–Blodgett to Self-Assembly*, Academic Press, Boston (1991). With permission.

Since their inception, numerous forms of thiol-based monolayer systems have been developed, tested, and applied. A typical SAM system consists of the following components:

- Substrate made of smooth silicon, glass, mica, plastic, or another metal
- Interstitial metal layer that provides better adhesion between the substrate and the noble metal
- Noble metal layer, usually an *fcc* metal with a (111) crystal face, deposited via thermal evaporation, RF sputtering, electrodeposition, physical vapor deposition, or electroless deposition
- Organo-sulfur compound that consists of (a) the head group (the sulfur, which has an affinity for metal), (b) the *n*-alkane shaft $(-CH_2)-_n$, and (c) the tail group (ranging in reactivity from inert to very reactive). The organosulfur compound is applied via solution or vapor deposition methods

The energetics of thiol layers formed of linear *n*-alkanethiols, particularly on Au surface, is determined by temperature programmed desorption (TPD) (**Fig. 12.10**) [18]. Two forces act in concert to construct a monolayer on Au and two peaks are apparent in the spectrum of a TPD experiment [18]. A lower energy peak is a physisorption peak that was found to increase linearly with increasing chain length, indicating a van der Waals type relationship. A higher temperature peak did not vary with alkane chain length and is due to the chemisorbed S–Au interaction. A third peak at higher energy appeared when

FIG. 12.10 *A temperature programmed desorption profile of hexanethiol layer deposited at 208°C after 2460 seconds. The lower peak represents physisorbed component of the monolayer. The higher energy peak represents the chemisorbed component. Chain length only affected the position of the lower energy peak.*

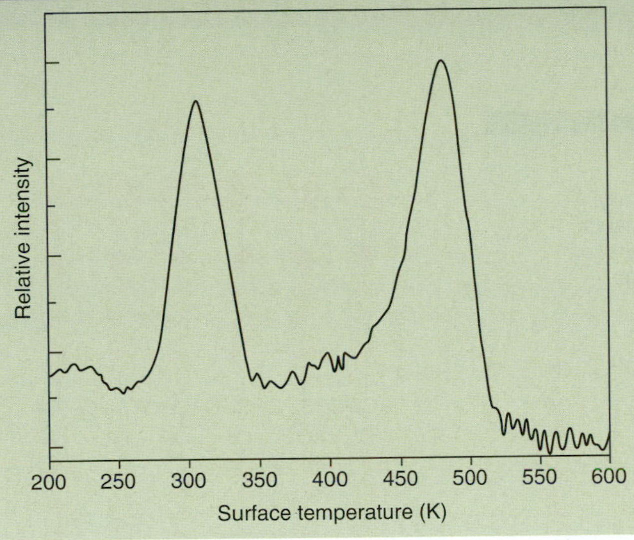

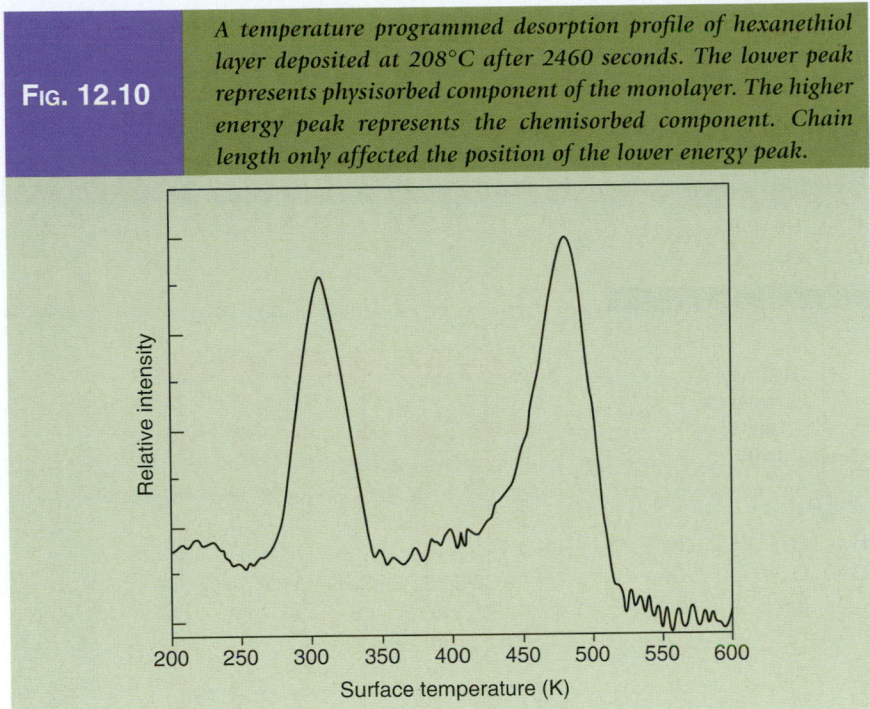

Source: Image redrawn with permission from Giacinto Scoles, Department of Chemistry, Princeton University.

chain length exceeded 14 methylene groups [18]. The overall relationships are, however, more complex.

12.1.3 Organosilanes

The chemisorption of alkysilanes to an activated (hydroxylated) silicon surface results in robust monolayers for a couple of reasons: (1) the covalent attachment to the surface and (2) lateral cross-linking with other silica groups as the monolayer grows (**Fig. 12.11**). The cross-linking is accomplished through a

FIG. 12.11 *A generalized version of acid catalyzed esterification—a nucleophilic addition–elimination reaction. The carboxylic ends of long-chain hydrocarbons interact with hydroxyl groups anchored to the surface of a substrate.*

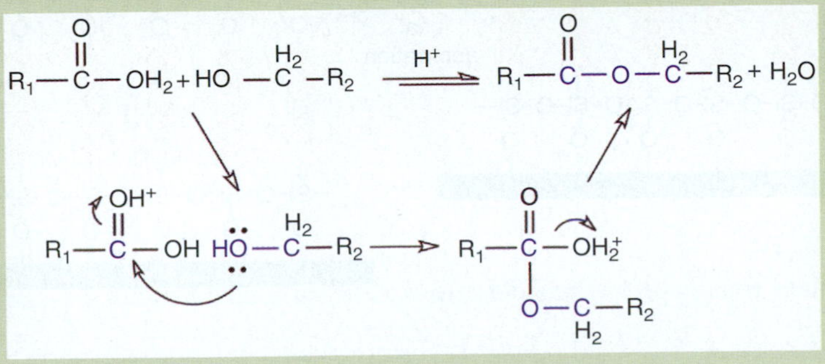

Fig. 12.12 *Counterclockwise from top left. Hydroxyl-terminated silane esters are added to an activated silicon surface to form the first monolayer. In the next step, 3-aminopropyl-dimethyl-ethoxysilane is reacted with the first monomer by an ester substitution reaction. The red dot represents the amine group. A thin layer of styrene monomers is applied by spin-coating onto the surface of the bilayer. The system is exposed to ultraviolet radiation, thereby initiating cross-linking of the polymer to form a thin robust coating.*

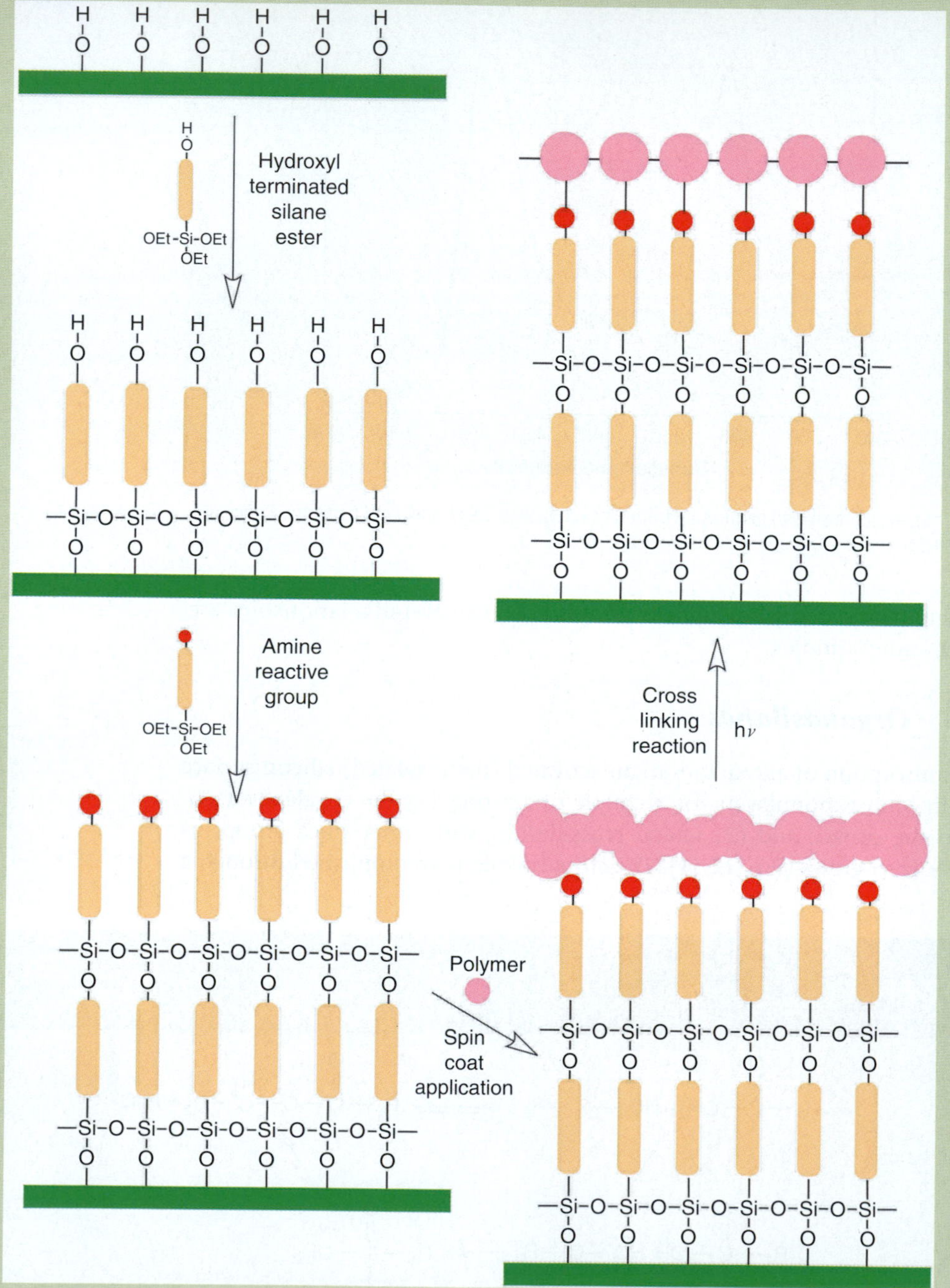

–Si–O–Si– covalent network. The reaction of the silyl ester groups proceeds in a manner similar to a standard esterification reaction between a carboxylic ester and a hydroxyl group but with more facility (**Fig. 12.12**):

$$K_{\mathrm{Eq}} = \frac{[\mathrm{R_1COOCR_2}][\mathrm{H_2O}]}{[\mathrm{R_1COOH}][\mathrm{R_2COH}]} = 3.38 \qquad (12.2)$$

These kinds of reactions, like the one for acetic acid + ethanol in the presence of an acid catalyst, have a relatively low equilibrium constant and therefore are reversible.

12.2 Synthesis and Chemical Modification of Nanomaterials

We touched upon several bottom-up chemical synthesis methods in chapter 4. Here we give more detailed examples of the synthesis and chemical modification of selected nanomaterials from the zero-, one-, and two-dimensional categories.

12.2.1 *Synthesis and Modification of Zero-Dimensional Materials*

Colloids, clusters, and quantum dots are members of the extended family of one-dimensional nanomaterials. We dedicate a special section to Au_{55} clusters—a truly versatile and important nanomaterial. The development of Au_{55}, its synthesis, and ligand decoration schemes are attributed to the work of Günter Schmid and his colleagues at University of Essen (now Essen-Duisberg) [19–27].

Au-55 Clusters. Synthesis of uniformly small particles proceeds best from the bottom up. One of the most popular methods to form metal clusters and colloids is through the reduction of metal cations. Common reducing agents include the base complements of organic acids such as sodium citrate, reducing alcohols, Na_2S, borohydrides [B_2H_6], sodium borohydride [$NaBH_4$], and even hydrogen gas. We understand why small clusters wish to agglomerate to form larger clusters via Ostwald ripening. In order to fabricate metal clusters of predetermined size (and we assume that nanoscale clusters are desired rather than colloids), special steps need to be taken. First of all, addition of a potential ligand species to the reaction mixture is required. The ligand serves to bind reduced metals and thereby modulate the growth of the embryonic clusters. Depending on relative concentrations of reactants (the metal salt, reducing agent and ligand), growth of clusters that are monodisperse with desired dimensions is possible. The generic process for synthesis of nanoclusters is summarized as

Reduction of metal cation → agglomeration prevention → ligand stabilization or ligand exchange → extraction from solvent → further surface modification

The synthesis of a triphenylphosphine ligand-stabilized gold cluster is shown in **Figure 12.13**.

Gold has a rich history. It has been part of our culture for over 5,000 years. With its malleability, corrosion resistance, and inherent luster, it is no wonder it has contributed to civilization's course—both positively and negatively. Gold leaf, for example, can be beaten into a layer that is 50 nm in thickness. Its inclusion into beautiful glasses is well documented. The chemistry of gold was problematic due its chemical inertness. With aqua regia and cyanide solutions the only means to dissolve gold, the evolution of its chemistry was limited in scope. Michael Faraday, of course, was well known for his chemical manipulation of gold species. Faraday reduced tetrachloroaurate [$AuCl_4^-$] with white phosphorus (the reducing agent) to generate colored solutions of gold colloids [19]. Ostwald was the first to bring attention to the property dependence of nanoclusters on their surface and, by that postulate, predicted that nanoparticles possess remarkable properties [19].

Gold colloid synthesis is straightforward. Of all the procedures available today, a method called the "Turkevitch route" is very popular [19,28]:

$$HAuCl_4 + (C_6H_5O_7)Na_3 \rightarrow Au^o + \text{oxidized products} \tag{12.3}$$

Fig. 12.13 *Reaction scheme of formation of Au_{55} ligand-stabilized cluster is depicted. At the top left, a solution containing dissolved metal cations is shown. The cations are converted into gold atoms after addition of a reducing agent like citrate. Once formed, the atoms nucleate and grow into aggregates that eventually stop at the cluster phase (depending on the reaction conditions). Ligands attach to the vertices of the Au_{55} cluster; there are 12 vertices in this structure. Not shown are the counter-anions, the chloride atoms of $Au_{55}[P(Ph)_3]_{12}Cl_6$.*

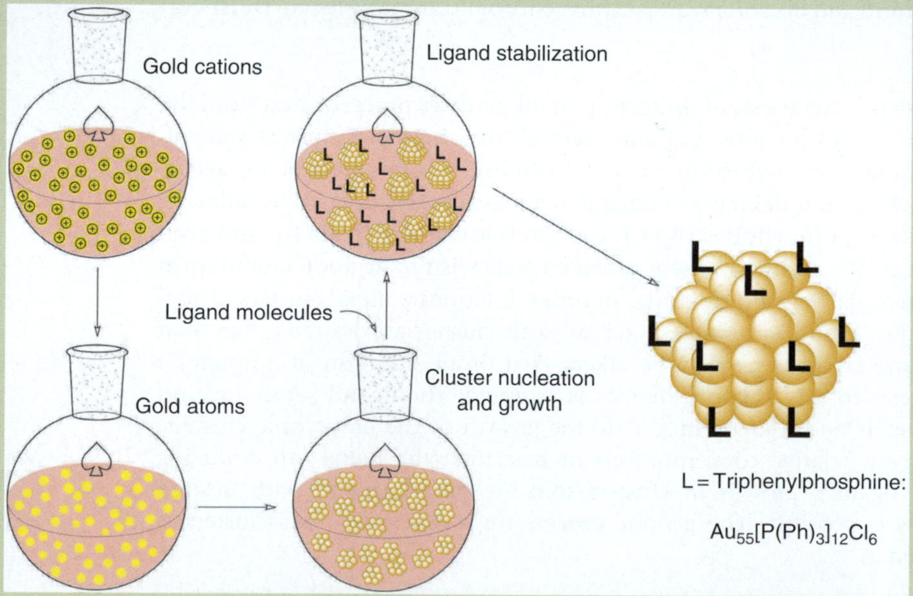

Source: Image redrawn with permission of Prof. Günter Schmid, Institut für Anorganische Chemie, University of Essen.

Approximately 5×10^{-6} mol of $HAuCl_4$ is dissolved in 19 mL of deionized water and heated to boiling; 1 mL of 0.5% sodium citrate is added with constant stirring for 30 min. The solution undergoes color changes from yellow to clear, gray, purple, deep purple, and finally to ruby-red. Water is added to maintain the level of solution to 20 mL. The "Brust route" is similar but employs sodium borohydride as the reducing agent:

$$HAuCl_4 + [CH_3(CH_2)_7]_4NBr\ (TOAB) + toluene + NaBH_4 \rightarrow Au^o \qquad (12.4)$$

This technique utilizes an emulsion layer made of water and toluene; 4.0×10^{-3} mol of tetraoctylammonium bromide (TOAB, surfactant, the phase transfer catalyst and the stabilizing ligand) is added to 80 mL of water and then 9.0×10^{-4} mol of $HAuCl_4$ in 30 mL water is added to the TOAB solution and stirred vigorously for 10 min. The aqueous phase is clear and the organic phase is orange. Sodium borohydride is added dropwise to the mixture and the color changes from orange to white to purple to dark red. To make sure that the product clusters are monodisperse with regard to size, the solution is stirred for an additional 24 h. The organic phase is washed with sulfuric acid to neutralize the solution. TOAB is not considered to be a strong ligand and will readily undergo ligand exchange with stronger ligands like thiols that bind covalently to the gold clusters [29].

Synthesis of gold clusters from $HAuCl_4$ by a sonolysis technique is used to create gold ribbons 20–30 nm in diameter. Glucose and hydroxyl radicals, formed at the cavities formed by sonication, serve as the reducing agents. When cyclodextrin is used, only spherical particles are synthesized [29]. The size of the clusters, their distribution, and stabilization are controlled by the following factors: reductant concentration, temperature, *p*H, stirring rate, rate of addition of reducing agent, and chemical composition of the solution.

G. Schmid et al. were the first to describe the physical properties of Au_{55} clusters [20]. Gold-55 (Au_{55}) is a magic-number cuboctahedral cluster. Stable forms of Au_{55} are synthesized by

$$AuCl[P(C_6H_5)_3] + B_2H_6 \rightarrow Au_{55}[P(C_6H_5)_3]_{12}Cl_6 + H_3B\text{–}P(C_6H_5)_3 \qquad (12.5)$$

Gaseous diborane is passed through a warm 150-mL solution of benzene containing 3.94 g of $AuCl[P(C_6H_5)_3]$. Diborane is the best reducing agent, but it also acts as a Lewis acid that binds phosphines. This process limits the amount of free ligand available at any time during the course of the reaction. Excess ligand concentration leads to smaller complexes and clusters, an undesirable outcome [20]. The temperature is raised to 50°C. After 40 min, the colorless solution turns dark brown. Upon cooling, a dark precipitate settles to the bottom of a now colorless solution. The precipitate is filtered, rinsed with dichloromethane, and filtered again through Celite to remove unwanted solids (e.g., colloidal gold). The product, $Au_{55}[P(C_6H_5)_3]_{12}Cl_6$, is re-precipitated slowly in 250 mL pentane to ensure that phosphine ligands saturate the Au_{55} cluster. The overall yield of the process is 29% [20]. The cluster is 2.1 nm in diameter (the cluster is 1.4 nm). An exact stoichiometric relationship between reactants and products has not been derived for this reaction [20].

The Au_{55} product is a dark brown powder that is soluble in dichloromethane and pyridine and insoluble in petroleum ether, benzene, and alcohols [20]. In air, the ligand-stabilized cluster decomposes to solid gold (agglomeration) and in solution reverts back to its precursor state. Mössbauer spectroscopy reveals the presence of four kinds of gold atoms: 13 central atoms, 24 uncoordinated peripheral atoms, 12 atoms coordinated to phosphine ligands, and 6 atoms coordinated to chlorine [20]. Three simulated renditions of $Au_{55}[P(Ph)_3]_{12}Cl_6$ are shown at the top of **Figure 12.14.** On the bottom, a molecular model of a gold cluster attached to DNA is shown.

The triphenylphosphine ligands of the cluster are labile and undergo ligand exchange readily in phase transfer reactions:

$$\begin{aligned} &Au_{55}[P(Ph)_3]_{12}Cl_6 \text{ (in dichloromethane)} + P(Ph)_2(C_6H_4SO_3)Na \text{ (in water)} \\ &\quad \rightarrow Au_{55}[P(Ph)_2(C_6H_4SO_3)Na]_{12}Cl_6 \\ &\quad \rightarrow Au_{55}[P(Ph)_2(C_6H_4SO_3)]_{12}(Cl_6)^{12-} + 12Na^{+} \end{aligned} \qquad (12.6)$$

A thiol derivative of an organo-silsesquioxanes (OSS) is one of the most intriguing ligands synthesized by G. Schmid et al. of the University of Essen in Germany [27]. Its synthesis is shown in **Figure 12.15.** Several changes in chemical reactivity and physical properties occur in substituted Au_{55} clusters. The

FIG. 12.14 *Left: various molecular models of $Au_{55}[P(Ph)_3]_{12}Cl_6$ are depicted. The last image shows the electron density, especially that of the phenyl rings. Right: molecular model of the interaction of Au_{55} clusters with DNA. The strong interaction of gold clusters with the major groove of DNA makes them extremely toxic to living things.*

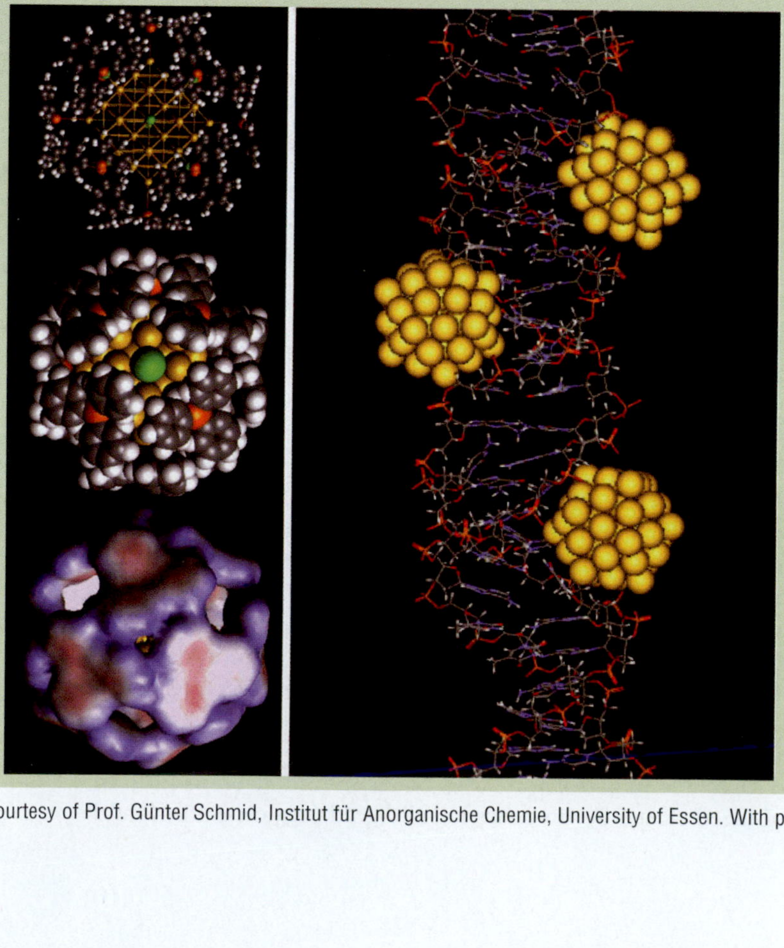

Source: Images courtesy of Prof. Günter Schmid, Institut für Anorganische Chemie, University of Essen. With permission.

FIG. 12.15 *Synthesis of T_8–OSS–SH is shown. The ultimate Au–T_8–OSS–SH complex is soluble in pentane, which is not possible for the triphenylphosphine derivative that is soluble only in dichloromethane type solvents. In addition to providing an impressive visual, the utility of these ligand-stabilized clusters in two-dimensional arrays shows promise.*

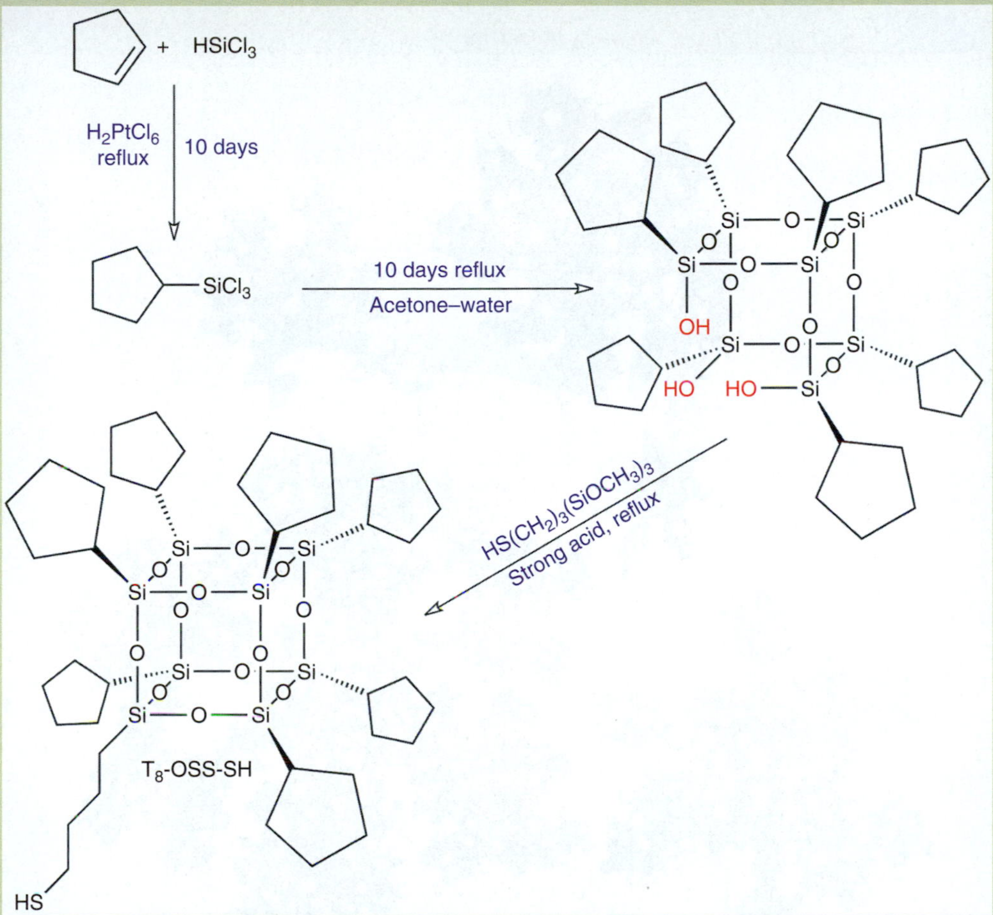

Source: Image redrawn with permission of Prof. Günter Schmid, Institut für Anorganische Chemie, University of Essen.

ligand-stabilized cluster increases in dimension from 2.1 to 4.4 nm upon exchange of the ligands. It is soluble in pentane as well as dichloromethane and is considerably more stable (due to its strong Au–S covalent bond) than its phosphine-substituted counterpart. An increase in the activation energy of electron tunneling increased from 0.16 *e*V for $Au_{55}[P(Ph)_3]_{12}Cl_6$ to 0.26 *e*V for the T_8–OSS–SH ligand system [27]. A ligand exchange chemical reaction scheme is shown in **Figure 12.16.** Ligand exchange occurs according to the following scheme:

$$Au_{55}[P(Ph)_2(C_6H_4SO_3)Na]_{12}Cl_6 + 12\ T_8\text{–OSS–SH} \rightarrow Au_{55}[T_8\text{–OSS–SH}]_{12}Cl_6 + 12\ PPh_3 \quad (12.7)$$

However, when $Au_{55}[P(Ph)_2(C_6H_4SO_3)Na]_{12}Cl_6$ ligand is exposed to T_8–OSS–SH, no ligand exchange takes place (equation 12.7), possibly due to phase transfer kinetic factors.

FIG. 12.16 *Ligand exchange reactions with $Au_{55}[P(Ph)_3]_{12}Cl_6$ are graphically depicted. Both the closo-dodecaborate and the OSS ligands are able to displace the triphenylphosphine. The utility of the thiol group in binding gold-based materials is once again demonstrated. $[Au_{55}(B_{12}H_{11}SH)_{12}]^{24-}$ is soluble in water and $[Au_{55}(T_8\text{–}OSS\text{–}SH)_{12}]$ is soluble in pentane. $Au_{55}[P(Ph)_3]_{12}Cl_6$ is soluble in dichloromethane: three ligand-stabilized clusters with three radically different solubility properties.*

Source: Image redrawn with permission of Prof. Günter Schmid, Institut für Anorganische Chemie, University of Essen.

We represent one more interesting ligand consisting of 12 boron atoms [23]. The exchange of the phosphine occurs thusly in dichloromethane:

$$Au_{55}[P(Ph)_3]_{12}Cl_6 + Na_2[B_{12}H_{11}SH] \rightarrow Au_{55}[(B_{12}H_{11}SH)Na_2]_{12}Cl_6 \qquad (12.8)$$

Na^+ can be exchanged by $(octyl)_4N^+$, thereby rendering the complex water soluble. The boron cluster is depicted in **Figure 12.16.** Both the OSS and boron ligand systems increase the size of the cluster and thereby increase the spacing of interdigitated arrays of two-dimensional ligand-stabilized quantum dots relative to the Au_{55} phosphine ligand-stabilized cluster.

Fig. 12.17 *A colorized TEM image of an array of ligand-stabilized $Au_{55}[P(Ph)_3]_{12}Cl_6$ clusters. Self-assembly in this case resulted in a square planar array of ligand-stabilized gold nanoparticles.*

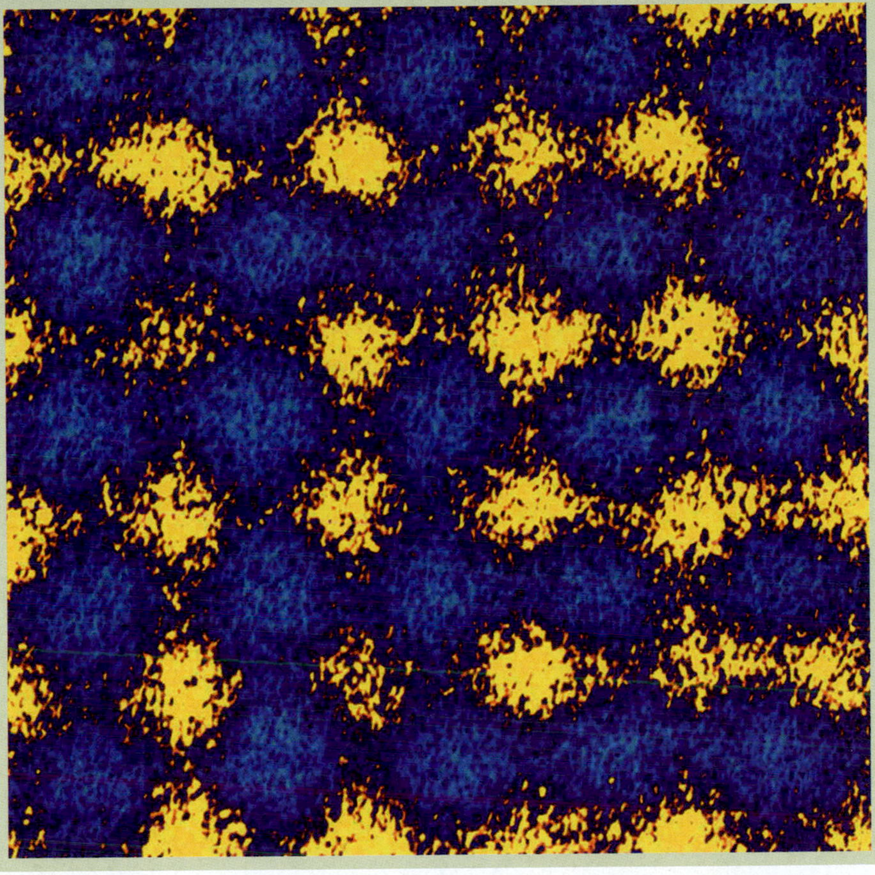

Source: Image courtesy of Prof. Günter Schmid, Institut für Anorganische Chemie, University of Essen. With permission.

Au_{55} clusters can be formed into arrays depending on the type of ligand and the concentration of clusters. After ligand-stabilized clusters are spread on a two-dimensional surface such as copper, configurations ranging from liquid (disordered) to solid de-interdigitated (similar in structure to a two-dimensional hexagonal close-packed structure) to "interdigitated" [30,31]. **Figure 12.17** shows a TEM image of ligand-stabilized Au_{55} clusters formed into a square array. The number of ligand systems that are able to interconnect quantum dot clusters is enormous. The clusters can be linked by covalent systems (e.g., sulfide linkages) or by intermolecular interactions such as π–π interactions, van der Waals, and electrostatic ones. A few interdigitated systems and other linking systems are shown in **Figure 12.18.** The size of cuboctahedron metal clusters (*sans* ligands) also affects the interlayer spacing of arrays: $[Au_{55}(PPh_3)_{12}Cl_6]$ (1.4 nm), $[Pt_{309}O_{30}]$ (1.8 nm), $[Pd_{561}(Phen_{36}O_{200})]$ (2.5 nm), $[Pd_{1415}(Phen_{60}O_{\sim1100})]$ (3.0 nm), and $[Pd_{2057}(Phen_{86}O_{\sim1600})]$ (3.6 nm).

FIG. 12.18 *Ligands interlinking clusters do so by several mechanisms ranging from covalent bonding (–R–S–R–) or disulfide (–S–R–S–S–R–S–) linkages that are covalent to aryl group interdigitation, and scores of other intermolecular interactions.*

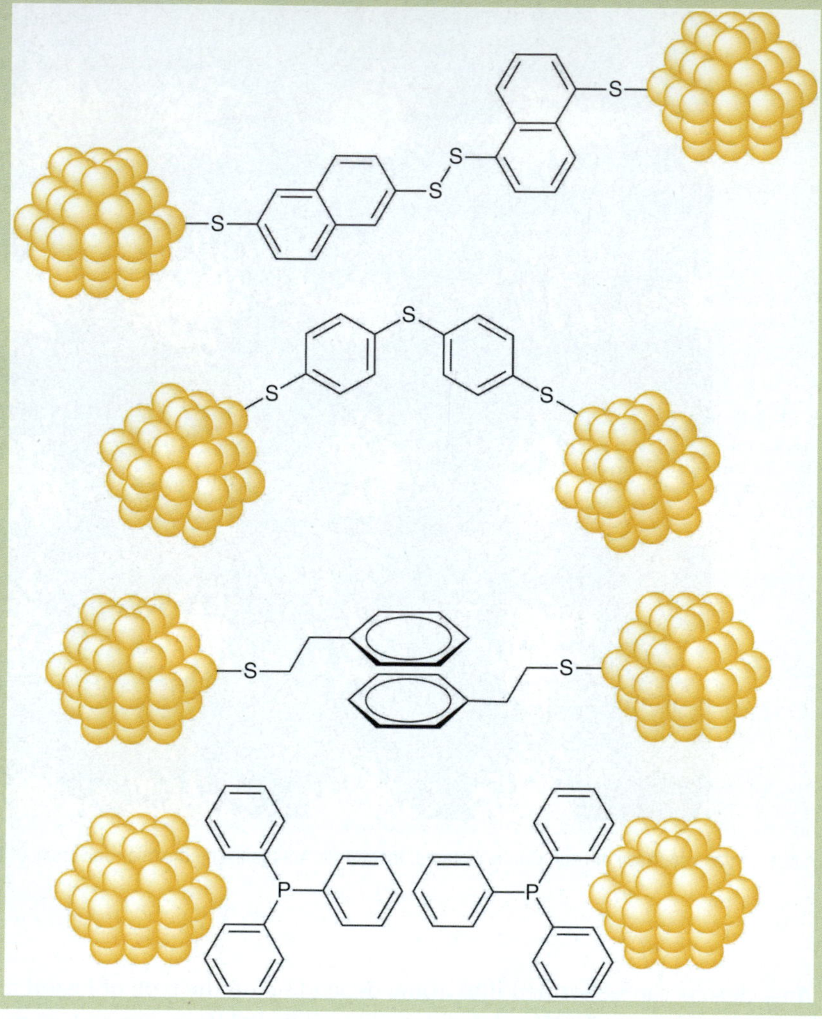

Semiconductor Quantum Dot Synthesis. There are several challenges facing the synthesis of quantum dots: (1) synthesis of dots that are monodisperse with regard to size, shape, and orientation and (2) fabrication of those dots into a uniform array. The inverse micelle-emulsion method of fabrication of semiconductor quantum dots has become increasingly popular. (Please take note that there are numerous methods to fabricate semiconductor *q*-dots from the top down that are not discussed in this chemical synthesis chapter.)

Semiconductors can be efficiently synthesized in reverse (inverse) micelles (emulsions of inverse micelles). Reverse micelles are formed in nonpolar solutions in which the hydrocarbon tails of the amphiphile project outward from the micelle into the nonpolar medium. Naturally, the polar groups of the

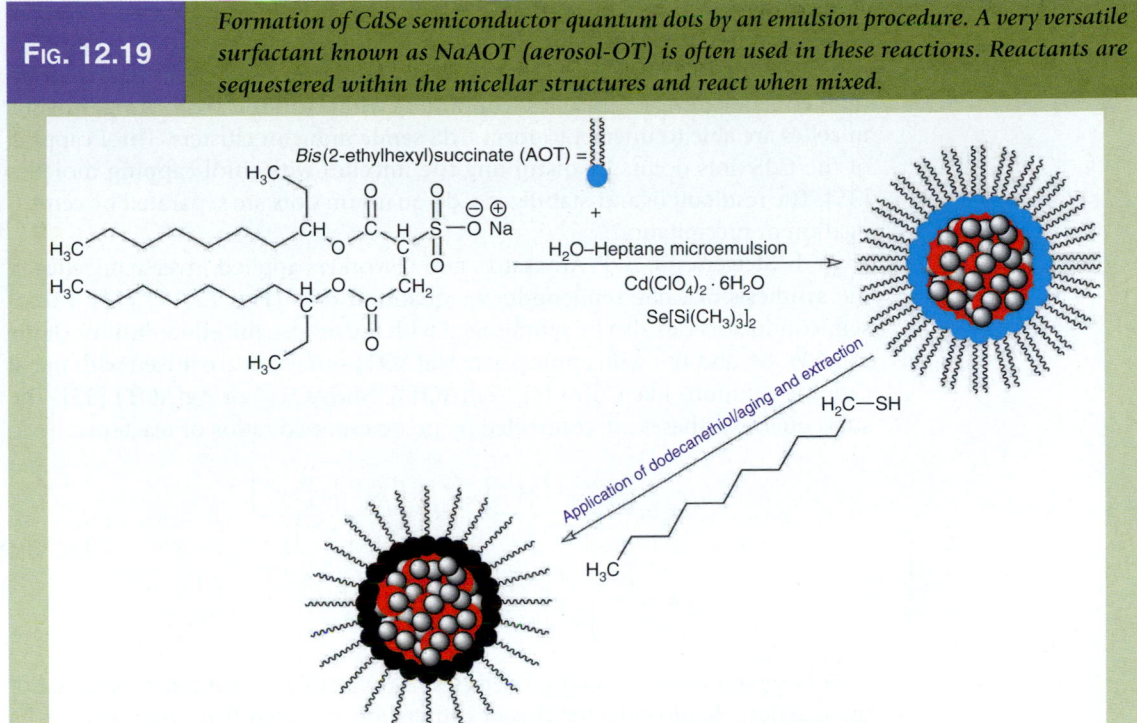

Fig. 12.19 *Formation of CdSe semiconductor quantum dots by an emulsion procedure. A very versatile surfactant known as NaAOT (aerosol-OT) is often used in these reactions. Reactants are sequestered within the micellar structures and react when mixed.*

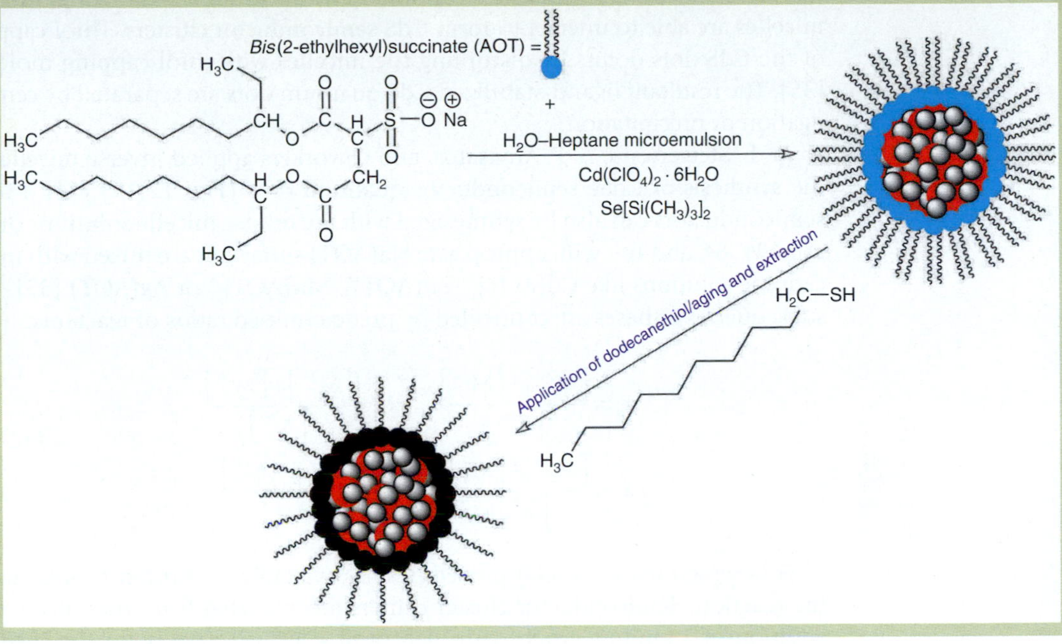

amphiphiles project into the micelles structure. An example of a CdSe quantum dots formed in inverse micelles from sodium *bis*(2-ethylhexyl)sulfosuccinate (a.k.a. NaAOT), $Cd(ClO_4)_2 \cdot 2H_2O$, and $Se[Si(CH_3)_3]_2$ ([*bis*(trimethylsilyl)Se) is shown in **Figure 12.19.** Emulsions can be created by vigorously mixing water with a nonpolar solvent like heptane.

Inverse micelles form in solutions in which the majority liquid component is nonpolar and the minority component is made of water. The polar core of the micelle is called the "water pool" and is characterized by the water-surfactant molar ratio *W*:

$$W = \frac{[H_2O]}{[S]} \tag{12.9}$$

where *S* is the concentration of the surfactant [32,33]. Large *W*s are characteristic of emulsions while reverse micelles have $W < 15$. The water pool radius, R_{Water}, is calculated from

$$R_{Water} = \frac{3V_{aq}[H_2O]}{\sigma[S]} \tag{12.10}$$

where V_{aq} is the volume of the water molecules and σ is the area of head of the polar group. Because Na(AOT) is used quite often, its R_{Water} is defined by the empirical relation [33]:

$$R_{Water} = 1.5W \tag{12.11}$$

In other words, the water pool radius is a linear function of the water content. Larger micelles can be achieved by increasing the water-surfactant ratio [3,34].

One group of reverse micelles containing S^{2-} anions is mixed into a solution that contains reverse micelles containing Cd^{2+} cations. Following mixing, micelles are able to interact to form CdS semiconductor clusters. Thiol capping of the CdS dots occurs by disrupting the micelles with thiol capping moieties [35]. The resultant ligand-stabilized CdS quantum dots are separated by centrifugation or precipitation.

M. L. Steigerwald, A. P. Alivisatos, and coworkers applied inverse micelles in the synthesis of CdSe semiconductor quantum dots (**Fig. 12.19**) [36]. Mixed semiconductors can also be synthesized with the reverse micelle solutions of the same W: S^{2-} and Te^{2-} with appropriate Na(AOT) surfactant are mixed with metal cationic solutions like $Cd(AOT)_2$, $Zn(AOT)_2$, $Mn(AOT)_2$, or Ag(AOT) [35]. The subsequent syntheses are controlled by predetermined ratios of reactants:

$$x_a = \frac{[M_1^{2+}]+[M_2^{2+}]}{[X_1^{2-}]}; \quad x_b = \frac{[M_1^{2+}]+[M_2^{2+}]}{[X_2^{2-}]}$$
$$x_c = \frac{[M_3^{2+}]+[M_4^{+}]}{[X_1^{2-}]}; \quad x_d = \frac{[M_3^{2+}]+[M_4^{+}]}{[X_2^{2-}]} \qquad (12.12)$$

Following mixing, several procedures are available for further processing of the reaction. Semiconductor cluster centers are extracted from the micelles by addition of a binding group with high (and selective) affinity for the clusters. Groups such as dodecanethiol are able to coat the semiconductor and are separated by a subsequent phase transfer reaction. The original surfactant is liberated from the clusters by addition of ethanol. The combination of dodecanethiol addition and extraction of the clusters can be accomplished by different procedures: (1) immediate addition–immediate extraction; (2) delayed addition (2 days)–immediate extraction; (3) immediate addition–delayed extraction; and (4) 90-min delay in addition of dodecanethiol [33]. Thermodynamic equilibrium, water pool radius, and other factors are allowed to exert influence over the size and quality of the semiconductor quantum dot clusters [33].

Aging (time before addition of dodecanethiol), time of extraction, oxidation, cation ratios, and cation–anion ratios, in addition to the solvent and surfactant, all affect the size and crystallinity of quantum dots. CdTe are made under conditions of excess Cd in order to prevent oxidation of Te: $x = [Cd^{2+}]/[Te^{2+}] = 2$ [33]. Procedure (1) resulted in dots 2.6–3.4 nm in diameter while procedure (2) allowed for larger particles: 3.4–4.1 nm [33]. The difference was readily observable from fluorescence spectra. CdS by procedure (1) showed no fluorescence. By procedure (2) and with an excess of CdS ($x = [Cd^{2+}]/[S^{2-}] = 2$), size-dependent, but otherwise weak, fluorescence was observed (e.g., red shift with increasing particle size). In other words, the exciton peak was visible when aging of the particle was allowed to proceed [33]. However, strong fluorescent peaks were observed in CdS dots when the cation–anion ratio favored the anion. The addition of excess S prevents the vacancies (e.g., $x = [Cd^{2+}]/[S^{2-}] = 1/2$) from forming in the crystal. Although aging enhances the crystallinity of the CdS dots, elimination of S vacancies by adding excess S^{2-} proved to enhance the peak intensity [33].

Triangular nanocrystals of CdS are formed from $Cd(AOT)_2$/isooctane/H_2O mixture with $W = 30$ [33]. Addition of H_2S and N_2 generates cadmium and sulfur coprecipitates. Dodecanethiol is added after 2 days. TEM analysis and computer simulation showed that the particles are pyramidal in shape. Alloyed semiconductor crystals of the form $Cd_{1-y}Zn_yS$ are made by procedure (1). The solid is made from precursors containing excess sulfur: $x = \{[Cd^{2+}] + [Zn^{2+}]\}/[S^{2-}] = 1/2$ [33].

M. Guglielmi et al. synthesized CdS semiconductor dots from cadmium acetate $[Cd(CO_2CCH_3) \cdot 2H_2O]$, thioacetamide $[CH_3CSNH_2]$, and the surfactant 3-mercaptopropyl-trimethoxysilane (MPTMS) in methanol and resulted in particles on the order of 8 nm in diameter [M. Guglielmi et al., *Journal of Sol–Gel Science Technology,* 8, 1017, 1992]. Other CdS dots formed by mixing cadmium dichlorate $[Cd(ClO_4)_2]$ with $[Na(PO_3)_6]$ complex with H_2S produced nanoparticles of differing size. The size was dependent on the *p*H of the starting solution. At *p*H 9.8, the CdS nanoparticle fluoresced blue color ($\lambda \sim 430$ nm); at *p*H 9.0, the color of the solution was turquoise ($\lambda \sim 480$ nm); and at *p*H 8.1, the solution appeared greenish ($\lambda \sim 500$ nm) [A. Henglein et al., *Journal of the American Chemical Society,* 109, 5649, 1987].

Of course, as with gold clusters, monolayers and more complex structures are possible with semiconductor quantum dots. AgS dots on a TEM grid showed hexagonal packing characterized by relatively large domains (>100 μm). Chain interdigitation, chain length, chain packing, and particle interaction all contribute to the formation of structures with higher order. AgS quantum dots decorated with alkyl chains—$(CH_2)_x$—have formed into *fcc* "supracrystals" with $x = 8$ [33]. **Figure 12.20** shows a two-dimensional hexagonal packed layer of a generic quantum dots.

Gold colloids were also be formed in inverse micelles [37]. The majority solvent used was *n*-heptane and the surfactant was NaAOT. Micelles consisting of $HAuCl_4$ and $N_2H_4 \cdot H_2SO_4$ were mixed to form gold colloids ranging in size from 30 to 100 nm (with aggregates of 200–400 nm). The size of the gold colloids increased as the $[H_2O]/[AOT]$ ratio was increased [37]. In another microemulsion synthesis, NaAOT-*Span80*-isooctane (aqueous $HAuCl_4$ and $N_2H_5 \cdot OH$) were mixed:

$$4HAuCl_4 + 3N_2H_5 \cdot OH \rightarrow 4Au^o + 16HCl + 3N_2 + 3H_2O \qquad (12.13)$$

The non-ionic cosurfactant *Span80* acted as a stabilizer and structural modifier for the gold colloids. W was 8.0 and spherical gold nanoparticles 3.5–8.6 nm were achieved [37]. A small amount of rod-like and cube-shaped particles were also synthesized (95 nm in length) [37]. When W approached 20, particle size increased in the range of 10–40 nm and rods and cubes became more prevalent. When cosurfactants NaAOT and tetra(ethylene glycol)dodecylether were used together with isooctane as the oil phase, pure *fcc* gold up to 80 nm in size was produced. The morphology (spherical, rod-like, or trigonal systems) of the gold crystals was dependent on experimental conditions [37].

Aqueous Incipient Wetness Impregnation. This method is used heavily in the field of catalyst preparation. Although the resulting catalyst particles are not considered to be quantum dots, they certainly are small enough to be included in that special family of materials. AIWI requires a support material that aids in

FIG. 12.20 *Hexagonal array of semiconductor quantum dots decorates a surface. Differing surface arrangements can be achieved depending on the size of the dot and its ligands (or if there are mixed ligand systems).*

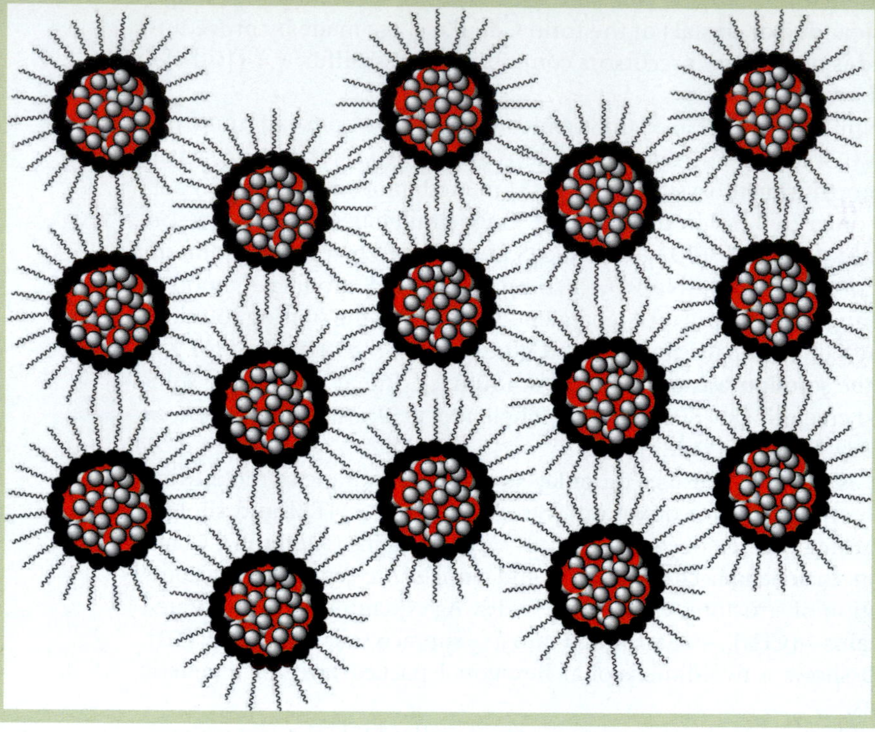

size control of the catalyst particles. The support is often in the form of fumed alumina (85–115 $m^2 \cdot g^{-1}$). One procedure is reviewed next to provide sufficient perspective about the procedure.

Fe–Mo catalyst for the formation of single-walled carbon nanotubes is prepared by mixing $Fe_2(SO_4)_3 \cdot 5H_2O$, $(NH_4)_6Mo_7O_{24} \cdot 4H_2O$ with fumed Al_2O_3 in deionized water at 80°C for 1 h with application of ultrasonication [38]. The mixture is dried overnight at 100°C under a nitrogen blanket. The resulting powder is then ground with mortar and pestle and calcined at 900°C for 10 min. After cooling, the catalyst-support powder is reground into a fine dispersion. The ratio of starting materials results in 2.7 mg of metal in a 6:1 Fe:Mo molar ratio on ~50 g of alumina support material [38]. Nanoparticles on the order of 1 nm to a few nanometers are produced by this method. The powder is then placed into a quartz boat in a reduction atmosphere to produce metal catalyst particles. CVD is then allowed to proceed and single-walled carbon nanotubes (SWNTs) are formed. Very simple.

Nb_2O_5–TiO_2 catalysts are prepared by AIWI of TiO_2 with aqueous solutions of niobium oxalate–oxalic acid. Nb loading of 0.48–8.5 wt.% was prepared by this process. Time of flight secondary ion mass spectrometry (ToF SIMS) experiments on the catalyst complex revealed the presence of monomeric, dimeric, and trimeric Nb_xO_y clusters, even at low loading fractions, and that the degree of aggregation of the niobia clusters increased with higher loading levels [39].

Once again, a very simple, low-cost, straightforward chemical procedure takes center stage in the arena of nanoscience. The power of nano!

Sol–Gel Synthesis. One of the oldest forms of nanotechnology comes in the form of sol–gel synthesis. Colloid chemists have developed this technology dating back numerous decades (or more), developing one of the simplest, inexpensive, low-temperature, and most effective bottom-up wet chemical synthesis of nanomaterials that are highly pure and monodisperse in size. Inorganic metal oxides as well as inorganic–organic hybrid materials are synthesized routinely by sol–gel methods. The *sol* is the homogeneous solution of molecular reactant precursors that are concerted into an infinite molecular weight three-dimensional polymer: the *gel* that forms an elastic-solid fill material with the same volume as the liquid [40]. Mixtures of precursors (different metals, oxides, and even organics) lead to binary, ternary, and higher order systems [40]. Colloid size, structure, and morphology are dependent on the solution *p*H, temperature, composition, concentrations, and the solvent. Transformation from the sol into the gel, for example, is stimulated by a change in the *p*H of the solution [40].

Sol–gel synthesis proceeds by hydrolysis and condensation of silicate precursors like triethylorthosilicate (TEOS) (**Fig. 12.21**). Hydrolysis is the process of breaking bonds by the action of H_2O. Condensation is the formation of a bond with the simultaneous release of a water molecule (similar to the synthesis of a peptide bond). Hydrolysis of a silicon precursor proceeds:

$$\mathrm{Si(OR)_4} + y\mathrm{H_20} \rightarrow \mathrm{Si(OR)_{4-y}(OH)_y} + y\mathrm{ROH} \tag{12.14}$$

where R is an ethyl group in TEOS. Subsequent condensation is illustrated by:

$$2\mathrm{Si(OR)_{4-y}(OH)_y} \rightarrow [\mathrm{(RO)_{4-y}(OH)_{y-1}}]\text{–Si–O–Si–}[\mathrm{(OR)_{4-y}(OH)_{y-1}}] + \mathrm{H_2O} \tag{12.15}$$

Fig. 12.21 *One of the most celebrated workhorses of sol–gel synthesis is tetraethylorthosilicate (TEOS). In water (or alcohol–water mixtures), TEOS undergoes dimerization and eventual polymerization. The main processes involved in sol–gel synthesis techniques are hydrolysis and condensation.*

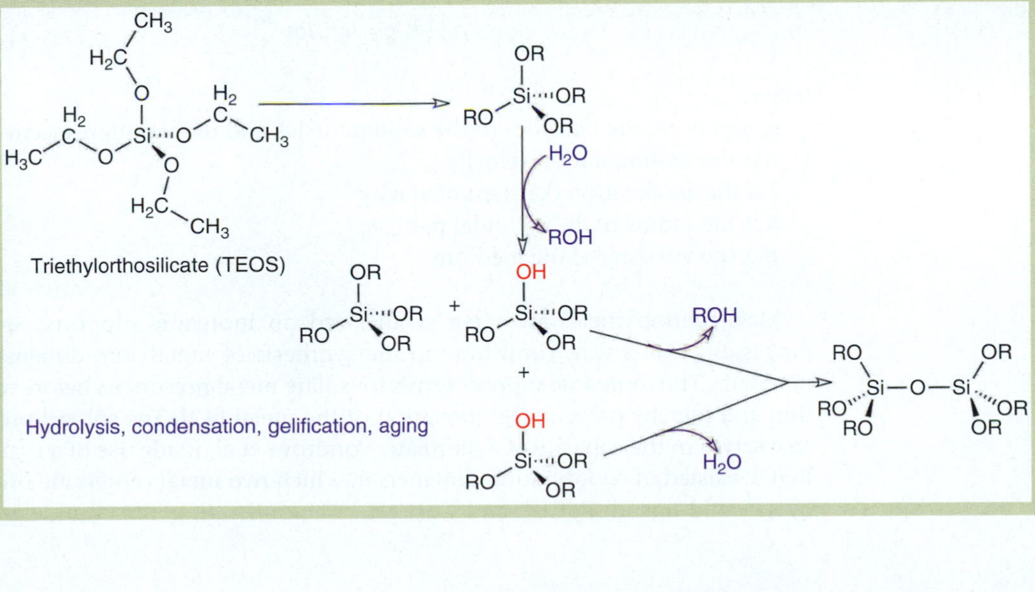

Although we have sol–gel synthesis in the section of zero-dimensional materials, its versatility is easily expanded to include one-dimensional, two-dimensional, and higher order structures and morphologies. Different metal groups and combinations of metals like aluminum, potassium, titanium, and others result in colloidal (and eventually ceramic) materials of great diversity, properties, and function.

Colloidal materials abound around us. Porous rocks, clays, mist, smoke, blood, opal, pearl, bones, milk, elgin, ice cream, and other foods are some of the natural materials that are made of colloidal materials [41]. Synthetic colloidal materials include paint, toothpaste, photographic silver iodide, polyurethane foam, and a multitude of other industrial materials. Typically, colloids range in size from 1 nm to 1 μm [41]. Because of high surface-to-volume ratios, colloids actually exist somewhere between physical phases—a separate and distinct colloidal phase. Colloidal particles are therefore sensitive to external forces such as electric fields and to intermolecular forces that we discussed earlier. Colloidal particles undergo Brownian motion, balancing attractive and repulsive forces by the interplay of electrostatic and steric factors. Reversible agglomeration is known as *flocculation* and irreversible agglomeration is called *coalescence* [41]. Agglomeration is prevented by the application of charge and steric stabilization—topics we discussed briefly in chapter 6.

Colloidal systems consist of a dispersed phase within a continuous phase. The colloidal system we are most interested in is that of a solid (dispersed) phase dispersed within an aqueous (continuous) phase. Solid–liquid dispersions contain a high solid content. Types of colloidal system dispersions (with examples in parentheses) include of liquid–gas (liquid aerosols), solid–gas (solid aerosols), gas–liquid (foams, froths), liquid–liquid (emulsions), gas–solid (solid foams), liquid–solid (ice cream), and solid–solid (opal). Colloids are characterized by *rheological* techniques (flow behavior) in which traits like thickening, shear thinning, and viscoelasticity are prominent. There are many ways to characterize particle size and shape (discussed briefly in chapter 3), such as light-scattering, small-angle x-ray scattering (SAXS) and small angle neutron scattering (SANS) analysis. Sedimentation (not discussed in chapter 3) relies on the following relationship:

$$(\rho_s - \rho_l)\tfrac{4}{3}R^3 g = 6\pi \upsilon R\eta \tag{12.15}$$

where

ρ_s and ρ_l are the densities of the solid particles and the liquid, respectively
υ is the sedimentation velocity
g is the acceleration constant of gravity
R is the radius of the colloidal particle
η is the viscosity of the medium

Metal nanoparticles are often synthesized on inorganic supports. Sol–gel processes, in this way, contribute to the synthesis of metal zero-dimensional materials. The inorganic support serves to isolate metal precursors before reduction and thereby prevent agglomeration of the metal [42]. The sol–gel product also serves in the capacity of a template. Vondrova et al. made use of a cyanogel that consisted of coordination polymers in which two metal centers are bridged by a cyanid ligand: $PdCl_4^{2-}$ and $Co(CN)_6^{3-}$. Upon heating, the cyanide groups are able to reduce the metals. In their experiments, silica gel was prepared from [SiO_2/Na_2O] and aqueous solutions of Na_2PdCl_4 and $K_3Co(CN)_6$ were added to the solution and allowed to form a gel. The co-gel was dried at 600°C and the cyanogel phase was reduced to metal [42]. The Pd/Co alloy metal particles had an average diameter of 7 nm [42].

Sol–gel synthesis demonstrates versatility and facility at low cost. It is truly and purely chemistry. Sol–gel methods are revisited in section 12.3.

12.2.2 Synthesis and Modification of One-Dimensional Materials

There are many techniques available from which to form one-dimensional materials, better known as nanowires. Obviously, it will not be possible to cover all the methods in this section. Selected methods are summarized in **Table 12.3.**

A broad class of substances can be synthesized in the form of one-dimensional nanomaterials (nanotubes, nanocylinders, whiskers, fibers, nanorods, and nanowires). The wires can be electrically conducting, semiconducting, or insulating. Metals, ionic compounds, ceramics, and organic materials can all be made into nanowires. Recently, biological materials such as DNA have been shown to be nanowires with great utility, versatility, and durability. The ultimate nanowire is a chain of atoms. We have already discussed carbon nanotubes in chapter 9 and will not present any topics related to them in this section.

Conductive nanowires 100 nm in diameter were fabricated by self-assembly of amyloid protein fiber biological templates (Step 1) and selective metal

TABLE 12.3 *Bottom-Up Chemical Nanowire Synthesis Methods*

Method	Description	Examples
Template-based methods	Electrodeposition	Gold nanocylinders in porous alumina materials. Electrodeposition is valid for all metals; some require nonaqueous solvents
	Electroless autocatalytic deposition	Electroless Au, Ag, Pd, Co, Ni, Cu
	Colloid, solution impregnation	Sol–gel or solution mixtures to make metals, semiconductors, ceramics, and other materials
	Microemulsions	Semiconductor rods and wires formed in micellular-based cavities
	Block copolymers	Formed in emulsions to form ordered materials with cavities, channels, and other template forms
	Biological materials	DNA and other biological materials with periodic structure serve as templates
Chemical vapor deposition	Decomposition of precursor materials	Carbon nanotubes Ge nanowires by diffusion of Ge_2H_6 vapor into mesoporous silica + heat [3,43]

deposition of Ag or Au (Step 2) [44]. In another study, block copolymers were first assembled to form within "lithographically defined" channels on a silicon substrate [45]. The block copolymers, 40 nm across, were introduced into the channels and self-assembled into long cylindrical molds after heat treatment at 230°C. Aqueous solutions of metal ions then were made to infiltrate the lines. The polymer was removed by application of oxygen plasma etching leaving the metal nanowire. The wires were 10 nm in diameter and over 50 μm in length. Buriak's group fabricated 25 parallel Pt wires. Jillian Buriak, the leader of the research team, stated:

> We've figured out a way to use molecules that will self-assemble to form the lines that can be used as wires. Then we use those molecules as templates and fill them up with metal, and then we have the wires that we want. You use the molecules to do the hard work for you.

Brown University researchers have found a straightforward way to synthesize Fe–Pt nanorods and nanowires by a technique that offers diameter, length, and composition control [46]. In their solution-based process, varying the ratio of the solvent (octadecene) and surfactant (oleylamine), nanowires ranging from 20 to 200 nm have been produced. Adding more surfactant resulted in longer nanowires, whereas more solvent yielded shorter rods. A 1:1 ratio produced 20-nm rods.

Solution synthesis of germanium nanowires was accomplished at 1 atm and less than 400°C [47]. The precursor molecule germanium 2,6,-dibutylphenoxide [$Ge(DBP)_2$] was dissolved in oleylamine surfactant and then immediately transferred into a 1-octadecene solution at 300°C. The resulting nanowires were single crystal of cubic phase and coated with oleylamine. The length of the wires ranged from 0.1 to 10 μm. A self-seeding mechanism and an aggregation mechanism are proposed to explain the growth process.

Organic nanowires are also produced by bottom-up methods. P. Nickels et al. showed that polyaniline (PAN) nanowires can be grown within a DNA template [47]. Polyaniline was grown on DNA templates immobilized on a surface by (1) oxidative polymerization with ammonium persulfate; (2)enzymatic oxidation with hydrogen peroxide (horseradish peroxidase); or (3) photo-oxidation using a Ru complex as the photo-oxidant [48].

Ultralong $Cd(OH)_2$ nanowires were synthesized according to a hydrothermal methods from $Cd(CH_3COO)_2 \cdot 2H_2O$ and $C_6H_{12}N_4$ aqueous solution at 95°C without the use of templates [49]. Wires with aspect ratio of several thousands were formed. The formation mechanism is attributed to the oriented attachment of small particles. Transformation into semiconductor material occurred by calcination at 350°C for 3 h [49].

Electrodeposition. DC electrodeposition of metals into templates is an effective method to form nanowires that are monodisperse with regard to diameter [50–63]. The length of the nanowire is dependent on the duration of the electrodeposition process. Both metal and semiconductor nanowires can be formed by electrodeposition. Electrodeposition was discussed briefly in chapter 4. The optical properties of nanogold rods synthesized in anodic porous alumina membranes are dependent on the size, shape, and orientation within the template material [50–63]. The optical response of nanogold–alumina membrane composites was discussed in chapter 5.

Electroless Methods. Electroless deposition plating is the autocatalytic, continuous, and chemical reduction of metal ions onto a substrate. The substrate does not have to be conducting. The required components of an electroless plating solution include an aqueous solution of metal ions of interest, a catalyst (usually a minute amount of the metal in reduced form), reducing agents, complexing agents (help monitor *p*H and control free metal ion concentration), and solution stabilizers (catalytic inhibitors to prevent out of control reactions that result in poorly structured products) operating in a specific range of *p*H, metal ion concentration, and temperature. No electrical current is required in this form of deposition—hence the name electroless. Whether forming wires or plating on a two-dimensional surface, the final metal layer or wire conforms to the contour of the template or the surface.

A gold electroless plating solution, for example, consists of: 0.01 M gold chloride-hydrochloride trihydrate; 0.014 M sodium potassium tartarate; 0.013 dimethylamine borate; 400 $mg \cdot L^{-1}$ sodium cynanide; $pH = 13.0$ and temperature 60°C. An example of the electroless deposition of copper is

$$Cu^{2+} + EDTA^{4-} \rightarrow Cu(EDTA)^{2-} \quad (12.16)$$

$$2H_2C{=}O + 4OH^- \rightarrow 2HCO_2^- + 2H_2O + H_2 + 2e^- \quad (12.17)$$

$$Cu^{2+} + 2e^- \rightarrow Cu^o \quad (12.18)$$

$$\begin{aligned} &Cu(EDTA)^{2-} + 2H_2C{=}O + 4OH^- \\ &\rightarrow Cu^o + 2H_2O + H_2 + 2HCO_2^- + EDTA^{4-} \end{aligned} \quad (12.19)$$

12.2.3 Synthesis and Modification of Two-Dimensional Materials

George Whitesides' rule for designing surfaces includes the following considerations, starting with the SAM [5]: (1) in general, functionalized alkane-thiols will form ordered SAMs with the terminal capping group pointed away from the substrate; (2) rule number 1 applies best for *n*-alkane thiols with small cap groups (e.g., –OH, –CN, $-CO_2H$, etc.); and (3) rule number 1 does not take into account steric, molecular free volume, lateral segregation (multicomponent assemblies), and neighbor and surface interactions. He goes on to say, "The development of guidelines for designing SAMs that predict the organization and composition of the monolayer and incorporate all these elements remains to be worked out."

In general, although we are getting better at supramolecular design, it is exceedingly more difficult to predict the physical and chemical properties of SAMs and we must rely on semi-empirical studies for at least a little while longer. However, modification of SAMs occurs by both covalent and intermolecular mechanisms. There is no limit to the variety of derivatization that can occur to receptive monolayers. Monolayer surfaces, especially, require modification if future design involves biological or biochemical applications [5].

Chemical Modification of SAMs. The chemistry of covalent chemical modification is influenced strongly by nucleophilic substitution reactions—e.g., basic organic chemistry. Bromine is a good leaving group and is used in many chemical modification procedures in nucleophilic substitution reactions.

FIG. 12.22 *A basic S_N2 nucleophilic substitution is shown. Bromine is a good leaving group in the presence of nucleophilic moieties. In SAM systems, due consideration is always allotted to designing the chemistry of the end group.*

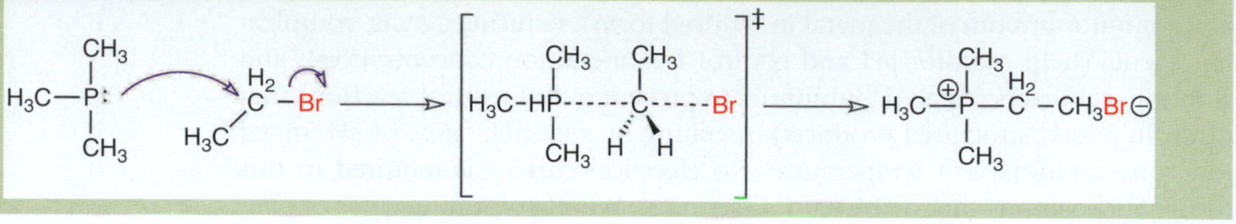

A basic nucleophilic substitution reaction is shown in **Figure 12.22.** If such a group is a terminal capping group on a SAM molecule, then the case for chemical modification is made possible. The strength of leaving groups depends on several parameters: size of the atom, electronegativity, electric charge, steric properties, and the solvent, to name a few. For example, F is not a good leaving group, whereas Br is a good leaving group. Inspection of some of the most basic SAM chemical procedures reveals that –OH is a good nucleophile and that –Cl and various alkoxides serve as good leaving groups. The conjugate bases of strong acids are good leaving groups. The sulfonate ion is an excellent leaving group and is employed in many kinds of substitution-chemical modification reactions of monolayer end groups.

The solvent, just like in supramolecular chemical reactions, is also an important consideration in chemical modification reactions and plays an important role in the rate of displacement reactions [64]. For example, the displacement of iodine in methyl iodide by an acetate anion takes place 10 million times faster in dimethylformamide (DMF) than it does in methanol [64]. If an S_N2 nucleophilic displacement reaction involves ions, then polar solvents are required. *Hydroxylic solvents,* for example, promote hydrogen bonding of reactants and products. Hydroxylic solvents include water and alcohols. *Polar aprotic solvents* include acetonitrile, DMF, dimethyl sulfoxide, and acetone. Nonpolar solvents consist of the family of low molecular weight alkanes, toluene, benzene, and other familiar solvents. The free energy of activation $\Delta G^{\ddagger}$ of reaction of nucleophiles with methyl iodide varies significantly from solvent to solvent. For example, the free energy barrier of $CN^- + CH_3I$ substitution is 59 $kJ \cdot mol^{-1}$ in DMF but as high as 92 $kJ \cdot mol^{-1}$ in methanol [64].

The other generic kind of chemical modification takes advantage of intermolecular interactions. A sample of such a reaction was illustrated by the golden molecular squares in chapter 11. Chemical modification by grafting, wrapping, associating, hydrogen bonding, electrostatic charge interactions, and all the other intermolecular interactions discussed so far occurs with high frequency in the literature. It all depends on what exactly the goal of the chemical treatment is: robust inert layers? biosensors? optical response? electrical conduction? The list goes on and on.

If the terminal group of the SAM molecule is a methyl group (not a good leaving group at all), then for all practical purposes the chemical portion of the synthesis is completed, unless one applies external stimuli by photo-excitation or an electrochemical potential [5]. The surface of the SAMs capped with methyl groups is hydrophobic; the surface is a low-energy surface (large contact angle) and is relatively nonreactive, if that is the place one wishes to stop.

More often than not, additional layers are desired to improve robustness and enhance properties. The most popular example of this process is the generation of additional layers by adding silane moieties to the previous layer. Our example of ALD shown in chapter 4 is analogous to this philosophy of layer building. During each step, a hydroxyl group acts as a nucleophile with silicon that has good leaving groups attached—for example, the $Si–O(Me)_3$ or $–SiCl_3$ type groups. Cross-linking of layers occurs via condensation reaction. The end result is a thick layer (dependent on the number of cycles employed) with lateral robustness.

Four advantages lend themselves to SAM modification following monolayer deposition: (1) application of commonly used procedures; (2) ligand incorporation into SAMs after monolayer deposition impossible a priori; (3) derivatization of many SAM samples by many kinds of ligands; (4) preservation of the molecular, short-, and long-range order of the parent monolayer; and (5) small amount of ligand (e.g., 10^{-9} mol) [5]. Disadvantages of SAM films are related to the surface coverage; the extent of the parent and modified structure is relatively unknown and heterogeneous coverage may be produced during chemical modification [5].

Table 12.4 lists several strategies of SAM chemical modification [5]. In addition to the chemical natures of the terminal group and the potential modifying moieties, the SAM itself is able to affect the chemical reaction by steric and geometric factors, restricting accessibility to functional groups, and modifying the nature of the solvent near the SAM–solution interface. The characteristics of the solution at the surface may be quite different from that of the bulk [5]. The SAM is a chemical entity, albeit a complex one, and as such exhibits properties that are based in its structure and composition. The organization of the chains (e.g., level of crystallinity or disorder), the surface density, orientation of functional groups and their distance from the substrate, and lateral steric effects (and van der Waals interactions) all contribute to the chemical and physical character of the monolayer.

Electron Beam Writing. Enhancement of an electron beam writing procedure was accomplished by attachment of silica-based SAMs capped with nitro-phenol [65,66]. Monolayers were formed on activated silicon surface substrate by application of nitrophenoxy-propyltrimethyloxysilane (NPPTMS). The terminal nitro groups were converted into amines by exposure to an electron beam (5–6 *k*eV). The nitrophenyl group is easy to reduce chemically to form amines but that would limit the scope of selectivity over surface patterning. How would one selectively mark off areas that require modification of the nitrophenyl groups by such a broad-base chemical treatment? Application of the e-beam solves this problem quite well (**Fig. 12.23**). Modification of the amine groups by altering the *p*H renders the amine groups (now in the ammonium state) receptive to reaction with citrate-stabilized gold nanoparticles. Gold in this form carries inherent negative charges. The process is similar to the formation of expanded monolayers of the golden molecular squares [67].

Covalent Chemical Modification. Covalent chemical modification of the distal layers of fluorinated self-assembled monolayers (F-SAMs) $[CF_3(CF_2)_7(CH_2)_2–S–Au]$

Table 12.4 *Functional Groups and Chemical Modification*

Type of modification	Mechanism	Examples Terminal group + ligand
Covalent reactions	Nucleophilic substitution, esterification, acylation, nucleophilic addition, cyclo-addition, metal-catalyzed olefin cross-metathesis	Maleimide + thiol (R = peptides, carbohydrates) Disulfide + thiol (R = DNA) Hydroxyl + siloxane Halogen + phosphine (lone pair donors) Amine + thiocyanate Vinyl group + acrylamide (Ru-catalyzed) Acetyl group + azide (of alkylthiolates) → triazole Alkyne + azide Ester + azide (formation of amide)
	Surface activation to form a reactive intermediate that is able to react with the ligand (e.g., formation of amide linkages by interchain anhydride intermediate) *Activation* by external stimuli (electric fields, photoexcitation) transforms unreactive groups into active ones.	Dehydrated carboxylic acid (via TFAA)‡ + amine → amide Activated carboxylic acid (via NHS)* + amine → amide Hydroquinone (via electrochemical oxidation) → quinone + diene
	Covalent bond breaking reactions of terminal cap group. In these reactions, there is release of a molecule bound to a surface group into solution.	Propionic ester modified quinones release ester and lactone during electrochemical reduction, resulting in intramolecular cyclation
	Surface-initiated polymerization is able to graft polymers to a SAM surface by (1) covalent linking of preformed polymer chains to reactive SAM terminal groups or (2) direct growth of the polymer from the terminal group.	(1) Poly(acrylic acid)/poly(ethylene glycol) (2) Poly(acrylonitrile) (by photoinitiated radical method) Block copolymers
Intermolecular interactions	*Nonspecific adsorption* of molecules from the solution phase by van der Waals forces, electrostatic interactions, combinations thereof, and hydrophobic effects	Surfactants, polymers, polyelectrolytes, proteins, organic dyes, and colloidal particles Nonspecific adsorption implies no control (or relatively little control) over orientation and thickness
	Fusion of vesicles: adsorption of lipid vesicles results in supported bilayers, single layers, or hybrid layers, the chemical nature of which depends on the chemistry of the functional group.	Supported bilayers of phospholipid vesicles yield layers that are defective. Hybrid bilayers serve as dielectric barriers with few pinholes. The combination of SAM + bilayer = good model for cell membranes.
	Selective deposition is a function of hydrophobicity and charge density. Electrostatic charge on SAM aids orientation of adsorbate molecules. Due to selectivity, the method is applicable to patterning techniques.	pH control of adsorption of polyelectrolytes, polyallylamine (PAA), and polyethyleneimine (PEI) onto carboxylic acid- or oligoethylene glycol (OEG)-terminated SAMs favors PAA on OEG at pH 4.8, while PEI deposits primarily on carboxylic acid groups. Positively charged surfaces adsorb antibodies. *Cytochrome c* adsorbs onto positive SAM surfaces with favorable electron-transfer orientation without losing its structure. Negatively charged surfaces do not produce the same effect.

Table 12.4 (Contd.)	Functional Groups and Chemical Modification	
Type of modification	Mechanism	Examples Terminal group + ligand
	Modifications by designed *molecular recognition*—the domain of supramolecular chemistry This kind of modification is reversible. Due to high selectivity, the precision of the method allows placement of ligands close together.	Use of hydrogen bonding networks, dative bonds between metals and ligands, electrostatic interactions, and/or hydrophobic interactions Concerted, cooperative host–guest relationships stabilize surface complexes. Reactions have small equilibrium constants and are therefore reversible (e.g., dissociation of adsorbates by addition of excess ligand).

Source: J. C. Love et al., *Chemical Reviews*, 105, 1103–1169 (2005).

can be attained by bombardment with low energy (>100 *e*V) polyatomic ions [68]. The projectile ion was $CH_2Br_2^+$ with an m/z ratio equal to 172. Detection of the species CH_2F^+, $CHBrF^+$, and CF_2Br^+ was proof that the F-SAM underwent chemical modification [68].

Corrosion Prevention. In an attempt to improve the corrosion resistivity of aluminum, a self-assembled monolayer of hexadecanedioic acid (HDDA) was applied to the surface of the metal (**Fig. 12.24**) [69]. To enhance the layer further, cross-linking within the HDDA layer was attempted by application of octyltrichlorosilane (OTS). Use of perfluorinated carboxylic acid as the base layer reveals that OTS is able to displace the 1° organic SAM and from Si–O (siloxane) linkages to the metal surface [69]. Carboxylic acids do not form the strongest links with metal surfaces and during any chemical modification, such considerations need to be included in the design phase of the monolayer experiment. If the result of chemical modification is the decomposition of the SAM, then another strategy needs to be considered.

12.3 Template Synthesis

Template synthesis is the fabrication of nanomaterials within porous materials and interstitial spaces. We focus primarily on porous materials formed by inorganic and organic–inorganic hybrid materials. According to the IUPAC definition (as before), there exist three classifications of porous materials: macroporous ($d > 50$ nm), mesoporous ($2 < d < 50$ nm), and microporous ($d < 2$ nm). An example of each kind of template is presented. Porosity ε is defined as the porous or void fraction of a material (as before). In more specific terms, it is also defined as the fraction of void volume available to an adsorbate or analyte. Obviously, this omits porous spaces that are inaccessible to adsorbates. The various kinds of porous materials are displayed in **Figure 12.25.**

Chemistry in Confined Spaces. Chemists are used to working within the framework of nonconfined systems such as the beaker and the flask. But how are

FIG. 12.23 *The conversion of the nitro group on the benzene ring into amines is accomplished with an electron beam. By this way, specific sectors are created on an otherwise homogenous monolayer coating. Further chemical modification is facilitated with the amine terminus of the molecule. Such patterning is possible by direct writing with the electron beam. Adjustments in pH convert the amine into an ammonium moiety that is receptive to electrostatic interaction with citrate-stabilized gold colloids that have negative charges. The latter part of this procedure was used effectively to produce the golden molecular squares encountered in chapter 11 [65–67].*

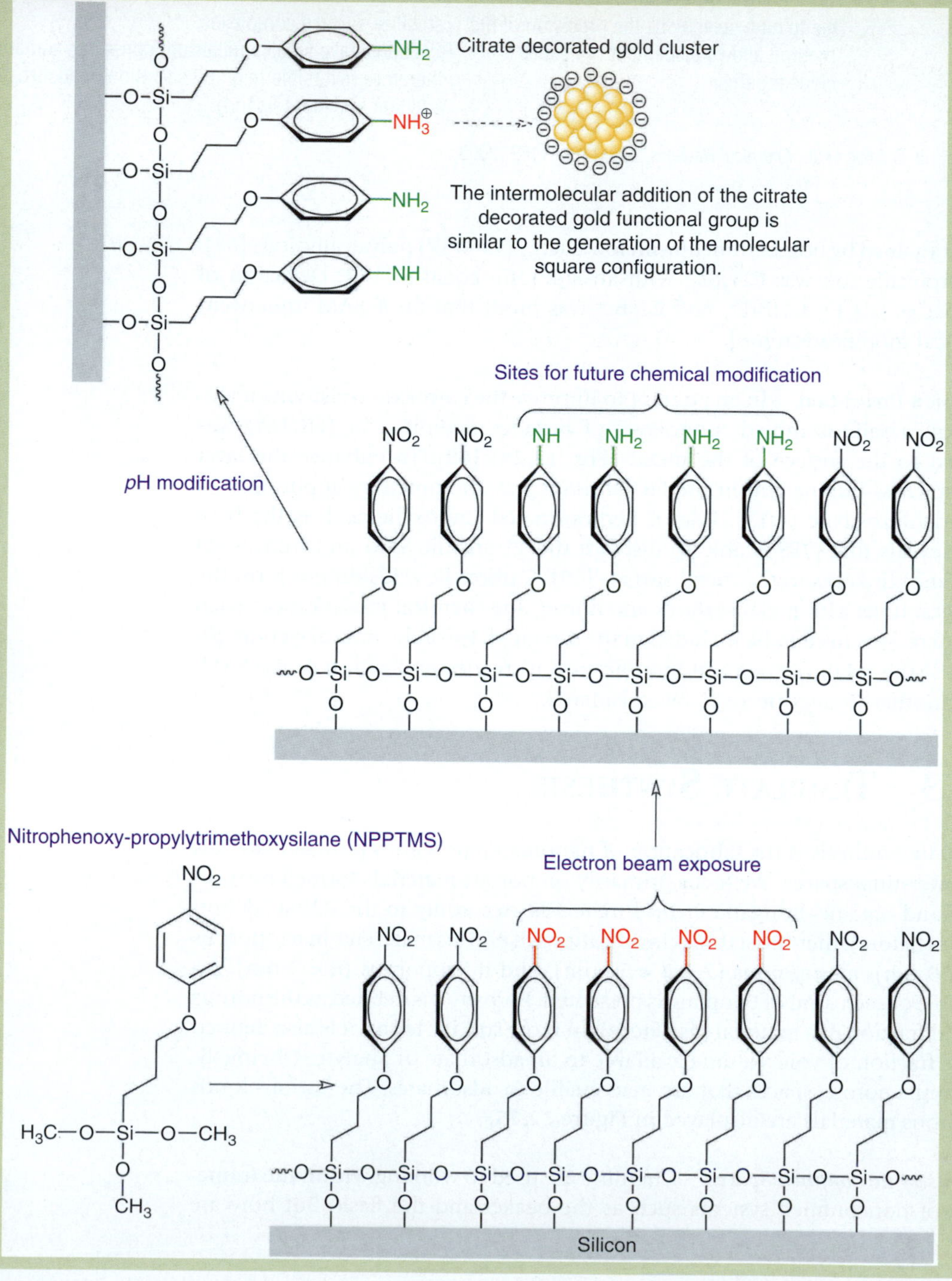

FIG. 12.24 *Hexadecanedioic acid (HDDA) was applied to the surface of the metal [69]. To enhance the layer further, cross-linking of the HDDA later was attempted by application of octyltrichlorosilane (OTS).*

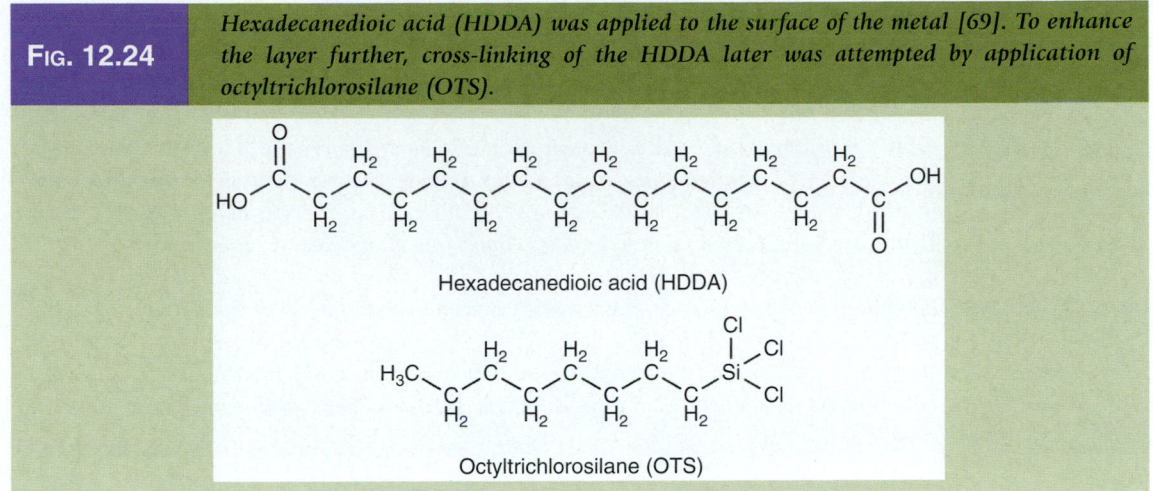

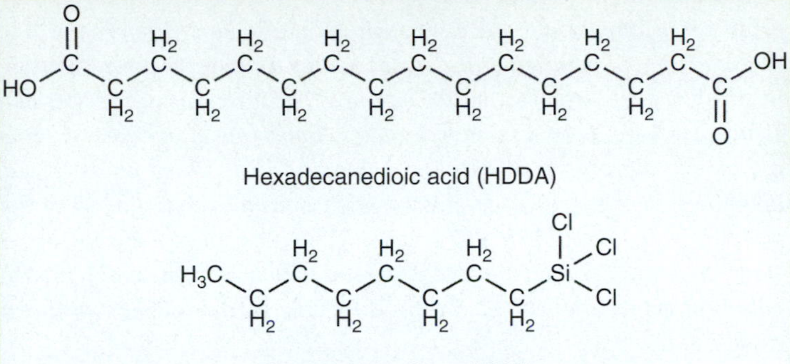

Source: L. D. Seger, J. P. Rasimas, R. Pesce-Rodriguez, and R. Fifer, Chemical modification and attempted polymerization of self-assembled monolayers of hexadecanedioic acid at aluminum surfaces, Army Research Lab, www.stormingmedia.us/21/2171/A217133.html (1997). With permission.

chemical reactions affected if that chemistry is accomplished in a space similar in scale to that of the process itself? Molecularly confined spaces are defined loosely as three-dimensional spaces less than 1 nm in diameter. In essence, we are talking about nanosized chemical reactors.

In one example, the pyrolysis of surface-linked 1,3-diphenylpropane (DPP) in MCM-41 zeolite was found to exhibit accelerated rates and alterations in product selectivity—both kinetic effects [70]. S. Polarz reported that the formation of electrochromic oxides like MoO_3 is limited to first-order kinetics in confined spaces and that the kinetics of transformation of precursors is also altered [71]. In another study, researchers found that chirality of cations can be controlled in confined spaces [72]. R. Anwander showed that surface-confined metal species are amenable to small-ligand chemistry of alkoxide metal centers. These species are not found in traditional solutions due to agglomeration [73]. Encapsulation and surface grafting contributed to nonagglomerated metal–ligand species as a result of steric effects—pathways originating from steric and highly unsaturated metal centers [73].

Diffusion Behavior in Porous Membranes. Diffusion of gases and liquids through porous materials is a major field of study in nanoscience. Purification of natural gas to remove toxic compounds and noncombustibles is a major undertaking by today's energy suppliers. Membranes that offer selectivity along with high throughput are highly sought after. Unfortunately, the two characteristics reside on opposite ends of the spectrum (e.g., you can have one but not the other). There are a few kinds of diffusion processes through porous membranes. The general categories are (**Fig. 12.26**): (1) bulk diffusion (nonselective, no separation); (2) Knudsen flow diffusion in which the mean free path of the molecule is on the order of the pore diameter; and (3) molecular or restricted diffusion (when pore size is on the order of molecular size). The Knudsen number K_n (the ratio between the mean free path and the system size) indicates which diffusion domain is applicable:

FIG. 12.25

Various kinds of porous materials are displayed. Top left: the "ink bottle" pore cavity is quite common and serves as an excellent nanosized beaker within which studies of chemical confinement are conducted. A structure is considered to be a pore if its width is less than its depth. Top middle: interstitial spaces are formed with hard spheres. The spaces are exploited in templating processes such as nanosphere lithography. Top right: interstitial spaces are found between cylinders formed by a sol–gel process. Such spaces are also found in single-walled carbon nanotube bundles. The interspatial cavity is defined by the van der Waals gap between and among bundles. Bottom left: enclosed cavities are not useful in the templating process due to their inaccessibility. These kinds of cavities, however, are able to affect the physical properties (e.g., thermal conductivity) of the material. Bottom middle: zeolites exhibit highly ordered structures that come in the form of channels or three-dimensional interconnected porous networks. Bottom right: the pore channels of porous alumina can be made to transverse the entire thickness of the membrane or be left capped at one end.

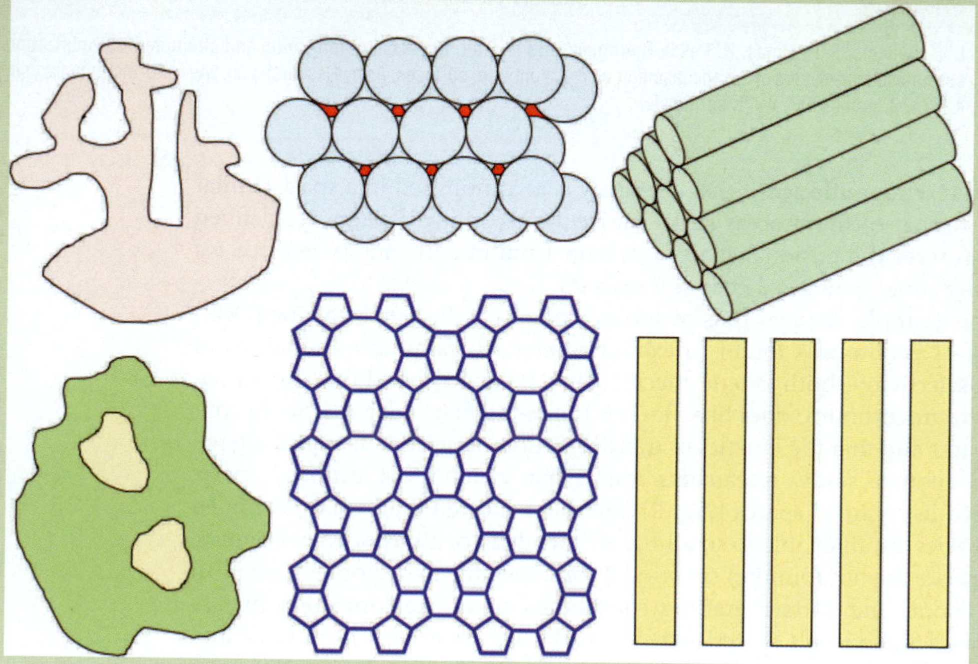

$$K_n = \frac{\lambda}{d} \tag{12.20}$$

where λ is the molecular mean free path and d is the pore diameter. When $K_n >> \lambda$, the condition of bulk diffusion prevails. Pores greater than 100 nm in diameter fall into this category ($D > 10^{-4}$ m$^2 \cdot$ s^{-1}). When $K_n \sim \lambda$, Knudsen flow predominates and diffusion behavior is described by Fick's law of diffusion. Pore diameter lies between 100 and 1 nm ($10^{-8} < D < 10^{-4}$ m$^2 \cdot$ s^{-1}). When $K_n \ll \lambda$, molecular or restricted flow (configurational diffusion and surface migration) is in effect. Pore diameter in the molecular flow is less than 1 nm ($10^{-8} < D < 10^{-16}$ m$^2 \cdot$ s^{-1}).

FIG. 12.26 *Diffusion through porous materials is important to many industrial processes. Various kinds of diffusion are depicted. Pressure differentials are important in bulk diffusion (far left) but lose their importance as pore size is diminished. In general, selectivity increases as pore size diminishes. However, throughput diminishes with decreasing pore diameter.*

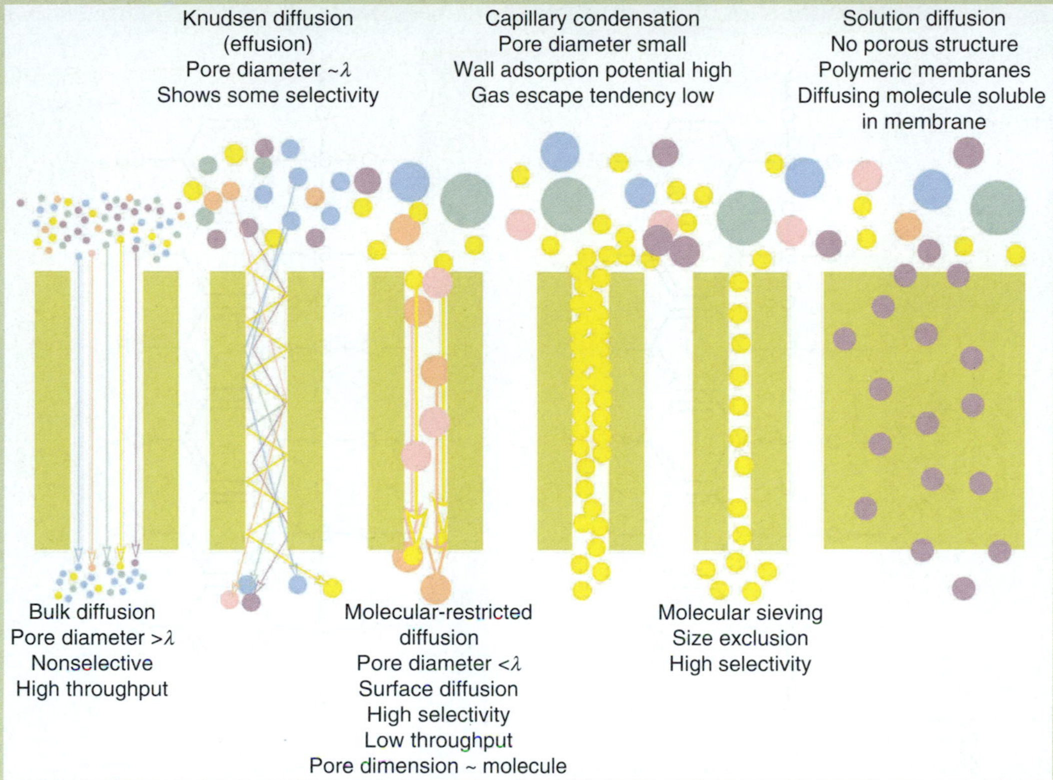

Four basic "selective" mechanisms of gas transport (mass transport) through porous materials are *Knudsen diffusion* (more akin to effusion), surface diffusion, capillary condensation, and molecular sieving [74]. The pore diameter under Knudsen diffusion conditions is on the order of the mean free path of the molecule. Mesoporous systems exhibit Knudsen flow. In other words, as molecules make their way through the pore channels, collisions with the wall are likely. Their rate of transport through the pore channel is inversely proportional to their molecular mass. The separation factor between two molecular species is the ratio of the squares of their respective masses. For optimal selectivity, the pore diameter should be less than the mean free path λ of the molecule. Therefore, pore size, surface chemistry, and the chemical nature of the molecule affect mass transport through the pore channels. *Surface diffusion* involves interaction of molecules with the pore surface; polar molecules interact with polar surfaces; and nonpolar molecules interact with nonpolar surfaces (van der Waals forces) **(Fig. 12.27)**. If the orifice is small enough to promote overlap of potential fields within the pore, then the case of *capillary condensation* results as molecules transit the pore channel. The propensity of escape is dependent on the molecule's ability to overcome the adsorption potential. Lastly, molecular sieving occurs

FIG. 12.27 *Chemical modification of the pore walls of porous materials affects both physical and chemical properties. Additional layers (or just the presence of one layer if the pore diameter is small enough) constrict the orifice and thereby enhance the size-exclusion ability of the material. Chemical modification of the surface allows for separation based on intermolecular chemical interactions.*

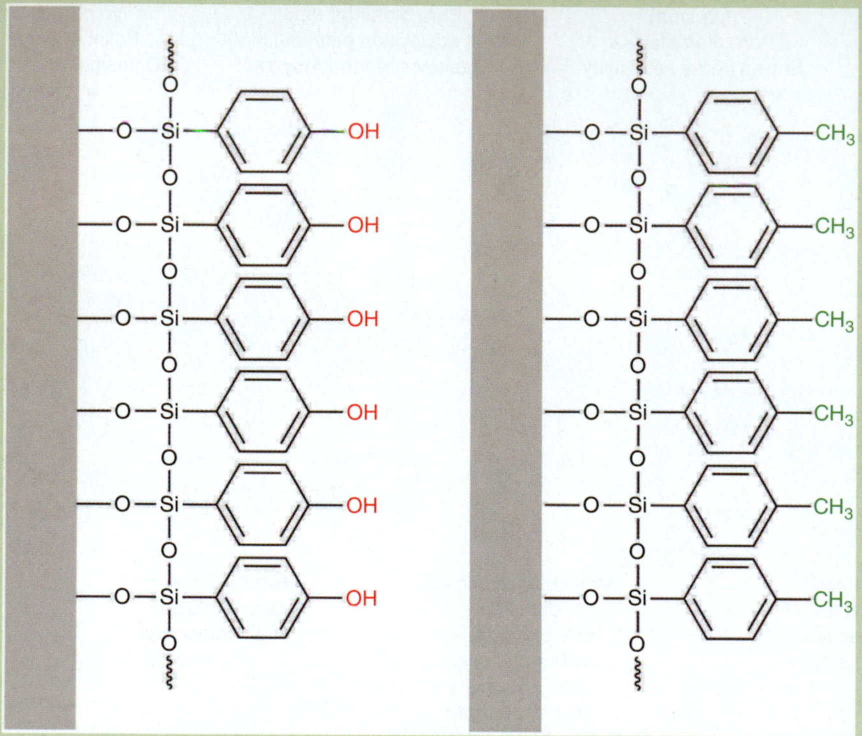

when the pore diameter is so small that size exclusion keeps large molecules out of the channel. Small molecules like CO_2 and H_2 that have kinetic diameter near 3 Å qualify for passage through very small pores.

We will not dwell much more on diffusion through porous materials but will make mention that chemical modification of pore channels is able to impact both the physical properties of the pore channel (e.g., the size and flow) and the chemical selectivity of potential substrates (e.g., traction with specific substrates that affect the kinetics of the flow processes). **Figure 12.27** illustrates the chemical modification of two surfaces of porous alumina formed by an anodize process. The shape, size, charge, and chemical affinity of the diffusing molecule play more important roles in smaller pore channels (also confinement effects). Gas permeation and separation, dialysis, electrodialysis, osmosis, and reverse osmosis are all membrane-based phenomena but detailed discussion beyond what was already presented is beyond the scope of this section.

The Periodic Minimal Surface. A minimal surface is a surface of a structure that results in the smallest possible area [75]. An example of a two-dimensional minimal surface is afforded by a film of soap formed within a wire frame [75].

The surface tension γ of the film of soap is proportional to the surface area (e.g., the as-formed film "is consistent with the shape of the frame and with the requirement that the mean curvature of the film be zero at all points" [75]. Of course with zeolites, we are interested in the *triply periodic minimal surface* (TPMS). These surfaces abound in nature, especially in the silicates, lyotropic colloids, detergent films, and lipid bilayers—all materials that are formed by self-assembly processes and that, interestingly, can serve the function of a template in synthetic processes to form nanomaterials. The skeleton of a sea urchin exhibits this phenomenon at the interface between calcite crystals and its soft-matter structure [76,77]. Atoms in zeolites lie on minimal surfaces [78,79].

Up until the discovery of template-based surfactant synthesized zeolites, minimal surfaces were difficult to quantify. It was found that these new zeolitic materials possessed structures that resemble liquid–liquid minimal surfaces. The synthetic zeolites have predetermined, tunable, and well-defined pore size and shape; excellent thermal and hydrolytic stability; and high order over macroscopic scales [75]. The correspondence between structure determined experimentally and theoretical calculations is excellent; the cubic mesophase of the H_2O–cetyltrimethylammonium bromide (CTAB) is the same as those predicted by calculation and the same as those found in the zeolite from which it was synthesized [80]. Ordered graphite foams from fullerenes and crystalline metal oxides also possess triply periodic minimal surfaces. We can never drift too far off the topic of energy and how it is always minimized in the structures that we study in nanoscience—or any other science, for that matter.

12.3.1 Macroporous Template Materials

There are many kinds of macroporous materials. Electrochemical etching of aluminum (anodizing) is capable of forming alumina structures with pores greater than 50 nm up to 200+ nm in diameter. In general, anodically formed porous aluminas are classified as mesoporous systems and will be discussed later in this section. Electrochemical etching of silicon, chemical etching of glass, ion track etching of polymers, and excimer laser micromachining are techniques capable of producing one-dimensional (pores oriented in one direction) macroporous materials [81]. Microemulsion chemistry and colloid sintering processes are capable of generating porous structures that range in two or three dimensions. Sacrificial scaffolds of crystalline materials have increasingly been used as templates to form porous materials [81].

The most popular method of forming macroscopic porous materials is from templates consisting of close-packed spherical polymer beads (a.k.a. latex beads) by an infiltration-sintering process. The process to form them is as follows: (1) synthesis of polymeric latex beads by emulsion polymerization; (2) packing of the beads into arrays by several techniques, such as evaporation, filtration, settling, and dip-coating; (3) infiltration of the interstitial volume with silicate, zirconia, titania, or aluminate sol–gel precursor solutions by capillary action or vacuum induction of the solution or by infiltration by solutions containing metal–anion components; and (4) treatment at high temperature (~500°C or more) to remove polymeric components and simultaneously sinter to form ceramic matrix. Filling scaffolds by electrochemical deposition and chemical vapor deposition are also effective methods of creating porous materials.

Close-packed open structures of 320–360 nm have been formed in a single-step process. The void spaces in stacked latex spheres, for example, were formed by vacuum-assisted induction (packing) that were covered with the metal alkoxide precursors. The assembly was calcined at 575°C [82]. Approximately 500-nm thick films that contain three-dimensional arrays of macropores titania, zirconia, or alumina microcrystalline material were formed.

Hybrid macroporous materials are also fabricated by colloidal crystal templating methods that yield porous thiol–metal oxide structures. [83]. Examples of hybrid macroporous materials include thiol-functionalized titania or zirconia with propylsiloxane, ethylsulfonate, or propylsulfonate linkages. The templating materials in this case are made of polymethyl methacrylate (PMMA). The purpose of using macro- rather than meso- or microporous materials is to improve the mass transport of a potential analyte through the porous material. The colloidal crystal system has built-in "synthetic flexibility" in that the selectivity, activity, and throughput characteristics of an adsorbent can be tailor made to fit an application. Hybrid macroporous materials consisting of zirconia and titania were chosen because those materials offer greater flexibility for chemical functionalization than do the silica-based substrates. Three types each of titania and zirconia macroporous structures are discussed and properties are displayed in **Table 12.5** for use in removal of heavy metals from solution by ion adsorption. An example of a template method that uses latex spheres is shown in **Figure 12.28a.**

12.3.2 Mesoporous Template Materials

Anodic Aluminum Oxide Membranes. Anodization of gate (refractory) metals leads to another class of mesoporous system. In anodizing, a metal is made the anode and oxidation of that metal in an electrolytic solution occurs. Anodizing aluminum has been an industry practice dating back to the first quarter of the twentieth century. Following the introduction of aluminum as a major industrial material, NPL of Teddington, United Kingdom, filed a British patent (no. 290,901) in 1926 that later was acquired by Alcoa in the United States in the 1930s. Anodizing is a process that creates an insulating oxide layer on a conductive metal anode, usually aluminum, in an electrolytic solution, usually a dilute polyprotic acid. By providing hexagonally packed pore channels that are simple to fabricate and the ability to manipulate pore diameter and length during and after anodizing, the anodic alumina oxide (AAO) offers a perfect template for nanoscale material synthesis. The utility of the AAO as a template is widespread.

Table 12.5 *Hybrid Inorganic–Organic Macroporous Structure Size and Surface Area*

Macropore material	Pore size (nm)	Template size (nm)	Surface area ($m^2 \cdot g^{-1}$)
TiO_2–O_3Si–Pr–SH	265	300	37
TiO_2–O_3S–Et–SH	270	300	12
TiO_2–O_3S–Pr–SH	275	300	12
ZrO_2–O_3Si–Pr–SH	415	480	18
ZrO_2–O_3S–Et–SH	450	480	20
ZrO_2–O_3S–Pr–SH	460	480	14

Metals and alloys, polymers, semiconductors, carbon nanotubes, and ceramics are just a few examples of the major classes of materials that can be fabricated or inserted into the pore channels of AAOs. The presence of hydroxyl groups on the surface of AAOs facilitates a wide range of chemical applications.

Robust porous oxides are formed on metals known as gate metals. These include aluminum, titanium, tantalum, and niobium. Zirconium and vanadium have also been anodized successfully. We limit our discussion, however, to

Fig. 12.28 *Macroscopic and mesoscopic materials are compared. Mesoscopic materials have pores that are "officially" under 50 nm. (a) Macroporous template synthesis: porous ceramic (titania or silicate) ordered arrays formed from spherical latex templates. The structure on the bottom is interconnected with channels similar to those found in zeolites. The size of the spheres can be several hundred nanometers.*

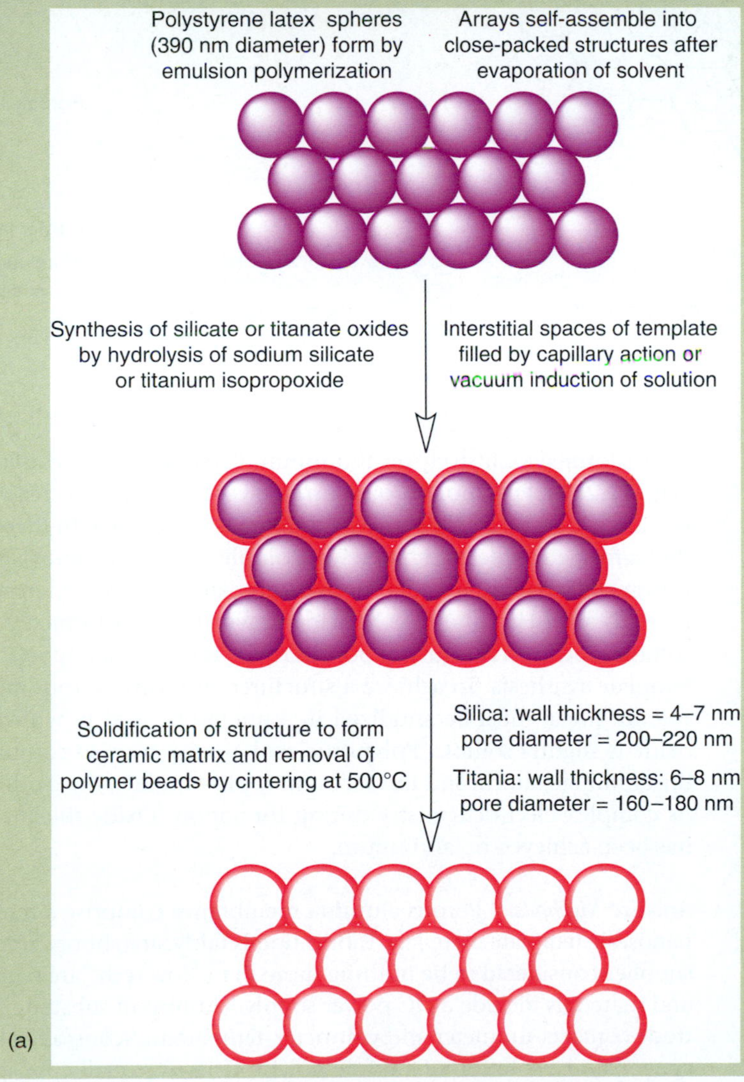
Polystyrene latex spheres (390 nm diameter) form by emulsion polymerization
Arrays self-assemble into close-packed structures after evaporation of solvent
Synthesis of silicate or titanate oxides by hydrolysis of sodium silicate or titanium isopropoxide
Interstitial spaces of template filled by capillary action or vacuum induction of solution
Solidification of structure to form ceramic matrix and removal of polymer beads by cintering at 500°C
Silica: wall thickness = 4–7 nm
pore diameter = 200–220 nm
Titania: wall thickness: 6–8 nm
pore diameter = 160–180 nm
(a)

(continued)

Fig. 12.28 (Contd.) *(b) Mesoporous template synthesis: porous aluminas are able to transcend all boundaries with regard to classification. In general, most anodically formed porous structures have pores in the range of 20–50 nm. Commercially available anodic films have pores that are much larger. The structure of anodically formed porous aluminum oxide is shown again, this time compared to a barrier layer oxide. The porous layer is formed on top of the barrier layer. The barrier layer thickness is the same throughout the process. The scalloped metal surface is shown in the figure. The size of the scallop depends on the applied voltage. By means of a double-anodize method, a perfect honeycombed structure can be formed. During the first anodization, usually consisting of an extended time period (e.g., 24 h), equilibrium positioning of the pore channels is achieved. All the pore channels have the same diameter with the same barrier layer thickness. After dissolution of the primary layer, re-anodization produces a new film that sprouts from the preexisting scalloped structure—already with the equilibrium size and distribution.*

(b)

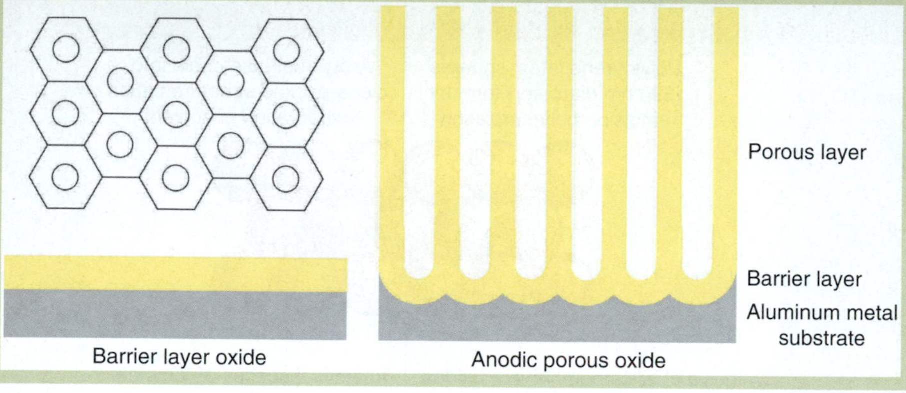

oxides formed exclusively on aluminum. Three general types of oxides are grown (**Fig. 12.28b**). First, nonporous barrier layer oxides are formed when aluminum is anodized in weak acids such as tartaric or boric acids. In these acid solutions, the as-formed oxide is completely insoluble and a thin layer (a few nanometers) is formed. Secondly, irregular nonporous oxides are formed in strong monoprotic acids such HCl and HNO_3, acids in which the as-formed oxides are highly soluble. The third category of porous AAOs is the most important with regard to template synthesis. To achieve a structured, uniform porous anodic oxide, aluminum metal must be anodized in a polyprotic acid in which the aluminum oxide is slightly soluble. Polyprotic acid anions are also comparable in size to the aluminate anion and this factor may play a role in the oxide's solubility and its complex electrochemistry during formation. Oxide thickness over 300 μm has been achieved on aluminum.

Anodize Methods. Porous alumina membranes comprise a remarkable class of nanoscale template materials. Fabrication is fairly straightforward. Equipment and supplies considered to be nothing more than "low tech" are required. Apparatus and materials include a DC power supply, aluminum substrate plate stock, electrode contacts, ammeter and voltmeter, temperature-controlled bath, an agitator system, and a solution of a dilute polyprotic acid (usually phosphoric, chromic,

TABLE 12.6 *Anodize Parameters for Common Polyprotic Electrolytes*

% Electrolyte (w/w)	Temp. (°C)	Potential (V)	Duration (h)	Thickness (μm)	Pore diameter (nm)	Pore density ($N \cdot cm^{-2}$)
4% H_3PO_4	0	130	8	50	200	1.3×10^8
1% H_2CO_4	–5	90	2	40	120	2.2×10^9
1% H_2CO_4	0	70	3	30	86	4.3×10^9
1% H_2CO_4	0	40	6	40	60	1.2×10^{10}
4% H_2CO_4	2	30	10	25	52	1.4×10^{10}
10% H_2SO_4	0	20	4	20	32	3.7×10^{10}
10% H_2SO_4	5	15	6	15	22	8.0×10^{10}
15% H_2SO_4	8	10	3	5	10	3.8×10^{11}
20% H_2SO_4	20	2	1	<1	~5	1.5×10^{12}

Source: G. L. Hornyak, *Characterization and optical theory of nanometal/porous alumina composite membranes*, Ph.D. dissertation, Colorado State University (1997).

sulfuric, or oxalic acids). Basic anodize formulations are listed in **Table 12.6.** The porosity ε for all types of membranes ranges from 20 to 30%. Porosity can be increased or decreased following anodizing and membrane detachment.

Control of pore diameter under voltage control conditions is accomplished by variation of applied voltage. The range in pore diameter varies from a few to several hundreds of nanometers. The thickness of the membrane is a function of the duration of anodizing. Thickness can be varied from a few nanometers to greater than 300 μm. Pore channel aspect ratio ($\mathcal{A}_{pc} = l/d$) ranges from less than one to greater than several thousand. An atomic force microscopy (AFM) image of a porous aluminum oxide made in oxalic acid is shown in chapter 1, **Figure 1.4**.

Two general strategies are applied during anodization. In the first, anodic potential is held constant for the duration of film growth while the current is allowed to seek its steady-state level. This configuration leads to pore channels that have the same diameter throughout the thickness of the membrane. In the second type, the applied current is held constant and the voltage is allowed to wander. In this case, pore channels form in conical configurations. **Figure 12.29** shows the current density time transient under voltage control on anodize time for two types of acid solutions: oxide slightly soluble (porous type, dashed red line) and oxide insoluble (barrier type, solid blue line). Initially, both transients appear to exhibit the same behavior due to the initial surge in current as the system strives to achieve the voltage according to Ohm's law:

$$V = I\Omega \tag{12.21}$$

Resistance Ω in the anodic circuit is provided by the newly formed barrier layer. At the onset, no oxide (insulating) coating is present and the current achieves its maximum value under specific voltage, electrolyte, and temperature conditions. However, as the oxide layer begins to form, a rapid decay in current occurs in both electrolytes. In the boric acid medium, the current decays exponentially without recovery. This results in a nonporous and thin barrier oxide

FIG. 12.29

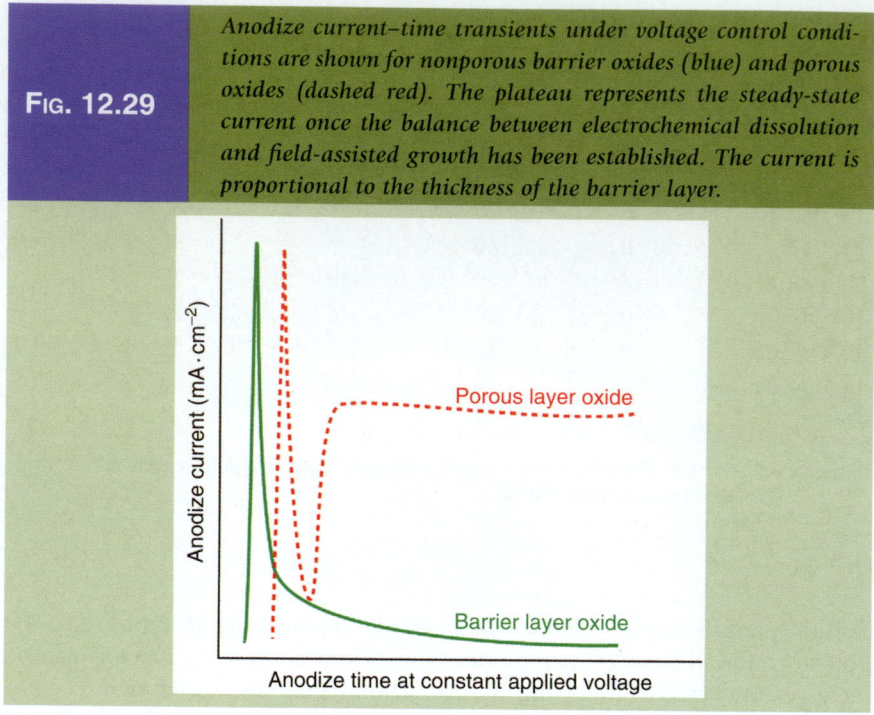

Anodize current–time transients under voltage control conditions are shown for nonporous barrier oxides (blue) and porous oxides (dashed red). The plateau represents the steady-state current once the balance between electrochemical dissolution and field-assisted growth has been established. The current is proportional to the thickness of the barrier layer.

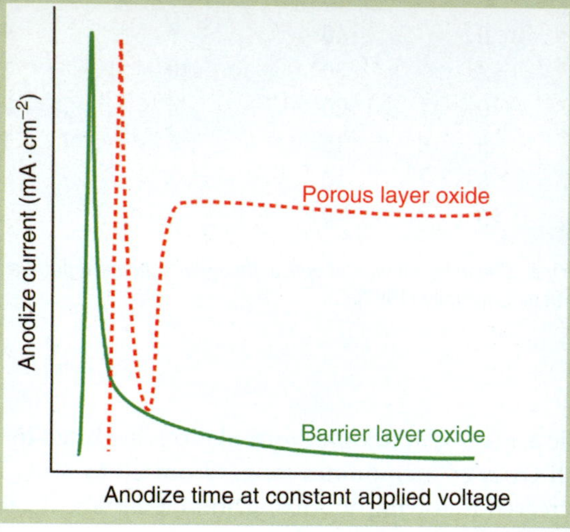

layer. In polyprotic acids, the current reemerges and achieves a steady-state level for the duration for the anodize process. A porous oxide layer is grown under these conditions.

A porous layer is formed due to a complex interplay between the chemical dissolution process and the electric field-assisted growth. This behavior is responsible for the steady-state plateau region of **Figure 12.29.** The steady-sate current is a function of the barrier layer thickness.

As-formed membranes are complex in composition and contain a wide range of amorphous, hydrated, and crystalline alumina species in addition to incorporated electrolyte. Three layers contribute to the as-formed membrane: the aluminum substrate, the barrier layer, and the porous layer. The barrier layer allows anodization to occur and is the place where chemical dissolution and electric field-assisted growth of the oxide exist in the steady state. The barrier layer thickness is a function of applied anodic voltage. The pore diameter, as a result, also depends on the applied potential. The barrier layer behaves much like an ohmic device. Ohm's law describes the electronic behavior of the barrier layer oxide:

$$V = \rho d i_+ \tag{12.22}$$

where ρ is the specific resistance of the oxide material during anodizing in terms of ohm centimeters, d is the thickness of the barrier layer in centimeters, and i_+ is the ionic current density at the anode in amperes per square centimeter.

Pore formation and spacing are explained by phenomenological reasoning. Aluminum metal plates of 99.999% purity are used to manufacture the best membranes. The plate is electropolished to form a shiny, smooth, and reflective surface. Electropolishing is similar to anodizing except that chemical dissolution of the as-formed oxide is favored over film growth: high temperature, high current, very high or very low *p*H conditions. Unless formed from single crystal aluminum, the surface is saturated with crevices, pits, ridges, and other irregularities. These areas are regions where current is focused, akin to a reversed lightning rod. As the potential is applied, a competition for available current density is initiated. An oxide layer is formed and dissolved, and pockets of lower resistance outcompete areas where thicker oxide layers were formed. Scalloping of the aluminum surface occurs as the current density vector increases in strength, favoring their survival over bigger or smaller cells. The cells with the appropriate barrier layer thickness survive and pack two-dimensionally into a hexagonal array. In membranes formed by this process, branched pore channels are commonplace as pore cells migrate to find equilibrium positions in the hexagonal array as a function of anodize time. Surface defects and imperfections prevent long-range order from happening in AAOs, although domains on the order of microns are possible.

Highly Ordered Membranes. Fabrication of highly ordered membranes involves a two-step anodize process [84]. In the first step, a porous membrane is formed according to the scheme described previously. After 24 h of anodizing, all scallops are the same size and equally distributed on the aluminum surface. At this time, the oxide is dissolved in acid, leaving an aluminum plate that is pretemplated for another round of anodizing. With the scallops preformed and in the equilibrium state, anodizing is resumed, but this time the oxide layer grows directly from the template scalloped layer. The result is an ordered array of nanochannels with no branching and no heptagonal or pentagonal channel lattices (e.g., hexagonal cells are prefered in the "equilibrium" structure). Hideki Masuda of the Tokyo Metropolitan University is credited with its development [H. Asoh, K. Nishio, M. Nakao, T. Tamamura, H. Masuda conditions for fabrication of ideally ordered anodic porous alumina using pretextured Al, *J. Electro. Chem. Soc.*, 148, B152–B156 (2001)].

Membrane Detachment. There are several ways to remove the AAO membrane from its aluminum metal substrate. The most efficient means is to apply the procedure of stepwise voltage reduction. We know that pore size is a function of applied potential. If we were to reduce the potential stepwise, then a network of smaller and smaller pore channels would form in the barrier layer. The detachment process current–time transient is similar in form to **Figure 12.29** except that it contains a series of steps (**Fig. 12.30**), each with less current intensity as voltage is turned down sequentially. The barrier layer is a highly permeable ionic membrane during anodizing. If an instantaneous 10% voltage reduction is applied, the current undergoes depletion and then recovers to the new steady-state condition dictated by the newly applied potential. Essentially, a new set of pores, proportionally smaller than the parent pore, are formed in the barrier layer. This process is repeated until the pore diameter is reduced along with a concomitant thinning of the barrier layer.

When 1 V is achieved (pore diameter $\approx$1.5 nm, barrier layer thickness $\approx$1 nm), the process is stopped and the aluminum and oxide still attached to the metal plate is placed in an acid detachment solution (25% H_2SO_4 for sulfuric

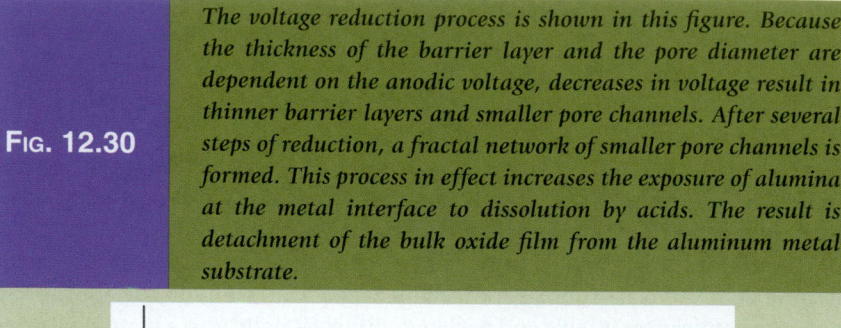

Fig. 12.30 *The voltage reduction process is shown in this figure. Because the thickness of the barrier layer and the pore diameter are dependent on the anodic voltage, decreases in voltage result in thinner barrier layers and smaller pore channels. After several steps of reduction, a fractal network of smaller pore channels is formed. This process in effect increases the exposure of alumina at the metal interface to dissolution by acids. The result is detachment of the bulk oxide film from the aluminum metal substrate.*

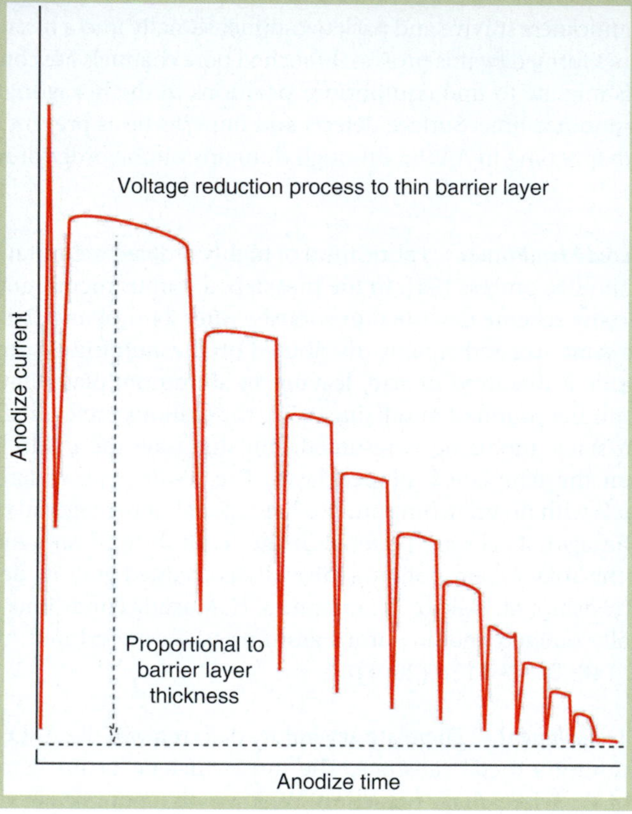

acid-formed films at room temperature, conditions that favor dissolution). The dissolution reaction is as follows:

$$2Al_2O_3 + 12H^+ \rightarrow 4Al^{3+} + 6H_2O + 3H_2(g) \tag{12.23}$$

$$2Al^o + 6H^+ \rightarrow 2Al^{3+} + 3H_2(g) \tag{12.24}$$

Hydrogen gas is produced between the alumina film and the metal substrate. Bubbles coalesce underneath the film indicating detachment and, after rinsing and drying, the film essentially pops off intact. Another method to remove the film that is practical for thin aluminum metal substrates is the application of saturated mercuric chloride. The mercuric chloride only attacks aluminum metal and not the oxide, and an amalgam between aluminum and mercury is formed.

In other words, treatment of the aluminum metal substrate with mercuric chloride removes the aluminum and leaves an intact barrier layer (unlike in the voltage reduction procedure). This is a good method to use if the barrier layer is deliberately left intact. In this way, the membrane is closed at one end of the pore channel—a perfect nanobeaker.

The AAO membrane has some ideal chemical, physical, thermal, and optical properties because it is made from alumina and, after drying, the membrane is insulating. Each cell could contain capacitors, resistors, conductors, or semiconductors and be insulated from its neighbors. Alumina is relatively inert chemically. Although it is an amphoteric material (able to dissolve in both acids and bases), at neutral *p*H, the material is inert to most chemical dissolution reactions. Optically, the membranes are highly transparent. Just like with colloids and quantum dots, light scattering (which causes diffusivity) does not occur until pore diameter is larger than 25 nm. That is why membranes made in sulfuric acid (with pore diameter <25 nm) are perfectly clear. Membranes with pores in the 50- to 100-nm range appear clear but slightly diffuse. Inclusion organic oxalates may contribute to the diffusivity in addition to the larger pore size. Membranes made in phosphoric acid, with pore diameter greater than 150 nm, appear opaque, grayish to white. AAOs can withstand thermal stress until phase changes occur. As we noted earlier, many phases of alumina, along with incorporated electrolyte, make up the matrix. At temperatures above 600°C, alumina undergoes a phase change into γ-Al_2O_3. At temperatures greater than 800°C, sulfates decompose into SO_2 and SO_3 gases and the matrix undergoes catastrophic loss of pore structure. At temperatures higher than 900°C, alumina is converted into γ-Al_2O_{3-}.

Chemical Derivatization of Anodic Surfaces. Chemical modification of the surfaces of the pore channels is quite straightforward. The alumina surface is a natural source of hydroxyl groups that are able to act as nucleophiles in ester exchange reactions. Alkylphosphonic acids and diethlybutylphosphonate are used to change the wetting character of the surface from hydrophilic to hydrophobic [85]. Silanes, carboxylates, phosphonates, and numerous other species are able to react with the pore channel walls of AAO membranes.

Template Synthesis with Porous Aluminas. Template synthesis with AAO hosts results in nanomaterials in the pore channels. This method is a general approach that allows for the synthesis of many kinds of nanomaterials by many types of synthetic methods [56,86,87]. Monodisperse materials are fabricated in highly ordered membranes. Two forms of synthetic strategies are employed: one in which the AAO membrane is detached from the aluminum metal substrate and another where the membrane is left on the aluminum with the barrier intact.

Synthesis with intact membranes requires thinning of the barrier layer by acid dissolution or by the voltage reduction method described previously so that electrical conductivity and ionic transport are facilitated through the barrier layer. The barrier layer, once thinned, allows for AC electrodeposition of metal and semiconductor materials. In another application, one that is gaining importance as a component in flat panel displays, carbon nanotubes are synthesized from Co, Fe, or Ni nanoparticles that are electrodeposited at the base of the pore channels by the AC method. The process comprises four steps following the initial anodizing process: barrier layer thinning, AC electrodeposition of metal catalyst nanoparticles, chemical vapor decomposition of carbon feedstock gas, and, lastly, the growth of carbon nanotubes.

Electroless deposition procedures to form honeycomb replicas or negative replicas of AAOs also require AC deposition to place a catalyst at the pore basin. In electroless deposition, a conductive material is deposited onto another material, conductive or nonconductive, by the autocatalytic reduction of metal ions in a chemical solution without using electrodes. H. Masuda of the Tokyo Metropolitan University has taken this form of plating to high levels in nanostructured material fabrication.

Electroplating metals and semiconductors into the pore channels of AAO straightforward. Following detachment and removal of the barrier layer, the freestanding membrane is coated with a layer of silver, usually applied by RF sputter procedure. The silver performs the function as a cathode and is easily removed following electroplating with HNO_3 treatment. Formation of gold nanostructures is depicted in **Figure 12.31.** Multitiered structures are also possible in concentric geometry (**Fig. 4.15,** if viewed from the top).

Alumina → derivatized walls → polymer coating → metal nanowire

Hybrid Inorganic–Organic Mesostructured Materials. Organically functionalized mesostructured materials were formed by the co-condensation of TEOS and organoalkylsilane with a neutral alkyl amine surfactant. Organosilane species included octyltriethoxysilane, mercaptopropyltrimethoxysilane, phenyltriethoxysilane, butyltrimethoxysilane, and propyltrimethoxysilane—all relatively long-chained alkylsilanes. At least 3-methylene units were required to interact with the hydrophobic core of the structure-directing micelle. This allowed for successful incorporation of the silane into the pore walls of the mesostructured material [88].

Most ordered mesoporous solids consisting of metal oxides, metals, and carbon have amorphous pore walls. Hierarchical ordered mesoporous solids with molecular-scale pore surface periodicity, however, certainly are not amorphous. Inagaki et al. managed to synthesize, with the aid of surfactant mediation, an ordered benzene–silica hybrid that contains hexagonally distributed mesopores with lattice constant equal to 5.25 nm and structural period of 0.76-nm spacing along the channel axis [89].

The tube was "self-assembled" by a combination of alternating hydrophilic silica and hydrophobic benzene layers with structural directing between the precursor molecules and the surfactants. The synthesis is as follows. An organosilane monomer $[(C_2H_5O)_3Si–C_6H_6–Si(C_2H_5O)_3]$ is heated in a basic solution of surfactant template material. Interactions between and among the surfactant, the organosilane, silica, and benzene organize to direct the self-assembly process to form into a tube. Average diameters range from 3 to 5 nm, and the diameter can be controlled by selection of the surfactant moiety. As the material was developed, sulfonation of the material was accomplished successfully without disruption of the mesoporous structure or periodicity [89] that proved to be stable at 500°C in air. Mesoporous structure synthesis can also be accomplished with emulsion chemistry [90].

The study of chemistry in confined spaces continues with the addition of each new kind of template. The overlap between "supramolecular chemistry in confined spaces" and cage-like structures continues as new insights into chemistry

Fig. 12.31 *Synthesis of gold nanorods by template synthesis/electrodeposition method. Top left: porous alumina was removed from alumina metal substrate by the process of voltage reduction in 25% oxalic acid. Pore channel diameter after anodizing is between 10 and 50 nm. Middle left: a layer of Ag is applied by RF plasma sputter process. The Ag layer is the cathode in all electrodeposition steps. Bottom left: Ag posts are electroplated into pore channels to provide platform deeper into the oxide. Bottom right: Au is plated on top of the Ag posts. The aspect ratio of the Au cylinders is controlled by the pore diameter and the time of electrodeposition. Middle right: the Ag is removed with a nitric acid wash. Top right: if required, the alumina is removed by reaction in NaOH solution. The red color of the gold cylinders resembles the actual color of the composite membrane*

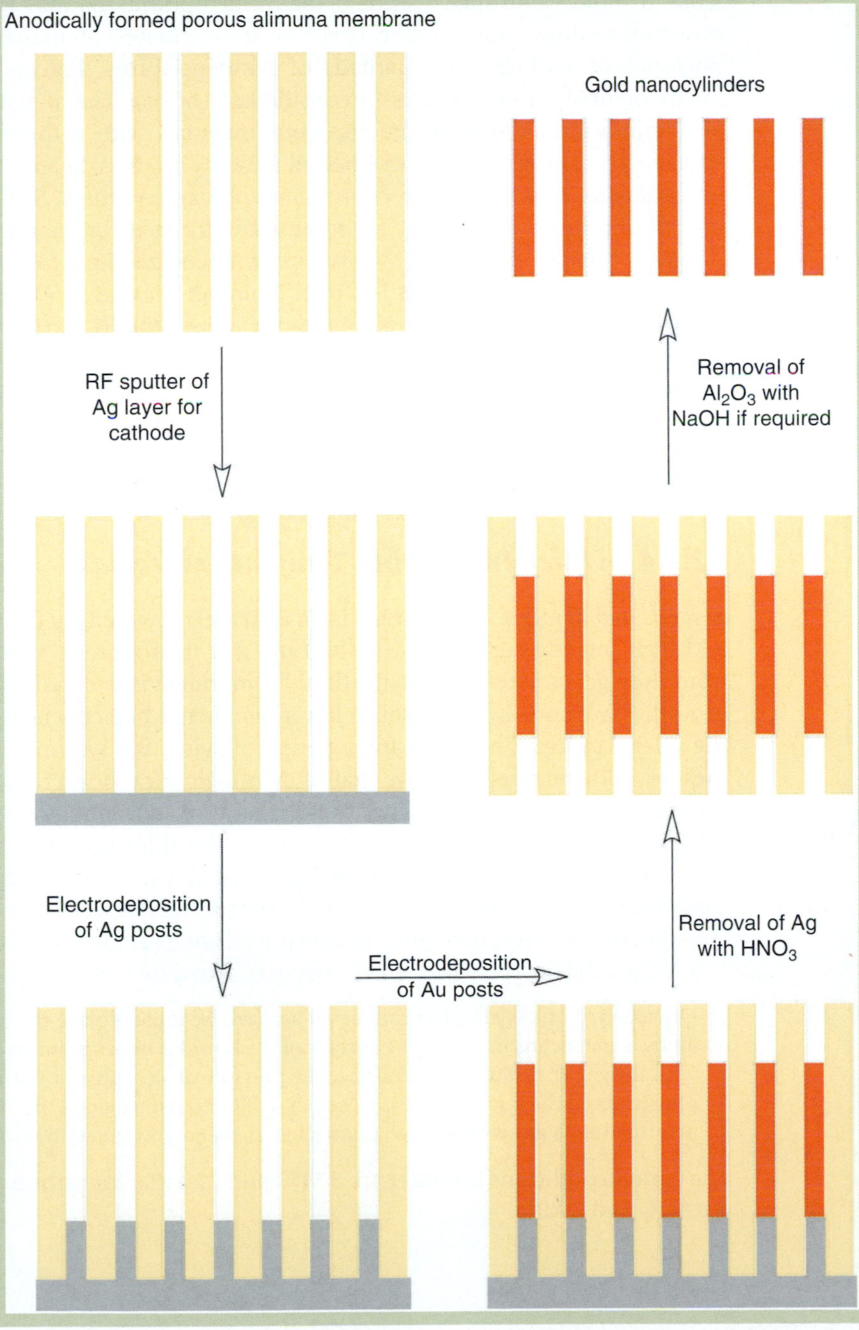

and kinetics continue to be uncovered [91]. These nanometer-ordered "minimal surface" structures continue to amaze.

12.3.3 Microporous Template Materials

Microporous templates have the smallest pore cavities and channels (<2 nm). The best known of these are the zeolites. Although zeolite pore channel size spans the microporous domain and into the meso range, we shall confine their discussion to this section (another boundary). Axel F. Cronstedt, the Swedish chemist who discovered nickel in 1751, coined the word zeolite (from the Greek *zein,* "to boil," and *lithos,* "stone") for a microporous natural material, after observing boiling chips in water. There are approximately 50 naturally occurring microporous zeolites and hundreds of synthetic forms that are classified as micro- or mesoporous. Zeolites are considered to be of a class of materials known as aluminosilicates that are microporous materials with a three-dimensional structure consisting of an interconnected network of pores. Si and Al atoms exist in tetrahedral crystal structure coordinated by oxygen atoms. Scolecites, natrolites, phillipsites, and analcime are some naturally occurring zeolites.

Mobil crystalline materials (MCMs) are synthetic zeolites. MCM zeolites are characterized by pore diameter less than 2 nm but as large as 50 nm, depending on synthesis conditions. The largest pores found in naturally occurring zeolites are 1–1.2 nm. The "liquid crystal template system" of Kresge et al. showed that pore size can be controlled by intercalation of layered aluminosilicates between ordered surfactant micelle species [92]. A calcination process follows chemical synthesis to form the zeolite. Uses of zeolites include molecular sieves, adsorption, shape-selective catalysis, membrane separations, nonlinear optics, and detergents.

12.3.4 Other Interesting Template Materials

Aerogels and Xerogels. An aerogel is an extremely low-density (porosity >90%) and high internal surface area (>1000 $m^2 \cdot g^{-1}$) microporous structure derived from a sol–gel process in which the liquid component is replaced by gas. Xerogels generally have porosity that is much lower (and hence higher density) [3]. During the sol–gel process and following a period of aging, the system is exposed to a pressure and temperature above that of the supercritical point of the solvent—a process called supercritical drying. This technique results in removal of the liquid vehicle without causing structural damage to the solid matrix of the material. Supercritical conditions are employed to mitigate the effect of capillary forces that would cause the collapse of the gel network otherwise. (Note: see Laplace's equation that involves the radius of curvature, contact angle, and surface energy.) S. S. Kistler, the first person to make aerogels, stated in 1931 [132]:

> Obviously if one wishes to produce an aerogel, he must replace the liquid with air by some means in which the surface of the liquid is never permitted to recede with the gel. If a liquid is held under the pressure always greater than the vapor pressure, and the temperature is raised, it will be transformed at the critical temperature into a gas without two phases having been present at any time.

The critical conditions for water are 374°C and 22 MPa; for carbon dioxide, they are 31°C and 7 MPa.

Aerogels are made from silicas, aluminum oxides, and carbons by a sol–gel process involving hydrolysis, condensation, and gelification. The synthesis is relatively expensive and requires long times for gelation—especially in the extreme low-density materials. Kistler reacted 1 mol of Na_2SiO_3 with 2 mol of HCl to form 1 mol of a silicon oxide complex $[SiO_2 \cdot H_2O]$ and 2 mol of NaCl. He treated the aged sol in an autoclave held at 374°C @ 221 bar. Kistler extracted the water with ethanol and thereby lessened the required temperature parameter from 374 to 243°C @ 64 bar. He produced aerogels with density ranging from 0.02 to 0.2 $g \cdot cm^{-3}$. The system was extended to other inorganic gels, various combinations of precursors, and to organic systems.

12.4 Polymer Chemistry and Nanocomposites

Polymers (from the Greek *polymers*, "having many parts," from *poly*, "many," and *meros*, "part") are materials that consist of a large number of similar units (monomers) that are bonded together. Swedish chemist J. J. Berzelius coined the word *polymer* in 1833, although his original intent was to describe materials with the same empirical formula that had different molecular weights. Polymers, a direct by-product of modern chemical engineering, have made a significant impact on our technological development by means of their versatility. Polymers are made of various mixtures of carbon and hydrogen, nitrogen, fluorine, or silicon. Molecular weights of bulk polymers are in the thousands to millions. Colloidal polymeric materials and nanocomposites are two areas that have received much attention recently.

12.4.1 *Introduction to Polymer Chemistry*

Polymeric materials exhibit many forms. Inorganic polymers include various strains of clay, sand, and fibers. Adhesives, fibers, coatings, plastics, and rubber

Example 12.1 *Degree of Polymerization of Polyethylene*

Calculate "n" for a polyethylene polymer that is 180 nm in length. Assume that the polyethylene is not coiled and that all the bonds are in one plane. The C–C bond length is 0.154 nm.

Solution:

The monomeric repeating unit of polyethylene is $-(C_2H_4)-n$

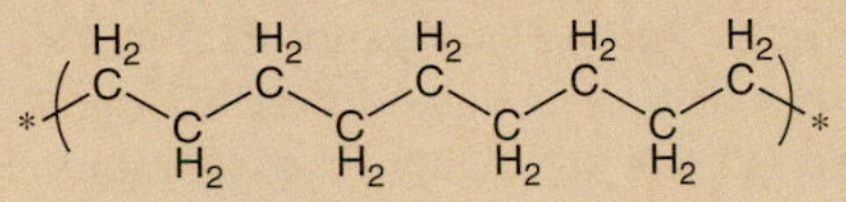

The C–C–C bond angle is 109.5° for a tetrahedral structure. The actual linear bond length is sin $[(1/2)\ 109.5°] \times$ *C–C bond length* $= 0.1258$ *nm*

Since every monomeric component has two bonds,

$n = 180/[(0.1258) \times 2] = 714$ *monomers*

constitute the class of synthetic organic polymers. Well-known biological polymers include polysaccharides, proteins, and naturally occurring gum rubber. Polymers are generally formed by addition (adding to a double bond) or condensation (e.g., elimination of water) reactions. Epoxies (cross-linked) and polyurethanes (urethane linkages) are members of other classes of polymers. A *copolymer* is a mixture of two polymers. A *block copolymer* consists of repeating units of blocks of two types of polymers. The degree of polymerization, represented usually by n, indicates the number of monomer units that have bonded to generate the polymer. The value of n ranges from a few hundred to several thousand. Molar mass is related to the degree of polymerization—roughly equal to the average molecular weight of the constituents times the number of monomeric units.

Types of Polymers. There are many kinds of monomers and hence many kinds of polymers. A few are listed in **Figure 12.32.**

Bonding in Polymers. Because polymers are complex structures containing large molecules, one can surmise that many possible types of forces exist between and among polymer chains. One does not expect polymers to be highly ordered but they do have considerable crystalline character. A stretched rubber band is the classic example of how some level of order is imparted to a polymer as it is elongated, freezing the components momentarily into a lattice. Although the order may not transcend the entire polymer (long-range order), ordered domains, called crystallites, exist within the structure of a polymer.

Some types of polymer chains are attached to each other by strong covalently bonded substituents. This is called cross-linking. Weak intermolecular forces discussed previously abound in polymers. Weak dispersion forces in poly(ethylene) and van der Waals forces in poly(propylene) exist in polymers and arise from synchronous attractions of electrons between large molecules, of which there is an abundance in polymers [93]. Medium to weak dipole–dipole forces are found in poly(vinylchloride) and poly(methylmethacrylate). Much stronger hydrogen bonding also plays a major role in the structure of condensation polymers such as polyamides and proteins and in cellulose. Strong electrostatic intermolecular forces are found in ionomeric compounds. Every type of intermolecular force is represented in polymers.

12.4.2 Polymer Synthesis

There are two general strategies available to fabricate polymeric materials. The first method is by the process of *addition*. There are three stages in the addition process: initiation, propagation, and termination. Addition processes involve addition to the double bond of an alkene monomer. Initiators can be free radicals, cations, anions, or organometallic. Organometallic catalysts called Ziegler catalysts consist of an organometallic (e.g., R_3Al) and a metal halide (e.g., $TiCl_4$). Because radicals are highly reactive, rapid chain reactions abound in the propagation step. All reactions are terminated when radical products combine with other radicals in the mixture. The second general mechanism is the *condensation* process. The term condensation implies that water is eliminated as bonds are formed between two monomers. Other small molecules, however, can also be eliminated, such as hydrochloric acid, methanol, and carbon dioxide.

FIG. 12.32 *Some of the most common polymers are shown in the figure. All can be applied to make nanomaterials.*

Polymerization method	Starting material (monomer)	Monomer structure
Free radical 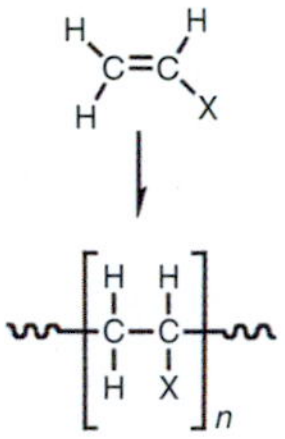	Poly(ethylene) PET	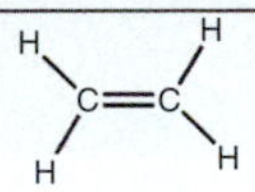
	Poly(propylene)	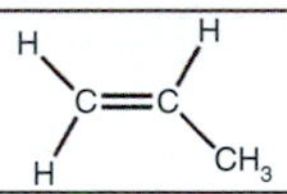
	Poly(vinylchloride) PVC	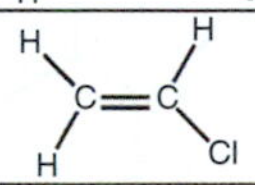
	Poly(styrene)	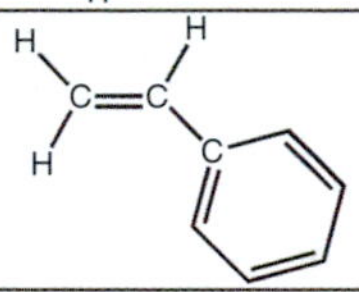
	Poly(acrylonitrile) Orlon	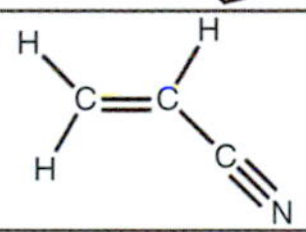
	Poly(tetrafluoroethane) Teflon	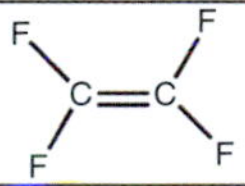
Hydrolysis	Poly(vinylalcohol) PVA	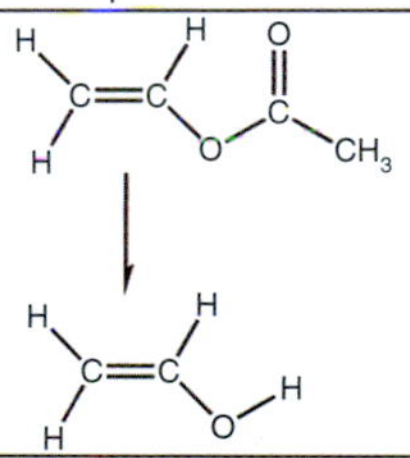
Anionic	Poly(methylmethacrylate) PMMA (and free radical)	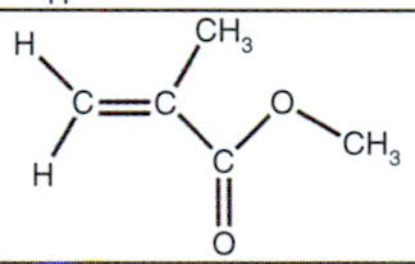
Cationic	Butyl rubber	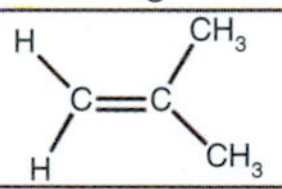
Condensation	Poly(butylenesuccinate) PBS	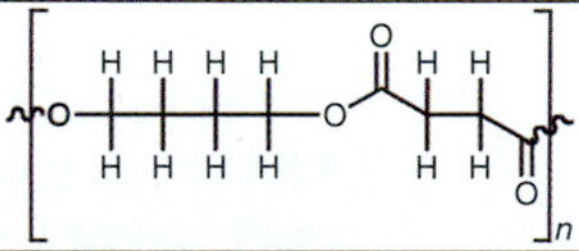
	Polyesters Combination of dicarboxylic acid or diester+diol	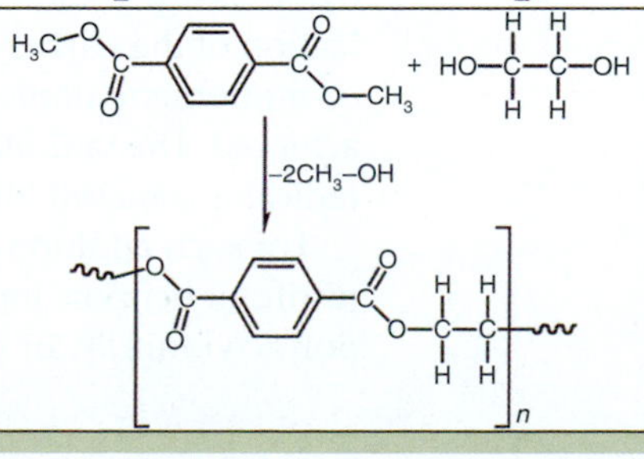

FIG. 12.33 *A free radical polymerization reaction is depicted. Following initiation, the polymer propagates until reactants are depleted by termination. The thermal decomposition of organic peroxides (e.g., R = benzoylperoxide) is one means of generating the radical initiator required for the polymerization of the ethylene monomer (Step 1). Step 2 involves consecutive addition of monomeric units until chain lengths of several hundreds to thousands are achieved. The radical process propagates in this way until reaction with another radical is achieved, thus terminating the polymerization (Step 3).*

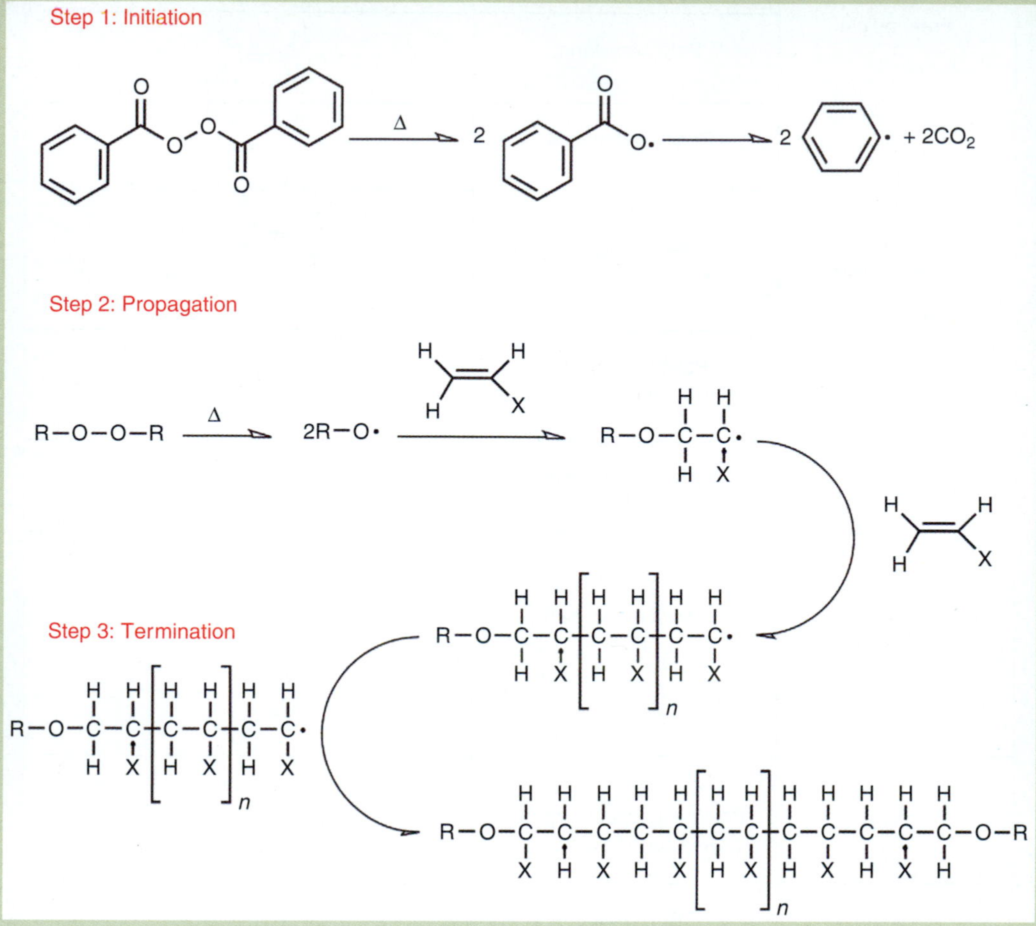

The *radical chain reaction process* is one way to form addition polymers (**Fig. 12.33**). The thermal decomposition of organic peroxides (e.g., R = benzoyl peroxide) is one means of generating the radical initiator required for the polymerization of the ethylene monomer (Step 1). Step 2 involves consecutive addition of monomeric units until chain lengths of several hundreds to thousands are achieved. The radical process propagates in this way until reaction with another radical is achieved, thus terminating the polymerization (Step 3).

For such addition polymer processes, if "X" is a hydrogen, methyl, chlorine, nitrile, or benzene moiety, then polyethylene, polypropylene, polyvinyl chloride, polyacrylonitrile, or polystyrene is formed, respectively. If all four constituents

FIG. 12.34 *An example of the anionic polymerization of ethylene is shown below. A nucleophilic reagent attacks a carbon member of the double bond to form a carbanion. A powerful base such as t-butyl-lithium can serve as the anionic initiator.*

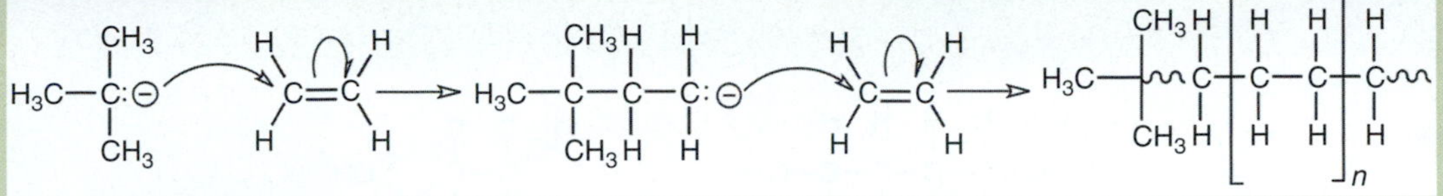

of the ethylene core are substituted with fluorine, then Teflon, poly(tetrafluoroethane), is the product of the polymerization process. There are many more possible types of polymers based on the free radical polymerization of the ethylene class of alkene monomers.

Anionic or cationic molecules are also able to initiate addition polymerization. An example of the anionic polymerization of ethylene is shown in **Figure 12.34.** A nucleophilic reagent attacks a carbon member of the double bond to form a carbanion. A powerful base such as *t*-butyl lithium can serve as the anionic initiator. Since primary carbanions are more stable than tertiary carbanions, the reaction proceeds to add more ethylene and thus propagates the polymer reaction to form poly(ethylene).

In cationic addition, an electrophilic carbocation or other cationic initiator attacks the double bond and creates cationic intermediates during the propagation step of the polymerization process.

Condensation reactions are commonly used to form polyamides like nylon and polyesters. Nylon-66 is formed by condensation between adipic acid and 1,6-hexamethyldiamine. Condensation reactions are responsible for the generation of the peptide linkage between the carboxylate and the amine of two amino acids to form a polypeptide. Poly(butylenesuccinate), or PBS, is formed by a condensation reaction in which water is eliminated (**Fig. 12.35**).

Polymers that are formed in this way, with water as a by-product, may be susceptible to future hydrolysis and subsequent degradation.

12.4.3 Block Copolymers

A copolymer is a macromolecular species that contains two or more monomers. Block copolymers are characterized by relatively long, ordered sequences of repeating units made of immiscible blocks of polymers. Block copolymers can be highly ordered nanostructures (e.g., high-density magnetic storage) or nanostructured without long-range order (e.g., active component in filters) or can form individual nanoscale materials such as vesicles (e.g., vehicles for drug delivery) [94]. Block copolymers are able to form a unique class of templates from soft-matter materials. Because they are able to function as a template, block copolymers are used in lithographic pattern transfer techniques. For polymers to be successful as nanotemplates, they must be ordered into molecular-scale assemblies [95]. Synthesis of diblock polymers requires pure reagents and strict control of reaction conditions. This is because there are no termination

FIG. 12.35 *Condensation reactions are commonly used to form polyamides like nylon and polyesters. Nylon-66 is formed by condensation between adipic acid and 1,6-hexamethyldiamine. Condensation reactions are responsible for the generation of the peptide linkage between the carboxylate and the amine of two amino acids to form a polypeptide. Poly(butylenesuccinate), or PBS, is formed by a condensation reaction in which water is eliminated.*

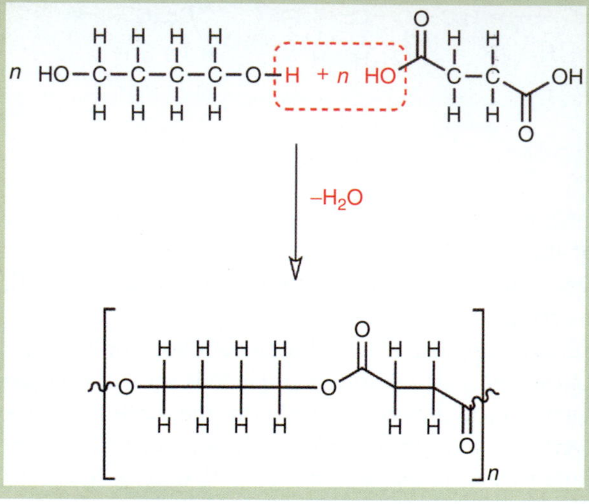

steps involved in the process and all processes (self-assembly, etc.) occur simultaneously [95].

An **A–B** diblock copolymer has the following sequence of monomer **A** and monomer **B** units:

A–A–A–A–A–A–A–A–A–A–A–A–B–B–B–B–B–B–B–B–B–B–B–B

An **A–B–C** triblock copolymer has the following sequence:

A–A–A–A–A–A–A–A–B–B–B–B–B–B–B–B–C–C–C–C–C–C–C–C

Other forms like gradient, graft, and star copolymers also exist. A–B block copolymers are useful templates in preparing spheres and cylinders [94,95]. A–B–C triblock copolymers are applied in synthesis of more complex nanostructures. Polymer chains, before assembly, tend to be on the order of 10 nm in size ($\sim 10^5$ g·mol^{-1})—for example, the radius of gyration, R_g [94,95]. Size and spacing of periodic structures can be controlled by altering the length of the block copolymer subunits or by swelling domains by adding more A or B (**Fig. 12.36**).

An important property of block copolymers is phase separation. There are two types: (1) macrophase separation and (2) microphase separation. Macrophase separation, as the name implies, is undesirable. At the nanoscale, however, microphase separation leads to new properties. Microphase separation occurs with polymers that are dissimilar (immiscible). The size of the microphases is modulated by the balance between enthalpic (larger the better) and entropic (smaller the better) considerations. In diblock systems, laminar structures are favored if **A** and **B** have $f_A/f_B = 50/50$; the gyroid form at $f_A/f_B \sim 40/60$; hexagonal cylinders between 40/60 and 20/80; and *bcc* structures at $f_A/f_B \sim 10/90$. Note that the ratios can be reversed.

FIG. 12.36 *Block copolymer morphology. Left: A–B (spherical, bcc); middle: A–B (cylindrical, hcp); right: A–B–C (lamellar). A third common form, cubic gyroid, is not pictured. A–B lamellar forms result if the packing fraction f is on the order of 50%. For cylindrical block copolymers with f_A = 32%, cylindrical structure is favored. If the temperature is raised, the structure transforms into the lamellar phase depicted in the right (except that it is a diblock copolymer and not the triblock system depicted in the figure).*

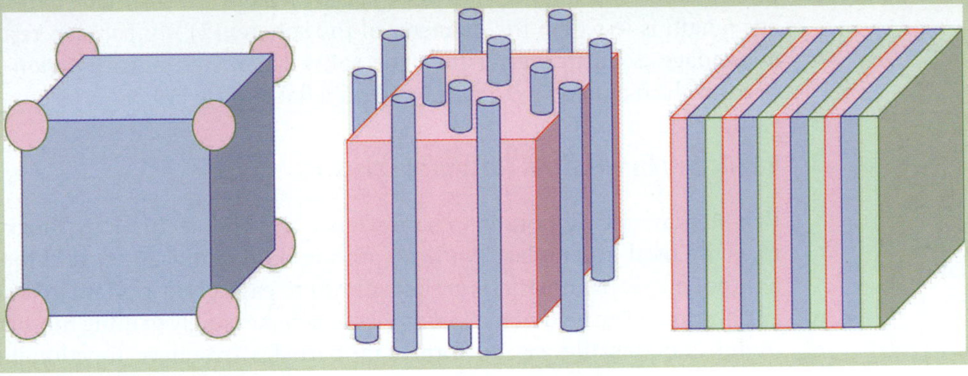

Sources: J. A. Elliott, *Mixtures of polymers*, Presentation, University of Cambridge, www.cus.cam.ac.uk/~jae1001/teaching/materials/M6_Lecture_4.pdf, 2004, and N. P. Balsara and H. Han, in *The chemistry of nanostructured materials*, P. Yang, ed., World Scientific Publishing Co., Ltd., Hackensack, NJ, 2003. With permission.

Block copolymers combine the physical properties of different polymers (homopolymers) and are excellent precursor materials for synthesizing a vast array of nanomaterials [94]. The morphology and properties of the product nanostructured polymeric material depend on polymer component molecular weight, distribution, and fraction present in the block copolymer. Diblock copolymers self-assemble into two phases after prolonged heating above the respective glass transition temperatures. For example, magnetic nanowires can be formed by biblock copolymer polystyrene (PS) and PMMA [96]. Block copolymers of styrene (f = 70%) and methyl methacrylate nanomers (MMA) yielded cylindrical PMMA of 20–40 nm. The diblock copolymer was applied by spin-coating with annealing at 170°C for 24 h to yield a film 100 nm in thickness. The PMMA phase was then decomposed by application of UV light (253.7 nm). Deposition of Co was accomplished by electrodeposition to form the nanowires.

Cobalt nanowires with 14 nm diameter were formed in diblock copolymer template consisting of PS and poly(butadiene) (PB) [94,97]. In this case, a thin film of PS + PB block copolymer was applied by spin-coating technique onto a SiN surface. After annealing, an array of 5-nm PB spheres formed in the PS matrix. Upon contact with ozone, the PB was removed, leaving an array of holes. Reactive ion etching (RIE) was applied to transfer the pattern to the SiN substrate, after which templating synthesis of nanomaterials could proceed by various methods.

Nanoscale vesicles similar to those found in cells can be formed from block copolymers. Poly(ethyleneoxide) (PEO) and poly(ethylethylene) combine to form biological-type vesicles in water. The hydrophobic PEE spontaneously forms the core of the center while the hydrophilic PEO extends into the aqueous medium [98]. Vesicles made of block copolymer are more durable and less permeable to water than those formed from phospholipids that are used in drug delivery [98]. Individual polymers are able to assemble into several forms depending on reaction conditions and polymer constitution. The hydrophobic core can be a single strand that folds in on itself and cross-links between cofolded sections of the strand. Vesicles with projections (appendages consisting of the hydrophilic component of the diblock polymer) from the surface of the hydrophobic core come in three varieties: (1) *hairy nanospheres* are made if the appendage length is less than the diameter of the sphere; (2) *star polymers* result if the appendage is much greater than the sphere diameter; and (3) *brush polymers* occur if short appendages are formed on a flat surface [99].

12.4.4 Emulsion Polymerization

Polymers are very important to nanoscience and nanotechnology. Block copolymers are used in template synthesis among other applications and biodegradable polymeric nanoparticles are becoming increasingly important in biomedical applications. The field of nanocomposites is increasingly gaining momentum. A useful and versatile way to form nanoparticles that have functionality is by emulsion polymerization. Couvreur et al. in 1979 introduced this method of polymerization to design nanoparticles with biodegradable polymers for the in vivo delivery of drugs [100]. Common monomers used in emulsion polymerization include esters like the methacrylates, alkyl-cyanoacrylates, butadiene, styrene, vinyl acetate, acrylic, and methacrylic acids [101]. Copolymers can also be grown by this method.

Latex (from the Latin *latex*, meaning "liquid, fluid") is a fluid that contains a dispersion of polymer particles in water. Latex solutions are traditionally associated with paints and coatings. Several natural latex solutions exist; natural rubber latex is a raw material used for the synthesis of rubber polymers. In 1835, white, milky fluid from plants was designated as latex. Latex is formed from polymerization of monomeric emulsions. Many kinds of polymers can be synthesized by emulsion polymer techniques: poly(vinyl acetate), poly(styrene), poly(chlorophene), and poly(alkyl cyanoacrylate) [102]. Polymerization to form latex particles occurs within dispersed particles suspended in an aqueous medium that are stabilized by surfactants. Latex particles are high molecular weight, spherical, and traditionally in the 100-nm diameter range, but they can be as large as 5 μm in diameter.

Emulsion polymerization involves a monomeric entity, a surfactant, and an aqueous phase that contains an initiator. The technique is a relatively simple process that requires mechanical stirring and temperature control (the beauty and elegance of bottom-up methods!). At the onset, an emulsion of monomer that is slightly soluble in the water medium is introduced [101]. A surfactant (a.k.a. emulsifying agent) is responsible for forming micelles that in turn are responsible for solubilizing monomers [101]. The emulsifying agent is responsible for the overall stability of the monomer, the embryonic polymer, and the polymer product. The polymerization reaction—initiation and the early stages of propagation—occurs inside the monomer-containing micelles.

Initiators form free radicals that seek out monomers for reaction as the polymer grows in size within the ever expanding micelles. Both initiation and propagation take place inside the micelles that provide safe haven for the slightly

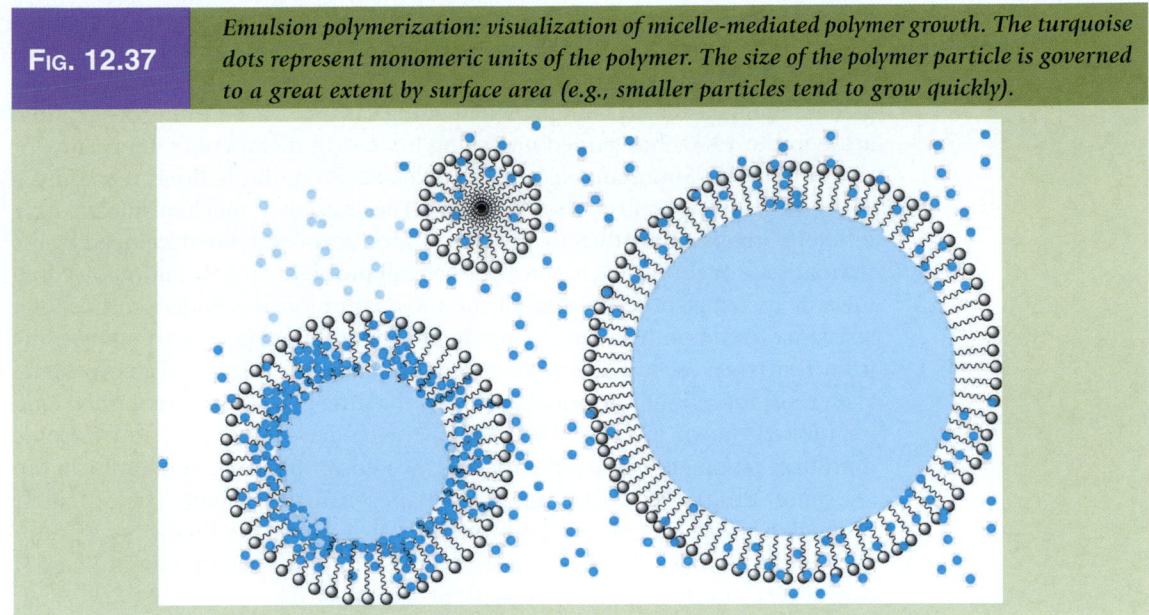

FIG. 12.37 *Emulsion polymerization: visualization of micelle-mediated polymer growth. The turquoise dots represent monomeric units of the polymer. The size of the polymer particle is governed to a great extent by surface area (e.g., smaller particles tend to grow quickly).*

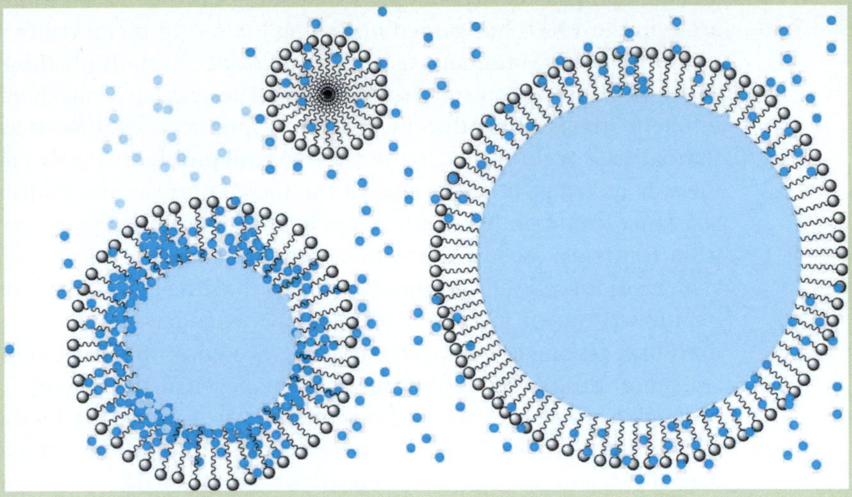

soluble monomer molecules. **Figure 12.37** shows the stages of growth of the polymer inside a micelle by emulsion polymerization. The mechanism proceeds as follows. Not much polymerization occurs in the large droplets (monomer-swollen latex particles) due to the high surface area advantage of smaller micelles. Thus, by statistical factors, droplet size is maximized at some predetermined level.

Polymeric cationic surfactants (a.k.a. polysoaps) are used in polymerization of styrene to form latex microspheres. It was found that the polymerization properties and the eventual properties of the latex balls depend sensitively on the characteristics of the emulsifier and the charge on the initiator [103]. With cationic initiators, for example, all emulsifiers guide the formation of stable monodisperse latexes [103]. In another study, it was found that critical micelle concentration (CMC; usually 10^{18} micelles · mL^{-1}) and average polystyrene particle size increased with decreasing alkyl chain length of the emulsifying agent [104]. As you might recall, the CMC is the equilibrium concentration of an amphiphilic component at which the formation of micelles is initiated. Anionic surfactants include alkyl benzene, sodium vinyl, alpha olefin, and alcohol sulfonates. Non-ionic surfactants (insensitive to pH) include alkyl polysaccharides and amines ethoxylates, alcohol, and alkylphenol, to name just a few. A common initiator used in emulsion polymerization is persulfate (a sulfate peroxide), a water-soluble species that decomposes at low temperatures (50°C) into sulfate radicals.

Poly(Alkylcyanoacrylate) Drug Delivery Vehicles. Poly(alkylcyanoacrylates) (PACCAs) were employed as polymers in the early 1980s. However, the corresponding monomers have been used since at least 1966 because of their excellent adhesive properties resulting from the bonds of high strength they are able to form with most polar substrates, including living tissues and skin. Therefore, the monomers have been used extensively as tissue adhesives for the closure of skin wounds and as surgical glue. More recently, one application of these biodegradable polymers is as nanoparticulate drug carriers. This area of research that arose in the 1980s has gained increasing interest in therapeutics, especially for cancer trea'ents. Simultaneous cellular resistance to multiple drugs represents a major problem in cancer chemotherapy. The resistance mechanism can have different origins: either directly linked to a specific mechanism developed by the tumor tissue or connected to the more general problem of distribution of a drug towards its targeted tissue. One of the most common mechanisms of cellular resistance to chemotherapeutic drugs has been attributed to an active drug efflux from resistance cells linked to the presence of transmembrane P-glycoprotein. With resistant cells, upon entering the cell, the drug binds to P-glycoprotein and is pumped out of the cell. PACA nanoparticles have been used to overcome multidrug resistance at the cellular level by co-encapsulation of doxorubicin (an anticancer drug) and cyclosporin A (a chemosensitizing agent). The extremely promising research in this field now makes PACA nanoparticles the most promising polymer colloidal drug delivery system and they are currently in Phase II clinical trials in the treatment of resistant cancers [131].

The synthesis of a PACA is shown in **Figure 12.38.** An anionic polymerization scheme is depicted. There are many ways to synthesize methacrylate polymers. H. Eerikäinen et al. synthesized methacrylic polymers and drugs by an aerosol flow reactor method [105]. PACA nanoparticles have applications in the treatment of intracellular infections, nonresistant and resistant cancers, delivery of oligonucleotides, peptides, proteins, and vaccines.

One of the most exciting applications of PACA nanoparticles is in the treatment of cancers. Intravenously administered drugs are distributed throughout the body as a function of the physicochemical properties of the molecule. The nonspecificity of the delivery of the drug means that the same pharmacologically active concentration of drug that reaches the tumor also affects the rest of the body. Colloidal drug carriers such as PACA nanoparticles offer a way to address this problem in cancer therapy. PACA nanoparticles are taken up by the liver (macrophages of the mononuclear phagocyte system (MPS), spleen, and, to a lesser extent, by bone marrow after intravenous injection [106]. The uptake of the PACA nanoparticles by the MPS is a nonspecific foreign body response by the host that is based on adsorption of blood proteins and complement activation [100,106]. Polymerization to form PACA follows an anionic mechanism initiated by nucleophiles such as OH^-, CH_3O^-, and CH_3COO^- leading to the formation of nanoparticles of low molecular mass due to rapid polymerization. In an aqueous medium, anionic polymerization can be controlled to produce higher molecular weight nanoparticles.

The *p*H of the solution plays an important part in the reaction; the formation of PACA nanoparticles above *p*H of 3 is not possible, most likely due to aggregation. Lowering the *p*H with a mineral acid such as HCl and adjusting the concentration of the anionic polymerization inhibitor (SO_2) in the monomer [106], along with using stabilizers (dextran-70, dextran-40, dextran-10, polyoxamer-18, -184, -237, etc.) and surfactants (polysorbate-20, -40, or -80), can be used to control particle size and molecular mass of the nanoparticles [107]. The concentration of the monomer and the speed of stirring are also important parameters in nanoparticle formation. The size of the nanoparticles varies from 50 to 300 nm [107–109].

Fig. 12.38

Poly(alkylcyanoacrylate) (PACA) can be used as tissue adhesives—drug carriers that are able to overcome multidrug resistance. Three different mechanisms for polymerization exist: free radical, anionic, and zwitterionic. Free radical mechanism is determined by high activation energy (125 kJ·mol^{-1}*). The polymerization is very slow and the reaction rate depends strongly on the temperature and the quantity of radicals. Anionic and zwitterionic techniques are more rapid and easier to handle and are better suited for biomedical applications. Propagation of polymerization occurs after formation of carbanions that are able to react with another monomer molecule leading to the formation of living polymer chains. Termination is due to the presence of action that leads to that end of the polymerization. Anionic chain terminators include* O_2*,* CO_2*,* H_2O*, and HCl.*

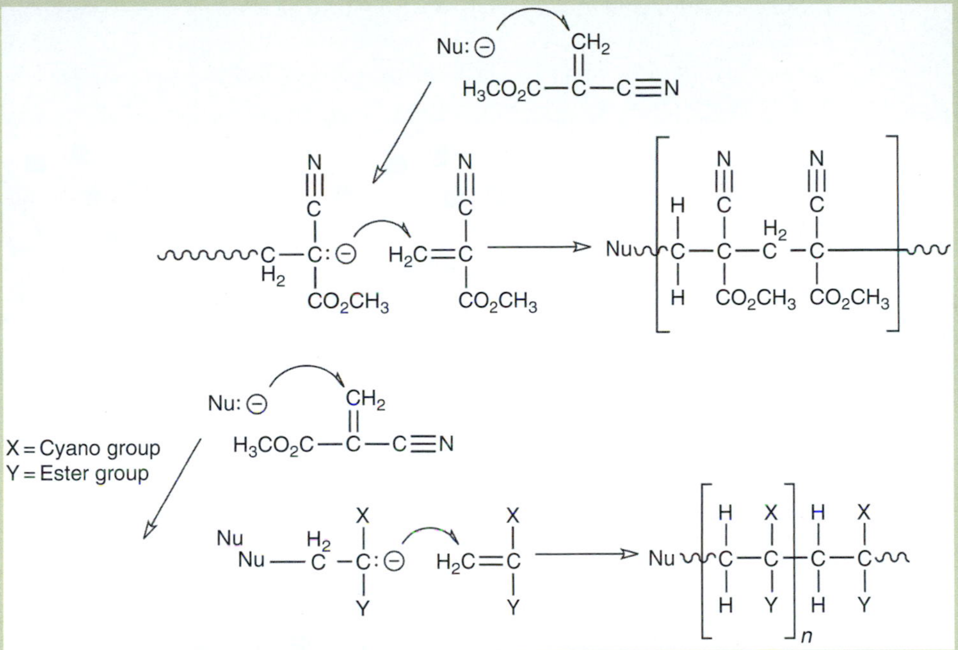

Source: R. D'Sa, *Poly(alkylcyanoacrylates): Biodegradable Polymeric Nanoparticles for Biomedical Applications*, U. of Colorado at Denver, 2004. With permisssion.

A generalization of this procedure involves: 100 µL of the monomer is dispersed in a *p*H 2.5 solution (adjusted with HCl) containing 10 mL of a 1% solution of dextran-70 [110]. The solution is allowed to polymerize for 3–4 h under strong magnetic stirring. The addition of drugs to this mixture will entrap them in the nanoparticle [111,112]. A generalized synthesis is shown in **Figure 12.39.**

Biodegradation of PACA Nanoparticles. In order for a polymeric drug delivery system to be suitable in vivo, the material must be biocompatible and biodegradable or easily excreted (e.g., by the kidneys). In nature, enzymes secreted by fungi, bacteria, and the digestive ingredients that reside within mammalian stomachs are able to effect biodegradation reactions. Many kinds of synthetic biodegradable polymers have been invented. The mechanism for PACA degradation depends on the local environment. Two main mechanisms exist and are shown in **Figure 12.40.** The first and predominant mechanism involves

FIG. 12.39

Emulsion polymerization is a very popular approach used to synthesize polymer colloids. Introduced in 1979 to design nanoparticles with biodegradable polymers for the in vivo delivery of drugs, polymerization is initiated by the hydroxyl ions of water and elongation of the polymer chain occurs according to an anion polymerization mechanism. To produce higher molecular mass as well as stable nanoparticles, polymerization must be carried out in an acidic medium (pH 1.0–3.5). After dispersing the monomer in an aqueous acidic medium containing surfactant and stabilizer, polymerization is continued for 3–4 h by increasing the pH of the medium to obtain the desired products. Particle size and molecular mass of nanoparticles depend upon the type and concentration of the stabilizer and/or surfactant used. The size and molecular mass of nanoparticles depend upon the pH of the polymerization medium, but nanoparticle production is not possible above a pH of 3.0, probably due to the aggregation and stepwise molecular mass increase at lower pH. Other factors that influence the formation of nanoparticles include the concentration of monomer and the speed of stirring.

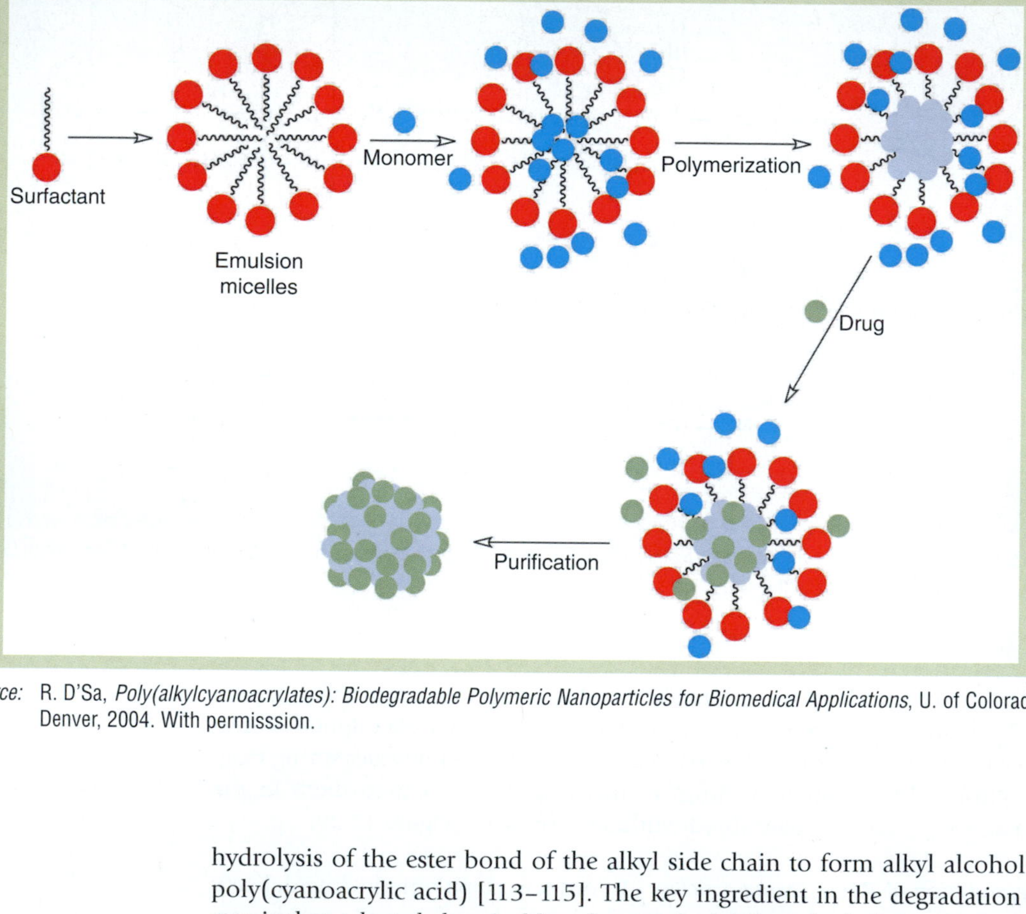

Source: R. D'Sa, *Poly(alkylcyanoacrylates): Biodegradable Polymeric Nanoparticles for Biomedical Applications*, U. of Colorado at Denver, 2004. With permisssion.

hydrolysis of the ester bond of the alkyl side chain to form alkyl alcohol and poly(cyanoacrylic acid) [113–115]. The key ingredient in the degradation process is that selected chemical bonds must be labile to hydrolysis. For example, base-catalyzed hydrolysis of polyester is initiated by the nucleophilic attack on a carbonyl group by a hydroxyl group. Both degradation products are water soluble and are therefore easily excreted by the kidneys. The esterases found in serum, lysosymes, and pancreatic juice [116,117] catalyze this reaction in vivo within a few hours, depending on the length of the alkyl chain [114,118,119].

FIG. 12.40 *Enzymatic degradation catalyzed by esterases leading to the production of alkylalcohol and poly(cyanoacrylic acids) is shown. One of the degradation mechanisms described in the literature consists of the hydrolysis of the ester bond of the alkyl side chain of the polymer. Degradation products consist of an alkylalcohol and poly(cyanoacrylic acid), which are soluble in water and can be eliminated in vivo via kidney filtration. This degradation has been shown to be catalyzed by esterases from serum, lysosomes, and pancreatic juice and is believed to occur as the major degradation pathway in vivo. According to this mechanism, nanoparticles are usually degraded within a couple of hours depending on the alkyl side chain length of the PACA forming the nanospheres.*

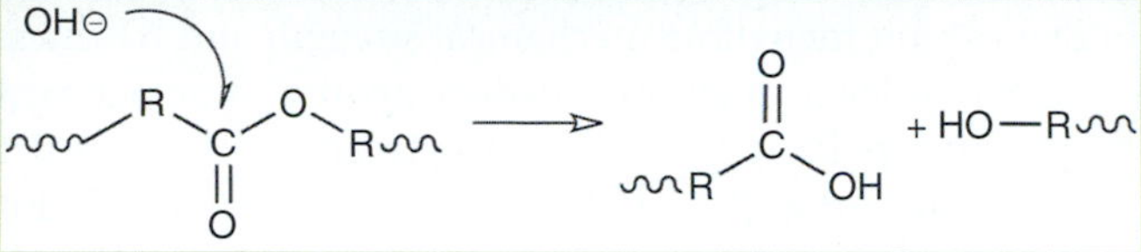

Source: R. D'Sa, *Poly(alkylcyanoacrylates): Biodegradable Polymeric Nanoparticles for Biomedical Applications*, U. of Colorado at Denver, 2004. With permisssion.

The second mechanism is an unzipping depolymerization followed by an immediate repolymerization to give a lower molecular weight polymer (oligomers). This base-catalyzed reaction [120] is thought to be initiated by amino acids of proteins. Since this reaction takes place very fast it is difficult to observe in complex biological systems and therefore has never been described. The alkylcyanoacrylate moieties are recovered, and this process continues until the polymer is consumed (**Fig. 12.41**).

12.4.5 Nanocomposites

A composite (from the Latin *compositus*, "to put together" (*com*) + *ponere*, "to place") is a material that is made up of various components. In other words, it is not a homogeneous material. The term is often applied to construction materials. Graphite fiber epoxy composites developed in the late 1970s revolutionized airframe and wing designs. Nanogold deposited in the pores of porous alumina membranes is a composite material.

There are many kinds of composites that contain nanomaterials. Concrete, for one, is an age-old bulk material composite that contains aggregate and

FIG. 12.41 *Unzipping depolymerization–repolymerization mechanism producing oligomers of PACA [110]. The alkylcyanoacrylate moieties are recovered, and this process continues until the polymer is consumed.*

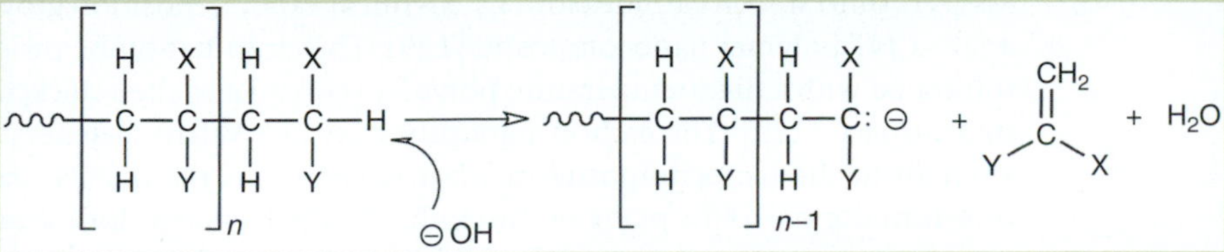

cement binder (10–100 μm). Cement is formed by grinding clay and calcined limestone into a fine powder. The performance of concrete has recently been shown to improve dramatically with the addition of SiO_2, fly ash, TiO_2, and Fe_2O_3 nanoparticles [121,122]. Addition of nanoparticles enhances the mechanical properties of concrete. Nanoparticles are able to occupy voids between cement grains, accelerate hydration and improve bonding between aggregates, improve toughness, shear, tensile and flexural strength, and enhance crack arrest. Experts are predicting that the concrete industry will be dramatically impacted by such nanotechnological enhancements.

Nature has developed complicated structures with nanostructured materials that exhibit incredible mechanical strength and hardness. For example, biominerals (calcite, aragontie, hydroxyapatite) interact with biomacromolecules (proteins, polysaccharides, proteoglycans) to form structures with hierarchical long-range order [123,124]. If you look inside an abalone shell you will see a pearly iridescent layer called nacre (mother of pearl), nature's incredible armor. Although the components are brittle (but deformable), the shell exhibits a robust structure [125]. Thin layers (5–20 nm) of elastic organic biopolymeric materials separate hexagonal platelets of aragonite ($CaCO_3$). The dimensions of the aragonite platelets are 10–20 μm by 200–500 nm thick. The lamellar composite has a 2-fold advantage in strength and 1000-fold advantage in toughness over the individual material components [125]. The iridescence is due to interference colors produced by the interaction of the 500-nm thick platelets with light. Nature puts this nanocomposite to use in the form of the mussel. Attempts to mimic this structure are ongoing [126]. Synthetic efforts that try to mimic the bricklike ultrastructure are in abundant supply.

Carbon Nanotube–Polymer Composites. Nanocomposite based on carbon nanotube reinforced polymers shows great potential. We now discuss the science of composites that utilize chemically modified carbon nanotubes for structural reinforcement and electrical, optical, and thermal enhancement. From chapter 9 we learned about several ways to chemically modify carbon nanotubes [127,128].

There are two generalized approaches for incorporating carbon nanotubes into a polymer matrix. One is to use aligned arrays, usually of the multi-walled carbon nanotube (MWNT) type, grown by chemical vapor deposition (CVD) that are millimeters in height [128]. The array can be impregnated with polymers in order to preserve the alignment. Secondly, use of loose, unaligned nanotubes of high purity, primarily of the SWNT type, can be applied towards multifunctional polymer composites, membranes, coatings, and fiber fabrication [128].

Polymer grafting discussed earlier is an effective way to place carbon nanotubes within a polymeric matrix. In a reversal of roles, researchers have grafted MWNTs onto silicon carbide cloth by chemical vapor deposition growth to form a SiC–CNT polymer nanocomposite [129]. The cloth-nanotube product can be infiltrated with a high-temperature polymer (epoxy) and then stacked and cured in a furnace [129]. The carbon nanotubes act like Velcro fasteners and afford strength to the nanocomposite in what is known as the *through thickness*, the direction normal to the plane of the cloth. The CNT-treated cloth shows fourfold improvement in fracture tests, fivefold improvement in energy dissipation by structural damping and structural energy, and a threefold increase in dimensional stability when compared to control materials [129]. Thermal and electrical conductivity performance was also significantly enhanced [129].

Single-walled carbon nanotubes have also been successfully incorporated within polymer matrices (Nylon-6,6, formed from two six-carbon precursors) to form nanocomposites by a process called interfacial in situ polymerization [130]. Interfacial polymerization of the polyamide Nylon-6,6 takes place, as the name indicates, between the interfacial layer of an aqueous phase and an organic (toluene) phase. Within the organic phase, the condensation monomer precursor is adipoyl chloride and within the aqueous phase the complementary precursor is 1,6-hexamethylene diamine, both in equimolar proportions to produce the neat nylon polymer.

Carboxylic acid-functionalized SWNTs are subjected to the same polymerization procedure as described before. Purified neat SWNTs, purified functionalized SWNTs, and NaDDBS-stabilized (sodium dodecylbenzene sulfonate surfactant) SWNTs were tested. The only SWNT species that showed good dispersion in both the solvent and the composite were the ones that were chemically functionalized.

References

1. T. Sugimoto, K. Sakata, and A. Muramatsu, Formation mechanism of monodisperse pseudocubic a-Fe_2O_3 particles from condensed ferric hydroxide gel, *Journal of Colloid Interface Science*, 159, 372–382 (1993).
2. V. R. Reddy, Gold nanoparticles: Synthesis and applications, *Synlett*, 11, 1791–1792 (2006).
3. G. Cao, *Nanostructures & nanomaterials: Synthesis, properties & applications*, Imperial College Press, London (2004).
4. G. M. Whitesides and M. Boncheva, Beyond molecules: Self-assembly of mesoscopic and nanoscopic components, *Proceedings of the National Academy of Science*, 99, 4769–4774 (2002)
5. J. C. Love, L. A. Estroff, J. K. Kriebel, R. G. Buzzo, and G. M. Whitesides, Self-assembled monolayers of thiolates on metals as a form of nanotechnology, *Chemical Reviews*, 105, 1103–1169 (2005).
6. J. L. Wilbur, A. Kumar, H. A. Biebuyck, E. Kim, and G. M. Whitesides, Microcontact printing of self-assembled monolayers: Applications in nanofabrication, *Nanotechnology*, 7, 452–457 (1996).
7. C. H. Giles, *The origins of surface film balance: Studies in the early history of surface chemistry*, Part 3, Society of Chemical Industry, January, 43–53 (1971).
8. Tapani Viitali, KSV Instruments, Inc., Finland.
9. N. Tillman, A. Ulman, and T. L. Penner, Layered metal phosphates and phosphonates: From crystals to monolayers, *Langmuir*, 5, 101–111 (1989).
10. N. K. Adam, *The physics and chemistry of surfaces*, 3rd ed., Oxford University Press, London (1941).
11. L. Cristofolini, T. Berzina, S. Erokhina, O. Konovalov, and V. Erokhin, Structural study of the DNA dipalmitoylphosphatidylcholine complex at the air–water interface, *Biomacromolecules*, 8, 2270–2275 (2007).
12. S. Pradhan, D. Ghosh, L.-P. Xu, and S. Chen, Interparticle charge transfer mediated by π–π stacking of aromatic moieties, JACS Communications, *Journal of the American Chemical Society*, 129, 10622–10623 (2007).

13. R. L. Donkers, Y. Song, and R. W. Murray, Substituent effects on the exchange dynamics of ligands on 1.6 nm diameter gold nanoparticles, *Langmuir*, 20, 4703–4707 (2004).
14. L. Salem, Attractive forces between long saturated chains at short distances, *Journal of Chemical Physics*, 37, 2100–2113 (1962).
15. A. Ulman, *An introduction to ultrathin organic films: From Langmuir–Blodgett to self-assembly*, Academic Press, Boston (1991).
16. R. G. Nuzzo, B. R. Zegarski, and L. H. L. Dubois, Fundamental studies of the chemisorption of organosulfur compounds on gold(111)—Implications for molecular self-assembly on gold surfaces, *Journal of the American Chemical Society*, 109, 733–740 (1987).
17. I. Engquist, Self-assembled monolayers, Department of Physics, Linköping University, Sweden, www.ifm.liu.se/applphys/ftir/sams.html, (1996).
18. G. Scoles, Physisorption and chemisorption of alkanethiols and alkylsulfides on Au(111), Princeton University, www.princeton.edu/~gscoles/theses/sean/chapter4.html (2006).
19. G. Schmid and B. Corain, Nanoparticulated gold: Synthesis, structures, electronics and reactivities, *European Journal of Inorganic Chemistry*, 3081–3098 (2003).
20. G. Schmid, R. Pfeil, R. Böse, F. Banderman, S. Meyer, G. H. M. Calis, and J. W. A. van der Velden, $Au_{55}[P(C_6H_5)_3]_{12}Cl_6^-$ ein Goldcluster ungewöhnlicher Größe, *Chemische Berichte*, 114, 3634 (1981).
21. G. Schmid, Metal nanoclusters as quantum dots. In *Encyclopedia of nanoscience and nanotechnology*, H. S. Nalwa, ed., American Scientists Publisher, 5, 387–398 (2004).
22. G. Schmid, E. Emmrich, J.-P. Majoral, and A.-M. Caminade, The behavior of Au_{55} nanoclusters on and in thiol-terminated dendrimer monolayers, *Small*, 1, 73–75 (2005).
23. G. Schmid, R. Pugin, W. Meyer-Zaika, and U. Simon, Clusters on clusters: Closo-dodecaborate as a ligand for Au55 clusters, *European Journal of Inorganic Chemistry*, 2051–2055 (1999).
24. G. Hornyak, M. Kröll, R. Pugin, T. Sawitowski, G. Schmid, J.-O. Bovin, G. Karsson, H. Hofmeister, and S. Hopfe, Gold clusters and colloids in alumina nanotubes, *Chemistry: A European Journal*, 3, 1951–1956 (1997).
25. G. L. Hornyak, S. Peschel, T. Sawitowski, and G. Schmid, TEM, STEM and AFM as tools to study clusters and colloids, *Micron*, 29, 183–190 (1998).
26. G. Schmid and G. L. Hornyak, Metal clusters: New perspectives in future nanoelectronics, *Current Opinions in Solar State and Materials Science*, 2, 204, (1997).
27. G. Schmid, R. Pugin, J.-O. Malm, and J.-O. Bovin, Silsesquioxanes as ligands for gold clusters, *European Journal of Inorganic Chemistry*, 1998, 813–817 (1998).
28. J. Turkevitch, P. C. Stevenson, and J. Hillier, Nucleation and growth process in the synthesis of colloidal gold, *Discussions of the Faraday Society*, 11, 55 (1951).
29. Colloidal gold, en.wikipedia.org/wiki/Colloidal_gold, October (2007).
30. A. Badia, S. Singh, L. Demers, L. Cuccia, G. R. Brown, and R. B. Lennox, Self-assembled monolayers on gold monolayers, *Chemistry: A European Journal*, 2, 359–363 (1996).
31. N. Sandhyarani, T. Pradeep, J. Chakrabarti, M. Yousuf, and H. K. Sahu, Distinct liquid phase in metal-cluster superlattice solids, *Physics Reviews B*, 62, R739 (2000).
32. M. P. Pileni, ed., *Reactivity in reverse micelles*, Elsevier, Amsterdam (1989).
33. M. P. Pileni, Semiconductor nanocrystals. In *Nanoscale materials in chemistry*, K. J. Klabunde, ed., Wiley-Interscience, New York (2001).
34. S. M. Emin, C. D. Dushkin, S. Nakabayashi, and E. Adachi, Growth kinetics of CdSe nanoparticles synthesized in reverse micelles using *bis*(trimethylsilyl) selenium precursor, *Central European Journal of Chemistry*, 5, 590–604 (2007).
35. A. N. Shipway, E. Katz, and I. Willner, Nanoparticle arrays on surfaces for electronic, optical and sensor applications, *ChemPhysChem*, 1, 18–52 (2000).

36. M. L. Steigerwald, A. P. Alivisatos, J. M. Gibson, T. D. Harris, R. Kortan, A. J. Muller, A. M. Thayer, T. M. Duncan, and L. E. Brus, Surface derivatization and isolation of semiconductor cluster molecules, *Journal of the American Chemical Society*, 110, 3046–3050 (1988).
37. D. Ganguli and M. Ganguli, *Inorganic particle synthesis via macro- and microemulsions: A micrometer to nanometer landscape*, Kluwer Academic Press/Plenum Publishers (Springer), New York (2003).
38. G. L. Hornyak, L. Grigorian, A. C. Dillon, P. A. Parilla, and M. J. Heben, A temperature window for chemical vapor deposition growth of single walled carbon nanotubes, *Journal of Physical Chemistry B*, 106, 2821–2825 (2002).
39. S. B. Bukallah, M. Houalla, and D. M. Hercules, Characterization of supported Nb catalysts by ToF-SIMS, *Surface Interface Analysis*, 29, 818–822 (2000).
40. R. Kelsall, I. Hamley, and M. Geoghegan, *Nanoscale science and technology*, John Wiley & Sons, Ltd., Chichester (2005).
41. I. W. Hamley, *Introduction to soft matter*, John Wiley & Sons, Ltd., Chichester (2000).
42. M. Vondrova, T. Klimczuk, V. L. Miller, B. W. Kirby, N. Yao, R. J. Cava, and A. B. Bocarsly, Supported superparamagnetic Pd/Co alloy nanoparticles prepared from silica/cyanogel Co-Gel, *Chemical Materials*, 17, 6216–6218 (2005).
43. R. Leon, D. Margolese, G. Stucky, and P. M. Petroff, Nanocrystalline Ge filaments in the pores of a mesosilicate, *Physics Review B*, 52, R2285–R2288 (1995).
44. T. Scheibel, R. Parthasarathy, G. Sawicki, X.-M. Lin, H. Jaeger, and S. L. Lindquist, Conducting nanowires built by controlled self-assembly of amyloid fibers and selective metal deposition, *Proceedings of the National Academy of Science*, 100, 4527–4532 (2003).
45. J. Chai, D. Wang, X. Fang, and J. M. Buriak, Assembly of aligned linear metallic patterns on silicon, *Nature Nanotechnology*, 2, 500–506 (2007).
46. C. Wang, Y. Hou, J. Kim, and S. Sun, A general strategy for synthesizing FePt nanowires and nanorods, *Angewandte Chemie International Edition*, 46, 6333–6335 (2007).
47. H. Gerung, T. J. Boyle, L. J. Tribby, S. D. Bunge, C. J. Brinker, and S. M. Han, Solution synthesis of germanium nanowires using a Ge^{2+} alkoxide precursor, *Journal of the American Chemical Society*, 128, 5244–5250 (2006).
48. P. Nickels, W. U. Dittmer, S. Beyer, J. P. Kotthaus, and F. C. Simmel, Polyaniline nanowire synthesis templated by DNA, *Nanotechnology*, 15, 1524–1529 (2004).
49. M. Ye, H. Zhong, W. Zheng, R. Li, and Y. Li, Ultralong cadmium hydroxide nanowires: Synthesis, characterization, and transformation in CdO nanostrands, *Langmuir*, 23, 9064–9068 (2007).
51. C. A. Foss, Jr., G. L. Hornyak, J. A. Stockert, and C. R. Martin, Optical properties of composite membranes containing arrays of nanoscopic gold cylinders, *Journal of Physical Chemistry*, 96, 7497–7499 (1992).
52. C. A. Foss, Jr., M. J. Tierney, and C. R. Martin, Template-synthesis of infrared-transparent metal microcylinders: Comparison of optical properties with the predictions of effective medium theory, *Journal of Physical Chemistry*, 96, 9001–9007 (1992).
53. C. J. Brumlik, C. R. Martin, and K. Tokuda, Microhole array electrodes based on microporous alumina membranes, *Analytical Chemistry*, 64, 1201–1203 (1992).
54. C. A. Foss, Jr., G. L. Hornyak, J. A. Stockert, and C. R. Martin, Template synthesis and optical properties of small metal particle composite materials: Effects of particle shape and orientation on plasmon resonance maxima, *Proceedings of Material Research Society Symposium on Nanometals*, Boston, MA, 286, 431–436 (1993).
55. C. A. Foss, Jr., G. L. Hornyak, J. A. Stockert, and C. R. Martin, Optically transparent nanometal composite membranes, *Advanced Materials*, 5, 135–136 (1993).

56. C. R. Martin, Nanomaterials—A membrane-based synthetic approach, *Science*, 266, 1961–1966 (1994).
57. C. A. Foss, Jr., G. L. Hornyak, J. A. Stockert, and C. R. Martin, Template-synthesized nanoscopic gold particles: Optical spectra and the effects of particle size and shape, *Journal of Physical Chemistry*, 98, 2963–2971 (1994).
58. G. L. Hornyak, K. L. N. Phani, D. L. Kunkel, V. P. Menon, and C. R. Martin, Fabrication, characterization and optical theory of aluminum nanometal/nanoporous membrane thin film composites, *Proceedings of the 2nd International Conference on Nanostructured Materials*, H.-E. Schaefer, R. Würschum, H. Gleiter, and T. Tsakalokos, eds., 6, 839–842 (1995).
59. G. L. Hornyak and C. R. Martin, Optical properties of a family of Au-nanoparticle-containing alumina membranes in which the nanoparticle shape is varied from needle-like (prolate) to spheroid, to pancake-like (oblate), *Thin Solid Films*, 300, 84–88 (1997).
60. G. L. Hornyak, C. J. Patrissi, C. R. Martin, J-C.Valmalette, J. Dutta, and H. Hofmann, Dynamical Maxwell-Garnett optical modeling of nanogold-porous alumina composites: Mie and kappa influence on absorption maxima, *Nanostructured Materials*, 9, 575–578 (1997).
61. G. L. Hornyak, C. J. Patrissi, and C. R. Martin, Fabrication, characterization and optical properties of gold-nanoparticle/porous-alumina composites: The non-scattering Maxwell-Garnett limit, *Journal of Physical Chemistry B*, 101, 1548–1555 (1997).
62. J. C. Hulteen, C. J. Patrissi, D. L. Miner, E. R. Crosthwait, E. B. Oberhauser, and C. R. Martin, Changes in the shape and optical properties of gold nanoparticles contained within alumina membranes due to low temperature annealing, *Journal of Physical Chemistry B*, 101, 7727–7731 (1997).
63. G. L. Hornyak, *Characterization and optical theory of nanometal/porous alumina composite membranes*, Ph.D. dissertation, Colorado State University (1997).
64. A. Streitwieser and C. H. Heathcock, *Introduction to organic chemistry*, Macmillan Publishing Co., New York (1985).
65. N. Tillman, A. Ulman, J. S. Schildkraut, and T. L. Penner, Incorporation of phenoxy groups in self-assembled monolayers of trichlorosilane derivatives: Effects on film thickness, *Journal of the American Chemical Society*, 110, 6136–6144 (1988).
66. J. A. Preece, The micro–nano device, presentation, University of Birmingham (2004).
67. T. Zhu, X. Fu, T. Mu, J. Wang, and Z. Liu, pH-dependent adsorption of gold nanoparticles on aminothiophenol-modified gold substrates, *Langmuir*, 15, 5197–5199 (1999).
68. N. Wade, T. Pradeep, J. Shen, and R. G. Cooks, Covalent chemical modification of self- assembled fluorocarbon monolayers by low-energy $CH_2Br_2^+$ ions: A combined ion/surface scattering and x-ray photoelectron spectroscopic investigation, *Rapid Communications in Mass Spectrometry*, 13, 986–993 (1999).
69. L. D. Seger, J. P. Rasimas, R. Pesce-Rodriguez, and R. Fifer, Chemical modification and attempted polymerization of self-assembled monolayers of hexadecanedioic acid at aluminum surfaces, Army Research Lab, www.stormingmedia.us/21/2171/A217133.html (1997).
70. A. C. Buchanan, III, M. K. Kidder, P. F. Britt, Z. Zhang, and S. Dai, Effects of pore confinement on the pyrolysis of 1,3-diphenylpropane in mesoporous silica, Oak Ridge National Laboratory, www.ornl.gov/~webworks/cppr/y2001/pres/115857.pdf (2001).
71. S. Polarz, Chemistry in confined spaces, presentation, ERA Chemistry, Workshop Mainz Bochum, www.erachemistry.net/index/file/158 (2005).
72. J. Sivaguru, H. Saito, M. R. Solomon, L. S. Kaanumalle, T. Poon, S. Jockusch, W. Adam, V. Ramanurthy, Y. Inoue, and N. J. Turro, The control of chirality by cations

in confined spaces: Photooxidation of enecarbamates inside zeolite supercages, *Photochemistry and Photobiology*, 86, 123–131 (2006).
73. R. Anwander, SOMC@PMS: Surface organometallic chemistry at periodic mesoporous silica, *Chemical Materials*, 13, 4419–4438 (2001).
74. Carbon nanotube membranes and adsorbants, private communication, M. J. Heben, National Renewable Energy Laboratory (2001).
75. Periodic minimal surfaces, University of Cambridge, Jacek Klinowski Group, www-klinowski.ch.cam.ac.uk/jkhome.htm (2002).
76. G. Donnay and D. L. Pawson, X-ray diffraction studies of echinoderm plates, *Science*, 166, 1147–1150 (1969).
77. H. U. Nissen, Crystal orientation and plate structure in echinoid skeletal units, *Science*, 166, 1150–1152 (1969).
78. S. Andersson, S. T. Hyde, K. Larsson, and S. Lidin, Minimal surfaces and structures: From inorganic and metal crystals to cell membranes and bio-polymers, *Chemical Reviews*, 88, 221–242 (1988).
79. V. Alfredsson, M. W. Anderson, T. Ohsuna, O. Terasaki, M. Jacob, and M. Bojrup, Cubosome description of the inorganic mesoporous structure MCM-48, *Chemical Materials*, 9, 2066–2070 (1997).
80. J. Charvolin and J. F. Sadoc, Cubic phases as structures of disclinations, *Colloid Polymer Science*, 268, 190–195 (1990).
81. Y. Xia, Y. Lu, K. Kamata, B. Gates, and Y. Yin, Macroporous materials containing three dimensionally periodic structures. In *The chemistry of nanostructured materials*, P. Yang, ed., 69–100, World Scientific Publishing Co. Pte. Ltd., Singapore (2003).
82. B. T. Holland, C. F. Blanford, and A. Stein, Synthesis of macroporous minerals with highly ordered three-dimensional arrays of spherical voids, *Science*, 281, 538–540 (1998).
83. R. C. Schroden, M. Al-Daius, S. Sokolov, B. J. Melde, J. C. Lytle, A. Stein, M. C. Carbajo, J. T. Fernandez, and E. E. Rodriguez, Hybrid macroporous materials for heavy metal ion adsorption, *Journal of Materials Chemistry*, 12, 3261–3267 (2002).
84. H. Asoh, K. Nishio, M. Nakao, T. Tamamura, and H. Masuda, Conditions for fabrication of ideally ordered anodic porous alumina using pretextured Al, *Journal of the Electrochemical Society*, 148, B152–B156 (2001).
85. P. G. Mingalyov, M. V. Buchnev, and G. V. Llisichkin, Chemical modification of alumina and silica with alkylphosphonic acids and their esters, *Russian Chemical Bulletin*, 50, 1693–1695 (200).
86. V. M. Cepak, J. C. Hulteen, G. Che, K. B. Jirage, B. B. Lakshmi, E. R. Fisher, and C. R. Martin, Fabrication and characterization of concentric tubular composite micro- and nanostructures using the template synthesis method, *Journal of Materials Research*, 13, 3070–80 (1998).
87. G. Che, B. B. Lakshmi, C. R. Martin, E. R. Fisher, and R. A. Ruoff, Chemical vapor deposition based synthesis of carbon nanotubules and nanofibers using a template method, *Chemical Matererials*, 10, 260–267 (1998).
88. L. Mercier and T. J. Pinnavaia, Direct synthesis of hybrid organic–inorganic nanoporous silica by a neutral amine assembly route: Structure–function control by stoichiometric incorporation of organosiloxane molecules, *Chemical Materials*, 12, 188–196 (2000).
89. S. Inagaki, S. Guan, T. Ohsuna, and O. Terasaki, An ordered mesoporous organosilica hybrid material with a crystal-like wall structure, *Nature*, 416, 304–307 (2002).
90. M. Yates, K. Ott, E. Birhbaum, and T. McCleskey, Hydrothermal synthesis of molecular sieve fibers: Using microemulsions to control crystal morphology, *Angewandte Chemie International Edition*, 41, 476–478 (2003).

91. N. J. Turro, Molecular structure as a blueprint for supramolecular chemistry in confined spaces, *PNAS,* 102, 10766–10770 (2005).
92. C. T. Kresge, M. E. Leonowicz, W. J. Roth, J. C. Vartuli, and J. S. Beck, Ordered mesoporous molecular sieves synthesized by a liquid crystal template mechanism, *Nature,* 359, 710–712 (1992).
93. J. D. Roberts, R. Stewart, and M. J. Casserio, *Organic chemistry,* W. A. Benjamin, Inc., New York (1971).
94. N. P. Balsara and H. Han, Block copolymers in nanotechnology. In *The chemistry of nanostructured materials,* P. Yang, ed., World Scientific Publishing Co., Ltd., Hackensack, NJ (2003).
95. P. Yang, ed., *The chemistry of nanostructured materials,* World Scientific Publishing Co., Pte., Ltd., Singapore (2003).
96. V. V. Warke, M. G. Bakker, D. E. Nikles, J. Mays, and P. Britt, Block copolymer nanolithography for the preparation of patterned magnetic recording media, poster, MINT Center, University of Alabama, Fall (2005).
97. A. Urbas, R. Sharp, Y. Fink, E. L. Thomas, M. Xenidou, and L. J. Fetters, Tunable block-copolymer–homopolymer photonic crystals, *Advanced Materials,* 12, 812–814 (2000).
98. B. M. Discher, Y. Won, D. S. Ege, J. C. Lee, F. S. Bates, D. E. Discher, and D. A. Hammer, Tough vesicles made from diblock copolymers, *Science,* 284, 1143–1146 (1999).
99. C. P. Poole and F. J. Owens, *Introduction to nanotechnology,* Wiley-Interscience, John Wiley & Sons, Inc., Hoboken, NJ (2003).
100. P. Couvreur, B. Kante, M. Roland, P. Guiot, P. Bauduin, and P. Speiser, Polycyanoacrylate nanocapsules as potential lysosomotropic carriers: Preparation, morphological and sorptive properties, *Journal of Pharmacy and Pharmacology,* 31, 331–332 (1979).
101. H. Y. Erbil, *Surface chemistry of solid and liquid interfaces,* Blackwell Publishing, Oxford (2006).
102. M. E. Karaman, L. Meagher, and R. M. Pashley, Surface chemistry of emulsion polymerization, *Langmuir,* 9, 1220–1227 (1993).
103. D. Cochin and A. Lascheqsky, Emulsion polymerization of styrene using conventional, polymerizable, and polymeric surfactants. A comparative study, *Macromolecules,* 30, 2278–2287 (1997).
104. S. Demharter, W. Richtering, and R. Mülhaupt, Emulsion polymerization of styrene in the presence of carbohydrate-based amphiphiles, *Polymer Bulletin,* 34, 271–277 (1995).
105. H. Eerikäinen, *Preparation of nanoparticles consisting of methacrylic polymers and drugs by an aerosol flow reactor method,* VTT Publications 563, Finland (2005).
106. F. Lescure, C. Zimmer, D. Roy, and P. Couvreur, Optimization of polycyanoacrylate nanoparticle preparation: Influence of sulfur dioxide and pH on nanoparticle characteristics, *Journal of Colloid Interface Science,* 154, 77–86 (1992).
107. S. J. Douglas, L. Illum, and S. S. Davis, Particle size and size distribution of poly(butyl 2-cyanoacrylate) nanoparticles. II. Influence of stabilizers, *Journal of Colloid Interface Science,* 103, 154–163 (1985).
108. M. J. Alonso, A. Sanchez, D. Torres, B. Seijo, and J. L. Vila-Jato, Joint effect of monomer and stabilizer concentrations on the physico-chemical characteristics of poly(butyl-2-cyanoacrylate) nanoparticles, *Journal of Microencapsululation,* 7, 517–526 (1990).
109. B. Seijo, E. Fattal, L. Roblot-Treupel, and P. Couvreur, Design of nanoparticles of less than 50 nm diameter: Preparation, characterization and drug loading, *International Journal of Pharmacy,* 62, 1–7 (1990).

110. C. Vauthier, C. Dubernet, E. Fattal, H. Pinto-Alphandary, and P. Couvreur, Poly(alkylcyanoacrylates) as biodegradable materials for biomedical applications, Advances in Drug Delivery Reviews, 55, 519–48 (2003).
111. K. S. Soppimath, T. M. Aminabhavi, A. R. Kulkarni, and W. E. Rudzinski, Biodegradable polymeric nanoparticles as drug delivery devices, *Journal of Control Release*, 70, 1–20 (2000).
112. C. Vauthier and P. Couvreur, Degradation of poly(alkylcyanoacrylates). In *Miscellaneous biopolymers and biodegradation of synthetic polymers handbook of biopolymers*, 9, J. P. Matsumara and A. Steinbuchel, eds., Wiley–VHC, New York (2002).
113. V. Lenaerts, P. Couvreur, D. Christiaens-Leyh, E. Joiris, M. Roland, B. Rollman, and P. Speiser, Degradation of poly(isobutyl cyanoacrylate) nanoparticles, *Biomaterials*, 5, 65–68 (1984).
114. L. Vansnick, P. Couvreur, D. Christiaens-Ley, and M. Roland, Molecular weights of free and drug-loaded nanoparticles, *Pharmacy Research*, 1, 36–41 (1985).
115. K. Langer, E. Seegmüller, A. Zimmer, and J. Kreuter, Characterization of polybutylcyanoacrylate nanoparticles: Quantification of PBCA polymer and dextran, *International Journal of Pharmceutics*, 110, 21–27 (1994).
116. R. H. Müller, C. Lherm, J. Herbort, and P. Couvreur, In vitro model for the degradation of alkylcyanoacrylate nanoparticles, *Biomaterials*, 11, 590–595 (1990).
117. D. Scherer, J. R. Robinson, and J. Kreuter, Influence of enzymes on the stability of polybutylcyanoacrylate nanoparticles, *International Journal of Pharmaceutics*, 101, 165–168 (1994).
118. F. Leonard, R. K. Kulkarni, G. Brandes, J. Nelson, and J. J. Cameron, Synthesis and degradation of poly(alkyl-cyanoacrylates), *Journal of Applied Polymer Science*, 10, 259–272 (1996).
119. C. Lherm, R. Muller, F. Puisieux, and P. Couvreur, II. Cytotoxicity of cyanoacrylate nanoparticles with different alkyl chain length, *International Journal of Pharmaceutics*, 84, 13–22 (1992).
120. B. Ryan and G. McCann, Novel sub-ceiling temperature rapid depolymerization—Repolymerization reactions of cyanoacrylate polymers, *Macromolecular Rapid Communications*, 17, 217–227 (1996).
121. K. Sobolev and M. Ferrada-Gutiérez, Nanotechnology of concrete, *Nano News* press release, The NanoTechnology Group, Inc. (2005).
122. K. Sobolev and M. Ferrada-Gutiérez, How nanotechnology can change the concrete world: Part 2, *American Ceramic Society Bulletin*, 11, 16–19 (2005).
123. G. Fu, S. Valiyaveettil, B. Wopenka, and D. E. Morse, $CaCO_3$ biomineralization: Acidic 8-kDa proteins isolated from aragoniteic abalone shell nacre can specifically modify calcite crystal morphology, *Biomacromolecules*, 6, 1289–1298 (2005).
124. X. Li, W.-C. Chang, Y. J. Chao, R. Wang, and M. Chang, Nanoscale structural and mechanical characterization of a natural nanocomposite material: The shell of red abalone, *Nano Letters*, 4, 613–617 (2004).
125. M. Berger, Nature's bottom-up nanofabrication of armor, NanoWerk, LLC, www.nanowerk.com/spotlight/spotid=870.php (2006).
126. P. Podsiadlo, Z. Liu, D. Paterson, P. B. Messersmith, and N. A. Kotov, Fusion of seashell nacre and marine bioadhesive analogs: High-strength nanocomposite by layer-by-layer assembly of clay and L-3,4-dihydroxyphenylalanine polymer, *Advanced Materials*, 19, 949–955 (2007).
127. P. Liu, Modifications of carbon nanotubes with polymers, *European Polymer Journal*, 41, 2693–2703 (2005).
128. D. B. Geohegen and H. M. Christen, Functional nanomaterials—Capabilities, presentation, Center for Nanoscale Materials Sciences, Oak Ridge National Laboratory, Tennessee (2005).

129. V. P. Veedu, A. Cao, X. Li, K. Ma, C. Soldano, S. Kar, P. M. Ajayan, and M. H. Ghasemi-Nejhad, Multifunctional composites using reinforced laminae with carbon-nanotube forests, *Nature Materials*, 5, 457–462 (2006).
130. R. Haggenmueller, F. Du, J. E. Fischer, and K. I. Winey, Interfacial in situ polymerization of single wall carbon nanotube/nylon 6,6 nanocomposites, *Polymer*, 47, 2381–2388 (2006).
131. R. D'Sa, *Poly(alkylcyanoacrylates): Biodegradable polymeric nanoparticles for biomedical applications*, U. of Colorado at Denver (2004).
132. S. S. Kistler, Coherent expanded aerogels and jellies, *Nature*, 127, 741–744 (1931).

Problems

12.1 What are the distinctions between intermolecular self-assembly and intramolecular self-assembly?

12.2 Describe the construction of a protein from amino acids all the way from 1 through 4° structure. Is all of it a self-assembly process or are other mechanisms involved?

12.3 List the generic forms of chemical functionalization.

12.4 Intermolecular interactions have advantages in nanosynthesis over traditional covalent procedures. Make a list.

12.5 What are so-called "full-shell clusters"? Why do they have a special stability? (Question courtesy of Günter Schmid, University of Essen.)

12.6 Why is a shell of ligand necessary to synthesize and to isolate metal nanoparticles? (Question courtesy of Günter Schmid, University of Essen.)

12.7 Porous alumina membranes are often employed by template synthesis methods. Ag nanopedestals are fabricated in the pore channels of anodic alumina by an electrodeposition process. How many coulombs of charge are required to form a nanocylinder of Ag 300 nm in length inside pores of a membrane that is 30% porous? The area of the alumina is 3.14 cm^2. Use:

$$l_{cyl} = \left(\frac{CM_{Ag}}{\mathcal{F}\mathcal{A}\varepsilon\rho} \right) \times 10^7$$

where

C is coulombs
M_{Ag} is the atomic mass of Ag
$\mathcal{A}$ is the area of the alumina film
ε is the porosity as before
$\mathcal{F}$ is the Faraday constant
ρ is the density of the Ag

From the preceding, are we able to calculate the diameter of the Ag cylinders? If the pore diameter is 50 nm, what is the aspect ratio of the Ag cylinders?

12.8 Pick a monomer and a polymer process (radical, anion, etc.) and lay out a polymerization process: initiation/growth and termination. Discuss the properties of your polymer.

12.9 The degree of polymerization δ is represented by:

$$\delta = \frac{\textit{Molecular weight of polymer}}{\textit{Molecular weight of monomer}}$$

For a polymer of poly(methylmethacrylate) (PMMA) that has 30,000 amu, what is the degree of polymerization?

12.10 What is the difference between a block copolymer and a regular copolymer?

12.11 Estimate the length of n-dodecane thiol, a molecule often used in self-assembly on gold surfaces. Two formulas are useful: (1) the root-mean-square length (left) and the extended length formula (right) (J. F. Shackelford, *Introduction to Materials Science for Engineers*, 4th ed., Prentice Hall, Englewood Cliffs, NJ, 1998):

$$\bar{L} = l\sqrt{m} \quad \text{and} \quad L_{ext} = ml \sin\frac{109.5°}{2}$$

where l is the length of a single bond and m is the number of bonds.

12.12 Estimate the thickness of a monolayer of olive oil (*triolein*; look it up) from 5 mL that is spread over 0.5 acres. Assume that the overall volume of the monolayer material remains the same (5 mL).

12.13 A tradition in general chemistry laboratory: Calculate the number of stearic acid molecules required to cover a watchglass. What preliminary conditions must be in place before the student can estimate Avogadro's number? How would you design such an experiment?

12.14 In an early experiment, Irving Langmuir determined that long chain fatty acid molecules have the same area (20 Å^2) per molecule regardless of the number of carbon atoms and that the polar end was stuck in the water (the subphase). What other important conclusion did Langmuir glean from this experiment?

12.15 Surface pressure π (a "two-dimensional" pressure) in a Langmuir trough is related to the difference between the surface tension of water (γ_o) and the surface tension of the subphase γ that is covered by the amphiphiles. Recall that in this scenario pressure is applied by the barrier laterally in the trough. The consequence of this action is to squeeze the monolayer together and form a tight assembly:

$$\Pi = \gamma_o - \gamma$$

If Π typically has a value of 10 mN·m^{-1}, after compression, what is the value of γ?

Relating the lateral two-dimensional pressure to the internal pressure of a bulk material in terms of newtons per square meter (e.g., the monolayer) requires the following relation (P. C. Hiemenz and R. Rajagopalan, *Principles of Colloid and Surface Chemistry*, 3rd ed., Marcel Dekker, New York, 1997.):

$$P = \frac{\Pi}{\Phi}$$

where Φ is the thickness of the monolayer. Assuming that the monolayer is 2 nm thick, what is the value of P?

Akin to the ideal gas laws ($P \propto V^{-1}$), is there an analogous inverse relationship between π and area (is $\pi \propto A^{-1}$)? Under what conditions is the preceding equation valid? Under what conditions do you expect it to break down?

12.16 When designing a biosensor using the LB technique, what fundamental considerations must be in place during the design process?

12.17 Research cantilevers on <veeco.com>. Calculate the vibrational frequencies of cantilevers of mass ___, dimensions, ______, made of: (1) silicon, (2) steel, (3) silica glass, (4) quartz, (5) carbon nanotube.

12.18 From the preceding examples, calculate the changed vibrational frequency if a single molecule of a protein with a mass of (1) 700, (2) 1200, and (3) 3000 Da is attached to the tip of the cantilever.

12.19 Design a simple circuit to stimulate and monitor resonant vibration of the preceding cantilevers.

12.20 For a cantilever of dimensions $500 \times 100 \times 0.5$ μm made of single crystal silicon, with an antibody of 14,000 Da attached to its tip, which binds to an antigen of 43,000 Da, calculate the rate of vibration with and without the attached antigen.

12.21 Use Stoney's equation to determine the relative deflections produced for a silicon cantilever of length 500 μm and thickness 0.5 μm versus a silicon nitride cantilever of length 600 μm and thickness 0.65 μm.

Section 5

Natural and Bionanoscience

Natural Nanomaterials

Nature's R&D with photonic systems that manipulate the flow of light and colour has been ongoing for at least 500 million years. She has developed methods and protocols for creating every manner of animal and plant appearance. These optical systems are self-assembled and have precisely specified optical functionalities. They serve diverse biological purposes and have been produced as a result of complex evolutionary selection pressures and developmental constraints. When it comes to photonics, nature truly is an experienced practitioner.

Where better, then, to look for inspiration…?

P. Vukusic, University of Exeter

Chapter 13

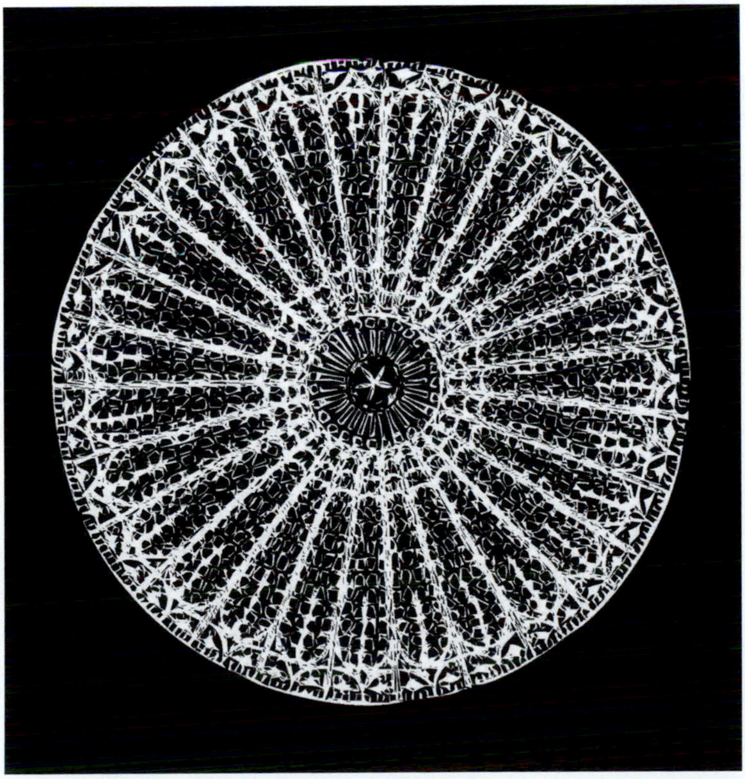

FIG. 13.0 *(a) Professor Peter Vukusic of the University of Exeter, School of Physics. (b) Tiles on a butterfly wing illustrate the complex hierarchy and beauty of natural materials interacting with light on the nanoscale.*

(a)

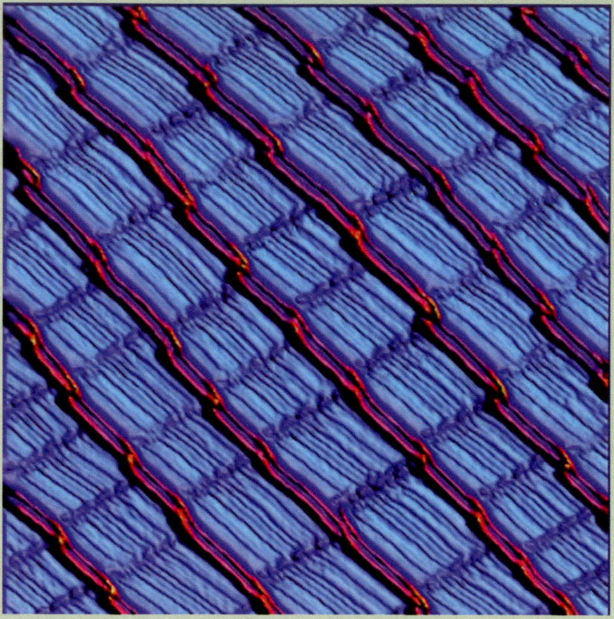

(b)

Source: Image supplied courtesy of P. Vukusic, University of Exeter. With permission.

THREADS

The chemistry and physics underlying nanoscience have been covered in whirlwind fashion in previous chapters. We now give an overview of naturally occurring materials. In this chapter we describe examples of interesting and remarkable natural nanomaterials, focusing on their structure and function and the role the nanoscale plays in macrosopic expression.

This chapter marks the first formal incursion into the biological division, although we have certainly touched upon related topics throughout the text. Many natural nanomaterials comprise inorganic materials; we discuss a few selected examples of those as well. Following this chapter, we delve in more detail into biochemical nanoscience in chapter 14 and take a look behind the structure and function of biochemical machinery from the top down as well as from the bottom up.

13.0 NATURAL NANOMATERIALS

In this chapter we give examples of natural nanomaterials and the nanoscale forces and effects that give rise to their macroscale properties. What do we mean by natural nanomaterials? Every material can be described at the nanoscale, but nanomaterials are materials that are remarkable because of their inherent nanostructure. We define natural nanomaterials as materials that occur in natural environments, without artificial modification or processing. This includes the biological domain as well as the materials found around it.

The chemical identity and properties of a substance depend upon its *molecular* structure. The nanostructure of biological material is due to its *supramolecular* level—arrangements of tens to hundreds of molecules into shapes and forms that can give the material striking properties through interactions with light, water, and other materials.

13.0.1 Nanomaterials All around Us

Nanomaterials abound in living systems and in inorganic rocks and minerals. The materials highlighted in this chapter are a small sample of the natural nanomaterials around us.

13.0.2 Aesthetic and Practical Value of Natural Nanomaterials

Natural materials are rich in texture and color created by specific nanoscale structures. The geometric patterns on seashells, the patterns and colors on butterfly wings, and the shape and texture of bird feathers are a few examples. The iridescence of feathers and insect bodies is also due to layered structures with nanoscale spacing. The striking and remarkable effects of these natural nanomaterials have inspired artists, decorators and engineers, and scientists.

13.0.3 Learning from Natural Nanomaterials

The effects of nanoscale properties are often striking and easily apparent to the casual observer. For this reason natural nanomaterials have attracted scientific curiosity. This interest has led to a greater understanding of nanoscale phenomena. By studying how nanomaterials produce the effects so readily observable on the macroscale, we can learn much about the effects that nanoscale structures have on the behavior of materials in general. Natural nanomaterials are a source of guidance for chemists, physicists, and engineers to create sophisticated synthetic materials with unique functions.

The physical origins of the remarkable properties of many biological materials are due to a complex, often hierarchical structure that can provide a model for designing radically improved artificial materials for engineering and medicine. These nano-inspired materials possess outstanding levels of adaptivity, multifunctionality, and mechanical performance.

Artificial materials based on principles of design from nature are called *biomimetic* materials. We will cover biomimetic materials and their applications in a later chapter on nanotechnology. In this chapter we will focus on the natural nanomaterials themselves and on the principles of nanoscience that we can learn from them.

13.0.4 The Nano Perspective

Natural nanomaterials provide an inspiring way of introduction to nanoscience. Their beauty and intriguing properties arouse curiosity and reward investigation. It can be enlightening to realize that common functional materials that we take for granted, such as paper, cloth, and clay, depend as much on their physical nanostructure as on their chemistry to give them their useful properties. In this chapter we call attention to aesthetic and human interest as we discuss natural nanomaterials.

Naturally occurring nanostructures are rich in patterns and properties adapted through natural selection over evolutionary time scales. Science has long recognized the value of natural chemical compounds; the emerging nanoscience is now examining the properties and potential uses of natural nanostructures.

13.1 Inorganic Natural Nanomaterials

A great variety of natural nanomaterials are formed by living organisms—by complex biochemical life processes. But there are also many natural nanomaterials that form spontaneously by much simpler chemical and physical processes. Among them are a number of simple self-assembling minerals and crystalline oxides of metals. Many complex natural inorganic substances, such as clays, carbonaceous soots, and natural inorganic thin films, exhibit properties based on their supramolecular nanoscale structure, and hence fall into our definition of natural nanomaterials. In this section we look at *inorganic natural nanomaterials*.

13.1.1 Minerals

Many minerals have complex nanostructures that give rise to optical, mechanical, absorbent, and catalytic properties. *Clays* are complex minerals composed of nanoscale particles derived from the weathering of silicate rocks containing oxides of the element silicon. Silicon can form many types of oxides, or silicate groups, that constitute anions to combine with different metals. Most of the outer crusts of the Earth and Moon are made up of silicate minerals.

Silicon. Silicon is to rocks what carbon is to organic material. It is abundant, combines readily with oxygen and hydrogen, and its tetravalent bond structure leads to a great variety of molecular geometries. Silicon, located in the periodic table below carbon, is a much larger atom, and has a complete stable octet shell of quantum shell $n = 2$ electrons in addition to four valence electrons in its $n = 3$ outer quantum shell. Whereas carbon compounds form thin flexible films in water or else dissolve in it, silicon compounds are denser and less soluble and more readily form three-dimensional structures. Silicon compounds tend to form crystals that may incorporate or exclude water molecules. In the organic world of carbon, water is a partner. In the inorganic world of silicon, water is incorporated into the silicate crystalline structure. For example, when water is incorporated into the crystal lattice of silicates, nanoscale dislocations are formed.

The mechanical and optical properties of two minerals with the same chemical composition depend on the supermolecular arrangement of material at the nanoscale. Silica minerals are excellent examples of materials of the same or very similar chemical composition that have dramatically different properties. Clear quartz crystal has a strikingly different appearance compared with opal, even though both are composed of silica; the difference is due not

FIG. 13.1 *Opal and quartz—differences based on nanostructure.*

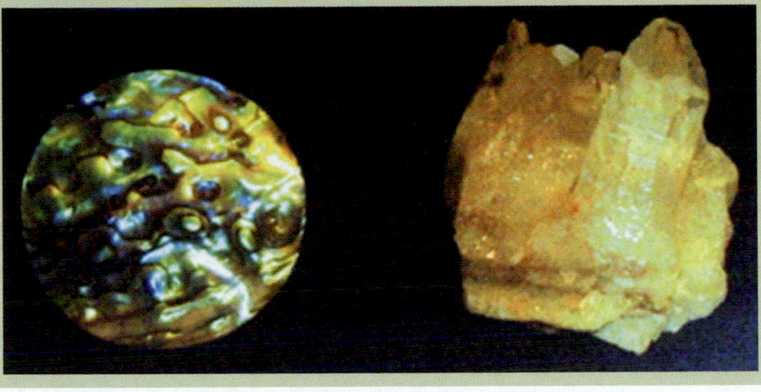

Source: Image courtesy of J. D. Gillean.

to chemical composition, or to crystal structure, but rather to nanoscale structures (**Fig. 13.1**).

Minerals were originally classified by crystalline structure and color, even as their chemical compositions were still being discovered. Silicate minerals crystallize with distinctive geometries based on the structure of their silicate ion group. For example, *tectosilicates*, or "framework silicates," have a three-dimensional crystal structure composed of silicate tetrahedra with SiO_2 in a 1:2 ratio. Tectosilicates comprise nearly 75% of the crust of the Earth. With the exception of the quartz group, tectosilicates are *aluminosilicates*, containing another abundant element, aluminum. Some of the interesting nanoscale-derived properties of silicates result from their interaction with metals and water.

Zeolites. *Zeolites* are hydrated aluminosilicate minerals made of tetrahedral building blocks of (AlO_4) and (SiO_4) linked by rings. These units form a rigid, three-dimensional crystalline structure with a network of interconnected tunnels and cages. More than 40 naturally occurring zeolites are known. An example is the mineral natrolite [$Na_2Al_2Si_3O_{10} \cdot 2H_2O$]. Zeolites differ in the number of aluminosilicate building blocks in the rings, the ratio of silicon to aluminum, and in how the building blocks are arranged. The rings are composed of six to a dozen or more tetrahedrally coordinated silicon (and/or aluminum) atoms linked by shared oxygen atoms.

The larger the aluminosilicate ring is, the larger is the size of the tunnels and cages within its superstructure that open into pores on the surface of the mineral. These tunnels pass completely through the crystal like a honeycomb enabling water, for example, to filter freely. In a given form of zeolite, the pore and channel sizes are nearly uniform, allowing the crystal to act as a nanoscale filter, or *molecular sieve*.

Molecular sieves are materials that can sort molecules based on their size and chemical or electronic affinity. The size of molecular or ionic species that can enter the pores of a zeolite is controlled by the diameters of the tunnels (e.g., size exclusion). The types of molecules that are allowed to pass through the pores are influenced by their electrical charges and chemical interaction with the sieve matrix (e.g., the chemistry of the surface).

The aluminosilicate crystalline structure is a negatively charged framework with pores that hold cations such as sodium, potassium, calcium, and magnesium (Na^+, K^+, Ca^{++}, Mg^{2+}), among others. These cations readily exchange with other alkali and alkaline earth ions, making zeolites natural *ion exchange systems*.

The crystal matrix of regularly spaced, precise, nanosized pores makes the zeolites natural molecular sieves. Due to the strong electron charge inside the pores (e.g., chemically active surfaces) and high surface energy, zeolites are able to act as catalysts. The internal surface of the pores is surrounded by the electron orbitals of the (AlO_4) and (SiO_4). The strongly electronegative environment is capable of rearranging the bonds of molecules trapped inside or passing through the openings in the matrix.

The size and electronic charge in the pores is also affected by the cation that is held in the pore bonding site. The pores in one form of zeolite are approximately 4 Å (0.4 nm) across when occupied by Na^+. With a larger K^+ ion, the pore opening is reduced to approximately 0.3 nm. A Ca^{2+} ion, with two charges, exchanges with two singly charged ions on the zeolite matrix. Thus, the pore opening increases to approximately 0.5 nm when Ca^{2+} is exchanged for sodium or potassium.

The nanoscale architecture of the silicate and alumina in zeolites gives them their remarkable properties. Because of their powerful properties as filters and catalysts, zeolites are invaluable scientific tools. Their practical usefulness makes them the basis for a valuable and important industry. Novel types of zeolites have been synthesized in laboratories since 1948, and new natural forms are still being discovered.

13.1.2 Clays

When aluminosilicates are weathered by the action of water containing dissolved carbonic acid or other chemicals, they break down into particles less than 2 μm in diameter to form clays. *Clays* are hydrous aluminum silicates rich in silicon and aluminum oxides and hydroxides. Clays have distinctive properties compared to other finely divided mineral particles due to their small particle size, layered structure, "stickiness" (self-adhesion), and plasticity, which is due to their ability to incorporate variable amounts of water into their structure. The layered crystalline structure of many clays (similar to a deck of cards) makes them shrink and swell as water is absorbed and removed between the layers.

Moderate drying by the sun removes water from clays and results in a hard material that is relatively strong and resistant to reabsorption of water. When clays are heated to high temperatures ("fired"), they lose water and form permanent bonds between particles, transforming them from a soft pliable substance to a hard durable material. From prehistoric times, clays have been used for art and ceremonial objects, as a building material, and as one of the earliest media for holding written records.

Montmorillonite Nanoclay. Clay comes in many different colors and exhibits differing properties depending on its proportions of silicon, aluminum, and other elements and their molecular combinations with oxygen and water. One especially important class of clays are the *montmorillonites* [1,2]. Montmorillonite is formed by the weathering of rocks that have a relatively low silica content. It is named after Montmorillon in France, where it is found in abundance. It is also an important soil constituent in many regions of North America. It is the main constituent of the volcanic ash weathering product, *bentonite*.

Montmorillonite's crystal structure has two tetrahedral sheets sandwiching a central octahedral sheet. The microscopic particles in montmorillonite clays are plate shaped with an average diameter of approximately 1 μm or less. In natural montmorillonite, particle size ranges from 50 to 500 nm. Its chemical composition is hydrated sodium calcium aluminum magnesium silicate hydroxide—$[(Na,Ca)_{0.33}(Al,Mg)_2Si_4O_{10}(OH)_2 \cdot nH_2O]$.

The sodium and calcium may be replaced by potassium, iron, and other cations, in different relative abundances, with resulting variations in physical properties. The structure unit of montmorillonite consists of two silica tetrahedral sheets with alumina octahedral sheets in between, as shown in **Figure 13.2.**

Like other clays, montmorillonite swells with the addition of water, but its volume changes with water content to an unusual extent. The degree of expansion depends on the exchangeable cation substituted in the molecule. With sodium as the predominant cation, the clay can swell to several times its original volume with the uptake of water. Montmorillonite swelling is a significant factor in soil stability and has to be taken into account in building roads, dams, and buildings where it is a major soil constituent. The expansion properties of sodium montmorillonite make it useful as a nonexplosive agent for splitting rock in stone quarries and for the demolition of concrete structures where the use of explosive charges is undesirable.

The absorbant properties of montmorillonite have led to its use in traditional medicines and cosmetics. Modern uses include anticaking additives for animal foods and other commodities, drilling muds for lubricating and expanding fissures in rocks during oil exploration and production, ion exchange agents for chemical processing, catalyst supports, binding agents for toxins and microorganisms, and water purification. Montmorillonites catalyze the self-assembly of certain organic molecules. Montmorillonite and other clays are truly versatile nanomaterials with many emerging uses and newly appreciated properties.

FIG. 13.2 *Schematic of montmorillonite structure—tetrahedral silica sheets sandwiching a central cation octahedral sheet.*

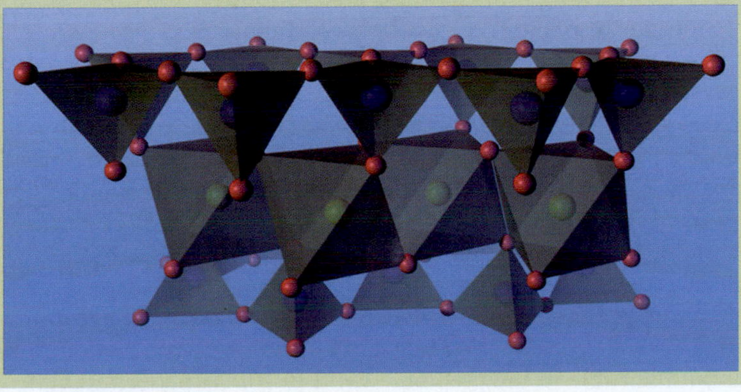

Source: Image by Anil K. Rao.

Montmorillonite is now commonly referred to as *nanoclay* in the emerging natural nanomaterials market, where new applications are being discovered each year. Commercial production may involve grinding to reduce particle size, sorting for uniform particle size, aggregation of particles into porous granules for easy handling, and chemical treatment to yield a desired mix of cations.

The remarkable properties of montmorillonite and other clays are due to their crystal structure and its particle size. These properties are a result not just of its chemical structure, but of the interaction of the clay nanoparticles with water and with each other. Although the diameter of natural montmorillonite clay particles varies, the thickness is on the order of nanometers. The large surface area of the clay particles enhances interaction with water and solutes. If the crystals were not divided into nanoparticles, the montmorillonite molecules would take the same form as other harder minerals with a similar crystalline symmetry. Instead, the exchange of cations produces disorder in the crystals, preventing the growth of large crystals and resulting in nanoparticles [3].

Catalyst for Life. Montmorillonite and other clays may have acted as catalysts in the evolution of life from prebiotic materials. During the 1990s, scientists discovered that clays catalyze the chemical reactions needed to construct RNA from *nucleotides* [4,5]. This discovery was bolstered by a recent finding that clays can catalyze the transformation of lipid micelles into vesicles. In previous chapters we discussed how lipids in water spontaneously form micelles, making a region surrounded by a single lipid layer. In living organisms, the membranes of vesicles and cells are composed of lipid bilayers.

Under certain conditions fatty acid micelles spontaneously form into vesicles with lipid bilayer membranes, similar to those in living cells, but the transition is very slow without a catalyst or nucleation site for the membranes [6,7]. In recent years, it was found that clays act as nucelation sites for the self-assembly of micelles into vesicles. In 2004 a research group at the Howard Hughes Medical Institute (HHMI) showed that small amounts of montmorillonite greatly accelerated the formation of vesicles from fatty acid micelles [8].

In addition to catalyzing RNA synthesis, montmorillonite catalyzes the formation of vesicles that may have been precursors to the evolution of the first living cells. The HHMI group was also able to induce vesicles to grow and to split into separate vesicles under laboratory conditions. By using montmorillonite with RNA attached, they demonstrated that some of the RNA was contained in the newly formed vesicles and remained there without leaking out. The researchers succeeded in inducing the vesicles to grow by incorporating additional micelles. When large vesicles were extruded through small pores, they divided into smaller vesicles containing RNA, with structure and properties similar to the precursors. They described these processes as "a proof of principle that growth and division is possible in a purely physical–chemical system," which is "vaguely analogous to biological cell division."

There are many more complicated processes necessary to get from vesicle replication with RNA to a living cell, but nanoclays are beginning to provide clues about their role in the prebiotic world [9,10]—a perfect example of one nanomaterial interacting with others.

13.1.3 Natural Carbon Nanoparticles

With its covalent bonding, carbon readily forms complex two- and three-dimensional structures with nanoscale features under a variety of conditions. Diamond and graphite are two naturally occurring forms of carbon that differ in bonding between the carbon atoms, and as a result have different crystal structures and physical properties.

Diamond Nanoparticles. Diamonds are networks of carbon atoms each linked by four bonds to their neighbors in a strongly bonded crystalline lattice. On the surface of the lattice, outer bonds are terminated by hydrogen atoms. The smallest diamond structure possible is the adamantane molecule (chapter 9), with a lattice of 10 carbon atoms linked in a tetrahedral lattice, each capped by a hydrogen atom. Large gemstone diamonds are formed only at very high pressures and temperatures, since these are the only conditions under which the diamond crystalline phase of carbon is more stable than graphite or amorphous carbon. The only natural conditions on Earth sufficient for the formation of large diamonds are more than 120 km (75 miles) deep within the Earth's mantle, where the high pressures and temperatures (40 kbar and 900°C) can compress carbon into the compact structure of diamond.

Until the emergence of nanoscience, it was generally taken for granted that diamond material could only be formed under extreme conditions of pressure and heat that existed uniquely deep within the Earth's molten depths. This is true of large, single crystalline diamonds, but very small nanodiamonds or even thin films of diamond can be formed under less rigorous conditions. Once scientists became aware of and interested in the structure of minerals and sediments on the nanoscale, they began to discover many occurrences of nanosized diamonds. The quest to understand how these nanodiamonds could have been formed has led to fascinating insights into the Earth's history.

Suitable conditions for formation of very small diamond crystals and films may occur close to and even on the surface of Earth. Nanoscale diamond particles may occur naturally as the result of rapid combustion of organic material or in meteorite impacts. Industrial researchers have developed methods to grow extremely thin diamond films at very low pressures, using gas phase plasmas and ion deposition, where the usual phase rules that govern materials in equilibrium do not apply. These are the types of conditions in flames, lightning discharges, radioactive decays, explosions, and high-energy impacts. It is interesting that ancient beliefs in the Indus civilization attributed the formation of diamonds to lightning strikes on rocks.

Nanodiamonds as Indicators of Impact Events. Micro- and nanosized diamonds found in recent years in nonigneous minerals may have originated from meteorite impacts, radioactive decay, or other processes not yet understood. Some tiny diamonds are strongly associated with known meteorite impacts. Some of the most dramatic examples are microscopic diamonds found in Meteor Crater, Arizona, formed when an iron-rich meteor crashed 50,000 years ago. The diamonds may have been formed from carbon in the impacted soil or from graphitic particles in the meteorite itself. Some of these diamonds have a unique hexagonal crystal structure only found in impact craters, which was named lonsdaleite after British mineralogist Dame Kathleen Lonsdale.

Canadian scientists have found diamond particles 3–5 nm across embedded in a layer of sediment 65 million years old that is linked by other evidence to the time when many scientists believe a giant meteor slammed into Earth and caused the mass extinction of dinosaurs and many other Cretaceous animal and plant species [11]. The nanodiamonds, along with iridium and other indicators in sediments, form a worldwide marker layer from which the nature and date of the impact have been inferred. Other research on the diamonds from this impact dust layer, found in areas of Montana and New Mexico, has shown that their isotope abundance is characteristic of diamonds formed from carbon on Earth rather than carbon particles formed in space and embedded in meteorites. This work further supports the hypothesis that these nanodiamonds were formed in the impact.

Recently, nanodiamonds have become the primary evidence for the existence of a massive meteoric impact 12,900 years ago. Scientists had suspected some sort of catastrophic event caused at least four major changes found in the fossil record. The first was a mini ice age, called the Younger Dryas cold phase, which started about 12,900 years ago and lasted for about 1,000 years. The second is the sudden disappearance of the Clovis culture, the earliest known human inhabitants of North America, in the same time period. The third is the disappearance of the large animals such as mastadons from North America. The fourth is the disruption of human settlements and disappearance of primitive Stone Age cultures in Europe during the same time period. The nanodiamond dust found over a wide area was the key piece of evidence tying other clues together. This is an example of how nanoscience has led to surprising breakthroughs in diverse fields of study. With new awareness of the existence of nanomaterials and the tools to identify them, scientists have obtained new insights into the history of the Earth.

Nanodiamonds from Radiation. Another unusual source of natural nanodiamonds on Earth is the action of radioactive isotopes in carbon-rich minerals. Scientists examining uranium-rich coal from Russia have discovered diamonds ranging in size from 2 to 40 nm and containing a few thousand to a few million carbon atoms. The diamonds appear to have formed when uranium atoms in the rock underwent fission, breaking up into high-energy ions. These fission fragments tear through the carbon structure, breaking the bonds and heating up a microscopic area around the fission track. As the carbon atoms move back together to form new bonds, they sometimes reconfigure into a diamond lattice structure [12].

Nanodiamonds from Petroleum Deposits. An even more surprising source of nanosized diamond particles is petroleum. These *diamondoids* had first shown up as deposits in crude oil equipment, and have now been shown to be present in petroleum deposits under the Gulf of Mexico. They are extremely small nanoparticles, typically containing up to 11 *adamantane* units (10 carbon atoms arranged in a diamond lattice). Although scientists have been able to synthesize similar diamondoid particles in the laboratory, they have not been able to make particles larger than four adamantane lattice cages. Some of the oil-based diamondoids have unique structures, such as one consisting of six adamantane units arranged in a disk, named cyclohexamantane [13,14].

It is not clear how the diamondoids found in oil are formed. One possibility is that clay minerals catalyze reactions between methane and other hydrocarbon precursors, adding carbon to adamantane seed crystals. One possibility is that these nanodiamonds were formed by the giant meteor impact in the Gulf of Mexico some 65 million years ago and migrated into the oil-bearing strata along with other organic material. This hypothesis may be supported in that the worldwide distribution of diamondoids in oil follows a pattern consistent with known meteor impact events. These particles are on the boundary between chemical molecules and nanoparticles, where many interesting and useful properties arise. Researchers are studying their use as building blocks for nanomachines, as platforms for antiviral drugs, and in other applications.

Diamond Nanoparticles of Unknown Origin. The sources of some micro- and nanodiamonds are still a mystery. Tiny diamond grains 20–80 μm in size have been found in metamorphic rocks (gneisses) from southwestern Norway and Asia. These are not the igneous rocks from the molten magma that is the source of gem diamonds. The metamorphic rocks have not yet been linked to any known meteor impacts. Diamonds that were weathered out of igneous rock might have been deposited in sediments that were the precursor of the metamorphic gneisses where they are now found. But sedimentary rocks underwent pressure and heat that transformed them into their present metamorphic phase. Diamonds are unlikely to have survived this transformation; instead, they would have been oxidized or converted to graphite. Although diamond is the hardest form of carbon and graphite is one of the softest minerals known, graphite is actually more stable than diamond [15–17].

The study of nanodiamonds is still very new, with many unknowns waiting to be resolved. Nanoscience has increased the awareness and attention given to small-scale particles in geoscience just as in other fields. As the basic scientific understanding of nanoscience grows, more knowledge and tools for solving mysteries in geochemistry, mineralogy, and archeology will be made available.

Graphite Nanoscience. Graphite is black and soft; it can be considered to be the purest form of coal. Diamond and petroleum deposits may be converted over geological time into graphite, as the most stable form of pure carbon. A single particle of carbon with a graphitic structure no longer has the properties associated with graphite, since these properties depend on the interaction between different nanolayers. Nevertheless, nanoparticles with basic graphitic structure do exist in the natural world.

The most familiar application of graphite is in the common pencil lead. Graphite powder is valued for its lubricating properties. Both applications depend on the nanostructure of the flat sheets of linked carbon atoms. For a long time, it was believed that the sheets were loosely coupled and could easily slice over each other, giving graphite its useful properties. However, it has been shown that in a vacuum (such as in space), graphite is a poor lubricant. This led to the discovery that the lubrication is due to the absorbed fluids between the layers, such as air and water. Recent studies suggest that an effect called *superlubricity* can also account for graphite's lubricating properties.

When a large number of crystallographic defects binds these planes together, graphite loses its lubrication properties and becomes what is known as *pyrolytic carbon*. This material is useful for blood-contacting implants such as prosthetic heart valves because it does not cause blood clotting. It is also highly diamagnetic; thus it will float in midair above a strong magnet.

Graphite forms *intercalation* compounds with some metals and small molecules. In these compounds, the host molecule or atom gets "sandwiched" between the graphite layers, resulting in compounds with variable stoichiometry. An example of an intercalation compound is potassium graphite, denoted by the formula KC_8.

Amorphous Carbon Nanoparticles. Amorphous carbon is a term applied to coal, soot, and other carbon compounds that are neither graphite nor diamond. However, these materials are not truly amorphous; they are polycrystalline or nanocrystalline materials of graphite or diamond within an amorphous carbon matrix.

Soot, or "black carbon," produced from burning organic material is ubiquitous in the environment. Soot and smoke contain a variety of carbon nanoparticles, which can serve as both an adsorbent and a catalytic site for chemical reactions. Soot from combustion where the amount of oxygen is limited has been found to contain significant amounts of carbon buckyballs.

13.1.4 Nanoparticles from Space

Nanoparticles with unusual properties can form in the exotic conditions that exist in interplanetary and interstellar space. Meteorites sometimes contain microscopic particles of diamond or other complex nanostructures. Diamonds could be formed by collisions between asteroids or with other planets, but they could also be formed in deep space. Some nanodiamonds embedded in meteorites contain an isotopic mixture of xenon not found in the solar system. It is believed that such diamonds were formed in collapsing stars as they died in supernova explosions. If this hypothesis is correct, stardust may contain tiny diamonds.

Astronomers studying interstellar space have determined that nanoparticles make up a large part of interstellar dust. Optical measurements indicate that carbonaceous particles ranging from soot to diamond exist in disks of gas and interstellar clouds. These particles may have formed directly from condensation of gases or from collision fragments. In the low density of space, exotic crystal structures may gradually coalesce into nanoparticles with unusual structures and properties. Astronomers have evidence that nanodust particles in space act as catalysts in formation of compounds in clouds of gas and dust, leading to building up of raw materials that may play a part in formation of life after the clouds coalesce into planets [18].

13.2 Nanomaterials from the Animal Kingdom

In this section we look at natural nanomaterials formed by animals. We start with organic structural fibers and matrices common to most forms of life. In this section we trace the progression of mineralization from shells to teeth and bones, and look at the polymers that make up scales, hair, and feathers. In later sections we will explore other directions that plants and insects took in producing their own unique nanomaterials.

13.2.1 Building Blocks of Biomaterials

In this section we look at some protein and biopolymer building blocks that illustrate some important principles and examples of nanostructure architecture in animals, plants, and single-celled organisms.

Multicellular organisms synthesize a protein matrix to hold cells together in a structured body. These *extracellular matrices* provide the scaffolding, support, and strength to tissues for all living organisms. Organisms arrange protein, mineral, and polymer building blocks from the bottom up, linking them in complex scaffolds and networks. They make an exquisite variety of composite materials whose organization on the nanoscale gives them strengths and features that far surpass the properties of their components. These *natural nanomaterials* have evolved far beyond the basic function of holding cells together: Natural nanomaterials protect, support, and serve an amazing variety of functions in the bodies of organisms.

We now look at some important structural proteins and some examples of the protein nanomachinery that controls the building of natural nanomaterials. Long, straight chain *collagen* fibers intertwine to provide strength; elastic *proteoglycans* and *elastins* give resilience; adhesive *glycoproteins* such as *fibronectin* and *laminin* bind the matrix and cells. These and other structural proteins are key building blocks for the nanostructured materials we will examine in this chapter. In addition to the building blocks, we will show a few examples of the proteins that act as nanomachines to guide and control the assembly of nanostructures. Nonprotein polymers are also important in natural structural materials. Three such polymers we will discuss are *cellulose*, *chitin*, and *keratin*. We encounter a few other interesting examples along the way. Cellulose is produced by plants, chitin is produced mainly by shellfish and insects, and keratin is produced mainly by vertebrates.

Most organisms combine protein and polymer building elements with inorganic minerals such as *carbonates*, *phosphates*, and *silicates*, ordering their nanostructure to create strong and rigid materials with unique optical and electrical properties. Some general principles of structural organization should emerge from this overview of natural nanomaterials. One recurring theme will be that of *ordered fibrous polymers embedded in a mixture of nonfibrous substances* to give a *composite* with superior properties to the single components. Where high strength and rigidity are required, nature generally selects *fiber-forming macromolecules*, usually ones that aggregate into *near-crystalline arrays*. Where elasticity and resilience are important, *highly branched polymers* such as *elastin* meet the requirement. In many natural materials from different sources and different compositions, we will find a *hierarchy of fiber assemblies* from the molecular level up to the macroscopic, with *interlocking linkages* to spread loads. An important theme is the importance of *calcium carbonates* and *phosphates* as components of skeletal structure, as well as serving as buffers or sinks for carbon dioxide and regulators of cellular metabolism.

A good introduction to fibers in connective tissues begins with the *collagens*, made up of *tropocollagen* building blocks arranged in a variety of near-crystalline aggregations into fibrils in a wide variety of organisms and tissues. The straight chain fibrils of collagen are supplemented by more branched polymers to provide elasticity and resilience. We encounter more details about collagens and similar polymer units, such as chitin and keratin, in many settings as we traverse through the nanoscale architecture of biomaterials in this and later sections [19].

Single-Celled Organisms, Plants, and Animals. Plants and animals have radically divergent life patterns and equally different body architectures and materials. The most important plant structural materials are nanostructured cellulose and lignin, familiar as paper, cotton, and wood. The highly important functional and structural nanodesign of plant leaves, flowers, seeds, and roots is beyond the scope of this section.

Before proceeding, we devote this section to the biological materials made by multicellular animals: nanostructures combining proteins and organic polymers with mineral crystals of carbonate and phosphate. These important nanostructures include *exoskeletons* of sponges, corals, and crustaceans (seashells); eggshells; cartilage; scales; teeth; bone; turtle shells; claws; hair; fingernails; feathers; and other tissues in which carbonate and phosphate crystals are layered in with organic polymers and proteins to provide increased hardness, rigidity, and strength. In the vertebrates, mineralized carbonates and phosphates are developed into *endoskeletons* of cartilage and bone.

Mineralized Biological Nanomaterials. Tissues that contain solidified inorganic salts are said to be *mineralized*. Mineralized tissues, such as shells and bones, are widespread among multicellular animals; sponges, corals, and sea urchins make mineralized tissues to support their cells. The origin of multicellular animals appears to be tied closely to carbonate-based chemistry. The most primitive mineralized materials made by multicellular animals include crystals of *calcium carbonate*, $CaCO_3$, as a component of their structure.

The first traces of biomineralization in the geological record are found in rocks dated to the *Ediacaran* era (which is when the first fossil evidence of multicellular animal life appears)—immediately following a period of high carbon dioxide content in the Earth's oceans. The Ediacaran was followed by the Cambrian epoch 545–490 million years ago. The "Cambrian explosion" was the era in which primitive life forms diverged into the major classes that make up life on Earth. The carbonate ion plays a central role in their metabolism [20].

Carbonate Chemistry. Let us look at the chemistry that is the foundation for the building of carbonate nanostructures. The carbonate ion, CO_3^{-2}, is formed when carbon dioxide, CO_2, dissolves in water:

$$CO_2 + H_2O \rightarrow H_2CO_3 \rightarrow H^+ + HCO_3^- \rightarrow 2H^+ + CO_3^{2-} \quad (13.1)$$

The neutral gas, carbon dioxide, readily dissolves in water where it is in equilibrium with the bicarbonate ion and the carbonate ion. This equilibrium buffers the acidity of the solution. Metal ions such as sodium, potassium, magnesium, calcium, and zinc can associate with the hydrated carbonate ion. The lighter and more active metals, sodium and potassium, form strongly hydroscopic and highly soluble salts, whereas magnesium and calcium and heavier metals are less soluble and tend to precipitate out of solution to form mineral crystals. Higher organisms use dissolved carbon dioxide as a source of carbon for both photosynthetic reduction of carbon into carbohydrates and for mineralization, as well as a means of controlling *p*H. This *sequestration* of carbon dioxide through photosynthesis and mineralization is an important factor through which life regulates the amount of carbon dioxide in the oceans and atmosphere.

Crystalline Structures of Carbonates. Calcium carbonate is the most abundant form of mineral carbonate, both in geological sediments and in biomaterials. Pure calcium carbonate can crystallize in several different forms, with the molecules stacked in different arrangements to form different crystal shapes and symmetries. The most common crystalline form of calcium carbonate is *calcite,* in which the carbonate ion groups lie in a single plane pointing in the same direction, leading to a *trigonal symmetry*. Another common carbonate mineral is *aragonite,* in which the carbonate ions lie in two planes, with adjacent carbonate groups pointing in opposite directions, giving an *orthorhombic* symmetry. Either of these forms of calcium carbonate can form by precipitation from solution depending on the concentration, temperature, *p*H, and presence of other salts. The formation of both can be catalyzed by protein enzymes.

Cellular Chemistry of Carbonate. In living cells, carbonate precipitation is catalyzed by specific protein enzymes, which determine the crystalline form of the mineralized carbonate. The family of enzymes called *carbonic anhydrases* catalyzes the conversion of dissolved carbon dioxide into carbonate ions. In bacteria these enzymes get rid of excess carbon dioxide and produce carbonate salt, which serves as a buffer in the cell cytoplasm, helping to regulate the *p*H in the cell by reacting with excess acid. In sponges and other multicellular organisms, this carbonate continues to serve its roles in regulating cell chemistry, but it can also be precipitated to form a mineralized skeleton. Special proteins bind to the carbonate and direct its precipitation in either calcite or aragonite form.

13.2.2 Shells

By exacting analysis of the nanostructure of seashells, scientists have found that the mantle of tissue next to the growth surface of the shell lays down layers of chitin coated on both sides by special proteins. The chitin determines the size and position of the calcium crystals, and the proteins control the deposition of calcium in specific crystalline forms for the plates or prisms.

Nanostructure of Shell Growth. In corals and shellfish, the development of carbonate-based skeletons has evolved as a support and protection for the organism. In animals that build carbonate exoskeletons, the carbonate minerals typically are precipitated as crystals, or *prisms*, embedded in layers or matrices of proteinaceous and polysaccharide material. In mollusks, the layer of the shell closest to the living organism consists of very smooth pearly nacre, whose mineral crystals are in the form of flattened aragonite platelets, while the thick body of the shell is built of calcite prisms.

The exoskeleton or shell is grown by a layer of cells that first lays down a coating of protein supported by polysaccharide polymers such as chitin. The proteins act like a nano-assembly mechanism to control the growth of the calcium carbonate crystals. The calcium and carbonate ions are also secreted by the mantle, but it is the tailored environment created by the organic matrix that determines how the crystals grow. The honeycomb-like matrix of protein and chitin remains as an encapsulating layer around each crystal. This envelope serves as a relatively flexible, crack-deflecting matrix for the mineral particles. The strong covalent bonding of the protein and chitin complements the ionic bonding of the brittle mineral to give the finished material a combination of toughness and hardness. The size of each crystal is typically around 100 nm.

Strength and Toughness Based on Nanostructure. As a result of the nanostructure biocomposite of polygonal carbonate crystals embedded in a polymer matrix, the nacre of mollusk shells has extraordinary physical properties. Nacre consists of vertical stacks of flat polygonal aragonite crystal tablets (or platelets). The argonite is arranged in parallel growth layers called *lamellae*. Each lamellae layer is built of several thousand layers of aragonite plates stacked in an interlocking offset pattern somewhat like a brick wall. The plates are mortared to each other by thin layers of protein and chitin glue. Platelets are typically hexagonal, with thicknesses between 0.3 and 1.5 μm and diameters ranging from 5 to 20 μm. Each plate is encased in a protein and chitin coating a few tens of nanometers thick. The matrix has pores that allow mineral bridges to connect the plates to each other. This structure makes nacre 3000 times as fracture resistant as crystalline mineral aragonite.

Methods for Investigating Shell Nanostructure. To investigate the microscopic structure of calcium carbonate crystals in snail shells and the interactions between proteins and the crystals on the nanoscale, researchers use atomic force microscopy (AFM). The AFM is guided and controlled by scanning electron microscopy (SEM), light microscopy (LM), and biochemical methods. For studies of nanoscale interactions of proteins with crystal faces, extracts of proteins taken from shells are mixed with solutions of calcium carbonate and calcium chloride.

Calcium Control by Protein Nanomachinery. A number of calcium-controlling proteins have been identified. Some of these proteins have active sites that orient calcium carbonate into seed crystals of either calcite or aragonite, a process called *crystal nucleation*. Other shell-building proteins bind onto specific faces and edges of calcium carbonate crystals, where they may promote or inhibit either calcite or aragonite formation. The inhibitory proteins may prevent the growth of the crystal in some directions while allowing growth in preferred directions. The inhibitory proteins have been shown to inhibit the dissolution of crystals once they are formed. Some of the calcium-controlling proteins found in shells have active sites that closely resemble portions of proteins found in milk whey, which are associated with the deposition of calcium in the bones of mammals. The complex mixture of calcium-building proteins is specific to each organism, resulting in growth of different shell types, patterns, and strengths.

The polysaccharide chitin is secreted by the mantle cells and acts as a scaffold to which the protein molecules are attached. Some proteins have been shown to act as adhesives between argonite tablets (the large protein lustrin A, found in abalone nacre). Other proteins, such as perlucin, nucleate new layers in calcium carbonate crystals. Other proteins, whose function is not yet fully understood, such as perlustrin, are similar to insulin- and insulin-like growth factor-binding proteins found in higher organisms. Structures for the calcium-controlling proteins from seashells are still being fully determined but their amino acid sequences are related to proteins that stimulate proliferation and differentiation of cells involved in building bone in vertebrates, such as fibroblasts, bone-marrow cells, and osteoblasts. Recently, the full structure has been worked out for ovocleidin-17, a protein similar to perlucin, which plays a similar role to build calcium layers in chicken eggs. Nature is very conservative in preserving useful nanotools once they are evolved, as evidenced by the close relationships between the calcium-controlling machinery across many organisms. Yet, as we find when we learn about other mineralized tissues, nature is infinitely inventive in the structures that organisms build to meet the needs of survival in their environments [21–28] (**Fig. 13.3**).

13.2.3 Exoskeletons

The mollusks adapted to a defensive lifestyle of building a hard protective shell into which the organism could retreat. Predators in turn evolved first teeth and claws, and then jaws, to crush and penetrate the shells, leading to an eons-long arms race in which the nanostructure of the shell was optimized. Other marine animals that originated during the Cambrian took on a more mobile and flexible form of mineralized exoskeleton. Crustaceans, such as crabs, lobsters, and shrimp, have a tough, flexible, chitin-rich exoskeleton divided into sections connected by flexible joints—ideal for supporting movement.

The crustacean's modular exoskeleton is divided into three parts: head, thorax, and abdomen—a body architecture that characterizes insects and vertebrates

Fig. 13.3 *Shells—variants of a mineralized nanostructure.*

Source: Image courtesy of J. D. Gillean.

as well. In addition to mobility, the lighter, more flexible exoskeleton of the crustaceans allows them to shed or *molt* their exoskeletons as they grow and then grow new ones. The thin exoskeleton is not only easy to break out of compared to shells, but also involves less of an investment in energy and nutrients to rebuild. The crustacean exoskeleton is also an excellent instructive example of how strength and flexibility are related to the nanostructure of a material with a minimal investment in material weight, volume, and components. Crustacean exoskeletons show a strong anisotropy in their properties, resulting from their hierarchical organization from the nanometer to the millimeter scale [29–31].

Crustacean exoskeletons have four layers, all relatively rich in proteins compared to mollusk shells. The outer layer is a soft chitin-free coating of lipids, proteins, and calcium salts, which seals and protects the exoskeleton. The exocuticle is a matrix of chitin and protein with embedded mineral particles, mainly calcite. The endocuticle is also a mineralized matrix of chitin and protein, but with less calcification and polymer cross-linking than in the exocuticle. The membranous layer is also composed of protein and chitin, but is not mineralized. The thicknesses of these layers vary in different types of crustaceans. They also vary on different parts of the body (claws, abdomen, eye covering, etc.) and depend on the stage of development and molting. In a typical crab, the epicuticle is on the order of 10 μm, the exocuticle 30 μm, the endocuticle several hundred micrometers, and the membranous layer 20–30 μm [32].

The crustacean exoskeleton is an outstanding example of a nanostructured natural material optimized to trade off high strength, light weight, and economy of fabrication. The protein and chitin in crustacean exoskeletons form fibrils arranged in ordered, three-dimensional, twisted honeycomb patterns. This structure, sometimes referred to as "twisted plywood," is formed by the helical stacking of plates composed of crystalline chitin–protein fibers. The stacked planes are gradually rotated about their normal axis, thereby creating complex structures that appear as arches when cut in cross-sections. This structure is also called the Bouligand pattern [29], named for the scientist who first deduced it from electron micrograph observations.

However, these twisted plywood planes are not simple arrays of parallel chitin–protein fibers. Interconnected fibers form a honeycomb-like network filled with pores in which calcite mineral particles are distributed. The different layers of the exoskeleton differ in density and hardness.

The finely woven twisted plywood structure of the crustacean exocuticle has a high stiffness (8.5–9.5 GPa). The hardness increases within the exocuticle between the surface region (130 MPa) and the region close to the interface to the endocuticle (270 MPa). The endocuticle is a much more coarsely twisted plywood structure, in which both the stiffness (3–4.5 GPa) and hardness (30–55 MPa) are much smaller than in the exocuticle. The different layers are important in absorbing mechanical loads on the exoskeleton and in allowing for staged growth [33].

Preventing Propogation of Crack Failure. The cuticles of spiny and slipper lobsters have a pattern of surface beads, or *tubercles*, and a system of pits that resist failure by blunting any cracks that start to form in the shell. The layered structure of the shell dissipates forces horizontally, while the tubercle and pit systems increase the surface area available to dissipate forces and interrupt the growth of cracks through the material. These systems may represent evolutionary solutions to predation, particularly by predators that strike, crush, or bite holes, rather than those that engulf.

13.2.4 Endoskeletons

Thus far we have discussed animals that develop hard exoskeletons to protect and support their bodies. Another group of animals began to build an internal structure for support. This strategy disencumbered the body of a restrictive heavy coating, and gave the advantages of mobility, unrestricted growth, and less investment in material for a given body size. In some of the simplest multicellular animals, the sponges, the extracellular matrix contains both mineral crystals and *collagen*. Collagen acts as a structural material in many animals, including *arthropods* (crustaceans and insects), where it plays a secondary role to the exoskeleton.

Cartilage. In *chordates* and *vertebrates,* strands of collagen are combined with other structural polymers to form an *endoskeleton* of *cartilage*.

The nanostructure of cartilage is important in giving it the strength to support the animal body and in providing a framework for mineralization that eventually leads to the development of *bone* in later descended vertebrates. Cartilage tissue can also be found in invertebrates such as horseshoe crabs, marine snails, and cephalopods, where its nanostructure is optimized to support muscular movement and attachment of tissue to the shell. Although cartilage is found in invertebrates, no invertebrate cartilage mineralizes in vivo. Mineralization of cartilage is a vertebrate innovation.

Cartilage forms a dense connective tissue that can withstand large shear and other stresses. The nanostructure of cartilage is very complex. A very important difference between cartilage and the materials of mollusk and insect exoskeletons is that cartilage has living cells embedded within the tissue, so it grows from within rather than from a boundary layer of cells. This difference opens the possibility of many more nano- and micro-architectures. This also enables cartilage to grow much more rapidly than if growth could occur only at the surface, with important implications for the life cycle of the organism.

Types of Cartilage. In vertebrates there are three main types of cartilage adapted to different functions. Each type differs in arrangement and concentration of collagen fibers in relation to the matrix and in the nanostructure of the fibrils themselves. For a given type of cartilage, the orientation of the fibers varies depending upon the loads that the cartilage bears during development.

Hyaline cartilage is a translucent cartilage found in growing tissue and in embryos, serving as a center of ossification or bone growth. The name hyaline comes from the Greek word *hyalos,* meaning glass. This refers to the translucent matrix or ground substance. Hyaline cartilage also forms the gristly *articular* cartilage lining bones in joints and is present inside bones as well. In hyaline cartilage, collagen molecules are arranged in cross-striated fibers, 15–45 nm in diameter.

Elastic cartilage is a dense yellow tissue that supports stiff organs such as the ear, nose, and larynx. Elastic cartilage is stiffer and more elastic than hyaline cartilage due to elastin bundles distributed in the matrix and attached to the collagen fibers.

Fibrocartilage is a white cartilage with high tensile strength, with larger and more densely concentrated collagen fiber bundles than in other types of cartilage. Fibrocartilage forms the tough tissue forming the discs separating vertebrae and connecting tendons or ligaments to bones.

Matrix. The matrix interacts with the fibers of cartilage to give its elastic properties. The matrix contains *proteoglycans,* which are large molecules with a protein backbone and *glycosaminoglycan* side chains. These molecules fill the spaces between the collagen fibers. The proteoglycans are hydroscopic, holding water that can be expelled and reabsorbed in response to loads, both compression and stretching. This gives cartilage resistance to compression and elastic resilience (ability to spring back into shape after load). An important example of a proteoglycan is *aggrecan,* whose glycosaminoglycan side chains contain *chondroitin sulfate* and *keratan sulfate*. Aggrecan is found in the tough gristly cartilage that supports bone joints. The matrix also contains the precursor building block molecules of collagen, *tropocollagen*.

Collagen: Self-Assembled Hierarchical Structure. The fibrous collagen strands in the matrix of cartilage have a hierarchical, self-assembling architecture. Tropocollagen building blocks bundle into *fibrils,* and collagen *fibrils* bundle into *fibers*.

Tropocollagen molecules have a typical molecular weight of about 300,000, but there are many variants of different chain lengths. Each tropocollagen building block is composed of three polypeptide strands, each of which is a left-handed helix. Three left-handed helices self-assemble into a right-handed triple helix coil about 300 nm long and 1.5 nm in diameter. The triple coil is not bound by chemical bonds, but is a *cooperative quaternary structure* stabilized by hydrogen bonding.

The bonding of tropocollagen subunits into self-assembled fibrils is controlled by a mixture of strong and weak intermolecular forces. There is some covalent cross-linking between tropocollagen units along the 300-nm long triple helices, and a variable amount of covalent cross-linking between neighboring tropocollagen helices in the fibrils. The result is a large number of different nanostructure combinations, giving rise to many different types of collagen found in tissues. These spontaneously bound structures with their large number of possible arrangements can have different mechanical and structural properties.

Quaisi-Crystalline or Near-Crystalline Structure. Unlike the highly ordered and regularly consistent bonding between molecules in a crystal, the aggregations between proteins are more variable, yet not random. This type of organization of matter is called *quaisi-crystalline, near-crystalline, or cooperative structure.* It is characteristic of very large molecules with both covalent and ionic sites for inter- and intramolecular interactions, both bondings and repulsions. In biomaterials it allows for great variety in nanostructure to produce materials with a great range of properties.

Types of Collagen. There are some 24 different classified types of collagen. In *hyaline* cartilage, collagen units are arranged in cross-striated fibers, 15–45 nm in diameter, called type II collagen. Type II collagen has tropocollagen units stacked in parallel with regularly staggered ends, which give this type of cartilage a characteristic banded appearance under high magnification under the electron microscope. Type II cartilage is the main type found in fully developed cartilage. Type I and type III collagens are less strong and are found in bone and its precursors [34–36].

Buehler has shown in a mathematical model that the force required to separate two fibers of collagen bonded along their length is dependent on the length of the fibers, and that there is an optimal length depending on the structure of the fibers [35,36].

Collagen and Cartilage as Precursors of Bone. Strong fibrous collagen molecules are not only the main structural protein in cartilage, but are also the main organic framework for bone and other supportive tissue in vertebrates. Cartilage is the precursor of bone in the evolutionary and embryonic development of vertebrates, forming the skeleton of the earliest types of vertebrates, of which lampreys, rays, and sharks are modern representatives [37].

The cartilage in sharks and rays consists primarily of a mesh of collagen fibers embedded in a gelatin-like matrix, along with a scattering of cartilage-generating cells called *chondrocytes*. (Sometimes the cartilage is surrounded by a thin layer of mineralized tissue that gives it a little extra stiffness.) The result is softness and flexibility.

Bone. Bone differs from shell in having a phosphorous mineral component that largely replaces calcium carbonate. The first step towards development of bone was the development of proteins, which fixed phosphorous as phosphate, so that the calcium mineralization was supplemented by the calcium phosphate mineral *hydroxyapatite*—$Ca_{10}(PO_4)_6(OH)_2$.

Types of Bone. Bone is formed with differing densities, microstructures, and nanostructures that give each type of bone its structural features. Each type developed to serve a different function in different animals.

Aspidine. The earliest appearance of bony tissue was in jawless fishes with bony mouth plates (similar to those of the hagfishes) and bony armor plating on the exterior of their bodies. These plates, with a characteristic structure called aspidine, had a hierarchical structure and composition like bone. Aspidine was composed mostly of hydroxyapatite; it had a characteristic microstructure, with three distinct layers. Each of the microstructure layers had a different nanostructure, with different densities and orientations of calcium phosphate crystals. The outer layer is a solid, highly mineralized array of dentine tubercles similar to the patterns on some shark teeth, underlain by a porous mineralized, honeycomb. These early animals probably had an endoskeleton made of cartilage, with a small percentage of mineralization, like the toothed but boneless jawed fishes.

Teeth. In the *sharks* and other *jawed fishes* with cartilage endoskeletons, hydroxyapatite serves mainly to provide the strong *enamel* and *dentin* layers of teeth.

Enamel is the hardest and most highly mineralized tissue in vertebrates, consisting mostly of flattened, needle-like crystallites of apatite supported by a small amount of protein, which adds toughness and resists cracking. *Dentin* is a softer, less mineralized, and honeycombed material that supports the enamel shell from within and below, much like the porous aspidine supported the hard outer layer.

Tooth enamel contains more than 90% mineral crystals, of which about 95% is phosphate. The rest of the mineral content is calcium carbonate with traces of magnesium and other alkali metals. Traces of fluoride anion are found in shark and other teeth, adding to hardness and making them less soluble in contact with corrosive agents such as acids and enzymes.

Unlike dentin and bone, tooth enamel does not contain collagen. Instead, it has two unique classes of proteins called *amelogenins* and *enamelins*. While the role of these proteins is not fully understood, it is believed that they serve as a framework in the growth of enamel layers, and they may add some strength by resisting crack propagation.

Because of its high mineral and low protein content, enamel is relatively brittle compared to dentin. For sharks this problem is solved by shedding teeth continually and growing new ones. Like seashells, teeth have a composite nanostructure, which gives them more strength than their simple mineral constituents. Unlike shells, teeth and other bones have a complex inner structure in which living cells continually replenish the mineral and structural protein content.

Nanostructure of Bone Growth. In growth and development from embryo to adult in vertebrates, bone is formed by mineralization of a cartilage scaffolding (**Fig. 13.4**). The evidence given by research is that the mineralization of tissue is associated primarily with the type I collagen fibril, that the fibril is formed first and then mineralized, and that mineralization replaces water within the fibril

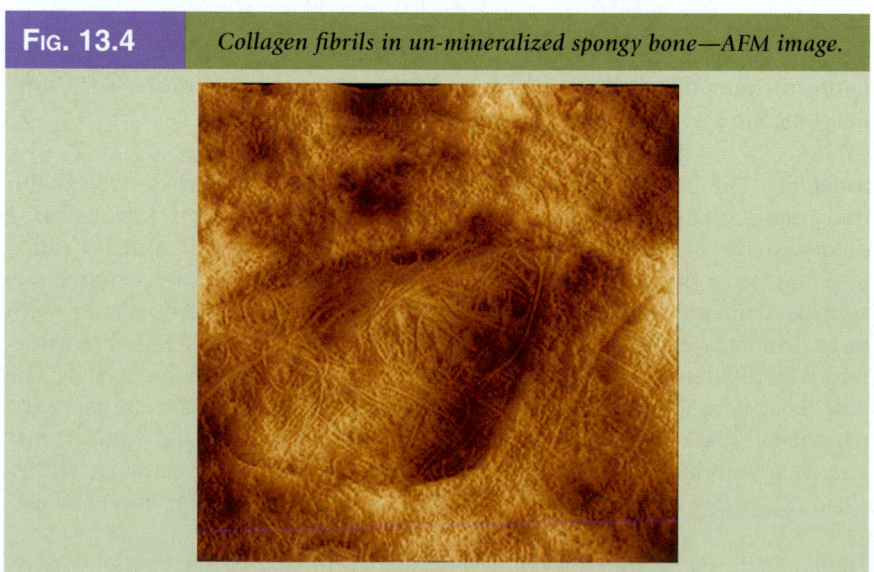

FIG. 13.4 *Collagen fibrils in un-mineralized spongy bone—AFM image.*

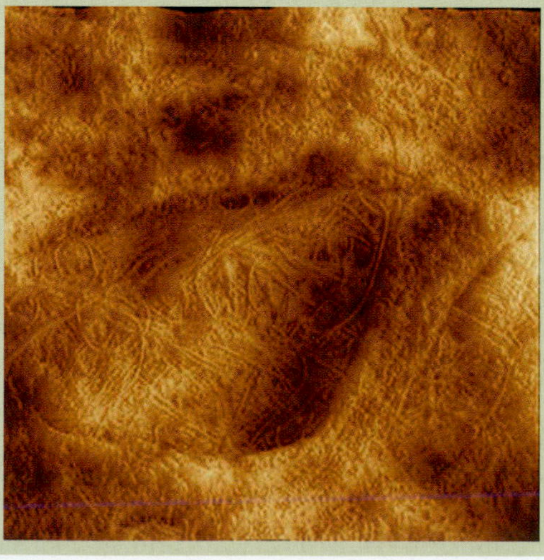

Source: Image courtesy of Veeco Instruments, Inc.

with mineral. The collagen fibril therefore plays an important role in mineralization, providing the aqueous compartment in which bone minerals grow. Collagen remains in the bone and contributes to its elasticity and fracture resistance [38].

Nanostructure of Collagen Mineralization Sites. In bone, the collagen triple helices lie in a parallel, staggered array, with gaps of 40 nm between the ends of the tropocollagen subunits. These spaces seem to serve as nucleation sites for the deposition of long, hard, fine crystals of the mineral component, mainly calcium hydroxyapatite, with some calcium phosphate.

Cortical and Trabecular Bone. Compact, or *cortical,* bone forms the hard dense outer shell of large bones; spongy or *trabecular* bone fills the interior spaces of bones.

Woven and Laminar Bone. Cortical bone is typically either woven or lamellar (layered). *Woven bone* has a small number of randomly oriented collagen fibers; it grows quickly without a preexisting cartilage scaffolding, but is relatively weak. *Lamellar bone* is stronger, formed of numerous stacked layers, or *lamellae,* and filled with collagen fibers. Woven bone is found mainly in areas of regrowth after a break. Most woven bone is gradually replaced by slower growing lamellar bone, on a scaffolding of calcified hyaline cartilage through a process known as *bony substitution*.

In laminar bone, the collagen fibers run parallel to other fibers in the same layer. The fibers run in opposite directions in alternating layers, assisting in the bone's ability to resist torsion forces. Plate-shaped carbonate apatite crystals are arranged in layers across the fibers; like the fibers, the orientations of these crystals are different in alternate layers. At the next higher level of organization in laminar bone, the laminar layers are grouped into cylindrical structures [39].

Mechanical Properties of Bone. Bone has two major characteristics: rigidity to support weight and toughness, the ability to withstand breaking. Bone has compressive strength some two times its torsional or tensional strength. This selectiveness is due to the arrangement of the apatite crystal and the collagen fibers. The apatite is arranged in small crystalline plates about $60 \times 30 \times 8$ nm in size found within and around collagen fibrils. The plates overlap and are aligned parallel to the long axis of the fibril. Also, the plates within each fibril have the same angular orientation, which differs from the orientation in adjacent fibrils. As compression is applied, the matrix material is squeezed together as the collagen deforms to absorb the force. The plates slide towards each other until they start to touch. This initial movement absorbs some of the applied force and prevents the bone from breaking. As the plates touch, frictional forces take over. The plates can no longer slide and the compression force is resisted. What happens when the forces are too great for the crystalline plates? They fracture, but because the plates are so small, large cracks cannot form. Also, because the plates of adjacent fibrils are not oriented at the same angle any small crack that does form cannot propagate easily from one plate to the next.

Piezoelectric Properties of Bone. Like many minerals, hydroxyapatite is piezoelectric; the ionic components of the crystals generate a small electric field in response to stress. Repeated stress, such as weight-bearing exercise or bone healing, results in the bone thickening at the points of maximum stress. It has been hypothesized that this is a result of bone's piezoelectric properties. This would imply a response of the collagen and matrix to the electric fields, which is still an area of active research. We will discuss some aspects of the response of bone to electromagnetic fields in the chapter on biomedical nanotechnology.

Micromechanical Models of Bone. A number of micromechanical mathematical models for the Young's modulus of bone have been developed to account for the strength and anisotropy of the material [40]. The models incorporate the platelet-like geometry of the load-bearing crystals, the alternating thin and thick lamellae, and the orientations of the crystal platelets in the lamellae. The thin and thick lamellae are modeled as orthotropic composite layers made up of thin rectangular apatite platelets within a collagen matrix, and classical orthotropic elasticity theory is used to calculate the Young's modulus of the lamellae. Bone is viewed as an assembly of such orthotropic lamellae bent into cylindrical structures, and having a constant, alternating angle between successive lamellae.

Bony Exoskeletons. The shells of turtles, armadillos, and some other vertebrates are outgrowths of the animal's skin (*dermus*) and endoskeleton. The individual plates are covered by scales called *scutes,* made of the tough protein keratin, like claws and fingernails. Underneath the scute is a layer of dermal tissue and calcified shell, or carapace, which is formed by fusion of vertebrae and ribs during development.

Unlike seashells and arthropods, turtle shells have living cells, blood vessels, and nerves, including a large number of cells on the calcareous shell surface and scattered throughout its interior. Bone cells that cover the surface and are dispersed throughout the shell secrete protein and mineral so that the bone can grow and reshape continuously.

13.2.5 Skin and Its Extensions

Keratin. *Keratin* is a polypeptide-based material with a complex hierarchical structure that is the major component of skin, scales, claws, beaks, feathers, horns, hair, and other hard, tough outgrowths of the dermal layer that covers the bodies of vertebrates. The properties that make structural proteins like keratins useful depend on their supermolecular aggregation. These depend on the properties of the individual polypeptide strands, which depend in turn on their amino acid composition and sequence.

Relation of Molecular Biology and Biochemistry to Nanostructure. The α-helix and β-pleated sheet motifs, and disulfide bridges, are crucial to the conformations of globular, functional proteins like enzymes, many of which operate semi-independently, but they take on a completely dominant role in the architecture and aggregation of keratins, just as in collagens.

Hair, wool, and related fibers all have keratins as their principal constituent. Their different physical properties are due to differences in their protein arrangement at the nanoscale level. The polypeptide chains of keratin are folded, but when stretched, the chain unfolds. The unfolding is reversible, giving the fibers their elastic properties (the stretched form of keratin is called β-keratin and the folded form is called α-keratin). Keratins can have different degrees of folding depending upon the side chains attached to the polypeptide. We will see more details of polypeptides in chapter 14, "Biomolecular Nanoscience".

Glycine and Alanine. Keratins contain a high proportion of the smallest of the 20 amino acids, glycine, whose "side group" is a single hydrogen atom, as well as the next smallest, alanine, with a small and uncharged methyl group. In the case of β-sheets, this allows sterically unhindered hydrogen bonding between the amino and carboxyl groups of peptide bonds on adjacent protein chains, facilitating their close alignment and strong binding. Fibrous keratin molecules can twist around each other to form helical intermediate filaments.

Limited interior space is the reason why the triple helix of the (unrelated) structural protein collagen, found in skin, cartilage, and bone, likewise has a high percentage of glycine. The connective tissue protein elastin also has a high percentage of both glycine and alanine. Silk fibroin can have these two as 75–80% of the total, with 10–15% serine, and the rest having bulky side groups. The chains are antiparallel, with an alternating C–N orientation. A preponderance of amino acids with small, unreactive side groups is characteristic of structural proteins, for which H-bonded close packing is more important than chemical specificity.

Disulfide Bridges. In addition to intra- and intermolecular hydrogen bonds, keratins have large amounts of the sulfur-containing amino acid cysteine, required for the disulfide bridges that confer additional strength and rigidity by permanent, thermally stable cross-linking—a role sulfur bridges also play in vulcanized rubber. Human hair is approximately 14% cysteine. The pungent smells of burning hair and rubber are due to the sulfur compounds formed. Extensive disulfide bonding contributes to the insolubility of keratins, except in dissociating or reducing agents such as urea.

The more flexible and elastic keratins of hair have fewer interchain disulfide bridges than the keratins in mammalian fingernails, hooves, and claws (homologous structures), which are harder and more like their analogs in other vertebrate classes. Hair and other keratins consist of α-helically coiled single protein strands (with regular intrachain H-bonding), which are then further twisted into superhelical ropes that may be further coiled. The keratins of reptiles and birds have β-pleated sheets twisted together, then stabilized and hardened by disulfide bridges.

Feathers. Feathers, produced by birds and their extinct relatives, are keratin structures, among the most complex structural organs found in vertebrates. The keratins in feathers, beaks, and claws—and the claws, scales, and shells of reptiles—are composed of protein strands hydrogen-bonded into β-pleated sheets, which are then further twisted and cross-linked by disulfide bridges into structures even tougher than the keratins of mammalian hair, horns, and hoof. The complex micro- and nanostructure of feathers controls the flow of air and heat around the surface of the animal, yielding exceptionally high performance in aerodynamics and insulation. An interesting aspect of the nanoscience of feathers is the aerodynamic interaction on the nanoscale between the complex surface of the finely divided feather structure and the surrounding air.

Feathers are attractive for humans for their beauty of form and color. A nanoscale phenomenon called the *Dyck texture* is what causes the colors blue and green in most parrots. This is due to an optical effect caused by the nanostructure of the feather itself, rather than pigment, or the *Tyndall effect*, as was previously believed (dispersion of shorter light wavelengths).

Hair. Hair, produced by mammals in many forms, has varied complex hierarchical keratin structures, which give hair it curliness and resilience. The nanostructure of the keratin fiber arrangement in hair has been studied using scanning microbeam small-angle x-ray scattering (SAXS). Scanning microbeam SAXS patterns of hair fibers can reveal the differences in patterns between the inner and outer sides of the curvature in curly hair. In very curly Merino and Romny wool, different types of cortices exist; the macroscopic curl of the hair fibers arises from inhomogeneous nanoscale distribution of two types of cortices. The same effect is observed for human hair [41].

Wools. Wool has two qualities that distinguish it from hair or fur: It has scales that overlap like shingles on a roof and it is crimped; in some fleeces the wool fibers have more than 20 bends per inch. Because of its economic importance, wool fiber keratins were one of the first proteins whose folding behavior was understood. X-ray crystallography was used to reveal differences in regular spacings of the protein structure between relaxed and stretched wool fibers [42]. The molecular conformation and the microstructure of wool fibers can also be studied by newer techniques, such as Raman spectroscopy and atomic force microscopy. Raman spectroscopy shows absorption by symmetrical vibrations, so it gives information about the C–C skeletal vibration region, and the S–S and C–S bond vibration regions or keratins. Raman spectroscopy experiments can reveal the secondary structural transformation from α-helical to β-pleated sheets, which takes place in wool fiber stretching. The stretching mechanism of wool fibers can be divided into two different mechanisms at different levels of the nanoscale structural hierarchy: (1) slippage of the folds of the polypeptide chain on the protein molecular level and (2) disordered conformation of fibrils and breaking of some portion of the S–S bonds linking keratin fibers at the supramolecular level [43].

13.2.6 Summary

In this section we have seen how the nanoscale architecture of natural biomaterials determines their properties. Polymers and proteins with strong covalent bonds provide tensile strength. Hydrogen bonding and cross-linking in proteins and complex polymers provide resilience and flexibility. And ionic bonded mineral crystals provide load-bearing and compression strength. Combinations of these materials in a hierarchy of organization from the molecular to nano- to micro- to macroscale create optimized material properties for different requirements. These materials provide examples of important principles of structure on the nanoscale.

An important theme is how proteins orchestrate the architecture of natural biomaterials on the nanoscale, sometimes with the aid of polymer scaffolds. Mineral crystals, micelles, the bilayer membranes of cells, and polymer scaffolds are self-assembling: They organize into predetermined shapes and networks based on the molecular structure. The proteins, by contrast, are synthesized by a bottom-up, serial mechanism and they perform their catalytic control of the building of biomaterials in a bottom-up fashion, intervening in the crystallization process of mineralization and in the extension of polymer meshes. The result is an intricate highly functional nanostructure.

In addition to strength and elasticity, the nanostructure of biomaterials gives rise to optical, electrical, hydrodynamic, and aerodynamic properties—for example, in diatoms, nautilus shell, and feathers. We will provide examples in the following sections on plant fibers, butterfly wings, lotus leaves, gecko feet, and others.

13.3 Nanomaterials Derived from Cell Walls

Plants and some bacteria differ from animals in having a cell wall surrounding their cellular membranes. In bacteria and most plants the cell wall is made of cellulose or a similar carbohydrate material. These cell walls have an intricate nanostructure. When the organisms die, their cellular cytoplasm and membranes are dissolved away, leaving the cell wall as a nanomaterial residue.

13.3.1 Paper

Paper is the most common familiar substance made from the cellulose residue of digested reeds, stems, wood, or fibers such as cotton. The oldest form of paper was made by the ancient Egyptians from the papyrus reed that grows on the Nile River, from which our word "paper" derives. Although paper is an artificial material, it consists of natural *cellulose* that has been purified and processed. Its properties that we value, such as absorbency and strength, depend on the nanostructure of the cellulose fibers. The microstructure and nanostructure of the fibers allows them to entangle and link by hydrogen bonding to form strong mats. The process of papermaking harnesses the natural self-assembling properties of the fibers. The nanostructure of the pores made by the plant cells makes paper fibers absorbent to water and inks, and makes it possible to hold in fillers such as clays and titanium dioxide to impart sheen and smooth finishes.

Ancient papermaking consisted essentially of gathering, pulping, and pressing the natural materials. Today, processing of natural cellulotic materials to make various types of papers is an extremely sophisticated technological field involving chemistry, mechanical engineering, and nanotechnology. Many types of fiber and fillers are added to modern papers. The appreciation of how nanoscience can be applied to paper is relatively new and is covered in a later section under nanotechnology.

At Louisiana Tech University, the potential for applying nanoscience to improved papermaking is being developed in a pioneering research and education program [44]. Canadian companies are also rapidly developing ways to improve paper through the application of nanoscience [45]. Similar developments are taking place in Europe, Israel, and the Far East.

13.3.2 Cotton

Cotton is a natural plant material whose nanostructure gives it highly valued properties. Cleaned cotton fibers consist of almost pure *cellulose*, the same material from which paper is made; however, the nanostructure of cotton makes it unique. Cotton has been prized for centuries for its absorbent wicking properties, which make it comfortable in warm climates, make excellent bandages and wound dressings, and allow cotton fabrics to be easily dyed with colorful pigments. These characteristics are due to the nanostructure of the fibers as produced by the cotton plant (genus *Gossypium*) in its seed pod.

The high strength, durability, and absorbency of cotton are due to the nanoscale arrangement of the fibers. Each fiber is made up of 20–30 or more concentric layers of cellulose, called *lamenae*, coiled in a spiral winding to form a series of natural springs. When the cotton boll is opened, the fibers dry into flat, twisted, ribbon-like shapes and become kinked together and interlocked. This interlocked form is ideal for spinning into a fine yarn.

Natural cotton fibers are covered with a cuticle layer of waxes and pectin, which is removed by cleaning. The outer wall of the fiber is composed of crystalline cellulosic fibrils. Inside this wall are three layers of closely packed spiraling parallel fibrils. The direction of the spiral is reversed in the middle layer relative to the two layers that enclose it, adding strength and elasticity to the fiber. The center of the fiber is the lumen, which contains the cell contents in the immature fiber and is empty when the cotton boll is fully ripe and open. The cross-section of the fiber is kidney or bean shaped when dry and may swell to a round shape when it absorbs moisture. The fibrils swell with absorption of moisture—more so in the immature fiber because it has less crystalline cellulose.

Inside the spiraling fibrils is a level of supramolecular nanostructure that optimizes the performance of the fiber. The cellulose molecules in cotton fibrils are organized into parallel formations called crystallites. Each elementary fibril is made up of 36 cellulose chains, which make up one basic crystalline unit of cotton cellulose. In the fiber, elementary fibrils are gathered into microfibrils and larger aggregates called macrofibrils. The diameters of elementary fibrils measure 3.5–10 nm wide, the microfibrils 10–40 nm, and the macrofibrils 60–300 nm. The exact arrangement of fibrils determines the properties of the fiber (**Fig. 13.5**).

The cellulose in the fibrils is laid down in a series of concentric growth rings, the lamellae, whose number and properties are affected by temperature fluctuations and moisture in the plant during fiber development. Up to 50 lamellae have been reported in fibers from some plants, but between 25 and 40 lamellae are more commonly found in a mature fiber. Different growing conditions, climate, and variety of plant can yield different qualities of fiber.

Various forms of microscopy have been used to elucidate the morphology and internal structure of cotton. With transmission electron microscopy (TEM), fibrils on the surface of cotton fibers and in thin transverse sections can be resolved, but at the cost of pretreatments that may alter the natural state or introduce artifacts.

Fig. 13.5 *Cotton fiber structure. This representation of a cotton fiber shows the following layers from outermost to innermost: cuticle (outer covering rich in pectins and waxes), outer crystallite celluloid layer, three layers of spiraling fibrils (directions of spirals alternates), and lumen.*

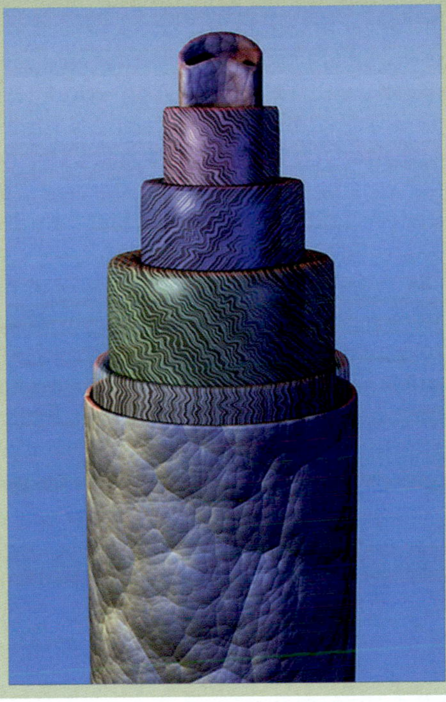

Source: Image by Anil K. Rao.

The scanning probe microscope (SPM) technique is nondestructive, and permits examining cotton fibers in ambient conditions in the presence of natural humidity. This technique uses a probe tip and measures the force changes between the tip and the sample as the tip is scanned. Information about the nanomechanical properties of the fibers can be obtained by operating SPM in force mode. In this mode, the force as a function of distance between the tip and the sample is recorded. The AFM and the scanning tunneling microscope (STM) have also been used to investigate the surface structure of cotton fibers [46,47].

In the following sections we will look at some natural nanomaterials synthesized by single-celled organisms, yeasts, fungi, and plants. Their nanomaterials are made from organic carbon polymers and from silicon.

13.3.3 Bacterial Fibers

Cellulosic nanomaterials are produced by single-celled organisms such as algae and bacteria as well as by plants. Some of these organisms produce useful materials that have familiar uses in many cultures, as well as promising new uses. In this section we show some examples.

Certain strains of *Acetobacter xylinum* bacteria produce fine cellulose fibrils with a diameter of 3–8 nm and a length of 50–80 nm. Traditionally, this bacterium is used in the Philippines to ferment coconut milk, producing a jelly-like food base known as *nata de coco* (Spanish, "cream of coconut"). This jelly is an ingredient used in Asian desserts. It can be prepared as a jelly, an agar, or a pudding with gelatin features. Nata de coco has also been used as a sunblock ingredient and moisturizer in lotions and cosmetics.

These bacterial fibers form an entangled mass, making a three-dimensional nanonetwork with extremely small pores. The cellulose produced by these bacteria has very different characteristics from most other naturally occurring cellulose. Bacterial cellulose fibrils are relatively insoluble because of their high molecular weight, stiffness, and extensive hydrogen bonding. The stiffness of the gels led to traditional use as a food additive when prepared as emulsions of 99% water and 1% cellulose.

Acetobacter xylinum bacterial cellulose can be pressed and dried into sheets with remarkable mechanical strength; the Young's modulus is as high as >15 GPa across the plane of the sheet. The high Young's modulus has been attributed to the unique supermolecular nanostructure in which fibrils of biological origin are preserved and bound tightly by hydrogen bonds. A team of scientists and engineers led by Sony in Japan have studied the mechanical properties of cellulose sheets and used the material in a commercial speakerphone. Because of their strength, stability, and optical transparency, these fibers have also been used as reinforcement material in flexible displays for televisions, personal computers, and portable phones developed in Japan by NTT, Mitsubishi, and others.

Researchers led by Dr. Mari Tabuchi in Japan have found new uses for this unique natural nanomaterial. Recognizing the potential of its polymer strength, insolubility, and nanoporiosity, they have used it to produce high-performance nanopore filters for a nanoscale gene detection system. The filters made from bacterial cellulose resisted up to 2 MPa of pressure without any clogging, whereas conventional membranes were capable of less than 0.5 MPa. The bacterial gel also gives improved performance as a separation medium for macromolecules and for light amplification in the system's photodetectors.

Pulps made from bacterial cellulose give a strong paper and are used for reinforcing conventional pulp papers and enabling paper-making from some fibrous materials, *Acetobacter xylinum*, cyanobacteria, and algae [48–50].

Biofilms. Bacteria can attach to surfaces where they grow in colonies. They secrete a complex matrix, which holds the colony in place, called a *biofilm*. Common examples of biofilms are the slippery coating in a stale puddle of water or in a vase of flowers, and dental plaque on teeth. Biofilms are important for health, sanitation, and their clogging of equipment in water. In this section we look at the nanostructure of the biofilm polymer matrix and how it is secreted.

Bacteria and other microorganisms that form biofilms synthesize an extracellular *polysaccharide* matrix, sometimes called EPS for *extracellular polymeric substance*. Polysaccharides are polymers made up of sugar units (or *saccharides*) bonded together. These large molecules are amorphous, adhesive, insoluble, and resistant to attack by enzymes that would normally attack sugars and starches. They make an ideal protective home for colonies of bacteria living in water.

This polymer matrix protects the cells in the biofilm. Channels within the matrix help distribute nutrients and signaling molecules. The tough biofilm matrix may persist after the bacteria die and may even become fossilized. Fossilized biofilm matrices are among the earliest evidence for primitive life on Earth.

How are bacteria and fungi able to stick to surfaces and form such strong films, even in flowing water? Cells use special protein molecules and hairlike nanostructures called *pili* (Latin for "hairs") to attach to surfaces and to each other. Each pilus (Latin for "hair" [singular]) is about 10 nm in diameter and may be 100+ nm in length. Pili are composed of proteins with binding sites specialized for receptors on the surfaces of other bacteria or on environmental surfaces. After attachment with the pili, cells begin to secrete the organized biofilm matrix. Once an attachment to a surface has been made, it is easier for other cells to attach to the first ones, and the biofilm can grow to macroscopic dimensions.

Cell cultures growing in water often form a surface biofilm called a *pellicle*. Complex pellicles can be formed on the surface of fermenting food preparations or on any liquid in which organic matter is decomposing. Cut flowers left in a water vase generate pellicles that are easy to harvest and study under the microscope. Some bacteria form stable pellicles with nanopores and other interesting properties.

13.3.4 Diatoms

One class of microorganisms, the *diatoms*, have cell walls composed of a silica skeleton with an overlay of pectin. The rigid silica cell walls support the pectin just like the cellulose walls of bacteria and most plants. But the silica is a permanent nondegradable material that accumulates in sediments after the cell dies. Like other microorganisms, diatoms are very abundant in the oceans. There can be anywhere from thousands to hundreds of millions of phytoplankton cells in 1 L of seawater, depending on nutrients, temperature, and light [51,52].

Sediments composed largely of diatom skeletons are found in some places and are known as *diatomaceous earths*. The fossil record of unique species preserved in the sediments makes diatoms a useful tool in dating sediments for modern ecological and evolutionary researchers.

The silica skeletons of diatoms, called *frustules*, have a complex micro- and nanostructure. There are more than 200 genera of diatoms; hundreds of thousand of species have been identified, with a great variety of intricate skeletal shapes and patterns. Diatom skeletons range in size from 1 or 2 μm to hundreds of microns. Their intricate substructures and pores are on the nanometer to micrometer scale. Pores can vary from 20 to 200 nm in diameter in different species. Some species have a finer network of secondary pores within the primary pores (**Fig. 13.6**).

The detailed structure of diatom frustules is below the resolving power of ordinary light microscopes. Nanoscale features of diatoms can be characterized using SEM, AFM, EDAX, and secondary ion mass spectrometry (SIMS).

Adhesive Nanofibers from Diatoms. Some species of diatom extrude adhesive nanofibers, which are used to anchor the cells in biofilm colonies. Researchers in Australia have examined the adhesive and mechanical properties of these

FIG. 13.6 *Diatom cells illuminated from the side display iridescence by trapping light. This is due to the interaction of their nanoscale structure with the wavelengths of visible light [51,52].*

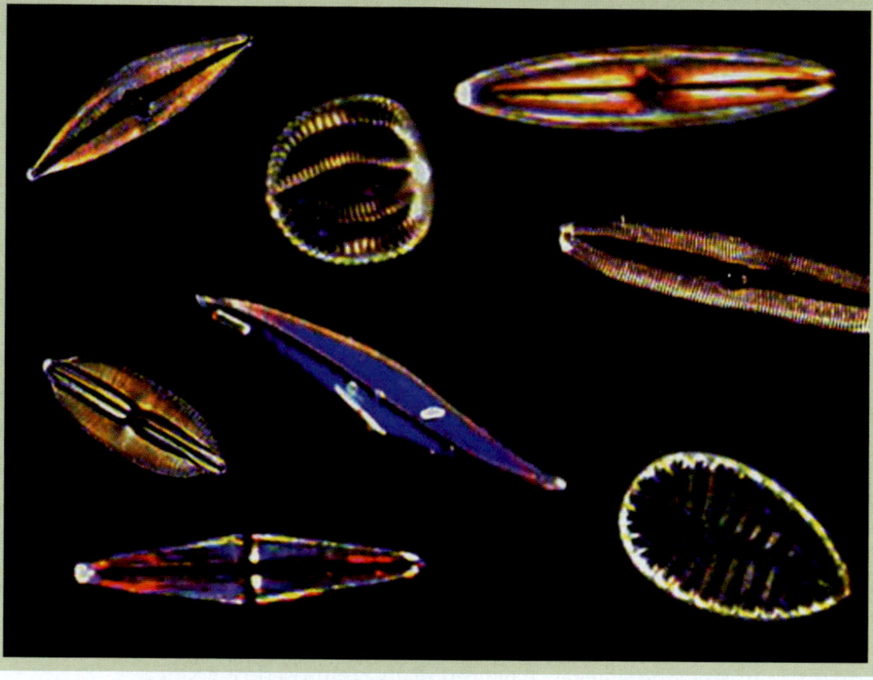

Source: Image courtesy of Ely Silk (www.viewsfromscience.com).

fibers using atomic force microscopy on the live diatom cells. The resulting force curves have a regular sawtooth pattern, the characteristic fingerprint of modular proteins. The protein fibers do not lose their elasticity even when stretched and relaxed for up to 600 repetitions. The high force required for bond rupture, high extensibility (~1.2 μm), and the accurate and rapid refolding upon relaxation together provide strong and flexible properties ideally suited for the cell-substratum adhesion of fouling diatoms. Studies of natural nanofibers like these allow us to understand the mechanism responsible for the strength of adhesion, and may one day lead to improved man-made adhesives and fibers [53].

13.3.5 Lotus Flower

The sacred lotus, *Nelumbo nucifera*, is an aquatic perennial, flowering plant. The plant roots itself to the bottom of ponds and rivers, with the leaves floating on the surface. Supported by thick stems, the lotus flowers rise several centimeters above the leaves. The sacred lotus is indigenous to a wide area, extending from eastern Asia to the Middle East. The plants were introduced to Europe in the eighteenth century and are now found worldwide, where they are typically raised in botanical green houses. *N. nucifera* flowers are used to supplement food as a garnish and add flavor to a variety of dishes. The petals are also used to make lotus tea and are considered an herbal remedy.

The sacred lotus is very important in practices of the Asian religions, Hinduism and Buddhism. It is revered due to the striking white flower that remains free of dirt even after having been submerged in muddy river water. It is a symbol of purity. From Bhagavad Gita 5.10, "One who performs his duty without attachment, surrendering the results unto the Supreme Lord, is unaffected by sinful action, as the lotus leaf is untouched by water."

The quote from the Bhagavad Gita illustrates a remarkable property of the lotus flower: its ability to easily shed water and dirt. This confers several survival benefits. By preventing plant pathogenic microorganisms (e.g., bacteria and virus) from gaining a foothold on the plant surface, the lotus can avoid some natural infections. Also, clean plant surfaces have been associated with decreased plant temperatures. Rates of diffusion through the stomata (pores in the plant epidermis) of water and respiratory gases are higher.

Apparently, surface adhesion to dirt is reduced by a combination of a waxy surface coating and microtopography of the plant surface. Closer examination reveals that the flower petals are essentially nonwettable.

The wettability of a surface is related to the *contact angle* (CA) between a droplet of liquid and the horizontal surface upon which it rests. The CA value results from the forces that develop between the interfaces of liquid, gas, and solid phases of the system. These *surface tension* forces are modified by the physical and chemical properties of the materials (liquid, gas, or solid) that comprise the system. If the angle is between 0° and 90°, the surface is considered wettable, with the former value denoting complete wettability. The result is that the liquid will spread out to form a monomolecular thin film on the horizontal surface.

If the contact angle of the system is between 90° and 180°, the surface is nonwettable; that is, the liquid drop is in contact with the surface at a single point and exhibits low interfacial tension. Although this state is difficult to achieve in real plant systems, they can get close. This condition is called *superhydrophobicity*. Because the CA between the liquid and the surface is very high (e.g., water drops on a waxed surface), the drops bead up and, if the surface is not normal to gravity, will roll off. Much as a snowball collects more snow as it rolls, the drops collect dust and dirt and carry them away, leaving the petal surface clean.

Wilhelm Barthlott and Christoph Neinhuis first noted that the combination of surface topography and chemistry of leaves and petals of the sacred lotus exhibited superhydrophobicity. The lotus cuticle consists of a combination of an insoluble polyester polymer called *cutin* impregnated with and covered by lipids called *epicuticular wax crystalloids*. The waxes consist of long-chain hydrocarbons that self-assemble to form the nanometer-sized crystalloids. These appear to be partially responsible for the surface hydrophobicity.

The surface of the lotus flower has an additional adaptation that increases the hydrophobicity effect. Scanning electronmicrographs show a very bumpy surface, consisting of a series of micrometer-sized projections called *epidermal papillae* or *trichomes*. Some plants have a combination of a waxy cuticle and low trichomes, which creates a relatively smooth surface. The droplets slide rather than roll over the surface. Dirt and dust are not picked up as efficiently. The sacred lotus has a very rough surface produced by underlying epidermal cells, which push upwards to create numerous trichomes. The spaces between the trichomes trap air and the hydrophobic epicuticular waxes exclude water. This reduces the contact area between the water drops and decreases wettability. As a result, water droplets ride on the tips of the projections and a cushion of air.

FIG. 13.7 *Water droplets bead on a lotus petal.*

Source: Lotus flower image courtesy of Gary B. Lawrence.

However, the wettability of dirt particles is increased and they adhere to the water droplet. The drops roll, trapping the particles and effectively cleaning the petal surface (**Fig. 13.7**).

The sacred lotus is an example of a surface that, due to the physical and chemical conditions at the micro- and nanometer scale, is able to produce a remarkable self-cleaning effect, which aids the survival of the plant. The lotus effect can be observed in other organisms, such as tulips and water cress; in the wings of butterflies and chitinous coverings; and in the feet of the gecko [54–57].

13.4 Nanomaterials in Insects

In this section we present a discussion on the nanostructure of some insects, including the truly remarkable and colorful nature of butterfly wings, and how their properties are due to their nanostructure.

13.4.1 Chitin

Chitin is a hard translucent natural material made by plants, insects, and other animals. Chitin is a major component of insect bodies and wings. It is also the main component of the exoskeletons of arthropods such as the crustaceans (crab, lobster, and shrimp) and of the beaks of cephalopods (squid and octopus).

Like cellulose, chitin is one of many naturally occurring polymers synthesized by organisms to give structure, strength, and protection to cells and bodies. Chitin and cellulose are closely related in molecular structure: Both are long, unbranched chains of glucose derivatives. Chitin is a long-chain *polysaccharide*, a polymer of *β-glucose* sugar units. The chemical formula of chitin is $(C_8H_{13}O_5N)_n$. Chitin differs chemically from cellulose by having one hydroxyl group on each glucose monomer substituted with an *acetylamine* group. The nitrogen in the amine group leads to stronger hydrogen bonding between adjacent polymer chains, making the chitin–polymer matrix stronger than cellulose.

The polysaccharide chitin polymer is translucent, pliable, strong, and resilient. In arthropods such as insects, the chitin exoskeleton may be made tougher by being embedded in a hardened proteinaceous matrix. The nanostructure of chitin polymer is organized into layers and interlocking structures whose growth is guided by special chitin-binding matrix proteins. These matrix proteins have chitin-binding domains that bind to saccharides and catalyze their polymerization into polysaccharide chitins. The amino acid sequences determine the folding of the proteins, which in turn determines the nanostructure of the chitin polymer material.

The larval stage of insects frequently has body walls with less protein than the adults. The strength given to the polysaccharide by intertwined protein chains can be seen by comparing the body walls of a caterpillar and a beetle. This is an example of supramolecular strengthening of materials on the nanoscale, as we saw in detail earlier when we discussed shell and bone. Next we will look at some interesting optical properties that are due to nanoscale structure in insect wings [58].

13.4.2 Chitin Structures in Insect Wings

Different types of organisms have different proteins that build structures adapted for the organism's environmental niche. A rich variety of structures have evolved over time and have been selected to perform functions that aid in species survival. Insects provide many examples of chitin nanostructures. By studying these nanostructures we can learn how insects do some of the remarkable and strange things of which they are capable. We may even be able to adapt these properties to artificial nanomaterials for new and useful applications.

Some insects such as the cicada (*Pflatoda claripennis*) and termite (family Rhinotermitidae) have ordered hexagonal packed array structures on their wings made of chitin. The spacings of the arrays range from 200 to 1000 nm. The structures tend to have a rounded shape at the apex and protrude some 150–350 nm out from the surface plane. Wing structures with spacings at the lower end of the range may be adapted to serve as an optimized antireflective coating (a kind of natural "stealth technology"). These close-spaced arrays may also act as a self-cleaning coating (another example of the *lotus effect* described in the previous section).

Structures with spacing at the upper end of the range might provide mechanical strength to bear loads imposed in flight. Arrays with these larger spacings might control the flow of air over the wing to improve the aerodynamic efficiency of the insect. The research team of Watson and Watson has demonstrated the multipurpose design of natural structures such as these in its studies [59].

13.4.3 Butterfly Wings

A butterfly's head and chest are covered with plates of hard chitin, while the abdomen is covered with soft chitin. The wings of a butterfly are a chitinous membrane. The brightly colored wings of some Lepidoptera species of butterfly, such as *Morpho rhetenor*, and moths, such as *hawkmoths*, are a consequence of the nanotopography of the wing's surface and its interaction with light [60]. The resulting colors are quite striking and visible at great distances, reportedly up to half a mile. There have been a number of hypotheses for this coloration. The most common explanation for why such a feature evolved is that the brilliant colors are used for communication and mate selection and have an obvious survival benefit for the species.

The wings also exhibit *iridescence*, which is defined as a shift in the color of an object when observed from different positions. This characteristic is easily demonstrated in optical storage media, such as compact discs (CDs). The very fine submicrometer scale repeating rows of pits, which encode digital information on the CD's surface, cause the light falling on them to be diffracted into the range of colors that comprise the visible spectrum. Each frequency of light is visible at a particular angle with respect to the surface of the disc. In effect, the CD acts as a diffraction grating that exhibits iridescence. Natural diffraction gratings appear to be reasonably common and are observed in a diversity of land and sea organisms, such as crustaceans (the seed shrimp, *Cyridinid ostracod)*, the scarab family Scarabaeidae, and other insects.

13.4.4 Color and Structure

The color of some materials is determined by a *pigment*—a chemical that absorbs certain frequencies of light and reflects others. This type of color is called *chemical color. Iridescence* or *physical color*, on the other hand, is a result of the interaction of light with the physical structure of the surface. The structures that interact with wavelengths of visible light are nanoscale in size. When light strikes a transparent surface some of the light is reflected. A few of the light rays penetrate into the material, reflect off the bottom surface, and pass upwards to join the light rays reflected from the upper surface.

If the two rays are *in phase* they combine to form a bright color. This is called *constructive interference*. If the rays are out of phase, the rays cancel and the light dims. This is *destructive interference*. The color, intensity, and angles of iridescence are dependent on the thickness and the refractive index of the substrate, and on the incident angle and frequency of the light striking the surface (**Fig 13.8**).

Natural iridescence is produced by several mechanisms. The most famous is that of the opal, the sedimentary mineral prized for its gem-like qualities. Here the iridescence is produced by packed silica spheres in the nanometer range. In opals that produce strong iridescence, the spheres are uniform in size and are arranged in layers. This provides the appropriate conditions for interference. Another example is the iridescence of diatoms, which also behave as photonic crystals due to their nanoscale structure (**Fig. 13.6**).

Butterflies and moths produce iridescence in their own unique way. A closer look at the wing surface shows rows of scale arranged much like tiles on a roof (**Fig. 13.9**). Each scale has smaller structures on its surface (**Fig. 13.10**). There are raised ribs that run the length of the scale and are spaced evenly part. There

FIG. 13.8 *Constructive and destructive interference of reflected light at a surface.*

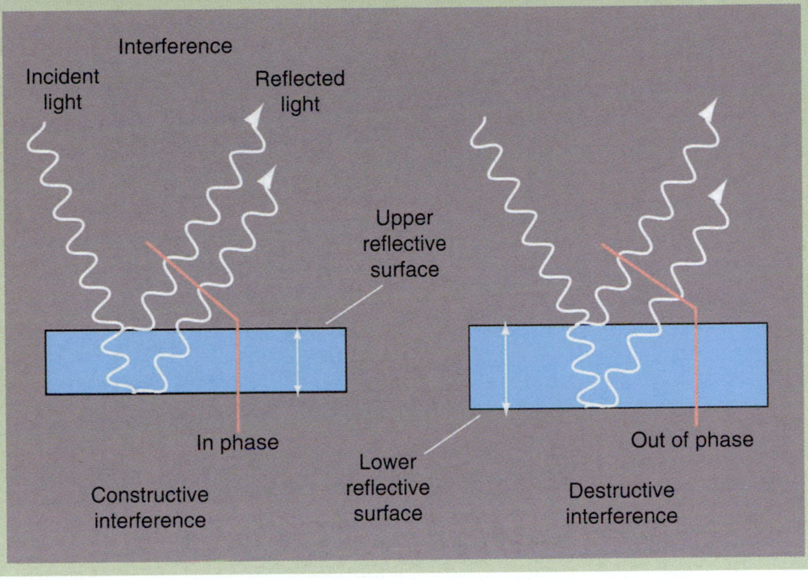

Source: Image by Anil K. Rao.

are even smaller, uniformly spaced "microribs" that interconnect the larger ribs (Fig. 13.11).

At a microscopic size the surface of the scale has a very intricate and highly ordered structure. The size of the spacing of the ribs and spars is comparable to the wavelength of light; about 400–700 nm. The spaces between the ribs and spars form natural photonic crystals or diffraction gratings that can generate constructive and destructive interference. A *photonic crystal* is a periodic nanostructure that can modify the passage of light rays. The refractive indices of the materials and the cavities restrict the frequencies of light that can propagate well. A particular

FIG. 13.9 *Butterfly wing with scales arranged like tiles on a roof.*

Source: Image courtesy of Dr. Peter Vukusic, University of Exeter.

Fig. 13.10 *Butterfly wing scales detail—the parallel ribs running the length of the scales are just resolved in this image.*

Source: Image courtesy of Dr. Peter Vukusic, University of Exeter.

system excludes the propagation of a particular set of frequencies; this is called a *photonic band gap*. The gaps will appear dark and represent the frequencies where the light rays are out of phase and undergo destructive interference.

Morpho Butterfly. The architecture of the surface of the scales varies between species. The male *Morpho rhetenor*, with its dazzling blue wings (capable of reflecting 80% of the incident light at the blue wavelengths), has a series of very small longitudinal ribs on the wing scale that in cross-section look like evergreen

Fig. 13.11 *Butterfly wing microribs are evenly spaced on scale surface like a diffraction grating with spacing of 400–500 nm.*

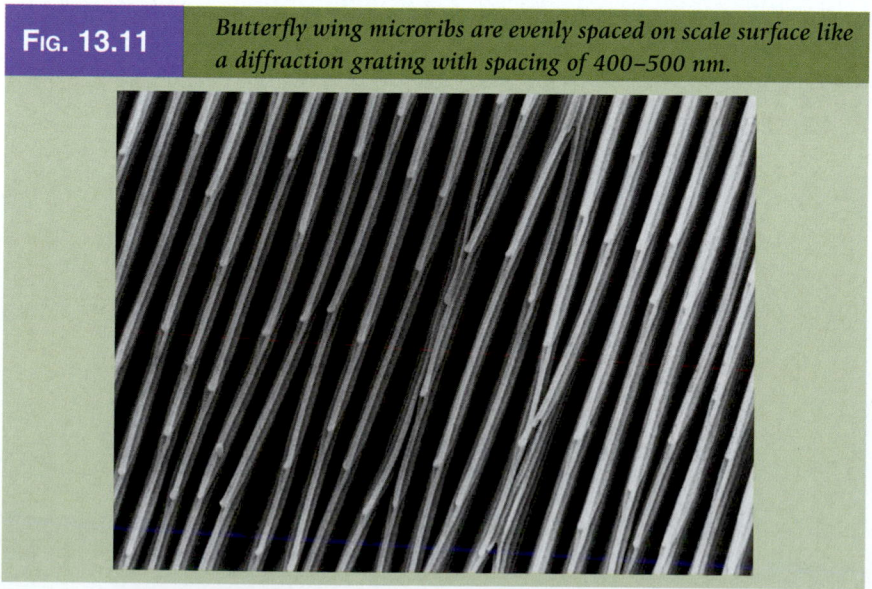

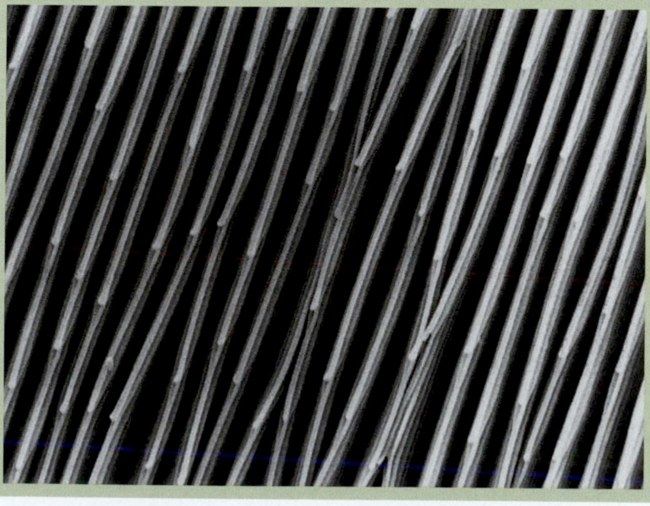

Source: Image courtesy of Dr. Peter Vukusic, University of Exeter.

Fig. 13.12 *Nanoscale image of cross section of butterfly wing microribs, showing ridged structure.*

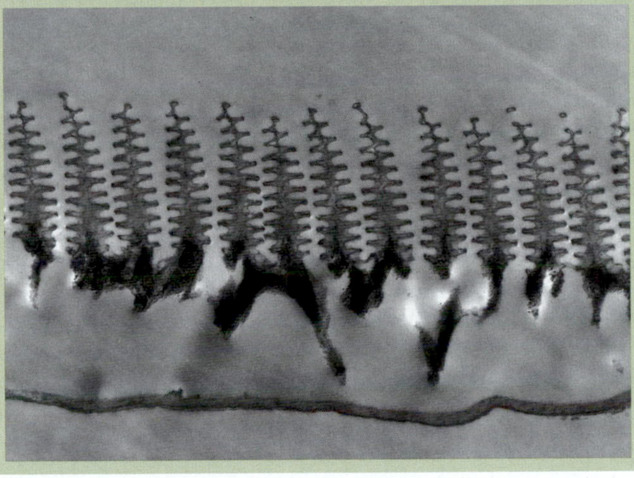

Source: Image courtesy of Dr. Peter Vukusic, University of Exeter.

trees. These are called *setae*. The ridges are made of a *cuticle* layer, composed of chitin polymerized from glucose (**Fig. 13.12**).

The lowest "branches" are the longest and they get smaller towards the tip of the setae. Each branch is transparent and separated from the next by an air-filled space. This alternating cuticle/air structure, with a refractive index ratio of 1.5:1, creates the right condition for interference to occur (**Fig. 13.13**). As light falls on

Fig. 13.13 *Nanoscale schematic of morpho microridge with light interface.*

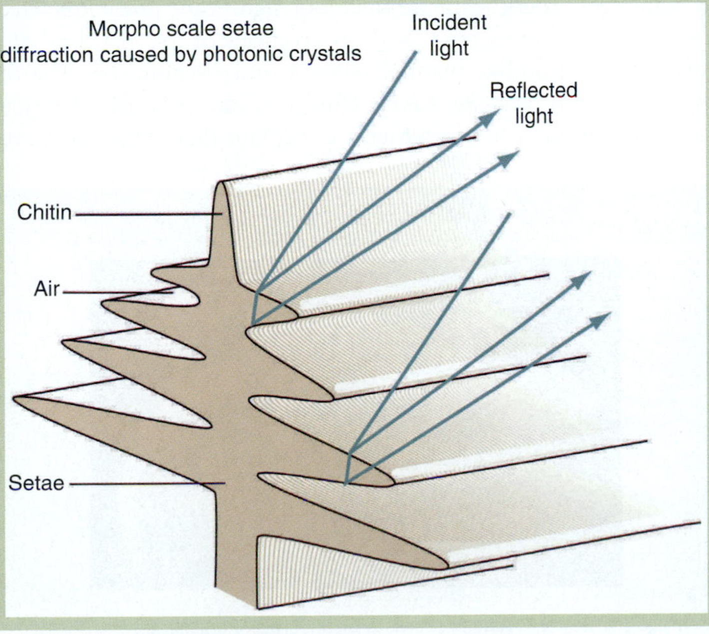

Source: Image courtesy of Anil K. Rao.

the setae it first strikes the upper branches and, as noted before, some of the light passes through the cuticle and some is reflected. Because the lower branches project further out than the higher ones, they also transmit, refract, and interfere with light. The cumulative effect is to produce constructive interference in the blue wavelengths and generate a strong blue color. This is the characteristic appearance of the wings of the *Morpho* butterfly.

In some moth species the nanotopography is different. Rather than surface projections, the series of transparent ribs, microribs, and the smaller nanostructures enclose subsurface cavities. This creates a spongy volume that has been referred to as the "pepper-pot structure." Again, these structures have a size range similar to that of the wavelength light and can generate interference. In many ways, this arrangement is similar to the iridescent opal.

The basic theme of objects that show iridescence is that diffraction and refraction of light is caused by the physical surface itself, not necessarily the chemical makeup of the underlying substrate. This characteristic has been utilized by a variety of systems, with varying architectures, to generate the brilliant and shifting colors associated with spectacular iridescence.

13.5 Gecko Feet: Adhesive Nanostructures

Many members of the family Gekkonidae, commonly called the gecko lizard, have the ability to cling to virtually any surface and at any orientation, even upside down on a glass surface (**Fig. 13.14**). Aristotle observed this in the fourth century B.C. This remarkable ability is independent of the surface type. They can adhere equally well to smooth or rough surfaces. The survival benefits are evident. A number of models had been proposed, ranging from suction cups to sticky adhesives. These prior models had significant problems. Gecko feet do not exude any sticky material and observations of the anatomy of the feet also showed no suction-like features, even at microscopic size. The feet have to adhere well, but also release easily. This gives the gecko the ability to move rapidly over any surface. It was difficult to explain this seemingly paradoxical

Fig. 13.14 *Gecko attached to a vertical glass surface.*

Source: Image courtesy of K. Autumn, Lewis & Clark College, Portland, OR.

FIG. 13.15 *Gecko foot pad showing scansors.*

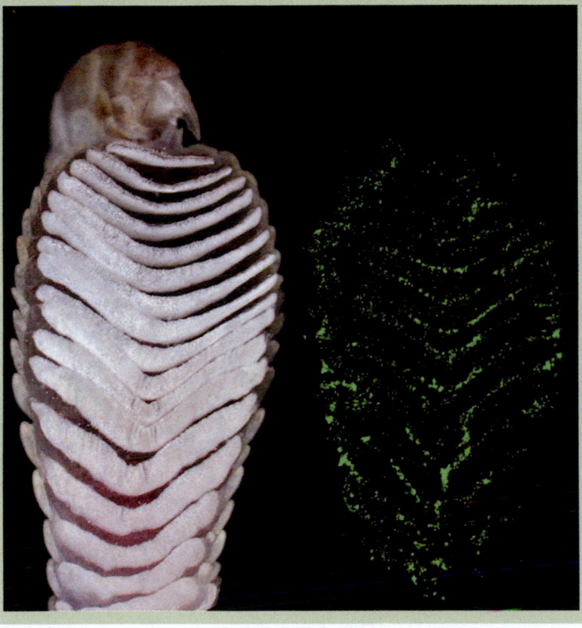

Source: Image courtesy of K. Autumn, Lewis & Clark College, Portland, OR.

situation. Iterative rounds of hypothesis testing have evolved the working model of how the gecko feet work to the present form, which focuses on the micro- and nanotopography of the gecko feet [61].

13.5.1 Gecko Feet

The gecko foot has a series of subdigital transverse ridges called *scansors* (**Fig. 13.15**), which contain numerous, 100-μm long projections called *setae* (**Fig. 13.16**). The setae are formed from epidermal extensions that were originally thought to assist in skin shedding but have evolved in some species to provide an adhesive function. There are approximately half a million of these setae on the feet of the Tokay gecko. Each seta is further subdivided into about a thousand 200-nm wide projections called *spatulae*. Each spatula has a flattened end, much like a spoon (hence the name spatula). As a result, the total surface area of the gecko feet is enormous and this provides a clue about the mechanism of the gecko's remarkable clinging abilities [62].

13.5.2 Mechanism of Adhesion

Although surfaces may appear to be smooth, at molecular dimensions these same surfaces are very rough; steep molecular mountain ranges rise up between deep valleys. So even when two objects appear to be closely touching, in reality only the rough tips touch and the surface area of contact is relatively small. The spatulae of the gecko's foot are able to project into the valleys between the mountains and increase the surface contact area tremendously. The normally

FIG. 13.16 *Gecko setae micrograph showing spatulae.*

Source: Image courtesy of K. Autumn, Lewis & Clark College, Portland, OR.

minute forces that do not come into play when the contact area is low now dominate. The force is called van der Waals attraction. Since van der Waals forces are physical rather than chemical, the gecko feet stick to hydrophobic and hydrophilic surfaces with the same force.

Van der Waals forces arise when the molecules have an uneven distribution of electrons across their surfaces. This creates a charge separation, which results in a molecule where one end is more negative than the other end. This is called a dipole. As unlike charges attract, the negatively charged area of the dipole is attracted to nearby positive charges and so generates a slight force. The force is typically very short range, within 2 nm or so. In order for strong adhesion to occur, you need a very close contact between the molecules of the two surfaces and a very large surface area. The gecko's spatulae are flexible enough to essentially mold themselves into the molecular nooks and crannies of any surface; this provides intimate contact and large surface area. The result is strong adhesion. Studies of the mechanics of adhesion of a single seta showed that it binds with a force of 200 μN. The theoretical maximum force that a single Tokay gecko with its 6.5 million setae can apply is on the order of 130 kg. In reality, not all of the setae can adhere simultaneously, so the practical maximum force is considerably lower [63].

13.5.3 Attachment and Release of Grip

How do the geckos apply and release their grip? The ability of the gecko to attach and detach its foot when it wants to is related to the manner in which the lizard applies its foot to a surface. The gecko places its toes on to a surface by first placing the base of the toe (the proximal end) to the surface. It then rolls the toe forward, towards the far (distal) end, incorporating more surface area, until the entire toe pad is in contact with the substrate. This begins adhesion, but the maximal force is reached when the toe is pulled slightly backward, about 5 μm (a force is applied parallel to the substrate surface). The movement changes the angle of the seta to the surface to below criticality, which in this case is about 30°. This apparently increases the contact area of the spatulae and surface to the practical maximum [64].

To release, the sequence of events is reversed. The distal end of the toe lifts and curls upward, increasing the contact angle to above 30°, at which point the spatulae detach. The ability of the gecko to rapidly move (up to 1 m sec^{-1}) across the surface is therefore a combination of the morphology of the foot, cycles of toe and foot movements, and the nanotopology of the scansors.

13.5.4 Self-Cleaning

Another interesting property of the gecko foot is its ability to stay clean even as it sticks to dirty walls. It appears to exhibit differential or competitive adhesion to dirt compared to a surface. Hansen and Autumn dusted the feet of the Tokay gecko with ceramic microspheres. These spheres essentially attenuated the van der Waals forces normally present between the spatulae and the wall. After the gecko had been walking over the surface for a while the spheres were shed. This suggested that the spheres were more likely to stick to the wall than the feet. The investigators have analyzed the interaction of the spatulae and the microspheres and concluded that about 25–60 spatulae were needed for the feet to cling to the spheres. Electronmicrography showed that far fewer fibers on average adhered to the spheres and the spheres were more likely to adhere to the wall. This is a form of self-cleaning that is unlike that of the lotus flower, but is nonetheless a result of the topography of the surfaces at the molecular level.

13.6 More Natural Fibers

13.6.1 Spider Silk

One of the more interesting examples of naturally engineered biomaterial is the spider's web that serves a vital role in the survival of its engineer. Its main function is predation; prey items are caught in the sticky component of the web and available to the spider for a leisurely meal. The web itself needs to survive a variety of assaults. It must withstand the constant buffeting by the elements and the force of impact of the insect. It is able to stretch and not break, but also exhibits great strength. Spider silks are comparable in strength to the best synthetic materials made by man. Depending on how the measures are taken (or whom you talk to), the strength of spider silk is some five times that of steel of the same weight. Also, one author noted that a 450-g strand of spider silk can stretch around the world. What is it about the structure and composition of the silk that allows it to exhibit such remarkable properties?

When discussing the qualities of spider silk (or any fibrous structure) it is necessary to evaluate how well that material can withstand stresses. Young's modulus of elasticity is a quantitative measure of that characteristic. It is defined as the amount of stretch a material can exhibit for a given stress or a measure of the stiffness of the material. Simply, it is the ratio between the stress applied and the resulting strain exhibited by the material. The stress is defined as the force (F) applied over the cross-sectional area (A) of the material and has the dimensions of pressure, F/A or newtons per square meter. This is called a pascal (Pa, or megapascal, MPa). The strain is the change in length over the original length. The value is dimensionless and is represented as a percentage, $\Delta L/L_{\text{original}}$. The slope of the curve is a measure of the stiffness of the material: the steeper the slope is, the stiffer the material is. As the stress increases, the material will stretch until it breaks; this is the maximum height of the curve and is called the tensile strength of the material. The area under that curve is the total energy required to break the material, as force × distance = work (energy). The curve shown here is linear. Most materials exhibit a nonlinear curve, which is characteristic of a particular material. Generally, Young's modulus varies over the course of the applied stress (**Fig. 13.17**).

Spider Silk Structure. The major superclass protein of spider silk is fibroin. From analysis of the gene and amino acid sequences the structure of the fibroin has been well characterized. Depending upon the degree of hydration and the tension on the fiber, the fibroin chains can be loosely folded (**Fig. 13.18a**) or aligned in a quasi-crystalline matrix (**Fig. 13.18b**). The primary amino acid sequences of the fibroin show several distinct repeated motifs. These motifs are responsible for the secondary structures found in the silk proteins. β-pleated sheet is a type of secondary structure formed by sequences of 6–12 alanine (Ala or polyAla) and glycine/alanine (Gly/Ala) residues. As the protein folds back on itself, each fold forms hydrogen bonds between the N–H and C=O of adjacent folds to create a structure called a β-pleated sheet [65].

Fibrous regions are formed by glycine-rich and glycine-/proline-repeat sequences. In some cases the amino acids hydrogen bond to form β-spiral and "nanospring" like structures. There is also some folding, although not to the extent seen in the β-pleated sheets. These regions are noncrystalloid, unlike the

FIG. 13.17 *Young's modulus.*

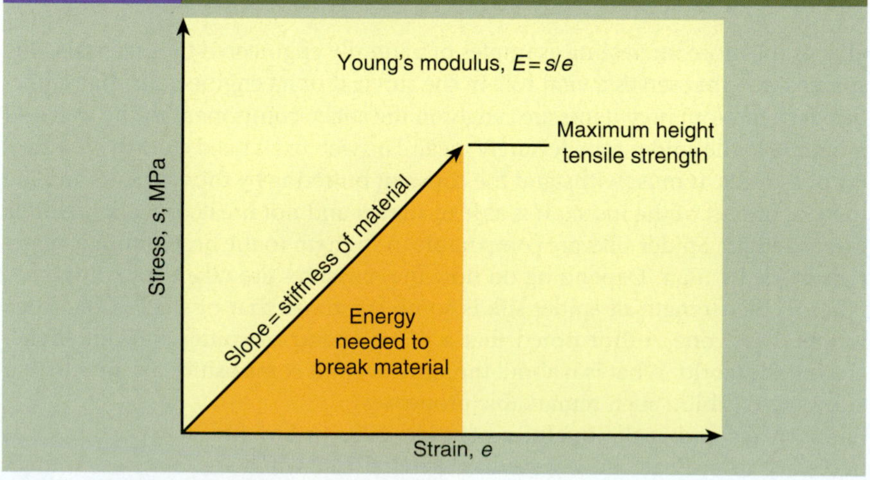

Source: Image by Anil K. Rao.

FIG. 13.18 *Nanostructure of spider silk fiber.*

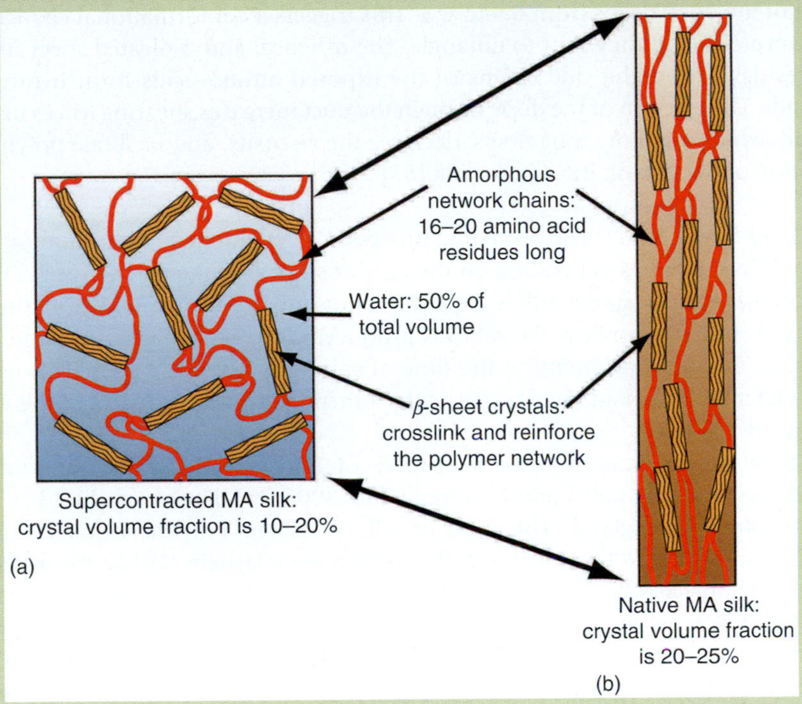

Source: Image courtesy of J. M. Gosline et al. (1999) *Journal Exp. Biol.*, 202, 3295–3303.

β-pleated sheet regions. As a result the fibers tend to be more elastic with lower tensile strength. Local β-pleated sheets can crystallize, forming stable regions that cross-link the extensible, noncrystalline fibers. This results in spider silks with both high stiffness and extensibility.

Changes in Young's modulus are not constant as a smoothly increasing force is applied to the silk fiber. Rather the slope of the curve can change abruptly, resulting in some discontinuities in the curve. These are called yield points. This is likely due to the sudden breaking of hydrogen bonds in both the fiber and β-pleated sheet regions of the protein, with the resulting change in the shape of the silk protein. The stress/strain curve behaves differently after this yield point when compared to the curve before, reflecting the change in the architecture of the silk fiber [66].

The silk is formed by polymerization of silk proteins. These proteins are stored in a liquid form in the glands located in the abdomen of the spider. Across the various arachnid species seven different gland and silk types have been identified. Not all species produce all seven of the silks. At most a spider may produce up to six types. The glands are connected via spinning ducts to one to four pairs of spinnerettes, which contain anywhere from 2 to 50,000 small tubes. The fibroin liquid is extruded through the tubes to produce the silk fibers.

In general, the concentrated liquid protein, called the spinning dope, is stored in the glands. The proteins in the dope are not aggregated into the typical fibrous spider silk, but are in a globular state. The material is quite viscous at this time, as high protein concentration (50% protein/water) causes the long polymers to become entangled. As the dope passes through the spinning duct, the pH of the dope drops from 6.9 to 6.3. This triggers a conformational change in the proteins and they start to untangle. The α-helical and β-pleated sheet structures develop as the side chains of the exposed amino acids form hydrogen bonds. The ejection of the dope through the duct increases shearing forces in the fluid, which align the long fibers, decrease the viscosity, and facilitate polymerization of the fibroin into spider silk [67].

Types of Spider Silk. The preceding process is a generalized description of silk formation. It varies depending on the type of silk being made. The production of the varieties of spider silk is dependent not only on the protein structure of the silk, but also on how the silk was produced, the diet of the spider, and the environmental condiments at the time of spinning. There are several varieties of spider silk, each adapted to particular functions in various parts of the web (**Fig. 13.19**) [68].

Dragline silk contains two subclasses of fibroin called major ampullate spidroin 1 and 2 (MaSP1 and 2, >several 200–300 kDa) that are produced by the major ampullate gland. This type of silk is rich in PolyAla sequences and Gly/Ala repeats and has a high tensile strength and elasticity. This is the first silk the spider makes when building a web. The dragline is extruded, the wind catches it, and it attaches (by chance) to another anchor point. The spider then moves back and forth across the line and produces more silk to reinforce it. The dragline eventually becomes part of the web frame. The fiber needs to be tough as it serves to support the bulk of the web and provides safety lines for the spider.

Minor ampullate glands produce another subclass of fibroin called the minor ampullate spidroin 1 and 2 (MiSP 1 and 2). These polymerize to form smaller

FIG. 13.19 *Types of spider webs.*

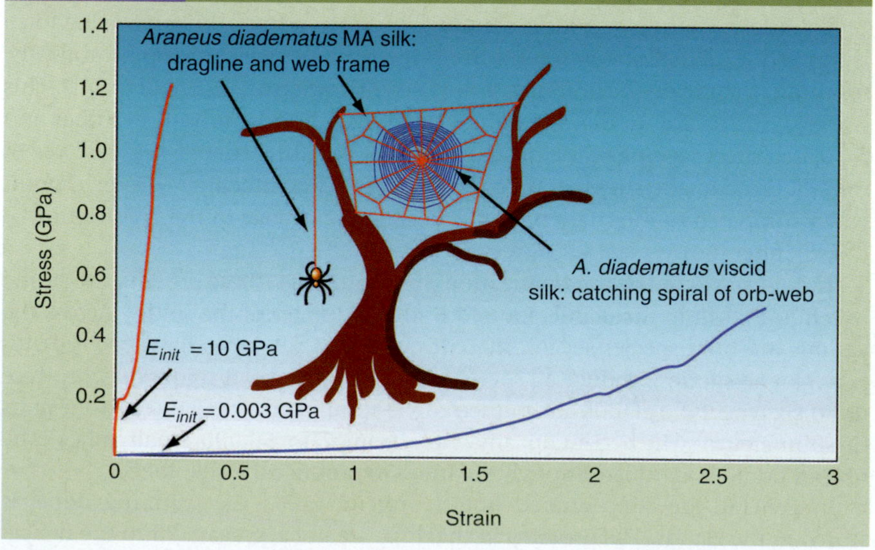

Source: Image courtesy of J. M. Gosline et al. (1999) *Journal Exp. Biol.*, 202, 3295–3303.

TABLE 13.1 *Spider Gland and Silk Types*

Glands	Silks
Major ampullate	Dragline and web frame
Minor ampullate	Web radii and capture
Flagelliform	Spiral capture and viscid
Aggregate	Web glue
Pyriform	Web attachment discs
Aciniform	Dragline and wrapping
Coronatae	Adhesive threads

diameter fibers than dragline silk. Although these proteins show fewer β-pleated sheet structures, they nonetheless have a comparable tensile strength to dragline silk, but lower elasticity. This is probably due to differences in protein cross-linking. The silks are used to reinforce the web with radial fibers and are also involved in producing the capture spiral. The flagelliform, aggregate, and aciniform glands produce web silk protein variants involved in the production of capture silks, web glues, and cocoon wrapping, respectively (summarized in **Table 13.1**).

From any perspective, the spider silk is a remarkable material. It is clear that a synthetic and easily manufactured version of spider silk would have great commercial implications. Researchers have attempted to produce such items with limited success. The complex chemical structure and the many different environmental conditions under which the silk can be produced suggest synthesis will be a difficult task. What the spider does so easily is not so easily mimicked [69].

13.6.2 Sponge Fibers

Sponges are a primitive form of multicellular life. Their cells secrete an intercellular matrix that supports and holds the cells together. Some sponges secrete a skeletal matrix that is high in silica content. The secretion of silica grains or *phytoliths* is not uncommon in sponges, sea invertebrates, and some grasses, but the optical properties of the skeleton of the deep-sea sponge *Euplectella*—also known as the glass sponge—are unusual. This sponge grows long fibers in its exoskeleton that guide light in a manner similar to man-made fiber-optic cables. But these fibers, made up of three layers of material, are tougher than artificial glass optical fibers [65].

Joanna Aizenberg and colleagues at Bell Labs in the United States and Tel Aviv University in Israel have been studying the *Euplectella* sponge and have found that the skeleton of the sponge is constructed from amorphous hydrated silica. The fibers are made of a network of *spicules*, composed of silica spheres between 50 and 200 nm in diameter assembled together to yield the final 100-μm fiber. In optical properties, the sponge fibers resemble the fibers used in telecom networks, with a high-refractive-index core and low-index cladding. When Aizenberg coupled light into them, she found that they acted like single- or few-mode waveguides.

The natural fiber has advantages over its man-made counterparts. It is tougher than brittle commercial fiber. The layers contained within the spicules' structure are connected by organic molecules that provide an effective crack-arresting mechanism. The fibers can be bent to a much tighter radius than man-made fibers.

The *spicules* grow in an ambient environment, rather than the high-temperature furnaces needed to make optical fiber. The natural, bottom-up growth of the fibers allows for the structure to become precisely doped with impurities that improve the fiber's performance.

The sea sponge fibers are multimode when surrounded by air or seawater and single mode when embedded in a material. Multimode fiber can carry multiple light waves over shorter distances, and single-mode fiber can carry a single light wave over longer distances. The free-standing spicules are multimode because the refractive index contrast is greater between the spicule shell and air than between its core and shell. This allows light to fill the whole fiber rather than just the core.

The fibers have a complex structure that incorporates a hierarchal set of elements that provide strength similar to the trusses in bridges or high-rise buildings. They are being studied to learn principles of optical and structural design that may improve artificial fibers.

13.7 Summary

In this brief overview of natural nanomaterials, some general principles of structural organization have been repeated in each type of material surveyed. A recurring theme is that natural materials are built from the bottom up, layer by layer, with incorporation of many heterogeneous atoms and molecules into the structure.

The bottom-up structures are not random, but highly ordered. They are, however, highly heterogeneous: typically, *fibrous polymers are embedded in a mixture of nonfibrous substances* to give a *composite* with superior properties to the single components. Where high strength and rigidity are required, we found *fiber-forming macromolecules*, aggregated into *ordered arrays*. Where elasticity and resilience are important, *highly branched polymers* such as *elastin* were found to meet the requirement.

In many natural materials, we found a *hierarchy of fiber assemblies* from the molecular level up to the macroscopic, with *interlocking linkages* to spread loads. In many different types of material, fibers are embedded in *layers with alternating orientations*. Another important theme is the importance of *calcium carbonates* and *phosphates* as components of skeletal structure as well as serving as buffers or sinks for carbon dioxide and regulators of cellular metabolism. A final important theme is the occurrence of sulfur linkages and silica compounds to meet unique optical and structural requirements.

References

1. J. K. Mitchell, *Fundamentals of soil behavior*, 3rd ed., John Wiley & Sons, New York, (2005).
2. K. Katti and D. Katti, Effect of clay–water interactions on swelling in montmorillonite clay. In *ASCE 16th engineering mechanics conference: Techniques for*

experimental analysis, instrumentation, and mechanics, University of Washington, Seattle, 1–9 (2003).
3. A. Viani, A. Gualtieri, and G. Artioli, The nature of disorder in montmorillonite by simulation of x-ray powder patterns, *American Mineralogist*, 87, 966–975 (2002); see web page at: webmineral.com/data/Montmorillonite.shtml
4. G. W. Beall and M. Goss, Self-assembly of organic molecules on montmorillonite, *Applied Clay Science*, 27, 179–186 (2004).
5. G. Ertem and J. P. Ferris, Template-directed synthesis using the neterogeneous templates produced by montmorillonite catalysis. A possible bridge between the prebiotic and RNA worlds, *Journal of the American Chemical Society*, 119, 7197–7201 (1997).
6. M. Franchi, E. Bramanti, L. M. Bonzi, P. L. Orioli, C. Vettori, and E. Gallori, Clay–nucleic acid complexes: Characteristics and implications for the preservation of genetic material in primeval habitats, Origins of Life and Evolution of the Biosphere, 29, 297–315 (1999).
7. J. Leng, S. U. Egelhaaf, and M. E. Cates, Kinetics of the micelle-to-vesicle transition: Aqueous lecithin–bile salt mixtures, *Biophysics Journal*, 85, 1624–1646 (2003).
8. M. M. Hanczyc, S. M. Fujikawa, and J. W. Szostak, Experimental models of primitive cellular compartments: Encapsulation, growth, and division, *Science*, 302, 618–622 (2003).
9. M. M. Hanczyc and J. W. Szostak, Replicating vesicles as models of primitive cell growth and division, *Current Opinions in Chemical Biology*, 8, 660–664 (2004).
10. L. E. Orgel, Prebiotic chemistry and the origin of the RNA world, *Critical Reviews in Biochemistry and Molecular Biology*, 39, 99–123 (2004).
11. D. B. Carlisle and D. R. Braman, Nanometre-size diamonds in the Cretaceous—Tertiary boundary clay of Alberta, *Nature*, 352, 708–709 (1991).
12. R. Monastersky, Radioactive alchemy: Diamonds from coal, *Science News*, 149, 133 (1996).
13. J. E. P. Dahl et al., Isolation and structural proof of the large diamond molecule, cyclohexamantane ($C_{26}H_{30}$), *Angewandte Chemie International Edition*, 42, 2040–2044 (2003).
14. J. E. P. Dahl, S. G. Liu, and R. M. K. Carlson, Isolation and structure of higher diamondoids, nanometer-sized diamond molecules, *Science*, 299, 96–99 (2002).
15. R. Monastersky, Microscopic diamonds crack geologic mold, *Science News*, 148, 22 (1995).
16. N. V. Sobolev and V. S. Shatsky, Diamond inclusions in garnets from metamorphic rocks: A new environment for diamond formation, *Nature*, 343, 742–745 (1990).
17. L. F. Dobrzhinetskaya, E. A. Eide, R. B. Larsen, B. A. Sturt, R. G. Trønnes, D. C. Smith, W. R. Taylor, and T. V. Posukhova, Microdiamond in high-grade metamorphic rocks of the Western Gneiss region, Norway, *Geology*, 23, 597–600 (1995).
18. T. Henning, H. Mutschke, S. Schlemmer, and D. Gerlich, Nanoparticles in space and the laboratory, NASA Laboratory Astrophysics Workshop, held May 1–3 2002 at NASA Ames Research Center, Moffett Field, CA 94035-100. Edited by F. Salama, NASA Conference Proceedings NASA/CP-2002-21186, 175 (2002).
19. R. Har-El and M. L. Tanzer, Extracellular matrix 3: Evolution of the extracellular matrix in invertebrates, *FASEB Journal*, 7, 1115–1123 (1993).
20. D. J. Jackson, L. Macis, J. Reitner, B. M. Degnan, and G. Wörheide, Sponge paleogenomics reveals an ancient role for carbonic anhydrase in skeletogenesis, *Science*, 316, 1302 (2007).
21. S. Blank, M. Arnoldi, S. Khoshnavaz, L. Treccani, M. Kuntz, K. Mann, G. Grathwohl, and M. Fritz, The nacre protein perlucin nucleates growth of calcium carbonate crystals, *Journal of Microscopy*, 212, 280–291 (2003).
22. K. Mann, F. Siedler, L. Treccani, F. Heinemann, and M. Fritz, Perlinhibin, a cysteine-, histidine-, and arginine-rich miniprotein from abalone (*Haliotis laevigata*) nacre,

inhibits in vitro calcium carbonate crystallization, *Biophysics Journal,* 93, 1246–1254 (2007).

23. J. P. Reyes-Grajeda, A. Moreno, and A. Romero, Crystal structure of ovocleidin-17, a major protein of the calcified *Gallus gallus* eggshell: Implications in the calcite mineral growth pattern, *Journal of Biological Chemistry,* 279, 40876–40881 (2004).
24. D. R. Lide, *CRC handbook of chemistry and physics,* 88th ed., CRC Press, Boca Raton, FL (2007).
25. J. L. Katz, Mechanics of hard tissue. In *The biomedical engineering handbook,* 3rd ed., CRC Press, Boca Raton, FL, Section VI: Biomechanics, chap. 47 (2006).
26. J. D. Bronzino, ed., *The biomedical engineering handbook,* 3rd ed., CRC Press, Boca Raton, FL (2006).
27. Y. Bar-Cohen, *Biomimetics: Biologically inspired technologies,* CRC Press, Boca Raton, FL (2005).
28. T. Vo-Dinh, *Nanotechnology in biology and medicine: Methods, devices, and applications,* CRC Press, Boca Raton, FL (2007).
29. D. M. Skinner, S. S. Kumari, and J. J. O'Brien, Proteins of the crustacean exoskeleton, *American Zoologist,* 32, 470–484 (1992).
30. D. Raabe, C. Sachs, and P. Romano, The crustacean exoskeleton as an example of a structurally and mechanically graded biological nanocomposite material, *Acta Materialia,* 53, 4281–4292 (2005).
31. P. Romano, H. Fabritius, and D. Raabe, The exoskeleton of the lobster *Homarus americanus* as an example of a smart anisotropic biological material, *Acta Biomaterialia,* 3, 301–309 (2007).
32. J. J. O'Brien, S. Kumari, and D. M. Skinner, Proteins of crustacean exoskeletons: I. Similarities and differences among proteins of the four exoskeletal layers of four brachyurans, *Biology Bulletin,* 181, 427–441 (1991).
33. S. F. Tarsitano, K. L. Lavalli, F. Horne, and E. Spanier, The constructional properties of the exoskeleton of homarid, palinurid, and scyllarid lobsters, *Hydrobiologia,* 557, 9–20 (2005).
34. R. E. Burgeson, New collagens, new concepts, *Annual Review of Cell Biology,* 4, 551–577 (1988).
35. M. J. Buehler, Nature designs tough collagen: Explaining the nanostructure of collagen fibrils, *PNAS USA,* 103, 12285–12290 (2006).
36. M. J. Buehler and S. Y. Wong, Entropic elasticity controls nanomechanics of single tropocollagen molecules, *Biophysics Journal,* 93, 37–43 (2007).
37. B. K. Hall, Consideration of the neural crest and its skeletal derivatives in the context of novelty/innovation, *Journal of Experimental Zoolology* (Mol. Dev. Evol.), 304B, 548–557 (2005).
38. D. Toroian, J. E. Lim, and P. A. Price, The size exclusion characteristics of type I collagen, *Journal of Biological Chemistry,* 282, 22437–22447 (2007).
39. H. D. Wagner and S. Weiner, On the relationship between the microstructure of bone and its mechanical stiffness, *Journal of Biomechanics,* 25, 1311–1320 (1992).
40. J. M. Crolet, M. Racila, R. Mahraoui, and A. Meunier, A new numerical concept for modeling hydroxyapatite in human cortical bone, *Computer Methods in Biomechanics and Biomedical Engineering,* 8, 139–143 (2005).
41. Y. Kajiuraa, S. Watanabea, T. Itoua, K. Nakamuraa, A. Iidab, K. Inouec, N. Yagic, Y. Shinoharad, and Y. Amemiya, Structural analysis of human hair single fibres by scanning microbeam SAXS, *Journal of Structural Biology,* 155, 438–444 (2006).
42. W. T. Astbury and H. J. Woods, The x-ray interpretation of the structure and elastic properties of hair keratin, *Nature,* 126, 913–914 (1930).
43. H. Liu and W. Yu, Microstructural transformation of wool during stretching with tensile curves, *Journal of Applied Polymer Science,* 104, 816–822 (2007).

44. See Websites for Nano Pulp and Paper Llc, http://www.nanopulpandpaper.com/, and Better Paper Technologies Llc, http://www.bptech.cc/
45. M. Koepenick, Paper innovation ramps up with nanochemistry, *Pulp and Paper Canada*, 102, 17–19 (2001).
46. J. M. Maxwell, S. G. Gordon, and M. G. Huson, Internal structure of mature and immature cotton fibers revealed by scanning probe microscopy, *Textile Research Journal*, 73, 1005–1012 (2003).
47. A. A. Baker, W. Helbert, J. Sugiyama, and M. J. Miles, Surface structure of native cellulose microcrystals by AFM, *Applied Physics*, A 66, S559–S563 (1998).
48. R. M. Brown, Jr., Algae as tools in studying the biosynthesis of cellulose, nature's most abundant macromolecule. In *Experimental phycology. Cell walls and surfaces, reproduction, photosynthesis*, W. Wiessner, D. G. Robinson, and R. C. Starr, eds., pp. 20–39, Springer-Verlag, Berlin (1990).
49. M. Tabuchi, Nanobiotech versus synthetic nanotech? *Nature Biotechnology*, 25, 389–390 (2007).
50. M. Tabuchi, F. Tomita, K. Kobayashi, S. Miki, T. Saijo, A. Shibata, and Y. Baba, Self-regulated bionanostructured membranes for MEMS, MEMS 2006 Istanbul, *19th IEEE International Conference on Micro Electro Mechanical Systems*, 514–517 (2006).
51. J. Bradbury, Nature's nanotechnologists: Unveiling the secrets of diatoms, *PLoS Biology*, 2, e306 (2004).
52. N. D. Crosbie and M. J. Furnas, Picocyanobacteria and nano-/microphytoplankton growth, *Aquatic Microbial Ecology*, 24, 209–224 (2001).
53. T. Dugdale, R. Dagastine, T. Chiovitti, P. Mulvaney, and R. Wetherbee, Single adhesive nano-fibers from a live diatom have the signature fingerprint of modular proteins, *Biophysics Journal*, 89, 4252–4260 (2005).
54. A. Summers, Biomechanics: Secrets of the sacred lotus, *Natural History*, 406, www.naturalhistorymag.com/master.html?http://www.naturalhistorymag.com/0406/0406_biomechanics.html (2006).
55. Nelumbo nucifera, Pacific Island Ecosystems at Risk (PIER) www.hear.org/pier/species/nelumbo_nucifera.htm (2007).
56. A. Agrawal, Wettability, nonwettability and contact angle hysteresis, MIT, Non-Newtonian Fluid Dynamics Research Group: web.mit.edu/nnf/education/wettability/wetting.html (2004).
57. W. Barthlott and C. Neinhuis, Purity of the sacred lotus, or escape from contamination in biological surfaces, *Planta*, 202, 1–8 (1997).
58. Z. Shen and Z. M. Jacobs-Lorena, Evolution of chitin-binding proteins in invertebrates, *Journal of Molecular Evolution*, 48, 341–347 (1999).
59. G. S. Watson and J. A. Watson, Natural nano-structures on insects—Possible functions of ordered arrays characterized by atomic force microscopy, *Applied Surface Science*, 235, 139–144 (2004) (8th European Vacuum Conference and 2nd Annual Conference of the German Vacuum Society).
60. A. R. Parker and H. E. Townley, Biomimetics of photonic nanostructures, *Nature Nanotechnology*, 2, 347–353 (2007).
61. K. Autumn, Properties, principles, and parameters of the gecko adhesive system. In *Biological adhesives*, A. Smith and J. Callow, eds., pp. 225–255, Springer-Verlag, Berlin, (2006).
62. K. Autumn, K., M. Sitti, A. Peattie, W. Hansen, S. Sponberg, Y. A. Liang, T. Kenny, R. Fearing, J. Israelachvili, and R. J. Full, Evidence for van der Waals adhesion in gecko setae, PNAS USA, 99, 12252–12256 (2002).
63. K. Autumn, A. Dittmore, D. Santos, M. Spenko, and M. Cutkosky, Frictional adhesion: A new angle on gecko attachment, *Journal of Experimental Biology*, 209, 3569–3579 (2006).

64. K. Autumn, Y. A. Liang, S. T. Hsieh, W. Zesch, W.-P. Chan, W. T. Kenny, R. Fearing, and R. J. Full, Adhesive force of a single gecko foot-hair, *Nature*, 405, 681–685 (2000).
65. J. M. Gosline, P. A. Guerette, C. S. Ortlepp, and K. N. Savage, The structure of spider silk, *Journal of Experimental Biology*, 202, 3295–3303 (1999).
66. V. Sundar, A. D. Yablon, J. L. Grazul, J. Aizenberg, and M. Ilan, Fibre-optical features of a glass sponge, *Nature*, 424, 899 (2003).
67. M. Beals, L. Gross, and S. Harrell, *Spider silk: Stress–strain curves and Young's modulus*, www.tiem.utk.edu/~mbeals/spider.html (1999).
68. X. Hu, K. Vasanthavada, K. Kohler, S. McNary, A. M. F. Moore, and C. A. Vierra, Molecular mechanisms of spider silk, *Cellular and Molecular Life Sciences*, 63, 1986–1999 (2006).
69. M. A. Garrido, M. Elices, C. Viney, and J. Pérez–Rigueiro, Active control of spider silk strength: Comparison of drag line spun on vertical and horizontal surfaces, *Polymer*, 43, 1537–1540 (2002).

Problems

13.1 The setae of geckos adhere to hydrophobic as well as hydrophilic surfaces by van der Waals forces. This implies that the size and shape of the tips of the setae are responsible for adhesion and chemistry—a truly nano phenomenon! Assume that each seta is capable of 200 μN of adhesion force and that there are 14,400 setae per square millimeter. What area of setae is required to lift a 180-lb human up from the floor?

13.2 Each seta consists of approximately 500 spatulae—the actual nanostructured material. How many spatulae are there per square millimeter? What is the average force load capability per spatulae?

If you were to use carbon nanotubes instead of spatulae, what would be the density per square millimeter (assume SWNTs with diameter = 1 nm)? If W (adhesion energy for van der Waals surfaces) equals 50 $mJ \cdot m^{-2}$, what is the lifting capability of such a synthetic adhesive?

13.3 In nature, as body mass is increased, not only does the number of setae increase, but also the structural components responsible for adhesion (the spatulae) get smaller (reduced diameter). This is evidenced in the progression in biological mass from a small beetle to a fly to a spider to a gecko. (H. Gao and H. Yao, Proceedings of the National Academy of Science, 101, 7851–7856, 2004.) Why do you think this trend occurs as it does?

13.4 The average mass of a diatom is approximately 60 pg. What is the maximum and minimum diatom biomass in a liter of ocean water, based on the population densities cited in this chapter?

13.5 Look up the properties of silkworm silk and compare them to those of spider silk. What are some of the difficulties in artificially making a fiber that compares to spider silk? Discuss and support your answers with references.

13.6 Would you expect a calcite crystal to be isotropic in Young's modulus? Why? Look up the values for Young's modulus for calcite, argonite, and rock salt (NaCl) and explain in terms of the crystal structure.

13.7 Go to the following Web pages for the interesting exercises using CD diffraction gratings to determine wavelength of incident light or measure the groove spacing:
<http://maxwell.physics.mun.ca/mpl/Physics1051_MM_F05/lab6.pdf>
<http://chem.lapeer.org/PhysicsDocs/GroovyCD.html>

13.8 Aragonite is a polymorph of calcite, which means that it has the same chemistry as calcite but it has a different structure and, more importantly, different symmetry and crystal shapes. Aragonite's more compact structure is composed of triangular carbonate ion groups (CO3), with a carbon at the center of the triangle and the three oxygens at each corner. Unlike in calcite, the carbonate ions do not lie in a single plane pointing in the same direction. Instead, they lie in two planes that point in opposite directions, thus destroying the trigonal symmetry that is characteristic of calcite's structure. To illustrate the symmetries of calcite and aragonite, draw or construct an equilateral triangle, a threefold rotation with three mirror planes that cross in the center. Now join two of these triangles together at their bases and you have a diamond-shaped figure with the symmetry of a twofold rotation with one mirror plane in the middle. This is the effect of the two carbonate planes with opposite orientations on the symmetry of this structure. Aragonite has an orthorhombic symmetry (2/m 2/m 2/m) instead of calcite's "higher" trigonal (bar 3 2/m) symmetry. A very rare mineral called vaterite is also a polymorph of aragonite and calcite—making them all trimorphs. Vaterite has a hexagonal symmetry (6/m 2/m 2/m). Draw or make models of each of these crystals. Draw or model the smallest unit from which you can construct the complete crystal in each case.

13.9 The sodium ion has a radius of ___. The potassium ion radius is ___, calcium _____, magnesium _____, and barium ______. The bond length of the Al–O bond in an Al–O–Al linkage is _________. The bond length of the Si–O–Si bond is ____. The bond length of the Al–O–SI bonds is ___. What size zeolite unit ring would be needed to give a pore size of at least 0.4 nm with each cation? Hint: Look up bond lengths in CRC Handbook of Chemistry and Physics.

13.10 An interesting set of exercises for learning about biofilms, titled "A Manual of Biofilm-Related Exercises," is available from the American Society of Microbiology at: <http://www.personal.psu.edu/faculty/j/e/jel5/biofilms/>.

13.11 What is the difference between single- and multimode optical fiber conductance of light?

Biomolecular Nanoscience

Some have said to me that "sequencing the human genome will diminish humanity by taking the mystery out of life." Nothing could be further from the truth. The complexities and wonder of how the inanimate chemicals that are our genetic code give rise to the imponderables of the human spirit should keep poets and philosophers inspired for millennia.

J. Craig Venter, Ph.D., CEO
Celera Genomics, Inc.

Chapter 14

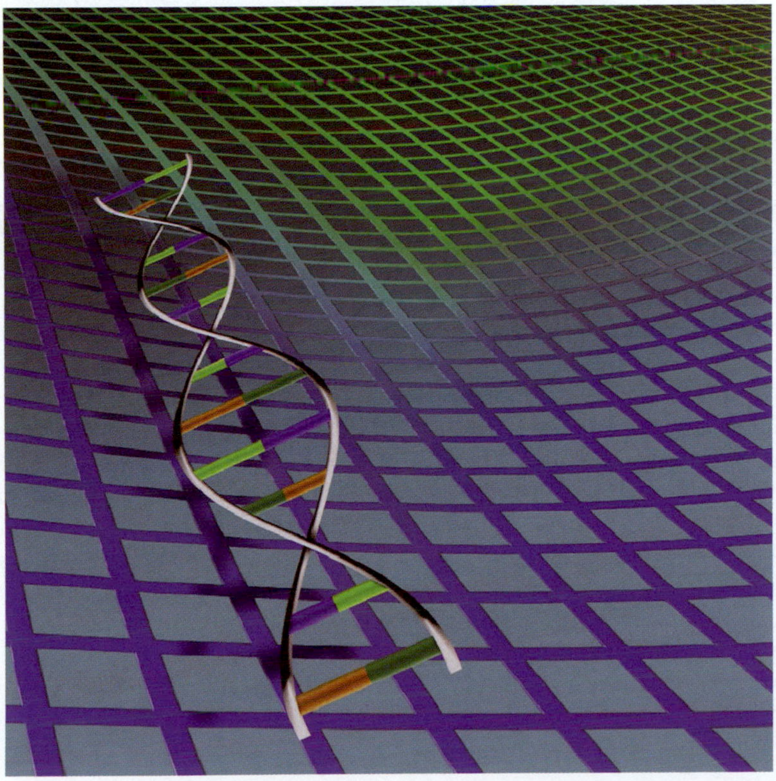

FIG. 14.0 *DNA strands.*

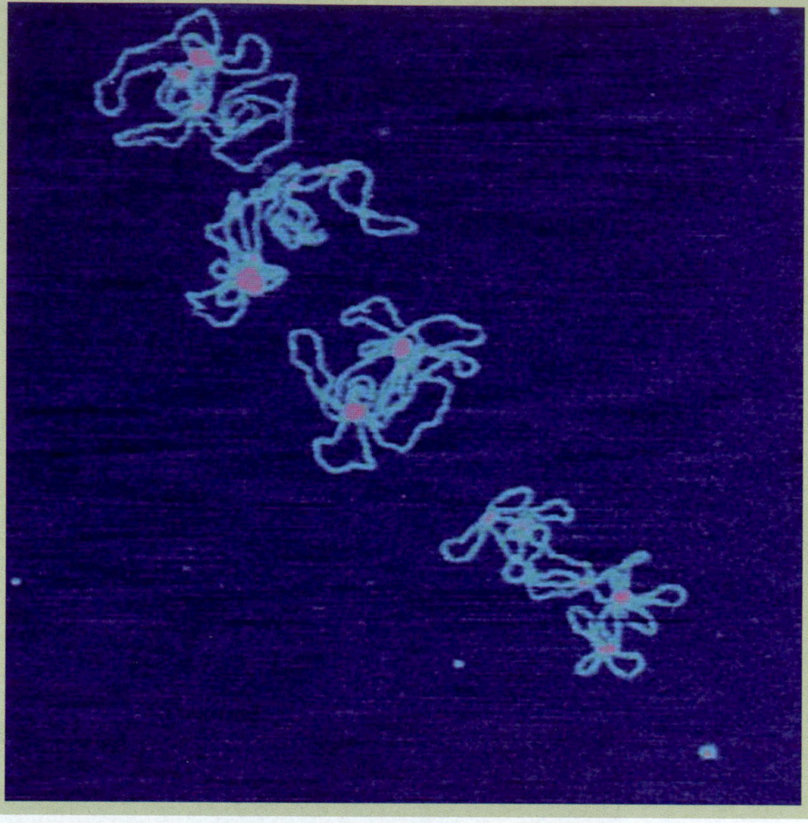

Source: Image courtesy of Veeco Instruments, Inc.

THREADS

Chapter 14 presents a short course in biochemistry and cellular biology, with emphasis on the nano perspective. Most physicists and engineers never took a formal course in biology. Because nanotechnology is highly interdisciplinary, knowledge of biochemical and cellular processes is essential. This chapter will hopefully bridge that gap. Chapter 13 introduced natural nanomaterials. Before that we studied the chemistry and physics of nanomaterials. With all that now in perspective, we look at living things and what makes them tick in a different light—emanating from the nanoscale.

This chapter introduces the organization of life as related to nanoscale properties of matter. From the intramolecular forces in water to the self-organizing forces of giant macromolecules such as proteins, life depends upon and is an emergent property of the energetics of matter on the nanoscale. Understanding how nanoscale forces orchestrate life will be exciting work for future scientists. How biological systems operate at the nanoscale and how nanoscale machinery integrates into larger systems are an important part of the broader discipline of nanoscience. Biological nanosystems, the product of millions of years of evolutionary development, will reward the nanoscientist with rich and intriguing examples of nanostructures, nanomachines, and principles of operation for nanosystems.

The nanoscience of living systems displays finely honed design, efficiency of material and energetics,

Threads (Contd.)

stability, and tolerance of errors and interferences that are the product of billions of years of exploration of the possibilities of structure and engine building on the molecular and cellular scale. The natural nanomachinery of living organisms has been rigorously optimized through natural experimentation and competitive selection to evolve into the elegant and finely tuned systems that our science has only begun to understand. Advances in understanding the basis of living systems will in turn be greatly empowered by the current progress in the fields of nanoscience and nanotechnology. Exposure to the challenge of biosystems on the nanoscale is part of a multidisciplinary approach that will inspire lateral thinking in approaching problems on the nanoscale. This is the motivation for a detailed component in the basics of life science on the nanoscale as a part of this treatise.

14.0 Introduction to Biomolecular Nanoscience

Life can be considered to be nanoscience in the wild native state. Living organisms are proof that self-replicating molecular nanoengines exist and function, performing work; building material structures of great strength and subtlety; reproducing and repairing themselves; and processing information to identify and control sources of materials and energy, ward off threats, and carry out survival strategies. In this chapter we will explore some of the nanoscale phenomena that underlie life processes.

14.0.1 *Definitions: Biomolecular Nanoscience*

Biomolecular nanoscience is the study of living systems on the nanoscale level. Biomolecular nanoscience studies how phenomena at the nanoscale underlie higher levels of organization in living things. Biomolecular nanoscience is an integrated transdisciplinary approach to the science of life. Biomolecular nanoscience includes aspects of biophysics, molecular biology, protein structure, and biochemistry; it aims to unify the branches of physics, biology, and biochemistry that deal with phenomena at the nanoscale and large-scale implications.

The large-scale aspects of life include how a person, an organism, or an organ is able to perform its functions without failure in the face of external threats and internal complications. These functions include metabolism, growth, development, reproduction, heredity, evolution of species, and intelligent behavior. Biomolecular nanotechnology provides a foundation upon which such phenomena can be related to underlying forces, structures, and organizations.

Research in nanoscience has opened new windows into the causes and mechanisms of previously mysterious biological phenomena, as shown by examples in the previous chapter such as the role of clay in catalyzing the synthesis of biomolecules, the relation of bone strength to nanostructure, the coloring of butterfly wings, and adhesion of gecko feet. There is every reason to expect that further investigations into biomolecular nanoscience will yield even more important insights.

14.0.2 Historical Origins

Modern nanoscience arose from a bottom-up approach, beginning with the physical aspects of manipulating atoms, as well as a top-down approach, working from the analysis of existing natural structures and substances in finer and finer detail down to the nano level. Biomolecular nanoscience is top down in its historical development, since whole organisms were studied long before their small components.

The story of how scientists and doctors, working over many centuries, explored ever finer details down to the nanoscale is a fascinating one. Since the first microscopes, there has been a drive to examine and study biological structures at an ever smaller level, seeking to understand the structure and basis of life. The earliest application of microscopy was the observation of microbiological specimens. Physical phenomena were applied to early experiments with life (e.g., the understanding of electricity was advanced by Galvani's study of frog legs). The study of x-rays was linked to their application for anatomical imaging from their first discovery. The study of proteins and molecular biology by physical chemists using x-ray diffraction laid the basis for understanding how proteins act as nanomachines. And now it is no wonder that nanoscience also includes life and living things.

Biochemistry in the nineteenth century made important progress in analyzing small molecules and elucidating important energy and material cycles in living systems. However, traditional chemistry dealt only with freely mixing combinations of molecules in bulk. In the highly structured organization of matter that exists in cells and at their boundaries, models of molecular interaction based on freely circulating and randomly mixed chemical molecules are no longer an adequate description. Great progress was made in the twentieth century in understanding macromolecules, catalysts, and the orchestration of molecular processing in the cell. The physical chemistry of catalysts, surface reactions, and polymers was useful in describing parts of the picture, but the biological environment proved even more complex—for example, in the processes by which proteins orchestrate cell division. In the twenty-first century, biomolecular nanotechnology develops on the foundations of the traditional sciences of biophysics, biochemistry, and molecular and cellular biology to meet the challenges of describing and understanding the nanoscale behind the life.

14.0.3 Biomolecular Nanoscience: Roots in Traditional Science

Nanoscience depends on traditional sciences of physics, biology, and chemistry. In this section we give an historical overview leading from bulk properties of matter and whole organisms to the understanding of matter on the nanoscale, as applied to biomolecular science.

Physics. The "integration of everything" allows us to move from one discipline to another. Understanding light and electricity from the perspective of physics enabled the application of spectroscopy chemistry and its insights into the structure of atoms and molecules. In the same way, understanding of how properties change with interactions at the nanoscale is opening surprising new insights and applications.

Chemistry. Classical chemistry, with strong input from physics, had begun to unravel matter at the molecular scale in the nineteenth century. By the middle of the twentieth century, chemists manipulated matter on the atomic scale, but largely only in bulk phase. By extremely subtle and intricate methods, chemists influenced the molecular structure by combining external influences such as heat, light, and electricity. By introducing specific reagents together in controlled conditions, they produced interactions that changed the organization of bonds and geometric connections (or stereochemistry) of molecules. This was accomplished in traditional chemistry with multiple trillions of molecules, all reacting at the same time and in the same way in a test tube with external control of bulk reaction conditions such as temperature and pressure.

When Smalley, Kroto, Curl, and coworkers discovered the structure of carbon buckyballs in the 1980s, they helped shift perspectives towards molecules as engineered structures that are produced directly in high yields by electronic vacuum processes and other engineering approaches. This viewpoint began to supplement and even supplant the predominant attitude that complex chemical molecules were destined to be the product of elegant, elaborate, carefully planned synthetic reactions involving many steps and much laboratory time by highly skilled chemists with intimate knowledge and subtle repertoires of reactions and source materials. This trend was already underway with polymer, catalyst, and ion complex ligand chemistry, including progress in the synthesis of conductive polymers.

In the meantime, life scientists and molecular biologists were working out how cell membranes, DNA, RNA, and protein structures brought about carefully controlled chemical and physical interactions in the cell. Biochemists and biotechnologists were thinking of the molecular components of cells as machinery performing tasks on the molecular level. Protein and membrane structures became as much a focus as enzyme and metabolite kinetics [1–6].

Molecular Biology. Armed with the concepts of chemistry, biochemists and molecular biologists began thinking of living structures as networks of chemical reactions [7,8]. With the discovery of DNA and the mechanism of transcription by Watson and Crick (and Franklin), foreshadowed by the discovery of helical protein structures by Pauling, molecular biologists began to think in terms of three-dimensional machinery that operates on the subcellular level [9–11]. In the meantime, the mechanism of cell-wall penetration by bacteriophages was elucidated, along with rotary flagellation engines propelling cilia, electrochemical and stereochemical mechanisms operated gates in cell membranes, DNA replicating and repairing machinery, cell division machinery, and the whole mechanism of life based on molecular mass transport and information exchange.

Medical scientists such as A. Gilman, J. P. Nolan, and Leonard A. Herzenberg began to map the complex network of molecular signaling pathways that control the lives of cells. In the process, they called on the work of physicists and engineers to develop new tools and techniques such as flow cytometry to separate cells based on differences in genome expression [12–15], statistical sampling microscopy to resolve details below the limits of optical microscopes [16], and statistical analysis of complex networks with many cross-talk interactions to identify relationships between pathways [17]. The knowledge of molecular control pathways casts new light on the understanding of evolution and development, including insights into physical and chemical constraints that limit and define what would otherwise be thought to be random selection of evolutionary pathways [18,19].

Engineering and Biotechnology. A number of initiatives accelerated the energetic exploration of nanobiology and nanomedicine [20,21]. Biological and microelectronic materials continued to be integrated into biochips and biosensors [22–27]. Groups at Los Alamos and Sandia National Laboratories made progress towards self-assembly of nanostructures to contain cells; the goal was to create a nanostructured environment in which cells would remain viable without external buffers or fluidic architecture. This work opened the proverbial door for cell-based sensors based on immobilized cells [28]. Protective implant capsules were synthesized for transplanted cells to deliver insulin or other complex substrates. Pores in the capsule are small enough to allow delivery of insulin while not the entry of rejection antibodies [29]. Other researchers developed ways to harness the replication machinery of DNA and synthesize useful proteins in bulk for medical applications. To obtain useful throughput quantity, the proteins are then purified robotically followed by automated analysis [30]. DNA of higher animals can also be modified; for example, human antibodies are being produced in the milk of genetically modified goats and cows [31,32].

DNA protein replicating machinery operating outside the cell has also been developed. This simplified the process of separating and purifying end products and opened the way for bulk production of proteins similar to conventional chemical processes. For example, at the Harvard Institute of Proteomics, at the company DNANO, cell-free extracts containing DNA, RNA, and substrates are currently being employed to synthesize proteins for biochemical and immunological products [33].

14.0.4 The Nano Perspective

In this section we see how the advancement of nanoscience benefited from insights and techniques shared with physics, chemistry, and cellular and molecular biology. Biomolecular nanoscience was inspired by naturally occurring molecular machinery in the cell, and molecular and genetic biologists found it increasingly useful to employ concepts and systematic analysis of molecular machinery in common with nanoscience. Thus, biomolecular nanoscience can be considered as the pursuit of nanoscience with relevance and application to biological science.

Biomolecular nanoscience draws upon engineering concepts and methods to contribute new models for understanding life based on complex chemical, physical, and structural machinery, with subtle control systems that involve detailed and highly specific interactions at the molecular level. This interdisciplinary approach, *biomolecular nanoscience*, provides a perspective from which we model and understand living processes at the level of molecular machinery.

14.1 Material Basis of Life

In this section we give an overview of the development of tools and methods for studying matter, starting historically with bulk properties of matter and whole organisms, and leading to matter on the nanoscale, as applied to biomolecular science.

Biomimetics and Bionanotechnology. Biomimicry is defined as the study of forms and functions in nature in order to apply them to engineering and manufacturing practice. The imitation of nature is as old as invention itself, but the understanding of biological systems and materials on the molecular scale has opened an immensely rich resource for designers and engineers [34]. Lessons learned from biology are being applied to nanotechnology and, in turn, are applied to biotechnology [35]. A compelling counterexample comes to mind: the elucidation of the structure of carbon nanospheres by the engineering designs of Buckminster Fuller led researchers to the discovery of the correct structure, at least in part, based on dodecahedron geometry [36].

To understand the domain of biomolecular nanoscience and to be able to use its tools, it is important to have a grasp of what is being measured, manipulated, and constructed and to have some appreciation for the marvelous and complex natural structures of life. This is true for the individual building blocks as well as the network of their interactions. These components and network pathways are the guiding architecture of cells and organisms for their development and metabolic life cycles.

14.1.1 Molecular Building Blocks—From the Bottom Up

In the following section we start with the smallest building blocks of matter that play a direct role in living systems and describe how they are combined into progressively larger components, up to the level of the living biological cell. We introduce and define some fundamental terms, such as DNA and RNA—important building blocks and mechanisms that we explore in more detail in later sections. The chemical building blocks that are significant in life processes, leaving aside the important aspect of electron and charge transfer, are the atomic species: C, H, O, N, P, and S—all elements that are coincidentally important to organic chemists. The simplest molecules of greatest importance are H_2O, CO_2, N_2, and O_2.

Water. Although we have discussed the importance of *water*, its structure, and chemical properties in previous sections, we reinforce its significance. Its ability to make hydrogen bonds (**Fig. 14.1**); dissociate into OH^- and H^+ (pH); and solvate essential minerals like Na^+, Cl^-, Ca^{+2}, K^+, Mg^{2+}, Fe^{3+}, Fe^{2+}, I^-, and numerous other metal cationic and anionic species cannot be understated. The interaction

FIG. 14.1 *Water structure—hydrogen bonds are continuously forming and rearranging. The two hydrogen and one oxygen atoms that make up the water molecule can associate into aggregates with other water molecules by forming hydrogen bonds. The hydrogen bonds between water molecules lead to exclusion of the hydrophobic organic portions of large biological molecules such as lipids and proteins. This exclusionary force drives the formation of micelles and double layer membranes. It is the main force determining the folding and shaping of proteins. Water molecules solvate with any hydrophilic portion of a protein or fatty acid through hydrogen bonds. The hydrophobic portion of the molecule is squeezed into a shape determined by the attraction of water for the hydrophilic portions.*

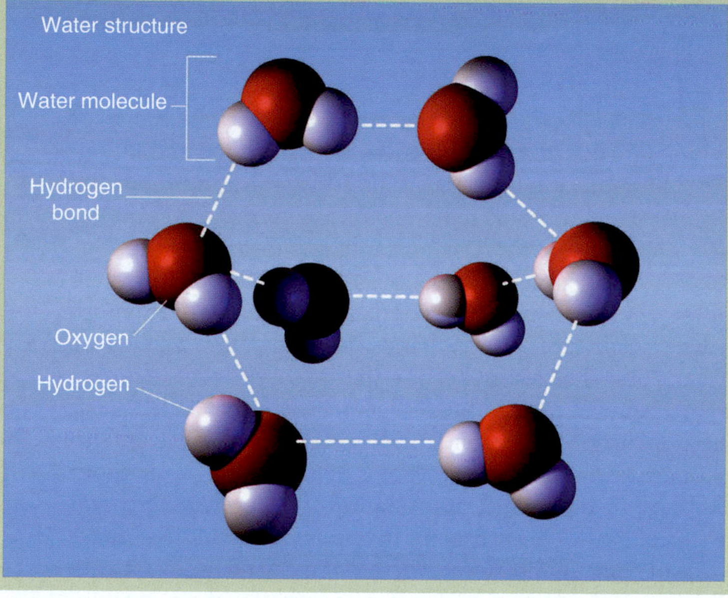

Source: Image by Anil Rao.

of hydrophilic and hydrophobic species by water forms the basis for structural order in cellular systems. It is no wonder that NASA is always looking for water on other planets.

Elements. *Carbon* is equally essential. The ability of carbon to form bonds with itself, hydrogen, oxygen, nitrogen, sulfur, and phosphate groups makes it the key building unit for life. The presence of heterocyclic atoms (nitrogen, oxygen, sulfur) in aromatic rings alters the electronic properties of the electron bond system and forms the basis for the macrocycle effect discussed in chapters 10 and 11. *Oxygen* is a key element that, in combination with hydrogen, gives rise to the unique properties of water; oxygen also modifies the properties of carbon compounds. Oxygen serves as a source of unpaired electrons that are available for dative binding of metals and is the central ingredient in aerobic respiration. *Nitrogen* is found in nucleotide purine and pyrimidine bases such as adenine, guanine, thymine, and cytosine (and uracil). It is a participant in the formation of peptide bonds in proteins. *Phosphorous* is a component of the backbone of DNA and RNA and of ADP and ATP (adenosine di- and triphosphate). *Sulfur* plays a major role in protein folding—a process that affords proteins their functionality.

Biological Molecules. Life depends on a variety of organic and inorganic compounds: food (metabolites including sugars, starches, fats, vitamins, and minerals); sensing and signaling materials (hormones, neurotransmitters, enzymes, regulator compounds such as nitric oxide); structural components (collagen fibers, shells, bones); chemical transport structures (hemoglobin); digestive entities (enzymes, stomach acids), and energy catalysts (ATP, chlorophyll). In **Table 14.1** we review a few essential biological molecules and their roles.

Amino Acids. A set of extremely important building blocks for life are the *amino acids*—organic compounds of carbon, hydrogen, oxygen, and nitrogen (see **Table 14.1**). The carboxylic oxygen component possesses acidic hydrogens and the amine nitrogen component is basic: Amino acids are *zwitterions* and, given the right solvent conditions, a positive charge from the acid part of the molecule can transfer to the base portion, leaving the two parts of the same molecule with opposite charges.

Amino acids in which the acidic carboxylic group and the basic amine group are attached to the same carbon atom are called *alpha amino acids*. Opposite electron affinities pulling on the same carbon make these alpha carbon atoms hungry for electrons, and thus they are very reactive. Alpha carbons readily form bonds to electron-rich nitrogen atoms in amino groups, linking amino acids together to form long chains.

In these chains, the amine group on one molecule links to the alpha carbon of a neighbor; in biochemistry this link is termed as a *peptide bond*. Peptide linkages are covalent metastable chemical bonds that are strong enough to make stable chains, but weak enough to be broken by catalysts in water so that chains can be rearranged under mild conditions. Out of all amino acids that are chemically possible, only alpha amino acids, with the carboxyl and amino groups located on the same carbon at the end of the molecule, form peptide linkages found in living organisms. Twenty specific alpha amino acids are the building blocks of proteins. Amino acids are precursor molecules with subnanometer dimensions. When they are assembled to form a peptide or a protein, at that stage, they become nanomaterials.

Proteins. Proteins are the primary building blocks for the molecular nanostructures and nanomachines that perform the work of cellular life and are bona fide nanomaterials. Proteins are long organic chains of chemically linked amino acid molecules (polyamides). The connecting links of the chain are the peptide bonds between alpha carbons and nitrogen atoms in an amine group. The backbone of the chain thus consists of carbon and nitrogen with bonds to hydrogen, oxygen, and organic groups and sometimes to phosphorous and sulfur. The minimum chain size for a protein to have a structural biological function seems to be about 40 amino acid groups. Shorter chains are referred to as peptides or polypeptides.

Peptides interact with proteins to serve as messengers and perform other roles in cells. The protein chains can be several thousand amino acid units in length that fold onto themselves to form stable shapes held together by intermolecular

Table 14.1 *Biological Molecules*

Biomolecule	Structure	Comments
Amino acids	Glycine 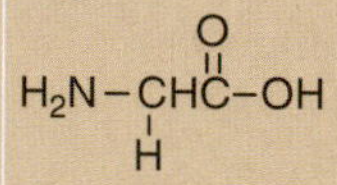 Tryptophan 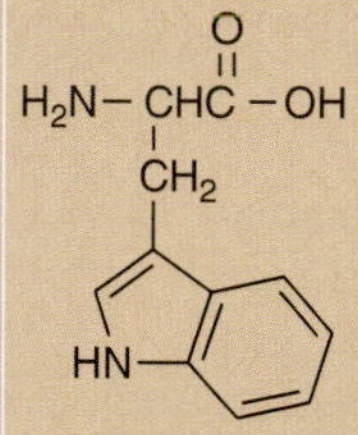	Glycine is the smallest amino acid: <0.5 nm, MW = 75 Tryptophan is the largest natural amino acid in proteins; <0.7 nm, MW = 246 All amino acids that make up life have the following basic structure: The **R** represents different groups in each amino acid; the amine group (red) and the carboxylic acid group (blue) are both attached to alpha carbon atoms that react with the nitrogen in other amino acid molecules to form peptide bonds. 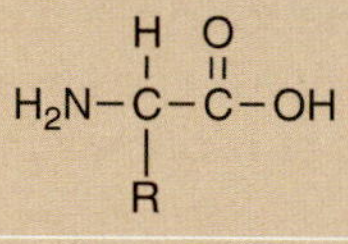
Carbohydrates	α-Furanose 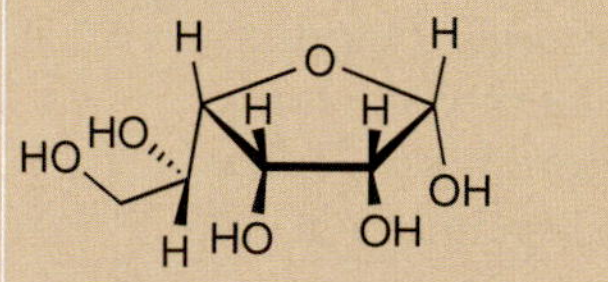 D-Glucose 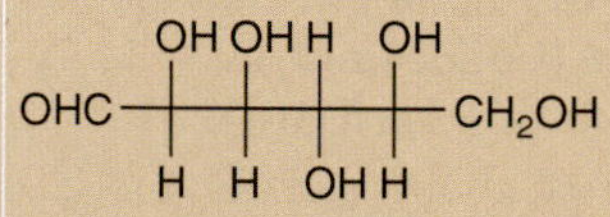	Carbohydrates (from the Greek *carbo,* "sugar," and *hydra,* "water") are aldehydes and ketones with hydroxyl groups. Carbohydrates function in the storage and transport of energy (starch, glycogen) and in structure (polysaccharides, cellulose, chitin). Monosaccharides include glucose, fructose, ribose, and deoxyribose. The size of simple sugars is on the order of amino acids. Polysaccharides can get relatively large. Glycogen is ~150 nm.
Lipids, fatty acids, and phospholipids	Myristoleic acid $CH_3(CH_2)_3CH{=}CH(CH_2)_7COOH$ Oleic acid $CH_3(CH_2)_7CH{=}CH(CH_2)_7COOH$ Cardiolipin (Diphosphatidylglycerol) 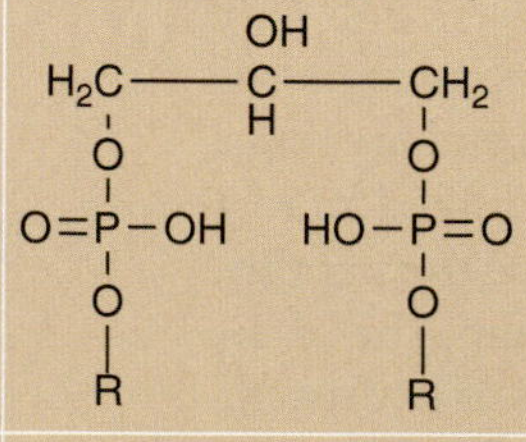	Many fatty acids are long-chain aliphatics with a carboxylic acid group at the terminus. Unsaturated fatty acids have no double bonds. Phospholipids contain a phosphate group and constitute a major component of cell membranes.
Nucleotide bases	Adenine 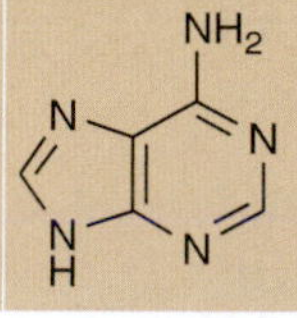 guanine 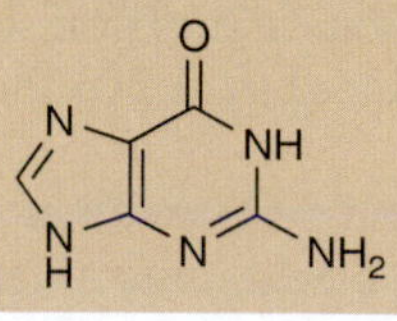	The four base nucleotides of DNA. The sizes and molecular weights are on the order of the amino acids. Thymine and adenine specifically complex together by hydrogen bonding, as do cytosine and guanine.

Table 14.1 (Contd.) *Biological Molecules*

Biomolecule	Structure	Comments
	Cytosine 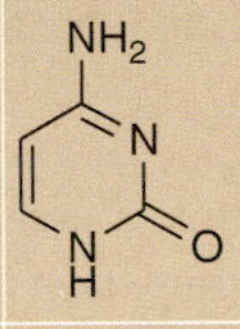 thymine 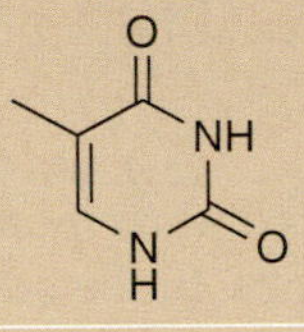	
Adenosine triphosphate	Adenosine-5′-triphosphate 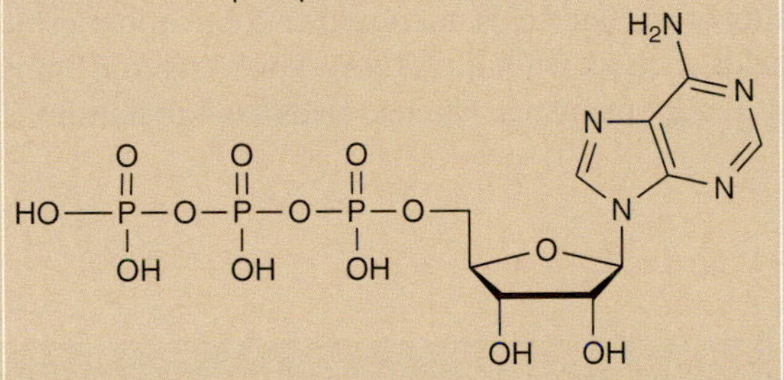	ATP serves the role of storing, transporting, and releasing energy to drive reactions in cells. All cells use ATP as their primary energy currency. Energy is stored in the phosphate–phosphate group bonds. These bonds are formed with energy consumed in respiration in the mitochondria and photosynthesis in plants, with the aid of special nanostructures that harvest and redirect electron energy. MW = 499; size = 0.95 nm
Proteins	1° Primary structure is the amino acid backbone 2° Determined by intramolecular hydrogen bonding. Two kinds: α-helix or β-sheet 3° α-helix + β-sheet; disulfide linkages 4° Quaternary structure consists of more than one protein (for hemoglobin)	Peptides formed from condensation reaction of amino acids to form peptide bonds Size ranges from 2 to 50 nm or greater; MW ranges from 6000 to >1,000,000 The simplest protein (peptide) is insulin. Hemoglobin, enzymes, fibrinogen, albumin, collagen
Cofactors Coenzymes Vitamins	Fat soluble: vitamins A, D, E, and K Water soluble: thiamin, riboflavin, nicotinic acid, pantothenic acid, ascorbic acid, folic acid, B_{12}, biotin NAD, FAD	The class of organic substances that function in trace amounts. Compounds that are essential to sustain the life are called cofactors. If an organism cannot synthesize a cofactor, it must obtain it from its environment.
Steroids	β-Estradiol (estrogen) 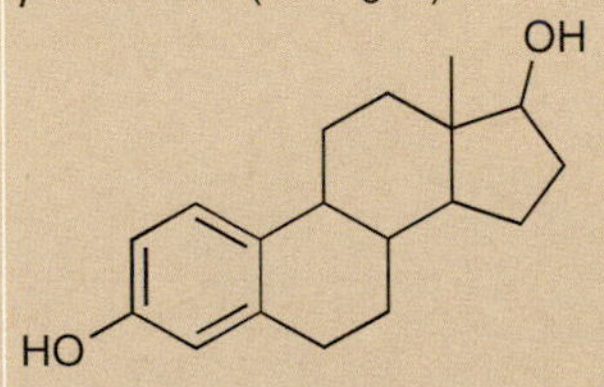 Testosterone 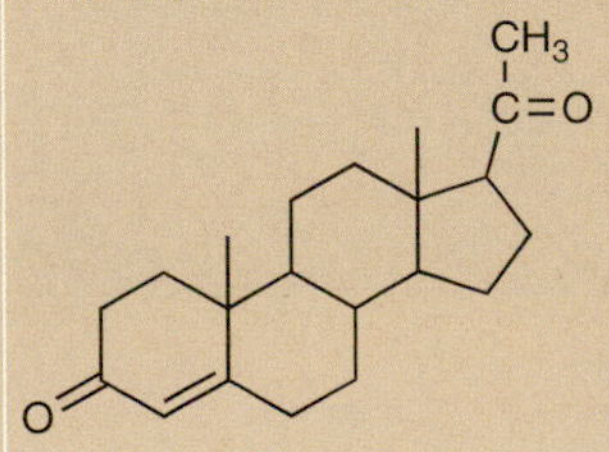	Includes cholesterol, vitamin D, and many hormones. Vitamin A (retinol) is an isoprenoid derivative, as is vitamin E (tocopherol). Terpenoids can be found in all classes of living things, and are the largest group of natural organic compounds. The flavors, scents, and colors of many plants are due to terpenoids, as are antibacterial and biologically active substances such as menthol, camphor, and the cannabinoids produced by the *Cannabis* plant. Steroid hormones include testosterone, progesterone, estradiol (an estrogen), cortisol, and others. Vitamin D is a steroid hormone precursor (cholecalciferol, also called calcitriol).

forces such as hydrogen bonding between water and hydrophilic portions of the chain (**Fig. 14.2**). The affinities of hydrophobic organic portions of the chain for each other and their repulsion by the surrounding aqueous environment play a strong role in the folding of proteins. Protein shape is highly correlated with their purported function: controlling access to chemically active reaction sites on the chains. Proteins serve as structural units but also act as catalysts and transport engines [37–43].

Nucleotides. Nucleotides are made up of three portions: a heterocyclic base, a sugar, and one or more phosphate groups. In the cell they have important roles in energy production, metabolism, and signaling. Nucleotides are linked in polymeric chains of nucleic acids anchored to a phosphate-sugar backbone to form nucleic acids such as DNA and RNA—encoding and transcribing genetic information in living organisms. Nucleic acids are the nanomachinery of DNA and RNA.

FIG. 14.2 ***Structure of proteins—the basic building blocks for life. Proteins are chains of amino acids: 20 common acids called "residues" linked by carbon nitrogen bonds, "peptide bonds." Chain length can range from 40 to several thousand, commonly several hundred. Chains fold: folding stays in place by intramolecular forces weaker than peptide bonds. Primary structure = residue sequence. Secondary structure = folding: folds are of different types classified as: α helix and β sheet. Tertiary structure = shape of a single protein or "conformation." Quaternary structure = fitting together of several proteins to form a complex.***

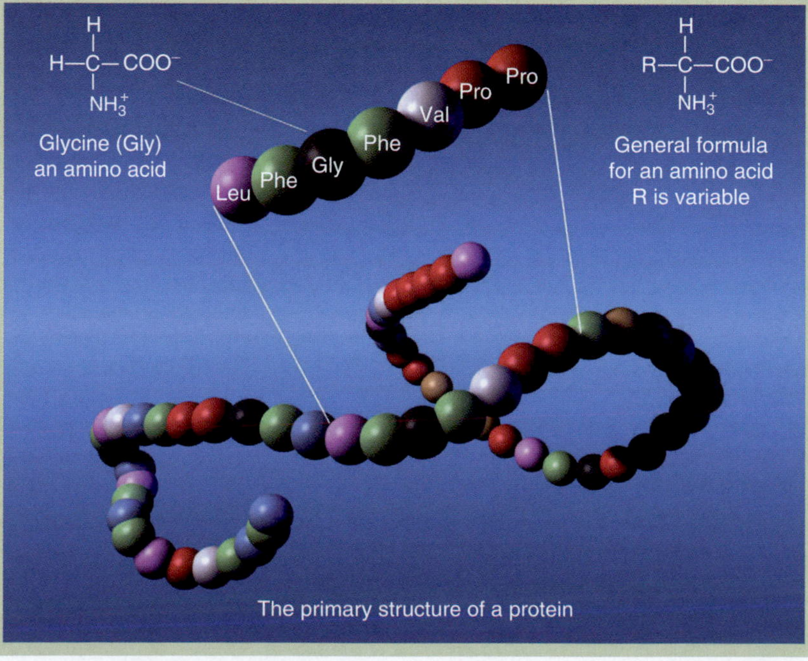

Source: Image by Anil Rao.

14.1.2 Cells and Organized Structures

Bottom Up Meets Top Down. In order to understand how proteins, DNA, and RNA function in living systems, we need to put them into the special context of increasing levels of organization. Starting with the organization of DNA into genes and then of folded proteins, structures such as pores and prions, the simplest life forms like viruses, phages, bacteria, and cells all show hierarchical development. Therefore, we now switch to a more traditional top-down approach and introduce a description of cellular anatomy. However, to maintain the nanobiology context, we place more emphasis on size and scale than is done in most traditional approaches.

Cells and Internal Structure. A brief overview of cellular biology is presented [44,45]. A typical cell in the human body is on the order of 10 μm (10,000 nm) in diameter—the size of a lymphocyte or red blood cell. A yeast cell is typically about 2–3 μm; a bacterium such as *Escherichia coli* is only 1/10 of that size and a flu virus is about 100 nm across. The human egg cell is about 35 μm in diameter. A typical cell is shown in **Figure 14.3.**

Living forms are classified into three categories based on their cell types. The oldest and simplest types of cells, the *archea* and the *bacteria,* are single celled life forms; although simple in structure, they are highly evolved. The more complex cells are called the *eukaryotes* and include all multicelled animals and plants as well as single-celled organisms that have a *cellular nucleus.* All cells have external *membranes* separating their *cytoplasm* from the environment. Most highly developed *eukaryotic* cells have a nucleus containing genetic material. Cells have distinct *organelles* such as *mitochondria* and *ribosomes, microtubules,* and a *cytoskeleton.*

The simpler and generally smaller cells characteristic of *prokaryotes* do not have a separate nucleus but do have DNA. Some bacteria form *capsules* that protect and anchor the cell and contain *spores* that store DNA in an inactive state. When conditions for normal cell growth are unfavorable, the spores are able to regenerate a new bacterial cell when the surrounding conditions change for the better. Bacteria have a *cell wall* made of *peptidoglycan,* a polymer consisting of sugars and amino acids that forms a mesh-like layer surrounding the outside of the plasma membrane. A typical bacterial cell is shown in **Figure 14.4.** Note the cell wall, the absence of a nucleus, and the generally less complex internal structure.

An *organelle* is a descriptive term for any discrete structure found inside a cell. Originally, organelles were identified and named because they showed up as visible regions and bodies under the microscope. Today, organelles are defined in terms of their roles in the material, energy, and information networks of the cell. There are many types of organelles with specialized functions, particularly in the eukaryotic cells of higher organisms. An organelle is to the cell what an organ is to the human body.

We focus especially on some of the molecular nanomachinery that performs essential functions in the cell such as maintaining the structure, sending and receiving signals to other cells, and synthesizing the building blocks used by the cell. Organelles are depicted inside the cell in **Figures 14.3** and **14.4.** Ribosomes are ~30 nm in size; mitochondria are cigar-shaped ellipsoids 500 × 900 × 300 nm; the chloroplast is ~4 μm, lysosomes are ~700 nm, and vacuoles are ~10 mm—all,

FIG. 14.3 *A typical eukaryote cell (animal kingdom). Cells are tiny chemical factories, using protein nanomachinery to produce and deliver materials according to instructions encoded in the DNA. In this figure of a typical animal cell we see the nucleus holding the DNA, surrounded by the nuclear envelope and communicating with the rest of the cell through the nuclear pores. The DNA produces RNA which interacts with the ribosomes to synthesize proteins. Ribosomes are concentrated in the nucleolus, but some are attached to sites on the rough endoplasmic reticulum and some float freely in the cytoplasm. The proteins are sorted and tagged for their destinations in the Golgi apparatus. From there they are transported in transport vesicles to their destinations in the cell machinery, either internally, in the cell membrane, or outside the cell as a secretion (either through a cell pore or a secretory vesicle). Mitochrondia produce energy to drive the processes. Lysosomes break down proteins and other substances that are no longer needed. Centrioles hold disassembled scaffolding units for erecting the apparatus that will be used when the cell divides. The membranes of the smooth endoplasmic reticulum serve as sites where various internal enzymatic reactions needed for the cell functions take place.*

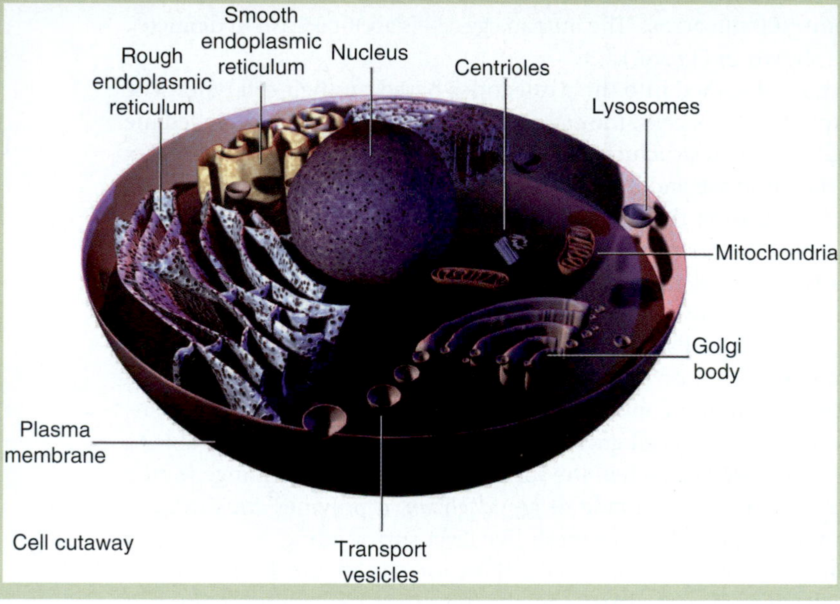

Source: Image by Anil Rao.

without any doubt, considered to be nanomaterials. In **Figure 14.3,** in a typical animal cell, the *nucleus* is central and prominent. The nucleus sequesters DNA and is surrounded by a *nuclear envelope*. Communication with the rest of the cell is conducted through pores in the nuclear envelope. DNA in the nucleus produces the template for *m*-RNA (messenger RNA) that in turn translates the blueprint to the ribosomal protein synthesis machinery.

Ribosomes are concentrated in the *nucleolus* (**Fig. 14.3**); some are attached to sites on the *rough endoplasmic reticulum* and some float freely in the cytoplasm. Proteins are sorted and tagged for their destinations in the *Golgi apparatus* and transported by vesicles to destinations in the cell, either internally, in the cell

FIG. 14.4 *A typical bacteria cell. In this figure of a typical bacterial (prokaryotes) cell, the plasma membrane is protected by a cell wall and capsule. There is no nucleus, and the DNA floats as a plasmid in the cytoplasm, along with ribosomes. A flagellum extends through the cell membrane and wall and is capable of rotary movement. Pili are extensions of the capsule and help attach the cell to its neighbors or to a stationary anchor point.*

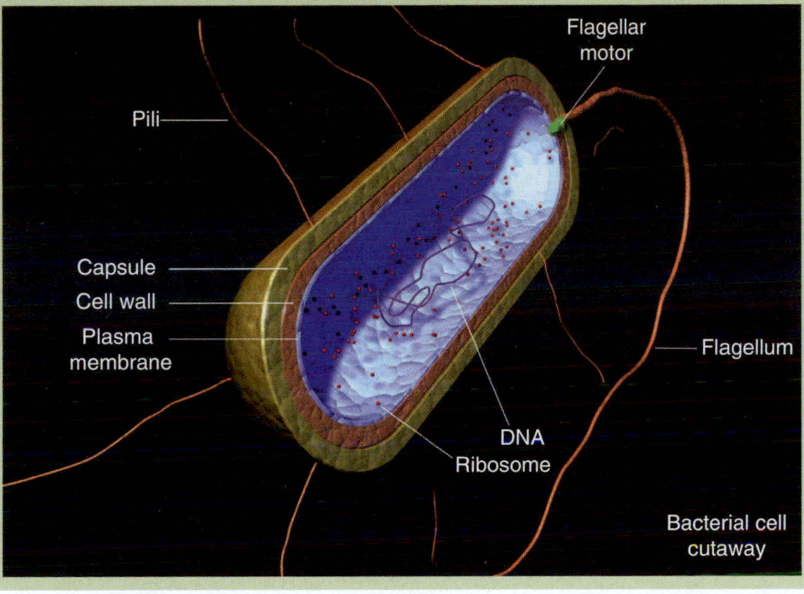

Source: Image by Anil Rao.

membrane or outside the cell via secretion (through a cell pore or a secretory vesicle). *Mitochondria* produce energy to drive cellular processes. *Lysosomes* break down proteins and other substances that are no longer needed. *Centrioles* hold disassembled scaffolding units for erecting apparatus for cell division. The membranes of the *smooth endoplasmic reticulum* serve as sites where various internal enzymatic reactions needed for cell functions take place.

We have made reference to cell membranes throughout the text. All cells have an outer *cellular membrane* (also called the *plasma membrane*) composed of a double layer of lipid molecules with a hydrophilic head and hydrophobic tail (**Fig. 14.5**). These amphiphilic molecules form a *lipid bilayer*, with the hydrophobic lipid tails lying together inward between the outer surfaces formed by the hydrophilic heads. This membrane is interspersed with proteins, carbohydrates, and a number of other large molecules (such as cholesterol) that are held in place by a combination of hydrophilic and lipophilic intermolecular forces.

The *cytoskeleton* is a macromolecular scaffolding or nanoskeleton that gives structure and support to the cytoplasm within cells. It is not present in most prokaryotes, but is an essential structure in all eukaryotic cells (**Fig. 14.6a**). The cytoskeleton maintains the shape and the internal organization of the cell but is not static or rigid; it is able to reconfigure within the cytoplasm depending on prevailing conditions. The cytoskeleton is composed of three types of protein

Fig. 14.5 *Phospholipid bilayer of the cell membrane. Note the inclusions and attachments, including peripheral, integral, and transmembrane proteins.*

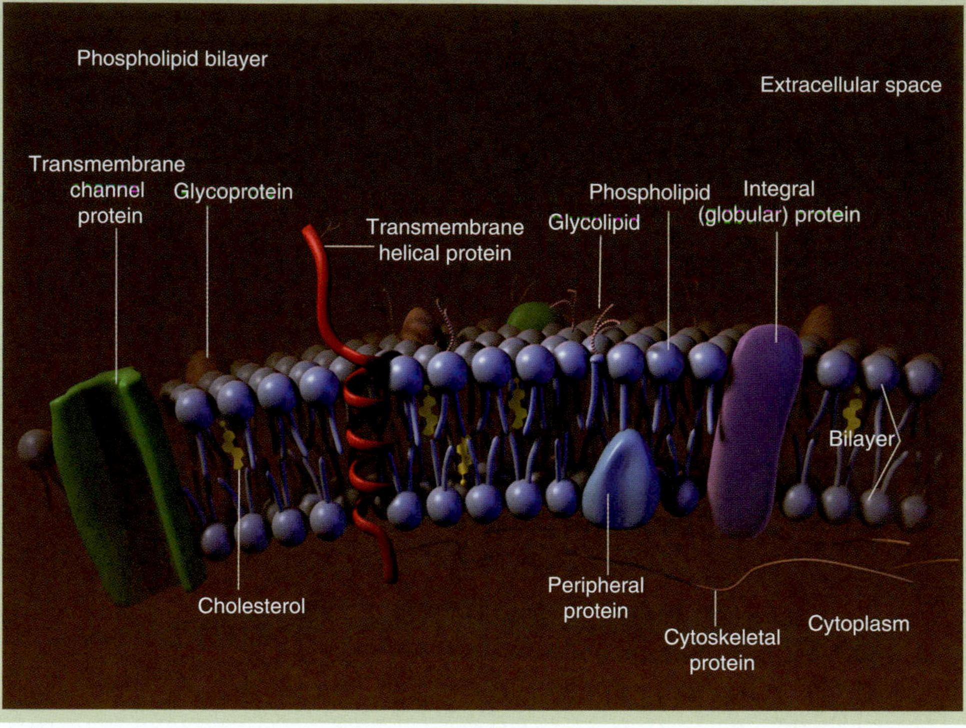

Source: Image by Anil Rao.

Fig. 14.6 *Cells showing endoplasmic reticulum. (a) Atomic force micrograph of cellular endoplasmic reticulum.*

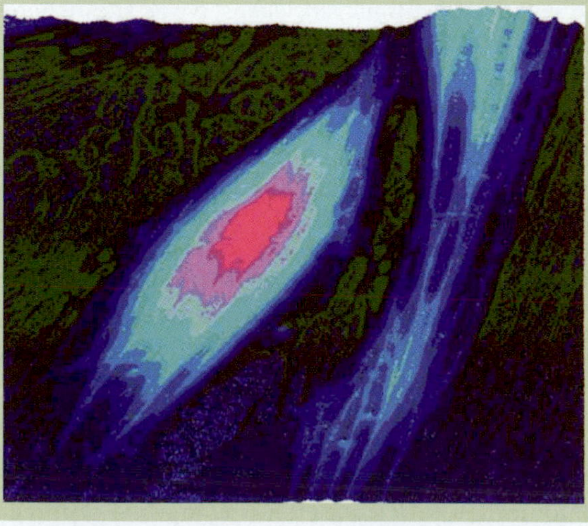

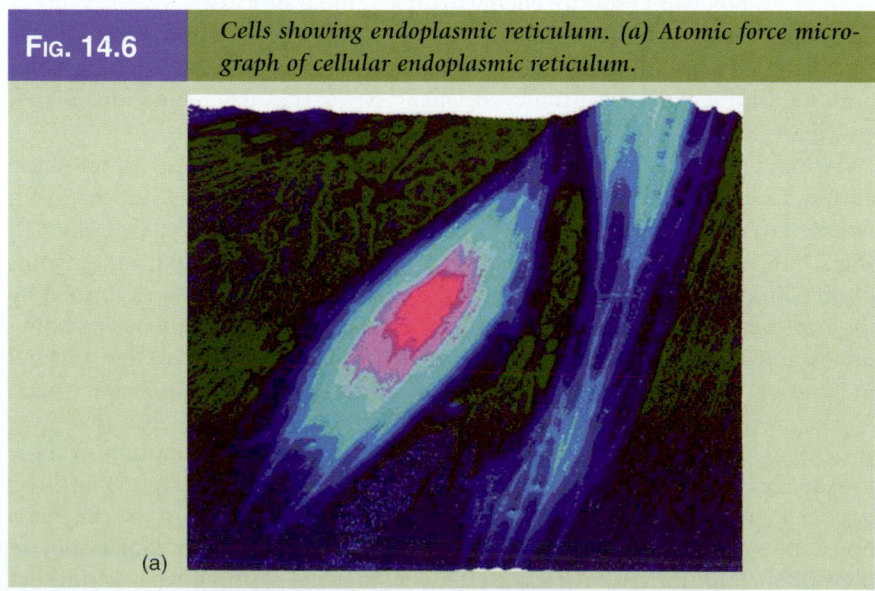

Source: Image courtesy of Veeco Instruments, Inc.

FIG. 14.6 (CONTD.) *(b) Actin filament. (c) Actin and Cadherin filaments support adhesion between cells at the cell membranes.*

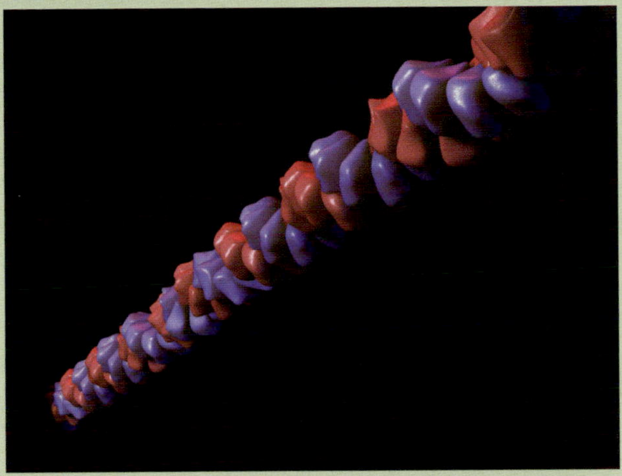

(b)

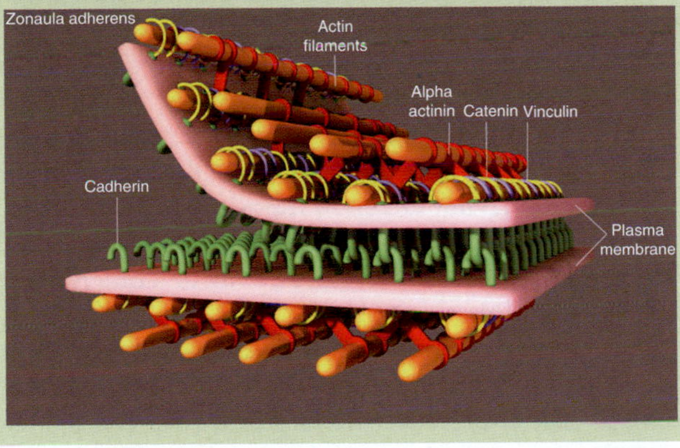

(c)

Source: Image by Anil Rao.

filaments, each organized in a different way and serving a different function: *actin filaments* (or microfilaments), *intermediate filaments*, and *microtubules*.

Actin filaments or microfilaments are composed of two chains of actin protein intertwined in an interlocking spiral to form the 7-nm diameter filaments (**Fig. 14.6b**). Actin filaments are linked by smaller protein units in a framework next to the cell membrane that supports the membrane and its junctions to other cells and the extracellular matrix (**Fig. 14.6c**). Actin filaments are very important in cell migration processes within an organism's body (*cytokinesis*) and anchor *myosin* for muscle contraction. Actin filaments are also distributed within the interior of many types of cells, where they control streaming of components suspended in the cytoplasm. The filamentous system is a remarkable example of a hierarchical nanostructured material.

Intermediate filaments consist of several different types of cytoskeleton elements that are components of supporting frameworks within the interior of cells. Intermediate filaments are generally larger (8–11 nm in diameter) and more stable than actin filaments. There are four types of intermediate filaments, classified according to their compositions and functions. *Vimentins* provide structural support to muscle and other cells. *Keratins* form a scaffold that supports the nucleus of *epithelial* (skin) cells. *Lamins* make up the nuclear lamina, a support framework within the nuclear envelope. Lastly, *neurofilaments* strengthen the long *axons* of *neurons*. Microtubles are hollow cylinders about 25 nm in diameter, made up of *protofilaments* composed of α- and β-tubulin proteins. The functions of different types of microtubules include: (1) intracellular transport of organelles like mitochondria and vesicles, (2) mitotic spindles in cell division, (3) inner core skeleton of cilia and flagella, and (4) scaffolds for synthesis of the cell wall in plants. Thus, the microtubule elements of the cytoskeleton act as tramways for movement of vesicles and organelles within the cell, and for separation of chromosomes during cell division. Portions of the cytoskeleton are anchoring points for cell movement mechanisms such as flagella and cilia, which have microtubule inner core skeletons.

Details of the Nucleus. The nucleus of eukaryotic cells has a double membrane enclosure holding the DNA material. The structure of the nucleus is complex; we will describe some of the main components in this section and in the section on DNA. The nucleus is enclosed in a *nuclear envelope* (or nuclear membrane), which separates the contents of the nucleus from the rest of the cell cytoplasm. The nuclear envelope is a double membrane composed of two separate lipid bilayer membranes. The *perinuclear space* between the two membranes that make up the nuclear envelope is about 20–40 nm wide. The outer membrane is continuous with the rough endoplasmic reticulum. The nuclear envelope contains pores that regulate the exchange of proteins, RNA, and other substances between the nucleus and the cytoplasm.

A *chromosome* is a single long chain molecule of DNA ca. 2 nm in diameter. In eukaryotic cells, the chromosome is packaged in the cell nucleus and includes a surrounding of special proteins, the lamins, which package and protect the DNA strand. In bacteria and other prokaryotic cells without nuclei, the chromosome is bare; it is attached to the cell membrane or floats in the cytoplasm as a closed loop of DNA called a plasmid. In eukaryotes, DNA undergoes several levels of folding before it is called a chromosome [101]. The 2-nm strands are much too long to fit into a several micron diameter cell. The first coiling is the double-stranded DNA. The second coiling event involves a 140-base pair sequence that coils around a histone protein to form a bead (histone-associated DNA is called *chromatin*) to form structures called *nucleosomes*, the diameter of which is 11 nm. These nucleosomes stack on top with linker regions in between (for flexibility). At this stage the protochromosome is 30 nm in diameter. Series of nucleosomes form a condensed region of the chromosome that is 300 nm in diameter that, in turn, condense to form 700-nm structures per strand. Two chromosomes then add to make a structure that is 1400 nm wide and clearly visible with an optical microscope.

The *nucleolus* is a dense body within the nucleus in which ribosomes, RNA, and proteins are concentrated; the ribosomes are assembled from their constituent protein and RNA components and protein synthesis takes place here. The nucleolus is not separated from the rest of the nucleus by a membrane and is a dynamic structure. The cell nucleus is surrounded by a double membrane called the nuclear envelope, which is supported by a framework of lamin protein fibers, called the nuclear lamina. The nuclear envelope separates the nuclear contents, the DNA, and nucleoli from the rest of the cytoplasm. The chromosomes are attached to the lamina; during mitosis, the lamina disintegrates prior to formation of the spindle fibers along which the chromosomes separate. The lamina connects with the rough endoplasmic reticulum adjacent to the nuclear envelope.

Plasmids. A plasmid is a circular strand of DNA separate from the chromosomal DNA that is capable of autonomous replication, generally at a higher rate than a chromosome found in the nucleus. Plasmids are important in bacteria where they are associated with rapid genetic adaptation and exchange of genetic material with other bacteria. In bacterial cells there may be little difference between a large plasmid and a small chromosome. Some plasmids are also found in eukaryotes. Mitochondria and chloroplasts resemble plasmids in structure. It is thought that eukaryotic organisms arose from ancient cells engulfing others to form symbiotic relationships.

Endoplasmic Reticulum. The endoplasmic reticulum is a network of tubules, vesicles, and sacs found in all eukaryotic cells. This interconnected network supports specialized functions in the cell, including protein synthesis, calcium sequestration, steroid production, storage and production of glycogen, and insertion of membrane proteins. Structure and composition of the membranes making up the endoplasmic reticulum are similar to those of the plasma membrane.

The *rough endoplasmic reticulum* is located near the nucleus. It holds ribosomes and other apparatus used in protein synthesis and transport. The *smooth endoplasmic reticulum* is located closer to the periphery of the cell. The membrane network of the smooth endoplasmic reticulum provides surface area for enzyme activity and storage and processing of enzyme products. In highly specialized cells, the smooth endoplasmic reticulum is adapted for specialized functions. In muscle cells, its vesicles and tubules store calcium, which is released by *calcium pumps* during the contraction process.

Lysosomes. Lysosomes are organelles surrounded by globular membranes within the cytoplasm. Lysosomes digest excess or damaged organelles and proteins, food particles, and engulfed viruses or bacteria. The membrane surrounding a lysosome prevents its digestive enzymes from destroying the cell. Lysosome structure resembles the lipid bilayer self-assembled structures discussed earlier.

Mitochondria and Chloroplasts. Mitochondria and chloroplasts are organelles (plasmids) within cells that have their own DNA and carry out special functions such as energy production and photosynthesis, respectively. They represent extremely evolved *symbionts* that have become highly integrated components of higher cells; they carry out essential molecular processing functions for the host cell but they retain their own DNA and replication cycle, and depend on the host for most essential life functions other than their specialized contribution. In many cells mitochondrial DNA codes for some of the many proteins that form the ribosomes. Thus, the cell nucleus and the mitochondria are mutually dependent for the production of proteins [46,47]. One only gets mitochondria, for example, from one's mother.

Ribosomes. A ribosome is a special type of organelle involved directly in the synthesis of proteins. Ribosomes contain the active ingredient of protein replication (ribosomal RNA or *r*-RNA) along with the associated proteins. Ribosomes are active in the nucleolus, bound to the endoplasmic reticulum, and floating freely in the cytoplasm, depending on the type and destination of the proteins they synthesize. The molecular recognition system involved in protein synthesis, starting with DNA, is indeed one of the most remarkable examples of living chemistry.

14.1.3 Viruses

Existing on the boundary between living and nonliving structures, viruses consist of a nucleic acid (DNA or RNA) genetic core enclosed in a protein coat. The virus is able to hijack the operation of specific host cells once the viral particle gains entrance. The protein coat, called the *capsid*, protects the core from attack by antibodies and provides a mechanism for penetrating the wall of the host cell—the simplest forms of survival. The core may also contain one or more protein enzymes used to aid replication in the cell, but the virus uses the molecular machinery of the host for most of its reproductive functions [48]. Outside the host cell, the virus particle with its protein coat is a dormant particle called a *virion* that may form crystals like a nonliving molecular material.

Viruses are of particular relevance for nanoscience because of their small size and the simplicity of their molecular machinery. A typical virus particle is between 10 and 300 nm in diameter. Many are roughly spherical in shape, but some are filaments with length up to 1400 nm. The diameter of the capsid is only about 80 nm. Most viruses are too small to view with a light microscope, but some are as large as or larger than the smallest bacteria and can be seen under high optical magnification.

14.1.4 Prions

Another type of subcellular disease factor, whose existence was controversial when first reported, is the prion [49]. Prions are misfolded proteins that appear to be able to catalyze misfolding of related proteins in living systems, resulting in dysfunction in cells and organisms. Like viruses, their existence was first inferred by the ability of tissue extracts to transmit disease in the absence of any known identifiable disease agent such as microbes or viruses. Their action is more biological in nature than mere toxins due to their ability to reproduce copies of themselves, leading to plaques (especially in the brain) and to disrupt normal metabolic functions. Thus, they are sometimes referred to as "quasi-bionts" like the viruses, but they do not have the structure of RNA or a protein coating like virions.

Prions were discovered after exhaustive research into the causes and cures for the devastating condition known as "mad cow disease" or BSE (bovine spongiform encephalitis). BSE is related to a similar disease in sheep and is transmitted to humans by consumption of contaminated meat products. Prions seem to be extremely refractory in that they are able to survive the digestive system and make their way to the brain, where they produce their damaging effects (a compelling reason to cook food thoroughly). All known prions are believed to infect and propagate by formation of an amyloid fold—polymerization of the protein into a fiber with a core consisting of tightly packed β-sheets [50].

14.1.5 Toxins and Disruptive Nanoparticles

Traditionally, toxins are defined as molecules that disrupt the operation of cells to cause damage or death by chemical action. Chemically inert particles, if sufficiently small, also are able to disrupt the molecular machinery of cells. Nonliving particles such as asbestos fibers and soot particles that lodge into the fine structures of the lung and some other tissues thereby have the capability to disrupt processes of cell growth and division, causing cancers and cell death. Immune cells are able to absorb and address many types and sizes of contaminating nonliving particles as well as bacteria [51]. Classical toxins range from heavy metal atoms that tie up and precipitate cytoplasm to complex biomolecules such as enzymes and proteins that disrupt and disorganize specific cellular metabolism pathways [52]. Toxic inert particles include asbestos, silica dust, metal oxides, and soot (although some particles may be chemically active as well and microscopically disruptive at the same time).

In general, toxins are nonliving chemical substances that interface with the complex nanomachinery of living systems and, in the worst case, cause blockage or disruption. Molecules that are toxic to one type of organism or cell may be harmless or useful to others. Many living organisms produce highly specific toxins as a defense mechanism. In a later section we explore how the immune systems of some organisms, including humans, produce extremely sophisticated toxins to protect the organism against harmful invaders. We also examine some concerns about possible toxic effects of new types of nanoparticles that are entering the environment as a result of nanotechnology.

14.1.6 Completing the Circle from Top Down to Bottom Up

When we started our exploration of the building blocks of life, we went from the simplest chemical constituents up to the complex structure of DNA and RNA. We are able to describe these constituents in terms of chemistry and the physical interactions between molecules up to the nanoscale. Above this level of organization, we begin to encounter structures on the order of hundreds of nanometers and higher. Above this point it becomes quite cumbersome to describe the components and their interactions at the detailed level of atoms and molecules; new levels of organization begin to emerge. Aggregates of molecules form units that act as identifiable separate entities and maintain their structures and identities. To the extent that such aggregates are stable and act as units, it is consistent with reality to consider them as objects on a higher emergent level of organization. Thus, we continue our description by starting with observable features under the microscope, such as cells, and work our way back down to molecular components in the context of cellular anatomy and functions.

Levels of Organization in Living Systems. Although it would be theoretically possible to describe the behavior of a complete living organism in terms of its chemical and molecular interactions, it is more efficient to employ larger aggregated units of organization as shorthand in order to simplify the description. This higher level of description not only is more efficient, but also makes for a better fit to our human ability to grasp complexity and to understand, model, plan, and manipulate. This description is analogous to grouping of lines of computer programming code into functions and subroutines; it helps us to deal with the complexity of large systems. We put assemblages of molecules into an aggregated conceptual box and use higher levels and features to describe living structure and behavior. In studying nanoscience or any complex system, there is always the possibility of an element of chaotic behavior in which new features emerge unexpectedly. This can be true even when we try to carefully control experiments, designs, and fabrications. Thus, the study of living systems is instructive for nanoengineers engaged in any complex design [53].

Emergent Properties: Complex Systems, Chaos, and Catastrophe Theory. We introduced emergent property theory briefly in chapter 8, "Nanothermodynamics." The emergence of new features and structures from apparently chaotic behavior is a widespread but constructive phenomenon. Stable, robust structures and performance can be based in underlying chaotic and random processes. This is the domain of chaos theory, fuzzy logic, and catastrophe theory, practical applications of which include complex robotic controls; simulated annealing for exploring optimal designs of circuits, polymers, and proteins; neural networks; signal processing; and track correlation. It is noteworthy that many early insights into the nature of chaos theory owe their origins to the study of natural living systems, such as ecological population dynamics and morphogenesis and embryonic development [54–58]. An important aspect of biological nanoscience is the insight that understanding of processes at the nanoscale can provide in revealing how large numbers of very small components can organize to produce effective behaviors without centralized direction or control.

Lower Size Limits for Life. What is the size of the smallest aggregate of molecules that is able to support all of the functions of life? As we have seen with viruses and prions, there is a continuum between living and nonliving aggregates of matter. Biomolecular nanoscience, aided by ever more powerful microscopy, continues to discover and investigate new naturally occurring life forms and life-related particles, as well as nonliving particles that can interact with living metabolisms, down to sizes in the nanoscale. With microscopy tools able to resolve and study structures at the nanoscale, an understanding is emerging of a rich complexity of life components and disease agents at the subcellular level.

The generally accepted lower limit for cellular life forms such as bacteria is 200 nm, which happens to be the lower limit of resolution of a standard light microscope. But noncellular quasi-bionts such as viruses have been found in sizes from 10 to 300 nm in diameter.

In 1999, a workshop convened by the U.S. National Academy of Sciences estimated that "the minimum diameter of a spherical cell compatible with a system of genome expression and a biochemistry of contemporary character would appear to lie somewhere between 200 and 300 nm, probably closer to the latter" [59]. For noncellular life forms such as viruses [60]:

> The lower size limits are seen in some RNA viruses like the Qß virus (which contains only three genes), and in certain animal viruses (e.g., poliovirus) that are in the range of 25 to 50 nm in diameter, while most others are in the range of 100 to 200 nm or even larger.

Symbiotic organelles or bacteria are also commonly found in the 200-nm range and are sometimes smaller. These include bacteria that can only grow inside cells of other organisms, intracellular organelles (e.g., mitochondria or chloroplasts), and the enigmatic nanobacteria smaller than 200 nm that have been reportedly found in kidney stones [61–63]. Thus, at the smallest and lowest number of components, organic units of matter can no longer have an independent existence; they can only carry out life functions by acting symbiotically with other life forms. It is somewhere on the boundary of independent living cells and dependent symbionts that the boundary between life and nonliving matter lies.

In addition to asking how small an organism can be in physical volume or number of molecules, it is of interest from the nano- and molecular biology perspective to ask how small the set of DNA machinery that programs the cell—the genome and the genes that it encodes—can be.

Upper and Lower Size Limits for Genomes in Cells. From a nanoscience and nanotechnology point of view, it is useful to learn how the simplest organisms are structured because they can provide a starting point for designing artificial nanomachinery for biochemical functions. For these reasons as well as fundamental scientific interest, it is of interest to ask, "What is the smallest number of genes needed to construct a viable organism?" [64].

Advances in the ability to create new types of organisms, either through modification of existing genomes or synthesizing DNA from components, have led to the new field of synthetic biology [65]. The emerging field of synthetic biology leads to a view of the cells as an assemblage of parts put together to build an organism with desired characteristics. This program has been approached theoretically and experimentally in the laboratories of Craig Ventnor, Francis Collins, and elsewhere [66,67]. To put the question in context, consider that the organism with the largest genome found thus far is a microscopic protozoan, *Amoeba dubia*, with a genome of 670 billion base pairs (670 Gb). A related microbe, *Amoeba proteus*, has a mere 290 billion base pairs, but that number is 100 times larger than the human genome. The 3-Gb-long human genome is thought to encode fewer than 30,000 genes, perhaps as few as 23,000 that are required to program the human body.

The free-living organism with the smallest genome found thus far is one of the most abundant proteobacteria in the world's oceans, *Pelagibacter ubique*, with a DNA sequence length of only 1.3 Mb. *P. ubique* is suited for a robust independent existence, with complete biosynthetic pathways for all 20 amino acids and all but a few cofactors [68]. Parasitic or symbiotic bacteria may have even smaller genomes because they depend on their host for essential life functions [69].

The mitochondria organelles in the cells of higher organisms may have originated as an independent symbiotic microbe that invaded host cells (or were engulfed) and gradually coadapted to integrate fully with the host. There is much genetic evidence for symbionts adapting to their host and becoming simpler and smaller in the process, notably in bacteria adapted to live inside the cells of insects [70,71]. The smallest such examples yet found have apparently undergone extreme genome reduction in adapting to intracellular life within their hosts and have many similarities to mitochondria. An alternative but less widely accepted explanation that has been posited is that extreme endosymbionts represent a stage in the evolution of an organelle that split off from the host nuclear genome [72].

Even smaller hypothetical nanobacteria may fragment into nongrowing entities that appear considerably smaller than the true, viable organisms, and these fragments may come together at a later time to form a viable organism. Some plant viruses exhibit just such a pattern. Each particle package separates RNA, and sometimes three separate particles are needed to establish an infection. While this strategy is known for a few RNA viruses, there are as yet no examples among the bacteria or other prokaryotes [60,73]. Other requirements for small cell size include: (1) reduction of the average size of proteins, (2) an RNA-world approach in which a single type of molecule accomplishes both catalytic and genetic functions, and (3) the use of overlapping genes and genes on complementary strands [59,60].

Genome Sizes in Viruses. Viruses, which need no genes for constructing and maintaining cellular structure and depend entirely on host cells for reproduction, may have as few as three genes. Most viruses encode from 3 to 400 genes. The largest known viral genome is found in the *mimivirus*: approximately 1.2 megabases that encode about 1000 genes. The discovery of this virus blurs the boundary between higher living organisms and viruses [74].

Continuing the Search for Life at the Nanoscale. The search at the nanoscale for new types of minimally sized life forms invites theoretical questions of extraterrestrial life and in studies of the origins of life on Earth. This search is also important for disease agents, including toxins, catalytic particles, and life forms. Determining their roles in human and animal disease remains an active and open area of investigation, with many open questions. Some evidence has been reported to implicate anobacteria or other nano life forms as agents in deposits of calcium as arterial plaque in blood vessels, kidney stones, and joints, as well as in tooth decay. However, alternative explanations have been presented, so the matter remains controversial [75,76].

14.2 Cellular Membranes and Signaling Systems

In the following section, we look into aspects of the life cycle of cells and how cells communicate with their surroundings.

> All organisms consist of cells that multiply through cell division. An adult human being has approximately 100,000 billion cells, all originating from a single cell, the fertilized egg cell. In adults there is also an enormous number of continuously dividing cells replacing those dying. Before a cell can divide it has to grow in size, duplicate its chromosomes and separate the chromosomes for exact distribution between the two daughter cells. These different processes are coordinated in the cell cycle. [77]

We discussed cell size briefly in chapter 5 and continue from another perspective here. Cells grow to a certain size and may divide to form new cells. In each type of tissue or in each type of single-celled organism, the optimal cell size is determined by the processes by which nutrients, oxygen, and waste products are brought into and out of the cell. Size matters because of the costs of transport; hence, the ratio of cell volume to surface area is a critical factor. The optimal size of a cell is related to its rate of metabolism, its shape (ratio of volume to surface), and the particular mechanism for transporting metabolic raw materials and products across the cell—as well as the particular specialized function that the cell is programmed to perform.

Cells in complex multicellular organisms organize and function as tissues and organs. Tissues in the human body have a very specific and intricately organized nanostructure, adapted to enhance their performance in load bearing; stress transport; fluid transport; resistance to penetration by hostile pathogens; and protection, balance, and containment of vital substances such as water, oxygen, salts, and nutrients. Organs have complex and specialized functions, which are interdependent and regulated by chemical and electrochemical signals in an interplay of nerve and chemical communications networks. All of the communication of cells with the surrounding fluid environment must involve the cell membrane. For organs to develop, function, and be maintained, cells must recognize and respond appropriately to their neighbors and environment.

14.2.1 Cell Membrane Function

Cells are contained and maintained by their membranes, both external and internal to the cytoplasm. The nucleus has its own membrane, as do other specialized structures such as mitochondria, organelles, etc. The external cell membrane contains specialized structures with sophisticated mechanisms for regulating what passes between the inside of the cell and the external environment. Each of these structures is a functional biological nanomachine, built of biological molecules such as proteins, glycoproteins, peptides, etc. [78,79].

Specialized ion channels and membrane domains act as gateways to control molecular events on the nanoscale that govern electrostatic and material balances, interactions with the external world, and communication with other cells of the same and different organs. Molecular biology, biophysics, and molecular genetics techniques for the study of ion channels are active areas of research on the nanoscale domain to understand topics such as the molecular origin of voltage dependence in ion channels, functioning of calcium-activated K^+ channels of large conductance, gap junction tunnels, the process of exocytosis, and other special aspects of cell membrane nano-anatomy [80]. More strongly ionic species—such as cations of sodium, calcium, and potassium ions; anions of chlorine; and most proteins—have low solubility and affinity for the lipid layer, which is a natural barrier between the cytoplasm and the exterior. In general, hydrophilic species can only pass across the lipid barrier by active expenditure of energy facilitated at special gateway sites (e.g., valinomycin).

Figure 14.5 showed a section of cell membrane composed of the phospholipid bilayer with several types of inclusions interspersed into the bilayer. The bilayer itself is composed of many pairs of phospholipids, with their hydrophilic tails forming a lipid-rich region sandwiched between the hydrophilic heads, which form two outer layers facing the external and internal environments. The outer environment will typically be filled with intercellular fluids that bathe the cells. The inner environment is the cell's cytoplasm. Both are generally based in aqueous media.

The total thickness of the lipid bilayer is only about 5–10 nm or less, including the water of hydration attached to the phosphate groups depending on proteins and lipids embedded within the bilayer. The core region is very hydrophobic and the phosphate surfaces are very hydrophilic: The effective concentration of water changes from nearly 0 inside the membrane (between the two layers of the lipid) to a concentration of around 2 *M* in the phosphate layers [81,83].

Cholesterol is an essential component of all cell membranes. The cholesterol molecule is amphiphilic, with a hydrophilic alcohol group coupled to a lipophilic steroid and hydrocarbon chain. With the alcohol anchoring the cholesterol molecule to the phosphate head layer of the membrane, the lipophilic portion inserts itself between the lipid chains, where it enhances membrane flexibility—somewhat like a lubricant. Cholesterol molecules also group together to form lipid rafts in the membrane that reduce the permeability of the membrane to hydrogen ions (acid), sodium, and other ions. Glycoproteins and glycolipids attach to the membrane and project out from the surface with specific markers that are recognized as tags to identify the cell, with hydrophilic sugar residues attached to the part of the polypeptide exposed at the surface of the cell. While the most common function of these projections is to allow recognition of the cell by other cells, enzymes, and antibodies, some types of "projecting molecules" serve as anchors to attach the cell to its surroundings or to other cells, in specialized types of cells. They play an important role in attaching cells to each other to construct tissues and organs.

Proteins that are embedded in the cell membrane are anchored in place by filaments of cytoskeletal protein (on the interior of the cell) or tethered to an extracellular matrix (composed of collagen and other polymers, which form scaffoldings between cells). Some embedded proteins and lipids are free to float through the membrane. Some types of proteins are embedded in the membrane to perform material transport functions. The attachment locations of these proteins can be *peripheral, integral,* or *transmembrane*. Peripheral proteins are inserted in or attached only to the phosphate surface or only to one half of the lipid bilayer. Integral proteins are embedded through both halves of the bilayer. Transmembrane proteins are integral proteins that cross through the membrane to make connection between the inner and outer environments of the cell.

Peripheral membrane proteins are associated with membranes but do not penetrate the hydrophobic core of the membrane. Peripheral membrane proteins typically have only weak electrostatic and temporary attachment to the membrane. They are often attached to integral proteins instead of or in addition to the phosphate layer, serving in a secondary attachment role in the process associated with the integral protein. Because they have weak electrostatic attachments, peripheral proteins can be separated from the membrane by salt solutions, whereas integral proteins can be disrupted from their membranes only with strong detergents [84,85].

14.2.2 Ion Pumps, Ion Channels, and Maintenance of the Cellular Environment

Membrane pores act as ion channels that help to control the concentration of ions inside the cell. This difference in ion concentration establishes an electronic potential difference across the cell membrane. Ion channels are present in all types of living cells. The controlled passage of ions through the potential gradient of the channels plays an important role in cell physiology for nerve and muscle cells.

Each type of cell can have many different kinds of ion channels controlling different types of ions. The most important ions for cell physiology are sodium, potassium, calcium, and chlorine (Na^+, K^+, Ca^{2+}, and Cl^-). A channel that can open and close is called a gated channel. A cell may have several different types of channels controlling the same ion in different ways in response to different conditions. Several types of mechanisms control gated channels. A channel can switch between an activated or open state to an inactivated or closed state in response to: (1) voltage changes (voltage-gated channels), (2) activation of receptors (receptor-gated channels), and (3) specific ions and chemicals (ligand-gated channels). In the sections that follow we will discuss some of the characteristics of ion channels and the functions that they enable in nerve and muscle cells.

14.2.3 Transmission of Neural Impulses: Action Potential and K Channel

Key Advances and Tools. Since the 1700s, people have known that nerve impulses are somehow associated with electricity. Electricity was first known as the buildup of static charge on pieces of amber. In 1771, Luigi Galvani discovered that frog legs can be made to contract by application of electric charge. This led to the direct experimentation of the effects of electricity on muscle and nerve tissue. By the mid-nineteenth century, Helmholtz measured the speed of propagation of the *action potential* (the membrane potential) along frog neurons. Helmholtz found that the propagation velocity was about 40 m per second—slower than would be the case if the action potential were due simply to electrical conduction through a wire. Thus, the action of nerves had to involve a more complex biochemical process. The term *synapse* was coined by Sherrington for these junctions in 1897.

Microscopic techniques revealed much about the structure of synapses in the following years, but it would be another 60 years before their mechanism of operation could be unraveled. This depended upon first understanding how electrical impulses were carried along the membranes of the nerve fibers, the *axons*, whose long processes stretched out from the cell body to connect with other neurons. At this time it was understood that the membrane potential was maintained by a difference in ion concentration gradients across the cell membrane. Using the microelectrode technique, it was determined that the potential gradient in nerve cells was due to an imbalance between sodium and potassium ions on the inside and outside of the membrane. The sodium and potassium gradients were in opposing directions. When the ion gates opened and the concentration gradient collapsed, the result was an electrical propagation along the nerve fiber. Somehow, the potential would have to be restored after each such discharge. The voltage clamp was an advance on the microelectrode, adding a second reference electrode and controlling the potential inside the cell with a feedback circuit [85].

In a famous series of experiments published in 1952, Alan Hodgkin and Andrew Huxley used the voltage clamp method to separate the components of the membrane currents due to potassium (K^+) and sodium (Na^+) ions. They were able to separate the components of the membrane conductance into three factors: Na^+ conductance, K^+ conductance, and Na^+ inactivation. They found that the Na^+ conductance is activated in response to depolarization but closed without repolarization. They concluded from this that there must be two control gates for the Na^+ channels. The K^+ channels, working in the reverse direction from the Na^+ channels, were activated by polarization but did not have a deactivation response. The K^+ channels work to restore the membrane potential to its resting level after a discharge. Since the K^+ channel gates work continuously to maintain the resting potential, they require only an activation mechanism. With advances in amplifiers, it is possible to measure extremely small currents corresponding to the passage of single ions through an ion channel [86].

Mathematical Description of the Cell Membrane. In the previous section we described in qualitative terms how the balance of chemical forces results in an electrical potential difference across the lipid bilayer membrane. In this section we will show the same concept in more precise mathematical terms. Molecules that are free to move, as in a gas or a solution, tend to diffuse from a region of higher concentration to a region of lower concentration (a *concentration gradient*). Because it requires expenditure of energy to concentrate a substance in solution or gas, we can describe the tendency to diffuse away from concentrated areas as a *force*.

If concentrated salt water is separated by such a membrane from pure water, the water will diffuse into the region of higher salt concentration until the salinity is equal on both sides of the membrane. This phenomenon is called *osmosis* (from the Greek *endo*, "inside" + *smos*, "push or thrust") and is observed when water is absorbed into plant tissues. It represents a real force because it can do the work of raising large amounts of water from roots of trees up to leaves, far in excess of the work that could be done by capillary action alone.

The Nernst Equation. The force associated with concentration diffusion is measured by the work done by the change in volume as water diffuses across a membrane. If a force in the form of pressure is applied to one side of the membrane, water is forced to cross the membrane from the side of high salt concentration to the other side. This is the basis for water purification by reverse osmosis. The equation analogous to the ideal gas law for the work due to differences in concentrations is

$$W_c = 2.303RT \log \frac{[A]_{out}}{[A]_{in}} \quad (14.1)$$

where W_c is the work, or energy; $[A_{out}]$ is the concentration in one region of solution (by convention, outside the cell); and $[A_{in}]$ is the concentration a different region of the solution (inside the cell). The concentration is expressed in terms of moles.

If, as in the cell, we have a membrane separating two unequal concentrations of ions, such as potasium (K^+) and chlorine (Cl^-), and the membrane is differentially permeable to the positive and negative ions, there is an electrical potential difference across the membrane if the concentrations of the two species differ on either side. The work required to separate the two ions is given by

$$W_E = \mathcal{F}\mathcal{E} \quad (14.2)$$

where W_E is the work or energy; $\mathcal{F}$ is Faraday's constant, a measure of the quantity of electrical charge per mole of substance; and $\mathcal{E}$ is the electrical potential difference across the membrane. Note that this work is independent of the distance between the charges and only depends on the potential difference, no matter over what distance the potential exists, independent of the membrane thickness.

When the chemical forces are in balance with the electrical forces, the system is at equilibrium; there is no net movement of ions across the barrier. The chemical force tending to move, say K^+ into the cell, will be equal to the electrical force tending to move it outside (or vice versa), and the work given by equation (14.1) will equal the work in equation (14.2), so

$$W_c = 2.303RT \log \frac{[K^+]_{out}}{[K^+]_{in}} \quad (14.3)$$

It follows that the membrane potential, $\mathcal{E}$, is given by

$$E = 2.303 \frac{RT}{F} \log \frac{[K^+]_{out}}{[K^+]_{in}} \quad (14.4)$$

This is the *Nernst equation*, first derived by Walther Nernst in 1888. The membrane potential E is also called the *Nernst potential* or the *diffusion potential*.

For any given ion species, such as K^+, Cl^-, etc. and given type of membrane, such as a squid neuron, the Nernst potential is the *equilibrium potential* at which there is no net flux for that ion across the membrane. This *membrane potential* is of fundamental importance to electrical activity in all cells and to the nanoscience of living things.

At 18°C, the constant 2.3 $RT/\mathcal{F}$ = 58 mV. For the giant squid axon, the ratio for concentration of potassium outside versus inside the cell is 1 to 20 [87], and the membrane potential is –75 mV. This is the equilibrium potential that the membrane tends to assume when membrane is permeable to potassium (i.e., when the potassium ion channel gates are open).

The Membrane Equivalent Circuit. A convenient way to represent the membrane potential is with an equivalent electronic circuit, in which the lipid membrane is represented by a capacitance element. The membrane potential due to each type of ion is represented by a battery connected across the membrane capacitor. Each battery has the voltage and polarity corresponding to the ion for which it is an analogue. For each type of ion, the ion gate is represented by a variable resistance in series with the battery. Ohm's law for electrical currents relates resistance and potential (voltage) to the current as

$$E = IR \tag{14.5}$$

where E is the electrical potential in volts (V), I is the current in amperes (A), and R is the resistance in ohms. The reciprocal of resistance is conductance (g) measured in siemens (S).

Conductance is an analogue for membrane permeability (P).

Current flowing across the giant axon membrane may be represented by the sum of conductive components (which we identify as ion channels) and capacitance (the cell membrane). The electrical properties can be represented by the circuit diagram in **Figure 14.7,** the classic *membrane equivalent circuit.*

The Hodgkin–Huxley Model Equations. Hodgkin and Huxley [88,89] discovered that the conductivity of different ions was dependent upon the potential difference across the cell membrane (V_m) and the equilibrium potentials (E) of the ions. They assumed the membrane potential was not equal to the equilibrium potential when there is a net flow of ions proportional to the difference between the membrane potential and the equilibrium potential. They represented the observed conductivity across the membrane as the sum of currents through the conductive elements of the equivalent circuit, or:

$$I_{\text{ion}} = g_{\text{ion}}\left(V_{\text{m}} - E_{\text{ion}}\right) \tag{14.6}$$

In the simple form of the Hodgkin–Huxley model represented by the preceding equivalent circuit diagram, there are only three types of channel: a sodium channel with conductivity g_{Na}, a potassium channel with conductivity g_{K}, and a weaker chlorine channel with conductivity g_{Cl}.

From their voltage clamp measurements, Hodgkin and Huxley found that g_{Na} and g_{K} change over time as well as with voltage, but the conductances of the

FIG. 14.7 *Membrane equivalent circuit.*

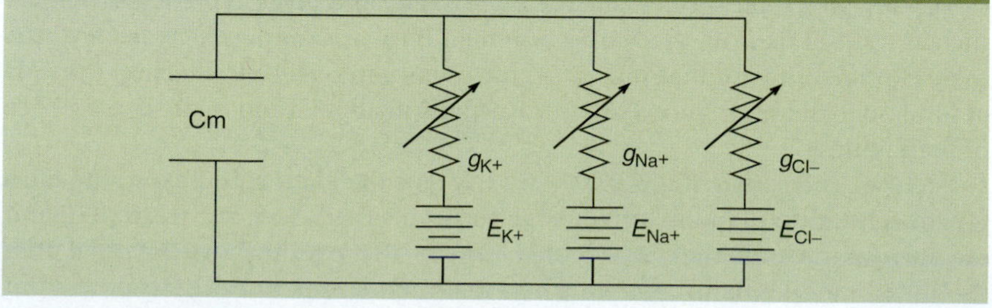

Source: Image by Mike Hammer, Lightspeed Llp.

other ions are constant. When the axonal membrane was depolarized, they observed a transient increase in Na^+ conductance and a slower noninactivating increase in K^+ conductance. In cells, conductance is caused by opening of individual ion channels, but each channel is either open or closed. In the Hodgkin–Huxley model, a probability function x (known as a gate variable) may be constructed to define what fraction of the channels are open as a function of the membrane voltage and time.

In the original simplified model, the time dependence of the conductance is represented by an activation coefficient x, which represents the probability of a gate in the channel being open. The conductance for a time-dependent channel can thus be written in terms of its activation coefficient x ($0 \leq x \leq 1$) and the maximum conductance $g_{\text{ion,max}}$ that is possible when all channel gates are open:

$$g_{\text{ion}} = g_{\text{ion,max}}(x) \tag{14.7}$$

where x is given by the equation:

$$\frac{dx}{dt} = \alpha_x x(1-x) - \beta_x x \tag{14.8}$$

where α_x and β_x are rate coefficients that are nonlinear functions of voltage (units are t^{-1}).

The activation coefficient can be raised to a higher power to represent the case in which a larger number of gates are present in a single ion channel. Additionally, each channel may have multiple activation coefficients to describe activation followed by, for example, voltage-dependent inactivation. With these refinements, the current is given by an equation with the following general form:

$$I_{\text{ion}} = g_{\text{ion,max}} xy(V_{\text{m}} - E_{\text{ion}}) \tag{14.9}$$

where x is the activation coefficient and y is the inactivation coefficient. This is the form of the equation for the Na^+ conductance responsible for the depolarization during the action potential because it inactivates without repolarization. Because the Na^+ conductance opened in response to depolarization but closed without repolarization, Hodgkin and Huxley reasoned that there must be two control gates. One is activated when a threshold depolarization is achieved; the second closes more slowly. The two gates are represented by two variables: m is the activation coefficient (substituted for x), and h is the inactivation coefficient (substituted for y). The time dependencies of the two gates are described by the differential equations:

$$\frac{dm}{dt} = \alpha_m x(1-m) - \beta_x m \tag{14.10}$$

$$\frac{dh}{dt} = \alpha_h x(1-h) - \beta_x h \tag{14.11}$$

where α and β are rate constants that are time-invariant functions of voltage. The Na^+ conductance may then be written:

$$g_{\mathrm{Na}} = g_{\mathrm{Na,max}}\left(m^3 h\right) \tag{14.12}$$

where $g_{\mathrm{Na,max}}$ is the maximum Na^+ conductance.

Expressions for rate constants were found by fitting curves to experimental data. When this was first done, the data from squid giant axons had to be numerically integrated by hand-cranked numerical calculators, an enormous effort described in Huxley's later paper [88].

The Na^+ current is described by

$$I_{\mathrm{Na}} = g_{\mathrm{Na}}\left(V_{\mathrm{m}} - E_{\mathrm{Na}}\right) \tag{14.13}$$

$$I_{\mathrm{Na}} = g_{\mathrm{Na,max}}\left(m^3 h\right)\left(V_{\mathrm{m}} - E_{\mathrm{Na}}\right) \tag{14.14}$$

where E_{Na} is the Na^+ reversal potential. The potassium current may be described similarly, although, instead of voltage-operated activation and inactivation gates, one need only include an activation gate.

The Hodgkin–Huxley model explained the mechanisms for initiation and propagation of action potentials in neurons and has remained the basis for computational modeling of nerve function. Hodgkin and Huxley received the Nobel Prize for physiology or medicine in 1963 in recognition of this accomplishment and related work on understanding neuronal mechanisms.

Potassium Channel: An Example of a Nano Ion Pump Engine. As an example and to give an appreciation of how ion channels work from a nanoscience point of view, we illustrate one type of gated channel for which the structure has been determined. Potassium channels are very important and common in many different types of cells, from bacterial to vertebrate nerve and muscle cells. There are a large number of different types of potassium ion (K^+) channels, but one of the most important is the voltage-gated family that opens and closes in response to changes in the membrane potential. This type of potassium channel is especially important for its role in the action potential of nerve cells. Voltage-gated potassium channels are found in almost all organisms including bacteria, where they maintain the ionic balance of the cell.

The molecular model shows a K^+channel from the bacterium *Streptomyces lividans* (**Fig. 14.8**). The channel is formed from a transmembrane protein group with a central pore through which the potassium ion is gated.

Functional Components: Filter and Gate Subunits. The K^+ channel consists of two major functional parts: the filter pore, which selectively allows potassium ions to pass, and the gate, which opens and closes the channel based on the membrane potential. The pore part of the channel both selectively filters and conducts ions through the membrane. The voltage-sensing gate portion of the channel changes its conformation in response to changes in transmembrane voltage, thus closing the pore. The filter region forms the extracellular end of the pore. The voltage activation gate lies at the opposite end of the channel, at its cytoplasmic mouth.

How the Subunits Are Constructed. The potassium channel is made up of four identical protein chains that link together to encircle the central pore space that

FIG. 14.8 *Potassium ion channel.*

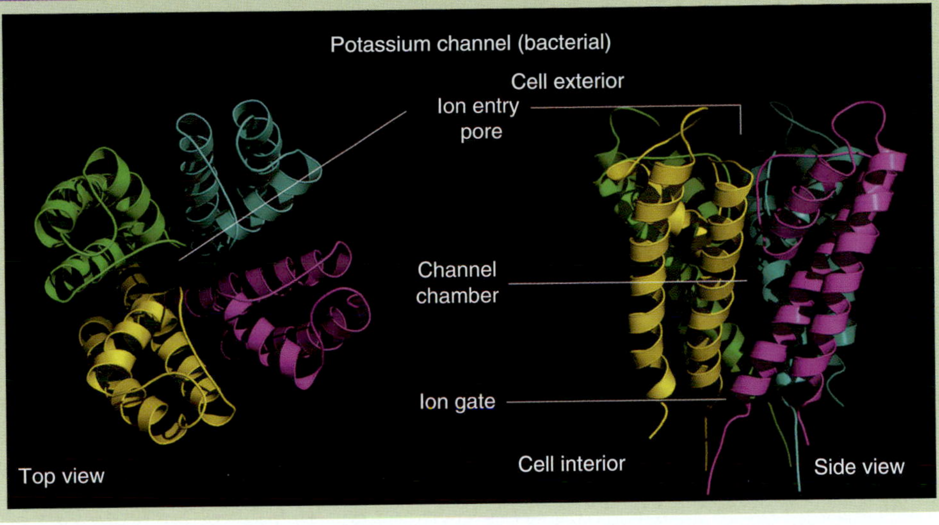

Source: Image by Anil Rao.

serves as the ion conduction pathway. The two functional domains of the channel, the filter and the gate, are formed by the opposite ends of the four grouped protein chains. Each protein chain is composed on six alpha helices. The first four helices of each chain form the voltage gate, which is thus composed of 16 alpha helices in all. The last two transmembrane helices of each chain, along with a loop of protein chain that circles back into the membrane (called a re-entrant P loop), are embedded inside the lipid portion of the membrane to form the pore walls. The P loop includes a short helix and a region of polypeptide chain that forms the ion selective filter.

The amino acid chain sequence of this potassium channel transmembrane protein is similar to all known K^+ channels, particularly in the pore region. X-ray analysis with data to 3.2 Å has shown that the four identical protein subunits create an inverted teepee, or cone, cradling the selectivity filter of the pore in its outer end. The narrow selectivity filter is only 12 Å (0.12 nm) long, whereas the remainder of the pore is wider and lined with hydrophobic amino acids. A large, water-filled cavity and helix dipoles are positioned so as to overcome electrostatic destabilization of K^+ ions in the pore in the interior of the membrane bilayer. Carbonyl oxygen atoms attached to the protein chain line the selectivity filter. The selectivity filter contains two K^+ ions about 7.5 Å apart that provide electrostatic repulsive forces to overcome attractive forces between K^+ ions and the polarizable covalent bonds in the protein chain making up the selectivity filter.

How It Works. The ion conduction pathway can switch between two main functional states: open and closed. When the cell membrane is polarized so that the interior of the cell is at a negative voltage relative to the exterior, the K^+ channels remain closed. When the membrane is depolarized, these channels open rapidly (in less than 1 ms), allowing ions to flow passively down their electrochemical gradients at near diffusion rates (up to one hundred million ions per second). K^+ channels have a selective permeability for potassium of at least 10,000 times greater than for sodium.

Comparing the x-ray structures of K^+ channels crystallized in a closed conformation to similar channels crystallized in an open conformation suggests that bending or kinking of the inner pore-lining helices plays a key role in pore gating. Studies of the potassium channel have identified two segments that contain several charged protein residues, and these charged residue regions apparently undergo conformational changes in response to the potential difference across the membrane; therefore, they could contribute to the gating mechanism.

Recent research is revealing that voltage-gated K^+ channels all have very similar pore structures and permeability mechanisms, although they use diverse mechanisms of gating (the processes by which the pore opens and closes). Certain features of potassium channels indicate a multi-ion conduction mechanism involving single-file queuing of ions inside a long, narrow pore. Because of these properties, K^+ channels are classified as "long pore channels." The pores of all the K^+ channels are blocked by tetraethylammonium (TEA) ions, making this and related compounds strong toxins.

Ball and Chain Mechanism: An Example of Nanomachinery in the Ion Gate. In many voltage-gated potassium channels, a protein ball and chain functions to open and close the ion gate, like a cork bottle stopper on a string. The amino-terminal domain of the protein chain, on the inside of the cell where the gate is located, forms the ball and chain. The ball region swings freely through the cytoplasm on its protein chain tether. The ball is the right size and shape to plug the pore when it swings in from the cytoplasm, thus deactivating the channel. The tether is long enough for the ball to reach the inside surface of the plasma membrane, where the positively charged, hydrophobic ball can be attracted and attached to anionic phospholipids of the lipid bilayer. A change in the membrane potential is apparently sufficient to dislodge the ball from the membrane and into the plugged position [90].

14.2.4 Synapses and Neurotransmitters

We have seen how the ion gating mechanism works for the propagation of the action potential along nerve cells. In the 1950s, Katz, Fatt, and Exxles demonstrated that ion channels were also fundamental to signal transmission across the *synapses*, the junctions between nerve cells. In the synapses, the ion channels are not gated by changes in membrane voltages as were the action potential gates. Instead, the synaptic ion channels were found to be regulated by small molecules, ligands like *acetylcholine*. In the 1960s and 1970s, working with large amounts of tissue to isolate very small amounts of ligands, researchers discovered many other small molecules and peptides that gate synapse channels, including glutamate, GABA, glycine, serotonin, dopamine, and norepinephrine. These became important in understanding and regulating overall brain and nerve activity.

Synapses. The junction between a neuron and another cell is called a synapse. Most commonly we think of synapses as being between two neurons, but synapses also connect nerve endings to muscles and sensory organs. Synapses connecting a nerve to muscle fibers are also called *neuromuscular junctions* or *myoneural junctions*.

Neurons. To make the function of the synapse clear requires an overview of the structure and function of *nerve cells*, the *neurons*. Neurons are the specialized cells that carry messages in the body. Neurons carry out specialized roles as motor neurons, sensory neurons, and highly specialized types of neurons found in the brain. A typical neuron has four main morphological regions: the *cell body*, *dendrites*, *axon*, and *presynaptic terminals*. The *cell body* is similar to that of other cells, with a nucleus and cytoplasm. The cell body is also referred to by the neurological terms *perikaryon* and *soma*. It is the base station that maintains the metabolic functions for the more specialized appendages, the *dendrites* and *axon*, collectively called *nerve processes*.

When action potentials travel down the axon to the presynaptic terminal, they initiate the release of neurotransmitters into the gap between the cell junctions (called the *synaptic cleft*). The action potential triggers the opening of calcium ion (Ca^{2+}) channels in the axon membrane, allowing calcium to enter the cell. The calcium ions in turn cause the opening of small vesicles bound to the membrane, containing neurotransmitters. The neurotransmitters are released into the narrow cleft and bind to ion channel receptors on the receiving cell membrane. Typically, the receptors are ligand-gated ion channels, which open in response to the neurotransmitter to change the membrane potential of the receiving cell (**Fig. 14.9**).

The potential change in the receiving cell may cause excitation or inhibition. Excitation usually is associated with the flow of positive ions into the cell and inhibition by negative ion flow. If enough excitation signals are received, they trigger an action potential in the dendrite of the receiving neuron, thus propagating the signal to the next cell. Synapses may release neuroinhibitors, which have the effect of lowering the activity of the receiving cell.

The receptors in the synapses are sensitive to any compound recognized by the gate in their ion channels. Toxins, anesthetics, and drugs that act on the nervous system may mimic the neurotransmitters, block the receptors, or instigate other effects to interfere with normal nerve function.

14.2.5 Hormones and Regulation of Cell Growth and Metabolism

In the previous section we saw how ion channels in the cell membrane work and how changes in electrical potential trigger the release of chemical neurotransmitters in the nervous system. Now we take a look at a more generalized system of intercellular communication: the regulation of growth and metabolism by chemical messengers acting in a less localized region than the synapses. As we go through some highlights of the system of hormone regulation, it is important to remember that the same fundamental mechanisms of communication through receptors embedded in the cell membrane are at work throughout all cells of the organism.

Endocrine System. Historically, the first cell signaling chemicals to be recognized were those associated with the *ductless glands* of the human body, which were

FIG. 14.9 *Neuronal synapse.*

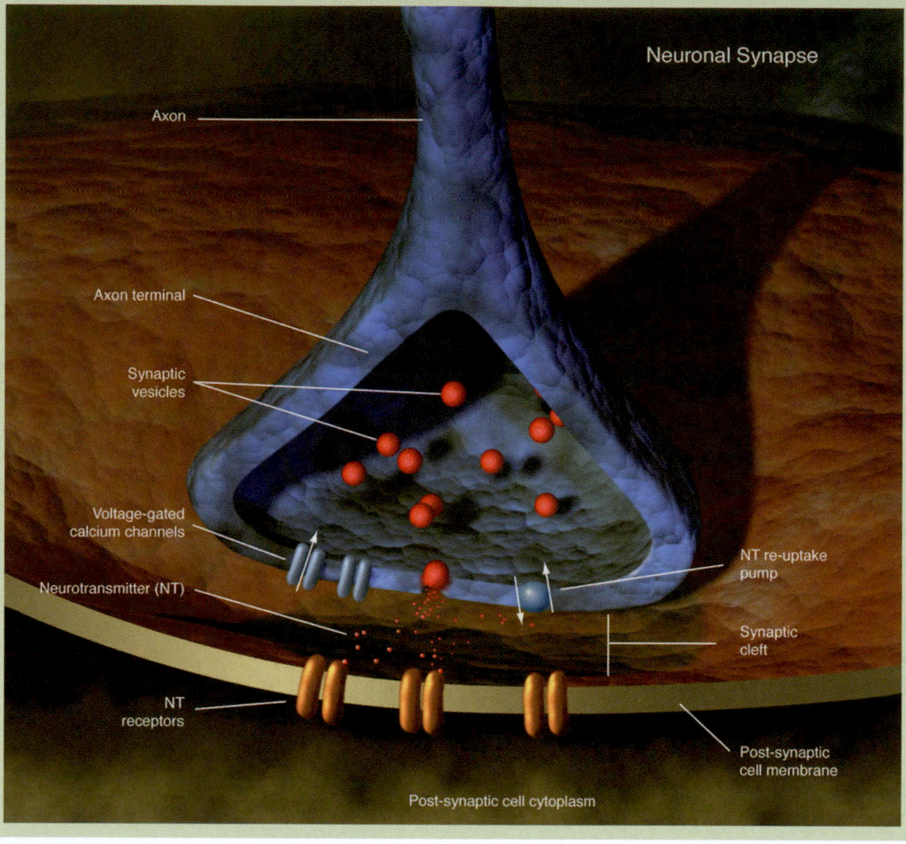

Source: Image by Anil Rao.

classified anatomically as the *endocrine glands*, in contrast to the *exocrine* glands, which had obvious visible ducts to deliver their secretions into the blood or digestive tract, like the sweat, salivary, and bile glands. The *endocrine system* is an anatomical and physiological classification for nine ductless glands that secrete important *hormones* [91,92].

Hormones are chemical regulators that are carried throughout the body, where they are recognized by specific cell receptors to regulate *growth*, *metabolism*, and *sexual development*. Over 100 different hormones have been identified and classified in humans. In insects, hormones control the growth and transition between different phases of development. Recent advances are leading to the realization that almost every organ of the body releases some type of chemical signaling compound. Thus, the endocrine system represents the most dramatic example of the phenomenon of cell signaling, which plays a much more generalized role in the body.

The Control of Growth. The growth of a cell or tissue is now recognized to be a complex process, controlled by a network of interactions among genes, metabolism, nutrition, and hormones. Different aspects of growth include developmental patterns, growth rate, body and tissue size, aging, and cancer. These interactions are under the control of *chemical*, *electrical*, and *physical* influences to which the individual cells respond. Because the cells are programmed to respond in specific ways to specific chemicals, we may think of these chemical messengers as *signals* that carry information to the cells. The signals are received and recognized at specific receptors.

Cell signaling requires recognition of the hormone signal by the receiving cell, which involves interaction of the hormone with a protein receptor. The *receptor protein* may be in the cell membrane or internal to the cell. Typically, the hormone fits into a steric structure in the receptor protein and produces a conformal or electronic change in the state of the protein. This process is similar to the way in which gating locations in cell membrane proteins interact with ions and other ligands, as we saw with the potassium ion channel.

Any regulatory interaction where the result causes the originating step to moderate is called a *negative feedback* loop. Negative feedback tends to stabilize a process by regulating the rate of the driving step. With negative feedback loops, any deviation from the normal or optimal control *set point* gives rise to a countering action that tends to return the system to the control state. In contrast, *positive feedback* leads to a runaway process that grows or speeds up to a breaking point or discontinuity, after which the cycle starts over [58].

Cellular Hormone Production Factory: Pancreas. An important example of a hormone system is the *insulin* and *glucogen* system, which regulates metabolism throughout the body. The complex of cells that interact to synthesize insulin and related hormones is an illustrative example of how the cellular signaling system interacts to control hormone production. Insulin up-regulates the metabolism of glucose, and glucogen down-regulates glucose metabolism in cells throughout the body. Insulin is produced in the *pancreas* gland by specialized *beta cells* located in regions of the pancreas called the *islets of Langerhans*.

14.3 DNA, RNA, and Protein Synthesis

A fundamental aspect of the structure of living organisms is the pair of special molecules that control protein formation. This molecular machinery generates proteins from a physical code stored in a polymer chain by means of an elegant and beautiful nanoprocessing engine that is inspiring and instructive for the design of molecular computers and actuators. Proteins are synthesized in living organisms by a catalytic chain building process directed by two special types of long chain molecular structures, called nucleic acids (or nucleosides) because they were first isolated in the cell nucleus [44,45,93].

14.3.1 DNA and RNA Function and Structure

The two types of nucleic acid play separate roles in protein synthesis: DNA (deoxyribonucleic acid) stores the information used to make a protein. DNA is the template that makes RNA (ribonucleic acid); the latter actually encodes the

FIG. 14.10 *The DNA double helix structure and the bases from which it is formed.*

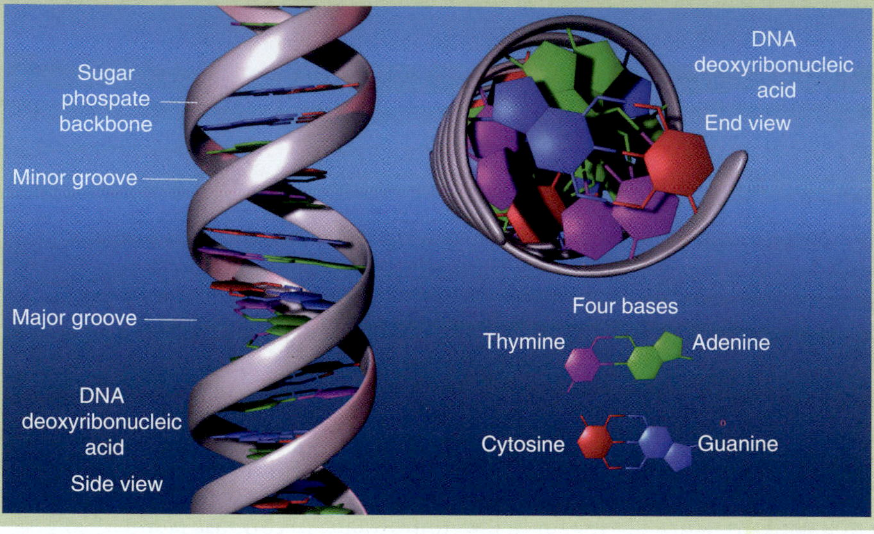

Source: Image by Anil Rao.

amino acid sequence and makes the protein chains with the aid of enabling catalysts.

DNA consists of pairs of nucleotides (also called bases) arranged in a double helix (like a ladder twisted into a spiral) (**Fig. 14.10**). Each nucleotide has three parts: a five-carbon deoxyribose sugar ring, a phosphate group composed of phosphorous and oxygen, and a nitrogen-containing base. The DNA nucleotide bases may be one of four compounds: adenine, cytosine, guanine, or thymine (**Fig. 14.10**). The four bases that are part of nucleotides in DNA are represented in code sequences by the letters A, C, G, and T. The sequences of these bases in a DNA molecular strand comprise the genetic code. The three-dimensional chemical structure of the bases is critical for allowing DNA to form helical chains and to self-replicate. Adenine can form two hydrogen bonds with thymine while cytosine can form three hydrogen bonds with guanine. A–T and C–G are called complementary pairs. This coupling property of the bases is the mechanism for the bonding process in DNA self-replication.

Bases in the DNA strand pair up with complementary bases to form a chain of RNA. RNA has the three bases—adenine, cytosine, and guanine—in common with DNA, but in RNA uracil replaces the DNA thymine. Each RNA nucleotide also has three parts: a five-carbon ribose sugar ring (instead of the deoxyribose in DNA), a phosphate group, and one of the nitrogen-containing bases (**Fig. 14.11**).

The RNA chain is usually arranged in a single helix. The bases in the RNA in turn pair up with amino acids to form a protein chain. The amino acids are chemically attracted onto the bases of the RNA strand and then linked by an enzyme into a protein polymer chain. Three bases in sequence on the RNA chain match three portions of each amino acid so as to absorb it in place during the chain-building process (**Fig. 14.12**).

Fig. 14.11 *Structure of RNA.*

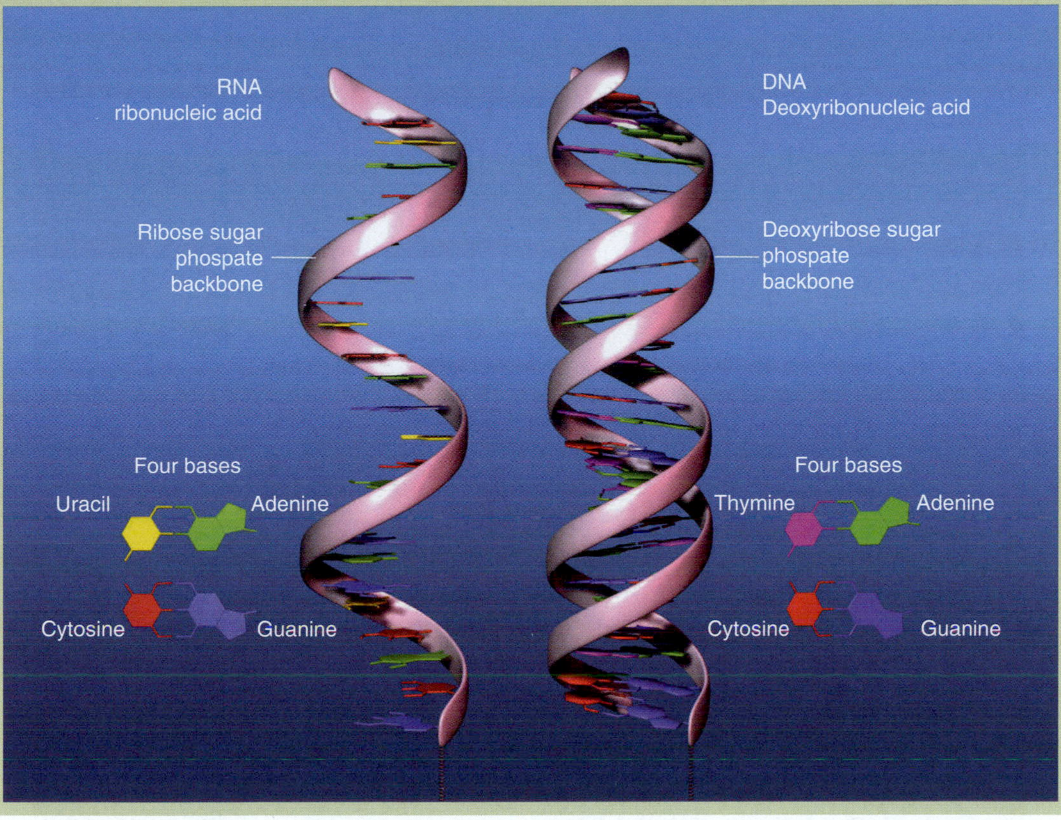

Source: Image by Anil Rao.

Thus, sets of three bases are said to code for each amino acid and are referred to as "codons." For example, the codon GCU uniquely attracts and binds to the amino acid alanine. The codon UAG does not match any of the 20 amino acids used in protein building, and causes the enzyme-linking process to stop. Recall from the discussion of amino acids that each amino acid has a base (amino) and acid (carboxyl) location on the molecule. These opposite chemical functional groups are aligned in neighboring amino acids by the RNA replicating process, ready to be joined by a polymerase catalyst. **Figure 14.12** illustrates how DNA synthesizes RNA, which in turn synthesizes proteins.

14.3.2 DNA Replication

DNA is replicated with the aid of enzymes called DNA polymerases. The DNA polymerases copy a template DNA chain by polymerizing nucleotides to form a DNA chain that is complementary to the original. Polymerases extend the DNA strand by adding nucleotides to the end of the growing chain. Different types of DNA polymerases carry out repair of damaged DNA strands. Polymerization of

FIG. 14.12 *DNA transcribes RNA which produces proteins DNA codes for synthesis of proteins in the cell by producing complementary RNA (messenger RNA, mRNA). Transfer RNA (tRNA) catalyzes the alignment of polypeptide chain protein constituents and their peptide bonding.*

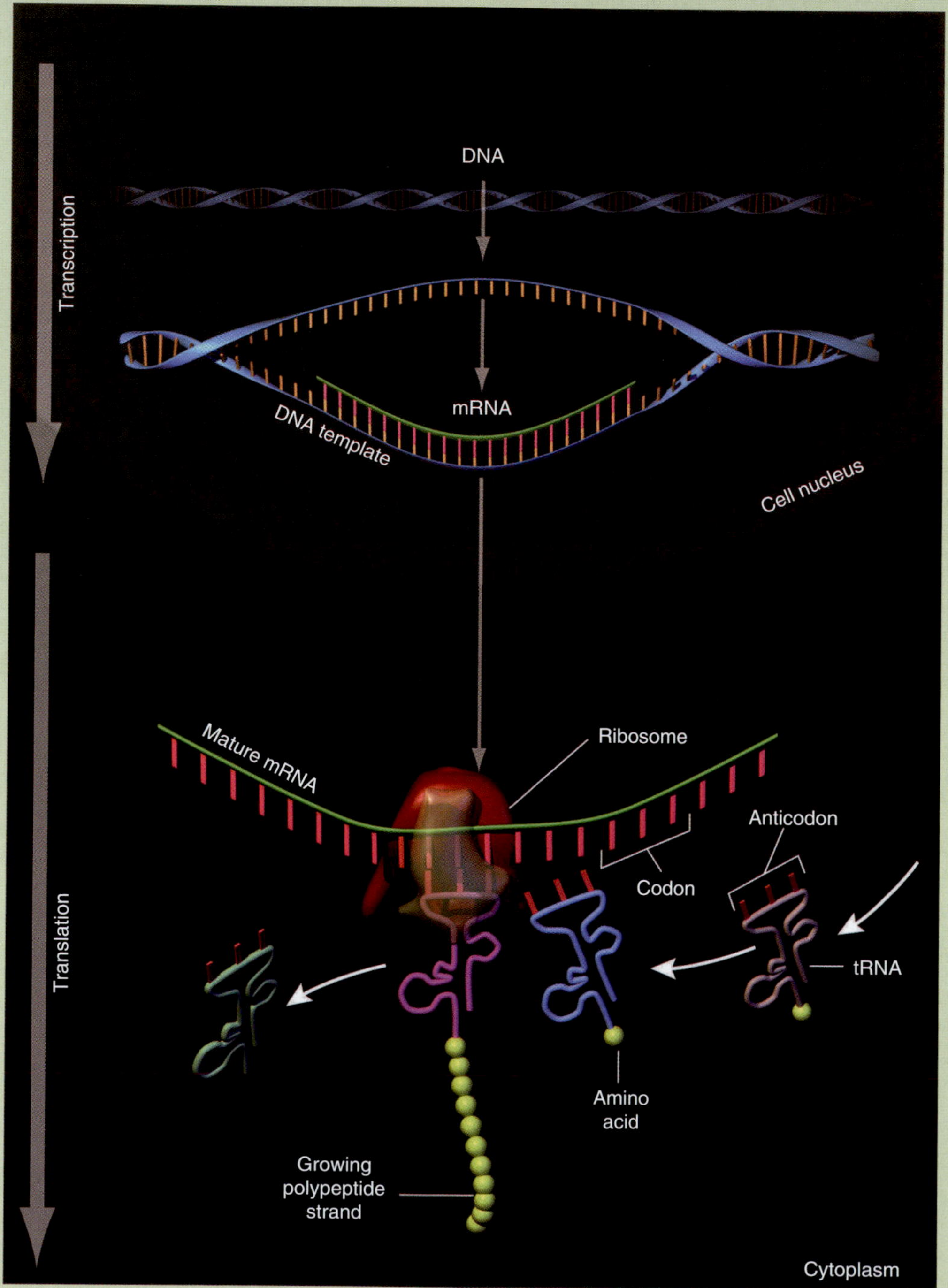

Source: Image by Anil Rao.

the DNA chain occurs by joining the hydroxyl group on a carbon of the sugar to the phosphate of an adjacent nucleotide. Other enzymes catalyze the building of RNA chains from DNA and the synthesis of proteins from RNA (**Fig 14.12**).

14.3.3 DNA as a Genetic Information Storage Material

DNA sequences preserve the information on how to program the synthesis of proteins. This information is not changed by the replication process, so it is available for reuse as many times as the living organism needs it, so long as the DNA strand is not damaged or altered. DNA is copied when organisms reproduce, so this information is passed down to subsequent generations. DNA is the material that encodes the genetic information of all living things. The units of information that encode for particular proteins or phenotypic traits are called genes.

DNA and Genes. A specific sequence of bases in the DNA chain will encode a sequence on the RNA chain that generates a protein. Most proteins are common to a large family of organisms and only a few are unique to a single species of life. The region on a chain of DNA that encodes a protein or any feature of inheritance is called a gene. Some proteins regulate the activity of others, and the complex mechanisms of the cell include regulation of the protein replication process itself, so the DNA and RNA by themselves do not provide a simple direct map to the phenotype of the organism.

The genome, defined as the total number of base pairs in the DNA of a particular type of cell, does not relate in a predictable fashion to the number of genes. DNA in many cases appears to have long sequences of base pairs whose function is not fully known, many of which are highly repetitive. Each gene codes for the synthesis of a protein indirectly by coding for an RNA codon sequence.

Exon and Intron. The portion of the DNA that codes for RNA that synthesizes a protein is called an exon. Some regions of the genome consist of long sequences in the DNA called introns, which are transcribed into RNA but not translated into protein. These may serve to separate the genes or regulate their transcription [94,95].

Genotype and Phenotype. Historically, starting with Mendel in the nineteenth century, genes were identified as the unit that carried any inherited characteristic, such as the color of a flower or shape of a leaf, beak, or wing. Genes were identified by the ability of observable features to be inherited. Mendel originated the concept of genes by counting the relative abundance of observable features in the offspring of crossed varieties of peas. His ideas were found to have universal significance long before the physical basis of the gene was understood.

When biologists began to understand the inner structure of the cell and the process of cell division, the gene was associated with the chromosomes located in the cell nucleus of higher cells. Once the role of DNA in encoding genes was understood, it was realized that the gene is not a simple physical entity. A gene can include segments of DNA that do not code for a protein directly but instead have sequences that bind other regulatory molecules involved in the protein synthesis [93].

The composition and configuration of the compounds making up an organism—from its molecular composition to the arrangement of its proteins and cells to the shapes and interconnections of its organs and appendages—are specific for each organism and are called its *phenotype*. The DNA sequence that encodes the genes and controls the development of the phenotype is called the *genotype*. It is of great theoretical and practical importance to be able to determine which genotypes give rise to the synthesis of particular products. By comparing the genotypes of different organisms and individuals we can understand relationships and the causes of inherited diseases and disorders.

DNA Sequencing to Determine Genotypes and Characterize Genes. A large body of knowledge has been developed by biologists and geneticists, starting late in the nineteenth century, on genetic inheritance based on observations of deformities and modification of the shapes and number of chromosomes in the cell—structures that can be made visible with staining and preparation under the microscope. By the middle of the twentieth century biologists had established that the number of chromosomes in each cell and the amount of nucleic acid were an invariant characteristic for normal members of any species, thus indicating strongly that genetic inheritance was carried by nucleic acid in the chromosomes. Some plants and animals have large chromosomes, making them useful subjects for experimental study. Higher animals such as mammals have similarities in their chromosomes—a state that leads to generalizations and understanding of the effects of chromosome abnormalities. Since chromosomes have specific shapes and numbers at certain stages of cell development, genetic biologists were able to map various inherited characteristics onto specific regions of the chromosomes, especially for species where chromosomes were easily observable, such as fruit flies, squid, and a number of plants. When the molecular mechanisms of DNA replication began to be worked out, it began to be possible to envisage correlating knowledge about genetic inheritance and chromosomes with detailed molecular sequences at various places on the DNA strand and, ultimately, with complete genotypes.

Genotypes can be determined in principle by analyzing the sequence of bases in the strands of DNA in the cell. Since there are millions of nucleotides in the genome of even the simplest organism, the analysis of their sequence is an enormous and very difficult accomplishment (the human genome contains about 3 billion base pairs, notated as 3 gigabases [3 Gb]). To perform a chemical analysis of each of the millions of nucleotides while keeping track of their sequence in the DNA chain requires complex and sophisticated strategies if the task is to be made feasible in even the simplest organisms.

The first methods developed to sequence DNA were termination sequencing developed by Frederick Sanger and Howard Chadwell and chemical sequencing developed by Allan Maxam and Walter Gilbert, both during the 1970s. Because of the difficulty and labor involved in performing sequencing, only short portions of genomes were sequenced at first. Since the award of the Nobel Prize in 1980 to Sanger and Gilbert, intensive and large-scale efforts have been made to extend these first steps. Worldwide coordinated projects were started to achieve rapid and inexpensive sequencing of complete genomes for any organism or individual, culminating in the first nearly complete draft sequences of the human genome in 2001 and 2002, spurred by the U.S. government-sponsored Human Genome Project [93–98].

The development of genome sequencing was a unique large-scale development involving many cooperative and competitive efforts, analogous for biology to space exploration programs or to the high-energy physics accelerator projects for discovery of elemental particles. A number of important steps marked the rapid progress in ability to decipher genome sequences. The first sequencing of a complete organism was carried out in 1983 by a team of scientists at the U.K. Medical Research Council for the Epstein–Barr virus, whose genome had approximately 170,000 base pairs (170 kb). The same year Mullis and coworkers developed a way to amplify replication of strands of DNA by repetitive cycling, which could be readily automated. This development, called variously the polymerase chain reaction (PCR), DNA thermocycling, or DNA amplification, made it feasible to obtain workable amounts of DNA from very small amounts of starting material and revolutionized DNA sequencing and cloning [99].

Initially, DNA sequencing reactions are only capable of analyzing the sequence of a few thousand nucleotides at a time. PCR and computer-assisted sequence matching programs developed in the 1990s made it possible to determine the sequence for much larger pieces of DNA. Since genomes typically contain millions of billions of base pairs, it was necessary to break up DNA into segments in order to sequence it. One had to somehow keep track of where the breaks in the DNA strand were made in order to reconstruct the entire sequence. One approach to this problem is called shotgun sequencing. In the shotgun sequencer method, the DNA is broken up and the fragments amplified by replicating them. Then the fragments are sequenced individually and, using computer matching programs, they are compared for overlaps to try to deduce the complete sequence of the original DNA.

Quite apart from progress in determining the sequence of entire genomes, DNA sequencing of shorter characteristic portions of the genome was very useful in genetic and forensic applications. Polymerase chain reaction (PCR), computer matching techniques, and other advances have pushed DNA sequencing from a small-scale, labor-intensive technology capable of reading several kilobase pairs of sequence per day into an automated, high-throughput, multibillion dollar industry. Nanotechnology has enabled sequencing technologies to be developed based on advances in electroanalytical chemistry and photosensors on the nanoscale that do not involve either electrophoresis or Sanger sequencing chemistries. For example, sequencing by synthesis (SBS) involves multiple parallel microsequencing addition events occurring on a surface where data from each round are detected by imaging [100].

By understanding and manipulating the synthetic molecular nanomachinery of protein synthesis we can produce useful protein products and modify the characteristics of organisms—for example, to eliminate genetic defects or synthesize desirable hormones, drugs, foods, or fuels in bioreactors, crops, or livestock. The ability to decipher and manipulate DNA sequences leads to the possibility of artificially making identical copies of individual organisms (called *cloning*) and inserting genes into bacteria, plants, or animals to alter their phenotypes (genetic engineering, making transgenic or chimeric organisms, performing gene surgery).

The protein products of DNA and RNA synthesis are the nanomachinery of the cell, determining the development and maintenance of the organism. Every aspect of shape, size, metabolism, and behavior of the organism is based on the structure and function of the proteins encoded by its DNA, and how they are expressed and energized.

14.3.4 RNA and DNA Nanoengines: Viruses and Phages

Now that we have given highlights of the functions of biomolecular nanomachinery, we will look at a very simple example of how DNA and protein structures work together as nanomachines. This example is easy to visualize because it is not confined inside the complex environment of the cell. In section 14.1.3 we introduced viruses in the survey of the sizes of organisms in relation to their biomolecular structure. Viruses fall below the minimum size possible to have all of the molecular machinery necessary for an independent reproduction cycle, but they can reproduce by invading and hijacking the machinery of cells. Now that we have surveyed how DNA and RNA work to replicate copies of themselves and to reproduce proteins, it is possible to understand how viruses can take over this machinery and alter it to make copies of viral DNA and protein coats.

In some viruses, called phages (from the Greek word meaning "to eat"), the protein coat has an active function: It attaches to the cell wall of a bacterial host and drives the viral DNA or RNA core through the wall into the body of the cell where it can replicate. Thus, we see a step in the continuum between nonliving molecular machinery and living organisms (**Fig. 14.13**).

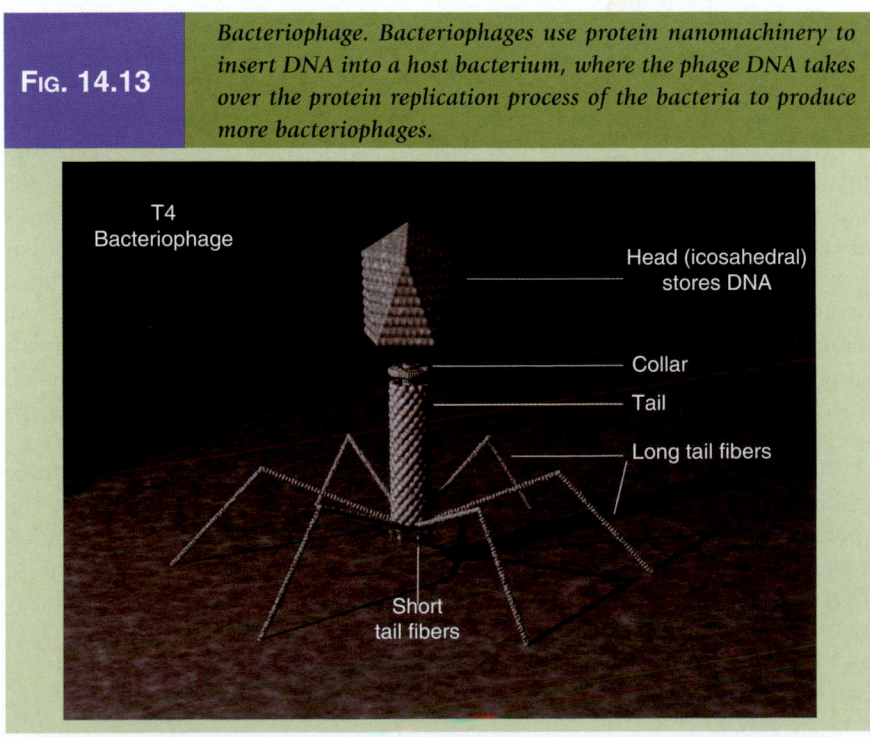

Fig. 14.13 *Bacteriophage. Bacteriophages use protein nanomachinery to insert DNA into a host bacterium, where the phage DNA takes over the protein replication process of the bacteria to produce more bacteriophages.*

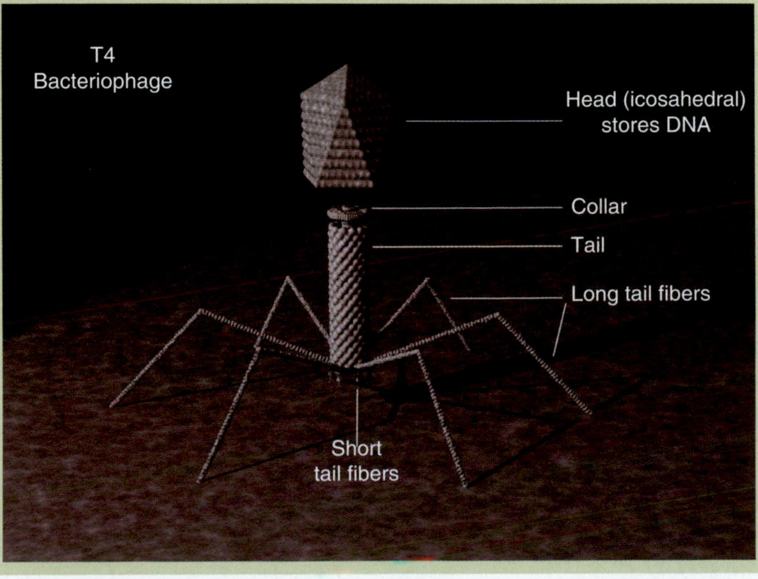

Source: Image by Anil Rao.

14.3.5 The Role of the Protein Environment

To understand how the nanomachinery of the cell works, it is necessary to take account of the role of the water and protein environment in which the process operates. As emphasized in earlier sections, water, hydration, and hydrogen bonding play a key role in biological processes at the nano level. As one example, we illustrate the role of "zinc sticky fingers" in shepherding the RNA transcription process (**Fig. 14.14**). The "sticky fingers" in this case are proteins bound to a zinc ion known as transcription factor IIIA. These finger-shaped proteins surround and stabilize the RNA backbone helix during the transcription process. The complex role of zinc finger proteins took many years of experimentation and separations to elucidate before it was isolated and its structure determined. We conclude this chapter with this example to remind potential entrants into the world of biomolecular nanoscience and nanotechnology that biological systems are extremely subtle and complex, with many surprises. In an introduction of this type there is space and time for only a brief introduction to some of them. Hopefully, this introduction will spark further curiosity about and appreciation for the amazing possibilities of biomolecular nanoscience.

FIG. 14.14 *Transcription actor IIIA zinc "Sticky Fingers" protein. Transcription factor IIIA is a protein that binds to RNA, shaping the rRNA backbone during the process of transcription.*

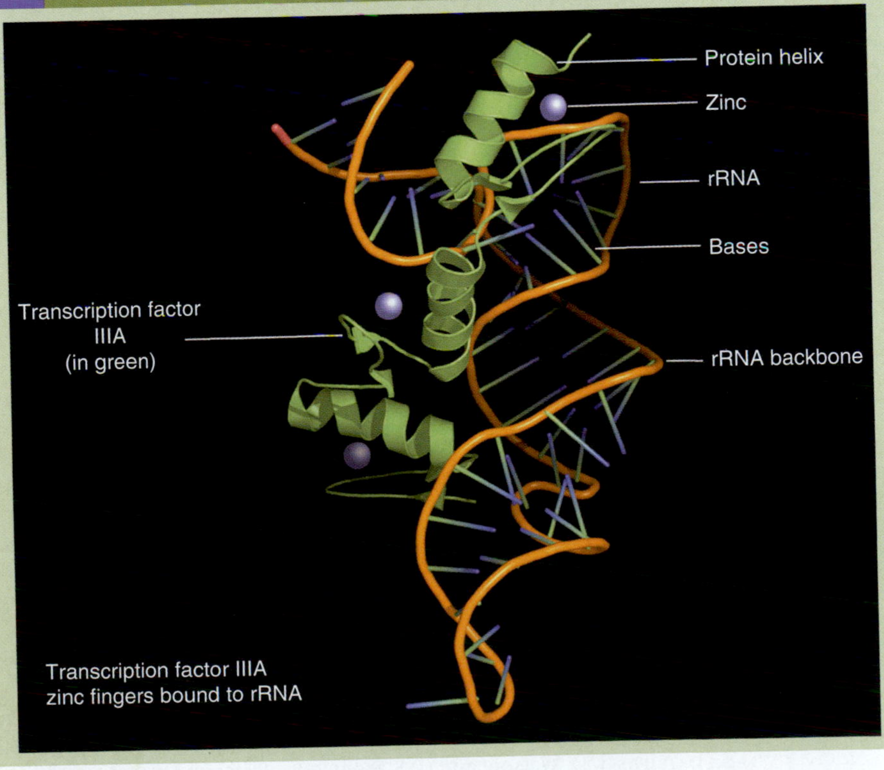

Source: Image by Anil Rao.

14.4 Concluding Remarks

14.4.1 Emerging Concepts and Developments

This chapter introduced life sciences from a nanoscience viewpoint. Approaching life as an emergent property of the energetics of matter on the nanoscale gives an integrated viewpoint for studying the self-organizing properties of macromolecules such as proteins, RNA, and DNA.

Biological nanosystems must be seen as the product of a long and complex process of evolutionary development, but understanding how they work should reveal intelligent short cuts for remedying disorders, exploiting biological machinery to produce new products, and designing soft nanomachines for energy, computation, and communication. Approaching the complex world of bioscience from the viewpoint of nanoscience is one way of introducing a field that is far too broad and intricate to master in an introductory survey. Since natural biology is very rich with nanoscale phenomena, it is not surprising that nanoscience intersects in many areas with cellular and molecular biology. As the field matures it is possible that nanoscience will contribute to the understanding of larger scale phenomena and unraveling of the complexities of signaling and immune systems.

Future Possibilities. Biomolecular nanoscience is a novel way of looking at the world of living systems. It should be a productive endeavor leading to new insights and knowledge about biology and life, as well as guiding the development of new types of artificial nanoengines. Biomolecular nanoscience is an integrated approach to the complex phenomena of living systems. As such, it lays a foundation for new approaches to biotechnology and biomedicine. We will explore some aspects of the possibilities for these fields in chapters in the next part.

References

1. J.-M. Lehn, Perspectives in supramolecular chemistry—From molecular recognition towards molecular information processing and self-organization, *Angewandte Chemie International Edition*, 29, 1304–1319 (2003).
2. F. Vogtle, *Fascinating molecules in organic chemistry*, John Wiley & Sons, Inc., New York, (1992).
3. T. Sekine, T. Niori, J. Watanabe, T. Furukawa, S. W. Choi, and H. Takezoe, Spontaneous helix formation in smectic liquid crystals comprising achiral molecules, *Journal of Materials Chemistry*, 7, 1307–1309 (1997).
4. H. Aldersey-Williams, *The most beautiful molecule: Discovery of the buckyball*, John Wiley & Sons, Inc., New York, (1997).
5. M. Meyyappan and L. Kelly, *Carbon nanotubes: Science and applications*, CRC Press, Boca Raton, FL, (2005).
6. S. Reich, C. Thomsen, and J. Maultzsch, *Carbon nanotubes*, John Wiley & Sons, Inc., New York, (2004).
7. A. Ullmann, ed., *Origins of molecular biology: A tribute to Jacques Monod*, ASM Press, Washington, D.C. (2003).
8. J. Cairns, J. D. Watson, and G. S. Stent, eds., *Phage and the origins of molecular biology*, CSHL, Woodbury, NY (2000).

9. R. B. Corey and H. R. Branson, The structure of proteins: Two hydrogen-bonded helical configurations of the polypeptide chain, by Linus Pauling, *Proceedings of the National Academy of Science, USA*, 37, 205–211 (1951). Available on JSTOR at www.jstor.org/view/00278424/ap001041/00a00030/0
10. L. Fredholm, The discovery of the molecular structure of DNA—The double helix: A scientific breakthrough, on The Official Web Site of the Nobel Foundation, at: http://nobelprize.org/educational_games/medicine/dna_double_helix/readmore.html
11. J. D. Watson and F. H. C. Crick, A structure for deoxyribose nucleic acid, *Nature*, 171, 737 (1953), Commentary by T. Zinnen on National Health Museum web site: www.accessexcellence.org/RC/AB/BC/casestudy2.htm
12. Honoring the lifetime achievements of Mack J. Fulwyler, special issue, *Cytometry*, A67, 53–179 (2005).
13. R. Kondratas, The history of the cell sorter, Smithsonian Institution Videohistory Collection, www.si.edu/archives/ihd/videocatalog/9554.htm
14. M. R. Loken, D. R. Parks, and L. A. Herzenberg, Two-color immunofluorescence using a fluorescence-activated cell sorter, *Journal of Histochemistry and Cytochemistry*, 25, 899–907 (1977).
15. M. B. Hale and G. P. Nolan, Phospho-specific flow cytometry: Intersection of immunology and biochemistry at the single-cell level, *Current Opinions in Molecular Therapy*, 8, 215–224 (2006).
16. A. J. Westphal et al., Kinetics of size changes of individual *Bacillus thuringiensis* spores in response to changes in relative humidity, *Proceedings of the National Academy of Science, USA*, 100, 3461–3466 (2003).
17. K. Sachs, O. Perez, D. Pe'er, D. A. Lauffenburger, and G. P. Nolan, Causal protein-signaling networks derived from multiparameter single-cell data, *Science*, 308, 523–529 (2005).
18. S. B. Carroll, *Endless forms most beautiful: The new science of Evo Devo and the making of the animal kingdom*, W. W. Norton & Co., New York (2005).
19. M. J. Denton, C. J. Marshall, and M. Legge, The protein folds as platonic forms: New support for the pre-Darwinian conception of evolution by natural law, *Journal of Theoretical Biology*, 219, 325–342 (2002).
20. E. D. Wolf, ed., Advanced submicron research and technology development at the national submicron facility, *Proceedings of the IEEE*, 71, 589–600 (1983). Available at IEEE archives website at: ieeexplore.ieee.org/iel5/5/31325/01456911.pdf?arnumber=1456911
21. NSF Nanobiotechnology Center, center website at: www.nbtc.cornell.edu/
22. R. E. Jenkins and S. R. Pennington, Arrays for protein expression profiling: towards a viable alternative to two dimensional gel electrophoresis, *Proteomics*, 1, 13–29 (2001).
23. M. J. Heller and A. Guttman, *Integrated microfabricated biodevices*, Marcel Dekker, New York (2002).
24. M. Schena, *Microarray analysis*, John Wiley & Sons, Inc., New York (2002).
25. J. S. Albala and I. Humphery-Smith, *Protein arrays, biochips and proteomics*, Marcel Dekker, New York (2003).
26. N. H. Malsch, ed., *Biomedical nanotechnology*, CRC Press, Boca Raton, FL (2005).
27. A. Guiseppi-Elie and T. Vo-Dinh, eds., *Biochips handbook*, CRC Press, Boca Raton, FL (2007).
28. H. K. Baca, C. Ashley, E. Carnes, D. Lopez, J. Flemming, D. Dunphy, S. Singh, Z. Chen, N. Liu, H. Fan, G. P. Lopez, S. M Brozik, M. Werner-Washburne, and C. J. Brinker, Cell-directed assembly of lipid–silica nanostructures providing extended cell viability, *Science*, 313, 337–341 (2006).
29. B. Gimi, T. Leong, Z. Gu, M. Yang, D. Artemov, Z. M. Bhujwalla, and D. H. Gracias, Self-assembled three-dimensional radio frequency (RF) shielded containers for cell encapsulation, *Biomedical Microdevices*, 7, 341–345 (2005).

30. P. Braun, Y. Hu, B. Shen, A. Halleck, M. Koundinya, E. Harlow, and J. LaBaer, Proteome-scale purification of human proteins from bacteria, *Proceedings of the National Academy of Science, USA*, 99, 2654–2659 (2002).
31. P. H. C. van Berkel, M. M. Welling, M. Geerts, H. A. van Veen, B. Ravensbergen, M. Salaheddine, E. K. J. Pauwels, F. Pieper, J. H. Nuijens, and P. H. Nibbering, Large scale production of recombinant human lactoferrin in the milk of transgenic cows, *Nature Biotechnology*, 20, 484–487 (2002).
32. E. Behboodi et al., Viable transgenic goats derived from skin cells, *Transgenic Research*, BW2118, 1–10 (2004).
33. T. V. Murthy, W. Wu, Q. Q. Qiu, Z. Shi, J. LaBaer, and L. Brizuela, Bacterial cell-free system for high-throughput protein expression and a comparative analysis of *Escherichia coli* cell-free and whole cell expression systems, *Protein Expression and Purification*, 36, 217–225 (2004).
34. J. Benyus, *Biomimicry: Innovation inspired by nature*, William Morrow and Co., New York (1997).
35. D. S. Goodsell, *Bionanotechnology: Lessons from nature*, Wiley-Liss, Inc., Hoboken, NJ (2004).
36. R. E. Smalley, Discovering the fullerenes, *Reviews of Modern Physics*, 69, 723–730 (1997).
37. C. Branden and J. Tooze, *Introduction to protein structure*, 2nd ed., Routledge, Abington, OX, UK (1999).
38. A. M. Lesk, *Introduction to protein architecture: The structural biology of proteins*, Oxford University Press, Oxford, UK (2001).
39. D. Whitford, *Proteins: Structure and function*, John Wiley & Sons, Inc., New York (2005).
40. A. Fersht, *Structure and mechanism in protein science: A guide to enzyme catalysis and protein folding*, 3rd ed., W. H. Freeman, New York (1998).
41. G. C. Howard and W. E. Brown, *Modern protein chemistry: Practical aspects*, CRC Press, Boca Raton, FL (2001).
42. G. A. Petsko and D. Ringe, *Protein structure and function*, New Science Press, London (2004).
43. H. Rehm, *Protein biochemistry and proteomics*, Elsevier, Burlington, MA (2006).
44. G. Karp, *Cell and molecular biology*, John Wiley & Sons, Inc., New York (2005).
45. H. Lodish, A. Berk, P. Matsudaira, C. A. Kaiser, M. Krieger, M. P. Scott, and L. Zipursky, J. Darnell, *Molecular cell biology*, 5th ed., W. H. Freeman, New York (2003).
46. J. Bereiter-Hahn, Behavior of mitochondria in the living cell, *International Review of Cytology*, 122, 1–63 (1990).
47. L. Grohmann, A. Brennicke, and W. Schuster, The mitochondrial gene encoding ribosomal protein S12 has been translocated to the nuclear genome in Oenothera, *Nucleic Acids Research*, 20, 5641–5646 (1992).
48. N. J. Dimmock, A. J. Easton, and K. N. Leppard, *Introduction to modern virology*, 6th ed., Blackwell Publishing, Inc., Malden, MA (2006).
49. J. Collinge, Prion diseases of humans and animals: Their causes and molecular basis, *Annual Review of Neuroscience*, 24, 519–550 (2001).
50. K. M. Pan, M. Baldwin, J. Nguyen, M. Gasset, A. Serban, D. Groth, I. Mehlhorn, Z. Huang, R. J. Fletterick, F. E. Cohen, and S. B. Prusiner, Conversion of alpha-helices into beta-sheets features in the formation of the scrapie prion proteins, *Proceedings of the National Academy of Science, USA*, 90, 10962–10966 (1993).
51. G. Oberdörster et al., Principles for characterizing the potential human health effects from exposure to nanomaterials: Elements of a screening strategy, *Particle and Fiber Toxicology*, 2, 8 (2005).
52. C. D. Klaassen, J. B. Watkins, and L. J. Casarett, eds., *Casarett & Doull's essentials of toxicology*, McGraw–Hill, New York (2001).

53. B. Onaral and J. P. Cammarota, *Complexity, scaling, and fractals in biomedical signals, Biomedical engineering handbook*, Chap. 58, CRC Press, Boca Raton, FL (2000).
54. R. Thom, *Structural stability and morphogenesis*, 2nd rev ed., Addison Wesley, Boston (1989).
55. G. Nicolis and I. Prigogine, *Self-organization in nonequilibrium systems: From dissipative structures to order through fluctuations*, John Wiley & Sons, Inc., New York (1977).
56. E. Ott, *Chaos in dynamical systems*, Cambridge University Press, Cambridge, UK (2002).
57. Z. Li, W. A. Halang, and G. Chen, eds., *Integration of fuzzy logic and chaos theory: Studies in fuzziness and soft computing*, Springer–Verlag, Berlin, (2006).
58. W. A. Halang and A. D. Stoyenko, *Constructing predictable real time systems*, Kluwer Academic Publishers, Dordrecht, the Netherlands (1991).
59. C. de Duve and M. J. Osborn (panel 1 moderators), Constraints on size of a minimal free-living cell. In *Size limits of very small microorganisms: Proceedings of a workshop* (1999), National Academy of Science, USA (2000), www.nap.edu/html/ssb_html/NANO/nanopanel1.shtml
60. K. Nealson (panel 2 moderator), Is there a relationship between minimum cell size and environment? In *Size limits of very small microorganisms: Proceedings of a workshop* (1999), National Academy of Science, USA (2000), www7.nationalacademies.org/ssb/nanopanel2.html
61. E. O. Kajander and N. Çiftçioglu, *Nanobacteria:* An alternative mechanism for pathogenic intra- and extracellular calcification and stone formation, *Proceedings of the National Academy of Science, USA*, 95, 8274–8279 (1998).
62. J. T. Hjelle, M. A. Miller-Hjelle, I. R. Poxton, E. O. Kajander, N. Ciftcioglu, M. L. Jones, R. C. Caughey, R. Brown, P. D. Millikin, and F. S. Darras, Endotoxin and nanobacteria in polycystic kidney disease, *Kidney International*, 57, 2360–2374 (2000).
63. J. O. Cisar, D.-Q. Xu, J. Thompson, W. Swaim, L. Hu, and D. J. Kopecko, An alternative interpretation of nanobacteria-induced biomineralization, *Proceedings of the National Academy of Science, USA*, 97, 11511–11515 (2000).
64. A. Mushegian, The minimal genome concept, *Current Opinion in Genetics & Development*, 9, 709–714 (1999).
65. D. Ferber, Synthetic biology: Microbes made to order, *Science*, 303 158–161 (2004).
66. J. I. Glass, N. Assad-Garcia, N. Alperovich, S. Yooseph, M. R. Lewis, M. Maruf, C. A. Hutchison, III, H. O. Smith, and J. C. Venter, Essential genes of a minimal bacterium, *Proceedings of the National Academy of Science, USA*, 103, 425–430 (2006).
67. H. O. Smith, C. A. Hutchison, III, C. Pfannkoch, and J. C. Venter, Generating a synthetic genome by whole genome assembly: X174 bacteriophage from synthetic oligonucleotides, *Proceedings of the National Academy of Science, USA*, 100, 15440–15445 (2003).
68. S. J. Giovannoni, H. J. Tripp, S. Givan, M. Podar, K. L. Vergin, D. Baptista, L. Bibbs, J. Eads, T. H. Richardson, M. Noordewier, M. S. Rappé, J. M. Short, J. C. Carrington, and E. J. Mathur, Genome streamlining in a cosmopolitan oceanic bacterium, *Science*, 309, 1242–1245 (2005).
69. V. Pérez-Brocal, R. Gil, S. Ramos, A. Lamelas, M. Postigo, J. M. Michelena, F. J. Silva, A. Moya, and A. Latorre, A small microbial genome: The end of a long symbiotic relationship? *Science*, 314, 312–313 (2006).
70. S. G. E. Andersson, Perspectives genetics: The bacterial world gets smaller, *Science*, 314, 259 (2006).
71. A. Nakabachi, A. Yamashita, H. Toh, H. Ishikawa, H. E. Dunbar, N. A. Moran, and M. Hattori, The 160-kilobase genome of the bacterial endosymbiont *Carsonella*, *Science*, 314, 267 (2006).
72. S. G. E. Andersson, A. Zomorodipour, J. O. Andersson, T. Sicheritz-Pontén, U. C. M. Alsmark, R. M. Podowski, A. K. Näslund, A.-S. Eriksson, H. H. Winkler,

and C. G. Kurland, The genome sequence of *Rickettsia prowazekii* and the origin of mitochondria, *Nature*, 396, 133–140 (1998).

73. J. Maniloff, Nanobacteria: Size limits and evidence, *Science*, 276, 1773–1776 (1997).
74. B. La Scola, S. Audic, C. Robert, L. Jungang, X. de Lamballerie, M. Drancourt, R. Birtles, J.-M. Claverie, and D. Raoult, A giant virus in amoebae, *Science*, 28, 2033 (2003).
75. J. Hogan, Are nanobacteria alive or just strange crystals? *New Scientist*, 182, 6–7 (2004).
76. M. R. Taylor, *Dark life: Martian nanobacteria, rock-eating cave bugs, and other extreme organisms of inner Earth and outer space*, Scribner, New York (1999).
77. Key regulators in the cell cycle, Nobel Assembly press release: Summary of award of Nobel Prize in physiology or medicine to Leland H. Hartwell, R. Timothy (Tim) Hunt, and Paul M. Nurst (2001).
78. G. B. Childs, Membrane structure (2003) University of Texas Medical Branch web page at: cellbio.utmb.edu/cellbio/memebrane.htm
79. Y. Yawata, *Cell membrane: The red blood cell as a model*, Wiley-VCH Verlag GmbH & Co., Weinheim, Germany (2003).
80. R. Latorre and J. C. Sáez, eds., *From ion channels to cell-to-cell conversations*, Springer-Verlag, Berlin (1997).
81. B. Hille, *Ion channels of excitable membranes*, 3rd ed., Sinaner Associates, Sunderland, MA (2001).
82. D. Marsh, Polarity and permeation profiles in lipid membranes, *Proceedings of the National Academy of Science, USA*, 98, 7777–7782 (2001).
83. D. Marsh, Membrane water-penetration profiles from spin labels, *European Biophysics Journal*, 31, 559–562 (2002).
84. J. Nagle and S. Tristram-Nagle, Structure of lipid bilayers, *Biochimica Biophysica Acta*, 1469, 159–195 (2000).
85. F. Goñi, Non-permanent proteins in membranes: When proteins come as visitors, *Molecular Membrane Biology*, 19, 237–245 (2002).
86. E. R. Kandel, J. H. Schwartz, and T. M. Jessell, *Principles of neural science*, 4th ed., McGraw–Hill, New York (2000).
87. G. M. Shepherd, *Neurobiology*, 3rd ed., Oxford University Press, Oxford, UK (1994).
88. A. Huxley, From overshoot to voltage clamp, *Trends in Neurosciences*, 25, 553–558 (2002).
89. A. L. Hodgkin and A. F. Huxley, A quantitative description of membrane current and its application to conduction and excitation in nerve, *Journal of Physiology (London)*, 117, 500–544 (1952).
90. D. A. Doyle, J. M. Cabral, R. A. Pfuetzner, A. Kuo, J. M. Gulbis, S. L. Cohen, B. T. Chait, and R. MacKinnon, The structure of the potassium channel: Molecular basis of K+ conduction and selectivity, *Science*, 280, 69–77 (1998).
91. C. A. Janeway, Jr., P. Travers, M. Walport, and M. J. Shlomchik, *Immunobiology*, 5th ed., Garland Publishing, Inc., New York (2001).
92. L. Du Pasquier and G. W. Litman, *Origin and evolution of the vertebrate immune system*, Springer-Verlag, Berlin (2000).
93. J. D. Watson, T. A. Baker, S. P. Bell, A. Gann, M. Levine, and R. Losick, *Molecular biology of the gene*, 5th ed., Benjamin Cummings Publishing Company, San Francisco, CA (2003).
94. P. J. Beurton, R. Falk, and H.-J. Rheinberger, *The concept of the gene in development and evolution (Cambridge studies in philosophy and biology)*, Cambridge University Press, Cambridge, UK (2003).
95. P. Portin, The concept of the gene: Short history and present status, *The Quarterly Review of Biology*, 68, 173–223 (1993).

96. J. Davis, *Mapping the code: The human genome project and the choices of modern science*, John Wiley & Sons, Inc., New York (1991).
97. M. A. Palladino, *Understanding the human genome project*, 2nd ed., Benjamin Cummings Publishing Company, San Francisco, CA (2005).
98. Human Genome Program Report, *Genomics and its impact on science and society*, U.S. Department of Energy, Washington, D.C. (2003).
99. P. Rabinow, *Making PCR: A story of biotechnology*, University of Chicago Press, Chicago, IL (1996).
100. K. Mitchelson, ed., *New high-throughput technologies for DNA sequencing and genomics 2*, Elsevier, Amsterdam, the Netherlands (2007).
101. C. P. Poole and F. J. Owens, *Introduction to nanotechnology*, Wiley Interscience, Hoboken, NJ (2003).

Problems

14.1 There are on the order of one hundred trillion cells (1×10^{14}) in a human body. Starting with a single fertilized egg cell, how many cell divisions are required to result in a full human body?

14.2 Cells—think nano:

a. Why are there so many cells in a mammalian body?
b. In general, why are cells the sizes that they are?
c. What advantages does a multicellular organism have compared to a single-celled organism?
d. What are the advantages as building blocks of a cell over a virus or macrophage?

14.3 Oxygen and nutrients must be transported into cells from their surfaces. Likewise, waste products must be transported out. For three spherical cells with radii of 1000, 2000, and 5000 nm, calculate their surface areas, volumes, and ratios of surface area to volume. What would you expect the relationship to be between cell volume and sustainable rate of metabolic activity?

14.4 Refering to Problem 14.1, and using the typical size for a human cell cited in Section 14.1.2 of this chapter, calculate the approximate total cellular surface area in a human body.

14.5 In a typical healthy human digestive tract, the number of bacteria is an order of magnitude higher than the number of cells in the body. How many bacteria are in a typical human digestive tract? Assuming that these bacteria are the size of *E. coli* (see Section 14.1.2), what is their total volume and surface area?

14.6 What is the difference between DNA and RNA? In the context of DNA and RNA, what is meant by "complementary?" Which of the following are complementary to each other?

a. Sugar and phosphate
b. Deoxyribose and ribose
c. Thymine and cytosine
d. Adenine and quinine

14.7 For the DNA segment below:

a. What is the RNA segment that you would expect to be transcribed corresponding to it?
b. Does it matter whether you transcribe from left to right or right to left to get the RNA?
c. For the protein translated by this sequence, does the direction of transcription matter?
d. What is a sequence of amino acids in the polypeptide chain that it would transcribe?

(If the direction matters, work out both peptides for both directions.)
DNA Segment: TAGTGCAAAGCTCAGCAT

14.8 Describe the structure of a cell membrane. Include polysaccharide chains, cross-linking, peptide chains, lipo-polysaccharides, lipid, proteins, and mesosomes in your discussion.

14.9 Ribosomes, ca. 18 nm in size, consist of two subunits (50S and 30S). Explain their role in protein synthesis. Why is this structure such a perfect nanomaterial? Relate your answer to size, structure, and function.

14.10 What is the function of the *cristae* found in mitochondria? Discuss the morphology and structure of mitochondria. What nanocomponent(s) is (are) exerting its (their) effects in mitochondria?

14.11 Explain the structure of the endoplasmic reticulum and the *cisternae*. What nanocomponent(s) is (are) exerting its (their) effects in the smooth ER? The rough ER?

14.12 Perhaps the pinnacle of molecular recognition is illustrated by the immune globulins—for example, the antibodies. These classes of proteins are not only able to recognize and neutralize antigenic materials that have invaded the body, but are also able to do so in a dynamic, versatile way. Antigens are considered to be divalent; antibodies are considered to be polyvalent. Explain what this means and relate your understanding to the overall solubility of antibody–antigen complexes.

14.13 Polarity, pH, solubility, and structure all play a role in amino acid chemistry. There are basically four kinds of amino acids: those with nonpolar R groups, uncharged polar R groups, positively charged basic R groups, and negatively charged acidic groups. In an ion-exchange column at high pH, what do you predict the order of elution would be for the following amino acids: glycine, phenylalanine, arginine, and glutamic acid? Which ones migrate towards the anode? The cathode? In an electrophoretic gel—also at high pH.

14.14 Polypeptides normally do not exist in linear form (extended). The α-helix is a form often adopted. What is the length in nanometers of a polypeptide chain that has 80 amino acids in the α-helix form? In the open linear extended form? Hint: The C–N–C–N–C backbone consists of single bonds with alternating carbons and nitrogens. The peptide bond containing the α-carbon exists in a plane. Determine the repeating unit, the pitch, and the rise—also bond lengths and angles—and calculate the lengths. Any biochemistry text will contain the relevant information with images.

14.15 Compare the structures of the α-helix and the β-sheet. Why the difference?

14.16 The behavior of proteins in solution is quite important to metabolism. List five characteristics of proteins that affect their solubility.

14.17 What is the length of all the unraveled human DNA in a human cell if set end to end? What assumptions did you use to make your calculation?

Part II

Nanotechnology

Section 6

Perspectives

INTRODUCTION: NANOTECHNOLOGY

I never think of the future—it comes soon enough.

ALBERT EINSTEIN

Chapter 15

Threads

Nanotechnology is about technology—the manufacturing of high-tech (and not so high-tech) goods. It is about converting nanoscience into products. It is not so much about science although we do not refrain from relating to scientific principles, concepts. Our focus is on topics associated with materials, manufacturing (fabrication), devices, and applications. The format of the text is that of a cooperative, interdisciplinary effort that is typical of nano—albeit with emphasis on engineering—where biology, chemistry, and physics (and computer science) come together to contribute to this very broad subject.

The chapter opens with a short course on how to start, maintain, and exit a business. Although it applies to nearly any kind of business, it most certainly applies to nanobusiness. This is an unprecedented approach, to include such a topic in a technology textbook, but it has to be done in order to keep pace with our changing world. We accomplish our due diligence by offering this material in this chapter as well as a section on education and workforce development. It is one thing to understand the technology and pass the course with flying colors; it is quite another thing to know what to do with it or plan a career centered on its impact.

Nanometrology, nanomanufacturing, and research have to be conducted somewhere. Where they take place is actually quite important. As one might gather from the prefix, nano implies things that are very small, and being small, working with nanomaterials, measuring them, and making them all require special facilities and considerations. We discuss nanometrology and nanomanufacturing in *chapter 16* but we also turn our attention to the buildings that house them—the buildings for advanced technology. This is a rather unusual section in that material of this ilk is not usually found in science and engineering texts—but we have already broken that rule by adding a short course in business development and operation. However, by understanding what it takes to make nano happen, the student should acquire an even broader appreciation of the subject matter—and perhaps interest in a career path along the lines of architecture and engineering buildings for advanced technology may just be a beneficial byproduct of this chapter.

Lastly, we provide a catalog of nanotechnology institutions and products. We highly recommend that you visit some of the Web site links provided so that a broader perspective of nanotechnology can be acquired. Following this chapter, a special chapter on nanometrology is presented.

15.0 Perspectives of Nanotechnology

We start this adventure in 1959, at the onset of the modern age of nanotechnology, with Richard Feynman's lecture "There's Plenty of Room at the Bottom." It was in this lecture that Feynman alerted the consciousness of the scientific community about the untapped potential of nanomaterials. In Part I, we discussed the important historical contributions to nanotechnology and the remarkable observations of nature, wittingly or unwittingly, that were made throughout the ages and the scientific aspects of nano. We now strive to maintain focus on applications, integrated structures, and devices. However, before we embark, a small amount of reorientation is required. We start by recycling a few core definitions.

15.0.1 Review of Definitions

In 2003, Dr. Rachel Brazil of the Royal Society of Chemistry stated quite succinctly her definition of a powerful but relatively nonspecific term—*nanotechnology* [1].

> At present, the term is used to encompass a wide spectrum of nanoscience, from nanoparticles in sunscreen to the production of 'nanobots' for in vivo medical applications. In defining nanotechnology, distinctions need to be made between 'science' and 'technology'. A narrower definition of the type of 'technology' covered by the term may also be considered, limiting nanotechnology to technology producing functional devices fabricated and operating on the scale of nanometres.

The National Science and Technology Council, Committee on Technology, Subcommittee on Nanoscale Science, Engineering, and Technology (NSET) formally established the following boundaries of nanotechnology in the year 2000 [2].

> Research and technology development at the atomic, molecular or macromolecular levels, in the length scale of approximately 1–100 nanometer range, to provide a fundamental understanding of phenomena and materials at the nanoscale and to create and use structures, devices and systems that have novel properties and functions because of their small and/or intermediate size. The novel and differentiating properties and functions are developed at a critical length scale of matter typically under 100 nm. Nanotechnology research and development includes manipulation under control of the nanoscale structures and their integration into larger material components, systems and architectures. Within these larger scale assemblies, the control and construction of their structures and components remains at the nanometer scale. In some particular cases, the critical length scale for novel properties and phenomena may be under 1 nm (e.g., manipulation of atoms at ~0.1 nm) or be larger than 100 nm (e.g., nanoparticle reinforced polymers have the unique feature at ~200–300 nm as a function of the local bridges or bonds between the nano particles and the polymer).

It is always a difficult matter to draw a hard line but one has to be drawn so that we can obtain the proper perspective and orientation. It also helps us navigate in the real world, specifically if funding and contractual issues are at stake. Let's move on and introduce (or reintroduce) the fundamental and defining characteristics of words and phrases that include the prefix *nano*.

In order to provide the necessary perspective, the following working definitions of nanotechnology, and its distinction from nanoscience, are listed below [3].

Nanoscale

The nanoscale, based on the nanometer (nm) or one-billionth of a meter, exists specifically between 1 and 100 nm. In the general sense, materials with at least one dimension below one micron but greater than one nanometer can be considered as nanoscale materials.

Nanoscience

Nanoscience is the study of nanoscale materials—materials that exhibit remarkable properties, functionality, and phenomena due to the influence of small dimensions.

Nanoscience is similar to materials science in that it is an integrated convergence of academic disciplines. There exist a couple of major distinctions between the two: size and biology. The size we understand by now but we also understand that materials science traditionally does not include biological topics.

Nanotechnology

Nanotechnology, based on the manipulation, control, and integration of atoms and molecules to form materials, structures, components, devices, and systems at the nanoscale, is the application of nanoscience, especially to industrial and commercial objectives.

Nanotechnology is a horizontal enabling convergent technology that cuts across all vertical industrial sectors while nanoscience is a horizontal integrating interdisciplinary science that cuts across all vertical science and engineering disciplines.

Nanotechnology is a disruptive technology with a high barrier of entry that will impact the development of enhanced materials and devices. Nanotechnology will require that a new genre of partnerships be formed among and between business, academe, and government. It will focus study and effort on potential societal implications of a new and certainly disruptive technology. Nanotechnology is predicted to significantly impact the wealth and security of nations. Nanotechnology is the next industrial revolution.

Nanotechnology is considered to be, more so than ever, a technology that will have great impact on all aspects of culture and society.

Nanotechnology is the application of nanoscience—plain and simple. You will see many different forms of the above definitions in the media and scientific literature, but essentially all the definitions, after distillation and purification, crystallize into a few key forms—in particular that nanotechnology consists of materials with small dimensions, remarkable properties, and great potential.

15.0.2 Technology Revolution or Evolution?

Technology (from the Greek *teknikos* meaning "art, artifice; to weave, build, join" and *tekton* meaning "carpenter") has played a major role in the history of civilization. There is not much question that technology is one of the pillars (and drivers) of civilization. After all, isn't it new technology that offers the developers of that technology an advantage in the game of life (survival)? However, a revolution implies rapid and dramatic change. What exactly are those technological advances that changed our civilization and in such a so-called revolutionary manner?

From agriculture, the *practice* of growing and harvesting food, sprung civilization. Indeed, one must agree that this is true. Early human lifestyle was altered forever as nomadic ways adapted by the hunter–gatherer gave way to the sedentary form of the farmer—truly a revolutionary change. However the processes of agriculture and irrigation were deliberately developed over an extended period of time—perhaps several thousand years. Innovations in agriculture continue as we speak—genetically modified organisms (GMOs) provide an excellent example of our high-tech foray into that arena. Although change in agricultural technology progressed quite slowly, even the gradual development of agriculture must be considered to be a revolutionary one based on its impact on civilization.

Other technological breakthroughs such as the advent of metal tools, implements, and weapons and the discovery of fire may have occurred rather spontaneously, for example, by accidental discovery (as in the case of fullerenes). These revolutionary breakthroughs are based on the discovery of a specific *type of material*—stone, copper, bronze, or iron. We can only speculate that the spread of these technologies, however, still proceeded rather slowly. There was no rapid-fire means of disseminating information (perhaps only by conquest) due to limitations in population and communication several millennia ago. Today no such barriers exist. High population density is a worldwide phenomenon, and information is spread globally in the blink of an eye. Revolutions can now be very fast. The *Industrial Revolution* of the 1800s ushered in the modern technological era. The emerging availability of hydrocarbon fuels in particular launched mechanization, mass production, transportation, and communication to levels never seen before. The telephone, television, computer, cell phone, and Internet have all changed our lives forever.

How does one then superimpose a scale of measure on a technological revolution? And what kinds of scales are relevant to nanotechnology? Revolutionary developments are measured in terms of speed, population, and impact (economic and cultural) or by weighing the overall flow of resources (in and out). For example, what percentage of a population is engaged directly in a technological revolution? "Engaged" in this case indicates the number of people working in the field, so to speak. With regard to agriculture, the percentage of population engaged in the field early on was quite low. As the practice spread over several thousands of years, greater than 90% of a population may have been rooted in the agrarian lifestyle, for example, in the Middle Ages. The pure technology sectors (R&D and manufacturing) of today may employ fewer that 10% of the total workforce but the impact (in terms of sales, use, and lifestyle) is quite enormous.

And what of nanotechnology? Is it really a revolution or is it just a natural progression in miniaturization? The advent of the transistor, integrated circuits, and the computer age certainly changed the way we accomplish our daily tasks. The *Biotechnological Revolution* added biological materials into the mix. The *Nanotechnology Revolution* is right on the back of this "revolution" inspired by biology. Many consider biotechnology to be a component of nanotechnology. All these pseudo-issues boil down to semantics and boundaries. We already have stated our opinion on boundaries. Within a generalized field such as nanotechnology that is highly interdisciplinary in nature, boundaries are consequently trampled and blurred—to a great extent.

Nanotechnology is expected to change the way we live. In that sense, it can be considered to be revolutionary. Nanotechnology is also the product of the natural evolutionary process of miniaturization. It is actually with the perception of nanotechnology (hype or not) that many are concerned. However, we should not belabor what is in essence a pseudo-dilemma—whether or not nanotechnology is the *Next Industrial Revolution*. Let's leave that discussion to the pundits, the politicians, and the omnipresent media. Rather, we should simply try and understand why these materials have remarkable properties—and within the context of this text, how nanomaterials are converted and integrated into applications. In this volume of course, we stress the applications of nanotechnology products and those enhanced (enabled) by nanotechnology. It is somewhat artificial to designate a technology to be revolutionary *before* it has run its course. **Table 15.1** outlines some of the salient features of a revolution.

TABLE 15.1 *The* Next Industrial Revolution *Compliance Check List*

Criterion	Yes	No	Comments
Basis	X		Nanotechnology is firmly grounded in reality—the physical properties and phenomena associated with nanotechnology are real and significant—they are the drivers behind the "revolution."
Speed	X		Although speed is not the only major component of a revolution, the faster it happens, the more revolutionary is its impact and the more profound its legacy. Nanotech has hit the mainstream since the mid-1990s—not that long ago. Changes nowadays are expected to occur quickly.
Track record		X	Outside of computers, pharmaceuticals, and nanoparticles, the glamorous and highly touted quantum-dots, carbon nanotubes, and drug delivery nanomaterials have not yet made a significant impact on the economy but are making headway as we speak.
Research papers, conferences	X		The scientific community has fully embraced nanoscience and nanotechnology. There is no single scientific conference in physics, chemistry, engineering, biology or medicine that does not have presentations, posters, sessions or the whole conference dedicated to nano.
Nanocompanies	X		There are more and more nanocompanies every year. Small Times Magazine lists 3500 nanocompanies just in the United States [4]. Nearly all Fortune 500 companies have some involvement in nanotechnology.
Patents	X		The number of patents (and the trend) is burgeoning—if anything, this trend indicates revolutionary proportions.
Stock market		X	Not big yet.
Institutions	X		About 500 institutions ranging from nonprofits, industry associations, research labs, economic development organizations, and university and educational institutions are registered with Small Times Magazine as nano groups [4].
% Workforce		X	Technology workforce contributes less than 10% of the total U.S. workforce in general. Nanotechnology, in its broadest sense (computers, pharmaceuticals, etc.), is already involved in many sectors. By 2015, 2 million more "nanotech" jobs are expected to emerge.
Education		X	Concerted efforts are underway to promote nanotechnology in K-12 and higher. Although a great proportion of the U.S. population knows about nanotechnology through movies, it is not considered to be a significant part of curricula at this time, but this too is changing.
Products	X	X	Nanosized transistors are already part of computer chips—although developed rather quietly without much fanfare via the natural evolutionary process of miniaturization. There is no doubt that more products will undergo enabling/enhancement from nanotechnology.
Evolutionary component		X	Evolutionary components are not considered to be revolutionary. In this case, nano is very evolutionary—emerging from micro- and biotech industries—a natural, nonrevolutionary process of miniaturization, ever since the first timepiece was developed.
Hype	X		Since the days of the "turbo," nothing has been hyped more than nanotechnology.
Upside	X		The reality-based upside and promise of nanotechnology is tremendous.

15.0.3 Outlook

What Is the Status Quo of Nanotechnology in the United States? Zyvex Labs, LLC, has recently spun out nanotube composites that are integrated into baseball bats, golf clubs, sailboat masts, ballistic armor, and radiation shielding [4]. The energy industry in collaboration with universities has developed new solar cells and new high-performance batteries. The medical industry in collaboration with universities has developed dendrimer-based viricides and advanced diagnostic instrumentation [4]. According to Jim Von Ehr, II (CEO of Zyvex), societal impacts need to be studied and carefully weighed before the release of products containing nanomaterials and devices. Encounters with luddites, nano-pretenders, and obstructionists must be sifted from those with legitimate environmental understanding and legitimate health and safety concerns. The U.S. government in the form of the National Nanotechnology Initiative (NNI) has taken a leadership role in addressing issues confronting the development of nanotechnology. Positives include the leadership role of the NNI, interdisciplinary cooperation, support of university research, and interagency cooperation between and among the agencies and departments of the federal government. Negatives include insufficient focus to the NNI's nine "focus areas"—and that *science and discovery*, although a wonderful couple, are not enough.

Invention and innovation are required to transform those discoveries into products. There is not enough focus on product development. The National Institute of Standards and Technology's (NIST) ATP opportunity (Advanced Technology Program) is one of the few government agency programs that is able to "fund the gap" between a new idea and a fundable prototype. Why is this important—because nano is a global competition—a competition with foreign governments (and their complementary economies) that do stress and fund commercialization.

According to Jim Von Ehr, II of the Zyvex Corporation, university funding is doing well but is unfocused; commercialization is not working as well as it should; the (costly) national lab system and its relationship to the economy should be redefined; and patent reform should occur sooner than later to avoid creating "patent thickets" that impede innovation [4]. The government's role in technology is to regulate and tax when required. He believes that the U.S. government is not supporting industry very well. Tax codes, accounting rules, financial reporting, liability procedures, and regulations should not be changed every year. The acquisition of talent by limiting foreign visas forces jobs overseas and losses to foreign competition. The U.S. industrial policy has become "come to school here, create new technology, go back to your home country, and commercialize it there."

In 2005, London outpaced New York in the number of initial public offerings (IPOs) and the United States received none of the largest 25. The Sarbanes–Oxley (SOX) Act of 2002 had a devastating effect on foreign investment in the United States. Before SOX, 90% of foreign funds were raised in the United States. Post-SOX, 90% of foreign funds are raised outside the United States. This demonstrates clearly how the government can impact the economy. The combination of the SOX policy and the inability to import high-tech talent drives jobs to other countries and companies out of business here. In summary: (1) move towards engineering, applications, and commercialization, (2) intensify focus on energy, healthcare, and nanomanufacturing; fund social impact studies when nanotech has a larger impact; and study environmental issues in a science-driven manner, and lastly, (3) industry will respond to incentives [4].

15.0.4 The Nano Perspective

We can safely say that without manufacturing there would be no nanotechnology—at least according to our definition. Upon implementation of the technology, we ask ourselves is it for the greater good? Will society as a whole benefit from this technology? Societal implications and technology have always been linked. We therefore should make things for everyone—at least in theory. In the perfect world, everybody should win. Nanotechnology is no different. We develop products for the greater good, for defense, for profit, for barter, and for numerous other reasons. Where does nanotechnology fit into this age-old relationship? The diversity of products, a few mentioned earlier, shows no signs of slowing. While reading, and hopefully studying, this text, please keep in mind how this new technology impacts products, develops new ones, and enhances our security and quality of life. Think also how this new technology can impact our fears as well as our hopes.

15.1 The Business of Nanotechnology

Although most will not admit it in public, and we do not indicate anything by stating this, most scientists are not particularly business savvy, and to be fair to scientists, most business folks do not know a hoot about science or technology—at least beyond the fundamentals. Although there certainly are exceptions to this general conundrum, do you all essentially agree? We therefore have decided to include a *short course in business* in this textbook. The age of the *scientist as entrepreneur* is upon us as more and more universities interact with the commercial sector. Biotechnology first and now nano have pushed the envelope of interaction between academia and industry. As a future scientist or engineer, one must start thinking in terms of a career in nanotechnology, and even better, start thinking in terms of starting a business.

There are companies and there are nanocompanies. A nanocompany is not necessarily a small company. It may actually be a big company, perhaps a Fortune 500 company. What kinds of companies are drawn to this new technology? What differences are there between creating a nanotechnology company from say a biotech one? Nanotechnology has a high barrier of entry—unlike that faced by the "dot-coms" of the 1980s and 1990s—where all that was needed were a couple of geniuses, a PC, and a garage. Nanotechnology requires expertise (e.g., PhD level), costly equipment (e.g., transmission electron microscopes, ultrahigh vacuum), clean rooms, and highly trained tech people. Development and manufacture of nanomaterials, more than ever, also requires partnerships between and among government, business, and academia. Because nanotechnology is a worldwide phenomenon, competition is intense on a global scale.

The people and entities that support nanotechnology—the intellectual property managers, the patent attorneys, the building designers, the consultants—are all faced with new challenges brought on by nanotechnology and the interdisciplinary nature of the subject matter. Hopefully we will be able to raise questions, pose challenges, and explain some of what surrounds this global phenomena in the following sections of the text.

15.1.1 Background

Throughout history, humans have sought improvement by innovative approaches to the various tasks of living. Although the invention of stone arrowheads and spear points, the wheel, and a host of other early innovations precede written record, they were in by some perspectives more impressive in their impact on society than the invention of the electric light and telephone in the late nineteenth century. One characteristic of each of those early inventions is that someone accomplished the equivalent of starting a business. With the wheel, for instance, the wheelwright developed his skills to be able to make a consistently round product of a standard size.* The specialists able to run with these innovations were the forerunners of modern businesses.

The modern concept of a business, (e.g., a company with employees, products, a board of directors, a CEO, and shareholders) dates back only a few hundred years. Throughout history, however, the purpose of business was to sell products and services to buyers—the customers. From the time that early humans graduated from a simple hunter–gatherer status to increased levels of stratification with diverse skill sets, the equivalent of business was established. After all, what is the difference between a flint-knapper making arrowheads and spear points in exchange for food and shelter and a modern multinational corporation making electronic components in exchange for money? In terms of the lowest common denominator, the answer is nothing—both are trading one item of value for another at an agreed rate of exchange.

Over the millennia, this simple form of proto-business, essentially a sole proprietorship, began to evolve and grow into defined organizations offering a range of goods—from a bread maker with a few employees and several different products to a modern multinational corporation with thousands of products and tens of thousands of employees. At the same time, various restrictions, requirements, and other hurdles were put into place to make certain types of businesses more difficult to enter than others. From early on, businesses began to need some sort of registration with civil authorities, if for no other reason than tax assessment. Today, a business must register with the appropriate governmental agency in the country or state that it operates and must have various tax account numbers for payment of a variety of taxes. It also is responsible for various reporting requirements, depending on its structure.

Nanoscience and nanotechnology have already and will continue to yield many new and wonderful discoveries and inventions. But, as has been the case throughout history, these inventions remain interesting curiosities unless they are applied to develop products and/or processes that make their way into the mainstream of commerce. Take the wheel, for example. It is easy to understand that many applications, from the simple wheel/axle wagon to more complex

* Imagine what it would be like to have a chariot with different-sized wheels. If it were somehow impossible to make wheels consistently, would it have ever developed?

applications like block-and-tackle, watches, and precision gearing were made possible by applying the basic principle of the wheel. These are merely direct applications of the wheel. Indirectly, the wheel has led to many support products and even whole industries. For example, because wheels require lubrication, someone started manufacturing, marketing, and distributing lubricants. Others developed new materials to make the wheel last longer and perform better. Today, it is difficult to imagine a product or process that doesn't have a wheel somewhere within its mechanism. Nanotechnology has the potential to do the same. Indeed, many observers have stated emphatically that nanotechnology-based companies will engender another Industrial Revolution that is expected to exert profound effects on the world economy and society.

The U.S. National Science Foundation (NSF) has estimated that the world nanotechnology industry will grow from approximately $35 billion in 2005 to $1 trillion in 2015, employing over 2 million workers.* Such rapid growth is unprecedented but reflects a conviction that nanotechnology will affect and become a part of nearly every segment of industry over the next decade.

15.1.2 Companies

What Makes a Company? A company is not just a good idea, but is also a fine starting point. A company is a provider of goods and/or services that customers are willing to buy. It is also a person or a group of people able to supply those goods or services. This implies that the company is meeting some need, either existing or created.

The concept of an existing need is obvious, but what is a *created need*? The history of invention is the history of created needs. Nearly every adopted technology is either superior to an existing technology or it can outperform the previous technology. For example, conventional stoves and ovens are quite adequate for food preparation but the microwave oven provides an additional feature—convenience and speed in food preparation. Its development spawned an industry of prepackaged foods ready to "be popped into the microwave" and ready to eat in minutes. Later models were developed with oscillating or spinning cooking surfaces. A series of secondary products including "microwavable" cookware and special splatter covers appeared. Today, roughly a half century after the first microwave oven was introduced, most home kitchens and break-rooms in the United States have one.

The relevance of the example above is that not only can an inventor create a product and then a need for it, but also the acceptance of that product can spur the creation of a whole new industry. The entrepreneur needs to open his or her mind to the myriad possibilities for products, and therefore businesses, that can manifest themselves with the advent of new technologies, and their support. It will help the entrepreneur to consider exploitable market niches and products that will appear as the field of nanotechnology develops and expands its impact

* These numbers appear in Roco, M.C., *Journal of Nanoparticle Research*, 5, 181–189 (2003) and other sources from the same author. While this is a generally accepted figure, other sources place the market by the middle of the second decade of the twenty-first century as high as $2.6 trillion.

on society. Conversely, it will help the entrepreneur to understand what is displaced (and even disappears) as a result of a new technology.*

Successful entrepreneurs do not simply invent wondrous gadgets. They look at the gadgets they (or others) have invented and ask what could be needed to improve and to fully exploit and support them.

What Is a Nanotechnology Company? This seemingly simple question is actually a very complex one. One definition of a nanotechnology company is that it manufactures a product less than 100 nm in at least two dimensions, or that the manufacturing process itself is controlled at that size or smaller. The two-dimensional requirement would rule out companies making, for example, graphite lubricant because while sheets of graphene are one carbon atom thick (less than 1 nm), the other two dimensions are generally much, much larger than 100 nm. If a company were manufacturing coatings with particle sizes of consistently less than 100 nm, they would also qualify. However, we can safely call companies that manufacture thin films nanocompanies regardless of the lateral dimensions of their product. The process control issue would exclude bulk chemical manufacturers from joining the ranks of nanocompanies. Although chemical synthesis, as represented on paper, consists of reaction between atoms and molecules, the reality is that it occurs in 10,000 L reactors. The only real controls of these processes are physical in nature, for example, temperature, pressure, rate of mixing, rate and order of reactant addition, etc. By contrast, manufacturing an integrated circuit with components that are 65 nm in size do qualify as a nanotechnology process.

Nanotechnology companies also arise from the service sector. For our purposes, we define the service sector as comprised of companies that are not directly involved in developing or manufacturing a technology. Examples include:

- Consulting engineers specializing in the development of processes that fit the above definition
- Technology transfer professionals who seek out nanotechnology clients
- Patent law firms or agents having experience in filing and prosecuting nanotechnology-related patents
- Toxicology laboratories with the ability to study the effects of nanotechnology products on living organisms and the environment
- Contractors who were experienced in building the specialized manufacturing facilities needed or indeed a host of other service providers.

15.1.3 Sources of Nanotechnology Inventions

Although many discoveries and applications arise from work in laboratories of companies, a significant proportion, probably a majority of them, will originate

* For example, what happened to the typewriter industry when personal computers with word processors, such as the one used to write this book, became affordable and nearly universal? What about the typewriter repairman or ribbon and carbon paper manufacturers?

in academic and government laboratories, including both basic discovery and product development. However, these venues are not equipped to develop and commercialize products. The federal government mandates that research funded by the U.S. government be commercialized by a U.S. company. As a result, and over the years, institutions and laboratories have established bureaucratic structures and procedures for transferring newly developed technologies and products to U.S. companies. The general approach starts with the filing of a patent.* The university or national lab normally covers the cost of the patent filing. The invention is then licensed either exclusively or nonexclusively, usually for some fee plus royalties based on product sales, to a promising company. The terms for licenses vary considerably.†‡

Of course, many entrepreneurs who receive licenses from academic or government labs are in partnership (or will be soon) with the inventors employed by the institutions. Although the inventors understand the potential of the invention, not many have a clear idea of how to go about founding, registering, funding, and operating a company. The remainder of this section will provide a general overview of these subjects.

15.1.4 Founding a Company—What to Do First?

Most of the details you encounter below is nothing new—it applies to any and all businesses. Although nanotechnology is revolutionary in many ways, the structure of the business and the process required to move forward are relatively

* When examining the patent status of the invention, the entrepreneur should keep in mind that nanotechnology is a worldwide effort and the filings need to be prepared such that they can be filed as geographically broadly as possible. A U.S.-only patent, commonly filed by universities and government labs, limits the value since anyone outside the United States can simply copy the technology and potentially produce products for sale in worldwide markets. The reason for limited filings by these institutions is not lack of understanding but lack of budget. The subject of patents is covered elsewhere in this book.

† License terms and negotiations are beyond the scope of this book and the reader is referred to the Association of University Technology Managers (AUTM, www.autm.org) and the Licensing Executives Society (LES, www.les.org) for more information about license terms and negotiations. Both offer courses on these subjects, for a fee.

‡ In the last decades of the twentieth century, many companies began to actively seek inventions and technologies from academic, government, and even other industrial sources and established professional positions devoted to the process of in-licensing of technology of potential value and secondarily out-licensing (or outright selling) of those no longer of internal interest. As a result of their experiences, a number of these technology transfer professionals have offered their services as consultants to companies seeking to acquire technologies. They can be found either by Web searches or through listings with the AUTM and LES. The entrepreneur with limited (or no) experience in this arena should seriously consider having professional help in negotiating terms to acquire rights to technologies and inventions. The role these professionals play is to help establish terms, generally financial and business terms (due diligence requirements, termination rights, and a host of other issues), but most of them are not attorneys and while they will also participate in contract negotiations they are not responsible for the actual contract drafting and language. An experienced commercial attorney is essential for this important task, with preference for one who has written contracts for closely related technologies.

ubiquitous and require a modicum amount of common sense. Students who receive degrees in science become good scientists but not necessarily good business people.

Before embarking on your entrepreneurial journey (always a wild ride), several questions need to be addressed. In doing so, the entrepreneur defines the company in order to decide what structure to use when registering it. These questions include:

- What type of company?
  - Manufacturing
  - R&D—contract or grant-based
  - Consulting
  - Service provider
- How much financing is required to launch the company?
  - Fees, including legal, to set up the company and obtain operating licenses
  - Salaries for entrepreneur, partners, and any initial employees
  - Facilities cost
  - Licensing fees for technology
  - Operational setup costs: supplies, Web site, office and laboratory equipment, and furnishings—everything that may be needed from paper clips and business cards to atomic force microscopes
  - Running costs until other sources of finance are available
- How will it be financed?
  - Entrepreneur alone
  - Insiders and family/friends
  - Angel/venture funding
  - Eventually through IPO
  - Product or service sales (with no need for outside financing)
- How will ownership be distributed?
  - Entrepreneur alone
  - Entrepreneur and partners or family members
  - Entrepreneur and financial backers
  - Employees through stock incentives
  - Large shareholder base through public offering
- Where will the company be registered?
  - If in the United States, which state?
  - If outside the United States, where?

All of these factors plus many others, including personal biases, play a role in the choice of structure. A good way to make certain these issues are considered is to write a *business plan*.

The principle purpose of a business plan, at least in the first draft form, is to focus the entrepreneur on defining the company in terms of products and markets, hurdles to success, financial needs, revenue and cash flow, personnel, facilities, and equipment needs. It will later be useful in raising money, whether from venture or angel sources or from banks but its greatest value to the entrepreneur is to his own understanding. It should include:

- Executive summary (actually written last)
- Company description—when, where, and by whom founded, assets, etc.

- Market served, including a market analysis
- Products
- R&D
- Manufacturing issues
- Management team
- Other personnel needs
- Hurdles and key milestones, with estimated dates and costs
- Financial analysis
  - Revenues and timing
  - Expenses and their timing
  - Break-even
  - Pre-revenue financing needs

It should be detailed enough to show that the potential entrepreneur has thought out the issues well but should not be too long (or no one will read it). As a rule of thumb it should be under 50 pages including financial tables and executive summary—that by itself should be a maximum of two pages.*

15.1.5 Business Structures

There is a wide variety of business structures in the United States. Since businesses are incorporated at the state rather than federal level, some variation in available structures that the entrepreneur in nanotechnology may consider are as follows:

- Various type of corporations
- Various forms of partnerships
- Limited liability companies
- Sole proprietorships

There are a few major characteristics that differentiate them. These include:

- Taxation status
- Liability to the owners
- Governance requirements (i.e., management and reporting requirements)
- Number of shareholders permitted
- Ease of raising capital

For instance, if the company requires a large capital budget to build the capacity to manufacture products, the most likely source would be through venture financing followed by a public offering of stock. This is best done with a General C Corporation. Venture capital funds, angel investors, and banks are familiar

* There are many sources of information and hints on writing business plans. Entering "business planning" into any search engine will turn up a large number of sites that will, sometimes with no fee, offer to help an entrepreneur with preparing business and marketing plans. Business plan software is available from several sources and can be found in many software stores. Reviews of business planning software can be found at www.homeofficereports.com/Business%20Plan.htm. There are also numerous books available from any bookseller. One that provides a good overview is T. Berry, *Hurdle: the Book on Business Planning,* Palo Alto Software (2002).

and comfortable with this structure. Investment bankers, for example, have experience with IPOs. Governance and reporting requirements for C Corporations provide a degree of confidence to investors that the company will be managed responsibly, or at least with shareholder and board of directors oversight, so as to provide the best possible return on the investments. The C Corporation fits the general image of a "company" in the minds of most: it has a CEO who reports to a board of directors, other corporate officers responsible for finance, R&D, manufacturing, HR, etc. To investors, it has two other important characteristics. The C Corporation can have an unlimited number of shareholders and a variety of classes of stock,* making it attractive as an investment assuming it also has products or potential products that could provide those investors a good return. Those products do not have to be ready to go to market in order to secure funding, but the company must produce sufficient data to show that they can be producing revenue in a timeframe acceptable to the investors.

A C Corporation also provides, in general, the most flexibility in offering ownership incentives to employees or outside service providers like consultants. Their compensation can include share grants either through outright sales or, as is more often the case, through so-called option grants to these individuals. This subject is an important one in light of recent U.S. Federal laws.†

If the company is a consulting business, the entrepreneur could establish a *sole proprietorship* or, if working with a partner a *general partnership*. These do not require registration but may require business licenses in some jurisdictions.‡ No special tax forms are needed because the income from the business flows into the owner's personal tax returns. The individual is personally responsible for all debts and obligations and therefore all personal property is at risk unless otherwise legally protected. Partnerships need to be structured very carefully since each partner is liable for all business debts and obligations incurred by the other partners, and in the case of sole proprietorships, all personal property is at risk unless protected by some means. In setting up a general partnership, care should be taken in putting together a partnership agreement (PA), a legal contract between the partners that governs the relationship. In general, this PA cannot

* In general, the class of common stock is what is traded on the various stock exchanges. Special classes of stock are often used when raising money before a public offering, and can have different values and voting rights, can even be interest bearing and include rights to convert to common stock at some fixed rate. When considering special classes of stock, legal advice is essential.

† For the interested reader, the Public Company Accounting Reform and Investor Protection Act of 2002, also known as the Sarbanes–Oxley Act, was passed in the wake of several corporate scandals, ostensibly to restore public confidence in corporate accounting practices. It has been the source of much controversy, with some claiming it interferes inappropriately with the operation of corporations, others claiming it does not go far enough, and yet others claiming that the arguments on both sides are overblown. This important act has had one effect that could be considered positive, the creation of a new line of business in software and consultation for compliance. Visit http://fl1.findlaw.com/news.findlaw.com/hdocs/docs/gwbush/sarbanesoxley072302.pdf for a full copy of the Act or use a search engine to find general articles that will provide some background.

‡ If the founder does not want to operate under his own name, he should register a trade name with his state authorities, the so-called Doing Business As or dba name.

limit an individual owner's liability for the business activities of other owners.*

Even though there is paperwork involved, instead of a sole proprietorship or a partnership, an entrepreneur should consider a structure that has the same tax consequences overall to limit their personal liability. Either a *Subchapter S Corporation* or a *Limited Liability Company* (LLC) will do that. The advice of an experienced business attorney should always be obtained before deciding the best type of structure for your business.

15.1.6 Registering a Company—Where?

In the United States, companies are incorporated (or registered) under state laws, not at the federal level. Each state has its own requirements for paperwork, reporting and taxation. While a company can register in any state, it generally needs to have an office, or *Registered Agent*, in the state it chooses. Many U.S. companies, especially those expecting to operate in several states, choose to register in a state that has advantageous tax or other corporate policies, often Delaware or Nevada, though many states are changing business laws to make themselves more attractive. Information about these and other potentially attractive states can be found on various Web sites to help the entrepreneur understand the basics of registration.†

Companies must also register in the state in which they are physically located or in which they have offices, employees, or other facilities, even if their primary registration is in one of the so-called business friendly states. For instance, a company with offices and manufacturing facilities in North Carolina, may choose to register in Delaware but then must register as a "foreign" company in North Carolina. If it establishes a research site in South Carolina or a sales representative with an office in California it will then register in each of those states as a foreign company as well.

An entrepreneur may initially decide to register the company in the state in which offices and laboratories are located. This may be done to reduce paperwork, including the filing of multiple tax returns and annual reports. If the business expands to several states, the primary registration can be relocated to one of the business friendly states. However, the list of these states is growing as more and more are competing for businesses, so such a move should be carefully investigated before any action is taken.

The business of nanotechnology is worldwide and although the above discussion has centered on U.S. companies and registrations, many of the same issues face the entrepreneur who wants to start a business elsewhere in the world. Of course, registration and corporate structures, though similar, vary from country to country.

---

* Except that the limited liability partnership can shield each partner from malpractice by other partners.

† Two such sites are www.mycorporation.com and www.corporate.com, though a search engine will also provide other sources. These sites not only provide information but also will aid the entrepreneur in registering his company in one of those states. Many business attorneys will assist in preparing registration documents, ownership contracts, bylaws, etc. and will act as registered agents for the business.

The nanotechnology entrepreneur may decide to establish subsidiary companies in countries other than the one in which he initially establishes his organization. Intergovernmental commercial treaties are in place to make the establishment of whole- or majority-ownership of corporate subsidiaries in countries other than the residence of the parent relatively easy—although various rules and restrictions may apply. In general, due diligence must be exercised if an entrepreneur wishes to establish a presence in multiple countries, for example, consulting legal entities for advice both in the home country and the foreign country is recommended. Information about the requirements and costs of registration in a wide range of countries is readily available on the Internet.*

Registering a Company—The Process. The Internet has changed the way businesses are registered. As late as the 1990s, a person wishing to register a company in the United States would have to obtain a form, fill it out manually, and either mail or carry it to the business registration office in his state, along with a check, sometimes certified, to pay the fees. Today, most states encourage registration via online forms and even charge a premium for using a paper form. Online registration has become streamlined so that someone who has registered several companies can usually complete the process in 15 min or less, and then pay the fees by credit card, never leaving his office.†

Once the company has been formed and registered, the entrepreneur needs to prepare the paperwork: company bylaws, employment contracts, quarterly (or even more often) and annual reporting to tax authorities, and many other details. Regular meeting of the board of directors is required as is an annual meeting of all shareholders. There are many more requirements that differ with the size of the company. Many of these details are both time consuming and tedious, but they have to be done or the company could be liable for various penalties. The entrepreneur, especially the scientist, is wise to either be prepared for paperwork or consult a business-oriented professional.

---

* A Web search can be made for individual country registration requirements by using a search engine and a simple search string such as "business registration (country name)." A World Bank Web site: http://www.doingbusiness.org/ExploreTopics/StartingBusiness/ contains a broad compilation of estimated costs and registration requirements for all countries. U.S. embassies around the world have an officer known as the commercial attaché on their staffs. Many other countries have similar positions. The job of the attaché is to assist companies in understanding the business climate in these locations and helping businesses make appropriate contacts and negotiating the paperwork. These individuals and their staff can be a valuable resource.

† Once the company has been registered in the United States, the next step is to obtain an employer identification number from the Internal Revenue Service (IRS). This, too, is done online by going to the IRS site, www.irs.gov, and filling out and submitting Form SS-4, Application For Employer Identification Number. The number is assigned at the time of completion and a copy is mailed to the filing address. It may be necessary to obtain other permits and licenses from the state or states in which the company is registered, such as a sales tax license or a tax exemption number. In some jurisdictions a city or county tax or other permit may be required. Most of the applications for these are now available online and many, if not most, of them can be filled out and filed electronically.

15.1.7 Finances

Financing is the single-most important aspect of a new business. While this may seem obvious, one of the biggest problems for the entrepreneur is determining, and then acquiring, the amount of money needed to sustain operations until the company has sufficient product sales to pull the load. Unless the entrepreneur is already wealthy and can either fund the company from his own pocket or has partners, friends or family with money, time needs to be spent, perhaps a great deal of time, soliciting interest in his company from groups that do have money and who may be willing to invest it. After investing their own money and that of partners, family, and friends, most entrepreneurs turn to angel investors, venture capitalists, and then public equities markets. Angel investors are individuals or groups who invest their own money in companies that interest them. Usually the investment is small, under one million dollars (and often much less), and the investor receives equity in the company, sometimes a board position.*

Venture capitalists (VCs) are professionals who seek out investments for large funds they manage. These funds tend to be very large and investments they make are measured in millions of dollars. Venture capital fund managers have a reputation of being very exacting in choosing companies for investment. They often will want to participate in governance via board positions and may also demand preferred stock.† VCs have extensive contacts in the investment banking industry and can provide invaluable assistance when seeking additional finance via other venture funds, loans, or public stock offerings.

Capital markets, or stock exchanges, were established to provide an avenue for raising capital to finance major manufacturing businesses. In the United States, many business owners and investors are familiar with companies trading on the New York Stock Exchange and the NASDAQ system. However, there are public equities markets in every country and trading shares, via the Internet, that are able to facilitate investment. There is no reason to consider only the major U.S. markets (there are other stock exchanges in the United States and globally) as sources of capital. Increasingly, companies are choosing to register shares for sale on markets in Canada, various European countries, Singapore, Hong Kong, and others.

15.1.8 Managing the Company

Management, especially in a technical setting, is a complex and often delicate task. Entire racks of books about management can be found on the shelf and in

* Some entrepreneurs will balk at this. The author's advice is not to do so, since the angel investor usually has been successful in business and can offer valuable help in managing the company. Also, he is likely to know other potential angel investors or venture capitalists. When seeking more capital, such contacts are indispensable, and will provide more credibility than anything else, with the possible exception of a marketed product.

† Preferred stock can come in many different forms, but is "superior" to common stock. This means that if the company is sold or liquidated, the holders of these types of shares receive preference in any distribution.

the catalog of any bookseller. Numerous case studies have been published about successful and unsuccessful management practices. The subject is too large to address in this text but from the author's point of view, the most successful approach to management is respect. Employees at all levels who feel they are treated with respect will help make a company succeed by working harder and longer hours. On the other hand, an atmosphere of disrespect often results in underperformance and, sometimes, outright sabotage.

The most important task of the senior managers of any company is understanding the strategic goals and plans of the company and keeping themselves and their employees focused on them. This is especially important in a small, growing company.

15.1.9 Developing and Manufacturing a Product

This complex subject will vary over a wide range depending on the product. Product development and process development go hand-in-hand: the world's finest mousetrap will not be a success if it cannot be manufactured consistently and at a cost that makes it affordable to the consumer while providing a profit to the manufacturer.

For certain products, the development cycle is quite short because of facilitated production. A good example is the manufacture of carbon nanotubes formed by chemical vapor deposition. If purification and other properties are not considered, the cost of production is very low and fairly well understood. Thus, a new company entering this field could be selling product within a few months.

For others, the cycle may be very long. An example may be a nanoparticle drug delivery system. Besides the extensive testing any pharmaceutical product requires, that is, years to show both safety and efficacy, there may well be additional testing of any effects on the environment that may arise as a result of its manufacture or use. Another such product could be a nanoparticulate coating for glass to reduce glare or transmission of a certain wavelength. Such nanoparticles are expected to enter the environment as a result of the use of coated glass and data on the effects of that exposure could take a significant effort to produce and analyze. Such testing should be carried out or the manufacturer could be held liable for any problems.*

Building a manufacturing facility will also vary greatly with the product. Integrated circuit manufacturers working at the nanoscale will invest billions of dollars in manufacturing facilities, equipment and training, as will manufacturers of bio-nanotechnology products. A company manufacturing the window coating mentioned above may well be able to do so for a much smaller investment.

Product development and manufacturing are areas that the entrepreneur in nanotechnology should be very careful when considering. They can be very expensive in both time and finances and under- or overestimating the time and cost is very easy to do. With the advent of numerous and ingenious bottom-up fabrication processes, the cost of manufacturing can be driven to reasonable levels for start-ups.

---

* The asbestos litigation is an example.

15.1.10 Marketing

The old adage that "if you build a better mouse trap the world will beat a path to your door" is only partly true. The world needs to know the mousetrap exists and where your door is located; and then it needs to know why it is a better mousetrap. That is the function of marketing.

For every shelf of management books at a bookseller there are at least four shelves of books about marketing, advertising, and sales. These books when taken collectively contain conflicting messages and methods about how to go about developing a marketing and advertising campaign and then selling the product. An entrepreneur with little or no marketing experience may not want to indulge in marketing strategies. When the company has a product identified and has it under development, the best approach is to hire a professional marketing person—let them develop and execute the marketing plan.

15.1.11 Exits

Once a nanotechnology company is successful, generally by having a product either ready for market or actually producing revenue, the founders are often faced with a choice: to stay on and manage a commercial operation or move on, allowing professional managers with the appropriate experience to take over. This is actually a difficult subject for many entrepreneurs. They have seen their company succeed and they often want to continue to grow the enterprise. The problem is that while a start-up company often needs the single-minded drive, enthusiasm, direction and, even, charisma of the founder, when it comes down to actually producing products and making a profit, management needs to evolve—a condition more evident if the company has become publicly held through a stock offering. In this case, the company comes under pressure from stock markets to increase revenue and profits. The type of executive manager who focuses exclusively on that issue, while still supporting new ideas and growth of product lines, is needed to manage the company in that environment.

At this point, the founders can choose to remain, perhaps in different roles, or leave the company. If they remain, they need to understand the differing needs and be willing to put the appropriate management into place: finance, marketing, R&D, manufacturing, HR, legal/patent etc. They then should act as enablers, helping and supporting these managers to do their jobs and make the company even more successful.

Leaving, although traumatic perhaps for both the founders and the associated staff, may be the best choice. Very often, founders leave and use the assets they have from their company to start other companies or to act as sources of funding for other entrepreneurs.* The successful exit is defined by another mechanism. Successful small companies (those that actually develop products and make a profit selling them) are very likely to be bought out by another company. If that small company is publicly-traded, the stock market sets a value

* In fact, the author has known a number of "serial entrepreneurs" who, having been successful with their first company, have gone on to found others based on different technologies or applications. Since they have a track record of success they often find it easier to attract financial backers in subsequent companies.

based on share price. How markets value traded companies is generally based on a multiple of earnings or potential earnings per share per year. The market capitalization is simply the per-share trading price at the close of any day multiplied by the number of shares actually issued to shareholders (as opposed to being held in reserve by the company) and, therefore, varies with the trading price. Usually, when a company is being sold, the total price is more than the market-based valuation, with the premium resulting from the negotiating acumen of the company being purchased.

If the company is not a public company, the sale price is determined by negotiations. If an entrepreneur is involved in selling a company with a revenue and profit stream it should be possible to calculate a value, taking into account future product revenues. If there is no revenue stream, the value is based on a model of future cash flows. It would be advisable to engage the services of a business evaluator with experience in high-technology companies to calculate a market value when revenues are either very low or not yet realized. Some of the best sources of such evaluators are the large accounting firms, who all have consulting arms prepared to do this kind of evaluation.

The above description of the founder being successful and able to exit the company with significant assets, that is, money or its equivalent, is the dream of all entrepreneurs. However, the reality of business is that many do not succeed. The technology often times cannot be developed in a commercially feasible form due to the lack of money, or that a competing product appears first, or for a wide range of other reasons. This situation does not necessarily cause bankruptcy. In fact, most businesses in the United States that close their doors do not owe money; they just do not succeed.

When the decision is made to "wind up" a company, that is, close it down, there are often assets left in terms of cash, equipment, intellectual property, etc. Many of these assets are sold or, as in the case of a lease of space, transferred to another company. the resulting cash is usually divided among the shareholders pro rata, with holders of preferred shares receiving the distribution first.

Many successful entrepreneurs have been through the closing of a company. It is not a reflection of failure on their part. They learn from the experience and then go on to found other companies.

15.2 Education and Workforce Development

Societal components collectively and integrally bend, mend, and mold our civilization. We now direct our attention to education and workforce development. With regard to the broader sense of societal implications, specifically that new technology brings along changes that are positive, negative, or indifferent, we will do our best to achieve a balanced presentation. We strive in this section to stay below and beyond the hype associated with nanotechnology, a technology and societal driver that pundits around the globe consider to be revolutionary.

As a student, you are standing at the threshold that opens into your future, your livelihood, and your potential contribution to society. The underlying purpose of this section is to inform, provoke thought, stimulate, and hopefully, to encourage direct action. One must seriously consider the following prediction by the NNI to understand the significance of what is to come. Specifically, that products impacted by nanotechnology are expected to contribute over $1 trillion to the global economy by 2015 [5]. The strategic plan released by the NNI in December 2004 opens with the following vision [2].

> The vision of the National Nanotechnology Initiative is a future in which the ability to understand and control matter on the nanoscale leads to a revolution in technology and industry.

This vision indicates a revolution (from the Latin *revolutus* to "turn, roll back," with the "general sense of great change in affairs") is in the making based on the promise of nanotechnology; a revolution that has the potential to radically transform both technology and society. We need to simply look back a few years to the twentieth century to understand the impact of new technology. The legacy of the automobile, the television, and the computer are well understood. Are nanoscience and nanotechnology capable of exerting such historical changes? The NNI strategic plan goes on to state [5]

> ...the NNI will expedite the discovery, development, and deployment of nanotechnology in order to achieve responsible and sustainable economic benefits, to enhance the quality of life, and promote the national security.

The expectations for nanotechnology are quite lofty. **Goal 3** of the strategic plan is centered on the development of education and workforce development. The introductory paragraph of **Goal 3** emphasizes the important link between education and workforce development and the all-important tangible infrastructure that must be in place, seemingly a priori:

> A well-educated citizenry, a skilled workforce, and a supporting infrastructure of instrumentation, equipment, and facilities are essential foundations of the initiative. Nanoscale science, engineering and technology education can help to (1) produce the next generation of researchers and innovators, (2) provide the workforce of the future with math and science education and technological skills they will need to succeed, and (3) educate the citizenry capable of making well-informed decisions in an increasingly technology-driven society.

For any responsible strategy, the vision, goals, objectives, and the timetable are identified well before any reasonable action is to take place. Our hope is to convince you that a career in nano or related fields will offer challenge, accomplishment, and most importantly, responsible change in society. We repeat once again the inspirational statement attributed to Nobel Prize winner Dr. Richard Smalley, *"Be a scientist—Save the world"* [6].

15.2.1 Technological Revolutions—The Workforce Point of View

Education, workforce development, infrastructure, and industry are intimately intertwined and have collectively defined much of our civilization for thousands of years. Once again we are confronted with a familiar conundrum—what is the exact meaning and potential impact of nanotechnology and why would it be considered revolutionary? In order to gain a relevant perspective, we must review some basic history. Several thousands of years ago, developments in agriculture and irrigation transformed the nomad into the farmer. Prehistory ended with the birth of civilization. Agriculture is a *practice* that is defined as the production and distribution of food. With the creation of sessile agrarian societies, the livelihood of the majority of the population was based on farming and affiliated occupations. With the accumulation of assets, and eventually wealth, the need for armies, rulers, merchants, and clerics followed suit as a division in classes ensued.

On the other hand, the Industrial Revolution was based on a single-minded *manufacturing philosophy*, more specifically, the mass production of products. The invention of the printing press by Johannes Gütenberg in 1447 was an incipient form of mass production. The Industrial Revolution happened relatively quickly compared to the fundamental one of agriculture, but there was still plenty of time to adapt. Although all offspring of the Industrial Revolution, products such as the television, automobile, and computer have sprung minirevolutions of their own. Throughout history, we have had revolutions based on a *practice* (agriculture), based on a *manufacturing philosophy* (industrial), based on a *fuel* (oil), based on *products* (steam engine, railroad, automobile, telephone, and computer), based on *nature* (biotechnology), and based on *size* (micro and nano).

The currently burgeoning revolution in biotechnology is so because of our greater understanding of the *disciplines* of biology, biochemistry, and genetics. And what of the contributions of micro and later on, nano? What brand do we place on the micro- and nano-revolutions? Micro and nano are of course prefixes that relate to size. We now actually have revolutions, micro earlier and now nano, based on *SIZE* and size alone! The drive to manufacture smaller and smaller components materialized in the first portable watch, created by Peter Heinlen in 1524. The integrated circuit became the embodiment of miniaturization that has now morphed into nanocircuits.

Because nanotechnology has the capability to affect every type of product made (to the best of our knowledge, everything is still made of atoms and molecules), new jobs are certain to follow suit once a great idea leaves the lab—at least to countries that value education and innovation (is that us?). Will we have to overhaul our compartmentalized approach to education to one that is fundamentally interdisciplinary? According to an accepted definition of revolution, these changes are supposed to represent radical and far-reaching consequences, both of a technological and social nature. With each and every revolution throughout our history, the time scale of change has been diminished from that of its predecessor—overall, from several thousands of years to a few decades today. Is a revolution in education and workforce development needed or should we plod on with the status quo?

15.2.2 The State of Education and Workforce Development

It is widely viewed that the status of science and engineering education in the United States, and perhaps the Western world for that matter, is in a state of decline. According to Nobel laureate Richard E. Smalley [6], the United States in particular is falling behind developing nations such as China, India, and industrialized nations such as Japan with regard to the generation of new scientists and engineers. There is not enough space in this section to offer proper treatment of the multifold and complex factors responsible for the apparent decline. Without the addition of tedious detail, it is generally accepted that (1) the United States is producing fewer engineers and scientists; (2) although still the leader, more and more intellectual property is being produced outside the United States than ever before; (3) more and more high-technology jobs are being shipped overseas; (4) the structure of K-12 education is not optimized to promote science and engineering; (5) higher education is generally considered to be under funded with increasingly higher tuition; (6) the influx of foreign intellectual talent is limited due to current emphasis on security and increased prosperity in other nations; and (7) significant commitment in time and effort with diminished employment expectation results in fewer home-grown graduate students in science and engineering. According to Richard Florida, the author of *Flight of the Creative Class*, global competition for creative talent will be the defining issue of the twenty-first century [7]. Please acknowledge that this brief list is not apocalyptic in its message, just a statement of accepted trends. Isaac Asimov is noted for stating

> Science can be introduced to children well or poorly. If poorly, children can be turned away from science; they can develop a lifelong antipathy; they will be in a far worse condition than if they had never been introduced to science at all.

At the juxtaposition of any revolution, especially one that is suddenly upon us, opportunity abounds for those that are prepared. There are many means of navigating the impending maelstrom based on two generalized factors: (1) acknowledgement that the world is changing and (2) adapting by necessity to that change just like our predecessors did thousands of years ago. Economic development groups across the country have three goals: (1) jobs, (2) jobs, and (3) jobs. A civilization without jobs is incomprehensible, intangible, and actually unimaginable. Without employment there is no prosperity, at least within the existing paradigm—the paradigm that defines our quality of life. Along with each revolution, job descriptions have changed, oftentimes drastically.

In 1958, Congress passed the National Defense Education Act following the launch of Sputnik. In 2005, Louis V. Gerstner, Jr., a former chairman and CEO of IBM, stated in a *Newsweek* article [8]

> America is no longer winning the skills race. South Korea, with one sixth of the U.S. population, graduates as many engineers as the United States. China graduates four times as many; India, five times as many. Just as more than half of America's current science and engineering work force is approaching retirement, the flow of foreign talent is starting to dry up. For the first time in my memory, we're at the wrong end of a brain drain, as foreign-born grads in science, technology and engineering either return home after getting U.S. degrees or stay home in the first place.

According to RAND Corporation reports, alarms concerning potential skilled labor shortfalls oftentimes are not always justified [9]. However, they go on to say that

> ... although previously anticipated STEM workforce shortages have not materialized in the economic sense, the implications of a shortage of skills critical to U.S. growth, competitiveness and security justify continued examination.

15.2.3 Current Workforce and Education Programs

Awareness about the current state of education and the workforce is at a high level, especially with the onset of nano. New programs available to attract or create scientists and engineers are springing up in every nation. Factors such as the age of the existing workforce, dependence on foreign nationals and outdated curriculum coupled with declining federal funding, and the accelerated pace of science and discovery bolster the case for urgency. There is a fierce global competition to commercialize nanotech-based products. The United States does not necessarily have the lead. The NNI has defined the needs for education and development of a workforce of the twenty-first century, and is backing up their proposal with funding initiatives. Emphasis on nanoscience and nanotechnology must begin in the K-12 school system and extend through community college, university, and vocational schools [10].

Programs developed by the Nanotechnology Institute (NTI) in Pennsylvania serve as excellent examples of advanced thinking and multilevel partnerships forged among business, academia, and government that promote nanotechnology education and workforce development. The organization is a collaboration of academic and research institutions with a state-funded economic development group, the Ben Franklin Technology Partners of Southeastern Pennsylvania. The following excerpt from their Web site summarizes the NTI workforce development mission (http://www.nanotechinstitute.org/nti/index.jsp).

> A highly-skilled, technically trained, nanotechnology workforce will be needed if the region is to gain full value from the commercialization opportunities that nanotechnology will generate. The NTI is partnering with regional community colleges to anticipate this need by building the partnerships and securing the tools and resources to develop and implement education, training and workforce development strategies.

Nanotechnology as we have learned has a high barrier of entry. Nanotechnology will require highly trained scientists and technicians, but those jobs need to be in place to attract these special graduates.

15.2.4 The Workforce of the Future

The *knowledge worker*, a phrase invented by Peter Drucker in 1959, identified the emergence of a changing workforce, one that relied more on brains than brawn [11]. Robert E. Kelley wrote an influential book in 1985 called *The Gold Collar Worker: Harnessing the Brainpower of the New Workforce* [12]. In it he writes that American business is suffering from a *brain drain* due to severe mismanagement of its most valuable resource—*brainpower*. If you recall, the 1980s was the decade of the information age (*a.k.a.* revolution). Kelley stated that the gold-collar worker should

> ... engage in complex problem-solving, not bureaucratic drudgery or mechanical routine; they are imaginative and original, not docile or obedient. Their work is challenging, not repetitious, and their results are rarely predictable or quantifiable especially if they're scientists or researchers. Gold-collar workers are everywhere...

According to Dr. Mary Ann Roe, the author of *Cultivating the Gold Collar Worker*, nanotechnology along with the other emerging technologies, will require a special workforce. She states that [13]

> Complex powerful currents swirl through the nation and the world, altering the economic landscape, while offering extraordinary opportunities for well-prepared individuals… Technology-driven, these currents propel dynamic change, force innovation and create new types of work in the private sector.

According to Dr. Roe, this belongs to the techno-professional who wears a gold-collar—an individual with a white-collar mind but blue-collar hands; a worker who is able to create as well as operate and think as well as do.

We try to envision the future and the worker therein. Within the lexicon of social prognosticators of this day and age, we are told often that the average employee changes job description, or the job itself, every five years. We predict that the new workforce, whether of white, blue, or gold collar, will have flexibility in the job market, a high level of training and education, an interdisciplinary background, advanced communication and computer skills, and perhaps an understanding of the international education and economic community. We boldly predict that the face of K-12 must change in order to accommodate the new age of technology. Instead of teaching to the test in a compartmentalized fashion, classrooms must open curricula to new ideas and efficiently teach interdisciplinary programs in the sciences. Partnership and close communication between educational institutions, the universities and community colleges in particular, and industry is necessary to ensure that there is balance between the needs of industry and a supply of educated and trained students. We are all aware of the shortage of trained technicians in the biotechnology sector.

15.2.5 Planning Ahead and Potential Career Paths

There is no guarantee of success. That key ingredient is always up to the individual. Contemplate for a moment the vast complexity of today's economy. Surely there are numerous paths to a fulfilling career. Make a list of potential paths for your future career. Then, try and integrate those paths with a nanotechnology theme.

The best approach begins with *planning* now. It is never too early to make plans. Therefore, start a *journal* to keep track of any ideas, thoughts, or references that happen your way. *Flexibility* is always a survival advantage. Omnivorous species like crows tend to out compete counterparts that rely on a limited food sources. Devour knowledge. Consider also that many of us end up doing work that we never intended to do. Nonetheless, a good strategy is to place your focus on one academic discipline such as chemistry but learn enough about physics, engineering, and biology to be able to understand your coworkers of the future. *Networking*, a good way to achieve that good strategy, is perhaps one of the most powerful tools that an individual can acquire. It really is "who you know" much of the time. Attend conferences—as many as physically possible. Open your horizons and *partnerships* to include business, academia, and government, the big three that are currently working together to pave the road for the nanotech revolution. Traditionally, business folks do not mingle much with academics. That tradition is changing drastically as universities are acquiring the entrepreneurial spirit and commercializing intellectual property like never before.

Research all available *funding sources*. Conduct investigations to find federal, state, local government, and private sources. For example, funding for women scientists and underrepresented groups are out there just waiting to be tapped. Stay abreast of *societal implications*, understand social and global trends, and track investments into nano-based businesses. Take a serious look into *careers that are tangential to the science and technology*. For example, there is a great need for a new generation of patent lawyers who possess interdisciplinary skills and knowledge. Health and safety, job sourcing and staffing, education at all levels, economic development, public policy, regulation, international commerce, and investment are but a few examples of jobs outside the technology. Lastly, and certainly of some great importance, make plans now to *start a nanobusiness*. Take care of your education (perhaps an MBA in the plan), do the research, conceive a novel product or process, make the contacts, establish the necessary partnerships, locate and secure funding and, yes, then you are in business. Although things "do come to those that wait," it is a much better policy to get informed and go after it. We wish the best of luck to you all.

15.3 Buildings for Nanotech

We now jump from creating and operating a business and finding and securing employment to buildings that house them. Nanotechnology along with biotechnology are placing exceedingly stringent demands on laboratory design, manufacturing strategy, and construction. This section presents a short introduction into another world—the domain of architecture and construction—where our new facilities try to keep pace with the *Nano Age*.

Old buildings are constantly being replaced due to wear and tear, but a new paradigm of building design, construction, and operation is upon us nonetheless and on many fronts. New buildings need to be designed and constructed to accommodate the rapid changes brought about by new technologies in the fields of biology and nanotechnology, new technologies that rely on interdisciplinary cooperation in a big way. New designs also include more efficient means to improve indoor air quality, electricity production, energy efficiency (heating), waste management, water conservation, and daylight lighting designs—some already enhanced by the new technologies.

New buildings also include environmentally approved technologies and practices that include use of recycled materials and reduction of toxic material components. Although we can spend a significant amount of time discussing new building philosophies and practices, we must focus on the relevant thrust of this chapter—how building and facility design and construction conform to the needs of nanotechnology and biotechnology—and conversely, how nanotechnology can contribute to new buildings. Now more than ever, the goals of research centers, are to conduct world class research, attract researchers and students, attract money, attract industry, provide jobs and, at the same time, be flexible [14].

Several "monuments to science and technology" have been constructed worldwide to support nanotech and biotech R&D. A few notable buildings are mentioned in this text: the Center for Integrated Nanotechnologies (CINT) of Sandia and the Los Alamos National Laboratories in Albuquerque, New Mexico; the Birck Nanotechnology Center at the University of Purdue in West Lafayette, Indiana, and the state-of-the-art leader amongst all buildings that support advanced technology—the NIST's Advanced Measurement Laboratory (AML) in

Gaithersberg, Maryland. We shall focus primarily on NIST's AML facility. Please refer to www.HDRInc.com, aml.nist.gov and www.nanobuildings.com for complete information concerning all of their advanced technology facilities.

15.3.1 Nanotechnology in Buildings—Environmental Aspects

Although we will present the strategy and design elements required in constructing buildings to support nanotechnology research, development, and manufacturing, it is only fair that we reserve this section to discuss the impact of the nanotechnology on building materials and construction in general. The construction business is estimated to be on the order of $1 trillion per year worldwide [15]. Many kinds of building materials already take advantage of nanomaterials [15]:

The Now

- Flexible solar panels
- Self-cleaning windows
- Self-cleaning concrete using catalytic TiO_2
- Wi-Fi paint (an additive mixed with paint that reduces transmission of radio waves through walls)
- Selective absorbing–reflecting solar windows
- Scratch-resistant flooring
- Antimicrobial steel surfaces
- TiO_2 and AgO antimicrobial coatings (public places)
- Nano-cement (enhanced physical properties)
- Nansulate coating (corrosion prevention, insulation)
- Translucent concrete (enables transit of light)
- Nanotech-enabled gypsum drywall (water resistant, durable, stronger, lighter)
- Nanotech-enabled steel (corrosion resistant, higher strength, high plasticity)
- Aerogel glasses (high R-value thermal insulation, acoustic insulation)
- Nanogel (Cabot Corp.) traps air at the molecular level → thin insulating layers

According to the USDA Forest Service Research Laboratory, nearly 2 million housing units were constructed in the United States in 2004 [16]. Wood comprised, by volume, 80% of all the building materials used. Half of the wood products are engineered wood composites. Nanotechnology is expected to contribute to the next generation of wood-based products that exhibit enhanced strength, properties, and endurance similar to those of carbon-based composite materials [16]. The new materials are designed to be biodegradable [16]. And of course, nanotechnology will promote the development of "intelligent wood"—biocomposite products with built-in arrays of nanosensors. By the way, wood is a natural nanomaterial. According to the USDA [16]:

> Building functionality into lignocellulosic surfaces at the nanoscale could open new opportunities for such things as self-sterilizing surfaces, internal self-repair and electronic lignocellulosic devices. The high strength of nanofibrillar cellulose together with its potential economic advantages will offer the opportunity to make lighter weight, strong materials with greater durability.

The U.S. Department of Energy's Twenty Year Plan identifies strategies for environmental friendly buildings—try and select the ones you believe involve nanotechnology. The United States has 5% of the world's population but consumes 25% of the world's resources and discards 40% of the world's waste, of which 50% is due to construction activities [17]. If we can at least optimize our buildings, perhaps a dent can be made on some of these statistics. We listed a few current innovative products enhanced by nanotechnology earlier. Let us continue by painting a picture of the near and far futures.

The Near Future

- Cellular building materials
- Disaster-resistant materials
- Intelligent materials (e.g., self-repairing, self-adjusting)
- Superior moisture barrier materials
- Nontoxic materials
- Resource-efficient materials
- Superior insulating materials
- Superior weather-resistant capability (e.g., low maintenance)
- Smart materials with sensors able to detect loads, temperatures, decay, fire, etc.
- Fiber-cement siding (self-cleaning, thin layer of silica)
- Anti-bacterial house fittings (Ag particles or light-activated particles)
- Fire retardant materials that incorporate clays
- Easy-clean water repellant surfaces
- UV-resistant paint coatings
- Thin film coatings for roofing materials

Other aspects of construction include household devices. Home and laboratory products enabled by nanotechnology include [18]:

- Water treatment systems (nanofiltration, light-activated nanoparticles)
- Energy-efficient lighting, LED, and electroluminescent lighting
- Solar energy generation and advanced energy storage

Buildings in the Not-So-Far Future [19]

- Earthquake-proof buildings using carbon nanotube reinforcement
- Nano-reinforced glass structural and enclosure elements
- Quantum dot lighting
- Next generation nanosensors

The interior lighting paradigm will also be impacted by nanotechnology—by solid-state lighting devices made with nanocomposite materials. Carbon nanotube-based organic composites, known as "ultra-low energy high brightness" (ULEHB) lights, are expected to produce the same quality light with a fraction of the energy. Many other materials are under development: self-healing concrete, UV-IR radiation blockers, smog-mitigating coatings, and LEWs and LECs (light-emitting walls and ceilings). According to George Elvin of Nanowerk.com, the "smart home" is also on the horizon. Smart homes are decorated with nanotechnology-based sensors that monitor temperature, humidity, and toxins as well as transmit medical information to your doctor. From larger structures to appliances, sensors would be able to monitor vibration, stress, crack propagation—the foundation for "intelligent buildings" [15,19].

There is no doubt that nanotechnology will impact (and has already impacted) the way buildings are constructed, the materials selected for building structure and function, and the way people interact within buildings. It is not hype. It is due to the remarkable properties of nanomaterials that there is the excitement and anticipation. Yes, there is hype out there and the nano-pretenders abound, but atoms and molecules do not spin tall tales—they just spin.

Through nanotechnology, we will be able to develop the building materials of the future—those that absorb and radiate heat, offer earthquake-proof stability, the electrical wires that conduct over long distances efficiently, windows that reflect or absorb radiation depending on the current need, solar cells that power our instruments, and cooling systems and lighting that do not damage our environment. Yes, it is possible to accomplish all of the above. Yes, much of it will, and already has, come from our knowledge of nanotechnology. The materials and processes just have to be cost-effective and competitive—complete with an analysis of long-range recovery of costs and comfort with environmental friendly interactions.

In the United States, buildings consume 39% of the total energy, 12% of the total water, 66% of the total electricity, and contribute to 38% of the total carbon dioxide emissions [20]. What can nanotechnology (and our own policies) do to help mitigate these numbers?

Environmental Protection Agency's (EPA) Elements of Green Building

- Energy efficiency and renewable energy
- Water stewardship
- Environmentally preferable building materials and specifications
- Waste reduction
- Toxic material reduction
- Indoor environment mitigation
- Smart growth and sustainable development

In addition to renewable energy sources, equipment and appliances should conform to EPA's *Energy Star* program—the Building Design Guidance [21,22]. Designers, architects, and builders should consider the following:

- *Include statement of energy design intent* (SEDI)—describes the energy performance outcome of your building in the bid package
- *Specify design team participation*—during construction to ensure that energy performance features are incorporated
- *Include approval process for change orders*—accountability!
- *Document construction methods*—include manufacturer's literature, summary of energy efficient features, and explanation of anticipated functions to assist construction team
- *Select qualified manufacturers*—rejection of unapproved alternatives and coordination of manufacturers to enhance compatibility
- *Seek incentives*—local utility company incentives and government incentives to offset costs
- *Communication of superior design intent*—label with Energy Star mark

What is a green building material [23–26]? A green material is made of renewable resources. The use of green building materials reduces the environmental impact of extraction, transportation, processing, fabrication, installation, reuse, and recycling and disposal [26]. Green building materials reduce maintenance-replacement costs over the life of the material, provide for energy conversation, improve occupant health and productivity, lower costs associated with changing space configurations, and offer greater design flexibility [26]. Material selection criteria, in addition to its suitability for the intended function, include resource efficiency, indoor air quality, energy efficiency, water conservation, and affordability [25,26].

The Leadership in Environmental Design. The Leadership in Environmental Design (LEED) is a green building rating system developed by the U.S. Green Building Council. The LEED concept sprung from the Natural Resources Defense Council (NRDC) in 1994 and a consortium of nonprofits, government agencies, architects, engineers, builders, developers, product manufacturers, and others [27]. Buildings are rated according to the following six criteria: sustainable sites, water efficiency, energy and atmosphere, materials and resources, indoor environmental quality, and innovation and design processes. Certification categories start at the baseline with "Certified" and then progress from "Silver" to "Gold" to "Platinum" with Platinum being the best possible rating.

15.3.2 The Needs of Scientists and Engineers (And Equipment and Instrumentation)

Architects and designers need to know what engineers and scientists require. There are, for example, several challenges that arise from integrating technologies. The transfer of energy across multiple length scales; optical amplification of quantum dots in a two-dimensional crystal; combining top-down with bottom-up fabrication; interfacing biological and synthetic systems; and interfacing mechanical force and fluid transport (e.g., nanomechanics and nanofluidics) at the nanoscale. Such integrated nanotechnologies are expected to impact our world, according to N.D. Shinn of Sandia National Laboratory in New Mexico, who states "... connecting scientific disciplines and multiple length-scales is the key to success," and require special buildings to house processes and procedures [28].

The CINT model is a prime example of an integrated building in which integrated science is accomplished. CINT makes use of a "core-gateway model" that emphasizes interaction between two national labs (Sandia and Los Alamos) and universities. More about CINT is presented later in this chapter.

Nanotechnology research requires observations of reactions in real time (at femtosecond time scales or lower). Nanotechnology research requires increased resolution—the capability to see atoms and gauge the level of material behavior at the molecular level without interference from the outside world [29]. The underlying goal of every new advanced technology building is to, of course, have built-in flexibility [14]. Collaboration, interaction, and integration are key components of any new design—and these ingredients must all coexist within one building or complex. A case in point is the interdisciplinary collaboration among multiple disciplines and interaction between academia, business, and government. In addition, new buildings, need to accommodate technology transfer like never before, evolve into a technology transfer user facility [14].

Interactive Spaces. A state-of-the-art research and manufacturing center is composed of many kinds of *spaces*—the result of forethought, adaptability, and an effective floor plan design. *Technical spaces* are places where characterization, laboratory work, and manufacturing are accomplished. These include clean rooms, imaging facilities, production (e.g., lithography), and metrology. *Nontechnical spaces* include offices, meeting rooms, libraries, conference rooms, and auditoriums. These spaces must be strategically placed around, within, or between technical spaces to enhance interaction between and among scientists, entrepreneurs, and managers.

A key element of advanced building design that is acquiring exponentially more importance with each passing year is the following—how do the architect and engineer accommodate the increasing level of interdisciplinary needs required for nanotechnology? We will need dry labs, wet labs, semiconductor clean rooms, bio-clean rooms, quiet labs, ultraquiet labs, labs for metrology and imaging, office spaces, interactive spaces, and spaces for support. Potential users may arise from numerous university departments—chemistry, physics, biology, chemical engineering, mechanical engineering, agronomy, mathematics, biophysics, computer sciences—and from industry and government.

Environmental Considerations [29]. Environmental stability is important in any laboratory or manufacturing facility. Factors that contribute to environmental stability include temperature, humidity, contamination, air velocity, and vibration. Environmental stability is defined over space (uniformity) and over time (drift) [29]. Uniformity is influenced by the amount of "in and out" of a room, the distribution scheme, and the room design. Drift is influenced by measurement and control systems, sensor locations, and equipment placement.

Temperature and Humidity. Temperature control for a room that houses a TEM is a challenging prospect. The room temperature surrounding a high-powered (+300 kV) TEM should conform to 20 ± 0.01 to 0.25°C. The drift of temperature should be no more than $0.5°C \cdot h^{-1}$ (fluctuation $0.05°C \cdot min^{-1}$) and air velocity less than $5\,m \cdot min^{-1}$. The heat generated from a generic TEM column and support equipment is 500 W and 800–1200 W respectively. In order to achieve ca. control, internal sources of heat gain need to be minimized (e.g., lighting), a high air exchange rate needs to be maintained ($>300\,AC \cdot h^{-1}$), and heat transfer through walls needs to be minimized [29]. Added features such as a heated floor pad aid in maintaining such a tight rein on temperature.

Humidity control, obviously, is vital for several highly sensitive measurement techniques. For example, the temperature and humidity requirement for JEOL's JXA-8200 electron probe microanalyzer is ±1°C and 60% rh or less so that dew does not condense on the cooling water hose. At NIST's AML, humidity control is exerted to a level of ±1% in metrological areas and ±5% in other laboratories.

Clean Rooms. A clean room (or cleanroom) is an isolated space within which a high level of particulate contaminant control is in effect. A clean room environment is designed to reduce environmental pollutants such as dust, microbes, aerosol particles, and chemical vapors [30]. According to ISO-14644 (The International Standards Organization)

TABLE 15.2 *United States Clean Room Standards*

US FED STD 209E	≥0.1 µm	≥0.2 µm	≥0.3 µm	≥0.5 µm	≥5 µm
1	35	7	3	1	
10	350	75	30	10	
100		750	300	100	
1,000				1,000	7
10,000				10,000	70
100,000				100,000	700

Source: US FED-STD-209E Cleanroom Standards. With permission.

> Cleanrooms and associated controlled environments provide for the control of airborne particulate contamination to levels appropriate for accomplishing contamination-sensitive activities. Products and processes that benefit from the control of airborne contamination include aerospace, microelectronics, pharmaceuticals, medical devices, healthcare, food, and others. Many factors besides airborne particulate cleanliness must be considered in the design, specification, operation, and control of cleanrooms and other controlled environments.

This is no easy feat to accomplish considering that the air outside in a typical city has on the order of 3.5×10^7 particles · m^{-3} 500 nm or larger. Microprocessors, for example, are assembled in a clean room. Other aspects of clean rooms such as temperature, humidity and pressure are also strictly regulated.

According to Federal Standard 209E, a *Class 10000* clean room should have no more than 10,000 particles larger than 0.5 µm in a cubic foot of air. A *Class 100* clean room should have no more than 100 particles larger than 0.5 µm in a cubic foot of air. A *Class 1* clean room should be essentially contaminant free (**Table 15.2**). Hard disk manufacturing requires a Class 100 clean room. The ISO, headquartered in Geneva, Switzerland, also recommends standards for clean rooms based on a logarithmic scale.

ISO-14644-1 Standards. Following normalization between cubic feet and square meters, clean room standards are converted accordingly: *Class 1* = ISO 3, *Class 10* = ISO 4, *Class 100* = ISO 5, *Class 1000* = ISO 6, and Class *10,000* = ISO 7 (**Table 15.3**). Other standards are also in existence.

TABLE 15.3 *International Standards Organization Clean Room Standards*

ISO-14644-1	≥0.1 µm	≥0.2 µm	≥0.3 µm	≥0.5 µm	≥1 µm	≥5 µm
ISO 1	10	2				
ISO 2	100	24	10	4		
ISO 3	1,000	237	102	35	8	
ISO 4	10,000	2,370	1,020	352	83	
ISO 5	100,000	23,700	10,200	3,520	832	29
ISO 6	1,000,000	237,000	102,000	35,2000	8,320	293
ISO 7				352,000	83,200	2,930
ISO 8				3,520,000	832,000	29,300
ISO 9				35,200,000	8,320,000	293,000

Source: International Organization for Standardization (ISO-14644). With permission.

Air entering a clean room is filtered for dust and the air inside is recirculated through high efficiency particulate air (HEPA) and ultra-low penetration air (ULPA) filters. Employees in a clean room facility must enter through airlocks that may contain an air shower and wear bunny suits that cover all of the person including hands and shoes. Any fibrous or particulate materials are prevented from entering the clean room (e.g., pencils, fabrics, soda pop). Clean rooms also maintain a positive pressure within to prevent egress of unfiltered air. The support equipment infrastructure and maintenance of a clean room (air conditioning, filter systems, etc.) may be enormous and complicated respectively.

Electricity. According to the IEEE (Institute of Electrical and Electronics Engineers) Std. 1100–1999

> Power quality is the concept of powering and grounding electronic equipment in a manner that is suitable to the operation of that equipment and compatible with the premise wiring system and other connected equipment.

Power disturbances arise from external or internal sources. External sources include lighting, faults, and utility switch surges such as voltage reduction and line maintenance. Internal sources arise from mechanical equipment (chillers, fans, pumps), elevators, shop equipment, and laboratory equipment. There are several means of reducing electrical disruptive effects by applying power conditioners, transient voltage surge suppressors reduction (dedicated circuits minimize line noise, transients), shielded-isolation transformers, uninterruptible power supplies (UPS), and standby generators [31].

Common and normal mode transients from line and load side sources, noise from and between lab equipment, stray ground currents, ELF (extra or extremely low frequency) and EMI (electromagnetic interference) sources, acoustic noise sources, vibration sources, irregular voltage and frequency, and sources of heat are some common problems facing architects and engineers who design any laboratory. With nano-capable equipment, such issues are exacerbated and special care must be taken to minimize electrical disruption so that they do not influence data acquisition and measurements. System wiring configurations also play an important role in reducing electrical disruptions. For example, grounding systems include lightning protection, safety wiring, communication system grounds, signal reference grounds, and instrument reference grounds. Isolated grounding of individual equipment is recommended [31]. For more information, please refer to www.HDRInc.com.

Vibration and Acoustics. Structural and mechanical design of advanced technology facilities must address vibration-sensitive issues [32,33]. Mechanical vibrations in general, depending on energy and their potential targets, can be detrimental to human health, comfort impairment, sensitive equipment, and structural components [34]. Many laboratory operations are sensitive to vibrations. These include metrology, high-end imaging (TEM, SEM), photolithography, and probe development [32,33]. Other methods and procedures are impacted by vibrations to some degree. These include optical microscopy, mass spectrometry, and other characterization methods. Low-end imaging, theory and modeling instruments, and nontechnical spaces, of course, are less sensitive to vibrations. The bottom-line, like for any requirement, must be balanced with cost-effectiveness, facility mission, aesthetics, facility operation costs, and future flexibility.

Sources of external vibration are numerous. Machinery, road traffic, and continuous construction activity are considered to be sources of *continuous vibration*. In this form, vibration continues uninterrupted for a defined period. *Impulsive vibrations* are infrequent and sporadic and are defined as "three distinct vibration events in an assessment period" (e.g., dropping heavy equipment and loading and unloading supplies). Rapid buildup to a maximum level followed by a damped decay of short duration is a characteristic of impulsive vibrations. *Intermittent vibrations* arise due to trains, passing heavy vehicles, and intermittent construction activity. Intermittent vibrations can be interrupted periods of continuous vibrations or repeated periods of impulsive vibrations that arise from continuous or repetitive sources.

Internal sources arise from periodic excitations (e.g. constant speed rotating equipment), footfall (e.g. from the scientists), AVAC, fluid transmission in pipes, and low-frequency airborne acoustic noise (e.g. chit-chatting) [34]. The effect of mechanical vibrations depends on the vibration amplitude, frequency range, duration, the predominant component, and time of day [34]. There are several standards that describe recommended vibration criteria (VC).

VC Curves. Generic vibration criteria curves are provided to architects and engineers who design housing for vibration-sensitive instruments [35]. With the onset of the microelectronics, medical, and biopharma industries in the 1980s, *criterion curves* were developed to lay down generic standards for vibration for a wide range of instrumentation, equipment, and tools.

The VC curve is represented in the form of a set of *one-third octave band velocity spectra* [34,35]. The curves are defined in terms of constant velocity (RMS) within the 8–80 Hz frequency range. The energy-averaged RMS velocity is calculated within proportional bandwidths, for example, a one-third octave bandwidth spectrum at each frequency range that lies within 23% of the center frequency (or 71% of the peak value) and is considered in all three orthogonal directions [34]. RMS velocity (as opposed to the "peak" or "peak-to-peak" criteria) is measured in terms of $\mu m \cdot s^{-1}$. The RMS velocity is related to the product of the frequency f and the wavelength λ (displacement) of the vibration. Displacement, velocity, and acceleration are all interrelated.

$$\upsilon = 2\pi f \lambda \tag{15.1}$$

VC curves extend from 4 to 100 Hz. Pneumatic isolation systems (e.g., for AFMs) may resonate with floor vibrations in the 1–3 Hz range. There is usually less concern for vibrations above 100 Hz. Relevant generic vibration criteria for advanced buildings are as follows: VC-A/B $\rightarrow$ 50 to 25 $\mu m \cdot s^{-1}$; VC-D/E = 6 to 3 $\mu m \cdot s^{-1}$; NIST-A $\rightarrow$ 0.025 μm displacement for $1 \leq f \leq 20$ Hz or 3 $\mu m \cdot s^{-1}$ for $20 \leq f \leq 100$ Hz; and NIST-A1 $\rightarrow$ 6 μm displacement for $f \leq 5$ Hz or 0.75 $\mu m \cdot s^{-1}$ for $5 \leq f \leq 100$ Hz [35]. Each instrument has a unique RMS υ versus f profile. For example, *Omicron AFM* VC criteria lie within 0.5 and 1 $\mu m \cdot s^{-1}$ ($1 \leq f \leq 10$ Hz); a *Mann 3696 Stepper* is 0.8–80 $\mu m \cdot s^{-1}$ ($1 \leq f \leq 100$ Hz); a *Hitachi S-4700ii SEM* range is 10–20 $\mu m \cdot s^{-1}$ ($1 \leq f \leq 10$ Hz); and a *JEOL 2010 HRTEM and 6400 and 5800 SEMs* are 6–20 $\mu m \cdot s^{-1}$ ($1 \leq f \leq 10$ Hz).

For offices and theory and modeling facilities, RMS velocities between 400–800 $\mu m \cdot s^{-1}$ are tolerable. General labs fall into the range of 50–100 $\mu m \cdot s^{-1}$ (VC-A±). Clean room levels depend on the designation of the clean room: Class 1000 $\rightarrow$ 25 $\mu m \cdot s^{-1}$ (VC-B); Class 100 $\rightarrow$ 6 $\mu m \cdot s^{-1}$ (VC-D); Class 10 $\rightarrow$ 3 $\mu m \cdot s^{-1}$ (VC-E). Metrology lab VC criteria range from 3–6 $\mu m \cdot s^{-1}$. AFM and atom pushing

≈3 μm · s^{-1} (VC-E or NIST-A) and nano-instrument development is defined under 1.25 μm · s^{-1} [32,33]. Vibration limits in general for nanolabs of 0.75–3 μm · s^{-1} are not unusual [36]. More information about vibrational concerns, criteria, and recommendations can be found on www.colingordon.com.

Electromagnetic Shielding. Electromagnetic field (EMF) shielding is the process of reducing the intensity of EM radiation between two areas. Placing a barrier made of a conducting material is able to accomplish this task. In particular, sensitive equipment needs to be shielded from radio frequency-electromagnetic sources (RF-shielding). Such shielding (e.g., Faraday cages) diminishes the interaction between radio waves and electrostatic fields. Magnetic shielding may also be required. TEMs, SEMs, E-beam writers, and semiconductor inspection systems are some examples of sensitive equipment that require shielding.

Mechanical Noise. Noise is measured in terms of decibels (dB). A *decibel* is a dimensionless unit of measure that gauges the intensity of sound. It is a measure that is used in electronics, signal transfer, and communications. The dB is the logarithm of a ratio—the ratio may be that of power, sound pressure (acoustics), voltage, or other indicators of intensity. The sound-intensity-level dB(SIL) decibel reference is standardized to the level of threshold intensity of hearing in humans in air—ca. 1 kHz with an intensity of $I_o = 10^{-12}\ \mathrm{W \cdot m^{-2}}$. One decibel is

$$\mathrm{dB} = 10\log\left(\frac{I}{I_o}\right) \tag{15.2}$$

or in other words, 1 dB equals ten *bels*. The exponent of the power of ten in the final log term is known as the bel.

The dB(SPL) is the decibel (sound-pressure-level) relative to 20 μPa (2×10^{-5} Pa), the minimum sound a human can hear in air (e.g., a mosquito at 3 m distance). Other symbolism is used depending on the frequency scale used for calibration. The dB(A), dB(B), and dB(C) are based on different frequency weighting. For example, dB(A) is based on the A-scale range. Some nighttime zoning restrictions call out a maximum of 45 bB(A). Ambient noise is not to exceed 55 dB (10 Hz < f < 10 kHz) within 0.5 m surrounding an electron microscope as specified by the manufacturer [36].

The primary source of mechanical noise arises from HVAC (heating, ventilation, and air conditioning) noise and vibration. Equipment operation (e.g., pumps) and air movement through ducts (aerodynamic noise) cause significant noise if not constructed properly. Air turbulence is the cause of high vibration levels at low frequencies (59 dB @ 31.5 Hz) in air ducts (lined plenum). The problem disappears by installing silencers, changing the duct size (made smaller), and reducing the air duct velocity. Adding acoustical tiles to the walls and acoustical transparent curtains help to mitigate the effect of the reflective electron microscope suite doors. Fan deficiencies (static pressure, airflow, and fan speed) should also be addressed. Most advanced technology buildings place as much mechanical equipment as possible outside the building. These include cooling towers, exhaust fans, scrubbers, and pumps. What is known for certain is that conventional noise control solutions don't necessarily work for advanced technology buildings [36].

15.3.3 Advanced Facilities That Support Nano and Biotech

National Institute of Standards Advanced Measurement Laboratory. The NIST has recently opened the doors to the most advanced measurement laboratory in the world—the NIST Advanced Measurement Laboratory. According to NSET (National Science and Technology Council, Executive Office of the President) in 2004

> NIST has recently constructed the most technologically advanced facilities in the world, the Advanced Measurement Laboratory, which will support industry in the conduct of this research with new ways to more accurately measure, quantify and calibrate important processes and properties.

The NIST's AML, costing \$235 million to construct, is a 49,843 m^2 (536,507 sq. ft) facility that is composed of five separate wings, two of which are buried 12 m (39 ft) underground [37,38]. The facility, under the auspices of the Department of Commerce, will provide

> ... sophisticated measurements and standards needed by U.S. industry and the scientific community for key 21st century technologies such as nanotechnology, semiconductors, biotechnology, advanced materials, quantum computing and advanced manufacturing. NIST research efforts planned for the new facility range from improved calibrations and measurement of fundamental quantities such as mass, length and electrical resistance to the development of quantum computing technology, nanoscale measurement tools, integrated micro-chip-level technologies for measuring individual biological molecules, and experiments in nanoscale chemistry.

The minimum standard criteria for air quality, temperature control, vibration, and humidity control are cutting edge (**Fig. 15.1**). Air quality is controlled to 3.5 particles per liter (100 particles per cu. ft) as compared to 3,500 particles per liter (100,000 particles per cu. ft) in most modern labs. In standard cu. ft. terms, this converts to 100 particles per cu. ft. Temperature is controlled from ±0.1 to 0.01°C as compared to ±2°C in most labs [37].

The NIST's AML consists of five interconnected units. The five buildings are united with above ground walkways and underground tunnels. The AML is environmentally stable with regard to humidity, temperature, vibration, electromagnetic interference, and contamination. Metrology activities are conducted in two underground facilities.

There are two single-floor analytical characterization laboratories (housing 187 laboratory modules). Most of the analytical labs are controlled to 20 ± 0.25°C, a few to ±0.1°C. Two underground metrology laboratories (housing 151 laboratory modules) are located 9 m (40 ft) below ground level. There are two types of labs: "quiet" metrology labs are dedicated to measurement, and "rotating or dynamic" labs involve some kind of moving equipment. Selected labs are isolated on concrete slabs (equipped with air springs that are able to cancel out even the slightest of vibrations) to reduce any level of vibration. A multi-floor nanofabrication facility serves as an ultra-clean room.

The Nanofab building has a 18,000 sq. ft "raised-floor" *Class 100* clean room—adaptable for upgrade to *Class 10* if required. A sub-fabrication area is located beneath the clean room, and an interstitial space separates the clean room floor from the mechanical suite. NIST has invested in a completely new set of equipment dedicated to producing 150 MM wafers. The equipment includes furnaces, LPCVD (liquid phase chemical vapor deposition), rapid thermal annealers, three reactive

Fig. 15.1 *The level of environmental control in NIST's new AML is compared to other laboratory facilities. The NIST facility has 338 reconfigurable modules. A 8.520 m² (91,700 sq. ft) nanofab facility* Class 100 clean room *is rated at <3.5 particles·L⁻¹.* Enhanced air quality *is achieved with HEPA (high efficiency particulate air) filters for general-purpose laboratories. Most laboratories have baseline temperature control within ±0.25°C and 48 laboratories have control to within ±0.1 or ±0.01°C.* Vibration *isolation is achieved to a level of 3 μm·s⁻¹ or less and down to 0.5 μm·s⁻¹ in 27 low-vibration modules.* Humidity *control is held at ±5% relative down to ±1% in special laboratory sections.* Electrical power filtering *provides institute-wide uninterruptible power and counter measures of voltage spikes, drop-outs and other dirty power problems that limit accuracy and precision, reduce analytical sensitivity and cause long-running experiments to crash.* Green building *features include natural daylighting, energy conservation, and recycling.*

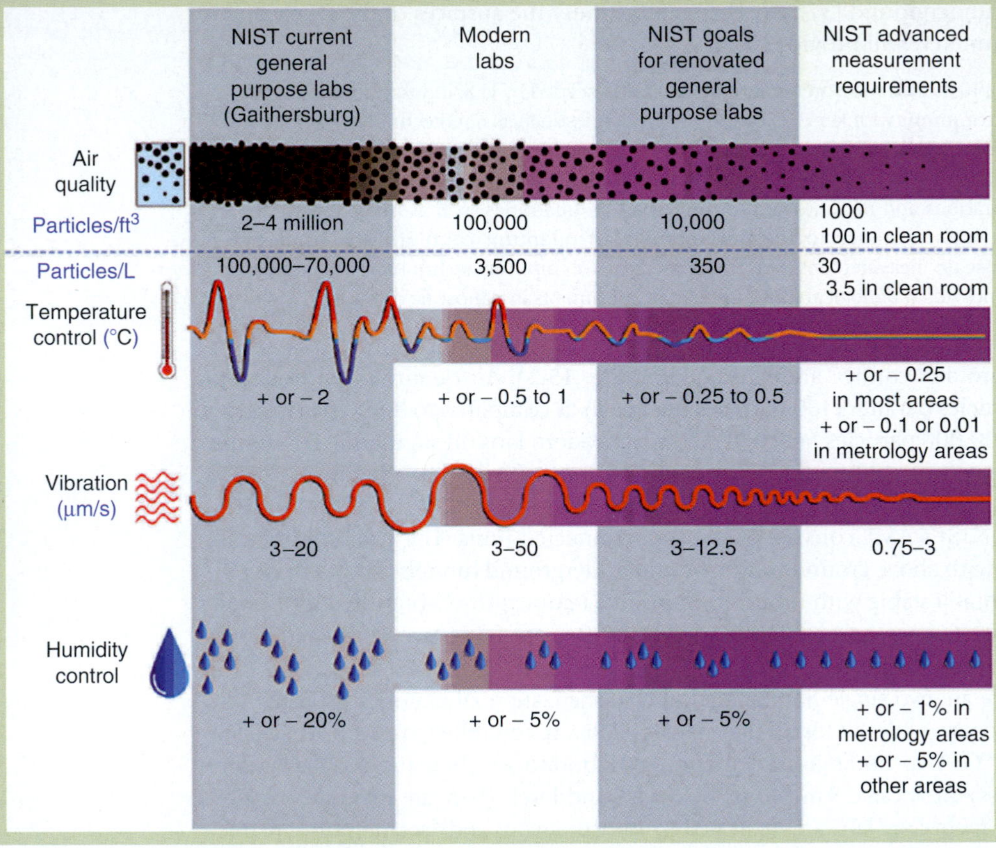

Source: Image courtesy of HDR, Inc. With permission.

ion etchers (RIEs), three metal deposition tools (thermal evaporator, electron beam evaporator and ion sputterer), contact lithography capability, electron beam lithography and focused ion-beam lithography. Field emission SEM, spectroscopic ellipsometers, contact profilometers, microscopes, and image capture instrumentation support product and research monitoring and metrology.

A selection of incredible images acquired at AML are displayed in **Figure 15.2**.

Fig. 15.2 *(a) Silicon step rulers that range in height from 10s to 100s of nanometers to a single monolayer measuring 0.3 nm. The microscope used to make the image sits on an isolated concrete slab equipped with air springs to cancel the smallest vibrations. (Image courtesy of J. Fu, NIST.) (b) 12 cobalt atoms in a circle on copper. The interference of electron waves produces a daisy pattern. An instrument autonomously picks up and places the atoms. (Image courtesy of J. Strocio and R. Celotta, NIST.) (c) A microheater is used to detect toxic gases. Variations in the thickness cause changes in color. NIST's ALM facility will produce arrays of these sensors. (Image courtesy of NIST.) (d) MgO cubes decorated with gold nanoparticles are imaged by a new 3-D chemical imaging method using a scanning TEM, a tilting stage, and sensitive detectors. (Image courtesy of J. Bonevich and, J.H. Scott, NIST.) (e) Magnetic domains of new generation logic devices are depicted. Changes in color correspond to changes in the direction of the magnetic field. This image was taken 12 m underground by the highest resolution magnetic imaging instrument in the United States. (Image courtesy of J. Unguris, NIST.) (f) Interference colors indicate the thickness of a clear organic film. Deep blue-brown indicates ~1 μm thickness that corresponds to the thickness required to embed a particle for analysis. The work was accomplished in NIST's ultraclean room. (Image courtesy of C. Zeissler, NIST.) (g) An infrared image illustrates temperature variations as a round swipe cloth is heated to detect the presence of explosives. Dark areas correspond to 40°C. Lightest areas are around 200°C. (Image courtesy of G. Gillen, NIST.)*

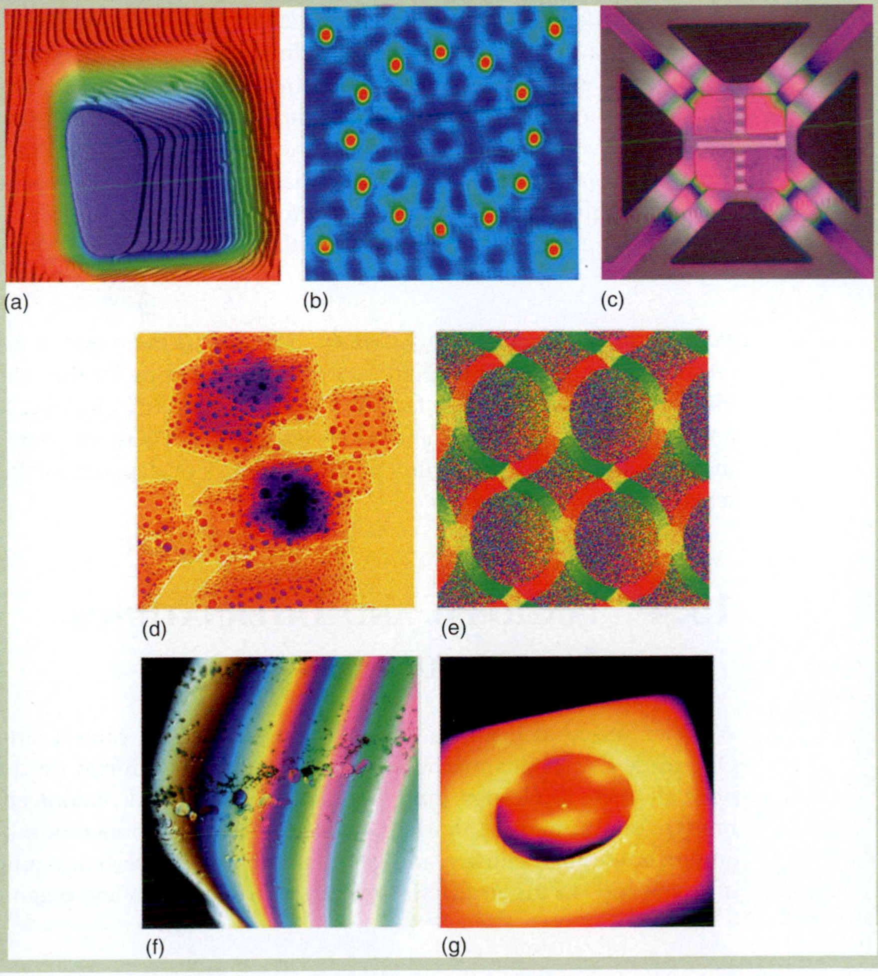

Center for Integrated Nanotechnologies (CINT). The CINT (also designed by HDR Architecture, Inc.) is located in Albuquerque, New Mexico and was constructed in 2006.

> The vision of CINT is to become a world leader in nanoscale science by developing the scientific principles that govern the design, performance, and integration of nanoscale materials.

CINT emphasis is to take scientific discovery and the integration of nanostructures into the micro- and macroscopic worlds.

CINT's core facility is a $9.8 million 35,600 sq. ft gateway located near two national laboratories: Sandia and Los Alamos. CINT is called a gateway facility because it is able to link two national laboratories with collaborative users from universities and industry. Nanophotonics, nanoelectronics, complex nanomaterials, nanomechanics, and nanoscale bio-micro interfaces are some of the divisions within the CINT facility. From the description of the focus areas given above, one can conclude that integration is indeed a prime directive at CINT—and revisiting the challenges defined earlier in this chapter, attempts to accommodate energy across multiple length scales, combination of top-down and bottom-up fabrication strategies, and interfacing biological and synthetic systems was certainly accomplished in good faith.

CINT has allocated 20,000 sq. ft for office suites, visitor accommodations, computer bays, and communication links. The synthesis wing consists of 15,000 sq. ft and houses chemical benches, hoods, equipment, and bench-top characterization tools. A 12,000 sq. ft integration wing possesses a *Class 100* clean room equipped with flexible fabrication. The characterization wing is a vibration-isolated 15,000 sq. ft facility that houses scanning probe microscopes, nanomechanics tools, laser optics, and microelectronics. Utilities, storage, and service space account for 20,000 sq. ft of the total space. The CINT overall is over 90,000 sq. ft.

Purdue's Birck Nanotechnology Center. We shall present one more facility, the recently constructed Birck Nanotechnology Center at Purdue University (by HDR Architecture, Inc.). The facility cost $45 million to build. It is 220,000 sq. ft and contains Class 10, 100, and 1000 clean rooms. There are over 100 labs and modules and nearly 100 faculty from 24 schools or departments participate in research at the facility.

15.4 National and International Infrastructure

We have reviewed business development, education, careers, and buildings. One more cog in this mechanism that requires introduction is a topic about how all these buildings (and people) are organized. Nanotechnology has brought about, more than ever before, the need for cooperation between and among academia, industry, and government, and between and among nations of our world. In the next few paragraphs, we review a few organizations and infrastructure that support advanced technology research, development, and societal implication awareness.

The logical step in the hierarchy is to create an infrastructure of user facilities (buildings, equipment, and instrumentation). There are several statewide and few nationally recognized infrastructure networks dedicated to nanoscience research and development. The new generation facilities in general support a collaborative environment and incorporates sustainable design solutions with flexible, modular designs.

15.4.1 Research and Development Organizations

The National Nanotechnology Infrastructure Network (NNIN). The NSF released a solicitation in 2003 to invite universities to submit proposals for membership in the NNIN. On March 1, 2004, the NNIN was officially launched. The NNIN is an integrated partnership of 13 user facilities that is dedicated to support nanoscale fabrication, synthesis, characterization, modeling, design, computation, and hands-on training [39]. The NNIN supports academia, small and large industry, and government in the capacity of a research facilitator with open access by providing leading edge tools, instrumentation, and expertise.

The member institutions include Cornell Nanoscale Facility, Stanford Nanofabrication Facility, University of Michigan Solid State Electronics Laboratory, Georgia Institute of Technology Microelectronics Research Center, University of Washington Center for Nanotechnology, Penn State Nanofabrication Facility, Nanotech at the University of California at Santa Barbara, the Minnesota Nanotechnology Cluster (MINTEC), Nanoscience at the University of New Mexico, the Microelectronics Research Center at the University of Texas, the Center for Imaging and Mesoscale Structures at Harvard, the Howard Nanoscale Science and Engineering Facility, and the Triangle National Lithography Center at North Carolina State University [40].

The NNIN, according to its Web site, "provides unparalleled opportunities for nanoscience and nanotechnology research." The NNIN provides extensive support in nanoscale fabrication, synthesis, characterization, modeling, design, computation, and hands-on training [39]. The NNIN embraces and promotes educational programs to inform the public, educate K-12 students, employ and train undergraduates, develop curriculum, develop technological workforce, train teachers, and numerous other programs and objectives.

The Department of Energy's National Science Research Centers (NSRC). The U.S. Department of Energy (DOE) has five centers for nanoscience (**Fig. 15.3**). (1) the *Molecular Foundry* (Lawrence Livermore National Laboratory) is involved in advanced light sources, electron microscopy, and the "nanowriter"; (2) advanced proton source, intense pulsed neutron source, and an electron microscope center for materials research are found at the *Center for Nanoscale Materials* (Argonne National Laboratory); (3) there is the *Center for Functional Nanomaterials* at Brookhaven National Laboratory; (4) the *Center for Nanophasic Materials* Sciences (Oak Ridge National Laboratory); and (5) Los Alamos and Sandia's Center for Integrated Nanotechnologies is involved with semiconductor research,

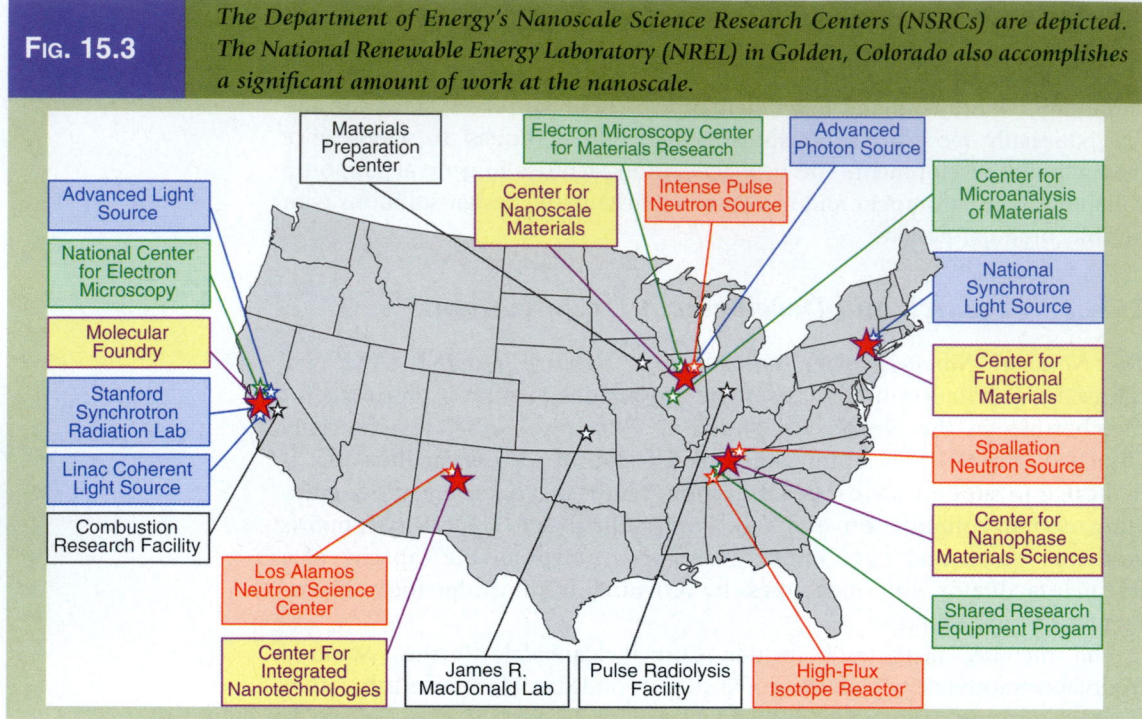

FIG. 15.3 *The Department of Energy's Nanoscale Science Research Centers (NSRCs) are depicted. The National Renewable Energy Laboratory (NREL) in Golden, Colorado also accomplishes a significant amount of work at the nanoscale.*

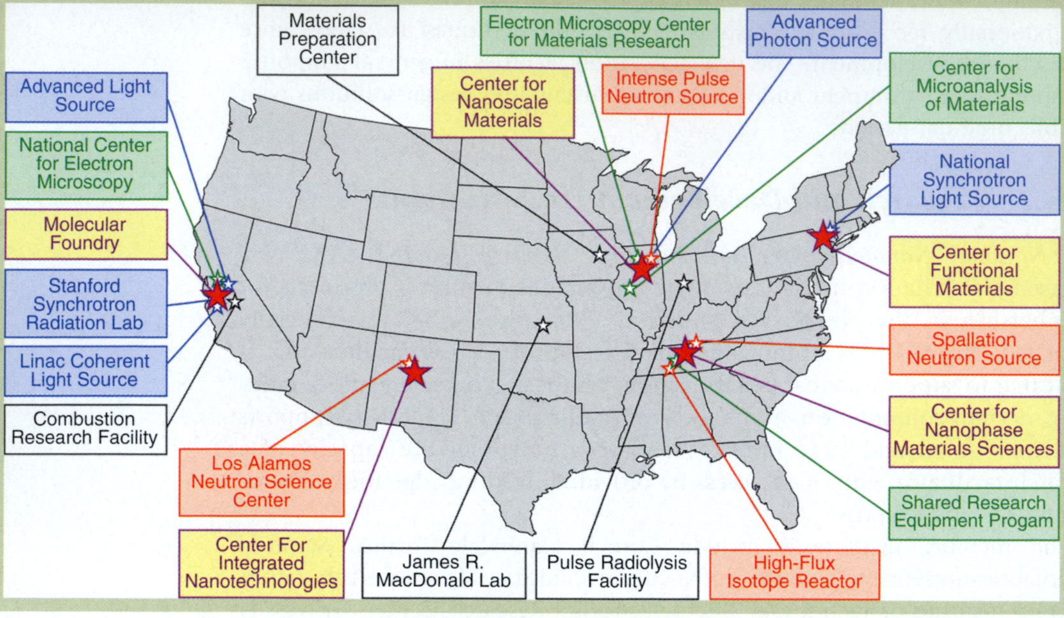

microelectronics, combustion research, and magnetic field laboratories. All of these facilities, and those of the NIST and other facilities in Europe and Asia, are concerned with integrating science and engineering across all scientific disciplines and across all size scales [28].

DOE's investment in nanotechnology is expected to exceed $500 million from 2007 to the present [40]. The Molecular Foundry, located near U.C. Berkeley, is a new center of high-powered nanotech R&D. The center contains clean room facilities, high-powered computing capability, highly sensitive equipment and instrumentation, and top-notch researchers [40].

15.4.2 Economic Development Organizations

The NanoBusiness Alliance (NBA). The NanoBusiness Alliance (www.nanobusiness.org), known as the "world's leading nanotechnology trade organization" made the national scene in nanotechnology business ca. 2002. The organization was founded by Mark Modzelewski and others who realized that nanotechnology was fast coming of age. The current executive director is Sean Murdock who recently testified before the House Science Committee in April 2008 in support of the National Nanotechnology Initiative Amendments Act of 2008.

15.4.3 Organizations Centered on Societal Implications

National Nanotechnology Initiative. The NNI (at www.nano.gov) was established in 2001. Mihail C. Roco is its executive director. The goals of the NNI are quite comprehensive and far-reaching:

- Advance a world-class nanotechnology research and development program
- Foster the transfer of new technologies into products for commercial and public benefit
- Develop and sustain educational resources, a skilled workforce, and the supporting infrastructure and tools to advance nanotechnology
- Support responsible development of nanotechnology

Although the NNI is involved in many other areas (e.g., the NNIN and other programs), its efforts have spearheaded recognition of the importance of societal implications of nanotechnology. A number of excellent resources are available from the NNI Web site. Please consult this Web site often and in detail. Intergovernmental agency and department efforts are under the auspices of the National Nanotechnology Coordination Office (NNCO).

Center for Responsible Nanotechnology (CRN) [41]. The Center for Responsible Nanotechnology is a nonprofit research and advocacy group that is focused on the major societal and environmental implications of nanotechnology. The group was formed in 2002 by Mike Treder and Chris Phoenix. They also stress the development, evaluation, and implications of molecular manufacturing (e.g., fourth generation nanotechnology). Their mission statement reads as follows.

> The mission of CRN is to: 1) raise awareness of the benefits, the dangers, and the possibilities for responsible use of advanced nanotechnology; 2) expedite a thorough examination of the environmental, humanitarian, economic, military, political, social, medical, and ethical implications of molecular manufacturing; and 3) assist in the creation and implementation of wise, comprehensive, and balanced plans for responsible worldwide use of this transformative technology.

Foresight Institute. The mission of the Foresight Institute is to "ensure the beneficial development of nanotechnology." The foresight institute is essentially a think tank and public interest institute, this according to their Web site at www.foresight.org. The group was founded in 1986 and lists as the foremost challenges (with solutions through nanotechnology):

- Providing renewable clean energy
- Supplying clean water globally
- Improving health and longevity
- Healing and preserving the environment (and maximizing productivity of agriculture)
- Making information technology available to all
- Enabling space development

In January of 2008, the Foresight Institute released a roadmap for nanotechnology development [42,43].

15.4.4 Nanotechnology News Services

A great way to keep abreast of nanotech news is to subscribe to a variety of free news services. Several notable services are listed below.

- Nanoscienceworks Newsletter (www.nanoscienceworks.org)
- Nanotechnology Now (www.nanotech-now.com)
- Nanotechweb.org, IOP Publishing (www.nanotechweb.org)
- Nature Nanotechnology (www.nature.com/nano/index.html)
- Nano Vip Newsletter (www.nanovip.com/nanotechnology-newsletter)
- Nanotechnology.com, the international small technology network (www.nanotechnology.com)
- Small Times Magazine (www.smalltimes.com and www.nanotechnews.com)
- Nanowerk Nanotechnology News (www.nanowerk.com)
- The A to Z of Nanotechnology (www.azonano.com)
- Nanodot (www.foresight.org/nanodot/)
- Nanotechnology News Network (www.nanonewsnet.com)
- Chemical & Engineering News (www.pubs.acs.org/nanofocus/)
- Nano Techwire (www.nanotechwire.com)
- Nano World News (www.nsti.org/news/)

The list keeps on growing and growing.

For business, the Forbes/Wolfe Nanotech Reports (www.newsletters.forbes.com), Scott Livingston's Axiom Capital Management (www.axiomcapital.com), and Lux Capital (www.luxcapital.com) should keep you updated about the investment aspects of nanotechnology.

15.4.5 International Organizations and Institutes

There are numerous organizations overseas. Europe has several that are involved in every aspect of nanotechnology. The list is by no means complete. Please find out what is going on in your world.

North America

- Nanotechnology—National Research Council Canada (www.nrc-cnrc.gc.ca)
- NRC National Institute for Nanotechnology (www.nint-innt.nrc.gc.ca)
- National Institute for Nanotechnology—University of Alberta (www.uofaweb.ualberta.ca/nint)
- Canadian NanoBusiness Alliance (www.nanobusiness.ca)

Europe

- Nanoforum.org: European Nanotechnology Gateway (www.nanoforum.org)
- Nanotechnology (cordis.europa.eu/nanotechnology)
- ENTA—European Nano Trade Alliance (www.euronanotrade.com)
- Institute of Nanotechnology (www.nano.org.uk)
- Swiss Federal Institute of Technology (Ecole Polytechnique Fédérale Lausanne, EPFL) (www.epfl.ch)

Asia

- Asian Institute of Technology (www.ait.ac.th)
- National Nanotechnology Center (NANOTEC) (www.asia-nano.org)
- Asia Nano Forum (www.asia-nano.org)

- The Australian Research Council Nanotechnology Network (www.ausnano.org)
- Nanotechnology Research Institute (NRI) (unit.aist.go.jp/nanotech)
- Nanotechnology India (www.indiannanotechnology.com)

Africa

- South African Nanotechnology Initiative (SANi) (www.sani.org)
- Focus Nanotechnology Africa, Inc. (www.fonai.org)

South and Central America

- Laboratorio Nacional de Nanotecnologia (LANOTEC) in Costa Rica (www.cenat.ac.cr/cenat)
- Brazilian Nanotechnology Networks
- Development of Nanoscience and Nanotechnology (Brazil)
- Nanoscience Millennium Institutes (Brazil)
- Fundacion Argentina de Nanotecnologia (FAN) (www.fan.org.ar)

15.5 Nanotechnology Products

Part II is about applications of nanoscience to commerce and industry. In other words, nanotechnology is about development and manufacturing of products enhanced by the remarkable properties of nanomaterials. We therefore present, to conclude the perspectives aspect of this text, a list of some products that you may or may not be aware that are associated with nanotechnology. Most of the products were found on the NNI's Web site, www.nano.gov. In addition, please consult www.nanotechproject.org/ for an extensive list of hundreds of products that have been enhanced with nanotechnology.

Automotive Industry

- Step assists, bumpers, paints, coatings, glare reduction, catalytic converters
- Cooling chips to replace compressors with no moving parts

Recreation

- Lighter stronger tennis racquets, long-lasting tennis balls, smart golf balls
- Nanotube reinforced masts for sailboats, new materials for hull and deck
- Golf shafts, skis, fog eliminators

Personal Use and Food

- Sunscreens, cosmetics, stain-free clothing
- Silver nanoparticle food storage containers, cutting boards, pans
- Nonstick bake ware
- Umbrella (based on lotus leaf)

Medicine, Therapeutics, and Hygiene

- Dental-bonding agents, burn and wound dressings
- Medical imaging with quantum dots
- Targeted drug delivery and gene therapy

- Water filters
- Lab-on-a-chip diagnosis
- Sanitized toilets

Structural Materials and Industrial Applications

- Stronger, lighter polymers; enhanced concrete, enhanced steel
- Wear-resistant nanoceramic coatings
- Catalysts
- Carbon nanotube-reinforced materials
- Various nano glues, nanoseal wood, nano-enhanced insulation
- Self-cleaning glass
- Exterior paint

Electronics and Computing

- Sub-100 nm transistors (old technology)
- Carbon nanotube triodes
- Organic LEDs and organic electroluminescent displays
- Cordless power tool batteries
- Carbon nanotube displays
- Protective self-assembling film layers for displays
- Cellular memory

Satisfy your own curiosity and research the amazing number of products out there that have already been enabled or enhanced with nanotechnology.

Acknowledgments

We would like to acknowledge Michael Burke, CEO & President of NanoThread, Inc. for his contribution to section 15.1. We are also indebted to the NNI for their central focused efforts to keep the United States abreast in the quest to commercialize nano.

References

1. R. Brazil, Nanotechnology—the issues: A response to the request for initial views on the Royal Society and the Royal Academy of engineering study on nanotechnology, Royal Society of Chemistry, London (2003).
2. M. C. Roco, J. Murday, C. Teague, S. Hays, and C. Merzbacher, The national nanotechnology initiative: Strategic plan, National Science and Technology Council Committee on technology, NSET (2004).
3. G. L. Hornyak, J. Dutta, H. F. Tibbals, and A. K. Rao, *Introduction to nanoscience*, CRC Press, Boca Raton, FL (2008).
4. J. Von Ehr, II, Comments at PCAST (President's Council of Advisors on Science and Technology) public meeting, Zyvex Labs, LLC (2007).
5. M. C. Roco and W. S. Bainbridge, eds., Societal implications of nanoscience and nanotechnology, Kluwer Academic Publishers, Boston, 3–4 (2001).
6. R. E. Smalley, Our energy challenge, Presentation, Columbia University, New York (2003).
7. R. Florida, *The flight of the creative class: The new global competition for talent*, HarperCollins, New York (2005).

8. L. V. Gerstner, Jr., Sputnik was nothing, *Newsweek*, November 28 (2005).
9. T. K. Kelly, W. P. Butz, S. Carroll, D. M. Adamson, and G. Bloom, The U.S. scientific and technical workforce: Improving data for decision making, Conference Proceedings, RAND Science and Technology, RAND Corporation (2004).
10. M. C. Roco and W. S. Bainbridge, *Nanotechnology: Societal implications—maximizing benefits for humanity*, Report of the National Nanotechnology Workshop, December 2–3, 2003, Arlington, Virginia; National Science Foundation, Springer Science & Business Media (2005).
11. P. F. Drucker, *Landmarks of tomorrow: A report on the new "post-modern" world*, Harper, New York (1959).
12. R. E. Kelley, *The gold-collar worker: Harnessing the brainpower of the new work force*, Addison-Wesley, Reading, MA (1985).
13. M. A. Roe, Cultivating the gold collar worker, *Harvard Business Review*, 79, 32–33 (2001).
14. M. Jamison, HDR, Inc., Buildings for nanotechnology research, Presentation, Colorado Nano/Micro Summit, Boulder CO, May 24 (2004).
15. G. Elvin, Risks in architectural applications of nanotechnology, Nanowerk Spotlight, www.nanowerk.com/spotlight/spotid=1007.php (2006).
16. T. H. Wegner, J. E. Winandy, and M. A. Ritter, Nanotechnology opportunities in residential and non-residential construction, USDA Forest Products Laboratory, 2nd International Symposium on Nanotechnology in Construction, Bilbao, Spain (2005).
17. A. Marsh, Nanotechnology, green building, sustainable design musts for world survival, profit, *Commercial Property News*, www.cpnonline.com/cpn/specialties/article_display.jsp?vnu_content_id=1003382018 (2006).
18. Opportunities for nanotechnology, www.nanovic.com.au/?a=industry_focus.building (accessed 2008).
19. G. Elvin, The nano revolution, A science that works on the molecular scale is set to transform the way we build, *The Magazine*, Architect OnLine, www.architectmagazine.com/industry-news.asp?sectionID=1006&articleID=492836&artnum=1 (2007).
20. Green building, en.wikipedia.org/wiki/Green_building (2008).
21. Green buildings, U.S. Environmental Protection Agency, www.epa.gov/greenbuilding/ (2007).
22. Building design guidance, U.S. Environmental Protection Agency, www.energystar.gov/index.cfm?c=new_bldg_design.new_bldg_design_guidance#construction (2007).
23. L. M. Froeschle, Environmental assessment and specification of green building materials, *The Construction Specifier*, October, 53 (1999).
24. D. M. Roodman and N. Lenssen, A building revolution: How ecology and health concerns are transforming construction, Worldwatch Paper 124, Worldwatch Institute, Washington, DC, March, 5 (1995).
25. R. Spiegel and D. Meadows, *Green building materials: A guide to product selection and specification*, John Wiley & Sons, Inc., New York (1999).
26. G. Dick, Green building materials, California Integrated Waste Management Board, www.ciwmb.ca.gov/GreenBuilding/Materials/ (2007).
27. Leadership in energy and environmental design, U.S. Green Building Council, LEED Rating Systems, www.usgbc.org/ (2008).
28. N. D. Shinn, CINT: A new model for a nanoscience research user facility, CINT-Sandia National Laboratories, Buildings for Advanced Technology II, Workshop, January 21–23, Mesa, AZ (2004).
29. N. Toussaint, Establishing and maintaining critical environments, HDR, Inc., Presentation, Buildings for Advanced Technology, Mesa, AZ, January 22 (2004).
30. Cleanroom, en.wikipedia.org/wiki/Clean_room#ISO_14644-1_cleanroom_standards (2007).

31. D. Bechtol, Electrical power and grounding, PE, HDR Architecture, Inc., Buildings for Advanced Technology II, Workshop, Mesa, AZ (2004).
32. H. Amick, Vibrations.... should i worry? What about? Colin Gordon & Associates, Buildings for Advanced Technology II, Workshop, Mesa, AZ (2004).
33. H. Amick, On generic vibration criteria for advanced technology facilities: With a tutorial on vibration data representation, *Journal of the Institute of Environmental Science*, 40, 35–44 (1997).
34. J. M. Proenca and F. Branco, Case studies of vibrations in structures, *Revue Européene de Génie Civil*, 9, 159–186 (2005).
35. C. G. Gordon, Generic vibration criteria for vibration sensitive equipment, International Society for Optical Engineering (SPIE), Conference on Current Developments in Vibration Control for Optomechanical Systems, Denver, CO (1999).
36. A. Yazdanniyaz, Mechanical system noise issues: Case studies, Arup, Presentation, Buildings for Advanced Technology II (BAT-II), Workshop, Mesa, AZ, January 22 (2004).
37. E. M. Vogel, NIST advanced measurement laboratory (AML) Nanofab, Presentation, National Institute of Standards and Technology (2004).
38. M. Baum, NIST launches advanced measurement laboratory: Research environment among the world's best for nanotech, leading-edge science, NIST News Release, www.nist.gov/public_affairs/releases/aml_dedication.htm (2004).
39. National Nanotechnology Infrastructure Network (NNIN), www.nnin.org (2004).
40. V. McCarthy, DOE's nano-centric research centers set to fuel new era in nano collaborative, Nano World News, Nano Science and Technology Institute (NSTI), www.nsti.org/news/item.html?id= 127 (2007).
41. Center for Responsible Nanotechnology, www.crnano.org/about_us.htm (2008).
42. K. E. Drexler, J. Randall, S. Corchnoy, A. Kawczak, and M. L. Steve, eds., Productive nanosystems: A technology roadmap, Battelle Memorial Institute and Foresight Institute (2007).
43. From here to there: Nanotechnology roadmap, Press Release, http://foresight.org/roadmaps, January 29 (2008).

Problems

15.1 What is the combined decibel level if you are talking in a restaurant at 70 dB and the adjacent noise from the kitchen contributes another 70 dB?

15.2 With regard to vibrations—acceleration, velocity, and displacement are all related by simple equations. How so?

15.3 Vibrations that influence equipment range from 1 to 100 Hz. Give examples of sources of 1-, 25-, 50-, and 100-Hz vibrations.

15.4 Clean rooms of Class 100 indicate that there are no more than 100 0.5-μm particles per cubic foot. What applications would require such tight control and why?

15.5 What products can you think of that cannot be enabled or enhanced by nanotechnology?

15.6 a. If someone is exposed to sound intensity of 1×10^{-12} W·m^{-2}, how does this translate into decibels? b. What is the intensity of sound in terms of W·m^{-2} of jet plane noise at takeoff if the recorded decibel level is 140?

15.7 What is your vision of a building (research center) of the future?

15.8 List all potential employers that involve nanotechnology but are not involved in science, technology, or manufacturing—for example, an intellectual property attorney.

15.9 List three reasons why it is not a good idea to place a TEM room next to a loading dock. (Hint: There is one subtle reason you might miss).

15.10 Is nanotechnology an industrial revolution in the making?

15.11 When starting a nano-based business, discuss the importance of partnerships.

15.12 Research the economic cluster model in Arizona. What do you think?

15.13 Do you agree that government should be heavily involved in funding nanotech? Why or why not?

Nanometrology: Standards and Nanomanufacturing

Whether you can observe a thing or not depends on the theory which you use. It is the theory which decides what can be observed.

I think that a particle must have a separate reality independent of the measurements. That is an electron has a spin, location and so forth even when it is not being measured. I like to think that the moon is there even if I am not looking at it.

Albert Einstein

Chapter 16

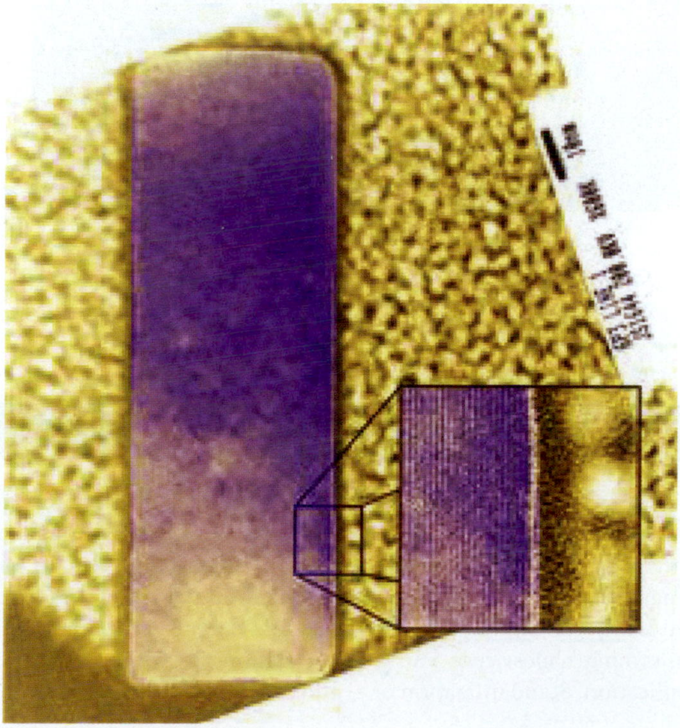

Fig. 16.0

James R. Von Ehr, II is a nano-entrepreneur. As the CEO and chairman of Zyvex Corporation, he was a major contributor and founder of the NanoTech Institute at the University of Texas at Dallas, and the founder of the Texas Nanotechnology Initiative in 2000. He was a major force behind the twenty-first century Nanotechnology Research and Development Act signed into law on December 3, 2003. Jim von Ehr is, and in no uncertain terms, one of the prime movers of nanotechnology in the United States.

Threads

Nanomanufacturing is where the "rubber meets the road." It is one manifestation of the proverbial "moment of truth." It is the point of "put up or shut up." It is the reality check of nanoscience. It is the discriminator between hype and reality. Nanomanufacturing is the process of transforming nanoscience into a useful material, an application. Standardization of practice, product performance and packaging specifications accompany any manufacturing process, and products that are made or will be made by nanomanufacturing are no exception to this rule. The first chapter introduced nanotechnology and its relations to business development, education and workforce development and provided some

Threads (Contd.)

insight into where nanotechnology is housed and how it is trekking across our globe. This chapter, the second chapter in the *Perspectives* division of the text, provides relevant background in metrology, standards and manufacturing.

One of the links between and among science, technology and commerce is metrology. Nanometrology is no exception to this rule and is the embodiment of measurement at the nanoscale. Nanometrology and nanomanufacturing go hand in hand. Industry standards forge a necessary link between the two and are therefore an integral part of any commercial process. In this chapter, we introduce all three and dwell on their interrelationship. Metrology at the nanoscale presents new operational paradigms—ones that link the macroscopic and quantum world like never before. Quantum standards of all things! Our science and technology now exist in a world where quantum standards are required. How far we have come indeed! Measurement of long held standards such as the meter and kilogram are now defined from the bottom-up, from quantum mechanics and quantum phenomena.

Following the *"Perspectives"* division, the next three chapters delve into the electromagnetic manifestations of nanotechnology that comprise the *"Electromagnetic Nanoengineering"* division of the text.

Nanomanufacturing is the next great threshold of nanotechnology. As our capabilities move beyond creating simple aggregations of "nano-stuff", we find ourselves wanting to control quality and repeatability, just as we do in conventional manufacturing. It is not enough to simply make small things; we need to control the process that makes those things so the things perform the right functions. Ideally, we want to control nanotechnology to the point where we can make large quantities of identical things, using the discreteness of matter at the atomic scale to achieve our goal of adaptable, affordable, atomically precise manufacturing. Living cells make proteins in this manner, getting every atom in the right place, using molecular-scale manufacturing factories called ribosomes. The flexibility of living systems in manipulating molecules to transform the sun's energy into useful chemical forms, and to use those chemicals to transform inert minerals into structures as varied as diatom shells, oak trees, or elephants, gives us an existence proof that this technology is both feasible and valuable.

Metrology is important to manufacturing at any scale. We must be able to measure what we're building, if we hope to control our manufacturing processes. With nanotechnology, this is more challenging than normal, due to the size scales. We can't use light-based observation, since the structures we want to observe are usually smaller than the wavelength of light. We either need exotic techniques like high-energy accelerators, or must build probe-based tools that reach into the nanoworld in order to touch, manipulate, and measure objects. As an action-oriented company, Zyvex has chosen the "reach into the nanoworld and manipulate it" approach.

The challenge for tools companies in the nano space is to survive and grow on what are still tiny markets. We will advance the field quickly and all thrive if we can innovate new tools and new markets. Innovative companies can quickly become market leaders. As we move to a world of "digital matter", innovation and speed of execution become increasingly important to a company's success.

Opportunities abound, and as we get better at nanomanufacturing, we will come to find that "imagination is our only limit." Physics and chemistry provide constraints, but human ingenuity lets us work within those constraints to achieve our aims, whether those are flying hundreds of people across the ocean at nearly the speed of sound, videoconferencing to someone on the other side of the world, or my personal favorite, eating fresh red raspberries in the middle of winter. We accomplished all those because creative people said "I can…"

James Von Ehr, II,
CEO and Chairman, Zyvex Corporation

16.0 The Transition, the Need

Many great ideas sit on shelves in university labs, offices, and libraries—great ideas that never make it to the next level due to a variety of factors: (1) the idea was ahead of its time, e.g. the manufacturing technology required was not available, (2) gap funding to develop *definitive proof-of-concept* was not acquired in a timely manner and the idea melded into oblivion, (3) the transfer of technology between the university and a commercial venture was not facilitative or effective, (4) although characterized as a valid realistic technology, its shelf life expired, or (5), the road to commercialization is too long (**Fig. 16.1**). Unavoidably, some ideas turn out to be not so great, not so practical. There are numerous examples.

Fig. 16.1 *An idea, a concept, a process, a device, or an invention is valued highly by the scientist or inventor (and the university) who developed it (red line). A businessperson, on the other hand, views that same idea, invention, etc., as a project that requires investment and development, complete with all the ancillary business trimmings. In other words, that same idea, concept, or invention is not worth much as much to the business developer as it is to the inventor—a disconnect—even though it may show great promise down the road (blue line). It is important that these two extremes meet somewhere at sometime—preferably early on. As time goes by and as the product is developed, it gains in value (blue line). If it sits on a university shelf, it loses value over time (red line).*

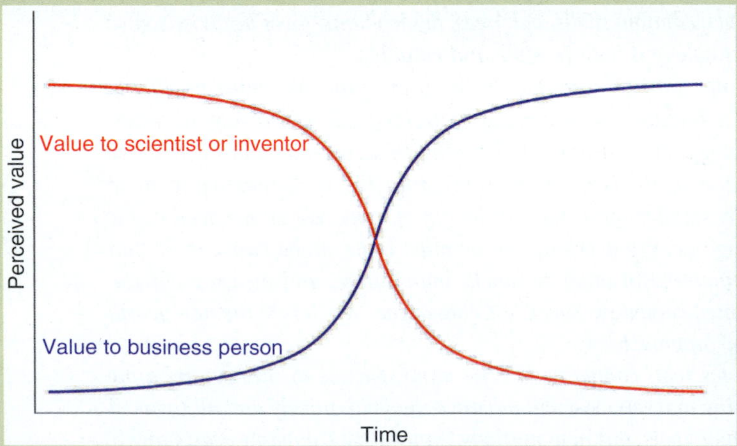

Intellectual property, technology transfer, and business development are just several facets of the process. There are also practical aspects of transferring a laboratory prototype into a real application that involve aspects such as measuring, standardizing, scale-up, manufacturing, reliability/testing, and packaging. For a nanoscience to become a nanotechnology, we need to understand metrology, and engineering and manufacturing practices as they apply to nanotechnology and how best to construct the equipment, instrumentation, facilities, and buildings that accommodate them. Numerous metrology and manufacturing challenges must be addressed before we are ready to place our stamp on the next industrial revolution.

16.0.1 Background to Nanometrology

Metrology has been with us forever. Our need to measure and standardize made civilization what it is today. Nanometrology is a natural extension of metrologies that sprung from microscale R&D and manufacturing. However, like never before, the face of metrology is changing dramatically—in a way that is driven by nanotechnology and the need to measure parameters that reside on the fringe of the limits of uncertainty and theory. Welcome to the *Nano Age*.

History. In antiquity, the length of one's forearm was known as a *cubit*. Many cultures adopted unique versions of the cubit. Day-to-day things were measured and recorded by comparing them to various parts of the body. The human experience is rich with examples of such whimsical measurement. The Greeks popularized the *foot* as a means of placing physical dimensions on paper. Of course, the measure depended on the size of an individual's foot—hardly a standardizable commodity. Horse height (at the shoulder) is still measured to this day in terms of *hands*—the height of which depends, however, on whose hand is abstracted for the measurement. One of the best systems developed to measure length (down to ~1.7 mm resolution) is credited to cultures inhabiting the Indus Valley several millennia ago. With the onset of commerce that went beyond simple barter, the need for standardization assumed greater importance.

The Romans popularized the *natural cubit*, about 1.5 ft. The Roman foot was divided into twelve *unciae* (inches), each about 25 mm in length [1]. The *mille passus* (or 1000 paces, each pace equal to 5 ft) and the mile (5000 ft) were adopted by the early Britons during the Roman occupation [1]. The *mille passus* was changed later into the statute mile (5280 ft) by edict of Queen Elizabeth I. Mass was measured first in terms of the *grain* (of wheat?). The Romans introduced the *libra* (one pound) from which the symbolism *lb* is obtained. The Roman pound was divided into twelve *uncia* (ounces). Imperial units became official following passage of the *British Weights and Measures Act of 1824*. To this day, with small modifications here and there, the Imperial system survives. The sexagesimal system of numbers dates back to the ancient Babylonians who divided the circle into 360°. Hence, we refer to degrees, minutes, and seconds much like cultures did so in antiquity.

Ironically, the metric system was first defined by John Wilkinson of England in 1668. The French adopted this system, based on the number 10, in 1791, with some influence from Benjamin Franklin two years after the onset of the French Revolution. The rest is history. Mixing the Imperial system with the metric system, however, has caused disasters. In 1983, metric conversion errors resulted in a Boeing 767 jetliner to run out of fuel. The pilots, according to the *New York Times*, were able to make a dead-stick landing [2]. In another notorious incident, NASA's Mars $125 million orbiter landed (crashed) prematurely due to metric-to-Imperial units conversion errors [3].

Metrology. Metrology is the science of weights and measures [4]. According to the U.S. Military Dictionary [5], metrology is

> ...the science of measurement, including the development of measurement standards and systems for absolute and relative measurements.

According to the *International Bureau of Weights and Measures, IBWM* (or *Bureau International des Poids et Mesures*)

> Metrology [from Greek metron 'measure', and logos 'study of'] is the science of measurement. Metrology includes all theoretical and practical aspects of measurement... the science of measurement, embracing both experiment and theoretical determinations at any level of uncertainty in any field of science and technology.

Metrology therefore has a scientific (fundamental) component as well as an industrial (applied) one. According to the IBWM [6]:

> Scientific or fundamental metrology concerns the establishment of measurement units, unit systems, the development of new measurement methods, realisation of measurement standards and the transfer of traceability from these standards to users in society. Applied or industrial metrology concerns the application of measurement science to manufacturing and other processes and their use in society, ensuring the suitability of measurement instruments, their calibration and quality control of measurements.

OK, we get the picture. We now define nanometrology. An excellent resource is available from Ref. [7], Nanometrology, IBWM, bipm.org/en/home/(2007) [7].

Nanometrology. Nanometrology, of course, is metrology at the nanoscale [8].

> ...or particles to self-assemble at the nanoscale to form new materials with unusual properties. Nanometrology, i.e., the ability to conduct measurements at these dimensions, to characterize the materials—that, together, will form the metrological infrastructure essential to the success of nanotechnology. The MSEL Nanometrology Program incorporates basic measurement metrologies to determine material properties, process monitoring at the nanoscale.

MSEL is the acronym for the *Materials Science and Engineering Laboratory* at NIST, the *National Institute of Standards and Technology* (formerly the *National Bureau of Standards, NBS*).

According to MSEL [8],

> No previous materials technology has shown so prodigiously a potential for concurrent advances in research and industry as does the field of nanomaterials in mechanical devices, electronic, magnetic and optical components, quantum

computing, tissue engineering and other biotechnologies, and as-yet unanticipated exploitations of as-yet undiscovered novel properties of nanoscale assemblies and particles.

The MSEL Nanometrology Program to develop nanometrology infrastructure is straightforward—to incorporate preexisting basic measurement metrologies applied for material properties, process monitoring, nanomanufacturing, analytical techniques, advanced imaging, and multiscale modeling [8].

The MSEL effort, obviously, is intended to cover a wide range of nanometrological domains: (1) *mechanical property measurement* such as elastic properties, plastic deformation, adhesion, friction, stiction, and tribological behavior; nanoindentation (thin films and nanostructures); the use of AFM, acoustic microscopy, surface acoustic wave spectroscopy, and Brillouin light scattering; development of micro- and nanoscale structures and test methods to measure strength and fracture behavior of interfaces and materials having very small volumes, (2) *chemical and structural characterization and imaging* with neutron and x-ray beams, scattering and spectroscopy methods, chemical bond identification and orientation, polyelectrolyte dynamics, and equilibrium structures; and development of electron microscopic techniques for 3-D imaging, advanced scanning probe microscopies, and physical probing of cell membranes with carbon nanotubes, (3) *fabrication and monitoring of nanomanufacturing processes* such as electrochemical and microfluidic fabrication methods; nanocalorimetrics to study interfacial reactions, in situ observations of nanoparticle and nanotube dispersion, and alignment and advanced instrumentation for nanotribological experimentation, and (4) *theory modeling* to predict material behavior from the nanoscale to the macroscale; development of finite-element analysis, multiscale Green's function methods, classical atomistic simulations, ab initio and quantum mechanical density functional theory (DFT); and interfacing of models at different length scales to ensure accuracy of physics of components and systems [8].

Terminology. There are several terms associated within the domain of metrology that one should become familiar. We list a few important ones.

1. A *measurand* is the quantity that is subjected to measurement—electric current, length, density, etc.
2. *Accuracy* is the ability to obtain the true value of a measurement. It is the agreement between measurement and the true value. It does not reflect upon the quality of an instrument [9].
3. *Precision* is the ability to repeat measurements in the same way. Precision reveals information about the quality of an instrument.
4. *Uncertainty* is a parameter associated with the result of a measurement—a confidence level within which the measured value is thought to reside. It is the interval around the value that is reinforced by further measurement. The combined distribution of measurements is assumed to take normal form. Experimental uncertainty due to random measurement effects is different from systematic errors manifested by equipment or independent sources. *Uncertainties* and *probabilities* go hand in hand, for example, the measured value has the stated probability to lie within the confidence interval [10].

5. *Error* is the disagreement between a measurement and the true value of a measurand. In some ways, it is the opposite of accuracy. Errors arise from a number of scenarios: Abbe error, thermal expansion, repeatability, human error, vibrations, and surface contacts. These are not the same as uncertainties. Do not confuse *error* with *uncertainty*, for example, a measurement with a large uncertainty may have negligible error.
6. *Calibration* is the process of determining and documenting the deviation of a measured result from that of an accepted standardized "true" value [10].
7. *Traceability* is a process whereby the measured result can be compared to a standard for the specific measurand at one or subsequent stages. The tolerance on a product is usually looser than that of the primary standard, shown below [10].

   Product → Manufacturer's Test Equipment → In-House Calibration Laboratory → Accredited Calibration Laboratory → National or International Standard

   Measurements derived with a scanning probe microscope provide an example of a *traceability chain*. The chain, starting with the definition of the meter and ending with SPM (scanning probe microscopy) measurements, proceeds as follows [11]:

   Definition of the SI unit, the Meter → Realization of the Meter → Calibration of the Laser Frequency → Interferometrically Traceable AFM → Calibration of Physical Transfer Standards (1-D pitch, 2-D pitch, flatness, and step-height standards) → SPM Calibration → Traceable SPM Measurements

8. *Accepted values* are values generally agreed upon as the standard. Accepted values may not be the true values. Question everything (within reason of course)!

We leave it up to the student to review other concepts and axioms of statistics such as *significant figures*, the *mean*, the *median*, *relative uncertainty*, and the *standard deviation*.

The number of tools involved in nanometrology is enormous and we will cover a few general kinds. Electron beam and scanning probe instruments, considered to be the workhorses of nanotechnology research and development, will be discussed in more detail of course. These instruments are in every way the purest methods of nanotechnology. Specialized spectroscopic techniques continue to serve nanotechnologists.

The field of nanometrology is enormous but of appropriate magnitude to support this apparently vaster field of nanotechnology. Although we are interested in measuring that which is very small, we are also interested in measuring things well beyond the smallness of nanoparticles—the single bond, the single molecule or atom, and the single electron transfer. We are also interested in how these collective atomic and molecular events come together to determine the behavior of bulk materials—from the bottom up. Metrology unconsciously takes place in all of nature. Metrology, for us humans, is required at ALL levels of complexity within the material continuum with which we deal on a daily basis.

16.0.2 Background to Nanomanufacturing

Nanofabrication and *nanomanufacturing* processes are, in many ways, the natural evolutionary extension to the nanoscale of preexisting manufacturing and micromanufacturing practices. The terms "synthesis" and perhaps "nanofabrication" are more appropriately applied in the laboratory setting, while the term "nanomanufacturing" resides as the exclusive domain of nanotechnology—in the factory and assembly line.

Several recently developed techniques are unique to nanoscale fabrication and manufacturing, and therefore must be considered to be innovative (if not revolutionary). Nanofabrication and nanomanufacturing are processes by which nanoscale materials or components are produced from the bottom up (from atoms and molecules), from the top down (from bulk materials) in small steps of high precision or by various mixtures of the two [12]. *Molecular manufacturing* (via the molecular assembler route) is a subset of the general field of bottom-up nanomanufacturing and is discussed in more detail later on in the text.

The National Science Foundation (NSF) established the *NanoManufacturing Program* in 2001 [13]. The NSF program objectives include (1) scale-up of nanotechnology for high rate production, reliability, robustness, yield, efficiency, and cost, (2) integration of nanostructures in functional microdevices and meso/macroscale systems and interfacing across dimensional scales, (3) interdisciplinary research, multi-functionality across energetic domains (mechanical, thermal, fluidic, chemical, biochemical, electromagnetic, and optical), (4) systems approach encompassing nanoscale materials, structures, fabrication and integration processes, production equipment and characterization, theory, modeling, simulation and control tools, biomimetic design, and (5), nanomanufacturing education and training of the workforce, involvement of socio-economic sciences, addressing the health, safety, and environmental implications, development of manufacturing infrastructure, as well as outreach and synergy of the academic, industrial, federal, and international communities. The program has a special interest in environmental, health, and safety aspects, as well as ethical, legal, and societal implications of nanomanufacturing [13].

Several centers for nanomanufacturing have cropped up around the globe (chapter 1). The mission of the NSF Center for High-Rate Nanomanufacturing (CHN) centered at Northeastern University in Boston, Massachusetts (plus the University of Massachusetts at Lowell and the University of New Hampshire) is to develop tools and processes that enable high-rate/high-volume bottom-up, precise, parallel assembly of nanoelements and polymer nanostructures [14,15]. Their fundamental purpose is to develop means of bridging the gap between nanoscience and nanotechnology and to provide education about nanomanufacturing and its surrounding concerns—environmental, economic, societal implications, and development of an emerging workforce [14,15]. The CHN has constructed a state-of-the-art facility within which to accomplish its twenty-first century manufacturing—the *George J. Kostas Facility for Micro and Nanofabrication* [16].

Metrology and manufacturing are bound tightly to one another—it would be impossible to manufacture any material or device without guidelines that are embedded in previous metrology and previous manufacturing. During the *Industrial Revolution* (ca. 1800 to 1920), a manufacturing process known as "mass production" was introduced. Clearly in order to mass produce a product, the product has to be the same size, shape, and function, all within some degree of uniformity. The low-limit tolerance specified by designers in those bygone days was in the neighborhood of 1 mil (25 μm) [17]. Clocks, firearms, automobiles, sewing machines, and numerous other machines of the industrial revolution were manufactured whole scale within such low tolerances. Manufacturing processes included machining, stamping, casting, forging, and other techniques that assembled interchangeable machine parts—all doable within a 1-mil tolerance [17]. The mechanical Vernier caliper was the measurement device of choice used to transfer design specification metrology to the stamp, gauge block, cast, and cutting tool [17].

The *Semiconductor Revolution* arrived on the scene in the early 1950s and its metrology was based on the micron, for example, the microcircuit. As we all know so well, the entire computer industry is based on semiconductor materials and integrated circuits. Manufacture of the integrated circuit requires planar multileveled lithographic processes that in turn require accurate patterning, transfer, and alignment [17]. Operational length scale down to 100 nm was made possible by optical laser interferometer technology. The optical-based laser early on and electron-based beam later were the vehicles of choice employed to transfer design specification metrology to the mask and the resist materials. Micromanufacturing has since employed more sophisticated pattern transfer technology that utilizes ion beams and higher energy optical sources. The evolution of these technologies helped prepare the way to the domain of the nanoworld.

The *Nanotechnology Revolution* has, in essence, just begun. Who can say with certainty that the nanotech revolution sprung into the scientific mainstream in 1990, 1995, or 2000 or whenever? We shall leave that debate to science and industrial historians. Regardless, we are on the verge of mass-scale production of nanosystems that consist of electronic, photonic, magnetic, mechanical, chemical, biological materials, and devices and all possible permutations and combinations thereof [17]. We have actually already breached that threshold with the sub-100-nm computer transistor since 2003 and as Intel's latest transistor is a remarkable 30 nm in size.

With nanotechnology, we no longer operate in the pure world of classical physics. The length scale is now that of the nanometer. Like with anything else, associated uncertainties exist at the nanometer level but also at that of the subnanometer. Accuracy is transferred to products by optical, electron, and atomic imaging; and microscopes capable of viewing the nanodomain. Our cutting tools have most certainly changed over the past hundred years or so—when exactly was the transition between one that was purely mechanical to one that is based on focused ion beams?

16.0.3 The Nano Perspective

No matter how one adheres to the evolutionary theme, traditional challenges assume special "revolutionary" significance as we attempt to interface more intimately with the nanoworld. According to the NNI (National Nanotechnology Initiative) report *Instrumentation and Metrology for Nanotechnology* [18], revolutionary advances are required in order to realize products and manufacturing processes based on nanotechnology. What form will these revolutionary developments take? Are there undiscovered analytical techniques beyond the current paradigms or will techniques and instrumentation evolve from preexisting technology with enhanced (revolutionary) capabilities? A combination of both is most likely.

Serious advances have contributed to nanocharacterization over the past few decades [19]. Atomic scale chemical sensitivity and mapping, structure and morphology determination (by tomography), electron holography (for electrical, magnetic, and thickness properties at high resolution), dual-characterization and fabrication tools, and higher data collection rates and sample analytical throughput are just a few of the numerous advances—above and beyond the development of the scanning and transmission electron microscopes and the scanning probe techniques.

There are a handful of key challenges facing nanomanufacturing and nanometrology [14,15,19]:

R&D Laboratory

- Development of communication between nano- and micro (and macro)-domains
- Assembly and connection between and among nano-elements and -devices
  - Made of similar or different materials
  - 2-D interfaces
  - 1-D electronic or mechanical connections
- Integration of top-down and bottom-up processes
- Development of higher-capacity computing systems and advanced modeling techniques
- Reproducibility and repeatability from lab to lab
- Blending of classical physics and quantum properties
  - Quantum entanglement
  - Heisenberg uncertainty issues
  - Scaling laws
- Characterization methods and instrumentation that require
  - Multifaceted instrumentation development
  - Enhanced 3-D characterization
  - Significantly enhanced characterization speed
  - Nondestructive interfacial characterization
  - In situ characterization
  - Quantitative measurement of dispersion of embedded nanomaterials
  - Complete in situ identification and tracking of cellular processes
  - Greater resolution, accuracy, and precision
  - Data recording without artifactual interference
  - Enhanced computer interfaces
- Physical properties
  - Ohmic contact
  - Heat transfer
  - Sticking and friction
  - Fluidics
  - Cooling
  - Surface area mitigation

Manufacturing

- Manufacture of nanocomponents and devices with high throughput
- Uniformity (bottom-up fabrication has high throughput but domain size is limited)
- Reliability (although progressive, relatively unknown at this time for many processes)
- High rate scale-up potential
- Contamination (do we need better than 100-class cleanrooms?)
- Efficient detection of defective elements and domains and their subsequent mitigation
- Packaging and stability
- Development of more nanometrology standards

Societal

- Technology transfer (systems need to be streamlined to keep up globally)
- Patent processing (new intellectual property paradigms with streamlined processing)
- Skilled workers? engineers and scientists? Where will they come from?
- Redefinition of partnerships and funding increases
- Occupational health and safety concerns
- Environmental safeguards
- Other societal elements

How does one make a device that is governed by quantum behavior, one that is susceptible to weak van der Waals forces, one that is influenced by thermal fluctuations in its immediate vicinity, or a device that may not comply with the second law of thermodynamics like we are accustomed? Does Nature possess inherent metrology that is able to provide quality inspection to all of its machinations? What happens to defective nanostructures in the biological domain? I think we are well aware of the answer to the last query. Nanotechnology has already made the scene in the academic realm and is contributing more and more to industrial applications. The precise control and transfer of materials and devices to nanodimensions is a key concern as sub-nanometer control is becoming more frequent.

The National Science and Technology Council in their report "Science for the 21st Century" stresses the need for nanometrology development and standards [20]. Along with this need, a parallel requirement to build laboratories and nanofabs is imperative. Nanometrology requires special facilities—facilities that have rooms without vibration, with enhanced electromagnetic shielding and thermal control to tolerances less than 0.1–0.01°C [21]. Nanotechnology is tagged with the moniker "revolutionary." This may or may not be true across the board, but it is most certainly setting the world of metrology on its head.

16.1 Nanometrology and Uncertainty

"The history of science is the history of measurement," so stated James M. Cattell, one of the founders of modern psychology [22,23].

> If Professor Cattell had the opportunity to read Porter (1995, 2003)[24], then he might have extended his statement by saying that the history of measurement is also the history of commerce and government as well as the history of many other significant aspects of modern social and political life.

... so stated George Englehard, Jr., author of *Historical views of the concept of invariance in measurement theory* [23]. Without measurement we have no quantitative information. We only have information based on purely qualitative descriptions. This is, however, not to understate the importance of qualitative observation and recording. Standardized quantitative measures are extremely important and are usually the gateway to in depth analysis of a phenomenon or property.

Metrology is divided into four general groups: (1) theory, (2) techniques, (3) instrumentation, and (4) legislation (standards) [25]. The demands on the range of knowledge and metrological skills of engineers and scientists have increased dramatically. Ironically, this was stated in 1986, more than 20 years ago [26]. One can just imagine the importance placed on accuracy and precision in acquiring measurements at the nanoscale.

SI Units. The seven SI base units are the meter (length, x, l, r, or m), the kilogram (mass, m or kg), the second (time, t or s), the ampere (electric current, I, i, or A), the Kelvin (temperature, T or K), the mole (amount of substance, n or mol), and the candela (luminous intensity, I_V or cd). These parameters, however, rely on one another in several ways and are therefore not entirely independent. The ohm (Ω), for example, is in terms of $m^2 \cdot kg \cdot s^{-3} \cdot A^{-2}$. All of these base SI units are directly relevant in nanometrology. We are now interested in how our perspective of them may become altered.

We are mostly familiar with the means and methods that such standards were set in the past, but new methods that radically differ from them are making the scene. These new methods, making an analogy with experimentation, represent proof-of-concept of measurement from an independent approach—a practice that is always good for good science. For example, the ohm can now be defined with extremely high accuracy from experiments involving the quantum Hall effect (QHE) and the value of the von Klitzing constant—that $R_{K\text{-}90}$ can be used to establish a reference standard of resistance with one standard deviation estimated to be 1×10^{-7} and a reproducibility that is significantly better than values obtained from traditional methods. The BIPM(Bureau International des Poidset Mesures) in 2000 has admitted as much. The QHE provides an "invariant reference" for resistance that is derived from physical constants with reproducibility better than two orders of magnitude than the best uncertainty derived from traditional SI units [27]—the power of the quantum domain!

There are several reasons for this kind of validation in addition to having roots in natural constants: (1) measurement results do not depend on external parameters (e.g., ambient conditions), (2) the values do not drift with time, and (3) the measures can be reproduced anywhere, a condition that "simplifies and improves traceability of any measurement to the primary standards" [27]. The discovery of the QHE and the Josephson effect (more about them later) made available to the scientific community two quantum standards from which the other basic seven standards can be derived [27]. Truly revolutionary! A bottom-up approach to metrology is making its mark.

16.1.1 Nanometrology

Norio Taniguchi, the person who coined the term nanotechnology, stated in 1974 that uncertainties between 100 and 1 nm would become the standard required for measurement and manufacturing [28]. According to J. B. Bryan in 1979, again somewhat prophetically, "sophisticated instrumentation and ultra-precise dimensional metrology" would be required [29]. From the days of mechanical machining to microlithography, the drive towards miniaturization, better precision, and better accuracy followed a steady drumbeat (**Fig. 16.2**). The need for accurate positioning along with precise carving tools led to the development of smaller and smaller actuators, motors, and sample stages that relied on the development of new materials and thin films and new methods of fabrication, but also now require the participation of disciplines such as chemistry and biology [30].

Traditional engineering materials (metals and alloys, ceramics, semiconductors, organics and polymers, and composites), all enhanced by nanotechnology, now include biological components. Nanoscale development has got us into intimate contact with natural biological processes—a domain where all materials and functions are based on nanoscale phenomena. Objects and devices have undergone miniaturization over the past several decades. Nanometrology evolved, was reinvented, then evolved further to keep up. Nanometrology is rapidly expanding its sphere of influence and will soon establish the rules of engagement at the nanoscale.

16.1.2 Uncertainty

A new age of metrology is upon us. Assignment of uncertainty to nanoscale materials and phenomena presents a new set of metrological challenges that has evolved from traditional challenges [31]. New models and procedures are needed to analyze and interpret new data.

Measurements are approximations of the true value of a physical quantity (the *measurand*), and therefore the result is reliable only when defined in terms of the uncertainty of the measurement [32]. Uncertainty can be statistical in nature (evaluated by statistical methods) or deterministic (evaluated by other means). Uncertainty arises from random effects. The student should review some of these terms by consulting any general statistics text. Standard uncertainty (standard deviation) μ_i is usually derived from the square root of the variance μ_i^2 [32]. A popular method to determine standard uncertainty is statistical analysis of independent observations that employ the method of least squares to fit a curve. Analysis of variance (ANOVA) can be applied to identify and quantify random effects.

The NIST relies on the *combined standard uncertainty* $\mu_c(y)$ in the definition of many measurement results (commercial, industrial, and regulatory). Combined uncertainties result from individual standard uncertainties (e.g., subject to the *law of propagation of uncertainty* and the method of *root sum squares*, RSS). Combined uncertainties are used in conjunction with the determination of fundamental constants, fundamental metrological research, and international comparisons and realizations of SI units [32]. However in cases of health and safety (or perhaps nanometrology?), *expanded uncertainty* is required—an extra confidence level U surrounding the measurement y of the *measurand* Y such that

FIG. 16.2 *The evolution of machining from the 1900s to the present day has changed dramatically. Achievable machining accuracy has been reduced from 100 μm to values well below that of the micron. The primary tools involved in high technology manufacture have evolved from those that are purely mechanical in nature, such as the lathe and milling machines, to tools that involve gratings and ion beams. The concomitant evolution of measurement devices occurred alongside the production tools—from the Vernier caliper to optical and electronic comparators to lasers and to the electron microscopes we use today.*

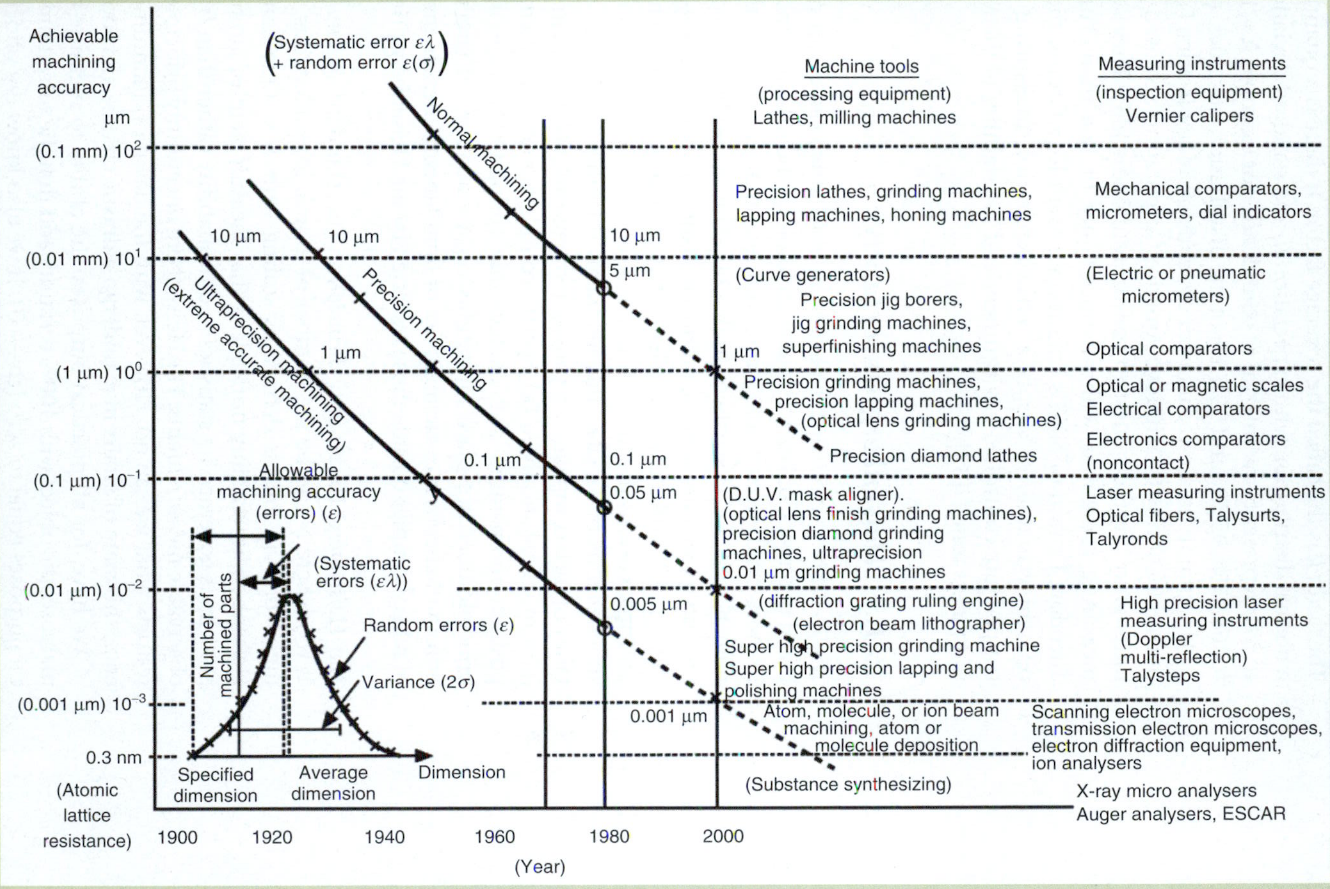

Source: N. Taniguchi, Current status in, and future trends of, ultraprecision machining and ultrafine materials processing, Tokyo Science University, Annals of the CRIP, 32, 573–582 (1983). CRIP is the College International pour la Recherche en Produtique. With permission.

$$U = k\mu_c(\gamma) \tag{16.1}$$

$$\gamma - U \leq Y \leq \gamma + U \quad \text{or} \quad Y = \gamma \pm U \tag{16.2}$$

The factor k is the called the coverage factor. If there is a quantity z described by a normal distribution with expectation μ_z and standard deviation σ, the interval $\mu_z \pm k\sigma$ represents the interval of confidence. For example, $k = 1$, $k = 2$, and $k = 3$ for 68.27%, 95.5% and 99.73% of the distribution of values, respectively [32].

Uncertainty in a measurement depends on many factors. For example, NIST in conjunction with Mitutoyo America Corporation achieved "remarkable low uncertainty" in their quest to control roundness [33]. They were successful in measuring roundness with an uncertainty of less than 3 nm. According to Theodore Doiron of NIST's Engineering Metrology Group, the heralded development was due to "a special error separation measuring method, a tight temperature controlled environment and the stability of the *Mitutoyo RA-H500* roundness tester" [33].

16.1.3 Heisenberg Uncertainty

The *Heisenberg uncertainty principle* limits what we can do in metrology. A new brand of physics, bolstered by nanotechnology and quantum mechanics, explores in more depth what can be done with quantum principles within the sphere of influence of the Heisenberg limit [34]. The Heisenberg uncertainty principle presents a lower limit on the product of standard deviations (e.g., for position and momentum and other complementary couples). The Heisenberg uncertainty principle governs the theoretical limitation of the combined accuracy of pairs of simultaneous but related measurements. In other words, the combined uncertainty of such complementary measurements cannot be less than Planck's constant h (or more specifically, $h/4\pi$). Specific complementary measurement pairs include position–momentum and energy–time couples. The Heisenberg uncertainty principle does not place limitations on the accuracy of single measurements, nonsimultaneous measurements, or of simultaneous measurements of other types of related couples outside those restricted by the Heisenberg principle [35].

The Heisenberg Uncertainty Principle. The standard *quantum limit* scales as $1/\sqrt{N}$, where N can be the number of atoms, ions, electrons, or photons, and the *Heisenberg limit* is defined as scaling with $1/N$. One consequence of the Heisenberg uncertainty principle is that measurement of atoms and subatomic particles is governed by statistics and probability rather than by actual physical, observable measurements. The Heisenberg uncertainty principle, in other words, restricts what we can do. However, with the dawn of quantum information science, this state of affairs may undergo a drastic change and soon.

We have, for all practical purposes, the ability to measure the position of a macroscopic object with 100% certainty and that the uncertainty is for all practical purposes equal to zero ($\Delta x \to 0$). If we also know the velocity (and mass) of the macroscopic object, we can usually predict where it is going, where it has been, and vice versa. Such is the case in classical physics [36]. In the domain of classical physics, exact simultaneous values can be assigned to all physical quantities [37]. These are tools that allow us to make predictions about the physical characteristics of a classically physical object, for example, the flight of a jet plane. As objects assume smaller dimensions, however, the Heisenberg limit assumes more importance. The question to ponder is at what point does the Heisenberg limit, or the standard quantum limit for that matter, become a serious consideration in metrology.

In order to observe (measure) the position and/or the momentum of an electron, the electron must undergo a perturbation—perhaps one caused by an interaction with a photon that *is* used for the measurement [36]. Therefore, by probing the position of the electron (e.g., so that $\Delta x \to 0$), more than likely you have (or rather the photon has) altered the electron's momentum, perhaps drastically and bumped that parameter into the domain of the unknown. The momentum of the photon (or electron, atom, or other materials) is known from

$$p = \frac{h}{\lambda} \tag{16.3}$$

and the change in the electron's momentum following collision results in an uncertainty

$$\Delta p \approx \frac{h}{\Delta \lambda} \tag{16.4}$$

where $\Delta\lambda$ is the uncertainty in the electron's position. The product of the uncertainty in position and the uncertainty in momentum results in the Heisenberg uncertainty principle*

$$(\Delta x)(\Delta p) = \left(\frac{h}{\lambda}\right)\lambda = h \to \frac{h}{4\pi} = \frac{\hbar}{2} \tag{16.5}$$

or, since the value of the uncertainty represents a minimum

$$(\Delta x)(\Delta p) \geq \frac{h}{4\pi} = \frac{\hbar}{2} \tag{16.6}$$

The *h*-bar version of Planck's constant is a more accurate rendition of the uncertainty principle. The exact value (with uncertainty) is

$$\hbar = \frac{h}{2\pi} = 1.054\ 571\ 628(53) \times 10^{-34}\,\mathrm{J \cdot s} = 6.582\ 118\ 99(16) \times 10^{-16} e\mathrm{V \cdot s} \tag{16.7}$$

In simple terms, the measurement process itself limits the ability to measure position and momentum simultaneously [36]. If, at the subatomic scale we were to know position, say of an electron, exactly (e.g., $\Delta x \to 0$), then the uncertainty in its momentum must approach infinity ($\Delta p \to \infty$).

Energy and time also form a complementary pair of uncertainties [36].

$$(\Delta E)(\Delta t) \geq \frac{h}{4\pi} \tag{16.8}$$

---

* *x*-space with a characteristic linewidth of Δx is also localized in *k*-space (e.g., the particle's wave number) with a characteristic width of $1/(2\ \Delta x)$ and then $\Delta x \Delta k \geq 1/2$. Since measurement of a particle's wave number is equivalent to measuring its momentum *p* and that $p = hk$, the uncertainty in *k* of order Δk translates to an uncertainty in *p* of order $\Delta p = h\Delta k$. Therefore, $\Delta x \Delta p \geq \hbar/2$ (or $h/4\pi$).

Example 16.1 *Measurement of an Electron's Position*

A free electron (such as one found in an electron beam) has kinetic energy equal to 25 *e*V. What is the velocity of the electron? If this velocity is known to within 0.5% accuracy, what is Δx of the position of the electron?

Solution:
Electron velocity

$$25\ eV = (25)1.6\times10^{-19}\,\mathrm{C}\left(\frac{\mathrm{J}}{\mathrm{C}}\right) = 4.0\times10^{-18}\,\mathrm{J}$$

$$\mathrm{KE}_e = \tfrac{1}{2}m_e\upsilon_e^2;\quad 4.0\times10^{-18}\,\mathrm{J} = \tfrac{1}{2}\left(9.1\times10^{-31}\ \mathrm{kg}\right)\upsilon_e^2$$

$$\upsilon_e = \sqrt{2\left(\frac{4.0\times10^{-18}\,\mathrm{J}}{9.1\times10^{-31}\ \mathrm{kg}}\right)} = 3.0\times10^{6}\,\frac{m}{\mathrm{s}}$$

The uncertainty in position is 0.5%.

Therefore, the uncertainty in momentum is

$$\Delta p = m_e\,\Delta\upsilon = 9.1\times10^{-31}\ \mathrm{kg}\left[(0.005)\ (3.0\times106\mathrm{m\cdot s^{-1}})\right] = 1.4\times10^{-26}\ \mathrm{kg\cdot m\cdot s^{-1}}$$

The uncertainty in position follows from

$$\Delta x \approx \frac{h}{\Delta p} = \frac{6.6\times10^{-34}\ \mathrm{J\cdot s}}{1.4\times10^{-26}\ \mathrm{kg\cdot m\cdot s^{-1}}} = 47\ \mathrm{nm}$$

Detection of an electron can be accomplished by bombarding the electron with a photon of velocity c, the speed of light. The position of an electron is known with an inherent uncertainty $\Delta x \approx \lambda$. The time it takes for this photon to traverse the distance equal to the uncertainty of the position of the electron is

$$\frac{\Delta x}{c} \approx \frac{\lambda}{c} = \Delta t \tag{16.9}$$

Example 16.2 *The Position of a High-Speed Projectile*

A 22-caliber bullet with mass 2.6 g is able to achieve a velocity of 390 m · s^{-1}. If we are able to determine its velocity to ±0.02% (υ = 390 ± 0.078 m · s^{-1}), what is the minimum uncertainty in its position? Is this value consequential?

How does this value compare to that of the electron in Example 16.1 above?

Solution:
Calculation of $\Delta\upsilon$

$$\upsilon = 390\pm0.078\ \mathrm{m\cdot s^{-1}},\quad \Delta\upsilon = 2\times0.078\ \mathrm{m\cdot s^{-1}} = 0.16\ \mathrm{m\cdot s^{-1}}$$

$$\Delta x = \frac{h}{2\pi\Delta p} = \frac{h}{2\pi(m_e\Delta\upsilon)} = \frac{6.6\times10^{-34}\,\mathrm{J\cdot s}}{2\pi\left(2.6\times10^{-3}\ \mathrm{kg}\right)\left(0.16\ \mathrm{m\cdot s^{-1}}\right)} = 2.5\times10^{-31}\,\mathrm{m}$$

The uncertainty in position of the electron is significant—on the order of several nuclear diameters. The uncertainty in position of the bullet is of little consequence to the measurement.

Also, since $E = h\nu = \frac{hc}{\lambda}$; $\Delta E = \frac{hc}{\lambda}$, the Heisenberg uncertainty with regard to energy and time reduces to

$$(\Delta E)(\Delta t) \approx \left(\frac{hc}{\lambda}\right)\left(\frac{\lambda}{c}\right) = h \tag{16.10}$$

What is the physical meaning of this expression? It means that in order to measure the energy of a particle exactly, an infinite amount of time is required [36,38]. The measurement of the frequency ν of a photon is a measure of its energy. A suitable experiment, for example, is the measure of the energy of an excited state of an electron. The energy (frequency) can only be determined during the lifetime of the excited state, and the observed emission line, therefore has a characteristic energy width, $\Delta E = h\Delta\nu$. The diffraction limit in optics corresponds to this Heisenberg uncertainty limit.

16.1.4 Quantum Entanglement

We decided to include a section on this interesting phenomenon although its rigorous understanding is beyond the scope of this text. Nonetheless, please try and get a feel for what is presented.

An interesting quantum phenomenon that has potential effects on metrology, and one that has assumed greater importance over the past few years, is the phenomenon of *quantum entanglement*. Entanglement is a quantum phenomenon in which two or more quantum states are intimately linked—however separated in space—that leads to correlations between observable physical properties of systems—some that are remote with regard to one another [39]. Quantum entanglement, from the universal perspective, is actually the vehicle that serves to link the quantum world to that of our macroscopic classical world of physical observables. We espouse the concept of continuum throughout this text, whether it be in the form of an energy continuum, a material continuum, or the artificial "academic discipline" continuum. Here we have yet another continuum to consider, that of a metrological quantum mechanical-classical physics continuum. Traditionally, we have not needed to make direct links between the quantum domain and macroscopic physical properties. We do today and the phenomenon of quantum entanglement helps supply that link, and uncoupling that entangled couple may just allow for a whole new generation of applications.

Thomas Durt of Vrije University in Brussels believes that entanglement is omnipresent [40].

> When you see light coming from a faraway star, the photon is almost certainly entangled with the atoms of the star and the atoms encountered along the way… and the constant interactions between electrons in the atoms that make up our body are no exception. … We are a mass of entanglements.

It was Erwin Schrödinger who coined the word entanglement [40]:

> When two systems, of which we know the states by their respective representatives, enter into temporary physical interaction due to known forces between them, and when after a time of mutual influence the systems separate again, then

> they can no longer be described in the same way as before, viz. by endowing each of them with a representative of its own. I would not call that one but rather the characteristic trait of quantum mechanics, the one that enforces its entire departure from classical lines of thought. By the interaction the two representatives [the quantum states] have become entangled.

Albert Einstein, although awarded a Nobel Prize for his work on the photoelectric effect, could not accept the probability-based nature of quantum mechanics, for example, "God does not play dice with the universe," so he said. Or, making his point with more aplomb, Einstein stated that

> I find the idea quite intolerable that an electron exposed to radiation should choose of its own free will, not only its moment to jump off, but also its direction. In that case, I would rather be a cobbler, or even an employee in a gaming house, than a physicist.

Einstein believed that something more deterministic existed underneath the probability, and hence, the concept of entanglement emerged in 1935 [41,42]. Einstein called quantum entanglement "spooky action at a distance." From a deterministic perspective, by evaluating the spin of one electron, only then does one know exactly the spin of the other electron. Before that point in time, electron spins exist, only within a range of probabilities. In other words, it is only after a direct observation that spins become fixed [41,42]. Entanglement does not minimize the principle of cause and effect, but it does underscore the importance of quantum particles, particles that are usually defined by probability values rather than fixed values [41,42].

The *Fano effect* is a characteristic of atomic spectra, bulk solids, and semiconductor heterostructures that offers clues about the relationship between the quantum and a bulk observable characteristic—the spectrum. The Fano effect describes the relationship between quantum interference of two competing optical pathways: the first pathway that couples the energy of a ground state with an excited state, and the second that connects the ground state with a continuum of energy states. It is actually a common phenomenon that occurs when a discrete quantum state (of an atom or molecule) interacts with a continuum state (of a surrounding host material) [43]. The effect is easily observed in spectra that display asymmetric lineshapes. In other words, the Fano effect influences the way an atom or molecule absorbs light and is expressed in its spectral response [43].

Experiments with small nanoscale systems, however, are made difficult by factors related to the Heisenberg uncertainty principle—in that the interaction of a nanoscale system and the continuum state that surrounds it may not be detected easily [43]. However, M. Kroner et al. have shown that photons scattered from a quantum dot with increasing laser intensity enabled them to observe weak interactions related to the Fano effect, for the first time, to a nanoscale phenomenon [43].

Two-Photon Correlation and Parametric Down Conversion. A means of generating a two-photon entangled system is shown in **Figure 16.3**. Quantum correlation of paired photons is made by an incident photon of higher energy impinging on a nonlinear optical crystal. A nonlinear optical material induces dielectric polarization P response in a nonlinear way to the applied electric field E. Usually a

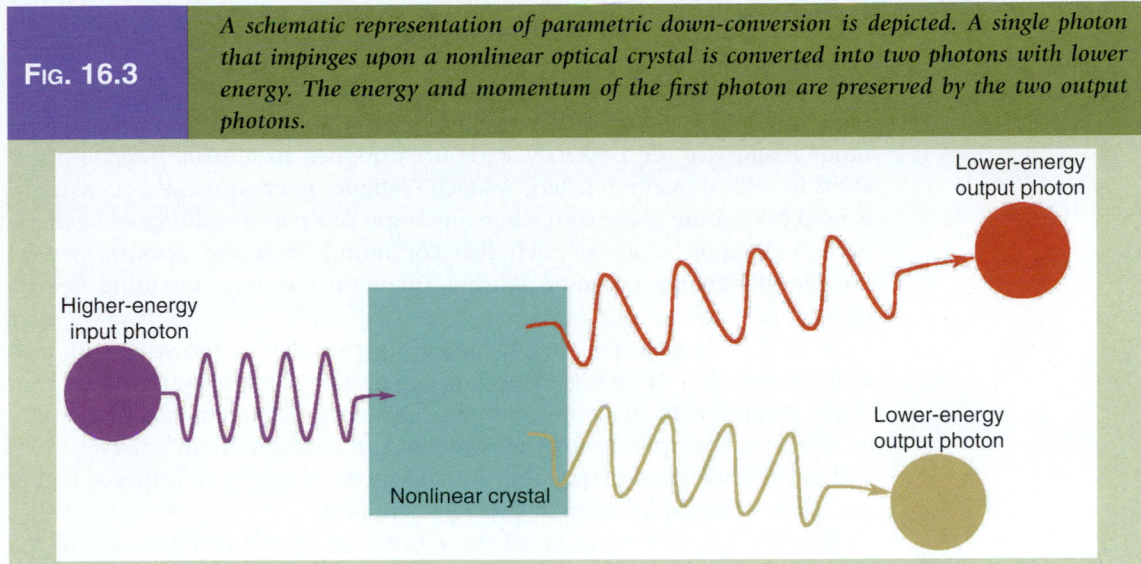

Fig. 16.3 *A schematic representation of parametric down-conversion is depicted. A single photon that impinges upon a nonlinear optical crystal is converted into two photons with lower energy. The energy and momentum of the first photon are preserved by the two output photons.*

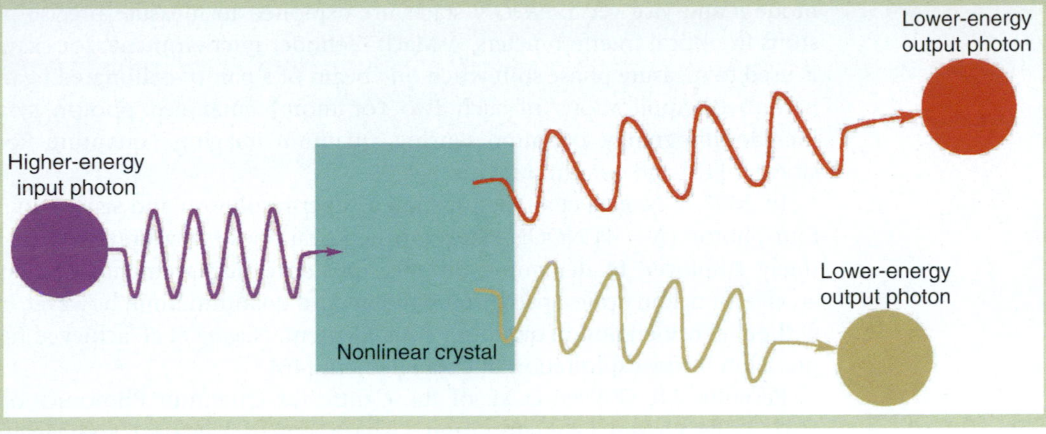

function of intense laser exposure, an example of a nonlinear optical phenomenon is second harmonic generation from such a crystal.

According to Alan Migdall, a physicist in the Optical Technology Division of the NIST in Gaithersberg, Maryland [44],

> ... such quantum correlation promises metrologists something of a free lunch: absolute measurements that don't require absolute standards.

As a matter of fact, quantum correlation techniques provide a method of determining the quantum efficiency of detectors without the need of calibration standards [44]. Migdall goes on to state that measurements of absolute infrared radiance without calibrated standards and without an infrared detector are also possible due to this phenomenon. This statement should cause same entanglement in your thinking.

Parametric down conversion is a process by which entangled photons are produced (**Fig. 16.3**). Input photons decay into entangled photons through the action of a nonlinear crystal. Therefore, if one knows the energy of the input laser and one of the output photons, then information about the entangled pair is known. And, even more interestingly, if measurements were made on the first photon of the entangled pair, it would be possible to predict the outcome of measurements on the second photon [44]. In other words, detection of one photon predicts the existence and properties of the second photon such as wavelength, emission time, direction, and polarization—all at a distance [44]. Measurement of the quantum efficiency of a detector is an important aspect of metrology (refer to chapter 14). By such methods, 150×10^{-18} s level of resolution has been achieved that approached the Heisenberg uncertainty limit on the simultaneous creation of photon pairs—and all this occurred in 1999, the Stone Age of nanotechnology.

We won't spend more time on this aspect but in general a quantum mechanical entangled state (photons) is referred to as a *NOON* state.

$$|\psi_{NOON}\rangle = |N\rangle_a |0\rangle_b + e^{i\theta} |0\rangle_a |N\rangle_b \tag{16.11}$$

One can see immediately from where the acronym *NOON* originates. The equation represents a superposition of N particles in mode a with zero particles in mode b and vice versa. *NOON* states are exploited to measure precise phase shifts in optical interferometers. A Mach–Zehnder interferometer, for example, is used to measure phase shift when one beam of a pair of collimated beams is perturbed. Applications of such two (or more) entangled photon systems include lithography, quantum sensing, quantum imaging, "quantum Rosetta stones," [45] and quantum metrology.

In 2007, T. Nagata et al. demonstrated super-resolution and sensitivity with four-photon ($N = 4$) *NOON*-states [46]. Optical phase measurements are routinely employed to determine distance, position, displacement, orientation, acceleration, and optical path [46]. The standard quantum limit, however, exists without consideration of quantum entanglement. Nagata et al. achieved higher precision by the exploitation of entanglement [46].

Recently, J.L. O'Brien et al. of the Centre for Quantum Photonics of the H.H. Wills Physics Laboratory and Department of Electrical and Electronic Engineering, University of Bristol in the United Kingdom reported control of single photons via an optically controlled NOT gate. A NOT gate is a gate based on digital logic that operates on the principle of a truth table, for example, input "A," output "NOT-A" [47,48]. Most photonic quantum technology relies on large-scale assemblies fixed to optical tables and complex interferometers (for subwavelength control). O'Brien and his team devised a milliscale device from silica waveguides on a silicon chip (an entire optical table). The characteristics of their apparatus are as follows, according to the authors [47,48]:

> … high visibility (98.5 ± 0.4%) classical interference, high-visibility (94.8 ± 0.5%) two-photon quantum interference, high-fidelity controlled-NOT entangled logic gates (logical basis fidelity, F = 94.3 ± 0.2%) and on-chip quantum coherence confirmed by high fidelity (>92%) generation of two-photon path entangled states…. We fabricated 100s of devices on a single wafer.

We add this information to demonstrate the relative simplicity of the setup. Quantum metrology measurements utilizing four-photon communication have already demonstrated by O'Brien and his colleague Takeuchi in 2003 (Hokkaido University, Japan) [47,48]. The team also claims that exotic quantum processes are not necessary to reach the ultimate limits of measurement [49].

Without nonlocalized entanglement, standard interferometers are limited by statistical shot-noise that scales with $1\sqrt{N}$ where N is the number of particles (photons) passing through the interferometer over time. With quantum correlations (entanglement), the interferometer sensitivity is improved by a factor of $\sqrt{N}$ to scale as $1/N$—the limit imposed by the Heisenberg uncertainty principle. Exploitation of quantum entanglement is used to overcome quantum projection noise (QPN) (noise due to fluctuations in population of a "fixed number" of atoms or, alternatively, QPN arising from statistical nature of projecting a superposition of two states into one state when a measurement is made). The use of cold atoms and precise spectroscopy measurement has enhanced the frequency measurement accuracy of atomic clocks, but the measurement is limited by QPN of ensembles of uncorrelated particles [50].

For example, noise can be reduced by a process called "squeezing" (amplitude or phase squeezing). Photons are the best choice for quantum metrologies, because they deliver a small amount of noise (decoherence). Photons are good candidates for quantum bits in quantum communication, quantum lithography, quantum computing, and of course, quantum metrology [46]. Other topics not discussed in detail can be elaborated upon in excellent references.

Unentangled Photon Metrology? According to B.L. Higgins et al. at the Center for Quantum Dynamics at the Griffith University in Brisbane, Australia, advances in precision measurement have always been the gateway to scientific discoveries [51]. He also states that entangled states are not necessary for precise and accurate measurements. For example, he replaces entangled input states with "multiple applications of the phase shift on entangled single-photon states." Measurement precision at the fundamental level, of the optical phases used in length metrology, for example, are difficult to achieve due to limitations of the number of quantum sources available, for example, photons [51]. He goes on to say that standard measurement techniques that apply each resource independently are subjected to phase uncertainty that scale as $1/\sqrt{N}$, also know as the standard quantum limit. The Heisenberg limit, on the other hand, scales to $1/N$, the true quantum limit and a $\sqrt{N}$ factor enhancement over the standard quantum limit. By interferometry methods alone, this limit cannot be beaten [52].

Conventional classical measurement relies on N independent physical systems that are separately prepared and detected. The experimental values (e.g., measurement) derived by such classical techniques rely on averaging the outcomes of N independent measurements. For quantum measurement, N physical systems are prepared in an entangled state. Measurement occurs over all N states simultaneously with a single delocalized measurement (of probability) that collectively considers all N components [52].

Nanomechanical Oscillators and the Uncertainty Principle. Nanoelectromechanical systems (NEMS) are the miniaturized version of microelectromechanical systems (MEMS) except with higher frequency ranges, higher quality factors Q, higher sensitivity, and lower power demands [53]. Q is a measure of the damping of oscillations in a mechanical system as a function of frequency—higher Q indicates more stable vibrations. For example, high Q gallium nitride (GaN) wires 30–500 nm in diameter and 5–20 μm in length grown at NIST were stimulated to vibrate between 400,000 to 2.8 million times per second [54,55]. Another feature of nano-oscillators is that they can be cooled close to their ground states under cryogenic temperatures.

Such nanomechanical oscillators are perfect tools to study quantum effects like back-action, coherent states, and superposition [53]. Quantum back-action is a phenomenon that reflects the Heisenberg uncertainty principle—for every measurement there is always a perturbation in the object subjected to measurement. In addition to gigahertz-level vibrations and resolution with atomic spacing, MEMS and NEMS systems are capable of detecting force in the zeptonewton range (10^{-21} N) and mass in the zeptogram range (10^{-21} g) [56]. We are now prepared to observe quantum behavior in a mechanical device.

Now consider this. If a superconducting single-electron transistor (SSET) were coupled to a resonating nanomechanical structure (e.g., a silicon nitride beam) and a voltage applied between them that corresponds to a quantized energy state of electrons passing through the SSET, what do you think would happen? According to Keith C. Schwab of Cornell University, Ithaca, New York, and his research team, "counter-intuitive" developments unfold [53,57]. Positions of the beam during oscillation alter SSET conductivity. Therefore, from the measurement of current in the SSET (from the observer's point of view), determination of the resonator position was made possible. Schwab et al. also determined that random charge fluctuations within the SSET influenced the frequency, position, and damping rate of the resonator [53,57]. The authors stated that damping in the resonator mode due to a back-action force caused the temperature to drop from 550 to 300 mK. In essence, the SSET acted more like an absorber of energy rather than a generator of heat. This phenomenon will no doubt generate new ways to cool down components of future NEMS mechanisms [53,57].

Keith Schwab et al. essentially investigated the relationship between a nanomechanical oscillator device and the uncertainty principle [58]. The purpose of the experiment was to probe the limitations placed by quantum mechanics on the resolution of position. In this case, the research team coupled a single-electron transistor (SET) and an ultrasensitive microwave detector network to a nanomechanical device (a 19.7-MHz resonator of high Q). The SET served as an appropriate amplifier for a radio frequency nanomechanical system [58]. The goal for the research was to conduct position detection approaching that set by the Heisenberg uncertainty principle, at mK temperatures [53]. Schwab et al. claim that "ultimate sensitivity should be limited by both the quantum fluctuations of the mechanical oscillator and the quantum fluctuations of the linear amplifying system" [58].

The nanomechanical device consisted of an aluminum layer on a silicon nitride resonator that was 8.7 μm in length and 200 nm wide. They were able to detect movement due to the stimulus of observation (e.g., back-action). He states:

> We made measurements of position that are so intense - so strongly coupled - that by looking at it we can make it move.... Quantum mechanics requires that you cannot make a measurement of something and not perturb it. We're doing measurements that are very close to the uncertainty principle; and we can couple so strongly that by measuring the position we can see the thing move.

Schwab is working on building a nanodevice that exhibits superposition quantum behavior—a nanodevice that can be in two places at the same time [53].

16.1.5 Applications

Please note that any application of quantum entanglement implies an integral relationship with metrology (nano or otherwise). Quantum entanglement phenomena are expected to contribute to metrology—some fairly far out. Certain government agencies, for example, believed that communication faster than the speed of light was possible. This was based on entanglement operating instantly at some distance [42]. However, theoretically, this is not possible because the message would in effect be read in the past. A more practical application of entanglement is that of encryption. If one half of a set of entangled pairs is sent to a complementary communications link, interception is rendered impossible due to disintegration of the entanglement.

Nanotechnology And? How does this all relate to applications and perhaps nanotechnology? According to Professor Ali Mansoori of the Departments of Bioengineering and Chemical Engineering and Physics at the University of Illinois at Chicago, the Heisenberg principle applies strictly to subatomic particles like electrons, positrons, and photons [59]. Nanotechnology, on the other hand, involves the position and momentum of atoms, molecules, for example, larger particles that do not come under the umbrella of the strict confines of the Heisenberg principle [59]. This does not imply that nanosystems are not affected by the Heisenberg uncertainty principle. As a matter of fact, it is through the application of nanomaterials that we are able to probe the interface between the quantum limit and the classical world of physics.

With regard to information technology, nanotechnology is expected to have great impact—in actuality, it already has. Future trends however indicate that smaller devices are inevitable. With their development, certain fundamental limits will be breached [60]. As fewer electrons, perhaps even single electrons, are switched to execute operational commands, at what point does this become physically feasible? Or, when the size of magnetic domains (e.g., bits) are reduced, at what point is stability compromised due to particle interactions with thermal fluctuations?

Quantum lithography and Two-Photon Microscopy. Quantum lithography and two-photon microscopy provide a fertile ground for applications of entangled quantum states. The resolution of objects smaller than the wavelength of the applied beam is difficult due to light scattering around objects (e.g., the Rayleigh limitation). The reduction of the wavelength, while keeping the wavelength of the radiation field constant, although paradoxical, is what needs to be done [52]. According to Vittorio Giovannetti et al. at the *National Enterprise for Nanoscience and Nanotechnology* of Italy's Istituto Nazionale per la Fisica della Materia (NEST-INFM) and Scuola Normale Superiore in Pisa, Italy, quantum effects can help. By using equipment that is sensitive to the de Broglie wavelength, using two or more photons, the wavelength can be reduced from one wavelength to half that wavelength without changing the source frequency

$$\lambda = \frac{2\pi\hbar}{p} = \frac{2\pi\hbar c}{E} = \frac{2\pi c}{\omega} \rightarrow \frac{2\pi\hbar c}{2E} = \frac{\lambda}{2} \tag{16.12}$$

In essence, two entangled photons (as a single entity!) enter an apparatus consisting of mirrors and beamsplitters. The entangled photons originate from a light wave that is split and then recombined on a surface. The recombined pattern is the same pattern that would have been generated from the original wave but at half the wavelength. For example, packets of 10 entangled photons originating from infrared radiation could end up as x-rays. More photons reduce the practical λ even further [52]. Beating the Rayleigh limit (e.g., best features limited by half the wavelength) is a piece of cake.

Quantum Computing. Although computers rely on quantum mechanics today, information is encoded (transmitted and processed) in accordance to classical principles (as a "0" or a "1"), e.g., two well-defined states. Another application is that of *quantum computing*, an application that has received significant attention over the past few years. In this case, the quantum information can exist as a superposition of different states (not so well-defined)—both in *0ñ* and *1ñ* states, for example. Interference in a double-slit experiment is an example of superposition of states. Quantum interference in principle is the same except that only one particle is required due to the potential of entanglement. Single-particle quantum interference has been demonstrated with photons, electrons, neutrons, and atoms [61].

In quantum computing, information is stored with quantum particles (photons, atoms, or electrons) called quantum bits (or *qubits*). The problem is, according to the Heisenberg uncertainty principle, that accessing the particles should destroy the information carried by the qubits. With quantum entanglement, interaction with qubits is possible without "frying your quantum memory," this according to Brian Clegg, author of the *God Effect: Quantum Entanglement, Science's Strangest Phenomenon* [41,42]. Clegg also believes that quantum computers are not likely to make the scene anytime soon.

The weirdest one—are you ready for this—is the prospect of *teleportation*. Within the bounds of the Heisenberg uncertainty principle, interacting with a quantum particle changes that particle, for example, one cannot input a particle (scan it) and send it somewhere and have it reassembled into an exact copy. However, if only half of an entangled pair is "scanned" and the other half of the pair is placed through a logic gate, it is possible to make an identical particle at some distance. The downside is that the original particle (or human) loses its identity. According to Brian Clegg, this has actually been accomplished with large molecules [41]. Nobel Prize winner Brian Josephson (of Josephson junction fame) stated that *telepathy* can be explained by quantum entanglement. Brian Clegg concludes

> What entanglement (and quantum theory in general) does do is remind us is that the real world is much stranger than we imagine. That's because the way things are in the world of the very small is totally different to large scale objects like desks and pens. We can't rely on experience and common sense to guide us on how things are going to work at this level. And that can make some of the effects of quantum physics seem mystical. In the end, this is something similar to science fiction writer Arthur C. Clarke's observation that "any sufficiently advanced technology is indistinguishable from magic."

Welcome to the *Nano Age*! On that note, we move onto the next section, *Quantum Metrology*.

16.2 Quantum Metrology

Quantum metrology is about time, frequency, and photonics; and fundamental constants and the path towards quantum detection. There is a concerted effort to standardize measurements at the quantum level, for example, to improve measurement accuracy, traceability, new nanoscale metrology solutions, and the development of quantum standards [62]. The measurement continuum extends from the classical domain of physical objects down to particles like atoms and single molecules. The electromagnetic domain, over increasing ranges of frequency, must also address quantum fields within which individual photons (*quanta*) are of interest to nanotechnologists and others—energy of single atoms, nanoparticles, photons, and phonons. The "continuum of measurement accuracy" is given by the following scheme:

> Classical Metrology → Classical-to-Nanometrology → Quantum Metrological Transition → Quantum Metrology

Determination of the length scale of the metrological transition between the classical to the quantum is challenging and depends on which physical property is under consideration and the lower limit of the energy scales [62]. For example, electric current is no longer a continuous property at the nanoscale (e.g., the Coulomb blockade). The same is true for ferromagnetism (at the superparamagnetic limit). Add to this that nanoparticles are susceptible to thermal fluctuations, and require cooling to ~100 nK for measurement acquisition. The challenge is quite formidable. The ultimate goal of quantum metrology is to provide measurements, based in quantum theory, with better accuracy than those provided by classical means. Quantum metrology is not as new as one might believe. For example, the redefinition of the meter has been based on the wavelength of light for quite some time. Ultimately, all SI unit based mechanical standards will be redefined by electrical units derived from quantum standards [63].

We discussed some of the obstacles, issues, and progress in development methods based on quantum entanglement and the relationship between measurements and the Heisenberg uncertainty principle. A new age of measurement is upon us; about that there is no uncertainty.

16.2.1 Atomic Clocks, the Meter, and Time

Time and Frequency. Measurement of time and frequency is a high priority in metrology. An oscillating clock is a device that is capable of measuring these critical parameters to high accuracy and precision. If one starts with N number of cold ions in the ground state that are excited by an electromagnetic pulse, then a superposition of the ground and excited states is created. A second identical pulse is applied to measure a phase factor φ between the two states, typically by measuring the intensity. From N independent ions, ω from the phase factor φ has an error of $\Delta\varphi = 1/\sqrt{N}$ and $\Delta\omega = 1/(\sqrt{N}\cdot t)$. For entangled states, ions not acting independently, $\Delta\varphi = 1/N$ and $\Delta\omega = 1/(N\cdot t)$, an enhancement of $\sqrt{N}$.

Much of nanometrology depends on time—reaction rates, stability functions, resonant frequencies, and other natural phenomena that exist at the nanoscale. The second, the nanosecond and smaller increments of time, therefore must also be standardized. A cesium (Cs) fountain atomic clock housed at NIST, Boulder, Colorado contributes to the international group of atomic clocks (Coordinated Universal Time, UTC) [64]. The uncertainty of this clock is 5×10^{-16} s. **Figure 16.4** shows the progress made with regard to uncertainty in standardized clocks since 1940.

Frequency and time intervals are measured during the up-down movement of a cesium cluster—hence the term fountain (**Fig. 16.5d**). With the assistance of several lasers, cesium atoms are pushed together and cooled (to near absolute zero). Another laser launches the cluster vertically ca. 1 m high as the others are shutoff. Gravity then pulls the cluster down into a microwave cavity that acts to

Fig. 16.4 *The progress made with timekeeping at NIST since the 1950s is shown. Improvement, similar in ways to Moore's law, is rather linear. The NIST-F1 clock, based on Cs oscillations, is accurate to one second every 80 million years. Most recently, a strontium lattice atomic clock (red dot) set a record for accuracy and precision. It is accurate to one second in every 200 million years.*

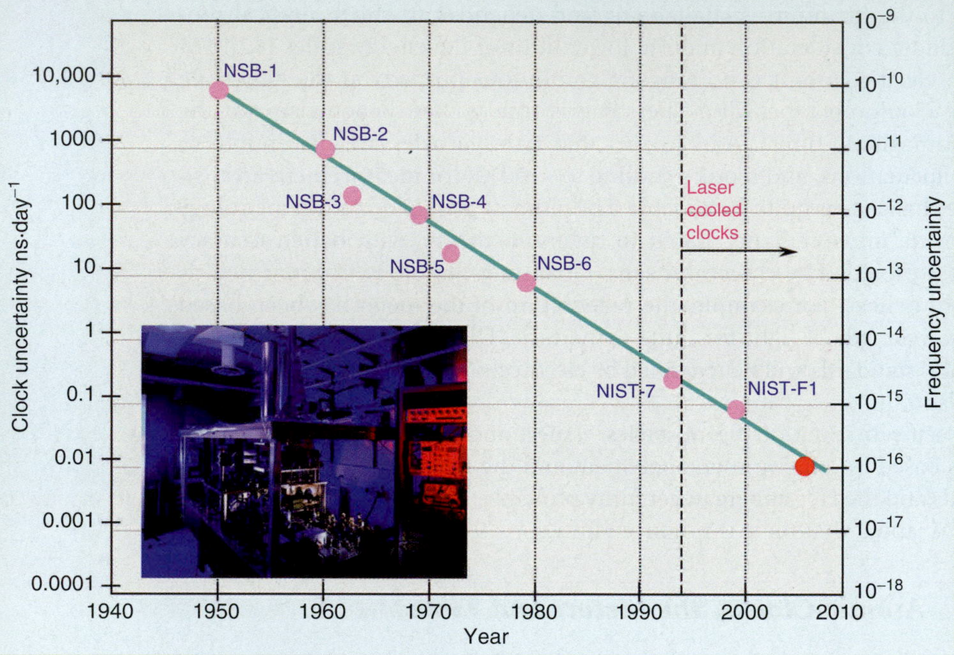

Source: Tom O'Brian, NIST Time and Frequency Division, Boulder, Colorado. With permission.

stimulate fluorescence, which is measured by a detector. After tuning to stimulate maximum fluorescence, the natural resonance frequency of Cs is attained, and it is that frequency that is used to define the second—9,192,631,770 Hz [64,65]. The Cs clock oscillator is a source of microwaves. For more information, please refer to http://tf.nist.gov/cesium/fountain.htm. The cooling and the mechanical motion allow for longer duration of observation of the Cs cluster, and it is from this extended observation that such accurate time intervals are determined [64,65].

Recently, advances have been made to improve the cesium fountain—better cooling techniques in particular that drive the temperature down to ca. 100 nK [62]. Enhanced cooling leads to reduced "Cs cold-collisional shifts" that contribute to the fountain systematic uncertainty [62]. Quieter laser sources have also improved the accuracy of the clock [62]. Newer atomic clocks based on optical (rather than microwave) measurements are expected to be 20 times more stable than the current atomic clocks and have the potential to be 100x to 1000x more stable, such as the mercury ion clock [62] An optical clock has been developed that oscillates at ~10^{15} Hz—arising from a single atom of mercury. The detector (also a vital factor without which no such clock is possible) is provided by a femtosecond (10^{-15} s) laser counter, called a comb [66]. Stephen Webster of the National Physical Laboratory (UK) states that

> The detector laser acts as a reduction gear that is capable of downshifting the optical frequency to a countable radio frequency without impairing the purity of the optical frequency.

In other words, the detector (the large gear) is able to reduce the oscillation of the mercury emission (the small gear) by a factor of 500,000 times—enough to reduce the signal to the microwave domain and be able to be counted electronically—with exceptional accuracy [66]. **Figure 16.5** displays various renditions of NIST atomic clocks.

Enhancing signal-to-noise ratio improves accuracy in frequency measurements. We mentioned quantum entanglement earlier in this chapter. *Entangled atomic clocks*, in quantum-entangled states (an instantaneous interaction between two

FIG. 16.5 *(a) NIST and the University of Colorado at Boulder teamed up to set the record for the most accurate operational atomic clock, based on thousands of strontium atoms trapped in grids formed by lasers. (b) An ultracold plasma of 26,000 Be ions fluoresce when excited by a laser. (c) The newest and most accurate atomic clock uses blue laser light to cool and trap strontium atoms in a vacuum chamber.*

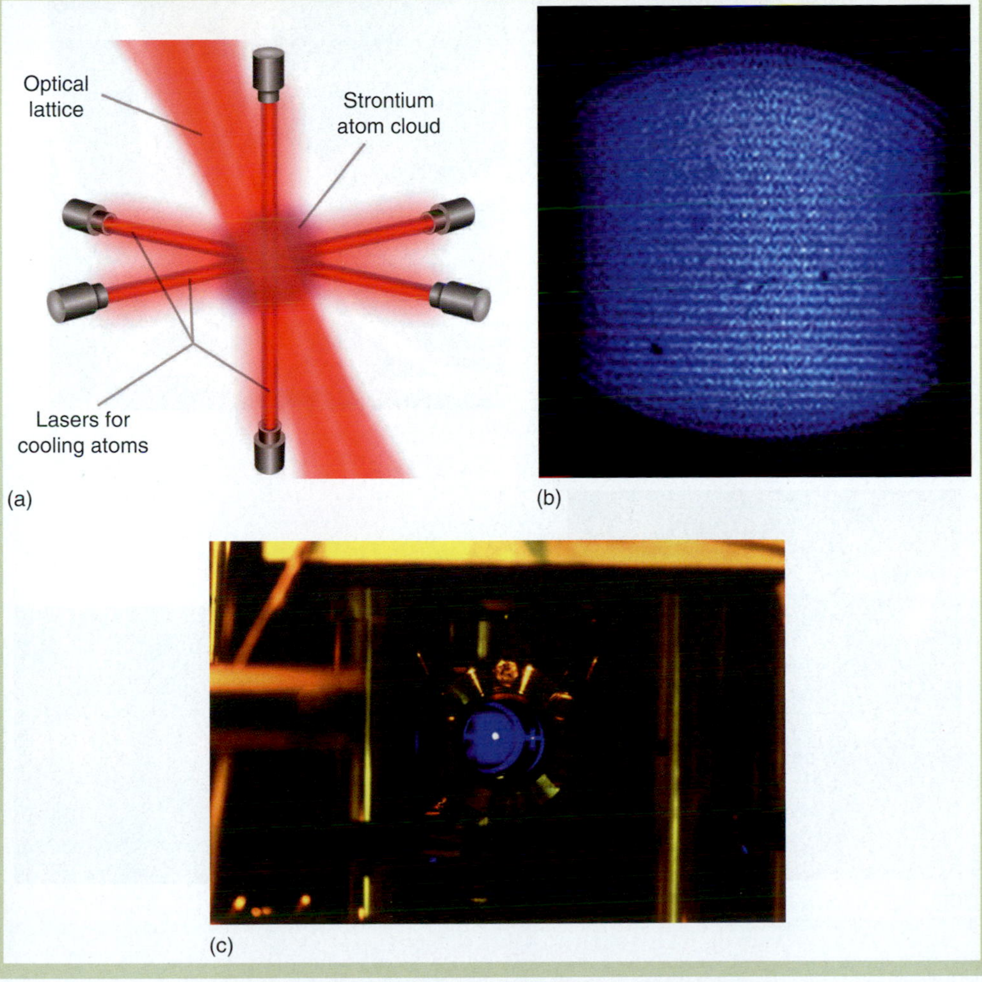

(continued)

Fig. 16.5 (Contd.) *(d) The cesium atomic fountain clock mechanism is displayed. (e) Aluminum and beryllium ions are trapped in this very small device—an extremely small atomic clock that could become the most accurate clock yet. (f) The NIST-chip-scale atomic clock includes a laser, a lens, an optical attenuator (that reduces laser power), a waveplate (that alters polarization), a cesium vapor cell, and a photodiode. (g) Ytterbium atoms are trapped within pancake-shaped wells. A yellow laser provides excitation between lower and higher energy levels.*

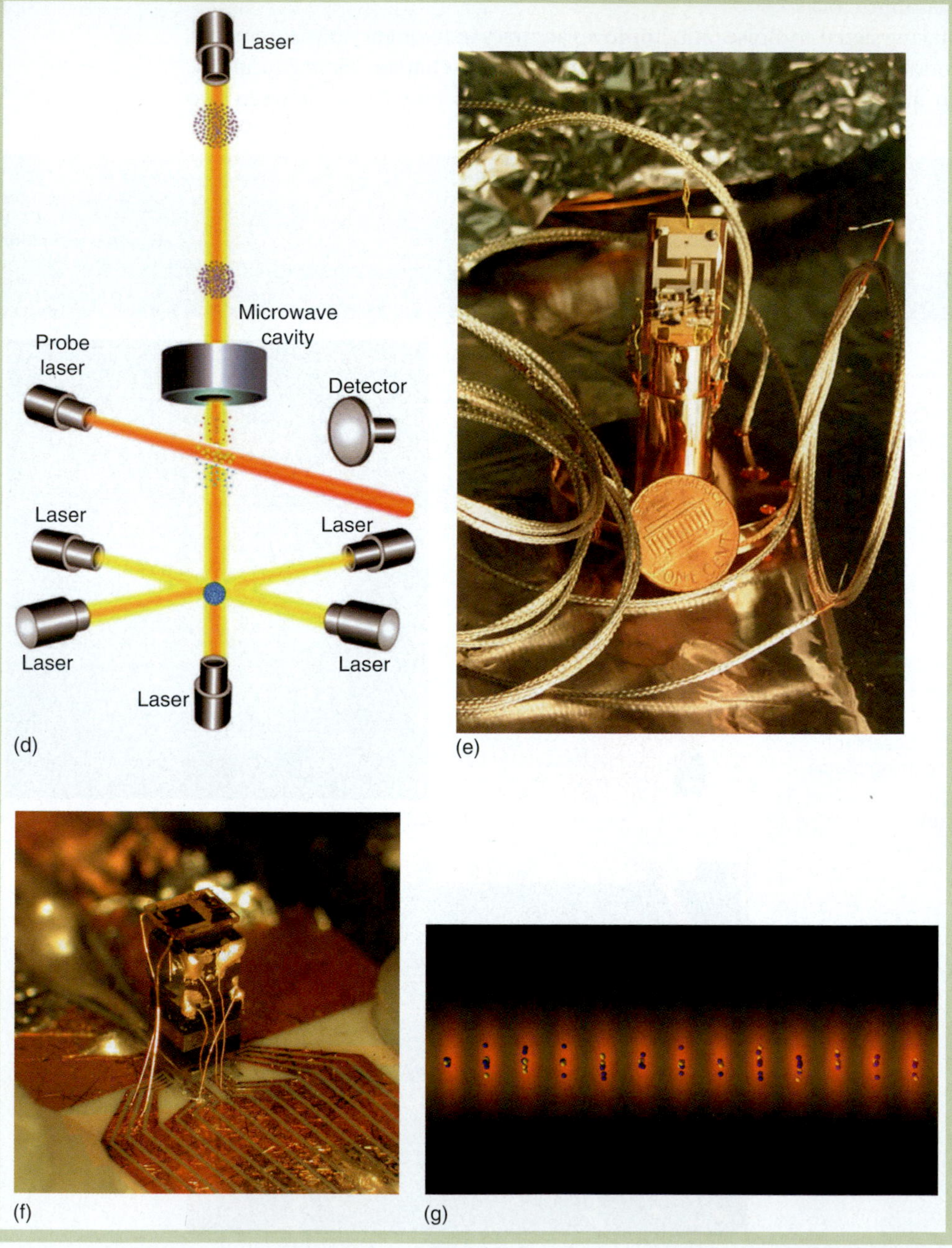

Fig. 16.5 (Contd.) *(h) An electromagnetic trap encloses one mercury ion. The optical frequency of the ion is the basis for this small atomic clock. Mercury ion clocks are the most accurate—losing 1 s every 400 million years. (i) Two magnetic coils and an optical lattice (the red laser beam) serve to trap ytterbium atoms. The blue-violet laser beams cool the clock's atomic timepiece and thereby slow its motion.*

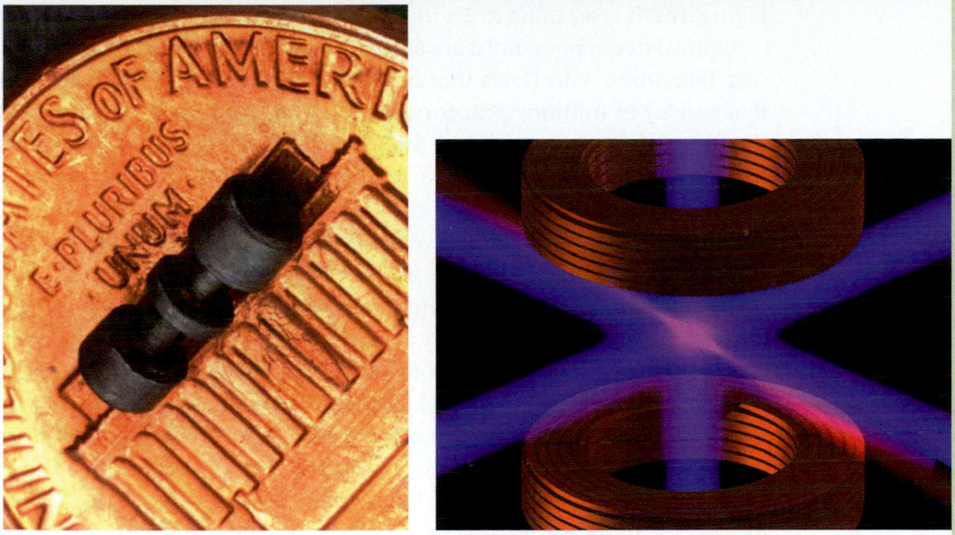

(h) (i)

Source: (a), (b), (d–g) and (i): National Institute of Standards and Technology; (c) M. Boyd and T. Ido, JILA (Joint Institute for Laboratory Astrophysics, NIST and University of Colorado at Boulder); and (h), J. Berquist and D. Wineland, NIST. With permission.

particles regardless of distance) from a suitably designed pair of atoms, should be able to enhance clock measurements in the presence of background noise [67].

> The development of an entangled atomic clock would be accurate and precise enough to test the value of the 'Fine Structure Constant', one of the fundamental constants of physics.

The Sommerfeld fine-structure constant is a measure of the strength of electromagnetic forces that govern the interaction between electrons and photons and was first used to explain the splitting or "fine structure" of the energy levels of the hydrogen atom [64,65]. It is a dimensionless ratio that involves four physical constants: Planck's constant, the elementary charge (the square of), vacuum permittivity (or vacuum permeability), and the speed of light.

$$\alpha = \frac{e^2}{\hbar c 4\pi\varepsilon_o} = \frac{\mu_o c e^2}{2h} = \frac{1}{137.035\,999\,07(09)} \tag{16.13}$$

The measurement has a precision of 0.70 ppb [68] and can be determined directly from quantum Hall effect measurements, the anomalous magnetic moment of the electron, or by direct substitution of constants [69].

Most recently, a Sr (strontium) lattice clock at 1×10^{-16} fractional uncertainty was reported by A. D. Ludlow et al., using a calcium (Ca) clock for remote optical evaluation. (*Science*, 319, 1805–1808 (2008)). The researchers evaluated clock performance comparisons over km-scale distances. They found that such lattice-confined neutral atom clocks outperformed the best efforts of Cs-based atomic clock primary standards [70]. The clock is accurate to within one second in every 200 million years. The current Cs standard clock is accurate to 1 s in every 80 million years or so [71]. The prototype mercury single-ion clock shown in **Figure 16.5h** is accurate to 1 s in 400 million years.

Optical frequency combs are accurate and precise tools that measure frequencies. Beginning with lasers that emit a continuous tandem of short pulses (fs) that consist of millions of colors, the characteristics of light are then converted into frequency numbers (lines) that resemble a comb in appearance. The timing between the pulses ordains the spacing between the teeth of the comb. According to NIST, the teeth of the comb can be used to measure light emitted by lasers, atoms, stars, and other sources with high precision. Why do we bring up optical combs in our discussion of atomic clocks? Such combs greatly simplify frequency metrology while simultaneously improving accuracy. New kinds of atomic clocks are expected with 100x the accuracy of Cs clocks ($f = 9 \times 10^9$ Hz, in the microwave domain) as a result of this technology. Obviously "ticks of a clock" that correspond to optical frequencies are expected to be more accurate—oscillations at frequencies approaching 500×10^9 Hz! Nothing exists at the time that is capable of counting such a high number of oscillations directly; therefore, optical frequencies must be stepped down via gears to the microwave region. For more information concerning this latest development of nanometrology, please consult www.nist.gov/public_affairs/newsfromnist_frequency_combs.htm.

16.2.2 The Quantum Triangle

The age of atomic resolution and manipulation is well developed, but another age is emerging rapidly and impacting metrology and other fields of science and technology—single-electron detection and manipulation. Are we able to manipulate a single electron to do our bidding? Are we able to see a single electron? Are we able to measure the charge of a single electron?

In 1879, Edwin Hall, at the age of 24, discovered that a transverse electrical field is formed in a solid if placed in a magnetic field that is perpendicular to the flowing current. He placed a thin film of gold connected to a battery in a strong magnetic field. He found that a small but measurable voltage, V_H, proportional to the magnetic field times the current was generated across the film, orthogonal to the magnetic field and the applied current (**Fig. 16.6**). It's as if the electrons behaved like a "cloud of mosquitoes in a cross-wind, in this case a magnetic wind that pushes the electrons towards one edge" [72]. The Hall potential arises in the direction of the wind [72].

The QHE is obviously the quantized version of the continuous bulk form of the Hall effect that relates Planck's constant h and the elementary charge e. Nobel laureate Klaus von Klitzing discovered in 1980 that two-dimensional electronic systems placed in a strong magnetic field at low temperatures showed plateau (rather than continuous) behavior as a function of the number of electrons. He was awarded the Nobel Prize in Physics in 1985 for his contribution. He showed that the quantized steps of the Hall effect can be expressed with high

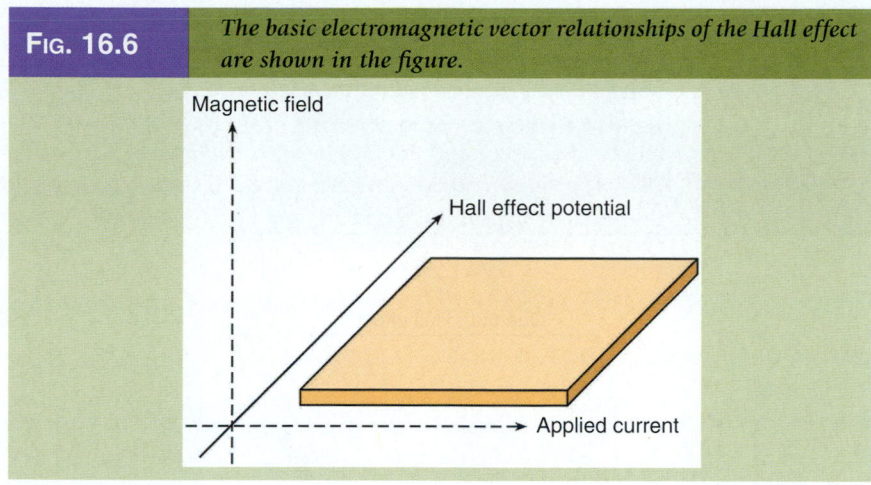

FIG. 16.6 *The basic electromagnetic vector relationships of the Hall effect are shown in the figure.*

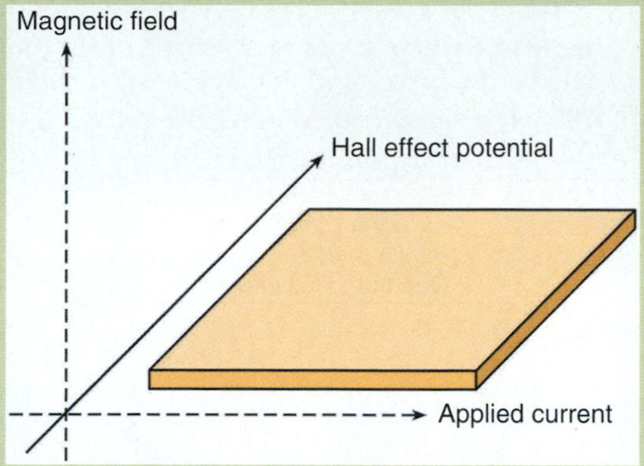

precision as integral steps proportional to the ratio between two fundamental physical constants: h and e. The von Klitzing constant R_K has become the new standard unit for electrical resistance.

Metrology, the Volt, and the Ohm. Impedance, temperature, strain, electrical, and power standards are mostly based on resistance standards, for example, the ohm [73,74]. Efforts are underway at NIST's AML (Advanced Measurement Laboratory) to redefine the legal ohm with quantum Hall resistance (QHR) measurements. With the use of cryogenic current comparators (CCC), direct scaling of the QHR up to 1 MΩ was possible in 2002. This level has since been amplified in 2004. High-ratio scaling is complemented by efforts to develop thin-film resistance standards for the *quantum metrology triangle* (QMT) (**Fig. 16.7**) [73,74]. The QMT relates voltage (V), current (I), and frequency (f) via a triangle consisting of the Josephson junction, quantum Hall, and single-electron tunneling effects (SET) that are important for nanometrologically precise electrical measurements [75]. Verification of the value of the elementary charge e and the fine structure constant α is possible by QMT experiments and calculations.

The challenge facing QMT experiments is to test the aforementioned Josephson voltage standard, the QHR, and the SET. Testing needs to be done in an extremely quiet high-ratio CCC combined with orders-of-magnitude larger SET sources. The ultimate goal is to produce useful levels of quantum current.

Single-electron metrology in the form of the single-electron electrometer has already become a viable technology [63]. The definition of single-electronics implies the controlled movement, positioning, and measurement of a single electron [63]. The potential effect of a single electron must not be understated, for example, a metallic nanosphere with radius 1 nm charged with one electron (charge, q or e = the elementary charge, 1.6022×10^{-19} C) produces an electric field on its surface in vacuum of $1.4\,\text{GV}\cdot\text{m}^{-1}$ [76]. This is enough to repel other electric charges. Please note that isolation of a single electron has not been accomplished—only that a single electron has been added to a preexisting electronic system. The "single-electron box" is a device that is capable of exhibiting single-electron charging effects. Consisting basically of a nanometal grain, a

FIG. 16.7 *The QMT: the relationships between frequency f, voltage V, and current I are functions of physical constants h and e, Planck's constant, and the value of the elementary charge, respectively. In this way, the current measure can be derived from measurement of frequency and not necessarily the kilogram (SI units) [63]. The current standard of the ampere is derived from the Josephson array voltage, the QHR, and Ohm's law [63].*

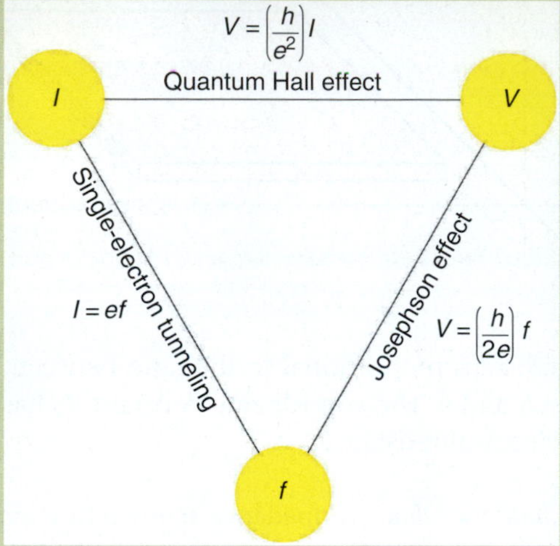

tunnel junction (a thin oxide coating) and a bias voltage energy source. Electrons can be added over a threshold voltage ($V_{\text{threshold}} = e/C$) to the particle according to the Coulombic energy relation

$$E_C = \frac{e^2}{2C} > k_B T \tag{16.14}$$

where C is the capacitance of the junction. Obviously the Coulombic energy must be greater than the ambient thermal energy $k_B T$ in order for quantum effects to be noticed [63].

As the bias voltage is raised, the electronic box can be filled with a predetermined number of electrons—resulting in a Coulomb blockade current (or charge) versus voltage behavior. Electrons can only be added or subtracted from a conducting island by tunneling—each time modifying the island potential by $\pm E_C$. Current is therefore inhibited (blocked) and is a classical case of Coulombic repulsion of charge carriers. Stepwise episodes exceeding the Coulombic blockade potential results in Coulomb staircase behavior. The current, therefore, through a tunneling junction is quantized. If a tunnel junction is under conditions of constant current I, charge accumulates on the tunnel junction similar to a capacitor and then single-electron tunneling oscillations appear with frequency

$$f = \frac{I}{e} \tag{16.15}$$

In superconductors, these are Cooper pairs that form Bloch oscillations.

$$f = \frac{I}{2e} \tag{16.16}$$

B. Josephson of Cambridge predicted that electrons should be able to tunnel in pairs, for example, Cooper pairs form between two superconductors separated by an insulating but very thin tunneling junction. A Cooper pair is a pair of electrons that can be coupled via lattice vibrations (phonons) over distances of hundreds of nanometers in superconductors (also condensation of electrons into a state that forms a bandgap). Are Cooper pairs an example of entangled electrons?

If DC voltage is applied to this junction, an AC current of frequency f is generated equal to $f = 2eV/h$. If an applied current of frequency f is applied, then a DC voltage equal to V_n would be generated: $V_n = nhf/2e$. The voltage measured across a Josephson junction V_J that is irradiated by a microwave EM field with frequency f_J is

$$V_J = n\frac{h}{2e}f_J \tag{16.17}$$

where n is the integral voltage step number [76].

The current generated by an electron pump is proportional to the charge q (an integral multiple of e)

$$I = qf \tag{16.18}$$

which is driven by an external frequency source (e.g., the clock frequency of the pump). Experiments have demonstrated accuracy of 1.5×10^{-8} at pA current levels [27]. From QHR and that $I = ef$, the Hall voltage is

$$V_H(i) = \left(\frac{R_K}{i}\right)ef \tag{16.19}$$

where R_K is the QHR (the von Klitzing constant).

The accuracy and precision of single-electron boxes depend on several factors [63]: (1) the Coulomb energy must exceed the ambient thermal energy as stated above, (2) capacitance C must be smaller than 12 aF to observe charging effects at liquid nitrogen temperatures and 3 aF at 300 K, (3) this requires that grain size be smaller than 15 and 5 nm, respectively, (4) for logic operations, the operating temperature has to be a factor of 50x lower and granules smaller than 1 nm, (5) quantum fluctuations must be negligible (number of electrons localized on any single island with no electron sharing between islands), and (6) all tunnel junctions must be opaque to electrons, for example, confined at a level of tunnel resistance [63]

$$R_K = \frac{h}{e^2} = 25812.807\ \Omega \tag{16.20}$$

where R_K is the integer QHE standard resistance that links voltage to current. Notice that the form of the QHE is similar to Ohm's law: $V = \Omega I$. The voltage standard (Josephson constant) is defined as

$$K_J = 483597.9\ \text{GHz}\cdot\text{V}^{-1} \tag{16.21}$$

The Josephson "superconductor–insulator–superconductor" junction when irradiated with microwave energy generates a DC potential determined by the frequency of the microwave radiation, the value of the elementary electronic charge, and Planck's constant. It is essentially a frequency-to-DC voltage converter [77,78].

$$e\left(R_K K_J\right) = in\left(\frac{f_J}{f}\right) \tag{16.22}$$

$$e\left(R_K K_J\right) \rightarrow a\left(\frac{f_J}{f}\right) = 2 \tag{16.23}$$

where f is the frequency of the SET pump cycle, f_J is the frequency of the microwave, a is a constant that represents a ratio of experimental integers such as the Josephson step number n, the Hall plateau number i, and the number of electrons pumped per cycle [77]. The dimensionless value is known to a combined relative uncertainty of 7.8×10^{-8} [77].

Hall resistance of semiconductors is measured in a magnetic field in which current flows in the x direction and the Hall voltage is measured in the y direction. The ratio of the two is the Hall resistance [63]. The AC Josephson effect implies that voltage applied across a Josephson junction (a superconducting tunnel junction separated by a thin barrier) produces an AC current with frequency proportional to $f = 2eV/h$. Conversely, irradiation of a Josephson junction with electromagnetic sources produces a voltage $V = hf/2e$ across the junction [63].

The accuracy of the QHR has been achieved to a few parts per 10^{10} and is established as a universal quantity DC standard for resistance that is independent of host material or device [76]. The reproducibility of the QHR is two orders of magnitude better than that of the ohm based on SI standards [76]. The von Klitzing constant R_K and the Josephson constant K_J have contributed significantly to a new age of electrical measurement, one quite well suited for nanoscale research and manufacturing. Work is underway to establish a primary AC resistance standard based on QHE [76].

16.2.3 The Single-Electron Transistor

A device that is able to combine atomic-scale resolution with ultrasensitive levels of charge detection is certainly desirable. Quantum mechanical tunneling and the Coulomb blockade are two fundamental phenomena responsible for the action of a SET [79]. The device is sensitive to nearby sources of electric fields, for example, electrons. N.B. Zhitenev and T.A. Fulton of Lucent Technologies in New Jersey have developed a scanning SET microscope that is capable of atomic resolution and detecting charge smaller than 0.001 e [79]. The limit of 0.001 e is the state-of-the-art value obtained at temperatures less than 4 K. The SETs of Zhitnenev et al. are based on the metal–metal tunnel junction developed by I. Giaever in 1960—two metal superconductors separated by a thin tunneling junction a few nanometers in thickness [79,80]. The device is shown in **Figure 16.8** below.

Recently, physicists in Finland and at Stony Brook University, New York, Jukka Pekola and Dmitri Averin, respectively, constructed a SET nanodevice that

FIG. 16.8

N.B. Zhitenev and T.A. Fulton explain the operation of the SET [79]. The input electrode is at a preset bias voltage. The output electrode is connected to the ground potential. The conducting island is insulated with two separate layers—thin films that serve as tunneling junctions. Current I flows from the input electrode to the output electrode. The gate with bias voltage V_G has capacitance C_G and is connected to the conducting island. Current is dependent on V, V_G, resistance R_J, and capacitance C_J of the junctions. On the voltage scale of e/C ($C = 2C_J + C_G$), the dependence of I is nonlinear. The current is a periodic function of V_G oscillation (as e/C_G) at constant V. A Coulomb blockade suppresses current over the range of V and VG at low V. Higher V demonstrates a dependence of current on 2R(V - e/2C).

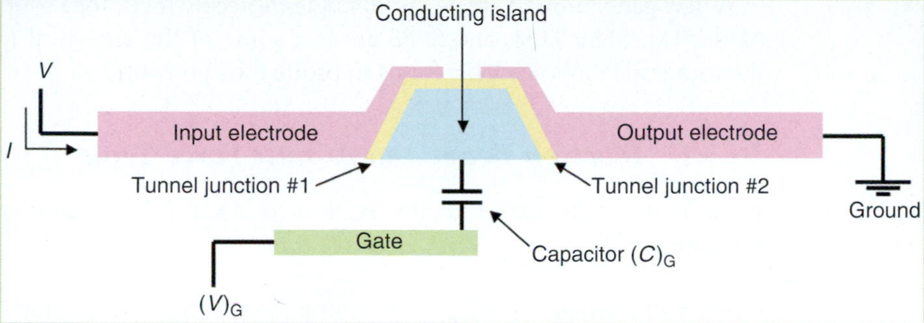

Source: N. B. Zhitenev and T. A. Fulton, Scanning single-electron transistor microscopy, In Encyclopedia of nanoscience and nanotechnology, Vol. 5, J. A. Schwarz, C. I. Contescu, and K. Putyera, eds., Marcel–Dekker (2004), Taylor & Francis (2005). With permission.

is able to measure current one electron at a time [81]. The device, cooled to 0.1 K, consists of a conducting island connected to two tunnel junctions—electrons flow in through one tunnel junction and out the other tunnel junction. A constant voltage was applied across the island junctions and an oscillating voltage applied to the gate electrode, similar to the scenario described above, to control the flow of electrons [82]. One electron is able to tunnel through the thin insulating junctions. Repulsion between electrons prevents more than one electron from tunneling through the junction at any one time [82]. Measurement of the amplitude and the mean value of the gate voltage allowed investigators to determine the current that flowed through the transistor per every cycle of oscillation

$$I = eN_e f \tag{16.24}$$

or the electron charge e times the number of electrons N_e times the frequency f of the gate voltage. The device has proven to be extremely precise but the accuracy needs to be improved—uncertainty is on the order of 10^{-2}. Accuracy could be improved, according to Pekola et al., if 10 such devices were placed in parallel. Such a configuration would produce 100 pA of current—enough to measure with greater accuracy [82]. This component would contribute to a complete "QMT" in which the current, voltage, and resistance are all related via the fundamental physical constants as described above, for example, Planck's constant h and the elementary charge e.

16.3 Nanometrology Tools

Pure nano analytical probe methods consist of electron beams and scanning microscopes. Both are capable of sub-nanometer resolution. Resolution of electron beam probes depends on the wavelength (or energy), the spot size of the electron beam and the sensitivity of the detector(s). Resolution of scanning probe microscope methods depends on the sharpness of the tip, the mechanical properties of the cantilever, the capability of the piezo-scanner, and the sensitivity of the input electronics and detector. Both methods offered the first glimpses into the nanoworld as they helped launch the *Nano Age*.

A new generation of single-atom/single-molecule metrology is upon us. The AFM, STM, SEM, TEM, and SERS are just some of the analytical (and hence, metrological) tools available for us to probe the nanoworld.

16.3.1 Electron Beam and Atomic Force Tools

We restrict our discussion in this section to electron beam and atomic force methods.

Electron Microscope Metrology. Scanning (SEM) and transmission (TEM) and scanning transmission (STM) electron microscopy are utilized in particle (or facet, hole) size determination and morphological characterization. Both techniques rely of the wavelength of the electron beam as the critical factor responsible for ultimate resolution. The wavelength of the beam for either technique is dependent on the de Broglie equation that we know so well by now

$$p = \frac{h\nu}{c} = \frac{h}{\lambda} \tag{16.25}$$

where

p is momentum
h is Planck's constant
λ is the wavelength
c is the speed of light
ν is the frequency

The classical form of the wavelength from energy is given by

$$\lambda = \frac{h}{\sqrt{2m_e eV}} \tag{16.26}$$

and for electron diffraction

$$n\lambda = 2d\sin\theta \tag{16.27}$$

where d is the interplanar distance.

Nanometrological inspection of nanocrystallites, nanotubes, quantum wells, and other sub-micrometer devices is often accomplished by SEM [83]. It is generally known that the size of nanometer structures appears larger and the distance between nanometer structures appears to be less than its real dimensions—when compared to TEM results [83]. The cause of the enlargement is linked to the radial distribution of secondary signal electrons, the effective resolution of the SEM, and the probe diameter *d*. Critical dimension (CD) SEM precision has evolved to 0.3 nm for the 45 nm technology node [84]. Recall that SEM images are two-dimensional projections of three-dimensional surfaces; therefore, it is difficult to provide accurate 3-D profile data, especially over large areas [85,86]. For this reason, SPMs (scanning proble methods) have greater potential than optical microscopes, SEMs, and surface roughness measure instruments for nanometrology of 3-D nanostructured surfaces [85,86].

SEM is one of the workhorses of the $200 billion semiconductor industry. The ITRS (International Technology Roadmap for Semiconductors) states that SEM will continue to provide at-line and inline imaging for characterization of cross-sectional samples, particle and defect analysis, and CD measurements [87]. Customers require effective CD (critical dimension) and defect review to transcend the 45-nm threshold. The semiconductor industry requires that SEM standard artifacts related to magnification, calibration, performance, and measurement of linewidth be addressed [87].

Scanning Probe Microscope Metrology. Nanoscale metrology with AFM is correlated with a unique set of issues. Errors such as overestimation of the size of positive features (e.g., facets) and underestimation of the size of negative features (e.g., pores) are a direct byproduct of the sharpness of the tip, contact angle with the surface, and the sensitivity and control of the force application system. There are four underlying sources of image artifacts in AFM methods: probes, scanners, image processing, and vibrations. AFM images, for example, are the result of convolution of the object of interest and the geometry of the probe [88]. A general rule of thumb is that if the probe is much smaller than the feature on the surface, probe-induced artifacts are minimized and the image is reproduced with good accuracy. For example, if a 100-nm feature is imaged with a 10-nm diameter probe, a good image should be acquired with a minimal number of artifacts [88]. On the other hand, if the probe tip is bigger than a protruding feature, images appear larger than the object are recorded (**Fig. 16.9**). On the flip side, if imaging of pores (holes) is desired, the pore diameter appears smaller if a larger probe tip is applied. Regardless, in most cases, although the depth profile may be inaccurate with a large probe, the pore diameter and the order in a pattern may still be accurately imaged. The shape of objects may be distorted as well with large probes. With some larger silicon probe tips, artifactual triangular patterns represent the geometry of the tip rather than the geometry of the feature [88].

Artifacts can be induced by other factors associated with AFM image acquisition and processing. Nonlinear motion, hysteresis, and the geometry of the ceramic piezoelectric scanner may distort images. The major sources of error in large-area measurements are due to the angular motions of the scanning actuator/stages [85,86]. Probe-sample angles less than nonperpendicular (or close to 12° off of normal) can be corrected in some AFMs. The artifact produced in this case is an image that may appear smaller on one side than the other. "Edge overshoot," in which one side of the feature is distorted, can be caused by hysteresis. Scanner drift, in which distortion is observed at initial contact with a feature, is caused

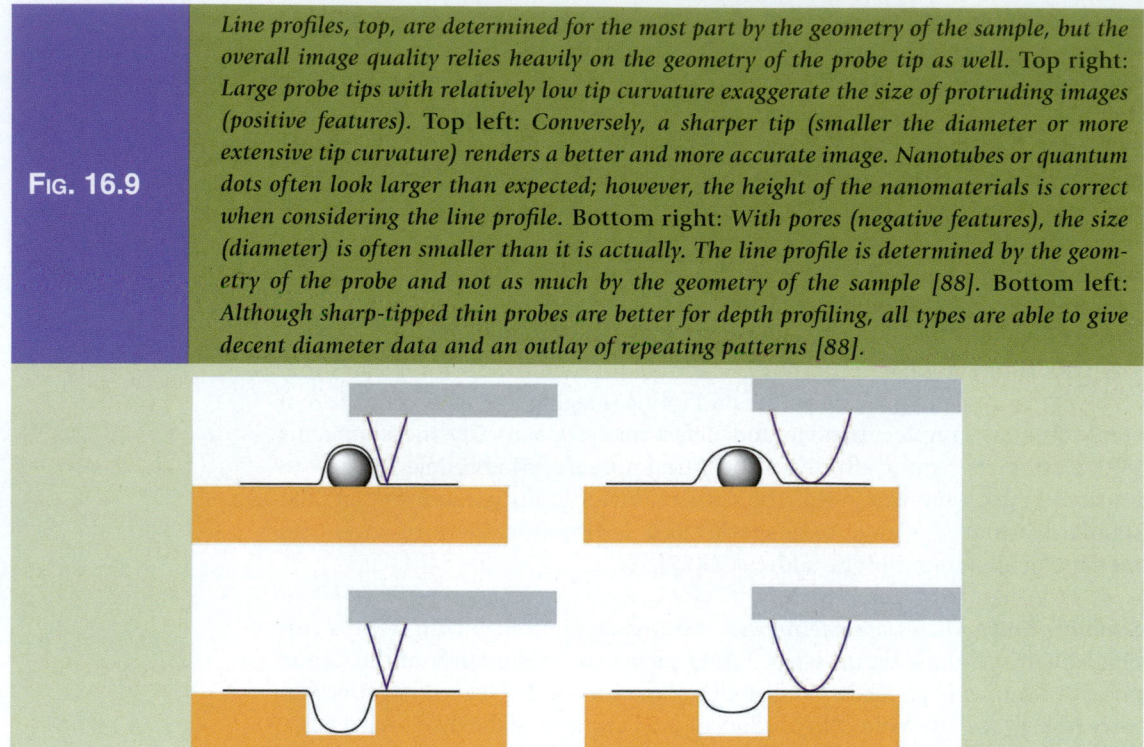

FIG. 16.9 *Line profiles, top, are determined for the most part by the geometry of the sample, but the overall image quality relies heavily on the geometry of the probe tip as well.* Top right: *Large probe tips with relatively low tip curvature exaggerate the size of protruding images (positive features).* Top left: *Conversely, a sharper tip (smaller the diameter or more extensive tip curvature) renders a better and more accurate image. Nanotubes or quantum dots often look larger than expected; however, the height of the nanomaterials is correct when considering the line profile.* Bottom right: *With pores (negative features), the size (diameter) is often smaller than it is actually. The line profile is determined by the geometry of the probe and not as much by the geometry of the sample [88].* Bottom left: *Although sharp-tipped thin probes are better for depth profiling, all types are able to give decent diameter data and an outlay of repeating patterns [88].*

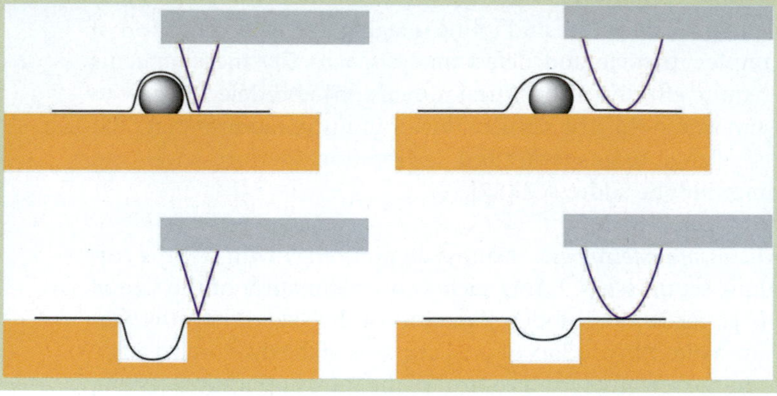

by thermal effects induced by external sources. Most of these kinds of distortions can be detected by imaging a calibration standard.

Image-processing artifacts are more rare. Too much application of low pass filters may cause steps in features to be distorted (too much smoothing)—too little may result in sharpness at steps. Synthetic periodic structures that appear to be the atomic structure may also be introduced by the use of Fourier filtering. Vibrations absolutely cannot be tolerated during image acquisition of atomic structure (or in actuality, any microscopic structure). A building floor acts as a "transducer" for many sources of vibrations, both external and internal. Acoustic vibrations (e.g., a voice) are not desired, especially during image capture. Other sources of AFM artifacts arise from surface contamination, faulty electronics, and vacuum leaks. An excellent article by Paul West and Natalia Starostina [88] of Pacific Nanotechnologies, Inc. (LOT-Oriel Gruppe Europa) can be found at www.lot-oriel.com/site/site_down/pn_artifacts_deen.pdf.

Analysis by scanning probe (SPM) techniques such as atomic force (AFM), chemical force (CFM), friction force (FFM), and atomic force acoustic (AFAM) microscopies rely on substrate surface chemistry (surface energy) for contrast. A new nanometrology, developed by the NIST in Gaithersburg, Maryland, is based on gradient micropatterned (∇μp) substrates intended to serve as reference substrates for SPM calibration [89]. The microscale chemical pattern-surface energy gradients provide a means of calibrating image contrast against traditional chemical measures such as contact-angle determination and surface spectroscopy [89]. The patterns were formed from a combined microcontact printing/vapor deposition of monochlorosilane self-assembled monolayer on silicon substrates [89].

Veeco Instruments, Inc. has recently introduced a new high-throughput AFM designed for quality assurance testing of semiconductor wafers [90]. The *InSight 3D Atomic Force Microscope* platform is capable of accuracy and precision for nondestructive, high-resolution three-dimensional measurements of 45- and 32-nm semiconductor features. Approximately 30 wafers can be processed per hour. According to Veeco, the new instrument is able to *"address critical dimension (mid-CD precision = 1.5 nm or 1% @ 3σ), depth (1 nm or 1% 3σ) and chemical mechanical planarization (CMP) metrology in a production environment."* The *InSight 3D AFM* is designed to compete with other on-line nanometrology techniques. *"CD-SEM and scatterometry are precise methods but not accurate enough, causing significant measurement issues. Veeco's InSight provides the lowest measurement uncertainty for CD metrology, which leads to improved process control."* The new instrument provides twice the accuracy of previous AFMs. Improvements in the X–Y stage, auto-focus laser pattern recognition system, new probe designs, and system reliability are responsible for the enhanced capability [90].

Atomic resolution is required in scanning probe devices. The ability to "see" single electrons and even smaller levels of charge is also important. The combination of a sharply refined scanning probe and a SET should be a powerful combination that enables us to achieve atomic resolution and minimal levels of charge [79].

16.3.2 Spectroscopic Tools

Single-Molecule Measurement and Imaging. Single-molecule measurement and imaging has acquired significant importance in the area of biomolecular study. In particular, the study of enzyme action offers a gateway into understanding the mechanisms of nanomachines [91]. Since the early 2000s, new methods have evolved to detect single biomolecules. For instance, study of biomolecules by using a dielectric bead with an attached biomolecule trapped by optical tweezers has been demonstrated recently [92–94].

Two new methods have gained prominence: super-resolution imaging using single-molecule fluorescence and single-molecule trapping by Brownian motion suppression [91]. Insights into analysis of single-molecule spectra have also made progress in the past few years [95].

Single-molecule spectra are easy to obtain in solid samples. However, to study biological molecules in situ is challenging due to Brownian motion effects exhibited at room temperature, for example, the diffusion of a 10-nm nanoparticle through a 0.3-μm laser spot area (or 100 fL) occurs in less than 1 ms [91]. Various means have been devised to circumvent this difficulty: Immobilization on a transparent glass surface, encapsulation within the aqueous environment of a gel pore or a vesicle, or application of the tethered bead-laser tweezer method mentioned earlier [91]. The ABEL (anti-Brownian electrokinetic) trap is able to constrain single molecules with nanoscale resolution [96]. The trap is able to monitor the Brownian motion of the particle by fluorescence microscopy, and then by applying a feedback voltage to control the electrical drift the Brownian motion is cancelled. The trap works well on nanoscale materials (molecules) that are able to provide an optical image and an electric charge [96]. A.E. Cohen et al. in 2005 have shown that the ABEL trap is capable of trapping fluorescent polystyrene nanospheres with diameter as little as 20 nm [96]. According to the authors, the ABEL trap can be applied to single-molecule spectroscopy, nanomanufacturing, and the detection of biological particles [96].

Single-molecule detection procedures have been applied to the analysis of bacterial neurotoxin (a protease) translocation across membranes [97]. In their 2007 paper, A. Fischer and M. Montal reported the analysis of the dynamics of *botulinum* neurotoxins by single-molecule assay with millisecond resolution. From this they were able to determine the mechanism of transport of the toxin across cell membranes. Botulin neurotoxins are composed of two protein chains linked by a disulfide bond: a 50-kDa light chain (LC, the cargo) and a 100-kDa heavy chain (HC, the channel/chaperone). They concluded that the neurotoxin transit occurred according to the following scheme: an insertion step (with LC unfolding), entry into the HC channel, conduction of the LC through the HC channel, release of the LC cargo (reduction of the disulfide bridge), and LC refolding into the cytosol [97]. The processes were monitored by way of single-channel conduction currents.

Short Wavelength Sources. Short wavelength sources include x-rays and free-electron lasers (FEL) that are able to produce short coherent photon pulses. The underlying micro- and nanostructure of materials plays an important role in the nano-, micro-, and macroscopic properties of novel materials that can be probed by short-wavelength techniques [98]. Characterization of such nanostructures is therefore a prerequisite to engineering design of devices, for example, that consider the degree of micro- and nanostructural properties like strain and stress, crystalline defects, and texture [98]. X-rays are sensitive to structural features on the scale of the chemical bond. In addition to standard static x-ray diffraction, structural changes triggered by infrared radiation can be measured by short x-ray pulses [99].

Synchrotrons produce stable and intense x-ray beams with low divergence. Picosecond pulses (10–400 ps) make this source suitable for the study of dynamic behavior with nanosecond time resolution [99]. Time-resolved x-ray diffraction is an excellent technique with which to measure phase transitions in nanomaterials. However analysis of living systems by synchrotron-based x-ray microscopes is limited to 20 nm due to the onset of radiation damage. Application of higher energy pulses is a means to overcome this limitation. Single-particle diffraction and flash imaging using wavelengths from 20 nm down to 0.2 nm allows for single-molecule resolution by x-rays [99]. To image noncrystalline samples, the photon pulses need to be shorter than the time it takes to damage the molecule (e.g., <100 fs and 10^{11} photons per 100-nm spot) [99]. Interest in solving the structure of noncrystallizable proteins is one of the driving forces behind this technique.

Larger structures require longer wavelength sources. Free-electron lasers produce photons that overlap with the soft-x-ray domain. The wavelength of such devices has been decreased from 32 nm down to 13 nm. By means of <6-nm wavelength sources, single-particle imaging by x-rays are on the horizon.

Scanning Nano-Raman Spectroscopy (SNRS). SNRS, also called tip-enhanced Raman spectroscopy (TERS), has been shown to be an effective means of analyzing nanoscale silicon-based structures [100]. SNRS or TERS with side-illumination optics have been shown to optimize near-field and far-field (background) Raman signals by optimizing the beam polarization [100]. The breakthrough yielded signals with an order of magnitude improvement in contrast with a lateral resolution of Raman images of ca. 20 nm [100].

Surface-Enhanced Raman Spectroscopy (SERS). An advanced form of Raman spectroscopy (SERS), (or SERRS—surface enhanced resonance Raman spectroscopy or scattering) is well known for attomole (10^{-18}) or better levels of detection with improvements since the discovery of the surface enhancement phenomomen in 1974 [102]. SERRS relies on the coincident overlap of the chromophore transition with the excitation frequency. The SERRS system includes silver nanoparticles upon which the analyte is able to adsorb the appropriate excitation energy [103]. From configurations like these, single-molecule detection has been achieved using SERRS [104,105]. According to A. Macaskill et al., SERRS is a more sensitive and specific detection method for the direct analysis of DNA compared to fluorescence measurements because SERRS is able to produce fingerprint spectra. This renders the technique better suited for multiplexed identification of target molecules without the need for separation. [103]. With fluorescence techniques, broad and overlapping spectra often limit its usefulness in that regard [103]. DNA by itself does not function as a classical chromophore, but attachment of an appropriate dye molecule (a fluorescent label) renders it so.

Single-Molecule (Atom) Spectroscopy and Detection. Single-molecule (atom) detection is easily obtainable via SPM methods. On the surface, it would seem that this goal would be more difficult to accomplish via spectroscopic methods, but, nonetheless, this too has also been achieved. Single-molecule spectroscopy and single-molecule detection provide important data especially when coupled with other scanning, imaging, and spectroscopic analysis [106]. Researchers in Genoa, Italy have managed to combine a commercial confocal (or multi-photon) scanning head, a mode-locked titanium–sapphire laser, and an inverted microscope to detect single-molecule fluorescence of indo-1, rhodamine-6G, fluorescein, and pyrene—the result of the ability to selectively collect signals from highly confined volumes [106]. Dilute solutions of the dyes were applied to glass slides by the process of spin coating.

Scatterometry. CD and other profile properties can be obtained from periodic structures by a procedure known as scatterometry. Scatterometry analyzes the light (usually extreme ultraviolet, $\lambda = 13.5$ nm) diffracted from such structures. According to the 2004 ITRS, 3-σ metrology of at least 0.2 nm is required for the 32-nm nodes by 2013. Currently, 0.7-nm precision is available for printed and physical isolated lines of 90-nm dimension [84].

Scatterometry is based on the analysis of collimated light reflected from a grating (or structured) surface. The resulting diffraction pattern relies on the wavelength of the incident light and the period (structure) of the features on a surface. For example, a perfectly flat surface would not produce a diffraction pattern of any order. Reflection depends on

$$R = \frac{I_{\text{out}}}{I_{\text{in}}} \tag{16.28}$$

where I_{out} and I_{in} are the intensities before and after reflectance. The quantity R depends on the grating pitch (the period), geometry, refractive index, number of multilayers, and base substrate properties [84]. The factors that influence R from the perspective of the incident beam include wavelength, angle of incidence, polarization, and azimuth angle [84]. Obviously the reflectance profile will be complicated. Analysis is therefore accomplished by comparing data to a reference profile derived from Maxwell's equations [84]. For example, if the method is required to deliver a measurement resolution of 0.1 nm, then the method must be able to *differentiate between the fits of two-signature curves 0.1 nm apart in the line or CD—within the constraints of the signal-to-noise ratio* [84]. Counting bin numbers over all pairs of spectra generates uniqueness scores. Hopefully, in all cases, apples are being compared to apples.

Manufacturing process control factors include (1) targets fitting within a prescribed area that are a function of the effective spot size of the incident probe, (2) scribe CD structures are employed to track processing shifts, and (3) characterization is required of two-dimensional and three-dimensional structures by providing grating average values for CD, sidewall profile, thickness, and optical properties of the stack [84]. The industry uses both angular- and wavelength-dependent scatterometry. Issues include (1) that the process works best with periodic structures, (2) relies on accurate values of optical properties of materials, (3) standards are needed (as indicated above) to validate the process, and (4) may become limited when analyzing structures with ever smaller dimensions [107].

16.3.3 Nanomechanical Tools

Challenges of Nanomechanical Metrology. Nanomechanical methods face, according to the NNI report on instrumentation and metrology [18], numerous challenges and barriers. It does seem like everything is a challenge in nanometrology, but the status quo represents, without question, the dynamic state of flux that nanometrology is undergoing and the importance of nanotechnology. **Table 16.1** lists an edited review of challenges and barriers facing nanomechanical metrology [18].

Mechanical behavior for many small, synthetic and biological systems is not well known. In order to get the most out of mechanical devices (e.g., performance and reliability), mechanical property measurements and testing need to be conducted. Properties such as localized strength and fracture toughness of materials and interfaces and the effect of surface phenomena need to be studied [108]. Depth-sensing nanoindenters have application as a universal testing machine. The Manufacturing Engineering Laboratory of NIST has devised a compression load test configuration using a nanoindenter that yields information on applied load and load-point displacement [108]. Such testing coupled with finite element analysis (FEM) should be able to mitigate the need for mechanical information about nanomaterials.

Nanoindentation. Quantitative nanoindentation methods are the most versatile mechanical testing systems applied to nanomaterials and nanodevices, along with AFMs. There are over 1000 commercial nanoindentation instruments available [109]. The nanoindentation technique yields information about hardness

TABLE 16.1 *Challenges Facing Nanomechanical Metrology*

Metrology	Challenges
Standards and calibration	Traceability of force and displacement, international collaboration on standards, nanoscale forces, and contact mechanics
Modeling	Computational power, data storage, and mining
Experimentation	Testing fixtures, sample development, positioning, and manipulation
Multiple technique integration	Atomic/molecular-scale resolution, single-event spatial resolution, time and position information synchronization
Automated measurements with high throughput	Speed, automation, yield, quality, size, and conditioning Robust probes, periodic reference standards Lack of wide range of testing environments (T, *f*), lack of models, high-speed methodologies, and well-characterized nanoscale probes
New instruments	Tip wear, control, cm to nm positioning, decoupled lateral and vertical force sensors and need for standards, quantitative mechanical property mapping
Real-time measurements	Contact area, surface treatment, robustness, real-time measurement, interfacial testing, thermal drift

and elastic properties from force, area, and displacement data acquired from indentations on the order of 5–10 nm [109]. However, worldwide, there exists a dearth of force calibration, standardized methods, and standard reference materials. Displacement is easily measured from interferometry but force measurements pose more problems due to reliance on artifactual SI standards (e.g., the kg mass). NIST is working on ways to trace force measurements to electronic and length SI units in the range between 1 mN and 10 nN [109].

Nanotribology and Surface Properties. There is a need to establish standards and methods to measure nanoscale adhesion, friction, and surface forces and characterization of surface properties (e.g., texture) at the nanoscale [110]. Normal and lateral forces of nanocontacts in particular need to be characterized at the nanonewton level. Understanding how meniscus and electrostatic forces affect friction measurements is an ongoing concern. Thin-film stress measurements are important to determine the robustness of electrodeposited thin films [111]. Mechanical stresses develop during nucleation and growth in such films and lead to loss of adhesion and the generation of bulk surface defects. Cantilever-based devices have been devised that are able to measure and resolve forces on the order of $0.03\,\mathrm{N \cdot m^{-1}}$ of films as they form. For example, profiles of stress formation during the deposition of 250 nm of Au on borosilicate glass revealed that there is a rapid rise in tensile stress during the first 20 nm of deposition [111]. It was also determined that the highest tensile stresses correlate with high nucleation densities obtained at more negative electrodeposition potentials.

Adhesion. Adhesion and mechanical properties characterization from axis-symmetric testing is a promising method to quantify adhesive performance while studying debonding mechanisms simultaneously [112]. Traditional "one-per" tests (serial testing) do not offer enough throughput, and as a result, combinatorial methods have become more attractive. The Johnson–Kendall–Roberts method (JKR) utilizes a hemispherical lens that is pressed into a sample. By monitoring the contact area between the lens and the sample, it is possible to describe the work of adhesion [112]. The Multilens Combinatorial Adhesion Tester (MCAT), equipped with an array of microlenses, is able to track the contact radii of 400 specimens simultaneously.

Peel tests, one of the primary methods applied by industry to evaluate adhesives, is a method that also suffers from the limitations of serial testing. Combinatorial and other high-throughput methods have also been applied to accelerate the rate of adhesive performance testing [113]. A thin polymer applied to a substrate is characterized by a surface energy gradient and simultaneously, an orthogonal temperature gradient. Failure maps can be constructed that reveal the transition from adhesion to failure as a function of surface energy, annealing time, and annealing temperature [113].

Computer Modeling of Nanomechanical Behavior. Nanoscale mechanical failure is difficult to measure [114]. Finite element method (FEM), classical atomistic simulations, and density functional theory (DFT) offer means of addressing this dilemma. FEM is an excellent procedure with which to model elastic behavior of bulk materials but comes up short with regard to failure analysis because failure mechanisms depend on atomic scale behavior. Classical atomistic simulations are able to handle thousands to millions of atoms but fall short at the chemistry end of the spectrum. DFT is rather accurate and describes the chemistry accurately. Unfortunately, even in this day and age, central processing unit (CPU) time is at a premium and DFT models are not able to correlate more than a few hundred atoms simultaneously. Use of all three models is often required to understand mechanical behavior from the atomic to the macroscopic level [114]. The results of computer modeling techniques are correlated with experimental nanoindentation and AFM data to provide quantitative predictions of material behavior.

Roundness. Roundness measurements, mentioned briefly earlier, determine the circularity of the contour of a 3-D object. Roundness is the quality or state of being round. Eccentricity is the deviation from roundness, but in general, out-of-roundness includes the presence of surface defects as well. Roundness has been traditionally evaluated according to NBSIR 79-1758, a document created but the National Bureau of Standards (now NIST). The world's highest accuracy for roundness measurements are obtained on a Mitutoyo RA-H5000 located in the AML of NIST in Gaithersberg, Maryland. Measuring uncertainty of 3 nm and long-term repeatability within 1 nm have been achieved with this instrument [115].

16.4 Nanometrology and Nanomanufacturing Standards

According Sarah Gale of *Small Times*, the magazine of the MEMS and nanoworlds, "*... the rush to develop standards for nanotechnology that the world will embrace is in* *full throttle*" as metrology and standards organizations worldwide are converging to establish standards for nanotechnology [116]. There are several key ingredients required to realize this ambitious goal including generalized components such as defining a common language, material dimensions and quality, and the foundation for nanomanufacturing practices [116].

16.4.1 Standards for Nanotechnology

Just like the NNI and the National Institutes of Health, other organizations have established plans of attack concerning the development of nanotechnology including several metrology organizations that are involved in the process of doing exactly that—creating a roadmap. Some of the most important standards organizations are listed in **Table 16.2**.

All of these organizations realize that global cooperation is imperative if universal standards regulating nanometrology, practices, and standards (and nanomanufacturing) are to play a viable role in the commercialization of nanotechnology. An excellent source of nanotechnology standards news can be found online at Nanotechnology Standards News (IHS Engineering), (http://engineers.ihs.com/news/topics/nanotechnology-standards-news.htm) [117].

The Institute of Electrical and Electronics Engineers (IEEE) has already released a standards document for nanotechnology and an overview of the roadmap to guide standardization efforts [118,119]. The IEEE roadmap reveals the complexity of developing standards for new materials, especially materials that exist at the nanoscale. The complexity and challenge of developing the IEEE roadmap can be seen by reference to the draft which was made available for the public in April, 2007 [120]. The roadmap identifies 25 key technologies, 74 standards opportunities, and 13 "high priority" technologies.

TABLE 16.2 *Standards and Testing Organizations of Note*

Organization	Acronym	Web Address
American Institute of Chemical Engineers	AIChE	www.alche.org
American National Standards Institute	ANSI	www.ansi.org
American Society of Mechanical Engineers	ASME	www.asme.org
American Society for Testing and Materials International	ASTM	www.astm.org
British Standards Institute	BSI	www.bsi-global.com
Institute for Electrical and Electronics Engineers	IEEE	www.ieee.org
International Organization for Standards	ISO	www.iso.org
International Union of Pure and Applied Chemis's	IUPAC	www.iupac.org
National Institute of Advanced Industrial Science and Technology	AIST	www.aist.go.jp
National Institute of Standards and Testing	NIST	www.nist.gov
National Sanitation Foundation International	NSF	www.nsf.org
Semiconductor Equipment and Materials International	SEMI	www.semi.org

Table 16.3 *Terminology for Nanotechnology (ASTM E 2456-06)*

Term	Definition	Term	Definition
Agglomerate	A group of particles held together by relatively weak forces (e.g., van der Waals or capillary) that may break apart into smaller particles upon processing.	Nanotechnology	A term referring to a wide range of technologies that measure, manipulate, or incorporate materials, and/or features with at least one dimension between approximately 1–100 nm. Such applications exploit the properties, distinct from the bulk/macroscopic systems, of nanocomponents.
Aggregate	A discrete group of particles in which the various individual components are not easily broken apart, such as in the case of primary particles that are strongly bonded together (e.g., fused, sintered, or metallically bonded particles).	Nanostructured	Containing physically or chemically distinguishable components, at least one of which is nanoscale in one or more dimensions.
Nanoscale	Having one or more dimensions from approximately 1–100 nm	Non-transitive particle	A nanoparticle that does not exhibit size-related intensive properties
Nanoscience	The study of nanoscale materials, processes and phenomena, or devices.	Transitive nanoparticle	A nanoparticle exhibiting size-related intensive properties that differ significantly from that observed in fine particles (100 nm < fine particle < 2.5 μm) or bulk materials.

IEEE. At the tail of 2005, the IEEE released a standards document that defines test methods for measurement of electrical properties of carbon nanotubes—*IEEE 1650-2005* [118]. It was the first formal effort to identify the minimal information required for reporting results for a nanomaterial [116]. The purpose of the document was to ensure that results can be reproduced in other labs and thereby proven—and on a large scale [121].

ASTM. ASTM (American Society for Testing and Materials) formed in 1898 by C.B. Dudley specifically to support the railroad industry develops and publishes voluntary consensus technical standards for technology and materials. It is considered by many to be the world's largest organization dedicated to standards. ASTM standards are developed from numerous volunteer consensus technical committees. ASTM developed one of the first nanotechnology standards in 2006 [122]. Logically, the first standard should revolve around "standard terminology related to nanotechnology." We have mentioned time and time again about the importance of language in science. Selected definitions originating from ASTM E 2456-06 are provided in **Table 16.3**.

16.4.2 NIST Efforts

The National Institute of Standards and Testing has undertaken efforts to develop metrology and standards for nanomaterials and processes. Part of the commitment is fully visible in the form of NIST's new AML in Gaithersburg, Maryland. NIST has also recently opened the Center for Nanoscale Science and Technology in 2006 that contains a nanofabrication facility and focuses on instrumentation research, metrology, and standards development. Standard reference materials are also being developed for the characterization of nanomaterials, for example, standards for gold nanoparticles ranging in diameter from 1 to 100 nm for instrument calibration, thin films, single-phase nanoscale particles for electron beam, and ion beam analytical imaging instrumentation. The Nanotechnology Characterization Laboratory (NCL) at the National Cancer Institute (NCI) is conducting tests on the efficiency and toxicity of nanoscale particles—in vitro and in vivo. The Department of Energy (DOE) laid the groundwork for five Nanoscale Science Research Centers (NSRC) across the country. Developing new tools and methods for nanomaterials include the transmission electron aberration corrected microscope (TEAM), dedicated to developing the next generation TEMs.

ANSI. Standards are in the process being developed to accommodate nanotechnology. In June 2004, the American National Standards Institute (ANSI) established a panel to help coordinate national efforts in standards development. The International Organization for Standardization (ISO) followed suit shortly thereafter. The international arm of ASTM, known as ASTM-International, is working with ANSI to develop and standardize nanotechnology terminology, characterization methods, health and safety protocols, intellectual property management, and international liaison and cooperation. ASTM has also formed a partnership with IEEE, the American Society for Mechanical Engineers, Semiconductor Equipment Manufacturers International, the International Institute for Chemical Engineers, and the National Sanitation Foundation to develop global terminology and standards.

Work is underway globally to develop (1) methods to detect nanomaterials within biological environments, (2) methods to understand heterogeneity in nanomaterials, (3) methods for measuring purity and metrology of particle size, (4) methods and standardized tools to assess nanomaterial shape, structure and surface area, (5) to detect and measure nanomaterial spatio-chemical composition, and (6) an inventory of nanomaterials and their use.

16.4.3 IEEE Roadmap for Nanoelectronics

The purpose of the IEEE Standards Association Nanoelectronics Standards Roadmap is to frame the big picture view of nanoelectronics standards, both in the near and far terms [120]. This includes creating standards for materials (nanoparticles such as nanopores, quantum dots, quantum-confined atoms, nanospheres, nanorods, nanoshells, complex nanoparticles, nanocrystals, and magnetic nanoparticles and ferrofluids); and nanoscale conductive interconnects like nanotubes and nanowires; and devices like sensors, storage devices, emitting, and switching.

All categories need to be identified and provided with standards—whether manufacturing or material—if commercialization is to proceed efficiently worldwide. Many of the standards will be extensions of preexisting standards that already support micron-level industrial practices developed over the past 20 years or so. Others are brand new and will have to be developed as we go along.

Equally impressive is another list provided in the IEEE roadmap that addresses "behaviors to exploit." The list is a "who's who" or more appropriately, a "what's what" of nanotechnology. It is a summary of the many properties and phenomena that are exhibited by nanomaterials and by itself is a valuable pedagogical tool.

16.5 Nanomanufacturing and Molecular Assembly

Nanomanufacturing is already a reality. Components for our computers already are stocked with nanosized transistors (<100 nm). However, bottom-up techniques are making headway into nanomanufacturing. Because nanotechnology is amenable to bottom-up synthesis, one would expect that chemical and biological methods of synthesis be applied on a larger scale. There are however, growing pains associated with whole-scale uses of bottom-up synthesis methods. The development of micro- and nanomanipulators offers another route to devices and products. And lastly, biomolecular assembly, a process that is based on natural assembly lines characteristic of living things, offers us a unique look at bottom-up assembly adapted to a synthetic world.

16.5.1 Lithographies

The accurate placement of patterns is a critical element in future nanomanufacturing practices. Current nanoelectronics, nanophotonics, and nanomagnetics all require the application of patterns during some part of their assembly. More accurate pattern gratings are needed for 2-D planar nanomanufacturing to become a reality. Solutions to pattern placement problems include eventual elimination of laser interferometers. M.L. Schattenburg of MIT, who states that current metrology tools are inadequate for nanomanufacturing, provides an intriguing overview of nanometrology in nanomanufacturing [17]. In general and historically, two basic criteria are required for viable manufacturing of 2-D systems: (1) CD metrology that regulates feature size and (2) accurate pattern placement (location and/or overlay of patterns on top of complementary structures). Rapid, reliable, and accurate inspection of products during assembly is another need. Currently, SEMs for CD evaluation have limited accuracy, AFMs equipped with nanotips are too slow, and laser interferometers are inadequate for metrology at this small scale [17]. The critical metrology component is the "metrology frame." The stable metrology frame consists of a length scale and a way to compare a manufactured component with that length scale. Microscopes, for example, traditionally are used to accomplish the latter.

According to Schattenburg, metrology frames "underpin all accurate metrology tools" that include lithography mask (reticle) writers, lithography scanners and steppers, CD metrology tools, pattern placement and overlay metrology tools, circuit and mask repair, and coordinate measuring tools [17]. Just to provide a perspective, according to the ITRS [17,123], metrology milestones indicate the direction the semiconductor industry is taking over a 15-year horizon [123]. From 2001 through 2016, CD is expected to decrease from 130 nm down to 22 nm; overlays from 45 nm down to 9 nm; mask image placement from 27 nm down to 6 nm. The corresponding frame errors, once again from 2001 through 2016, are expected to decrease from 11 nm down to 2.3 nm; and the length scale error from 2.8 nm down to 0.6 nm (based on MIT research efforts) [17]! From 2005 on or so, solutions are in the works and some issues, the overlay parameters in particular, have no working solutions at this time.

Now, let's discuss an amazing (little) device called the MIT Nanoruler. C.G. Chen and M.L. Shattenburg et al. of MIT in 2004 reported the development of a ruler based on a new form of interference lithography capable of large-scale work that is fast, accurate, and precise. The nanoruler was designed to pattern large gratings. For example, a 400-nm period grating was etched onto a 300-mm wafer in ca. 20 min [124]! The grating pattern lower limit of the 2004 device was 150 nm. The tool is housed in a cleanroom (housed in a larger cleanroom) that holds temperature fluctuations to ±0.005°C. Scanning beam interference lithography (SBIL), a concept conceived by Schattenburg, possesses precision "better than 1 nm" across a 300-mm wafer. SBIL consists of two narrow (2 mm diameter) ultraviolet laser beams that generate interference pattern fringes. A spot formed by the fringes is scanned linearly across the wafer's surface. The SBIL writes the grating with thousands of parallel fringe spots.

16.5.2 Nanomanipulators and Grippers

Manufacturing that is accomplished at the macroscale, like automobile assembly lines, is driven to a great part by automated components, for example, robots. We now explore their micro- and nano-counterparts. Manipulators with high positioning accuracy and stages with nanometer sensitivity are well represented in today's marketplace.

Nanomanipulators. Nanomanipulators are often coupled with scanning electron microscopes to provide a view of field operations for research, development, and production applications. Some models, like the *S100 nProber* (sub-100-nm) manufactured by the Zyvex Corporation, have four positioners that grasp, move, test, and position nanocomponents to accommodate four axes of movement. Transfer of samples and interfaces with preexisting laboratory equipment are possible with such devices. These nanomanipulators are equipped with a joystick and keypad to provide tight control over manipulation. **Figure 16.10** displays images of selected nanomanipulators and grippers, courtesy of the Zyvex Corporation. Zyvex also manufactures high aspect ratio *NanoEffector Probes* (tip radius <50 nm) that are able to probe small features with 50-nm contact and small geometries (four 50-nm contacts within 100 nm of each other). Eight probe tips can be utilized in a 500-nm workspace and is utilized by Zyvex's *nProber*. For more information, please consult Zyvex's website at www.zyvex.com.

Fig. 16.10 *(a) A Zyvex nProber is capable of sub-100-nm manipulation. The semiautomated system consists of 8 encoded positioners, an xyz-encoded center stage (for step and repeat capability), and vision feedback for point and click positioning. (b) A field-emission gun scanning electron microscope (FEI Quanta 200 FEG SEM), a parametric analyzer, and advanced contamination system is shown. The FEI Quanta 200 provides optimized resolution, video rates, beam shift, vacuum technology and user control for nanoprobing integrated circuit and failure analysis. (c) A close up of the eight encoded positioners that enable advanced probing and increased throughput. (d) A Zyvex microgripper is shown.*

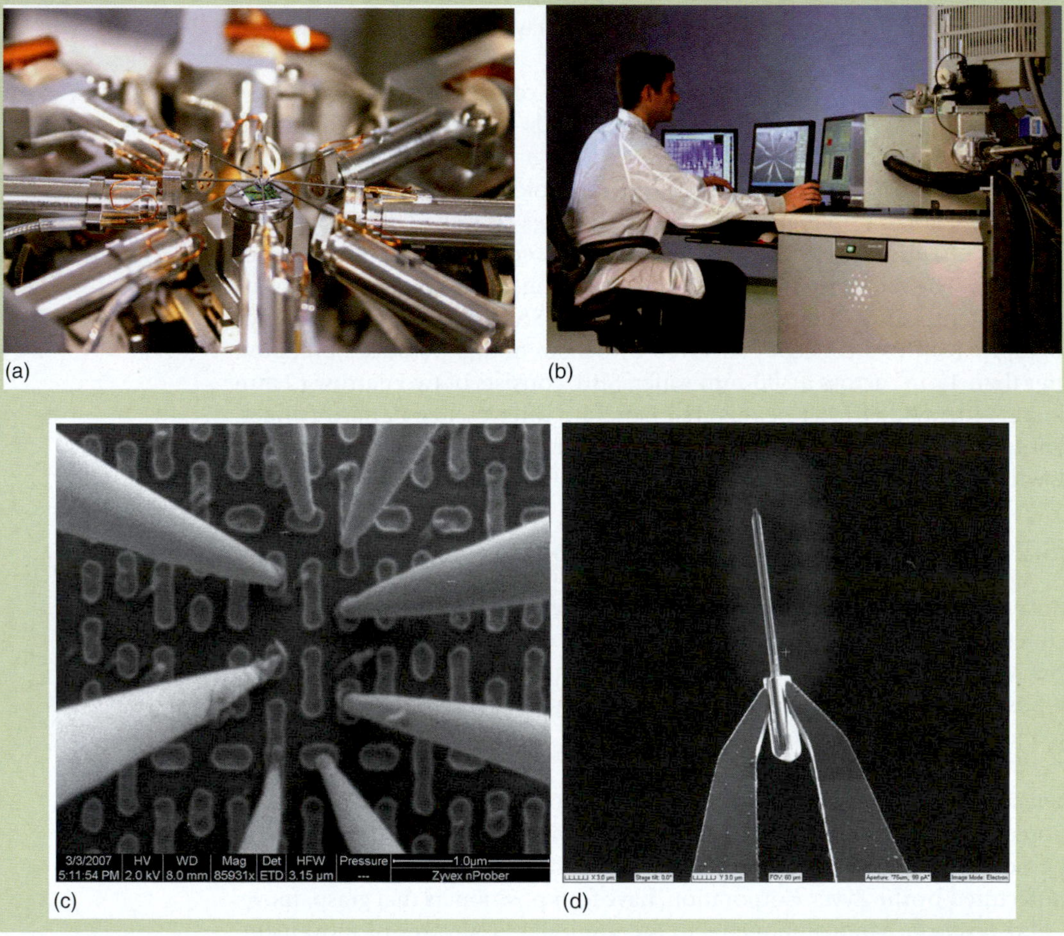

Source: Images courtesy of Jim Von Ehr, II; Zyvex Instruments, Richardson, Texas. With permission.

Nanorobotics? The field of nanorobotics, although infringing on the realm of science fiction, is pursued quite vigorously by several groups. A recent call for papers (expected publication in fall 2008) by the *International Journal of Robotics Research* is serious about expanding the world of robotics down to the nanoscale. Please refer to www.ijrr.org. The current state of the art and future challenges in nanorobotics are scheduled to be discussed. Their participation in nanomanufacturing is inevitable. Nanorobots are defined as

> ... intelligent systems with overall dimensions at or below the micrometer range that are made of assemblies of nanoscale components with individual dimensions ranging between 1 and 100-nm. Nanorobots will be able to perform at least one of the following actions: actuation, sensing, signaling, information processing, intelligence, swarm behavior at the nanoscale.

Such nanobots face critical design, fabrication, and control challenges, according to the journal. In particular, nanorobots would be exposed to unusual microenvironments that are vastly altered from macroscopic scenarios. Solicited paper topics include design and manufacture of nanobots, kinematic and dynamic modeling, robotic control, AFM/SPM-based assembly and manipulation, molecular self-assembly and swarm behavior, nanosensors, bio-nano robots, and applications (medical, biological, and industrial manufacturing) [125]. We shall wait and see. Crossing this threshold would indeed be an amazing milestone.

16.5.3 Bottom-Up Manufacturing

Bottom-up chemical methods of nanomanufacturing are expected to deliver a major contribution to manufacturing at the nanoscale. Although far from being realized, bottom-up techniques in general will provide inexpensive means to form nanostructures (dots, wires, and thin films) with high throughput. We shall review a few major kinds of bottom-up manufacturing techniques. Detailed descriptions of these techniques are found in Part I of this book [101]. Challenges facing self-assembly on a larger scale include attainment of long-range order, overall robustness, development of methods to mitigate contaminants, errors and defects; and control of microenvironments.

Microfabrication. Microfabrication techniques are essentially top-down "machining" processes. By subtraction from a bulk substrate material, the following repeating process takes place to form circuits, components, and devices:

> Substrate → Application of Thin Resist Film → Mask and Lithography → Etch → Liftoff of Mask → Characterization and Release

Lithography involves cleaning (following each step), application of the mask, exposure to pattern, development, curing, and inspection. Lithography is usually an optically driven process but direct writing is also accomplished by electron or ion beams and AFM techniques.

Nanofabrication with Soft Lithography. In the soft lithography process, polydimethylsiloxane (PDMS) is poured into a predetermined pattern (formed by traditional lithography). Following curing, the PDMS is peeled out as a negative replica of the primary mask. This rubbery material is then attached to a substrate and becomes a stamp for the pattern. The stamp's surface is coated with thiols and then pressed against a surface consisting of a thin film of gold. The thiols form a self-assembled monolayer on the gold. George Whitesides and his group at Harvard University developed this process known as micro-contact printing (μCp).

Nanofabrication with Direct Manipulation. STM, AFM, spin-polarized STM, and dip-pen nanolithography are a few methods that create 2-D hierarchies by direct writing. STM, for example, utilizes a sharp, metal-conducting tip that is brought into close contact (within 1 nm) of a sensitized surface. The surface can then be "etched" by the STM tip that is rastered to produce a pattern. Dip-pen lithography is capable of depositing a wide variety of component species ranging from carbon nanotubes, colloidal particles, self-assembled monolayers, reactive species, and many more. An AFM tip immersed in a liquid meniscus (that is a reservoir containing the targeted delivery materials) is rastered across a surface according to a predetermined pattern.

Nanosphere Lithography. Nanosphere lithography employs spherical latex particles (or other suitable materials) that are assembled into a close-packed pattern. By methods such as evaporation or sputtering, materials are allowed to fill the interstices of the spherical matrix to form a negative pattern based on the original template. The process was developed by Richard P. van Duyne of Northwestern University.

Other Template Methods. Templates include porous templates made of metal oxides and various silicas. Aluminum oxide templates, formed via an anodized process, contain nearly perfect arrays of parallel pore channels with diameters a few nanometers to larger than 250 nm. The anodic process to form the films is extremely "low tech" and requires some highly polished aluminum, one of four kinds of mineral acids (sulfuric, oxalic, chromic, or phosphoric), a cooling apparatus, and a DC-power supply. The diameter of the pores formed during anodizing is directly proportional to the applied voltage. Highly ordered arrays are formed by performing the anodizing twice. Removal of the first oxide layer after a 24 h period leaves a templated surface of perfectly ordered scallops. The array of hexagonally packed scallops have attained the steady-state equilibrium with regard to diameter and distribution. Anodizing over this layer yields an equivalent distribution of pore channels, already in an ideal, ordered state. Pore channels can be filled with whatever material required. For example, carbon nanotubes can be grown in the pore channels to form substrates for flat panel displays. Through-hole anodic membranes have utility as maskants for patterning surfaces.

Self-Assembly. Supramolecular host–guest chemistry, molecular self-assembly, and inorganic–organic hybrid materials are just some processes and/or materials that rely on self-assembly in order to form nanostructured materials. Self-assembly is the spontaneous formation of a material from preexisting components that is reversible (to some degree—of course, in manufacturing, we want things to be nonreversible), controllable, and is based on stochastic chemical processes. There is static self-assembly, dynamic self-assembly, and programmed self-assembly. More information about self-assembly can be gleaned from *Introduction to Nanoscience*.

Micelles, the meniscus effect, crystallization, and polymerization occur in nature unabated. The action of protein folding, the self-assembly of the tobacco mosaic virus, and the molecular engine called the ribosome are all based on intermolecular binding forces and substrates held together with strong covalent bonds. The environment is very important in self-assembly. Other drivers such as energy minimization, nucleation, and the effect of templates also influence self-assembly. Nature is a good tutor. Applications of self-assembly to synthetic systems is dependent on the degree of control, for example, controlling the local environment, external fields, particle dispersion, and timing.

16.5.4 Molecular Scale Assembly Lines

Richard Feynman first proposed in 1959 the concept of "working with atomic precision" to fabricate nanosystems [126]. Most of the discussion, however, in this section is gleaned from Eric Drexler's essay *Productive nanosystems: The physics of molecular fabrication* (*Physics Education*, 40, 339–346 (2005)). Drexler is one of the leading proponents of molecular manufacturing, and he states the following [126]:

> Fabrication techniques are the foundation of physical technology and are thus of fundamental interest. Physical principles indicate that nanoscale systems will be able to fabricate a wide range of structures, operating with high productivity and precise molecular control. Advanced systems of this kind will require immediate generations of system development, but their components can be designed and modeled today.

Nature, after all, does precisely this—fabricate materials bottom up from molecular assembly systems that deliver precise control over structure—with a few glitches here and there. Ribosomes are biological molecular machines that are directed by digital codes to fabricate proteins—a testament to a functional nanosystem that has demonstrated incredible success (and reliability) over eons that use inexpensive materials that are "biodegradable." The yearly output of such molecular machines is on the order of billions of tons of exactly the same product(s). Drexler goes on to state that synthetic systems should be able to outperform biological counterparts. An excellent example that supports this claim is the contrast between our avian friends and modern day aircraft. To provide balance, however, we add that synthetic systems are far inferior to their biological counterparts in many other ways.

Drexler envisions the assembly of precise, intricate structures by enjoining molecules with direct mechanical control. He claims that this is possible if tight control over molecular motion, bonding transformations that exclude unwanted molecular encounters while simultaneously ensuring positional accuracy down to 1 Å. Enzymes, after all, are able to do just that—precise and accurate control down to the atomic level. Reliability would be a beneficial outcome of this manufacturing methodology. The secret to achieving the successes achieved by biological systems is dependent on the nature of scaling laws and their applicability to nanosystems [126].

Productive Nanosystems. Nature perhaps will be our greatest tutor—via biomimetics—copying nature should certainly give us a head start over developing such systems from scratch. An interesting juxtaposition has been reached. We emulate natural systems to learn how to build nanosystem molecular assemblies in the long term yet our technology is mostly "very synthetic." For example, most of our technology does not emulate nature at all. In accordance, Drexler's plan, actually, intends to move away from assembly in fluid media (as in the case of biological materials). Due to "unnecessary drag" caused by fluid and gaseous environments to media and restrictions imposed by biomolecules that are essentially large monomers with low bond density and stiffness, performance can be

Fig. 16.11

A molecular planetary gear is shown. The structure is held together with strong covalent bonds and is able to rotate at a frequency of 1 GHz while delivering power density of 10^{15} $W \cdot m^{-3}$. A molecular machine such as this resembles a conventional machine; but do these molecular analogues really perform like conventional machines? And, what factors exist at the nanoscale that serve to disrupt this notion? According to molecular dynamic models, at temperatures much less that k_BT, molecular models have shown that "bearing interfaces" of strong covalent solid molecular machines show no static friction. Dynamic friction, however, is likely through phonon interactions. Another aspect is thermal fluctuations. Thermal fluctuations could provide a real barrier to accelerated use of nanomaterials. Thermal fluctuations, as one might imagine, contribute to increasing uncertainty in position and reduce energy barriers. One solution to reduce thermal fluctuations is to operate at extremely low temperatures.

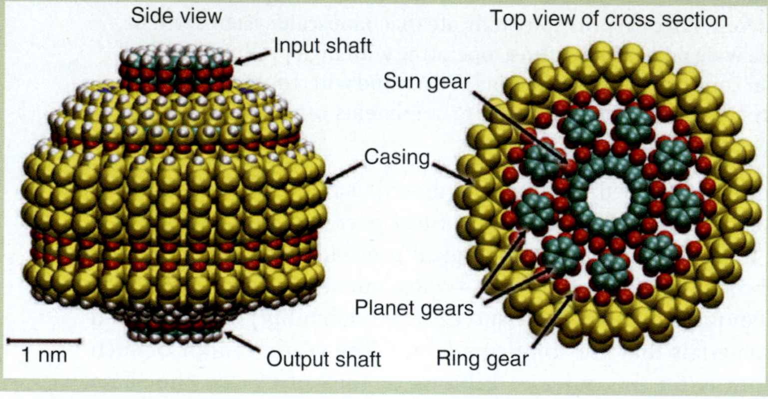

Source: K. E. Drexler, Productive nanosystems: The physics of molecular fabrication, *Physics Education*, 40, 339–346 (2005). With permission.

enhanced by moving out of the liquid/gaseous environment and employing materials that are stronger, harder, and stiffer [126]. In other words, nanosystem productivity should be enhanced by extrapolation of our macroscopic machines down to the nanoscale in the form of a modern assembly line. This type of philosophy has already been accomplished at the microscale. An example of a futuristic molecular planetary gear is shown in **Figure 16.11**. A molecular mill mechanism is shown in **Figure 16.12**. The figure represents a repetitive mechanism that is able to add one atom to a cluster.

Nanotribology. Friction is another parameter that requires more study at the nanoscale. Friction (from the Latin *frictionem* "a rubbing down", from *fricare* "to rub") is an age-old engineering issue. It was from a frictional process after all that some of our ancestors were able to make fire. The ancient Egyptians figured out that moving monumental masses on wooden rollers were easier pulled on wet sand than dry [127]. Pork fat was used in the Middle Ages to lubricate axels of wagons [127]. Leonardo Da Vinci in the late 1400s and early 1500s, by way of his inclined plane experiments, introduced friction in the context of modern engineering by sliding a rectangular block down a planar surface [128,129]. He found that the magnitude of friction had a linear dependence on load but was independent of geometrical area of contact and the type of material (the universal coefficient of friction ≈0.25) [127]. Guillaume Amontons in the late 1600s

FIG. 16.12 *In this artist's rendition, the concept of atomic precision is depicted by a mechanical system that forces components to react, for example, guiding molecular reactions. In the image, a molecular machine attaches a hydrogen atom to a cluster. The machine components are shown undersized and without detail.*

Source: Image courtesy of Eric Drexler. With permission.

reaffirmed Da Vinci's results and found that frictional force is directly proportional to the perpendicular force and independent of the type of material (universal coefficient of friction ≈0.33). Leonhard Euler related frictional force with gravitational force and was the first to distinguish between static and kinetic friction. Later on, Charles Augustin Coulomb in the mid-1700s studied the independence of friction from velocity.

With perhaps a peek into the nanodomain, Coulomb suggested that micrometer roughness was responsible for friction. It was later shown that highly polished surfaces demonstrated more friction than roughened surfaces. John Theophilus Desaguliers in the early 1700s implied that a molecular component to friction was at work—adhesion that was dependent on geometrical contact area [127]. Interestingly, it is rather difficult to derive classical friction laws from fundamental atomic principles [128]. Philip Bowden and David Tabor of Cambridge University proposed an adhesion bonding mechanism between atoms to friction [128]. G.A. Tomlinson in 1929 proposed that phonons were responsible for the mechanism of friction. With the advent of the quartz crystal microbalance (QCM) and the lateral force microscope (a type of AFM), the mechanisms of friction were explored with greater frequency.

The phononic mechanism was proven by J. Krim et al. by showing how krypton monolayers demonstrated measurable sliding friction over a gold surface [129]. Phonon effects were measured by a QCM. In another study, Abdelmaksoud et al. showed that trace levels of interfacial slippage between tricresylphosphate (TCP) and metal or metal oxide yielded low macroscopic coefficients of friction while rigidly attached TCP showed high macroscopic coefficients of friction [130]. Static friction, as opposed to sliding friction, requires more force to overcome. At the nanoscale, when considering monolayers sliding over atomically smooth substrates, there is no static friction—ca. 10^5 times less than for macroscopic counterparts [128,129]. Coefficients of friction derived from nanotribological experiments tend to disagree by several orders of magnitude with their macroscopic counterparts [128,129].

Ernst Meyer at the University of Basel in 2004 studied friction with a lateral force microscope sliding over the surface of an atomically smooth sodium chloride surface. He showed that there is a transition from stick slip to sliding due to atomically modulated friction phenomena. Once overcoming the stick-slip regime, an ultra-low domain of friction is encountered, for example, superlubricity [127]. MEMS friction testers (tens of contact regions) capable of detecting forces as small a 5 μN have been developed to bridge the gap between macroscopic (numerous contact regions) and nanoscale friction (single contact region).

At the macroscale, over $100 billion is lost in the United States annually due to friction [128]. Much of this kind of friction depends on the interaction with gravity. Gravity has no effect at the nanoscale, rather electromagnetic forces are dominant. Material shear at the macroscale is relatively insignificant. This shear, however, of a single atomic layer of a nanomachine can be terminal to its operation [128]. Superlubricity, nonetheless, is expected to help solve some of the frictional issues encountered today. In order for this to occur, according to J. Krim, understanding of friction at the atomic-, nano-, and mesoscale is imperative. The future of friction, she goes to say, will be part of the system design process and reduced or eliminated "before the fact" rather than application of lubricants "after the fact."

Self-Assembly. The competition to such mechanized "mechanically guided systems" is the natural process of Brownian motion-driven self-assembly—a process that takes place usually in aqueous media. Biology exploits self-assembly via the action of programmable machines to make precursor building blocks [126]. However, Drexler believes that directed mechanical assembly will replace biologically based self-assembly that produces interfaces that are "soft and weak." Although Brownian motion-based self-assembly is prevalent today, in the long run, it will not survive the nanomanufacturing demands required by future technology, according to Drexler.

Scaling Laws. Classical scaling laws are mathematical laws that predict how variation in one quantity affects variations in other quantities. Scaling laws are commonly referred to as power laws. The Stefan–Boltzmann law, for example, is a power law. Scaling laws are able to relate one function to that function at different size regimes. For example, ants are able to lift ca. 20× their own weight in mass while humans are not quite able to achieve that level of perfection. Put another way, if there were no scaling laws, then we should be able to lift 2 tons of weight (assuming a 200-lb person) in proportion to what an ant is able to lift (e.g., a linear transposition).

The simplest way to get a feel for scaling laws is to study mechanical systems. In mechanical systems, mass is proportional to volume. However, if a linear dimension of an object is reduced by a factor of x, then the volume of that object is reduced by a factor of x^3. At the nanoscale (ca. 10^{-8} m), several classical scaling laws retain a sense of relevance—especially with regard to mechanical functions [126]. These laws do break down at smaller scales where atomic structure and quantum effects exert more influence. Please refer to **Table 16.4**. Classical

TABLE 16.4 *Physical Properties: Scaling and Magnitude*

Quantity	Scaling	Magnitude	
Magnetic force	L^4	10^{-19}	N
Volume	L^3	10^{-24}	m^3
Mass	L^3	10^{-21}	kg
Electrostatic energy	L^3	10^{-19}	J
Torque	L^3	10^{-15}	m N
Gravitational force (weight)	L^3	10^{-20}	N
Area	L^2	10^{-16}	m^2
Force (at working stress)	L^2	10^{-7}	N
Mechanical power	L^2	10^{-7}	W
Electrostatic force (constant field)	L^2	10^{-11}	N
Electric current	L^2	10^{-6}	A
Length	L^1	10^{-8}	m
Deformation (constant stress)	L^1	10^{-11}	m
Motion time	L^1	10^{-8}	s
Stiffness	L^1	10^{4}	$N\ m^{-1}$
Voltage (constant field)	L^1	10^{0}	V
Gravitational stress	L^1	10^{-5}	$N\ m^{-2}$
Mechanical working stress	L^0	10^{9}	$N\ m^{-2}$
Modulus of elasticity	L^0	10^{12}	$N\ m^{-2}$
Electrostatic stress (constant field)	L^0	10^{5}	$N\ m^{-2}$
Adhesive strength (dispersion forces)	L^0	10^{9}	$N\ m^{-2}$
Strain	L^0	10^{-3}	—
Density	L^0	10^{3}	$kg\ m^{-3}$
Speed	L^0	10^{0}	$m\ s^{-1}$
Current density	L^0	10^{10}	$A\ m^{-2}$
Electric field	L^0	10^{8}	$V\ m^{-1}$
Amplitude of thermal vibrations	$L^{-1/2}$	10^{-12}	m
Acceleration	L^{-1}	10^{8}	$m\ s^{-2}$
Spring stiffness	L^{-1}	10^{4}	$N\ m^{-1}$
Deformation (constant force)	L^{-1}	10^{-11}	m
Mechanical power density	L^{-1}	10^{17}	$W\ m^{-3}$
Electrical resistance	L^{-1}	10^{0}	Ω
Motion frequency	L^{-1}	10^{8}	s^{-1}
Relative productivity (scaled parts)	L^{-1}	10^{5}	s^{-1}
Relative productivity (atomic parts)	L^{-4}	10^{3}	s^{-1}

Source: K. E. Drexler, Productive nanosystems: The physics of molecular fabrication, *Physics Education*, 40, 339–346 (2005). With permission.

continuum scaling laws as they apply to magnitudes down to 10^{-8} m break down at smaller dimensions where atomic structure and, if smaller, quantum effects need to be considered [126].

For a more detailed discussion, please refer to Drexler's paper mentioned earlier in this text.

Impediments to Mechanical Nanosystem Fabrication. Scale-down of a macroscopic machine to the nanoscale is not just a matter of making it smaller. Other factors need to be considered. For example, forces that are irrelevant at the macroscale such as van der Waals forces can be significant at the nanoscale. There are several aspects that could contribute to the failure of a molecular system: (1) thermal fluctuations that allow energy barriers to be breached along a path to an undesirable state, (2) structures rearrange or become fragmented, and (3) displacement of a nanotool during a molecular operation causes unwanted bonding transformations (threshold displacement of $\Delta x_i \approx 0.1$ nm). Drexler believes that many of these challenges can be overcome by designing molecular tools using stiff materials positioned by 100-nm scale devices [126].

16.6 Concluding Remarks

We come to the end of the *Perspectives* division of the text. We have experienced a brief foray into the wondrous world of nanometrology and have briefly touched upon nanomanufacturing. Great efforts are expended on roadmaps describing just how to go about tackling nanometrology [131]. Many believe that indeed nanometrology is the next big thing in measurement [132]. Are we really interested in quantifying the surface coverage of conjugated molecules on functionalized nanoparticles [133] or understanding the transition between stick-slip friction and continuous sliding in atomic friction [134]? Based on what you have soaked up so far, I hope that your answer is in the affirmative. The federal government and the international community do understand the importance of nanometrology because it is metrology, after all is said and done, that allows us to practice commerce, conduct meaningful research, and make decisions—some of the fundamental bases of our civilization [135]. Will we ever be able to quantify and standardize self-assembly processes [136]? And, how has nature gone about metrology and manufacturing all these years?

We hope that the student has gained a sense of importance of quantum metrology and quantum phenomena and how nanomaterials are capable of playing a major role in the modulation of quantum measurements and properties.

References

1. History of measurement, en.wikipedia.org/wiki/History_of_measurement, modified 12 (2007).
2. Jet's fuel ran out after metric conversion errors, *New York Times*, July 30 (1983).
3. NASA's metric confusion caused mars orbiter loss, www.cnn.com/TECH/space/9909/30/mars.metric/ (1999).

4. Metrology, Computer Encyclopedia, computerlanguage.com/ (2007).
5. Metrology, *The Oxford essential dictionary of the U.S. military,* Oxford University Press, Oxford (2007).
6. Metrology, International Bureau of Weights and Measures, IBWM, bipm.org/en/home/ (2007).
7. Nanometrology, International Bureau of Weights and Measures, IBWM, bipm.org/en/home/ (2007).
8. Nanometrology, Materials Science and Engineering Laboratory, www.msel.nist.gov/Nanometrology.pdf (2007).
9. Accuracy, precision, and uncertainty, Science Division, Bellevue Community College, scidiv.bcc.ctc.edu/Physics/Measure&sigfigs/Measure&sigfigsintro.html (accessed 2007).
10. Introduction to metrology, MIKES, Mittatekniikan Keskus Centre for Metrology and Accreditation, Finnish National Standards Laboratories, cs.utu.fi/rlahdelma/ADMSSIM/L01_Metro.pdf (2005).
11. A. Lassila, Traceable dimensional measurements on nanometer range, Annual Report, *MIKES,* Mittatekniikan Keskus Centre for Metrology and Accreditation, Finnish National Standards Laboratories (2004).
12. Nanomanufacturing, en.wikipedia.org/wiki/Nanomanufacturing (2007).
13. Nanomanufacturing (NM), Solicitation, Civil, Mechanical and Manufacturing Innovation, National Science Foundation, nsf.gov/funding/pgm_summ.jsp?pims_id=13347 (2008).
14. Center for high-rate nanomanufacturing, nano.neu.edu/industry/industry_showcase/industry_day/ (2007).
15. A. Busnaina, Center for high-rate nanomanufacturing (CHN), Presentation, nsec.neu.edu (viewed 2007).
16. The George J. Kostas facility for micro and nanofabrication, kostas.neu.edu/ (2007).
17. M. L. Schattenburg, Nanometrology in nanomanufacturing, Presentation, Massachusetts Institute of Technology, NASA Tech Briefs, Cambridge, MA snl.mit.edu/papers/presentations/2003/Schattenburg/MLS-NASA-Tech-2003.pdf (2003).
18. M. T. Postek and R. J. Hocken, *Instrumentation and metrology for nanotechnology,* Report of the National Nanotechnology Initiative Workshop, Gaithersburg, MD (2004).
19. R. Cavanaugh, Instrumentation and metrology for nanocharacterization, chap. 2 In *Instrumentation and metrology for nanotechnology,* M. T. Postek and R. J. Hocken, eds., Report of the National Nanotechnology Initiative Workshop, Gaithersburg, MD (2004).
20. J. Marburger and R. R. Colwell, Science for the 21st century, National Science and Technology Council, Executive Office of the President; Washington, DC, July (2004).
21. E. M. Vogel, NIST advanced measurement laboratory (AML) Nanofab, Presentation, National Institute of Standards and Technology (2004).
22. J. M. Cattell, Mental measurement, *Philosophical Review,* 2, 316–332 (1893).
23. G. Englehard, Historical views of the concept of invariance in measurement theory, In *Objective measurement: Theory into practice,* M. Wilson, ed., Vol. 2, pp. 73–79, Ablex, Norwood, NJ (1994).
24. T. M. Porter, *Trust in numbers: The pursuit of objectivity in science and public life,* Princeton University Press, Princeton, NJ (1995).
25. A. J. Fiok, J. M. Jaworski, R. Z. Morawski, J. S. Oledzki, and A. C. Urban, Theory of measurement in teaching metrology in engineering faculties, Warsaw University of Technology, sierra.iem.pw.edu.pl/~jsol/pedeefy/art34.pdf (accessed 2007).
26. A. J. Fiok, J. M. Jaworski, R. Z. Morawski, J. S. Oledzki, and A. C. Urban, Theory of measurement in teaching metrology in engineering faculties, *Measurement,* 6, 63–67 (1988).
27. B. Jeckelmann and B. Jeanneret, The quantum hall effect as an electrical resistance standard, *Measurement Science and Technology,* 14, 1229–1236 (2003).

28. N. Taniguchi, Current status in, and future trends of, ultra-precision machining and ultrafine materials processing, *Annals of the CIRP*, 32, 573–582 (1983).
29. J. B. Bryan, The Abbe principle revisited, *Precision Engineering*, 1, 129–132 (1979).
30. D. J. Whitehouse, Nanometrology, chap. 8, In *Handbook of surface and nanometrology*, Institute of Physics (IOP) Publishing, a subsidiary of Taylor & Francis, Bristol (2003).
31. A. Weckenmann, T. Wiedenhöfer, and K.-D. Sommer, Determining measurement uncertainty in nanometrology with the example of nanoclusters, *tm-Technisches Messen*, 71, Part 2, 93–100 (2004).
32. J. W. Lyons, Guidelines for evaluating and expressing the uncertainty of NIST measurement results, National Institute of Standards and Technology, physics.nist.gov/Pubs/guidelines/sec6.html (1993).
33. J. Salsbury, Nanometrology—NIST and Mitutoyo achieve world's highest accuracy roundness measurements, *Mitutoyo News*, mitutoyo.com/pressrelease/NISTPR%20.pdf (2004).
34. C. M. Caves, Quantum-limited measurements: One physicists crooked path from quantum optics to quantum information, Department of Physics presentation, University of New Mexico (viewed 2008).
35. Uncertainty Principle, *The Columbia Electronic Encyclopedia*, www.cc.columbia.edu/cu/cup/ through encyclopedia2.thefreedictionary.com/Heisenberg+limit (2007).
36. J. D. Wilson and A. J. Buffa, *College physics*, Prentice-Hall, Upper Saddle River, NJ (2000).
37. J. Hilgevoord, The Uncertainty Principle, *Stanford Encyclopedia of Philosophy*, plato.stanford.edu/entries/qt-uncertainty/ (2006).
38. D. Halliday, R. Resnick, and J. Walker, *Fundamentals of physics*, 7th ed., John Wiley & Sons, Inc., New York (2005).
39. Quantum entanglement, en.wikipedia.org/wiki/Quantum_entanglement (2008).
40. M. Brooks, Entanglement: The weirdest link, *New Scientist*, 181, 2440–2427 (2004).
41. P. Comstock, The strange world of quantum entanglement, Interview of Brian Clegg, *California Literary Review*, calitreview.com/topics/science/51/ (2007).
42. B. Clegg, *The god effect: Quantum entanglement, science's strangest phenomenon*, St. Martin's Press, New York (2006).
43. M. Kroner, A. O. Govorov, S. Remi, B. Biedermann, S. Seidl, A. Badolato, P. M. Petroff, W. Zhang, R. Barbour, B. D. Gerardot, R. J. Warburton, and K. Karrai, The nonlinear Fano effect, *Nature*, 451, 311–314 (2008).
44. A. Migdall, Correlated photon metrology without absolute standards, *Physics Today*, 52, 41–46 (1999).
45. H. Lee, P. Kok, and J. P. Dowling, A quantum Rosetta stone for interferometry, *Journal of Modern Optics*, 49, 2325–2338 (2002).
46. T. Nagata, R. Okamoto, J. L. O'Brien, K. Sasaki, and S. Takeuch, Beating the standard quantum limit with four-entangled photons, *Science*, 316, 726–7294 (2007).
47. A. Politi, M. J. Cryan, J. G. Rarity, S. Yu, and J. L. O'Brien, Silica-on-Silicon Waveguide Quantum Circuits, *Science Express*, www.sciencemag.org/sciencexpress/recent.dtl (2008).
48. Tiny photonics logic circuit achieves quantum entanglement. *R&D Magazine*, March 28 (2008).
49. J. L. O'Brien, Precision without entanglement, Perspectives, *Science*, 318, 1393–1394 (2007).
50. G. Santarelli, P. Laurent, P. Lemonde, A. Clairon, A. G. Mann, S. Chang, A. N. Luiten, and C. Salomon, Quantum projection noise in an atomic fountain: A high stability cesium frequency standard, *Physics Review Letters*, 82, 4619 (1999).

51. B. L. Higgins, D. W. Berry, S. D. Bartlett, H. M. Wiseman, and G. J. Pryde, Entanglement-free Heisenberg-limited phase estimation, *Nature*, 450, 393–396 (2007).
52. V. Giovannetti, S. Lloyd, and L. Maccone, Quantum-enhanced measurements: Beating the standard quantum limit, *Science*, 306, 1330 (2004).
53. A. Naik, O. Buu, M. D. LaHaye, A. D. Armour, A. A. Clerk, M. P. Blencowe, and K. C. Schwab, Cooling a nanomechanical resonator with quantum back-action, *Nature*, 443, 193–196 (2006).
54. S. M. Tanner, J. M. Gray, C. T. Rogers, K. A. Bertness, and N. A. Sanford, High-Q GaN nanowire resonators and oscillators. *Applied Physics Letters*, 91, 203117 (2007).
55. High-Q NIST nanowires may be practical oscillators, *Nanotech-Now*, www.nanotech-now.com/news.cgi?story_id=26600 (2007).
56. K. C. Schwab and M. L. Roukes, Putting mechanics into quantum mechanics, *Physics Today*, 58, 36–42 (2005).
57. D. Nikbin, Quantum back-action has cooling effect, *Physics World*, physicsworld.com/cws/article/news/25905 (2006).
58. K. C. Schwab, Nanomechanics coupled to SET, Homepage, Cornell University, www.kschwabresearch.com/articles/detail/7 (2008).
59. G. A. Mansoori, *Principles of nanotechnology*, World Scientific Publishing Co. Pte. Ltd., Singapore (2005).
60. G. Aeppli, P. Warburton, and C. Renner, Will nanotechnology change IT paradigms? *BT Technology Journal*, 24, 163–169 (2006).
61. A. Zellinger, Fundamentals of quantum information, *Physics World*, physicsworld.com/cws/article/print/1658 (1998).
62. Science & technology programme in quantum metrology, Department of Trade and Industry (DTI), National Measurement System Directorate (NMSD), April 2004 to March 2007, London (2007).
63. C. Wasshuber, Computational single-electronics, In *Computational Electronics*, S. Selberherr, ed., Springer-Wein, New York (2001).
64. NIST-F1 Cesium fountain atomic clock: Primary and frequency standard for the United States, National Institute of Standards and Testing (NIST), Boulder, CO, http://tf.nist.gov/cesium/fountain.htm (2008).
65. S. Jefferts, D. Meekhof, T. Heaver, and E. Donley, NIST-F1 Cesium fountain atomic clock, National Institute of Standards and Technology, Boulder, CO, tf.nist.gov/cesium/fountain.htm (2007).
66. J. Roach, New clock will lead to more accurate measure of time, National Geographic News (2001).
67. C. F. Roos, M. Chwalla, K. Kim, M. Riebe, and R. Blatt, "Designer atoms" for quantum metrology, *Nature*, 443, 316–319 (2006).
68. Fine structure constant, en.wikipedia.org/wiki/Fine-structure_constant (2007).
69. Current advances: The fine structure constant and Quantum Hall effect, NIST Reference on Constants, Units and Uncertainty, physics.nist.gov/cuu/Constants/alpha.html (2000).
70. A. D. Ludlow, T. Zelevinsky, G. K. Campbell, S. Blatt, M. M. Boyd, M. H. G. de Miranda, M. J. Martin, J. W. Thomsen, S. M. Foreman, J. Ye, T. M. Fortier, J. E. Stalnaker, S. A. Diddams, Y. Le Coq, Z. W. Barber, N. Poli, N. D. Lemke, K. M. Beck, and C. W. Oates, Sr lattice clock at 1×10^{-16} fractional uncertainty by remote optical evaluation with a Ca clock, *Science*, 319, 1805–1808 (2008).
71. New atomic clock super-accurate, Denver and the West, *The Denver Post*, February 15 (2008).
72. J. Kinaret, M. Jonson, A. Bárány, P. Sylwan, and S. Björn-Rasmussen, Mosquitoes in a cross-wind, Poster, Nobel Prize in Physics 1998, The Royal Swedish Academy

of Sciences; nobelprize.org/nobel_prizes/physics/laureates/1998/illpres/practice.html (1998).
73. Physics of Quantum Hall resistance standards, National Institute of Standards and Testing, Boulder CO, www.boulder.nist.gov/div814/div817b/whatwedo/qhall/qhall.htm (2002).
74. Metrology of the Ohm, Electronics and Engineering Laboratory (EEEL), National Institute of Standards and Testing, www.eeel.nist.gov/817/whatwedo-817g/ohm/ohm.html (2004).
75. S. Liou, W. Kuo, Y. W. Suen, W. H. Hsieh, C. S. Wu, and C. D. Chen, Shapiro steps observed in a superconducting single electron transistor, *Chinese Journal of Physics*, 45, 230 (2007).
76. B. Doucot, B. Duplantier, V. Pasquier, and V. Rivasseau, eds., *The Quantum Hall effect: Poincaré seminar 2004*, Birkhäuser Verlag, Basel (2005).
77. T. Yang, N. B. Belecki, and J. F. Mayo-Wells, A practical Josephson voltage standard at 1 V, NVL-NIST, nvl.nist.gov/pub/nistpubs/sp958-lide/html/315-318.html
78. R. E. Elmquist, M. E. Cage, Y. Tang, A-M. Jeffery, J. R. Kinard, R. F. Dziuba, N. M. Oldham, and E. R. Williams, The ampere and electrical standards, *Journal of Research of the National Institute of Standards and Technology*, January-February (2001).
79. N. B. Zhitenev and T. A. Fulton, Scanning single-electron transistor microscopy, In *Encyclopedia of nanoscience and nanotechnology*, Vol. 5, J. A. Schwarz, C. I. Contescu, and K. Putyera, eds., Marcel-Dekker, New York (2004); Taylor & Francis (2005).
80. I. Giaever, Energy gap in superconductors measured by electron tunneling, *Physics Review Letters*, 5, 147–148 (1960).
81. J. P. Pekola, J. J. Vartianen, M. Möttönen, O.-P. Saira, M. Meschke, and D. V. Averin, Hybrid single-electron transistor as a source of quantized current, *Nature Physics*, Published Online, www.nature.com/nphys/journal/vaop/ncurrent/abs/nphys808.html, 9 December (2007).
82. B. Dumé, Ampere could be defined one electron at a time, Nanotechweb.org, www.nanotechweb.org/cws/article/tech/32295 (2008).
83. J. Cazaux, Errors in nanometrology by SEM, *Nanotechnology*, 15, 1195–1199 (2004).
84. J. Allgair and B. Bunday, A review of scatterometry for three-dimensional semiconductor feature analysis, *Future Fab International*, 19, future-fab.com/documents.asp?grID=216&d_ID=3390 (2005).
85. J. Aoki, W. Gao, and S. Kiyono, A high-precision AFM for nanometrology of large area micro-structured surfaces, *EUSPEN Glasgow*, 254–255 (2004).
86. J. Aoki, W. Gao, S. Kiyono, and T. Ono, A high-precision AFM for nanometrology of large area micro-structured surfaces, *Key Engineering Materials*, 295–296, 65–70 (2005).
87. Scanning electron microscope-based metrology, In Office of microelectronics programs: Programs, activities and accomplishments, NISTIR-7367, Electronics and Electrical Engineering Laboratory (EEEL), National Institute of Standards and Technology, January (2007).
88. P. West and N. Starostina, *AFM image artifacts*, Pacific Nanotechnologies, LOT-Oriel Gruppe Europa, www.lot-oriel.com/site/site_down/pn_artifacts_deen.pdf (viewed 2008).
89. D. Julthongpiput, M. J. Fasolka, W. Zhang, T. Nguyen, and E. J. Amis, Gradient chemical micropatterns: A reference substrate for surface nanometrology, *Nano Letters*, 5, 1535–1540 (2005).
90. Veeco introduces new insight 3D atomic force microscope: Unparalleled accuracy for in-line CD, depth and CMP metrology applications, Veeco.com (2007).
91. W. E. Moerner, New directions in single-molecule imaging and analysis, *Proceedings National Academy of Sciences*, 104, 12596–12602 (2007).

92. C. Bustamante, Z. Bryant, and S. B. Smith, Ten years of tension: Single-molecule DNA mechanics, *Nature*, 421, 423–427 (2003).
93. K. C. Neuman and S. M. Block, Optical trapping, *Review of Scientific Instruments*, 75, 2787–2809 (2004).
94. G. Hummer and A. Szabo, Free energy surfaces from single-molecule force spectroscopy, *Accounts of Chemical Research*, 38, 504–513 (2005).
95. I. Gopich and A. Szabo, Fluorophore-Quencher distance correlation functions from single-molecule photon arrival trajectories, *Journal of Physical Chemistry B*, 109, 6845–6848 (2005).
96. A. E. Cohen and W. E. Moerner, Method for trapping and manipulating nanoscale objects in solution, *Applied Physics Letters*, 86, 93–109 (2005).
97. A. Fischer and M. Montal, Single molecule detection of intermediates during Botulinum neurotoxin translocation across membranes, *Proceedings of National Academy of Science*, 104, 10447–10452 (2007).
98. T. A. Siewert, and D. Balzar, Metrology for nanoscale properties: X-Ray methods, In *Nanometrology*, p. 8, National Institute of Standards and Testing, www.msel.nist.gov/Nanometrology.pdf (viewed 2008).
99. C. Caleman, Towards single-molecule imaging: Understanding structural transitions using ultrafast x-ray sources and computer simulations, Acta Universitasis Upsaleinsis, Uppsala (2007).
100. N. Lee, R. D. Hartchuh, D. Mehtani, A. Kisliuk, J. F. Maguire, M. Green, M. D. Foster, and A. P. Sokolov, High contrast scanning Nano-Raman spectroscopy of silicon, *Journal of Raman Spectroscopy*, 38, 789–796 (2007).
101. G. L. Hornyak, J. Dutta, H. F. Tibbals, and A.K. Rao, *Introduction to nanoscience*, CRC Press, Boca Raton, FL (2008).
102. C. L. Haynes, C. R. Yonzon, X. Zhang, and R. P. Van Duyne, Surface-enhanced Raman sensors: Early history and the development of sensors for quantitative biowarfare agent and glucose detection, *Journal of Raman Spectroscopy*, 36, 471–484 (2005).
103. A. Macaskill, A. A. Chemonosov, V. V. Koval, E. A. Lukyanets, O. S. Federova, W. E. Smith, K. Faulds, and D. Graham, Quantitative surface-enhanced resonance Raman scattering of phthalocyanine-labeled oligonucleotides, *Nucleic Acid Research Advance Access*, 1–6 (2007).
104. K. Kneipp, Y. Wang, H. Kneipp, L. T. Perelmean, I. Itzkan, R. R. Dasari, and M. S. Filed, Single molecule detection using surface-enhanced Raman scattering, *Physical Review Letters*, 78, 1667–1670 (1997).
105. S. Nie and S. Emory, Probing single molecules and single nanoparticles by surface-enhanced Raman scattering, *Science*, 275, 1102–1106 (1997).
106. F. Cannone, G. Chirico, and A. Diaspro, Two-Photon interactions a single fluorescent molecule level, *Journal of Biomedical Optics*, 8, 391–395 (2003).
107. Optical grating scatterometry, Optical Technology Division, Physics Laboratory, National Institute of Standards and Technology; physics.nist.gov/Divisions/Div844/facilities/scatterometry/scatterometry.html (2005).
108. E. R. Fuller and G. D. Quinn, Mechanical metrology for small-scale structures, In *Nanometrology*, p. 4, National Institute of Standards and Testing, www.msel.nist.gov/Nanometrology.pdf (viewed 2008).
109. D. T. Smith, Nanoindentation methods and standards, In *Nanometrology*, p. 5, National Institute of Standards and Testing, www.msel.nist.gov/Nanometrology.pdf (viewed 2008).
110. S. M. Hsu, Nanotribology and surface properties, In *Nanometrology*, p. 6, National Institute of Standards and Testing, www.msel.nist.gov/Nanometrology.pdf (viewed 2008).

111. G.R. Stafford and O. Kongstein, Nanostructure fabrication processes: Thin film stress measurements, In *Nanometrology*, p. 17, National Institute of Standards and Testing, www.msel.nist.gov/Nanometrology.pdf (viewed 2008).
112. A. M. Forster and S.-H. Moon, Combinatorial adhesion and mechanical properties: Axisymmetric adhesion testing, In *Nanometrology*, p. 19, National Institute of Standards and Testing, www.msel.nist.gov/Nanometrology.pdf (viewed 2008).
113. C. M. Stafford and Y. M. Chiang, Combinatorial adhesion and mechanical properties: Innovative approaches to peel tests, In *Nanometrology*, p. 20, National Institute of Standards and Testing, www.msel.nist.gov/Nanometrology.pdf (viewed 2008).
114. L. E. Levine and A. M. Chaka, Nanomechanics: Coupling modeling with experiments, In *Nanometrology*, p. 18, National Institute of Standards and Testing, www.msel.nist.gov/Nanometrology.pdf (viewed 2008).
115. J. Salsbury, Nanometrology: NIST and Mitutoyo achieve world's highest accuracy roundness measurements, *Mitutoyo News*, NISTPR.pdf (viewed 2008).
116. S. F. Gale, The state of standards: Nano, *Small Times*, 8, 21–23 (2008).
117. IEEE Completes Standards Roadmap for Emerging Nanoelectronic Applications, *Nanotechnology Standards News*, electronics.ihs.com/news/ieee-nanoelectric-roadmap.htm (2007).
118. IEEE 1650–2005, IEEE Standard test methods for measurement of electrical properties of carbon nanotubes, grouper.ieee.org/groups/1650, December (2005).
119. P. W. Brazis, IEEE nanoelectronics standards roadmap (NESR) overview, Institute of Electrical and Electronics Engineers, andards.ieee.org/getieee/nano/nanoelectronics_roadmap_v1.pdf (2005).
120. Nanoelectronics standards roadmap, IEEE nanotechnology standards roadmap initiative, IEEE Standards Association (IEEE-SA), April 17 (2007).
121. D. Gamota, in The state of standards: Nano, article by, S.F. Gale, *Small Times*, 8, 21 (2008).
122. Standard terminology relating to nanotechnology, *ASTM E 2456-06*, P. Picariello, Staff Manager of Committee E56 on Nanotechnology, Subcommittee E56.01 on Terminology & Nomenclature, ASTM International, West Conshocken, PA, www.astm.org (2006).
123. A. Allan, D. Edenfeld, W. H. Joyner, A. B. Khang, M. Rodgers, and Y. Zorian, 2001 Technology roadmap for semiconductors, *Computer*, 35, 42–53 (2002).
124. P. Konkola, C. Chen, R. K. Heilmann, C. Joo, J. Montoya, C.-H. Chang, and M. L. Schattenburg, Nanometer-level repeatable metrology using the Nanoruler, *Journal of Vacuum Science and Technology B*, 21, 3097–3101 (2003).
125. C. Mavroidis and A. Ferriera, eds., *Current state of art and future challenges in nanorobotics*, International Journal of Robotics Research (2008).
126. K. E. Drexler, Productive nanosystems: The physics of molecular fabrication, *Physics Education*, 40, 339–346 (2005).
127. E. Meyer, R. M. Overnay, K. Dransfeld, and T. Gyalog, *Nanoscience: Friction and rheology at the nanoscale*, World Scientific Publishing Co., Pte. Ltd., Singapore (2002).
128. J. Krim, Friction at the nanoscale, *Physics World*, physicsworld.com/cws/article/print/21309, IOP Publishing (2005).
129. J. Krim, D. H. Solina, and R. Chiarello, Nanotribology of a Kr monolayer: A quartz-crystal microbalance study of atomic-scale friction, *Physical Review Letters*, 66, 181–184 (1991).
130. M. Abdelmaksoud, J.W. Bender, and J. Krim, Bridging the gap between macro- and nanotribology: A quartz crystal microbalance study of tricresylphosphate uptake on metal and oxide surfaces, *Physical Review Letters*, 92, 176101–176103 (2004).

131. M. Désenfant and M. Priel, Road map for measurement uncertainty evaluation, *Measurement*, 39, 841–848 (2006).
132. D. Graham, Nanometrology—Is it the next big thing in measurement? *The Analyst*, 132, 95–96 (2007).
133. L. F. Pease, III., D.-H. Tsai, R. A. Zangmeister, M. R. Zachariah, and M. J. Tarlov, Quantifying the surface coverage of conjugate molecules on functionalized nanoparticles, *Journal of Physical Chemistry C*, 111, 17155–17157 (2007).
134. Instrumentation, metrology, and analytical methods, chap. 2 In *Environmental, health, and safety research needs for engineered nanoscale materials*, The National Nanotechnology Initiative, NSET Subcommittee, Office of Science and Technology Policy, Washington DC (2006).
135. A. Socoliuc, R. Bennewitz, E. Gnecco, and E. Meyer, Transition from stick-slip to continuous sliding in atomic friction: Entering a new regime of ultralow friction, *Physical Review Letters*, 92, 134301–134303 (2004).
136. J. A. Pelesko, *Self assembly: The science of things that put themselves together*, Chapman & Hall/CRC, Taylor & Francis Group, Boca Raton, FL (2007).

Problems

16.1 Explain, in your own words, the meaning of the Heisenberg uncertainty principle.

16.2 Calculate the uncertainty in the position of the moon if its velocity is $1.01\,\text{km}\cdot\text{s}^{-1}$ and if we can measure its velocity to 10^{-8} km. Use $\Delta x \Delta p \approx h$. Is the value of the uncertainty consequential?

16.3 What is the uncertainty in position of an electron of an atom if there is $2.0 \times 10^7\ \text{m}\cdot\text{s}^{-1}$ uncertainty in its velocity?

16.4 The modern meter is defined as the length of the path traveled by light in vacuum during the interval of 1/299 792.458th of a second. From this data, calculate the speed of light?

16.5 The uncertainty in calculating the length of the meter has dropped five orders of magnitude since the early 1900s. What is the wavelength of the sources and why were they chosen?

Source	Uncertainty	Wavelength
Pt-Ir Bar	2×10^{-6}	N/A
λ Cd Lamp	7×10^{-8}	?
λ^{86}Kr Lamp	4×10^{-9}	?
λ HeNe Laser	2.5×10^{-11}	?

16.6 What source was used to provide the timing element in the HeNe laser measurement of the meter given above?

16.7 From 1904–1960, the reproducibility of measurements against the prototype meter bar (Pt-Ir) was on the order of $\pm 0.25\,\mu\text{m}$. What is this in terms of uncertainty?

16.8 Does the Heisenberg uncertainty principle place an ultimate limit on the measurement of position?

16.9 Devise a scenario for molecular assembly based on a natural model.

16.10 What is the difference between accuracy and precision?

16.11 Which excited state has a broader spectral signature—one with a lifetime of 10^{-10} s or one with a lifetime of 10^{-8} s? How much broader is the peak?

16.12 Three groups of four students each are asked to make measurements of the diameter of spherical nanoparticles (in nm) by TEM. The nanoparticles were purchased from a vendor. The students compared the diameter to a scale bar placed in the corner of the image.

Student group	#1	#2	#3	#4	#5	#6
Nano	20.1	20.935	17.0	22.23	16	18.11
Micro	19.011	19.113	19.201	19.166	19.205	19.022
Milli	21	20	21	20	21	20
Pico	18.23	18.26	18.20	18.23	18.24	18.25
Centi	22.1	20.5	18.5	21.0	19.0	19.5

Rank the groups according to the following criteria:

a. Uncertainty (highest to lowest)
b. Accuracy (best to worst)
c. Precision (best to worst)
d. Error (most to least)
e. Significant Figures (most to least)

The professor managed to find the specification sheet provided by the vendor for the nanoparticles and announced to the groups that the diameter is 20 ± 1.0 nm. Rerank accuracy and error as stipulated below.

a. Accuracy (best to worst)
b. Error (most to least)

Do more significant figures help you improve accuracy?

16.13 If an instrument is capable of attomole level detection, how close is that to single-molecule detection? How many molecules are there in an attomole? At what mole-level is a single molecule?

16.14 In a laser-tweezer system, the object trapped by the optical laser tweezer is the large dielectric polystyrene bead ($d \approx 1$ mm). A biomolecule is tethered chemically to the surface of the latex bead. Can the laser tweezer trap a single molecule without the assistance of the polystyrene sphere? Estimate the laser power required if 10 mW is required to trap a 1 μm diameter polystyrene sphere and that the gradient optical forces in the tweezer system are proportional to the polarizability α of the trapped object that is in turn proportional to the volume d^3 of the potential trapped molecule. Hint: consult W.E. Moerner, New directions in single-molecule imaging and analysis, *Proceedings of National Academy of Science*, 104, 12596–12602 (2007).

16.15 If the uncertainty of the cesium fountain atomic clock at NIST in Boulder, Colorado is 5×10^{-16} s, how many years would it take for the clock to gain or lose 1 s?

Section 7

Electromagnetic Nanoengineering

Nanoelectronics

... there is plenty of room to make [computers] smaller .... nothing that I can see in the physical laws ... says the computer elements cannot be made enormously smaller than they are now. In fact, there may be certain advantages.

Richard Feynman, 1959

Chapter 17

THREADS

Chapter 16 introduced some of the basics of quantum metrology while also introducing, albeit not in a direct fashion, some electronics. *Chapter 17*, a primer on nanoelectronics, is the first chapter in the *Electromagnetic Nanoengineering division* of the text. The division consists of three chapters that among them address three aspects of electromagnetics: electronics, optics, and magnetism. Following this division, things become quite mechanical as we move on to the three chapters of the *Mechanical Nanoengineering division* of the book.

Nanoelectronics is at the heart of any nanosensor or nanodevice. As with other properties, we expect electronic behavior to behave differently at the nanoscale than it does at the bulk scale. It is a domain in which a metal conductor, if reduced enough in size, now becomes a semiconductor. It is a domain in which electrical conduction, if forced through a small enough device, now exhibits staircase behavior rather than a continuous smooth curve.

17.0 ELECTRONICS AND NANOELECTRONICS

Electronics has a rich history. Most of us are familiar with the great events and discoveries and we therefore will not dwell upon them for too long. By the early nineteenth century, most physicists and chemists, including Michael Faraday, were convinced that the flow of charged particles were responsible for electrolysis and other electrical phenomena. Nanoelectronics is an extension of electronics down to the nanoscale. Macroscopic phenomena are all altered at this scale as new properties emerge. Distinctions between nanoscale and quantum phenomena all begin to emerge.

17.0.1 Basic Electronic Terminology and Symbols

Not all of us had the "blessed good fortune" of taking electrical engineering courses while in college. Therefore, for the benefit of those of us who missed out on that special opportunity, a brief review of electronics is given below for the purpose of perspective and orientation (**Table 17.1**).

17.0.2 Fundamental Types of Electronic Materials (and Nanomaterials)

Electronics, as the name implies, is a field of engineering that involves electrons—moving and storage. Electronics is an applied engineering discipline and industrial sector that involves the shuffling of electrons back and forth in order to perform designated tasks, to perform work. Electronics is a branch of physics that is concerned with the behavior of electrons in devices—whether in the vacuum, gas, liquid or solid phases.

There are several kinds of electronic materials—dielectric, optical, ferroelectric, magnetic, semiconductor, metal, superconductor, organic, polymer, and most recently, biomolecular. **Table 17.2** summarizes the different kinds of electronic materials and provides a representative example of each.

Table 17.1 *Basic Electronic Definitions and Relations*

Name	Form		Units, comments
Ampere	A	(17.1)	The "old" SI definition of the ampere: A force (generated between two parallel, straight wires of infinite length with negligible cross section in vacuum and carrying current in the same direction) equals to 2×10^{-7} N · m^{-1}
Coulomb	C	(17.2)	
Elementary charge	e	(17.3)	Elementary charge $e = 1.6022 \times 10^{-19}$ C. 1 C $= 6.2415 \times 10^{18}$ charge carriers. 1 A = 1 C · s^{-1}
Current	I or i	(17.4)	Current is in terms of Amperes and is symbolized by I or i.
Current density	$\mathrm{J} = \frac{\mathrm{A}}{\mathrm{m}^2}$	(17.5)	Current density is in terms of amperes per square meter
Volt	$\mathrm{V} = \frac{\mathrm{J}}{\mathrm{C}}$	(17.6)	1 V equals 1 J · C^{-1}
Power	W	(17.7)	Electric power is given in terms of watts: J · s^{-1}. The kilowatt · hour is a measure of total energy used. Also W = i · V = V · A, the rate of electrical energy transfer
Charge	$q = it$ $i = \frac{dq}{dt}$	(17.8) (17.9)	Total charge in coulombs, (C · s^{-1}, current) × (t, time). Charge is quantized: $q = N_e$, where N_e is the number of electron charges
Coulomb's law	$F = \frac{1}{4\pi\varepsilon_o}\left(\frac{q_1 q_2}{r^2}\right)$	(17.10)	Force in newtons, N, where ε_o is the permittivity of free space and r is the radial distance between the two charges in m ($\varepsilon_o = 8.854 \times 10^{-12}$ C^2 · N^{-1} · m^{-2} or C · V^{-1} · m^{-1}). $1/4\pi\varepsilon_o = 8.99 \times 10^9$ N · m^2 · C^{-2}
Electrostatic energy	$E = \frac{1}{4\pi\varepsilon_o}\left(\frac{q}{r^2}\right)$	(17.11)	Taking a test charge from a distance r from a point charge, the magnitude of the electric field E is F/q_1
Electrostatic potential	$\mathrm{V} = \frac{1}{4\pi\varepsilon_o}\left(\frac{q}{r}\right)$	(17.12)	The electric potential V due to a test charge q relative to a particle the zero potential taken to infinity. r is the radial distance
Charge density	$D = \varepsilon_o E$ $D = \varepsilon_o \kappa E$ $n_{p,n,e,h} = \frac{N_{p,n,e,h}}{\mathrm{m}^3}$	(17.13) (17.14) (17.15)	κ is the relative permittivity—a dimensionless number commonly known as the *dielectric constant*. E is the strength of the electric field in V · m^{-1} n is the charge carrier density in m^{-3}. N is the number of charge carriers where p, n, e, h are positive, negative, electrons, holes, or other forms of charge carriers, respectively ($n_{Cu} = 8.49 \times 10^{28}$ · m^{-3} and $\rho_{Cu} = 1.69 \times 10^{-8}$ Ω · m; $n_{Si} = 1 \times 10^{16}$ · m^{-3} and $\rho_{Si} = 2.5 \times 10^3$ Ω · m)
Ohm's law and current density	$\mathrm{V} = IR$ $J = \sigma E$	(17.16) (17.17)	Voltage (in volts, V) equals current (I, in amperes A or C · s^{-1}) times resistance (R or Ω, in Ohms). Current density J is in terms of A · m^{-2} or C · s^{-1} · m^{-2}. Conductivity σ is in terms of reciprocal ohm · cm (usually as Ω^{-1} · cm^{-1}). E is the strength of the electric field: E = V · L^{-1} (usually as V · cm^{-1})
Current density	$J = n_e \upsilon e$	(17.18)	Current density as a function of electron carriers in terms of charge per unit area. n_e is the number of electrons per unit volume; υ is the velocity of the electrons in m · s^{-1}. J is in terms of A · m^{-2} or C · m^{-2} · s^{-1}

(*continued*)

TABLE 17.1 (CONTD.) *Basic Electronic Definitions and Relations*

Name	Form	Units, comments
Farad	$$F=\frac{A\cdot s}{V}=\frac{C}{V}=\frac{C^2}{J}=\frac{C^2}{N\cdot m}=\frac{s^2\cdot C^2}{m^2\cdot kg}=\frac{s^2\cdot A}{m^2\cdot kg} \quad (17.19)$$	The SI unit for capacitance. Derivation of the SI units of F is shown at the left [1]
Resistivity and conductivity	$$\rho=\frac{\Omega A}{m} \quad (17.20)$$ $$\sigma=\frac{1}{\rho} \quad (17.21)$$	*Resistivity* is resistance times area divided by length in terms of $\Omega\cdot m$ *Conductivity* is the reciprocal of resistivity in terms of $\Omega^{-1}\cdot m^{-1}$. Also called siemens (or *mhos* or reciprocal ohms, $A\cdot V^{-1}$)
Carrier mobility	$$\mu=\frac{\bar{\upsilon}}{E} \quad (17.22)$$ $$\bar{\upsilon}=\frac{J}{eN_e} \quad (17.23)$$	Where $\bar{\upsilon}$ is the drift velocity (in $m\cdot s^{-1}$), the average velocity of electrons due to an external electric field. It is assumed to be constant in a crystal (or changes vary slowly) [3]
Conductivity/carriers	$$\sigma=nqu \quad (17.24)$$ $$\sigma=n_eq_eu_e \quad (17.25)$$ $$\sigma=n_nq_n\mu_n+n_pq_p\mu_p \quad (17.26)$$	Conductivity is dependent on the charge volume density of carriers n in m^{-3} (negative and positive) $\times$ the total charge $q\times$ the mobility. For multivalent ions, the magnitude of q depends on the valency Z_i of the ion. For monovalent ions and electrons, $q=e$. The equation for a pure electronic condition is shown by equation (17.26) [2]

17.0.3 Fundamental Kinds of Electronic Devices

During your day-to-day experience, you will most likely run across an array of electronic devices—from your personal devices like cell phones, digital cameras, GPS trackers, computers, and DVDs to sophisticated devices found outside your home such as ATMs, cash registers, security equipment, sophisticated laboratory equipment, and automobile gadgets. All of these devices share similar components—transistors, capacitors, voltage regulators, rectifiers, converters, and many more. Many of these already are marketed as nano-versions—many are totally new kinds of devices (can you think of any?). There are many kinds of electronics that are unique to the nano and quantum domains. Some of them were already described in chapter 16.

The *transistor* is the heart of modern day electronic devices and regulates the flow of current or voltage. Therefore, it is a solid-state device capable of amplification, switching, and signal generation. Semiconductor transistors consist of p-type silicon and n-type silicon to form the p/n-junction, commonly known as the depletion zone—where positive and negative charges cancel each other (**Fig. 17.1**). Applications of transistors include their use in rectifying junctions, amplifiers, and switches. A *rectifying junction* is a junction that allows current to flow in just one direction. Most transistors in today's devices consist of *complementary metal oxide semiconductors* (CMOS), devices that employ two complementary transistors per gate in microprocessors. *Gates* control output that is dependent on the combination of inputs (e.g., logic gates). *Field effect transistors* (FET) are used to amplify weak signals. MOSFETs are metal oxide semiconductor FETs.

Batteries are centers of electric potential derived from chemical energy that provide static charge potential, or electromagnetic force, required to power devices. Batteries provide the necessary static charge potential to drive reactions,

TABLE 17.2 *Electronic Materials*

Material	Description	Examples	Applications
Conductors	Materials that have a surplus of "free" charge carriers. Electrons in the conduction band behave like an electron gas. Materials with relatively low electrical resistance and no bandgap. The measure of electrical conductivity is σ, the ratio of the current density J to the electric field strength E in terms of siemens per meter ($S \cdot m^{-1}$ or $\Omega^{-1} \cdot m^{-1}$). Conductivity between $10^4 < \sigma < 10^8\ \Omega^{-1} \cdot m^{-1}$	Metals like Ag, Cu, Au, Pt, Al, metal alloys, ionic SLNs, tap water, graphite, and some kinds of carbon nanotubes	Charge carriers in integrated circuit devices, transmission wires, magnetic coils, single-electron devices, low resistance connectors between devices
Super-conductors	Materials with incredibly low electrical resistivity with exclusion of an interior magnetic field (Meissner effect). The resistance is essentially zero (ca. 10^{16} times less than room temperature values) when below some critical temperature T_c (usually between 0.01 and 125 K) [3]. There are two types of superconductors: Type I can be destroyed by a magnetic field and Type II only gradually so. Type I is characterized by the presence of Cooper pairs	27 elements, many alloys, Cu_xO_y, ceramics and Se- or S-based organic materials [3]	Super-efficient electronic circuits that generate no heat—lossless power transmission lines, high-speed levitated trains, faster computers and switching devices called cryotrons [3], also NMR, MRI, and Josephson junctions. Superconducting digital electronics, quantum computing, microwave communication systems
Ballistic conductors	Crystalline character of highly pure materials allows electrons to transit without scattering—different from superconductivity due to lack of Meissner effect. Conductivity given by Landauer's equation. Current density 100 to 1000x better than Al	Multiwalled nanotubes are ballistic conductors at room temperature [4]	Energy-efficient electron transport and information processing
Semi-conductors	Materials that have a bandgap with energy in the IR and UV-visible range. Conductivity: $10^{-4} < \sigma < 10^4\ \Omega^{-1} \cdot m^{-1}$. The bandgap energy E_g is <2 *e*V [2]. Electrons (*e*) and holes (*h*) are charge carriers	*Intrinsic semiconductors* are pure elemental materials (Si) where $N_e = N_h$. *Extrinsic semiconductors* are doped with impurities	Transistors, solar cells, rectifier junctions
Insulators	Conductivity is very low: $10^{-10} < \sigma < 10^{-16}\ \Omega^{-1} \cdot m^{-1}$. Characteristic bandgap energy >2 *e*V [2]. Low density of electron charge carriers (N_e)	Industrial ceramic materials—Al_2O_3 ($\kappa = 10.1$), BeO ($\kappa = 6.7$), polyester ($\kappa = 3.6$), borosilicate glass, polyethylene	Insulators, capacitors, ferroelectric materials, piezoelectric materials

(*continued*)

Table 17.2 (Contd.) *Electronic Materials*

Material	Description	Examples	Applications
Dielectric materials	Insulating materials (see above) that are able to store charge via process of polarization by external field. Electrons are bound to atoms. There is very limited or no mobility of charge carriers. Insulators have completely filled valence bands and empty conduction bands for example, no intraband transitions. Bandgaps are typically >5 *e*V, and therefore, limited IR or visible transitions (except phonons)	NaCl and other alkali halides, fused quartz, polymeric materials	Capacitors, insulators
Ferroelectric materials	Capacitors below T_c with the ability to instantaneously switch polarity by the application of an external electric field—a condition called spontaneous polarization	Ceramic materials like $BaTiO_3$ due to its tetragonal microstructure	Capacitors, filters, resonant tunneling devices, ferro-electric RAM's, holographic 3-D storage, photonic networks
Piezoelectric materials	Piezoelectric materials are able to convert electricity into mechanical energy. Conversely, when pressure is applied, piezos generate electricity. The fraction of mechanical energy converted to electrical energy is *k*. Piezoelectric materials are considered ferroelectrics	Quartz or SiO_2, ($k = 0.1$); $BaTiO_3$, ($k = 0.1$) and combinations of $PbTiO_3$ and $PbZrO_3$ (PZT) [Shack]	Transducers convert one kind of energy into another kind of energy. AFM and STM raster mechanics, quartz crystal microbalances, radios, electronic devices
Organics and polymers	Electronic materials made of organic (carbon-based) materials rely on charge transfer complexes and conjugated π-systems. Mechanisms include: mobility gaps, tunneling, phon-assisted hopping; resonance stabilization, delocalization of π-electrons. Organic photoelectric complexes often involve a chromophore and a semiconductor	Conductive polymers: polyaniline (PANI), polypyrrole (PPY), polyacetylene (PA) Chromophores: ruthenium bipryridine (Rubpy), chlorophyll	Molecular electronics, solar cells, quantum devices Photosynthesis
Biomolecular materials	Use of biological materials in electronic circuits. Proteins, lipids, etc., are able to transfer charge, transfer charged molecules, undergo color changes	DNA Bacteriorhodopsin Self-assembled monolayers Proteins/lipids	Neuroelectronic interfaces Biological motors Storage elements Electronic switches, gates, biosensors, and biological transistors [5]
Nano-electronic materials	Analogous to macroscopic counterparts but possess nanoscale dimensions. Biomaterials can produce cascades. Electronic properties of nanomaterials may be quantized	Carbon nanotubes Quantum dots Organic thin films Inorganic nanowires Small molecules (e.g., rotaxanes and catenanes) Charge transfer complexes	Single-electron transistors Nanowires NEM systems Unimolecular rectifiers Computer transistors Field-effect transistors Nanosensors

Fig. 17.1 *Schematic rendition of a prototype silicon electronic device. When forward bias is applied to the p-type semiconductor terminal (as depicted in the figure), the holes are pushed away from the positive contact and electrons away from the negative contact towards the p/n junction and electric current is made to flow. When no potential is applied, the depletion zone consists of canceled hole–electron pairs, a process known as recombination. When increasing bias potential, the depletion zone is thinned so that charge carriers are able to tunnel through the junction as electrical resistance is decreased.*

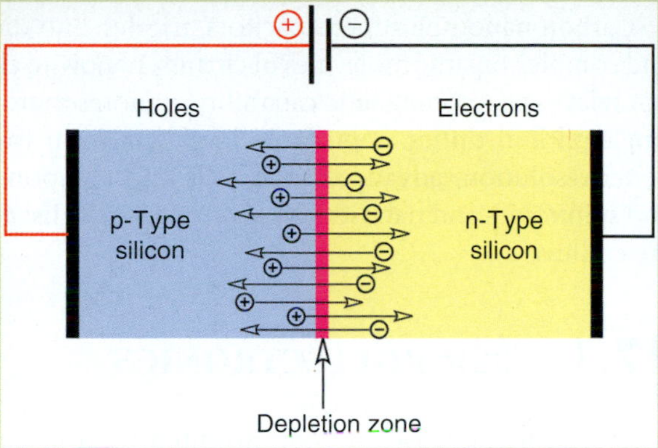

devices, or processes. Nanotechnology has already created a new generation of batteries that outperform anything on the market today. Battery components include the *anode*, the *cathode*, and the *electrolyte* (solid or liquid).

A *capacitor* is a passive electronic component that is able to store energy. Its mode of energy storage is not like that of a battery. Capacitors, consisting of a dielectric material sandwiched between two conducting plates, store energy in the form of charge. *Capacitance* is directly proportional to the surface area of the electrical contact with the dielectric and measured in terms of farads (**Table 17.1**). Capacitors are used in conjunction with transistors and form two core elements in integrated circuits. A *resistor* is another passive component that provides a source of resistance in an electronic circuit, for example, resisting the flow of electrons. Its function is to produce a voltage drop across its terminals. Resistors also produce heat when activated and are usually made of carbon particulates with a binder material. A *converter* is a device that processes AC power into DC power or vice versa. *Transducers* are devices that convert a physical signal into one that is electronic. The stimulus may be physical (e.g., pressure), chemical, or biological. These devices all have nanoscale analogs that operate in similar ways. Nanoscale sensors, transducers, transistors, rectifying junctions, resistors, and capacitors all group together in nano-integrated circuits to perform predetermined functions.

17.0.4 The Nano Perspective

The first level of function in the biological world is at the nanoscale—as atoms and molecules react stochastically without any perceived function or objective except to find the lowest energy state. Nature has optimized the nanoworld—optimized its *special nanoelectronics* and has shown an incredible track record for several hundreds of millions of years—one that humbles us for sure. We are, however, at the stage where atoms and molecules can be manipulated by us with due intent—to do our bidding as it were. Yes, we have controlled atoms, molecules, and electrons before in our chemical reactions, but we have never controlled them singly. Nanotechnology is the application of nanoscience to industry and commerce. We take nanoscience and turn it into devices, materials, and applications. We take an atom, a molecule, or an electron and tell it to "stay" or "go."

Carbon nanotube supercapacitors, triodes, and diodes; the use of biological and complex organic molecules in circuits, nanowire electrical conduits, quantum dot relays, gigantomagnetic capability, nanoresonators, single-electron transfer, single-photon optics, zeptomole level analytical capability, atomic level and better resolution, advanced solar cells and components, quantum computing and spintronics and nanoscale field emitters—the list goes on and the possibilities are endless.

17.1 Microelectronics

In 1904, Sir Ambrose Fleming invented the thermionic valve (diode or two electrode rectifier). The age of microelectronics officially began with the invention of the germanium transistor, used initially to amplify electrical currents, in the late 1940s by J. Bardeen. W. Brittain, and W. Shockley, all of Bell Labs. It was smaller, cheaper, and more efficient than the vacuum tube it replaced. Use of silicon grew popular due to its superior properties and the ability to grow it as a single crystal with high purity. Silicon also readily forms a thin oxide layer (better than germanium)—a layer that serves as a gate oxide and demonstrates facility during fabrication and function. Then William Shockley left Bell Labs to form Shockley Semiconductor after he developed the junction (sandwich) transistor.

In 1957, eight colleagues, known as the "Fairchild Eight," left Shockley's company to form Fairchild Semiconductor, and two of the eight, R.N. Noyce and G. Moore, left to form Intel [6]. The planar process to form integrated circuits was underway [7]. The process makes use of silicon's thin oxide layer through which impurities are able to diffuse. The concept of the integrated circuit was developed in 1958 when R.N. Noyce (Fairchild and the Intel) and J. St. Clair Kilby (Texas Instruments) independently placed transistors and circuits on a silicon wafer or chip [6]. Noyce and Kilby did not know each other at the time of their breakthrough, but their two respective companies engaged in a lawsuit that lasted well into the 1960s [8]. According to Gordon Moore of Intel (the author of Moore's Law), 10^{17} transistors are manufactured every year—an amount equivalent to the number of ants in the world [6]. That figure breaks down to about three billion transistors manufactured per second!

17.1.1 Introduction to Band Structure

Charge carriers and their mobility are responsible for electrical conduction. Charge carriers can be electrons, holes, or positive and negative ionic species. Semiconductors differ from metals in that electrical conductivity increases as a function of temperature—whereas in metals conductivity decreases with increasing temperature due to increasing resistance. The average velocity of the charge carriers is the product of the carrier mobility μ times the strength of the electric field E.

The Electronic Structure of Atoms and Molecules. According to the *Pauli exclusion principle,* no two electrons can be in the same exact state, for example, two electrons in the same orbital must have opposite spins to form a closed shell. Each electron of an atom is defined by four quantum numbers that form a unique set: n, l, m, and s (principal, two orbital angular momentum, and spin, respectively). The electrons fill consecutive levels of atomic orbitals. The s orbital can accommodate 2 electrons, the p 6, the d 10, and the f 14, and so on and so forth.

Types of Solid Materials. When metal atoms come together to form molecules or clusters, a reshuffling of electrons and energy levels among the valence electrons takes place (core shell electrons that exist in filled shells remain unchanged). If these atoms have unfilled valence shells, electrons become mobile and contribute to a phenomenon known as an electron cloud. In other words, valence electrons donated by atoms become delocalized and spread over the entire structure—an extension of molecular orbital theory. The luster, high electrical and thermal conductivity, and malleability of metals are due to this electron cloud. Why is a metal conducting and why is an insulator non-conducting?

Ionic solids are formed by strong nondirectional Coulombic attractions between ions and have large bond ene gies on the order of 2–4 *e*V per atom. Low electrical conductivity (no free electrons), water solubility, and visible light transparency (visible photons do not have enough energy to excite electrons) are properties associated with ionic solids. Covalent solids (Group IV elements like carbon and silicon and Group III–V molecules like gallium arsenide) have strong localized, directional bonds with high bond energies on the order of 4–7 *e*V per atom. Covalent solids have stable closed shell configurations. Higher melting points, low electrical conductivity, and solubility in nonpolar media (hydrocarbons) are characteristic of covalent solids. Metallic solids are characterized by Coulombic attractions between the positive atom cores and the negatively charged and delocalized valence electron gas (or electron sea). The cohesive bond energy of metals is less, between 1–4 *e*V and metals therefore are able to absorb visible light. The bonds in metals are nondirectional and allow for free electron movement throughout the crystal structure of the metal.

Band Theory. Sodium serves as a good example for the formation of electronic bands in metals. Sodium has one electron in its valence shell, the $3s^1$ electron. If two sodium atoms form a bond, two molecular orbitals are formed: A σ bonding orbital and a σ^* antibonding orbital. The energy of an occupied bonding orbital is lower due to constructive orbital overlap (constructive interference) between the two atomic orbitals and the resulting enhanced amplitude in the internuclear region [9]. This gives rise to bond strength between the two atoms. The antibonding orbital has higher energy due to destructive interference between two electrons that cancels their respective amplitude and generates a node. These electrons do not reside in the internuclear region [9].

FIG. 17.2 *Band structure of sodium (Na) is depicted (only even numbered clusters are shown for ease of illustration—for Na_3, there would be an orbital midway in energy between the bonding and antibonding orbitals—a nonbonding orbital). The black dots represent electrons. The electron configuration of sodium is $1s^2 2s^2 2p^6 3s^1$. There is one valence electron. When a molecular orbital is formed between two Na atoms, splitting occurs to form HOMO and LUMO molecular orbitals. As more atoms are added to the cluster, a band is formed. In metals, the bandgap is small and can be overcome by thermal energy and electrons can be easily promoted into the conduction band. The progression is described by the following:* atom → molecule → cluster → nanoparticle → bulk material. *In each case, going from atoms to the bulk, the bandgap energy decreases as more states become populated.*

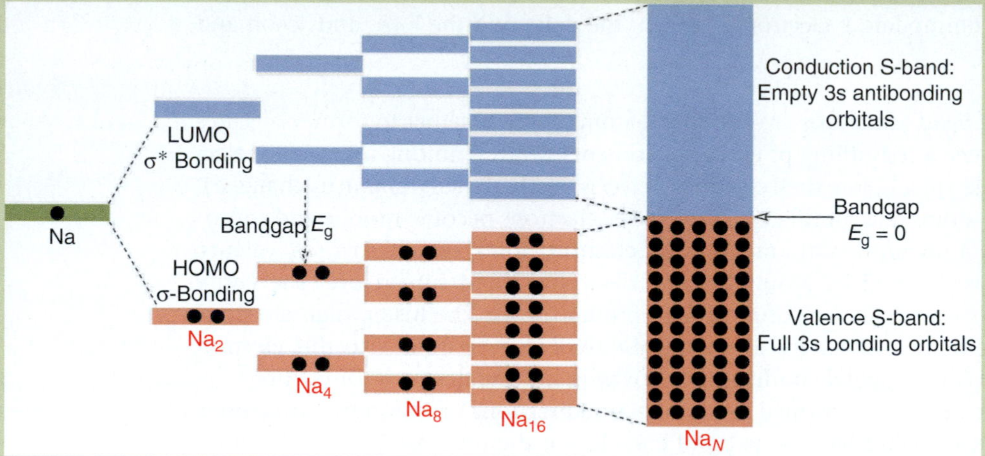

Electrons in a metal (or crystal) become free when excited and move into unoccupied molecular orbitals (LUMO) that lie just above the occupied molecular orbitals of highest energy (HOMO). In a metal, this requires little energy, which is provided by thermal energy $k_B T$ (~25 m*e*V). From molecular orbital theory, electronic bands are formed when many atoms come together to form a solid. **Figure 17.2** depicts the change in the electronic orbital structure of sodium as more atoms are added. Compare the electronic structures of sodium (11 electrons, $1s^2 2s^2 2p^6 3s^1$) with neon (10 electrons, $1s^2 2s^2 2p^6$). Sodium is conducting and neon has a band gap of ~20 *e*V—the energy between the closest filled and empty bands. The determination of electrical conduction relies on two parameters: whether the band is full or not and the degree of separation between the bands.

Bands made of just s-orbitals are called s-bands. Bands that include p-orbitals also form if the energy gap between them is not too great. Many metals that have the capability to form d-bands are constructed from overlap of many d-orbitals [9]. Bandgap energies exist between orbital energies (e.g., s, p, and/or d) or between HOMO and LUMO bands. Magnesium (Mg), for example, has two electrons in the 3s shell but three low-lying 3p orbitals. In this case, the valence band (the 3s band) is full and the conduction band, the 3p band, is empty. Due to the overlap between the 3s and 3p bands, called an sp-band, there is no bandgap and therefore, Mg is a conductor.

The Fermi Energy. A Fermi–Dirac distribution describes the population of electrons (or Fermions, particles with 1/2 spins, that obey the Pauli exclusion principle) as a function of the Fermi energy, the Boltzmann constant, and temperature. At absolute zero, electrons populate all available energy states below the Fermi energy E_F (**Fig. 17.3**). Due to thermal excitation at higher temperatures, electrons start to occupy the conduction band and deplete the valence band. Electrons that occupy states at the higher energy states of the valence band (the HOMO) become mobile with thermal stimulation. The Fermi–Dirac behavior is described by the following relation.

$$f(E) = \frac{1}{e^{\left[(E-E_F)/k_B T\right]}+1} \tag{17.27}$$

where
$f(E)$ is the Fermi function
E_F is the Fermi level
k_B is the Boltzmann constant

FIG. 17.3 *The Fermi function at T = 0 K (left) and T > 0 K (right). The probability (between 0 and 1) that an electron will populate an energy state is a function of the band structure and the temperature. As temperature increases, thermal excitation of electrons promotes high-lying electrons from the valence band into the conduction band. The Fermi level is defined at 0 K and represents the energy of the highest filled state in the valence band. At $f(E_F)$, the value is exactly 0.5. When $E > E_F$, enough electrons are promoted that conduction of electricity is possible.*

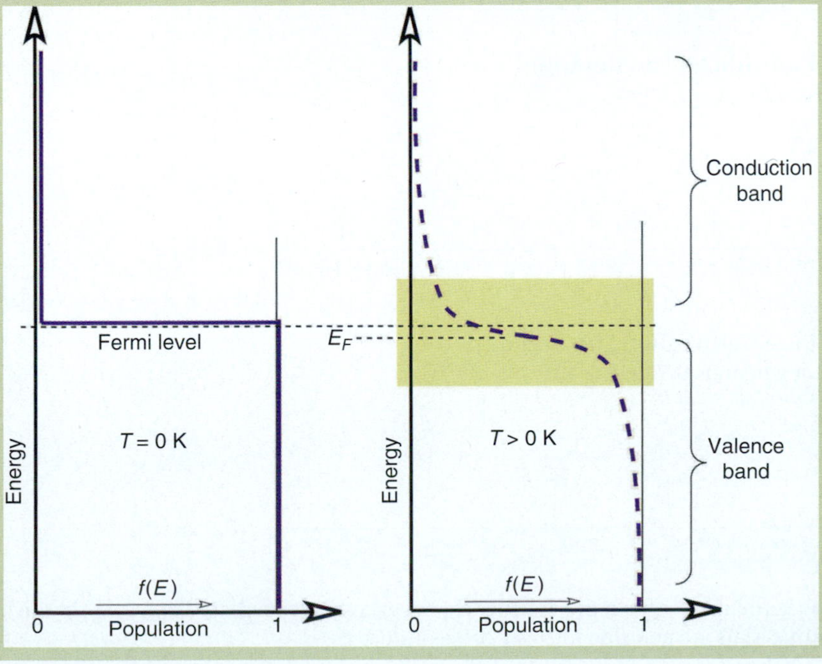

The distribution describes the population of electrons in the band structure as a function of energy and temperature—the probability that an energy level is occupied by an electron.

There are two terms that you will encounter from time to time: *intraband transitions* and *interband transitions*. When a photon excites an electron to a higher level within the same band, it is called an intraband transition—a transition from occupied to empty states within the same band with the participation of a phonon [3]. Transitions within the conduction band for example are called

Example 17.1 ***Probability of Electron Promotion***

What is the room temperature (300 K) probability of electron promotion from the valence band into the conduction band of four common materials: diamond, graphite, germanium, and silicon? (Hint: Assume that the Fermi level lies halfway between the conduction and valence bands).

Solution for Diamond:
The bandgap for diamond is 5.5 *e*V.

$$E_F = \frac{E_g}{2} = \frac{5.5\,eV}{2} = 2.75\ eV;\ \text{and that } k_B = 1.381\times10^{-23}\,\text{J}\cdot\text{K}^{-1} = 8.617\times10^{-5}eV\cdot\text{K}^{-1}$$

$$f(E)_{\text{Diamond}} = \frac{1}{e^{\left[(2.75\,eV)/\left(8.617\times10^{-5}\,eV\cdot\text{K}^{-1}\cdot 300\text{ K}\right)\right]}+1} = \frac{1}{e^{106.4}+1} = \frac{1}{1.58\times10^{46}} = 6.31\times10^{-47}$$

Solution for Graphite:
The bandgap for graphite is 0.0 *e*V. It is a good conductor of electricity. The band is a p-band that consists of abutting π and π* molecular orbitals in the solid.

$$f(E)_{\text{Graphite}} = \frac{1}{e^{\left[(0\,eV)/\left(8.617\times10^{-5}\,eV\cdot\text{K}^{-1}\cdot 300\text{ K}\right)\right]}+1} = \frac{1}{e^{0}+1} = \frac{1}{2} = 0.50$$

Solution for Pure Silicon (no doping):
The bandgap for silicon is 1.09 *e*V.

$$E_F = \frac{E_g}{2} = \frac{1.09\,eV}{2} = 0.545eV$$

$$f(E)_{\text{Silicon}} = \frac{1}{e^{\left[(0.545\,eV)/\left(8.617\times10^{-5}\,eV\cdot\text{K}^{-1}\cdot 300\text{ K}\right)\right]}+1} = \frac{1}{e^{21.43}+1} = \frac{1}{1.43\times10^{9}+1} = 6.98\times10^{-10}$$

Solution for Pure Germanium (no doping):
The bandgap for silicon is 0.72 *e*V.

$$E_F = \frac{E_g}{2} = \frac{0.72\,eV}{2} = 0.36eV$$

$$f(E)_{\text{Germanium}} = \frac{1}{e^{\left[(0.36\,eV)/\left(8.617\times10^{-5}\,eV\cdot\text{K}^{-1}\cdot 300\text{ K}\right)\right]}+1} = \frac{1}{e^{13.9}+1} = \frac{1}{1.12\times10^{6}+1} = 8.95\times10^{-7}$$

Both silicon and germanium offer a much better chance at conduction than diamond. Although not in doped states, the two materials are genuine intrinsic semiconductors.

FIG. 17.4 *Electrical conductivities of some common materials.*

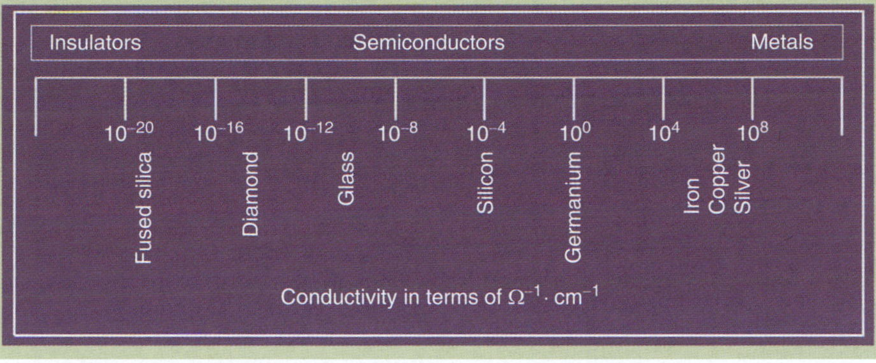

intraband transitions. Intraband transitions are not quantized and occur only in metals with maximum energy correlating to the minimum and maximum band edge [3]. If more energetic radiation is applied, interband transitions become stimulated. The mechanism of absorption depends on the energy difference between the bands. Interband transitions occur in both metals and semiconductors and are similar to optical resonance excitation of bound electrons [3]. A transition from the valence band to the conduction band is an example of an interband transition.

17.1.2 Basic Conductor and Semiconductor Physics

Conductors. **Figure 17.4** lists materials that traverse the electrical conductivity continuum.

In metals, valence and conduction bands overlap. There are then no forbidden energy gaps. Metals, therefore, are excellent conductors. At the temperature of absolute zero, electrons are in the lowest energy levels with the highest filled level known as the Fermi level E_F. At ambient temperatures, some electrons from these highest energy levels are excited by thermal stimuli to empty levels above. The E_F represents the energy at which the energy levels are half filled or half empty.

Conductors or insulators are characterized by the degree of separation of valence and conduction bands and whether the bands are filled or empty. If N atoms make up a solid, each band may consist of $2N$ electrons, in the sense relating to filling of electronic orbitals. Let us consider an example consisting of two elements, Ne and Na. The number of electrons in sodium is odd, therefore the last band up is only half filled. Ne on the other hand has an even number of electrons and the last "band" is therefore full (**Table 17.3**).

TABLE 17.3 *Electronic Configurations of Neon and Sodium*

Name	No. of electrons	Configuration
Neon	10	$1s^2\ 2s^2\ 2p^6$
Sodium	11	$1s^2\ 2s^2\ 2p^6\ 3s^1$ or [Neon] $3s^1$

For a solid material to conduct electricity, conduction electrons must be induced to flow by addition of some energy from an external applied voltage. The energy level of conduction electrons is approximated by the thermal energy k_BT, equal to ca. 0.025 *e*V, at room temperature. The relative separation of bands therefore is a consideration, and we should also consider the energy of the next unfilled band. Thus in this example, we have:

Neon

- The energy separation between the nearest empty band and the closest filled band is around 20 *e*V.
- For conduction to take place in neon, the electrons in the upper most band, completely filled, would have to "jump" across the band gap of 20 *e*V, very unlikely with only 25 m*e*V of energy k_BT (thermal energy)!

Sodium

- The band is partially filled, thus a "little bit" of energy can be added to the electrons—no quantum jump is involved.
- The final electrons "added" to make sodium, the $3s^1$ electrons, are very easy to strip from the "neon" ion core. They become the conduction electrons, often known as the *free electrons* in the metal.

The electronic configurations of the following elements that happen to be excellent conductors as solids are listed in **Table 17.4**:

This is not to say that all conductors have one electron in an unfilled band, but certainly the best conductors do!

Insulators. An insulator has a wide forbidden energy gap as we have noted earlier. The valance band is completely filled, and the conduction band is completely empty. Since there are no free electrons present, an insulator should not be able to conduct electricity. An electrical insulator, therefore, resists the flow of electricity. Application of a voltage difference across a good insulator results in negligible electrical current. In comparison, a conductor allows current to flow readily. Electrical wiring and electronic circuits require both insulators and conductors. For example, wires typically consist of a current-carrying metallic core sheathed in an insulating coating.

Resistivity. Resistivity is the measure of a material's effectiveness in resisting current flow. Materials with resistivities higher than $10^8\ \Omega \cdot m$ are considered as

TABLE 17.4 *Electronic Configurations of Conductors*

Name	No. of electrons	Configuration
Aluminum	13	[Neon] $3s^2\ 3p^1$
Copper	29	[Argon] $3d^{10}\ 4s^1$
Silver	47	[Krypton] $4d^{10}\ 5s^1$
Gold	79	[Xenon] $4f^{14}\ 5d^{10}\ 6s^1$

good insulators. Glass, rubber, and many plastics are well-known examples of good conductors. Resistivities as high as $10^{16}\ \Omega \cdot m$ can are attained in some exceptional insulating materials. Conductors on the other hand have resistivity values as low as $10^{-8}\ \Omega \cdot m$.

Charge carriers are responsible for transporting current. One kind of charge carrier is the electron. In most conductors, electrons move relatively freely, while in insulators electrons do not move freely. When atoms of simple metals combine to form a solid, the outer valence electrons become free for conduction. In an ideal insulator, all electrons stay tightly bound to the atoms, so there are no electrons that can be readily moved through the material for conduction.

In order to understand better the mechanism of conduction (and hence, insulation), consideration of electronic band structure is required. The electrons in an isolated atom possess discrete energies, a consequence of quantum mechanics. These discrete levels evolve into bands of allowed energies when the atoms condense into a solid. Forbidden regions separate the allowed bands, as schematically displayed in **Figure 17.2**. The electrons in a solid fill in the bands, from lower to higher energy.

The distinction between an insulator and a conductor is how the electrons fill in the allowed bands. For a simple metal, the highest band containing electrons will be only half full. The thermal energy (at ordinary temperatures) will be sufficient to generate conduction electrons—electron states of slightly higher energy are available in the incompletely filled band.

In comparison, the highest energy band containing electrons is completely full in a good insulator. The thermal energy of the electrons is not sufficient for promotion from this band, known as the *valence band* (VB), to the next band with available energy states, known as the *conduction band* (CB). The gap between the valence and conduction band, known as the band gap, is at least several electron volts (*e*V) wide in an insulator—thermal electron energies are 100 times smaller.

Semiconductors. The energy band model of a semiconductor is similar to that of an insulator except that the forbidden energy gap is relatively narrow in comparison. At very low temperatures, semiconductors behave like insulators. However, at higher temperatures, a few electrons are promoted from the VB to the CB by simple thermal excitation. Electrons promoted to the CB now are able to conduct electricity in the presence of an applied electric field. Corresponding electron vacancies, or holes, created in the VB contribute to over all conductivity. Two commonly used semiconductor materials are silicon (bandgap = 1.1 *e*V) and germanium (bandgap = 0.7 *e*V).

The topic of semiconductors is enormous and we will provide only a cursory overview. Since the invention of the transistor ca. 40 years ago, three single-component, ten binary, and a few tertiary semiconductor materials (e.g., mercury cadmium telluride [HgCdTe] and aluminum gallium arsenide [AlGaAs]) have been developed. These single and binary semiconductors, ordered by the magnitude of their bandgap, are listed in **Table 17.5**. Others not listed include ones that have significant potential like mercury telluride (HgTe), manganese selenide (MnSe), gallium antimonide (GaSb), indium nitride (InN), scandium nitride (ScN), aluminum nitride (AlN), zinc selenide (ZnSe), and boron nitride (BN).

TABLE 17.5 Common Semiconductor Materials

Material	Bandgap (eV)	Bandgap type
InSb	0.230	Direct
InAs	0.354	Direct
Ge	0.664	Indirect
Si	1.124	Indirect
InP	1.344	Direct
GaAs	1.424	Direct
CdTe	1.475	Direct
AlAs	2.153	Indirect
GaP	2.272	Indirect
ZnTe	2.394	Direct
SiC	2.416	Indirect
GaN	3.503	Direct
C	5.5	Indirect

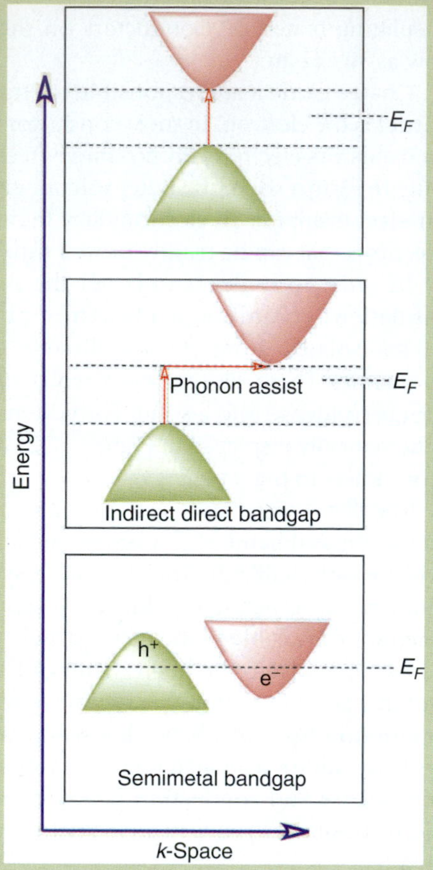

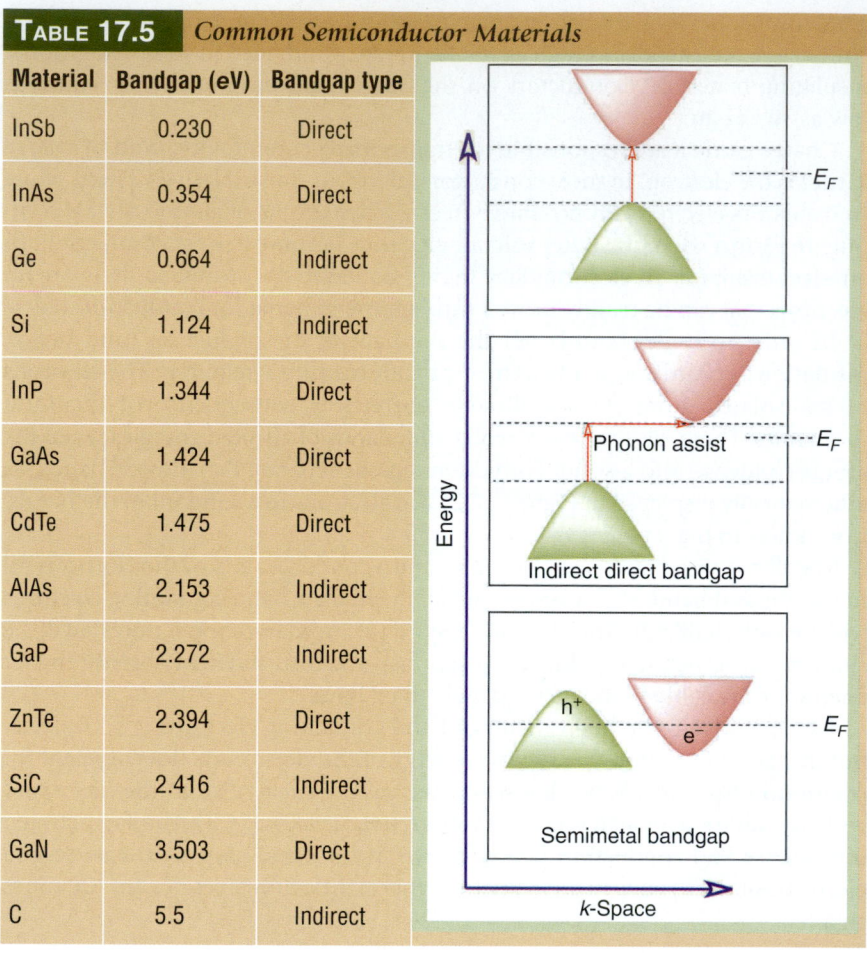

Table 17.5 also indicates the nature of the bandgap, that is, whether it is direct or indirect gap material, as this determines to a large extent just what kinds of applications the semiconductor may be suited for. Direct bandgap semiconductors are those in which the minimum energy of the CB lies directly on top of the maximum energy of the VB (in momentum or *k*-space). Therefore, in direct bandgap semiconductors, electrons in the conduction band have a high probability to combine directly with holes that exist in the valence band without any loss of momentum. When recombination does occur, a photon is released that corresponds to the bandgap energy (radiative or spontaneous recombination) [10]. Gallium arsenide and copper indium diselenide ($CuInSe_2$) are good examples of direct bandgap semiconducting materials. Direct bandgap materials find applications in light-emitting diodes and lasers [10].

In crystalline silicon, recombination does not occur as directly. In "momentum space," the VB minimum and CB maximum are not aligned perfectly. Silicon is an example of an indirect bandgap semiconducting material. In this case, direct transition is forbidden (e.g., does not conserve momentum). Recombination does occur however but it is in a form that is not mediated by the emission of a photon. Energy from excitation is released in the form of phonons. *Semimetals*, another class of semiconductors, are those in which the lowest energy of the conduction band is actually lower in energy than the highest state of the VB. In this scenario, however, the lowest state of the VB is shifted in momentum space from the highest state of the VB. Arsenic, antimony, bismuth, HgTe, and graphite are examples of semimetal materials. Boron nitride (white), although having the same structure (isoelectronic) as graphite (black), is a semiconductor. Graphical representations of direct, indirect semiconductors, and semimetals are shown in the image in **Table 17.5**.

Intrinsic Semiconductor: Group IV Elements (Si and Ge). An intrinsic semiconductor is a pure elemental semiconducting material (ideal) that has no chemical impurities. No atoms are displaced from their proper sites in the structure of the material. This state is highly unlikely however. For such an ideal material at absolute zero, the valence band is filled completely and the conduction band is absolutely empty. At temperature, $T > 0\,\text{K}$, two processes occur: (1) Due to thermal vibration of the lattice, electron–hole pairs (EHPs) are generated due to break up of some covalent bonds and (2) recombination of EHPs to produce covalent bonds also occurs in thermal equilibrium. The number of electrons per unit volume (n_i) in CB is equal to the number of holes per unit volume (p_i) in VB. This means that the Fermi level, E_F, is located in the middle of the forbidden energy gap. Hence, at $T > 0$ charge carriers are obtained by breakdown of some covalent bonds.

Extrinsic Semiconductors. The electrical properties of a semiconductor are drastically altered when doped by "impurity atoms." Such a solid is called an extrinsic semiconductor. There are two types of extrinsic semiconductors: *n-type semiconductors* and *p-type semiconductors*.

n-Type Semiconductors. When a group IV element (Si/Ge) is doped with a group V element (P/As/Sb), an n-type semiconductor is formed (**Fig. 17.5**). At temperatures

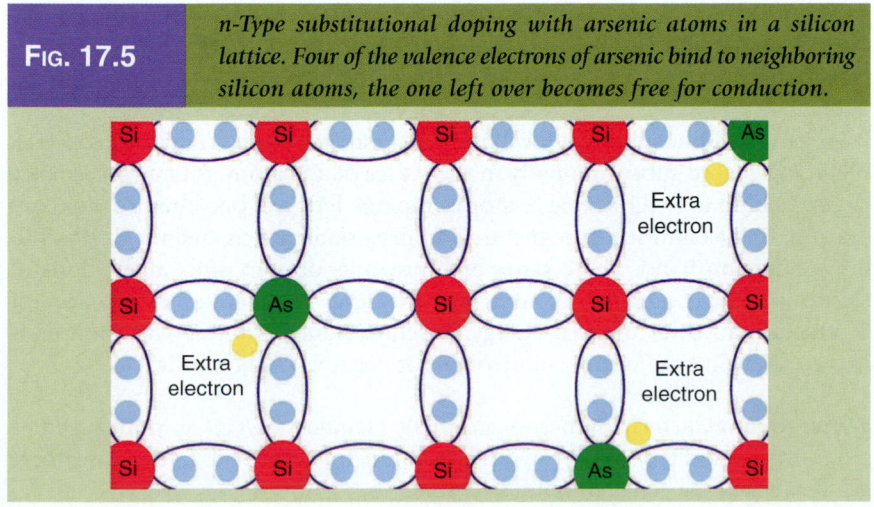

Fig. 17.5 *n-Type substitutional doping with arsenic atoms in a silicon lattice. Four of the valence electrons of arsenic bind to neighboring silicon atoms, the one left over becomes free for conduction.*

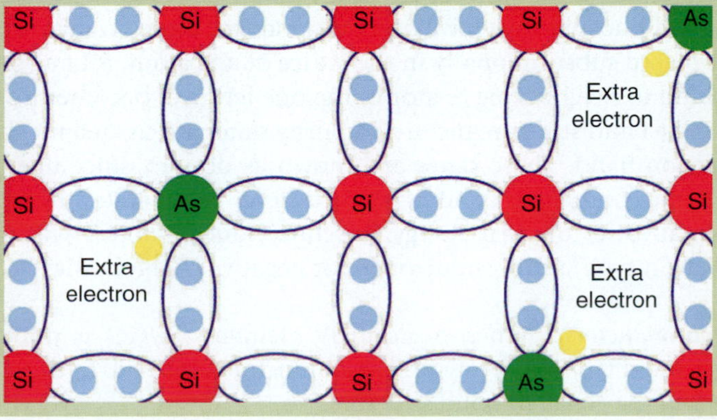

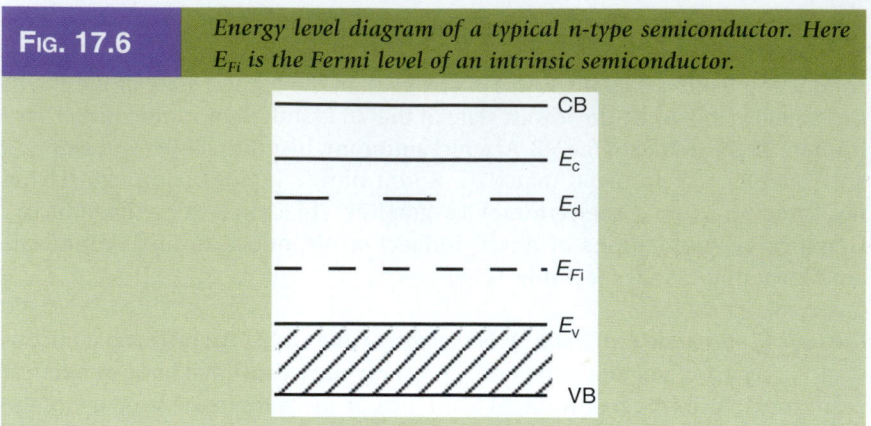

FIG. 17.6 *Energy level diagram of a typical n-type semiconductor. Here E_{Fi} is the Fermi level of an intrinsic semiconductor.*

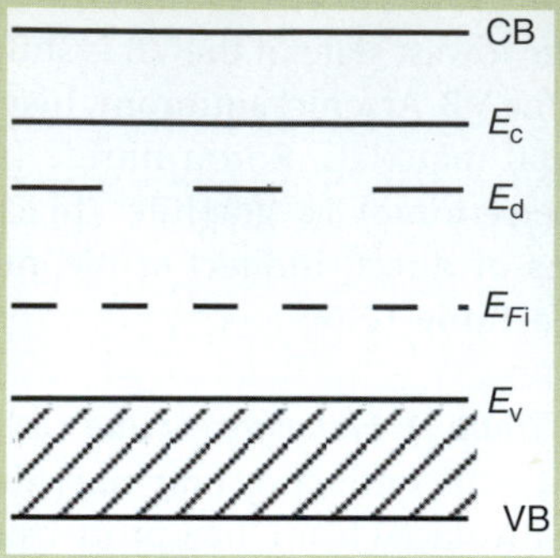

above absolute zero, four of the impurity atom electrons play the same role as the four valence electrons of the VB. The fifth valence electron is unbonded and is therefore considered to be free. Thus, addition of donor atoms provides for additional allowed energy states to the semiconductor (**Fig. 17.6**).

At very low temperatures, $T < 100\,\text{K}$, electrons are attached to the donor atoms, that is, they occupy the E_d state (see **Fig. 17.6**). At around 100 K, the thermal energy enables the extra impurity electron to shift into one of the many empty states of the nearby conduction band, where it has effective mass m_e^*, mobility μ_e, and can carry current. Note that in an n-type material, there are two types of charge carriers from three processes:

Doping—produces electron as charge carriers (majorities carriers)
Thermal vibration—produces electrons and holes
Recombination—takes away electron and holes
$n \equiv$ defined as the electron density
$p \equiv$ defined as the hole density

In an extrinsic semiconductor in thermal equilibrium, N_d is the donor concentration. The product of concentrations of majority and minority carriers in thermal equilibrium is independent of the doping impurity concentration and is a function of only of the temperature and the semiconductor material, that is, $n_p = n_i 2$ (constant) for every T.

Here, $n \gg n_i \gg p$.

Arsenic and phosphorous have one more valence electron than silicon (Si). When it is placed substitutionally in the lattice of a Si atom, four of the valence electrons bind to neighboring Si atoms, the one left over becomes free for conduction. In the band structure picture, impurity states are created much closer to the conduction band. These states are known as donors, since they "donate" electrons to the conduction band. It is much easier to "liberate" the electron for conduction at lower thermal energy. A semiconductor doped with donors is known as an "n-type" semiconductor (n for negative charge carrier).

p-Type Semiconductors. When a group IV element (Si/Ge) is doped with a group III element (In/Ga), a p-type semiconductor is formed. In this case, there is a vacancy in the covalent bond structure.

FIG. 17.7 *Energy level diagram of a typical p-type semiconductor. Here E_{Fi} is the Fermi level of an intrinsic semiconductor.*

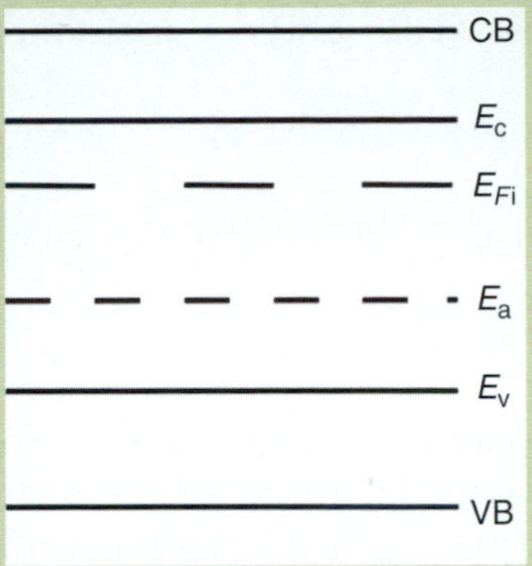

Addition of the impurity adds allowed energy states (E_a) that are empty at absolute zero. At higher temperature ($T > 100$ K) electrons from VB can jump to E_a and leave a vacancy behind (**Fig. 17.7**). In a p-type material, the following processes occur:

Doping—produces holes as charge carriers (majority carriers)
Thermal vibration—produces electrons and holes
Recombination—takes away electrons and holes
Here, again $n_p = n_i 2$. Here, $p \gg n_i \gg n$.

Gallium or boron has one fewer valence electron than silicon—when it is placed substitutionally in the lattice for a Si atom only three of the neighboring Si atoms are successfully bound, a hole remains, which also aids conduction since electrons may now move more freely through the lattice, "jumping" into a hole and leaving a hole behind to be filled (**Fig. 17.8**).

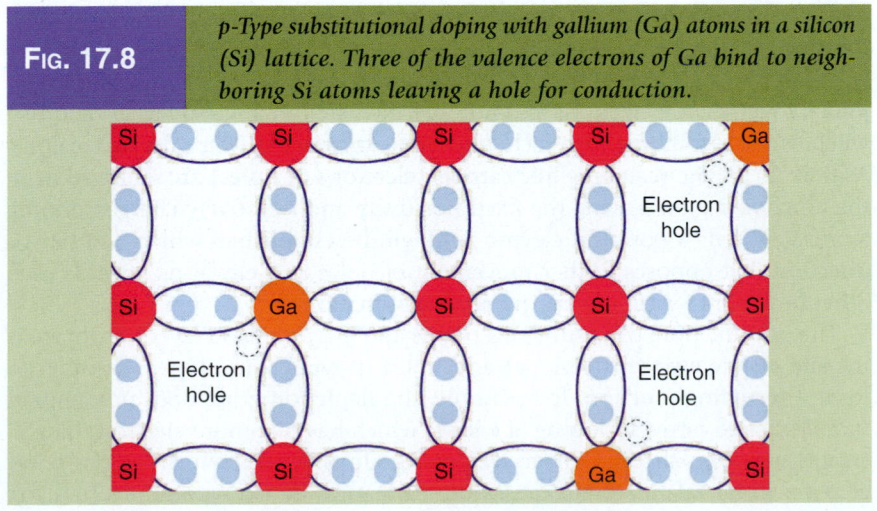

FIG. 17.8 *p-Type substitutional doping with gallium (Ga) atoms in a silicon (Si) lattice. Three of the valence electrons of Ga bind to neighboring Si atoms leaving a hole for conduction.*

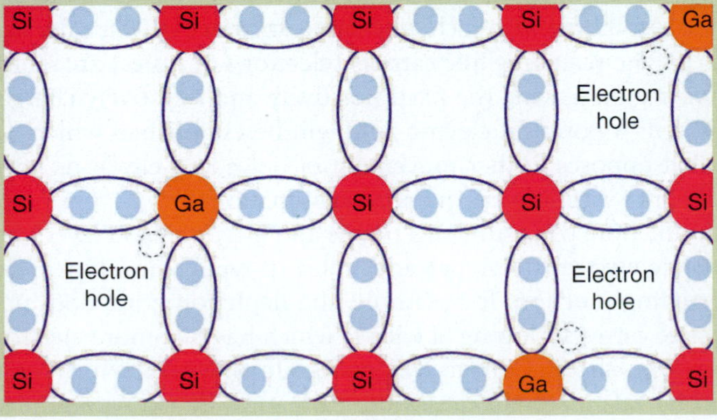

Table 17.6 *Resistivity of Antimony-Doped Germanium Compounds*

Donor concentration (μ %)	Resistivity (Ω·cm)
5	109
50	105
150	10
300	0.1

In the band structure picture, impurity states are created much closer to the valence band. These states are known as acceptors, since they can accept electrons from the valence band. Again, it is much easier to "liberate" the electron for conduction at lower thermal energy. A semiconductor doped with acceptors is known as a "p-type" semiconductor (p for positive charge carrier).

Doping changes the resistivity of semiconductors drastically. A typical example of antimony-doped germanium with different antimony concentrations and its relative resistivity is shown in **Table 17.6** above.

A diode is created when a p-type material is joined to an n-type material to create a p/n-junction (**Fig. 17.1**). The p-type material will have a high concentration of holes and few electrons. The n-type material will have a high concentration of electrons and few holes. It is obvious that there will be a large concentration gradient of both free electrons and free holes. Therefore there must be a large electric field in the p/n-junction region, which will oppose the tendency of the electron and hole concentrations to even themselves out. As previously mentioned this electric field arises when the carriers diffuse down their concentration gradients. As they do so they uncover (remove the screening carriers from) the fixed doping atoms, which have positive and negative charges for donors (n-type) and acceptors (p-type), respectively.

The formation of a p/n-junction is schematically shown in **Figure 17.9**, in which two pieces of uniformly and oppositely doped semiconductors are brought together. The p/n junction is not actually created in this way, but it is useful to imagine such a process in order to illustrate the basic physics. Initially the two pieces of semiconductors consist of fixed-charge doping atoms surrounded by an equal number of oppositely charged free carriers. When they are joined together, the electrons and holes (which are moving around at random) will begin to recombine with each other at the interface between the two regions. As they do so, the screening free carriers (electrons or holes) are removed from the junction region leaving the fixed positively and negatively charged doping atoms exposed. A powerful electric field will be established which will have a direction that opposes further movement of holes and electrons towards each other (and hence inhibit subsequent recombination).

The p/n-junction, consisting of a p-type and n-type silicon layer in contact, is the site of free electrons (n-type) and holes (p-type), canceling each other to form a rectifying junction, for example, the depletion zone does not contain mobile charge carriers. Doping Si with P, which has one more electron than Si, creates an n-type silicon. Conversely, doping silicon with B, which has one fewer electron in its valence shell, causes the formation of a p-type semiconductor.

FIG. 17.9 *Formation of a p/n-junction (schematic only). The doping atoms are shown with a circle around the charge sign, while the free electrons do not have a circle. When the n and p regions contact each other there is rapid recombination of electrons and holes at the interface. A depletion region forms, containing a strong electric field. Very soon, the field strength is sufficient to prevent further migration of electrons and holes towards each other, and equilibrium is reached. The depletion region width is greatly exaggerated in the figure.*

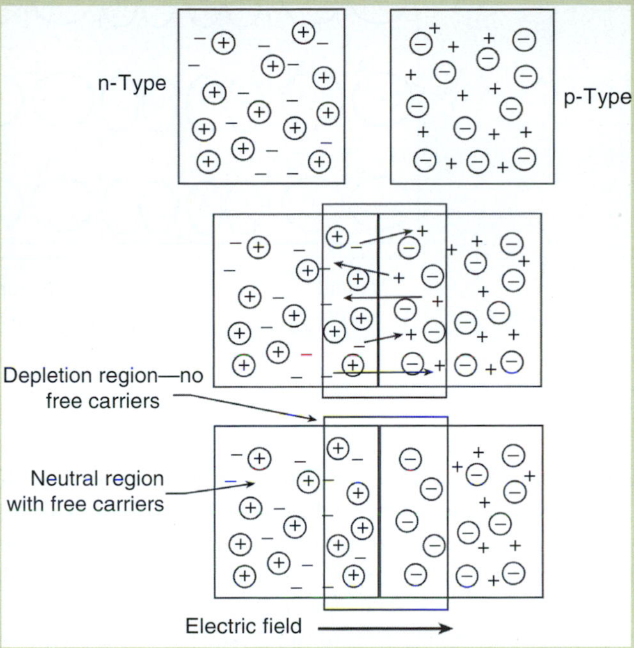

When a positive bias voltage (positive pole, forward bias) is applied to the p-type side, charge flows across the interface (**Fig. 17.9**). If polarity is reversed (reverse bias), very little charge can flow.

When equilibrium is established (after a few microseconds) the silicon wafer will consist of three regions: a neutral n-type region; a neutral p-type region; and a narrow-charged depletion region. The depletion region is typically 0.1–1 μm wide and has a very low concentration of free electrons and holes (hence the name). The electric field is confined to the depletion region, and is nearly zero in the uniformly doped n-type and p-type regions that comprise the bulk of the wafer (which is typically 500 μm thick). The electric field strength will rise from near zero at the two edges of the depletion region to a maximum in the center, and can reach a strength of $106\,V \cdot cm^{-1}$.

Of course, if the n-type or p-type regions are not uniformly doped (e.g., from a solid-state diffusion of dopants) then the electric field will be nonzero outside the depletion region as well. **Figure 17.10** shows in cross section the main regions of a particular diode together with the electric field strength. The electric field strength is nonzero to the left of the depletion region because that region has been diffused, and has a nonuniform doping concentration. The electric field

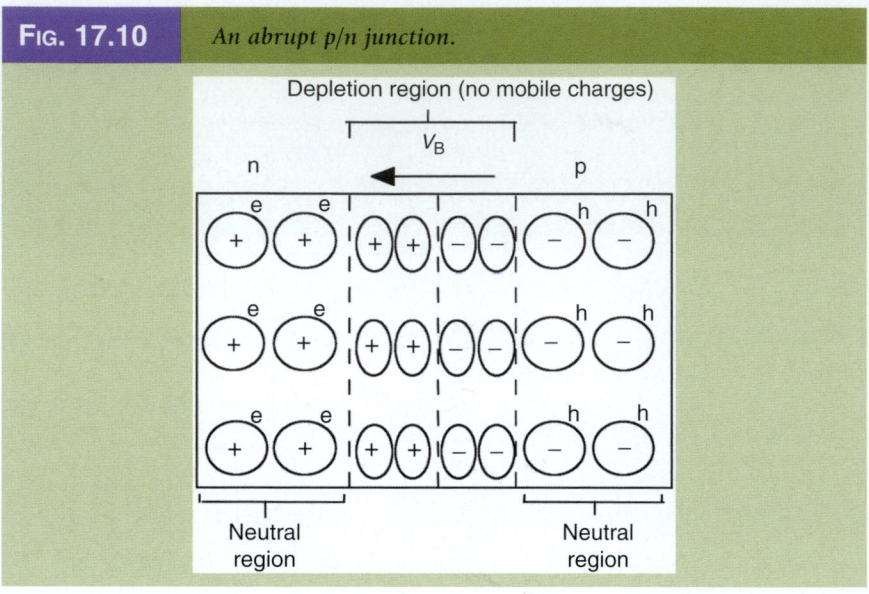

FIG. 17.10 *An abrupt p/n junction.*

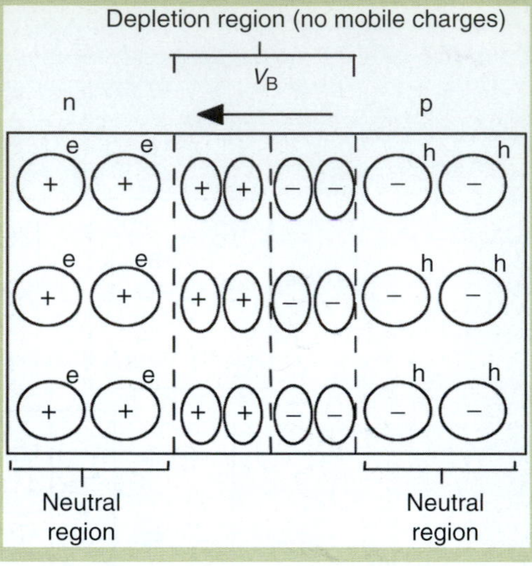

strength is zero in the right-hand p-type region because the doping concentration is uniform. The electric field strength in the depletion region has a triangular shape and rises to a maximum in the center of the depletion region.

The direction of the electric field (conventional field, + to −) is from left to right because the n-type region has fixed positive charges (the phosphorus atoms that donated an electron to the crystal) and the p-type region has fixed negative charges (the boron atoms that accepted an electron from the crystal). The electric field opposes further movement of holes down the concentration gradient from the p-type silicon to the n-type silicon and similarly for electrons. At equilibrium, the electric field exactly balances the concentration gradient.

Because of the concentration gradient, electrons diffuse from n to p and holes from p to n. Hence a potential barrier VB builds up across the junction. VB causes drift currents which exactly balance the diffusion currents (in equilibrium), that is, there is no net flow of electrons or holes across the junction.

Barrier height VB in absence of external bias

Consider **Figure 17.11**.

(a) Fermi level is shifted due to doping.
In general,

$$E_F = E_{Fi} + \frac{KT}{2} \ln \frac{\overline{n}}{\overline{p}} \tag{17.28}$$

E_{Fi} is the Fermi level in an intrinsic semiconductor

$\overline{n}$ is the electron concentration, $\overline{p}$ is the hole concentration } at temperature T in thermal equilibrium

(i) n-type:

$$\bar{n} \simeq N_d$$

$$\bar{p} = \frac{n_i^2}{N_d} \tag{17.29}$$

N_d is the donor concentration

Substituting equation (17.29) in equation (17.28)

$$E_F = E_{Fi} + KT \ln \frac{N_d}{n_i}$$

$$\text{i.e., } E_F > E_{Fi} \tag{17.30}$$

$$E_F - E_{Fi} = q_N^V = KT \ln \frac{N_d}{n_i}$$

(ii) p-type:

$$\bar{p} \simeq N_A$$

$$\bar{n} = \frac{n_i^2}{N_A} \tag{17.31}$$

N_A is the acceptor concentration

$$\therefore E_F = E_{F_i} - KT \ln \frac{N_A}{n_i}$$

$$\text{i.e., } E_F < E_{F_i} \tag{17.32}$$

$$E_F - E_{F_i} = qVp = -KT \ln \frac{N_A}{n_i}$$

From equations (17.30) and (17.32) we see that Fermi level in n-type is shifted up as compared to intrinsic Fermi level and is shifted down in p-type due to doping.

(b) At the junction, Fermi level must be the same on both sides. Thus, the electron energy levels are higher on the p-side as compared to the n-side (**Fig. 17.11**).

$$V_B = V_N + |V_P|$$

$$= \frac{KT}{q} \ln\left(\frac{N_d N_A}{ni^2}\right) \tag{17.33}$$

Forward bias (**Fig. 17.12**):

V_A is the applied voltage and V_t is the total voltage.

$$V_t = V_B - |V_A| \tag{17.34}$$

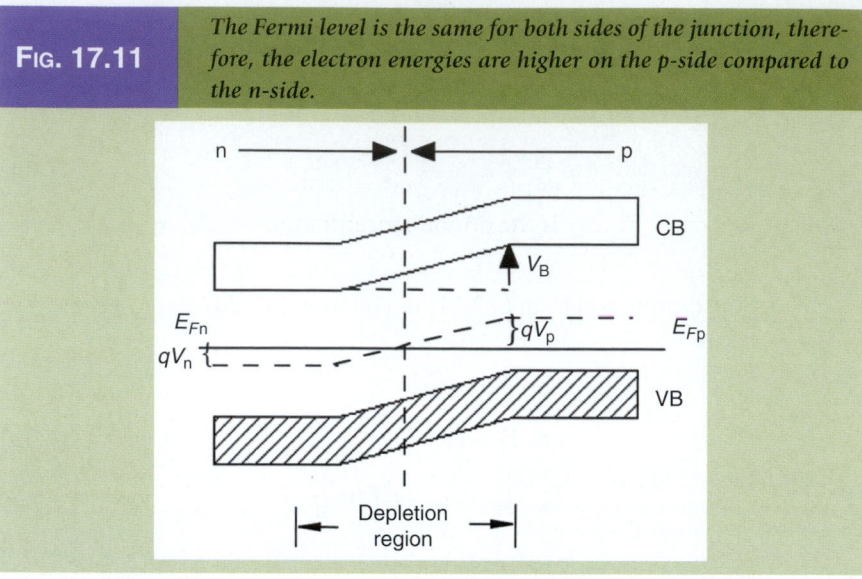

FIG. 17.11 *The Fermi level is the same for both sides of the junction, therefore, the electron energies are higher on the p-side compared to the n-side.*

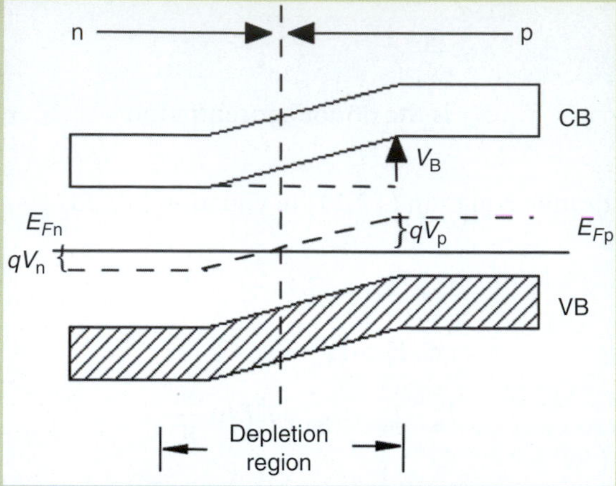

FIG. 17.12 *Forward bias is depicted. As the applied voltage is increased, the V_t and the barrier width both decrease.*

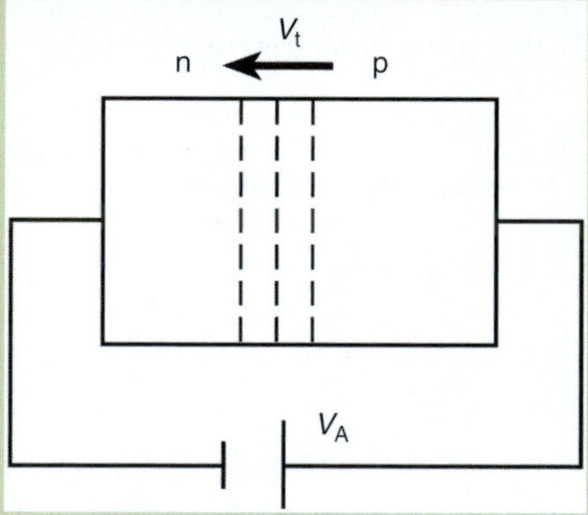

As V_A is increased, V_t and barrier width both decrease.
Reverse bias (**Fig. 17.13**):

$$V_t = V_B + |V_A| \tag{17.35}$$

As V_A increases, V_t decreases.
The Poisson equation in one dimension is given by

$$\frac{d^2V(x)}{dx^2} = -\frac{dE(x)}{dx} = -\frac{Q}{\varepsilon_r\varepsilon_o} \tag{17.36}$$

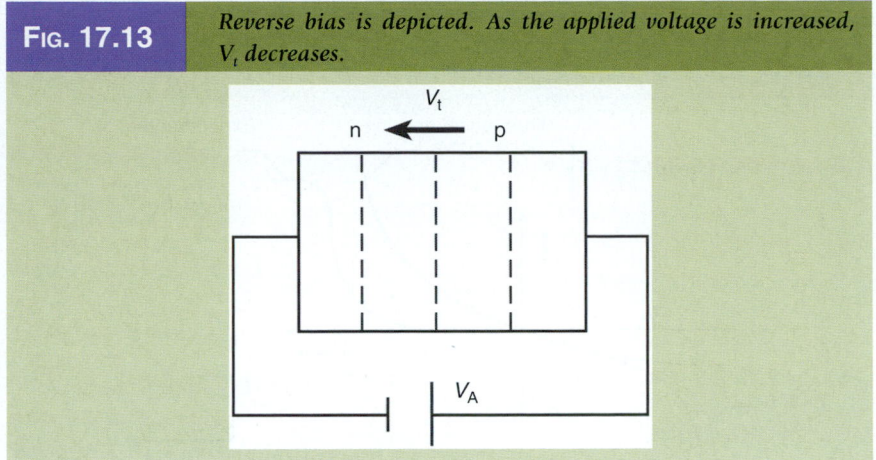

FIG. 17.13 *Reverse bias is depicted. As the applied voltage is increased, V_t decreases.*

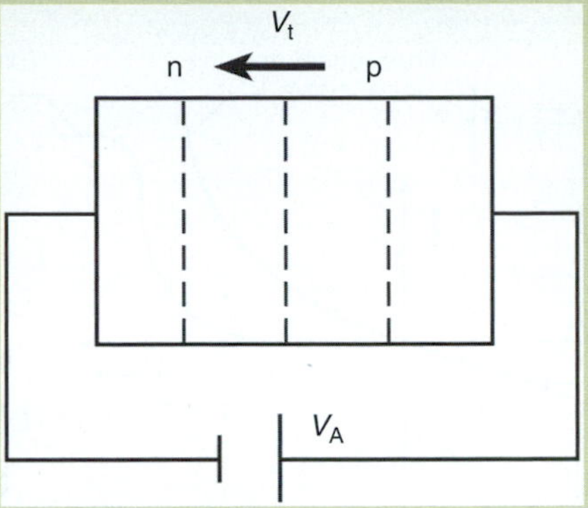

where

ε_o is the permittivity of vacuum
ε_r is the dielectric constant of the material
Q is the charge density
E is the electric field
V is the potential.

Solving under the abrupt junction approximately, we get

$$V_t = V(-d_n) - V(d_p)$$
$$= \frac{q}{2\varepsilon_r\varepsilon_o}\left(N_d d_n^2 + N_A d_p^2\right) \tag{17.37}$$

where

$Q = qN_d$ on the n-side, and $Q = -qN_A$ on the p-side.
d_n is the barrier width on the n-side
d_p is the barrier width on the p-side

Now,

$$N_d d_n = N_A d_p \tag{17.38}$$

$$\therefore\ d_n^2 = \frac{2\varepsilon_r\varepsilon_o}{q}\frac{N_A}{N_d(N_A + N_d)}V_t$$

Similarly,

$$d_p^2 = \frac{2\varepsilon_r\varepsilon_o}{q}\frac{N_A}{N_d(N_A + N_d)}V_t$$

∴ Total barrier width

$$d = d_n + d_p$$
$$= \left[\frac{2\varepsilon_r\varepsilon_o}{q}V_t\frac{(N_A + N_d)}{N_A N_d}\right]^{1/2} \tag{17.39}$$

FIG. 17.14 (a) Plot of the I–V characteristics of a diode (V_r is the cut-in voltage, V_{br} the breakdown voltage; (b) I–V characteristics of a pure conductor.

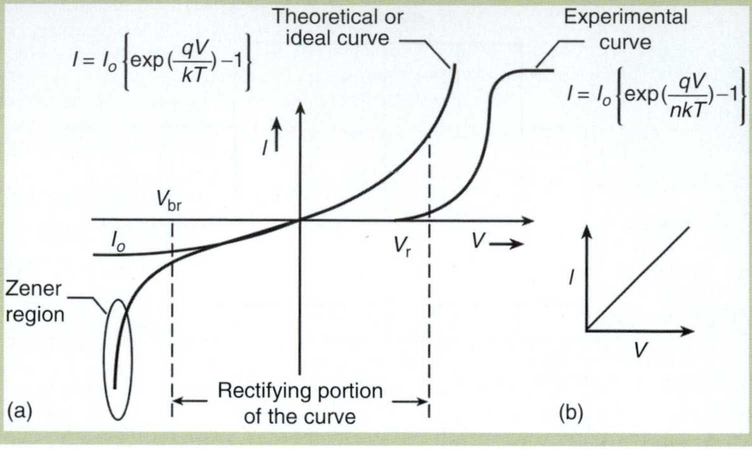

I–V Characteristics. Current through the p/n junction (**Fig. 17.14**) is given by

$$I = I_o\left[\exp\left(\frac{qV}{KT}\right) - 1\right]$$

where

V is the applied voltage
K is the Boltzman constant
T is the absolute temperature

1. When the junction is forward biased, V is positive
   If

$$V \gg \frac{4KT}{q} \quad \text{i.e.,} \quad V \gg 100 \text{ mV}$$

   then

$$\exp\left(\frac{qV}{KT}\right) \gg 1$$

   then

$$I = I_o \exp\left(\frac{qV}{KT}\right)$$

2. When the junction is reverse biased, V is negative
   Then, if

$$\exp\left(\frac{qV}{KT}\right) \ll 1, \quad I = -I_o \tag{17.40}$$

   I_o is thus called the saturation value for a reverse bias condition.

Fig. 17.15 *(a) NPN-type; (b) NPN-transistor symbol; (c) PNP-type; (d) PNP-transistor symbol.*

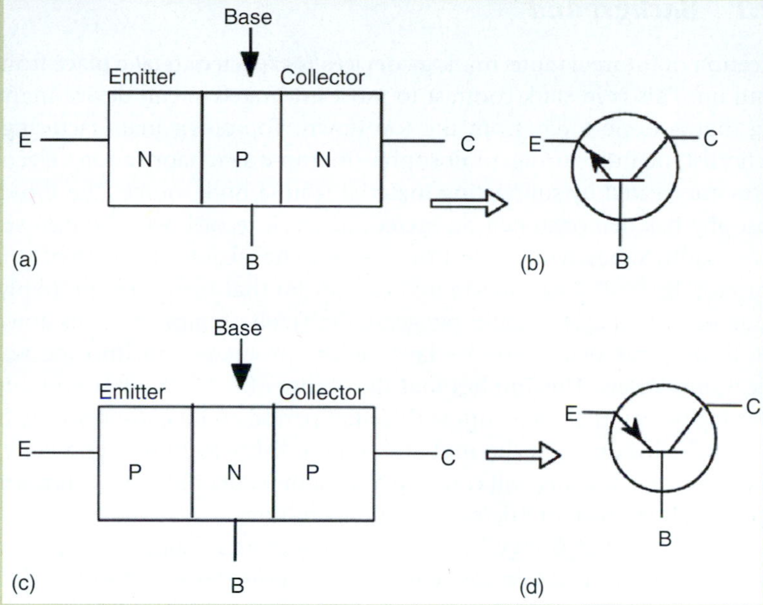

17.1.3 Transistors

The transistor is a three-layer semiconductor device consisting of either two n-type and one p-type layers of material (referred to as npn transistors) or two p-type and n-type layers (referred to as pnp transistors) (**Fig. 17.15**). The middle range is called the base and the two outer regions are called the emitter and the collector. In most transistors, the collector region is made larger than the emitter region since it is required to dissipate more heat. The emitter is heavily doped, the base is lightly doped and is thin, while the doping of the collector is in between the other two. The emitter is so called since it emits electrons (holes in case of a pnp transistor) into the base. The base passes most of these electrons (hole for pnp) onto the collector. The collector gathers these electrons (holes for pnp) from the base.

A transistor has two p/n junctions, the one between the emitter and base is called the emitter junction while the one between the base and collectors is called the collector junction. There are four possible ways of biasing these two junctions (**Table 17.7**).

Table 17.7 *Junction Bias*

Condition	Emitter junction	Collector junction	Region of operation
FR	Forward biased	Reverse biased	Active
FF	Forward biased	Forward biased	Saturation
RR	Reverse biased	Reverse biased	Cut-off
RF	Reverse biased	Forward biased	Inverted

17.2 Nanoscale Electronics

17.2.1 Background

Fabrication of future nanotechnology devices is expected to take place from the bottom up. This is in stark contrast to most integrated circuit device manufacturing that is done today from the top down. Top-down manufacturing is a reductionist manufacturing philosophy. In top-down fabrication, electronic features are created by subtracting material from a bulk source. The top-down philosophy has demonstrated an incredible track record over the past several decades. Sadly, Si-based microelectronics cannot travel down the path of Moore's law forever. In 1997, Gordon Moore pointed out that there were probably five more generations (at 18 months per generation) left of improvements down the trackfollowing his own "Moore's law" before processes run into the wall of physical limitations. This implies that the ability of the industry to double the computing power of a microchip within that period could conservatively begin a noticeable slowdown in the mid-2000s. [11]. Fabrication of computer chips, and in no uncertain terms, will come up against the wall of economic unfeasibility as further scaling down of microelectronics continues.

Higher and higher energy lithography sources and support equipment are required to achieve the resolution required for nanometer-scale features. Because of the different functions they perform, logic chips are a more complicated design than memory chips are. As a result, the design complexity is expected to increase exponentially with the number of transistors to be integrated, and this in turn impacts the design-cycle time, causing a tendency for the actual integration level to fall short of the scales potentially attainable with current technology. This expenditure in investment is expected to be economically impractical as some juxtaposition between economics, smaller features, and complexity is attained along somewhere soon down the line.

17.2.2 The Current State of Microelectronics and Extensions to the Nanoscale

There are many kinds of physical limitations that would hinder our "staying the course" of traditional manufacturing practice. Stray signals and heat-dissipation limitations will be encountered as more transistors are packed onto a chip, and the increasing magnitude of difficulty of fabricating ever-smaller devices will severely impact future progressive steps. Experts expect such challenging issues to dramatically express themselves as transitions approach the 100-nm landmark. Because of this and other factors, the explosive increase in transistor densities and processing rates of improved integrated circuits projected by Moore's law is countered by mounting costs required for facilities to manufacture transistors, chips, and wafers. Inevitably, market equilibrium will be reached when the continually decreasing physical scale of the microelectronics reaches the rapidly increasing economic factors that impact production. This impasse could be reached as early as 2015. The cost of a facility, for example, that is required to fabricate high technology devices and components is projected to be nearly

$200 billion. That event will signal the end of the long and remarkable advances in CMOS computer-chip processing power.

We all are aware how storage capacity and computer speed increased significantly over the years. We too are well aware how the concomitant cost of manufacturing integrated microcircuits, transistors, chips, and devices has been reduced. The primary bearer of the load of the electronic industry has been, and still is, photolithography—in particular, the use of visible and ultraviolet sources. Photolithography, however, has a fundamental resolution limit—70 nm [11]. Even with the advent of shorter wavelength sources such as extreme ultraviolet (EUV), x-rays, and electron beams, obstacles in the form of resolution limits, radiation damage, and increasing costs to develop the next generation of high-vacuum, high-energy lithography equipment have inspired investigators to develop other means.

Much of modern IC R&D and manufacturing processes is devoted to optimization of chip size, wafer size, and detection of defectivity and interconnects—all of this just to keep pace with Moore's law. We have to, however, ask this question. Can we fabricate circuits <1 μm by simply extending practices associated with current device production technologies? The answer is an unequivocal no! What then is required for us to get down to that remarkable level of properties and phenomena in components and devices that are manufactured whole scale? One of the first steps to take is to develop an interdisciplinary "off the beaten path" approach because the limitations of current manufacturing options are on the horizon. For example, ArF lasers with a wavelength 193 nm, a current technology, is able to achieve critical dimensions down to 100 nm. To resolve smaller structures would require light of smaller wavelength (higher energy). If we go beyond the extreme ultraviolet, yes Martha, we are in the domain of the x-ray. Consequences of taking such an "energetic step" is to search for new materials that are able to stand up against x-ray bombardment. This will only add to the already astronomical R&D expenses involved in modem IC manufacture.

17.2.3 Nanotechnology-Based Strategies: Single-Electron Tunneling

Bottom-up strategies are nanotechnology-based methods that offer real solutions to the dilemma described in the previous section. Strategies using molecules, biological entities (proteins/DNA), particle deposition, etc., are straightforward techniques to synthesize, fabricate, and explore new component and device hierarchies. Self-organization is a fabrication strategy that allows for building patterns, processes, and structures at a higher level(s) through multiple and diverse interactions among components at lower level(s), for example, molecules. The self-organization process is inspired from nature. As we state over and over in this text, nature has perfected these processes over millions of years, evolved many complex structures that function as sensors, actuators, memory, and logic units. Synthetic sensors have a linear response because they are easily connected to the next stage of processing. In comparison, biological sensors usually have nonlinear characteristics, for example, they are "compensated by feedback systems" within which they play an integral role.

The dynamic range and overload capability, as a result, are often spectacular. As examples of memory devices, nature stores biological information in DNA where approximately 50 atoms are used for one bit of information about the cell—currently our level of "one bit" far exceeds this modest number. Modern hard disks can have an areal density of 150 GB · in.$^{-2}$, a level that corresponds to more than a million atoms. However, all these changes will take more than a decade from making an impact in the marketplace at the same level as today's IC technology!

Electron Tunneling and the Single-Electron Transistor. Once again we delve into the fundamentals of single-electron transfer. In 1985, Dmitri Averin and Konstantin Likharev of the University of Moscow proposed the idea of a new three-terminal device called a single-electron tunneling (SET) transistor that was eventually fabricated by Theodore Fulton and Gerald Dolan at Bell Labs in the United States in the following two years. Unlike FETs, single-electron devices are based on the tunneling effect that has its fundamentals in quantum mechanical phenomenon. The tunneling effect of electrons is observed if two metallic electrodes are separated by an insulating barrier about 1-nm thick (the length of 10 hydrogen atoms). Electrons at the Fermi energy level are able to "tunnel" through the insulator, even though in classical terms their energy would be too low to overcome the potential barrier.

The electrical behavior across a tunnel junction depends on how effectively the barrier transmits electron waves (that decrease exponentially as a function of barrier thickness) and on the number of electron-wave modes that impinge on the barrier (given by the area of the tunnel junction divided by the square of the electron wavelength). A single-electron transistor takes advantage of the quantum transfer of charge through the barrier. Such charge transfer becomes quantized when the junction is made sufficiently resistive.

If a tunnel junction interrupts an ordinary conductor, electric charge will move through the system by both a continuous and a discrete process. Since only discrete electrons can tunnel through junctions, charge will accumulate at the surface of the electrode against the isolating layer, until a high enough bias has built up across the tunnel junction (see right side of **Fig. 17.1**). Then one electron will be transferred. Likharev coined the term "dripping tap" as an analogy of this process. In other words, if a single-tunnel junction is biased with a constant current I, the so-called Coulomb oscillations will appear with frequency

$$f = \frac{I}{e} \tag{17.41}$$

where e is the charge of an electron (**Fig. 17.16**). Notice that equation (17.41) is identical to equation (16.15) presented in chapter 16 and describes SET oscillations.

The charge continuously accumulates on the tunneling junction barrier until it is energetically favorable for an electron to tunnel. This discharges the tunnel junction by an elementary charge e. Similar effects are observed in superconductors.

FIG. 17.16 *Current-biased tunnel junction showing Coulomb oscillations.*

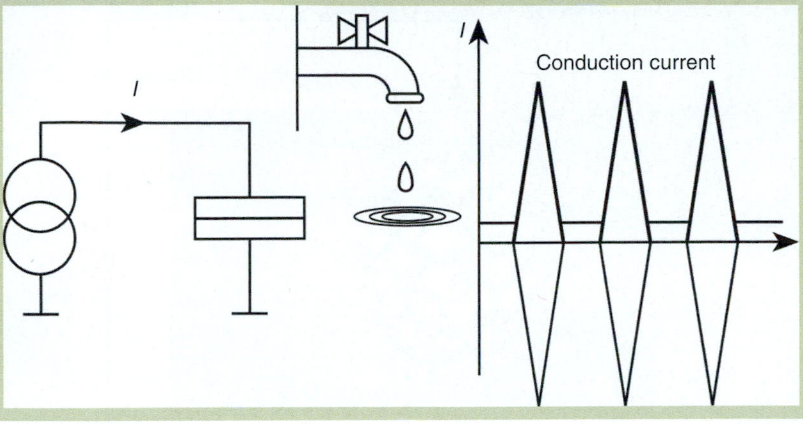

There, charge carriers are Cooper pairs, and the characteristic frequency becomes $f = I/2e$; (equation (16.16)) is related to what are known as the *Bloch oscillations* (**Fig. 17.17**).

The current-biased tunnel junction is one very simple circuit that is able to control the transfer of electrons. Another one is the electron-box (see **Fig. 17.2**). A particle is only on one side connected by a tunnel junction. On this side electrons can tunnel in and out. Imagine for instance a metal particle embedded in oxide, as shown in **Figure 17.18**.

The top oxide layer is thin enough for electrons to tunnel. To transfer one electron onto the particle, the Coulombic energy $E_c = e^2/2C$, where C is the capacitance of the particle, is required. Neglecting thermal and other forms of energy, the primary energy source is the bias voltage V_b. So long as the bias voltage is small, smaller than a threshold $V_{th} = e/C$, no electron is able to tunnel because

FIG. 17.17 *The electron-box can be filled with a precise number of electrons.*

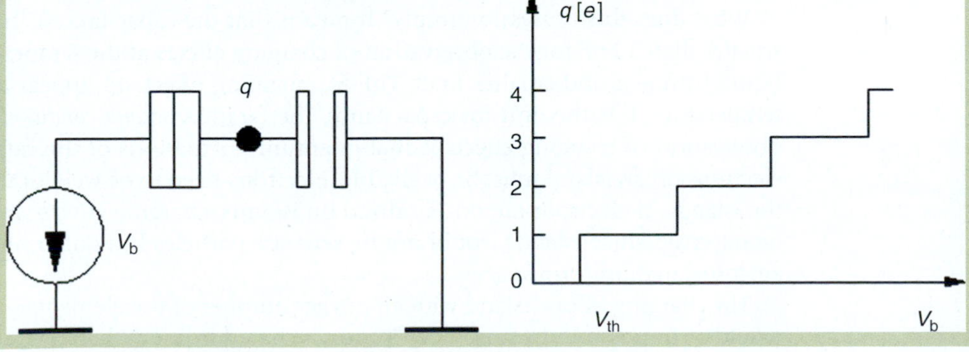

Fig. 17.18 *Metal particle embedded in oxide. Tunneling is only possible through the thin top layer of oxide.*

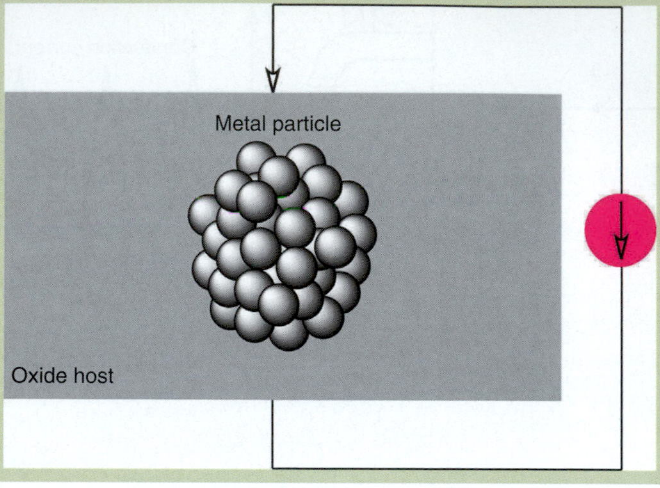

not enough energy is available to charge the island. This behavior is known as the *Coulomb blockade* behavior. Raising the bias voltage will populate the particle with one, then two, and three electron, and so on and so forth, leading to a staircase-like characteristic.

It is easily understandable that these single-electron phenomena, such as *Coulomb oscillations* and *Coulomb blockade*, only matter if the Coulomb energy exceeds that of the thermal energy. Otherwise thermal fluctuations will disturb the motion of electrons and will wash out the quantization effects. The necessary condition is

$$E_c = \frac{e^2}{2C} > k_B T \tag{17.42}$$

where

k_B is Boltzmann's constant
T is the absolute temperature

What does this expression imply? It means that the capacitance C has to be smaller than 12 nF for the observation of charging effects at the temperature of liquid nitrogen and smaller than 3 nF for charging effects to appear at room temperature. F is the unit for capacitance, the farad. *A second condition* for the observation of charging effects is that quantum fluctuations of the number of electrons on an island must be negligible. Electrons need to be well localized on the islands. If electrons are not localized on islands, charging effects would not be observed since islands would not be separate particles but rather one larger agglomerated uniform space.

The charging of one island with an integer number of the elementary charges would be impossible, because one electron is shared by more than one island. In this case, the Coulomb blockade would vanish because no longer would a lower limit of charge exist. This leads to the requirement that all tunnel junctions must be opaque to electrons in order to confine electrons on islands. This tunnel junction "transparency" is described by the tunnel resistance R_T and must fulfill the following condition for discrete charging effects to be observable. This should be understood as an order-of-magnitude measure, rather than an exact threshold.

$$R_T > \frac{h}{e^2} = 25{,}813\ \Omega \qquad (17.43)$$

where h is Planck's constant.

Therefore, these effects are experimentally verifiable for very small high-resistance tunnel junctions like those associated with small particles with small capacitances and/or at very low temperatures. Advanced fabrication techniques such as the production of granular films with particle sizes down to 1 nm and a deeper physical understanding allow today the study of many charging effects at room temperature.

In summary, if a voltage source charges a capacitor, through an ordinary resistor, the charge on the capacitor is strictly proportional to the voltage and shows no sign of charge quantization. But if the resistance is provided by a tunneling junction, the metallic area between the capacitor plate and one side of the junction forms a conducting "island" surrounded by insulating materials. In this case the transfer of charge onto the island becomes quantized as the voltage increases, leading once again to the so-called Coulomb staircase (**Fig. 17.17**). The Coulomb staircase is seen only under certain conditions. The energy of the electrons due to thermal fluctuations must b significantly smaller than the Coulomb energy, which is the energy needed to transfer a single electron onto the island when the applied voltage is zero. This Coulomb energy is given by $e^2/2C$, where e is the charge of an electron and C is the total capacitance of the gate capacitor and the tunnel junctions. Secondly, the tunnel effect itself should be weak enough to prevent the charge of the tunneling electrons from becoming delocalized over the two electrodes of the junction, as happens in chemical bonds.

SET Function. So how does a SET transistor work? The key point is that charge passes through an island in quantized units. For an electron to jump onto an island, its energy must equal the Coulomb energy $e^2/2C$. When both the gate and bias voltages are zero, electrons do not possess enough energy to enter an island, and as a result current does not flow. As the bias voltage between the source and drain is increased, an electron passes through the island when the energy in the system reaches the Coulomb energy, for example, the Coulomb blockade. The critical voltage needed to transfer an electron onto the island, equal to e/C, is called the Coulomb gap voltage.

Now imagine that the bias voltage is set under the Coulomb gap voltage. If the gate voltage increases, the energy of the initial system (with no electrons on the island) gradually increases concomitantly while the energy of the system with one excess electron on the island gradually decreases. At the gate voltage corresponding to the point of maximum slope on the Coulomb staircase, both of these configurations equally qualify as the lowest energy states of the system. This lifts the Coulomb blockade, thereby generating electrons to tunnel into and out of an island.

The Coulomb blockade is lifted when the gate capacitance is charged with exactly minus half an electron, which is not as surprising as it may seem. The island is surrounded by insulators, which means that the charge on it must be quantized in units of e, but the gate is a metallic electrode connected to a plentiful supply of electrons. The charge on the gate capacitor merely represents a displacement of electrons relative to a background of positive ions.

Researchers have long considered whether SET transistors could be used for digital electronics. Although the current varies periodically with gate voltage (in contrast to the threshold behavior of the FET), the SET would still form a compact and efficient memory device. However, even the latest SET transistors suffer from "offset charges." That means that the gate voltage needed to achieve maximum current varies randomly from device to device. Such fluctuations make it impossible to build complex circuits.

One way to overcome this problem might be to combine the island, two tunnel junctions, and the gate capacitor that comprise a single-electron transistor in a single molecule. An intrinsically quantum behavior of SET transistors should not be affected at the molecular scale. In principle, the reproducibility of such futuristic transistors would be determined by chemistry, and not by the accuracy of the fabrication process per se.

It is not yet clear whether electronics based on individual molecules and single-electron effects will replace conventional circuits based on scaled-down versions of FETs. Only one thing is certain: That if the pace of miniaturization continues unabated, the quantum properties of electrons will become crucial in determining the design of electronic devices before too long—most likely before the end of the next decade.

Single-electron transistors are not so different from traditional MOSFETs (**Fig. 17.19**).

Fig. 17.19 *Comparison of SET and MOSFET.*

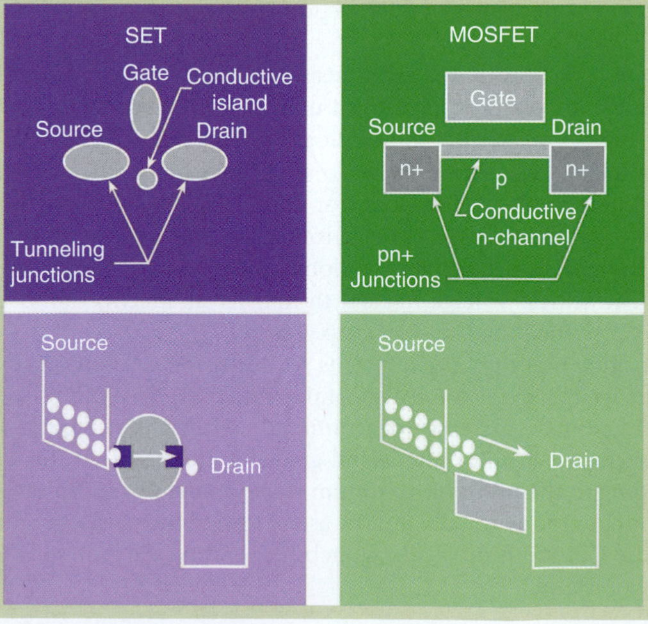

1. In the SET, electron conduction takes place one electron at a time while in a MOSFET, many electrons simultaneously take part in conduction.
2. In the SET, the drain gate controls Coulomb blockade behavior while in MOSFETs the gate controls the channel.
3. The SET requires an opaque junction where $R_T > R_Q \sim 26\,k\Omega$ while MOSFETs require highly transparent junctions.

17.2.4 Nanotechnology-Based Strategies: Molecular Wires

The concept of fabricating nanoscale devices with molecules was initiated in the early 1970s by chemists Ari Aviram of the IBM Thomas J. Watson Research Center in Yorktown Heights, New York, and Mark Ratner, a professor at Northwestern University, Chicago, Illinois. The two started working on the idea of synthesizing molecular electronic elements and devices and proposed the concept of a rectifier consisting of a single molecule. [12]. Excellent reviews on molecular electronics research is available in the literature [13–15]. In 1997, Professor Robert Metzger and his colleagues at the University of Alabama produced the molecule named *hexadecylquinolinium tricyanoquinodimethanide* [16]. They used this molecule to measure the bi-directional current flow through it between two aluminum electrodes. The rectification ratio of these two currents, the ratio indicating the preferential flow direction, ranged from 2.4 to 26.4 for different room temperature samples. These results demonstrated a definite preference for the direction of the electron flow. The chemical configuration of this molecule is shown in **Figure 17.20**.

Fig. 17.20 *The DC electrical conductivity of g-(n-hexadecyl)quinolium tricyanoquinodimethanide (or $C_{16}H_{33}$-Q-3CNQ) assembled in Langmuir–Blodgett multilayers demonstrated unimolecular rectification behavior. In other words, the conductivity of electrons from one of this impressive molecule is not the same as conductivity originating from the other end. Monolayers with the negative end of the zwitterion on the substrate and the tail (the hydrocarbon chain) up and away from the substrate conducted electrons away from the substrate.*

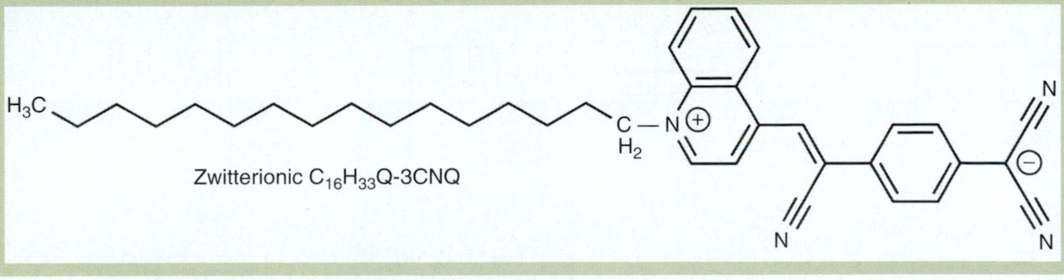

Source: R. M. Metzger, B. Chen, U. Höpfner, M. V. Lakshmikantham, D. Vuillaume, T. Kawai, X. Wu, H. Tachibana, T. V. Hughes, H. Sakurai, J. W. Baldwin, C. Hosch, M. P. Cava, L. Brehmer, and G. J. Ashwell, *Journal of American Chemical Society*, 119, 10455–10466 (1997). With permission.

There are several compelling reasons to consider that molecular electronics will play a major role in the future of nanoelectronics. Essentially all electronic processes in nature, from photosynthesis to signal transduction, are mediated by molecular structures. There are four major advantages for molecular circuits:

1. ***Size***. The size scale of molecules is between 1 and 100 nm, which allows to make functional nanostructures with accompanying advantages in cost, efficiency, and power dissipation.
2. ***Assembly and Recognition***. Intermolecular interactions can be utilized to form structures by nanoscale self-assembly. Molecular recognition can be used to modify electronic behavior, to obtain both switching and sensing capabilities on a single-molecule scale.
3. ***Dynamical Stereochemistry***. Many molecules have multiple distinct stable geometric structures or isomers (an example is the rotaxane molecule in **Figure 17.21d**, in which a rectangular slider has two stable binding sites along a linear track). Such fixed geometry isomers have very distinct optical and electronic properties. For example, the retinal molecule (in eyes) switches between two stable structures, a process that transduces light into a chemoelectrical pulse that allows vision.

FIG. 17.21 *Examples of molecular transport junctions. The top panel depicts molecules with various localized, low-energy molecular orbitals (colored dots) bridging two electrodes L (left) and R (right). In the middle panel, the black lines are unperturbed electronic energy levels; the red lines indicate energy levels under an applied field. The bottom panel depicts representative molecular structures. (a) A linear chain, or alkane. (b) A donor–bridge–acceptor (DBA) molecule, with a distance l between the donor and acceptor and an energy difference EB between the acceptor and the bridge. (c) A molecular quantum dot system. The transport is dominated by the single metal atom contained in the molecule. (d) An organic molecule with several different functional groups (distinct subunits) bridging the electrode gap. The molecule shown is a rotaxane [2], which displays a diverse set of localized molecular sites along the extended chain. Two of those sites (red and green) provide positions on which the sliding rectangular unit (blue) can stably sit. A second example of a complex molecule bridging the electrodes might be a short DNA chain.*

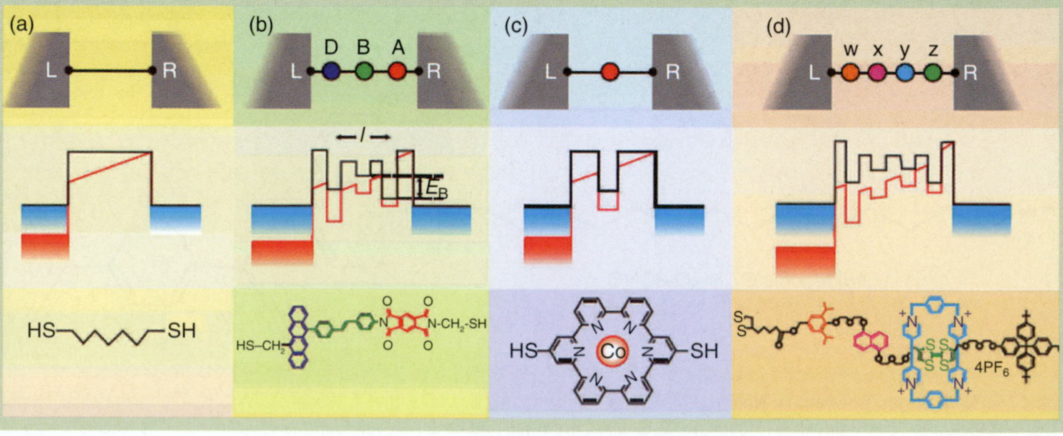

Source: Adapted from R. E. Hummel, *Electronic properties of materials*, Springer-Verlag, Berlin (1985).

FIG. 17.22 *Schematic representation of the self-assembly of molecular devices.*

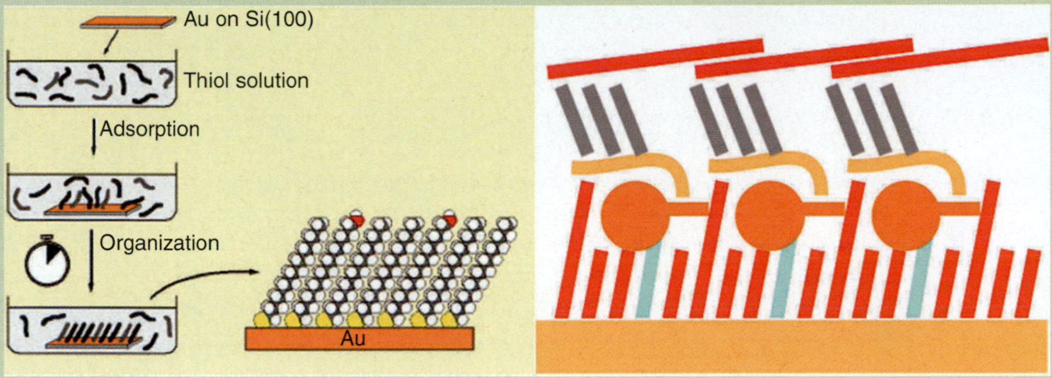

4. ***Synthetic Tailorability.*** The molecules can be tailor made to the expected composition and geometry to control a molecule's transport, binding, optical, and structural properties. The tools for such molecular synthesis are highly developed **(Fig. 17.22)**

But what are these devices specifically? To begin with, there are two generalized classes of molecules that have demonstrated possession of the characteristics acceptable for molecular-scale electronic devices: (1) carbon nanotubes and (2) polyphenylene-based chains.

Polyphenylene Molecular Wires. The polyphenylene-based chains, which are made of chains of aromatic benzene rings, are much smaller molecules than carbon nanotubes are. It then seems more likely to provide an immediate source of molecular-scale rectifiers and switches. As a result, these materials have established a base for much of the recent research into molecular electronic wires and devices. An example of such a device is shown in **Figure 17.23**, conducting molecular wire assembly, demonstrated by Tour [17]. The basic structure of this molecule is shown in the figure.

To understand the basics of what is involved in producing a current carrier of such a small magnitude, let us first look at how one would produce such a "simple" conductor or wire.

If we remove two hydrogen atoms from a benzene ring (C_6H_6), then we obtain a phenylene structure (C_6H_4), which is a ring with free sites or what we call two bonding sites. Now if we have many phenylene groups together, the missing hydrogen atoms will act as attraction sites; a chain of phenylene groups can be formed as depicted in **Figure 17.23 (Top)**. The chain is terminated with a phenyl-group molecule, which is a benzene ring with only one hydrogen removed (C_6H_5). Since this single strand of nanofilament can be of arbitrary length, it is called a "polyphenylene" chain, where "poly" means "more than one." Other molecular groups can be adapted to this chain to specialize it for specific properties or characteristics that may be desired.

For example, triply bonded ethynyl or acetylenic links are often inserted as spacers to eliminate interference from adjacent hydrogen atoms in order to

Fig. 17.23 *(Top): "Tour wires." A conductive polyphenylene molecular chain comprised of a series of modified benzene rings (Middle): A wire with an insulating island is depicted. (Bottom): Molecular rectifying diode, originally proposed by Aviram and Ratner, uses chemically doped polyphenylene molecular wires for the conduction of electrons. The dopants are intramolecular modifiers. Two intramolecular doping agents, X and Y, are applied to the polyphenylene backbone structure [18]. "X" is an electron donating group (e.g., "n-type") and "Y" is an electron accepting group (e.g., "p-type"). The donor–acceptor complex is separated within the molecular diode by a semi-insulating bridge structure "R" (e.g., dimethylene groups) that provides a potential energy barrier to preserve the potential drop between the Tour wires and X and Y coplanar assemblies to allow for tunneling. The thiol linkages, "S," also provide potential barriers to the Au-electrodes that serve to maintain isolation of the molecular wire structure. None of the barriers have potential that is insurmountable when a proper bias is applied [18].*

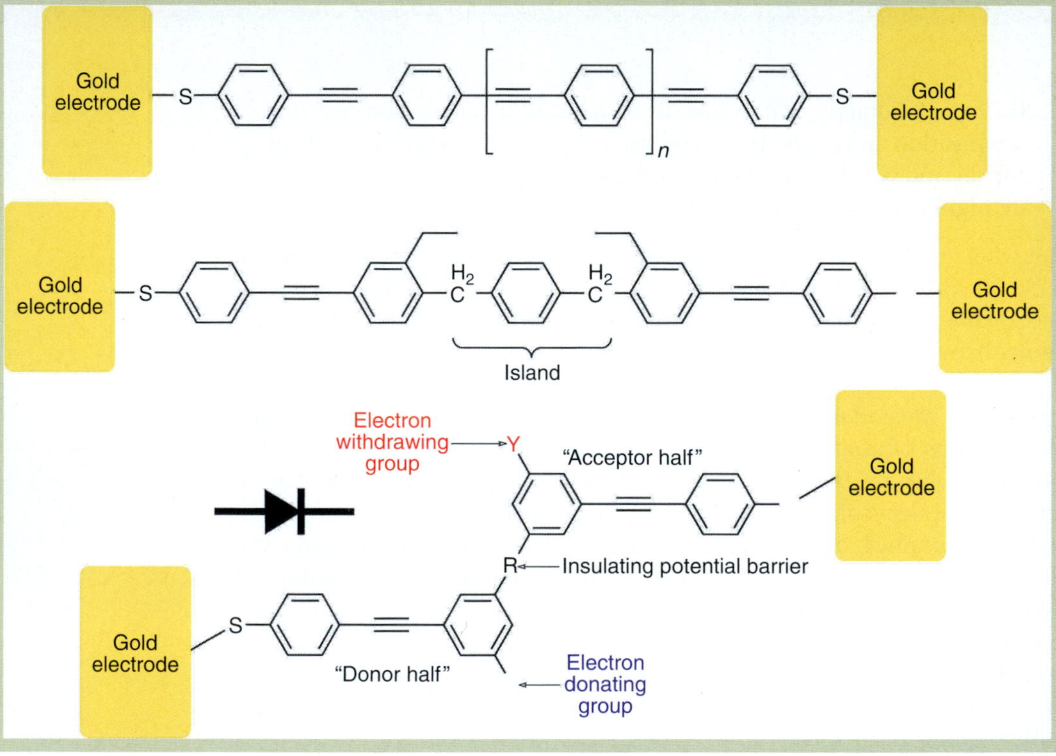

Source: Image redrawn with permission from CRC Press.

enhance the conductivity of the molecular wire. A three-ring polyphenylene chain demonstrated approximately 30×10^{-9} A of current passing through the molecule. The technique for synthesizing conductive polyphenylene-based chains has been refined by James Tour so well with great repeatability over the past few years that they have come to be known as "Tour wires." [17]. Adding aliphatic methylene groups to the chain makes the molecule an insulator, and depending on where the polyphenylene-based molecular wire is added, aliphatic methylene could function either as a "resistor" or as an "insulator."

Both rectifying diodes and resonant-tunneling diodes have been recently demonstrated. A molecular-rectifying diode, based on the work by Aviram and Ratner, has been successfully demonstrated by teams led by Metzger [16] at the University of Alabama and Reed at Yale University.[17]. Although other configurations of polyphenylene-based molecular rectifiers are being designed, molecular-resonant tunneling diodes, which are quantum devices that employ quantum effects in their simplest form by utilizing energy quantization for the "on"-"off" switching of an electric current, have been found to integrate rather nicely. The energy quantization "regulates" the variation of voltage bias across the source and drain contacts of the diode. However, current can pass equally well in either direction for the resonant-tunneling diode, unlike that of the rectifier.

In order to control the desired direction of the electron flow, a set of "insulators" acting as tunnel barriers are placed very close together to create a potential-energy "well" or "island" similar to the case explained in single-electron transistors. The probability that electrons can tunnel depends upon the difference between the energy level of the incoming electrons and that of the electrons within the potential well. If the energy quantum state of the electrons arriving in the device is different from the energy level allowed inside the potential well, current flow is not allowed and the device is switched "off." But if the energy quantum state of the arriving electrons matches one of the energy levels of the island barrier, a condition said to be "in resonance," then the state of the device is "on" and current flows.

Tour and Reed produced the first working "prototype," as shown schematically in **Figure 17.24**. By inserting two aliphatic methylene groups into the polyphenylene-based chain "wire" on either side of a single aliphatic ring, they were able to demonstrate a resonant-tunneling diode. **Figures 17.23 (Middle)** and **17.24** show that the "island" is formed by the CH_2 or methylene group located on either side of the benzene ring while the insulating property of the

FIG. 17.24 *Polyphenylene molecule* (Fig. 17.23 Middle) *with resonant-tunneling diode configuration.*

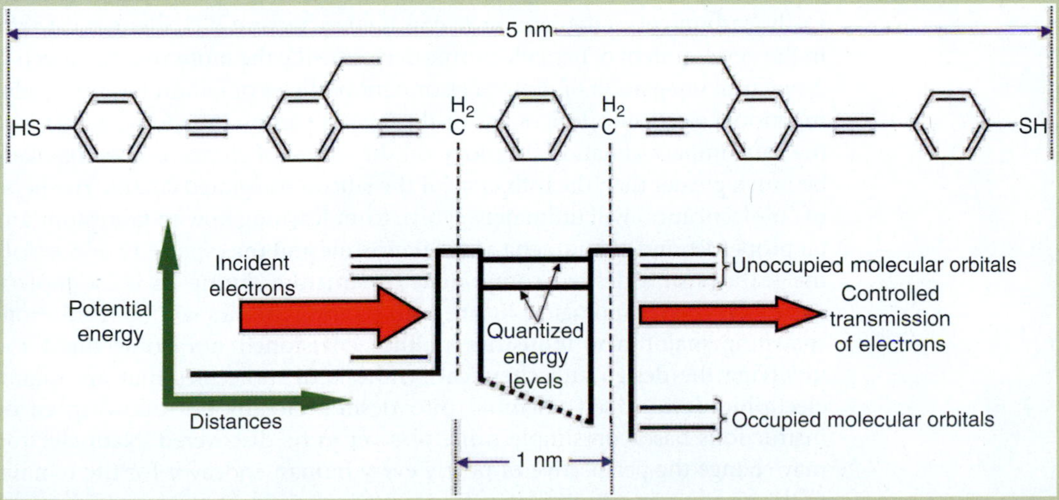

aliphatic group, as mentioned earlier, acts as potential-energy barriers to the flow of electrons. The aromatic ring between the aliphatic groups provides a very narrow gap of approximately 0.5 nm, which becomes a region of lower potential energy that the electrons must pass through.

Because of the possibility of multiple quantum levels implicit with resonant-tunneling devices, the additional advantage of multistate switching behavior is permitted. A continuously increasing bias voltage can produce multiple energy levels that come into resonance with the incident electrons. This multiple switching behavior represents additional logic states for each device.

Carbon Nanotubes. Similar progress has been experienced in the research of carbon nanotubes, also known as tubular fullerenes. A carbon nanotube, as depicted in **Figure 17.9**, is a cylinder of carbon atoms comprising a single molecule that measures from 1 to 20 nm (nominally 10 atoms) in diameter. Carbon nanotubes can be either conductors or semiconductors—depending upon their "chiral" or twist angles and diameters. In 1997, Cees Dekker and his associates at the Delft University of Technology in the Netherlands were the first to demonstrate the capability of nanotubes to act as wires [19,20]. The following year this same team demonstrated a transistor using a nanotube as one of its components. They then expanded their research into finding ways to adapt nanotubes as electronic devices and discovered that a kink in a nanotube caused it to act like a rectifier diode. Through manipulation of individual molecules, researchers at IBM Corporation then successfully produced an array of transistors out of carbon nanotubes. [20]. While a technique for the self-ordering of both the conducting and semiconducting nanotubes like in the case of organic molecules has so far eluded researchers, the IBM team has discovered a technique for producing only semiconducting nanotubes through the destruction of the conductors. This is a significant step towards the practical application of nanotubes in computer chips.

Application of molecular electronics requires flexible reaction chemistry capable of producing long strings with desired branch structures, and a high electrical conductivity for these large structures is a necessity. While polyphenylenes provide the more flexible reaction chemistry, and can be self-assembled more easily, carbon nanotubes are more conductive. Technical challenges yet remain in the development of nanoelectronic devices using the molecular approach [21]. A practical integration of molecules or nanoparticles or nanotubes into scalable, functional electronic devices will still take some more time to be achieved. But the total impact of nanotechnology on the future of electronics seems likely to be much greater than the influence of the silicon integrated circuit. The benefits of "molectronics" will ultimately evolve from learning how to tailor fundamental properties and phenomena at the nanoscale and the capability of controlling the features of a desired component or structure during its basic molecular assembly. The technological future with nanoelectronics will be revolutionary, spawning major new industries neither envisioned nor conjectured today, involving the design and chemical synthesis of molecules that are quantum electronic devices, self-assembly into desired circuits by following encoded instructions based on simple principles yet to be discovered. Nanoelectronics may change the perspective of nearly every human endeavor for the remainder of the twenty-first century.

References

1. Farad, en.wikipedia.org/wiki/Farad (2008).
2. J. F. Shackelford, *Introduction to materials science for engineers*, 4th ed., Prentice-Hall, Upper Saddle River, NJ (1996).
3. R. E. Hummel, *Electronic properties of materials*, Springer-Verlag, Berlin (1985).
4. C. Berger, Y. Yi, Z. L. Wang, and W. A. van de Heer, Multiwalled carbon nanotubes are ballistic conductors at room temperature, *Applied Physics A-Materials Science & Processing*, 74, 363–365 (2004).
5. S. Kornguth, Nanotechnology and biomolecular electronics, National Institute of Standards and Technology, www.atp.nist.gov/clso/nano_tech.htm (2005).
6. Transistors and chips, IEEE Virtual Museum, www.ieee-virtual-museum.org (2008).
7. M. Di Ventra, S. Evoy, and J. R. Heflin, *Introduction to nanoscale science and technology*, Kluwer Academic Publishers, Boston (2004).
8. J. Redin, A tale of two brains, www.xnumber.com/xnumber/kilby.htm (2007).
9. D. F. Shriver, P. W. Atkins, and C. H. Langford, *Inorganic chemistry*, W.H. Freeman & Co., New York (1990).
10. Direct bandgap, en.wikipedia.org/wiki/Direct_bandgap (2009).
11. C. G. Huang, S. I. Lee, and Y. D. Hong, Driving forces of future semiconductor technology, In *Future trends in microelectronics*, S. Luryi, J. Xu, and A. Zaslavsky, eds., p. 13, Wiley-Interscience publication, New York (1999).
12. A. Aviram and M. A. Ratner, Molecular rectifiers, *Chemical Physics Letters*, **29**, 277 (1974); R. M. Metzger et al., Electrical rectification in a Langmuir–Blodgett monolayer of dimethyanilinoazafullerene sandwiched between gold electrodes, *Journal of Physical Chemistry B*, **107**, 1021 (2003).
13. A. Nitzan, Electron transmission through molecules and molecular interfaces, *Annual Review of Physical Chemistry*, **52**, 681 (2001).
14. V. Mujica and M. A. Ratner, Molecular conductance junctions: A theory and modeling, In *Handbook of nanoscience, engineering, and technology*, W. A. Goddard III et al., eds., CRC Press, Boca Raton, FL (2002); C. Joachim, J. K. Gimzewski, and A. Aviram, Electronics using hybrid-molecular and mono-molecular devices, *Nature*, **408**, 541 (2000).
15. J. C. Ellenbogen, A brief overview of nanoelectronic devices. presented at the Proceedings of the 1998 Government Microelectronics Applications Conference (GOMAC98), Arlington, VA, 13–16 March; www.mitre.org/technology/nbanotech/GOMAC98_article.html (1998).
16. R. M. Metzger, B. Chen, U. Höpfner, M. V. Lakshmikantham, D. Vuillaume, T. Kawai, X. Wu, H. Tachibana, T. V. Hughes, H. Sakurai, J. W. Baldwin, C. Hosch, M. P. Cava, L. Brehmer, and G. J. Ashwell, Unimolecular electrical rectification in hexadecylquinolinium tricyanoquinodimethanide, *Journal of the American Chemical Society*, 119, 10455–10466 (1997).
17. M. A. Reed, C. Zhou, C. J. Muller, T. P. Burgin, and J. M. Tour, Conductance of a molecular junction, *Science*, 278, 252–254 (1997).
18. J. C. Ellenbogen and J. C. Love, Architecture for molecular electronic computers, chap. 7 In *Handbook of nanoscience, engineering and technology*, W. A. Goddard III, D. W. Brenner, S. E. Lyshevski, and G. J. Iafrate, eds., pp. 7-1–7-64, CRC Press, Boca Raton, FL (2003).
19. P. Avouris, P. Collins, and M. Arnold, Engineering carbon nanotubes and nanotube circuits using electrical breakdown. *Science*, 292(5517), (2001).
20. S. Tans, A. Verschueren, and C. Dekker, Single nanotube-molecule transistor at room temperature. *Nature*, 393, 49–51 (1998); Yao, et al. Carbon nanotube intramolecular junctions, *Nature*, 402, 273 (1999).
21. J. Tour, M. Kozaki, and J. Seminario, Molecular scale electronics: A synthetic/computational approach to digital computing. *Journal of American Chemical Society*, 120, 8486–8493 (1998).

Problems

17.1 Why are semiconductor quantum dots not very good in classical microelectronic applications? Give at least two reasons.

17.2 Discuss your understanding of tunneling of electrons in single-electron nanostructures?

17.3 Define Fermi energy. Describe the Fermi energy in a metal. Describe the Fermi energy for a semiconductor. What is its importance? How does Fermi energy relate to nanoparticles? Do semiconductors have a true Fermi energy?

17.4 What is the basic difference between classical microelectronics and single "electron" electronics?

17.5 What are compound semiconductors? Give some examples. What are some of their uses? Are there nanoparticulate compound semiconductors? If yes, give some examples.

17.6 What are charge carriers? Identify the charge carriers in metallic substances, semiconducting substances, and conductive liquids. Can nanoparticles be used as charge carriers?

17.7 Explain from the figure below the carrier recombination mechanisms in semiconductors?

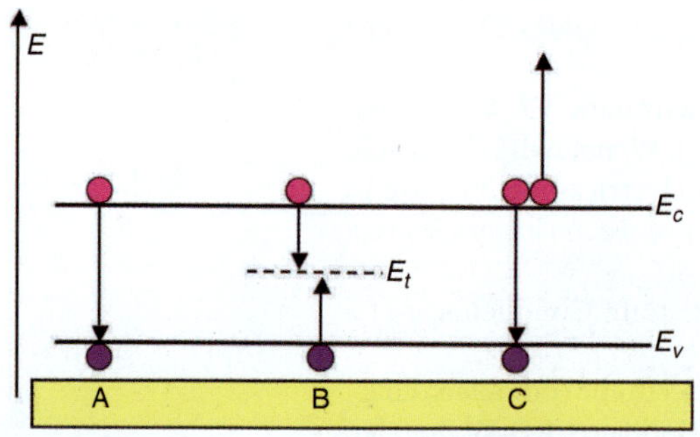

17.8 Explain the Coulomb blockade in your own words.

17.9 Calculate the intrinsic carrier density in silicon at 300 K.

17.10 When $E_C - E_F > 3k_BT$

$$n = N_C e^{(E_C - E_F)/k_BT} \quad \text{and} \quad N_C = 2\left(\frac{2\pi m_p^* k_B T}{h^2}\right)^{3/2}$$

Find the concentration of electrons and holes.

17.11 Show that for intrinsic semiconductors, the correct alternative expression for n and p are

$$n = n_i e^{(E_f - E_i)/k_BT} \quad \text{and} \quad p = n_i e^{(E_i - E_f)/k_BT}$$

17.12 Use classical Coulomb potential energy expressions in finding the lowest potential energy of two electrons that occupy four quantum dots in a quantum automata cell. Show the energy of the electrons is lower when the diagonal configuration is occupied than when the electrons occupy dots that are adjacent. The dimensions of the 2-D box is a × a × a × a.

17.13 Reconsider Problems 17.8–17.11. Is there anything that we need to modify to account for diminished size (as in the case of nanoparticles)?

17.14 Consider a molecular wire. How would you expect its electrical conductivity to behave? Does an electron in a conjugated linear molecule behave like a "particle in a box"? Plot what you would expect in terms of its I–V behavior.

17.15 What is an ohmic junction? Is there an issue with such junctions at the nanoscale? Why or why not?

17.16 Is overheating a problem in nanoscale circuitry?

17.17 Explain the role of electron tunneling in a single-electron transfer device. Have we been able to transfer a single electron?

17.18 What is ballistic conduction? How does it relate to the "mean free path" of an electron? What nanoscale material has potential to serve as a ballistic conductor? Is ballistic conduction related to superconduction?

17.19 What is a unimolecular rectifier? Nanoionics?

17.20 Describe the action of a carbon nanotube field emission device.

17.21 Why is it difficult to dope a nanoscale semiconductor?

NANO-OPTICS

Some physicists may be happy to have a set of working rules leading to results in agreement with observation. They may think that this is the goal of physics. But it is not enough. One wants to understand how Nature works.

PAUL DIRAC

Chapter 18

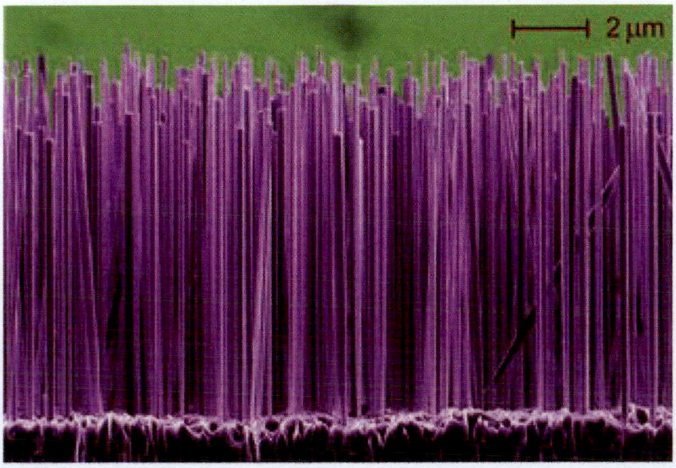

Threads

Chapter 18 represents the second part of the *Electromagnetic Nanoengineering* section, following *chapter 17* on *Nanoelectronics*. It is interesting to note that electronics, optics, and magnetism are essentially inseparable. They are all part of the electromagnetic continuum. All involve electrons and photons. Nonetheless, we draw the boundary for our convenience. In *chapter 18*, we study the basics of light and its interaction with matter. From there, we go on to investigate light's interaction with nanomaterials. Once again, we find that the size of matter has a great influence on the optical response of the material.

Chapter 18 is sandwiched in between electronics (*chapter 17*) and magnetism (*chapter 19*). The order of the chapters should make no difference with regard to the subject matter at hand.

18.0 Introduction to Optics

Optics (from the Greek *opticus* "of sight or seeing"; *optos* "seen visible," related to *ops* "eye") is the study of the interactions of light with matter and other optical phenomena. Infrared, visible, and ultraviolet are ranges of EM (electromagnetic) radiation that we are most familiar. X-rays, microwave, and radar are included in the extended family of EM phenomena. The field of quantum physics provided us a fundamental understanding of optics from the "bottom up." Classical optics (geometric optics) described the propagation of light through materials like prisms. Concepts such the light ray, refractive index, diffraction, and polarization were developed through the work of early scientists in the field. **Figure 18.1** depicts a page taken from the *Cyclopaedia*, compiled by Ephraim Chambers, published in 1728 London [1].

There are many pioneers in this field, too numerous to list and discuss, but we shall recount a few notable contributors [1–3]. The first lenses were derived from quartz that was polished dating to eighth century B.C. Democritus attempted to explain color and claimed it was due to the "roughness of constituent atoms." Euclid in the 4th–3rd century B.C. was one of the first to study optics from a geometrical sense. In his seven axioms, he defines visual rays, cones, and "seeing as a result of lines falling on objects"; angles and their relation to size and clarity; and left, right, and high rays. Archimedes (5th–4th century B.C.) studied *catoptrics*, reflections from surfaces and refraction. The Roman Seneca during the reigns of Caligula, Claudius, and Nero studied the properties of water with regard to magnification. Hero of Alexandria (first century) considered the

Fig. 18.1 Table of Opticks, *published in* Cyclopaedia *in 1728 London. The Cyclopaedia (or Universal Dictionary of Arts and Sciences) was compiled by Ephraim Chambers, an apprentice to a globe craftsman. His epitaph read* "Heard of by many, Known to few, Who led a Life between Fame and Obscurity, Neither abounding nor deficient in Learning, Devoted to Study, but as a Man, Who thinks himself bound to all Offices of Humanity, Having finished his Life and Labours together, Here desires to rest."

Fig. 18.1 *See caption on page 966.*

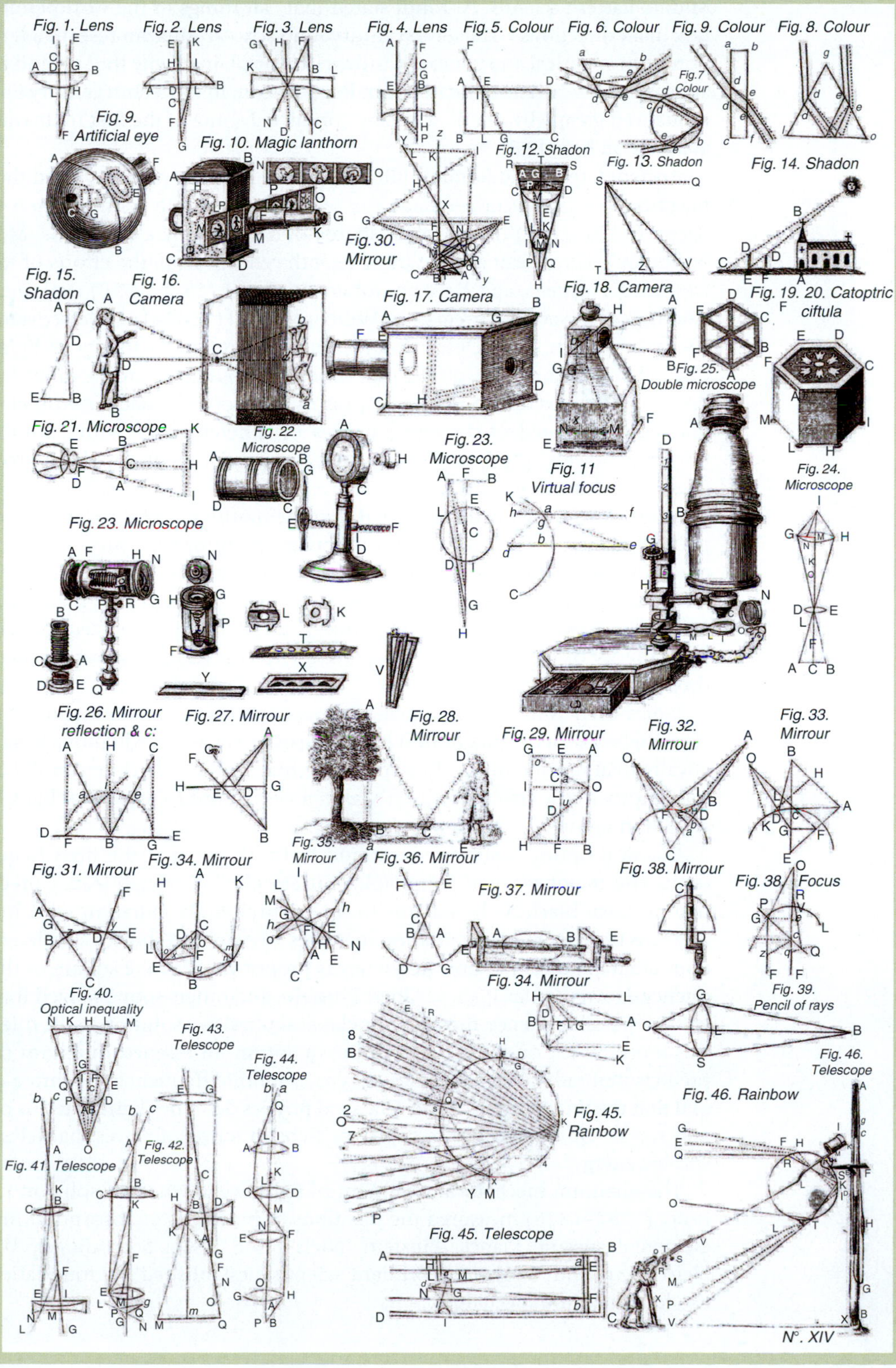

phenomena of reflection, smooth versus porous surfaces (the beginning of nanophotonics?), and that light must travel at high speeds. The ninth century Middle Eastern scientist Al-Kindi stated that "all things in the world emit light rays in all directions." The tenth century Arab scholar Alhazan researched reflection from spherical and parabolic mirrors and explained why the sun and moon appear larger when near the horizon. Roger Bacon, in thirteenth century Europe, influenced greatly by Arabic scholars, conducted some of the first mathematical analyses on light [2,3].

Ancient Greeks and Romans filled glass spheres with water and used them to magnify objects. Spectacles made the scene ca. thirteenth century when Roger Bacon (1220–1292) used broken shards of a glass sphere for a lens. Modern optics took form, beginning in the sixteenth century, from the efforts of Kepler, Descartes, Huygens, and Newton. Johannes Kepler (1571–1630) explained the workings of the eye and eyeglasses. Willebrod Snell (1580–1626) discovered the law of refraction. Rene Descartes promoted a corpuscular theory of light and that light consists of "tiny globes that travel and bounce according to the laws of optics—color was due to different spins of the globes." Isaac Newton claimed that light consisted of discrete particles of different sizes with "immutable refracting powers" [3]. Interestingly, such corpuscular theories of light preceded those based on wave mechanics.

Thomas Young in 1801 proposed that the nature of light was wavelike and particulate per se. He found that interference patterns were produced by light emerging from adjacent pinholes. Interestingly, his idea was ridiculed in his homeland England [3]. Augustin-Jean Fresnel developed a mathematical treatment of wave optics. Laplace, Fourier, and Poisson supported the corpuscular theory of light. Young, Fresnel, and Francois Arago supported the wave theory.

James Clerk Maxwell explained EM theory and laid the foundations for the classical behavior of light. Rudolf Luneburg used Maxwell's equations to systematically treat ray and diffraction optics. Heinrich Hertz, discoverer of the quantum photoelectric effect, also used Maxwell's equations to prove that light waves are EM in nature.

Classical optics, however, was found to break down at the high frequency limit. The major tenet of "ultraviolet catastrophe" (Rayleigh-Jeans) predicted that an ideal black body radiator (a 3-D cavity) emits radiation with infinite power as the energy of the radiation increases (shorter and shorter wavelength)—that radiated power per unit frequency is proportional to the square of the frequency. However Max Planck, Albert Einstein, and others soon realized that this was not possible. Hence, through nonclassical physics, a solution to the dilemma was found based on the concept of the quantum. In essence, quantum theory predicts that radiated energy goes to zero at infinite frequencies within a cavity and that the total power is finite. Classical physics describes light in terms of rays and waves. Quantum physics describes light in terms of waves, particles, and specific energy.

The quantum mechanical treatment of light began when Joesph von Fraunhofer (1787–1826) measured the positions of hundreds of absorption lines in elemental spectra. Planck, Einstein, Niels Bohr, Erwin Schrodinger, Werner Heisenberg, and many more brilliant scientists established the foundation for modern quantum mechanics.

Modern Developments. Vacuum tubes and transistors in the 1960s evolved into integrated circuits in the 1980s and on into VLSI (very large-scale integration) in which ICs were created by combining thousands of transistor-based circuits into a single chip. In the 2000s, we are on the verge of molecular electronics. In an analogous way, crossing over from vacuum tubes and transistors, discrete fiber optic components were developed in the 1970s that evolved into planar optical waveguides in the 1980s and on into integrated optical circuits, until finally we are in the developmental period of optical photonic crystals. The frequency ranges and velocities of the two primary components, the electron and the photon, differ significantly: $f_{electron} \sim 10^{10}$ Hz versus $f_{photon} \sim 10^{15}$ Hz, $\upsilon_{electron} \sim 10^5$ $m \cdot s^{-1}$ versus $\upsilon_{photon} \sim 10^8$ $m \cdot s^{-1}$.

Optical techniques have several advantages: sensitivity (single photons, single molecules), selectivity (fluorescence, filters), spectroscopy, (chemical analysis) and speed (femtosecond timescales).

History of Photonics. Photonics is the interaction of light with matter. It is the technology of light. Optical fibers and waveguides, semiconductor physics, light-emitting diodes (LEDs), laser diodes, p/i/n-photodiodes, avalanche photodiodes (APD), organic photonic devices, photonic crystal fibers, and many more devices are based on the interactions between light and matter. Photonics as a field began in the 1960s with the invention of the laser, and subsequently the laser diode and optical fibers and the Erbium-doped fiber amplifier in the 1970s. Some of the first applications of photonics were devoted to telecommunications. Whereas the optics industry focused on cameras and light sources, photonics pursued the development of optical fibers, laser LEDs, compact discs, digital videos, and LEDs.

The first visible spectrum LED was invented by N. Holonyak of GE in 1962 and he is considered to be the "father of the LED." In the 1990s, Shuji Nakamura of Nichia Corporation developed the first high-brightness blue LED based on InGaN (p-doped GaN). The first white LED used a $Ce{:}Y_3Al_5O_{12}$ or YAG (yttrium–aluminum–garnet) phosphor coating that mixed yellow (down-converted) light with blue to produce white light.

18.0.1 Interactions of Light with Matter

The propagation of EM radiation is mediated by the materials that it encounters. In vacuum, EM radiation travels at ca. 10^8 $m \cdot s^{-1}$. The absorption of light makes an object dark or opaque to the wavelengths or colors of the incoming wave. Some materials are opaque to some wavelengths of light but transparent to others. Glass and water are opaque to ultraviolet light, but transparent to visible light. The wavelengths of light absorbed by a material depend on the material composition, and hence properties derived from them. If a material or matter absorbs light of certain wavelengths or colors of the spectrum, an observer will not see these colors in the reflected light. On the other hand if certain wavelengths of colors are reflected from the material, an observer will see them and see the material in those colors. For example, the leaves of green plants contain chlorophyll, a pigment that absorbs the blue and red colors of the spectrum and reflects the green. Leaves therefore appear green. The fundamental interactions of matter with light are reviewed in **Figure 18.2**.

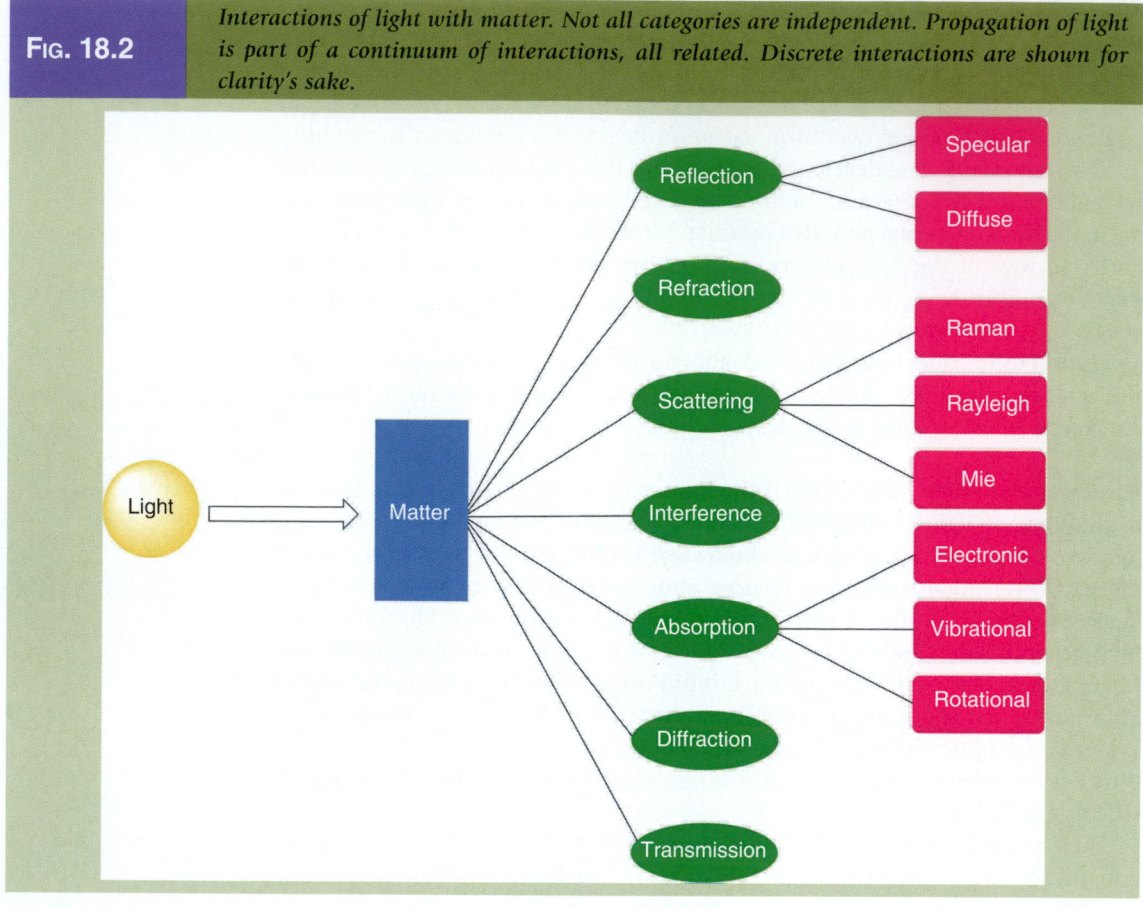

FIG. 18.2 *Interactions of light with matter. Not all categories are independent. Propagation of light is part of a continuum of interactions, all related. Discrete interactions are shown for clarity's sake.*

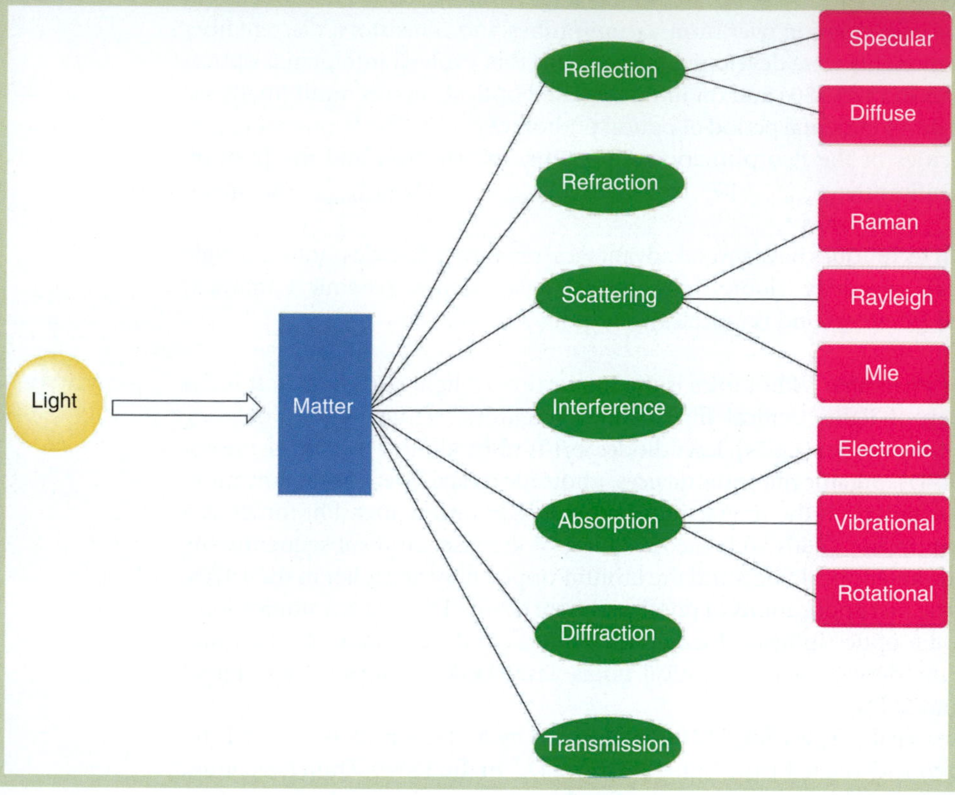

Light (EM radiation) can be viewed classically or from the quantum mechanical point of view. Since light is an EM phenomenon that consists of oscillating electric and magnetic fields and that electrons and protons are charged particles with motion that generate oscillating electric and magnetic fields, interaction between them is inevitable. The energy of electrons is quantized and the energy of EM radiation is quantized as well. We review some basic terminology and relate them to materials without going into too much detail (**Table 18.1**). Keep in mind that all interactions between light and matter can be described by the sum of all absorption and emission of photons and the most probable paths that photons take. All EM radiation exists within a continuum with matter related through $E = mc^2$. All photon and matter interactions are the same; it is just their expression that we are able to distinguish, catalog, and measure. Additionally, it is also with the size of the matter that makes a difference in those properties that we call observables.

Optical properties of nanomaterials are interesting to study due to the variety of new possible applications, some of which have already been demonstrated or have been predicted theoretically. These interesting properties arise either due to the intrinsic absorption or scattering processes. When light is incident on a material, the intensity of the incoming light onto the material reduces by either

Table 18.1 *Interactions between Radiation and Matter*

Phenomena	Description	Examples
Reflection	Change in the wavefront at an interface between two material media in which the incident wave is directed back into the medium of its origin. A reflected wave that preserves the geometrical structure of the incident wave to produce a mirror image is called *specular reflection*. The angle of incidence is equal to the angle of reflection: $\theta_i = \theta_r$. *Diffuse reflection* occurs when light strikes a roughened surface and is the macroscopic embodiment of scattering. Smooth surfaces reflect and rough surfaces scatter. In specular reflection, momentum of the light (an elastic process) is considered to be preserved.	Shiny mirror surfaces reflect light. Metals are able to reflect light due to the presence of mobile electrons on their surface. The electrons are able to generate an opposing field that deflects that of the impinging radiation.
Scattering	When radiation is not reflected, other mechanisms account for its distribution. Light impinging on a roughened surface may be deflected in all directions with loss of coherence and phase. This is called scattering. The degree and kind of scattering is dependent on the size of the particles or surface facet and the wavelength of the impinging light. Scattering is the microscopic form of reflection—in the general sense.	Molecules, small and somewhat larger particles in the nano- and micro-domains are responsible for scattering light. Scattering is highly size-dependent—thereby, it is very relevant to nano-optics. SERS is based on Rayleigh scattering—both elastic and inelastic ramifications thereof.
Refraction	Refraction is related to reflection. Light incident on a surface is either reflected (or scattered) or refracted—usually both. It depends, like all EM phenomena, on the wavelength and the nature of the material (e.g., the refractive index). Refraction is described by Snell's law: $$\frac{n_1}{n_2} = \frac{\sin\theta_2}{\sin\theta_1} \quad (18.1)$$	Transparent materials and liquids are able to refract (bend) light. The refractive index of air (or vacuum) is equal to 1.
Absorption	Absorption is a molecular phenomenon with mechanisms of electronic transitions, vibrations, and rotations.	Chromophores and fluorophores are examples of organic materials that have specific electronic transitions. Raman and IR spectroscopy are based on the absorption of wavelengths that induce vibrations, both symmetrical (Raman) and asymmetrical (IR).
Transmission	Transmission is the ability of light to pass through a material. Transmission is, in a general way, complementary to absorption. Transmission is a wavelength and material-dependent process.	Transmission of light is what is left over after reflection, scattering, and absorption have occurred.
Diffraction	Diffraction is the change in direction and intensity of waves after interaction with an obstacle (e.g., an aperture) that is of size on the order of the wavelength. It is the bending of waves around a feature or spreading of them through an aperture.	Diffraction is important in optical imaging techniques. The ultimate resolution of EM wavelength based optics is limited by the Heisenberg principle. The practical resolution of optical methods is limited by $1/2\lambda$. Diffraction gratings make use of this phenomenon in spectrometers.

the absorption in the material, transmission through the material, or reflection and scattering from the material as shown in **Figure 18.2**.

Intrinsic absorption of an optical material occurs by three fundamental processes: electronic absorption, lattice or phonon absorption, and free-carrier absorption. *Electronic absorption* occurs due to the interaction between the incident radiation and the motions of charged particles in a material (observed towards the higher frequency end of the infrared spectrum). Only EM radiation with sufficient energy to cause an electron to transfer between the valence band (VB) and conduction band (CB) will be absorbed by this mechanism. *Lattice absorption* characteristics are observed at the lower frequency regions, in the middle to far-infrared wavelength range, define the long wavelength transparency limit of the material, and are the result of the interactive coupling between the motions of thermally induced vibrations of the constituent atoms of the substrate crystal lattice and the incident radiation. Hence, all materials are bound by limiting regions of absorption caused by atomic vibrations in the far-infrared and motions of electrons and/or holes in the short-wave visible regions.

In the interband region, the frequency of the incident radiation has insufficient energy ($E = h\nu$) to transfer electrons to the CB and cause absorption; here the material is essentially loss free. In addition to the fundamental electronic and lattice absorption process, *free-carrier absorption* in semiconductors can be present. This involves electronic transitions between initial and final states within the same energy band (*intraband transitions*).

The absorbance of an object is quantified by the *Beer–Lambert law*. In absorption, the frequency of the incoming light wave is at or near the energy levels of the electrons in the matter. The electrons will absorb the energy of the light wave and change their energy state. After that, the electron returns to the ground state emitting a photon of light or the energy is dissipated internally. Beer–Lambert's law is usually expressed as follows:

$$A = -\log\left(\frac{I_T}{I_o}\right) \tag{18.2}$$

where

I_T is the intensity of the transmitted light through material

I_o is the intensity of the impinging light.

18.0.2 The Nano Perspective

Nano-optics is the study of optical phenomena and techniques near or beyond the diffraction limit [4]. Phenomena associated with nano-optics were some of the first nanotechnology to be controlled by humans. Nature also abounded with examples of nano-optical phenomena albeit unbeknownst to us in both synthetic and natural cases. Optical nano persisted throughout our history in the form of stained glass, photography, and quantum dots now. With the advent of the scanning electron microscope (SEM) and its application, we were able to decipher the colors of the Lycurgus cup. With transmission electron microscope (TEM) and atomic force microscope (AFM), we were able to see quantum dots for the first time. Nano-optics is evolving alongside nanotechnology and new instrumentation designed to see them better.

18.1 The Surface Plasmon

There are several challenges facing optics in general as miniaturization continues unabated to the nanoscale. For example, spatial resolution is limited by diffraction. The diffraction limit is described by the Heisenberg principle:

$$\Delta x \cdot \Delta p_x \geq \frac{\hbar}{2} \tag{18.3}$$

and for photons

$$\Delta x \cdot \Delta k_x \geq \frac{1}{2} \tag{18.4}$$

where k is the wave vector. In order to shrink or overcome this limit, confocal microscopy and near-field microscopy (NSOM) have been developed recently. In addition, Raman spectroscopy and multiphoton fluorescence are capable of spatial resolution less than 200 nm, which is the highest optical resolution. Quantum dots of course take center stage in this section. Their versatility, stability, and overall expectations are true representatives of things to come and of the future of nano-optics.

18.1.1 *The Surface Plasmon Resonance*

Since clusters are an ensemble of atoms, we can consider two types of effects that are related to the size of these clusters: *intrinsic effects* are related to the changes due to the surface-to-volume ratio of the cluster, and *extrinsic effects* that are related to the size of the cluster which varies according to the external constraints that are generated due to an applied field. The field of technology that involves the surface plasmon is called *nanoplasmonics*.

The interface between the cluster and the *surrounding media* (usually a dielectric material) or host plays an extremely important role in the electrical and optical properties. Numerous theories have been developed to explain the properties of clusters as a function of their nature, size, shape, and the surrounding matrix. No single theory however can explain all the effects in entirety.

One of the notions to be developed in this section is the definition of the surface plasmon. Since a large number of atoms of the metallic nanoparticle are actually on the surface, the neighboring electrons form a sort of an electron gas since they are in continuous interaction with their neighbors. [5]. The surface plasmons are thus collective excitation of free electrons on the surface of the clusters. In other words, plasma oscillations on metal surfaces are collective longitudinal excitations of the conduction electron gas. The plasmon is a quantum of a plasma oscillation. When the light is incident on nanoparticles, its electric field perturbs this electron cloud as electrons are excited into the CB. This creates surface charge separation in a small particle ($r \ll \lambda$) and is called dipolar resonance (**Fig. 18.3**). Higher-order resonances exist for larger particles that are best described by Mie scattering theory, a topic beyond the scope of this chapter.

The dipole moment per unit volume is known as the polarization. The dipolar oscillation of all electrons will have the same phase and when the frequency

FIG. 18.3 *The surface plasmon of two types of metal particles are depicted. The direction of propagation of the light is perpendicular to the horizontal axis (as depicted) of the elongated shape. Therefore, the electric field of the light runs parallel to the horizontal axis of the elongated shape. In both the sphere and wire, the plasmon generates an opposing electric field to counter that of the light.*

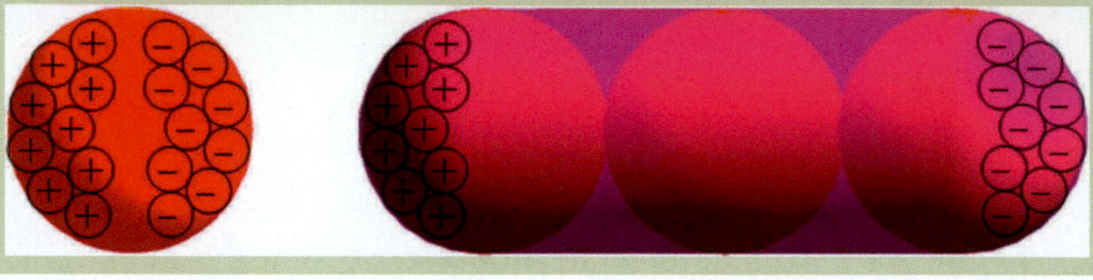

of the EM field resonates with the coherent electron motion, there is a strong absorption in the optical spectrum. It is this plasmon oscillation that is responsible for the ruby color imparted to gold colloids with small diameter (<20 nm). The frequency and width of the surface plasmon absorption depend on the size and shape of metal nanoparticles, the dielectric constant of the metal itself (e.g., the composition of the particle), and the dielectric constant of the surrounding medium.

The classical Drude model of electrical conduction, developed in the 1900s by Paul Drude, explains the transport properties of electrons in materials (especially metals). The Drude model is the application of the "kinetic theory of gases" to model the "electron gas" of a metal. Assumptions include that the material contains immobile positive ions and an "electron gas" of classical, noninteracting electrons of density n, with motion damped by a frictional force due to collisions of the electrons with the core nuclei themselves, and importantly with the surface boundary. Relaxation time is symbolized by τ. From the Drude model, we can see that if a metal nanoparticle has the size and shape in which its CB electrons have a higher possibility to polarize, it is easier for the surface plasmon resonance to occur with a lower frequency and sharp bandwidth. We also can use the Drude model to explain the surface plasmon resonance observed in noble metals, such as gold, silver, and copper, and that size and shape play a role in the development and maintenance of the surface plasmon (e.g., higher polarizability at lower frequency and sharper bandwidth than that exhibited in the bulk). Therefore, the plasmon resonance of the metal nanoparticles (noble metal particles in particular) are strongest and shifted towards the visible part of the EM spectrum. A metal colloid and clusters surrounded by dielectric media are shown in **Figure 18.4**.

The electric field strength of spherical metal nanoparticles (dielectric constant ε_m) surrounded by an insulating medium (dielectric constant ε_o) exposed to light is described by

$$E_{\text{sphere}} \propto \left(\frac{\varepsilon_m - \varepsilon_o}{\varepsilon_m + 2\varepsilon_o} \right)_{\lambda} \tag{18.5}$$

FIG. 18.4 Top: *Metal colloids are embedded collectively within a monolithic dielectric surrounding medium.* Bottom: *Ligand-stabilized clusters are embedded within a localized dielectric. In each case, the surrounding medium affects the position of λ_{max} of the surface plasmon dipolar resonance. Bulk material complex dielectric functions adequately describe the optical response of colloidal metals but caution must be used when applying them to smaller entities such as clusters. Metal clusters with diameter <10 nm start to exhibit quantum properties.*

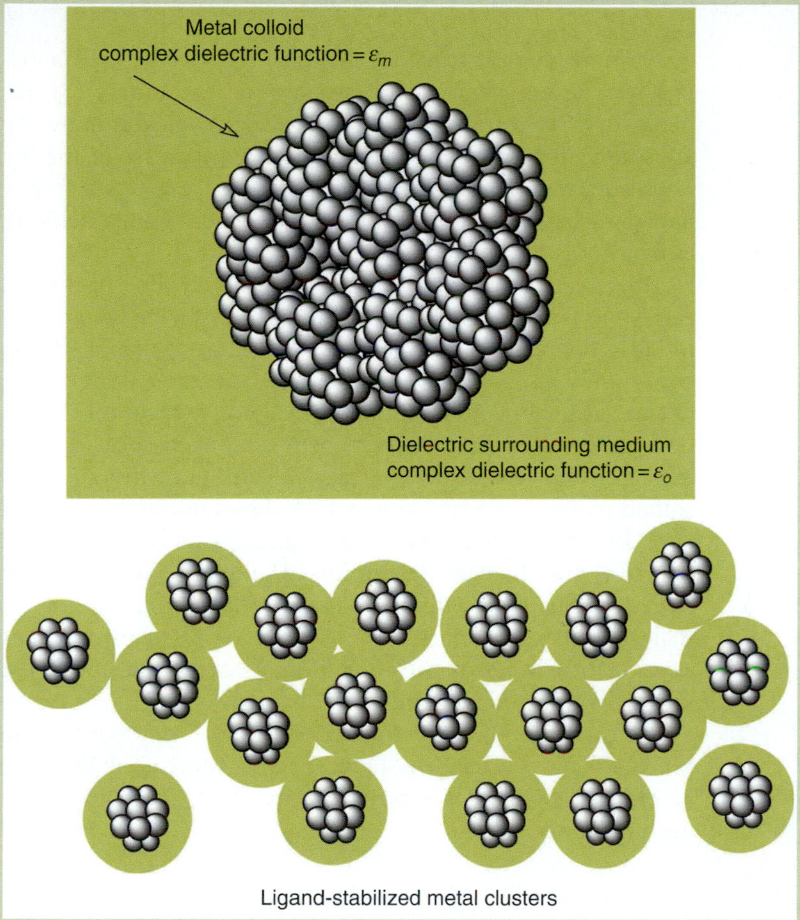

The factor "2" in the denominator is indicative of spherical shapes. From this equation, we can readily see that the localized electric field strength is maximized when

$$[\varepsilon_m = -2\varepsilon_o]_\lambda \tag{18.6}$$

the condition for the dipolar resonance. For nanoparticles that are greater than ca. 20 nm, other factors such as higher-order resonance absorption modes and scattering phenomena need to be considered. For example, minor differences in the dimensions of lithographically produced Ag particles shown by Jensen et al. [6] showed a significant effect on the resonance of the plasmon.

At the resonance, the extinction of Ag particles can be very large and increases linearly with particle volume.

We can use this characteristic of surface plasmon resonance (depending on the size and shape of nanoparticles) for sensor applications described in Ref. [6]. Moreover, for larger particles, the light cannot polarize the nanoparticles homogenously, and retardation effects lead to the excitation of higher-order modes or multipoles. Therefore, several resonances are generated leading to a broad extinction profile, or a few other low-energy peaks in the extinction spectra.

For nonspherical particles, the localized electric field induced by a surface plasmon is not evenly distributed around the surface [7] like nanorods and nanowires. The distribution of the local electric field is more complex, and there are some effects resulting in the significant increase of the local electric field. Thus, surface plasmon resonance in this case is different from that of spherical nanoparticles. For example, cylindrical or oblate nanoparticles are often described as nanorods with reference to their aspect ratio.

The plasmon resonance of a nanorod is divided into two bands. The first corresponds to the oscillation of the electrons being perpendicular to the major rod axis, often referred as the transverse surface plasmon absorption. The other corresponding to the oscillation of the electrons along the major rod axis is referred as the longitudinal surface plasmon absorption. The separation of two peak absorption frequencies depends on the aspect ratio; when this ratio increases, the separation increases. In addition, the $\lambda_{\max}$ of the nanorods is shifted to lower energies. **Figure 18.5** shows results from the report of El Sayed et al. [8]. The inset shows the peak absorption of transverse surface plasmon (squares) and the peak absorption of longitudinal surface plasmon (spheres) corresponding to the different aspect ratios. From this inset, we can observe that the transverse surface plasmon resonance is nearly unchanged for different aspect ratios, whereas the longitudinal surface plasmon resonance changes linearly with the aspect ratio.

The oscillation between the surface and the center of the nanoparticle (x_o) is generated by the incident light. The movement of the electrons generates an induced dipole that is periodic (**Fig. 18.6**). The equation that describes the movement of electrons (e.g., the induced electric field) is given by

$$E_{\text{induced}} = E_o \exp^{-iwt} \tag{18.7}$$

If the field is created by an incident EM wave of frequency ω, then the field experience by the electrons is represented by this time-dependent expression

$$m_e\left(\frac{\partial^2(x)}{\partial(t)^2}\right) + m_e\gamma\frac{\partial(x)}{\partial(t)} = e \cdot E_o \exp^{-iwt} \tag{18.8}$$

where m_e is the electron's rest mass, and γ is an absorption factor that takes into account the damping forces arising from friction discussed earlier.

The dipole oscillation for a nanoparticle between the center and the surface x_o is expressed as

$$p = ex_o \tag{18.9}$$

FIG. 18.5 Top: *The transverse and longitudinal plasmon resonance modes are depicted for two Au nanoparticle shapes. The sphere, of course, is isotropic with regard to the plasmon resonance and is therefore expected to display just one absorption peak.* Bottom: *Absorption spectrum of a nanorod (solid line) with aspect ratio = 3. The two peaks represent the transverse mode (left in the spectrum) and the longitudinal mode (at the right in the spectrum). The dotted line represents the absorption profile for spherical particles. The box in the upper corner plots the behavior of λ_{max} as a function of aspect ratio. As the particles become longer, the λ_{max} of the longitudinal mode (circles) shifts to lower energies. The λ_{max} of the transverse mode (squares) remains relatively unchanged [8].*

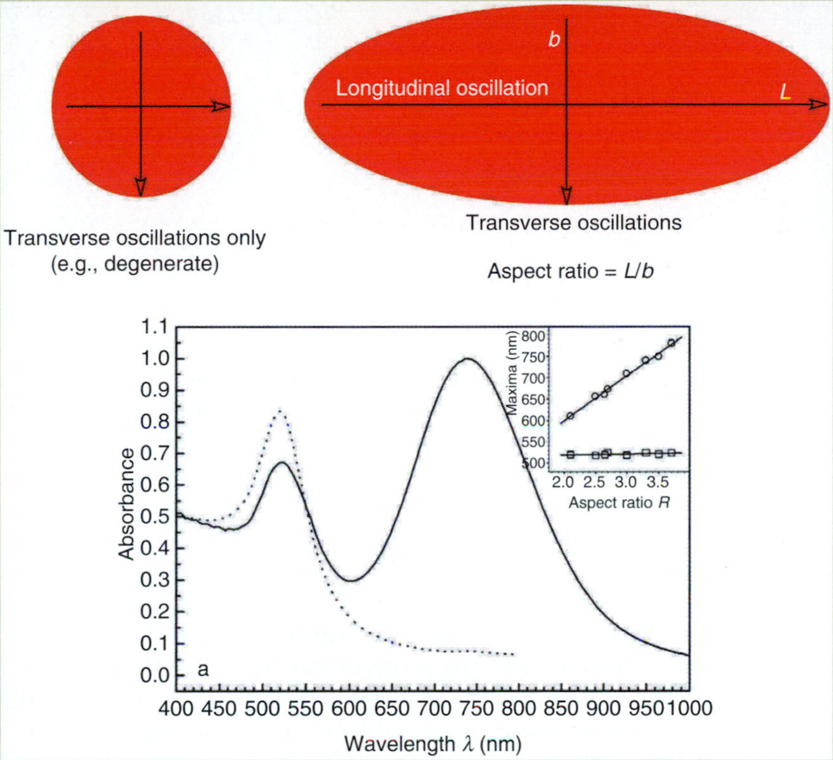

where p is the polarization due to one electron with charge e (similar to the more recognized expression for the dipole moment for a molecule: $\mu = \delta d$). With respect to the total number of electrons n, this becomes

$$P = np \tag{18.10}$$

The polarization is also defined from the perspective of the electric field:

$$P = n\alpha \cdot E \tag{18.11}$$

where α is the polarizability (remember your basic Raman spectroscopy).

Colloidal dispersions of metals behave differently from the bulk. If we consider a single representative spherical particle of diameter d with complex dielectric constant ε_m [32] embedded in a medium of dielectric constant ε_o in an oscillating

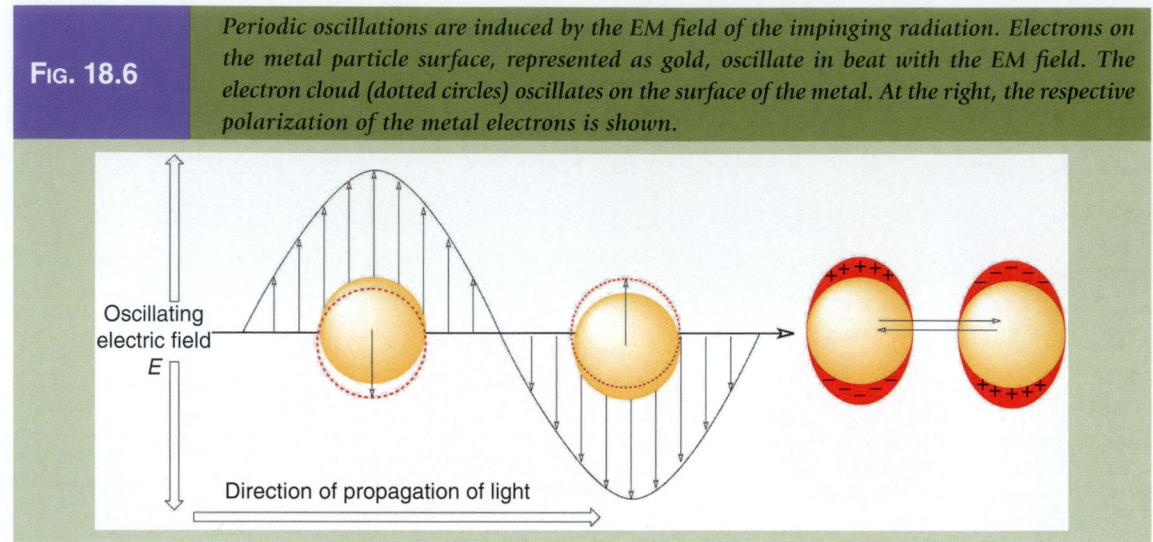

Fig. 18.6 *Periodic oscillations are induced by the EM field of the impinging radiation. Electrons on the metal particle surface, represented as gold, oscillate in beat with the EM field. The electron cloud (dotted circles) oscillates on the surface of the metal. At the right, the respective polarization of the metal electrons is shown.*

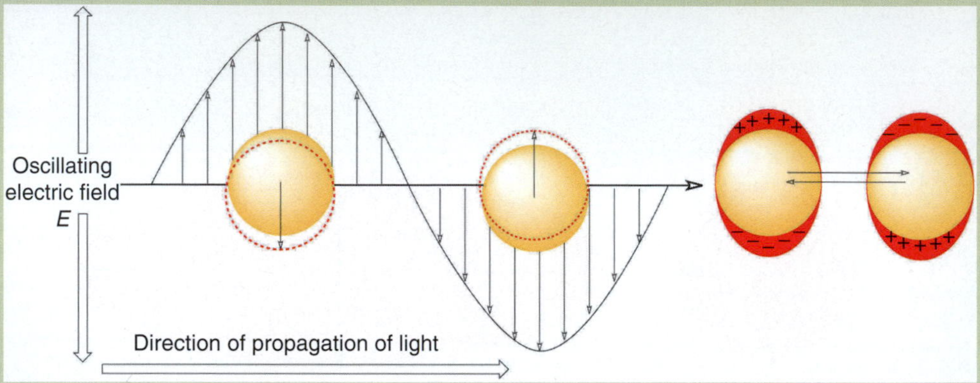

electric field, a density oscillation is built up. This is known as the surface plasma mode. The dielectric function under this mode is found proportional to $1/\tau^2$ where τ is the average size-limited scattering time—a main factor for the optical properties for small particles. So much for classical EM treatment.

A new effect, that is, the quantum size effect, plays a considerable role for tiny particles when the classical scattering concept breaks down. In this case the surface plasmon mode is treated quantum compared to the classically derived. Considering all such effects, silver is found to be the most effective metal with sharp response to surface plasmon resonance. Subsequent research in this field reveals that plasmon peak shift sensitivity depends on the interaction of the cluster with surrounding atoms or molecules and the medium. The more electrons that are spilled out of the geometrical surface of the metal cluster the more a red shift is favored. On the other hand the more electrons that are injected into the metal cluster through the surface, the more a blue shift is observed.

Nanoparticle EM response is adequately described by the quasi-static approximation. The major assumption is that the nanoparticle must be much smaller than the wavelength of light. If this is the case, then the electrodynamic condition becomes one of electrostatics. Under these conditions, the oscillations of the time-dependent external field occur slowly relative to the motions of the free electrons comprising the plasmon, for example, the opposing field is easily set up. As a result, only the dipolar resonance of the plasmon contributes to the optical response. There are no phase shifts (retardation effects), multipole phenomena (e.g., quadrupole scattering and extinction), radiation damping, and dynamic depolarization. This condition lends itself to treatment by *electrostatic* principles, a far simpler method to predict absorption than if an electrodynamic treatment is required. The physical particle size regime that qualifies for a quasi-static domain is $r < 10\,\text{nm}$ or so. The Maxwell–Garnett, equation 18.12, and Bruggeman (not shown) effective medium theories are used successfully to describe the optical response of small metal clusters

$$f_m\left(\frac{\tilde{\varepsilon}_m-\tilde{\varepsilon}_o}{\tilde{\varepsilon}_m+\kappa\tilde{\varepsilon}_o}\right)=\left(\frac{\tilde{\varepsilon}_c-\tilde{\varepsilon}_o}{\tilde{\varepsilon}_c+\kappa\tilde{\varepsilon}_o}\right) \tag{18.12}$$

where
- f_m is the inclusion fraction (usually a metal)
- ε_m, ε_o, and ε_c are the dielectric constants of the metal, host oxide, and composite, respectively.

The factor κ in the denominator is indicative of spheres if equal to 2.

18.1.2 Scattering

Mie theory offers a means of describing the optical response of particles larger than those that reside within the quasi-static limit—particles 20–50 nm and larger in diameter—or when the ratio of particle circumference to wavelength is larger than 10 [9]. As depicted in **Figure 18.2**, there are several kinds of scattering phenomena. Apart from absorption, scattering phenomena gives rise to novel applications of nanomaterials.

When we discuss about extinction in optical properties, for small nanoparticles (diameter much smaller than the wavelength of the incident light), only the dipole absorption contributes to the extinction of the nanoparticles and scattering is negligible. However, when the size of the nanoparticles increases (for large nanoparticles), the extinction is not only dependent on the higher-order multipole modes but also the light scattering of light can be thought of as the redirection of light that takes place when an EM wave (i.e., an incident light ray) encounters an obstacle or nonhomogeneity, let us say a scattering particle.

As the EM wave interacts with the discrete particle, the electron orbits are perturbed periodically with the same frequency as the electric field of the incident wave—similar to dipolar plasmon resonance we discussed earlier. The oscillation or perturbation of the electron cloud results in a periodic separation of charge within the molecule called an *induced dipole moment*. The oscillating induced dipole moment becomes a source of EM radiation. Scattered light is the result. The majority of light scattered by the particle is emitted at the identical frequency of the incident light, a process referred to as *elastic scattering*. In summary, the above comments describe the process of light scattering as a complex interaction between the incident EM wave and the molecular/atomic structure of the scattering object; hence, light scattering is not simply a matter of incident photons or EM waves "bouncing" off the surface of an encountered object [9].

Rayleigh Scattering. Rayleigh scattering is a molecular phenomena and the limit in size can be extended to ca. $d = 0.10\lambda$, for example, where photons do not lose energy. The intensity of Rayleigh scattering is proportional to $1/\lambda^4$. This means that Rayleigh scattering is more intense for the blue wavelength light—hence the blue color of the sky. Rayleigh scattering is elastic. SERS (surface enhanced Raman spectroscopy) relies on some of the equations we presented earlier in this section: the one describing the polarization and the ones describing the resonance condition. SERS utilizes scattered photons induced by a laser source to detect changes in vibrational states of analyte molecules. For example, Campion et al. state that for an analyte signal

$$G = \chi^{12}\left(\frac{\varepsilon_m - \varepsilon_o}{\varepsilon_m + 2\varepsilon_o}\right)_\lambda^2 \left(\frac{\varepsilon_m - \varepsilon_o}{\varepsilon_m + 2\varepsilon_o}\right)_s^2 \tag{18.13}$$

where

$\chi = r/(r + d)$ with r equal to the radius of the spherical metal nanoparticle

d is the distance of the analyte molecule from the surface of the metal particle.

The λ and s represent the laser and Stokes fields, respectively. Obviously the SERS signal is enhanced significantly as the relative denominators approach zero [10]. SERS is a true nanoscale phenomenon.

Mie Scattering. Scattering from molecules and very tiny particles ($<0.10\lambda$) is predominantly Rayleigh type of scattering. For particle sizes larger than this limit, Mie scattering predominates. This scattering produces a pattern like an antenna lobe, with a sharper and more intense forward lobe for larger particles. Mie scattering is not as strongly wavelength dependent and produces an almost white glare around the sun when a lot of particulate material is present in the air. It also gives us the white light we see from mist and fog. Mie scattering is compared to Rayleigh scattering in **Figure 18.7**.

In Mie theory, if nanoparticle size is less than the incident wavelength, the Mie size parameter is less than unity.

$$x = \frac{2\pi n_o a}{\lambda} \tag{18.14}$$

In this case, scattering is defined by Rayleigh scattering. The Mie scattering cross section is reduced to the Rayleigh scattering cross section.

$$C_{\text{scattering}} = \frac{8\pi}{3}\left(\frac{2\pi n_o a}{\lambda}\right)^4 a^6 \left(\frac{m^2 - 1}{m^2 + 2}\right)^2 \tag{18.15}$$

Mie scattering encompasses the general spherical scattering problem (absorbing or nonabsorbing) and places no limitations on particle size. $C_{\text{scattering}}$ is inversely

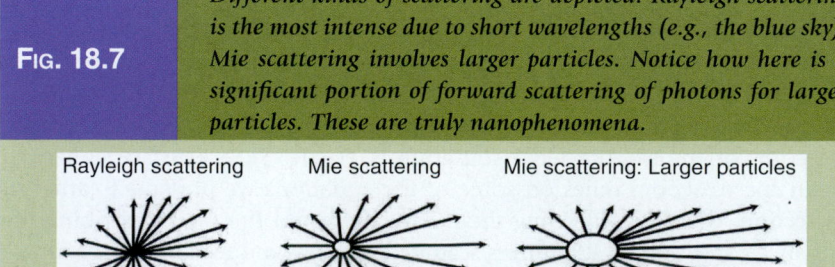

FIG. 18.7 *Different kinds of scattering are depicted. Rayleigh scattering is the most intense due to short wavelengths (e.g., the blue sky). Mie scattering involves larger particles. Notice how here is a significant portion of forward scattering of photons for larger particles. These are truly nanophenomena.*

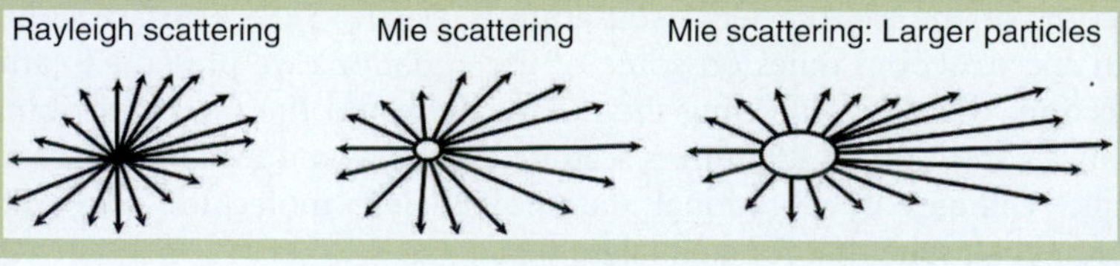

proportional to λ^4. Therefore, for a same-sized particle, the scattering efficiency $Q_{scattering}$ decreases as the wavelength becomes longer where

$$Q_{scattering} = \frac{C_{scattering}}{\pi a^2} \tag{18.16}$$

Conversely, as the particle diameter a becomes larger, scattering efficiency is expected to improve.

Mie scattering theory therefore converges to the limit of geometric optics for large particles and may be used for describing most spherical particle scattering systems, including Rayleigh scattering. As you can see, we can create a continuum between quantum optics, quasi-static optics, and Rayleigh scattering, Mie scattering, and geometric optics of the classical world.

18.1.3 Color Generation from Nanoparticles and Nanostructures

Color Due to Interference. This mode of color is quite familiar to us—from the vivid colors displayed by bubbles made of water and surfactants. The color is based on interference (constructive) of light wavelengths as they traverse the thickness of the bubble film. Depending on the thickness of the bubble, different colors are made visible. Relative thickness of microtomed sections for TEM analysis is determined by interference colors. Bubble films are on the order of several hundred nanometers in thickness.

Jewelry made of anodized titanium often display brilliant colors due to the thin refractive oxide coating. Colors such as bronze ($\lambda \approx 300$ nm), blue ($\lambda \approx 400$ nm), yellow ($\lambda \approx 600$ nm), and purple ($\lambda \approx 700$ nm) are some colors available. Optimal oxide thickness is equal to ca. half the wavelength of the color desired. Of course, incidence angle and refractive index also play a role in the resultant color.

Color Due to Diffraction. The best example of diffraction colors is a CD (**Fig. 18.8**).

Color Due to Scattering. We have discussed above the phenomenon of scattering. Different colors are generated by different kinds of scattering, different particle sizes, and different wavelengths. The sky is blue because short wavelengths are scattered by molecules. The sky is red because long wavelengths (e.g., reds) are scattered by larger particles. The deficiency in scattering intensity is compensated by longer path lengths.

Color Due to the Surface Plasmon. Perhaps the most famous example of metal plasmon color is provided by the Lycurgus cup, housed in the British Museum. The cup, made in fifth century Rome, consists of Au and Ag nanoparticles housed in a soda lime NaO glass matrix. SEM analysis in the 1990s showed that noble metal nanoparticles were responsible for the color. A remarkable dichroism is also exhibited by the cup. Upon reflection, the cup looks green. Upon transmission, the cup is a crimson red. Gold-55 quantum dots occupy the smallest limit

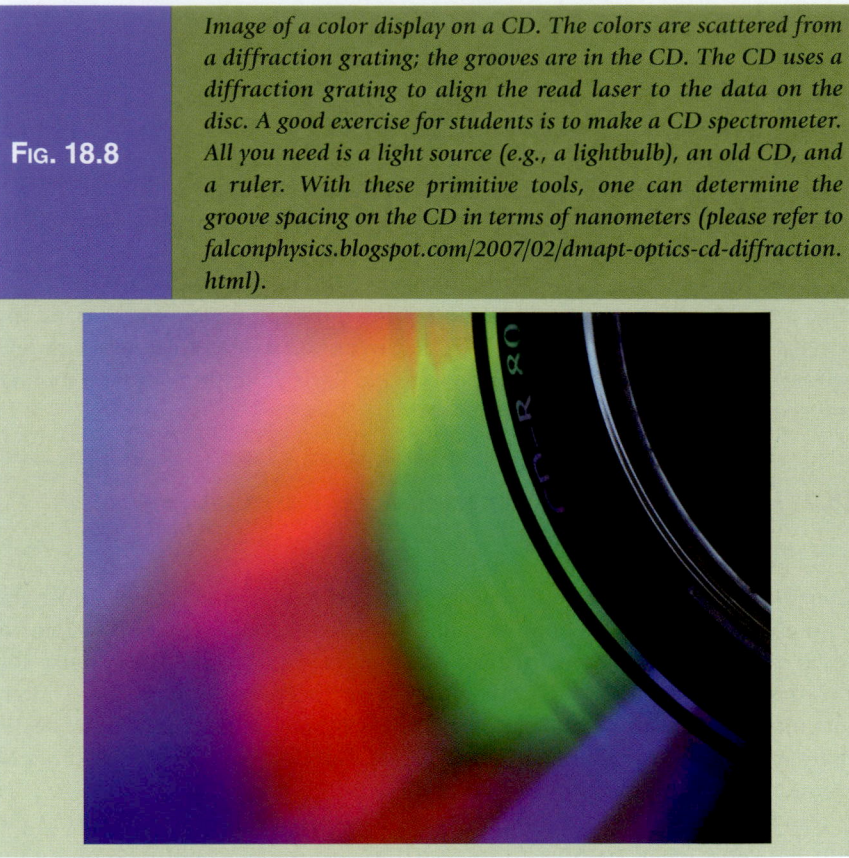

Fig. 18.8 *Image of a color display on a CD. The colors are scattered from a diffraction grating; the grooves are in the CD. The CD uses a diffraction grating to align the read laser to the data on the disc. A good exercise for students is to make a CD spectrometer. All you need is a light source (e.g., a lightbulb), an old CD, and a ruler. With these primitive tools, one can determine the groove spacing on the CD in terms of nanometers (please refer to falconphysics.blogspot.com/2007/02/dmapt-optics-cd-diffraction.html).*

for the plasmon resonance. They are, with their protecting ligand shell, around 4 nm in diameter. The color is a ruby red with λ_{max} at ca. 520 nm.

Color Due to Quantum Fluorescence. Semiconductor quantum dots are known for their intense fluorescent colors. Although made of exactly the same material, different colors are generated due simply to the difference in size of the quantum dots (QDs) (**Fig. 18.9**).

18.1.4 Applications of Nanoplasmonics

In nanophotonics and information technology, nanoscale metal structures embedded in organic devices could also potentially be beneficial in improving the light absorption, as well as the charge separation and charge collection processes. In biomedical applications, by linking specific antibodies to the metal surface, tumor cells can be targeted and imaged before pathologic changes occur at the anatomic level. HIV-I virus can be detected with Ag nanoparticles, and super paramagnetic nanoparticles used in MRI can detect disease in soft tissues (e.g., liver). Au nanoparticles applied on electrode surfaces improve electrode chemical analysis. In sensing, detecting low concentrations of molecules or

FIG. 18.9 *Fluorescence of CdSe quantum dots decreasing in size from left to right are depicted. More will be said about QDs in the next section. The wavelength of fluorescence emission of CdSe quantum dots is related to the size of the nanoparticle. The emission of longer wavelengths is indicative of larger size.*

Source: Image reprinted with permission from Professor Zhiqun Lin, Department of Materials Science and Engineering, Iowa State University.

biological species exploits the dependence of the surface plasmon resonance frequency on the dielectric constant of the surrounding medium. The sensitivity of single, small metal particles could approach the single-molecule detection limit for large biomolecules.

Waveguides. Waveguides based on surface plasmon resonance are also possible. Arrays of closely spaced metal nanoparticles are able to interact by establishing coupled plasmon modes. The resonance creates a dipole field due to the oscillations of the plasmons as we discussed earlier. If a periodic array (a line or plane or other structure) of metal nanoparticles is created, electrodynamic interactions between the adjacent metal nanoparticles can combine to form a waveguide. The phenomena is based on the near-field coupling of adjacent metal particles. The propagation of light along the array can be forced inside such a pipe on a scale well below the diffraction limit and be made to take 90° turns that are significantly lower than the wavelength of the light [11].

The screening parameter utilized by effective medium theories (e.g., Maxwell–Garnett) incorporates a term known as the screening parameter κ (described briefly above). If parallel Au nanorods are placed within the pore channels of anodically formed porous alumina, they too have the ability to "screen" light into the transparent host dielectric medium. This concept forms the basis for development of transparent metal nanostructured materials [12]. The screening effect of a spherical particle (e.g., $\kappa = 2$) is not as strong as screening by higher aspect ratio parallel nanorods (e.g., $\kappa = 1$) [12].

SERS. Surface enhanced Raman spectroscopy (SERS) has exploited the phenomenon of surface plasmon for several decades. SERS techniques have developed to the point of single-molecule detection. The surface plasmon of the nano–noble–metal particle is able to enhance analytical signals several orders of magnitude. Illumination with a fiber optic source of a SERS substrate at the tip of the probe is all that is required to detect analyte samples at incredibly low levels.

18.2 Quantum Dots

A quantum dot is a semiconductor nanostructure that is very small along all three spatial dimensions. What is very small? Small in this case implies that the motion of CB electrons, VB holes, and excitons are restricted in the three spatial directions. More clarification is required. Quantum dots (a.k.a. zero-dimensional structures) are nanostructures where all dimensions are comparable to the exciton Bohr radius. QDs generally have a diameter that is less than 10 nm. An electron–hole pair created when an electron leaves the VB and enters the CB is called an exciton.

18.2.1 The Bohr Exciton Radius

Excitons have a natural physical separation between the electron and the hole which is dependant on the material. This average distance is called the *exciton Bohr radius*. In large semiconductor crystals, the exciton Bohr radius is small compared to the size of the crystal, and the exciton is then relatively free to move around in the crystal. However, in nanocrystals, *quantum confinement* occurs that serves to restrict the motion of electrons, holes, and excitons in 1, 2, or 3 dimensions, respectively. Thus, the exciton Bohr radius may additionally be defined as the natural physical separation in a crystal between an electron in the CB and the hole it leaves behind in the VB. The size of this radius controls how large a crystal must be before its energy bands can be treated as continuous, which distinguishes a semiconductor bulk crystal from a quantum dot. Mathematically, the exciton Bohr Radii is described by

$$a_B = \frac{4\pi\varepsilon_o\varepsilon_r\hbar^2}{m_o e^2}\left[\frac{1}{m_e^*} + \frac{1}{m_h^*}\right] \tag{18.17}$$

where

ε_o and ε_r are the absolute and relative dielectric constants of the medium respectively
m_o is the rest mass of an electron
e is the fundamental electronic charge
m_e^* and m_h^* are the effective masses of an electron and hole exciton pair, respectively

In your spare time, provide a physical interpretation of this equation.

The difference in energy between the lowest energy state of the CB and the highest energy state of the VB of a semiconductor is the energy bandgap. In bulk semiconductors, CB and VB have continuous energy states separated by the bandgap that allows us to easily control by doping and other methods.

In semiconducting quantum dots, on the other hand, motion of charge carriers is confined due to the particle dimensions approaching the exciton Bohr radius. The CB and VB energy states split up and become *discretized*. Charge transfer occurs within these discrete levels, resulting in photoemission. Mathematically, these discrete energy levels can be expressed by the formula in equation

$$E_n = \left[\frac{h^2}{8ma^2}\right] n^2 \tag{18.18}$$

where

E_n is the energy of the quantum dot at some discrete energy level n
m is the mass of the QD
a is the diameter of the QD

Such discretization of the bandgap shifts the emission of the QD to a higher energy, for example, a blue shift. Comparison of the bandgap of bulk semiconductors, QDs, and molecules are shown in **Figure 18.10**.

18.2.2 Tuning the Gap

The discrete energy states of QDs are similar to those of organic molecules. Electrons excited from VB to CB may lose energy radiatively or nonradiatively by discrete jumps within these levels. The optical properties of quantum dots have strong size dependence. Excitation as well as emission bands can be tailored by

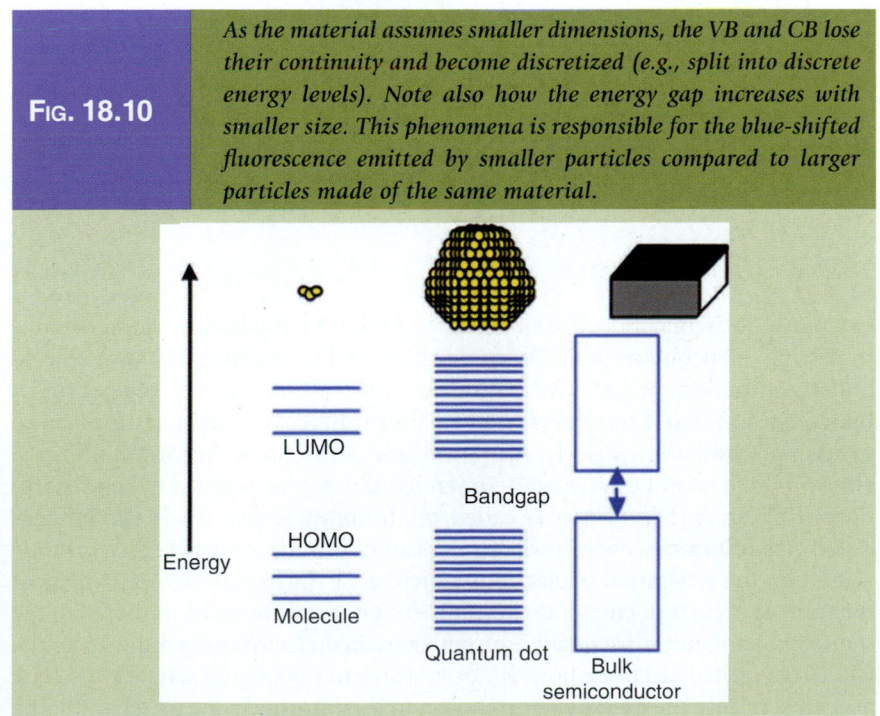

Fig. 18.10 *As the material assumes smaller dimensions, the VB and CB lose their continuity and become discretized (e.g., split into discrete energy levels). Note also how the energy gap increases with smaller size. This phenomena is responsible for the blue-shifted fluorescence emitted by smaller particles compared to larger particles made of the same material.*

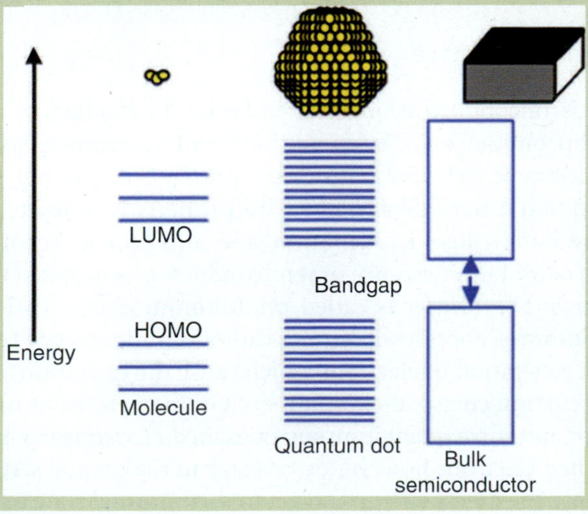

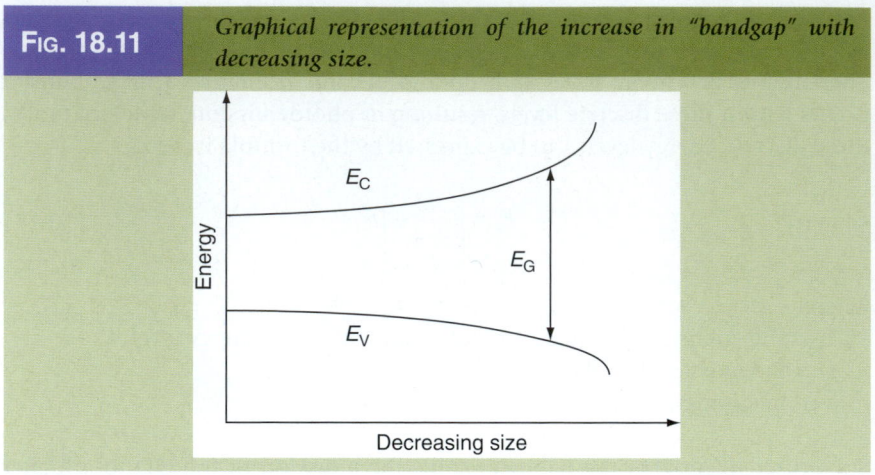

FIG. 18.11 *Graphical representation of the increase in "bandgap" with decreasing size.*

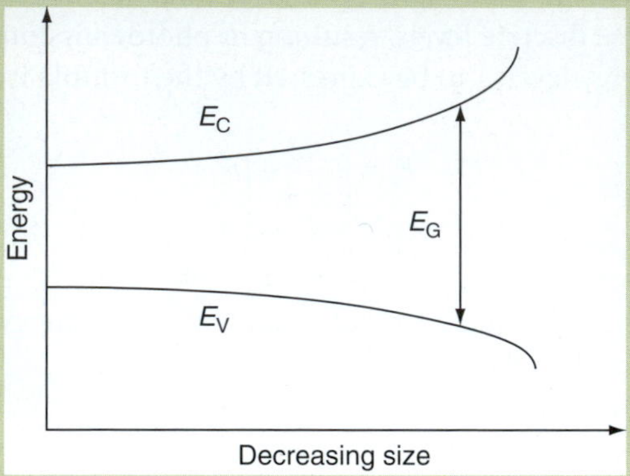

varying the size of QDs. In general, the smaller the diameter of the QD, the greater is its exciton binding energy and the higher its emission energy (**Fig 18.11**). The emission spectrum of a QD can be tuned from ultraviolet past into the infrared. The gap of ZnS, for example, is 3.6 *e*V for bulk material but it can be increased to about 4.5 *e*V if in the form of a QD of diameter between 1 and 4 nm [13]. The bandgap energy E_g is related to the diameter a of the QD from

$$E_g = E_{g(bulk)} + \frac{3\varepsilon_o \varepsilon_r \hbar^2 \pi^2}{2a^2}\left[\frac{1}{m_e^*} + \frac{1}{m_h^*}\right] \tag{18.19}$$

and

$$\Delta E_g = \frac{3\varepsilon_o \varepsilon_r \hbar^2 \pi^2}{2a^2}\left[\frac{1}{m_e^*} + \frac{1}{m_h^*}\right] \tag{18.20}$$

18.2.3 Luminescence

Quantum dot semiconductors, just like in bulk semiconductors, are amenable to doping. Concomitantly, QDs, in addition to fluorescence, also are able to undergo luminescent radiative pathways. Luminescence is the generation of light emission that is not fluorescence. This property is characteristic of semiconductors and the cause is excitation and subsequent recombination of electron–hole pairs. Luminescence in semiconductors is initiated by the absorption of photons. The former is called photoluminescence while the latter is called electroluminescence or sometimes cathodoluminescence (e.g., a cathode is used for the generation of electrons which cause the excitation). Regardless of the form of excitation energy, the result is excitation of electrons in the VB of the semiconductor and subsequent jumping of excited electrons from the VB to the CB. These excited electrons however do not stay in the excited state for too long and must release the excess energy and get to their normal state by recombining

Fig. 18.12 *Different luminescent recombination processes for electrons and holes in doped semiconducting QD (not to scale).*

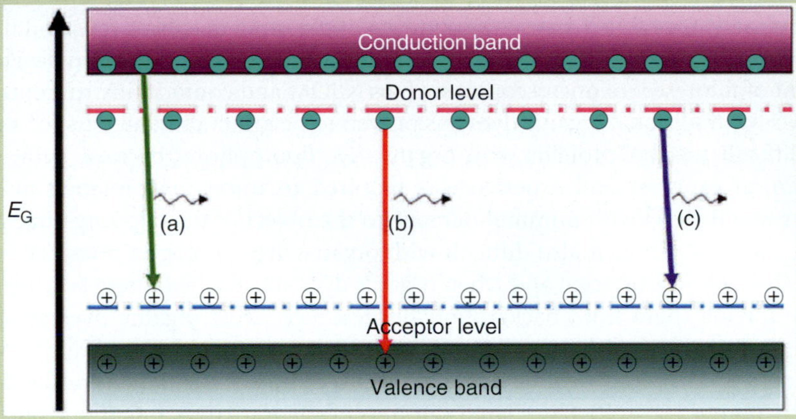

with holes. This happens of course by emission of energy either in the form of heat (via phonons) or light.

When a recombination results in emission of light it is called radiative recombination, otherwise it is nonradiative recombination. Radiative recombination is favored if donor/acceptor energy levels are introduced in the energy band of a semiconductor through doping. Dopants that reside within the crystal in between lattice sites are called interstitial dopants. Interstitial dopants do not give rise to luminescent recombinations. Dopant ions that reside within a semiconductor QD (e.g., by substituting lattice atoms) are called substitutional dopants, and this form results in luminescent recombinations that occur in three different ways (**Fig. 18.12**).

18.2.4 Applications

Crystalline semiconducting QDs have many applications: (1) as components in nonlinear optical devices, (2) in photoluminescent and electroluminescent materials, and most importantly (3) as fluorophores in biolabeling, medical imaging, and other types of image applications. ZnS QDs in particular have received a significant amount of attention over the past decade or so. ZnS has excellent optical properties and is chemically quite stable. ZnS has potential for use in optoelectronic devices such as high-field electroluminescence, cathodoluminescence, and field emission displays (FED).

Fluorophores. Fluorophores are molecules or functional organic groups that can emit light or fluoresce after absorbing light at a particular wavelength. In other words, the component of a molecule which makes it fluorescent is named a fluorophore. Fluorophores absorb energy at specific wavelengths and emit at other specific wavelengths. This emitted energy depends on the type of the chemical composition, the immediate environment (a ligand stabilized shell, for example), and functional groups. Fluorophores are particularly useful for bacterial, protein, and cell studies and in other biologically related investigations.

For example they can be incorporated into other biological proteins to produce fluorescence. This process of fusion is known as biological labeling or biolabeling in short. Biological labeling of living bacteria or cells, etc., has helped immensely in getting better understanding of the inner details of cells and their functions. This is relatively easy to accomplish because of the advanced development of fluorescence microscopies such as NSOM and confocal instruments.

Use of traditional organic dyes has proven to be rather difficult. First of all, it is difficult to label proteins with organic dye fluorophores because sufficient technical expertise and experience is required to inject the biolabels at the microscopic level with minimal damage to the object of the tag. Targeting specific proteins in vivo is also difficult with organic dyes. Biological materials can also be highly fluorescent and when organic dyes are attached, there is difficulty in resolving them from background fluorescence. Most organic dyes are not chemically stable (to light and heat), and this results in a decay of fluorescent properties. Therefore, organic dyes are usually stored in the dark at refrigerator temperatures (ca. 4°C). Since large amounts of chemical dyes are often required, poisoning of the cell may result.

In QDs, the absorption and emission lines are much farther apart than that of organic dyes. This is illustrated in **Figure 18.13** where the fluorescent isothiocyanate (FITC or fluorescein) absorption–emission profile is compared to that of a generic QD. In **Figure 18.13 (Middle)**, the degradation of luminescence efficiency between hybridized EUB338-conjugated QDs and 4′-6-diamidino-2-phenylindole (DAPI) organic fluorescent labels on *E. coli* is depicted.

Most organic dyes used for biolabeling have a narrow excitation band and often the emission band overlaps with it (**Fig. 18.13 Top**). Semiconducting QDs on the other hand can be excited by any wavelength above its bandgap with built-in tunability. The emission wavelength can be tuned by manipulation of the size, use of dopant materials, and their specific placement in the energy band structure of the semiconductor QD. Semiconductor fluorescent QDs can be synthesized for multicolor luminescence.

QDs have a high number of photons and can give high luminescent intensities [14]. QDs therefore are capable of a broad range of emission with narrow emission spectral bandwidth [14]. For example, ZnS-capped CdSe QDs demonstrated 20x more intensity and 100x better stability than the common organic dye rhodamine 6G [15].

QDs have been used in combination with bacteriophages, and subsequently tag infectious bacteria [16]. There is a need to identify infectious bacteria quickly and with great amplification of signal to increase sensitivity. A "quick and easy" method to do just that has been developed by research groups at the National Cancer Institute (NCI) and the National Institute of Standards and Testing (NIST). The team developed a phage-based method that utilized QDs for tagging (**Fig. 18.14**). Organic fluorophores like GFP and luciferase have disadvantages as discussed earlier. Specifically, they suffer from low signal-to-noise ratio (due to background autofluorescence) and low photostability (e.g., susceptible to photobleaching).

New fluorescent materials that incorporate CdSe QDs exhibit broadband absorption with narrow emission and with size-dependent local λ_{max} of absorption. ZnS QDs, due to an outer shell of a few atomic layers, further enhance the photostability of the QD fluorophore [16]. The method developed by the

Fig. 18.13 Top: *Generic spectra emphasizing the difference between organic dye (FITC) and a generic QD fluorescence. The organic fluorescence* (top left) *is within range of its excitation (absorption) wavelength whereas in the case of the QD, the absorption and emission regions are quite separated.* Middle: *Comparison between organic and QD fluorescence lifetime is depicted. Green relates to EUB338 QD fluorophores, blue to an organic dye DAPI.* Bottom: *Biolabeled* E. coli *with chitosan-capped Mn-doped ZnS QDs [31].*

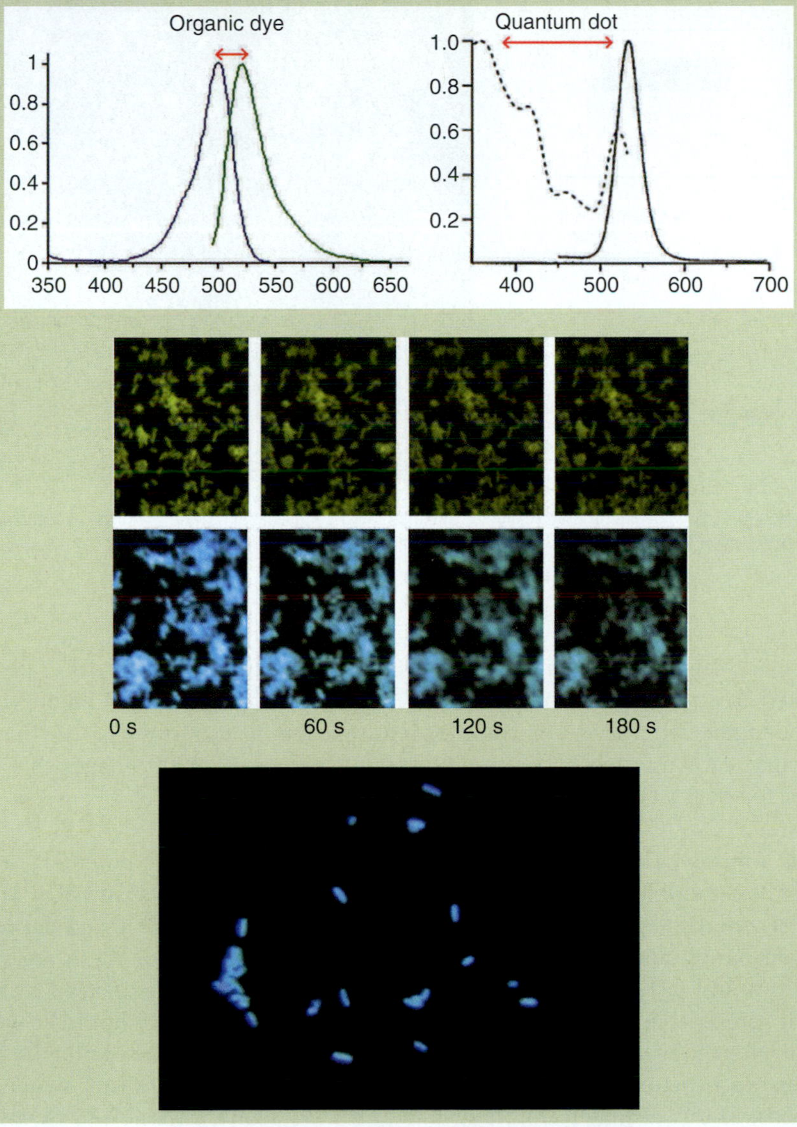

Fig. 18.14 *The process of tagging microorganisms with nanoengineered QD complex to detect infectious bacteria is shown. (a) First, the phage is genetically engineered to assemble a small surface peptide (a biotinylation peptide) that is fused to a capsid protein. Biotin (vitamin H) is then attached to the lysine residue of the tagged peptide. Streptavidin-functionalized QDs are then attached. (b) Western blot analysis of control (T7-myc phage) versus experimental (T7 biophage) results with streptavidin–horseradish peroxidase is shown. (c) Fluorescence micrograph of* E. coli *with 100-fold excess of biotinylated phage is depicted. (d) Bright field TEM of the same region is shown. Scale bars are 1 and 2 μm, respectively.*

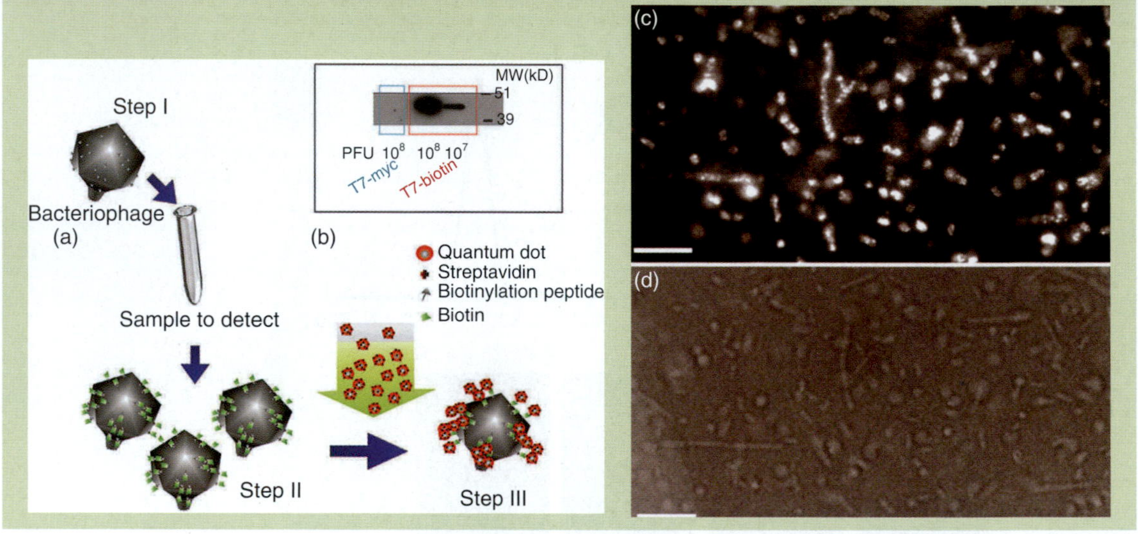

Source: R. Edgar, M. McKinstry, J. Hqang, A. B. Oppenheim, R. A. Fekete, G. Giulan, C. Merril, K. Nagashima, and S. Adhya, *Proceedings of the National Academy of Science*, 103, 4841–4845 (2006). With permission.

NCI–NIST team utilizes in vivo biotinylation of genetically engineered T7 phage linked to streptavidin-modified QDs. Amazingly, this genetically engineered, nanoengineered phage–QD complex reduces amplification time to 20–45 min because each infected bacterium is able to produce 10–1000 phages that are detected by the QD complex.

Field Emission Displays. Fluorescent quantum dots have the potential to be used as cathodoluminescent phosphors, and hence can be used in FEDs. Most flat screen displays make use of LCDs that are electronically switched between a transparent and opaque state. However, viewing angles for these are very narrow, and faster switching speed is also a problem. Hence, instead of this light "on/off technology," light emitters as the picture elements should be used, equivalent to a miniature version of old CRT technology. This configuration gives much better viewing angles with full color display and faster switching speeds. In this display technology, like of that in a CRT, emission of electrons and corresponding emission of color is due to electrons impinging the picture elements. This technology is known as a *field emission display* (FED). In FEDs, electrons are emitted from cold cathodes unlike in CRTs and the most challenging task is to develop low-voltage phosphors with cathodoluminescent

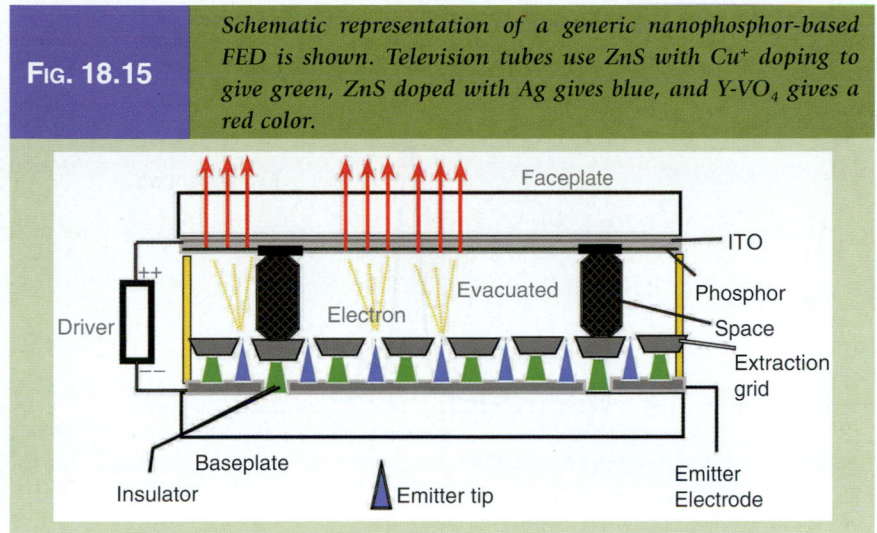

FIG. 18.15 *Schematic representation of a generic nanophosphor-based FED is shown. Television tubes use ZnS with Cu^+ doping to give green, ZnS doped with Ag gives blue, and Y-VO_4 gives a red color.*

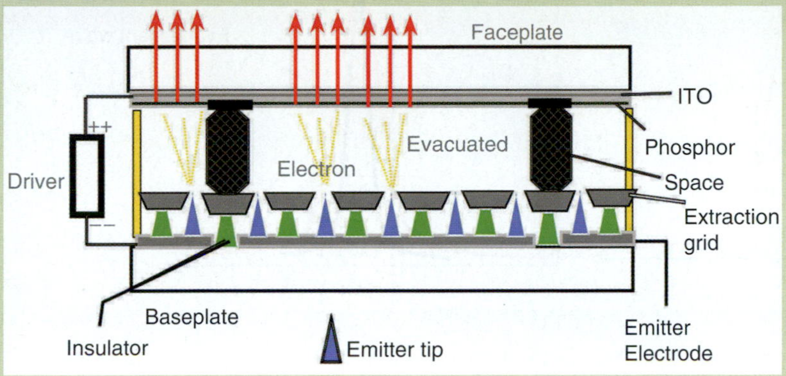

efficiency to get good resolution, stability, and brightness [17]. Good FEDs should have low current saturation and the nanophosphors showed less current saturation than the microphosphors [18].

There are more advantages to this nanoscale FED technology. The processing temperature of the nanophosphors are hundreds of degrees lower than the commercial microphosphors that are fabricated by mechanical milling with sizes of the order of 2 μm. These micron-sized particles pose a problem for screen efficiency of very high-resolution displays [19]. Nanophosphors, on the other hand, are fabricated at lower temperatures and are very small: 2–100 nm in diameter. These small particles, not surprisingly, can be made to luminesce at different wavelengths for the reasons we described before. This gives QD-based FEDs great flexibility. What do you think that the resolution of such a device could be, potentially? A schematic of a typical nano-emitter is shown in **Figure 18.15**.

18.3 Near-Field Microscopies

18.3.1 The Diffraction Limit

Measurement and standards need to be extended to the nanoscale (e.g., nanometrology). One of the areas where this is required is nanoscale optics. NSOM has made incredible progress in this area—to the point of single-molecule spectroscopy. The ability to detect, measure, and manipulate large, single biological molecules like proteins and RNA is leading to a greater understanding of folding and other mechanisms. Efforts are underway to accomplish all of this—and within living cells [20].

Far-field optics is limited by diffraction (**Fig. 18.16**).

$$\Delta x = 0.61\left(\frac{\lambda}{\mathrm{NA}}\right) \quad \text{and} \quad \frac{\lambda}{\mathrm{NA}} \rightarrow \frac{4\pi}{\Delta k} \rightarrow \frac{\Delta p}{\hbar} \tag{18.21}$$

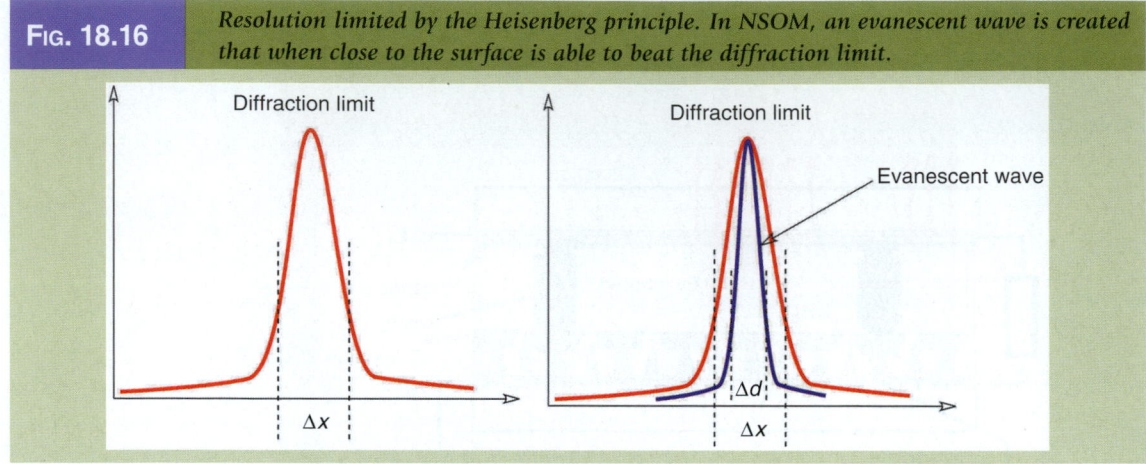

FIG. 18.16 *Resolution limited by the Heisenberg principle. In NSOM, an evanescent wave is created that when close to the surface is able to beat the diffraction limit.*

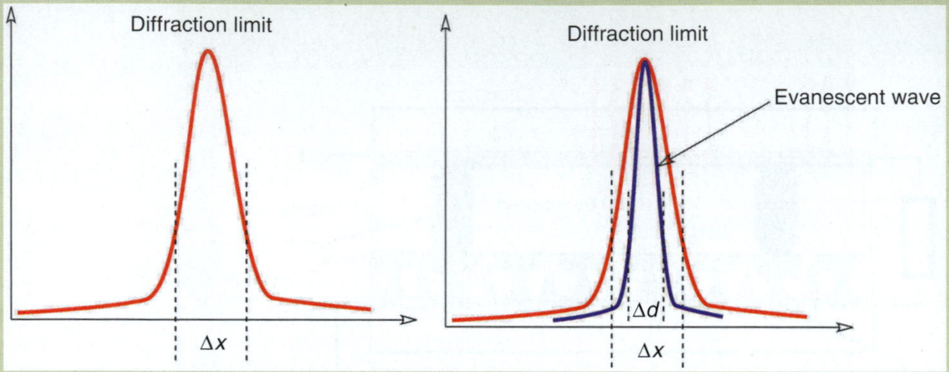

where
NA is the numerical aperture.

$$(\Delta x \cdot \Delta p) = 1.22 \cdot h > \left(\frac{\hbar}{2}\right) \quad (18.22)$$

$$\text{Resolution limit} \approx \frac{\lambda}{2} \quad (18.23)$$

Optical resolution is therefore restricted by the diffraction limit defined by the uncertainty principle. Depending on the wavelength, for most standard optical (and UV) microscopes this limit is ca. 150–300 nm. Resolution limit of 30 nm with far-field microscopes was attained with nonlinear materials and solid immersions [4]. This is not possible using conventional UV-visible light microscopy because it has a wavelength range of 300–700 nm. Thus conventional optical microscopy fails because the resolution is restricted to half the wavelength used.

18.3.2 Near-Field Microscopy

The diffraction limit can be overcome by adding a nanoscale object in the near-field and resolution is improved significantly. The invention of NSOM overcomes the major problem of resolution limits. In order to obtain an illumination in the nano range, we need optical apparatus that is able to produce a small spot size with the beam in the near-field of the probe (with diameter between 10 and 100 nm). To overcome this problem a small aperture that illuminates a region on the sample that is a few nanometers in diameter is needed. We should keep in mind that the character of light changes as it passes through small apertures. Thus the localization of the light waves result in the formation of *evanescent waves*. As the distance from the aperture increases, the intensity of the evanescent waves decreases rapidly. Therefore the aperture has to be close to the object, only a fraction of wavelength away. That is why it is called near-field microscopy [21].

Much more research and progress is going on to design electronics devices with feature size on the nanometer scale, in which the electron is confined in the nanometer scale structure called quantum dot. In these structures the spatial confinement of electron approaches the de Broglie wavelength, so the matter wave properties of electron changes rigorously. Therefore, along with the electronic properties their optical properties also vary from that of bulk materials. With current techniques it is a challenging task to observe optical transition between the electronic states in quantum dots. But by using the nanometer scale light source (obtained from the conventional laser light from an aluminum-coated fiber tip), Guest et al. have now succeeded in exciting and detecting single optical transition on the nanometer scale in a solid material [21].

In NSOM, a subwavelength aperture (e.g., an aperture that is smaller than the wavelength of the illuminating source) is used as the scanning probe. The probe is scanned a few nanometers above the surface of the sample. The confined light transmits through the sample and is collected by traditional optical apparatus [22]. The probe tip is a small aperture at the end of a tapered optical fiber that is coated with aluminum. With such a probe, even though the original light source may be 500-nm wavelength, it is possible to beat the diffraction limit. Resolution down to 50 nm and better is achievable.

18.3.3 Applications

Near-field polarimetry was used to investigate the structure of isotactic polystyrene with subdiffraction limit resolution [22]. The crystallites of ultrathin polymer films (<100 nm) crystallites were imaged by this technique. The objective was to investigate the radial strain, local tilt of the crystal axis, and strain in the amorphous layer above and below the growth planes of the crystallites of the polymer by optically characterizing the birefringence profiles of the material [22]. Characterization of chain conformation near the growth front, amorphous layers (positioned over and under crystallites), and the orientation of folded chains was not possible by traditional diffraction limited methods. Use of the NSOM technique was applied to reveal the structure during early growth stages of the crystallites (**Fig. 18.17**) [22].

In anisotropic crystals, two rays of light produced by a double refraction have different velocities in the crystal (e.g., it takes longer time for the slow ray to traverse the crystal). The faster ray travels a distance ΔR beyond the surface of the crystal before the slow ray reaches the surface. ΔR is known as the retardation. Birefringence is the difference in refractive index within a material. Without retardation, for example, light emerging from a crystal can recombine by interference. If there is no retardation, the resultant ray is identical to the incident ray. If there is retardation, then the resultant ray is altered. If the light source is monochromatic, then the crystal will appear light or dark depending on the degree of retardation. Retardation is a function of crystal orientation and thickness.

Fig. 18.17 *Retardance and topographical images of PS crystallites acquired with NSOM polarimetric technique. (a) Retardance, (b) topography, (c) retardance with overlaid fast axis orientation, and (d) topography with overlaid fast axis orientation, and (e and f), high resolution scans. For a more detailed interpretation, see Ref. [22].*

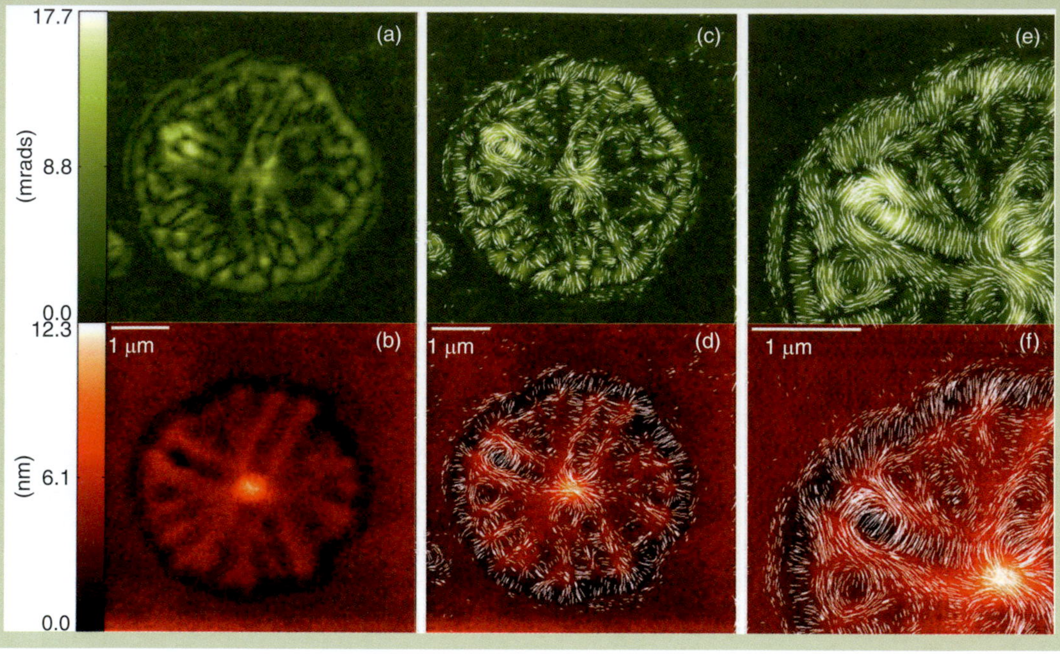

18.4 Nanophotonics

18.4.1 Photonics

According to the Photonics Dictionary [23]

> The technology of generating and harnessing light and other forms of radiant energy whose quantum unit is the photon. The science includes light emission, transmission, deflection, amplification and detection by optical components and instruments, lasers and other light sources, fiber optics, electro-optical instrumentation, related hardware and electronics, and sophisticated systems. The range of applications of photonics extends from energy generation to detection to communications and information processing.

Photonics is the study of the interactions of light with matter. Photonics is not just the study of light, it is more appropriately a technology that happens to be based on the interactions of light with matter. The goal of photonics is to develop components and devices. Photonics first began to emerge with the invention of the laser in 1960 and the laser diode a decade later. The invention of the optical fiber as a means of transmitting information via light, and therefore information, formed the basis for optical telecommunications. The field is now quite enormous and consists of a variety of subdisciplines and applications that includes laser technology, biological and chemical sensing, medical diagnostics and therapeutics, display technology, optical computing, fiber optics, metrology, holography, photodetection, and photonic crystals.

In 1987, Eli Yablonovitch at Bell Communications Research Center in New Jersey and Sajeev John of the University of Toronto created an array of 1-mm holes in a material with refractive index equal to 3.6 (a.k.a. Yablonovite) [24,25]. They found that the array prevented microwave radiation from propagating in any direction. It took nearly a decade to fabricate photonic crystals that do the same in the near-IR and visible range. They based their work on the premise that "photonic bandgap" (PBG) behavior would be similar to the behavior of electron waves in natural crystals, for example, that phenomena such as reciprocal space, Brillouin zones, dispersion relations, Block wave functions, and Van Hove singularities must also be applicable for photonic waves [26].

18.4.2 Photonic Structures in Living Systems

There are numerous examples of photonic systems that are displayed by living things. The nanostructure of the wing of the butterfly genus *Morpho* is composed of intricate photonic structures on the order of 500–600 nm that interact with light to produce a beautiful iridescent blue color. Other butterflies also have what are known as structural colors—interference colors that do not depend on chemical means or a pigment to become visible. Professor Peter Vukusic, a member of the Thin Film Photonics Group in the School of Physics at the University of Exeter, United Kingdom discussed the "optical tricks that have given butterflies, beetles and other creatures an evolutionary advantage" [27]. It was Isaac Newton and Robert Hooke who first connected photonic structures to the iridescent colors of a peacock's feather, the Abalone's shell (mother of pearl), and natural opals.

There exist two generalized mechanisms that produce color in butterfly wings. The first is due to the presence of pigment molecules that are able to absorb certain wavelength light and selectively transmit or reflect other wavelengths. The second mechanism is more interesting. It leads to remarkable iridescent colors and is based on structure, for example, a structural color. Interference colors are due to the thin-film interference or diffraction phenomenon. Interference in thin film as discussed above is due to the difference in the refractive indices of the two mediums, which form the interface for the light to pass through [26]. Wave mechanics has shown that the color of butterfly scales is due to interference phenomena. When light is incident on thin films some of the light reflects from the outer surface while the rest reflects from the inner surface of the thin film. One sees interference colors when the film thickness is on the order of the wavelength of visible light. These "reflected" wavelengths interfere with each other either constructively or destructively and that causes increase or decrease of amplitude of the original wavelength. Constructive interference is responsible for the color production.

The structural hierarchy of the blue *Morpho* butterfly wing is shown in **Figure 18.18**. Starting with scales (not shown), there exist veins (or ridges) that consist of photonic structures. If viewed from the side (**Fig. 18.18 Left**), one can see that the tiered structures correspond to light that interferes with the structure.

Fig. 18.18 *The ultrastructure of the blue* Morpho *butterfly is depicted. On the left, the veins of the structure are depicted. On the right, the transverse section of the veins is shown. The tiered structure is ca. 600 nm in height.*

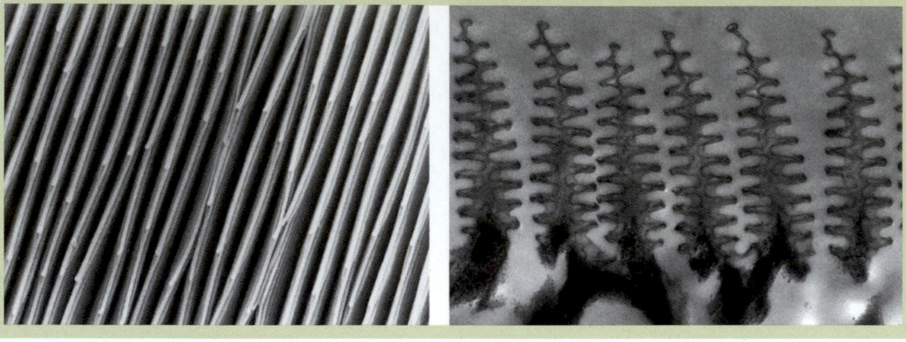

Source: Images are courtesy of Professor Pete Vukusic, School of Physics, University of Exeter, UK. With permission.

18.4.3 Photonic Crystals

Photonic crystals are periodic structures made of dielectric materials that interact with resonance with radiation at wavelengths consistent with the periodicity of their lattice network. One of the goals of nanophotonic crystals is to serve as waveguides and to confine light without experience losses. Photonic crystals, called "semiconductors of light" by Marion Florescu of NASA's Jet Propulsion Laboratory and Cal Tech, can be compared to semiconductors [28]. Semiconductors consist of periodic arrays on the atomic scale that control the flow of electrons. Photonic crystals consist of periodic arrays on the scale of wavelength of dielectric materials that control the propagation of light waves. Both material regimes have a characteristic bandgap. We understand the nature of the bandgap in semiconductors. We shall discuss PBGs later in this section. Photonic crystals are designed to confine, manipulate, and control the propagation of photons in three dimensions. Photonic crystals can be one-, two-, or three-dimensional (**Fig. 18.19**).

Applications of photonic crystals include their use in devices to control the flow of radiation, dielectric mirrors for antennas, microresonators, low-threshold nonlinear devices, microlasers and amplifiers, controlled miniaturization, and pulse sculpting (Florescu). PCs can act as a filter regardless of polarization or direction of motion of photons to localize photons, to inhibit the spontaneous emission of chromophores, to modulate stimulated emission, and to serve as waveguides [29]. Control of these phenomena will enable us to make LEDs, zero-threshold semiconductor diode lasers, and enhance performance of quantum optical devices as well as all other kinds of optical devices [29]. PCs and applications of photonics reach across the EM spectrum—from UV to radio waves.

The major characteristics of synthetic photonic crystals is that dielectric networks retain connectivity, the active elements have the same average optical path, and that there be a high ratio of dielectric indices. Even if the crystal is heterogeneous, the average optical path should be the same regardless of the media. There should be an overlap of frequency gaps along different directions. The beauty of photonics is that PCs can be fabricated to specifically overlap with the radiation range of interest.

FIG. 18.19 *Generalized photonic crystal structures depicting 1-, 2-, and 3-D configurations are shown.*

Source: C. P. Poole and F. J. Owens, *Introduction to Nanotechnology*, John Wiley & Sons, Inc., New Jersey (2003). With permission.

The Photonic Bandgap. The photonic bandgap (PBG) is a range in wavelength within which there is no absorption or propagation of light by a material. In other words, the PBG is a wavelength range in 3-D dielectric structures in which EM waves are forbidden regardless of the direction of propagation of the incident beam [26]. The PBG is the key to the way light might be controlled by a material structure. For example, by introducing defects in the structure or by doping, certain pathways can be closed or diverted. Creating periodic structures out of materials with different dielectric properties (high and low refractive index materials) serve as waveguides—similar to the way electrons are "guided" through a semiconductor via doped domains. The goal of photonic crystals is to exclude light transmission in all directions for specific wavelengths—once again, similar to semiconductors that exclude the flow of electrons for specific energy bands. It was actually the de Broglie model for electronic wave propagation in a crystalline solid that inspired research on the PBG.

The PBG is determined by the radius of the holes (or other features like dielectric rods), the periodicity of the holes (or rods), the lattice structure, the thickness of the material, and the refractive index. In essence, optical properties can be absolutely under control by designing the photonic device with proper attributes. Periodic structures along the direction of propagation of light restrict or allow modes or limit the density of states of photons. For example, the structures shown in **Figure 18.20** are two different kinds of photonic structures.

The behavior of light in a photonic crystal is described by Maxwell's equations for periodic dielectrics. Dispersion relations (dependence of energy on the wavelength or *k*-vector) are accurate because there is little interaction between photons in such a crystal (**Fig. 18.21**) [30].

If some features were removed from the lattice, for example, a couple of rows of rods or holes removed in a nonrandom way, that region becomes devoid of structure and would act as a waveguide. In such a configuration, an "allowed frequency" is now allowed to traverse the bandgap—similar in effect to doping a semiconductor material crystal [30].

In semiconductors, p or n-doping creates energy levels in the bandgap. The same is true in photonic crystals except that the PBG is activated to accommodate the passage of light.

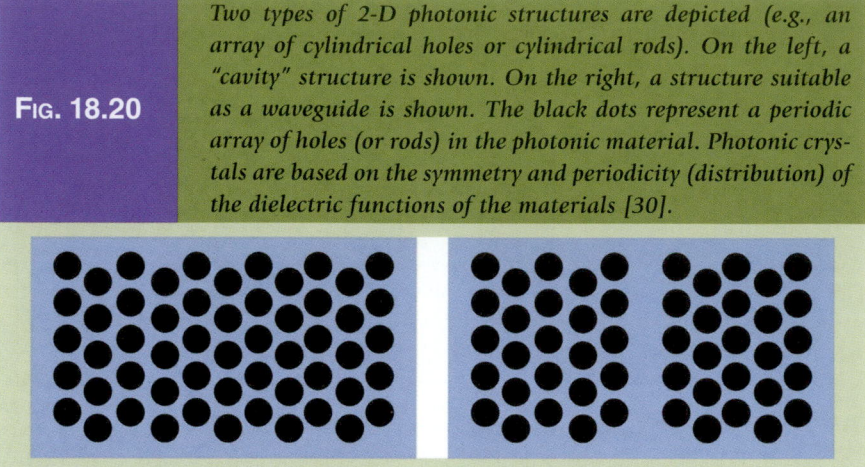

FIG. 18.20 *Two types of 2-D photonic structures are depicted (e.g., an array of cylindrical holes or cylindrical rods). On the left, a "cavity" structure is shown. On the right, a structure suitable as a waveguide is shown. The black dots represent a periodic array of holes (or rods) in the photonic material. Photonic crystals are based on the symmetry and periodicity (distribution) of the dielectric functions of the materials [30].*

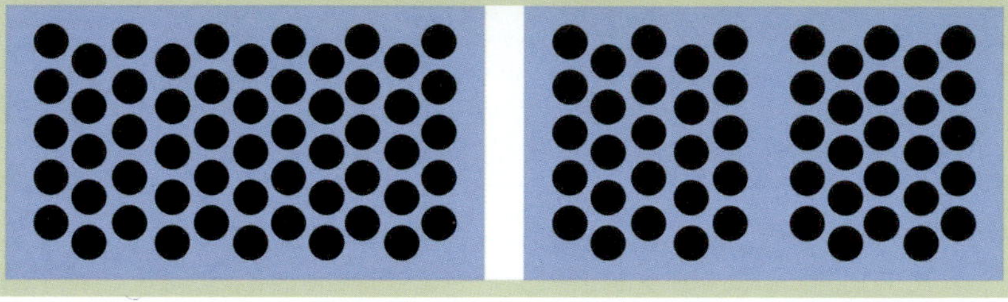

Source: C. P. Poole and F. J. Owens, *Introduction to Nanotechnology*, John Wiley & Sons, Inc., New Jersey (2003). With permission.

Light is actually able to turn sharp corners in such a structure. Because the light is traveling in what was once forbidden territory, it has no way to escape the forbidden zone back and make its way back into the crystal's periodic array crystal [30].

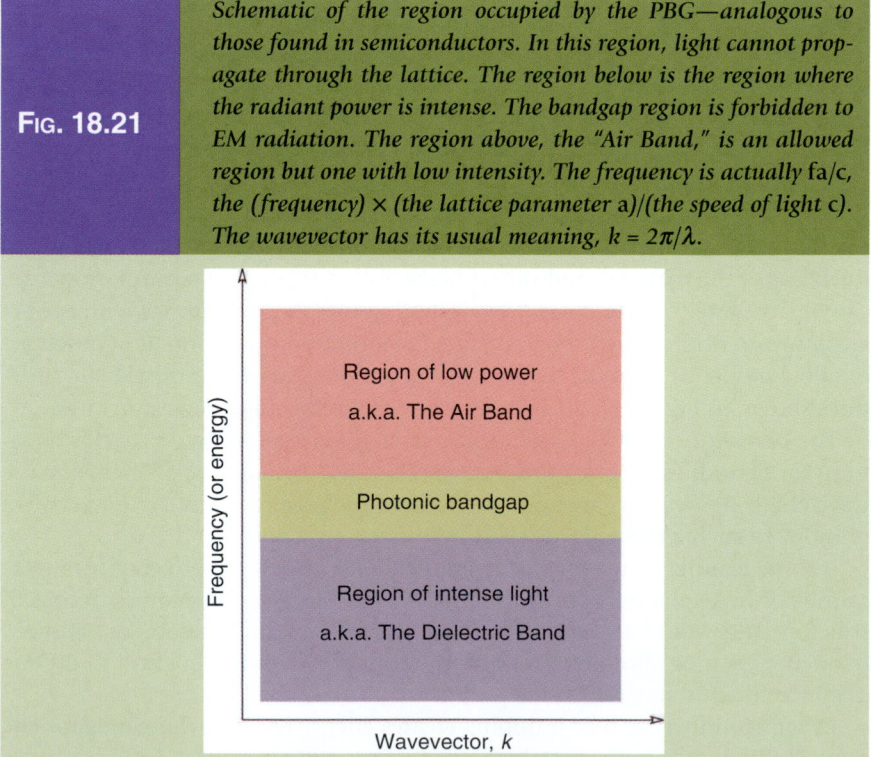

FIG. 18.21 *Schematic of the region occupied by the PBG—analogous to those found in semiconductors. In this region, light cannot propagate through the lattice. The region below is the region where the radiant power is intense. The bandgap region is forbidden to EM radiation. The region above, the "Air Band," is an allowed region but one with low intensity. The frequency is actually* fa/c, *the (frequency) × (the lattice parameter* a*)/(the speed of light* c*). The wavevector has its usual meaning,* $k = 2\pi/\lambda$.

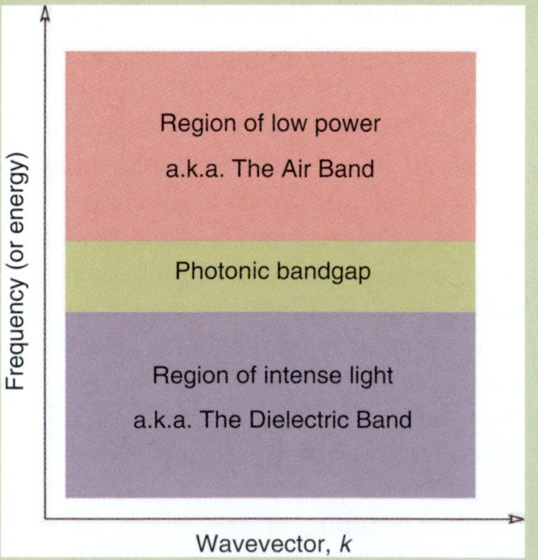

By creating such defects or altering the geometry of its members, resonant cavities can be produced in the crystal. This allows characteristics of photonic crystals to be tunable [30].

18.4.4 Fabrication of Nanophotonic Crystals

The ability to tune the properties of a photonic crystal (e.g., material selection, component geometry, size and orientation, and spacing) is the key to photonic applications. Use of photonic crystals as filters and laser components. For example, in spontaneous emission, coupling a photonic crystal to a laser allows for independent control of the rate of decay and the coupling between atoms and photons in the crystal [30].

Fabrication and Application of 1-D and 2-D Photonic Materials. 1-D and 2-D photonic crystals are quite easy to fabricate [29].

Physical or chemical deposition techniques layer by layer are able to generate 1-D structures (see **Fig. 18.10**). 2-D systems are a little more difficult but overall are still relatively easy to obtain with today's lithographic technology. Pattern transfer and selective etching allow for the creation of 2-D structures. Applications of these materials have already achieved commercial level status in the form of optical notch filters, dielectric mirrors, optical resonance cavities, and Bragg gratings on optical crystal [29].

Fabrication and Application of 3-D Photonic Materials. Three-dimensional photonic crystals are more problematic to fabricate. In 1994, K.M. Ho conceptualized a simple cubic lattice by stacking dielectric rods (**Fig. 18.22**), similar to

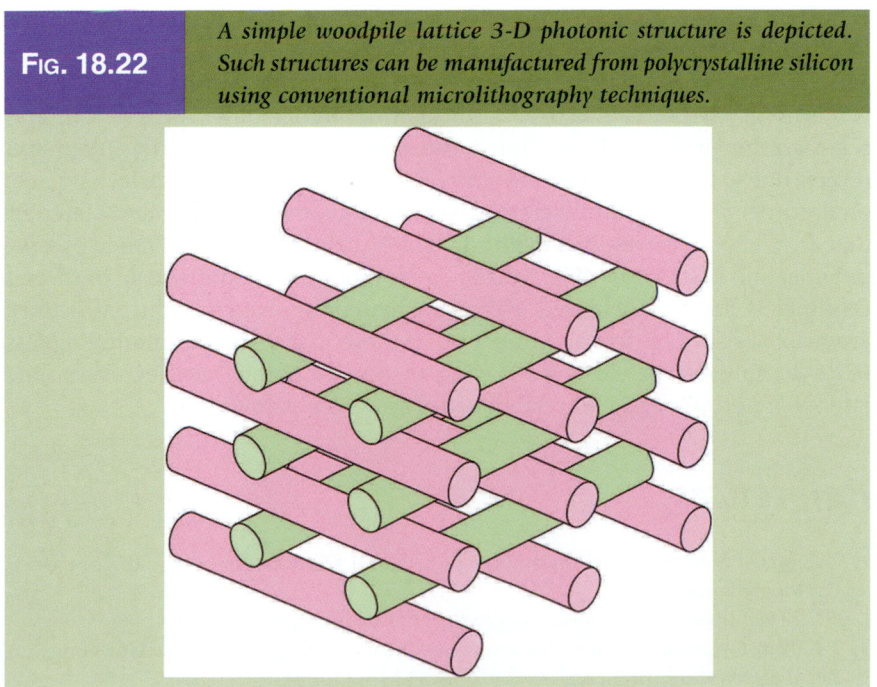

Fig. 18.22 *A simple woodpile lattice 3-D photonic structure is depicted. Such structures can be manufactured from polycrystalline silicon using conventional microlithography techniques.*

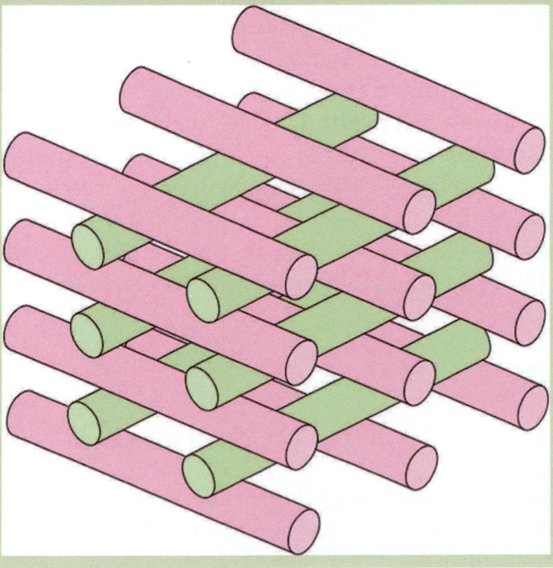

Fig. 18.23 *A photonic structure made of latex beads is shown.*

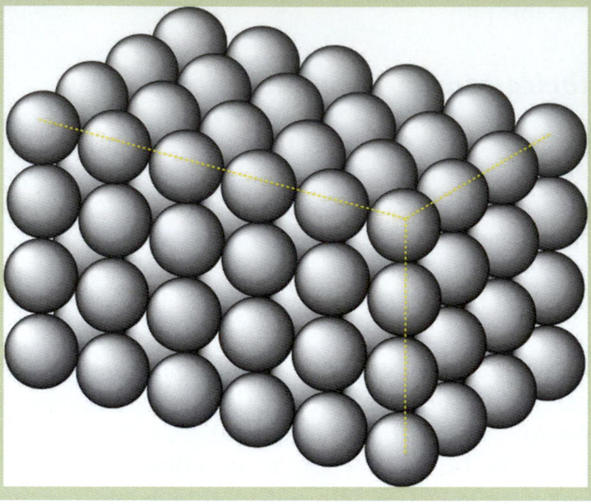

stacking popsicle sticks that we made structures out of in elementary school. A thin film of poly-Si (polycrystalline silicon) was first deposited on a silicon wafer and patterned by a standard photolithography technique. Features (the wood of the woodpiles) formed were accomplished by reactive ion etching. The gaps between the individual lumber were filled with SiO_2 and the process was repeated after another layer of poly-SI was deposited on top of the first layer. This time the pattern was placed 90° to the longitudinal direction of the first layer. This process was repeated until the desired configuration was achieved. Selective etching by hydrofluoric acid (HF) removed the interstitial SiO_2.

Photonic crystals can also be fabricated by holographic, two-photon, and self-assembly processes. Crystalline lattices of spherical colloids are varied and diverse due to the flexibility of the template starting materials. Particle diameters on the order of a few nanometers to several microns are available today. Materials that are compatible with this method include dielectric materials made of latex polymer and silica beads (**Fig. 18.23**). Long-range order on the scale of several centimeters has been achieved with these materials. Inverse opaline structures are also possible by this method but in these cases, the templating agents (e.g., the spheres) are dissolved leaving an array of porous cavities. Crystalline lattices of nonspherical colloidal particles are also possible.

References

1. Cyclopaedia or Universal dictionary of arts and sciences, /en.wikipedia.org/wiki/Cyclopaedia (2008).
2. History of optics, en.wikipedia.org/wiki/History_of_optics (2008).
3. T. Ohm, Optics highlights: An anecdotal history of optics from Aristophanes to Zernike, www.ece.umd.edu/~taylor/optics3.htm (2008).

4. B. Goldberg, Nanophotonics and nanoplasmonics: Control of light at one-hundreth of the wavelength, Presentation, NANOMAT 2007, Bergen, Norway (2007).
5. Optics, en.wikipedia.org/wiki/Optics (2008).
6. T. R. Jensen, M. D. Malinsky, C. L. Haynes, and R. P. Van Duyne, Nanosphere lithography: Tunable localized surface plasmon resonance spectra of silver nanoparticles, *Journal of Physical Chemistry*, 104, 10549–10556 (2000).
7. P. C. Andersen and K. L. Rowlen, Brilliant optical properties of nanometric noble metal spheres, rods, and aperture arrays, *Applied Spectroscopy*, 56 (2002).
8. S. Link and M. A. El-Sayed, Shape and size dependence of radiative, non-radiative and photothermal properties of gold nanocrystals, *International Reviews in Physical Chemistry*, 19, 409–453 (2000).
9. C. N. Davies, Survey of scattering and absorption of light by particles, *British Journal of Applied Physics*, 5, S64-S65 (1954).
10. A. Campion, Raman spectroscopy of molecules adsorbed on solid surfaces, *Annual Reviews of Physical Chemistry*, 36, 549–572 (1985).
11. G. Cao, *Nanostructures & nanomaterials: Synthesis, properties and applications*, Imperial College Press, London (2004).
12. G. L. Hornyak, Optical characterization and optical theory of nanometal/porous alumina composite membranes, Dissertation, Colorado State University, Fort Collins, CO (1997).
13. A. A. Khosravi, M. Kundu, B. A. Kuruvilla, G. S. Shekhawat, R. P. Gupta, A. K. Sharma, P. D. Vyas, and A. K. Kulkarni, *Applied Physics Letters*, 67, 2506 (1995).
14. W. Chan, D. Maxwell, X. Gao, R. Bailey, M. Han, and S. Nie, Luminescent quantum dots for multiplexed biological detection and imaging, *Current Opinion Biotechnology*, 13, 40 (2002).
15. W. C. W. Chan and S. M. Nie, Quantum dot bioconjugates for ultrasensitive nonisotopic detection, *Science*, 281, 2016 (1998).
16. R. Edgar, M. McKinstry, J. Hqang, A. B. Oppenheim, R. A. Fekete, G. Giulan, C. Merril, K. Nagashima, and S. Adhya, High-sensitivity bacterial detection using biotin-tagged phage and quantum-dot nanocomplexes, *Proceedings of the National Academy of Science*, 103, 4841–4845 (2006).
17. H. Weller, U. Koch, M. Gutierrez, and A. Henglein, Photochemistry of colloidal metal sulfides. 7. absorption and fluorescence of extremely small ZnS particles (the world of neglected dimensions), *Berichte Der Bunsen-Gesellschaft-Physical Chemistry Chemical Physics*, 88, 649 (1984).
18. A. D. Dinsmore, D. S. Hsu, H. F. Gray, S. B. Qadri, Y. Tian, and B. R. Ratna, Mn-doped ZnS nanoparticles as efficient low-voltage cathodoluminescent phosphors, *Applied Physics Letters*, 75(6), 802 (1999).
19. R. O. Peterson, FED phosphors: Low or high voltage, *Information Display-Journal of the Society for Information Display*, 3, 22 (1997).
20. L. S. Goldner, Single molecule spectroscopy, National Institute of Standards and Testing, physics.nist.gov/Divisions/Div844/facilities/smspec/sm.html (2008).
21. ftp://203.159.21.137/dutta/Ruhullah-Optics%20in%the%20Nano-World.pdf
22. L. S. Goldner, G. N. Goldie, M. J. Fasolka, F. Ranaldo, J. Hwang, and J. F. Douglas, Near-field polarimetric characterization of polymer crystallite, *Applied Physics Letters*, 85, 1338–1340 (2004).
23. Photonics, *Photonics Dictionary*, www.photonics.com/directory/(2007).
24. E. Yablonovitch, Inhibited spontaneous emission in solid state physics and electronics, *Physical Review Letters*, 58, 2059–2062 (1987).
25. S. John, Strong localization of photons in certain disordered dielectric superlattice, *Physical Review Letters*, 58, 2486–2489 (1987).
26. E. Yablonovitch, Photonic bang-gap structures, *Journal of the Optical Society of America B*, 10, 283–295 (1993).

27. Natural photonics, *Physics World*, physicsworld.com/cws/article/print/18931 (2004).
28. M. Florescu, Photonics crystals: A new frontier in modern optics, Presentation, NASA Jet Propulsion Laboratory & California Institute of Technology, www.mcc.uiuc.edu/nsf/ciw_2006/talks/Florescu.ppt (2006).
29. M. Di ventra, S. Evoy, and J. R. Heflıln Jr., *Introduction to nanoscale science and technology*, Kluwer Academic Press, Boston (2004).
30. C. P. Poole and F. J. Owens, *Introduction to nanotechnology*, Wiley-Interscience, John Wiley & Sons, Inc., Hoboken, NJ (2003).
31. J. Dutta et al., Unpublished results from the AIT Center of Excellence in Nanotechnology, Bangkok, Thailand (2008).
32. G. Mie, Beiträge zur optik trüber Medien, speziell kolloidaler Metallösungen, *Ann Physics (Leipzig)*, 25, 377 (1908).

Problems

18.1 Why is the color of 20-nm gold particles red? What happens when two or three gold particles agglomerate together?

18.2 What is the Lycurgus cup and how does it relate to nanotechnology?

18.3 What is near-field scanning optical microscopy (NSOM)?

18.4 What is a Bohr radius? What does it have to do with electronic transitions in semiconductor materials?

18.5 What are excitons?

18.6 What led Bohr to his radical proposal of "quantum leaps" as an alternative to Rutherford's "planetary" model?

18.7 In solitary atoms, electrons are free to inhabit only certain, discrete energy states. However, in solid materials where there are many atoms in close proximity to each other, bands of energy states form. Where do these "bands" come from?

18.8 What is Fermi energy? What is its importance?

18.9 What is a photonic bandgap material? For visible bandgap, what particle sizes would you choose?

18.10 Niels Bohr in 1913 hypothesized that electrons in hydrogen were restricted to certain discrete levels. This comes about because the electron waves can have only certain wavelengths, that is, $n\lambda = 2\pi r$, where r is the orbit radius. Based on this, one can show that the energy of hydrogen atom for $n = 2$ is 3.4 eV

18.11 How much energy is required for taking out an electron from a Si atom?

18.12 Calculate the energy difference between the ground state and the first excited state for an electron in a quantum dot of size 10^{-8} cm (mass of electron = 9.1×10^{-31} kg and Planck's constant = 6.626×10^{-34} J·s)

18.13 The probability that an electron will occupy a state at the energy E_C is the same as the probability that a hole will occupy a state at the energy E_V. What is the energy E_F of the Fermi level?

18.14 Assuming that $n = 1$ states for both electrons and holes participating in an optical transition, estimate the emitted photon energy from a GaAs (Bandgap = 1.42 eV) single-quantum well laser with a well length of 10 nm. Assume the hole mass to be 0.62 times the free space mass and the hole mass to be 0.067 times the free space mass of electrons. (Given: Free space electron mass = 0.511×10^6 eV/C^2; $\hbar = 6.58 \times 10^{-16}$ $eV \cdot s$; $c = 3.0 \times 10^{10}$ $cm \cdot s^{-1}$.)

18.15 Suppose an electron accelerating at 5 kV strikes a copper target. What type of x-rays will be emitted?

Nanomagnetism

Here again it was the quantum theory which came to the rescue.

Walther Nernst

Chapter 19

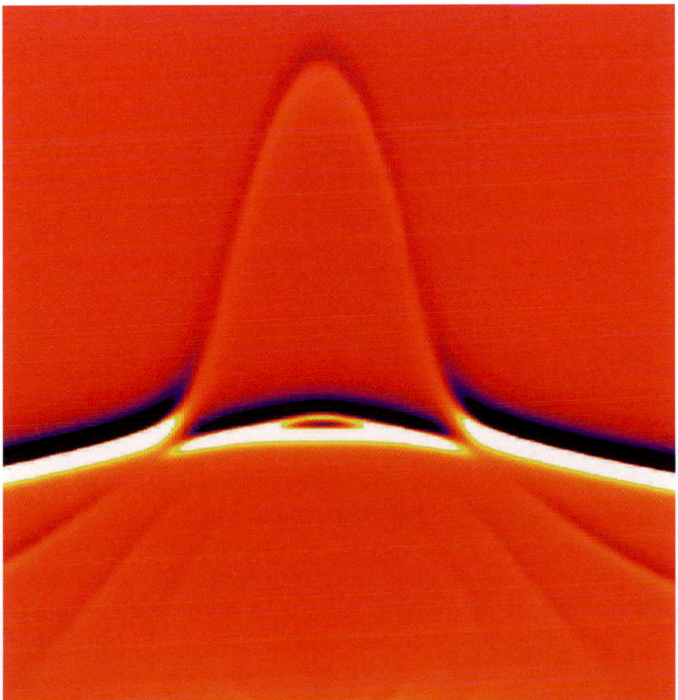

Threads

We conclude the *Electromagnetic Nanoengineering* section with *chapter 19* that deals with magnetic properties of nanomaterials. Once again, size is a critical parameter with regard to the magnetic response of nanomaterials. For several decades now, information has been stored via the action of magnetic materials.

Mechanical Nanoengineering is the next section of the book. In this section, mechanical properties and the properties of thin films and nanocomposites are discussed in *chapters 20* through *22*. This is followed by a section on *Chemical Nanoengineering* and applications of nanomatrials and the final section deals with *Biological and Environmental Nanoengineering* aspects of nanotechnology.

19.0 Introduction

Magnetism is one of the phenomena by which materials can exert attractive or repulsive forces on other materials. Some well-known materials that exhibit easily detectable magnetic properties are nickel, iron, cobalt, and their alloys. All materials are influenced to a greater or lesser extent by the presence of a magnetic field. Magnetism also does have other manifestations, particularly as one of the two components of electromagnetic waves such as light or the cell phone signals often known as electromagnetic waves.

What Is a Magnet? A magnet is normally known as a piece of iron, steel, or magnetite that has the property of attracting iron or steel. The most well-known magnet is probably the lodestone, also often called magnetite, which is a naturally occurring rock that was first discovered in a region known as magnesia and was hence named after this region. Magnetism may be naturally present in a material or the material may be artificially magnetized and magnets can either be permanent or temporary. After being magnetized, a permanent magnet retains the properties of magnetism indefinitely while a temporary magnet (e.g., a magnet made of soft iron) is usually easy to magnetize but it loses most of its magnetic properties when the magnetizing cause is discontinued. Permanent magnets are usually more difficult to magnetize, but they remain magnetized practically forever. These materials which can be magnetized are called *ferromagnetic materials*.

19.0.1 History

The term "magnetism" was introduced by a shepherd by the name of Magnés (from the city of Magnesia in modern day Turkey) who found that his iron-tipped cane was attracted to magnetic rock deposits. The first scientific discussion on magnetism was enumerated by Thales (625 B.C. to about 545 B.C.) according to Aristotle. The earliest literary reference to magnetism lies in a fourth century B.C. book called *Book of the Devil Valley Master,*"The lodestone makes iron come or it attracts it," that was published in China, where also the first compass was reported by Shen Kuo (1031–1095), which is reported to have been first used by Alexander Neckham, by 1187, in Europe for navigation.

In 1269 Peter Peregrinus wrote the *Epistola de Magnete*, the first extant treatise describing the properties of magnets. The modern understanding of the interrelation between electricity and magnetism was triggered in 1819 by Hans

Christian Oersted, from Denmark, who accidentally discovered that an electric current could influence a compass needle. This landmark experiment is popularly known as Oersted's Experiment. Andre-Marie Ampere, Carl Friedrich Gauss, Michael Faraday, and others carried out separately experiments to enumerate the links between magnetism and electricity in greater detail.

Possibly one of the best contribution to the understanding of the electromagnetic nature of waves was clarified by James Clerk Maxwell who presented the now famous *"Maxwell's equations,"* unifying electricity, magnetism, and optics into a new field called *electromagnetism*. 1921–1928 was probably the best-known years where active minds joined to explain the origin of magnetism that changed physics and magnetism. Notable amongst them were the introduction of spins in atoms by Pauli, Uhlenbeck, Goudsmidt, and Dirac; the quantum theory of magnetism propounded by Heisenberg, Schrödinger, and Dirac; and the magnetic exchange interaction in materials proposed by Heisenberg.

19.0.2 Magnetic Phenomena and Their Classical Interpretation

Magnetic Flux. A group of magnetic field lines emitted outward from the north pole of a magnet is called *magnetic flux*. The symbol for magnetic flux is Φ (phi). The SI unit of magnetic flux is the weber (Wb). One *weber* is equal to 1×10^8 magnetic field lines.

Magnetic Field (H). A Current through a coil produces a magnetic field (H). If N be the number of turns of the coil of turn length L and I be the current passed through the coil, the relation for magnetic field (H) is given by

$$H = NI/L \text{ ampere-turns/m}$$

The above equation can be related as $B = (1 + \chi)\mu_o H$.

We stated earlier that different types of magnetism exist and that they are characterized by the magnitude and the sign of the susceptibility. Since various materials respond so differently in a magnetic field, we suspect that several fundamentally different mechanisms must be responsible for the magnetic properties. In the first part of this chapter we shall attempt to unfold the multiplicity of the magnetic behavior of materials by describing some pertinent experimental findings and giving some brief interpretations (**Fig. 19.1**).

Diamagnetism. In a diamagnetic material, there are no unpaired electrons, so the intrinsic electron magnetic moments cannot produce any bulk effect and the magnetization arises solely from the electrons' orbital motions. When a material is put in a magnetic field, the electrons circling the nucleus will experience a Lorentz force from the magnetic field, in addition to their Coulomb attraction to the nucleus. Depending on which direction the electron is orbiting, this force may increase the centripetal force on the electrons, pulling them in towards the nucleus, or it may decrease the force, pulling them away from the nucleus. This effect systematically increases the orbital magnetic moments that were aligned opposite the field, and decreases the ones aligned parallel to the field (in accordance with Lenz's law). This results in a small, bulk magnetic moment, with an opposite direction to the applied field.

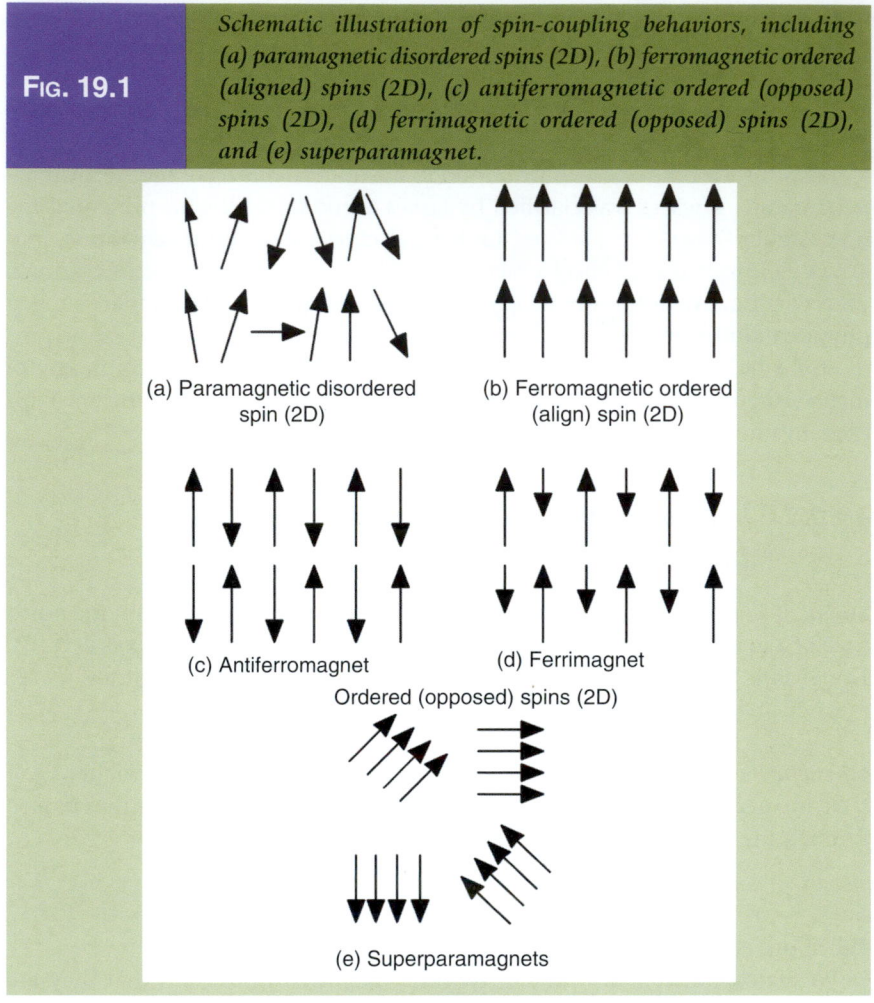

FIG. 19.1 *Schematic illustration of spin-coupling behaviors, including (a) paramagnetic disordered spins (2D), (b) ferromagnetic ordered (aligned) spins (2D), (c) antiferromagnetic ordered (opposed) spins (2D), (d) ferrimagnetic ordered (opposed) spins (2D), and (e) superparamagnet.*

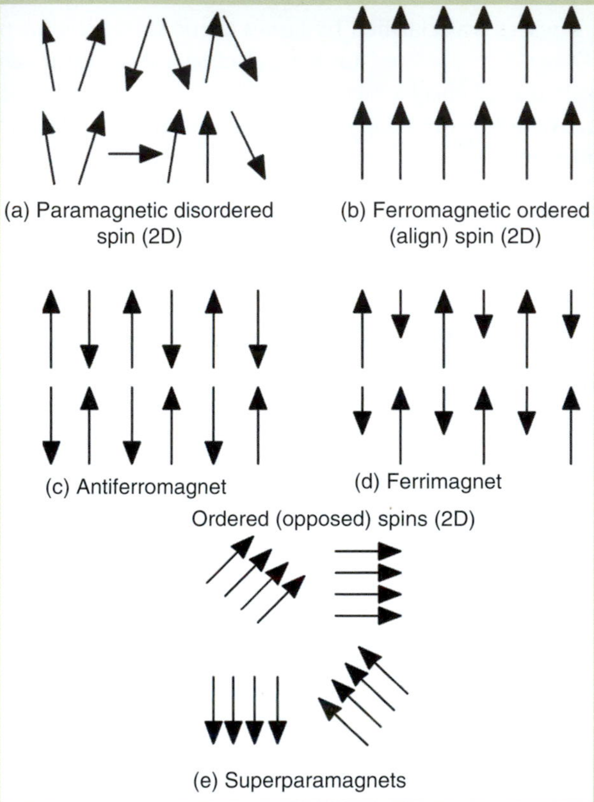

Paramagnetism. In a paramagnet there are unpaired electrons. When an external magnetic field is applied, these magnetic moments will tend to align themselves in the same direction as the applied field, thus reinforcing it.

Ferromagnetism. A ferromagnet has unpaired electrons. In addition to the electrons' intrinsic magnetic moments wanting to be parallel to an applied field, there is also in these materials a tendency for these magnetic moments wanting to be parallel to each other due to exchange interaction between electrons. Thus, even when the applied field is removed, the electrons in the material can keep each other continually pointed in the same direction. Every ferromagnet has its own individual temperature, called the Curie temperature, or Curie point, above which it loses its ferromagnetic properties. This is because the thermal tendency to disorder overwhelms the energy lowering due to ferromagnetic order. In ferromagnetic rare earth metals (e.g., Gd, Dy, Sm), the magnetic moments (e.g., the Si) are strongly localized at the atomic positions.

Antiferromagnetism. In an antiferromagnet, there is a tendency for the intrinsic magnetic moments of neighboring valence electrons to point in opposite directions. When all atoms are arranged in a substance so that each neighbor is

FIG. 19.2 *Magnetic susceptibility of various kinds of magnetic material.*

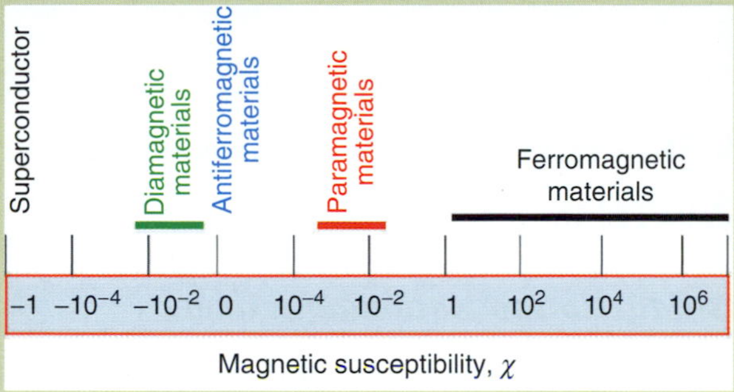

"anti-aligned," the substance is antiferromagnetic. Antiferromagnets have a zero net magnetic moment. In varying temperatures, antiferromagnets can be seen to exhibit diamagnetic and ferrimagnetic properties. In antiferromagnetic materials, the local spin densities are nonzero, yet the overall spin (and magnetic moment) vanishes. Examples of antiferromagnets are MnO, FeO, CoO, and NiO. Chromium is a peculiar ferromagnet with nonvanishing local spin densities but with a periodicity which is incommensurate with the Cr lattice.

Ferrimagnetism. Ferrimagnets retain their magnetization in the absence of a field. However, like antiferromagnets, neighboring pairs of electron spins like to point in opposite directions but there is more magnetic moment from the sublattice of electrons which point in one direction, than from the sublattice which points in the opposite direction. We can distinguish various kinds of magnetic properties using magnetic susceptibility, ratio of magnetization, and applied magnetic field (**Fig. 19.2**).

The interaction of localized moments can also be described using the molecular field approach where it is assumed that the interaction of spin S with all the other spins can be approximated with an effective magnetic field B_{mf} due to those neighboring spins. In a metal which is constructed with paramagnetic atoms, the bonding is mediated by the valence electrons, which are often also responsible for the atomic paramagnetism. In Na for example, the single 2s valence electron ($S = 1/2$) gives rise to the conduction band, which is know to be almost free electron like. This means that these electrons hardly feel the atomic potentials but instead an almost featureless average potential. In this state, the local spin density vanishes everywhere. In other words the bulk is nonmagnetic. This is not true for itinerant ferromagnetic metals such as the 3d transition metals (Fe, Co, Ni). The bulk is ferromagnetic where each atom contributes $0.54\mu_B$. This is due to an imbalance in the up-down spin densities. The local spin densities at the Fermi level of ferromagnetic metals do not vanish. This is important in the characterization of magnetic materials.

Several exchange-type interactions are responsible for magnetism in metals. One is the intra-atomic exchange, which causes atoms to attain a net spin that underlies Hund's rules. Exchange is also responsible for spin order. That interaction is the interatomic exchange coupling can be either indirect or direct.

19.0.3 The Nano Perspective

Why the Interest in Nanoscale Magnetic Materials? There is a dramatic change in magnetic properties when the critical length governing some phenomenon (magnetic, structural, etc.) is comparable to the nanoparticle or nanocrystal size. Since a large amount of surface is exposed, effects due to surfaces or interfaces are stronger.

19.1 Characteristics of Nanomagnetic Systems

19.1.1 Introduction to Nanomagnetism

In magnetic materials, a magnetic field is produced because of the movement of electrons within the material, which produces a field around the material and a magnetization effect exists within an atom, as shown in **Figure 19.3**. Because of the charge of an electron, magnetic moment appears just like magnetic field is generated, when current flows through solenoid coils. A magnetic moment can appear even due to the orientation of the spins of an electron. The magnetic moment generated due to the orbital motion or spin motion of a single electron is called *Bohr magneton*, which is the smallest unit of magnetic moment of solids. This magnetic moment can interact with the magnetic field as that in the current loop.

$$\text{Bohr magneton} = \frac{qh}{4\pi m_e} = 9.27 \times 10^{24}\ [\text{A} \cdot \text{m}^2]$$

where

q is the charge of an electron
h is the Planck's constant
m_e is the rest mass of an electron

In most materials, the electron spin will have its pair, up and down spin, and the spin will cancel each other resulting in a zero net spin, hence rendering these materials immune to any effects of an applied magnetic field (no magnetic moment). Following Hund's rule, certain materials such as the transition metals have unbalanced spins that lead to a nonzero net spin rendering a magnetic moment to the atoms (**Table 19.1**).

FIG. 19.3 *Electron orbit in an atom.*

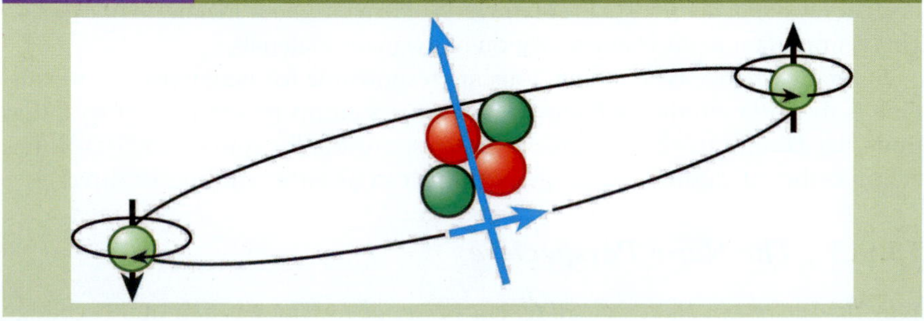

TABLE 19.1 *Electron Configuration of Some Transition Metals*

Metal	3d					4s
Sc	↑					↑↓
Ti	↑	↑				↑↓
V	↑	↑	↑			↑↓
Cr	↑	↑	↑	↑	↑	↑
Mn	↑	↑	↑	↑	↑	↑↓
Fe	↑↓	↑	↑	↑	↑	↑↓
Co	↑↓	↑↓	↑	↑	↑	↑↓
Ni	↑↓	↑↓	↑↓	↑	↑	↑↓
Cu	↑↓	↑↓	↑↓	↑↓	↑↓	↑

Magnetism at nanometric sizes deals more with the newly discovered phenomena attributed to nano-sized materials such as spin-polarized tunneling, oscillatory exchange coupling between magnetic–nonmagnetic multilayers, magnetoresistivity etc., which take place mostly in low dimensional materials. When an object becomes so small that the number of surface atoms is a sizable fraction of the total number of atoms, then surface effects tends to be important. A typical size of a classically expected magnetic domain is of the order of 1 μm. As the dimensions of magnetic materials decrease down to nanometer scales, these materials start to exhibit new and very interesting physical properties mainly due to quantum size effects. Magnetic behavior is changed, the magnetization of noble metals is considerably induced, electron interference patterns are observed, oscillatory exchange interactions occur between adjacent layers separated by a nonmagnetic spacer, and magnetoresistivity is enhanced many orders of magnitude. All these new phenomena are affected by the imperfections of nanostructures. At the nanoscale the intrinsic properties of the materials become extrinsic and can be adjusted by size in the nanoscale, and thus devices virtually having any characteristic can be made realistic. The homogeneity and purities in chemical composition, crystalline structures, external morphology, etc., determine the physical properties. These new materials are important for magnetic sensors; giant magnetoresistance (GMR) reading heads are used to read magnetically stored data. The recording industry desires for ever denser and more reliable recording media and hence nanosize properties are of distinct interest.

Thus, in magnetic materials, electrons' orbital angular motion around the nucleus and the electrons' intrinsic magnetic moment are the most important sources of magnetization. The magnetic behavior of a material depends on the structure of the materials as well as the temperature (at high temperatures, random thermal motion makes it more difficult for the electrons to maintain alignment).

A qualitative, as well as a quantitative distinction between different types of magnetism, can be achieved in a relatively simple way following a method proposed by Faraday. The magnetic material to be investigated is suspended from

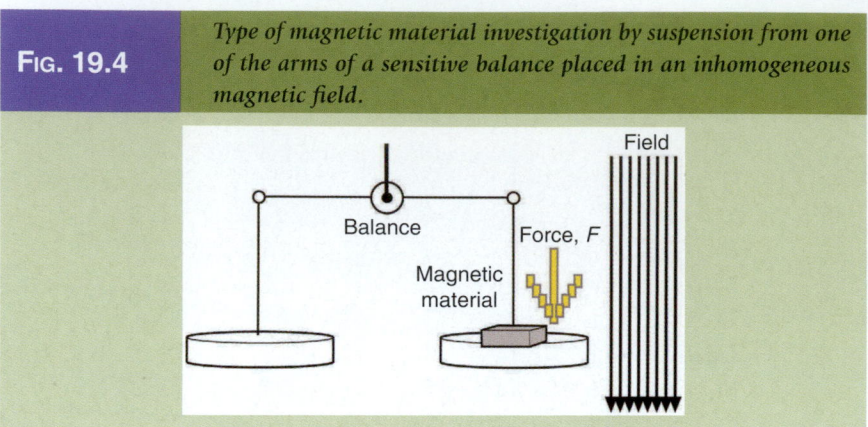

FIG. 19.4 *Type of magnetic material investigation by suspension from one of the arms of a sensitive balance placed in an inhomogeneous magnetic field.*

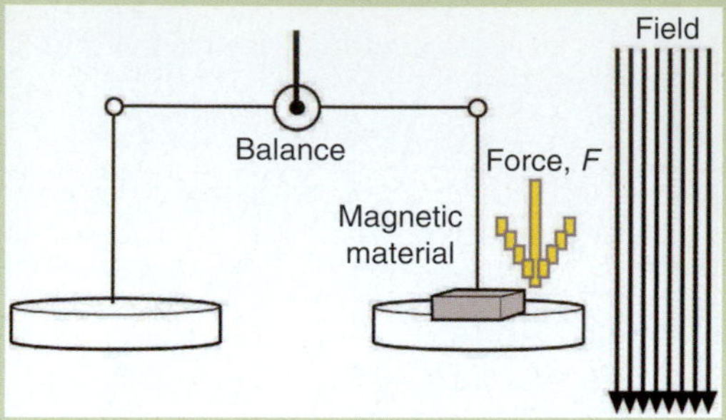

one of the arms of a sensitive balance and is allowed to reach into an inhomogeneous magnetic field (**Fig. 19.4**).

Diamagnetic materials are expelled from this field, whereas para-, ferro-, antiferro-, and ferrimagnetic materials are attracted in different degrees. It has been empirically found that the apparent loss or gain in mass, that is, the force F on the sample exerted by the magnetic field is

$$F = V\chi H \frac{dH}{dx}$$

where

dH/dx is the change of the magnetic field strength $|\mathbf{H}|$ in the direction of the field

V is the volume of the sample

The magnetic material can be characterized by a material constant χ, called the susceptibility which expresses how responsive this material is to an applied magnetic field. Frequently, a second material constant, permeability μ, is used. This constant is related to the susceptibility by

$$\mu = 1 + 4\pi\chi$$

For empty space, and for all practical purposes, also for air, χ is 0 and thus $\mu = 1$. For diamagnetic materials one finds χ to be small and negative, and thus μ slightly less than 1. For para- and antiferromagnetic materials χ is again small, but positive. Thus, μ is slightly larger than 1. Finally, χ and μ are large and positive for ferro- and ferrimagnetic materials. The magnetic constants are temperature dependent, except for diamagnetic materials.

The magnetic field parameters at a given point in space are defined to be the magnetic field strength **H**, which we introduced above, and the magnetic flux density or magnetic induction **B**. In free space **B** and **H** are identical. Inside a magnetic material, the induction **B** consists of the free-space component (**H**), plus a contribution to the magnetic field which is due to the presence of matter.

$$\mathbf{B} = \mathbf{H} + 4\pi\mathbf{M}$$

where **M** is called the magnetization of the material. For a material in which the magnetization is thought to be proportional to the applied field strength we define

$$\mathbf{M} = \chi\mathbf{H}$$

Combining the above two equations we get

$$B = H(1 + 4\pi\chi) = \mu H$$

Finally, we need to define the magnetic moment, μ_m, through the following equation

$$M = \frac{\mu_m}{v}$$

that is the magnetization is the magnetic moment per unit volume.

It needs to be noted that in magnetic theory several unit systems are commonly in use. Scientific and technical literature on magnetism is still widely written in electromagnetic cgs (emu) units. The magnetic field strength in cgs units is measured in oersted (Oe) and the magnetic induction in gauss.

19.1.2 Characteristics of Nanomagnetic Materials

Size Dependence. In general, the magnetic moment of transition metals decreases with the increase of number of atoms (size) in the clusters. However, the moment of different transition metal clusters are found to depend differently on the size of clusters. In the lower size limit, the clusters have a high-spin majority configuration and the behavior is more like an atom, but with the increase of size, the moment moves towards that of the bulk with slow oscillations.

Surface Magnetism. The magnetic properties at the surfaces of ferromagnets are significantly altered from the bulk for several reasons. For example, since the number of nearest neighbors is reduced, a valence electron tends to spend more time at each ionic site, due to reduced coordination compared with the bulk. Weaker bonding to neighboring sites causes the ions to have a more isolated atomic character.

Magnetic Anisotropy and Domains in Small Particles. Small magnetic particles therefore are typically monodomain, since the energy cost to form a domain wall outweighs the reduction in magnetic energy. Typically, critical sizes for monodomain particles are in the range of 20–2000 nm and depend on the ferromagnetic material under consideration.

19.1.3 Magnetization and Nanostructures

Changes in magnetization of a material occur via activation over an energy barrier. Each physical mechanism responsible for an energy barrier has an associated length scale. The fundamental magnetic lengths are the crystalline anisotropy length l_k, the applied field length l_H, and the magnetostatic length l_S, which are defined as follows:

$$l_k = \sqrt{J/k}$$

$$l_H = \sqrt{2J/HM_S}$$

$$l_S = \sqrt{J/2\pi M_S^2}$$

Here, k is an anisotropy constant of the bulk material due to the dominant anisotropy and J is the exchange within a grain. If more than one type of barrier is present, then magnetic properties are dominated by the shortest characteristic length. For most common magnetic materials, these lengths are of the order of 1–100 nm. For example, nickel at 1000 Oe and room temperature has lengths $l_S \cong 8$ nm, $l_k \cong 45$ nm, and $l_H \cong 19$ nm.

When a sufficiently large magnetic field is applied, the spines within the material align with the field. The maximum value of magnetization achieved in this state is called the saturation magnetization, M_S. As the magnitude of the field decreases, spins cease to be aligned with the field and the total magnetization decreases. In ferromagnets, a residual magnetic moment remains at zero field. The value of the magnetization at zero field is called the remanent magnetization, M_R. The ratio of the remanence magnetization to the saturation magnetization M_R/M_S is called the remanence ratio and varies from 0 to 1. The magnitude of the field that must be applied in the negative direction to bring the magnetization of the sample back to zero is called the coercive field. H_C magnetic recording applications require a large remnant magnetization, moderate coercivity, and (ideally) a square hysteresis loop. The different regimes of nanomaterials in the realm of nanomagnetism is summarized in **Figure 19.5**.

Classification of Magnetic Nanomaterials. The magnetic behavior of most experimental realizable systems is a result of contributions from both interaction and size effects. The correlation between nanostructure and magnetic properties suggests a classification of nanostructure morphologies. The following classification is designed to emphasize the physical mechanisms responsible for the magnetic behavior that **Figure 19.6** schematically represents. At one extreme are systems of isolated particles with nanoscale diameters, which are denoted by type A. These noninteracting systems derive their unique magnetic properties strictly from the reduced size from the components, with no contribution from interparticle interactions. At the other extreme are bulk materials with nanoscale structure denoted by type D, in which a significant fraction (up to 50%) of the sample volume is composed of grain boundaries and interfaces. In contrast to type A systems, magnetic properties here are dominated by interactions. The length of

Fig. 19.5 *Different regimes of magnetism with respect to the number of atoms in the material.*

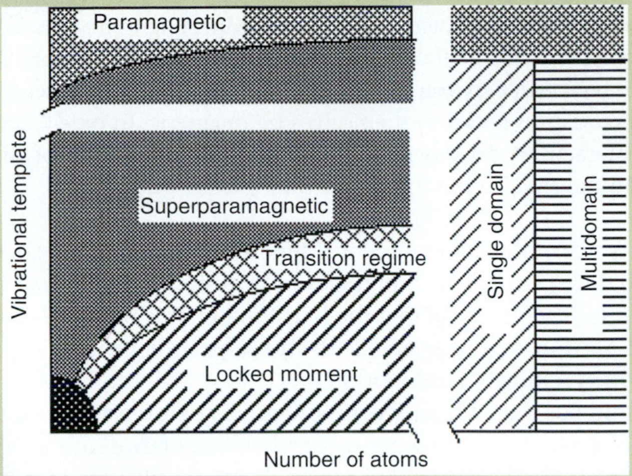

Fig. 19.6 *Schematic representation of the different types of magnetic nanostructures. Type A materials include the ideal ultrafine particle system, with interparticle spacing large enough to approximate the particles as noninteracting. Ferrofluids, in which magnetic particles are surrounded by a surfactant, preventing interactions, are a subgroup of type A materials. Type B materials are ultrafine particles with a core-shell morphology. Type C nanocomposites are composed of small magnetic particles embedded in a chemically dissimilar matrix. The matrix may or may not be magnetoactive. Type D materials consists of small crystallites dispersed in a noncrystalline matrix. The nanostructure may be two phase, in which nanocrystallites are a distinct phase from the matrix, or the ideal case in which both the crystallites and the matrix are made of the same material.*

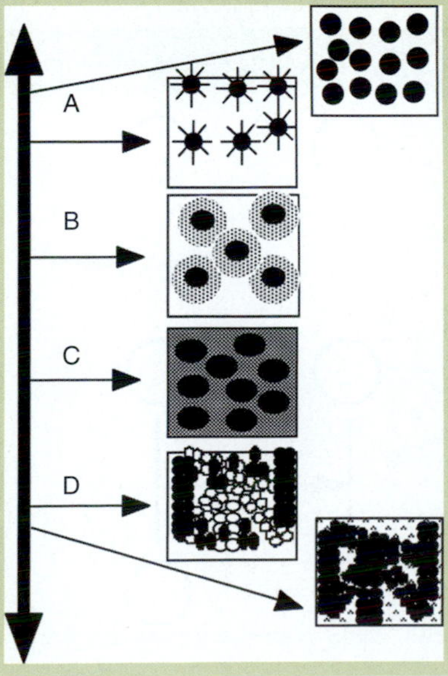

the interactions can span many grains and is critically dependent on the character of interface. Due to this dominance of interaction and grain boundaries, the magnetic behavior of type D nanostructures cannot be predicted simply by applying theories of polycrystalline materials with reduced length scale. In type B particles, the presence of a shell can help prevent particle interactions, but often at the cost of interaction between the core and the shell. In many cases the shells are formed via oxidation and may themselves be magnetic. In type C materials, the magnetic interactions are determined by the volume fraction of the magnetic particles and the character of the matrix they are embedded in.

Ferrofluids. A ferrofluid is a synthetic liquid that holds small magnetic particles in a colloidal suspension, with particles held aloft due to thermal energy. The particles are sufficiently small that the ferrofluid retains its liquid characteristics even in the pressure of a magnetic field, and substantial magnetic forces can be induced, which results in fluid motion.

A ferrofluid has three primary components. The carrier is the liquid element in which the magnetic particles are suspended. Most ferrofluids are either water based or oil based. The suspended materials are small ferromagnetic particles such as iron oxide, on the order of 10–20 nm in diameter. The small size is necessary to maintain stability of the colloidal suspension, as particles significantly larger than this will precipitate. A surfactant coats the ferrofluid particles to help maintain the consistency of the colloidal suspension.

The magnetic properties of the ferrofluid is strongly dependent on particle concentration and on the properties of the applied magnetic field. With an applied field, the particles align in the direction of the field, magnetizing the fluid. The tendency for the particles to agglomerate due to magnetic interaction between particles is opposed by the thermal energy of the particles. Although particles vary in shape and size distribution, an insight into fluid dynamics can be gained by considering a simple spherical model of the suspended particles (**Fig. 19.7** and **Table 19.2**).

FIG. 19.7 *Representative model of a typical ferrofluid.*

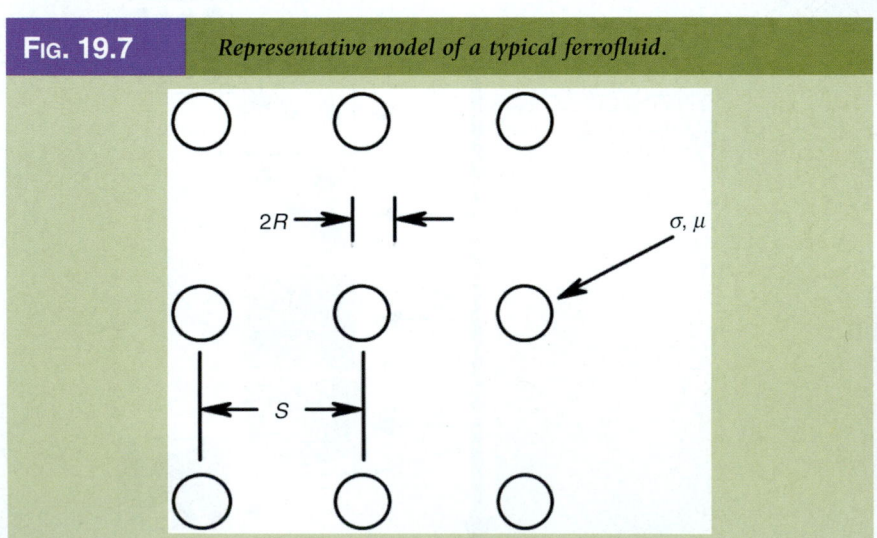

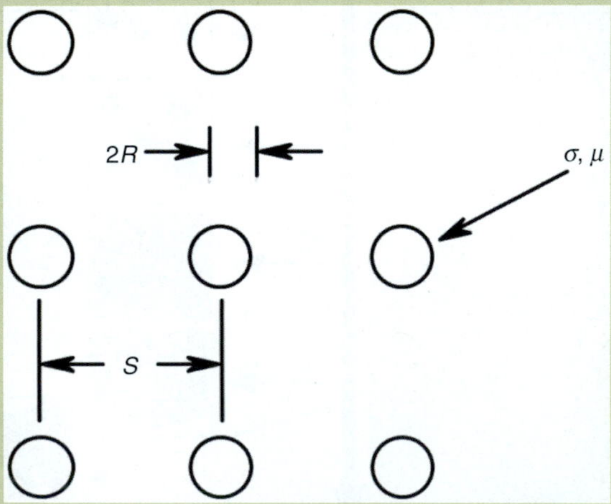

TABLE 19.2 *Material Parameters of a Typical Ferrofluid [1]*

Sample volume	1.7×10^{-6} m^3
Electrical conductivity of particles	$\sigma_f = 3 \times 10^6$ $(\Omega \cdot m)^{-1}$
Electrical conductivity of ferrofluid	$\sigma_{fluid} < 10^{-7}$ $(\Omega \cdot m)^{-1}$
Volume percentage of particles	3% by volume
Initial magnetic permeability of particles	$\mu \approx 100\mu_0$
Particle mean radius	$R \approx 10^{-8}$ m
Density of magnetic particles	$\rho_{FE} = 7.8$ g $\cdot$ cm^{-3}
Fluid density	$\rho_f = 1$ g $\cdot$ cm^{-3}
Fluid viscosity of carrier fluid	$\eta = 1$ cp $= 0.01$ g $\cdot$ cm^{-1} $\cdot$ s^{-1}

The particles are free to move in the carrier fluid under the influence of an applied magnetic field, but on average the particles maintain a spacing S to nearest neighbor. In a low density fluid, the spacing S is much larger than the mean particle radius $2R$, and magnetic dipole–dipole interactions are minimal.

Applications for ferrofluids exploit the ability to position and shape the fluid magnetically. Some applications are

- Rotary shaft seals
- Magnetic liquid seals, to form a seal between regions of different pressures
- Cooling and resonance damping for loudspeaker coils
- Printing with magnetic inks
- Inertial damping, by adjusting the mixture of the ferrofluid the fluid viscosity may be changed to critically damp resonances accelerometers
- Level and attribute sensors
- Electromagnetically triggered drug delivery

Single-Domain Particles. Groups of spins all pointing in the same direction and acting cooperatively are separated by domain walls, which have a characteristic width and energy associated with their formation and existence. The motion of a domain wall is a primary means of reversing magnetization. The dependence of *coercivity* on particle size is similar to that schematically illustrated in **Figure 19.8**.

In large particles, energetic considerations favor the formation of domain walls. Magnetization reversal thus occurs through the nucleation and motion of these walls. As the particle size decreases towards some critical particle diameter $D_{critical}$, the formation of domain walls becomes energetically unfavorable and the particles are called single domain. Changes in the magnetization can no longer occur through domain wall motion and instead require the coherent rotation of spins, resulting in larger coercivity. As the particle size continues to decrease below the single domain value, the spins get increasingly affected by thermal

FIG. 19.8 *Qualitative illustration of the behavior of the coercivity in ultrafine particle systems as the particle size changes.*

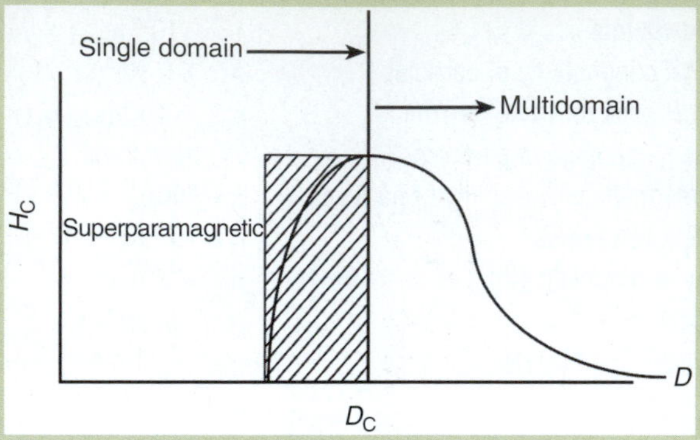

fluctuations and the system becomes superparamagnetic. Particles with sufficient shape anisotropy can remain single domain to much larger dimensions than their spherical counterparts.

Single-Domain Characteristics. In a granular magnetic solid with a low volume fraction, one has a collection of single-domain particles, each with a magnetic axis along which all the moments are aligned. In **Figure 19.9**, the typical nanoparticle sizes for some common ferromagnetic materials (single-domain size, D_{critical}) and the stability of the magnetic properties or the superparamagnetic limit at room temperature ($D_{\text{superparamagnetic}}$) is shown in **Figure 19.9**.

In the absence of a magnetic field, parallel and antiparallel orientations along the magnetic axis are energetically equivalent but separated by an energy barrier

FIG. 19.9 *Single domain size, $D_{critical}$ and magnetic stability size or the superparamagnetic limit at room temperature, $D_{superparamagnetic}$ for some common ferromagnetic materials [2].*

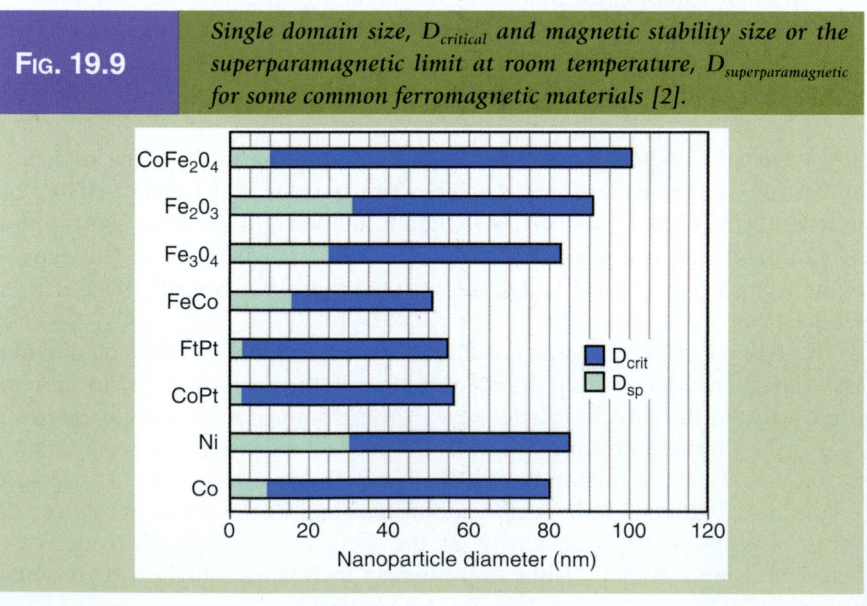

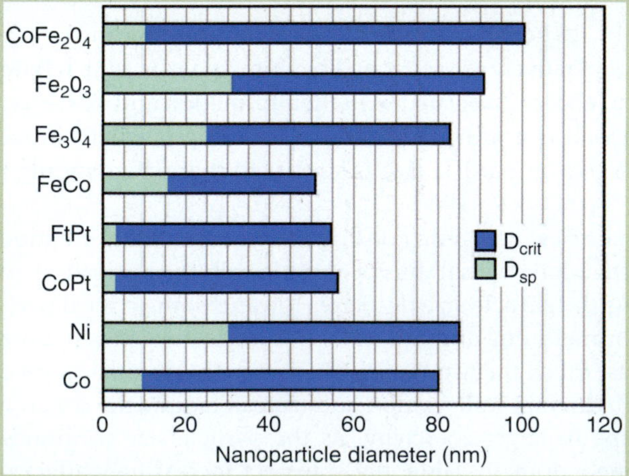

of CV, where C is the total anisotropy per volume, and V is the particle volume. Since the size of each single domain remains fixed, under an external field, only the magnetic axes rotate. Thus the measured magnetization (M) of a granular magnetic field solid with a collection of single-domain particles is the global magnetization

$$M = \frac{\langle \vec{M}\vec{H} \rangle}{H} = M_S \langle \cos\theta \rangle$$

where θ is the angle between the magnetic axes of a particle. M_S is the saturation magnetization, $\vec{H}$ is the external field, and the average $\langle \cos\theta \rangle$ is taken over many ferromagnetic particles. The hysteresis loop of a granular solid is thus a signature of the rotation of the magnetic area of the single-domain particles. This should be contrasted with the hysteresis loop of a bulk ferromagnet, in which the sizes and the direction of the domains are altered drastically under an external field.

Superparamagnetic Materials. Superparamagnetic materials consist of individual domains of elements that have ferromagnetic properties in bulk. Their magnetic susceptibility is between that of ferromagnetic and paramagnetic materials. **Figure 19.10** illustrates the effect of a superparamagnetic material (gray circle) on the magnetic field flux lines.

At sufficient high temperatures, the energy barrier (CV) is overcome by the thermal energy. Consequently, the magnetic moments within a particle rotate rapidly in unison, exhibiting the superparamagnetic relaxation phenomenon. The simplest form of the relaxation time can be described by the Arrhenius relation

$$\tau = \tau_0 \exp\left(\frac{CV}{k_B T}\right)$$

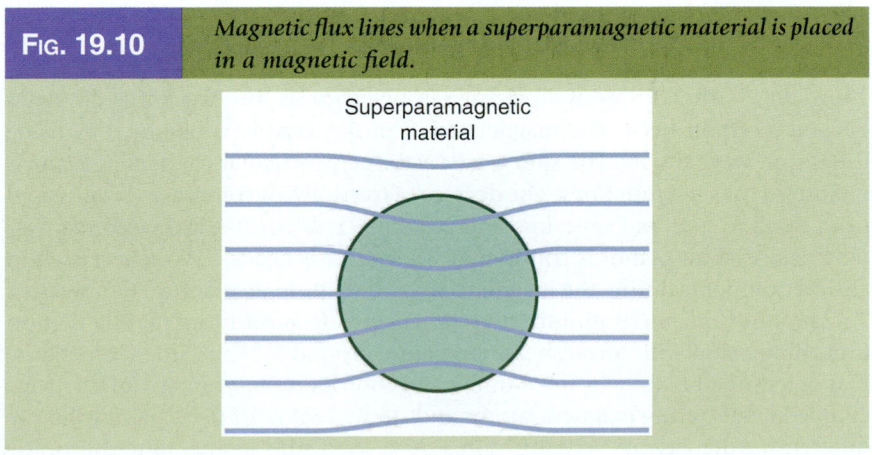

FIG. 19.10 *Magnetic flux lines when a superparamagnetic material is placed in a magnetic field.*

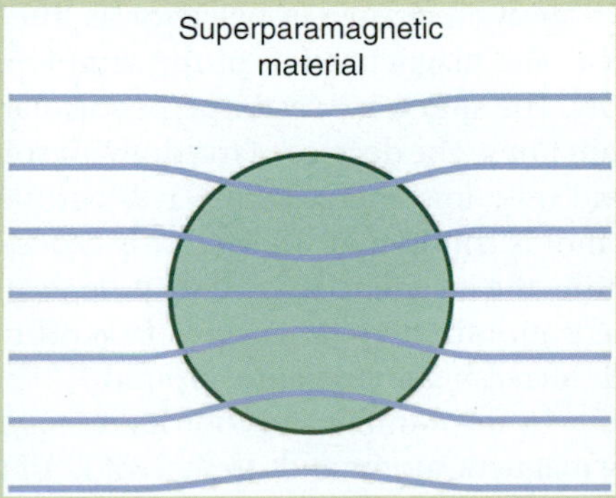

where

τ is the relaxation time
τ_0 is the characteristic time
CV is the total anisotropy energy

The behavior of τ is dominated by the exponential argument. Assuming representative valves ($\tau_0 = 10^{-9}$ s, $k_B = 10^6$ erg/cm^3, and $T = 300$ K), a particle of diameter 11.4 nm will have a relaxation time of 0.1 s. Increasing the particle diameter to 14.6 nm increases τ to 10^8 s. For an instrument that can measure certain magnetic characteristics (e.g., magnetometry, AC susceptibility, Mössbauer spectroscopy) with a measurement time of τ_i, one can define a blocking temperature.

$$T_B = \frac{CV}{k_B \ln(\tau_i/\tau_0)}$$

At $T < T_B$, τ_i is less than τ and the instrument detects the ferromagnetic nature (e.g., a hysteresis loop) of the system. However, at $T > T_B$, because τ_i is longer than τ, the time-averaged value of the ferromagnetic characteristics (e.g., magnetization and coercivity) vanish within the measuring time of τ_i. Then the system is in an apparent paramagnetic or superparamagnetic state, even though within each particle the magnetic moments remain ferromagnetically aligned. Because of the superparamagnetic relaxation, the value of remnant magnetization (M_R) and coercivity (H_c) decrease with increasing temperature and vanish at the blocking temperature (T_B). Above T_B, all apparent ferromagnetic characteristics disappear. The blocking temperature in a superparamagnetic system decreases with increasing measuring fields, being proportional to $H^{2/3}$ at large magnetic fields and proportional to H^2 at lower fields.

19.2 Magnetism in Reduced Dimensional Systems

19.2.1 Two-Dimensional Systems

Thin Films. There is always a change in behavior as film thickness decreases beyond a certain limit. The magnetization of the sample is obtained by using usual spin wave theory. The spin wave vector perpendicular to the film plane is quantized for such thin films. The degrees of freedom for spin waves is decreased from three to two and only low energy spin waves with a wave vector in two dimensions (film plane) is considered. This effect is expected to reflect itself in critical exponents during the magnetic phase transition near critical temperature T_c. As a result magnetization in two dimensions falls off more rapidly than in three dimension with increasing temperature around T_c. The Curie temperature T_c is determined mainly by the number and coordination symmetry of exchange coupled neighboring magnetic atoms and strength of J. All these parameters are different in the case of very thin films. And the Curie temperature drastically changes in the thin-film case. T_c is found to depend on the number of layers.

T_c even vanishes when we go to a one-dimensional system; the magnetization can be enhanced for ultrathin films, especially for monolayers. The atoms try to maximize their spins as per Hund's rule and also obey Pauli's exclusion principle, which favors the increase of individual atomic moments at the surfaces. Also the electrons want to be far apart from each other to minimize the Coulomb energy.

Monolayers. For monolayers, magnetic moments are very large. Even a nonmagnetic metal in its bulk form can become spontaneously ferromagnetic in its monolayer form though there are difficulties in growing ideal monolayers. For a high quality Cr monolayer on Fe(100) the atomic magnetic moment is observed to rise up to $3\,\mu_B$ instead of $0.4\,\mu_B$ for antiferromagnetic Cr in its bulk form.

Quantum Wells. With the decrease of the film thickness, the film can behave like a quantum well for spin carriers. These carriers are reflected by the walls of the wells and the wave functions interfere to form a standing wave with discrete energy levels, and the reflection coefficient is spin dependent. This effect influences almost all the physical properties, such as inverse photoemission and photoemission, magnetic anisotropy, magneto-optic response, electrical conductivity, Hall effect, and superconductivity.

19.2.2 One-Dimensional Systems

A one-dimensional magnetic system on an atomic level occurs as a chain of magnetic atoms even in the bulk form of some crystalline materials. These magnetic chains are magnetically decoupled from neighboring chains by nonmagnetic intermediated atoms. Artificial one-dimensional magnetic structures are becoming more attractive for their magnetoresistance. Though the geometry is experimentally difficult for a continuous thin-film case however, it can be realized by growing parallel stripes separated by nonmagnetic very narrow spacer wires on a substrate. The current in the substrate plane can be applied to get high magnetoresistance. One dimensional system may contain many parallel magnetic stripes and many properties of the interface are changed. For instance, continuous spectra split into discrete energy levels; magnetization direction in the sample plane may be switched between perpendicular and parallel to the stripes.

19.2.3 Zero-Dimensional Systems

The most remarkable changes are observed in ultrafine magnetic particles since the surface-to-volume ratio is highest for "zero-dimensional" systems. The surface magnetism increases or decreases, or even can be disordered and/or dead for some surface-treated (e.g., Ni) particles at low temperatures. The smallest magnetic particles are magnetic molecules and clusters. The size of artificial magnetic particles may be reduced down to even 1 nm. The fine particles behave as a monodomain magnet because the particle diameter would be even smaller than domain wall thickness. When the particle size shrinks further the superparamagnetic limit is reached. When the size of the spin in the case of a single domain ferromagnetic particle without crystalline anisotropy can be many orders of the magnitude greater than for a single atom it is known as *superparamagnetism*. The fine particle system on a substrate can be in a distinctly different phase from the bulk, depending on the particle density and size in the nanometer range. The ultimate goal is to achieve particle array for magnetic recording. Thus both signal-to-noise ratio and storage capacities would be increased a few orders of magnitude. If an assembly of identical uniaxial superparamagnetic particles is initially polarized along the easy axis, then the magnetization will reduce with increasing time. An important property of superparamagnetic particle systems is that they are nonhysteretic and their magnetization curves scale with H/T.

19.3 Physical Properties of Magnetic Nanostructures

19.3.1 Substrate Effects on Structures and Related Properties

At nanoscale the magnetic phases are dramatically changed and new magnetic phases arise in ultrathin films of epitaxially grown metallic Fe, for instance, on a single-crystal Cu(100) substrate. The magnetic properties, phase boundaries, and critical temperatures depend on the film thickness and on different crystal structures of the film (fcc) that is imposed by the substrate through adhesion at the interface. The most important parameter determining the magnetic structure is the exchange overlap integral of electronic wave functions on neighboring atoms. This overlap depends on neighboring atomic distances determined by crystal lattice parameters.

The major contribution to the magnetism comes from electronic spins while the magnetic anisotropies originate from the interaction between spins and orbit; exchange interactions are changed because of lack of neighbors. A significant magneto-elastic energy due to lattice mismatch is induced as well. Anisotropies can overcome the magnetostatic demagnetizing energy to give a perpendicular magnetization for ultrathin-film cases.

19.3.2 Oscillatory Exchange Coupling

Oscillatory exchange coupling is an important phenomena observed in ferromagnetic thin films separated by a nonmagnetic spacer. In this phenomena one of the layers polarizes the conduction electrons of the nonmagnetic metallic spacer. If the lifetime of the polarization is long enough, these polarized electrons carry this information to the layer across the spacer. Thus the second layer is coupled with the first one through polarized conduction electrons, which is also known as RKKY.

19.3.3 Spin-Polarized Tunneling

Quantum mechanical effects may be important in particular for very small particles at low temperatures, when quantum mechanical reorientation of the magnetization may occur. This process proceeds through a tunneling mechanism. When two ferromagnetic materials are separated by a very thin insulator the tunneling current from one layer to the other depends on the relative orientation of magnetization of the layers and the potential barrier, which in turn, depends on the insulator thickness and the type of the layer materials. Parallel orientation of the adjacent layer corresponds to lower resistivity. Indeed tunneling rate depends on the junction quality. The tunneling current is also affected by the temperature and voltage.

The magnetocrystalline anisotropy in the particle define a preferred direction along which the magnetization will be oriented at very low temperatures. Equal and opposite directions are energetically degenerate. If there is only one anisotropy axis, then the ground state will be degenerate, and the magnetization will be either parallel or antiparallel to the magnetization axis. However, in the more general case, if there are two or more anisotropy axes, then the ground state will be a superposition of states where the magnetization is oppositely aligned (compare this with the NH_3 molecule). Hence, if the magnetization is initially aligned along the easy axis (for example, by cooling it in a strong magnetic field), then it will oscillate at the tunneling frequency between the spin-up and spin-down state.

19.3.4 Magnetoresistivity

Magnetoresistivity is an important phenomena observed at the nanoscale, which is a change of electrical resistivity when an external magnetic field is applied to the substance. The magnetoresistivity can be varied by the relative orientation of magnetization in neighboring ferromagnetic layers separated by a nonmagnetic metallic thin layer. The prototype system is layers of Co separated by spacer layers of Cu. The in-plane resistance of this system depends sensitively on the magnetic field applied to the layers. The origin of the effect is a strong exchange coupling between the ferromagnetic layers. The magnetoresistivity also depends on the direction of the current with respect to the film normal. The parameters affecting the magnetoresistivity for multilayer films are spin-scattering parameters for outer surfaces and interfaces, effective masses, inner potentials, and relaxation times for magnetic and nonmagnetic films. All of these parameters are strongly influenced by imperfections of the samples. This property is ideal for a magnetic sensor as it directly relates the applied magnetic field and the electrical resistivity. The resistance in normal metals is increased in the presence of an external magnetic field and depends on the relative orientation of the current with respect to the field. However, the resistivity decreases with the applied field for magnetic–nonmagnetic composite systems when the size of the nonmagnetic metallic component separating magnetic components becomes smaller than the mean free path of charge carriers. This is called GMR and can be much larger than AMR.

19.3.5 Magnetic Moments of 3d Transition Metal Clusters

An overall decreasing trend of the magnetic moment with increasing cluster size is observed in all three cases, with some oscillating fine structure features. The

Fig. 19.11 *Magnetic moment per atom in units of Bohr magneton for iron, cobalt, nickel, and rhodium clusters as a function of cluster size.*

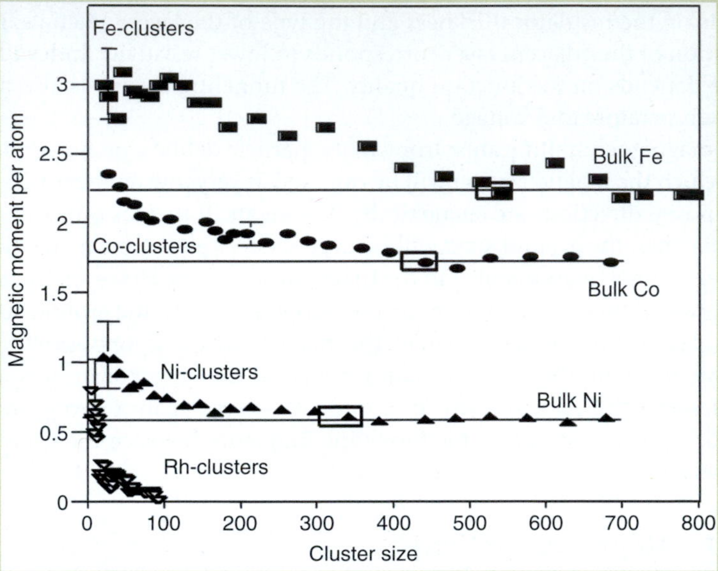

magnetic moment of an atom at the surface is generally larger than in the bulk for small sizes where the magnetic moments per atom for Fe, Co, and Ni are approximately $3\mu_B$, $2\mu_B$, and $1\mu_B$ respectively (Co is slightly higher, probably due to orbital effects.) These values correspond to the maximum spin that can be obtained with 7, 8, and 9 electrons in a d orbital. For example a nearly free iron atom has 8 valence electrons of which 1 is in the 4s orbital and 7 in the 3d orbital. Of those 7 and 5 are in spin-up states (forming the majority spin band) and 2 in spin-down states (the minority spin band). Hence the net magnetic moment per atom due to the 3d orbitals is $3\,\mu_B$. In fact for Fe, $\mu_{bulk} = 2.2\,\mu_B$; for Co, $\mu_{bulk} = 1.6\,\mu_B$, for Ni, $\mu_{bulk} = 0.6\,\mu_B$. Clusters with as few as 600 atoms already appear to have bulk-like magnetic moments (**Fig. 19.11**). It is interesting to note that measurements of Cr clusters ($N > 10$) have demonstrated that these clusters do not deflect. This indicates that they are nonmagnetic or more likely that they are antiferromagnetic.

19.3.6 The Temperature Dependence of Magnetic Moments

Loss of ferromagnetic order occurs at the Curie temperature where thermal motion overcomes the order imposed by the interatomic exchange interaction. In itinerant magnetism there are two distinct pictures: in the band picture, the magnetic moment reduction is caused by thermally induced electronic excitations from the top of the majority spin band to the Fermi level of the minority spin band, which reduces the total moment. In the localized moment picture, the global moment is reduced through local misalignments. The molecular beam method favors measurements of magnetic moments as a function of temperature over a wide range of temperature, ranging from 80 to 1000 K. In this way the ferromagnetic to paramagnetic phase transition can be probed. The magnetic moments have been measured as a function of temperature for several sizes and it is clear that they decrease with increasing temperature. It is also observed that for the magnetic moments of Ni and Co clusters the magnetization curves appear to converge to their respective bulk behaviors. Fe is anomalous and no obvious trend can be discerned. Small clusters usually prefer icosahedral structures.

19.4 Recent Progress in Nanoscale Sample Preparation

19.4.1 Epitaxial Methods

Epitaxial layer-by-layer growths in ultrahigh vacuum systems are the most useful and common methods of preparation of nanosized samples, since lattice symmetry and size of the crystal strongly influence the magnetic properties. However, due to their relatively higher surface free energy, the magnetic materials are difficult to grow on any substrate. Also any particular ferromagnetic material can be deposited in more than one crystal structure. Moreover some artificial solids are possible to grow, exhibiting new physical properties.

The most critical parameters in magnetic multilayer growth are lattice and relative surface free energies mismatch between the substrate and overlayer film. The interdiffusion, chemical reactions, and alloying between the substrate and overlayer atoms give additional problems. For epitaxial multilayer film preparations, noble metals, semiconductors, and some oxides are used as single-crystal substrates. Many tricks have been developed to overcome problems such as low temperature and high growth rate. Each molecule has to stick at a nearest place with an energy minimum of the previous layer, and thus, growth symmetry follows the substrate structure. Electrochemical deposition techniques and self-organization are other useful and promising methods.

Magnetic materials are deposited on cylindrical pores to get nanowires perpendicular to the surface of the substrate. The beam of magnetic materials can be focused by interference field of an intense laser onto the substrate to form parallel wires.

19.5 Nanomagnetism Applications

19.5.1 Overview

Nanosized magnetic iron oxide particles have been studied extensively due to their wide range of applications in ferrofluids, high-density information storage, magnetic resonance imaging (MRI), biological cell labeling, sorting and separation of biochemicals, targeting, and drug delivery [3]. For many of these applications surface modification of nanosized magnetic particles were accomplished by physical/chemical adsorption or surface coating of desired molecules, depending on the specific applications. A silica coating on the surface of nanosized iron oxide particles, for example, have also been shown to prevent their aggregation in liquids and improve their chemical stability.

For example, poly(1-vinylimidazole) polymer-grafted nanosized magnetic particles have been used in applications as magnetic carriers in a wide range of disciplines. Poly(1-vinylimidazole) is chosen to graft on nanosized magnetic particles, as the resultant organic–inorganic hybrid magnetic materials are anticipated to expand the sorbent-based separation technology to a multiphase complex system, ranging from biological cell sorting to industrial effluent detoxification and recovery of valuables. Metal ion binding properties of imidazole and poly(1-vinylimidazole) have been reported by many researchers for the removal of various metal ions from aqueous solutions.

MRI detectable and targeted quantum dots have also been developed. Quantum dots were coated with paramagnetic and pegylated lipids. The quantum dots are usually functionalized by covalently linking RGD peptides. With recent developments in chemistry and the synthesis of powerful, innovative, specific, and multimodal contrast agents, for example, by introducing fluorescent properties as well, MRI is becoming increasingly important for molecular imaging. Quantum dots, semiconductor nanocrystals in a size range of 2–6 nm, have gained much interest in the past few years for biological imaging purposes, especially because of their bright fluorescence, their photostability, and their narrow and tunable emission spectrum. The in vivo use requires the quantum dots to be water soluble and biocompatible. Efforts have been undertaken to achieve these goals, and quantum dots have been used successfully for imaging studies of live cells and animal models, mainly in combination with two-photon fluorescence microscopy.

The semiconductor currently used in integrated circuits, transistors, and lasers such as silicon and gallium arsenide (GaAs) are nonmagnetic. The carrier is almost independent of the spin direction. The miniaturization becomes more difficult in nanostructures. Magnetic semiconductors, alternative semiconductors which have both properties of magnetic materials and semiconductors, exchange interactions that can give rise to pronounced spin-related phenomena not just in nanostructures but also in more conventionally sized devices.

Semiconductor spin electronics can be divided into two fields such as semiconductor magneto-electronics and semiconductor quantum spin electronics. Semiconductor magneto-electronics are used to implement new functions by using semiconductor materials that are also magnetic or combinations of semiconductors and magnetic materials. For example, semiconductors such as optical isolators, magnetic sensors, and nonvolatile memory are possible to be developed if the magnetism can be controlled by light and electric field.

The other field that is referred as semiconductor quantum spin electronics is mainly focused on using the quantum mechanical nature of spin in semiconductors. For example, since the various types of spins in nonmagnetic semiconductors

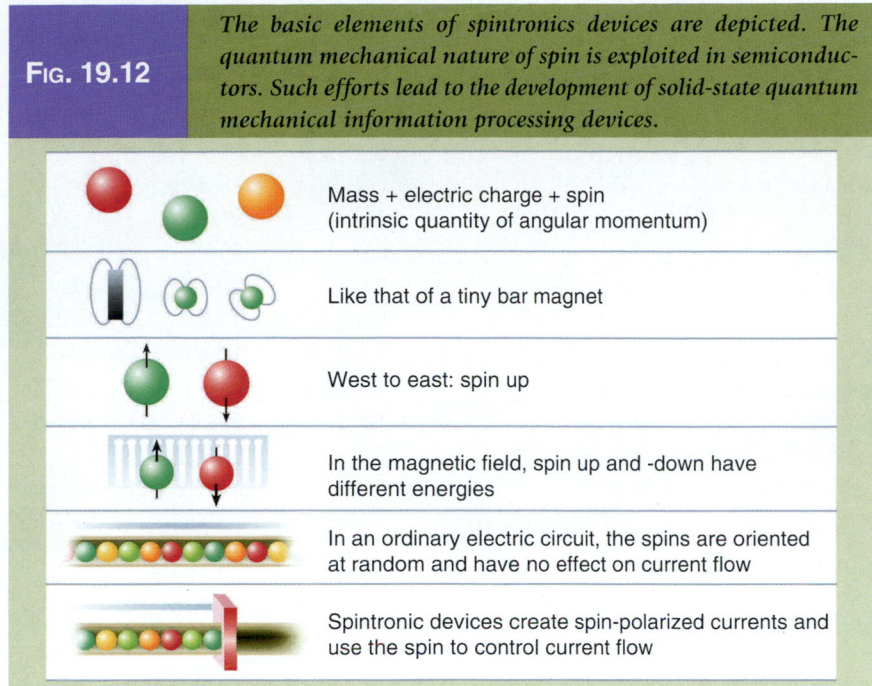

FIG. 19.12 *The basic elements of spintronics devices are depicted. The quantum mechanical nature of spin is exploited in semiconductors. Such efforts lead to the development of solid-state quantum mechanical information processing devices.*

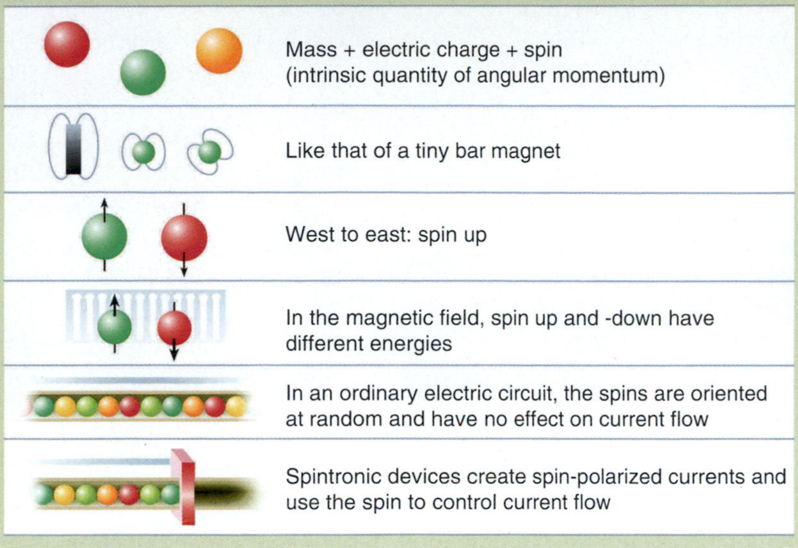
Mass + electric charge + spin
(intrinsic quantity of angular momentum)
Like that of a tiny bar magnet
West to east: spin up
In the magnetic field, spin up and -down have different energies
In an ordinary electric circuit, the spins are oriented at random and have no effect on current flow
Spintronic devices create spin-polarized currents and use the spin to control current flow

have a much longer coherence time than electrical polarization and can be controlled by light or electric fields, it is easy to manipulate spin as a quantum mechanical entity. Such properties lend themselves to the development of solid-state quantum information processing devices; in this way, spin in semiconductors is heralding a new era both in classical and quantum physics and technology (**Fig. 19.12**).

The spin of the electron was ignored in mainstream charge-based electronics. A technology has emerged called spintronics (spin transport electronics or spin-based electronics) [4]. It is not the electron charge but the electron spin that carries information. Spintronics offers opportunities for a new generation of devices. This technology combined standard microelectronics with spin-dependent effects that arise from the interaction between spin of the carrier and the magnetic properties of the material. In an ordinary electric circuit the spins are oriented at random and have no effect on current flow. Spintronic devices create spin-polarized currents and use the spin to control current flow. Devices that rely on an electron's spin to perform their functions form the foundations of *spintronics*.

The advantages of these alternative techniques would be nonvolatility, increased data processing speed, decreased electric power consumption, and increased integration densities compared with current semiconductor devices. The field of spintronics is addressed by the experiment and theory of the optimization of electron spin, the detection of spin coherence in nanoscale structures, the transport of spin-polarized carriers across relevant length scales and

Table 19.3 *Comparison of Memory Technologies for the Year 2011 [5]*

	CMOS			
Technology	**DRAM**	**Flash**	**SRAM**	**MRAM**
Reference	SIA 1999	SIA 1999	SIA 1999	
Generation at introduction	64 Gb	64 Gb	180 Mb/cm^2	64 Gb
Circuit speed (MHz)	150	150	913	>500
Feature size (nm)	50	50	35	<50
Access time (ns)	10	10	1.1	<2
Write time	10 ns	10 μs	1.1 ns	<10 ns
Erase time	<1 ns	10 μs	1.1 ns	N/A
Retention time	2–4 s	10 years	N/A	Infinite
Endurance cycles	Infinite	10^5	Infinite	Infinite
Operating voltage (V)	0.5–0.6	5	0.5–0.6	<1
Voltage to switch state	0.2 V	5 V	0.5–0.6 V	<50 mV
Cell size	2.5 F^2[a]/bit 0.0005 μm^2	2F^2/bit	12F^2/bit	2F^2/bit

[a] F = minimal lithographic feature size.

heterointerfaces, and the manipulation of both electron and nuclear spins on sufficiently fast timescales.

Merging of electronics, photonics, and magnetics will ultimately lead to new spin-based multifunctional devices such as spin–FET (field effect transistor), spin–LED (light-emitting diode), spin RTD (resonant tunneling device), optical switches operating at very high frequency in the terahertz range, modulators, encoders, decoders, and quantum bits for quantum computation and communication. A summary of the future memory technologies involving nanomagnetism is shown in **Table 19.3**.

19.5.2 Current Status of Spin-Based Electronics Devices

There are a number of effects that couple magnetization to electrical resistance. These include

- Ordinary magnetoresistance (OMR)
- Anisotropic magnetoresistance (AMR)
- Giant magnetoresistance (GMR)
- Tunneling magnetoresistance (TMR)
- Ballistic magnetoresistance (BMR)
- Colossal magnetoresistance (CMR)

The GMR is perhaps the major innovation in spin-based electronics. GMR is observed in very thin-film materials constructed of alternate ferromagnetic and nonmagnetic layers. The resistance of the material is lowest when the magnetic domains in ferromagnetic layers are aligned and highest when they are anti-aligned (**Fig. 19.13**).

A spin valve (**Fig. 19.14a**) is a GMR-based device that comprises of two ferromagnetic layers (e.g., alloys of nickel, iron, and cobalt) isolated within a

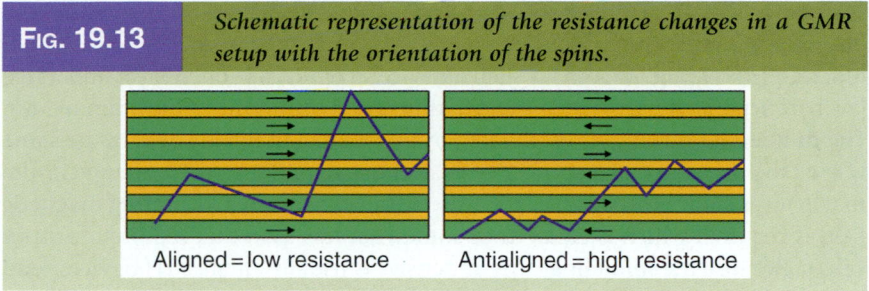

FIG. 19.13 *Schematic representation of the resistance changes in a GMR setup with the orientation of the spins.*

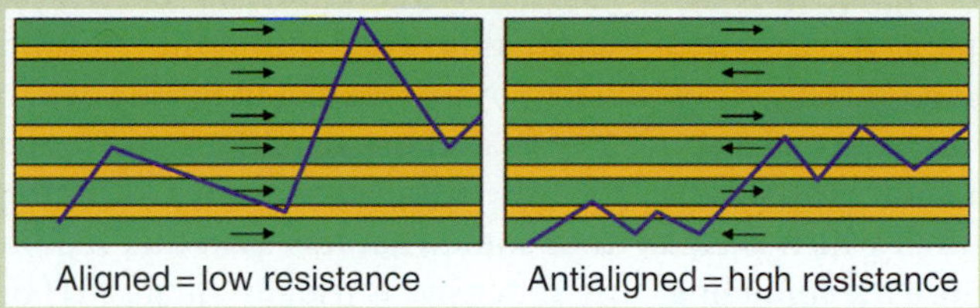

thin nonmagnetic metal (usually copper), with one of the two magnetic layers being aligned (called *pinned*) whereby the magnetization in the pinned layer is relatively insensitive to any external magnetic fields. The other magnetic layer is called the *free* layer, and its magnetization can be changed by the application of a relatively small magnetic field. As the magnetizations in the two layers change from a parallel to a antiparallel alignment, the resistance of the spin valve increases typically from 5% to 10%. Antiferromagnetic layer (called "*pinning*" layer) is usually constructed by using an antiferromagnetic material that is grown on top of the pinned layer so that they can exchange their spins. The two films form an interface that acts to resist changes to the pinned magnetic layer's magnetization.

In alternative structures, the pinned layer was replaced with a synthetic antiferromagnet: Two magnetic layers separated by a very thin (~10 Å) nonmagnetic conductor (usually ruthenium metal is used). The magnetizations in the two magnetic layers are strongly antiparallely coupled and are thus effectively immune to outside magnetic fields. This structure improves both stand-off magnetic fields and the temperature of operation of the spin valve. Another way of making these structures is by growing a nano-oxide layer (NOL) at the outside surface of the soft magnetic film. This layer reduces resistance due to surface

FIG. 19.14 *Spin-dependent transport structures. (a) Spin valve and (b) magnetic tunnel junction.*

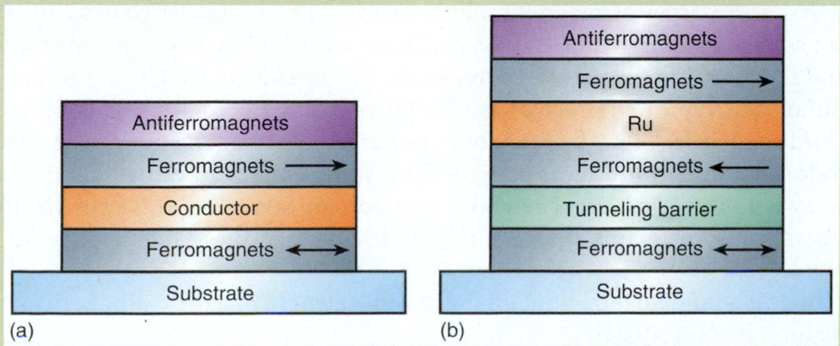

scattering, thus reducing background resistance and thereby increasing the percentage change in magnetoresistance of the structure. A magnetic tunnel junction (MTJ) (**Fig. 19.14b**) is such a device in which a pinned layer and a free layer are separated by a very thin insulating layer (commonly aluminum oxide). The tunneling resistance is modulated by an external magnetic field in the same way as the resistance of a spin valve is, and usually 20% to 40% change in the magnetoresistance can be observed. Also for its operation a saturating magnetic field is required which is equal to or somewhat less than that required for spin valves. Because the tunneling current density is usually small, MTJ devices tend to have very high resistances.

Applications of Spin-Based Electronics Devices. The GMR-based galvanic isolator which is a combination of an integrated coil and a GMR sensor on an integrated circuit chip is perhaps a good example of the current application. GMR isolators introduced in 2000 eliminate ground noise in communications between electronics blocks, thus performing a function similar to that of optoisolators providing electrical isolation of grounds between electronic circuits. The GMR isolator is ideally suited for integration with other communications circuits and the packaging of a large number of isolation channels on a single chip.

Magnetic Data Storage. MRAM (magnetic random access memory) that uses magnetic hysteresis to store data and magnetoresistance to read data is another application of the GMR effect. GMR-based MTJ or pseudospin valve memory cells are integrated on an integrated circuit chip and function like a static semiconductor RAM chip with the added feature that data are retained with the power off. Potential advantages of the MRAM compared with silicon electrically erasable programmable read-only memory (EEPROM) and flash memory are 1000 times faster write times, no wear out with write cycling (EEPROM and flash wear out with about 1 million write cycles), and lower energy for writing. MRAM data access times are about 1/10,000 that of hard disk drives.

Magnetic nonstructures are starting to play a role in technology, particularly in "non-volatile magnetic data storage." New combinations with semiconductor technology are developing such as MRAMs where the storage capacitor of a traditional semiconductor in replaced by a nonvolatile magnetic dot. The storage media to today's magnetic hard disk drives may be viewed as an array of magnetic nanoparticles. The magnetic coating of the disk consists of a ternary Co–Pt–Cr mixture, which segregates into magnetic Co–Pt grains. These grains are magnetically separated by Cr at the grain boundaries. Typical grain sizes are 10–20 nm using about 10^3 grains/bit at a recording density of 1 Gbit/in.^{2} for commercial devices. The grains segregate randomly, which introduces statistical noise into the read out signal due to the variation in grain size, coercivity, and domain structure. This explains the large number of grains that are required to reduce these fluctuations in a device.

There is a fundamental limit on the improvement of magnetic storage density. The limit is controlled by

1. Thermal flipping of the bits as the grains become smaller.
2. Energy barrier between the two stable magnetizations along the easy axis becomes comparable with kT. Eventually the superparamagnetic limit is reached where individual grains stay magnetized, but their orientation fluctuates thermally.

For typical magnetic storage media, the superparamagnetic limit imposes a minimum particle size of about 10 nm that is a maximum recording density of several terabits per square inch. This is almost four orders of magnitude higher than the density of 1 Gbit/in.^{2} found in top-of-the-line disc drives today. While the current particle size is already close to the superparamagnetic limit the number of particle/bits is still more than 10^3. There are many signal-to-noise issues on the way towards reducing this number and reaching the theoretical limit. Inconsistent switching of different particles and an irregular domain structure require averaging over many particles.

Controlling coercivity, size, orientation, and position of magnetic nanoparticles will be essential for reducing the number of particles needed to store a bit. For example, a large crystalline anisotropy can produce a higher switching barrier than the shape anisotropy above. Single-domain nanoparticles with high saturation magnetization and coercivity are being optimized for this purpose. The orientation of segregated grains can be controlled using multilayered structures where the first layer acts as a seed for small grains and subsequent layers shape the crystalline orientation for the desired anisotropy. A further improvement would be the move from longitudinal to perpendicular recording where the demagnetizing field does not destabilize the written domains.

The ultimate goal in magnetic storage is single-particle-per-bit on quantized recording. It is aimed at producing single-domain particles close to the superparamagnetic limit with uniform switching properties. Lithography is currently the method of choice for producing regular arrays of uniform magnetic dots. Dot arrays with a density of 65–250 Gbit/in.^{2} have been produced by electron beam lithography. The performance and parameters for various types of volatile and nonvolatile memory chip alternatives are shown in **Table 19.4**.

19.5.3 Sensors

Nanostructured material has also been used in the development of some reading head devices. The traditional inductive pick up of the magnetic signal is replaced by a magnetoresistive sensor in state-of-the-art devices.

Permalloy/(Cu/Co) multilayers are utilized in reading heads. Currently, reading heads in high-end disc drives are based on the 2% AMR of permalloy. The resistance is highest for the current parallel to the magnetization and lowest perpendicular to it, producing a sinusoidal orientational dependence. The magnetic stray field between adjacent bits with opposite orientation rotates the magnetization in the permalloy film with respect to the current and thus induces a resistance change. That is directly connectible into a read-out voltage.

Present activities with GMR are directed towards lowering the switching field while keeping a large magneto-resistance. To obtain the best of both characteristics one obtains soft permalloy layer for easy switching with a high-spin co-layer that enhances the magnetoresistance.

Table 19.4 *Comparison of Performance and Parameters for Various Types of Volatile and Nonvolatile Memory Chip Alternatives [6]*

Parameter	DRAM	SRAM	NOR flash	NAND flash	FeRAM	MRAM
Read cycles	$>10^{15}$	$>10^{15}$	$>10^{15}$	$>10^{15}$ before cycling	10^{12}–10^{15}	$>10^{15}$
Write cycles	$>10^{15}$	$>10^{15}$	10^4–10^5	10^6	10^{12}–10^{15}	$>10^{15}$
Write voltage (V)	2.5–5	3.3–5	10–10	18	0.8–5	0.8–5
Cell write time (ns)	10–100	1–50	6×10^3	2×10^5	10–50	10
Write energy (pJ)	Few 10^{-2}		9000	1	1	10–100
Random access time (ns)	40–70	6–70	150	~10,000	40–70	40–70
Cell size (F^2)	8	~100	12	4.6	9–13	6–10
Retention (years)	None	None	10	10	10	10
Scaling issues	Charge		Tunnel oxide → read current → access time	Erase voltage tunnel oxide scaling.	3D + material texture	Switching field increases with scaling and uniformity
Status/forecast	256 Mb/1 Gb	4–16 Mb	32 Mb/128 Mb	256 Mb/1 Gb	1 Mb/4 Mb	Few kb/1 Mb (?)
Applications	PC memory	Cache memory	Program code and data	Data files (camera, MP3)	Contactless smartcard	Envisaged: embedded (SOC) and mass storage

Another type of a nonvolatile magnetic storage device avoids moving parts altogether at the expense of having to pattern the storage medium. This is a combination of magnetic memory elements with semiconductor circuits that sense and amplify the magnetic state (MRAM).

Further into the future are logic devices based on magnetic nanostructures. A bipolar spin switch has been demonstrated that acts like a transistor.

19.5.4 Nanomagnetism for Biomedical Applications

The use of magnetic nanoparticles for biological and medical applications has been developed recently. The size of the particles can range from a few nanometers to several micrometers and thus is compatible with biological entities ranging from proteins (a few nm) to cells and bacteria (several μm).The combination of biology and magnetism is useful, because the biochemistry enables a selective binding of the particles, while the magnetism enables easy manipulation and detection. Using magnetic field gradients, the magnetic particles can be subjected to significant forces even when they are embedded in a biological environment. The absence of ferromagnetism in most biological systems, which typically have only weak dia- or paramagnetism, means that the magnetic moment from the ferromagnetic particles can be detected with little noise in a biological environment.

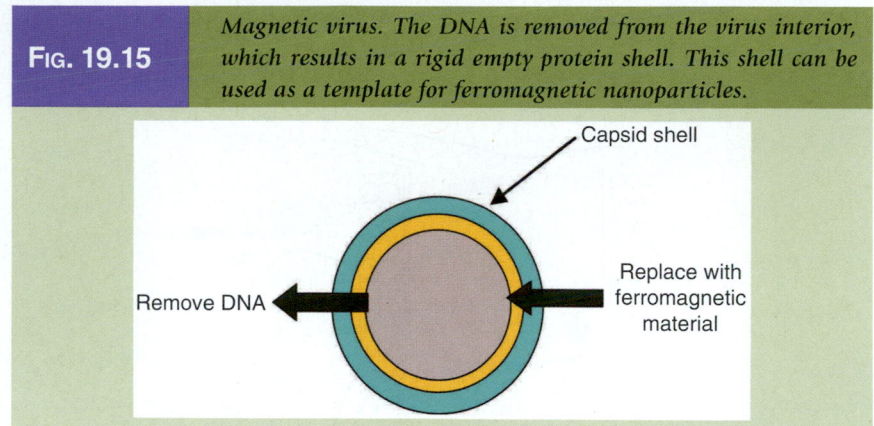

Fig. 19.15 *Magnetic virus. The DNA is removed from the virus interior, which results in a rigid empty protein shell. This shell can be used as a template for ferromagnetic nanoparticles.*

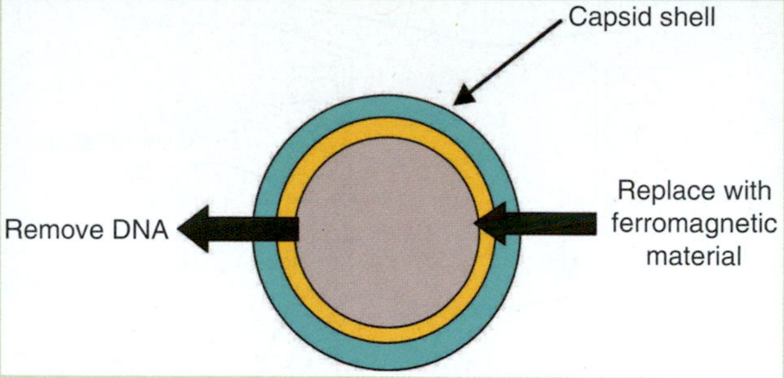

Based on these ideas, a variety of nanomagnetism applications in the biomedical field have emerged. A simple application is to bind magnetic particles to the interested biological systems, which then allows manipulating the biological material via the magnetic field gradient. This has been already applied to several problems, such as separating red blood cells from blood, cancer cells for bone marrow, drug delivery, and hypothermal treatment.

Applications in biology and medicine normally require that the particles are stable in aqueous solution at neutral pH values. For this reason, magnetite (Fe_3O_4) and maghemite (γ-Fe_2O_3) are most commonly utilized for biofunctionalized magnetic nanoparticles.

The ability to tag biological molecules with functionalized magnetic particles has been already exploited for biological sensors. Traditionally many biological sensors, like DNA microarray, use fluorescent markers for the detection of specific biological molecules. For example, the DNA can be bound to a fluorescent molecule, then DNA is exposed to an array with many different well-defined DNA strands and it will only stick to complementary matching DNA. The position of the light signal from the fluorescent marker then indicates which are the right DNA strands. Similarly it has been demonstrated that magnetic particles can be used as tags and the binding can be identified by detecting the stray magnetic field of the particles (**Fig. 19.15**). Key advantages of using magnetic particles versus fluorescent molecules have been argued to be the following:

- Magnetic particles typically have an unlimited shelf life compared to fluorescent markers which deteriorate with time (note: quantum dots circumvent this problem).
- The use of magnetic particles together with magnetoelectronic sensors allows for a complete electronic readout of the sensors.
- The sensitivity of the sensors based on magnetic versus fluorescent tagging are comparable.
- Magnetic tags allow for manipulation of the target molecules such that they can be moved towards the magnetic field sensor using magnetic field gradient

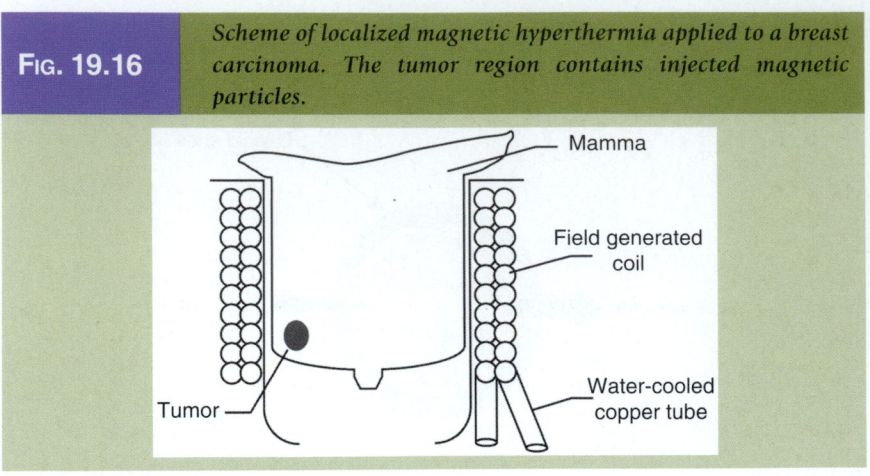

FIG. 19.16 *Scheme of localized magnetic hyperthermia applied to a breast carcinoma. The tumor region contains injected magnetic particles.*

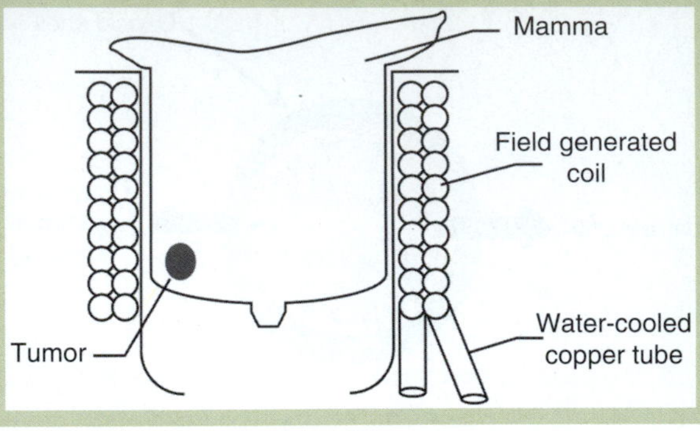

Hyperthermia. There are several reports regarding the efficacy of hyperthermia for the treatment of musculoskeletal tumors [7–10]. Currently, the principal methods of hyperthermia for clinical use are microwaves or radiofrequency waves to generate heat in the tumor (**Fig. 19.16**). Using these methods, however, it is difficult to heat deeply seated tumors effectively and selectively [11]. New types of ferromagnetic thermoseeds are being developed experimentally to solve this problem, but they have never been used clinically because of their unreliable rise in temperature. Application of magnetic materials for hyperthermia of biological tissue with the goal of tumour therapy is known in principle for more than four decades. Much empirical work was done in order to manifest a therapeutic effect on several types of tumors by performing experiments with animals or using cancerous cell cultures. However, routine medical applications are not known till now and there is a demand for a more profound understanding of the related material properties to render that method reliably for tumour therapy of human beings. The heating of oxide magnetic materials with low electrical conductivity in an external alternating magnetic field is due to loss processes during the reorientation of the magnetization. If thermal energy $k_B T$ is too low to facilitate reorientation, hysteresis losses dominate which depend on the type of the remagnetization process (wall displacement or several types of rotational processes). With decreasing particle size, thermal activation of reorientation processes leads to a dependence on temperature and measurement frequency to superparamagnetic behavior of the particle ensemble and the occurrence of the so-called Néel losses [12]. In the case of ferrofluids, losses related to the rotational Brownian motion of magnetic particles may arise too, and hence care should be taken to analyze the results, depending on the temperature and measurement frequency.

Ocular. The anterior segment of the eye is bounded by the cornea and the lens–iris diaphragm, and contains the aqueous humor. The posterior segment begins behind the lens–iris diaphragm and includes, from inside outwards the vitreous (gel/fluid), retina (neurosensory tissue), and choroid (heavily muscular).

FIG. 19.17 *Internal tamponade using silicone magnetic fluids—magnetic fluid for use in eye surgery [13].*

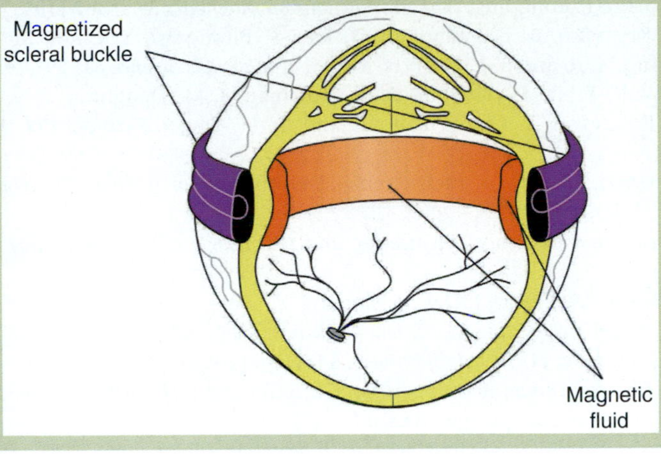

The retinal photoreceptors are supported by the choroid (**Fig. 19.17**). The retina and the choroid stay attached to each other with a suction pump, which keeps the subretinal space dry.

Retinal detachment is a major cause of vision loss in adults. It occurs when the retina separates from the choroid, resulting in eventual death of the retina and subsequent loss of vision. As one ages, the vitreous gel normally undergoes liquefaction, collapses, and separates from the retina. Separation of the vitreous gel may result in the formation of a tear in the retina at a site of vitreoretinal adhesion. The tear provides a pathway for the vitreous fluid to pass through and underneath the retina, overcoming the suction attraction of the retina to the choroid, thus detaching the retina from the underlying choroid.

The goal of surgery is to close the holes in the retina, preventing further fluid flow into the subretinal space, allowing for reattachment of the retina. Efforts are going on to develop an internal tamponade from modified silicone fluid containing spherically stabilized 4–10 nm sized metal particles, which could be held in place with an external magnetized scleral buckle. With an appropriate magnetic fluid inside the vitreous cavity, a stable, 360° internal tamponade might be achieved. The enriching magnetized scleral buckle and magnetic fluid would produce a ring of silicone oil in opposition to the retinal periphery. The central vitreous cavity would be free of the magnetic fluid and the lens, anterior structures, or macula, thus avoiding the complications of currently available treatment modalities.

There are still other applications on magnetic nanoparticles, such as; drug delivery, hyperthermia treatment. Key challenges for nanoparticles for biological application include the modification of nanoparticles for enhanced aqueous solubility, biocompatible or bio-recognition, optimization of their magnetic properties. Looking further ahead into the future there are still many possibilities for the magnetic nanoparticles in biomedical field. Many possible applications of magnetic nanoparticles in biology and medicine clearly offer plenty of opportunities for future research and development.

References

1. Z. Jibin and L. Yongping, *IEEE Transactions on Magnetics* 28, 3367 (1992).
2. K. M. Krishnan, A. B. Pakhomov, Y. Bao, P. Blomqvist, Y. Chun, M. Gonzales, K. Griffin, X. Ji, and B. K. Roberts, *Journal of Materials Science* 41, 793–815 (2006).
3. S. A. Wolf, D. D. Awschalom, R. A. Buhrman, J. M. Daughton, S. von Molnar, M. L. Roukes, A. Y. Chtchelkanova, and D. M. Treger, *Science*, 294, 1488–1494 (2001).
4. J. F. Gregg, I. Petej, E. Jouguelet, and C. Dennis, *Journal of Physics D: Applied Physics*, 35, R121–R155 (2002).
5. *Handbook of nanoscience, engineering and technology*, CRC Press, Boca Raton, FL (2003).
6. J. De Boeck, W. V. Roy, J. Das, V. Motsnyi, Z. Liu, L. Lagae, H. Boeve, K. Dessein, and G. Borghs, *Semiconductor Science and Technology*, 17, 342–354 (2002).
7. R. K. Gilchrist et al., *Annals of Surgery*, 146, 596 (1957).
8. C. Streffer and D. van Beuningen, in: J. Streffer (Ed.), *Hyperthermia and the therapy of malignant tumors*, Springer, Berlin (1987).
9. A. Jordan, R. Scholz, J. SchuKler et al., *International Journal of Hyperthermia*, 13, 83 (1997).
10. P. Burgman, A. Nussenzweig, and G. C. Li, in: M. H. Seegen-Schmiedt, P. Fessenden, and C. C. Vernon (Eds.), *Thermoradiotherapy and thermochemotherapy, Vol. 1: Biology, Physiology, Physics*, Springer, Berlin, 1995.
11. D. C. F. Chan, D. B. Kirpotin, and P. A. Bunn Jr., *Journal of Magnetism and Magnetic Materials*, 122, 374 (1993).
12. L. Néel and C. R. Hebd, *Seances Acad. Sci.*, 228, 664 (1949).
13. J. P. Dailey et al., *Journal of Magnetism and Magnetic Materials*, 194, 140 (1999).

Problems

19.1 What is the origin of magnetic moment in magnetic materials?

19.2 What is a Bohr magneton?

19.3 Which type of magnets have higher coercivity? Soft or hard?

19.4 What are the three different types of material response to an applied electromagnetic field called?

19.5 Hund's rule states that the lowest energy configuration for an atom is the one having the maximum number of unpaired electrons allowed by the Pauli principle in a particular set of degenerate orbitals. Show the spins in the various levels in the table below:

	Configuration	1s	2s	$2p_x$	$2p_y$	$2p_z$
C	$1s^2\ 2s^2\ 2p^2$					
N	$1s^2\ 2s^2\ 2p^3$					
O	$1s^2\ 2s^2\ 2p^4$					

19.6 We've discussed how *M* affects *J*, and the ability to transport charge, as manifested through magnetoresistive effects. One can also consider the converse: can a current *J* of carriers with a net spin polarization affect *M*?

19.7 Consider that the molar volume of magnetite is $4.4 \times 10^{-5}\ m^{-3}$ and magnetic moment of Fe_3O_4 is 9.27×10^{-24}, then calculate the magnetization of magnetite (Fe_3O_4) assuming that the magnetization is only due to the six 3d electrons of the Fe^{2+} ions and that only the spin angular momentum of the electrons contributes to the magnetic moment.

19.8 Determine the values of $\hat{S}$, L, and $\hat{J}$ for Cr^{3+} which has three electrons in the 3d subshell. All lower energy shells are filled.

19.9 Calculate the magnetization of Fe_3O_4 assuming that only the six 3d electrons of the Fe^{2+} ions contribute to the magnetization and that only the spin angular momentum of the electrons contribute to the magnetic moment. Consider the molar volume of magnetite as 4.4×10^{-5} m^{-3}.

19.10 What is giant magnetoresistance? What are magnetic multilayered monolayers?

19.11 How would you fabricate a magnetic nanowire?

19.12 What future applications do you see arising from such giant magnetoresistance?

Section 8

Mechanical Nanoengineering

Nanomechanics

Masood Hasheminiasari and John J. Moore

I am not afraid to consider the final question as to whether, ultimately in the great future we can arrange atoms the way we want; the very atoms, all the way down!... The principles of physics, as far as I can see, do not speak against the possibility of maneuvering things atom by atom. It is not an attempt to violate any laws... but in practice, it has not been done because we are too big... At the atomic level, we have new kind of forces and new kind of possibilities, new kind of effects.

Richard Feynman, 1960

Chapter 20

Fig. 20.0 *Richard P. Feynman (1918–1988), scientist, physicist, and teacher who was a joint recipient of the Nobel Prize in Physics in 1965.*

Source: http://www.feynmangroup.com/company/whos_feynman.cfm, 2008.

Threads

Chapter 20, Nanomechanics, is the first chapter in the *Mechanical Nanoengineering* section of the text. We are now ready to define, or attempt to embrace, the mechanics of nanomaterials. This chapter is about how small size can influence the mechanical behavior of materials and its impact on the bulk properties. There are two chapters in the *Mechanical Nanoengineering* section. In *chapters 21* and *22*, thin films and nanocomposites are discussed. The sections all provide examples of applications of this nanotechology.

Following the mechanical-oriented chapters, we embark on the chemical engineering aspects of nanotechnology—specifically, the domain of the catalyst and the composite.

20.0 Introduction

This chapter is designed to develop an understandable description of what is called "nanomechanics," the mechanical behavior of an object in the nanometer level, that is, objects for which at least one dimension is in the nanometer range. Our main concern is not to focus on the rigid body dynamics or on the quantum mechanics in detail. Instead we will focus on the static deformation of solid objects, both with and without external forces in one, two, and three dimensions. Then, we will discuss about nanomechanical devices and their applications in the nanotechnology area.

In the first part of this chapter, the main goal is to build the foundations required to understand the concept of nanomechanics and provide some examples in order to show how the small size scale can impact and vary the bulk properties. In the second part of the chapter, we will describe some aspects of nano-microelectromechanical systems (NEMS/MEMS) and illustrate other nanomechanical devices.

This chapter begins by discussing a very basic mechanical problem, in which two atoms of a molecule bond together by their mutual interactions. The motion of these atoms is restricted to one dimension. Next we move to three atoms and then expand our previous discussion of a two-atom chain to a more general problem.

20.0.1 Two-Atom Chain Mechanics

Let's consider a molecule consisting of only two atoms. We assume that the atoms can only move horizontally towards or away from one another. We presume that there is an attractive force between the atoms; due to an electrostatic attraction, if the atoms have opposite electric charges, to a covalent bonding or due to an attraction known as the van der Waals force, generated by induced dipole moments in each atom. So, we assume that there is a net force $f(r)$ between these two atoms, which is only a function of distance between two atoms, r. If the force is attractive, then the atoms will accelerate and combine with each other. This does not happen, because when the two atoms get too close to one another, the electron clouds of each atom repel one another through their electrostatic repulsion, and are furthermore limited to the law of quantum mechanics from occupying the same volume of space. Thus, the attractive force becomes repulsive as the atoms approach each other.

It is usually more useful to work with the interaction potential energy $\phi(r)$ rather than force $f(r)$, which is defined by the relation

$$f(r) \equiv -\frac{d\phi}{dr} \tag{20.1}$$

The potential energy is also described as the negative of the work done by the force for a displacement $r - r_0$ from the point of zero potential energy r_0

$$\phi(r) = -W = -\int_{r_0}^{r} f(r)\, dr \tag{20.2}$$

which is equivalent to equation (20.1).

20.0.2 Interaction Potentials

One basic characteristic of atoms and molecules is the electric charge, e. Electric charge is related to the property of particles to exert forces on each other by means of electric field. Pair of particles with electric charges e_1 and e_2 exert repulsive (at $e_1e_2 > 0$) or attractive (at $e_1e_2 < 0$) forces on one another which are described by

$$F_1 = -F_2 = -\nabla_i V_c(r_1, r_2), \quad i = 1, 2 \tag{20.3}$$

where

r_1 and r_2 are radii of the particles

$V_c(r_1, r_2)$ is the electrostatic Coulomb potential

$$V_c(r_1, r_2) = \frac{1}{4\pi\varepsilon_0}\frac{e_1e_2}{r}, \quad r = |r_1 - r_2| \tag{20.4}$$

where $\varepsilon_0 = 8.854188 \times 10^{-12}$ C/Vm is the permittivity constant in vacuum. The Coulomb potential is a long-range interaction and would be noticeable even at large separation distances, because the potential decays slowly with distance. But as atoms get closer the electrostatic field of the positively charged atomic nuclei or ion is neutralized by the negatively charged electron clouds surrounding the nuclei, which is called a short-range interaction potential. In general, the interaction potential can be based on purely theoretical calculations or phenomenological considerations.

Phenomenological interaction potential functions are often a more realistic view of atomic interaction than potentials which are derived exclusively from theoretical calculations. Phenomenological atomic interactions are in most cases based on a simple analytical expression which may or may not be justified from theory and contains one or more parameters adjusted to the experimental results. There are many interaction potentials developed for a two-atom system, pair-wise potentials, of which some are briefly mentioned in the next section.

Lennard–Jones Potentials. The general form of the Lennard–Jones potential is [1]:

$$\phi(r) = \frac{\lambda_n}{r^n} - \frac{\lambda_m}{r^m} \tag{20.5}$$

This potential was developed to treat inert gases, but it is often used to describe metals and other materials. The most common form is called Lennard–Jones (6-12) potential, where $n = 12$ and $m = 6$ and has the following form:

$$\phi(r) = 4\varepsilon\left[\left(\frac{\sigma}{r}\right)^{12} - \left(\frac{\sigma}{r}\right)^{6}\right] \tag{20.6}$$

Due to its simple form, this potential is often used to treat the cross-interaction of two different materials, as indicated in **Table 20.1**. The properties of the noble gases are estimated by 10% accuracy. Since this potential is designed for noble gases, we cannot expect to gain adequate results from the Lennard–Jones potential in metallic systems [2–4].

Buckingham Potentials. The Buckingham potential has both inverse 6th and inverse 8th power dependency, which make this potential complicated [5]. A simpler form of this potential which has eliminated the inverse 8th power functionality is called the modified version of the Buckingham potential and has the form:

TABLE 20.1 *Lennard–Jones Potential Parameters for Different Materials*

Symbol	Mass ($\times 10^{-7}$ kg)	ε ($\times 10^{-21}$ J)	σ ($\times 10^{-10}$ m)
Ne	33.51	0.5315	2.786
Ar	66.34	1.6539	3.405
Kr	139.16	2.2075	3.639
Xe	218.02	3.0497	3.962
Cu	105.52	65.626	2.338
Ag	179.13	55.276	2.644

TABLE 20.2 *Modified Buckingham Potential Parameters for Different Nonbonded Materials*

Symbol	Type	Mass ($\times 10^{-7}$ kg)	r_m ($\times 10^{-10}$ m)	ε ($\times 10^{-21}$ J)	α
C	sp,sp^2	19.925	3.88	0.357	12.5
H	Hydrocarbon	1.674	3.00	0.382	12.5
O	Carbonyl	26.565	3.48	0.536	12.5
N	sp^3	23.251	3.64	0.447	12.5
F	Flouride	31.545	3.30	0.634	12.5
Cl	Chloride	58.064	4.06	1.950	12.5
Br	Bromide	131.038	4.36	2.599	12.5
I	Iodide	210.709	4.64	3.444	12.5
S	Sulfide	53.087	4.22	1.641	12.5
Si	Silane	46.454	4.50	1.137	12.5
P	Phosphine	51.464	4.36	1.365	12.5

$$\phi(r) = \frac{\varepsilon}{1-\frac{6}{\alpha}}\left[\frac{6}{\alpha}e^{\alpha\left(1-\frac{r}{r_m}\right)} - \left(\frac{r}{r_m}\right)^{-6}\right] \tag{20.7}$$

where there are three independent parameters $(\varepsilon, r_m, \alpha)$ with ε as the depth of the energy minimum and r_m as the corresponding value of the distance r between two atoms. The steepness of the exponential is measured by α.

This potential is often used to describe the attractive and repulsive forces experienced by the pairs of uncharged, nonbonded atoms. **Table 20.2** illustrates a set of parameters for different materials [6].

Moreover, we can treat the interaction of atoms from different materials by the introduction of the mean value procedure:

$$\varepsilon_{12} = \frac{\varepsilon_1 + \varepsilon_2}{2}, \quad r_{m12} = \frac{r_{m1} + r_{m2}}{2} \tag{20.8}$$

Barker Potentials. Barker determined potentials for ground-state krypton–krypton and xenon–xenon interactions [7], which are in good agreement with a wide range of experimental data including second virial coefficients, gas transport properties, solid state data, long-range interactions, and measurements of differential scattering cross sections.

While the many-body interactions have been neglected, the third-order triple dipole three-body interactions have been considered. Therefore, Barker potentials can be considered as effective pair potentials

$$\phi(r) = \varepsilon\left[\phi_0(r) + \phi_1(r) + \phi_2(r)\right] \tag{20.9}$$

where

$$\phi_0(r) = e^{\alpha\left(1-\frac{r}{r_m}\right)} \sum_{i=0}^{5} A_i \left(\frac{r}{r_m} - 1\right)^i - \sum_{i=0}^{2} \frac{C_{2i+6}}{\left(\frac{r}{r_m}\right)^{2i+6} + \delta},$$

$$\phi_1(r) = \begin{cases} e^{\beta\left(1-\frac{r}{r_m}\right)}\left[P\left(\frac{r}{r_m}-1\right)^4 + Q\left(\frac{r}{r_m}-1\right)^5\right] & r \geq r_m \\ 0 & r < r_m \end{cases} \quad (20.10)$$

$$\phi_2(r) = \begin{cases} e^{\gamma\left(1-\frac{r}{r_m}\right)^2}\left[R\left(\frac{r}{r_m}-1\right)^2 + S\left(\frac{r}{r_m}-1\right)^3\right] & r \geq r_m \\ 0 & r < r_m \end{cases}$$

Again, ε is the depth of the potential at its minimum, where the value of the interatomic distance is $r = r_m$. Parameters for krypton and xenon are listed in **Table 20.3**.

In the next figure, **Figure 20.1**, the Barker potential for krypton is compared with the corresponding Lennard–Jones and Buckingham potentials. The Lennard–Jones potential matches nearly exactly with the Buckingham potential, but the Barker potential shows a deeper minimum [8].

In conclusion, of all the pair potentials reviewed in this section, the *Barker* potentials are the most accurate with the broadest range of validity. Especially the independence of temperature predestines these functions as model potentials within molecular dynamics (MD) calculations for nanosystems. There are many

Table 20.3 *Barker Potential Parameters for Krypton and Xenon*

Parameters	Krypton	Xenon
ε	2.787	3.898
r_m	4.0067	4.3623
α	12.5	12.5
A_0	0.23526	0.2402
A_1	-4.78686	-4.8169
A_2	-9.2	-10.9
A_3	-8	-25
A_4	-30	-50.7
A_5	-205.8	-200
C_6	1.0632	1.0544
C_8	0.1701	0.1660
C_{10}	0.0143	0.0323
δ	0.01	0.01
β	12.5	12.5
P	-9	59.3
Q	68.67	71.1
γ	0	-50
R	0	2.08
S	0	-6.24

FIG. 20.1 *A comparison between the Barker, modified Buckingham, and Lennard–Jones potentials for krypton.*

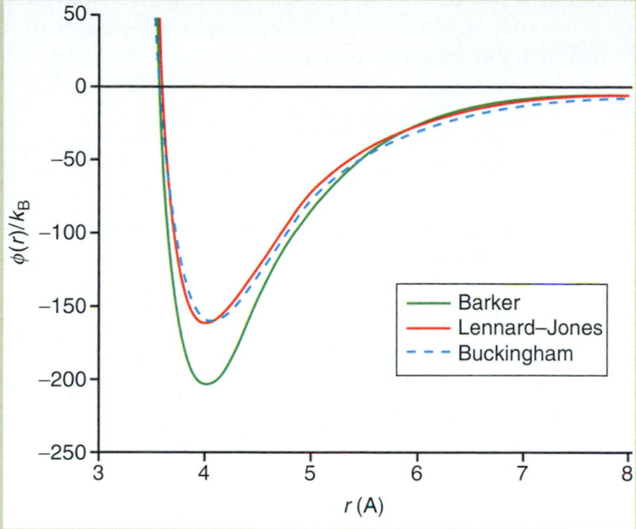

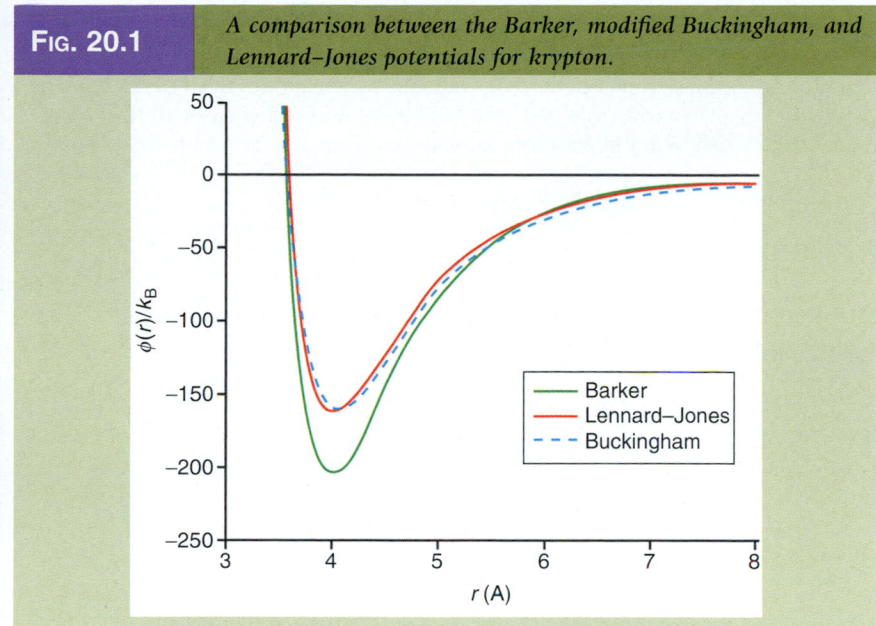

Source: M. Reith, World Scientific (2003).

other kinds of pair-wise interaction potentials such as the Morse potential, the Schommers potential, and many more potentials that are beyond the scope of this section. Among the different many-body potentials, the embedded atom potential (EAM) is one of the most used. Now, we briefly mention the EAM, which is usually applied to metallic systems.

Embedded Atom Potential. This multibody potential is specifically designed to treat metallic systems. One appealing aspect of the EAM potential is its physical picture of metallic bonding, where each atom is embedded in a host electron gas created by all neighboring atoms [9]. The atom–host interaction is more complicated than the simple pair-wise models. This interaction is described in terms of an empirical embedding energy function. The embedding energy function incorporates some important many-atom effects by providing the amount of energy required to insert an atom into the electron gas of a given density. The total potential energy, ϕ, includes the embedding energies, G, of all the atoms in the system, and an electrostatic Coulomb interaction V_c:

$$\phi = \sum_i G_i\left(\sum_{i \neq j} \rho_j^a(r_{ij})\right) + \sum_{i,j>i} V_c(r_{ij}) \tag{20.11}$$

Here, ρ_j^a is the averaged electron density for a host atom j, viewed as a function of the distance between this atom and the embedded atom i. Therefore, the host electron density is employed as a linear superposition which is assumed to be spherically symmetric. More information about the shape of the functions G, ρ, and V_c can be gathered from Clementi and Roetti [10] and Foiles et al. [11]. For a comprehensive review of the EAM, readers are referred to Daw et al. [12].

EXAMPLE 20.1 *Lennard–Jones Potential for Two Argon Atoms in an Ar_2 Molecule*

The Lennard–Jones potential is one of the easiest pair potentials available, which applies to atoms interacting through the van der Waals interaction. Therefore, this pair potential gives an excellent description of the interactions between inert gas atoms, such as argon, krypton, and xenon. A simple form of Lennard–Jones (6-12) potential is assumed:

$$\phi(r) = -\frac{A}{r^6} + \frac{B}{r^{12}}$$

with the parameters A determining the strength of the attractive interaction, and B the repulsive interaction. The potential energy has a minimum at equilibrium spacing ($r = r_0$), so we have:

$$\left.\frac{d\phi}{dr}\right|_{r=r_0} = 0$$

$$\therefore\ 6\frac{A}{r_0^7} - 12\frac{B}{r_0^{13}} = 0 \rightarrow (r_0)^6 = \frac{2B}{A}$$

$$r_0 = \left(\frac{2B}{A}\right)^{\frac{1}{6}}$$

The binding energy E_b, the difference between the potential energy minimum and that when the two atoms are infinitely far apart, is given by:

$$E_b = \phi(\infty) - \phi(r_0)$$

$$\phi(r_0) = -\frac{A^2}{2B} + \frac{A^2}{4B} = -\frac{A^2}{4B}$$

$$\therefore\ E_b = \frac{A^2}{4B}$$

In argon, the equilibrium spacing is found to be r_0 = 0.38 nm, and the binding energy is E_b = 10.4 meV = 1.7 × 10^{-21} J [13]. So, we can calculate the constants to be

$$A = 2r_0^6 E_b = 63\ \text{Å}^6\ eV$$

and

$$B = r_0^{12} E_b = 9.4 \times 10^4\ \text{Å}^{12}\ eV$$

So the final form of pair-wise potential can be written as

$$\phi(r) = \frac{9.4 \times 10^4}{r^{12}} - \frac{63}{r^6}\,(eV)$$

The binding energy of an argon molecule is less than the thermal energy at ambient temperature, k_BT = 26 meV. So, solid argon forms only at quite low temperatures, below 100 K. But, the binding energy for the much stronger ionic, metallic, and covalent interactions in typical solids is in the range of several tens of electron volts, rather than a few meV. Therefore, these types of bonds cannot simply be modeled with Lennard–Jones potentials.

20.0.3 External Forces

We assume that equal and opposite external forces f_{ext} are applied to each atom. The atoms will move apart until they reach a new equilibrium point. The potential energy associated with the external force is given by $\phi_{ext}(r) = -f_{ext}r$. So, the total potential energy is then $U_{tot} = \phi(r) + \phi_{ext}(r)$. For small f_{ext} the minimum for the total potential $U_{tot}(r)$ will shift to the new equilibrium position; for f_{ext} that is too large, no minimum occurs and there will be no equilibrium point, then the atoms will unbind.

It is often useful to understand how two atoms in a solid will respond to very weak forces, such that the atoms only displaced a very small amount from their equilibrium. We can use the Lennard–Jones interaction potential to model this phenomenon. For a very weak force, the very small shift in the equilibrium allows us to expand the interaction potential by using a Taylor series expansion.

$$\begin{aligned}\phi(r) &= \phi(r_0) + \left.\frac{d\phi}{dr}\right|_{r_0}(r-r_0) + \left.\frac{1}{2!}\frac{\partial^2\phi}{\partial r^2}\right|_{r_0}(r-r_0)^2 \\ &\quad + \left.\frac{1}{3!}\frac{\partial^3\phi}{\partial r^3}\right|_{r_0}(r-r_0)^3 + \cdots \\ &\approx \phi(r_0) + \left.\frac{1}{2}\frac{\partial^2\phi}{\partial r^2}\right|_{r_0}(r-r_0)^2\end{aligned} \tag{20.12}$$

We have used the fact that $d\phi/dr(r_0) = 0$, and we have neglected the higher-order terms in the Taylor expansion. So, we are just left with a *harmonic potential approximation* for the interaction which depends on the square of displacement $u = r - r_0$ from equilibrium. For the Lennard–Jones potential, the curvature is given by equilibrium spacing and binding energy:

$$\left.\frac{\partial^2\phi}{\partial r^2}\right|_{r_0} = 72\frac{E_b}{r_0^2} \tag{20.13}$$

The approximation works very well for very small displacements from equilibrium, but it fails as one moves far from the equilibrium. Furthermore, in the presence of a weak external force, the equilibrium point shifts to where $dU_{tot}/dr = 0$; using the expansion (equation 20.12) for interaction potential, this leads to

$$-f_{ext} + \left.\frac{\partial^2\phi}{\partial r^2}\right|_{r_0}(r-r_0) = 0 \tag{20.14}$$

or

$$u \equiv r - r_0 = \frac{1}{\frac{\partial^2\phi}{\partial r^2}}f_{ext} = \frac{1}{k}f_{ext} \tag{20.15}$$

Thus, we find that the displacement u from equilibrium for small forces f_{ext} is linear with respect to an external force. The linear response for small displacement u is a generic property of most of the materials.

20.0.4 Dynamic Motion

In this section, we allow the atoms to move and have a kinetic energy in addition to their potential energy. We assume that the center of mass for our system remains at rest. And we introduce some parameters such as $r = r_2 - r_1$ and center of mass location:

$$r_{cm} = \frac{M_1 r_1 + M_2 r_2}{M_1 + M_2} \tag{20.16}$$

The positions of atoms can be rearranged in terms of r_{cm} and r as

$$\left.\begin{aligned} r_1 &= r_{cm} - \frac{M_2}{M_1 + M_2} r \\ r_2 &= r_{cm} + \frac{M_1}{M_1 + M_2} r \end{aligned}\right\} \tag{20.17}$$

If the center of mass is at rest, so $dr/dt = 0$, then the velocities satisfy

$$M_1 \dot{r}_1 = -M_2 \dot{r}_2 \tag{20.18}$$

So, the kinetic energy can be written as

$$\begin{aligned} K &= \frac{1}{2} M_1 \dot{r}_1^2 + \frac{1}{2} M_2 \dot{r}_2^{\,2} \\ &= \frac{1}{2} \mu \dot{r}^2 \end{aligned} \tag{20.19}$$

Using the reduced mass $\mu = M_1 M_2 / (M_1 + M_2)$ and with the momentum $p = \mu \dot{r}$, the kinetic energy is $K = p^2/2\mu$ and the Hamiltonian for the system, $H = K + U$ is then

$$H = \frac{1}{2\mu} p^2 + \phi(r) \tag{20.20}$$

and Hamilton's equations of motion yield

$$\mu \ddot{r} = -\frac{d\phi}{dr}(r) = f(r) \tag{20.21}$$

If we consider very small displacement, $u = r - r_0$, from the equilibrium spacing, then using the Taylor expansion of interaction potential (equation 20.12) reveals

$$\mu \ddot{u} = -\left.\frac{d^2\phi}{dr^2}\right|_{r_0} u \tag{20.22}$$

which is the equation of motion for a *simple harmonic oscillator*, and has the general solution of the form

$$u(t) = u_0 \cos(\omega_0 t + \varphi) \tag{20.23}$$

Example 20.2 *Natural Resonance Frequency of Argon Atoms Calculated by Lennard–Jones Potential*

For the Lennard–Jones interaction potential with two argon atoms, and with masses $M_1 = M_2 = 2\mu = 6.6 \times 10^{-23}$ g

$$k = \left.\frac{\partial^2 \phi}{\partial r^2}\right|_{r_0} = 52\ \text{meV/Å}^2 = 0.83\ \text{N/m}$$

We know find the natural frequency $\omega_0/2\pi = 0.8$ THz. This is low for a mechanical atomic resonance frequency and is due to the weak van der Waals forces in the argon molecule. The typical resonance frequency for covalently or ionically bonded atoms is on the order of 10 THz [14].

where the resonance frequency ω_0 is given by

$$\omega_0 = \sqrt{\frac{1}{\mu}\frac{\partial^2 \phi}{\partial r^2}} \tag{20.24}$$

and amplitude u_0 and phase φ are defined by the initial conditions.

20.1 Three-Atom Chain

Now we add a third atom to the system, and restrict the motion to one dimension, and also assume that all the atoms are identical, with mass M. Then we consider equal and opposite external forces applied at two ends as shown in **Figure 20.2**. The two end atoms 1 and 3 will be displaced symmetrically and the middle atom does not move

$$\begin{aligned} u_1 &= r_2 - r_1 - r_0 = f_{\text{ext}}/k \\ &= \frac{1}{\frac{\partial^2 \phi}{\partial r^2}} f_{\text{ext}} \end{aligned} \tag{20.25}$$

with an identical expression for u_3.

Moreover, we now briefly describe the dynamic behavior of this three-atom chain as we have done for the two-atom system. Again, we assume that the center of mass remains at rest, so

Fig. 20.2 *Schematic representation of three-atom chain connected with springs.*

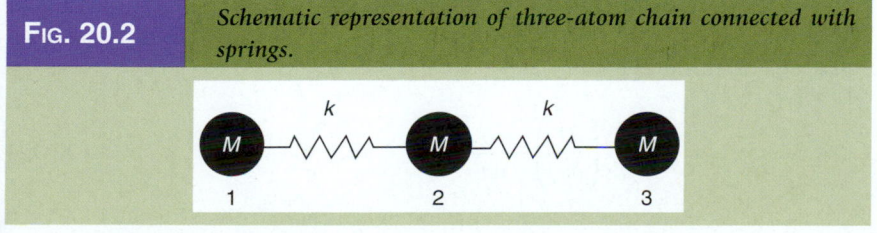

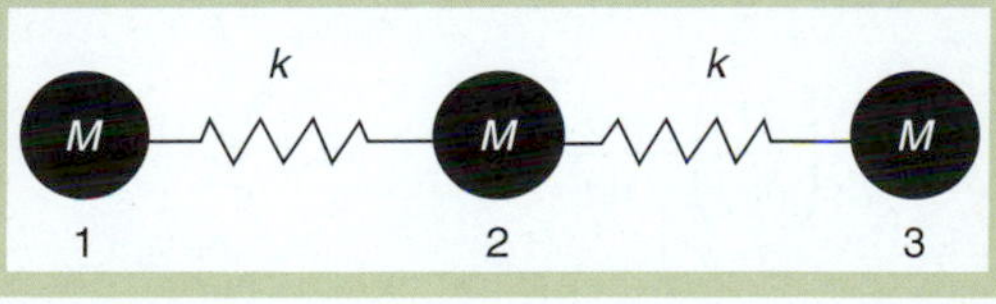

$$\dot{u}_1 + \dot{u}_2 + \dot{u}_3 = 0 \tag{20.26}$$

The Hamiltonian for the system is defined by

$$H = K + U = \frac{1}{2M}(p_1^2 + p_2^2 + p_3^2) + \frac{k}{2}(u_1 - u_2)^2 + \frac{k}{2}(u_2 - u_3)^2 \tag{20.27}$$

The corresponding equations of motion are

$$\left.\begin{aligned} M\ddot{u}_1 &= k(u_2 - u_1) \\ M\ddot{u}_2 &= k(u_1 - 2u_2 + u_3) \\ M\ddot{u}_3 &= k(u_2 - u_3) \end{aligned}\right\} \tag{20.28}$$

The solutions where all degrees of freedom have the same harmonic and time dependence possess the form

$$u_n = A_n e^{-i\omega t} \quad \text{where } n = 1, 2, 3 \tag{20.29}$$

Inserting these solutions to equation (20.28), we find the linear system of equations

$$\left.\begin{aligned} -M\omega^2 A_1 &= k(A_2 - A_1) \\ -M\omega^2 A_2 &= k(A_1 - 2A_2 + A_3) \\ -M\omega^2 A_3 &= k(A_2 - A_3) \end{aligned}\right\} \tag{20.30}$$

We can rearrange and write the above system of equations in a matrix form, where $\omega_0 = \left(\frac{k}{M}\right)^{\frac{1}{2}}$ is the frequency of the system. So, the system of equations can be written as an eigenvalue–eigenvector equation

$$\left(\frac{\omega}{\omega_0}\right)^2 \begin{bmatrix} A_1 \\ A_2 \\ A_3 \end{bmatrix} = \begin{bmatrix} 1 & -1 & 0 \\ -1 & 2 & -1 \\ 0 & -1 & 1 \end{bmatrix} \begin{bmatrix} A_1 \\ A_2 \\ A_3 \end{bmatrix} \tag{20.31}$$

This approach for the three-atom chain can be expanded to the n-atom chain in a similar way and readers are advised to go to Ref. [14] for a more detailed understanding.

20.2 LATTICE MECHANICS

A dynamic solution for the system of particles that comprise a stable lattice structure can be described by the Lagrangian formalism, which is briefly mentioned as follows. An arbitrary system of n particles can be described by the Lagrange equations of motion [15]

$$\frac{d}{dt}\left(\frac{\partial T}{\partial \dot{q}_j}\right) - \frac{\partial T}{\partial q_j} = Q_j, \quad j = 1, 2, \ldots, s \tag{20.32}$$

where

s is the number of independent degrees of freedom

Q_j are the generalized forces.

In order to write the lattice equation of motion, we require the Lagrangian in terms of the particle displacement vectors, u_n. The kinetic energy of lattice particles can be written in a matrix form

$$T = \frac{1}{2}\sum_n \dot{u}_n^T M \dot{u}_n \tag{20.33}$$

where M is a diagonal matrix of particle masses written for one unit cell. This gives the lattice Lagrangian in the general form of

$$L = \frac{1}{2}\sum_n \dot{u}_n^T M \dot{u}_n - U(u) \tag{20.34}$$

Here, U is the lattice potential energy, and u is a formal notation for all the displacement vectors, u_n in a given lattice. Substituting equation (20.34) into the Lagrange equation of motion (20.32), written for one unit cell with s degrees of freedom

$$\frac{d}{dt}\frac{\partial L}{\partial \dot{u}_{n,s}} - \frac{\partial L}{\partial u_{n,s}} = f_{n,s}^{\text{ext}}, \quad s = 1, 2, \ldots, S \tag{20.35}$$

We obtain the lattice equation of motio ı

$$M\ddot{u}_n + \frac{\partial U}{\partial u_n} = f_n^{\text{ext}}, \quad n = (n_1, n_2, n_3) \tag{20.36}$$

where f_n^{ext} is the vector of external forces exerted on the current unit cell n. In equation (20.35) $u_{n,s}$ and $f_{n,s}^{\text{ext}}$ are individual components of the vector u_n and f_n^{ext}, respectively. One special form of the lattice equation of motion can be obtained within the *harmonic approximation*, which consists of expanding the potential energy in Taylor series about the equilibrium

$$U(u) = U(0) + \sum_{n,s}\left.\frac{\partial U}{\partial u_{n,s}}\right|_0 u_{n,s} + \frac{1}{2}\sum_{n,n',s,s'}\left.\frac{\partial^2 U}{\partial u_{n,s}\partial u_{n',s'}}\right|_0 u_{n,s}u_{n',s'} + \cdots \tag{20.37}$$

and ignoring the second-order and higher terms. The zero-order term can be ignored in equation (20.37), since a constant shift of the Lagrangian does not alter the equations of motion. Therefore, the harmonic approximation leads to the lattice Lagrangian

$$L = \frac{1}{2}\sum_n \dot{u}_n^T M \dot{u}_n + \frac{1}{2}\sum_{n,n'} u_n^T K_{n-n'} u_n \tag{20.38}$$

where the superscript T indicates a transposed vector and the equation of motion reaches the final matrix form

$$M\ddot{u}_n(t) - \sum_{n} K_{n-n'} u_{n'}(t) = f_n^{ext}(t) \tag{20.39}$$

Here, K are the lattice stiffness matrices, which represent linear elastic properties of the lattice structure. These matrices are composed of the atomic force constants, according to

$$K_{n-n'} = \left. \frac{\partial^2 U(u)}{\partial u_n \partial u_{n'}} \right|_{u=0} \tag{20.40}$$

For one-, two-, or three-dimensional lattice with nearest unit cell interactions, there are up to 3, 9, or 27 nontrivial K-matrices, respectively. The lattice equation of motion (20.39) is an ordinary differential equation with a finite difference. The effective solution of these equations involve transform methods, such as Fourier and Laplace transforms, which convert the operation of convolution into an ordinary matrix multiplication in the transform domain.

In order to derive the K-matrices, the potential energy of the atomic interactions needs to be written for one associate cell only; this gives

$$U = V(u_n - u_{n-1} + \rho) + V(u_{n+1} - u_n + \rho) \tag{20.41}$$

where V is a pair-wise potential such as the Lennard–Jones with defined equilibrium distance ρ. According to equation (20.40), the K-matrices yield

$$K_{-1} = \frac{-\partial^2 U}{\partial u_n \partial u_{n+1}} = k, \quad K_0 = \frac{-\partial^2 U}{\partial u_n^2} = -2k, \quad K_1 = \frac{-\partial^2 U}{\partial u_n \partial u_{n-1}} = k \tag{20.42}$$

Here, the linear force constant k depends on the parameters of the potential

$$k = 36 \cdot 2^{\frac{2}{3}} \varepsilon/\sigma^2 (\text{Lennard–Jones}) \tag{20.43}$$

EXAMPLE 20.3 *Monatomic Chain Lattice (Equation of Motion)*

The monatomic chain lattice, **Figure 20.3**, is the simplest example that can be described with previous understandings. We assume that the interaction exists only between the nearest lattice atoms; this can be viewed so that the cutoff radius of the potential function is close to 1.5 ρ, where ρ is the equilibrium distance for the potential.

FIG. 20.3 *Monatomic chain lattice interacting with its nearest neighbors.*

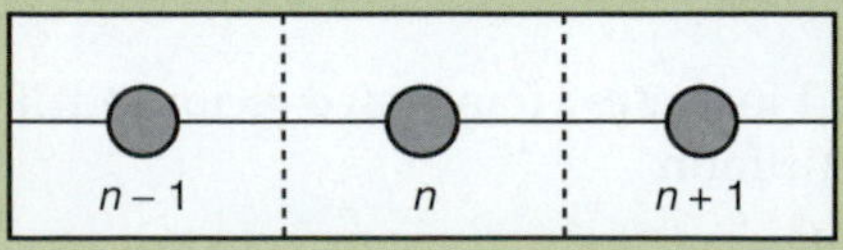

Substituting the K-matrices into the general equation of motion (20.39) yields

$$M\ddot{u}_n - k(u_{n-1} - 2u_n + u_{n+1}) = f_n^{\text{ext}} \tag{20.44}$$

20.3 Stress and Strain

In the first part of this chapter, we have introduced a brief description of solids from the atomic point of view and then broadened our knowledge to two-dimensional objects which can be expanded to three-dimensional objects. Classical dynamics assumes that all objects are infinitely rigid. But, the bonds between the atoms that make up the solid are not infinitely rigid, and many thermodynamic properties of insulators are based on the flexibility of these bonds. Therefore, rigid body dynamics is an approximation to the actual motion of solids.

Now, we consider a displacement vector $u(r)$ at a given position and then add a small Δr to the previous position; therefore, the new position vector is at $u(r+\Delta r)$. Then the relative displacement vector is defined as $\Delta u = u(r+\Delta r) - u(r)$. We can also expand each component u_i of displacement $u(r + \Delta r)$ in a vector Taylor series about the point r

$$u_i(r+\Delta r) = u_i(r) + \sum_{j=1}^{3} \frac{\partial u_i}{\partial x_j}\Delta r_j + \text{Higher-order terms} \tag{20.45}$$

Dropping the higher-order terms, we can write the relative displacement, Δu, with respect to the initial position in component form as

$$\Delta u_i = \sum_{j=1}^{3} \frac{\partial u_i}{\partial x_j}\Delta r_j \tag{20.46}$$

These derivatives can be assembled into a tensor D, where components are given by $D_{ij} = \partial u_i/\partial x_j$

$$D = \begin{pmatrix} \dfrac{\partial u_1}{\partial x_1} & \dfrac{\partial u_1}{\partial x_2} & \dfrac{\partial u_1}{\partial x_3} \\ \dfrac{\partial u_2}{\partial x_1} & \dfrac{\partial u_2}{\partial x_2} & \dfrac{\partial u_2}{\partial x_3} \\ \dfrac{\partial u_3}{\partial x_1} & \dfrac{\partial u_3}{\partial x_2} & \dfrac{\partial u_3}{\partial x_3} \end{pmatrix} \tag{20.47}$$

The tensor D is the basis for the definition of the strain tensor, which can be split into two species: symmetric and antisymmetric

$$\begin{aligned} D &= S + \Omega \\ S &= \frac{1}{2}(\nabla u + (\nabla u)^T) \\ \Omega &= \frac{1}{2}(\nabla u - (\nabla u)^T) \end{aligned} \tag{20.48}$$

where T here indicates the transpose. The tensor S is the strain tensor, and Ω is the rotation tensor. Written out in component form, the strain tensor is

$$S_{ij} = \frac{1}{2}\left(\frac{\partial u_i}{\partial x_j} + \frac{\partial u_j}{\partial x_i}\right) \tag{20.49}$$

and in tabular form

$$S = \begin{pmatrix} \frac{\partial u_1}{\partial x_1} & \frac{1}{2}\left(\frac{\partial u_1}{\partial x_2} + \frac{\partial u_2}{\partial x_1}\right) & \frac{1}{2}\left(\frac{\partial u_1}{\partial x_3} + \frac{\partial u_3}{\partial x_1}\right) \\ \frac{1}{2}\left(\frac{\partial u_1}{\partial x_2} + \frac{\partial u_2}{\partial x_1}\right) & \frac{\partial u_2}{\partial x_2} & \frac{1}{2}\left(\frac{\partial u_2}{\partial x_3} + \frac{\partial u_3}{\partial x_2}\right) \\ \frac{1}{2}\left(\frac{\partial u_1}{\partial x_3} + \frac{\partial u_3}{\partial x_1}\right) & \frac{1}{2}\left(\frac{\partial u_2}{\partial x_3} + \frac{\partial u_3}{\partial x_2}\right) & \frac{\partial u_3}{\partial x_3} \end{pmatrix} \tag{20.50}$$

The rotation tensor is

$$\Omega_{ij} = \frac{1}{2}\left(\frac{\partial u_i}{\partial x_j} - \frac{\partial u_j}{\partial x_i}\right) \tag{20.51}$$

and in tabular form

$$\Omega = \begin{pmatrix} 0 & \frac{1}{2}\left(\frac{\partial u_1}{\partial x_2} - \frac{\partial u_2}{\partial x_1}\right) & \frac{1}{2}\left(\frac{\partial u_1}{\partial x_3} - \frac{\partial u_3}{\partial x_1}\right) \\ \frac{1}{2}\left(\frac{\partial u_2}{\partial x_1} - \frac{\partial u_1}{\partial x_2}\right) & 0 & \frac{1}{2}\left(\frac{\partial u_2}{\partial x_3} - \frac{\partial u_3}{\partial x_2}\right) \\ \frac{1}{2}\left(\frac{\partial u_3}{\partial x_1} - \frac{\partial u_1}{\partial x_3}\right) & \frac{1}{2}\left(\frac{\partial u_3}{\partial x_2} - \frac{\partial u_2}{\partial x_3}\right) & 0 \end{pmatrix} \tag{20.52}$$

The tensor Ω gives the local rotation of volume element at the point r. Note that Ω does not represent the body rotation of the solid as a whole. Both S and Ω behave as second-order tensors, which under a rotation of the coordinate system given by transformation R, the strain tensor S', in the new coordinate system, have components which are given in terms of the old coordinate system

$$S'_{ij} = \sum_{m=1}^{3}\sum_{n=1}^{3} R_{im}S_{mn}R_{nj} \tag{20.53}$$

The displacement vector that gives the strain shown in figure is

$$u = (0, sx_1, 0) \tag{20.54}$$

The strain tensor is then given by

Example 20.4 *Pure Shear Strain*

We now consider the case of pure shear strain, where the solid is strained as depicted in **Figure 20.4**. The two plane surfaces perpendicular to x_1 are shifted by Δl along x_2.

Fig. 20.4 *Schematic representation of pure shear strain applied to a cube.*

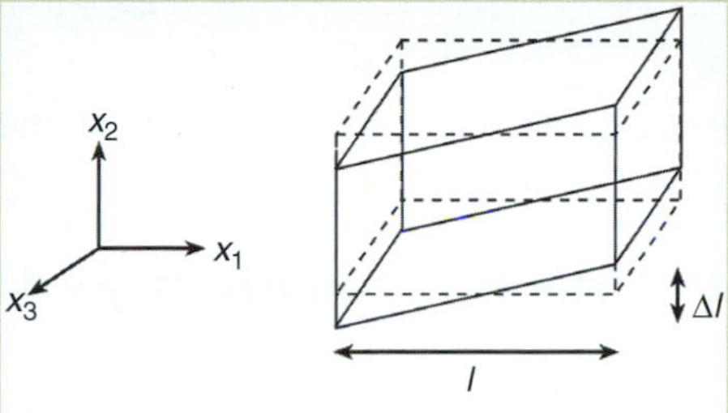

$$S = \begin{pmatrix} 0 & \frac{s}{2} & 0 \\ \frac{s}{2} & 0 & 0 \\ 0 & 0 & 0 \end{pmatrix} \tag{20.55}$$

Thus, the rotation tensor is given by

$$\Omega = \begin{pmatrix} 0 & -\frac{s}{2} & 0 \\ -\frac{s}{2} & 0 & 0 \\ 0 & 0 & 0 \end{pmatrix} \tag{20.56}$$

The motion depicted in **Figure 20.4** consists of both shear strain and rotation tensor. A slight change in displacement vector $u = (-\frac{1}{2}sx_2, \frac{1}{2}sx_1, 0)$ can make the rotation tensor to be zero while the strain tensor remains unchanged. The readers are encouraged to try to sketch what shape the cube would be under this displacement vector.

20.4 Linear Elasticity Relations

We now focus on linear elastic materials and their responses to any external forces. The most general linear elastic relation that relates stress T to strain S is defined as below

$$T_{ij} = \sum_{k=1}^{3}\sum_{l=1}^{3} \alpha_{ijkl} S_{kl} \tag{20.57}$$

where the elastic moduli α_{ijkl} are constants. We assume that the material is homogeneous and the elastic moduli are independent of position within the solid. There are $3^4 = 81$ distinct components in the elastic moduli tensor. But, due to the symmetry of strain and stress tensors, this reduces to 36 components. Using the definition above, we can rewrite the elasticity relation in a 6×6 elasticity matrix

$$\tau_i = \sum_{j=1}^{6} c_{ij}\varepsilon_j \tag{20.58}$$

The constants $c_{ij}\varepsilon_j$ are called the elastic stiffness coefficients, which have the dimensions of N/m^2 in SI units.

20.4.1 Orthotropic and Isotropic Materials

A material that has a mirror symmetry about all three planes, $x_1 - x_2, x_2 - x_3$, and $x_1 - x_3$, is known as an orthotropic material, and these special types of symmetry can reduce the number of independent components of elastic moduli to 12 distinct constants as shown below:

$$c = \begin{bmatrix} c_{11} & c_{12} & c_{13} & 0 & 0 & 0 \\ c_{21} & c_{22} & c_{23} & 0 & 0 & 0 \\ c_{31} & c_{32} & c_{33} & 0 & 0 & 0 \\ 0 & 0 & 0 & c_{44} & 0 & 0 \\ 0 & 0 & 0 & 0 & c_{55} & 0 \\ 0 & 0 & 0 & 0 & 0 & c_{66} \end{bmatrix} \tag{20.59}$$

Moreover, in the case of isotropic materials, three mirror symmetries for independent 90° rotations about the axes are added and also an additional 45° rotation about one axis is found, which can reduce the number of independent components to two distinct elastic constants

$$c = \begin{bmatrix} c_{11} & c_{12} & c_{12} & 0 & 0 & 0 \\ c_{12} & c_{11} & c_{12} & 0 & 0 & 0 \\ c_{12} & c_{12} & c_{11} & 0 & 0 & 0 \\ 0 & 0 & 0 & c_{44} & 0 & 0 \\ 0 & 0 & 0 & 0 & c_{44} & 0 \\ 0 & 0 & 0 & 0 & 0 & c_{44} \end{bmatrix} \tag{20.60}$$

where $c_{44} = (c_{11} - c_{12})/2$, the two elastic constants required to describe isotropic materials are traditionally referred to as Lame constants (1798–1870) λ and μ.

$$\begin{aligned} \lambda &= c_{12} \\ \mu &= c_{44} = (c_{11} - c_{12})/2 \end{aligned} \tag{20.61}$$

20.4.2 Crystalline Materials

Many materials are neither isotropic nor orthotropic, but display more restrictive symmetries. Here, we cite the symmetries and the corresponding form of elastic

Table 20.4 *Lattice Constants, Density, and Elastic Moduli of Diamond Structure Materials*

Material	a_0 (Å)	ρ (g/cm³)	c_{11} (GPa)	c_{12} (GPa)	c_{44} (GPa)
Silicon (Si)	5.4307	2.330	165	64	79.2
Germanium (Ge)	5.6200	5.323	129	48	67.1
Diamond (C)	3.5670	3.515	1040	170	550

stiffness matrix for a few common semiconductor materials. A more comprehensive and detailed discussion can be found in Landolt-Bbornstein [16].

Diamond Structure. Crystals formed from elements in group IV of the periodic table typically have diamond structures. The diamond structure consists of two interpenetrating face-centered cubic lattices with origins offset by ¼ of the cubic diagonal. The form of elastic stiffness coefficients is that given for isotropic materials (equation 20.60) except that there are three independent coefficients. The lattice constants, density, and elastic moduli for silicon, germanium, and diamond are tabulated in **Table 20.4** [16].

Wurtzite Structure. Gallium nitride (GaN), aluminum nitride (AlN), indium nitride (InN), and zinc oxide (ZnO) are binary compounds having what is commonly called a wurtzite structure. The wurtzite structure is formed from two interpenetrating hexagonal close-packed lattices, each lattice filled with one type of atom. The form for the elastic constants matrix is given by equation (20.62).

$$c = \begin{bmatrix} c_{11} & c_{12} & c_{13} & 0 & 0 & 0 \\ c_{12} & c_{11} & c_{13} & 0 & 0 & 0 \\ c_{13} & c_{13} & c_{33} & 0 & 0 & 0 \\ 0 & 0 & 0 & c_{44} & 0 & 0 \\ 0 & 0 & 0 & 0 & c_{44} & 0 \\ 0 & 0 & 0 & 0 & 0 & \dfrac{c_{11}-c_{12}}{2} \end{bmatrix} \qquad (20.62)$$

The lattice constants, density, and elastic stiffness for GaN, AlN, InN, and ZnO are shown in **Table 20.5** [16,17].

Table 20.5 *Materials Characteristics of Crystals with Wurtzite Structure*

Material	a_0 (Å)	c_0 (Å)	ρ (g/cm³)	c_{11} (GPa)	c_{33} (GPa)	c_{44} (GPa)	c_{12} (GPa)	c_{13} (GPa)
GaN	3.189	5.185	6.095	374	379	101	106	70
AlN	3.112	4.982	3.255	345	395	118	125	120
InN	3.540	5.705	6.880	190	182	10	104	121
ZnO	3.249	5.207	5.675	209	218	44.1	120	104

20.5 Molecular Dynamics

Molecular dynamics is a simulation technique in which the time evolution of a set of interacting particles is obtained by integrating the equations of motion, which are derived from Newton's equations of motion

$$m_i\ddot{r}_i = m_i a_i = -\nabla_i U + F_i, \quad i = 1, 2, \ldots, n \tag{20.63}$$

applied to each atom i in a system containing n atoms. Here, m_i is the atomic mass, $\nabla_i U$ the first derivative (gradient) of potential energy, and F_i the force acting on atom i due to the interaction with other atoms. In order to solve the second-order differential equations of motion (20.63), several algorithms have been introduced. Here, we just mention these which are mostly applied in nanosystems.

20.5.1 Verlet Algorithms

As a direct solution to the second-order differential equations (20.63), the Verlet algorithm, a time-integration method, is widely used to solve the equations of motion. This method uses the current position r_i and acceleration a_i as well as the previous position r_{i-1} of an atom to derive the position r_{i+1} for the next time step in the following manner

$$r_{i+1} = 2r_i - r_{i-1} + a_i\Delta t^2 \tag{20.64}$$

with

$$r_i = r(t_i), \quad a_i = \frac{F(t_i)}{m_i}, \quad t_i = i\Delta t, \qquad i = 0, 1, 2, \ldots, N \tag{20.65}$$

where the interaction forces F_i and acceleration a_i have to be calculated for each particle according to the following equations

$$\begin{aligned} F_i &= -\nabla_{r_i} U(r_1, \ldots, r_N) \\ a_i &= -\frac{1}{m_i}\nabla_{r_i} U(r_1, \ldots, r_N) \end{aligned} \tag{20.66}$$

Since the velocities do not appear directly they can be obtained by applying the central difference method

$$v_i = \frac{r_{i+1} - r_{i-1}}{2\Delta t} \tag{20.67}$$

The time step here is denoted by Δt. Since the atomic vibrations are on the order of approximately 100 fs, a time step smaller than that is required, typically of 2 fs. Several modifications have been proposed to improve the numerical precision of the basic Verlet algorithm [18]. One of these is the so-called half-step leap-frog scheme [19]:

$$v_{i+1/2} = v_{i-1/2} + \Delta t\, a_i$$
$$r_{i+1} = r_i + \Delta t\, v_{i+1/2} \tag{20.68}$$

Here the current velocities have to be calculated from the mid-step values

$$v_i = \frac{1}{2}[v_{i-1/2} + v_{i+1/2}] \tag{20.69}$$

Another derivative, the velocity Verlet algorithm [20], works without a mid-step at the cost of additional storage for a_i

$$r_{i+1} = r_i + v_i\,\Delta t + \frac{1}{2}a_i\,\Delta t^2$$
$$v_{i+1} = v_i + \frac{1}{2}[a_i + a_{i+1}]\Delta t \tag{20.70}$$

There are further derivatives [18], but basically all Verlet methods produce the same error and generate identical position trajectories. So, there seems to be no need to implement a more complicated Verlet algorithm than is given by equation (20.70).

20.5.2 Nordsieck/Gear Predictor–Corrector Methods

Nordsieck [21] and Gear [22] developed an integration technique on the basis of Taylor expansions of the positions, velocities, accelerations, and further derivatives

$$r(t+\Delta t) = r(t) + v(t)\Delta t + \frac{1}{2}a(t)\Delta t^2 + \cdots + \frac{1}{k!}q_k(t)\Delta t^k$$
$$v(t+\Delta t) = v(t) + a(t)\Delta t + \frac{1}{2}q_3(t)\Delta t^2 + \cdots + \frac{1}{(k-1)!}q_k(t)\Delta t^{k-1}$$
$$a(t+\Delta t) = a(t) + q_3(t)\Delta t + \frac{1}{2}q_4(t)\Delta t^2 + \cdots + \frac{1}{(k-2)!}q_k(t)\Delta t^{k-2}$$
$$q_i(t+\Delta t) = q_i(t) + q_{i+1}(t)\Delta t + \frac{1}{2}q_{i+2}(t)\Delta t^2 + \cdots + \frac{1}{(k-i)!}q_k(t)\Delta t^{k-i}, \quad i = 3, 4, \ldots, \tag{20.71}$$

where

$$q_k(t) = \frac{\partial^k}{\partial t^k} r(t) \tag{20.72}$$

Now for the position $r^{(0)}$ and its scaled derivatives $r^{(k)}$ with

$$r^{(0)} = r, \quad r^{(1)} = v\Delta t, \quad r^{(2)} = \frac{1}{2}a\,\Delta t^2, \quad r^{(k)} = \frac{1}{k!}q_k\Delta t^k \tag{20.73}$$

a simple Taylor series predictor becomes

$$\begin{bmatrix} r_{i+1}^{\sim(0)} \\ r_{i+1}^{\sim(1)} \\ r_{i+1}^{\sim(2)} \\ \vdots \end{bmatrix} = P \begin{bmatrix} r_i^{(0)} \\ r_i^{(1)} \\ r_i^{(2)} \\ \vdots \end{bmatrix} \tag{20.74}$$

where P is the Pascal triangle matrix with the binomial coefficients in its columns

$$P = \begin{bmatrix} 1 & 1 & 1 & 1 & 1 & 1 & \cdots \\ 0 & 1 & 2 & 3 & 4 & 5 & \cdots \\ 0 & 0 & 1 & 3 & 6 & 10 & \cdots \\ 0 & 0 & 0 & 1 & 4 & 10 & \cdots \\ \vdots & \vdots & \vdots & \vdots & \vdots & \vdots & \ddots \end{bmatrix} \tag{20.75}$$

The predictor does not generate the exact values for the position and its derivatives. But, with help of the predicted position $r_{i+1}^{\sim(0)}$ the forces of the time step $i+1$ can be calculated, and therefore the correct accelerations, a_{i+1}. The comparison with the predicted accelerations $r_{i+1}^{\sim(2)}$ from equation (20.74) gives a measure of the error corresponding to the predictor step:

$$\varepsilon_{i+1} = \frac{1}{2} a_{i+1} \Delta t^2 - r_{i+1}^{\sim(2)} \tag{20.76}$$

Then this error is used to improve the predicted values in a corrector step which has this form

$$\begin{bmatrix} r_{i+1}^{(0)} \\ r_{i+1}^{(1)} \\ r_{i+1}^{(2)} \\ \vdots \end{bmatrix} = \begin{bmatrix} r_{i+1}^{\sim(0)} \\ r_{i+1}^{\sim(1)} \\ r_{i+1}^{\sim(2)} \\ \vdots \end{bmatrix} + \begin{bmatrix} c_0 \\ c_1 \\ c_2 \\ \vdots \end{bmatrix} \varepsilon_{i+1} \tag{20.77}$$

Usually the Nordsieck/Gear algorithm works with $i = 3, \ldots, 8$ values for which the corrector vector can be found. If accuracy and long periods are not important for the simulations, then the Verlet algorithms have to be preferred. But, for high accuracy problems or long-time simulations the 6-value Nordsieck/Gear predictor–corrector yields better results, though at the cost of decreased step sizes. From equation (20.66), it becomes obvious that the problem of modeling a material is essentially that of finding the potential U, which reproduces the behavior of the material under the simulation conditions. Depending on the origin of the potential, there are three different MD techniques: empirical, tight-binding, and first principles.

The empirical methods employ classical potentials, which can be given by different techniques, for example, the dependence of the energy on the nuclei position can be extracted from the first principle description. Another choice is to fit the potential to experimental data. The simplest form of the many-body potential is in the form of a sum of pairwise terms, with the energy of a pair only depending on their relative distance, r_{ij}

$$U(r_1, \ldots, r_N) = \frac{1}{2} \sum_{i \neq j} \varphi(r_{ij}) \tag{20.78}$$

Unfortunately, the types of materials that can be realistically modeled using this approach are limited to the noble gases, where electronic bonding is absent and atoms are interacting through the weak van der Waals forces.

The potential for metals and other materials must incorporate the quantum mechanical effect of bond weakening, a consequence of the Pauli principle [23]. Several schemes were developed based on the analytical form

$$U = \frac{1}{2}\sum_{i\neq j}\varphi(r_{ij}) + \sum_{i}\phi(n_i) \tag{20.79}$$

As before, φ is a two-body interaction potential part whereas ϕ is a function giving the energy of atom as a function of its coordination n_i. In general, the classical potentials for metals and semiconductors are designed from the start with a cutoff radius, which limits the interaction to only the nearest neighbor atoms. If an abrupt truncation in the term of a step function is employed, then the energy and its derivatives are not continuous functions of atomic coordinates, which can disrupt a minimization process or lead to the unwanted effects in a dynamic simulation. In order to resolve this issue, a smoothing function can be introduced which tapers the interaction to zero at a given distance. For example, the Brenner potential for carbon has incorporated a switch type function as described below [24]

$$U(r) = \begin{Bmatrix} 1 & r < r_1 \\ \dfrac{1}{2}\left\{1 + \cos\left[\dfrac{\pi(r - r_1)}{r_2 - r_1}\right]\right\}, & r_1 \le r \le r_2 \\ 0 & r > r_2 \end{Bmatrix} \tag{20.80}$$

This has the property of leaving the interaction unchanged for distances less than the inner cutoff distance, $r_1 = 0.17$ nm, and decreases to zero at the second cutoff, $r_2 = 0.2$ nm. Additionally, the first derivative is continuous on the full range, which avoids problems in minimization and dynamic simulations.

All methods discussed above assume that the electronic system is in the ground state and follows the nuclear motion. This approximation is valid in most cases, but there are some physical situations, where this approximation is no longer appropriate. For instance, response of matter under intense laser pulses or behavior of materials in the plasma state where some of the species are in the excited or ionized states cannot be described using basic MD. In these cases, more complicated MD has to be considered. However, these physical phenomena are beyond the scope of nanomechanics discussed in this chapter.

20.5.3 Molecular Dynamics Applications

In this section, we review applications of the Newtonian equations of motion to some typical MD simulations in the field of nanomechanics and nanomaterials.

Inelasticity and Failure of Gold Nanowires. Nanowires are found to have great potential as structural reinforcements, as elements in electronic circuitry, and in many other applications [25,26]. The examples depicted here are MD simulations of the tensile failure of gold nanowires [27]. The wire size was initially 16 nm in length with a square cross section of length 2.588 nm. The wire was first relaxed to a minimum energy configuration with free boundaries everywhere, and then

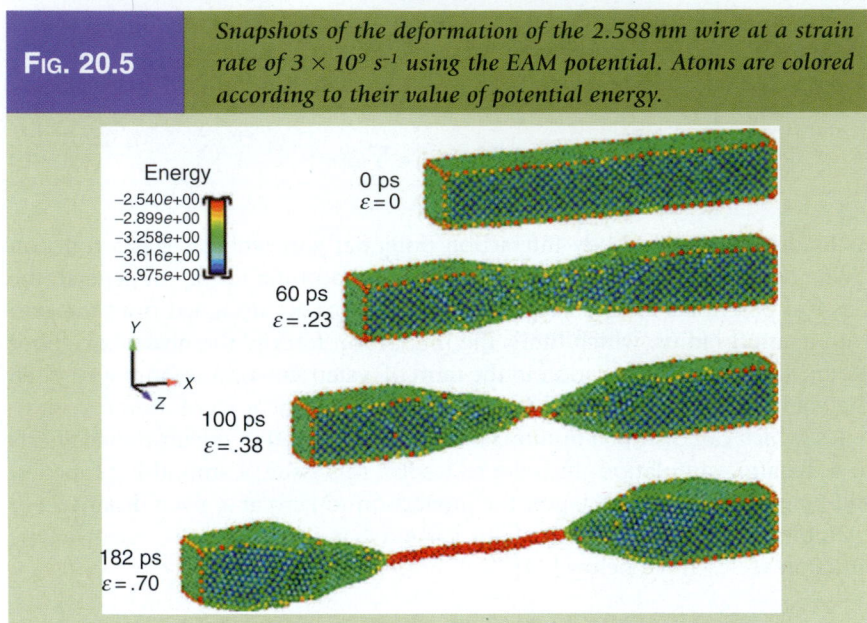

FIG. 20.5 *Snapshots of the deformation of the 2.588 nm wire at a strain rate of 3×10^9 s^{-1} using the EAM potential. Atoms are colored according to their value of potential energy.*

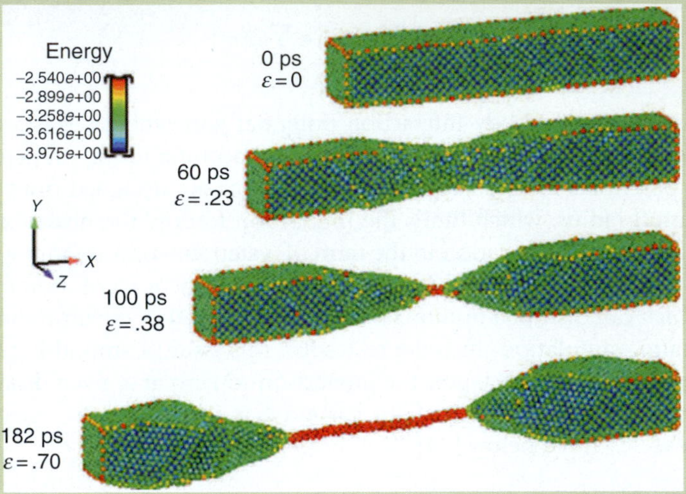

Source: S. M. Foiles, M. L. Baskes, and M. S. Daw, *Physical Review B*, 33(12), 7893–7991 (1986).

thermally equilibrated at a fixed length to 300 K. Finally, a ramp velocity was applied to the nanowire ranging from zero at one end to a maximum value at the loading end. So, one end of the nanowire was fixed while the other end was elongated at a constant velocity at each time step corresponding to an applied strain rate of 3.82×10^9 s^{-1}.

The gold nanowire depicted in **Figure 20.5** resembles the same failure mechanisms as a macroscopic tensile specimen, such as necking and yielding. However, one very interesting quality of the gold nanowire is its incredible ductility, which is manifested in the elongation of extremely thin nanobridges, as seen in the last snapshots in **Figure 20.5**.

Thermal Stability of Nanosystems. Mechanical stability of nanosystems due to temperature change is an example of material properties that can be altered extremely when we go from the macroscopic to the microscopic realm. Thermal stability and melting temperature of macroscopic systems are well defined and usually are stable up to the melting point. But in contrast to macroscopic systems, the melting point of nanosystems depends on the number of particles and is also a function of the shape of the system. For example, the melting temperature of macroscopic aluminum (Al) is about 933 K and the structure is stable close to the melting point. But, this is not the case for Al systems of nanometer size. Reith and Schommers [28] have performed a MD simulation on the basis of the Schommers pair potential for an Al nanosystem. The thermal stability of a three-dimensional object (F-shape) standing on the surface is studied. The structures are usually unstable far below the melting point and even dissolve. In **Figure 20.6**, a three-dimensional F-shaped structure consisting of aluminum atoms is shown. **Figure 20.6a** corresponds to the initial configuration. It corresponds to the crystalline structure of aluminum at zero absolute temperature.

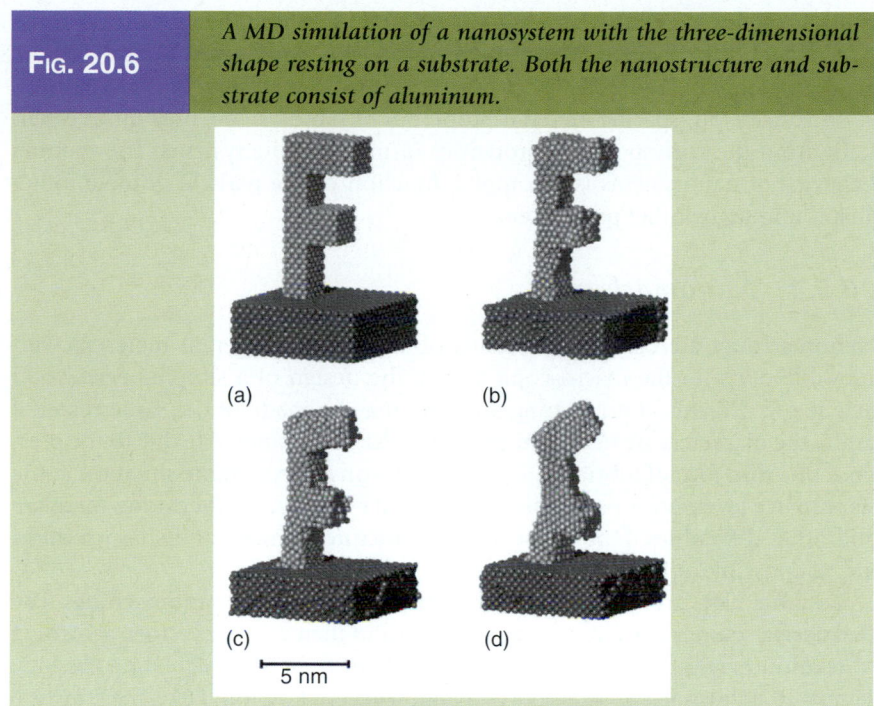

FIG. 20.6 *A MD simulation of a nanosystem with the three-dimensional shape resting on a substrate. Both the nanostructure and substrate consist of aluminum.*

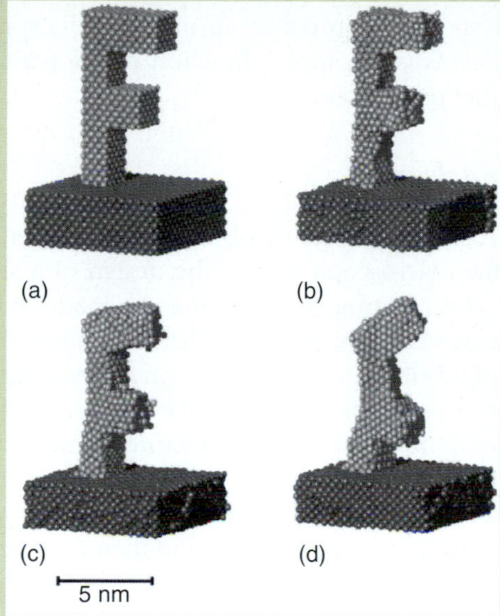

Source: M. Reith, W. Schommers, and S. Baskouts, *Mod. Phys. Lett. B*, 14, 621 (2000).

The time step used in this calculation is 5×10^{-15} s; the particle number n of the nanosystem is $n = 1660$ (without substrate). After 2000 time steps the system has reached a temperature of 250 K (**Fig. 20.6b**), also after 6000 time steps (**Fig. 20.6c**) the temperature is 270 K, and the nanosystem remains at this temperature for a further step, **Figure 20.6d**. Though the melting point of Al is 933 K, it can be clearly seen from **Figure 20.6d** that this configuration is structurally disturbed already at 270 K, which is significantly below the melting temperature. Furthermore, it is typical for the behavior of nanostructures that a tiny change in the initial condition can lead to different final shapes as depicted in **Figure 20.7**. In the case of **Figure 20.7a**, the temperature reaches 400 K after 10^4 time steps. In **Figure 20.7b**, the temperature is 500 K after 5000 time steps. So, as we can see

FIG. 20.7 *The effect of initial condition variations on the final shape of the nanostructures. (a) The temperature reaches 400 K after 10^4 time steps and (b) the temperature is 500 K after 5000 time steps.*

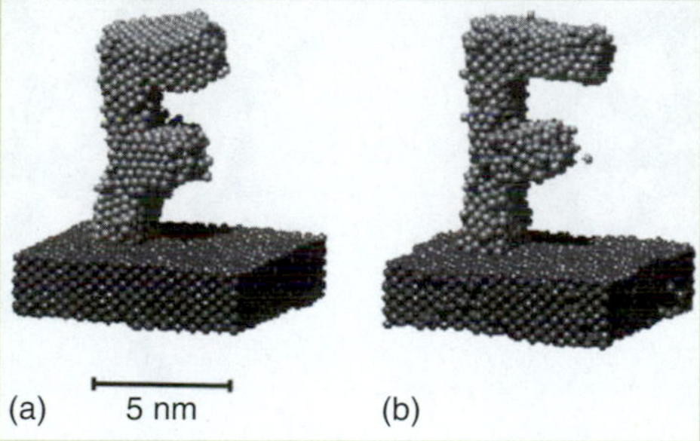

Source: M. Reith, W. Schommers, and S. Baskouts, *Mod. Phys. Lett. B*, 14, 621 (2000).

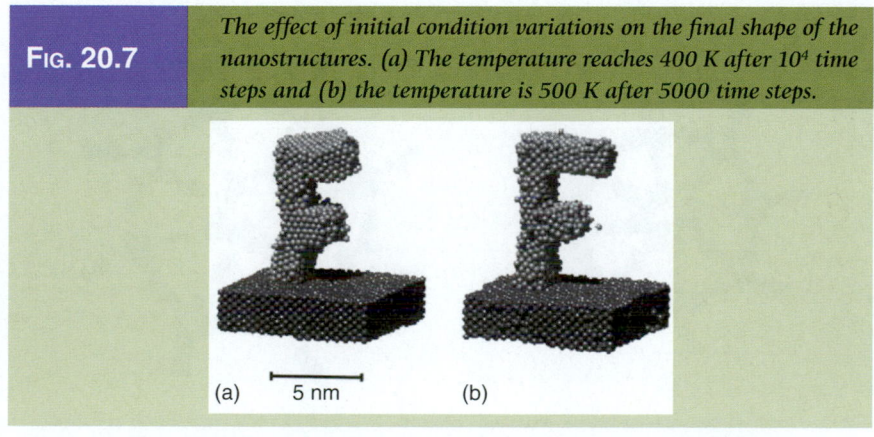

from the figure, the time steps have an important effect on the final shape of the nanosystem. The system in **Figure 20.7b** is at a higher temperature but still is more stable than that in **Figure 20.7a** because it has a smaller number of time steps.

In conclusion, specific material properties of nanosystems may differ essentially from the corresponding properties of macroscopic systems. The thermal behavior of nanosystems is a complex function of the particle number, outer shape, and many other parameters.

20.5.4 Nanomachines

Nanomachines are defined as systems of at least two different materials with movable parts. In the macroscopic world, the design of a simple bearing and axle is not difficult. Both bearing and axle may be made of the same material and if the diameters are nearly the same, the axle is still movable due to the presence of a thin film of lubricant [8]. But, if we consider the same situation at the nanometer level, both parts will stick together forever. Furthermore, there are still other difficulties. Due to the atomic structure, smooth and sharp surfaces are not possible at the nanometer size.

Rotating parts are of particular focus in connection with nanomachines. The first design setup is to create a static structure and then velocity vectors, according to revolution speed, which are added to each atom before calculating the MD. **Figure 20.8** illustrates an example of a nanowheel of krypton. Here, the structure disintegrates because the centrifugal forces are too strong [8].

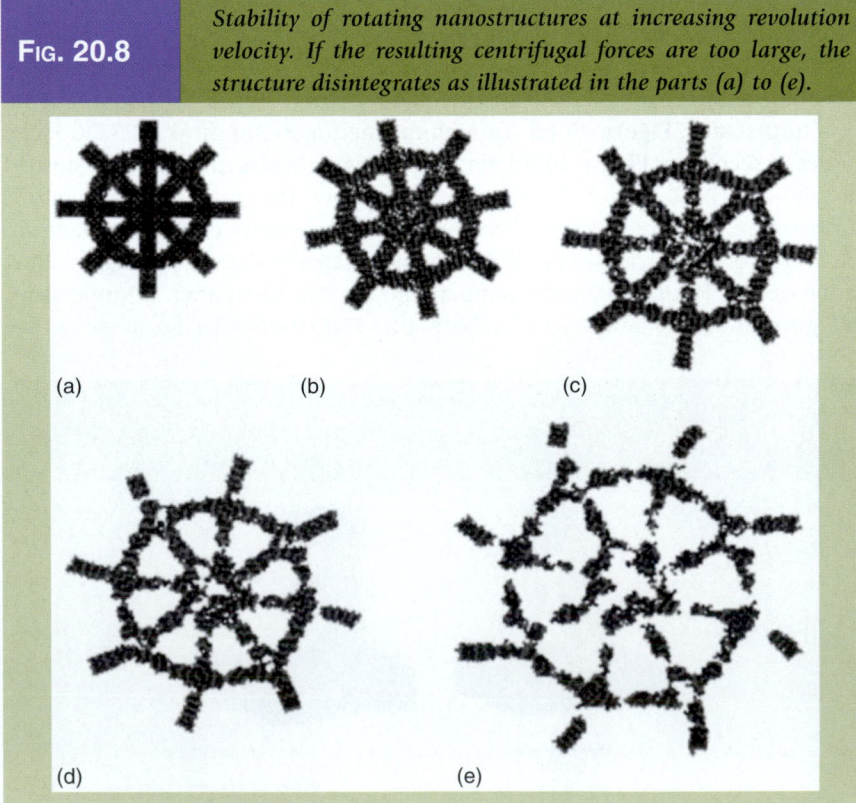

FIG. 20.8 *Stability of rotating nanostructures at increasing revolution velocity. If the resulting centrifugal forces are too large, the structure disintegrates as illustrated in the parts (a) to (e).*

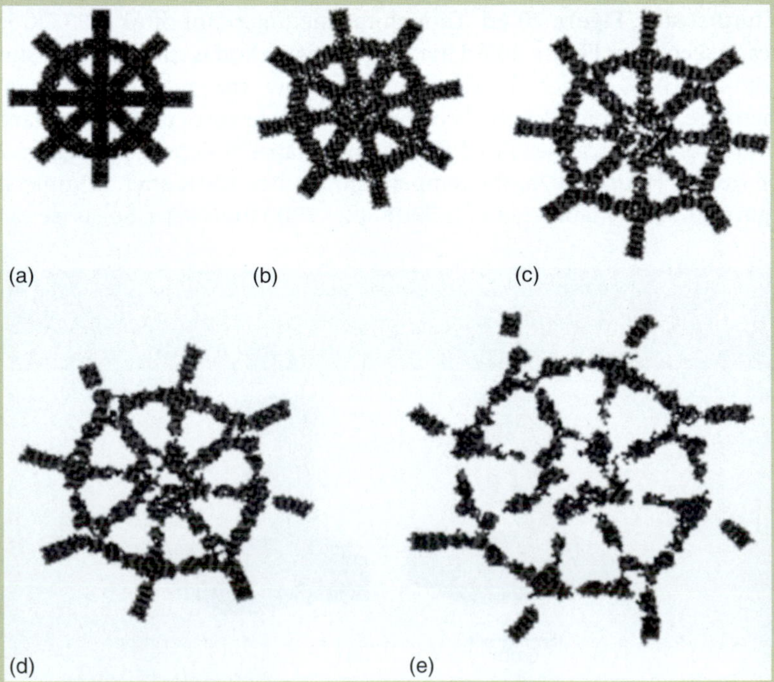

Source: M. Reith, World Scientific (2003).

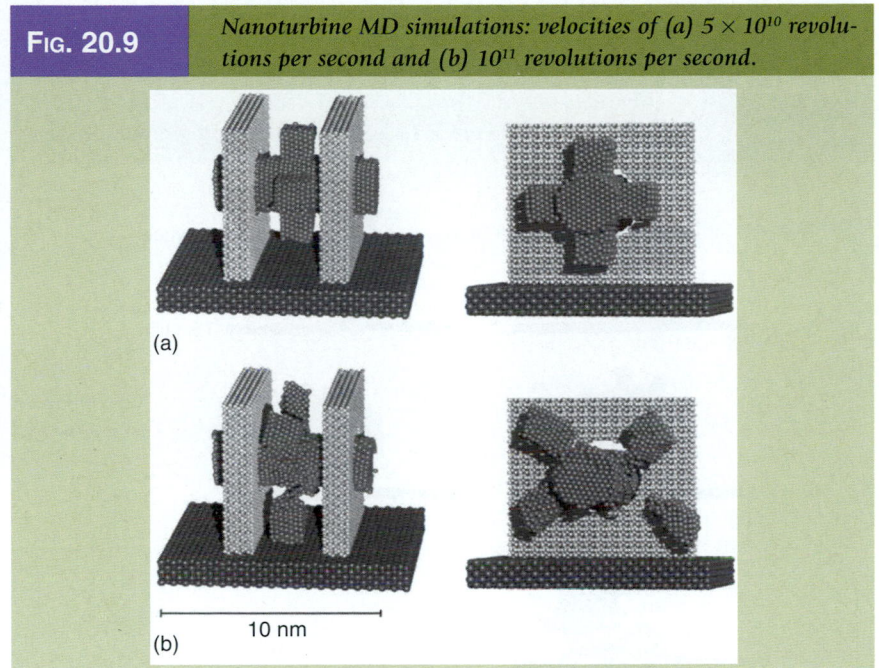

FIG. 20.9 *Nanoturbine MD simulations: velocities of (a) 5×10^{10} revolutions per second and (b) 10^{11} revolutions per second.*

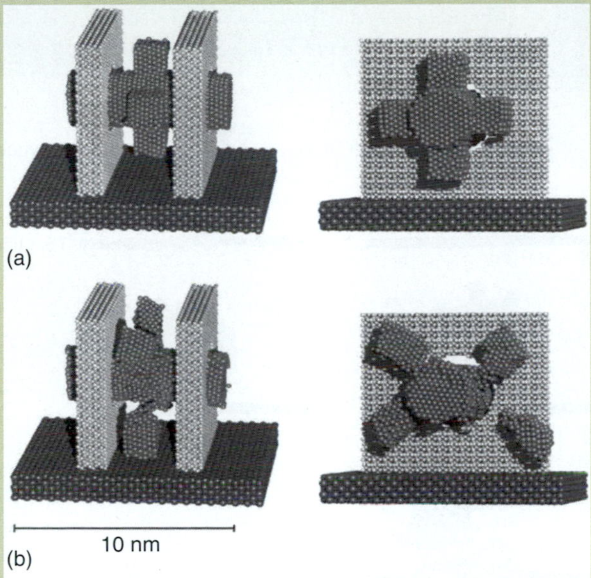

Source: M. Reith, World Scientific (2003).

Moreover, a more complicated example of a nanomachine and its integrity is depicted in **Figure 20.9** [8]. This is a model of a nanoturbine that consists of two bearings and an axle standing on a substrate. As you can see from the figure, the fit of axle and bearing is rather loose. However, MD studies have shown that such a nanoturbine can remain stable up to 5×10^{10} revolutions per second, but it ruptures at 10^{11} revolutions per second [8].

20.5.5 Wear at the Nanometer Level

Within the frame of MD systems, friction in the macroscopic domain is not defined. At the microscopic MD level, the forces are derived as quantities which are dependent on the structural configuration (particle position) but not on the particle velocities. Therefore, a force which is proportional to the velocity cannot be introduced at the microscopic level and thus a friction constant in the macroscopic sense is not definable as a constant in the microscopic world. At this level, wear is described by complex processes. To understand this phenomenon, an example is provided here. **Figure 20.10** shows a MD model for a spinning wheel moving towards a thin film [8]. The wheel rotates at about 10^{12} revolutions per second and is kept at a constant temperature of 300 K and its diameter is approximately 10 nm.

When the wheel approaches the surface and contacts the surface, friction effects emerge and, depending on the magnitude of the vertical force, the wheel may be destroyed as depicted in the figure. Therefore, the friction at the microscopic level is a complex process and may not be described by one constant only. The wear at the microscale level depends on the specific structure of the surface and additionally on the shape and motion of the wheel.

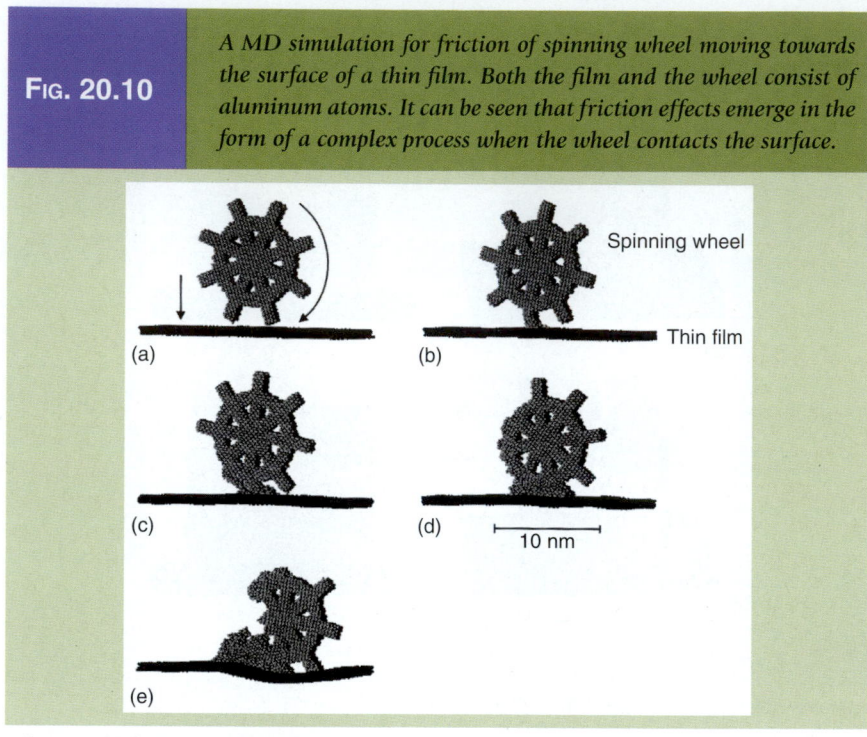

Fig. 20.10 *A MD simulation for friction of spinning wheel moving towards the surface of a thin film. Both the film and the wheel consist of aluminum atoms. It can be seen that friction effects emerge in the form of a complex process when the wheel contacts the surface.*

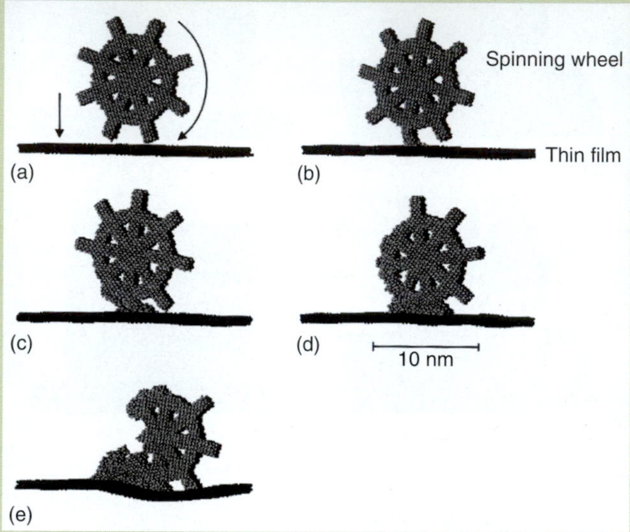

Source: M. Reith, World Scientific (2003).

20.6 Structure and Mechanical Properties of Carbon Nanotubes

Carbon nanotubes (CNTs) have been synthesized for a long time from the action of a catalyst over the gaseous environment originating from the thermal decomposition of hydrocarbons. There are several other techniques such as sputtering or evaporation that can be used to produce CNTs, but thermal decomposition is the most widely used method to synthesize CNTs. The accidental discovery of single-wall carbon nanotubes (SWCNTs) by Iijima et al. [29] and Bethune et al. [30] at NEC in 1990 made a huge impact on science and technology. CNTs consist of honeycomb lattices of carbon rolled into cylinders nanometer in diameter and micrometer in length. CNTs have incredible properties.

- 1/6 the weight of steel; 5 times its Young's modulus; 100 times its tensile strength; 6 orders of magnitude higher in electrical conductivity than copper and can be strained up to 15% without fracture
- Metallic or semiconductor depending on its chirality
- 10 times smaller than the smallest silicon tip in a scanning tunneling microscope (STM)

20.6.1 Structure of Carbon Nanotubes

To simplify the understanding of a carbon nanotube's shape, a perfect graphene sheet (single atomic layer made of a hexagonal display of sp^2 hybridized carbon

Fig. 20.11 *Schematic representation of graphite sheet and CNTs that are formed by rolling.*

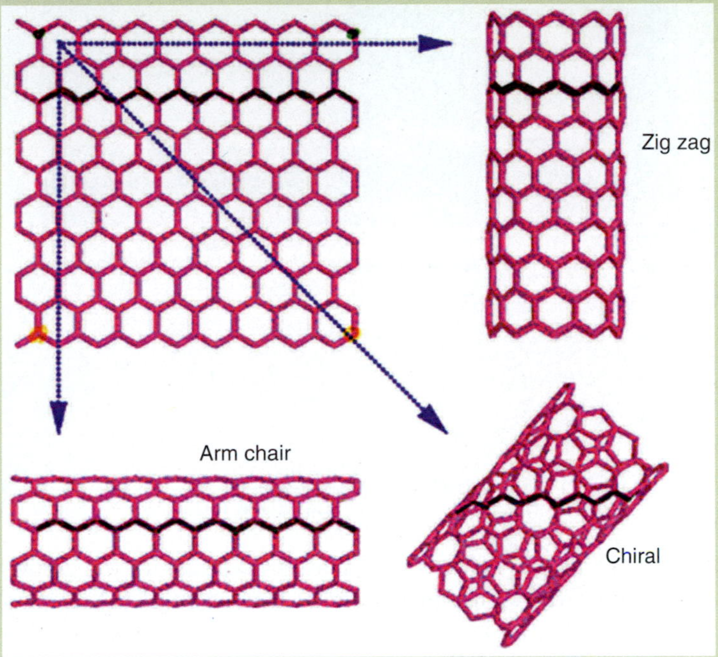

atoms) is depicted in **Figure 20.11**, and is then rolled into three different cylinders [31].

Though carbon atoms are involved in aromatic rings, the C = C bond angles are no longer planar as they should ideally be. This means that the hybridization of carbon atoms is no longer pure sp^2 but involves some percentage of the sp^3 character. This angle change results in a more active surface of the carbon nanotube compared to its graphitic sheet. As illustrated in **Figure 20.11**, there are three different ways to roll a graphitic sheet into a SWCNT; some of the resulting nanotubes enable symmetry mirrors both parallel and perpendicular to the nanotube axis (armchair and zigzag configurations). The other way of forming CNTs is shown in **Figure 20.11** (chiral nanotube), which does not have any symmetry mirrors that were mentioned above. The various ways to roll graphene into tubes are therefore mathematically defined as the vector of helicity C_h and the angle of helicity θ as defined below:

$$C_h = na_1 + ma_2 \tag{20.81}$$

with

$$a_1 = \frac{a\sqrt{3}}{2}x + \frac{a}{2}y \quad \text{and} \quad a_2 = \frac{a\sqrt{3}}{2}x - \frac{a}{2}y \tag{20.82}$$

where $a = 0.246\,\text{nm}$

Fig. 20.12 *Image of two neighboring chiral single-wall carbon nanotubes as produced by high-resolution tunneling microscopy.*

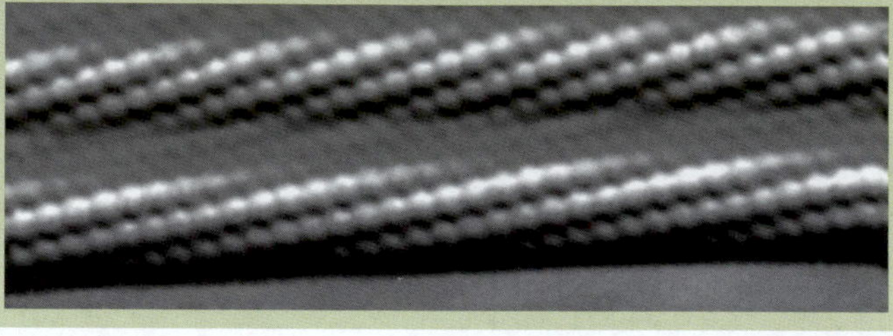

Source: Image courtesy of Prof. Yazdani, University of Illinois at Urbana. With permission.

and

$$\cos\theta = \frac{2n+m}{2\sqrt{n^2+m^2+nm}} \tag{20.83}$$

where n and m are integers of the vector of helicity considering the unit vectors a_1 and a_2. For example the zigzag type nanotubes have an angle of helicity of 0° (as shown in **Figure 20.11**), whereas armchair type CNTs have an angle of helicity of 30°. It is clear from **Figure 20.11** that having the vector of helicity perpendicular to any of the three overall C=C bond directions will produce the zigzag type nanotube $(n, 0)$, while having the vector of helicity parallel to one of the three C=C bond directions will provide armchair type structures (n, n). Because of the six-fold symmetry of the graphene sheet, the angle of helicity for the chiral (n, m) nanotubes is such that $0 < \theta < 30°$ [32]. **Figure 20.12** indicates two examples of what chiral single-wall carbon nanotubes look like, as seen by means of a high-resolution STM.

20.6.2 Mechanical Properties of Carbon Nanotubes

Considerable progress has been made in investigating the mechanical properties of SWCNTs and multiwall carbon nanotubes (MWCNTs). The theoretical predictions and experimental measurements are very promising and motivate further studies of future applications for lightweight and high-strength products. CNTs are very strong through the three-folded bonding of the curved graphene sheet, which is stronger than in diamond due to their differences in C–C bond length (0.142 versus 0.154 nm for graphene and diamond, respectively) [32].

The tensile strength of SWCNTs can be 20 times that of steel [33] and has actually been measured equal to about 45 GPa [34]. One important application of CNTs is in composite materials, which are reinforced by the introduction of SWCNTs or MWCNTs. Mechanical properties of CNTs can be measured by atomic force microscopy (AFM); the tests involve measurements of deformations under controlled forces in lateral [35] or normal [36] directions, and a tensile test can be done by incorporating two AFM tips at both ends of CNTs [37]. In order to apply continuum mechanics, we need to define the thickness of the nanotube or a graphene sheet for the continuum beam approximation.

Most researchers working in this field use a value of 0.34 nm close to the interlayer separation in graphite sheets. One important point in the measurements of elastic properties, such as Young's modulus and shear modulus, is the huge amount of precision that must be done in order to minimize any further errors. Even a very small error in the measurement of deflection enters into equations of beam deflection as d^4, which leads to a large uncertainty in the results [32]. Several techniques have been employed to measure the mechanical properties of nanotubes including bar, beam, and shell models. The bar model has been used in the experiment by Lourie and Wagner [38], in which the compressive response was calculated using micro-Raman spectroscopy. The reported values for Young's modulus are 2.3–3.6 TPa for SWCNTs, and 1.7–2.4 TPa for MWCNTs. Another test was performed by Yu et al. [33,37] using tensile loading of SWCNTs and MWCNTs. The Young's modulus obtained ranged from 320 to 1470 GPa for SWCNTs and from 270 to 950 GPa for MWCNTs.

A cantilever beam model was used in an experiment conducted by Wong et al. [35] in which individual MWCNTs were bent using an AFM tip. After fitting the measured data points to the analytical solution, a Young's modulus of 1.28 ± 0.59 TPa was obtained. In another model simulated by Salvetat et al. [39,40], the deflection of a simple-supported beam was modeled and a Young's modulus of ~1 TPa for MWCNTs, grown by arc discharge, was reported, whereas CNTs grown by the catalytic decomposition of a hydrocarbon gas showed a modulus of 1–2 orders of magnitude smaller.

There are numerous experimental measurements [41–45] and theoretical calculations [46–54] for mechanical properties of CNTs that due to the limited space in this chapter are just mentioned in the reference section. Readers are strongly recommended to review these references to get a more comprehensive understanding of how the mechanical properties of CNTs are obtained.

20.7 Nanomechanical Measurement Techniques and Applications

Experimental tools are used to measure the mechanical properties of materials at the nanometer level. The most prominent measurement technique that evaluates the mechanical properties is scanning probe microscopy, which is a broad term for different techniques that can be performed in both contact and noncontact modes. Scanning probe microscopes (SPM) are tools that scan a sharp probe tip across the specimen and provide nanoscale information about the sample. One of the very common SPM and STM was invented by Binnig and Rohrer in the 1980s. A sharp tip is positioned very close to the surface of the specimen in an STM instrument and the tip is scanned through the sample.

One major drawback of the STM is that it is unable to image nonconducting surfaces, since electronic current is involved in the measurements. Another measurement technique, AFM, was proposed by Binnig et al. [55], not long after the invention of STM. The probe was allowed to contact the surface directly and the resulting probe deflection can be measured by optical methods to generate an image. In this technique, the interaction force between the tip and the sample surface is recorded. AFM provides information about the mechanical properties of the surface, but not of electronic properties.

In some AFM instruments, the static mode is used to obtain force–distance curves, which then can be analyzed to produce the local, reduced Young's modulus. In practice, these values are not measured directly and the raw data must be calibrated and converted to produce real displacement and force data. Although absolute values of mechanical properties are difficult to measure, relative measurements are still promising. Another difficulty is that the movement of the tip on the sample can occur not only vertically but also along the cantilever axis, which limits the validity of applied contact mechanics [32]. Because of these limitations, the static AFM method is not widely used especially in the case of stiff surfaces. Using acoustic vibrations of the AFM sample surface, we can access local elastic and inelastic properties of the specimen.

20.7.1 AFM Measurements: Mechanical Properties of CNTs

Mechanical properties of solids such as elasticity, inelasticity, and plasticity are usually measured on a macroscopic scale for materials. But recently new techniques such as AFM and STM have been developed to study these properties on the nanoscale region. CNTs are one of the promising nanomaterials that are widely used nowadays. Mechanical measurements on CNTs performed with the AFM have confirmed theoretical expectations [51] of their superior mechanical properties. As we mentioned previously, there are several ways to measure the mechanical properties of CNTs. One method is to bend the CNT by the AFM tip and calculate the Young's modulus from the force–displacement curve. Another method, which is a more direct measurement of the elastic properties of CNTs, has been performed on MWCNTs [33] and SWCNT ropes [37] under axial strain. Two AFM tips were used to hold MWCNTs or SWCNT ropes and then tensile strain was applied to them portioned inside the scanning electron microscope (SEM). The AFM tips were integrated with different cantilevers, one rigid with a spring constant of 20 N/m and the other compliant with a spring constant of 0.1 N/m, as depicted in **Figure 20.13**.

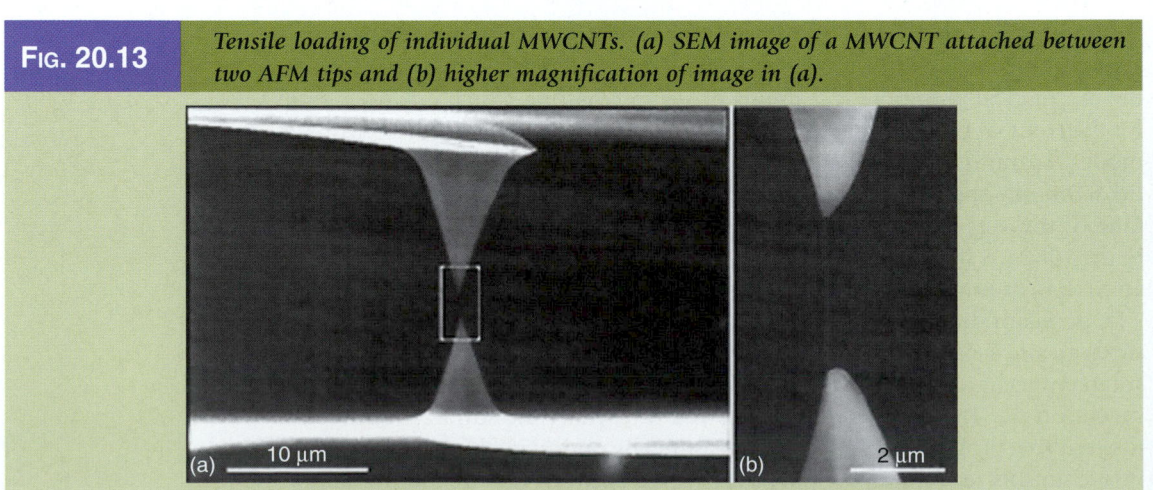

Fig. 20.13 *Tensile loading of individual MWCNTs. (a) SEM image of a MWCNT attached between two AFM tips and (b) higher magnification of image in (a).*

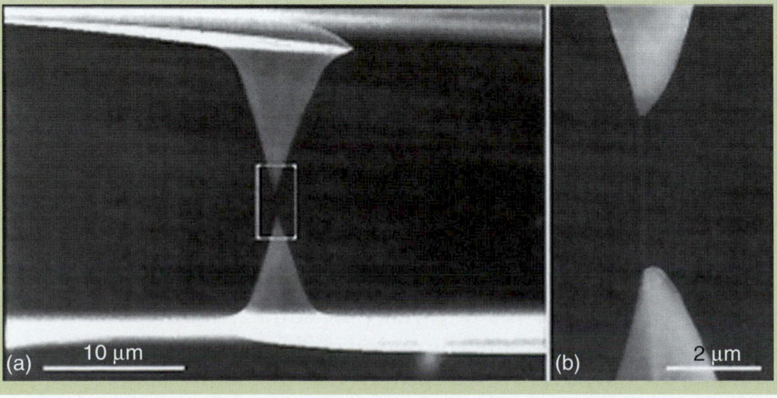

Source: M. F. Yu, O. Lourie, M. J. Dyer, K. Moloni, T. F. Kelley, and R. S. Ruoff, *Science*, 287, 637–640 (2000). With permission.

Fig. 20.14 *(a) Schematic representation of two AFM tips straining the MWCNTs and (b) Plot of stress versus strain for different individual MWCNTs.*

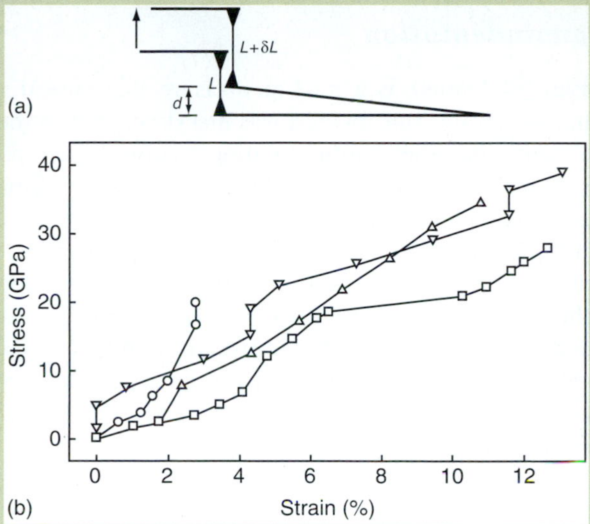

Source: M. F. Yu, O. Lourie, M. J. Dyer, K. Moloni, T. F. Kelley, and R. S. Ruoff, *Science*, 287, 637–640 (2000). With permission.

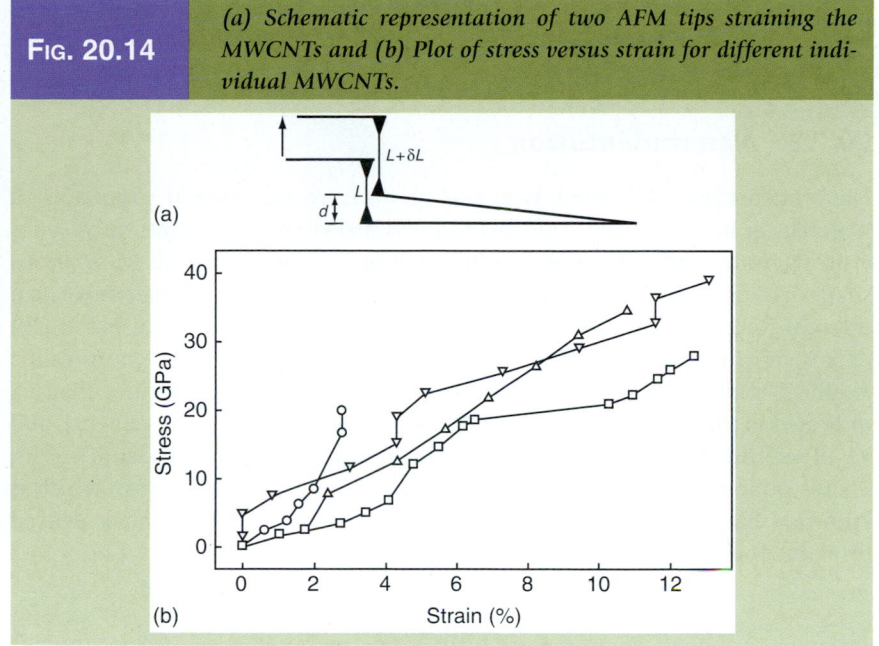

The rigid cantilever was driven using a linear piezomotor. On the other end, the compliant lever was bent due to the applied tensile force. The force is calculated as $F = kd$ where k is the spring constant of the flexible AFM cantilever and d its displacement in the vertical direction. The strain of the nanotube is $\delta L/L$, which is shown in **Figure 20.14**. The stress–strain curve was derived from the force–displacement data points and then the Young's modulus was obtained. For this setup configuration, Young's modulus values ranging from 270 to 950 GPa were found.

All these measurements for the mechanical properties of CNTs that were mentioned in this and previous sections are summarized in **Table 20.6**.

Table 20.6 *Summary of the Mechanical Properties of CNTs Measured Using SPM Methods*

Young's modulus E (GPa)	Tensile strength σ (GPa)	Shear modulus G (GPa)	Nanotube type	Method of testing	Ref.
1300 ± 600	—	—	MWCNTs arc grown	Lateral bending	[35]
1000 ± 600	—	—	SWCNTs	Normal bending	[36]
1000 ± 600	—	~1	SWCNT ropes	Normal bending	[40]
1020	30	—	SWCNT ropes	Tensile loading	[37]
870 ± 400	—	—	MWCNT arc grown	Normal bending	[36]
270–950	11–63	—	MWCNTs arc grown	Tensile loading	[33]
400	—	—	SWCNT rope	Normal bending	[56]
12±6	—	—	MWCNT catalytic	Normal bending	[36]
—	45±7	—	SWCNT rope	Lateral bending	[34]

It should be noted that the absolute values of mechanical properties have large uncertainties due to the huge influence of the precision of tube diameters and lengths used in the experiments.

20.7.2 Nanoindentation

Nanoindentation is defined as a tool to measure mechanical properties of materials at the nanoscale. This method was first developed in the early 1980s evolving from traditional Vicker hardness testing. Nanoindenter tips with various shapes are used in this technique to analyze the resistance of materials to an external force. A schematic representation of continuous load–displacement data is shown in **Figure 20.15**, which consists of three different regions called loading, holding, and unloading regions. Some important quantities that are indicated in this figure are P_{max} (peak load), h_{max} (maximum displacement), and $S = dP/dh$ (the slope of unloading curve at maximum indentation depth).

It is noticeable that S has the dimensions of force per unit length, which is known as elastic contact stiffness. The elastic modulus of materials can be derived using the following equations:

$$E_r = \frac{(\sqrt{\pi} \cdot S)}{2\beta\sqrt{A}} \quad (20.84)$$

where

β is a constant that depends only on the geometry of the indenter
A is the projected contact area

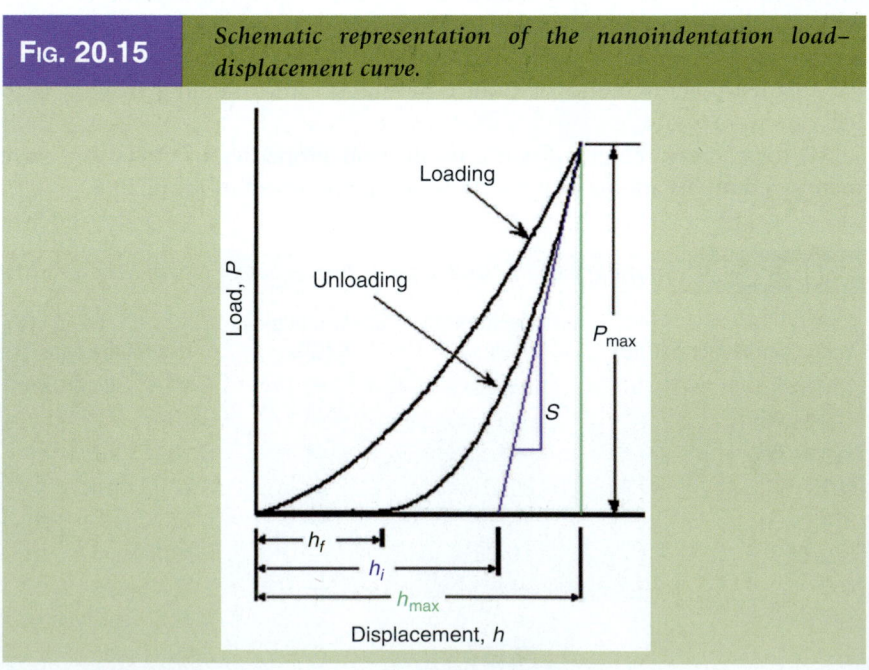

FIG. 20.15 *Schematic representation of the nanoindentation load–displacement curve.*

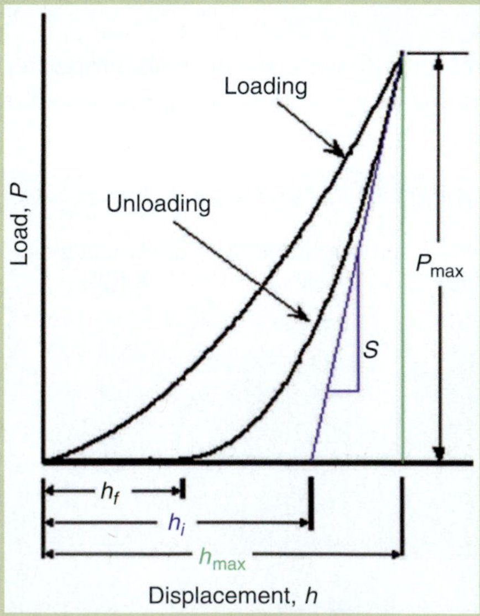

E_r is called the reduced modulus, a parameter that considers the effect of a non-rigid indenter on the load–displacement behavior. The elastic modulus E of the test material is calculated using the following expression:

$$\frac{1}{E_r} = \frac{1-\nu^2}{E} + \frac{1-\nu_i^2}{E_i} \tag{20.85}$$

where

ν is the Poisson's ratio of the test material

E_i and ν_i are the elastic modulus and Poisson's ratio of the indenter, respectively.

It may seem inappropriate that we have to know the material's Poisson's ratio in order to calculate its modulus. But, even a rough number, say $\nu = 0.25 \pm 0.1$, produces only about a 5% error in the calculation of elastic properties of most materials. For different types of nanoindenter tips, a specific β constant is used for calculations. For indenters with square cross sections such as the Vickers pyramid, $\beta = 1.012$; for triangular cross sections such as the Berkovich and the cubic corner indenters, $\beta = 1.034$.

An additional mechanical property that is usually obtained from nanoindentation is hardness H as defined below:

$$H = \frac{P}{A} \tag{20.86}$$

where P is the applied load and A is the projected contact area of indentation at load P as a function of contact depth. Therefore, we must obtain the elastic contact stiffness (S) and projected contact area (A) in order to be able to derive the elastic modulus and hardness. There are two distinct methods that measure the stiffness and projected contact area, that is, the continuous stiffness measurement (CSM) method and the unloading stiffness measurement (USM) method.

In the CSM method, a small oscillating force is applied either to the sample or indenter during the indentation period and the contact stiffness value is calculated from the displacement response against the depth of indentation. Once the stiffness of contact S is defined, the elastic modulus and hardness can be obtained using equations (20.84)–(20.86).

In the USM method, there are several ways to calculate the contact stiffness. The method of Oliver and Pharr [57] is the most widely used. According to the method, data analysis procedure begins by fitting the load–displacement data acquired during unloading to the power–law relation:

$$P = B(h - h_f)^m \tag{20.87}$$

where

P is the applied load to the test surface

h is the resulting penetration

B and m are empirically determined fitting parameters

h_f is the final displacement after complete unloading

The contact stiffness S is then obtained using

$$S = \frac{dP}{dh} = Bm(h_{max} - h_f)^{m-1} \tag{20.88}$$

It is worth noting that in nanoindentation the projected contact area is not obtained by optical imaging. Rather it is computed as a function of the contact depth h_c, applying an empirical relationship given by

$$A = f(h_c) = C_0 h_c^2 + C_1 h_c + C_2 \sqrt{h_c} \tag{20.89}$$

The constants C_0, C_1, and C_2 are determined prior to the experiment by indenting a sample of known properties such as fused silica. The contact depth h_c is different from the total penetration depth h, and is estimated using

$$h_c = h - \varepsilon \frac{P}{S} \tag{20.90}$$

where ε is a constant that depends only on the geometry of the indenter. For cones, $\varepsilon = 0.72$ and for spheres, $\varepsilon = 0.75$. There is empirical justification for using $\varepsilon = 0.75$ for Berkovich and Vickers tips as well.

20.8 Nano-Microelectromechanical Systems (NEMS/MEMS)

MEMS are based on devices that have a characteristic length of 1 mm or less but more than 100 nm and consist of electrical and mechanical parts. The similar term commonly used in Europe is microsystem technology (MST) and in Japan it is called micromachines.

NEMS refer to nanoscopic devices that have a characteristic length of 100 nm or less and combine the electrical and mechanical parts. **Figure 20.16** compares MEMS and NEMS in the size range and their applications. These devices (NEMS/MEMS) are referred to as intelligent miniaturized systems that consist of sensing, processing, and/or actuating parts and combine mechanical and electrical components and operations.

20.8.1 MEMS Fabrication Techniques

In this section, we will discuss various MEMS fabrication techniques that are commonly used to fabricate different microdevices such as sensors and actuators. Micromachining is one of the most important steps in fabricating MEMS, and is described in more detail in the following sections.

Bulk Micromachining. Bulk micromachining is the oldest MEMS technology available, which is currently by far the most successful in manufacturing MEMS devices, such as pressure sensors and ink-jet printer heads. The basic concept behind micromachining is to remove materials selectively. This can lead to the

FIG. 20.16 *Comparison of NEMS and MEMS dimensionality.*

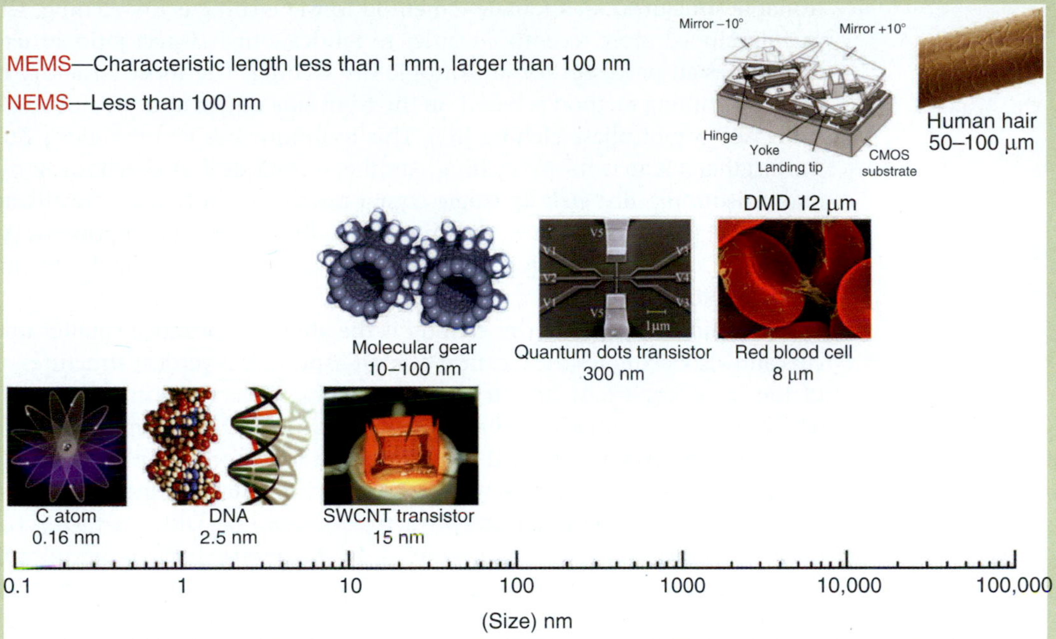

Source: A. P. Graham, G. S. Duesberg, R. Seidel, M. Liebau, E. Unger, F. Kreupl, and W. Hoenlein, *Diamond and Related Materials*, 13, 1296–1300 (2004); W. G. van der Wiel, S. De Franceschi, J. M. Elzerman, T. Fujisawa, S. Tarucha, and L. P. Kouwenhoven, *Reviews of Modern Physics*, 75, 1–22 (2003); Texas Instruments DLP Products, Plano, TX, http://www.dlp.com. With permission.

creation of different micromechanical components such as beams, plates, or membranes that are used to fabricate movable parts in MEMS devices.

The use of anisotropic wet etchants to remove silicon is the basic micromachining technique used in the semiconductor industry. Backside wet etching is used in order to create movable parts, as depicted in **Figure 20.17**.

One of the basic ways to control the etching process is to remove the etchant from the substrate at a specified time or thickness. But this technique is unable to produce features that are thinner than 20 μm. Therefore, new etch stop techniques

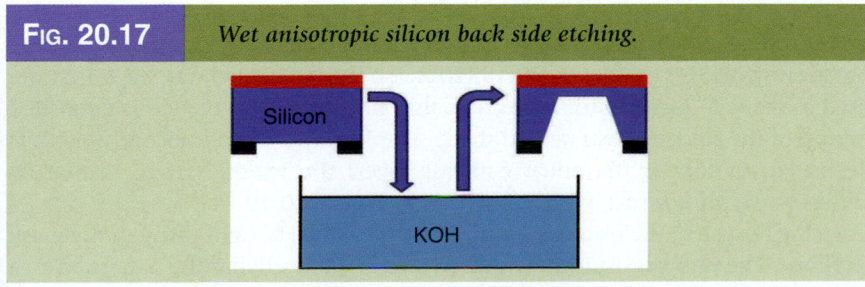

FIG. 20.17 *Wet anisotropic silicon back side etching.*

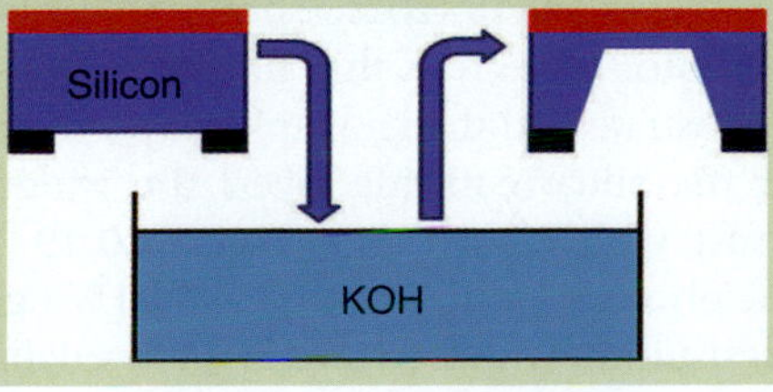

Source: B. Bhushan, *Springer handbook of nanotechnology*, Springer-Verlag, Berlin, Germany (2004). With permission.

were introduced to have more precise results. Doped regions and electrochemical bias are used to slow down or stop the etch process, and hence create more controllable structures. An alternative method to wet etching is dry etching, which was developed more recently in order to fabricate high-aspect-ratio structures and to design processes for anisotropic dry etching. The most basic dry bulk micromachining method is based on the front side undercut of microstructures using XeF_2 vapor phase etching [61]. This technique has its limitations due to the fact that it is an isotropic etching. Another technique that uses both isotropic and anisotropic dry etch is single-crystal-reactive etching and metallization (SCREAM) [62], which can create structures with suspended components. Most of the dry etching techniques are plasma based and have several advantages compared to wet etching.

The main advantage of dry etching is the ability to produce smaller undercuts and also facilitates the creation of high-aspect-ratio vertical structures. One of the most important dry etching techniques is reactive ion etching (RIE), which combines both physical and chemical processes. In this method, the surface of the material is activated by the incident ions from the plasma and then reactive species react with the material producing faster etching in the vertical direction. Another technique, deep reactive ion etching (DRIE), which is based on ion etching, was recently introduced. In this method, the passivation and etching steps are performed in a two-step cycle sequence. The DRIE technique is capable of achieving an aspect ratio of 30:1 and silicon etching rates of 2–3 μm/min [32].

Surface Micromachining. Another important new fabrication technique that is widely used to create movable parts on top of a silicon substrate is surface micromachining [63]. This technique is based on the deposition of thin films of polysilicon and other structural films on top of a sacrificial layer that is subsequently removed by etching. This process creates a movable micromechanical structure that can be integrated with on-chip electronics to produce MEMS devices. One of the major advantages of this technique is that extremely small sizes can be obtained.

The basic surface micromachining process is depicted in **Figure 20.18** [32]. As can be seen from the figure, the process starts with depositing a sacrificial layer on top of the silicon substrate and then the sacrificial layer is patterned. Furthermore, the structural layer is deposited and then patterned, which would be anchored to the substrate through the opening created during the previous step. Finally, the sacrificial layer is removed by etching and a movable part is obtained.

One of the most widely used techniques in industry is called multiuser MEMS processes (MUMPs). **Figures 20.19–20.26** show the sequential steps that are used to make a micromotor [64,65]. A thin film of silicon nitride is deposited on top of the silicon substrate and then a blanket layer of polysilicon (poly 0) is deposited on top of the silicon nitride layer; the wafer is then coated by UV-sensitive photoresist, which is shown in **Figure 20.19**.

In the next step, the photoresist is patterned using UV light through the mask level one. The photoresist in the exposed area is removed and the patterned area is left behind, as shown in **Figure 20.20**.

Fig. 20.18 *Surface micromachining fabrication process.*

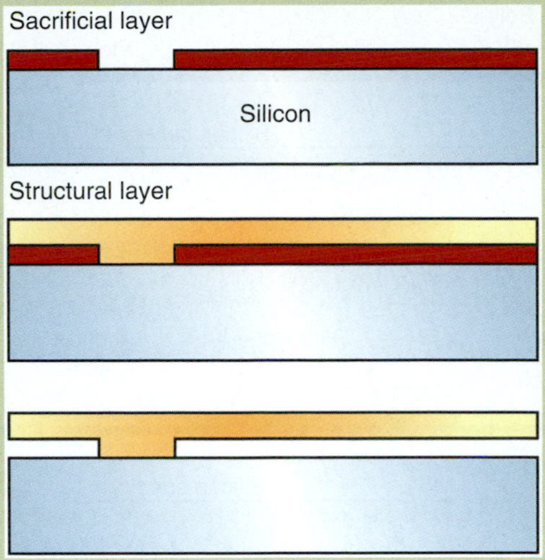

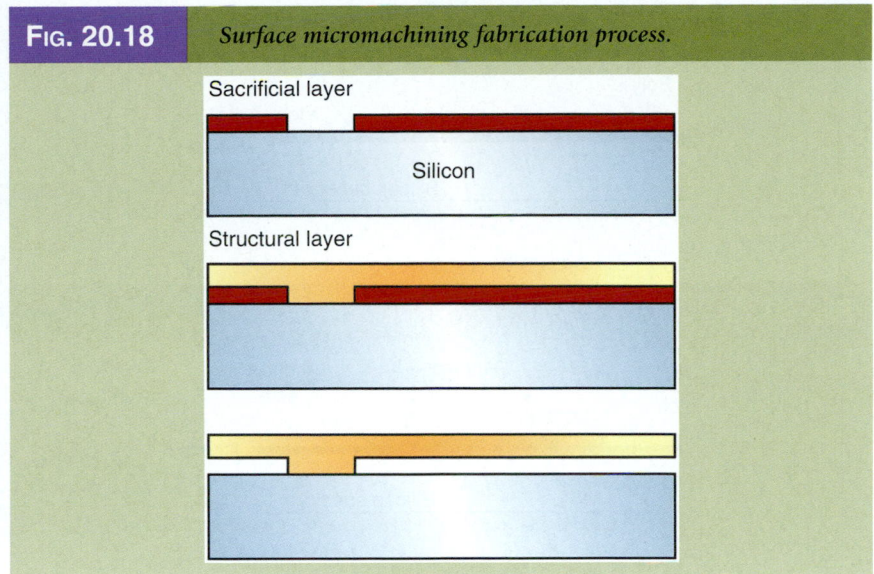

The unwanted part of (poly 0) is etched using RIE and, after this step, the photoresist layer is totally etched away by putting the structure in a solvent bath (**Fig. 20.21**). This method of patterning the wafers by photoresist and then etching the remaining photoresist is a commonly used method in MUMPs.

The first sacrificial layer is deposited on top of the previous layers and then is patterned lithographically using its photo mask. The unwanted oxide layer is removed applying RIE and the remaining photoresist is stripped, see **Figure 20.22**.

Fig. 20.19 *Schematic of micromotor deposition steps, step 1.*

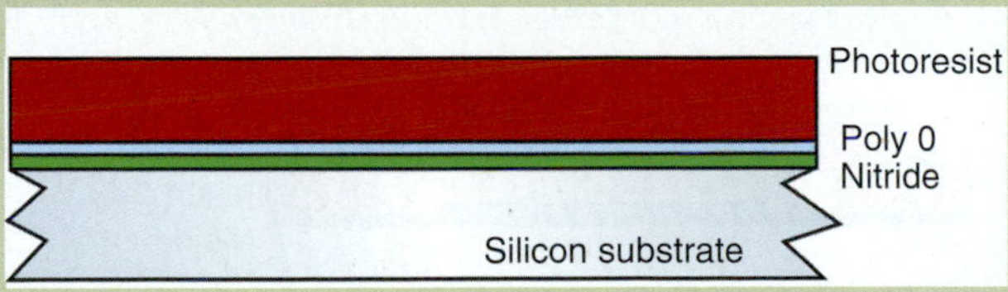

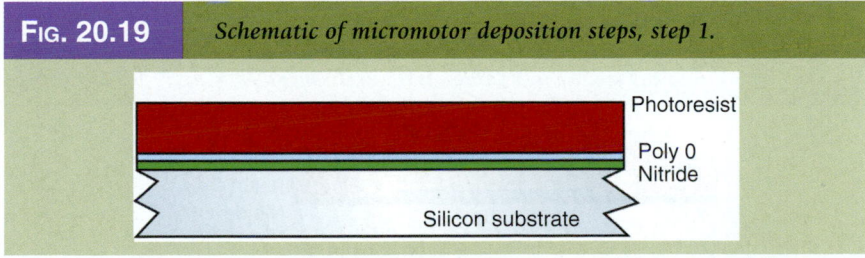

Fig. 20.20 *Schematic of micromotor deposition steps, step 2.*

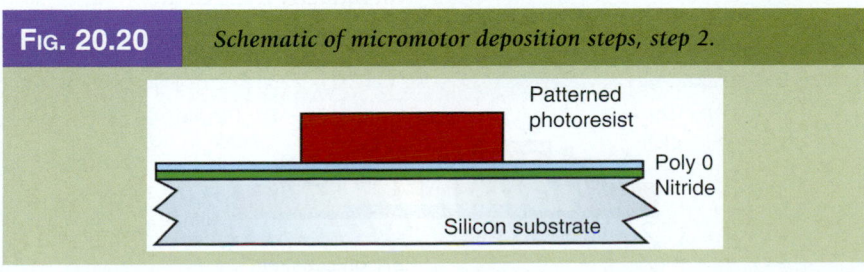

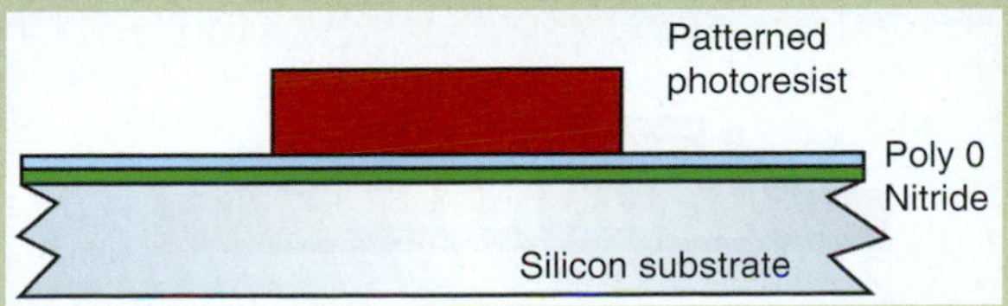

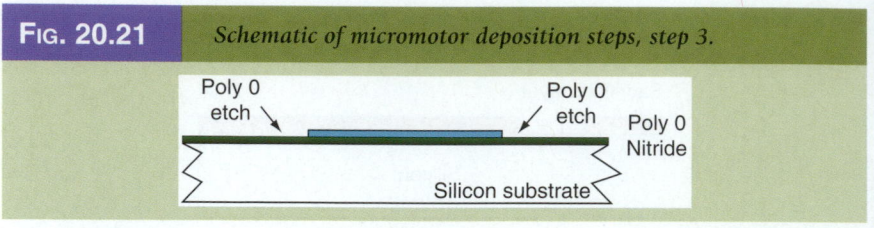

FIG. 20.21 *Schematic of micromotor deposition steps, step 3.*

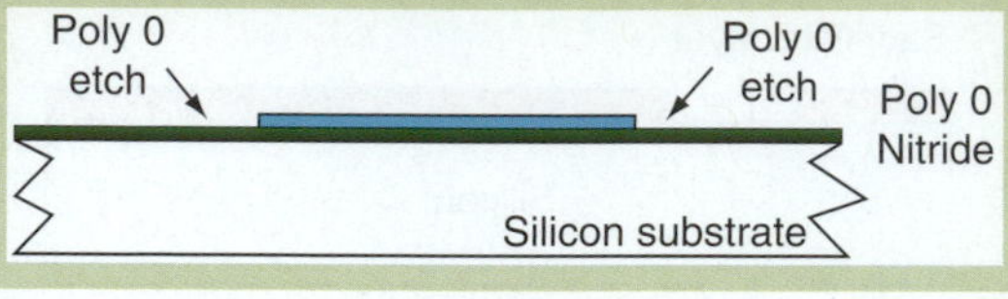

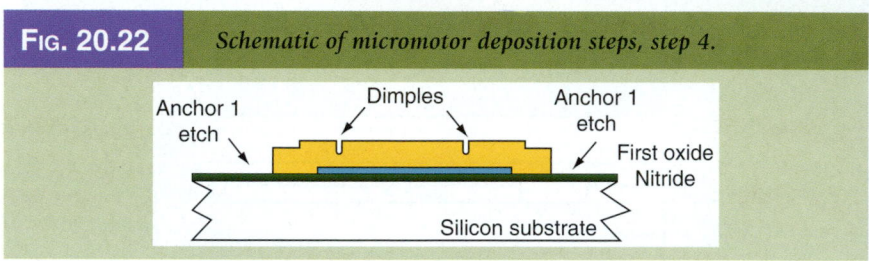

FIG. 20.22 *Schematic of micromotor deposition steps, step 4.*

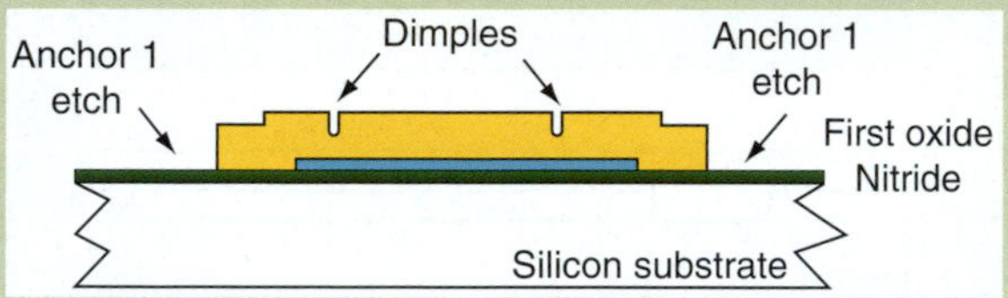

Another layer of polysilicon (poly 1) is coated on top of the existing layers and then lithographically patterned using its specific mask. In the next step, another oxide layer (sacrificial layer) is deposited on top of the layers as depicted in **Figure 20.23**.

The second oxide layer is patterned twice to obtain a contact through the poly 1 and also the substrate layers. The wafer is patterned again and the unwanted oxide layer is removed using an appropriate etchant, as shown in **Figure 20.24**.

Next, the last polysilicon layer (poly 2) is deposited on top of the structure and is patterned through its photoresist mask and the metal layer is deposited on top of poly 2, as shown in **Figure 20.25**.

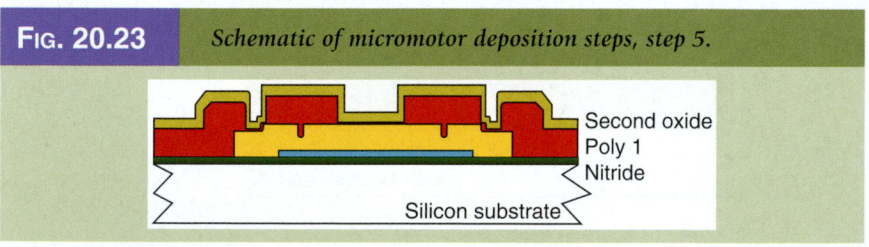

FIG. 20.23 *Schematic of micromotor deposition steps, step 5.*

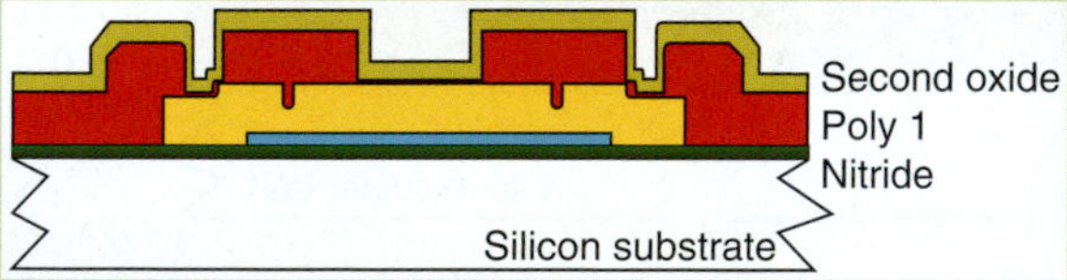

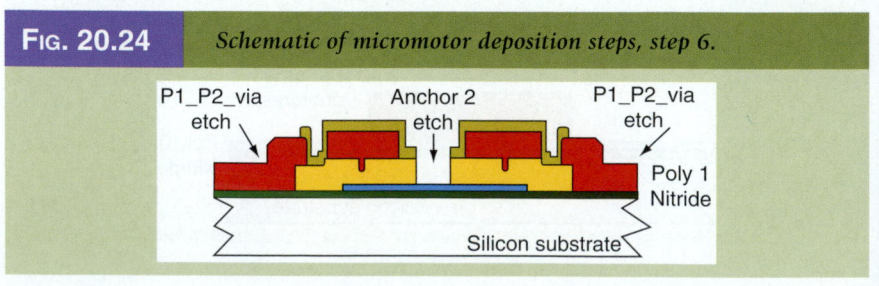

FIG. 20.24 *Schematic of micromotor deposition steps, step 6.*

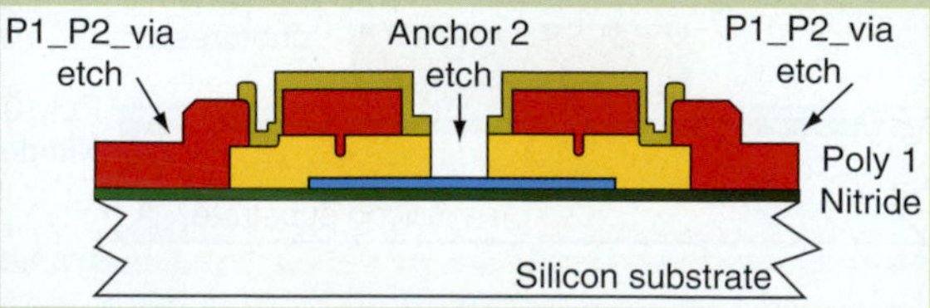

FIG. 20.25 *Schematic of micromotor deposition steps, step 7.*

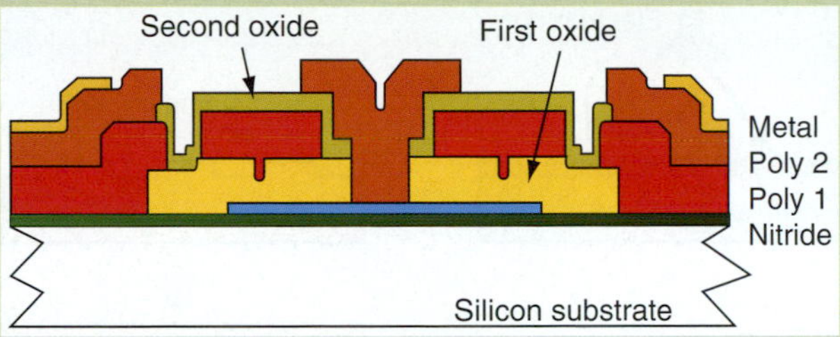

FIG. 20.26 *Schematic of micromotor deposition steps, final step.*

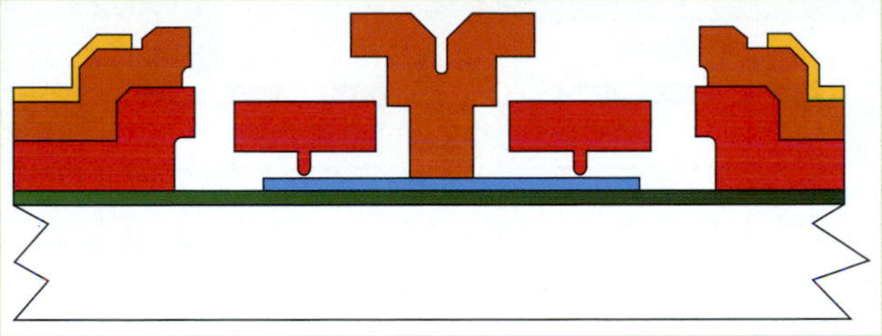

Finally, the structures are released by immersing the chip into 49% HF solution. The (poly 1) "rotor" can be seen around the fixed (poly 2) hub, which can be moved electrostatically as illustrated in **Figure 20.26**.

20.8.2 NEMS Fabrication Techniques

Recent developments in nanotechnology have enabled the production of nanoscale devices such as NEMS and other instruments at the nanometer level. These improvements create new possibilities for next-generation communication and electromechanical devices. For example, as the dimensions shrink to the nanometer level in sensor technology, the sensitivity and accuracy of devices are improved.

NEMS are produced by nanomachining in a top-down approach (from large to small sizes) and bottom-up approach (from small to large scale) [66–70]. The top-down approach is based on fabrication methods that produce nanostructures (similar to micromachining used in MEMS technology) including electron beam lithography and STM writing which removes an atom at a time from the surface. The bottom-up approach includes chemical synthesis, thin film deposition techniques, molecular beam epitaxy (MBE), and various plasma techniques that could be physical or chemical.

As discussed in the previous section, UV and x-rays are two major lithographic techniques that are commonly used in MEMS fabrications but UV lithography does not provide nanometer resolution, and x-ray masks are difficult to make, while lithography by x-rays has its own safety problems. Therefore, more precise techniques such as *e*-beam lithography are mostly used in patterning the

Fig. 20.27 *Schematic illustration of nanoimprint fabrication.*

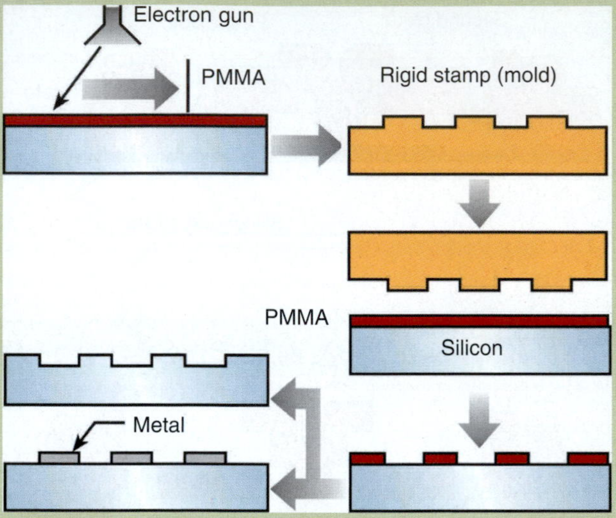

Source: B. Bhushan, *Springer handbook of nanotechnology*, Springer-Verlag, Berlin, Germany (2004). With permission.

layers [71]. This method uses an electron beam to expose an electron sensitive resist such as polymethylmethacrylate (PMMA) dissolved in trichlorobenzene (positive) or polychloromethylstyrene (negative) [72]. The *e*-beam gun is usually a component part in SEM and TEM instruments, but can obtain resolutions of only 10 nm. Another interesting technique which is being developed is called nano-imprint [73]. This technique uses an *e*-beam to fabricate hard material (mold) to stamp and deform polymeric resist and then these steps are followed by a RIE step to transfer the pattern into the substrate, as depicted in **Figure 20.27**. This technique is very economical, since a large number of nanostructures can be fabricated by a single stamp.

Another useful technique that can be applied in the fabrication of nanostructures is SPM. Electrons emitted from a biased SPM tip can be used in order to expose a resist (the same way as *e*-beam) [74]. Different instruments such as constant current STM, noncontact AFM, and AFM with constant current can be used to obtain the lithographic patterns.

20.8.3 NEMS/MEMS Motion Dynamics

In this section, the theory of beams and cantilevers and their motion is briefly described. Here, we assume that the beam is straight, untwisted, and has a constant cross section, as shown in **Figure 20.28** [75]. Moreover, the beam thickness (d) and width (w) are small compared to its length (l), which reduces the system to a one-dimensional problem. Furthermore, the normal stresses in lateral directions are negligible [76]. So with these assumptions, the only remaining normal stress σ_z can be written as

FIG. 20.28 *Schematic representation of a cantilever and its cross section.*

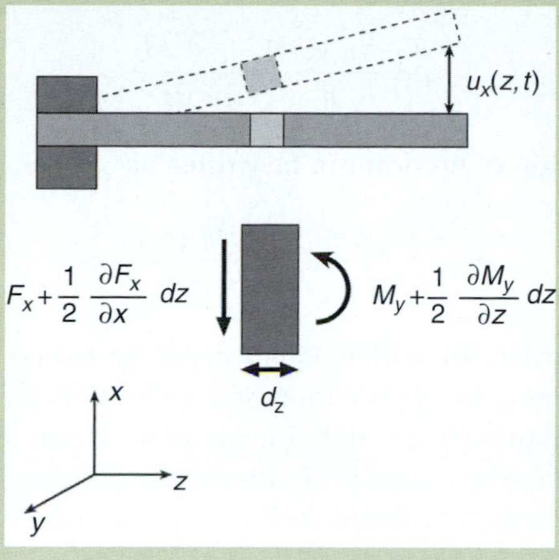

$$\sigma_z = kx \tag{20.91}$$

where k is constant and $x = 0$ is in the center of the beam. With no external momentum applied, the total bending momentum is defined as

$$M = M_y = \int_A x\sigma_z dA = k\int_A x^2 dA, \quad I_y = \int_A x^2 dA \tag{20.92}$$

From previous equations, one can see that $k = M_y/I_y$ and the cross section stress is derived as

$$\sigma_z = E\varepsilon_z = \frac{M_y x}{I_y} \tag{20.93}$$

where E is Young's modulus. If the deflection in the x direction is small, the second derivative of the deflection is the inverse of radius of the curvature r:

$$\frac{\partial^2 u_x(z,t)}{\partial z^2} \approx r^{-1} \tag{20.94}$$

Also the strain can be obtained by $\varepsilon = -x/r$. Therefore, we obtain the Euler–Bernoulli beam theory:

$$M_y = -EI_y \frac{\partial^2 u_x(z,t)}{\partial z^2} \tag{20.95}$$

As we know, the total momentum has to be zero and the equation of motion becomes

$$m\frac{d^2 u_x(z,t)}{dt^2} = \sum F_{\text{int}} = \frac{\partial f_x}{\partial z}dz \tag{20.96}$$

With the mass of the beam given by $m = \rho A dz$ (where ρ is the density of the beam and A is its cross section), the equation of motion becomes

$$\rho A \frac{d^2 u_x(z,t)}{dt^2} = -\frac{\partial^2 M_y}{\partial z^2} \tag{20.97}$$

And the final equation of motion can be written as

$$\rho A \frac{d^2 u_x(z,t)}{dt^2} + EI_y \frac{d^4 u_x(z,t)}{dz^4} = 0 \tag{20.98}$$

This linear fourth-order differential equation can be solved using separation of variables [76–78]. Here, we are not interested in the complete solution; instead we want to find the natural resonant frequency of the beam. This can be obtained by using a Fourier transformation [75]. After solving the Fourier transformation, one can obtain the natural resonant frequency as follows:

$$\omega_i = \frac{\beta_i^2}{l^2}\sqrt{\frac{EI_y}{\rho A}} \tag{20.99}$$

Here, β_i is constant, which depends on the boundary conditions used to solve the differential equation of the cantilever (clamped–free cantilever) or the beam (clamped–clamped beam). The moment of inertia of a beam with different cross sections can be found in the appropriate tables.

20.8.4 MEMS Devices and Applications

MEMS devices are inherently small and thus can offer such attractive characteristics as reduced size, weight, and less power dissipation compared to macroscopic systems. Moreover, these miniaturized systems can obtain better precision and also work at higher speeds than their macroscopic-sized components. In this section, we will review some of the devices that are commercially available nowadays and demonstrate their applications.

Pressure Sensor. Pressure sensors are commercial devices that are widely used in various industrial and biomedical applications. These sensors can be based on four different mechanisms such as piezoelectric, piezoresistive, capacitive, and resonant sensing [32]. In this section two different types of piezoresistive and capacitive pressure sensors are mentioned.

A pressure that is applied to the sensor will deform the silicon band structure, thus altering the resistivity of the material. The device consists of a silicon diaphragm suspended over a vacuum cavity to form a pressure sensor. An external pressure applied to the diaphragm would introduce stress on the sensing resistors, resulting in a resistance change based on the external pressure. These resistors are temperature dependent and consume direct current (DC), which make them less attractive compared to capacitive pressure sensors.

Capacitive pressure sensors are more attractive than piezoresistive pressure sensors since they are temperature independent and do not consume any DC power and are very stable over time. Moreover, complementary metal oxide semiconductor

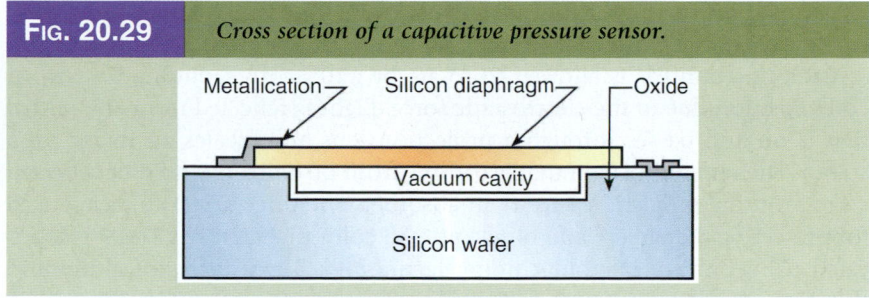

Fig. 20.29 *Cross section of a capacitive pressure sensor.*

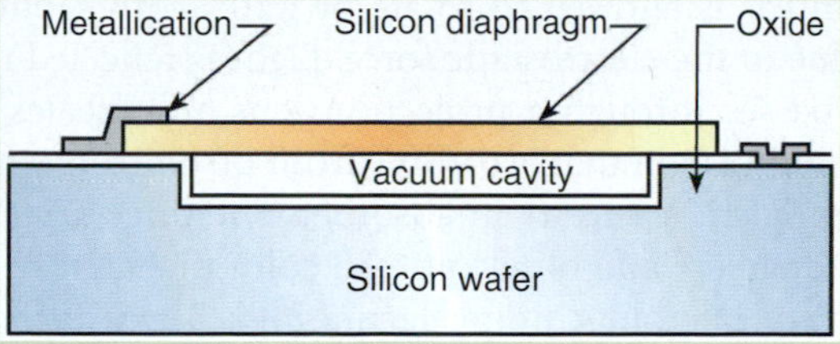

(CMOS) microelectronic circuits can be easily interfaced with these sensors to improve the overall performance of the system. A schematic representation of a capacitive pressure sensor is depicted in **Figure 20.29**.

This device consists of a suspended silicon diaphragm over a vacuum cavity. As the pressure is applied to the outer surface of the device, the diaphragm deflects towards the cavity, resulting in an increase in the capacitance value. This change in capacitance is monitored with electronic devices, and the sensor capacitance values are converted to an output voltage corresponding to the diaphragm position. This voltage is used to generate a feedback signal to the top electrode to maintain the diaphragm at its nominal position.

Digital Displays. The digital micromirror device (DMD) was introduced by Texas Instruments (TI) in 1987. The DMD is an integral part of TIs digital light processing (DLP) technology. This DMD technology can achieve higher resolution and brightness, and produce lightweight projection displays that generate images with higher fidelity and stability [79,80]. This technology has various applications such as computer projectors, high definition television (HDTV), and movie projectors (DLP cinema) [80]. A DMD consists of up to 2.07 million aluminum micromirrors with a typical area of 16×16 μm as illustrated in **Figure 20.30**.

Fig. 20.30 *Two pixels (two micromirrors) of a DMD chip.*

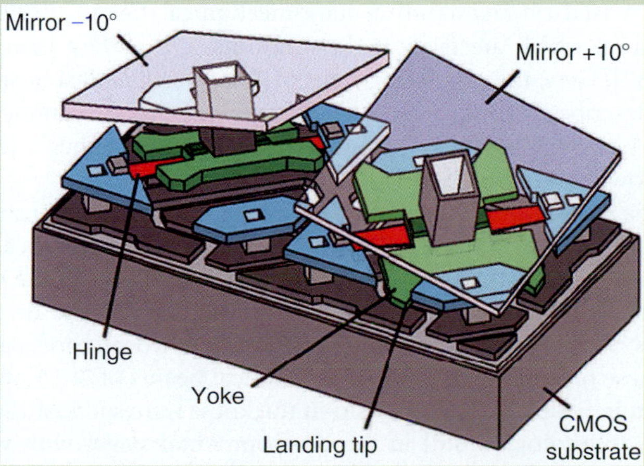

Source: L. J. Hornbeck, *MRS Bulletin*, 26, 325 (2001). With permission.

These micromirrors switch forward and backward thousands of times per second by electrostatic attraction.

Each micromirror is allowed to rotate by ±10°, corresponding to "on" or "off" position due to the electrostatic force. Light is reflected from any mirror that is on and passes through a projection lens and creates an image on a screen. The remaining light that is coming from off-mirrors would be reflected away from the projection lens to an absorber. The three DMD chips are used for projecting red, green, and blue colors in color displays. The DMD is fabricated using surface micromachining technology. Three layers of aluminum thin films are deposited and patterned to form the mirror and its suspension structure.

Polymeric material is used as a sacrificial layer and is removed by plasma etching to produce the suspended micromirror structures. This fabrication process is also compatible with CMOS fabrication, which enables the manufacturer to achieve a higher yield and lower cost compared to the fabrication of these DMD devices without any underlying circuit technology.

20.8.5 NEMS Devices and Applications

NEMS devices are fabricated using either top-down or bottom-up approaches. The top-down approach incorporates more precise lithographic techniques, such as electron beam lithography, which give better accuracy compared to the MEMS lithographic techniques, and the device can be fabricated by employing etching and lithography steps on the bulk material. This top-down approach is widely used to manufacture NEMS devices because it is based on existing technology that is used to produce integrated circuits (ICs) and MEMS structures. The only significant difference between the nanoscale and microscale processing steps is the patterning method used for various features. In contrast, the bottom-up approach follows the same path that nature constructs objects, by assembling atoms and molecules on top of one another.

In this section, the goal is to introduce some of the early developments in the field on NEMS devices and their applications. The first generation of NEMS devices are based on freestanding nanomechanical beams, oscillators, and tethered plates, which are fabricated using bulk and surface nanomachining processes [32]. Here, the processing steps of a nanomechanical beam of silicon are briefly described. Carr et al. [81] fabricated a submicron clamped–clamped mechanical beam and suspended plates with a nanometer tether. The processing steps are shown in **Figure 20.31**.

In the first step, PMMA is deposited on top of the silicon-on-insulator (SOI) substrate, and then the PMMA layer is patterned using electron beam lithography. An aluminum film is then deposited and patterned into the silicon etch mask. The nanomechanical beam is then patterned by RIE and the underlying SiO_2 layer is removed by immersing the device in a hydrofluoric acid solution. Applying these processing steps, nanomechanical beams of 7–16 μm in length, 120–200 nm in width, and 50–200 nm in thickness were successfully fabricated [81]. NEMS technology is still in the developmental stages with very limited commercial success. Nevertheless, NEMS devices have been used for precision measurements [82] and for probing the material's properties at the nanometer level [83,84].

Fig. 20.31 *Cross-sectional schematics of a process to fabricate nanomechanical structures using silicon on insulator substrates.*

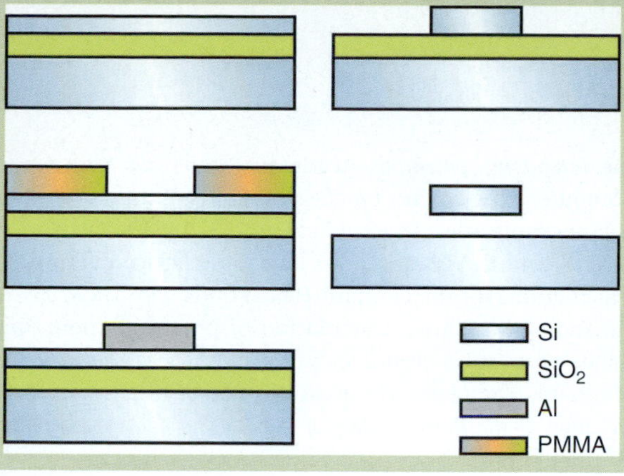

Source: B. Bhushan, *Springer handbook of nanotechnology*, Springer-Verlag, Berlin, Germany (2004). With permission.

20.9 Summary

In the first part of the chapter, the foundations of nanomechanics were introduced. A simple two-atom chain of a molecule was briefly discussed in order to derive the interaction potentials associated with a specific material and being able to calculate the force generated due to this potential. Also this approach was expanded to three and n-atom chains to study the static and dynamic behavior of materials at the nanoscale.

MD, a simulation technique, was reviewed and its applications on different systems reveal that failure mechanisms of materials at the nanoscale can be different from macroscopic failures. The tensile failure of a gold nanowire specimen shows an incredible ductility during its elongation, which is manifested in the elongation of extremely thin atom chains. Furthermore, the thermal stability of a nanosystem was modeled which indicated that these nanosystems are very sensitive to their initial conditions and number of particles used for the simulation.

CNTs exhibit fascinating mechanical properties which, compared to steel, have 1/6 the weight of steel, 5 times its Young's modulus, 100 times its tensile strength, and can be strained up to 15% without fracture. Nanomechanical measurement devices such as the AFM, STM, and nanoindenters can be used to obtain the mechanical properties of material experimentally at the nanometer level.

NEMS/MEMS devices and their applications are briefly mentioned in the last section of this chapter. Their different fabrication techniques were introduced and some commercial products that are based on this technology were reviewed.

Acknowledgments

We are especially grateful to Professor Ivar Reimanis, who has reviewed this chapter and provided valuable feedback. We are also particularly indebted to the following authors who provided some of the figures and plots: M. Reith,

S. M. Foiles, M. L. Baskes, M. S. Daw, W. Schommers, S. Baskouts, M. Meyyappan, Ali Yazdani, M. F. Yu, O. Lourie, M. J. Dyer, K. Moloni, T. F. Kelley, R. S. Ruoff, B. Bhushan, M. D. Ventra, S. Evoy, J. R. Heflin, G. Genta, and L. J. Hornbeck.

References

1. I. M. Torrens, *Interatomic potentials*, Academic Press, New York, London (1972).
2. L. Verlet, Computer experiments on classical fluids. I. thermodynamical properties of Lennard-Jones molecules, *Physical Review*, 159, 98 (1967).
3. R. G. Della Valle and E. Venuti, Quasi harmonic lattice dynamics and molecular dynamics calculations for the Lennard-Jones solids, *Physical Review B*, 58, 1 (1998).
4. T. Halicioglu and G. M. Pound, Calculation of potential energy parameters form crystalline state properties, *Physica Status Solidi A-Applied Research*, 30, 619 (1975).
5. R. A. Buckingham, The classical equation of state of gaseous helium, neon and argon, *Proceedings of the Royal Society of London Series A-Mathematical and Physical Sciences*, 168, 264 (1938).
6. K. E. Drexler, *Nanosystems: Molecular machinery, manufacturing, and computation*, John Wiley & Sons, Inc., New York (1992).
7. J. A. Barker, R. O. Watts, J. K. Lee, T. P. Schafer, and Y. T. Lee, Interatomic potentials for krypton and xenon, *Journal of Chemical Physics*, 61, 8 (1974).
8. M. Reith, *Nano-engineering in science and technology*, World Scientific Publishing Co. Pte. Ltd., Singapore (2003).
9. M. S. Daw and M. I. Baskes, Semiempirical, quantum mechanical calculation of hydrogen embrittlement in metals, *Physical Review Letters*, 50, 1285 (1983).
10. E. Clementi and C. Roetti, Roothaan-hartree-fock atomic wavefunctions, *Atomic Data and Nuclear Data Tables*, 14(3–4), 177–478 (1974).
11. S. M. Foiles, M. L. Baskes, and M. S. Daw, Embedded atom method functions for the fcc metals Cu, Ag, Au, Ni, Pd, Pt, and their alloys, *Physical Review B*, 33(12), 7893–7991 (1986).
12. M. S. Daw, S. M. Foiles, and M. L. Baskes, The embedded atom method: A review of theory and applications, *Materials Science Reports*, 9, 251–310 (1993).
13. N. W. Ashcroft and N. D. Mermin, *Solid state physics*, Saunders College, Philadelphia, PA (1976).
14. A. N. Cleland, *Foundations of nanomechanics*, Springer, Berlin, Germany (2003).
15. H. Goldstein, *Classical mechanics*, Addison-Wesley, Reading, MA (1980).
16. Landolt-Bornstein, *Numerical data and functional relationships in science and technology*, Vol. 11, Springer, Berlin (1979).
17. O. Ambacher, Growth and applications of group-III nitrides, *Journal of Physics D: Applied Physics*, 31, 2653–2710 (1998).
18. M. P. Allen and D. J. Tildesley, *Computer simulation of liquids*, Oxford Science Publications, New York (1990).
19. R. W. Hockney, The potential calculation and some applications, *Methods in Computational Physics*, 9, 136 (1970).
20. W. C. Swope, H. C. Andersen, P. H. Berens, and K. R. Wilson, WA computer simulation method for the calculation of equilibrium constants for the formation of physical clusters of molecules: Application to small water clusters, *Journal of Chemical Physics*, 76, 637 (1982).
21. A. Nordseick, On numerical integration of ordinary differential equations, *Mathematics of Computation*, 16, 22 (1962).
22. C. W. Gear, *Numerical initial value problems in ordinary differential equations*, Englewood Cliffs, NJ, Prentice Hall, NJ (1971).
23. V. M. Harik and M. D. Salas, *Trends in nanoscale mechanics*, Kluwer Academic Publishers, the Netherlands (2003).

24. D. W. Brenner, Empirical potential for hydrocarbons for use in simulating the chemical vapor deposition of diamond films, *Physical Review B*, 42, 9458–9471 (1990).
25. C. M. Lieber, Nanoscale science and technology: Building a big future from small things, *MRS Bulletin*, 28(7), 486–491 (2003).
26. P. Yang, The chemistry and physics of semiconductor nanowires, *MRS Bulletin*, 30(2), 85–91 (2005).
27. H. S. Park and J. A. Zimmerman, Modeling inelasticity and failure in gold nanowires, *Physical Review B*, 72, 054106 (2005).
28. M. Reith, W. Schommers, and S. Baskouts, Thermal stability and specific material properties of nanosystems, *Modern Physics Letters B*, 14, 621 (2000).
29. S. Iijima and T. Ichihashi, Single-shell carbon nanotubes of 1 nm diameter, *Nature*, 363, 603–605 (1993).
30. D. S. Bethune, C. H. Kiang, M. S. de Vries, G. Gorman, R. Savoy, J. Vazquez, and R. Bayers, Cobalt-catalysed growth of carbon nanotubes with single-atomiclayer walls, *Nature*, 363, 605–607 (1993).
31. M. Meyyappan, *Carbon nanotubes: Science and applications*, CRC Press, Boca Raton, FL (2004).
32. B. Bhushan, *Springer handbook of nanotechnology*, Springer-Verlag, Berlin, Germany (2004).
33. M. F. Yu, O. Lourie, M. J. Dyer, K. Moloni, T. F. Kelley, and R. S. Ruoff, Strength and breaking mechanism of multiwalled carbon nanotubes under tensile load, *Science*, 287, 637–640 (2000).
34. D. A. Walters, L. M. Ericson, M. J. Casavant, J. Liu, D. T. Colbert, K. A. Smith, and R. E. Smalley, Elastic strain of freely suspended single-wall carbon nanotube ropes, *Applied Physics Letters*, 74, 3803–3805 (1999).
35. E. W. Wong, P. E. Sheehan, and C. M. Lieber, Nanobeam mechanics: Elasticity, strength and toughness of nanorods and nanotubes, *Science* 277, 1971–1975 (1997).
36. J. P. Salvetat, J. M. Bonard, N. H. Thomson, A. J. Kulik, L. Forró, W. Benoit, and L. Zuppiroli, Mechanical properties of carbon nanotubes, *Applied Physics A-Materials Science and Processing*, 69, 255–260 (1999).
37. M. F. Yu, B. S. Files, S. Arepalli, and R. S. Ruoff, Tensile loading of ropes of single wall carbon nanotubes and their mechanical properties, *Physical Review Letters*, 84, 5552–5555 (2000).
38. O. Lourie and H. D. Wagner, Evaluation of Young's modulus of carbon nanotubes by micro-Raman spectroscopy, *Journal of Materials Research*, 13 (9), 2418–2422 (1998).
39. J. P. Salvetat, A. J. Kulik, J. M. Bonard, G. A. D. Briggs, T. Stockli, K. Metenier, S. Bonnamy, F. Beguin, N. A. Burnham, and L. Forro, Elastic modulus of ordered and disordered multiwalled carbon nanotubes, *Advanced Materials*, 11(2), 161–165 (1999).
40. J. P. Salvetat, G. A. D. Briggs, J. M. Bonard, R. R. Bacsa, A. J. Kulik, T. Stockli, N. A. Burnham, and L. Forro, Elastic and shear moduli of single-walled carbon nanotube ropes, *Physical Review Letters*, 82(5), 944–947 (1999).
41. M. M. J. Treacy, T. W. Ebbesen, and J. M. Gibson, Exceptionally high Young's modulus observed for individual carbon nanotubes, *Nature*, 381(6584), 678–680 (1996).
42. A. Krishnan, E. Dujardin, T. W. Ebbesen, P. N. Yianilos, and M. M. J. Treacy, Young's modulus of single-walled nanotubes, *Physical Review B*, 58(20), 14013–14019 (1998).
43. P. Poncharal, Z. L. Wang, D. Ugarte, and W. A. de Heer, Electrostatic deflections and electromechanical resonances of carbon nanotubes, *Science*, 283(5407), 1513–1516 (1999).
44. M. F. Yu, M. J. Dyer, J. Chen, and K. Bray, Multiprobe nanomanipulation and functional assembly of nanomaterials inside a scanning electron microscope, in: International Conference IEEE-NANO2001, Maui (2001).
45. D. A. Dikin, X. Chen, W. Ding, G. J. Wagner, and R. S. Ruoff, Resonance vibration of amorphous SiO_2 nanowires driven by mechanical or electrical field excitation, *Journal of Applied Physics*, 93, 226 (2003).

46. G. Overney, W. Zhong, and D. Tomanek, Structural rigidity and low-frequency vibrational-modes of long carbon tubules, *Zeitschrift fur Physik D-Atoms Molecules and Clusters*, 27(1), 93–96 (1993).
47. G. G. Tibbetts, Why are carbon filaments tubular, *Journal of Crystal Growth*, 66(3), 632–638 (1984).
48. G. H. Gao, T. Cagin, and W. A. Goddard, Energetics, structure, mechanical and vibrational properties of single-walled carbon nanotubes, *Nanotechnology*, 9(3), 184–191 (1998).
49. B. I. Yakobson, C. J. Brabec, and J. Bernholc, Nanomechanics of carbon tubes: Instabilities beyond linear response, *Physical Review Letters*, 76(14), 2511–2514 (1996).
50. S. Timoshenko and J. Gere, *Theory of elastic stability*, McGraw-Hill, New York (1988).
51. J. P. Lu, Elastic properties of carbon nanotubes and nanoropes, *Physical Review Letters*, 79(7), 1297–1300 (1997).
52. N. Yao and V. Lordi, Young's modulus of single-walled carbon nanotubes, *Journal Applied Physics*, 84(4), 1939–1943 (1998).
53. E. Hernandez, C. Goze, P. Bernier, and A. Rubio, Elastic properties of C and BxCyNz composite nanotubes, *Physical Review Letters*, 80(20), 4502–4505 (1998).
54. X. Zhou, J. J. Zhou, and Z. C. Ou-Yang, Strain energy and Young's modulus of single-wall carbon nanotubes calculated from electronic energy-band theory, *Physical Review B*, 62(20), 13692–13696 (2000).
55. G. Binnig and Heinrich, Scanning tunneling microscopy—from birth to adolescence, *Reviews Modern Physics*, 59, 615 (1987).
56. G.-T. Kim, G. Gu, U. Waizmann, and S. Roth, Simple method to prepare individual suspended nanofibers, *Applied Physics Letters*, 80, 1815–1817 (2002).
57. W. C. Oliver and G. M. Pharr, An improved technique to determining hardness and elastic modulus using load and displacement sensing indentation experiments, *Journal of Materials Research*, 7, 1564–1583 (1992).
58. A. P. Graham, G. S. Duesberg, R. Seidel, M. Liebau, E. Unger, F. Kreupl, and W. Hoenlein, Towards the integration of carbon nanotubes in microelectronics, *Diamond and Related Materials*, 13, 1296–1300 (2004).
59. W. G. van der Wiel, S. De Franceschi, J. M. Elzerman, T. Fujisawa, S. Tarucha, and L. P. Kouwenhoven, Electron transport through double quantum dots, *Reviews of Modern Physics*, 75, 1–22 (2003).
60. Texas Instruments DLP Products, Plano, TX, http://www.dlp.com.
61. B. Eyre, K. S. J. Pister, and W. Gekelman, Multi-axis microcoil sensors in standard CMOS, *Proceedings of the SPIE Conference on Micromachined Devices and Components*, Austin, pp. 183–191 (1995).
62. K. A. Shaw, Z. L. Zhang, and N. C. MacDonnald, SCREAM: A single mask single-crystal silicon process for microelectromechanical structures, *Proceedings of the IEEE Workshop Micro Electro Mechanical Systems*, Fort Lauderdale, pp. 155–160 (1993).
63. J. M. Bustillo and R. S. Muller, Surface micromachining for microelectromechanical systems, *Proceedings of the IEEE*, 86(8), 1552–1574 (1998).
64. MCNC, Make a micromotor, http://mems.mcnc.org/smumps/Mumps.html.
65. http://www.ece.wis.edu/~priasmor/mem.html.
66. B. Bhushan, *Handbook of micro/nanotribology*, 2nd ed., CRC press, Boca Raton, FL (1999).
67. G. Timp, ed., *Nanotechnology*, Springer, New York (1999).
68. E. A. Rietman, *Molecular engineering of nanosystems*, Springer, New York (2001).
69. H. S. Nalwa, ed., *Nanostructured materials and nanotechnology*, Academic press, San Diego, CA (2002).
70. W. A. Goddard, D. W. Brenner, S. E. Lyshevski, and G. J. Iafrate, *Handbook of nanoscience, engineering, and technology*, CRC press, Boca Raton, FL (2003).
71. P. Rai-Choudhury, ed., *Handbook of microlithography, micromachining and microfabrication*, SPIE, Bellingham, WA (1997).

72. L. Ming, C. Bao-qin, Y. Tian-Chun, Q. He, and X. Qiuxia, The sub-micron fabrication technology, *Proceedings of 6th International Conference on Solid-State and Integrated-Circuit Technology*, IEEE, pp. 452–455 (2001).
73. S. Y. Chou, Nano-imprint lithography and lithographically induced self-assembly, *MRS Bulletin*, 26, 512–517 (2001).
74. H. T. Soh, K. W. Guarini, and C. F. Quate, *Scanning probe lithography*, Kluwer, Boston, MA (2001).
75. M. D. Ventra, S. Evoy, and J. R. Heflin, *Introduction to nanoscale science and technology*, Springer, New York (2004).
76. G. Genta, *Vibration of structure and mechanics*, Springer (1999).
77. H. A. C. Tilmans, M. Elwenspoek, and J. H. J. Fluitman, Micro resonant force gauges, *Sensors and Actuators A-Physical*, 30, 35 (1992).
78. A. A. Shabana, *Vibration of discrete and continuous systems*, Springer-Verlag, New York (1997).
79. L. J. Hornbeck, Digital light processing update: Status and future applications, *Proceedings of the Society for Imaging and Photooptical Engineering*, 3634, 158 (Projection Displays V) (1999).
80. L. J. Hornbeck, The DMD™ projection display chip: A MEMS-based technology, *MRS Bulletin*, 26, 325 (2001).
81. D. W. Carr and H. G. Craighead, Fabrication of Nanoelectromechanical systems in single crystal silicon using silicon on insulator substrates and electron beam lithography, *Journal of Vacuum Science and Technology B*, 15, 2760–2763 (1997).
82. A. N. Cleland and M. L. Roukes, A nanometre-scale mechanical electrometer, *Nature*, 392, 160–162 (1998).
83. K. Schwab, E. A. Henriksen, J. M. Worlock, and M. L. Roukes, Measurement of the quantum of thermal conductance, *Nature*, 404, 974–977 (2000).
84. S. Evoy, A. Olkhovets, L. Sekaric, J. M. Parpia, H. G. Craighead, and D. W. Carr, Temperature-dependent internal friction in silicon Nanoelectromechanical systems, *Applied Physics Letters*, 77, 2397–2399 (2000).
85. L. Anand and Y. Wei, Mesoscopic modeling of the deformation and fracture of nanocrystalline metals, *IUTAM Symposium*, Beijing, China, 3-10 (2005).
86. L. A. Girifalco, R. A. Lad, Energy of cohesion, compressibility and the potential energy functions of the graphite system, *Journal of Chemical Physics*, 25(4) 693–697 (1956).

Problems

20.1 Calculate the electrostatic coulomb potentials of Li^+ ions in vacuum at the distances of 10, 5, and 1 nm. Also, explain why this potential fails as the particles get closer to one another. (Permittivity of vacuum, $\varepsilon_o = 8.854 \times 10^{-12}$ $C^2/(J \cdot m)$ and $e = 1.602 \times 10^{-19}$ C).

20.2 Determine the equilibrium spacing and binding energy of two Ne atoms using the Lennard–Jones (6-12) potential.

20.3 Plot a graph comparing Barker and Lennard–Jones (6-12) potentials for Xe atoms. Then, calculate the maximum relative error between these two potentials for xenon atoms.

20.4 Calculate the potential energy of an HCl molecule using modified Buckingham potentials. Assume that the spacing is ($r = 2r_m$). Furthermore, evaluate the amount of force that exists between these two atoms at the above spacing.

20.5 Define the natural resonance frequency of the system depicted in **Figure 20.32** using Lennard–Jones (6-12) potential. Assume binding energy of 1.5×10^{-21} J, $r_0 = 0.35$ nm, and $m = 3 \times 10^{-23}$ g.

20.6 Find the eigenvalue $(\omega/\omega_0)^2$ and eigenvector $[A_1\ A_2\ A_3]$ for the system of three argon atoms using the system of equations (equation 20.31).

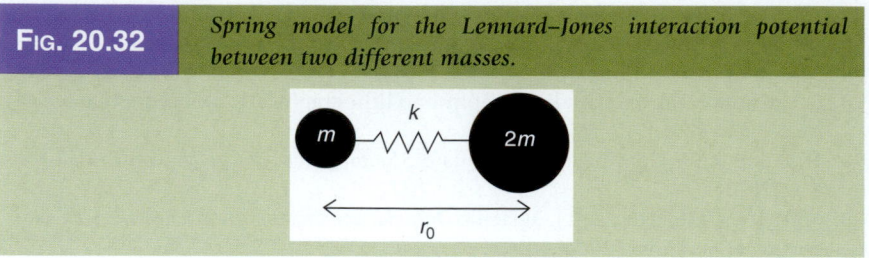

FIG. 20.32 *Spring model for the Lennard–Jones interaction potential between two different masses.*

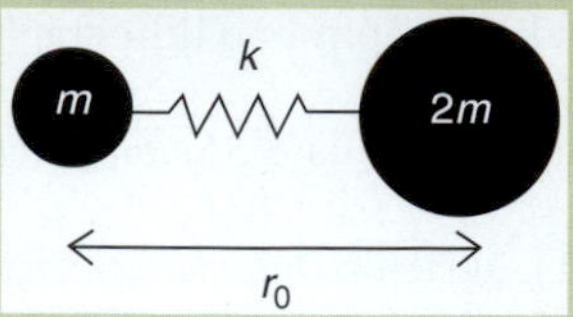

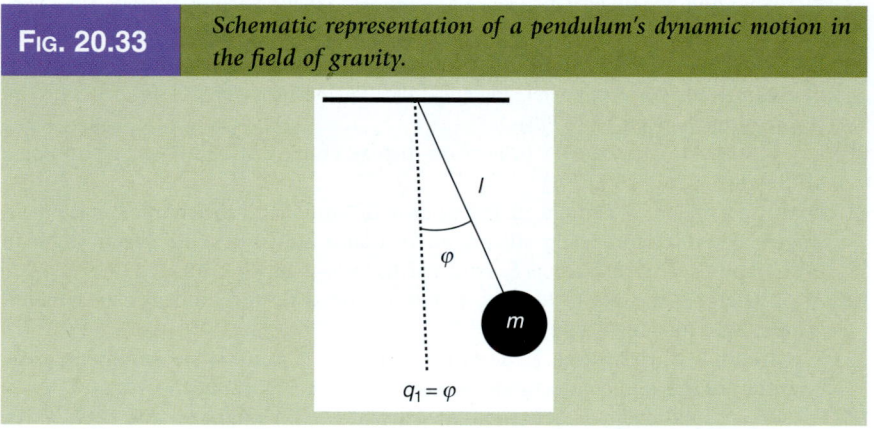

FIG. 20.33 *Schematic representation of a pendulum's dynamic motion in the field of gravity.*

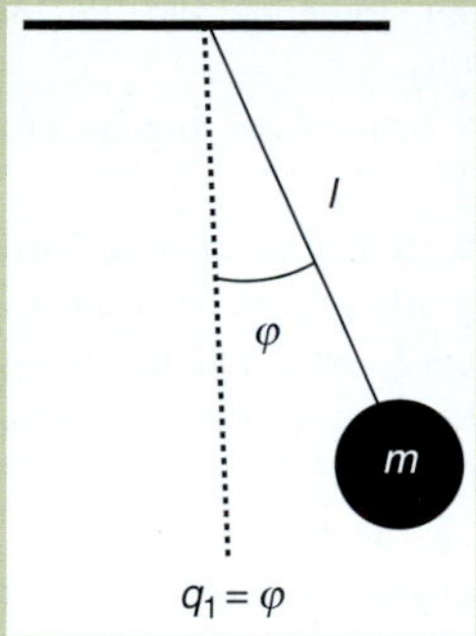

FIG. 20.34 *Schematic representation of a graphene sheet and its vector of helicity.*

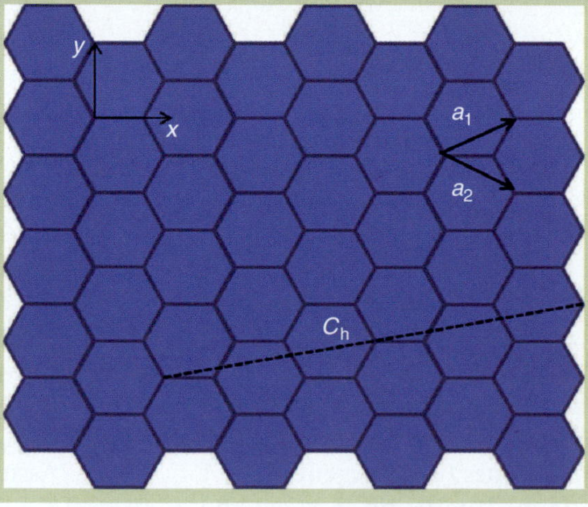

20.7 Derive the Lagrangian and equation of motion for a pendulum depicted in **Figure 20.33** using the generalized coordinate q_i.

20.8 Given the displacement vector $u = (x_1x_2, x_3^2, x_1x_3)^T$ find the expression for the strain tensor S and rotation tensor Ω.

20.9 Elastic stiffness coefficients of an isotropic nanocrystalline nickel are found to be $c_{11} = 247$ GPa, $c_{12} = 147$ GPa [85]. Define the Lame constants for this material and then calculate its Young's modulus E, and shear modulus G. (Hint: The Young's modulus and shear modulus in terms of the Lame constants are given as follows: $E = \frac{\mu(3\lambda + 2\mu)}{\lambda + \mu}$ and $G = \frac{1}{\mu}$.)

20.10 Compare the level of accuracy between Verlet algorithms and Nordsieck/Gear predictor–corrector methods.

20.11 Explain the failure mechanism of a nanowire shown in **Figure 20.5** and then contrast its differences from macroscopic failure mechanisms.

20.12 The melting point of bulk aluminum is about 933 K. However, as mentioned in this chapter, the nanostructured Al can melt at about 270 K. Explain why this is the case and what other parameters influence the melting point of materials at nanometer scales.

20.13 Describe the common difficulties associated with constructing nanomachines which are absent at the macroscopic level.

20.14 Compare the wear mechanisms in macroscopic and microscopic domains by contrasting their differences.

20.15 Compute the angle of helicity for the zigzag and armchair type nanotubes. Also, explain why the angle of helicity ranges between 0° and 30° for all chiralities.

20.16 Define the vector of helicity C_h and the angle of helicity θ for **Figure 20.34.**

20.17 A carbon nanotube's diameter is defined as $d = |C_h|/\pi$. Prove that the nanotube's diameter has the form of $d = (a/\pi)\sqrt{n^2 + m^2 + nm}$, where $a = 0.246$ nm. Then, calculate the nanotube diameter for the picture shown in problem 20.16.

20.18 The Lennard–Jones potential energy of a carbon–carbon system is given by: $\phi = \frac{A}{\sigma^6}\left[\frac{1}{2}y_0^6\left(\frac{\sigma}{r}\right)^{12} - \left(\frac{\sigma}{r}\right)^6\right]$, [86]. Show that the equilibrium spacing r_0 is defined as $r_0 = \sigma y_0$, and then calculate the binding energy of this carbon–carbon system assuming that $A = 24.3 \times 10^{-79}$ (J·m^6), $\sigma = 0.142$ nm, and $y_0 = 2.7$.

20.19 Describe two different SPM methods that are used for nanomechanical measurements and indicate their advantages/disadvantages compared to each other.

20.20 A silica sample was tested by a CSM nanoindentation instrument and a reduced modulus of 71 GPa was obtained. Assume that the Berkovich diamond tip has a modulus of 1140 GPa, and Poisson ratios of the diamond tip and silica sample are 0.07 and 0.2, respectively. Calculate the modulus of this silica specimen applying these data points.

20.21 Explain the main differences between wet and dry etching and contrast their advantages/disadvantages upon one another.

20.22 Describe the functionality of a capacitive pressure sensor and its advantages compared to piezoresistive pressure sensors.

Nanostructure and Nanocomposite Thin Films

John J. Moore, Jianliang Lin, and In-Wook Park

What this tells us is that if you're building nanostructures, the surface is what's really important.

Paul Evans

Chapter 21

Threads

Chapter 21 brings on some substance and detail about nanostructured and nanocomposite thin films. It is the second chapter in the *Mechanical Nanoengineering* division of the text. Following *chapter 21*, we provide applications of these thin films and with this chapter, we round out the division. In *chapter 24*, although part of the *Chemical Nanoengineering* division of the text, mechanical properties as they apply to nanocomposites are reviewed.

21.0 Introduction

Nanostructured coatings have recently attracted increasing interest because of the possibilities of synthesizing materials with unique physical–chemical properties [1,2]. A number of sophisticated surface-related properties such as optical, magnetic, electronic, catalytic, mechanical, chemical, and tribological can be obtained by advanced nanostructured coatings [3,4]. There are many types of design models for nanostructured coatings, such as three-dimensional nanocomposite coatings [2,5], nanoscale multilayer coatings [6,7], functionally graded coatings [1,4], etc. The optimized design of nanostructured coatings needs to consider many factors, for example, ion energy and ion flux of the depositing species, interface volume, crystallite size, single-layer thickness, surface and interfacial energy, texture, epitaxial stress and strain, and overall coating architecture, all of which depend significantly on materials selection, deposition methods, and process parameters [2,8].

In particular, pulsed reactive magnetron deposition techniques have been investigated, more recently, since it is possible to conduct reactive sputtering without arcing during deposition. Pulsed reactive sputtering can also change and control the plasma constituents, increase the ion energy and ion flux, and control microstructural growth of the thin film through ion bombardment [8]. The applications of pulsing in reactive magnetron sputtering opens up considerable opportunities for the control of ion energy and ion flux to optimize the deposition process and tailor the as-deposited coating structure and properties.

The focus of this chapter is to introduce the relationships between processing, structure, properties, and functionality of nanostructured coatings using various deposition processes, such as unbalanced magnetron sputtering (UBMS), hybrid coating system of cathodic arc evaporation (CAE) and magnetron sputtering (MS), pulsed closed-field unbalanced magnetron sputtering (P-CFUBMS), and high-power pulsed magnetron sputtering, as shown in **Figure 21.1**.

21.1 Classification of Nanostructured, Nanocomposite Tribological Coatings

21.1.1 *Nanoscale Multilayer Coatings*

Research on using nanoscale multilayers (i.e., "superlattices") to increase the hardness and toughness of coatings has provided significant advancements in

Fig. 21.1 *Tetrahedron representing the relationship among processing, structure, properties, and functionality for nanostructured, multifunctional tribological coatings.*

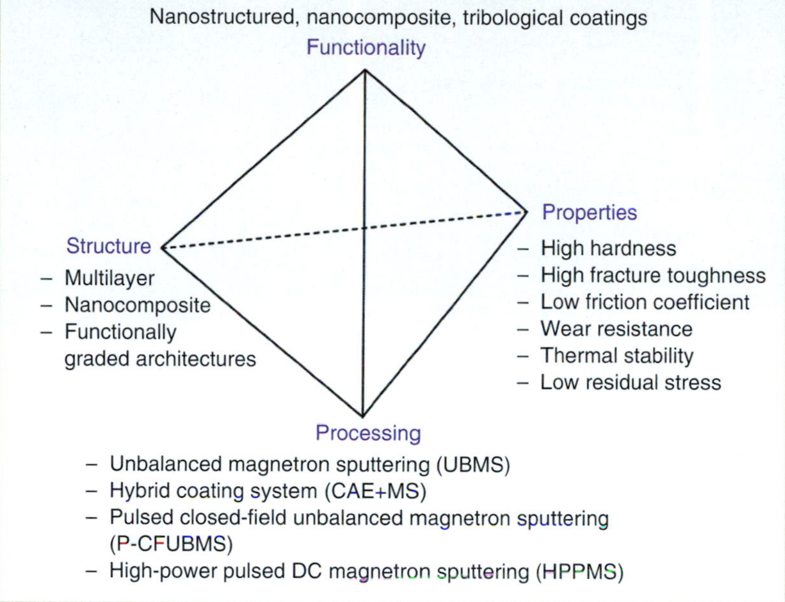

understanding the advantages of employing this type of coating architecture. Early research by Palatnik with multilayers of metals showed that significant improvements in strength were achieved when layer thickness was decreased below 500 nm [6,9]. In early modeling, Koehler [7] predicted that high shear strength coatings could be produced by alternating layers of high and low elastic modulus. Key elements of the concept are that very thin layers (≤10 nm) inhibit dislocation formation, while differences in elastic modulus between layers inhibit dislocation mobility. Lehoczky demonstrated [47] these concepts on metallic Al/Cu and Al/Ag multilayers and showed that a Hall–Petch type equation could be used to relate hardness to $1/(\text{periodicity})^{1/2}$, where periodicity is a minimum periodic length between layers in the multilayer coating architecture. Springer and Catlett [10], and Movchan et al. [11] reported on mechanical enhancements in metal/ceramic (e.g., Ti/TiN, Hf/HfN, W/WN, etc.)[12] and ceramic/ceramic (e.g., TiN/VN [13], TiN/NbN [14,15], TiN/$V_xNb_{1-x}N$ [16,17], etc.) laminate structures that followed a Hall–Petch relationship. These pioneering works were followed by intensive research in multilayers [18,19], which has produced coatings significantly harder than the individual components making up the layers. To achieve increased hardness, the layers must have sharp interfaces and periodicity in the 5–10 nm range.

The multilayer architectures, as shown in **Figure 21.2**, exhibiting high hardness are frequently called superlattices [20]. The different design architectures have been classified and some reports have formalized the multilayer design [4,21].

FIG. 21.2 *Cross-sectional TEM images and selected area diffraction patterns (SADP) of nanoscale multilayer coatings: (a) (Ti, Al)N/SiN, (b) (Ti, Al)N/WN, and (c) CrN/BCN.*

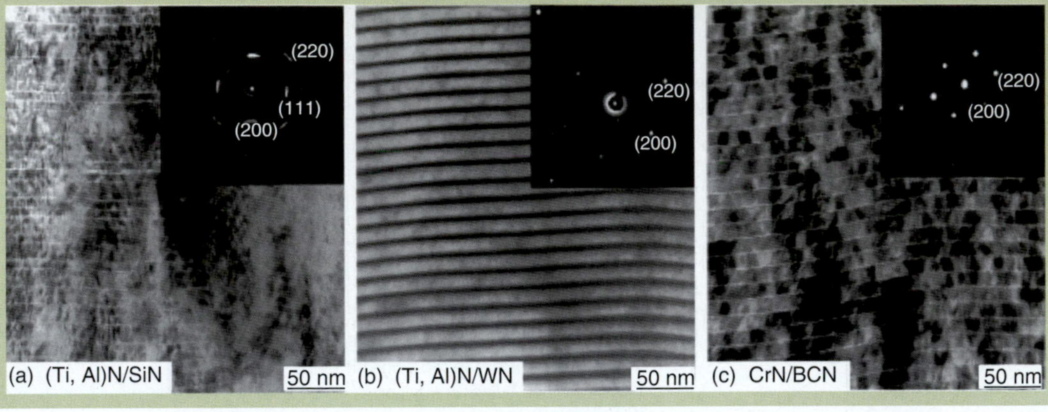

Source: K. Yamamotoa, S. Kujimeb, and K. Takahara, *Surface and Coatings Technology*, 200, 435–439 (2005). With permission.

Multilayer architectures clearly increase coating hardness and have commercial applications, especially in the tool industry. However, they can be difficult to apply with uniform thickness on three-dimensional components and rough surfaces. If the layers are not of the correct periodicity, the superlattice effect is lost. Another relatively new technology, nanocomposites, offers the same advantages as multilayers (plus other benefits) and their properties are not critically dependent on thickness or substrate geometry.

21.1.2 Nanocomposite Coatings

Nanostructured composite (i.e., "nanocomposite") coatings are usually formed from ternary or higher-order systems and comprise at least two immiscible phases: two nanocrystalline phases and, more commonly, an amorphous phase surrounding nanocrystallites of a secondary phase. The most interesting and extensively investigated nanocomposite coatings are ternary, quaternary, or even more complex systems, with nanocrystalline (nc-) grains of hard transition metal nitrides (e.g., TiN, CrN, AlN, BN, ZrN, etc.), carbides (e.g., TiC, VC, WC, ZrC, etc.), borides (e.g., TiB_2, CrB_2, VB_2, WB, ZrB_2, etc.), oxides (e.g., Al_2O_3, TiO_2, SiO_2, MgO, TiO_2, Y_2O_3, ZrO_2, etc.), or silicides (e.g., $TiSi_2$, $CrSi_2$, $ZrSi_2$, etc.) surrounded by amorphous (*a*-) matrices (e.g., Si_3N_4, BN, C, etc.). The physical, mechanical, and thermal properties of these hard materials are summarized in **Table 21.1** [22]. The synthesis of such nanocomposite (nc-/*a*-) coatings critically depends on the ability to co-deposit both the nanocrystalline and amorphous phases, such as Ti–Si–N (nc-TiN/nc-and *a*-$TiSi_2$/*a*-Si_3N_4) [2], Ti–Al–Si–N (nc-TiAlN/*a*-Si_3N_4) [5], W–Si–N (nc-W_2N/*a*-Si_3N_4) [23], Cr–Si–N (nc-CrN/*a*-Si_3N_4) [24], Ti–B–C–N (nc-TiB_2 and TiC/*a*-BN) [25], TiC/DLC (nc-TiC/*a*-C) [26], WC/DLC (nc-WC/*a*-C) [27], etc., as schematically presented in **Figure 21.3a**. A variety of hard compounds can be used as the nanocrystalline phases, including nitrides, carbides, borides, oxides, and silicides. Veprek [28] suggested that the

Table 21.1 *The Physical, Mechanical, and Thermal Properties of Hard Materials*

Phase	Crystal structure	Lattice parameters (nm)	Density (g · cm^{-3})	Melting point (°C)	Linear thermal expansion, α (10^{-6} · K^{-1})	Thermal conductivity, λ (W · m^{-1} · K^{-1})	Electrical resistivity (10^{-6} Ω · cm)	Enthalpy at 298 K (kJ · mol^{-1})	Young's modulus (10^5 N · mm^{-2})	Microhardness (10 N · mm^{-2})	Oxidation resistance (×100 °C)
Nitrides											
AlN	hex	0.311/0.498	3.05	2200	6	10	10^{11}	288.9	3.15	1200	13
BN	hex	0.251/0.669	2.25	3000	3.8	284.7	3×10^{14}	252.5	0.9	4400 HV	10
CrN	*fcc*	0.415	6.1	1050	2.3	11.72	640	118–124	4	1800–2100	7–7.5
	cub-B1	0.4149	5.39–7.75	1450	2.3	11.72	640	123.1	3.236	1100	
Cr_2N	hex	0.4760/0.4438	5.9	1500	9.4			30.8	3.138	2250 HV	86.3–110.3
HfN	*fcc*	0.452	13.8	3310	6.9	11.3	26	369.4	3.33–4.8	1700–2000	
Si_3N_4	hex	0.78/0.56	3.44	1900	2.4	20–24′	10^{18}	750.5	2.1	1410 HV	12–14
TaN	*hcp*	0.52/0.29	13.6–13.8	3000	3.6	8.58	128	225.7	5.756	3240	5–8
Ta_2N	hex	0.30/0.493	15.8	3000		10.05	263	270.9		3000	
TiN	cub-B1	0.423	5.21	3220	9.35	30	21.7	336.2	2.512	2400 HV	5
VN	*fcc*	0.41	6.13	2050	8.1	11.3	85–100	147.8	4.6	1520	5–8
ZrN	*fcc*	0.46	6.93	3000	6	16.75	13.6	365.5	5.1	2000	12
Carbides											
B_4C	rhom	0.5631/1.2144	2.52	2450	6	27.63	10^6	72	4.5	3700	11–14
Cr_3C_2	ortho	1.146/0.552/ 0.2821	6.68	1900	10.3	18.8	75	88.8	4	1500–2000	12
NbC	*fcc*	0.45	7.78	3490	6.65	14.24	35–74	139.8	3.4	2400	11
SiC	α:hex	β:0.4360	3.2	2200	5.68	15.49	10^5	71.6	4.8	3500	14–17
	β:*fcc*	α:0.3–7.3/1–1.5	3.17	2700	5.3	63–155	10^5	73.3	3.9–4.1	1400 HV	13–14
TaC	cub-B1	0.4454	14.65	3877	6.04	22.19	25	159.5	2.91	1490	11–14
TiC	cub-B1	0.429–0.433	4.93	3150	7.4	17–23.5	68	179.6	3.22	3200 HV	11–14
VC	cub-B1	0.4173	5.36	2770	6.55	4.2	156	105.1	4.34	2760 HV	8–11
WC	hex	0.29/0.28	15.7	2600	5.2–7.3	121.42	17	35.2	7.2	2080	8
ZrC	cub-B1	0.4989–0.476	6.51	3400	6.93	20.5	42	181.7	4	2600 HV	12

(*continued*)

Table 21.1 (Contd.) *The Physical, Mechanical, and Thermal Properties of Hard Materials*

Phase	Crystal structure	Lattice parameters (nm)	Density ($g \cdot cm^{-3}$)	Melting point (°C)	Linear thermal expansion, $\alpha (10^{-6} \cdot K^{-1})$	Thermal conductivity, $\lambda (W \cdot m^{-1} \cdot K^{-1})$	Electrical resistivity ($10^{-6}\ \Omega \cdot cm$)	Enthalpy at 298 K ($kJ \cdot mol^{-1}$)	Young's modulus ($10^5\ N \cdot mm^{-2}$)	Microhardness ($10 N \cdot mm^{-2}$)	Oxidation resistance (×100 °C)
Borides											
AlB_2	hex	0.3006/0.3252	3.17	1975				67			
CrB	ortho	0.2969/0.7858/ 0.2932	6.05	1550				75.4			14–18
CrB_2	hex	0.279/0.307	5.6	2200	11.1		56	94.6	2.15	2250	
HfB_2	hex	0.3141/0.3470	11.01	3200	5.3	430	10	336.6		2800	11–17
MoB_2	hex	0.3/0.31	7.8	2100			45	96.3		1380 HV	11–14
NbB_2	hex	0.31/0.33	6.8	3000	7.1–9.6	16.75	32	150.7	2.6	2600	11–14
SiB_6	ortho	1.4470/ 1.8350/0.9946	2.43	1950	8.3		10^7		3.3	1910	
TaB_2	hex	0.31/0.33		3150	5.1	21.35	68	209.3	2.62	2200	11–14
TiB_2	hex	0.3/0.32	4.5	2900	6.39	25.96	9	150.7	3.7	3840	11–17
Ti_2B	Tet	0.61/0.46		2200						2500	
VB_2	hex	0.3/0.31	4.8	2400	5.3		16	203.9	5.1	2080	13
WB	Tet	0.31/1.7	15.5	2860						3750	
W_2B	Tet	0.56/0.47	16.5	2770	4.7		21.43			2350	8–14
ZrB	*fcc*	0.47	6.5	3000				163.3		3600	
ZrB_2	hex	0.32/0.35	6.1	3000	6.83	23.03	9.2	326.6	3.5	2200	11
Oxides											
Al_2O_3-α	hex	0.5127	3.99	2043	8	30.1	10^{20}	1580.1	4	2100 HV	17
	rhom	0.513	3.9	2030	7.2–8.6	4.2–16.7	10^{20}	1678.5	3.6	2100 HV	20
BeO	hex	0.2699/0.4401	3	2450	9	264	10^{23}	569	3	1230–1490 HV	17
CrO_2	Tet	0.441/0.291	4.8					582.8			
CrO_3	ortho	0.573/0.852/0.474	2.81	170–198				579.9			
Cr_2O_3	rhom	0.536	5.21	2440	6.7		10^{13}	1130.4		1000 HV	
	hex	0.495876/1.35942	5.21	2400	5.6			1130.4		2300 HV	
HfO_2	mono	0.512	9.7	2900	10	3	5×10^{15}	1053.4		900	

MgO	cub	0.4208	3.6	2850	11.2	36	10^{12}	568.6	3.2	745 HV	17
SiO_2	quartz	0.4093/0.5393	2.33	1703–1729	0.4	1.38	10^{22}	911	0.5–1.0	1130–1260	
	trigonal	0.421/0.539	2.2	1713	0.5–0.75	1.2–1.4	10^{21}	911.5	1.114	1200	
ThO_2	*fcc*	0.5859	10.05	3250	10	10	10^{16}	1173.1	1.38	950 HV	17
TiO	cub-B1	0.417	4.88	1750	7.6	11		520		1300	
TiO_2	Tet	0.4593/0.2959	4.19	1900	4.21–4.25	8	1.2×10^{10}	945.4	2.05–2.80	767–1000 HK	
	cub-B1	0.45933/0.29592	4.25	1867	9				0.8–2.0	1000 HV	
Ti_2O_3	rhom	0.5454	4.6	2130			10^{19}–10^{24}	1433.1		980 HV	
Ti_3O_5	mono	0.9828/0.3776/0.9898		1780				2461			
ZrO_2	cub	0.511	5.6	2750	7.5–10.5	0.7–2.4	10^{16}	1035	1.63	1200	17
Silicides											
CrSi	cub	0.4629	5.38	1550				53.2		1000	
$CrSi_2$	hex	0.442/0.655	4.91	1630				100.5		1100	14–18
Cr_3Si	cub	0.455	6.52	1710				105.5		900–980	
$MoSi_2$	Tet	0.32/0.786	6.3	2050	8.4	221.9	21	108.9	3.84	1290	17
$NbSi_2$	hex	0.48/0.66	5.5	1950	8.4		6.3	50.2		700	8–11
$TaSi_2$	hex	0.4773/0.6552	9.2	2200	8.9/8.8		38	150.7		1410	11
$TiSi_2$	ortho	0.8236/0.4773/0.8523	4.39	1520	11.5		18	134.4	2.556	892	11
VSi_2	hex	0.46/0.64	4.5	1650	11		9.5	95		960	
WSi_2	Tet	0.321/0.788	9.5	2165	6.5	45	12.5	92.1	5.3	1090	16
$ZrSi_2$	ortho	0.372/1.416/0.367	4.87	1700	9.7		161	159.4	2.348	1030 HV	8–11

cub: cubic, cub-B1: cubic NaCl-type, *fcc*: face-centered cubic, *hcp*: hexagonal closed-packed, hex: hexagonal, HK: Knoop hardness, HV: Vickers hardness, mono: monoclinic, ortho: orthorhombic, rhom: rhombohedral/trigonal, tet: tetragonal, tri: triclinic

Fig. 21.3 *(a) Schematic diagram of a nanostructured nanocomposite coating proposed by Veprek [28] and (b) HRTEM image and selected area diffraction pattern (SADP) of the nanocomposite Ti–B–N (nc-TiN/a-BN).*

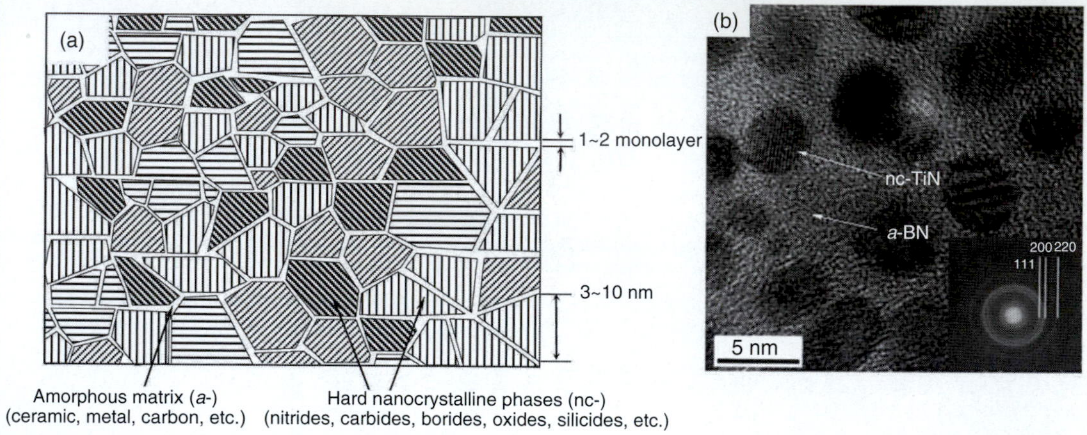

Source: Y. H. Lu, P. Sit, T. F. Hung, H. Chen, Z. F. Zhou, K. Y. Li, and Y. G. Shen, *Journal of Vacuum Science and Technology, B*, 23(2), 449 (2005). With permission.

nanocrystalline grains must be 3~10 nm in size and separated by 1~2 monolayers within an amorphous phase as shown in **Figure 21.2a**. For example, the Ti–B–N nanocomposite, which consists of nanocrystalline TiN (~5 nm in size) in an amorphous BN matrix, has been synthesized and observed by Lu [29], as shown in **Figure 21.3b**.

21.1.3 Functionally Graded Coatings

In order to counteract brittle failure and improve fracture toughness, two concepts have been explored. The first involves the use of graded interfaces between the coating and substrate and between layers in the coating. For example, a WC–TiC–TiN (outside layer) graded coating for cutting tools was reported by Fella et al. [30], which showed considerably less wear than single-layer hard coatings used in the cutting of steels. This type of coating is functionally and chemically graded to achieve better adhesion, oxidation resistance, and mechanical properties. One example of how functionally graded architectures improve coating performance is the adhesion of diamond-like carbon (DLC) to steels. DLC, and especially hydrogen-free DLC, has a very high hardness and generally has a large residual compressive stress. The coatings are relatively inert, and adhesion failures of coated steel surfaces were a roadblock to success. This problem was solved through designing and implementing a graded interface between the coating and the substrate. Examples of effective gradient compositions are Ti–TiN–TiCN–TiC–DLC for hydrogenated DLC [31] and Ti–TiC–DLC for hydrogen-free DLC [32]. In the development of the latter composition, the importance of a graded elastic modulus through the substrate coating/interface was highlighted as shown in **Figure 21.4**. The gradual build-up in material stiffness from the substrate with $E = 220$ GPa to the DLC layer with $E = 650$ GPa avoids sharp interfaces

FIG. 21.4 *Schematic diagram of a functionally gradient Ti–TiC$_x$–DLC coating, where chemistry and elastic moduli are transitioned from the metallic substrate to a hard DLC top layer.*

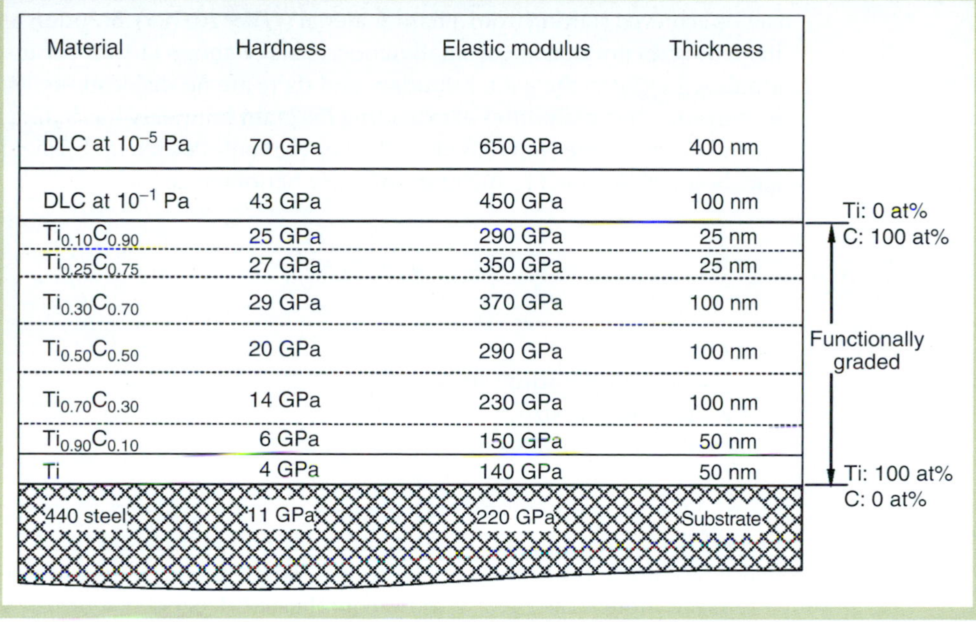
Material
Hardness
Elastic modulus
Thickness
DLC at 10^-5 Pa
70 GPa
650 GPa
400 nm
DLC at 10^-1 Pa
43 GPa
450 GPa
100 nm
Ti0.10C0.90
25 GPa
290 GPa
25 nm
Ti0.25C0.75
27 GPa
350 GPa
25 nm
Ti0.30C0.70
29 GPa
370 GPa
100 nm
Ti0.50C0.50
20 GPa
290 GPa
100 nm
Ti0.70C0.30
14 GPa
230 GPa
100 nm
Ti0.90C0.10
6 GPa
150 GPa
50 nm
Ti
4 GPa
140 GPa
50 nm
440 steel
11 GPa
220 GPa
Substrate
Ti: 0 at%
C: 100 at%
Functionally graded
Ti: 100 at%
C: 0 at%

Source: A. A. Voevodin, M. A. Capano, S. J. P. Laube, M. S. Donley, and J. S. Zabinski, *Thin Solid Films*, 298, 107–115 (1997). With permission.

that can provide places for crack initiation, good chemical continuity, and load support for the hard DLC top-coat. This functionally graded approach can be combined with multilayer and nanocomposite architectures to further enhance tribological properties.

21.2 Background of Nanostructured Super-Hard Coatings

Hardness is defined as the resistance to plastic deformation. Plastic deformation of crystalline materials occurs predominantly by dislocation movement under an applied load. Therefore, a higher resistance to dislocation movement of a material will generally enhance its hardness. One approach to obtain high resistance to dislocation movement and plastic deformation is to preclude the formation of stable dislocations. *Super-hard* coatings, with a hardness value in excess of 40 GPa, have attracted significantly increasing interest during the past 10–15 years [33]. A concept for super-hard nanocomposite coatings was suggested by Veprek [34]. The strength and hardness of engineering materials are orders of magnitude smaller than the theoretically predicted values. They are determined mainly by the microstructure, which has to be designed in such a way as to efficiently hinder the multiplication and movement of dislocations and the growth of microcracks. This can be achieved in various ways known from metallurgy, such as solution, work, and grain boundary hardening [35,36]. In this way, the strength and hardness of a material can be increased by a factor of 3–7 times, that is, super-hard materials should form when such an enhancement can be achieved starting from a hard material ($HV > 20\,\mathrm{GPa}$). Solution and work hardening do not operate in small nanocrystals of about <10 nm because solute atoms segregate to the grain boundary and there are no dislocations. Therefore, we consider the possibilities of extending the grain boundary hardening in poly- and microcrystalline materials, described by the Hall–Petch relationship [37,38], equation (21.1), down to the range of a few nanometers:

$$\sigma_C = \sigma_0 + \frac{k_{gb}}{\sqrt{d}} \tag{21.1}$$

where

σ_C is the critical fracture stress
d is the crystallite size
σ_0 and k_{gb} are constants

Many different mechanisms and theories describe Hall–Petch strengthening [37,38]. Dislocation pileup models and work hardening yield the $d^{-1/2}$ dependence but different formulas for σ_0 and k_{gb}, whereas the grain–grain boundary composite models also give a more complicated dependence of σ_C on d. The strength of brittle materials, such as glasses and ceramics, is determined by their ability to withstand the growth of microcracks. Brittle materials do not undergo any plastic deformation up to their fracture. Their strength or hardness is proportional to the elastic modulus. The critical stress, which causes the growth of a microcrack of size a_0, is given by the general Griffith criterion (equation 21.2).

$$\sigma_C = k_{crack}\sqrt{\frac{2E\gamma_s}{\pi\, a_0}} \propto \frac{1}{\sqrt{d}} \tag{21.2}$$

Here E is the Young's modulus, γ_s is the surface cohesive energy, and k_{crack} is a constant which depends on the nature and shape of the microcrack and on the kind of stress applied [35]. Thus, the $d^{-1/2}$ dependence of the strength and hardness in a material can also originate from the fact that the size a_0, of possible flaws such as voids and microcracks that are formed during the processing of the material, also decreases with decreasing grain size. For these reasons, the Hall–Petch relationship, equation (21.1), should be considered as a semiempirical formula which is valid down to a crystallite size of 20–50 nm (some models predict an even higher limit [39,40]). With the crystallite size decreasing below this limit, the fraction of the material in the grain boundaries strongly increases which leads to a decrease of its strength and hardness due to an increase of *"grain boundary sliding"* [40,41]. A simple phenomenological model (i.e., rule of mixtures) describes the softening in terms of an increasing volume fraction of the grain boundary material f_{gb}, with the crystallite size decreasing below 10–6 nm (equation 21.3) [42].

$$H(f_{gb}) = (1 - f_{gb})H_c + f_{gb}H_{gb} \tag{21.3}$$

Here $f_{gb} \propto (1/d)$. Due to the flaws present, the hardness of the grain boundary material H_{gb} is smaller than that of the crystallites H_c. Thus the average hardness of the material decreases with d decreasing below 10 nm. The first report of a *reverse (or negative) Hall–Petch relationship* was by Chokshi et al. [43]. Later on it was the subject of many studies, with controversial conclusions regarding the critical grain size where a *normal Hall–Petch relationship* changes to a reverse one [39,44]. Various mechanisms of grain boundary creep and sliding were discussed and are described by deformation mechanism maps in terms of temperature and stress [45,46]. Theories of grain boundary sliding are critically reviewed in Ref. [39]. Recent computer simulation studies [41] confirm that the negative Hall–Petch dependence in nanocrystalline metals is due to the grain boundary sliding that occurs due to a large number of small sliding events of atomic planes at the grain boundaries without thermal activation. Although many details are still not understood, there is little doubt that grain boundary sliding is the reason for softening in this crystallite size range. Therefore, a further increase of the strength and hardness with decreasing crystallite size can be achieved only if grain boundary sliding can be blocked by appropriate design of the material. This is the basis of the concept for the design of super-hard nanocomposites, suggested by Veprek [34].

As mentioned in section 21.1.1, another possible way to strengthen a material is based on the formation of nanoscale multilayers consisting of two different materials with large differences in elastic moduli, sharp interface, and small bilayer thickness (lattice period) of about 10 nm [7]. Because this design of nanostructured super-hard materials was suggested and experimentally confirmed before super-hard nanocomposites were developed, we will deal with superlattice coatings in section 21.2.1.

21.2.1 Nanoscale Multilayer Coatings

In a theoretical paper published in 1970, Koehler suggested [7] a concept for the design of strong solids, which are nowadays called superlattices. Originally he suggested depositing epitaxial multilayers of two different metals, A and B, having as different elastic constants as possible $E_A < E_B$ but similar thermal expansion and strong bonds. The thickness of the layers should be so small that no dislocation source could operate within the layers. If under applied stress a dislocation, which would form in the softer layer A, would move towards the A/B interface, elastic strain induced in the second layer B with the higher elastic modulus would cause a repulsing force that would hinder the dislocation from crossing that interface. Thus, the strength of such multilayers would be much larger than that expected from the rule of mixture. Koehler's prediction was further developed and experimentally confirmed by Lehoczky who deposited Al/Cu and Al/Ag superlattices and measured their mechanical properties [47]. According to the rule of mixtures, the applied stress which causes elastic strain is distributed between the layers proportional to their volume fractions and elastic moduli. Lehoczky has shown that the tensile stress–strain characteristics measured on multilayers consisting of two different metals displayed a much higher Young's modulus and tensile strength than that predicted by the rule of mixtures, and both of which increased with decreasing thickness of the double layer (lattice period). For layer thicknesses <70 nm, the yield stress of Al/Cu multilayers was 4.2 times larger and the tensile fracture stress was 2.4–3.4 times higher than the values given by the rule of mixture [47]. This work was followed by the work of a number of researchers who confirmed the experimental results on various metal/metal, metal/ceramic, and ceramic/ceramic multilayer systems (see section 21.1.1). In all these cases, an increase in hardness by a factor of 2–4 was achieved when the lattice period decreased to about 5–10 nm. For a large lattice period, where the dislocation multiplication source can still operate, the increase of the hardness and the tensile strength (most researchers measured the hardness because it is simpler than the measurement of tensile strength conducted by Lehoczky) [47] with decreasing layer thickness is due to the increase in the critical stress, which is dependent on multiplicate dislocations such as a Frank–Read dislocation source (equation 21.4):

$$\sigma_C = \frac{Gb}{l_{pp}} \propto \frac{1}{l_{pp}} \tag{21.4}$$

where
- G is the shear modulus
- b is the Burgers vector
- l_{pp} is the distance between the dislocation pinning sites [35]

Usually one finds strengthening dependence similar to the Hall–Petch relationship but with a somewhat different dependence on the layer period (λ^{-n}), instead of the crystallite size d in equation (21.1), with $n = 1/2$ for layers with different slip systems and $n = 1$ for layers with similar slip systems [48]. A more recent theoretical discussion of the Hall–Petch relationship for superlattices was published by Anderson and Li [49]. According to their calculations, strong deviations from continuum Hall–Petch behavior occur when the thickness of the layers is so small that the pileup contains only one to two dislocations.

In a remark added in the proof, Koehler mentioned that the ideas described in his paper would also be valid if one of the layers is amorphous. Recently, several papers have appeared in which one of the layers consists of an amorphous CN_x and a transition metal nitride such as TiN [50] or ZrN [51]. However, with decreasing layer thickness the layered structure vanished and a nanocrystalline composite (i.e., "nanocomposite") structure appeared [50,51]. Such films also exhibit a high hardness of 40 GPa or more.

21.2.2 Single-Layer Nanocomposite Coatings

Using similar ideas for restricting dislocation formation and mobility as used in multilayer approaches to "hardening," nanocomposite coatings can also be super-hard [25,28]. These composites have 3–10 nm crystalline grains embedded in an amorphous matrix and the grains are separated by 1–3 nm. This designs an *architecture* which leads to ultrahard (hardness above 100 GPa) coatings as reported by Veprek and co-authors most recently [2]. The nanocrystalline phase may be selected from the nitrides, carbides, borides, and oxides, as shown in **Table 21.1**, while the amorphous phase may also include metals and DLC as shown in **Figure 21.1a**. The initial model proposed by Veprek to explain hardness in nanocomposites is that dislocation operation is suppressed in small grains

(3–5 nm) and that the narrow space between them (1 or 2 monolayer separation) induces incoherence strains [34,52]. The incoherence strain is likely increased, when grain orientations are close enough to provide interaction between matched but slightly misoriented atomic planes.

In the absence of dislocation activity, Griffith's equation, as shown in equation (21.2), for crack opening was proposed as a simple description of the nanocomposite strength. This equation suggests that strength can be increased by increasing the elastic modulus and surface energy of the combined phases, and by decreasing the crystalline grain sizes. It is noted that elastic modulus is inversely dependent on grain sizes that are in the nanometer range due to lattice incoherence strains and the high volume of grain boundaries [2]. In practice, grain boundary defects always exist, and a 3-nm grain size was found to be close to the minimum limit. Below this limit, a reverse Hall–Petch effect has been observed and the strengthening effect disappears because grain boundaries and grains become indistinguishable and the stability of the nanocrystalline phase is greatly reduced [34,43,53].

Nanocomposites with metal matrixes are in a special category for this discussion. They have been demonstrated to increase hardness, but also have good potential for increasing toughness. Mechanisms for toughening within these systems are discussed in the next section, while mechanisms for hardening are discussed here. The composite strength of ceramic/metal nanocomposites may be described by the following form of the Griffith–Orowan model (equation 21.5) [54] when the dimensions of the metal matrix permit operation of dislocations:

$$\sigma_C = \sqrt{\frac{2E(\gamma_s + \gamma_p)}{\pi a} \frac{r_{\text{tip}}}{3d_a}} \tag{21.5}$$

where

γ_p is the work done during plastic deformation
r_{tip} is the curvature of the crack tip
d_a is the interatomic distance

It is noted that the crack tip blunting and the work of plastic deformation considerably improve material strength, while the lower elastic moduli of metals cause a reduction in strength as compared to ceramics. However, in nanocomposites, dislocation operation may be prohibited because the separation of grains is very small. For example, the critical distance l_{pp} for a Frank–Read dislocation source is very small. Matrix dimensions in hard nanocomposite coatings are from 1 to 3 nm, which is well below the critical size for dislocation source operation, even in very soft metal matrixes. Therefore, the mechanical behavior of such nanocomposites can be expected to be similar to that of ceramic matrix composites.

Composite designs that increase elastic modulus and hardness do not necessarily impart high toughness. First, dislocation mechanisms of deformation are prohibited and crack opening is the predominant mechanism for strain relaxation when stresses exceed the strength limit. Second, Griffith's equation does not take into account the energy balance of a moving crack, which consists of the energy required to break bonds and overcome friction losses, potential energy released by crack opening, and kinetic energy gained through crack motion [55]. From crack energy considerations, a high amount of stored stress dictates a high rate of potential energy release in the moving crack. In such conditions, a crack can achieve the self-propagating (energetically self-supporting) stage sooner, transferring into a macrocrack and causing brittle fracture. However, nanocomposites contain a high volume of grain boundaries between crystalline and amorphous phases. This type of structure limits initial crack sizes and helps to deflect, split, and blunt growing cracks.

21.3 New Directions for Nanostructured Super-Tough Coatings

While super-hard coatings are very important, quite notably for protection of cutting tools, most tribological applications for coatings either require or would receive significant benefit from increased toughness and lower friction. In particular, *high fracture toughness* is necessary for applications where high contact loads and, hence, significant substrate deformations are encountered [27]. A material is generally considered tough if it possesses both high strength and high ductility. High hardness (H) is directly related to high elastic modulus (E) and high yield strength, but it is very challenging to add a measure of ductility to hard coatings. For example, the super-hard coating designs, as stated earlier, prevent dislocation activity, essentially eliminating one common mechanism for ductility. Therefore, designs that increase ductility must also be considered to provide tough tribological coatings. Pharr [56] has suggested that an indication of fracture toughness (i.e., H/E ratio) can be obtained by examining the surface radial cracks created during indentation, described by equation (21.6):

$$K_c = \alpha_1 \left(\frac{E}{H} \right)^{1/2} \left(\frac{P}{c^{3/2}} \right) \tag{21.6}$$

where

P is the peak indentation load
c is the radial crack length
α_1 is an empirical constant related to the indenter geometry

K_c describes the "critical stress intensity" for crack propagation, but it is not an intrinsic parameter that can be used to measure fracture toughness directly. However, it is inversely proportional to fracture toughness. Thus, fracture toughness of coatings would appear to be improved by both a high hardness and a *low elastic modulus*. In this work, the H/E values were calculated and discussed relatively for each coating [57].

21.3.1 *Functionally Graded Multilayer Coatings*

An effective route for improving toughness in multilayers is the introduction of ductile, low elastic modulus alternate layers into the coating structure to relieve

FIG. 21.5 *A multilayer coating with multiple Ti/DLC pairs on top of a functionally graded layer for an optimum combination of cohesive and adhesive toughness: (a) design schematic and (b) cross-sectional photograph of the coating produced with 20 Ti/DLC pairs.*

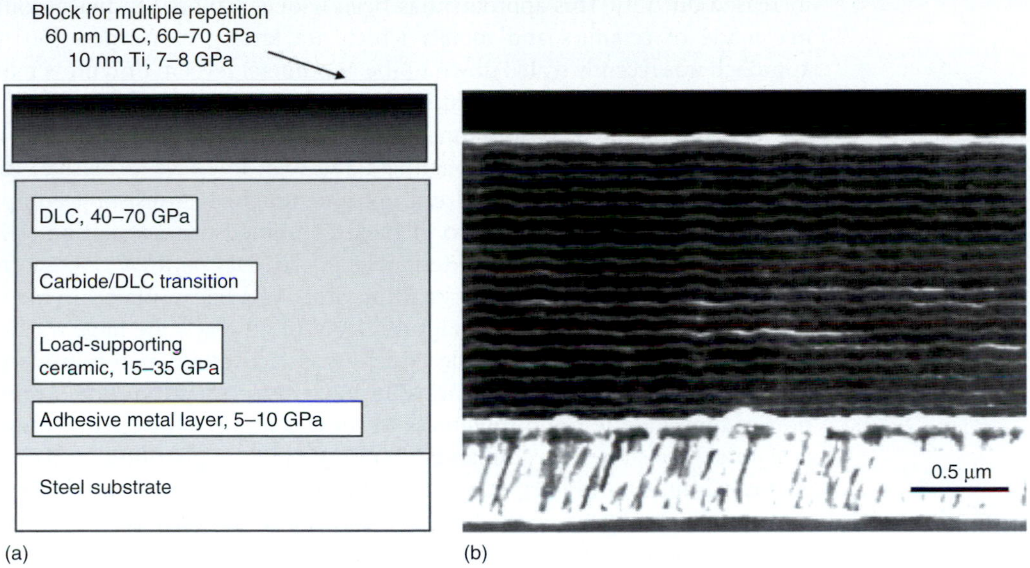

Source: A. A. Voevodin, S. D. Walck, and J. S. Zabinski, *Wear*, 203–204, 516–527 (1997). With permission.

stress and allow crack energy dissipation by plastic deformation in the crack tip. This approach will result in a decreased coating hardness, but the gain in the fracture toughness improvement may be more important in many tribological applications, excluding coatings for the cutting tool industry. For example, $[Ti/TiN]_n$ multilayer coatings on cast iron piston rings relaxed interface stress and improved combustion engine performance [58]. **Figure 21.5a** shows a schematic of a multilayer $[Ti/DLC]_n$ coating on a graded load support foundation, where the ductile Ti layers in the multilayer stack were graded at every DLC interface to avoid brittle fracture [4]. A cross-sectional photograph of this coating with 20 [Ti/DLC] pairs is shown in **Figure 21.5b**. The ductile Ti layers reduced the composite coating hardness to 20 GPa as compared to a single layer DLC coating, which has a hardness of about 60 GPa. However, due to dramatic improvement in toughness the multilayer coating design permitted operation during sliding friction at contact pressures as high as 2 GPa without fracture failure compared to 0.6–0.8 GPa for a single-layer DLC.

In general, the combination of multilayer and functionally graded approaches in the design of wear protective coatings produces exceptionally tough wear protective coatings for engineering applications. One potential drawback slowing the widespread use of new coatings was the need for reliable process controls to ensure that the correct compositions, structures, and properties are implemented during growth. However, modern process instrumentation and control technologies are able to meet the challenge and permit successful commercialization [59]. Thus, functionally graded and multilayer designs are commonly utilized in the production of modern tribological coatings.

21.3.2 Functionally Graded Nanocomposite Coatings

An alternative to employing multilayers to toughen coatings is embedding grains of a hard, high-yield strength phase into a softer matrix, allowing for increased ductility. This approach has been widely explored in macrocomposites made of ceramics and metals which are known as cermets [60]. This approach was recently scaled down to the nanometer level in thin films made of hard nitrides and softer metal matrixes [28]. A combination of the nanocrystalline/amorphous designs with a functionally graded interface, as shown in **Figure 21.6a**, provides high cohesive toughness and high interface (adhesive) toughness in a single coating. Several examples of tough wear-resistant composite coatings have been reported. Two of them combined nanocrystalline carbides with an amorphous DLC matrix designated as TiC/DLC and WC/DLC composites. In another example, nanocrystalline MoS_2 was encapsulated in an Al_2O_3 amorphous matrix as shown in **Figure 21.6b**. In all cases, the large fraction of the grain boundary phase provided ductility by activating grain boundary slip and crack termination by nanocrack splitting. This provided a unique combination of high hardness and toughness in these coatings. Novel nanocomposite designs suggested by Voevodin and Zabinski [61] for tough tribological coatings are very promising and provide a very attractive alternative to multilayer architectures. One of these novel designs incorporates the *chameleon* approach in which components are added to the coating that provide a low friction coefficient under extreme environmental conditions such as low and high humidity and low and high temperatures [61]. Nanocomposite coatings are more easily implemented, since they do not require precise control in the layer thickness and frequent cycling of the deposition parameters, as is required for fabrication

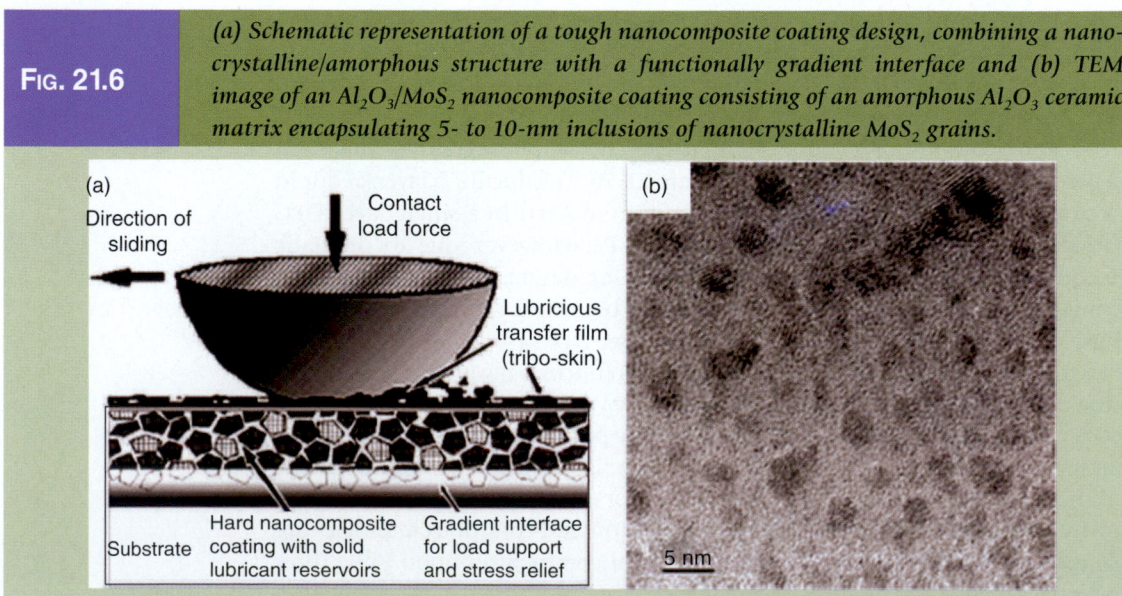

FIG. 21.6 *(a) Schematic representation of a tough nanocomposite coating design, combining a nanocrystalline/amorphous structure with a functionally gradient interface and (b) TEM image of an Al_2O_3/MoS_2 nanocomposite coating consisting of an amorphous Al_2O_3 ceramic matrix encapsulating 5- to 10-nm inclusions of nanocrystalline MoS_2 grains.*

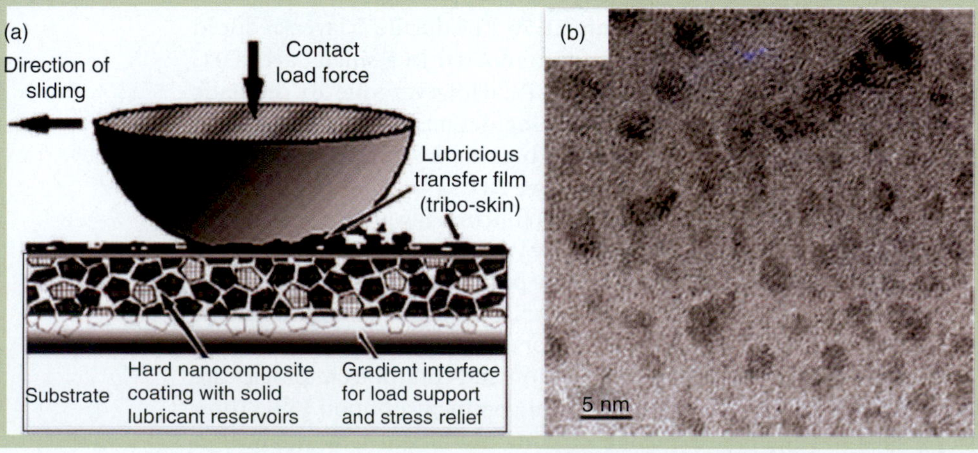

Source: A. A. Voevodin and J. S. Zabinski, *Composite Science Technology*, 65, 741–748 (2005). With permission.

of multilayer coatings. They are however relatively recent developments, and suitable scale-up of deposition techniques is currently under intense study.

To prevent tribological failures, there are additional requirements related to the normal (load) and tangential (friction) forces. In general terms, a tough *wear-resistant* coating must support high loads in sliding or rolling contact without failure by wear, cohesive fracture, and loss of adhesion (delamination). A *low friction coefficient* reduces friction losses and may increase load capability. Tribological coatings where a low friction coefficient is also required may be obtained by producing nanocomposite coatings with a mix of hard and lubricating phases, in which a hard primary phase (e.g., nitrides, carbides, or borides, etc.) provides wear resistance and load-bearing capability and a lubricating secondary phase (e.g., *a*-C, *a*-Si_3N_4, *a*-BN, etc.) reduces the friction between two contacting components. Finally, *thermal stability* is required to optimize coating performance and lifetime. The amorphous phases in grain boundaries can act as diffusion barriers (e.g., *a*-Si_3N_4, *a*-SiO_2, etc.) for improved thermal stability. For instance, nc-TiN/*a*-Si_3N_4 coatings with an amorphous Si_3N_4 matrix did not show grain growth at temperatures up to 1050°C as well as super-hardness of about 45 GPa [62]. Moreover, silicon nitride acts as an efficient barrier against oxygen diffusion at the grain boundaries and also by forming an oxidation-resistant SiO_2 surface layer, thus resulting in excellent thermal stability.

In summary, in addition to high hardness, other aspects such as high toughness, low friction coefficient, and high thermal stability are decisive characteristics of nanostructured coatings for their potential as protective tribological coatings.

21.4 Processing Techniques and Principles

Thin-film deposition is a process in which elemental, alloy, or compound thin films are deposited onto a bulk substrate. The deposition of the thin film may also be coupled with a previous surface treatment modification of the substrate in order to provide the required properties of the total coating or thin-film system. Thin-film deposition of metallic, insulating, conductive, and dielectric materials plays an important role in a large number of manufacturing, production, and research applications. There are a wide range of deposition processes that can be used to produce nanostructured and nanocomposite coatings, based on nitrides, carbides, and oxides, onto different substrate surfaces. The deposition process can be broadly classified into (i) physical vapor deposition (PVD) and (ii) chemical vapor deposition (CVD). These processes are used to deposit a broad range of thin-film materials, including semiconductors, superconductors, insulators, barrier layers, magnetic, optical films, and tribological and wear-resistant coatings, metals, compound, and organics, which play an important role in a large number of manufacturing, production, and research applications, such as protective coatings, optical coatings, microelectronic and optoelectronic devices, and decorative coatings [63]. This section will provide an introduction to the various important deposition techniques as well as some of the current day applications of the films produced.

21.4.1 Plasma Definition

Many vapor deposition techniques take advantage of conducting the process in a plasma medium. Plasma is often referred to as the fourth state of matter. Like the other three states of matter (solid, liquid, and gas), plasma has its own unique properties. A plasma is a gas containing charged and neutral species in varying degrees of excitation, including some or all of the following: electrons, positive ions, negative ions, atoms, and molecules. On average a plasma is electrically neutral, because any charge imbalance would result in electric fields that would tend to move the charges in such a way as to eliminate the imbalance [64].

To generate a plasma, a specific amount of energy needs to be added to separate the gas component molecules (gas breakdown) into a collection of ions, electrons, neutral atoms, etc., such as by applying an electric field or substantially increasing the temperature. Depending on the amount of energy added, the resulting plasma can be characterized as thermal or nonthermal ("cold" plasma). A nonthermal plasma is one in which the mean electron temperature (energy), which usually is of 1–10 *e*V, is higher than that of the bulk gas molecules. This *nonequilibrium* plasma can be generated at low pressures and is used in sputtering, etching, etc. **Figure 21.7a** shows a typical nonthermal plasma used in MS. On the other hand, if the energy is high enough, the ions and electrons are in local thermal (thermodynamic) equilibrium (at the same temperature). This *equilibrium* plasma is usually generated at near and above atmospheric pressures and is used in thermal plasma processing (spraying, heating, melting, etc.). **Figure 21.7b** shows a typical thermal plasma used in thermal plasma spraying.

To sustain a plasma requires that the rate of ionization must balance the loss of ions and electrons from the plasma volume by recombination and diffusion or convection to the plasma boundary. An important parameter of a plasma is the degree of ionization, which is the fraction of the original neutral species which has become ionized. The ionization process will create excited, dissociated, and/or ionized reactive species that are involved in the chemical reactions on the substrate. There are a number of important ionization mechanisms involved in the plasma deposition process. Typical ionization processes and reactions are

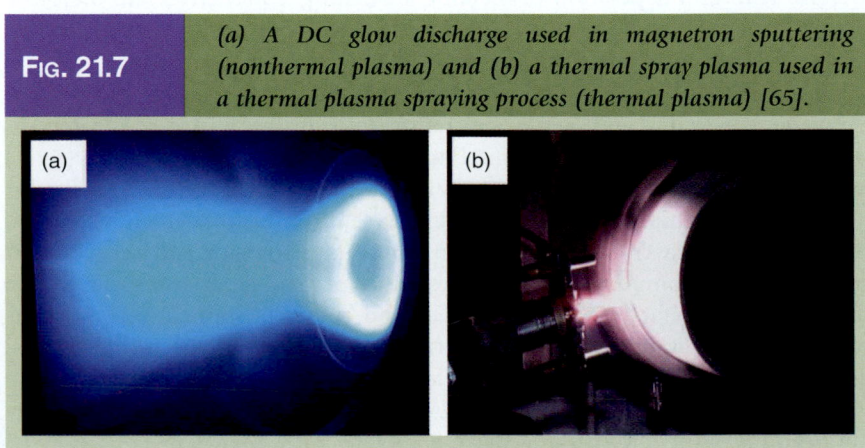

FIG. 21.7 *(a) A DC glow discharge used in magnetron sputtering (nonthermal plasma) and (b) a thermal spray plasma used in a thermal plasma spraying process (thermal plasma) [65].*

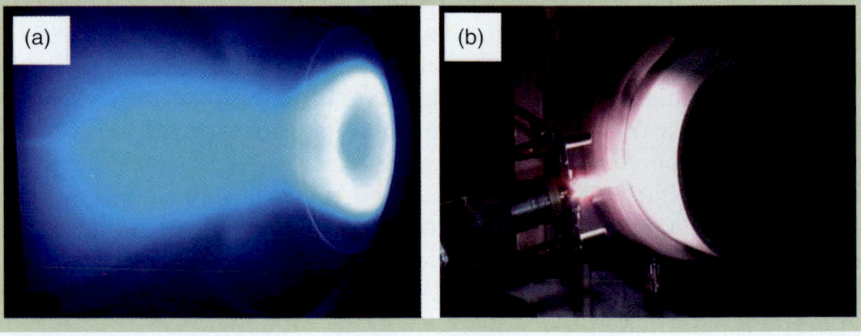

TABLE 21.2 *Summary of the Main Ionization Processes in a Sputter-Discharged Plasma*

Ionization process	Reactions	Comments
1. Direct ionization	$e + A \rightarrow A^+ + 2e$	Ions and additional electrons are created.
2. Penning ionization and excitation	$e + A \rightarrow A^m + e$ $A^m + B_2 \rightarrow A^+ + B_2^+ + e$	A metastable species A^m is created by electronic excitation; ions are created when metastable species A^m collides with species B.
3. Dissociative ionization	$e + AB \rightarrow A^+ + B + 2e$	An electron collides with a molecule AB to create ions and radicals.
4. Charge excitation	$A^+ + B \rightarrow A + B^+$	An ion collides with an atom to transfer the charge, where A and B can be the same species.
5. Ion–electron recombination	$A^+ + e \rightarrow A$	Ions and electrons combine to form a neutral species.

summarized in **Table 21.2**. All these processes will increase the ionization rates and excitation rates in the plasma, and this increase is one reason why inert gases like helium and argon are added to process the plasma discharge: they are relatively easy to ionize, while argon is also relatively inexpensive.

21.4.2 Chemical Vapor Deposition

The chemical vapor deposition (CVD) process is a popular thin-film deposition technology used to produce a wide range of metal, ceramic, and polymers coatings. In CVD, the material being deposited is generated from chemical vapor precursor species that are decomposed by reduction or thermal decomposition and come into contact with a heated substrate surface, and where they react or decompose forming a solid phase on the hot surface [66]. Reduction is normally accomplished by hydrogen at an elevated temperature, while decomposition is accomplished by thermal activation. A basic CVD process consists of the following steps: (i) diffusion of reactants to the substrate, (ii) adsorption onto the surface, (iii) surface chemical reaction(s) leading to deposition of solid, (iv) gaseous by-products desorption from the surface, and (v) gaseous by-products diffusing into the stream (**Fig. 21.8**).

FIG. 21.8 *Schematic diagram of the CVD process.*

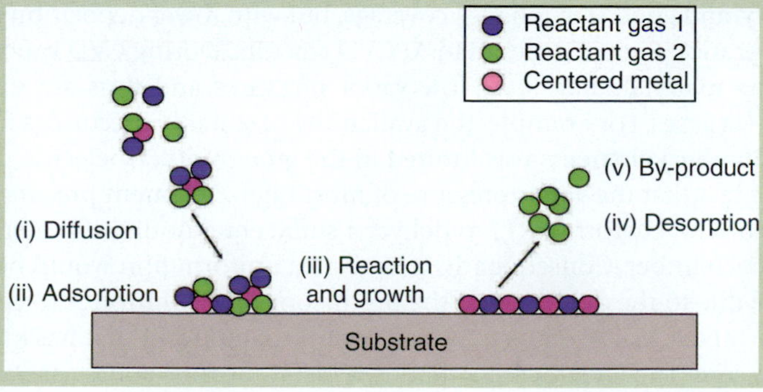

FIG. 21.9 *Schematic drawing of a hot-wall CVD reactor.*

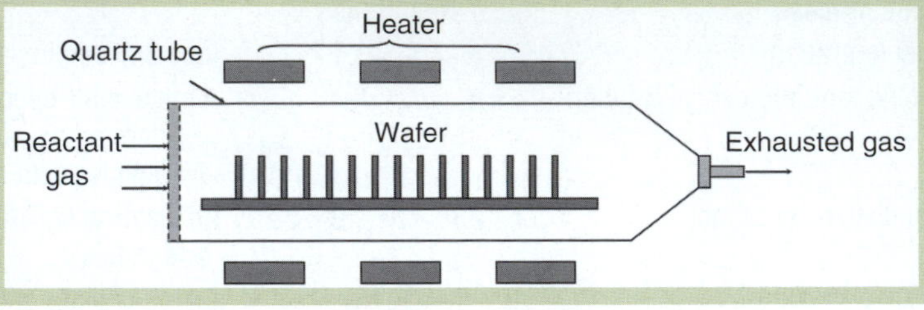

Figure 21.9 shows a typical configuration of a hot-wall CVD deposition system. In CVD, vapor supersaturation affects the nucleation rate of the film whereas substrate temperature influences the rate of film growth. These two factors together influence the extent of epitaxy, grain size, grain shape, and texture. Low gas supersaturation and high substrate temperatures promote the growth of single-crystal films. High gas supersaturation and low substrate temperatures result in the growth of less coherent and possibly amorphous films [67].

CVD encompasses a wide range of reactor and process types. The choice of process/reactor type is determined by the application via the requirements for substrate material, coating material and morphology, film thickness and uniformity, availability of precursors, and cost [66]. Several main CVD reactors that are widely being used are introduced here, and the reader should refer to other references [68–70] for detailed information on other systems.

There are a number of CVD processes that include (i) atmospheric pressure CVD (APCVD), (ii) low-pressure CVD (LPCVD), (iii) plasma-enhanced or -assisted CVD (PECVD or PACVD), and (iv) metal–organic CVD (MOCVD). Atmospheric pressure CVD (APCVD) operates at atmospheric pressure. APCVD has a very high deposition rate and is the simplest in design. The main disadvantages of APCVD are the low purity, poor uniformity of the deposited films and poor step (shape) coverage. In the mid-1970s, it was realized that LPCVD processing operating at medium vacuum (e.g., 1–10 mTorr) could have significant advantage over APCVD systems. By reducing the pressure, it was found that the diffusion coefficient was sufficiently enhanced and that deposition became surface controlled. With the hot-wall system, as shown in **Figure 21.9**, the deposition temperature could be maintained very uniformly, thereby achieving excellent film uniformity [69]. LPCVD processing produces films with excellent purity, uniformity, and good step (shape) coverage, but with lower deposition rates and higher operation temperatures than APCVD reactors. During CVD vapor deposition, many materials have very low vapor pressures and thus are difficult to transport via gases. For example, the availability of suitable precursors for high-Z (atomic number) elements was limited in the growth of ferroelectric thin films due to the fact that the vapor pressure of most high-Z element precursors is too low (e.g., below 1 mTorr at RT) to deliver a sufficient amount of material to the deposition chamber. Consequently, a large-area uniform film would be difficult to achieve due to the depletion of the precursors. One solution is to chemically attach the metal (Ga, Al, Cu, etc.) to an organic compound that has a very high vapor pressure. The CVD process that uses metal–organic source gases is called

MOCVD [71,72]. For instance, MOCVD may use tantalum ethoxide ($Ta(OC_2H_5)_5$), to create tantalum pentoxide (Ta_2O_5), or tetradimethylamino titanium (TDMAT) to create titanium nitride (TiN). The organic–metal bond is very weak and can be broken via thermal means on wafer substrates, thereby depositing the metal while the high vapor pressure organic is pumped away.

CVD is not a "line of sight" process and offers some distinct advantages such as uniform deposition over complex geometries ("conformal" deposition) and large areas, good conformal step coverage, and relatively easy control of stoichiometry of deposited films. However, CVD suffers from limitations in growth of thin films [73,74]. CVD processing is generally accompanied by volatile reaction by-products and unused precursor species. In addition special safety precautions must be taken to insure a minimum of the organic by-products are released into the atmosphere. For example, in MOCVD, as the human body absorbs organic compounds relatively easily, the metal organics are easily absorbed by humans. Once in the body, the weak metal–organic bond is easily broken, thus poisoning the body with heavy metals that often cannot be easily removed by normal bodily functions. Another limitation of APCVD and LPCVD processing is the high deposition temperature. In general, the typical substrate temperature is in the range of 800–1200°C; therefore, the substrates used are limited to a narrow range, such as cemented carbides to minimize the risk of dimensional and microstructural change. The high operation temperature also limits the variety of the materials that can be produced without changing the film properties; for example, a deposition temperature less than 400°C is needed for the deposition of certain silicon thin films used in microelectronic applications.

With respect to overcoming the high process temperature limitations, several modifications of the conventional CVD process have been developed. In recent years, the use of a plasma to dissociate and ionize the gaseous precursors and deposit the coatings can be carried out at much lower temperatures than in conventional CVD. This process is called plasma-enhanced (or -assisted) chemical vapor deposition (PECVD) and provides considerable potential to be utilized in the deposition of semiconductive coatings [75–77]. PECVD reactors also operate under low pressure, but do not depend completely on thermal energy to accelerate the reaction processes. They transfer energy to the reactant gases by using a "glow discharge." The glow discharge used by a PECVD reactor is generally created by applying a radio frequency (RF) field to a low-pressure gas, creating free electrons within the discharge region. The electrons are sufficiently energized by the electric field that gas phase dissociation and ionization of the reactant gases occur when the free electrons collide with the gas species. Energetic species are then adsorbed on the film surface, where they are subjected to ion and electron bombardment, rearrangements, reactions with other species, new bond formation, and film formation and growth. The typical deposition temperature in PECVD can be reduced below 500°C. The relatively low deposition temperature makes PECVD suitable for a wide range of substrates including tool steel and hot work tool steel and Si wafers used in microelectronics so that the deposition temperature is below the tempering temperature of tool steels, therefore maintaining the required microstructure and properties of the substrate and minimizing distortion of the substrate [78].

CVD is capable of producing a wide range of coating materials, including typical nitrides, carbides, and oxides, such as TiN [79,80], TiCN [81], titanium carbide (TiC) [82,83], DLC [84–86] TiB_2 [87], and Al_2O_3 [88]. CVD/PECVD has also been used in a multitude of semiconductor wafer fabrication processes, including the production of amorphous and polycrystalline thin films, for example, amorphous silicon for solar cell [89,90], polycrystalline silicon for gate contact [91], deposition of SiO_2 [92] and silicon nitride [93], and growing single-crystal epitaxial layers [94,95]. It has also been used to produce thick oxides used for isolation between metal interconnects, doped oxides for global planarization, and dielectrics for isolation and capacitors.

21.4.3 Physical Vapor Deposition

Compared to the CVD process, PVD can be carried out at lower deposition temperatures and without corrosive products. The substrate deposition temperature in the PVD process is in the range of 150–550°C thereby greatly bordering the substrate selection range. Almost all kinds of tool steel, for example, hot working tool steel and high speed steel, can be used as substrates in the PVD process, and no heat treatment is required after coating deposition due to the low process temperature. A range of desired compositional coatings can be readily produced by PVD that exhibit fine grain size, dense structure, and improved wear properties. In the PVD process, the coating material is vaporized or sputtered in a chamber to produce a flux of atoms or molecules which condenses on the substrate to be coated. Chemical compounds can be deposited using a composite target or by introducing a reactive gas (nitrogen, oxygen, or simple hydrocarbons) containing the desired reactants, which dissociate and ionize in the plasma to form reactive species (e.g., N, O, C) that react with metal atoms sputtered from the PVD source or target [68]. In this section, variations of PVD processes including cathode arc evaporation (CAE), electron beam evaporation, plasma sputtering, ion beam sputtering (IBS), pulsed laser deposition (PLD), and thermal plasma processing will be introduced. Some important developments in plasma sputter deposition including the UBMS, pulsed magnetron sputter deposition, and high-power pulse magnetron sputtering are emphasized later for their important role in enhancing the chemical and/or structural nature of the deposited films. Each of these systems and techniques will be described in this section, as well as some of the current applications of the films produced.

Electron Beam Evaporation. Electron beam evaporation falls in the catalog of general evaporation, in which a block of the source material to be deposited is transformed into vapor form by means of high-energy electron beam bombardment, and then it is allowed to condense on the substrate [96]. The electron beam evaporation process typically involves the following steps: (1) Generation of an electron beam by an electron gun, which uses thermal or plasma emission of electrons produced by an incandescent filament cathode, (2) emitted electrons are accelerated towards an anode, which usually is the crucible itself holding the evaporation material, by a electric field (kilovolts), (3) beside the electric field, a magnetic field is used to bend the electron trajectory and move (scan) the electron beam across the treated surface allowing the electron gun to be positioned below the evaporation line, (4) vaporization of the material to the volatile state in the form of vapors, and (5) deposition of the vaporized material onto th

FIG. 21.10 *A diagram of the electron beam evaporation equipment.*

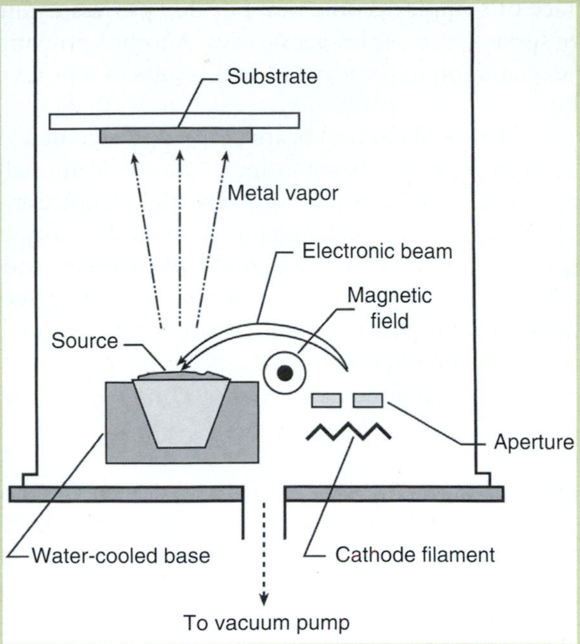

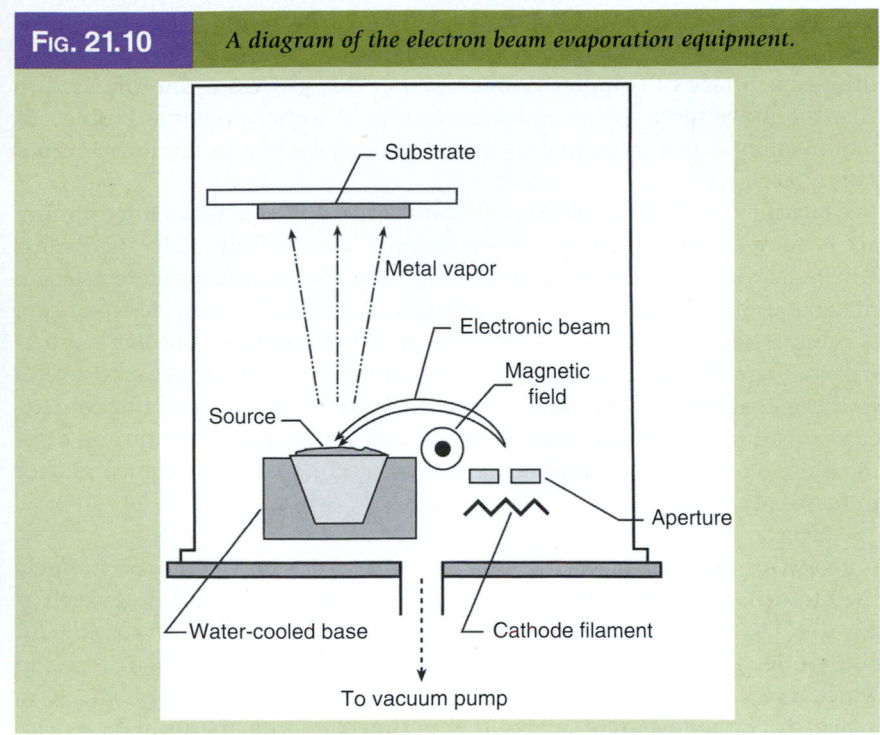

substrate. A diagram of the electron beam evaporation equipment is shown in **Figure 21.10**.

Electron beam evaporation offers several advantages over competing processes. By adjusting the value of the accelerating voltage and the combination of different electric fields, the electron beam can be localized or unfocused, accelerated or retarded, and pulsed [97], thus it is possible to obtain localized heating on the material being evaporated with a range of high densities of evaporation power (several kilowatts). This ability to vary the evaporation power allows precise control of the evaporation rates, from as low as one nanometer per minute to as high as a few micrometers per minute. This technique is an extremely versatile means of depositing uniform high-purity thin films. The materials used for evaporation are available in near limitless shapes and forms, for example, the pellets, slugs, and disks. In addition, elevated temperatures in excess of 3500°C can be used allowing the production of thin-film coatings from pure metals, including materials with high melting point, such as refractory metals W, Ta, C, etc., as well as numerous alloys and compounds. Cooling the crucible avoids contamination problems from heating and degasification. Electron beam evaporation also offers excellent material utilization to other methods, co-deposition and sequential deposition systems, precise film composition, structural and morphological control, uniform low temperature deposition, and freedom from contamination.

However, electron beam evaporation is a "line-of-sight" deposition process, necessitating a two- or three-axis rotation of the components to be coated—and thus special rotational rigs have to be designed within the deposition chamber. The two-axis translational and rotational motion of the shaft helps for coating the outer surface of complex geometries, but this process cannot be used to coat the inner surface of complex geometries. Another potential problem is that filament degradation in the electron gun results in a nonuniform evaporation rate.

Since the introduction of electron beam evaporation in the 1950s, thin-film applications requiring electron beam evaporation are continually increasing. Due to its widespread material availability, very high deposition rate, efficient material utilization, and unmatched film purity and uniformity, electron beam evaporation is employed in the production of wear resistant and thermal barrier coatings (TBC) in aerospace, hard coatings for cutting tools, corrosion resistant coatings, optical films for coating lenses and mirrors, infrared detectors, costume jewelry and filters with anti-reflection, and insulating and resistor films on electronic components for nanotechnology and semiconductor industries [98–102].

Cathodic Arc Evaporation (CAE). For more than a decade, CAE has been widely used to deposit various kinds of refractory and wear-resistant coatings such as Ti–Al–N [103] and its variants [104] for cutting tools, and CrN [105] coatings for automotive applications. CAE deposition is another evaporation-like process in which a very high current (hundreds of amperes) direct current arc is struck on a metallic cathode surface, where the arc interacts with the cathode surface vaporizing the cathode materials with a high-power density at the contact point [106]. Due to a high-power density on the arc electrodes, the CAE process is characterized by a combination of a high deposition rate and a high degree of ionization (≥90%) of evaporated species with high ion energies (20–150 *e*V), which makes this process a versatile deposition technology for producing well adherent and dense metal and compound films. Due to the nature of the high-current vacuum arc discharge, however, only target materials with good electro-conductivity can be used as evaporation sources. Also materials with a too high or low melting point or poor mechanical strength cannot be used.

The main disadvantage of the cathodic arc deposition is the production of microdroplets (macroparticles) due to the high power density on the cathode. These macroparticles will also become embedded in the films with a typical size from 0.2 μm to several micrometers. These macroparticles are undesirable since they will degrade the uniformity of the film surface and consequently functions of the film, especially for thin films used for microelectronics and electro-optics.

The macroparticle emission from the cathode spot can be, to a certain extent, reduced by [107] (i) decreasing the arc current, (ii) decreasing the cathode surface temperature, (iii) using a pure cathode material without gas, and (iv) contamination of the cathode surface by a reaction product with a higher melting point. The macroparticle can also be greatly reduced or removed from the arc plasma by several approaches during plasma transport to the substrate, of which the magnetic filter has been the most successful. There are many designs for macroparticle filters and the most popular design is based on the work by Aksenov et al. in the 1970s [108]. They used a curved magnetic filter with a positive bias of approximately 20 V applied between the filter and anode. The magneti

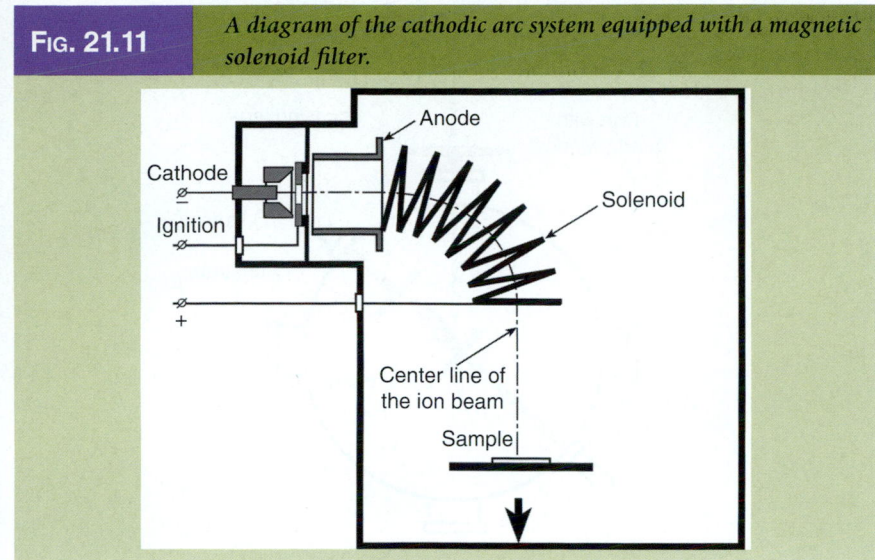

Fig. 21.11 *A diagram of the cathodic arc system equipped with a magnetic solenoid filter.*

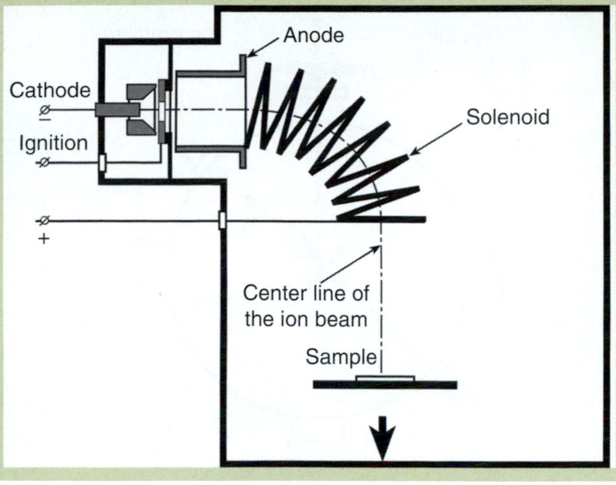

Source: Y. H. Liu, J. L. Zhang, D. P. Liu, T. C. Ma, and G. Benstetter, *Surface and Coatings Technology*, 200(7), 2243–2248 (2005). With permission.

filter can prevent line-of-sight of the macroparticles and guide the vacuum arc plasma by the curved magnetic field to the substrate. A typical cathodic vacuum arc system equipped with a magnetic solenoid filter is shown in **Figure 21.11** [109].

Pulsed Laser Deposition (PLD). Since Dijkkamp and Venkatesan [110] successfully deposited a high-temperature superconductive $YBa_2Cu_3O_7$ film using PLD in 1987, PLD has gained great attention in the past two decades for its ease of use and success in depositing materials of complex stoichiometry that are normally difficult to deposit by other methods, especially multielement oxides.

PLD uses short and high-power laser pulses (typically ~108 W · cm^{-2}) to evaporate and ablate material from the surface of a solid target in an ultrahigh vacuum chamber. The primary ablation mechanisms involve many complex physical phenomena such as collisional, thermal, and electronic excitation, and exfoliation and hydrodynamics [111]. The ablated species expand into the surrounding vacuum in the form of a plume containing many energetic species including atoms, molecules, electrons, ions, clusters, particulates, and molten globules, before depositing on the typically hot substrate. Then, the ablated material is deposited through the plasma plume onto the heated substrate surface with nucleation and growth of a thin film. A diagram of the PLD equipment is shown in **Figure 21.12**.

The PLD can occur in the presence of a wide variety of gases, which makes it an extremely versatile technique for preparing a wide range of ceramic oxides, nitrides, metallic multilayers, and various superlattice thin films [112]. Unlike thermal evaporation, which produces a vapor composition dependent on the vapor pressures of elements of the target material, the PLD generates entire/congruent evaporation of the target irrespective of the evaporating point of the constituent elements or compounds of the target that is facilitated by an extremely

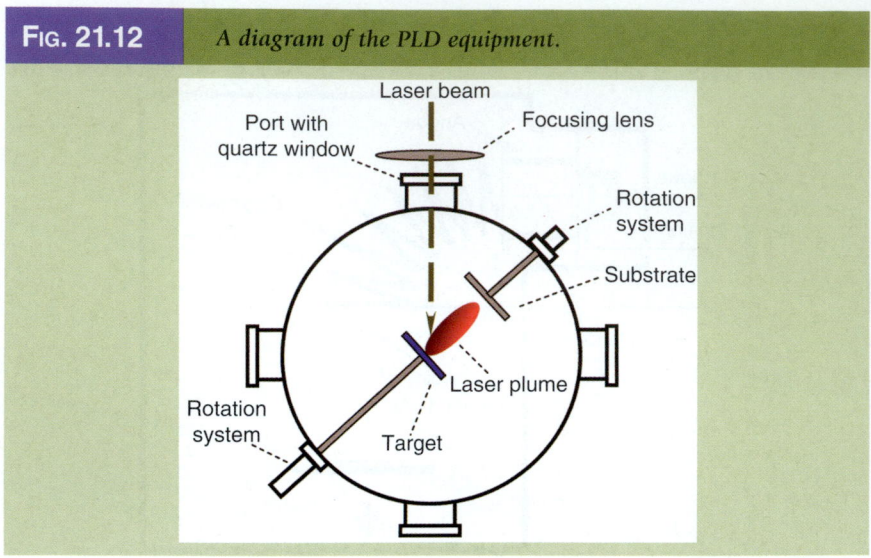

FIG. 21.12 *A diagram of the PLD equipment.*

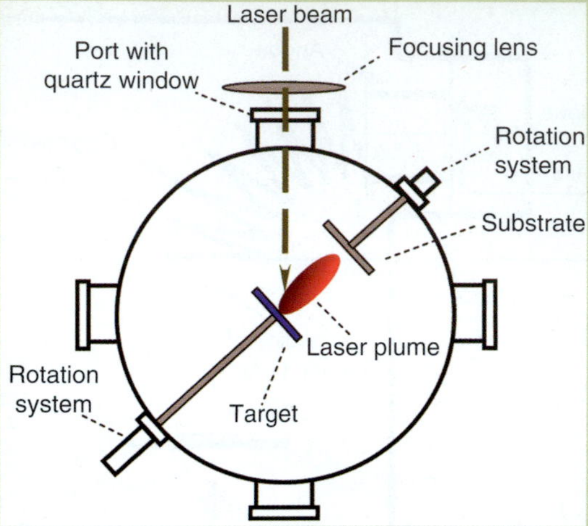

high heating rate of the target surface (10^8 K · s^{-1}). Materials are dissociated from the target surface and ablated out with the same stoichiometry as the target. Therefore, it is generally easier to obtain complex film stoichiometry for multi-element materials using PLD than with other deposition technologies. The conceptually simple design and the versatility of PLD make it highly cost-effective, in that complex multilayer films are relatively straightforward to be produced within a single system by moving various targets into and out of the laser beam focal point. In addition, by using mirrors to change the beam path, several deposition systems can be clustered around a single laser [111]. In the PLD process, due to the short laser pulsed duration (~10 ns) and hence the small temporal spread (<10 ms) of the ablated materials, the deposition rate can be extremely fast (~10 mm · s^{-1}). Consequently a layer-by-layer nucleation is favored and ultrathin and smooth films can be produced.

In spite of the mentioned advantages of PLD, some disadvantages have been identified in use of this deposition technique. One of the major problems is the splashing or the particulate (macroparticles) deposition on the film. The physical mechanisms leading to splashing include surface boiling, expulsion of the liquid layer by shock wave recoil pressure, and exfoliation. The size of particulates may be as large as a few microns. Such particulates will greatly affect the growth of the subsequent layers as well as the electrical properties of the film and therefore should be eliminated. The spot size of the laser and the plasma temperature has significant effects on the deposited film uniformity. The target-to-substrate distance is another parameter that governs the angular spread of the ablated materials.

Another problem is the narrow angular distribution of the ablated species, which is generated by the adiabatic expansion of the laser-produced plasma plume and the pitting of the target surface. These features limit the use of PLD in producing a large-area uniform thin film such that PLD has not been fully deployed in industry. Recently, remedial measures such as inserting a shadow mask to block off the particulates and rotating both target and substrate in order to produce a larger uniform film have been developed to minimize some of the PLD problems.

Applications of the PLD technique range from the production of high-temperature superconducting films, for example, $YBa_2Cu_3O_{7-x}$, $NdBa_2Cu_3O_7$, ferroelectric $BaTiO_3$ and other perovskite thin films [113–116], certain magnetic materials, for example, yttrium iron garnet (YIG) [117,118] and ferromagnetic shape-memory (FSM) alloy Ni–Mn–Ga [119], biocompatible coatings for medical applications, for example, protein [120], pepsin [121], to TBC for turbine blades. In spite of this widespread usage, the fundamental processes occurring during the transfer of material from target to substrate are not fully understood and consequently need further research.

Thermal Plasma Processing of Thin Films. Thermal plasma processing is identified as utilizing high-temperature plasma to deposit metallic and nonmetallic materials in a molten or semimolten state on a substrate [122–124]. Thermal plasma processing may also be referred to as injection plasma processing (IPP), in which different types of processes were developed, such as thermal plasma CVD, plasma flash evaporation, and plasma spray [125]. The plasma spray process is a well-established commercial process that can be operated in normal atmospheric conditions and is referred to as atmospheric plasma spray (APS), or in a vacuum chamber (VPS), or with a protective gas at low pressure (LPPS). However, the basic processing principles are similar in that firstly a high-temperature plasma is initiated from a plasma gas (Ar, N_2, H_2, or He) by applying a high voltage between a cathode and anode (as shown in **Fig. 21.13a**), and material in the form of powder is injected into the high-temperature plasma flame, where it is rapidly heated, melted, and accelerated to a high velocity. The hot material impacts on the substrate surface and rapidly cools forming a coating. A photo showing a thermal plasma spray process at work is presented in **Fig. 21.13b**. Similar to the thermal spray process, plasma flash evaporation also uses the raw

FIG. 21.13 *(a) Schematic diagram of a solution precursor thermal plasma spray process [126] and (b) a photo showing a thermal plasma process at work [127].*

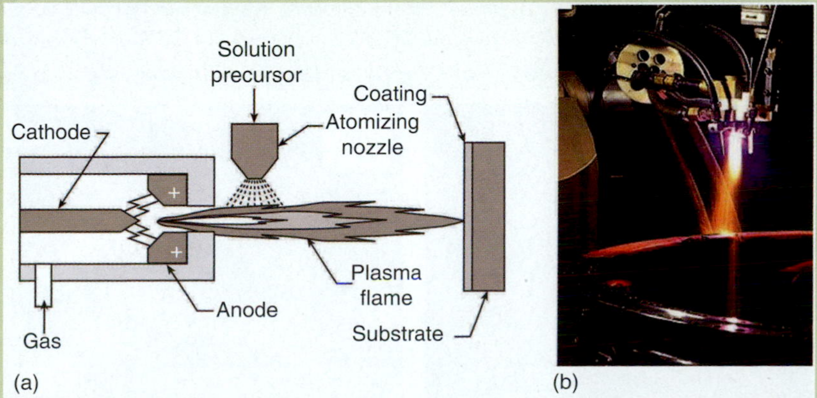

powder material as the reactant. However, the powder particle size needs to be fine enough (<10 μm) in an effort to achieve evaporation, instead of only melting the reactant powders.

Plasma spraying has the advantage that it can spray very high-melting point materials such as refractory metals like tungsten, and ceramic, for example, zirconia, which are not suitable for combustion processes. Plasma-sprayed coatings are generally much denser, stronger, and cleaner than the other thermal spray processes. Plasma spray coatings probably account for the widest range of thermal spray coatings and applications and thus making this process the most versatile. Due to the high deposition rate, typical coating thickness ranges between 50 μm to a few millimeters.

However, thermally plasma-sprayed coatings normally contain inhomogeneities such as inter-splat porosity, unmelted particles, and micro-cracks, as shown in **Figure 21.14a** [123]. The presence of interconnected porosity is of prime concern as it limits the effectiveness of the barrier provided by the coating. Another disadvantage of thermal spray is that it is hard to coat complex shaped substrates since it is a line-of-sight process. Postspray polishing is usually needed to obtain the desired surface finish on the surface. LPPS coating can provide coatings with improved microstructure and decreased porosity (**Fig. 21.14b**). The individual Al_2O_3 splats are in the LPPS Al_2O_3 coating shown in transmission electron microscope (TEM) (**Fig. 21.14b**). The splats are seen to have a nanocrystalline Al_2O_3 phase embedded in an ordered amorphous Al_2O_3 phase.

Recently, another new development in thermal plasma processing, plasma chemical vapor deposition (PCVD) process, has been developed. Unlike the previous two-spray process, the reactants fed into the high-energy-density plasma are gaseous and liquid precursors in the process of PCVD [128–131]. With the high temperatures, the precursor material that is injected into the plasma is rapidly vaporized and dissociated and accelerated towards the substrate. The thermal plasma in most thermal plasma processes is initiated by a high voltage DC discharge due to its high energy density, high velocity, and ease of use. However, the steep temperature and velocity gradients occurring in the plasma generally result in nonuniformity in terms of heating, trajectory, and reactions

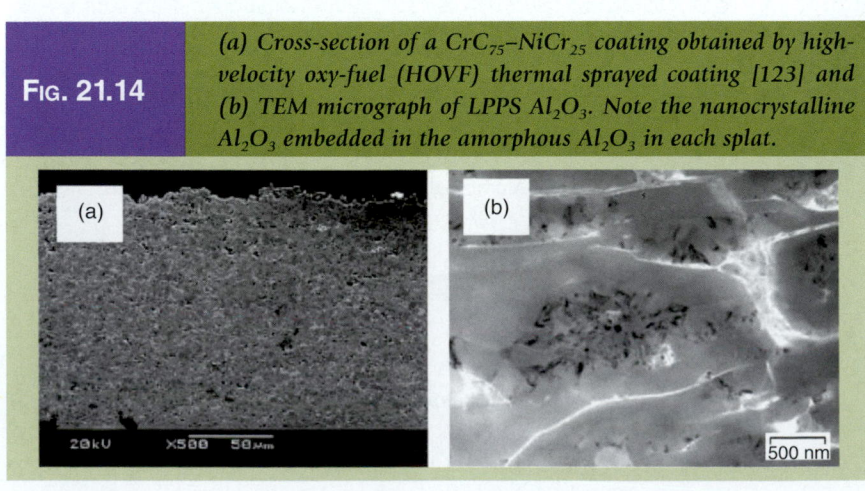

Fig. 21.14 *(a) Cross-section of a CrC_{75}–$NiCr_{25}$ coating obtained by high-velocity oxy-fuel (HOVF) thermal sprayed coating [123] and (b) TEM micrograph of LPPS Al_2O_3. Note the nanocrystalline Al_2O_3 embedded in the amorphous Al_2O_3 in each splat.*

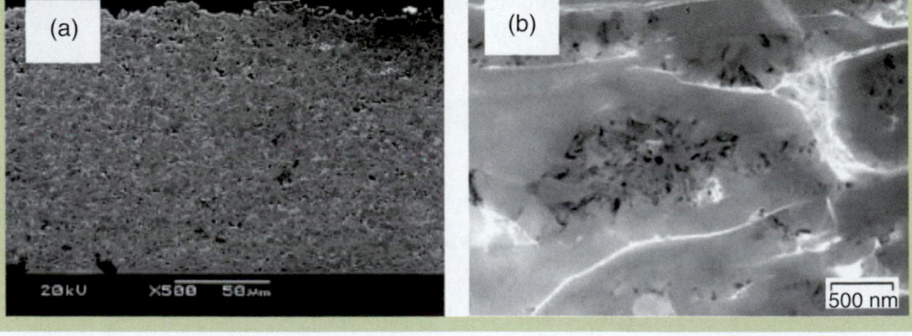

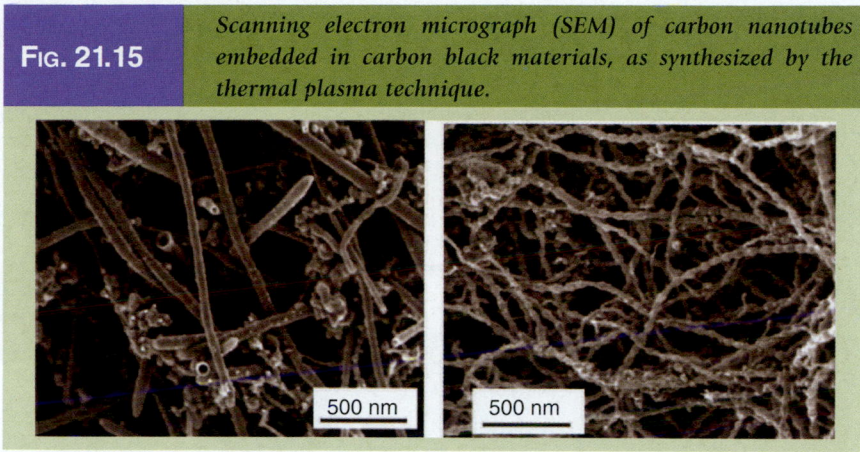

Fig. 21.15 *Scanning electron micrograph (SEM) of carbon nanotubes embedded in carbon black materials, as synthesized by the thermal plasma technique.*

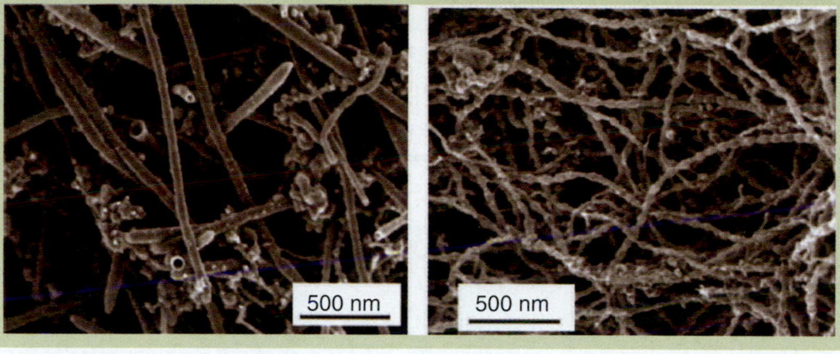

Source: H. Okuno, E. Grivei, F. Fabry, T. M. Gruenberger, J. Gonzalez-Aguilar, A. Palnichenko, L. Fulcheri, N. Probst, and J.-C. Charlier, *Carbon*, 42(12–13), 2543–2549 (2004). With permission.

in addition to the high heat flux to the substrate, which leads to difficulties in controlling the substrate temperature. In recent years, the use of RF and hybrid plasmas has developed rapidly since they offer better control of the plasma to minimize trajectory and reaction variations [125].

Thermal plasma processing is considered to be one of the prime candidates for producing high-temperature thermal barrier coatings for aircraft engines, industrial gas turbine blades, and antiwear and anticorrosion coatings for high-temperature applications due to its relatively low cost, high deposition rate, and high durability [132]. In recent years, thermal plasma processing has been reportedly used to synthesize carbon nanotube and nano-necklaces at relatively low operation temperatures as shown in **Figure 21.15** [124].

Sputter Deposition. The PVD deposition techniques discussed so far are based fundamentally on thermally evaporating materials at elevated (above melting point) temperatures. Another important PVD deposition mechanism is sputtering. Sputter deposition is the deposition of atoms that are vaporized from a solid target or source by bombarding the surface with energetic ions [133]. Since Grove first observed sputtering in a DC gas discharge tube in 1852, sputtering has been widely used for surface cleaning and etching, thin-film deposition, surface and surface layer analysis, sputter ion sources, and for the modification of the properties of thin films (implantation) depending on the ion energy range (1 *e*V to 10,000 k*e*V) involved in the deposition [134]. The incident particles are usually inert gas ions, but any ion, neutral atom, molecule, or even a photon can be used if the energy is sufficient [135].

As schematically portrayed in **Figure 21.16**, in sputter deposition, the impact of the energetic particles on the target surface kinetically knocks one or more of the surface or near-surface atoms off their equilibrium sites. These atoms, which have received considerable kinetic energy from the initial particles, move deeper into the target material and undergo further collisions. This process continues until eventually causing the ejection of atoms (sputtering) from the target surface [136].

FIG. 21.16 *A schematic representation of the physical sputtering process.*

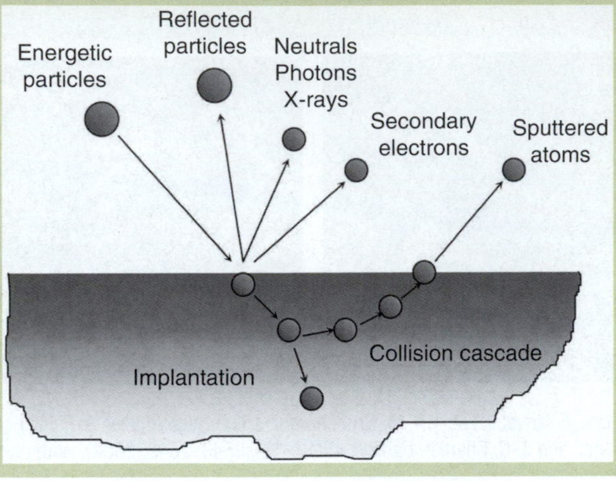

The interaction of energetic ions with surfaces will also create a variety of interactions besides sputtering, such as generation of secondary electrons, neutrals, photons, x-rays, and implantation of atoms into the substrate (**Fig. 21.16**). These interactions play an important role in the field of surface science, such as the ionization of atomic particles at/near surfaces, the modification of surfaces, and surface analysis. For example, the emission of secondary electrons is essential for maintaining the discharged plasma. Optical emission from a glow discharge during sputtering can be used for process control and chemical analysis in the optical emission spectrometer (OES) [183].

The use of energetic ions in film deposition has numerous advantages in that controlled ion bombardment can lead to significant micro/nanostructural modifications and properties. Ion-assisted deposition of materials can produce coatings with superior properties since the ion assistance provides incident atoms with additional energy. A typical kinetic energy of sputter deposition is 1–10 *e*V, which is orders of magnitude above that of typical evaporated particles. This added energy can modify the nucleation process, improve film adhesion, increase film density, change the film texture, trigger phase changes, influence film stress, and change film microstructure. All these factors can be utilized to produce coatings with improved corrosion resistance, hardness, and optical, physical, and wear properties.

Sputter Deposition I: Ion Beam Sputtering (IBS). Sputter deposition is typically practiced in either a plasma (**Fig. 21.17a**) or an ion beam (**Fig. 21.17b**) configuration. In plasma systems (**Fig. 21.17a**), a cathode is used which is bombarded by ions from the plasma. If the energy and pressure conditions are appropriate, the cathode (target) material is then sputtered off and can deposit on nearby surfaces. In IBS (**Fig. 21.17b**), a beam of ions is generated using a remote source. This ion beam is then directed onto the target, and atoms are sputtered from the target onto a nearby sample [106].

FIG. 21.17 *Schematic drawing of (a) conventional plasma sputtering and (b) dual ion beam sputtering system.*

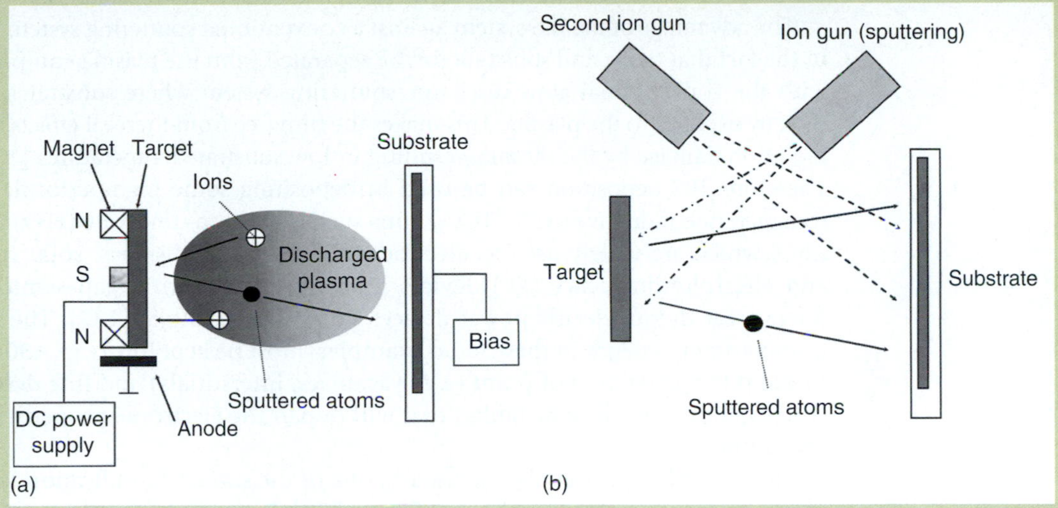

During the IBS process, the high-intensity Kaufman ion source is routinely used to generate a high-energy ion beam. The element, alloy, or compound targets are sputtered by the ion beam in a prescribed direction controlled by a magnetic or electrical filed. The sputtered atoms are deposited on a heated or biased substrate to form the thin film. The growing film may also be irradiated with ions from a second ion source, as indicated in **Figure 21.17b**. This technique is termed dual ion beam sputtering and has been used extensively both to modify the film structure and to synthesize compounds. The second ion source may operate with inert or reactive ions with energies ranging from 20 *e*V to 10 k*e*V.

The IBS process can also be operated in the plasma sputter mode. After a pure ion beam has been extracted from an ion source, electrons may be added to the ion beam to form a plasma beam (equal number of positive ions and electrons) which will not diverge and not cause a charge build-up on the target surface. In the Kaufman source these electrons are normally generated from a hot filament ("neutralizer filament"), such as tungsten, that easily generates electrons due to thermal emission. The beam is volumetrically neutral due to the addition of the electrons. Plasma beams have the advantage that the electrons can easily be deflected (steered) by a magnetic or electrostatic field and the ions will generally follow. They can be operated in vacuum and at a high pumping speed. Therefore contamination can easily be controlled. Also the flux and energy of the bombarding particles can easily be monitored and controlled, and insulating surfaces can be sputtered.

The IBS technique offers several advantages. The process of ion beam sputter deposition produces sputtered atoms with high average energy, for example, 3–10 *e*V (compared to 0.1 *e*V or so in thermal evaporation). Films made from these atoms show improved properties and adhesion when compared to conventional deposition techniques such as thermal evaporation. Sputter deposition of compound and alloy films is feasible since stoichiometry is normally preserved in the growing film. During the IBS process, the source materials are directly sputtered onto the substrate and the ion beams can be directionally controlled, the films can be deposited with extremely high accuracy and repeatability.

The advantage of the IBS system against a conventional sputtering system lies in the fact that target and substrate can be separated from the plasma compared with the conventional glow discharge sputtering system where substrates are directly exposed to the plasma. This makes the film free from thermal effects and radiation damage by the plasma, resulting in low substrate temperatures [137]. Therefore, IBS deposition can be used in depositing some high-performance transparent conductive oxide (TCO) films such as indium–tin oxide (ITO) and ZnO, which are widely used as electrodes for flat panel displays, solar cells, and electroluminescence (EL) devices [138,139], and narrow-gap semiconductors for thermoelectric power devices, for example, Mg_2Si [137]. The ion bombardment energy, in these latter examples, must be kept low (e.g., <30 *e*V) to avoid the generation of point (e.g., vacancies, interstitials) and line defects (e.g., dislocations, stacking faults) that will impair the electronic properties of these films.

The IBS technique can also be used as one of the surface modification techniques to reduce surface roughness of materials by selectively detaching atoms and nanoparticles from the surface with the bombarding energetic ions from 1–10 to a few tens of kiloelectron volts onto the film surfaces. This technique can be applied to a surface that needs to have submicrometer surface roughness. Ion beam machining is used for ultraprecision machining of high-melting point and hard, brittle materials where machining depth needs to be precisely controlled. For ultraprecision machining of a large region in IBS processes, a high current density ion beam and a large-area ion source with a uniform beam extraction unit are necessary [140]. However, the small ion beam size (~100 mm diameter) in the IBS limits the substrate size for the film deposition. Therefore, the equipment required is somewhat more expensive when measured in terms of rate and coated area per unit cost.

Sputter Deposition II: Balanced and Unbalanced Magnetron Sputtering. More frequently, the sputter deposition is practiced in a plasma configuration (**Fig. 21.17a**). The simplest system still in use today is the diode sputtering deposition, which consists of the target and anode facing each other. In many cases, the anode is the chamber wall which is grounded, and the cathode is then biased negatively. With the appropriate gas density and an adequate electric field between the anode and cathode, a plasma can be formed by gas breakdown into ions within the chamber. Even though planar diode sputtering deposition is still used today due to its simplicity and the relative ease of fabrication of targets for a wide range of materials, it has several disadvantages. In diode sputtering, not all of the electrons escaping the target contribute to the ionized plasma glow area, thereby resulting in a low deposition rate. The wasted electrons are accelerated away from the target causing radiation and other problems, for example, the heating of the substrate.

Sputtering first achieved widespread use in research and industrial application with the introduction of magnetic assistance [141]. The use of magnetic fields to enhance the sputtering rate led to the term "magnetron sputtering." A MS source addresses the electron problem by placing magnets behind, and sometimes, at the sides of the targets. These magnets capture the escaping electrons and confine them to the immediate vicinity of the target (**Fig. 21.18a**). The ion current (density of ionized atoms hitting the target) is increased by an order of magnitude over conventional diode sputtering systems, resulting in faster deposition rates at lower pressure. The lower pressure in the chamber also helps create a cleaner film and a lower target temperature, enhancing the deposition of high-quality films. Additional modifications to physical sputter deposition have been made to enhance the chemical and/or structural nature of the deposited films.

As mentioned above, the magnetic field behind the MS targets is designed to trap the electrons, and hence, the plasma in the vicinity of the substrate. In traditional planar-balanced magnetron sputtering, the north and south magnets behind the targets are balanced (i.e., equal strength) (**Fig. 21.18a**) [142]. One of the disadvantages of the balanced magnetron source is that the plasma is effectively trapped near the surface of the sputtering target resulting in lower plasma density over large coating volumes (**Fig. 21.18b**). Normally, ion current density (ICD) at the target (cathode) (the current drawn per unit area of a negatively biased substrate) in conventional balanced magnetron sputtering is less than $1\,mA \cdot cm^{-2}$ [142]. The problem was partially solved by adding auxiliary ionization sources or using RF. The other disadvantage is the poor deposition rate compared to thermal evaporation methods.

In recent years many researchers have tried to overcome the disadvantages of MS and increase the plasma density. These developments include the unbalanced magnetron, closed-field configuration of magnetrons, pulsed magnetron sputter deposition, and high-power pulsed DC magnetron sputtering (HPPMS).

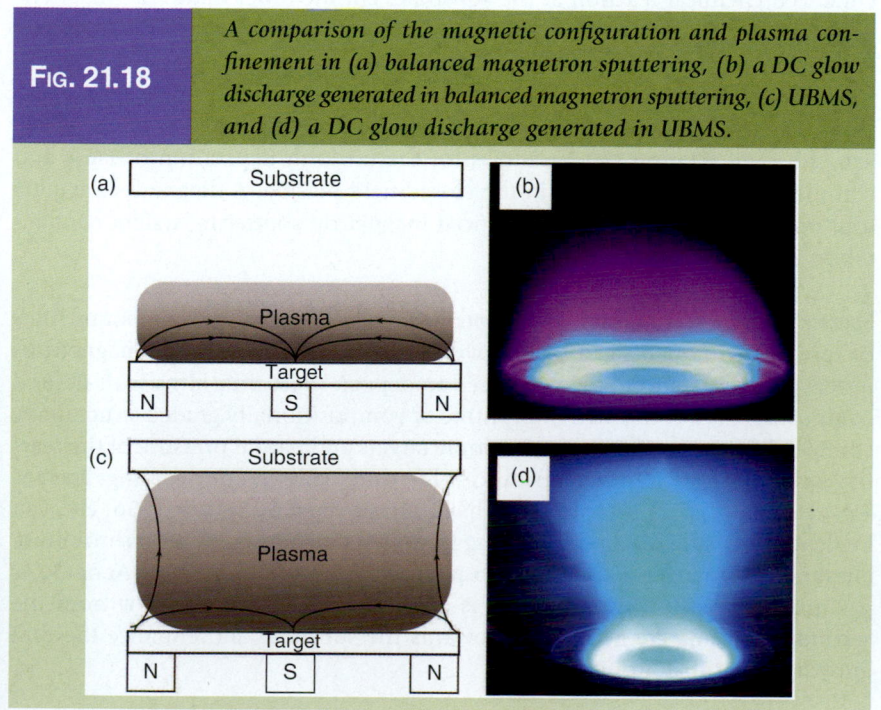

FIG. 21.18 *A comparison of the magnetic configuration and plasma confinement in (a) balanced magnetron sputtering, (b) a DC glow discharge generated in balanced magnetron sputtering, (c) UBMS, and (d) a DC glow discharge generated in UBMS.*

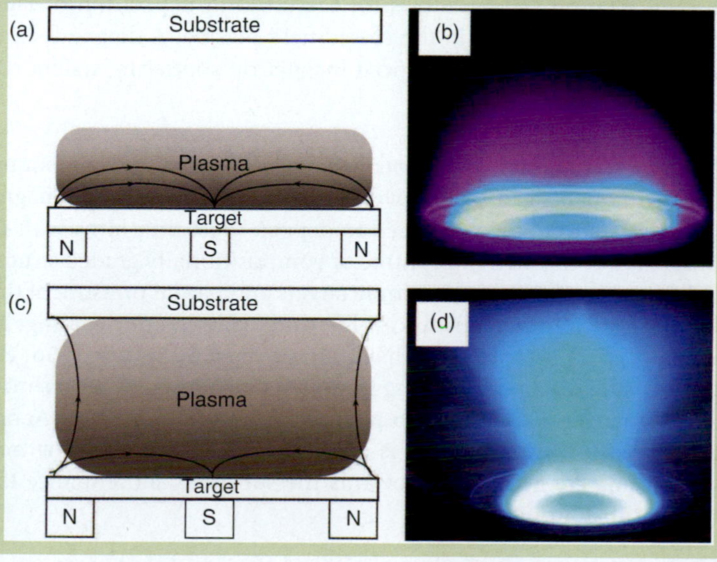

The invention of the unbalanced magnetron by Windows and Savvides in 1986 offered a better solution to enhance the plasma density [143,144]. An unbalanced magnetron uses stronger magnets on the outside than the center resulting in the expansion of the magnetron field lines, electrons, and plasma away from the surface of the target towards the substrate (**Fig. 21.18c**). The effect of the unbalanced magnetic field is to trap fast-moving secondary electrons that escape from the target surface. These electrons undergo ionizing collisions with neutral gas atoms at locations away from the target surface and produce a greater number of ions and further electrons in the region of the substrate, thereby considerably increasing the plasma density and substrate ion bombardment (**Fig. 21.18d**). The ICD in UBMS is increased to 2–10 $mA \cdot cm^{-2}$ [142].

Sputter Deposition III: Closed-Field Unbalanced Magnetron Sputtering (CFUBMS). In the late 1980s an important improvement in MS was developed—termed the closed-field unbalanced magnetron sputtering. The purpose of CFUBMS is to enhance ionization and increase the ICD in MS as proposed by Sproul et al. [145,146] and Tominaga [147]. The ICD can be further increased to 5–20 $mA \cdot cm^{-2}$ compared to the UBMS (2–10 $mA \cdot cm^{-2}$). [142].

A comparison of a mirrored and a closed-field magnetron configuration [145] is shown in **Figure 21.19**. The major feature in the CFUBMS system was the idea of using unbalanced magnetrons in an arrangement whereby neighboring magnetrons are of opposite magnetic polarity. Using this arrangement, the deposition zone in which the substrates are located is surrounded by linking magnetic field lines (**Fig. 21.19a**). This traps the plasma region, prevents loss of ionizing electrons escaping to the chamber walls resulting in much higher plasma density (ICDs) (**Fig. 21.19b**), and dense, hard, well-adhered coatings by enhanced chemical reaction at the substrate. On the other hand, the magnetic field lines are not closed in the mirrored configuration (**Fig. 21.19c**), thereby resulting in a poor plasma density (**Fig. 21.19d**).

Closed-field systems can be configured using any even number of magnetrons. The use of multiple magnetrons in conjunction with a two-axis or three-axis substrate rotation system allows for the uniform deposition of large and complex shaped components in the closed-field plasma. A diagram showing a four magnetron closed-field unbalanced magnetron sputtering system configuration is presented in **Figure 21.20**.

Sputter Deposition IV: Pulsed Magnetron Sputtering. Ceramic and compound films of various components can be reactively deposited using multiple magnetrons in MS. Reactive sputtering can be used to deposit most thin films with desired composition as a controlled monolithic or compositionally graded structure by control of the power density on multiple targets and partial pressure of the reactive gas [148–150]. Various nitride, carbide, and oxide ceramic coatings such as TiN [151], CrN [152], TiC [153], TiAlN [154], CrAlN [155], Al_2O_3 [156] etc., can be deposited using reactive sputtering from metal targets (titanium, chromium, aluminum, etc.) in a reactive gas atmosphere, for example, N_2/Ar, CH_3/Ar or O_2/Ar gas mixture. When a negative bias is applied to the substrate, ions from the secondary plasma are accelerated towards the substrate and enhance the film growth at the substrate.

FIG. 21.19 *A comparison of the magnetic field configurations and plasma photographs in (a) the "mirrored" magnetic field configuration, (b) a DC glow discharge generated in the "mirrored" magnetron sputtering, (c) "closed-field" magnetic field configuration, and (d) a DC glow discharge generated in "closed-field" magnetron sputtering.*

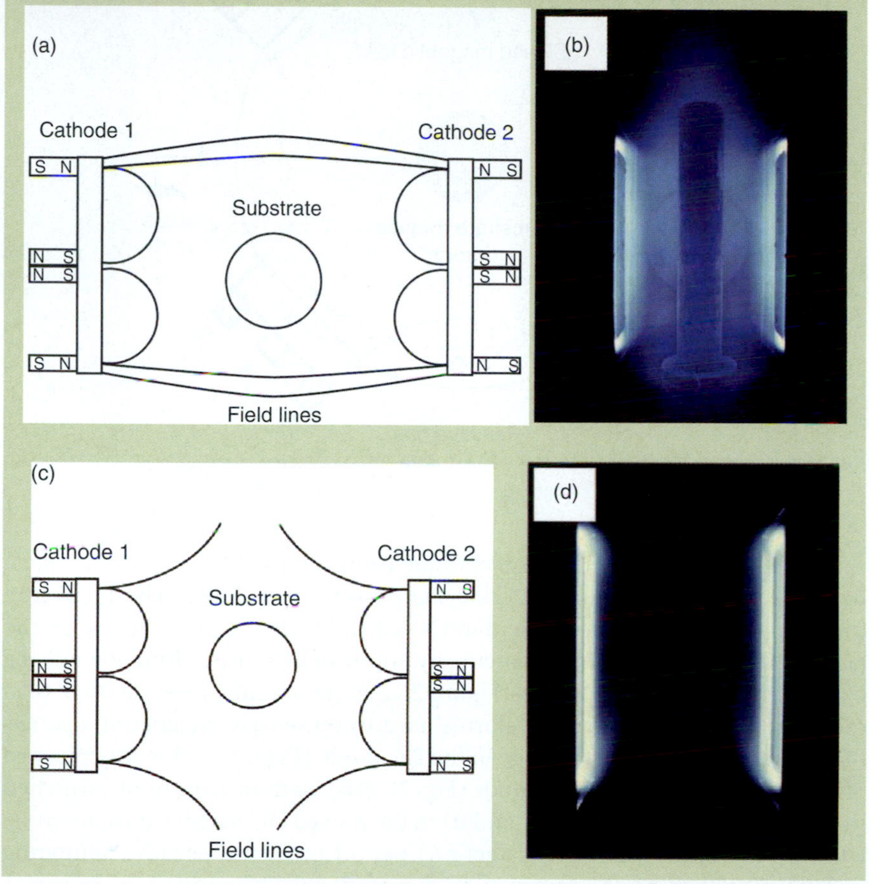

However, for deposition of insulating films, the insulating film builds up on the surface of the chamber and/or anode. When the insulating layer on the anode becomes thick, the sputtering discharge becomes unstable. This phenomenon is called "disappearing anode." The target can also charge up quickly due to a nonconductive layer formed on the target at which point the target is said to be "poisoned," which makes sputtering more difficult, decreases the sputtering rate, and increases the power on the target. The charges will generate arcing problems in which microparticles will be ejected from the target and incorporated into the deposited coatings. This condition leads to nonuniform deposition, and inhomogeneities and defects in the coatings, as well as a reduced deposition rate [157].

To overcome this problem, the use of RF sputtering was developed in the 1960s. However, RF sputtering is not used extensively for commercial MS due to its low deposition rate, high cost, and the generation of a high temperature from the self-bias voltage associated with the RF power.

FIG. 21.20 *Diagram showing a four magnetron closed-field unbalanced magnetron sputtering system configuration.*

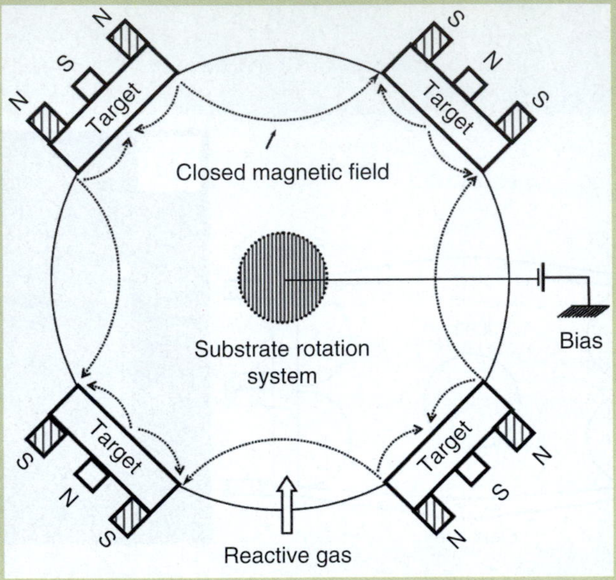

In recent years, an alternative technique using a pulsed DC plasma has been developed and is called pulsed DC magnetron sputtering. Pulsed DC magnetron sputtering utilizes a pulsed potential to neutralize the positive charge on the target surface and eliminate the arcing by appropriately controlling the pulsing parameters of the target potential. **Figure 21.21** schematically presents some typical target voltage–potential waveforms used in pulsed DC magnetron sputtering. The continuous target voltage in the DC mode (**Fig. 21.21a**) is either turned off periodically in the unipolar mode (**Fig. 21.21b**), or more commonly, switched to a positive voltage (**Fig. 21.21c** and **d**) in the pulsed DC magnetron sputtering. During the normal pulse-off (target-on) period (τ_{on}), the negative sputtering voltage is applied to the target as in conventional DC sputtering. However, the target negative potential is periodically interrupted by a positive pulse voltage with a period of τ_{rev}. The reversed positive voltage is variable depending on the power supply allowance, which is either reversed to a smaller positive voltage than the nominal negative pulse voltage in the asymmetric bipolar mode (**Fig. 21.21c**) or reversed to the same magnitude of positive voltage as the nominal negative pulse voltage in the symmetric bipolar mode (**Fig. 21.21d**). The duty cycle is defined as the negative pulse time divided by the period of the pulsing cycle (τ_{cycle}) as shown in equation (21.7):

$$\text{Duty cycle} = \frac{\tau_{cycle} - \tau_{rev}}{\tau_{cycle}} \quad (21.7)$$

For a given positive pulse width, the full range of frequencies may not be available due to the power supply limitation, for example, for an Advanced Energy

FIG. 21.21 *The target voltage waveforms when operated in (a) continuous DC, (b) unipolar pulsed mode, (c) asymmetric bipolar pulsed mode, and (d) symmetric bipolar pulsed mode, (τ_{rev}: the reversed positive pulse period, τ_{on}: the normal negative target potential period, and τ_{cycle}: the whole pulse period).*

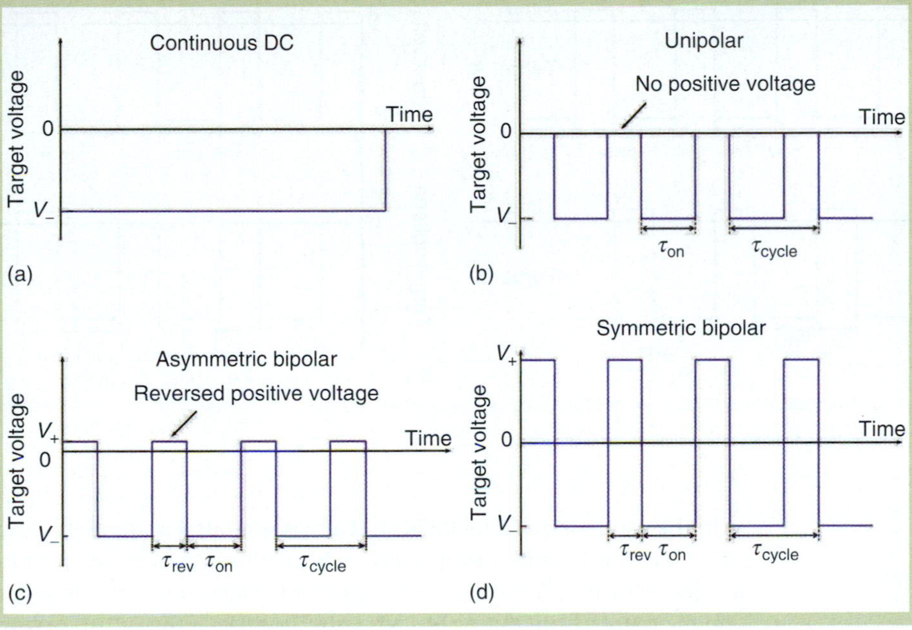

Pinnacle Plus power supply, the smallest duty cycle is 50% [158]. In practice, the waveforms are virtually never as intended due to nonlinearities of either the plasma or the power supply circuitry. Therefore, the shapes of the resulting power waveforms are complex.

In the closed-field configuration, two or more magnetrons are commonly used for reactive sputtering. All magnetrons can be run in a pulsing condition, thereby resulting in different combinations of pulsing modes. Normally, the magnetron potential can be pulsed in either an asynchronous bipolar mode (**Fig. 21.22a**), in which the two target voltage waveforms are out of phase, or a in synchronous bipolar mode (**Fig. 21.22b**), in which the two target voltage waveforms are in phase. The degree of out of phase in the asynchronous mode is totally dependent on the frequencies and reverse time on each magnetron.

During the reversed positive pulse, the charge built-up on the insulting material during the negative pulse period is discharged to eliminate breakdown and arcing. Therefore, a stable deposition process and smooth coating structure can be obtained. Moreover, with precise control of the partial pressure of reactive gas, high deposition rates can be achieved in depositing insulating films. A comparison of the microstructure of IrO_2 films deposited by P-CFUBMS and normal DC magnetron sputtering is shown in **Figure 21.23**. A dense and smooth surface IrO_2 film was produced by P-CFUBMS. On the other hand, the film deposited by DC magnetron sputtering exhibited considerable microparticles covering the film surface.

FIG. 21.22 *Different target voltage waveforms in a dual magnetron pulsed mode: (a) asynchronous bipolar pulsing and (b) synchronous bipolar pulsing.*

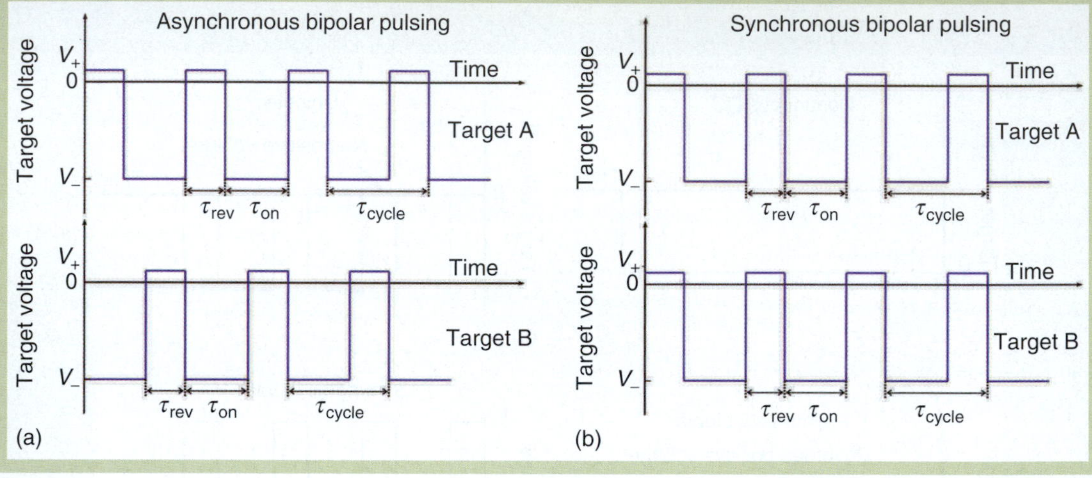

In the sputtering of a metal film, if a bias potential is applied to the substrate, one can control the film properties such as adhesion, texture, morphology, and density of the film. This control is achieved because one can extract ions to bombard the growth front during film growth. However, if the substrate is an insulator, it can get charged up very quickly and one would lose the benefit of ion bombardment, particularly during the initial stages of growth. RF potential can be used to neutralize the substrate. However, the ion energy distribution generated by a RF power is not uniform and is not well controlled. Recently, it has been recognized that using a pulsed bias potential on the substrate can overcome this difficulty and can produce better controlled microstructures and a superior film quality [159].

FIG. 21.23 *SEM photomicrographs of fracture sections of iridium oxide coatings deposited by (a) DC reactive sputtering and (b) pulsed reactive sputtering.*

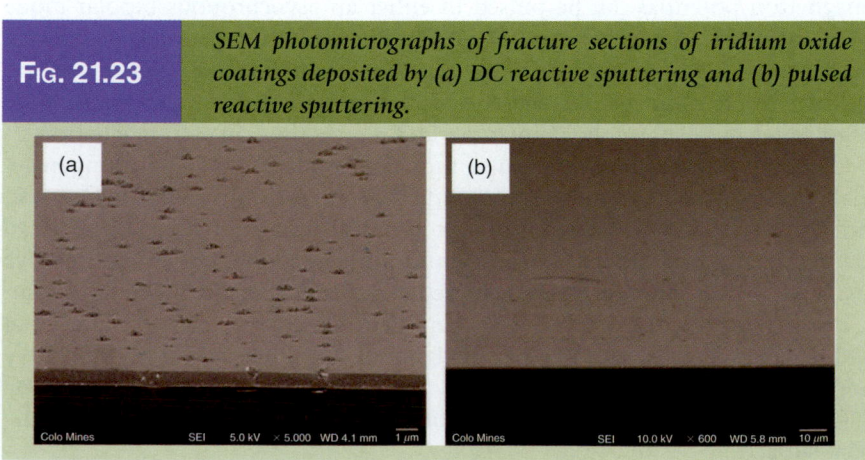

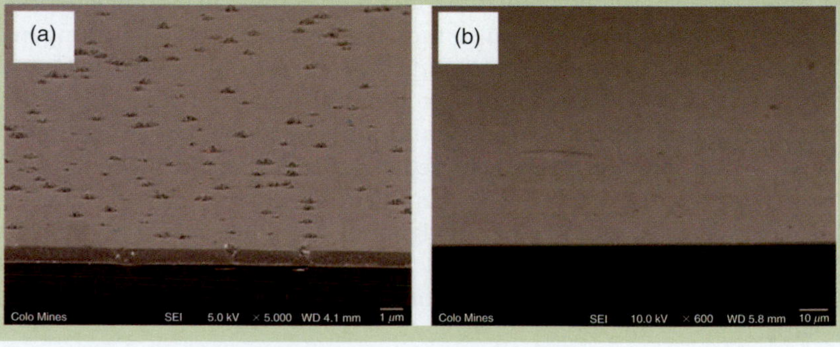

Sputter Deposition V: High-Power Pulsed DC Magnetron Sputtering (HPPMS). Recently, a new high-power pulsed DC magnetron sputtering (HPPMS), operating in a unipolar mode, has attracted wide attention. In this new method, a high-density plasma is created in front of the sputtering source by using pulsed high-target power density, ionizing a large fraction of the sputtered atoms [160–162].

In normal DC and pulsed DC magnetron sputtering, the degree of ionization is low (typically less than 10%), which is due to the low power density (e.g., $3\,W \cdot cm^{-2}$) limited by the target overheating from the ion bombardment [163]. However, in HPPMS, the average thermal load on the target is low by operating the target at high power density (e.g., $3000\,W \cdot cm^{-2}$) in a pulsed condition; consequently a considerably large fraction of ionized species can be created by the high probability for ionizing collisions between the sputtered atoms and energetic electrons. It was reported that the fraction of ionized target species in high-power pulsed DC magnetron sputtering can be considerably increased up to 70% or higher compared with less than 10% for conventional DC magnetron sputtering [164].

The high degree of ionized ions with controlled energy makes it possible to control the film growth behavior and produce high quality films since energetic condensation of the film can be easily achieved. The potential of this emerging new high-power pulsed magnetron sputtering technique includes depositing fully dense films with equiaxed structure, controlling the orientation of the film, and depositing thick (maybe up to 20 μm or more) films with low residual stress. As the deposited species are largely ions, it is possible to further control metal ion energies and their trajectories by biasing the substrate. Alami synthesized Ta films on a Si substrate placed along the wall of a 2-cm deep and 1-cm wide trench using a mostly neutral Ta flux by conventional DC magnetron sputtering and a mostly ionized Ta flux by HPPMS [162]. Drastic structural changes were achieved in the films simply by changing the power source, as shown in **Figure 21.24**. The Ta thin film grown by HPPMS exhibited a smooth surface and a dense crystalline structure with grains oriented perpendicular to the substrate surface (**Fig. 21.24a** and **c**), whereas the film grown by DC magnetron sputtering exhibited a rough surface, pores between the grains, and an inclined columnar structure (**Fig. 21.24b** and **d**). The improved homogeneity achieved by HPPMS is a direct consequence of the high ion fraction of sputtered species being controlled by the bias on the substrate. However, this new technique is in its developing stage, and the deposition rate and cost of power supply are the most important challenges that need to be considered.

21.5 General Considerations and Practical Aspects of Sputtering Deposition

Sputtering is nowadays considered as a flexible and effective coating method for surface engineering applications. There are numerous papers and textbooks on the fundamentals, phenomena, and applications of sputter deposition. The information contained within this section is not intended to be an all-inclusive

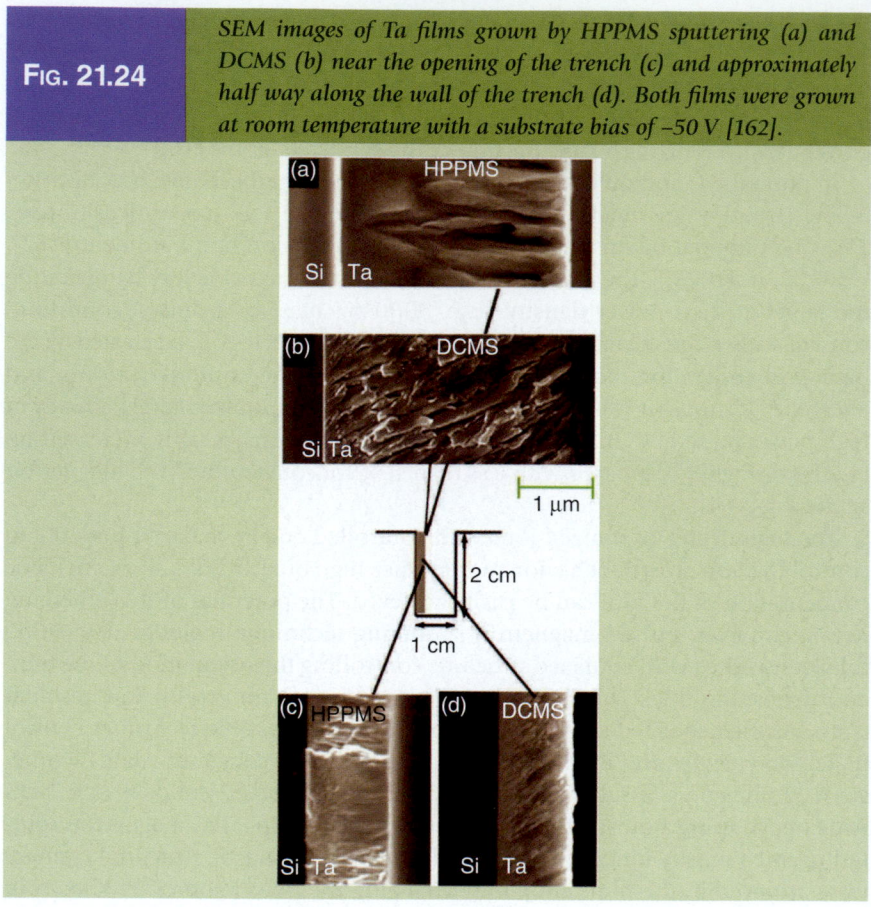

Fig. 21.24 *SEM images of Ta films grown by HPPMS sputtering (a) and DCMS (b) near the opening of the trench (c) and approximately half way along the wall of the trench (d). Both films were grown at room temperature with a substrate bias of −50 V [162].*

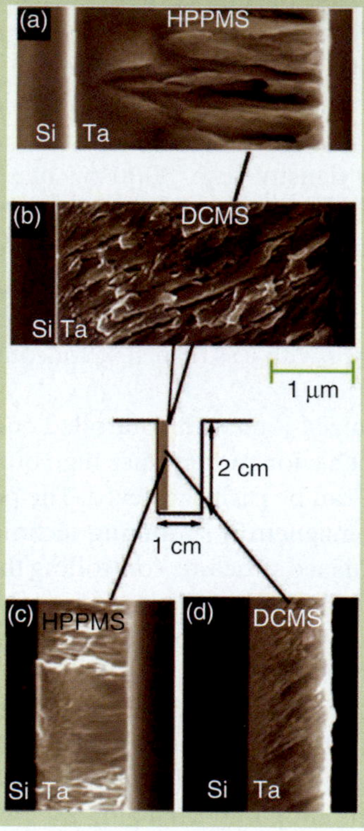

survey of the field. Instead, several aspects which are important practical aspects for the successful deposition of nanostructure and nanocomposite coatings, including reactive gas control with process stability, film structure control, discharged plasma properties and monitoring, and energy-enhanced deposition, will be briefly introduced.

21.5.1 Reactive Sputtering Deposition Process Stability

Compound thin films can be deposited either by direct deposition of a compound source in thermal evaporation, electron beam evaporation, CAE, PLD, sputtering, or by reactive sputtering. However, it should be noted that thermal evaporation of a compound source is often not possible due to noncongruent melting of the compound, while the compound must be conducting for CAE of a compound target. More recently, reactive sputtering has become the deposition of choice of compound thin films. In reactive sputtering deposition, atoms and small molecules are ejected (sputtered) from a target surface under particle bombardment and travel through the plasma discharge to the substrate, where they react with the gas to form a wide variety of compound films on the substrate. Careful control of the reactive gas is important in reactive sputtering. Too low a supply of the reactive gas will cause high rate metallic sputtering, but may give rise to an understoichiometric compound composition of the deposited film. Too high a supply of the reactive gas will allow for stoichiometric composition of the deposited film, but will cause poisoning of the target surface, which may reduce the deposition rate significantly [165].

In reactive sputtering, the reaction takes place both at the cathode/target and substrate surface during the deposition. The reaction at the cathode is usually related to target poisoning and the reaction at the substrate is called a metallic mode [166]. Normally, the sputter yield of the poisoned target (compound material) is substantially lower than the sputter yield of the elemental target material, which causes the deposition rate to exhibit a hysteresis curve with the variation of the flow rates of the reactive gas [165–167].

A typical hysteresis curve is shown in **Figure 21.25** [168]. (a) At low O_2/Ar pressures, the metallic Zr + Y thin films are deposited below the critical oxygen pressure to provide the required oxide stoichiometry. (b) When the partial oxygen pressure increases, the YSZ thin films are deposited through the metallic mode. In the metallic mode, the deposition rates of the yttrium stabilized zirconia (YSZ) are higher than those of Y/Zr metal thin films. (c) Above the critical oxygen pressure, the deposition rate abruptly decreases. (d) The target is poisoned (oxide formation on the metallic target) and the sputtering mode moves to the oxide mode with low deposition rate. (e) When the partial oxygen pressure decreases, the sputtering is still in the oxide mode for a certain oxygen flow rate. (f) The sputtering deposition will return to the original metallic mode at an oxygen pressure below the critical oxygen pressure, when the oxide layer of the target surface is fully removed.

The presence of the hysteresis hampers the process control of the reactive sputtering process. As such several researchers have tried to eliminate the hysteresis

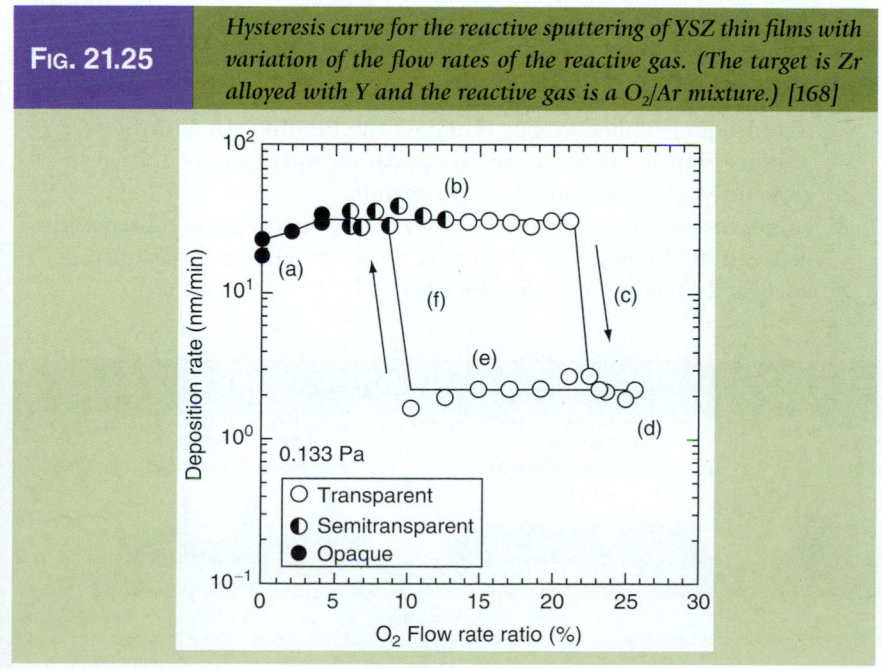

FIG. 21.25 *Hysteresis curve for the reactive sputtering of YSZ thin films with variation of the flow rates of the reactive gas. (The target is Zr alloyed with Y and the reactive gas is a O_2/Ar mixture.) [168]*

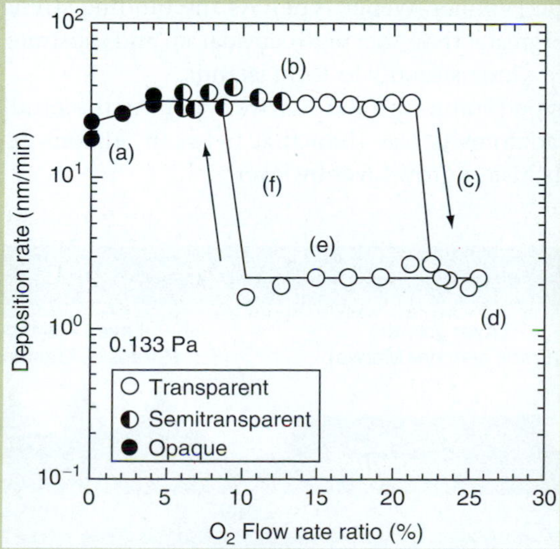

effect. Kadlec et al. [169] suggested that the hysteresis effect can be avoided if the pumping speed of the pumping system is greater than a critical value. The use of very high-speed pumping systems has proved to provide a solution to the problem of reactive sputtering of a metal titanium target to provide a reasonable rate for the preparation of large areas of the titanium oxide. The pumping speed of a system has been shown to affect the process of reactive sputtering of indium–tin oxide [170]. The simple solution requires large pumps, with arrangements of the system to give high gas conductances, which can be difficult to arrange and the cost of the extra pumps is high. For these reasons, the high pumping speed approach is not used very often. Recently, Depla et al. [171] studied the hysteresis behavior during reactive magnetron sputtering of Al_2O_3 using a rotating cylindrical magnetron. They found the hysteresis shifts towards lower oxygen flows when the rotation speed of the target is increased. Sproul et al. [167] demonstrated that using partial pressure control of the reactive gas allows deposition of films in the transition region between the elemental and poisoned states of the target, which leads to higher deposition rates compared to flow control and better film properties.

21.5.2 Film Structure Control (Structure Zone Models)

Thin-film formation is a continuous process in which atoms absorb onto the substrate, nucleate, and grow to form a dense film structure. In a vapor deposition process without plasma assistance, the mode of nucleation and film growth is mainly determined by the thermochemistry of the substrate and film materials and the temperature of the system [172]. There are three steps involved in a thin-film deposition process: (i) production of the appropriate atomic, molecular, or ionic species; (ii) transport of these species to the substrate through a medium; and (iii) condensation on the substrate, either directly or via a chemical reaction to form a solid deposit.

The film nucleation and initial growth can be described by three growth types, as shown in **Figure 21.26** [173]:

a. Island type (Volmer–Weber type): As the binding strength between adatoms is greater than that between adatom and substrate, the clusters grow three dimensionally to form islands.
b. Layered type (Frank–van der Merwe type): As the binding strength between adatoms is less than that between adatom and substrate, nucleated clusters grow layer by layer.

FIG. 21.26 *Three modes of thin-film nucleation and initial growth processes.*

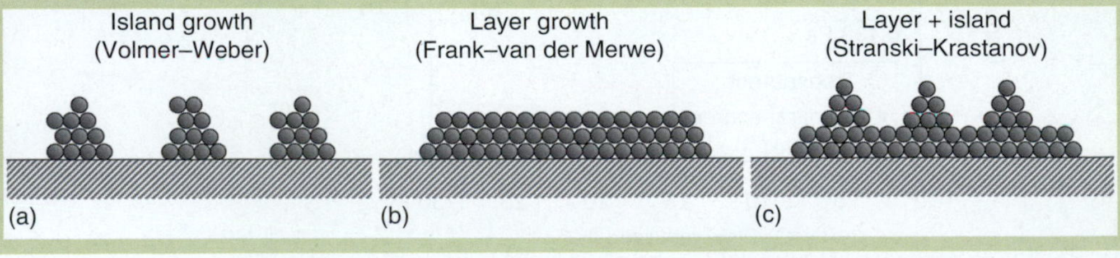

c. Mixed type (Stranski–Krastanov type): A mixed growth mode combines layer growth and island growth.

In this case, after a few monolayers have grown, subsequent layer growth becomes unfavorable and islands are formed. The reasons for the thin-film growth in this mode remain unclear. Factors such as the release of strain energy may trigger the formation of islands. This growth mode is fairly commonly observed in many kinds of thin-film growth systems.

In general, an external input energy is usually necessary to facilitate the formation of the thin films and to improve the quality of the film, which can be divided into three categories: (i) chemically induced mobility, (ii) thermal energy, and (iii) kinetic energy. The chemically induced mobility comes from the nature of the vapor species. The thermal energy originates from heating of the substrate, and the kinetic energy can be obtained from the energetic ion bombardment and the momentum transfer of ion species and neutral adatoms during the deposition of the film. All types of energy assist the migration of the adatoms on the substrate surface and supply the formation energy of the thin films. Generally, most plasma-related deposition systems require a moderate substrate temperature of about 300~400°C to enhance the quality of the thin film. However, thin films with acceptable quality can be prepared at a relatively low substrate temperature, even at room temperature, using high-energy ion bombardment deposition systems. By the addition of a plasma and high-energetic ion species and ion flux (concentration), the film will grow far from its thermodynamic equilibrium state, which makes the nucleation and growth processes much more complicated.

In the past two decades, different structure zone diagrams have been proposed to correlate the microstructure of thin films with the deposition conditions. In the 1960s, efforts were undertaken to generalize the effect of PVD deposition parameters on thin-film structure and properties. The melting temperature T_m of the film was assumed as the basic material parameter and the substrate temperature T was assumed as the main process parameter. Different zone models were developed based on their ratio T/T_m (the homologous temperature consideration), and typical examples include the Movchan–Demchishin model [174] for vacuum vapor deposition, the Grovenor model for thermal evaporation [175], the Thornton model [176] for cathode and magnetron sputtering, and the Messier model [177] for ion beam deposition.

Grovenor et al. [175] examined the microstructure of metal films grown by thermal evaporation. The microstructures were classified into four zones according to their homologous temperatures as depicted in **Figure 21.27a**. At a substrate temperature below $0.15\,T_m$, the film consists of porous columnar grains (Zone I). When the substrate temperature is between 0.15 to $0.3\,T_m$, the film exhibits a transitional structure (Zone T) between Zone I and Zone II. Zone II is a dense columnar structure, resulting from deposited atoms having sufficient surface mobility to diffuse and to increase the grain size. Zone III structure is controlled by bulk diffusion as the substrate temperatures are higher than $0.5\,T_m$.

The most well-known and important zone model is the Thornton model for cathode and magnetron sputtering [176]. In the Thornton model, the sputtering gas pressure is used as a process parameter to determine the film structure

Fig. 21.27 *(a) Grovenor [175], (b) Thornton [176], and (c) Messier [177] structure zone diagrams.*

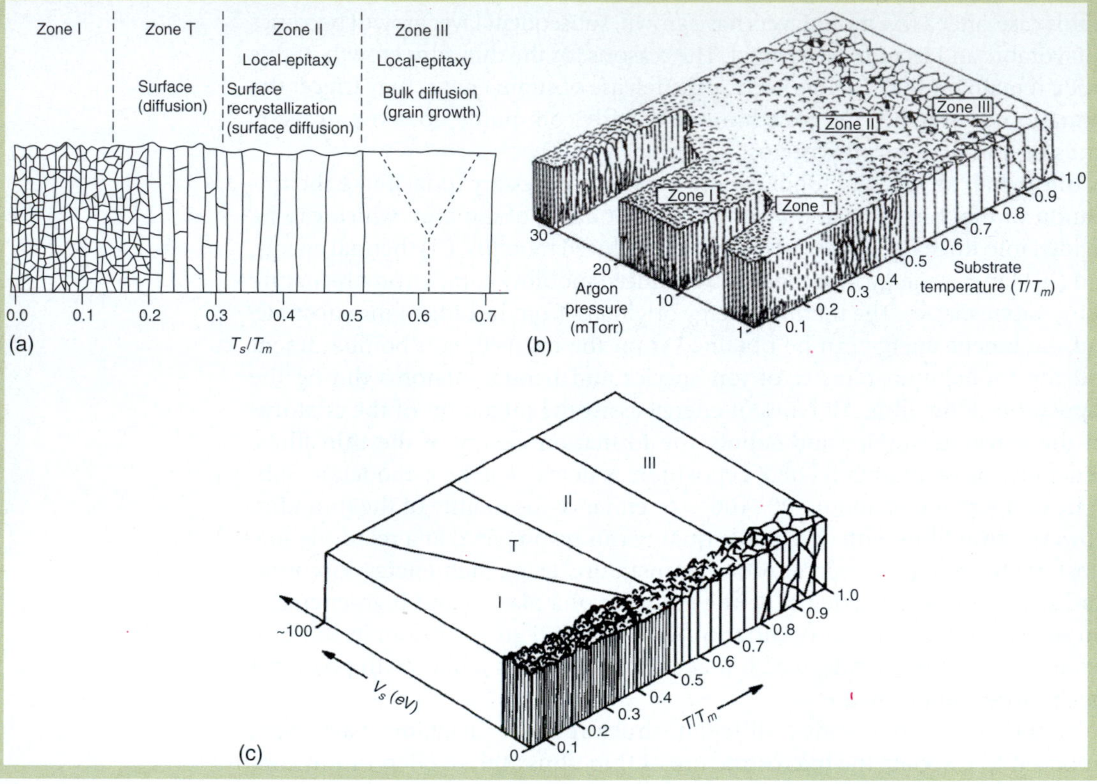

(Fig. 21.27b). Four different zones, including Zone I, Zone T, Zone II, and Zone III were proposed. Zone I ($T/T_m < 0.3$) is a columnar structure with pronounced pores and open columnar boundaries. This formation is due to the weak surface diffusion of atoms in combination with shadowing effects. Zone T is a transition structure that translates the Zone I structure to the Zone II structure, with fine, dense fibrous grains, changing with a rise of temperature to dense columnar grains. Zone T films are denser and have a much smoother surface morphology than the two surrounding zones (Zone I and Zone II). The film structure moves to Zone II when the substrate temperature is higher than about $0.5\,T_m$. Zone II consists of columnar grains separated by distinct boundaries and results from surface diffusion-controlled growth. Lattice diffusion dominates at high substrate temperature, giving rise to the large equiaxed grains of Zone III. For the entire temperature range, a rise in gas pressure causes a shift in the ranges of occurrence of different zones to a higher value of T/T_m ratio. A reduction in sputtering gas pressure during deposition increases the mean free path for elastic collisions between vapor species and the sputtering gas atoms. This behavior leads to higher kinetic energy of the vapor species impinging on the substrate surface, thereby producing a denser microstructure.

For ion beam techniques of coating deposition, Messier et al. [177] modified the Thornton model by replacing gas pressure in the sputtering zone with the energy of ions reaching the substrate surface V_s (*eV*) (**Fig. 21.27c**). As the bombardment energy increases, the width of Zone T increases at the expense of Zone I. It was also found that inside the columnar structure of Zones I and T, the intrinsic structure can be either polycrystalline or amorphous.

The Thornton and Messier models are widely used today on account of their simplicity and good agreement with industrial practice. A combination of these two models can successfully predict the microstructure of deposited thin films in many cases. However, the effect of ion bombardment rate such as ion/metal flux ratio is not considered or incorporated, which is becoming an increasingly important parameter in plasma-assisted sputtering deposition.

21.5.3 Sputtering Glow Discharges

Plasma-assisted material processing includes such techniques as etching, film deposition, and modifications of solid surfaces. Most of the modern PVD processes are operated with additional ion or plasma enhancement to achieve coatings with good adhesion, and mechanical and tribological properties [178]. In this section the plasma refers to a glow discharge that is used for sputter deposition of thin films.

The most common methods for generation of a plasma (breakdown and ionization of a gas) include the DC discharge, pulsed DC discharge, RF and microwave discharge, dielectric barrier discharges, and using electron and laser beams [179]. Here, to describe the plasma architecture we use the simplest DC discharge that is generated in a vacuum chamber between two counter electrodes by applying a DC power across the electrodes to break down the gas (e.g., Ar), as illustrated in **Figure 21.28**. In general, a neutral gas always contains a few electrons and ions due to ionization, for example, produced by cosmic radiation [180]. When a DC potential, V_c, is applied between the cathode and anode, these free charge carriers are accelerated by the electric field and new charged particles may be created when these charge carriers collide with atoms and molecules in the gas or with the surfaces of the electrodes (reactions 1, 2, and 4 in **Table 21.2**). This leads to an avalanche of charged particles that is eventually balanced by charge carrier losses, so that a steady-state plasma develops [179].

The plasma between the cathode and anode consists of different distributions of potential, space charge, and current density, as shown in **Figure 21.28** [181,182]. The edge of the plasma in contact with other surfaces is significantly different from the bulk plasma regions. A dark space or sheath is usually observed adjacent to all surfaces in contact with the plasma. As shown in **Figure 21.28**, the cathode will repel secondary electrons at high velocity away from the cathode, leaving behind the ions with slow mobility. Therefore, the high net positive space charge near the cathode will form a sheath, which is called a "cathode dark space." The cathode dark space is a region of the discharge where the electrical potential drops suddenly between the cathode and the edge of the negative glow and, as a result, there is an extremely low electron density. This lack of electrons results in low levels of excitation of the gas species in the region, and hence the area appears dark. The negative glow region is the brightest part of the discharge where the electric field is close to zero and most ionization collisions

FIG. 21.28 *Spatial distribution of a DC glow discharge and the potential profile.*

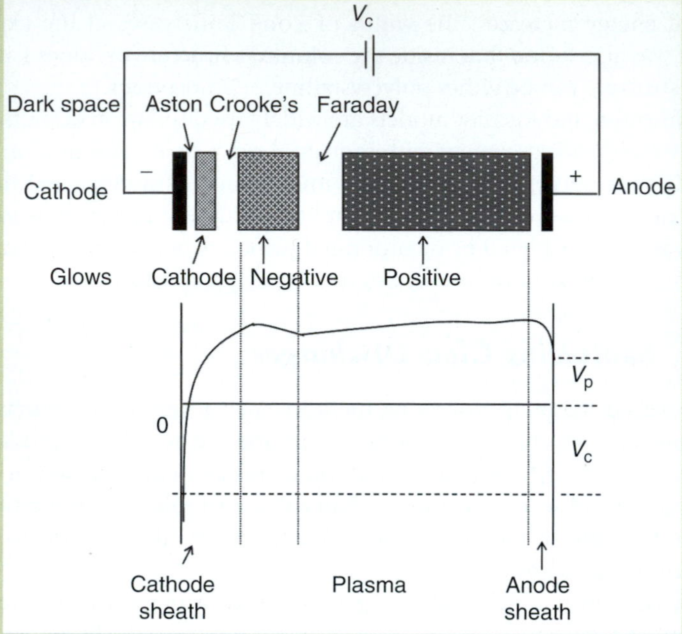

take place in this region. Following the negative glow region is the Faraday dark space and the positive glow. In the homogenous positive glow, a constant longitudinal electrical field is maintained. The electrons gain energy in this field and form an electron energy distribution with an appreciable number of energetic electrons for the formation of a sufficiently large number of ions and electrons to balance the charge carrier losses to the wall [179]. However, in a sputter deposition system, the interelectrode separation needs to be small in order to increase the deposition rate. If the anode (substrate) is located in the negative glow region, the positive glow and the Faraday dark space do not exist [180].

In sputter deposition, if an unbiased or grounded substrate is placed in the plasma, it will rapidly charge negatively due to the fact that electrons have greater velocities than ions. The negative surface will attract ions and retard the electrons until the escaping fluxes of ions and electrons are equal. In this steady state, the potential on this surface is known as the floating potential and it is typically negative of the plasma potential and is given by equation (21.8):

$$V_f = \frac{-KT_e}{2e} \ln\left(\frac{m_i}{2\pi m_e}\right) \tag{21.8}$$

where m_e and m_i are the electron and ion masses and T_e is the electron temperature. Since the ion mass is 3–4 orders of magnitude higher than that of an electron, the floating potential will have a (negative) value several times the electron temperature in volts.

In a vacuum system, the electrons would leave the plasma at a faster rate than ions, ending up at the vacuum chamber walls. The result for the plasma would be a slow increase in the net positive charge. As the plasma charges positively, it becomes less energetically possible for the electrons to leave, because now the walls of the chamber are more negative than the plasma. Eventually, a steady-state condition is reached in which the plasma potential is high enough that the loss rate of electrons is reduced to the same level as the loss rate of ions. In this way the plasma will retain its overall neutrality. The plasma potential V_p, which is now the average potential of the bulk plasma with respect to the chamber, will be on the order of several volts more positive than the chamber potential. Therefore, the plasma will remain the most positive body during a deposition process [64]. As a result of this plasma potential, ions that reach the edge of the plasma are then accelerated with the same voltage to the chamber wall. In the sputter deposition process, a magnetron field is usually applied behind the cathode in an effort to confine the secondary electrons near the cathode region to increase the plasma density, as discussed in more detail in section 21.4.3 under *Sputter Deposition*.

The primary energy loss mechanism for ions in the cathode fall is quantum mechanical through resonant charge exchange collisions. In simple diode discharges, ions move only a relatively small fraction of the cathode fall distance (cathode sheath) before they experience a charge exchange collision and are no longer accelerated by the field. Since this type of reaction does not require physical collision, the neutralized particle continues towards the target while the newly created ion is itself accelerated over a small fraction of cathode sheath. As a result, the average energy of ions incident at the target is much less than 1 *e*V and the glow discharge sputter yields are less than expected. However, this effect is partially mediated by sputtering due to fast neutrals [183].

21.5.4 Energetic Enhanced Deposition

The microstructure and properties of a thin film can be tailored and controlled by incorporating kinetic energy during film deposition. In general, the energy involved in energetic-enhanced deposition includes thermal energy and energetic particle bombardment. Both energies can be used to enhance the adatom mobility and reduce void formation during the film growth stages. For example, for most thin-film applications, a porous Zone I structure is undesirable because it degrades the film's mechanical, tribological, electronic, and optical properties. The film structure can be moved towards the dense, void-free Zone T structure by the use of energy-enhanced deposition [96]. Higher thermal energy can be achieved by increasing the substrate temperature during film deposition. Under raised substrate temperature, the migration rate of adatoms to shadowed regions can be large enough to surpass the rate of void incorporation during film growth [96], thereby increasing the film density. The substrate temperature also has influence on the film grain size, phase change, the formation of crystalline films, and texture alignment. However, many thin-film deposition applications and substrate materials need to be processed under relatively low temperatures. Therefore, the kinetic energy transferred from energetic bombardment in plasma-assisted depositions is widely used.

Various thin-film processing techniques that are based on ion- or plasma-assisted deposition were reviewed in section 21.4. All of these methods have the significant advantage that the structure and properties of thin films may be usefully modified by means of suitably considering the energy of ions/neutral species ranging from a few electron volts up to hundreds of kiloelectron volts impingement on the substrate surface. Some of the particles are not just assisting but they may condense and thereby become part of the growing film. In this section, emphasis will be focused on low energy plasma film deposition techniques, including magnetron sputtering, pulsed magnetron sputtering, and HPPMS, in which the substrate is immersed in a high density plasma to accomplish the deposition with the assistance of energetic bombardment plasma ions with ion energies up to hundreds of electron volts. The means of increasing the ion energy and plasma density, in particular, the generation of high ion energy and ion flux in a pulsed plasma, and the effect of ion bombardment on the growing film structure and properties will be introduced here. As to other high-energy ion/plasma characteristics and processing methods, such as plasma/ion implantation and ion beam deposition, readers are referred to Refs. [68,96].

Effect of Ion-Assisted Deposition. The ion energy and ion flux within the plasma have been proved to be important factors influencing the growing film structure and properties. A certain level of ion bombardment energies are critical for obtaining high-quality thin films. The impingement of a large amount of energetic ions or atoms upon a substrate surface produces a wide variety of effects during film deposition, such as film densification, decreased grain size, increased hardness and Young's modulus, high compressive stress, reduced roughness, resputtering, and film texture/orientation change depending on the ion energy range.

In general, the ion energies used in magnetron sputtering range up to a few hundred electron volts. The physical structure of a thin film can be changed by the mobility of the adatoms during growth. The enhanced ion energy and ion flux bombardment can transfer kinetic energy to other atoms as the collision sequence develops in time on the surface of the growing film, allowing the surface atoms to move around on the surface (adatom mobility) and find energetically favorable sites. Increasing ion bombardment produces resputtering and forward sputtering of the surface atoms, which fills the voids that would naturally occur along the columnar grains from shadowing effects [148,184,185]. The incident ion transfers kinetic energy to other atoms as the collision sequence develops in time, thereby densifying the film without a need of thermal diffusion.

Figure 21.29 provides a 2D molecular dynamics simulation of structure development versus the kinetic energy carried by atoms perpendicularly incident on a film held at 0K substrate temperature. The lowest of the three ratios of incident energy E_t to the adatom potential-well depth E_c corresponds to thermal deposition. The lowest normalized impinging energy E_t/E_c is 0.02eV, which corresponds to a thermal deposition process. The highest normalized impinging energy E_t/E_c of 5eV is characteristic of sputtered energy-enhanced deposition. The effect of added process energy on void filling is obvious [96].

Petrov et al. [186] studied the substrate bias effect on the sputter-deposited TiN film structure. They found a closing of intercolumnar porosity when increasing the substrate bias V_s, from 80 to 120V at substrate temperature T_s = 300°C, as a result of forward sputtering and ion irradiation enhanced adatom mobility **(Fig. 21.30)**.

Ion bombardment of the growing film can restrict the grain growth, increase nucleation sites, and permit the formation of nanocrystalline films. The high

FIG. 21.29 *Two-dimensional molecular dynamics simulation of the deposition of energetic atoms impinging perpendicularly onto a substrate held at 0 K. The horizontal line is the substrate interface. Normalized impinging energy E_t/E_c is (a) 0.02, (b) 0.5, and (c) 1.5, where E_t is the incident energy and E_c is adatom potential-well depth [96].*

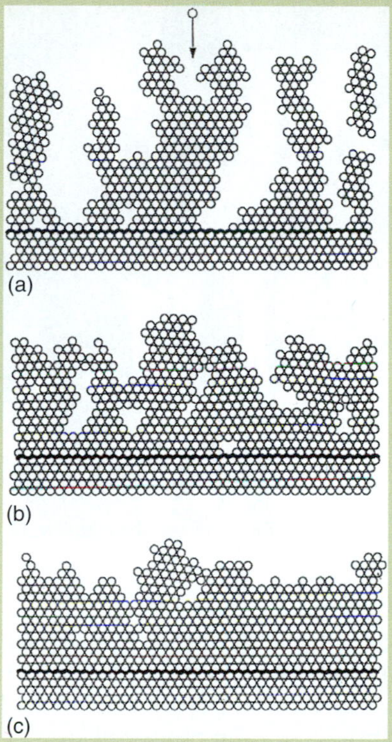

surface adatom mobility and diffusivity can increase the surface smoothness of the films by filling in the voids. Forward sputtering places atoms in interstitial sites, which results in compressive residual stresses in the PVD hard coatings [187]. This stress has both a positive effect, increasing the effective hardness of the coating [188], and a negative effect, causing film delamination as the film thickness increases. Energetic ions during ion bombardment can knock off some metal atoms from the substrate or may penetrate the substrate lattice to angstrom levels in addition to cleaning and heating the substrate. This bombardment leads to defects and roughness on the substrate at an atomic level, and this atomic level of roughness is believed to be responsible for the improved adhesion of the coating [189].

Furthermore, the crystallographic orientation of grains can also be influenced by the energy and flux of bombarding ions [190–192]. The ion-assisted reactive sputtering film orientation depends strongly on processing conditions, such as substrate temperature, substrate orientation, substrate bias, film thickness, ion flux, and ion energy. Theoretical modeling of ion energy effects on densification and texture evolution needs to take into account thermodynamic properties and

Fig. 21.30 *Cross-sectional TEM photomicrograph showing the evolution of microstructure in a TiN film sputter-deposited at T_s = 300°C. The negative substrate bias V_s was changed stepwise without interrupting the growth. The voided region along column boundaries (indicated by arrows) becomes dense when increasing V_s from 80 to 120 V [186].*

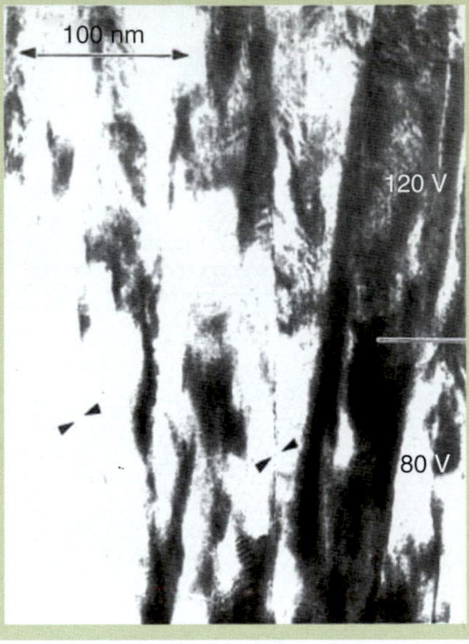

stress and strain energies. In general, it is recognized that the texture evolution observed in PVD thin coatings could be related to the minimization of free energy of the coatings, which is composed of surface and strain energies [192]. In the normal film growth model, the grains will grow on low, rather than high energy surfaces to minimize the surface energy. However, when additional kinetic or thermal energy is incorporated into the deposition process, the adatoms will receive higher momentum transfer and gain higher mobility to move around. At the same time, higher strain energy will be developed in the films, resulting in the change of the film orientations.

However, molecular dynamic computer simulations also predict that there should be optimum ion energy for film densification. Excessive ion energy will result in deeper penetration in the lattice leaving vacancies which cannot be filled by arriving vapor species [193]. Ion bombardment of crystalline films will cause heavy atomic structure damage and, in extreme cases, full amorphization [194] for some materials. **Figure 21.31** shows the defect incorporation as a function of substrate bias V_s in a multilayered TiN film bombarded by argon ions accelerated by V_s = 120, 80, and 40 V at T_s = 700°C [186]. As the substrate bias is increased, the projected range of the penetration of ions becomes larger giving rise to a defect generation at an increasingly larger distance below the growth surface, thereby resulting in dislocation networks and point defect aggregates.

At even higher energies, for example >400 *e*V, the rate of sputtering (resputtering) of the film surface will exceed the rate of deposition, effectively prohibiting film growth. This range can be characterized as an etch mode, which is important for surface preparation and patterning processes. In the case of polymer

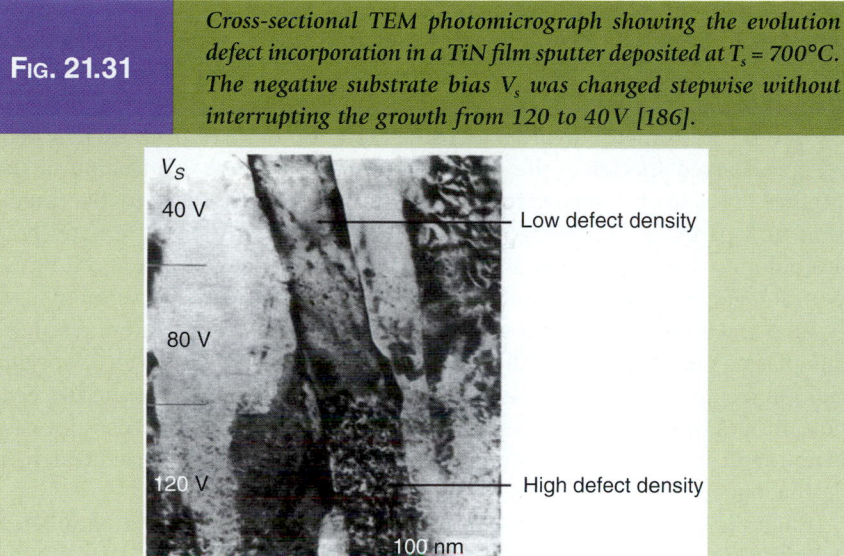

FIG. 21.31 *Cross-sectional TEM photomicrograph showing the evolution defect incorporation in a TiN film sputter deposited at T_s = 700°C. The negative substrate bias V_s was changed stepwise without interrupting the growth from 120 to 40 V [186].*

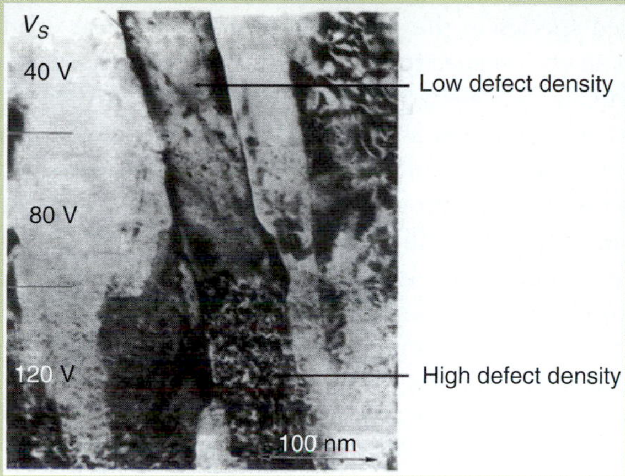

substrates, ions of too high energy and/or too high dose lead to substrate damage by chain-scissoring [195]. Particle energies of a few kiloelectron volts are used to remove material and to produce shallow implant profiles, which is important to some semiconductor processing. Another example of a process in this energy range is plasma nitriding, in which nitrogen ions (N^+) are implanted into the substrate surface forming hard and inert nitride phases in the near-surface region of steel or aluminum alloys [196]. Finally, at the very high end of the ion energy scale, ion implantation at hundreds of kiloelectron volts or some megaelectron volts is conducted routinely, but these processes and applications are beyond the scope of this chapter and will not be considered further.

Energy-Enhanced Deposition Using a Substrate Bias and a Pulsed Plasma. The energetic bombardment in magnetron sputtering can come from the flux of gas ions from the plasma impinging on the biased substrate, from the self energy of the depositing atoms, or from the flux of energetic neutrals that result when using high target power [197]. In general, the threshold ion energy of conventional DC magnetron sputtering is about 10 *e*V. Since ion energy is the difference between the plasma potential and the potential of the surface that ions bombard, an effective way to enhance the ion bombardment is to apply a negative bias to the substrate to extract ions from the plasma to bombard the growing film during film growth. Extensive studies have demonstrated that substrate bias plays an important role on film growth and microstructural evolution and properties [198–200].

It is important to understand the behavior of plasmas in order to gain a better understanding of the relationship between the deposition process and the film structure and properties [201]. In general, plasma diagnostics refers to the techniques used to gather information about the nature (properties) of the plasma used in a deposition process. Diagnostics help us obtain information about plasma properties in the sputter deposition glow discharge, such as the plasma chemical compositions and species, temperature, plasma density, ion/electron energy distributions, and nonelectronic species. The most commonly used plasma diagnostic techniques can be categorized as optical and electrostatic spectrometry.

Optical spectrometry involves focusing radiation emitted from excited neutral and charged species in the plasma through an optical window on to the entrance slit of an optical spectrometer. A photomultiplier is used to detect radiation of a particular wavelength at the spectrometer exit slit. Alternatively, a photodiode array is mounted at the exit port of the spectrometer and a broad spectral region is detected simultaneously [202,203]. Optical spectrometry uses a spatial and temporal resolution instrument which allows measurement of the bulk plasma properties. It is difficult for optical spectrometry to detect a specific position inside the plasma, for example, to measure the plasma properties near the cathodes or close the substrate. However, the relatively low cost and easy operation of an optical spectrometer make it a useful tool, and is used widely in plasma diagnostics applications.

Langmuir probes are used for monitoring discharge or plasma parameters including spatial distribution of potential, electron density, and electron temperature. The Langmuir probe consists of molybdenum or tungsten electrodes inserted into the plasma. The plasma parameters are estimated by the current–voltage curve of these electrodes. More detailed information on determining the electron density and temperature can be found in Refs. [204,205].

Mass spectrometry is another powerful technique for identifying unknown species, studying ionic species, and probing the fundamental principles of chemical reactions in the plasma. Mass spectrometry is based upon the motion of a charged particle, that is, ion, in an electric or magnetic field. The mass to charge ratio (m/z) of the ion affects this motion. Since the charge of an electron is known, the mass to charge ratio will indirectly measure an ion mass. Mass spectrometry is operated under high vacuum condition. A sample (preferably a gas) is introduced and broken down into charged fragments by electron impact or chemical ionization. The fragments, accelerated by applying a voltage, pass through a mass selector which separates them by their ratio of mass to charge (m/z). The separate fragments are detected and measured as ion current. Under constant conditions, a molecule will break up in the same number of ways, giving a reproducible mass spectrum. Compared to optical spectrometers, mass spectrometers can be placed at any position inside a plasma, thereby providing flexible and specific measurement of the plasma parameters as a function of location in the plasma.

As described in the previous section, pulsed DC magnetron sputtering has been used to eliminate arcing problems during the reactive sputter deposition of insulating films, stabilize the discharge, and reduce the formation of defects in the film. Besides its primary goal of eliminating arcing, it has been found that applying pulsing in DC magnetron sputtering also has benefits on coating structure and properties due to changes in ion energy and ion flux caused by changes in plasma parameters [206–211]. In recent years, in order to understand the nature of the pulsed plasma, time- and space-resolved Langmuir probe and electrostatic quadrupole plasma mass spectrometer measurements have been intensively used to investigate the plasma condition in pulsed DC magnetron sputter deposition [207,211–213]. These investigations have been conducted on various DC magnetron sputter configurations, such as one planar cathode, two planar cathodes facing each other in a mirror configuration, and so on. Observations of some interesting plasma properties and their relationship to the pulsing conditions have been revealed.

One important approach in pulsed plasma examination is using a time-resolved mass spectrometer to elucidate the dynamics of energetic ions in a pulsed DC magnetron plasma and related ion energy distributions (IED) in the plasma. It was found that pulsing provides a wide range of ion energies and ion fluxes in the plasma, and the energy of the energetic species can be up to hundreds of electron volts.

Three different ion energy regions in a pulsed plasma have been well documented [211,212]. **Figure 21.32a** shows a typical time averaged $^{29}N_2^+$ ion energy distribution in a discharged plasma for CrAlN film deposition when the Cr and Al targets were pulsed synchronously at 100 kHz and 5.0 μs. The Al target voltage waveform is shown in **Figure 21.32b**. It can be seen that pulsed ion energies generally consist of three regions, which are also correlated to distinct phases of the discharge voltage.

The low ion energy region "A" in **Figure 21.32a** usually is in the range of 0–10 *e*V, which corresponds to the negative pulse period (sputtering period) in the target voltage waveform (**Fig. 21.32b**). The plasma potential is several volts above the grounded chamber wall surface potential during this period. The "B" middle ion energy region (20 ~ 50 *e*V) is the energy gained from the target potential in the reverse positive pulse period (**Fig. 21.32b**), which has an average positive value of a fixed percentage of the nominal sputtering (negative) voltage. The "C" ion energy tail region is the kinetic energy gained from the fast and high positive voltage overshoot at the beginning of the positive pulse period. When the fast and steep positive voltage overshoot is developed on the target, a large electron current is extracted from the bulk plasma towards the target. Due to the slow ion movement, compared with that of electrons, the equilibrium cannot be reached in a short time period, therefore a negative charge density gradient

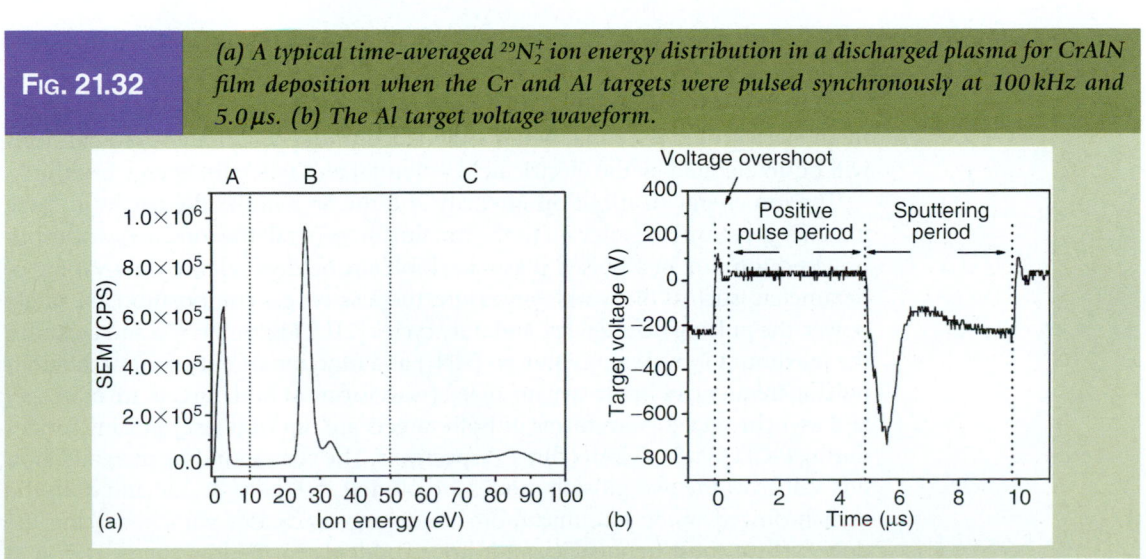

FIG. 21.32 *(a) A typical time-averaged $^{29}N_2^+$ ion energy distribution in a discharged plasma for CrAlN film deposition when the Cr and Al targets were pulsed synchronously at 100 kHz and 5.0 μs. (b) The Al target voltage waveform.*

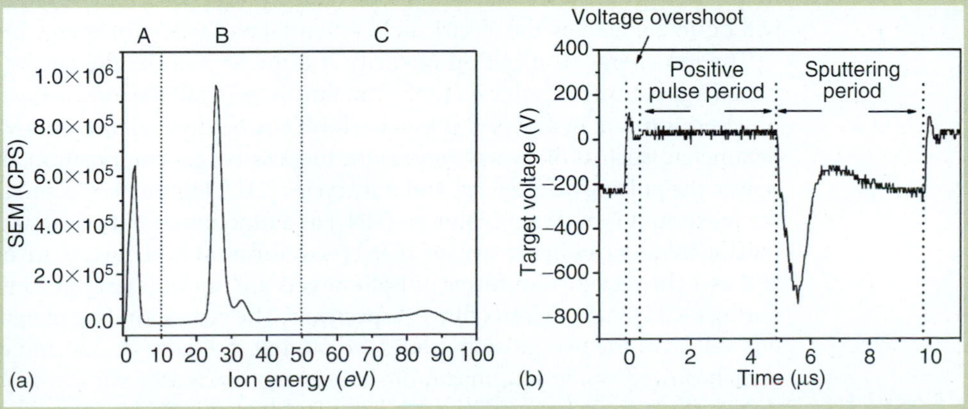

Fig. 21.33 *The maximum ion energy ($^{29}N_2^+$) as a function of the duty cycle at different pulsing frequencies in (a) synchronized mode and (b) asynchronized mode. The integrated ion flux as a function of the duty cycle at different pulsing frequencies in (c) synchronized mode and (d) asynchronized mode [211].*

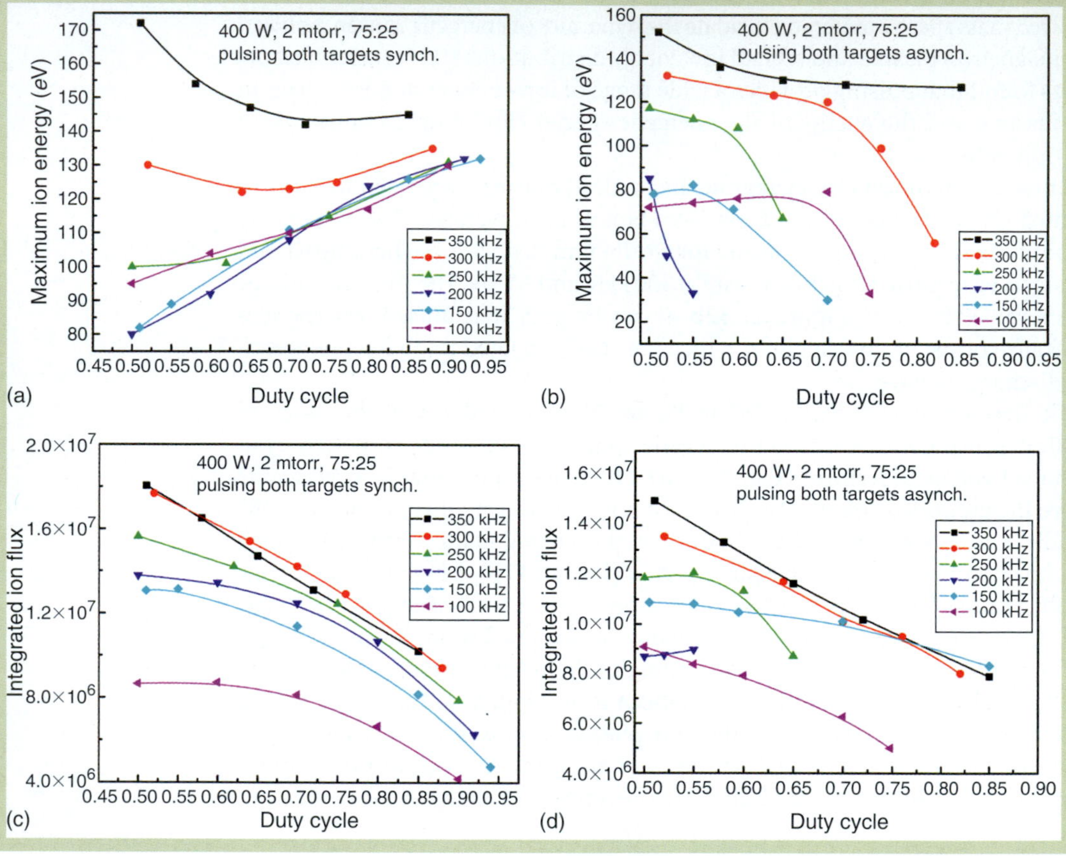

and a positive charge density gradient will be created in the bulk plasma resulting in an electric field potential being established under this charge gradient. Ions will be accelerated by the electric field potential and will gain energy from it.

The ion energy distribution intensity and the area under the curves in these ion energy regions are relevant to the ion flux. In general, the ion energy distribution and ion flux in a pulsed plasma exhibit strong dependence on a variety of parameters, such as the working pressure, the reactive gas composition, the target power, the pulsing frequencies, and duty cycles [211]. **Figure 21.33a** and **b** exhibit the maximum ion energy evolution ($^{29}N_2^+$) as a function of duty cycle at different pulsing frequencies in the synchronized (waveforms of both targets are in phase) and asynchronized (waveforms of both targets are out of phase) pulsing modes during P-CFUBMS of CrAlN films, respectively. The corresponding integrated ion flux values in the two pulsing modes are plotted in **Figure 21.33c** and **d**. In the asynchronized mode, the maximum ion energy decreases with increasing the duty cycle at most frequencies and drops suddenly to the low energy region at certain duty cycles due to the total separation of the two target positive pulse periods (**Fig. 21.33b**) [211]. The ion flux decreases with an increase in the duty cycle at all frequencies in both the synchronized and asynchronized pulsing modes (**Fig. 21.33c** and **d**). As the duty cycle decreases (the positive pulse time increases), the cathode is switched proportionally to a positive voltage for a longer period of time, which in turn provides more time for positive ions to stream away from the target. This increased escape time results in a higher number of ions gaining the additional kinetic energy available from the positive potential switch, thereby increasing the flux of higher energy ions at the substrate.

The increased ion energy and ion flux in a pulsed DC plasma will strongly affect the structure and properties of the films. The high ion energy and ion flux will apply additional ion bombardment, increase the mobility of the atoms on the substrate, and reduce the shadowing effect of the columnar structure, thereby changing the film growth microstructure. Muratore and Moore [206] compared the IED of pulsed DC magnetron sputtering and conventional DC magnetron sputtering deposition of TiO films. An increased ion energy up to 140 *e*V was observed in pulsed DC magnetron sputtering compared with 17 *e*V in the DC discharge. Increase of crystallographic texture and an 11% increase in hardness were observed in the TiO films processed with the pulsed magnetron sputtering. Backer et al. [214] calculated the ion energy impinging on the pulsed DC magnetron sputtering of Ti and TiO_2 films from the measurements of plasma parameters using a planar Langmuir probe and found that the ion energy values during the positive pulse period increased more than ten-fold the time-averaged value. The high ion energy has an important effect on the film surface roughness. It was found that the titanium films prepared at pulsing frequencies between 100 and 275 kHz exhibited low surface roughness. Higher ion energy at high pulsing frequencies will increase the titanium film surface roughness, which is not desirable. The effect of pulsed ion energy and ion flux bombardment on thin-film microstructure and properties will be illustrated in more detail using the technological example in section 22.3.

References

1. A. A. Voevodin, J. S. Zabinski, and C. Muratore, Recent advances in hard, tough, and low friction nanocomposite coatings, *Tsinghua Science and Technology,* 10(6), 665–679 (2005).
2. S. Veprek, A. Niederhofer, K. Moto, T. Bolom, H.-D. Mannling, P. Nesladek, G. Dollinger, and A. Bergmaier, Composition, nanostructure and origin of the ultrahardness in nc-TiN/a-Si_3N_4/a- and nc-$TiSi_2$ nanocomposites with H_V = 80 to ≥105 GPa, *Surface and Coatings Technology,* 133–134, 152–159 (2000).
3. D. Pilloud, J. F. Pierson, and L. Pichon, Influence of the Silicon concentration on the optical and electrical properties of reactively sputtered Zr–Si–N nanocomposite coatings, *Material Science and Engineering: B*, 131, 36–39 (2006).
4. A. A. Voevodin, S. D. Walck, and J. S. Zabinski, Architecture of multilayer nanocomposite coatings with super-hard diamond-like carbon layers for wear protection at high contact loads, *Wear,* 203–204, 516–527 (1997).
5. I.-W. Park, K. H. Kim, J. H. Suh, C.-G. Park, and M.-H. Lee, Role of amorphous Si_3N_4 in the microhardness of Ti-Al-Si-N nanocomposite films, *Journal of the Korean Physical Society,* 42(6), 783 (2003).
6. L. S. Palatnik, A. I. Il'inskii, and N. P. Sapelkin, *Soviet Physics-Solid State,* 8, 2016 (1967).

7. J. S. Koehler, Attempt to design a strong solid, *Physical Review B*, 2, 547–551 (1970).
8. P. J. Kelly, R. Hall, J. O. Brien, J. W. Bradley, P. Henderson, G. Roche, and R. D. Arnell, Studies of mid-frequency pulsed dc biasing, *Journal of Vacuum Science and Technology, A*, 19, 2856 (2001).
9. L. S. Palatnik and A. I. Il'inskii, Doklady, *Soviet Physics-Technical Physics*, 9, 93 (1964).
10. R. W. Springer and D. S. Catlett, Structure and mechanical properties of Al/Al_xO_y vacuum-deposited laminates, *Thin Solid Films*, 54, 197–205 (1978).
11. B. A. Movchan, A. V. Demchishin, G. F. Badilenko, R. F. Bunshah, C. Sans, C. Deshpandey, and H. J. Doerr, Structure-property relationships in microlaminate TiC/TiB_2 condensates, *Thin Solid Films*, 97, 215–219 (1982).
12. K. K. Shih and D. B. Dove, Ti/Ti-N Hf/Hf-N and W/W-N multilayer films with high mechanical hardness, *Applied Physics Letters*, 61, 654–656 (1992).
13. U. Helmersson, S. Todorova, S. A. Barnett, J.-E. Sundgren, L. C. Markert, and J. E. Greene, Growth of single-crystal TiN/VN strained-layer superlattices with extremely high mechanical hardness, *Journal of Applied Physics*, 62, 481 (1987).
14. M. Larsson, P. Hollman, P. Hedenqvist, S. Hogmark, U. Wahlström, and L. Hultman, Deposition and microstructure of PVD TiN–NbN multilayered coatings by combined reactive electron beam evaporation and DC sputtering, *Surface and Coatings Technology*, 86–87, 351–356 (1996).
15. K. M. Hubbard, T. R. Jervis, P. B. Mirkarimi, and S. C. Barnett, Mechanical properties of epitaxial $TiN/(V_{0.6}Nb_{0.4})N$ superlattices measured by nanoindentation, *Journal of Applied Physics*, 72, 4466–4468 (1992).
16. P. B. Mirkarimi, L. Hultman, and S. A. Barnett, Enhanced hardness in lattice-matched single-crystal $TiN/V_{0.6}Nb_{0.4}N$ superlattices, *Applied Physics Letters*, 57, 2654–2656 (1990).
17. P. B. Mirkarimi, S. A. Barnett, K. M. Hubbard, T. R. Jervis, and L. Hultman, Structure and mechanical-properties of epitaxial $TiNV_{0.3}NB_{0.7}N(100)$ superlattices, *Journal of Materials Research*, 9, 1456 (1994).
18. H. Holleck and H. Schulz, Advanced layer material constitution, *Thin Solid Films*, 153, 11–17 (1987).
19. H. Holleck and V. Schier, Multilayer PVD coatings for wear protection, *Surface and Coatings Technology*, 76–77, 328–336 (1995).
20. K. Yamamotoa, S. Kujimeb, and K. Takahara, Properties of nano-multilayered hard coatings deposited by a new hybrid coating process: Combined cathodic arc and unbalanced magnetron sputtering, *Surface and Coatings Technology*, 200, 435–439 (2005).
21. P. Robinson, A. Matthews, K. G. Swift, and S. Franklin, A computer knowledge-based system for surface coating and material selection, *Surface and Coatings Technology*, 62, 662–668 (1993).
22. R. Riedel, *Handbook of ceramic hard materials*, John Wiley & Sons, Inc., New York (2000).
23. A. Cavaleiro and C. Louro, Nanocrystalline structure and hardness of thin film, *Vacuum*, 64(3), 211–218 (2002).
24. E. Martinez, R. Sanjinés, A. Karimi, J. Esteve, and F. Lévy, Mechanical properties of nanocomposite and multilayered Cr–Si–N sputtered thin films, *Surface and Coatings Technology*, 180–181, 570–574 (2004).
25. I.-W. Park, K. H. Kim, A. O. Kunrath, D. Zhong, J. J. Moore, A. A. Voevodin, and E. A. Levashov, Microstructure and mechanical properties of superhard Ti–B–C–N films deposited by dc unbalanced magnetron sputtering, *Journal of Vacuum Science and Technology, B*, 23(2), 588 (2005).
26. M. Stüber, H. Leiste, S. Ulrich, H. Holleck, and D. Schild, Microstructure and properties of low friction TiC-C nanocomposite coatings deposited by magnetron sputtering, *Surface and Coatings Technology*, 150, 218 (2002).

27. A. A. Voevodin, J. P. O'Neill, S. V. Prasad, and J. S. Zabinski, Nanocrystalline WC and WC/a-C composite coatings produced from intersected plasma fluxes at low deposition temperatures, *Journal of Vacuum Science and Technology, A*, 17(3), 986 (1999).
28. S. Veprek, P. Nesladek, A. Niederhofer, F. Glatz, M. Jilek, and M. Sima, Recent progress in the superhard nanocrystalline composites: Towards their industrialization and understanding of the origin of the superhardness, *Surface and Coatings Technology*, 108–109, 138 (1998).
29. Y. H. Lu, P. Sit, T. F. Hung, H. Chen, Z. F. Zhou, K. Y. Li, and Y. G. Shen, Effects of B content on microstructure and mechanical properties of nanocomposite $Ti–B_x–N_y$ thin films, *Journal of Vacuum Science and Technology, B*, 23(2), 449 (2005).
30. R. Fella, H. Holleck, and H. Schulz, Preparation and properties of WC-TiC-TiN gradient coatings, *Surface and Coatings Technology*, 36, 257 (1988).
31. A. A. Voevodin, J. M. Schneider, C. Rebholz, and A. Matthews, Multilayer composite ceramic metal-DLC coatings for sliding wear applications, *Tribology International*, 29, 559 (1996).
32. A. A. Voevodin, M. A. Capano, S. J. P. Laube, M. S. Donley, and J. S. Zabinski, Design of A Ti/TiC/DLC functionally gradient coating based on studies of structural transitions in Ti–C thin films, *Thin Solid Films*, 298, 107–115 (1997).
33. L. Shizhi, S. Yulong, and P. Hongrui, Ti-Si-N films prepared by plasma-enhanced chemical vapor deposition, *Plasma Chemistry and Plasma Processing*, 12(3), 287 (1992).
34. S. Veprek and S. Reiprich, A concept for the design of novel superhard coatings, *Thin Solid Films*, 268, 64 (1995).
35. R. W. Hertzberg, *Deformation and fracture mechanics of engineering materials*, 3rd ed., John Wiley & Sons, Inc., New York (1989).
36. A. Kelly and N. H. MacMillan, *Strong solids*, 3rd ed., Clarendon, Oxford (1986).
37. E. O. Hall, The deformation and ageing of mild steel: III discussion of results, *Proceedings of the Physical Society of London Section B*, 64, 747–753 (1951).
38. N. J. Petch, The cleavage strength of polycrystals, *Journal of the Iron and Steel Institute*, 174, 25 (1953).
39. A. Lasalmonie and J. L. Strudel, Influence of grain size on the mechanical behaviour of some high strength materials, *Journal of Material Science*, 21, 1837 (1986).
40. E. Arzt, Size effects in materials due to microstructural and dimensional constraints: A comparative review, *Acta Materialia*, 46, 5611–5626 (1998)
41. J. Schiotz, F. D. Di Tolla, and K. W. Jacobsen, Softening of nanocrystalline metals at very small grain sizes, *Nature (London)*, 391, 561 (1998).
42. J. E. Carsley, J. Ning, W. W. Milligan, S. A. Hackney, and E. C. Aifantis, A simple, mixtures-based model for the grain size dependence of strength in nanophase metals, *Nanostructured Materials*, 5, 441 (1995).
43. A. H. Chokshi, A. Rosen, J. Karch, and H. Gleiter, On the validity of the Hall-Petch relationship in nanocrystalline materials, *Scripta Metallurgica*, 23, 1679 (1989).
44. R. W. Siegel and G. E. Fougere, Mechanical properties of nanophase metals, *Nanostructured Materials*, 6, 205–216 (1995).
45. M. F. Ashby, A first report on deformation-mechanism maps, *Acta Metallurgica*, 20, 887–897 (1972).
46. S. C. Lim, Recent developments in wear-mechanism maps, *Tribology International*, 31, 87–97 (1998).
47. S. L. Lehoczky, Strength enhancement in thin-layered Al-Cu laminates, *Journal of Applied Physics*, 49, 5479 (1978).
48. S. A. Barnett, Deposition and mechanical properties of superlattice thin films, In *Physics of Thin Films, Vol. 17 Mechanics and Dielectric Properties*, M. H. Francombe and J. L. Vossen, eds., p. 2 Academic Press, Boston, MA (1993).

49. P. M. Anderson and C. Li, Hall-Petch relations for multilayered materials, *Nanostructured Materials*, 5, 349 (1995).
50. W. D. Sproul, Reactive sputter deposition of polycrystalline nitride and oxide superlattice coatings, *Surface and Coatings Technology*, 86–87, 170 (1996).
51. M. L. Wu, X. W. Lin, V. P. Dravid, Y. W. Chung, M. S. Wong, and W. D. Sproul, Preparation and characterization of superhard CN_x/ZrN multilayers, *Journal of Vacuum Science and Technology, A*, 15, 946–950 (1997).
52. S. Veprek, New development in superhard coatings: The superhard nanocrystalline-amorphous composites, *Thin Solid Films*, 317, 449–454 (1998).
53. W. D. Sproul, Multilayer, multicomponent, and multiphase physical vapor deposition coatings for enhanced performance, *Journal of Vacuum Science and Technology, A*, 12(4), 1595–1601 (1994).
54. G. E. Dieter, *Mechanical metallurgy*, McGraw Hill, New York (1976).
55. M. Marder and J. Finberg, How things break, *Physics Today*, 49, 24 (1996).
56. G. M. Pharr, Measurement of mechanical properties by ultra-low load indentation, *Materials Science Engineering, A*, 253, 151(1998).
57. A. Leyland and A. Matthews, On the significance of the H/E ratio in wear control: A nanocomposite coating approach to optimised tribological behaviour, *Wear*, 246, 1 (2000).
58. V. V. Lyubimov, A. A. Voevodin, A. L. Yerokhin, Y. S. Timofeev, and I. K. Arkhipov, Development and testing of multilayer physically vapour deposited coatings for piston rings, *Surface and Coatings Technology*, 52, 145 (1992).
59. A. A. Voevodin, P. Stevenson, J. M. Schneider, and A. Matthews, Active process control of reactive sputter deposition, *Vacuum*, 46, 723 (1995).
60. J. Musil, P. Zeman, H. Hruby, and P. H. Mayrhofer, ZrN/Cu nanocomposite film—A novel superhard material, *Surface and Coatings Technology*, 120–121, 179 (1999).
61. A. A. Voevodin and J. S. Zabinski, Nanocomposite and nanostructured tribological materials for space applications, *Composite Science Technology*, 65, 741–748 (2005).
62. S. Veprek, Conventional and new approaches towards the design of novel superhard materials, *Surface and Coatings Technology*, 97, 15 (1997).
63. J. E. Crowell, Chemical methods of thin film deposition: Chemical vapor deposition, atomic layer deposition, and related technologies, *Journal of Vacuum Science and Technology, A*, 21(5), S88–S95 (2003).
64. S. M. Rossnagel, J. J. Cuomo, and W. D. Westwood, eds., *Handbook of plasma processing technology–Fundamentals, etching, deposition, and surface interactions*, William Andrew Publishing/Noyes, Park Ridge, New Jersey (1990).
65. http://www.gordonengland.co.uk/ps.htm
66. J. R. Creighton and P. Ho, Introduction to chemical vapor deposition (CVD), In *Chemical vapor deposition*, J.-H. Park, ed., ASM International, Materials Park, OH (2001).
67. L. B. Freund and S. Suresh, *Thin film materials: Stress, defect formation and surface evolution*, Cambridge University Press, Cambridge (2003).
68. D. M. Mattox, *Handbook of physical vapor deposition (PVD) processing: Film formation, adhesion, surface*, William Andrew Inc., Westwood New Jersey (1998).
69. A. Sherman, *Chemical vapor deposition for microelectronics: Principles, technology, and applications*, William Andrew Inc., Westwood, New Jersey (1987).
70. J. B. Fortin and T.-M. Lu, *Chemical vapor deposition polymerization: The growth and properties of parylene thin films*, Springer, Norwell, Massachusetts (2003).
71. K. F. Jensen, Chemical vapor deposition, In *Microelectronics processing: Chemical engineering aspects*, D. W. Hess and K. F. Jensen, eds., American Chemical Society, Washington, DC (1989).
72. M. Razeghi, *The MOCVD challenge, volume 2: A survey of GaInAsO-GaAs for photonic and electronic device applications*, CRC Press, Boca Raton, FL (1989).

73. A. Eishabini-Riad and F. D. Barlow, *Thin film technology handbook*, McGraw-Hill, New York (1998).
74. C. Cao, Ferroelectric thin films and applications, In *Ferroelectric thin films and applications in functional thin films and functional materials: New concepts and technologies*, D. Shi, ed., Springer, New York (2003).
75. M. Stoiber, J. Wagner, C. Mitterer, K. Gammer, H. Hutter, C. Lugmair, and R. Kullmer, Plasma-assisted pre-treatment for PACVD TiN coatings on tool steel, *Surface and Coatings Technology*, 174–175, 687–693 (2003).
76. E. Vassallo, A. Cremona, F. Ghezzi, F. Dellera, L. Laguardia, G. Ambrosone, and U. Coscia, Structural and optical properties of amorphous hydrogenated silicon carbonitride films produced by PECVD, *Applied Surface Science*, 252(22), 7993-8000 (2006).
77. N. Martins, P. Canhola, M. Quintela, I. Ferreira, L. Raniero, E. Fortunato, and R. Martins, Performances of an in-line PECVD system used to produce amorphous and nanocrystalline silicon solar cells, *Thin Solid Films*, 511–512, 238–242 (2006).
78. D. Heim, F. Holler, and C. Mitterer, Hard coatings produced by PACVD applied to aluminium die casting, *Surface and Coatings Technology*, 116–119, 530–536 (1999).
79. S. Shimada, Y. Takada, and J. Tsujino, Deposition of TiN films on various substrates from alkoxide solution by plasma-enhanced CVD, *Surface and Coatings Technology*, 199(1), 72–76 (2005).
80. J. Bonitz, S.E. Schulz, and T. Gessner, Ultra thin CVD TiN layers as diffusion barrier films on porous low-k materials, *Microelectronic Engineering*, 76(1–4), 82–88 (2004).
81. K.-T. Rie and J. Wöhle, Plasma-CVD of TiCN and ZrCN films on light metals, *Surface and Coatings Technology*, 112(1–3), 226–229 (1999).
82. Y.-G. Jung, S.-W. Park, and S.-C. Choi, Effect of CH_4 and H_2 on CVD of SiC and TiC for possible fabrication of SiC/TiC/C FGM, *Materials Letters*, 30(5–6), 339–345 (1997).
83. F. Ding and Y. Shi, The study of diamond/TiC composite film by a DC-plasma–hot filament CVD, *Surface and Coatings Technology*, 201(9–11), 5050–5053 (2007).
84. M. Noda and M. Umeno, Coating of DLC film by pulsed discharge plasma CVD, *Diamond and Related Materials*, 14(11–12), 1791–17 (2005).
85. H. Hanyu, S. Kamiya, Y. Murakami, and Y. Kondoh, The improvement of cutting performance in semi-dry condition by the combination of DLC coating and CVD smooth surface diamond coating, *Surface and Coatings Technology*, 200(1–4), 1137–1141 (2005).
86. S. Yoon, H. Yang, R. J. Ahn, and Q. Zhang, Preparation of a-C and DLC-C films using electron cyclotron resonance CVD and RF Bias, *Vacuum*, 49(1), 67–74 (1998).
87. S. H. Lee, K. H. Nam, S. C. Hong, and J. J. Lee, Low temperature deposition of TiB_2 by inductively coupled plasma assisted CVD, *Surface and Coatings Technology*, 201(9–11), 5211–5215 (2007).
88. A. Osada, E. Nakamura, H. Homma, T. Hayahi, and T. Oshika, Wear mechanism of thermally transformed CVD Al_2O_3 layer, *International Journal of Refractory Metals and Hard Materials*, 24(5), 387–391 (2006).
89. H. Sonobe, A. Sato, S. Shimizu, T. Matsui, M. Kondo, and A. Matsuda, Highly stabilized hydrogenated amorphous silicon solar cells fabricated by triode-plasma CVD, *Thin Solid Films*, 502(1–2), 306–310 (2006).
90. S. Faÿ, L. Feitknecht, R. Schlüchter, U. Kroll, E. Vallat-Sauvain, and A. Shah, Rough ZnO layers by LP-CVD process and their effect in improving performances of amorphous and microcrystalline silicon solar cells, *Solar Energy Materials and Solar Cells*, 90(18–19), 2960–2967 (2006).
91. B. Y. Moon, J. H. Youn, S. H. Won, and J. Jang, Polycrystalline silicon film deposited by ICP-CVD, *Solar Energy Materials and Solar Cells*, 69(2), 139–145 (2001).

92. K. Saito, Y. Uchiyama, and K. Abe, Preparation of SiO_2 thin films using the cat-CVD method, *Thin Solid Films*, 430(1–2), 287–291 (2003).
93. J. Perez-Mariano, S. Borros, J. A. Picas, A. Forn, and C. Colominas, Silicon nitride films by chemical vapor deposition in fluidized bed reactors at atmospheric pressure (AP/FBR-CVD), *Surface and Coatings Technology*, 200(5–6), 1719–1723 (2005).
94. M. J. Thwaites and H. S. Reehal, Growth of single-crystal Si, Ge and SiGe layers using plasma-assisted CVD, *Thin Solid Films*, 294(1–2), 76–79 (1997).
95. A. Tallaire, A. T. Collins, D. Charles, J. Achard, R. Sussmann, A. Gicquel, M. E. Newton, A. M. Edmonds, and R. J. Cruddace, Characterisation of high-quality thick single-crystal diamond grown by CVD with a low nitrogen addition, *Diamond and Related Materials*, 15(10), 1700–1707 (2006).
96. D. L. Smith, *Thin-film deposition: Principles and practice*, McGraw-Hill, New York (1995).
97. K. Oczos', The shaping of materials by concentrated fluxes of energy (in Polish). Publications of the Rzeszo'w Technical University, Rzeszo'w (1988).
98. J. Yao, J. Shao, H. He, and Z. Fan, Optical and electrical properties of TiO_x thin films deposited by electron beam evaporation, *Vacuum*, 81(9), 1023–1028 (2007).
99. S.-W. Hsu, T.-S. Yang, T.-K. Chen, and M.-S. Wong, Ion-assisted electron-beam evaporation of carbon-doped titanium oxide films as visible-light photocatalyst, *Thin Solid Films*, 515(7–8), 3521–3526 (2007).
100. C. Rebholz, M. A. Monclus, M. A. Baker, P. H. Mayrhofer, P. N. Gibson, A. Leyland, and A. Matthews, Hard and superhard TiAlBN coatings deposited by twin electron-beam evaporation, *Surface and Coatings Technology*, 201(13), 6078–6083 (2007).
101. A. Lotnyk, S. Senz, and D. Hesse, Epitaxial growth of TiO_2 thin films on $SrTiO_3$, $LaAlO_3$ and Yttria-stabilized Zirconia Substrates by electron beam evaporation, *Thin Solid Films*, 515(7–8), 3439–3447 (2007).
102. R. Al Asmar, J.-P. Atanas, Y. Zaatar, J. Podlecki, and A. Foucaran, Characterization and ellipsometric investigation of high-quality ZnO and $ZnO(Ga_2O_3)$ thin alloys by reactive electron-beam co-evaporation technique, *Microelectronics Journal*, 37(10), 1080–1085 (2006).
103. F.-R. Weber, F. Fontaine, M. Scheib, and W. Bock, Cathodic arc evaporation of (Ti,Al)N coatings and (Ti,Al)N/TiN multilayer-coatings-correlation between life-time of coated cutting tools, structural and mechanical film properties, *Surface and Coatings Technology*, 177–178, 227–232 (2004).
104. Y. Tanaka, N. Ichimiya, Y. Onishi, and Y. Yamada, Structure and properties of Al–Ti–Si–N coatings prepared by the cathodic arc ion plating method for high speed cutting applications, *Surface and Coatings Technology*, 146–147, 215 (2001).
105. A. M. Merlo, The contribution of surface engineering to the product performance in the automotive industry, *Surface and Coatings Technology*, 174, 21–26 (2003).
106. S. M. Rossnagel, Thin film deposition with physical vapor deposition and related technologies, *Journal of Vacuum Science and Technology, A*, 21(5), S74 (2003).
107. J. Vyskočil and J. Musil, Cathodic arc evaporation in thin film technology, *Journal Vacuum Science and Technology, A*, 10(4), 1740–1748 (1992).
108. I. I. Aksenov, V. A. Belous, V. G. Padalka, and V. M. Khoroshikh, Transport of plasma streams in a curvilinear plasma-optics system, *Soviet Journal of Plasma Physics*, 4, 425–428 (1978).
109. Y. H. Liu, J. L. Zhang, D. P. Liu, T. C. Ma, and G. Benstetter, A triangular section magnetic solenoid filter for removal of macro- and nano-particles from pulsed graphite cathodic vacuum arc plasma, *Surface and Coatings Technology*, 200(7), 2243–2248 (2005).
110. D. Dijkkamp, T. Venkatesan, X. D. Wu, S. A. Shaheen, N. Jisrawi, Y. H. Min-Lee, W. L. McLean, and M. Croft, Preparation of Y-Ba-Cu oxide superconductor thin

films using pulsed laser evaporation from high T_c bulk material, *Applied Physics Letters*, 51(8), 619–621 (1987).
111. D. B. Chrisey and G. K. Hubler, *Pulsed laser deposition of thin films*, John Wiley & Sons, Inc., New York (1994).
112. Z. L. Wang, Y. Liu, and Z. Zhang, *Handbook of nanophase and nanostructured materials*, Springer-Verlag, New York (2002).
113. S.-M. Kim, S. C. Song, and S. Y. Lee, Effect of CeO_2, $BaTiO_3$ and $CeO_2/BaTiO_3$ double buffer layers on the superconducting properties of $Y_1Ba_2Cu_3O_{7-x}$ grown on metallic substrates by pulsed laser deposition, *Physica C: Superconductivity*, 351(4), 379–385 (2001).
114. V. Beaumont, B. Mercey, and B. Raveau, Performant superconducting $NdBa_2Cu_3O_7$ films grown by pulsed laser deposition at "low temperature" in an argon rich atmosphere, *Physica C: Superconductivity*, 340(2–3), 112–118 (2000).
115. N. Scarisoreanu, M. Dinescu, F. Craciun, P. Verardi, A. Moldovan, A. Purice, and C. Galassi, Pulsed laser deposition of perovskite relaxor ferroelectric thin films, *Applied Surface Science*, 252(13), 4553–4557 (2006).
116. W. Biegel, R. Klarmann, B. Stritzker, B. Schey, and M. Kuhn, Pulsed laser deposition and characterization of perovskite thin films on various substrates, *Applied Surface Science*, 168(1–4), 227–233 (2000).
117. N. B. Ibrahim, C. Edwards, and S. B. Palmer, Pulsed laser ablation deposition of yttrium iron garnet and cerium-substituted YIG films, *Journal of Magnetism and Magnetic Materials*, 220(2–3), 183–194 (2000).
118. Y. Nakata, T. Okada, M. Maeda, S. Higuchi, and K. Ueda, Effect of oxidation dynamics on the film characteristics of Ce:YIG thin films deposited by pulsed-laser deposition, *Optics and Lasers in Engineering*, 44(2), 147–154 (2006).
119. A. Hakola, O. Heczko, A. Jaakkola, T. Kajava, and K. Ullakko, Ni–Mn–Ga films on Si, GaAs and Ni–Mn–Ga single crystals by pulsed laser deposition, *Applied Surface Science*, 238(1–4), 155–158 (2004).
120. Y. Tsuboi, M. Goto, and A. Itaya, Pulsed laser deposition of silk protein: Effect of photosensitized-ablation on the secondary structure in thin deposited films, *Journal of Applied Physics*, 89(12), 7917–7923 (2001).
121. G. Kecskeméti, N. Kresz, T. Smausz, B. Hopp, and A. Nógrádi, Pulsed laser deposition of pepsin thin films, *Applied Surface Science*, 247(1–4), 83–88 (2005).
122. L. Pawlowski. *The science and engineering of thermal spray coatings*, John Wiley & Sons, Inc., New York (1995).
123. J. A. Picas, A. Forn, and G. Matthäus, HVOF coatings as an alternative to hard chrome for pistons and valves, *Wear*, 261(5–6), 477–484 (2006).
124. H. Okuno, E. Grivei, F. Fabry, T. M. Gruenberger, J. Gonzalez-Aguilar, A. Palnichenko, L. Fulcheri, N. Probst, and J.-C. Charlier, Synthesis of carbon nanotubes and nano-necklaces by thermal plasma process, *Carbon*, 42(12–13), 2543–2549 (2004).
125. T. Yoshida, The future of thermal plasma processing for coating, *Pure and Applied Chemistry*, 66(6), 1222–1230 (1994).
126. M. Gell, L. Xie, X. Ma, E. H. Jordan, and N. P. Padture, Highly durable thermal barrier coatings made by the solution precursor plasma spray process, *Surface and Coatings Technology*, 177–178, 97–102 (2004).
127. http://www.plasma-group.co.uk/
128. P. R. Strutt, B. H. Kear, and R. Boland, US Patent No. 6025034 (2000).
129. S. D. Parukuttyamma, J. Margolis, H. Liu, C. P. Grey, S. Sampath, H. Herman and J. B. Parise, Yttrium aluminum garnet (YAG) films through a precursor plasma spraying technique, *Journal of American Ceramic Society*, 84, 1906 (2001).
130. J. Karthikeyan, C. C. Berndt, J. Tikkanen, S. Reddy, and H. Herman, Plasma spray synthesis of nanomaterial powders and deposits, *Materials Science and Engineering, A*, 238, 275–286 (1997).

131. N. P. Padture, K. W. Schlichting, T. Bhatia, A. Ozturk, B. Cetegen, E. H. Jordan, Towards durable thermal barrier coatings with novel microstructures deposited by solution-precursor plasma spray, *Acta Materialia*, 49(12), 2251–2257 (2001).
132. X. Ma, F. Wu, J. Roth, M. Gell, and E. H. Jordan, Low thermal conductivity thermal barrier coating deposited by the solution plasma spray process, *Surface and Coatings Technology*, 201, 4447–4452 (2006).
133. J. L. Vossen and W. Kern, eds., *Thin film processes II*, Academic Press, London (1991).
134. E. G. Spencer and P. H. Schmidt, Ion-beam techniques for device fabrication, *Journal of Vacuum Science and Technology*, 8(5), S52 (1971).
135. J. J. Cuomo, S. M. Rossnagel, and H. R. Kaufman, *Handbook of ion beam processing technology: Principles, deposition, film modification, and synthesis*, Noyes Publications, Park Ridge, New Jersey (1989).
136. P. Sigmund, Sputtering by ion bombardment: Theoretical concepts, In *Sputtering by particle bombardment*, I, R. Behrisch, ed., Springer-Verlag, Berlin (1981).
137. T. Serikawa, M. Henmi, T. Yamaguchi, H. Oginuma, and K. Kondoh, Depositions and microstructures of Mg–Si thin film by ion beam sputtering, *Surface and Coatings Technology*, 200, 4233–4239 (2006).
138. S. Iwatsubo, Temperature effect on structure and surface morphology of indium tin oxide films deposited by reactive ion-beam sputtering, *Vacuum*, 80(7), 708–711 (2006).
139. E. Horváth, A. Németh, A. A. Koós, M. C. Bein, A. L. Tóth, Z. E. Horváth, L. P. Biró, and J. Gyulai, Focused ion beam based sputtering yield measurements on ZnO and Mo thin films, *Superlattices and Microstructures*, 42(1–6), 392–397 (2007).
140. B. Y. Kim, J. S. Lee, K.-R. Kim, B. H. Choi, and B. S. Park, Development of ion beam sputtering technology for surface smoothing of materials, *Nuclear Instruments and Methods in Physics Research Section B: Beam Interactions with Materials and Atoms*, 261(1–2), 682–685 (2007).
141. A. Matthews, Plasma-based PVD surface engineering processes, *Journal of Vacuum Science and Technology, A*, 21(5), S224 (2003).
142. R. D. Arnell and P. J. Kelly, Recent advance in magnetron sputtering, *Surface and Coatings Technology*, 112, 170–176 (1999).
143. B. Windows and N. Savvides, Charged particle fluxes from planar magnetron sputtering sources, *Journal of Vacuum Science Technology, A*, 4(2), 196–202 (1986).
144. B. Windows and N. Savvides, Unbalanced dc magnetrons as sources of high ion fluxes, *Journal of Vacuum Science and Technology, A*, 4(3), 453–456 (1986).
145. W. D. Sproul, P. J. Rudnik, M. E. Graham, and S. L. Rohde, High-rate reactive sputtering in an opposed cathode closed-field unbalanced magnetron sputtering system, *Surface and Coatings Technology*, 43, 270 (1990).
146. S. L. Rohde, W. D. Sproul, and J. R. Rohde, Correlations of plasma and magnetic field characteristics to TiN film properties formed using a dual unbalanced magnetron system, *Journal of Vacuum Science and Technology, A*, 9(3), 1178 (1991).
147. K. Tominaga, Preparation of AIN films by planar magnetron sputtering system with facing two targets, *Vacuum*, 41, 1154–1156 (1990).
148. W. D. Sproul, Physical vapor deposition tool coatings, *Surface and Coatings Technology*, 81, 1–7 (1996).
149. W. D. Sproul, M. E. Graham, M.-S. Wong, and P. J. Rudnik, Reactive d.c. magnetron sputtering of the oxides of Ti, Zr, and Hf, *Surface and Coatings Technology*, 89,10–15 (1997).
150. W. D. Sproul, High-rate reactive DC magnetron sputtering of oxide and nitride superlattice coatings, *Vacuum*, 51, 641 (1998).

151. O. Sánchez, M. Hernández-Vélez, D. Navas, M. A. Auger, J. L. Baldonedo, R. Sanz, K. R. Pirota, and M. Vázquez, Functional nanostructured titanium nitride films obtained by sputtering magnetron, *Thin Solid Films*, 495(1–2), 149–153 (2006).
152. S. Inoue, F. Okada, and K. Koterazawa, CrN films deposited by rf reactive sputtering using a plasma emission monitoring control, *Vacuum*, 66(3–4), 227–231 (2002).
153. E. Kusano, A. Satoh, M. Kitagawa, H. Nanto, and A. Kinbara, Titanium carbide film deposition by DC magnetron reactive sputtering using a solid carbon source, *Thin Solid Films*, 343–344, 254–256 (1999).
154. P. W. Shum, W. C. Tam, K. Y. Li, Z. F. Zhou, and Y. G. Shen, Mechanical and tribological properties of titanium-aluminum-nitride films deposited by reactive closed-field unbalanced magnetron sputtering, *Wear*, 257, 1030–1040 (2004).
155. J. Lin, B. Mishra, J. J. Moore, and W. D. Sproul, Microstructure, mechanical and tribological properties of $Cr_{1-x}Al_xN$ films deposited by pulsed-closed field unbalanced magnetron sputtering (P-CFUBMS), *Surface and Coatings Technology*, 201, 4329–4334 (2006).
156. J. O'Brien and P. J. Kelly, Characterisation studies of the pulsed dual cathode magnetron sputtering process for oxide films, *Surface and Coatings Technology*, 142–144, 621–627 (2001).
157. A. Anders, Fundamentals of pulsed plasmas for materials processing, *Surface and Coatings Technology*, 183, 301–311 (2004).
158. Advanced Energies Inc., *PinnacleTM plus user manual* (2002).
159. E. V. Barnet and T.-M. Lu, *Pulsed and pulsed bias sputtering*, Kluwer Academic Publishers, Norwell, MA (2003).
160. V. Kouznetsov, K. Macak, J. M. Schneider, U. Helmersson, and I. Petrov, A novel pulsed magnetron sputter technique utilizing very high target power densities, *Surface and Coatings Technology*, 122, 2)0 (1999).
161. K. Macak, V. Kouznetsov, J. M. Schneider, U. Helmersson, and I. Petrov, Ionized sputter deposition using an extremely high plasma density pulsed magnetron discharge, *Journal of Vacuum Science and Technology, A*, 18, 1533 (2000).
162. J. Alami, P. O. Å. Persson, D. Music, J. T. Gudmundsson, J. Bohlmark, and U. Helmersson, Ion-assisted physical vapor deposition for enhanced film properties on nonflat surfaces, *Journal of Vacuum Science and Technology, A*, 23(2), 278 (2005).
163. C. Christou and Z. H. Barber, Ionization of sputtered material in a planar magnetron discharge, *Journal of Vacuum Science and Technology, A*, 18, 2897 (2000).
164. A. P. Ehiasarian, R. New, W.-D. Munz, L. Hultman, U. Helmersson, and V. Kouznetsov, Influence of high power densities on the composition of pulsed magnetron plasmas, *Vacuum*, 65, 147 (2002).
165. S. Berg and T. Nyberg, Fundamental understanding and modeling of reactive sputtering processes, *Thin Solid Films*, 476, 215–230 (2005).
166. S. Berg, T. Nyberg, H.-O. Blom, C. Nender, D. A. Glocker, and S. I. Shah, eds., *Handbook of thin film process technology*, p. A5.3:1 Institute of Physics Publishing, Bristol (1998).
167. W. D. Sproul, D. J. Christie, and D. C. Carter, Control of reactive sputtering process, *Thin Solid Films*, 491, 1–17 (2005).
168. T. Hata, S. Nakano, Y. Masuda, K. Sasaki, Y. Haneda, and K. Wasa, Heteroepitaxial growth of YSZ films on Si (100) substrate by using new metallic mode of reactive sputtering, *Vacuum*, 51, 583 (1998).
169. S. Kadlec, J. Musil, and H. VyskoEil, Hysteresis effect in reactive sputtering: A problem of system stability, *Journal of Physics D: Applied Physics*, 19, 187–190 (1986).
170. R. P. Howson and H. A. Ja'fer, Reactive sputtering with an unbalanced magnetron, *Journal of Vacuum Science and Technology, A*, 10, 1748 (1992).

171. D. Depla, J. Haemers, G. Buyle, and R. De. Gryse, Hysteresis behavior during reactive magnetron sputtering of Al_2O_3 using a rotating cylindrical magnetron, *Journal of Vacuum Science and Technology, A*, 24, 934–938 (2006).
172. A. Anders, Plasma and ion sources in large area coating: A review, *Surface and Coatings Technology*, 200, 1893–1906 (2005).
173. K. Wasa, M. Kitabatake, and H. Adachi, *Thin film materials technology: Sputtering of compound materials*, 1st ed., William Andrew Publishing, Norwich, New York (2004).
174. V. A. Movchan and A. V. Demchishin, Investigation of structure and properties of thick vacuum condensates of nickel, titanium, tungsten, aluminum oxide and zirconium dioxide (in Russian), *Fizika Metallov i Metallovedenye*, 28(4), 83–86 (1969).
175. C. R. M. Grovenor, H. T. G. Hentzell, and D. A. Smith, The development of grain structure during growth of metallic films, *Acta Metalliallia*, 32, 773 (1984).
176. J. A. Thornton, High rate thick film growth, *Annual Review of Materials Science*, 7, 239–260 (1977).
177. R. Messier, A. P. Giri, and R. A. Roy, Revised structure zone model for thin film physical structure, *Journal of Vacuum Science and Technology, A*, 2(2), 500–511 (1984).
178. R. Boxman, P. Martin, and D. Sanders, *Handbook of vacuum arc science and technology*, Noyes Publications, Park Ridge, NJ (1995).
179. H. Conrads and M. Schmidt, Plasma generation and plasma sources, *Plasma Sources Science and Technology*, 9, 441 (2000).
180. M. Venugopalan and R. Avni, Analysis of glow discharges, In *Thin films from free atoms and particles*, K. J. Klabunde, ed., Academic Press, Orlando, FL (1985).
181. A. von Engel, *Ionized gases*, Oxford University Press, London and New York (1965).
182. E. Nasser, *Fundamentals of gaseous ionization and plasma electronics*, John Wiley & Sons, Inc., New York (1971).
183. J. E. Greene, Sputter deposition, AVS short course program, Rocky Mountain Chapter, Golden, CO (2006).
184. I. Petrov, F. Adibi, J. E. Greene, L. Hultman and J.-E. Sundgren, Average energy deposited per atom: A universal parameter for describing ion-assisted film growth? *Applied Physics Letters*, 63, 36–38 (1993).
185. I. Petrov, P. B. Barna, L. Hultman, and J. E. Greene, Microstructure evolution during film growth, *Journal of Vacuum Science and Technology, A*, 21(5), S117–S128 (2003).
186. I. Petrov, L. Hultman, U. Helmersson, J.-E. Sundgren, and J. E. Greene, Microstructure modification of TiN by ion bombardment during reactive sputter deposition, *Thin Solid Films*, 169, 299–314 (1989).
187. H. Ljungcrantz, L. Hultman, and J.-E. Sundgren, Ion induced stress generation in arc-evaporated TiN films, *Journal of Applied Physics*, 78(2), 832 (1995).
188. D. T. Quinto, Mechanical property and structure relationships in hard coatings for cutting tools, *Journal of Vacuum Science and Technology, A*, 6, 2149 (1988).
189. S. PalDey and S. C. Deevi, Single layer and multilayer wear resistant coatings of (Ti,Al)N: A review, *Materials Science and Engineering, A*, 342, 58–79 (2003).
190. J. Pelleg, L. Z. Zevin, and S. Lungo, Reactive-sputter-deposited TiN films on glass substrates, *Thin Solid Films*, 197, 117–128 (1991).
191. J. E. Greene, J.-E. Sundgren, L. Hultman, I. Petrov, and D. B. Dergstrom, Development of preferred orientation in polycrystalline TiN layers grown by ultrahigh vacuum reactive magnetron sputtering, *Applied Physics Letters*, 67, 2928 (1995).
192. J. P. Zhao, X. Wang, Z. Y. Chen, S. Q. Yang, T. S. Shi, and X. H. Liu, Overall energy model for preferred growth of TiN films during filtered arc deposition, *Journal Physics D: Applied Physics*, 30, 5–12 (1997).
193. T. Mueller, A. Gebeshuber, R. Kullmer, C. Lugmair, S. Perlot, and M. Stoiber, Minimizing wear through combined thermochemical and plasma activated diffusion and coating process, *Materiali in Tehnologije*, 38, 6 (2004).

194. A. R. Gonzalez-Elipe, F. Yubero, J. P. Espinos, A. Caballero, M. Ocana, J. P. Holgado, and J. Morales, Amorphisation and related structural effects in thin films prepared by ion beam assisted methods, *Surface and Coatings Technology*, 125, 116–123 (2000).
195. R. Rank, T. Wuensche, M. Fahland, C. Charton, and N. Schiller, Adhesion promotion techniques for coating of polymer films, 47th Annual Technical Conference, Society of Vacuum Coaters, Dallas, TX, p. 632 (2004).
196. W. Möller, S. Parascandola, T. Telbizova, R. Günzel, and E. Richter, Surface processes and diffusion mechanisms of ion nitriding of stainless steel and aluminium, *Surface and Coatings Technology*, 136, 73 (2001).
197. B. Windows, Issues in magnetron sputtering of hard coatings, *Surface and Coatings Technology*, 81, 92–98 (1996).
198. S. K. Karkari, A. Vetushka, and J. W. Bradley, Measurement of the plasma potential adjacent to the substrate in a midfrequency bipolar pulsed magnetron, *Journal of Vacuum Science and Technology*, 21, L28 (2003).
199. M. Audronis, A. Leyland, P. J. Kelly, and A. Matthews, The effect of pulsed magnetron sputtering on the structure and mechanical properties of CrB_2 coatings, *Surface and Coatings Technology*, 201, 3970 (2006).
200. J.-W. Lee, S.-K. Tien, and Y.-C. Kuo, The effects of pulse frequency and substrate bias to the mechanical properties of CrN coatings deposited by pulsed DC magnetron sputtering, *Thin Solid Films*, 494, 161 (2006).
201. J. E. Greene, Optical spectroscopy for diagnostics and process control during glow discharge etching and sputter deposition, *Journal of Vacuum Science and Technology*, 15, 1718 (1978).
202. A. Belkind, A. Freilich, J. Lopez, Z. Zhao, W. Zhu, and K. Becker, Characterization of pulsed DC magnetron sputtering plasmas, *New Journal of Physics*, 7, 90 (2005).
203. N. B. Dahotre and T. S. Sudarshan, eds., *Intermetallic and ceramic coatings*, Marcel Dekker, Inc., New York (1999).
204. I. Langmuir and H. M. Mostt-Smith, *General Electric Review*, 27(449), 583 (1924).
205. M. A. Lieberman and A. J. Lichtenberg, *Principles of plasma discharges and materials processing*, John Wiley & Sons, Inc., New York (1994).
206. C. Muratore, J. J. Moore, and J. A. Rees, Electrostatic quadrupole plasma mass spectrometer and langmuir probe measurements of mid-frequency pulsed DC magnetron discharges, *Surface and Coatings Technology*, 164, 12 (2003).
207. J. W. Bradley, H. Bäcker, P. J. Kelly, and R. D. Arnell, Space and time resolved langmuir probe measurements in a 100 kHz pulsed rectangular magnetron system, *Surface and Coatings Technology*, 142–144, 337–341 (2001).
208. J. W. Bradley, H. Bäcker, P. J. Kelly, and R. D. Arnell, Time-resolved langmuir probe measurements at the substrate position in a pulsed mid-frequency DC magnetron plasma, *Surface and Coatings Technology*, 135, 221 (2001).
209. P. S. Henderson, P. J. Kelly, R. D. Arnell, H. Bäcker, and J. W. Bradley, Investigation into the properties of titanium based films deposited using pulsed magnetron sputtering, *Surface and Coatings Technology*, 174–175, 779 (2003).
210. M. Mišina, J. W. Bradley, H. Bäcker, Y. A. Gonzalov, S. K. Karkari, and D. Forder, Investigation of the pulsed magnetron discharge by time- and energy-resolved mass spectrometry, *Vacuum*, 68, 171 (2003).
211. J. Lin, J. J. Moore, B. Mishra, W. D. Sproul, and J. A. Rees, Examination of the pulsing phenomena in pulsed-closed field unbalanced magnetron sputtering (P-CFUBMS) of Cr–Al–N thin films, *Surface and Coatings Technology*, 201, 4640 (2007).
212. J. W. Bradley, H. Bäcker, Y. Aranda-Gonzalvo, P. J. Kelly, and R. D. Arnell, The distribution of ion energies at the substrate in an asymmetric bi-polar pulsed DC magnetron discharge, *Plasma Sources Science and Technology*, 11, 165–174 (2002).

213. J. Lin, J. J. Moore, B. Mishra, M. Pinkas, W. D. Sproul, and J. A. Rees, Effect of asynchronous pulsing parameters on the structure and properties of CrAlN films deposited by pulsed closed field unbalanced magnetron sputtering (P-CFUBMS), *Surface and Coatings Technology,* 202, 1418–1436 (2007).
214. H. Backer, P. S. Henderson, J. W. Bradley, and P. J. Kelly, Time-resolved investigation of plasma parameters during deposition of Ti and TiO2 thin films, *Surface and Coatings Technology,* 174–175, 909–913 (2003).

Problems

21.1 List at least five reasons and corresponding examples to employ surface engineering and coatings in industrial applications.

21.2 List typical microstructure designs involved in nanostructured and nanocomposite coatings. Explain the key features and critical requirements for each structural design.

21.3 Define the Hall–Petch and reversed Hall–Petch relationships. Explain briefly (i) why the hardness of nanostructured films will increase when the grain size is reduced, (ii) why the hardness may start to decrease when the grain size is below 10 nm in nanocrystalline films, and (iii) why the hardness can be further enhanced in the nanocomposite coating design even though the nanocrystalline size is below 10 nm?

21.4 Explain the relationship between the bilayer period and the properties (e.g., hardness) of the nanoscale multilayer coatings. List at least three possible mechanisms of hardness enhancement in the superlattice coatings.

21.5 Explain the nanocomposite coating design. List at least five nanocomposite coating systems based on the information in the text and other research papers, and identify the nanocrystalline and amorphous phases in each coating system.

21.6 Explain why high toughness is as important as hardness for the industrial application of nanostructured coatings. Explain the relationship between the hardness and fracture toughness of nanostructured coatings.

21.7 Distinguish between physical and chemical vapor deposition using technical examples. Prepare a table to summarize the advantages and disadvantages of different PVD and CVD techniques, for example, electron beam evaporation, cathodic arc evaporation, PLD, thermal plasma processing and sputter depositions.

21.8 Define thermal and cold plasmas. Use cartoon drawings to explain the main ionization mechanisms in a sputter-discharged plasma.

21.9 Compare the characteristics of balanced, unbalanced, and closed-field unbalanced magnetron sputtering techniques. Explain why a higher plasma density and ICD can be achieved if the magnetrons are designed as unbalanced and in closed-field configuration.

21.10 Describe the mechanism of arc formation during DC reactive sputtering of Al in an Ar and O_2 atmosphere. Explain the principles of pulsed magnetron sputtering for suppressing the arc problem. Consider an Al_2O_3 layer of area A and thickness h. Calculate the frequency of target pulsing required for arc suppression given the target current density (J = 10 mA·cm^{-2}), dielectric constant of Al_2O_3 film (ε_r = 10), and dielectric breakdown electric field (E = 10^5 V·cm^{-1}). (Hint: The capacitance C and the electric field E that build up across the layer are $C = \varepsilon_r\varepsilon_0 A/h$ and $E = q/(Ch)$, respectively.)

21.11 Calculate the duty cycles and draw the voltage waveforms for the following symmetrically pulsed plasma using a sputtering voltage of 400 V: (1) 100 kHz and 1.0 μs, (2) 100 kHz and 5.0 μs, (3) 300 kHz and 1.0 μs, and (4) 300 kHz and 1.4 μs.

21.12 Explain the principles of the high-power pulsed DC magnetron sputtering (HPPMS) technique. Prepare a table showing the comparisons between HPPMS, conventional DC, and pulsed DC magnetron sputtering under various aspects, for example, the degree of ionization, plasma density, target power density, etc.

21.13 Assuming you are producing Al_2O_3 coatings using a magnetron sputtering system in Ar and O_2 gas mixture, list potential procedures to minimize the target poisoning effect in the hysteresis curve.

21.14 The Zone structure model proposed by Thornton established relationships between the sputtered coating structure and surface morphology to the pressure and substrate temperature, respectively. Describe the key features of Zone I, Zone T, Zone II, and Zone III structures of the sputtered coatings, and explain why and how a change of the pressure and substrate temperature will change the coating structure and morphology.

21.15 In a DC planar magnetron sputtering system, the cathode fall distance (L) can be estimated from Child's law:

$$L^2 = \left(\frac{4\varepsilon_0}{9}\right)\left(\frac{2e}{m}\right)^{1/2}\left(\frac{V^{3/2}}{J}\right)$$

where

ε_0 (8.85×10^{14} $\mathrm{F \cdot cm^{-1}}$) is the permittivity of vacuum

J is the target current density in $\mathrm{mA \cdot cm^{-2}}$

e/m is the charge/mass ratio of the extracted ions

V/J a target operation voltage of 1000 V and a target current density of $1\,\mathrm{mA \cdot cm^{-2}}$

i. Use Child's law to estimate the cathode fall distance (L) for Ar DC sputtering.
ii. If the cathode to anode spacing is 10 cm, determine the magnetic field that needs to be applied to trap electrons within 0.5 cm of the target?

21.16 What is a floating substrate potential? Explain how the floating potential is calculated in a DC discharged plasma.

21.17 What is the ion energy range in a conventional DC discharged plasma? Explain how the ion bombardment on the growing film in DC magnetron sputtering may be increased. Explain why a wide range of ion energies will be produced in a pulsed DC discharged plasma in reference to the features in a pulsed target voltage waveform.

21.18 Explain why ion bombardment is important in a magnetron sputtering process. Explain the effect of substrate bias and/or pulsed ion energetic bombardment on the following structure and properties of the films: (a) composition, (b) texture, (c) density, (d) grain size, (e) surface roughness, (f) residual stress, and (g) defect.

Applications of Thin Films

John J. Moore, Jianliang Lin, and In-Wook Park

In thinking about nanotechnology today, what's most important is understanding where it leads, what nanotechnology will look like after we reach the asssembler breakthrough.

K. Eric Drexler

Chapter 22

THREADS

Chapter 22 is a continuation of *chapter 21* and provides examples of applications of the thin film technology gleaned from it. With this chapter, we finish the *Mechanical Nanoengineering* division of the text and move on to the *Chemical Nanoengineering* division.

22.0 Technological Applications of Thin Films

In the previous chapter, the concept, classification, and technical properties of nanocomposite and nanostructured thin films and coatings have been introduced. In addition, important and popular thin-film deposition techniques and principles have been reviewed. For a variety of applications, different structures and properties of a nanostructured and nanocomposite coating system are required, leading to specific coating design and correlated deposition process parameter control. In general, nanocomposite coatings can demonstrate different mechanical, electrical, optical, electrochemical, catalytic, and structural properties than those of each individual component [1]. This multifunctional behavior is closely connected to the film structure. The structure, however, depends on the phases, chemical composition, and the arrangement of the phases in the material and this in turn is strongly governed by the deposition process.

In this section, some technological examples of nanostructured and nanocomposite coatings will be illustrated, and which been used widely in different areas, such as high-temperature self-lubricating coatings for aerospace applications, high hard and tribological coatings for pressure die casting die protection, and high hard and toughness tribological coatings for bearing protection.

22.1 Unbalanced Magnetron Sputtering of Ti–Al–Si–N Coatings

Nanocomposite coatings based on nanocrystalline hard transition metal carbide, for example, nc-TiN imbedded in a solid amorphous carbon matrix, a:Si_3N_4, have been shown to enhance the film hardness and toughness while maintaining low sliding friction coefficients. These coatings have significant applications as a protective layer for roller or sliding bearings and gears in the automotive industry.

Quaternary Ti–Al–Si–N coatings have been prepared by a hybrid coating system, where cathodic arc evaporation (CAE) was combined with a magnetron sputtering technique. Various analyses (e.g., high resolution transmission electron microscopy [HRTEM], x-ray photoelectron spectroscopy [XPS], x-ray diffraction [XRD]) revealed that the synthesized Ti–Al–Si–N coatings exhibited nanostructured composite microstructures consisting of solid-solution (Ti, Al, Si)N crystallites and amorphous Si_3N_4. The Si addition caused the grain refinement of (Ti, Al, Si)N

FIG. 22.1 *Dark-field TEM and SEM images of Ti–Al–Si–N coatings containing (a) 0, (b) 4, (c) 9, and (d) 19 at% Si, respectively.*

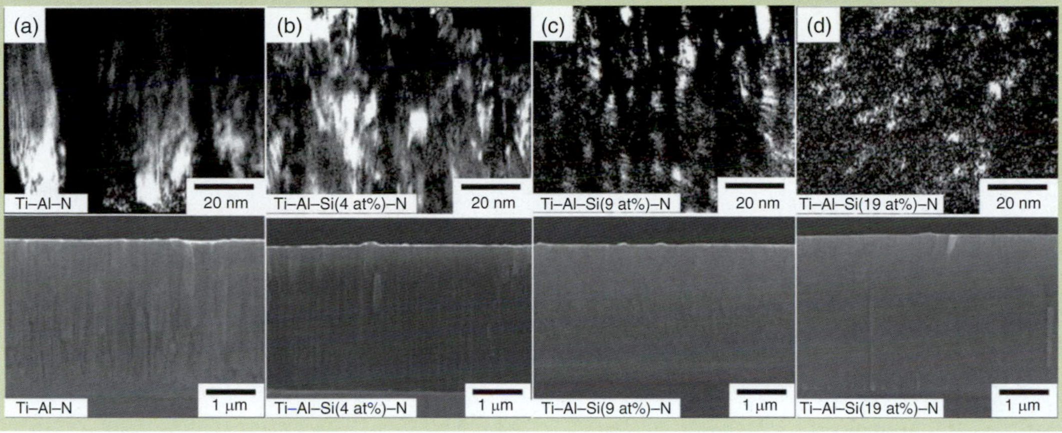

Source: J. Takadoum, H. Houmid-Bennani, and D. Mairey, *Journal of European Ceramic Society*, 18, 553 (1998). With permission.

crystallites and their uniform distribution with percolation phenomenon of amorphous silicon nitride similar to that of the Si effect in TiN films [2]. **Figure 22.1** shows dark-field transmission electron microscope (TEM) images of Ti–Al–Si–N coatings containing different amounts of Si. The (Ti, Al)N crystallites (**Fig. 22.1a**) appear to be large grains with a columnar structure. The (Ti, Al)N crystallites became finer with a uniform distribution as the Si content was increased.

In **Figure 22.1b**, the crystallites were embedded in an amorphous matrix. These crystallites were revealed to be solid-solution (Ti, Al, Si)N phases of typical face-centered cubic (*fcc*) crystal structure from the electron diffraction patterns. The (Ti, Al, Si)N crystallites showed a partly aligned microstructure penetrated (*percolated*) by an amorphous phase, but were not distributed homogeneously in the amorphous matrix. The microstructure, however, changed to that of a nanocomposite, having much finer (Ti, Al, Si)N crystallites (approximately 10 nm) and uniformly embedded in an amorphous matrix as the Si content in films increased to 9 at% (**Fig. 22.1c**). Such a microstructure as shown in **Figure 22.1c** was in agreement with the concept of a nanocomposite architecture suggested by Veprek et al. [3]. Therefore, the Ti–Al–Si–N coatings with Si content of 9 at% exhibited maximum hardness among the experimental conditions. On the other hand, at a higher Si content of 19 at% (**Fig. 22.1d**), (Ti, Al, Si)N crystallites decreased (~3 nm) and the film consisted mainly of the amorphous phase.

Figure 22.2a shows the nanohardness and Young's modulus of Ti–Al–Si–N coatings with various Si contents and average grain sizes. As the Si content increased, the nanohardness and Young's modulus of the Ti–Al–Si–N coatings steeply increased, and reached maximum values of ~55 and 650 GPa at Si content of 9 at%, respectively, and then dropped again with further increase of Si content. The reason for large increases in hardness and Young's modulus of Ti–Al–N with Si addition is due to the grain boundary hardening both by strong cohesive energy of interphase boundaries and by Hall–Petch strengthening derived from crystal size refinement, as mentioned in section 21.3, which were

FIG. 22.2 *(a) Nanohardness and Young's modulus and (b) friction coefficients of nanocomposite Ti–Al–Si–N [nc-(Ti, Al)N/a-Si_3N_4] coatings as a function of Si content.*

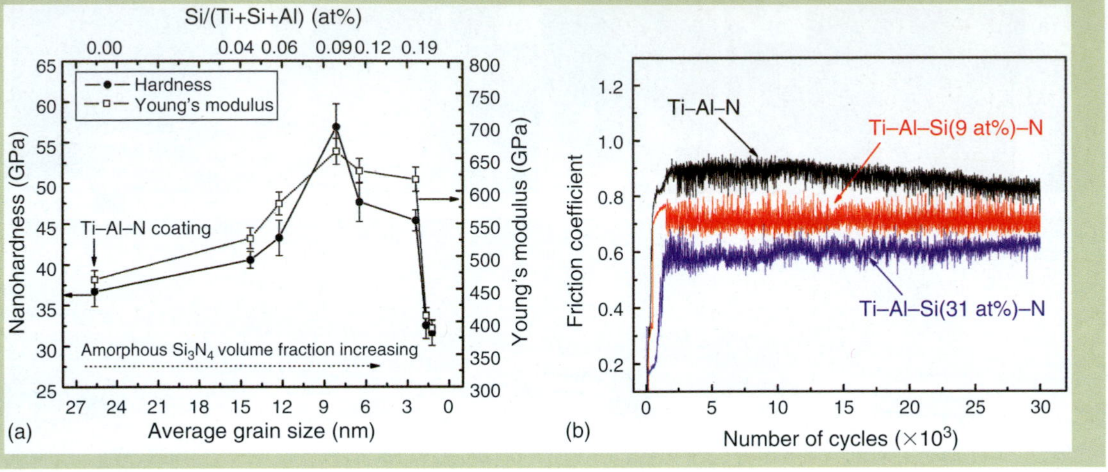

simultaneously caused by the percolation of amorphous Si_3N_4 (H = ~22 GPa, E = ~250 GPa) [2] into the Ti–Al–N film. Another possible reason would be due to solid-solution hardening of crystallites by Si dissolution into Ti–Al–N. The maximum hardness value at the silicon content of about 9 at% results from the nanosized crystallites and their uniform distribution embedded in the amorphous Si_3N_4 matrix as shown in **Figure 22.2**. On the other hand, the hardness reduction with further increase of Si content after maximum hardness as shown in **Figure 22.2a** has been explained with the thickening of amorphous Si_3N_4 phase with increase of Si content [4]. When the amorphous Si_3N_4 is increased, the ideal interaction between nanocrystallites and the amorphous phase is lost, and the hardness of the nanocomposite becomes dependent on the property of the amorphous phase. On the other hand, Young's modulus, which must be related with density and atomic structure of the film, also largely increased from 470–670 GPa with Si addition. This latter result was attributed to the densification of Ti–Al–Si–N films by filling the open structure of the Ti–Al–N grain boundaries with amorphous silicon nitride. Young's modulus reduction with further increase of Si content above 9 at% was explained by the increase of volume fraction of the amorphous Si_3N_4 phase, which has a lower atomic density than the crystalline (Ti, Al, Si)N phase.

Figure 22.2b shows the friction coefficients of the Ti–Al–N, Ti–Al–Si(9 at%)–N, and Ti–Al–Si(31 at%)–N films against a steel ball counterpart. The average friction coefficient of the film decreased from 0.9 to 0.6 with increasing Si content. This result is caused by tribo-chemical reactions, which often take place in many ceramics [5], for example, Si_3N_4 reacts with H_2O to produce a SiO_2 or $Si(OH)_2$ tribo-layer [6]. The products of SiO_2 and $Si(OH)_2$ are known to play the role of a self-lubricating layer.

Figure 22.3 shows the surface morphologies of the wear track and composition analyses for the wear debris after a sliding test. The surface morphology of the wear track for the Ti–Al–N film was rough, and the width of the wear track

Fig. 22.3 *SEM morphologies of wear track and composition analyses for the wear debris after the sliding test: (a) Ti–Al–N, (b) Ti–Al–Si(9 at%)–N, and (c) Ti–Al–Si(31 at%)–N film.*

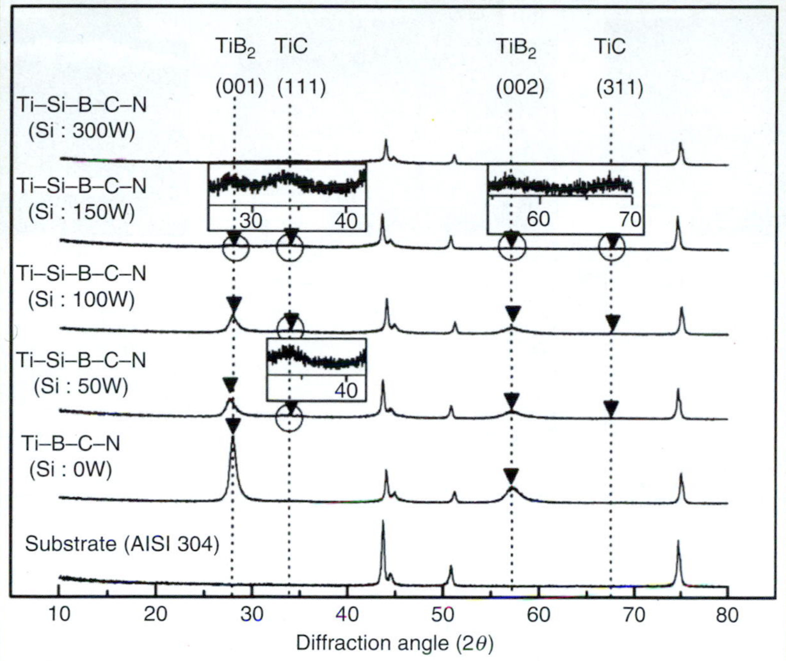

was relatively narrow as shown in **Figure 22.3a**, whereas, the surface morphology for the Ti–Al–Si(9 at%)–N film was relatively smooth and the width of wear track was wide (**Fig. 22.3b**). This result is due to the adhesive wear behavior between the hard film (~55 GPa) and the relatively soft steel (~700 $H_{v0.2}$). Thus, the steel ball is worn and smeared onto the Ti–Al–Si(9 at%)–N film having higher hardness (~55 GPa). On the other hand, the surface morphology of wear track for the Ti–Al–Si(31 at%)–N film was very smooth, and the width of the wear track narrowed again as shown in (**Fig. 22.3c**). This reflects that the formation of a self-lubricating tribo-layer such as SiO_2 or $Si(OH)_2$ was activated on increasing the Si content. From energy dispersive x-ray spectroscopy (EDS) analyses of the wear debris (**Fig. 22.3a** and **b**), the peak intensity of Fe for the Ti–Al–Si(9 at%)–N film was higher than those for the Ti–Al–N film, and the peak intensities of Ti and Al were reversed for the two films. Similar EDS results were found between Ti–Al–Si(9 at%)–N and Ti–Al–Si(31 at%)–N films. This indicates that the harder film was more wear resistant against the steel counterpart.

22.2 Unbalanced Magnetron Sputtering of Ti–Si–B–C–N Coatings

Figure 22.4 shows the x-ray diffraction patterns of Ti–B–C–N and Ti–Si–B–C–N films on AISI 304 stainless steel substrates with various Si target powers at a

FIG. 22.4 *XRD patterns of Ti–B–C–N and Ti–Si–B–C–N films on AISI 304 stainless steel substrates with various Si target powers at a fixed TiB_2–TiC composite target power of 700 W.*

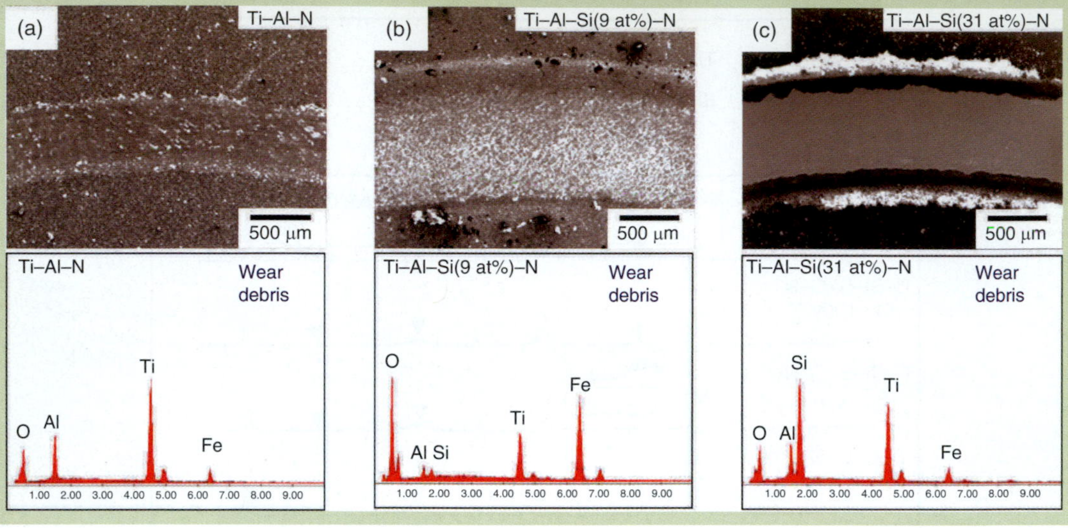

fixed TiB_2–TiC composite target power of 700 W. At a Si target power of 0 W, the diffraction pattern of the Ti–B–C–N film exhibited a crystalline hexagonal TiB_2 phase with preferred orientations of (001) or (002) crystallographic planes. Any XRD peaks corresponding to the crystalline TiC or TiN phase were not observed from the Ti–B–C–N diffraction pattern. As the Si target power was increased, the diffraction peak intensities of TiB_2 (001) and (002) gradually reduced and completely disappeared at the Si target power of 300 W. At the Si target power of 50 and 100 W, the TiB_2 peaks corresponding to the same (001) and (002) planes as well as small TiC peaks for (111) and (311) crystallographic planes were present. And, at a Si target power of 150 W, very small diffraction TiB_2 peaks for (001) and (002) as well as TiC peaks for (111) and (311) were observed. Furthermore, at the highest Si target power of 300 W, the XRD pattern presented no diffraction peaks for the film, indicating that the film is comprised mainly of an amorphous phase. The gradual changes in the XRD patterns of Ti–Si–B–C–N films with Si additions into Ti–B–C–N are similar to the case of N addition into Ti–B–C, as previously reported by the authors for the Ti–B–C–N nanocomposite system [7].

In the report, it was revealed that the crystallites in Ti–B–C–N films were composed with solid-solution (Ti, C, N)B_2 and Ti(C, N) crystallite (~10 nm in size). Addition of nitrogen into the Ti–B–C film led to grain refinement of (Ti, C, N)B_2 and Ti(C, N) crystallites, and their distribution is coupled with a percolation phenomenon of amorphous BN and carbon phase. In order to determine the chemical composition and to investigate the bonding status of the Ti–Si–B–C–N coating, x-ray photoelectron spectroscopy (XPS) was performed on Ti–B–C–N and Ti–Si–B–C–N coatings deposited by unbalanced magnetron sputtering from a TiB_2–TiC composite target and a Si target at different Si target powers. **Figure 22.5a** provides the content of each element in the Ti–Si–B–C–N coating

FIG. 22.5 *XPS data for (a) content of Ti, Si, B, C, N, and O, and XPS spectra of (b) Si 2p, (c) C 1s, and (d) O 1s for Ti-B-C-N with Si target power of 0 W and Ti-Si-B-C-N coating with Si target power of 300 W(14.2 at% Si in film).*

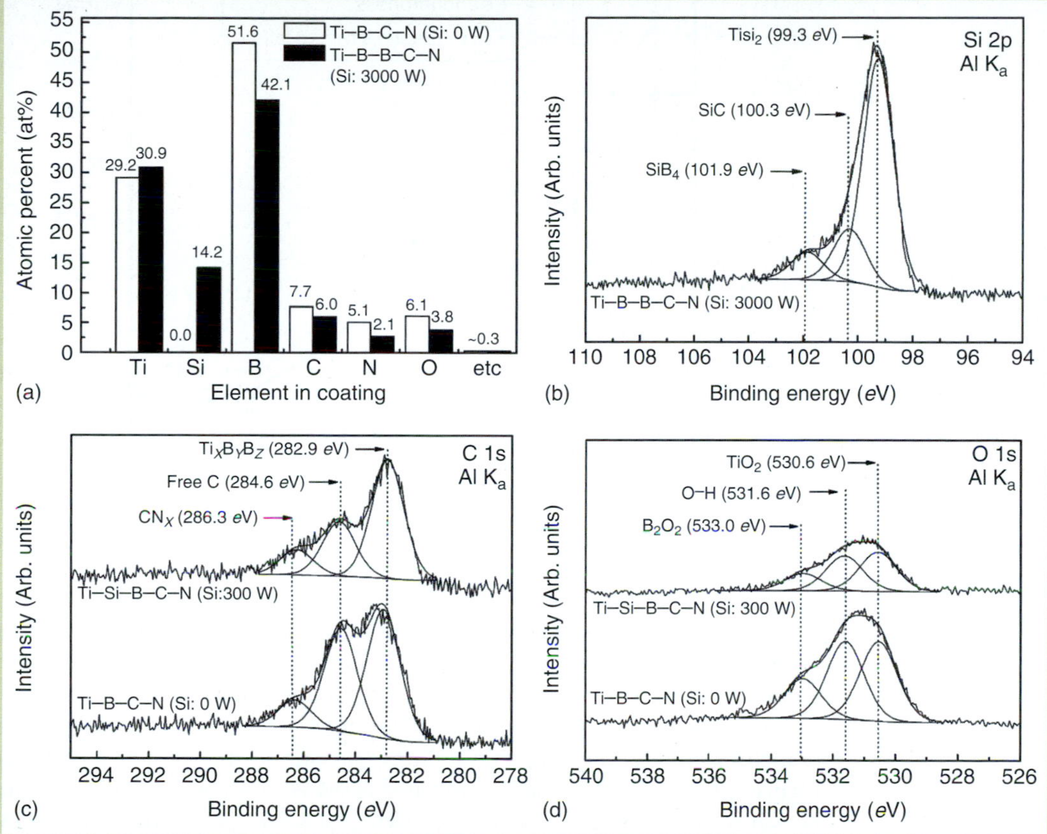

with two different Si target powers and a fixed TiB_2–TiC target power of 700 W. As the Si target power was increased, the Si content was increased in the Ti–Si–B–C–N film from 0 to 14.2 at%. The boron content steeply decreased. **Figure 22.5b–d** present the XPS spectra binding energies of Si, C, and O for the Ti–B–C–N and Ti–Si(14.2 at%)–B–C–N coatings. For the Si 2p region (**Fig. 22.5b**), $TiSi_2$ is clearly present with smaller amounts of SiC and SiB_4. The C 1s region (**Fig. 22.5c**) confirms the presence of $Ti_xB_yC_z$ components in amorphous free carbon and CN_x. With the addition of Si into the Ti–B–C–N coating, the free-carbon peak intensity significantly decreased. The large decrease in the free-carbon peak intensity of Ti–B–C–N with 14.2 at% Si addition is most likely the result of formation of SiC as shown in **Figure 22.5b**.

Based on the results from the XRD and XPS analyses, it is concluded that the Ti–Si–B–C–N coatings are nanocomposites consisting of nanosized $(Ti,C,N)B_2$ and Ti(C,N) crystallites embedded in an amorphous $TiSi_2$ and SiC matrix including some carbon, SiB_4, BN, CN_x, TiO_2, and B_2O_3 components.

Figure 22.6 presents the nanohardness and *H*/*E* values of the Ti–Si–B–C–N coatings as a function of Si target power. The hardness of the Ti–Si–B–C–N

FIG. 22.6 *Nanohardness and H/E values of Ti–Si–B–C–N coating as a function of Si target power.*

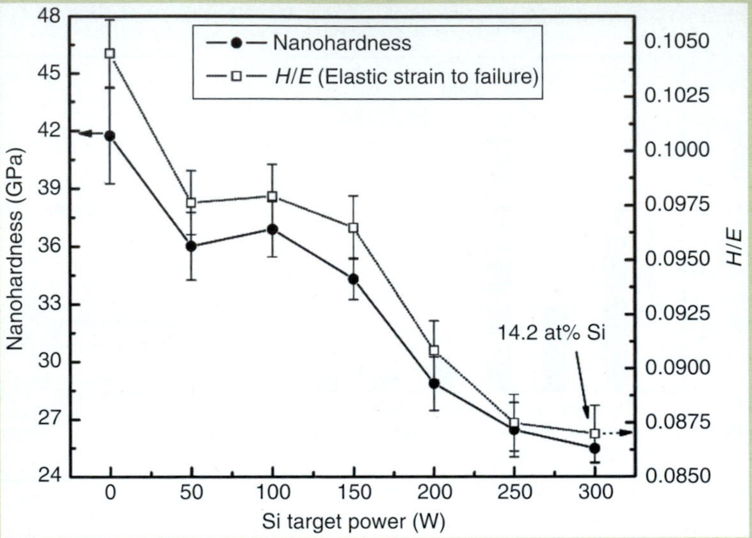

coating decreased from ~42 GPa at 0 W Si target power to ~36 GPa at 50 W Si target power. The hardness was constant at about 35 GPa from 50 to 150 W Si target power and decreased again with a further increase in Si target power to about 25 GPa at a Si target power of 300 W(14.2at% Si in coating). The decrease in hardness of Ti–B–C–N with 50 W Si target power is most likely due to a reduction in hard TiB_2-based crystallites. The supporting evidence is shown in **Figure 22.5a**, where boron content decreases with increasing silicon content. However, the Ti–Si–B–C–N coating with a Si target power from 50 to 150 W exhibited a high hardness of about 35 GPa. The reason for maintaining the high hardness (~35 GPa) in Ti–Si–B–C–N coatings with a small amount of Si is most likely due to percolation of amorphous $TiSi_2$ and SiC in the grain boundaries. Veprek et al. [8] have found that ultra-hardness ($80\,GPa \leq H_v \leq 105\,GPa$) is achieved when the nanosized and/or amorphous $TiSi_2$ precipitate in the grain boundaries in their Ti–Si–N (nc-TiN/*a*-Si_3N_4/*a*- and nc-$TiSi_2$) nanocomposites. On the other hand, the hardness reduction with further increase in Si target power above 200 W can be explained by either an increase in the soft amorphous $TiSi_2$, SiC, and SiB_4 phases or reduction of hard TiB_2-based crystallites.

When the amount of amorphous phase is increased the ideal interaction between nanocrystallites and the amorphous phase can be lost, and the hardness of the nanocomposite becomes dependent on the property of the amorphous phase [9]. In addition, *H/E* values, the so-called *elastic strain to failure criterion*, were calculated from the obtained hardness (*H*) and Young's modulus (*E*). Recently, Leyland and Matthews [10] have suggested that a high *H/E* value is often a reliable indicator of good wear-resistance. In **Figure 22.6**, the *H/E* values exhibited a similar tendency with hardness. As the Si target power increased, the *H/E* value of Ti–Si–B–C–N coatings decreased from ~0.105 to 0.087. From the results of hardness and *H/E*, it can be suggested that the Ti–Si–B–C–N coating

FIG. 22.7 *(a) Friction coefficients and (b) wear rates of Ti–Si–B–C–N coating against WC-Co ball as a function of Si target power.*

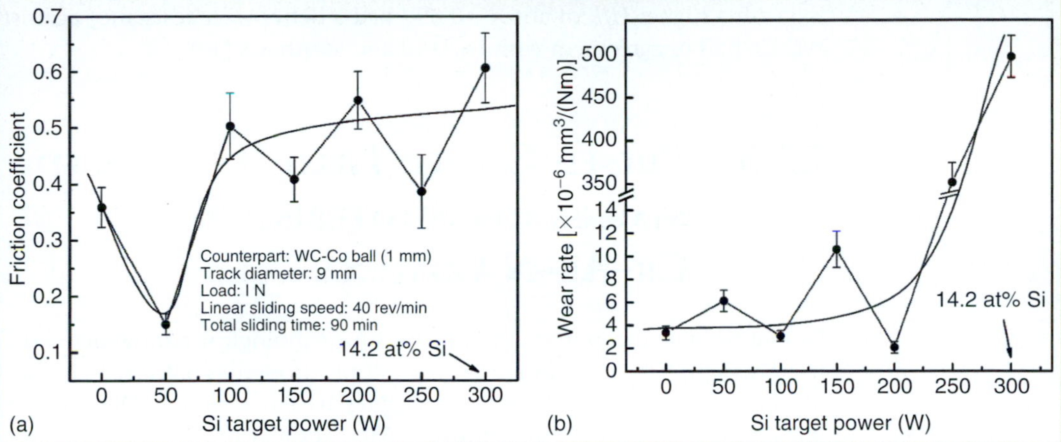

with Si target power up to 150 W could provide a better wear-resistance with higher fracture toughness than that of Ti–Si–B–C–N coatings with Si target power above 200 W.

Figure 22.7a shows the friction coefficients of the Ti–Si–B–C–N coating against a WC-Co ball as a function of Si target power. The average friction coefficient of the Ti–Si–B–C–N coating rapidly decreased by increasing the Si target power and showed a minimum value of ~0.15 at a Si target power of 50 W, and then rebounded with further increase in Si target power above 100 W. This large decrease in the friction coefficient of the Ti–Si–B–C–N coating with 50 W Si target power is most likely due to the formation of smooth solid-lubricant tribo-layers formed by tribo-chemical reactions during the sliding test. For example, silicon compounds such as $TiSi_2$, SiC, or SiB_4 in the coating react with ambient H_2O and oxygen to produce SiO_2 or $Si(OH)_2$ tribo-layers. These by-products of SiO_2 and $Si(OH)_2$ have been known [5] to play the role of a self-lubricating layer. This Si effect on tribological behavior in nanocomposites has also been found [11] by other authors. In addition, the carbon and hydroxide phases in the coating, as shown in **Figure 22.5c** and **d**, with small Si content may also contribute to the minimum friction coefficient value.

However, the friction coefficient slightly increased with further increase of Si content in the coating. The increase in friction coefficient with increased target power is most likely due to a reduction in the free-carbon and hydroxide phases, which are self-lubricating phases, in the Ti–Si–B–C–N coating as shown in **Figure 22.5c** and **d**. **Figure 22.7b** represents the wear rates of the Ti–Si–B–C–N coating on AISI 304 substrates as a function of Si target power. The wear rate of the Ti–Si–B–C–N coating slightly increased from $\sim 3 \times 10^{-6}\ mm^3 \cdot N^{-1} \cdot m^{-1}$ at 0 W to $\sim 10 \times 10^{-6}\ mm^3 \cdot N^{-1} \cdot m^{-1}$ at a Si target power around 200 W. These very low wear rates would be due to the adhesive wear behavior between the hard coating (~35 GPa) and relatively soft WC-Co (~22 GPa) ball. On the other hand, at the Si target power above 250 W, the wear rate of the Ti–Si–B–C–N coating steeply increased to about $500 \times 10^{-6}\ mm^3 \cdot N^{-1} \cdot m^{-1}$ at a Si target power of 300 W. This large increase in wear rate would be due to the abrasive wear behavior between the relatively soft coating (~25 GPa) and the WC-Co ball. Combining the results of the *H/E* (**Fig. 22.6**) values and wear rates (**Fig. 22.7b**), the Ti–Si–B–C–N coating with a higher *H/E* of above ~0.090 had a better wear-resistance against the WC-Co ball in agreement with Leyland and Matthews [10].

22.3 Pulsed Closed Field Unbalanced Magnetron Sputtering of Cr–Al–N Coatings

One of the most important applications for tribological nanostructured coatings is as a protective layer to improve multiple properties of the working surface of a bulk material, or tool, used in an aggressive environment. Development of transition metal nitride thin films (e.g., TiN, TiAlN, CrN, CrAlN, etc.) has been widely documented as a means to increase productivity and tool life in material-forming processes, such as die casting, metal forming, plastic molding, glass forming tool, and machining/cutting applications. In these applications, however, the coating material often experiences extreme mechanical, thermal, and chemical loading conditions. For example, in the modern high-pressure die casting process, a molten aluminum alloy at temperatures ranging from 670 to 710°C is injected into the die cavity at high velocities of the order of 30–100 $m \cdot s^{-1}$. The injection pressures are of the order of 50–80 MPa, with a temperature gradient of around 1000°C$\cdot cm^{-2}$ [12]. In the machining of steels, stainless steels, and cast irons with coated cemented carbide tools, the cutting edges are worn according to different wear mechanisms at high cutting speeds; the amount of heat generated in the cutting zone is considerable and the temperature at the cutting edge of coated cutting tools may exceed 1000°C in an ambient oxidation environment [13].

The demand for continual improvement of hard coatings leads to the need to develop multifunctional hard tribological coatings, which can provide a wide range of properties. The successful application and improvement of these tribological hard coatings strongly depends on the microstructure (nanostructure) design and deposition process control.

The CrAlN ternary compound film is a very promising die coating candidate that shows high toughness and hardness, good wear resistance, and excellent oxidation resistance combined with corrosion resistance [14–17]. It was found that the aluminum content in the film plays a significant role in determining the structure and properties of the Cr–Al–N coatings. The incorporation of aluminum into the B1 cubic CrN lattice will lead to the precipitation of a B4 hexagonal AlN phase if the solubility limit of AlN in the coating is exceeded. The formation of the wurtzite hexagonal structure is not desired due to its low hardness and poor ductility [18]. It has been predicted that CrN shows the highest solubility of 77.2% for aluminum among the transition nitrides with a B1 cubic structure [19,20]. Therefore, it is possible to add a large amount of aluminum into CrAlN films without changing the cubic phase, thereby extending the CrAlN film oxidation resistance temperature while maintaining good mechanical and

FIG. 22.8 *The GIXRD patterns for $Cr_{1-x}Al_xN$ films with different aluminum concentrations.*

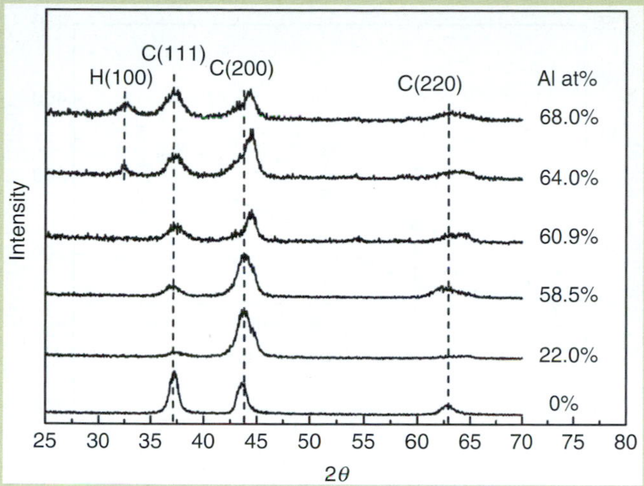

tribological properties. **Figure 22.8** shows the change in the crystal structure of $Cr_{1-x}Al_xN$ deposited by CFUBMS as a function of aluminum content. The B4 hexagonal AlN structure was observed when the aluminum content is at 64 at% [14]. The coexistence of cubic and hexagonal phases is observed when the aluminum concentration in the film is at or beyond 64.0 at%.

Figure 22.9 provides the cross-sectional scanning electron microscope (SEM) photomicrographs of CrN and $Cr_{0.415}Al_{0.585}N$ films. The columnar grain size of CrN is about 100 nm (**Fig. 22.9a**). A significant decrease in the grain size was observed with an increase in the aluminum concentration in the films (**Fig. 22.9b**). The $Cr_{1-x}Al_xN$ films' hardness and Young's modulus values are plotted as a function of aluminum concentration in **Figure 22.10**. The results show that the CrN film has a hardness of 25.0 GPa. As aluminum is incorporated into the film, the

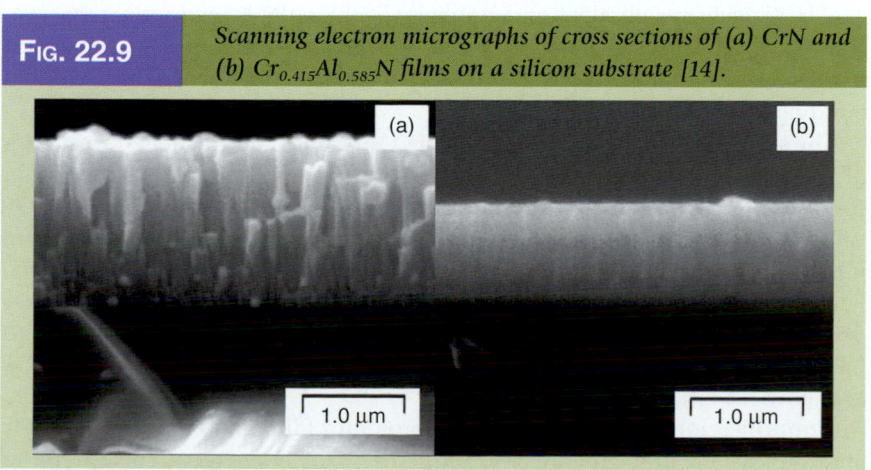

FIG. 22.9 *Scanning electron micrographs of cross sections of (a) CrN and (b) $Cr_{0.415}Al_{0.585}N$ films on a silicon substrate [14].*

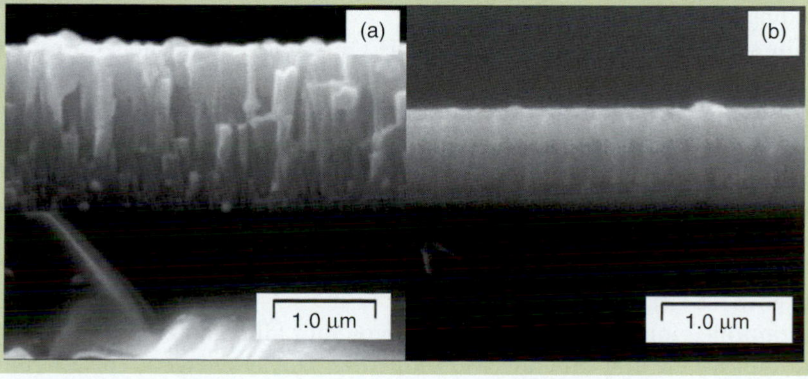

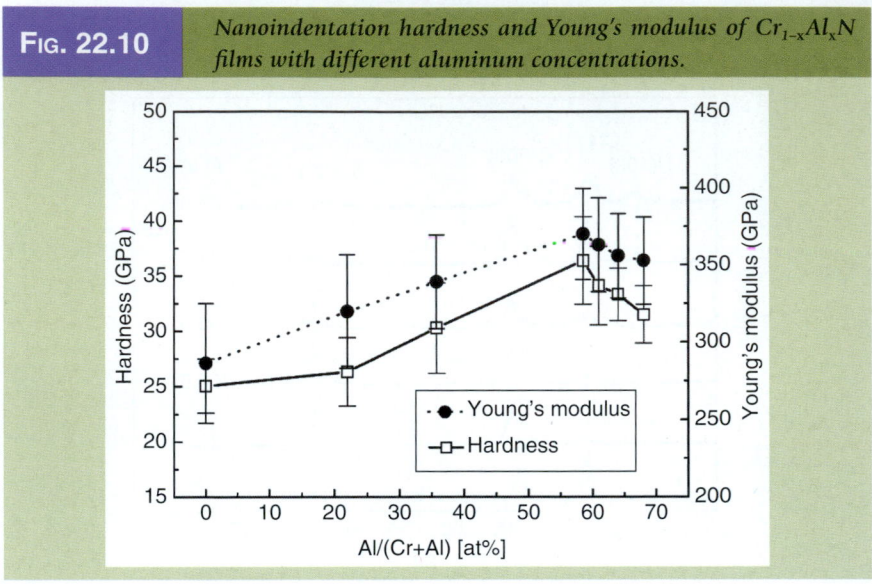

FIG. 22.10 *Nanoindentation hardness and Young's modulus of $Cr_{1-x}Al_xN$ films with different aluminum concentrations.*

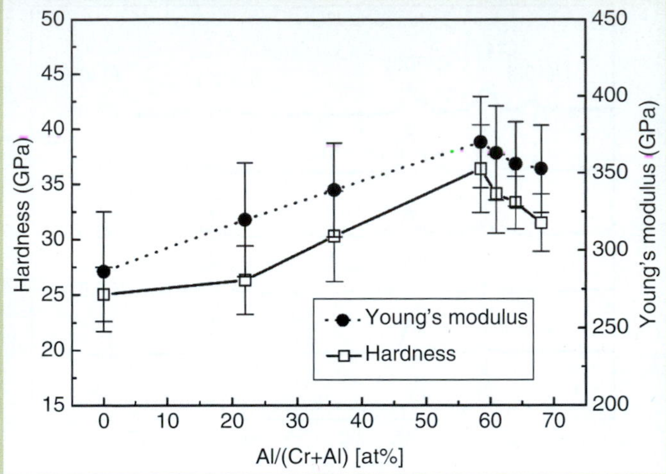

hardness and Young's modulus values of the $Cr_{1-x}Al_xN$ films increase, and both reach the maximum values of 36.3 and 370 GPa, respectively, at 58.5 at% aluminum concentration. A further increase in aluminum concentration results in a decrease in both hardness and Young's modulus. The increase of the film hardness is possibly related to the decrease in the grain size and a denser film structure (**Fig. 22.9b**).

According to the "Hall–Petch" relationship (equation 21.1), the hardness of the material increases with decreasing crystallite size, especially prominent for grain sizes down to tens of nanometers. In addition, the incorporation of aluminum into the CrN lattice will decrease the lattice parameter because of the smaller atomic size of aluminum atoms compared with the Cr atom. This behavior will increase the covalent energy in the films, because the interatomic distance *d*, is related to the covalent bandgap E_h according to the expression $E_h = Kd^{-2.5}$. Furthermore, addition of aluminum in CrN increases the covalent bonding, as CrN is a largely metallically bonded material, while AlN is predominantly covalently bonded [21]. Therefore, the increase in hardness of $Cr_{1-x}Al_xN$ films with increasing aluminum content is probably related to the above grain size and bonding energy effect.

The steady-state coefficient of friction (COF) values and calculated wear rates of the $Cr_{1-x}Al_xN$ nanostructured films are plotted as a function of aluminum content in **Figure 22.11**. It was found that the CrN film exhibits a lower COF value (0.28) than all of $Cr_{1-x}Al_xN$ films. Adding a small amount of aluminum into $Cr_{1-x}Al_xN$ films ($x = 0.22$) resulted in a sudden increase in COF to 0.55. Further increasing the aluminum content in the film, the COF of $Cr_{1-x}Al_xN$ films decreased to a low value of 0.37 at $x = 0.585$ and then started to increase up to 0.45 at $x = 0.68$. The wear rate exhibits a similar trend compared with the COF change in $Cr_{1-x}Al_xN$ films.

Controlled ion bombardment of growing thin films can be used to modify and improve the film structure and properties. Higher energetic species (up to

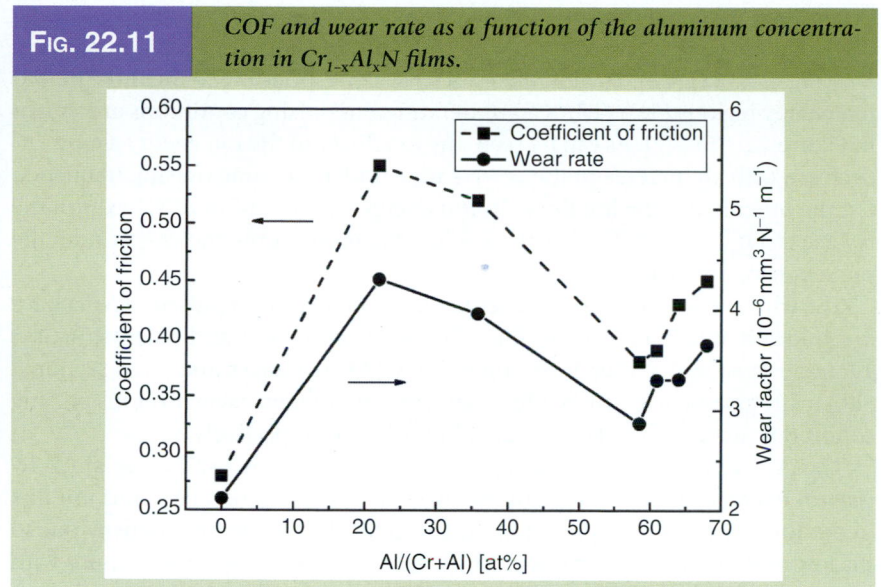

FIG. 22.11 *COF and wear rate as a function of the aluminum concentration in $Cr_{1-x}Al_xN$ films.*

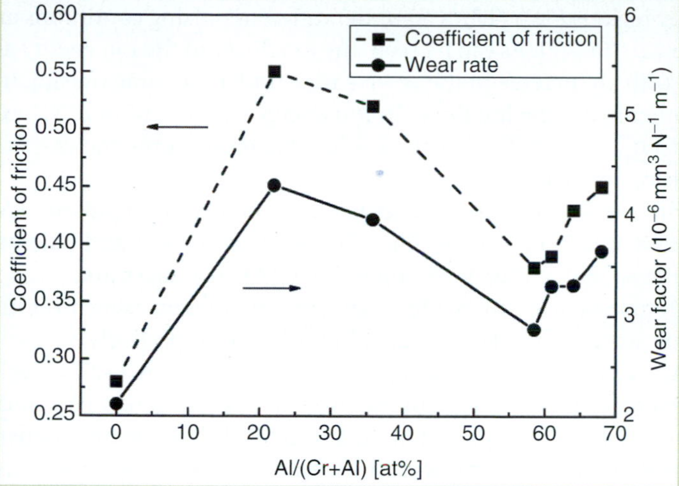

hundreds *e*V) found in the plasma by pulsing the target(s) in magnetron sputtering has been discussed in section 21.5.4. In this section, the effect of the pulsed ion energy and ion flux on the P-CFUBMS-deposited CrAlN film structure and properties will be illustrated.

Figure 22.12a shows the $^{29}N_2^+$ IEDs for pulsing both Cr and Al targets asynchronously at 100 and 350 kHz with different reverse times, respectively. Different

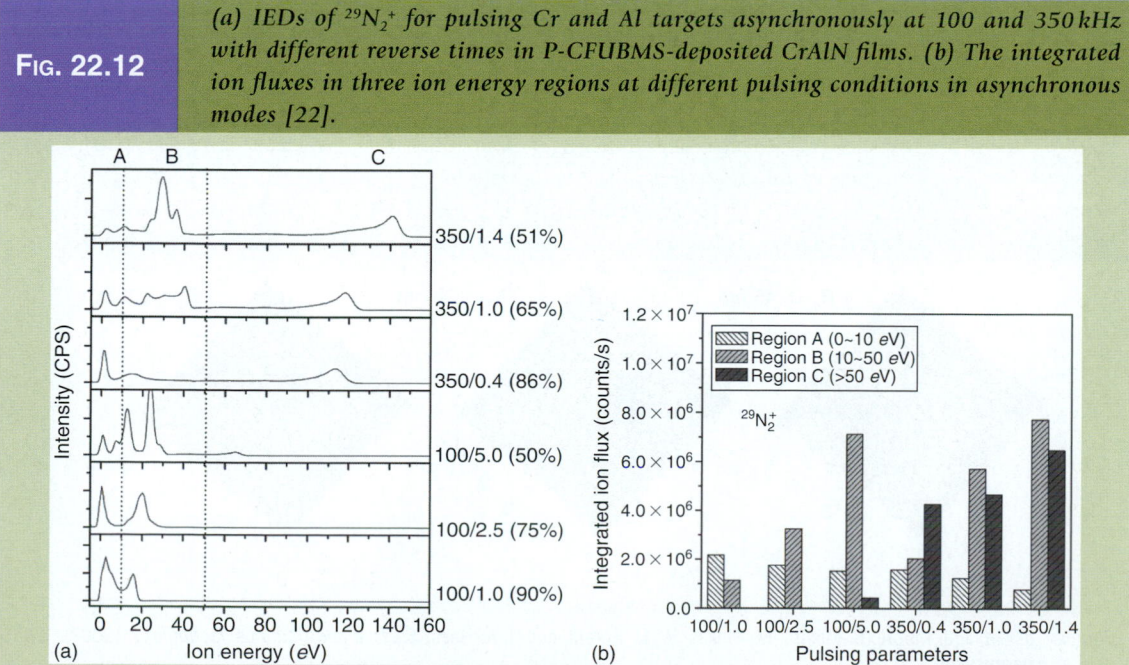

FIG. 22.12 *(a) IEDs of $^{29}N_2^+$ for pulsing Cr and Al targets asynchronously at 100 and 350 kHz with different reverse times in P-CFUBMS-deposited CrAlN films. (b) The integrated ion fluxes in three ion energy regions at different pulsing conditions in asynchronous modes [22].*

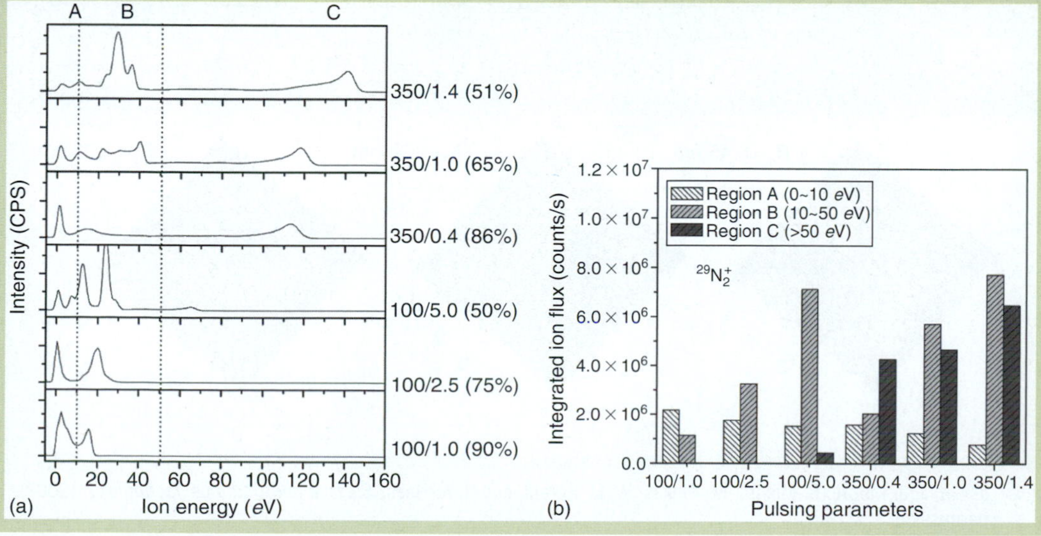

maximum $^{29}N_2^+$ energies with a range from 22 to 150 *e*V were observed in the discharged plasma in accordance with different asynchronous pulsing parameters (**Fig. 22.12a**) [22]. The ion fluxes of $^{29}N_2^+$ corresponding to the three pulsed ion energy regions ("A", "B", "C") under different pulsing conditions are shown in **Figure 22.12b** [22]. As can be seen, the ion fluxes in the ion energy region "A" decrease with an increase in the reverse time under the same pulsing frequency. On the other hand, the ion fluxes in ion energy regions "B" and "C" exhibit the reverse trend, in which they increase when the reverse time increases under the same pulsing frequency.

The wide range of pulsed ion energies and ion fluxes in a pulsed plasma have a significant influence on the film surface morphology and microstructure. Three-dimensional atomic force microscopy (AFM) images and cross-sectional SEM micrographs of Cr–Al–N films deposited at different asynchronous pulsing conditions are shown in **Figures 22.13** and **22.14,** respectively.

The change in film surface roughness and morphology can be explained by the ion energy/ion flux change in the plasma. The low ion energy and ion flux in the 100 kHz and 1.0 µs pulsed plasma lead to a low nucleation density due to the low mobility of the adatoms on the substrate, developing coarser grains with a high density of sub-grains and large cell boundaries (**Fig. 22.13a**). This is consistent with an open columnar structure, corresponding to zone 1 structure in the Thornton Zone model [24] as seen in the SEM image (**Fig. 22.14a**). This film exhibited a mean surface roughness of 5.45 nm.

The film that was deposited in the 100 kHz and 5.0 µs condition exhibits a very dense grain structure and the smoothest surface with a surface roughness of 1.04 nm (**Figs. 22.13b** and **22.14b**). This may be attributed to the significant increase of ion flux in the "B" (middle) ion energy region, and also the total ion energy reached 122 *e*V. In this case, a large flux of ions with middle energy level bombard the substrate surface, significantly enhancing the adatom mobility and diffusivity by the momentum transfer from the impingement. The highly

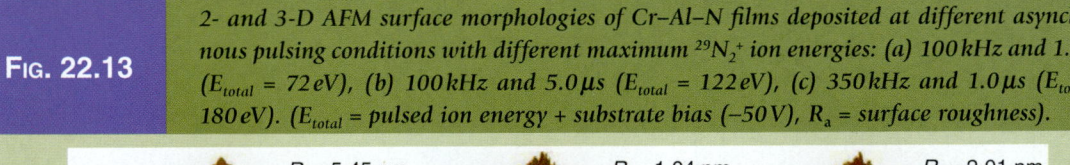

FIG. 22.13 *2- and 3-D AFM surface morphologies of Cr–Al–N films deposited at different asynchronous pulsing conditions with different maximum $^{29}N_2^+$ ion energies: (a) 100 kHz and 1.0 µs (E_{total} = 72 eV), (b) 100 kHz and 5.0 µs (E_{total} = 122 eV), (c) 350 kHz and 1.0 µs (E_{total} = 180 eV). (E_{total} = pulsed ion energy + substrate bias (−50 V), R_a = surface roughness).*

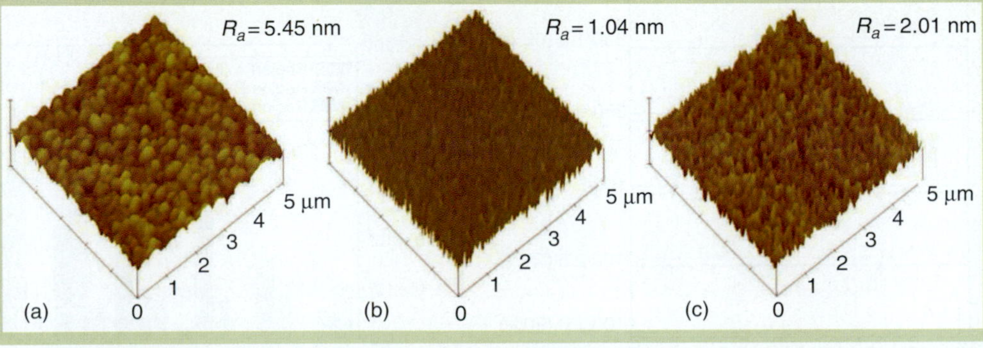

Source: J. Lin, J. J. Moore, B. Mishra, M. Pinkas, W. D. Sproul, and J. A. Rees, *Surface and Coatings Technology*, (2007). With permission.

FIG. 22.14 *Cross-sectional SEM micrographs of Cr–Al–N films deposited at different asynchronous pulsing conditions with different maximum $^{29}N_2^+$ ion energies: (a) 100 kHz and 1.0 μs (E_{total} = 72 eV), (b) 100 kHz and 5.0 μs (E_{total} = 122 eV), and (c) 350 kHz and 1.0 μs (E_{total} = 180 eV).*

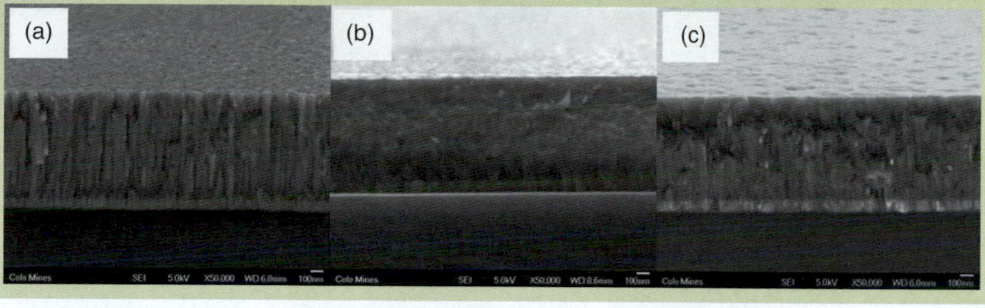

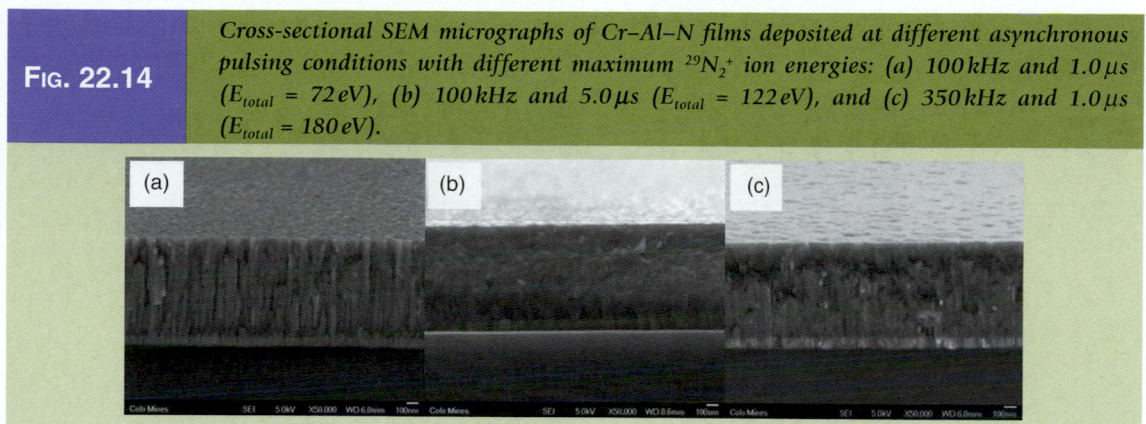

Source: J. Lin, J. J. Moore, B. Mishra, M. Pinkas, W. D. Sproul, and J. A. Rees, *Surface and Coatings Technology*, (2007). With permission.

mobile adatoms can move or diffuse into the inter-grain voids under high energy ion bombardment, and a denser structure is attained.

The pulsed ion energy and ion flux is further enhanced in the 350-kHz pulsed plasma (**Fig. 22.12**). However, it is noted that the ion energy increase is largely from the "C" high ion energy region while the contribution from the "A" low ion energy region is decreased compared to the 100-kHz conditions. In the cross-sectional field emission scanning electron microscopy (FESEM) micrographs of the 350-kHz and 1.0-μs films (**Fig. 22.14c**), a different microstructure compared to those in 100-kHz pulsing conditions is revealed. Columnar grains are still observed but "renucleation" (interruption of film growth due to high ion energy bombardment followed by localized growth) on the individual column is seen throughout the film. The incoming high energy (>180 *e*V) ion bombardment can disrupt the continued growth of the columnar grains. Thus, only few grains grew throughout the film thickness and short columnar grains were formed (**Fig. 22.14c**), while the film surface roughness increased to 2.25 nm by kinetic roughening (**Fig. 22.13c**).

A comparison of the cross-sectional TEM micrographs and corresponding selected area electron diffraction (SAED) patterns of Cr–Al–N films deposited at 100 kHz and 5.0 μs and 350 kHz and 1.0 μs are presented in **Figure 22.15a** and **b**, respectively. In both films the energetic-enhanced deposition conditions resulted in the formation of a dense nanocrystalline structure. The average grain size in the 100-kHz and 5.0-μs film is 20–50 nm (**Fig. 22.15a**). The SAED pattern displays a typical form of a nanocrystalline material consisting of the *fcc* (Cr, Al) N phase. No amorphous rings are observed in the SAED pattern. In the film deposited at 350 kHz and 1.0 μs, the arc- and spot-shaped SAED pattern indicate that a bi-model grain size is formed in this high-energy bombarded film. The small grains are formed on renucleation sites along the larger columnar grains, as shown in **Figure 22.15b**. However, this high ion energy may be excessive for the film growth, in that it can cause increased point (vacancy) and line (dislocations) defect density and intensive resputtering and a consequent decrease in nucleation sites. The intergranular residual damage is the most prevalent defect

Fig. 22.15 *Cross-sectional TEM photomicrographs of Cr–Al–N films deposited at different asynchronized pulsing conditions. (a) 100 kHz, 5.0 μs and (b) 350 kHz, 1.0 μs.*

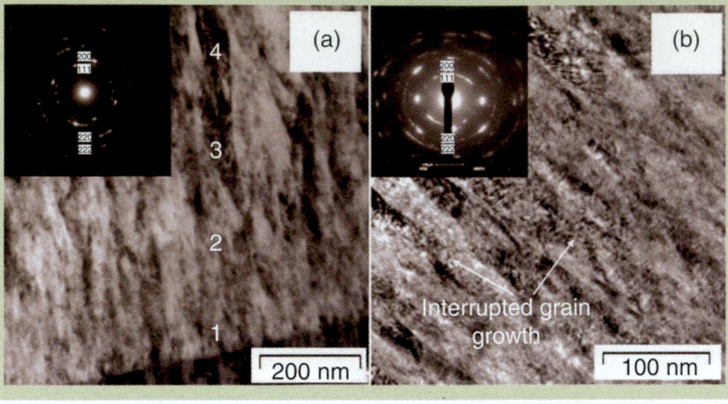

in the 350-kHz and 1.0-μs Cr–Al–N film, which was subjected to excessive ion bombardment as shown in **Figure 22.16a**. The grains contain an appreciable amount of lattice defects visible by the speckled contrast within the columnar grains. Edge dislocations that compensate for the high strain within the columnar grains are revealed in **Figure 22.16b**, which is a high-resolution TEM micrograph of the distorted lattice of the Cr–Al–N film.

High-resolution lattice images of Cr–Al–N films deposited at 100 kHz and 5.0 μs and 350 kHz and 1.0 μs are shown in **Figure 22.17a** and **c**, respectively. Fourier transform filtered images of the same areas are shown in **Figure 22.17b** and **d**, respectively. From the filtered lattice images, the film deposited at 100 kHz

Fig. 22.16 *Cross-sectional TEM photomicrograph showing (a) high dislocation density within the columnar grains and (b) high-resolution TEM photomicrograph showing the distorted lattice and low-angle grain boundaries, in the Cr–Al–N film deposited in asynchronized mode at 350 kHz and 1.0 μs.*

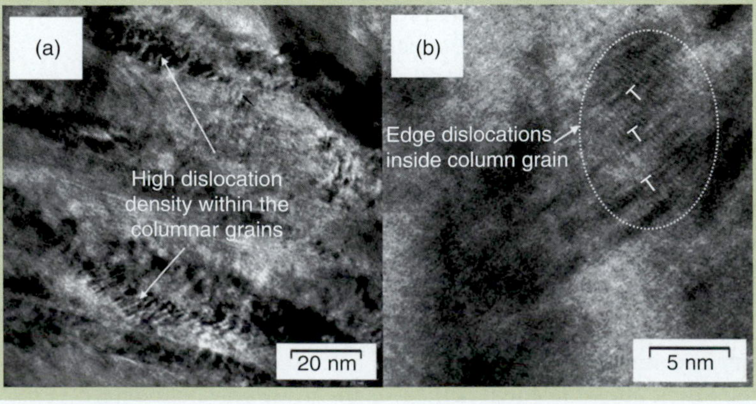

Fig. 22.17 *(a) High-resolution TEM lattice images of Cr–Al–N film deposited at 100 kHz and 5.0 μs, (b) the fast Fourier transform (FFT) filtered image of (a), (c) high-resolution TEM lattice images of Cr–Al–N film deposited at 350 kHz and 1.0 μs, and (d) the FFT filtered image of (c).*

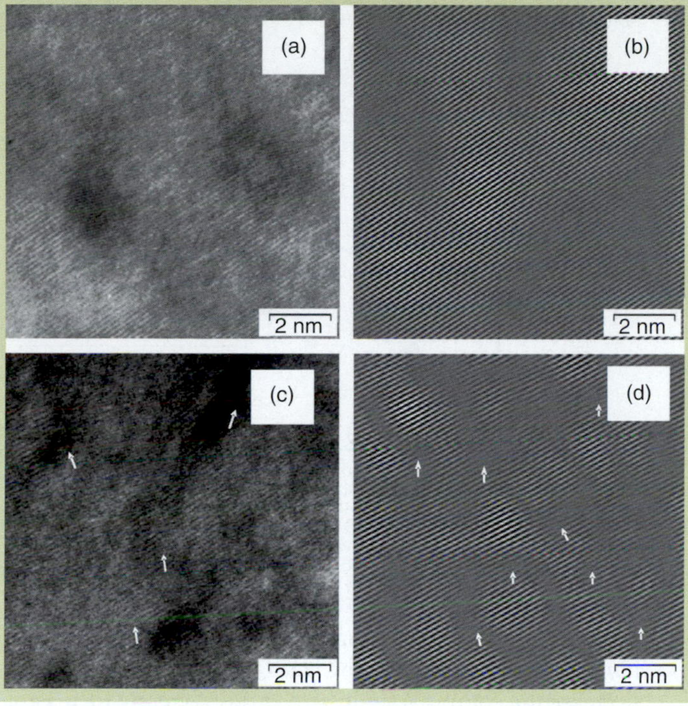

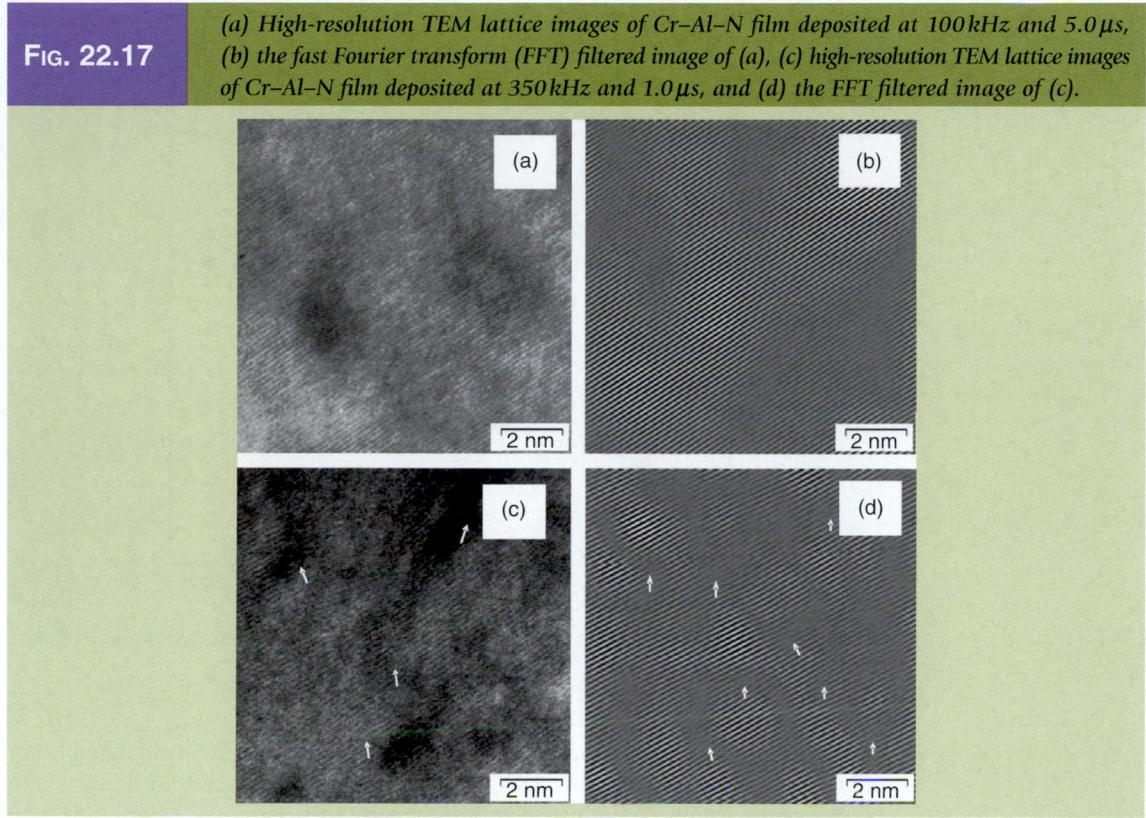

Source: J. Lin, J. J. Moore, B. Mishra, M. Pinkas, W. D. Sproul, and J. A. Rees, *Surface and Coatings Technology*, (2007). With permission.

and 5.0 μs exhibits a uniform lattice with few defects. On the other hand, many defects in the form of edge dislocations (indicated by arrows) are observed in the 350-kHz and 1.0-μs film.

The evolution of the film microstructure at controlled ion bombardment by pulsing the targets has significant influence on the film mechanical and tribological properties. The hardness and Young's modulus of Cr–Al–N films deposited at different asynchronous pulsing conditions are plotted in **Figure 22.18**. The film hardness increased from 34 to 48 GPa when the total ion energy in the discharged plasma increased from 72 to 200 *e*V accordingly. The Young's modulus of the films exhibits a similar trend. The increased hardness of the films deposited with pulsed ion energy and ion flux bombardment can be explained by two features: (i) improved density and decreased (nanocrystalline) grain size and (ii) development of internal residual stress and large defect densities.

The maximum ion energies were increased from 72 to 122 *e*V when the reverse time was increased from 1.0 to 5.0 μs in asynchronized mode at 100 kHz. At the same time, the main ion flux contributed from the "B" ion energy region

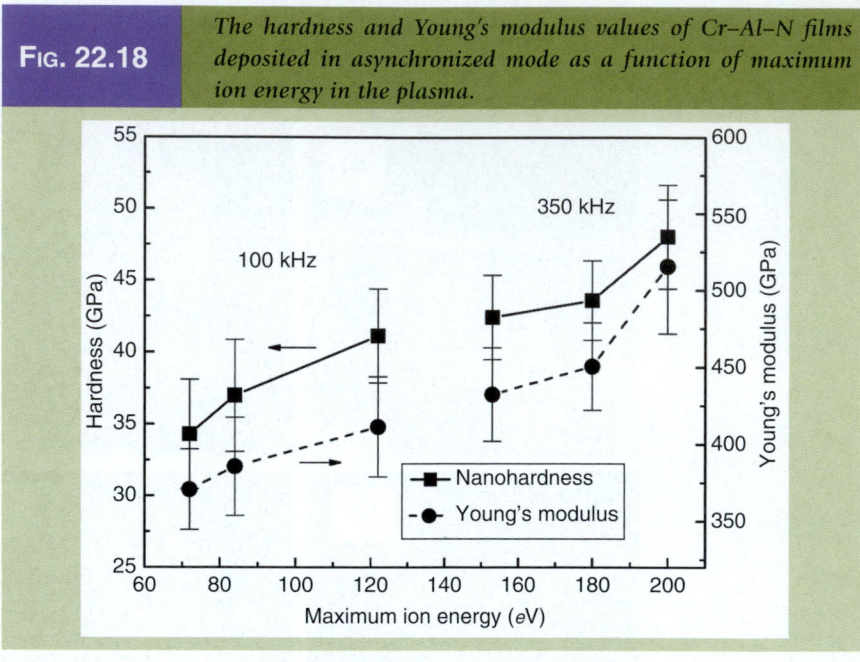

Fig. 22.18 *The hardness and Young's modulus values of Cr–Al–N films deposited in asynchronized mode as a function of maximum ion energy in the plasma.*

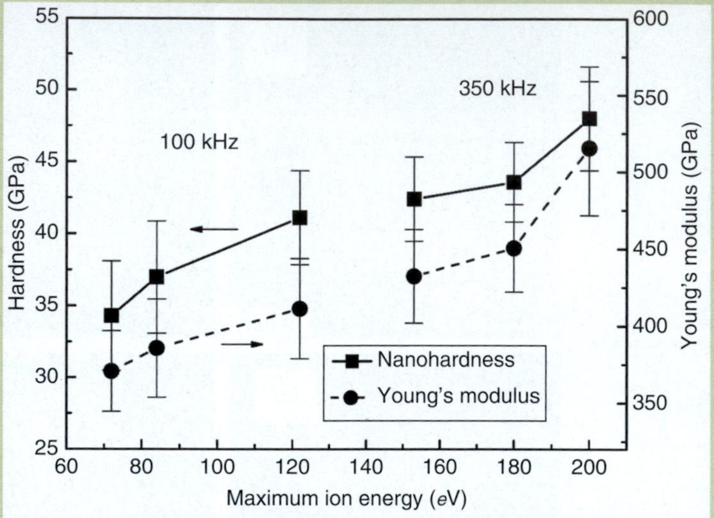

(10~30 *e*V) increases correspondingly. This controlled middle ion energy bombardment can effectively increase adatom mobility, enhance film density, and decrease the columnar grain size without greatly increasing the defect density in the film (**Fig. 22.15a**). The increased hardness values from 34 to 41 GPa in these films mainly result from the film structural improvements. Nevertheless, when the Cr–Al–N films were deposited in an asynchronized mode at 350 kHz, the excessive ion energy bombardment from the "C" high ion energy region will result in high defect incorporation as well as high residual stress in the films. In general, a high defect concentration in a compressively stressed material will restrict the plastic flow, and thus be a contributing factor in enhancing the hardness [23]. Therefore, further increase in the hardness (41~48 GPa) in these excessively ion bombarded Cr–Al–N films is possibly related to the strain hardening and a high defect density.

The steady-state COF values and calculated wear rates of the Cr–Al–N films deposited in the asynchronized mode with different maximum ion energies are plotted in **Figure 22.19**. As can be seen, the COF values and wear rate of Cr–Al–N films increased with an increase in the maximum ion energy. The films deposited at 100-kHz pulse frequency exhibit low COF values in the range of 0.38 to 0.46. However, the high point and line defect densities and high residual stress incorporated into the films deposited at 350-kHz conditions can decrease the toughness and increase the brittleness of the film. Cr–Al–N films deposited at 350-kHz pulse frequency exhibit decreased wear resistance, as shown in **Figure 22.19**.

The technological example of P-CFUBMS of CrAlN films demonstrates the importance of film composition design and the deposition process control for

FIG. 22.19 *COF and wear rate of Cr–Al–N films deposited in asynchronized mode with different maximum ion energy.*

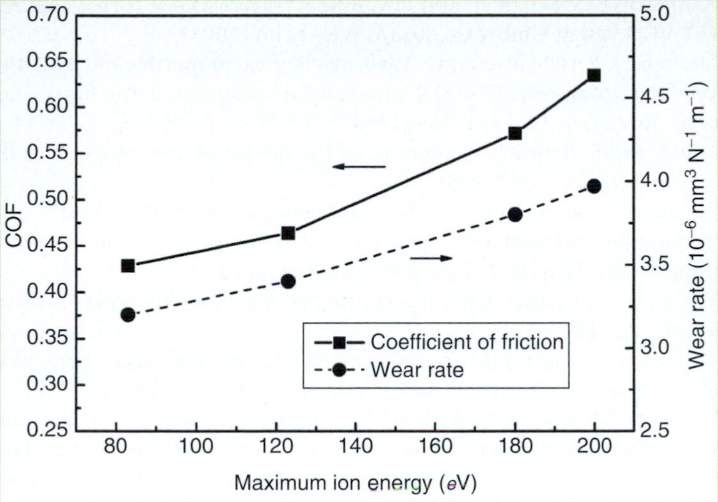

obtaining high quality films. For the pulsed magnetron sputtering, there is an advantage to maintain the maximum pulsed ion energy less than 120 *e*V and increase the ion flux in the "A" and "B' middle ion energy regions (10–70 *e*V) for obtaining improved film nanostructure and tribological properties. If the pulsed ion energies are excessive (>120 *e*V), there will be an increase in point and line defects, introduced in the form of high residual stress in the crystalline structure.

22.4 Concluding Remarks

The main purpose of chapters 21 and 22 is to provide a basic level of understanding of the concept, background, and processing of nanostructure and nanocomposite thin films. The chapter discusses how nanostructure and nanocomposite thin films can result in improved performance to meet the required applications, such as high wear resistance, low friction coefficient, self-lubrication, high oxidation, and/or corrosion resistance. These chapters also provide a review of some of the thin-film deposition techniques which are widely used to process nanostructure and nanocomposite thin films. Approaches to control the film chemistry and ion energy (ion flux) in tailoring the structure and properties of the films to meet specific tribological applications are emphasized. A number of multicomponent, nanostructured coating examples processed using various deposition processes were given to demonstrate the relationship among processing, structure, properties, and performance: Ti–Al–Si–N, Ti–Si–B–C–N, and Cr–Al–N.

References

1. P. M. Ajayan, L. S. Schadler, and P. V. Braun, *Nanocomposite science and technology*, WILEY-VCH Verlay, GmbH Co. KGaA, Wienheim (2003).
2. M. Diserens, J. Patscheider, and F. Lévy, Mechanical properties and oxidation resistance of nanocomposite TiN–SiN_x physical-vapor-deposited thin films, *Surface and Coatings Technology*, 120–121, 158 (1999).
3. S. Veprek and S. Reiprich, A concept for the design of novel superhard coatings, *Thin Solid Films*, 268, 64 (1995).
4. S. H. Kim, J. K. Kim, and K. H. Kim, Influence of deposition conditions on the microstructure and mechanical properties of Ti–Si–N films by DC reactive magnetron sputtering, *Thin Solid Films*, 420–421, 360 (2002).
5. S. Wilson and A. T. Alpas, Tribo-layer formation during sliding wear of TiN coatings, *Wear*, 245, 223 (2000).
6. J. Takadoum, H. Houmid-Bennani, and D. Mairey, The wear characteristics of silicon nitride, *Journal of European Ceramic Society*, 18, 553 (1998).
7. I.-W. Park, K. H. Kim, A. O. Kunrath, D. Zhong, J. J. Moore, A. A. Voevodin, and E. A. Levashov, Microstructure and mechanical properties of superhard Ti–B–C–N films deposited by dc unbalanced magnetron sputtering, *Journal of Vacuum Science and Technology, B*, 23(2), 588 (2005).
8. S. Veprek, P. Nesladek, A. Niederhofer, F. Glatz, M. Jilek, and M. Sima, Recent progress in the superhard nanocrystalline composites: Towards their industrialization and understanding of the origin of the superhardness, *Surface and Coatings Technology*, 108–109, 138 (1998).
9. J. Patscheider, T. Zehnder, and M. Diserens, Structure–performance relations in nanocomposite coatings, *Surface and Coatings Technology*, 146, 201 (2001).
10. A. Leyland and A. Matthews, On the significance of the H/E ratio in wear control: A nanocomposite coating approach to optimised tribological behaviour, *Wear*, 246, 1 (2000).
11. J. Xu and K. Kato, Formation of tribochemical layer of ceramics sliding in water and its role for low friction, *Wear*, 245, 61 (2000).
12. S. Gopal, A. Lakare, and R. Shivpuri, Evaluation of thin film coatings for erosive-corrosive wear prevention in die casting dies, In *Surface modification technologies XII*, T. S. Sudarshan, K. A. Khor, and M. Jeandin, eds., ASM International, Materials Park, Ohio (1998).
13. H. O. Gekonde and S. V. Subramanian, Tribology of tool–chip interface and tool wear mechanisms, *Surface and Coatings Technology*, 149, 151 (2002).
14. J. Lin, B. Mishra, J. J. Moore, and W. D. Sproul, Microstructure, mechanical and tribological properties of $Cr_{1-x}Al_xN$ films deposited by pulsed-closed field unbalanced magnetron sputtering (P-CFUBMS), *Surface and Coatings Technology*, 201, 4329–4334 (2006).
15. J. C. Sánchez-López, D. Martínez-Martínez, C. López-Cartes, A. Fernández, M. Brizuela, A. García-Luis, and J. I. Oñate, Mechanical behaviour and oxidation resistance of Cr–Al–N coatings, *Journal of Vacuum Science and Technology, A*, 23, 4 (2005).
16. J. Lin, B. Mishra, J. J. Moore, W. D. Sproul, and J. A. Rees, Effects of the substrate to chamber wall distance on the structure and properties of CrAlN films deposited by pulsed-closed field unbalanced magnetron sputtering (P-CFUBMS), *Surface and Coatings Technology*, 201, 6960 (2007).
17. C. Brecher, G. Spachtholz, K. Bobzin, E. Lugscheider, O. Knotek, and M. Maes, Superelastic (Cr,Al)N coatings for high end spindle bearings, *Surface and Coatings Technology*, 200 1738 (2005).

18. Y. Sun, Y. H. Wang, and H. P. Seow, Effect of substrate material on phase evolution in reactively sputtered Cr–Al–N films, *Journal of Materials Science*, 39, 7369–7371 (2004).
19. A. Sugishima, H. Kajioka, and Y. Makino, Phase transition of pseudobinary Cr–Al–N films deposited by magnetron sputtering method, *Surface and Coatings Technology*, 97, 590 (1997).
20. Y. Makino and K. Nogi, Synthesis of pseudobinary Cr–Al–N films with B1 structure by RF-assisted magnetron sputtering method, *Surface and Coatings Technology*, 98, 1008 (1998).
21. H. Holleck, Material selection for hard coatings, *Journal of Vacuum Science Technology*, A, 4, 2661 (1986).
22. J. Lin, J. J. Moore, B. Mishra, M. Pinkas, W. D. Sproul, and J. A. Rees, Effect of asynchronous pulsing parameters on the structure and properties of CrAlN films deposited by pulsed closed field unbalanced magnetron sputtering (P-CFUBMS), *Surface and Coatings Technology*, 202, 1418 (2008).
23. L. Karlsson, A. Horloing, M. P. Johansson, L. Hultman, and G. Ramanath, The influence of thermal annealing on residual stresses and mechanical properties of arc-evaporated TiC_xN_{1-x} (x = 0, 0.15 and 0.45) thin films, *Acta Materialia*, 50, 5103 (2002).

Problems

22.1 Give three examples of nanostructured tribological coatings used in industry. Select one of your interest to prepare a case study under the following items from the literature to study the process–structure–property relationship: (a) deposition technique and parameters, (b) chemical composition, (c) phase structure, (d) grain size and microstructure, (e) properties (e.g., hardness, adhesion, tribological, stability with temperature, corrosion resistance, surface roughness, stress and defect, etc.), and (f) on-going research.

22.2 The ball-on-disk wear test is a widely used technique to evaluate the wear properties of tribological films. Calculate the wear rate of the film expressed in $mm^3 \cdot N^{-1} \cdot m^{-1}$, given the following test parameters: tests were carried out along a circular track of 12-mm diameter under a load of 5 N and at a constant sliding speed of 40 $mm \cdot s^{-1}$, for sliding distances up to 1000 m. After the wear tests, the average cross-sectional area of the wear track in the film was determined to be 150 μm^2 using a surface profilometer.

22.3 Briefly describe the possible effects of the following deposition parameters on the deposition process and the structure and properties of films: (a) working pressure, (b) deposition temperature, (c) substrate bias, (d) target power, and (e) the gas flow rate.

22.4 List at least two characterization techniques which can be utilized to determine each of the following properties of nanostructured coatings, and explain the advantages and limitations of each technique: (a) grain size, (b) chemical compositions, (c) crystalline phase, (d) stress, (c) surface morphology, (d) cross-sectional morphology, (e) texture, and (f) hardness.

Section 9

Chemical Nanoengineering

Nanocatalysis

Scott W. Cowley

I hence will name it the catalytic force of the substance, and I will name decomposition by this force catalysis. The catalytic force is reflected in the capacity that some substances have, by their mere presence and not by their own reactivity, to awaken activities that are slumbering in molecules at a given temperature.

Jöns Jacob Berzelius, 1836

Chapter 23

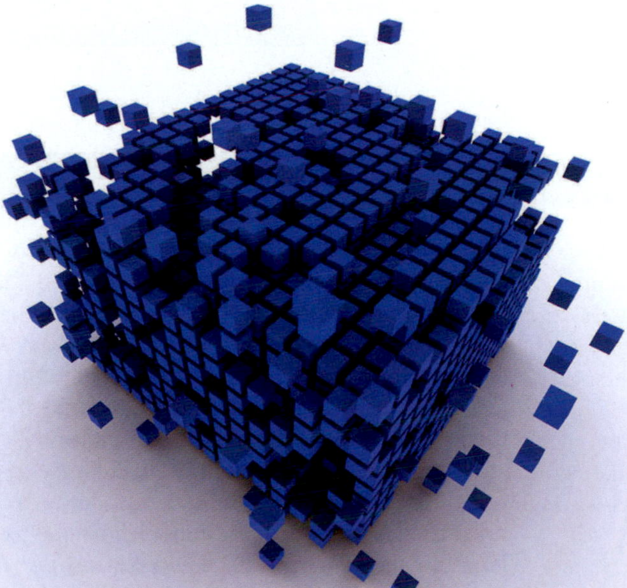

Fig. 23.0

Jöns Jacob Berzelius was one of the pioneers of modern chemistry. He is responsible for elucidating the "law of definite proportions" and for developing a table based on atomic weights. He was the first to use the phrase "organic chemistry." He is known for stating the following about catalysis: "Thus it is certain that substances have the property of exerting an effect quite different from ordinary chemical affinity, in that they promote the conversion without necessarily participating in the process with their own component parts."

Threads

Chapter 23 presents a generalized discussion about catalysis by nanoparticles. This chapter is the first chapter in the *Chemical Nanoengineering* division of the book. Catalysts have been with us for a long time. Over the past 50 years, catalysts have made many industrial processes successful. Catalysts are expected to be enabled and enhanced by applications of nanotechnology—whether in the form of new syntheses, better characterization, or new applications.

The next chapter, *chapter 24*, in this division explores the domain of nanocomposites. We have already had a discourse about nanocomposites in *chapters 21* and *22*, especially as they apply to thin films. Polymer chemistry and the chemical modification of carbon nanotubes in particular team up to present an entirely new and innovative chapter in our already rich experience with composite materials.

Following *chapter 24*, *chapters 25–27* delve into biological aspects of nanotechnology and applications. *Chapter 28* is a member of the *Biological Nanoengineering* division and presents discussions about the environment.

23.0 Introduction to Catalytic and Nanocatalytic Materials

23.0.1 *The Importance of Catalysis in a Modern Society*

Catalysts have an enormous impact on the social and economic structure of our world today. It is projected that by the year 2010, approximately $12 billion worth of catalyst materials will be used each year to produce hundreds of billions of dollars worth of goods on a worldwide basis [1]. Catalysts are used to make fertilizers, fuels, chemicals, medicines, textiles, and many other important products. For example, the Haber process, developed by Fritz Haber in 1913, uses an iron-based catalyst to convert nitrogen and hydrogen gases into ammonia. When ammonia is combined with nitric or sulfuric acid, then ammonium nitrate or ammonium sulfate fertilizers are produced. The utilization of fertilizers results in an increased production of food, which supports a much larger population. A similar example can be given for the catalytic conversion of petroleum products into the various hydrocarbon fuels needed for our farming, manufacturing, and transportation industries, thus bringing people, food, and goods to the market place. Catalysts also play a major role in improving our quality of life from an environmental perspective. For example, a catalyst containing platinum, palladium, rhodium, cerium, and other compounds is used to control the emission levels of toxic gases emitted from our automobiles.

23.0.2 *What Is a Catalyst?*

The term "catalyst" was first used by the Swedish chemist Jöns Jacob Berzelius in his published work on the catalytic decomposition of hydrogen peroxide in the *Edinburg Philosophical Journal* in 1836. He observed that a catalytic substance increased the rate of reaction without being changed or being consumed itself. However, the exact function of the catalyst remained a mystery to Berzelius for he described it as "an inherent force whose nature is still unknown."

Advances in scientific methods and analytical techniques have greatly improved our understanding of how a catalyst works. Today we know that a catalyst increases the rate of reaction by lowering the activation energy (E_a) required to convert reactants into products. The catalyst may actually be consumed during the reaction, but it is always regenerated by the end of the reaction cycle. Once regenerated, the catalyst is available to participate in another reaction cycle. Consequently, only a small amount of catalyst is needed to convert a large amount of reactant into product. Thus catalytic materials become very important when considering if a chemical process is "green" or "sustainable." Consider the thermal decomposition of formic acid (HCO_2H) into water (H_2O) and carbon monoxide (CO), or into hydrogen (H_2) and carbon dioxide (CO_2) as shown in equation (23.1).

$$CO + H_2O \leftarrow HCO_2H \rightarrow H_2 + CO_2 \tag{23.1}$$

At room temperature, ~21°C, the reaction rate is nearly immeasurable, because only a very small population of molecules has sufficient energy to react, that is, they have insufficient energy to overcome the activation energy (E_a) barrier that is required for a chemical change, as shown in **Figure 23.1**. The formic acid molecules must collide with sufficient energy to go through the transition states

FIG. 23.1 *Schematic of the thermal and catalytic decomposition of formic acid into carbon dioxide and hydrogen or into carbon monoxide and water.*

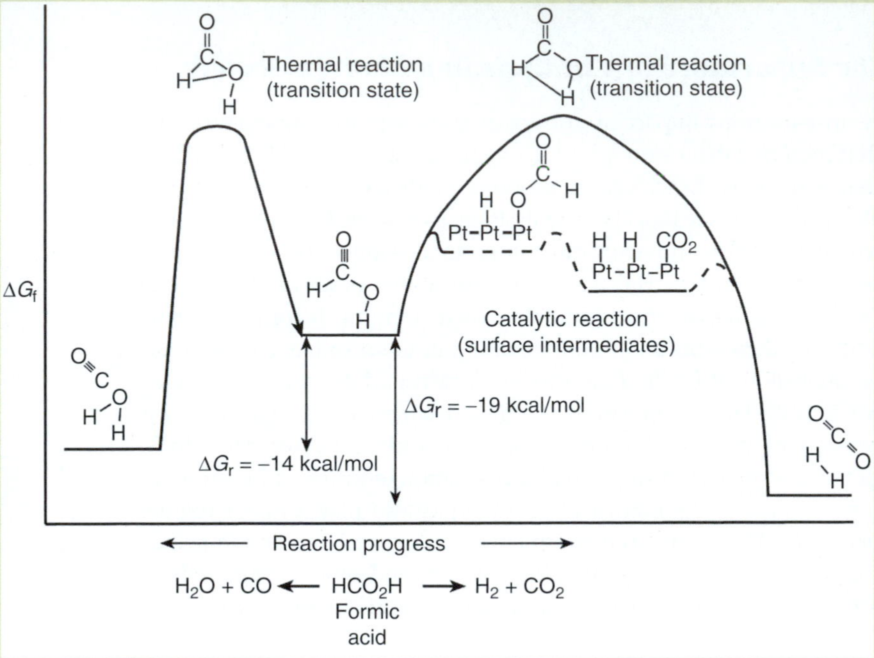

that are shown at the top of the energy curve in **Figure 23.1**, and form the products given in equation (23.1). If the temperature is increased, a larger population of molecules will have enough energy to overcome the activation energy barrier and the reaction will proceed at a measurable rate. However, now there is enough energy to cross the reaction barriers in both directions resulting in a nonselective process. The use of a catalyst can dramatically alter this situation by selectively lowering the activation energy for one pathway but not for the other. Platinum selectively decomposes formic acid into hydrogen and carbon dioxide, while aluminum oxide gives carbon monoxide and water as the major products. In summary, a catalyst accomplishes the following:

1. It lowers the activation energy of the reaction, thus it speeds up the rate of the reaction. A catalyst does not change the thermodynamic equilibrium of a reaction; it only increases the speed for which equilibrium is reached. In addition, a lower activation energy means that a lower reaction temperature can be used.
2. It participates in the reaction chemistry, but it is always regenerated and is available for other reaction cycles. Therefore, only a small amount of catalyst is required for the overall reaction process.
3. It can increase the selectivity for a given reaction by lowering the activation energy of one pathway over that of another.

23.0.3 The Nano Perspective

Only a portion of the catalysts used today can be classified as nanocatalysts. The term nanocatalyst is defined as a material that has catalytic properties on at least one nanoscale dimension. The concept of a "nanocatalyst" is not new to scientists and engineers working in this field. Although the term "nano" was not commonly used until more recent times, researchers traditionally focused on producing very small particles of active catalytic agents in order to maximize the reaction efficiency and to reduce the overall cost of the chemical process. Recent advances in synthetic methods for the production of nanomaterials has produced new nanocatalysts with novel properties and reactivity. Traditional commercial nanocatalysts such as enzymes, zeolites, and transition metal nanocatalysts represent about 98% of the global nanocatalyst market. Newer nanocatalyst materials account for the other 2%. The global market for nanocatalysts is forecast to be $5 billion by the year 2009, with the newer novel nanocatalyst materials expanding to about 7% of the market.

A simple schematic of a commercial platinum catalyst is shown in **Figure 23.2**. The functioning catalyst consists of small nanosized (1–5 nm) metal crystallites supported on a porous metal oxide support. In this case, the active catalyst is platinum and the porous support is aluminum oxide (alumina). The support plays several important roles. Since platinum is a very expensive metal, it is important that every platinum atom be involved in the reaction in order to recover the cost of the platinum in reasonable time. Since the chemical reaction occurs only at the surface of the metal crystallite, any atom in the interior of the crystallite is inaccessible and provides no value to the reaction process. In other words, they cost the user money but don't pay for themselves. Therefore, it is beneficial to have as many atoms on the surface of the crystallite as possible to increase the benefit-to-cost ratio of the catalyst. The porous support provides an inexpensive, but high surface area, base on which to disperse the platinum crystallites. In some cases the support is inert and in other cases it plays an important role in the chemistry as well. Support materials commonly consist of silica (SiO_2), alumina (Al_2O_3), activated carbon, amorphous silica–alumina (SiO_2–Al_2O_3), and crystalline zeolites (SiO_2–Al_2O_3), and have surface areas of 100–1000 $m^2 \cdot g^{-1}$. In addition the high surface area of the support permits the catalyst to be quite compact, but able to

FIG. 23.2 *Schematic of a supported platinum metal catalyst.*

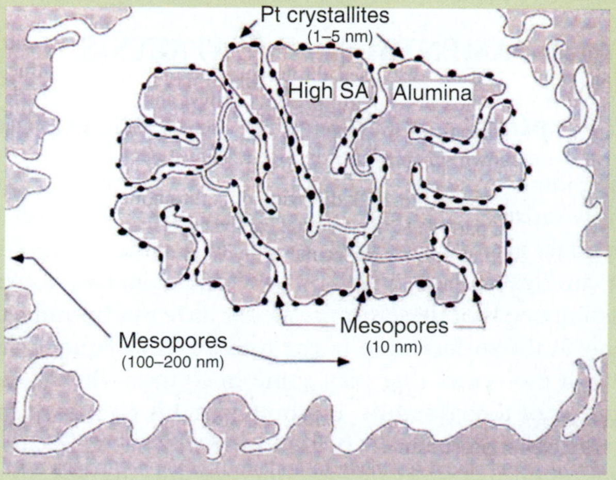

Source: R. Farrauto and C. Bartholomew, *Fundamentals of industrial catalytic processes*, John Wiley & Sons, (2006). With permission.

FIG. 23.3 *A transmission electron microscope (TEM) image of rhodium metal nanocrystallites on a nonporous silica microsphere.*

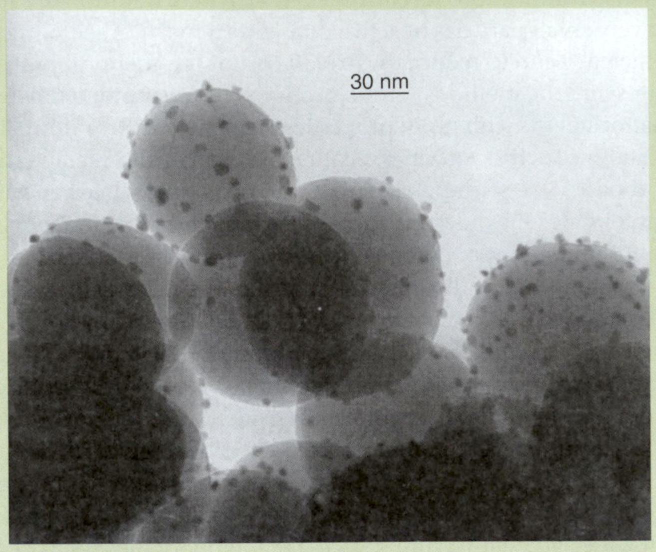

Source: S. Chakraborti, A. K. Datye, and N. J. Long, *Journal of Catalysis*, 108, 444–451 (1987). With permission.

handle a large volume of reactant. A TEM image of rhodium metal nanocrystallites on a nonporous silica microsphere is shown in **Figure 23.3** [3].

There are other benefits to making the metal crystallite smaller. The material properties, such as optical, magnetic, and surface chemistry of the crystallite, change dramatically once they are in the nanocrystallite range. For example, gold is a very inert metal and consequently makes a poor catalyst. Gold particles, which are smaller than 20 nm, are purple in color rather than the traditional yellow color, and are capable of oxidizing carbon monoxide into carbon dioxide at room temperature.

23.1 Fundamentals of Catalysis

23.1.1 *Adsorption of a Molecule on a Catalyst Surface*

An atom in the interior of a crystallite may have as many as 12 nearest neighbors and is fully satisfied in terms of its bonding needs. However, as shown in **Figure 23.4**, surface atoms have fewer nearest neighbors. A terrace surface atom will have nine, an edge atom six, and a corner surface atom four nearest neighbors. Thus the bonding needs of these surface atoms have not been met, resulting in a higher energy at the surface than in the interior. Consequently, surfaces will form bonds with molecules that they come in contact with in order to lower the overall energy of the crystallite. If a bulk metal is cleaved to expose a new active surface, it will immediately bond to the molecules in air to satisfy its bonding needs. **Figure 23.5** [4] shows a high-resolution transmission electron microscope image of Pd nanocrystallites on a silica surface.

FIG. 23.4 *Illustration of metal crystallites on a support surface. Top views of crystalline structures on a substrate support material. The lack of nearest neighbors results in higher surface energy that translates into catalysis. Terrace surface atoms have nine nearest neighbors. Edge atoms have six and corner atoms are extremely undersaturated with only four nearest neighbors. Other forms of surface species also exist.*

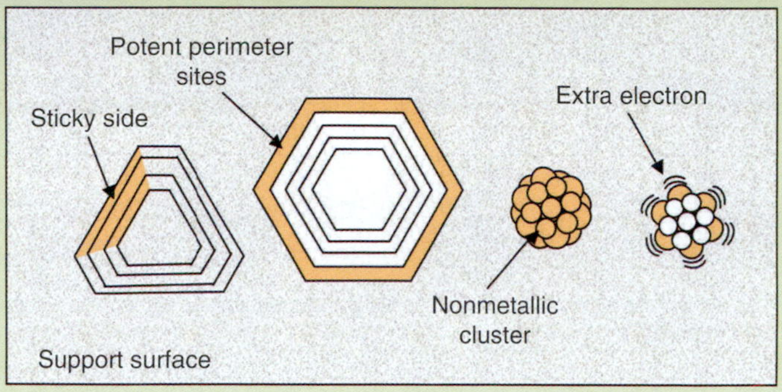

Since oxygen forms a strong bond with most metals, it reacts faster with the metal surface than the other molecules present in air, and soon a monolayer of oxygen atoms are bonded to the surface. This adsorption process is called chemical adsorption or chemisorption. The speed at which chemisorption occurs depends on the metal–oxygen bond strength for the various metals. As one might imagine, the fewer neighbors that a surface atom has, the more reactive it will be. Thus, the strength of bonding to the metal surface occurs in the order of corner atoms > edge atoms > terrace atoms. We can take advantage of this phenomenon

FIG. 23.5 *A HRTEM image of palladium nanocrystallites on a silica support.*

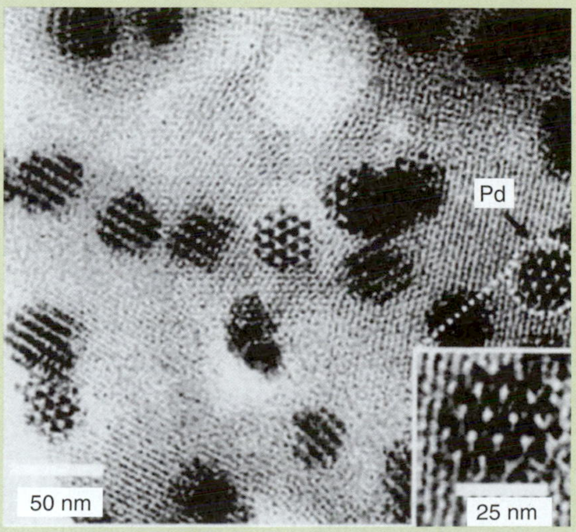

Source: H.-Y. Lee, S. Ryu, H. Kang, Y.-W. Jun, and J. Cheon, *Chemical Communication*, 1325–1327 (2006). With permission.

FIG. 23.6 *Illustration of the dissociative adsorption of hydrogen molecules into hydrogen atoms on a nickel metal surface.*

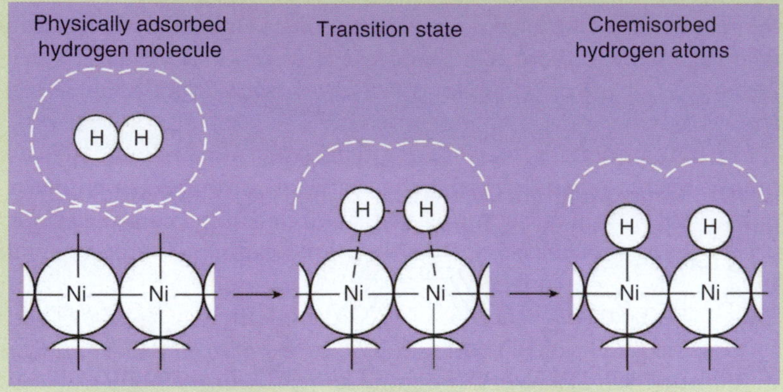

Source: G. C. Bond, *Heterogeneous catalysis:* Principles and applications, Oxford Press, Oxford, (1974). With permission.

by generating an active metal surface in the presence of reactant molecules in a closed system, rather than air. **Figure 23.6** [2] illustrates the adsorption of hydrogen onto a nickel surface. As hydrogen molecules approach the surface the atoms in the molecule begin to form hydrogen–metal bonds and simultaneously break the hydrogen–hydrogen bond, resulting in the dissociation of the hydrogen molecule into hydrogen atoms on the nickel surface, as shown in equation (23.2)

$$H_2 + 2Ni_{(S)} \rightarrow 2HNi_{(S)} \tag{23.2}$$

where
$Ni_{(S)}$ represents a surface nickel atom
$HNi_{(S)}$ represents a hydrogen atom bonded to a nickel surface atom

23.1.2 Adsorption Theory

In order for a chemical reaction to occur on the surface of a catalyst, all the reactants must adsorb onto the surface and all the products must desorb. In 1916, Irving Langmuir published his theoretical work on the adsorption of gases on solid surfaces, and in 1936 he received the Nobel Prize for his "discoveries and investigations in surface chemistry." His theory was based on the following four precepts:

1. The surface of the solid catalyst is uniform, thus all surface bonding sites are equivalent and interact with each adsorbing gas molecule in the same way.
2. The mechanism for bonding the gas to the surface is the same for every surface site.
3. Each adsorbed molecule behaves as an independent species and does not interact with other adsorbed molecules on the surface.
4. As the partial pressure of the adsorbate gas (P_A) is increased, the population of gas molecules adsorbed on the surface is also increased

until a maximum coverage is obtained. This maximum coverage is referred to as a monolayer. Thus, the moles of gas, which are adsorbed on the surface, are directly proportional to the number of adsorption sites on the surface.

Langmuir's precepts are idealistic and exceptions to the precepts are well known. Although more complex theories have evolved to address these discrepancies, his theory serves as a good model for introducing the student to the concepts of adsorption. Let's assume that a clean metal surface is placed into a closed vessel containing only the adsorbate gas and the system is allowed to reach equilibrium, and according to equation (23.3)

$$\mathrm{A_{(g)}} + \mathrm{S} \underset{k_2}{\overset{k_1}{\longleftrightarrow}} \mathrm{AS} \tag{23.3}$$

where

$A_{(g)}$ represents the adsorbate gas
S represents an unoccupied surface bonding site
AS represents a surface site occupied by or chemically bound to an adsorbate molecule

When the partial pressure of A is increased, the equilibrium is shifted to the right, resulting in a higher population of adsorbed species. The rate of adsorption (r_a) and rate of desorption (r_d) can be expressed in terms of the partial pressure of adsorbate (P_A), the adsorption rate constant k_1, the number of unoccupied sites (S), the desorption rate constant k_{-1}, and the number of occupied sites (AS), as given in equations (23.4) and (23.5).

$$r_a = k_1 P_A S \tag{23.4}$$

$$r_d = k_{-1} AS \tag{23.5}$$

At equilibrium, the rate of adsorption equals the rate of desorption and

$$k_1 P_A S = k_{-1} AS \tag{23.6}$$

or

$$AS = K_A P_A S, \qquad \text{where } K_A = k_1 / k_{-1} \tag{23.7}$$

The total number of surface sites (S_T) is equal to the number of unoccupied sites (S) plus the number of occupied sites (AS).

$$S_T = S + AS = S + K_A P_A S \tag{23.8}$$

Thus

$$\theta_A = AS / S_T = K_A P_A S / (S + K_A P_A S) \tag{23.9}$$

$$\theta_A = K_A P_A / (1 + K_A P_A) \tag{23.10}$$

Equation (23.10) is known as the Langmuir isotherm equation, and θ_A represents the fraction of surface sites occupied by adsorbent A. When θ_A is plotted

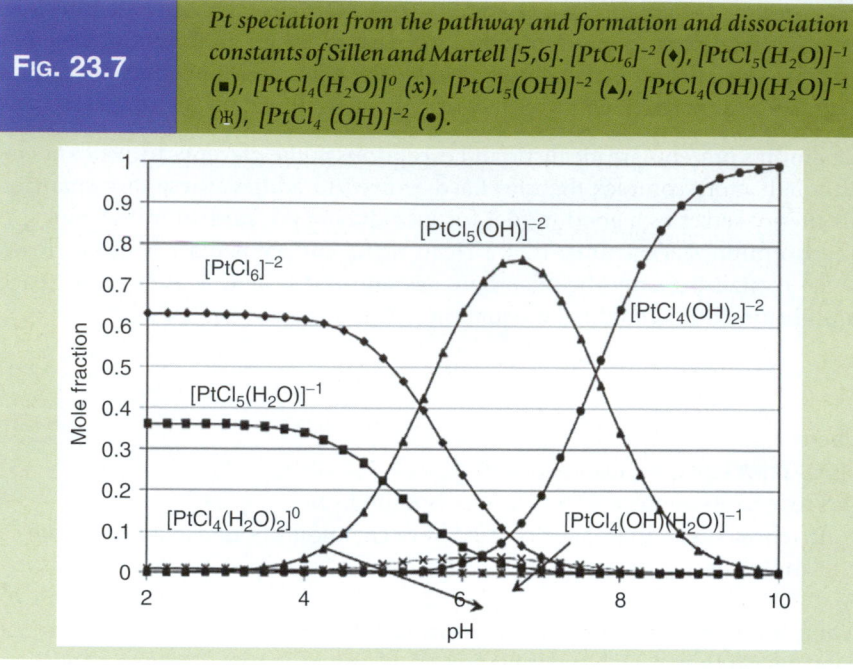

FIG. 23.7 *Pt speciation from the pathway and formation and dissociation constants of Sillen and Martell [5,6].* $[PtCl_6]^{-2}$ (♦), $[PtCl_5(H_2O)]^{-1}$ (■), $[PtCl_4(H_2O)]^{0}$ (x), $[PtCl_5(OH)]^{-2}$ (▲), $[PtCl_4(OH)(H_2O)]^{-1}$ (✱), $[PtCl_4(OH)]^{-2}$ (●).

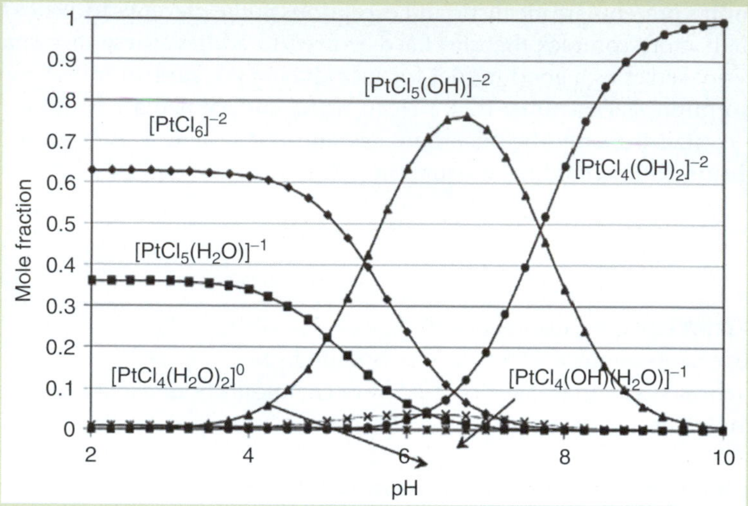

versus the P_A, a Langmuir isotherm is obtained, as shown in **Figure 23.7**. If Langmuir's precepts are valid, then a monolayer of A is formed on the surface when the P_A is large enough to make θ_A equal to 1.0. At this point no further adsorption is possible. Information about the bond strength and the effect of temperature on the adsorption equilibrium are contained in K_A. If the temperature is kept constant, then the magnitude of the K_A value reflects the strength of the adsorbate–surface bond.

In **Figure 23.7**, the shape of the adsorption curves correlates to progressively weaker adsorbate–surface bonding as $K_{A1} > K_{A2} > K_{A3}$. For example, the relative K_A values for the adsorption of hydrogen onto nickel, iron, or copper are $K_{A(Ni)} > K_{A(Fe)} > K_{A(Cu)}$. If one is interested in the hydrogenation of ethene ($H_2C{=}CH_2$) to ethane ($H_3C{=}CH_3$), as shown in equation (23.11), then nickel metal would be a good candidate and copper a poor one. Because Ni is the catalyst, it is not included in the overall reaction equation, but written over the arrow, indicating it is not consumed during the reaction.

$$H_2C = CH_2 + H_2 \xrightarrow{Ni} H_3C - CH \tag{23.11}$$

If the surface is kept constant, then the shape of the curve represents the effect of temperature on the adsorption equilibrium, that is, as the temperature is increased the adsorption equilibrium shifts further to the left, thus K_{A1} would represent the lowest temperature and K_{A3} the highest. Since most adsorptions are nearly always exothermic, the adsorbate tends to desorb from the surface as the temperature is increased.

In reality the four precepts suggested by Langmuir are not often valid. Crystallite surfaces tend to be nonuniform as shown in **Figure 23.4**. The mechanism for adsorption can vary depending on the type of sites available on the surface. For example, carbon monoxide can adsorb in a linear fashion to a single-surface metal atom or it can form a bridge across two metal atoms. In many cases, adsorbate molecules can interact with each other. Depending on the temperature and the strength of the adsorbate–surface bond versus adsorbate–adsorbate interactions, multiple layers of adsorbate can form. This kind of multiple layer adsorption is best described by the BET equation published by Stephen Brunauer, Paul Emmett, and Edward Taylor in 1938.

23.1.3 Surface Reactions

Once reactant molecules adsorb onto a surface, they must undergo a reaction to form products. Eventually the products desorb and the catalyst surface is restored to its initial chemical state (regenerated) and is ready for the next reaction cycle. The reaction conditions that produce the optimum product yield are determined by varying the reaction temperature, pressure, and catalyst contact time with the reactant. The catalyst contact time can be changed by changing the flow or agitation rate in the vessel. One of the objectives of a researcher in this field is to understand the reaction mechanism. If some knowledge of the reaction mechanism is available, then the catalyst formulation can be improved using sound scientific principles, rather than relying on a trial and error approach. Details of the surface reaction chemistry can be obtained by doing a kinetic study combined with a variety of surface and bulk analytical methods. This section focuses on the kinetic approach and section 23.3 focuses on the most common analytical methods that are used.

In his early career, Cyril Hinshelwood published his work (1921) on molecular kinetics at surfaces, using the precepts of Langmuir. Today it is referred to as the Langmuir–Hinshelwood mechanism. The overall reaction given in equation (23.12) is used to illustrate this mechanism, where $A_{(g)}$ and $B_{(g)}$ represent the gaseous reactants and $C_{(g)}$ represents the gaseous product. Although this equation represents the overall stoichiometric reaction, it does not include all the intermediate steps of the reaction.

$$A_{(g)} + B_{(g)} \rightarrow C_{(g)} \tag{23.12}$$

Hinshelwood assumed that following reaction sequence took place on the surface of the catalyst:

$$A_{(g)} + S \rightarrow AS \text{ (associative adsorption of } A \text{ onto vacant site } S\text{)} \tag{23.13}$$

$$B_{(g)} + S \rightarrow BS \text{ (associative adsorption of } B \text{ onto vacant site } S\text{)} \tag{23.14}$$

$$AS + BS \rightarrow CS + S \text{ (surface reaction)} \tag{23.15a}$$

$$CS \rightarrow C_{(g)} + S \text{ (desorption of } C \text{ from } S \text{ to regenerate vacant site)} \tag{23.15b}$$

and that the reactants A and B were associatively adsorbed, that is, they were adsorbed without the molecules breaking apart on the surface. If we assume that the adsorption steps are fast and in equilibrium and that the surface reaction is the slow or rate-limiting step, a rate expression can be written as follows:

$$\text{Rate} = k_3(AS)(BS) \tag{23.16}$$

In order for this rate expression to be meaningful, it must be written in terms that can be measured experimentally. The values AS and BS are very difficult, if not impossible, to measure under experimental conditions; therefore, the rate expression must be rewritten in terms of quantities that can be measured, such as the concentration or partial pressures of reactants and products, that is, C_A, C_B, and C_C, or P_A, P_B, or P_C. If adsorption is fast and reversible, then it reaches equilibrium quickly and can be written as follows:

$$K_A = \frac{(AS)}{(P_A S)} \text{ or } (AS) = K_A P_A S \tag{23.17}$$

$$K_B = \frac{(AS)}{(P_B S)} \text{ or } (AS) = K_B P_B S \tag{23.18}$$

$$K_C = \frac{(AS)}{(P_C S)} \text{ or } (AS) = K_C P_C S \tag{23.19}$$

Substituting for AS and BS, we can express the rate expression in terms of partial pressures of A (P_A) and B (P_B):

$$\text{Rate} = k_3(K_A P_A S)(K_B P_B S) = k_3(K_A P_A)(K_B P_B)S^2 \tag{23.20}$$

The number of vacant sites (S) is another experimental value that is very difficult to measure, but the total number of surface sites (S_T) can be measured using adsorption techniques. At any point during the reaction, S_T is equal to the sum of the vacant and occupied sites as given in equation (23.21):

$$S_T = S + AS + BS + CS = S(1 + K_A P_A + K_B P_B + K_C P_C) \tag{23.21}$$

$$S = \frac{S_T}{(1 + K_A P_A + K_B P_B + K_C P_C)} \tag{23.22}$$

$$\text{Rate} = \frac{k_3(K_A P_A)(K_B P_B)S_T^2}{(1 + K_A P_A + K_B P_B + K_C P_C)^2} \tag{23.23}$$

$$\text{Rate} = \frac{k'(P_A P_B)S_T^2}{(1 + K_A P_A + K_B P_B + K_C P_C)^2} \tag{23.24}$$

It is important to note that under the reaction conditions normally used, the adsorbed species AS and BS are mobile and move from site to site. When AS and BS are on adjacent sites and collide with sufficient energy to overcome the E_a of the reaction, then a surface reaction occurs and the product CS is formed. The example above is of course very idealized. For real reactions the molecules may actually dissociate into smaller fragments on the surface. These fragments can recombine to form new products. The structure and relative ratio of these fragments on the surface is controlled by the strength of bonding on the catalyst, the partial pressure of the reactants, and the reaction temperature. The hydrogenation of carbon monoxide serves as a good example of this. When copper is used as a catalyst material, it is unable to dissociate carbon monoxide into carbon and oxygen atoms, thus carbon monoxide adsorbs as an intact molecule on the surface. Hydrogen on the other hand dissociates into hydrogen atoms. When adsorbed hydrogen atoms collide with an adsorbed carbon monoxide molecule, a reaction occurs and methanol is produced. The net methanol synthesis reaction is shown in equation (23.25):

$$CO + 2H_2 \xrightarrow{Cu} CH_3OH \text{ (net reaction with Cu catalyst)} \quad (23.25)$$

A simplified mechanism is given below:

$$CO + S \rightarrow COS \text{ (associative adsorption of CO)} \quad (23.26)$$

$$H_2 + 2S \rightarrow 2HS \text{ (dissociative adsorption of } H_2\text{)} \quad (23.27)$$

$$COS + 2HS \rightarrow H_2COS + S \text{ (surface reaction)} \quad (23.28)$$

$$H_2COS + 2HS \rightarrow CH_3OHS + S \text{ (surface reaction)} \quad (23.29)$$

$$CH_3OHS \rightarrow CH_3OH + S \text{ (desorption of methanol product)} \quad (23.30)$$

On the other hand, if nickel is used as the catalyst then a different outcome is obtained and carbon monoxide and hydrogen are converted into methane. Nickel can dissociate (break apart) the CO molecule on the surface and copper cannot. Both metals can dissociate hydrogen into hydrogen atoms. Hydrogen is so strongly adsorbed onto nickel that each time a carbon monoxide molecule adsorbs and breaks apart it is surrounded by a very large population of hydrogen atoms and the most probable reactions are multiple collisions with hydrogen atoms to eventually form methane. The adsorbed oxygen atom shares the same fate and is quickly converted into water. The net methane synthesis reaction is shown in equation (23.31) and a simplified mechanism is presented in equations (23.32)–(23.37).

$$CO + 3H_2 \xrightarrow{Ni} CH_4 + H_2O \text{ (net reaction with Ni catalyst)} \quad (23.31)$$

$$CO + 2S \rightarrow CS + OS \text{ (dissociative adsorption of CO)} \quad (23.32)$$

$$H_2 + 2S \rightarrow 2HS \text{ (dissociative adsorption of } H_2\text{)} \quad (23.33)$$

$$CS + 2HS \rightarrow H_2CS + S \text{ (surface reaction)} \quad (23.34)$$

$$H_2CS + 2HS \rightarrow CH_4 + 2S \text{ (formation and desorption of } CH_4\text{)} \quad (23.35)$$

$$2HS + OS \rightarrow H_2OS + S \text{ (surface reaction)} \quad (23.36)$$

$$H_2OS \rightarrow H_2O + S \text{ (desorption of water)} \quad (23.37)$$

Iron can also dissociate carbon monoxide into carbon and hydrogen atoms, but it does not adsorb hydrogen nearly as well as nickel. Therefore, carbon species have an opportunity to collide with each other as well as with hydrogen atoms. This process is known as the Fisher–Tropsch synthesis of hydrocarbons and is currently being used to synthesize gasoline and other hydrocarbon products at the Sasol plant in South Africa.

$$n\text{CO} + (n+2)\text{H}_2 \xrightarrow{\text{Fe}} (\text{CH}_2)_n + \text{H}_2\text{O} \text{ (net reaction with Fe catalyst)} \quad (23.38)$$

$$\text{CO} + 2\text{S} \rightarrow \text{CS} + \text{OS} \text{ (dissociative adsorption of CO)} \quad (23.39)$$

$$\text{H}_2 + 2\text{S} \rightarrow 2\text{HS} \text{ (dissociative adsorption of H}_2\text{)} \quad (23.40)$$

$$\text{CS} + 2\text{HS} \rightarrow \text{H}_2\text{CS} + \text{S} \text{ (surface reaction)} \quad (23.41)$$

$$\text{H}_2\text{CS} \rightarrow \text{SCH}_2\text{CH}_2\text{S} \text{ (surface reaction)} \quad (23.42)$$

$$n\text{H}_2\text{CS} \rightarrow \text{S(CH}_2)_n\text{S} + (n-2)\text{S} \text{ (surface reaction)} \quad (23.43)$$

$$\text{S(CH}_2)_n\text{S} + 2\text{HS} \rightarrow \text{CH}_3(\text{CH}_2)_{(n-2)}\text{CH}_3 \text{ (product formation and desorption)} \quad (23.44)$$

As one can see, the proper selection of catalyst materials and reaction conditions is critical in determining the reaction pathway and thus the end product.

23.2 Synthesis

23.2.1 *Synthesis Requirements*

The primary synthetic objectives, for a commercially viable catalyst, are to make a catalyst that is highly active, highly selective for the desired product, mechanically durable, thermally stable, long lived, and cost-effective. It can be quite a challenge to meet all of these requirements. To illustrate this philosophy, let's examine the components of a catalytic converter on an automobile to reduce exhaust emissions. The catalytic converter consists of a ceramic monolith or honeycomb, as shown in **Figure 23.6**, that is coated with a fine powder of highly porous (~150 $m^2 \cdot g^{-1}$) gamma aluminum oxide, which is also known as gamma-alumina (γ-Al_2O_3). The ceramic honeycomb gives mechanical strength to withstand the vibrations and thermal variations experienced inside the catalytic converter. Its open structure provides uniform flow and good contact with the combustion gases exiting from the engine. However, it has a very low surface area and surface features that don't permit good dispersion or adhesion of the active catalyst components, which are platinum, palladium, and rhodium. The alumina is added because it bonds well to the ceramic and provides a surface that gives good binding and dispersion of the active catalyst metals. Dispersion means the metal or metal oxide crystallites of the active catalytic material are small and is defined by the following equation:

$$\%D = \frac{M_S}{M_T} \times 100 \quad (23.45)$$

where

$\%D$ represents the percent dispersion

M_S represents moles of metal atoms on the particle surface

M_T is the total moles of metal atoms in the particle

For nanocatalyst particles the value of $\%D$ approaches 1, that is, nearly every atom in the particle is exposed to the surface. Since only atoms on the surface can participate in a catalytic process, a high metal dispersion corresponds to a high catalytic activity.

In this example, the active catalytic components are platinum, palladium, and rhodium. These are very expensive elements. If a company wishes to recover the cost of the initial expense of the catalyst, it is important that as many platinum atoms as possible contribute to making the desired product. For a 200-μm-sized platinum metal particle, only the atoms on the surface contribute to the reaction chemistry. However, in this case there are more platinum atoms in the interior of the particle than there are on the surface. The interior atoms cost money, but provide no catalyst benefit to the user. The smaller the catalyst particle the higher its effectiveness, that is, the more atoms that are on the surface participating in the reaction, and the catalyst pays for itself in a shorter time.

One of the objectives of catalyst synthesis is to generate active catalyst particles that have diameters in the nanometer range. Such small particles will have nearly every atom exposed to the surface. Although this is the desired outcome, it is difficult to achieve because atoms on the surface are higher in energy than those on the interior. The more atoms that are exposed to the surface, the higher the overall energy of the particle. Thermodynamics dictates that larger particles are more stable than smaller ones. Thus thermodynamics drives the synthesis in the direction of larger particles, which is opposite to the desired objective. The challenge is to prepare a catalyst with nanosized particles that are stable under the conditions used during the reaction. The next section will discuss how that is done.

23.2.2 Example of a Conventional Synthetic Technique

There are many ways to prepare a catalyst with nanosized particles. This section will focus on the more common methods for making a commercial catalyst material. There are four traditional ways for preparing a catalyst. The first and most common method is called the impregnation or incipient wetness technique. In this method, an aqueous solution containing the soluble metal salt of the active catalyst, such as $(NH_3)_4Pd(NO_3)_2$, $PtCl_4$, $PlCl_2$, $Ni(NO_3)_2$, $HAuCl_4$, etc., is coated (impregnated) onto the surface of a highly porous support material, such as alumina (Al_2O_3), silica (SiO_2), or silica–alumina (SiO_2–Al_2O_3). The catalyst is dried at ~120°C to remove excess water and then heated to ~500°C in air, also known as calcining, to decompose the metal salt into its active metal or metal oxide form. For example, platinum and gold decompose directly into the metal form under calcining conditions, while palladium and nickel form a metal oxide. Different forms of the catalyst are used for different types of reactions. For example, a hydrogenation reaction requires the metal form and an oxidation reaction usually requires the oxide form. If the metal form of the catalyst is required, then it can be reduced to the metal form in a subsequent step using a reducing agent, such as hydrogen.

The key to forming metal or metal oxide particles in the nanometer range lies in the details of the preparation. The chemistry of the support surface plays a very important role in generating such small catalyst particles. When exposed to an aqueous solution, a porous metal oxide support has both positive and negative charges on its surface. The relative population of positive or negative charges depends upon the pH of the solution and the isoelectric point (IEP) of the support. The IEP is defined as the pH of the solution where the population of positive and negative charges is equal and the net charge on the surface is zero. The IEP for silica is zero at a pH of approximately 2, while that of γ-Al_2O_3 is zero at a pH of approximately 7. If one is making a catalyst that requires platinum, then a soluble platinum salt, such as $(NH_3)_4Pt(NO_3)_2$ or H_2PtCl_6, is used. When the salt is dissolved into water, ions as shown in equations (23.46) and (23.47) are formed.

$$(NH_3)_4Pt(NO_3)_2 \rightarrow (NH_3)_4Pt^{2+} + 2(NO_3)^- \qquad (23.46)$$

$$H_2PtCl_6 \rightarrow [PtCl_6]^{2-} + 2H^+ \qquad (23.47)$$

In the first example, the platinum ions have a positive (+2) charge and in the second they have a negative (–2) charge. The chloroplatinic acid (H_2PtCl_6) solution has a pH around 2.5, which means the silica surface is slightly negatively charged. Consequently, the negatively charged platinum compound $[PtCl_6]^{-2}$ tends to aggregate with itself rather than be dispersed on the surface, resulting in metal particles that are larger than the desired nanoparticle range. When alumina is used as the support, the surface is predominantly positively charged at a pH of 2, and thus the platinum ions are highly dispersed on the alumina surface, rather than aggregating into metal salt crystallites on the surface. As the catalyst is dried and calcined very small metal particles result. A similar result can be obtained for a silica support by modifying it with a small amount of lanthanum oxide (lanthana). The resulting surface has an IEP value of zero at a pH of approximately 6. In the case of $PtCl_4$, the nature of the dissolved species and its charge are pH dependent (see **Fig. 23.7**).

It is easily seen that the IEP of the support surface and the charge of the metal ion in solution are factors that can be manipulated to optimize the dispersion of the metal on the surface during preparation of the catalyst and can lead to the generation of nanosized particles. Other preparation methods include coprecipitation, ion exchange, and chemical vapor deposition. Details of these synthetic methods are presented in other references.

23.2.3 Nontraditional Methods for Preparing Nanocatalysts

The problem with conventional catalyst preparations is the lack of uniformity of the catalyst particles produced. One obtains a range of particle sizes that is solely dependent on the random aggregation of metal atoms on the support surface. Newer methods of catalyst preparation permit the formation of nanometal clusters of a well-defined size and composition, which results in a material which demonstrates a highly size-dependent catalyst activity and selectivity. The control of particle size introduces a new dimension of control in the catalyst performance that is not achievable using traditional synthetic methods. These techniques involve atomic layer deposition, laser ablation, and high vacuum techniques. For example, nanosized clusters of metals or metal alloys can be generated using vacuum techniques and mass spectrometry to selectively deposit metal clusters of a uniform size. These clusters demonstrate unique adsorption and catalytic properties and provide the basis for obtaining a better understanding of how nanosized particles affect the fundamental catalytic performance.

Recent work at the University of Georgia and the Technical University of Munich have discovered that gold nanoparticles are negatively charged. This phenomenon is thought to be responsible for aiding the low temperature adsorption and oxidation of carbon monoxide into carbon dioxide. A conventional catalytic converter must be heated by the engine exhaust to a temperature, called the light-off temperature, where the temperature is sufficient to sustain the oxidation reactions needed for pollution control, especially hydrocarbons and carbon monoxide being oxidized to less toxic substances, such as carbon dioxide and water. A significant amount of pollution occurs between the cold start-up of the engine and the catalyst light-off time. A low temperature catalyst would make a tremendous impact on this problem.

Although new nanocatalyst materials have tremendous economic potential, there are still many problems to be resolved. Uniform nanocatalyst particles tend to be expensive to produce and are often unstable at the reaction conditions normally employed in many catalytic processes. New nanomaterials have provided the scientific community with a much improved understanding of the fundamental workings of a catalyst. Incorporating this new knowledge into an economically viable catalytic material is the challenge to be met by current and future scientists.

23.3 Catalyst Characterization

23.3.1 Overview

It is important to know the bulk and surface composition of a catalyst once it has been prepared. For a new catalyst formulation, this important information allows the researcher to accurately reproduce the synthesis and to better understand how the catalyst actually performs its catalytic function. With such knowledge, the researcher can use scientific principles to design a better catalyst in the future, rather than relying on trial and error methods of discovery. Some industrial catalysts have gradually evolved over decades. However, the introduction of modern analytical methods has greatly shortened the development time for new catalyst formulations that are more active, selective, and cost-effective.

Catalysts usually have a limited lifetime and begin to show a gradual or even rapid deactivation under certain reaction conditions. Analytical methods can provide valuable information about the cause of this deactivation. If the problem is identified, then new formulations can be designed to prolong the catalyst lifetime or reaction conditions can be selected to avoid or slow down the deactivation process. For example, platinum or palladium nanoparticles can react with trace impurities in the reaction feedstock, such as hydrogen sulfide, to form a stable platinum or palladium sulfide surface that is catalytically inactive. Surface analysis can identify the chemical culprit and then the researcher can take appropriate action to minimize or eliminate this problem. Today there are numerous analytical techniques available to scientists and engineers in this field. In the interest of brevity, only a few of the more commonly used methods will be discussed here.

23.3.2 Bulk Characterization Techniques

Although all of the chemistry takes place at the solid–gas or solid–liquid interface of the catalyst surface and the reaction medium, it is useful to know the bulk composition of the catalyst as well. There are a number of important questions that need to be answered. What is the bulk formula of the catalyst? Is the solid crystalline or amorphous? If it is crystalline, what is the crystalline phase? Is a particular crystalline phase necessary for the catalyst to express its unique activity? What effect does the size of the nanoparticle has on the catalyst performance? As mentioned previously, gold has always been considered a poor catalyst. It is used as jewelry and in many scientific applications where an inert surface is required, because its surface is so stable. However, nanoparticles of gold are able to participate in chemical reactions that larger particles cannot.

Three important bulk characterization techniques will be discussed here. They are inductively coupled plasma (ICP), x-ray diffraction (XRD), and high-resolution transmission electron microscopy (HRTEM).

Inductively Coupled Plasma. ICP is a method where the sample is dissolved in acid or base to make a solution of known concentration. The solution is fed into a plasma flame and atomized. The excited atoms or ions emit radiation that is characteristic of that element. A detector identifies the emission frequencies, a computer data collection system identifies the elements that produced those frequencies, and the signal intensity is used to determine the amount of each element present in the original sample. This information is used to determine the amount of each element present in the catalyst sample and to identify any trace elements that may be present. The technique can detect most elements in parts per million or parts per billion amounts.

X-ray Diffraction. XRD is used to determine the crystal structure of the bulk material and in some cases can be used to show changes in catalyst particle size. It has a limit of about 5 nm, so it is not useful for determining the size of nanoparticles. However, it is still a useful tool in observing the success of a preparation. **Figure 23.8** [11] shows the XRD pattern for a catalyst prepared using $(NH_3)_4Pt(NO_3)_2$ and $PlCl_4$. The SiO_2 support has been treated with La_2O_3 to change the IEP of the surface from approximately 2.0 to 5.3. This means the SiO_2 support surface has a net positive charge for the pH of the solution used during the catalyst preparation. In the case of the $(NH_3)_4Pt(NO_3)_2$ compound, the platinum species has a positive charge and is poorly adsorbed onto the support surface, resulting in aggregation and the production of large Pt particles in the finished catalyst. The XRD pattern shows two peaks that are characteristic of Pt metal on the support surface and with an average particle diameter of

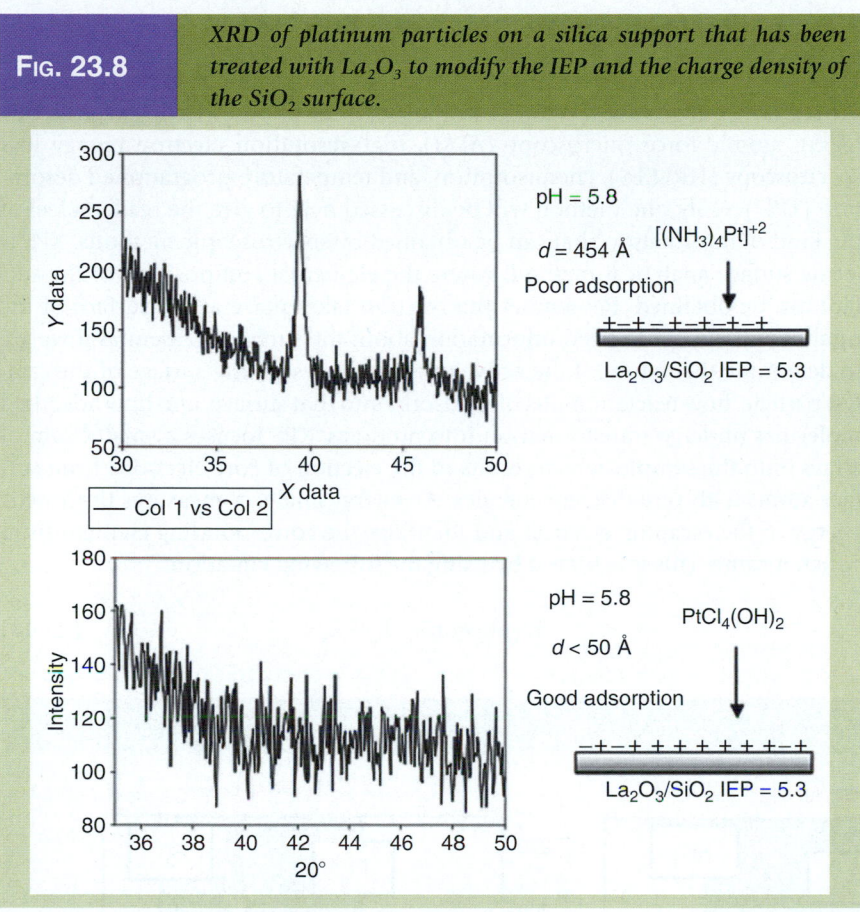

Fig. 23.8 *XRD of platinum particles on a silica support that has been treated with La_2O_3 to modify the IEP and the charge density of the SiO_2 surface.*

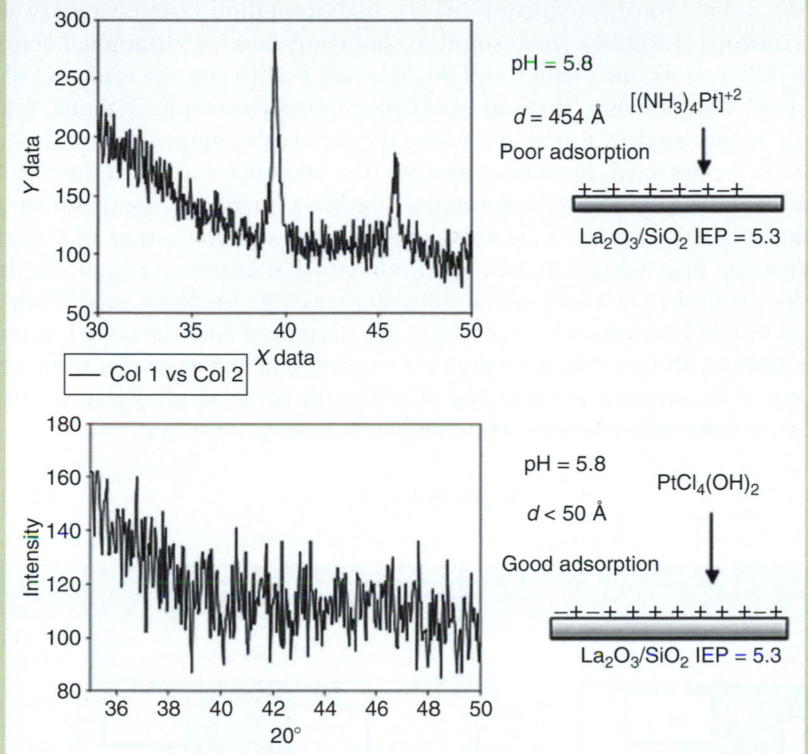

Source: S. W. Cowley et al., Unpublished results (2007). With permission.

about 450 Å. The use of $PtCl_4$ results in a negatively charged platinum species in solution, resulting in good adsorption on the surface, and generates Pt metal particles smaller than 5 nm, which cannot be observed by XRD. Notice the absence of Pt peaks in the XRD data.

High-Resolution Transmission Electron Microscopy. HRTEM can be used to observe metal particles on a variety of metal oxide supports. The dark spots in **Figure 23.3** represent small particles of Rh metal on a silicon oxide support. By measuring the diameter of the metal particles, as compared to the 30 nm scale in the micrograph, one can obtain a good idea of the average particle diameter in the sample. This technique is employed to obtain size information about nanosized particles. By using a higher-energy electron beam, much higher resolution is obtained and details of the crystal structure or arrangement of the atoms in the nanoparticles can be seen. **Figure 23.5** is an example of a HRTEM image. The nanoparticles appear as rafts of atoms on the support surface rather than a sphere-like particle. This shape presents a highly active surface and explains the higher activity often observed for nanosized metal particles.

23.3.3 Surface Characterization Techniques

There are a large number of surface techniques that can be used to characterize the surface of the catalyst. A few examples are x-ray photoelectron spectroscopy (XPS), atomic force microscopy (AFM), high-resolution electron energy loss spectroscopy (HREELS), chemisorption, and temperature-programmed desorption (TPD). Only one method will be discussed here to give the reader a feel of the kind of information that can be obtained by spectroscopic methods. XPS is a true surface analytical method, where the elemental composition of the surface can be obtained. Remember the reaction takes place at the surface of the catalyst particle, so detailed information about the surface is essential if we are to determine the nature of the active catalyst species on the surface of the catalyst particle, how reactant molecules adsorb onto that surface, and how adsorbed molecules undergo transformation into products. XPS focuses a small beam of x-rays onto the sample, which results in the ejection of core electrons from surface atoms with very discrete energies. An energy analyzer measures the kinetic energy of the escaping electron and identifies the corresponding element from which it came. This is achieved by using the following equation:

$$h\nu(\text{photon}) = E_B + E_K \tag{23.48}$$

FIG. 23.9 *(a) XPS spectra of $(NH_3)_4Pt(NO_3)_2$ adsorbed onto SiO_2 and (b) La_2O_3-treated SiO_2.*

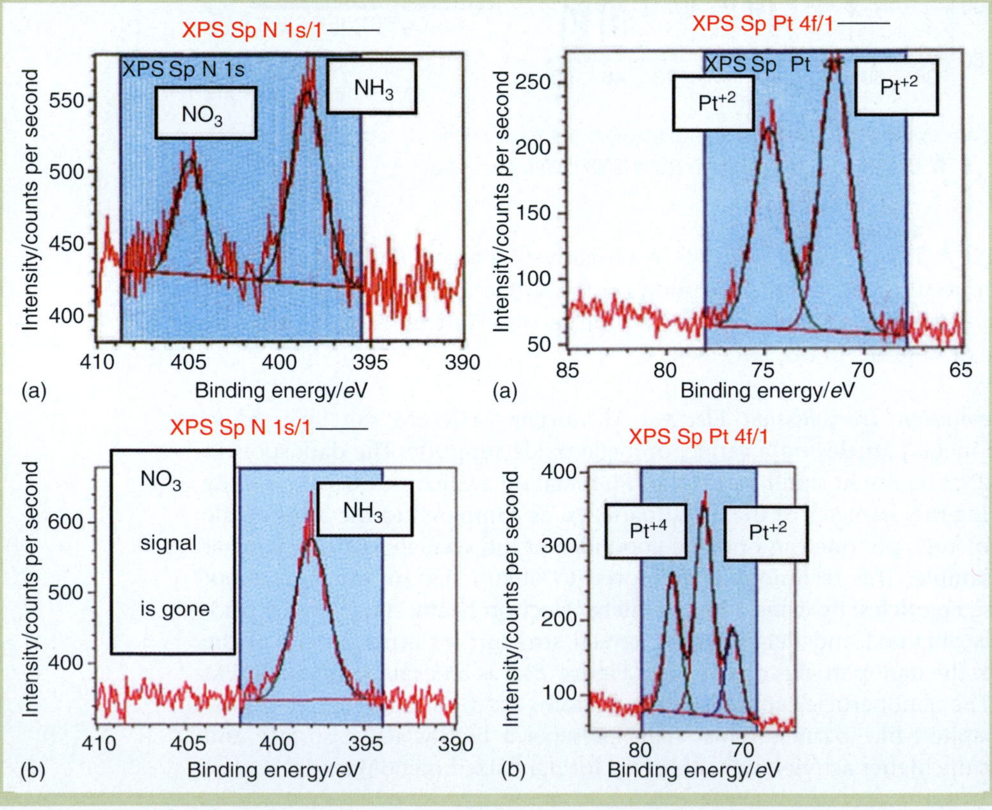

where

$h\nu$ is the x-ray photon energy that is transferred to a core electron in a surface atom

E_B is the binding energy of the atom

E_K is the kinetic energy of the electron after it leaves the atom

E_B is the energy required to remove a core electron from an atom to a distance where the electron can no longer feel the attractive force of the nucleus. E_K is the residual kinetic energy of the electron after it has broken free of the binding force of the atom. Only electrons located in the top few layers of the sample can escape their host atoms and reach the surface without undergoing collisions with other atoms and loosing some of the original photon energy transferred to them by the x-ray beam. The discrete kinetic energy (E_K) of the escaping electrons is measured and quantified by the XPS instrument. Since $h\nu$ and E_K are now known, the E_B value can be calculated. Each atom has a unique E_B value and in this fashion the elements present on the surface and their amounts are determined. In addition, small changes in the E_B give clues as to the oxidation state of the metal. **Figure 23.9** shows important chemical changes that occur during the synthesis of platinum nanoparticles on a SiO_2 surface and a La_2O_3-treated SiO_2 surface.

This information is used to further our understanding of the chemical transformation responsible for the formation of nanosized catalyst particles and how those particles function as catalysts. For more detail concerning catalysts, please refer to the excellent sources listed in the references section of this chapter [7–10].

Acknowledgments

Scott W. Cowley is an associate professor in the Department of Chemistry & Geochemistry at the Colorado School of Mines in Golden, Colorado. His research interestes include heterogeneous catalysis, surface science and analysis, surface analysis, solid sorbents for environmental applications, oxide film growth and corrosion, xerography and laser printer materials, and photoconducting films. Dr. Cowley was program co-chair of the North American Catalysis Society in 1995, chairman of the Rocky Mountain Fuel Society (1991–1995), and chairman of the Western States Catalysis Club (1996–2000).

Scott Cowley received a BS in chemistry from Utah State University in 1967, and a MS in physical organic chemistry from Utah State in 1972, under the direction of Grant Gill Smith, studying the "High Temperature Thermal Decomposition of Organic Compounds." He received a PhD in 1975 from Southern Illinois University in physical organic chemistry, under the direction of Gerard V. Smith studying the "Mechanism for the Catalytic Hydrodesulfurization of Thiophene." Dr. Cowley then served in a postdoctoral position at the University of Utah working on catalysis with Frank Massoth studying the "Mechanism of Catalytic Hydrodesulfurization and Catalytic Fischer-Tropsch Synthesis of Olefins." During 1976–1979 he was an assistant professor of fuels engineering at the University of Utah. He has been at CSM in the Department of Chemistry since 1979.

References

1. R. J. Farrauto and C. H. Bartholomew, *Fundamentals of industrial catalytic processes*, John Wiley and Sons, New York (2005).
2. G. C. Bond, *Heterogeneous catalysis: Principles and applications*, Oxford Press, Oxford (1974).
3. S. Chakraborti, A. K. Datye, and N. J. Long, Oxidation reduction treatment of rhodium supported on non-porous silica spheres, *Journal of Catalysis*, 108, 444–451 (1987).
4. H.-Y. Lee, S. Ryu, H. Kang, Y.-W. Jun, and J. Cheon, Selective catalytic activity of ball-shaped Pd@MCM-48 nanocatalysts, *Chemical Communications*, 1325–1327 (2006).
5. L. G. Sillen and A. E. Martell, The stability constant of metal ion complex, Special Publication No.25 (Suppl.1), The Chemical Society, Burlington House, London (1971).
6. W. A. Spieker, J. Liu, J. T. Miller, A. J. Kropf, and J. R. Regalbuto, An EXAFS study of the coordination chemistry of hydrogen hexachloroplatinate(IV): 1. Speciation in aqueous solution, *Applied Catalysis A: General*, 232, 219–235 (2002).
7. W. A. Spieker, J. Liu, X. Hao, J. T. Miller, A. J. Kropf, and J. R. Regalbuto, An EXAFS study of the coordination chemistry of hydrogen hexachloroplatinate(IV): 2. Speciation of complexes absorbed onto alumina, *Applied Catalysis A: General*, 243, 53–66 (2003).
8. J. C. Vickerman, ed., *Surface analysis—The principle techniques*, John Wiley & Sons, Chi Chester, New York, Weinheim, Brisbane, Singapore, Toronto (1997).
9. D. T. Wickman, B. W. Logsdon, S. W. Cowley, and C. D. Butler, A TPD and XPS investigation of palladium on modified alumina supports used for the catalytic decomposition of methanol, *Journal of Catalysis*, 128, 198–209 (1991).
10. S. Youngwilai, Ph.D. Thesis, The chemical modification of silica surfaces to enhance the dispersion and activity of platinum metal crystallites, Department of Chemistry, Colorado School of Mines, Golden (2003).
11. S. W. Cowley et al., Unpublished results (2007).

Nanocomposites and Fibers

Make the workmanship surpass the materials.

Ovid
Metamorphoses, 1st Century B.C.

Human wisdom is the aggregate of all human experience, constantly accumulating, selecting and reorganizing its own materials.

John Story

Chapter 24

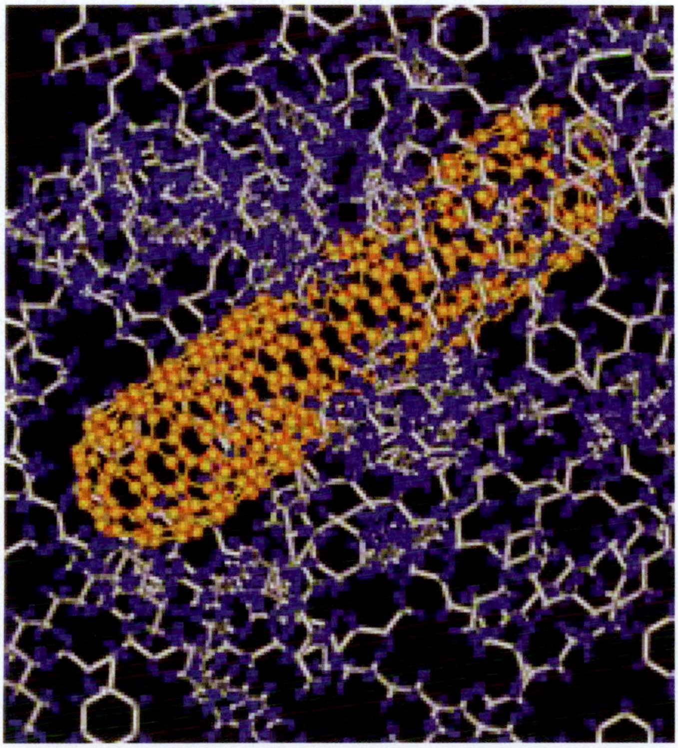

Threads

Chapter 24 is the second chapter in the *Chemical Nanoengineering* division of the text, following our discussion of catalysis in *chapter 23. Nanocomposites* are a great part of the new wave of nanomaterials—inorganic, organic, biological, and any combination and permutation thereof. Most materials in nature are composites. Composite materials permeate our synthetic products as well. The reason for composites is straightforward: combinations of pure materials and phases result in a material with superior performance. The reason for nanomaterials is also straightforward: inclusion of nanoscale materials in bulk material matrices result in composites with superior properties. We provide a brief review of engineering materials, engineering properties, and mechanical, electrical, and thermal test methods. All of these categories apply to nanomaterials and to nanomaterials embedded within a matrix.

The importance of composite materials transcends our technology. There is not much that isn't a composite material. Extending this theme further, there is not much in biology that is not made of composite materials, more specifically, nanocomposite materials. Following this chapter, the biologically inclined should feel at home in the next four chapters. Bio-nano (or nano-bio) technology and chemistry, biomimetics, and nanomedical topics are presented. The final chapter, *chapter 28*, addresses environmental applications of nanotechnology to round off the text.

24.0 Nanocomposites and Fibers

It will not be through our development of new materials that lift our technology to the next level but rather through combinations of new materials. Composite materials are able to dodge the tradeoffs that restrict the performance of our traditional engineering materials. The addition of nanoscale materials to composites furthers this trend by enabling new materials that are able to dodge the tradeoffs that restrict the performance of our traditional composite engineering materials. Polymeric composites influenced our technology (and way of life) decades ago just as nanocomposites are expected to influence our technology and way of life for decades to come.

Nanocomposites consist of bulk material matrix with a reinforcing phase of one or more nanomaterials. In general, mechanical testing of nanocomposites has shown mixed results [1]. Because nanomaterials are small, the high surface-to-volume ratio and concomitant high surface energy (e.g., high interfacial energy) should promote facilitative bonding with the matrix material. Such enhanced bonding should afford the composites filled with nanomaterials superior performance over those filled with micrometer-sized materials [2]. The interphase layer plays a critical role in the mechanical property performance of the nanocomposite. If, for example, in the interphase region, the polymer chain has limited capacity to bond with the nanoparticles, then the density of the interphase region is expected to be lower than that of the bulk matrix. The condition is exacerbated if a high volume fraction of filler is required. One of the prime driving forces behind the use of nanomaterial fillers is the overall reduction in the amount of filler required. Micrometer-sized fillers, for example, require V_f above 20% in order to be effective. Therefore, the objective of composite researchers is two-fold: (1) reduction of the amount of filler, and (2) enhancement of bonding between the matrix material and the nanomaterial inclusion. Both numbers one and two are obtainable with nanotechnology.

Nanotechnology impacts both the inclusion and the host. What is it about nanoparticles (zero-, one-, and two-dimensional) that make them excellent inclusion materials in composites? Then, why do polymers, metals, or ceramics made of nanoscale building blocks and then transformed into a macroscopic material demonstrate superior properties? How do nanomaterials affect the structure–property relationship between the filler and matrix? Traditional polymer chemistry has taught us about the relationship between molecular weight, chain length, architecture, fillers, ordering, functional groups, and scaffolds with solubility, rheology, mechanical properties, and other chemical behavior. Also the effects of solvent, size, chemistry, and fabrication conditions are relatively well understood for polymers. What then happens to such baseline polymers when nanomaterials are inserted? How do mechanical, physical, and chemical properties respond to nanometer-sized inclusions of varying size, shapes, compositions, and orientations—whether in the form of particles, tubes, porous materials, discs, wires, or blocks?

Polymer nanocomposites by definition are multicomponent systems in which the major component is a polymer and the minor component is a nanomaterial with one dimension below 100 nm [3]. Properties of nanocomposites filled with small levels of nanoparticle fillers (1%–5%) are comparable or better than materials that contain a larger volume fraction of similar but conventional-sized materials (15%–40% or more) [3]. Smaller amounts of fillers in nanocomposites enhance processing and reduce overall weight of the composite. Other properties of nanomaterial-filled composites include enhanced optical clarity, self-passivation, and increased flame retardation, oxidation, and ablation resistance [3]. Automotive parts, coatings, and flame retardants are some commercial avenues already utilizing the unique properties of these nanocomposites [3]. Nanoscale building blocks are individually remarkable because of several reasons: they are free of defects, reactive, and have unusual physical properties. It is however not a simple task to keep these remarkable properties intact in a macroscale composite [4].

Please note that we will not focus on nanocomposite materials in general. In this chapter, for example, we do not intend to discuss SAMMS (self-assembled monolayers on mesoporous substrates) or OPVs (organic photovoltaic solar cells) although these are most certainly considered to be nanocomposite materials. We do focus on traditional polymer matrices that incorporate nanoscale materials. As a second note, we limit our discussion to composite mechanical properties although brief excursions are made into thermal, electronic, and optical domains, especially when presenting literature and data concerning carbon nanotubes.

Most polymers are either thermoplastics (e.g., heavily cross-linked long-chain molecules like moldable polyethylene that softens on heating), thermosets (e.g., epoxy resins that are 3-D covalent bonded network, light-weight, rigid, thermally stable composites that harden upon application of heat), elastomers and rubbers (e.g., linear memory polymers, occasionally cross-linked), and natural polymers (e.g., cellulose, lignin, and proteins).

24.0.1 Background

A composite material (from the Latin *compositus* "to put together," from *com* "together" + *ponere* "to place") is an intimate combination of two or more materials that differ in composition, structure, and form. A composite material is a heterogeneous engineering material with superior properties in which one phase is dispersed within another—the phases bonded mechanically or chemically. For example, the tempering of steel with highly dispersed finely divided carbides imparts an unusual combination of strength and toughness to the steel—resulting in an in situ composite with fine structure [5]. Although the constituents of a composite act together, they act together to perform unique mechanical, chemical, thermal, optical, or electrical tasks that individually would not be possible. Structural concrete, for example, consists of steel reinforcement embedded within the concrete matrix—the steel bars are responsible for mitigating tension loads while the concrete is responsible for mitigation of compression loads.

A composite material exhibits a combination of functions and materials working in concert. Most composites have a reinforcing element, (or filler) dispersed within a matrix (or binder). In traditional composites, distinct identifiable interfaces exist that retain the chemical identity of each component and do not merge into one another. In other words, the phases of a composite do not react with or dissolve into each other. The reinforcing element is usually a fiber (e.g., made of glass or carbon). The direction of the reinforcement or property vector is also a factor in the design of the material. Reinforcement, for example, can also be made to be isotropic or anisotropic. Thermal or electrical conductivity can be made to be isotropic or anisotropic. The same is true for optical properties.

Constituents of a composite technically should not dissolve or blend within each other. In other words, each constituent within a composite retains its original identity—steel remains steel, concrete remains concrete, carbon fibers remain as carbon fibers. Composites, therefore, are distinctly different from the pure classes of engineering materials like metals and semiconductors, ceramics and glasses, and polymers and plastics. Composite materials consist of one or more material that is integrated to form a unique material—one that demonstrates enhanced properties above and beyond those of its individual components. In addition to the preexisting properties of the dispersed component and the matrix, the physical properties of the composite depend on the geometry, size, shape, roughness, and orientation of the dispersed component and the nature of the interface between the minority inclusion and the host matrix material.

Brief History of Synthetic Composites. Perhaps one of the first synthetic composite materials was the straw-reinforced mud brick developed by early humans. The straw added tensile properties while the bricks provided mass and thermal and compression resistance—an excellent combination for a dwelling. Modern reinforced concrete is based on the same concept although component material physical properties have certainly been improved over several millennia. The Mongols of the twelfth century developed small and powerful bows by combining several natural products: tendons (for tension) and horn sheaths from cattle (to prevent compression), bamboo (core structure), and silk (soaked in resin to wrap the structure). The strength of the Mongol bows compared favorably to modern composite bows.

In the mid to late 1800s, chemists developed the first polymers and invented cross-linked synthetics such as celluloid, melamine, and Bakelite. Polyester arrived on the scene in the 1930s and glass-reinforced polymer composites soon followed. Carbon black and fumed silica have been added to components in plastics since the invention of tires. Embedding nanoparticles in tires has improved their performance. Starting with larger embedded particles and proceeding to smaller inclusions, tensile strength was improved. The following particle inclusion size with tensile strength (in parentheses) demonstrates this trend. The tensile strength is given in parentheses: 250–350 nm (10 MPa), 70–100 nm (15 MPa), 30–35 nm (22 MPa), and 20–25 nm (25 MPa).

Roger Bacon in 1958 at the Parma Technical Center of Union Carbide near Cleveland, Ohio was the first to fabricate a high-performance carbon fiber [6]. He experimented with arc-discharge fabrication techniques to produce carbon whiskers a few to greater than 5 μm in diameter and ca. 3 cm in length. The structure of the whiskers revealed a hierarchy of concentric tubes, each in the form of a rolled up scroll (e.g., a multiwalled carbon nanotube). Fibers made of the whiskers demonstrated tensile strength of 20 GPa and Young's modulus on the order of 700 GPa. The fibers consisted of continuous graphite sheets rolled into scrolls to make a filament. The whiskers were grown under conditions of 92 atm of Ar at 3900 K in a DC electric arc. The electrical room temperature resistivity of the tubes was 65 μΩ·cm—a value typical for crystalline graphite [7]. Although T.A. Edison baked cotton and other organic filaments at high temperature to produce carbon filaments back in the late 1870s and that the U.S. Navy produced carburized fabric from rayon and cotton for missile heat shields in 1957, Bacon was the first person to unwittingly produce carbon nanotubes.

Heat treatment of rayon at temperatures as high as 3000 K yielded high performance fibers and cloths. Stretching the polymer fabric at high temperatures (e.g., hot stretching) resulted in orienting the graphitic layers along the long axes of the fibers. By 1965, fibers (like Thornel 25) with Young's modulus of ca. 170 GPa were attained. Carbon fiber reinforced plastic (CFRP) composites consist of layers of carbon fiber cloth in a mold that is in the shape of a structure (e.g., an aircraft wing component). The orientation and weave of the carbon fiber cloth is selected to optimize mechanical properties of the final product. The mold is then filled with an epoxy polymeric material and cured in an autoclave (to outgas) at a predetermined temperature. The resulting product is a lightweight material with good stiffness and tensile properties.

The fuselage of Boeing's new *787 Dreamliner* is composed of CFRP. The new fuselage design is lighter in weight and is expected to show better mechanical performance (fatigue resistance). The 787 is the first airliner to use composite materials for most of its construction—over $6 billion in carbon fiber is expected to go into the new planes. The plane consists by weight of 50% composite, 20% aluminum, 15% titanium, 10% steel, and the remainder other materials—over 35 tons of CFRP of which 23 tons are carbon fibers. By volume, the plane will consist of 80% composite materials [8].

Natural Composites. Natural composites surround us every day—mainly in the form of wood, clays, and paper. Biological nanocomposites dwell within us and within all of life. Bone, spider silks, sea urchin skin, cellulose, nacre (mother of pearl), teeth, and a plethora of other natural products made of combinations of proteins, lipids, carbohydrates, nucleotides, and all other biological subunits form, assemble, and merge to create some of the most remarkable materials to be found. Yes, all biological materials are manufactured from the bottom up and we challenge you to name structural materials in living things that are not nanocomposites. We shall discuss a few types later in the chapter.

24.0.2 Overview of Engineering Materials

Keep in mind that nanomaterials, and nanocomponents within composites, are nanosized versions of the basic classes of engineering materials: metals, alloys, and semiconductors; ceramics and glasses; and polymers and inorganic carbon-based materials. The world of nano also includes a vast array of biological materials. Composites usually consist of a majority (matrix) material and a minority (inclusion or dispersed phase) material. Basic engineering materials are reviewed below.

Ceramics. Early humans made use of ceramics in the form of crystalline stone tools and eventually pottery well before the use of metals [9]. Hardness, temperature resistance, compression resistance, and strength are parameters used to characterize ceramics. However, brittleness and the tendency to fracture in a brittle way under tensile, shear, or impact stresses are undesired qualities in modern materials. It is these exact traits however that helped us get the edge over animals and kickoff the Stone Age. The first tools with sharp edges were formed by cleavage of appropriate "hard" materials, for example, flint, a sedimentary cryptocrystalline form of quartz [10]. Quartz is crystalline SiO_2 that is harder than flint and therefore served better as early hammers and grinders.

Ceramics are composed of combinations of metallic and nonmetallic elements. Examples include aluminum oxides (Al_2O_3), titania (TiO_2), zirconia (ZrO_2), silica (SiO_2), and sodium oxide (Na_2O) [9,11]. Other kinds of ceramics include various metal carbides like silicon carbide (SiC), titanium carbide (TIC), boron carbide (BC), and tungsten carbide (WC); nitrides like silicon nitride (Si_3N_4), titanium nitride (TiN), and boron nitride (BN); and borides like titanium boride (TiB_2) [9]. Ceramic materials are able to serve either as the matrix host material in a composite or as the minority inclusions in nanocomposites—whether in a metal or polymer matrix. Traditional uses of ceramics include their applications as refractory materials.

Recently, ceramics have made inroads into electrical, magnetic, piezo, optical, thermal, and high-temperature superconducting (YBa_2CuO_7) devices in the form of magnetic insulators, capacitors, high-frequency circuits, sensors, and transducers that convert optical, thermal, mechanical, and electrical stimuli into analytical signals [9]. Nanoscale ceramic materials are used as catalysts, photoconductors (charge carrier and separation), and template materials that are revolutionizing device fabrication.

Metals and Alloys. Metals are highly reflective, thermally conducting, and malleable due to the presence of highly mobile surface electrons. Advantageous characteristics of metals include high electrical conductivity, thermal conductivity, tensile strength (strength), high modulus of elasticity (stiffness), and toughness (fracture resistance) [9,11]. Fabrication is facilitated by metals' inherent ductility and their tendency towards plastic deformation before failure. The average density of metals is ca. $7.5\,g\cdot cm^{-3}$ [9], considered to be a disadvantage in the aerospace industry.

Nanometals are traditionally exploited in a variety of catalytic applications, for example, catalytic converters in automobiles and the various steam reforming and polymerization processes that abound in today's polymer technology. Nanometals are inserted into a composite if layer-specific optical, electrical, or thermal properties are desired. Metals are often mixed with ceramics to produce high-performance composite materials. Metals made of nano rather than micrograins have shown superior performance. Descriptions of metal structure and properties, along with those of other engineering materials, are found in Ref. [12].

Semiconductors. We are all quite familiar with the impact made on today's technology by semiconductors. Ranging from pure materials like silicon to oxides like titania and complex semiconductors like CIGS (copper–indium–gallium–selenide), highly integrated and complex electronic applications all contain semiconductor materials. Nanosized semiconductors already have had major impacts on the computer industry. The ultimate nano-version of semiconductors is the quantum dot. Semiconductors have found their way into nanocomposites with optical, electronic, or sensing functions. Gold–semiconductor nanocomposite hybrids can be formed by citrate stabilization of Au from $HAuCl_4$ with borohydride reduction and coupling to CdS, CuS, NiS, or PbS semiconductors [13]. Other examples include Au nanometal clusters embedded in an amorphous Si matrix (1-nm layers) to generate a unique optical response based on the surface plasmon [14], metal–semiconductor nanocomposite layered optical fibers [15], and next generation solar cells based on semiconductor nanocomposites [16]. In this chapter, we do not focus on optical properties are based on these technologies solar cells are discussed in more detail in chapter 28.

Polymers. Polymers are lightweight materials (long-chain hydrocarbons) that have found applications in every major industry. Polymers, consisting mainly of carbon and light elements like N, O, P, S, F, and Cl, have a low average density of $\sim 1.5\,g\cdot cm^{-3}$ [9]. There are three general classes of polymers: thermosets (polyurethanes, phenolic resins, epoxy resins, polyesters, and vinyl esters), thermoplastics (polyethylene, nylon, ABS, PI, PP, PC, and PEEK), and elastomers (silicones and EPDM). Thermoset polymers are stronger than thermoplastic polymers (polyethylene or polyvinyl chloride) [2].

Organic polymers are relatively easy to fabricate and have a high strength-to-mass ratio. One disadvantage of polymers is long-term instability under loads. Polymers, however, form the perfect matrix for a composite. Polymers can be easily mixed to form new materials. The incorporation of carbon or glass has created materials that overcome disadvantages like lack of stiffness, lack of hardness, and lack of tensile strength inherent to most organic polymers. At the nanoscale, the block-co-polymer has become an indispensable nanomaterial template and matrix element. The carbon nanotube-reinforced polymer is just one reason why nanocomposites are such a hot area today.

Inorganic Carbons. Inorganic carbon materials like graphite, carbon blacks, fullerenes, carbon nanotubes, and diamondoids have made important contributions to the composite materials industry—especially from the contribution of the carbon reinforced epoxy resin composite. The spectrum of inorganic carbons ranges from the humble pencil lead, to the strongest material known to science, the carbon nanotube. It is no wonder that researchers are fervently trying to incorporate carbon nanotubes into composites with all the major classes of engineering materials as host. A special section in the chapter is devoted to carbon fibers, whiskers, multiwalled, and single-walled carbon nanotubes.

Materials from Biology. Diatoms have been used as filler materials for centuries. Biologically based materials such as nucleotides, carbohydrates, proteins, and lipids have shown value as components in composites. There is great effert underway to copy nature's materials. For example, conductive self-assembled DNA composite membranes supported on nanoporous polycarbonate track-etched substrates showed enhanced mechanical properties, proton conductivity (8.0×10^{-2} $S \cdot cm^{-1}$), and decreased methanol permeability (10 wt% @ 5.0×10^{-7} $cm^2 \cdot s^{-1}$) for 1.6 mg·cm^2 DNA loading [17]. Water soluble composites were prepared with C_{60} and saccharose, fructose, or dextrans by mechanochemical procedures [18]. Lipid-based nanocomposites are formed by a simple beaker immersion method [19]. Formation of nanoparticle–lipid hybrid films is accomplished with electrostatic control, starting with the deposition of a lipid on a solid substrate [19]. Ordered nanocomposites can be formed from proteins. Proteins are nontoxic, don't require organic solvents to disperse, can be poured into molds, and are easily functionalized [20].

24.0.3 Types of Composite Materials and Generic Structures

We all are exposed to many kinds of composites on a daily basis. We can also classify them in numerous ways—by composition, by inclusion size, by inclusion morphology, and so on and so forth. Particulate zero-dimensional (nanosized) and three-dimensional (macroscopic) composite products are represented by caulking compounds (latex, acrylic, silicone, calcium carbonate, talc, and others). Fishing lines consist of hollow glass spheres or tungsten embedded in poly(vinyl chloride). Composites with fibrillar or filamentous (one-dimensional) composite products are found in surf boards, automobile skin (polyesters, epoxides impregnated with glass fibers), and sporting equipment such as tennis racquets, bicycles, and golf clubs. Laminate composites (two-dimensional) include chemically retardant safety gear and armored windows made of glass–ceramics or ceramic–organic polymers. Just take a look around and you will find that many of our products contain some kind of composite material.

They can be classified according to compositional variables: by the type of majority component (major engineering materials), by the type of minority components (once again, members of the basic types of engineering materials), by the number of minority components, by the amount of majority component, by the amount of minority component, by the form of the minority component (fibrillar, particulate, laminate), and finally, by the dispersion of minority components (bulk, agglomerate, surface enriched, etc.) [5].

They are also classified according to type and inclusion size as did Brooks et al: (1) natural composites (wood, bone, bamboo, and muscle tissue), (2) microcomposite materials (metal alloys, toughened and reinforced thermoplastics, sheet and molding compounds, aligned or random continuous fibers, particulates), and (3) macrocomposite materials (reinforced concrete, tennis racquets, skis etc.) [5]. To this we add a fourth category, nanocomposite materials (carbon nanotube reinforced polymers).

Others categorize composites by the nature of the matrix material (e.g., metal, ceramic, or polymer). **Table 24.1** lists selected synthetic composites. Nature gives us abundant examples of nanocomposite materials. Some have been discussed in Part I [21] and more will be discussed in chapter 26.

Types of Nanocomposites. Nanocomposites consist of traditional matrix materials that happen to have nanomaterial fillers. The nanomaterial inclusions

TABLE 24.1 *Synthetic Composite Materials*

Synthetic composite	Description
Concrete	A ceramic + ceramic composite, ceramic (quicklime, CaO) cement (usually Portland cement) + ceramic (fly ash, aggregate), hardens after a third component (water) is added; reinforced concrete is a higher level composite in which steel bars (rebar) are added to provide tensile strength. The composite has the best of both worlds: compression strength is provided by the cementious material and tensile properties by the steel reinforcement.
Carbon fiber reinforced plastic (CFRP)	Graphitic carbon threads consisting of thousands of carbon filaments a few micrometer in diameter; plastic material is usually thermosetting materials like epoxy resin but others like polyester, vinyl esters, and nylon are used; carbon fibers can be made from other polymers like polyacrylonitrile that is heated to form graphitic fibers; tensile strength greater than 5 GPa and modulus of elasticity greater than 500 GPa are common.
Fiberglass	Fiber-reinforced polymer (FRP); glass fiber + polymeric support; silica-based fibers mostly $[(SiO_2)_n]$ but also contains Al_2O_3, Fe_2O_3, CaO, MgO, Na_2O, TiO_2, ZrO_2; formed by extrusion through nozzles to form an amorphous solid (25 μm diameter fibers); polymer matrix materials include epoxies (high strength), polyesters (for generalized structures), phenolics (high temperature), and silicones (electrical applications). Composites of continuous fibers, discrete fibers, or woven fabric offer different mechanical properties.
Other fiber-matrix composites	Para-aramid (Kevlar) + epoxy or polyester Metal-matrix: B + Al; Al_2O_3 + Al; SiC + Al; SiC + Ti Ceramic matrix: SiC + Al_2O_3; SiC + Si_3N_4

can be zero-dimensional (metal oxides, carbon black, polymeric colloids, carbides like SiC and WC, boron and silicon nitrides, and block copolymers), one-dimensional (carbon nanotubes, metal oxide whiskers and nanotubes, boron nitride nanotubes, and linear block copolymers), two-dimensional nanolayered materials (phyllosilicates, nanoclays, hydrotalcites, and 2-D block copolymers) and three-dimensional networks (block copolymers).

Basic Composite Structure. In order to understand the mechanical and physical properties of composite materials, knowledge of macro-, micro-, and nanostructure is required beforehand. Mechanical properties include tensile strength, ductility, toughness, hardness, impact resistance, resilience, fatigue resistance, and creep and many more—not all are considered to be independent of one another. Physical properties include the glass transition temperature (T_g), and permeability and dielectric, optical, thermal, and electrical properties. Chemical properties include reactivity, corrosion resistance, cross-linked structure, intermolecular bonding, polymerization, and biochemical compatibility.

Average properties of traditional composites can be estimated similar to the way electrical resistance is evaluated, for example, in series. With $x \leq 1$ representing the inclusion fraction

$$\frac{1}{\text{Composite property}} = \frac{(x)(\text{Inclusion})}{\text{Inclusion property}} + \frac{(1-x)(\text{Host})}{\text{Host property}} \tag{24.1}$$

This relation, however, may not apply appropriately to nanocomposites for several reasons: filler and polymer dimensions are on comparable size scales, interfacial areas are immense, and distances between filler and polymer phases are separated by mere nanometers. All these factors imply that the properties of the interfacial-bound polymer regions would most likely dominate the properties of the composite rather than any weighted average of bulk properties of the two phases. In addition, nanometer properties most likely differ drastically from those of bulk material counterparts. Nanometer-scale inclusions would also, due to their intimate relationship with polymers at the atomic scale, impart effects on chain morphology and conformation. The conclusion? It would be dangerous to evaluate nanocomposites based on traditional composite theory alone.

Inclusion Configurations in Polymer Hosts. Simplistically speaking, physical properties of composite materials are either isotropic (independent of direction) or anisotropic (dependent on the direction of the applied force, thermal stress, or optical probe—a.k.a. orthotropic). Most metals and ceramic composites have isotropic mechanical properties. Filler materials can occupy several predetermined orientations within a polymeric host material (**Fig. 24.1**). Physical properties depend on not only the type of material of the inclusion but also its aspect ratio, size, shape, concentration, the presence of interpenetrating networks, porosity (if applicable), and orientation.

24.0.4 The Nano Perspective

We have repeated over and over how nanomaterials possess unique and even remarkable properties. What happens to these unique properties and phenomena

FIG. 24.1 *Various inclusion configurations in polymer matrices are depicted. (a) Randomly orientated group of small fibers (e.g., multiwalled carbon nanotubes) is dispersed throughout the polymer matrix. (b) Long strands of fibers are dispersed randomly. (c) Composite material with parallel fibers oriented perpendicular to the plane of the composite afford good compression resistance. (d) Fibers oriented along the load-bearing (tensile) axis of the material enhance tensile properties. Long, uniform, unbundled, evenly dispersed single-walled carbon nanotubes embedded in such a polymer would consist of a holy grail for composite researchers. (e) Micrometer or nanofibers crosshatched in the form of a textile offer strengthening properties along more directions in the composite. (f) Composite with nanoparticle inclusion. Depending on the concentration and type of inclusion, the composite may exhibit unique electrical conductivity or optical properties.*

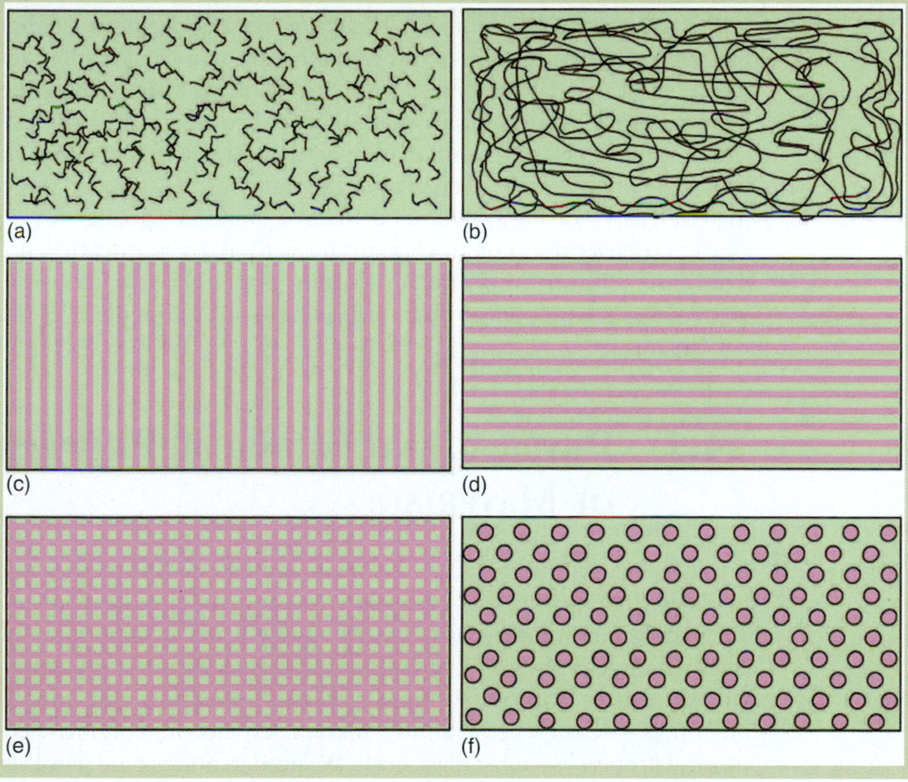

when a nanomaterial is embedded within a bulk material? Or when nanomaterials interact with other nanomaterials to form a bulk material, what are the properties of the resulting homogeneous material or composite? What happens when nanomaterials are woven together to make a fiber and then serve as a structural element in a composite? As is a usual theme in these texts, we refer to a lesson from nature.

Why Nanomaterials? Nanomaterials have remarkable, and oftentimes unusual, physical and chemical properties as well as increased surface area and interfaces. Composites, therefore, that contain such nanomaterials are in turn expected to outperform composites made of bulk materials. Composites that contain nanomaterials are called nanocomposites. The possibility that nanomaterials combine in ways that are unknown to either parent material in the bulk phase is also likely. We have learned that in composites, the merged materials retain their identity and properties. In nanocomposites, materials remaining separate and distinct in the nanocomposite may not adhere to the fundamental definition given earlier. By binding a nanoparticle or nanofiber, do you change it? Are we confronted with a Heisenberg-like dilemma consisting of a complementary pair of binding and properties? The problem faced by engineers is how to incorporate such materials without losing those amazing characteristics. How does one bind carbon nanotubes into a polymer matrix without compromising tensile strength, stiffness, and electrical and thermal conductivity?

We restrict the scope of our discussion in this chapter to nanocomposite materials defined by one or more nanoscale materials embedded within a matrix material. To further squeeze the definition, let's continue to define nanocomposites as materials that are intended to serve as a structural material first but may have enhanced electrical and thermal properties. In the broadest sense, nearly every nanomaterial, beyond that of the standalone material, is in one way, shape, or form a nanocomposite. A monolayer of silica particles on top of an ITO surface is then, by application of this broad definition, a nanocomposite. A quantum dot tethered to a carbon nanotube is then also a nanocomposite, but let's not split filaments over this arbitrary demarcation.

24.1 Physical and Chemical Properties of Materials

A brief review of mechanical, electrical, and thermal properties and testing is provided. Engineering materials undergo engineering testing; nanocomposites are no different in that regard. We just need to learn more about them to understand why exactly physical properties are different and enhanced. Nanocomposites are expected to improve stiffness, strength, toughness, and other mechanical properties—density, permeability, thermal expansion, conductivity, and other physical properties—all at lower costs. We need to understand why this prediction is true or why it will prove to be true.

24.1.1 Mechanical Properties

Mechanical properties of a bulk material are the sum total of physical and chemical properties measured at the macroscale. A physical device pulls, compresses, pushes, indents, or impacts to determine the mechanical performance of a material. Failure analysis, however, is very much a microscopic process in which the root cause of poor performance or failure is rooted out at the atomic level. As it turns out, all mechanical properties valid at the macro level transpose directly to the nano level—via some scaling effect of course. Let's review a few mechanical properties and associated terminology.

Stress. Stress is the measure of average force exerted per area: $F \cdot m^{-2}$. The units of stress are commonly given in terms of pascal (usually associated with mega or giga although now nano and tera are making headway). Stress, σ, is defined as

$$\sigma = \frac{F}{A_o} \tag{24.2}$$

where
F is force in newtons
A_o is the initial area of the material cross section in m^2 [22]

Linear stress is a rank-two tensor and is represented by a [3 × 3] matrix. There are several kinds of stress: tensile, compressive, shear, bending, and torsional, to name a few. A tensile stress situation is shown in **Figure 24.2**.

Strain. Tensile strain is caused by tensile stress. Tensile strain is a dimensionless geometrical expression of deformation—the extent to which a body is distorted after application of a deforming stress. It is the ratio of the extension or compression (in terms of length, area or volume) of a material to its original dimensions. Linear strain is measured as a change in length along a line or the change in angle between two lines. Other forms of strain are plane, shear, and volumetric strain. The formula for linear strain is given below:

$$\varepsilon = \frac{\Delta L}{L_o} \tag{24.3}$$

where
L_o is the initial length
ΔL is the change in length under stress

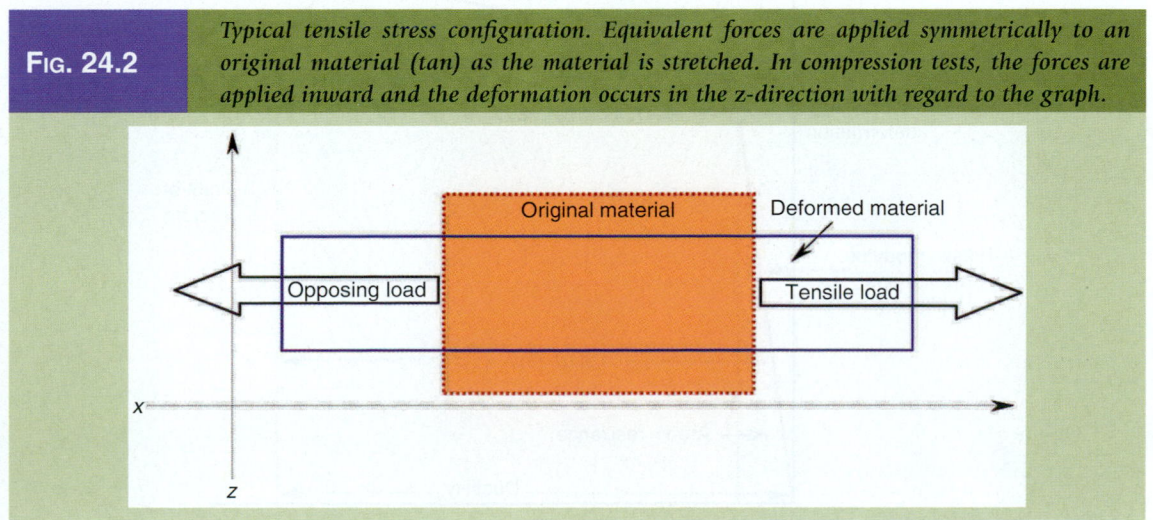

FIG. 24.2 *Typical tensile stress configuration. Equivalent forces are applied symmetrically to an original material (tan) as the material is stretched. In compression tests, the forces are applied inward and the deformation occurs in the z-direction with regard to the graph.*

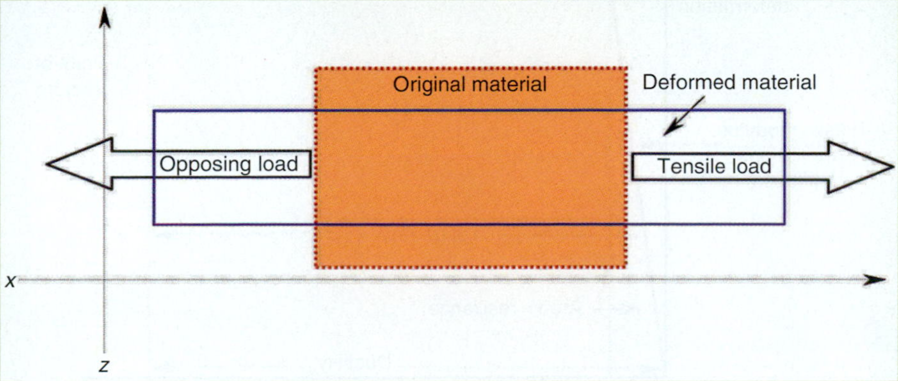

Tensile Strength. Tensile strength is an intensive property and is the measure of the level of tensile stress required to pull a material apart [23]. Yield strength is the point at which elastic deformation gives way to permanent, or plastic, deformation. The ultimate strength is the maximum stress that the material is subjected, and breaking strength is the point where rupture occurs in the material (**Fig. 24.3**).

Young's Modulus, E. Young's modulus (a.k.a. modulus of elasticity or tensile modulus) is a measure of stiffness. It is the measure of the rate of change of stress

FIG. 24.3

A simple tensile test is able to reveal mechanical properties of a generic metal. The modulus of elasticity *(Young's Modulus) is determined from the initial slope of the stress–strain curve.* Elastic deformation *is temporary deformation and the linearity of this region reflects Hooke's law. Young's modulus is also referred to as the* stiffness *of the material. The region past the elastic region is the point of* plastic deformation *or permanent deformation. The original shape of the material cannot be recovered after it is mechanically stressed beyond this region. The* yield strength *represents the minimum amount of stress necessary to generate a small amount of the permanent deformation [11]. The* ultimate tensile strength *is reflected by the highest point of the curve, for example, the point of maximum stress.* Toughness *is defined as the total area under the curve and is a reflection of a combination of mechanical properties—overall reflecting the ability of a material to absorb energy until fracture [24].* Ductility *is the percent elongation at fracture.* Resilience *is the ability of a material to absorb energy and is calculated from the area under the curve to the elastic limit point [24]. A curve such as this is also dependent upon the rate at which the test is conducted.* Hardness, *not shown in the graph, is the ability of a material to withstand local plastic deformation.*

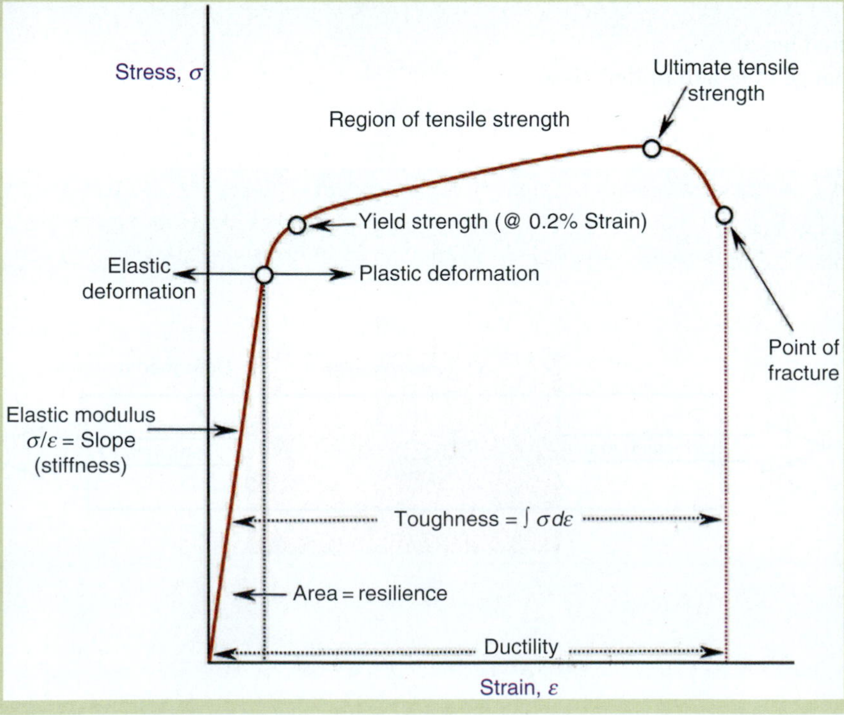

with strain and is determined from the slope of stress–strain curves extracted during tensile tests. Young's modulus is a numerical evaluation of Hooke's law, the ratio of stress to strain or the measure of resistance to elastic deformation. Stress and strain are related through an expression related to Hooke's law

$$\sigma = E\varepsilon \tag{24.4}$$

where E is a proportionality constant equal to $FL/A\Delta L$

$$E = \frac{\sigma}{\varepsilon} = \frac{F \cdot A_o^{-1}}{\Delta L \cdot L_o^{-1}} = \frac{F \cdot L_o}{A_o \cdot \Delta L} \tag{24.5}$$

where E is the Young's modulus in terms of force per unit area or pressure ($F \cdot m^{-2}$ or Pa).

Young's modulus is based on Hooke's simple spring law where

$$F = \left(\frac{E \cdot A_o}{L_o}\right)\Delta L = kx \tag{24.6}$$

Young's modulus is used to determine the length that a material will stretch under tension or the load under which a material will buckle under compression. The Young's modulus (in GPa) of selected common materials is given in **Table 24.2**.

A typical stress–strain curve for a typical metal was shown in **Figure 24.3**. A standardized test coupon (usually in the shape of a dog bone) is clamped at both ends while a tensile (pulling) stress is applied. The cylindrical cross-sectional area of the tensile specimen is recorded. Significant kinds of data can be extracted from such a simple test: Young's modulus (elastic deformation, stiffness), plastic deformation, tensile strength, resilience, ductility, toughness, yield strength, and point of fracture.

Loss Modulus. Loss modulus is an engineering term that describes the dissipation of energy into heat as a material deforms under tensile or shear mechanical testing in viscoelastic solids. It is often referred to as the damping factor (in Pa). Although beyond the scope of this chapter, the loss modulus is the imaginary component of the storage modulus. The storage modulus measures the level of stored energy (the elastic portion) and the dissipation of heat (of the viscous portion). Loss modulus is often correlated with shear modulus (modulus of rigidity or the shear stress over shear strain) from which the viscosity of material can be determined. Opposing forces are often applied to a material upon testing. Elastic deformation under shear stress is shown in **Figure 24.4**.

Poisson's Ratio. Materials stretched along the z-axis tend to contract along the x- and y-axis (**Fig. 24.5**). Poisson's ratio is the ratio of the relative contraction strain over that of the relative extension (axial) strain. For example

$$\nu = -\frac{\varepsilon_x}{\varepsilon_z} \tag{24.7}$$

Table 24.2 *Young's Modulus of Common Materials*

General engineering material	Example	Young' modulus (GPa)
Metal	Steel	200
	Cu	115
	Brass	110
	Al	69
Ceramics	Glass	60–90
	Tungsten carbide	550
	Silicon carbide	450
Carbon materials	Diamond	1100
	Graphite	
	Single-walled carbon nanotubes	1000+
Fibers	C-glass	69
	E-glass	72.4
	S-glass	85.5
	Graphite fiber	340–380
	Al_2O_3 whisker	430
	SiC fiber	430
	Boron filament	410
Polymers	Epoxy	6.9
	Polyester	6.9
	Low-density polyethylene	<0.2
	High-density polyethylene	1.4
	Nylon	3–7
Composites	Tendon	1.5
	Bone (compression)	9.4
	Bone (tensile)	16
	Pine wood (along grain)	9.0
	Oak wood (along grain)	11
	Plywood	4.0
	Rubber	<0.1
	CFRP	125–150
	Concrete (compression)	41
	W-filament ($V_f = 0.50$) strengthened Cu	260
	W-particle ($V_f = 0.50$) strengthened Cu	190
	E-glass fibers ($V_f = 0.73$) in epoxy	56
	Al_2O_3 whiskers ($V_f = 0.14$) in epoxy	41
	C fiber ($V_f = 0.67$) in epoxy	221
	Kevlar fibers ($V_f = 0.82$) in epoxy	86
	Boron filaments ($V_f = 0.70$) in epoxy	210–280

FIG. 24.4 *Elastic deformation along the x-axial direction due to shear stress by opposing forces is depicted.*

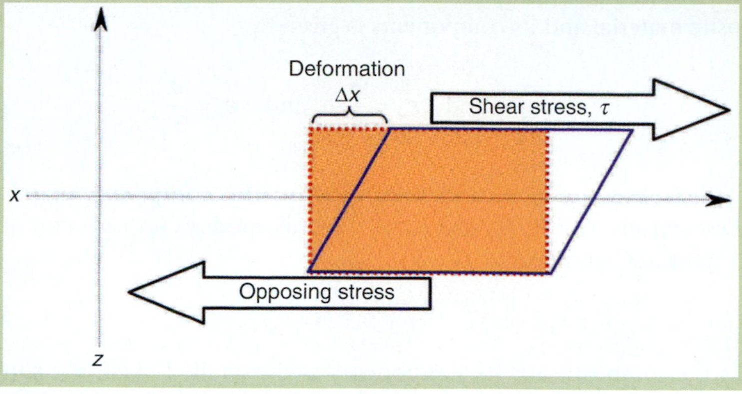

FIG. 24.5 *From Poisson's ratio, the characteristic contraction perpendicular to the tensile stress is defined.*

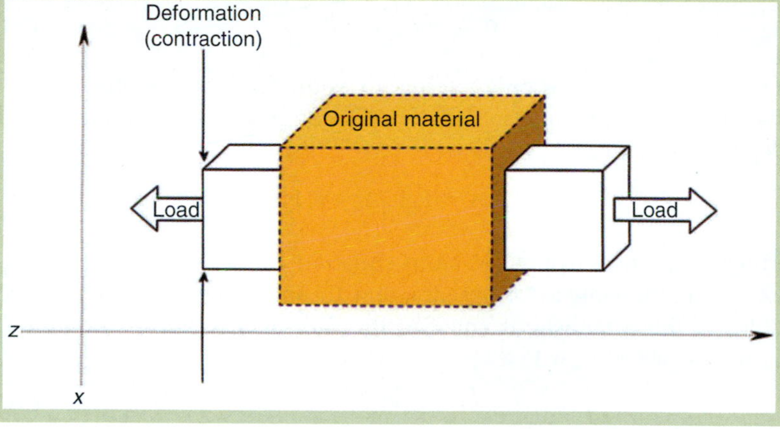

where ε_x and ε_z are the strains in the x and z directions. Poisson's ratio is a fundamental engineering relationship that describes the elastic behavior of materials.

Average Properties. Composites consist of more than one material. How then does one predict the mechanical properties of the composite from the known mechanical properties of the individual components? The mechanical properties of the composite are also dependent upon the geometric configuration of the structural materials—from the micro to the nanostructure—and how well the inclusions are bound to the matrix. Assuming tight binding, simple averaging of component mechanical properties yields a good first approximation of the mechanical properties of the composite.

The simplest case is that of a circular composite consisting of intimately bonded fibers running parallel through the length of the material [11]. Shackelford provides a good definition of average mechanical properties [11]. If a load is applied to a cylindrical composite parallel to the fibers (uniaxial stressing), an isostrain condition is appropriate for consideration. First of all, the strain for the composite material and its components is given by

$$\varepsilon_c = \frac{\sigma_c}{E_c} \quad \text{and} \quad \varepsilon_m = \frac{\sigma_m}{E_m} \quad \text{and} \quad \varepsilon_f = \frac{\sigma_f}{E_f} \tag{24.8}$$

where the subscripts c, m, and f correlate to the composite, matrix, and fiber, respectively, and E_c, E_m, *and* E_f are the bulk modulii for each component. For the isostrain condition

$$\varepsilon_c = \varepsilon_m = \varepsilon_f \tag{24.9}$$

because the composite and its components are integrally linked. The sum total of the load F_c is equal to the sum of the loads on the individual constituents

$$F_c = F_m + F_f \tag{24.10}$$

Since $\sigma = F/A$ (from equation 24.2, page 419)

$$\sigma_c A_c = \sigma_m A_m + \sigma_f A_f \tag{24.11}$$

This formulation is commonly referred to as the "rule of mixtures." The ROM works well for calculating the density of a composite. Combining equations (24.7), (24.8), and (24.9) yields

$$E_c \varepsilon_c A_c = E_m \varepsilon_m A_m + E_f \varepsilon_f A_f \tag{24.12}$$

Dividing by A_C and ε_c (obtained from equation 24.12) and noting that A_m/A_C and A_f/A_C are the same as volume fraction V_m and V_f, respectively, the Young's modulus of the composite is equal to the sum of the Young's moduli of the components (where $V_m + V_f = 1$)

$$E_c = V_m E_m + V_f E_f \tag{24.13}$$

For the *isostress* condition, in which the fibers are loaded perpendicular to the axis of the cylindrical composite, the following relationship applies

$$\frac{1}{E_c} = \frac{V_m}{E_m} + \frac{V_f}{E_f} \tag{24.14}$$

A good question to ask at this time is the following: "Does elastic modulus scale with size as implied by the above formulae to the nanomaterials?"

Composite mechanical properties are also analyzed via micromechanics in which analysis of individual phases on a multi-axial level takes place. Two kinds, the "method of cell" and the Mori–Tanaka averaging scheme, are used to predict mechanical behavior of composites. Fiber interaction effects relating inclusion strain to average matrix strain can be predicted by the Mori–Tanaka method.

Hardness Testing. Hardness is a test applied to materials in the solid phase. Engineering parameters such as elasticity, plasticity, viscosity, viscoelasticity, strength, strain, brittleness, ductility, and toughness can be gleaned from hardness testing [25]. Hardness values are dependent on the type of test applied and the geometry of the probe. Hardness is the ability of a material to resist permanent deformation induced by an applied force. There are several types of hardness testing: Rockwell, Knoop, Vickers, and Brinell are just a few of the most common ones. Scratch, indentation, and rebound hardness are some types of engineering definitions associated with applied hardness. In scratch hardness, resistance to fracture and plastic deformation due to friction applied via a sharp probe are measured. Indentation hardness measures the resistance to plastic deformation due to a persistent load delivered via a sharp probe. Rebound hardness measures elasticity from the height of recoil of an object dropped from a predetermined distance.

Hardness can be defined at the macro-, micro-, or nanoscales according to the level of applied force and the extent of the displacement. Macrohardness measurements are facilitative means of obtaining engineering values, especially in the case of metals. Microindenters function by continuous measurement of penetration depth induced by an applied load. Nanoindenters apply forces on the order of 1 nN. The indenter is driven into the sample by the action of a magnetic coil fixed to an indenter assembly. A capacitance displacement gauge monitors penetration depth.

The Vickers hardness test utilizes a diamond probe with a 136° angle between opposite faces of the indenter (or 22° between the indenter face and the surface) (**Fig. 24.6**) [26].

The area of the indentation is determined from

$$A = \frac{d^2}{2\left(\frac{\sin 136°}{2}\right)} \approx \frac{d^2}{1.854} \tag{24.15}$$

$$HV = \frac{F}{A} \approx \frac{1.854F}{d^2} \tag{24.16}$$

where HV is in terms of $\mathrm{kg \cdot F \cdot mm^{-2}}$ ($1\,\mathrm{kg \cdot F \cdot mm^{-2}} = 9.807\,\mathrm{MPa}$). Vickers hardness and other physical and mechanical properties are listed in **Table 24.3**.

Good correlation was achieved between nanoindentation tests and the standard Vickers hardness tests conducted on model ferritic alloys [27]. Nanoindenter loads of 0.05 g (Nanoindenter-II, Nano Instruments, Inc., Oak Ridge) were compared to conventional Vickers microhardness derived from 200 to 500 g applied loads [27]. Two methods of testing were used: one in which constant displacement depth was measured and the other in which a constant load was applied to all specimens. Nanohardness data was corrected to mitigate the differences between projected and actual indenter area. Overall, good correlation was achieved between the two techniques for hardness values in the range of 0.7 and 3 GPa. Tensile property measurements were also conducted on all specimens. The linear correlation between Vickers hardness and yield strength is apparently valid for data derived by the nanoindentation technique [27].

Rheology and T_g. Differential scanning calorimetry (DSC, a good way to determine the glass transition temperature, T_g) and rheological methods are

Fig. 24.6 *A Vickers hardness tip is depicted. The tip is made of diamond in the shape of a square-based pyramid. Consistency is obtained in that impressions are similar regardless of the load applied. The Vickers Pyramid Number (HV) is determined from the ratio F/A where F is the applied force and A is the surface area of the indentation.*

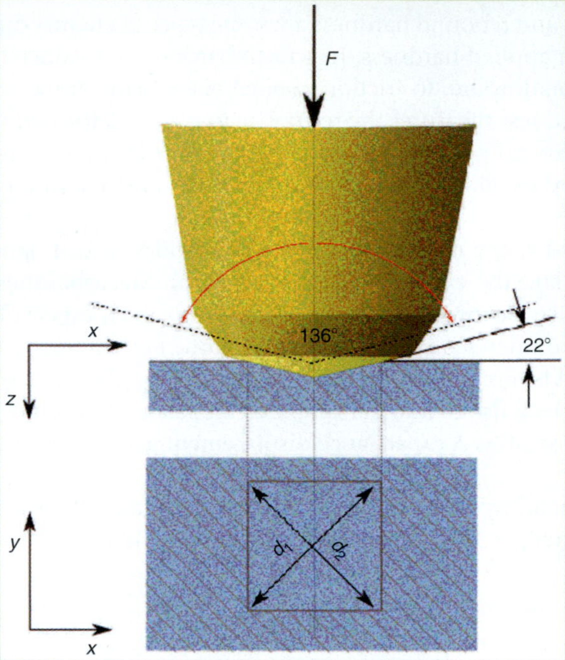

Source: Image from Wikipedia commons, en.wikipedia.org/wiki/Vickers_hardness_test (2008). With permission.

Table 24.3 *Engineering Properties of Elemental Materials*

	Vickers hardness (MPa)	Mohs hardness (MPa)	Young's modulus (GPa)	Poisson ratio	Thermal conductivity $[W \cdot (m \cdot K)^{-1}]$	Thermal expansion $(10^{-6}\ K^{-1})$	Electrical conductivity $(10^7\ S \cdot m^{-1})$	Density $(g \cdot cm^{-3})$
Fe	608	4	211	0.29	80	11.8	1.0	7.874
Al	167	2.75	70	0.35	235	23.1	3.8	2.7
Au	216	2.5	78	0.44	320	14.2	4.5	19.3
Ag	251	2.5	83	0.37	430	18.9	6.2	10.49
Cu	369	3	130	0.34	400	16.5	5.9	8.92
Bi	N/A	2.25	32	0.33	8	13.4	7.7×10^5	9.78
Ca	N/A	1.75	20	0.31	200	22.3	2.9	1.55
Sn	N/A	1.5	50	0.36	67	22	9.1×10^6	7.31
Be	1670	5.5	287	0.032	190	11.3	2.5	1.848
Mo	1530	5.5	329	0.31	139	4.8	2.0	10.28
W	3430	7.5	411	0.28	170	4.5	2.0	19.25

Note: Vickers hardness of the elements: periodictable.com/Elements/004/data.html.

good means of probing the interfacial area. The T_g is sensitive to the level of nanoparticles that have either an attractive (decreased mobility) or repulsive (increased mobility) reaction with the polymer matrix [28]. Rheology is the study of materials that deform or flow. With more filler, the polymer mobility and T_g (relaxation) are influenced.

Tribology. Tribological behavior is also affected by nanofillers. Dry lubrication of bearings is accomplished with homogeneous polymers like poly(tetrafluoroethylene), a.k.a. PTFE (e.g., they are nonabrasive with low friction), but they present issues with wear resistance. When micrometer-sized hard fillers are added to increase wear resistance, hardness, modulus, and strength, abrasion factors begin to emerge [28]. What is required is an engineered material that provides the best of both worlds—enter the nano-solution. PTFE does not wear well. If 15-wt% micrometer-sized fillers are added, two orders of magnitude improvement are observed but at a cost—an increase in the dry sliding coefficient [29]. Tests were conducted on four kinds of 304-stainless steel surfaces: electropolished, wet-sanded, lapped, and dry-sanded [30]. PTFE (2–20 μm grain size) + alumina (44, 80, or 500 nm) composites were prepared. Filling rates were varied from 0, 1, 5, and 10 wt%. The lapped surface showed the best results of the four. If up to 10 wt% of γ-alumina nanoparticles are added to the polymer, three orders of magnitude (OOM) reduction in wear rate is observed and 4-OOM reduction is observed if 1-wt% γ-alumina nanoparticles are mixed with the PTFE [30]. Other tests using α-alumina showed even better performance. Apparently α-alumina leads to stronger interfacial interactions with the PTFE polymer than does γ-alumina [31]. Apparently nanofillers (especially the 4-OOM filler at 1 wt%) stabilize the phase of PTFE that is tougher and fibrillates more easily and these conditions promote the formation of "well-adhered transfer films," a phase that is not normally available at room temperature [28]. Fibrillation in this case fills cracks better thereby reducing the formation of wear debris. DSC analysis revealed that nanofilled PTFE had a higher melting temperature than unfilled PTFE.

Mechanical Testing at the Nanoscale. Failure in materials usually involves some kind of fracture, and as a result, most failure analysis involves examination of the results of mechanical loading tests—tensile overload, fatigue, or creep [5]. However, we do not focus on failure analysis in this chapter. Regardless, most failures can be traced to atomic or molecular phenomena—phenomena now that are closer to the nanoscale in size. The question to be asked at this stage is how exactly does one test for nanomaterial mechanical properties? Apparently an entire new arsenal of nanomechanical testers has been invented that involve the use of AFM probes and nanoindenters, for example. Although containing nanomaterials, a nanocomposite is realized as a macroscopic, graspable material and is therefore amenable to the same kinds of mechanical testing that is perpetuated on traditional composites.

Transposition of bulk properties to nanoscale materials may be appropriate in some cases, depending on size, but the practice of doing so may be fraught with pitfalls as well. Bulk optical properties, for example, are used to describe the optical response of colloidal particles (ca. $10 < d < 100$ nm). However, for particles below 10 nm in size where surface effects and quantum effects start to exert influence, caution must be inserted into the equation. In addition, nanomaterials are purported to have fewer defects. This too is expected to affect their material properties.

24.1.2 Thermal Properties

Material thermal properties of concern to engineers come in the form of thermal conductivity, the ability of a material to conduct heat, and a parameter known as the coefficient of thermal expansion (CTE), the microscopic expansion of a material (strain) under thermal stress.

Coefficient of Thermal Expansion. CTE is the change in dimension (linear, area, or volume) of a material in response to a change in temperature. The degree of expansion (or the fractional increase in strain) divided by the change in temperature is known as the coefficient of thermal expansion. Thermal expansion for fibers, rods, or cables is expressed by the ratio of strain along their length

$$\varepsilon_{\text{Thermal}} = \frac{L - L_o}{L_o} = \frac{\Delta L}{L_o} \tag{24.17}$$

Thermal expansion is proportional to the change in temperature

$$\varepsilon_{\text{Thermal}} \propto T - T_o = \Delta T \tag{24.18}$$

Heat is stored in materials during heat transfer in several ways—in the form of translational, rotational, vibrational, or electronic energies. Carbon materials, in general, are characterized by low coefficients of thermal expansion. This is one of the reasons such materials perform well in spacecraft that is exposed to temperature extremes. Polymers, on the other hand, have higher coefficients of thermal expansion than do metals. Ceramic materials possess some of the lowest thermal coefficients of expansion (**Table 24.4**).

Thermal expansion can be isotropic or anisotropic. Crystals, for example, may expand along one crystalline direction more so than another. Graphite expands more along one direction (x, y) of its plane than along the other, the z-direction. The thermal coefficient of expansion is expressed in linear, areal, or volumetric forms as

$$\alpha_L = \frac{\Delta L}{L_o \Delta T}, \quad \alpha_A = \frac{\Delta A}{A_o \Delta T}, \quad \text{and} \quad \beta_V = \frac{\Delta V}{V_o \Delta T} \tag{24.19}$$

CTE is measured via interferometry, a procedure that analyzes the changes in the interference pattern of monochromatic light. Each shift in the fringe pattern corresponds to a change in L of 0.5λ the laser wavelength. Plots of strain versus temperature, and the slope of the strain–temperature curve yield the instantaneous value of CTE. Quartz dilatometry is used to determine CTE of materials that are expected to have large changes in dimension under thermal stress. CTE mismatches lead to mechanical failure of materials even if CTE values have apparent small values. Therefore, careful selection of materials with regard to CTE must be done during the design phase of an engineering material.

Table 24.4 *Linear Coefficients of Thermal Expansion (CTE) of Common Materials*

General engineering material	Example	CTE ($10^{-6} \cdot K^{-1}$ or $10^{-6} \cdot C^{-1}$)
Metals	Carbon steel A36	11.7
	Stainless steel alloy 304	17.2
	Cast iron G1800	11.4
	Stainless steel alloy 440A	10.2
	W (refractory metal)	4.5
	Mo (refractory metal)	4.9
	Ta (refractory metal)	6.5
	Au (noble metal)	14.2
	Ag (noble metal)	19.7
	Pt (noble metal)	9.1
	Ni (nonferrous metal)	13.3
	Zn (nonferrous metal)	23.0
	Zr (nonferrous metal)	5.9
	Al 7075 (nonferrous metal)	23.4
	Bronze (nonferrous metal)	18.0
Semiconductors	GaAs	5.9
	Si	2.5
Ceramics	Aluminum oxide	7.0
	SiC	4.6
	Zirconia	9.6
	Soda lime glass	9.0
	Si_3N_4	2.7–3.1
Carbon materials	Graphite	2.0–2.7
	Diamond	0.11–1.23
Polymers	Butadiene–acrylonitrile, styrene–butadiene, and silicone (elastomers)	220–270
	Phenolics (thermosets)	122
	Polyesters (thermosets)	100–180
	Epoxies (thermosets)	80–117
	Low-density polyethylene (LDPE)	180–400
	Polystyrene (PS)	90–150
	Polymethyl methacrylate (PMMA)	90–162
	Polyvinyl chloride (PVC)	90–180
	Teflon (PTFE)	126–216
Fibers	Aramid (Kevlar 49)	−2.0 (longitudinal)
		60 (transverse)
	Carbon fiber	−0.6 (longitudinal)
	Polyacrylonitrile (PAN)	10.0 (transverse)
	E-glass	5.0

(*continued*)

Table 24.4 (Contd.) *Linear Coefficients of Thermal Expansion (CTE) of Common Materials*

General engineering material	Example	CTE ($10^{-6} \cdot K^{-1}$ or $10^{-6} \cdot C^{-1}$)
Composites	Concrete	10.0–13.6
	Aramid fiber/epoxy matrix $V_f = 0.6$	−4.0 (longitudinal)
		70 (transverse)
	High modulus carbon fiber/epoxy	−0.5 (longitudinal)
		32 (transverse)
	E-glass fibers/epoxy resin	6.6
	Douglas fir	3.8–5.1 (grain parallel)
		25.4–33.8 (grain perpendicular)
	Oak	4.6–5.9 (grain parallel)
		30.6–39.1 (grain perpendicular)
	SiC–Metal	0.4 [32] (significantly less than pure metal)
	AlN fiber-reinforced polystyrene	0.3 times less than pure PS

Source: www.stormcable.com/uploads/Thermal_expansion_data_table_tb06.pdf (viewed 2008).

Thermal conductivity. Thermal conductivity is the ability of a material to conduct heat—the specific heat flux that flows through a material if a certain temperature gradient is in place. Heat conduction is defined by Fourier's law:

$$k = -\frac{dQ/dt}{A(dT/dx)} = \frac{\Delta Q/\Delta t}{A(\Delta T/\Delta x)} \tag{24.20}$$

where

$dQ/\Delta dt$ is the rate of heat flux (transfer) across an area A in $J \cdot s^{-1}$
κ is the thermal conductivity coefficient of a specific material
A is the area
x is the thickness of the conducting surface

For a flat slab under steady-state heat conduction, the differentials become average values (RHS of equation 24.20). The units of k are $J \cdot (s \cdot m \cdot K)^{-1}$ or more commonly, $W \cdot m^{-1} \cdot K^{-1}$ ($1\,W = 1\,J \cdot s^{-1}$). In general, metals are good conductors of heat and polymers are good insulators (**Table 24.5**).

24.1.3 Electronic Properties

Nanomaterials cover the entire gamut of electronic phenomena. Nanomaterials can act as conductors, semiconductors, and insulators. Carbon nanotubes in particular offer engineers a wide range of properties that are amenable to use in advanced materials—composites in particular, whether in the form of shielding materials or those designed to conduct electrically in an anisotropic fashion. We shall review a few of the basic engineering parameters that we need to be aware when discussing electronic and magnetic properties.

Table 24.5 *Thermal Conductivities of Common Materials*

General engineering material	Example	Thermal conductivity ($W \cdot m^{-1} \cdot K^{-1}$)
Metals	Carbon steel	54
	Stainless steel	16
	Al	250
	Co	69
	Au	310
	Ag	429
	Cu	401
Ceramics	Quartz	3
	Aluminum oxide	30
	Granite	1.7–4.0
	Mica	0.71
	Glass	1.05
	Plaster board gypsum	0.17
	TiC (100°C)	25
	Silica glass (100°C)	2.0
	Soda lime glass (100°C)	1.7
	ZrO_2 (100°C)	2.0
	Porcelain (100°C)	1.7
Carbon materials	Diamond	900–2320
	Graphite (100°C)	180
	Carbon nanotubes	1800–6000
Polymers	Nylon	0.25
	High-density polyethylene (HDPE)	0.42–0.51
	Polystyrene (PS)	0.033
	Polypropylene (PP)	0.1–0.22
	Polyvinyl chloride (PVC)	0.19
Composites	Leather	0.14
	Balsa wood	0.048
	Cement	0.29
	Cork	0.043
	Cotton	0.03
	Fire-clay brick	1.4
	Fiberglass insulation	0.04
	Oak	0.17
	Vermiculite	0.058

Resistivity and Conductivity. Electrical resistance depends on the composition and geometry of a material. For example, resistance Ω increases with sample length L and decreases with sample area A [11]. Resistivity ρ, on the other hand, is a parameter that is independent of sample geometry with units of $\Omega \cdot \text{m}$

$$\rho = \frac{\Omega A}{L} \tag{24.21}$$

Conductivity σ is the reciprocal of resistivity

$$\sigma = \rho^{-1} \tag{24.22}$$

Another term that is encountered frequently when discussing nanocomposite electrical properties is the *percolation limit*, or the correlation between geometrical and electrical connectivity. The percolation limit is defined as the minimum concentration of filler required to support electrical conduction in a composite. Particle contact and percolation require a large volume fraction when the metal particles and the polymer grains are of comparable size. When the conducting particles are small relative to the grain size of the polymer, they are forced into interstitial (interfacial) sites and in contact with one another within a lower volume. This configuration constitutes a "low percolation limit," another advantage of small size. The geometrical and electrical onset of connectivity may not occur simultaneously in some cases but their lack of correlation may be explained by the tunneling phenomena. **Table 24.6** lists the conductivity of selected metals, semiconductors, and insulators in terms of $\Omega^{-1} \cdot m^{-1}$ (siemens per meter, $S \cdot m^{-1}$, also indicated as $S \cdot cm^{-1}$) [11,33,34].

24.1.4 Chemical Properties

The chemistry of nanocomposites takes place within and during its manufacture, not after its introduction to its application environment. Structural nanocomposites are designed to be inert. We do not want composite materials that react with radiation, humidity, erosion, acids, corrosive agents, toxic gases, sequestered liquids, or any other chemical or physical process. We want our structural materials to retain their integrity under a variety of conditions. If there is any surface chemistry applied to nanocomposites, it is in the form of protective materials like sealants, paints, and other coatings. These treatments are not discussed in this chapter.

This chapter is placed in the *Chemical Nanoengineering* division of the text. Although some mechanical, electrical, thermal, and optical properties are discussed, we are specifically interested in the interfacial chemistry between the nanophase and the bulk matrix phase of the composite. What chemistry is required, for example, to render a single-walled carbon nanotube reactive to potential matrix bonding?

Chemical Modification. Nanomaterials have high surface energy. In general, due to this excess surface energy, nanomaterials seek stabilization by many methods. Chemical stabilization is one means by which nanomaterials achieve this goal. Gold-55 clusters, for example, require a ligand shell to help stabilize its geometry or otherwise agglomeration results. Many kinds of nanoparticulate fillers in composites have reactive surfaces, like silicates, that react readily with the polymer matrix. Others, such as nanogold, require a ligand shell in order to interface strongly with the matrix material of the composite. Ligands may provide specially designed functional groups that bind with side-chain counterparts of a polymer. These processes are highly chemical in nature and will be discussed throughout the remainder of the chapter.

TABLE 24.6 *Electrical Conductivity of Selected Engineering Materials*

General engineering material	Example	Conductivity ($S \cdot m^{-1}$ or $\Omega^{-1} \cdot m^{-1}$)
Metals	Al	3.54×10^7
	Cu	5.8×10^7
	Fe	1.03×10^7
	Au	4.26×10^7
	Steel	$5.7–9.4 \times 10^6$
Semiconductors	Ge	2.0
	Si	0.40×10^{-3}
	PbS	38.4
	Indium–tin oxide (ITO)	$\sim 1 \times 10^4\ S \cdot cm^{-1}$
Carbon materials	Graphite	1.28×10^5
Ceramics	Aluminum oxide	10^{-10} to 10^{-12}
	Borosilicate glass	10^{-13}
Polymers	Polyethylene	10^{-13} to 10^{-15}
	Nylon 66	10^{-12} to 10^{-13}
	Doped conductive polymers: polyaniline (PANI), polypyrrole (PPy)	10^{-5} to 10^2
	Polypyrrole doped with AsF_3	$\sim 10^3\ S \cdot cm^{-1}$
	Polypyrrole doped with I_2	$\sim 10^2\ S \cdot cm^{-1}$
	Polyacetylene doped with AsF_3	$\sim 10^5\ S \cdot cm^{-1}$
	Polyacetylene doped with I_2	$\sim 10^4\ S \cdot cm^{-1}$
Composites	K-intercalated graphite (10%)–polystyrene [35]	$1.3 \times 10^{-1}\ S \cdot cm^{-1}$
	Graphite (10%)–polystyrene [35]	$5.0 \times 10^{-3}\ S \cdot cm^{-1}$
	Br-intercalated graphite fiber–epoxy [36]	$1.1 \times 10^{-1}\ S \cdot cm^{-1}$
	Graphite fiber–epoxy [36]	$2.2 \times 10^{-3}\ S \cdot cm^{-1}$
	Graphene–silica ($V_f = 0.11$) spun composite	$0.45\ S \cdot cm^{-1}$ [37]
	MWNT–silica ($V_f = 0.093$) spun composite	$0.57\ S \cdot cm^{-1}$ [37]

Note: Data from *Materials Properties Tables: Electrical Conductivity and Resistivity*, NDT Resource Center, Iowa State University, www.ndt-ed.org/GeneralResources/MaterialProperties/ET/et_matlprop_index.htm.

Intermolecular Interactions. Intermolecular interactions play a major role in nanoinclusion–matrix interfacial chemistry. The entire gamut of these "noncovalent, nonmetallic, and nonionic bonding" participates in interfacial bonding with matrix polymeric materials. Hydrogen bonds, van der Waals, dative bonds, π-stacking, and aromatic dipole–dipole, dipole-induced dipole, and induced dipole–induced dipole; capillary, and hydrophobic interactions all are capable of binding to polymers. Some nanoparticles actually are able to catalyze or direct the formation of the polymer during synthesis of the nanocomposite.

Chemical Modification of Carbon Nanotubes. Some kinds of nanomaterials do not require stabilization by chemical or other means. Single-walled carbon nanotubes, for example, are kinetically stable under room ambient conditions.

However, they do not make for good fillers in nanocomposites unless some chemical modification takes place on their surface. The modification is required so that integral contact with a host matrix material, usually a polymer, is possible. The chemical bonds thus formed are able to conduct load transfer from the host bulk majority component to the carbon nanotubes—the whole idea behind fiber reinforcement in a composite.

24.2 Natural Nanocomposites

Biomimetic technology is discussed in detail in chapter 26 and we do not wish to steal any of its thunder; however, we cover a few notable examples of natural nanocomposites found in nature that serve as excellent models for our technology. A few selected examples of natural nanocomposites are listed and described in **Table 24.7**. Every hierarchical structure found in living things is a nanocomposite. Perhaps we can extend this expression to include "every structure in nature is a nanocomposite." What do you think?

24.2.1 Skin of the Sea Cucumber

Lessons from the Humble Sea Cucumber. Recently, S.J. Rowan and C. Weder of Case Western Reserve University have developed a material (with a low-modulus matrix) that can harden or soften depending on its immediate environment—a process known as chemoresponsive mechanic adaptability [39]. When threatened, the sea cucumber, an echinoderm that is able to rapidly and reversibly control the stiffness of its inner skin membrane (dermis), is able to transform its outer skin into a hard shell-like protective material in seconds—ten-times harder than the relaxed version. The reaction is due to enzymes that are able to bind protein fibers together. Another set of enzymes allows the skin of the sea cucumber to return to its relaxed state [39]. The humble sea cucumber serves as the model for the new nanopolymeric material.

The material developed by Rowan et al. is able to harden to a level that is greater than 2500 times its softer state—reversal of which occurs after soaking in water. The hardening of the material is due to the collective action of cellulose fiber matrix held together with hydrogen bonds. When wetted, water is able to hydrolyze bonds between fibers and the material undergoes relaxation. When dried, the material reforms hydrogen bonds and that process induces stiffening of the membrane. The elastic host polymer and stiff cellulose nanofibers demonstrates reversible reduction in tensile modulus by a factor of 40—from 800 to 20 MPa—during cycling of the material in tests. Polymers in which thermal transitions corresponded to simulated physiological conditions exhibited more dramatic tensile modulus swings—4.2 to 1.6 MPa, a factor of 2625 [39]. The material, once commercialized, could be used in brain implants that reduce inflammation or into clothing that can transform into armor.

24.2.2 Hard Natural Nanocomposites

The More Humble Mollusk. The shells of mollusks protect their soft bodies from predators and other traumas inflicted by nature. The composite, generally

TABLE 24.7 *Natural Composite Materials*

Natural composite	Description
Cellulose (wood)	Polysaccharides of linear polymers of D-glucose subunits linked by $\beta(1\rightarrow 4)$ glycosidic bonds; found in cell walls of plants; composite formed from flexible cellulose + stiff lignin.
Chitin	Nitrogenous polysaccharides $[(C_8H_{13}O_5N)_n]$ in linear configuration (*N*-acetyl-D-glucose-2-amine subunits) that form $\beta(1\rightarrow 4)$ like cellulose; chitin embedded in a proteinaceous matrix forms the exoskeleton composite material of insects; hydrogen bonding between adjacent polymers gives the material increased strength.
Bone	Parallel collagen triple helices in staggered array; 40-nm gaps at ends of tropocollagen serve as nucleation centers for hydroxyapatite, $[Ca_{10}(PO_4)_6(OH)_2]$, forming a composite material that is very hard (due to the mineral part) but elastic and fracture resistant (due to the collagen part).
Silk	β-Pleated sheet proteinaceous materials made up of gly–ser–gly–ala–gly (gly at every other position) monomer subunits; fibers have high tensile strength and are resistant to stretching. Spider silk is a composite of a gel core + surrounding solid casing of aligned molecules, usually consisting of alternating gly-ala or just ala amino acids. Spider silk (2–4 μm diameter) tensile strength is on the order of high-grade steel but a better strength-to-weight ratio; capable of 40% stretch without rupture which makes the material ductile with high toughness; β-sheets stack to form crystalline and amorphous domains, and it is the relationship between these two domains in the composite that give spider silk its remarkable mechanical properties.
Collagen	Modified α-helical protein materials consisting of repeating monomers of glycine–proline–hydroxyproline found in triple-helical structure; glycine at every third position; chains are stabilized by steric repulsion due to pyrrolidone rings of proline and hydroxyproline; three chains are hydrogen bonded to each other. Collagen + keratin gives skin its integrity and elasticity.
Nacre	Aragonite (ortho-calcium carbonate, $CaCO_3$) + biological macromolecules; a.k.a. mother of pearl; organic–inorganic composite material that is strong and resilient. Composition is of hexagonal platelets of aragonite 10–20 μm in width and 0.5-μm thick in parallel lamellar arrangement; the sheets are separated by elastic biopolymers made of complex polysaccharides like chitin and lustrin and β-sheet proteins found in silks; due to this configuration, transverse crack propagation is inhibited.
Diatoms	Cell walls consist of nanostructured polymerized silicates and biomolecules (frustulins on surface and silaffin polypeptides embedded within the silica) for extraordinary mechanical stability; the diatom cell wall consists of crystalline and amorphous regions that gives the composite excellent mechanical properties [38].

known as nacre, consists of calcium carbonate (the ceramic phase, polymorphs of $CaCO_3$, aragonite, or calcite) and an organic binder. One kind of organic constituent is the highly ordered layered β-chitin proteins arranged in parallel arrays within interlamellar sheets. Another kind is similar in structure to silk in the form of an amorphous gel phase. The organic phase is responsible for stabilizing the metastable aragonite and for directing the morphology and orientation of the crystals [40].

Comparison of Hard Natural Materials. A comprehensive study was conducted to verify the mechanical properties of some natural materials. J. D. Currey et al.

in 2000 found that nacre outperformed highly mineralized bone because of its extremely well-ordered microstructure. The organic material in nacre forms a nearly continuous enclosure around all the aragonite platelets (~500 nm thick), and is designed for toughness. This is surprising because normal bone contains more organic material (mostly collagenacious) than does nacre (~1%). The rostral bone of the whale, however, does have similar organic phase content within its bony structure. The compositions were as follows: bovine bone (65% mineral, 25% organic, and 10% water), rostrum bone (96% mineral, 1% organic, and 3% water), and nacre (98% mineral, 1% organic, and 1% water) [41]. Young's modulus for the three materials was found to be 20, 40, and 34 GPa respectively; bending strength of 220, 55, and 210 MPa, respectively, and hardness of 55–70, 227, and 200 $kg \cdot m^{-2}$, respectively [41].

The highly mineralized content of the rostrum bone resulted in greater stiffness (increases Young's modulus) and stronger in bending but loss of strength and toughness compared to bovine bone. Although nacre has a highly mineralized content, the loss of strength and toughness was not observed. The organic layer acts as a toughening device. According to Currey et al., there is a mystery associated with nacre and it is the following: why was the material never modified by the pressure of natural selection to have a larger component of organic materials that would impart more strength and toughness while still remaining quite stiff?

Nanostructured Inorganic Framework/Polymer Nanocomposites. Inspired by the structure of natural nacre and bone—natural composite materials that are composed of alternating layers of soft and hard materials in a periodic array—mechanically robust, multifunctional silica/polymer nanocomposites were prepared by a straightforward self-assembly/evaporation process [42]. Depending on process parameters, Y. Yang et al. formed nanostructured, conjugated polymers of hexagonal, cubic, or lamellar structures consisting of an inorganic framework that protects, stabilizes, and orients the polymer while providing mechanical and chemical stability [42]. Polymerizable amphiphilic diacetylene (PDA) surfactant molecules served as structure-directing agents and as the monomeric precursor material. By varying the size of the headgroup of the diacetylene surfactant (oligoethylene glycol, EO), it was possible to control the mesostructure of the resultant polymer. Addition of a silica inorganic host influenced the polymerization of the PDA. The process is generally known as a hybrid organic/inorganic self-assembly process.

The synthesis began with coupling DA with diethylene glycols to form DA-OE_n where $n = 3, 4, 5$, or 10. Tetraethyl orthosilicate (TEOS, [$Si(OC_2H_5)_4$]) provided the source of the silica. HCl catalyst in tetrahydrofuran (THF) and water was all that was required to generate the nanocomposite. The size of the head group played a major role in directing the final morphology of the composite: $n = 3$ produced lamellar composites, $n = 5$ produced hexagonal structured composites, and $n = 10$ formed cubic composites [42].

Mechanical properties were determined by the nanoindentation procedure. The Young's modulus ranged from approximately 9–5 GPa and the hardness from 0.6 to 0.4 GPa—both values comparable to calcined mesoporous silica films that are used routinely in microelectronic devices. The membranes were also impermeable to gases indicating that there is no significant porosity within its structure.

24.3 Carbon Fibers and Nanotubes

The importance of carbon fibers was mentioned briefly in section 24.0.1. Although, carbon fibers, whiskers, rods, filaments, and nanotubes have been fouling catalytic processes for decades, the discovery of nanosized carbon tubes has opened the doors to a whole new generation of advanced materials.

Carbon Fibers. A fiber (from the Latin *fibra* "a fiber, filament," related to *filum* "thread") is a high aspect ratio material ($L >> D$). There are many classes of carbon filamentous materials and others based on inorganic materials like alumina, boron nitride, and silicon carbide. Conventional carbon fibers range in diameter from a few to 10 μm (and larger depending on the application). Relatively solid carbon nanofibers (CNF) have diameters on the order of 50–200 nm. MWNTs, as we know well by now, have diameters between 10 to 20 nm and are as large as 50 nm. The diameter of SWNTs is limited to a few nanometers. The structure of CNFs, MWNTs, and SWNTs varies depending on fabrication conditions. Fiber nanostructure, for example, is demonstrated in the form of cones, cups, or plates.

Although we have implicitly touched upon fiber technology in previous sections in this chapter, we go ahead to provide a more complete treatment of the subject matter, of nanofibers and nanotubes based on carbon in particular. A nanofiber or nanotube is defined as a filament (hollow or solid) with diameter less than 100 nm—although we can safely extend this boundary to elongated materials with diameter less than 1 μm. Large diameter filaments are characterized by solid morphology (e.g., no central canal).

Nanofibers are produced in several ways—interfacial polymerization, electrospinning, and weaving to name a few. Nanotubes are most efficiently produced by CVD techniques. Nanofibers have multiple functions and applications beyond enhancing mechanical and physical properties of polymer composites—as filters, in medical devices, composites, garments, insulation, and energy storage to name a few. We define fibers as those materials that are not necessarily carbon nanotubes but can contain carbon nanotubes. Graphitic nanofibers range in microstructure from solid herring-bone to stacked graphitic platelets—parallel or perpendicular to the fiber axis.

Carbon Nanotubes. Carbon nanotubes and their ilk once again are the headliners in a chapter of a nanotechnology text. That should be of no surprise by now. They are amazing materials with amazing properties and why not place them within a polymeric matrix with the intent of fabricating a supercomposite. Carbon fibers have been used to reinforce polymers for several decades—with the earliest efforts linked to the aerospace industry in the mid- to late-twentieth century. One of the holy grails of aerospace design is to incorporate materials that are stronger (higher tensile strength) but lighter, for example, higher strength to weight ratio. Carbon-reinforced resins and carbon nanotube reinforced resins offer an optimistic route to achieve that goal.

Carbon nanotubes have certainly changed the landscape with the advent of nanotube-reinforced composites, creating a new generation of excitement. According to E.V. Barrera of Rice University in Texas, advanced applications of composites are expected in many areas: radiation protection, heat dissipation, static discharge capacity, high strength-lightweight parts and housings, heat engine components, deicing coatings, lightning protection, stress sensors, organic LEDs, electrically conductive ceramics, paintable polymers, antifouling paints, UV protection coatings, and overall corrosion protection [43].

Most research today is limited by the availability of carbon nanotubes. Berrara et al. go on to say that, and we quote indirectly, that a dearth of research materials limits the concentration of NTs in composites to low volume fraction—ca. <10% and that research has focused on nanotube dispersion, untangling, alignment, bonding, molecular distribution, and retention of nanotube properties [43]. However, it is known from theoretical studies that NT fractions of 40%–50% show "broader promise" [43].

Superior NT properties take form in many configurations. For example, the percolation limit (i.e., the minimal concentration of inclusion materials to maintain electrical conduction) can be as low as 1wt% because of the small dimension of the nanotube [43,44,45]. In addition to concentration, alignment of carbon nanotubes is also problematic at this time.

If you were to sell carbon nanotubes, what would be the content of your pitch? Perhaps tout that CNTs have tensile strength 10–100 times and an elastic modulus 5–10 times that of the best steel at one-fifth the weight; or that the electrical current capacity of CNTs can be 1000 times better than copper wire; or its thermal conductivity twice that of diamond; or how about its thermal stability up to 2800°C in vacuum; or that it can be a conductor, semiconductor, or insulator depending solely on its structure without doping or chemical modification. Is this too good to be true? Yes indeed, carbon nanotubes are the strongest, stiffest, and lightest materials known to science.

Performance of Advanced Carbon Composite Materials. More and more demand is placed on developing advanced materials that conform to the extreme environments encountered by tomorrow's aircraft. The development of nanocomposites that can greatly improve the strength, ablation resistance, stealth, thermal performance, and radiation hardening characteristics of materials is imperative for the next generation of aircraft. Radiation-hardened materials (and devices) have the ability to withstand damage or malfunctions caused by high-energy particles (cosmic rays, high energy protons and electrons, solar particle events, high-energy electrons from the Van Allen radiation belt, etc.) and electromagnetic radiation (ultraviolet, infrared, and x-rays) for applications in space, high altitude flight, or in or near nuclear reactors. Radiation and high-energy particles cause electron ejection, lattice displacements, depletion of minority carriers, recombination, and ionization. Neutrons, for example, can cause atomic displacement that disrupts the structure, and thereby reduces performance.

24.3.1 *Types of Fibers, Whiskers, and Nanotubes*

Fibers. The fiber filament offers one-dimensional reinforcement to a composite- or, it can be manufactured as a stand-alone material. It is with the carbon nanotube that we are most interested.

Non-Wovens. Another class of engineered fibrous materials is known as *non-wovens*. Non-wovens are materials that are not specifically woven or knit. The most prominent example of a non-woven material is paper; another is felt. Non-wovens, therefore, are engineered fabrics that take form as sheets or webs. Entanglement in non-wovens is generated mechanically, thermally, or chemically. Non-wovens act as absorbents, liquid repellants, flame retardants, and filters. Although non-wovens are capable of being bound internally by the action of adhesives, serrated mechanical needles, binders, or polymer melts, the strength of non-wovens does not compare to other fiber-reinforced materials [46].

Carbon Nanotubes. We are all familiar with the exceptional properties exhibited by the legendary eighth century Damascus swords—swords that possessed characteristic wavy banding patterns (called *damask*), incredibly sharp blades, and extraordinary mechanical properties [47,48]. Recent electron micrographic studies have revealed that the sword contained an ultrahigh content of carbon, nanowires (iron-based cementite), and carbon nanotubes [47,48]. The bundles of nanotubes and nanotube-encased cementite run parallel to the blade's surface. Softer steel is found between the nano-inclusions. The result is a blade that is strong and flexible. Following an etching process, wavy lines, consisting of nanostructured components sticking out from the blade's edge (e.g., sawtooth-like) became apparent.

Although Sumio Iijima of NEC Corporation is formally credited with the discovery of carbon nanotubes in 1993, two relatively unknown Soviet scientists, L. V. Radushkevich and V. M. Lukyanovich, were the first to formally describe hollow, nanometer dimension carbon tubes (**Fig. 24.7**) [49]. Since then, carbon nanotubes have been discovered and rediscovered until their "formal discovery" was finally attributed to S. Iijima et al. in 1991 [50,51]. Interestingly, the objective of carbon nanotube work over the past 50 years or so just before the 1990s and the advent of the *Nano* Age was specifically to reduce nanotube formation in order to prevent coking and fouling of catalysts used in other processes.

S. Iijima, and simultaneously D. Bethune et al. of IBM, discovered the first SWNTs in 1993 [52,53].

Elastic Modulus of Carbon Nanotubes. The theoretical Young's modulus is predicted to be from 1 to 5 TPa (that's terapascal!) whereas the best stainless steel and Kevlar values are in the range of 200 and 250 GPa, respectively. SWNTs show ca. 24% elongation of the materials at breaking as compared to 50% maximum for steel and around 2% for Kevlar. Because carbon nanotubes have such a low density, 1.3–1.4 $g \cdot cm^{-3}$, the specific strength of CNTs can be as high as 4.8×10^7 $N \cdot m \cdot kg^{-1}$, much better than high carbon steel (1.54×10^5 $N \cdot m \cdot kg^{-1}$).

Fig. 24.7 *The abstract from a seminal paper in the 1952 edition of the* Journal of Physical Chemistry of Russia *describing carbon nanotubes with 50-nm diameter from CO at 600°C shown. The discovery by L.V. Radushkevich and V.M. Lukyanovich remained obscure due to the Cold War climate of the time. Sumio Iijima of NEC Corporation of Japan went on to formally describe multiwalled carbon nanotubes in 1991 [49,50,51].*

1952 *ЖУРНАЛ ФИЗИЧЕСКОЙ ХИМИИ* *т. XXVI, вып. 1*

О СТРУКТУРЕ УГЛЕРОДА, ОБРАЗУЮЩЕГОСЯ ПРИ ТЕРМИЧЕСКОМ РАЗЛОЖЕНИИ ОКИСИ УГЛЕРОДА НА ЖЕЛЕЗНОМ КОНТАКТЕ

Л. В. Радушкевич и В. М. Лукьянович

Данная работа возникла в связи с электронно-микроскопическим изучением структуры различных адсорбентов, главным образом активных углей, графита и т. п. При исследовании препаратов углерода мы обратили внимание на сажу, получающуюся при разложении окиси углерода на железном контакте при температуре около 600° С. Так как сажа из окиси углерода изучалась адсорбционными методами и для нее была определена удельная поверхность по изотерме адсорбции, то представлялось интересным проверить эти результаты путем непосредственного измерения размеров частиц. Но уже первые наблюдения, сделанные нами [1], показали, что образующийся из CO углерод имеет весьма своеобразную структуру, до настоящего времени никем не описанную, и поэтому, естественно, наше внимание было перенесено на систематическое изучение этой структуры, а также на условия ее образования.

Young's modulus was determined from thermal vibration amplitude under TEM observation. T. Ebbesen of NEC Corporation measured the deflection of the tips of SWNTs as temperature was increased from 300 to 600°C. The amplitude of the vibration provided a means to assess the elasticity of the nanotubes—demonstrating high bending stiffness. Young's modulus is also calculated by an AFM method (**Fig. 24.8**) [54]. In this case, the AFM probe tip applies pressure to the distal end of a secured carbon nanotube and a correlation between applied force and material deflection is ascertained. Application of Euler's small deflection equation yields the Young's modulus

$$E = \frac{FL^3}{3\delta I} \tag{24.23}$$

where δ is the deflection of the cantilever (the carbon nanotube) at the point of contact, L is the length of the nanotube (between the fixed position and the point of force), F is the force applied by the AFM tip, and E is the Young's modulus as before. I is the areal moment of inertia of the cross section of the nanotube about its axis

$$I = \frac{\pi(r_o^4 - r_i^4)}{4} \tag{24.24}$$

where r_o and r_i are the outer and inner radii of the elastic cylinder, respectively [55].

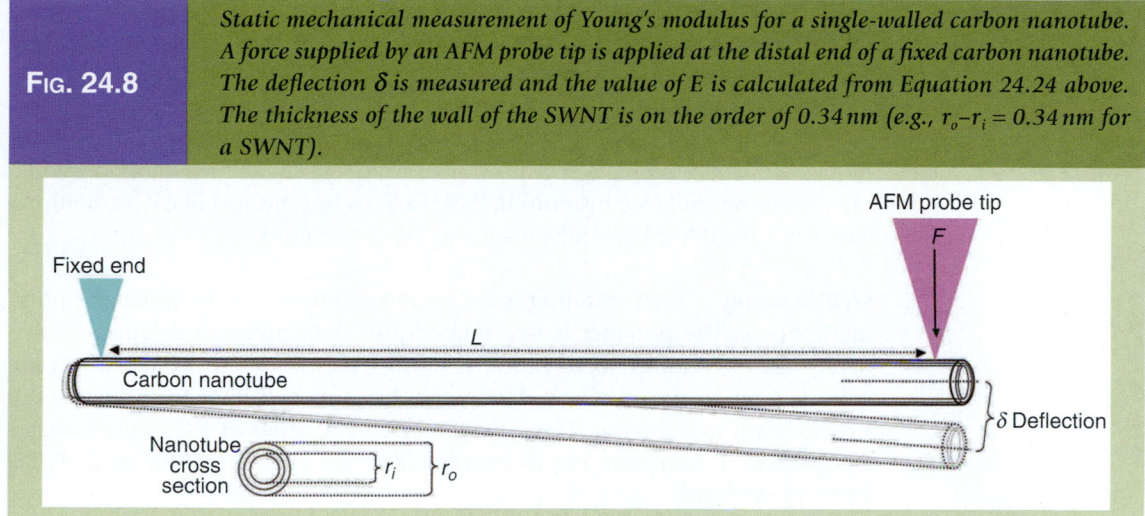

Fig. 24.8 *Static mechanical measurement of Young's modulus for a single-walled carbon nanotube. A force supplied by an AFM probe tip is applied at the distal end of a fixed carbon nanotube. The deflection δ is measured and the value of E is calculated from Equation 24.24 above. The thickness of the wall of the SWNT is on the order of 0.34 nm (e.g., $r_o - r_i = 0.34$ nm for a SWNT).*

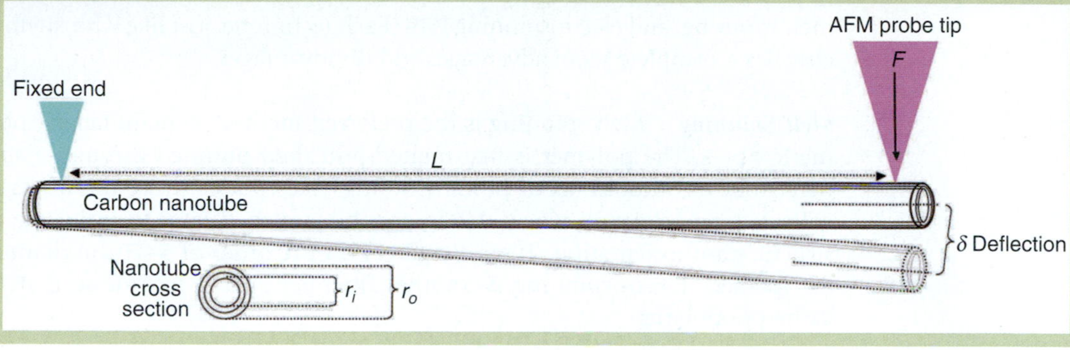

Why do carbon nanotubes have such extraordinary stiffness and strength? The strength is due to the strength and adaptability of the carbon–carbon bond, one of the strongest in nature. Stiffness is due to the phenomenon of hybridization. The carbon–carbon bonds in the hexagonal carbon rings are a blend of sp^2 and sp^3 character. When stressed, the bonds can rehybridize to accommodate the stress (increased bond angle by assuming more sp^3 character). Once the stress is removed, recovery to the original configuration occurs. The degree of rehybridization is a function of the degree of applied stress. How do these remarkable materials perform inside a polymer matrix? The answer depends on how well the CNTs are linked to the matrix elements.

Thermal Properties. *Carbon allotropes* show a variety of thermal properties. The thermal conductivity for diamond ranges from 900 to 2320 $W \cdot m^{-1} \cdot K^{-1}$, a value five times better than silver. Thermal conductivity in carbon nanotubes depends on the temperature and the large phonon mean free path. In 1999, Hone et al. determined that the thermal conductivity of a single SWNT rope at room temperature could vary from 1800 to 6000 $W \cdot m^{-1} \cdot K^{-1}$ [56]. Goddard et al. determined the thermal conductivity of a (10,10) tube approached 2980 $W \cdot m^{-1} \cdot K^{-1}$ [57]. How would a carbon nanotube fiber behave in a polymer composite?

24.3.2 Synthesis of Fibers and Nanotubes

Carbon nanotubes can be fabricated by laser vaporization (quartz tube at 1200°C with 1.2% CO–Ni alloy graphite target), carbon arc method (500 torr He, 20–25 V dc @ 50–120 A; no catalysts for MWNTs but Co, Ni, and Fe nanocatalysts are required for SWNT graphite targets), and chemical vapor deposition

(600–1000°C, argon atmosphere, carbon gas sources like methane, propylene, or acetylene and supported catalysts Co, Ni, or Fe on alumina or floating ferrocene and thiophene).

Polymer-Based Fibers. Polymer-based nanofibers are fabricated by several techniques: drawing, template synthesis, phase separation, self-assembly, weaving, melt spinning, and electrospinning [58]. Each technique, just like with anything else, has a complete set of advantages and disadvantages.

Melt Spinning. Melt spinning is the preferred method to manufacture polymeric fibers. The polymer is first melted and then pumped through a spinneret that has thousands of holes. Cooling takes place before the fibers are collected on a take-up wheel. Fibers can be stretched later to influence the structure and orientation. Typically, fibers on the order of 5–70 μm diameter are formed. Electrospinning is another fiber-generating technique that has gained popularity.

Drawing. Drawing involves the use of a pipette (a few micrometers in diameter) dipped into a viscoelastic solution (e.g., a droplet containing citrate molecules) and withdrawn rapidly (@100 $\mu m \cdot s^{-1}$). A nanofiber is pulled during withdrawal and deposited on a collector surface. Minimum equipment is required but drawing is a discontinuous process and scale-up would therefore become somewhat problematic [58].

Template Synthesis. Template synthesis is a generic means to produce nanowires from nearly any engineering material or combination of materials. Advantages include the manufacture of monodisperse fibers with specific diameter. Fiber length, however, is limited by the aspect ratio of the template pocket or template tube.

Phase Separation. In phase separation, nanofibrous poly(L-lactic)acid fibers (PLLA) are formed in a five-step process: (1) polymer dissolution of PLLA in tetrahydrofuran solvent to form a 1% w/v mixture, (2) gelation at −18°C, (3) solvent extraction with distilled water, (4) water removal and freezing, and lastly (5) freeze drying to produce a porous nanofibrous structure [59]. Advantages include tailored design of mechanical properties by manipulation of the polymer concentration, but a major disadvantage is that the process is limited to specific polymers.

Self-Assembly. *Self-assembly* is becoming exceedingly popular and is an excellent way to fabricate really small fibers but it often involves rather complex chemistry [58].

Electrospinning. Electrospinning provides an effective alternative to nanofiber manufacture and is exceedingly popular. Electrospun fibers with diameters of 250 nm have been a part of industrial materials for more than 20 years. Electrospinning employs electrostatic and mechanical force to spin fibers from the tip of a finely tuned orifice called a spinneret. The spinneret is held at a positive or negative charge by means of a DC power supply. The key principle involves the balance between the electrostatic charge and the surface tension of the polymer solution. Heat may be applied to keep the polymer in liquid form. When the electrostatic repelling force overcomes the surface tension, the viscoelastic liquid essentially spills forth from the spinneret to form finely divided but continuous filament. The filament is subsequently collected onto a rotating or stationary collector of opposite charge. The distance between the spinneret tip and the rotating collector is between 15 and 30 cm. Fibers ranging from 10 nm to 1 μm can be produced by this process [46].

The following polymer–solvent pairs (solvent in parentheses) are employed in electrospinning processes: nylon-6 and nylon-66 (formic acid), polyacrylonitrile (dimethylformamide, DMF), polyester (PET), (trifluoroacetic acid/dimethyl chloride), polyvinyl alcohol (PVA) (water), polystyrene (PS) (DMF/toluene), polyaramide (sulfuric acid), and polyimides (phenol) [46].

Commercial Synthesis of Carbon Fibers. Carbon fibers (a.k.a. graphite fibers or carbon whiskers) can be formed under pyrolytic conditions that utilize relatively large catalytic particles or by chemical and physical modification of polymers such as polyacrylonitrile (PAN). For example, long chains of PAN are drawn and subsequently aligned during a drawing process to form continuous filaments. The filaments are then heated to temperatures above 300°C in air and oxidized to disrupt the hydrogen bonding [6]. The oxidized form of PAN is then placed in an inert atmosphere (e.g., argon) and heat treated at temperatures 1500–3000°C to stimulate graphitization. Carbon precursor polymeric materials heated between 1500 and 2000°C possess high tensile strength because of the turbostratic carbon structure (due to carbonization). Carbon precursors heated to temperatures between 2500 and 3000°C have a highly graphitic structure and exhibit Young's modulus near 500 GPa [6].

Synthesis of Multiwalled Carbon Nanotubes. Carbon nanotubes (CNT) form under a variety of conditions—specifically, any environment in which there is a carbon source, anaerobic (pyrolytic) conditions, high temperature, and/or the presence of catalytic materials. Methods to fabricate carbon nanotubes include electric arc-discharge, laser ablation, solar furnace, and various forms of chemical vapor deposition (CVD). Arc-discharge and laser ablation methods are limited with regard to practical upscale as both are relatively energy intensive. The most practical means to generate nanotubes is afforded by chemical vapor deposition. Flame synthesis has become a most profitable commercial synthetic method.

The use of MWNT fillers to strengthen polymers and ceramics has become a $200 million dollar or more industry worldwide. Extremely effective means of producing pure multiwalled carbon nanotubes by a flame process are currently in use. For example, catalyst particles are placed on a support or seeded into a flame. Catalyst particles of Ni are generated by a thermal evaporation/condensation process with subsequent entrainment into a flame. The carbon source gas in many procedures consists of an ethylene/H_2 mixture to produce MWNTs [60]. Nanofibers are produced with a ternary gas mixture of $CO/C_2/H_2$ with Ni at 700°C [60]. Flame synthesis is capable of producing MWNTs in large quantities at a reduced cost [61]. A thermal evaporation technique is used to create the catalyst nanoparticles of Fe or Ni through gas condensation followed by entrainment into the flame. Each system yields consistent results, with CO/H_2 mixtures generally yielding single-walled nanotubes (SWNTs) with Fe while C_2H_2/H_2 mixtures usually produce multiwalled nanotubes (MWNTs) with Ni. A ternary gas mixture of $CO/C_2/H_2$ produces a better yield of nanofibers than either a CO/H_2 or C_2H_2/H_2 mixture at 700°C with a Ni catalyst [61]. A combination or perhaps a synergy between thermal—plus adsorbate-induced restructuring and adsorbate–particle steric factors affect particle structure and reactivity [61].

Commercial Synthesis of SWNTs. Several processes are available to manufacture SWNTs at levels viable for commercialization. Two of the most effective include the flame process and the HiPco process.

Flame Synthesis of SWNTs. A current commercially viable process to synthesize SWNTs is similar to the flame method discussed earlier for MWNT production. However, in this case, in place of a Ni catalyst with ethylene/H_2, Fe is used in a CO/H_2 gas mixture to generate SWNTs [60]. SWNTs have been detected in the postflame region (height-above-burner in millimeters, HAB) of a $C_2H_2/O_2/Ar$ flame at 50 torr with iron pentacarbonyl vapor. The iron pentacarbonyl decomposes in the flame to form the metal catalyst particles 5–10 nm in diameter in the 10- to –40-mm region of the flame [61]. After ca. 30 ms, incipient tubes were observed at 30-mm HAB, and nanotubes were detected after 40-mm HAB. Cluster (bundling) occurred between 40- and 70-mm HAB, and the growth rate was determined to be on the order of $10\,\mu m \cdot s^{-1}$ [61].

SWNTs are also produced in oxy-fuel inverse diffusion flames of high fuel-rich stoichiometric mixture fraction [62]. Fuel rich regions are devoid of soot and polyaromatic hydrocarbons (PAH). In an inverse diffusion flame, oxygen is introduced into the center of the flame and fuel around the periphery of the oxidizer (**Fig. 24.9**). As a result, carbon nanotubes are formed in the periphery of the flame and are not exposed to oxygen. The SWNTs on the order of 1 μm in length have been produced under such conditions.

Companies such as Nano-C, Inc., of Westwood, Massachusetts have developed processes to produce SWNTs by flame technology without the need for expensive gases and external energy supplies with facilitative scalability [63]. The flame, generated by premixed benzene–oxygen, can be adjusted with high specificity to produce SWNTs with low levels of contamination (no amorphous carbons, spherical fullerenes, and MWNTs) [63]. A catalyst precursor, iron pentacarbonyl $Fe(CO)_5$, is added to the nonsooting flame to form SWNTs [63].

If no catalyst is added to the mix, fullerenes are produced. Equivalence ratios φ are defined as the fuel-oxidizer ratio divided by the stoichiometric fuel-oxidizer ratio corresponding to the conversion of all hydrocarbons into CO_2 and H_2O. Acetylene and ethylene, for example, form SWNTs in the equivalence ratio range of 1.7–3.8. Good quality nanotubes are formed under conditions of acetylene fuel and O_2 oxidizer ($\varphi = 1.6$), Ar diluent at 18 mol%, $Fe(CO)_5$ catalyst precursor forming metal concentration of 6000 ppm, 50 torr, $30\,cm \cdot s^{-1}$ gas flow, HAB at 50 mm, and temperature profile at 1800 K @ 10 mm and 1500 K @ 80 mm [61].

HiPco Process. Large-scale production of single-walled carbon nanotubes is demonstrated by the HiPco process—short for "high pressure carbon

FIG. 24.9 Left: *Inverse diffusion flame is depicted. The flow of fuel is introduced around an oxidizer (usually air) core. In this configuration, SWNTs do not pass through the oxidation zone of the flame. SWNTs are collected downstream on a device such as a cold finger.* Right: *HAB versus φ shows regions of metal formation, nanotube conception, nanotube aggregation, and soot formation.*

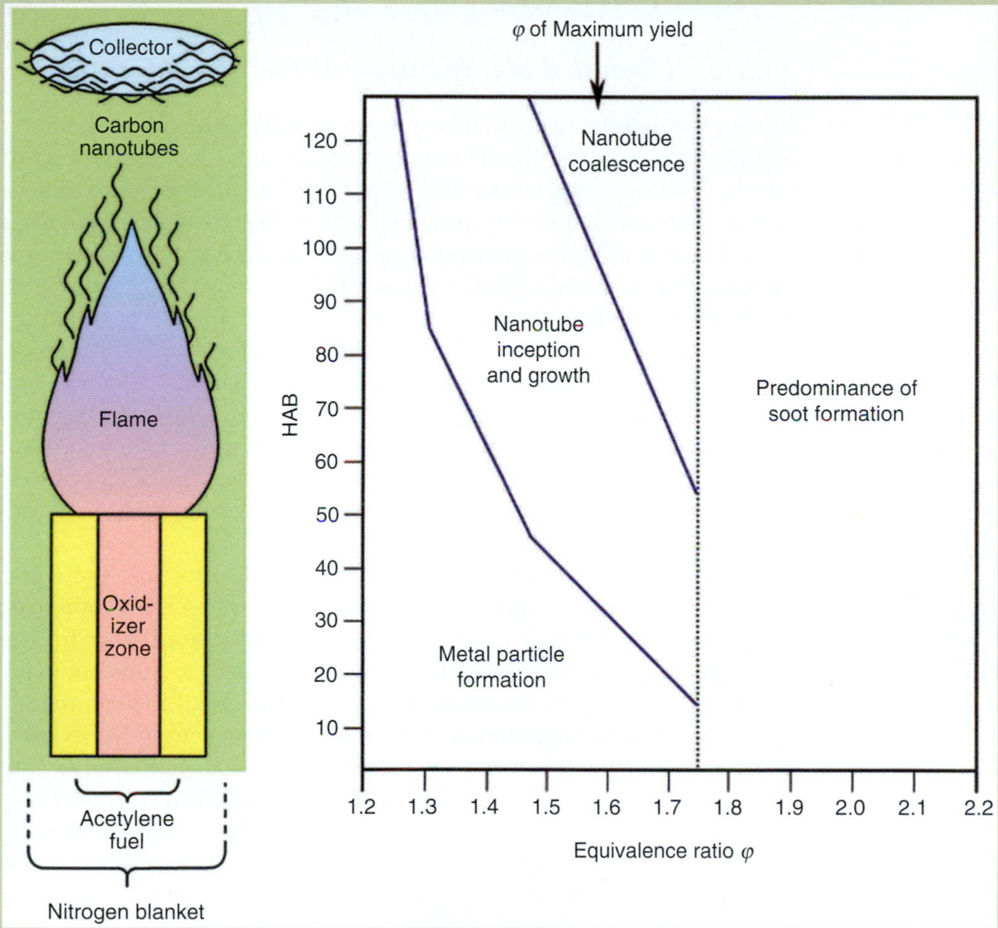

Source: Image adapted from M. J. Height, J. B. Howard, and J. B. Vandersande, Method and apparatus for synthesizing filamentary structures, US Patent 7335344, Publication Date: February 26 (2008).

monoxide." HiPco is a gas-phase CVD process operating under 30–50 atm and 900–1100°C. The carbon source material is flowing carbon monoxide (CO) and iron pentacarbonyl is the catalyst precursor material. Fe clusters are formed in situ from the decomposition (above 300°C) of the $Fe(CO)_5$ and subsequent condensation of reduced iron atoms in the gas stream. CO disproportionation occurs according to the following reaction (similar to the industrial Boudouard reaction)

$$2CO \rightarrow CO_2 + C_{SWNT} \tag{24.25}$$

Purity on the order of 97-mol% SWNTs and 3-mol% Fe is achieved by this process [64]. Standard production conditions are as follows: 30 atm, 1050°C, CO volume flow rate @ 8.4 $L \cdot min^{-1}$ (or 250 standard liters per min mass flow; slm), and 0.25 torr $Fe(CO)_5$ @ 1.4 8.4 $L \cdot min^{-1}$ (42 slm) for 24–72 h. The rate of production of SWNTs under these conditions was ca. 450 $mg \cdot h^{-1}$ or 10.8 $g \cdot day^{-1}$ [64].

24.3.3 Chemical Modification of Carbon Nanotubes

Although nanotube fillers in metals seem to work quite well, the same cannot be said for nanotubes in polymer matrices. In order for carbon nanotubes to become effective components in a composite (as we know, pure tubes are not easy to integrate within the polymer matrix due to their apparent "atomic smoothness"), the tubes must be chemically modified to provide sites that are capable of cross-linking with the polymer. There exist two generalized approaches to chemical modification of carbon nanotubes: (1) covalent bonding to the structure of the nanotube in which the alteration of nanotube properties occurs due to disruption of the resonance-stabilized conjugate structure, and (2) modification of nanotubes by weak molecular forces—in which the basic structure and properties of the nanotube are not significantly perturbed. Numerous chemical procedures of both kinds are available. Tubes are either synthesized in pure form or purified post de facto, then chemically modified, and integrated within the polymer matrix.

Depending on the application, tubes of monodisperse size and chirality are ideally desired (much progress has been made in recent years to synthesize, purify, and/or separate tubes to achieve this objective). Applications range from anticorrosion paints (high purity and uniformity not required) to thin conductive polymer films (higher level of purity and orientation required) to potential structural materials exposed to high stress situations (e.g., aircraft wings) that require longer, uniform, and highly pure tubes. One technological challenge faced by composite engineers is addressing the solubility issues associated with nanotube bundles—in general, the larger the bundle, the worse the solubility. Bundled nanotubes, in any event, are not desirable due to the "sword-in-the-sheath" effect, for example, that tubes found in the middle of bundles would easily slide out of the engineering position because the only force acting to hold them in place are van der Waals forces, not strong enough to overcome high mechanical external shear stresses. The same issues confront the incorporation of MWNTs because internal tubes are not held as strongly (only by van der Waals forces).

Chemical modification. Chemical modification techniques are summarized and discussed more intensely in chapter 9. Covalent derivatization occurs by well-known organic chemical methods: addition that includes cycloaddition, nucleophilic substitution, radical addition, electrophilic addition, hydrogenation, etc. Derivatization via intermolecular interactions occurs by well-known intermolecular reactions such as van der Waals, π-interactions, hydrophobic interactions, hydrogen bonding (once derivatized), and electrostatic methods, etc.

Covalent procedures include *oxidation* of terminal ends or sidewalls by treatment with strong acid at elevated temperature under reflux conditions. Addition of $-CO_2H$ groups activate CNTs for further substitution. *Halogenation* occurs with F_2 by 1,2-addition or 1,4-addition at 150–400°C. Further derivatization is achieved with other well-known chemical processes such as alkylation with Grignard and organolithium agents as well as diols and diamines. Substitution with terminal amino groups render aminoalkylated CNTs soluble in water. Cycloaddition with Cl_2 occurs with application of CH_2Cl_2. *Hydrogenation* is achieved with Li metal and methanol in liquid NH_3, glow discharge, or atomic bombardment. Nanotube walls appear corrugated following hydrogenation. Stoichiometry of hydrogenated nanotubes is $C_{11}H$. Hydrogenated tubes are stable to 400°C. *Cycloaddition* occurs readily with carbenes, nitrenes, azomethine ylids (that form pyrrolidine fused rings), and nitrile imines. Pyrrolidine rings are useful centers for further reactions. Examples include Diels–Alder and dipolar cycloaddition. Radical addition methods by thermochemical, photochemical methods are also utilized for functionalization. Ozonolysis, mechanochemical functionalization, and plasma activation are other means to modify the chemistry of nanotubes.

Noncovalent nanotube functionalization methods include polymer wrapping. MWNT and SWNT polystyrene (PS), poly(vinyl alcohol) (PVA), polyhydroxyaminoether (PHAE), and epoxy thermoset composites are produced following solvent evaporation. Some problems include aggregation of tubes limits solubility. Low loading due to saturation at 1%–2% results. Other noncovalent methods include polymer grafting "To" and polymer grafting "From." Grafting to oxidized tubes is also practiced.

Chemical Modification of SWNTs. Although SWNTs have superior mechanical, thermal, and electrical properties, realization of their advantages in composites requires special preparation of the nanotubes. By chemical functionalization, the interfacial binding between the polymer host and the SWNT becomes enabled—hopefully without compromising the properties of the nanotube. Without such chemical preparation, the SWNT is not able to transfer exterior loads encountered by the composite.

Functionalization has other advantages. Neat SWNTs are not very soluble. Depending on the chemical nature of the functional group, SWNTs can be dispersed in aqueous as well as organic solvents. Homogenious dispersal of SWNTs within a polymer results in better overall mechanical performance of the composite.

Why is this important? SWNTs usually exist in the form of tightly bound ropes—aggregations of van der Waals bonded SWNTs producing bundles that can be on the order of micrometers in thickness. Such ropes are virtually insoluble in any solvent. The mechanical properties of polymers that contain SWNT ropes are not enhanced to the degree theoretically possible. For example, SWNTs located within the rope will have a tendency to slide over one another, similar to graphite planes. Chemical modification occurs by linking to SWNTs with strong covalent bonds or by linking with intermolecular interactions.

FIG. 24.10 Left: *Chemical modification of SWNT: (1) oxidation with strong acids, reflux, temperature > 100°C, to produce carboxylic acid reaction sites –COOH, (2) chemical reaction with acetyl amide to form reactive terminal group, (3) and (4) addition of free radical initiator and styrene monomer to initiate polymerization process, (5) propagation and subsequent termination of the polymerization process to form a polystyrene.* Right: *Chemically modified SWNTs aligned in an electric field are cross-linked by polystyrene chains. Anisotropic mechanical properties result from such alignment [43].*

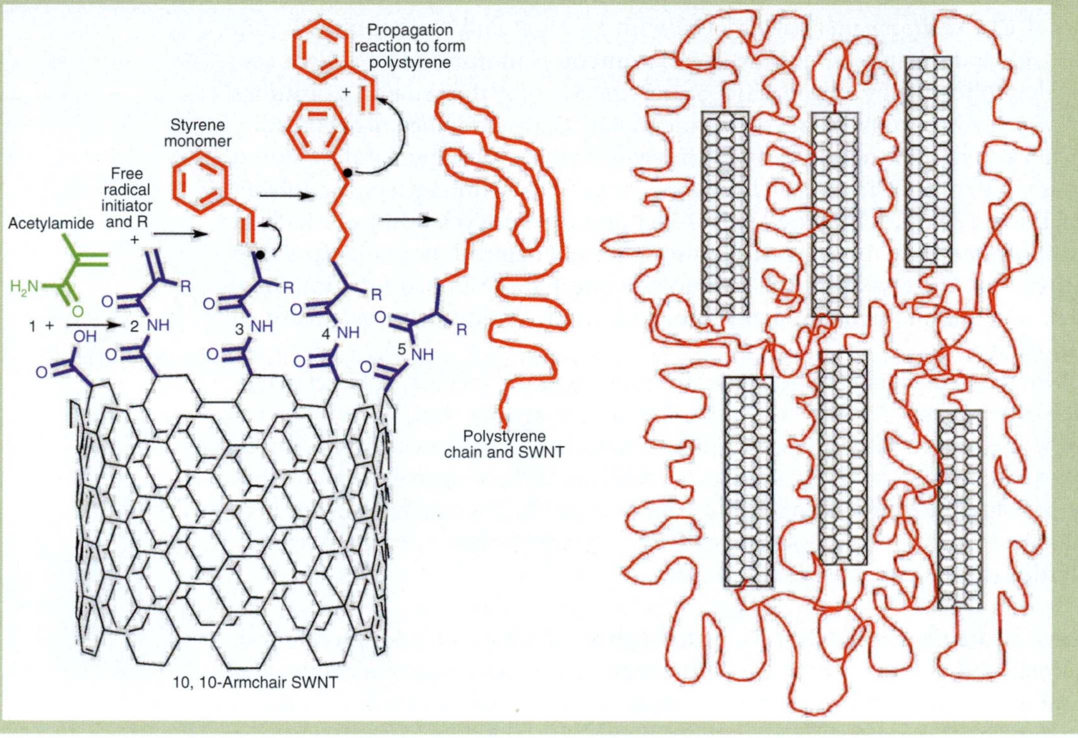

Source: Image adapted from E. V. Barrera and K. Lozano, *Journal of Materials*, 32, 38–42 (2000).

Oxidation of SWNTs by treatment in strong acids at elevated temperatures under reflux conditions yields open-ended SWNTs stabilized by carboxyl groups (Fig. 24.10).

24.3.4 Carbon Nanotube Applications

From yarns to "beds of nails," there seems to be no limit in the applications of carbon nanotubes.

SWNT Bed of Nails. SWNTs have been arranged in a configuration known as a "bed of nails" [65]. The SWNTs were packed in a hexagonal close-packed 2-D triangular lattice structure. The length of the tubes (membrane thickness) was on the order of 75 nm and the density of tubes in the membrane was ca. $10^{14} \cdot cm^{-2}$. The diameter of the tubes is tunable between 0.4 to 3 nm, with applications in batteries (intercalation of Li^+), Pt nanoparticle dispersion, and molecular transport with high sensitivity and flux [65]. The membranes were formed by application of the FIB (focused ion beam) mill perpendicular to a previously prepared neat fiber of SWNTs.

Carbon Nanotube Woven Fibers. Perhaps one of the most amazing breakthroughs in nanotechnology (and there have been many) occurred in the laboratory of R.H. Baughman et al. at the University of Texas at Dallas' NanoTech Center (**Fig. 24.11**). Baughman and his team have managed to weave 100 m (that's meters!) of nanotube composite fibers that exhibit better toughness than

FIG. 24.11 *Woven nanotube yarns from the Baughman group at the NanoTech Center at the University of Texas at Dallas. (a) Fibers/yarns are woven in a way similar to the way early humans did it—by rolling random strands until a fiber is formed. (b) Several weaving and braiding steps are required to make the yarn seen at the left. 10-nm diameter agglomerated MWNTs are woven into 30-nm diameter bundles. The 30-nm bundles form a yarn that is 1 μm in diameter. The next step is to form a plied yarn that is 6 μm in diameter. And finally, the yarn is braided to form the image in the figure to the left—40 μm or 0.04 mm in diameter. (c) A fabric is woven from the yarns, representing a higher level in the nanotube hierarchy. These accomplishments are absolutely amazing.*

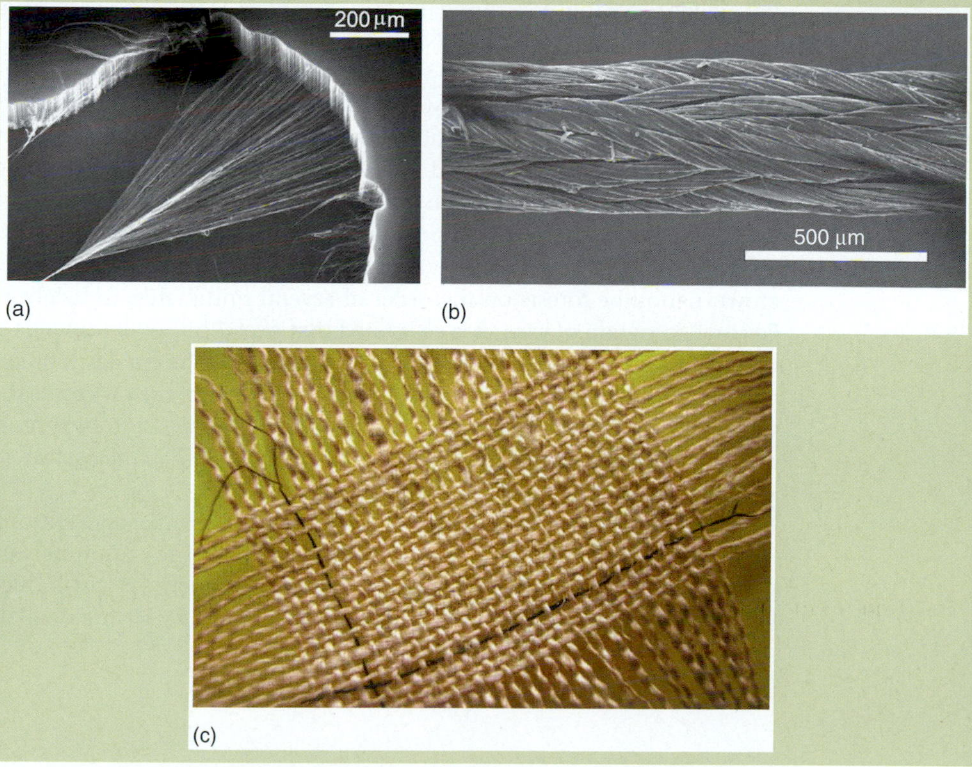

Sources: (a, b) M. Zhang, K. R. Atkinson, and R. H. Baughman, *Science*, 306, 1358–1361 (2004); (c) A. B. Dalton, S. Collins, E. Munoz, J. M. Razal, V. H. Ebron, J. P. Ferraris, J. P. Coleman, B. G. Kim, and R. H. Baughman, *Nature*, 423, 703 (2003). With permission.

any natural or synthetic organic fiber [66]. The composites consist of carbon nanotubes in a PVA (polyvinylalcohol) matrix. Toughness is the energy required, in this case, to rupture a fiber. Spider silk, for example, is five times tougher than steel of the same mass. These carbon nanotube composite fibers are 20x tougher than steel, 17x as tough as Kevlar, and 4x as tough as spider silk.

Kaili Jiang et al. at the Department of Physics at Tsinghua University in Beijing developed a method to "self-assemble" yarns made of carbon nanotubes by drawing them out from super-aligned planar arrays [67]. The process is actually similar to the way wool was drawn and spun since antiquity. String was made in Paleolithic times (ca. 20,000 years ago) by rolling tufts of animal hair up and down one's thigh into a fiber. The length of the fiber was determined by the amount of hair added. The process also works with grass. Just take some grass and roll it back and forth in your hands. You'll end up getting a strand in which the individual blades of grass are intertwined.

Ray Baughman embellished the method by introducing a twist during the spinning to make them "multi-plied" and torque stabilized [68]. MWNT sheets can be spun at a rate of $7\,\mathrm{m \cdot min^{-1}}$ from "nanotube forests." The rate can be increased to $30\,\mathrm{m \cdot min^{-1}}$ if lower quality (wool) is required. The MWNT sheets initially form as an aerogel that eventually condense into sheets as thin as 50 nm. A process called densification is required to impart strength to the sheet. Densification is accomplished by twisting the yarn. In the aerogel phase, the material has no strength. Densification, in addition to strengthening the material, aligns the nanotubes along the axial direction of the spun yarn. The process can be applied to MWNTs, DWNTs, and SWNTs.

According to Baughman, the yarns possessed some quite remarkable properties: strength greater than 460 MPa, hysteretical deformation with 48% energy damping, tough as fibers in bullet-proof vests, and no degradation at 450°C following immersion into liquid nitrogen. Yarn strength increased with polymer infiltration as demonstrated by high creep resistance and high electrical conductivity test results [68]. The physical properties as one might imagine, would be enhanced if longer nanotubes can be utilized in the yarn. Baughman et al. have grown nanotube forests on the order of several millimeters in height (Zhang). For each kilogram of yarn, it is estimated that over 3 billion kilometers of nanotubes must be incorporated into the yarn. Apparently the rate-limiting step in the process is not the weave step. The rate-limiting aspect of the process is the rate of CVD growth of carbon nanotubes. Scale-up is not expected to be a problem for this nanomaterial. The strength of such sheets is better than the best strength that steel can afford. The power of nano! Truly incredible.

The nanotube yarns can be configured into knots. Unlike conventional yarns, knots in nanotube yarns do not lose strength [69]. The nanotube yarns (and yarns in general) reversibly dissipate mechanical energy. Twist-induced reinforcement of the strength of the yarn is approximated by Hearle's equation:

$$\frac{\sigma_Y}{\sigma_F} \approx \cos^2 \alpha (1 - k \operatorname{cosec} \alpha) \tag{24.26}$$

$$k = \sqrt{\frac{dQ}{\mu}} \Big/ 3L \tag{24.27}$$

σ_Y/σ_F is the ratio of yarn strength to the strength of the fiber; α is the angle of the helix with respect to the axis of the yarn; d is the fiber diameter; μ is the coefficient of friction, L is the fiber length and Q is the fiber migration length. Strength is less for a yarn than a fiber making up the yarn. According to this equation, the $(1-k\,\text{cosec}\,\alpha)$ term describes the transfer of tensile stress to transverse stress, for example, the locking of fibers [69]. To get the best performance, decreasing the nanofiber diameter d, increasing the fiber length L, and/or increasing the coefficient of friction μ need to be done.

Applications of the yarn are straightforward as structural reinforcement elements, but applications of the textile (sheet) are potentially quite unusual. These include polarized incandescent light, microwave plastic welding, electrodes (transparent, elastic), conducting appliques (decorative fabrics that are optically transparent, electrically conducting, and microwave absorbing), and flexible OLEDs [69].

Cellular Foams. Condensed carbon nanotube free-standing foams with shock absorbing structural reinforcement behavior were synthesized by capillary-driven assembly [70]. Vertically aligned MWNT arrays were grown on patterned silica Si(100) substrates by CVD at 800°C. The nanotube arrays were then exposed to oxygen plasma in a glow-discharge chamber for 10 min. The MWNTs did not degrade in the chamber. The product was then immersed in solvents such as acetone, toluene, DMF, THF, or methanol. Evaporation of various liquids occurred at room temperature from the interstices to form the foams. Annealing at 800°C did not change the conformation. Foams with intricate open cellular structure were formed by the collapse and condensation of the nanotubes. These advanced materials demonstrated flexibility and good mechanical strength [70].

24.4 Organic Polymer Nanocomposites

What is it about nanomaterials (zero-, one-, and two-dimensional) that make them excellent inclusion materials in composites? How do nanomaterials, as opposed to fillers with larger size, affect the structure–property relationship between filler and matrix, for example, the interfacial area? Traditional polymer chemistry has taught us about the relationship between and among molecular weight, chain length, architecture, fillers, ordering, and functional groups and scaffolds with solubility, rheology, mechanical properties, and other chemical behavior. Also the effects of solvent, size, chemistry, and fabrication conditions are well understood. What then happens to such baseline polymers when nanomaterials are inserted? How are mechanical, physical, and chemical properties impacted by nanometer-sized inclusions of varying size, shapes, compositions, and orientations—whether in the form of particles, tubes, porous materials, discs, wires, or blocks?

Nanoscale building blocks are individually remarkable because of several reasons. They are free of defects, reactive, and have unusual physical properties. It is however, not a simple task to keep these remarkable properties intact in a macroscale composite.

24.4.1 Introduction to Polymers

We start this section on nanocomposites with polymers based on organic carbon. Polymer chemistry was introduced in Part I [21]. There are many kinds of polymers, but our discussion early on is limited to those based on organic carbon. Many inorganic materials are capable of polymeric reactions that connect building blocks into chain and network configurations. Examples of inorganic polymers include silicates, silicones, and glasses [11].

Brief History of Organic Polymers. The ancient Mayans used extract of rubber trees to make balls. Goodyear in 1839 improved the performance of rubber products by treating the polymer at 132°C in the presence of sulfur, a process called vulcanization that results in cross-linking of polymer molecules. In 1907, the first synthetic plastic, called Bakelite, was used as an insulating material due to its hardness and high heat resistance properties. In 1917, the molecular structure of cellulose was determined by an x-ray diffraction technique. In 1920, Nobel Laureate Hermann Staudinger revealed that polymers consisted of long chains made of molecular repeating units called monomers. Poly(vinyl chloride), or PVC, made the scene in 1927, polystyrene in 1930, nylon in 1938, and poly(ethylene) in 1941. Moldable high-temperature polymers were developed by 1970 and Kevlar was invented in 1971. Kevlar is an example of a fiber-reinforced polymer with extreme hardness and excellent temperature resistance (<300°C). By 1976, polymers and plastics overtook steel as the most widely used material in the United States.

Polymers offer an easily attainable, commercially viable route to stronger, lighter, less expensive, and versatile materials. Polymers are relatively easy to work with and do not require extreme conditions during their synthesis. Raw materials are readily available (at least until oil runs out). It is also relatively easy to embed inclusion materials within the matrix of a polymeric host. The rich chemistry of polymers allows us nearly an unlimited store of reactions to draw from to fabricate a composite.

Selection of the monomer, modification of side chains, various cross-linking strategies, materials concentration, and external control parameters such as temperature and pressure yield nearly an unlimited store of polymer materials. It is no wonder that polymers (composites made from them), and nanocomposites in particular, will establish a new class of materials with future promise in the field of high-performance materials.

Devising the order of presentation in this section was somewhat challenging. We decided that presenting polymer nanocomposites (PNCs) first, in a general sense, was a good idea because so many classes of nanocomposites are based on organic polymeric host matrices. There are, for example, clay–polymer nanocomposites, metal-polymer nanocomposites, and carbon fiber-reinforced polymer nanocomposites, just to name a few. In addition, several characteristics of polymers need to be understood before we proceed into sections describing other types of nanocomposites, and that terms associated continually with polymers need to be defined.

Polymerization. Although we will not spend too much time in this area, the basics are reviewed briefly. Polymerization is a process by which small chemical units (called monomers) are linked together by means of a chemical reaction. There are two basic fabrication schemes: addition (or chain) growth and step (or condensation) growth. Chain growth relies on the carbon–carbon double bond across which addition occurs readily, and condensation growth relies on molecules with bifunctional active groups that condense (react) to form water. Proteins are assembled via a condensation process to form peptide bonds. Chain growth proceeds in the presence of free radical, anionic, or cationic catalysts. Initiation, propagation, and termination (terms that are relatively self-explanatory) are the three general phases of this kind of polymerization.

If different kinds of monomers exist within the polymeric soup, a *copolymer* is produced that is similar in concept to a metal alloy [11]. *Block copolymers* (materials extremely important in nanotechnology) consist of ordered domains of repeating units that arise from different monomers. Polymeric templates and other nanostructures are based on block copolymer materials. *Polymer blends* consist of mixtures of preexisting polymers. *Linear polymers*, as the term implies, are made of long-chain hydrocarbons that can be linked. The degree of polymerization n is derived from the ratio between the molecular weight of the polymer and monomer, respectively. Chain length is approximated by the following relation:

$$L_{\text{chain}} = N_{\text{CC}} L_{\text{bond}} \sin\left(\frac{109.5°\text{C}}{2}\right) \tag{24.28}$$

where

N_{CC} is the number of carbon bonds

L_{bond} is the bond length (0.154 nm)

Placement of side groups can be *isotactic* (all on one side), *syndiotactic* (alternating), or *atactic* (placed irregularly) [11]. The kind of placement influences the overall structure of the polymer as well as the potential for cross-linking. *Branching* is a consequence of reaction conditions, monomer type, and side group steric factors.

Thermoplastic polymers consist of high molecular weight linear chains that are branched (no cross-linking) or unbranched that soften when heated. Conversely, thermoplastics freeze into a brittle state when cooled sufficiently. The chains of thermoplastic polymers interact primarily by van der Waals forces (polyethylene), dipole–dipole interactions, and hydrogen bonds (nylon) or aromatic stacking (polystyrene). Other examples include addition polymers such as polypropylene. Thermoplastic polymers are characterized by a parameter known as the glass transition temperature T_g, the temperature at which flexibility is attained. Thermoplastic polymers, in general, can be remelted.

Thermoset polymers, on the other hand, exhibit better mechanical properties due to higher cure temperatures and extended cross-linking. Most thermoset plastics contain a 3-D network of covalently bonded molecules. Curing of thermosets, therefore, is an irreversible process. Thermosets are used in structural applications and as adhesives. Bakelite (insulators), vulcanized rubber (tires), epoxy resins (graphite reinforced plastics), melamine resin (hard coatings), urea-formaldehyde (plywood), and polyimides (printed circuit boards) are some examples of thermoset plastic polymers. Thermosets exhibit

Fig. 24.12 *Organic polymers can be:* Top: *layered (laminate configuration);* Middle: *branched; or* Bottom: *highly cross-linked.*

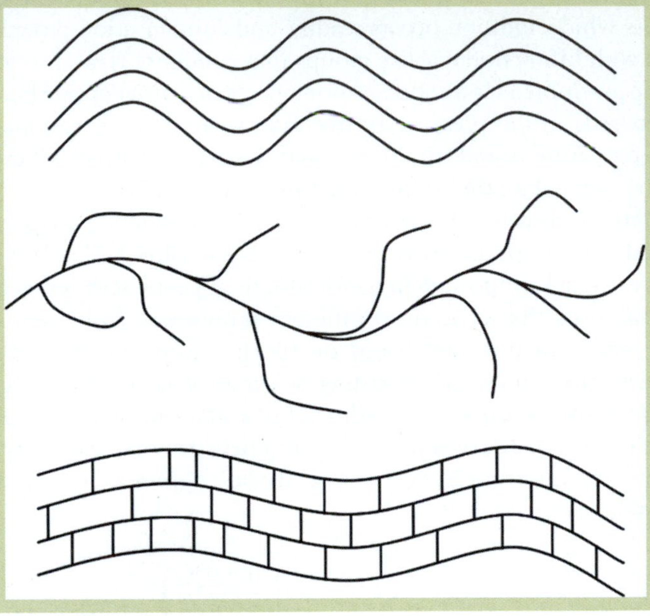

Source: C. R. Brooks and A. Choudry, *Failure analysis of engineering materials*, McGraw-Hill, New York (2002). With permission.

high thermal stability, rigidity, and dimensional stability, are resistant to creep and deformation under load, have high strength-to-mass ratios when compared to metals, and demonstrate high electrical and good thermal resistance properties.

Basic Types of Polymers. The basic types of polymer structure are shown in **Figure 24.12**.

24.4.2 Interfacial Area

A significant resource that helped place this section into focus was provided by an excellent article by Linda S. Schadler et al. of the Rensselaer Polytechnic Institute (RPI) [28]. By definition, polymer nanocomposites contain fillers with sizes between 1 and 100 nm—along one or more dimensions. Such a small size is a major advantage to nanocomposites due to a dramatic increase in interfacial contact area over composites with larger fillers or fibers (please refer to example 24.1). The result is a composite matrix that has a significant percent of its volume directly associated with the interfacial area. Then, as stated earlier, the mechanical properties of the composite are dominated by the tightly bound interfacial phase. There are many other compelling reasons that structural perfection is achieved as the reinforcing elements approach smaller dimensions. Interestingly, the creation of such a large interfacial volume requires less filler material and not more (e.g., lower loading). For example, composites with micrometer-sized

EXAMPLE 24.1 *Surface Area and Fill Volume*

Conduct a first approximation of specific surface area of a spherical gold nanoparticle filling material of (a) 1 µm, (b) 100 nm, and (c) 1 nm diameter in a generic polymer matrix of $V_{polymer} = 1\ cm^3$ if the fill factor ε is 4.5 vol%. The density ρ of gold is $19.3\,g \cdot cm^{-3}$.

Solution:

Surface area in m^2, $A_p = 4\pi r_p^2$

Specific surface area in $m^2 \cdot g^{-1}$, $S = \dfrac{A_p}{\rho V_p}$

Total surface area in m^2, $A_T = \sum N_p A_p$

Particle volume in m^3, $V_p = \frac{4}{3}\pi r_p^3$

Volume of fill factor, $V_{ff} = V_{polymer}\ \varepsilon = 1\ cm^3\ 0.045 = 4.5 \times 10^{-8}\ m^3$

(a) Number of 1-µm particles → $N_{1\mu m} = \dfrac{\varepsilon}{V_{1\mu m}} = \dfrac{4.5 \times 10^{-8} m^3}{5.2 \times 10^{-19} m^3} = 8.6 \times 10^{10}$ particles

Specific surface area: $S_{1\mu m} = \dfrac{3.1 \times 10^{-12} m^2}{19.3\ g \cdot cm^{-3} \left(\dfrac{100\ cm}{m}\right)^3 \left(5.2 \times 10^{-19} m^3\right)} = 0.31\ m^2 \cdot g^{-1}$

Total surface area: $A_{T-1\mu m} = (8.6 \times 10^{10})(3.1 \times 10^{-12}\ m^2) = 0.27\,m^2$

(b) Number of 100-nm particles → $N_{100\,nm} = 8.6 \times 10^{13}$ particles
Specific surface area: $S_{100\,nm} = 3.1\,m^2 \cdot g^{-1}$
Total surface area: $A_{T-100\,nm} = 2.7\,m^2$

(c) Number of 1-nm particles → $N_{1\,nm} = 8.6 \times 10^{19}$ particles
Specific surface area: $S_{1\,nm} = 313\,m^2 \cdot g^{-1}$
Total surface area: $A_{T-1\,nm} = 277\,m^2$

What would the specific surface area be if the nanoparticle was not gold but rather titania ($\rho = 4.23\,g \cdot cm^{-3}$)?

fillers require loadings on the order of 60 vol% as opposed to < 5 vol% with nanoparticles [28]. This allows for great flexibility in material design but simultaneously makes predictions about mechanical properties more difficult than for traditional composites.

Nanomaterials, nanoparticles in particular, have several inherent advantages due to their small size over their larger counterpart materials. First, collective surface area can be enormous leading to enormous interfacial linking; second, nanoparticles, due to their small size, have access to sites not available to larger particles; third, nanoparticles have higher surface-to-volume ratio leading to increased reactivity; lastly, nanoparticles can be more soluble. Therefore, the interfacial polymer chain structure, mobility, kinetics, and local chemistry all differ from the bulk polymer material. The structure of the interphasic region is different from the both polymer and the filler material. Obviously, if the interfacial region is homogeneous and covers a large surface area, the properties of the interphase region control the behavior of load transfer in the composite. Because nanomaterials have large surface area, the interphase, depending on the amount of filler material, is capable of dominating the mechanical behavior of the nanocomposite. There is still much to be learned about the interphase.

FIG. 24.13 *The interfacial region is highlighted in the image. As a consequence of nanoparticle surface area, the interfacial region can be quite large in a nanocomposite. The nanoparticle is also capable of inducing crystallization of the matrix within its vicinity.*

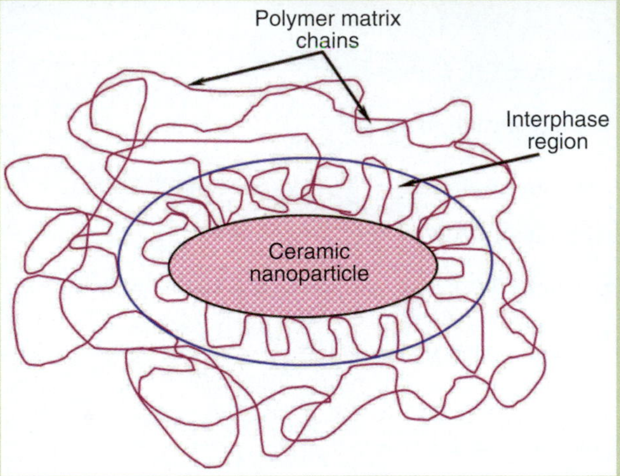

As stated previously, nanoscale fillers are relatively free of defects. Micrometer-sized fillers, on the other hand, scale with critical crack sizes that are known to cause mechanical failure [28]. Nanosized fillers would not be subjected to this form of mechanical failure. Therefore, a large volume of interfacial area that is radically different from that of the matrix host polymer material leads to better properties in this sense [28,71,72]. Crystallinity, mobility, chain conformation, molecular weight, chain entanglement density, charge distribution, the phases present, and cross-link density due to migration of small molecules where appropriate are some characteristics of the polymer at the interface that are impacted by the presence of the nanofiller, according to Schadler of RPI [28] (**Fig. 24.13**).

24.4.3 Nanofilled Composite Design, Synthesis, and Properties

Composite design requires consideration of the filler and the matrix. What feature do you wish to improve? Elongation? Toughness? Storage modulus? Tensile strength? Impact resistance? How does one go about fabricating a composite with enhanced properties? What is your choice of matrix? Choice of filler? Choice of chemical modification? Fiber length? Fiber diameter? Percent inclusion? All in all, chemical engineers must be ecstatic when confronted with all these options, permutations, and possibilities. It would indeed seem that anything one tries would result in a new material. We provide a few examples of nanocomposite design, synthesis, and performance. Although SWNTs, MWNTs, and carbon nanofiber fillers possess superior mechanical, thermal, and electronic properties than those of polymer hosts, their addition to polymers have not always improved the mechanical properties to the extent possible [66]. Many factors have to be considered.

Nanoparticles. There are many examples of nanofillers with enhanced interaction zone (large interfacial, larger interfacial volume, and better particle–polymer bonding) that show improved material properties. Addition of 1.5-vol% nanosilica to cross-linked polyethylene nanocomposites improved electrical endurance strength (breakdown strength and voltage endurance) by an order of magnitude compared to that of micrometer-sized fillers due to better local conduction and electron scattering [28,73]. The dielectric permittivity, space charge distribution, and dynamics also were significantly altered in the nanopolymer [28,73].

Nanofibers. A strong correlation between polymer ordering at the interface and reinforcement (Young's modulus) was determined for nanocomposites with multi-walled nanotubes in PVA [28,74]. The reinforced polymers showed significant increase in Young's modulus over controls. PVA is a polymer that is capable of "nanotube-induced" ordering [28,74]. The experimental data exceeded FOM and Mori–Tanaka expectations (ideal input parameters of straight aligned nanotubes, nanotube-matrix bonding, and nanotube properties) [28,74]. Interfacial effects are not included in these models. According to Jonathan N. Coleman et al. [74], the results are due to enhanced interfacial phenomena:

> Rather than acting as intrinsically stiffer reinforcing agents, our results suggest that the major role played by the nanotubes in improving the mechanical properties of composites is to nucleate an ordered polymer coating. It is the presence of this stiff ordered phase that dominates the reinforcement mechanism

The extent of the interface and other characteristics can be tuned by specific chemical modification of the carbon nanotubes. As a matter of fact, it was shown that the addition of MWNTs in semicrystalline polymers resulted in fibers with a higher specific energy to fracture than even Kevlar [75,76].

γ-Alumina Fillers. In poly(ethylene terephthalate) (PET) thermoplastic composites, γ-alumina nanofiller was able to alter the crystalline content of the PET matrix. At 1-wt% (<1-vol%) loading, the spherical γ-alumina was dispersed efficiently and nucleation and growth were inhibited relative to PET controls. With increased levels of filler and subsequent agglomeration, the γ-alumina stimulated nucleation and growth of PET, controlled the lamellar structure, and decreased the tendency of PET fibers to fibrillate (the tendency via abrasive action to develop smaller fibers) [28]. Although surface fibrillation is a good thing in adhesives, longitudinal fibrillation in composites may result in low composite strength under applied transverse stress [29].

What's the bottom line you ask? The structure and properties of the interface are different from the bulk but also are able to dominate the behavior of the nanocomposite leading to improved mechanical behavior [28].

24.4.4 Enhanced Polymer Nanocomposites

Interfacial Friction Damping. Interfacial shear, a mechanical phenomenon that is detrimental to high stiffness and strength in a composite, is beneficial to applications that require mechanical damping. Interfacial slippage may result between nanotubes in bundles, concentric nanotubes in MWNTs, and between nanotubes and the coordinating polymer chain [77,78]. Damping materials are required for acoustic (noise) and vibration suppression in dynamic systems [77]. Viscoelastic polymer-based damping systems suffer from compactness issues, unreliability, a high weight hit, low thermal conductivity and poor performance at elevated temperatures [77]. Nanotubes in polymers offer increased interfacial area contact (better load-mass transfer), high aspect ratio, and low mass density. These features allow for interfacial sliding of nanotubes for enhanced energy dissipation, seamless integration, and low-weight penalty without losing the mechanical properties of the nanotubes or structural integrity of the composite [77].

Tube-within-a-tube sliding was demonstrated by MWNT nanocomposites, but the required loading fraction was too high, ~50% to induce sliding. Apparently weak van der Waals forces between concentric tube layers are an ineffective means of dissipating energy [77,78]. Direct shear testing of epoxy thin films with dense packing of MWNTs revealed strong viscoelastic behavior with up to a 1400% increase in damping (ratio) compared to the baseline epoxy material. The mechanical properties (strength and stiffness) of the polymer remained intact. The interfacial sliding in MWNT composites was due to nanotube–nanotube interfaces [77].

J. Suhr et al. of Rensselaer Polytechnic Institute demonstrated the effect of temperature on SWNT inclusion/polycarbonate composites. They found that the NT-polymer interface activated at low dynamic strain levels of ~0.35% if the temperature was raised to 90°C. The increased mobility of the polymer chain backbone and thermal relaxation of "radial compressive stresses at the NT-polymer interface" were responsible for the enhanced damping, according to researchers [77]. Composite beams showed under dynamic cycling load tests greater than 1000% increase in the loss modulus of the nanocomposite with 2-wt% fraction SWNT filler. The damping increase was due to frictional sliding at the nanotube–composite interfacial contact areas without sacrificing the mechanical strength and stiffness of the composite [77].

Synthesis of SWNT-Polycarbonate Nanocomposites. HiPco SWNT bundles were first oxidized by sonication in nitric acid to generate carboxylic acid groups on the SWNT walls. Carboxylic acid groups enhance exfoliation of SWNT bundles. The exfoliation, however, was not complete as 35-nm diameter rope bundles were detected after treatment with nitric acid. Modified SWNTs were transferred to tetrahydrofuran (THF), sonicated, and then mixed with polycarbonate (also in THF) in a predetermined proportion (to a maximum of 1.5 wt%) required for the nanocomposite (**Fig. 24.14**). Dipole–dipole intermolecular interactions between the carboxylic acid groups on SWNTs and the carbonate groups of the polymer initiated good dispersion within the matrix [77].

The composite mixture precipitated immediately after resonication and drop-wise introduction into anhydrous methanol (the antisolvent for polycarbonate) in a 1:5 THF:methanol ratio. Following filtration and drying under vacuum, the material was placed in a standard tensile test mold and hot-pressed (205°C). Baseline polycarbonate control specimens (without the nanotube filler materials) were made by the same procedure [77]. Uniaxial dynamic cyclic loading tests were conducted from −60 to 90°C, maintained for 10 min at each ten degree increment of temperature within that range. The storage and loss modulus of the nanocomposite test coupons and controls were calculated.

FIG. 24.14 *SWNT–polycarbonate component is depicted. The sliding occurs at the interface between the two materials of the nanocomposite.*

Source: Image adapted from J. Suhr, W. Zhang, P. Ajayan, and N. Koratkar, *Nano Letters*, 6, 219–223 (2006). With permission.

Interfacial Slip Measurements. A novel "nano means" to measure interfacial slipping by AFM was accomplished by J.D. Whittaker et al. in 2006 [79]. A suspended single nanotube across a lithographically formed trough was used to study the nanotube–silicon dioxide adhesion forces (**Fig. 24.15**). Trenches, 300–400 nm in width and 40–50 nm in depth, were made of silica fabricated by e-beam lithography and dry etching techniques. Tubes were embedded in oxide

FIG. 24.15 *The Veeco probe used in this study possessed a nominal tip curvature of 20 nm. The spring constant of the probe was 0.24 $N \cdot m^{-1}$. The tip curvature and spring constant of a MikoMasch probe were 10 nm and 0.14 $N \cdot m^{-1}$, respectively. The probe was pushed onto the nanotube surface. Deflection of the tube and slippage from the anchor points on the trough were monitored.*

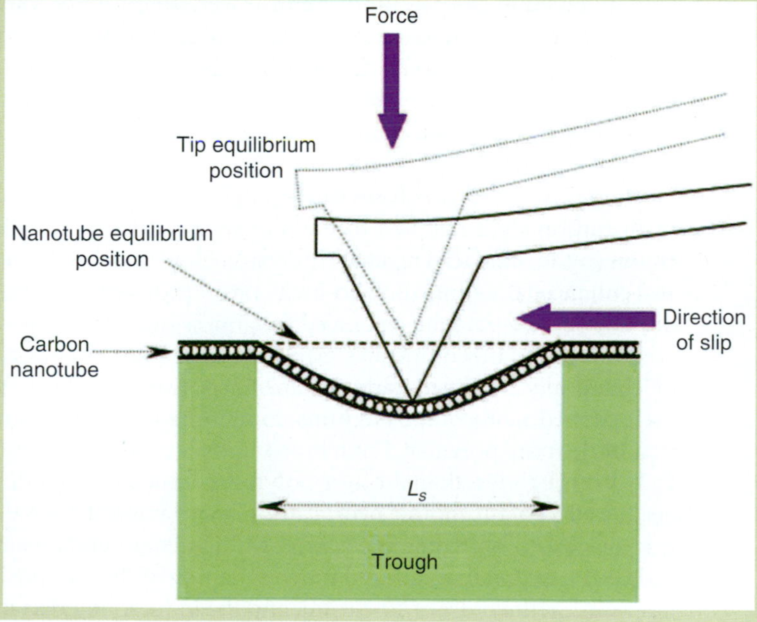

Source: Image adapted from J. D. Whittaker, E. D. Minot, D. M. Tannenbaum, P. L. McEuen, and R. C. Davis, *Nano Letters*, 6, 953–957 (2006). With permission.

by a CVD technique. Measures were taken to ensure that no oxide was applied to the suspended portion of the nanotube. Pressure was applied via an AFM tip to the nanotube (**Fig. 24.15**). The slipping force depends on the geometry of the nanotube–oxide interface by measuring the degree of slipping of different nanotube–oxide contact lengths. Measurements were conducted on nanotubes on bare silica and those embedded in silica.

Apparently, 7–8 nN applied tension are required to cause slippage along the SiO_2 surface, but an embedded nanotube requires 10 nN of applied tension before slippage occurs [79]. Also, the researchers found that nanotubes stretch up to 10%, confirming results obtained by other laboratories obtained by other means. A truly nanometer-scale method to measure interfacial structures!

Composite Toughening Mechanisms with CNFs. Modified carbon nanofibers (MCNFs) were chosen as the filler material due to their facilitated mass production and anisotropic shape. CNF diameter (50–200 nm) lies between SWNTs (a few nm), MWNTs (20–40 nm), and conventional carbon graphitic fibers (1–10 μm). It was found that although CNFs did not influence the modulus and strength of the matrix, they most certainly were able to improve the toughness of the composite (the elongation-to-break ratio) [80].

Ultrahigh molecular weight polyethylene (UHMWPE) was chosen as the polymer matrix because it is made of very long chains with MW in the millions, has a high entanglement density, and has high modulus and tensile strength. Long chains improve the load transfer to the polymer backbone by strengthening intermolecular interactions. As a result, UHMWPE has the highest impact strength of any thermoplastic. UHMWPE is also known as high-modulus or high-performance polyethylene.

Modification of CNFs was accomplished by oxidation of the surface with potassium chlorate and sulfuric acid to produce carboxylic acid and hydroxyl groups. The oxidized CNFs were then dispersed in octadecylamine [$CH_3(CH_2)_{18}NH_2$] at 200°C for 20 h [80]. Bonds are formed between the amine group of octadecylamine and the carboxylic acid via a condensation reaction. These long hydrocarbon chains are then able to interact with the long chains of the polymer matrix.

The MCNF acted as solvent carriers in the stiff polymer matrix, and the short hydrocarbon chains grafted to the fibers were able to plasticize the UHMWPE chains in the interfacial region. This resulted in "interfacial flow under stretching and enhanced the elongation-to-break ratio" [80]. Untreated CNFs in UHMWPE did not demonstrate the enhanced toughness qualities [80].

UHMWPE nanocomposites exhibited increased toughness due to inclusion of chemically modified carbon nanofibers (MCNFs) [80]. The toughness of melt-pressed nanocomposite films containing 0.2 and 5-wt% MCNF was 10× that of the pure polymer. The nanocomposite at 118°C also showed a factor of 2× better toughness than the pure polymer composite. According to the researchers, the effect of the mobile hydrocarbon layers at the UHMWPE/MCNF interface was enough to overcome the barrier of high chain entanglements of the solid UHMWPE matrix that induced the toughening of the composite. The interfacial layer was on the order of 10–20 nm, and the induced interfacial flow resulted in a large elongation-to-break ratio (>500%) [80].

Enhanced Storage Moduli of SWNT-PS Nanocomposites. Storage modulus is the measure of energy stored in a sample during frequency cycle testing and the loss modulus is the energy lost. We stay with carboxylic acid–octadecylamine modification but this time choose SWNTs as our reinforcing material and change the polymer matrix to polystyrene—all part of the process of chemical engineering design and exploitation of numerous options. In this case, octadecylamine (ODA) or amino-terminated polystyrene were grafted onto oxidized SWNTs [81]. The SWNTs were purified in HNO_3 and their lengths shortened (why?). Following modification, the SWNTs were rendered soluble in dichloromethane and aromatic organic solvents. The chemically modified tubes were then mixed with polystyrene.

SWNTs were modified in $SOCl_2$ in DMF at 70°C for 24 h. Amino-terminated PS or ODA was mixed for 5 h at 90°C. The acid chloride groups on the modified SWNTs acted as linkers to the PS or ODA. The grafted SWNTs (1–3 wt%) were then mixed with PS in benzene, the mixture cast as a test coupon and dried at 90°C for 7 days [81].

The storage modulus of the polymer composite containing unmodified but cut SWNTs was smaller than that of the pure polymer. This is due to the lack of adequate dispersion of SWNTs in the PS-polymer and that interaction between the PS matrix and SWNT was poor. This resulted in slippage. The PS-SWNT showed better storage modulus than did the ODA-SWNT due to better compatibility between PS moieties. Both modified filler composites showed nonterminal behavior characteristic of good dispersion [81].

Flexible Electronic Applications. Pulickel Ajayan of the Department of Physics at Rensselaer Polytechnic Institute is one of the major contributors to nanotube composite research. One example, of many, of his work involves the development of a flexible hybrid composite structure using MWNT arrays in a poly(dimethylsiloxane) (PDMS) matrix [82]. The PDMS matrix (excellent conformal filling properties) with the dense aligned and patterned nanotube network retains robustness under high stress conditions (tensile and compressive strain) while maintaining electrical conductivity. SWNTs are first grown on a patterned SiO_2–Si substrate by CVD of ferrocene and xylene at 800°C. The array consists of circular domains of vertically aligned MWNTs 500 μm in diameter ~1 mm apart spread in a triangular 2-D array. PDMS is poured over the array and cured at 100°C for 10 h. The PDMS-composite is then peeled off the substrate to form a free-standing film ca. 100-μm thick [82]. Use of these materials is expected in strain gauges, field emission devices, and tactile and gas sensors. This is a different kind of composite—one that has minimal interfacial component compared to others we have discussed. It also has a different mission, different design parameters, different materials, different synthesis, etc.

Alumina/Magnetite and PMMA/PS Nanocomposites. Mechanical properties of alumina (Al_2O_3) and magnetite (Fe_3O_4) nanoparticles embedded in poly(methylmethacrylate) (PMMA) and polystyrene (PS) polymer matrices were analyzed by tensile testing, dynamic mechanical analysis (DMA), and nanoindentation. Overall, the mechanical tests showed that the nanocomposite systems possessed worse mechanical properties than the respective pure polymer systems. The cause of the poor mechanical performance of the nanocomposites was credited to poor interaction of the nanoceramic materials with the polymer matrix. In this case, it is not enough just to dump nanoparticles into a polymer matrix. Some consideration must be invested into chemical preparation of the inclusion.

Ceramic-Polymer Nanocomposites. The mechanical properties of poly(vinyl alcohol) (PVA) matrix nanocomposites with SiC or Al_2O_3 nanowires have demonstrated remarkable increase (to a level of 90%) in elastic modulus and increased strength above the native polymer at levels as low as 0.8 vol% [83]. S. R. C. Vivekchand et al. claim that the enhanced stiffness was due to the nanowire-induced crystallization of the polymer due to high aspect and surface-to-volume ratios of the nanowires and possible in-plane alignment of the nanowires. Increased strength was due to "significant pull-out of the nanowires" and the corresponding stretching of the matrix as a function of complete wetting by the polymer of the nanowires [83].

24.5 Metal and Ceramic Nanocomposites

24.5.1 Metal Nanocomposites

Metal Alloys. The formation of a bulk metal from nano-sized constituents improves the mechanical properties of the metal [84]. These materials are not necessarily nanocomposites. They are simply pure materials that are comprised of the nanoparticulate metal. Nanostructured copper films (27-nm grain size) formed by electrodeposition showed enhanced yield strength at 119 MPa [84]. Steel, on the other hand is hardened by the addition of carbon. In this case, nanoparticulate iron carbide filling the interstitial spaces between grains is able to prevent the mobility of dislocations and thereby prevent mechanical failure. In general, materials that contain smaller grains, that is, nanograins, have more grain boundaries that are able to block the propagation of mobile dislocations. The result of nanograin-embedded materials is increased strength with loss of ductility (e.g., materials are more brittle under tension) [84].

Alloys are composite materials—although the boundary is stretched somewhat because both components are metals—but they are composite materials nonetheless. In one example, nanoparticulate Fe and Cu fabricated by the process of ball milling were compacted and consolidated by the action of a tungsten-carbide die at 1 GPa for 24 h and then heat-treated at 400°C at a pressure of 870 MPa (sinter-forging) [84,85]. The grains of the alloy ($Fe_{85}Cu_{15}$ or $Fe_{60}Cu_{40}$) ranged from 20 to 70 nm. Overall, alloying at the atomic level obtained a homogeneous microstructure of nanocrystalline grains in the consolidated product [85]. Although its elastic modulus was similar to that of pure iron of grain size 50–150 μm, fracture occurred at a level of stress five times higher in the nanostructured alloy. The materials also exhibited very high strength under compression indicating low flaw populations with the attainment of nanoscale grain structure [85].

Metal Matrix Composites. These kinds of composite materials are complementary to cermets. In cermets, the matrix material is a ceramic and the filler is a metal. In metal matrix composites, the opposite case is in effect: the matrix is a metal and the filler is a ceramic. Aluminum–carbon fiber hybrids are examples of a metal matrix composites. Reinforcement can be continuous or discontinuous. Discontinuous matrix elements result in a nanocomposite with isotropic mechanical, thermal, or electronic properties. Monofilamental wires of SiC or carbon fiber are embedded within the metal oriented along one or another axis to provide anisotropic mechanical properties.

24.5.2 Inorganic Nanofibers

Silicon Carbide. SiC nanostructures have many potential applications ranging from high-strength, temperature resistance, and extremely hard materials. The Si–C bond ($318\,kJ \cdot mol^{-1}$) is nearly as strong as the C–C bond ($346\,kJ \cdot mol^{-1}$). The most stable form of SiC is the β-form that adopts a diamond-like structure.

SiC nanotubes and SiC-coaxial nanotubes can be formed from carbon nanotube (CNT) templates by a process called "shape memory synthesis," a template method that exploits a gas–solid phase reaction. The thermal decomposition of gaseous SiO results in the decomposition of Si on the carbon nanotube. Following heating, the Si-CNT is converted into a tube of SiC while releasing CO gas [86]. Similar structures were obtained by reacting carbon nanotubes with Si powder at 1200°C [87]. Sun et al. prepared β-SiC, biaxial SiC–SiO_2, and multiwalled SiC with intertube spacing between 3.5 and 4.5 Å [87]. Concentric SiC tubes were formed layer-by-layer via Si diffusion [87]. SWNTs were also used as the carbon source and template material to form SiC nanotubes. The thermally induced template reaction consisted of vaporized Si in N_2 or NH_3 carrier gas to form SiC nanofibers and nanotubes [88]. Depending on the duration, altered forms of SiC were generated [88].

SiC nanofibers can be synthesized from polymer blend precursors and then subjected to a melt spinning technique to form them into filaments. Polycarbosilane (PCS) with chemical structure [–SiH(CH_3)CH_2–], the SiC precursor, is first finely dispersed in phenol–formaldehyde (PF) resin, the carbon precursor, in the proportion of 3:7 (PCS:PF). Fibers were formed by continuous melt-spinning process. After soaking in acid for stabilization (e.g., to render them infusible) and heating at 1000°C, fibers were embedded in a carbon matrix (from the phenolic resin) and nanofibers were derived from the PCF precursor. Treatment with nitric acid removed the carbonized matrix and released nanofibers that were collected after filtration. The fibers were of amorphous structure with a large concentration of oxygen compounds—ca. 100 nm in diameter and over 100 μm in length. Further heat treatment at 1500°C transformed the amorphous oxygenated material into pure β-SiC [89].

Boron. Boron nanofibers 20- to 100-nm diameter form another class of important engineering fibers. Catalyst-free growth of α-tetragonal B tubes was achieved by pyrolysis of diborane at 630–750°C in 200 mTorr [90]. Boron is attractive because it has a high melting point (2200°C), low density ($2.34\,g \cdot cm^{-3}$), high hardness (2900 Knoop hardness), and high Young's modulus (380–400 GPa). Boron imparts stiffness, toughness, and strength to nanocomposites [90].

Boron Nitride Nanotubes and Fibers. BN is isoelectronic with carbon, and therefore one would expect BN compounds to be flexible in terms of configurations.

For example, there are diamond and graphite analogs with BN. BN fibers can be prepared by decomposition of borazine fibers and boron oxide in nitrogen at 1800°C or from the thermal decomposition of cellulose fibers in the presence of boric acid at 1000°C. BN fibers find great utility in metal matrix composites.

24.5.3 Cermets

A cermet is a composite material composed of a ceramic and a metal. Ultimately, the objective of cermets is to generate a material with high temperature resistance, chemical and oxidation resistance and hardness (typical of ceramics) that also demonstrates some level of plastic deformation and high thermal conductivity typical of metals. In cermets, the metal (e.g., Ni, Co, or Mo, usually $V_f < 20\%$) serves as the filler material. Ceramic oxides often used for cermet matrices include alumina, tungsten carbide, borides, and various oxides like MgO and BeO [91]. Cermets also include a class of materials in which the outermost coating of a metal is modified to resemble a ceramic like layer. In the 1950s in Siberia, such layers were accidentally formed on drill bits during deep drilling operations. The cermet layer afforded the bit reduced friction, renewability and enhanced hardness, and smoothness—all due to nanoparticulate ceramic materials embedded at the metal surface. Common applications of cermet materials are found in turbine blades and other components exposed to high temperature in jet engines and ceramic to metal joints and seals.

A good example of a cermet is the tungsten carbide–cobalt cutting tool—a high hardness carbide ($0.6 < V_f < 0.9$) embedded in a ductile metal matrix like cobalt. The function of the carbide moiety is to cut hardened steel while the metal provides toughness and prevents crack propagation caused by particle-to-particle contact by the brittle carbide phase [11]. Both materials possess refractory properties and can therefore withstand high temperatures [11]. In WC-Co nanocomposites in general, the hardness of the cermet is inversely proportional to the size of the grain and fracture toughness is inversely proportional to the hardness—finer grain size leads to lower toughness. However, in nanostructured cermets, the mechanisms of strengthening are different due to the large volume fractions of grain boundaries. Therefore, it is expected that the fracture toughness of WC-Co composites will actually improve as grain size approaches nanodimensions [92]. Dramatic reduction in flaw size is expected. For example, with grain size less than 30 nm, flaws are expected to be a few nanometers in size, a condition beneficial for fracture toughness that is independent of the hardness of the material. Secondly, the increased amount of interfaces between WC nanograins and the Co metal enhances overall toughness by increasing the direction of crack path through the metal–ceramic interfaces rather than through conventional pathways. Lastly, an enhanced number of interfaces reduce the probability of conventional crack propagation (via dislocations) to that of sliding and short-range diffusion and other interface-dependent mechanisms [92]. These factors become more pronounced as both the ceramic and metal phase approach nanoscale [92].

Strengthened cermets ($V_f < 0.15$) consist of nano- to micrometer-sized particulates of oxide particles embedded within a metal. The oxides serve to strengthen the metal by preventing dislocation mobility—as discussed earlier.

The tensile strength, for example, can be increased by a factor of four times in aluminum if aluminum oxide ($V_f = 0.10$) is dispersed within the matrix of the metal [11].

24.5.4 Concrete

Concrete is an isotropic composite material that is comprised of cement. Cement is a siliceous binding material that is moldable when wet and hard when dried and a material that is able to bind other materials together. Cement (from the Latin *caementa* "stone chips for mortar"; from *caedra* "to cut down, chop, hew, fell") was a term used by the Romans to characterize materials that looked like small stones—*opus caementium*. Early concrete consisted of crushed stones held together with lime as a binder. Volcanic ash, pulverized rocks, and burnt lime were mixed together to form cement [93]. Hydraulic cements harden after mixing with water due to chemical reactions. Portland cement, the most common form of the material, consists of limestone, clay minerals, and gypsum treated at high temperatures to alter its chemistry.

Carbon Fiber and Nanotube-Reinforced Concrete. Because carbon fibers, and now carbon nanotubes, have such unique mechanical properties, why not use them in concrete? Carbon fiber reinforced concrete (CFRC) has been in existence for several decades (**Figure 24.16**). CFRCs contain 3 to 15 vol% short carbon fibers embedded in concrete. The carbon fibers originate from either pitch-based or PAN-based carbon fiber. The application of carbon nanotubes in concrete is a natural evolutionary progression of CFRC technology. The modulus of elasticity (theoretically > 1 TPa), yield stress (between 20 and 60 GPa), and elastic strain properties (ca. 10%) of CNTs make them perfect candidates to bolster the already good mechanical properties of concretes in existence today [94].

FIG. 24.16 *SWNT bundles are apparent in this SEM image of unhydrated cement powder.*

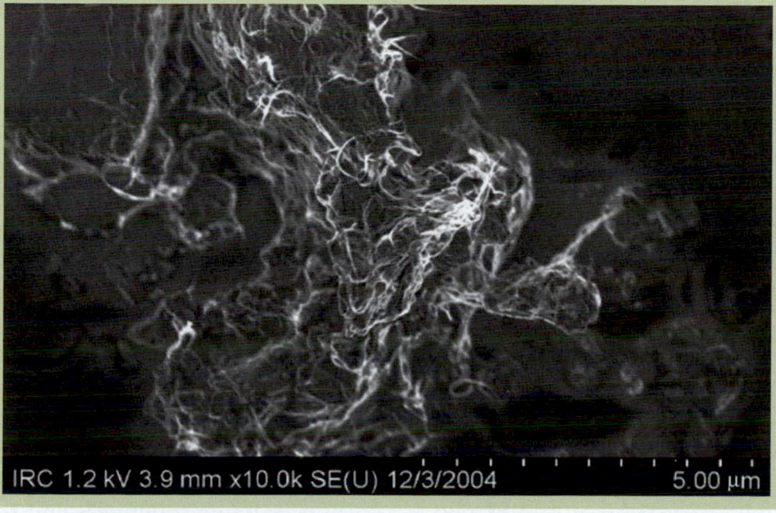

Source: J. Makar et al., Canadian Research Council. With permission.

Commercial grade SWNTs were dispersed by application of sonication in isopropanol. Portland Type 10 was added under constant sonication to produce a 0.02-wt% SWNT-cement ratio. Following 4h of mixing, the isopropanol was evaporated and the resultant mixture was ground into fine particles. No hydration was allowed during these stages of preparation. Mixtures of water, CNT-concrete, and superplasticizer were prepared and placed in elastic moulds and allowed to completely hydrate to form concrete nanocomposites. The unhydrated cement powder is shown in **Figure 24.16** [94].

SEM images show how SWNT fibers (actually bundles) were able to bridge cracks ca. 500nm in width in the matrix. Do SWNTs actually reinforce the concrete? It appeared that the SWNTs were pulled into the cracks (pull-out). Such pulling may be an indication of poor adhesion of SWNTs to the cement matrix. However a fivefold increase in fracture toughness and corresponding flexural toughness indicates otherwise. The cured concrete nanocomposites were subjected to Vickers hardness testing and showed higher microhardness values. Vickers hardness, as indicated earlier, can be directly correlated to elastic modulus and the compressive strength of the cement [94].

Nanocomposite Cements. Material performance can be enhanced with nanotechnology. I think we all agree that this is fast becoming a true statement. Concrete, as we now begin to understand, is very much a ceramic–ceramic nanocomposite. The addition of TiO_2 to cement, another ceramic, although not affording mechanical strength per se, is able to reduce the level of pollutants like NO_x, aromatics, ammonia, and aldehydes in the atmosphere near and around concrete structures like bridges and buildings [95]. The reduction in NO_x percentage over a period is absolutely dramatic—from 100% (normalized) NO_x to 0% in about 6h time [95].

There are many ways to improve the performance of cements and concrete with nanotechnology. Some include enhancements during the curing cycle (e.g., hydration efficiency), and other ways utilize the incorporation of reinforcing elements (as given in the example given above with carbon nanotubes) and enhancement of bonding between preexisting components of the cement. Some of the most significant developments are expected to be in the areas of superplasticity and high-strength fibers for energy-absorbing capability [96]. The addition of nanosilica is part of a trend to incorporate smaller and smaller particles (from fly ash to fumed silica and now nanosilicas) into ocncretes [96].

Nanoparticles lead to increased flexural and compressive strength in various formulations of cement and concrete because of the following: (1) Excellent dispersion of nanoparticles increases the viscosity of the mixture that helps suspend cement grains and aggregates allowing for better "workability;" (2) nanoparticles serve to fill voids between grains that immobilize "free-water" (known as the filler effect); (3) well-dispersed nanoparticles act as centers of crystallization for cement hydrates and this process accelerates the hydration mechanism; (4) nanoparticles catalyze the formation of smaller crystals of $Ca(OH)_2$, calcium silicate hydrates (a.k.a. CSH); (5) nanosilicates promote the *pozzolanic reaction* that consume $Ca(OH)_2$ to form more CSH; (6) nanoparticles enhance the "contact zone" resulting in better bonds between aggregate and cement phases (e.g., enhanced surface area effects); and (7) nanoparticles add crack arrest and interlocking effects between slip planes. This results in better toughness and shear, tensile, and flexural strength [96].

Nanotechnology is expected to stimulate research in many areas: accelerated hydration, mechanico-chemical activation of cement ingredients, nanoparticle-reinforced binders, nanoengineered internal bonding, nacre-like structures, superplasticizers for better workability, binders with humidity-sensitive moisture delivery to avoid cracking, reduced level of binder materials, self-healing materials, and self-cleaning, purifying materials [96].

24.6 Clay Nanocomposite Materials

Nanocomposites composed of clays and polymers demonstrated improved mechanical and thermal properties over control samples [97,98]. The enhanced properties are related to the degree of dispersion and the degree of exfoliation (dispersion of platelets) of the tactoids (clay platelet stacks) in the polymer matrix [99]. The degree of exfoliation is dependent on the coating of the platelet (usually an organic monolayer) and the solvent. Sodium montmorillonite is a crystalline aluminosilicate with platelet morphology. In this section, we apply clays as the minority inclusion material. We could have easily placed clay-polymer nanocomposite text in either the ceramic section or the polymer section. Because they form such unique nanocomposites, we decided to place them in a special section dedicated to them.

Clays are naturally occurring silicate nanomaterials composed of fine particles that interact dramatically with water—giving it a variable range of plasticity. When hardened, clay becomes a ceramic. There are a few classes of natural clays of importance to scientists: kaolinite, montmorillonites, illites, bentonites, and chlorites. Laponite is a synthetic smectite clay that resembles the natural clay hectorite. Synthesis is accomplished by combining salts of Na, Mg, and Li with sodium silicate. The resultant powder consists of nanoparticulates 0.92-nm thick and 25-nm across. Laponite forms good composites with carboxymethylcellulose and is used in making a variety of glazes.

In clays, the bonds between atoms in a layer are very strong but the bonds between layers are much weaker. The clay layer consists of three subunits consisting of an octahedral central layer (Al^{3+}) sandwiched between two tetrahedral layers consisting of silica (silicon and oxygen). The layer may be charged if other cations are substituted in its structure (e.g., Mg^{2+} for Al^{3+}) to generate a negative charge or substitution of Si^{4+} with Al^{3+} to yield an overall positive charge. Negative charges can be neutralized by the introduction of hydrated Na^+, K^+, or Ca^{2+} in the interlayer region of the clay [100] (**Fig. 24.17**).

Clay tactoids, stacks of platelets, in contact with monomers are able to form three distinct nanocomposites (**Fig. 24.18**). In the first configuration, the tactoid remains intact because it is coordinated all around by the polymer. In the second one, intercalation occurs in which the polymer is able to squeeze in between the nanoscale thin layers of clay. In the third configuration, complete exfoliation of the tactoid occurs as individual platelets are coordinated within the polymer matrix.

24.6.1 Polypropylene–Clay Nanocomposites

In order for clay platelets to disperse in a hydrophobic polymer, the surface of the clay must be modified. Clays naturally are water soluble polar polymeric

Fig. 24.17

A schematic rendition of the structure of a generic clay with polymer intercalation is depicted. Clays are known to intercalate water and swell to enormous dimensions. For a sodium montmorillonite clay, the primary structure consists of a layer of aluminum hydroxide between two layers of silicate. The overall chemical structure is $Na_{0.33}[(Al_{1.67}Mg_{0.33})Si_4O_{10}(OH)_2] \cdot nH_2O$. The platelets have a net negative charge and are weakly bound via electrostatic forces with an interlayer of hydrated Na^+, Li^+, Ca^{2+}, Fe^{2+}, or Mg^{2+} cations. The thickness of each plate is 1 nm but laterally can achieve dimensions of 1 μm [99]. Stacks of platelets are called tactoids. The space between layers is called the gallery. The particles are hydrophilic but the surface can be altered through chemical substitution to render the particles hydrophobic (via an organic cation substitution reaction) [99]. Following surface modification, the tactoids are completely exfoliated in chloroform or trichloroethylene. The surface modified layer is not shown in the image.

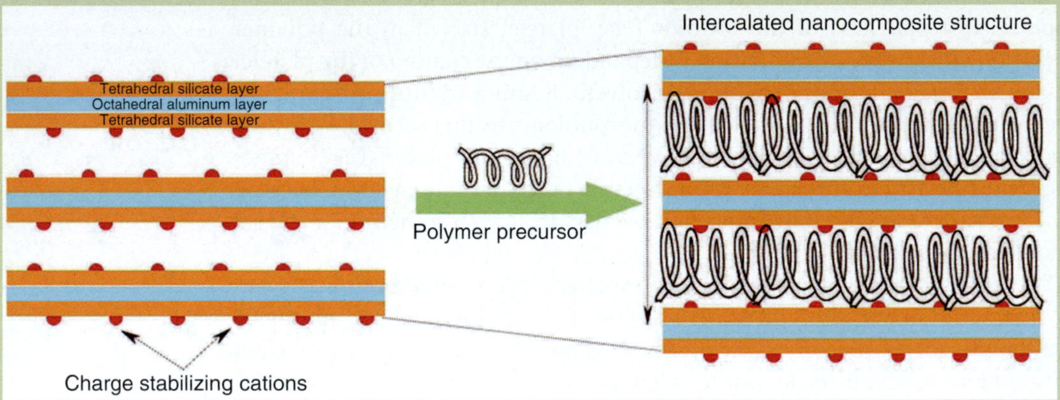

materials. Modification of the surface can occur with application of alkyl ammonium chlorides in which there are anywhere from a few to 18 carbons in the chain. The intercalated chains are able to expand the gallery spacing to 2.4 nm [101]. C_{18} hydrocarbon derivatized montmorillonite is called C_{18}-mmt. Addition of maelic anhydride modified polypropylene (PP-MA) compatibilizer to C_{18}-mmt results in intercalated clays with gallery spacing greater than 3 nm. Exfoliation of the clay was accomplished with addition of the PP-homopolymer. The best-case scenario requires that the tactoids become completely exfoliated (e.g., no correlation between platelets). If this is the case, interfacial area interaction with the polymer is maximized. The chemistry involved in this process is obviously complex. Physical parameters such as temperature also influenced the microstructure of the nanocomposite.

Inorganic fillers such as talc, calcium carbonate, glass fibers, glass beads, and mica are traditional means of improving mechanical properties of polymers. Larger particles act as stress concentrators that initiate cracks and degrade strength and toughness. The idea behind the use of nanoparticles is to enhance stiffness and strength while simultaneously improving toughness [101]. Clay platelets appeared to be a perfect solution because of their small size, large aspect ratio, and ability to be oriented within the polymer. In addition, enhanced

FIG. 24.18 *A basic clay–polymer nanocomposite is displayed. Three possible configurations are possible:* Top: *a conventional nanocomposite in which the polymer envelopes the entire clay tactoid;* middle: *intercalation of the polymer between the lamellae of the clay offers a new set of enhanced properties; and* bottom: *exfoliation of the tactoid occurs in which randomly oriented clay platelets are individually enveloped in the polymer matrix. Each configuration is expected to exhibit different mechanical properties.*

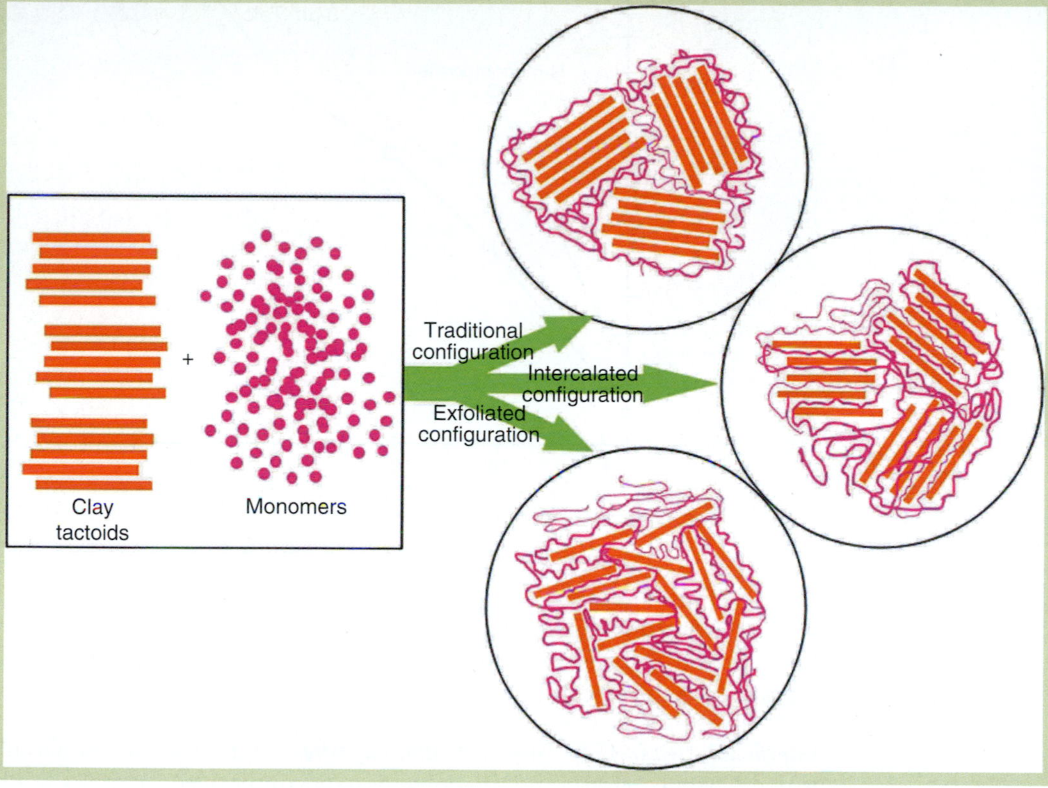

properties could be achieved with a lower volume fraction of filler material. This was demonstrated by Young's modulus of polypropylene-composite materials (**Fig. 24.19**).

PP-MA-mmt nanocomposites have also demonstrated a flame-retardant character. The reason for this remarkable attribute is the presence of a char layer on the surface. The char layer was found to have a high concentration of mmt [101].

24.6.2 Montmorillonite Clay Nanocomposites

Organically modified nanometer scale layered magnesium aluminum silicate platelets are added to polymeric materials to improve mechanical and physical properties. The silicate platelets are 1 nm in thickness and 70–150 nm across. The platelet surface is hydrophobically modified with a monolayer that allows for complete dispersion and miscibility within thermoplastic matrix polymeric materials such as acrylics, styrene/acrylics, and vinyl acetate copolymers. Addition

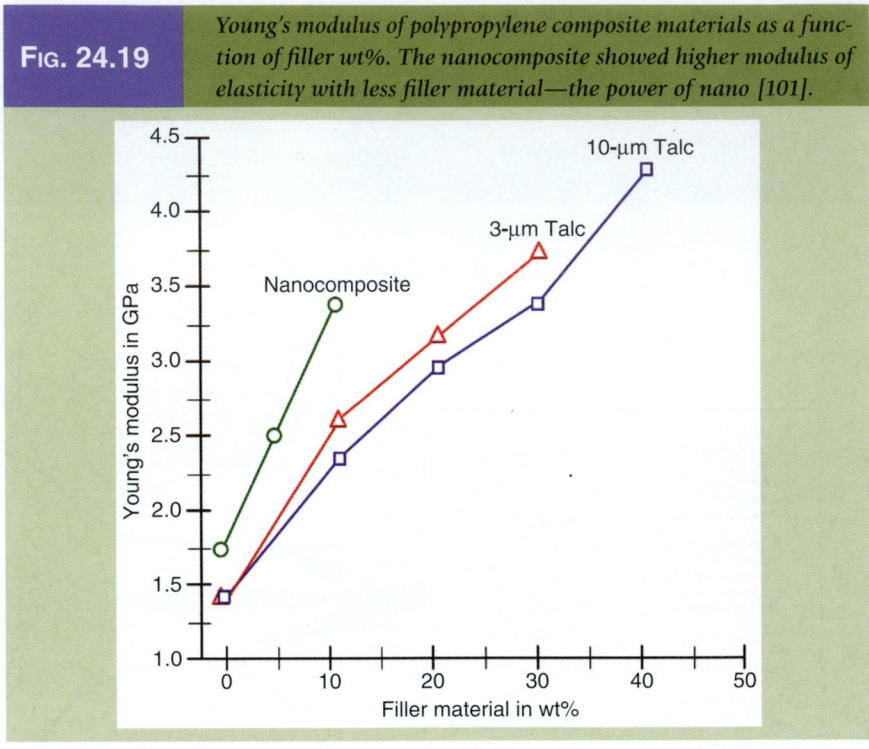

Fig. 24.19 *Young's modulus of polypropylene composite materials as a function of filler wt%. The nanocomposite showed higher modulus of elasticity with less filler material—the power of nano [101].*

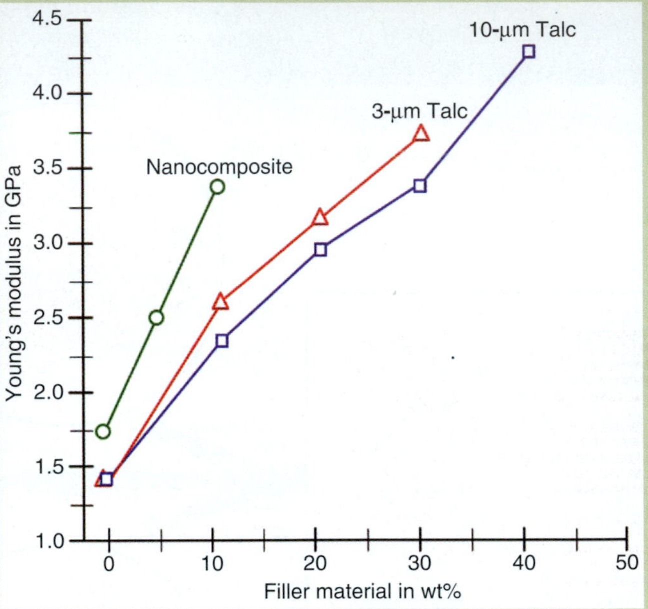

of the clay enhances flexural and tensile modulus, lowering the coefficient of linear thermal expansion of the polymer while increasing the flame-retardant capacity of the composite.

Interfacial Area in Clay–Polymer Nanocomposites. Advantage of interfacial area is demonstrated clearly in clay–polymer nanocomposites. Layered silicates in polymers have interfacial surface area on the order of 700 $m^2 \cdot g^{-1}$. Add to this that the distance between 1-nm thick plates is a mere 7 nm—and this achieved with a volume loading of 7 vol%. The morphology and resultant physical properties of the composite are significantly influenced, if not dominated, by the platelets.

24.6.3 Halloysite Nanotube Clay Composites

Halloysites nanotubes are another form of clay with the chemical composition of $Al_2Si_2O_5(OH)_4$. The tubes have diameter less than 100 nm (range from 40 to 200 nm) and length ranging from 500 nm to 1.2 μm. Halloysites are formed by the surface weathering of aluminosilicate minerals [102]. According to Paul Schroeder, in the figure, images of halloysites are depicted. Well-formed halloysite tubes occur naturally in surface hot springs (hydrothermal) environments. The mineral is made from a 0.7 nm structure comprised of silica tetrahedra and water sheets. The sheet misfits cause the layers to curl into tube-shaped morphology. The inside diameter averages 150 nm. The samples depicted in the figure were acquired from northwestern Turkey (**Fig. 24.20**). In 2007, halloysite nanotubes (HNT) were successfully functionalized in polypropylene at levels of 3%–5%. The modulus of the nanocomposite was measured to be twice that of the polymer alone [102]. In addition to an ingredient in polymers, halloysites can also be coated with metal for electronic applications.

In another study, halloysite nanotubes were made to react with 2,5-bis (2-benzoxazoyl)thiophene (or BBT) for a role as reinforcing elements in

Fig. 24.20 *Natural halloysite clays are depicted. Notice the hollow features at the end of the nanotubes.*

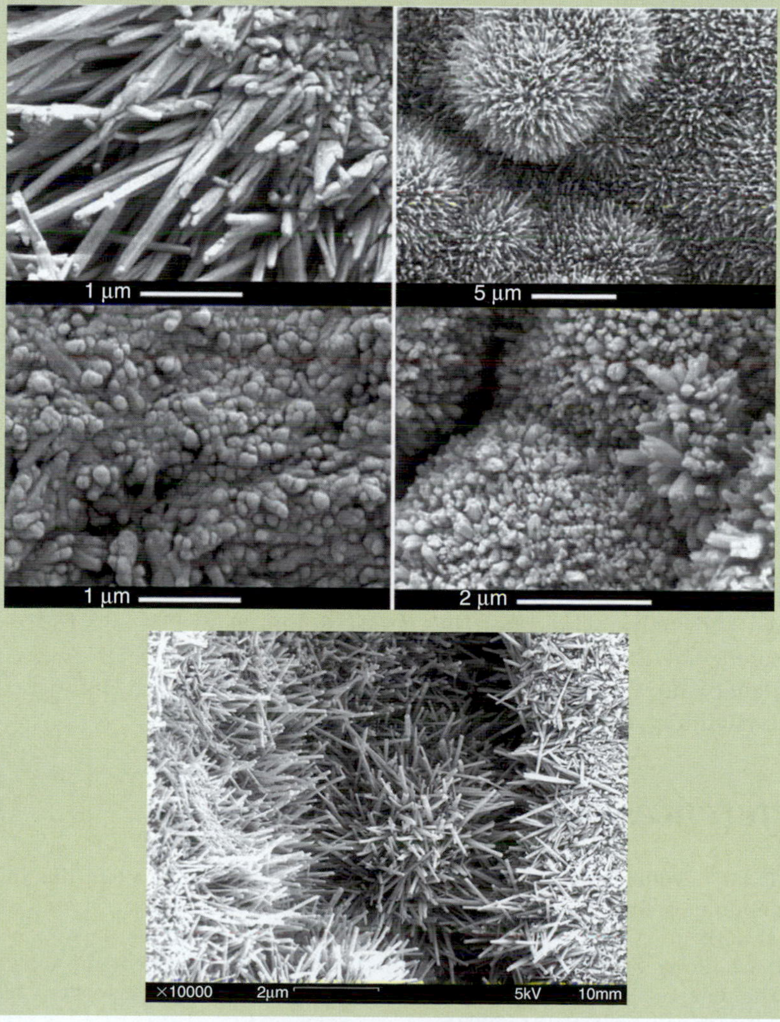

Source: Paul A. Schroeder, Department of Geology, University of Georgia. With permission.

Fig. 24.20 (Contd.)

polypropylene (PP) nanocomposites [103]. Clays tend to interact with organic materials via electron transfer mechanisms. This was confirmed in the BBT–HNT couple. Because of this, it is possible to enhance the interfacial properties of clay–polymer composites. BBT is capable of donating electrons to the positively charged HNTs and is therefore the interfacial modifier of the halloysite. Fibril formation of BBT is induced in the presence of HNTs in the nanocomposite. BBT fibrils of high crystallinity are formed under melt shearing. BBT-HNT-PP nanocomposites showed substantially increased tensile and flexural properties due to the presence of the crystallites [103]. HNTs were also complexed with polyvinyl alcohol (PVA [104]. Solutions of PVA and HNTs were prepared with ultrasonication. Particle size and distribution of HNTs was independent of the ratio between HNT and PVA; however, too much HNT depressed the crystallinity temperature (at T_g).

References

1. P. H. T. Vollenberg and D. Heikens, Particle size dependence of the Young's modulus of filled polymers: I. Preliminary experiments, *Polymer*, 30, 1656–1662 (1989).
2. M. Z. Rong, M. Q. Zhang, S. L. Pan, B. Lehmann, and K. Friedrich, Analysis of the interfacial interactions in polypropylene/silica nanocomposites, *Polymer International*, 53, 176–183 (2003).
3. K. I. Winey and R. A. Vaia, eds., Polymer Nanocomposites, *MRS Bulletin*, 32, Single Issue, April (2007).

4. P. Podsiadlo, A. K. Kaushik, E. M. Arruda, A. M. Waas, B. S. Shim, J. Xu, H. Nandivada, B. G. Pumplin, J. Lahann, A. Ramamoorthy, and N. A. Kotov, Ultrastrong and stiff layered polymer nanocomposites, *Science*, 318, 80–83 (2007).
5. C. R. Brooks and A. Choudry, *Failure analysis of engineering materials*, McGraw-Hill, New York (2002).
6. Carbon Fiber, en.wikipedia.org/wiki/Carbon_fiber (2008).
7. R. Bacon, Growth, structure and properties of graphite whiskers, *Journal of Applied Physics*, 31, 283–290 (1960).
8. Boeing 787, en.wikipedia.org/wiki/787_Dreamliner (2008).
9. M. Ohring, *Materials science of thin Films: Deposition and structure*, Academic Press, San Diego, CA (2002).
10. Flint, en.wikipedia.org/wiki/Flint#cite_note-0 (2008).
11. J. F. Shackelford, *Introduction to materials science for engineers*, 4th ed., Prentice-Hall, Upper Saddle River, NJ (1996).
12. J. F. Shackelford, *Introduction to materials science for engineers*, 6th ed., Prentice Hall, Upper Saddle River, NJ (2004).
13. M. S. Bakshi, P. Thakur, S. Sachar, and T. S. Banipal, Synthesis of nanocomposite gold-semiconductor materials by seed-growth method, *Materials Letters*, 61, 3762–3767 (2007).
14. K. S. Lee, I. H. Kim, T. S. Lee, B. Cheong, J. G. Park, J. G. Ha, and W. M. Kim, Dielectric confinement and the surface plasmon damping in Au: semiconductor nanocomposite thin films, *Surface and Coatings Technology*, 198, 51–54 (2005).
15. A. K. Sharma and B. D. Gupta, Metal-semiconductor nanocomposite layer based optical fibre surface plasmon resonance sensor, *Journal of Optics A: Pure and Applied Optics*, 9, 180–185 (2007).
16. M. Valdésa, M. A. Frontini, M. Vázquez, and A. Goossens, Low-cost 3D nanocomposite solar cells obtained by electrodeposition of $CuInSe_2$, *Applied Surface Science*, 254, 303–307 (2007).
17. J. Won, S. K. Chae, J. H. Kim, H. H. Park, Y. S. Kang, and H. S. Kim, Self-assembled DNA composite membranes, *Journal of Membrane Science*, 249, 113–117 (2005).
18. L. S. Litvinova, A. V. Gribanov, M. V. Mokeev, and V. N. Zgonnik, Physicochemical properties of water-soluble fullerene C60-carbohydrate composites, *Russian Journal of Applied Chemistry*, 77, 438–440 (2004).
19. V. M. Rotello, ed., Nanoparticles: Building blocks for nanotechnology, Plenum Press, New York (2003).
20. N. K. Cygan, A. Rice-Ficht, and K. Carson, Fabrication of protein nanocomposites, Presentation, Texas Institute for Intelligent Bio-Nano Materials and Structured for Aerospace Vehicles, TiiMS 3rd Annual Meeting, August (2005).
21. G. L. Hornyak, J. Dutta, H. F. Tibbals, and A. K. Rao, *Introduction to nanoscience*, CRC Press, Boca Raton, FL (2008).
22. Stress, en.wikipedia.org/wiki/Stress_(physics) (2008).
23. Tensile strength, en.wikipedia.org/wiki/Tensile_strength (2008).
24. R. Kelsall, I. Hamley, and M. Geoghegan, *Nanoscale science and technology*, John Wiley & Sons, Ltd., Chichester (2005).
25. Hardness, en.wikipedia.org/wiki/Hardness_%28materials_science%29 (2008).
26. Vickers hardness, en.wikipedia.org/wiki/Vickers_hardness_test (2008).
27. R. E. Stoller and P. M. Rice, Correlation of nanoindentation and conventional mechanical properties, *Materials research Society Symposium–Proceedings*, MRS (2000).
28. L. S. Schadler, L. C. Brinson, and W. G. Sawyer, Polymer nanocomposites: A small part of the story, *Journal of Materials*, 59, 53–70 (2007).
29. D. L. Burris and W. G. Sawyer, Improved wear resistance in alumina-PTFE nanocomposites with irregular shaped nanoparticles, *Wear*, 260, 915–918 (2006).

30. D. L. Burris and W. G. Sawyer, Tribological sensitivity of PTFE/alumina nanocomposites to a range of traditional surface finishes, *Tribology Transactions*, 48, 147–153 (2005).
31. W. G. Sawyer, K. D. Freudenberg, P. Bhimaraj, and L. S. Schadler, A study on the friction and wear behavior of PTFE filled with alumina nanoparticles, *Wear*, 254, 573–580 (2003).
32. S. Yu, P. Hing, and X. Hu, Thermal expansion behaviour of polystyrene-aluminium nitride composites, *Journal of Physics D: Applied Physics*, 33, 1606–1610 (2000).
33. C. A. Harper, ed., *Handbook of materials and processes for electronics*, McGraw-Hill, New York (1970).
34. J. K. Stanley, *Electrical and magnetic properties of metals*, American Society for Metals, Park, Ohio (1963).
35. H. Kim, H. T. Hahn, L. M. Viculis, S. Gilje, and R. B. Kaner, Electrical conductivity of graphite/polystyrene composites made from potassium intercalated graphite, *Carbon*, 45, 1578–1582 (2007).
36. D. A. Jaworske, R. D. Vannucci, and R. Zinolabedini, Mechanical and electrical properties of graphite fiber-epoxy composites made from pristine and bromine intercalated fibers, *Journal of Composite Materials*, 21, 580–592 (1987).
37. S. Watcharotone, D. A. Dikin, S. Stankovich, R. Piner, I. Jung, G. H. B. Dommett, G. Guennadi, S.-E. Wu, S.-F. Chen, C.-P. Liu, S. T. Nguyen, and R. S. Ruoff, Graphene-silica composite thin films as transparent conductors, *Nano Letters*, 7, 1888–1892 (2007).
38. E. Baeuerlein, P. Behrens, and M. Epple, eds., *Handbook of biomineralization: Biomimetic and bioinspired chemistry*, Wiley-VCH, Weinheim (2007).
39. J. R. Capadona, K. Shanmuganathan, D. J. Tyler, S. J. Rowan, and C. Weder, Stimuli-responsive polymer nanocomposites inspired by the Sea Cucumber Dermis, *Science*, 319, 1370–1374 (2008).
40. B. Pokroy, E. Zolotoyabko, and N. Adir, Purification and functional analysis of a 40kD protein extracted from the *Strombus decorus persicus* mollusk shell, *Biomacromolecules*, 7, 550–556 (2006).
41. J. D. Currey, P. Zioupos, P. Davies, and A. Casinos, Mechanical properties of nacre and highly mineralized bone, *Proceedings of the Royal Society of London Series B-Biological Science*, 268, 107–111 (2000).
42. Y. Yang, Y. Lu, M. Lu, J. Huang, R. Haddad, G. Xomeritakis, N. Liu, A. P. Malanoski, D. Sturmayr, H. Fan, D. Y. Sasaki, R. A. Assink, J. A. Shelnutt, F. van Swol, G. P. Lopez, A. R. Burns, and C. Jeffrey, Brinker functional nanocomposites prepared by self-assembly and polymerization of diacetylene surfactants and silicic acid, *Journal of American Chemical Society*, 125, 1269–1277 (2003).
43. E. V. Barrera, M. L. Shofner, and E. L. Corral, Applications: Composites, In *Carbon nanotubes: Science and applications*, M. Meyyappan, ed., pp. 253–275, CRC Press, Boca Raton, FL (2005).
44. V. A. Davis et al., Phase behavior and rheology of SWNTs in superacids, *Macromolecules*, 37, 154–160 (2004).
45. J. Fournier et al., Percolation network of polypyrole in conducting polymer composites, *Synthetic Metals*, 84, 839–840 (1997).
46. L. C. Wadsworth and K. Duckett, *Nonwovens science and technology II*, www.engr.utk.edu/mse/Textiles/Nanofiber%20Nonwovens.htm (2004).
47. M. Reibold, P. Paufler, A. A. Levin, W. Kochmann, N. Pätzkel, and D. C. Meyer, Materials: Carbon nanotubes in an ancient Damascus sabre, *Nature*, 444, 286 (2006).
48. M. Inman, Legendary swords' sharpness, strength from nanotubes, study says, *National Geographic News*, news.nationalgeographic.com/news/2006/11/061116-nanotech-swords.html (2006).

49. L. V. Radushkevich and V. M. Lukyanovich, O strukture ugleroda obrazujucegosa pri termiceskom razlozeni okisi ugleroda kontakte, *Zhurnal Fizicheskoi Khimii*, 26, 88–95 (1952).
50. S. Iijima, Helical microtubules of graphite carbon, *Nature*, 354, 56–58 (1991).
51. M. Monthioux, Who should really get the credit for the discovery of carbon nanotubes? *Carbon*, 44, 1621–1624 (2006).
52. S. Iijima and T. Ichihashi, Single-shell carbon nanotubes of 1 nm diameter, *Nature*, 363, 603–605 (1993).
53. D. S. Bethune, C. H. Kiang, M. S. De Vries, G. Gorman, R. Savoy, J. Vazquez, and R. Beyers, Cobalt catalysed growth of carbon nanotubes with single-atomic-layer walls, *Nature*, 363, 605–607 (1993).
54. E. W. Wong, P. E. Sheehan, and C. M. Lieber, Nanobeam mechanics: elasticity, strength, and toughness of nanorods and nanotubes, *Science*, 277, 1971–1975 (1997).
55. R. Saito, G. Dresselhaus, and M. S. Dresselhaus, *Physical properties of carbon nanotubes*, Imperial College Press, London (1998).
56. J. Hone, M. Whitney, C. Piskoti, and A. Zettl, Thermal conductivity of single-walled carbon nanotubes, *Physical Review B*, 59, R2514–R2516 (1999).
57. J. Che, T. Cagin, and W. A. Goddard III, Thermal conductivity of carbon nanotubes, *Nanotechnology*, 11, 65–69 (2000).
58. S. Ramakrishna, K. Fujihara, W.-E. Teo, T.-C. Lim, and Z. Ma, *An introduction to electrospinning and nanofibers*, World Scientific Publishing Co., Pte., Ltd., London (2005).
59. P. X. Ma and R. Zhang, Synthetic nano-scale fibrous extracellular matrix, *Journal of Biomedical Materials Research*, 46, 60–72 (1999).
60. R. L. Vander Wal and T. M. Ticich, Flame and furnace synthesis of single-walled and multi-walled carbon nanotubes and nanofibers, *Journal Physical Chemistry B*, 105, 10249–10256 (2001).
61. M. J. Height, J. B. Howard, and J. B. Vandersande, Method and apparatus for synthesizing filamentary structures, US Patent 7335344, Publication Date: February 26 (2008).
62. C. J. Unrau, R. L. Axelbaum, P. Biswas, and P. Fraundorf, Synthesis of single-walled carbon nanotubes in oxy-fuel inverse diffusion flames with online diagnostics, *Proceedings of the Combustion Institute*, 31, 1865–1872 (2007).
63. H. Richter, P. M. Jardim, J. B. Vander Sande, and J. B. Howard, Synthesis of fullerenic materials by controlled premixed combustion, *Nano-C Poster Symposium*, Boston, MA (2006).
64. M. J. Bronikowski, P. A. Willis, D. T. Colbert, K. A. Smith, and R. E. Smalley, Gas-phase production of carbon single-walled nanotubes from carbon monoxide via the HiPco process: A parametric study, *Journal of Vacuum Science and Technology, A*, 19, 1800–1806 (2001).
65. Y. Wang, S. Da, M. J. Kim, K. F. Kelly, W. Guo, C. Kittrell, R. H. Hauge, and R. E. Smalley, Ultrathin "bed of nails" membranes of single-wall carbon nanotubes, *Journal of American Chemical Society*, 126, 9502–9503 (2004).
66. A. B. Dalton, S. Collins, E. Munoz, J. M. Razal, V. H. Ebron, J. P. Ferraris, J. N. Coleman, B. G. Kim, and R. H. Baughman, Super-tough carbon-nanotube fibres, *Nature*, 423, 703 (2003).
67. K. Jiang, Q. Li, and S. Fan, Nanotechnology: Spinning continuous carbon nanotube yarns, *Nature*, 419, 801 (2002).
68. M. Zhang, K. R. Atkinson, and R. H. Baughman, Multifunctional carbon nanotube yarns by downsizing an ancient technology, *Science*, 306, 1358–1361 (2004).
69. R. H. Baughman, M. Zhang, S. Fang, A. A. Zakhidov, M. Kozlov, S. B. Lee, A. E. Aliev, C. D. Williams, and K. R. Atkinson, Solid state processing, structure and

multifunctional applications of carbon nanotube yarns and transparent sheets, Presentation, Science and technology without boundaries, Baughman.pdf (viewed 2008).

70. N. Chakrapani, B. Wei, A. Carrillo, P. M. Ajayan, and R. S. Kane, Capillarity-driven assembly of two-dimensional cellular carbon nanotube foams, *Proceedings of National Academy of Science*, 101, 4009–4012 (2004).
71. P. M. Ajayan, L. S. Schadler, and P. V. Braun, *Nanocomposite science and technology*, Wiley-VCH Verlag, Weinheim, Germany (2004).
72. R. A. Vaia, Polymer nanocomposites open a new dimension for plastics and composites, *AMPTIAC Newsletter*, 6, 17–24 (2002).
73. M. Roy, J. K. Nelson, R. K. MacCrone, L. S. Schadler, C. W. Reed, and R. Keefe, Polymer nanocomposite dielectrics-the role of the interface, *IEEE Transactions on Dielectrics Electrical Insulation*, 12, 629–643 (2005).
74. J. N. Coleman, M. Cadek, K. P. Ryan, A. Fonseca, J. B. Nagy, W. J. Blau, and M. S. Ferreira, Reinforcement of polymers with carbon nanotubes: The role of an ordered polymer interfacial region. Experiment and modeling, *Polymer*, 47, 8556–8561 (2006).
75. S. Ruan, P. Gao, and T. X. Yu, Ultra-strong gel-spun UHMWPE fibers reinforced using multiwalled carbon nanotubes, *Polymer*, 47, 1604–1611 (2006).
76. F. T. Wallenberger and N. E. Weston, *Natural fibers, plastics and composites*, Kluwer Academic Publishers, Dordrecht, the Netherlands (2004).
77. J. Suhr, W. Zhang, P. Ajayan, and N. Koratkar, Temperature activated Interfacial friction damping, in carbon nanotube polymer composites, *Nano Letters*, 6, 219–223 (2006).
78. N. A. Koratkar, J. Suhr, A. Johsi, R. S. Kane, L. S. Schadler, P. M. Ajayan, and S. Bertolucci, Characterizing energy dissipation in single-walled carbon nanotube polycarbonate composites, *Applied Physics Letters*, 87, 063102 (2005).
79. J. D. Whittaker, E. D. Minot, D. M. Tannenbaum, P. L. McEuen, and R. C. Davis, Measurement of the adhesion force between carbon nanotubes and a silicon dioxide substrate, *Nano Letters*, 6, 953–957 (2006).
80. X. Chen, K. Yoon, C. Berger, I. Sics, D. Fang, B. S. Hsiao, and B. Chu, In-situ x-ray scattering studies of a unique toughening mechanism in surface-modified carbon nanofiber/UHMWPE nanocomposite films, *Macromolecules*, 38, 3883–3893 (2005).
81. H. T. Ham, C. M. Koo, S. O. Kim, Y. S. Choi, and I. J. Chung, Chemical modification of carbon nanotubes and preparation of polystyrene/carbon nanotube composites, *Macromolecular Research*, 12, 383–390 (2004).
82. Y. J. Jung, S. Kar, S. Talapatra, C. Soldano, G. Viswanathan, X. Li, Z. Yao, F. S. Ou, A. Avadhanula, R. Vajtai, S. Curran, O. Nalamasu, and P. M. Ajayan, Aligned carbon nanotube-polymer hybrid architectures for diverse flexible electronic applications, *Nano Letters*, 6, 413–418 (2006).
83. S. R. C. Vivekchand, U. Ramamurty, and C. N. R. Rao, Mechanical properties of inorganic nanowire reinforced polymer–matrix composites, *Nanotechnology*, 17, S344–S350 (2006).
84. C. P. Poole and F. J. Owens, *Introduction to nanotechnology*, Wiley-Interscience, John Wiley & Sons, Inc., Hoboken, NJ (2003).
85. L. He, L. F. Allard, K. Breder, and E. Ma, Nanophase Fe alloys consolidated to full density from mechanically milled powders, *Journal of Materials Research*, 15, 904–912 (2000).
86. N. Keller, C. Pham-Huu, G. Ehret, V. Keller, and M. Ledoux, Synthesis and characterisation of medium surface area silicon carbide nanotubes, *Carbon*, 41, 2131–2139 (2003).

87. X.-H. Sun, C.-P. Li, W.-K. Wong, N.-B. Wong, C.-S. Lee, S.-T. Lee, and B.-K. Teo, Formation of silicon carbide nanotubes and nanowires via reaction of silicon (from disproportionation of silicon monoxide) with carbon nanotubes, *Journal of American Chemical Society*, 124, 14464–14471 (2002).
88. M. H. Rümmeli, E. Borowiak-Palen, T. Gemming, M. Knupfer, K. Biedermann, R. J. Kalenczuk, and T. Pichler, On the formation process of silicon carbide nanophases via hydrogenated thermally induced templated synthesis, *Applied Physics A*, 80, 1653–1656 (2005).
89. Z. Correa, H. Murata, T. Tomizawa, and A. Oya, Preparation of SiC nanofibers by using the polymer blend technique, *Advanced Science and Technology*, 51, 60–63 (2006).
90. T. T. Xu, J.-G. Zheng, N. Wu, A. W. Nicholls, J. R. Roth, D. A. Dilkin, and R. S. Ruoff, Crystalline boron nanoribbons: Synthesis and characterization, *Nano Letters*, 4, 963–968 (2004).
91. Cermet, en.wikipedia.org/wiki/Cermet (2008).
92. P. Seegopaul and Z. Fang, Tungsten-carbide-cobalt nanocomposites: Production and mechanical properties, In *Dekker Encyclopedia of nanoscience and nanotechnology*, Vol, 5, J. A. Schwartz, C. I. Contescu, and K. Putyera, eds., pp. 3943–3952, CRC Press, Boca Raton, FL (2004).
93. Cement, en.wikipedia.org/wiki/Cement (2008).
94. J. Makar, J. Margeson, and J. Luh, Carbon nanotube/cement composites—early results and potential applications, National Research Council of Canada, 3rd International Conference on Construction Materials: Performance, Innovations and Structural Applications, Vancouver, BC, August (2005).
95. L. Cassar, Presentation, CTG Italcementi Group, Italy, 2nd International Symposium on Nanotechnology in Construction, Bilboa, Spain, November (2005).
96. K. Sobolev and M. Ferrad-Gutiérrez, How nanotechnology can change the concrete world: Part 2, *American Ceramic Society Bulletin*, 11, 16–19 (2005).
97. A. Mourchid, E. Lecolier, H. Van Damme, and P. Levitz, On viscoelastic, birefringent, and swelling properties of laponite clay suspensions: Revisited phase diagram, *Langmuir*, 14, 4718–4723 (1998).
98. J. W. Gilman, C. L. Jackson, A. B. Morgan, R. Harris, Jr., E. Manias, E. P. Giannelis, M. Wuthenow, D. Hilton, and S. H. Phillips, Flammability properties of polymer-layered-silicate nanocomposites: Polypropylene and polystyrene nanocomposites, *Chemistry of Materials*, 12, 1866–1873 (2000).
99. D. L. Ho and C. J. Glinka, Effects of solvent solubility parameters on organoclay dispersions, *Chemistry of Materials*, 15, 1309–1312 (2003).
100. C. Greenwell, *The clay-polymer project*, www.exclaim.org.uk/CP_Intro.html (2005).
101. F. M. Mirabella, Jr., Polypropylene and thermoplastic olefin nanocomposites, In *Dekker Encyclopedia of nanoscience and nanotechnology*, Vol. 4, J. A. Schwartz, C. I. Contescu, and K. Putyera, eds., pp. 3015–3030, CRC Press, Boca Raton, FL (2004).
102. Halloysite nanotubes, *Natural Nano*, www.naturalnano.com (2008).
103. M. Liu, B. Guo, Q. Zou, M. Du, and D. Jia, Interactions between halloysite nanotubes and 2,5-bis(2-benzoxazolyl) thiophene and their effects on reinforcement of polypropylene/halloysite nanocomposites, *Nanotechnology*, 19, 205709–205718 (2008).
104. M. Liu, B. Guo, M. Du, and D. Jia, Drying induced aggregation of halloysite nanotubes in polyvinyl alcohol/halloysite nanotubes solution and its effect on properties of composite film, *Applied Physics A: Materials Science and Processing*, 88, 391–395 (2007).

Problems

24.1 What is the maximum strain energy that can be stored in a tendon with $E_o = 1.5$ GPa and density equal to 1120 kg·m^{-3}?

24.2 Everything else being equal and occupying equal volumes (volume fractions), how much more surface area do nanotubes with diameter 2 nm have compared to 2-μm diameter tubes? or 20-μm diameter fibers?

24.3 Derive Equation 24.14, $\frac{1}{E_c} = \frac{V_m}{E_m} + \frac{V_f}{E_f}$, assuming cylindrical geometry and that the load is perpendicular to the axis of the fiber composite. What other formulas do the isostrain and isostress relationships resemble? (Hint: think electricity).

24.4 The coefficient of thermal expansion for a thermoplastic sheet is $7.56 \times 10^{-5}\cdot C^{-1}$. How much will a 4 × 8 foot sheet of the material expand in the long dimension when taken out of a freezer at 30°F to room temperature (70°F)?

24.5 Determine the specific surface area of a clay that is 1 nm in thickness and 1 μm in lateral dimensions. The density of the clay is 2.65 g·cm^{-3}.

24.6 Averaging of properties for composites seems to work well for macroscopic to micrometer-sized materials. Do the formulae apply to nanomaterials? Why or why not?

24.7 Boeing's new *Stratoliner 787* is made primarily of composites based on carbon fibers. What advantages does this material provide over conventional aluminum materials used in planes for many decades?

Section 10

Biological and Environmental Nanoengineering

NANOBIOTECHNOLOGY

Nanotechnology allows us to do things we are unable to do otherwise. Nanotechnology gives us a 'set of tools.'

DR. OMID FAROKHZAD
Harvard Medical School

Chapter 25

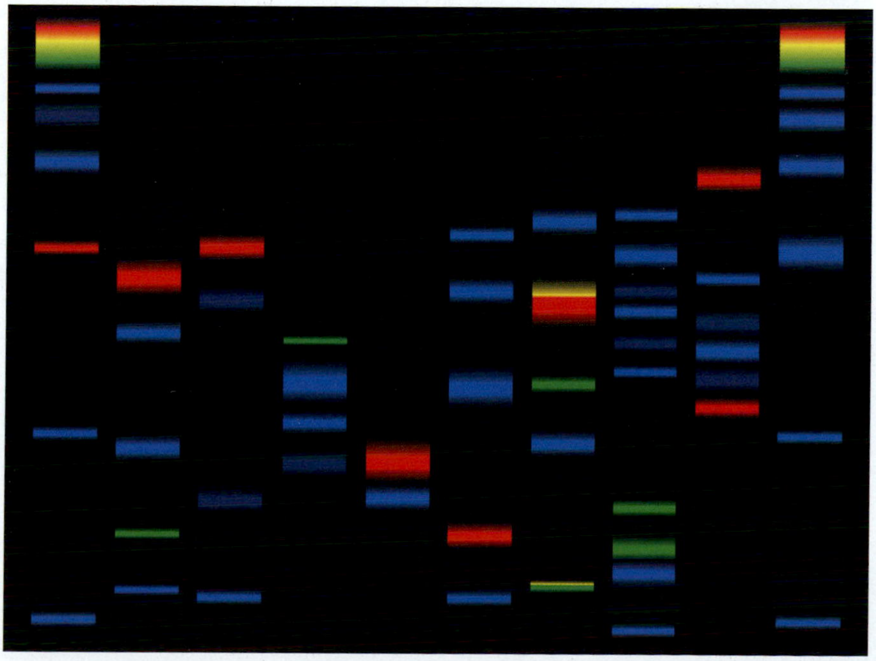

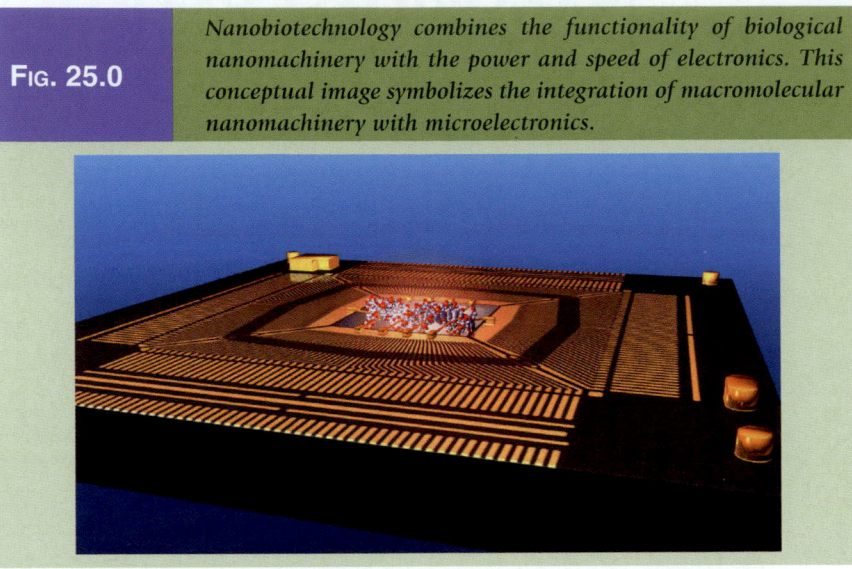

Fig. 25.0 *Nanobiotechnology combines the functionality of biological nanomachinery with the power and speed of electronics. This conceptual image symbolizes the integration of macromolecular nanomachinery with microelectronics.*

Threads

Chapter 25 presents a short introduction to nanobiotechnology, which is the application of nanotechnology to biological systems to achieve practical engineering results. These applications include separation and identification of biological products, modification of biological materials to improve their useful properties, and the use of biological molecules to create sensors. Nanoscale materials, sensors, and devices can be and are being designed based on almost any biological system, and inorganic nanodevices can be applied to many biological applications. In this selective introduction to a very broad and rapidly expanding field, we focus on a few examples which strongly illustrate the role of nanoscale phenomena in creating new types of devices and materials. To distinguish this focus area, we use the term nanobiotechnology.

As in instructive example, in this chapter we take one biological system and one nanoscale physical system and show how they can be integrated to produce a nano-biosensor. From the many possibilities, we chose the biological immune system and the physical nanoscale cantilever. In the course of this discussion, we cover considerably more detail on the immune system than is immediately needed for application to the sensors, in order to give an understanding of the depth and complexity of biological systems, and to give the student some resources for exploring other possibilities. In addition, we briefly survey other types of sensors and some other nanoscale biodevices and materials.

25.0 Introduction to Nanobiotechnology

Nanobiotechnology is an emerging technology on the interface between biotechnology and nanotechnology. Biotechnology manipulates the molecules and processes of biological systems to gain practical improvements, for example, in producing harvestable yields of useful drugs and antibodies, or in developing indicators and sensors based on the interaction of antibodies with substrates. Nanobiotechnology uses nanoscale mechanical, electrical, optical, and magnetic effects in biological systems to create new types of devices and applications which are beyond the scope of previous molecular biology- and biochemistry-based biotechnologies. As a result, the ability to sense and detect the state of biological systems and living organisms is being radically transformed.

25.0.1 Definitions

Because of the rapid rise of new developments in the interface between nanotechnology and biology, it is useful to go over some definitions of terms. Besides the well established but not always clearly defined term "biotechnology," new terms are coming into use, including "bio-nanotechnology," "biomolecular nanotechnology," "biomedical nanotechnology," and others.

25.0.2 Biotechnology

Biotechnology generally refers to the manipulation of key biological systems such as the DNA–RNA encoding and synthesis of proteins and the targeting of antigens by antibodies [1–3]. The understanding of the molecular basis of these fundamental biological processes led to the ability to modify and exploit their operation. Biotechnology utilizes enzymes, viruses, biochemistry, and biophysics to achieve its results. DNA can be modified to produce useful proteins; antibodies can be modified to include indicators or "reporters" such as fluorescent chemical groups, which change to indicate the presence of a selected antigen. DNA can be interfaced to photochemical or electrochemical sensors to detect the presence of complementary strands of DNA or RNA, thus providing a means of detecting and analyzing genetic material in unknown samples, or identifying a disease agent by its DNA. Biotechnology includes the modification of DNA to produce unique genotypes for animals and plants, and cloning of organisms to reproduce their genotypes. Biotechnology is being impacted by nanotechnology, as seen by many examples elsewhere in this book.

25.0.3 Bio-Nanotechnology

Bio-nanotechnology has emerged as the manipulation and exploitation of proteins, the natural nano-engines of biological systems. This is a vast field which is growing naturally out of the discipline of protein chemistry, and has many exciting possibilities [4].

25.0.4 Biomolecular Nanotechnology

Biomolecular nanotechnology is an emerging field which is being defined more broadly than protein bio-nanotechnology to include the most sophisticated application of biotechnology, nanofabrication, materials science, nanoelectronics, biochemistry, and biophysics to design useful and interesting structures on the nanoscale. Biomolecular nanotechnology includes using DNA as a template for molecular-scale computational engines, and adapting muscle and flagella mechanisms to make artificial organic nano-engines, for example [5].

25.0.5 Biomedical Nanotechnology

Biomedical nanotechnology is the application of nanotechnology to biomedical applications such as diagnostics and high-throughput screening, drug delivery, artificial tissues for implants and prostheses, and medical imaging. We will survey some of these areas in the next chapter of this book. Other areas of interface between nanotechnology and biology, such as modification of natural nanomaterials and biomimetic design of nanodevices, are covered elsewhere in this and the companion books [6–17].

25.0.6 Nanobiotechnology

This brings us to *nanobiotechnology*, the subject of this chapter, which we define as the application of nanotechnology to the above fields in ways which exploit phenomena that are unique to the nanoscale. As an example of this class of applications, we will discuss cantilever sensors, whose principles of operation depend upon the ratio of surface to volume that is characteristic of the nanoscale.

In order to make a nano-biosensor with a nanoscale cantilever, we utilize the biological immune system and integrate a selected portion of it with a nanoscale physical mechanism. The main example in this chapter will be to show how we integrate the molecular action of the biological immune system with the nanoscale physics and chemistry of absorption on the surfaces of nanoscale cantilevers, while using a variety of electronic and optical means to translate the results into a measurable form. This example is just one of a number of ways that nanoscale phenomena can be exploited to create sensors capable of responding to very small inputs. In the next section, we briefly describe the biological immune system before proceeding to the physical description of nanoscale cantilevers.

25.1 The Biological Immune System

The biological immune system is a primary example of how molecular recognition works at the nanolevel. The macromolecules of the immune system are directly harnessed in nanotechnology devices and as models for making similar artificial molecular recognition nanomachinery.

Living organisms fight off invasions of disease agents with an array of complex cellular and molecular defenses called the immune system [18–20]. Biological immunity works through molecular pathways which control recognition and signaling in cells. We have introduced some basic aspects of gene expression and molecular biology in chapter 14. We will now go further into the details of how these nanoscale systems work in the immune system.

25.1.1 Natural Molecular Recognition

In the chapter in Part I that discusses molecular nanobiology, we discussed how DNA and RNA act as molecular templates to reproduce and transmit information.

In this section we examine another example of molecular recognition, the immune system. The immune system, especially of vertebrates, is a set of highly developed nanoscale mechanisms of molecular recognition, which can be utilized to create drugs, probes, and diagnostic devices on the molecular and nanoscale levels. Humans and higher vertebrates have two levels of immunity which work together to defend against infectious pathogens: the *innate* and *adaptive* immune systems.

Vaccines and the Discovery of the Immune System. Long before the birth of scientific medicine, there was some cultural awareness of protection by inoculation. In Thucydides's historical account of the Athenian plague in 430 B.C., he noted that those who survived became immune. (His account is also noteworthy as one of the early instances of careful and systematic descriptions of symptoms.) It was also known that exposure to some types of illness did not confer future protection, for example the bubonic plague. Classical medicine recognized the effects on the body of illness, and the main symptoms are still a standard part of medical diagnosis taught to every medical student today: *rubor* (redness), *calor* (warmth), *tumor* (swelling), and *dolor* (pain). These are now recognized as inflammation due to the body's immune responses.

Various forms of inoculation (defined as the introduction of an infectious agent into the body, or into a culture medium) through the skin or nose were part of the ancient medical practices of India, China, Persia, and primitive folk medicines. The disastrous effects of the smallpox epidemics of the seventeenth and eighteenth centuries led to trials of inoculation (also called variolization in the case of smallpox) by infecting a scratch on the skin with material from another person with the disease. This risky procedure became obsolete after the late eighteenth century when the English physician Edward Jenner demonstrated that serum from the weaker, related disease cowpox could be used to inoculate successfully against the much more dangerous smallpox. Widespread use of vaccination after a long period of gaining acceptance eventually led to the eradication of smallpox late in the twentieth century.

By the late nineteenth century Robert Koch showed that infectious diseases are caused by microorganisms, and Louis Pasteur developed vaccines against cholera and rabies. In 1890 Emil von Behring and Shibasaburo Kitasato discovered *antibodies* that immobilized specific pathogens in the serum of vaccinated individuals. In the meantime Elie Metchnikoff discovered that many microorganisms could be engulfed and digested by phagocytic cells, which he called macrophages. Much to the astonishment of the scientific world of the day, these amoeba-like cells were shown to circulate in the bloodstream, seeking out and attacking pathogenic microbes, while not harming the sister cells of their own organism. Like amoebas, the macrophages demonstrate chemotaxis, motion directed by a chemical trail or gradient.

During the next century these and other white blood cells were studied and characterized microscopically, and antibodies were studied chemically and in their effects on cell cultures, as answers were sought to how the body's immune cells could recognize invaders and distinguish them from itself. Advances were made in vaccines and in using white blood cells to diagnose disease. Studies of rejection of blood transfusions in humans, and tissue grafts among inbred, genetically homogeneous mice helped understand the mechanisms for generation of antibodies and their relation to the genetic machinery of the cells. With the advent of molecular biology on the nanoscale, the beauty and complexity of the immune system has become apparent as the workings of intricate biomolecular nanomachinery [21].

25.1.2 The Innate Immune System

The *innate immune system* reacts to a wide range of innately recognized foreign bodies with a rapid cascade of reactions that requires no learning period or prior exposure to the pathogens. The first response to injury or invasion is inflammation, turned on by chemical signals released by damaged cells.

The most striking instruments in the armory of innate immunity are specialized amoeba-like white blood cells called *phagocytes* that engulf and digest microorganisms and foreign materials in the body. In the human immune system these phagocytic cells are called *macrophages*. In general, the phagocytes attack anything that is recognized as not being part of the organism ("not self").

The innate immune system includes a number of other types of specialized cells that attack pathogens using different strategies, and are recruited to the site of attack or inflammation by chemical signaling. Some of these cells also coordinate responses with the second, adaptive, level of defense.

The innate immune system works through many mechanisms. The most interesting aspect of innate immunity in terms of nanoscale molecular recognition is the way in which white blood cells sense and recognize pathogens. Receptors on the surface of white blood cells recognize repetitive molecular patterns characteristic of bacterial cell walls. These receptors are called Toll-like receptors, named after the type of genes that express them. The first example of this gene was found in studies of the fruit fly, *Drosophila*, and related genes are found in a wide range of organisms. The Toll-like proteins recognize molecular patterns that are common to many pathogens, such as the spacing of sugar units in the polymeric polysaccharide cell walls of bacteria. The receptor proteins attach firmly only when the pattern spacing is a match, leaving normal cell sugar metabolism undisturbed. These antigen proteins occur both as freely secreted molecules, which mark bacteria for destruction by macrophages, and as trans-membrane proteins, which bind defensive cells to the invading bacteria to surround it and in some cases to signal for the initiation of adaptive responses.

25.1.3 The Adaptive Immune System

The second part of the immune system is *adaptive immunity*, which is launched as an *immune response* to a particular pathogen. Adaptive immune response is mediated by the generation of *antibodies* which bind to a specific pathogen. The production of antibodies can only be induced by exposure to and infection by a pathogen, after an incubation period in which the adaptive immune system recognizes the pathogen and sets in motion the molecular machinery to build specifically modified antibody molecules. The adaptive immune system can generate other responses besides antibodies, and the term *antigen* is used for any substance that can produce an adaptive immune response. The adaptive immune system inserts and retains a molecular memory in its section of the DNA, which can provide *protective immunity* against repeated infection by a specific pathogen.

FIG. 25.1 *Structure of an antibody: the antigen-binding fragment binds to a specific antigen with a high degree of selectivity.*

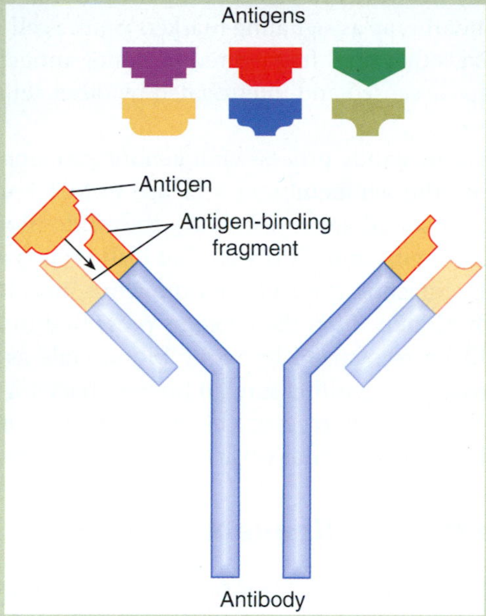

This is the basis for *vaccines*, substances to which the adaptive immune system can be exposed in order to generate antibodies without creating an infection (**Fig. 25.1**).

Biomolecular Specificity of Antibodies. The field of immunology is relatively young in terms of the amount to be learned at the molecular level, but an enormous amount of information has been laboriously elucidated on cellular recognition through complex signaling pathways. This research has discovered the role of the structure of the cell membrane and how sensors imbedded in the membrane interact with the expression of genes by regulating DNA in the cell nucleus to produce antibodies and stimulate the generation of phagocytes in response to infection. In this section we will give some examples of the antibody nanomachinery involved in the immune processes.

Protein–peptide complexes in the immune system are an important example of cellular signaling, trafficking, and targeting mechanisms in the body that are mediated by interactions between proteins and peptides. This sample covers only the barest introduction to the full richness of the immune system, but it is illustrative to anyone asking how assemblages of molecules can collectively recognize self and nonself, attack invaders, learn and remember threats, and organize perfectly matched tools to eliminate them. In effect the immune system displays a kind of molecular machine intelligence that is intimately connected with the genetic molecular machinery at the heart of life.

Antibodies in the Adaptive Immune System. Antibodies and antigens are produced by the protein-synthesizing machinery of the cell in a special region of the DNA genome called the *major histocompatibility complex* (MHC). This large genomic region exists only in vertebrates, as a very highly evolved and intricate defense mechanism. The proteins expressed by the DNA in the MHC are positioned in the cell membrane as signaling markers on the cell surface. The MHC proteins are templates that hold fragmented pieces of antigens on the cell surface, where they can be sensed and interrogated by other cells, especially white blood cells.

Mammalian cells constantly process endogenous proteins and present their peptide fragments on the cell membrane attached to MHC proteins. As long as the proteins being processed are endogenous, the white blood cells recognize the displaying cell as "self," and all is well. But if the cell is processing foreign protein expressed by infection from a bacterium, virus, or cancerous disruption of the molecular machinery, then the proteins presented on its surface by the MHC template will be recognized by white blood cells as "nonself." White blood cells are constantly engulfing foreign bodies (bacteria, viruses, particles, macromolecules), breaking them apart using lysozymes, and displaying the fragments using their own particular version of the MHC proteins.

25.1.4 White Blood Cells and Antibodies

In order to survive in a competitive world, all organisms have evolved defense mechanisms to protect against competitors, predators, and pathogens. These defenses first evolved on the nanoscale: they are so basic an aspect of life that they evolved before life diverged into plants and animals at the single-cell level. Thus all cells have a heritage of molecular defense mechanisms to suppress competitors, deter predators, fight pathogens, and respond to pathogen virulence factors. In multicellular organisms, these molecular defenses must interplay with cellular signaling pathways that orchestrate cooperation between the organism's own cells. And in multicellular organisms, there is the opportunity of some cells to specialize in fighting invaders and toxins that threaten the whole organism—these are the white blood cells.

The innate and adaptive immune systems work together in vertebrates through a coordinated mechanism at the cellular and molecular level, involving white blood cells and biomolecular-selective protein antigens and antibodies.

White Blood Cells. The number and appearance of white blood cells under the microscope, their separation from the much more numerous red blood cells by centrifugation, and their importance to the practical diagnosis of infection play a large and important part in the practice of medicine and study of immunology. The specificity of stains was discovered by trial and error in the early days of microscopy. Early biochemists and medical researchers speculated that if a chemical stain would selectively attach itself to a pathogen, there might be substances which could attach to pathogens and selectively kill or disable them without harming other cells. This suggested the concept of a "magic bullet" that would seek out and destroy disease germs, even long before the molecular mechanisms of antigens were understood, or before the concept of germ-destroying nanobots was proposed.

In humans and vertebrates the *leukocytes* or white blood cells play the major role in both innate and adaptive immune response. A number of different types of leukocytes are produced by special stem cells in various organs of the immune system such as the bone marrow and spleen. They are released into the blood at a rate of billions per day. The number and type of white blood cells circulating in the body varies in response to infection. Under the microscope they can be distinguished and classified according to their staining properties as agranulocytes and granulocytes; the latter have granules that show up with staining under the microscope. The granulated structures are due to specialized lysosomes associated with the digestion of engulfed particles, so that the staining differences correspond to different cell structures and functions.

Innate leukocytes include the macrophages, dendritic cells, and the granulocytes: neutrophils, basophils, and eosinophils. The macrophages, dendritic cells, and neutrophils are phagocytic, and engulf and digest pathogens. Other innate nonphagocytes are mast cells and natural killer cells. The nonphagocytes disable pathogens by contact or by attachment to mark them for attack by the adaptive immune system.

Molecular signaling pathways control production of leukocytes and direct them to find, identify, and fight sources of infection or antigens. Each of the innate leukocyte cell types plays a specialized role in the immune response:

1. Macrophages move through the body engulfing pathogens that they encounter and recognize or are summoned to in response to chemical signals. Macrophages migrate within tissues and release enzymes, proteins, and other factors which act on pathogens. Macrophages ingest and destroy worn out and dead cells of their own organisms, in addition to pathogens.
2. Neutrophils are carried through the bloodstream and follow chemical signals by chemotaxis to the site of infections where they concentrate as part of the inflammation response. Dendritic cells are phagocytes active mainly in tissues close to the external environment, such as skin, respiratory passages, and digestive tract. Their name refers to root-like appendages, similar in appearance but unrelated to some nerve cells. Dendritic and macrophage innate cell types play a role in activating the adaptive immune system, signaling the presence of new antigens by presenting them to special adaptive immune cells.
3. Natural killer (NK cells) cells are leukocytes that attack and destroy tumor cells, or cells that have been infected by viruses.
4. Basophils and eosinophils are related to neutrophils. They release chemical signals in response to parasites and play a role in the inflammations associated with allergic reactions.
5. Mast cells are very similar to basophils. In response to injury or antigens their granules are released and play an important part in inflammation.

The white blood cells of the adaptive immune system are special types of leukocytes, called lymphocytes. B cells and T cells are the major types of lymphocytes. B cells are involved in the humoral immune response, whereas T cells are involved in the cell-mediated immune response. When the MHC in precursor white blood cells are presented with an antigen, the genetic mechanisms produce new lines of adaptive white blood cells with new specific receptors for the antigen. This process takes time, so there is an incubation period before immunity takes hold.

25.2 Using Antibodies in Biosensors: Immunoassays

Antibodies have evolved to be molecular detectors playing a vital function in the maintenance of the organism. They possess the properties that are most desirable in a sensor: sensitivity, specificity, and discrimination. Antibodies work at the level of cell membranes and macromolecules, the nanoscale. Thus they are extremely sensitive, capable of detecting and latching onto a single molecule of antigen. Antibodies are highly specific: each one is custom designed to recognize and capture a specific molecular shape, electronic bonding, and pattern. Antibodies are highly discriminating: they can distinguish between antigens that are very similar to each other, and distinguish pathogens that are similar to cells that belong to their own organism (self).

Antibodies can be used in bulk chemistry as analytical indicators. Historically they were employed in serum assays where a precipitate or other visible change indicated the presence of a pathogen. Such tests are called *immunoassays*. With micro- and nanotechnology, antibodies can be harnessed at the molecular level to make very sophisticated and sensitive immunoassays. The design of ways to immobilize antibodies onto microchips and combine them with optical or electrical indicators has become a major area of research and development with the goal of producing analytical tools for research and medical diagnosis. In this section we give a brief introduction to this field and define some key terminologies.

25.2.1 Antibodies in Molecular Recognition Sensors

Use of Antibodies in Molecular Recognition Devices. Antibodies can be removed from blood and other tissues, and used for vaccines or other purposes. Like DNA and RNA, antibodies can be used as a macromolecular template to recognize and bind to other selected molecules. DNA segments can be fixed onto biochip substrates and used to search for matching complementary segments with sensitivity, specificity, and discrimination, even in complex mixtures of similar molecules. Readout of the biochip can be with optical, electronic, or chemical indicators.

25.2.2 Production of Antibodies

Making Antibodies. Most antigens have a variety of features that generate more than one antibody. So exposing whole animals to antigens produces a serum vaccine with heterogeneous mixtures of antibodies. This has advantages for fighting pathogens, as it gives more than one way for the body to localize and eliminate the pathogens. But for research and diagnostic purposes a single, specific antibody is required. And for safety and precision of therapy, serum antibodies have been replaced by *monoclonal antibodies* for vaccines and other medical uses [22].

25.2.3 Monoclonal Antibodies

Monoclonal antibodies, derived from a single cloned line of cells, are produced by molecular biotechnology techniques. In 1975 Köhler and Milstein, after many years of work, produced a technique for hybridizing myeloma cells with antibody-secreting cells from mice. The myeloma cancer cells could be grown indefinitely in culture, unlike normal cells, thus providing a way to manufacture antibodies. (Köhler and Milstein were awarded a Nobel Prize for this work.) Monoclonal antibodies produced in culture or in mice can be bonded to radioactive atoms, magnetic particles, fluorescent molecules, or other additives for use against cancer cells.

25.2.4 Reverse Transcriptase

Reverse Transcriptase. Genetic engineering provides a means of producing proteins by modifying the DNA sequence of cells so that they replicate the desired amino acid sequence. It was made possible by the discovery in 1970 that certain viruses use an enzyme—*reverse transcriptase*—to transcribe DNA from RNA, the opposite of the normal direction for nucleotide synthesis. (H. Temin and D. Baltimore received a Nobel Prize in 1975 for this discovery.) The RNA viruses that reproduce using reverse transcriptase are called *retroviruses*, and genetic material produced using *reverse transcriptase* is called retroviral DNA. Retroviral DNA techniques can be used to produce many proteins in culture or in transgenic animals or plants, providing another source for peptides, including antibodies.

25.2.5 Recombinant DNA

Transfer of Genes between Organisms. To produce a bioengineered product, gene sequences that code for the product must be inserted into organisms that can be easily cultured or harvested. Plasmids, small circles of DNA found in bacteria, can introduce operational DNA into bacterial cells. Plasmids used to deliver DNA are called *vectors*. Stitching a sequence of DNA into a plasmid ring is carried out by cut and paste enzymes, the restriction endonucleases and ligating enzymes. DNA altered in this manner is called *recombinant DNA*. Once in the bacteria, the plasmid is reproduced along with the products coded for by the recombinant DNA, so bacterial cultures can be used to produce genetically engineered products like human insulin, growth hormone, and interferon. Recombinant DNA technology can also be applied to plants and animals, producing genetically engineered organisms.

Recombinant DNA technology has advanced and been enhanced by refinements so that it is now used for DNA sequencing and large-scale production of hormones and vaccines. It employs a combination of traditional chemical, microbiological techniques along with molecular biology and genetics. Its chief relevance to nanotechnology is as a source for special macromolecules with specific properties, which can be used in nanomachinery and sensors. In turn, nanotechnology is contributing new abilities to analyze and manipulate DNA and other molecules in the cell.

25.2.6 Antibodies as Selection Tools for Biosensors

Antibodies are one of the biomolecules that can be used as highly specific selection and capture tools at the molecular level. Antibodies are produced by the adaptive immune system of mammals in response to all kinds of proteins and macromolecules, making them much more versatile than DNA and RNA. While DNA and RNA are very important for genetic and medical research and diagnosis, antibodies can be adapted for many more diverse purposes. In the next section we look at how antibodies can be attached to micromolecular devices to make extremely sensitive analytical detectors, and how nanoscale phenomena make the devices work.

25.3 Cantilevers as Nano-Biosensors

Sensors are devices which can be used to read and report information on their environment. Sensors typically change state in the presence of a targeted chemical or physical condition such as pressure or temperature.

Biosensors use a biological molecule or structure to effect the function of the sensor. Nano-biosensors are biosensor devices whose function takes advantage of some properties or phenomena unique to the nanoscale of interactions between matter and its environment.

Small size alone may be the chief attribute of a nanosensor, but often when we explore devices that are scaled down into the nano region, we find new behaviors that were not present at the macroscale. These phenomena can be due to changes in surface-to-volume ratios, internal material stress, heat and electrical conductivity, ratio of light wavelengths to size of material, and other phenomena. These behaviors, unique to the nanoscale, are what make nanosensors interesting and especially useful.

Natural biological sensors work at the nanoscale, and are sensitive to very small amounts of material and small changes in their environment. But they generally do not offer us a ready and quick reporting mechanism that would make them useful in an engineered device. Use of biological sensors without nanotechnology usually involves waiting for an incubation period to see the results in terms of a biological outcome, or at best involves a number of separation and amplification steps.

Integrating biomolecules into nanodevices gives us a way to combine the sensitivity and selectivity of biomolecular sensors with the speed and versatility of interfacing and readout that we require of our electronic and mechanical systems. Micro- and nanoscale cantilevers provide a good example of this paradigm.

25.3.1 *Sensing Physical Properties*

Physical parameters include mass, pressure, temperature, electrical potential, force, acceleration, tension, electric charge, etc., which can be related to biological measurements such as membrane potential, metabolism rate, and concentration of specific species (such as oxygen, hydrogen ions = pH). Physical measurements on the nanoscale can be made by exploiting physical, optical, and/or electronic properties of materials and structures on the scale of a few thousand atoms, the nanoscale. One especially important type of sensor is based on measuring the nanoscale motion of microscopic cantilevers with nanoscale coatings interacting with nanoscale quantities of substances in their environment.

25.3.2 Cantilevers and Selective Binding

Cantilever Sensors. Microscopic cantilevers are widely used tools that have become a mainstay of nanotechnology; they can be etched from silicon and other materials, as described elsewhere in this book. A cantilever or microprobe is the basis for the atomic force microscope, and arrays of cantilevers have been adapted to fabricate microelectronic memory devices and a variety of actuators and resonators [23–26]. Microchip cantilevers can also be used as mirrors, resonators, and capacitors to fabricate electro-optical and electromechanical devices that can be adapted as sensors for biomedical assays [27–29]. We will discuss some of the biomedical applications for this technology in a later section. First, we will look at how the properties of cantilevers on the nanoscale have made it possible to adapt them as sensors that have demonstrated potential for nanomedicine. In recent years this has been one of the most promising directions for a practical application in medical diagnosis of cancer, identification of proteins, and elucidation of molecular signaling pathways in molecular biology.

25.3.3 Active Cantilever Sensors

Active Cantilever Sensing—Vibration Resonance. A tiny cantilever etched from silicon or other substrates will have a characteristic resonant vibration frequency, like the tine of a microscopic tuning fork. The physics of vibrations of bulk materials applies well to submicroscopic cantilevers provided that their composition is uniform, without defects that lead to loss of elasticity. [30,31] A carbon nanotube or fiber is composed on bonded atoms, so it can be used as a cantilever that is more perfectly elastic than a deflecting or resonating structure etched from even the best single-crystal silicon, but nonlinear effects at the nanoscale must be taken into account. [32]

Nanoscale cantilevers can be used to measure density and viscosity of fluids, because their rate of vibration is damped by collisions with molecules in the fluid. The rate of vibration of a cantilever can be monitored electronically by a capacitive circuit design with the cantilever as an element. The deflection of a cantilever can also be measured optically, by changes in the angle of reflection or by interferometry. Cantilevers can be used to detect and determine the mass of molecules that are absorbed or bonded to their surface, since the resonant frequency of oscillation of a vibrating object depends on its mass as well as its elasticity. At the nanoscale, the difference in mass made by the addition of just a few large molecules such as proteins or DNA is large enough in relation to the mass of the cantilever to make it possible to detect minute quantities of biological substances [33].

25.3.4 Passive Cantilever Sensors

Passive Cantilever Sensing—Measuring Deformation. Nanoscale cantilevers can be distorted passively, without vibrating, due to differences in the amount and strength of absorption on opposite sides of the sliver of material. This is perhaps the most striking example of how things are different at the nanoscale with regard to cantilevers. This phenomenon follows the same laws of surface chemistry and physics as for larger-scale bulk materials, but at the nanoscale, the surface forces dominate over the internal properties of the material.

For a cantilever on the order of 500-μm long, 100-μm wide, and 0.5-μm thick, the free-energy differences between a surface covered with absorbed molecules and a clean surface, although small, are enough to force a displacement in the shape of the cantilever which can be detected when one side of a cantilever is coated with an agent that binds with an absorbate. The absorption of the analyte may be highly specific, as between an antibody and an antigen, or may be merely weakly selective. In the latter case, combinations of coatings on groups of cantilevers can be used to obtain patterns for recognition, as in the artificial nose device [34].

25.3.5 Surface Effects on Nanocantilevers

Absorption Strength and Surface Crowding: Absorption Affinity. To envisage the effect of absorbed molecules on the cantilever, consider that as bonds are formed with the absorbate on one specially coated or prepared side of the sliver of material, the absorbed molecules will tend to push their way into the matrix of molecules that make up the surface of the cantilever. This crowding into the surface will tend to distend the surface ever so slightly. The stronger the affinity or attractive force that binds the absorbed molecules to the substrate, the greater the crowding force will be and the more it will tend to push the molecules apart near that surface of the cantilever. If no absorption takes place on the other side, there will be a slight differential expansion on the absorption side, and this expansion will cause a deflection, with a convex curvature on the absorption side, and a concave curvature on the opposite side.

25.3.6 Steric Effects

Surface Crowding: Steric Effects. An additional force for expansion on the absorption side will occur if the absorbed molecules repel each other or push against each other; in that case they will tend to pull apart the molecules in the cantilever to which they are bonded. These two forces are sufficient to produce a measurable deflection if the cantilever is small enough that the effects on the surface overpower the forces holding the internal molecules in their preferred places relative to each other in the bulk material matrix.

Complexity due to Interplay between Affinity and Steric Effects. Some of the leading researchers in the field of cantilevers have pointed out that understanding the molecular causes of the deformation is critical, because the observed bending depends on both the strength of absorption affinity and the nanomechanical response of the cantilever. These two factors may be convoluted such that a strong binding affinity will not necessarily produce a large surface stress. Hans Peter Lang, Martin Hegner, and Christoph Gerber in Switzerland have shown that steric crowding of the absorbed molecules may be more important than free-energy changes on the surface, although the two factors are interlinked. [34,35] What is certain is that as the size of the cantilever becomes smaller, classical models for describing the interaction with its surroundings begin to break down, and care must be taken in designing a device and interpreting the results. [36,37]

Even with classical elastic behavior of the cantilever, thermal effects may bias the degree of displacement, for example. Thus, in designing a device for practical application in biomedicine, reference cantilevers are placed alongside the elements that are coated to bind with the analyte, so that any purely thermal effects or other environmental influences can be compensated.

25.3.7 Surface Free Energy at the Nanoscale

Cantilever motion can be explained by changes in the surface free energy of one surface of the cantilever relative to the opposite side. The strong intermolecular forces associated with specific binding between molecules, such as between an antibody and an antigen, result in much higher free-energy change than for nonspecific bonds such as physical absorption layers. Cantilever deflections, large enough to be useful for sensors, should be a result of specific binding.

An estimate for the magnitude of a cantilever deflection Δh is given by Stoney's equation:

$$\Delta h = 3\sigma(1 - v) / E \cdot (L / d)^2 \qquad (25.1)$$

where

σ is the change in surface free-energy density (or surface stress) due to specific binding

E is the elastic modulus (or Young's modulus) of the cantilever material

v is its Poisson ratio

L and d are the length and the thickness of the cantilever, respectively

Longer and thinner cantilevers will result in larger deflections for the same degree of surface stress. Note that the above formula does not take account of steric crowding forces [35,38,39].

25.4 Micro- and Nanosensors and Applications

Many cantilever and similar devices for detecting molecules have been described and developed. In principle, a string or wire of nanodimensions could be used in a similar way, provided that a means of inducing and measuring vibrations were devised [35]. A number of books and reviews are available on the subject of nanoscale cantilever molecular sensors [34,35]. Among the many types of cantilever devices used to measure biomolecules, we present a few examples that have particular potential or interest for nanomedicine.

25.4.1 Biomedical Cantilever Applications

To make a cantilever sensor with a medical application, one may coat one surface with a reactant such as an antibody specific for an antigen. When the surface is exposed to solutions of the antigen, it becomes coated with molecules of the antigen, producing a small but detectable deformation of the cantilever due to

FIG. 25.2 *Absorption and steric crowding on an antibody coated cantilever.*

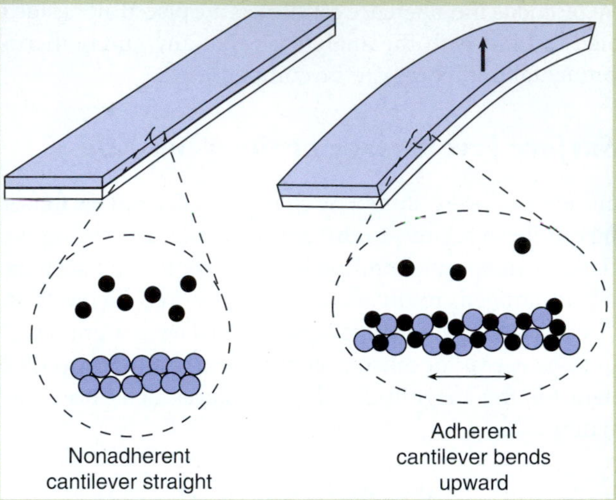

surface forces. This deformation can be detected by optical reflection. Only for devices of nanoscale proportions are the intermolecular nanomechanics at the surface sufficient to produce a detectable deflection usable as a sensor (**Fig. 25.2**).

25.4.2 Cantilever Sensor for Cancer Screening

An interdisciplinary team of coworkers at University of California Berkeley, University of Southern California, and Oak Ridge National Laboratory developed an application of this technique to detect the prostate cancer antigen (PSA) [38]. Using 600-μm-long and 0.65-μm-thick silicon nitride cantilevers, it was feasible with their technique to detect PSA concentrations of $0.2\,\mathrm{ng \cdot mL^{-1}}$. Since this antigen is produced by the body in response to cancer cells, this sensor could be useful in a blood screening test for early detection of cancer in humans. This technique may also lead to devices for high-throughput label-free analysis of protein–protein binding, DNA hybridization, and DNA–protein interactions, as well as drug discovery.

25.4.3 Biotechnology Applications of Cantilevers

In actual medical use, diagnosis for complex diseases such as cancer would require quantitative detection of multiple proteins. Therefore, current issues in research are how to improve sensitivity, specificity, and detection of real-world diagnostic concentrations of antigen in the presence of interferences under clinically relevant conditions. Many designs for using micro- and nanoscale cantilevers have been developed for biomedical detectors.

The Quartz Crystal Microbalance. The quartz crystal microbalance is a special case of the use of nanoscale effects in a cantilever. It is used to measure mass by the change in frequency of a piezoelectric quartz crystal when a small mass such as a virus or a large biomolecule is added. The frequency change is proportional to the added mass, so long as the added mass has elastic properties similar to those of the quartz itself. The mass change is given by the *Sauerbrey equation*:

$$\Delta f = \left[\frac{-2 f_{ro}}{A\sqrt{\rho_q G_q}} \right] \Delta m_q \qquad (25.2)$$

where

Δf is the change in the resonance frequency due to the added mass
f_{ro} is the resonance frequency of the unloaded resonator
ρ_q and G_q are the density and shear modulus of the quartz
A is the surface area of the resonator [40–43].

Arrays of Cantilevers. As we mentioned earlier, arrays of microelectronic cantilevers have been developed for use as semiconductor memory and digital light-processing devices [23–26,44,45]. Digital versions of cantilever arrays can be adapted not only to sensitize patterns for making biochips, in maskless photolithography [26–29], but also to read out the status of microchip sensors by reflecting precisely focused laser beams onto sensors in an array. These devices can be used to prepare and interrogate arrays of antibodies, and also with DNA and RNA segments to bind to complementary sequences, making arrays that can select and identify genetic information [46–48]. Microarrays can even use a mixture of DNA and antibody sensors to detect genetic and protein information at the same time.

In the case of the digital micromirror array, each element is not a freely suspended cantilever, but is designed to switch between two states. Thus the mirror elements are not used directly as sensors, but to prepare and interrogate arrays of sensors by directing light to precise addresses in a microarray. The use of cantilevers as digitally controlled mirror arrays has had a large impact on the efficiency of DNA sequencing, perhaps one of the more direct examples of the interplay between nanoscale fabrication technology and molecular biology [49–52].

25.4.4 Surface Acoustic Wave Nanosensors

Another surface interface phenomenon which can be exploited for nanoscale detectors is the *surface acoustic wave* (SAW). Surface acoustic waves are compression vibrations traveling over the surface of an elastic material. SAW waves have a longitudinal and a vertical shear component. This type of wave propagation is distinct from *body waves* which travel through the interior of a material. The amplitude of a surface acoustic wave decays exponentially with depth into the material.

In dielectric materials, surface acoustic waves can be generated by application of electric fields, and detected by the transduction of the waves into electromagnetic fields. This behavior is used in electronic circuit devices. SAW devices are designed to act as filters, oscillators, and transformers in circuits, using one transducer to excite the wave and another to detect it at opposite ends of a strip of piezoelectric material.

SAW waves exhibit strong coupling with adsorbed material on the device surface. This effect allows SAW devices to be used as sensors by measuring changes in frequency and amplitude of the wave in response to contact with external substances.

Coating a SAW device with a specific antigen or other compound that bonds to a biological target molecule turns the device into a sensitive and specific detector that can be used to monitor for known antigens or even for bacteria [53]. SAW devices for biological applications have not yet seen widespread application, but they are an interesting example of the use of a nanoscale phenomenon which may in future have significant applications.

Conventional SAW devices use piezoelectric material. Researchers at Pennsylvania State University have developed SAW devices based on *magnetoelastic* materials. These have the advantage that a coil can be used to detect changes in the frequency of the waves, thus providing for wireless operation [54]. The resonant frequency of a magnetoelastic sensor changes in response to temperature, pressure, ambient flow rate, and liquid viscosity and density, as well as to absorbed mass, making this type of sensor very versatile.

25.4.5 Electrochemical Nanosensors

Electrochemical sensors convert chemical reaction energy into electrical potential or current. They are extremely sensitive, but must be in electrical contact with a tissue or fluid in order to be effective. Gases can be detected by electrochemical sensors if a moistened membrane is maintained on the surface of the sensor, in which the gas is absorbed.

Electrochemical probes measure the *electronic potential difference* at the interface between the surface of two materials, usually a solid and a liquid. Every chemical element has an inherent tendency to attract or donate electrons to its surroundings. Any time that two different materials are in contact with one another, an electronic potential exists across the interface. The greater the difference in the chemical tendency to donate or accept electrons, the larger the potential difference. If there are no barriers to chemical reactions at the interface, the reaction proceeds with exchange of electrons until the potential is neutralized and equilibrium is reached.

The difference in potential between two substances can be measured when both are in contact with a conductive medium, an *electrolyte*, through which electrons can be exchanged. In an *electrochemical cell*, two dissimilar electrodes are in contact with an electrolyte. An electronic potential is created by the donation or acceptance of electrons through chemical reactions which take place between the electrodes and the electrolyte.

If the two electrodes are connected by an electronic conducting path in parallel with the path through the electrolyte, a circuit is formed over which electrons can flow. Electronic charge flows over the conductor from the electrode with the greater tendency to accept electrons from the electrolyte towards the electrode which donates electrons in reactions with the electrolyte. This process accelerates the chemical reactions at the electrodes by providing a path for electron transfer that is of lower resistance than the path through the electrolyte. The energy of the chemical reaction is converted into electrical current, which is harnessed in batteries and fuel cells, and used in electrochemical sensors [55].

The potential difference at the interface between electrode and electrolyte is affected by adsorbed substances at the electrode surface. This provides a sensitive means for measuring the nature and concentration of molecules which interact with the surface of an electrode [56,57].

Electrochemical Detection of Free Radicals and Nitric Oxide in Tissues. An example of the usefulness of nanoprobes for electrochemical sensing is the detection of nitric oxide (NO) in tissues. Nitric oxide is a simple free radical that has been found to act as a cell signaling messenger playing a number of roles in the body. One of its most important functions is to signal the muscles in the wall of blood vessels to relax, allowing more blood to circulate. Nitroglycerin's effectiveness in relieving cardiovascular distress is due to the release of nitric oxide. This was one of the clues which led to nitric oxide being identified as a key endothelium-derived relaxing factor. Nitric oxide has been found to play a number of roles in different cell signaling cycles; its absence or excess can lead to tissue damage in heart, kidneys, and muscle.

Researchers at the Oak Ridge National Laboratory have developed an electrochemical probe to track concentrations of NO in tissue, using a functionalized carbon nanotube. Important free radicals and other species which can be monitored with the nanoprobe include NO, H_2O_2, O_2, H_2S, ATP, glucose, and superoxides [58,59]. The nanoprobe uses an electrochemical sensor principle and is based on functionalized carbon nanotubes with a sensor diameter of 100 nm (**Fig. 25.3**).

The advantages of an electrochemical nanoprobe compared with luminescence assays and other spectroscopic methods are

1. It is real-time rather than based on a derivative product.
2. It provides improved spatial resolution.
3. It does not require luciferin or other expensive bioluminescence or fluorescence reagents.

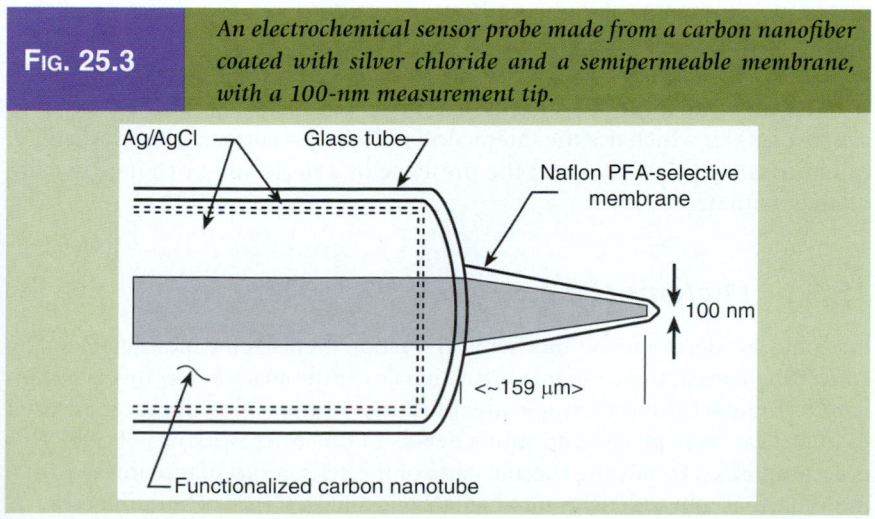

FIG. 25.3 *An electrochemical sensor probe made from a carbon nanofiber coated with silver chloride and a semipermeable membrane, with a 100-nm measurement tip.*

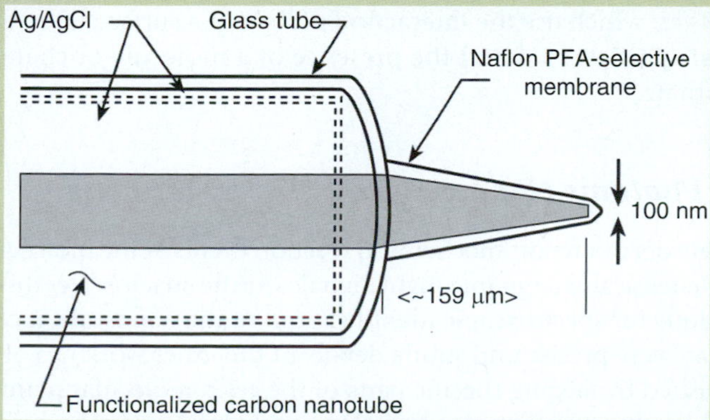

4. It can be applied to direct measurement of cells in a culture or tissue without use of a luminometer or spectrometer, and requires no shielding from external light sources.

A Carbon Fiber Nanoelectrode for Glucose Detection. Electrochemistry is a good method for detecting glucose and other simple compounds in blood and tissue. As one example, a glucose micro-biosensor has been developed by researchers in Austria using electrochemical co-deposition of glucose oxidase (GOx) along with MnO_2 as a mediator, onto a single carbon fiber microelectrode with a tip diameter of 100 nm [60]. A thin silver/silver chloride film is applied as an immobilization and interference-free protective layer. This type of micro-biosensor can be used as an amperometric glucose detector probe in blood or tissue. Improvements in glucose detection are important for research in diabetes, obesity, and metabolism.

Significance for Nanobiotechnology. Nanoscale electrochemical sensors work on the same principles as macroscale sensors; what makes them significant for use in nanobiotechnology is the newly developed ability to fabricate such sensors in nanoscale sizes for detection of very small amounts of substances in complex mixtures typically found in biological systems [61–63]. The ability to use nanoscale electrochemical probes is also due to advances in electronics, which enables the precise measurement of extremely small potentials and currents. Arrays of electrochemical microsensors are being fabricated with both DNA and antibody probes, in order to analyze DNA and proteins on a single platform for proteomics and medical applications [64]. The importance of interfaces, chemical boundaries, and electrical forces in biological systems makes electrochemical measurement an important route to many biological measurements [65,66].

25.5 Optical Nanosensors

In this section we look at some examples of optical sensors, which allow direct measurement of light without depending on cantilevers or other mechanical nanodevices, and some of their biomedical applications. These include *photonic sensors*, which use the interaction of light with molecules in spectroscopic absorption, reflectance, fluorescence, or luminescence; and *plasmon sensors*, which use the interaction of light at a surface layer, waveguide, or nanoscale particle to detect the presence of a molecule or change in state of an absorbate.

25.5.1 Photonic Nanosensors

Photonic sensors can report biochemical reaction events at the molecular level by converting chemical energy into light signals. At the macroscale, this is a standard technique for spectroscopic identification. At the nanoscale, it can be used to interrogate very precise and subtle details of the inner workings of cells. This is accomplished by tagging specific parts of the cell's molecular pathway with a molecule that converts energy into light through fluorescence or bioluminescence, which can be detected by a nanoprobe or by nanoscale-capable microscopy techniques [67–72].

Photonic Sensor for Antibody Detection Inside a Single Cell. Tuan Vo-Dinh and colleagues in a research team at the Oak Ridge National Laboratory have developed an optical nanoprobe for measurements inside a single cell [73]. The nanoprobe uses a fluorescent antibody that attaches to a molecule produced when the carcinogen benzo[a]pyrene (BaP) attaches to DNA to form a mutation-causing adduct that can lead to cancer. Detection of this biomarker is significant in monitoring for DNA damage due to BaP exposure and for possible precancer diagnosis. BaP is a polycyclic aromatic hydrocarbon of serious environmental and toxicological interest because of its mutagenic/carcinogenic properties and its widespread presence in the environment [74]. The measurements were performed on model cells from a rat liver cell line which contained the carcinogen metabolite.

The nanoprobes were fabricated by the same technique used for cell patch clamp experiments (described in chapter 14) [75]. A glass tube was heated and drawn out until its cross section at the tip was on the order of 50 nm. A reflective coating of silver was deposited on the outside of the glass probe, whose average diameter with the coating was 250–300 nm. The nanoprobes, which required great care to be made consistently, acted as fiberoptic waveguides which could be inserted through the cell membrane to take laser light into individual cells. The nanoprobes were coated with the fluorescent antibodies specific for the carcinogenic antigen, inserted into individual cells, incubated 5 min to allow antigen–antibody binding, and then removed for fluorescence detection. A concentration of 9.6×10^{-11} M for the carcinogen in the individual cells was determined.

Fiberoptic chemical sensors and biosensors offer important advantages for in situ monitoring applications because of the optical nature of the detection signal. The application of submicron fiberoptic chemical probes has been pioneered by Kopelman and coworkers, who developed probes for monitoring pH [76,77] and nitric oxide [78]. The use of submicron tapered optical fibers has also been demonstrated and used to investigate the spatial resolution that is possible using near-field scanning optical microscopy [79,80].

This example of an antibody-based nanoprobe for measurements of chemicals inside a single cell is being followed by further improvement of the technique for applications in biotechnology, molecular biology research, and medical diagnostics. This technique could eventually be used to develop advanced biosensing systems to study intracellular signaling and gene expression inside single cells.

25.5.2 Surface Plasmon Nanosensors

A *surface plasmon* is a charge density wave that is induced by light striking an interface between a thin film and another medium. *Surface plasmon resonance* (SPR) is the coupling of energy from incident light with the charge density in the surface, resulting in energy being transferred into a thin layer on the surface instead of being reflected. Resonance occurs only when the energy of the incident light is of the right frequency and angle of incidence for coupling.

SPR has been introduced in earlier chapters, as the phenomenon which gives rise to intense colors in butterfly wings, diatoms, nanodots, and other materials with thin layers on the nanoscale, corresponding to resonant frequencies with incident light. It is a useful phenomenon for optical detection and measurement where thin films can vary in composition and thickness. The waveform and intensity of the plasmon depend strongly on the thickness of the film and the type of material on which the film is deposited.

SPR is used to study the formation of self-assembled layers and structures of organic molecules formed at the surfaces of electrodes [81]. SPR can also be used to detect biological substances, such as DNA and proteins. A surface can be coated in patterns with different specific antigens or DNA complementary test strands to make an array in which a combination of biological target molecules can be detected (**Fig. 25.4**).

Dual Polarization Interferometry. Dual polarization interferometry (DPI) is an extension of the application of SPR which divides focused laser beams into two waveguides, one, the "sensing" waveguide, with an exposed surface, and the other to serve as a reference beam. When light that has passed through both waveguides is combined, an interference pattern is created. Measurement of the interference pattern from two different polarizations of light permits the refractive index and the thickness to be determined for an adsorbed layer on the surface of the sensing waveguide. Because the polarization can be switched rapidly, real time measurements can be obtained for chemical reaction or flows taking place on the surface of the sensing waveguide. The DPI technique is used to obtain

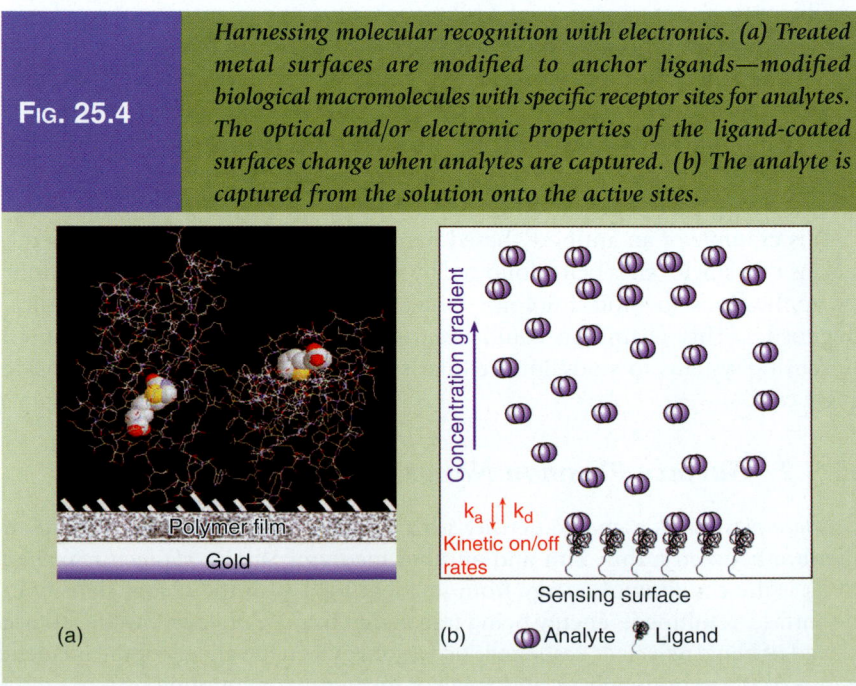

Fig. 25.4 *Harnessing molecular recognition with electronics. (a) Treated metal surfaces are modified to anchor ligands—modified biological macromolecules with specific receptor sites for analytes. The optical and/or electronic properties of the ligand-coated surfaces change when analytes are captured. (b) The analyte is captured from the solution onto the active sites.*

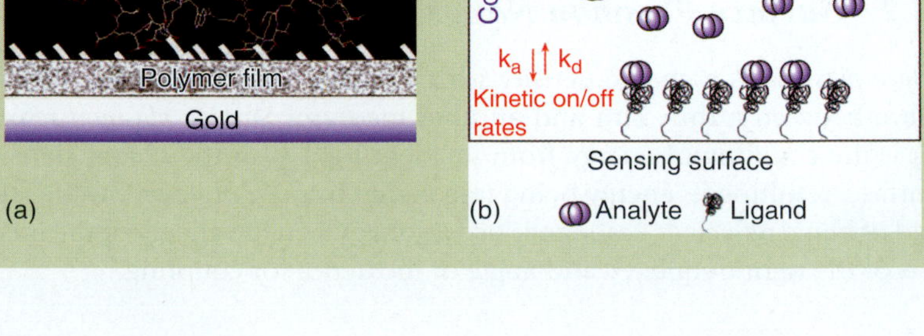

structural information on proteins and other molecules adsorbed on the waveguide interface [82,83].

Plasmon Waveguide Resonance. Plasmon waveguide resonance (PWR) spectroscopy is a highly sensitive extension of the surface plasmon technique which is especially suited for characterizing the kinetics, affinities, and conformational changes involved in receptors in cell membranes, without the need for radioactive or other labeling techniques.

PWR spectroscopy utilizes a thin dielectric coating (e.g., silica) on a thin metal film (e.g., silver) deposited onto a prism. The dielectric functions as a waveguide allowing plasmon excitation by polarized light. Plasmon resonance in the waveguide provides narrow resonance and high sensitivity. The use of light with different polarizations to probe immobilized biological films provides detailed structural information on oriented molecular arrays.

Lipid bilayers can be deposited onto the resonator surface with their incorporated membrane proteins, and changes in resonance monitors ligand binding to the proteins. Lipid bilayers containing microdomains whose refractive index anisotropies and thicknesses differ from those of the bulk membrane will display multiple resonances and thus can be mapped.

PWR can measure differences in ligand-binding constants, and protein properties that result from differences in the microenvironment. Thus, this technology provides useful insights into the structural and functional consequences of microdomains in cell membranes [84].

25.5.3 Nanoscale Optical Resonance Grids—Using the Butterfly Wing Effect

At the nanoscale, there are new types of opportunities for photonic effects, due to resonance, as we have seen elsewhere in this volume. Now we consider a new type of nanoscale sensor which works on the same principle as the butterfly wing photonics, taking advantage of the nanoscale effects of guided mode optical surface resonance to detect very small amounts of material trapped by absorption between vanes of a nanoscale optical resonance grid. This and similar nanoscale optical phenomena are leading to many new types of biological sensors. These new systems are capable of sensing at extremely sensitive levels and are suitable for parallel integration for detection of multiple signals.

25.5.4 Guided-Mode Resonance Sensors

Leaky optical resonance modes arise on periodic films when an incident light beam couples to the waveguide-grating layer system. This results in generation of a *guided-mode resonance* (GMR) field response in the spectrum. The resonance effect leads to dramatic redistribution of the diffracted energy and may manifest as sharp reflection and transmission peaks radiating from the structure. Application of this fundamental effect to sensors was first proposed in 1992 [85].

The sensor is based on the high parametric sensitivity inherent in the basic GMR effect [85,86]. Figure 25.5 schematically illustrates a generic GMR sensor and its operation. As an attaching biomolecular layer changes the parameters of the resonance element, including thickness, fill factor, and refractive index, the resonance frequency changes [87]. Thus, incident light is efficiently reflected in a narrow spectral band whose central wavelength (and resonance angle) is highly sensitive to chemical reactions occurring at the surface of the sensor element. A target analyte interacting with a bio-selective layer on the sensor can thus be identified without additional processing or use of foreign tags. Computed examples and experimental results have been presented that illustrate key sensor properties and chart paths for establishing useful sensor technology based on this effect [88].

25.5.5 Applications of Guided Mode Sensors

These photonic resonance concepts enable a new class of highly sensitive and selective biosensors and chemical sensors with applications in medical diagnostics, drug development, environmental monitoring, homeland security, and others.

Guided-Mode Resonance Sensor for Protein Detection. **Figure 25.5** illustrates the measurement of protein binding to a surface in air environment utilizing a double-layer resonant element illuminated at normal incidence. The clean grating surface is first chemically modified with amine groups by treating with a 3% solution of aminopropyltrimethoxysilane in methanol (**Fig. 25.6 Top, Left**). The device is then washed in a solution of bovine serum albumin (BSA, 100 mg · mL^{-1}) and a deposited 38 nm thick layer of BSA results in a reflected resonant peak spectral shift of 6.4 nm (**Fig. 25.6 Bottom, Left**). It is significant that minimal signal degradation occurs as the biomaterial attaches to the sensor surface with reflectance remaining at ~90% before and after BSA attachment. The biomaterial layer thickness is determined by fitting the measured data with theoretically computed values. The observed shift in resonance wavelength of 6.4 nm is easily detected with economic spectrum analyzers.

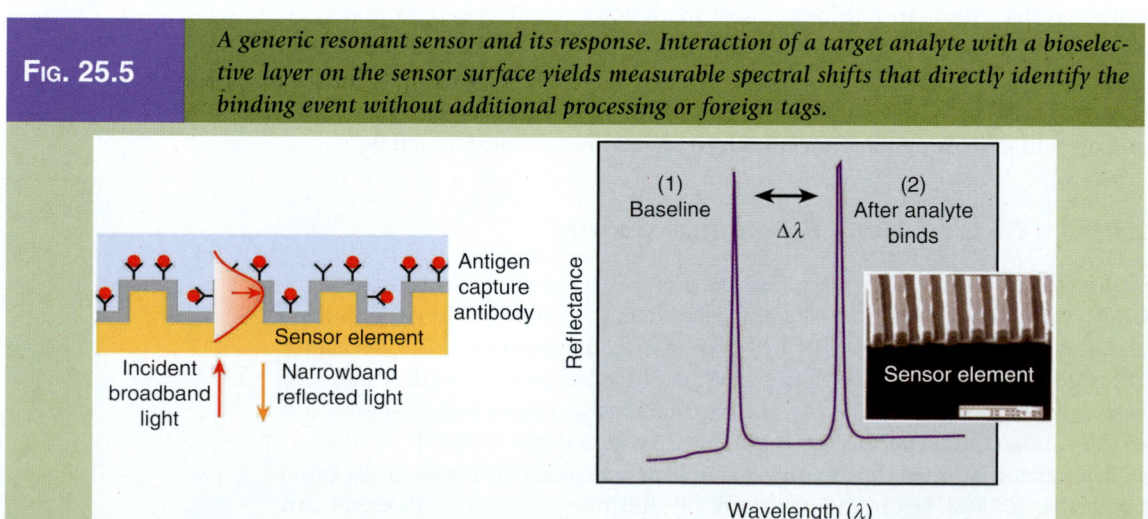

Fig. 25.5 *A generic resonant sensor and its response. Interaction of a target analyte with a bioselective layer on the sensor surface yields measurable spectral shifts that directly identify the binding event without additional processing or foreign tags.*

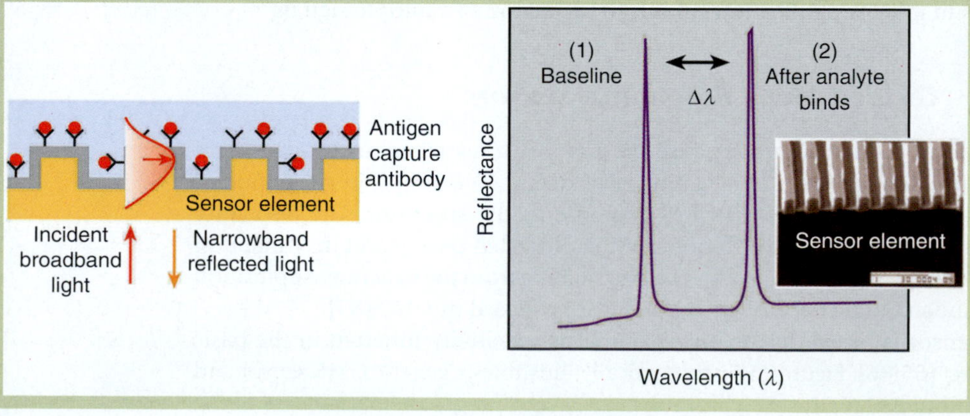

FIG. 25.6 *Measured leaky-mode resonance sensor spectral response in air and comparison with theory* (Top, Left). *The sensor is illuminated by a TE-polarized laser beam and the sensor surface is modified with a silane chemical linker; a 38-nm layer of attached bovine serum albumin (BSA) results in a 6.4-nm peak shift* (Bottom, Left). *Sensor parameters are* $n_C = 1.0$, $n_1 = 1.454$ (SiO_2), $n_2 = 1.975$ (HfO_2), $n_S = 1.454$, $d_1 = 135\,nm$, $f = 0.58$, $d_2 = 208\,nm$, $\Lambda = 446\,nm$, *and* $\theta = 0°$. *A scanning electron micrograph (SEM) of the sensor surface is also shown* (Top, Right).

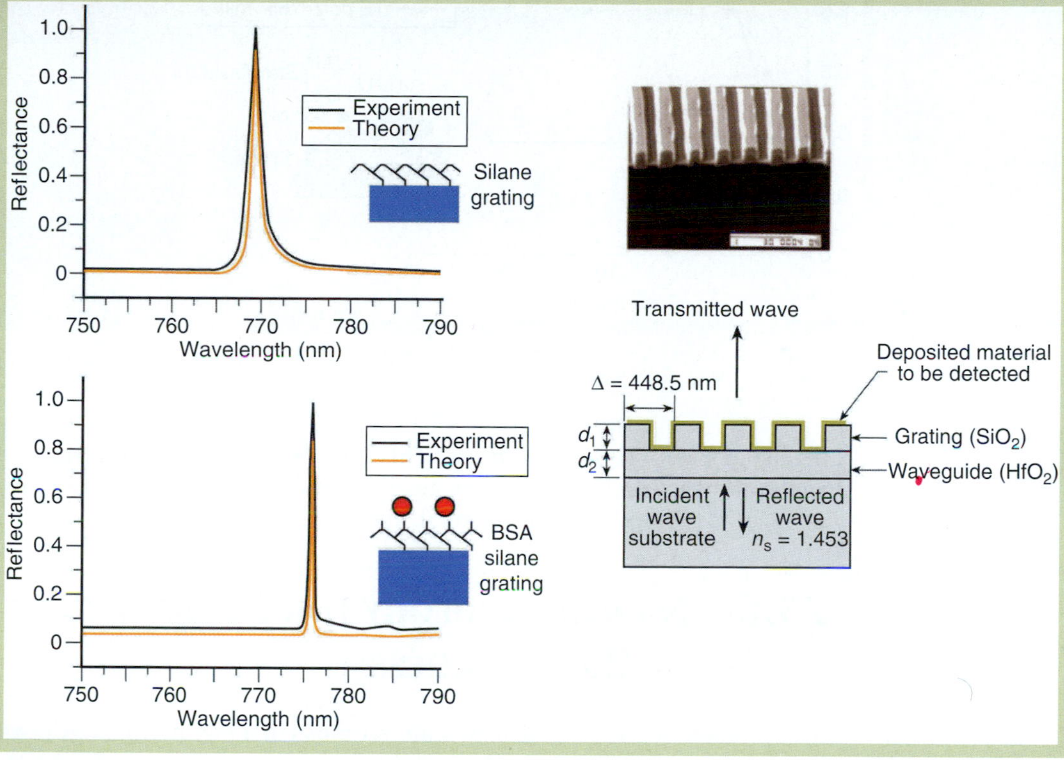

Significantly, polarization diversity permits monitoring of thickness and refractive index variations simultaneously. Occurrence of multiple leaky modes at resonance contributes to sculpting diverse spectral bands that are useful for new sensor types and many other optical elements. **Figure 25.7** shows an example application of a GMR sensor where both polarization states simultaneously monitor the reaction, providing added accuracy and precision in the quantitative determination of the biomolecular response.

Other Applications of Guided Mode Optical Sensors. Nanoscale optical sensors based on the butterfly wing resonance principle are leading to many new types of biological sensors. These new systems are capable of sensing at extremely sensitive levels and are suitable for parallel integration for detection of multiple signals [89–91]. Researchers are using butterfly wings as templates to fabricate optical grids and nanotubes for use in sensors [92,93]. The sensor systems being produced based on the butterfly wing principle dramatically outperform existing

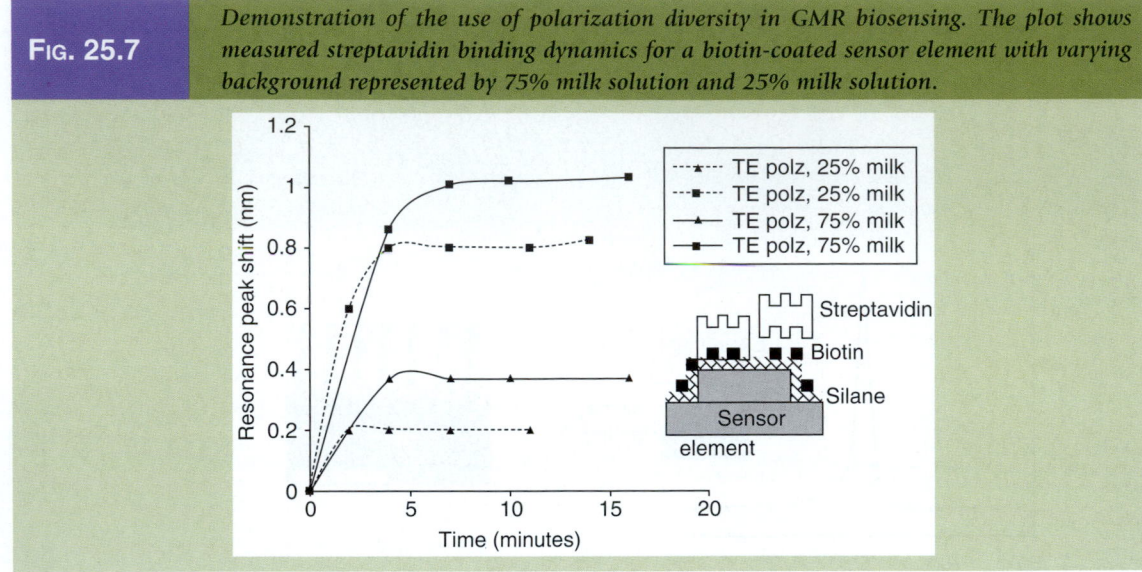

FIG. 25.7 *Demonstration of the use of polarization diversity in GMR biosensing. The plot shows measured streptavidin binding dynamics for a biotin-coated sensor element with varying background represented by 75% milk solution and 25% milk solution.*

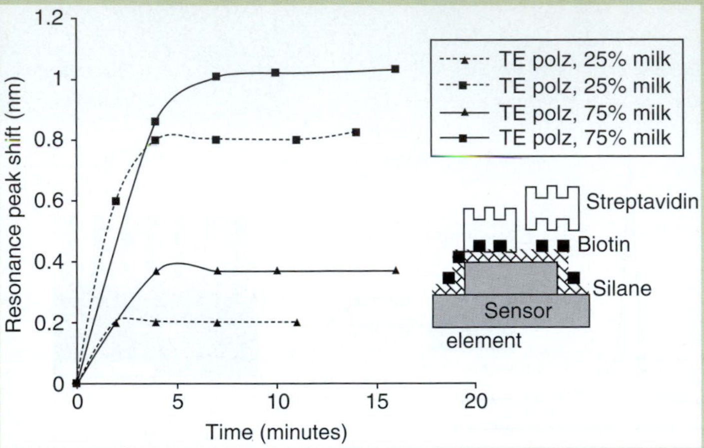

nanoengineered photonic sensors. They are being studied to develop new applications for detection of water, alcohols, and chlorinated hydrocarbons for a variety of biological, environmental, and medical applications.

25.6 Nanotechnology for Manipulation of Biomolecules

In addition to sensing on the nanoscale, biomolecular nanotechnology involves ways of manipulating cells and macromolecules using nanoscale forces. Some of the most important nanoscale manipulation tools are the atomic force microscope and atomic force microprobe, which are discussed elsewhere in this book. Two other techniques for nanoscale manipulation are especially useful for biological studies: dielectrophoresis and optical tweezers. These techniques depend on the electromagnetic field phenomena acting on very small scales.

25.6.1 *Optical Tweezers*

Light consists of oscillating electronic and magnetic fields, a beam of light can exert electromagnetic forces when it interacts with matter. Light also has momentum, so when it is refracted, reflected, or absorbed, it produces a force on an object. On the macroscale, these forces are so small relative to the mass of an object that they are hardly detectable. But on the micro- and nanoscale, these forces become significant. Random scattering of suspended microscopic particles by incident light can be observed with a microscope. When light is highly coherent, as in a laser beam, these forces are aligned and have very directed and controllable effects.

Optical Trapping. A tightly focused laser beam which converges through a pinched focal point creates an optical trap for a charged or polarizable particle small enough to fit inside the focal area of the beam. This trapping effect is only possible on the microscale and smaller, where the electromagnetic forces on the particle are large relative to gravity and inertia, and where the wavelength of the laser light is on the same order as the size of the particle. The density gradient of the electromagnetic fields combines with the momentum transfer to create the trap geometry. This effect was discovered by Askin and coworkers in 1986 [94,95]. Since then it has been applied using laser focusing arrangements called *optical tweezers* to manipulate objects on the micro- and nanoscale. It is especially applicable to biological cells, since living cells are highly polarizable.

Optical Tweezers. Optical tweezers can manipulate particles as small as a single atom. In the biological sciences these instruments have been used to apply forces in the pN-range and to measure displacements in the nm range of objects ranging in size from 10 nm to over 100 mm. Optical tweezer instruments are now available as sophisticated computer controlled systems devices that are able to measure displacements and forces with high precision and accuracy.

Applications of Optical Tweezers. Optical tweezers have been used to trap and manipulate dielectric inorganic particles, viruses, bacteria, living cells, organelles, proteins, and DNA segments. They can be used for confinement and sorting of cells, tracking of movements of bacteria and motile cells, measurement of small forces, and manipulation and rearrangement of cell membranes and structures.

Because visible light forms ideal traps at the micron scale, smaller objects are usually studied under the microscope with visible wavelength lasers by attaching specimens to micron-sized beads, which are then trapped, where the effects of light forces can be studied. Forces required to rearrange DNA or proteins can be measured using this technique. Molecular motors such as kinesin, myosin, and RNA polymerase can be studied to determine the size and continuity of motion steps and forces required to produce movements [96–99].

25.6.2 Dielectrophoresis

Electrophoresis is the phenomenon of movement, or *electrokinetics*, of a particle induced by an electrical field gradient, known since the early 1800s. An electrical field induces a force on a polarizable particle. As in the case of optically induced forces, the electrophoresis forces become significant at the micro- and nanoscale. Electrophoresis is used to separate colloidal particles and biological macromolecules such as proteins and DNA fragments, and has become a highly useful and refined technique in molecular biology and pathology [100–103].

Dielectrophoresis is a more complex phenomenon in which polarizable particles move in response to alternating electromagnetic fields [104]. Complex standing waves and gradients can be created around alternating current electrodes, which can trap and move polarizable particles such as biological cells. Dielectrophoresis has become a refined and useful technique in molecular biology and nanotechnology with the availability of sophisticated electronics and computer analysis to direct the patterns of fields [105]. Dielectrophoresis can be employed as the active principle for submicron scale instruments to (1) separate and isolate specific cells of interest, (2) concentrate and amplify cell concentrations, (3) accurately measure indicators for different cells, and (4) measure a quantitative result based on cell differences. Dielectrophoresis is being designed in microfabricated devices to separate cells for research, diagnosis, and preparation of purified fractions.

In dielectrophoresis an alternating current is applied to a volume containing cells to create a nonuniform temporal and geometric pattern of electromagnetic fields. A dielectrophoretic force acting on the polarized cells causes them to move along the electric field gradient created by the nonuniform electrode geometry. At specific radio frequencies, cells are moved differently depending on their polarizability and geometry. Polarizibility can distinguish genetically similar but distinct cells, which will have different movement responses at given frequencies.

Dielectrophoresis can be used like the more conventional methods of cell concentration and separation, such as centrifugation, filtration, fluorescence-activated cell sorting in flow cytometry, or optical tweezers. Dielectrophoresis does not require the chemical binding of labels or tags to the cell; it is based on the inherent polarizability of the cell's molecular contents. Thus it can operate directly on native, unlabeled cells, without the expense of labeling, or development of labels and tags. A single dielectrophoretic instrument can isolate and analyze a wide range of particle types (cells, bacteria, viruses, DNA, and proteins) using the same basic procedure.

Dielectrophoresis does not generally harm cells, so they remain viable for further study, processing, or therapy. Research is being undertaken to find whether, under certain conditions, dielectrophoresis could be capable of selectively destroying cells such as bacteria and cancer cells [106].

Development is also being done to refine dielectrophoresis separation and analysis techniques to analyze intact cells to provide detailed selective information on their DNA makeup, and to electrically lyse cells to break out their contents for processing and analysis.

25.6.3 Some Dielectrophoresis Applications

Separation of Cancer Cells from Blood. Dielectrophoresis can be used to separate cancer cells from normal cells, based on differences in their polarizability. A measure of the dielectric properties of cells is the plasma membrane capacitance, which has been found to differ greatly between normal and cancer cells. For example, MDA-231 human breast cancer cells have been found to have a mean plasma membrane specific capacitance of $26\,\mathrm{mF \cdot m^{-2}}$, more than double the value ($11\,\mathrm{mF \cdot m^{-2}}$) observed for resting T lymphocytes.

Researchers at the M. D. Anderson Cancer Center have used a dielectrophoresis affinity column to separate several different cancerous cell types from blood. The affinity column consists of a series of thin, flat chambers with microelectrode arrays on the bottom wall, through which the blood flows. Dielectrophoresis forces generated by the application of alternating current fields to the electrodes influence the rate of elution of cells from the chamber. Dielectric properties were measured for the various cell types in the mixtures to be separated, and theoretical modeling was used to design the optimal flow column and applied field parameters. Various ratios of cancerous to normal cells at different concentrations separated in test runs in the column. The results included a 100% efficiency for purging MDA-231 cells from blood at a ratio of about 1 cancer cell to 100,000 normal cells. Cell viability was not compromised, and separation rates were at least 103 cells $\cdot$ s^{-1} [107]. These results could lead to a diagnostic test for very early detection of cancer.

Concentrating Bacteria for Water Analysis. One of the problems that arise with nanoscale detection and analysis is finding the molecule or cell of interest when it is hidden in a large volume of material. It does little good to have a sensor that is capable of detecting a single molecule if the sensor is unlikely to ever encounter the molecule of interest in the midst of billions of others. Hence ways of concentrating or replicating the molecule or cell to be detected are sought.

The traditional way of amplifying the bacteria count so that an assessment of bacterial contamination can be made is to allow a small sample of bacteria to grow for hours or days in a culture medium, until the colony, held in place by a gel, is sizeable enough to be seen and evaluated. DNA can be amplified by a cyclical reaction technique, the *polymerase chain reaction* (PCR), by which a single strand of DNA is allowed to replicate repeatedly in a medium of base components and replication enzymes, through temperature cycles that cause the DNA chains to replicate and separate, doubling the amount of DNA on each cycle [108]. Centrifugation, filtration, evaporation, and other concentration techniques are not practical for detecting small numbers of bacteria in large volumes of water for a number of reasons.

Because it can selectively and efficiently separate different types of cells, dielectrophoresis can be used to concentrate cells into a small volume for efficiency of counting and analysis.

Insulator-Based Dielectrophoresis for Concentration of Bacteria. Researchers at Sandia National Laboratories have developed a novel form of dielectrophoresis, called *insulator-based dielectrophoresis* (iDEP), to selectively—and very quickly—concentrate live pathogenic bacteria from large water samples. iDEP can enable detection of small amounts of biological material in large sample volumes, without bacterial cell culturing. This significantly lowers costs, speeds analysis, and reduces risks of contamination [109].

iDEP could be used in medical diagnostic tests where a few anomalous cells result from diseases, such as cancer, sickle cell anemia, and leukemia. It could also be useful for protein isolation and concentration, and mass spectrometry sample preparation for proteomics and drug discovery.

Conventional dielectrophoretic sorters have electrodes within a sampling chamber and use the nonuniform electric field adjacent to electrodes to induce dielectrophoretic motion of cells. Dielectrophoretic electrodes typically require precision microfabrication, produce bubbles, and electrolyze products that can harm device operation, and can damage cells with strong field gradients.

In contrast, iDEP uses electrodes located on the outside of the sample chamber. Current from the electrodes conducts through a particle-bearing liquid into the device where patterned walls or insulating obstacles produce the required nonuniform electric field. This arrangement eliminates many of the disadvantages of conventional devices: insulating structures can be replicated economically, produce no electrolytic effect, and can be contoured to be gentle on cells.

Integration of Dielectrophoresis into Highly Parallel Devices for High Throughput. Dielectrophoresis is well suited for integration into highly parallel devices for rapid screening and analysis of multiple types of cells or macromolecules in complex mixtures [110].

Dielectrophoresis Summary. Dielectrophoresis makes use of the effects of small amplitude electromagnetic forces interacting with very small particles. At the scale of microsized particles and cells, the nanoscale geometric landscapes produce effects that are not intuitive or familiar to us from experience with macroscale objects and forces. Understanding and using these nanoscale phenomena to design new types of dielectrophoresis devices requires mastery of the descriptive equations and modeling techniques for predicting the behavior of alternating electromagnetic fields, which is beyond the scope of this introductory chapter. Further aspects of electromagnetics, physical chemistry, biochemistry, molecular and cell biology, and other disciplines are found in other chapters of this book and in the references.

25.6.4 Micro- and Nanofluidics

For large bodies of liquid, shape and flow characteristics can be understood with negligible consideration for electric charge and surface tension effects. As the volume of a stream or pool of liquid decreases, surface tension and evaporation become increasingly important factors in the behavior of the liquid, leading to the formation of droplets. At the microscale and below, electric charge becomes a very important force in the behavior of the droplets. These phenomena lead to the behaviors of aerosols and clouds, and have been exploited in ink-jet printers and similar technologies. They have also been exploited in biological and chemical applications, a few of which we will explore here.

Ink-Jet Printing. Ink-jet printing ejects ink under pressure through very small capillary nozzles in a series of droplets. In some versions of the technology the droplets are electrically charged, and can be accelerated and focused by electric fields. A similar technique is used in biology to inject minute amounts of material into cells. This procedure is called ballistic injection, or variously biolistic, particle bombardment, or gene-gun injection. It is used to transfer genetic DNA material into plants and animal cells.

Ballistic Injection. To transfer genes into cells, plasmid DNA encoding the gene of interest is coated onto 1- to 3-μm-sized gold or tungsten microbeads, and these particles are then accelerated by one of several motive force techniques to penetrate cell membranes. Ballistic DNA injection has been successfully used to transfer genes to a wide variety of mammalian cell lines in culture, and in vivo to the epidermis, muscle, liver, or other organs which can be exposed surgically. Ballistic DNA injection can deliver precise DNA dosages, but genes delivered by this method tend to be expressed only transiently and considerable cell damage may occur at the discharge site [111].

Ink-Jet Coating to Functionalize Cantilever Sensors. In an earlier section we saw how a nanomechanical cantilever could be functionalized to serve as a chemical or biochemical sensor. Ink-jet printing can be used to rapidly and accurately direct controlled deposition of functional layers onto cantilevers for sensor fabrication. Alkanethiols can be deposited onto gold-coated cantilever surfaces to make sensors for pH or ion concentrations in liquids. Single-stranded DNA oligomers can be bonded onto gold surfaces by thiol linkages to make detectors for gene fragments. Chemical gas detectors can be made by printing thin layers of selected polymers from dilute solutions onto cantilevers. The ink-jet method is noncontact, and is simpler, faster, and more accurate than applying coatings with microcapillaries or pipettes; it is more easily scalable to large arrays and can coat arbitrary patterns [112].

Electrospray Ionization for Mass Spectrometry Ion Sources. An ink-jet type system is used in mass spectrometry as a way to produce gas phase ions of large molecules such as proteins and DNA fragments, which otherwise tend to break up if volatized and ionized by electron bombardment or photolysis. This technique, along with matrix-assisted laser desorption from surfaces, makes mass spectrometry a useful technique for the analysis of large biomolecules. In electrospray ionization, a liquid is forced from a very small charged metal capillary opening. The liquid contains a dilute solution of the substance to be analyzed in the mass spectrometer, in a volatile solvent mixture. The analyte molecules are in solution in the ionic state, either as anions or cations. The liquid exits the capillary and forms an aerosol mist of droplets about 10 μm across. As the solvent evaporates, the analyte molecules are forced closer together, repel each other, and break up the droplets in a process called coulombic fission. Driven by repulsive coulombic forces between the charged molecules, the process repeats itself until the analyte is free of solvent, leaving isolated ions. The ions are separated in the mass analyzer stage of the spectrometer [113,114]. This is a very important application of nanoscale science which was recognized with a Nobel Prize in Chemistry to John Bennett Fenn in 2002.

Other Applications for Capillary Jet Injection. Ink-jet-like spray techniques are used to create a variety of unique particles and structures, taking advantage of intermolecular forces that become significant in microdroplets of mixed molecules as solvents evaporate and charges interact [115]. Ink-jet techniques are used to produce microencapsulation and nanoencapsulation of particles, as micelles and similar structures are formed in droplets composed of mixtures of a drug and an encapsulating polymer or other coating [116]. Ink-jet and electrospray techniques can be used to spray coatings with controlled nanosurface structures for biocompatible microdevices and materials [117]. Microfluidic spray droplets are used in surgery and dentistry, to coat patterns of functionalization in all kinds of biochips, and many other biotechnology applications [118].

25.6.5 Biochips, Labs on Chips, and Integrated Systems

Any of the sensor modalities surveyed in this section can be used to create arrays with multiple sensors, thus yielding very powerful screening and analysis tools. In diagnostic and chemical analysis applications in the clinic or laboratory, an array with a large number of sensors will be preferred, each primed to detect a specific analyte of interest. The various types of "biochips" that have been developed are based on printing a pattern of sensors in an array on a chip which can be interrogated for the presence or absence of the analyte to be detected [119].

Biochips. The engineering of biochips represents an integration of molecular biology and biochemistry with microelectronics and digital control. Demanding engineering challenges continue to exist in making such arrays smaller, more efficient, and more reliable, and exploration of the disciplines and technologies that go into the making of biochips is beyond the scope of this chapter. The future application of this technology will be involved with higher degrees of integration with electronics and digital wireless communications. We will describe a few examples by way of introduction.

Labs on Chips. Microfluidics can play a role in conjunction with nanosensors to create a "lab on a chip" in which extremely small samples are delivered to the sensing area for analysis by a nanosensor, an array of sensors, or a series of sensors. In some cases micro- and nanodot technologies are used to deliver samples, using injection techniques similar to those used for ink-jet printers, precision microelectronics soldering systems, and nanospray ion sources used in mass spectrometry. Many biochips rely on wired connections to the sensors to monitor the result of interaction with the analyte, but optical or wireless methods will be preferred where possible [120].

Integration of Cells with Biochips. One example of how microfluidics supports the use of nanosensors in cell research is seen in the use of a microfluidic device that allows neurons to be grown at low density. Brain researchers seek to analyze the signals between neurons with the goal of finding new treatments for brain damage and disease. Elucidation of neuron signaling is made difficult because neurons die without communication with networks of other neurons. Thus they can only be kept alive in vitro in dense cultures of many neurons, making it difficult to distinguish the signaling patterns.

A research group at the University of Illinois at Urbana-Champaign has developed a method of culturing neurons at low density, using a microfluidic chamber. The group has optimized conditions in microfluidic devices so that neurons can survive for around 11 days as opposed to 3 or 4. They used a material commonly employed in microfluidics but by heating the chip and washing it with buffer have made the nanoliter channels a suitable environment to grow neurons. This work was reported in a journal titled *Lab on a Chip*, which as the name implies is a good source for further information on this subject. [121]

Carbon Nanofibers for Inserting DNA into Cells. Carbon nanofibers have been coated with plasmid DNA and inserted into cells, where the DNA was viable in expressing RNA and proteins. This represents a radically new way of modifying the genetic behavior of cells in culture. If it were able to be applied in the human body, it could lead to "gene therapy on a stick," with DNA plasmids containing genes for expressing insulin, growth hormone, or neurotransmitters, for example, could be inserted into cells to repair deficiencies or alter organ and brain functions [122].

In experiments cells have survived for more than three weeks with the plasmid coated carbon nanotubes continuing to express their DNA. If the DNA coating is only physically absorbed onto the surface of the carbon nanofibers, it is transferred into the cell and is passed on to daughter cells when the cells divide. But if the DNA is chemically bonded to the nanofiber, it is not transferred into the cell's endoplasm domain and thus is not inherited by progeny cells upon cell division. This could provide a way of controlling genes that turn on cell growth, in order to repair bone or nerves or grow new skin or tendons, but without unlimited growth that would lead to tumors or cancer.

This development is just one example of cellular engineering on the molecular level using probes that control biomolecular reactions with far more precision than could be possible without nanoscale probes. The possibilities of these technologies have only just begun to be realized.

25.7 Summary

Biomolecular nanotechnology is a rapidly emerging field, with many more aspects than can be covered in a short introductory study. The fields of biomolecular motors, DNA computation engines, artificial muscles and organs, implantable medical devices, and many others have only been mentioned or not covered at all, and will be left to future chapters and sections. Many aspects of what we have called biomolecular nanotechnology overlap strongly with medical nanotechnology, which is the subject of the next chapter.

In selecting topics we have tried to provide continuity with the material covered in Part I, by illustrating how natural nanomaterials such as the *Morpho* butterfly and nanoscale biomolecular systems such as the dielectric potential of cellular membranes can be harnessed in technical applications. The examples selected are intended to give a broad overview of the field, and the references provided will point the reader to many additional applications and techniques.

Acknowledgments

We wish to acknowledge the significant contributions from the authors who graciously gave permission for use of their work and illustrations, citation in the references, especially Resonant Sensors Incorporated, for giving Robert Magnusson's and Debra Wawro's permissions for contributions and use of figures and data.

References

1. D. Springham, G. Springham, V. Moses, and R. E. Cape, *Biotechnology: The science and the business*, Taylor & Francis, Boca Raton, FL (1999).
2. P. Rabinow, *Making PCR: A story of biotechnology*, University of Chicago Press, Chicago, IL (1996).
3. C. Ratledge and B. Kristiansen, *Basic biotechnology*, Cambridge University Press, Cambridge, UK (2006).

4. D. S. Goodsell, *Bionanotechnology: Lessons from nature*, Wiley-Liss, Inc., Hoboken, NJ (2004).
5. M. Koehler and S. Diekmann, Biomolecular nanotechnology, *Reviews in Molecular Biotechnology*, 82, 1–2 (2001).
6. N. H. Malsch ed., *Biomedical nanotechnology*, CRC Press, Boca Raton, FL (2005).
7. R. R. H. Coombs and D. W. Robinson, *Nanotechnology in medicine and the biosciences*, CRC Press, Boca Raton, FL (1996).
8. T. Vo-Dinh, ed., *Nanotechnology in biology and medicine: Methods, devices, and applications*, CRC Press, Boca Raton, FL (2007).
9. J. D. Bronzino, ed., *Biomedical engineering handbook*, IEEE Press/CRC Press, Boca Raton, FL (2005).
10. J. Moore and G. Zouridakis, *Biomedical technology and devices handbook*, CRC Press, Boca Raton, FL (2004).
11. A. Guiseppi-Elie and T. Vo-Dinh eds., *Biochips handbook*, CRC Press, Boca Raton, FL (2007).
12. M. J. Heller and A. Guttman, *Integrated microfabricated biodevices*, Marcel Dekker, New York (2002).
13. J. S. Albala and I. Humphery-Smith, *Protein arrays, biochips and proteomics*, Marcel Dekker, New York (2003).
14. M. Schena, *Microarray analysis*, John Wiley & Sons, Inc., New York (2002).
15. K. Mitchelson, ed., *New high throughput technologies for DNA sequencing and genomics 2*, Elsevier, Amsterdam, the Netherlands (2007).
16. C. M. Niemeyer and C. A. Mirkin, *Nanobiotechnology*, Wiley-VCH, Weinheim, Germany (2004).
17. J. Benyus, *Biomimicry: Innovation inspired by nature*, William Morrow and Co., New York (1997).
18. G. Virella, *Medical immunology*, 7th ed., Informa Healthcare, New York (2007).
19. C. A. Janeway Jr., P. Travers, M. Walport, and M. J. Shlomchik, *Immunobiology*, 5th ed., Garland Publishing, Inc., New York (2001).
20. L. Du Pasquier and G. W. Litman, *Origin and evolution of the vertebrate immune system*, Springer Verlag, Berlin, Germany (2000).
21. R. Latorre and J. C. Sáez, eds., *From ion channels to cell-to-cell conversations*, Springer Verlag, Berlin, Germany (1997).
22. J. Haurum and S. Bregenholt, Recombinant polyclonal antibodies: Therapeutic antibody technologies come full circle, *IDrugs*, 8, 404–409 (2005).
23. H. G. Craighead, Nanoelectromechanical systems, *Science*, 290, 1532–1535 (2000).
24. C. A. Savran et al., Micromechanical detection of proteins using aptamer-based receptor molecules, *Analytical Chemistry*, 76, 3194–3198 (2004).
25. R. Berger et al., Micromechanics: A toolbox for femtoscale science, *Microelectronic Engineering*, 35, 373–379 (1997).
26. P. Vettiger, J. Brugger, M. Despont, U. Drechsler, U. Dürig, W. Häberle, M. Lutwyche, H. Rothuizen, R. Stutz, R. Widmer, and G. Binnig, Ultrahigh density, high-data-rate NEMS-based AFM data storage system, *Microelectronic Engineering*, 46, 11–17 (1999).
27. S. Singh-Gasson, R. D. Green, Y. Yue, C. Nelson, F. Blattner, M. R. Sussman, and F. Cerrina, Maskless fabrication of light-directed oligonucleotide microarrays using a digital micromirror array, *Nature Biotechnology*, 17, 974–978 (1999).
28. R. McKendry, J. Zhang, Y. Arntz, T. Strunz, M. Hegner, H. P. Lang, M. K. Baller, U. Certa, E. Meyer, H.-J. Guntherodt, and C. Gerber, Multiple label-free biodetection and quantitative DNA-binding assays on a nanomechanical cantilever array, *The Proceedings of the National Academy of Science USA*, 99, 9783–9788 (2002).
29. Y. Arntz, J. D. Seelig, H. P. Lang, J. Zhang, P. Hunziker, J. P. Ramseyer, E. Meyer, M. Hegner, and C. Gerber, Label-free protein assay based on a nanomechanical cantilever array, *Nanotechnology*, 14, 86–90 (2003).

30. I. G. Main, *Vibrations and waves in physics*, Cambridge University Press, Cambridge, UK (1993).
31. A. P. French, *Vibrations and waves*, W. W. Norton, New York (1971).
32. S. I. Lee, S. W. Howell, A. Raman, R. Reifenberger, C. V. Nguyen, and M. Meyyappan, Nonlinear tapping dynamics of multi-walled carbon nanotube tipped atomic force microcantilevers, *Nanotechnology*, 15, 416–421 (2004).
33. H. P. Lang, M. Hegner, and C. Gerber, Cantilever array sensors, *Materials Today*, 8, 30–36 (2005).
34. J. Gardner and P. N. Bartlett, eds., *Sensors and sensory systems for an electronic nose*, Kluwer Academic Publishers, Dordrecht, the Netherlands (1992).
35. J. Fritz, M. K. Baller, H. P. Lang, H. Rothuizen, P. Vettiger, E. Meyer, H. -J. Güntherodt, Ch. Gerber, and J. K. Gimzewski, Translating biomolecular recognition into nanomechanics, *Science*, 288, 316–318 (2000).
36. A. K. Gupta, P. R. Nair, D. Akin, M. R. Ladisch, S. Broyles, M. A. Alam, and R. Bashir, Anomalous resonance in a nanomechanical biosensor, *The Proceedings of the National Academy of Science USA*, 103, 13362–13367 (2006).
37. R. Kamalian, Y. Zhang, and A. M. Agogino, *Microfabrication and characterization of evolutionary MEMS resonators*, Proceedings of the 7th Symposium on Micro- and Nano-Mechatronics for Information-based Society, IEEE Robotics and Automation Society, pp. 109–114 (2005).
38. G. Wu, R. H. Datar, K. M. Hansen, T. Thundat, R. J. Cote, and A. Majumdar, Bioassay of prostate-specific antigen (PSA) using microcantilevers, *Nature Biotechnology*, 19, 856–860 (2001).
39. G. Wu, H. Ji, K. Hansen, T. Thundat, R. Datar, R. Cote, M. F. Hagan, A. K. Chakraborty, and A. Majumdar, Origin of nanomechanical cantilever motion generated from biomolecular interactions, *The Proceedings of the National Academy of Science USA*, 98, 1560–1564 (2001).
40. P. E. Sheehan and L. J. Whitman, Detection limits for nanoscale biosensors, *Nano Letters*, 5, 803–807 (2005).
41. G. Sauerbrey, Verwendung von Schwingquarzen zur Wägung dünner Schichten und zur Mikrowägung, *Zeit Physics*, 155, 206–222 (1959).
42. K. K. Kanazawa and J. G. Gordon, Frequency of a quartz microbalance in contact with liquid, *Analytical Chemistry*, 57, 1770–1771 (1985).
43. B. D. Vogt, E. K. Lin, W. Wu, and C. C. White, Effect of film thickness on the validity of the Sauerbrey equation for hydrated polyelectrolyte films, *Journal of Physical Chemistry B*, 108, 12685–12690 (2004).
44. D. R. Collins, J. B. Sampsell, L. J. Hornbeck, J. M. Florence, P. A. Penz, and M. T. Gately, Deformable mirror device spatial light modulators and their applicability to optical neural networks, *Applied Optics*, 28, 4900–4904 (1989).
45. L. J. Hornbeck, Digital light processing and MEMS: An overview, Advanced Applications of Lasers in Materials Processing/Broadband Optical Networks/Smart Pixels/Optical MEMs and Their Applications: IEEE/LEOS 1996, 7–8 (1996).
46. G. H. McGall, A. D. Barone, M. Diggelmann, S. P. A. Fodor, E. Gentalen, and N. Ngo, The efficiency of light-directed synthesis of DNA arrays on glass substrates, *Journal of American Chemical Society*, 119, 5081–5090 (1997)
47. A. C. Pease, D. Solas, E. J. Sullivan, M. T. Cronin, C. P. Holmes, and S. P. A. Fodor, Light-generated oligonucleotide arrays for rapid DNA sequence analysis, *The Proceedings of the National Academy of Science USA*, 91, 5022–5026 (1994).
48. G. McGall, J. Labadie, P. Brock, G. Wallraff, T. Nguyen, and W. Hinsberg, Light-directed synthesis of high-density oligonucleotide arrays using semiconductor photoresists, *The Proceedings of the National Academy Science USA*, 93, 13555–13560 (1996).
49. E. Maier, S. Meier-Ewert, D. Bancroft, and H. Lehrach, Automating array technologies for gene expression profiling, *Drug Discovery Today*, 2, 315–324 (1997).

50. K. M. O'Brien, J. Wren, V. K. Dave, D. Bai, R. D. Anderson, S. Rayner, G. A. Evans, A. E. Dabiri, and H. R. Garner, ASTRAL, A hyperspectral imaging DNA sequencer, *Review of Scientific Instruments*, 69, 2141–2146 (1998).
51. H. R. Garner, R. P. Balog, and K. L. Luebke, The evolution of custom microarray manufacture, *IEEE Engineering in Medicine and Biology Magazine*, 21, 123–125 (2002).
52. S. E. Lyshevski, *Nano and molecular electronics handbook, series: Nano- and microscience, engineering, technology and medicine volume: 9*, CRC Press, Boca Raton, FL (2007).
53. D. Branch and S. M. Brozik, Low-level detection of a *Bacillus anthracis* simulant using love-wave biosensors on 36° YX $LiTaO_3$, *Biosensors and Bioelectronics*, 19, 849–859 (2004).
54. M. K. Jain, Q. Cai, and C. A. Grimes, A wireless micro-sensor for simultaneous measurement of pH, temperature, and pressure, *Smart Materials and Structures*, 10, 347–353 (2001).
55. V. S. Bagotsky, *Fundamentals of electrochemistry*, 2nd ed., Wiley, New York (2006).
56. J. Wang, *Analytical electrochemistry*, 3rd ed., Wiley-VCH, Hoboken, NJ (2006).
57. H. B. Girault, *Analytical and physical electrochemistry*, Dekker, New York (2004).
58. X. Zhang, L. Cardosa, M. Broderick, H. Fein, and J. Lin, An integrated nitric oxide sensor based on carbon fiber coated with selective membranes, *Electroanalysis*, 12, 1113–1117 (2000).
59. X. Zhang, Y. Kislyak, J. Lin, A. Dickson, L. Cardosa, M. Broderick, and H. Fein, Nanometer size electrode for nitric oxide and S-nitrosothiols measurement, *Electrochemistry Communications*, 4, 11–16 (2002).
60. S. B. Hocevar, B. Ogorevc, K. Schachl, and K. Kalcher, Glucose microbiosensor based on MnO_2 and glucose oxidase modified carbon fiber microelectrode, *Electroanalysis*, 16, 1711–1716 (2004).
61. W. Lorenz and W. Plieth, *Electrochemical nanotechnology: In situ local probe techniques at electrochemical interfaces*, Wiley-VCH, Weinheim, Germany (1998).
62. R.-I. Stefan, J. F. van Staden, and H. Y. Aboul-Enein, *Electrochemical sensors in bioanalysis*, Dekker, New York (2001).
63. A. Brajter-Toth and J. Q. Chambers, *Electroanalytical methods for biological materials*, Dekker, New York (2002).
64. J. C. Harper, R. Polsky, D. R. Wheeler, S. M. Dirk, S. M. Brozik, Selective immobilization of DNA and antibody probes on electrode arrays: Simultaneous electrochemical detection of DNA and protein on a single platform, *Langmuir*, 23, 8285–8287 (2007).
65. J. W. Schultze, T. Osaka, and M. Datta, *Electrochemical microsystem technologies (New trends in electrochemical technology, Vol. 2)*, Taylor and Francis, New York (2002).
66. E. Palecek, F. Scheller, and J. Wang, *Electrochemistry of nucleic acids and proteins: Towards electrochemical sensors for genomics and proteomics*, Elsevier, Boston, MA (2005).
67. O. S. Wolfbeis, *Fiber optic chemical sensors and biosensors, Vol 1*, CRC Press, Boca Raton, FL (1991).
68. G. Boisde and A. Harmer, *Chemical and biochemical sensing with optical fibers and waveguides*, Artech House, Boston, MA (1996).
69. S. Donati, *Electro-optical instrumentation: Sensing and measuring with lasers*, Prentice Hall, Upper Saddle River, NJ (2004).
70. G. Orellana, *Frontiers in chemical sensors: Novel principles & techniques, series on chemical sensors & biosensors, Vol. 3*, Springer Verlag, Berlin, Germany (2005).
71. R. M. Nakamura, Y. Kasahara, and G. A. Rechnitz, eds., *Immunochemical assays and biosensor technology*, American Society for Microbiology, Washington, DC (1992).
72. T. Vo-Dinh, M. J. Sepaniak, G. D. Griffin, J. P. Alarie, Immunosensors: Principles and applications, *Immunomethods*, 3, 85–92 (1993).

73. T. Vo-Dinh, J.-P. Alarie, B. M. Cullum, and G. D. Griffin, Antibody-based nanoprobe for measurement of a fluorescent analyte in a single cell, *Nature Biotechnology*, 18, 764–767 (2000).
74. T. Vo-Dinh, ed., *Chemical analysis of polycyclic aromatic compounds*, Wiley, New York (1989).
75. B. Hille, *Ion channels of excitable membranes*, 3rd ed., Sinauer, Sunderland, MA (2001).
76. W. Tan, Z.-Y. Shi, and R. Kopelman, Development of submicron chemical fiber optic sensors, *Analytical Chemistry*, 64, 2985–2990 (1992).
77. T. Tan, Z.-Y. Shi, S. Smith, D. Birnbaum, and R. Kopelman, Submicrometer intracellular chemical optical fiber sensors, *Science*, 258, 778–781 (1992).
78. S. L. R. Barker, Y. D. Zhao, M. A. Marletta, and R. Kopelman, Cellular applications of a sensitive and selective fiber optic nitric oxide biosensor based on a dye-labeled heme domain of soluble guanylate cyclase, *Analytical Chemistry*, 71, 2071–2075 (1999).
79. E. Betzig, J. K. Trautman, T. D. Harris, J. S. Weiner, and R. L. Kostelak, Breaking the diffraction barrier: Optical microscopy on a nanometric scale, *Science*, 251, 1468–1470 (1991).
80. V. Deckert, D. Zeisel, R. Zenobi, and T. Vo-Dinh, Near-field surface-enhanced Raman imaging of dye-labeled DNA with 100-nm resolution, *Analytical Chemistry*, 70, 2646–2650 (1998).
81. L. L. Norman and A. Badia, Electrochemical surface plasmon resonance investigation of dodecyl sulfate adsorption to electroactive self-assembled monolayers via ion-pairing interactions, *Langmuir ASAP*, 10, 1021 (2007).
82. G. H Cross, A. Reeves, S. Brand, M. J. Swann, L. L Peel, N. J. Freeman, and J. R Lu, The metrics of surface adsorbed small molecules on the Young's fringe dual-slab waveguide interferometer, *Journal of Physics D: Applied Physics*, 37, 74–80 (2004).
83. N. J. Feeman, L. L. Peel, M. J. Swann, G. H. Cross, A. Reeves, S. Reeves, and J. R. Lu, Real time, high resolution studies of protein adsorption and structure at the solid–liquid interface using dual polarization interferometry, *Journal of Physics: Condensed Matter*, 16, S2493–S2496 (2004).
84. G. Tollin, Z. Salamon, S. Cowell, and V. J. Hruby, Plasmon-waveguide resonance spectroscopy: A new tool for investigating signal transduction by G-protein coupled receptors, *Life Sciences*, 73, 3307–3311 (2003).
85. R. Magnusson and S. S. Wang, New principle for optical filters, *Applied Physics Letters*, 61, 1022–1024 (1992).
86. S. S. Wang and R. Magnusson, Theory and applications of guided-mode resonance filters, *Applied Optics*, 32, 2606–2613 (1993).
87. D. Wawro, S. Tibuleac, R. Magnusson, and H. Liu, Optical fiber endface biosensor based on resonances in dielectric waveguide gratings, *Biomedical, Diagnostic, Guidance, and Surgical-Assist Systems II, Proceedings of SPIE*, 3911, 86–94 (2000).
88. R. Magnusson, Y. Ding, K. J. Lee, P. S. Priambodo, and D. Wawro, Characteristics of resonant leaky mode biosensors, *Nanosensing: Materials and Devices II, Proceedings of SPIE*, 6008, 60080U pp. 1–10 (2005).
89. P. Alivisatos, The use of nanocrystals in biological detection, *Nature Biotechnology*, 22, 47–52 (2004).
90. A. R. Parker and H. E. Townley, Biomimetics of photonic nanostructures, *Nature Nanotechnology*, 2, 347–353 (2007).
91. R. A. Potyrailo, H. Ghiradella, A. Vertiatchikh, K. Dovidenko, J. R. Cournoyer, and E. Olson, Morpho butterfly wing scales demonstrate highly selective vapour response, *Nature Photonics*, 1, 123–128 (2007).
92. J. Huang, X. Wang, and Z. L. Wang, Controlled replication of butterfly wings for achieving tunable photonic properties, *Nano Letters*, 6, 2325–2331 (2006).

93. W. Zhang, D. Zhang, T. Fan, J. Ding, Q. Guo, and H. Ogawa, Fabrication of ZnO microtubes with adjustable nanopores on the walls by the templating of butterfly wing scales, *Nanotechnology*, 17, 840–844 (2006).
94. A. Ashkin, J.M. Dziedzic, J. E. Bjorkholm, and S. Chu, Observation of a single-beam gradient force optical trap for dielectric particles, *Optics Letters*, 11, 288–290 (1986).
95. A. Ashkin, History of optical trapping and manipulation of small-neutral particle, atoms, and molecules, *IEEE Journal of Selected Topics in Quantum Electronics*, 6, 841–856 (2000).
96. A. Pralle, M. Prummer, E. L. Florin, E. H. Stelzer, and J. K. Horber, Three-dimensional high-resolution particle tracking for optical tweezers by forward scattered light, *Microscopy Research and Technique*, 44, 378–386 (1999).
97. Y. Ishii, A. Ishijima, and T. Yanagida, Single molecule nanomanipulation of biomolecules, *Trends in Biotechnology*, 19, 211–216 (2001).
98. S. C. Kuo, Using optics to measure biological forces and mechanics, *Traffic*, 2, 757–763 (2001).
99. M. J. Lang, C. L. Asbury, J. W. Shaevitz, and S. M. Block, An automated two-dimensional optical force clamp for single molecule studies, *Biophysical Journal*, 83, 491–501 (2002).
100. R. Westermeier, *Electrophoresis in practice: A guide to methods and applications of DNA and protein separations*, Wiley VCH, Weinheim, Germany (2005).
101. M. G. Khaledi, ed., *High-performance capillary electrophoresis: Theory, techniques, and applications*, Wiley, New York (1998).
102. K. D. Altria, *Capillary electrophoresis guidebook: Principles, operation, and applications*, Humana Press, Totowa, NJ (1996).
103. J. P. Landers, ed., *Handbook of capillary electrophoresis*. 2nd ed., CRC Press, Boca Raton, FL (1996).
104. W. A. Goddard, *Handbook of nanoscience, engineering, and technology*, CRC Press, Boca Raton, FL (2003).
105. T. B. Jones, *Electromechanics of particles*, Cambridge University Press, New York (2005).
106. J. Vykoukal and P. R. C. Gascoyne, Particle separation by dielectrophoresis, *Electrophoresis*, 23, 1973–1983 (2002).
107. P. R. C. Gascoyne, X.-B. Wang, Y. Huang, and F. F. Becker, Dielectrophoretic separation of cancer cells from blood, *IEEE Transactions on Industry Applications*, 33, 670–678 (1997).
108. J. M. S. Bartlett and D. Stirling, *A short history of the polymerase chain reaction, in PCR protocols, 2nd ed., Methods in molecular biology, Vol. 226*, Humana Press, Totowa, NJ (2003).
109. B. H. Lapizco-Encinas, B. A. Simmons, E. B. Cummings, and Y. Fintschenko, Dielectrophoretic concentration and separation of live and dead bacteria in an array of insulators, *Analytical Chemistry*, 76, 1571–1579 (2004).
110. E. Cummings, Streaming dielectrophoresis for continuous-flow microfluidic devices, *IEEE Engineering in Medicine and Biology Magazine*, 22, 75–84 (2003).
111. T. M. Klein, R. Arentzen, P. A. Lewis, and S. Fitzpatrick-McElligott, Transformation of microbes, plants and animals by particle bombardment, *Nature Biotechnology*, 10, 286–291 (1992).
112. A. Bietsch, J. Zhang, M. Hegner, H. P. Lang, and C. Gerber, Rapid functionalization of cantilever array sensors by inkjet printing, *Nanotechnology*, 15, 873–880 (2004).
113. J. B. Fenn, M. Mann, C. K. Meng, S. F. Wong, and C. M. Whitehouse, Electrospray ionization for mass spectrometry of large biomolecules, *Science*, 246, 64–71 (1989).
114. P. Kebarle, A brief overview of the present status of the mechanisms involved in electrospray mass spectrometry, *Journal of Mass Spectrometry*, 35, 804–817 (2000).

115. K.-H. Roh, D. C. Martin, and J. Lahann, Biphasic Janus particles with nanoscale Anisotropy, *Nature Materials*, 4, 750–763 (2005).
116. I. G. Loscertales et al., Micro/nano encapsulation via electrified coaxial liquid jets, *Science*, 295, 1695–1698 (2002).
117. C. Berkland, W. Daniel, D. W. Pack, and K. Kim, Controlling surface nano-structure using flow-limited field-injection electrostatic spraying (FFESS) of poly(D,L-lactide-co-glycolide), *Biomaterials*, 25, 5649–5658 (2004).
118. V. Farkas, L. Daniel, R. C. Leif, and D. V. Nicolau, Imaging, manipulation, and analysis of biomolecules, cells, and tissues, *Proceedings of SPIE*, 6441, 64410Z (2007).
119. P. C. H. Li, *Microfluidic lab-on-a-chip for chemical and biological analysis and discovery*, CRC Press, Boca Raton, FL (2005).
120. J. Melin and S. R. Quake, Microfluidic large-scale integration: The evolution of design rules for biological automation, *Annual Review of Biophysics and Biomolecular Structure*, 36, 213–31 (2007).
121. L. J. Millet, M. E. Stewart, J. V. Sweedler, R. G. Nuzzo, and M. U. Gillette, Microfluidic devices for culturing primary mammalian neurons at low densities, *Lab Chip*, 7, 987 (2007).
122. T. E. McKnight et al., Intracellular integration of synthetic nanostructures with viable cells for controlled biochemical manipulation, *Nanotechnology*, 14, 531–556 (2003).

Problems

25.1 Why is immunology important in nanotechnology? What are some of the ways that the immune system can be used to make sensors?

25.2 What other biological molecular systems can be used to make sensors? Which is the most versatile?

25.3 How would you expect the vibrational frequencies of nanocantilevers to change with increasing (1) mass; (2) dimensions: a. thickness, b. length, c. width; (3) density; (4) modulus of elasticity; (5) bond strength of chemical constituents; (6) amount of material absorbed on the surface? Justify your explanation in terms of physical laws.

25.4 How would you rank the expected vibrational frequencies of nanocantilevers with the same physical dimensions to compare if made from the following substances: (1) silicon, (2) steel, (3) silica glass, (4) quartz, (5) carbon nanotube, (6) copper? Justify your ranking (Hint: look up the physical properties.)

25.5 On the same basis as the preceding examples, calculate the changed vibrational frequency if a single molecule of a protein with a mass of (1) 700 Da, (2) 1200 Da, or (3) 3000 Da is attached to the tip of the cantilever.

25.6 Design a simple circuit to stimulate and monitor resonant vibration of the preceding cantilevers.

25.7 For a cantilever of dimensions $500 \times 100 \times 0.5\,\mu m$ made of single-crystal silicon with an antibody of 14,000 Da attached to its tip, which binds to an antigen of 43,000 Da, calculate the rate of vibration with and without the attached antigen.

25.8 Use Stoney's equation to determine the relative deflections produced for a silicon cantilever of length 500 μm and thickness 0.5 μm versus a silicon nitride cantilever of length 600 μm and thickness 0.65 μm.

25.9 Perhaps the pinnacle of molecular recognition is illustrated by the immune globulins—for example, the antibodies. These classes of proteins are not only able to recognize and neutralize antigenic materials that have invaded the body, but are also able to do so in a dynamic, versatile way. Antigens are considered to be divalent; antibodies are considered to be polyvalent. Explain what this means and relate your understanding to the overall solubility of antibody–antigen complexes?

25.10 What is the piezoelectric effect? Why does a layer of absorbed molecules change the surface acoustic resonance of an electrically excited vibrating circuit element?

25.11 What are the relative merits of optical and electrochemical (potentiometric) nanosensors?

25.12 How do optical tweezers work? How does dielectrophoresis differ from the optical tweezers effect, and how it is similar?

BIOMIMETICS

Nature works for maximum achievement at minimum effort. We have much to learn.

GEORGE JERONIMIDIS
Director, Centre for Biomimetics at the University of Reading, United Kingdom

Chapter 26

Fig. 26.0 *Photosynthesis is a splendid example of natural molecular nanomachinery from which valuable lessons for technology can be drawn.*

Threads

We introduced nanobiotechnology in *chapter 25*. Nanobiotechnology is the application of nanoscale biological materials, structures, and processes to synthetic goals. Biomimetics is a related topic that deserves special attention—biomimetics is the process of copying nature into the design of artificial devices.

Four chapters (Nanobiotechnology, Biomimetics, Medical Nanotechnology, and Environmental Nanotechnology) are rooted strongly in biology. Nature is the supreme wizard of nanotechnology—so why not copy its materials, structures, and processes? If "imitation is the sincerest form of flattery" then let's flatter nature to the best of our abilities. Following this chapter and the one on bionanomedicine, we wrap up the book with environmental aspects of nanotechnology—an area with growing applications of nanoscience and nanotechnology.

26.0 The Bio Sciences and Technologies

In this chapter, we examine the important role that biological science plays in nanotechnology development. A list of the biologically inspired influences that have been transmuted into human technology would be very long [1–14]. We give examples, starting with simple nanoparticles and proceeding through surfaces, membranes, and more complex structures and nanomachinery. Before we explore biologically inspired nanotechnologies, we will give some definitions that clarify the background and aims of biomimetics in general.

26.0.1 *Biomimetics, Bioengineering, and Other Bioengineering Fields*

The use of biological models in other fields is described by many terms—biomimetics, bionics, biomechanics, biophysics, bioengineering, etc., terms that we will define to avoid confusion. Historically many aspects of bio-inspired engineering and technology originated separately, but their overlap and convergence is being accelerated by nanotechnology.

Biomimetics. *Biomimetics* (from the Greek *bios* meaning "life, course or way of living" + *mimetikos* "imitative") has roots in the beginnings of civilization. Biomimetics is science or technology that copies or imitates nature to find new methods and applications. *The Encyclopedia of Biomaterials and Biomedical Engineering* gives the following definition [15].

> Biomimetics is the study of how Nature, building atom by atom, through eons of time, developed manufacturing methods, materials, structures, and intelligence. These studies are inspiring engineering and design of manmade miniature objects.... It is in the nanoworld that Nature is way ahead of human engineering as it has learned to work with much smaller, more versatile building blocks and master the self-assembly of those building blocks.... We believe that, just as in the macroworld, blindly mimicking Nature will be unproductive, but studying Nature and taking hints as to how to build nanostructures from the bottom-up ... will be productive.

The term biomimetics was coined in the 1950s by the American engineer and biophysicist Otto Herbert Schmitt. His investigations into the nervous system of the squid led to the development of the "Schmitt trigger," a device used so often in modern electronic devices that its biomimetic origins are often forgotten [16]. Advances in understanding of how nature works on the nanoscale led to new opportunities for biomimetic nanotechnology. Many significant applications are emerging, with huge room for growth [17].

Bionics. Bionics (from the Greek *bios* meaning life + the suffix *-ic* "like") is the application of biology to engineering, especially when used to enhance human capabilities directly, as in spacesuits or prosthetics. Bionics usually refers to augmentation of human performance. Bionics is the *transfer of technology* between living organisms and synthetic devices created by humans. The iron lung (P. Drinker, L.A. Shaw in 1928), the artificial heart (R. Jarvik in 1982), prostheses, cochlear ear implants (G. von Békésy in 1973), electronically stimulated limbic systems (K.D. Wise, D.J. Edell), artificial eyes (R. Birge in the 1990s), and electronic retinas (C. Mead of Cal Tech) are just a few examples of bionic applications to the human condition. The term bionics was coined by J.E. Steele in 1958, a scientist at Wright-Patterson Air Force Base in Dayton, Ohio who worked in aerospace medical systems [11,18].

Biomechanics. *Biomechanics* is the study of the mechanics of living things [19–24]. Biomechanics historically deals with the macroscale, and it is still heavily associated with functional anatomy, physical therapy, and sports medicine. However, biomechanics involves materials and phenomena that begin at the molecular level and extend up the biological hierarchy. *Biodynamics* and kinematics are subsets of this discipline that relate to movement [7]. *Biomaterials* and nanomaterials, and their properties, are important factors in biomechanical research and analysis. Metabolic and energetic factors are also important to biomechanics. Biomechanics was the one of the earliest of the bio-disciplines. Books describing the motion of animals were written by Giovanni Alfonso Borelli of Italy during the Renaissance—*De Motu Animalium I* and *De Motu Animaliam II* [25]. Leonardo da Vinci further advanced biomechanics with studies of anatomy and the design of machines based on biological actions. In the nineteenth century, understanding of biodynamics was advanced by photographic studies of animal and human motion, pioneered by Musgrave and others.

Bioengineering. This field is the application of engineering principles and practices to medicine and biology—increasingly vested in nanotechnology. It overlaps biomedical engineering (engineering applied to medicine) and genetic engineering (bioengineering that uses biotechnology to manipulate and make use of genetic coding) [26–28]. The National Institutes of Health defines bioengineering as follows [29]:

> Bioengineering is rooted in physics, mathematics, chemistry, biology, and the life sciences. It is the application of a systematic, quantitative, and integrative way of thinking about and approaching the solutions of problems important to biology, medical research, clinical proactive, and population studies. The NIH Bioengineering Consortium agreed on the following definition for bioengineering research on biology, medicine, behavior, or health recognizing that no definition could completely eliminate overlap with other research disciplines or preclude variations in interpretation by different individuals and organizations.
>
> Bioengineering integrates physical, chemical, or mathematical sciences and engineering principles for the study of biology, medicine, behavior, or health. It advances fundamental concepts, creates knowledge for the molecular to the organ systems levels, and develops innovative biologics, materials, processes, implants, devices, and informatics approaches for the prevention, diagnosis, and treatment of disease, for patient rehabilitation, and for improving health.

The NIH is placing increasing emphasis on nanoscience and nanomaterials in its bioengineering initiatives with the emergence of *biomedical nanotechnology* and the application of nanotechnology to the design and development of biomedical devices, for analysis, diagnosis, and medical therapeutics [30,31].

Biophysics. Biophysics is an interdisciplinary field that applies the theories and methods of physics and related sciences to biology. Biophysics is interdisciplinary: it overlaps with biomechanics and bioengineering [32–34]. It tends to be oriented towards basic science, and is very involved with nanoscale phenomena in biology.

Biotechnology and Bio-Nanotechnology. The term biotechnology was coined to describe genetic and protein engineering technologies. These use molecular biology, cell biology, and biochemistry for practical applications such as genetic modification of organisms for the production of drugs and development of DNA assays. As nanotechnology has been increasingly applied, terms such as *bio-nanotechnology* and *nanobiotechnology* have emerged and perhaps converged [35–38]. Bio-nanotechnology refers to nanoscale phenomena such as protein and organelle structure and function, and supramolecular mechanisms in proteins and cellular systems [39]. *Nanobiotechnology* focuses on the application of nanotechnologies such as nano-cantilevers, quantum dots, and nanomechanical structures to applications such as biosensors and biochips for diagnostics and

DNA screening. Nanobiotechnology was discussed extensively in chapter 25, and additional examples are given in this chapter [40,41].

Biomimicry. The term *biomimicry* implies copying nature. Originally it was used to describe convergent adaptations or behaviors found in nature (one organism copying another). An example is the viceroy butterfly, with similar coloration to the monarch butterfly (whose taste is repellent to birds). In recent times biomimicry has been used to describe a general approach to engineering and economics with emphasis on sustainability and environmental compatibility [3]. According to the Biomimicry Institute [42]

> Biomimicry (from the Greek bios, "life" + mimesis, "imitate") is a new science that studies nature's best ideas and then imitates these designs and processes to solve human problems.

Biognosis is a related term, used to refer to systems of knowledge and practice based on nature and natural principles.

Biomimetic Nanotechnology: An Interdisciplinary Field. Biomimetics and nanotechnology are both interdisciplinary, new, and rapidly growing. For the purpose of this chapter, we define *biomimetic nanotechnology* as the design of materials and devices based on nanoscale biological structure and function. The purpose of biomimetics is to pattern useful materials and devices after biological blueprints. As stated by Claus Mattheck at the Forschungszentrum Karlsruhe in Germany [43]:

> Modern biomimetics is a systematic approach for researchers who know that new developments and insights can only be achieved in transdisciplinary collaborations. If this is true for examples of co-working between physicists or chemists and engineers, why not networking with biologists?

Table 26.1 summarizes the frequency of the bio-terms discussed above and their relation to nano.

TABLE 26.1 *Google Hits of Bio + Nano Overlap and Place*

	Nanotechnology (14,400,000)	Nanoscience (810,000)	Nanomaterials (1,350,000)	Cumul. score
Biotechnology	520,000	215,000	466,000	4
(49,300,000)	2	1	1	
Bioengineering	1,280,000	39,600	31,500	5
(3,330,000)	1	2	4	
Biomimetics	310,000	14,100	140,000	10
(1,130,000)	3	5	2	
Biophysics	203,000	38,500	139,000	11
(4,540,000)	4	3	3	
Biomechanics	72,000	16,200	17,000	14
(3,370,000)	5	4	5	
Bionics	21,000	2,400	3,300	18
(908,000)	6	6	6	

The clear winner of the nanotechnology sweepstakes is *biotechnology*—not surprisingly because biotechnology is clearly a technological discipline and a very broad one at that. Bioengineering is next with an average score of five—another fairly broad category. *Biomimetics* has more specificity and it is placed third overall with a high overlap with nanomaterials. Biophysics and biomechanics have been around for quite a while. *Bionics* rounds out the table. Bionics, based on the findings of this table, is not as nano-driven as the others. Interestingly biotechnology, biomimetics, and biophysics appear to have the strongest nanomaterial component. Clearly, the "field" of nanotechnology by itself is a powerful force in today's science and technology—at least when considering the number of Google hits.

26.0.2 Biomimetics as an Emerging Science and Engineering Discipline

Nature has many solutions to problems that intrigue scientists and engineers. How does bone adapt to growth, concentrating material resources where they are needed? How does the anisotropic structure of bird feathers contribute to its toughness? How can an apple hold its firmness and shape when it consists of 97% water?

The Biomimetic Dialogue. Biological and synthetic designs are dramatically different. Nature does not utilize steel, flat surfaces, or sharp corners—nature uses proteins, curves, and adaptable shapes. As our knowledge of biology at the cellular and subcellular level increases dramatically, biomimetics is becoming a major player in nanobiotechnology. Nanotechnology enables scientists to understand the mechanisms that allow basilisk lizards to walk on water, penguins and sharks to reduce drag during swimming, and insects to fly and hover. The nanostructure of the lotus plant leaf is the key to understanding how it stays clean in the muddy environment that it inhabits. Swarm intelligence and artificial neural networks are another outgrowth of biologically inspired thinking [44].

26.0.3 Biomimetic Systems

The study of life on the nanoscale fosters appreciation for the self-organizing and evolutionary properties of biosystems: how they process and store energy, materials, and information, with hierarchical organization from the macro to the nanoscale [45,46].

Macroscopic Biomimetics. Velcro is a familiar biomimetic product which originated with an observation from nature—the clinging mechanism of *cockleburs* to dog hair and to fabric. In 1948, a Swiss engineer named George de Mestral, inspired by the microscopic hook-like structures in cockleburs, designed the first Velcro fastener (from the French *velour* and *crochet*)—a simple two-component device with hooks on one side and soft loops on the other. The result—Velcro Industries N.V. became a multimillion dollar industry [8].

Another macroscopic example of biomimetics is architecture based on termite mounds—structures that are able to maintain constant temperature and humidity regardless of ambient conditions that vary from near 0°C to over 40°C. A high-rise office complex in Zimbabwe called the Eastgate Centre is modeled after the internal structure of the termite mound, with energy consumption 10% that of a conventional building [9,47].

Microscopic Biomimetics. Nature provides us with numerous models of hierarchically structured materials. Bones, teeth, shells, skeletal components, and other materials are biocomposites with micro- and nanostructure composed of lipids, proteins, and polysaccharides and inorganic mineral materials such as hydroxyapatite, calcium carbonate, and silica. In the microscopic structure of natural materials we see a hierarchy from the bottom up, starting with the basic building blocks of biology—proteins, lipids, carbohydrates, and the nucleic acids. Complex architectures are then assembled that contain different types of building blocks held together by combinations of different bonding forces [45].

Molecular Biomimetics. The design of molecules is the domain of synthetic chemistry rather than nanotechnology. In earlier chapters we saw examples of how small molecules regulate nanoscale machinery in the cell, as key players in neurotransmission, hormonal regulation, and immunity. Because small molecules are so important in the regulation of cellular nanomachinery, it is worth examining some instances of biomimetic design of molecules where the structure and activity of the small molecule produces profound supramolecular effects. This is the most common mode of action for molecules used as drugs and pesticides. In such cases, molecular biomimetics has implications for the nano- and macroscale, as we shall see.

Industrialization brought the production of tremendous quantities of synthetic detergents and pesticides, which had no natural breakdown pathways and thus begin to build up in people, animals, and the environment, along with their toxic derivatives. One of the goals of biomimetic chemistry, or molecular biomimetics, is to find and create molecules that are closer to natural products, so that their fate in the environment and food chain is to break down without harmful effects. Another goal is to find more effective and lower cost drugs by looking for elegant molecules in natural systems as models.

26.0.4 The Nano Perspective

Most biomimetics has occurred on the macroscale to microscale. Now, a confluence of biomimetics with nanotechnology is in full force—the mimicking of nature at the molecular level. Organisms have exploited nanotechnology for nearly 4 billion years—DNA, RNA, proteins, and inorganic nanomaterials all contribute to the structures in living cells that perform a variety of functions at the nanoscale. Nature's nanotechnology is optimized and efficient and has undergone, via evolutionary pressures, the most rigorous product development and testing in the most demanding laboratory over the longest periods of time. It is no wonder then that we scientists seek to duplicate nature's marvelous materials and devices in order to create our own nanomaterials and machines from the bottom up.

So, what clues does nature provide to help us with fabrication of our biomimetic devices? According to P. Ball from his book *Made to Measure* [10], biomimetic construction (nanochemistry) should follow these guidelines (from nature):

- Use of composite materials that are made of alternating layers of aragonite ($CaCO_3$) and biopolymer rather than monolithic materials. Examples include bone, nacre (e.g., abalone shell, mother of pearl)—in which the composite outperforms either component separately
- Liquid phase manufacturing
- Materials based on carbon (and not silicon or metals)
- Parallel processing for high throughput
- Hierarchical organization for high strength-to-density ratios and multifunctional performance
- Self-repair
- Soft, flexible materials (not hard stiff materials with flat surfaces, sharp corners)
- Self-assembly (molecular recognition) and self-replication (intelligence) via weak intermolecular forces
- Template synthesis (genetic replication)
- Compartmentalization: tissues → cells → organelles → nanostructured biological materials

Biomimetic systems arising from the collaboration of physicists, chemists, molecular and cellular biologists, and bioengineers are expected to have numerous applications in bioengineering, pharmacology, and medicine. We will start our exploration of biomimetics with a look at design of molecules, and then proceed to the nanoscale.

26.1 Biomimetic Design of Molecules

Biomimetic principles of design can be applied on any size scale from the macroscopic down to the molecular level. Molecular biomimetics is the design of novel molecules based on structures and functions of natural products from plants and animals. Molecular biomimetics may use the chemical structure of a natural chromophore to produce improved dyes and pigments, or use other properties, as in artificial sweeteners or synthetic flavor compounds. Even using formic acid, the active ingredient in ant venom, as a starting point for making Formica polymer could be considered a type of biomimicry. But deeper biomimetics involves understanding how simple molecules produce large biological effects, and designing synthetic routes to achieve similar effects for drugs.

On the chemical level, synthetic drugs and insecticides act by blocking or duplicating the stereochemistry and functional activity of natural hormones, antibodies, and other biochemical substances. Plants, bacteria, and marine creatures, with their interactions in biodiverse ecosystems, offer a natural laboratory for discovery of biological interactions. Bacteria and archaea, plants and sea creatures, and other primitive life forms have interacted and developed over billions of years of evolution, to produce an enormous library of biochemical and nanomachinery processes, which can be studied and harnessed for applications to produce subtle and advanced medicines, foods, materials, catalysts, pesticides, and energy and growth regulators.

Centers of biodiversity such as tropical rainforests and coral reefs represent especially rich storehouses of biochemical information and materials that are only beginning to be tapped, even as they are disappearing due to ignorance of their value. Sources of biodiversity are not limited to tropical rainforests and reefs: boreal and alpine environments, and semidesert regions that have remained undisturbed by glaciation for millions of years, such as southwest Australia, also hold unique highly developed communities of plants and animals. Temperate forests, plains, savannahs, and wetlands also possess abundant networks of species with their own finely developed diverse biochemistry.

Learning from these natural biochemical networks is a more efficient way of discovering biochemical nanomachinery pathways than attempting to deduce them from basic chemistry and molecular biology, simply because the number and complexity of all the biological possibilities is so vast. The study of natural compounds synthesized by plants and animals yields valuable insights into substances and mechanisms that act on cellular pathways. Once these effects are discovered in nature, biochemical and molecular biology can then elucidate their mechanisms, and can guide the design of drugs with similar effects.

Ethnobotany, the study of the lore for traditional uses of plants in ancient indigenous cultures, has been found to accelerate the discovery of new drugs from plants, animals, and marine life, because cultural traditions can represent the tested results of thousands of years of trial and error experience. Plants produce large numbers of allelopathic chemicals which act as attractants, repellents, growth inhibitors, and poisons on competitors, pests, and predators. Knowledge of these compounds and how they work can improve the odds in the search for new and more effective drugs and pesticides. Currently, it is estimated that fewer than 5 in every 10,000 compounds investigated in the effort to develop new pharmaceuticals results in an effective approved new medication.

26.1.1 Design and Discovery of Drugs

The drug discovery process can be top down or bottom up. The top-down approach observes an effect produced by a substance and then works to establish the mechanism and chemical basis of the action. This traditional approach relies on evaluating compounds from traditional remedies or from exhaustive screening of classes of chemicals.

The bottom-up approach is more targeted, based on knowledge of chemical signaling pathways obtained from biochemical and molecular biology research: drugs are designed and synthesized to precisely match receptors, to block or enhance metabolic and signaling paths that control functions in cells, to regulate the homeostasis of the body, or to enhance the performance of the body's natural immune response mechanisms. When design moves on to production, biotechnology is used to manufacture drugs in quantity by redirecting the natural DNA–RNA molecular synthesis process in yeasts, bacteria, or genetically modified plants and animals. Modern drug discovery employs both top-down and bottom-up approaches.

The same processes apply to the discovery and production of pesticides and plant growth regulators as to pharmaceuticals. The traditional screening of compounds for pesticide activity is now supplemented by targeted design of compounds aimed at blocking metabolic pathways. New knowledge of how receptors and cytoplasmic reticulum structure operate in cellular processes is leading to drugs that act on the macromolecular and nanostructural levels.

As we shall see in this section, the chemicals produced naturally by animals and plants to regulate their internal and external interactions can yield a wealth of information for the discovery of new drugs, pesticides, growth regulators, and other useful molecules. Learning the modes and mechanisms by which these molecules act can reveal surprising aspects of the internal nanomachinery of cells which we might otherwise have never suspected or discovered in millions of years—but then, that is how long it took plants and insects to develop them.

26.1.2 Targeting with Magic Bullets

The path to synthesis of new drugs historically developed from use of simple inorganic poisons to highly sophisticated targeting based on knowledge of the nanoscale functioning of cells. Early medicines relied heavily on toxic metals such as mercury and arsenic to kill pathogens in the body—the cure offered by a quacksalver (from quicksilver → mercury + salve → healing potion or lotion) could be worse than the disease.

In 1908, Paul Ehrlich, a brilliant bacteriologist who developed new staining methods for classifying bacteria, set up a systematic search for improved drug molecules. Stains had been used in microscopy for nearly 100 years, first simply to enhance microscopic images, and later to identify microorganisms. Certain dyes were highly specific in binding to different types of bacteria. Ehrlich theorized that by combining aniline dye stains with arsenic he could produce a "magic bullet" that would bind specifically to bacteria in preference to human cells, delivering the poison to the pathogen with minimal harm to the body.

Previously, in 1859, the French chemist Antoine Béchamp had reacted aniline and arsenic acid to produce an organic preparation of arsenic which was hoped to be less toxic than the inorganic acid. The new compound, named atoxyl, though still highly poisonous, was indeed less toxic than free arsenic alone in the treatment of skin conditions. In 1905, the British physicians H.W. Thomas and A. Breinl discovered that atoxyl was active against the spiral trypanosome microorganism which causes sleeping sickness. Ehrlich teamed with the organic chemist Alfred Bertheim who synthesized new derivatives of the atoxyl molecule that were then systematically tested in Ehrlich's laboratory by his colleague Sahachiro Hata (**Fig. 26.1**).

Hata went through hundreds of trials of different aniline arsenic derivatives in an attempt to find one that was more specifically toxic to disease causing organisms than to humans; eventually the 606th compound, azobenzene, was effective against the syphilis pathogenic spirochete *Treponema pallidum* (**Fig. 26.2**). The new compound, although crude and still toxic, was a great improvement over previous drugs. Named salvarsan, or "Ehrlich 606," it was hailed as a magic bullet against the feared disease. It marked the first instance of targeted chemotherapy design, and set the pattern for drug research for most of the next century: systematic evaluation of derivatives of a lead compound by teams of bacteriologists, chemists, and clinicians. Ehrlich received the Nobel Prize for Medicine together with Ilya Ilyich Mechnikov in 1908.

The work of Ehrlich, Bertheim, and Hata was a great advance over random trials of chemicals for drug effectiveness, but at the same time other, remarkably

FIG. 26.1 *Paul Ehrlich, originator of the "magic bullet" concept, with his colleague Sahachiro Hata, who carried out the testing that discovered the effect against the syphilis microbe.*

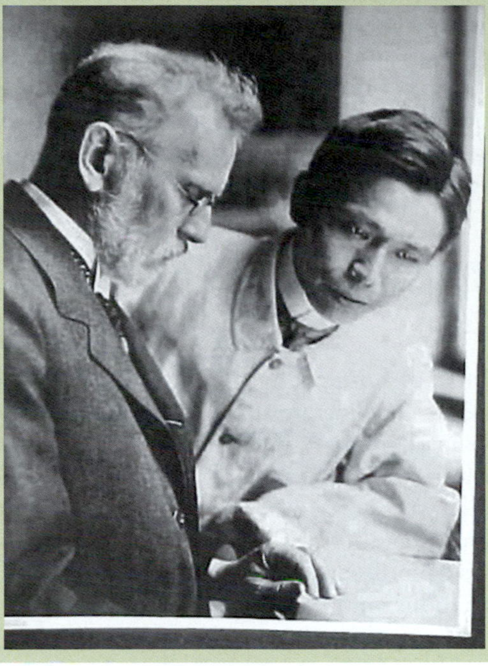

FIG. 26.2 *The syphilis pathogenic spirochete* Treponema pallidum.

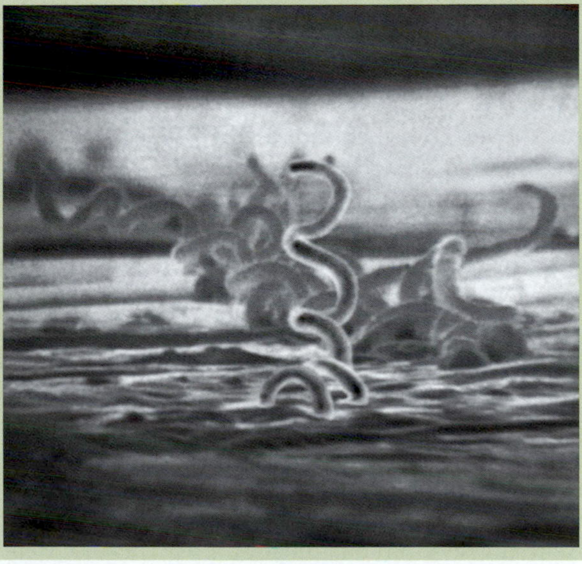

effective drugs were being inspired by natural biomimetic phenomena other than staining dyes: compounds from natural and traditional remedies led to more effective and safer medicines [48,49].

Many natural plant compounds are found to act specifically against devastating disease organisms, with very low toxicity to humans. Quinine was isolated from the bark of a tropical tree as a remedy for malaria. When its widespread use for a century or more led to resistant strains of pathogens, it was another natural product from traditional medicine, artemisin, which provided an alternative. In the next section we look at a natural plant substance which acts not by inhibiting disease organisms but by regulating the metabolism of our own bodies.

26.1.3 Aspirin: Signaling Pathways Revealed by the Willow

A remarkable example of a drug derived from plants is aspirin, one of the first drugs to be designed by a deliberate attempt to imitate the molecular structure and function of a natural product—the first biomimetic drug. Aspirin was developed as a safer version of willow bark extract, a traditional remedy for aches and fever which had the drawback of releasing strong acids into the stomach, producing digestive distress and even ulcers with continued use.

Willow bark has been used since ancient times for pain relief in arthritis and other inflammations, and to reduce fever. In the nineteenth century the active compound was isolated from willow bark, and named salicylic acid (from Latin *salix* = willow) (**Fig. 26.3**). Salicylic acid is a derivative of a more complex compound found in many related trees, the β-glycoside salicin (**Fig. 26.4**). Salicin breaks down into D-glucose (**Fig. 26.5a**) and salicylic acid (**Fig. 26.5b**), a ubiquitous plant hormone which acts as a signaling agent in some very basic and ancient biochemical pathways involving nonspecific immune responses [50].

The carboxylic group in the salicylic acid extracted from willow bark is made more acidic by the presence of a phenol radical, leading to its harmful effects on the stomach lining. In 1897, Felix Hoffmann, working at the German chemical company Bayer, developed a synthetic method for modifying salicylic acid, substituting the phenol to make it less harmful to the stomach, creating aspirin (**Fig. 26.6**). (Anecdotally, Hoffman supposedly was seeking a remedy to aid his father, who suffered from stomach ulcers from taking salicylic acid for arthritis over a long period.)

Aspirin's recognized therapeutic uses, as well as its phenomenal success as an over the counter remedy, generated significant profits, and much research effort was expended seeking improvements and alternatives. It was realized that research efforts into the mechanism of aspirin's many remarkable effects would open the door to a major class of new drugs; but for many years the mechanism of action of aspirin was not fully understood. Eventually, aspirin was found to inhibit the *arachidonic acid pathway* that leads to the synthesis of *eicosanoids*, potent mediators of pain and inflammation. In 1971, researchers led by the British pharmacologist John Robert Vane, at the Royal College of Surgeons in London, showed that aspirin suppresses the production of the prostaglandins and thromboxanes involved in inflammation [51]. This discovery led to major

Fig. 26.3 *The willow tree, source of salicin.*

Fig. 26.4 *Salicin molecule. (a) Three-dimensional model and (b) chemical bond structure.*

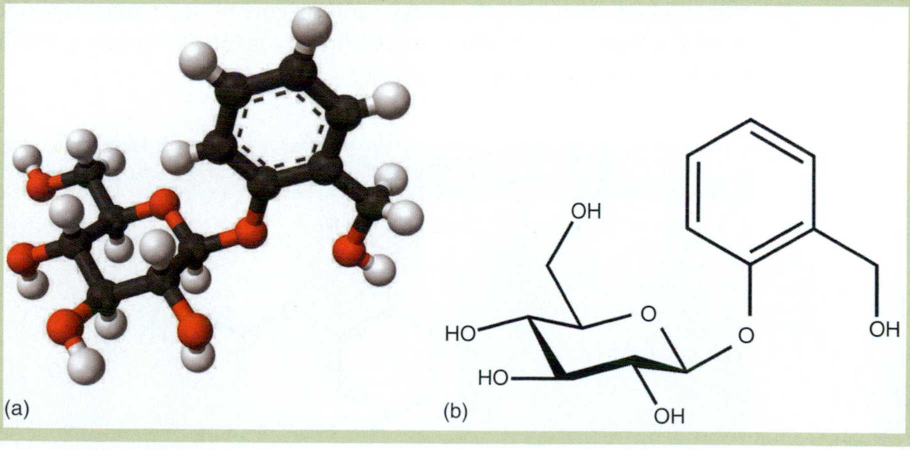

FIG. 26.5 *Salicin component molecular structures. (a) Glucose and (b) salicylic acid.*

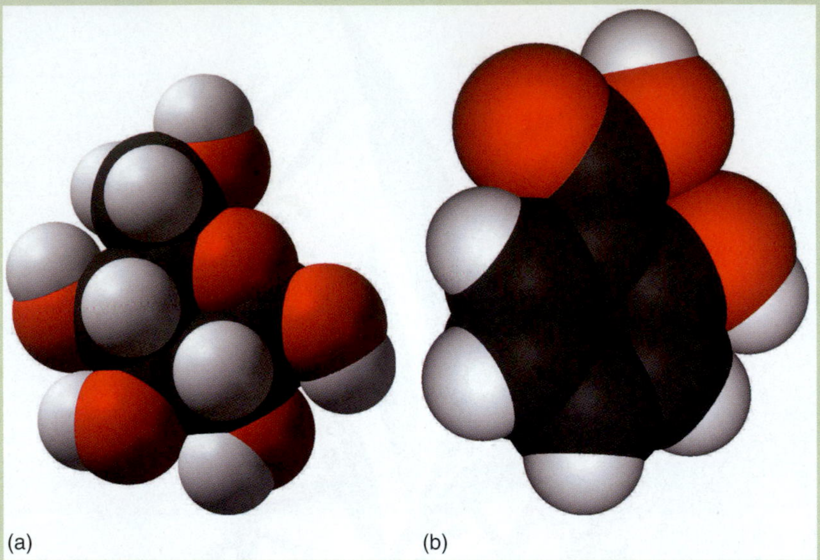

breakthroughs in understanding molecular signaling pathways between cells. The Nobel Prize for Medicine for 1982 was awarded to Professors Bengt Samuelsson, John Vane, and Sune Bergstrom for this discovery.

Aspirin was the model for a new class of pharmaceutical agents known as nonsteroidal antiinflammatory drugs (NSAIDs). Many but not all NSAIDs are derivatives of salicylates; all have similar effects—most act by nonselective inhibition of the enzyme cyclooxygenase, needed to synthesize prostaglandin and thromboxane. *Prostaglandins* are local (*paracrine*) hormones whose diverse effects include transmission of pain information to the brain, modulation of the hypothalamic thermostat, and regulating inflammation. Thromboxanes are involved in aggregation of platelets that form blood clots. Aspirin can irreversibly block the formation of thromboxane A2 in platelets, producing an inhibitory effect on platelet aggregation. This is the mechanism of aspirin's anticoagulant effects used to reduce the incidence and severity of heart attacks. A side-effect is a general reduction in the ability of the blood to clot, which may result in excessive bleeding with the use of aspirin [52].

FIG. 26.6 *Aspirin: Acetylsalicylic acid molecule chemical structure.*

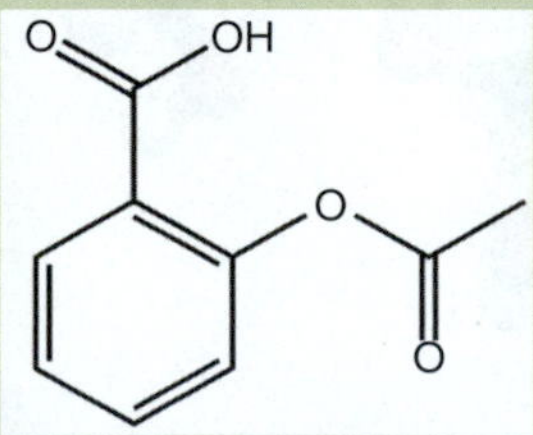

Fig. 26.7 *Action of aspirin on the structure of COX-2 (prostaglandin H synthase). The COX-2 molecule is a dimer; the blue and green halves are identical. In each monomer, the active serine site has been acetylated by aspirin, inactivating it.*

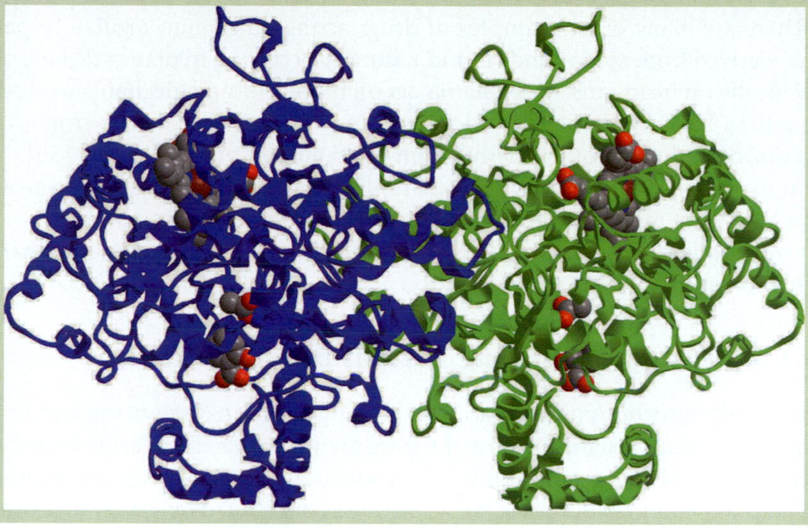

Source: Image courtesy of Jeff Dahl.

Aspirin suppresses the production of prostaglandins and thromboxanes by its *irreversible* inactivation of the cyclooxygenase (COX) enzyme, which is required for prostaglandin and thromboxane synthesis (**Fig. 26.7**). Aspirin acts as an acetylating agent, covalently bonding an acetyl group to a serine residue in the active site of the COX enzyme. This *irreversible* mode of action is different from other NSAIDs (such as diclofenac and ibuprofen), which are *reversible* inhibitors. Recently, *reversible* blocking of COX-2 by synthetic NSAID drugs has been found to lead to harmful effects, showing that we still have some surprising biomimetic lessons to learn from the subtle mode of action of the natural product.

In addition to inactivation of prostaglandin and thromboxane production, aspirin has two additional modes of action, contributing further to its strong analgesic, antipyretic, and antiinflammatory effects. Aspirin buffers and transports protons across membranes involved in energy release by ATP in the mitochondria, where it uncouples oxidative phosphorylation and disrupts energy release. As a weak acid, aspirin can carry protons, diffusing from the inner cell membrane space into the mitochondrial matrix, where it ionizes to release protons.

Aspirin stimulates the formation of NO radicals that enable the body's white blood cells (leukocytes) to fight infections more effectively. Dr. Derek W. Gilroy was awarded Bayer's International Aspirin Award in 2005 for his research revealing the effects of aspirin on NO production [53].

The study of aspirin has contributed to our understanding of inflammation and how it has evolved as a protective response to insult or injury, a primordial response that eliminates or neutralizes foreign organisms and materials. Salicylic acid and its derivatives have been found to modulate signaling through a number of transcription factor complexes that play central roles in many biological processes, including inflammation. Inflammation involves many mechanisms that protect us against tissue injury and promote the restoration of tissue after damage: our well-being and survival depend upon its efficiency and carefully balanced control [54,55].

There are many other examples of drugs acting on human regulatory pathways, derived from compounds found naturally occurring in plants: *digitalis* acts specifically on heart muscle; *coumarins* act on blood clotting mechanisms. Today more than 120 important and widely used medicines derive directly from plant precursors, and many others derive from fungi and marine organisms [56].

In the next section we look at an example of a molecule whose action would have been even more difficult to predict without its existence in nature; it works not by modulating a signaling pathway, but by blocking a fundamental nanoscale process involved in cell replication.

26.1.4 *Taxol: Novel Drug Actions on the Nanolevel*

Use of a biomimetic approach can be especially useful in discovering nanoscale effects, providing insights into interactions between macromolecules involved in cellular metabolism. One example of a nanoscale drug discovered by examining natural plant metabolites is taxol. Taxol was discovered by screening extracts from yew trees (**Fig. 26.8**). The yew is important in European folk traditions

FIG. 26.8 *Pacific yew tree, source of taxol.*

Source: Figure courtesy of Botanical Research Institute of Texas, Fort Worth, Texas.

dating back to the Druids and earlier. Yew berries are poisonous, and extracts of yew leaves and bark were used in Celtic medicines and ceremonies.

From the 1960s the United States National Institutes of Health (NIH) sponsored research, screening thousands of natural product extracts for potential use as cancer drugs. By the 1980s tests of extracts from the bark of the Pacific yew tree yielded promising results on cancer-prone mice strains. Further tests found that a compound in the Pacific yew extract, taxol, was especially effective in halting the growth of certain types of human breast cancer which were resistant to other treatments. The structure and mechanism of taxol were elucidated by further research sponsored by the NIH and pharmaceutical companies (**Fig. 26.9**). It was found that taxol disrupted the functioning of the microtubules involved in guiding chromosomes as they separated in mitosis during cell division. Rapidly growing tumor cells were preferentially affected compared to normal cells, as they were blocked from reproducing by the action of taxol. Taxol, also known by the generic name paclitaxel, has been found effective in treatment of lung, ovarian, breast, head and neck cancers, and advanced stages of Kaposi's sarcoma, and is used for the prevention of *restenosis* in patients with blocked blood vessels.

Research work developed ways to synthesize taxol and related compounds using precursors and biotechnology without the need to harvest large amounts of raw materials from the trees. Thus the value of the trees was not material, but informational. Taxol is an especially interesting drug from the nano perspective, because it acts by absorbing on to specific sites in the microtubule structures that act as scaffolding and guideways in the separation of chromosomes during cell division. Taxol stabilizes these structures, making them rigid and unable to function—it "gums up the works." It acts in the domain between the chemical and the mechanical—the nanoworld.

FIG. 26.9 *Taxol molecule chemical structure.*

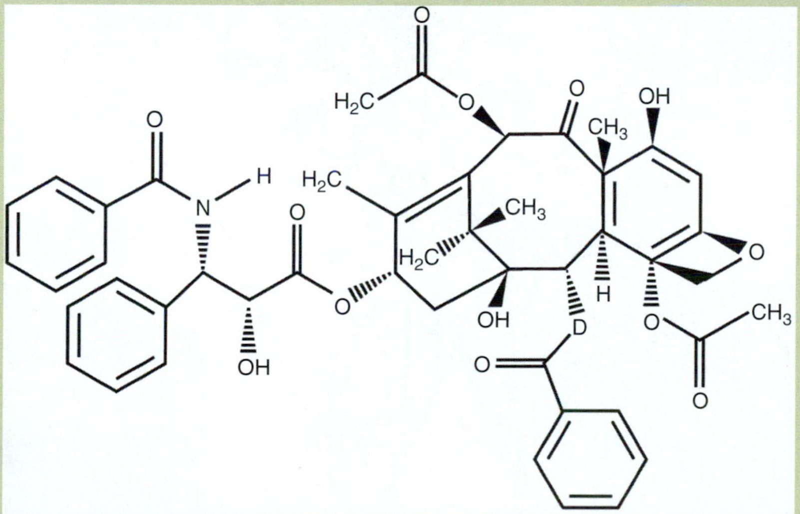

Modern biomimetic science relies on natural models like the yew tree to find examples of specific biological activity at the macromolecular level that would otherwise be unexpected. The action of taxol is at the nanoscale level of microtubules formed in mitosis, thus it lies between the domains of biochemistry and cellular biology. Taxol-derived drugs are examples of discovery of drug design and action by observations from nature. Taxol is also a cautionary tale of the costs of ignoring the importance of preserving the irreplaceable information contained in ancient biodiverse ecological systems [57,58].

26.1.5 Pyrethrum: Learning from the Daisy

The pyrethrum daisy is a flowering plant of the chrysanthemum family, native to the Caucasus mountains (**Fig. 26.10**). Use of the flowers as an insecticide and insect repellent dates back to antiquity; the dried flowers have long been prized and traded across the world. The plant produces natural organic compounds called *pyrethrins* with potent insecticidal activity. The pyrethrins are secreted in the seed cases in varying concentrations in different varieties. One species, *Chrysanthemum cinerariaefolium*, is grown commercially as a source of the insecticides.

The chemical structure of pyrethrins was first published by Hermann Staudinger and Lavoslav Ružička in 1924 (**Fig. 26.11**) [59]. The different variations of natural pyrethrins are all structurally related esters with a cyclopropane core [60,61]. Natural pyrethrins are viscous lipophilic liquids. They are nonpersistent, biodegradable, and break down relatively quickly on exposure to light or oxygen. Pyrethrins are neurotoxins that attack the nervous systems of all insects, being especially effective against the muscles of insects such as mosquitoes,

Fig. 26.10 Pyrethrum chrysanthemum *daisy, source of pyrethrin.*

Source: Figure courtesy of Botanical Research Institute of Texas, Fort Worth, Texas.

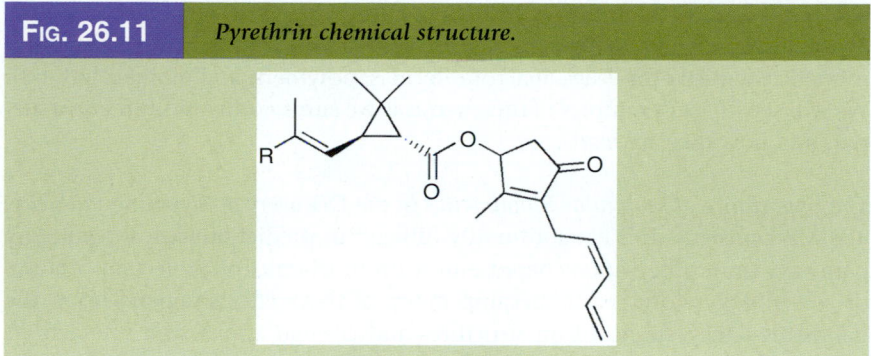

FIG. 26.11 *Pyrethrin chemical structure.*

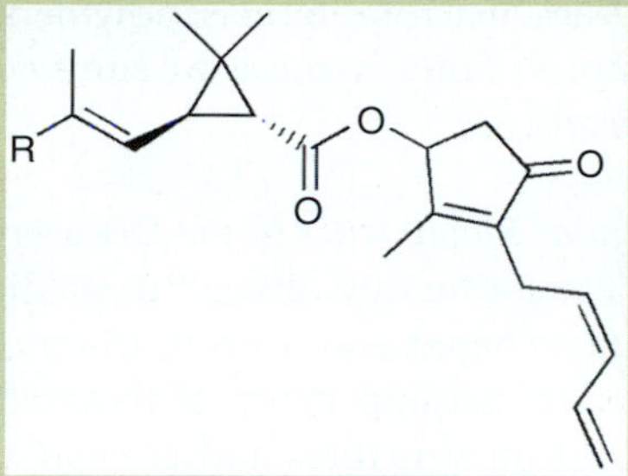

fleas, and houseflies. Low concentrations can have an insect repellent effect. Because of their powerful "knock-down" effect, insects are slow to develop into resistant strains. They are harmful to fish, but have very low toxicity to mammals and birds. In humans they can irritate eyes, skin, and respiratory systems but are not toxic to the human nervous system or metabolic pathways. They are considered to be amongst the safest insecticides for use around food [62].

The chemical structure of pyrethrins has served as a model for a variety of synthetic insecticides called pyrethroids such as permethrin and cypermethrin, which share their specificity, low mammalian toxicity, and nonpersistence in the environment, while being even more effective against insects than the natural substances. Thus the natural molecule has served as a source of information for the design of effective insecticides, not only for their action on insect muscle, but also to show how to design molecules that do not build up in the food chain and environment. Pyrethroids act on muscle, which is itself a very interesting example of biological nanomachinery. Studies of how pyrethrins work to disrupt muscle function can shed light on mechanisms of paralysis caused by diseases such as tetanus, polio, and muscular dystrophy. In a later section we will examine the structure of muscle and how it works, and how it serves as a model for some biomimetic engines.

Other Biomimetic Pesticides. Many other plants produce pesticides that have revealed new chemical paths for fighting crop and human pests. The Mexican marigold produces thiols, which are effective against nematodes that attack plant roots in valuable crops. Other plants produce allelopathic growth inhibitors against competing plants. Natural plant compounds can be used as models for safe, specific, and effective weed killers that do not build up in the environment, causing unintended harm to beneficial insects and animals, and creating chemical resistance. For these reasons plant scientists and agronomists are looking for natural and biomimetic substances as alternatives to costly petrochemical-based synthetics.

Biomimetic Methods of Molecular Synthesis. Natural processes can serve as models as well as natural materials. Organisms make complex molecules and link them together in polymers and nanostructures, working at ambient temperatures without high energy expenditure or large amounts of wasted by-products. The enzymes and catalytic proteins that make this possible are being understood and harnessed by biomimetic synthetic methods being developed by chemists and engineers. Chemo-enzymatic synthetic methods are a growing area of biomimetic chemistry for drugs, macromolecules, polymers, and biofuels [63–68]. We will see further examples of this theme as we survey other biomimetic materials and how they are made.

The Importance of Molecular Biomimetics in the Discovery of Signaling Pathways and Mechanisms. It is extraordinarily difficult to predict biological regulatory pathways from observations based purely on biochemistry, molecular biology, or cell biology alone, partly because many of these effects depend upon the nanoscale—they are based on structures and configurations, and interactions of chemistries, macromolecular structures, and nanomachinery involved in cell biology. Observation of interactions involving compounds produced by the organisms with which we share our natural environment can reveal effects, which when investigated using the tools of nanotechnology and the scientific methods of biochemical, cellular, and molecular biology can be unraveled down to their nanoscale mechanisms.

Areas for Further Study. As a hint of the interesting rewards of further study in this field, we give just two examples. Many models for useful macromolecules come from marine organisms. For example diazonamide A, a model for a family of anticancer drugs with a unique type of architecture and activity, was isolated from the marine acidian *Diazona angulata,* a rare invertebrate found in the Philippine sea [69].

Other sources for molecular biomimicry come from insects, spiders, and mites. In the earlier discussion of aspirin, we mentioned that aspirin acts upon the important arachidonic acid pathway. You may have wondered how this pathway got its name, which comes from the Greek word for spider. Arachidonic acids are found in the venom of many spiders, ticks, and other arachnid arthropods. These animals use the toxic versions of these acids to trigger violent inflammation responses in the victims of their bites, to help break down and digest their meals.

In a particularly interesting instance of biomimicry, researchers have isolated a highly effective class of macromolecular peptide pesticides that work very selectively against disease-carrying ticks, which have become resistant to other more conventional pesticides. The peptides come from a web-weaving spider which preys on other arachnids [70].

It is particularly satisfying to be able to turn the nanomolecular tools of these voracious bloodsuckers against such troublesome pests of their own family, thanks to their own indiscriminate predatory adaptations. It is more significant to consider this as an instructive example of our interconnectedness with the web of life—and sources for biomimetics.

26.2 Biomimetic Nanomaterials

In this section, we will discuss how natural materials inspire advanced nanomaterials with unprecedented performance characteristics: high strength, light weight, self-cleaning, and unique optical and electronic behaviors. We will survey some of these emerging materials, their properties, and their applications. We will also look for emergent themes and design principles that natural materials demonstrate [71–77].

Smart materials mimicking biological functions are increasingly closer to nature's technology. These include tough artificial shell and bone, artificial muscle made with electroactive polymers, nanoscale molecular actuators, smart membranes, nanoengineered encapsulation, and nanoadhesives. All of these nanotechnologies incorporate insights into how nature works. Their development is helping us cure diseases, save energy and water, and improve efficiency of food and materials production. By understanding nature on the nanoscale, we will be able to live in our environment more intelligently, sustainably, and elegantly.

26.2.1 Biomimetic Mineral Nanoparticles

The control of size and morphology of microcrystals is an area in which biological organisms excel, but which has until recently been beyond the capabilities of human engineers and scientists. Living organisms not only replicate organic structures such as DNA and cell membranes but are also capable of building inorganic structures with precise control of form and properties. As we saw in the previous book, these include shells, teeth, bone, and silica structures.

Control of Crystal Growth by Mimicking Natural Biological Processes. One of the pioneers of early studies which explained precise crystal structures of mineral micro- and nanoparticles grown by biological organisms is Professor Stephen Mann, who observes that [78]

> Inorganic building blocks play a fascinating and crucial role in self-organised assembly for many biological structures including bone, teeth and shell. Biomineralisation uses a limited number of solid inorganic materials such as calcium carbonate, silica and iron oxides to form new materials that bear no relation to underlying structures. Biological systems have developed an exquisite control of inorganic processes. There are species of magnetotactic bacteria that produce nanoscale magnets in their cells. Not just depositing iron oxide but producing crystals that are both the perfect size and shape and aligned in a chain.

Biomimetic Synthesis of Novel Inorganic Crystalline Structures. Mann's group has explored the use of protein-based micellular structures to grow bioinorganic nanocomposites with magnetic and quantum resonance properties, surfactant molecule assemblies as templates for inorganic crystal growth, and microemulsions to create minireaction chambers in micelles. They use biological microstructures to synthesize inorganic complexes, adapting bacterial filaments to fabricate ordered silica macrostructures and tobacco mosaic virus to make inorganic nanotubes. They also recently developed a new process using inorganic nanoparticles to make magnetic spider silk.

Nanoparticle Architectures. How to direct and control the self-assembly of nanoparticles is a fundamental question in nanotechnology. The success, growth, and application of nanotechnology depend upon our ability to manipulate nanoscale objects. A group led by Nicholas A. Kotov is examining three critical questions in the growth of nanoparticle structures:

1. What are the methods of organization of nanocolloids in more complex structures?
2. What kind of structures do we need for different applications?
3. What are the new properties appearing in nanocolloid superstructures?

Professor Kotov at the University of Michigan studies how to organize nanoparticles into a useful variety of larger and more complex systems, making new materials utilizing biomimetic models. His team developed a technique to make nanocrystals in a fluid assemble into free-floating sheets the same way some structures form in living organisms. This fluid process lends itself to automated production that builds materials one nanoscale layer after another. A robotic arm applies nanolayers onto substrates such as glass or silicon wafers, alternating compositions from different liquid source materials, to make composite nanoengineered plastics [79,80].

26.2.2 Shell as a Biomodel

In a previous volume, we looked at the natural structure of shells, based on layered, hierarchical nanostructures of mineral crystals, polymers, and proteins. Shells combine proteins and mineral crystals in a nanoscale architecture that produces high strength and toughness, using very little energy and wasting no material in the process. These natural materials and processes are being used as models to design new generations of strong and resilient materials.

Abalone Shell as a Model for Armor. Researchers led by Marc A. Meyers at the University of California, San Diego (UCSD), are pursuing a number of biomimetic projects, using the shell of the abalone as a model. Meyer describes his team's basic research on new materials in biomimetic terms:

> We have turned to nature because millions of years of evolution and natural selection have given rise in many animals to some very sturdy materials with surprising mechanical properties. In our search for a new generation of armors, we have exhausted the conventional possibilities, so we've turned to biology-inspired, or biomimetic, structures.

Mollusk shells, bird beaks, and other natural biocomposites are based on a hierarchy of structures from the molecular level to the macroscale. At the nanoscale, shell is made of thousands of layers of calcium carbonate "tiles," about 10-μm across and 0.5-μm thick. The layered stacks of thin tiles refract light to yield the characteristic luster of the mother of pearl. The shell nacre's nanostructure of interlocking calcium carbonate tiles and shock-absorbing protein adhesive give it the ability to absorb heavy blows without breaking (**Fig. 26.12**).

The main constituent of the abalone shell is calcium carbonate. Only about 3% of the shell is made up of organic components, but the fracture resistance of the nanocomposite nacre is about 3000 times higher than for the pure carbonate mineral. A key to the strength of shell is a positively charged protein adhesive that binds to the negatively charged top and bottom surfaces of the calcium

Fig. 26.12 *Abalone shell tiles: (a) Schematic representation of stacked layers of aragonite tiles. (b) Arrangement of tiles on inner surface of 10-mm shell; back-scattered image (SEM). (c) Schematic drawing of stacking of abalone tiles and their separation under tension. Abalone shell composite is some 3000 times tougher than the calcium carbonate crystals from which it is made. When an impact force is applied, the organic layer deforms slightly. This absorbs some of the shock; the rest causes the tiles to slide until frictional forces oppose the movement. The aragonite tiles can fracture, but the fracture is limited to the individual tiles. As each layer is offset laterally from the other, cracks have a difficult time propagating from one tile layer to the next. If a crack should spread it generally has a difficult time passing through the organic, elastic layer to an adjacent mesolayer.*

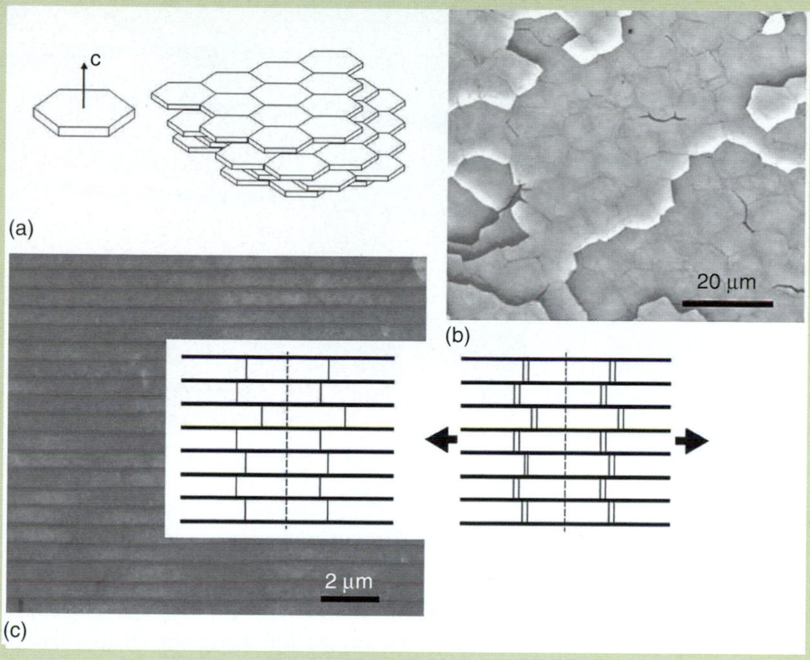

Source: Image adapted from A. Lin and M. A. Meyers, *Materials Science and Engineering A*, 390, 27–41 (2005).

carbonate tiles. The glue holds layers of tiles firmly together, but is soft enough to permit the layers to slip apart, absorbing the energy of a heavy blow in the process.

Meyers discovered previously unknown details about the nanoscale assembly of the nacre building blocks that help to explain their resilience and provide a guide to the design of stronger biomimetic materials.

> Contrary to what others have thought, the tiles abutting each other in each layer are not glued on their sides, rather they are only glued on the top and bottom, which is why adjacent tiles can separate from one another and slide when a strong force is applied.

The elastic adhesive and free edges in the shell's interior allow it to yield to impacts without breaking, unlike conventional laminated materials (**Figs. 26.13** and **26.14**).

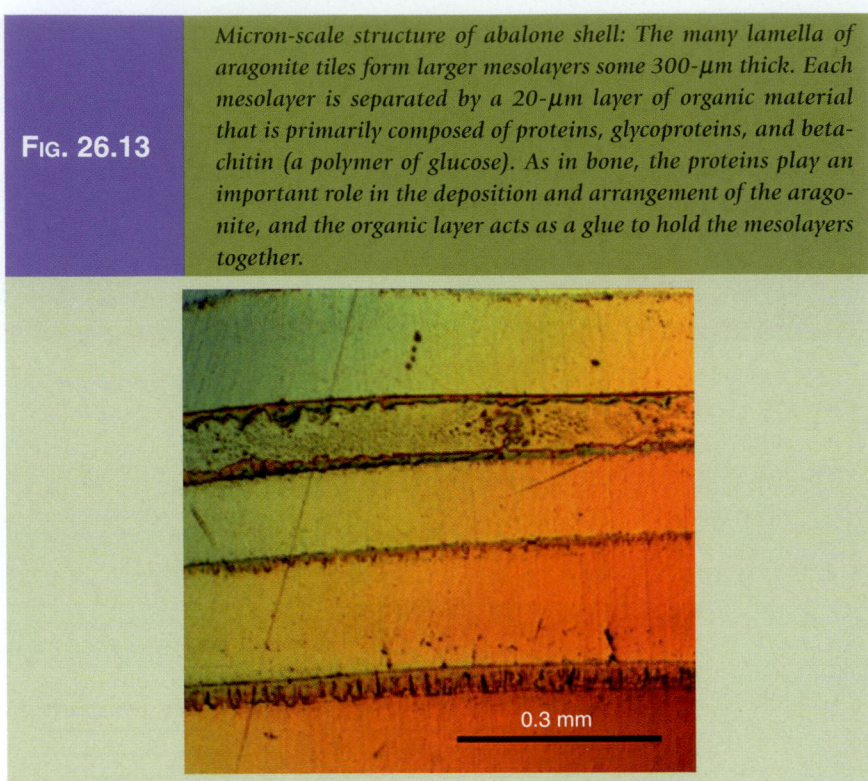

FIG. 26.13 *Micron-scale structure of abalone shell: The many lamella of aragonite tiles form larger mesolayers some 300-μm thick. Each mesolayer is separated by a 20-μm layer of organic material that is primarily composed of proteins, glycoproteins, and beta-chitin (a polymer of glucose). As in bone, the proteins play an important role in the deposition and arrangement of the aragonite, and the organic layer acts as a glue to hold the mesolayers together.*

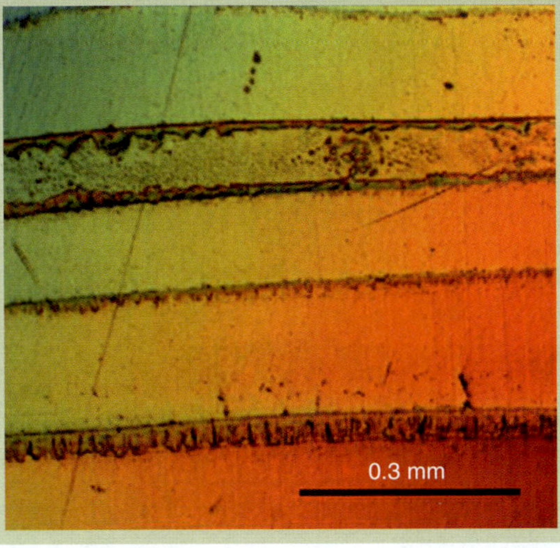

Source: Electron micrograph image from A. Lin and M. A. Meyers, *Materials Science and Engineering A*, 390, 27–41 (2005). With permission.

Abalones fill fissures within their shells that form due to impacts; they deposit "growth bands" of organic material alternating with mineral deposition. Meyer's group inserted glass slides beneath the mantle of abalone grown in a laboratory aquarium. When they withdrew the slides at various time intervals they were able to observe the pearly growth layers on the slide with a transmission electron microscope. These growth layers showed that the abalone mantle seeded calcium carbonate crystal precipitates at intervals separated by about 10 μm. Tiles began to form from the seed locations, growing 0.5-μm thick and spreading slowly outward. The tiles formed a hexagonal shape as individual tiles in each layer gradually grew to abut a neighboring tile. Microscopic imaging showed the growth surface of the shells with a Christmas-tree appearance as abalones add layers of tile faster than each layer is filled in. The UCSD group is developing a mathematical description of the growth process which will serve as a systematic design tool for biomimetic materials [81].

Other Biomimetic Nanocomposite Materials Modeled on Shell. Meyer's group is not alone in using natural shell structures as biomimetic models for strong materials. At the Particle Technology Laboratory of the Eidgenössische Technische Hochschule (ETH) in Zurich, researchers led by Ludwig J. Gauckler are pursuing bioinspired design principles, making strong nanocomposites through the

FIG. 26.14 *(a) Fracture mechanisms and (b) estimated tensile strength of platelet-reinforced composites. The aspect ratio of the platelets (s) determines whether the composite fails under the platelet fracture mode ($s > s_c$) or platelet pull-out mode [$s < s_c$(a)]. The tensile strength of polymer matrix composites reinforced with $CaCO_3$ and Al_2O_3 platelets is estimated in (b) (black and blue surfaces, respectively) on the basis of a shear-lag mechanical model. The limit between the platelet fracture and pull-out modes is indicated by the black and blue full lines for $CaCO_3$ and Al_2O_3 platelets, respectively [82].*

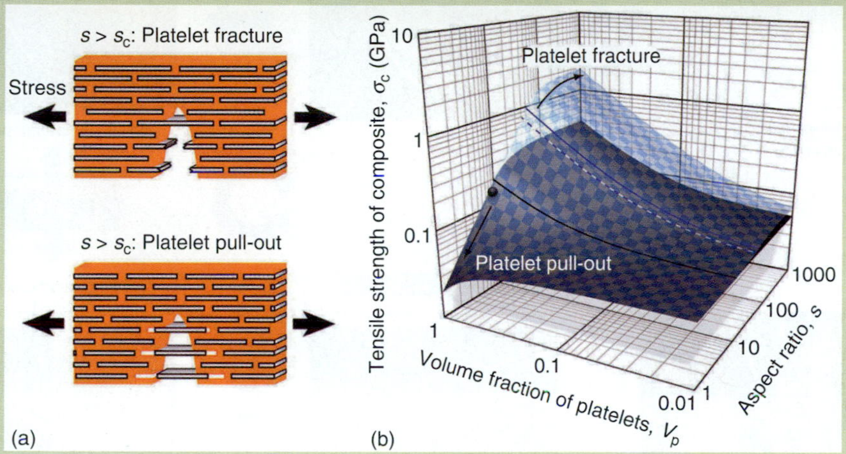

Source: Image from L. J. Bonderer, *Science*, 319, 1069–1073 (2008). Reprinted with permission from AAAS.

bottom-up colloidal assembly of sub-micrometer-thick ceramic platelets within a ductile polymer matrix [82] (**Figs. 26.14** and **26.15**). In 2007, a research group led by Hansma in California reported how bioinspired optimized adhesives combined with carbon nanotubes or graphene sheets can yield strong, lightweight, damage-resistant nanocomposite materials [83,84]. A common feature of biological nanocomposites is that only a small amount of adhesive is needed—excess adhesive actually weakens the material. The optimal amount of adhesive is just enough to transfer the load to the strong elements.

Mimicking the Self-Organizing Power of Proteins to Make Biomimetic Nanocomposites. At the University of Michigan, Professor Kotov is applying automated production technology to make biomimetic materials using alternating layers of clay nanoparticles and adhesive [79,80]. It takes 300 layers of adhesive polymer alternating with clay nanosheets to create a piece of this material as thick as a piece of plastic wrap. The glue-like polymer used in this nanomaterial is polyvinyl alcohol, which forms cooperative hydrogen bonds with the clay nanosheets. This "nanoglue" gives rise to what Kotov called "the Velcro effect." When hydrogen bonds are broken, they can reform easily in a new place. The dense hydrogen bonds allow for effective load transfer between the nanoparticle and polymer layers.

Biomimetic Materials Based on Wood Nanostructure. The microfibril structure of wood inspired one of the earliest successful biomimetic applications based on micro- and nanostructure of natural materials. R. Gordon, C.R. Chaplin, and

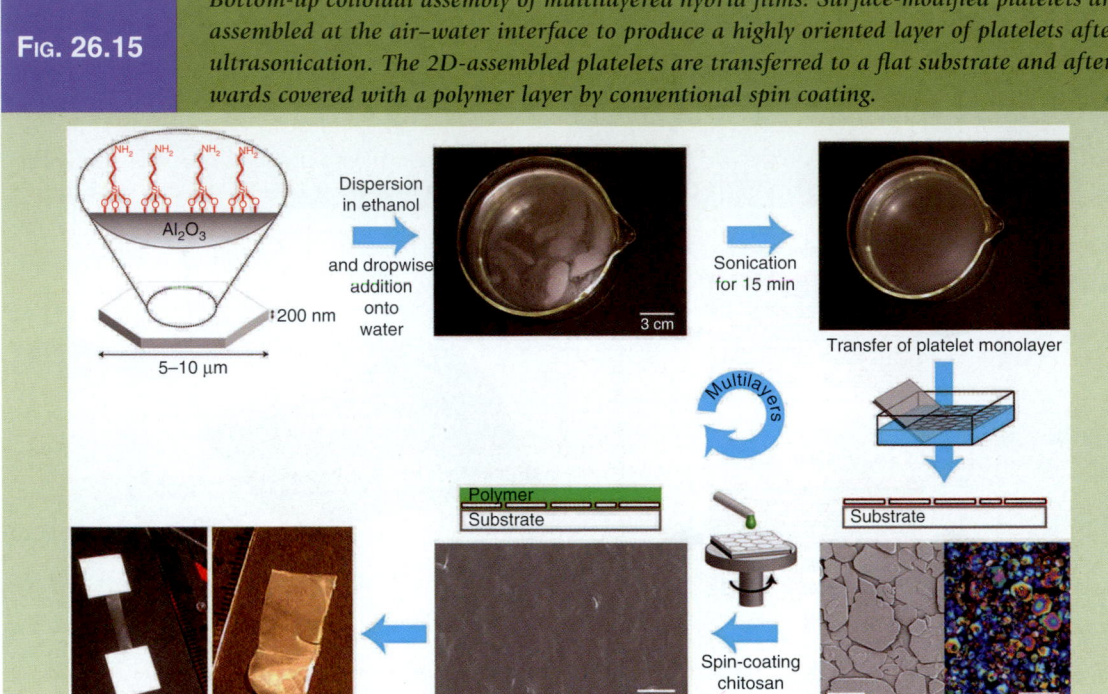

Fig. 26.15 *Bottom-up colloidal assembly of multilayered hybrid films. Surface-modified platelets are assembled at the air–water interface to produce a highly oriented layer of platelets after ultrasonication. The 2D-assembled platelets are transferred to a flat substrate and afterwards covered with a polymer layer by conventional spin coating.*

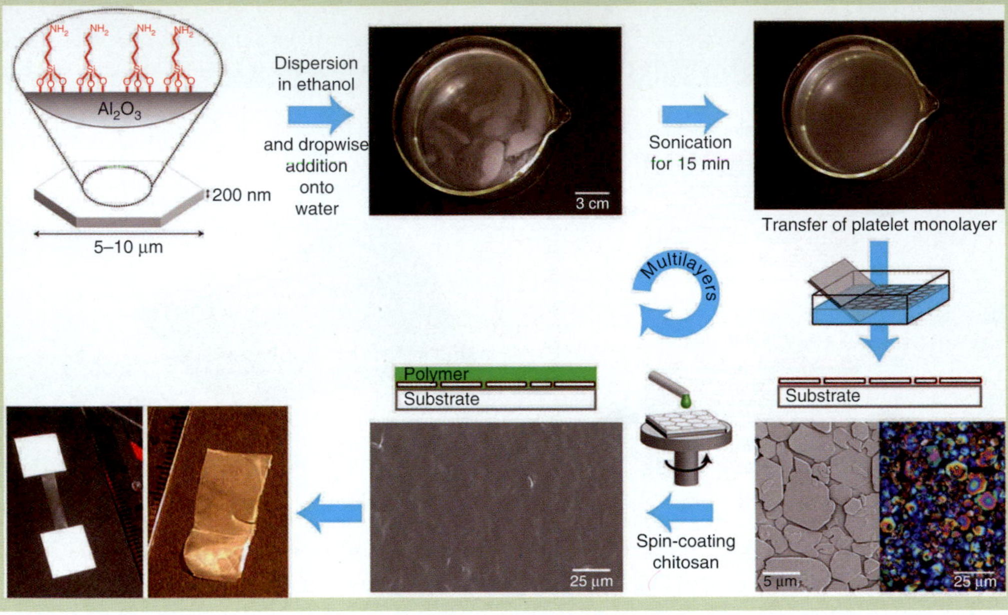

Source: Image from L. J. Bonderer, *Science*, 319,1069–1073 (2008). Reprinted with permission from AAAS.

G. Jeronimidis at Reading University patented a bio-inspired composite structural panel with high strength and toughness based on ultrastructural features in wood [85–87]. The orientation of the fibers in this biomimetic composite is based on angles found in microfibrils of wood tracheids. Reading is among a number of centers around the world that are investigating the micro- and nanostructure of wood, feathers, and other materials for biomimetic applications [88–96].

Modern lightweight composite structures are manufactured with so-called gradient textile techniques. As in nature, every single fiber strand is exactly laid within the structure in the direction necessary to neutralize outer forces so that no unnecessary fibers or weight are incorporated. Manufacturing of these ultralight composites was made possible by the development of adequate "finite element" computing methods for the calculation of forces in curved and irregular shapes. Biomimetic manufacturing technologies based on plant models are the special focus for physicists, biologists, and engineers led by Claus Mattheck at the Forschungszentrum Karlsruhe in Germany, where they are studying the biological design rules that govern the development of optimized shapes in growing plant structures [43,90].

26.2.3 Nanoengineering Bone

Crustaceans and other shell makers create their exoskeletons primarily as protection from external threats; so the shell structure is an excellent adaptation to protect against crushing and cracking forces applied from the outside. Vertebrates rely on bony endoskeletons which must perform a much more varied and complex set of functions; bones and teeth have to bear complex loads of moving bodies, provide a protective cage for vital organs, anchor tendons and muscles, and act as joints, fulcrums, and levers. Because of this functional complexity, the nano- and microstructure of bone are more complex than the carbonate nanostructure of shells. Different parts of the same skeleton must have different directional, compression, and tension strengths to prevent fracture and distortion under stresses, loads, impact, and fatigue.

Because of its medical importance, research into the nanostructure of bone has focused mainly on artificial bone for replacement, scaffolding for bone augmentation and growth, and interfaces between bone and implants. With the rapid growth of biomimetics, the structure of bone is serving as a model for materials development. Efforts have been made both to mimic the bony material itself as well as to mimic the process by which bone forms, but such efforts are still in their early stages compared to the biomimetics of shell-like materials [78].

Bone is an interesting natural material in its own right, due to its combination of mechanical properties. Its combination of strength and resilience with light weight is the result of its nanocomposite structure composed of mineral and hydrated organic macromolecules (see chapter 13). Bone consists of layers of mineral crystals (hydroxylapatite) interspersed with protein and biopolymers that add resilience to the compression-bearing strength of the mineral component **(Fig. 26.16)**. Teeth have additional strength from enamel and fluorides **(Fig. 26.17)**.

Biomimetic Composites for Artificial Bone. One approach to promoting the growth of bone for healing is to provide scaffold material into which bone-forming cells can migrate, leading to fusion of new bone growth with the scaffold. Early versions of scaffolding involved coatings of hydroxylapatite applied to fractures. More recently sol-gel nanocoating techniques have been shown to improve bone grafting. Bone forms by mineralization of precursor cartilage tissues. Successful growth, maintenance, and healing of bone depend on each portion of the structure matching the types of forces to which it is subjected in the body. Much work is being done to understand how precursor cells are programmed and influenced to promote mixtures of mineralization and organic fibers that optimally match the compressing and tension loads that must be borne by the skeleton, and how this can be mimicked and encouraged by scaffolds, growth-promoting substances, and electromagnetic stimulation.

F.Z. Cui and coworkers in Beijing, China have developed a bone scaffold by biomimetic synthesis of nano-hydroxyapatite and collagen assembled into mineralized fibrils [97]. This material shows some features of natural bone in composition and hierarchical microstructure, with three-dimensional porous scaffold materials that mimic the microstructure of cancellous bone. In cell culture and animal tests, bone-forming osteoblast cells from rats adhered, spread, and proliferated throughout the pores of the scaffold material within a week. This scaffold composite has promise for the clinical repair of bone defects. This and similar work being done elsewhere shows how biomimetic structures can integrate with natural tissues [98,99].

Fig. 26.16 *Bone resists tension, torsion, and compressive forces due to its nanostructure, made of hydroxyapatite mineral and collagen protein fibers. (a) The hydroxyapatite is arranged in concentric layers (called lamella) within a cylindrical, functional unit called an osteon. The apatite is arranged in small crystalline plates about $60 \times 30 \times 8$ nm in size found within and around collagen fibrils (bundles of collagen filaments.) The plates overlap and are aligned parallel to the long axis of the fibril. The plates within each fibril have the same angular orientation; adjacent fibrils are oriented at different angles to each other. Compression deforms the collagen, allowing the plates to slide towards each other until they start to touch. This initial movement absorbs some of the applied force and prevents the bone from breaking. As the plates touch, fictional forces take over and the plates can no longer slide. What happens when the forces are too great for the crystalline plates? They fracture, but because the plates are so small large cracks cannot form. Because the plates of adjacent fibrils are oriented at different angles small cracks cannot propagate easily from one plate to the next. (b) Micrograph of osteoblath reveals layered bone structure.*

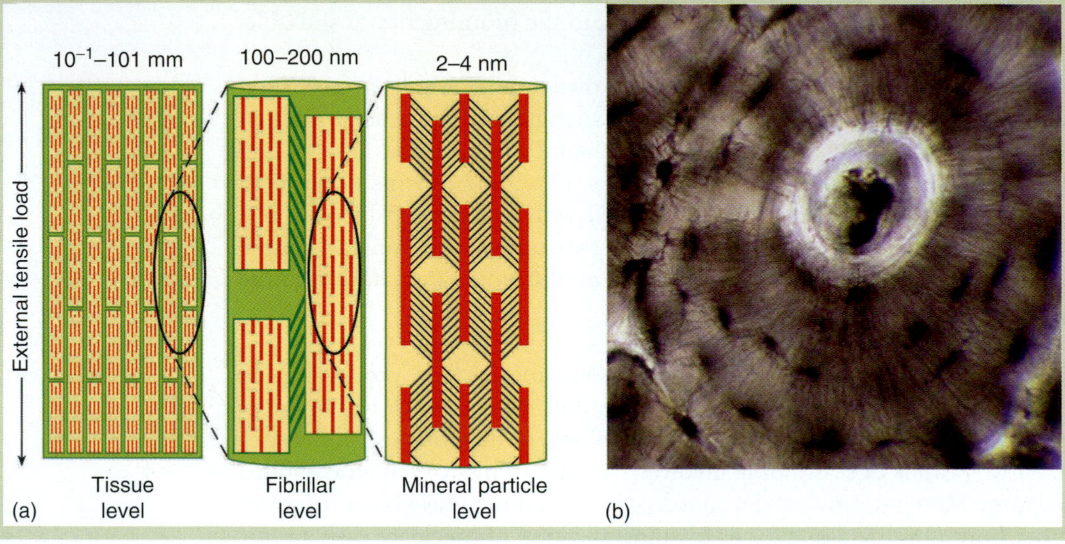

Source: Image courtesy of Max Planck Institute of Colloids and Interfaces. With permission.

26.2.4 Sponge Fiber Photonics

In the previous sections we looked at natural nanomaterials whose main functions are structural and load bearing; their nanostructures are optimized for interaction with forces, tensions, loads, and pressures. In this section we look at interactions with light—in photonic nanomaterials—and the biomimetic inventions that they are inspiring. The natural optical fibers found in the sea sponge *Euplectella* are tougher than current man-made glass fibers, and contain a unique mixture of dopants that enhance their optical properties. Fiber-optic engineers are studying the nanostructure of these natural silica fibers to learn how to improve the engineering of fiber-optic cables.

Euplectella is commonly called the "glass sponge." It is also known as the "Venus flower-basket" or "wedding basket" because its silica skeleton forms an intricate cage, which often houses a pair of mating shrimp. The skeleton of the

Fig. 26.17 *Teeth are adapted for withstanding pressure and wear. Extreme examples are the teeth of coral-eating fish. The enameloid outer layer of parrot fish teeth consists of fluoroapatite crystals and collagen fibers, where the latter only comprises about 3% of the enameloid. The high mineral content makes a very hard surface, withstanding both compressive and shearing forces. The orientation of the fluoroapatite crystals determines the hardness of the enameloid. The crystals, 100 nm in diameter and several microns in length, are bundled together to form large fibrils. At places where the strongest shearing forces are present (where the grinding action takes place) the fibrils are perpendicular to the surface. This presents a rather small cross-sectional area to the shearing forces and a lower risk for crystal fracture. Should a fracture occur, propagation of the fracture is limited because of the small crystal size. (a) Collagen fibers, (b) mineral plates.*

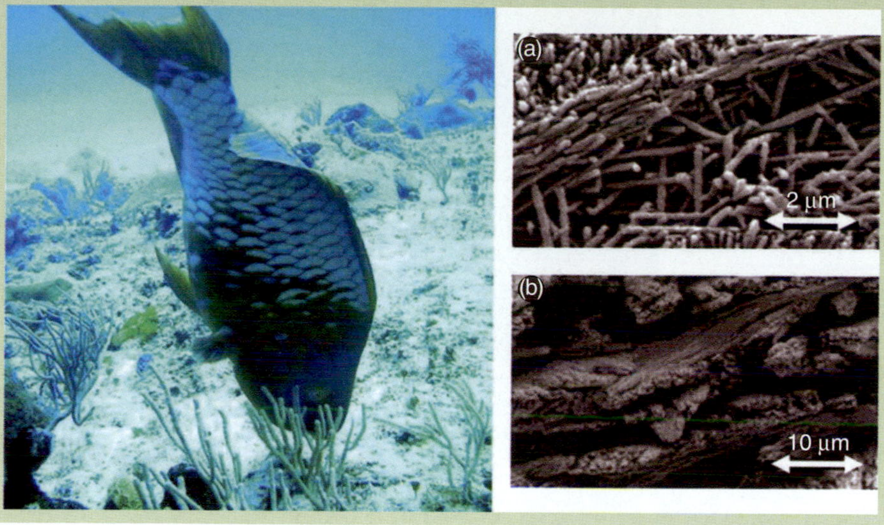

sponge is supported by a network of amorphous hydrated silica fibers, called spicules, with especially long and clear fibers growing from the base of the animal. These silica fibers resemble optical glass fibers used as waveguides in communication networks, of similar size and with a highly refractive core surrounded by a lower refractive index cladding. Researchers found that the sponge fibers carried light like single-mode or low-mode waveguide fibers (**Fig. 26.18**).

The natural spicules are tougher than man-made glass fibers. They can be bent without breaking, whereas telecommunications fibers are brittle and have to be protected. The natural fibers are composed of nanoscale layers of silica separated by organic molecules that form a crack-arresting elastic buffer. The fibers contain silica spheres between 50 and 200 nm in diameter, assembled together between many organic layers to yield a 100-μm fiber.

The natural fibers grow by self-assembly at the ambient temperatures, compared to the high energy and hard-to-control glass furnaces from which optical fibers are drawn. Low-temperature self assembly would allow engineers to dope glass fibers with precise concentrations of alkaline earth ions such as sodium, calcium, and magnesium in definite layers to selectively control the refractive index of the glass for improved optical transmission performance.

FIG. 26.18 *Micro- and nanostructure of a silica fiber from a glass sponge.*

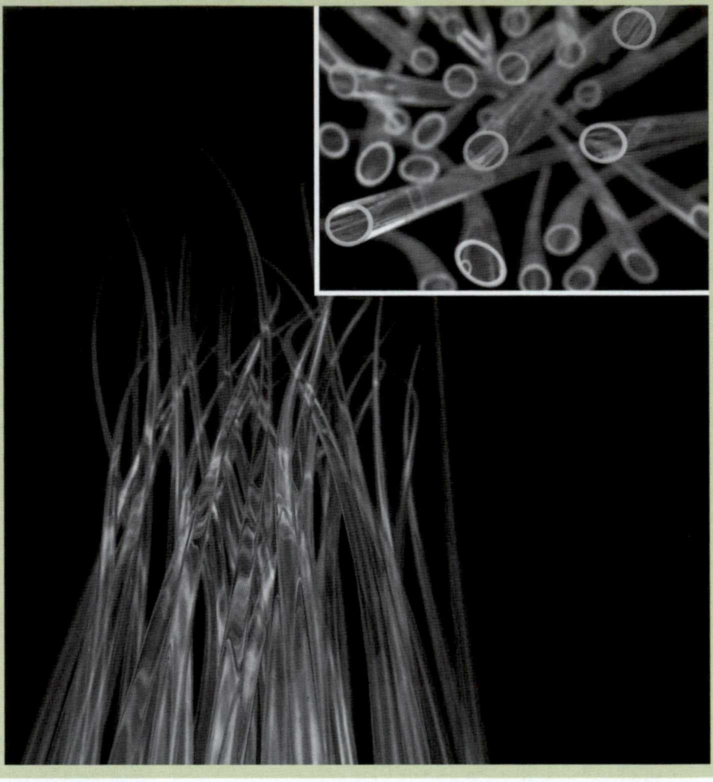

Source: Image courtesy of Jeff Christiansen, Halff Associates, Visual Science and Technology, Dallas, Texas. With permission.

Further biological studies to see how fibers in different sponges are adapted to depth, temperature, and available light could suggest ways to optimize the optics of man-made fibers. The sponge fiber has stimulated thinking about new ways to design and construct optical fibers for communications, and should "shed light on low-temperature, biologically inspired processes that could give rise to better fiber-optical materials and networks" [100].

Biomimetic Silica Nanofabrication Processes. A research group headed by Dr. Dan Morse at the University of California, Santa Barbara, is investigating the biomolecular mechanisms that direct the nanofabrication of silica in living organisms such as sponges, diatoms, and grasses. Mimicking these living organisms could lead to ways to direct the synthesis of photovoltaic and semiconductor nanocrystals of titanium dioxide, gallium oxide, and other materials. Applying natural synthetic methods to materials with which nature has never built structures before is a way to expand the limits of optical, electronic, and mechanical performance. Using natural examples could lead to production of semiconductors and photovoltaic materials in more energy efficient and environmentally benign ways.

Morse summed up the advantages of using natural templates: "Sponges are abundant right here off-shore and they provide a uniquely tractable model system that opens the paths to the discovery of the molecular mechanism that governs biological synthesis from silicon. This sponge produces copious quantities of fiberglass needles made from silicon and oxygen." Silicon, one of the most abundant elements on earth, is the basis for semiconductor technologies for computers, telecommunications devices, fiber optics, and other high-tech applications.

Morse and his group discovered that the silicon fiber of the sponge contains a protein filament that controls the synthesis of the spicules. They cloned and sequenced the DNA of the gene that codes for the protein filament, to determine how it acts as a catalyst to seed growth of silica. This was the first discovery of a protein catalyst that could control the growth of a rock-hard silica biomaterial. The protein actively promotes silicon deposition from low concentrations in sea water while simultaneously serving as a template to guide the nanostructure of the growing glass fiber [101–103].

Morse extended this work to develop synthetic mimics of the natural protein, and show that the same type of catalytic activities can be applied to the nanostructured synthesis of oxides of metals such as titanium and gallium with valuable photovoltaic and semiconductor properties. Low temperature, biomimetic catalytic fabrication of valuable materials used in electronics, photonics, and energy conversion devices could eventually replace high-energy, high-temperature processes that require vacuums, caustic chemicals, hazardous waste products, difficult control regimes, and consequent low yields and lost efficiencies.

Other Photonic Biomimetic Nanomaterials. Sponges and diatoms are not the only natural models for photonic nanomaterials, nor is silicon the only natural element on which such materials are built. In chapter 13, we saw the photonic properties of opals and butterfly wings, and in chapter 25 of this book, we see how the photonic nanostructure of the butterfly wing can be adapted to make chemical nanosensors.

Insect chitins, feathers, diatom shells, and many other natural substances have photonic nanostructures. Reflectors, diffraction gratings, and two- and three-dimensional photonic crystals have been found in nature, including some designs not encountered previously in physics [104–107]. Some optical biomimetic engineering methods make direct analogues of the reflectors and antireflectors found in nature [108]. However, recent nanotechnology ventures beyond merely mimicking in the laboratory what happens in nature, leading to thriving new areas of research.

26.2.5 The Lesson of the Lotus—Nanocontrol of Surfaces

The lotus has been a symbol of purity for thousands of years because its leaves and petals shed contamination. Electron microscopy enabled researchers to resolve the mechanism by which the lotus and other plants and animals repelled water and dirt. It took decades to understand the full details of the effect, through in-depth studies of extreme water repellency (superhydrophobicity), at centers like the Nees Institute for Biodiversity of Plants at the University of Bonn by Boris F. Striffler, Zdenek Cerman, and Wilhelm Barthlott and others [109–110].

The "Lotus Effect" is now understood to be associated with very high contact angles for water on surfaces, based on low-energy hydrophobic surfaces and surface micro- and nanostructures that decrease contact and increase effective hydrophobicity [111–118].

Other Plant and Insect Surfaces. Variations of the lotus effect have been found in many other plants and animals. Superhydrophobicity based on nanofibers has been found in gecko feet, water bugs, caterpillars, spiders, spider silk, wool, feathers, and leaves of many types [119–124]. Nanostructured superhydrophobic surfaces of many aquatic and semiaquatic plants and animals allow them to retain a layer of air under water. Besides keeping them clean and dry, the trapped air layer aids buoyancy, reduces drag, and allows for respiration [125–127].

Control of Surface Interactions by Nanosurfaces. Studies of the shapes of natural surfaces on the nanolevel—*nanosurfaces*—have led to understanding the interactions of liquids with surfaces [128,129]. On rough surfaces, depending on the interplay between surface roughness and surface chemistry, different modes of wetting can occur. Surface wetting phenomena are classified into three categories:

1. Drops suspended on top of hydrophobic roughness features, with air trapped underneath ("fakir" drops). This situation is frequently referred to as "Cassie" or "composite" wetting because the liquid rests on a composite surface composed of solid and air.
2. Wenzel wetting, where drops are impaled on roughness features.
3. Superwetting, where drops are sucked into the surface structure, ultimately spreading to a contact angle of 0°.

The category where drops rest on top of the rough nanofeatures is the Lotus leaf case; there the wetting contact angle is reduced, allowing drops to roll around easily in response to very small forces. When the surface tilts, the drops roll off, tending to take along particles lying on the surface, and leaving the surface dry and clean. The other extreme, superwetting, is found in the surfaces of insects that absorb moisture from the air in desert climates [130].

Lotus Effect Nanotechnology. Active research continues into enhanced biomimetic superhydrophobic and superhydrophilic materials. Studies reveal how nanoscale geometry, chemistry, and topology affect surface behavior [131–134]. Active switching between superhydrophilicity and superhydrophobicity of nanosurfaces can be effected by photoswitching, pH, temperature, and other methods [135–140]. The effect of plasma etching and interactions with surfactants on nanosurfaces has also been studied [141,142]. A range of fabrication methods in different kinds of materials has been developed [143–155].

Self-Cleaning Surfaces. The Lotus Effect (trademarked as "Lotus-Effect") has generated many successful, eco-friendly products such as self-cleaning paints, and fabrics that require less washing, detergents, and dry cleaning than conventional surfaces [156–159].

Transportation and Lighting. Another application is for windows and surfaces in vehicles and buildings—for prevention of icing, lubricating, and slowing corrosion. Automobile makers are investing in lotus research for self-cleaning surfaces and antifogging windows and lights. Compound insect eyes have been mimicked for water repellent, nonreflective, antiglare properties [160–165]. Nanoengineered surfaces mimic the ability of aquatic and semiaquatic plants and animals to retain a layer of air under water that can reduce drag on ships [166].

Biomedical Applications. Surface properties are vital in processes such as blood clotting and cell adhesion. Biomimetic nanosurfaces have been designed as anticlotting surfaces for medical implants, wetting of pharmaceutical and food powders, and providing cell-friendly surfaces for tissue culture. Nanosurface structure is a factor in biocompatibility of materials, and is crucial for attachment of cells to growth scaffolding, implants, and culture vessels [167–170].

Technological Applications. Nanosurface control is employed by chemical analysis, for mixing and laminar flow in microfluidic systems, for high-speed computer disk drive heads, and to protect surfaces of nano-manipulator instrumentation [171–174]. Biomimetic nanosurfaces are being directed to the development of papers and printing techniques for publications, signs, displays, and paper money for both improved wettability and self-cleaning applications [175].

Significance for Soil and Agriculture. Hydrophobicity of nanosurfaces is important for the efficiency of photosynthesis, plant metabolism, and the storage of fruits and vegetables. Research has shown how this effect increases the efficiency of photosynthesis, leaf respiration, and protection from pathogens. It has also been studied for the application of pesticides and herbicides to leaves and insect cuticles [176,177].

One of the most important areas of nanosurface research is revealing factors that affect hydrophobicity of soils and productivity of agriculture, especially with increased droughts and global temperatures. Understanding absorption of water by soil and plants, reducing runoff and surface evaporation, and erosion, reducing irrigation requirements and improving productivity of lands, could be among the most important outgrowths gained from the lotus leaf [178–184].

Technical and Economic Significance. Commercial products modeled on lotus leaf superhydrophobicity continue to grow in importance. The Lotus effect has become one of the best examples of how basic research in biological nanoscience produces practical applications through biomimetics. It shows that biomimetics is not simply a matter of imitating nature but of understanding the principles involved and applying them in an intelligent scientific manner.

Biomimetic Synthesis and Surface Modification of Natural Materials. Using natural substances like insect chitin and wood as starting material, chemists, materials scientists, and nanoengineers are developing new materials with improved properties. In many cases surface properties are modified to create superhydrophobicity or hydrophilicity for applications such as water-resistant fabrics, paints, or coatings. Other new materials are based on synthesis of biomimetic versions of natural materials with properties not found in nature [185–194].

Nanoscale interaction of solid surfaces with layers of molecules in liquids and gases surrounding them has inspired biomimetic materials with unprecedented properties and capabilities. They may even lead to technologies like "Spiderman" suits with advanced camouflage, self-cleaning, and adaptable superadhesive materials [195].

26.2.6 Gecko Glue and Other Biomimetic Nanoadhesives

Nanostructures can prevent adhesion of water and other substances, or nanostructures can create strong adhesion effects, as we saw in the gecko foot. In this section we will review how natural surfaces like the gecko foot provide adhesion without glues, and how those same surfaces can be manipulated to turn adhesion on and off as needed—it all depends on the nanostructure.

Geckos' toes are covered with millions of tiny hairs called setae; each seta branches out into billions of nanoscale spatulae. By studying geckos, spiders, flies, and other animals, researchers have learned how their fibrillar, self-cleaning feet can control adhesion by conforming to the nanoscale contour of surfaces, without using glue or leaving residues following detachment. This form of adhesion is due to forces that become important at the nanoscale; when van der Waals and capillary forces act at millions of gecko hairs in close contact to a surface, they add up to a bond that is a thousand times stronger than the force geckos need to hang onto a wall. The gecko can adhere strongly without having to press down on the surface, and it is able to detach and reattach rapidly without having to exert force [196,197].

Biomimetic Gecko Adhesives. In 2002, an interdisciplinary research team found that the network of gecko setae forms intermolecular bonds with surfaces by means of van der Waals forces [198]. The team later experimentally demonstrated the mechanism for gecko adhesion with synthesized biomimetic gecko hair tips. Researchers have continued developing artificial biomimetic adhesives based on the gecko effect, and we give some examples of their work here.

Gecko Adhesive Using Hard Polymer Microfibers. Researchers at the University of California, Berkeley, have developed adhesive gecko foot surfaces for use with climbing robots. They fabricated patches of microfiber arrays with 42 million polypropylene microfibers per cm^2. The patches can support $9\,N \cdot cm^{-2}$, with preloading of just $0.1\,N \cdot cm^{-2}$—a patch $2\,cm^2$ can support a 400 g load. Like the natural gecko foot, the Berkeley patches do not adhere when pressed down, but only when they slide over a surface, producing a shear force. The nanofiber adhesive surface is "smart" in that greater load increases the contact area and causes more fibers to engage, leading to greater adhesion strength [199,200].

Professor Kellar Autumn of Lewis and Clark College (Oregon) has identified seven key benchmark properties for gecko adhesives [201]. The Berkeley polypropylene gecko adhesive was able to demonstrate five of these seven properties in a single material:

1. Anisotropic directional attachment—the microfibers do not attach by being pressed into the surface, and instead require a sliding motion parallel to the surface for the fibers to bend and attach.

2. High pulloff force to preload ratio—a preload of less than 0.1 N is sufficient to engage the fibers, and after the preload is removed, the patch can sustain a shear load of 4 N.
3. Low detachment force—the polypropylene gecko adhesive is directional and shows adhesion in the direction parallel to the surface, and the patch can be easily removed with a force of less than 0.001 N.
4. Anti-self-matting—the polypropylene fibers do not stick to one another, and do not clump even with 100s of loading/unloading cycles.
5. Nonsticky default state—the polymer used in the synthetic adhesive, polypropylene, is almost as hard as the keratin used by natural gecko, and the surface of the patch is nonsticky.

Achievement of Keller Autumn's gecko properties six and seven continues to be the goal of further research efforts:

6. Topography independence (sticking to rough surfaces) and material independence (adhesion by van der Waals rather than chemical forces)
7. Self-cleaning

Eventually researchers hope to build arrays incorporating the necessary features to approach or surpass the adhesion achieved by geckos, which is about $10\,N \cdot cm^{-2}$.

Effects of Geometry and Materials on Adhesive Fiber Arrays. Low detachment force, self-cleaning, and nonsticky default states are best provided by hard polymers, rather than the soft polymers typically used in pressure sensitive adhesives. Researchers at Carnegie Mellon University (CMU), using the gecko and spider as models, have developed microfibers made of polyurethane. The adhesion strength is dependent on the material properties as well as the tip shape and fiber size [202–204]

Theoretical and experimental work on the adhesive mechanisms of geckos, flies, spiders, and other animals has also been actively pursued at the Max Planck Institute in Germany. Researchers there have produced the first artificial gecko adhesives with some degree of self-cleaning ability [205,206]. The work at Berkeley and elsewhere shows that hard plastic microfibers adhere when bent over by sliding forces at the surface, but are not sticky when pressed down normal to the surface. The adhesive force increases with sliding distance, which provides a natural automatic braking effect as the gecko runs over the surface. The Berkeley group found that their synthetic patches became stronger the more they were used [207].

Polymer and Carbon Nanotube Composite Tape. Researchers at the Rensselaer Polytechnic Institute and the University of Akron created biomimetic "gecko tape" using polymer surfaces covered with carbon nanotube hairs, which can stick and unstick on a variety of surfaces, including Teflon. By imitating the nanopatterned microtubes on the gecko foot, and varying the shape, pattern, and compliance of the nanofibers, the team fabricated adhesive tapes from carbon nanotubes with high performance [208].

Geckel Nanoadhesives—Combining Wet and Dry Adhesive Strategies. Dr. Philip Messersmith and other researchers at Northwestern University have merged two opposite adhesion strategies found in nature, combining gecko-inspired dry adhesive spatulae with mussel-inspired glue to obtain a nanostructured surface that works better than either alone. The researchers call the hybrid material a geckel nanoadhesive.

Messersmith has done research sponsored by the NIH/National Institute of Dental and Craniofacial Research on the underwater adhesion of mussels. He coated each synthetic gecko-inspired microfiber with a synthetic adhesive protein inspired by the mussel. This double biomimicry opens a potentially superior route to the design of temporary adhesive materials.

The unusual protein compound used to coat the artificial setae mimics the reversible bonding action of a mussel adhesive protein that Messersmith's group previously isolated and studied over several years. The wet adhesive force of each pillar increased nearly 15 times when coated with the mussel mimetic. The dry adhesive force of the pillars also improved when coated with the compound [209].

Gecko Adhesive for Medical Applications. Dr. Robert Langer at the Harvard-MIT Division of Health Sciences and Technology, and Jeff Karp at MIT and Brigham and Women's Hospital and Harvard Medical School, used gecko nano-adhesives coated with a thin layer of glue to help bandages stick in wet environments, such as heart and lung tissue. Their group included team members working at MIT, Draper Laboratory, Massachusetts General Hospital, and the University of Basel, Switzerland, fabricating materials and performing animal experiments.

The tape is designed for medical and surgery applications. The material can be biodegradable, to dissolve over time without having to be removed. The tape can fold or roll up to be inserted through a small incision, and then unfurled for application in minimally invasive or natural orifice transluminal surgery procedures where suturing is particularly difficult. There are significant design challenges for adhesives used in medical applications. For use in the body, adhesives must hold fast in a wet environment and be constructed of biocompatible materials: they must not cause inflammation; they must be biodegradable, meaning they decompose over time without producing toxins; and they must be elastic, to conform and stretch with tissue. To meet these requirements, the MIT team developed a new "biorubber" polymer. Pillars and holes were etched using micropatterning technology adapted from computer chip production tools to create different hill and valley profiles at nanoscale dimensions. The stickiest profile determined by tests on pigs was one with pillars spaced just wide enough to grip and interlock with the underlying tissue.

A very thin layer of sugar-based glue was coated onto the profile, which created a strong bond even on a wet surface. In this application, low lift-off force is not required, since the biodegradable bandages are left in place. The effects of patterning with and without glue were tested on intestinal tissue from pigs and in living rats. Nanopatterned adhesive bonds were twice as strong as unpatterned adhesives, and coating nanopatterned adhesive with glue increased adhesive strength more than 100%.

The biorubber can be infused with drugs which are released as the polymer degrades. The elasticity and degradation rate of the biorubber are tunable, as is the pillared landscape, so that the adhesives can be customized to have the right elasticity, resilience, and grip for different medical applications. Despite the differences between the biorubber medical adhesive and the gecko foot, Karp said that his team was inspired by the gecko to create a patterned interface to enhance the surface area of contact and thus the overall strength of adhesion. Nanostructures inspired by nature can be adapted to create new types of materials to fit the complex needs of new applications [210].

There are many other research groups around the world developing adhesive systems based on the gecko foot model, using many different materials and biomimetic designs. Novel gecko-inspired dry adhesives have been produced by a silicon dioxide based MEMS process at the University of California, Santa Barbara [211]. A nanorod dry adhesive has been fabricated at the University of Manchester in the United Kingdom [212]. We leave as a useful exercise for the student the project of coming up with other novel designs and comparing them with those in the literature.

Biomimetic Tribology—Coming Unglued with Nanoscale Surface Control. Precise control of surfaces at the nanoscale can produce repellent, absorbent, adhesive, and lubricating effects. Study of natural systems has revealed that friction between surfaces can be viewed as a continuous phenomenon ranging from adhesion at one end to lubrication at the opposite extreme. Thus tribology—the science of lubrication and friction—can be a beneficiary of biomimetics. Biotribologists gather information about biological surfaces in relative motion, their friction, adhesion, lubrication, and wear, and apply this knowledge to technological innovation. Biological systems excel at optimizing interactions between matter at the micro- and nanometer scale. The miniaturization of devices such as hard-disk drives, MEMS, and biosensors increases the opportunities for applications of nanoscale tribological phenomena [213–219].

Living organisms have many tribological challenges on the nanoscale. Moving surfaces contact each other in joints, muscles, and blood vessels. Examples of moving micromechanical systems can be found in diatoms with hinges and interlocking devices on the micrometer-scale and below. The immune system produces molecules that can switch states from lubricating to adhesive; these glycoproteins control the movement of white blood cells as they move between endothelial cells and adhere to foreign particles. As we saw earlier, protein macromolecules play key roles in absorbing stress in strong self-healing adhesives in bone and shells.

Wear-Resistant Joints and Self-Healing Adhesives in Diatoms. Diatoms are a type of algae with nanostructured amorphous silica surfaces (see chapter 13). Diatoms have evolved interconnected junctions and self-healing adhesives that prevent wear between rigid surfaces in relative motion. Their silica cell coverings grow efficiently at ambient conditions to produce an amazing variety of sizes, shapes, and nanostructured patterns, many of which exhibit photonic resonance with visible light [220,221].

Diatoms serve as model organisms for nanotribological biomimetics and as templates for novel three-dimensional microelectromechanical systems. Some diatom species have evolved strong, self-healing underwater adhesives; others have elastic linkages between the halves of their silica coverings. Some species secrete viscous mucilage which binds colonies together while protecting the silica shells from wear as they rub against each other. Hinges and interlocking devices in diatoms are very stable and can be seen under the microscope preserved intact in fossil deposits that are millions of years old. The lubrication mechanisms used by diatoms have been studied as models for nanolubricants to reduce stiction for MEMS [222].

Some diatoms are free floating, but others have evolved adhesives for stable and strong attachment in water. Some diatom species found in Antarctic seas synthesize special proteins that bind to ice, and also prevent recrystallization of water. These may be templates for biomimetic cryoprotective substances [223]. Diatom adhesives show opportunities to tailor new synthetic adhesives, lubricants, and protective coatings for specific applications such as de-icing.

Immuno-adhesives as Models for Biomimetic Switchable Adhesives. As we saw in the previous chapter (chapter 25, this book), the body's immune system includes white blood cells which flow through vessels and enter the extracellular matrix in tissues, sticking to and engulfing foreign particles. The immune system employs exquisitely sophisticated adhesives to deploy antigens to their targets, including the adhesive portions of the specific antibody macromolecules targeted to bind to recognized antigens.

These adhesives are highly selective—they use molecular and nanopattern templates to distinguish between self and nonself, allowing free flow in the first case and adhering tightly in the other. Their adhesive function can also be turned on by signaling substances produced by parts of the immune system that detect invaders and initiate the process of inflammation. Understanding how these nanoadhesives work is an important goal for biomimetic nanoengineering, both for the development of drugs and the design of artificial switchable adhesives for technological applications [224]. In addition, this is an important research area for medicine, for the prevention and treatment of immune-related disorders, the management of infection and healing, and in pathology and biotechnology applications.

Adhesive Processes in the Immune Response. Before we discuss adaptive adhesion molecules and how they function in the immune system, let us review the steps that take place in the nonspecific immune response—the *inflammatory response*. The inflammatory response begins when tissue damage or metabolic disruption triggers mast cells to release histamines stored in granules into the neighboring tissue. The histamines stimulate dilation of capillaries, allowing plasma and leukocytes to penetrate into surrounding tissue (The characteristic swelling and redness results). Damaged or infected cells release signaling molecules which attract phagocytes, which adhere to and engulf dead cells, bacteria, and foreign particles. At the same time, activation of antibodies or other triggers may initiate a cascade of complement proteins which attract phagocytes and penetrate the membranes of damaged cells, leading to lysis and acceleration of the inflammation process. The tissue returns to its normal state as histamine and complement protein release ceases [225].

Rolling and Sticking Behavior of White Blood Cells. When the inflammatory response is observed through the microscope, granulocytes and other leukocytes can be seen to adhere to and release from capillaries and small venules in the inflamed area. Leukocytes are observed to roll slowly along the endothelium, tethering and detaching as they move along in the direction of blood flow. The rolling velocity is typically 10–100 times lower than a nonadherent white blood cell moving next to the vessel wall. Adhesion molecules on the white blood cells and the endothelium regulate their interactions. The molecular mechanism underlying this leukocyte-endothelial interaction is of great interest for understanding switchable and adaptive adhesives [226–228]. The adaptive adhesion of white blood cells onto the endothelium takes place in a cascade of stepwise events [229]. These are classified as:

1. Initial tethering
2. Rolling adhesion (involving probing for signs of inflammation)
3. Firm adhesion
4. Escape from blood vessels into tissue

The Molecular Mechanisms for Adaptive Adhesion. It is known that phagocyte adhesion is mediated by specific biological macromolecules, which include cell membrane receptors, extracellular ligands, and cytoskeletal components (see chapter 14). Interdisciplinary researchers have analyzed adhesion between cells to gain an understanding of the underlying mechanical and molecular properties that govern the processes. Progress has been made towards modeling biochemical and biophysical cellular nanoadhesive processes in terms of quantitative parameters which could be useful in guiding the biomimetic design of tailored synthetic adhesives and lubricants [230,231].

Switching the Permeability of the Endothelium. One question addressed by this research is how white blood cells interact with the endothelium—the thin layer of cells that line the interior walls of blood vessels. Normally, the endothelium must resist interactions with blood cells, to allow the free flow of blood; but white blood cells may stop at particular sites and pass through the endothelium into the underlying subendothelial matrix—the layers of smooth muscle cells, structural proteins, and fibroblasts that make up the blood vessel wall.

White blood cells may pass through the matrix into surrounding tissue when cells of the endothelium are signaled to relax and open; this retraction of the blood vessel barrier may be induced by signaling compounds in the arachidonic acid pathway, which are produced by platelets as part of the clotting process. This signaling mechanism is important to understand how white blood cells contribute to clotting and build-up of plaques in the arteries, and how certain cancer cells migrate through the blood vessel walls during metastasis [232,233].

Separate Molecular Adhesive Mechanisms for Rolling and Migration. Researchers have studied white blood cell adhesion to endothelial cells under well-defined flow conditions in a variety of systems. They have been able to distinguish separate mechanisms for the initial adhesion and rolling steps and the firm adhesion and migration of white blood cells through the endothelium [234,235].

Selectin. Initial adhesion and rolling appear to be controlled by interactions between *selectin* and *carbohydrates* in the cell membranes. Selectins are a special type of lectins, the glycoproteins that bind sugar polymers.

Integrins. The firm adhesion and white blood cell migration steps are mediated primarily by interactions between *integrins* and *peptides* at inflammation sites. Integrins are integral membrane proteins that bind cells to the intracellular matrix and other cells, as well as acting as signal receptors.

Signaling and Switching of Integrins. Integrins play important roles in the cell membranes of all animals, and especially in white blood cells. Unlike most transmembrane proteins that form one-way cell membrane receptors, integrins can carry signals in both directions from outside and inside the cell. In the case of inflammation, signals from other receptors on the white blood cell are transmitted to its integrin transmembrane sites. These signals induce the extracellular domains of these sites to undergo conformational movements (changes in their molecular arrangement) that enable calcium ligand binding—switching from a nonadhesive to an adhesive state. In this state integrins rapidly stabilize contacts between white blood cells in the bloodstream and the endothelium at sites of inflammation [236,237].

Adaptive Adhesion and Molecular Properties. Integrins are the most sophisticated adhesion molecules known, and as such have aroused much interest for nanobiomimicry as well as for their medical importance. Through conformational changes, integrins can mediate both firm and transient types of adhesion. In order to design adaptive nanoadhesives based on integrins, we must first understand the functional properties of these molecules that control the dynamics of adhesion. There is evidence that adhesion depends on physical chemistry in addition to mechanical features such as deformability, morphology, or signaling conformations [238,239]. Possible physicochemical properties that affect dynamic states of adhesion are reaction rate, affinity, mechanical elasticity, kinetic response to stress, and molecular size and length.

Adhesion Dependence on Applied Forces. The activity of adhesion molecules may also be regulated by the force distribution present in blood vessels. The specifics of molecular adhesion can vary with the local wall shear stress—the force required to produce a certain rate of flow of a viscous liquid. Levels of venous and arterial shear stresses range between 0.1–0.5 Pa and 0.6–4 Pa, respectively. Adhesion of leukocytes to blood vessel walls may depend on shear force in a manner not unlike the adhesion of gecko fibrils to surfaces [240,241].

Integrins as a Biomimetic Model for Nanotools. The molecular properties that enable integrins to switch their transient adhesion states are of interest for their medical implications [242–246] as well as for technological applications, such as nano-grippers and sensors [247]. The complex synergistic interactions of the large molecules involved in cellular adhesion provide models for biomimetic nanoscale adhesion and manipulation [248–250]. The molecular conformations and forces involved in cellular adhesion have been investigated experimentally as well as theoretically [251–256]. This active and ongoing research provides models and inspirations for sophisticated biomimetic nanobiotechnology that utilizes the full potential of nanoscale mechanical, physical, and chemical forces in an integrated fashion. For example, cellular adhesion molecules have been adapted as the active selective elements for field flow fractionation for a flow cytometry device, giving highly selective cell separation which is aimed at identification and diagnosis of blood disorders [257].

Adaptive Protein Grippers as Antiparasite Weaponry. An example of an adaptive gripper from the immune system as a model for biomimetic drugs comes from the mosquito. An international team discovered this molecular gripper mechanism in research for treatment and prevention of malaria. They unraveled the mechanism for natural resistance shown by some mosquitoes to malarial parasites. This resistance was known to be associated with thioester proteins (TEPs) in the innate immune system of insects, which have some similarities with the complement factor glycoproteins of the mammalian immune system.

TEP1 proteins are activated by a biochemical reaction triggered by the presence of malarial parasites in the mosquito. The reaction cleaves a thioester bond, opening up the protein into an active state that locks covalently onto the parasite's surface, targeting it for elimination. The protein acts in a very specific way as an adaptive adhesive, similar to those we saw earlier in the targeted immune systems associated with white blood cells in vertebrates.

The team used x-ray crystallography to determine TEP1's three-dimensional structure. They found that the genetic differences between mosquitoes that are resistant and those that are susceptible to the parasite mostly manifest in a region of the TEP1 protein they called "the warhead," the portion that grabs the malarial parasite. The senior researcher of the group is Dr. Johann Deisenhofer, who was awarded the 1988 Nobel Prize in chemistry for using x-ray crystallography to describe the structure of a protein involved in photosynthesis.

Another lead researcher in the group is postdoctoral fellow, Dr. Richard Baxter, who described the role of TEP1 as a scout that finds the enemy, then plants a homing signal, and calls in the air strike. Other members of the research team were at UT Southwestern and the Institut de Biologie Moléculaire et Cellulaire in Strasbourg, France. The French group previously determined that the gene for TEP1 occurs in two forms, or alleles. One is found in mosquitoes that are resistant to malarial infection. Detailed analysis of the x-ray crystallography structure showed that the differences cluster around the warhead area, reinforcing the theory that it is a key element of the binding to the malaria parasites, and suggested a gripping type of mechanism similar to that of complement proteins in the immune system of higher organisms (**Fig. 26.19**). Understanding how mosquitoes fight off malarial parasites with macromolecular protein weapons could provide a model for effective drug and control strategies [258].

Tough Underwater Adhesives. In an earlier section we discussed the nanostructure of shell, with emphasis on the mineral platelets and how they were linked by elastic glue. In the immune system we have seen the complex macromolecular mechanisms involved in natural adhesion. In the shell these types of molecules are used to hold mineral building blocks, and in the immune system they are adapted to bind to foreign bodies. The macromolecular glues in shells are adapted for toughness—to allow the shell to absorb and redistribute shock forces. This is a very useful property for engineering, if its mechanism can be understood and imitated [259].

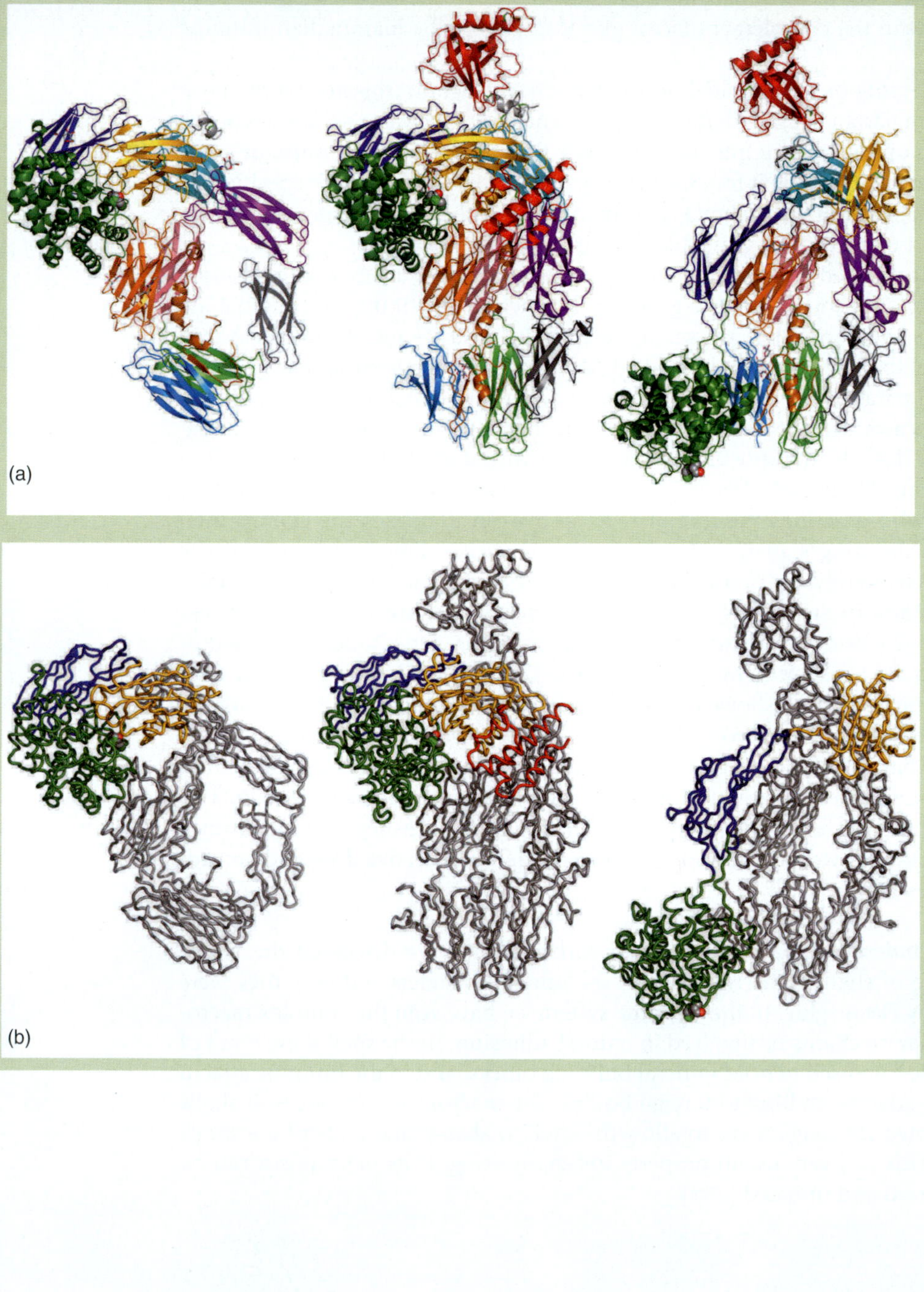

FIG. 26.19 *(a) Thioester protein structures: TEP and human immune complement factor C3, left to right: TEP1r, human C3 inactive state, and human C3b activated state with the reaction center exposed. The thioester is shown as spheres and the remaining protein domains as ribbon cartoons. The thioester-containing domain (TED) is in green, the complement C1r/C1s, Uegf, Bmp1 domain (CUB) is in navy blue, and the macroglobulin-8 domain (MG8) is in yellow. The anaphylotoxin domain (ANA), present in complement factor but not TEP1, is in red. (b) Thioester proteins as grippers: The thioesters and the domains that are active in gripping the antigens are shown in color for clarity, with the remaining parts of the protein shown as grey tubing.*

Fig. 26.19 (Contd.) *(c) Thioester protein genetics: a portion of the sequence alignment for alleles of the TEP1 gene from the two laboratory strains of mosquito that are susceptible (S) and refractory (R) to infection by the malarial parasite* Plasmodium berghei. *Differences between the two alleles are concentrated in the thioester domain and surroundings, and on the right of the figure the pink spheres locating these differences illustrate that they cluster in space about the domains most important for the proteins' function.*

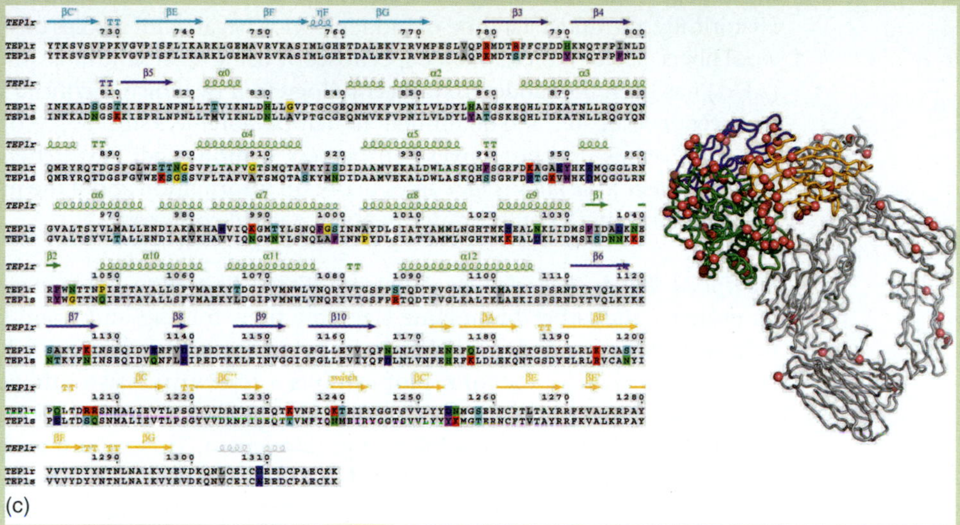

Source: Images courtesy of Dr. Richard Baxter and Professor Johann Deisenhofer, UT Southwestern Medical Center at Dallas. Images produced with PyMOL (http://www.pymol.org) for the proteins and ESPript (http://espript.ibcp.fr) for the sequence alignment.

The binding proteins in shells and strong underwater glues produced by mollusks and diatoms share two important features in their nanoscale structure and mechanism of reacting to stress. These are *self-healing* and *sacrificial bonds*.

Self-Healing Adhesives. The natural organic adhesive in nanocomposite nacre is composed of large, long chain molecules with sacrificial bonds and hidden lengths (unfoldable modules). Sacrificial bonds are weak bonds (such as Coulomb, van der Waals, or hydrogen bonds) between molecular segments. The sacrificial bonds allow the molecule to be reversibly stretched by their breaking and rebonding, dissipating large amounts of energy, and thereby preventing the covalent bonds making up the backbone of the molecular chain from breaking.

Sacrificial Bond Energetics on the Nanoscale Makes Natural Adhesives Self-Healing. Sacrificial bonds and hidden length in the folded conformations of long-chain natural adhesives provide an energy-absorbing mechanism that protects the molecule from being torn apart. This reversible, molecular-scale energy-dissipation greatly increases the fracture toughness of biomaterials. The sacrificial bonds themselves are relatively weak (van der Waals, electrostatic, Coulomb, hydrogen bonds) compared to the covalent bonds linking the backbone of the polymer chain. But breaking the multiple sacrificial bonds also involves the energy needed to reduce entropy and increase enthalpy as molecular segments are stretched after being released from their folded attachments. This energy is on the order of 100 *e*V—large compared to the energy needed to break the polymer backbone, on the order of a few electron volts. In many but not all cases, the breaking of sacrificial bonds is reversible, adding a "self-healing" property to the material [260].

Identifying and Measuring Sacrificial Bonds with Nanotechnology Tools. Historically, the understanding of hidden bonds began with studies of stretching wool fibers. Single-molecule force spectroscopy using an atomic force microscope (AFM) has been the modern nanotechnology tool by which sacrificial bonding has been investigated and quantified. In natural polymers such as glycoproteins, the force versus distance curves can be very complicated. Researchers closely analyze *atomic force microscopy* curves to obtain information about the molecules and their bonding to other components of natural composites.

Sacrificial Bonds Revealed by Sawtooth Force Patterns. AFM investigation of proteins and other biopolymers reveals their folding and bond-breaking behavior under applied stress. When individual strands of polymer fibers are stretched using AFM, the force plot exhibits a repetitive "sawtooth" response, rising and falling as the material is extended. Each tooth in the saw represents an unfolding of the long, complex molecular chain. This pattern is characteristic of material with sacrificial bonds connecting modular structures. This provides direct evidence of the effects of the hydrogen bonds, ionic attractions, covalent bonds, and steric rearrangements into less favorable energetic conformations. The ability of the bonds to re-form when the force is relaxed can be determined by repeating the stretch cycle. The force plots show the action mechanism of the multimodular nanostructure that gives natural adhesives self-healing behavior.

Researchers have observed this sacrificial bond behavior in protein-based fibers from many sources including wool and adhesives from mussels, diatoms, and algae. Dugdale and coworkers at the University of Melbourne, Australia, showed that single adhesive nanofibers produced by certain diatoms to attach to surfaces have the signature fingerprint of modular proteins: their force–extension curves have regular sawtooth patterns [261].

Amyloid Fibrils in Natural Adhesives. Researchers Anika Mostaert and Suzanne Jarvis at Trinity College Dublin studied AFM curves of algal adhesives and found highly ordered amyloid structures in the fibrils. Amyloids are normally associated with neurodegenerative diseases such as Alzheimer's where they are a primary component of plaques in brain tissue.

But the amyloid sheets found by the Dublin researchers provide sacrificial bonds in natural adhesives. They found evidence for similar material in parasitic worms. By exploring the differences between types of amyloid proteins found in different organisms, they seek to find potential templates for the design of biomimetic adhesives. Their work may also shed light on the origins, causes, and treatments for neurodegenerative diseases [262–264].

Future Opportunities for Biomimetic Nanocomposites. Synthetic adhesives and lubricants still have a long way to go to match the performance of many natural systems. Using the lessons learned from nature and armed with the tools of nanotechnology, biotribologists and other biomimeticists draw upon natural designs that have been optimized for millions of years to realize adaptive, self-healing and environmentally friendly lubricants and adhesives.

In this section we have looked at surfaces and their interactions. Next we add one more dimension to our exploration of biomimetic materials: biological membranes.

26.2.7 Biomimetic Membranes and Nanocapsules

Through the cell membrane with its gates and chemically selective lipid and protein domains, the cell senses and communicates via chemical signals with its complex biochemical matrix. We will see some biomimetic ways of making artificial lipid bilayers for practical applications and some of the artificial membrane structures that have been made to mimic useful functions such as selective permeability and molecular sensing.

Biomimetic Membranes. The bilipid membrane of cells has served as a biomimetic model for decades. Engineers duplicate the function and structure of natural membranes in more durable and versatile materials to control molecular transport and fabricate sensitive and selective sensors. There are a number of ways to create biomimetic membranes. Lipid vesicles (artificial liposomes) can be formed by agitation of phospholipids in water, using sonication (high-frequency sound waves) or rapid mechanical mixing. Planar-supported bilayers can also be formed from phospholipids and other bipolar macromolecules. The classic method was developed by Langmuir and Blodget in their studies of the surface forces of polar organic molecules in water. (**Fig. 26.20**).

There are many other ways of producing membranes with nanoscale organization. As we saw in section 26.2.2, membranes of proteins and inorganic crystals

Fig. 26.20 *A Langmuir–Blodget instrument for formation of lipid bilayers from thin layers of lipids on the surface of water. (a) Macromolecular monolayer on liquid surface is entrained as support is withdrawn, (b) coated wafer is withdrawn from apparatus.*

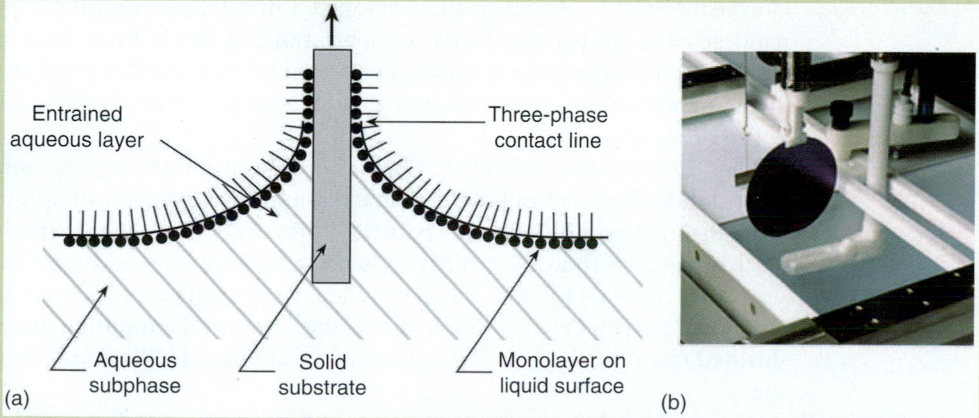

Source: Image courtesy of KSV Instruments SA. With permission.

are being formed in biomimetic processes to make analogues of shell and other biocomposites.

Artificial Membrane Mazes. Artificial lipid bilayers are very useful for investigating material transport and signaling mechanisms in natural cells, but making practical sensors requires a more durable material with long-term stability in many types of environment. Synthetic membranes have been fabricated using nanopores in aluminum, glass, silicon, polyester, and polycarbonate polymers. A simple way of producing a material with extremely small pores is by repeated drawing, bundling, and re-stretching of glass capillaries containing an etchable core, then removing the cores. Such artificial membranes, however formed, can be used to mimic some of the material transport functions of the natural cell membrane.

Forests of vertically aligned carbon nanofibers can control molecular transport in a manner analogous to natural cell membranes. Dense arrays of carbon nanofibers are used as membranes integrated within fluidic structures. Size-dependent transport, perpendicular to the orientation of the fibers, can be controlled based on the wall-to-wall spacing of the individual fibers. When biomolecules, such as DNA and proteins, are attached to the sides of the nanofibers, such membranes can function as sensors, using the resulting combination of size fractionation and chemical specificity to select analytes. Integrating electrically addressable fibers adds another dimension to controlled transport as well as electrochemically based sensing. The biomimetic modification of carbon nanofiber structures is useful for performing chemical separations and for mimicking the properties of natural membranes. [265–270]

Mimicking Natural Encapsulation. We have progressively covered nanostructures and how they interact with forces, optical materials and light, nanoengineered surfaces and their interaction with liquids and nanoparticles, and membranes and their interactions with molecules. Now we will look at a special case of membranes: nanoencapsulation.

Drug Delivery by Nano-Dumplings. Biomimicry has been used to inspire encapsulation in self-assembled structures similar to liposomes, globular proteins, and other nanostructures in cells. A team led by Dr. Karen Wooley at Washington University Saint Louis, Missouri, developed a novel class of synthetic polymer nanocapsules, similar to globular proteins, ranging in size from 10 to 100 nm. Dr. Tomasz Kowalewski, a member of the team who used AFM to obtain images of the nanoparticles, suggested that they be called "knedels"—after a type of Polish dumpling.

Shell-crosslinked knedels (or SCKs) are spheres in which a hydrophobic core is surrounded by a hydrophilic shell that resembles the structure of lipoproteins. With the right conditions, the particles form by self-assembly: amphiphilic block copolymers (polymer chains with separate water-soluble and water-insoluble segments) nucleate to form a micelle made up of from tens to several hundred individual polymer chains. The hydrophobic cores of the resulting micelles are shielded from the surrounding aqueous environment by the solubilized outer shell.

These "nano-dumplings" can be made in different sizes and chemical compositions for drug delivery, gene therapy, and immunology. Their stability, release rates of encapsulated contents, and attraction to different types of external molecules can be manipulated by varying the polymers in the shells. Possible applications range from removing hydrophobic contaminants from aqueous solutions to use as recording materials [271].

Nanoscale Bioreactors. In a demonstration of a multifunctional, smart nanoscale drug delivery system, researchers at the University of Basel have created a drug-loaded nanocontainer that targets specific cells and releases its payload in response to a specific physiological signal. Smart nanocontainers could become anticancer drug delivery vehicles that target tumors and release their contents only when they receive a tumor-specific biochemical signal [272].

Bacteria and Viruses as Templates for Nanocapsules. Diatoms, bacterial cell walls, and other natural capsules are used as templates for nanoencapsulation—these natural structures can be adapted for active drug delivery [273]. This biomimetic technology takes the natural mechanisms used by microbes to invade the body and adapts them for research and medical purposes. For example, the dendrimer plug from a bioengineered virus was adapted as a drug delivery capsule. Researchers developed viral coat nanoparticles that incorporate receptors in their outer shells acting on biological effectors inside cells. The receptor and effector together act to detect a specific biochemical signal that then affects the nanocontainer and its contents. The effects can include drug release or the generation of a diagnostic signal [274].

The viral nanocontainers are loaded with an enzyme that converts substrate molecules into a light-emitting fluorescent form. The substrate molecules are specifically transported by bacterial membrane pore proteins engineered into the viral nanocontainer, where they react to produce light that can be seen using fluorescence microscopy, demonstrating that the delivery capsule is working. This method could insert an enzyme that converts an inactive drug into its active form for release only inside a diseased cell.

26.2.8 Some Other Biomimetic Materials

Nature is rich in examples of nanostructured materials. There are so many new biomimetic nanotechnologies from which we can only select a few examples in this chapter. The reader will want to pursue further reading about many other interesting and informative examples of biomimetic nanotechnology. We close this section with a few more interesting examples.

Biomimetic Water Harvesting. Certain plants and insects possess the ability to capture water from fog. The desert cockroach (*Arenivaga investigata*) can harvest water from apparently dry air [130]. The Namibian desert dwelling beetle Stenocara has bumps on its wing scales with superhydrophilic nanostructured surfaces. The peaks of the bumps are glassy-smooth and hydrophilic; the slopes of the bumps and troughs in between are covered with hydrophobic wax. As droplets accumulate in size, they roll from the tops into the waxy channels to a spot on the beetle's back that supplies its mouth [275].

QinetiQ, Ltd. of the United Kingdom developed sheets that capture water vapors from cooling towers and industrial condensers based on the beetle wing design. They capture 10 times more water than the preexisting ethno-biomimetic technology—the fog catching nets used by inhabitants of remote mountain towns in Peru and Chile (inspired by the capture of fog by cloud forest plants). The sheets are being tested for water collection on tents and roofs in arid regions.

Biomimetic Food and Nutrition Science. In the introduction to this chapter, we raised questions about how apples store fluids. The nanostructured gels in which nutrients are stored and protected in fruits, as well as how nanoencapsulation affects biological activity of their antioxidative macromolecules, are studied by agricultural scientists and nutritionists [276,277]. Food scientists are exploring the nanoscale structure of fruits and how they are affected by freezing and drying to seek improved hardiness and preservation of freshness [278].

Biomimetic lessons also come from feathers, hair, wool, skin, claws, beaks, eggshells, teeth, corals, sponges, crustacean and insect chitins, spider and silkworm silks, insect eyes, pollens, seeds, grass stems, wood, celluloses from plants and bacteria, bacterial cell walls, and many others [279,280].

26.3 Biomimetic Nanoengineering

Up to now we have looked at biomimetic materials with interesting properties due to their nanostructures. Now we will look at more active nanoscale devices. There is a wealth of natural sources from which to draw inspiration for nanoscale devices. These include muscle that exerts force, viral mimetic batteries for high power density, photosynthetics to capture and store light energy in chemical form, rotors for cilia, gates for ionic transport, and DNA itself, whose storage and processing of information has been adapted in biomimetic molecular computing engines.

26.3.1 *Artificial Muscles*

Muscle is a natural engine that uses chemical energy to contract nanostructures, thereby exerting force which animals use for movement and mechanical work. Muscle generates macroscale forces by a hierarchical organization of millions of nano-actuators acting in parallel. Muscle produces undulatory motion (as in a slug or mollusk foot) or is harnessed to shells and bones to form levers for more complex movements.

Nanostructure of Natural Muscle. Muscle exhibits a hierarchical organization when viewed microscopically at progressively greater magnification. Typical muscle is made up of fiber bundles connected to bony joints via ligaments and tendons. The active portion of the muscle is composed of fibers called fascicles. Each fascicle is composed of smaller fibers, consisting of individual cells. Each cell contains bundles of myofibrils, the active engine of muscle—bundles of contractile protein filaments aligned in parallel with the ends overlapping.

The filaments are composed of two types of protein: actin and myosin. When seen at high magnification, the overlapping portions of actin and myosin chains form visible bands, or striations. The overlapping nanostructures making up the striations are called sacromeres. A sacromere consists of a parallel bundle of

Fig. 26.21 *Schematic diagram of muscle structure.*

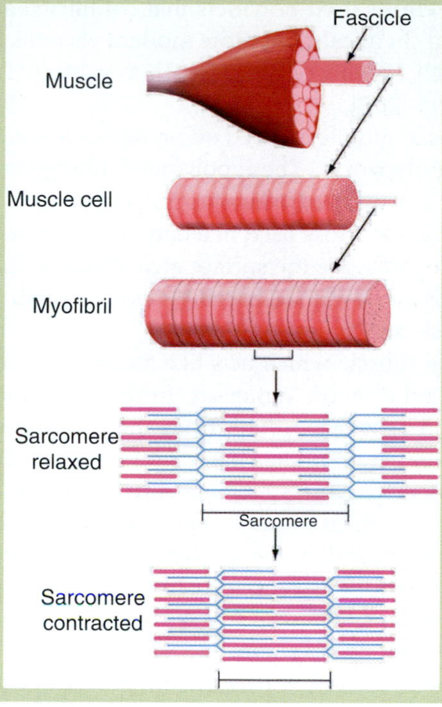

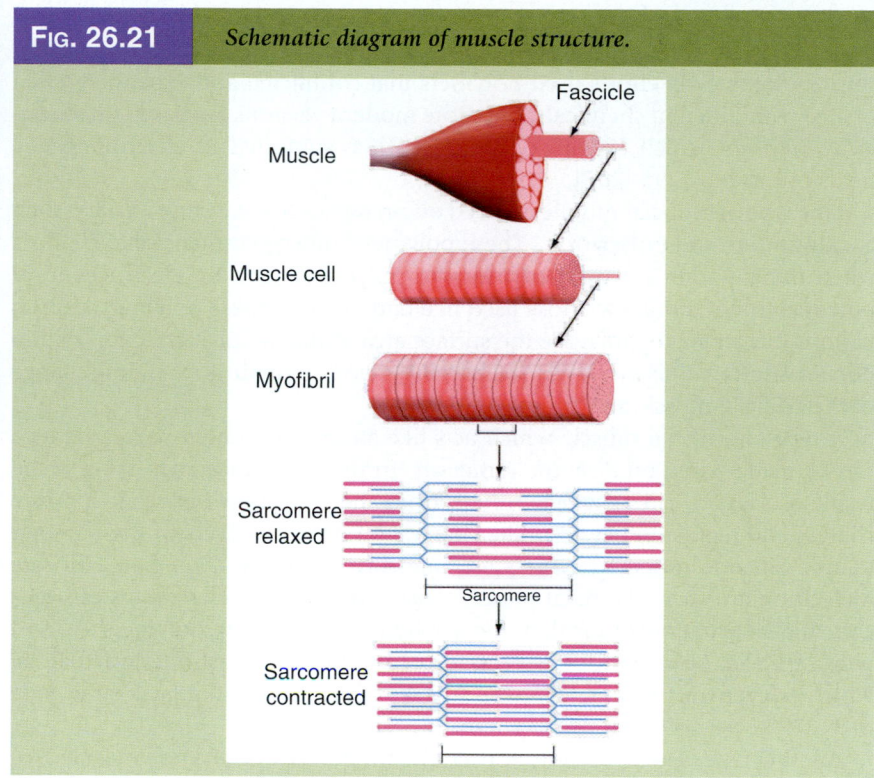

Source: Image courtesy of A. Rao.

myosin filaments interlaced on each end with actin filaments. The actin filaments are linked together so that their long chains extend between the fibrin filaments like fingers on a hand (**Fig. 26.21**) [281–287].

Action of Natural Muscle. When a muscle is stimulated by a nerve impulse, the actin filaments move across fibrin filaments, producing a lengthwise contraction in the sacromere. The energy for the contraction comes from the chemical bonds of ATP, driving a reaction in which the myosin and actin bond to form actomyosin, shortening the chain. This process can be repeated cyclically to result in a significant shortening of the entire bundle.

Nanolevel biomolecular investigations have revealed the details of the contraction mechanisms and energetics, in which voltage-gated calcium channels play an important role. The interactions of thin and thick filament proteins of the contractile apparatus are driven by troponin and calcium binding [288–291].

Natural muscles can repeat millions of work cycles before tiring, in part because the work load is shared and distributed over many fibers, some working while others recover. Natural muscles can contract by more than 20% or their length, at rates exceeding 50% per second. The efficiency of conversion of glucose to energy is nearly 40%, more than double the energy efficiency of artificial mechanical engines. Thus natural muscle has been the starting model for many efforts to duplicate some or all of these qualities by artificial means.

Biomimetic Artificial Muscle. Biomimetic versions of nanoscale-based actuators have been designed to mimic the smooth and efficient action of natural muscle. Some early efforts used polymers that contracted and expanded when treated with different chemicals, but more modern versions are based on *electroactive polymers*, which expand or contract when subjected to electromagnetic fields or currents [292–297].

One type of artificial muscle is based on *electronically conducting polymers* such as polyaniline and polypyrrole. These polymers undergo dimensional changes when dopant ions are inserted into the polymer lattice by electrochemical potential forces, similar to those used in a battery. With these polymers, carbon nanotubes can greatly increase the surface area available for interaction with a dopant solution, thus increasing the effective speed, strength, and volume change of the artificial muscle actuation.

A type of artificial muscle which acts like a capacitor, rather than a battery, can be made based on *dielectric elastomers*. In these polymers, actuation is the result of *Maxwell stress*, which results from the attraction between opposite charges and repulsion between like charges, on different layers of the polymer matrix. Silicone rubber polymers are one class of material used to fabricate dielectric elastomers. Artificial muscles based on dielectric elastomers can generate higher actuator strains than those based on conducting polymers. A company, Artificial Muscle Inc., has been formed to produce elastomeric actuators, which can generate strains of 120%, stresses of 3.2 MPa, and a peak strain rate of 34,000% per second for 12% strain.

Another type of artificial muscle uses the volume change of a polymer electrolyte caused by electrostatic repulsion when ions are absorbed. One version, called the *ionic polymer/metal composite actuator*, is manufactured by Environmental Robots Inc. These artificial muscles amplify low strains using the cantilever effect; they consist of two metal-nanoparticle electrodes filled with and separated by layers of a solid electrolyte. The actuators act as supercapacitors when an applied interelectrode potential injects electronic charge into the high-surface-area electrodes. This charge draws solvated ions to migrate between the electrodes. The volume of the solvated ions causes one electrode to expand relative to the other, thereby bending a cantilever actuator strip. Actuation can be produced by electrolysis, which causes a local pH change and transport of hydrated ions and water into the nanostructure.

Actuation rates and efficiencies for ionic systems are generally lower than for dielectric polymer systems. But a robot based on this system made artificial muscle history in 2005 as the best arm wrestling robot in competition with a human athlete during the first human–robot arm wrestling competition in San Diego, California. This marked a milestone in what Dr. Y. Bar-Cohen of the NASA Jet Propulsion Laboratory calls the "grand challenge" for robotics [296].

Other artificial muscles are based on shape-memory alloys, which can generate strains of up to 8%, but require conversion of electrical to thermal energy, followed by conversion of the latter to mechanical energy, with heat as a by-product. Actuators that use electrochemically generated gases confined in carbon nanotubes have produced strains of 300%, but with low energy conversion efficiency and cycle life.

The Nanotech Institute of the University of Texas at Dallas, led by Dr. R. Baugham, is making low-voltage, low-strain actuators that use electrochemical charge injection into nanostructured electrodes, resulting in electrostatically driven electrode expansion. One device type, carbon nanotube artificial muscles, can generate 100 times the stress of natural muscle and provides comparable actuation rates (20% per second), but the actuator strain is at best 2%. Baugham has suggested improving performance by replacing the carbon nanotubes by nanostructured elastomeric conducting polymers that would have comparable surface areas but would be more easily deformable (by a factor of 100).

Baugham recently reviewed approaches that could radically improve ionic polymer/metal composite actuators by replacing cantilever actuators that operate by bending with actuators that operate in tension. This more biomimetic design could be made by separating the opposing electrodes by a liquid electrolyte, which would provide the solvated ions for actuation [294].

The capacitance of current ionic polymer/metal composite actuators is less than one-tenth that of other supercapacitors. Increasing capacitance by increasing electrode surface area could increase actuator strains, as long as the electrode did not restrict movement in the actuator stroke direction. Baugham proposed that highly twisted carbon nanotube yarns, with their high electrical conductivities and high surface area, would be ideal for this type of electrode.

New directions for improved artificial muscle more closely mimic natural muscle: more elastic material is closer to muscle protein, liquid electrolyte is closer to intercellular plasma, action by tension instead of bending, and the nanoscale structure of the surfaces that provides the actuation force—these design directions are all moving closer to the sacromere structure. Thus we see that biomimetics continues to be a productive approach for design of nanoactuators.

26.3.2 Viral Energy Storage

A key theme in biomimetic nanotechnology is the exploitation of high surface areas created by nanoscale structures. This is true for the lotus and gecko effects, and for artificial muscles. This effect is exploited for energy storage in a very direct biomimetic manner by researchers at MIT, who patterned battery electrodes with self-assembling nanostructures grown by genetically engineered viruses [298,299].

Using genetic engineering to alter a few viral genes, the MIT team made the viruses bond to conductive metal nanoparticles of gold and cobalt oxide. They altered nucleotide sequences in the viral DNA which directed the outer coat proteins to add an amino acid that binds to cobalt ions, which react with water to form cobalt oxide, an advanced battery material with higher storage capacity than the carbon-based materials now used in lithium-ion batteries.

To improve the conductivity of the electrodes, the genetic code was further modified to express an additional strand of virus proteins that bind to gold. The viruses then assembled into nanowires with both cobalt oxide and gold particles. The virus used to make the wire, the M-13 bacteriophage, is about 6 nm in diameter and 880 nm in length.

The conductive viral nanoparticles were layered between oppositely charged polymers to form thin, flexible sheets. To make electrodes, support sheets are dipped into a solution of engineered viruses. The viruses, with hydrophobic and hydrophilic ends, assemble into a uniform layer. The coated sheet is then dipped into a solution containing metal ions. The viruses arrange the ions into an ordered crystal structure which is ideal for high-density batteries.

Used as electrodes in thin lithium-ion batteries, the high surface area sheets improve performance in smaller volume. By increasing surface area and eliminating inert supporting material, energy density is raised by a factor of three. Equally important, viral nanowires are grown at normal room temperature, leading to an energy-efficient manufacturing process.

Exploiting the abilities of microbes to recognize the correct molecules and assemble them where they belong, the MIT team is able to precisely control the nanostructure of electronic materials. The team led by Dr. A. M. Belcher is working towards building faster, cheaper, and environmentally friendly transistors, batteries, solar cells, diagnostic materials for detecting cancer, and semiconductor devices [299].

> My dream is to have a DNA sequence that codes for the synthesis of materials, and then out of a beaker to pull out a device. And I think this is a big step along that path.
>
> **Dr. A. M. Belcher**
> *MIT [299]*

26.3.3 Photosynthesis

Photosynthesis, the conversion of light into stored chemical energy in plants, is one of the crowning achievements of nature's biological nanotechnology [300,301]. Its fundamental importance to life on earth has been recognized by the award of 10 Nobel Prizes for unraveling its mechanisms. One of these Nobel laureates, P. D. Bower, used the term: "a splendid molecular machine," to describe ATP synthase, a key component of the reaction chain [302]. Increased understanding of this process, and advances in chemical synthesis and nanotechnology have made it possible to create biomimetic devices and semibiological hybrids capable of many of the functions of natural photosynthesis [303–306].

In this section, we describe the basics of the natural photosynthetic process and some biomimetic systems. To begin, photosynthesis requires light-harvesting antennas, linked to reaction centers that convert photoexcitation energy to chemical potential in the form of long-lived electrochemical charge separation. With these two steps, engineers can create molecular-level optoelectronic devices with a variety of uses, but natural photosynthesis goes further—it converts electrochemical energy into chemical bonds where the energy can be stored for use as fuel or to build biological polymer materials. This step involves light-driven proton pumps embedded in the lipid membranes of cells, which build up electrochemical potential to drive the synthesis of adenosine triphosphate (ATP), the natural energy storage and transport substance common to all bacterial, plant, and animal cells [307–314].

Photosynthesis and the Development of Natural Photocells. Photosynthesis (from the Greek *photos* "light" and *syntheses* "put together, combine") is the conversion of solar energy into chemical energy (in the form of carbohydrates) from raw materials CO_2 and H_2O, with O_2 as a by-product. Photosynthesis—the ultimate achievement of natural bio-nanotechnology—harnesses photochemical and biochemical energy to fuel life's functions (**Fig. 26.22**).

Photosynthesis Overview. Sometime over three billion years ago, blue-green algae declared their energy independence by developing the capability to split water into hydrogen and oxygen. The process is called photosynthesis. Perhaps one of the most ambitious directives of biomimetics is to duplicate nature's grand method of energy conversion—to design the next generation of energy production on a photosynthetic model.

Chemical reactions associated with photosynthesis are shown below.

The creation of carbohydrates:

$$6CO_{2(g)} + 12H_2O_{(aq)} + h\nu \rightarrow C_6H_{12}O_{6(aq)} + 6O_{2(g)} + 6H_2O_{(aq)} \qquad (26.1)$$

FIG. 26.22 *Photosynthesis in plant leaf cells.*

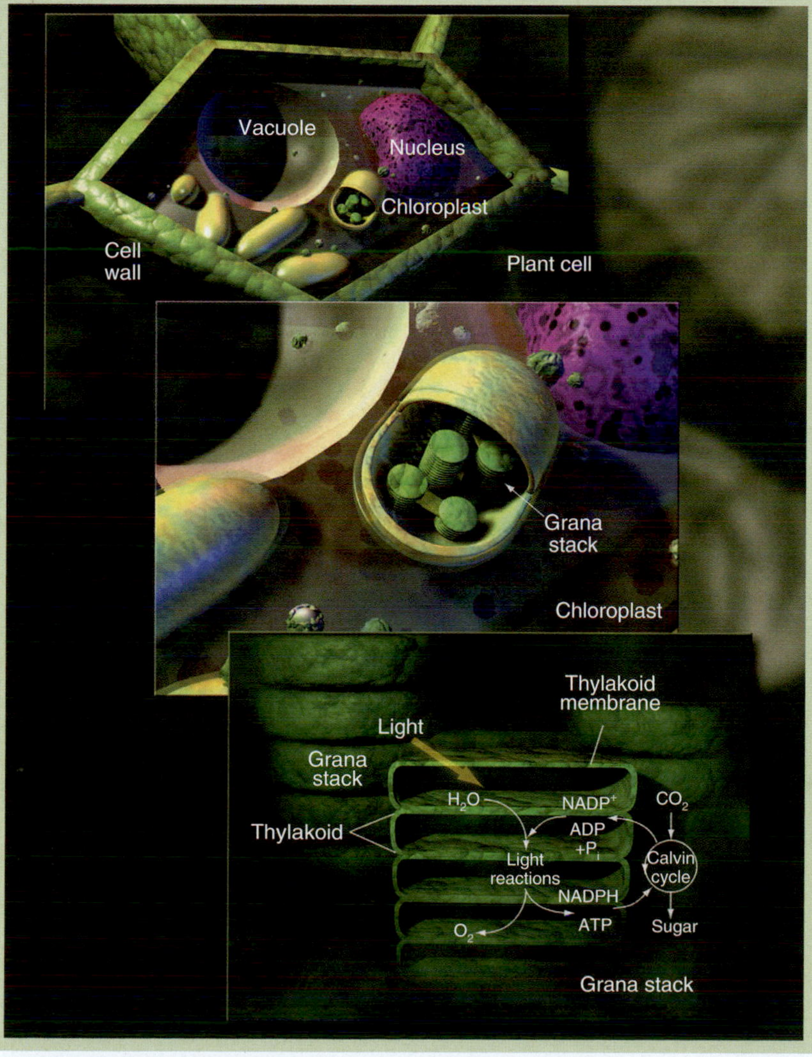

(*continued*)

Fig. 26.22 (Contd.)

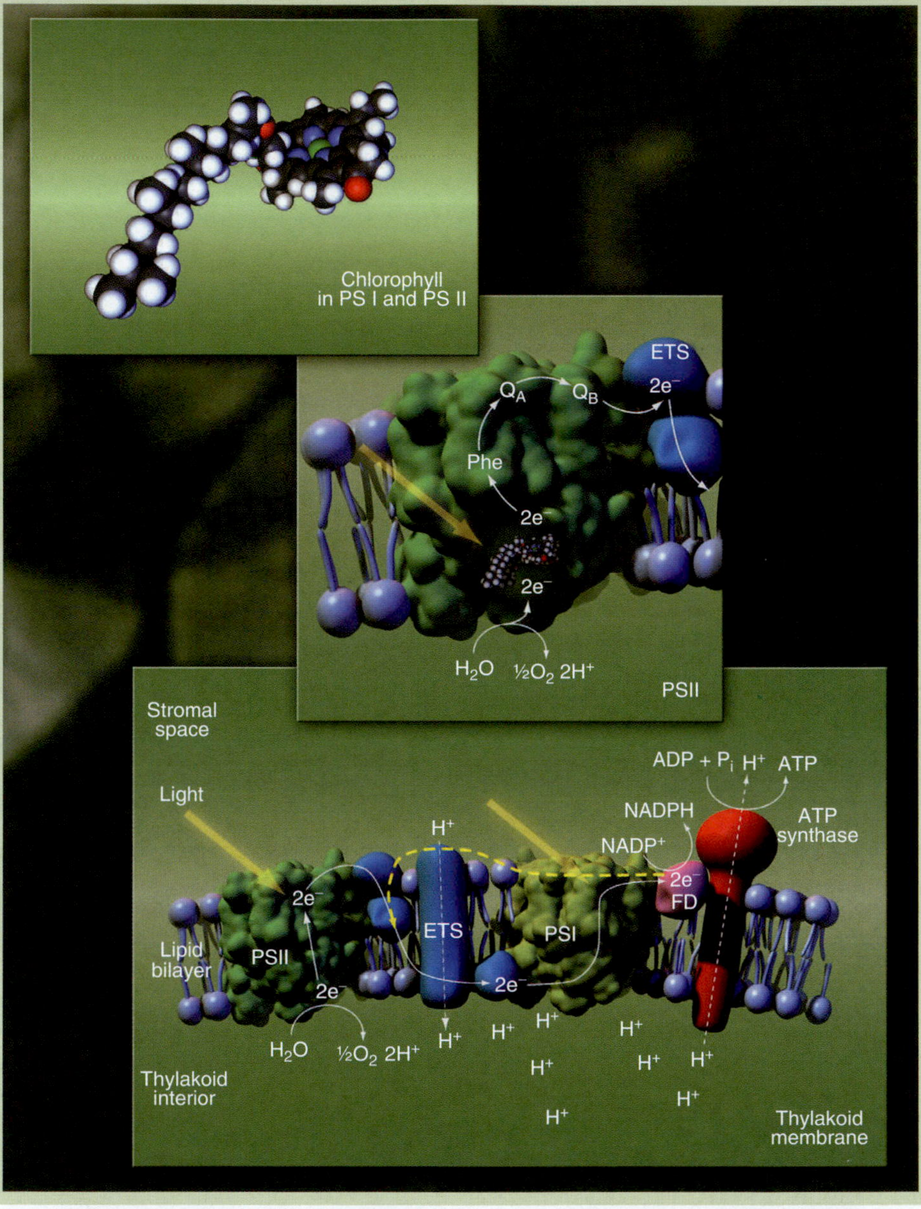

Source: Image courtesy of A. Rao.

The production of oxygen from water:

$$2H_2O + 2NADP^+ + 2ADP + 2P_i + h\nu \rightarrow 2NADPH + 2H^+ + 2ATP + O_2 \qquad (26.2)$$

where P_i is inorganic phosphorous. Carbon fixation is given by

$$3CO_2 + 9ATP + 6NADPH + 6H^+ \rightarrow C_3H_6O_3 \sim PO_4 + 9ADP + 8P_i + 6NADP^+ + 3H_2O \quad (26.3)$$

For this to occur, ATP provides an oxidizer stronger than oxygen. The process of respiration proceeds in the reverse direction.

Plant pigments like chlorophyll, xanthophylls, and carotenes absorb strongly in most of the visible region of the solar spectrum. When the pigments absorb light, photons raise some of their electrons from the ground state to a higher energy. The excited electron can transfer its energy in several ways: (1) it can reemit the energy at the same wavelength (reflectance or emission) or at a longer wavelength (fluorescence), (2) the energy can be dissipated as heat, or (3) the energy can be used to fuel biochemical processes (e.g., photosynthesis). If the process were confined to one pigment molecule (chlorophyll), then the energy would dissipate as the electron regained its ground state. However, an antennae complex consists of several pigment molecules, which allows the excited electron energy to be delocalized from its original site, making it available for charge transfer reactions which can drive the process further.

Significant photocatalytic products produced (in the thylakoids discussed below) include O_2, NADPH + H^+, and ATP—the last two of which provide for the reduction of CO_2 in the surrounding stroma of the thylakoids [315]. The reduction process is modulated by the nanomaterial *ferredoxin*, an iron–sulfur containing protein nanomaterial. Overall, water is oxidized and oxygen released by the plant. Most of the chemical functions are accomplished by five large protein complexes: photosystem I (with bound antennae), photosystem II (with bound antennae), light-harvesting complex II, cytochrome b_6f, and ATP synthase [316].

The photoreaction center consists of ligated chlorophylls that are oxidized to a cation radical following transfer of exciton energy from the antenna pigments—electron transfer (charge separation) reactions occur on the order of picoseconds (corresponding to molecular vibrations). Charge separation induces the creation of a large redox potential between the oxidized chlorophyll a and potential acceptors—the largest known in biology: ~1.5–1.8 V for photosystems I and II, respectively. Electron transfer reactions that follow are energetically downhill. The highly oxidizing chlorophyll reaction centers are prevented from reacting with highly reducing acceptor species—a tribute to ingenious molecular nanodesign. In photosystem I, Fe–S clusters, soluble ferrodoxin and flavoproteins catalyze the reduction of $NADP^+$. In photosystem II, a tetranuclear Mn cluster serves the four-electron oxidation of H_2O to oxygen. The redox reactions are catalyzed by metalloproteins containing either iron or Mn.

The Leaf. Surface nanostructure is a prime driver in biological function. The structure of the leaf is geared to maximize the photosynthesis process—the phenotypical expression of billions of years of development. **Figure 26.22** gives an illustration of the nano and microstructures that make a leaf what it is—a lean photosynthetic nanomachine.

The Chloroplast. Chloroplasts are semiautonomous micron-sized structures that house many kinds of structural and functional nanomaterials. The chloroplast in higher plants is an elongated vesicle filled with an aqueous matrix material called *stroma*. The chloroplast is surrounded by two enveloping membranes.

Thylakoid Membranes. Thylakoid membranes, the chlorophyll-sequestering laminar systems contained within the chloroplast, are stacked pancake-like vesicles (**Fig. 26.22**) found in photosynthetic prokaryotes and eukaryotes. The thylakoid vesicles contain most of the proteins required for the light reactions of photosynthesis and are made primarily of lipids with membrane-embedded proteins. All of the light-harvesting and energy-transducing aspects of photosynthesis take place in the thylakoid membranes.

The thylakoids are ca. 500 nm in size and are stacked to form structures called *grana*. Approximately every other thylakoid possesses an appendage that extends into the *stroma* (the interstitial spaces in chloroplasts) to form a three-dimensional interconnected network. These extensions are known as *stroma lamellae*. The physically contiguous membrane encloses an aqueous phases known as the *thylakoid lumen*. The thylakoid membrane is in a unique class compared to membrane bilayers of other organelles and the cell plasma membrane [315].

Chlorophyll. Chlorophyll pigments are tethered to thylakoid protein components of photosystem I and photosystem II. Chlorophylls trap solar energy and convert it into usable chemical energy to conduct oxidation–reduction reactions. Trapped energy is stabilized by transfer to other chlorophylls, pigments, and secondary redox reactions in protein complexes within the thylakoid membrane [315]. *Chlorophyll* a absorbs energy from the violet-blue, orange red, and some from green-yellow-orange wavelengths of visible light. Chlorophyll a reflects green light (hence the color of most leaves.)

Oxidation and Reduction Processes. Chlorophyll acts as a light antenna in which photons are absorbed, exciting electrons to higher energy states. The energized electrons drive a series of photochemical electron transfer reactions involving *quinones*, which carry the energy in proton bonds to a reaction center where it is stored in the bonds of the ATP molecule, which reacts further to form *nicotine adenine dinucleotide phosphate* ($NADP^+$). In the final step of this chemical process, hydrogen is taken from water to form NADPH, releasing oxygen as a by-product. The NADPH stores the energy until it is used in the next step to energize the formation of carbon–carbon bonds, consuming carbon dioxide in the process (**Fig. 26.23**). The end products are carbohydrates (The general formula of carbohydrates is $[CH_2O]_n$.) Thus the overall process consumes water and carbon dioxide, two greenhouse gases, produces fixed carbon which is the food base for all animal life, and releases oxygen into the atmosphere [300–302].

Photosynthesis and Artificial Nano. Photosystem I (PSI) complex has been isolated from spinach and used to power electronic devices—to fabricate the first biomimetic solid-state photosynthetic solar cell [317]. Creating the interface is nontrivial because it is quite a leap from a biological system that requires water and salts to function—both materials not desired in solid-state electronic devices. Researchers at MIT extracted PSI (10–20 nm in size) from chloroplasts and stabilized them with surfactants in a solid-state device [318]. The PSI systems are relatively large; the tightly coupled antennae complexes contain 175 chlorophyll molecules per PSI. The efficiency of charge separation is close to 100% when in its natural state and the terminal electron-accepting moieties are stable with a low electrical potential (<0.6 V) [319].

The device consists of a bottom transparent layer coated with a conducting material (a thin layer of gold that assists in the self-assembly of the PSI units). A semiconducting layer is evaporated on the biological materials to prevent electrical shorts and then another conducting layer is applied. About 12% of the photonic energy in incident light is converted into electrical charge. Higher levels of charge efficiency (~20%) could be attained by fabricating multiple layers of these PSI sandwiches [317,318].

FIG. 26.23 *(a) Schematic of the photosynthesis reaction center in plant leaf showing the essential membrane-bound elements that comprise the light reactions of photosynthesis. Light is absorbed by protein-bound pigments within a light-harvesting complex (green); the energy is transferred to a reaction center (yellow) in which the energy is used to separate charge across the membrane, leading to reduction of a membrane-soluble quinine Q_B; the quinone migrates to a second protein reaction center (red) that couples its reoxidation ($Q_0 \Leftrightarrow Q_i \Leftrightarrow Q$) to transfer of protons across the membrane; the backflow of protons across the membrane drives ATP synthesis at another transmembrane protein element (purple). (b) Structure of the photosynthesis reaction center: the light activated reaction center from* Rhodopseudomonas viridis *was the first of the integral membrane proteins involved in photosynthesis whose structure was determined to atomic resolution (by Deisenhofer, Epp, Miki, Huber and Michel in 1984).*

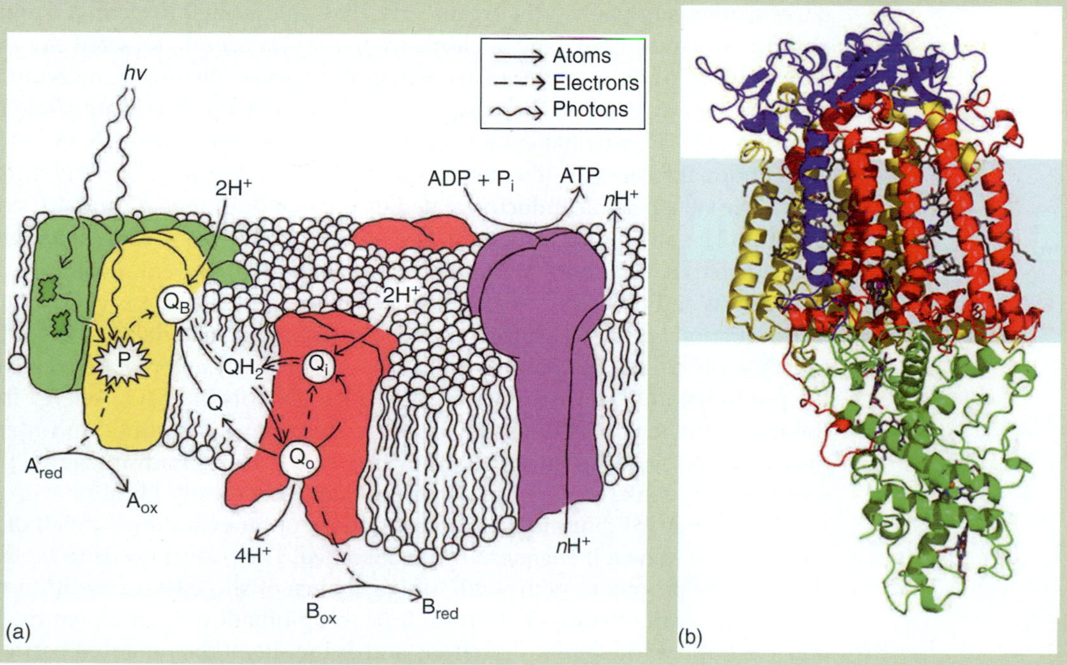

(*continued*)

FIG. 26.23 (CONTD.) *(c) Laue diffraction pattern of the bacterial reaction center of* R. viridis. *Resolution at detector edge 2.9 Å, temporal resolution 2 ms.*

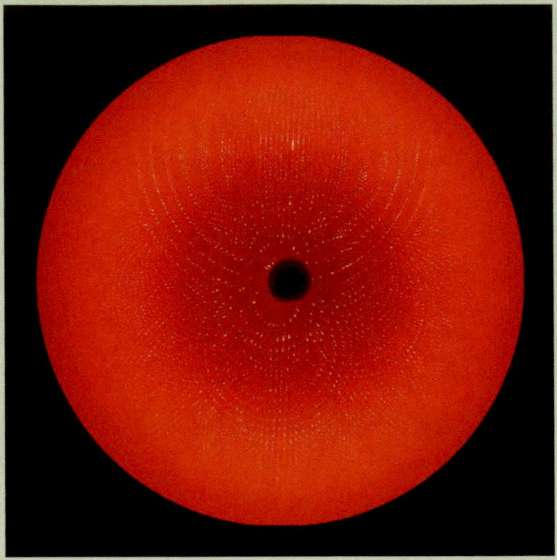

(c)

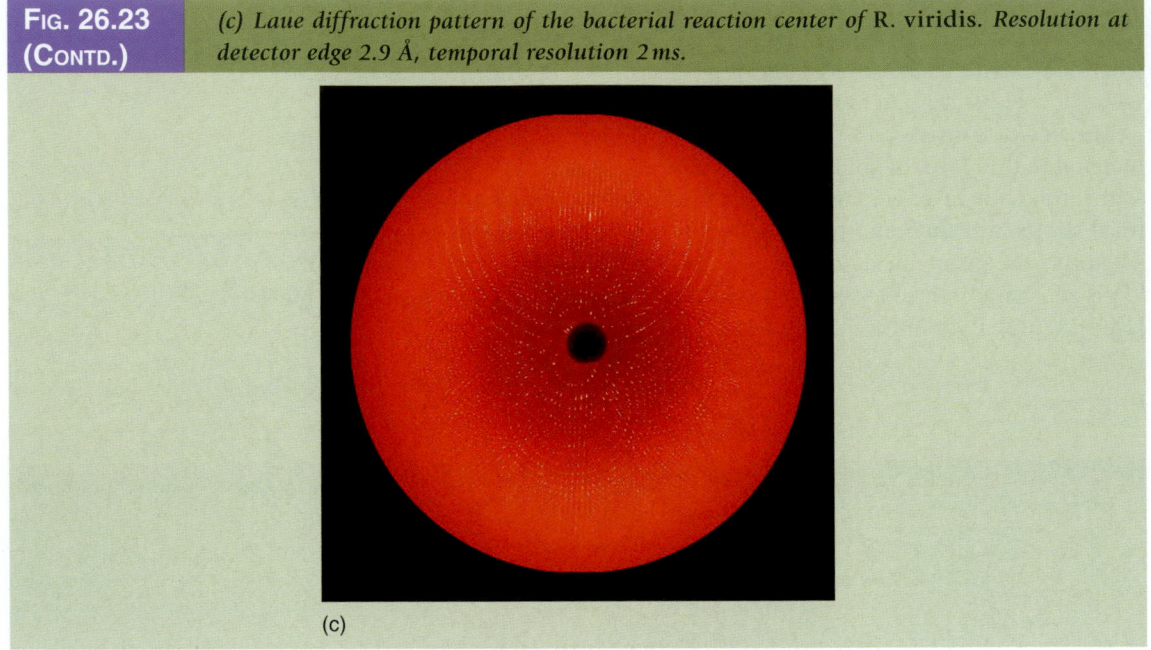

Source: Images courtesy of Dr. Richard Baxter, UT Southwestern Medical Center, Dallas. With permission.

The Grätzel Cell. Michael Grätzel of the Swiss Federal Institute of Technology (EPFL) developed the dye-sensitive solar cell in 1991 (known as the *Grätzel cell*) [320,321]. The mechanism of the Grätzel cell is similar to several aspects of photosynthesis in plants—an organic dye (like chlorophyll in plants) captures photons and transfers energy to electrons. In the Grätzel cell, tethered dye molecules absorb photons and transfer energy to titanium dioxide semiconductor nanoparticles [322,323]. The basic concepts of the Grätzel cell are shown in **Figure 26.24**. The schematic flow of electrons is shown in **Figure 26.25**.

Traditional solar cells, like those described earlier, behave like transistors in which the silicon semiconductor materials provide both the "n" and "p" components. The silicon absorbs light and is responsible for charge separation. These materials, as a consequence, must be very pure to prevent recombination of electrons and holes at defects. In Grätzel cells, this dilemma is overcome by differentiation between absorbers and charge separator materials [324–330].

In the photoexcitation process, electron–hole pairs (excitons) are created in the p/n-junction of semiconductors. In semiconductors like TiO_2, energy from the ultraviolet region of the solar spectrum (<400 nm) is required in order to overcome the bandgap energy (~3.2 *e*V for TiO_2). One disadvantage to pure semiconductors like silicon and titanium oxide is that only photons with the required energy ($>E_g$) or more are able to induce photoexcitation—much of the energy is lost as heat if energies $>E_g$ are absorbed. The p-layer needs to be fairly thick for the process to occur with a high chance of success—a condition that also promotes the chance of electron–hole recombination. In other words, silicon acts both as the source of excitons and as the site of the potential barrier of charge separation. Dye-sensitized solar cells resolve this problem. Dye molecules

Fig. 26.24 *An electronic flow scheme of a Grätzel cell is depicted. A dye molecule tethered to a TiO_2 nanoparticle traps solar energy similar to the action of chlorophyll—photoexcitation of electrons. The excited electron of the dye molecule is in close proximity to the TiO_2 semiconductor. By the process of charge transfer, excited electrons enter the conduction band of the TiO_2 and accumulate at the anode and flow in an external circuit. Electrons from the counter electrode are injected into the iodide/tri-iodide redox cycle from which iodide (the reduced form) relinquishes its electron by transferring it to the dye molecule. In this way, the electronic state of the dye molecule is back to square one and ready for another photoexcitation episode.*

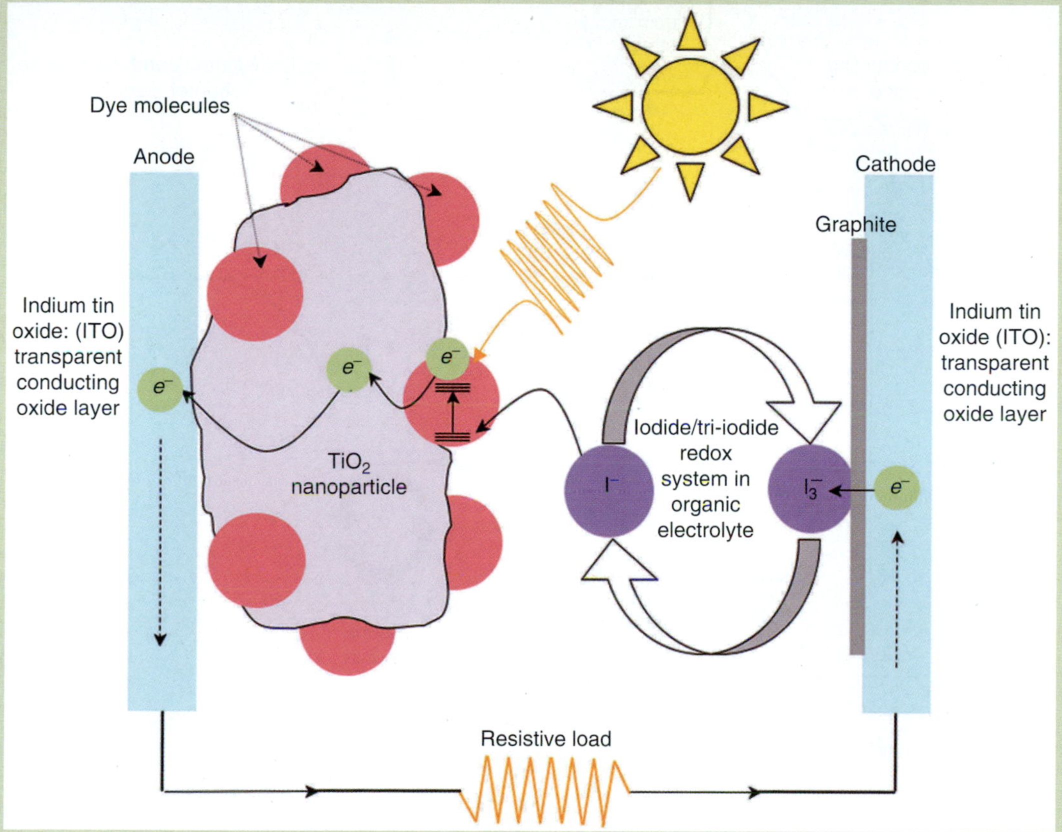

absorb lower energy regions of the spectrum. Since the solar spectrum consists of ca. 46% visible light, use of dye molecules allows the cell to capture more energy, as opposed to semiconductors like TiO_2 that absorb ultraviolet radiation (ca. 3% of the solar spectrum).

In a dye-sensitized solar cell, the function of the semiconductor (e.g., the titanium oxide nanoparticles) is to serve as a charge carrier for transfer of electrons from the dye molecules. Visible light absorption is accomplished by a thin unimolecular layer of dye molecules tethered to the semiconductor surface by organic ligands (**Fig. 26.26**).

The efficiency of the Grätzel cell improved to 12.3% by 2004 [331]. Modifications of the concept developed by Michaël Grätzel continue to this day—especially

Fig. 26.25 *A schematic version of the Grätzel cell depicting electronic states of the tethered dye molecule and n-type TiO_2 semiconductor conduction and valence bands is depicted. The voltage, against a standard calomel electrode, ranges from +2.5 V to −0.9 V (bottom-to-top). An electron in a transition metal complex excited by a photon transfers the excited electron to the conduction band of the semiconductor by the process of injection. The maximum voltage this device is able to deliver is the difference between the potential of the I^-/I_3 redox couple and the Fermi level of the semiconductor.*

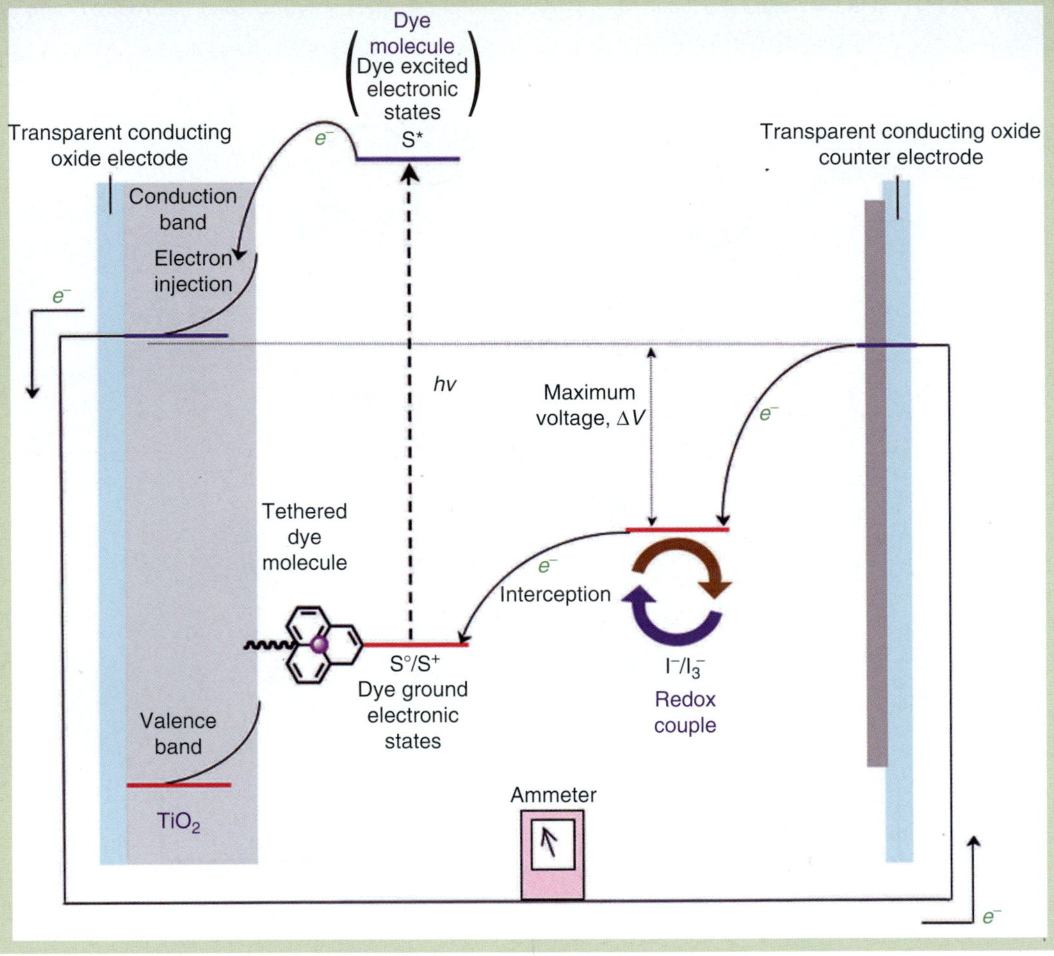

in the electrolyte and dye materials. For example, K.G.U. Wijayantha et al. of the University of Bath in the United Kingdom replaced the organic dye molecules with CdS quantum dots [332–334]—a nanotechnological adaptation of the Grätzel cell (**Fig. 26.27**). CdS quantum dots were self-assembled on the surface of dispersed nanocrystalline TiO_2 by tethering with organic ligands as bifunctional linking molecules—3-mercaptopropionic acid $HO_2C–CH_2–CH_2–SH$ (MPA). As with the organic dye depicted in **Figure 26.26** above, the carboxylic acid groups serve as links between the TiO_2 and the quantum dots. 3-MPA, a material used often in biosensors, promotes electron transfer reactions between cytochrome *c* and gold surfaces and is known as a promoter molecule [335].

FIG. 26.26 *A more realistic rendition of the Grätzel cell is depicted. All dye-decorated semiconductors are connected to the transparent conducting anode. The semiconductor particles form a compact layer with very high surface area, which is a few microns in thickness. The thin-layer (5–10 μg·cm^{-2}) Pt mirror also serves to catalyze the cathodic reduction of tri-iodide to iodide. The electrolyte I^-/I_3^- mediator is soluble in a nitrile organic solvent. Once all components are in place, the cell is sealed from the ambient environment. More advanced cells utilize a dry solid-state hole-transporting polymer called PVK (poly(N-vinyl-carbazole). Efficiency of Grätzel polymer glass cells is on the order of 5% while that of the liquid cell is on the order of 10%.*

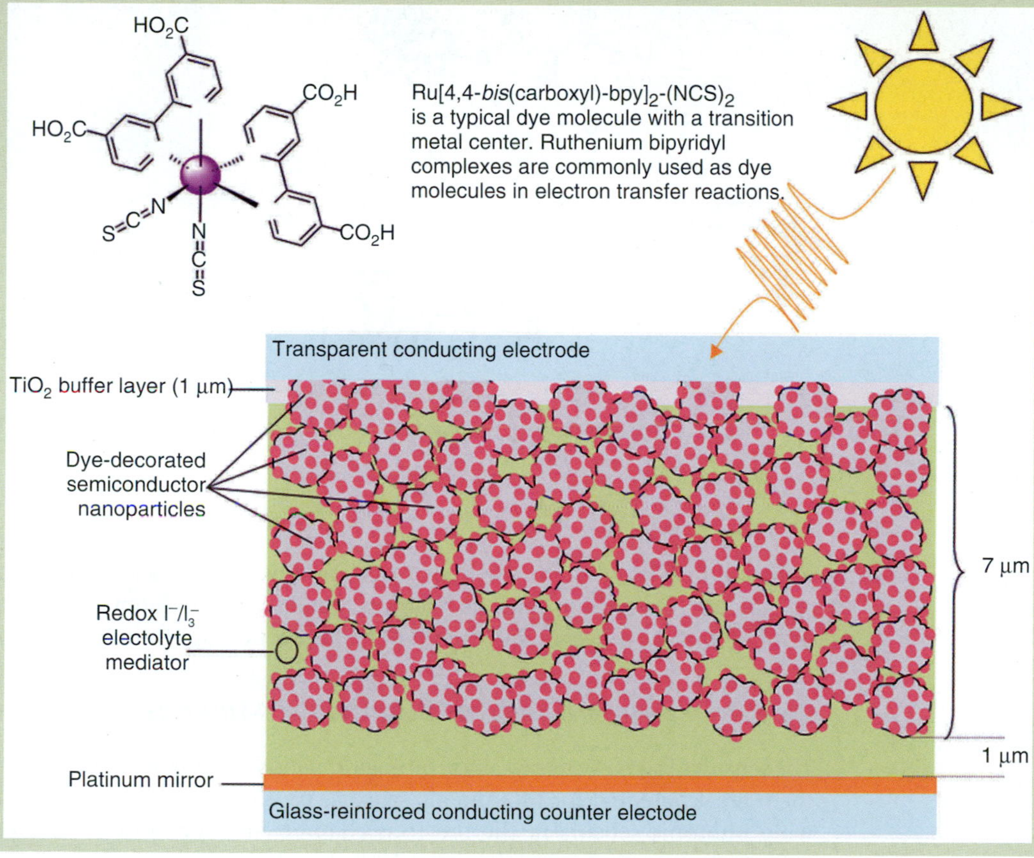

Future Directions for Electronic and Energy Transfer in Supramolecular and Biomolecular Systems. Biomimetic photosynthesis, photoelectricity production, and photonic production of hydrogen offer possible alternatives to conventional sources of fuel and electricity. Research in biomimetic photonics has the goal of providing low-cost renewable energy and fuels without polluting or global warming by-products. Highly focused and skilled research groups around the world are pursuing this goal by various routes. One route would use plants, bacteria, and other cells, enhancing their photosynthetic productivity by genetic engineering and hybrid biomimetics [336–338]. Other approaches include splitting water into hydrogen and oxygen by biomimetic photocatalytic systems.

FIG. 26.27 *Generic rendition of a TiO_2 nanoparticle decorated with CdS quantum dots. The tethering ligand is 3-mercaptopropionic acid. Use of quantum dots allows for nearly 100% capture of the solar spectrum. Although organic dyes function well in the visible range, they are expensive and are relatively unstable over the long term. The use of quantum dots pose several advantages: (1) they are easy (and relatively inexpensive) to fabricate (e.g., spin casting), (2) their optical properties are size dependent, therefore tunable, (3) they are relatively robust, and (4) they are able to absorb light over the IR, near-IR, visible, and UV range of the solar spectrum.*

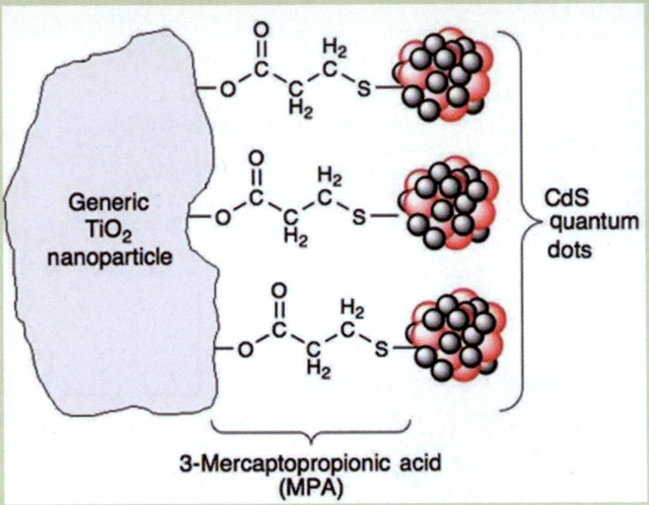

Others utilize biomimetic methods to create novel and efficient photovoltaic and photoelectric cells to generate electric power directly. No doubt a combination of all of these approaches will eventually find practical application [339–340].

26.3.4 Sensors Based on Biomimetic Moieties

The sensitive interactions of natural biological nanostructures with molecules give them great potential for biomimetic sensors. We saw examples in the butterfly wing photonic sensors in the previous chapter. And the adhesive molecules discussed earlier in this chapter offer many possibilities for use as selective biosensors [341]. We will discuss two additional examples that are illustrative in their biomimicry and principle of operation. These are gas sensors using the high surface area of diatom shells, and biomimetic nanoscale temperature sensors.

Silicon Diatom Model for Gas Sensors. A nitric oxide sensor based on the nanopatterned diatoms shells was designed by Sandhage et al. [342]. They started by making pure silicon replicas of the silica (silicon dioxide) diatom frustules. Unlike insulating silica, silicon is a semiconductor which can carry electrical current. The result was micro- and nanostructured silicon shells with high surface area ($>500\ m^2 \cdot g^{-1}$) which readily absorb gases, with resulting rapid changes in electrical impedance. The impedance changes can be measured by resistance to the flow of electrical current or by wireless coupling in a radiofrequency field.

When exposed to gaseous nitric oxide in concentrations as low as 1 ppm, the sensitivity and response speed of the diatom-patterned nanosensor are much greater than for conventional porous silicon NO sensors. The applied bias voltage can be as low as 100 mV—considerably smaller than that needed in other devices. Also, the silicon frustules luminesce strongly in UV light, which could provide a route to other types of sensor design.

Nanoscale Biomimetic Temperature Sensors. Lee and Kotov have reviewed the design of thermal sensors on the nanoscale [343]. This type of measurement is a good paradigm for how everything changes at the nanoscale. Temperature is straightforward in the macroscale. But it ceases to be simple when we leave large-scale molecular statistics for the nanoscale, where random variations of molecular motion and energy are not averaged out so smoothly. This is evident when we consider Brownian motion.

Temperature measurement at the nanoscale poses challenges, where we need to measure heat with high spatial resolution. Nano- and biotechnology require precise thermometry down to the nanoscale regime if temperature of nanodevices is to be calibrated. The development of a nanoscale thermometer is not merely a matter of size—it requires materials with novel physical properties, because all physicochemical and thermodynamic properties are drastically altered at the nanoscale. Progress on nanoscale thermal sensors will require use of molecular and biological moieties, as well as nanoscale superstructures, such as nanosprings and cantilevers.

Biological systems have a number of temperature-dependent molecules and processes which can be adapted for nanoscale thermal measurement. Use of thermo-transformable responsive entities borrowed from biology can lead to high spatial resolution and enhanced biocompatibility because of the moieties' reduced size and direct applicability to biomedical or clinical sensing and imaging. (For example, sensing very small cancer tumors, which have a higher local temperature in the body.)

Examples are temperature-dependent changes in double-stranded DNA structure from B- to Z-DNA, which has been investigated as a possible molecular nanothermometer. Differences in the electronic properties of the two structures and the charge-transfer process from fluorescent probes result in marked changes in optical emission. Certain messenger RNAs (mRNAs) change conformation with temperature. Areas of RNA chains undergo temperature-dependent conformal changes that can be monitored by ultraviolet and nuclear magnetic resonance spectroscopy.

26.3.5 Biomimetic Molecular Nanoengines

The design of macromolecular engines is an active area of nanotechnology, and biology is a very rich source of models and ideas for such designs. We briefly mention two topics in the area to call attention to its importance, although it is an advanced topic that is beyond the scope of this introductory chapter on biomimetic nanotechnology.

Biologically Inspired Nanoengines. Much work has been done on design of nanostructures using natural molecular motors such as the rotators in the flagella of bacteria, spermazoa, and similar cellular biomotive engines [344–350]. It is interesting to note that when mankind invented the wheel, no one could have known that nature had already developed highly sophisticated rotary engines. There was a natural precedent that no one would be able to see for thousands of years.

Another very interesting and semibiomimetic area is the adaptation of the information processing and storage properties of the genetic code machinery to design molecular computing engines. This falls somewhat outside what is usually considered biomimetic, but can be considered to be inspired by nature [351–354].

DNA for Parallel Processing. Leonard Adelman proposed using DNA to solve complex mathematical problems in 1994. He mapped the traveling salesman problem (a difficult to solve mathematical formulation also known as the Hamiltonian path problem) whose solution is a path through a set of points from start to end, going through all points once and only once. This problem becomes combinatorically difficult as the number of points increases.

Adelman represented each city by a unique DNA nucleotide code sequence; replication created new sequence combinations. DNA as a computing engine maps naturally to combinatoric problems where parallel processing can create many solutions at the same time for comparison. Most conventional computing architectures have to solve problems one step at a time (linear processing). DNA has the added benefits of being a cheap, energy-efficient, and extremely small for high information-holding density. The main disadvantages are setting up the program and the random errors that can occur in biological systems.

Molecular Penrose Tiling. Molecular tiles can be designed with capillary interactions that correspond to simple logic operations—AND, XOR, etc., and allowed to self-assemble to determine whether they organize into assemblies according to mathematical and logical rules, as in Penrose tilings. Interesting results have been obtained by researchers, including Adleman and Paul Rothemund. Rothemund designed experiments to see whether one could make tiles that obey the Penrose tile matching rules and to see how well they could make a Penrose tiling. There has been much interest in whether Penrose-type tilings can self-assemble with few errors. Experiments have shown that one could make sufficiently complex matching rules using capillary force interactions, and that the energetics of hydrophilic and hydrophobic bonds on the tile elements could skew the distribution of structures [353] (**Fig. 26.28**).

Neural Networks and Swarm Computing. The coordinated firing patterns of neurons inspired abstract mathematical analysis which found deep connections with quantum state phenomena, leading to the development of neural network algorithms for parallel computing and pattern recognition. This field underwent intense development from the 1970s through the present, and has moved into many practical applications [355,356]. The cooperative, independent but coordinated movements of flocks of birds and swarms of ants have inspired studies which found an underlying relationship to distributed computing models and

Fig. 26.28 *Penrose tile patterns produced by molecular computing elements.*

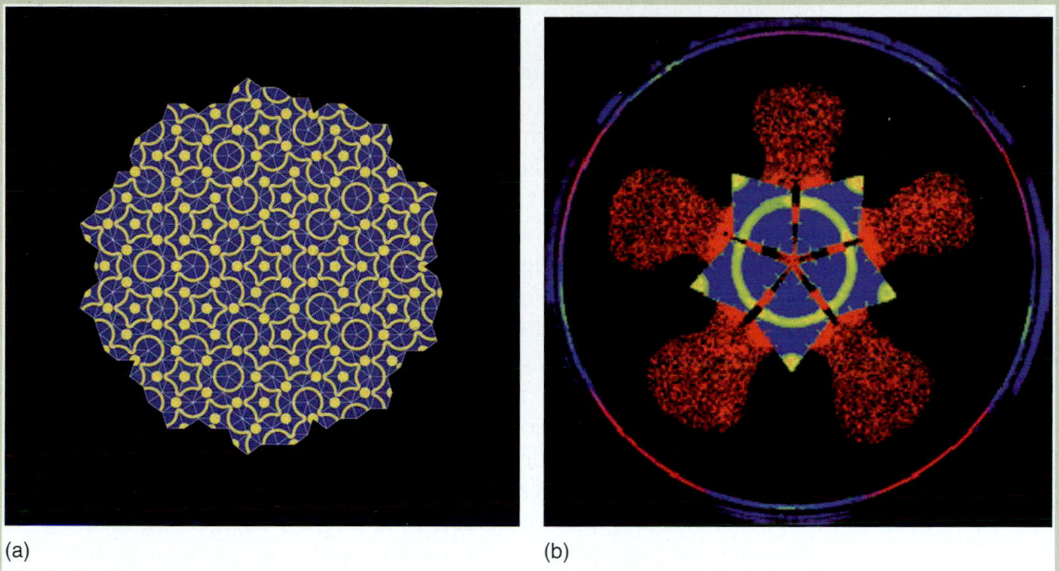

Source: Image from P. W. K. Rothemund, *Proceedings of the National Academy of Sciences USA*, 97, 984–989 (2000). With permission.

to statistical mechanics. This has led to the development of swarm computing algorithms for parallel processing which are finding application in many combinatoric and exponentially difficult problems in graphics, searches, and other areas, including nanotechnology [357].

DNA for Diagnosis. DNA computing is most naturally applied to biomedical applications, where its ability to recognize, match, and replicate trillions of combinations can be used for diagnosing and treating cancers, virus infections, and other disorders that involve genetic expression. Since the object of the tests is also DNA, there is no input/output interface problem. As a demonstration, researchers at the NSF National Center for Chemical Bonding built an automaton using DNA encoding simple logic to play a programmed game of Tic Tac Toe, always obtaining a win or tie, as any computer does. The automaton is called MAYA (Molecular Array of YES and AND logic gates) and its second generation (MAYA-II) has more than 100 DNA logic circuits. This technology could be used in the future to develop instruments that can simultaneously diagnose and treat cancer, diabetes, or other diseases.

DNA Nanotube Computer Circuits. DNA can serve as a template for digital and quantum computing [358–360]. IBM has used self-replicating and self-organizing properties of DNA to arrange carbon nanotube into nanowire circuitry [361]. Much DNA computing tends to leave the biomimetic domain once results are pursued—just as early airplanes ceased to resemble birds. This perhaps reflects our lack of real understanding, not of DNA and biomolecules, but of how the brain really works on a much deeper level than is addressed by nanotechnology.

TRIZ—Biomimetic Problem Solving Methodologies. Julian Vincent at the University of Bath has developed a methodology for solving engineering design problems which attempts to emulate the decision-making processes used by nature—perhaps the ultimate application of biomimetics. Vincent and his colleagues adopted a formalism called the "theory of inventive problem solving." Known by its Russian acronym Triz, it was developed by Genrich Altshuller as a way to systemize engineering and economic decision making. It is similar to operational analysis methods, and to design methodologies such as the Boothroyd–Dewhurst design for manufacturing method. Vincent put biology into the matrix to highlight the striking contrasts between the technical solutions favored in engineering and those found in nature [362].

Data is loaded into the matrix for design solutions used in engineering and those found in nature, using variables such as composition, structure, spatial arrangement, time, energy and information, and then the contradictions and agreements are computed. The conclusions are striking: there was only a 12% degree of similarity between the two approaches, suggesting that nature usually has quite different problem-solving strategies from those of engineers. The matrix revealed that engineering solutions varied with size scale. For manipulation at small scales, from molecular to micro, our engineering methods nearly always rely on high use of energy—reflecting that we use heat to drive random recombinations of molecules in chemical- and energy-generating processes.

At larger scales, structure, which involves directed manipulation of matter, becomes gradually more significant than energy for engineers. But in biology, the proportions of each solution class stay more or less constant at all sizes, from nanometers to kilometers. Crucially, this implies that biology manipulates information, with the saving of energy and material, at the nanoscale. This reflects that we can apply information more effectively at large scales where we are more familiar with visualizing the manipulation of the problem elements. But now nanotechnology, especially with the direction provided by insights from nature, is giving engineers the ability to manipulate matter intelligently on the nanoscale.

Vincent is planning to make a systematic version of Triz biomimetic design methodology available to engineers, designers, and planners on the Internet. In time it may become a widely used design tool, leading indirectly to a global infusion of biomimetic technology.

26.4 Conclusion

> The new technology we build in the future should be recyclable and sustainable, reliable and energy efficient. By elucidating the delicate and intricate assembly of living organisms, it will be possible to create new materials and systems. Markus Milwich et al. Competence Network Biomimetics [187].

By applying the principles learned from nature, biomimetics contributes to realizing "smart," dynamic, complex, environmentally friendly, self-healing, and multifunctional artificial structures, machines, lubricants, and adhesives. By illustrating how mechanical, physical, chemical, and entropic nanoscale forces can work in concert to provide strength, resilience, adaptivity, self-healing, and functionality, biomimetics plays a key role in the development of a mature and integrated nanotechnology that like natural systems, applies the appropriate tools to execute nanoscale tasks.

> The laws of biomimicry are obvious and straightforward–Nature runs on sunlight, uses only the energy it needs, fits form and function, recycles everything, rewards cooperation, banks on diversity, demands local expertise, curbs excesses from within and taps the power limits J. M. Benyus, Biomimicry: Innovation Inspired by Nature [3].

Life itself is still a miracle to us. Organisms are complex open systems through which energy and materials flow on a dynamic homeostatic trajectory far from thermodynamic equilibrium. Engineers, scientists, economists, and managers can learn by understanding and generalizing natural approaches to the challenges of life. Human engineers will need to be able to apply creativity and disciplined thought to solve the many new problems that we encounter and make for ourselves in the built environment of an industrialized world. Understanding natural methods is a powerful starting point, but biomimetics implies taking the natural examples and improving and expanding upon them in new materials and combinations.

References

1. Y. Bar-Cohen, ed., *Biomimetics: Biologically inspired technologies*, CRC Press, Boca Raton, FL (2005).
2. Y. Bar-Cohen, Biomimetics—using nature to inspire human innovation, *Bioinspiration and Biomimetics*, 1, P1–P12 (2006).
3. J. M. Benyus, *Biomimicry: Innovation inspired by nature*, HarperCollins, New York (1998).
4. J. F. Vincent et al., Biomimetics: Its practice and theory. *Journal of Royal Society Interface*, 3, 471–482 (2006).
5. T. Speck and C. Neinhuis, Bionik, biomimetik, *Naturwissenschaftliche Rundschau*, 57, 177–191 (2004).
6. S. Vogel, *Cats' paws and catapults–mechanical worlds of nature and people*, Norton, New York (1998).
7. G. Jeronimidis, Biodynamics, *Architectural Design*, 74, 90–96 (2004).
8. R. Hooper, Ideas stolen right from nature, *Wired Magazine*, www.wired.com/science/discoveries/news/2004/11/65642 (2004).
9. B. Poole, Biomimetics: Borrowing from biology, on *Trendy Science Blogsite*, http://trendyscience.blogspot.com/2007/08/biomimetics-borrowing-from-biology.html (2007).
10. P. Ball, *Made to measure*, Princeton University Press, New York (1997).
11. M. H. Dickinson, Bionics: Biological insight into mechanical design, *Proceedings of the National Academy of Science USA*, 96, 14208–14209 (1999).
12. Centre for Biomimetics, University of Reading, www.rdg.ac.uk/biomim/home.htm (2007).
13. BIONIS: The Biomimetics Network for Industrial Sustainability, Newsletter Number 23, Link at: www.biomimetics.org.uk November (2005).
14. What is Biomimetics? University of Reading, http://www.rdg.ac.uk/biomimetics/about.htm (2007).
15. J. Kikuchi, A. Ikeda, and M. Hashizume, Biomimetic Materials, In *Encyclopedia of biomaterials and biomedical engineering*, G. L. Bowlin and G. Wnek, eds., Informa/Routledge, New York, 96–102 (2004).

16. E. R. Ritneour, *Dr. Otto H. Schmitt*, The Schmitt Charitable Foundation, www.otto-schmitt.org (2008).
17. E. J. Lerner, Biomimetic nanotechnology, *The Industrial Physicist*, Aug-Sep, 18–21 (2004).
18. Bionics, U.S. History Encyclopedia, www.answers.com/topic/bionics?cat = health (2008).
19. Y. C. Fung, *Biomechanics: Mechanical properties of living tissue*, 2nd ed., Springer, Berlin, Germany (1993).
20. C. R. Ethier and C. A. Simmons, *Introductory biomechanics: From cells to organisms*, Cambridge University Press, Cambridge, UK (2007).
21. J. D. Humphrey, Continuum biomechanics of soft biological tissues, *Proceedings of the Royal Society London A: Mathematical, Physical, and Engineering Science*, 459, 3–46 (2002).
22. J. D. Humphrey and S. DeLange, *An Introduction to biomechanics: Solids and fluids, analysis and design*, Springer, Berlin, Germany (2004).
23. J. D. Humphrey, *Cardiovascular solid mechanics: Cells, tissues, and organs*, Springer, Berlin, Germany (2002).
24. S. Vogel, *Comparative biomechanics: Life's physical world*, Princeton University Press, Princeton, NJ (2003).
25. A. Thurston, Giovanni Borelli and the study of human movement: An historical review, *Australian and New Zealand Journal of Surgery*, 69, 276–288 (1999).
26. J. D. Bronzino, ed., *Biomedical engineering handbook*, CRC Press, Boca Raton, FL (2007).
27. S. A. Berger, W. Goldsmith, and E. R. Lewis, *Introduction to bioengineering*, Oxford University Press, Oxford, UK (1996).
28. Y. C. Fung, *Introduction to bioengineering*, World Scientific Publishing, Singapore, Malaysia (2001).
29. NIH Working Definition of Bioengineering, National Institutes of Health, 1997, www.becon.nih.gov/bioengineering_definition.htm (2008).
30. N. H. Malsch, ed., *Biomedical nanotechnology*, CRC Press, Boca Raton, FL (2007).
31. M. P. Hughes, *Nanoelectromechanics in engineering and biology*, CRC Press, Boca Raton, FL (2002).
32. R. M. J. Cotterill, *Biophysics: An introduction*, John Wiley and Sons, New York (2002).
33. J. A. Tuszynski and M. Kurzynski, *Introduction to molecular biophysics*, CRC Press, Boca Raton, FL (2003).
34. R. Glaser, *Biophysics*, 1st ed., Springer, Berlin, Germany (2004).
35. D. Springham, G. Springham, V. Moses, and R. E. Cape, *Biotechnology: The science and the business*, Taylor & Francis, London (1999).
36. K. L. Lerner, B. W. Lerner, and B. Wilmoth, eds., *Biotechnology: Changing life through science*, Gale Cengage Learning, Andover, UK (2007).
37. M. El-Mansi, A. L. Demain, and C. F. Bryce, eds., *Fermentation microbiology and biotechnology*, CRC Press, Boca Raton, FL (2006).
38. K. Fukui and T. Ushiki, *Chromosome nanoscience and technology*, CRC Press, Boca Raton, FL (2007).
39. D. S. Goodsell, *Bionanotechnology: Lessons from nature*, Wiley-Liss, New York (2004).
40. O. Shoseyov and I. Levy, eds., *Nanobiotechnology: Bioinspired devices and materials of the future*, Springer Verlag, Berlin, Germany (2007).
41. C. M. Niemeyer and C. A. Mirkin, eds., *Nanobiotechnology: Concepts, applications and perspectives*, Wiley-VCH, Weinheim, Germany (2004).
42. The Biomimicry Institute, Missoula, Montana, USA, website: http://www.biomimicryinstitute.org/ (2007).
43. C. Mattheck, *Design in nature—learning from trees*, Springer Verlag, Heidelberg, Germany (1998).

44. S. Olariu and A. Y. Zomaya, eds., *Handbook of bioinspired algorithms and applications*, CRC Press, Boca Raton, FL (2005).
45. R. Lipowsky, Biomimetic systems and transport systems, *European Whitebook of fundamental research in materials science*, pp. 78–82, Max Planck Institut für Metallforschung, Stuttgart, Germany (2001).
46. G. E. Wnek and G. L. Bowlin, eds., Biomimetics, In *Encyclopedia of biomaterials and biomedical engineering*, Marcel Dekker, New York, 1009–1016 (2004).
47. R. Webb, Offices that breathe naturally, *New Scientist*, 1929, 38 (1994).
48. K. Brown, *The Pox: The life and near death of a very social disease*, Sutton, Stroud, UK (2006).
49. B. Witkop, Paul Ehrlich and his magic bullets—Revisited, *Proceedings of the American Philosophical Society*, 143, 540–557 (1999).
50. C. C. Mann and M. L. Plummer, *The aspirin wars: Money, medicine, and 100 years of rampant competition*, Harvard Business School Press, Boston, MA (1991).
51. J. R. Vane and R. M. Butting, The mechanism of action of aspirin, *Thrombosis Research*, 110, 255–258 (2003).
52. H. Tohgi, et al., Effects of low-to-high doses of aspirin on platelet aggregability and metabolites of thromboxane A2 and prostacyclin, *Stroke*, 23, 1400–1403 (1992).
53. T. Morris, M. Stables, and D. W. Gilroy, New perspectives on aspirin and the endogenous control of acute inflammatory resolution, *Scientific World Journal*, 6, 1048–1065 (2006).
54. T. Lawrence and D. W. Gilroy, Chronic inflammation: A failure of resolution? *International Journal of Experimental Pathology*, 88, 85–94 (2007).
55. M. Paul-Clark, et al., 15-epi-lipoxin A 4-mediated induction of nitric oxide explains how aspirin inhibits acute inflammation, *Journal of Experimental Medicine*, 200, 69–78 (2004).
56. C. R. McCurdy and S. S. Scully, Analgesic substances derived from natural products (natureceuticals), *Life Sciences*, 78, 476–484 (2005).
57. J. Goodman and V. Walsh, *The story of Taxol: Nature and politics in the pursuit of an anti-cancer drug*, Cambridge University Press, Cambridge, UK (2001).
58. Y. Ji, J.-N. Bi, B. Yan, and X.-D. Zhu, Taxol-producing fungi: A new approach to industrial production of taxol, *Chinese Journal of Biotechnology*, 22, 1–6 (2006).
59. J. E. Casida, ed., *Pyrethrum: The natural insecticide*, Academic Press, New York (1973).
60. T. J. Class, et al., Pyrethroid metabolism: Microsomal oxidase metabolites of (S)-bioallethrin and the six natural pyrethrins, *Journal of Agriculture and Food Chemistry*, 38, 529–537 (1990).
61. I. T. Baldwin, M. J. Karb, and P. Callahan, Foliar and floral pyrethrins of *Chrysanthemum cinerariaefolium* are not induced by leaf damage, *Journal of Chemical Ecology*, 19, 2081–2087 (1993).
62. G. Vettorazzi, *International regulatory aspects for pesticide chemicals: Toxicity profiles*, Vol. 1, CRC Press, Boca Raton, FL (1979).
63. J. S. Dordick, Enzymatic and chemoenzymatic approaches to polymer synthesis and modification, *Annals of the New York Academy of Sciences*, 672, 352–362 (1992).
64. K. M. Koeller and C.-H. Wong, Complex carbohydrate synthesis tools for glycobiologists: enzyme-based approach and programmable one-pot strategies, *Glycobiology*, 10, 1157–1159 (2000).
65. J. W. Peeters, et al., Chemoenzymatic synthesis of branched polymers, *Macromolecular Rapid Communications*, 26, 684–689 (2005).
66. H. Yu, et al., One-pot three-enzyme chemoenzymatic approach to the synthesis of sialosides containing natural and non-natural functionalities, *Nature Protocols*, 1, 2485–2492 (2006).
67. N. Anand, et al., Stereoselective chemoenzymatic process for the preparation of optically enriched phenylglycidates as precursors of taxol side chain, US Patent

7060471, Issued June 13, (Council of Scientific and Industrial Research, India) (2006).

68. T. Tanaka, et al., A novel glycosyl donor for chemo-enzymatic oligosaccharide synthesis: 4,6-dimethoxy-1,3,5-triazin-2-yl glycoside, *Chemical Communications*, DOI: 10.1039/b801090k (2008).
69. E. Wilson, Total synthesis surprise: Scientists revise structure of coveted anticancer marine natural product, *Chemical and Engineering News*, 79, 11 (2001).
70. A. K. Mukherjee, B. L. Sollod, S. K. Wikel, and G. F. King, Orally active acaricidal peptide toxins from spider venom, *Toxicon*, 47, 182–187 (2006).
71. R. S. Greco, F. B. Prinz, and R. L. Smith, *Nanoscale technology in biological systems*, CRC Press, Boca Raton, FL (2004).
72. M. Stoneham, How soft materials control harder ones: Routes to bioorganization, *Reports on Progress in Physics*, 70, 1055 (2007).
73. N. A. Kotov, *Nanoparticle superstructures - Nanoparticle assemblies and superstructures*, CRC Press, Boca Raton, FL (2005).
74. N. Yui, *Supramolecular design for biological applications*, CRC Press, Boca Raton, FL (2002).
75. J. B. Park and J. D. Bronzino, eds., *Nano- and microscience, engineering, technology and medicine, Volume: 4: Biomaterials: Principles and applications*, CRC Press, Boca Raton, FL (2002).
76. M. J. Schulz, A. D. Kelkar, and M. J. Sundaresan, *Nanoengineering of structural, functional and smart materials*, CRC Press, Boca Raton, FL (2005).
77. J. A. Schwarz, C. I. Contescu, and K. Putyera, eds., *Dekker Encyclopedia of nanoscience and nanotechnology*, (5 Vols), Marcel Dekker, Inc., New York (2004).
78. D. Green, et al., The potential of biomimesis in bone tissue engineering: lessons from the design and synthesis of invertebrate skeletons, *Bone*, 30, 810–815 (2002).
79. P. Podsiadlo, et al., Ultrastrong and stiff layered polymer nanocomposites, *Science*, 318, 80–83 (2007).
80. B. S. Shim, et al., Nanostructured thin films made by dewetting method of layer-by-layer assembly, *Nano Letters*, 7, 3266–3273 (2007).
81. A. Lin and M. A. Meyers, Growth and structure in abalone shell, *Materials Science and Engineering A*, 390, 27–4 (2005).
82. L. J. Bonderer, A. R. Studart, and L. J. Gauckler, Bioinspired design and assembly of platelet reinforced polymer films, *Science*, 319, 1069–1073 (2008).
83. A. K. Geim and K. S. Novosolov, The rise of Graphene, *Nature Materials*, 6, 183–191 (2007).
84. P. K. Hansma, P. J. Turner, and R. S. Ruoff, Optimized adhesives for strong, lightweight, damage-resistant, nanocomposite materials: New insights from natural materials, *Nanotechnology*, 18, 044026 (2007).
85. R. Gordon, C. R. Chaplin, and G. Jeronimidis, Composite material, US Patent 4409274, WestVaco Corp. (1983).
86. G. Jeronimidis, Wood, one of nature's challenging composites, In *The mechanical properties of biological materials*, J. F. V. Vincent and J. D. Currey, eds., Vol. 34, pp. 169–182, Symposia of the Society for Experimental Biology, Cambridge University Press, Cambridge, UK (1980).
87. J. R. Barnett and V. A. Bonham, Cellulose microfibril angle in the cell wall of wood fibres, *Biological Reviews*, 79, 461–472 (2004).
88. I. Burgert, et al., Structure-function-relationships of four compression wood types—Micromechanical properties at the tissue and fiber level, *Trees-Structure and Function*, 18, 480–485 (2004).
89. I. Burgert, N. Gierlinger, and T. Zimmermann, Properties of chemically and mechanically isolated fibres of spruce (*Picea abies* [L.] Karst.). Part 1. Structural and chemical characterization. *Holzforschung* 59, 240–246 (2005).

90. C. Mattheck, *Trees: The mechanical design*, Springer Verlag, Heidelberg, Germany (1996).
91. K. J. Niklasm, *Plant biomechanics. An engineering approach to plant form and function*, University of Chicago Press, Chicago, IL (1992).
92. A. Reiterer, et al., Experimental evidence for a mechanical function of the cellulose microfibril angle in wood cell walls, *Philosophical Magazine A*, 79, 2173–2184 (1999).
93. I. Burgert, Exploring the micromechanical design of plant cell walls, *American Journal of Botany*, 93, 1391–1401 (2006).
94. S. H. Li, et al., Biomimicry of bamboo bast fiber with engineering composite materials, Biomolecular and biomimetic materials: Materials Research Society Fall Meeting Symposium. S3, 125–130 (1995)
95. J. F. V. Vincent, Ideas from skins, *Interdisciplinary Science Reviews*, 24, 52–57 (1999).
96. R. H. C. Bonser, L. Saker, and G. Jeronimidis, Toughness anisotropy in feather keratin, *Journal of Material Science*, 39, 2895–2896 (2004).
97. S. S. Liao, et al., Hierarchically biomimetic bone scaffold materials: Nano-HA/ collagen/ PLA composite, *Journal of Biomedical Materials Research B: Applied Biomaterials*, 69B: 158–165 (2004).
98. G. Heness and B. Ben-Nissan, Innovative Bioceramics, *Materials Forum*, 27, 3–21 (2004).
99. D. Vashishth, K. E. Tanner, and W. Bonfield, Experimental validation of a microcracking-based toughening mechanism for cortical bone, *Journal of Biomechanics*, 36, 121–124 (2003).
100. V. C. Sundar, et al., Fibre-optical features of a glass sponge, *Nature*, 424, 899–900 (2003).
101. M. M. Murr and D. E. Morse, Fractal intermediates in the self-assembly of silicatein filaments, *Proceedings of the National Academy of Sciences*, 102, 11657–11662 (2005).
102. G. Fu, et al., $CaCO_3$ biomineralization: Acidic 8-kDa proteins isolated from aragonitic abalone shell nacre can specifically modify calcite crystal morphology, *Biomacromolecules*, 6, 1289–1298 (2005).
103. G. Fu, et al., Acceleration of calcite kinetics by abalone nacre proteins, *Advanced Materials*, 17, 2678–2683 (2005).
104. A. R. Parker, 515 Million years of structural colour. *Journal Optics A: Pure and Applied Optics*, 2, R15–28 (2000).
105. A. R. Parker, et al., Aphrodite's iridescence, *Nature*, 409, 36–37 (2001).
106. A. R. Parker, et al., An opal analogue discovered in a weevil, *Nature*, 426, 786–787 (2003).
107. A. R. Parker, Z. Hegedus, and R. A. Watts, Solar-absorber type antireflector on the eye of an Eocene fly (45 Ma). *Proceedings of the Royal Society of London B: Biological Sciences*, 265, 811–815 (1998).
108. A. R. Parker and H. E. Townley, Biomimetics of photonic nanostructures, *Nature Nanotechnology*, 2, 347–353 (2007).
109. W. Barthlott and C. Neinhuis, Purity of the sacred lotus or escape from contamination in biological surfaces, *Planta*, 202, 1–7 (1997).
110. A. Solga, et al., The dream of staying clean: Lotus and biomimetic surfaces, *Bioinspiration and Biomimetics*, 2, S126–S134 (2007).
111. M. Callies and D. Quéré, On water repellency, *Soft Matter*, 1, 55–61 (2005).
112. D. Quéré, Non-sticking drops, *Reports on Progress in Physics*, 68, 2495 (2005).
113. J. Bico, C. Marzolin, D. Quere, Pearl drops, *Europhysics Letters*, 47, 220 (1997).
114. X.-M. Li, D. Reinhoudt, and M. Crego-Calama, What do we need for a superhydrophobic surface? A review on the recent progress in the preparation of superhydrophobic surfaces, *Chemical Society Reviews*, 36, 1350 (2007).

115. Y.-J. Sheng, S. Jiang, and H.-K. Tsao, Effects of geometrical characteristics of surface roughness on droplet wetting, *Journal of Chemical Physics*, 127, 234704 (2007).
116. H. Zhang, R. N. Lamb, and D. J. Cookson, Nanowetting of rough superhydrophobic surfaces, *Applied Physics Letters*, 91, 254106 (2007).
117. R. D. Narhe and D. A. Beysens, Water condensation on a super-hydrophobic spike surface, *Europhysics Letters*, 75, 98 (2006).
118. J. Hyvaluoma, et al., Droplets on inclined rough surfaces, *European Physics Journal E*, 23, 289–293 (2007).
119. K. Autumn and W. Hansen, Ultrahydrophobicity indicates a non-adhesive default state in gecko setae, *Journal of Comparative Physiology A*, 192, 1205–1212 (2006).
120. W. R. Hansen and K. Autumn, Evidence for self-cleaning in gecko setae, *Proceedings of the National Academy of Sciences*, 102, 385–389 (2005).
121. G. S. Bakken, et al., It's just ducky to be clean: The water repellency and water penetration resistance of swimming mallard *Anas platyrhynchos* ducklings, *Journal of Avian Biology*, 37, 561 (2006).
122. Y. Fang, et al., Hydrophobicity mechanism of non-smooth pattern on surface of butterfly wing, *Chinese Science Bulletin*, 52, 711 (2007).
123. Y. Zheng, X. Gao, and L. Jiang, Directional adhesion of superhydrophobic butterfly wings, *Soft Matter*, 3, 178 (2007).
124. H. I. Hima, et al., Novel carbon nanostructures of caterpillar-like fibers and interwoven spheres with excellent surface super-hydrophobicity produced by chemical vapor deposition, *Journal of Material Chemistry*, 18, 1245 (2008).
125. X. Gao, and L. Jiang, Water-repellent legs of water striders, *Nature* 432, 36 (2004).
126. W. H. Thorpe, Plastron Respiration in Aquatic Insects, *Biological Reviews*, 25, 344 (1950).
127. P. J. P. Goodwyn, D. Voigt, and K. Fujisaki, Skating and diving: Changes in functional morphology of the setal and microtrichial cover during ontogenesis in *Aquarius paludum fabricius* (Heteroptera, Gerridae), *Journal of Morphology*, DOI: 10.1002/jmor.10619 (2008).
128. N. J. Shirtcliffe, et al., Plastron properties of a superhydrophobic surface, *Applied Physics Letters*, 89, 104106 (2006).
129. C. Dorrer and J. Ruhe, Wetting of silicon nanograss: From superhydrophilic to superhydrophobic surfaces, *Advanced Materials*, 20, 159 (2008).
130. M. J. O'Donnell, Hydrophilic cuticle - the basis for water vapour absorption by the desert burrowing cockroach, *Arenivaga investigate*, *Journal of Experimental Biology*, 99, 43–60 (1982).
131. Y. B. Gerbig, et al., Effect of nanoscale topography and chemical composition of surfaces on their microfrictional behavior, *Tribology Letters*, 21, 161 (2006).
132. X. Zhang, et al., Effect of pattern topology on the self-cleaning properties of textured surfaces, *Journal of Chemical Physics*, 127, 014703 (2007).
133. B. D'Urso, J. T. Simpson, and M. Kalyanaraman, Emergence of superhydrophobic behavior on vertically aligned nanocone arrays, *Applied Physics Letters*, 90, 044102 (2007).
134. N. Zhao, et al., A novel ultra-hydrophobic surface: Statically non-wetting but dynamically non-sliding, *Advanced Functional Materials*, 17, 2739–2745 (2007).
135. N. Verplanck, Y. Coffinier, V. Thomy, and R. Boukherroub, Wettability Switching Techniques on Superhydrophobic Surfaces, *Nanoscale Research Letters*, 2, 577 (2007).
136. P. Roach, N. J. Shirtcliffe, and M. I. Newton, Progress in Superhydrophobic Surface Development, *Soft Matter*, 4, 224–240 (2008).

137. X. Yu, et al., Reversible pH-responsive surface: From superhydrophobicity to superhydrophilicity, *Advanced Materials*, 17, 1289 (2005).
138. A. Tuteja, et al., Designing superoleophobic surfaces, *Science*, 318, 1618 (2007).
139. F. DiBenedetto, et al., Photoswitchable organic nanofibers, *Advanced Materials*, 20, 314 (2008).
140. W. Sun, et al., Reversible switching on superhydrophobic TiO_2 nano-strawberry films fabricated at low temperature, *Chemical Communications*, 2008, 603–605 (2008).
141. M. Morra, E. Occhiello, and F. Garbassi, Surface characterization of plasma-treated PTFE, *Surface and Interface Analysis*, 16, 412 (1990).
142. M. Ferrari, et al., Surfactant adsorption at superhydrophobic surfaces, *Applied Physics Letters*, 89, 053104 (2006).
143. A. Nygard, et al., A Simple approach to micro-patterned surfaces by breath figures with internal structure using thermoresponsive amphiphilic block copolymers, *Australian Journal of Chemistry*, 58, 595–599 (2005).
144. G. R. J. Artus, et al., Silicone nanofilaments and their application as superhydrophobic coatings, *Advanced Materials*, 18, 2758 (2006).
145. D. Kim, et al., Superhydrophobic nano-wire entanglement structures, *Journal of Micromechanics and Microengineering*, 16, 2593–2597 (2006).
146. P. van der Wal, and U. Steiner, Super-hydrophobic surfaces made from Teflon, *Soft Matter*, 3, 426 (2007).
147. T. Ishizaki, et al., Fabrication and characterization of ultra-water-repellent alumina-silica composite films, *Journal of Physics D: Applied Physics*, 40, 192 (2007).
148. M. O. Gallyamov, et al., Formation of superhydrophobic surfaces by the deposition of coatings from supercritical carbon dioxide, *Colloid Journal*, 69, 411–424 (2007).
149. I. A. Larmour, et al., Remarkably simple fabrication of superhydrophobic surfaces using electroless galvanic deposition, *Angewandte Chemie*, 46, 1710 (2007).
150. M. Qu, et al., Fabrication of superhydrophobic surfaces on engineering materials by a solution-immersion process, *Advanced Functional Materials*, 17, 593 (2007).
151. J. Li, et al., Carbon nanofibers "spot-welded" to carbon felt: A mechanically stable, bulk mimic of Lotus leaves, *Advanced Materials*, 20, 420 (2008).
152. T. Mizukoshi, et al., Control over wettability of textured surfaces by electrospray deposition, *Journal of Applied Polymer Science*, 103, 3811 (2007).
153. M. Motornov, et al., Superhydrophobic surfaces generated from water-borne dispersions of hierarchically assembled nanoparticles coated with a reversibly switchable shell, *Advanced Materials*, 20, 200 (2008).
154. X.-J. Huang, et al., A one-step route to a perfectly ordered wafer-scale microbowl array for size-dependent superhydrophobicity, *Small*, 4, 211 (2008).
155. Y. Liu, et al., Superhydrophobic behavior on transparency and conductivity controllable ZnO/Zn films, *Journal of Applied Physics*, 103, 056104 (2008).
156. J. E. Ruckman, Water vapour transfer in waterproof breathable fabrics: Part 3: under rainy and windy conditions, *International Journal of Clothing Science Technology*, 9, 141 (1997).
157. I. P. Parkin and R. G. Palgrave, Self-cleaning coatings, *Journal of Materials Chemistry*, 15, 1689–1695 (2005).
158. J.-T. Yeh, C.-L. Chen, and K.-S. Huang, Preparation and application of fluorocarbon polymer/SiO_2 hybrid materials, part 2: Water and oil repellent processing for cotton fabrics by sol-gel method, *Journal of Applied Polymer Science*, 103, 3019 (2007).
159. T. Wang, X. Hu, and S. Dong, A general route to transform normal hydrophilic cloths into superhydrophobic surfaces, *Chemical Communications*, 2007, 1849–1851 (2007).

160. H. Saito, K. Takai, and G. Yamauchi, Water- and ice-repellent coatings, *Surface Coatings International*, 80, 168 (1997).
161. A. Marmur, Super-hydrophobicity fundamentals: Implications to biofouling prevention, *Biofouling*, 22, 107 (2006).
162. Y. T. Cheng, et al., Effects of micro- and nano-structures on the self-cleaning behaviour of lotus leaves, *Nanotechnology*, 17, 1359–1362 (2006).
163. Y. C. Chang, et al., Design and fabrication of a nanostructured surface combining antireflective and enhanced-hydrophobic effects, *Nanotechnology*, 18, 285303 (2007).
164. X. Gao, et al., The dry-style antifogging properties of Mosquito compound eyes and artificial analogues prepared by soft lithography, *Advanced Materials*, 19, 2213–2217 (2007).
165. J. A. Howarter and J. P. Youngblood, Self-cleaning and next generation anti-fog surfaces and coatings, *Macromolecular Rapid Communications*, DOI: 10.1002/marc.200700733 (2008).
166. Y. Zhang, S. Sundararajan, Superhydrophobic engineering surfaces with tunable air-trapping ability, *Journal of Micromechanics and Microengineering*, 18, 035024 (2008).
167. H. Schott, Contact angles and wettability of human skin, *Journal of Pharmaceutical Science*, 60, 1893–1895 (1971).
168. W.-C. Liao and J. L. Zatz, Critical surface tensions of pharmaceutical solids, *Journal of Pharmaceutical Science*, 68, 488494 (1979).
169. M. O. Riehle, Biocompatibility: Nanomaterials for cell- and tissue engineering, *Nanobiotechnology*, 1, 308–309 (2005).
170. T. Sun, et al., No platelet can adhere - Largely improved blood compatibility on nanostructured superhydrophobic surfaces, *Small*, 1, 959 (2005).
171. G. McHale, N. J. Shirtcliffe, and M. I. Newton, Super-hydrophobic and super-wetting surfaces: Analytical potential? *The Analyst*, 129, 284–287 (2004).
172. J. Ou, G. R. Moss, and J. P. Rothstein, Enhanced mixing in laminar flows using ultrahydrophobic surfaces, *Physical Review E*, 76, 016304 (2007).
173. K. Fukuzawa, et al., Conformation and motion of monolayer lubricant molecule on magnetic disks, *IEEE Transactions on Magnetics*, 41, 3034 (2005).
174. K. Gjerde, et al., On the suitability of carbon nanotube forests as non-stick surfaces for nanomanipulation, *Soft Matter*, 4, 392 (2008).
175. C.-M. Tøg, et al., Influence of surface structure on wetting of coated offset papers, *Holzforschung*, 61, 516 (2007).
176. P. J. Holloway, Surface factors affecting the wetting of leaves, *Pesticide Science*, 1, 156 (1970).
177. A. N. Round, et al., The influence of water on the nanomechanical behavior of the plant biopolyester cutin as studied by AFM and solid-state NMR, *Biophysical Journal*, 79, 2761–2767 (2000).
178. J. Poulenard, et al., Water repellency of volcanic ash soils from Ecuadorian paramo: Effect of water content and characteristics of hydrophobic organic matter, *European Journal of Soil Science*, 55, 487–496 (2004).
179. R. W. McDowell, The effectiveness of coal fly-ash to decrease phosphorus loss from grassland soils, *Australian Journal Soil Research*, 43, 853–860 (2005).
180. G. McHale, M. I. Newton, and N. J. Shirtcliffe, Water-repellent soil and its relationship to granularity, surface roughness and hydrophobicity: A materials science view, *European Journal of Soil Science*, 56, 445–452 (2005).
181. N. J. Shirtcliffe, et al., Critical conditions for the wetting of soils, *Applied Physics Letters*, 89, 094101 (2006).
182. G. McHale, et al., Implications of ideas on super-hydrophobicity for water repellent soil, *Hydrological Processes*, 21, 2229–2238 (2007).

183. D. A. L. Leelamanie and J. Karube, Effects of organic compounds, water content and clay on the water repellency of a model sandy soil, *Soil Science and Plant Nutrition*, 53, 711–719 (2007).
184. F. Bartoli, A. J. Poulenard, and B. E. Schouller, Influence of allophane and organic matter contents on surface properties of Andosols, *European Journal of Soil Science*, 58, 450–464 (2007).
185. B. Chen and J. Fan, Microstructures of chafer cuticle and biomimetic design, *Journal of Computer-Aided Materials Design*, 11, 1573–4900 (2004).
186. T. G. Rials and W. G. Glasser, Engineering plastics from lignin. XIII. Effect of lignin structure on polyurethane network formation, *Holzforschung*, 40, 353–360 (2006).
187. M. Milwich, et al., Biomimetics and technical textiles: Solving engineering problems with the help of nature's wisdom, *American Journal of Botany*, 93, 1455–1465 (2006).
188. J. Gravitis, Nano level structures in wood cell wall composites, *Cellulose Chemistry and Technology*, 40, 291–298 (2006).
189. J. Cao, R. Wijaya, and F. Leroy, Unzipping the cuticle of the human hair shaft to obtain micron/nano keratin filaments, *Biopolymers*, 83, 614–618 (2006).
190. N. Kohli, et al., Direct transfer of preformed patterned bio-nanocomposite films on polyelectrolyte multilayer templates, *Macromolecular Bioscience*, 7, 789–797 (2007).
191. T. T. Teeri, H. Brumer 3rd, G. Daniel, and P. Gatenholm, Biomimetic engineering of cellulose-based materials, *Trends in Biotechnology*, 25, 299–306 (2007).
192. Y. Kaneko, S. Matsuda, and J. Kadokawa, Chemoenzymatic syntheses of amylose-grafted Chitin and Chitosan, *Biomacromolecules*, 8, 3959–3964 (2007).
193. Technical Research Centre of Finland,. Water Repellent Wood Fiber Products Developed. ScienceDaily 28 January 2008, website at: http://www.sciencedaily.com-/releases/2008/01/080123163554.htm (2008)
194. Z. Lin, S. Renneckar, and D. P Hindman, Nanocomposite-based lignocellulosic fibers 1: Thermal, *Cellulose*, 15, 333–346 (2008).
195. N. M Pugno, Towards a Spiderman suit: Large invisible cables and self-cleaning releasable superadhesive materials, *Journal of Physics: Condensed Matter*, 19, 395001 (2007).
196. K. Autumn and N. Gravish, Gecko adhesion: Evolutionary nanotechnology, *Philosophical Transactions of the Royal Society of London A: Mathematical, Physical, and Engineering Science*, 366, 1575–1590 (2008).
197. E. Arzt, Biological and artificial attachment devices: Lessons for materials scientists from flies and geckos, *Materials Science and Engineering C: Biomimetic and Supramolecular System*, 26, 1245–1250 (2006).
198. K. Autumn, M. Sitti, Y. A. Liang, A. M. Peattie, and W. R. Hansen, Evidence for van der Waals adhesion in gecko setae, *Proceedings of National Academy of Sciences*, 99, 12252–12256 (2002).
199. J. Lee, C. Majidi, B. Schubert, and R. Fearing, Sliding induced adhesion of stiff polymer microfiber arrays: 1. Macroscale behaviour, *Journal of the Royal Society Interface*, (10.1098/rsif.2007.1308) (2008).
200. B. Schubert, et al., Sliding induced adhesion of stiff polymer microfiber arrays: 2. Microscale behaviour, *Journal of the Royal Society Interface*, (10.1098/rsif.2007.1309) (2008).
201. K. Autumn, Gecko adhesion: Structure, function, and applications, *MRS Bulletin*, 32, 473–478 (2007).
202. S. Kim, et al., Effect of soft backing layer thickness on adhesion of single-level elastomer fiber arrays, *Applied Physics Letters*, 91, 161905–161907 (2007).

203. S. Kim, B. Aksak, and M. Sitti, Enhanced friction of polymer microfiber adhesives with spatulate tips, *Applied Physics Letters*, 91, 221913–221915 (2007).
204. M. Murphy, et al., Adhesion and anisotropic friction enhancement of angled heterogeneous micro-fiber arrays with spherical and spatula tips, *Journal of Adhesion Science and Techology*, 21, 1281–1296 (2007).
205. E. Arzt, S. Gorb, and R. Spolenak, From micro to nano contacts in biological attachment devices, *Proceedings of National Academy of Science USA*, 100, 10603–10606 (2003).
206. R. Spolenak, S. Gorb, and E. Arzt, Adhesion design maps for bio-inspired attachment systems, *Acta Biomaterialia*, 1, 5–13 (2005).
207. C. Majidi, et al., High friction from a stiff polymer using microfiber arrays, *Physics Review Letters*, 97, 076103 (2006).
208. L. Ge, et al., Carbon nanotube-based synthetic gecko tapes, *Proceedings of the National Academy of Sciences*, 104, 10792–10795 (2007).
209. H. Lee, B. P. Lee, and P. B. Messersmith, A reversible wet/dry adhesive inspired by mussels and geckos, *Nature*, 448, 338–341 (2007).
210. A. Mahdavi, et al., A biodegradable and biocompatible gecko-inspired tissue adhesive, *Proceedings of the National Academy of Sciences*, 105, 2307–2312 (2008).
211. A. K. Geim, et al., Microfabricated adhesive mimicking gecko foot-hair, *Nature Materials*, 2, 461–463 (2003).
212. M. T. Northen and K. L. Turner, A batch fabricated biomimetic dry adhesive, *Nanotechnology*, 16, 1159–1166 (2005).
213. I. C. Gebeshuber, Biotribology inspires new technologies, *Nano Today*, 2, 30–37 (2007).
214. B. Bhushan, ed., *Handbook of micro/nanotribology*, CRC Press, Boca Raton, FL (1999).
215. B. Bhushan, ed., *Modern tribology handbook, Vol. 1 - Principles of tribology, Section II*, CRC Press, Boca Raton, FL (2001).
216. B. Bhushan, Tribology: Friction, wear, and lubrication, In *The engineering handbook*, R. C. Dorf, ed., p. 210, CRC Press, Boca Raton, FL (2000).
217. M. Scherge and S. Gorb, *Biological micro- and nanotribology – Nature's solutions*, Springer Verlag, Berlin, Germany (2001).
218. D. Qur, et al., Slippy and sticky microtextured solids, *Nanotechnology*, 14, 1109–1112 (2003).
219. P. K. Hansma, et al., Optimized adhesives for strong, lightweight, damage-resistant, nanocomposite materials: New insights from natural materials, *Nanotechnology*, 18, 044026 (2007).
220. I. C. Gebeshuber, et al., Diatom bionanotribology—Biological surfaces in relative motion: Their design, friction, adhesion, lubrication and wear, *Journal of Nanoscience and Nanotechnology*, 5, 79–87 (2005).
221. R. Gordon, ed., A special issue on diatom nanotechnology, *Journal of Nanoscience Nanotechnology*, 5, 1–4 (2005).
222. Y. X. Zhuang and A. Menon, On the stiction of MEMS materials, *Tribology Letters*, 19, 111 (2005).
223. J. A. Raymond and C. A. Knight, Ice binding, recrystallization inhibition, and cryoprotective properties of ice-active substances associated with Antarctic sea ice diatoms, *Cryobiology*, 46, 174–181 (2003).
224. C. E. Orsello, et al., Molecular properties in cell adhesion: A physical and engineering perspective, *Trends Biotechnology*, 19, 310–316 (2001).
225. R. L. Thurmond, et al., The role of histamine H1 and H4 receptors in allergic inflammation: The search for new antihistamines, *Nature Reviews Drug Discovery*, 7, 41–53 (2008).
226. O. Abbassi, et al., Canine neutrophil margination mediated by lectin adhesion molecule-1 in vitro, *Journal of Immunology*, 147, 2107–2115 (1991).

227. D. A. Jones, et al., P-selectin mediates neutrophil rolling on histamine-stimulated endothelial cells, *Biophysical Journal*, 65, 1560–1569 (1993).
228. K. Ley, et al., Lectin-like cell adhesion molecule 1 mediates leukocyte rolling in mesenteric venules in vivo, *Blood*, 77, 2553–2555 (1991).
229. T. A. Springer, Traffic signals for lymphocyte recirculation and leukocyte emigration: The multistep paradigm, *Cell*, 76, 301–314 (1994).
230. B. R. Alevriadou, et al., Real-time analysis of shear-dependent thrombus formation and its blockade by inhibitors of von Willebrand factor binding to platelets, *Blood*, 81, 1263–1276 (1993).
231. M. B. Lawrence, et al., Effect of venous shear stress on CD18-mediated neutrophil adhesion to cultured endothelium, *Blood*, 75, 227–237 (1990).
232. H.-W. Denker, Molecular approaches to cell-cell adhesion: From leukocyte extravasation to embryo implantation, *Cells Tissues Organs*, 172, 150–151 (2002).
233. K. V. Honn, et al., Enhanced tumor cell adhesion to the subendothelial matrix resulting from 12(S)-HETE-induced endothelial cell retraction, *FASEB Journal*, 3, 2285–2293 (1989).
234. R. P. McEver, K. L. Moore, and R. D. Cummings, Leukocyte trafficking mediated by selectin-carbohydrate interactions, *Journal of Biological Chemistry*, 270, 11025–11028 (1995).
235. C. V. Carman and T. A. Springer, Integrin avidity regulation: Are changes in affinity and conformation underemphasized? *Current Opinion in Cell Biology*, 15, 547–556 (2003).
236. H. Ait-Oufella, E. Maury, B. Guidet, and G. Offenstadt, The endothelium: A new organ (L'endothélium: un nouvel organe), *Reanimation*, 17, 126–136 (2008).
237. J. G. Lock, B. Wehrle-Haller, and S. Strömblad, Cell-matrix adhesion complexes: Master control machinery of cell migration, *Seminars in Cancer Biology*, 18, 65–76 (2008).
238. D. K. Brunk, D. J. Goetz, and D. A. Hammer, Sialyl Lewis(x)/E-selectin-mediated rolling in a cell-free system, *Biophysical Journal*, 71, 2902–2907 (1996).
239. D. K. Brunk and D. A. Hammer, Quantifying rolling adhesion with a cell-free assay: E-selectin and its carbohydrate ligands, *Biophysical Journal*, 72, 2820–2833 (1997).
240. D. F. J. Tees and D. J. Goetz, Leukocyte adhesion: An exquisite balance of hydrodynamic and molecular forces, *News in Physiological Sciences*, 18, 186–190 (2003).
241. S. Reboux, G. Richardson, and O. E. Jensen, Bond tilting and sliding friction in a model of cell adhesion, *Proceedings of the Royal Society of London A: Mathematical, Physical and Engineering Science*, 464, 447–467 (2008).
242. M. Shimaoka and T. A. Springer, Therapeutic antagonists and conformational regulation of integrin function, *Nature Reviews Drug Discovery*, 2, 703–716 (2003).
243. S. Huveneers, H. Truong, and E. H. J. Danen, Integrins: Signaling, disease, and therapy, *International Journal of Radiation Biology*, 83, 743–751 (2007).
244. S. Choi, et al., Small molecule inhibitors of integrin α2β1, *Journal of Medicinal Chemistry*, 50, 5457–5462 (2007).
245. F. G. Giancotti, Targeting integrin β4 for cancer and anti-angiogenic therapy, *Trends in Pharmacological Sciences*, 28, 506–511 (2007).
246. C. Coisne, et al., Therapeutic targeting of leukocyte trafficking across the blood-brain barrier, *Inflammation and Allergy - Drug Targets*, 6, 210–222 (2007).
247. A. R. Aricescu and E. Y. Jones, Immunoglobulin superfamily cell adhesion molecules: Zippers and signals, *Current Opinion in Cell Biology*, 19, 543–550 (2007).
248. R. P. McEver, Adhesive interactions of leukocytes, platelets, and the vessel wall during hemostasis and inflammation, *Thrombosis and Haemostasis*, 86, 746–756 (2001).

249. D. Vestweber, Adhesion and signaling molecules controlling the transmigration of leukocytes through endothelium, *Immunological Reviews*, 218, 178–196 (2007).
250. M. R. Morgan, et al., Synergistic control of cell adhesion by integrins and syndecans, *Nature Reviews Molecular Cell Biology*, 8, 957–969 (2007).
251. R. P. McEver and R. D. Cummings, Cell adhesion in vascular biology. Role of PSGL-1 binding to selectins in leukocyte recruitment, *Journal of Clinical Investigation*, 100, 485–491 (1997).
252. T. A. Springer and J.-H. Wang, The three-dimensional structure of integrins and their ligands, and conformational regulation of cell adhesion, *Advances in Protein Chemistry*, 68, 29–63 (2004).
253. S. Zhuang, et al., Multiple α subunits of integrin are involved in cell-mediated responses of the Manduca immune system, *Developmental and Comparative Immunology*, 32, 365–379 (2008).
254. J. Takagi, Structural basis for ligand recognition by integrins, *Current Opinion in Cell Biology*, 19, 557–564 (2007).
255. A. S. Popel and R. N. Pittman, Mechanics and transport in microcirculation, In *The biomedical engineering handbook*, 2nd ed., J. D. Bronzino, ed., pp. 31–101, CRC Press, Boca Raton, FL (2000).
256. C. W. Patrick, et al., Fluid shear stress effects on cellular function, In *The biomedical engineering handbook*, 2nd ed., J. D. Bronzino, ed., pp. 114–201, CRC Press, Boca Raton, FL (2000).
257. J. Li and W. Zhong, A two-dimensional suspension array system by coupling field flow fractionation to flow cytometry, *Journal of Chromatography A*, 1183, 143–149 (2008).
258. R. H. G. Baxter, et al., Structural basis for conserved complement factor-like function in the antimalarial protein TEP1, *Proceedings of the National Academy of Sciences*, 104, 11615–11620 (2007).
259. B. N. J. Persson, On the mechanism of adhesion in biological systems, *Journal of Chemical Physics*, 118, 7614–7621 (2003).
260. G. E. Fantner, et al., Sacrificial bonds and hidden length: Unraveling molecular mesostructures in tough materials, *Biophysical Journal*, 90, 1411–1418 (2006).
261. T. M. Dugdale, et al., Single adhesive nanofibers from a live diatom have the signature fingerprint of modular proteins, *Biophysical Journal*, 89, 4252–4260 (2005).
262. A. S. Mostaert and S. P. Jarvis, Beneficial characteristics of mechanically functional amyloid fibrils evolutionarily preserved in natural adhesives, *Nanotechnology*, 18, 044010 (2007).
263. A. S. Mostaert, et al., Nanoscale mechanical characterisation of amyloid fibrils discovered in a natural adhesive, *Journal of Biological Physics*, 32, 2887–2893 (2006).
264. A. S. Mostaert, T. Fukuma, and S. P. Jarvis, Explanation for the mechanical strength of amyloid fibrils, *Tribology Letters*, 22, 233–237 (2006).
265. J. H. Fendler, *Membrane-mimetic approach to advanced materials*, (*Advances in polymer science)*, Vol. 113, Springer Verlag, Berlin (1994).
266. C. R. Martin, Nanomaterials: A membrane based synthetic approach, *Science*, 266, 1961–66 (1994).
267. X.-Y. Zhang, et al., Synthesis of ordered single crystal silicon nanowire arrays, *Advanced Materials*, 13, 1238–1241 (2001).
268. M. A. Guillorn, et al., Individually addressable vertically aligned carbon nanofiber-based electrochemical probes, *Journal of Applied Physics*, 91, 3824–2828 (2002).
269. M. J. Doktycz, et al., Nanofiber Structures as Mimics for Cellular Membranes, *Nanotechnology*, 3, 420–423 (2003).

270. T. A. Desai, et al., Microfabricated immunoisolating biocapsules, *Biotechnology and Bioengineering*, 57, 118–120 (1998).
271. K. Senior, "Nano-dumpling" with drug delivery potential, *Molecular Medicine Today*, 4, 321 (1998)
272. P. Broz, et al., Toward intelligent nanosize bioreactors: A pH-switchable, channel-equipped, functional polymer nanocontainer, *Nano Letters*, 6, 2349–2353 (2006).
273. D. Akin, et al., Bacteria-mediated delivery of nanoparticles and cargo into cells, *Nature Nanotechnology*, 2, 441–449 (2006).
274. E. R. Ballister, et al., Nanotubes from biomimetically bioengineered viruses for drug delivery: In vitro self-assembly of tailorable nanotubes from a simple protein building block, *Proceedings of the National Academy of Sciences*, 105, 3733–3738 (2008).
275. A. R. Parker and C. R. Lawrence, Water capture by a desert beetle, *Nature*, 414, 33–34 (2001).
276. M. L. Fishman, P. H. Cooke, and D. R. Coffin, Nano structure of native pectin sugar acid gels visualized by atomic force microscopy, *Biomacromolecules*, 5, 334–341 (2004).
277. A. Hentschel, S. Gramdorf, R. H. Müller, and T. Kurz, Beta-Carotene-loaded nanostructured lipid carriers, *Journal of Food Science*, 73, N1-6 (2008).
278. J. Fava, et al., Structure and nanostructure of the outer tangential epidermal cell wall in *Vaccinium corymbosum* L. (Blueberry) fruits by blanching, freezing-thawing and ultrasound, *Food Science and Technology International*, 12, 241–251 (2006).
279. R. H. J. Hannink and A. J. Hill, eds., *Nanostructure control of materials*, Woodhead Publishing Limited, Abington, Cambridge, UK (2006).
280. J. Dyck, The evolution of feathers, *Zoologica Scripta*, 14, 137 (1985).
281. J. A. Tuszynski, *Molecular and cellular biophysics*, CRC Press, Boca Raton, FL (2007).
282. D. Jones, J. Round, and A. de Haan, *Skeletal muscle: From molecules to movement*, Elsevier, New York (2004).
283. R. Bartlett, *Introduction to sports biomechanics*, Taylor and Francis, London (1996).
284. M. M. Dewey, et al., Structure of limulus striated muscle: The contractile apparatus at various sarcomere lengths, *The Journal of Cell Biology*, 58, 574–593 (1974).
285. B. M. Millman, The filament lattice of striated muscle, *Physiological Reviews*, 78, 359–391 (1998).
286. A. M. Herrera, et al., Sarcomeres' of smooth muscle: Functional characteristics and ultrastructural evidence, *Journal of Cell Science*, 118, 2381–2392 (2005).
287. M. Reconditi, et al., Structure-function relation of the Myosin motor in striated muscle, *Annals of the New York Academy of Sciences*, 1047, 232–247 (2005).
288. O. M. Hernandez, et al., Plasticity in skeletal, cardiac, and smooth muscle, Invited review: Pathophysiology of cardiac muscle contraction and relaxation as a result of alterations in thin filament regulation, *Journal of Applied Physiology*, 90, 1125–1136 (2001).
289. J. Arikkath and K. P. Campbell, Auxillary subunits: Essential components of the voltage-gated calcium channel complex, *Current Opinion in Neurobiology*, 13, 298–307 (2003).
290. M. Kang and K. P. Campbell, The gamma subunit of voltage-activated calcium channels, Mini Review, *Journal Biological Chemistry*, 278, 21315–21318 (2003).
291. D. Michele and K. P. Campbell, Cardiomyopathy in muscular dystrophies, In *Molecular mechanisms for cardiac hypertrophy and failure*, R. A. Walsh, ed., pp. 541–567, Taylor & Francis, London (2005).
292. M. Shahinpoor, et al., *Artificial muscles: Applications of advanced polymeric nanocomposites*, Taylor & Francis, London (2007).

293. W. Yim, J. Lee, and K. J. Kim, An artificial muscle actuator for biomimetic underwater propulsors, *Bioinspiration and Biomimetics*, 2, S31–S41 (2007).
294. R. H. Baughman, Playing Nature's game with artificial muscles, *Science*, 308, 63–65 (2005).
295. J. Ayers, J. L. Davis, and A. Rudolph, eds., *Neurotechnology for biomimetic robots. Based on a conference held in May 2000*, MIT Press, Cambridge, MA (2002).
296. Y. Bar-Cohen, ed., Electroactive polymer (EAP) actuators as artificial muscles - Reality, potential and challenges, In *Electroactive polymer actuators and devices conference: Smart structures and materials symposium*, JPL, San Diego, CA (2005).
297. M. Shahinpoor, Ionic polymer–conductor composites as biomimetic sensors, robotic actuators and artificial muscles—a review, *Electrochimica Acta*, 48, 2343–2353 (2003).
298. P. J. Yoo, et al., Spontaneous assembly of viruses on multilayered polymer surfaces, *Nature Materials*, 5, 234–240 (2006).
299. C. Mao, et al., Virus-based toolkit for the directed synthesis of magnetic and semiconducting nanowires, *Science*, 303, 213–217 (2004).
300. D. W. Lawlor, *Photosynthesis*, Routledge, Andover, UK (2004).
301. M. Pessarakli, ed., *Handbook of photosynthesis*, 2nd ed., CRC Press, Boca Raton, FL (2005).
302. P. D. Boyer, The ATP synthase—a splendid molecular machine, *Annual Review Of Biochemistry*, 66, 717–49 (1997).
303. R. Govindjee, J. T. Beatty, H. Gest, and J. F. Allen, eds., *Discoveries in photosynthesis, advances in photosynthesis and respiration*, Vol. 20, Springer, Berlin, Germany (2006) (Reprinted from *Photosynthesis Research*, 73, 76, 80).
304. B. R. Selman and S. Selman-Reimer, eds., *Energy coupling in photosynthesis*, Elsevier, New York (1981).
305. S. Tanaka and R. A. Marcus, Electron transfer model for the electric field effect on quantum yield of charge separation in bacterial photosynthetic reaction centers, *Journal of Physical Chemistry B*, 101, 5031 (1997).
306. J. Deisenhofer, et al., Structure of the protein subunits in the photosynthetic reaction centre of *Rhodopseudomonas viridis* at 3 Å resolution, *Nature*, 318, 618–624 (1985).
307. J. J. Katz and M. R. Wasielewski, Biomimetic approaches to artificial photosynthesis, *Biotechnology And Bioengineering Symposium*, 8, 423–452 (1978).
308. D. Gust and T. A. Moore, Mimicking photosynthesis, *Science*, 244, 35–41 (1989).
309. D. Gust, T. A. Moore, and A. L. Moore, Molecular mimicry of photosynthetic energy and electron transfer, *Accounts of Chemical Research*, 34, 40–48 (2001).
310. L. Hammarstrom, et al., A biomimetic approach to artificial photosynthesis: Ru(II)-polypyridine photo-sensitisers linked to tyrosine and manganese electron donors, *Spectrochimica Acta Part A: Molecular and Biomolecular Spectroscopy*, 57, 2145-2160 (2001).
311. L. Hammarström, L. Sun, B. Åkermark, and S. Styring, A biomimetic approach to artificial photosynthesis: Ru(II)–polypyridine photo-sensitisers linked to tyrosine and manganese electron donors, *Spectrochimica Acta Part A: Molecular and Biomolecular Spectroscopy*, 57, 2145–2160 (2001).
312. D. A. LaVan and J. N. Cha, Approaches for biological and biomimetic energy conversion, *Proceedings of National Academy of Science USA*, 103, 5251–5255 (2006).
313. M. A. Baldo, Photosynthetic photovoltaic cells, Final Report US DTI ADA469444 (2006).
314. A. F. Collings and C. Critchley, eds., *Artificial photosynthesis: From basic biology to industrial application*, Wiley-VCH Verlag, Weinheim, Germany (2005).
315. D. R. Ort and C. F. Yocum, eds., Electron transfer and energy transduction in photosynthesis: An overview, chap. 1, In *Oxygenic photosynthesis: The light reactions,*

advances in photosynthesis, Vol. 2, Kluwer Academic Publishers, the Netherlands, 1–9 (1996).

316. L. A. Staehelin and G. W. M. van der Staay, Structure, composition, functional organization and dynamic properties of thylakoid membranes, chap. 2, In *Oxygenic photosynthesis: The light reactions, advances in photosynthesis*, Vol. 2, D. R. Ort and C. F. Yocum, eds, Kluwer Academic Publishers, the Netherlands, 11–30 (1996).
317. P. J. Kiley, et al., Self-assembling peptide detergents stabilize isolated photosystem I on a dry surface for an extended time, *PloS Biology*, 3, 230–237 (2005).
318. R. Das, et al., Integration of photosynthetic protein molecular complexes in solid-state electronic devices, *Nano Letters*, 4, 1079–1083 (2005).
319. B. D. Bruce, M. A. Baldo, and S. Zhang, Integration of photosynthetic complexes into novel biomolecular electronic devices, 2005 NSF Nanoscale Science and Engineering Grantees Conference, December 12–15, Arlington, VA (2005).
320. B. O'Regan and M. Grätzel, A low-cost, high efficiency solar cell based on dye-sensitized colloidal TiO_2 films, *Nature*, 353, 737–740 (1991).
321. M. Grätzel, Applied physics: Solar cells to dye for, *Nature*, 421, 586–587 (2003).
322. P. Wang, et al., A stable quasi-solid-state dye-sensitized solar cell with an amphiphilic ruthenium sensitizer and polymer gel electrolyte, *Nature Materials*, 2, 402–407 (2003).
323. D. Di Censo, et al., Synthesis, characterization, and DFT/TD-DFT calculations of highly phosphorescent blue light-emitting anionic Iridium complexes, *Inorganic Chemistry*, 47, 980–989 (2008).
324. U. Bach, et al., Solid-state dye-sensitized mesoporous TiO_2 solar cells with high photon-to-electron conversion efficiencies, *Nature*, 395, 583–585 (1998).
325. M. K. Nazeeruddin and M. Graetzel, Transition metal complexes for photovoltaic and light emitting applications, *Structure and Bonding*, 123, 113–175 (2007).
326. J.-H. Yum, et al., Efficient co-sensitization of nanocrystalline TiO_2 films by organic sensitizers, *Chemical Communications*, 44, 4680–4682 (2007).
327. C. Lee, et al., Phenomenally high molar extinction coefficient sensitizer with donor-acceptor, *Inorganic Chemistry* 47, INOCAJ ISSN:0020-1669. AN 2007:997170 (2008).
328. M. Grätzel, A high molar extinction coefficient charge transfer sensitizer and its application in dye-sensitized solar cell, *Journal of Photochemistry and Photobiology A: Chemistry*, 185, 331–337 (2007).
329. H. Choi, et al., A highly efficient and thermally stable organic sensitizers for solvent free electrolyte based dye-sensitized solar cells, *Angewandte Chemie*, 47, 327–330 (2008).
330. R. Buscaino, et al., A mass spectrometric analysis of sensitizer solution used for dye-sensitized solar cell, *Inorganica Chimica Acta*, 361, 798–805 (2008).
331. J.-J. Lagref, et al., Artificial photosynthesis based on dye-sensitized nanocrystalline TiO_2 Solar Cells, *Inorganica Chimica Acta*, 361, 735–745 (2008).
332. L. M. Peter, et al., Photosensitization of nanocrystalline TiO_2 by self-assembled layers of CdS quantum dots, *Chemical Communications*, 10, 1030–1031 (2002).
333. K. G. U. Wijayantha, et al., Fabrication of CdS quantum dot sensitized solar cells via a pressing route, *Solar Energy Materials and Solar Cells*, 83, 363–369 (2004).
334. L. M. Peter, et al., Transport and Interfacial transfer of electrons in dye-sensitized nanocrystalline solar cells, *Journal of Electroanalytical Chemistry*, 127, 524–525 (2002).
335. I. L. Medintz, et al., Quantum dot bioconjugates for imaging, labelling and sensing, *Nature Materials*, 4, 435–446 (2005).
336. T. Kuritz, et al., Molecular photovoltaics and the photoactivation of mammalian cells, *IEEE Transactions on Nanobioscience*, 4, 196–200 (2005).
337. B. R. Evans, et al., Enhanced photocatalytic hydrogen evolution by covalent attachment of Plastocyanin to Photosystem I, *Nano Letters*, 10, 1815–1819 (2004).

338. H. M. O'Neill and E. Greenbaum, Spectroscopy and photochemistry of Spinach photosystem I entrapped and stabilized in a hybrid organosilicate glass, *Chemistry of Materials*, 17, 2654–2661 (2005).
339. University of Leiden, Harnessing Solar Energy for the Production of Clean Fuels, EC ESF Task Force White Paper, at http://www.ssnmr.leidenuniv.nl/content_docs/cleansolarfuels.pdf (25 March 2008)
340. X.-G. Zhu, A. R. Portis, and S. P. Long, Would transformation of C3 crop plants with foreign Rubisco increase productivity? A computational analysis extrapolating from kinetic properties to canopy photosynthesis, *Plant Cell and Environment*, 27, 155–165 (2004).
341. H. S. Sakhalkar, et al., Leukocyte-inspired biodegradable particles that selectively and avidly adhere to inflamed endothelium in vitro and in vivo, *Proceedings of National Academy of Science USA*, 100, 15895–15900 (2003).
342. Z. Bao, M. R. Weatherspoon, S. Shian, C. Ye, P. D. Graham, S. M. Allan, G. Ahmad, M. B. Dickerson, B. C. Church, Z. Kang, H. W. Abernathy III, C. J. Summers, M. Liu, and K. H. Sandhage, Chemical reduction of three-dimensional silica micro-assemblies into microporous silicon replicas, *Nature*, 446, 172–175 (2007).
343. J. Lee and N. A. Kotov, Thermometer design at the nanoscale, *Nano Today*, 2, 48–51 (2007).
344. C. J. Brokaw, Molecular mechanism for oscillation in flagella and muscle, *Proceedings of the National Academy of Science USA*, 72, 3102–3106 (1975).
345. L. M. Godsel and D. M. Engman, Flagellar protein localization mediated by a calcium–myristoyl/palmitoyl switch mechanism, *The EMBO Journal*, 18, 2057–2065 (1999).
346. G. M. Whitesides, The Once and Future Nanomachine: Biology outmatches futurists' most elaborate fantasies for molecular robots, Nanotechnology web site at: http://www.mtmi.vu.lt/pfk/funkc_dariniai/nanostructures/nano_robots.htm (2008) (Keynote paper, Symposium on Functional Combinations in Solid States, Finland, 2002).
347. K. Namba, Revealing the mystery of the bacterial flagellum - A self-assembling nano-machine with fine switching capability, *Japan Nanonet Bulletin*, 11, 5th Feb. (2004).
348. M. Manghi, X. Schlagberger, and R. R. Netz, Propulsion with a rotating elastic nanorod, *Physics Review Letters*, 96, 068101 (2006).
349. G. Jensen, A nanoengine for gliding motility, *Molecular Microbiology*, 63, 4–6 (2007).
350. Z. Wang, Synergic mechanism and fabrication target for bipedal nanomotors, *Proceedings of the National Academy of Sciences*, 104, 17921–17926 (2007).
351. M. R. Diehl, K. Zhang, H. J. Lee, and D. A. Tirrell, Engineering cooperativity in biomotor protein assemblies, *Science*, 311, 1468–1471 (2006).
352. L. M. Adleman, Computing with DNA, *Scientific American*, Aug. 54–61 (1998).
353. P. W. K. Rothemund, Using lateral capillary forces to compute by self-assembly, *Proceedings of the National Academy of Science USA*, 97, 984–989 (2000).
354. M. Amos, *Theoretical and experimental DNA computation*, Springer, New York (2005).
355. L. Medskar and L. C. Jain, eds., *Recurrent neural networks (CRC Press International Series on Computational Intelligence)*, CRC Press, Boca Raton, FL (1999).
356. L. C. Jain and V. R. Vemuri, Industrial applications of neural networks, CRC Press, Boca Raton, FL (1998).
357. J. B. Waldner, *Nanocomputers and swarm intelligence*, ISTE, London (2007).
358. J. Macdonald, et al., Medium scale integration of molecular logic gates in an automaton, *Nano Letters*, 6, 2598–2603 (2006).
359. E. Braun and K. Keren, From DNA to transistors, *Advances in Physics*, 53, 441–496 (2004),

360. K. Keren, et al., DNA-templated carbon nanotube field-effect transistor, *Science*, 302, 1380–1382 (2003).
361. C. M. Lieber, The incredible shrinking circuit, In *The rise of nanotechnology*, Scientific American, New York, 285, 58–64 (2001).
362. J. F. V. Vincent, et al., Putting biology into TRIZ: A database of biological effects, *Creativity and Innovation Management*, 14, 66–72 (2005).

Problems

26.1 Which of the following technological developments were most likely based on observations of biological analogs? (a) Airplane, (b) wheel, (c) axe, (d) steam engine, (e) artificial intelligence, and (f) jet engine.

26.2 Why do you think that biomimetics and bionics have become increasingly more important (and achievable)?

26.3 Do some research and determine how many kinds of Velcro fasteners are on the market. Are there any that approach the nanoscale with regard to working components?

26.4 Research, define, and draw the hierarchical structure of rope. Is its structure related to any biological structure(s) that you know?

26.5 Explain the meaning of "hidden structure" and "sacrificial bonds." How do they contribute to strength of materials? Draw a diagram illustrating sacrificial bonds.

26.6 What percentage of glue in relation to mineral bricks is found in shells? Compare the structure of shells to the structures of different types of brick walls. Would the wall be stronger if less mortar is used? If more?

26.7 What is the function of steel bars used in concrete? What are some analogous structural features in nanofabrications?

26.8 What is meant by "rolling" in white blood cells? What are the steps in the process? What types of materials mediate each step?

26.9 Why can discovery of an adhesive molecule in a mosquito help prevent malaria in humans?

26.10 What are some of the practical commercial applications learned from the lotus leaf?

26.11 Do humans use more energy to make materials and power their engines than animals and plants? Explain the reason for your answer.

26.12 What is the difference between biomimetics and hunting and gathering natural materials?

Medical Nanotechnology

While some may dream of nanorobots circulating in the blood, the immediate applications in medicine will occur at the interfaces among ... nanotechnology, micro-electronics, microelectromechanical systems (MEMS) and microopticalelectro-mechanical systems (MOEMS). ... The bounty will not be realized until those trained in these new paradigms begin to ... address basic medical and scientific questions.

D.A. LaVan and R. Langer, MIT,
NSF Symposium 2001

Chapter 27

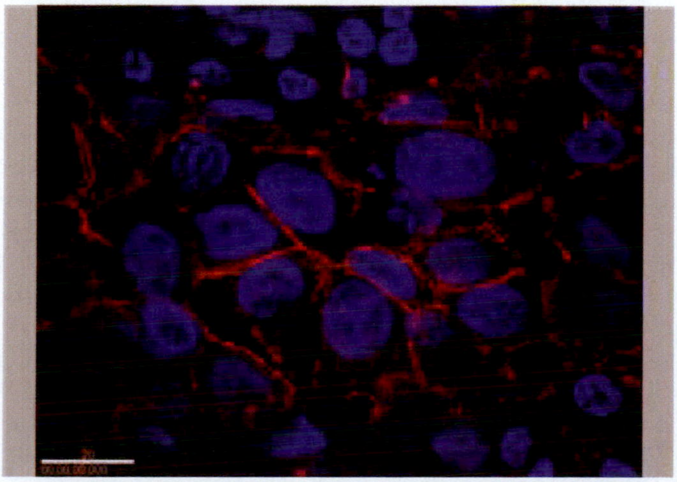

FIG. 27.0 *What if doctors could reach out and correct diseases at the macromolecular level? Nanotechnology provides tools to help bring that dream closer to reality.*

THREADS

Chapter 27 presents an introduction to nanotechnology as applied to medicine. This is a rapidly developing area, with many impacts. This chapter emphasizes device and materials nanotechnology rather than molecular and protein engineering. The impact of nanotechnology on treatment is emphasized: keeping in view that since this is about medicine, an integrated approach focused on the patient rather than the technology will be taken. This introduction focuses on some examples of current applications which are active areas of research and development in biomedicine, with an overview of bio-nanoengineering developments that have promise for the practice of medicine or which point to new directions for research and development. Several areas are discussed which are currently more microtechnology than nanotechnology; in these cases the path to feasible benefits through application of nanotechnologies that are in development is evident. Hopefully this will serve to motivate engineering and biomedical students to become a participant in making those benefits a reality.

As the practice of medicine is an applied art as much as a science, this chapter is placed as part of the book on nanotechnology rather than nanoscience. Biomedical science draws heavily upon biomolecular nanoscience, chemistry, physics, and engineering, which are covered elsewhere in the books on nanoscience and nanotechnology. This chapter is intended to be useful to those in medicine as an introduction to nanotechnology and its relevance to their field. Emphasis is therefore placed on current and near-term applications and developments rather than future long-term possibilities. This introductory survey should help clinicians understand the potential of nanotechnology-based innovations for those in their care. In keeping with evidence-based and peer-mentoring approaches in medicine, this chapter includes extensive references to medical and bioscience sources for further examination and study.

27.0 Introduction to Medical Nanotechnology

27.0.1 Definitions: Medicine and Medical Nanoscience

Medicine is the knowledge and practice of maintaining and restoring health. Health is the state of a person, an organism, or an organ in which its systems are able to perform their functions without failure in the face of external threats and internal complications. Living systems are constantly meeting challenges such as stress; injury; malformation in development; genetic errors; invasions by viral, bacterial, and parasitic agents; cancer; degeneration; and challenging normal life events such as pregnancy and delivery; puberty and menopause; aging; and death. It is the goal of medicine to support, maintain, and restore productive functioning of life while minimizing suffering and doing no harm. The roots of medicine lie in our empathy for our fellow creatures, starting with our fellow human beings. Medicine is essentially a human art which is supported by observation, evidence, recording and passing on of knowledge and experience, training, and standards. "There is no one division of medicine by which we know and another by which we act" [1]. Medical science is inseparable from medical practice, as the ultimate significant observation is the outcome for a patient.

Medical science is the application of scientific methods to the study of living systems with the goal of improving medical practice. Medical science is based on any scientific or technological discipline that can contribute knowledge and techniques that advance the practice and understanding of medicine. These have historically included anatomy, physiology, chemistry, physics, engineering, and other disciplines. The development of medical science is inextricably involved with the other sciences. The student who aspires to work in medical practice or research must be prepared with a solid base in multiple disciplines relevant to human health. Increasingly, these disciplines will include aspects of nanoscience and nanotechnology as applications to medicine emerge.

Health and Molecular Signaling. Modern understanding of health is based on the concept of regulation of metabolism by a complex network of molecular-based communication mechanisms known as cell signaling that governs basic cellular activities and coordinates cell actions. Cells in the body perform their life cycle functions in part by genetic programming, but also by responding to molecular signals generated within the cell and received through receptors on the cell membranes. These networks respond to, are controlled by, and can be disrupted by processes that take place on the electrical, molecular, macromolecular, and supramolecular scales. The latter are the domain of nanotechnology, where current advances are offering applications for medicine.

Homeostasis. Healthy organisms tend to maintain homeostasis, from the Greek words meaning "like or same" and "still or static." Homeostasis is defined as the stable state controlled by a system of feedback: the system reacts to changes sensed in its state and/or environment to counter influences that tend to destabilize it or divert its development from the normal path. For example, body temperature, blood pressure, levels of carbon dioxide in the lungs and tissues, the osmotic pressure within cells are homeostatically regulated. Living organisms and systems as a whole are not static: they undergo growth, development, and death in their normal life cycle. Health must be considered as a dynamic rather than a static process, by which a healthy cell or organism responds appropriately to environmental and developmental challenges. Medical science advances the understanding of how these responses are regulated through a complex network of molecular and supramolecular interactions.

Medical science draws upon engineering concepts and methods to create its own unique models for understanding biological networks as not only chemical but also physical and structural—as complex machinery with subtle control systems acting through specific detailed interactions at the macromolecular and nanoscale level. This approach underlies medical nanoscience, a perspective which gives us a framework to model, understand, and intervene in living processes at the level of supramolecular machinery with selectivity and precision.

27.0.2 Historical Origins: Medical Breakthroughs

Historically, medical science originates with empirical observations of outcomes of surgical and pharmaceutical applications. Prior to development of written records and a system of critical evaluation of outcomes, primeval societies had oral traditions regarding herbs and other remedies. Archeological evidence for primitive surgical operations has been found. The development of written records accelerated the communication of medical remedies and the development of a long-term body of experience and outcomes. Stone Age sites in Baluchistan have revealed evidence of dental drilling with flint tools in an early farming culture [2]. Some of the oldest known medical writings are Egyptian papyri from 1600 to 1700 B.C., which record surgical cases and outcomes, and document-established practices which may predate the papyri by several thousand years. [3]

Early civilizations held healers in high esteem; their reputations became legendary and they were elevated to god-like status. In ancient Egypt, Imhotep (meaning "one who comes in peace"), was a historical figure who rose from common beginnings to become the vizier to the Pharaoh Djoser (reigned 2630–2611 B.C.). In ancient cultures, the roles of doctor, priest, scribe, sage, poet, astrologer, architect, vizier, and chief minister were all intertwined. There was a degree of interaction between the Egyptian culture and the civilizations of the Tigris and Euphrates valley, such as the Persians. After his death Imhotep (or Greek: Imouthes) was elevated to demigod status and became a cult figure for healing and a focal point for collections of writings and medical practices in temples and courts in many cultures [4,5,6]. He became identified with the Greek god of healing, Aslepius, and the Greek and Arabian civilizations continued the medical heritage of the Egyptians and Asians.

The most famous Greek doctor was Hippocrates, who was also elevated to cult status, with an attributed genealogy descending from Aslepius. Modern Western science-based medicine reveres Hippocrates as a physician who applied philosophical enquiry to medicine and encoded ethical standards for practice. The admonition in the Hippocratic oath: "First, do no harm," is still the foundation of medical practice.

In the Indus civilization, the Ayurvedic tradition of medicine (meaning "life knowledge") is documented by the works of Agnivesha, whose writings were later revised by Charaka, about 300 B.C. Charaka held that health is not predetermined and life may be prolonged by human effort. This was a major advance in outlook for prescientific societies. He defined the goals of medicine as to cure the diseases of the sick, protect the healthy, and to prolong life. Medical students in the Charaka tradition were given a code of practice with parallels to the Hippocratic oath, including honesty, devotion to learning, sober living, and respect for patient confidentiality. The Indus tradition was very systematic with division of medicine into categories such as toxicology, obstetrics, etc., and a rigorous qualifying exam for practitioners.

In China, medical knowledge was documented in the classical period in texts with parallels to the other ancient traditions, basing illness on a balance of the primary elements and astrological influences. A unique aspect of Chinese medicine was the practice of acupuncture, which is systematically treated in the early treatises, with highly developed anatomical descriptions.

In large part because of the care to avoid doing harm, the practice of medicine has traditionally been experience based and very conservative when adopting new practices and theories. The germ theory of disease was famously resisted by medical practitioners in the nineteenth century until the evidence gained from actual practice became overwhelming, notably due to the work of Louis Pasteur, Claude Bernard, and Robert Koch [7]. At least as much harm has been done in the history of medicine by adherence to theories as doctrines as by rational experimentation with new techniques. That said, the goal of minimizing harm is best served by a skeptical approach to new developments.

Medical Breakthroughs. In 2000, the *New England Journal of Medicine* reviewed major medical progresses of the past 500 years [8], presenting a number of breakthrough medical milestones in chronological order. Besides such major advances in practice such as anesthesia and rigorous statistical clinical trials, the scientific breakthroughs included a series that trends toward understanding of life science on a broadly smaller scale, from the gross to the sub-microscopic:

Elucidation of human anatomy and physiology, beginning in the sixteenth century

Accurate anatomies, circulation, pulse, blood pressure, electrical nerve stimulus, control of muscle by nerves

Discovery of cells and their substructures, from the sixteenth to the twentieth centuries

Microscope, discovery of bacteria and protozoa, complex inner structures of plant and animal tissues, plant cells, animal cells, cell division, nucleus, etc., electron microscope, isolation of mitochondria

Development of biochemistry, from the seventeenth to the twentieth centuries

Concept of active "fermet" replaces idea of passive "humours," discovery of oxygen, other gases, development of organic and physical chemistry, enzyme chemistry, pathways of metabolism, Krebs cycle, role of calcium, sodium, potassium, magnesium, etc., chlorophyll, hemoglobin, hormones, neurotransmitters, cell signaling

Discovery of the relation of microbes to disease beginning in the nineteenth century

Displacement of spontaneous generation theory of microbial life by continuous inheritance principle, association of microbes with diseases, discovery of viruses, pasteurization, vaccination with weakened microbes and viruses, isolation of bacteria in pure culture, antiseptics, recognition of importance of sanitation and sterilization, disease vectors

Elucidation of inheritance and genetics beginning in the nineteenth century

Mendelian laws of inheritance, inheritance of errors of metabolism, genes, chromosomes, mapping of genes on chromosomes, specification of enzymes by genes, identification of DNA as genetic material, base pairing rules in DNA, isolation and x-ray diffraction of DNA molecules, identification of double helix structure of DNA, role of RNA in transcription of proteins coded in DNA, messenger RNA, methodology for decoding sequences of bases in DNA, discovery of reverse transcriptase which converts RNA into DNA, polymerase chain reaction method for DNA amplification, establishment of a relationship between a molecular mutation and a specific disease, "molecular disease" concept

Knowledge of the immune system

Discovery of first known antibodies (in diphtheria antitoxin), identification of role of phagocytes engulfing foreign bodies, cellular theory of immunity, major advances in vaccines, killed virus vaccines, first vaccine produced by DNA technology (hepatitis B)

Medical imaging and biomarkers for diagnosis and research

Discovery of x-rays, use to image bone and hard tissue structure, radioactive tracers for imaging, ultrasound, development of contrast agents; use of power of electronic controls and digital computation to generate detailed multilayer and dimensional images of organs and vessel structure: magnetic resonance imaging, computed tomography, Doppler ultrasound, functional magnetic resonance imaging, image enhancement agents, radiotracers for metabolic and cellular pathways, photochromic markers, bioluminescent markers of molecular activity within cells; use of medical imaging to guide surgery and minimally invasive procedures, use of antibodies labeled with photonic and magnetic targets to identify cancer cells

Discovery and development of antimicrobial agents

First antibiotic compound: Salvarsan, discovery of antibiotic activity of dyes, Prontosil for strep infections, sulfa drugs, penicillin, discovery of antibiotics in soil organisms, development of streptomycin, understanding the ability of microbes to develop resistance to antibiotics

Development of molecular pharmacotherapy

Concept of chemotherapy, extension of pharmaceutic application from microbes to cancer cells, removal or deactivation of hormone-producing glands (ovaries, testes, etc.) to treat cancer of organs regulated by hormones (breast cancer, prostate cancer, etc.), treatment of lymphomas by nitrogen mustard, methotrexate for leukemia, cis-platinum for cancer, beta-blockers; design of drugs for specific molecular targets, explanation of genetic-based variability of drug response

When we review the above catalog of medical advances, there is a trend from general treatment of whole organs and bodies to specific focus on detailed molecular mechanisms in both diagnosis and treatment. Over the past 500 years advances in chemistry, physics, and electrical engineering went hand in hand with advances in medicine to apply therapy more precisely and efficiently. Advances in science and technology have led to new concepts and tools, such as chemical reactions and chemical analysis and synthesis, which have been applied in the life sciences with resulting advances in medical understanding and practice.

In many cases, studies and observations from medicine and the life sciences have led to new scientific breakthroughs in chemistry and physics. The discovery of oxygen by Priestley [9], which led to the replacement of the phlogiston theory of combustion, was based on his experiments with fermentation, respiration, and the ability of plants to restore the life-supporting properties of air in confined spaces. Priestley's interactions with people like Benjamin Franklin accelerated the advance of medical and scientific knowledge, part of a newly emerging network of organizations which spurred and supported science, sometimes in the face of political and religious opposition. The activities of these early societies, supported by the wealth and protection of enlightened patrons, led to the meetings between Priestley and Lavoisier, who completed the conceptual development of the theory of combustion, a fundamental breakthrough in the understanding of matter and energy [10,11].

In an equally important breakthrough, the discovery of electrical stimulation of movement in frog legs by Galvani led to understanding of electric potential and the development of electric batteries by Volta, as well as understanding of how nerve impulses are transmitted. The story of Galvani and Volta and their disagreements provides an instructive example of tensions produced between the medical and technological outlooks [12].

In the twentieth century, the elucidation of *electronic charge transfer* in the process of photosynthesis and energetics of cellular metabolism has led to breakthroughs in photoelectric engineering and solar power generation. The study of human speech and hearing pathologies and development had impacts on computer speech synthesis and recognition. There is an inseparable relationship between advances in science and technology and advances in medicine. New ways of understanding, interpreting, and investigating the world based on nanotechnology will create new understanding of health and disease, and will generate new technical tools for study and application to medical practice.

27.0.3 Medical Nanoscience: Roots in Medical Science

The field of nanoscience is essentially a new perspective on science and technology, with new possibilities opened by vastly more powerful tools for examination and manipulation of matter on the sub-micron scale. This new paradigm is rapidly opening possibilities in medical research and practice. The impact of nanoscience on medicine parallels those made by microscopy, chemistry, physics, electronics, and computing which led to new theories of disease, more effective approaches to treatment, and powerful new imaging and surgical tools.

Early medical diagnosis and treatment was based on what could be deduced from appearance and feel of the body, pain and sensitivity, fever, examination of bodily fluids, external appearance of symptoms, and course of the disease. For example, early physicians diagnosed diabetes by the sweetness of urine. With the development of chemistry and the microscope, diagnoses could be made on the basis of the appearance of cells and composition of bodily fluids; for example, the relative number of different types of blood cells (the blood count) became a diagnostic yardstick; fever was quantified by use of the thermometer; clinical chemistry analyzed the pH and other chemical composition of the body fluids; microscopic examination showed the presence of identifiable microbes in sputum. With the development of x-rays and ultrasound, the anatomy of the body could be examined and the presence of fractures and foreign bodies could be diagnosed accurately.

Without advanced tools to directly observe molecules at the nanoscale, these discoveries were made by a combination of careful observation and experiment, insightful hypotheses, and skillful deduction. For example, viruses were first postulated by their ability to convey disease after passing through nanopore filters, whose extremely small pore sizes blocked the passage of previously known cellular disease agents. This early example of nanotechnology used the ceramic filter developed by Charles Chamberland. Viral structure was later elucidated with the electron microscope and x-ray diffraction. Clinical chemistry, histology, and molecular biology produced more subtle and precise windows into biology, with determination of multiple antibody assays and/or genotypes conducted at the patient's bedside or in the clinic.

Drugs are now designed to target specific metabolic pathways and cell membrane receptors or cell processes. For example, the taxols such as paclitaxel disrupt the tubulin process involved in the configuration of microtubules essential in cell division (mitosis), thus preferentially interfering with the growth of cancer cells [13].

27.0.4 Future Possibilities for Medical Nanotechnology: Nanomedicine

In this section, we take a quick look at some of the more far-reaching proposals and research areas for nanomedicine. The spirit of these proposals is summed up in the following quotation from the U.S. National Institutes of Health (NIH) Roadmap for Medical Research [14]:

> What if doctors could search out and destroy the very first cancer cells that would otherwise have caused a tumor to develop in the body? What if a broken part of a cell could be removed and replaced with a miniature biological machine? What if pumps the size of molecules could be implanted to deliver life-saving medicines precisely when and where they are needed? These scenarios may sound unbelievable, but they are the long-term goals of the NIH Roadmap's Nanomedicine initiative that we anticipate will yield medical benefits as early as 10 years from now.

The NIH defines nanomedicine as the highly specific application of nanotechnology to medical intervention at the molecular scale for curing disease or repairing damaged tissues, such as bone, muscle, or nerve. The nanometer size scale—about 100 nm or less—is the scale on which biological molecules and structures inside living cells operate.

Medical science has powerful tools to examine the parts of cells in detail down to the molecular level. It is the goal of nanomedicine to understand further how intracellular structures operate, in order to build "nano" structures or "nano" machines that are compatible with living tissues and can safely operate inside the body. The ultimate goal is to design diagnostic tools and engineer structures for highly specific and precise treatments of disease and repair of tissues.

Nanomedicine Research Programs. Nanoscience is being developed in conjunction with advanced medical science for further precision in diagnosis and treatment. Multidisciplinary biomedical scientific teams including biologists, physicians, mathematicians, engineers, and computer scientists are working to gather information about the physical properties of intracellular structures upon which biology's molecular machines are built. New emphasis is being given to moving medical science from the laboratory to the bedside and the community.

As researchers gain knowledge of the interactions between molecules and larger structures, patterns will emerge, and we will have a greater understanding of the intricate operations of processes and networks inside living cells. Mapping these networks and understanding how they change over time will, in turn, enable researchers to use this information to correct biological defects in unhealthy cells [15]. New tools that will work at the nanoscale and allow scientists to build synthetic biological devices, such as nanosensors to scan for the presence of infectious agents, or metabolic imbalances that could spell trouble for the body, and miniature devices to destroy infectious agents or fix the "broken" parts in cells.

The NIH Nanomedicine Roadmap. The NIH has developed a series of roadmaps planning research to develop new tools to intervene at the nanoscale or molecular level [16]. These include the *National Technology Centers for Networks and Pathways*, a network of centers which will create new tools to describe the dynamics of protein interactions. The centers will develop instruments, methods, and reagents for quantitative measurements at subcellular resolution and very short timescales. In addition, the NIH is creating *Nanomedicine Development Centers* (see boxes) that will focus on the engineering of new tools for medical interventions and diagnosis at the nanoscale or molecular level. The Nanomedicine Centers' research will support the development of synthetic biological devices, such as miniature, implantable pumps for drug delivery and implantable or mobile sensors to scan for signs of infectious agents, metabolic imbalances, or other biomarkers to detect disease.

Emerging medical nanoscience is targeted to impact the following areas:

1. New medical materials for cell growth scaffolding and tissue repair
2. Enhancement of diagnosis and imaging
3. Enhancement of drug delivery
4. Understanding and control over biomolecular mechanisms
5. Discovery of properties and medical effects of smaller units of life and nanoparticles

Areas of medical care that will benefit from the above nanoscience advances are

1. Plastic surgery and wound healing using nanogels and nanoengineered scaffolding materials
2. Repair of cut nerves using nanofabricated growth channels with cell growth coating patterns
3. Improved healing and fusion of bone fractures using nanopatterned porous implants
4. Selective image enhancement of diseased cells with antibody-coated nanoparticles
5. Reduction of MRI interaction with surgical and sensor probes by nanoengineering coatings
6. Targeted drug delivery with surface-modified and -coated nanoparticles
7. Drug delivery across the blood–brain barrier with "smart" nanoparticles
8. Custom-designed molecular enzymes to selectively switch cell functions on or off
9. Custom-designed phage-like molecular machines to deliver drugs or kill cancer cells
10. Custom-designed molecular enzyme machines to diagnose and repair subcellular structures
11. A suite of molecular enzyme machines to selectively initiate cell death in cancer cells
12. Artificial molecular agents to engulf and deactivate prions, attack viruses, and digest refractory plaques

Center for Nucleoprotein Machines. For example, the NIH-funded National Nanomedicine Center for Nucleoprotein Machines based at Georgia Tech, in collaboration with Emory University and the Medical College of Georgia, will take a biomedical engineering design approach to the repair of DNA, focusing on a model nanomachine that carries out nonhomologous end joining (NHEJ) of DNA double-strand breaks. This and other DNA repair machines have relatively simple structures (<20 components) and significant biological and clinical relevance, and thus are promising as feasible models for nanoscience engineering approaches. DNA repair is vitally important to human health, as both normal metabolic activities and environmental factors can cause DNA damage, resulting in as many as 100,000 individual molecular lesions per cell per day. If allowed to accumulate without repair, these lesions interfere with gene transcription and replication, leading to premature aging,

The First Eight NIH Nanomedicine Centers and Their Research Areas

Summary of the work at the Nanomedicine Development Centers funded to date (2006) through the National Eye Institute and their goals for diseases and medical research. Each of the centers is highly multidisciplinary and involves multiple institutions such as schools of biomedical science, medicine, engineering, and hospitals. For a guide and links to further information see the NIH Web site. Source: http://nihroadmap.nih.gov/nanomedicine/fundedresearch.asp (accessed May 2007).

1. **Nanomedicine Center for Mechanical Biology**—Columbia University leading partnership of six institutions
   Research on the roles of force, rigidity, and form in regulating cell functions signaling pathways and gene expression, and their influence on diseases such as cancers, immune disorders, genetic malformations, and neuropathies, using tools of nanotechnology and modern cell biology.

2. **UCSF/UCB Center for Engineering Cellular Control Systems**—University of California, San Francisco, and University of California, Berkley
   Work on engineering "grand challenges" to develop modified cells or cell-like molecular assemblies that perform intelligently guided precision therapeutic functions, such as tissue repair or "search and deliver" treatment of microscopic tumors or cardiovascular lesions, by focus on reengineering cell guidance, cell force generation, and cell motility systems.

3. **National Center for Design of Biomimetic Nanoconductors**—University of Illinois, Urbana-Champaign
   Research to design synthetic arrays of ion transport channels based on biological ion channels and other ion transport proteins, inserted in arrays of pores on substrates to study cell signaling, energy transport, and generation of osmotic pressures and flows, in order to gain insight into biological processes and disease targeting, and to develop practical applications such as biosensors, osmotic pumps, and power generation.

4. **Center for Protein Folding Machinery**—Baylor College of Medicine with Stanford University
   An interdisciplinary program to define the basic chemical and physical principles used by molecular chaperones in the folding of proteins, in order to engineer protein machines to assist the folding of any protein of interest, and to develop strategies to alleviative or prevent protein misfolding associated with human diseases.

5. **Nanomedicine Development Center for the Optical Control of Biological Function**—University of California Lawrence Berkeley National Laboratory
   Developing methods for rapidly turning select proteins in cells on and off with light, developing chemical and molecular toolkits for integration of optical control into proteins, viral delivery of photo-switchable proteins into cells, and light delivery systems to address these nanodevices in vivo, with the aim of treating retinal and cardiac pathologies by gaining optical control over the signaling and enzymatic activity of cells.

6. **The Center for Systemic Control of Cyto-Networks**—University of California, Los Angeles
   The goal for this center is to use engineering principles to develop global system control methods to investigate and manipulate the complex cell signaling network governing homeostasis of cells, in order to control and correct perturbations in the network by invading organisms, accumulation of pathologic substances, and uncontrolled cell growth that are the hallmarks of most morbid and mortal illness, especially conditions like cancers, infectious diseases, and stem cell related disorders.

7. **Nanomotor Drug Delivery Center**—Purdue University
   Creation of biocompatible membranes and arrays with embedded phi29 in vitro viral packaging biomotors for DNA insertion applications in medicine, by reverse engineering the phi29 motor; incorporating the active nanomotor into lipid bilayers; and developing active nanomotor arrays that enable drug delivery and diagnostics.

8. **Nanomedicine Center for Nucleoprotein Machines**—Georgia Tech, with Emory University and the Medical College of Georgia
   Using nanotechnology and biomolecular approaches, elucidate the structure–function relationships within and among DNA repair nanomachines, for precise modification of DNA and RNA, leading to therapeutic strategies for a wide range of diseases.

apoptosis, or unregulated cell division. The nucleoprotein machine engineering approach is

1. Develop protein tags and fluorescence probes including quantum dot bioconjugates for nanomachine targeting
2. Decipher structure–function relationship for the NHEJ reaction
3. Characterize the dynamics of nanomachine assembly and disassembly in the repair process
4. Determine the dimensions and structure of repair foci at high resolution in fixed cells
5. Establish engineering design principles for DNA double-strand break repair

The Georgia Nucleoprotein Machines center will complement the other NDCs that focus on filaments, membranes, and protein-folding enzymes. The probes, tools, and methodologies developed in these centers will be useful as tools for biological and disease research, and may ultimately provide genetic cures for common human diseases based on the ability to manipulate the somatic human genome using nanomedicine.

To match the Nanomedicine Initiative, many of the NIH Institutes that focus on specific medical specializations and diseases are supporting programs to find areas of application for the tools being developed at the nanomedicine centers. The goal is to improve diagnosis and treatment with nanotechnology-based techniques and materials in the areas of cancer, radiology, and others. For example, the National Cancer Institute (NCI) has a nanotechnology plan that is distinct from, but complementary to, the NIH Nanomedicine Roadmap. The NCI plan

> focuses on using knowledge from basic research discoveries and translating that into clinical oncology applications. The endpoints of this effort will be technology platforms in the context of diagnostics and therapeutics. [17]

The goals of nanomedicine are extrapolations of advances that have been made in the past by application of new science and technology to medicine. Not all will be successful in their application. Many will be controversial. This is not new in the history of medicine.

27.0.5 Putting Medical Nanoscience into Practice: Medical Nanotechnology

In this chapter we explore how nanoengineered materials and devices are being applied in clinical medicine in many areas: radiology, oncology, endocrinology, neurology, orthopedics, cardiology, otology, ophthalmology, emergency care, obstetrics and gynecology, gastroenterology, surgery, and others. (We discuss nanotechnology applications to laboratory medicine in chapter 11. These include biomolecular nanotechnology and nanodevices such as microfluidic biochips, diagnostic lab-on-a-chip devices, nanodroplet dispensers, cell manipulation and separation chips and micromanipulators, and DNA/RNA nanotechnology, with applications in pathology, diagnostics, cytology, genetics, forensic medicine, etc.)

Clinical nanotechnology applications are being applied to enhance or enable new diagnostic and therapeutic methods, including

1. Enhancement agents for medical imaging
2. Finding and destroying cancer cells
3. Delivering drugs deep into tumors (cancer)
4. Delivering insulin through novel routes (diabetes)
5. Delivering drugs through the blood–brain barrier (Alzheimer's, Parkinson's, etc.)
6. Guiding and stimulating nerve regeneration (spinal cord injury, paralysis)
7. Improving neural stimulation (cardiac pacemakers, neuroprosthetics)
8. Noninvasive, sensitive detection of nerve activity (ECG, brain–machine prosthetics)
9. Less invasive hearing and vision prosthetics (hearing and vision loss)
10. Improved remote medical monitoring (preventive, postoperative, recuperative, etc.)
11. Wearable and minimally invasive wireless physiological sensors (GI, Ob-Gyn, etc.)
12. New tissue scaffolds and artificial tissues and organs (surgery, wound care, etc.)
13. Advanced minimally invasive and effective surgical tools and techniques

In the following sections of this chapter, we give examples of the above applications of nanotechnology to medicine that are currently being developed and put into practice, along with discussion of some microscale technologies that are leading to future, improved, nanoscale techniques. Ultimately these families of technologies will merge across the micro-, nano-, and molecular scales into an integrated medical science supporting more powerful and effective medical practice, made more accessible and cost-effective by the availability of tools that operate on the scale of the machinery of the cell.

27.1 Nanoparticles and Nanoencapsulation for Medical Applications

Nanoparticles made of metal, carbon nanotubes, polymers, or other materials can be used in a variety of medical applications, especially when combined with antigen-specific coatings or functional groups on their surfaces. In this section we look at some examples.

The therapeutic and diagnostic usefulness of inorganic nanoparticles depend on their size and physical properties in addition to their chemical composition. They must be small enough to circulate through the bloodstream and tissues without becoming lodged in capillaries or other microanatomies. But they must be larger than atomic size in order to lend enhancement to images or separation techniques.

27.1.1 Nanoparticles for Medical Imaging

Nanoparticles have been used as contrast and image enhancement agents for x-ray and computed tomography imaging. Conventional image contrast agents are molecules such as iodinated benzoic acid derivatives, but these have risk factors and side effects associated with intravenous iodine injection. These chemicals are typically of low molecular weight, and they clear from the human body rapidly, making it difficult to target these agents to disease sites. Iodinated molecules have been encapsulated into liposomes to make a nanoscale particle, but the stability and concentration of agent delivered to the imaged site by this means is low [18].

Experimental image enhancement agents containing gadolinium (Gd) and radioisotopes have been developed for CT imaging, based on dendritic conjugation compounds [19]. However, in such systems, only a relatively small number of gadolinium atoms may be delivered to/in the vicinity of the target tissues. Both approaches deliver, at most, a couple of hundred heavy atoms (i.e., iodine or gadolinium). Another approach is to encapsulate solid nanoparticles of iodine compounds (sized from 200- to 400-nm diameter) in a polymer coating [20], but this involves risk of iodine toxicity should the coating break down.

In order to enhance an x-ray image, an agent must deliver a detectable number of heavy atoms into the imaged tissue without toxic effects. Nanoparticles of elemental heavy metals have the highest density (number of heavy metal atoms/volume), but they must be biologically inert and stable. Nanoparticles of inert metals such as gold are not very cost-effective. To overcome these issues, researchers at General Electric developed nanoparticles made of heavy metal compounds encapsulated in gold shells [21].

Gold-coated nanoparticles can be made by vapor or electrodeposition onto nanoparticles; gold can also be deposited into nanoscale mold templates of silicon, carbon, alumina, or other material with nanosize pores or wells. In the case of silicon, the wells can be created artificially in a silicon wafer using nanofabrication techniques. In the case of carbon nanotubes or alumina the templates are made by controlling the synthesis or electrodeposition of the material.

Organic compounds with sulfide (–S–H) groups (thiols) can be used to coat gold particles with uniform organic monolayers. The sulfide group attaches to the metal surface leaving the organic portion of the molecules exposed as an organic layer. By functionalizing proteins with thiol groups, gold particles can be coated with selectively binding antigens, antibodies, or target compounds for receptors on the surfaces of cells. By targeting receptors unique to certain types of cancer cells, gold nanoparticles can be made to enhance an x-ray image to increase the ability to detect the cancer cells by many orders of magnitude.

Metal and silicon nanoparticles can be used to enhance MRI. Silicon particles fabricated into shapes and coated with conductive layers can have enhanced magnetic resonance interactions with an imaging field. Such specially fabricated nanoparticles are being developed and evaluated at Johns Hopkins and Chicago [22,23].

Coatings and RF filters made from nanoengineered particles can reduce image artifacts and enhance the visibility of many biomedical devices, both implantable and interventional, that today are difficult to image due to eddy currents and other problems that interfere with MRI fields. Special coatings can also improve the ability to image guidewires and devices used in many surgical procedures.

Thin-film nanomagnetic particle coatings have been developed that can shield conductive wires and surgical instruments from radiofrequency (RF)-induced fields in MRI instruments. The high magnetic fields used for MRI, plus the RF signals, normally prevent use of conductors inside the field space. This is a problem for pacemaker wires and other devices. But specially engineered coatings based on nanoparticles have been able to shield such devices, allowing their unimpeded use with MRI.

These magnetic nanoparticle coatings can enable devices contraindicated for MRI due to safety concerns—devices such as pacemakers, defibrillators, neurostimulators, guidewires, endoscopes, etc.,—to be used in the MRI. Safety problems usually involve device heating and, in some cases, induced voltages that can cause very rapid heartbeats. An MRI safe pacemaker and ECG lead using an RF filter developed at Johns Hopkins has been successfully tested on pacemaker leads and licensed to Biophan Technologies Inc. for pursuit of applications and Federal Drug Administration (FDA) approvals for clinical use. Biophan also have a license from Nanoset LLC of Rochester, for medical rights to thin-film nanomagnetic particle coatings that provide a magnetic shield without electrical conductivity, and exclusive rights to a carbon composite material developed at the University of Buffalo. These developments will open the way to increased and simplified use or MRI for guiding surgical procedures.

Nanoshell particles with optical resonances in the infrared have been functionalized and used to enhance imaging of cancer cells. Metal nanoshells are composite spherical nanoparticles consisting of a dielectric core covered by a thin metallic shell, which is typically gold. By varying the relative dimensions of the core and the shell, the optical resonance of nanosize particles can be tuned from the near-UV to the mid-infrared. Work on this type of nanoshell for cancer treatment is being carried out by research groups at Rice and Arizona universities [24].

Nanoshells can destroy attached cells by absorbing infrared light at a frequency that is not absorbed by tissue. The plasmon resonance absorption heats the particles and destroys cells selectively bound to the nanoshell particle [25,26]. The nanoshells consist of a dielectric core and a gold shell, whose core–shell ratio determines their optically resonant frequency. Because these nanoparticles show intense absorption, light scattering, and emission properties in the "water window" of the near infrared (800–1300 nm), they are optimally suited for bioimaging and biosensing applications. In addition, the Rice group has developed rare earth nanoemitters, which are brightly luminescent rare earth ions incorporated into a silica nanoparticle matrix.

27.1.2 Nanoparticles for Targeting Cancer Cells

In this section, we look at a few of the many ways in which nanoparticles are being used to devise new therapies for cancer. In this rapidly advancing field, we will take some examples that show the directions and possibilities among the many new applications [27].

Ferromagnetic micro- and nanoparticles can be functionalized with antibodies, allowing cancer cells to be separated out of tissue samples such as blood and concentrated manyfold for diagnostic analysis. This is an important promising technique because cancer cells are released into the bloodstream in large numbers by microscopic tumors too small to be detectable by imaging modalities. If the circulating cancer cells can be concentrated and detected from a blood test, it would provide a means for early detection of cancer, with greatly improved prognosis for treatment versus detection after the tumor has grown to a size detectable by imaging. This technique is being developed and evaluated in a number of research centers and is being introduced into therapeutic use by more than one medical device company [28–32].

In principle, separation could be based on mass or charge rather than magnetic susceptibility, with centrifugation; or use of nanoparticles with electrically polarizable or charged functionalities with electrophoretic separations. In practice, separation by strong permanent magnets is simpler. In some cases, separation and concentration are enhanced by a combination of techniques, as in magnetic separation followed by concentration from suspensions by centrifugation. A system has been developed which uses 20-nm diameter luminescent/magnetic nanocomposite particles composed of superparamagnetic particle cores coated with CdSe/ZnS quantum dot shells, for ease of quantitative measurement of separated cells [33].

Nanomagnetic particles or ferrofluids can be used for magnetically controlled drug targeting. This technology is based on binding established anticancer drugs with ferrofluids that concentrate the drug in the area of interest (tumor site) by means of magnetic fields. Then, the drug desorbs from the ferrofluid and acts against the tumor. Nanoparticles are one option along with magnetic liquids for magnetically controlled anticancer chemotherapy [34,35]. Magnetic particle separations can also be used to separate cancer cells from bone marrow and other tissues, and for the isolation, identification, and genetic analysis of specific DNA sequences [36,37].

27.1.3 Nanoencapsulation for Drug Delivery to Tumors

In the previous section we described how nanoparticles can be coated with selective compounds to adhere to cancer cells for imaging and for delivering killing blows of energy. In a similar manner, nanoparticles can be filled with absorbed or encapsulated drugs and targeted onto cancer cells or disease agents. The most straightforward way of using nanoparticles to attack cancer is to embed them into tumors. This approach has advantages over simply circulating the drug through the body, because it enables release of the concentrated drug on the site of the tumor, with minimal effect on healthy cells [38].

Drug-laden nanoparticles can be injected into tumors with minimally invasive procedures. Their effect can be enhanced by using drugs that are further activated by radiation that can be directed onto the tumor. Drugs can be infused into the tumor which may be very potent, but difficult to deliver selectively through the circulatory system because of toxicity, insolubility, or reaction with enzymes or other compounds [39].

Nanoparticle Delivery of β-Lapachome. An example of a drug that can be enhanced by nanoparticle delivery is the promising compound *β-lapachone*, an *o*-naphthoquinone found in the bark of the South American lapacho tree.

β-Lapachone is known to induce cytotoxic effects in a wide variety of malignant human cell types including colon, lung, prostate, breast, pancreatic, ovarian, and bone cancers, as well as some blood cancers and retinoblastoma [40].

Dr. David Boothman and his colleagues at the University of Texas Southwestern Medical Center found that β-lapachone interacts with an enzyme called NQO1 (NAD(P)H:quinone oxidoreductase), which is present at high levels in certain types of solid cancer tumors. In tumors, the compound is metabolized by NQO1 and produces cell death but does not initiate apoptosis in noncancerous tissues, since they normally do not express this enzyme. In the tumor cells, β-lapachone induces a novel apoptotic pathway dependent on NQO1, which reduces β-lapachone to an unstable hydroquinone that rapidly undergoes a two-step oxidation back to the parent compound, perpetuating a self-sustaining redox cycle. A deficiency or inhibition of NQO1, such as is the case in normal cells, protects them from the effects of β-lapachone—but when β-lapachone interacts with NQO1 in the tumor cell, the cell kills itself [41].

Thus, β-lapachone has great potential for the treatment of specific cancers with elevated NQO1 levels. (e.g., breast, non-small cell lung, pancreatic, colon, and prostate cancers). Dr. Boothman's team is developing β-lapachone mono(arylimino) prodrug derivatives, specifically a derivative converted in a tumor-specific manner (i.e., in the acidic local environment of the tumor tissue), in order to reduce normal tissue toxicity while eliciting tumor-selective cell killing by NQO1 bioactivation [42].

In order to ensure delivery of β-lapachone and its derivatives to the local environment of the tumor without losing or diluting them in the body, one could simply inject them into the tumor. But experiments showed that the drug is carried away by the blood circulation relatively rapidly, before it has time to fully react with a large number of tumor cells. Dr. Boothman and his group are developing a variety of polymer implants that can be placed in the tumors to slowly release the anticancer drug in an effective manner. The implants include nanoscale polymer plugs molded from nanocells derived from a number of natural nanomaterials [43].

27.1.4 Nanoencapsulation for Penetration of the Blood–Brain Barrier

Delivery of drugs across the blood–brain barrier is another area where nanotechnology gives important new routes of access. The existence of the blood–brain barrier was discovered in the nineteenth century when Paul Ehrlich and his student Edwin Goldman found that dyes used to stain tissues would not pass between the central nervous system and the other tissues of the body. In the central nervous system, the epithelial cells lining the walls of blood vessels overlap in tight junctions, unlike those in the rest of the body. This closes off easy transport of large molecules (greater than molecular weight around 500 Da) between the blood and the brain. This helps protect the sensitive and vital central nervous system from disturbance by chemicals and pathogens (for example, viruses) that are tolerated by the more robust tissues of the body.

The tight epithelial barrier is highly lipophilic. Small lipophilic molecules can dissolve through the lipid bilayer and penetrate the barrier. Other molecules needed for the brain to function make use of specific natural transport mechanisms in the cell membranes (See chapter 14). Small polar molecules, such as glucose and amino acids, and larger proteins, like insulin and transferrin, are transported through the blood–brain chemical traffic by "gatekeeping" processes. Each of the required small molecules has its own transporter protein that carries it through the cell membranes—this process is called carrier-mediated transport. For proteins, specific cell membrane receptors bind the large molecules and pull them across the barrier in a mechanism called receptor-mediated transcytosis. In addition, some ionic proteins (e.g., cationic albumin) bind to and penetrate the blood–brain barrier using electrostatic interactions, in a process called absorptive-mediated transcytosis [44].

Many of the mechanisms that mediate transport across the blood–brain barrier are unknown; elucidating them is an active area of research in genomics, proteomics, and molecular biology. In the meantime, pharmaceutical research is seeking to exploit the pathways that are known. Some success has been made in modifying drugs, linking them to molecules that have transporter proteins, and thus hitching a ride across the barrier. For example, nipecotic acid, which has potential for treating Parkinson's disease, has been conjugated to ascorbic acid, which has access to ascorbate transporters, and has been delivered across the blood–brain barrier in rats, while the unconjugated nipecotic acid is barred. Other pathways have also been exploited using these "Trojan horse" and "chimeric peptide" techniques [45].

Potential drugs for treating Alzheimer's disease, Huntington's disease, stroke, and brain cancers often have molecular weights from 10,000 or 100,000 da or even greater. Many new high molecular weight peptides are being identified with potential use for central nervous system therapy. It has been estimated that up to 98% of potential drugs for the brain are not usable because of the blood–brain barrier, but this area has until recently been underdeveloped in the neurosciences. One reason is the difficulty of identifying each transport mediator path and synthesizing a chimeric or Trojan horse version of each drug to fit it.

Nanotechnology is beginning to offer a possible alternative for transport through the blood–brain barrier that is more generally applicable to a wide range of drugs. Drugs can be encapsulated in biodegradable polymers to make artificial liposomes; the coatings contain active sites to which antibodies can be attached. The antibodies are recognized by the brain-capillary receptors, which mediate their passage through the blood–brain barrier. Once inside the central nervous system, the liposomes release their contents.

These biodegradable polymeric nanoparticles, with appropriate surface modifications that can deliver drugs of interest through the blood–brain barrier, are being formulated with various physicochemical properties. Different surfactant concentrations, stabilizers, and amyloid-affinity agents are being evaluated to determine how they influence the transport mechanism.

Recently, the radiolabeled Cu^{2+} or Fe^{3+} metal chelator clioquinol, which has a high affinity for amyloid plaques which are a factor in neurodegenerative disease, has been encapsulated within small, spherical, lipophilic drug carriers capable of crossing the blood–brain barrier [46]. This and similar nanoencapsulation formulations have the potential to deliver many drugs to the central nervous system, opening new possibilities for therapy.

27.1.5 Nanoparticles and Nanoencapsulation for Insulin Delivery

One of the requirements for effectiveness of a therapeutic drug is delivery—getting the right amount in the right form to the right place. Proteins make up the nanomachinery of cells, and therefore proteins or peptides would make the ideal drug for many diseases, but proteins can be broken down and modified by many enzymes in the body. Thus, most successful pharmaceuticals are not peptides; many are small molecules, such as aspirin, which act to regulate complex biological networks. Some drugs, such as pacitoxol, are larger molecules (but still small compared to proteins), which inhibit or disrupt a precise part of the cellular machinery.

Insulin. One of the first peptides to be a successful therapeutic agent was insulin, which was discovered and applied to treat diabetes early in the twentieth century [47–51]. This represented an enormous breakthrough for one of the oldest documented diseases. The main cause of type I (insulin-dependent) diabetes mellitus is degeneration of insulin-producing cells in the islets of Langerhans, located in the pancreas. Named after the German pathologist Paul Langerhans, who discovered them in 1869, the islets are clusters of specialized cells which produce a number of hormones, including insulin. The islets contain five types of cells: alpha cells that make glucagon, which raises the level of glucose in the blood; beta cells that make insulin, which lowers blood sugar and is needed by cells to metabolize it; delta cells that make somatostatin, which inhibits the release of numerous other hormones in the body; and PP cells and D1 cells, about which little is known.

Insulin used to control diabetes has to be injected because like other peptides it is broken down into amino acids by digestive enzymes if taken orally. Soon after its discovery, researchers began to seek alternatives to subcutaneous injection with hypodermic needles. Besides the inconvenience and discomfort of injection, there are risks of improper dosage and rates of release. Insulin is released in controlled amounts by the pancreas in response to changes in blood sugar and other network stimuli, which are difficult to simulate by injections. And accidental injection of insulin directly into the bloodstream results in insulin shock, a dangerous and potentially fatal condition of hypoglycemia.

One approach to the insulin delivery problem is to synthesize or find compounds that have similar activity to insulin, but would survive modification in the digestive tract, passing into the bloodstream in an active form [52]. Other approaches seek injectable forms of insulin that have a more controlled release, and/or which are suitable for alternative forms of delivery, such as inhalation or intravenous or transcutaneous administration by micropumps. Progress has been slow on most fronts.

Nanoencapsulation of Insulin. A widely explored path is to encapsulate insulin in a protective coating that would allow its release after passing through the digestive system. A variant of this technique that has been extensively developed is to encapsulate insulin for injection into the soft tissues, in order to achieve a gradual, controlled rate of release into the bloodstream for a basal level of insulin, which can be supplemented as needed by injections. Many methods of encapsulation for injection have been tried, including liposomes, which can be administered intravenously [53].

The encapsulation of insulin and other drugs into liposomes, microcapsules, and nanocapsules is an example of practical nanoengineering to which much effort has been devoted over many years with the goal of seeking improved treatments for diabetes. A related approach is to infuse insulin into porous or absorbent polymer particles for gradual release [54]. The greatest benefits would come from a noninjected delivery mechanism, so most effort has been directed towards usable oral insulin formulations [55–59].

In a typical method for preparing nanoparticles of insulin or other peptide, the peptide is dissolved in an aqueous solution; then a nonsolvent such as a low molecular weight (C1 to C6) alcohol is stirred in with the aqueous solution. The alcohol absorbs up to 100% of its weight of water, causing the peptide to precipitate out of solution, with particles having diameters in the range of about 100–200 nm. If the mixture contains a suitable polymer, the particles are spontaneously coated as they precipitate; this process is called phase inversion nanoencapsulation.

Zinc insulin is a slowly released form of insulin that has been encapsulated in various polyester and polyanhydride nanosphere formulations using phase inversion nanoencapsulation. The encapsulated insulin maintains its biological activity and is released from the nanospheres over a span of hours. Some formulations have been shown to be active orally. These formulations typically have about 10% of the efficacy of intraperitoneally delivered zinc insulin, but they are able to control plasma glucose levels when faced with a simultaneously administered glucose challenge. The key properties that make such formulations promising for oral administration are size of dosage, release kinetics, bioadhesiveness, and ability to traverse the gastrointestinal epithelium.

As colloidal and nanofabrication techniques have advanced, more sophisticated forms of encapsulation have been developed and tested, using new types of polymer and inorganic coatings and matrices. With the advent of genetic engineering and biotechnology on a large scale, numerous bioactive peptides besides insulin are available in large quantities. Administering these substances by the oral route remains a formidable challenge due to their insufficient stability in the gastrointestinal tract and their poor absorption pattern. This has given new impetus to investigating new approaches to improve their oral bioavailability. The use of polymeric microparticles and nanoparticles is an actively pursued concept. Encapsulating or incorporating peptides in particles should at least protect these substances against degradation and, in some cases, also enhance their absorption.

Chitosan Nanoencapsulation. Nanoencapsulation with polymeric nanoparticles is offering new drug delivery routes for therapy and diagnosis of a number of diseases. An example of one promising encapsulation starting material is chitosan [60], a derivative of the chitin polysaccharide which we encountered in chapter 13. Extensive research is taking place on many formulations for encapsulating insulin, including detailed models of absorption and desorption rates [61].

A typical method developed for encapsulating nanoparticles with chitosan is the polyionic coacervation fabrication process, used for protein encapsulation and subsequent release. This process has been systematically manipulated and studied in a number of laboratories in efforts to obtain predictable effectiveness [62,63]. Bovine serum albumin (BSA) is widely used as a model protein, which is encapsulated using the polyanion tripolyphosphate (TPP) as the coacervation cross-link agent to form chitosan–BSA–TPP nanoparticles.

The BSA-loaded chitosan–TPP nanoparticles are characterized for particle size, morphology, zeta potential (colloidal electrokinetics), BSA encapsulation efficiency, and subsequent release kinetics. These properties have been found to be dependent on chitosan molecular weight, chitosan concentration, BSA loading concentration, and chitosan/TPP mass ratio. Protein-loaded nanoparticles can be prepared under varying conditions in the size range of 200–580 nm, with a high positive zeta potential. An advantage of chitosan over some other encapsulation materials is that later stage particle degradation and disintegration does not yield a substantial follow-on release, as the remaining protein molecules, with adaptable 3-D conformation, seem to be tightly bound and entangled with the cationic chitosan chains.

The polyionic coacervation process for fabricating protein-loaded chitosan nanoparticles offers simple preparation conditions and a useful range for manipulation of physiochemical properties of the nanoparticles (e.g., size and surface charge). A weakness of chitosan nanoparticle encapsulation is typically with difficulties in controlling initial burst effects which can release large quantities of protein molecules [64].

Insulin Pumps. Another approach to diabetes therapy is the use of external and implantable pumps to deliver insulin in response to fluctuations in blood sugar. An early external full size pump is the artificial beta cell (patented in 1979), which regulates blood glucose concentration by continuously analyzing blood from the patient and deriving a computer output signal to drive an insulin infusion pump. Advances in electronics and integration led to wearable pumps with small reservoirs of insulin formulation, but there are many problems with any type of artificial insulin pump. The pump must react to changes in blood sugar, so it must have an accurate and timely measurement device to drive its feedback loop. Any error or delay in reading or delivery is a potentially serious problem. And completely implanting a pump leaves the problem of how to maintain the supply of insulin. A challenging problem for insulin pumps and any system implanted into the body is the buildup of plaques and bacterial infections. Following a period in which further miniaturization and integration of insulin pumps looked promising; attention is now turning to other methods. Nevertheless, there will be a role for pumps in many therapeutic and evaluation situations; research to solve the challenges is continuing on silicon micropumps and minimally invasive microneedle designs [65–67].

Nanoparticles for Inhalation Therapy. Shortly after the discovery of insulin, efforts began to investigate the possibility of delivery by inhalation. The surface area of the interior of the lungs in an adult human is roughly the size of a tennis court, so if insulin could be delivered to the capillaries and alveoli of the lung, there is high potential for direct absorption into the bloodstream, by-passing the problems of digestion and absorption in the gastrointestinal tract and avoiding the problems of injection [68,69]. The fact that no successful inhalation therapeutics for insulin have emerged since the first work is testimony to the challenges of understanding the processes by which micro- and nanoparticles are processed by the cilia and alveoli in the lungs.

Much work has been done on pulmonary diseases, airborne bacteria and virus infections, smoking, and air pollution, which provides insights that may be useful in formulating nanoparticles for inhalation therapies. Nanotechnology has much to contribute to solving lung diseases as well as finding new effective inhalation drug delivery methods. Inhalation routes to drug delivery are an important and growing area or biomedical research in many areas besides diabetes, and are likely to be an area where nanotechnology will make a large impact [70].

27.1.6 Nanoencapsulation for Protection of Implants from the Immune System

Another area where nanotechnology is showing promise is encapsulation of living cells for implantation. One option for treatment of diabetes is transplantation of healthy pancreas beta cells to the patient, but rejection of the foreign cells by the host immune system is a major problem. For several decades medical researchers have tried various attempts to encapsulate or shield the transplanted tissue with a barrier that would protect it from immune attack. Only in recent years, with advances in nanotechnology for fabricating nanostructured porous biocompatible materials, has this approach been brought closer to feasibility.

Nanoscale capsules are being fabricated to contain living cells. Pores in the sides of the nanoparticle cages allow small molecules such as nutrients, oxygen, and carbon dioxide to pass through, but can be sized to keep out antibodies and protect the enclosed cells from attack by macrophages. Assemblies of encapsulated cells, enclosed in silica gel, silicon, and other materials, are being used to make bioartificial organs, which are being tested for effectiveness in various types of tissue implants ranging from pancreatic beta cells to bone marrow [71].

In one of the most striking examples of the transfer of silicon-based nanoengineering to biomedical use, a group at Johns Hopkins have developed self-assembling silicon nanocubes which can be used for cell encapsulation. Living cells have been successfully enclosed in the containers as they fold from their flat silicon lithography state into closed cubes with windows in their walls for circulation [72].

Most work on transplanting encapsulated cells has been in the area of insulin-producing pancreatic beta cells. After years of having experiments that result in the implants being smothered by plaques and invaded by the immune system of the host, promising results are beginning to appear, using new nanoengineered encapsulation materials and techniques [73].

In a recent example of work in this field, researchers in Germany achieved the first successful transplantation of functioning microencapsulated islets of Langerhans. They used a novel alginate-based microencapsulation formulation to implant human islets into immunocompetent diabetic mice [74]. This and perhaps other approaches, which must be subjected to many rigorous trials in animals and humans, may soon lead to a treatment for diabetes that restores normal blood sugar regulation.

27.2 Guiding and Stimulating Tissue Function and Growth

Besides encapsulating cells and tissues, nanostructures can be used to guide and stimulate the growth of cells, serving as scaffolding for growing new tissues. Tissue scaffolding, or tissue engineering, is an emerging technique in surgery and wound healing, which is being given new options and opportunities by the development of new nanomaterials and nanostructures. Tissue scaffold materials must be biocompatible; in some cases biodegradability is desirable.

Scaffolding should have good properties for cell adhesion and binding to connective tissue and, if appropriate, bone material. Depending on the application, porosity allowing cells and cell extensions to penetrate the material is needed. Major areas for application of tissue scaffolding are treatment of burns for regrowth of skin, guiding and stimulating regrowth of bone after surgery or injury, reconstruction and growth of ligaments and connective tissues, and guiding regrowth of nerves. This is a large field with many materials being developed and applied to many tissue types [75,76]. We will discuss some of the structural and surgical uses for tissue scaffolding in a later section, but first we will look at the very important special case of systems for support of nerve growth and nerve activity.

27.2.1 *Nanoguides for Neural Growth and Repair*

The technology of silicon microchips has been in use for several decades to make structures to guide cell growth on the microscale. The technique of growing cells on micropatterned and nanopatterned glass or silicon substrates has become important for evaluating nerve growth [77,78]. Nanotechnology is now available to fabricate more detailed and finer structures for use in sensor functionalization and cell growth patterning, by fabricating microchannel cell growth guides with specific degrees of surface roughness and coatings nano-bioengineered to promote cell adhesion [79] (**Fig. 27.1**).

Guiding and Monitoring Nerve Growth. Micropatterned plates have been developed for use as a template for evaluating the growth of neurons in the presence of growth stimulators and inhibitors. This type of growth guide is used in evaluating the effects of different neuronal growth stimulants and inhibitors in the laboratory, with the goal of understanding how to promote regeneration of severed or damaged nerves [80,81]. Fabricated cell growth surfaces with patterns and coatings, and printing equipment to apply cells, adhesion agents, and reagents to plates is now readily available from a number of sources [82]. This type of work is relevant to treatment of spinal cord injuries and other forms of paralysis.

Technology for neural growth guidance is now being taken from the research and diagnostic laboratory into the clinical laboratory, where experiments are

Fig. 27.1 *A template for molding experimental nerve growth guides.*

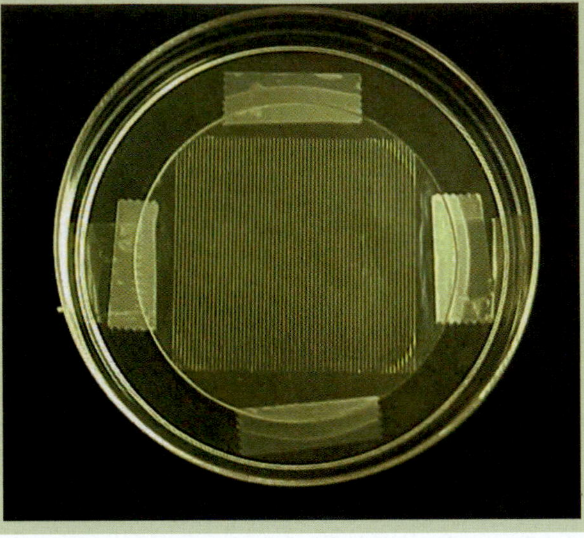

being conducted on promoting repair and growth of nerves, including the spinal cord. After injury, axonal regeneration occurs across short gaps in the peripheral nervous system, but regeneration across larger gaps remains a challenge. Cellular channels during development and after peripheral nerve injury have been shown to provide guidance cues to growing axons. In the following section we look at some examples of technologies aimed at bridging the nerve regeneration gap, which may one day be applied to restore mobility to patients with spinal nerve column damage.

Promoting Nerve Repair. A number of research groups are obtaining promising results in animal studies using various types of micro- and nanoporous guides for nerve regeneration. In the complex biological system of nerve repair, early physiological intervention to minimize the spread of injury will always be the first line of defense in treating nerve damage. Devices and techniques based on nanotechnology may eventually contribute to such intervention.

Basic research in prevention and treatment of permanent nerve injury, especially of the spinal cord, include (1) reduction of edema and free-radical production, (2) rescue of neural tissue at risk of dying in secondary processes such as abnormally high extracellular glutamate concentrations, (3) control of inflammation, (4) rescue of neuronal/glial populations at risk of apoptosis, (5) repair of demyelination and conduction deficits, (6) promotion of neurite growth through improved extracellular environment, (7) cell growth and replacement therapies, (8) transplantation approaches, (9) gene therapy to activate expression of growth factors, (10) rehabilitation to retrain and relearn motor tasks, (11) restoration of lost function by electrical stimulation, and (12) relief of chronic pain syndromes [83,84].

Nanotechnology will impact many of these areas, but it must be integrated into the entire therapeutic regime. Possible benefits from nanotechnology capabilities will be in (1) rapid, efficient, and minimally invasive surgical repair, (2) improved automation of rehabilitation through sensors and smart materials to provide feedback, (3) improved electrical stimulation devices for prostheses and pain relief, and (4) nanoengineered microdevices to promote cell growth.

Certain cell, molecular, and bioengineering strategies for repairing the injured spinal cord are showing encouraging results (either alone or in combination) in animal models. The most promising route is application of nanoengineered nerve growth guidance matrices in combination with (1) seeding of neuronal support cells such as glial and Schwann cells, and (2) molecular coatings and growth factors embedded into the matrix material.

To understand why this nanoengineering approach is being pursued so actively, consider the status of the alternatives. Transplantation of stem cells faces many difficulties and is in an early stage of investigation. Gene therapy is another potentially promising approach. Research using cell cultures and gene-altered mice has shown that switching on just two genes can induce considerable regeneration of damaged nerve fibers in the spinal cord, suggesting that genetic therapy or drugs that activate perhaps only a handful of genes might be enough to induce regeneration of spinal cords in humans. This result was surprising but promising because of the large number of different genes known to be involved in nerve growth.

In experiments at Duke University, using cell cultures and gene-altered mice, researchers have found that switching on just two genes can induce considerable regeneration of damaged nerve fibers in the spinal cord. Inserting the genes that expressed the two nerve growth regulatory proteins GAP-43 and CAP-23 increased spinal cord regeneration by as much as 60 times in transgenic mice as compared to controls [85].

Further research has been stimulated by these discoveries, but several obstacles remain to using gene therapy for nerve regeneration. Transgenic animals used in the experiments expressed the axon growth promoter genes throughout life, whereas normally they turn off after development of the spinal cord is complete. For the process to be used therapeutically, there would have to be some means of turning the genes on after an accident (and turning off other genes that suppress neuron growth) in a safe and rapid procedure. Other questions that would have to be answered would be how long the genes need to be expressed to get an effect, and what would be the side effects (such as neural cancer) of leaving them turned on in adults, and how to turn them off at the right time. These are challenging research problems that require much new effort in molecular and cell biology, genomics, and proteomics.

In the meantime, rapid progress is being made in developing implantable devices for neural growth promotion and support, some of which may be useful in delivering future gene therapies. The size scale for fabricating nerve growth conduits is in micrometers, but features such as patterning on the inside of the growth channels, porosity, and cross-linking of polymer structures involve nanoscale materials engineering. This is a challenging and productive area with many multidisciplinary teams actively pursuing the goal of nerve repair with promising results. Here are some examples of the different materials and designs for neural conduits.

Polylactide Foams. A research team at the University of Liege in Belgium has made macroporous polylactide foams and assessed the ability of dorsal root ganglion (DRG) derived neurons to survive and adhere in vitro [86]. (Polylactides are biodegradable, aliphatic polyester thermoplastic polymers made from lactic acids in one of several chiral forms.) The foams were fabricated using a thermally induced polymer–solvent phase separation. Two types of pore structures were obtained, oriented or interconnected pores. The foams were coated with polyvinyl alcohol to improve the wettability for cell culture. Microscopic observations of the cells seeded onto the polymer foams showed that the interconnected pore networks were more favorable to cell attachment than the anisotropic ones.

The Liege group investigated the capacity of the highly oriented foams to support in vivo peripheral nerve regeneration in rats. A sciatic nerve gap of 5-mm length was bridged with a polymer implant showing macrotubes of 100-μm diameter. At 4 weeks postoperatively, the polymer implant was still present and well integrated anatomically. An abundant cell migration was observed at the outer surface of the polymer implant, but not within the macrotubes. This dense cellular microenvironment was found to be favorable for axogenesis.

Polylactide Filaments. A research project conducted jointly between the University of Texas at Arlington, the University of Texas Southwestern Medical Center, and the University of Kentucky is producing nerve guidance channels made from laminin-coated poly(L-lactide) filaments to induce directional axonal growth and to enhance the rate of axonal growth after injury [87]. Dorsal root ganglia grown on these filaments in vitro extend longitudinally oriented neurites in a manner similar to native peripheral nerves. The extent of neurite growth is significantly higher on laminin-coated filaments compared with uncoated and poly-L-lysine-coated filaments. Schwann cells were found to grow on all types of filaments, and were associated with greater neurite growth.

To improve regeneration across extended nerve defects, the team fabricated wet-spun microfilaments of different fiber densities, with the capability for drug release to support cellular migration and guide axonal growth across a lesion. In bundles that were not loaded with drug release, after 10 weeks, nerve cable formation increased significantly in the filament bundled groups when compared to empty-tube controls. At lower packing densities, the number of myelinated axons was more than twice that of controls or the highest packing density. In a consecutive experiment, PLLA bundles with lower filament-packing density were examined for nerve repair across 1.4- and 1.8-cm gaps. After 10 weeks, the number of successful regenerated nerves receiving filaments was more than twice that of controls. These initial results demonstrate that PLLA microfilaments enhance nerve repair and regeneration across large nerve defects, even in the absence of drug release. Ongoing studies are examining nerve regeneration using microfilaments designed to release neurotrophins or cyclic AMP.

Polylactide Tubules. A group at the University of Iowa has developed biodegradable conduits that provide a combination of physical, chemical, and biological cues at the cellular level to facilitate peripheral nerve regeneration [88]. The conduit consists of a porous poly(D,L-lactic acid) tubular support structure with a micropatterned inner lumen. Schwann cells were pre-seeded into the lumen to provide additional trophic support.

In evaluation experiments, tubular conduits with micropatterned inner lumens seeded with Schwann cells were compared with three types of conduits used as controls: M (conduits with micropatterned inner lumens without pre-seeded Schwann cells), NS (conduits without micropatterned inner lumens pre-seeded with Schwann cells), and N (conduits without micropatterned inner lumens, without pre-seeded Schwann cells).

The conduits were implanted in rats with 1-cm sciatic nerve transections and the regeneration and functional recovery were compared in the four different cases. The number or size of regenerated axons did not vary significantly among the different conduits. The time of recovery and the sciatic function index, however, were significantly enhanced using the MS conduits, based on qualitative observations as well as quantitative measurements using walking track analysis. This and other experiments indicated that the micropatterning and the Schwann cells provide a combination of physical, chemical, and biological guidance cues for regenerating axons at the cellular level. The patterned and seeded conduits performed significantly better than conventional biodegradable conduits.

Biosynthetic Nerve Implants. Another type of conduit for promoting nerve regeneration, the biomimetic biosynthetic nerve implant (BNI), was developed by a group based at the Texas Scottish Rite Hospital for Children [89]. The BNI is a hydrogel-based, transparent, multichannel matrix designed as a 3-D substrate for nerve repair. Polymer scaffold casting devices were designed for reproducible fabrication of grafts containing several microconduits. A number of different polymers were evaluated for making the grafts, including cellulose, hydroxymethyl cellulose, hydroxyethyl cellulose, carboxymethyl cellulose, carboxymethyl chitosan, poly-2-hydroxyethyl-meth-acrylate, poly(*R*-3-hydroxybutyric acid-co-(R)-3-hydroxyvaleric acid)-diol (PHB), collagen, gelatin, glycinin, both neat and as mixtures.

The grafts have been tested in vivo using a sciatic nerve animal model for repair of the adult hemitransected spinal cord. At 16 weeks postinjury of the sciatic nerve, empty tubes formed a single regenerated nerve cable. In contrast, animals that received the multiluminal BNI showed multiple nerve cables within the available microchannels, better resembling the multifascicular anatomy and ultrastructure of the normal nerve. In the injured spinal cord, the BNI loaded with genetically engineered Schwann cells were able to demonstrate survival of the grafted cells with robust axonal regeneration through the implant up to 45 days after repair.

Carbon Nanotubes Enhance Cell Adhesion Surfaces. In a related series of experiments, the Texas Scottish Rite group tested electrodeposited, photolithographic, and micromachined gold microelectrodes for nerve cell stimulation [90]. The gold microprobe interface was modified by addition of conductive polymers and carbon nanotubes. It was observed that the addition of carbon nanotubes favors the formation of nodules and increases the surface roughness. Also, electrochemical impedance spectroscopy revealed that conductive polymer composites lower the impedance of gold microelectrodes by three orders of magnitude. The carbon nanotube/polymer composite coated electrodes maintain intimate contact with axons, enabling high-quality nerve spike signals and electrical stimulation of neurons.

Carbon Nanotube Sheets as Neuron Growth Support. In cooperation with the University of Texas Southwestern Medical Center and the University of Texas at Dallas NanoTech Institute, the Texas Scottish Rite group used sheets and yarns made from multiwalled carbon nanotubes to support the long-term growth of a variety of cell types ranging from skin fibroblasts and Schwann cells to postnatal cortical and cerebellar neurons [91]. The study found that the carbon nanotube sheets stimulate fibroblast cell migration compared to plastic and glass culture substrates, entice neuronal growth to the level of those achieved on polyornithine-coated glass, and can be used for directed cellular growth. The carbon nanotube yarns were recently developed at the NanoTech Institute [92]. These findings have positive implications for the use of this type of material in applications such as nerve growth channels, as well as for tissue engineering, wound healing, neurostimulation, and biosensors.

Biocompatibility Issues with Carbon Nanotubes. The published studies on biocompatibility of carbon nanotube materials are contradictory [93]. A number of recent studies found that neural cells adhere to multiwall carbon nanotubes [94,95]. Studies with cardiac cells found some short-term effects attributed to physical rather than chemical interactions, but no long-term toxicity [96]. Studies on toxicity in the lung have produced some ambiguous results; these appear to be related to absorption of other nanoparticles on the carbon nanotubes [97,98]. Biocompatibility issues for nonspecific protein absorption onto single-walled carbon nanotubes can be circumvented by co-adsorption of a surfactant and poly(ethylene glycol), whereas specific binding is achieved by co-functionalization of nanotubes with biotin and protein-resistant polymers, thus making the nanotubes essentially inert to most forms of interaction with biological systems, and providing a substrate which can be used as a base for preparing protein-specific molecular recognition systems [99].

The Texas Scottish Rite studies used multiwalled carbon nanotubes produced with a minimal residual content of catalytic transition materials to obtain good cell growth. The sheets were found to stimulate fibroblast cell migration, and neuronal growth was enticed to the level achieved on polyornithine-coated glass, which is the standard used for directed cellular growth.

Natural Material Scaffolds from Agarose and Laminin. One way to avoid issues of biocompatibility is to use a well-characterized natural material and coat it with laminin (for a description of laminin, see chapter 14). Researchers at the Cell and Tissue Engineering Laboratory of Case Western Reserve University made hydrogels from agarose and loaded the gel structure with laminin and nerve growth factors to create a three-dimensional scaffold for neurite growth [100]. The agarose hydrogel scaffolds were engineered to stimulate and guide neuronal process extension in three dimensions in vitro. The extracellular matrix protein laminin was covalently bound to agarose hydrogel using the bifunctional cross-linking reagent 1,1′-carbonyldiimidazole. Compared to unmodified agarose gels, laminin-modified gels significantly enhanced neurite extension in chick dorsal root ganglia cells. The Case Western team used inhibitors to study which types of receptors on the surfaces of the ganglia cells were active in the adhesion and growth process on the laminin. They also embedded nerve growth factors into the hydrogels. The resulting trophic factor gradients stimulated directional neurite extension. As a result of this and similar research, agarose hydrogel scaffolds may find application as biosynthetic 3-D bridges that promote regeneration across severed nerve gaps.

Natural Material Scaffolds with Collagen. Another natural connective material used for nanoengineered nerve growth guides is collagen (see chapter 13). For this purpose collagen polymer can be cross-lined chemically or with microwave radiation. Collagen polymers thus made can incorporate peptides to promote nerve growth. A number of research groups around the world have fabricated collagen tube and fiber microdevices with nanoengineered substructures and have demonstrated their ability to support nerve regeneration.

The Institute for Frontier Medical Science, at Kyoto University in Japan, made tubeless grafts with 2000 collagen filaments in each, to bridge 20-mm defects in rat sciatic nerve. Effective growth of myelinated axons was observed in the collagen filament nerve guides [101,102].

At the Bio-Organic and Neurochemistry Laboratory of the Central Leather Research Institute in Chennai, India, researchers fabricated multilayered collagen tubes by a lamellar evaporation technique and successfully used 14-mm tubules for regeneration of 10-mm nerve gaps in a rat model. Fourier transform infrared spectra of the collagen films showed that the native triple helicity of collagen was unaltered during the multilayered preparation process. Several different means of inducing cross-linking in the fibers were studied, including treatment with glutaraldehyde and microwave radiation [103,104].

Scanning electron microscopy of cross-linked tubes showed porous, fibrillar structures of collagen filaments in the matrices. Microscopic histology analysis showed that the tubule surfaces provide for good adherence and proliferation for the sprouting axons from the cut proximal nerve stumps. Among the two types of cross-linking, the microwave-irradiated collagen conduits results in ample myelinated axons compared with the GTA group, where more unmyelinated axons were observed. Solute diffusion studies on the tubes indicated that they are highly porous to a wide range of molecular sizes during regeneration.

Functional evaluations of the regenerated nerves were performed by measuring the sciatic functional index (SFI), nerve conduction velocity (NCV), and electromyography (EMG). The conduction velocity and recovery index improved significantly after 5 months, reaching the normal values in the autograft and microwave-induced cross-linked collagen groups compared to glutaraldehyde and uncross-linked collagen tubes.

Studies were conducted with nerve growth promoting peptides incorporated into the collagen matrix. Immunofluorescence studies demonstrated the staining of S100 proteins in the peripherally located cells indicating the proliferation of Schwann cells in the early days of regeneration. The staining pattern of integrin-αV was observed mostly in the perineural regions in close proximity to the peptide-incorporated collagen tubes. Evaluation of the sciatic functional index and conduction velocity at 90 days postoperatively showed regeneration of lesioned nerves with the peptide incorporated collagen implants [105,106].

Extensive evaluations were carried out for different cross-linking methods, including microwave, glutaraldehyde, di-tertiary butyl peroxide, and dimethyl suberimidate. The physical properties of collagen-based biomaterials are profoundly influenced by the method and extent of cross-linking. Cross-linking density, swelling ratio, thermomechanical properties, stress–strain characteristics, and resistance to collagenase digestion were determined to evaluate the physical properties of cross-linked matrices. The spatial orientation of amino acid side chain residues on collagen plays an important role in determining the cross-linking density and consequent physical properties of the collagen matrix. The microwave cross-linked matrices gave the best result for nerve regeneration [107].

Summary of Progress in Nerve Regeneration. Nerve regeneration research being conducted around the world offers possible solutions for the need to sacrifice a healthy nerve to make a graft, and for the shortage of graft material available, for the repair of severed nerves [108]. This work is progressing towards clinical treatments for the repair of spinal cord injuries and cures for paralysis. More than 50 clinical trials are in progress worldwide on various treatments for spinal cord injury. Consequently, in this millennium, unlike in the last, no spinal cord injury patient will have to hear "nothing can be done" [83]. Nanoengineering of tissues will have played a significant part in making such cures possible.

27.2.2 Neuronal Stimulation and Monitoring

The cardiac pacemaker is one of the best-known and most widely used neuroprostheses. Since 2001, the number implanted in the United States has approached 200,000 per year. Other pacemakers are not implanted permanently but are used during cardiac catheterization and other procedures [109–112]. Other types of electronic stimulators include cardiac defibrillators, cochlear implants, bone growth stimulators, and neural stimulators for the deep brain and spinal cord, and vagus, sacral, and other nerve stimulators [113].

Nanotechnology advances these devices with improved battery technologies, biocompatible materials and surface treatments for enclosures and leads, electrode miniaturization and efficiency improvements, and smaller-sized integrated circuits for control and power, while speed and processing capabilities increase. These improvements made possible the accessible, inexpensive emergency cardiac defibrillators in public places and on transportation such as aircraft, with control systems safe to be used by nonspecialists.

These indirect benefits, like battery life and power, will not be covered in this chapter, although they are an important result of nanotechnology innovations. The design and use of each of the above types of stimulator involves specialist knowledge in cardiology, neurology, and other medical fields. In this chapter we have selected applications in nerve repair, stimulation, and neuroprosthetics as examples which illustrate challenges and opportunities for nanotechnology. Advances discussed, such as improvements in electrodes, nanoengineered biocompatible materials, stimulation methods, etc., are applicable to cardiac pacemakers and other types of implantable neuroprosthesis which use electrodes.

An area in which nanotechnology is likely to benefit cardiac pacemakers directly is in improvements in electronic leads and electrode biocompatibility and durability. The leads have been a weak link for cardiac pacemakers while size, power, and battery life have improved, thus requiring longer life service for the wires and electrodes. Cardiac pacemaker leads must be biocompatible with tissue, but unlike surgical implants, the internal portion near the heart must not promote tissue adhesion and growth which would make them difficult to remove. In addition they must heal to skin and surface tissues to seal to prevent becoming a channel for infection. They must resist the growth of bacterial and fungal biofilms. These are challenges worthy to be taken up by nanoengineering of materials. Bioengineering must also be advanced to understand long-term factors affecting cell growth and adhesion to leads.

27.2.3 Neurostimulation for Pain and Nervous Disorders

Nanofabrication is increasing the resolution and capabilities of neurostimulation devices. Neurostimulation is used medically for cardiac pacemaking, deep brain stimulation to control tremors in Parkinson's disease, management of chronic pain, stimulation of tissue healing, prevention and reversal of nerve degeneration, and other conditions and therapies, including chronic neuropathy, diabetic neuroarthropathy, and cardiomyoplasty [114].

Integrated micro- and nanoscale devices make it possible to apply much finer resolution with many more electrode stimulation points, which can be dynamically programmed. Every advance in computer and signal processing power resulting from electronics nanoengineering contributes directly to the power and sophistication of programmable medical devices such as neurostimulation systems. These advantages may appear in new generations of cochlear implants, cardiac pacemakers, deep brain stimulation, and in new types of devices [115].

An important aspect of the development of cortical prostheses is the enhancement of suitable implantable microelectrode arrays for chronic neural recording. A promising approach to this function is the use of implantable silicon-substrate micromachined probes. Work on these probes has improved their reliability and signal quality. In rodent models the probes provide high-quality spike recordings over extended periods of time lasting up to 127 days. More than 90% of the probe sites consistently record spike activity with signal-to-noise ratios sufficient for amplitudes and waveform-based discrimination. Histological analysis of the tissue surrounding the probes generally indicated the development of a stable interface sufficient for sustained electrical contact [116]. Surface treatments, new electronic circuit materials, and other advances contributed by nanotechnology will result in continual improvements in making such probes resistant to the challenging environment of implantation. In addition to the general improvements in size, power, and mobility made possible by nanotechnology, we have seen in a previous section how nanoengineering of the surfaces of electrodes, including use of carbon fiber nanotubes, is contributing to improved interfaces between neurons and electrostimulation devices [90].

Electrical stimulation of neural tissue by surgically implanted neuroelectronic devices is now an approved modern therapy. Reduction of the size and power requirements with integrated microelectronic devices makes it feasible in many cases to energize an implanted device by RF electromagnetic transmission of power, eliminating wires and batteries [117–120]. Implanted neurostimulation devices such as pacemakers are already available that receive power from RF energy, thus eliminating transcutaneous wires that are a source of infection and complications [121,122]. The question of RF interferences becomes an important design consideration for such devices (see section 27.2.6).

Improvements in energy storage through nanoengineered supercapacitors and hypercapacitors, aerocapacitors, and conductive polymers [123], coupled with lower power requirements for nanoengineered electronics, allow room for great improvements in size and capability of embedded devices, thus allowing very small implantable devices to perform electrostimulation in selected points of the nervous, sensory, and neuromuscular systems. Such devices may make it practical for increased use of implanted electrostimulation for bone and tissue grafts, and to stimulate function in the endocrine system and other organs.

27.2.4 Neuroprosthetics

Neural interfaces to nano- and microelectronic devices open new opportunities to design more powerful neurostimulators for prosthetics. Broadly defined, a neural prosthesis is a device implanted to restore a lost or altered neural function [124,125]. The Greek word "*prosthesis*" originally refers to the addition of a syllable to the beginning of a word. In classical medicine it means an artificial replacement for a missing part of the body [126]. The term "prosthetics" denotes the medical art of providing prostheses to improve the life of patients.

Assistive Devices. Neuroprostheses are distinct from "assist devices" such as heart ventricular assist pumps, which do not interface to the voluntary nervous system [127]. Neuroprostheses are also distinguished from assistive devices which translate or amplify movement, for example, systems which enable paralyzed persons to control computers or devices by eye, tongue, or small muscle movement [128]. Nanotechnology is a key driver advancing the state of the art for assistive devices, and most dramatically for neuroprosthetics. As we have seen, nanotechnology is improving the electrodes that interface to nerves, and it is providing smaller and more powerful sensors, actuators, and distributed control systems to make prosthetics more natural and effective.

Scalable, Distributed Networks of Nanoelements

Nanotechnology will have its greatest impact on medicine with assemblies of cooperating interconnected networks of computing, communicating, and sensing nanoprocessors driving assemblies of modular interworking nanoactuators to make up a micro- or macrodevice like a powerful but subtle motor neuroprosthesis.

Types of Neural Prostheses. Neural prostheses are of two types—motor and sensory. Sensory neuroprostheses are devices that translate external stimuli such as sound or light into signals that are interfaced to the brain either directly or via neural pathways, to restore lost or damaged perception ability. Glasses and external hearing aids are prostheses, but a sensory neuroprosthesis is an active device that delivers electrical stimulus to the nervous system, such as a cochlear implant or artificial retina.

Motor Neuroprostheses. Motor neuroprosthetic devices take signals from the brain or motor nerve pathway and convert that information into control of an actuator device to execute the user's intentions. Motor neuroprosthetics work in one of two ways, either by (1) translating motor nerve impulses to electrical stimulation that excites or inhibits neuromuscular paths to paretic or paralyzed organs and limbs (functional electrical stimulation), or (2) picking up electricity generated by the brain or nerves and interpreting it to control prostheses or assistive devices (device control). In both cases, nerve signals can be interfaced to the neuroprosthesis by recording electrical impulses externally through the surface of the skin (myoelectric control) or through implanted electrodes.

Functional stimulation enables neural command signals to control muscle movement where native motor nerve function is paretic or impaired. Device control collects and maps nerve impulses to control electronic or electromechanical aids—actuated braces, artificial limbs or hands, synthetic speech generators, devices to allow control of bowel or bladder sphincter function, etc. Nerve signals interfaced to an active neuroprosthesis, or brain–computer interface, can be used to drive assistive technologies such as computer-based communication programs, environmental controls, and assistive robots [129,130].

For a motor neuroprosthesis to restore function, it must be integrated with the human owner's nervous system. The signals picked up by the electrodes from the nerves or brain must be translated into smooth and controlled actuation of the prosthesis [131]. This involves some combination of learning and adapting by the user and some sophisticated electromechanical control strategies to decode signals from either (1) the remaining peripheral nerves, or (2) the brain, and translate them into actions [132].

The design and implementation of a device capable of complex motor tasks—such as grasping, manipulation, and walking with smooth gait, coordinating movement with vision and balance, responding with appropriate force and velocity—all require that the prosthesis have a sophisticated distributed network of control, actuation, and feedback [133]. Limb and hand prostheses also need a high degree of fidelity to mechanical and dynamic properties of the natural limb. Otherwise the task of learning to use the device will be difficult [134,135]. Therefore to be effective, the nanotechnical design of nanoactuators and nanosensors must be fully integrated with the control systems from design to implementation, including the very special ergonomic interface between patient and device [136].

There is a high degree of overlap between prosthetics, robotics, virtual reality, design of space suits, and other human augmentation technologies [137–145]. These highly multidisciplinary fields present many opportunities for application of nanotechnologies in materials as well as electronics. Nanotechnology is already enabling more natural prostheses by providing (1) smaller (and more affordable) sensors, processor elements, and the wiring and interconnections to network them into a distributed control systems [146–151], (2) smaller, more powerful, efficient, and responsive scalable actuators whose mode of movement is natural and smooth because it is based on molecular forces, similar to those in natural muscle [152,153], and (3) engineered materials that match the strength/weight ratios, elasticity/rigidity, and mechanical energy storage characteristics of key components of natural extremities [154–157].

Nanomaterials such as conductive polymers and carbon nanotubes offer routes to nanosensors [158], and nanoscale-distributed computing elements [159], as well as to modular, lightweight, and strong artificial muscles that can be ganged in parallel to match force requirements [160–163]. Nanoscale magnetometers, accelerometers, pressure sensors, and gyroscopic devices will be able to more precisely detect even minute movements and angle changes; these will support the design of internal device rotation mechanisms that feature smooth, accurate movement as well as accurate transmittance of control and feedback information for human operation.

Nanotechnology will have its greatest impact on medicine not with tiny mobile robots, but with assemblies of cooperating interconnected networks of computing, communicating, and sensing nanoprocessors, driving assemblies of modular interworking nanoactuators to make up a micro- or macrodevice like a powerful but subtle motor neuroprosthesis. To see the current state of the art in prosthetics for limbs, and how integrated nanotechnology can be harnessed by distributed control to enable their design, consider some examples of artificial limbs.

Leg, Knee, and Foot Prostheses. Loss of a foot or leg is one of the most common amputations, due to war, civilian encounters with land mines, accidents, and complications of diabetes. Before the 1990s artificial legs, knees, and feet were difficult to use, and required more energy for movement than normal walking [164].

Without a control and feedback system that reproduces the natural interaction of the limb with the body of the wearer and the external environment, amputees walking with a leg prosthesis consume more energy than a non-amputee at comparable walking velocities. The elastic tendons of the foot and leg store and release energy with each step, and artificial limbs have been designed to reproduce this spring mechanism, with improved results [165–167].

Designs with distributed microprocessor control in a complex prosthesis have given improved usability [168], but until recently, processors, sensors, and actuators have been too large, power-hungry, and expensive to offer a practical solution. The most challenging function to reproduce was the proprioception or kinesthetic sense that provided virtually unconscious and autonomous feedback to the body, enabling us to keep our balance through all types of movement and terrain. This requires low power, rugged, fast sensors, processors, and actuators that have only recently become available through advances in micro- and nanotechnology.

Advanced strong and lightweight materials, including smart materials with built-in sensors, are also necessary to achieve the full potential of the electromechanical control system. As these prerequisites have become available, being able to build and test prototypes has led to better understanding of how the limb interacts with its physical environment as well as with the body and the nervous system.

Laboratories and companies around the world working on the leading edge of nanotechnology and control have now produced highly advanced artificial feet, knees, and legs that are giving life-like restoration of function to amputees [169–171]. A walk through a few step cycles with a leading commercial prosthesis shows how it uses highly miniaturized but powerful accelerometers, processors, and actuators with an optimized control strategy that reproduces essential features of natural gait and posture. A prosthesis like the Proprio Foot [172], for example, might use adaptive neural network algorithms to learn from the user's movements and stride to optimize its response [173]. Such a system would use fuzzy logic feedback algorithms to ensure smooth reactions and avoid overshooting responses [174]. Distributed parallel processing enables the system to plan the positioning of parts of the limb for the next movement in real time while executing a step. Force sensors and accelerometers provide feedback for adjustment to slope, speed, and sudden off-balance movements by making thousands of measurements per second.

Distributed and embedded artificial intelligence and electrophysical feedback control the autonomous functions involved in walking that we do not have to think about—like keeping the leg positioned relative to the center of gravity of the body, rotating the ankle, and controlling the angle of the foot to the ground, with compensation for forward speed, slope, and other factors. At the same time, the prosthesis interfaces to the user's nervous system to respond to motor commands and execute voluntary movements. A requirement for success of the system is that it should interface, interact, and adapt to the forces and patterns of movement from the physical environment and the rest of the body, as well as to the nerve impulses, just like a natural limb would do. Thus we see that our definition of a motor neuroprosthesis was somewhat limited: it has to interact appropriately with its entire environment, not just with the nervous system.

Two Important Concepts for Design of Motor Prostheses

Proprioception. The sum of all the tactile information constantly being fed to your brain that tells you where you are and what you're doing.

Kinesthesia. The ability of the brain and body to precisely and reproducibly know where limbs and fingers, etc., are positioned, how fast they are moving, and how much force is being applied. Kinesthetic memory enables musicians and athletes to achieve high performance.

Hand Prostheses. The hand has one of the highest densities of motor nerves and sensory nerve endings of any part of the human body, serving its highly developed tactile and sensory capabilities with 22 degrees of freedom in its four fingers and thumb. Prostheses to replace a lost hand are one of the oldest forms of prosthetics: the surgeon Ambroise Paré designed an anthropomorphic hand for wounded soldiers in the sixteenth century. Modern motor neuroprosthetics provide considerable functionality, thanks to advances in microelectronics, microactuators, and robotics [175].

Like the lower extremity, the hand interacts with its environment as well as with the nervous system, but the hand is much more intimately and intricately connected with the brain in the performance of voluntary planned and executed tasks. A number of prosthetic hands have been developed with embedded controls [176–179]. Much has been learned about hand movement from development of robotic hands for industry and space, as well as prosthetic use [180–182]. As in the case of the artificial foot, the most recent and advanced designs use distributed control and sensors with local feedback rather than requiring the user to decide and communicate how much force is exerted by each finger at a given time [183].

Artificial hands with advanced capabilities can be used for robotics or human augmentation aids as well as prosthetics [184–186]. Devices for robotics and prosthetics have been developed with 20 degrees of freedom, using shape memory alloy actuators [187,188]. Some advanced designs implement the complex degrees of freedom of the fingers and thumb and employ sensitive microphones to detect vibration when a grasped object is slipping [189–191].

Ongoing progress towards more natural artificial hands is making use of MEMS accelerometers, as well as force and pressure sensors (and hence more nanotechnology). Designs have been proposed for smart artificial skin with embedded nanosensors, which will give more options for designers of prostheses.

Prosthetics, like other medicine, is human oriented, seeking to adapt assistive devices to the unique needs of each patient, within inevitable constraints of cost, technical feasibility, and rehabilitative learning [192]. With the advance of nanotechnology making devices adaptable as well as affordable, this ideal becomes more realizable. The range of options for fitting a pirate hook was much narrower than with a modern nanosensor and processor-enhanced prosthesis.

For example, when a patient has some functioning tendons leading to a missing hand, these can be harnessed to give a more natural interface to the prosthesis. The loss of motor function suffered by many patients is not complete. For these patients who still possess residual functions, modular, more naturally controllable systems for supporting these functions are needed rather than complex systems to replace them [193].

Neuromotor devices that are one-way, sending signals collected from the brain side of the gap to the muscles, leave the patient relying primarily on visual feedback for control (or auditory feedback in the case of voice prostheses) [194]. Such control can be learned with therapy [195,196]. Indeed, the brain is very plastic; patients are able to learn to activate and use prostheses via a number of different neural pathways, regardless of the nature of the function [197].

An ideal prosthetic hand would replicate sensory–motor capabilities of the amputated limb. The special feedback that the body, and especially the hand, provides for such tasks is referred to as the haptic sense. Haptics involves primarily the sense of touch and perception of resistance to applied force, with some elements of kinesthetics. Haptics is important in the skilled performance of dexterous tasks such as surgery.

A number of research centers are working on biomechatronic hands which aim to approach natural dexterity, speed, and strength with normal weight and appearance. A challenge is to minimize power requirements, operate noiselessly, and be resistant to water, oils, food, and be easily cleanable. The goal is to have an intelligent, adaptable self-programming control system that makes it easy for the user to develop skill. Some current designs have excellent dexterity, but are lacking in some of the other requirements.

This is a very active and exciting area for application of new nanotechnology. In the case of actuators, miniature electronic motors or pneumatic muscles have yet to be replaced by nanoengineered artificial muscles in an integrated design for an artificial hand. Lifelike, durable, soil-resistant skin coverings have yet to be developed and evaluated (perhaps using some of the soil- and water-shedding features learned from the lotus petal) (see chapter 12).

Nanotechnology Is Bringing the Bionic Man Closer to Reality—The Brain–Machine Interface

"Abstract systems that allow a brain to control a computer are inching ever closer to reality—but their most important applications may be different from those envisaged by science fiction."

Is this the bionic man? *Nature* (2006) [198]

The Brain–Machine Interface. The control of physical objects by the power of thought alone has long captured the human imagination. Using our thoughts to control a computer or robot used to be the realm of science fiction writers. Now with the aid of new technology and years of study of neural activity in the brain, the control of machines and computers by the brain is becoming a reality. Systems are being made in which patterns of neuronal firing in the brain can be translated into electronic controls to support communication, mobility, and independence for paralyzed people [198–200]. Nanoengineered electronic and magnetic detection devices are helping brain–machine interfaces reach a level of speed and responsiveness that will make a brain interface a usable prosthesis that does not require long and difficult concentration and training for the user [201].

Motor and Sensory Interfaces. To fully interface with the brain, a neuroprosthesis must not only receive signals from the brain but must also return sensory information for feedback. This feedback is visual, auditory, kinesthetic, and haptic. In later sections we will discuss sensory neuroprostheses, but first we will examine aspects of motor control.

Promising Breakthroughs in Brain–Machine Prostheses. Many research groups have been working for decades to integrate sensors, computers, and knowledge of the patterns of nerve firing with which the brain controls movement, in order to help people who have brain or spinal cord damage to communicate and interact with the outside world. The goal is to use prostheses to replace or restore lost motor functions in paralyzed humans by routing movement-related signals from the brain, around damaged parts of the nervous system, to external effectors.

Much remains to be done before neuroprostheses that can respond to the brain become a clinical reality, but in experiments, paralyzed patients with electrodes implanted in the part of the brain that controls movement, the cerebral motor cortex, have been able to control computers and televisions, open e-mail, and move objects using a robotic arm [202]. Although control using the system was rather slow, it did not require complete focused attention to operate: the patient was able to engage in conversation while doing tasks with the neuroprosthesis. Experiments using implants in monkeys that were not paralyzed have demonstrated the potential for faster response times, with techniques to speed up the brain–machine interface [203].

Challenges to Be Overcome. It is possible to control prosthetic devices by extracellular recordings from the cortical neurons in the brain, where movement control originates [204,205]. Extensive preclinical experimentation with implanted neurosensor electrodes is laying the foundation for design of such systems; the pattern of neuron firing has to be at least approximated in order to map signals onto movement in the prosthesis. The development of smaller and faster neuroelectrodes is important for future success in patients.

To make a brain–machine interface requires the estimation of a mapping from neural spike trains collected in motor cortex areas onto the kinematics of a normal limb—but fortunately, this mapping does not have to be absolutely accurate—otherwise the task would be impossible [206]. Imagine that someone ripped the dashboard and steering column out of a car, or the cockpit out of an airplane, and handed it to you and asked you to reconnect all of the wires and controls back together so that it worked again. That would be an extremely simple task compared to mapping the connections between the brain and muscles after a spinal cord injury. That a brain–machine neuroprostheses can work is as much a tribute to the adaptive and regenerative power of the brain and nervous system as it is to the years of patient experimentation and intricate deciphering by neuroscientists. If the brain is presented with an interface that is at all workable and predictable, the cortex can adapt in most cases to learn how to associate patterns with movement [207].

There are purely clinical factors and obstacles for success of a brain neuroprosthesis. Firing patterns and maps from brain to motor effector neurons must be measured in intact animals and human subjects, to establish a baseline for translation from brain to machine. Movement signals in the motor cortex of the brain may not persist after nerve paths to a limb are cut off. ("Use it or lose it" holds true in the brain.)

But given the clinical obstacles, early experiments give indications that workable prostheses may be possible. There are nanoengineering challenges to be solved in both materials and control system architectures. The systems for recognizing and interpreting brain signals to movement must be robust and adaptable enough to deal with changes in pattern presented after injury, during learning, and due to individual differences. Current experimental systems typically sample ensembles of 100–200 cortical neurons. Advances in nanotechnology will allow denser and larger arrays to be implemented. The neural interfaces must be compatible with nerve cells and their environment, so that the transmission of signals does not fade over time due to buildup of plaques on the electrodes or withering away of the neurons in contact with the surfaces. In the opinion of many scientists, "most of these difficulties are now engineering challenges, rather than problems of principle" [198].

Design of Control Algorithms. Much work by neuroscientists has been done to tell us what part of the cortex controls which limbs, and how control spreads out down the neural networks [208,209]. If a control system could simply be "hardwired" to the brain to talk to a prosthesis, the paradigm would be to interface electrodes to the appropriate part of the brain's motor cortex or motor pathway, and analyze the relationship between the cortical activity and measured arm or hand movements; this relationship would then map cortical activity to similar prosthetic arm movements. However, the pathway to the brain is not so simple, and there is the clinical problem that measured limb movements are not feasible for amputees or patients with physical mobility limitations.

The firing patterns of the nerves controlling natural neuromuscular systems are very complex, and are modulated by chemical neurotransmitters and inhibitors, so a purely electronic interface can never be completely natural. The motor signals spread out, and sensory feedback returns, over a network of dendritic paths in patterns that can be mapped experimentally to the musculature and sensory surfaces of the body [210,211]. The areas of the cortex that control particular motor movements set up predictable patterns prior to initiating motor commands [212]. For a neuroprosthesis we merely want to provide the brain a tool to work with, rather than try to duplicate its operation.

We know from neuroscience that, to a large degree, the brain is adaptable and can learn new control tasks by response to feedback presented to it. If we can present the cortex cells with a coding system of responses that activate the prostheses, then the brain can learn to exercise control [213–215]. Hopefully, this learning can be automatic, implemented by feedback loops within the cortex that are similar to natural control mechanisms for the original limb. Otherwise control of the prosthesis will require extra effort and conscious attention to execute. Thus getting the control coding right is as important as the weight, power, and agility of the robotics in the prosthesis. Fortunately, progress is being made in designing workable control schemes that enable the brain to successfully map intent onto movements of the prostheses. And the brain has shown itself to be remarkably adaptable in learning to control devices by neural stimulation.

Adaptive Coding and Training. Recent research has shown that adaptive control strategies shorten the learning time and increase the effectiveness of brain–machine interfaces. A number of coding and training methods have been used to interface such neuroprostheses [216]. These coding techniques compute statistical scores to match patterns in the cortex with movements of the limb or prosthesis. Adaptive methods can automatically improve their performance with practice, by iterative computation that optimizes the correlation between nerve firing patterns and movements with practice.

Systematic investigation of several linear (Wiener filter, LMS adaptive filters, gamma filter, and subspace Wiener filters) and nonlinear models (time-delay neural network and local linear switching models) are applied to experiments in monkeys performing motor tasks (reaching for food and target hitting), in order to map cortical function onto motor movements [217]. Other correlation models that have been used include Bayesian graphical networks, Bayesian quadratic, Fisher linear and hidden Markov model classifiers [218] as well as neural networks, Kalman filters [219], and Kalman filters with smoothing [220], and other statistical component analysis techniques [221,222]. The timing and synchronization of firing between groups of neurons have been analyzed as a means of predicting cortex output based on coincidence and synchronization of firings [223].

Distributed Control Networks of Nanocomputers. Because the brain is more universal than any specialized branch of the neural pathways connecting it to the extremities, and is the starting point for voluntary control signals, one might assume that the neuroprosthetic interface could be controlled largely and directly by the brain, by-passing the functions of the spinal cord, ganglia, motor nerves, etc. But distribution of function, especially motor functions, serves to prevent overload of the brain; much preprocessing of stimuli takes place along the pathways.

In the case of the leg, a great many of the control feedback loops are automatic, and can be implemented entirely within the limb. But for the hand or for speech, a complex mixture of voluntary initiation and automatic cascades of control must take place, with the option for voluntary control to override the automatic function at any point playing a much greater role. To match the learning and adaptive power of neural networks that control complex tasks like manipulation or speech, the control systems must be adaptive [224].

Analysis of such control systems reveals that they are best implemented with distributed, communicating networks of processing elements [225–227]. Thus a network of nanofabricated small processors with communication links, sensors, and actuators will be an optimal platform for implementing agile and responsive control systems for prosthetics (as well as robotics) which can adapt to external conditions and to the stimuli from the user's brain. Such a network also has the advantage of requiring less energy for the same computation capability of one or two fast processors for controlling prostheses.

Noninvasive Interfaces for Brain–Machine Interfaces. Noninvasive or minimally invasive methods of communicating between brain and prosthesis would be highly desirable—eliminating surgery, electrodes, and wires. Some obvious minimally invasive techniques, such as using finger, toe, or eye movements, have the disadvantage of requiring a high degree of the patient's attention, even if muscle movement is possible [228]. One way to avoid the disadvantages of wires through the skull and skin would be to communicate with an implanted electrode via wireless RF signals—but this raises many technical problems and still requires an implant.

In principle, it is also possible to control computer cursors through noninvasive electrodes monitoring the brain from outside the head. The electrical potential of neurons collectively firing is detectable on the scalp. Its recording is termed the *electroencephalogram* (EEG), whereas the pattern of signals measured by electrodes on the surface of the cerebral cortex is called the *electrocortiogram* (ECoG). The EEG shows an average of many neurons firing in a broad region of cortex, filtered through the skull and scalp, but a number of patterns can be detected nevertheless. Considerations for control programming are similar to those for ECoG, but mapping is more arbitrary [229,230]. The lower interface resolution of the EEG makes it difficult to provide a wide range of subtle and distinct movement controls, and requires long training periods. EEG-based systems have generally been too slow for controlling rapid and complex sequences of movements, so this type of interface remains a less desirable option.

Some direct comparison studies have been conducted between EEG and ECoG, in order to measure the relative information-processing capacity that can be achieved using brain–computer interfaces with the two interface modes. Methods of translating brain computer interface data between noninvasive EEG and invasive ECoG are sought with the goal of further improving EEG control [231,232].

Researchers have used asynchronous EEG analysis and machine learning techniques to implement brain control of advanced robots with impressive results. Using an EEG-based brain–machine interface that recognized three mental states, they achieved mental control comparable to manual control on the same task, with a performance ratio of 0.74 [233].

EEG readings have been combined with other physiological measurements in multimodal monitoring to give an enhanced picture of the status and activity in the brain. Other data integrated with EEG include hemodynamics, functional MRI, and others [234].

Magnetic Neural Stimulation and Monitoring. Advances in micro- and nanotechnology are on the cusp of giving an alternative to low-performance EEG interfaces and invasive electrode implants. With advances in materials and electronic nanosensors, it is becoming feasible to use magnetic fields to stimulate and monitor the brain. Magnetic stimulation is noninvasive; focused magnetic fields can stimulate nerves deep within the body and brain without implanted electrodes or shocks through the skin [235].

Basic Principles. The firings of neurons and the travel of ion currents along axon membranes generate magnetic fields. A steady current induces a static magnetic field, and any change in current creates a change in the magnetic field. Thus, magnetism is a second-order effect of the movement of an electrical charge: the magnetic field is proportional to the velocity of the charge and the change in magnetic field is proportional to the acceleration or deceleration of the charge. This indirectness makes magnetic fields more difficult to use, but it also creates the possibility of noninvasive communication with nerves, without implanted electrodes or painful transcutaneous shocks.

High-intensity changes or pulses in strong magnetic fields can induce electrical fields in the nervous system, which can exceed the firing threshold for neurons (*the action potential*). The induced electrical currents are proportional to the rate of change of magnetic field (dB/dt). Thus magnetic stimulation can induce electrical currents in the neuron cell membranes like those induced by implanted electrodes, but without physical contact [236] (**Fig. 27.2**).

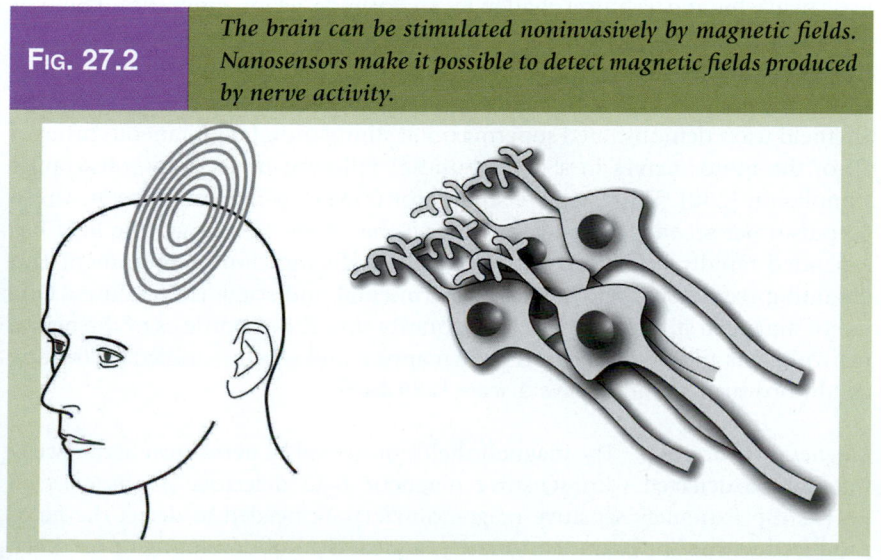

Fig. 27.2 *The brain can be stimulated noninvasively by magnetic fields. Nanosensors make it possible to detect magnetic fields produced by nerve activity.*

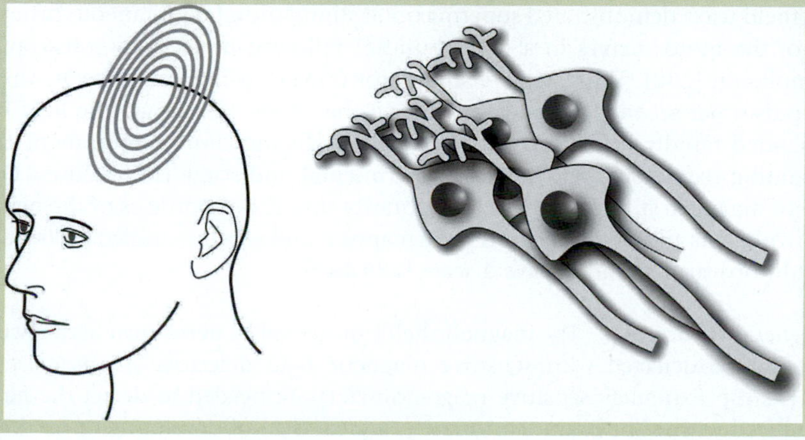

Magnetic Stimulation. The induction of nerve firing by application of strong focused magnetic fields is a new medical technique used to stimulate motor nerves in the limbs (*functional stimulation*) or neurons in the brain (*transcranial magnetic stimulation*). Magnetic stimulation requires strong magnetic fields, which must vary or pulse in order to generate an electric field—generation of an electrical field requires movement of an electrical charge relative to a magnetic field. Electrical stimulation of nerves could be obtained with a simple voltaic jar or electrostatic spark, but magnetic stimulation requires equipment that can generate short, intense, pulsed and focused magnetic fields of very high strength, about 0.5 T at the surface of the cortex, which typically requires a 2 T magnetic coil outside of the body.

To localize the stimulation, arrays of magnetic coils are positioned outside of the head to focus the combined fields inside the brain. Relatively low-frequency magnetic fields can be used (typically a few kilohertz) which are not absorbed by the skull or spinocerebral fluid. The electromagnets are controlled so that the strength of the magnetic pulse exceeds the excitation threshold only at the focal point. The resolution that can be achieved is less than with electrodes, but the noninvasive advantage is spurring efforts to increase the precision of the magnetic focus [237].

Development of Medical Applications. Electric and magnetic fields are complementary: electrical currents generate magnetic fields, and changes in magnetic fields induce electrical fields. The relationship between electricity and magnetism was demonstrated experimentally by Faraday in 1831 and explained theoretically by Maxwell in the latter half of the nineteenth century. Applications of the phenomena were developed by Edison, Marconi, and many others since, forming the basis of our electronic economy.

Electrical stimulation of nerves was known since the days of Galvin and Volta, but the idea of magnetic stimulation had to wait for Faraday and Maxwell to demonstrate the relationship between electricity and magnetism. There were a number of early attempts to stimulate the brain and motor nerves using magnetic fields, but the technical challenges are formidable for either stimulation or monitoring with magnetism.

The first successful development of technology for magnetic stimulation was conducted at the University of Sheffield in England starting in 1976. In 1982 the Sheffield team demonstrated supermaximal stimulation (simultaneous firing of all of the motor nerves in a nerve bundle) followed in 1985 by transcranial stimulation [238]. The repetitive stimulator (rTMS), which can generate up to 30 pulses per second, became available in the 1990s. Since then the field has expanded rapidly, with several companies producing clinical equipment and obtaining regulatory approvals for experimental and some clinical uses. One use of magnetic stimulation is to temporarily shut down portions of the neural network. This allows connections to be mapped, and is being studied for treating conditions such as Parkinson's disease [239,240].

Magnetic Monitoring. The magnetic fields produced by nerve currents are weak but can be detected with sensitive magnetic field detectors (*magnetometers*) [241–243]. Extremely sensitive magnetometers are needed to detect the fields produced within the brain. Fortunately, a great deal of development has been invested toward high-performance magnetic sensors for many applications such as computer disk drives, oil exploration, and security detection. The current state-of-the-art *magnetoencephalography* (MEG) can map brain activity on a one millimeter grid or less. With powerful signal processing and statistical analysis, MEG images can be co-registered with MRI scans with good accuracy. Because magnetic fields are induced perpendicular to the direction of current flow, MEG gives orthogonal information to EEG in terms of the types of neural tissue and direction of nerve impulses that are revealed.

The first generation of MEG equipment was typically bulky, requiring shielded rooms, high power consumption, cryogenic cooling of detectors, and significant processing times to deconvolute data from relatively few sensors. Thus MEG has until recently been limited to research and highly specialized diagnostic applications for life-threatening conditions [244]. This is a rapidly developing area which is being accelerated by applying nanofabrication to existing types of magnetic sensors and entirely new concepts made possible by nanotechnology. The results are 1000-fold improvements in sensitivity and reductions in size and power requirements by factors of 10 to 100.

Devices for Magnetic Stimulation and Monitoring. The improvements in performance necessary to make noninvasive magnetic communication with the brain a practical reality are already being delivered by nanotechnology. For stimulation, the impacts of nanotechnology are largely indirect; nanofabrication of interstitial compounds and alloys is producing better high-temperature superconducting materials to reduce the size and cryogenic environmental constraints for high-performance superconducting magnets, and nanoparticle thin films are being used to fabricate magnetic shielding materials.

Historically three classes of magnetic sensors have been developed: mechanical, electronic, and quantum. Researchers are re-examining magnetic sensing to find opportunities for enhancement based on phenomena that appear at the nanoscale, with very small masses and volumes. Many older types of magnetic measurement devices can be sub-miniaturized with nanofabrication, but nanotechnology is also making new designs possible based on previously inaccessible physical phenomena. Both types of development are producing concrete results, and applications are expanding.

Mechanical magnetic sensors include geometric magnetometers, where the sensor is moved or deformed by interaction with the magnetic field, and resonance sensors, whose vibration rate is influenced by field forces [245]. Electronic sensors include Hall effect sensors [246], which measure the resistance to flow of electrons caused by their deflection in a magnetic field; and magnetoresistive, giant magnetoresistive, and colossal magnetoresistive sensors, based on thin film conduction effects (the 2007 Nobel Prize in Physics was awarded to the discoverers of giant magnetoresistance, Albert Fert and Peter Grünberg) [247–250], and flux-gate devices, which compare the difference in current required to magnetize a coil in two directions. Some sensor designs utilize more than one physical effect in the same device for enhanced performance. Quantum sensors include the *superconducting quantum interference device* (SQUID), based on *Josephson junction* currents—the magnetically sensitive tunneling of electrons through a thin insulating barrier separating two superconductors [251].

The SQUID is the highest sensitivity magnetometer commercially available. Magnetic scanning systems approved for mapping neural activity are based on SQUID sensors. Although the first generation was bulky, it has been used successfully for brain and cardiac imaging. A second generation design has been optimized with highly sophisticated software and good engineering to increase resolution and reduce the size and weight of the cooling and shielding systems. Applications include diagnostic imaging for neonatal brain assessment, liver susceptometry, and gastric ischemia, and for difficult to diagnose and serious conditions [252,253].

Advances in Magnetic Sensor Design. A number of new nanoscale magnetometer designs are being developed which approach or exceed the sensitivity of SQUID, without cryogenic cooling, and with less power consumption, lower cost, and smaller size. One, the *optical atomic magnetometer*, developed by the U.S. National Institute of Standards and Technology with the University of Colorado and Sandia Laboratories, is based on the interaction of laser light with atoms oriented in a magnetic field in the gas phase [254]. Workers at Princeton University and at the University of California, Berkeley, and elsewhere are also developing optical atomic magnetometers, and improvements in performance continue to be published [255,256]. Another promising new magnetometer is a nanoscale cantilever design developed at Lucent Technologies' Bell Labs [257]. These and other designs may open new possibilities for magnetic medical imaging.

Optical Atomic Magnetometers. The NIST optical atomic magnetometer measures the change in alignment when atoms with a magnetic spin moment interact with a beam of laser light. In the absence of an external magnetic field, the atoms will align with the laser beam's crossed electric and magnetic fields. Any perturbation by a magnetic field will disorient the alignment with the beam, reducing the amount of light transmitted through the gas. Magnetic shielding is used to make the detector selective and directional. The fabrication of a cell containing the gas, a small solid-state laser, and a detector for the transmitted light can be scaled down to microchip form, with nanoscale geometries, to make an extremely small, sensitive, and economical sensor element.

At NIST a prototype, millimeter-scale microfabricated rubidium vapor cell with a low-power laser, was able to detect the heartbeat of a rat. In Berkeley researchers used the atomic magnetometer to detect magnetic particles flowing through water. Princeton researchers using a high-sensitivity atomic magnetometer based on potassium vapor performed MEG experiments [258]. Physicists at the University of Wisconsin and elsewhere are refining optical atomic magnetometer designs to reduce noise for biomedical applications [259].

The millimeter-scale prototype at NIST contains about 100 billion atoms of rubidium gas in a vial the size of a grain of rice. The change in spin alignment was easily detectable, and scalable down to much smaller sizes. The atomic magnetometer is about 1000 times more sensitive than previous devices of a similar size. With sensitivity below 70 femtotesla (fT) per root Hertz, it is comparable to, or even exceeds SQUID sensors. It can be made much smaller than a SQUID, and operates at much higher temperatures, at around 150°C.

Currently the complete NIST device is a few millimeters on each side. Developers predict that with the small size and high performance such sensors could lead to magnetocardiograms that provide similar information to an electrocardiogram (ECG), without requiring electrodes on the patient's body, even from outside clothing. The current versions of the atomic detector are sensitive enough to detect alpha waves from the human brain, which produce magnetic fields of about 1000 fT just outside the skull.

To pick up the full range of magnetic fields emanating from the human head, the atomic optical devices would need to be more sensitive—down to 10 fT or less, which is projected to be feasible. The thermal magnetic noise level generated by the human brain is on the order of 0.1 fT. A sensitivity of 0.2 fT is projected for the Princeton potassium-based atomic magnetometer, if supercooled shielding were used to reduce the noise level at the detector. This would enable imaging of individual cortical modules in the brain, which have a size of 0.1–0.2 mm. This could provide an alternative to MRI and PET imaging, without injection of contrast enhancement agents or tracers.

Previous atomic magnetometers and SQUIDs are larger and require much more power than the gas laser design. Even with the laser and heating components, the new devices use relatively low power and can be extremely small. Thus, they could be used in high-resolution arrays of distributed sensors. The small size allows the sensor to get close to the heart or brain for magnetic measurements. Developers project that the sensors could even be used to make portable MEG helmets for brain–machine interfaces. They could also be used to identify markers for specific chemicals by measuring nuclear quadrupole resonance of excited atoms, opening further possibilities for monitoring and research.

Nanoscale Electromechanical Resonator Magnetometers. Mechanical magnetometers can be made using nanoscale cantilevers or bridges, coated or implanted with magnetic materials to harness the sensitivity of nanoscale resonance vibrations. Fundamental breakthroughs in nanotechnology made in the past few years by Bell Labs and the New Jersey Nanotechnology Center (NJNC) have led to a new nanomechanical magnetometer design with performance that is potentially 100- to 1,000-times greater than existing commercial devices, at extremely low cost based on silicon lithographic fabrication [257].

The new Bell Labs MEMS magnetometers employ a silicon resonator carrying an electric circuit. Oscillation of the resonator in a magnetic field generates a current around a closed loop circuit damped by a resistor. Variations in magnetic field strength alter the amplitude and frequency of the resonator. This mechanical sensitivity can be used to measure the magnetic field by coupling the mechanical motion of a silicon bar or paddle to the ambient magnetic field.

In order to give a sensitive measure of a magnetic field, this nonlinear resonator must have negligible internal damping—a high Q-factor. Nanoscale crystalline oscillators made from quartz or silicon can be made with much higher Q numbers than all but superconducting electronic oscillators.

Magnetometers that use electronic detection (Hall, magnetoresistance, or flux-gate devices) have sensitivity limited by their electronic Q-factor, which depends on the resistance to electrons traveling through the metal in the circuit;

it is difficult to reduce this factor (and increase sensitivity) without resorting to superconducting materials (which is why SQUIDs remain the ultimate purely electronic detector). A tuning fork resonator made from single-crystal silicon (with less internal friction than that of the hardest metal) will vibrate almost a thousand times longer than the best room temperature electronic oscillators.

Researchers are working to optimize resonator magnetometer designs to achieve substantial improvements in sensitivity by modifying the microscale geometries with nanoscale features. Nanoscale mechanical resonators with mechanical Q-factors approaching 10,000 or more at room temperature can be made from semiconductor-grade silicon and similar single-crystal materials. This is a huge improvement over electronic detectors, without cryogenic cooling for superconductivity. Electromechanical resonator magnetometers should be up to 100- to 1,000-times more sensitive than existing commercial devices.

In the meantime, improved designs using nanotechnology continue to be applied to optical atomic magnetometers, as well as to devices based on the Hall effect, magnetoresistance, and SQUID [260–263]. Advances in signal processing are being applied to provide capacity to extract, analyze, and efficiently present magnetic sensing data for medical use [264].

A challenge in the design and application of MEG nanosensors is that the forces measured and signals generated by nanodevices can be many orders of magnitude smaller than the environmental magnetic noise. This is a general problem for all nanosensors, so it is illustrative to see how this significant metrological challenge is solved in the case of magnetic brain sensing. The environmental noise can be attenuated by a combination of shielding, primary sensor geometry, and synthetic analytical methods (signal processing). One of the most successful applications of computer power for noise cancellation is the use of synthetic higher-order gradiometers [265].

The MEG signals measured on the scalp surface must be interpreted and converted into information about the distribution of currents within the brain. This task is complicated by the fact that such inversion is nonunique. Additional mathematical simplifications, constraints, or assumptions must be employed to obtain useful source images, along with sophisticated signal extraction algorithms [266]. Beam-forming techniques such as synthetic aperture magnetometry beamformers can also reduce extraneous signals and focus the detector into the body [267].

Future Opportunities in Medical Magnetic Sensing. Imaging systems based on the newest sensors have not yet been built, but they will undoubtedly bring wider use of MEG, perhaps taking the portable noninvasive, efficient brain–machine interface closer to reality. In the meantime, other techniques for brain stimulation are being explored, such as stimulation of neural tissue by light [268,269], and vibrotactile or acoustic stimulation [270]. And the possibilities of feedback with MEG and other modalities for brain–machine interfaces are being explored [271]. So there are entirely new paths that the development of neuroprostheses could take. Whatever they are, nanotechnology will play a key role in implementing them (**Fig. 27.3**).

Sensory Neuroprosthetics: Haptics. Haptics refers to the sense of touch, and more generally to the sense of pressure and force feedback from the body to the

FIG. 27.3 *Nerve cells can be stimulated by light, even if they are not photoreceptors, if the light pulse depolarizes the membrane potential (experimental setup for cells in culture).*

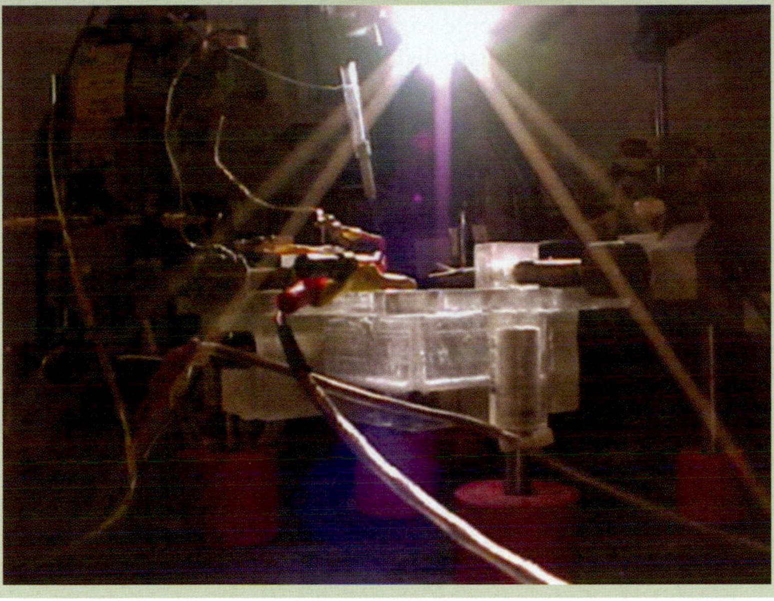

brain [272]. This brings us to the area of sensory neuroprosthetics, of which haptic sensors are a special case, since they are usually an integral part of motor prostheses. We discussed the importance of haptics and kinesthetics in the sections above on control and feedback of motor prostheses, and for brain–machine interfaces in general.

Haptics, along with vision, hearing, and balance, is an essential component of the stream of feedback that the nervous system sends to be brain [273]. A deficit in the haptic sense is usually the result of loss of the sensory nerve endings due to injury or amputation, but in rare and more difficult cases it can be due to brain injury. Haptics is a diffuse sense, so there is no one prosthetic device that satisfies the brain's requirement. It has been less well understood than more obvious senses, even taste and smell, which are more localized and less intuitive.

Because tactile feedback is so important to the fine motor control necessary for dexterous and delicate tasks, as well as athletic performance, it is becoming an important research area. Nanotechnology-based haptic sensors are having an impact on robotics, design of spacesuits, and in medicine—not only for neuroprosthetics, but in the design of robotics-assisted surgery systems [274,275]. It is extremely important in design of prosthetic strategies and in rehabilitation. Surgical connection of sensory as well as motor nerves to a prosthesis can give dramatically improved results [276,277]. So-called smart materials and embedded nanosensors will provide many options for implementing haptics in medical devices and surgical tools, and will assist in improving the quality of telemedicine [124,129,147].

Cognitive prosthetics. The concept of a *cognitive prosthesis*, a system developed to support and augment the cognitive abilities of its user, has only been made possible recently with the advances in computing and interface technologies [278]. To the extent that cognitive prosthetics includes augmentative and alternative communication for people with impaired communication, and virtual reality systems, there is not a sharp dividing line between the term and the functions provided by other sensory-motor augmentation modes such as implants to relieve seizures, hearing prostheses, and brain–computer interfaces in general [124,279]. However the concept does raise the possibilities of powerful augmentation of human capabilities, as well as treatment for mental deficits, both small and profound [280]. The idea raises many psychological, social, ethical, and medical concerns, as well as research issues.

Medical, Social, and Ethical Issues. Typically neuroprosthetics are resorted to only after pharmacologic and neurosurgical options have been exhausted. Bioengineers and other designers of systems to augment human capabilities are cognizant that their role should be to assist the body's adaptation and compensation for a deficit, rather than replace any remaining function. Systems that surpass natural capabilities can be intimately interfaced to the human body, in "bionic human" or cyborg scenarios [281,282]. Nanotechnology is making such capabilities more feasible and affordable, obliging us to confront the social, medical, and ethical consequences [283].

Future Directions for Brain–Machine Interfaces. The progress that has been made is remarkable, but many obstacles must still be overcome. Without invasive implants, current experimental brain–machine prostheses require the patient to be tethered to bulky equipment, which needs tuning and maintaining by a team of technicians. Prototype implants have wires that penetrate the skull and skin, with the risk of infection. Wireless signal transmission for brain implants is still in the future, along with wearable magnetic brain–machine interfaces.

A more difficult obstacle is that the performance of microelectrodes recording from neurons tends to fall off over time; better engineering of interfaces using nanoengineered materials is needed to improve biocompatibility and allow lower stimulation potentials. Even if implants are supplanted by noninvasive magnetic communication, there is still research to be done in neurocognition and how to interface learning systems.

Although patterns of control for hands and arms have been mapped, individual differences are not fully explored—some experimental patients are able to control prostheses much more easily than others. Concerted efforts are developing adaptive learning systems that require less effort from the patient by embedding neural networks and adaptive filters into the control system of the prosthesis, with impressive results.

Paths and mechanisms for feedback is another area where more research is needed—to succeed in duplicating or restoring limb function instead of merely controlling machines, researchers have to work out how the body tells the brain where its limbs are positioned in space—proprioception. Better pressure and vibration sensors, accelerometers, actuators, and force sensors are needed in order to develop improved artificial proprioception, haptics, and kinesthetics.

All of these pieces need to be integrated seamlessly in intelligent control systems that respond precisely and adapt over time to changes in their environment. Nanotechnology is playing an important role in sensors, actuators, communications, and computing elements to make the brain–machine interface a clinical reality. Most of the difficulties for motor neuroprosthetics are now engineering challenges, rather than problems of principle. They will be solved by closely knit interdisciplinary teams that include doctors, engineers, rehabilitation specialists, and patients.

27.2.5 Neuroprosthetics for the Ear

Restoring sensory pathways is as important for neural prosthetics as restoring motor function. Sight and hearing are valuable not only because of their role in performing tasks, but because they are the bridges for social communication. Hearing loss is the most common form of sensory impairment. Electronic aids for hearing have a long history intertwined with the telecommunications inventions that shape the modern world. Alexander Graham Bell was an audiologist; the telephone was a by-product of his interest in making an electronic hearing aid [284–286].

The study of how the human ear distinguishes sounds is important in design and optimization of large-scale telephone networks sending usable speech over long distances. Voice compression, recognition, and synthesis are modeled largely on an understanding of how the ear and brain process speech. The study of how hearing works—how information is encoded and decoded in sounds by the brain—has been essential in developing technology, and in turn has helped to develop aids to ameliorate hearing disorders [287]. Research to optimize telecommunications led to signal processing technology that made advanced hearing aids possible. Artificial stereocilia MEMS, modeling the resonators in the cochlea, are being fabricated in nanotechnology laboratories in order to understand the mechanisms of hearing [288–294].

Cochlear implants for hearing disorders are one of the most mature and best established areas of any electronic neuroprostheses. Nanotechnology has enhanced microelectronics, batteries, and micromechanical transducers in cochlear implant devices and thus has contributed significantly to the quality of life of persons with hearing impairment. Patients with cochlear implants benefit from improved understanding of speech in noise, sound quality, and localization of sounds compared to patients with acoustic hearing aids, without the ear canal occlusion, acoustic feedback, and inconvenience and cosmesis of external devices [295–297].

Cochlear prostheses are implanted in the middle ear, where they stimulate the ossicles electromechanically, rather than acoustically, through either electromagnetic or piezoelectric transducers. Sound signals transduced by an externally worn microphone are sent to the implant by wireless transmission. Cochlear implants achieve an average threshold improvement of 10 dB from 500 to 4000 Hz. At 6000 Hz the gain is about 20 dB compared with conventional fitted acoustic hearing aids [298].

Another type of hearing neuroprosthesis, the auditory brainstem implant, uses electrodes placed over the cochlear nucleus or inserted into the brainstem. This direct interface is capable of restoring some residual hearing in many patients who have lost both hearing nerves [299,300].

Because the organ to which the prosthesis connects the brain is inside the skull, neuroprostheses for hearing can bypass many of the problems faced in getting signals from the organ and its prosthesis to the relevant part of the brain. Cochlear implants (like cardiac pacemaker or spinal stimulators or deep brain stimulators) can be self-contained, with power transmitted through the skull by electromagnetic induction. Thus the main problems are long-term biocompatibility (build up of biofilms and plaques) and refinement of the signal processing to improve performance by attempting to match the natural function of the ear. Challenges include recognition of pitch for understanding and enjoyment of music, and preventing bacterial and fungal infections. The latter is an area that has prospects of being improved through nanomaterials [301–305].

Neuroprosthesis for Balance. Besides the sensory organ for hearing, the inner ear contains the vestibular arches, filled with fluid and cilia which can sense microscopic inertial fluid flows caused by head and body movements. Disruption of this function can cause severe dizziness, loss of balance, inability to walk or even sit upright, and sensations of sea-sickness or air-sickness. Some research is being pursued to develop an implantable MEMS neuroprostheses that could restore or compensate for loss or disturbance of this sensory organ [306–310].

Neuroprosthesis for Tinnitus. Tinnitus is a condition which results in a constant sensation of sound, regardless of whether an audial stimulus is present. It is difficult to treat, and can be very troubling. Most treatment seeks to modulate the patient's response, rather that treating the tinnitus itself. One alternative development is an implantable device to deliver electrical stimulation in the middle ear, close to the cochlea; the goal is to turn off the nerve pathway, similar to stimulators for relief of pain and tremor [311].

Anatomy of the Ear. In order to see the relevance of nanotechnology to cochlear implants, consider the neuroanatomy of the inner ear and how it is stimulated.

Mechanoacoustical pressure is delivered to the membrane that seals off the inner ear, the cochlea, by an extremely delicate linkage from the eardrum to the malleus, incus, and stapes (hammer, anvil, and stirrup) bones. Sound is collected by the eardrum, causing the stapes to vibrate against the membrane which separates the cochlea from the outer ear, transmitting pressure waves to fluids in the interior. This linkage is a powerful mechanical amplifier—the human ear can detect motions of the eardrum on the order of a picometer—smaller than the diameter of an atom.

The cochlea is a hollow tapering helix supported by a bony spiral shelf, the *osseous spiral lamina*, which winds around a central core, the *modiolus*. The cochlea's spiral cone geometry, like a French horn or conch shell, acts as a mechanical acoustical transform to select for different vibration frequencies along its interior. The interior of the cochlea is separated into two fluid-filled chambers (the *scala vestibuli* and *scala tympani* or upper and lower ducts) by a thin sac, called the *cochlear duct*, filled with gelatinous material. The large end of the spiral is sealed from the outer ear by two membranes, the oval and round windows, on either side of the cochlear duct. The duct separates the two chambers all the way up the spiral to its apex, where there is a small opening between them. The sensory hair cells are inside the cochlear duct adjacent to a thin layer of tissue (the *tectorial membrane*). Each hair cell has a group of stereocilia projecting into the viscous gelatin, which resonate with sound [312,313].

The chambers on either side of the cochlear duct are filled with an electrolyte solution which conducts sound from the oval window, to which the stapes is attached, into the scala vestibuli. When pressure waves travel through the upper side of the cochlear duct to the apex of the spiral, and down the lower side, the duct is compressed by the fluid, causing movements of the cell hairs. The shearing movement of the hair cells opens potassium ion channels in hair cell membranes, depolarizing the cells and initiating an electrical signal. (See chapter 14, for a review of potassium ion channels in nerve cells.)

The mechanical stress is transmitted to the ion channels in hair cells via tiny filaments that connect neighboring hairs in a bundle. **Figure 27.4** shows a schematic of the hair cells in the ear. The hairs are about 500 nm in diameter and tip links are on the order of 2 nm in diameter.

> Each hair cell is a micromachine that is constructed from hundreds of much smaller components (e.g. ion channels and tip links). But hair cells do not operate in isolation. Hair cells are part of a larger system: the inner ear.
>
> Thus the inner ear represents VERY LARGE SCALE INTEGRATION of hundreds of thousands of biological microelectromechanical devices.
>
> **A. J. Aranyosi**
> *MIT Micromechanics Group*

The stereocilia are linked in a complex network, with an inner and outer layer. The cochlea contains about 28,000 hair cells in humans. The absolute number of cells does not directly relate to auditory acuity; cats have 39,000, bats, rats, and dolphins about 15,000. In the human, the interior linear extent of the cochlea is about 3.5 cm. So the density of sense cells is about 800 per millimeter (spaced at about 1.25 μm along the cochlea).

Behind the hair cells is a layer of neurons, the spiral ganglion, which contains four or five times more cells than the sensory cell layers. The network of auditory neurons in the spiral ganglion lies close to the interior or modiolar wall of the cochlear duct. The spiral ganglion neurons control the selection and organization of stimuli that are sent to the auditory cortex in the brain.

The hair cells are extremely sensitive to acoustical vibrations, and vulnerable to damage. Up to one-third of the sensory hair cells typically die with age and damage by intense or chronic sound overload, chemicals, and disease. Unlike disorders in the mechanical acoustical path, nerve damage cannot be compensated by external devices.

Design of Cochlear Implants. Cochlear implants serve to bypass damaged inner ear hair cells and transfer auditory information to the brain. A cochlear implant takes sound from an external microphone and converts it into electrical impulses to stimulate the cochlea. The microphone and the signal processing module are worn outside the head, and the stimulator is implanted behind the ear, with wires inserted into the cochlea. Stimulation is usually applied to the ganglion layer, which is accessed by inserting electrodes into the scala tympani, the lower cochlear chamber next to the modiolus where it can be placed close to the spiral ganglion.

Fig. 27.4 *Hair cells in the cochlear duct have stereocilia (microvilli) linked to potassium ion channels. Movements induced in the stereocilia by mechanical stimulus open the ion channels to depolarize the cell, initiating nerve firing.*

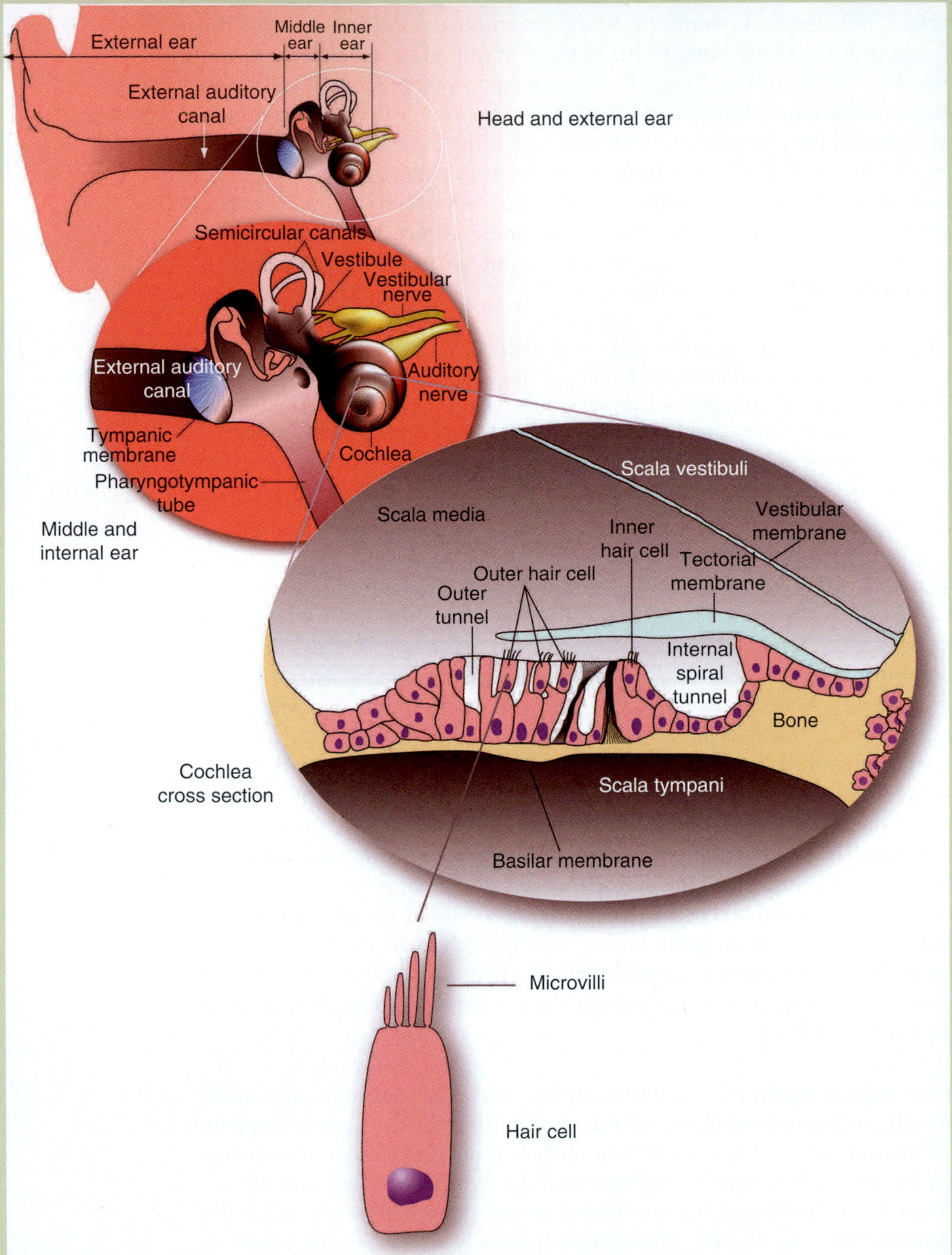

Coding of sound and mapping stimulation onto the cochlea are guided by neurocognitive research in how the cochlea receives and processes sound. Early cochlear implants used a single electrode and encoded sounds by converting sound frequencies into electronic pulse frequencies, which are perceived as sound when applied to the ganglion. This encoding is oversimplified and does not reproduce fine details of sound perception. Newer cochlear implants map different frequencies spatially along the cochlea, from lower to higher frequencies. This improved mapping is closer to the natural perception of sound by the ganglion, and is the coding used by most current cochlear implant devices.

Current devices have between 16–100 electrodes. While it may never be necessary to approximate the number of discrete stimulation points on the spiral ganglion that are presented by the 20,000 or more hair cells, clearly there is plenty of room at the bottom for delivery of sound coded impulses with higher resolution along the cochlear duct.

Researchers and developers are working to overcome a number of challenges in what is called the *electroneural bottleneck*. Currently, the voltage required to stimulate the neurons in the spiral ganglion is not localized—it stimulates a relatively large area. The electrodes cannot be placed very close together because it would result in cross-talk. The electrodes cannot be too conforming or embedded in the cochlea because the device may have to be removed without damage to the tissues. The electrodes are currently handmade assemblies; integrated device electrode assemblies have been proposed and used experimentally in the lab, but performances in electrode resistance, durability, reliability, and biocompatibility are still not sufficient for clinical use. Currently the signal-processing power available to fit into a low-power wearable package is not sufficient to process received sound into many more channels for discrete delivery, and a large number of channels would present a wiring challenge.

Work has been done on designs using nested wiring with electrode contacts that can be de-insulated by laser, which may resolve some obstacles. Electrode coatings and treatment with platinum, iridium, gold, platinum black, and alloys have been studied to lower resistance and reduce corrosion. Shape metal alloys have been proposed to produce better conformance to the modiolus and spiral ganglion. Designs with quadrupolar electrodes, with contact points on various places in the modiolus have been proposed, which might help to focus the excitation area. Coating of electrodes with brain-derived growth factors has been studied experimentally in an effort to reduce long-term atrophy of spiral ganglion cells [314–316].

Long-term wear of cochlear implants has given rise to biofilm contamination and infection; the ear is particularly vulnerable to infection since it is open to the mouth and throat.

Nanotechnology may in future offer solutions to these challenges with better materials, circuits, signal-processing arrays, and stimulation methods. Application of nanotechnology may even come up with convenient, low-cost nanoengineered smart acoustical materials that could filter out damaging frequencies and noise levels when inserted or injected into the ear as an expanding foam, while allowing the wearer to hear normal sounds in comfort, thus preventing hearing loss.

27.2.6 Vision Prosthetics

Loss of sight has profound psychological as well as social and physical consequences; some 30% of the sensory input to the brain comes from the eye [317,318]. The sense of sight involves highly parallel processing of image data from light focused onto a surface. Vision involves the process by which the eye and the optical nerves gather light, extract image data by sampling areas on the projection, detect features in the image, and send the information to the brain for further processing, recognition, and analysis.

The brain is very plastic, especially with respect to the pathways to the visual cortex; with a prosthesis that maps pixels onto the surface of a suitable area of skin, with dense nerve receptors, it is possible for the brain to map the signals to the visual cortex so that the patient learns to visualize from the haptic inputs. The adaptability of the brain gives a good prognosis for the development of a number of visual prostheses types, whether mapping light images to haptic nerve endings or stimulating some level in the layers of processing that lead from the eye to the brain, all the way up to direct stimulation of the visual cortex.

Any of these routes is a formidable undertaking from both technical and neurological viewpoints. If the ear is the foot of sensory prosthetics, then the eye is the hand. Both are difficult, but hearing is a mapping onto a one-dimensional sensory cell space—the linear array of cilia—whereas vision is a mapping onto a two-dimensional space of rods and cones in the retina. Additional dimensionality for both hearing and sight is added by parallelism in time—sensation from millions of cells is simultaneous over the sensing space rather than serial. One of the jobs of the layers of neurons behind the retina is to organize a sampling scheme for transmitting images to the brain. Vision is not totally asynchronous—somewhere on the path to the brain the incoming data is organized into scans, which is why the visual cortex can be satisfied to generate images from strobed motion picture and television screens, so long as their output rate exceeds the sampling rate of the brain's optical system. The latter is not a simple scanning process, but in practice a frame rate of about 30 frames per second is perceived as continuous [319–321].

The Retina. The retina is the point at which light is converted into neural impulses, which are processed into images by networked layers of neurons carrying information to the visual cortex. The retina samples images with photoreceptor cells—rods and cones—whose overall size is on the order of microns. Cones are color sensitive. Rods have a sharper acuity than cones, but do not discriminate colors [322,323].

The human retina contains approximately 120 million rod and 1 million cone cells. The densely packed cones in the center of the retina where vision is most acute, the fovea, have a center to center spacing of about 2.5 μm. Cone density in the fovea is between 100,000 and 300,000/mm^2, but rods and cones are both present in surrounding periphery of the retina. Rods are absent in the fovea, and are packed at a density of 80,000–100,000/mm^2 in the periphery.

Fovea. The area of the fovea, where rods are absent, has a radius of 200–300 μm; the central part of the fovea where cones are packed most densely is only about

50 × 50 μm. The total number of cones in the fovea is approximately 200,000. The total number of cones in the entire retina is approximately 6,400,000. The total number of rods in the retina is 110,000,000–125,000,000.

The peak rod density is located in a ring around the fovea between 1.5 and 5 mm from the center, where the rod density is between 100,000 and 160,000 rods/mm^2. In terms of the radius of the field of vision, the peak is between 5 and 18° from the center. The rod-free area of the fovea is only 1 or 2° of the visual field.

The photoreceptors communicate with ganglion neurons located in a tissue layer 150–300 μm behind the retina, and separated from it by a layer of support cells, the retinal pigment epithelium. There are 1 or 2 cones and about 20 rods for each ganglion cell. A network of interconnected neurons process the information from more than six million receptors in the retina down to where it is carried by approximately one million axons in the optic nerve to the brain.

By-Passing the Retina. Retinal degeneration or detachment is one of the main causes of vision loss. Therefore most efforts to develop a visual neuroprosthesis have attempted to stimulate the ganglia cell layer behind the retina, as the simplest strategy to interface to the visual signal processing that is in place in the optical nerve path. Some prostheses have been designed and tested that stimulate the visual cortex directly, producing a low resolution pattern of visual sensation, and some have mapped digitized imaging onto touch sensors in the back or other skin areas, in a kind of transposed Braille that delivers images rather than encoded letters.

Artificial Retinas. Most recent research in visual neuroprosthetics has been focused on artificial retina replacements or bypasses converting visual information into patterns of electrical stimulation onto inner retinal neurons. More than a dozen projects around the world are aimed at using advances in nanotechnology and high-density integrated microelectronics to develop an implant analogous to the cochlear implant for the ear, to restore lost vision. Like the cochlear implant, an eye implant is not anticipated to fully restore all lost function. Research devices are expected to provide enough visual perception of contours, outlines, and shades of light to allow a blind person to move freely in unfamiliar environments (**Fig. 27.5**).

Treatment of diseases such as retinitis pigmentosa is focused on growing knowledge of affected biochemical pathways, development of animal models, and possible gene therapy, especially for genetically defined subsets of patients, based on newly identified genes [324]. As is the case with many other parts of the nervous system, vision, its development, and its degeneration are controlled by a large number of different genes, and this approach is still in the early stages of development. Another possible approach being investigated is transplantation of cells to the retina [234]. Visual neuroprostheses will probably still be needed, even when other therapies become available, because of the large number and diversity of causes of vision loss.

In the remainder of this section, we look at some projects developing prostheses that can be implanted in the visual cortex, around the optic nerve, or in the eye. These approaches have shown promise for useful perception to patients with visual impairments [325–330].

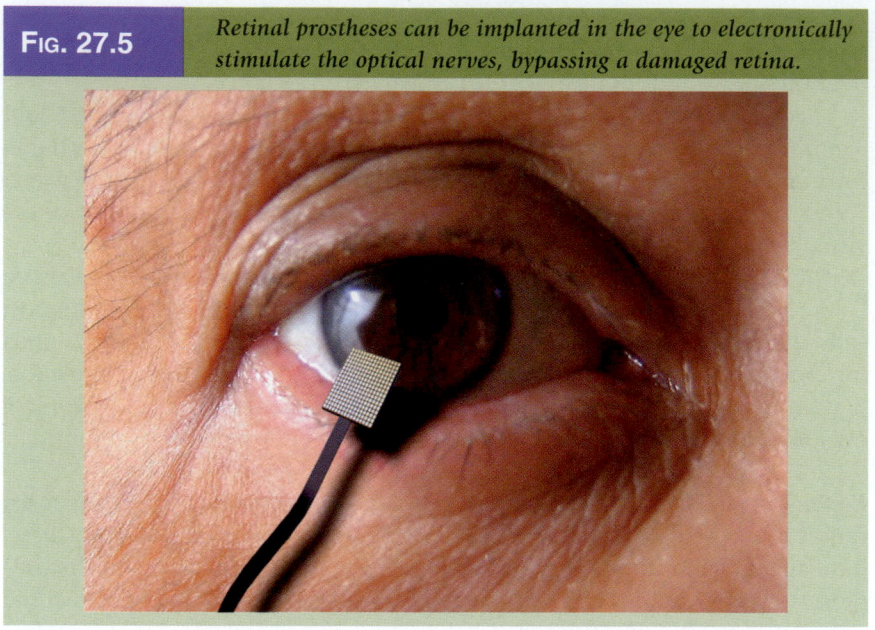

Fig. 27.5 *Retinal prostheses can be implanted in the eye to electronically stimulate the optical nerves, bypassing a damaged retina.*

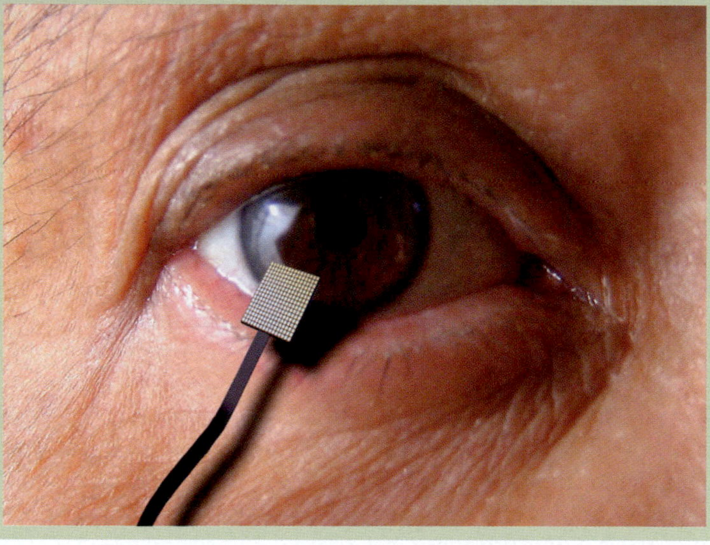

A large-scale project has been undertaken for a number of years by the U.S. Department of Energy National Laboratories and the National Science Foundation, with a team that includes a number of universities, institutes, and private industry. The artificial retina device bypasses nonfunctioning retinal cells to transmit signals directly to the optic nerve. The device consists of a tiny camera and microprocessor mounted in eyeglasses, a receiver implanted behind the ear, and an electrode-studded array that is tacked to the retina. Power is provided by a wireless battery pack worn on the belt.

A microprocessor converts the camera image and transmits information to an implanted wireless receiver. The receiver sends the signals through a tiny cable to the electrode array to generate stimulating pulses. The pulses are perceived as patterns of light and dark spots corresponding to the electrodes stimulated. Patients have learned to interpret the visual patterns produced, enabling them to detect when lights are on or off, describe an object's motion, count individual items, and locate objects. To evaluate the long-term effects of the retinal implant, five devices have been approved for home use (**Fig. 27.6**).

The DOE project has produced three successively more sophisticated models, which are being tested and evaluated. Surgical time was reduced from the 6 h required for the first model to 2 h for the second version, which has 60 electrodes. The third model will have more than 200 electrodes, and will use more advanced materials than previous ones. A special coating, only a few microns thick, will replace the bulky sealed package used in previous models. The new model will use flexible conductive materials for the electrode array so that it will conform to the shape of the inner eye. The latest model will be many times smaller than earlier models. Engineering goals include enhancing the resolution with more electrodes, and decreasing the size of the device and complexity of the surgical procedure [331–335].

FIG. 27.6 *An artificial retina system consists of an external camera with a signal processor which transmits encoded image stimuli to an electrostimulator array in the eye, which can trigger nerve firings behind the retina that can produce visual perceptions. A cochlear implant works much the same way with an external microphone and a stimulator next to the nerves in the ear. Experiments are underway to stimulate the visual nerve network with chemical neurotransmitters using microfluidic devices with nanoscale membranes.*

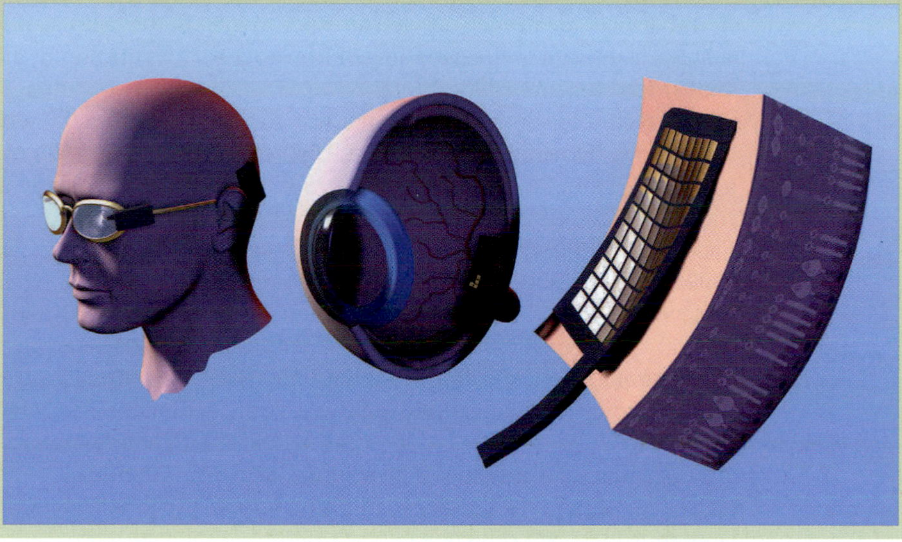

Other retinal prostheses systems have been developed including subretinal versions, and versions with high pixel density, in Australia, Europe, Japan, and elsewhere including international teams [336–341].

Researchers face numerous challenges in developing retinal prosthetic devices that are effective, safe, and durable enough to last for the lifetime of the individual. Materials must be biocompatible with delicate eye tissue, yet able to withstand corrosion. The device must remain fixed to a precise area of the retina and not compress or pull the tissue. The apparatus also needs to deliver enough power to stimulate electrodes, without generating excess heat to damage the remaining functional nerve tissue. Image processing needs to be performed in real time, so there is no delay in interpreting an object in view. In addition, effective surgical approaches are critically important to ensure a successful implant.

A number of interdisciplinary research teams are looking at the effect of implant surgery on the retina, the sensitivity of the retina to electronic charge, spatial resolution in relation to stimulation, the best patterns and locations for nerve stimulation, and evaluating the learning and adaptation of patients with early versions of the devices [342–367].

The possibility of electrode or tissue damage limits excitation schemes to those that may be employed with electrodes that have relatively low charge densities. The excitation thresholds that have been required to achieve vision have been found to be relatively high. This may result in part from poor apposition between neurons and the stimulating electrodes and is confounded by the effects of the photoreceptor loss, which initiates other pathology in the surviving retinal tissue. The combination of these and other factors imposes a restriction on the pixel density that can be used for devices that actively deliver electrical stimulation to the retina. The resultant use of devices with relatively low pixel densities presumably will limit the degree of visual resolution that can be obtained with these devices. Further increases in pixel density, and therefore increased visual acuity, will necessitate either improved electrode–tissue biocompatibility or lower stimulation thresholds. To meet this challenge, innovations in materials and devices are needed, as well as experiments on the factors and functional parameters relevant to the designing of implants, such as thresholds and electrical point spread functions [342].

Nanotechnology offers a potential way to avoid the obstacles with electrical stimulation, by using nanofabricated microfluidics to inject chemical neurotransmitter stimulants into the subretinal ganglion. Researchers at Stanford University have developed a prototype test system to study the possibility, by characterizing the stimulus produced with a microfluidics system fabricated on a 500-nm thick silicon nitride membrane, with a single 5- or 10-μm aperture overlaying a microfluidic delivery channel in a silicone elastomer. Controlled excitation based on picoliter amounts of neurotransmitter delivered was obtained in rat neurons grown in culture on the surface of the apparatus. In the experiments the stimulation radius was as small as 10 μm, comparable to what has been achieved in electrical stimulation experiments with a micrometer-size electrode. This experimental study shows how future neurostimulation systems, scaled up with arrays of delivery channels, might be possible with advances in integrated nanofabrication and cell engineering [368].

27.3 Summary

Most of the medical devices which we covered in this chapter are on a borderline between micro- and nanotechnology, but there is a clear trend towards more subtle nanoscale mechanisms. The same is true for other types of devices which we did not have the scope to cover, such as robotic surgery, and implantable sensors for blood pressure, gastric reflux, and other monitors [369–371].

Nanomedicine will have finally come of age when design of devices has fully merged with design of drugs—when the devices that are used for medicine may be made of protein and DNA and macromolecules as easily as silicon or polymers, all designed to fit their work in a natural way in cells and biological systems—and when the drugs and pharmaceuticals are atomic- and molecular-scale biochemical, genomic, and proteomic devices, designed with a full understanding of how they will fit into and do their work in cells and organisms—and when sensors and nanosurgical tools no longer have a clear distinction from genetically engineered bacteria and macrophages. This is the direction that the science and technology are going, and it will be reached only when a much deeper and fuller understanding of the fundamentals of biology are understood by engineers, and engineering by biomedical researchers, and the principles of physics, chemistry, electronics, and systems organization understood by both.

Even if that day comes, it will not bring a health utopia. There will always still be many challenges, as life is continually adapting and evolving. Solving one metabolic pathway or gene expression disorder, or finding a cure for one disease, will not eliminate the eternal work of survival and adaptation. New disorders and new organisms will arise to take advantage of any possible pathway that is open to feed energy, sustenance, and propagation. Living systems will always need to be changing to maintain their dynamic existence, which is like a flame—a structure maintained by a complex of dynamic chemical reactions as fuel and oxidants pass through. Though each organism may appear to us as a solid, stable body, we are dynamic equilibrium structures, apt to be disintegrated by a sufficient disturbance. The wonder of life is how stable and resilient it is, how all the systems work together to compensate for disturbances and counter multifold attacks, and how such complexity is all able to work together smoothly and beautifully in a harmonious whole. So the conclusion is that there will always be plenty of work to be done by nanoscientists and nanoengineers, so study diligently and apply yourself intelligently.

Conclusion. We do not have to wait for the future to find ways to use nanotechnology. To advance sensors in performance, size, cost, and power, some designs are moving into new nanoscale territory, such as the interaction between gas phase atomic magnetic resonance and laser optics, while others go down in the nanoscale to the mechanical principles of cantilevers. At the nanoscale, the electromechanical rules yield different results from the micro- or atomic scale, so re-examination of the tuning fork magnetometer or the gas phase atomic clock at the nanoscale gives new opportunities.

What is so fascinating and exciting about the nanoscience field at this time is that it is eminently probable that an even better design for some important medical device or process, improved by orders of magnitude, simpler, easier to build and assemble into arrays and interface into imaging systems, will be invented by someone who is now just now learning about nanoscience and nanotechnology—perhaps even someone who is studying this book.

Acknowledgments

The author wishes to thank the many colleagues who provided material, comments, and review of the drafts. Much material has been drawn from experiences in projects at the University of Texas Southwestern Medical Center, the University of Texas at Dallas, the University of Texas at Arlington, and the Texas Scottish Rite Hospital for Children. Any errors remain his own. Special thanks to the staff of the UT Southwestern Library.

References

1. H. J. Cook, What stays constant at the heart of medicine, *British Medical Journal*, 333, 1281–1282 (2006).
2. A. Coppa, L. Bondioli, A. Cucina, D. W. Frayer, C. Jarrige, J. -F. Jarrige, G. Quivron, M. Rossi, M. Vidale, and R. Macchiarelli, Palaeontology: Early Neolithic tradition of dentistry, *Nature*, 440, 755–756 (2006).
3. B. Morris, Surgery on Papyrus, *Student British Medical Journal*, 12, 309–348 (2004).

4. M. Kennedy, *A brief history of disease, science, and medicine*, Asklepiad Press, Mission Viejo, CA (2004).
5. I. Shaw, *The Oxford history of ancient Egypt*, Oxford University Press, Oxford, UK (2000).
6. J. F. Nunn, *Ancient Egyptian medicine*, University of Oklahoma Press, Norman, OK (1996).
7. R. Porter, *The greatest benefit to mankind: A medical history of humanity from antiquity to the present*, HarperCollins, New York (1997).
8. The Editors, Looking back on the millennium in medicine, *New England Journal of Medicine*, 342, 42–49 (2000).
9. K. Conley, *Joseph Priestley and the discovery of oxygen*, Mitchell Lane Publication Inc., Hockessin, DE (2005).
10. American Chemical Society, National Historical Chemical Landmarks website: http://acswebcontent.acs.org/landmarks/landmarks/priestley/discovery_O2.html (2008).
11. B. Jaffe, *Crucibles: The story of chemistry from ancient alchemy to nuclear fission*, 4th ed., Dover Publications, Mineola, NY (1976).
12. M. Pera, trans. by J. Mandelbaum, *The ambiguous Frog: The Galvani-Volta controversy on animal electricity*, Princeton University Press, Princeton, NJ (1992).
13. H. Xiao, P. Verdier-Pinard, N. Fernandez-Fuentes, B. Burd, R. Angeletti, A. Fiser, S. B. Horwitz, and G. A. Orr, Insights into the mechanism of microtubule stabilization by Taxol, *Proceedings of the National Academy of Science USA*, 103, 10166–10173 (2006).
14. J.-M. Lehn, Perspectives in supramolecular chemistry—From molecular recognition: Towards molecular information processing and self-organization, *Angewandte Chemie*, 29, 1304–1319 (2003).
15. NIH Office of Portfolio Analysis and Strategic Initiatives, website at: http://nihroadmap.nih.gov/nanomedicine/ (2008).
16. NIH Roadmap webpage at http://nihroadmap.nih.gov/initiatives.asp (2008).
17. NCI Nanotechnology Plan webpage at http://www.cancer.gov/researchfunding/NIHRoadmapFAQs (2008).
18. J. U. Leike, A. Sachse, and K. Rupp, Characterization of continuously extruded iopromide-carrying liposomes for computed tomography blood-pool imaging, *Investigative Radiology*, 36, 303–308 (2001).
19. A. T. Yordanov, et al., Novel iodinated dendritic nanoparticles for computed tomography (CT) imaging, *Nano Letters*, 2, 595–599 (2002).
20. J. R. Swanson, H. W. Bosch, K. J. Illig, D. M. Marcera, and R. L. Mueller, Process of preparing x-ray contrast compositions containing nanoparticles, US Patent 5543133, NanoSystems L.L.C., Collegeville, PA (1996).
21. P. J. Bonitatebus Jr., O. H. E. Axelsson, O. H. Erik, A. M. Kulkarni, B. C. Bales, D. J. Walter, A. S. Torres, and C. Treynor, Nanoparticle-based imaging agents for x-ray/computed tomography, US Patent 20070098641, General Electric Company, Fairfield, CT (2007).
22. J. W. M. Bulte and D. L. Kraitchman, Iron oxide MR contrast agents for molecular and cellular imaging, *NMR in Biomedicine*, 17, 484–499 (2004).
23. S. Eroglu, B. Gimi, L. Leoni, B. Roman, G. Friedman, T. Desai, and R. L. Magin, NMR imaging of biocapsules for monitoring the performance of cell and tissue implants, *IEEE-EMB Special Topic Conference on Microtechnology in Medical Biology, 2nd Annual International*, pp. 193–198 (2002).
24. C. Loo, A. Lin, L. Hirsch, M.-H. Lee, E. Chang, J. West, N. Halas, and R. Drezek, Gold nanoshell bioconjugates for molecular imaging in living cells, *Optics Letters*, 30, 1012–1014 (2005).
25. C. Loo, A. Lin, L. Hirsch, M.-H. Lee, J. Barton, N. Halas, J. West, and R. Drezek, Nanoshell-enabled photonics-based imaging and therapy of cancer, *Technology In Cancer Research and Treatment*, 3, 33–40 (2004).

26. C. Loo, A. Lowery, N. Halas, J. West, and R. Drezek, Immunotargeted nanoshells for integrated cancer imaging and therapy, *Nano Letters*, 5, 709–711 (2005).
27. M. M. Amiji, *Nanotechnology for cancer therapy*, CRC Press, Boca Raton, FL (2006).
28. I. Safarik and M. Safarik, Use of magnetic techniques for the isolation of cells, *Journal of Chromatography B*, 722, 33–53 (1999).
29. M. Lewin, N. Carlesso, C.-H. Tung, X.-W. Tang, D. Cory, D. T. Scadden, and R. Weissleder, Tat peptide-derivatized magnetic nanoparticles allow in vivo tracking and recovery of progenitor cells, *Nature Biotechnology*, 18, 410–414 (2000).
30. Z. M. Saiyed, S. D. Telang, and C. N. Ramchand, Application of magnetic techniques in the field of drug discovery and biomedicine, *Biomagnetic Research and Technology*, 1, 1–2 (2003).
31. B. Molnara, F. Siposb, O. Galamb, and Z. Tulassay, Molecular detection of circulating cancer cells: Role in diagnosis, prognosis and follow-up of colon cancer patients, *Digestive Diseases*, 21, 320–325 (2003).
32. Q. A. Pankhurst, J. Connolly, S. K. Jones, and J. Dobson, Applications of magnetic nanoparticles in biomedicine, *Journal of Physics D: Applied Physics*, 36, R167–R181 (2003).
33. D. Wang, J. He, N. Rosenzweig, and Z. Rosenzweig, Fe_2O_3 beads-CdSe/ZnS quantum dots core-shell nanocomposite particles for cell separation, *Nano Letters*, 4, 409–413 (2004).
34. A. S. Lübbe, C. Alexiou, and C. Bergemann, Clinical applications of magnetic drug targeting, *Journal of Surgical Research*, 95, 200–206 (2001).
35. U. Häfeli, et al., *Scientific and clinical applications of magnetic carriers*, Springer Verlag, Berlin, Germany (1997).
36. M. Shinkai and A. Ito, Functional magnetic particles for medical application, *Advances in Biochemical Engineering/Biotechnology*, 91, 191–220 (2004).
37. B.-I. Haukanes and C. Kvam, Application of magnetic beads in bioassays, *Biotechnology*, 11, 60–63 (1993).
38. C. G. Thanos and D. F. Emerich, The pinpoint promise of nanoparticle-based drug delivery and molecular diagnosis, *Biomolecular Engineering*, 23, 171–184 (2006).
39. K. K. Jain, Targeted drug delivery for cancer, *Technology in Cancer Research and Treatment*, 4, 311–454 (2005).
40. P. Merritt and L. A. Snyder, Pharmacology of β-Lapachone and Lapachol, *Cyberbotanica*, URL: http://biotech.icmb.utexas.edu/botany/beta.html, BioTech Resources, Austin, TX (1997).
41. D. A. Boothman and A. B. Pardee, Inhibition of Radiation-Induced Neoplastic Transformation by β-lapachone, *Proceedings of the National Academy of Science*, 86, 4963–4967 (1989).
42. K. E. Reinicke, E. A. Bey, M. S. Bentle, J. J. Pink, S. T. Ingalls, C. L. Hoppel, R. I. Misico, G. M. Arzac, G. Burton, W. G. Bornmann, D. Sutton, J. Gao, and D. A. Boothman, Development of β-Lapachone prodrugs for therapy against human cancer cells with elevated NAD(P)H:Quinone oxidoreductase 1 levels, *Clinical Cancer Research*, 11, 3055–3064 (2005).
43. D. Sutton, S. Wang, N. Nasongkla, J. Gao, and E. E. Dormidontova, Doxorubicin and β-lapachone release and interaction with micellar core materials: Experiment and modeling, *Experimental Biology and Medicine*, 232, 1090–1099 (2007).
44. D. Filmore, Breaching the blood-brain barrier, *Modern Drug Discovery*, 5, 22–27 (2002).
45. K. S. Soppimath, T. M. Aminabhavi, A. R. Kulkarni, and W. E. Rudzinski, Biodegradable polymeric nanoparticles as drug delivery devices, *Journal of Controlled Release*, 70, 1–20 (2001).
46. C. Roney, P. Kulkarni, V. Arora, P. Antich, F. Bonte, A. Wu, N. N. Mallikarjuana, S. Manohar, H. F. Liang, A. R. Kulkarni, H. W. Sung, M. Sairam, and T. M. Aminabhavi,

Targeted nanoparticles for drug delivery through the blood–brain barrier for Alzheimer's disease, *Journal of Controlled Release*, 108, 193–214 (2005).
47. J. Rafuse, Seventy-five years later, insulin remains Canada's major medical-research coup, *Canadian Medical Association Journal*, 155, 1306–1308 (1996).
48. M. Bliss, *The discovery of Insulin*, McClelland and Stewart, Toronto (1982).
49. R. D. Simoni, R. L. Hill, and M. Vaughan, The discovery of Insulin: The work of Frederick Banting and Charles Best, *Journal of Biological Chemistry*, 277, 26 (2002).
50. L. Rosenfeld, Insulin: Discovery and controversy, *Clinical Chemistry*, 48, 2270–2288 (2002).
51. L. Rosenfeld, Margaret and Charley: The personal story of Dr. Charles Best, the co-discoverer of Insulin, *Journal of the American Medical Association*, 291, 1903–1904 (2004).
52. B. Zhang, G. Salituro, D. Szalkowski, Z. Li, Y. Zhang, I. Royo, D. Vilella, M. T. Díez, F. Pelaez, C. Ruby, R. L. Kendall, X. Mao, P. Griffin, J. Calaycay, J. R. Zierath, J. V. Heck, R. G. Smith, and D. E. Moller, Discovery of a small molecule Insulin mimetic with antidiabetic activity in mice, *Science*, 284, 974–977 (1999).
53. R. S. Spangler, Insulin administration via liposomes, *Diabetes Care*, 13, 911–922 (1990).
54. S. Furtado, D. Abramson, L. Simhkay, D. Wobbekind, and E. Mathiowitz, Subcutaneous delivery of insulin loaded poly(fumaric-co-sebacic anhydride) microspheres to type 1 diabetic rats, *European Journal of Pharmaceutics and Biopharmaceutics*, 63, 229–236 (2006).
55. E. Allémann, J.-C. Leroux, and R. Gurny, Polymeric nano- and microparticles for the oral delivery of peptides and peptidomimetics, *Advanced Drug Delivery Reviews*, 43, 171–189 (1998).
56. M. Aboubakar, F. Puisieux, P. Couvreur, M. Deyme, and C. Vauthier, Study of the mechanism of insulin encapsulation in poly(isobutylcyanoacrylate) nanocapsules obtained by interfacial polymerization, *Journal of Biomedical Materials Research*, 47, 568–576 (1999).
57. S. Watnasirichaikul, N. M. Davies, T. Rades, and I. G. Tucker, Preparation of biodegradable Insulin nanocapsules from biocompatible microemulsions, *Pharmaceutical Research*, 17, 684–689 (2000).
58. G. P. Carino, J. S. Jacob, and E. Mathiowitz, Nanosphere based oral insulin delivery, *Journal of Controlled Release*, 65, 261–269 (2000).
59. J. S. Jacob, Y. S. Jong, D. T. Abramson, E. Mathiowitz, C. A. Santos, M. J. Bassett, and S. Furtardo, Nanoparticulate therapeutic biologically active agents, US Patent 20050181059, Spherics, Inc. (2004).
60. S. Salmon and S. M. Hudson, Crystal morphology, biosynthesis, and physical assembly of cellulose, chitin, and chitosan, *Polymer Reviews*, 37, 199–276 (1997).
61. S. A. Agnihotri, N. N. Mallikarjuna, and T. M. Aminabhavi, Recent advances on chitosan-based micro- and nanoparticles in drug delivery, *Journal of Controlled Release*, 100, 5–28 (2004).
62. Y. Zheng, Y. Wu, W. Yang, C. Wang, S. Fu, and X. Shen, Preparation, characterization, and drug release in vitro of chitosan-glycyrrhetic acid nanoparticles, *Journal of Pharmaceutical Sciences*, 95, 181–191 (2005).
63. Q. Gan and T. Wang, Chitosan nanoparticle as protein delivery carrier—Systematic examination of fabrication conditions for efficient loading and release, *Colloids and Surfaces B: Biointerfaces*, 59, 24–34 (2007).
64. C. Pinto Reis, R. Neufeld, A. Ribeiro, and F. Veiga, Nanoencapsulation II. Biomedical applications and current status of peptide and protein nanoparticulate delivery systems, *Nanomedicine: Nanotechnology, Biology, and Medicine*, 2, 53–65 (2006).
65. E. Renard, G. Costalat, and J. Bringer, From external to implantable insulin pump, can we close the loop? *Diabetes and Metabolism*, 28, S19–S25 (2002).

66. J. D. Zahn, Y.-C. Hsieh, and M. Yang, Components of an integrated microfluidic device for continuous glucose monitoring with responsive insulin delivery, *Diabetes Technology and Therapeutics*, 7, 536–545 (2005).
67. S. L. Tao and T. A. Desai, Microfabrication of multilayer, asymmetric, polymeric devices for drug delivery, *Advanced Materials*, 17, 1625 (2005).
68. Y. Y. Huang and C.H. Wang, Pulmonary delivery of insulin by liposomal carriers, *Journal of Controlled Release*, 113, 9–14 (2006).
69. G. T. McMahon and R. A. Arky, Inhaled insulin for diabetes mellitus, *New England Journal of Medicine*, 356, 497–502 (2007).
70. Z. T. Bloomgarden, Insulin treatment and type 1 diabetes topics, *Diabetes Care*, 29, 936–944 (2006).
71. E. J. A. Pope, K. Braun, and C. M. Peterson, Bioartificial organs I: Silica gel encapsulated pancreatic islets for the treatment of Diabetes Mellitus, *Journal of Sol-Gel Science and Technology*, 8, 635–639 (1997).
72. B. Gimi, T. Leong, Z. Gu, M. Yang, D. Artemov, Z. M. Bhujwalla, and D. H. Gracias, Self-assembled three dimensional radio frequency (RF) shielded containers for cell encapsulation, *Biomedical Microdevices*, 7, 341–345 (2005).
73. P. de Vos, A. F. Hamel, and K. Tatarkiewicz, Considerations for successful transplantation of encapsulated pancreatic islets, *Diabetologia*, 45, 159–173 (2002).
74. S. Schneider, P. J. Feilen, F. Brunnenmeier, T. Minnemann, H. Zimmermann, U. Zimmermann, and M. M. Weber, Long-term graft function of adult rat and human islets encapsulated in novel alginate-based microcapsules after transplantation in immunocompetent diabetic mice, *Diabetes*, 54, 687–693 (2005).
75. W. T. Godbey and A. Atala, In vitro systems for tissue engineering, *Annals of the New York Academy of Sciences*, 961, 10–26 (2002).
76. L. G. Griffith and G. Naughton, Tissue engineering—Current challenges and expanding opportunities, *Science*, 295, 1009–1014 (2002).
77. G. W. Gross, W. Wen, and J. Lin, Transparent indium-tin oxide patterns for extracellular, multisite recording in neuronal culture, *Journal of Neuroscience Methods*, 15, 243–252 (1985).
78. U. Egert, B. Schlosshauer, S. Fennrich, W. Nisch, M. Fejtl, T. Knott, T. Muller, and H. Hammerle, A novel organotypic long-term culture of the rat hippocampus on substrate-integrated multielectrode arrays, *Brain Research Protocols*, 2, 229–242 (1998).
79. M. Bani-Yaghoub, R. Tremblay, R. Voicu, G. Mealing, R. Monette, C. Py, K. Faid, and M. Sikorska, Neurogenesis and neuronal communication on micropatterned neurochips, *Biotechnology and Bioengineering*, 92, 336–345 (2005).
80. Editors' Choice, Ephrins: From axon guidance to neurite inhibitor to viral receptor, *Science STKE*, 2005 tw281(2005).
81. M. D. Benson, M. I. Romero, M. E. Lush, Q. R. Lu, M. Henkemeyer, and L. R. Parada, Ephrin-B3 is a myelin-based inhibitor of neurite outgrowth, *Proceedings of the National Academy of Sciences*, 102, 10694–10699 (2005).
82. Application Note 111, *Versatile biomolecular printing on a variety of surface types*, BioForce Nanosciences, Inc., Ames, IA 50010 USA (2007).
83. C. E. Hulsebosch, Recent advances in pathophysiology and treatment of spinal cord injury, *Advances in Physiology Education*, 26, 238–255 (2002).
84. M. E. Schwab, Repairing the injured spinal cord, *Science*, 295, 1029–1031 (2002).
85. H. M. Bomze, K. R. Bulsara, B. J. Iskandar, P. Caroni, and J. H. P. Skene, Spinal axon regeneration evoked by replacing two growth cone proteins in adult neurons, *Nature Neuroscience*, 4, 38–43 (2001).
86. V. Maquet, D. Martin, B. Malgrange, R. Franzen, J. Schoenen, G. Moonen, and R. Jérôme, Peripheral nerve regeneration using bioresorbable macroporous polylactide scaffolds, *Journal of Biomedical Materials Research*, 52, 639–651 (2000).

87. T.-T. B. Ngo, P. J. Waggoner, A. A. Romero, K. D. Nelson, R. C. Eberhart, and G. M. Smith, Poly(L-lactide) microfilaments enhance peripheral nerve regeneration across extended nerve lesions, *Journal of Neuroscience Research*, 72, 227–238 (2003).
88. G. E. Rutkowski, C. A Miller, S. Jeftinija, and S. K. Mallapragada, Synergistic effects of micropatterned biodegradable conduits and Schwann cells on sciatic nerve regeneration, *Journal of Neural Engineering*, 1, 151–157 (2004).
89. M. Romero-Ortega and P. Galvan-Garcia, A biomimetic synthetic nerve implant, US Patent 20070100358, Texas Scottish Rite Hospital for Children (2007).
90. T. Kmecko, G. Hughes, L. Cauller, J.-B. Lee, and M. Romero-Ortega, Nanocomposites for neural interfaces, In *Electrobiological interfaces on soft substrates*, J. P. Conde, B. Morrison III, and S. P. Lacour, eds., Materials Research Society Symposium Proceedings, 926E, 926-CC04-06 (2006).
91. P. Galvan-Garcia, E. W. Keefer, F. Yang, M. Zhang, S. Fang, A. A. Zakhidov, R. H. Baughman, and M. I. Romero, Robust cell migration and neuronal growth on pristine carbon nanotube sheets and yarns, *Journal of Biomaterials Science, Polymer Edition*, 18, 1245–1261 (2007).
92. M. Zhang, K. R. Atkinson, and R. H. Baughman, Multifunctional carbon nanotube yarns by downsizing an ancient technology, *Science*, 306, 1358–1361 (2004).
93. S. K. Smart, A. I. Cassady, G. Q. Lua, and D. J. Martin, The biocompatibility of carbon nanotubes, *Carbon*, 44, 1034–1047 (2006).
94. J. Chłopek, B. Czajkowska, B. Szaraniec, E. Frackowiak, K. Szostak, and F. Béguin, In vitro studies of carbon nanotubes biocompatibility, *Carbon*, 44, 1106–1111 (2006).
95. M. A. Correa-Duarte, N. Wagner, J. Rojas-Chapana, C. Morsczeck, M. Thie, and M. Giersig, Fabrication and biocompatibility of carbon nanotube-based 3D networks as scaffolds for cell seeding and growth, *Nano Letters*, 4, 2233–2236 (2004).
96. S. Garibaldi, C. Brunelli, V. Bavastrello, G. Ghigliotti, and C. Nicolini, Carbon nanotube biocompatibility with cardiac muscle cells, *Nanotechnology*, 17, 391–397 (2006).
97. A. Magrez, S. Kasas, V. Salicio, N. Pasquier, J. W. Seo, M. Celio, S. Catsicas, B. Schwaller, and L. Forró, Cellular toxicity of carbon-based nanomaterials, *Nano Letters*, 6, 1121–1125 (2006).
98. J. M. Wörle-Knirsch, K. Pulskamp, and H. F. Krug, Oops they did it again! Carbon nanotubes hoax scientists in viability assays, *Nano Letters*, 6, 1261–1268 (2006).
99. M. Shim, N. W. S. Kam, R. J. Chen, Y. Li, and H. Dai, Functionalization of carbon nanotubes for biocompatibility and biomolecular recognition, *Nano Letters*, 2, 285–288 (2002).
100. X. Yu, G. P. Dillon, and R. V. Bellamkonda, A laminin and nerve growth factor-laden three-dimensional scaffold for enhanced neurite extension, *Tissue Engineering*, 5, 291–304 (1999) doi:10.1089/ten.1999.5.291.
101. S. Yoshii and M. Oka, Peripheral nerve regeneration along collagen filaments, *Brain Research*, 888, 158–162 (2001).
102. S. Yoshii, M. Oka, M. Shima, A. Taniguchi, and M. Akagi, 30 mm regeneration of rat sciatic nerve along collagen filaments, *Brain Research*, 949, 202–208 (2002).
103. M. R. Ahmed and R. Jayakumar, Peripheral nerve regeneration in RGD peptide incorporated collagen tubes, *Brain Research*, 993, 208–216 (2003).
104. M. R. Ahmed, U. Venkateshwarlu, and R. Jayakumar, Multilayered peptide incorporated collagen tubules for peripheral nerve repair, *Biomaterials*, 25, 2585–2594 (2004).

105. V. Charulatha and A. Rajaram, Crosslinking density and resorption of dimethyl suberimidate-treated collagen, *Journal of Biomedical Materials Research*, 36, 478–486 (1997).
106. V. Charulatha and A. Rajaram, Influence of different crosslinking treatments on the physical properties of collagen membranes, *Biomaterials*, 24, 759–767 (2003).
107. M. R. Ahmed, S. Vairamuthu, M. Shafiuzama, S. H. Basha, and R. Jayakumar, Microwave irradiated collagen tubes as a better matrix for peripheral nerve regeneration, *Brain Research*, 1046, 55–67 (2005).
108. S. Yoshii, M. Oka, N. Ikeda, M. Akagi, Y. Matsusue, and T. Nakamura, Bridging a peripheral nerve defect using collagen filaments, *Journal of Hand Surgery*, 26A, 52–59 (2001).
109. K. A. Ellenbogen and M. A. Wood, *Cardiac pacing and ICDs*, 4th ed., Blackwell, Malden, MA (2005).
110. R. E. Klabunde, *Cardiovascular physiology concepts*, 4th ed., Lippincott Williams & Wilkins, Philadelphia, PA (2005).
111. D. E. Mohrman, *Cardiovascular physiology*, McGraw-Hill, New York (2002).
112. R. M. Berne and M. N. Levy, *Cardiovascular physiology*, 8th ed., Mosby, St. Louis, MO (2001).
113. P. J. Rosch and M. S. Markov, *Bioelectromagnetic medicine*, Informa Health Care, London (2004).
114. K. Mullett, State of the art in neurostimulation, *Pacing and Clinical Electrophysiology*, 10, 162–175 (1987).
115. A. L. Benabid, G. Deuschl, A. E. Lang, K. E. Lyons, and A. R. Rezai, Deep brain stimulation for Parkinson's disease, *Movement Disorders*, 21(Suppl. 14), S168–S170 (2006).
116. R. J. Vetter, J. C. Williams, J. F. Hetke, F. A. Nunamaker, and D. R. Kipke, Chronic neural recording using silicon-substrate microelectrode arrays implanted in cerebral cortex, *IEEE Transactions on Biomedical Engineering*, 51, 896–904 (2004).
117. N. M. Neihart and R. R. Harrison, Micropower circuits for bidirectional wireless telemetry in neural recording applications, *IEEE Transactions on Biomedical Engineering*, 52, 1950–1959 (2005).
118. F. Silveira and D. Flandre, *Low power analog CMOS for cardiac pacemakers design and optimization in bulk and SOI technologies*, Springer Verlag, Berlin, Germany (2004).
119. M. Ghovanloo and G. Lazzi, *Transcutaneous magnetic coupling of power and data, in Wiley Encyclopedia of biomedical engineering*, John Wiley & Sons, New York (2006).
120. C. M. John and V. John, Device for neuromuscular peripheral body stimulation and electrical stimulation (ES) for wound healing using RF energy harvesting, US Patent 2000161216, San Francisco, CA (2006).
121. M. Glikson and P. Friedman, The implantable cardioverter defibrillator, *Lancet*, 357, 1107–1117 (2001).
122. R. Allen, *Medtronic sets the pace with implantable electronics*, Electronic Design Online ID #5951 (2003).
123. R. Kötz and M. Carlen, Principles and applications of electrochemical capacitors, *Electrochimica Acta*, 45, 2483–2498 (2000).
124. W. E. Finn and P. G. LoPresti, *Handbook of neuroprosthetic methods*, CRC Press, Boca Raton, FL (2002).
125. K. W. Horch and G. S Dhillon, eds., *Neuroprosthetics, theory and practice*, World Scientific Publishing, Singapore (2004).
126. P. Rebelo and M. Van Walstijn, Designing acoustic thresholds, In *Les journées du design sonore*, IRCAM, Paris (2004).

127. D. H. Delgado, V. Rao, H. J. Ross, S. Verma, and N. G. Smedira, Mechanical circulatory assistance: State of art, *Circulation*, 106, 2046–2050 (2002).
128. H. H. Hu, P. Jia, T. Lu, and K. Yuan, Head gesture recognition for hands-free control of an intelligent wheelchair, *Industrial Robot: An International Journal*, 34, 60–68 (2007).
129. H.-N. Teodorescu and L. C. Jain, *Intelligent systems and technologies in rehabilitation engineering*, CRC Press, Boca Raton, FL (2001).
130. D. Taylor, Neural control of assistive technology, In *Wiley Encyclopedia of biomedical engineering*, John Wiley & Sons, Inc., New York (2006).
131. C. M. Light, P. H. Chappell, B. Hudgins, and K. Engelhart, Intelligent multifunction myoelectric control of hand prosthesis, *Journal of Medical Engineering and Technology*, 26, 139–146 (2002).
132. G. M. Friehs, V. A. Zerris, C. L. Ojakangas, M. R. Fellows, and J. P. Donoghue, Brain–machine and brain–computer interfaces, *Stroke*, 35, 2702–2705 (2004).
133. J. D. Weingarten, G. Lopes, R. E. Groff, M. Buehler, and D. E. Koditschek, Automated gait adaptation for legged robots, IEEE International Conference on Robotics Automation (ICRA), New Orleans, LA (2004).
134. R. Rupp and H. J. Gerner, Neuroprosthetics of the upper extremity: Clinical application in spinal cord injury and future perspectives, *Biomedizinische Technik, (Berlin)*, 49, 93–98 (2004).
135. R. Gailey, Rehabilitation of a traumatic lower limb amputee, *Physiotherapy Research International*, 3, 239–243 (2006).
136. M. Lowe, A. King, E. Lovett, and T. Papakostas, Flexible tactile sensor technology: Bringing haptics to life, *Sensor Reviews*, 24, 33–36 (2004).
137. H. Atmani, F. Merienne, D. Fofi, and P. Trouilloud, Computer aided surgery system for shoulder prosthesis placement, *Computer Aided Surgery*, 12, 60–70 (2007).
138. M. Ciocarlie, C. Goldfeder, and P. Allen, Dimensionality reduction for hand-independent dexterous robotic grasping, presentation, IROS: IEEE/RSJ International Conference on Intelligent Robots and Systems, San Diego, CA (2007).
139. P. S. Blaer and P. K. Allen, Data Acquisition and View Planning for 3-D Modeling Tasks, presentation, IROS: IEEE/RSJ International Conference on Intelligent Robots and Systems, San Diego, CA (2007).
140. R. D. Howe and Y. Matsuoka, Robotics for surgery, *Annual Review of Biomedical Engineering*, 1999, 1, 211–240 (1999).
141. S. J. Weghorst, K. S. Morgan, and H. B. Sieburg, *Medicine meets virtual reality: Health care in the information age*, IOS Press, Incorporated, Fairfax, VA (1996).
142. J. D. Westwood, *Medicine meets virtual reality 2001: Outer space, inner space, virtual space (Studies in health technology and informatics)*, IOS Press, Incorporated, Fairfax, VA (2001).
143. B. Beckwith, Medicine meets virtual reality 2001 (Review), *Clinical Chemistry*, 47, 2190 (2001).
144. S. Canright, ed., *Amy Ross, space suit designer*, NASA Education Home, STS-118, Resources for Educators, webpage at http://www.nasa.gov/audience/foreducators/stseducation/stories/Amy_Ross_Profile.html (2007).
145. P. Danaher, K. Tanaka, and A. R. Hargens, Mechanical counter-pressure vs. gas-pressurized spacesuit gloves: Grip and sensitivity, *Aviation, Space, and Environmental Medicine*, 76, 381–384 (2005).
146. S. E. Lyshevski, *Nano- and micro-electromechanical systems: Fundamentals of nano- and microengineering*, CRC Press, Boca Raton, FL (2005).
147. R. Zurawski, *Embedded systems handbook*, CRC Press, Boca Raton, FL (2006).
148. M. Rieth, *Nano-engineering in science and technology: An introduction to the world of nano-design*, World Scientific Publishing, Singapore (2003).

149. M. Gad-el-Hak, *The MEMS handbook*, CRC Press, Boca Raton, FL (2002).
150. B. G. Lipták, *Instrument engineers' handbook: Process control and optimization*, CRC Press, Boca Raton, FL (2005).
151. D. Hristu-Varsakelis, W. S. Levine, R. Alur, K.-E. Arzen, J. Baillieul, and T. A. Henzinger, *Handbook of networked and embedded control systems*, Birkhäuser, Boston, MA (2005).
152. M. Shahinpoor, K. J. Kim, and M. Mojarrad, *Artificial muscles: Applications of advanced polymeric nanocomposites*, Taylor & Francis, New York (2007).
153. Y. Osada and D. E. De Rossi, *Polymer sensors and actuators (Macromolecular systems—Materials approach)*, Springer Verlag, Berlin, Germany (1999).
154. Y. Gogotsi, *Nanomaterials handbook*, CRC Press, Boca Raton, FL (2006).
155. H.-J. Fecht and Y. Champion, *Nano-architectured and nanostructured materials: Fabrication, control and properties*, Wiley-VCH, Weinheim, Germany (2006).
156. H. S. Nalwa, *Handbook of nanostructured biomaterials and their applications in nanobiotechnology*, (2 Vols), American Scientific Publishers, Stevenson Ranch, CA (2005).
157. B. D. Ratner, A. Hoffman, F. Schoen, and J. Lemons, *Biomaterials science: An introduction to materials in medicine*, 2nd ed., Academic Press, Burlington, MA (2004).
158. R. W. Bogue, Nanotechnology: What are the prospects for sensors? *Sensor Reviews*, 24, 253–260 (2004).
159. W. Lu and C. M. Lieber, Nanoelectronics from the bottom-up, *Nature Materials*, 6, 841–850 (2007).
160. K. Bullis, Ultrastrong carbon-nanotube muscles: Artificial muscles made from carbon nanotubes are 100 times stronger than human muscles, *MIT Technology Review*, Dec. 8, 1–2 (2006).
161. University of Texas at Dallas, Nano technologists demonstrate artificial muscles powered by highly energetic fuels, *Science Daily*, March 17, webpage at: http://www.sciencedaily.com-/releases/2006/03/060317110801.htm (2006).
162. Y. Bar-Cohen, Electroactive polymers as artificial muscles: Reality, potential and challenges, EAP Actuators and Devices Conference: Smart Structures and Materials Symposium, San Diego, CA, March 6, NASA Jet Propulsion Laboratory, Pasadena, CA (2005).
163. NASA/Jet Propulsion Laboratory, Scientists "muscle" Sci-Fi into reality, *Science Daily* June 11, webpage at: http://www.sciencedaily.com-/releases/2002/06/020611071940.htm (2002).
164. D. H. Nielsen, D. G. Shurr, J. C. Golden, and K. Meier, Comparison of energy cost and gait efficiency during ambulation in below-knee amputees using different prosthetic feet—A preliminary report, *Journal of Prosthetics and Orthotics*, 1, 24–31 (1989).
165. S. E. Irby, K. R. Kaufman, R. W. Wirta, and D. H. Sutherland, Optimization and application of a wrap-spring clutch to a dynamic knee-ankle-foot orthosis, *IEEE Transactions on Rehabilitation Engineering*, 7, 130–134 (1999).
166. B. S. Farber and J. S. Jacobson, An above-knee prosthesis with a system of energy recovery: A technical note, *Journal of Rehabilitation Research and Development*, 32, 337–348 (1995).
167. J. Dewen, W. D. Pilkey, J. Zhang, and W. A. Gruver, Analytical evaluation of an energy-storing foot prosthesis, *System Man Cybernetics: IEEE International Conference on Intelligent Systems 21st Century*, Vol. 1, Waseda University International Conference Center, Tokyo, Japan, pp. 513–517 (1995).
168. K. R. Kaufman, B. K. Iverson, D. J. Padgett, R. H. Brey, J. A. Levine, and M. J. Joyner, Paired outcome assessment of a microprocessor controlled knee versus a mechanical prosthesis, *Gait and Posture*, 24, S57–S59 (2006).
169. G. Chamberlain, Artificial foot and knee designed jointly by U.S., Russian labs, *Design News*, September (1999).

170. Sandia National Laboratories, American and Russian nuclear labs work with prosthetics company to develop artificial knee for landmine victims, *Science Daily*, August 12, On website at http://www.sciencedaily.com-/releases/1999/08/990812081041.htm (1999).
171. *Ossur Proprio Foot*, Ossur hf, 110 Reykjavik, Iceland, web page at: http://bionics.ossur.com/Home/INTRODUCTION/LAWS-OF-BIONICS (2008).
172. The New York Times, *Electronic brain in an artificial foot*, October 3, web animation at:http://www.nytimes.com/packages/html/science/20061003_FOOT_GRAPHIC/index.html (2008).
173. P. Chen, Y. Hasegawa, and M. Yamashita, Grasping control of robot hand using fuzzy neural network, *Advances in Neural Networks*, 1178–1187 (2006).
174. J. M. Winters, Commentary: A case for soft neurofuzzy control interfaces for humans with disabilities, In *Biomechanics and neural control of movement*, J. M. Winters, and P. E. Crago, eds., chap. 39, pp. 548–550, Springer Verlag, New York (2000).
175. G. Lundborg, Tomorrow's artificial hand, *Scandinavian Journal of Plastic and Reconstructive Surgery and Hand Surgery*, 34, 97–100 (2000).
176. J. Yang, E. P. Pitarch, K. Abdel-Malek, A. Patrick, and L. Lindkvist, A multi-fingered hand prosthesis, *Mechanism and Machine Theory*, 39, 555–581 (2004).
177. M. C. Carrozza, B. Massa, S. Micera, R. Lazzarini, M. Zecca, and P. Dario, The development of a novel prosthetic hand-ongoing research and preliminary results, *IEEE/ASME Transactions on Mechatronics*, 7, 108–114 (2002).
178. P. J. Kyberd, C. Light, P. H. Chappell, J. M. Nightingale, D. Whatley, and M. Evans, The design of anthropomorphic prosthetic hands: A study of the Southampton hand, *Robotica*, 19, 593–600 (2001).
179. P. Kyberd, The intelligent hand, *IEE Review*, 46, 31–35 (2000).
180. F. L. Lewis, D. M. Dawson, and C. T. Abdallah, *Robot manipulator control: Theory and practice*, 2nd ed., CRC Press, Boca Raton, FL (2003).
181. T. R. Kurfess, *Robotics and automation handbook*, CRC Press, Boca Raton, FL (2004).
182. M. C. Carrozza, G. Cappiello, L. Beccai, F. Zaccone, S. Micera, and P. Dario, Design methods for innovative hand prostheses, *IEMBS04: 26th Annual International Conference of the IEEE Engineering Medicine and Biology Society*, San Francisco, Vol. 6, pp. 4345–4348 (2004).
183. P. H. Chappell, A. Cranny, D. P. J. Cotton, N. M. White, and S. P. Beeby, Sensory motor systems of artificial and natural hands, *International Journal of Surgery*, 6, doi:10.1016/j.ijsu.2006.06.028, 436–440 (2008).
184. M. C. Carrozza, B. Massa, S. Micera, M. Zecca, and P. Dario, A "wearable" artificial hand for prosthetics and humanoid robotics applications, *Proceedings of the 2001 IEEE -RAS International Conference on Humanoid Robots*, 22–24 (2001).
185. M. C. Carrozza, F. Vecchi, S. Roccella, L. Barboni, E. Cavallaro, S. Micera, and P. Dario, The ADAH project: An astronaut dexterous artificial hand to restore the manipulation abilities of the astronaut, 7th ESA Workshop on Advanced Space Technologies for Robotics and Automation "ASTRA 2002" ESTEC, Noordwijk, the Netherlands, November 19–21 (2002).
186. K. J. De Laurentis, Design of a rapidly fabricated, smart material actuated robotic manipulator for space applications, 55th International Astronautical Congress, International Astronautical Federation, International Academy of Astronautics, International Institute of Air and Space Law IAC-04-I.4.11, Vancouver, Canada (2004).
187. K. J. De Laurentis and C. Mavroidis, Actuators for artificial limbs, *Technology and Health Care*, 10, 91–106 (2002).

188. W. Craelius, R. L. Abboudi, N. A. Newby, and J. Flint, Control of a mutli-finger prosthetic hand, *IEEE Transactions on Rehabilitation Engineering*, 7, 121–129 (1999).
189. M. Harris, P. Kyberd, and W. S. Harwin, Design and development of a dextrous manipulator, *Transactions of the Institute of Measurement and Control*, 27, 137–152 (2005).
190. A. Poulton, P. J. Kyberd, and D. Gow, Progress of a modular prosthetic arm, In *Universal access and assistive technology*, S. Keates, P. Langdon, J. Clarkson, and P. Robinson, eds., pp. 193–200, Springer Verlag, Berlin, Germany (2002).
191. P. J. Kyberd, N. Mustapha, F. Carnegie, and P. H. Chappell, A clinical experience with a hierarchically controlled myoelectric hand prosthesis with vibro-tactile feedback, *Prosthetics and Orthotics International*, 17, 56–64 (1993).
192. J. M. Winters and M. F. Story, eds., *Medical instrumentation: Accessibility and usability considerations*, CRC Press, Boca Raton, FL (2006).
193. R. Rupp and H. J. Gerner, *Neuroprosthetics of the upper extremity-clinical application in spinal cord injury and future perspectives, Biomedizinische Technik, (Berlin)*, 49, 93–98 (2004).
194. F. W. J. Cody, ed., *Neural control of skilled human movement*, Portland Press Ltd., Colchester, UK (1995).
195. M. Kuttiva, G. Burdea, J. Flint, and W. Craelius, Manipulation practice for upper-limb amputees using virtual reality, *Presence: Teleoperators and Virtual Environments* 14, 175–182 (2005).
196. J. M. Winters and P. E. Crago, eds., *Biomechanics and neural control of movement*, Springer Verlag, New York (2000).
197. R. K. Shields, Muscular, skeletal, and neural adaptations following spinal cord injury, *Journal of Orthopaedic and Sports Physical Therapy*, 32, 65–74 (2002).
198. Editorial, Is this the bionic man? *Nature*, 442, 164–171 (2006).
199. W. Craelius, The bionic man: Restoring mobility, *Science*, 295, 1018–1021 (2002).
200. S. H. Scott, Neuroscience: Converting thoughts into action, *Nature*, 442, 164–171 (2006).
201. T. M. Vaughan and J. R. Wolpaw, The third international meeting on brain-computer interface technology: Making a difference, *IEEE Transactions on Neural Systems and Rehabilitation Engineering*, 14, 126–127 (2006).
202. L. R. Hochberg, M. D. Serruya, G. M. Friehs, J. A. Mukand, M. Saleh, A. H. Caplan, A. Branner, D. Chen, R. D. Penn, and J. P. Donoghue, Neuronal ensemble control of prosthetic devices by a human with tetraplegia, *Nature*, 442, 164–171 (2006).
203. G. Santhanam, S. I. Ryu, B. M. Yu, A. Afshar, and K. V. Shenoy, A high-performance brain–computer interface, *Nature*, 442, 195–198 (2006).
204. M. A. Lebedev and M. A. L. Nicolelis, Brain–machine interfaces: past, present and future, *Trends in Neurosciences*, 29, 536–546 (2006).
205. P. D. Cheney, J. Hill-Karrer, A. Belhaj-Saïf, B. J. McKiernan, M. C. Park, and J. K. Marcario, Cortical motor areas and their properties: implications for neuroprosthetics, *Progress in Brain Research*, 1, 135–160 (2000).
206. V. Brezina, I. V. Orekhova, and K. R. Weiss, The neuromuscular transform: The dynamic, nonlinear link between motor neuron firing patterns and muscle contraction in rhythmic behaviors, *Journal of Neurophysiology*, 83, 207–231 (2000).
207. M. A. Lebedev, J. M. Carmena, J. E. O'Doherty, M. Zacksenhouse, C. S. Henriquez, J. C. Principe, and M. A. L. Nicolelis, Cortical ensemble adaptation to represent actuators controlled by a brain machine interface, *Journal of Neuroscience*, 25, 4681–4693 (2005).
208. W. O. Friesen and R. J. Wyman, Analysis of Drosophila motor neuron activity patterns with neural analogs, *Journal of Biological Cybernetics*, 38, 41–50 (1980).
209. A. E. Lindsay, K. A. Lindsay, and J. R. Rosenberg, New concepts in compartmental modeling, *Computing and Visualization Science*, 10, 79–98 (2007).

210. K. A. Lindsay, J. M. Ogden, and J. R. Rosenberg, Dendritic subunits determined by dendritic morphology, *Neural Computation*, 13, 2465–2476 (2001).
211. D. J. Weber, R. B. Stein, D. G. Everaert, and A. Prochazka, Decoding sensory feedback from firing rates of afferent ensembles recorded in Cat dorsal root ganglia in normal locomotion, *IEEE Transactions on Neural Systems and Rehabilitation Engineering*, 14, 240–245 (2006).
212. J. Wessberg, C. R. Stambaugh, J. D. Kralik, P. D. Beck, M. Laubach, J. K. Chapin, J. Kim, S. J. Biggs, M. A. Srinivasan, and M. A. Nicolelis, Real-time prediction of hand trajectory by ensembles of cortical neurons in primates, *Nature*, 16, 361–365 (2000).
213. J. P. Donoghue, Connecting cortex to machines: Recent advances in brain interfaces, *Nature Neuroscience*, 5, 1085–1088 (2002).
214. S. G. Mason, Z. Bozorgzadeh, G. E. Birch, The LF-ASD brain-computer interface: On-line identification of imagined finger flexions in subjects with spinal cord injuries, *Proceedings of the ASSETS 2000*, ACM Press, Washington, USA, 108–109 (2000).
215. J. M. Carmena, M. A. Lebedev, R. E. Crist, J. E., O'Doherty, D. M. Santucci, D. F. Dimitrov, P. G. Patil, C. S. Henriquez, and M. A. L. Nicolelis, Learning to control a brain-machine interface for reaching and grasping by primates, *PLoS Biology*, 1, 193–208 (2003).
216. S.-P. Kim, J. C. Sanchez, Y. N. Rao, D. Erdogmus, J. M. Carmena, M. A. Lebedev, M. A. L. Nicolelis, and J. C. Principe, A comparison of optimal MIMO linear and nonlinear models for brain–machine interfaces, *Journal of Neural Engineering*, 3, 145–161 (2006).
217. S. Rezaei, K. Tavakolian, A. M. Nasrabadi, and S. K. Setarehdan, Different classification techniques considering brain computer interface applications, *Journal of Neural Engineering*, 3, 139–144 (2006).
218. M. Bogdan, M. Schröder, and W. Rosenstiel, Artificial neural net based signal processing for interaction with peripheral nervous system, *Proceedings of the 1st International IEEE EMBS Conference on Neural Engineering 2003*, pp. 134–137 (2003).
219. G. J. Gage, J. K. Otto, K. A. Ludwig, and D. R. Kipke, Co-adaptive Kalman filtering in a naive rat cortical control task, *IEMBS04: 26th Annual International Conference of IEEE Engineering, Medical and Biological Society*, Vol. 6, pp. 4367–4370 (2004).
220. W. Wu, A. Shaikhouni, J. R. Donoghue, and M. J. Black, Closed-loop neural control of cursor motion using a Kalman filter, *IEMBS-4: 26th Annual International Conference of IEEE Engineering, Medical and Biological Society*, Vol. 6, pp. 4126–4129 (2004).
221. R. Tomioka and K. Aihara, Classifying matrices with a spectral regularization, *Proceedings of 24th International Conference on Machine Learning*, ICML07, pp. 895–902, ACM Press, Washington, USA (2007).
222. N. J. Hill, M. Schröder, T. N. Lal, and B. Schölkopf, Comparative evaluation of Independent Components Analysis algorithms for isolating target-relevant information in brain-signal classification, *Brain-Computer Interface Technology*, 3, 95 (2005).
223. H. Parikh, G. Gage, T. C. Marzullo, D. Kipke, Real-time detection of unitary events for cortical control, *EMBS05: 27th Annual International Conference of IEEE Engineering, Medical and Biological Society*, Shanghai, China, pp. 2122–2125 (2005).
224. M. C. Lovett, C. D. Schunn, C. Lebiere, and P. Munro, *Sixth International Conference on Cognitive Modeling: ICCM-2004*, Psychology Press, Routledge, London (2004).
225. M. R. Jane, D. Marini, and A. De Gloria, *Transputer Applications and Systems '94: Proceedings of the 1994 World Transputer Congress*, Cernobbio, Italy, IOS Press, Amsterdam, the Netherlands (1994).

226. R. R. Burridge, A. A. Rizzi, and D. E. Koditschek, Sequential composition of dynamically dexterous robot behaviors, *International Journal of Robotics Research*, 18, 534–555 (1999).
227. G. A. D. Lopes and D. E. Koditschek, Visual servoing for nonholonomically constrained three degree of freedom kinematic systems, *International Journal of Robotics Research*, 26, 715–736 (2007).
228. J. R. Wolpaw, N. Birbaumer, D. J. McFarland, G. Pfurtscheller, and T. M. Vaughan, Brain-computer interfaces for communication and control, *Clinical Neurophysiology*, 113, 767–791 (2002).
229. M. Schröder, T. N. Lal, T. Hinterberger, M. Bogdan, N. J. Hill, N. Birbaumer, W. Rosenstiel, and B. Schölkopf, Robust EEG channel selection across subjects for brain computer interfaces, *EURASIP Journal on Applied Signal Processing, Special Issue: Trends in Brain Computer Interfaces*, 19, 3103–3112 (2005).
230. M. Schröder, M. Bogdan, W. Rosenstiel, T. Hinterberger, and N. Birbaumer, Automated EEG feature selection for brain computer interfaces, *Proceedings of the 1st International IEEE EMBS Conference on Neural Engineering, 2003*, pp. 626–629 (2003).
231. J. A. Wilson, E. A. Felton, P. C. Garell, G. Schalk, and J. C. Williams, ECoG factors underlying multimodal control of a brain–computer interface, *IEEE Transactions on Neural Systems and Rehabilitation Engineering*, Vol. 14, pp. 246–249 (2006).
232. E. A. Felton, J. A. Wilson, J. C. Williams, and P. C. Garell, Electrocorticographically controlled brain–computer interfaces using motor and sensory imagery in patients with temporary subdural electrode implants: Report of four cases, *Journal of Neurosurgery*, 106, 495–500 (2007).
233. D. M. Taylor, S. I. Tillery, and A. B. Schwartz, Direct cortical control of 3D neuroprosthetic devices, *Science*, 296, 1829–1832 (2002).
234. M. S. Hämäläinen, Progress and challenges in multimodal data fusion, *CARS 2007—Proceedings of the International Congress*, 1300, 15–18 (2007).
235. A. Pascual-Leone, N. Davey, J. C. Rothwell, E. M. Wassermann, and B. K. Puri, *Handbook of transcranial magnetic stimulation*, CRC Press, Boca Raton, FL (2002).
236. V. Walsh and A. Pascual-Leone, *Transcranial magnetic stimulation*, MIT Press, Cambridge, MA (2003).
237. P. B. Fitzgerald, S. Fountain, and Z. J. Daskalakis, A comprehensive review of the effects of rTMS on motor cortical excitability and inhibition, *Clinical Neurophysiology*, 117, 2584–2596 (2006).
238. A. T. Barker, R. Jalinous, and I. L. Freeston, Non-invasive magnetic stimulation of human motor cortex, *Lancet*, 1(8437), 1106–1107 (1985).
239. T. Kujirai, Corticocortical inhibition of the motor cortex, *Journal of Physiology*, 471, 501–509 (1993).
240. A. Pascual-Leone, D. Bartres-Faz, and J. P. Keenan, Transcranial magnetic stimulation: studying the brain-behaviour relationship by induction of virtual lesions, *Philosophical Transactions of the Royal Society of London B: Biological Science*, 354, 1229–1238 (1999).
241. S. Tumanski, *Thin film magnotoresistive sensors*, CRC Press, Boca Raton, FL (2001).
242. J. R. Brauer, *Magnetic actuators and sensors*, IEEE/Wiley Interscience, New York (2006).
243. S. L. Mouaziz, *Micro and nano tools for magnetic field imaging—Series in microsystems*, P. A. Besse, J. Brugger, M. Gijs, R. S. Popovic, and P. Renaud, eds., Vol. 21, Hartung-Gorre Verlag Konstanz, Germany (2007).
244. S. H. Allos, D. J. Staton, L. A. Bradshaw, S. Halter, J. P. Wikswo, Jr., and W. O. Richards, Superconducting quantum interference device magnetometer for diagnosis of Ischemia caused by mesenteric venous thrombosis, *World Journal of Surgery*, 21, 173–178 (1997).

245. D. King, A resonant MEMS magnetometer, *IEE Seminar and Exhibition on MEMS Sensor Technology, 2005*, pp. 1–12 (2005).
246. C. Schott, F. Burger, H. Blanchard, and L. Chiesi, Modern integrated silicon Hall sensors, *Sensor Reviews*, 18, 252–257 (1998).
247. A. P. Ramirez, Colossal magnetoresistance, *Journal of Physics: Condensed Matter*, 9, 8171–8199 (1997).
248. E. Paperno and B.-Z. Kaplan, Sub-nano-tesla in-plane vector magnetometer employing single magnetoresistor, *Nineteenth Convention on Electrical and Electronics Engineering in Israel*, 1996, pp. 188–191 (1996).
249. S. R. Brankovic, X. M. Yang, T. J. Klemmer, and M. Seigler, Electrodeposition of 2.4T Co37Fe63 alloys at nanoscale for magnetic recoding application, *IEEE Transactions on Magnetics*, 42, 132 (2006).
250. S. A. Solin, T. Thio, D. R. Hines, and J. J. Heremans, Enhanced room-temperature geometric magnetoresistance in inhomogeneous narrow-gap semiconductors, *Science*, 289, 1530–1532 (2000).
251. J. Clarke, Principles and applications of SQUIDs, *Proceedings of IEEE*, 77, 1208–1223 (1989).
252. Y. Okada, K. Pratt, C. Atwood, A. Mascarenas, R. Reineman, J. Nurminen, and D. Paulson, BabySQUID: A mobile, high-resolution multichannel magnetoencephalography system for neonatal brain assessment, *Review of Scientific Instruments*, 77, 024301 (2006).
253. L. A. Bradshaw, A. Irinia, J. A. Sims, M. R. Gallucci, R. L. Palmer, and W. O. Richards, Biomagnetic characterization of spatiotemporal parameters of the gastric slow wave, *Neurogastroenterology and Motility*, 18, 619–631 (2006).
254. V. Shah, S. Knappe, P. D. D. Schwindt, and J. Kitching, Subpicotesla atomic magnetometry with a microfabricated vapour cell, *Nature Photonics*, 1, 649–652 (2007).
255. S. Xu, M. H. Donaldson, A. Pinesb, S. M. Rochester, D. Budkerc, and V. V. Yashchuk, Application of atomic magnetometry in magnetic particle detection, *Applied Physics Letters*, 89, 224105 (2006).
256. I. K. Kominis, T. W. Kornack, J. C. Allred, and M. V. Romalis, A subfemtotesla multichannel atomic magnetometer, *Nature*, 422, 596–599 (2003).
257. D. S. Greywall, Sensitive magnetometer incorporating a high-Q nonlinear mechanical resonator, *Measurement Science Technology*, 16, 2473–2482 (2005).
258. H. Xia, A. B.-A. Baranga, D. Hoffman, and M. V. Romalis, Magnetoencephalography with an atomic magnetometer, *Applied Physics Letters*, 89, 211104 (2006).
259. Z. Li, R. T. Wakai, and T. G. Walker, Parametric modulation of an atomic magnetometer, *Applied Physics Letters*, 89, 134105 (2006).
260. L. Balcells, E. Calvo, and J. Fontcuberta, Room-temperature anisotropic magnetoresistive sensor based on manganese perovskite thick films, *Journal of Magnetism and Magnetic Materials*, 242–245, 1166–1168 (2002).
261. M. M. Raja, R. J. Gambino, S. Sampath, and R. Greenlaw, Thermal sprayed thick-film anisotropic magnetoresistive sensors, *IEEE Transactions on Magnetics*, 40, 2685–2687 (2004).
262. G. B. Donaldson, SQUIDs—ultimate magnetic sensors, *Physica Status Solidi*, 2, 1463–1467 (2005).
263. A. Candini, G. C. Gazzadi, A. di Bona, M. Affronte, D. Ercolani, G. Biasio, and L. Sorba, Hall nano-probes fabricated by focused ion beam, *Nanotechnology*, 17, 2105–2109 (2006).
264. J. Vrba and S. E. Robinson, Signal Processing in magnetoencephalography, *Methods*, 25, 249–271 (2001).
265. A. A. Fife, J. Vrba, S. E. Robinson, G. Anderson, K. Betts, M. B. Burbank, D. Cheyne, E. Cheung, S. Govorkov, G. Haid, V. Haid, C. Hunter, P. R. Kubik, S. Lee,

J. McKay, E. Reichl, C. Schroyen, I. Sekachev, P. Spear, B. Taylor, M. Tillotson, and W. Sutherling, Synthetic gradiometer systems for MEG, *IEEE Transactions on Applied Superconductivity*, 9, 4063–4068 (1999).

266. J. Vrba and S. E. Robinson, Linearly constrained minimum variance beamformers, synthetic aperture magnetometry, and MUSIC in MEG applications, *Signals, Systems and Computers, 34th Asilomar Conference 2000*, Vol. 1, pp. 313–317 (2000).
267. R. Frostig, ed., *In vivo optical imaging of brain function (Methods and new frontiers in neuroscience)*, CRC Press, Boca Raton, FL (2002).
268. J. Wells, C. Kao, K. Mariappan, J. Albea, E. D. Jansen, P. Konrad, and A. Mahadevan-Jansen, Optical stimulation of neural tissue in vivo, *Optics Letters*, 30, 504–506 (2005).
269. J. Wells, C. Kao, P. Konrad, T. Milner, J. Kim, A. Mahadevan-Jansen, and E. D. Jansen, Biophysical mechanisms of transient optical stimulation of peripheral nerve, *Biophysical Journal*, 93, 2567–2580 (2007).
270. G. Caetano and V. Jousmäki, Evidence of vibrotactile input to human auditory cortex, *Neuro Image*, 29, 15–28 (2005).
271. T. N. Lal, M. Schröder, J. Hill, H. Preissl, T. Hinterberger, J. Mellinger, M. Bogdan, W. Rosenstiel, T. Hofmann, N. Birbaumer, and B. Schölkopf, A Brain Computer Interface with Online Feedback based on Magnetoencephalography, In *Proceedings of the 22nd International Conference on Machine Learning*, L. De Raedt and S. Wrobel, eds., pp. 465–472, ACM Press, Washington, USA (2005).
272. D. Hecht and M. Reiner, Field dependency and the sense of object-presence in haptic virtual environments, *Cyber Psychology and Behavior*, 10, 243–251 (2007).
273. C. R. Wagner and R. D. Howe, Mechanisms of performance enhancement with force feedback, *First Joint Eurohaptics Conference and Symposium on Haptic Interfaces for Virtual Environment and Teleoperator Systems WHC05*, pp. 21–29 (2005).
274. A. M. Okamura, Methods for haptic feedback in teleoperated robot-assisted surgery, *Industrial Robot: An International Journal*, 31, 499–508 (2004).
275. G. A. Calvert, C. Spence, and B. E. Stein, *The handbook of multisensory processes*, MIT Press, Cambridge, MA (2004).
276. L. Hochberg and D. Taylor, Intuitive prosthetic limb control, *Lancet*, 369, 345–346 (2007).
277. T. A. Kuiken, L. A. Miller, R. D. Lipschutz, B. A. Lock, K. Stubblefield, P. D. Marasco, P. Zhou, and G. A. Dumanian, Targeted reinnervation for enhanced prosthetic arm function in a woman with a proximal amputation: A case study, *Lancet*, 369, 371–380 (2007).
278. J. Arnott, N. Alm, and A. Waller, Cognitive prostheses: communication, rehabilitation and beyond, *IEEE International Conference on Systems Man and Cybernetics IEEE SMC 1999*, Vol. 6, pp. 346–351 (1999).
279. W. Barfield and T. Caudell, *Fundamentals of wearable computers and augmented reality*, Lawrence Erlbaum Associates/CRC Press, Boca Raton, FL (2001).
280. S. K. Rosahl, Neuroprosthetics and neuroenhancement: Can we draw a line? *Virtual Mentor*, 9, 132–139 (2007).
281. M. E. Clynes and N. S. Kline, Cyborgs and space, *Astronautics*, September, 26–27 (1960).
282. R. Clarke, Human-artefact hybridisation: Forms and consequences, *Ars Electronica 2005 Symposium on Hybrid - Living in Paradox*, Linz, Austria, September 2–3 (2005).
283. F. Jotterand, Nanomedicine: How it could reshape clinical practice? *Nanomedicine* 2, 401–405 (2007).
284. US NIDCD: National Institute on Deafness and Other Communication Disorders, *Cochlear Implants*, NIH Publication No. 00-4798, Website at: http://www.nidcd.nih.gov/ (2007).

285. US NIDCD: National Institute on Deafness and Other Communication Disorders, Statistics about hearing disorders, ear infections, and deafness, Website at: http://www.nidcd.nih.gov/health/statistics/hearing.asp (2008).
286. R. D. Kent, *The MIT Encyclopedia of communication disorders*, MIT Press, Cambridge, MA (2004).
287. D. E. Ingber, Cellular mechanotransduction: Putting all the pieces together again, *FASEB Journal*, 20, 811–827 (2006).
288. D. I. Margolin, *Cognitive neuropsychology in clinical practice*, Oxford University Press, New York (1992).
289. P. R. Cook, *Music, cognition, and computerized sound: An introduction to psychoacoustics*, MIT Press, Cambridge, MA (2001).
290. V. K. Madisetti and D. Williams, *The digital signal processing handbook*, CRC Press, Boca Raton, FL (1997).
291. M. Kahrs and K. Brandenburg, *Applications of digital signal processing to audio and acoustics*, Springer Verlag, Berlin, Germany (1998).
292. B. Wilson, Digital signal processing applications for hearing accessibility, *IEEE Signal Processing Magazine*, 20, 14–18 (2003).
293. J. Tierny, M. A. Zissman, and D. K. Eddington, Digital signal processing applications in cochlear-implant research, *Lincoln Laboratory Journal*, 7, 31–62 (1994).
294. G. J. M. Krijnen, M. Dijkstra, J. J. van Baar, S. S. Shankar, W. J. Kuipers, R. J. H. de Boer, D. Altpeter, T. S. J. Lammerink, and R. Wiegerink, MEMS based hair flow-sensors as model systems for acoustic perception studies, *Nanotechnology*, 17, S84–S89 (2006).
295. G. Clark, *Cochlear implants: Fundamentals and applications*, Springer Verlag, Berlin (2003).
296. US FDA: Food and Drug Administration, *Cochlear Implants*, website at: http://www.fda.gov/cdrh/cochlear/index.html (2008).
297. M. Valente, H. Hosford-Dunn, and R. J. Roeser, *Audiology: Treatment*, Thieme Medical Publishers, New York (2000).
298. J. W. Hall, *New handbook for auditory evoked responses*, Allyn & Bacon, Boston, MA (2006).
299. G. Miller, *Sensory organ replacement and repair*, Morgan & Claypool Publishers, San Rafael, CA (2006).
300. A. R. Moller, *Cochlear and brainstem implants (Advances in otorhinolaryngology)*, Karger, Basel, SZ (2006).
301. K. S. Pawlowski, Anatomy and physiology of the Cochlea, In *Ototoxicity*, P. S. Roland and J. A. Rutka, eds., BC Decker Inc, London (2004).
302. K. S. Pawlowski, D. Wawro, and P. S. Roland, Bacterial Biofilm Formation on a Human Cochlear Implant, *Otology and Neurotology*, 26, 972–975 (2005).
303. T. A. Johnson, K. A. Loeffler, R. A. Burne, C. N. Jolly, and P. J. Antonelli, Biofilm formation in cochlear implants with cochlear drug delivery channels in an in vitro model, *Otolaryngology–Head and Neck Surgery*, 136, 577–582 (2007).
304. R. Cristobal, C. E. Edmiston Jr., C. L. Runge-Samuelson, H. A. Owen, J. B. Firszt, and P. A. Wackym, Fungal biofilm formation on cochlear implant hardware after antibiotic-induced fungal overgrowth within the middle ear, *The Pediatric Infectious Disease Journal*, 23, 774–777 (2004).
305. B. Gold and N. Morgan, *Speech and audio signal processing: Processing and perception of speech and music*, John Wiley & Sons, New York (1999).
306. G. P. Jacobson, C. W. Newman, and J. M. Kartush, *Handbook of balance function testing*, Springer Verlag, Berlin, Germany (1997).
307. J. M. Goldberg and C. Fernandez, Vestibular mechanisms, *Annual Review of Physiology*, 37, 129–162 (1975).

308. K. W. Lindsay, T. D. Roberts, and J. R. Rosenberg, Asymmetric tonic labyrinth reflexes and their interaction with neck reflexes in the decerebrate cat, *Journal of Physiology*, 261, 583–601 (1976).
309. M. F. Reschke, J. M. Krnavek, J. T. Somers, and G. Ford, A brief history of space flight with a comprehensive compendium of vestibular and sensorimotor research conducted across the various flight programs, NASA/ SP-2007-560, National Center for AeroSpace Information, Hanover, MD (2007).
310. A. M. Shkel and F.-G. Zeng, An electronic prosthesis mimicking the dynamic vestibular function, *Audiology and Neurotology*, 11, 113 (2006).
311. A. A. Maltan and T. K. Whitehurst, Stimulation using a microstimulator to treat tinnitus, US Patent 20070021804, Advanced Bionics Corporation, Valencia, CA (2007).
312. S. A. Gelfand, *Essentials of audiology*, 2nd ed., Thieme, New York (2001).
313. V. P. Eroschenko and M. S. H. di Fiore, *Di Fiore's atlas of histology*, 10th ed., Lippincott Williams & Wilkins, Philadelphia, PA (2004).
314. S. S. Corbett, III, J. W. Swanson, J. Martyniuk, T. A. Clary, F. A. Spelman, B. Clopton, A. H. Voie, and C. N. Jolly, Multi-electrode cochlear implant and method of manufacturing the same, US Patent 5630839, PI Medical Corporation (Portland, OR); University of Washington, Seattle, WA (1997).
315. C.-P. Richter and S. Ho, Cochlear implant including a modiolar return electrode, US Patent 7194314, Northwestern University (2007).
316. D. Rejalia, V. A. Leec, K. A. Abrashkina, N. Humayuna, D. L. Swiderskia, and Y. Raphaela, Cochlear implants and ex vivo BDNF gene therapy protect spiral ganglion neurons, *Hearing Research*, 228, 180–187 (2007).
317. D. Seybold, The psychosocial impact of acquired vision loss, *Vision 2005—Proceedings of the International Congress*, 1282, 298–301 (2005).
318. S. H. Schwartz, *Visual perception*, 3rd ed., McGraw-Hill, New York (2004).
319. J. S. Glaser, *Neuro-ophthalmology*, 3rd ed., Lippincott Williams & Wilkins, Philadelphia, PA (1999).
320. P. L. Kaufman and A. Alm, *Adler's Physiology of the eye*, 10th ed., Harcourt/Mosby, London, UK (2002).
321. J. L. Smith, *Neuro-ophthalmology*, Vol. 4, C. V. Mosby Co., St. Louis, MO (1968).
322. B. Cassin and M. L. Rubin, eds., *Dictionary of eye terminology*, 5th ed., Triad Publications, Gainesville, FL (2006).
323. M. E. Brezinski, *Optical coherence tomography: Principles and applications*, Academic Press, Boston, MA (2006).
324. D. T. Hartong, E. L. Berson, and T. P. Dryja, Retinitis pigmentosa, *Lancet*, 386, 1795–1809 (2006).
325. T. Suzuki, M. Akimoto, H. Imai, Y. Ueda, M. Mandai, N. Yoshimura, A. Swaroop, and M. Takahashi, Chondroitinase ABC treatment enhances synaptogenesis between transplant and host neurons in a mouse model of retinal degeneration, *Cell Transplantation*, 16, 493–503 (2007).
326. W. Roush, Envisioning an artificial retina, *Science*, 268, 637–638 (1995).
327. R. R. Lakhanpal, D. Yanai, Douglas, J. D. Weiland, G. Y. Fujii, S. Caffey, R. J. Greenberg, E. de Juan, Jr., and M. S. Humayun, Advances in the development of visual prostheses, *Current Opinion in Ophthalmology*, 14, 122–127 (2003).
328. J. D. Weiland, W. T. Liu, and M. S. Humayun, Retinal prosthesis, *Annual Review of Biomedical Engineering*, 7, 361–401 (2005).
329. P. Hossain, I. W. Seetho, A. C. Browning, and W. M. Amoaku, Science, medicine, and the future—Artificial means for restoring vision, *British Medical Journal*, 330, 30–33 (2005).
330. M. Javaheri, D. S. Hahn, R. R. Lakhanpal, J. D. Weiland, and M. S. Humayun, Retinal prostheses for the blind, *Annals in Academy of Medicine Singapore*, 35, 137–144 (2006).

331. G. Dagnelie, Visual prosthetics 2006: Assessment and expectations, *Expert Reviews in Medical Devices*, 3, 315–326 (2006).
332. D. E. Casey, ed., Envisioning sight for the blind: The DOE artificial retina project, *Artificial Retina News*, 1, 1–3 (2006).
333. J. D. Weiland and M. S. Humayun, A biomimetic retinal stimulating array, *IEEE Engineering in Medicine and Biology Magazine*, 24, 14–21 (2005).
334. J. D. Weiland and M. S. Humayun, Intraocular retinal prosthesis—Big steps to sight restoration, *IEEE Engineering in Medicine and Biology Magazine*, 25, 60–66 (2006).
335. C.-Y. Wu, F. Cheng, C.-T. Chiang, and P.-K. Lin, A low-power implantable Pseudo-BJT-based silicon retina with solar cells for artificial retinal prostheses, *Proceedings of 2004 International Symposium. on Circuits and Systems*, ISCAS 04. IV, pp. 37–40 (2004).
336. G. J. Suaning and N. H. Lovell, CMOS neurostimulation system with 100 electrodes and radio frequency telemetry, Conference of IEEE EMBS (Vic), Melbourne (1999).
337. D. C. Rodger and Y.-C. Tai, Microelectronic packaging for retinal prostheses, *IEEE Engineering in Medicine and Biology Magazine*, 24, 52–57 (2005).
338. F. Paillet, D. Mercier, and T. M. Bernard, Second generation programmable artificial retina, *Proceedings of the 12th Annual IEEE International ASIC/SOC Conference 1999 XII*, pp. 304–309 (1999).
339. F. Gekeler and E. Zrenner, Status of the subretinal implant project. An overview, *Ophthalmologe*, 102, 941–52 (2005).
340. E. Funatsu, Y. Nitta, Y. Miyake, T. Toyoda, J. Ohta, and K. Kyuma, An artificial retina chip with current-mode focal plane image processing functions, *IEEE Transactions on Electron Devices*, 44, 1777–1782 (1997).
341. J. Ohta, et al., Silicon LSI-based smart stimulators for retinal prosthesis—A flexible and extendable microchip-based stimulator, *IEEE Engineering in Medicine and Biology Magazine*, 25, 47–59 (2006).
342. A. Stett, A. Mai, and T. Herrmann, Retinal charge sensitivity and spatial discrimination obtainable by subretinal implants: Key lessons learned from isolated chicken retina, *Journal of Neural Engineering*, 4, S7–S16 (2007).
343. T. Schanze, H. G. Sachs, C. Wiesenack, U. Brunner, H. Sailer, Implantation and testing of subretinal film electrodes in domestic pigs, *Experimental Eye Research*, 82, 332–340 (2006).
344. M. T. Pardue, M. J. Phillips, H. Yin, B. D. Sippy, S. Webb-Wood, A. Y. Chow, and S. L. Ball, Neuroprotective effect of subretinal implants in the RCS rat, *Investigative Ophthalmology and Visual Science*, 46, 674–682 (2005).
345. M. T. Pardue, M. J. Phillips, B. Hanzlicek, H. Yin, A. Y. Chow, and S. L. Ball, Neuroprotection of photoreceptors in the RCS rat after implantation of a subretinal implant in the superior or inferior retina, *Retinal Degenerative Diseases*, 572, 321–326 (2006).
346. M. T. Pardue, S. L. Ball, M. J. Phillips, A. E. Faulkner, T. A. Walker, and A. Y. Chow, Status of the feline retina 5 years after subretinal implantation, *Journal of Rehabilitation Research and Development*, 43, 723–732 (2006).
347. A. P. Fornos, J. Sommerhalder, B. Rappaz, A. B. Safran, and M. Pelizzone, Simulation of artificial vision, III: Do the spatial or temporal characteristics of stimulus pixelization really matter? *Investigative Ophthalmology and Visual Science*, 46, 3906–3912 (2005).
348. S. I. Fried, H. A. Hsueh, and F. S. Werblin, A method for generating precise temporal patterns of retinal spiking using prosthetic stimulation, *Journal of Neurophysics*, 95, 970–978 (2006).

349. L. Johnson, et al., Impedance-based retinal contact imaging as an aid for the placement of high resolution epiretinal prostheses, *Journal of Neural Engineering*, 4, S17–S23 (2007).
350. E. Margalit and W. B. Thoreson, Inner retinal mechanisms engaged by retinal electrical stimulation, *Investigative Ophthalmology and Visual Science*, 47, 2606–2612 (2006).
351. T. Yagi and Y. Hayashida, Artificial retina implantation, *Nippon Rinsho*, 57, 1208–1215 (1999).
352. H. A. Hassan, S. R. Montezuma, and J. F. Rizzo, In vivo electrical stimulation of rabbit retina: Effect of stimulus duration and electrical field orientation, *Experimental Eye Research*, 83, 247–254 (2006).
353. W. H. Dobelle, Artificial vision for the blind by connecting a television camera to the visual cortex: State of the art, *American Society for Artificial Internal Organs Journal*, 46, 3–9 (2000).
354. W. H. Dobelle, J. Turkel, D. C. Henderson, and J. R. Evans, Mapping the representation of the visual field by electrical stimulation of human visual cortex, *American Journal of Ophthalmology*, 88, 727–735 (1979).
355. J. F. Doorish, A wireless photovoltaic Mini epi-Retinal Prosthesis (MeRP) 1: Concept and design, *Journal of Modern Optics*, 53, 1267–1285 (2006).
356. A. Y. Chow, V. Y. Chow, K. H. Packo, J. S. Pollack, G. A. Peyman, and R. Schuchard, The artificial silicon retina microchip for the treatment of vision loss from *Retinitis pigmentosa*, *Archives of Ophthalmology*, 122, 460–469 (2004).
357. K. Hungar, M. Gortz, E. Slavcheva, G. Spanier, W. C. Weidig, and W. Mokwa, Production processes for a flexible retina implant (Eurosensors XVIII, Session C6.6), *Sensors and Actuators A-Physical*, 123, 172–178 (2005).
358. P. Walter, Z. F. Kisvárday, M. Görtz, N. Alteheld, G. Rossler, T. Stieglitz, and U. T. Eysel, Cortical activation via an implanted wireless retinal prosthesis, *Investigative Ophthalmology and Visual Science*, 46, 1780–1785 (2005).
359. J. Ohta, T. Tokuda, K. Kagawa, S. Sugitani, M. Taniyama, A. Uehara, Y. Terasawa, K. Nakauchi, T. Fujikado, and Y. Tano, Laboratory investigation of microelectronics-based stimulators for large-scale suprachoroidal transretinal stimulation (STS), *Journal of Neural Engineering*, 4, S85–S91 (2007).
360. J. S. Pezaris and R. C. Reid, Demonstration of artificial visual percepts generated through thalamic microstimulation, *Proceedings of the National Academy of Science USA*, 104, 7670–7675 (2007).
361. R. J. Jensen, O. R. Ziv, and J. F. Rizzo, Thresholds for activation of rabbit retinal ganglion cells with relatively large, extracellular microelectrodes, *Investigative Ophthalmology and Visual Science*, 46, 1486–1496 (2005).
362. L. B. Merabet, J. F. Rizzo, A. D. Amedi, D. C. Somers, and A. Pascual-Leone, What blindness can tell us about seeing again: Merging neuroplasticity and neuroprostheses, *Nature Reviews Neuroscience*, 6, 71–77 (2005).
363. F. Duret, M. E. Brelen, V. Lambert, B. Gerard, J. Delbeke, and C. Veraart, Object localization, discrimination, and grasping with the optic nerve visual prosthesis, *Restorative Neurology and Neuroscience*, 24, 31–40 (2006).
364. D. W. Chun, J. S. Heier, M. B. Raizman, and B. Michael, Visual prosthetic device for bilateral end-stage macular degeneration, *Expert Review of Medical Devices*, 6, 657–665 (2005).
365. H. L. Hudson, S. S. Lane, J. S. Heier, R. D. Stulting, L. Singerman, P. R Lichter, P. Sternberg, and D. F. Chang, Implantable miniature telescope for the treatment of visual acuity loss resulting from end-stage age-related macular degeneration: 1-year results, *Ophthalmology*, 113, 1987–2001 (2006).

366. J. O. Winter, S. F. Cogan, and J. F. Rizzo, Retinal prostheses: Current challenges and future outlook, *Journal of Biomaterials Science Polymer Edition*, 18, 1031–1055 (2007).
367. M. C. Peterman, D. M. Bloom, C. Lee, S. F. Bent, M. F. Marmor, M. S. Blumenkranz, and H. A. Fishman, Localized Neurotransmitter Release for Use in a Prototype Retinal Interface, *Investigative Ophthalmology and Visual Science*, 44, 3144–3149 (2003).
368. W. H. Ko, A review of implantable sensors, *Pacing and Clinical Electrophysiology*, 6, 482–487 (1983).
369. R. Bogue, MEMS sensors: Past, present and future, *Sensor Review*, 27, 7–13 (2007).
370. M. Norris, Design considerations for wireless implants, *Design News*, May 14 (2007).
371. B. Sarmentoa, A. Ribeirob, F. Veiga, and D. Ferreira, Development and characterization of new insulin containing polysaccharide nanoparticles, *Colloids and Surfaces B: Biointerfaces*, 53, 193–202 (2006).

Problems

27.1 Is nanotechnology relevant to medicine? How does medicine depend on technology in general? What are the considerations medical researchers and caregivers take when evaluating a new technology for patient care?

27.2 What are the nine major medical breakthroughs identified in the text? What three, if any, do you consider the most important? (Discuss the reasons for your choices.)

27.3 What role does the NIH consider nanotechnology will play in medical research and healthcare, and why?

27.4 What is the earliest application of nanotechnology in medicine of which you are aware?

27.5 What are the typical size ranges of (a) human cells, (b) bacteria, (c) viruses, and (d) DNA?

27.6 How are nanoparticles used to enhance medical imaging? What advantages do they have over other image contrast agents? Can you think of any potential disadvantages? (Discuss for specific types of contrast agents and nanoparticles.)

27.7 What radiation frequencies are used with nanoparticles to destroy cancer cells? How can the nanoparticles be made selective for cancer cells?

27.8 What are three ways that nanotechnology can be used to treat diabetes?

27.9 How can nanoencapsulation be used to deliver drugs through the blood–brain barrier?

27.10 How can nanotechnology be used to stimulate nerve growth? What are the factors needed for successful application of this technique?

27.11 Is nanotechnology applicable to (a) motor prosthetics, (b) sensory prosthetics, (c) neither, and (d) both? Explain.

27.12 What are some potential improvements that nanotechnology presents for (a) neurostimulation for pain, (b) hearing aids, and (c) prosthetics for the blind? What are some challenges faced in improving these types of devices and where is nanotechnology relevant?

27.13 What are the factors that limit the density of placement of electrodes in a cochlear implant?

27.14 What are the factors that limit the density of stimulation electrodes in a retinal implant?

Environmental Nanotechnology

Environmentally friendly cars will soon cease to be an option ... they will become a necessity.

Fujio Cho
President of Toyota Motors, 2004

Chapter 28

Threads

We save this chapter on environmental nanotechnology for the last—the one to leave the student, hopefully, with an important perspective of things to come and, unequivocally, to place value on the place that we all live. We need to know how nanotechnology will impact our environment, our health and our safety, and our society in general. There are numerous positive attributes to nanotechnology in this regard. The environmental footprint of nano is expected to be three orders of magnitude less than that observed for today's technology, the microtechnology. There are also potentially negative effects of this wondrous technology—as you might imagine, ones that are able to tap into the very fiber of nature itself. Which nanomaterials are environmentally neutral? Which ones are beneficial? Which ones are potentially dangerous to human health?

The nanotechnology we have presented and discussed so far included topics as varied as nanometrology, nanomanufacturing, optical-electromagnetic devices, thin films and interfaces, catalysis, polymers and fibers, and those associated with bionanotechnology and nanomedicine—truly a vast and diverse assembly of materials, devices, and applications. Once manufactured, used or disposed, what happens to the nanomaterials from which they originated? Because nanomaterials possess remarkable properties, how are we to know that these remarkable properties keep on expressing themselves once we are through with the technology? Will these remarkable properties work against us and against the environment? Or will the balance between the positive and the negative be in favor of the positive? We certainly hope the latter is emphasized in all possible scenarios.

We limit our discussion to the nanotechnology of environmental devices and materials (and natural nanotechnological devices and materials) that are designed to measure, mitigate, or provide power—the sensors, filters, practices, and other commercial and industrial (and natural) applications. We need to know more how our materials, by-products, and devices affect the environment, and thereby us. What do we do and how do we do it? Although nanotechnology will help us become better informed, better equipped, and better defended, unfortunately, the answer to this question transcends solutions rendered to them through nanotechnology.

28.0 The Environment (and Technology)

Nanotechnology is expected to play a critical role in environmental-related issues—from sensing, monitoring, mitigation, and power to perhaps generating its own brand of pollution, contamination, and infection. The stages upon which all these actors play are the traditional ones that we are all familiar—the air, the soil, and the water—and of course, the workplace and home and the human body itself. Pollutants, contaminants, and pathogens come in a variety of forms and are released from an incredible number of diverse sources—ranging from trash, industrial wastes and spills, seepage, combustion to more subtle expressions of human activity like radiation (e.g., cell phones), noise, congestion, recycled air, and daily consumer activities.

Nanotechnology is expected to impact all industrial sectors—from aerospace to energy and from construction to medicine. Nanotechnology's impact on the environment is also expected to be significant—its vector remains to be determined. There is a need for sensors that have better detection limits, versatility, and durability. For example, since 1992, there are over two million storage tanks under surveillance by the Environmental Protection Agency (EPA) that sequester toxic and/or volatile materials—and what of storage tanks that are not included in the monitor registry [1,2]? With an adequate arsenal of micro- and nanosensors, we should be able to detect low-level leaks from unregistered tanks and determine their source before any serious environmental damage ensues.

Where can nanomaterials and devices take us? Use of nanomaterials in instruments, devices, equipment, and other products and the energy required for their manufacture and operation both have direct and indirect impact on our environment—in both cases, the less the better. In addition, detection and mitigation of low-level pollutants can be accomplished with better nanoscale catalysts, inline and remote detectors, and nanochemical reactors. Facilitative and affordable field analysis technologies have and will continue to spring from advances in nanotechnology. For example, surface-enhanced Raman fiber-optic probes have been around for quite some time [3]. The surface-enhanced resonance Raman scattering (SERRS) method is capable of attomole level sensitivity [4]—no big deal anymore since single-molecule detection has already been achieved with this technique [5]. SERS is without question a nanotechnology-enhanced analytical tool that relies on nanosized silver or gold particles to generate a gigantic signal.

Current environmental procedures rely on a significant investment in time, cost, and energy—including sample acquisition, preparation, and laboratory analysis [6]. Better, more accurate methods able to scan larger sample sizes with higher throughput and greater sensitivity need to be developed. The methods and equipment must be robust, reliable, and reusable with the capability for remote sensing and operation in real time—all with enhanced facility and cost-effectiveness [6]. Quite the wish list! We present a broad spectrum of environmental issues and technology in this chapter. We divide the chapter into three major categories: water and soil, air, and energy. With this chapter, we round out *Introduction to Nanoscience and Nanotechnology*.

28.0.1 Background

Agriculture, villages, and cities have brought on pollution and resource exploitation—upsetting Earth's natural balance, like never before, in order to provide food, shelter, and energy for burgeoning populations. Environmental insults have occurred since *Homo sapiens'* opposable thumb and frontal lobe started working together. For example, evidence of the use of fire to clear forests was found at an archeological dig in Tanzania dating back 60,000 years [7]—an early example of anthropomorphic alteration of the natural environment. As technology developed and populations continued to explode, the concomitant levels of pollution and resource exploitation increased proportionally—air pollution from dust, wood, smoke, tanneries, and animal manure; and water pollution from large cities in the form of sewage. Specific toxins found in lead-sweetened wine ("sugar of lead," lead acetate) and lead piping resulted in poisoning among the Romans [7]. It is a well known that ancient Babylonians, Minoans, Phoenicians, and Romans stripped local forests of timber to supply their cities.

Although the technology of the age was not sophisticated, the devastation was widespread nonetheless. This is not to say that all this early form of environmental insults was conducted without some consciousness. The Sumerians for example passed laws as early as ca. 2700 B.C. to protect remaining forests [7]. From the human health viewpoint, acid mists in copper mines were known to endanger the lungs of miners—so observed the Greek physician Galen in ca. 200 B.C. [8]. In first century Rome, workers in zinc smelting operations used animal bladders as respirators—so observed Pliny the Elder [8]. Plutarch recommended that only criminal slaves should work in the lead and mercury mines—an early societal issue associated with a technology. Yes, to the doubters out there, anthropomorphic generated activities do impact the environment—and yes, as global populations continue to swell, these impacts are manifesting themselves on a global scale.

The effects of fossil fuel exploration, drilling, mining, and combustion are well known. In thirteenth century England, the combustion of coal fouled the city—forcing "royal personages" to flee to cities with cleaner air [9]. The combustion of coal and other fossil fuels, exacerbated by population increases worldwide, has contributed most to pollution and other forms of environmental anxiety and aesthetics—simply because with the advent of the "Oil Age," humankind was able to alter the energy balance by tapping energy resources manufactured in the distant past. Tapping ancient resources enabled us to produce more food than we could have otherwise thereby shifting the natural equilibrium of things [10]. The subsequent usurping of the world's food supply, greater than 40% of NPP (net primary production) [11], resulted in our current dilemma—overpopulation, insatiable energy demand, and rampant resource exploitation and waste.

We have certainly upset the natural balance of things. People will comment readily that "the smog is really bad today"; "the water tastes terrible"; or "look at all the trash everywhere." Most people will agree that all pollution is bad, ugly, and unhealthy and some believe in science and unfortunately, some don't—but that's what makes our world go around. What is real and tangible is that nanotechnology is here to stay. Nanotechnology is ready to enhance and enable preexisting technologies and to develop new ones that will help the environment. Both the aware and the unaware would benefit and/or pay the price—perhaps perceptibly or more likely, perhaps imperceptibly.

28.0.2 Traditional Methods of Detecting Environmental Contaminants

Many traditional methods exist to detect environmental pollutants, contaminants, and pathogens. Since most of us are quite familiar with most of those methods, much detail and discussion is not allocated to them in this section. Analytical instruments and procedures require the following: (1) rapid response, (2) portability, (3) measurements in real time, (4) multiparameter capability, (5) simple design, (6) low detection limits, (7) high throughput, and (8) large working range [12]. Quite the wish list! How can nanotechnology help achieve these goals? A brief catalog of traditional laboratory analytical techniques is provided in **Table 28.1**. In the third column of the table, improvements due to

Table 28.1 *Traditional Laboratory Analytical Techniques Used to Measure Environmental Pollutants*

Technique	Acronym	Samples	Nano-enablement and nano-uses
Gas chromatography	GC	Organic pollutants, halogenated organic compounds	Thinner capillary columns filled with nanomaterial supports; enhanced detection systems
GC/mass spectrometry	GC/MS	Organic samples	Microfluidic samples containing nanogram quantities of material [13]. In situ analysis of carbon nanotubes by MS
High-performance liquid chromatography	HPLC	Persistent pesticides, herbicides, and polychlorinated biphenyls (PCBs)	Postcolumn derivatization; nanospray sample introduction (emitters), nanoflow separation (eliminating dead volume); HPLC-chip/MS systems, biocompatible HPLC-chips [14]; and micro-nano HPLC columns (75 μm diameter—ca. 4000x increase in sensitivity) [15]
Fourier transform infrared spectroscopy	FTIR		Live cell FTIR [16]
Atomic absorption/emission spectroscopy	AAS/AES	Heavy metals	Nanospray enhancements
Inductively coupled plasma spectroscopy	ICP	Heavy metals	ICP/MS with ppt to ppq (parts per quintillion) level of detection, nanoparticle introduction systems; ICP-MS to analyze gold nanoparticles [17]
Ion and ion-exclusion chromatography	IC and IEC	Waterborne cations and anions, carbon nanotubes	Ion exchange chromatography to separate single-walled carbon nanotubes (SWNTs) based on electrical properties [18]
Electrode methods		Heavy metal ion pollutants	Separation based on redox potentials + point-contact spectroscopy and conductance; in situ nano-contact sensor for heavy metal detection—15 pairs of nanoelectrodes on Si chip [19];
Fluorescence	FS	Pathogens	Nanophotonic light sources for fluorescence spectroscopy and cellular imaging [20]

nanotechnology of various aspects of the technique and some nano-applications are provided—whether for the detector, the mechanism that delivers the sample or other associated features of the technique.

28.0.3 *Types of Environmental Sensors*

For any environmental assessment or risk analysis, it is good to have data upon which to base conclusions (obviously!). In order to acquire data, detectors, sensors, collectors and analysis are required. Generic classes of sensors are listed and described in **Table 28.2**. Natural sensors are able to detect single molecules (e.g., the pheromone bombykol by the *Bombyx* moth). The Defense Advanced research Projects Agency (DARPA) wants to create a "dog's nose" to detect land mines with greater sensitivity (e.g., for amyl acetate, 1–10 ppb level of detection is possible by a dog's nose).

TABLE 28.2 *General Kinds of Sensors Used in Environmental Analysis*

Category of sensor	Basic mode of operation	Examples of nano-applications
Biosensors	Biosensors consist of a working enzyme, nucleotide, or biomolecule tethered to a detector surface. The working enzyme must show a strong affinity for a type of molecule or pathogen Biosensors using lanthanide oxides have large Stokes shift, sharp emission spectra, long luminescence lifetimes, and photostability [21]	Acetylcholinesterase (for organophosphates) [6]; Luciferase (for Hg^{2+} detection) [6] GABA (γ-aminobutyric acid for pyrethroid and bicyclophosphate detection) [6] Antibodies (that bind with insecticides, herbicides, or microbes) [6] Immunosensors and DNA sensors on the surface of magnetic La- or Eu-doped Gd_2O_3 phosphor nanoparticles to measure pesticides in water, toxins in food, and MTBE (methyl *t*-butyl ether) remediating bacterial DNA in soil [21]
Electrochemical sensors	Detection of changes in electrical resistance induced by contact with substrate Electric current passed between or on electrodes that interact with chemicals Electrodes can be coated with nanometer layers of substrate-selective materials	Analytical improvements in detection limits, fabrication, and remote communication [22]. Ultrasensitive electrochemical sensors using carbon nanotubes—able to detect trace concentrations of Pb in the ppb range—best yet for an electrochemical sensor [23]
Mass sensors	Measure change in mass associated with substrate reaction or sample collection	Cantilever sensor arrays on a chip that rapidly detect proteins; measurement of mass loading effects, stress, and charge simultaneously; integration with microfluidic channels capable of picoliter volumes and mass detection down to a few attograms [24] Quartz crystal microbalance (QCM) and surface acoustic wave (SAW) are refined nano-processes that have nearly achieved their theoretical limits [25]
Optical sensors	Measures change in visible light (energy or flux) following chemical or physical reaction of substrate	Semiconductor photocatalysis of dye molecules SERS detection of adsorbed species
Thermal sensors	Measures temperature changes in gaseous, aqueous, or solid-state material or environment; collect sensitive data with high accuracy from many kinds of environment	Monitoring of global warming Systems-on-a-chip containing nanosensors and actuators equipped for remote sensing

Biosensors. One of the first biosensors was the canary in the mine—a small creature with a high metabolic rate sensitive to toxic gases. Biosensors abound in nature and nearly any function can be traced to some form of biosensing. From the perspective of *Homo sapiens*, Leland C. Clark, Jr. is considered to be the father of the synthetic biosensor [26]. In the 1954, Clark invented the *Clark electrode*, an electrochemical sensor that is capable of measuring oxygen in blood [27]. Biosensor history began in 1962 with the glucose sensor [26–28]. The glucose sensor is derived from the Clark electrode [26–28]. The electrochemical reduction of oxygen occurs on a catalytic platinum surface according to

$$O_2 + 2e^- + 2H_2O \rightarrow H_2O_2 + 2OH^- \tag{28.1}$$

Clark encased the working electrode (and the counter electrode) in nonconductive polyethylene—a material with limited permeability to oxygen. Later in the 1960s, Clark experimented with trapping an enzyme that reacted to oxygen against an electrode with a dialysis membrane. He believed that by monitoring enzymatic activity the concentration of oxygen could be determined. Thus, the first chemobiosensor was created—equipped with the first enzyme transducer electrode [27,28]. The trapped enzyme was glucose oxidase (GOD). A decrease in measured oxygen (or an increase in H_2O_2) was proportional to the glucose concentration in solution. Thus, the first glucose sensor was based on the amperometric detection of hydrogen peroxide [27,28].

A biosensor combines a biological moiety with a physicochemical sensor to detect a targeted analyte [29]. The analyte may be biological, organic, or inorganic. The substrate-sensitive component may be an enzyme, a nucleic acid, an antibody, a cell receptor, or other biologically based materials. The biosensing element is linked to a transducer that is able to translate physical or chemical changes based on optical, electrical, electrochemical, piezoelectric, thermal, or other parameters. When an appropriate substrate interacts with the biological component of the sensor, a signal is generated that is converted by a *transducer* into a signal that is easy to measure, for example, electrical resistance, current, or voltage change. In the case of the first glucose sensor, the measured material was hydrogen peroxide, the result of an electrochemical transformation of oxygen and water at a Pt electrode. According to Martin Chaplin of the London South Bank University Faculty of Engineering, Science, and the Built Environment, a successful biosensor must have the following [30]:

- The enzyme (or biocatalyst) must be highly specific, be stable during storage and repetitive analytical applications.
- The analytical reaction should be independent to the best degree possible of physiochemical conditions such as stirring, *p*H, and temperature.
- The response of the biosensor should be accurate, precise, and reproducible over the analytical range without dilution or concentration.
- The sensor must be small, sterilizable, and compatible with in vivo samples and not be prone to fouling or proteolysis.
- The biosensor should be relatively inexpensive, portable, and user friendly (and easy to fabricate?).
- A preexisting market should be in place for the biosensor.

There are many kinds of biosensors—calorimetric, potentiometric, amperometric, optical, piezoelectric, and immunobiosensors—just about all the kinds of sensor types mentioned earlier except with an added biological twist.

Biosensors are typically based on the phenomenon of molecular recognition—recognition between an enzyme and a substrate for example. Many biosensors, therefore, are enzyme based. For example, an oxidoreductase-type enzyme (the sensor), anchored to a detector system electrode (the transducer), is placed in a natural water stream. An environmental toxin that happens along is oxidized in the active pocket of the enzyme. The substrate is oxidized yielding two electrons that are able to reduce nearby FAD (flavine adenine dinucleotide, a redox coenzyme, or prosthetic group) to $FADH_2$ (**Fig. 28.1**). The electrode subsequently oxidizes the $FADH_2$ back to FAD thereby releasing two electrons and a signal is

FIG. 28.1 *Basic mechanism of an oxidoreductase enzyme-based biosensor is depicted. A substrate makes contact with the active pocket of the oxidoreductase. The substrate is oxidized and two electrons are generated that transform FAD into $FADH_2$. The $FADH_2$ is oxidized back to FAD by a mediator yielding the two electrons. The electrons are scooped up by an amperometric or potentiometric transducer (an electrode) and converted into a signal for processing.*

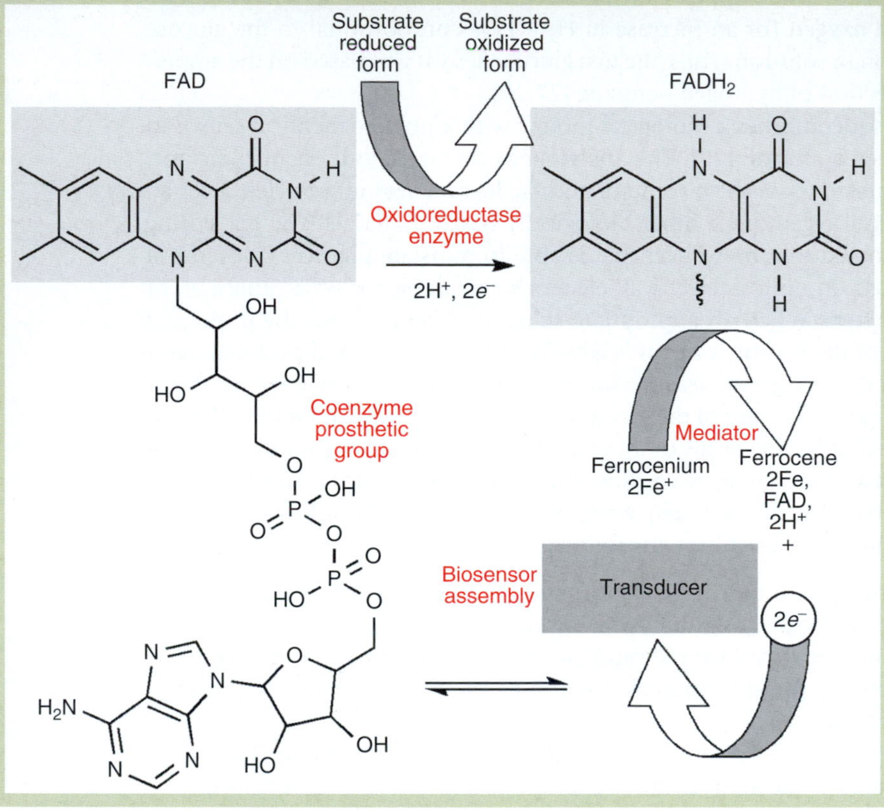

generated. The size of the current ideally is proportional to the concentration of the analyte.

Aptamer-based sensors are a new class of ultrasensitive biosensors called "smart materials" that employ oligonucleotides like double- or single-stranded DNA and single-stranded RNA or peptides to bind molecular targets such as proteins or metabolites with high affinity and specificity. Aptamers (from the Latin *aptus* "to fit") are engineered through iterations of in vitro selection (called SELEX, systematic evolution of ligands by exponential enrichment) [32]. Aptamers, often referred to as DNAzymes, are segments of DNA (or RNA) that possess enzymatic function, and can be coupled with fluorophores or gold nanoparticles to selectively detect targeted analytes [31]. Nucleic acid aptamers, for example, have been shown to inhibit the replication of HIV-I [33].

The range of substrates that aptamers can selectively bind is enormous—well on the order of antibodies. However, advantages of aptamers over antibodies are multifold: (1) they can be engineered in laboratories by chemical synthesis methods (do not require biological hosts), (2) are easily tagged or cross-linked with thiols, (3) have desirable storage properties and longevity while immobilized, and (4) engender little or no immune response [32].

Juewen Liu and Yi Lu of the Department of Chemistry at the University of Illinois at Urbana-Champaign recently developed aptamers linked to nanomaterials that are sensitive to cocaine, adenosine, and K^+ [31]. In one system, two types of DNA-functionalized 13-nm diameter gold nanoparticles were produced: one functionalized with the 3′-end thiol and the other with the 5′-end thiol (**Fig. 28.2**). Already a built-in templating component is installed in the system as the 3′ or 5′ ends of the nucleotides direct further modification by self-assembly. The two types of precursors were then linked to an adenosine aptamer and a cocaine aptamer, respectively, that self-assembled into an agglomerated form (the color of the aggregate is blue). In the presence of adenosine, the adenosine aptamer adjusted its structure to accommodate two adenosine molecules. In the presence of cocaine, the appropriate aptamer adjusted its structure to

Fig. 28.2 *An example of an aptamer sensor that simultaneously targets cocaine and adenosine. The presence of adenosine and cocaine are necessary in order to disperse the gel nanoparticle aggregation. This is evidenced by a change in color from blue (the aggregate phase) to red (the dispersed phase). Different permutations of adenosine–cocaine systems can be generated: a highly cooperative system is depicted in the figure, systems with three kinds of links to gold nanoparticles, and lastly, systems with no cooperativity (respond individually to bind the substrates). This aptamer scheme is quite an ingenious tribute to nanotechnology—depending on the aptamer, the nanoparticles, and the way they are linked.*

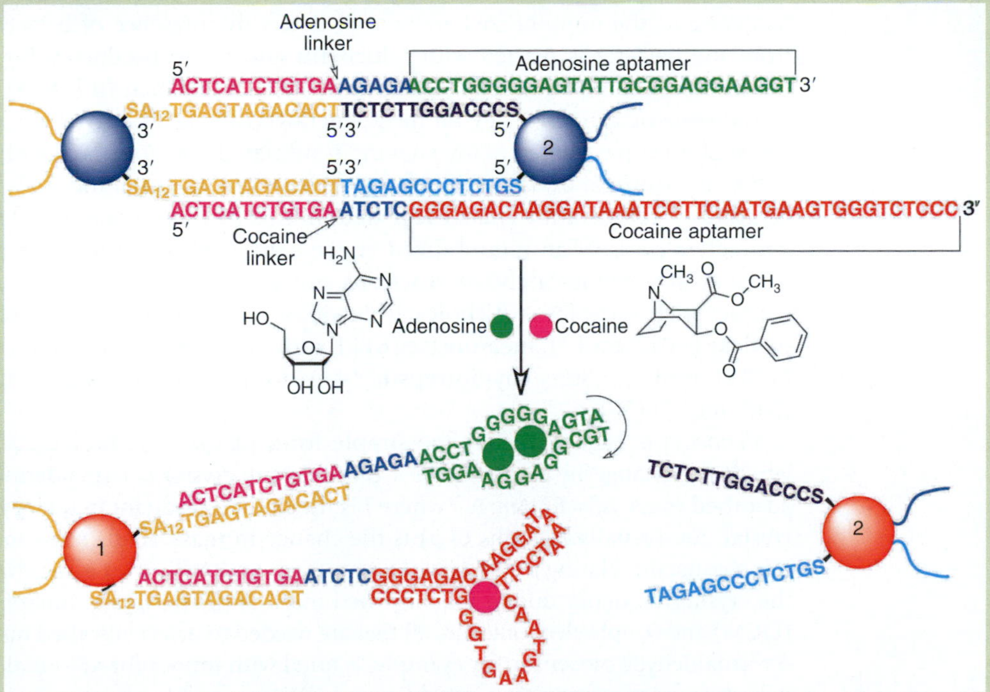

Source: J. Liu and Y. Lu, *Advanced Materials*, 18, 1667–1671 (2006). With permission.

accommodate one molecule of cocaine. Complete dispersion of the aggregate occurred only when both cocaine and adenosine were present—resulting in a red color indicative of dispersed gold colloids [31].

Due to design generalities among the aptamers, sensors that respond to any combination of the three can be developed, a process called controlled cooperativity. According to J. Liu et al., the gold nanoparticles can be replaced with other nanomaterials such as quantum dots, nanotubes, or polymers that would open the door to a diverse range of applications [31]—quite an amazing display of nanoengineering!

Potentiometric biosensors are able to transfer information generated from a biological reaction into an electronic signal. Ion-selective electrodes resemble a modified *p*H probe. They consist of an outer immobilized enzyme encased in a semipermeable membrane around an active glass membrane that encloses a *p*H probe (an internal Ag/AgCl electrode in dilute HCl). An electric potential is formed between the probe and an external reference electrode. The probe detects changes in the hydrogen ion concentration, whether by absorption or generation of H^+, due to catalytic activity of the enzyme. There are three generic kinds of ion-selective electrodes that are exploited in ion-selective biosensors—glass electrodes for cations (with hydrated glass membrane) based on cation competition; glass *p*H electrode coated with gas-permeable membrane selective for CO_2, NH_3, or H_2S or other gases; and solid-state electrodes that utilize specific ion conductors [30]. Each kind can be enhanced with nanotechnology.

Optical biosensors invoke one of two mechanisms for the purposes of detection. In the first, a change in light absorption following a chemical reaction is quantified, and in the second, light output generated by a luminescent process is measured. An example of an optical biosensor is one that incorporates luciferase as the immobilized enzyme to detect the presence of bacteria. ATP from the bacteria is reacted with D-luciferin and O_2 to produce yellow light (ca. 560 nm) in high quantum yield with sensitivity less than 10^{-12} M ATP [30].

Calorimetric biosensors rely on heat generated during a catalytic reaction as the analytical parameter. Many enzyme-modulated reactions are exothermic. Therefore, the heat produced in a reaction is an indication of the analyte concentration. Temperature changes are determined by thermistors at the ingress and egress ports of an immobilized enzyme array. Heats of reaction (ΔH_{rxn}) for several enzyme-catalyzed reactions are known—for substrate–enzyme couples such as cholesterol/cholesterol oxidase (53 kJ · mol^{-1}), glucose/glucose oxidase (80 kJ · mol^{-1}), urea/urease (61 kJ · mol^{-1}), hydrogen peroxide/catalase (100 kJ · mol^{-1}), esters/chymotrypsin (4–16 kJ · mol^{-1}), and starch/amylase (8 kJ · mol^{-1}) [30].

Piezoelectric biosensors obey the simple form of the Sauerbrey equation in which the change in frequency of a piezoelectric crystal is dependent on the adsorbed mass: $\Delta f = Kf^2 \Delta m \cdot A^{-1}$ where f is the natural resonant frequency of the crystal, Δm (usually in terms of g) is the change in mass of adsorbed material, K is a constant relating to the crystal parameters, and Δf (in Hz) is the change in the crystal frequency due to the adsorbed mass. A quartz crystal microbalance (QCM) and simple electronics are all that are needed to detect adsorbed materials. A formaldehyde biosensor, for example, is fitted with immobilized formaldehyde dehydrogenase coating on a quartz crystal. QCMs are able to detect nanolayers of adsorbed material.

Immunosensors employ an antibody–antigen couple as the basis of detection. Biosensors are able to incorporate enzyme-linked immunosorbent assays (ELISA) within the sensing mechanism [30]. In ELISA, an antibody specific for a target antigen is immobilized on the surface of a transducer. Then, a mixture of a known concentration of antigen–enzyme complex plus an unknown amount of antigen is flowed over the transducer—a competition-based detection scenario. After time for equilibration, the antigen and antigen–enzyme conjugates become distributed between two states: either bound or free depending on their relative concentration. The unbound material is rinsed away and the level of antigen–enzyme complex that is bound is determined by the rate of the enzymatic reaction (or the transducer signal) [30].

Inorganic Electrochemical Sensors. Adsorption of gases and subsequent heterogeneous catalytic activity (oxidation or reduction) at the surface of semiconducting metal oxides results in concomitant changes in the electrical resistance of the sensor material, for example, a signal that is characteristic of a specific gas–surface interaction [6]. The magnitude of the change in electrical resistance is dependent on the type of gas, it's partial pressure, and ambient conditions such as temperature. For nanocrystalline sensor materials, crystallite size is also an important consideration [6]. Selectivity in gas sensors is problematic and a number of strategies have been applied to help with this issue.

Tethered DNA electrochemical sensors were fabricated by M. Curreli et al. to produce functionalized In_2O_3 nanowire electrodes for the purposes of biosensing, identification, and quantification of infectious agents [34]. Considering the properties of nanomaterials (e.g., high surface-to-volume ratio), low detection limits can be achieved due to miniscule changes in the conductivity of the nanosensor. Recognition of the analyte is afforded by analyte-specific molecular recognition groups. Indium oxide nanowires provide an alternative to silicon nanowires and carbon nanotubes. One advantage over silicon nanowires is that In_2O_3 has no oxide layer—therefore it can act as a gas sensor with ppb level detection [34]. Immobilization of DNA-based sensors on the nanowire is accomplished by tethering tailored single-stranded DNA receptors to In_2O_3 nanowires with HQ-PA [4-(1,4-dihydrobenzene)butyl phosphonic acid] linking groups [34].

HQ-PA is an electrochemically active hydroquinone that readily undergoes redox reactions at low potentials [35]. The oxidized form (**Fig. 28.3**) is able to react with many kinds of functional groups that can be spliced into biomolecules. Phosphonic acids are known to interact with In_2O_3. The reversible redox behavior of HQ-PA on indium–tin oxide (ITO) and In_2O_3 is similar. Cyclic voltammetry studies conducted on ITO showed that the oxidation wave was centered at +330 mV and the reduction wave at −200 mV [35].

Mass Sensors. MEMs and NEMs sensors that rely on microcantilever transducers are also used to detect the presence of chemicals, pathogens, explosives, and ionic species. Once again a tried and true chemical procedure is applied—the functionalization of a surface in order to bind target molecules or entities. In this case, it is the surface of the silicon (or silicon oxide, silicon nitride) microcantilever, and a small and sensitive force or mass detector that is chemically modified. Analytical signals are generated following molecular adsorption of

FIG. 28.3 *A functionalized electrochemical cell made of In_2O_3 nanowires and tethered, DNA-functionalized hydroquinone electroactive groups is shown. In_2O_3 nanowires and electrodes were formed by photolithography and metal deposition on SiO_2/Si to form an electrode array. A self-assembled monolayer of HQ-PA was applied to the indium oxide nanowires. Thiol-terminated C_{12} hydrocarbons (dodecane-1-thiol) formed a self-assembled monolayer on the gold electrodes to prevent thiol-terminated DNA from interacting with the gold electrodes. The DNA sequence is generic and shown only for image clarity and is not provided by the authors of the paper.*

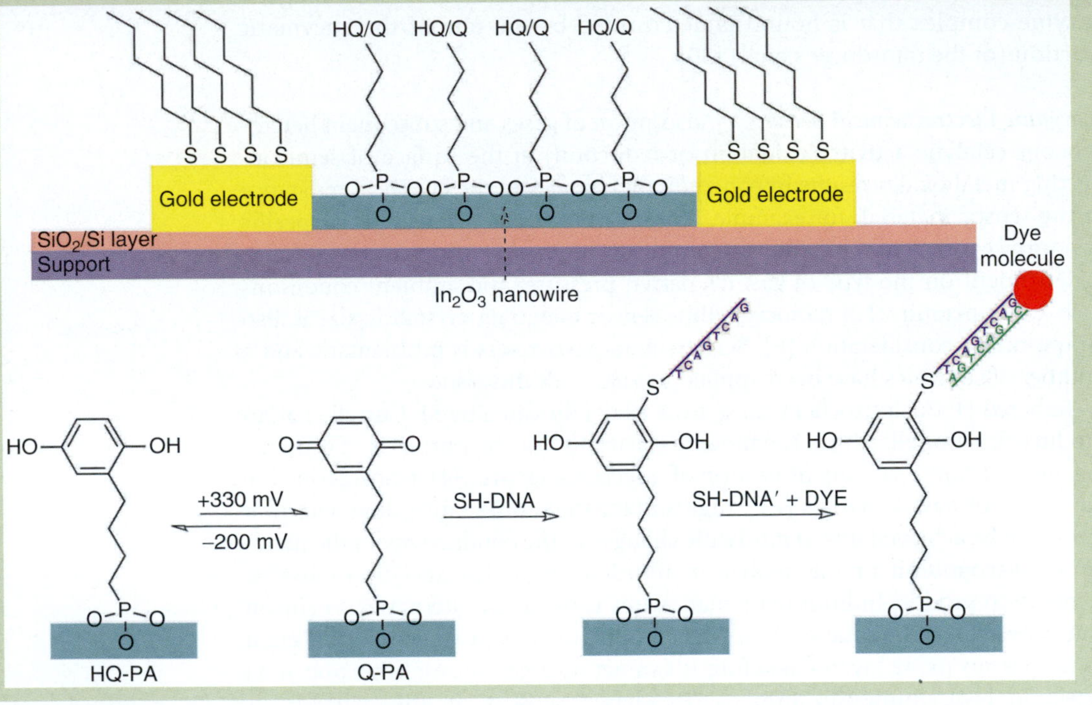

Source: M. Curreli, C. Li, Y. Sun, B. Lei, M. A. Gundersen, M. E. Thompson, and C. Zhou, *Journal of American Chemical Society*, 127, 6922–6923 (2005). With permission.

target molecules by monitoring bending and changes in resonant frequency or the position of the cantilever [6]. Several techniques such as optical beam deflection (the principle behind AFM) (**Fig. 28.4**), piezoresistivity or electricity, interferometry, capacitance, and electron tunneling are some of the most applied [36]. For example, forces and surface stresses in the micronewton to nanonewton range allow for quantification of the concentration of adsorbed analyte [37].

An array of silicon cantilevers is coated with sensitive substrate-specific layers. Each cantilever is capable of sensing a unique pollutant. Physical or chemical processes induced by adducts are transduced into mechanical motion due to changes in surfaces stresses like bending. Surface stress leads to bending (static mode) and mass increase leads to lower resonance frequency of the cantilever (dynamic mode). Surface stress is not necessarily related to mass but rather to molecular forces such as steric hindrance, electrostatic repulsion, or conformational

Fig. 28.4 *Mass sensor based on cantilever detection is displayed. Cantilevers are typically 10–500 μm in length and as thin as a few microns or less. The sensitivity of a cantilever depends on its "spring constant": the lower the constant, the higher the sensitivity. Laser light is shined on a specific area of the surface and the reflected beam is captured by a position-sensitive photodetector. Polymeric materials coating cantilever surfaces serve to differentiate among VOCs in air. At the lower left corner of the image, a schematic of a cantilever array is depicted. Each cantilever is uniquely derivatized in order to identify and bind a specific substrate.*

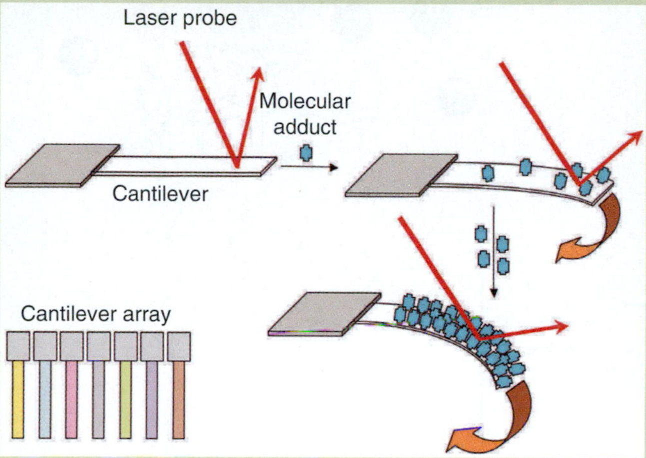

changes within the functionalized layer. Therefore, two complementary sets of data can be acquired from a single measurement [6].

Cantilever sensors have been developed to detect VOCs, TNT, heavy metal ions and anions like chromate, Salmonella bacteria, viruses, fungal spores, and pesticides. Calixarene-crown ether complexes on cantilevers were employed to detect Cs^+ in the range of 10^{-7} to 10^{-11} M [38]. In this case, a selective cesium ion sensor based on an ion-selective SAM-coated microcantilever was capable of detecting low concentrations of Cs^+ in the presence of high concentrations of K^+ and Na^+ cations. A thin gold layer was applied to the surface of a microcantilever and a subsequent thiol layer as well (**Fig. 28.5**).

The sensitivity of this nanotechnologically enhanced electrode is several orders of magnitude better than the best available ion-selective electrode (ISE) [38].

Optical Sensors. Optical sensing is usually accomplished with surface plasmon resonance spectroscopy, interferometry, and luminescent spectroscopy. One class of optical sensing is based on the photoluminescent properties of nanocrystalline porous silicon. The photoluminescent response of this material depends on the chemical nature of its surface [6]. Adduct metal ions and organic materials are able to quench the photoluminescence of porous silicon.

28.0.4 Introduction to Environmental Mitigation

We are all quite familiar with many kinds of environmental pollutants and pathogens and therefore will not list them. There are thousands of kinds of hazardous

Fig. 28.5 *Functionalized cantilever specific for cesium cations is shown. The cesium recognition agent is 1,3-alternate-25,27-*bis*(11-mercapto-1-undecanoxy)-26,28-calix[4]benzocrown-6 (highlighted in black). This molecular recognition system was able to discriminate cesium ions in high concentrations of sodium and potassium cations.*

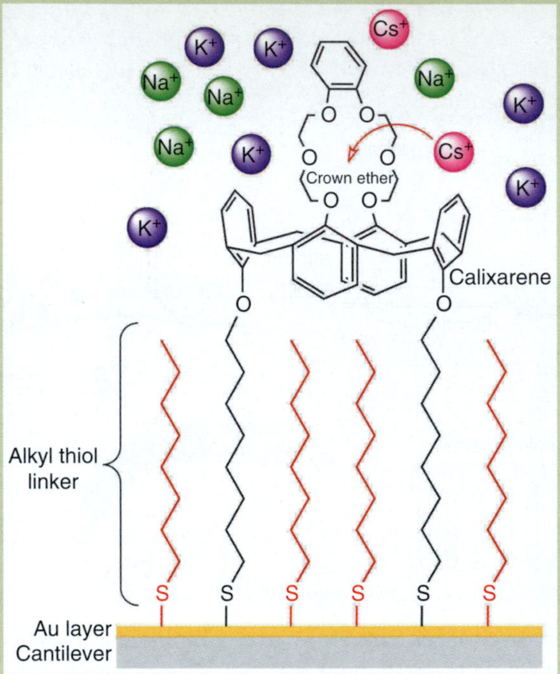

Source: H.-F. Ji, R. Dabestani, G. M. Brown, and P. F. Britt, *Chemical Communication*, 6, 457–458 (2000). With permission.

substances, and they exist as particulates, organics, inorganics, and biological materials—in water (including groundwater), soil (and sediments), and air (indoor and outdoor)—and as toxins in our bodies. Volatile organic compounds form the biggest class of hazardous pollutants. Inorganic elements such as heavy metals and radionuclides are also prevalent. Polyaromatic hydrocarbons (PAHs), halogenated pesticides, organophosphates, nitrosamines, and other organic chemicals have drawn much attention over the years.

There are three generalized types of mitigation: (Type I) *post de facto* treatment of pollutants released into the environment; (Type II) treatment of pollutants before whole-scale release into the environment—a.k.a., "end-of-pipe" management and cleanup; and (Type III) elimination of pollutant from a process by substitution with another process or by alteration of further the process—a.k.a., prevention. Post de facto treatments are (were?) the most common. Treating the pollutant before release obviously reduces the amount of further post de facto remedial applications in which the pollutant may achieve a highly widespread and dispersed state. The best option is, of course, to remove the pollutant from the process in the first place. This can be accomplished by substituting with alternative sources, modifying processes in such a way as to eliminate the detrimental material or by releasing less material. Selected examples of mitigation, and remedial or preventive processes are listed in **Table 28.3**.

TABLE 28.3 *Environmental Remediation, Mitigation, or Preventative Processes Using Nanotechnology*

Pollutant	Nano-mitigation
Automobile exhaust gases	Supported (on alumina) precious metal (Pt-Pd, Au) nanocatalysts of precious metals to remove CO, NO_x, and HCs [39,40] NO_x reduction by Rh supported on alumina or vanadium oxides
Mercury	Coal burning plant and industrial wastewater. High surface area nanoporous silica-based ceramic material with tailored pore size coated with attached monolayer specific for attracting Hg. 99.9% Hg in simulations captured [43,44]
Asbestos	Stopped using asbestos
Volatile organic compounds	5–100 nm MnO nanoparticle catalysts—removal to ppb levels [46] Catalytic destruction of indoor VOCs by nanoparticle scrubbers at room temperature Photocatalytic destruction of VOCs in air scrubbers Manufacture of nanotechnology-based paints without VOCs but also to act as an antimicrobial and antifungal surface Ethylene oxidation by Ag on support
Heavy metals	Granular microscale Fe > 50 μm replaced by nanoscale Fe and Fe/Pd that are more reactive in reduction of inorganics [48]
Halogenated organic compounds	Complete and rapid dechlorination of aqueous species by nanoscale Fe and Fe–Pd (99.9% Fe) [41]
Dense nonaqueous phase liquids (DNAPLs)	Emulsified zero-valent iron (EZI) nanoparticles act like DNAPL in water (sinks to bottom in organic phase), then attract pollutant into hydrophobic sphere and reduces the chlorinated hydrocarbon [42] Removal of lindane (γ-hexachloro-cyclohexane) by FeS nanoparticles
Perchlorate	Complete reduction of perchlorate in contaminated groundwater to chloride by Fe nanoparticles
Pathogens	Viruses, bacteria, endotoxins, and DNA filtered by nanofilters that remove 200-nm particles with 99.9% efficiency [45]
Nuclear waste	Copper canisters have been used for nuclear cleanup in the past. Now "super-sponges" made of inorganic–organic SAMM (self-assembled monolayer on mesoporous support) with channel diameter 20–200 nm can reduce concentrations to 1 ppb [47]
Water treatment	Capacitative deionization technology (CDT) based on carbon aerogels—to remove chloride, nitrate, silica at the positive electrode, and calcium, magnesium, and sodium at the negative electrode [49]. No chemicals are required

Application of Catalysts. Nanotechnology has been applied for many years to reduce the amount of toxins produced by the combustion of gasoline. Rhodium, platinum, and palladium nanosized catalysts have been utilized in catalytic converters for decades—but they are very expensive! The use of transition nanoparticle metal carbides and oxycarbides may provide a lower cost alternative [6]. However, numerous nanotechnology-enhanced contaminant mitigation techniques have been developed recently. Bio- and genetic engineering have provided recombinant DNA to create artificial protein polymers (tunable biopolymers) coupled with molecular recognition capability to target the removal of specific toxic wastes [6,50]. Nanoparticle materials ranging from 1 to 10 nm in size offer routes to chemical remediation of many potential environmental pollutants. The key to this success of the new generation of nanocatalysts is the incredible surface to volume ratio, enhanced surface activity of such small particles, and in many cases, modified surfaces that have the ability to recognize targeted pollutant materials.

Catalytic Gold. Gold, not known as a catalyst, is quite a good one when reduced in size [51]. As a matter of fact, research and development of gold catalysts has increased such that from 1991 to 2001 over one-third of the patents awarded for pollution control using catalysts have involved gold nanoparticles [52,53]. Gold catalysts on clay particles (magnesium silicate hydrate) coupled with ozone destroyed odors [54]. Gold nanoparticles were very effective in controlling trichloroethylene (TCE) levels, a groundwater contaminant. Michael Wong of Rice University found that Au–Pt nanoparticle co-catalysts broke down TCE 100 times faster than did catalysts made of the traditional material, palladium, and more economically [55]. Detection systems using gold nanoparticles developed at the Indian Institute of Technology found that nanogold was able to attract and then bind pesticide residues in flowing water [55]. Coal-fired power plants are one of the primary sources of anthropogenic mercury. Since gold has a strong affinity for mercury, a pilot study showed that gold catalytic systems were effective in oxidizing mercury in power plant effluent streams from the Lower Colorado River Authority Fayette Power Plant [55].

Nanoparticulate gold on oxide supports, in particular, showed versatility in many kinds of pollution control studies. For example, supported nanogold was able to remove CO from room air at ambient conditions and from fuel cell hydrogen feed gases [52]. The oxidation of methane, propane, and the removal of nitrogen oxides from the atmosphere have also been accomplished at relatively low temperatures with gold catalysts (as opposed to traditional catalyst methods) [56]. More recently, Nanostellar Inc. (a leader in nanoengineered catalyst design and manufacture) announced that gold oxidation catalysts reduced diesel hydrocarbon emissions by as much as 40% over commercially available materials in tests [39]. Since there are over 14 million light-duty and 2 million heavy-duty diesel vehicles produced annually worldwide, the environmental impact provided by such newly developed nanocatalyst systems is expected to be, and in no uncertain terms, enormous [39].

Photocatalysts. Photocatalytic-driven reactions are gaining widespread acceptance in purifying indoor air, among other applications. Nanosized semiconductor materials such as TiO_2 and ZnO are capable of interfacial charge transfer to organic pollutants situated on the nanoparticle surface. When TiO_2 nanoparticles are irradiated with ultraviolet radiation (<400 nm), electron–hole pairs are created (**Fig. 28.6**). Recombination of surface charges is reduced by the presence of contaminant species. Both electrons and holes are able to create reactive species that can destroy toxic species at the surface. The surface reactions of both are shown in **Figure 28.6**. Holes are capable of creating reactive species that can destroy detrimental organic species (e.g., halogenated hydrocarbons) and transform them into CO_2, H_2O, and HCl while electrons are able to reduce harmful chemicals by unleashing O^{2-} species that accomplish similar reactions.

If noble metals like gold and platinum are chemisorbed to the TiO_2 or ZnO, photocatalytic activity is accelerated due to enhanced electron attraction. The presence of metal also helps to keep electrons and holes from recombining in the semiconductor and thereby increase the efficiency of the photocatalyst [6]. Many methods have been developed to overcome the dependence on UV radiation (less than 3% of the solar spectrum). These include doping the

Fig. 28.6 *Schematic illustration of the photocatalytic effect of nanoparticulate titanium dioxide. Energetic light in the ultraviolet regime of solar radiation generates an electron–hole pair in TiO_2. Electrons and holes either recombine on the surface (neutralizing the photocatalytic action) or react with surface-adsorbed species to create reactive species that are able to oxidize organic compounds or kill bacteria. If chlorinated organics are degraded, mineral acids like HCl can form. By this way, pollutants are converted into nontoxic or environmentally benign chemicals without input of artificial sources of energy. One major disadvantage of TiO_2 and other semiconductors is that UV light is required—a wavelength range that comprises only a few percent of the solar spectrum.*

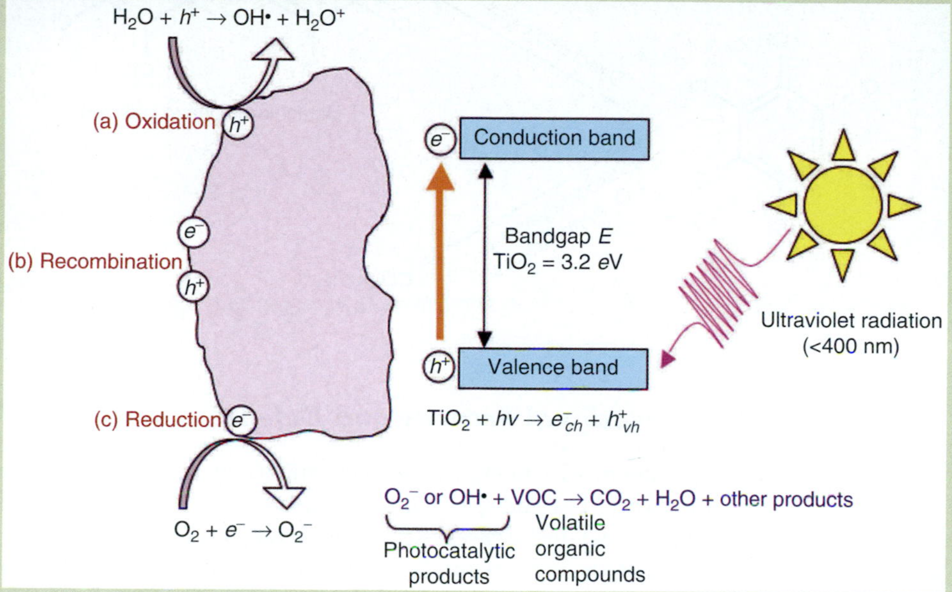

relatively wide-bandgap semiconductor with manganese or other metals and semiconductors to produce bandgap states with lower energy or to attach organic dye molecules that are able to transfer excited electrons to the photocatalyst. In this way, the doped or otherwise altered oxide makes possible the absorption of visible light, e.g., light that comprises just less than 50% of the solar spectrum—more effective than UV radiation that contributes only about three percent of the solar light energy reaching the Earth's surface.

There are many kinds of photocatalysts. The most popular is TiO_2. Other photocatalytic oxides include SnO_2, ZnO, Fe_2O_3, and V_2O_5. Many sulfides are also able to act as photocatalysts: FeS, ZnS, MoS_2, CdS, and PbS; and selenides like CdSe. Semiconductor system duos utilize TiO_2 with CdS, SnO_2, ZnO, or CdSe. ZnO/CdS couples have also been studied. **Figure 28.7** illustrates the products of catalytic oxidation versus direct exposure.

In order for photocatalysts to be effective, however, immobilization onto a solid substrate is recommended. This action not only conserves catalytic material resources (by regeneration) but also leads to higher reaction rates. In particular, polypeptide-cased membranes that sequester nanoscale metals have become more important in recent years [59].

FIG. 28.7 *Two modes of pentachlorophenol degradation is depicted. The pathway utilizing the photocatalyst produces CO_2 and HCl (the process of total mineralization), much preferred over the production of octachlorodibenzo-p-dioxin, a species that is more toxic than its precursor, by direct photolysis [57]. Microbial decomposition of 4-chlorophenol proceeds very slowly—the reaction has a half-life of 500 days [58].*

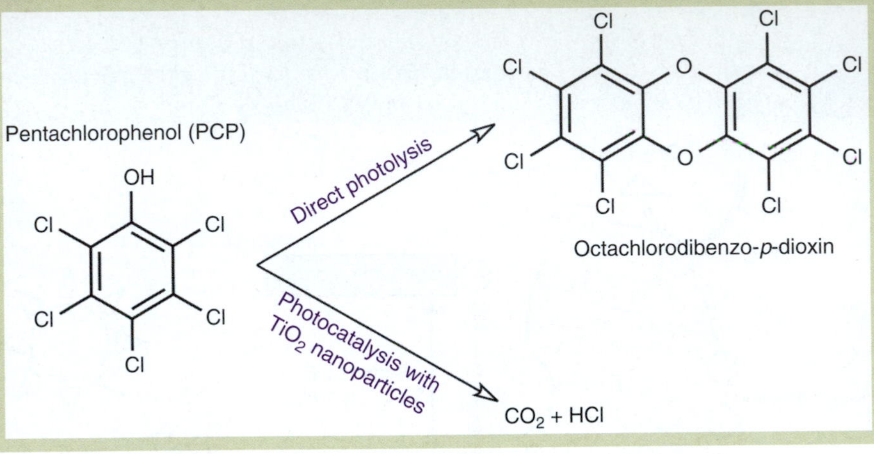

28.0.5 National Security and Defense

Security issues involve many aspects of the environment, health, and safety. Water poisoning (e.g., cyanides), radiation exposure (e.g., dirty bombs), explosives (e.g., land mines), nerve gases (e.g., sarin), bioagents (e.g., anthrax), drugs (e.g., cocaine)—and yes, and perhaps nanotechnology-enhanced materials themselves—are some well-known materials known to terrorists. Security programs focus on detection, preparedness, response, and recovery (mitigation). We will focus on detection and mitigation in this section. Toxic materials come in all forms and shapes. Bacterial toxins such as botulin, diphtheria, and anthrax require, in terms of $g \cdot kg^{-1}$ of body weight, 0.001, 0.10, and 0.004–0.02 $g \cdot kg^{-1}$, respectively, to deliver a lethal dose that kills 50% of subjects tested (a.k.a. LD-50) [60]. In this section, we will review several aspects of defensive techniques applied to mitigate potential chemical, biological, and radioactive agents.

Nanosensors. Room temperature CdZnTe sensors have been developed to detect gamma rays emitted by radionuclides from Cs and Co [61]. Gold nanowire sensors able to detect ppb levels of mercury in air or water have been developed [62]. Robust and regenerable sensors have shown superior performance over thin-film devices [62]—most likely due to enhanced surface area. A research team at the Lawrence Livermore National developed compact, low-power piezoresistive cantilever-based sensor arrays for chemical detection of many kinds of organic vapors over a wide concentration range [63]. Gas fingerprinting using carbon nanotube transistor arrays has been demonstrated by Frenchmen P. Bondavalli et al. [64]. Carbon nanotube field-effect transistors (CNT-FET) for gas sensing rely on the sensitive response of the Schottky junction between CNTs and numerous kinds of drain/source metal electrodes [64]. The mechanism is based on the change induced in the work function of the contact metal by a specific gas (e.g., sarin), thereby modifying the CNT-FET electrical response [64].

Nanomaterials as Toxic Agents. Nanomaterials themselves can pose formidable hazards to national security. A study by Y. Zhang et al. at the Arizona State University showed that nanomaterials (CdSe quantum dots, hematite, and six commercial metal oxides) in water possessed varied stability and that conventional water treatment methods were not effective in removing them completely [48]. Toxicity studies showed that nanomaterials flatten the microvilli in intestine and can pass easily through epithelial cells [48]. S.L. Harper et al. of Oregon State University conducted in vivo studies (zebra fish embryos) of the biodistribution and toxicity of nanomaterials [65]. Fullerenes and gold nanoparticles were used to investigate the effects of size and surface functionalization, and 11 kinds of nano-metal oxides were used to study the effects of chemical composition. They determined that toxicity was dependent on surface functionalization and chemical composition. Core size and functionalization also influenced the toxicity of gold nanoparticles. Polystyrene and CdSe fluorescent nanomaterials were employed in the investigation of biodistribution [65].

T. Knight et al. of the University of Southern Maine investigated the cytotoxicity of 15-nm diameter gold nanoparticles [66]. They found that significant cytotoxicity was developed in Jurkat human T cells by gold nanoparticles [66]. The gold nanoparticles had differential effects on other kinds of cell types [66]. J. Chen et al. of the Harvard Medical School evaluated the potential toxic effects of metal oxide nanoparticles (Fe_2O_3, CuO, ZnO) and their metal ion counterparts (Fe^{3+}, Cu^{2+}, Zn^{2+}) on human neuroglioma and neuroblastoma cells [67]. They found that CuO caused significant dose-dependent episodes of cellular death while the cytotoxicity of the other Fe_2O_3 and ZnO nanoparticles was marginal when compared to control results [67]. Of the ionic renditions, Cu^{2+} demonstrated a significant effect on cell toxicity [67].

Carbon nanotubes are proving to be useful materials in drug delivery and carbon nanocages filled with metals in MRI applications [68]. However, the potential toxicity of carbon nanotubes has received scrutiny over the past few years. R. Sharma et al. of Florida State University showed that carbon nanotubes (30-nm diameter) penetrated skin tissue quickly (within 2–3 min) and were able to damage the epidermal layer—at the level of the membrane and nucleus [68].

Surface-Enhanced Raman Probes. Land mines can be detected quite well by metal detectors (sensitivity to 0.5 g), but metal detectors respond to anything made of metal—tin cans, old utensils, and automobile parts. Therefore, new generation detectors are required that sense organic constituents of explosives and not specifically the metal components. Surface-enhanced Raman spectroscopy (SERS) analytical tools have been recently developed to detect parts per billion levels of explosives and degraded explosives from land mine vapors emanating from the soil [69]. Many types of explosives contain nitro groups attached to an aromatic moiety (**Fig. 28.8a** and **b**). These include picric acid, TNT (2,4, 6-trinitrotoluene), DNT (2,4-dinitrotoluene, an impurity), and degradation products (by the action of bacteria) 1,3-dinitorbenzene and 4-amino-2,6-dinotrotoluene vapors. All can be detected by SERS [70]. The diagnostic signal is due to the nitrate-bending modes at 820 cm^{-1} and the symmetrical stretching modes at

Fig. 28.8 *(a) Schematic representation of 2,4,6-TNT and 2,4-DNT detection by a SERS sensor. The nitrate moiety of TNT and DNT makes contact with the roughened gold substrate. When laser light is shined on the surface (785 nm, in the near-IR range), the symmetrical vibrational modes of the nitrate are modulated by the scattering. (b) The Raman peak at ca. 1384 cm^{-1} is diagnostic for TNT vapors. It can easily be seen above the background signals. Since natural nitrated aromatic compounds are scarce, there is little chance of interference. Therefore, the peak for adsorbed nitrate is diagnostic. Naturally occurring nitrogen dioxides undergo catalytic conversion to NO_3^- at the surface. A sharp peak at 1035 cm^{-1} is characteristic of these species [70–72].*

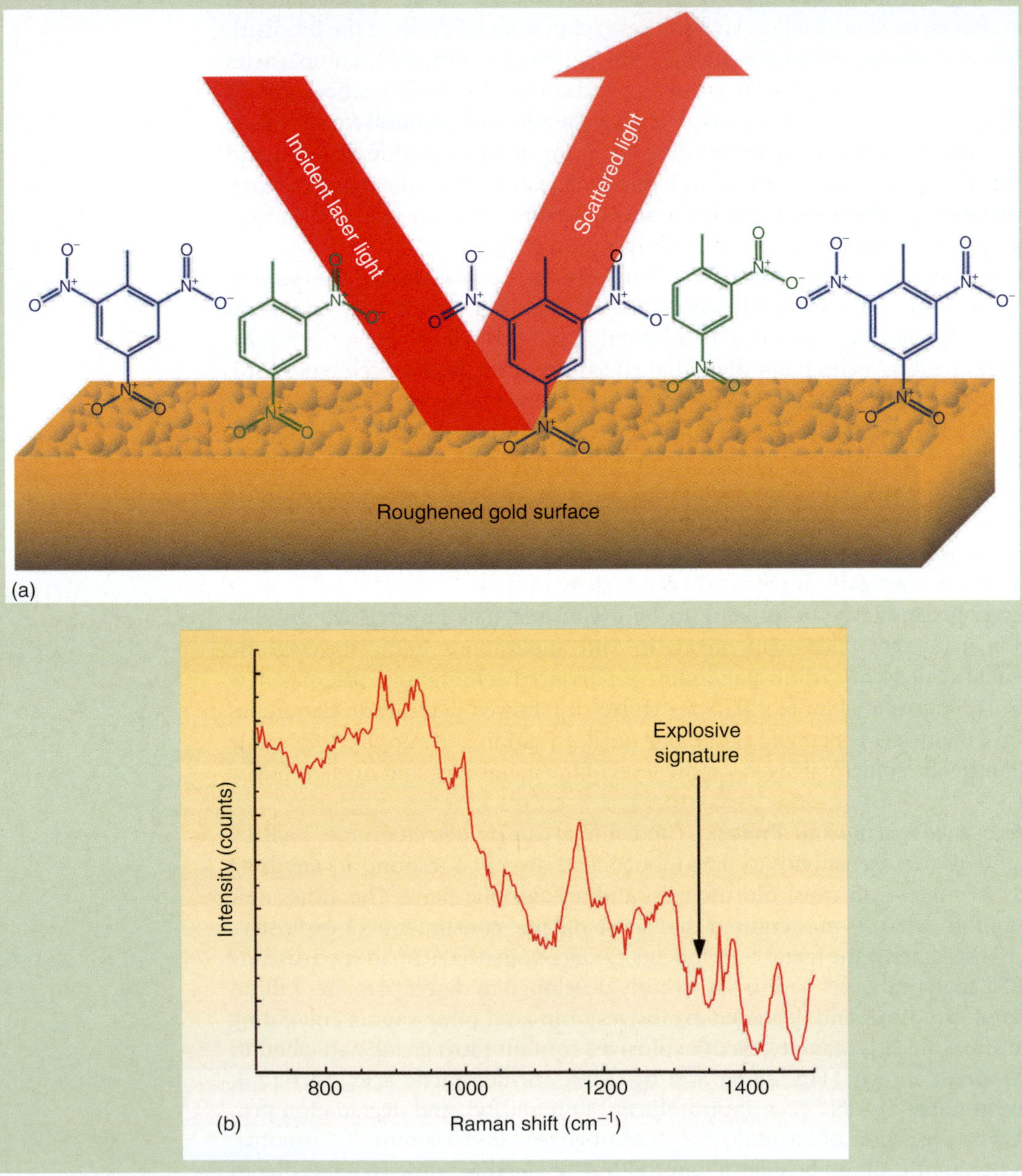

Source: Images courtesy of Kevin M. Spencer, EIC Laboratories, Inc. With permission.

ca. 1337 cm^{-1}. The nitrate signature is strongly enhanced and shows that adsorption does not take place by the aromatic ring. TNT and DNT have been detected down to 1 ppb level in concentration (1 fg or 4.4×10^{-18} mol, the attomole level) by SERS detectors used in the field [71]. Interference from the background in field tests was not significant [72].

The surface-enhanced Raman substrate consists of metallic (gold, sliver, copper) micro- to nanospherical or ellipsoidal particulates (or roughened surfaces with micro- or nanofacets) upon which analyte substrates are adsorbed. SERS is capable of 10^{12}–10^{14} enhancement of detection limits [72]. The wavelength of the excitation laser matches the resonance of the surface plasmon of the metal nanoparticles. For the detection of TNT, wavelength in the near-IR (785 nm) provided the probing beam. The metal plasmon is tunable by adjusting the size, shape, and orientation of the metal nano or microparticles with respect to the probing radiation [73]. The SERS surface is oriented towards a commercial fiber-optic probe. A small fan directs the ambient vapor over the substrate [72]. SERS requires no sample preparation and with regard to spectral acquisition, sample evaluation takes place in ca. 30 s. SERS is also able to detect many kinds of suspected chemical agents as long as a diagnostic peak exists for that material.

Cyanide found in drinking water is also detectable by SERS. Cyanide has a very strong Raman band at ca. 2200 cm^{-1} and levels down to 2 ppb have been recorded. The detection of cyanide has been shown to be quantitative at ppm levels of concentration [72]. While SERS is perfect for chemical agents, resonant Raman spectroscopy (RRS) has shown utility for detecting biological species [72]. In this case ultraviolet radiation is used to excite biological molecules—especially proteins and nucleotides. Because Raman signal intensity is proportional to v^4 (e.g., the higher the energy of probing radiation, the more intense the Raman signal), nucleotides at concentrations on the order of 500 nM have been successfully detected [72]. This is a new area of national security study and there is still a way to go before a viable system is on the market.

Immunosensor Detection of TNT. Environmental detection and subsequent remediation of hazardous pollutants such as explosives is needed. On the detection front and in addition to SERS methods, portable fluorescence immunoassay biosensors have been developed by the Naval Research Laboratory (NRL) to meet the need [74]. The biosensor is based on a competitive fluoroimmunoassay in which a fluorescent molecule similar in structure to the analyte competes with the analyte for binding sites on antibodies immobilized on the surface of an optical probe [74]. In other words, by this method, fluorescent signal intensity is inversely proportional to the amount of analyte in the sample.

A fiber-optic biosensor (FOB) makes use of molecular recognition and evanescent wave sensing to detect many kinds of analytes. Optical fibers are able to excite fluorescent molecules that are near to the core of the fiber. An evanescent wave, the electric field generated from internally reflected light, has a penetration depth of 125 nm. Fluorescent molecules that enter the evanescent wave (bound by the antibodies on the surface of the fiber-optic probe) are excited and emit light at longer wavelengths (e.g., fluoresce). A portion of the fluorescence is transmitted by the fiber to a detector. Molecules out of range of the evanescent wave are not detected. This is an example of a competition-type immunoassay. A schematic of a fiber-optic detector is shown in **Figure 28.9**.

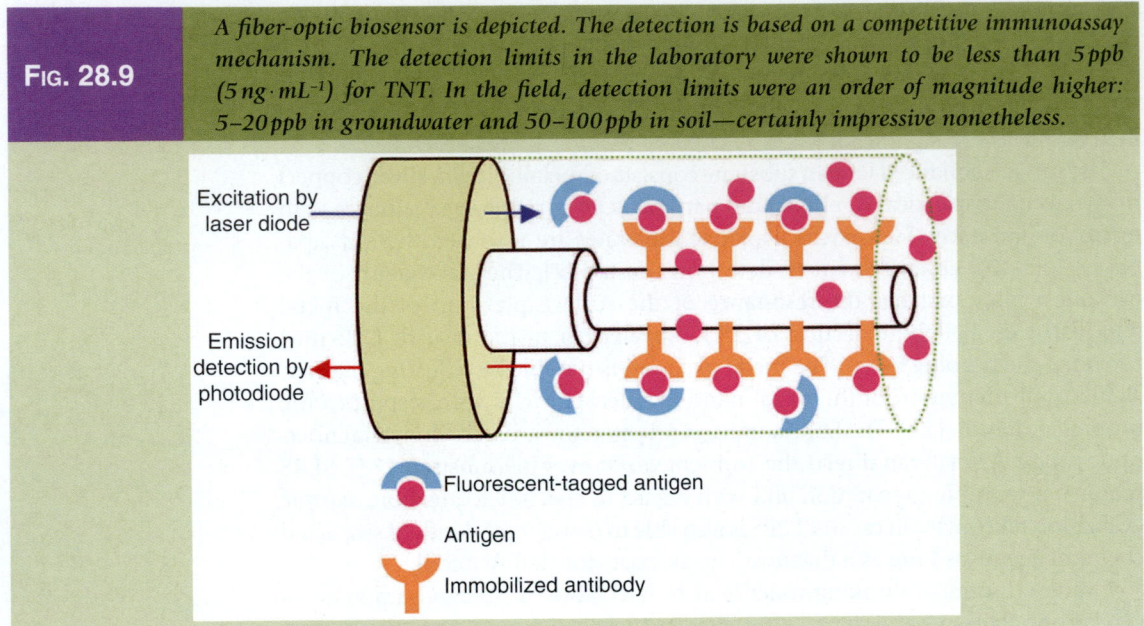

FIG. 28.9 *A fiber-optic biosensor is depicted. The detection is based on a competitive immunoassay mechanism. The detection limits in the laboratory were shown to be less than 5 ppb (5 ng·mL⁻¹) for TNT. In the field, detection limits were an order of magnitude higher: 5–20 ppb in groundwater and 50–100 ppb in soil—certainly impressive nonetheless.*

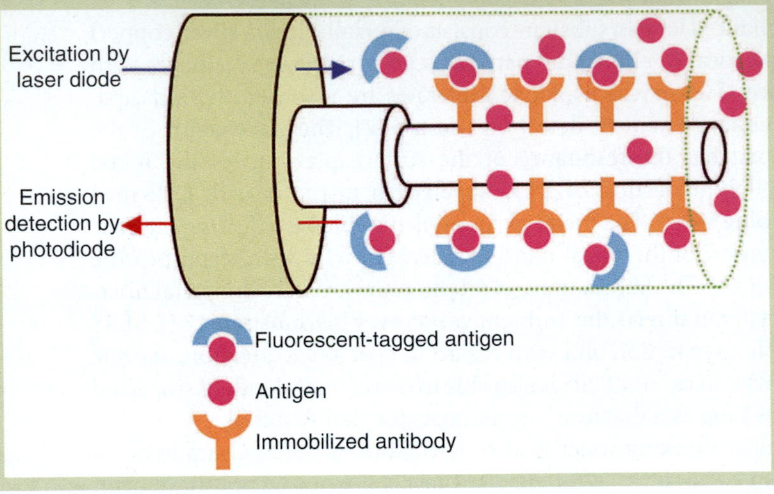

Source: Image courtesy of Naval Research Laboratory.

A continuous flow immunosensor (CFI) also utilizes immobilized antibodies and fluorescent dye marked signal molecules. Signal molecules similar to the analyte are tagged with a fluorophore (an organic dye). Analyte molecules are detected as they displace the fluorophore, thereby reducing the fluorescent signal. As opposed to the competitive type of immunoassay, the CFI technique is an example of a displacement-type immunoassay [74].

Chemical and Biological Agents and Nanomaterials. Chemical warfare *nerve agents* include VX (methylphosphonothiolic acid: LD-50 = $15\,mg \cdot kg^{-1}$ body weight), GB (sarin: LD-50 = $0.01\,mg \cdot kg^{-1}$ body weight; 26x more deadly than cyanide, 0.5 mg lethal dose for adults), GD (soman), GF (cyclosarin), and GA (tabun). *Blister or vesicant agents* include HD (sulfur mustard yperite), HN (nitrogen mustard), L (lewisite), and CX (phosgene oxamine). *Choking agents* include CG and DP (phosgene and diphosgene), Cl (chlorine), and PS (chloropicrin). *Blister agents* like mustard gas irreversibly alkylate selected amines in proteins, enzymes, and DNA that leads to cell dysfunction and eventual death, depending on the level, of the organism [75]. Most nerve agents are based on phosphonic acid. Nerve gases bind to active centers of phosphatase enzymes such as acetylcholinesterase that are necessary for nerve impulse transmission [75]. Selected chemical structures are shown in **Figure 28.10**.

Successful mitigation of chemical agents requires the following: (1) that formation of nontoxic by-products, (2) that decontamination is accomplished quickly, (3) that procedures and materials are applied easily, (4) that detox materials are not corrosive or hurtful to nearby surfaces, (5) that mitigation be accomplished quickly, and (6) that waste products be disposed of safely and efficiently [75]. Mitigation proceeds by physical methods such as adsorption or catalytic reactions that involve hydrolysis or oxidation.

FIG. 28.10 *Chemical structures of various warfare agents.*

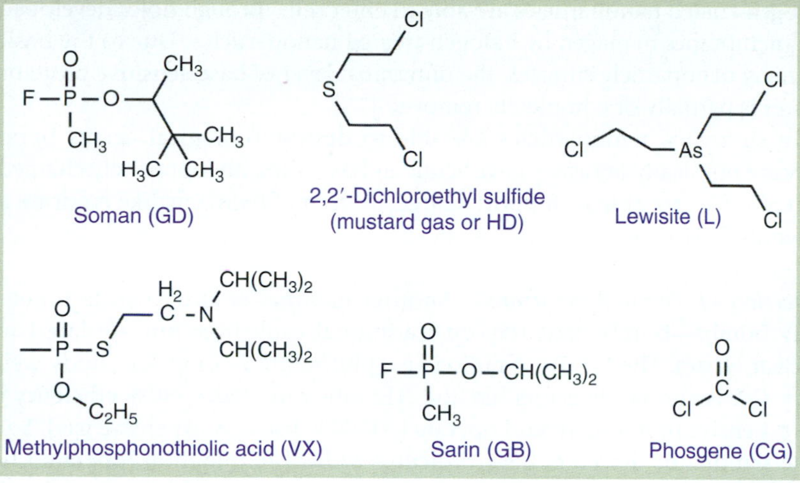

Traditionally, activated carbons adsorb warfare agents quite well although the adsorbate remains in a toxic form. Engineered nanoparticles offer the same surface advantage as do activated carbons but in addition are able to neutralize the toxins via chemical action. VX, GB, DP, and HD have been shown to hydrolyze at room temperature after contact with nanosized Al_2O_3 [76]. The by-products of the alumina-catalyzed reaction are harmless surface-bound nontoxic phosphonates or aluminophosphonates like $Al[OP(O)(CH_3)OR]_3$ [76].

Magnesium oxide nanoparticles possess basic sites that interact with HD by hydrohalogen elimination.

Nerve gases like VX can be neutralized by nanosized MgO to form hydrolysis products. Technological challenges facing catalytic procedures involve keeping the catalyst surface from clogging and aggregation of the catalyst. Aluminas possess a high level of Lewis acid sites that enhance hydrolytic processes. In this case however, the surface of the alumina can be eroded by the chemical agents during reaction. The erosion behavior keeps the surface fresh and active, and this is one of the reasons nanoscale aluminas are expected to contribute favorably to warfare agent mitigation [75].

Biological warfare agents include botulin toxin (LD-50 = 0.001 $mg \cdot kg^{-1}$ body weight), diphtheria toxin (LD-50 = 0.10 $mg \cdot kg^{-1}$ body weight) and anthrax toxin (LD-50 = 0.004–0.02 $mg \cdot kg^{-1}$ body weight) [60]. Traditional methods to address biological toxins include disinfection with chlorine and bleach (chloramine solutions). These materials, however, are quite corrosive and quite toxic themselves. Oil-in-water emulsions containing antitoxins are effective against bacteria but are difficult to dispose and are ineffective against airborne pathogens [75].

One of the best ways to mitigate biological pathogens is by application of nanopowders like MgO, ZnO, and CaO [75]. Nanoparticles have incredibly high adsorption capacity toward potential bactericides like chlorine bromine, or iodine and are able to store them on the surface until required. Halogen-loaded MgO, for example, becomes positively charged in solution—a perfect scenario for attracting bacteria that are negatively charged [77]. Laser confocal microscopy data showed that Br- and Cl-decorated MgO nanoparticles coagulate spontaneously with bacteria to form clumps [77]. In the case of anti-fungal treatment, halogen-coated nanoparticles are able to enter cells through holes developed in the membranes damaged by halogen-treated nanoparticles. Due to the basicity of many nanoparticle surfaces, the outermost layer of base-sensitive proteins of spores is partially or completely removed [75].

In summary, nanoparticles are able to destroy biological agents because they are physically abrasive, have acidic or basic sites, are oppositely charged to bacteria, and are able to support a large amount of oxidizers like chlorine and bromine [75].

Detection of Thermal Neutrons. Another national security risk is posed by dirty bombs—bombs that rely on traditional explosives but are laced with nuclear wastes. The LD-50 of radioactive plutonium is $1\,mg \cdot kg^{-1}$ body weight [60]. Conventional detectors like the ^{3}He-tube can attain >80% efficiency but are expensive to maintain and operate (>1000 V, leakage, expensive gas). Solid-state scintillator detectors have difficulty with discrimination, operate under high voltages, and longevity of the photomultiplier tube is questionable. Another kind of detector consists of polycrystalline boron nitride materials and is able to convert neutrons to alpha particles and generate an electrical signal. The problem is that it is not single crystalline. A recently developed detector employs ^{10}B pillars embedded in silicon—P–I–N diode pillars (doped-p/intrinsic undoped layer/doped n-layer) that are 2-μm width and spacing and 50-μm in height [78]. Detector efficiencies on the order of 65% have been reported and better results are expected with smaller features [78]. Advantages over other kinds of detectors are that the pillar detector is three-dimensional and that the interstitial converter material is in very close proximity. Theoretical efficiency can be 70% with the current configuration but better results can be obtained if the device geometry is rendered in the nanoscale [78].

28.0.6 The Nano Perspective

There are always positive factors associated with any technology. There are always negative factors. There are always factors that lie somewhere in between in the grey area. Positive impacts of nanotechnology are expected to be numerous and varied. According to T. Masciangoili et al. of the Environmental Protection Agency (EPA, 2004), nanodevices will require less material upfront, during production and as a by-product of manufacturing processes [79]—the case for dematerialization (reduction of the need for raw and manufactured materials), minimization of industrial wastes and effluents, and reduction of toxins entering the environment. Nanotechnology is expected to exert positive societal changes that benefit the environment by enhancement of vehicular transportation systems, urban planning and development, and information management. A significant impact of nanotechnology on the environment is expected to come in the form of green manufacturing, pollution prevention, treatment, mitigation and remediation, and the development of sensors [79].

Pollution can be mitigated a priori by a few general common sense practices: (1) source and resource reduction and conservation, for example, the less material one works with, the less potential there is for pollution; (2) process efficiency, for example, the more efficiency that is built into a process with regard to water use, energy input, raw materials, etc., the less material is wasted or produced that is polluting in nature; (3) effective postprocess pollution mitigation, for example, highly sensitive and efficient "end-of-pipe" means of reducing pollutants before they enter environmental conduits; and (4) substitution of toxic substances (i.e., substances that exert detrimental effects at low levels) with those that are benign to the environment, health, and safety. These factors are relevant for any kind of material regardless of size and any kind of technology regardless of kind. Two of the grand challenges issued by the National Nanotechnology Initiative (NNI, at nano.gov) involve environmental themes: (1) efficient energy and storage, and (2) nanoscale processes for environmental improvement.

The Extreme Importance of Gold and Silver Nanoparticles. According to the American financier Bernard M. Beruch, in the early 1900s

> Gold has worked down from Alexander's time. When something holds good for two thousand years I do not believe it can be so because of prejudice or mistaken history.

Although gold is a coveted commodity worldwide, its value in the nanoworld is ever increasing exponentially. Gold's nanoscale properties have proven to be quite remarkable, especially in its application as a catalyst. David T. Thompson of the World Gold Council [51–53]

> Many of the applications of gold are based on its unique properties, and this uniqueness can be rationalized in terms of the large relativistic contraction of its 6s orbitals resulting in a very small atomic radius compared with that which might be expected from the position of gold in the periodic table…

Although we focus on environmental issues in this chapter, gold's importance in all of nanotechnology cannot be understated. Christopher W. Corti, consultant to the World Council, states that [53]

> The gold industry wants to see increasing demand for gold in industrial and medical applications and has recognised that it must make a commitment to new science and its exploitation if that objective is to be achieved. We also live in an age where protection of the environment and the deleterious impact of global warming are issues of international importance. These two issues are central to the gold mining industry's aim to develop important new industrial applications for gold.

And guess what? Much of this new science is centered on the applications of nanogold. Gold is yet again proving that it is one of the most remarkable (and valuable) of materials—whether in bulk or nanophase! Nanosilver materials have made major environmental impacts in air purification, disinfection, and odor mitigation technology. The importance of nanosilver will be discussed in more detail later on in the text.

Nanotechnology Impact on the Environment. We are beginning to realize how nanotechnology and nanomaterials will be able to mitigate environmental insults and reduce the amount of industrial materials, but what of the nanomaterials themselves? New technologies bring along new benefits and new consequences. Nanotechnology is no different. Research must stay ahead of potential negative environmental impacts of nanotechnology—especially with regard to human health and safety where a few parts per million could exert toxic or even lethal effects. We must understand the transport, transformation, and fate of nanomaterials in various environmental media. We must learn from the impacts of earlier technologies and understand how nanomaterials would be different with regard to environmental mobility, reactivity, and accumulation from earlier technological materials [80].

Nanomaterials are smaller and are therefore able to penetrate regions otherwise forbidden to larger particles and are quite reactive due to enhanced surface-to-volume ratio. Many nanomaterials, for example, are able to penetrate skin. The size and surface (and surface chemistry) of nanomaterials allow for facilitated transport, reactivity, and bio-uptake (e.g., endocytosis, membrane penetration, or transmembrane channels) [80]. Nanomaterials can be hydrophilic or hydrophobic.

Environmental transport of nanomaterials may occur in air, surface water, groundwater, or soils. Gravity, inertial forces, and buoyant forces (like turbulence) affect the dispersion of large particles (>100 nm). The transport of nanoparticles, on the other hand, may occur primarily due to diffusion processes [81]. Nanomaterials, therefore, can mix and disperse quite rapidly and be subjected to the movement of air and water. It also must be kept in mind that smaller particles are subject to forces originating from electrical charge, van der Waals attraction, and diffusion (as stated before). Agglomeration is always a possibility with nanomaterials. Fullerenes like C_{60}, typically insoluble in water, form colloids that remain suspended in water, and therefore are able to persist [82].

28.1 Water and Soil Quality, Monitoring, and Mitigation

Water covers over 70% of the Earth's surface of which 0.3% is fresh [83]. Seventy percent of the fresh water is used for agricultural purposes. Energy demand contributes to ca. 50% of the cost of desalination. There is nothing more critical to maintain life processes than water. The percent of water in our bodies ranges from 60% to 80% depending on our age. We get our daily requirement, at a level that varies greatly from person to person, from water directly, from foods and from drinks like tea, coffee, and juices. Where do we get our water? Well, we get it from the environment—from rivers, rainwater, snow, groundwater, lakes, and the oceans. All these sources are susceptible to pollution from human activity.

Pollution (from the Latin *pollutionem* "defilement" based on the Greek *lyma* "filth, dirt, disgrace," *lymax* "rubbish, refuse") is the introduction of materials into the environment that cause harm to human health, other living organisms, and the environment itself. Pollution occurs when substances exceed natural levels and have detrimental effects—in the form of aesthetics, and health and sustainability of ecosystems. Water pollution comes in many forms. Pollutants arise from industrial processes, mining, agricultural runoff, domestic sources, hospitals and just about every other anthropogenic source. Pollution also arises from natural sources like volcanic eruptions, floods, windstorms, and biological phenomena. Pollutants take the form of metal cations, particulates, viruses, organic chemicals, radiation sources, or other species that simply do not belong in natural or treated water supplies. Once in water, pollutants are exposed to great environmental mobility, depending on their size, concentration, solubility, and chemical characteristics.

So long as populations were relatively disperse, water pollution was not an issue. With the onset of civilization, pollution soon followed wherever cities were founded. Accumulation of sewage was one of the first water pollution concerns to confront humanity. Some cultures dealt with it efficiently like the Hindus and Israelis (due to adherence to strict religious codes) and the Romans (due to efficient engineering capability). Indus River valley civilizations developed city-wide sanitation and codes of hygiene as early as 2500 B.C. The Romans built *cloaca maxima* (the great sewer) in 500 B.C. Pollution from waterborne heavy metal contaminants were known to the ancient Romans. In 1388, the British Parliament passed an act forbidding the disposal of garbage into ditches and rivers. In 1690, Paris becomes the first city to install an extensive sewer system. Before the advent of carbon nanotube water filters [83,84], humans resorted to less sophisticated methods to cleanse their water.

28.1.1 Traditional Water Treatment

Understanding the factors involved in drinking water quality was not well understood by the ancients although there is evidence that water treatment in the form of filtering through charcoal, exposure to sunlight, and boiling and straining were accomplished by ancient Sanskrit cultures 6000 years ago and the ancient Greeks [85]. The driving force behind water treatment in early times was to reduce the level of turbidity. Although the historical record shows that aesthetic issues such as water odor, appearance, and taste have been noted, the notion of water quality took thousands of years to develop [85]. The Egyptians used alum hydrated aluminum potassium sulfate, $KAl(SO_4)_2 \cdot 12H_2O$ to force suspension of solids from solution. The practice of filtration reemerged in the 1700s and 1800s and it was not uncommon to filter water slowly through sand [85]. Fine sand can be 60 μm or less in diameter—not quite a nanomaterial. The link between disease and waterborne microbes was made in the mid-1800s, and it was not until Louis Pasteur linked germs to disease that major sanitation operations were launched. Filtration was used to reduce turbidity and microbial contaminants such as cholera, typhoid, and dysentery [85].

In 1908, chlorine was used as a disinfectant in New Jersey as was ozone in Europe. The U.S. Public Health Service set drinking water standards in 1914. By 1974, the Safe Drinking Water Act was passed by Congress but new kinds of pollutants made the scene by the 1960s: industrial, agricultural, and synthetic chemicals and new treatments that involved aeration, flocculation, and activated carbon adsorption were reinvigorated. More than 800 organic and inorganic chemical species originating from industrial, agricultural, and municipal discharge have been identified in common drinking water [86]. Other estimates raise the level to 1500 or more [86]. Modern techniques designed to remove chemical species, many of them carcinogenic, include aeration, flocculation (coagulation and sedimentation), ion exchange, and activated carbon filtration. Application

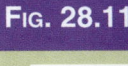

Fig. 28.11 *Traditional water treatment process is shown. Nanomaterials have already contributed to such methods. In the future, more solutions are expected to emerge from nanotechnology.*

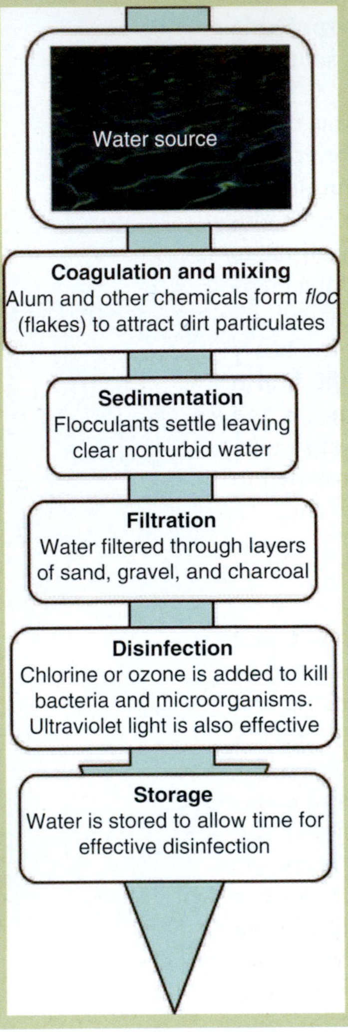

Water source
Coagulation and mixing
Alum and other chemicals form *floc* (flakes) to attract dirt particulates
Sedimentation
Flocculants settle leaving clear nonturbid water
Filtration
Water filtered through layers of sand, gravel, and charcoal
Disinfection
Chlorine or ozone is added to kill bacteria and microorganisms. Ultraviolet light is also effective
Storage
Water is stored to allow time for effective disinfection

Nanoscale considerations of water treatment

Colloid chemistry plays a major role in the coagulation and mixing phase of water purification. Flocculation is a process by which particles of sub-micron-sized clay aggregate into larger macrosized particles. Flocculation occurs as a result of a chemical reaction between the clay particles (the dirt) and flocculants (e.g., salt water or alum). Flocculants cause colloids, otherwise too small to be filtered, to aggregate. Flocculants usually contain multivalent cations—aluminum sulfate, calcium oxide, iron(II) sulfate, iron(III) chloride, sodium silicate, sodium aluminate, and aluminum chlorohydrate—or can be found in the form of polymers. Under the appropriate *p*H, temperature, and salinity conditions, positively charged flocculants, interact with negatively charged colloids thereby removing electrostatic barriers to aggregation. Flocculation (reversible aggregation?), coagulation (irreversible aggregation?), and aggregation are terms that are often used interchangeably. Van der Waals and stronger electrostatic forces play major roles in the flocculation process. According to Malvern Instruments, Ltd., suspended colloids in a stable colloidal phase can be removed from water by flocculation, coagulation (aggregation), sedimentation, and finally phase separation—in that order.

Nanomaterials play a role in the filtration process. Sand particles range in size from submicron to several micron dimensions. Sand particles may be coated with a monolayer to enhance separation properties. Charcoal consists of a porous structure of micro-, meso-, and/or macropores. Die to its special surface chemistry, charcoal has a natural affinity for organic materials and is able to remove them from aqueous media. Activated charcoal (or activated carbon) is usually reserved for the next level of water purification—found in the home or laboratory. GACs and PACs (granular and powered activated carbons, respectively) are effective at removing particulates and organics from water. PAC can be removed from water by coagulation, flocculation, and sedimentation. Removal or organics such as *p*-chlorophenol occurs via adsorption and diffusion into pores and onto surfaces of activated carbons mentioned earlier [87].

of activated carbon is the "best broad spectrum control technology available today" [86]. **Figure 28.11** illustrates a generalized scheme of water purification adopted by most municipalities in the United States.

The Clean Water Act. The Clean Water Act (CWA) of 1972 established guidelines to clean up America's waterways. The CWA (and specifically the Federal Water Pollution Control Amendments of 1972) is the primary federal law that regulates the quality of water in the United States. The act identified point sources such as industrial facilities, municipalities, and agriculture as targets for regulation. It set the standards for levels of toxic substances with the goal of eliminating additional water pollution by 1985. Over the years, several modifications to CWA were passed by a number of congressional sessions.

28.1.2 Nanomaterial Contamination in Aqueous Environments

Groundwater is water found beneath the surface of the Earth. Approximately one-third of all freshwater supplies originated from some form of groundwater in the United States. In Asia, the level of use is increased to over 50%, and in Europe the percentage is even higher. Groundwater sources are usually cleaner than surface water sources. Aquifers, the source of water for much of agricultural use, are replenished as part of the hydrologic cycles: rainfall, percolation, or directly by surface water recharging. Seepage from landfills, septic tanks, underground fuel storage tanks, road salts (ice melt), hazardous waste sites (over 200,000 abandoned sites in the U.S. alone), mine runoffs, agricultural runoffs, and poorly managed municipal treatment sites all combine to contaminate groundwaters. What are the contaminants? We are quite familiar with most of them by now in this section: VOCs, SOCs (synthetic organic compounds), HOCs (halogenated organic compounds), heavy metals, inorganic salts (anions and cations), radioactive materials, particulates and biological pathogens like bacteria, protozoans, and viruses, and associated biological toxins.

Natural Nanoparticles in Soil and Water. Natural soils consist of aggregations of particles composed of silicates (sand and silt), aluminosilicates (zeolites and clays), iron oxides, organics, and an assortment of biological materials—a veritable plethora of nanomaterials, aggregates, and conglomerations. Nanophase materials found in the soil are responsible for water percolation, purification, and water content in general. What happens to the natural nanomaterials in soils as we introduce synthetic ones?

Synthetic C_{60} and Carbon-Based Nanoparticles. According to researchers at Purdue University, C_{60} nanoparticles have little impact on the structure and function of the soil microbial community [88]. Naturally occurring oil samples were treated with 1 μg $C_{60} \cdot g^{-1}$ soil in aqueous suspension or 1000 μg $C_{60} \cdot g^{-1}$ soil in granular form and incubated for 180 days. Soil respiration byproducts and soil community structure were monitored [88]. Neither the fullerene or the application solvent caused any measurable decrease in the amount of organisms found in the soil. Neither were the levels of oxygen and enzyme activity affected, that is, under laboratory conditions [88]. In real world waterways and soils, exposure pathways may affect toxicity [88].

John D. Fortner et al. at Georgia Tech have compiled significant level of research concerning C_{60} and carbon nanotube effects in soils and water [89–93]. They showed that fullerenes tend to form a stable clump (as rectangular solids—**Fig. 28.12**) with the size of the clump in the range of 25–500 nm (called nano-C_{60}) and recommended further environmental modeling be conducted on not only C_{60} but the C_{60} aggregates as well. The size of the clumps is affected by *p*H, rate of addition to water, and the level of dissolved salts (that cause C_{60} to form a solid mass that settles on the bottom) [89].

Fortner et al. also found that aggregates inhibited prokaryote growth and decreased aerobic respiration [89]. In subsequent studies with *Escherichia coli* (Gram-negative) and *Bacillus subtilis* (Gram-positive) bacteria, they demonstrated that C_{60} associated strongly with both kinds of bacteria and displayed antimicrobial

FIG. 28.12 *Agglomerated C_{60} in water-soil samples.*

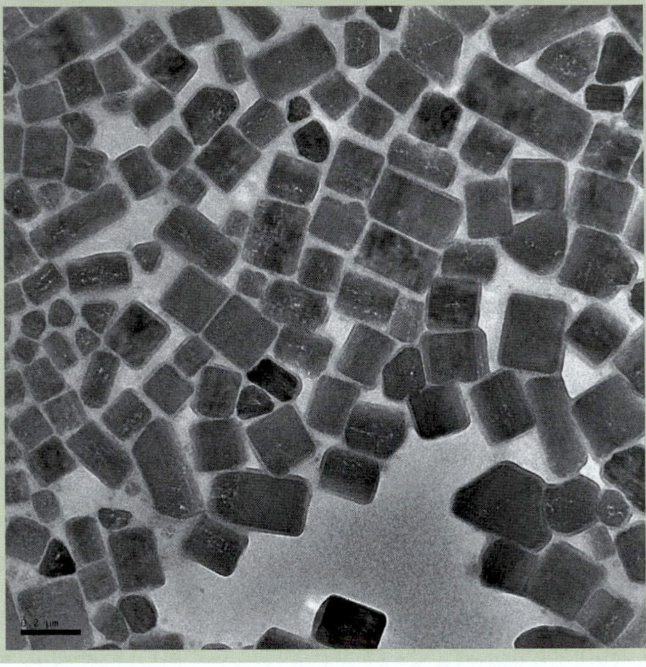

Source: Reprinted with permission from Georgia Institute of Technology: Professor Joseph Hughes, Dr. John Fortner, Rice University: Dr. Joshua Falkner and Professor Vicki Colvin.

behavior [90]. As before, higher salt concentration inhibited or eliminated the antimicrobial properties of C_{60} [90].

Aggregation dissolution occurs with the formation of fullerene oxides—to the tune of an average of 29 oxygens per fullerene molecule arranged in repeating hydroxyl-hemiketal moieties [93]. C_{60} in aqueous phase was also shown to produce reactive singlet oxygen in the presence of UV irradiation [91]. These results suggest that such aqueous phase reactivity to a strong oxidizer and ultraviolet radiation found in the environment should be considered when studying the transport, fate, and risk analysis of C_{60} materials [93]. Studies of the behavior of multiwalled carbon nanotubes in the presence of natural organic matter (NOM) revealed that MWNTs are readily dispersed. The individually dispersed MWNTs remained stable for over a month [92]. The authors concluded that the dispersal of carbon-based nanomaterials in natural, aqueous environments occurs better than expected and that this too must be considered when future transport, fate, and risk analyses are conducted [92].

Silver Nanoparticles. The effects of silver nanoparticles in soils and waterways are relatively unknown although nanosilver is a proven bactericide that is found in more and more consumer products—food packages, odor-resistant socks and clothing, household appliances, and now in wound dressings like Band-Aid [94]. Certain heavy metals in colloidal form and metal salts are known to be toxic to humans and environmentally hazardous to beneficial soil species and organisms such as fish, algae, crustaceans, some plants, fungi, and bacteria [95–99]. Silver nanoparticles are currently used in combination with TiO_2 (nanosilver-titanium dioxide coating, NS-TDC) in an effort to disinfect surfaces at train stations in Hong Kong—escalators, handrails, machines, elevator buttons, etc. [100]. Its manufacturers claim that the NS-TDC spray kills a variety of bacteria, mold, and viruses like the H_1N_1 flu virus [100]. It is not known how much silver is introduced into waterways due to application and erosion of NSTDC materials.

A new kind of energy-efficient washing machine, manufactured by Samsung, was introduced to Swedes in 2005. The machine makes use of silver nanoparticles at low water temperatures [94]. However, the Swedish version of the EPA is concerned that nanosilver will cause damage to water organisms and that it will be costly to remove silver particles from wastewater effluents. The Stockholm Water Authority claims that the washing machines are expected to dump two to three times more silver into waterways than the current level. Another potential problem facing the use of silver nanoparticles is that they may destroy essential bacteria required to treat sewage.

28.1.3 Activated Carbon—A Simple Traditional Nanotechnology

Activated carbon is a nanomaterial—a material with an enormous surface area per gram—on the order of 400–2000 $m^2 \cdot g^{-1}$. Activated carbon is composed of an "amorphous material" with microcrystalline structure—quasi-graphitic particulates—that consists of highly developed porosity and extended interparticulate surface area [86]. Quasi-graphite consists of stacked graphene sheets cross-linked in a random manner. The irregular arrangement generates interstitial spaces that give rise to a network of polydisperse pores of nonuniform size and shape [86]. The interlayer *d*-spacing between individual graphene sheets is larger (ca. 0.35 nm) than that of graphite (0.335 nm). ESR analysis revealed that the aromatic structures in activated carbon contain free-radical elements, and hence unpaired electrons at unsaturated edge carbons. The material, although stabilized by the resonance structure typical of graphene materials, renders the material reactive to heteroatoms such as hydrogen, oxygen, nitrogen, and sulfur. Therefore, typical activated carbons contain 88% C, 0.5% H, 0.5% N, 1% S, and ~7% O and the rest inorganic ash residues [86]. E.G. Furuya et al. showed the action of interparticle mass transport of phenol in activated carbons [87].

Activated carbon is manufactured by a two-step process: (1) pyrolysis of carbon-containing materials at 600–900°C in the absence of oxygen and (2) activation by oxidation in CO_2, O_2, or H_2O at 600–1200°C or higher. Pyrolysis of organic materials above 1800°C forms turbostratic graphite. Turbostratic graphite possesses only short-range order on the scale of a few to tens of nanometers. Activation by chemical means also occurs—in acids, bases, or materials like $ZnCl_2$ at 450–900°C. Activation in acidic media produces carbons with affinity for heavy metals but minimally so for chlorinated hydrocarbons—that are preferred if activation takes place in basic media (and heavy metals are not favored). Materials most often adsorbed in water industrial and municipal treatment facilities are free available chlorines (FACs), halogenated organics (HOCs like DDT, endrin, lindane, TCE and others), aromatics (benzene, toluene, dioxins, PCBs, and plasticizers), heavy metals (Pb, Cd, Hg as dissolved ions, colloidal oxides, or carbonates), and taste and odor compounds (T&O) produced by microbes [99].

Nanosized pores and pockets are formed in carbon particles that contribute to its tremendous overall surface area. The pores, spaces, or cavities depending on pretreatment conditions range from less than 2 nm (micropores that comprise ~95% of the surface area) to mesopores with diameter <50 nm and to macropores, in which the pore diameter is greater than 50 nm. Powdered activated carbon (PAC) is ca. 40 μm in diameter. Granulated activated carbon (GAC) is 400–600 μm in size.

Activated carbon, due to the small size of its features, adsorbs organic pollutants by means of van der Waals forces—a characteristic of nanomaterials and due to the random interactions of electron clouds and unsaturated valencies with target contaminants. X-ray studies have shown that heteroatoms are most likely found adsorbed onto edge sites [86]. The overall reactivity and catalytic properties of activated carbons are directly related to their nanostructure (adsorption capacity) and chemical nature (adsorptive species). Active sites, active centers, and attached heteroatoms determine the material's propensity for polar or non-polar adsorption. Regeneration of activated carbon is energy intensive.

Adsorption of contaminants takes place by two well-known processes—physisorption (van der Waals, 10–20 kJ · mol^{-1}) and chemisorption (sharing or exchange of electrons, 40–400 kJ · mol^{-1}). The adsorption profile is described by traditional isotherm formulae: Langmuir, Freundlich, or linear. In drinking water treatment, the most commonly used isotherm application is the Freundlich isotherm—a Type I isotherm in which only surface adsorption is considered and there is no interaction between one adsorbed molecule with another (e.g., no capillary condensation like in BET analysis).

$$\theta = K \cdot (C_{\text{equil}})^{n^{-1}} \tag{28.2}$$

where

K and n are empirical, experimentally determined constants (usually with $0 < n < 1$)

C_{equil} is the equilibrium concentration of the solute (equal to p_{equil} if in gaseous form at solid–gas interface)

θ is a measure of surface coverage in terms of

$$\theta = m_{\text{solute}} / m_{\text{adsorbant}} \tag{28.3}$$

where

m_{solute} is the mass of solute adsorbed

$m_{\text{adsorbant}}$ is the mass of the adsorbing material

If $n = 1$, then $\theta = K \cdot C_{\text{equil}}$; if $n \to \infty$, then $\theta \to 1$ as the limits. Plots of log θ versus log C_{equil} yield a linear relationship. In soil analysis, the values of K and n are dependent on soil properties such as pH, organic matter content, clays, aluminum, and iron oxides.

Adsorption capacities can be determined by preparing identical bottles of powdered activated carbon suspended in water. Different levels of a potential contaminant are added to the bottles and mixed until equilibrium is achieved. The aqueous adsorbent–solute mixtures are filtered and the concentration of remaining unadsorbed contaminant is determined. A plot of log θ versus log C_{equil} is then derived. The average line is the Freundlich isotherm. The Freundlich isotherm for a specific contaminant is then applied to determine the amount of contaminant adsorbed in terms of $mg \cdot g^{-1}$ of activated carbon. By convention,

EXAMPLE 28.1 *Freundlich Isotherm Adventures*

Data were collected from different bottles of granular activated carbon (GAC). Varying concentrations of trichloroethylene (TCE) were added to each, C_{equil}. After filtering, the remaining TCE, the unadsorbed concentration was calculated and the adsorbed amount was determined in $mg \cdot g^{-1}$ carbon capacity. The resulting data is shown in **Table 28.4** below. Derive a Freundlich plot (**Fig. 28.13**). Is it economically reasonable to remove TCE from water using GAC adsorbents?

TABLE 28.4 *Isotherm Data*

	1	2	3	4	5	6
C_{equil} TCE ($mg \cdot L^{-1}$)	0.0010	0.0030	0.030	0.30	0.50	1.0
log C_{equil}	-3.00	-2.52	-1.52	-0.523	-0.301	0.00
θ, Capacity ($mg \cdot g^{-1}$)	0.40	0.75	3.4	15	20	30
log θ	-0.400	-0.125	0.531	1.18	1.30	1.48

FIG. 28.13 *Freundlich isotherm derived from data in* Table 28.4.

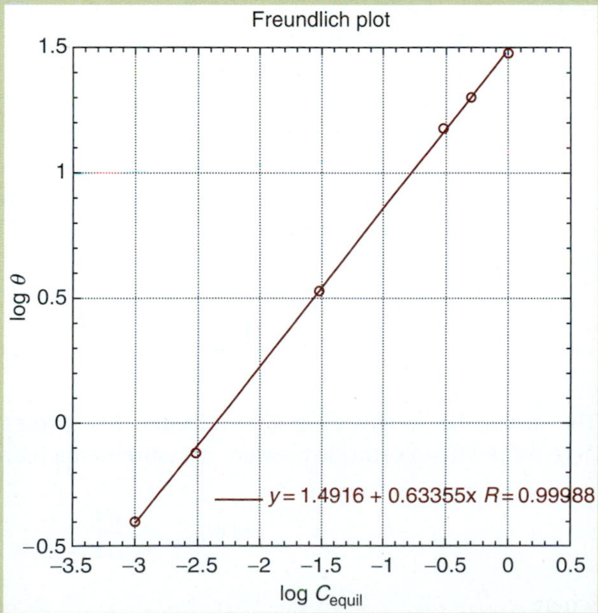

TCE is much better adsorbed than is chloroform. Therefore, TCE can be removed economically from water streams by GAC adsorbents [99].

capacity values ($mg \cdot g^{-1}$) corresponding to $1.0\,mg \cdot L^{-1}$ are used to compare adsorptions of different contaminants [101].

In industry practice, chloroform is one of the least absorbable materials by activated carbons ($2.6\,mg \cdot g^{-1}$ @ $1.0\,mg \cdot L^{-1}$). Therefore, any material with a Freundlich isotherm value less than that of chloroform is considered to be unadsorbable by activated carbon [101]. In industrial practice, activated carbons are used as prefilters in many applications, specifically in reverse osmosis.

28.1.4 Membranes and Separation Technology

Water purification, desalination in particular, is accomplished by a few tried and true techniques—*reverse osmosis* (RO), electrodialysis, and distillation. Before we describe reverse osmosis, let us describe the phenomenon of osmosis first. Osmotic pressure is a colligative property of a solution and is a function of the number of solute particles per unit volume. Osmotic pressure, however, is independent of the molecular composition or shape of the solute. Osmotic pressure is often used to determine the molecular weight of macromolecules. Osmosis is the diffusion of water through a cell membrane or semipermeable membrane (only allows water to transport) from a solution of low solute concentration to that of a higher solute concentration [102]. Osmotic pressure is the hydrostatic pressure that is generated by differential concentrations across a semipermeable membrane [102,103]. The process of osmosis is pressure driven. Osmotic pressure Π is defined (in the ideal limiting case) as

$$\Pi = iMRT \tag{28.4}$$

where

i is the van't Hoff factor (the number of moles of solute in solution)
M is the molarity (concentration) of the solute
R is the gas constant equal to $0.082\,L \cdot atm \cdot mol^{-1} \cdot K^{-1}$
T is temperature in Kelvin

In cells, isotonic solutions have $\Pi_{sln} = \Pi_{cell}$; in hypertonic solutions, $\Pi_{sln} > \Pi_{cell}$ (e.g., cells shrink), and in hypotonic solutions, $\Pi_{sln} < \Pi_{cell}$ (cells swell). Another form of this equation is often used

$$\Pi_{osmosis} = \frac{n_{solute}RT}{V} \tag{28.5}$$

The molecular weight of macromolecules like proteins and polymers can be determined from osmotic pressure measurements (osmometry) and applying

$$MW_{solute} = c\frac{RT}{\Pi} \tag{28.6}$$

where

MW_{solute} is the molecular weight of the solute in $g \cdot mol^{-1}$
c is the concentration in $g \cdot L^{-1}$.

Reverse Osmosis. Reverse osmosis (RO) is a process by which water is purified (e.g., increasing the concentration of solute on one side) against the natural chemical potential of the solvent/solute system. It requires energy to do so [104]. The nanotechnology involved in RO processes is confined to the chemistry and structure of the semipermeable membrane. Some RO membranes make use of specially coated 20-nm polyamide nodules (with polysulfone as a porous layer support) under high pressure to overcome natural osmotic pressure. Although adapted by industry to treat waters with a wide range of salinity and brackishness, the process is energy intensive and RO membranes have relatively low throughput. Reverse osmosis processes require significant energy, on the order of 1.5–2.5 kW-h of electricity to produce one cubic meter of fresh water. *Electrodialysis* employs an electric potential to force ions through a highly resistive membrane but does not work well with seawater substrates. *Distillation*, the most energy-intensive method, is one we are all quite familiar ($1000\ kW \cdot AF^{-1}$ to separate H_2O from its salts and $800{,}000\ kW \cdot AF^{-1}$ just to overcome the heat of vaporization, AF = acre · foot, 326,000 gallons). When measured in terms of $MJ \cdot cm^{-3}$ (not considering distillation which has prohibitively high energy demand), desalinating water with 20,000 ppm salt content requires $\sim 20\ MJ \cdot cm^{-3}$ by electrodialysis and $\sim 15\ MJ \cdot cm^{-3}$ by RO. An optimal level of energy use considered to be is in the neighborhood of $\sim 10\ MJ \cdot cm^{-3}$.

Mass transport of a solvent through a semipermeable membrane driven by a difference in chemical potential on either side of the membrane is called *osmosis* (from the French *endosmose* "inward passage" on the Greek *osmo* "pushing"). The chemical potential of a solvent decreases with solute but increases with pressure. The process of osmosis strives to create a balance between the two. The solvent flows from the solution with the lower solute concentration (greater μ_{slv}) into the one with the higher solute concentration (lesser μ_{slv}).

$$\mu_{\text{solvent}} = \left(\frac{\partial G}{\partial n_{\text{solvent}}} \right)_{T,P,n_{\text{solute}}} = \mu^{o}_{\text{solvent}}(T,P) - k_B T \left(\frac{n_{\text{solute}}}{n_{\text{solvent}}} \right) \tag{28.7}$$

In reverse osmosis, the solvent is forced to flow from the solution with higher solute concentration into the solution with low or no solute concentration. A reverse osmosis configuration is shown in **Figure 28.14**.

RO is a separation process in which solutes are retained and concentrated on one side of a semipermeable membrane while fresh water is forced through the membrane on the other side. Problems associated with RO membranes in addition to energy demand and low throughput include fouling by bacteria and colloidal particles; compaction due to exposure to high pressure that rearranges the polymeric elements; and chemical decomposition, membrane defects and poor sealing, and oxidation and degradation of matrix elements—polyamide by Cl^- and cellulose acetate from *p*H below 3.5 or higher than 7.5. Categories of RO systems depend on pore size and composition: *particle filtration* removes particles >1 μm, *microfiltration* for particles >50 nm, *ultrafiltration* for particles >3 nm, *nanofiltration* for particles >1 nm, and finally *hyperfiltration* for the removal of sub-nanometer particles [104].

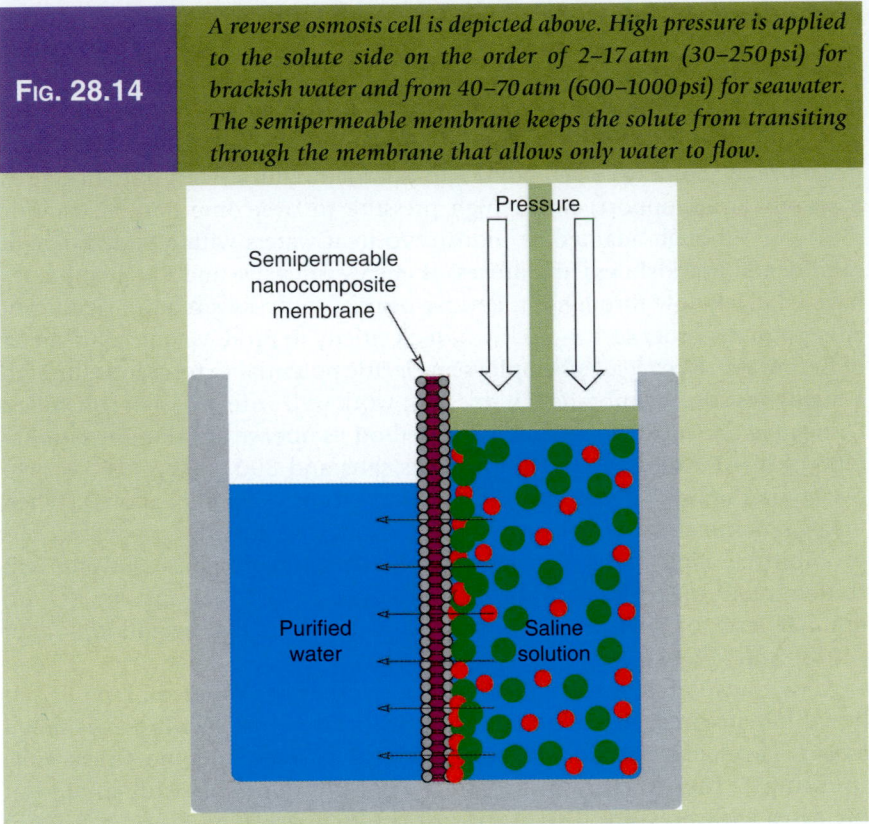

FIG. 28.14 *A reverse osmosis cell is depicted above. High pressure is applied to the solute side on the order of 2–17 atm (30–250 psi) for brackish water and from 40–70 atm (600–1000 psi) for seawater. The semipermeable membrane keeps the solute from transiting through the membrane that allows only water to flow.*

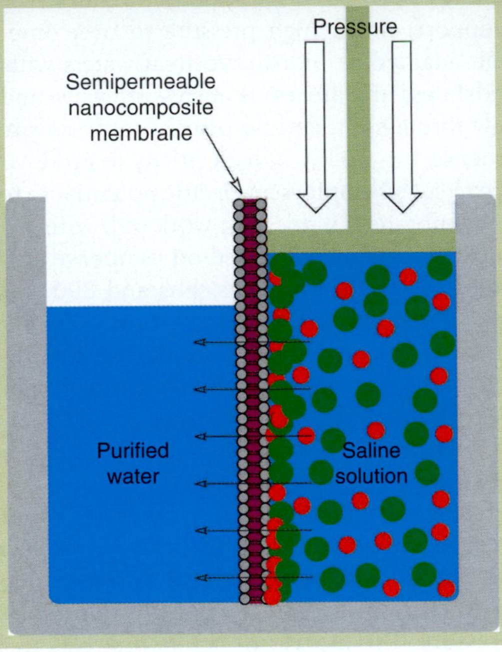

RO and Nanoparticles, Proteins, and Colloids. Oncotic pressure is a form of osmotic pressure and is a characteristic of the circulatory system. The walls of capillaries in the circulatory system are semipermeable—permeable to water but not to plasma proteins. Therefore, an osmotic pressure is generated. There is also a tendency for proteins (mostly negatively charged) to attract cations

EXAMPLE 28.2 *Desalination and Reverse Osmosis*

What pressure is required to desalinate water with sodium chloride concentration of 0.08 M at 25°C?

Solution:

Apply $\pi = iMRT$

Sodium chloride dissociates into sodium cations and chloride anions in solution

$$NaCl \rightarrow Na^{+} + Cl^{-}$$

Therefore, $i = 2$.

Concentration is equal to 0.08 M, $T = 298\,K$, and $R = 0.082\,L \cdot atm \cdot mol^{-1} \cdot K^{-1}$.

$\pi = iMRT = (2)(0.08\,M)(0.082\,L \cdot atm \cdot mol^{-1} \cdot K^{-1})(298\,K) = 3.9\,atm$ pressure required to purify water with NaCl concentration of 0.08 M.

(the Gibbs–Donnan effect) that contributes to the osmotic gradient between the plasma and the interstitial fluids. Because of these two effects, water is drawn from the interstitial regions into the plasma. The resulting pressure is called *colloid oncotic pressure*. The pressure is proportional to the difference in protein concentration between the plasma and the interstitial fluid. The hydrostatic pressure in the capillary causes fluid to leave the plasma (e.g., flow) and the oncotic pressure tends to pull it back into the capillary.

Semipermeable Nanocomposite Membranes. A new semipermeable nanocomposite membrane developed by Erik M.V. Hoek et al. of UCLA is capable of producing highly purified water by high-pressure reverse osmosis [105]. The membrane is comprised of specially synthesized nanoparticles dispersed throughout a porous polymeric matrix. The membrane is designed to enhance water flow at the expense of contaminants such as dissolved salts, organics, and bacteria. The surfaces of the nanoparticles are rendered hydrophilic. Repelling organic materials and bacteria prevent clogging of the membrane, a problem associated with current technology.

Carbon Nanotube Water Filters. Researchers at Lawrence Livermore National Laboratory have developed a membrane on a silicon chip that is the size of a quarter that has potential applications in the areas of desalination, demineralization, and purification of water [83,84]. The membrane consists of single- and double-walled carbon nanotubes 1–2-nm in diameter and a silicon chip (**Fig. 28.15**). Flow of contaminants is restricted through the hollow pipeline of the nanotube while the flow of liquids and gases is allowed. Molecular dynamic simulations have predicted that fast flow through carbon nanotubes is likely [83]. According to Olgica Bakajin and Alexandr Noy, membranes containing millions of aligned carbon nanotubes, formed by chemical vapor deposition (CVD) on a silicon chip, have incredibly smooth interior walls [83]. The challenge lies in the fabrication of the membrane—specifically, to fill the gaps between the carbon nanotubes so that seepage is prevented. In order to overcome this obstacle, the nanotubes were coated with silicon nitride. Gap-free membranes were produced. Excess silicon nitride was removed and ends opened by reactive ion etching.

The performance of the membrane showed promise—the membrane was able to block ingress of 2-nm diameter gold nanoparticles while water was allowed through the nanotubes. This is a nano-example of a size-exclusion process. Remarkably, measured gas flow exceeds predictions of a Knudsen diffusion model by ca. 100x [83,84]. And more remarkably, water flow exceeds values calculated by continuum hydrodynamic models by three to four orders of magnitude—comparable to flow rates calculated by molecular dynamic simulations [83,84]. The membranes also outperformed commercially available polycarbonate membranes by several orders of magnitude despite having pore channels an order of magnitude less in diameter [83,84]. Enhanced water flow is explained by few theories: (1) the hydrophilic–hydrophobic relationship between the water molecules and the internal walls of the carbon nanotube results in frictionless flow and (2) confinement-induced ordering that results in the formation of "water wires of extremely long length" [83].

FIG. 28.15 *Carbon nanotube water filter membrane is depicted. Carbon nanotubes are shown at the top left. The diameter distribution of single-walled carbon nanotubes is shown in the bottom left. The membrane with filler is shown in the center of the image.*

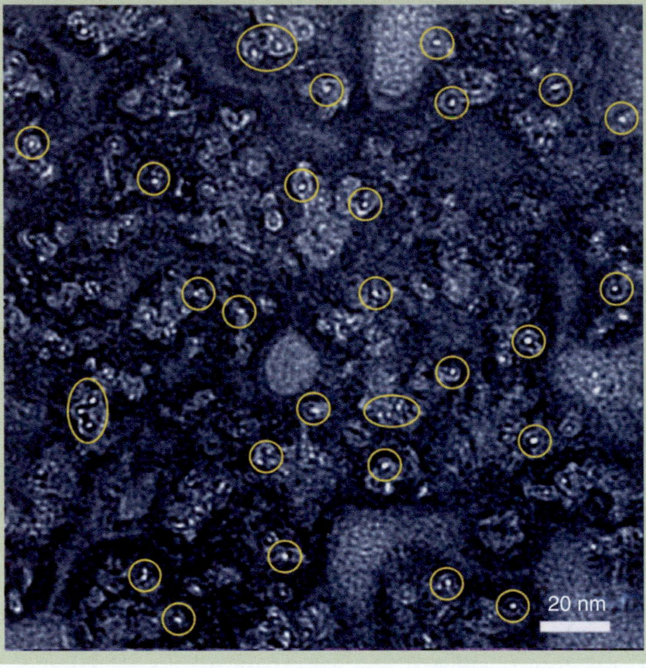

Source: J. K. Holt, H. G. Park, Y. Wang, M. Stadermann, A. B. Artyukhin, C. P. Grigoropoulos, P. Costas, A. Noy, and O. Bakajin, *Science*, 312, 1034–1037 (2006). With permission.

Carbon nanotube filters are expected to reduce the consumption of energy by as much as 75% compared to reverse osmosis [106]. H.G. Park et al. at the Lawrence Livermore National Laboratory also have developed an ion-exclusion method to purify water using carbon nanotubes that should offer high throughput along with cost-effective manufacturing [107]. The principal mechanism of synthesis is similar to the size-exclusion system described above in which open carbon nanotubes were fixed within a silicon nitride matrix except that in this case, the openings of the carbon nanotubes were modified chemically—functionalized with negative chemical moieties that repel anions but allow cations to pass. The permeability of such membranes has been shown to exceed that of a conventional RO membrane by a factor of 100x [107].

Chemically Modified Silica Particles. Peter Majewski and Chiu Ping Chan of the Ian Wark Research Institute at the University of Southern Australia have developed a water purification system based on specially coated silica nanoparticles, e.g., functionalized self-assembled monolayers, nanometers in thickness. The active nanometer coatings were shown to eliminate biological molecules, viruses, oocysts, proteins, bacteria, and waterborne parasites like *Cryptosporidium parvum* [108]. The pathogenic agents were removed at the *p*H of drinking water by electrostatic attraction and subsequent immobilization between the surface-engineered-silica (SES) and the pathogens. Stirring for 1 h and filtration remove the pathogen–silica complexes from the water [108,109].

Bacteria and Nanofilters. Bioremediation is the process of water purification by bacterial action. Treated water is filtered through a porous membrane with pore size ranging from 10 μm–1 nm [110,111]. N. Hilal et al. reported in 2008 that the new bioremediation membranes can be cleaned without removal from the medium and the waste products used for fuel—a definite improvement over conventional membranes that are easily fouled.

28.1.5 Oil Spills

There are many means of mitigating the effects of an oil spill via nanotechnologies. *Bioremediation methods*, a procedure that follows mechanical skimmers and other physical methods of cleanup, consist of "microbial cultures, enzyme additives, or nutrient additives" that are designed to operate in wetlands or shallow water [112]. The purpose of the additives is to boost the natural nanotechnology of the microbial community to decompose oil materials. *Chemical dispersants* made of surfactants and other supramolecular precursors keep oil from settling onto shores and sensitive habitats by emulsifying oil and rendering it soluble in water. The oil particulates remain afloat in a dispersed state and are made available for accelerated biodegradation [112].

Another method that is gaining acceptance is the use of *specially modified aerogels* decorated with hydrophobic moieties to enhance oil interaction. Such hydrophobic aerogels possess extremely high surface area, a property characteristic of nanomaterials, and are therefore able to absorb sixteen times their weight of oil [113]. The inorganic–organic hybrid aerogels incorporate trifluoromethylpropyl groups into the silicate backbone of the cross-linked aerogel during synthesis.

$$(CH_3O)_4Si + CF_3CH_2CH_2Si(OCH_3)_3 \rightarrow \text{aerogel} - CH_2CH_2CF_3 \qquad (28.8)$$

Once the oil has been absorbed, the solid oil-soaked aerogel can be plucked from the surface. Synthesis of aerogels however is not cost-effective and alternatives that utilize charcoal are under consideration [113]. *Magic sand*, another surface-modified silica material, is synthesized from sand coated with an organosilane monolayer—trimethylsilane. In this case, the oil adsorbs to the hydrophobic surface and the sand-monolayer-oil particles sink to the bottom where the oil can be recovered by dredging [114]. Although the sand particle is on the order of microns in size, the monolayer, depending on the silane, is less than a nanometer to a few nanometers in thickness.

The use of nanoparticulates in conjunction with a self-assembled monolayer (SAM) material perhaps shows the most promise. Interface Scientific Corporation developed such a material in 2005 but details about its structure are not available. The developers claim that the nanoparticles absorb 40 times its weight in oil—a level that exceeds any method to date—and the oil can be recovered. The combination of the nanoparticle and SAM produces a hydrophobic environment that excludes all water. The properties can be altered by altering the length and functional groups of the monolayer—and perhaps by the size of the nanoparticle—typical techniques involved in nanochemistry.

28.1.6 Chemical and Biological Sensors and Detectors

We have introduced chemical and biological detectors in an earlier section but we reiterate at this time their importance in environmental mitigation strategies—in order to mitigate a problem one needs to know how what kind and how much of a problem it is. Many kinds of pollutants exist in water supplies [6]: Pathogens that arise from human activity and natural sources; heavy metals like arsenic generated by geological conditions, mining runoff and industrial use; pesticides from aerial spraying and agricultural runoff; algal and bacterial sources of neurotoxins, hepatotoxins, and cytotoxins that originate in waters with high levels of nutrients (due to synthetic chemical fertilizers); nitrates from natural and agricultural sources; fluorides from water supply purification practices and geological sources; and organic compounds and halogenated organic materials from industry and transportation. There are many kinds of waterborne pollution that are toxic to living things—quite the challenge for nanotechnologists developing sensors to help detect them!

28.2 Air Quality, Monitoring, and Mitigation

Several factors and sources affect global and local air quality. Many sources are anthropogenic; others are naturally occurring. The U.S. Clean Air Act (CAA) was passed in 1963 followed by the Air Quality Act in 1967, the Clean Air Act Extension of 1970, and the Clean Air Act amendments in 1977 and 1990 [115]. The 1990 revision added statutes concerning emissions trading, acid rain, ozone depletion, and newly classified toxic air pollutants [115]. The CAA sets standards for emission of fine particulates and regulates hazardous chemicals such as volatiles and heavy metal vapors. The EPA's Office of Air and Radiation is currently assessing the impact of engineered nanomaterials and is in the process of determining whether or not nanomaterials should be classified as "potential criteria pollutants" under CAA Section 110, as "potential hazardous air pollutants" under section 112 or pursue other implications [116–118].

The U.S. Toxic Substances Control Act (TSCA) was originally passed in 1976 and is administered by the U.S. EPA. The law authorizes the EPA to secure information on new chemical substances—nanotechnology; however, it is not specifically a chemical substance in the traditional sense. According to the American Bar Association meeting in May 2007, TSCA needs to address the following [116–118]:

- What guidance should EPA provide to nanomaterial manufacturers to determine if a nanomaterial is a new chemical under TSCA?
- What regulatory alternatives are there for nanomaterials that do not qualify as new chemicals?
- Should EPA amend its regulations to address nanomaterials specifically?

- What materials should manufacturers provide to the EPA to help it conduct risk assessments?
- How does TSCA relate to international standards for nanomaterials?

Atmospheric Nanoparticles. Particles found in the atmosphere come in a variety of sizes—ranging from particles as large as 10 μm to 1 nm with the most common sizes detected near 5, 50, and 300 nm [119]. Both natural and synthetic processes are responsible for the production of atmospheric nanoparticles, but anthropogenic sources contribute significantly more to fine particle fractions [120]. Combustion processes produce fly ash. Atmospheric particles impact air quality by reducing visibility, and catalyzing cloud formation and radiation forcing—a process that affects the ratio of absorbed and reflected sunlight. The impact of nanoparticles is both mitigated and acerbated, depending on your local and the prevailing winds, by traveling long distances from one continent to another [119]. Particles are removed from the atmosphere by gravity (via coalescence), diffusion, and rain but residence times can last from minutes to days or longer.

Nanoparticles are capable of reacting chemically or catalyzing chemical reactions while airborne, e.g., the conversion of SO_2 and NO_2 gases into sulfuric and nitric acids, respectively, or depleting ozone by catalyzing the formation of chlorine compounds [119].

28.2.1 Gas Separation: Advanced Membrane Technology

Membrane-based gas transport and separation are accomplished by several kinds of membranes that possess altered chemistry and geometric configuration. Depending on the prevailing pressure (from vacuum to high pressure), mean free path factors determine the different types of mass transport through membranes.

Viscous Flow. Viscous flow occurs when the mean-free path λ_{mpf} of the gas atom or molecule is significantly less than the diameter of the pore channel. Flow patterns are determined by the collisions of the molecules. No effective separation occurs in this scenario and bulk flow of gas proceeds very much like it does for a liquid (turbulent or laminar), relatively unimpeded through pore channels. The pressure domain characteristic of viscous flow is between 10^{-1} and 10^3 torr. In *molecular (Fickian) diffusion*, the mean-free path is much less than 0.01 d_{pore} and follows Fick's law of diffusion and the pressure domain lies between 10^{-6} and 10^{-3} torr. The transport diffusivity relates the macroscopic flux of the gas species to the concentration gradient. This type of diffusion is like Brownian motion in which the movement of the gas molecules is random and not a function of its previous motion.

Knudsen (or Transition) Flow. This is based on the inverse square root ratio of molecular weights of gas species—in this case, the mean-free path of the gas molecules is greater than 1/100th of the diameter of the pore ($0.01\ d_{\text{pore}} < \lambda_{mpf} < 1.00\ d_{\text{pore}}$). Collisions with the pore wall occur frequently. Knudsen flow dominates when the pore diameter is between ca. 2 and 50 nm. The Knudsen number Kn is the ratio of the mean-free path to the pore diameter

$$\text{Kn} = \frac{\lambda_{mfp}}{d_{\text{pore}}} \tag{28.9}$$

The transport equation of Knudsen diffusion of species i is given by J_i:

$$J_i = \left(\frac{\varepsilon}{d_{\text{mem}}\tau}\right)\left(\frac{D_K}{RT}\right)\Delta P_i \tag{28.10}$$

where

ε is the porosity (volume pores/volume membrane)
d_{mem} is the thickness of the membrane
D_K is the Knudsen diffusivity
R is the gas constant
T is the absolute temperature
ΔP is the pressure differential across the membrane
τ is the tortuosity

$$D_K = \frac{2}{3} r_{\text{pore}} \sqrt{\frac{8RT}{\pi M_i}} \tag{28.11}$$

where

r_{pore} is the radius of the pore channel
M_i is the molecular weight of the gas in terms of $g \cdot mol^{-1}$

The enrichment factor of the mass transport between two gases is the ratio of the squares of their respective molecular masses.

$$\text{Enrichment}_{ij} = \frac{J_i}{J_i} = \sqrt{\frac{M_i}{M_j}} \tag{28.12}$$

Ultra-Microporous Molecular Sieving. This depends on diffusion rates of molecules, for example, with small molecules diffusing quicker than larger ones, but also depends on specific adsorption differences between similar-sized molecules like O_2 and N_2. These membranes may have a tortuous but continuous network of channels.

Solution-Diffusion Separation. This mode of separation occurs through membranes that are essentially solid (with no continuous passages—nonporous polymeric membranes like PDMS, PS, etc.) and is based on solubility and mobility factors of gas molecules [121]. Passage of gas through such membranes occurs due to thermally agitated motion of polymer chain segments that form transient penetrant-scale gaps—driven by diffusion phenomena [121]. Pressure and concentration gradients provide the driving force with working pressures in the neighborhood of 100 atm. Metallic membranes (Pd and Ag alloys) break down H_2 on one side into H atoms that diffuse through the metal film and reassemble on the other side.

Traditionally Established Industrial Methods. Gases in purified form are desired for many applications in research and industry [121]. Separation of methane, propane, and other hydrocarbons from H_2S, H_2O, CO_2, and N_2 and other impurities found in natural gas streams and biogas is required in order to tap new energy reserves. Compression of the gases for storage also requires energy investment. Recovered hydrogen from industrial product streams generated by steam reforming and other processes are required in ammonia synthesis and gasification reactions of oil refinery processes. High-purity hydrogen is required to power fuel cells. The effort to separate CO_2 from air has gathered significant momentum over the recent past in order to mitigate the ingress of greenhouse gases into the atmosphere. Traditional gas separation methods are energy intensive. These include cryogenic-distillation methods, amine absorption/regeneration, and pressure swing adsorption (PSA).

PSA is used to remove CO_2 during the commercial synthesis of hydrogen for applications in oil refineries, ammonia production, and to elevate methane content in biogas [122]. For example, a nitrogen purification PSA process passes 95%–99.5% pure N_2 (at 6–8 atm.) over a sorbent material (activated carbon molecular sieve) that removes oxygen and water, the *retentates*, from the input gas stream by adsorption [123]. The purified nitrogen *raffinate* is not adsorbed and the product stream proceeds over the bed to collection (also energy intensive) [123]. If higher purity is required, catalytic processes combine H_2 with O_2 to produce water. Cooling or an additional adsorption process is applied to remove accumulated water [123]. Regeneration occurs by depressurization to atmospheric pressure. Oxygen PSA exploits alumina to remove water vapor. Zeolite molecular sieves are used to remove water, CO_2, and other gases [123]. Application of PSA to capture CO_2 produced by coal-fired plants is currently under consideration [122]. Another version of PSA is adapted to provide carbon dioxide and humidity control to portable life support systems (PLSS) for NASA. A sorbent is coupled to the PSA system and is regenerated by exposure to vacuum in space during extravehicular activity (EVA) [124]. Such microporous membranes offer a nonenergy intensive alternative to gas separation—using the built-in vacuum of space to perform the purging function.

Zeolite Membranes. Zeolites are microporous materials (according to IUPAC convention) comprised of aluminosilicates. Zeolites are natural polymorphic minerals, but many synthetic versions have been developed for experimental and industrial use. Applications of zeolites fall into these major categories—separation via sieving, ion-exchange, size-exclusion etc., and catalysis and template synthesis. Zeolites are chemically resistant, can tolerate high temperatures and pressures, are able to form into pinhole-free membranes, and the porous structure can be tailor-made to suit end user criteria [125]. Low-cost facilitated synthesis, highly ordered structures, and flexibility in design have made zeolites a leading nanomaterial for membrane separation technology. Pore, or channel, diameter in six and eight-member ring zeolites range from 0.3 to 0.4 nm; 10-member ring zeolites have characteristic pore diameter ranging from 0.5 to 0.6 nm; and larger pore zeolites having 12-member rings have pores 0.7–0.8 nm [125]. Membranes can be supported on alumina or silica substrates, called composite membranes, or fabricated as free-standing contiguous films—limited in size to a few square centimeters. Membrane-based separation, as opposed to conventional techniques such as pressure swing adsorption, should reduce facility and equipment costs and energy consumption and allow for continuous operation [125].

Hydrogen is a clean burning fuel and developmental efforts are underway to power cars, electrical devices, and household appliances with hydrogen. In order to reach these goals, one of the holy grails of renewable fuel development, in addition to efficient production and storage, is to find an efficient way to separate hydrogen from other light gas by-products of industrial processes like steam-reforming and gasification reactions.

Kanna Aoki of the Nanoelectronics and Collaborative Research Center at the University of Tokyo and colleagues prepared a highly hydrophilic A-type zeolite by hydrothermal synthesis on the outer surface of a porous α-alumina tube [126]. The membrane possessed two types of pores: 0.4- to 0.43-nm zeolitic pores and nonzeolitic pores. Molecules such as ethylene were not able to penetrate the zeolitic pore system. Separation was dependent on the hydrophobic nature and molecular size of permeates, and affinity to the pore walls by the permeate gases. Combinations of gases also affected the permeation rates. The presence of water molecules reduced the permeation character of hydrophobic gases [126].

28.2.2 CO_2 Mitigation

Established industrial methods of removing CO_2 from combustion flue gases include amine absorption recovery and PSA. Sequestration (capture and storage) or conversion of CO_2 is desired for several reasons, topics upon which we shall not elaborate. Addition of a separation technology to fossil fuel conversion systems would help mitigate anthropogenic sources of CO_2 from reaching the atmosphere. Experimental carbon-based membranes, functionalized mesoporous oxide membranes, and sieving zeolitic membranes are members of a new family of regulated nanostructured materials that have all shown promise in selectively removing CO_2 from complex gas streams.

Y. Ohta et al. utilized dynamic Monte Carlo simulations to estimate the permselectivity of a binary gas mixture consisting of CO_2/N_2 in zeolite-like porous membranes with seven-member ring structure and 0.34-nm pore diameter [127]. The separation factor and permeability of CO_2 were shown to be superior, and simulation methods should prove to be useful in further investigations [127].

Kiochi Yamada, Shingo Kazama, and Katsunori Yogo of the Research Institute for Innovative Technology for the Earth (RITE) in Japan have developed several kinds of gas separation membranes.

T. Koutetsu et al. devised a dendrimer composite membrane on porous substrates for CO_2 separation. Poly(amidoamine), or PAMAN, dendrimers demonstrated superior in situ CO_2/N_2 gas separation [128]. The surfaces of commercially available ultrafiltration membranes were modified to produce a gas-selective layer. Chitosan layer (200 nm) formed the base layer to attach PAMAN dendrimers (total thickness = ca. 300 nm with 20%–40% dendrimer). Separation characteristics were determined under pressure differential of 100 kPa and 40°C. The analyte mixture consisted of 5%-vol. CO_2 and 95-vol% N_2. The selectivity of CO_2 over N_2 was determined to be 400 and the permeance at 40°C was 1.6×10^{-7} $m^2 \cdot s^{-1} \cdot kPa^{-1}$ [128].

N. Hiyoshi et al. in 2005 reported that in the aminosilane-modified mesoporous SBA-15 (a silaceous zeolite) membranes resulted in an increase in CO_2 per nitrogen atom via aminosilane adsorption increased as the surface density of the amine was increased [129]. The authors claim that amine pairs onto which CO_2 adsorbed formed alkylammonium carbamates—making a case for a chemisorption mechanism. The adsorption also depended on the structure of the amine [129]. S. Duan et al. in 2006 also developed PANAM dendrimer composite membrane for CO_2 separation [130]. Under similar conditions, Duan et al. achieved better permeance but less selectivity: 230 CO_2/N_2 and 1.6×10^{-7} $m^2 \cdot s^{-1} \cdot kPa^{-1}$, respectively [130]. Kovvali et al. also accomplished work with dendrimers for CO_2 separation [131].

There is great incentive to develop CO_2 capture via innovative membranes with sub-nanoscale materials control [132]. Objectives include separating CO_2 from other gases present in flue gases or synthetic processes as is means of storing CO_2. The strategy is, in general, to develop low-cost polymeric and inorganic membranes for the purposes of separation and storage. Current systems that are not permselective enough require complex cascading structures to obtain pure streams of carbon dioxide.

28.2.3 *Hydrogen Production and Purification*

From the environmental viewpoint, hydrogen is the premier fuel of combustion processes.

$$H_{2(g)} + \frac{1}{2}O_{2(g)} \rightarrow H_2O_{(g)} \tag{28.13}$$

Catalytic reforming produces hydrogen with 40%–92% purity. The effluent is channeled into a PSA treatment system to achieve 99.95% grade hydrogen. Petrochemical off-gases produce 10%–20% pure hydrogen that is funneled into a cryogenic system to achieve 99.95% H_2 that is separated from various hydrocarbon gases. Cryogenic separation is based on the difference in boiling point of H_2 from feedstock impurities—the method is also used to recover CO and other gases. Refrigeration is accomplished by a process based on Joule–Thompson expansion. Water and carbon dioxide are usually removed before application of the cryogenic distillation process. Hydrogen with ~40% purity is generated from petrochemical off-gases, but in this case the effluent is channeled to membrane or PSA purification processes [133].

Bifunctional PSA systems are able to produce H_2 at purity levels 99.95%–99.9999% from 40 to 90 + %H_2 feedstock. An activated carbon layer (a great traditional nanomaterial) removes CO, CH_4, CO_2, H_2O, and H_2S. Molecular sieves (another traditional nanomaterial) remove N_2, Ar, O_2, and the remaining CH_4 and CO [133]. PSA is widely used in H_2 purification for fuel cell applications and is the principal method for hydrogen purification in use today [133]. Operational issues include extensive maintenance due to wear and tear of valves and their inherent complexity. Air Products and Chemicals, Inc. in 2006, produced over 1.75 million tons per year of hydrogen [133].

Polymeric membranes operate on the phenomena of partial pressure difference and the degree of gas–polymer interaction. The rate of diffusion is inversely proportional to the thickness of the membrane. Polymer membrane methods have the following issues: they require high feed pressures (20–200 atm) irreversible membrane damage occurs from impurities, physical defects like cracks that reduce efficiency, and permeability to other gases. Polymeric membranes are therefore not used for purification of hydrogen for fuel cell applications.

28.2.4 Chemical Sensing and Detection

We continue again with a section on gas sensing to supplement previous material. Nanotechnology, of course, is expected to play a major role in this area.

Carbon Nanotube–Nanocomposite Gas Sensors. The detection capability of gas sensors (hydrogen, ammonia, and acetone in experiments) has been recently enhanced with the application of conductive polymers and single-walled carbon nanotubes [134,135]. One type of sensing material is a composite consisting of 1-wt% SWNTs in poly(3,3′-dialkyl-quarterthiophone). According to Sara Vieire et al. at the University of Cambridge, doping with gas at levels of 10 ppb yielded exponential changes in the overall conductivity profile of the nanocomposite sensors. Conducting polymers are good materials to use for sensing due to inherent physical properties such as conductivity, lightness, ease of manufacture, and cost-effectiveness. However, the materials lack specificity in distinguishing among different kinds of gases, for example, lack of selectivity. Semiconducting SWNTs, on the other hand, are highly selective but are expensive to manufacture. Just like for fiberglass, a composite material that outperforms either component by itself, the combination of conductive polymers with carbon nanotubes makes good sense. The device consists of a commercially available silicon-on-insulator complementary metal-oxide semiconductor (SOI-CMOS).

Gas Sensing by Thin Metal/Metal Oxide Films. Tin oxide (SnO_2) is one of the most prevalent of gas-sensing materials. It is used to detect AsH_3, H_2S, NO, NO_2, H_2, NH_3, CO, CCl_4, C_3H_6, CH_4, O_2, MeOH, EtOH, CO_2, and numerous other hydrocarbon gases. SnO_2 is often supported on a wide variety of materials: alumina, Pt, Pd, Ag, ZrO_2, CuO, and La_2O_3.

The use of nanoparticles has significantly enhanced the performance of gas-sensing devices [136]. H. Ogawa in 1982 showed that enhanced sensitivity was obtained with SnO_2 nanoporous films rather than dense nanoparticulate films of the same material [137]. He explained his results by applying Hall measurements—specifically, that a space–charge region develops in the nanocrystallite when its size is approximately twice the Debye length (the distance over which significant charge separation can occur). This condition results in higher sensitivity to adsorbed gas species [6], and in concert with high, specific surface area of the porous film ($\varepsilon \sim 98\%$ porosity), fast transport of the analyte gas adsorbing and desorbing species is enhanced. This explains why carrier mobility is strongly dependent on the concentration of the substrate in porous films while no such dependence is found for dense thin films [137].

One of the greatest challenges facing gas sensing is specificity (e.g., selectivity)—a challenge where nanotechnology can certainly help.

28.3 Energy

The late Nobel laureate Richard E. Smalley toured the country in the early 2000s giving his "Our Energy Challenge" lectures [138]. These lectures summarized the state of humanity and listed the 10 most important challenges that our species will face in the next 50 years! He also made significant mention of nanotechnology and how many of the solutions will arise from that wondrous technology. We will summarize some of his points and discuss the ones relevant to environmental issues.

According to Smalley, humanity's "top ten" problems for the next 50 years are [138]: (1) energy, (2) water, (3) food, (4) environment, (5) poverty, (6) terrorism and war, (7) disease, (8) education, (9) democracy, and (10) population. You may wish to reorder this list to suit your priorities, but most will agree that it does cover some of the most critical issues that we as a species will face (and are already facing) in the coming years. Population is actually at the root of all of those issues and could easily displace energy as the No. 1 issue. For example, the global population is expected to exceed 10 billion by the year 2050 [138]. Energy consumption, however, does not scale linearly with population growth. Although human population has quadrupled over the twentieth century, energy consumption has increased by a factor of 16 [138].

In 2004 (14.5 TW or 220 MBOE $\cdot$ day^{-1}), oil contributed nearly 35% to our total energy resources; coal 25%, gas 20%, fission 5%, biomass 10%, hydroelectric 5%, and solar wind, and geothermal a mere 0.5%. In 2050 (expected 30–60 TW or 450–900 MBOE $\cdot$ day^{-1}), these ratios are expected to undergo a drastic change: oil <5%, coal ~5%, gas >10%, fusion–fission ~15%, biomass >10%, hydroelectric ~5%, and solar, wind, and geothermal ~50%. The acronym MBOE $\cdot$ day^{-1} stands for "millions of barrels of oil equivalents per day." "TW" is of course the terawatt. When reviewing the list of potential energy sources for the future, conservation efficiency, hydroelectric (maxed out), biomass, wind, and wave and tide cannot possibly provide enough energy. Chemical sources such as natural gas and clean coal are faced with extraction and purification costs. Nuclear sources and others such as geothermal, solar terrestrial, solar satellite, and lunar–solar are burdened with high cost, engineering difficulties, and minimal potential or waste disposal issues.

The sun delivers 165,000 TW daily to the Earth—of which we can potentially capture 20 TW by placing six solar farm grids of 10,000 sq. mi. each on the continents [138]. By way of a proposed *Distributed Store-Gen Grid*, energy is transported by wire (rather than rail) on a continental scale that interconnects 100 million local storage and generation sites, electricity can be provided to most of the civilized world. The network includes local storage and generation capability as well as a centralized power grid. Highways in the United States cover about 0.2% of the land area. If we were to cover that amount of area with solar cells that deliver 50% efficiency, we would be able to generate 1 TW of electricity [139].

What can nanotech do to support this master plan? To begin with, the cost of photovoltaics could drop 10-fold due to the use of less material, lowered processing costs, and revolutionary manufacturing breakthroughs. Nano-enhanced materials and devices would be capable of photocatalytic reduction of CO_2 to methanol and direct photoconversion of H_2O to H_2 fuel sources. A likely 10- to 100-fold improvement of batteries, supercapacitors, and flywheels for automotive (plug-in hybrid vehicles) and Store-Gen Grid applications is expected. Power cables made of superconducting or quantum materials would enable worldwide transmission of electricity and replace aluminum and copper in electrical motors.

Carbon nanotube quantum wires would exhibit negligible current loss with one-sixth the weight of copper. Hydrogen storage on carbon nanotubes and other suitable nanomaterials and advanced materials for high-pressure storage of liquid hydrogen will be enabled by new nanomaterials. If the cost of reversible, low temperature fuel cells were dropped 10- to 100-fold, widespread use of portable and reliable energy sources would become reality [138].

28.3.1 Solar Energy and Nano

In 1839, 19-year old French scientist Alexandre E. Becquerel discovered the photoelectric effect while illuminating metal electrodes in an electrolyte. English engineer Willoughby Smith in 1873 was the first to observe the phenomenon of photoconductivity while working with a new material with high resistance called selenium (that happened to be photoactive)—a material used to check for flaws in transoceanic cables. In 1877, W.G. Adams and R.E. Day, Willoughby's students, verified that selenium produces a current when exposed to light. C.E. Fritts a few years later in 1883 constructed the first photovoltaic cell consisting of a sandwich of Se between a gold leaf and a brass plate. In 1887, J. Moser described dye-sensitized solar cells. Albert Einstein described the photoelectric effect in 1904. In 1918, Polish scientist J. Czochralski devised a means of fabricating monocrystalline Si and the first Si solar cell was manufactured in 1941. PV cell efficiency started out at 2%. In 1959 and 1960, 9% and 14% efficiencies were achieved, respectively. In July 2007, 42.8% efficiency was achieved with an experimental VHESC (yes, a "very high efficiency solar cell") multijunction by a University of Delaware team. In August 2008, NREL created a reliable solar cell with 40.8% efficiency (www.NREL.gov/news/press/2008/625.html).

There are many kinds of solar cells, all of which have useful environmental potential with regard to efficiency or reduced costs: multijunction concentrators (>40%), silicon cells (single crystal, polycrystalline, and thin film with efficiencies <30%, ~20%, and ~16%, respectively); thin-film technologies such as $Cu(In,Ga)Se_2$ or CIGS (~18%), CdTe (~16%) and amorphous stabilized Si:H (~12%); and emerging PV materials such as dye-cells (~14%) and organic cells (~5%). An overview of solar cell efficiency is presented in **Figure 28.16**.

Nanotechnology offers numerous advantages to solar cell design and manufacture: (1) large surface area collectors, (2) antennae consisting of quantum dots that are able to capture 90% or better of the solar spectrum, (3) tunable solar material interfaces that demonstrate enhanced efficiency, (4) multielectron stimulation, and (5) reduced manufacturing costs due to revolutionary nanoprocesses such as nano-imprint lithography that operate at lower temperatures with higher throughput.

The groundwork for a new solar plant, owned and operated by Nanosolar, Inc., was announced in 2006. The plant is expected to manufacture paper-thin flexible solar materials that are expected to produce over 430 MW of power per year [140]. The solar active component of the thin-film nanotechnology is a copper–indium–diselenide ink that absorbs light and converts it into energy. In December 2006, Boeing-Spectrolab developed a solar cell that is capable of 41% efficiency and earlier in the year, researchers at Lawrence Berkeley National Laboratory developed a cell from a new semiconducting material made of zinc–manganese–tellurium doped with a few atoms of oxygen that demonstrated 45% efficiency [141].

Fig. 28.16 *The development of more efficient solar cells from 1975 to the present is depicted. Efficiency greater that 40% has been achieved by the National Renewable Energy Laboratory. The theoretical maximum efficiency of photosynthesis is estimated to be 11%. The efficiency of photosynthesis is limited by the wavelength range (400–700 nm). This range accounts for only 45% of the total solar spectrum available for photosynthesis. Then, there are issues with quantum efficiency per CO_2. Therefore, only 25% of the 45% is actually considered to be photosynthetically active radiation (PAR). Other reasons such as reflection, respiration, and lack of optimal solar radiation levels also serve to reduce efficiency. The comparison with the NREL cell is not direct due to variable circumstances.*

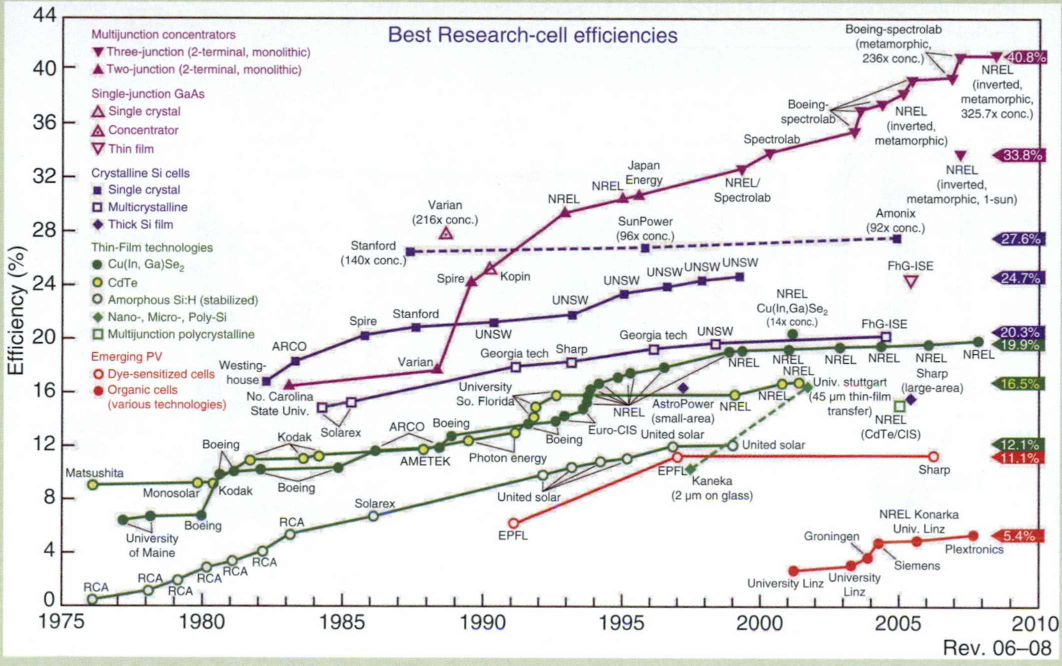

Source: Larry Kazmerski, National Renewable Energy Laboratory (NREL), Golden, Colorado. With permission.

Conventional solar cells are made of an *encapsulant*, a *front contact grid*, an *antireflective coating*, *n*-type-silicon, *p*-type-silicon, and a *back electrical contact* (**Fig. 28.17**).

During our description, think of ways that material components can be enhanced with nanotechnology. The encapsulating material serves to seal the cell and keep the external environment out. It is usually made of glass or clear plastic. The antireflective coating effectively channels photons into the matrix of the solar cell (reduces solar loss). A grid made of conducting material forms close contact with the n-doped silicon as is a metal contact with the *p*-doped silicon.

The ultimate goals of any photoelectric device is to generate free charge carriers from interaction with photons and to separate the charge carriers to generate an electric current before the excited states can decay or recombine [142]. Electron charge carriers exist in the conduction band of the semiconductor—electrons that were promoted due to energy input. The vacancies left by the excited electrons are called holes and are known as virtual charge carriers. The simultaneous process forms *excitons* and the basis for photovoltaic mechanisms. In the

FIG. 28.17 *A generic silicon-based solar cell is depicted. The photocell has three major functions: (1) harvest a maximum amount of photons, (2) convert photons into electrical charge, and (3) generate electrical charge from the device.*

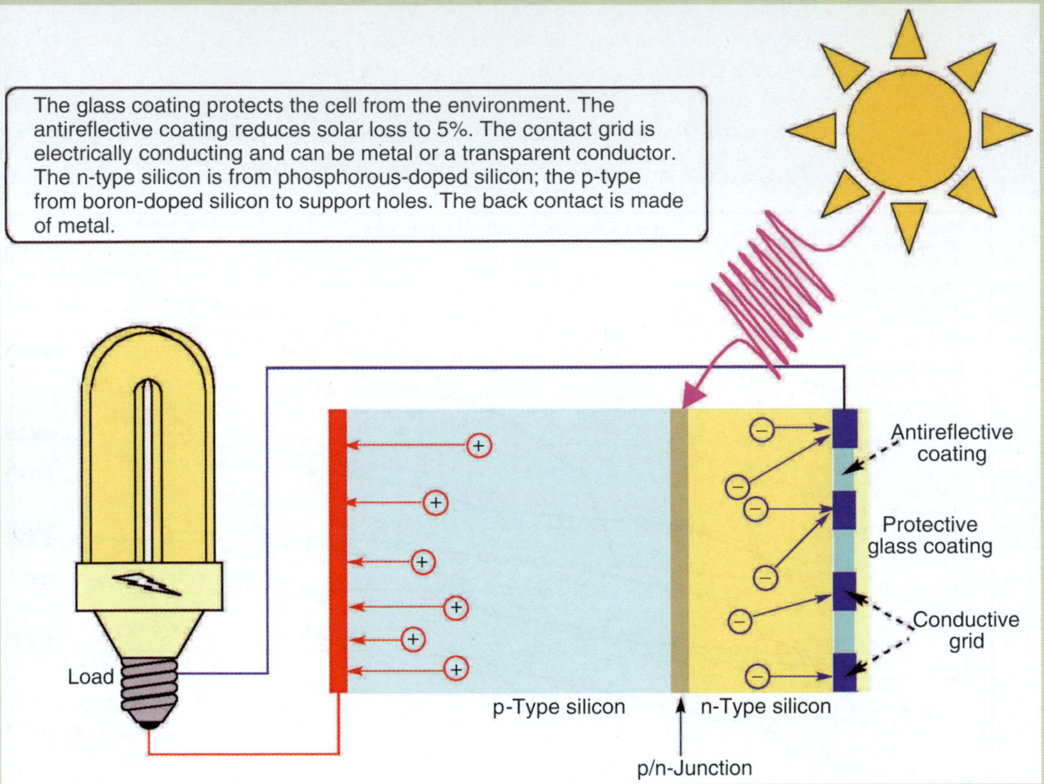
The glass coating protects the cell from the environment. The antireflective coating reduces solar loss to 5%. The contact grid is electrically conducting and can be metal or a transparent conductor. The n-type silicon is from phosphorous-doped silicon; the p-type from boron-doped silicon to support holes. The back contact is made of metal.
Antireflective coating
Protective glass coating
Conductive grid
Load
p-Type silicon
n-Type silicon
p/n-Junction

broadest sense, excitons are mobile combinations of an electron and a virtual particle called a hole in an excited semiconductor. In the narrower sense, excitons are strongly coupled electrons and holes. Wannier excitons are typical of covalent semiconductors and insulators, are separated by a distance that exceeds the lattice spacing of the solid-state material, and move like unbound particles. Frenkel excitons occur in molecular crystals, are separated by a distance on the order of the lattice spacing, and are usually localized to one site.

Doping alters the electronic properties of semiconductors by introducing excess positive or negative charges. In most bulk silicon solar sells, a thin n-type Si layer is deposited on top of a much thicker p-type Si layer. The *p*/*n* junction interface establishes a bias that results in recombination of electrons and holes, for example, depleted zone of paired charges. When electromagnetic radiation is applied to the surface, a higher population of charge carriers is induced creating a stronger bias across the interface. Charge carriers then move towards their respective conducting electrode terminals and are made available to do work.

Efficiency is the ratio of input to output energy. In 1961, W. Shockley and H.J. Queisser determined that a conventional PV cell is capable of ~32% efficiency, the Shockley–Queisser limit [143]. Only a small fraction of the solar spectrum matches the bandgap of Si. Impinging photonic energy above or below the bandgap, therefore, will produce heat in the semiconductor. The natural (theoretical) efficiency of commonly used bulk solar materials ranges from ca. 23% to 30%.

There are inherent limits to solar collection efficiency due to a variety of loss mechanisms: (1) spectral range of absorption limitations due to bandgap energy, (2) transformation of surplus light energy into heat, (3) optical losses due to reflection and shadowing of the cell surface, and (4) losses due to electrical resistance and other material factors (contamination, surface effects, crystal defects, etc.) [144]. In bulk semiconductors, one photon creates a single electron–hole pair regardless of photon energy—whether a high-energy photon or a low-energy photon.

Energy conversion efficiency (ECE) η_{ECE} is a parameter used to measure the efficiency of solar cells. *ECE* is the percentage or radiant power (incident irradiance $E_o = \text{W} \cdot \text{m}^2$) converted to electrical energy.

$$\eta_{\text{ECE}} = \frac{P_{\text{max}}}{E_o A_{\text{cell}}} \tag{28.14}$$

where

P_{max} is the maximum power output of the solar cell in watts (W)
E_o is the incident irradiance
A_{cell} is area of the cell

Quantum efficiency (*QE*), a subset of the greater category of *quantum yield* (the ratio of any quantum induced phenomena to input photons) should not be confused with *energy conversion efficiency*. Quantum efficiency is the percentage of photons of a select wavelength that produce electron–hole pairs. Good examples of devices that are measured by QE are charge-coupled devices (CCD, QE ≈ 90%) and photographic films (QE ≤ 10%) [145]. QE is often measured as a function of wavelength. If QE is integrated over the whole spectrum, the potential electric current that a certain cell produces can be estimated.

$$\eta_{\text{QE}} = \frac{N_e}{N_p} = \frac{\phi_o / hv_o}{\phi_{dl} / hv_o} = \frac{\phi_o}{\phi_{dl}} \tag{28.15}$$

where

N_e is the number of electrons
N_p is the number of photons
ϕ_o and ϕ_{dl} are the incident optical power (radiant flux, in watts or $\text{J} \cdot \text{s}^{-1}$) and the absorbed optical power, respectively

Optical power is absorbed in the depletion layer. For solar cells, the external quantum efficiency (*EQE*) is an important parameter. *EQE* is the current measured outside the solar cell device and represents efficiency in both the absorption of photons as well as the collection of electrical charge.

$$EQE = \frac{\text{electrons} \times \text{s}^{-1}}{\text{photons} \times \text{s}^{-1}} = \left(\frac{i}{e}\right)\left(\frac{hv_o}{\phi_o}\right) \tag{28.16}$$

Quantum Dot Solar Cells. In cells that contain quantum dots, the number of charge carrier pairs produced depends on the energy of the photon. In 2001, 14% efficiencies were achieved by photocells fabricated from nanoparticles in conducting polymer films [146]. The shape of the quantum dot also is expected to play a role in its operational efficiency [147,148].

In 2006, I. Robel et al. at Notre Dame University fabricated solar cells based on CdSe quantum dots (QDs) tethered with the bifunctional surface linker HS-R-CO_2H onto mesoscopic TiO_2 films [147]. The TiO_2–CdSe QD composite (acting as the photoanode) was able to exhibit a photon-to-charge-carrier efficiency of 12%. The function of the CdSe QDs (ca. 3 nm in diameter) was similar to that of a dye molecule. Following visible excitation, CdSe QDs inject electrons into TiO_2 nanocrystals (ca. 40–50 nm in diameter). The process was verified with femtosecond transient absorption as well as emission quenching experiments. In 2007, the research team found that the rate constant for electron transfer from the thermalized *s*-state of CdSe increased with decreasing particle size: 7.5-nm diameter particle ($E_g = 1.92$ *eV*) $\rightarrow 10^7\ s^{-1}$ and the 2.4-nm diameter particle ($E_g = 2.4$ *eV*) $\rightarrow 10^{10}\ s^{-1}$ [148]. The researchers claimed that the significant level of electron loss was due to scattering, charge recombination at the TiO_2/CdSe interface, and internal TiO_2 grain boundaries [147]. If a liberated electron collides with a nearby atom, it loses energy by the creation of atomic vibrations that end up heating the semiconductor.

Due to the size dependence of absorption by quantum dots, the solar cell can be tuned to absorb wavelengths ranging from 650 to 400 nm by decreasing the diameter of the CdSe dot. In this way, a greater range of visible light can be harvested for energy conversion. The thickness of a photocell based on QDs is able to influence the degree of absorption of light energy. Tenfold increase in the thickness, from 0.2 to 2 μm, resulted in significant darkening of the solar cell. This phenomenon is obviously due to the availability of more sites for CdSe binding with concomitant increase in efficiency of the solar cell.

Quantum dots also have the ability to generate multiple charge carriers from a single photon, unlike bulk semiconductors that generate at the most one exciton pair when exposed to an energetic near ultraviolet photon [149–151]. The remainder of the photon energy is converted into heat. QDs are able to generate more electron–hole pairs by a process called impact ionization—a mechanism first proposed by A.J. Nozik of the National Renewable Energy Laboratory in Golden Colorado [152–155]. Impact ionization is a process whereby high-energy charge carriers are able to distribute energy by the creation of more charge carriers. When in refined form, theoretical photon conversion efficiencies greater than 100% are possible [149].

In 2004, researchers at the Los Alamos National Laboratory discovered that three excitons each were generated when 5-nm lead selenide quantum dots were exposed to high-energy blue light [150,151]. In these experiments, in which quantum dots consisting of 1000 atoms were mixed in a liquid and sealed inside a glass sheath, laser pulses over a wide range of energies were directed at the mixture. Recently, 8-nm diameter quantum dots made of lead sulfide were shown to release seven electrons per dot when exposed to ultraviolet radiation [150,151]. This apparent leap in photon conversion efficiency should be able to benefit other types of solar conversions systems such as water splitters that produce hydrogen for fuel cells [150,151].

Organic Solar Cells. The photoactive elements of organic photovoltaics (OPVs) are made of polymers and/or nanocrystalline phases. Most OPVs are called bulk heterojunction devices in which *p*-type (electron donating) and *n*-type (electron accepting) organic materials are blended into a polymer matrix [156]. Power efficiencies (η) near 3% have been achieved with conjugated organic polymers like PCBM (phenylbutyric acid methyl ester) with electron-accepting substituted fullerenes [60] and soluble donors like polythiophene or poly(phenylenevinylene) derivatives [157]. Organic heterojunctions are defined by four processes: absorption, exciton diffusion, charge transfer, and charge collection. Charge transfer and collection are highly efficient processes with $\eta_{EXE} \to$ 100%. The major difference among OPVs is the efficiency of absorption ($\eta_A \to 50\%$, optical absorption path ~100 nm) and exciton diffusion ($\eta_{ED} \to 10\%$, diffusion length ~5 nm, HOMO–HOMO, LUMO–LUMO). Since absorption and exciton diffusion are limiting, the overall efficiency of the heterojunction OPV is expected to be around 10%.

C.W. Tang in 1975 conducted some of the first investigations to duplicate nature's ultimate solar cell—the one that exploits the amazing properties of chlorophyll-*a* in photosynthesis [158]. Tang also created a chlorophyll-*a*-PV cell in 1975. Monomers of chlorophyll-*a* were prepared on a chromium electrode by electrodeposition to form a microcrystalline array. The structure has a strong absorption band in the far red range of the spectrum at 740–745 nm [158]. The power conversion efficiency of the sandwich structure (Cr-Chl-*a*-Hg) was reported to be 0.01%—the highest for an organic PV cell of the day. [158]. Later on in 1986, he developed one of the first organic photocells, called the Tang cell [159]. Via lessons learned from nature, a new generation of solar cell development will most likely be modeled on biological materials like chlorophyll.

Advantages of organic solar materials are multifold: (1) the solar "cell" materials can be applied by painting on roofs, walls, or other surfaces that are targeted for solar collection—a incredibly low-tech solution to a high tech problem, (2) offer great potential for facilitated upscale (high throughput and can cover large areas), (3) are flexible and lightweight, (4) offer an economical solution as the cost-to-efficiency ratio for silicon is teetering towards the former, (5) can be applied on any kind of surface, especially large area flexible plastic surfaces [160], (6) ultra-fast optoelectronic response and charge carrier generation at organic donor–acceptor interface, (7) continuous tunability of optical energy bandgaps via molecular design, synthesis, and processing steps, and (8) integrability into textiles, packaging, and consumer products [161]. Underlying these advantages is the ability to synthesize organic molecules with great flexibility—flexibility that allows for tunability for adoption to specific applications. The major disadvantage is that OPVs promise, at best, only moderate solar conversion efficiencies. Low efficiencies are due to photon loss, exciton loss, and carrier loss.

Photoreactive polymers serve as the basis of OPV technology. In standard solar cells, both the donor and acceptors are basically the same material—doped silicons. On the other hand, in OPVs the donor material may not be anything like the acceptor material. To start, donor and acceptor molecules must be able to absorb light. Many donors contain conjugated molecules with a electron-donating character. Acceptor molecules need to be able to stabilize electrons originating from the donor molecules. Stabilizing molecules include electron-withdrawing groups. Carbon nanotubes and fullerenes, for example, are known for their electron stabilization ability.

C.W. Tang's first cell consisted thin films of copper-phthalocyanine (Cu-PC) donor and perylene tetracarboxylic acid derivative acceptor [159]. Cu-Pc is a porphyrin-like ring that hosts a Cu cation. A power conversion efficiency of 1% was achieved with this cell. In this new concept, bias voltage does not control the charge-separation efficiency, and therefore cells with fill-factor up to 65% can be achieved. In addition, the interface between the two organic materials and not the organic material–electrode contact is the primary factor that determines the PV properties of the cell [159].

In 2006, Klaus Müllen et al. of the Max Planck Institute for Polymer Research in Mainz, Germany developed a bulk heterojunction cell consisting of poly(2, 7-carbazole) (PCz) and perylene tetracarboxydiimide dye (PDI) donor–acceptor pair for organic solar cell applications [156]. PCz, in a highly soluble state, served as the electron donor material and the PDI was the electron acceptor material. The pair demonstrated complementarity within a broad range of the solar spectrum. The reported EQE efficiency was 16% and the power efficiency was 0.6% under solar conditions. The bandgap of PCz is ~3 *e*V, well into the ultraviolet range of the spectrum. To make this material practical in solar cells, it must be paired with a material that has a lower characteristic bandgap—enter PDI (bandgap ~2 *e*V). Previously, the Müllen group has shown that PDI in combination with discotic liquid crystalline hexa-*peri*-hexabenzocoronone (HBC) displayed a maximum external quantum efficiency (EQE) of 34% with η = 1.95% at 490 nm [162]. Müllen et al. stated that photoinduced charge transfer occurred between the HBC and the perylene complement in addition to efficient charge transport through vertically segregated perylene and HBC π-systems—all constructed with simple bottom-up nanofabrication solution techniques [162]. The cell is shown in **Figure 28.18**.

Müllen et al. showed that conjugated polymers with high bandgap can be used for solar applications if a proper electron acceptor is selected [156]. Cyclic voltammetry (Ag/Ag^+, calibrated against ferrocene) of PCz/PDI on ITO revealed that the HOMO of PDI was −5.8 *e*V, below that of PCz (−5.6 *e*V). The estimated LUMOs of each, determined by bandgap calculations from absorption spectra, were −2.6 *e*V for PCz and −3.8 *e*V for PDI. According to Müllen, PDI are able to form excitons, and hole migration from excited PDI states to PCz is driven by the 0.2 *e*V difference in HOMO energies. With regard to the corresponding LUMO energies, the 0.8 *e*V difference in energy is able to drive electron migration from PCz to PDI. The group is working on mixing PCz with other dyes that have broader absorption spectra [156].

Fullerene-modified polymers are increasingly popular for use in OPVs. In such cells, charge separation mechanisms in bulk heterojunction PV cells are based on electron transfer from an absorbing polymer to a fullerene-modified material such as (6,6)-phenyl-C_{61}-butyric acid methyl ester (PCBM). Liu et al. have shown that resonant energy transfer from a red-emitting organic chromophore (Nile red) to PCBM

Tae Wan Kim and his group of the Department of Physics at Hongik University is Seoul, South Korea studied the photovoltaic effects of the Cu-Pc layer thickness—10- to 50-nm [163]. The group studied ITO/Cu-Pc/Al and ITO/CuPc/C_{60}/BCP/Al cells. P. Peumans et al. in 2001 developed an ITO/Cu-Pc/C_{60}/BCP/Al with power conversion efficiency of 3.6% at the AM-1.5 solar spectrum [164]. Excitons generated in OPV semiconductors are characterized by excitons

Fig. 28.18

The PCz-PDI organic solar cell is shown. Thin films of PDI and PCz were prepared by a spin-coating application onto activated (oxygen plasma) ITO surface. A thin layer of silver (100 nm) was evaporated through a mask to form the cathode. Monochromatic light ranging from 300 to 800 nm was supplied with a maximum intensity of 6 W·M^{-2} at 600 nm. Solar light at 100 W·m^{-2} was supplied by a solar simulator. The PDI films yielded a charge-carrier mobility of 0.2 cm^2·V^{-1}·s^{-1} in its crystalline phase. PCz was synthesized from 2,7-dibromocarbazole by alkylating with 2-decyltetradecylbromide (R = decyltetradecyl) and subsequent polymerization with bis-(1,5-cyclooctadiene)Ni, *a.k.a. Ni(COD)$_2$. Characterization by electrochemistry was conducted with a voltammeter, a three-electrode cell with a working electrode of ITO, a silver quasi-reference electrode (AgQRE calibrated against Fc/Fc$^+$ redox couple with E^o = −4.8 eV), and a Pt counter electrode. The electrolyte was 0.1. M tetrabutylammonium perchlorate. Independently, PDI showed absorption maxima at 490 and 530 nm, PCz at 395 and 270 nm. When blended, a broad absorption between 300 and 600 nm was displayed—perfect for a solar cell application.*

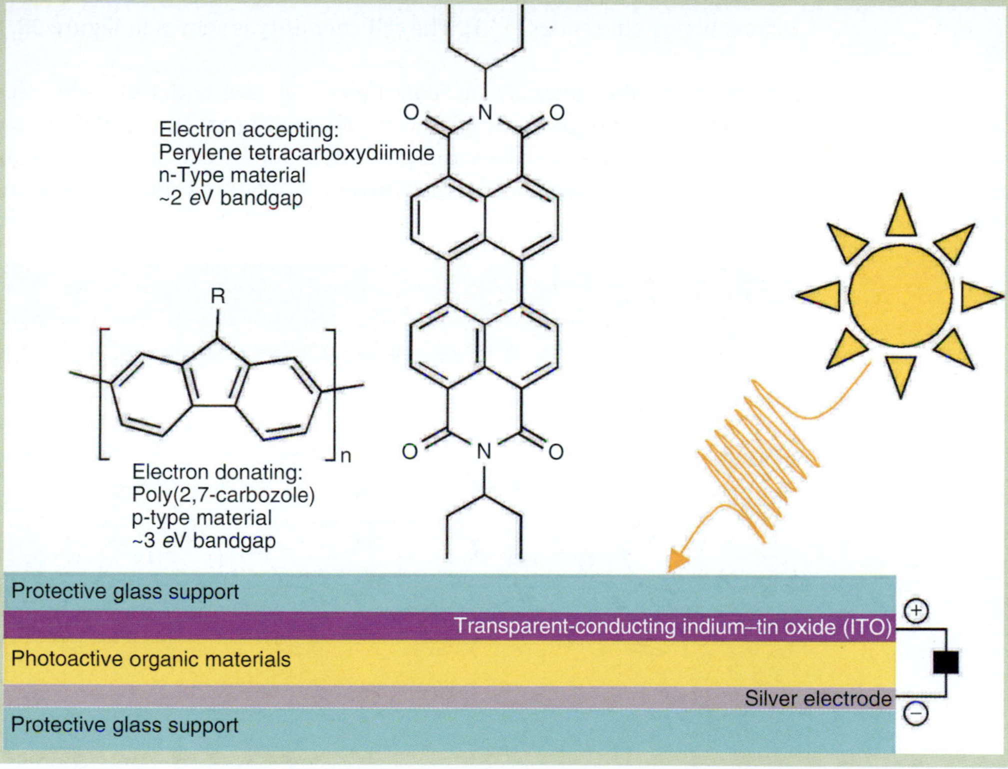

Source: Image courtesy of Klaus Müllen, Max-Planck Institute for Polymer Research in Mainz. With permission.

with higher binding energy than those found in inorganic material counterparts. Therefore, an electric field is applied to help separate the excitons into holes and electrons. The exciton diffusion length is approximately 10 nm in organic PV materials, therefore, according to the authors, the photoactive layer thickness should impact the efficiency of the cell [163].

A 5-mm wide strip of ITO (sheet resistance = 15 Ω·sq^{-1}) was fabricated with selective etching in an HCl:HNO_3 (3:1) solution and placed on a glass support.

Cu-Pc was deposited by thermal evaporation to thicknesses ranging from 10- to 50-nm in samples. In other specimens, C_{60} and BCP and then a 150-nm thick layer of aluminum were also deposited via thermal evaporation—quite simple straightforward processes [163]. Under dark conditions, nonlinear current density-voltage behavior was noted as Cu-Pc layers became thicker. Rectifying behavior was exhibited by films thicker than 30 nm indicating that exciton diffusion length is a factor in OPV materials—a trait desirable in solar cells. Therefore, layers of Cu-Pc less than 30 nm in thickness are not suitable for OPV cells. Open-current voltage (VOC, the voltage measured without the presence of current) depends on the Fermi energy difference between the electrodes and the energy levels (HOMO/LUMO) of the organic materials.

The current flow in ITO/Cu-Pc(20 nm)/C_{60}(20 nm)/BCP(15 nm)/Al exhibited more than 100x the current flow measured in single layer cells [163]. The group demonstrated that the efficiencies of heterojunction cells outperformed those of the single-layered Cu-Pc materials. The multilayered cells were more stable and yielded higher efficiencies [163]. The cell chemistry is shown in **Figure 28.19**.

Carbon Nanotube Materials in Solar Cells. In standard TiO_2 cells, electrons undergo a circuitous path, jumping from TiO_2 nanoparticles until they reach the electrode. Many do not achieve this goal, and thereby effectively reduce the efficiency of the cell. Researchers at the University of Notre Dame, P.V. Kamat et al.,

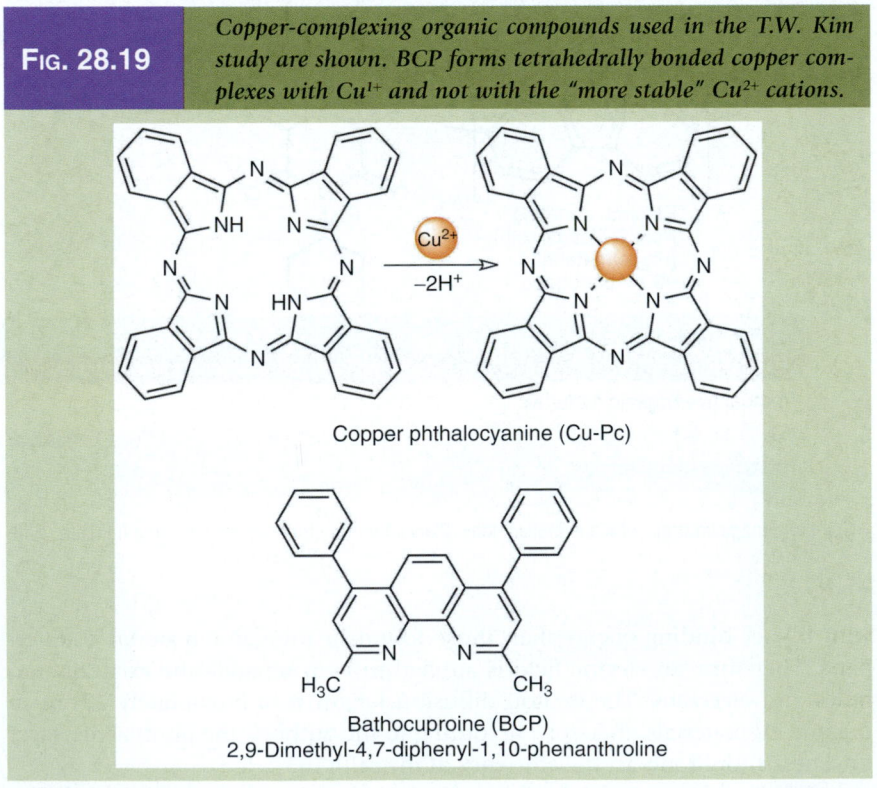

FIG. 28.19 *Copper-complexing organic compounds used in the T.W. Kim study are shown. BCP forms tetrahedrally bonded copper complexes with Cu^{1+} and not with the "more stable" Cu^{2+} cations.*

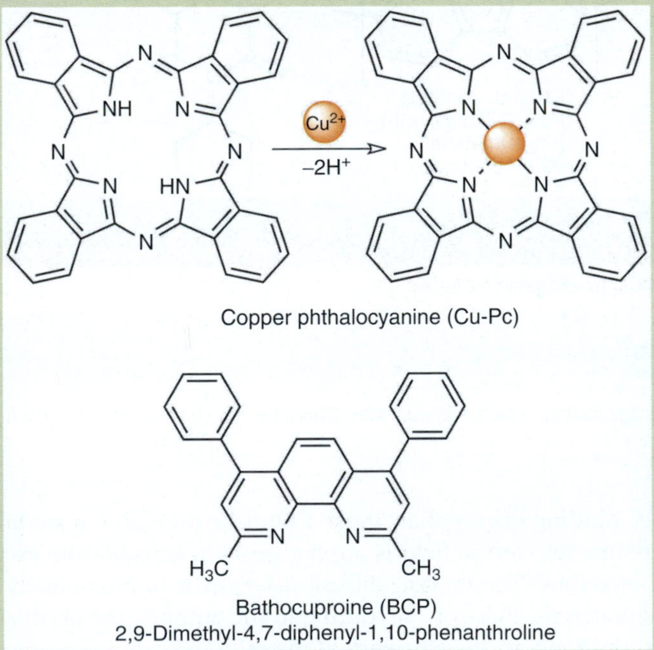

Source: S. W. Hur, T. W. Kim, and J. W. Park, *Journal of Korean Physics Society*, 45, 627–629 (2004). With permission.

have adapted single-walled carbon nanotubes to solar cells [165–169]. P.V. Kamat et al. have developed a cell that utilizes single-walled carbon nanotubes as conduits for electrons between the tethered TiO_2 and the electrode [165]. The SWNT-enhanced cells demonstrated twice the efficiency over cells comprised only of TiO_2 nanoparticles. An ultraviolet light lamp served as the source of the radiation. Carbon nanotubes are known to be good electron acceptors. A shift of approximately 100 mV of SWNT-TiO_2 system when compared to unsupported TiO_2 suggests that Fermi level equilibrium between the two systems was achieved [165].

The efficiency of solar cells using SWNTs decorated with TiO_2 can be enhanced if a one-molecule thick layer of organic dyes is adsorbed onto the semiconductor surface (as before). The dyes are able to expand the range of adsorption by the cell into the visible region of the solar spectrum. Another strategy involves the tethering quantum dots to the ends of the SWNTs. In this case, not only would the cell be able to convert visible range energy into electricity but also the ability to convert high-energy photons into multiple excitons.

Double-Walled Carbon Nanotube Solar Cells. J. Wei et al. in 2007 showed how double-walled carbon nanotubes (DWNTs) can be applied to solar cell technology [170]. The DWNTs served as both photogeneration sites and electron collection-charge carriers. A semitransparent layer of DWNTs was applied on an *n*-type crystalline Si substrate to create high-density *p*–*n* heterojunctions in which charge separation takes place by means of electron extraction by the *n*-Si and holes through nanotubes [170]. A power conversion efficiency of >1% was achieved by these cells [170]. The main difference from other cells that use organic PV materials and CNTs is that in those cells the conjugated polymers generate the excitons and the CNTs act as a conduit for electrons. In this cell, the CNTs play a role in charge separation [170].

Ultimate Black Materials—The Super Black Object. Ideal absorbers absorb all wavelengths without reflection (e.g., like blackbody cavities)—as even graphite and black paint reflect 5%–10% at the air–dielectric interface [171]. The major roadblock to developing such a material has been the inability to reduce the material's index of refraction to unity, so that optical reflection is totally eliminated. [171]. According to the authors, simulations do predict low index of refraction ($1.01 < n < 1.10$) by low-density, vertically aligned, multiwalled carbon nanotubes (VA-CNTs, 8–11 nm in diameter) in an array (e.g., a structure with high aspect ratio). The thickness of the CNT layer was between 10–800 μm and spacing between tubes was ca. 50 nm. These CNTs showed low refractive index, super low reflectance, diffuse reflection, birefringence, and strong absorption in the visible range of light [171]. The optical properties of such a surface is tunable by varying nanotube diameter, spacing, and length—all pretty standard features of nanotechnological engineering. Applications of such materials could impact pyrolytic detectors and solar energy conversion [171].

28.3.2 Batteries

The battery is one of the most important inventions. Benjamin Franklin in 1748 called an array of charged glass plates a battery. Luigi Galvani determined that electricity was involved in the mechanism of nerve impulses. Alessandro Volta invented the first real wet-cell battery in 1800—the voltaic pile—and found a means to conduct the charge over some distance. The Daniel cell (a cell that employs two electrolytes) was developed in 1836 by J.F. Daniels. In 1839, the first fuel cell was developed by W.R. Grove by combining oxygen and hydrogen. Frenchman G. Leclanche in 1866 invented the fundamental lead acid battery. The first commercially available dry cell was invented by C. Gassner in 1881. The alkaline storage battery was developed in 1901 by T.A. Edison. 1n 1954, the first solar battery was developed by G. Pearson, C. Fuller, and D. Chapin. Of course, there are many scientists and inventors who have made significant contributions but were left off this list. Nanomaterials should be able to provide battery technology with several advantages that include enhanced surface area, high electrical conductivity, electrical conducting networks, and perhaps even flexibility [172].

Some battery terms are listed and defined in this paragraph. A *primary battery* is not rechargeable. In other words, the electrochemical reaction that powers the cell is not reversible and the battery is usually discarded after it is drained. A dry cell alkaline battery is an example of a primary battery. A *secondary battery* is a battery that can be recharged (its electrochemical reaction is reversible) by passing a current through the battery opposite in direction to that of discharge during use. The lead acid battery found in most cars is an example of a secondary battery. An *ampere-hour* ($A \cdot h$) is equal to a current of one ampere over an hour's time. Battery capacity is measured in terms of $A \cdot h$. Specific battery capacity is measured in terms of $A \cdot h \cdot g^{-1}$ (or more appropriately, $mA \cdot h \cdot g^{-1}$). The *anode* in a battery is the site of electron release; the *cathode*, the site of electron absorption. More electrolyte and more electrode material translate into greater capacity for a battery. Smaller cells made of the same material as a larger cell, therefore, have less capacity although generating the same voltage [173]. The specific capacity of standard AAA through D batteries is ca. 100–140 $mA \cdot h \cdot g^{-1}$. The standard graphite anode battery is capable of ca. 372 $mA \cdot h \cdot g^{-1}$ capacity [174].

New enhanced batteries are expected to power automobiles that can operate for 200 miles without the need for recharging—or more modestly, at least be able to run for 24 h without having to plug into a power supply.

Next Generation Batteries. Desirable characteristics of the next generation of batteries include large discharge capacity at potentials near that of Li, maintenance of high currents, and facile reversability to ensure rechargeability. Silicon actually has one of the highest theoretical capacity because it can store more ions (~4200 $mA \cdot h \cdot g^{-1}$), but during Li insertion and extraction, Si-based insertion anodes (materials with highest known intercalation ability and low discharge potential) undergo a 400% change in volume for Li-ion secondary cells. The swelling results in decrepitation, pulverization, and subsequent loss of electrical contact [175]. Si nanowires, rather than bulk materials, improved strain relaxation (without pulverization) and conductivity. Si nanowires were grown directly on the current collector substrates. Improvement was due to good electrical conduction between Si nanowires and the current collector, short Li-insertion distances, high interfacial contact area with electrolyte, good electrical conductivity along the nanowires, and good material durability [176]. Discharge capacities

75% of the maximum were achieved with little diminishment during recycling [176].

High-Capacity Lithium Ion Batteries with Germanium Nanowires. Maximum lithium capacity germanium–lithium ($Ge_{4.4}Li$) batteries are expected to deliver $1600\,mA \cdot h \cdot g^{-1}$ with 370% volume change. Because the room temperature diffusion of Li in Ge is 400x better than that in Si, Ge is attractive as an anode component in cells that intend to deliver high power rate ratios [175]. C.K. Chan et al. fabricated Ge nanowires by vapor–liquid–solid growth on metallic current collector substrates. An initial discharge capacity of $1140\,mAh \cdot g^{-1}$ was shown to be stable over 20 cycles [174]. Fabrication of Ge nanowire anodes is straightforward. Ge nanowires 50- to 100-nm diameter, 20- to 50-μm long are formed by VLS (vapor–liquid–solid) growth with CVD of GeH_4 from gold catalysts on a metal substrate (the electrical collector, usually a stainless steel foil). The nanowires were single crystalline and required no further processing—a one-step nanotechnology fabrication method [174]. The electrode is shown in **Figure 28.20**.

Carbon Nanotube Based Batteries. Batteries have been recently fabricated from a random network of carbon nanotubes that serve as charge carriers at the interface with active components [172]. Highly purified SWNTs formed by laser synthesis were applied as anode materials for thin-film lithium ion batteries. The specific surface area and lithium capacity was compared to other traditional carbon materials like graphite, carbon black, and MWNTs [177]. BET-specific surface was found to be $915\,m^2 \cdot g^{-1}$, much greater than the other carbonaceous materials mentioned. The discharge capacity of the SWNT-Li ion battery was $1300\,mA \cdot h \cdot g^{-1}$ after 30 charge–discharge cycles with a current density of $20\,\mu A \cdot cm^{-2}$ [177]. Lithium-ion battery anodes made of SWNT paper showed

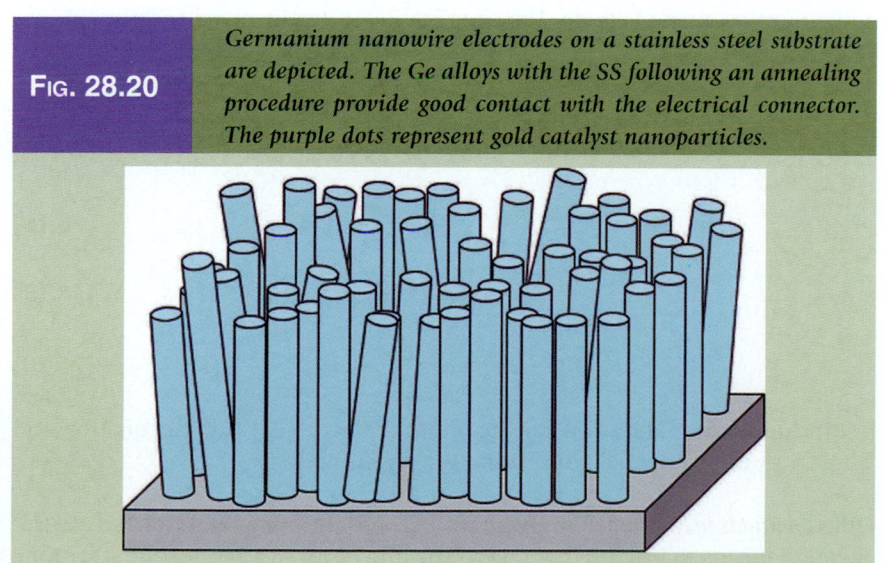

FIG. 28.20 *Germanium nanowire electrodes on a stainless steel substrate are depicted. The Ge alloys with the SS following an annealing procedure provide good contact with the electrical connector. The purple dots represent gold catalyst nanoparticles.*

Source: C. K. Chan, X. F. Zhang, and Y. Cui, *Nano Letters*, 8, 307–309 (2008). With permission.

lower overall weight due to lack of a binder material and extremely simple preparation [178]. The performance of the paper anode battery was slightly less than the performance of conventional SWNT electrodes [178].

Researchers at Rensselaer Polytechnic Institute (RPI) and MIT have developed a new fully flexible and robust material that eliminates the need for bulky multi-layered batteries (e.g., the kind you find in your cell phone). The electrodes, separator, and electrolyte were integrated into a single, flexible nanocomposite unit made of nanoporous cellulose paper embedded with aligned carbon nanotubes and electrolyte [179]. The team grew carbon nanotubes on a Si substrate and plugged the gaps between the tubes with cellulose, thereby making a flexible membrane. Two of these sheets were abutted with the cellulose sides placed inward and saturated in electrolyte to form a supercapacitor material. A 100 g sheet of the paper battery demonstrated 1300 mAh discharge capacity ($130\,mAh \cdot g^{-1}$) [179]. The use of room temperature ionic liquid (RTIL) electrolytes like 1-butyl, 3-methylimidazolium chloride is able to dissolve cellulose with the assistance of microwave energy and allows for its use as an electrolyte in potential supercapacitor applications [179]. Potential applications of this technology include its use in Li-ion batteries, supercapacitors, and hybrids [179].

SnO_2 Electrodes in Lithium Batteries. Tin oxide nanoparticles ranging in size from 3 to 8 nm were tested for battery capacity and cyclability. The 3-nm diameter particles showed the best characteristics of the group—ca. $740\,mAh \cdot g^{-1}$ with negligible capacity fading [180].

28.3.3 Hydrogen Production and Storage

Photocatalysis. Splitting water into hydrogen and oxygen is accomplished regularly and for over 3.5 billion years by photosynthetic organisms. However, how can we accomplish such a fundamental ancient task? In 1972, A. Fujishima and K. Honda achieved the photoelectrolysis of water with a TiO_2 photoanode with a platinum counter electrode. The chemical reactions involved with the photocatalytic splitting of water are given below [181]. The cell of Honda and Fujishima is shown in **Figure 28.21**.

$$TiO_2 + 2h\nu \rightarrow 2e^- + 2h^+ \tag{28.17}$$

$$H_2O + 2h^+ \rightarrow \tfrac{1}{2}O_{2(g)} + 2H^+ \tag{28.18}$$

$$2H^+ + 2e^- \rightarrow H_{2(g)} \tag{28.19}$$

$$H_2O + 2h\nu \rightarrow \tfrac{1}{2}O_{2(g)} + H_{2(g)} \tag{28.20}$$

not including the expression for the photocatalyst in the overall equation and where $h\nu$ = ultraviolet light applied to the photocatalyst.

Other Methods to Manufacture Hydrogen. There are many chemical and electrochemical means to produce hydrogen. The process of steam methane reforming

Fig. 28.21 *The original ultraviolet (<415 nm) photoelectrochemical cell devised by A. Fujishima and K. Honda in 1972 is depicted [181,182]. Water is electrolyzed without the application of an external voltage. The photoanode consisted of an n-type TiO_2 semiconductor material. The cathode was made of platinum. Modern day cells have reached an efficiency of 24% [182]. Charge separation is possible due to the presence of a potential that is provided ultimately from solar energy. In photosynthesis, such a gradient is established by the presence of a 5-nm thick lipid membrane that separates two different aqueous solutions in chloroplasts.*

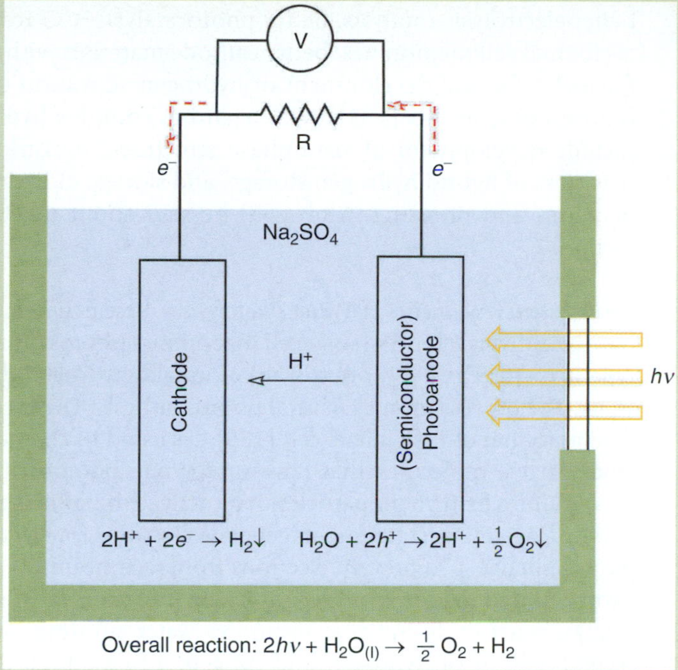

(SMR) at ca. 800°C is currently the most economic process to synthesize hydrogen. By an ancillary process of chemical shift, carbon monoxide and water react to form more hydrogen [183].

$$CH_4 + H_2O \rightarrow 3H_2 + CO \tag{28.21}$$

$$CO + H_2O \rightarrow H_2 + CO_2 \tag{28.22}$$

The process of partial catalytic oxidation (POX) of methane, natural gas, landfill gas, industrial wastes, solvents, oils, biomass, agricultural wastes, sewage sludge, kerosene, biodiesel, and other hydrocarbon sources proceeds at high temperatures (1100–1350°C) and can achieve ca. 50% conversion efficiency. Gasification occurs with the addition of steam and subsequent desulfurization to produce syngas. Shift reactions and purification results in the production of H_2 and CO_2 [183]. Coal gasification proceeds at 48% efficiency and exploits coal, biomass, and residual oil resources. Pure oxygen is required and the fuel must be in a pulverized state. The operating temperature is problematic: 1100–1300°C [183]. Nuclear-powered H_2 production is capable of driving thermochemical splitting of water but the specter of nuclear wastes and the dropping costs of advanced technologies make the nuclear option less attractive [183].

DOE goals include the development of novel synthetic techniques to achieve >40% efficiency at modest cost, development of materials to generate 10% efficiency solar to stored hydrogen energy, enhance nanocatalyst design to be more efficient, development of carbon-resistant reforming catalysts, better electrolysis catalysts, better photocatalysts, CO-resistant anode materials for fuel cell membranes, better cathode materials with lower overpotentials for fuel cells, and development of hydrogen activation catalysts that depend less on noble metals [183]. With regard to complex hydrides, DOE priorities include development of nanophase structures, transition metal dopants, a new class of hybrid hydrogen storage, and storage of hydrogen at various temperatures and pressures. More will be said about the last item in the next section.

Photosystem II Mimetics by Water Photolysis. Researchers at Penn State University have recently developed a solar cell that mimics photosynthesis [184]. The prototype is inefficient (~1%) but is able to make hydrogen directly from splitting water. The process mimics natural photosynthesis. The mechanism of the cell is similar to that of the Grätzel cell [185] discussed in chapter 26 except that electrons in dye molecules that have undergone photoexcitation are transferred onto iridium oxide nanoparticles that catalyze the splitting of water into oxygen and hydrogen ions (rather than generate electrons for electricity). The technological challenge is to prevent electrons from recombining with the dye and from hydrogen and oxygen from recombining into their most stable state—water. A catalyst must be found that promotes water splitting but does not promote recombination of hydrogen and oxygen. T.E. Mallouk and W.J. Youngblood recently presented their results at the American Association for the Advancement of Science Annual Meeting, at Boston in February 2008 [184].

Iridium oxide nanoparticles in direct contact with orange-red dye molecules (absorb in blue range, a high energy domain) formed clusters approximately 2 nm in diameter—the catalyst and the dye molecule were approximately the same size. The catalyst was impregnated into a TiO_2 anode. Pt served as the cathode. The electrodes were immersed in separate compartments in a salt solution. Each surface Ir complex is able to convert water into O_2 when receiving an excited electron from a dye molecule. The researchers claim that each surface iridium atom is able to catalyze the water-splitting reaction approximately 50 times per second—a level comparable to the turnover rate exhibited by PSII in green plants.

Tunability can be achieved according to Mallouk by improving the efficiency of the dye, improving the catalyst, and making adjustments to the geometry of the system [184]. The distance between molecules is one parameter that may be adjusted to direct electrons better—lengthening some paths and shortening others [184]. Mallouk et al. also reported about the ability of trivalent rhodium hydroxide nanoparticles on semiconducting layered calcium niobate for catalytic hydrogen evolution [186].

Hydrogen Storage (and a Real Scientific Debate). Hydrogen is the most abundant element in the universe and is the cleanest burning fuel. The combustion of hydrogen is straightforward with no detrimental by-products.

$$H_{2(g)} + \tfrac{1}{2}O_{2(g)} \rightarrow H_2O \tag{28.23}$$

What could be better than producing water from a combustion product or use of hydrogen in a fuel cell that also produces water? Hydrogen is easily manufactured from renewable energy sources [187], plentiful, and is lightweight. This is the perfect scenario, right? The only problem with the use of hydrogen is finding a safe, efficient, compact, and economical way to store it [188]. There are two issues facing both compression and liquefaction scenarios—two traditional methods to store hydrogen. First, hydrogen is explosive and will ignite at relatively low concentrations, for example, recall the fates of the Hindenburg airship that used compressed hydrogen or the space shuttle Challenger that used liquid hydrogen. Most hydrogen storage requires "hydrogen boil-off," venting that is most certainly a potential safety hazard. Second, compressing or cooling hydrogen is an energy-intensive process—not desirable from an economic perspective. Development of materials strong enough for use in storage canister walls would also present technological challenges. Quantum Technologies has developed storage canister equipped with composite fiber walls capable of withstanding 10,000 psi.

Other traditional methods of storing hydrogen are by gas-on-solid adsorption (using activated carbon or zeolites) and chemical/metal hydride materials (pure or alloyed metals react with H_2 to store ca. 2–10 wt%) [183].

Scientists are in general agreement that alternative methods of storing hydrogen by materials such as activated carbons and metal hydrides have not lived up to the goals established by the Department of Energy's *FreedomCar Program* roadmap—that on-board mobile hydrogen storage system should provide 6.5 wt% of H_2 in order to be effective as a transportation fuel [189]. The FreedomCar roadmap, by 2010, asks for onboard hydrogen storage systems to achieve $2\,kWh \cdot kg^{-1}$ (6 wt%), $1.5\,kWh \cdot L^{-1}$, and \$4/kWh. By 2015, the bar is raised to $3\,kWh \cdot kg^{-1}$ (9 wt%), $2.7\,kWh \cdot L^{-1}$, and \$2/kWh [189]. On carbon nanotubes or graphite, one hydrogen molecule per carbon atom yields 14 wt% of hydrogen. Other ways to store hydrogen need to be developed if the Freedom (from oil?) Car is to become a reality. Will the solution(s) come through nanotechnology?

Hydrogen Storage with Carbon Nanotubes? In 1997, Dillon et al. at the National Renewable Energy Laboratory (NREL) in Golden, Colorado claimed that single-walled carbon nanotubes were able to store hydrogen gas (at high density) inside the channels of the tubes by a condensation capillary mechanism [190]. The NREL group reported that single-walled carbon nanotubes, found in arc-produced soots, were able to store 8 wt% by weight of hydrogen at room temperature and moderate pressures [190]. The paper generated an incredible level of international interest and a whirlwind of controversy over the next decade (and now beyond). For example, a group at the Max Planck Institute in Germany, led by Michael Hirscher, disputed the claim by reporting that SWNTs adsorb less than 1 wt% H_2. Hirscher et al. reported that the metal alloy contaminants found in the arc-generated soot were responsible for H_2 adsorption reported by Dillon et al. [191].

Most scientists agreed in the years following that a physisorption mechanism mediated by the outside walls (e.g., interstitially between nanotubes that make up a bundle) of the tubes were able to attract H_2 to nanotubes—and that the mechanism of adsorption was not based on capillarity or condensation inside the tubes. Computational methods have since demonstrated that simple physisorption mechanisms cannot account for 6 wt% of hydrogen storage on pristine nanotubes [192]. According to J. Karl Johnson of the University of Pittsburgh, the fact that physisorption by itself is not able to account for the adsorption of hydrogen does not rule out other mechanisms [193]. Johnson showed that only 1–2 wt% hydrogen storage is possible on simulated pristine nanotubes at room temperature and reasonable pressures [188].

The shape of the thermally programmed desorption curve (TPD) also presented a forum for controversy. The curve obtained initially resembled that of a titanium alloy. As it turned out, an ultrasonication system utilized a Ti alloy (Al and V) during a "cutting" process [188]. Cutting is a process by which nanotube caps can be opened with the intent of allowing ingress of gas during exposure to H_2. From further study it was determined that the Ti alloys were able to adsorb 2.5–3.4wt% H_2 on their own.

Most recently, researchers at the Stanford Synchrotron Radiation Laboratory (SSRL) demonstrated that >7-wt% H_2 storage is possible with saturated (chemically modified with hydrogen) carbon nanotubes [194–196]. Theoretical analyses and total energy density functional theory (DFT) calculations predict that 7.7wt% storage is possible by chemisorption mechanisms by saturation of the carbon–carbon double bonds of the nanotube (**Fig. 28.22**) [197]. The mechanism of hydrogen adsorption is by the chemisorption process while the contribution from physisorption mechanisms was shown to be negligible at room temperature [194–196]. One likely problem associated with hydrogen storage on saturated carbon nanotube is the potential for slow kinetics during desorption due to the strength of the C–H bond [197].

A. Nikitin et al. claim that chemisorbed hydrogen can be desorbed and recombined at temperatures as low as 200–300°C, not entirely an unreasonable temperature in an automobile system [194–196]. The desorption temperature would be controlled by the desorption kinetics and the nanotube diameter—a characteristic that can be manipulated to tune the strength of the C–H bond—a true nanotechnology in action! The challenge then is to find an economic and efficient means of hydrogenating the SWNTs with molecular hydrogen. The *spillover process* is one such way [198]. H_2 molecules dissociate over the surface of a catalysts deposited on carbon nanotubes. Upon breakdown, hydrogen radicals spill over from the catalyst particle to the carbon nanotube and add across the double bonds to form C–H bonds—one nanotechnology helping out another [194–196]. Y.-W. Lee et al. have shown that Pt nanoparticles inside carbon nanotubes enhance the uptake of hydrogen fivefold [198].

Another aspect of the Stanford technology is that *non-bundled nanotubes* of the *same diameter* are required to provide maximum adsorption. Nanotubes, as we know so well by now, prefer to exist in a bundled state due to strong van der Waals attractions, and to date, no one has been able to fabricate nanotubes

Fig. 28.22 *An illustration of the partial hydrogenation of single-wall carbon nanotubes. Hydrogenation proceeds by exposing the nanotubes to atomic hydrogen and the hydrogen radicals add to the double bond of the SWNT. The hydrogenation process is reversible at 500–600°C [199]. Others claim reversibility at lower temperatures [194–196]. Smaller SWNTs are less stable against hydrogenation than larger tubes. Following hydrogenation, SWNTs undergo structural deformations, reduced electrical conductance, and increased semiconductor behavior [199]. The diameter of SWNTs increased from 1.0 to 1.3 nm or from 1.8 to 2.1 nm, depending on the starting material, following hydrogenation.*

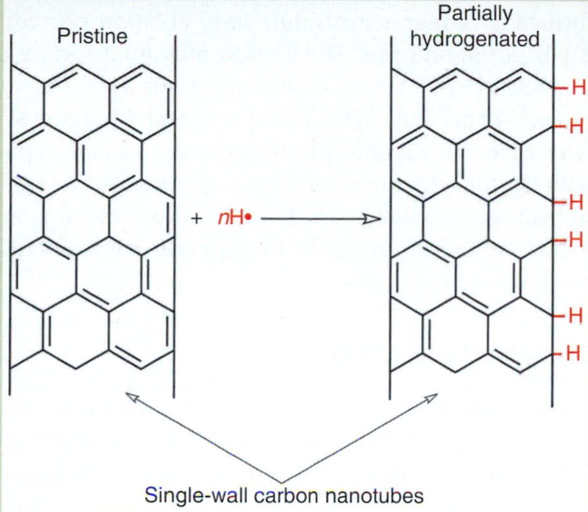

Source: G. Zhang, P. Qi, X. Wang, Y. Liu, D. Mann, X. Li, and H. Dai, *Journal of the American Chemical Society*, 128, 6026–6027(2006).

with the same diameter, electrical properties, and chirality without extensive postsynthesis separation.

Meanwhile, the investigation into Ti-decorated carbon nanotubes continues [200]. Recent theoretical calculations showed that light transition metal atoms that decorate one- and two-dimensional carbon-based nanostructures (carbon linear chain, graphene and nanotubes) were capable of hydrogen uptake. Interestingly, the researchers claim that even graphene would serve as a potential high-capacity H_2 storage material [200].

In October 2006, the DOE Hydrogen Program decided to pull the plug on research funding to continue work with pure, undoped, single-walled carbon nanotubes for vehicular hydrogen storage. The decision was based on the inability of pure SWNTs to store 6wt% hydrogen at close to room temperature and low pressure—experimental data show that adsorption of 0.6wt% hydrogen—an order of magnitude less than required. At cryogenic temperatures (77°C), only 3–6wt% was achievable at 20-bar pressure. The DOE report, however, affirms that certain areas of carbon nanotube research with metal-doped hybrid materials may warrant further investment by DOE [201]. The NREL group, for example, has achieved 3wt% storage at room temperature and nominal pressure on metal-doped SWNTs [202]. Perhaps the Ti-alloy impurity is promoting a spillover process that feeds hydrogen radicals to the nanotubes. Regardless of current obstacles, it seems likely that hydrogen storage by saturated or metal-doped nanotubes is not too far off in the future.

Storage with Single-Walled Carbon Nanohorns. David B. Geohegan et al. at Oak Ridge National Laboratory have a different approach to hydrogen storage and catalyst-assisted hydrogen storage—by application of single-walled carbon nanohorns (SWCNHs) with tunable pore size [203,204]. The SWCNHs can be therefore optimized, by control of synthesis parameters and postprocess treatment, for hydrogen adsorption and metal-catalyzed spillover. SWCNHs are formed by a high-temperature laser ablation procedure capable of generating $9\,g \cdot h^{-1}$ of nanohorns. The highest measured surface area of SWCNHs is equal to $1892\,m^2 \cdot g^{-1}$. Levels of hydrogen storage of 2.5 wt% were achieved with opened-Pt decorated SWCNHs [203,204]. Charging SWCNHs in an electric field is expected to significantly increase the storage capacity of hydrogen. Studies with C_{82} fullerenes revealed that a charge of $6e^-$ was able to increase the capacity of hydrogen storage to 8 wt% [205]. Single metal atoms and organic molecules inside fullerenes are able to generate sufficient electric fields to enhance hydrogen storage [205].

28.3.4 Fuel Cells

The fuel cell offers one of the panaceas to the energy challenge posed by Professor Smalley. Fuel cells are best described as a renewable electrochemical battery (or better, a reusable battery). The technology is developed enough so that potential consumer goods are likely on a large scale—products in the form of battery replacement devices, remote power generators, and automobile power. The fuel cell is based on the electrochemical conversion of hydrogen and oxygen to produce water. The result of the electrochemical process results in maximum 1.2 V and $1\,W \cdot cm^{-2}$ of power [206]. Nanostructured catalysts and membranes are integral in the design and operation of a fuel cell. A generic fuel cell is depicted on **Figure 28.23** [206].

Sir William Grove conceived the concept of the fuel cells as early as 1839. Grove thought that by reversing the process of electrolysis, in which hydrogen and oxygen are generated from the electrolysis of water, electricity could be produced. His first fuel cell was called the "gas voltaic battery." Ludwig Mond and Charles Langer coined the term fuel cell later on before the turn of the century. There are many kinds of fuel cells—the polymer exchange membrane (the one we focus on in this section), solid oxide fuel cell (high operating temperatures—generates heat and electricity), the alkaline fuel cell (requires pure oxygen and hydrogen and used in the space program), and the direct methanol fuel cell among others. The efficiency of a fuel cell powered by pure hydrogen can be as high as 80%—when converted into mechanical power, the overall efficiency is still quite high at 64%. The primary reactions of each are shown in **Table 28.5**.

The polymer (or proton) electrolyte membrane (PEM) serves as a conduction medium for protons. Nafion is a sulfonated tetrafluoroethylene copolymer ionomeric membrane invented by Du Pont in the 1960s and developed in the 1970s. Nafion is able to operate at relatively low temperatures—ca. 85°C. Nafion

Fig. 28.23 *A generic rendition of a fuel cell is depicted on the top. Below, a detailed description of the active region is shown.*

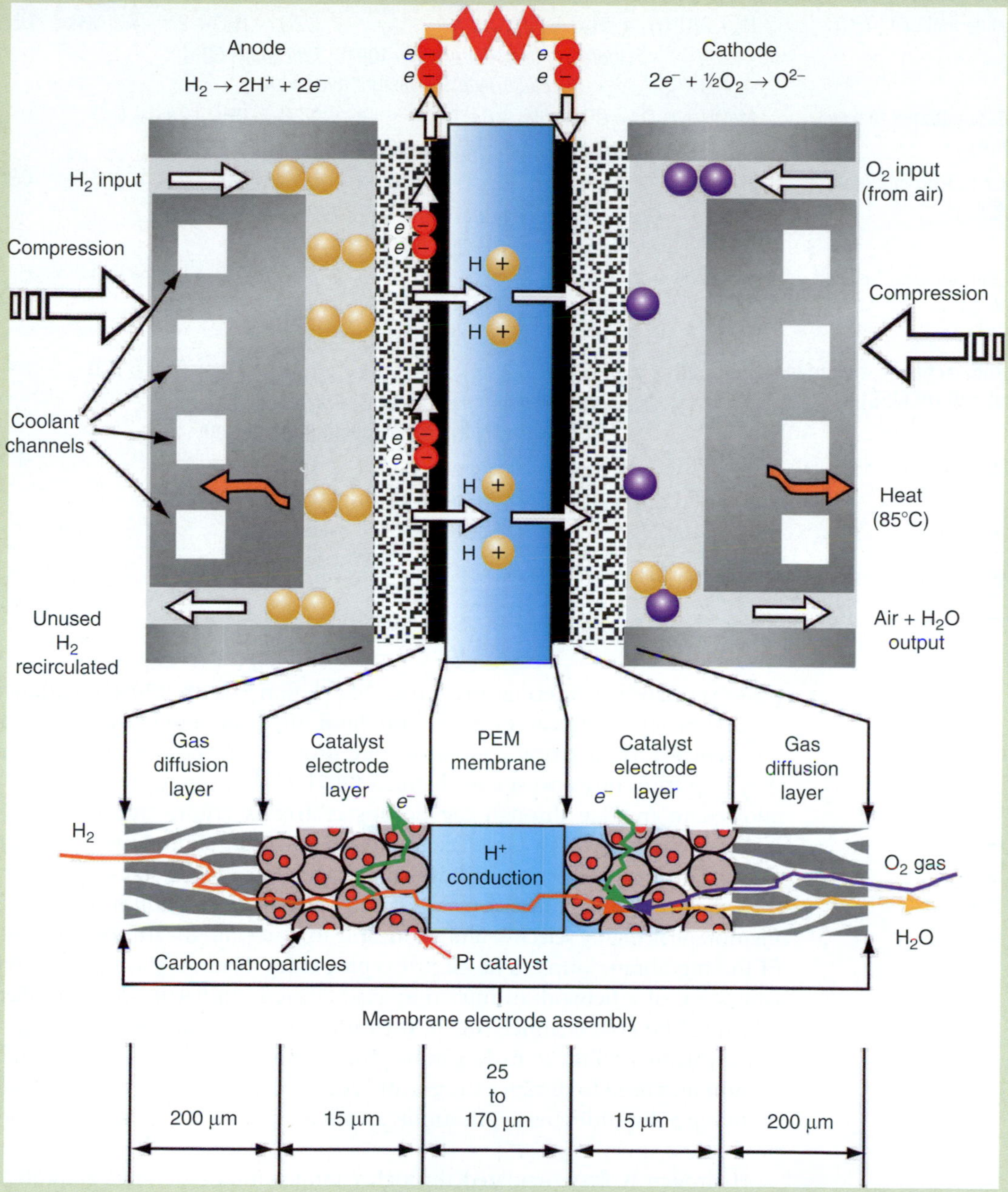

Source: Image courtesy of Dr. David Jacobson, National Institute of Standards and Technology. With permission.

TABLE 28.5 *Types of Fuel Cells and Anode/Cathode Reactions*

Type of fuel cell	Anode	Cathode	Ion
Alkaline fuel cell (AFC)	$H_2 + 2(OH)^- \rightarrow 2H_2O + 2e^-$	$1/2O_2 + H_2O + 2e^- \rightarrow 2(OH)^-$	OH^-
	Operating temperature: 80–100°C; *Efficiency*: 60%; *Electrolyte*: Potassium hydroxide		
Direct methanol fuel cell (DMFC)	$MeOH + H_2O \rightarrow CO_2 + 6H^+ + 6e^-$	$1/2O_2 + 2H^+ + 2e^- \rightarrow H_2O$	H^+
	Operating temperature: 60–90°C; *Efficiency*: 25%		
Molten carbonate fuel cell (MCFC)	$H_2 + CO_3^{2-} \rightarrow CO_2 + H_2O + 2e^-$	$CO_2 + 1/2O_2 + 2e^- \rightarrow CO_3^{2-}$	CO_3^{2-}
	Operating temperature: 600–650°C; *Efficiency*: 45%–60% *Electrolyte*: Carbonate melt		
Phosphoric acid fuel cell (PAFC)	$H_2 \rightarrow 2H^+ + 2e^-$	$1/2O_2 + 2H^+ + 2e^- \rightarrow H_2O$	H^+
	Operating temperature: 200–220°C; *Efficiency*: 40%–45% *Electrolyte*: Phosphoric acid		
Proton exchange membrane fuel cell (PEMFC)	$\mathbf{H_2 \rightarrow 2H^+ + 2e^-}$	$\mathbf{1/2O_2 + 2H^+ + 2e^- \rightarrow H_2O}$	$\mathbf{H^+}$
	***Operating temperature*: 70–80°C; *Efficiency*: 35%–45% *Electrolyte*: Hydrogen conducting membrane**		
Solid oxide fuel cell (SOFC)	$H_2 + 2O^{2-} \rightarrow H_2O + 2e^-$	$1/2O_2 + 2e^- \rightarrow O^{2-}$	O^{2-}
	Operating temperature: 800–1000°C; *Efficiency*: 50%–65% *Electrolyte*: Solid oxide		

Note: The proton exchange fuel cell system is in bold in the table.

possesses excellent thermal properties like Teflon (up to 190°C), conductive properties, and chemical resistance and durability (does not release fragments). The vehicles for electrical conduction are protons on the sulfonate groups as they hop from one acid site to the next [207]. Only cations (and protons) are allowed to migrate through the porous matrix as anions and electrons are blocked. It is known as a super-acid catalyst because the sulfonic groups attached to the fluorinated (electron withdrawing) backbone and the stabilizing effect of the extended polymer matrix make Nafion a strong proton donor ($pK_a < 0$). The membrane is highly selective and permeable to water and the degree of hydration of the membrane impacts its electrical properties. Nafion is a nanomaterial—consisting of a network of interconnected clusters similar in form to inverted micelles. The pore channels are on the order of 4 nm in diameter although the exact structure is difficult to determine [207]. Nafion is also used in ion-exchange membranes used to produce Cl_2 gas and NaOH during the electrolysis of water, in drying or humidifying applications, and as a super-acid catalyst in the production of fine chemicals [208].

Hydrogen is first circulated through channels made of solid graphite and then passed into a porous mesh layer called the "gas diffusion media" [206]. Once well into the media, the hydrogen gas contacts graphitic nanoparticle supported platinum catalyst and is stripped of electrons. The electrons are swept into a current emanating from the anode. Meanwhile, the remaining protons make their way through the PEM to the cathode and encounter another layer of carbon-supported platinum catalysts. The protons interact with oxygen and incoming electrons from the external circuit to form liquid water that is drained from the system [206].

28.3.5 *Solar Heating and Power Generation*

Although not as exciting, exotic, or sophisticated as its photovoltaic cousins, solar heating systems, as a general rule, are able to convert 25%–40% of solar radiation into usable heat [144]. For a small hot water heating system (50 L per person at 45°C per day), the recommended dimensions of a solar collector are on the order of 1.2–1.5 m^2 [144]. The principle of solar heating is straightforward. The process is based on a material that absorbs sunlight energy and releases it in the form of heat whether directly to a water source or to a heat exchange element (heat pump). There are two general types of solar collectors: the *flat plate collector* in which solar energy absorbing coated pipes are enclosed within a glass paneled box, and a solar water heater that employs a *batch collector* that combines the collection and storage systems within one device. Any material that is able to enhance surface area or absorption properties are sure to benefit this aspect of the solar energy applications—enter nanotechnology.

Solar Concentration. The efficiency of solar cells can be increased by external means as well—by the use of concentrating solar collectors that apply Fresnel lenses and parabolic mirrors to concentrate and focus solar energy onto solar cells. Advantages include the use of smaller areas for solar cells and increasing the level of solar flux impinging on the cells. Another advantage is afforded by the installation of solar tracking systems [209].

Concentrating solar energy for heating by an array of reflective mirrors is a means of providing heat to power a generator that is used to create electricity. Such systems require a large amount of space and generous helpings of direct sunlight. Parabolic trough collectors consist of a parabolic mirror with an absorber tube placed at its center, thereby heating the liquid within the tube. The heated effluent is used to boil water that consequently produces steam to run a generator.

Geothermal Energy. Nanotechnology contributions to extracting energy by geothermal mechanisms include development of new materials that exhibit enhanced thermal conductivity or corrosion resistance [210]. Nanoscale materials are expected to better manage geothermal reservoirs, acquire subsurface information, and enhance thermoelectric materials that can detect small thermal gradients. Transport of thermal energy (100–200°C) to within 1 km is practical if geothermal energy sources are to prove useful. Strength enhancement of geothermal drilling materials is also required as higher temperatures are encountered.

28.4 Epilogue

We have reached the end of our adventure in nanotechnology with this last chapter—the discussions about the environmental aspects. What we have observed over the course of this book, and perhaps from Part I as well, is that nanotec nology is causing a paradigm shift with regard to properties, capability, fabrication and applications. For example, the micro-age was based on litho phy. Although lithography is still a viable and thriving top down, not m mention was made of this technique in the latter chapters—rather emphasis placed on bottom-up self-assembly methods. The shift of fabrication techniq from the engineer to the chemist is in full sway as more amazing and incredi devices are created. Another trend in nanotechnology is also making itsel known—the relationship between nature and a new generation of devices. Nature also makes things from the bottom up—from the basic amino acids, carbohydrates, and lipids.

New methods to clean water and store hydrogen, among many other projects, are constantly under development [211–224]. We take No. 1 and 2 from Smalley's list of the grand challenges that face humanity and add a few more examples to take with you: that of the bottom-up assembly of materials that remove heavy metals to purify water (by SAMMS) and the storage of hydrogen with metal organic frameworks (MOFs)—both examples of true chemical synthesis bottom-up nanotechnology.

28.4.1 SAMMS

SAMMS are self-assembled monolayers on mesoporous supports—the combination of organic and inorganic materials to form a device. SAMMS serve as the perfect example of where nanotechnology will take us—better detection, sensing and overall capability, higher throughput, and ultimately more economical and environmental sense.

SAMMS. Self-assembled monolayers on mesoporous supports (SAMMS) provide a new generation of SAM technology applied to removal of waterborne contaminants—a pure and true nanotechnology! Novel nanocomposite sorbents based on copper ferrocyanide immobilized within a mesoporous ceramic matrix have shown great promise in removing Cs from water streams—complete removal of Cs in the presence of competing metal ions under a variety of conditions for solutions in which the [Cs] = 2 ppm [211–213]. The Pacific Northwest Laboratory group reported loading capacities of more than 1.35 mmol $Cs \cdot g^{-1}$ adsorbent [211–213]. The system displayed fast binding kinetics that was due to the rigid pore structure and the extremely high surface area of the ceramic sorbent material. DOE has put into place some goals designed to promote the reduction of vitrified wastes—specifically that just the radionuclide-laden SAMMS need to be vitrified and not the whole volume of all solid wastes [211–213]. Since actinides and other heavy metals form insoluble hydroxides, separation methods usually require alteration of pH to less than 4—a process that doubles the waste volume [211–213]. SAMMS, on the other hand, display solution-solid rations of ca. 100 (and even higher)—a process that actually reduces the amount of vitrified waste by a factor of 50 [211–213]. An image depicting a SAMMS molecular sponge adsorbent is shown in **Figure 28.24**.

SAMMS technology is based on three generations of self-assembled monolayer work. First, the development of surfactant molecules to create micellular templates was accomplished; second, aggregation of silica-coated micelles into

. 28.24

A nanostructured adsorbant SAMMS is depicted. Starting with micelles that served as the original template material, a silaceous gel was generated. Upon high-temperature treatment, the organic component was removed leaving the silica skeleton—a highly porous, uniform, and interconnected system. After activation of the silica surface, monolayers were attached by application of specially modified silanes or other linking groups. Quite a bit of bottom–up nanotechnology is involved in its fabrication and operation. Depending on the functional groups, the size, and configuration of the pore channels, a wide range of selectivity can be achieved.

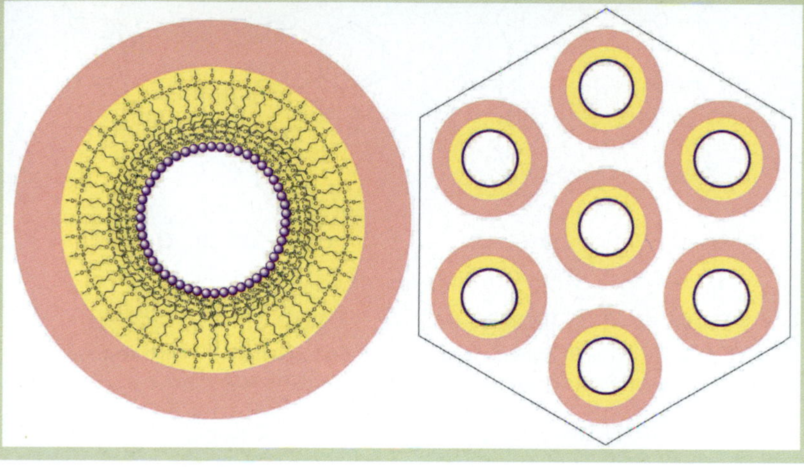

Source: Image courtesy of Pacific Northwest National Laboratory. With permission.

a mesostructured material occurred with high surface area (~1000 $m^2 \cdot g^{-1}$); and finally, self-assembly of silanes into an ordered monolayer on the surface of the silicate pore structure with a high concentration of binding groups [211–213]. Many kinds of SAMMS can be prepared, for example, glycinyl-urea (Gly-UR) and salicyclamide (**Fig. 28.25**).

28.4.2 One More Pass at Hydrogen Storage

We bring up hydrogen storage one more time because it is one of the most important challenges facing modern nanotechnology. Carbon nanotubes (inconsistency) and metal hydride (cost, low specific uptake, unfavorable kinetics and contamination) storage systems have significant limitations [214]. Metal-organic frameworks (MOFs) provide for a new medium of hydrogen storage [214–217]. Dissociative methods of storing hydrogen (discussed earlier) are hindered by large H–H bond dissociation thresholds. Associated adsorption, the case in which H_2 is adsorbed as a molecule, does not suffer from such an impediment [217]. MOFs are crystalline solids composed of metal clusters and organic ligands. A 7.5wt% level of storage was demonstrated with a type of MOF called MOF-177 at 60 bar and 77 K—conditions that are not particularly practical [217]. Recently, by molecular simulation methods, it was demonstrated that Li-doped MOF (called Li-MOF-C30) showed significantly improved results at much higher H_2 saturation temperatures of –30°C @ 100-bar [217]. The Li-doped material was able to adsorb 6.0wt% H_2. One of the DOE FreedomCar

Fig. 28.25 *Various types of SAMMS ligands are depicted. Depending on the end group, a high level of versatility can be achieved to targets specific materials. HOPOs are hydroxy pyridinones. Thiol-derivatized SAMMS are good for cleaning up Pd, Ag, Cd, Pt, Au, Hg, Tl, and Pb; Cu-FC-EDAs are tailor made to extract Cs; CU-EDAs target V, Cr, Mo, Tc, Ge, As, Se, Br, and I (EDA, ethylene diamines form a system of three bidentate ligands to bind Cu^{2+}; HOPOS and Prop-phos work for Th, Pa, U, Np, Pu, and Am mitigation) [211–213].*

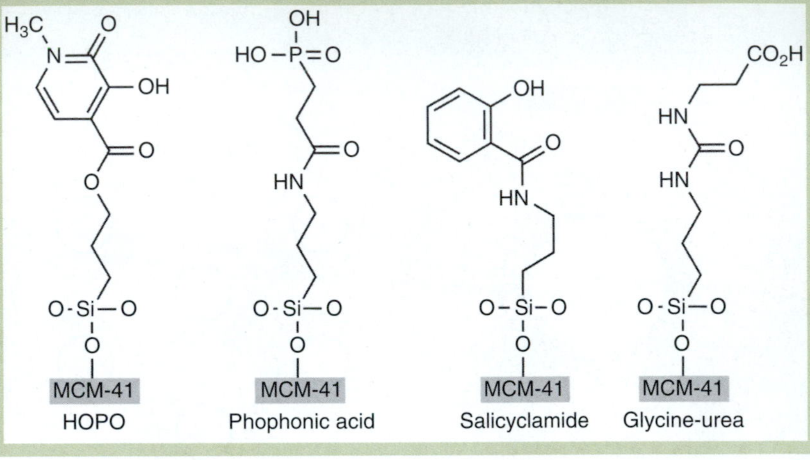

criteria is to make H_2 storage practical between −30 and 80°C. This material, therefore, shows high compatibility for hydrogen storage.

Highly porous metal-organic frameworks (**Fig. 28.26**) have uniform-sized cavities within a crystalline-metal-organic frame consisting of $[OZn_4]$ conjoined in an octahedral array by $[O_2C–C_6H_4–CO_2]^{2-}$ (a.k.a. 1.4-benzenedicarboxylate or BDC). Other kinds of ligands, for example, the tetrazolates, have four nitrogens at the disposal of metals wishing to become coordinated. The adsorption

Fig. 28.26 *Example of MOF cage structures. The yellow sphere represents the scope of the enclosed volume (and not hydrogens). The zinc oxide (blue tetrahedrons and orange dots respectively) are linked with organic ligands. Black dots represent carbon atoms. Very clever—very nano!*

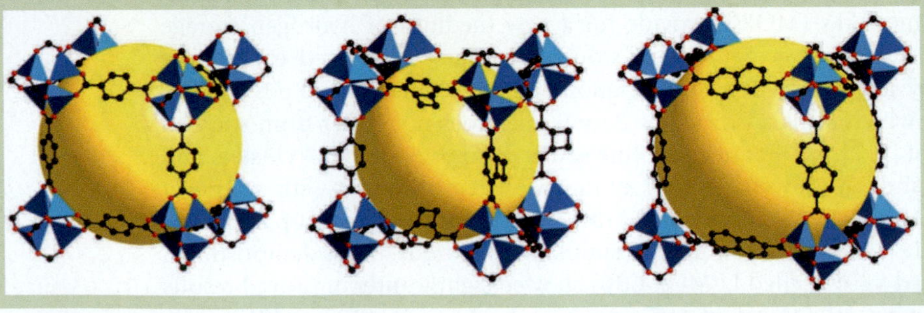

Source: N. L. Rosi, J. Eckert, M. Eddaoudi, D. T. Vodak, J. Kim, M. O'Keefde, and O. M. Yaghi, *Science*, 300, 1127–1129 (2003). With permission.

enthalpy of this MOF was as low as 10 kJ · mol^{-1} and an adsorption capability of 6.9 wt% (60 g · L^{-1}) at 77 K and 90 bar—one of the best demonstrated so far—this according to Professor Jeffery R. Long of the Department of Chemistry at U.C. Berkeley (alchemy.cchem.berkeley.edu/hydrogen.html).

28.4.3 Concluding Thoughts

We explored, among many things, the mechanisms, materials, and devices of organic photocells [218], practical applications of bimetallic nanoparticles for groundwater treatment [219], the mechanisms of activated carbon [220], how cadmium chalcogenide quantum dots are anchored onto stable oxide semiconductors to make QD-sensitized solar cells [221,222], how to make SAMMS for specific environmental remediation of heavy metals [223], and fabrication of ordered SAMs onto gold for the purposes of sensing [224]. There is so much more. Sadly, it is impossible to cover all of nanotechnology in a single textbook. The purpose of the book is to expose the student to broad aspects of this wondrous new area and perhaps, hopefully, inspire that student to apply his or her newly acquired knowledge to help protect our environment.

Please keep in mind Smalley's ten grand challenges and how you, perhaps, could be the scientist that discovers ways to mitigate them. As the late Richard Everett Smalley stated "Be a scientist. Save the World!" Best of luck to all of you.

Acknowledgments

We wish to thank with great appreciation the Department of Energy, the National Renewable Energy Laboratory and the National Institute for Standards and Technology for providing information that was vital to the content of this text.

References

1. EPA, Measurement and analysis of Adsistor and Figaro gas sensors used for underground storage tank leak detection, U.S. Environmental Protection Agency (EPA), Report #EPA/600/R-92/219 (1992).
2. EPA, Measurement and analysis of vapor sensors used at underground storage tank sites, U.S. Environmental Protection Agency (EPA), Report #EPA/600/R-95/078 (1995); T. M. Masciangioli, N. Savage, and B. P. Karn, Environmental implications of nanotechnology, Poster, www.environmentalfutures.org/nanotech.htm (2004).
3. K. I. Mullen and K. T. Carron. Surface enhanced Raman spectroscopy with abrasively roughened fiber optic probes, *Analytical Chemistry*, 63, 2196–2199 (1991).
4. J. C. Cook, C. M. P. Cuypers, B. J. Kip, R. J. Meier, and E. Koglin, Use of colloidal SERS for the trace detection of organic species: A survey based on pyridine as the probe molecule, *Journal of Raman Spectroscopy*, 24, 609–619 (2007).
5. C. J. L. Constantino, T. Lemma, P. A. Antunes, and R. Aroca, Single molecular detection of a perylene dye dispersed in a Langmuir–Blodgett fatty acid monolayer using surface-enhanced resonance Raman scattering, *Spectrochimica Acta Part A: Molecular and Biomolecular Spectroscopy*, 58, 403–409 (2002).
6. H. Krug, ed., *Nanotechnology, Volume 2: Environmental aspects*, Wiley-VCH, Weinheim (2008).

7. R. H. Grove, *Green imperialism: Colonial expansion, tropic island edens and the origins of environmentalism*, Cambridge University Press, Cambridge, UK (1995).
8. Ancient Civilizations, *Environmental History Timeline*, www.runet.edu/~wkovarik/envhist/1ancient.html (accessed 2007).
9. A. Markham, *A brief history of pollution*, St. Martin's, New York (1994).
10. J. Ewing and D. Quinn, *Monkeys are made of chocolate*, PixyJack Press, LLC., Masonville, CO (2005).
11. The flow of energy: Primary production to higher trophic levels, University of Michigan, www.globalchange.umich.edu/globalchange1/current/lectures/kling/energyflow/energyflow.html (2005).
12. S. Morgan, Nanotechnology applications in water management, Presentation, Nanotechnology Victoria, Nanotechnology_Water_Management_July06.pdf (2006).
13. T. Hatakeyama, D. L. Chen, and R. F. Ismagilov, Microgram-scale testing of reaction conditions in solution using nanoliter plugs in microfluidics with detection by MALDI-MS, *Journal of American Chemical Society*, 128, 2518–2519 (2006).
14. Helping you explore, be novel, be first in nanotechnology, www.agilent.com/find/nano (2006).
15. Introduction to Micro/Nano-HPLC, www.innosep.com/2.htm (2007).
16. S. K. Sundaram and T. J. Weber, New nano-canary in the nanotoxicology coalmine: The body itself, Pacific Northwest National Laboratory, www.physorg.com/news10971.html (2006).
17. A. Scheffer, C. Engelhard, M. Sperling, and W. Buscher, ICP-MS as a new tool for the determination of gold nanoparticles in bioanalytical applications, *Analytical and Bioanalytical Chemistry*, 390, 249–252 (2007).
18. X. Li, X. Tu, S. Zaric, K. Welsher, W.-S. Seo, W. Zhao, and H. Dai, Selective synthesis combined with chemical separation of single-walled carbon nanotubes for chirality selection, *Journal of American Chemical Society*, 129, 15770–15771 (2007).
19. N. Tao, A nanocontact sensor for heavy metal ion detection, Proceedings: EPA Nanotechnology and the environment: Applications and implications, STAR (Science to achieve results) Progress Review Workshop, Arlington, VA (2002).
20. O. Hayden and C. K. Payne, Nanophotonic Light Sources for Fluorescence Spectroscopy and Cellular Imaging, *Angewandte Chemie International Edition*, 44, 1395–1398 (2005).
21. D. Dosev, M. Nichkova, M. Liu, B. Guo, G.-Y. Liu, B. D. Hammock, and I. M. Kennedy, Application of luminescent Eu:Gd_2O_3 nanoparticles to the visualization of protein micropatterns, *Journal of Biomedical Optics*, 10, 064006–064012 (2005).
22. G. Hanrahan, D. G. Patil, and J. Wang, Electrochemical sensors for environmental monitoring: design, development and applications, *Journal of Environmental Monitoring*, 6, 656–664 (2004).
23. Y. Tu, Y. Lin, W. Yantasee, and Z. Ren, Carbon nanotube based nanoelectrode arrays: fabrication, evaluation, and application in voltammetric analysis, *Electroanalysis*, 17, 79–84 (2005).
24. M. Watari and J. Ransley, Label-free nanomechanical protein detection on sensor arrays, *IRC in nanotechnology*, Cambridge Nanoscience, www.nanoscience.cam.ac.uk/irc/characterisation/22.html (2006).
25. P. Datskos, Chemical sensors based on nanomechanical resonators, *Oak Ridge National Laboratory-fact sheet*, www.ornl.gov/sci/engineering_science_technology/sms/Hardy%20Fact%20Sheets/Chemical%20Sensors.pdf (viewed 2008).
26. L. C. Clark, R. Wolf, D. Granger, and Z. Taylor, Continuous recording of blood oxygen tensions by polarography. *Journal of Applied Physiology*, 6, 189–193 (1953).
27. L. C. Clark, Jr. and C. Lyons, Electrode systems for continuous monitoring in cardiovascular surgery, *Annals of New York Academy of Science*, 102, 29–45 (1962).

28. L. P. Clark, people.clarkson.edu/~ekatz/scientists/clark_leland.htm (2007).
29. Biosensor, en.wikipedia.org/wiki/Biosensor (2008).
30. M. Chaplin and C. Bucke, *Enzyme technology*, Cambridge University Press, Cambridge, UK (1990).
31. J. Liu and Y. Lu, Smart nanomaterials responsive to multiple chemical stimuli with controllable cooperativity, *Advanced Materials*, 18, 1667–1671 (2006).
32. Aptamer, en.wikipedia.org/wiki/Aptamer (2008).
33. D. M. Held, J. D. Kissel, J. T. Patterson, D. G. Nickens, and D. H. Burke, HIV-1 inactivation by nucleic acid aptamers, *Frontiers in Bioscience*, 11, 89–112 (2006).
34. M. Curreli, C. Li, Y. Sun, B. Lei, M. A. Gundersen, M. E. Thompson, and C. Zhou, Selective functionalization of In_2O_3 nanowire mat devices for biosensing applications, *Journal of American Chemical Society*, 127, 6922–6923 (2005).
35. E. W. L. Chan, M. N. Yousaf, and M. Mrksich, Understanding the role of adsorption in the reaction of cyclopentadiene with an immobilized dienophile, *Journal of Physical Chemistry A*, 104, 9315–9320 (2000).
36. N. V. Lavrik, M. J. Sepaniak, and P. G. Datskos, Cantilever transducers as a platform for chemical and biological sensors, *Review of Scientific Instruments*, 75, 2229–2253 (2004).
37. J. Fritz, M. K. Baller, H. P. Lang, H. Rothuizen, P. Vettiger, E. Meyer, H.-J. Guntherodt, C. Gerber, and J. K. Gimzewski, Translating biomolecular recognition into nanomechanics, *Science*, 288, 316–318 (2000).
38. H.-F. Ji, R. Dabestani, G. M. Brown, and P. F. Britt, A novel self-assembled monolayer (SAM) coated microcantilever for low level caesium detection, *Chemical Communication*, 6, 457–458 (2000).
39. Nanostellar introduces gold in oxidation catalyst that can reduce diesel hydrocarbon emissions by as much as 40 percent more than commercial catalysts, www.nanotech-now.com/news.cgi?story_id= 21935 (2007).
40. Rational catalyst design, Nanostellar, Inc., www.nanostellar.com/technology.htm (2007).
41. C.-B. Wang and W. X. Zhang, Synthesizing nanoscale iron particles for rapid and complete dechlorination of TCE and PCBs, *Environmental Science and Technology*, 31, 2154–2156 (1997).
42. C. A. Dunn, NASA announces 2005 invention of the year: Environmental cleanup technology earns top honors, *NASA technology innovation*, 13 (2006).
43. S. V. Mattigod, G. E. Fryxell, and K. E. Parker, A thiol-functionalized nanoporous silica sorbent for removal of mercury from actual industrial waste, In *Environmental applications of nanomaterials: Synthesis, sorbents and sensors*, G. E. Fryxell and G. Cao, eds, chap. 11, Imperial College Press, London, 275–284 (2007).
44. S. V. Mattigod, G. E. Fryxell, X. Feng, K. E. Parker, and E. M. Pierce, Removal of mercury from aqueous streams of fossil fuel power plants using novel functionalized nanoporous sorbents, In *Coal combustion byproducts and environmental issues*, K. S. Sajwan, I. Twardowska, T. Punshon, A. K. Alva, eds, chap. 10, Springer, New York, 99–104 (2006).
45. Patented electro-static filter cartridges designed for environmental water purification, industrial water clean-up, RO protection and ultra-pure water, Argonide Advanced Filtration Technologies, www.argonide.com (2006).
46. Manganese oxide nanopowder, American Elements Products, www.americanelements.com (2008).
47. Nanoengineered sorbents, *Chemical and Engineering News*, 79, 32–38 (2001).
48. Y. Zhang, B. A. Koeneman, Y. Chen, P. Westerhoff, D. G. Capco, and J. Crittenden, Fate, transport, and toxicity of nanomaterials in drinking water, Nanoscience and Technology Institute, Symposium: Micro/Nano-Technology for National Security Applications, NSTI Nanotech 2007, Santa Clara, May 20–24 (2007).

49. D. Talley, Aerogel process nears market, CDT Systems, Inc., cdtwater.com/press/0401_H2Onews.php (2004).
50. G. Prabhukumar, M. Matsumoto, A. Mulchandani, and W. Chen, Cadmium removal from contaminated soil by tunable biopolymers, *Environmental Science and Technology*, 38, 3148–3152 (2004).
51. G. C. Bond and D. T. Thompson, Catalysis by gold, *Catalysis Review-Science and Engineering*, 41, 319–388 (1999).
52. C. W. Corti, R. J. Holliday, and D. T. Thompson, Developing new industrial applications for gold: Gold nanotechnology, *Gold Bulletin*, 35, 111–118 (2002).
53. C. W. Corti, Going green, *Annual Report 2006*, Rand Refinery Limited, pp. 38–41 (2006).
54. Toyota Chuo Kenkyushu K. K., Japanese Patent, 9150033 (1997).
55. R. Holliday, Evolving industrial uses of gold, *Annual Report 2006*, Rand Refinery Limited, pp. 34–37 (2006).
56. D. Andreeva, V. Idakiev, T. Tabakova, and A. Andreev, Low temperature water-gas shift reactions over Au/Fe_2O_3, *Journal of Catalysis*, 158, 354–355 (1996).
57. G. Mills and M. R. Hoffmann, Photocatalytic degradation of pentachlorophenol on titanium dioxide particles: Identification of intermediates and mechanism of reaction, *Environmental Science and Technology*, 27, 1681–1689 (1993).
58. M. A. Fox and M. T. Dulay, Heterogeneous photocatalysis, *Chemical Reviews*, 93, 341–357 (1993).
59. K. Venkatachalam, V. G. Gavalas, S. Xu, A. C. de Leon, D. Bhattacharyya, and L. G. Bachas, Poly (Amino Acid)-facilitated electrochemical growth of metal nanoparticles, *Journal of Nanoscience and Nanotechnology*, 6, 2408–2412 (2006).
60. D. Ratner and M. A. Ratner, *Nanotechnology and homeland defense: New weapons for new wars*; Pearson Educations, Inc., Prentice-Hall, NJ (2004).
61. J. Indusi, Advanced sensors and detectors, Brookhaven National Laboratory, http://www.bnl.gov/homeland/sensors.asp (2006).
62. S. Keebaugh, W. J. Nam, and S. J. Fonash, Manufacturable, highly responsive nanowire mercury sensors, Nanoscience and Technology Institute, Symposium: Micro/Nano-Technology for National Security Applications, NSTI Nanotech 2007, Santa Clara, May 20–24 (2007).
63. A. Loui, T. V. Ratto, T. S. Wilson, E. V. Mukerjee, Z.-Y Hu, T. A. Sulchek, and B. R. Hart, A compact, low-power cantilever-based sensor array for chemical detection, Nanoscience and Technology Institute, Symposium: Micro/Nano-Technology for National Security Applications, NSTI Nanotech 2007, Santa Clara, May 20–24 (2007).
64. P. Bondavilli, P. Legagneux, and D. Pribat, Gas fingerprinting using carbon nanotubes transistor arrays, Nanoscience and Technology Institute, Symposium: Micro/Nano-Technology for National Security Applications, NSTI Nanotech 2007, Santa Clara, May 20–24 (2007).
65. S. L. Harper, B. Maddux, J. Hutchison, and R. L. Tanguay, Biodistribution and toxicity of nanomaterials in vivo: Effects of composition, size, surface functionalization and route of exposure, Nanoscience and Technology Institute, Symposium: Micro/Nano-Technology for National Security Applications, NSTI Nanotech 2007, Santa Clara, May 20–24 (2007).
66. T. Knight, S. S. Wise, M. D. Mason, J. P. Wise, Sr., and A.-K. Ng, Cell-based assays for cytotoxic and pro-inflammatory effects of gold nanoparticles, Nanoscience and Technology Institute, Symposium: Micro/Nano-Technology for National Security Applications, NSTI Nanotech 2007, Santa Clara, May 20–24 (2007).
67. J. Chen, J. Zhu, H.-H. Cho, K. Cui, F. Li, X. Zhou, J. T. Rogers, S. T. C. Wong, and X. Huang, Differential cytotoxicity of metal oxide nanoparticles, Nanoscience and

Technology Institute, Symposium: Micro/Nano-Technology for National Security Applications, NSTI Nanotech 2007, Santa Clara, May 20–24 (2007).

68. R. Sharma, K. Shetty, R. Liang, and C. J. Chen, Interaction of carbon nanotube material with rat skin by 21 T MRI, Nanoscience and Technology Institute, Symposium: Micro/Nano-Technology for National Security Applications, NSTI Nanotech 2007, Santa Clara, May 20–24 (2007).
69. K. Spencer, Detection of landmines with SERS, Application Summary, EIC Laboratories, Inc. (2004).
70. J. M. Sylvia, J. A. Janni, J. D. Klein, and K. M. Spencer, Surface-enhanced Raman detection of 2,4-dinitrotoluene impurity vapor as a marker to locate landmines, *Analytical Chemistry*, 72, 5834–5840 (2000).
71. N. T. Kawai and K. M. Spencer, Raman Spectroscopy for Homeland Defense Applications, Application Note #18, InPhotonics, Inc. and EIC Laboratories (2004).
72. K. M. Spencer, J. M. Sylvia, J. A. Janni, and J. D. Klein, Advances in land mine detection using surface-enhanced Raman spectroscopy, *Proceedings of SPIE*, Vol. 3710 (2003).
73. G. L. Hornyak, C. J. Patrissi, and C. R. Martin, Surface enhanced Raman simulations of prolate and oblate nanoparticles, *Nanostructured Materials*, 9, 705–710 (1997).
74. Explosives detecting immunosensors, ESTCP Cost and Performance Report, CU-9713, U.S. Department of Defense (2000).
75. P. K. Stoimenov and K. J. Klabunde, Biological and chemical weapon decontamination by nanoparticles, In *Dekker encyclopedia of nanoscience and nanotechnology*, J. A. Schwartz, C. I. Contescu, and K. Putyera, eds., Vol. 1, pp. 241–245, CRC Press, Boca Raton, FL (2004).
76. G. W. Wagner, L. R. Procell, R. J. O'Connor, S. Munavalli, C. L. Carnes, P. N. Kapoor, and K. J. Klabunde, Reactions of VX, GB, GD, and HD with Nanosize Al_2O_3: Formation of aluminophosphonates, *Journal of American Chemical Society*, 123, 1636–1644 (2001).
77. P. K. Stoimenov, R. L. Klinger, G. L. Marchin, and K. J. Klabunde, Metal oxide nanoparticles as bactericidal agents, *Langmuir*, 18, 6679–6686 (2002).
78. R. J. Nikolic, C. L. Cheung, C. E. Reinhardt, and T. F. Wang, Future of semiconductor based thermal neutron detectors, Symposium: Micro/nano technologies for national security applications—explosive detection & radiological defense, Nanoscience and Technology Institute, NSTI Nanotech 2006, Boston, May 7–11 (2006).
79. T. M. Masciangioli, N. Savage, and B. P. Karn, Environmental implications of nanotechnology, Poster, www.environmentalfutures.org/nanotech.htm (2004).
80. V. Colvin, Nanotechnology: Environmental impact, Presentation, Nanotechnology and the environment: Background and resources, www.environmentalfutures.org/nanotech.htm (date unknown).
81. Nanomaterials and the Environment, In *Environmental, health, and safety research needs for engineered nanoscale materials*, chap. 4, The National Nanotechnology Initiative, NSET Subcommittee, Office of Science and Technology Policy, Washington D.C. (2006).
82. G. V. Andrievsky, V. K. Klochkov, E. L. Karyakina, and N. O. Mchedlov-Petrossyan, Studies of aqueous colloidal solutions of fullerene C-60 by electron microscopy, *Chemical Physics Letters*, 300, 392–396 (1999).
83. J. K. Holt, H. G. Park, Y. Wang, M. Stadermann, A. B. Artyukhin, C. P. Grigoropoulos, P. Costas, A. Noy, and O. Bakajin, Fast mass transport through sub-2-nanometer carbon nanotubes, *Science*, 312, 1034–1037 (2006).

84. J. K. Holt, Fast water transport through carbon nanotubes and implications for water treatment, Desalinating Water Cheaply—Exploring Technologies, Australian Academy of Sciences, Canberra, Australia (2006).
85. The history of drinking water, Environmental Protection Agency, Office of Water, Report EPA-816-F-00-006 (2000).
86. R. C. Bansal and M. Goyal, *Activated carbon adsorption*, Taylor & Francis, Boca Raton, FL (2005).
87. E. G. Furuya, H. T. Chang, Y. Miruya, H. Yokomura, S. Tajima, S. Yamashita, and K. E. Knoll, Interparticle mass transport mechanism in activated carbon adsorption of phenols, *Journal of Environmental Engineering*, 122, 909–916 (1996).
88. Z. Tong, M. Bischoff, L. Nies, B. Applegate, and R. F. Turco, Impact of fullerene (C60) on a soil microbial community, *Environmental Science and Technology*, 41, 2985–2991 (2007).
89. J. D. Fortner, D. Y. Lyon, C. M. Sayes, A. M. Boyd, J. C. Falkner, E. M. Hotze, L. B. Alemany, Y. J. Tao, W. Guo, K. D. Ausman, V. L. Colvin, and J. B. Hughes, C_{60} in water: Nanocrystal formation and microbial response, *Environmental Science and Technology*, 37, 3382–3391 (2003).
90. D. Y. Lyon, J. D. Fortner, C. M. Sayes, V. L. Colvin, and J. B. Hughes, Bacterial cell association and antimicrobial activity of a C_{60} water suspension, *Environmental Science and Technology*, 39, 4307–4316 (2005).
91. J. Lee, J. D. Fortner, J. B. Hughes, and J.-H. Kim, Photochemical production of reactive oxygen species by C_{60} in the aqueous phase during UV irradiation, *Environmental Science and Technology*, 41, 179–184 (2007).
92. H. Hyung, J. D. Fortner, J. B. Hughes, and J.-H. Kim, Natural organic matter stabilizes carbon nanotubes in the aqueous phase, *Environmental Toxicology Chemistry*, 24, 2757–2762 (2005).
93. J. D. Fortner, D.-I. Kim, A. M. Boyd, J. C. Faulkner, S. Moran, V. L. Colvin, J. B. Hughes, and J.-H. Kim, Reaction of water-stable C60 aggregates with ozone, *Environmental Science Technology*, 41, 2529–2535 (2007).
94. A. Navrotsky, Nanomaterials in the environment, agriculture and technology (NEAT), *Journal of Nanoparticle Research*, 2, 321–323 (2000).
95. A. Navrotsky, Environmental nanoparticles, In *Encyclopedia of nanoscience and nanotechnology*, Marcel-Dekker, New York, 1147–1156 (2004); Taylor & Francis (2005).
96. R. Senjen, Nanosilver—A threat to soil, water and human health? Friends of the Earth Australia, www.nano.foe.org.au (2007).
97. R. Eisler, A review of silver hazards to plants and animals, In *4th International Conference Proceedings: Transport, fate and effects of silver in the environment*, A. W. Andred and T. W. Bober, eds., pp. 143–144, Madison, Wisconsin (1996).
98. L. J. Albright and E. M. Wilson, Sub-lethal effects of several metallic salt-organic compound combinations upon heterotrophic microflora of a natural water, *Water Research*, 8, 101–105 (1974).
99. L. Braydich-Stolle, S. Hussain, J. J. Schlager, and M. Hofmann, In vitro cytotoxicity of nanoparticles in mammalian germline stem cells, *Toxicological Science*, 88, 412–419 (2005).
100. J. Fraser, Colloidal silver antibacterial liquid sprayed on Hong Kong subways as public health measure, NewsTarget.com, www.newstarget.com/020851.html (2006).
101. W. H. Beauman, Water purification and treatment, Short Course, Everpure, Inc., Westmont, IL (2001).
102. Osmosis, en.wikipedia.org/wiki/Osmosis (2008).
103. Osmotic Pressure, en.wikipedia.org/wiki/Osmotic_pressure (2008).
104. Reverse Osmosis, en.wikipedia.org/wiki/Reverse_osmosis (2008).

105. M. Abraham, Today's seawater is tomorrow's drinking water: UCLA engineers develop nanotech water desal membrane, Henry Samuelli School of Engineering and Applied Sciences, www.engineer.ucla.edu/news/2006/Desal_Membrane.htm (2006).
106. A. Risbud, Cheap drinking water from the ocean, *Technology Review*, MIT, www.technologyreview.com/Nanotech/16977/ (2006).
107. A. Parker, Tiny tubes make the flow go, *Research Highlights*, Lawrence Livermore National Laboratory, S & TR, February (2007).
108. Cheap, clean drinking water purified through nanotechnology, *Science News*, www.sciencedaily.com/releases/2008 (2008).
109. P. J. Majewski and C. P. Chan, Water purification by functionalised self-assembled monolayers on silica particles, *International Journal of Nanotechnology*, 5, 291–298 (2008).
110. Bacteria and Nanofilters: Future of clean water technology, *Science News*, www.sciencedaily.com/releases/2008/02/080222095403.htm (2008).
111. N. Hilal, Bacteria and Nanofilters: Future of clean water technology, *University of Nottingham News*, February 12 (2008).
112. P. S. Zurer, Countering Oils Spills, *Chemical and Engineering News*, 81, 32–33 (2003).
113. J. G. Reynolds, in P.S. Zurer, Countering oils spills, *Chemical and Engineering News*, 81, 32–33 (2003).
114. D. Robson, Magic sand, *ChemMatters Chemistry*, April, 8–9 (1994).
115. Clean air acts, en.wikipedia.org/wiki/Clean_Air_Act (2008).
116. The clean air act and nanotechnology, Nanotechnology Teleconference Series, Section of Environment, Energy and Resources, American Bar Association; www.abanet.org/environ/programs/teleconference/nanotech/session3/ (2007).
117. Nanotechnology and the toxic substances control act, Nanotechnology Teleconference Series, Section of Environment, Energy and Resources, American Bar Association; www.abanet.org/environ/programs/teleconference/nanotech/tsca/ (2007).
118. Nanotechnology and the clean water act, Nanotechnology Teleconference Series, Section of Environment, Energy and Resources, American Bar Association' www.abanet.org/environ/programs/teleconference/nanotech/cwa/ (2007).
119. A. Navrotsky, Nanomaterials in the environment, agriculture and technology (NEAT), *Journal of Nanoparticle Research*, 2, 321–323 (2000).
120. C. Anastacio and S. T. Martin, Atmospheric nanoparticles, In *Nanoparticles and the environment*, J. F. Banfield and A. Navrotsky, eds., Vol. 44, pp. 293–349 Reviews in Mineralogy and Geochemistry; Mineralogical Society of America, Washington D.C. (2001).
121. W. J. Koros, Gas separation, In *Membrane separation systems: Recent developments and future directions*, R. W. Baker, E. L. Cussler, W. Eykamp, W. J. Koros, R. L. Riley, H. Strathmann, eds., William Andrew Publishing, Inc., New York, 349–355 (1991).
122. Pressure Swing Adsorption, en.wikipedia.org/wiki/Pressure_swing_adsorption (2008).
123. Non-cryogenic air separation processes: Nitrogen or oxygen generation using pressure swing adsorption (PSA), Universal Industrial Gases, Inc., www.uigi.com/noncryo.html (2007).
124. G. Alptekin, An advanced rapid recycling CO_2 and H_2O control system for PLSS, NASA SBIR-NNJ05JB96C, TDA Research (2005).
125. W. Xing, W. C. D. da Costa, and Z. F. Yan, Environmental separation and reactions: Zeolite membranes, In *Encyclopedia of nanoscience and nanotechnology*, Vol. 2, Marcel-Dekker, New York (2004); Taylor & Francis (2005).

126. K. Aoki, K. Kusakabe, and S. Morooka, Separation of gases with an A-type Zeolite membrane, *Industrial and Engineering Chemistry Research*, 39, 2245 (2000).
127. Y. Ohtaa, H. Takabab, and S.-I. Nakaoa, A combinatorial dynamic Monte Carlo approach to finding a suitable Zeolite membrane structure for CO_2/N_2 separation, *Microporous and Mesoporous Materials*, 101, 319–323 (2007).
128. T. Kouketsu, S. Duan, T. Kai, S. Kazaman, and K. Yamada, PAMAM dendrimer composite membrane for CO_2 separation: Formation of a chitosan gutter layer, *Journal of Membrane Science*, 287, 51–59 (2007).
129. N. Hiyoshi, K. Yogo, and T. Yashima, Adsorption characteristics of carbon dioxide on organically functionalized SBA-15, *Microporous and Mesoporous Materials*, 84, 357–365 (2005).
130. S. Duan, T. Kouketsu, T. Kai, S. Kazaman, and K. Yamada, Development of PAMAM dendrimer composite membrane for CO_2 separation, *Journal of Membrane Science*, 283, 2–6 (2006).
131. A. S. Kovvali, H. Chen, and K. K. Sirkar, Dendrimer membranes: A CO_2-selective molecular gate, *Journal of American Chemical Society*, 122, 7594–7595 (2000).
132. K. Yamada, S. Kazama, and K. Yogo, Development of innovative gas separation membranes through sub-nanoscale control materials control, Stanford University Global Climate & Energy Project (GCEP), CO_2 Capture, gcep.stanford.edu/research/factsheets/membrane_subnanoscale.html (2005).
133. D. Guro, Fuel cell grade hydrogen purity requirements: Imapct on purification, analysis and cost, USFCC Fuel Cell Seminar, Air Products (2006).
134. B. Dumé, Conductive polymers and CNTs in the mix, Nanotechweb.org, nanotechweb.org/cws/article/tech/32863 (2008).
135. S. M. C. Vieira, P. Beecher, I. Haneef, F. Udrea, W. I. Milne, M. A. G. Namboothiry, D. L. Carroll, J. Park, and S. Maeng, Use of nanocomposites to increase electrical "gain" in chemical sensors, *Applied Physics Letters*, 91, 203111 (2007).
136. M. K. Kennedy, F. E. Kruis, H. Fissan, B. R. Mehta, S. Stappert, and G. Dumpich, Tailored nanoparticle films from monosized tin oxide nanocrystals: Particle synthesis, film formation, and size-dependent gas-sensing properties, *Journal of Applied Physics*, 93, 551–560 (2003).
137. H. Ogawa, M. Nishikawa, and A. Abe, Hall measurement studies and an electrical conduction model of tin oxide ultrafine particle films, *Journal of Applied Physics*, 53, 4448–4455 (1982).
138. R. E. Smalley, Our energy challenge, Presentation, 27th Illinois Junior Science & Humanities Symposium, smalley.rice.edu/smalley.cfm?doc_id= 4862, April 3 (2005).
139. P. D. Persans in "Busy old fool, unruly sun.....," The Energy Challenge, AIP Industrial Physics Forum, blogs.physicstoday.org/industry07/2007/10/busy_old_fool_unruly_sun_1.html, Seattle, WA (2008).
140. B. K. Miller, Largest solar factory coming to bay area, GlobeSt.Direct, www.nanosolar.com/cache/GlobeSt100MM.htm (2006).
141. M. Kanelios, Solar cell breaks efficiency record, CNET News.com (2006).
142. J. Nelson, *The physics of solar cells*, Imperial College Press, London (2003).
143. W. Shockley and H. J. Queisser, Detailed balance limit of efficiency of p-n junction solar cells, *Journal of Applied Physics*, 32, 510–519 (1961).
144. Photovoltaics: Solar electricity and solar cells in theory and practice, *Solar Magazine*, www.solarserver.de/wissen/photovoltaik-e.html (2007).
145. Quantum efficiency, en.wikipedia.org/wiki/Quantum_efficiency (2008).
146. M. Grätzel, Photoelectrochemical cells, *Nature*, 414, 338–344 (2001).
147. I. Robel, R. Subramanian, M. Kuno, and P. V. Kamat, Quantum dot solar cells: Harvesting light energy with CdSe nanocrystals molecularly linked to mesoscopic TiO_2 film, *Journal of American Chemical Society*, 128, 2385–2393 (2006).

148. I. Robel, M. Kuno, and P. V. Kamat, Size-dependent electron injection from excited CdSe quantum dots into TiO_2 nanoparticles. *Journal of American Chemical Society*, 129, 4136–4137 (2007).
149. G. Overton, Photovoltaics: Quantum dots promise next generation solar cells, *Laser Focus World*, www.laserfocusworld.com/articles/252486 (2006).
150. P. Weiss, Photon double whammy: Careening electrons may rev up solar cells, *Science News Online*, 165, 259 (2004).
151. P. Weiss, Quantum dot leap: Tapping tiny crystals' inexplicable light-harvesting talent, *Science News Online*, 169, 344 (2006).
152. A. J. Nozik, Exciton multiplication and relaxation dynamics in quantum dots: applications to ultrahigh-efficiency solar photon conversion, *Inorganic Chemistry*, 44, 6893–6899 (2005).
153. R. J. Ellingson, M. C. Beard, J. C. Johnson, P. Yu, O. I. Micic, A. J. Nozik, A. Shabaev, and A. L. Efros, Highly efficient multiple exciton generation in colloidal PbSe and PbS quantum dots, *Nano Letters*, 5, 865–871 (2005).
154. J. E. Murphy, M. C. Beard, A. G. Norman, S. P. Ahrenkiel, J. C. Johnson, P. Yu, O. I. Micic, R. J. Ellingson, and A. J. Nozik, PbTe colloidal nanocrystals: Synthesis, characterization, and multiple exciton generation, *Journal of American Chemical Society*, 128, 3241–3247 (2006).
155. R. J. Ellingson, A. J. Nozik, M. C. Beard, J. Johnson, J. Murphy, K. Knutsen, K. Gerth, J. Luther, M. Hanna, O. Micic, A. Shabaev, and A. Efros, Nanocrystals generating >1 electron per photon may lead to increased solar cell efficiency, *Solar and Alternative Energy*, SPIE, spie.org/x8735.xml (2008).
156. J. Li, F. Dierschke, J. Wu, A.C. Grimsdale, and K. Müllen, Poly(2,7-carbazole) and perylene tetracarboxydiimide: a promising donor/acceptor pair for polymer solar cells, *Journal of Materials Chemistry*, 16, 96–100 (2006).
157. M. M. Wienk, J. M. Kroon, W. J. H. Verhees, J. Knol, J. C. Hummelen, P. A. van Hal, and R. A. J. Janssen, Efficient methano[70]fullerene/MDMO-PPV bulk heterojunction photovoltaic cells, *Angewandte Chemie International Edition*, 42, 3371–3375 (2003).
158. C. W. Tang and A. C. Albrecht, Chlorophyll-*a* photovoltaic cells, *Nature*, 254, 507–509 (1975).
159. C. W. Tang, Two-layer organic photovoltaic cell, *Applied Physics Letters*, 48, 183–185 (1986).
160. S. Mitra and C. Li, Unique nanotube composites constructed for organic solar cells, *Solar and Alternative Energy*, SPIE, spie.org/x8735.xml (2008).
161. S.-S. Sun and N. S. Sariciftci, eds, *Organic photovoltaics: Mechanisms, materials, and devices (Optical engineering)*, CRC Press, Boca Raton, FL (2005).
162. L. Schmidt-Mende, A. Fechtenkötter, K. Müllen, E. Moons, R. H. Friend, and J. D. MacKenzie, Self-organized discotic liquid crystals for high-efficiency organic photovoltaics, *Science*, 293, 1119–1122 (2001).
163. S. W. Hur, T. W. Kim, and J. W. Park, Organic photovoltaic effects depending on CuPc layer thickness, *Journal of Korean Physics Society*, 45, 627–629 (2004).
164. P. Peumans and S. R. Forrest, Very-high-efficiency double-heterostructure copper phthalocyanine/C60 photovoltaic cells, *Applied Physics Letters*, 79, 126–128 (2001).
165. A. Kongkanand, R. M. Domínguez, and P. V. Kamat, Single wall carbon nanotube scaffolds for photoelectrochemical solar cells: Capture and transport of photogenerated electrons, *Nano Letters*, 7, 676–680 (2007).
166. T. Hasobe and P. V. Kamat, Photoelectrochemistry of stacked cup carbon nanotube films. Tube-Length dependence and charge transfer with excited porphyrin, *Journal of Physical Chemistry C*, 111, 16626–16634 (2007).

167. A. Kongkanand and P. V. Kamat, Electron Storage in single wall carbon nanotubes: Fermi level equilibration in semiconductor–SWCNT suspensions, *ACS Nano*, 1, 13–21, NDRL 4722 (2007).
168. A. Kongkanand and P. V. Kamat, Interactions of single wall carbon nanotubes with methyl viologen radicals: Quantitative estimation of stored electrons, *Journal of Physical Chemistry C*, 111, 9012–9015 (2007).
169. P. V. Kamat, Meeting the clean energy demand: Nanostructure architectures for solar energy conversion, *Journal of Physical Chemistry C*, 111, 2834–2860, NDRL 4697 (2007).
170. J. Wei, Y. Jia, Q. Shu, K. Wang, D. Zhuang, G. Zhang, Z. Wang, J. Luo, A. Cao, and D. Wu, Double-walled carbon nanotube solar cells, *Nano Letters*, 7, 2317–2321 (2007).
171. Z.-P. Yang, L. Ci, J. A. Bur, S.-Y. Lin, and P. M. Ajayan, Experimental observation of an extremely dark material made by a low-density nanotube array, *Nano Letters*, 8, 446–451 (2008).
172. A. Kiebele and G. Grüner, Carbon nanotube based battery architecture, *Applied Physics Letters*, 91, 144104 (2007).
173. Battery (Electricity), en.wikipedia.org/wiki/Battery_(electricity) (2008).
174. C. K. Chan, X. F. Zhang, and Y. Cui, High capacity Li ion battery anodes using Ge nanowires, *Nano Letters*, 8, 307–309 (2008).
175. U. Kasavajjula, C. Wang, and A. J. Appleby, Nano- and bulk-silicon-based insertion anodes for lithium-ion secondary cells, *Journal of Power Sources*, 163, 1003–1039 (2007).
176. C. K. Chan, H. Peng, G. Liu, K. McIlwrath, X. F. Zhang, and Y. Cui, High-performance lithium battery anodes using silicon nanowires, *Nature Nanotechnology*, 3, 31–35 (2007).
177. R. P. Raffaelle, T. Gennett, J. Maranchi, P. Kumta, A. F. Hepp, M. J. hebern, A. C. Dillon, and K. C. Jones, Carbon nanotube anodes for lithium ion batteries, *Materials Research Society Symposium Proceedings*, 706, Z10.5.1–Z10.5.7 (2002).
178. S. H. Ng, J. Wang, Z. P. Guo, J. Chen, G. X. Wang, and H. K. Liu, Single wall carbon nanotube paper as anode for lithium-ion battery, *Electrochimica Acta*, 51, 23028 (2005).
179. V. L. Pushparaj, M. M. Shaijumon, A. Kumar, S. Murugesan, L. Ci, R. Vajtai, R. J. Linhardt, O. Nalamasu, and P. M. Ajayan, Flexible energy storage devices based on nanocomposite paper, *Proceedings of the National Academy Science*, 104, 13574–13577 (2007).
180. C. Kim, M. Noh, M. Choi, J. Cho, and B. Park, Critical size of a nano SnO_2 electrode for Li-secondary battery, *Chemistry of Materials*, 17, 3297–3301 (2005).
181. A. Fujishima and K. Honda, Electrochemical photolysis of water at a semiconductor electrode, *Nature*, 238, 37–38 (1972).
182. I. Okura and M. Kaneko, *Photocatalysis: Science and technology, biological and medical physics Series*, Springer-Verlag, Berlin, Germany (2002).
183. Journey to sustainable energy: The H_2 solution, Presentation, Clean Energy Research Center, University of Southern Florida, cerc.eng.usf.edu/CERC_Overview_Dec04.pdf (2004).
184. A. E. Messer, Solar cell directly splits water for hydrogen, Eureka Alert, www.eurekalert.org/pub_releases/2008-02/ps-scd021108.php (2008).
185. B. O'Regan and M. Grätzel, A low-cost, high-efficiency solar cell based on dye-sensitized colloidal TiO_2 films, *Nature*, 353, 737–740 (1991).
186. H. Hata, Y. Kobayashi, V. Bojan, W. J. Youngblood, and T. E. Mallouk, Direct deposition of trivalent rhodium hydroxide nanoparticles onto a semiconducting layered calcium niobate for photocatalytic hydrogen evolution, *Nano Letters*, 8, 794–799 (2008).

187. L. Becker, Hydrogen storage, Review Article, www.csa.com (2001).
188. R. Dagani, Tempest in a tiny tube, *Chemical Engineering News*, 80, 25–28 (2001).
189. Hydrogen storage technologies roadmap, FreedomCar Fuel Partnership (2005).
190. A. C. Dillon, K. M. Jones, T. A. Bekkedahl, C. H. Kiang, D. S. Bethune, and M. J. Heben, Storage of hydrogen in single-walled carbon nanotubes, *Nature*, 386, 377–379 (1997).
191. M. Hirscher, M. Becher, M. Haluska, U. Dettlaff-Weglikowska, A. Quintel, G. S. Duesberg, Y.-M. Choi, P. Downes, M. Hulman, S. Roth, I. Stepanek, and P. Bernier, Hydrogen storage in sonicated carbon materials, *Applied Physics A*, 72, 129–132 (2001).
192. V. V. Simonyan, P. Diep, and J. K. Johnson, Molecular simulation of hydrogen adsorption in charged single-walled carbon nanotubes, *Journal of Chemical Physics*, 111, 9778–9783 (1999).
193. V. V. Simonyan and J. K. Johnson, Hydrogen storage in carbon nanotubes and graphitic nanofibers, *Journal of Alloys and Compounds*, 330–332, 654–658 (2002).
194. A. Nikitin, X. Li, Z. Zhang, H. Ogasawara, H. Dai, and A. Nilsson, Hydrogen storage in carbon nanotubes through the formation of stable C–H bonds, *Nano Letters*, 8, 162–167 (2008).
195. A. Nikitin, H. Ogasawara, D. Mann, R. Denecke, Z. Zhang, H. Dai, K. Cho, and A. Nilsson, Hydrogenation of Single-Walled Carbon Nanotubes, *Physical Review Letters*, 95, 225507 (2005).
196. A. Nikitin, New carbon nanotube hydrogen storage results surpass Freedom Car requirements, *Nanowerk*,://www.nanowerk.com/spotlight/spotid= 4154.php (2008).
197. T. Furuta, H. Goto, T. Ohashi, Y. Fujiwara, and S. Yip, Theoretical evaluation of hydrogen storage capacity in pure carbon nanostructures, *Journal of Chemical Physics*, 119, 2376–2385 (2003).
198. Y.-W. Lee, R. Bhowmick, and B. M. Clemens, Hydrogen storage in palladium and platinum-doped SWNTs by spill-over mechanism, Materials Research Society 2006 Fall Meeting, Z6.6, Boston, MA (2006).
199. G. Zhang, P. Qi, X. Wang, Y. Lu, D. Mann, X. Li, and H. Dai, Hydrogenation and hydrocarbonation and etching of single-walled carbon nanotubes, *Journal of American Chemical Society*, 128, 6026–6027 (2006).
200. E. Durgun, S. Ciraci, and T. Yildirim, Functionalization of carbon-based nanostructures with light transition-metal atoms for hydrogen storage, *Physical Review B*, 77, 085405–085413 (2008).
201. Go/No-Go decision: Pure, undoped single walled carbon nanotubes for vehicular hydrogen storage, U.S. Department of Energy Hydrogen Program, October (2006).
202. M. J. Heben, A. C. Dillon, K. E. H. Gilbert, P. A. Parilla, T. Gennett, J. L. Alleman, G. L. Hornyak, and K. M. Jones, Assessing the hydrogen storage adsorption capacity of single-wall carbon nanotube/metal composites, In *Hydrogen materials and vacuum systems*, G. R. Myneni and S. Chattopadhyay, eds., pp. 77–89, American Institute of Physics, Conference Proceedings, No. 671 (2003).
203. D. B. Geohegan, H. Hu, A. A. Puretzky. B. Zhao, D. Styers-Barnett, and I. Ivanov, Synthesis and processing of single-walled carbon nanohorns for hydrogen storage and catalyst supports, *FY 2006 Annual Report*, pp. 473–475 (2006).
204. D. B. Geohegan, H. Hu, A. Purtzky, M. Yoon, B. Zhao, and C. M. Rouleau, Single-walled carbon nanohorns for hydrogen storage and catalyst support, DOE Hydrogen Program Review, May 15 (2007).
205. M. Yoon, S. Yang, E. Wang, and Z. Zhang, Charged fullerenes as high capacity hydrogen storage media, *Nano Letters*, 7, 2578 (2007).

206. D. L. Jacobson, PEM Fuel Cells, National Institute for Standards and Testing (NIST), Gaithersburg, MD., physics.nist.gov/MajResFac/NIF/pemFuelCells.html (2006).
207. Nafion, en.wikipedia.org/wiki/Nafion (2007).
208. Nafion: Physical and Chemical Properties, Perma Pure, LLC., www.permapure.com (2006).
209. A. F. Hepp and S. G. Bailey, Inorganic photovoltaic materials and devices: Past, present and future, In *Organic photovoltaics: Mechanisms, materials, and devices (Optical Engineering)*, chap. 2, S.-S. Sun and N. S. Sariciftci, eds., CRC Press, Boca Raton, FL, Chapter 2, pp 19–36 (2005).
210. S. L. Gillett, Nanotechnology: Clean energy and resources for the future, White Paper for Foresight Institute: Preparing for Nanotechnology (2002).
211. G. E. Fryxell, T. S. Zemanina, H. wu, S. Kelly, Y. Lin, K. Kremmer, and K.N. Raymond, Actinide-specific interfacial chemistry of monolayer coated mesoporous ceramics, Final Report, U.S. Department of Energy, Project No. 65370 (2001).
212. Y. Lin, G. E. Fryxell, H. Wu, and M. Engelhard, Selective sorption of cesium using self-assembled monolayers on mesoporous supports, *Environmental Science and Technology*, 35, 3962–3966 (2001).
213. G. E. Fryxell, S. V. Mattigod, K. Parker, and R. Skaggs, Heavy metal sequestration using functional nanoporous materials, U.S. EPA Workshop on Nanotechnology for Site Remediation, es.epa.gov/ncer/publications/workshop/pdf/10_20_05_nanosummary.pdf, October 20–21 (2005).
214. N. L. Rosi, J. Eckert, M. Eddaoudi, D. T. Vodak, J. Kim, M. O'Keefde, and O. M. Yaghi, Hydrogen storage in microporous metal-organic frameworks, *Science*, 300, 1127–1129 (2003).
215. S. S. Kaye, A. Dailly, O. M. Yaghi, and J. R. Long, Impact and preparation and handling on the hydrogen storage properties of Zn_4O(1,4,-benzene dicarboxylate)$_3$(MOF-5), *Journal of American Chemical Society*, 129, 14176–14177 (2007).
216. M. Dinca and J. R. Long, High-enthalpy adsorption in cation-exchanged variants of the microporous metal-organic framework $Mn_3[(Mn_4Cl)_3(BTT)_8(CH_3OH)_{10}]_2$, *Journal of American Chemical Society*, 129, 1172–11176 (2007).
217. S. S. Han and W. A. Goddard, III, Lithium-doped metal-organic frameworks for reversible H2 storage at ambient temperature, *Journal of American Chemical Society*, 129, 8422–8423 (2007).
218. J. Perlin, The story of photocells, In *Organic photovoltaics: Mechanisms, materials, and devices (Optical Engineering)*, chap. 1, S.-S. Sun and N. S. Sariciftci, eds., CRC Press, Boca Raton, FL, 3–18 (2005).
219. D. W. Elliott and W.-X. Zhang, Field assessment of nanoscale bimetallic particles for groundwater treatment, *Environmental Science and Technology*, 35, 4922–4926 (2001).
220. D. Engber, How does activated carbon work? www.slate.com/id/2131430/ (2005).
221. H. J. Lee, D.-Y. Kim, J.-S. Yoo, J. Bang, S. Kim, and S.-M. Park, Anchoring Cadmium chlacogenide quantum dots (QDs) onto stable oxide semiconductors for QD sensitized solar cells, *Bulletin of Korean Chemical Society*, 28, 1–6 (2007).
222. Y.-J. Shen and Y.-L. Lee, Assembly of CdS quantum dots onto mesoscopic TiO_2 films for quantum dot-sensitized solar cell applications, *Nanotechnology*, 19, 045602–045608 (2008).
223. R. Dagani, Silica sorbents with a taste for metals: Organic-inorganic nanocomposites selectively take up cesium, mercury, or other pollutants, *Chemical and Engineering News*, 79, 28–30 (2001).
224. T. Sawaguchi, Y. Sato, and F. Mizutani, Ordered structures of self-assembled monolayers of 3-mercaptopropionic acid on Au(111): In situ scanning tunneling microscopy study, *Physical Chemistry Chemical Physics*, 3, 3399–3404 (2001).

Problems

28.1 Reorder the list generated by Richard Smalley to suit your perspective of the 10 most pressing challenges facing our species over the next 50 years.

28.2 What are some advantages of solar cells that employ quantum dots? List some disadvantages.

28.3 Why do you think groundwater would require less treatment than surface water? What kinds of contaminants would you expect to find in groundwater?

28.4 Why do you think natural groundwaters (or spring waters) taste so clean and fresh? Do you believe wetlands play a role in water purification?

28.5 Is single molecule detection important in environmental analysis?

28.6 List some important differences between fuel cells and photovoltaic cells. Which one is an electrolytic cell? Which one a voltaic cell? Which on uses batteries to store energy? Which one was designed to replace batteries?

28.7 What are the challenges affecting the widespread use of photovoltaic cells and that of fuel cells?

28.8 How many other means can you devise to measure the levels of glucose (hint: think of all aspects of the glucose reaction)?

28.9 Why do you think that C_{60}-agglomerated structures look rectangular?

28.10 The equation for osmotic pressure is similar to the gas law $PV = nRT$. Show how this is so.

28.11 What is the osmotic pressure of 10 g of sucrose in 100 mL of solution (MW = 180.16 g·mol^{-1})?

28.12 (a) 2.00 g of an unknown polymer are placed in 100 mL of solution. The measured osmotic pressure Π is 5.0×10^{-3} atm at 25°C. Find its molecular weight (MW). (b) Calculate MW for a 1% solution of a protein (ideal, isoelectric) that gives an osmotic pressure of 55 mm H_2O at 0°C. (Hint: pressure is in terms of mm H_2O and not mm Hg).

28.13 Are 10-nm diameter Au nanoparticles in solution able to cause osmotic pressure?

28.14 Why is a semipermeable membrane necessary for osmosis to occur? What about osmotic pressure? Is a semipermeable membrane necessary for osmotic pressure?

28.15 Calculate the total surface area of an electrode containing Ge nanowires that are 20 nm in diameter and 50 μm in length. The electrode surface area is 25 cm^2 and the wire surface density is 80%.

Index

A

B

C

D

E

F

G

H

I

J

K

L

M

N

O

P

Q

R

S

T

U

V

W

X

Y

Z